The number of permutations of n objects, of which n_1 are of (kind, . . . , and n_k are of a kth kind, is

$$\frac{n!}{n_1!\, n_2! \cdots n_k!}$$

where $n_1 + n_2 + \cdots + n_k = n$.

Combinations An arrangement, without regard to order and without repetition, of r objects selected from n distinct objects, where $r \leq n$. The number of such arrangements is

$$C(n, r) = \binom{n}{r} = \frac{n!}{r!(n - r)!}$$

BINOMIAL THEOREM

For $n > 1$ a positive integer, $(x + y)^n = \binom{n}{0} x^n + \binom{n}{1} x^{n-1} y + \cdots + \binom{n}{k} x^n y^k + \cdots + \binom{n}{n} y^n$

PROBABILITY

Equally Likely Events If the sample space S of an experiment has n equally likely outcomes and the event E in S occurs in m ways, then

$$P(E) = \frac{m}{n} = \frac{c(E)}{c(S)}$$

Additive Rule For any two events E and F of a sample space S

$$P(E \cup F) = P(E) + P(F) - P(E \cap F)$$

Complement For any event E in a sample space S,

$$P(E) = 1 - P(\overline{E})$$

Conditional Probability If E and F are two events in a sample space S and if $P(F) \neq 0$,

$$P(E|F) = \frac{P(E \cap F)}{P(F)}$$

Product Rule If E and F are two events in a sample space S,

$$P(E \cap F) = P(F)\, P(E|F)$$

Independent Events Two events E and F of a sample space S are independent if and only if

$$P(E \cap F) = P(E)\, P(F)$$

Bayes' Formula Let S be a sample space partitioned into n events, $A_1, A_2, \ldots, A_n$. If E is an event in S for which $P(E) > 0$, then

$$P(A_j|E) = \frac{P(A_j)\, P(E|A_j)}{P(E)} = \frac{P(A_j)\, P(E|A_j)}{P(A_1)\, P(E|A_1) + P(A_2)\, P(E|A_2) + \cdots + P(A_n)\, P(E|A_n)}$$

for $j = 1, 2, \ldots, n$.

Binomial Probabilities In a Bernoulli trial, the probability of exactly k successes in n trials is

$$b(n, k; p) = \binom{n}{k} p^k q^{n-k} = \frac{n!}{k!(n - k)!} p^k q^{n-k}$$

where p is the probability of success and $q = 1 - p$ is the probability of failure.

STATISTICS

Sample Data The data used are a sample $x_1, x_2, \ldots, x_n$ taken from the population of N items, $n < N$.

Sample Mean $\overline{X} = \dfrac{x_1 + x_2 + \cdots + x_n}{n}$

Sample Standard Deviation $S = \sqrt{\dfrac{(x_1 - \overline{X})^2 + (x_2 - \overline{X})^2 + \cdots + (x_n - \overline{X})^2}{n - 1}}$

Population Data The data used are from the entire population $x_1, x_2, \ldots, x_N$ of N items,

Population Mean $\mu = \dfrac{x_1 + x_2 + \cdots + x_N}{N}$

Population Standard Deviation $\sigma = \sqrt{\dfrac{(x_1 - \mu)^2 + (x_2 - \mu)^2 + \cdots + (x_N - \mu)^2}{N}}$

Z = score $Z = \dfrac{x - \mu}{\sigma}$

Eighth Edition

Mathematics
An Applied Approach

Michael Sullivan Chicago State University
Abe Mizrahi Indiana University Northwest

JOHN WILEY & SONS, INC. WILEY

ACQUISITIONS EDITOR	Michael Boezi
ASSOCIATE PUBLISHER	Laurie Rosatone
FREELANCE DEVELOPMENTAL EDITOR	Anne Scanlan-Rohrer
EXECUTIVE MARKETING MANAGER	Julie Lindstrom
SENIOR PRODUCTION EDITOR	Norine M. Pigliucci
SENIOR DESIGNER	Harry Nolan
COVER DESIGN	Howard Grossman
INTERIOR DESIGN	Jerry Wilke Design
ILLUSTRATION EDITOR	Sigmund Malinowski
ELECTRONIC ILLUSTRATIONS	Techsetters, Inc.
PHOTO EDITOR	Lisa Gee
ASSISTANT EDITOR	Jennifer Battista
PROGRAM ASSISTANT	Kelly Boyle
PRODUCTION MANAGEMENT SERVICES	Suzanne Ingrao/Ingrao Associates
COVER PHOTOS	(Peach) © Tim Turner/FoodPix/Getty Images (Apple) © Brian Hagiwara/FoodPix/Getty Images (Plum) © Richard Kolker/The Image Bank/Getty Images
INSET COVER PHOTOS	© Getty Images and Digital Vision

This book was set in Minion by Progressive Information Technologies and printed and bound by Von Hoffmann Corporation. The cover was printed by Von Hoffmann Corporation.

This book is printed on acid free paper. ∞

ISBN 0-471-32784-0

WIE ISBN 0-471-65664-X

Printed in the United States of America

10 9 8 7 6 5 4 3 2 1

To Our Families

About the Authors

Michael Sullivan

is Professor Emeritus in the Department of Mathematics and Computer Science at Chicago State University where he taught for 35 years before retiring a few years ago. Dr. Sullivan is a member of American Mathematical Society, the Mathematical Association of America, and the American Mathematical Association of Two Year Colleges. He is President of the Text and Academic Authors Association and represents that organization on the Authors Coalition of America. Mike has been writing textbooks in mathematics for over 30 years. He currently has 13 books in print: 3 texts with John Wiley & Sons and 10 with Prentice-Hall. Six of these titles are co-authored with his son, Michael Sullivan III.

Mike has four children: Kathleen, who teaches college mathematics; Michael, who teaches college mathematics, Dan, who is a Prentice-Hall sales representative, and Colleen, who teaches middle-school mathematics. Nine grandchildren round out the family.

Abe Mizrahi

enjoyed an active career in mathematics before his untimely passing in 2001. He received his doctorate in mathematics from the Illinois Institute of Technology in 1965, and was a Professor of Mathematics at Indiana University Northwest. Dr. Mizrahi was a member of the Mathematics Association of America, and wrote articles that explored topics in math education and the applications of mathematics to economics.

Dr. Mizrahi served on many CUPM committees and was a panel member on the CUPM Committee on Applied Mathematics in the Undergraduate Curriculum. Dr. Mizrahi was the recipient of many NSF grants and served as a consultant to a number of businesses and federal agencies.

Preface to the Instructor

Since the publication of the Seventh Edition of *Mathematics, An Applied Approach* in 2000, my co-author Abe Mizrahi unexpectedly passed away. Abe and I began writing together in 1970 and published *Finite Mathematics* in 1973 with John Wiley and Sons. *Mathematics: An Applied Approach* was published in 1976. In 1982, Abe and I published *Calculus* with Wadsworth Publishing Company, an engineering calculus text that saw three editions. Abe enjoyed a long and successful teaching career at Indiana University Northwest, from which he had retired just prior to his untimely death. While Abe did not participate in this revision, his influence remains.

The Eighth Edition

The Eighth Edition of *Mathematics, An Applied Approach*, builds upon a solid foundation by integrating new features and techniques that further enhance student interest and involvement. The elements of previous editions that proved successful remain, while many changes have been made. Virtually every change is the result of thoughtful comments and suggestions from colleagues and students who have used previous editions. I am sincerely grateful for this feedback and have tried to incorporate changes that improve the flow and usability of the text.

New to the Eighth Edition

Chapter Projects

Each chapter begins with a student-oriented essay that previews a project involving the mathematics of the chapter. The Chapter Project appears at the end of the chapter and is designed for an individual or collaborative experience. Each project builds on the mathematics of the chapter but also requires students to stretch their understanding of the concepts.

A Look Back, a Look Forward

At the beginning of each chapter, a connection is made between previously studied material and material found in the current chapter.

Preparing for This Section

Most sections now open with a referenced list (by section and page number) of key items to review in preparation for the section ahead. This provides a just-in-time review for students.

Objectives

At the beginning of every section is a numbered list of objectives. The objective is cited at the appropriate place it is encountered in the section, and marked with a numbered 1 icon.

Chapter Reviews

Each chapter concludes with a variety of features to synthesize the important ideas of the chapter.

Things to Know

A review of the important definitions, formulas, and equations from the chapter.

Objectives

The section objectives are listed again, with references to the review exercises that relate to them.

True/False Items; Fill in the Blanks

Short, quick-answer questions to test vocabulary and concepts.

Review Exercises

These now reflect the objectives of each section. Blue problem numbers can be used by the student as a Practice Test.

Mathematical Questions from Professional Exams

Where appropriate, questions from CPA, CMA, and Actuary Exams have been reproduced.

Exercises and Examples

Sourced Problems

Many new examples and exercises that contain sourced data or sourced facts have been added to each problem set.

Conceptual Problems

These new problems ask the student to verbalize or to write an answer to a problem that may have multiple solutions or require some research. These problems are clearly marked with an icon and green color to make them easy to identify.

Technology

As an optional feature, examples and exercises have been included that utilize a graphing utility or spreadsheets (EXCEL). These examples and exercises are clearly identified using an icon or an icon, and blue problem numbers.

Design

This edition has a fully integrated, pedagogically based, utilization of color in both the text and the illustrations. Just look at it to see the effect.

Using the Eighth Edition Effectively and Efficiently with Your Syllabus

To meet the varied needs of diverse syllabi, this book has been organized with flexibility of use in mind. The illustration shows the dependencies of chapters on each other.

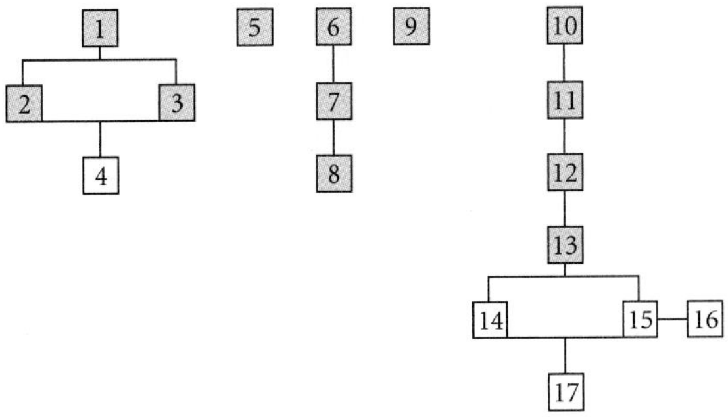

The following chapter descriptions along with the Table of Contents will help you match the topics of this book with your syllabus. Please do not hesitate to contact me (through the publisher John Wiley & Sons) if you have any questions about the content and organization of this book.

Chapter 1 Linear Equations
This chapter serves as the basis for Chapters 2, 3, and 4. It can be used as the starting point of a course.

Chapter 2 Systems of Linear Equations; Matrices
This chapter uses information from Chapter 1.

Chapter 3 Linear Programming: Geometric Approach
This chapter also uses information from Chapter 1 and is independent of Chapter 2.

Chapter 4 Linear Programming: The Simplex Method
This chapter requires Sections 2.1 and 2.2 of Chapter 2 and Sections 3.1 and 3.2 of Chapter 3. Appendix B: Using LINDO to Solve Linear Programming Problems may be used in conjunction with Chapter 4.

Chapter 5 Finance
This chapter can be used as the starting point of a course.

Chapter 6 Sets; Counting Techniques
This chapter can be used as the starting point of a course.

Chapter 7 Probability
This chapter requires the information presented in Chapter 6.

Chapter 8 Additional Probability Topics
This chapter requires information found in Chapter 7.

Chapter 9 Statistics
While not absolutely necessary, a coverage of Section 7.1 in advance may prove helpful.

Chapter 10 Functions and Their Graphs
This chapter can be used as the starting point of a course. While this chapter reviews material studied in College Algebra or Precalculus, it also introduces many of the concepts needed later in calculus. For example, while the student is reviewing function notation, the difference quotient is introduced. When the graph of a function is discussed, the concepts of increasing and decreasing functions, local maximum and minimum, and average rate of change are introduced.

Chapter 11 Classes of Functions
Here quadratic, power, polynomial, rational, exponential, and logarithmic functions are introduced in the context of how they will be utilized in calculus. Section 11.6, Continuously Compounded Interest, may be omitted without a loss of continuity.

Chapter 12 The Limit of a Function
While intuitive in approach, the techniques introduced here will be used later in the discussion of the derivative and the integral of a function. Uses of limits as they apply to continuity, end behavior, and asymptotes are included.

Chapter 13 The Derivative of a Function
The derivative and its applications to geometry (tangent line), economics (marginal analysis), and physics (velocity and acceleration) are discussed in an organized, student-friendly manner. The subsection on Velocity and Acceleration may be omitted without a loss of continuity.

Chapter 14 Applications: Graphing Functions; Optimization
All the information required to obtain the graph of a function is given here using methodology from both algebra and calculus, with emphasis on the calculus. Additional applications are given to optimization (Section 14.4), elasticity of demand (Section 14.5), related rates (Section 14.6), and linear approximations using differentials (Section 14.7). These sections are independent of each other and may covered in any order or be omitted without any loss of continuity.

Chapter 15 The Integral of a Function and Applications
The indefinite integral and the definite integral are introduced here with various applications. Sections 15.5, 15.6, and 15.7 are independent of each other and may be covered in any order or be omitted without any loss of continuity.

Chapter 16 Other Applications and Extensions of the Integral
This chapter contains optional material. Section 16.3 depends on Section 16.1.

Chapter 17 Calculus of Functions of Two or More Variables
This chapter is optional. Sections 17.1 – 17.4 follow in sequence. Sections 17.4 and 17.5 are optional and may be covered in any order or omitted without loss of continuity.

Appendix A: Review
This material serves as a just-in-time review of Intermediate Algebra topics. When applicable, its content is referenced in the text in the Preparing for this Section feature.

Appendix B: LINDO
This material presents an introduction to the software package LINDO, for solving linear programming problems.

Appendix C: Graphing Utilities
This appendix provides an overview of some common uses of a graphing calculator.

Other Books in this Series

Also available are *Finite Mathematics: An Applied Approach, 9/e* (ISBN 0-471-32899-5), for a one-term course in finite mathematics and *Brief Calculus: An Applied Approach, 8/e* (ISBN 0-471-45202-5), for the one-term course in calculus.

The Faculty Resource Network

The *Faculty Resource Network* is a peer-to-peer network of academic faculty dedicated to the effective use of technology in the classroom. This group can help you apply innovative classroom techniques, implement specific software packages, and tailor the technology experience to the specific needs of each individual class. Ask your Wiley representative for more details.

Acknowledgments

There are many colleagues I would like to thank for their input, encouragement, patience, and support. They have my deepest thanks and appreciation. I apologize for any omissions.

Contributors

- Bill Ardis and Neil Wigley for working the odd-numbered problems
- Michael Divinia of San Jose City College and Ann Ostberg for sourced problems
- Walter Hunter of Montgomery County Community College for the Excel examples
- Thomas Polaski of Winthrop University for the projects for Chapters 1–7 and 10–18
- David Santana-Ortiz of Rand Corporation for the projects for Chapters 8 and 9
- Tim Comar of Benedictine University, Kathleen Miranda of SUNY Old Westbury, and Kurt Norlin of Laurel Technical Services for working on the answer sections.
- Ken Brown of Southeastern Louisiana University, Mike Divinia of San Jose City College, and Henry Smith of River Parishes Community College for providing new applications and problems using real data
- Kathleen Miranda for accuracy checking the text

Reviewers

Maula Allen	California State University, Hayward
Mark Ashbrook	Arizona State University
Chunsheng E. Ban	Ohio State University
Satish C. Bhatnagar	University of Nevada, Las Vegas
Adel Boules	University of North Florida
Michael Divinia	San Jose City College
Morteza Ebneshahrashoob	California State University, Long Beach
Chris Edwards	University of Wisconsin, Oshkosh
Stephanie L. Fitch	University of Missouri–Rolla
Rutger Hangelbroek	Western Illinois University
David Harpster	Minot State University
Christine Keller	Southwest Texas State University
Richard Leedy	Polk Community College
Tsun-Zee Mai	University of Alabama
Michael A. Nasab	Long Beach City College
Bette Nelson	Alvin Community College
Judith Nowalsky	University of New Orleans
Krish Revuluri	Harper College
William H. Richardson, Jr.	Francis Marion University
Evan Siegel	John Jay College of Criminal Justice

Katie Stables	Western Washington University
Mary Jane Sterling	Bradley University
Bob Stong	University of Virginia
Jeff Stuart	Pacific Lutheran University
Patrick Sullivan	Valparaiso University
Stephen J. Tillman	Wilkes University
Mike Trapuzzano	Arizona State University
Helene Tyler	Manhattan College
Cheryl Whitelaw	Southern Utah University
Yangbo Ye	The University of Iowa

Recognition and thanks are due particularly to the following individuals for their valuable assistance in the preparation of this edition:

- Michael Boezi, for his enthusiasm and support
- Kelly Boyle, for her skill at coordinating a complicated project
- Lisa Gee, for identifying representative photos
- Suzanne Ingrao, for her organizational skills as production manager
- Bonnie Lieberman, for taking a chance
- Julie Lindstrom, for her innovative marketing skills
- Sigmund Malinowski, for his dedication to quality illustrations
- Harry Nolan, for creating a beautifully designed book
- Kurt Norlin of Laurel Technical Services for checking all the answers and especially for his high regard for accuracy
- Norine Pigliucci, for making it all happen on time
- Laurie Rosatone and Bruce Spatz, for their forthright and candid viewpoints and strong encouragement
- Anne Scanlan-Rohrer, for her loyalty and professionalism as development editor

And, to Kathleen Miranda, who read page proofs to ensure accuracy, worked problems for the Solutions Manuals, offered many useful suggestions, and listened so often.

I also want to thank the Wiley sales staff for their continued support and confidence over the years.

Finally, I welcome comments and suggestions for improving this text. Please do not hesitate to contact me through the publisher, John Wiley & Sons.

Sincerely,

Michael Sullivan

Preface to the Student

As you begin this course in Finite Mathematics and Calculus, you may feel overwhelmed by the number of topics that the course contains. Some of these may be familiar to you, while others may be new to you. Either way, I have written this text with you, the student, in mind.

I have taught courses in Finite Mathematics and Calculus for over 30 years. I am also the father of four college graduates who, while in college, called home from time to time frustrated. I know what you are going through and have written a text that doesn't overwhelm or unnecessarily complicate.

This text was designed and written to help you master the terminology and basic concepts of in Finite Mathematics and Calculus. Many learning aids are built into the format of the text to make your study of the material easier and more rewarding, helping you focus your efforts to get the most from the time and effort you invest.

How to Use this Book Effectively and Efficiently

First, and most important, this book is meant to be read! Please, read the material assigned to you. You will find that the text has additional explanations and examples that will help you.

Many sections begin with *Preparing for this Section,* a list of concepts that will be used in the section. Take the short amount of time required to refresh your memory. This will make the section easier to understand and will actually save you time and effort.

A list of *Objectives* is provided at the beginning of each section. Read them. They will help you recognize the important ideas and skills developed in the section.

After a concept has been introduced and an example given, you will see Now Work Problem xx. Go to the exercises at the end of the section, work the problem cited, and check your answer in the back of the book. If you get it right, you can be confident in continuing on in the section. If you don't get it right, go back over the explanations and examples to see what you might have missed. Then rework the problem. Ask for help if you miss it again.

If you follow these practices throughout the section, you will find that you have probably done many of your homework problems. In the exercises, every Now Work

Problem is in yellow with a pencil icon . All the odd-numbered problems have answers in the back of the book and worked-out solutions in the Student Solutions Manual. Be sure you have made an honest effort before looking at a worked-out solution.

At the end of each chapter, there is a Chapter Review. Use it to be sure you are completely familiar with the definitions, formulas, and equations listed under Things To Know. If you are unsure of an item here, use the page reference to go back and review it. Go through the Objectives and be sure you can answer 'Yes' to the question 'I should be able to . . . ' Review exercises that relate to each objective are listed to help you.

Lastly, do the problems in the Review Exercises identified with blue problem numbers. These are my suggestions for a Practice Test. Do some of the other problems in the review for more practice to prepare for your exam.

Please do not hesitate to contact me, through the publisher of this book, John Wiley and Sons, with any suggestions or comments that would improve this text. I look forward to hearing from you.

Best wishes.

Michael Sullivan

Supplements

The following ancillary materials are designed to support the text.

Student Solutions Manual
The Student Solutions Manual contains worked-out solutions to all of the odd-numbered problems. ISBN: 0-471-33379-4

Instructor's Solutions Manual
This manual contains worked-out solutions for all problems in the text.
ISBN: 0-471-44823-0

TI 83 Technology Resource Manual
This manual contains basic instructions for using technology with the text. Students get suggestions for using their calculators and a description of the steps used to solve particular problems from the text.
ISBN: 0-471-44824-9

Test Bank
The test bank contains a wide range of problems and their solutions, which are keyed to the text and exercise sets.
ISBN: 0-471-44825-7

Computerized Test Bank
Available in both IBM and Macintosh formats, the Computerized Test Bank allows instructors to create, customize, and print a test containing any combination of questions from the test bank. Instructors can also edit the questions or add their own.
ISBN: 0-471-44822-2

eGrade
eGrade is an online assessment system that contains a large bank of skill-building problems, homework problems, and solutions. Instructors can automate the process of assigning, delivering, grading, and routing all kinds of homework, quizzes and tests while providing students with immediate scoring and feedback on their work. Wiley eGrade "does the math" . . . and much more. For more information, visit http://www.wiley.com/college/egrade or contact your Wiley representative.

Applications Index

Life and Physical Sciences

Contents

CHAPTER 1

Linear Equations

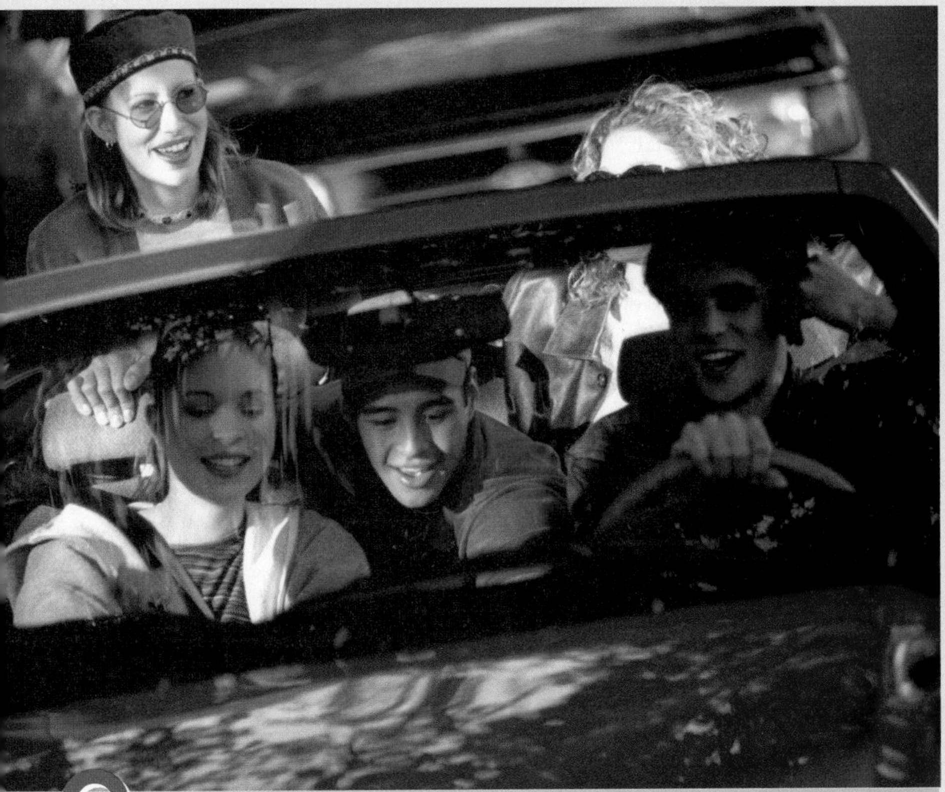

OUTLINE

On the way back to college, you and a friend decide to stop off in Charlotte, North Carolina. Because you have only one full day to see the sights, you decide that renting a car is the best way to see the most. But which car rental company should you use? Naturally, the cheapest! But what is the cheapest? Is it the one with unlimited mileage or the one with a better daily rate and limited miles? The mathematics of this chapter provides the background for solving this problem. The Chapter Project at the end of the chapter will help you understand how to make the best decision.

A LOOK BACK, A LOOK FORWARD

In Appendix A, we review algebra and geometry skills from earlier courses. In this chapter we make the connection between algebra and geometry through the rectangular coordinate system. The idea of using a system of rectangular coordinates dates back to ancient times, when such a system was used for surveying and city planning. Apollonius of Perga, in 200 B.C., used a form of rectangular coordinates in his work on conics, although this use does not stand out as clearly as it does in modern treatments. Sporadic use of rectangular coordinates continued until the 1600s. By that time, algebra had developed sufficiently so that René Descartes (1596–1650) and Pierre de Fermat (1601–1665) could take the crucial step, which was the use of rectangular coordinates to translate geometry problems into algebra problems, and vice versa. This step was important for two reasons. First, it allowed both geometers and algebraists to gain critical new insights into their subjects, which previously had been regarded as separate but now were seen to be connected in many important ways. Second, the insights gained made possible the development of calculus, which greatly enlarged the number of areas in which mathematics could be applied and made possible a much deeper understanding of these areas.

1.1 Rectangular Coordinates; Lines

PREPARING FOR THIS SECTION *Before getting started, review the following:*

> Real Numbers (Section A.1, pp. A-1–A-14)

> Algebra Review (Section A.2, pp. A-15–A-25)

OBJECTIVES

1 Graph linear equations
2 Find the equation of a vertical line
3 Calculate and interpret the slope of a line
4 Graph a line given a point on the line and the slope
5 Use the point–slope form of a line
6 Find the equation of a horizontal line
7 Find the equation of a line given two points
8 Use the slope–intercept form of a line

We locate a point on the real number line by assigning it a single real number, called the *coordinate of the point*. For work in a two-dimensional plane, we locate points by using two numbers.

We begin with two real number lines located in the same plane: one horizontal and the other vertical. We call the horizontal line the **x-axis**, the vertical line the **y-axis**, and the point of intersection the **origin O**. We assign coordinates to every point on these number lines as shown in Figure 1, using a convenient scale. In mathematics, we usually use the same scale on each axis; in applications, a different scale is often used on each axis.

FIGURE 1

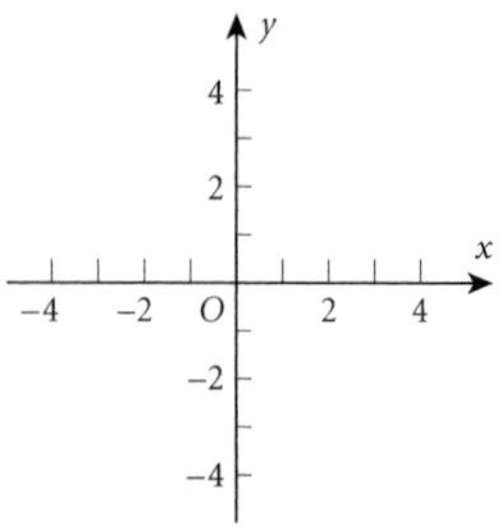

The origin O has a value of 0 on both the x-axis and the y-axis. We follow the usual convention that points on the x-axis to the right of O are associated with positive real numbers, and those to the left of O are associated with negative real numbers. Points on the y-axis above O are associated with positive real numbers, and those below O are associated with negative real numbers. In Figure 1, the x-axis and y-axis are labeled as x and y, respectively, and we have used an arrow at the end of each axis to denote the positive direction.

FIGURE 2

FIGURE 3

Quadrant II	Quadrant I
$x < 0, y > 0$	$x > 0, y > 0$
Quadrant III	Quadrant IV
$x < 0, y < 0$	$x > 0, y < 0$

FIGURE 4

The coordinate system described here is called a **rectangular** or **Cartesian*** **coordinate system.** The plane formed by the x-axis and y-axis is sometimes called the **xy-plane**, and the x-axis and y-axis are referred to as the **coordinate axes.**

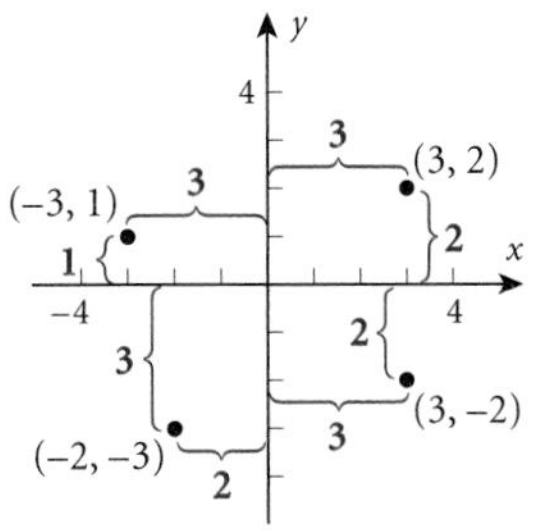

Any point P in the xy-plane can then be located by using an **ordered pair** (x, y) of real numbers. Let x denote the signed distance of P from the y-axis (*signed* in the sense that, if P is to the right of the y-axis, then $x > 0$, and if P is to the left of the y-axis, then $x < 0$); and let y denote the signed distance of P from the x-axis. The ordered pair (x, y), also called the **coordinates** of P, then gives us enough information to locate the point P in the plane.

For example, to locate the point whose coordinates are $(-3, 1)$, go 3 units along the x-axis to the left of O and then go straight up 1 unit. We **plot** this point by placing a dot at this location. See Figure 2, in which the points with coordinates $(-3, 1)$, $(-2, -3)$, $(3, -2)$, and $(3, 2)$ are plotted.

The origin has coordinates $(0, 0)$. Any point on the x-axis has coordinates of the form $(x, 0)$, and any point on the y-axis has coordinates of the form $(0, y)$.

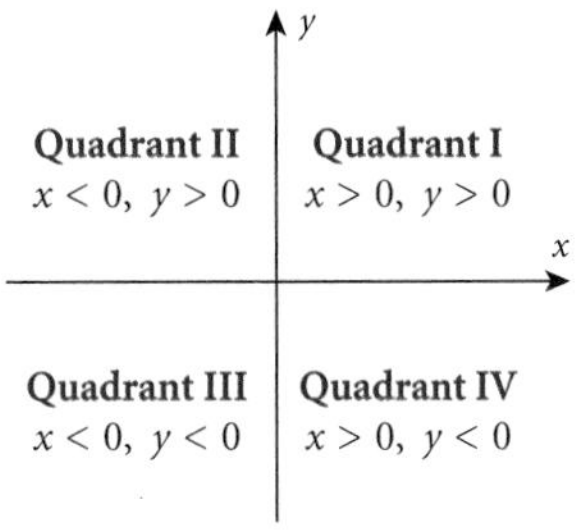

If (x, y) are the coordinates of a point P, then x is called the **x-coordinate,** or **abscissa,** of P; and y is the **y-coordinate,** or **ordinate,** of P. We identify the point P by its coordinates (x, y) by writing $P = (x, y)$, referring to it as "the point (x, y)," rather than "the point whose coordinates are (x, y)."

The coordinate axes divide the xy-plane into four sections, called **quadrants,** as shown in Figure 3. In quadrant I, both the x-coordinate and the y-coordinate of all points are positive; in quadrant II, x is negative and y is positive; in quadrant III, both x and y are negative; and in quadrant IV, x is positive and y is negative. Points on the coordinate axes belong to no quadrant.

 NOW WORK PROBLEM 1.

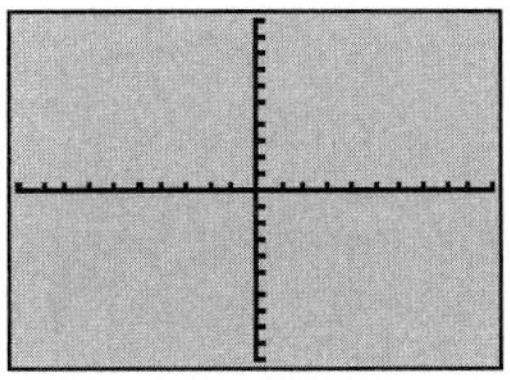

COMMENT: On a graphing utility, you can set the scale on each axis. Once this has been done, you obtain the **viewing rectangle.** See Figure 4 for a typical viewing rectangle. You should now read Section C.1, The Viewing Rectangle, in Appendix C.

Graphs of Linear Equations in Two Variables

A **linear equation in two variables** is an equation of the form

$$Ax + By = C \qquad (1)$$

where A and B are not both zero.

Examples of linear equations are

$$3x - 5y - 6 = 0 \qquad \text{This equation can be written as}$$
$$3x - 5y = 6 \qquad A = 3, B = -5, C = 6$$

Named after René Descartes (1596–1650), a French mathematician, philosopher, and theologian.

$-3x = 2y - 1$ This equation can be written as
$-3x - 2y = -1$ $A = -3, B = -2, C = -1$
or as
$3x + 2y = 1$ $A = 3, B = 2, C = 1$

$y = \frac{3}{4}x - 5$ Here we can write
$-\frac{3}{4}x + y = -5$ $A = -\frac{3}{4}, B = 1, C = -5$
or
$3x - 4y = 20$ $A = 3, B = -4, C = 20$

$y = -5$ Here we can write
$0 \cdot x + y = -5$ $A = 0, B = 1, C = -5$

$x = 4$ Here we can write
$x + 0 \cdot y = 4$ $A = 1, B = 0, C = 4$

The **graph** of an equation is the set of all points (x, y) whose coordinates satisfy the equation. For example, $(0, 4)$ is a point on the graph of the equation $3x + 4y = 16$, because when we substitute 0 for x and 4 for y in the equation, we get

$$3 \cdot 0 + 4 \cdot 4 = 16 \qquad 3x + 4y = 16, x = 0, y = 4$$

which is a true statement.

It can be shown that if A, B, and C are real numbers, with A and B not both zero, then the graph of the equation

$$Ax + By = C$$

is a **line**. This is the reason we call it a **linear equation.**

Conversely, any line is the graph of an equation of the form $Ax + By = C$.

Since any line can be written as an equation in the form $Ax + By = C$, we call this form the **general equation** of a line.

Graph linear equations 1

Given a linear equation, we can obtain its graph by plotting two points that satisfy its equation and connecting them with a line. The easiest two points to plot are the *intercepts*. For example, the line shown in Figure 5 has the intercepts $(0, -4)$ and $(3, 0)$.

FIGURE 5

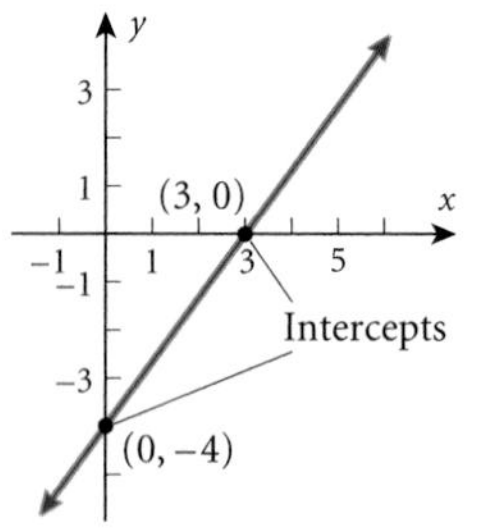

Intercepts

The points at which the graph of a linear equation crosses the axes are called **intercepts.** The ***x*-intercept** is the point at which the graph crosses the x-axis; the ***y*-intercept** is the point at which the graph crosses the y-axis.

Steps for Finding the Intercepts of a Linear Equation

To find the intercepts of a linear equation $Ax + By = C$, with $A \neq 0$ or $B \neq 0$, follow these steps:

STEP 1 Let $y = 0$ and solve for x. This determines the x-intercept of the line.
STEP 2 Let $x = 0$ and solve for y. This determines the y-intercept of the line.

EXAMPLE 1 Finding the Intercepts of a Linear Equation

Find the intercepts of the equation $2x + 3y = 6$. Graph the equation.

SOLUTION **Step 1** To find the x-intercept, we need to find the number x for which $y = 0$. We let $y = 0$ in the equation and proceed to solve for x:

$$\begin{aligned} 2x + 3y &= 6 \\ 2x + 3(0) &= 6 && y = 0 \\ 2x &= 6 && \text{Simplify} \\ x &= 3 && \text{Solve for } x. \end{aligned}$$

The x-intercept is $(3, 0)$.

Step 2 To find the y-intercept, we let $x = 0$ in the equation and solve for y:

$$\begin{aligned} 2x + 3y &= 6 \\ 2(0) + 3y &= 6 && x = 0 \\ 3y &= 6 && \text{Simplify} \\ y &= 2 && \text{Solve for } y. \end{aligned}$$

The y-intercept is $(0, 2)$.

Since the equation is a linear equation, its graph is a line. We use the two intercepts $(3, 0)$ and $(0, 2)$ to graph it. See Figure 6.

FIGURE 6

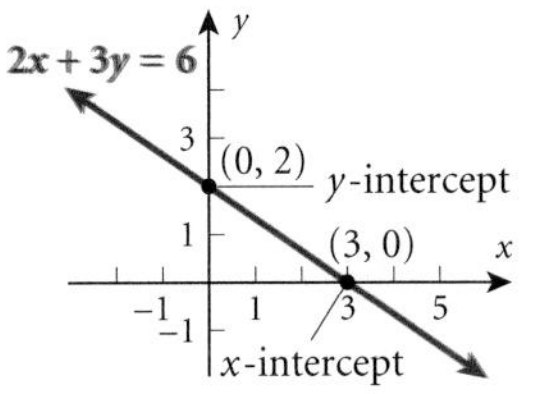

EXAMPLE 2 Graphing a Linear Equation

Graph the equation: $y = 2x + 5$

SOLUTION This equation can be written as

$$-2x + y = 5$$

This is a linear equation, so its graph is a line. The intercepts are $(0, 5)$ and $(-\frac{5}{2}, 0)$, which you should verify. For reassurance we'll find a third point. Arbitrarily, we let $x = 10$. Then $y = 2x + 5 = 2(10) + 5 = 25$, so $(10, 25)$ is a point on the graph. See Figure 7.

x	y
0	5
$-\frac{5}{2}$	0
10	25

FIGURE 7

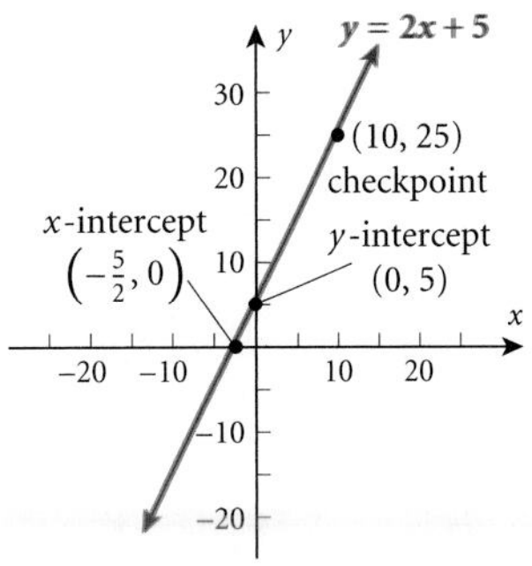

NOW WORK PROBLEM 5.

When a line passes through the origin, it has only one intercept. To graph such lines, we need to locate an additional point on the graph.

EXAMPLE 3 Graphing a Linear Equation

Graph the equation: $-x + 2y = 0$

FIGURE 8

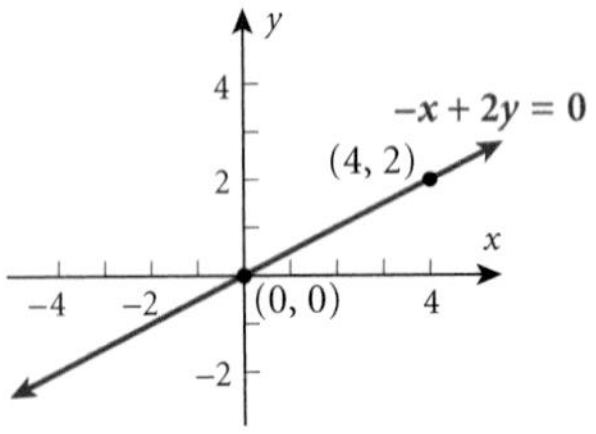

SOLUTION This is a linear equation, so its graph is a line. The only intercept is (0, 0). To locate another point on the graph, let $x = 4$. (This choice is arbitrary; any choice of x other than 0 could also be used). Then,

$$\begin{aligned} -4 + 2y &= 0 \qquad -x + 2y = 0, x = 4 \\ 2y &= 4 \\ y &= 2 \end{aligned}$$

So, $y = 2$ when $x = 4$ and (4, 2) is a point on the graph. See Figure 8.

Next we discuss linear equations whose graphs are vertical lines.

EXAMPLE 4 Graphing a Linear Equation (a Vertical Line)

Graph the equation: $x = 3$

FIGURE 9

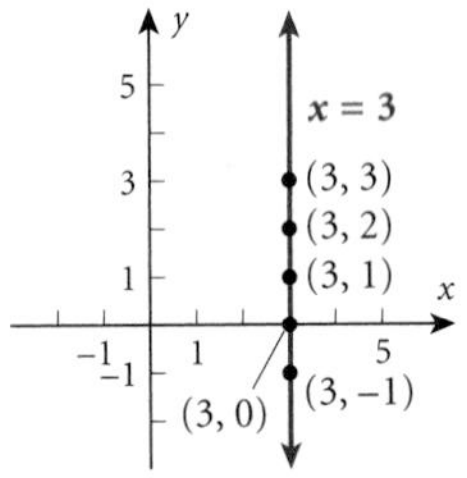

SOLUTION We are looking for all points (x, y) in the plane for which $x = 3$. Since $x = 3$, no matter what y-coordinate is used, the corresponding x-coordinate always equals 3. Consequently, the graph of the equation $x = 3$ is a vertical line with x-intercept (3, 0) as shown in Figure 9.

As suggested by Example 4, we have the following result:

Equation of a Vertical Line

A vertical line is given by an equation of the form

$$\boxed{x = a}$$

where $(a, 0)$ is the x-intercept.

2

EXAMPLE 5 Finding the Equation of a Vertical Line

Find an equation for the vertical line containing the point $(-1, 6)$.

SOLUTION The x-coordinate of any point on a vertical line is always the same. Since $(-1, 6)$ is a point on the vertical line, its equation is $x = -1$.

NOW WORK PROBLEM 9(a).

Slope of a Line

Calculate and interpret the slope of a line 3

An important characteristic of a line, called its *slope*, is best defined by using rectangular coordinates.

Slope of a Line

Let $P = (x_1, y_1)$ and $Q = (x_2, y_2)$ be two distinct points. If $x_1 \neq x_2$, the **slope** m of the nonvertical line L containing P and Q is defined by the formula

$$m = \frac{y_2 - y_1}{x_2 - x_1} \qquad x_1 \neq x_2 \tag{2}$$

If $x_1 = x_2$, L is a vertical line and the slope m of L is **undefined** (since this results in division by 0).

Figure 10(a) provides an illustration of the slope of a nonvertical line; Figure 10(b) illustrates a vertical line.

FIGURE 10

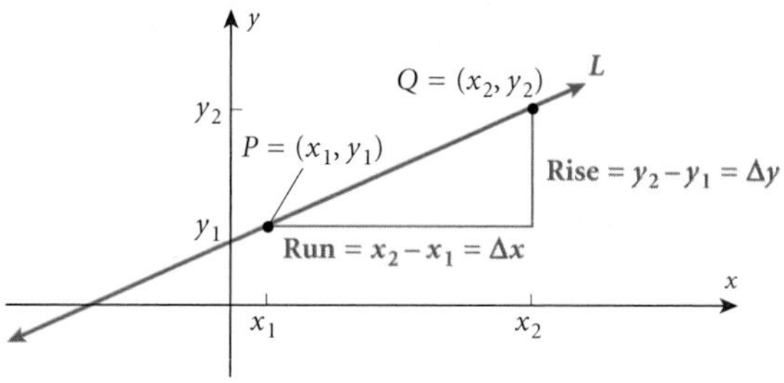

Slope of L is $m = \frac{y_2 - y_1}{x_2 - x_1}$

(a)

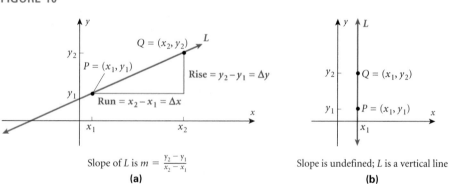

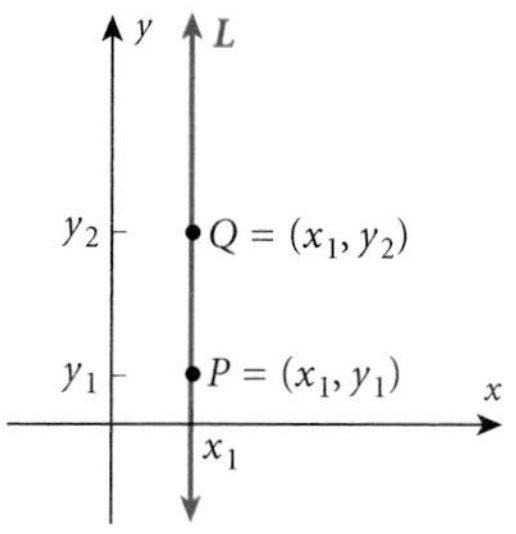

Slope is undefined; L is a vertical line

(b)

As Figure 10(a) illustrates, the slope m of a nonvertical line may be given as

$$m = \frac{y_2 - y_1}{x_2 - x_1} = \frac{\text{Rise}}{\text{Run}} = \frac{\text{Change in } y}{\text{Change in } x}$$

The change in y is usually denoted by Δy, read "delta y," and the change in x is denoted by Δx.

The slope m of a nonvertical line L measures the amount y changes, Δy, as x changes from x_1 to x_2, Δx. This is called the **average rate of change of y with respect to x.** Then, the slope m is

$$m = \frac{\Delta y}{\Delta x} = \text{average rate of change of } y \text{ with respect to } x$$

EXAMPLE 6 Finding and Interpreting the Slope of a Line

The slope m of the line containing the points $(3, -2)$ and $(1, 5)$ is

$$m = \frac{\Delta y}{\Delta x} = \frac{5 - (-2)}{1 - 3} = \frac{7}{-2} = \frac{-7}{2} = -\frac{7}{2}$$

We interpret the slope to mean that for every 2-unit change in x, y will change by -7 units. That is, if x increases by 2 units, then y decreases by 7 units. The average rate of change of y with respect to x is $-\frac{7}{2}$.

NOW WORK PROBLEMS 13 AND 17.

Two comments about computing the slope of a nonvertical line may prove helpful:

1. Any two distinct points on the line can be used to compute the slope of the line. (See Figure 11 for justification.)
2. The slope of a line may be computed from $P = (x_1, y_1)$ to $Q = (x_2, y_2)$ or from Q to P because

$$\frac{y_2 - y_1}{x_2 - x_1} = \frac{y_1 - y_2}{x_1 - x_2}$$

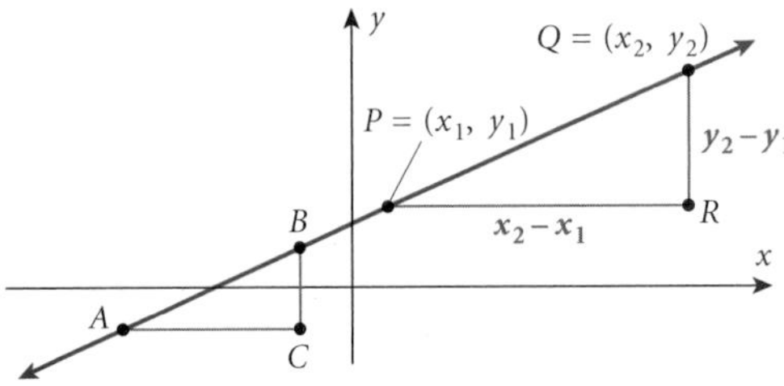

FIGURE 11 Triangles *ABC* and *PQR* are similar (they have equal angles). So ratios of corresponding sides are proportional. Then:

Slope using P and $Q = \frac{y_2 - y_1}{x_2 - x_1} =$

Slope using A and $B = \frac{d(B, C)}{d(A, C)}$

where $d(B, C)$ denotes the distance from B to C and $d(A, C)$ denotes the distance from A to C.

To get a better idea of the meaning of the slope m of a line L, consider the following example.

EXAMPLE 7 Finding the Slopes of Various Lines Containing the Same Point (2, 3)

Compute the slopes of the lines L_1, L_2, L_3, and L_4 containing the following pairs of points. Graph all four lines on the same set of coordinate axes.

$$\begin{aligned} L_1&: P = (2, 3) & Q_1 &= (-1, -2) \\ L_2&: P = (2, 3) & Q_2 &= (3, -1) \\ L_3&: P = (2, 3) & Q_3 &= (5, 3) \\ L_4&: P = (2, 3) & Q_4 &= (2, 5) \end{aligned}$$

SOLUTION Let m_1, m_2, m_3, and m_4 denote the slopes of the lines L_1, L_2, L_3, and L_4, respectively.

FIGURE 12

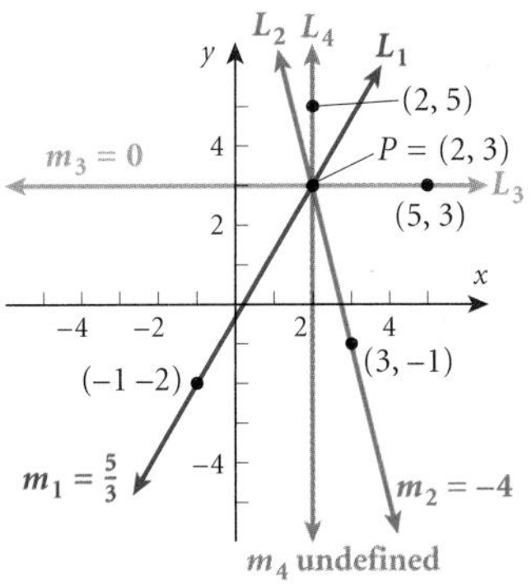

Then

$$m_1 = \frac{-2-3}{-1-2} = \frac{-5}{-3} = \frac{5}{3}$$ A rise of 5 divided by a run of 3

$$m_2 = \frac{-1-3}{3-2} = \frac{-4}{1} = -4$$ A rise of −4 divided by a run of 1

$$m_3 = \frac{3-3}{5-2} = \frac{0}{3} = 0$$ A rise of 0 divided by a run of 3

m_4 is undefined — The x coordinates of P and Q_4 are equal ($x_1 = x_2 = 2$)

The graphs of these lines are given in Figure 12.

As Figure 12 illustrates,

1. When the slope m of a line is positive, the line slants upward from left to right (L_1).
2. When the slope m is negative, the line slants downward from left to right (L_2).
3. When the slope m is 0, the line is horizontal (L_3).
4. When the slope m is undefined, the line is vertical (L_4).

COMMENT: Now read Section C.3, Square Screens, in Appendix C.

SEEING THE CONCEPT: On the same square screen, graph the following equations:

FIGURE 13

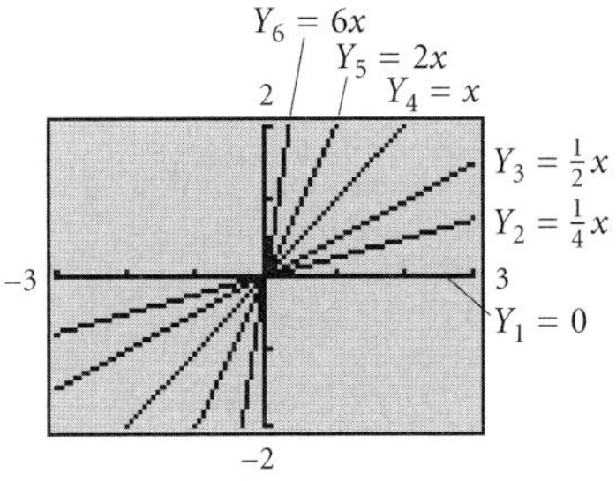

$Y_1 = 0$	Slope of line is 0.
$Y_2 = \frac{1}{4}x$	Slope of line is $\frac{1}{4}$.
$Y_3 = \frac{1}{2}x$	Slope of line is $\frac{1}{2}$.
$Y_4 = x$	Slope of line is 1.
$Y_5 = 2x$	Slope of line is 2.
$Y_6 = 6x$	Slope of line is 6.

See Figure 13.

SEEING THE CONCEPT: On the same square screen, graph the following equations:

FIGURE 14

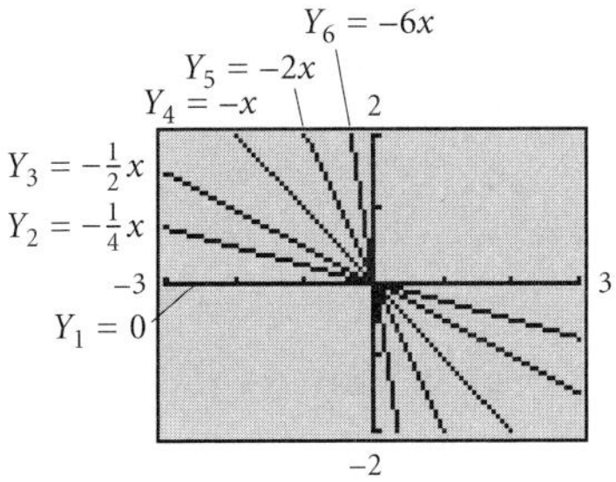

$Y_1 = 0$	Slope of line is 0.
$Y_2 = -\frac{1}{4}x$	Slope of line is $-\frac{1}{4}$.
$Y_3 = -\frac{1}{2}x$	Slope of line is $-\frac{1}{2}$.
$Y_4 = -x$	Slope of line is −1.
$Y_5 = -2x$	Slope of line is −2.
$Y_6 = -6x$	Slope of line is −6.

See Figure 14.

Figures 13 and 14 illustrate that the closer the line is to the vertical position, the greater the magnitude of the slope.

The next example illustrates how the slope of a line can be used to graph the line.

4 **EXAMPLE 8** **Graphing a Line When its Slope and a Point Are Given**

Draw a graph of the line that contains the point (3, 2) and has a slope of

(a) $\frac{3}{4}$ **(b)** $-\frac{4}{5}$

SOLUTION **(a)** Slope $= \frac{\text{rise}}{\text{run}}$. The fact that the slope is $\frac{3}{4}$ means that for every horizontal movement (run) of 4 units to the right, there will be a vertical movement (rise) of 3 units. If we start at the given point (3, 2) and move 4 units to the right and 3 units up, we reach the point (7, 5). By drawing the line through this point and the point (3, 2), we have the graph. See Figure 15(a).

(b) The fact that the slope is $-\frac{4}{5} = \frac{-4}{5}$ means that for every horizontal movement of 5 units to the right, there will be a corresponding vertical movement of -4 units (a downward movement). If we start at the given point (3, 2) and move 5 units to the right and then 4 units down, we arrive at the point $(8, -2)$. By drawing the line through these points, we have the graph. See Figure 15(b).

Alternatively, we can set $-\frac{4}{5} = \frac{4}{-5}$ so that for every horizontal movement of -5 units (a movement to the left), there will be a corresponding vertical movement of 4 units (upward). This approach brings us to the point $(-2, 6)$, which is also on the graph shown in Figure 15(b).

FIGURE 15

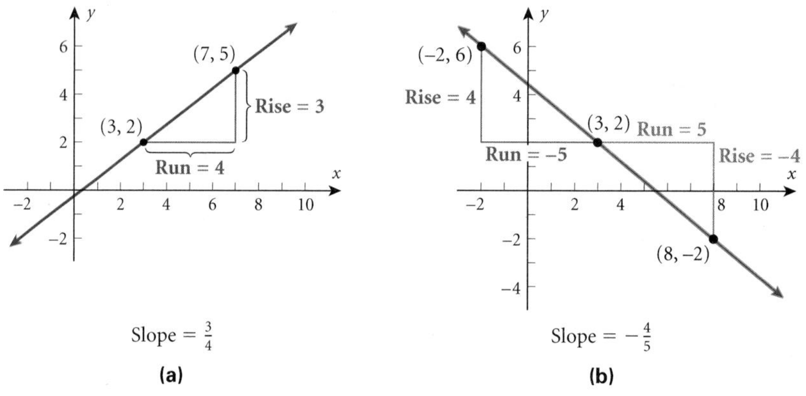

NOW WORK PROBLEM 25.

FIGURE 16

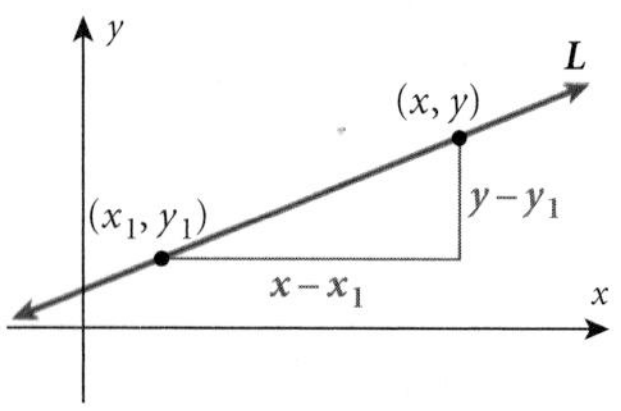

Other Forms of the Equation of a Line

Let L be a nonvertical line with slope m and containing the point (x_1, y_1). See Figure 16. Since any two distinct points on L can be used to compute slope, for any other point (x, y) on L, we have

$$m = \frac{y - y_1}{x - x_1} \quad \text{or} \quad y - y_1 = m(x - x_1)$$

Point–Slope Form of an Equation of a Line

An equation of a nonvertical line with slope m that contains the point (x_1, y_1) is

$$y - y_1 = m(x - x_1) \tag{3}$$

5 **EXAMPLE 9 Using the Point–Slope Form of a Line**

FIGURE 17

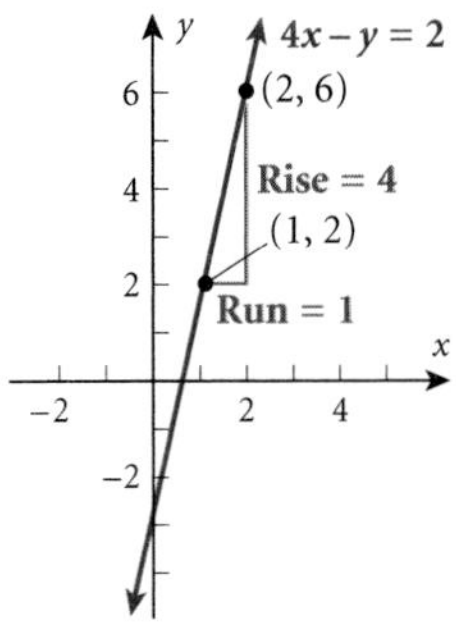

An equation of the line with slope 4 and containing the point (1, 2) can be found by using the point-slope form with $m = 4$, $x_1 = 1$, and $y_1 = 2$:

$$
\begin{aligned}
y - y_1 &= m(x - x_1) && \text{Point–slope form.}\\
y - 2 &= 4(x - 1) && m = 4, x_1 = 1, y_1 = 2.\\
y - 2 &= 4x - 4 &&\\
4x - y &= 2 && \text{General equation}
\end{aligned}
$$

See Figure 17.

NOW WORK PROBLEM 37.

6 **EXAMPLE 10 Finding the Equation of a Horizontal Line**

Find an equation of the horizontal line containing the point (3, 2).

FIGURE 18

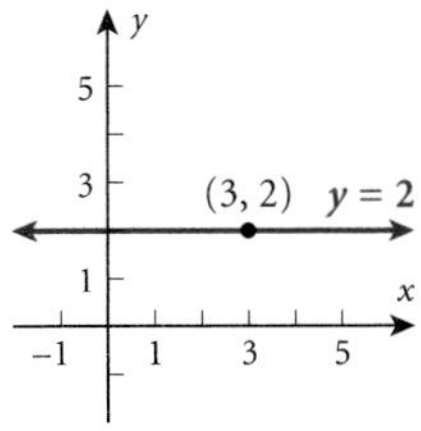

SOLUTION The slope of a horizontal line is 0. To get an equation, we use the point–slope form with $m = 0$, $x_1 = 3$, and $y_1 = 2$:

$$
\begin{aligned}
y - y_1 &= m(x - x_1) && \text{Point–slope form.}\\
y - 2 &= 0 \cdot (x - 3) && m = 0, x_1 = 3, y_1 = 2.\\
y - 2 &= 0 &&\\
y &= 2 &&
\end{aligned}
$$

See Figure 18 for the graph.

As suggested by Example 10, we have the following result:

Equation of a Horizontal Line

A horizontal line is given by an equation of the form

$$y = b$$

where $(0, b)$ is the y-intercept.

NOW WORK PROBLEM 9(b).

7 **EXAMPLE 11 Finding an Equation of a Line Given Two Points**

Find an equation of the line containing the points (2, 3) and (−4, 5). Graph the line.

SOLUTION Since two points are given, we first compute the slope of the line:

$$m = \frac{5-3}{-4-2} = \frac{2}{-6} = \frac{1}{-3} = -\frac{1}{3}$$

We use the point (2, 3) and the fact that the slope $m = -\frac{1}{3}$ to get the point–slope form of the equation of the line:

$$y - 3 = -\frac{1}{3}(x - 2)$$

See Figure 19 for the graph.

FIGURE 19

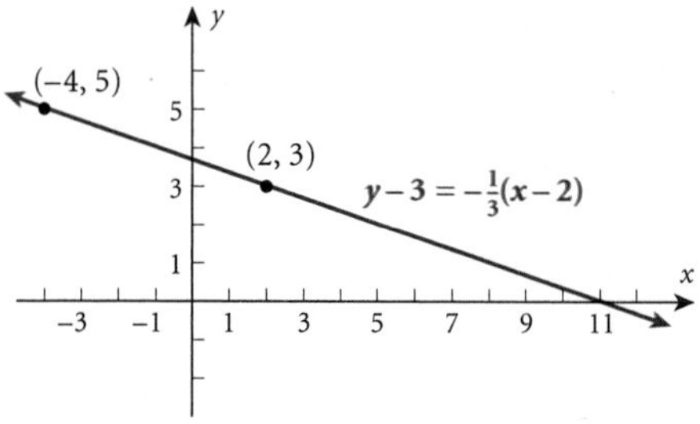

NOW WORK PROBLEM 41.

In the solution to Example 11 we could have used the point (−4, 5) instead of the point (2, 3). The equation that results, although it looks different, is equivalent to the equation we obtained in the example. (Try it for yourself.)

The general form of the equation of the line in Example 11 can be obtained by multiplying both sides of the point–slope equation by 3 and collecting terms:

$$\begin{aligned} y - 3 &= -\frac{1}{3}(x-2) && \text{Point–slope equation.} \\ 3(y-3) &= 3\left(-\frac{1}{3}\right)(x-2) && \text{Multiply by 3.} \\ 3y - 9 &= -1(x-2) && \text{Simplify} \\ 3y - 9 &= -x + 2 && \text{Simplify} \\ x + 3y &= 11 && \text{General equation} \end{aligned}$$

This is the general form of the equation of the line.

Use the slope–intercept form of a line 8

Another useful equation of a line is obtained when the slope m and y-intercept $(0, b)$ are known. In this case we know both the slope m of the line and a point $(0, b)$ on the line. Then we can use the point–slope form, Equation (3), to obtain the following equation:

$$\begin{aligned} y - y_1 &= m(x - x_1) && \text{Point-slope form} \\ y - b &= m(x - 0) && x_1 = 0, y_1 = b \\ y &= mx + b && \text{Simplify and solve for } y \end{aligned}$$

Slope–Intercept Form of an Equation of a Line

An equation of a line L with slope m and y-intercept $(0, b)$ is

$$\boxed{y = mx + b} \qquad (4)$$

SEEING THE CONCEPT: To see the role that the slope m plays in the equation $y = mx + b$, graph the following lines on the same square screen.

$$Y_1 = 2$$
$$Y_2 = x + 2$$
$$Y_3 = -x + 2$$
$$Y_4 = 3x + 2$$
$$Y_5 = -3x + 2$$

FIGURE 20

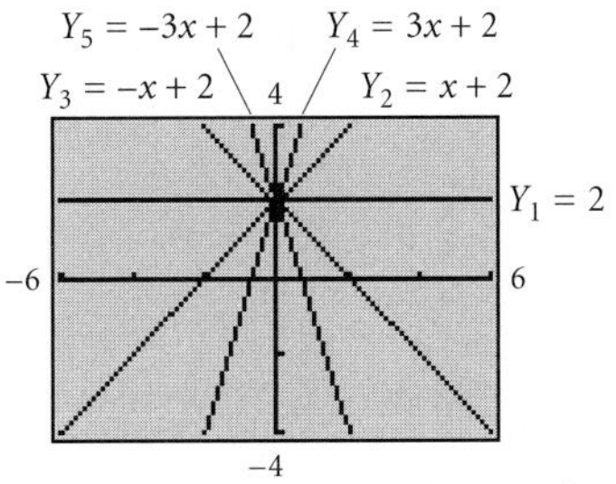

See Figure 20. What do you conclude about the lines $y = mx + 2$?

SEEING THE CONCEPT: To see the role of the y-intercept b in the equation $y = mx + b$, graph the following lines on the same square screen.

$$Y_1 = 2x$$
$$Y_2 = 2x + 1$$
$$Y_3 = 2x - 1$$
$$Y_4 = 2x + 4$$
$$Y_5 = 2x - 4$$

FIGURE 21

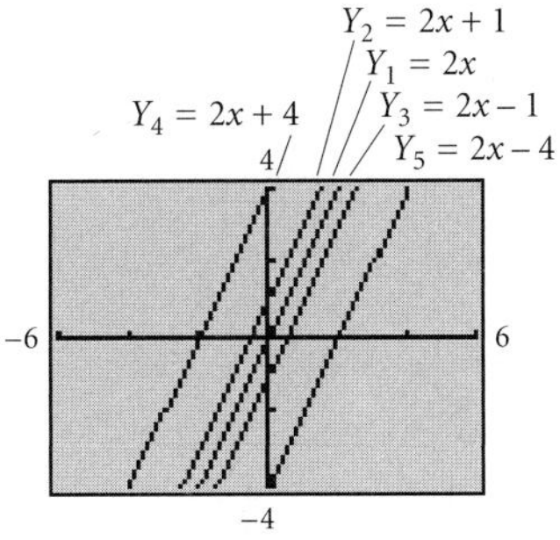

See Figure 21. What do you conclude about the lines $y = 2x + b$?

When an equation of a line is written in slope–intercept form, it is easy to find the slope m and y-intercept $(0, b)$ of the line. For example, suppose the equation of the line is

$$y = -2x + 3$$

Compare it to $y = mx + b$:

$$\begin{aligned} y &= -2x + 3 \\ &\quad\ \uparrow \quad\ \ \uparrow \\ y &= \ \ mx + b \end{aligned}$$

The slope of this line is -2 and its y-intercept is $(0, 3)$.

Let's look at another example.

EXAMPLE 12 Finding the Slope and y-Intercept of a Line

Find the slope m and y-intercept $(0, b)$ of the line $2x + 4y = 8$. Graph the line.

SOLUTION To obtain the slope and y-intercept, we transform the equation into its slope–intercept form. To do this, we need to solve for y:

$$\begin{aligned} 2x + 4y &= 8 \\ 4y &= -2x + 8 \\ y &= -\frac{1}{2}x + 2 \end{aligned}$$

The coefficient of x, $-\frac{1}{2}$, is the slope, and the y-intercept is $(0, 2)$.

We can graph the line in either of two ways:

1. Use the fact that the y-intercept is $(0, 2)$ and the slope is $-\frac{1}{2}$. Then, starting at the point $(0, 2)$, go to the right 2 units and then down 1 unit to the point $(2, 1)$. Plot these points and draw the line containing them. See Figure 22.
2. Locate the intercepts. The y-intercept is $(0, 2)$. To obtain the x-intercept, we let $y = 0$ and solve for x. When $y = 0$, we have

$$\begin{aligned} 2x + 4 \cdot 0 &= 8 \\ 2x &= 8 \\ x &= 4 \end{aligned}$$

The intercepts are $(4, 0)$ and $(0, 2)$. Plot these points and draw the line containing them. See Figure 22.

FIGURE 22

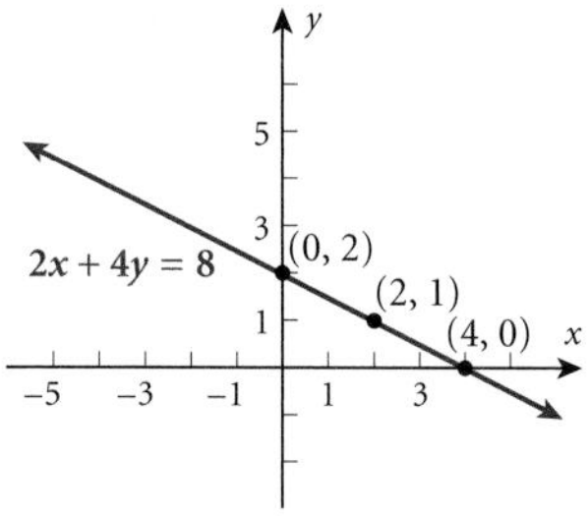

[*Note*: The second method, locating the intercepts, only produces one point when the line passes through the origin. In this case some other point on the line must be found in order to graph the line. Refer back to Example 3.]

NOW WORK PROBLEM 59.

EXAMPLE 13 Daily Cost of Production

A certain factory has daily fixed overhead expenses of \$2000, while each item produced costs \$100. Find an equation that relates the daily cost C to the number x of items produced each day.

SOLUTION The fixed overhead expense of \$2000 represents the fixed cost, the cost incurred no matter how many items are produced. Since each item produced costs \$100, the variable cost of producing x items is $100x$. Then the total daily cost C of production is

$$C = 100x + 2000$$

FIGURE 23

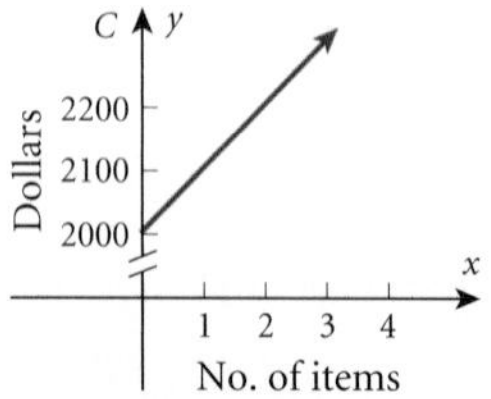

The graph of this equation is given by the line in Figure 23. Notice that the fixed cost \$2000 is represented by the y-intercept, while the \$100 cost of producing each item is the slope. Also notice that, for convenience, a different scale is used on each axis.

SUMMARY The graph of a linear equation, $Ax + By = C$, where A and B are not both zero, is a line. In this form it is referred to as the general equation of a line.

1. Given the general equation of a line, information can be found about the line:
   (a) Place the equation in slope–intercept form $y = mx + b$ to find the slope m and y-intercept $(0, b)$.
   (b) Let $x = 0$ and solve for y to find the y-intercept.
   (c) Let $y = 0$ and solve for x to find the x-intercept.
2. Given information about a line, an equation of the line can be found. The form of the equation to use depends on the given information. See the table below.

Given	Use	Equation
Point (x_1, y_1), slope m	Point–slope form	$y - y_1 = m(x - x_1)$
Two points $(x_1, y_1), (x_2, y_2)$	If $x_1 = x_2$, the line is vertical If $x_1 \neq x_2$, find the slope m: $m = \dfrac{y_2 - y_1}{x_2 - x_1}$ Then use the point–slope form	$x = x_1$ $y - y_1 = m(x - x_1)$
Slope m, y-intercept $(0, b)$	Slope–intercept form	$y = mx + b$

EXERCISE 1.1 Answers to Odd-Numbered Problems Begin on Page AN-1.

1. Give the coordinates of each point in the following figure. Assume each coordinate is an integer.

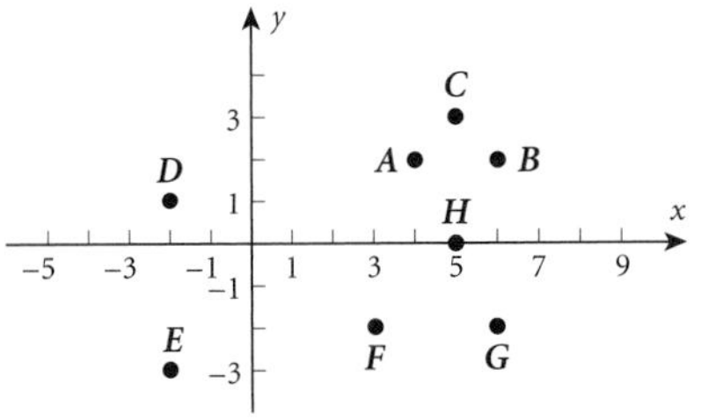

2. Plot each point in the xy-plane. Tell in which quadrant or on what coordinate axis each point lies.
   (a) $A = (-3, 2)$ (b) $B = (6, 0)$
   (c) $C = (-2, -2)$ (d) $D = (6, 5)$
   (e) $E = (0, -3)$ (f) $F = (6, -3)$

3. Plot the points $(2, 0)$, $(2, -3)$, $(2, 4)$, $(2, 1)$, and $(2, -1)$. Describe the collection of all points of the form $(2, y)$, where y is a real number.

4. Plot the points $(0, 3)$, $(1, 3)$, $(-2, 3)$, $(5, 3)$, and $(-4, 3)$. Describe the collection of all points of the form $(x, 3)$, where x is a real number.

In Problems 5–8, use the given equations to fill in the missing values in each table. Use these points to graph each equation.

5. $y = 2x + 4$

x	0		2	-2	4	-4
y		0				

6. $y = -3x + 6$

x	0		2	-2	4	-4
y		0				

7. $2x - y = 6$

x	0		2	-2	4	-4
y		0				

8. $x + 2y = 8$

x	0		2	-2	4	-4
y		0				

In Problems 9–12: (a) find the equation of the vertical line containing the given point; (b) find the equation of the horizontal line containing the given point.

9. $(2, -3)$ **10.** $(5, 4)$ **11.** $(-4, 1)$ **12.** $(-6, -3)$

In Problems 13–16, find the slope of the line. Give an interpretation of the slope.

13.
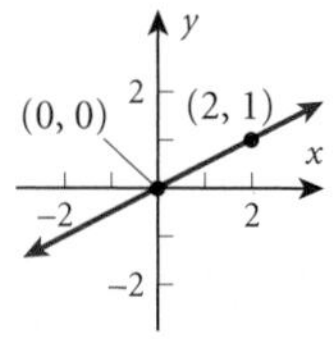

14.
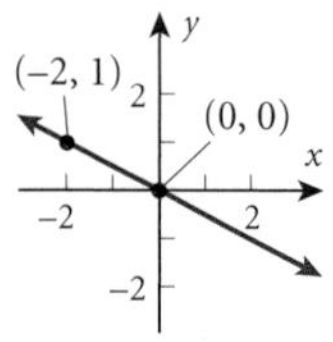

15.
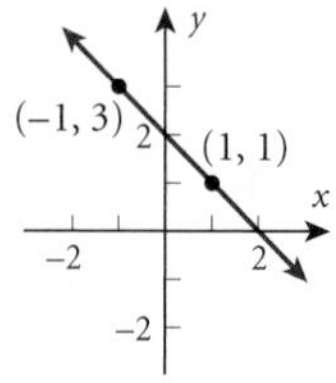

16.
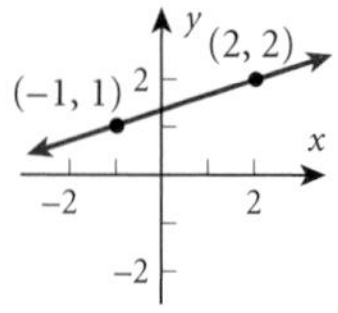

In Problems 17–24, plot each pair of points and find the slope of the line containing them. Interpret the slope and graph the line.

17. (2,3); (1,0) **18.** (1,2); (3,4) **19.** $(-2, 3)$; $(2, 1)$ **20.** $(-1, 1)$; $(2, 3)$

21. $(-3, -1)$; $(2, -1)$ **22.** $(4, 2)$; $(-5, 2)$ **23.** $(-1, 2)$; $(-1, -2)$ **24.** $(2, 0)$; $(2, 2)$

In Problems 25–32, graph the line containing the point P and having slope m.

25. $P = (1, 2)$; $m = 2$ **26.** $P = (2, 1)$; $m = 3$ **27.** $P = (2, 4)$; $m = -\frac{3}{4}$ **28.** $P = (1, 3)$; $m = -\frac{2}{3}$

29. $P = (-1, 3)$; $m = 0$ **30.** $P = (2, -4)$; $m = 0$ **31.** $P = (0, 3)$; slope undefined **32.** $P = (-2, 0)$; slope undefined

In Problems 33–36, write the equation of each line in the form $Ax + By = C$.

33.
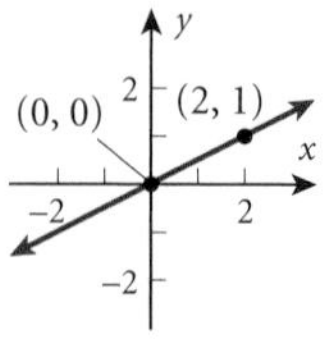

34.
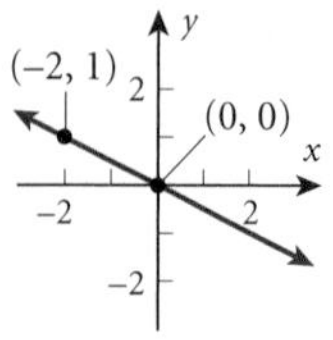

35.
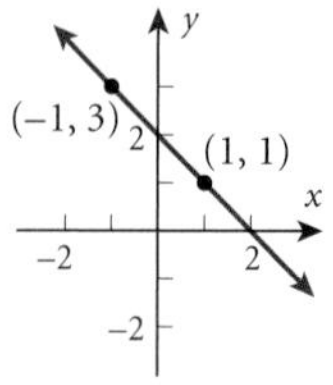

36.
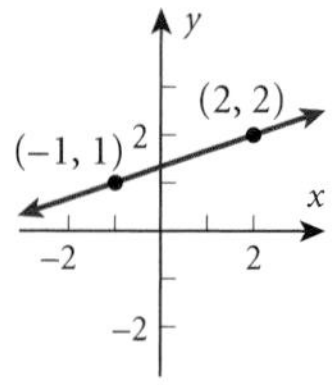

In Problems 37–54, write the equation of each line in the form $Ax + By = C$.

37. Slope $= 2$; containing the point $(-4, 1)$

38. Slope $= 3$; containing the point $(-3, 4)$

39. Slope $= -\frac{2}{3}$; containing the point $(1, -1)$

40. Slope $= \frac{1}{2}$; containing the point $(3, 1)$

41. Containing the points $(1, 3)$ and $(-1, 2)$

42. Containing the points $(-3, 4)$ and $(2, 5)$

43. Slope $= -2$; y-intercept $= (0, 3)$

44. Slope $= -3$; y-intercept $= (0, -2)$

45. Slope $= 3$; x-intercept $= (-4, 0)$

46. Slope $= -4$; x-intercept $= (2, 0)$

47. Slope $= \frac{4}{5}$; containing the point $(0, 0)$

48. Slope $= \frac{7}{3}$; containing the point $(0, 0)$

49. x-intercept $= (2, 0)$; y-intercept $= (0, -1)$

50. x-intercept $= (-4, 0)$; y-intercept $= (0, 4)$

51. Slope undefined; containing the point $(1, 4)$

52. Slope undefined; containing the point $(2, 1)$

53. Slope $= 0$; containing the point $(1, 4)$

54. Slope $= 0$; containing the point $(2, 1)$

In Problems 55–70, find the slope and y-intercept of each line. Graph the line.

55. $y = 2x + 3$

56. $y = -3x + 4$

57. $\frac{1}{2}y = x - 1$

58. $\frac{1}{3}x + y = 2$

59. $2x - 3y = 6$

60. $3x + 2y = 6$

61. $x + y = 1$

62. $x - y = 2$

63. $x = -4$

64. $y = -1$

65. $y = 5$

66. $x = 2$

67. $y - x = 0$

68. $x + y = 0$

69. $2y - 3x = 0$

70. $3x + 2y = 0$

---

71. Find the equation of the horizontal line containing the point $(-1, -3)$.

72. Find the equation of the vertical line containing the point $(-2, 5)$.

73. Cost of Operating a Car According to the American Automobile Association (AAA), the average cost of operating a standard-sized car, including gasoline, oil, tires, and maintenance increased to \$0.122 per mile in 2000. Write an equation that relates the average cost C of operating a standard-sized car and the number x of miles it is driven.

Source: *AAA Traveler Magazine*

74. Cost of Renting a Truck The cost of renting a truck is \$280 per week plus a charge of \$0.30 per mile driven. Write an equation that relates the cost C for a weekly rental in which the truck is driven x miles.

75. Electricity Rates in Illinois Commonwealth Edison Company supplies electricity to residential customers for a monthly customer charge of \$7.58 plus 8.275 cents per kilowatt-hour for up to 400 kilowatt-hours.

(a) Write an equation that relates the monthly charge C, in dollars, to the number x of kilowatt-hours used in a month, $0 \le x \le 400$.
(b) Graph this equation.
(c) What is the monthly charge for using 100 kilowatt-hours?
(d) What is the monthly charge for using 300 kilowatt-hours?
(e) Interpret the slope of the line.

Source: Commonwealth Edison Company, December, 2002.

76. Electricity Rates in Florida Florida Power & Light Company supplies electricity to residential customers for a monthly customer charge of \$5.25 plus 6.787 cents per kilowatt-hour for up to 750 kilowatt-hours.

(a) Write an equation that relates the monthly charge C, in dollars, to the number x of kilowatt-hours used in a month, $0 \le x \le 750$.
(b) Graph this equation.
(c) What is the monthly charge for using 200 kilowatt-hours?
(d) What is the monthly charge for using 500 kilowatt-hours?
(e) Interpret the slope of the line.

Source: Florida Power & Light Company, January 2003.

77. Weight–Height Relation in the U.S. Army Assume the recommended weight w of females aged 17-20 years in the U. S. Army is linearly related to their height h. If an Army female who is 67 inches tall should weigh 139 pounds and an Army female who is 70 inches tall should weigh 151 pounds, find an equation that expresses weight in terms of height.

Source: http://www.nutribase.com/nutrition-fwchartf.htm.

78. Wages of a Car Salesperson Dan receives \$375 per week for selling new and used cars at a car dealership in Omaha, Nebraska. In addition, he receives 5% of the profit on any sales he generates. Write an equation that relates Dan's weekly salary S when he has sales that generate a profit of x dollars.

79. Cost of Sunday Home Delivery The cost to the *Chicago Tribune* for Sunday home delivery is approximately \$0.53 per newspaper with fixed costs of \$1,070,000. Write an equation that relates the cost C and the number x of copies delivered.

Source: Chicago Tribune, 2002.

80. Disease Propagation Research indicates that in a controlled environment, the number of diseased mice will increase linearly each day after one of the mice in the cage is infected with a particular type of disease-causing germ. There were 8 diseased mice 4 days after the first exposure and 14 diseased mice after 6 days. Write an equation that will give the number of diseased mice after any given number of days. If there were 40 mice in the cage, how long will it take until they are all infected?

81. Temperature Conversion The relationship between Celsius (°C) and Fahrenheit (°F) degrees for measuring temperature is linear. Find an equation relating °C and °F if 0°C corresponds to 32°F and 100°C corresponds to 212°F. Use the equation to find the Celsius measure of 68°F.

82. Temperature Conversion The Kelvin (K) scale for measuring temperature is obtained by adding 273 to the Celsius temperature.

(a) Write an equation relating K and °C.
(b) Write an equation relating K and °F (see Problem 81).

83. Water Preservation At Harlan County Dam in Nebraska, the U.S. Bureau of Reclamation reports that the storage content of the reservoir decreased from 162,400 acre-feet (52.9 billion gallons of water) on November 8, 2002 to 161,200 acre-feet (52.5 billion gallons of water) on December 8, 2002. Suppose that the rate of loss of water remains constant.

(a) Write an equation that relates the amount A of water, in billions of gallons, to the time t, in days. Use $t = 1$ for November 1, $t = 2$ for November 2, and so on.
(b) How much water was in the reservoir on November 20 ($t = 20$)?
(c) Interpret the slope.
(d) How much water will be in the reservoir on December 31, 2002 ($t = 61$)?
(e) When will the reservoir be empty?
(f) Comment on your answer to part (e).

Source: U.S. Bureau of Reclamation.

84. Product Promotion A cereal company finds that the number of people who will buy one of its products the first month it is introduced is linearly related to the amount of money it spends on advertising. If it spends \$400,000 on advertising, 100,000 boxes of cereal will be sold, and if it spends \$600,000, 140,000 boxes will be sold.

(a) Write an equation describing the relation between the amount spent on advertising and the number of boxes sold.
(b) How much advertising is needed to sell 200,000 boxes of cereal?
(c) Interpret the slope.

In Problems 85–92, use a graphing utility to graph each linear equation. Be sure to use a viewing rectangle that shows the intercepts. Then locate each intercept rounded to two decimal places.

85. $1.2x + 0.8y = 2$

86. $-1.3x + 2.7y = 8$

87. $21x - 15y = 53$

88. $5x - 3y = 82$

89. $\frac{4}{17}x + \frac{6}{23}y = \frac{2}{3}$

90. $\frac{9}{14}x - \frac{3}{8}y = \frac{2}{7}$

91. $\pi x - \sqrt{3}\,y = \sqrt{6}$

92. $x + \pi y = \sqrt{15}$

In Problems 93–96, match each graph with the correct equation:

(a) $y = x$; (b) $y = 2x$; (c) $y = \frac{x}{2}$; (d) $y = 4x$.

93.

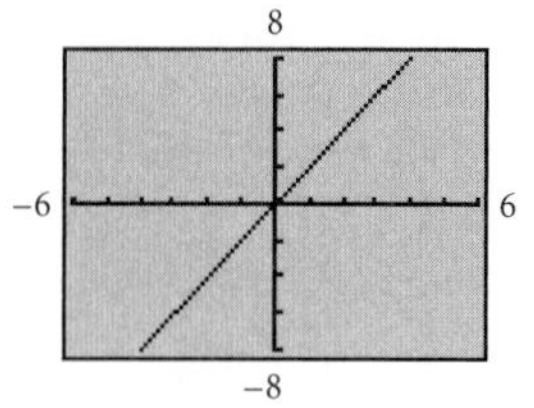

94.

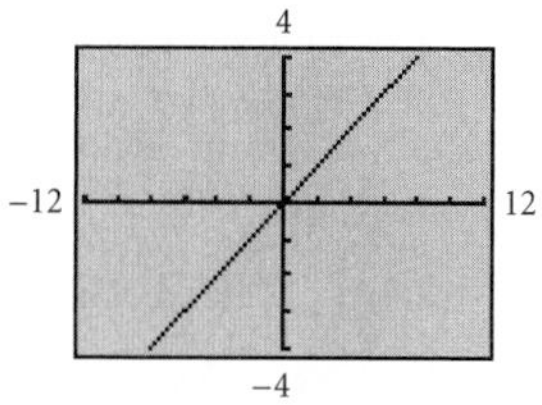

95.

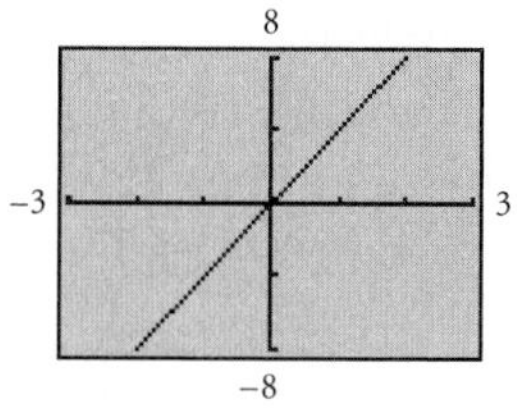

96.

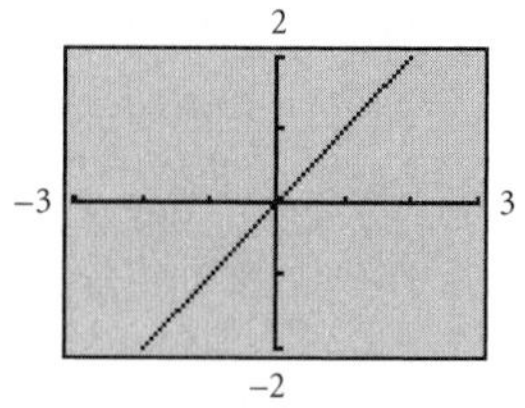

In Problems 97–100, write an equation of each line. Express your answer using either the general form or the slope-intercept form of the equation of a line, whichever you prefer.

97.

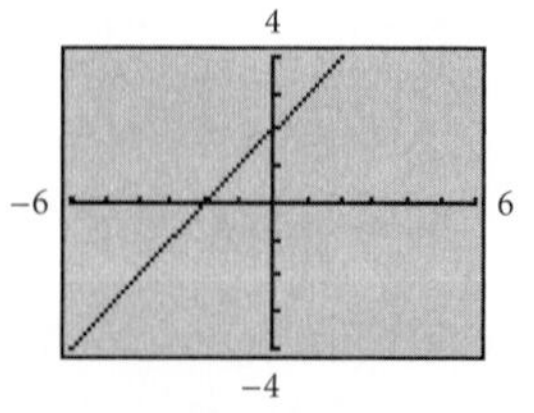

98.

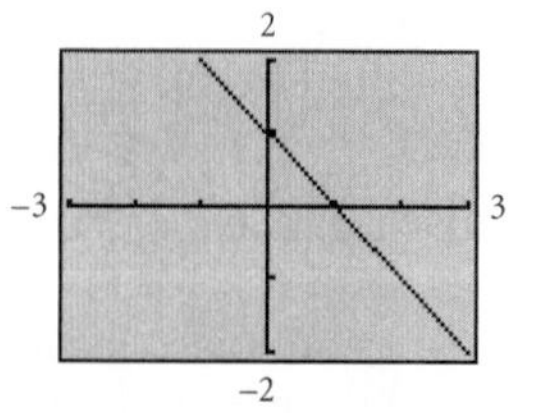

99.

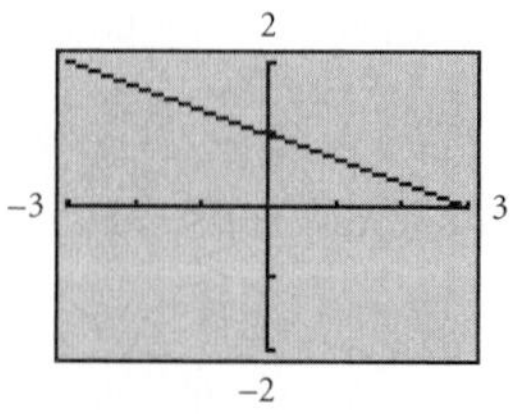

100.

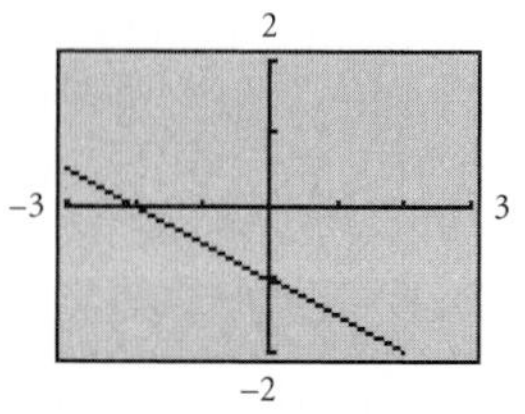

101. Which of the following equations might have the graph shown. (More than one answer is possible.)

(a) $2x + 3y = 6$
(b) $-2x + 3y = 6$
(c) $3x - 4y = -12$
(d) $x - y = 1$
(e) $x - y = -1$
(f) $y = 3x - 5$
(g) $y = 2x + 3$
(h) $y = -3x + 3$

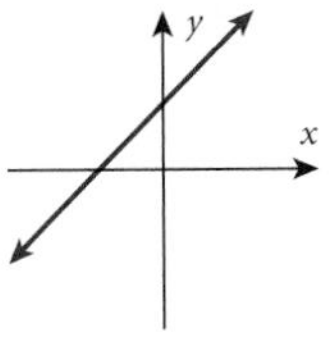

102. Which of the following equations might have the graph shown. (More than one answer is possible.)

(a) $2x + 3y = 6$
(b) $2x - 3y = 6$
(c) $3x + 4y = 12$
(d) $x - y = 1$
(e) $x - y = -1$
(f) $y = -2x + 1$
(g) $y = -\frac{1}{2}x + 10$
(h) $y = x + 4$

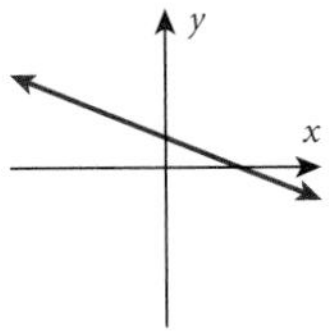

103. Write the general equation of the x-axis.

104. Write the general equation of the y-axis.

105. Which form of the equation of a line do you prefer to use? Justify your position with an example that shows that your choice is better than another. Have reasons.

106. Can every line be written in slope–intercept form? Explain.

107. Does every line have two distinct intercepts? Explain. Are there lines that have no intercepts? Explain.

108. What can you say about two lines that have equal slopes and equal y-intercepts?

109. What can you say about two lines with the same x-intercept and the same y-intercept? Assume that the x-intercept is not $(0, 0)$.

110. If two lines have the same slope, but different x-intercepts, can they have the same y-intercept?

111. If two lines have the same y-intercept, but different slopes, can they have the same x-intercept? What is the only way that this can happen?

112. The accepted symbol used to denote the slope of a line is the letter m. Investigate the origin of this symbolism. Begin by consulting a French dictionary and looking up the French word *monter*. Write a brief essay on your findings.

1.2 Pairs of Lines

OBJECTIVES
1 Show that two lines are coincident
2 Show that two lines are parallel
3 Show that two lines intersect
4 Find the point of intersection of two intersecting lines
5 Show that two lines are perpendicular

Consider two lines L and M in the same plane. Then exactly one of the following three circumstances must hold:

1. They have no points in common, in which case they are **parallel.**
2. They have one point in common, in which case they **intersect.**
3. They have two points in common, in which case they are **coincident** or **identical,** and all the points on L are the same as the points on M.

Figure 24 illustrates the three circumstances.

FIGURE 24

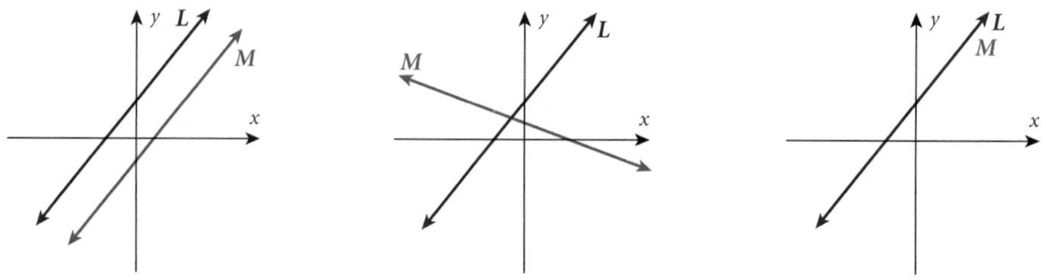

Parallel Lines: No points in common

(a)

Intersecting Lines: One point in common

(b)

Coincident Lines: All points on L are the same as the points on M.

(c)

Figure 25 illustrates coincident lines that are vertical and nonvertical.

FIGURE 25

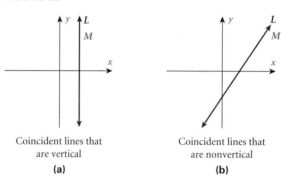

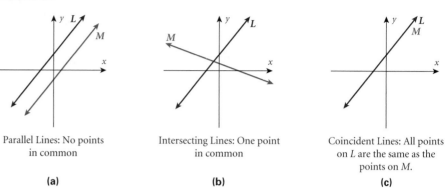

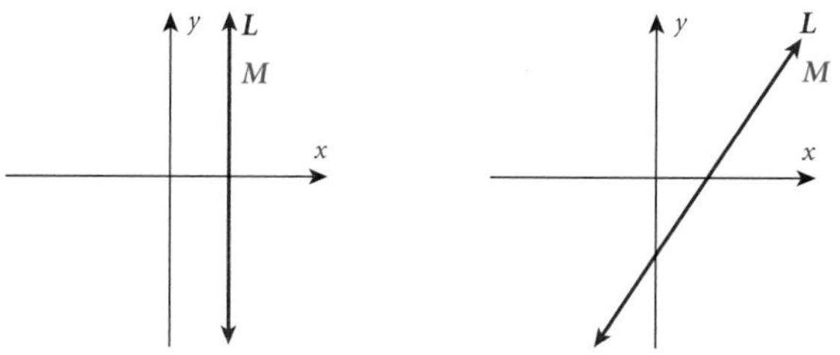

Coincident lines that are vertical

(a)

Coincident lines that are nonvertical

(b)

We are led to the following result.

Coincident Lines

Coincident lines that are vertical have undefined slope and the same x-intercept. Coincident lines that are nonvertical have the same slope and the same intercepts.

Show that two lines are coincident 1

To show that two nonvertical lines are coincident only requires that you show they have the same slope and the same y-intercept. Do you see why?

EXAMPLE 1 Showing that Two Lines Are Coincident

Show that the lines given by the equations below are coincident.

$$L:\ 2x - y = 5 \qquad M:\ -4x + 2y = -10$$

SOLUTION We put each equation into slope–intercept form:

$$L:\ 2x - y = 5 \qquad\qquad M:\ -4x + 2y = -10$$
$$-y = -2x + 5 \qquad\qquad 2y = 4x - 10$$
$$y = 2x - 5 \qquad\qquad y = 2x - 5$$

FIGURE 26

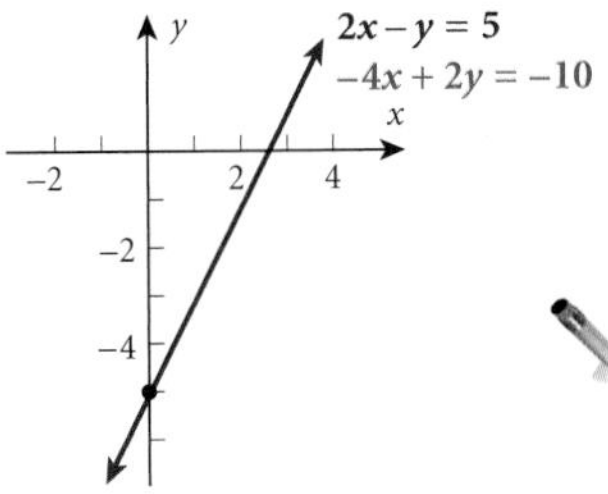

The lines L and M have the same slope 2 and the same y-intercept $(0, -5)$ so they are coincident. See Figure 26.

NOW WORK PROBLEM 5.

Next, consider the parallel lines in Figure 27.

FIGURE 27

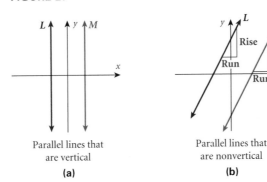

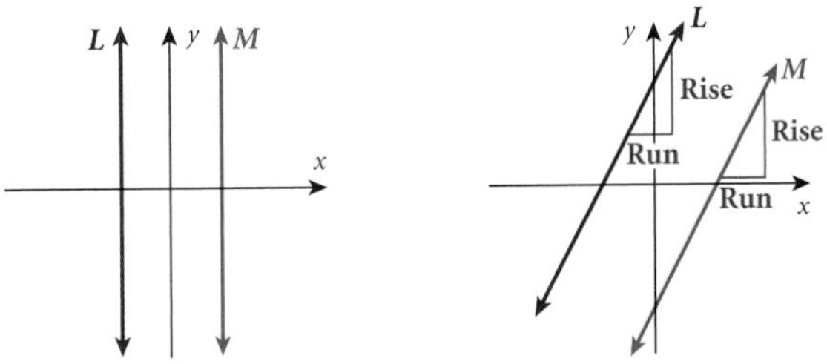

We see in Figure 27(a) that the two vertical parallel lines have different x-intercepts. For the two nonvertical parallel lines in Figure 27(b), equal runs result in equal rises. As a result, nonvertical parallel lines have the same slope. Since they also have no points in common, they will have different x- and y-intercepts.

Parallel Lines

Parallel lines that are vertical have undefined slope and different x-intercepts. Parallel lines that are nonvertical have the same slope and different intercepts.

Show that two lines are parallel 2

To show that two nonvertical lines are parallel only requires that you show they have the same slope and different y-intercepts. Do you see why?

EXAMPLE 2 Showing that Two Lines Are Parallel

Show that the lines given by the equations below are parallel.

$$L:\ 2x + 3y = 6 \qquad\qquad M:\ 4x + 6y = 0$$

SOLUTION To see if these lines have equal slopes, we put each equation into slope–intercept form:

$$L:\ 2x + 3y = 6 \qquad\qquad M:\ 4x + 6y = 0$$
$$3y = -2x + 6 \qquad\qquad 6y = -4x$$
$$y = -\frac{2}{3}x + 2 \qquad\qquad y = -\frac{2}{3}x$$
$$\text{Slope} = -\frac{2}{3} \qquad\qquad \text{Slope} = -\frac{2}{3}$$
$$y\text{-intercept} = (0, 2) \qquad\qquad y\text{-intercept} = (0, 0)$$

FIGURE 28

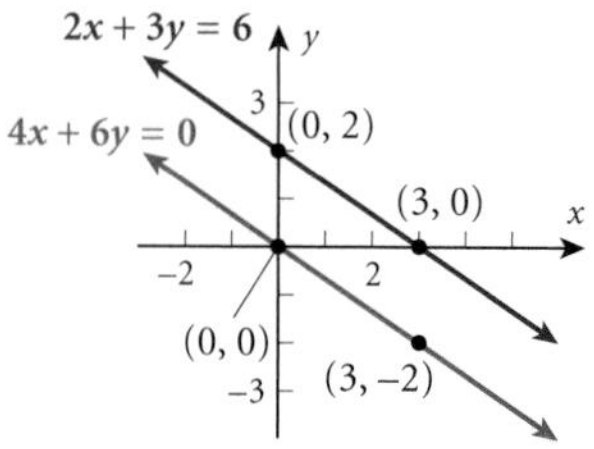

Since each has slope $-\frac{2}{3}$ but different y-intercepts, the lines are parallel. See Figure 28.

NOW WORK PROBLEM 1.

If two lines L and M have exactly one point in common, the common point is called the **point of intersection.** The slopes of intersecting lines are unequal. Do you see why?

Intersecting Lines

Intersecting lines have different slopes.

3 EXAMPLE 3 Showing that Two Lines Intersect

Show that the lines given by the equations below intersect.

$$L:\ 2x - y = 5 \qquad M:\ x + y = 4$$

SOLUTION We put each equation into slope–intercept form.

$$L:\ 2x - y = 5 \qquad\qquad M:\ x + y = 4$$
$$-y = -2x + 5 \qquad\qquad y = -x + 4$$
$$y = 2x - 5$$

The slope of L is 2 and the slope of M is -1, so the lines intersect.

NOW WORK PROBLEM 3.

4 EXAMPLE 4 Finding the Point of Intersection

Find the point of intersection of the two lines

$$L:\ 2x - y = 5 \qquad M:\ x + y = 4$$

SOLUTION The slope–intercept form of each line, as found in Example 3, is

$$L:\ y = 2x - 5 \qquad M:\ y = -x + 4$$

If (x_0, y_0) denotes the point of intersection, then (x_0, y_0) is a point on both L and M. As a result, we must have

$$y_0 = 2x_0 - 5 \qquad \text{and} \qquad y_0 = -x_0 + 4$$

We set these equal and solve for x_0.

FIGURE 29

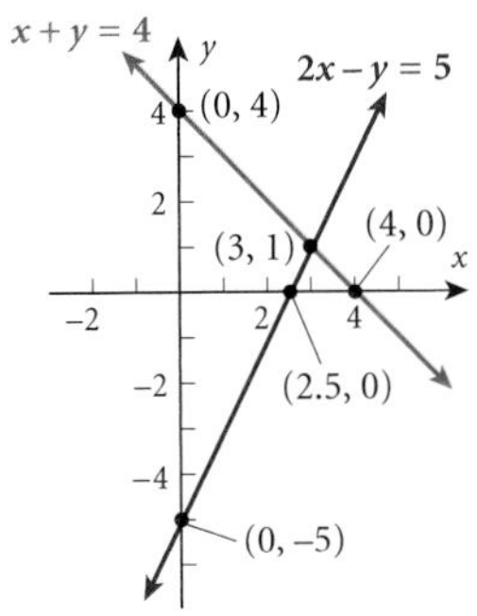

$$\begin{aligned} 2x_0 - 5 &= -x_0 + 4 \\ 3x_0 &= 9 \\ x_0 &= 3 \end{aligned}$$

Substituting $x_0 = 3$ in $y_0 = 2x_0 - 5$ (or in $y_0 = -x_0 + 4$), we find

$$y_0 = 2x_0 - 5 = 2(3) - 5 = 1$$

The point of intersection of L and M is (3, 1). See Figure 29.

Check: We verify that the point (3, 1) is on both L and M.

$$L\colon\ 2x - y = 2(3) - 1 = 6 - 1 = 5 \qquad M\colon\ x + y = 3 + 1 = 4$$

NOW WORK PROBLEM 13.

Perpendicular Lines

When two lines intersect and form a right angle, they are said to be **perpendicular.** For example, a vertical line and a horizontal line are perpendicular.

> **Perpendicular Lines**
>
> Two nonvertical lines L and M with slopes m_1 and m_2, respectively, are perpendicular if and only if $m_1m_2 = -1$, that is, if and only if the product of their slopes is -1.

5 **EXAMPLE 5** **Showing that Two Lines Are Perpendicular**

Show that the lines given by the equations below are perpendicular.

$$L\colon\ x - 2y = 6 \qquad M\colon\ 2x + y = 1$$

SOLUTION To see if these lines are perpendicular, find the slope of each:

$$\begin{aligned} L\colon\ x - 2y &= 6 \qquad & M\colon\ 2x + y &= 1 \\ -2y &= -x + 6 & y &= -2x + 1 \\ y &= \frac{1}{2}x - 3 & & \end{aligned}$$

$$\text{Slope} = m_1 = \frac{1}{2} \qquad \text{Slope} = m_2 = -2$$

Since $m_1m_2 = \frac{1}{2} \cdot (-2) = -1$, the lines are perpendicular.

NOW WORK PROBLEM 25.

EXAMPLE 6 **Finding the Equation of Two Lines: One Parallel to and the Other Perpendicular to a Given Line**

Given the line $x - 4y = 8$, find an equation for the line that contains the point (2, 1) and is

(a) Parallel to the given line **(b)** Perpendicular to the given line

SOLUTION First find the slope of the line $x - 4y = 8$ by putting it in slope–intercept form $y = mx + b$:

$$x - 4y = 8$$
$$-4y = -x + 8$$
$$y = \frac{1}{4}x - 2 \qquad y = mx + b;\ m = \frac{1}{4};\ b = -2$$

The slope of the line is $\frac{1}{4}$.

(a) We seek a line parallel to the given line that contains the point (2, 1). The slope of this line must be $\frac{1}{4}$. (Do you see why?) Using the point–slope form of the equation of a line, we have

$$y - y_1 = m(x - x_1)$$
$$y - 1 = \frac{1}{4}(x - 2) \qquad m = \frac{1}{4}, x_1 = 2, y_1 = 1$$
$$y = \frac{1}{4}x + \frac{1}{2}$$

(b) We seek a line perpendicular to the given line, whose slope is $\frac{1}{4}$. The slope m of this line obeys

$$m \cdot \frac{1}{4} = -1 \qquad \text{Product of the slopes is } -1$$
$$m = -4 \qquad \text{Solve for } m.$$

Since the line we seek contains the point (2, 1), we have

$$y - y_1 = m(x - x_1)$$
$$y - 1 = -4(x - 2) \qquad m = -4;\ x_1 = 2, y_1 = 1$$
$$y = -4x + 9$$

Figure 30 illustrates these solutions.

FIGURE 30

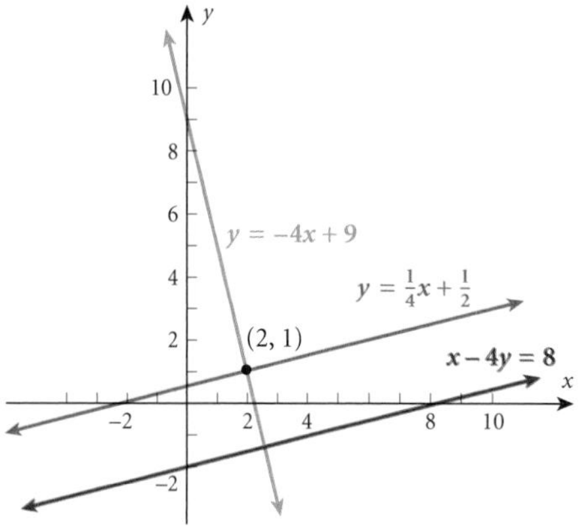

NOW WORK PROBLEMS 35 AND 41.

SUMMARY

Pair of Lines	*Conclusion: The lines are*
Both vertical	(a) Coincident, if they have the same x-intercept (b) Parallel, if they have different x-intercepts

One vertical, one nonvertical	(a) Intersecting (b) Perpendicular, if the nonvertical line is horizontal
Neither vertical	Write the equation of each line in slope–intercept form: $$y = m_1x + b_1, \quad y = m_2x + b_2$$ (a) Coincident, if $m_1 = m_2$, $b_1 = b_2$ (b) Parallel, if $m_1 = m_2$, $b_1 \neq b_2$ (c) Intersecting, if $m_1 \neq m_2$ (d) Perpendicular, if $m_1m_2 = -1$

EXERCISE 1.2 Answers to Odd-Numbered Problems Begin on Page AN-3.

In Problems 1–12, determine whether the given pairs of lines are parallel, coincident, or intersecting.

1. L: $x + y = 10$; M: $3x + 3y = 6$
2. L: $x - y = 5$; M: $-2x + 2y = 8$
3. L: $2x + y = 4$; M: $2x - y = 8$
4. L: $2x + y = 8$; M: $2x - y = -4$
5. L: $-x + y = 2$; M: $2x - 2y = -4$
6. L: $x + y = -4$; M: $3x + 3y = -12$
7. L: $2x - 3y = -8$; M: $6x - 9y = -2$
8. L: $4x - 2y = -7$; M: $-2x + y = -2$
9. L: $3x - 4y = 1$; M: $x - 2y = -4$
10. L: $4x + 3y = 2$; M: $2x - y = -1$
11. L: $x = 3$; M: $y = -2$
12. L: $x = 4$; M: $x = -2$

In Problems 13–24, the given pairs of lines intersect. Find the point of intersection. Graph each pair of lines.

13. L: $x + y = 5$; M: $3x - y = 7$
14. L: $2x + y = 7$; M: $x - y = -4$
15. L: $x - y = 2$; M: $2x + y = 7$
16. L: $2x - y = -1$; M: $x + y = 4$
17. L: $4x + 2y = 4$; M: $4x - 2y = 4$
18. L: $4x - 2y = 8$; M: $6x + 3y = 0$
19. L: $3x - 4y = 2$; M: $x + 2y = 4$
20. L: $4x + 3y = 2$; M: $2x - y = 1$
21. L: $3x - 2y = -5$; M: $3x + y = -2$
22. L: $4x + y = 6$; M: $4x - 2y = 0$
23. L: $x = 4$; M: $y = -2$
24. L: $x = 0$; M: $y = 0$

In Problems 25–30, show that the lines are perpendicular.

25. $-x + 3y = 2$; $6x + 2y = 5$
26. $2x + 3y = 4$; $9x - 6y = 1$
27. $x + 2y = 7$; $-2x + y = 15$
28. $4x + 12y = 3$; $15x - 5y = -1$
29. $3x + 12y = 2$; $4x - y = -2$
30. $20x - 2y = -7$; $x + 10y = 8$

In Problems 31–34, find an equation for the line L. Express the answer using the general form or the slope–intercept form, whichever you prefer.

31.
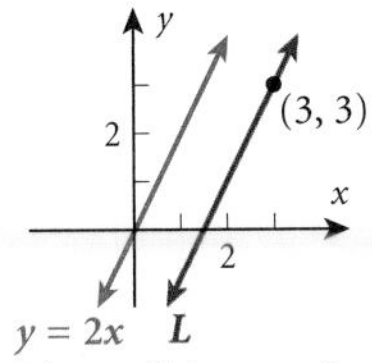

L is parallel to $y = 2x$

32.
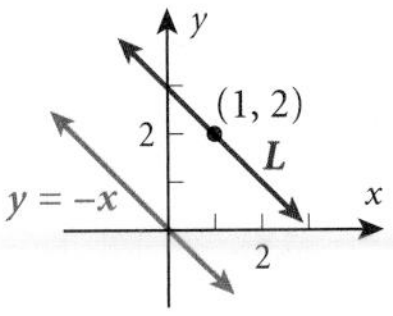

L is parallel to $y = -x$

33.
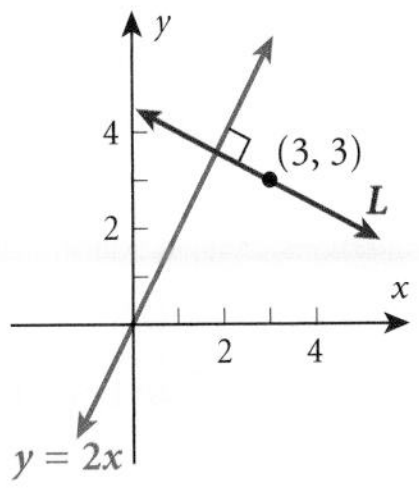

L is perpendicular to $y = 2x$

34.
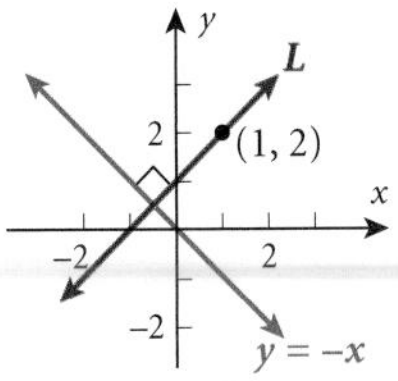

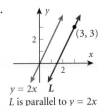

L is perpendicular to $y = -x$

In Problems 35–44, find an equation for the line with the given properties. Express the answer using the general form or the slope–intercept form, whichever you prefer.

35. Parallel to the line $y = 4x$; containing the point $(-1, 2)$

36. Parallel to the line $y = -3x$; containing the point $(-1, 2)$

37. Parallel to the line $2x - y = -2$; containing the point $(0, 0)$

38. Parallel to the line $x - 2y = -5$; containing the point $(0, 0)$

39. Parallel to the line $x = 3$; containing the point $(4, 2)$

40. Parallel to the line $y = 3$; containing the point $(4, 2)$

41. Perpendicular to the line $y = 2x - 5$; containing the point $(-1, -2)$

42. Perpendicular to the line $6x - 2y = -5$; containing the point $(-1, -2)$

43. Perpendicular to the line $y = 2x - 5$; containing the point $\left(-\frac{1}{3}, \frac{4}{5}\right)$

44. Perpendicular to the line $y = 3x - 15$; containing the point $\left(-\frac{2}{3}, \frac{3}{5}\right)$

45. Find t so that $tx - 4y = -3$ is perpendicular to the line $2x + 2y = 5$.

46. Find t so that $(1, 2)$ is a point on the line $tx - 3y = -4$.

47. Find the equation of the line containing the point $(-2, -5)$ and perpendicular to the line containing the points $(-2, 9)$ and $(3, -10)$.

48. Find the equation of the line containing the point $(-2, -5)$ and perpendicular to the line containing the points $(-4, 5)$ and $(2, -1)$.

49. The figure below shows the graph of two parallel lines. Which of the following pairs of equations might have such a graph?

(a) $x - 2y = 3$
$x + 2y = 7$

(b) $x + y = 2$
$x + y = -1$

(c) $x - y = -2$
$x - y = 1$

(d) $x - y = -2$
$2x - 2y = -4$

(e) $x + 2y = 2$
$x + 2y = -1$

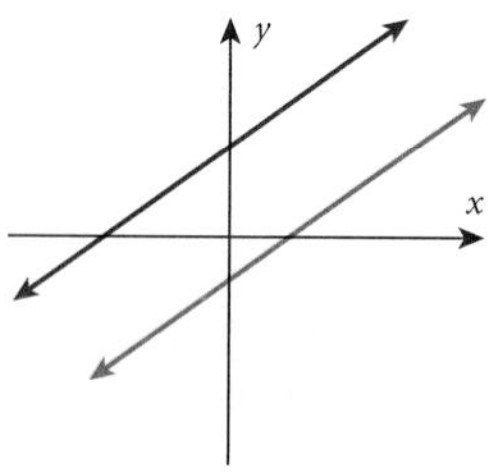

50. The figure below shows the graph of two perpendicular lines. Which of the following pairs of equations might have such a graph?

(a) $y - 2x = 2$
$y + 2x = -1$

(b) $y - 2x = 0$
$2y + x = 0$

(c) $2y - x = 2$
$2y + x = -2$

(d) $y - 2x = 2$
$x + 2y = -1$

(e) $2x + y = -2$
$2y + x = -2$

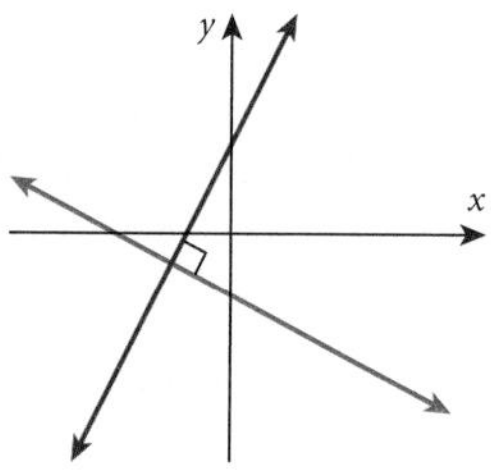

1.3 Applications: Prediction; Break-Even Point; Mixture Problems; Economics

OBJECTIVE 1 Solve applied problems

Solve applied problems 1

In this section we present four applied situations in which linear equations are used. These subsections are independent of each other.

Prediction

Linear equations are sometimes used as predictors of future results. Let's look at an example.

EXAMPLE 1 Predicting the Cost of a Home

In 1999 the cost of an average home in Chicago was \$170,100. In 2000 the cost was \$173,600. Assuming that the relationship between time and cost is linear, develop a formula for predicting the cost of an average home in 2004.

Source: National Association of Realtors.

SOLUTION We agree to let x represent the year and y represent the cost. We seek a relationship between x and y. Two points on the graph of the equation relating x and y are

$$(1999, 170100) \qquad \text{and} \qquad (2000, 173600)$$

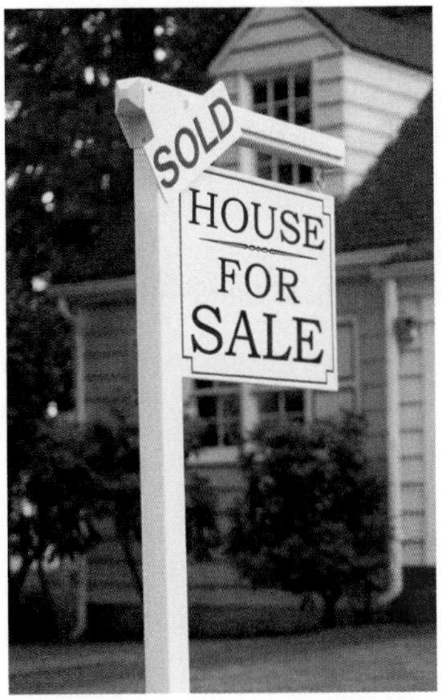

The assumption is that the equation relating x and y is linear. The slope of this line is

$$\frac{173{,}600 - 170{,}100}{2000 - 1999} = 3500$$

Using this fact and the point (1999, 170100), the point–slope form of the equation of the line is

$$y - 170{,}100 = 3500(x - 1999)$$
$$y = 170{,}100 + 3500(x - 1999)$$

For $x = 2004$ we predict the cost of an average home to be

$$\begin{aligned} y &= 170{,}100 + 3500(x - 1999) \\ &= 170{,}100 + 3500(2004 - 1999) \\ &= 170{,}100 + 3500(5) \\ &= \$187{,}600 \end{aligned}$$

Figure 31 illustrates the situation.

FIGURE 31

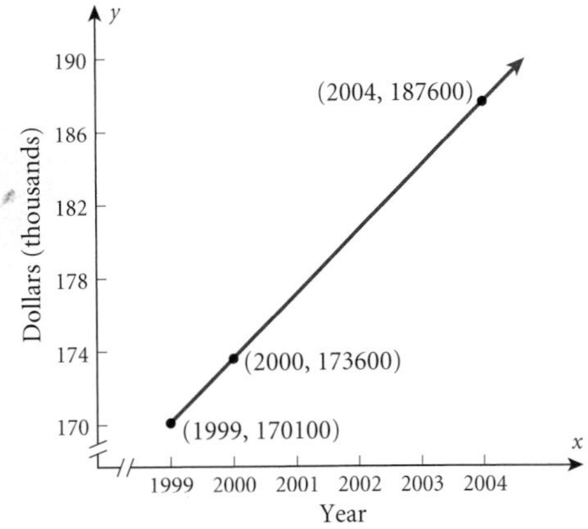

This prediction of future cost is based on the assumption that annual increases remain the same. In this example, the assumption is that each year the cost of a house will go up \$3500 (the slope of the line). Of course, if this assumption is not correct, the predicted cost will also not be correct.

NOW WORK PROBLEM 1.

Break-Even Point

In many businesses the cost C of production and the number x of items produced can be expressed as a linear equation. Similarly, sometimes the revenue R obtained from sales and the number x of items produced can be expressed as a linear equation. When the cost C of production exceeds the revenue R from the sales, the business is operating at a loss; when the revenue R exceeds the cost C, there is a profit; and when the revenue R and the cost C are equal, there is no profit or loss. The point at which $R = C$, that is, the point of intersection of the two lines, is usually referred to as the **break-even point** in business.

EXAMPLE 2 Finding the Break-Even Point

Sweet Delight Candies, Inc., has daily fixed costs from salaries and building operations of \$300. Each pound of candy produced costs \$1 and is sold for \$2.

(a) Find the cost C of production for x pounds of candy.
(b) Find the revenue R from selling x pounds of candy.
(c) What is the break-even point? That is, how many pounds of candy must be sold daily to guarantee no loss and no profit?
(d) Graph C and R and label the break-even point.

SOLUTION **(a)** The cost C of production is the fixed cost of \$300 plus the variable cost of producing x pounds of candy at \$1 per pound. That is,

$$C = \$1 \cdot x + \$300 = x + 300$$

(b) The revenue R realized from the sale of x pounds of candy at \$2 per pound is

$$R = \$2 \cdot x = 2x$$

(c) The break-even point is the point where $R = C$. Setting $R = C$, we find

$$2x = x + 300$$
$$x = 300$$

That is, 300 pounds of candy must be sold to break even.

(d) Figure 32 shows the graphs of C and R and the break-even point. Notice that for $x > 300$, the revenue R always exceeds the cost C so that a profit results. Similarly, for $x < 300$, the cost exceeds the revenue, resulting in a loss.

FIGURE 32

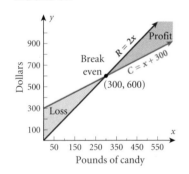

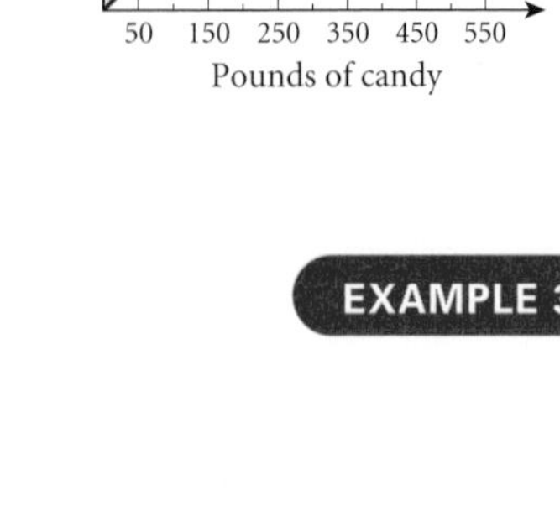

EXAMPLE 3 Analyzing Break-Even Points

After negotiations with employees of Sweet Delight Candies and an increase in the price of sugar, the daily cost C of production for x pounds of candy is

$$C = \$1.05x + \$330$$

(a) If each pound of candy is sold for \$2.00, how many pounds must be sold daily to make a profit?
(b) If the selling price is increased to \$2.25 per pound, what is the break-even point?

(c) If it is known that 325 pounds of candy can be sold daily, what price should be charged per pound to guarantee no loss?

SOLUTION **(a)** If each pound is sold for \$2.00, the revenue R from sales is

$$R = \$2x$$

where x represents the number of pounds sold. When we set $R = C$, we find that the break-even point is the solution of the equation

$$\begin{aligned} 2x &= 1.05x + 330 \\ 0.95x &= 330 \\ x &= \frac{330}{0.95} = 347.37 \end{aligned}$$

If 347 pounds or less of candy are sold, a loss is incurred; if 348 pounds or more are sold, a profit results.

(b) If the selling price is increased to \$2.25 per pound, the revenue R from sales is

$$R = \$2.25x$$

The break-even point is the solution of the equation

$$\begin{aligned} 2.25x &= 1.05x + 330 \\ 1.2x &= 330 \\ x &= \frac{330}{1.2} = 275 \end{aligned}$$

With the new selling price, the break-even point is 275 pounds.

(c) If we know that 325 pounds of candy will be sold daily, the price per pound p needed to guarantee no loss (that is, to guarantee at worst a break-even point) is the solution of the equation

$$\begin{aligned} R &= C \\ 325p &= (1.05)(325) + 330 \\ 325p &= 671.25 \\ p &= \$2.07 \end{aligned}$$

We should charge \$2.07 per pound to guarantee no loss, provided at least 325 pounds will be sold. ▶

EXAMPLE 4 Analyzing Break-Even Points

A producer sells items for \$0.30 each.

(a) Determine the revenue R from selling x items.

(b) If the cost for production is

$$C_1 = \$0.15x + \$105$$

where x is the number of items sold, find the break-even point.

(c) If the cost can be changed to

$$C_2 = \$0.12x + \$110$$

would it be advantageous?

(d) Graph R, C_1, and C_2 together.

SOLUTION **(a)** The revenue R from selling x items is

$$R = \$0.30x$$

(b) If the cost for production is $C_1 = \$0.15x + \$105$, then the break-even point is the solution of the equation

$$\begin{aligned} 0.3x &= 0.15x + 105 \quad R = C_1 \\ 0.15x &= 105 \\ x &= 700 \end{aligned}$$

The break-even point is 700 items.

(c) If the revenue received remains at $R = \$0.3x$, but the cost for production changes to $C_2 = \$0.12x + \$110$, then the break-even point is the solution of the equation

$$\begin{aligned} 0.3x &= 0.12x + 110 \quad R = C_2. \\ 0.18x &= 110 \\ x &= 611.11 \end{aligned}$$

The break-even point for the cost in (a) was 700 items. Since the cost in (b) will require fewer items to be sold in order to break even, management should probably change over to the new cost.

(d) Figure 33 shows the graphs of R, C_1, and C_2.

FIGURE 33

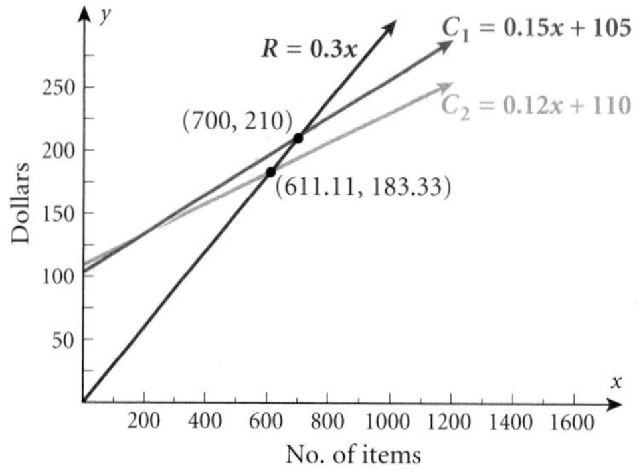

NOW WORK PROBLEM 15.

Mixture Problems

Oil refineries sometimes produce gasoline that is a blend of two other fuels; special blend coffees are created by mixing two other flavors of coffee. These problems are referred to as **mixture problems** because they combine two or more quantities to form a mixture.

EXAMPLE 5 Mixing Peanuts

A store that specializes in selling nuts sells cashews for \$5 per pound and peanuts for \$2 per pound. At the end of the month the manager finds that the peanuts are not selling well. In order to sell 30 pounds of peanuts more quickly, the manager decides to mix the 30 pounds of peanuts with some cashews and sell the mixture of peanuts and cashews for $3 a pound. How many pounds of cashews should be mixed with the peanuts so that the revenue remains the same?

SOLUTION There are two unknowns: the number of pounds of cashews (call this x) and the number of pounds of the mixture (call this y). Since we know that the number of pounds of cashews plus 30 pounds of peanuts equals the number of pounds of the mixture, we can write

$$y = x + 30$$

Also, in order to keep revenue the same, we must have

$$\begin{pmatrix}\text{Price}\\\text{per}\\\text{pound}\\\$5\end{pmatrix}\cdot\begin{pmatrix}\text{Pounds}\\\text{of}\\\text{cashews}\\x\end{pmatrix}+\begin{pmatrix}\text{Price}\\\text{per}\\\text{pound}\\\$2\end{pmatrix}\cdot\begin{pmatrix}\text{Pounds}\\\text{of}\\\text{peanuts}\\30\end{pmatrix}=\begin{pmatrix}\text{Price}\\\text{per}\\\text{pound}\\\$3\end{pmatrix}\cdot\begin{pmatrix}\text{Pounds}\\\text{of}\\\text{mixture}\\y\end{pmatrix}$$

That is,

$$5x + 2(30) = 3y$$

$$\frac{5}{3}x + 20 = y \qquad \text{Divide both sides by 3}$$

We now have two equations

$$y = x + 30 \qquad \text{and} \qquad y = \frac{5}{3}x + 20$$

FIGURE 34

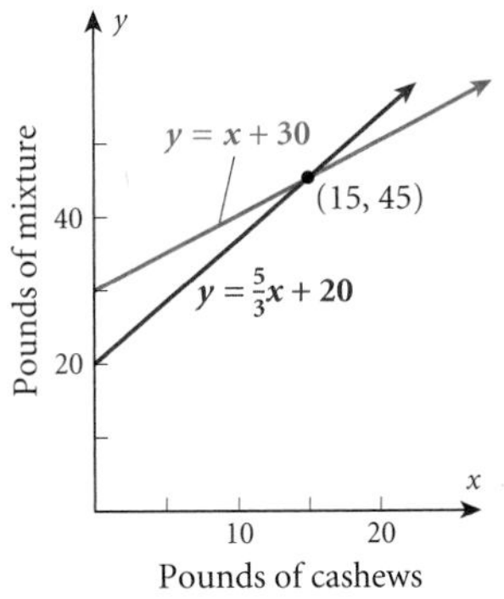

Since the number of pounds of the mixture is the same in each case, we have

$$\frac{5}{3}x + 20 = x + 30$$

$$\frac{2}{3}x = 10$$

$$x = 15$$

The manager should mix 15 pounds of cashews with 30 pounds of peanuts. See Figure 34. Notice that the point of intersection (15, 45) represents the pounds of cashews (15) in the mixture (45 pounds).

NOW WORK PROBLEM 17.

Certain types of problems that involve investments in two interest-bearing accounts can be solved as a mixture problem.

EXAMPLE 6 Financial Planning

Kathleen has $80,000 to invest and requires an overall rate of return of 5% per year. She can invest in a safe, government-insured Certificate of Deposit, but it only pays 3% per year. To obtain 5%, she agrees to invest some of her money in noninsured corporate bonds paying 6% per year. How much should be placed in each investment to achieve her goals?

SOLUTION The question is asking how Kathleen should split her money between the two investments in order for her to realize her goal of earning a 5% rate of return on her money.

We need two dollar amounts: the amount (called the principal) to invest in corporate bonds (call this x) and the principal to invest in the Certificate of Deposit (call this y). Since we know that the amount she invests in corporate bonds plus the amount she invests in the Certificate of Deposit equals \$80,000, we can write

$$x + y = \$80{,}000.$$

Solving this for y, we get

$$y = \$80{,}000 - x$$

which is the amount that will be invested in the Certificate of Deposit. See Table 1.

TABLE 1

	Principal \$	Rate	Time yr	Interest \$
Bonds	x	$6\% = 0.06$	1	$0.06x$
Certificate	$80{,}000 - x$	$3\% = 0.03$	1	$0.03(80{,}000 - x)$
Total	80,000	$5\% = 0.05$	1	$0.05(80{,}000) = 4000$

Since the total interest from the investments is equal to $0.05(\$80{,}000) = \$4000$, the equation relating the interest earned on the accounts is given as:

interest earned on the bonds + interest earned on the certificate = total interest earned

$$0.06x + 0.03(80{,}000 - x) = 4000$$

(Note that the units are consistent: the unit is dollars on both sides of the equation.) Now we solve the equation for x, the amount invested in corporate bonds.

$$0.06x + 2400 - 0.03x = 4000$$
$$0.03x + 2400 = 4000$$
$$0.03x = 1600$$
$$x = 53{,}333.33$$

Kathleen should invest \$53,333.33 in corporate bonds and $\$80{,}000 - \$53{,}333.33 = \$26{,}666.67$ in the Certificate of Deposit.

FIGURE 35

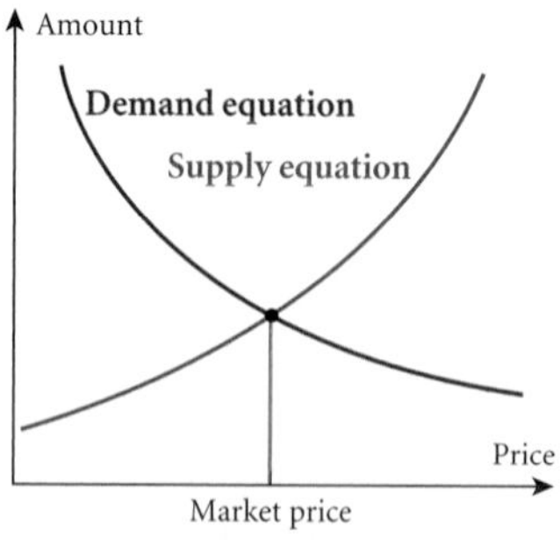

Economics

The **supply equation** in economics is used to specify the amount of a particular commodity that sellers are willing to offer in the market at various prices. The **demand equation** specifies the amount of a particular commodity that buyers are willing to purchase at various prices.

An increase in price p usually causes an increase in the supply S and a decrease in demand D. On the other hand, a decrease in price brings about a decrease in supply and an increase in demand. The **market price** is defined as the price at which supply and demand are equal (the point of intersection).

Figure 35 illustrates a typical supply/demand situation.

EXAMPLE 7 Supply and Demand

The supply and demand for flour have been estimated as being given by the equations

$$S = 0.8p + 0.5 \qquad D = -0.4p + 1.5$$

FIGURE 36

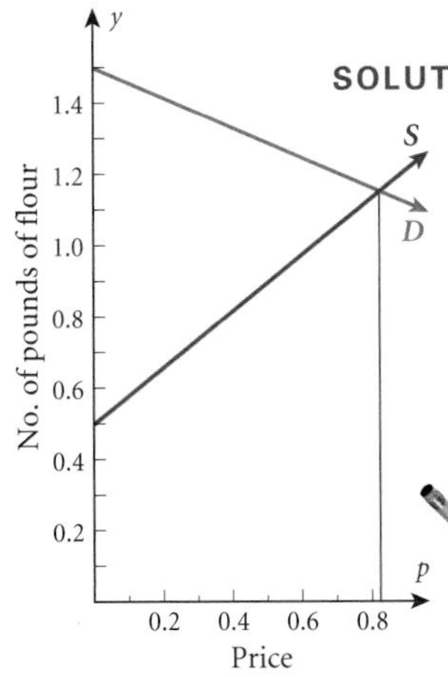

where p is measured in dollars and S and D are measured in pound units of flour. Find the market price and graph the supply and demand equations.

SOLUTION The market price p is the solution of the equation

$$S = D$$
$$0.8p + 0.5 = -0.4p + 1.5$$
$$1.2p = 1$$
$$p = 0.83$$

At a price of \$0.83 per pound, supply and demand for flour are equal. The graphs are shown in Figure 36.

NOW WORK PROBLEM 27.

EXERCISE 1.3 Answers to Odd-Numbered Problems Begin on Page AN-4.

Problems 1–8 involve prediction.

1. **Predicting Sales** Suppose the sales of a company are given by

$$S = \$5000x + \$80{,}000$$

where x is measured in years and $x = 0$ corresponds to the year 2002.

(a) Find S when $x = 0$.
(b) Find S when $x = 3$.
(c) Find the predicted sales in 2007, assuming this trend continues.
(d) Find the predicted sales in 2010, assuming this trend continues.

2. **Predicting Sales** Rework Problem 1 if the sales of the company are given by

$$S = \$3000x + \$60{,}000$$

3. **Predicting the Cost of a Compact Car** In 2000, the cost of a compact car averaged \$12,500. In 2003, the cost of a compact car averaged \$14,450. Assuming that the relationship between time and cost is linear, develop a formula for predicting the average cost of a compact car in the future. What do you predict the average cost of a compact car will be in 2005? Interpret the slope.

4. **Oil Depletion** The Yates Oil Field, discovered in 1926 in Texas, through 1999 has produced 1.4 billion barrels of oil. In 1999, the field had an estimated reserve of 600 million barrels. Based on previous production levels, in 2002 the field should have had an estimated reserve of 540 million barrels. Now assume the rate of depletion is constant.

(a) Write an equation that relates the amount A, in millions of barrels, of oil left in the field at any time t, where t is the year.
(b) If the trend continues, when will the oil well dry out?
(c) Interpret the slope.

Source: Energy Information Administration, Department of Energy.

5. **SAT Scores** The average score on the mathematics portion of the SAT has been steadily increasing for South Carolina students over the past 15 years. In 1987 the average SAT mathematics score was 468, while in 2001 the average SAT mathematics score was 488. Assume the rate of increase is constant.

(a) Write an equation that relates the average SAT mathematics score S at any time t, where t is the year.
(b) If the trend continues, what will the average SAT mathematics score of South Carolina students be in 2004?

Source: Digest of Education Statistics, 2001, National Center for Education Statistics.

6. **Financial Statements** In March 2001, Kellogg Co. purchased Keebler Foods Co. After the purchase, revenue increased from \$1.99 billion at the end of the first financial quarter in 2002 to \$2.13 billion at the end of the second financial quarter in 2002. Now assume the rate of increase is constant.

(a) Write an equation that relates the revenue R, in billions of dollars, and the time t. Use $t = 1$ for the first quarter of 2002, $t = 2$ for the second quarter, and so on.
(b) What is the projected revenue for the fourth quarter?
(c) Interpret the slope.

Source: Kellogg Company Annual Report, 2002.

7. **Percent of Population with Bachelor's Degrees** In 1990 the percent of people over 25 years old who had a bachelor's degree or higher was 20.3%. By 2000, the percent of people over 25 with a bachelor's degree or higher was 25.6%. Assume the rate of increase is constant.

(a) Write an equation that relates the percent P of people over 25 with a bachelor's degree or higher at any time t, where t is the year.
(b) If the trend continues, estimate the percentage of people over 25 who will have a bachelor's degree or higher by 2004.
(c) Interpret the slope.

Source: Digest of Education Statistics, 2001, National Center for Education Statistics.

8. **College Degrees** In 1993 1,159,931 bachelor's degrees were conferred by colleges and universities in the United States. In 1997 1,168,023 bachelor's degrees were awarded. Suppose we assume the relationship between time and degrees conferred is linear.

(a) Write an equation that relates the number N of bachelor's degrees awarded to the year t.
(b) If the trend continues, estimate the number of bachelor's degrees that were awarded in 2003.
(c) Interpret the slope.

Source: Digest of Education Statistics, 1999. National Center for Education Statistics.

Problems 9–16 involve break-even points.
In Problems 9–12, find the break-even point for the cost C of production and the revenue R. Graph each result. Indicate the break-even point and where a profit results and where a loss results.

9. $C = \$10x + \$600$ $\quad R = \$30x$

10. $C = \$5x + \$200$ $\quad R = \$8x$

11. $C = \$0.20x + \$50$ $\quad R = \$0.3x$

12. $C = \$1800x + \$3000$ $\quad R = \$2500x$

13. **Break-Even Point** A manufacturer produces items at a daily cost of \$0.75 per item and sells them for \$1 per item. The daily operational overhead is \$300. What is the break-even point? Graph your result.

14. **Break-Even Point** If the manufacturer in Problem 13 is able to reduce the cost per item to \$0.65, but with a resultant increase to \$350 in operational overhead, is it advantageous to do so? Give reasons. Graph your result.

15. **Profit from Sunday Home Delivery** For \$1.79 per copy, the Chicago Tribune will deliver the Sunday newspaper to your front door. The cost to the Tribune for Sunday home delivery is approximately \$0.53 per newspaper with fixed costs of \$1,070,000.

(a) Determine the revenue R from delivering x newspapers.
(b) Determine the cost C of delivering x newspapers.
(c) Determine the profit P of delivering x newspapers.
(d) Determine the break-even point.
(e) Graph R and C together and label the break-even point.
(f) Graph P and label its x-intercept.
(g) Comment on the relationship between the break-even point and the x-intercept.

Source: Chicago Tribune, 2002.

16. **Wages of a Car Salesperson** In 2000 median earnings, including commissions, for car salespersons was \$17.81 per hour or \$712.40 per week. Dan receives \$375 per week for selling new and used cars. In addition, he receives 5% of the profit on any sales he generates.

(a) Write an equation that relates Dan's weekly salary S when he has sales that generate a profit of x dollars.
(b) If Dan has sales that generate a profit of \$4000, what are his weekly earnings?
(c) To equal the median earnings of a car salesperson, Dan would have to have sales that generate a profit of how many dollars?

Source: U.S. Department of Labor, Occupational Outlook Handbook 2002–2003 Edition.

Problems 17–24, involve mixture problems.

17. **Mixture Problem** Sweet Delight Candies sells boxes of candy consisting of creams and caramels. Each box sells for \$8.00 and holds 50 pieces of candy (all pieces are the same size). If the caramels cost \$0.10 to produce and the creams cost \$0.20 to produce, how many caramels and creams should be in each box for no profit or loss? Would you increase or decrease the number of caramels in order to obtain a profit?

18. **Mixture Problem** The manager of Nutt's Nuts regularly sells cashews for \$6.50 per pound, pecans for \$7.50 per pound, and peanuts for \$2.00 per pound. How many pounds of cashews and pecans should be mixed with 40 pounds of peanuts to obtain a mixture of 100 pounds that will sell for \$4.89 a pound so that the revenue is unchanged?

19. **Investment Problem** Mr. Nicholson has just retired and needs \$10,000 per year in supplementary income. He has \$150,000 to invest and can invest in AA bonds at 10% annual interest or in Savings and Loan Certificates at 5% interest per year. How much money should be invested in each so that he realizes exactly \$10,000 in extra income per year?

20. **Investment Problem** Mr. Nicholson finds after 2 years that because of inflation he now needs \$12,000 per year in supplementary income. How should he transfer his funds to achieve this amount? (Use the data from Problem 19.)

21. Mixture Problem California Coffee Roasters sells Kona coffee for \$22.95 per pound and Colombian coffee for \$6.75 per pound. Suppose they offer a blend of those two coffees for a price of \$10.80 per pound. What amounts of Kona and Colombian coffees should be blended to obtain the desired mixture? *Hint:* Assume that the total weight of the blend is 100 pounds.

Source: California Coffee Roasters.

22. Theater Attendance The Star Theater wants to know whether the majority of its patrons are adults or children. During a week in July 2600 tickets were sold, and the receipts totaled \$16,440. The adult admission is \$8 and the children's admission is \$4. How many adult patrons were there?

23. Mixture Problem One solution is 15% acid and another is 5% acid. How many cubic centimeters of each should be mixed to obtain 100 cubic centimeters of a solution that is 8% acid?

24. Investment Problem A bank loaned \$10,000, some at an annual rate of 8% and some at an annual rate of 12%. If the income from these loans was \$1000, how much was loaned at 8%? How much at 12%?

Problems 25–32 involve economics.
In Problems 25–28, find the market price for each pair of supply and demand equations.

25. $S = p + 1$ $D = 3 - p$

26. $S = 2p + 3$ $D = 6 - p$

27. $S = 20p + 500$ $D = 1000 - 30p$

28. $S = 40p + 300$ $D = 1000 - 30p$

29. Market Price of Sugar The supply and demand equations for sugar have been estimated to be given by the equations

$$S = 0.7p + 0.4 \qquad D = -0.5p + 1.6$$

(a) Find the market price.
(b) What quantity of supply is demanded at this market price?
(c) Graph both the supply and demand equations.
(d) Interpret the point of intersection of the two lines.

30. Supply and Demand Problem The market price for a certain product is \$5.00 per unit and occurs when 14,000 units are produced. At a price of \$1, no units are manufactured and, at a price of \$19.00, no units will be purchased. Find the supply and demand equations, assuming they are linear.

31. Supply and Demand Problem For a certain commodity the supply equation is given by

$$S = 2p + 5$$

At a price of \$1, there is a demand for 19 units of the commodity. If the demand equation is linear and the market price is \$3, find the demand equation.

32. Supply and Demand Problem For a certain commodity the demand equation is given by

$$D = -3p + 20$$

At a price of \$1, four units of the commodity are supplied. If the supply equation is linear and the market price is \$4, find the supply equation.

1.4 Scatter Diagrams; Linear Curve Fitting

OBJECTIVES
1 Draw and interpret scatter diagrams
2 Distinguish between linear and nonlinear relations
3 Use a graphing utility to find the line of best fit

Scatter Diagrams

Draw and interpret scatter diagrams 1

A **relation** is a correspondence between two sets. If x and y are two elements in these sets and if a relation exists between x and y, then we say that x **corresponds to** y or that y **depends on** x and write $x \rightarrow y$. We may also write $x \rightarrow y$ as the ordered pair (x, y). Here, y is referred to as the **dependent** variable and x is called the **independent** variable.

Often we are interested in specifying the type of relation (such as an equation) that might exist between two variables. The first step in finding this relation is to plot the ordered pairs using rectangular coordinates. The resulting graph is called a **scatter diagram.**

EXAMPLE 1 Drawing a Scatter Diagram

The data listed in Table 2, represent the apparent temperature versus the relative humidity in a room whose actual temperature is 72° Fahrenheit.

TABLE 2

Relative Humidity (%), x	Apparent Temperature, y	(x, y)	Relative Humidity (%), x	Apparent Temperature, y	(x, y)
0	64	(0, 64)	60	72	(60, 72)
10	65	(10, 65)	70	73	(70, 73)
20	67	(20, 67)	80	74	(80, 74)
30	68	(30, 68)	90	75	(90, 75)
40	70	(40, 70)	100	76	(100, 76)
50	71	(50, 71)			

(a) Draw a scatter diagram by hand.

(b) Use a graphing utility to draw a scatter diagram.*

(c) Describe what happens to the apparent temperature as the relative humidity increases.

SOLUTION **(a)** To draw a scatter diagram by hand, we plot the ordered pairs listed in Table 2, with the relative humidity as the x-coordinate and the apparent temperature as the y-coordinate. See Figure 36(a). Notice that the points in a scatter diagram are not connected.

(b) Figure 36(b) shows the scatter diagram using a graphing utility.

FIGURE 36

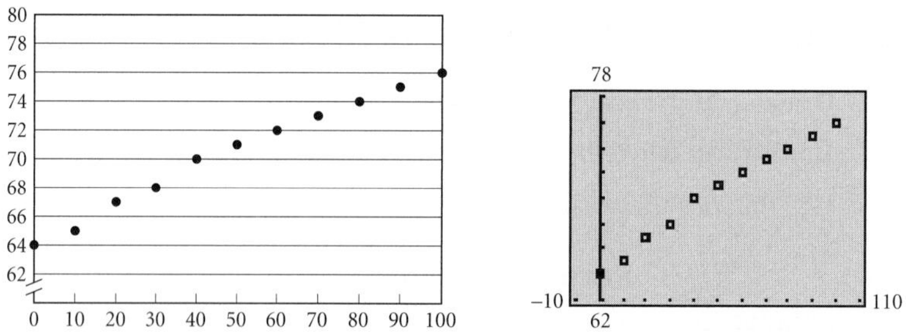

(a) (b)

(c) We see from the scatter diagrams that, as the relative humidity increases, the apparent temperature increases.

NOW WORK PROBLEM 7(a).

Consult your owner's manual for the proper keystrokes.

Distinguish between linear **2**
and nonlinear relations

Curve Fitting

Scatter diagrams are used to help us see the type of relation that may exist between two variables. In this text, we concentrate on distinguishing between linear and nonlinear relations. See Figure 37.

FIGURE 37

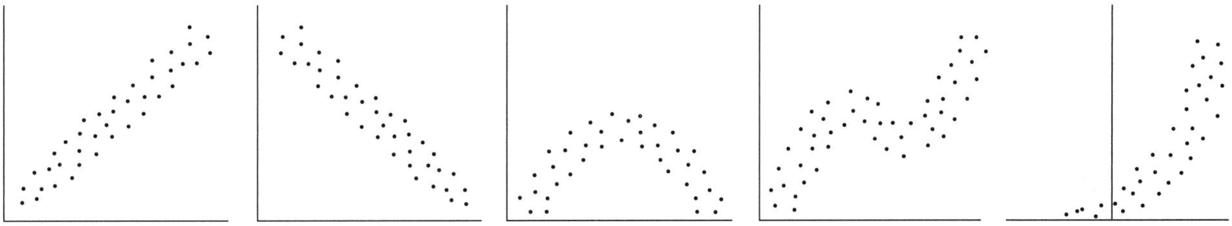

(a) Linear
$y = mx + b, m > 0$

(b) Linear
$y = mx + b, m < 0$

(c) Nonlinear

(d) Nonlinear

(e) Nonlinear

EXAMPLE 2   **Distinguishing between Linear and Nonlinear Relations**

Determine whether the relation between the two variables in Figure 38 is linear or nonlinear.

FIGURE 38

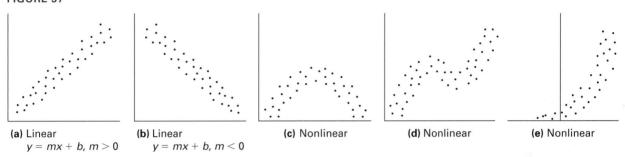

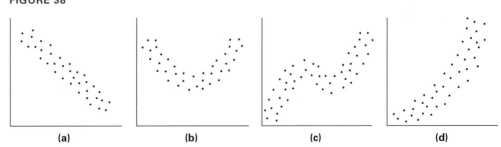

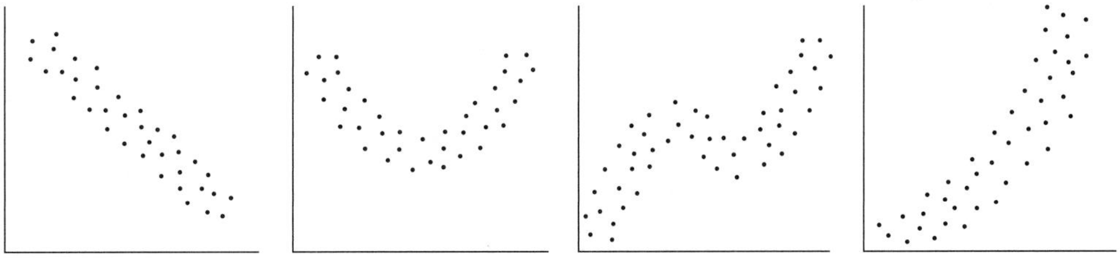

(a)   (b)   (c)   (d)

SOLUTION   (a) Linear   (b) Nonlinear   (c) Nonlinear   (d) Nonlinear   ▶

    NOW WORK PROBLEM 1.

In this section we will study data whose scatter diagrams imply that a linear relation exists between the two variables.

Suppose that the scatter diagram of a set of data appears to be linearly related as in Figure 37(a) or (b). We might wish to find an equation of a line that relates the two variables. One way to obtain an equation for such data is to draw a line through two points on the scatter diagram and find the equation of the line.

EXAMPLE 3 Find an Equation for Linearly Related Data

Using the data in Table 2 from Example 1, select two points from the data and find an equation of the line containing the points.

(a) Graph the line on the scatter diagram obtained in Example 1(a).
(b) Graph the line on the scatter diagram obtained in Example 1(b).

SOLUTION Select two points, say (10, 65) and (70, 73). (You should select your own two points and complete the solution.) The slope of the line joining the points (10, 65) and (70, 73) is

$$m = \frac{73 - 65}{70 - 10} = \frac{8}{60} = \frac{2}{15}$$

The equation of the line with slope $\frac{2}{15}$ and passing through (10, 65) is found using the point–slope form with $m = \frac{2}{15}$, $x_1 = 10$, and $y_1 = 65$.

$$y - y_1 = m(x - x_1)$$

$$y - 65 = \frac{2}{15}(x - 10)$$

$$y = \frac{2}{15}x + \frac{191}{3}$$

(a) Figure 39(a) shows the scatter diagram with the graph of the line drawn by hand.
(b) Figure 39(b) shows the scatter diagram with the graph of the line using a graphing utility.

FIGURE 39

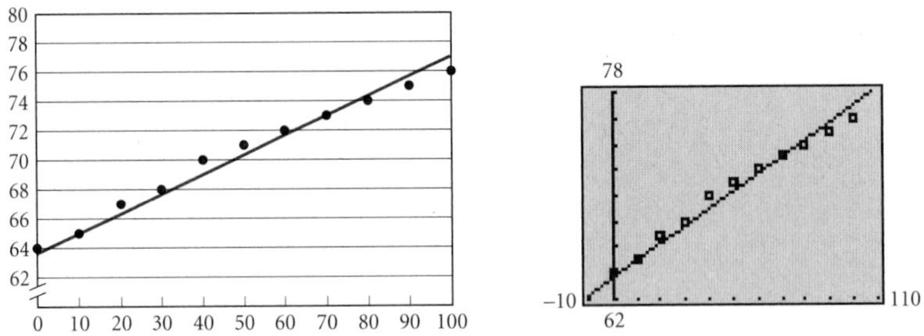

(a) (b)

NOW WORK PROBLEMS 7(b) AND (c).

Use a graphing utility to find the line of best fit 3

LINE OF BEST FIT

The line obtained in Example 3 depends on the selection of points, which will vary from person to person. So the line that we found might be different from the line that you found. Although the line that we found in Example 3 appears to "fit" the data well, there may be a line that "fits better." Do you think your line fits the data better? Is there a line of *best fit*? As it turns out, there is a method for finding the line that best fits linearly related data (called the *line of best fit*).

EXAMPLE 4 Finding the Line of Best Fit

Using the data in Table 2 from Example 1:

(a) Find the line of best fit using a graphing utility.
(b) Graph the line of best fit on the scatter diagram obtained in Example 1(b).
(c) Interpret the slope of the line of best fit.
(d) Use the line of best fit to predict the apparent temperature of a room whose actual temperature is 72°F and relative humidity is 45%.

SOLUTION **(a)** Graphing utilities contain built-in programs that find the line of best fit for a collection of points in a scatter diagram. (Look in your owner's manual under Linear Regression or Line of Best Fit for details on how to execute the program.) Upon executing the LINear REGression program on a TI-83 Plus we obtain the results shown in Figure 40. The output that the utility provides shows us the equation $y = ax + b$, where a is the slope of the line and b is the y-intercept. The line of best fit that relates relative humidity to apparent temperature may be expressed as the line $y = 0.121x + 64.409$.

(b) Figure 41 shows the graph of the line of best fit, along with the scatter diagram.

FIGURE 40

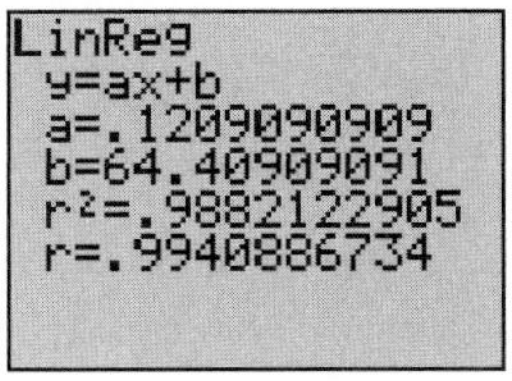

FIGURE 41

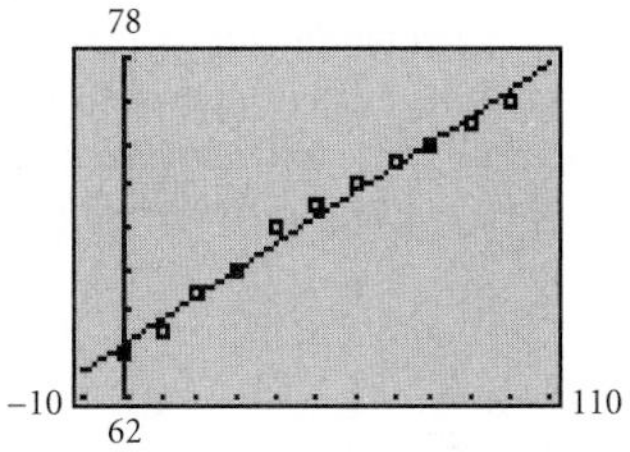

(c) The slope of the line of best fit is 0.121, which means that, for every 1% increase in the relative humidity, apparent room temperature increases 0.121°F.

(d) Letting $x = 45$ in the equation of the line of best fit, we obtain $y = 0.121(45) + 64.409 \approx 70$°F, which is the apparent temperature in the room.

NOW WORK PROBLEMS 7(d), (e) AND (f).

Does the line of best fit appear to be a good fit? In other words, does the line appear to accurately describe the relation between temperature and relative humidity?

And just how "good" is this line of best fit? The answers are given by what is called the *correlation coefficient.* Look again at Figure 40. The last line of output is $r = 0.994$. This number, called the **correlation coefficient, r,** $-1 \leq r \leq 1$, is a measure of the strength of the *linear relation* that exists between two variables. The closer that $|r|$ is to 1, the more perfect the linear relationship is. If r is close to 0, there is little or no *linear* relationship between the variables. A negative value of r, $r < 0$, indicates that as x increases y decreases; a positive value of r, $r > 0$, indicates that as x increases y does also. The data given in Example 1, having a correlation coefficient of 0.994, are indicative of a strong linear relationship with positive slope.

EXERCISE 1.4 Answers to Odd-Numbered Problems Begin on Page AN-5.

In Problems 1–6, examine the scatter diagram and determine whether the type of relation, if any, that may exist is linear or nonlinear.

1.
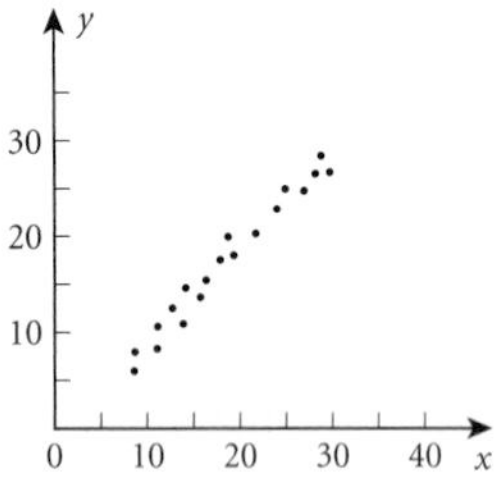

2.
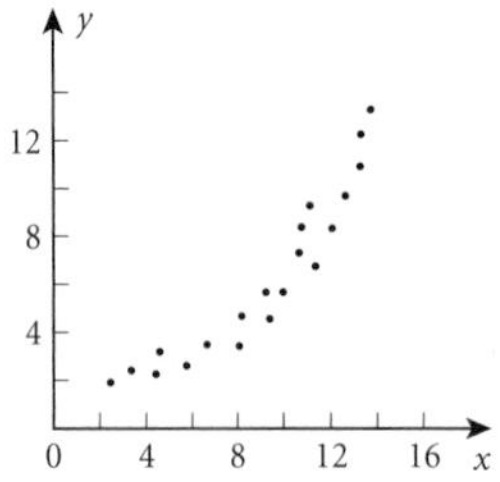

3.
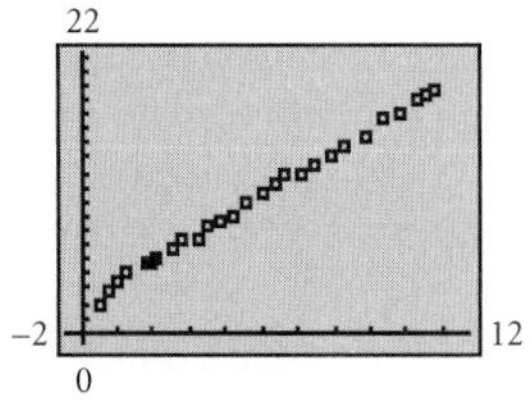

4.
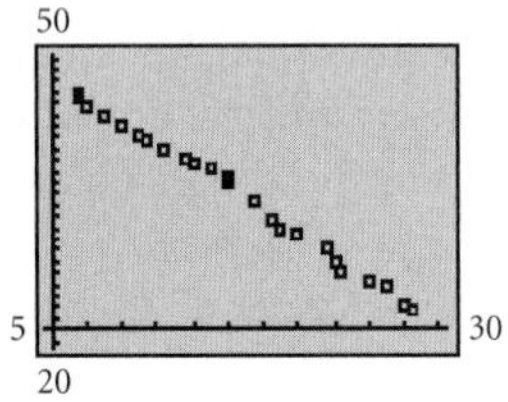

5.
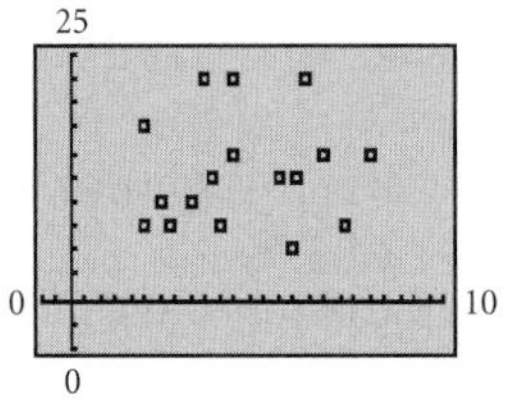

6.
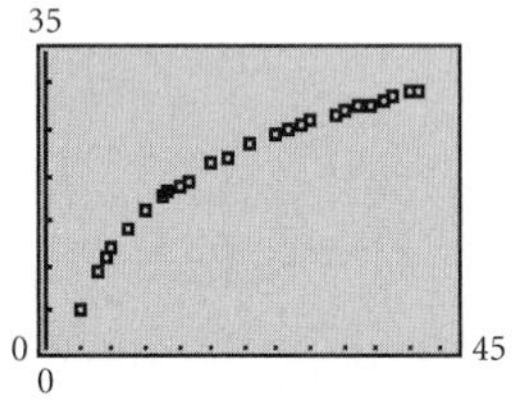

In Problems 7–14:

(a) *Draw a scatter diagram by hand.*
(b) *Select two points from the scatter diagram and find the equation of the line containing the points selected.**
(c) *Graph the line found in part (b) on the scatter diagram.*
(d) *Use a graphing utility to draw a scatter diagram.*
(e) *Use a graphing utility to find the line of best fit.*
(f) *Use a graphing utility to graph the line of best fit on the scatter diagram.*

7.

x	3	4	5	6	7	8	9
y	4	6	7	10	12	14	16

8.

x	3	5	7	9	11	13
y	0	2	3	6	9	11

9.

x	−2	−1	0	1	2
y	−4	0	1	4	5

10.

x	−2	−1	0	1	2
y	7	6	3	2	0

11.

x	20	30	40	50	60
y	100	95	91	83	70

Answers will vary. We will use the first and last data points in the answer section.

12.

x	5	10	15	20	25
y	2	4	7	9	11

13.

x	−20	−17	−15	−14	−10
y	100	120	118	130	140

14.

x	−30	−27	−25	−20	−14
y	10	12	13	13	18

15. **Consumption and Disposable Income** An economist wishes to estimate a line that relates personal consumption expenditures C and disposable income I. Both C and I are in thousands of dollars. She interviews eight heads of households for families of size 3 and obtains the data below.

I(000)	20	20	18	27	36	37	45	50
C(000)	16	18	13	21	27	26	36	39

Let I represent the independent variable and C the dependent variable.

(a) Draw a scatter diagram by hand.
(b) Find a line that fits the data.*
(c) Interpret the slope. The slope of this line is called the **marginal propensity to consume.**
(d) Predict the consumption of a family whose disposable income is $42,000.
(e) Use a graphing utility to find the line of best fit to the data.

16. Marginal Propensity to Save The same economist as in Problem 15 wants to estimate a line that relates savings S and disposable income I. Let $S = I - C$ be the dependent variable and I the independent variable.

(a) Draw a scatter diagram by hand.
(b) Find a line that fits the data.
(c) Interpret the slope. The slope of this line is called the **marginal propensity to save.**
(d) Predict the savings of a family whose income is $42,000.
(e) Use a graphing utility to find the line of best fit.

17. Mortgage Qualification The amount of money that a lending institution will allow you to borrow mainly depends on the interest rate and your annual income. The following data represent the annual income, I, required by a bank in order to lend L dollars at an interest rate of 7.5% for 30 years.

Annual Income, I($)	Loan Amount, L($)
15,000	44,600
20,000	59,500
25,000	74,500
30,000	89,400
35,000	104,300
40,000	119,200
45,000	134,100
50,000	149,000
55,000	163,900
60,000	178,800
65,000	193,700
70,000	208,600

Source: Information Please Almanac, 1999

Let I represent the independent variable and L the dependent variable.

(a) Use a graphing utility to draw a scatter diagram of the data.
(b) Use a graphing utility to find the line of best fit to the data.
(c) Graph the line of best fit on the scatter diagram drawn in part (a).
(d) Interpret the slope of the line of best fit.
(e) Determine the loan amount that an individual would qualify for if her income is $42,000.

18. Mortgage Qualification The amount of money that a lending institution will allow you to borrow mainly depends on the interest rate and your annual income. The following data represent the annual income, I, required by a bank in order to lend L dollars at an interest rate of 8.5% for 30 years.

Annual Income, I($)	Loan Amount, L($)
15,000	40,600
20,000	54,100
25,000	67,700
30,000	81,200
35,000	94,800
40,000	108,300
45,000	121,900
50,000	135,400
55,000	149,000
60,000	162,500
65,000	176,100
70,000	189,600

Source: Information Please Almanac, 1999

Let I represent the independent variable and L the dependent variable.

(a) Use a graphing utility to draw a scatter diagram of the data.
(b) Use a graphing utility to find the line of best fit to the data.
(c) Graph the line of best fit on the scatter diagram drawn in part (a).
(d) Interpret the slope of the line of best fit.
(e) Determine the loan amount that an individual would qualify for if her income is $42,000.

19. Apparent Room Temperature The data on page 42 represent the apparent temperature versus the relative humidity in a room whose actual temperature is 65° Fahrenheit. Let h represent the independent variable and T the dependent variable.

(a) Use a graphing utility to draw a scatter diagram of the data.
(b) Use a graphing utility to find the line of best fit to the data.
(c) Graph the line of best fit on the scatter diagram drawn in part (a).
(d) Interpret the slope of the line of best fit.
(e) Determine the apparent temperature of a room whose actual temperature is 65°F if the relative humidity is 75%.

Relative Humidity, h(%)	Apparent Temperature, T (°F)
0	59
10	60
20	61
30	61
40	62
50	63
60	64
70	65
80	65
90	66
100	67

Source: National Oceanic and Atmospheric Administration

Relative Humidity, h(%)	Apparent Temperature, T (°F)
0	68
10	69
20	71
30	72
40	74
50	75
60	76
70	76
80	77
90	78
100	79

Source: National Oceanic and Atmospheric Administration

20. Apparent Room Temperature The following data represent the apparent temperature versus the relative humidity in a room whose actual temperature is 75° Fahrenheit. Let h represent the independent variable and let T be the dependent variable.

(a) Use a graphing utility to draw a scatter diagram of the data.
(b) Use a graphing utility to find the line of best fit to the data.
(c) Graph the line of best fit on the scatter diagram drawn in part (a).
(d) Interpret the slope of the line of best fit.
(e) Determine the apparent temperature of a room whose actual temperature is 75°F if the relative humidity is 75%.

Chapter 1 Review

OBJECTIVES

Section	You should be able to	Review Exercises
1.1	1 Graph linear equations	1–8, 23–26
	2 Find the equation of a vertical line	12, 13
	3 Calculate and interpret the slope of a line	5–8(a), 45(b–e, g, h), 46(b–e, h), 47(c, d)
	4 Graph a line given a point on the line and the slope	9–11
	5 Use the point–slope form of a line	9–11
	6 Find the equation of a horizontal line	11, 14
	7 Find the equation of a line given two points	5–8(b), 15–18
	8 Use the slope–intercept form of a line	23–26
1.2	1 Show that two lines are coincident	28, 31
	2 Show that two lines are parallel	19, 20, 27, 32
	3 Show that two lines intersect	29, 30

	4 Find the point of intersection of two intersecting lines	33–38
	5 Show that two lines are perpendicular	21, 22
1.3	1 Solve applied problems	39–42
1.4	1 Draw and interpret scatter diagrams	43, 44, 45(a), 46(a), 47(a)
	2 Distinguish between linear and nonlinear relations	43, 44
	3 Use a graphing utility to find the line of best fit	45(f), 46(f), 47(e)

IMPORTANT FORMULAS

Linear Equation, General Form (p. 3)
$Ax + By = C$ A, B not both zero

Vertical Line (p. 6):
$x = a$ $(a, 0)$ is the x-intercept

Slope of a Line (p. 7)

$$m = \frac{y_2 - y_1}{x_2 - x_1} \text{ if } x_1 \neq x_2; \text{ undefined if } x_1 = x_2$$

Point–Slope Form of the Equation of a Line (p. 11)
$y - y_1 = m(x - x_1)$ m is the slope of the line; (x_1, y_1) is a point on the line

Horizontal Line (p. 11):
$y = b$ $(0, b)$ is the y-intercept

Slope–Intercept Form of the Equation of a Line (p. 13)
$y = mx + b$ m is the slope of the line; $(0, b)$ is the y-intercept

Pair of Lines (pp. 19–25)	*Conclusion: The lines are*
Both vertical	(a) Coincident, if they have the same x-intercept (b) Parallel, if they have different x-intercepts
One vertical, one nonvertical	(a) Intersecting (b) Perpendicular, if the nonvertical line is horizontal
Neither vertical	Write the equation of each line in slope–intercept form: $y = m_1x + b_1, y = m_2x + b_2$ (a) Coincident, if $m_1 = m_2, b_1 = b_2$ (b) Parallel, if $m_1 = m_2, b_1 \neq b_2$ (c) Intersecting, if $m_1 \neq m_2$ (d) Perpendicular, if $m_1m_2 = -1$

TRUE–FALSE ITEMS Answers are on page AN-7.

T F **1.** In the slope–intercept equation of a line, $y = mx + b$, m is the slope and $(0, b)$ is the x-intercept.

T F **2.** The graph of the equation $Ax + By = C$, where A, B, C are real numbers and A, B are not both zero, is a line.

T F **3.** The y-intercept of the line $2x - 3y = -6$ is $(0, 2)$.

T F **4.** The slope of the line $2x - 4y = -7$ is $-\frac{1}{2}$.

T F **5.** Parallel lines always have the same intercepts.

T F **6.** Intersecting lines have different slopes.

T F **7.** Perpendicular lines have slopes that are reciprocals of each other.

T F **8.** A linear relation between two variables can always be graphed as a line.

T F **9.** All lines with equal slopes are distinct.

T F **10.** All vertical lines have positive slope.

FILL IN THE BLANKS Answers are on page AN-7.

1. If (x, y) are rectangular coordinates of a point, the number x is called the ____________ and y is called the ____________.

2. The slope of a vertical line is ____________; the slope of a horizontal line is ____________.

3. If a line slants downward from left to right, its slope will be a ____________ number.

4. If two lines have the same slope but different y-intercepts, they are ____________.

5. If two lines have the same slope and the same y-intercept, they are said to be ____________.

6. Lines that intersect at right angles are said to be ____________ to each other.

7. Distinct lines that have different slopes are ____________ lines.

REVIEW EXERCISES

Answers to odd-numbered problems begin on page AN-7.
Blue Problem numbers indicate the author's suggestions for use in a practice test.

In Problems 1–4, graph each equation.

1. $y = -2x + 3$
2. $y = 6x - 2$
3. $2y = 3x + 6$
4. $3y = 2x + 6$

In Problems 5–8, (a) calculate and interpret the slope of the line containing each pair of points; (b) find an equation for the line containing each pair of points. Write the equation using the general form or the slope–intercept form, whichever you prefer. (c) Graph each line.

5. $P = (1, 2) \quad Q = (-3, 4)$
6. $P = (-1, 3) \quad Q = (1, 1)$
7. $P = (-1, 5) \quad Q = (-2, 3)$
8. $P = (-2, 3) \quad Q = (0, 0)$

In Problems 9–22, find an equation of the line having the given characteristics. Write the equation using the general form or the slope–intercept form, whichever you prefer. Graph each line.

9. Slope $= -3$; containing the point $(2, -1)$
10. Slope $= 4$; containing the point $(-1, -3)$
11. Slope $= 0$; containing the point $(-3, 4)$
12. Slope undefined; containing the point $(-3, 4)$
13. Vertical; containing the point $(8, 5)$
14. Horizontal; containing the point $(5, 8)$
15. x-intercept $= (2, 0)$; containing the point $(4, -5)$
16. y-intercept $= (0, -2)$; containing the point $(5, -3)$
17. x-intercept $= (-3, 0)$; y-intercept $= (0, -4)$
18. Containing the points $(3, -4)$ and $(2, 1)$
19. Parallel to the line $2x + 3y = -4$; containing the point $(-5, 3)$
20. Parallel to the line $x + y = 2$; containing the point $(1, -3)$
21. Perpendicular to the line $2x + 3y = -4$; containing the point $(-5, 3)$
22. Perpendicular to the line $x + y = 2$; containing the point $(1, -3)$

In Problems 23–26 find the slope and y-intercept of each line. Graph each line.

23. $9x + 2y = 18$
24. $4x + 5y = 20$
25. $4x + 2y = 9$
26. $3x + 2y = 8$

In Problems 27–32 determine whether the two lines are parallel, coincident, or intersecting.

27. $3x - 4y = -12$, $6x - 8y = -9$
28. $2x + 3y = -5$, $4x + 6y = -10$
29. $x - y = -2$, $3x - 4y = -12$
30. $2x + 3y = 5$, $x + y = 2$
31. $4x + 6y = -12$, $2x + 3y = -6$
32. $-3x + y = 0$, $6x - 2y = -5$

In Problems 33–38, the given pair of lines intersect. Find the point of intersection. Graph the lines.

33. L: $x - y = 4$; M: $x + 2y = 7$
34. L: $x + y = 4$; M: $x - 2y = 1$
35. L: $x - y = -2$; M: $x + 2y = 7$

36. $L: 2x + 4y = 4$
$M: 2x - 4y = 8$

37. $L: 2x - 4y = -8$
$M: 3x + 6y = 0$

38. $L: 3x + 4y = 2$
$M: x - 2y = 1$

39. **Investment Problem** Mr. and Mrs. Byrd have just retired and find that they need \$10,000 per year to live on. Fortunately, they have a nest egg of \$90,000, which they can invest in somewhat risky B-rated bonds at 12% interest per year or in a well-known bank at 5% per year. How much money should they invest in each so that they realize exactly \$10,000 in interest income each year?

40. **Mixture Problem** One solution is 20% acid and another is 12% acid. How many cubic centimeters of each solution should be mixed to obtain 100 cubic centimeters of a solution that is 15% acid?

41. **Attendance at a Dance** A church group is planning a dance in the school auditorium to raise money for its school. The band they will hire charges \$500; the advertising costs are estimated at \$100; and food will be supplied at the rate of \$5 per person. The church group would like to clear at least \$900 after expenses.

(a) Determine how many people need to attend the dance for the group to break even if tickets are sold at \$10 each.
(b) Determine how many people need to attend in order to achieve the desired profit if tickets are sold for \$10 each.
(c) Answer the above two questions if the tickets are sold for \$12 each.

42. **Mixture Problem** A coffee manufacturer wants to market a new blend of coffee that will cost \$6.00 per pound by mixing \$5.00 per pound coffee and \$7.50 per pound coffee. What amounts of the \$5.00 per pound coffee and \$7.50 per pound coffee should be blended to obtain the desired mixture? [*Hint:* Assume the total weight of the desired blend is 100 pounds.]

In Problems 43 and 44, draw a scatter diagram for each set of data. Then determine whether the relation, if any, that may exist is linear or non-linear.

43.

x	0	1	2	3	4	5	6
y	90	45	21	12	5	3	2

44.

x	3	5	7	9	11	13
y	74	70	67	58	55	51

45. **Concentration of Carbon Monoxide in the Air** The following data represent the average concentration of carbon monoxide in parts per million (ppm) in the air for 1987–1993.

Year	Concentration of Carbon Monoxide (ppm)
1987	6.69
1988	6.38
1989	6.34
1990	5.87
1991	5.55
1992	5.18
1993	4.88

Source: U.S. Environmental Protection Agency.

(a) Treating the year as the x-coordinate and the average level of carbon monoxide as the y-coordinate, draw a scatter diagram of the data.
(b) What is the slope of the line joining the points (1987, 6.69) and (1990, 5.87)?
(c) Interpret this slope.
(d) What is the slope of the line joining the points (1990, 5.87) and (1993, 4.88)?
(e) Interpret this slope.
(f) Use a graphing utility to find the slope of the line of best fit for these data.
(g) Interpret this slope.
(h) How do you explain the differences among the three slopes obtained?
(i) What is the trend in the data? In other words, as time passes, what is happening to the average level of carbon monoxide in the air? Why do you think this is happening?

46. **Housing Costs** The data on page 46 represent the average price of houses sold in the United States for 1994–2001.

(a) Treating the year as the x-coordinate and the price of the houses as the y-coordinate, draw a scatter diagram of the data.
(b) What is the slope of the line joining the points (1994, 154500) and (1998, 181900)?
(c) Interpret this slope.
(d) What is the slope of the line joining the points (1998, 181900) and (2001, 213200)?
(e) Interpret this slope.
(f) Use a graphing utility to find the slope of the line of best fit for this data.

Year	Price (Dollars)
1994	154,500
1995	158,700
1996	166,400
1997	176,200
1998	181,900
1999	195,600
2000	207,000
2001	213,200

Source: U.S. Census Bureau, 2002.

(g) Interpret this slope.

(h) How do you explain the differences among the three slopes obtained?

(i) What is the trend in the data? In other words, what is happening to the average price of a home in the United States? Why do you think this is happening?

47. Value of a Portfolio The following data represent the value of the Vanguard 500 Index Fund for 1996–1999.

Year	Value per Share
1996	$ 69.17
1997	90.07
1998	113.95
1999	135.33

Source: Vanguard 500 Index Fund, Annual Report 2001.

(a) Treating the year as the x-coordinate and the value of the Vanguard 500 Index Fund as the y-coordinate, draw a scatter diagram of the data.

(b) Do the data appear to be linearly related?

(c) What is the slope of the line connecting (1996, 69.17) and (1999, 135.33)?

(d) Interpret the slope.

(e) Use a graphing utility to find the line of best fit for these data.

(f) Assuming the line of best fit truly represents the trend in the data, predict the value of a share of Vanguard 500 Index Fund in the year 2001.

48. Value of a Portfolio Investment ads warn that "past performance is no guarantee of future performance." Vanguard's Annual Report shows the value of a share was $121.86 in 2000 and $105.89 in the year 2001.

(a) Add these data to the chart in Problem 47 and draw a revised scatter diagram representing the years 1996–2001.

(b) Do the data appear to be linearly related?

(c) What would you say about the prediction you made in Exercise 47 (f)?

49. Make up four problems that you might be asked to do given the two points $(-3, 4)$ and $(6, 1)$. Each problem should involve a different concept. Be sure that your directions are clearly stated.

50. Describe each of the following graphs in the xy-plane. Give justification.

(a) $x = 0$

(b) $y = 0$

(c) $x + y = 0$

Chapter 1 Project

CHOOSING A RENTAL CAR COMPANY*

You have a decision to make. You will be flying to Charlotte, North Carolina for a one-day stay. You will need to rent a midsize car while you are there, and you want to do this in the most economical way possible. The cost to rent a car sometimes depends on the number of miles it is driven, but sometimes the cost includes unlimited mileage. You need to figure out which rental company gives you the best deal. As it turns out, the equations involved are linear, so the problem can be analyzed using techniques discussed in this chapter.

**All rates quoted have been taken from quotes provided by car rental company web sites in February 2003. There are other fees and taxes which are later added to each quote; these fees and taxes are ignored in our analysis.*

You begin by contacting two car rental agencies: Avis and Enterprise. Avis offers a midsize car for $64.99 per day with unlimited mileage. This means that for this rental the number of miles driven does not impact on the cost. Enterprise offers a midsize car for $45.87 per day with 150 miles free. But each mile beyond 150 that the car is driven costs $0.25. It is clear that Enterprise offers the better deal if the car is driven fewer than 150 miles. But what if the car is driven more than 150 miles? At what point will the Avis rental become a better deal?

Let's analyze the situation. Let x denote the number of miles the car is driven.

1. Suppose A is the cost of renting at Avis. Find a linear equation involving A and x.

2. Now let E be the cost of renting at Enterprise. Find a linear equation involving E and x, if $x \leq 150$. Find a linear equation involving E and x if $x > 150$.

3. Graph the linear equations found in parts (1) and (2) on the same set of coordinate axes. Be careful about the restrictions on x for the equations found in part (2).

4. Find the mileage beyond which the Avis rental is more economical by finding the point of intersection of the two graphs. Label this point on your graph.

5. Explain how you can use the solution to part (4) to decide on which car rental is more economical.

6. In an effort to find an even better deal, you contact AutoSaveRental. They offer a midsize car for $36.99 per day with 100 miles free. Each mile driven over 100 miles costs $0.25. Find equation(s) that involve the cost S at AutoSave and the miles x driven.

7. Graph this equation on the same graph found in part (3).

8. It should be clear that AutoSave is the best choice among the three if you are driving less than 100 miles. For what values of x, if any, do Avis or Enterprise offer a better deal?

9. Still not sure you have the best deal, you call Usave Car Rental. They offer a midsize car for $35.99 per day with 200 free miles. Each mile over 200 costs $0.25. Find an equation(s) for the cost U at Usave and graph it on the same graph found in part (7). For what values of x, if any, is Usave the best deal?

10. Discuss your choice of car rental companies if you think you will drive somewhere between 125 and 175 miles.

MATHEMATICAL QUESTIONS FROM PROFESSIONAL EXAMS*

1. **CPA Exam** The Oliver Company plans to market a new product. Based on its market studies, Oliver estimates that it can sell 5500 units in 1992. The selling price will be $2 per unit. Variable costs are estimated to be 40% of the selling price. Fixed costs are estimated to be $6000. What is the break-even point?

   (a) 3750 units (c) 5500 units
   (b) 5000 units (d) 7500 units

2. **CPA Exam** The Breiden Company sells rodaks for $6 per unit. Variable costs are $2 per unit. Fixed costs are $37,500. How many rodaks must be sold to realize a profit before income taxes of 15% of sales?

   (a) 9375 units (c) 11,029 units
   (b) 9740 units (d) 12,097 units

3. **CPA Exam** Given the following notations, what is the break-even sales level in units?

   SP = Selling price per unit VC = Variable cost per unit
   FC = Total fixed cost

   (a) $\dfrac{SP}{FC \div VC}$ (c) $\dfrac{VC}{SP - FC}$

   (b) $\dfrac{FC}{VC \div SP}$ (d) $\dfrac{FC}{SP - VC}$

4. **CPA Exam** At a break-even point of 400 units sold, the variable costs were $400 and the fixed costs were $200. What will the 401st unit sold contribute to profit before income taxes?

   (a) $0 (c) $1.00
   (b) $0.50 (d) $1.50

5. **CPA Exam** A graph is set up with "depreciation expense" on the vertical axis and "time" on the horizontal axis. Assuming linear relationships, how would the graphs for straight-line and sum-of-the-year's-digits depreciation, respectively, be drawn?

   (a) Vertically and sloping down to the right
   (b) Vertically and sloping up to the right
   (c) Horizontally and sloping down to the right
   (d) Horizontally and sloping up to the right

The following statement applies to Questions 6–8:
In analyzing the relationship of total factory overhead with changes in direct labor hours, the following relationship was found to exist: $Y = \$1000 + \$2X$.

6. **CPA Exam** The relationship as shown above is

   (a) Parabolic (d) Probabilistic
   (b) Curvilinear (e) None of the above
   (c) Linear

7. **CPA Exam** Y in the above equation is an estimate of

   (a) Total variable costs (d) Total direct labor hours
   (b) Total factory overhead (e) None of the above
   (c) Total fixed costs

8. **CPA Exam** The $2 in the equation is an estimate of

   (a) Total fixed costs
   (b) Variable costs per direct labor hour
   (c) Total variable costs
   (d) Fixed costs per direct labor hour
   (e) None of the above

CHAPTER 2

Systems of Linear Equations; Matrices

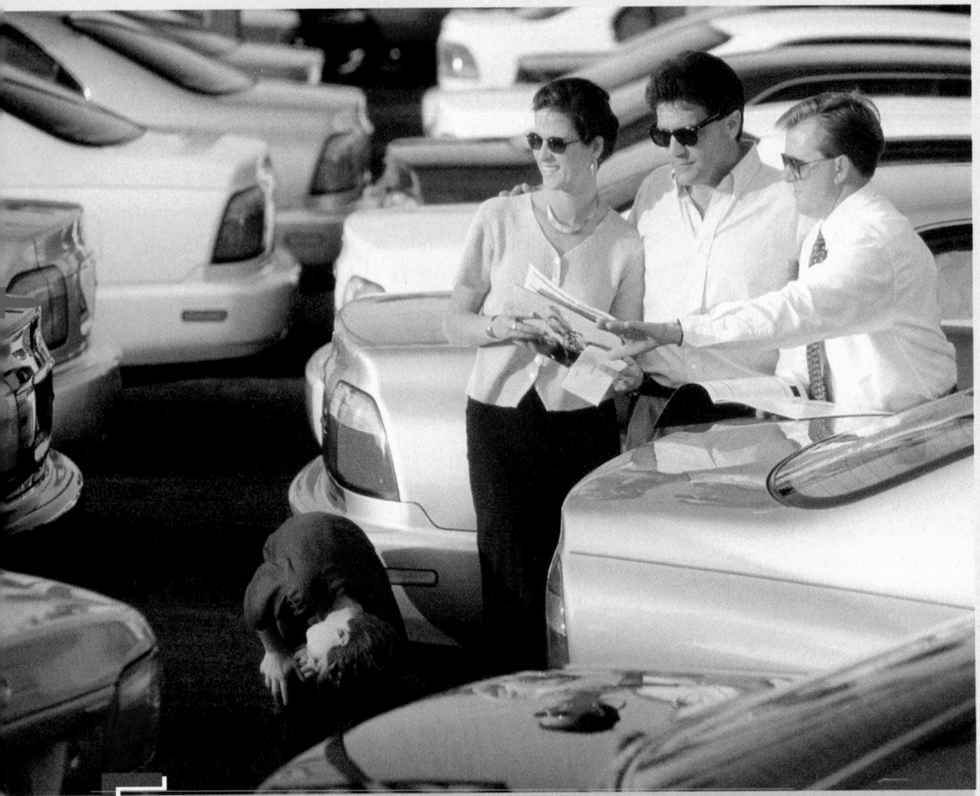

OUTLINE

Economists are always talking about the influence of consumer spending on the economy. What do they mean? Suppose you have $5000 in disposable income and could spend it on a new bathroom for your house or on a plasma TV or on a dream vacation in Fiji. Would your decision impact the economy? What if you decided to spend the $5000 on the construction of the new bathroom? This would help the construction industry but would not help the consumer electronics industry or the travel industry. What if 200 people were in the same position as you and each of them made the decision to spend $5000 on construction, resulting in increased demand for construction of $1,000,000? How would this impact other segments of the economy? Which ones does it help? Which ones does it hurt? A famous economic model, the Leontief Model, was constructed to answer such questions. We study this model in Section 2.7 and answer some of the questions listed here in the Chapter Project at the end of the chapter.

A LOOK BACK, A LOOK FORWARD

In Section 1.2 of Chapter 1, we discussed pairs of lines: coincident lines, parallel lines, and intersecting lines. Each line was given by a linear equation containing two variables. So a *pair* of lines is given by *two* linear equations containing two variables. We refer to this as a *system of two linear equations containing two variables.*

In this chapter we take up the problem of *solving* systems of linear equations containing two or more variables. As the section titles suggest, there are various ways to do this. The *method of substitution* for solving equations in several unknowns goes back to ancient times. The *method of elimination,* though it had existed for centuries, was put into systematic order by Karl Friedrich Gauss (1777–1855) and by Camille Jordan (1838–1922). This method led to the *matrix method* that is now used for solving large systems by computer.

The theory of *matrices* was developed in 1857 by Arthur Cayley (1821–1895), though only later were matrices used as we use them in this chapter. Matrices have become a very flexible instrument, invaluable in almost all areas of mathematics.

2.1 Systems of Linear Equations: Substitution; Elimination*

PREPARING FOR THIS SECTION *Before getting started, review the following:*

> Pairs of Lines (Section 1.2, pp. 19–23)

OBJECTIVES

1. Solve systems of equations by substitution
2. Solve systems of equations by elimination
3. Identify inconsistent systems of equations containing two variables
4. Express the solutions of a system of dependent equations containing two variables
5. Solve systems of three equations containing three variables
6. Identify inconsistent systems of equations containing three variables
7. Express the solutions of a system of dependent equations containing three variables

We begin with an example.

EXAMPLE 1 Movie Theater Ticket Sales

A movie theater sells tickets for \$8.00 each, with seniors receiving a discount of \$2.00. One evening the theater took in \$3580 in revenue. If x represents the number of tickets sold at \$8.00 and y the number of tickets sold at the discounted price of \$6.00, write an equation that relates these variables.

SOLUTION Each nondiscounted ticket brings in \$8.00, so x tickets will bring in $8x$ dollars. Similarly, y discounted tickets bring in $6y$ dollars. Since the total brought in is \$3580, we must have

$$8x + 6y = 3580$$

**Based on material from* Precalculus, *6th ed., by Michael Sullivan. Used here with the permission of the author and Prentice-Hall, Inc.*

The equation found in Example 1 is an example of a **linear equation containing two variables.** Some other examples of linear equations are

$$2x + 3y = 2 \qquad 5x - 2y + 3z = 10 \qquad 8x_1 + 8x_2 - 2x_3 + 5x_4 = 0$$

2 variables   3 variables   4 variables

In general, an equation containing n variables is said to be **linear** if it can be written in the form

$$a_1x_1 + a_2x_2 + \cdots + a_nx_n = b$$

where $x_1, x_2, \ldots, x_n$ are n distinct variables*, $a_1, a_2, \ldots, a_n, b$ are constants, and at least one of the a_i's is not 0.

In Example 1, suppose that we also know that 525 tickets were sold that evening. Then we have another equation relating the variables x and y.

$$x + y = 525$$

The two linear equations

$$8x + 6y = 3580$$
$$x + y = 525$$

form a *system* of linear equations.

In general, a **system of linear equations** is a collection of two or more linear equations, each containing one or more variables. Example 2 illustrates some systems of linear equations.

EXAMPLE 2   **Examples of Systems of Linear Equations**

(a) $\begin{cases} 2x + y = 5 \\ -4x + 6y = -2 \end{cases}$ (1)   Two equations containing
   (2)   two variables, x and y

(b) $\begin{cases} x + y + z = 6 \\ 3x - 2y + 4z = 9 \\ x - y - z = 0 \end{cases}$ (1)   Three equations containing
   (2)   three variables, x, y, and z
   (3)

(c) $\begin{cases} x + y + z = 5 \\ x - y = 2 \end{cases}$ (1)   Two equations containing
   (2)   three variables, x, y, and z

(d) $\begin{cases} x + y + z = 6 \\ 2x + 2z = 4 \\ y + z = 2 \\ x = 4 \end{cases}$ (1)   Four equations containing
   (2)   three variables, x, y, and z
   (3)
   (4)

(e) $\begin{cases} x_1 - 2x_2 + x_3 - x_4 = 5 \\ 3x_1 + x_2 - x_3 - 5x_4 = 2 \end{cases}$ (1)   Two equations containing four
   (2)   variables x_1, x_2, x_3, and x_4

We use a brace, as shown above, to remind us that we are dealing with a system of linear equations. We also will find it convenient to number each equation in the system.

A **solution** of a system of linear equations consists of values of the variables that are solutions of each equation of the system. To **solve** a system of linear equations means to find all solutions of the system.

---

*The notation x_n is read as "x sub n." The number n is called a **subscript** and should not be confused with an exponent. We use subscripts to distinguish one variable from another when a large or undetermined number of variables is required.*

For example, $x = 2, y = 1$ is a solution of the system in Example 2(a) because

$$\begin{cases} 2x + y = 5 & (1) \\ -4x + 6y = -2 & (2) \end{cases} \qquad \begin{cases} 2(2) + 1 = 4 + 1 = 5 & (1) \\ -4(2) + 6(1) = -8 + 6 = -2 & (2) \end{cases}$$

A solution of the system in Example 2(b) is $x = 3, y = 2, z = 1$, because

$$\begin{cases} x + y + z = 6 & (1) \\ 3x - 2y + 4z = 9 & (2) \\ x - y - z = 0 & (3) \end{cases} \qquad \begin{cases} 3 + 2 + 1 = 6 & (1) \\ 3(3) - 2(2) + 4(1) = 9 - 4 + 4 = 9 & (2) \\ 3 - 2 - 1 = 0 & (3) \end{cases}$$

Note that $x = 3, y = 3, z = 0$ is not a solution of the system in Example 2(b).

$$\begin{cases} x + y + z = 6 & (1) \\ 3x - 2y + 4z = 9 & (2) \\ x - y - z = 0 & (3) \end{cases} \qquad \begin{cases} 3 + 3 + 0 = 6 & (1) \\ 3(3) - 2(3) + 4(0) = 3 \neq 9 & (2) \\ 3 - 3 - 0 = 0 & (3) \end{cases}$$

Although these values satisfy Equations (1) and (3), they do not satisfy Equation (2). Any solution of the system must satisfy *each* equation of the system.

When a system of equations has at least one solution, it is said to be **consistent;** otherwise, it is called **inconsistent.**

NOW WORK PROBLEM 3.

Two Linear Equations Containing Two Variables

Based on the discussion in Section 1.2, we can view the problem of solving a system of two linear equations containing two variables as a geometry problem. Because the graph of each equation in such a system is a line, a system of two linear equations containing two variables represents a pair of lines. The lines either (1) are parallel or (2) are intersecting or (3) are coincident (that is, identical).

1. If the lines are parallel, then the system of equations has no solution, because the lines never intersect. The system is **inconsistent.**
2. If the lines intersect, then the system of equations has one solution, given by the point of intersection. The system is **consistent** and the equations are **independent.**
3. If the lines are coincident, then the system of equations has infinitely many solutions, represented by the totality of points on the line. The system is **consistent** and the equations are **dependent.**

Based on this, a system of equations is either

(I) Inconsistent; has no solution

or

(II) Consistent; with

(a) One solution (equations are independent)

or

(b) Infinitely many solutions (equations are dependent)

Figure 1 illustrates these conclusions.

FIGURE 1

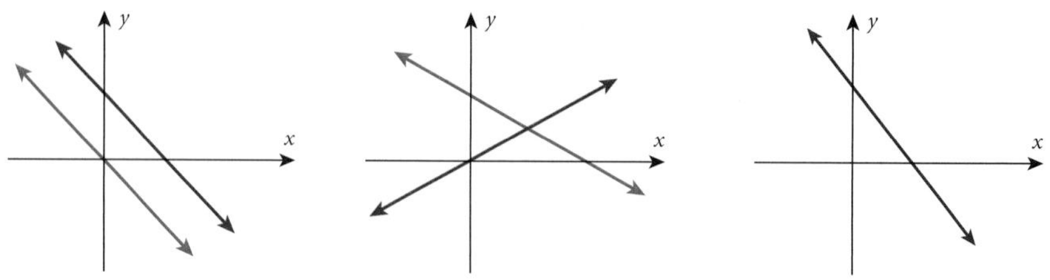

(a) Parallel lines; system has no solution and is inconsistent

(b) Intersecting lines; system has one solution and is consistent; the equations are independent

(c) Coincident lines; system has infinitely many solutions and is consistent; the equations are dependent

EXAMPLE 3 Graphing a System of Linear Equations

Graph the system: $\begin{cases} 2x + y = 5 & (1) \\ -4x + 6y = 12 & (2) \end{cases}$

SOLUTION Equation (1) is a line with x-intercept $(\frac{5}{2}, 0)$ and y-intercept $(0, 5)$. Equation (2) is a line with x-intercept $(-3, 0)$ and y-intercept $(0, 2)$.

Figure 2 shows their graphs.

FIGURE 2

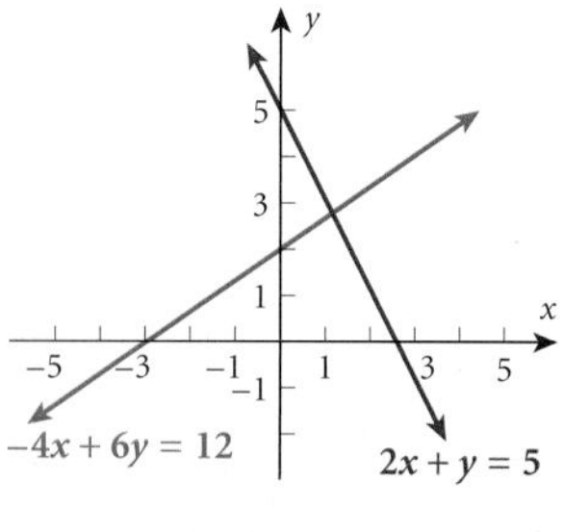

From the graph in Figure 2 we see that the lines intersect, so the system is consistent and the equations are independent. We can also use the graph as a means of approximating the solution. For this system the solution would appear to be close to the point $(1, 3)$. The actual solution, which you should verify, is $(\frac{9}{8}, \frac{11}{4})$.

To obtain the exact solution, we use algebraic methods. The first algebraic method we take up is the *method of substitution.*

Method of Substitution

Solve systems of equations by substitution 1

We illustrate the method of substitution by solving the system of Example 3.

EXAMPLE 4 Solving a System of Equations Using Substitution

Solve: $\begin{cases} 2x + y = 5 & (1) \\ -4x + 6y = 12 & (2) \end{cases}$

SOLUTION We solve the first equation for y, obtaining

$$2x + y = 5 \qquad (1)$$
$$y = -2x + 5 \qquad \text{Subtract } 2x \text{ from each side}$$

We substitute this result for y in the second equation. This results in an equation containing one variable, which we can solve.

$$-4x + 6y = 12 \qquad (2)$$
$$-4x + 6(-2x + 5) = 12 \qquad \text{Substitute } y = -2x + 5 \text{ in (2)}$$
$$-4x - 12x + 30 = 12 \qquad \text{Remove parenthesis.}$$
$$-16x = -18 \qquad \text{Combine like terms; subtract 30 from each side.}$$
$$x = \frac{-18}{-16} = \frac{9}{8} \qquad \text{Divide each side by } -16.$$

Once we know that $x = \frac{9}{8}$, we can find the value of y by **back-substitution**, that is, by substituting $\frac{9}{8}$ for x in one of the original equations.

We will use the first equation.

$$2x + y = 5 \qquad\qquad\qquad (1)$$

$$2\left(\frac{9}{8}\right) + y = 5 \qquad\qquad \text{Substitute } x = \frac{9}{8} \text{ in (1).}$$

$$\frac{9}{4} + y = 5 \qquad\qquad \text{Simplify}$$

$$y = 5 - \frac{9}{4} \qquad\qquad \text{Subtract } \frac{9}{4} \text{ from each side.}$$

$$= \frac{20}{4} - \frac{9}{4} = \frac{11}{4}$$

The solution of the system is $x = \frac{9}{8} = 1.125$, $y = \frac{11}{4} = 2.75$.

✓ **CHECK:**

$$\begin{cases} 2x + y = 5: & 2\left(\frac{9}{8}\right) + \frac{11}{4} = \frac{9}{4} + \frac{11}{4} = \frac{20}{4} = 5 \\[2mm] -4x + 6y = 12: & -4\left(\frac{9}{8}\right) + 6\left(\frac{11}{4}\right) = -\frac{9}{2} + \frac{33}{2} = \frac{24}{2} = 12 \end{cases}$$ ▶

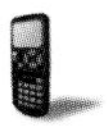

COMMENT: We can also verify our algebraic solution in Example 4 using a graphing utility.

First, we solve each equation for y. This is equivalent to writing each equation in slope–intercept form. Equation (1) in slope–intercept form is $Y_1 = -2x + 5$. Equation (2) in slope-intercept form is $Y_2 = \frac{2}{3}x + 2$. Figure 3 shows the graphs using a graphing utility. From the graph in Figure 3, we see that the lines intersect, so the system is consistent and the equations are independent. Using INTERSECT, we obtain the solution $(1.125, 2.75)$, which is equivalent to $\left(\frac{9}{8}, \frac{11}{4}\right)$. ▶

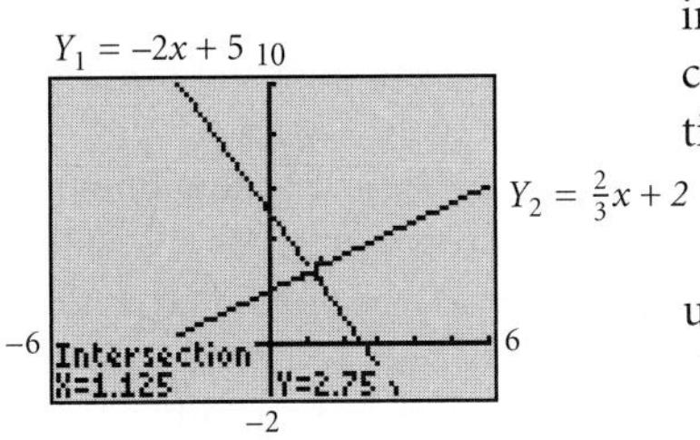

The method used to solve the system in Example 4 is called **substitution.** The steps used are outlined in the box below.

FIGURE 3

$Y_1 = -2x + 5$

$Y_2 = \frac{2}{3}x + 2$

Steps for Solving by Substitution

STEP 1 Pick one of the equations and solve for one of the variables in terms of the remaining variables.

STEP 2 Substitute the result in the remaining equations.

STEP 3 If one equation in one variable results, solve this equation. Otherwise, repeat Steps 1 and 2 until a single equation with one variable remains.

STEP 4 Find the values of the remaining variables by back-substitution.

STEP 5 Check the solution found.

EXAMPLE 5 Solving a System of Equations Using Substitution

Solve: $\begin{cases} 3x - 2y = 5 & (1) \\ 5x - \ \ y = 6 & (2) \end{cases}$

SOLUTION **STEP 1** After looking at the two equations, we conclude that it is easiest to solve for the variable y in Equation (2):

$$5x - y = 6 \qquad (2)$$
$$y = 5x - 6 \qquad \text{Add } y \text{ and subtract 6 from each side.}$$

STEP 2 We substitute this result into Equation (1) and simplify:

$$3x - 2y = 5 \qquad (1)$$
$$3x - 2(5x - 6) = 5 \qquad y = 5x - 6$$

STEP 3

$$-7x + 12 = 5 \qquad \text{Simplify}$$
$$-7x = -7 \qquad \text{Simplify}$$
$$x = 1 \qquad \text{Solve for } x$$

STEP 4 Knowing $x = 1$, we can find y from the equation

$$y = 5x - 6 = 5(1) - 6 = -1 \qquad x = 1$$

STEP 5 *Check:* $\begin{cases} 3(1) - 2(-1) = 3 + 2 = 5 \\ 5(1) - \ \ (-1) = 5 + 1 = 6 \end{cases}$

The solution of the system is $x = 1, y = -1$.

NOW WORK PROBLEM 13 USING SUBSTITUTION.

Method of Elimination

Solve systems of equations by elimination 2

A second method for solving a system of linear equations is the *method of elimination.* This method is usually preferred over substitution if substitution leads to fractions or if the system contains more than two variables. Elimination also provides the necessary motivation for solving systems using matrices (the subject of the next section).

The idea behind the method of elimination is to replace the original system of equations by an equivalent system so that adding two of the equations eliminates a variable. The rules for obtaining equivalent equations are the same as those studied earlier. However, we may also interchange any two equations of the system and/or replace any equation in the system by the sum (or difference) of that equation and any other equation in the system.

Rules for Obtaining an Equivalent System of Equations

1. Interchange any two equations in the system.
2. Multiply (or divide) each side of an equation by the same nonzero constant.
3. Replace any equation in the system by the sum (or difference) of that equation and a nonzero multiple of any other equation in the system.

An example will give you the idea. As you work through the example, pay particular attention to the pattern being followed.

EXAMPLE 6 Solving a System of Linear Equations Using Elimination

Solve: $\begin{cases} 2x + 3y = 1 & (1) \\ -x + y = -3 & (2) \end{cases}$

SOLUTION We multiply each side of equation (2) by 2 so that the coefficients of x in the two equations are opposites of one another. The result is the equivalent system

$$\begin{cases} 2x + 3y = 1 & (1) \\ -2x + 2y = -6 & (2) \end{cases}$$

If we now replace Equation (2) of this system by the sum of the two equations, we obtain an equation containing just the variable y, which we can solve.

$$\begin{cases} 2x + 3y = 1 & (1) \\ -2x + 2y = -6 & (2) \end{cases}$$

$$\begin{aligned} 5y &= -5 && \text{Add (1) and (2).} \\ y &= -1 && \text{Solve for } y. \end{aligned}$$

We back-substitute this value for y in Equation (1) and simplify to get

$$\begin{aligned} 2x + 3y &= 1 && (1) \\ 2x + 3(-1) &= 1 && \text{Subsitute } y = -1 \text{ in (1).} \\ 2x &= 4 && \text{Simplify.} \\ x &= 2 && \text{Solve for } x. \end{aligned}$$

The solution of the original system is $x = 2$, $y = -1$. We leave it to you to check the solution.

The procedure used in Example 6 is called the **method of elimination.** Notice the pattern of the solution. First, we eliminated the variable x from the second equation. Then we back-substituted; that is, we substituted the value found for y back into the first equation to find x.

Steps for Solving by Elimination

STEP 1 Select two equations from the system and eliminate a variable from them.

STEP 2 If there are additional equations in the system, pair off equations and eliminate the same variable from them.

STEP 3 Continue Steps 1 and 2 on successive systems until one equation containing one variable remains.

STEP 4 Solve for this variable and back-substitute in previous equations until all the variables have been found.

NOW WORK PROBLEM 13 USING ELIMINATION.

Let's return to the movie theater example (Example 1).

EXAMPLE 7 Movie Theater Ticket Sales

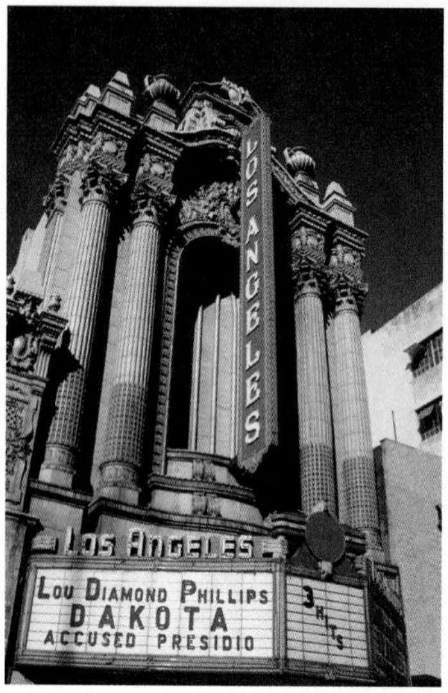

A movie theater sells tickets for \$8.00 each, with seniors receiving a discount of \$2.00. One evening the theater sold 525 tickets and took in \$3580 in revenue. How many of each type of ticket were sold?

SOLUTION If x represents the number of tickets sold at \$8.00 and y the number of tickets sold at the discounted price of \$6.00, then the given information results in the system of equations

$$\begin{cases} 8x + 6y = 3580 & (1) \\ x + y = 525 & (2) \end{cases}$$

We use elimination and multiply equation (2) by -6 and then add the equations.

$$\begin{cases} 8x + 6y = 3580 & (1) \\ -6x - 6y = -3150 & (2) \end{cases}$$

$$2x = 430 \qquad \text{Add (1) and (2).}$$

$$x = 215 \qquad \text{Solve for } x.$$

Since $x + y = 525$, then $y = 525 - x = 525 - 215 = 310$. We conclude that 215 nondiscounted tickets and 310 senior discount tickets were sold.

Identify inconsistent systems of equations containing two variables 3

The previous examples dealt with consistent systems of equations that had one solution. The next two examples deal with two other possibilities that may occur, the first being a system that has no solution.

EXAMPLE 8 An Inconsistent System of Linear Equations

Solve: $\begin{cases} 2x + y = 5 & (1) \\ 4x + 2y = 8 & (2) \end{cases}$

SOLUTION We choose to use the method of substitution and solve equation (1) for y.

$$2x + y = 5 \qquad (1)$$

$$y = -2x + 5 \qquad \text{Subtract } 2x \text{ from each side.}$$

Now substitute $y = -2x + 5$ for y in equation (2) and solve for x.

$$4x + 2y = 8 \qquad (2)$$

$$4x + 2(-2x + 5) = 8 \qquad \text{Substitute } y = -2x + 5 \text{ in (2).}$$

$$4x - 4x + 10 = 8 \qquad \text{Remove parentheses.}$$

$$0 \cdot x = -2 \qquad \text{Subtract 10 from each side.}$$

This equation has no solution. We conclude that the system itself has no solution and is therefore inconsistent.

Figure 4 illustrates the pair of lines whose equations form the system in Example 8. Notice that the graphs of the two equations are lines, each with slope -2; one line has y-intercept $(0, 5)$, the other has y-intercept $(0, 4)$. The lines are parallel and have no point of intersection. This geometric statement is equivalent to the algebraic statement that the system is inconsistent and has no solution.

FIGURE 4

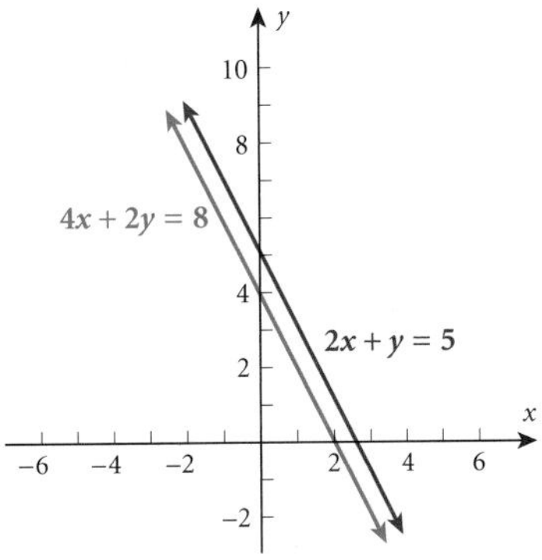

NOW WORK PROBLEM 19.

The next example is an illustration of a system with infinitely many solutions.

4 **EXAMPLE 9** **Solving a System of Linear Equations with Infinitely Many Solutions**

Solve: $\begin{cases} 2x + y = 4 & (1) \\ -6x - 3y = -12 & (2) \end{cases}$

SOLUTION We choose to use the method of elimination:

$$\begin{cases} 2x + y = 4 & (1) \\ -6x - 3y = -12 & (2) \end{cases}$$

$$\begin{cases} 6x + 3y = 12 & (1) \\ -6x - 3y = -12 & (2) \end{cases}$$ Multiply each side of equation (1) by 3.

$$\begin{cases} 6x + 3y = 12 & (1) \\ 0 = 0 & (2) \end{cases}$$ Replace equation (2) by the sum of equations (1) and (2).

The original system is equivalent to a system containing one equation, so the equations are dependent. This means that any values of x and y for which $6x + 3y = 12$ (or, equivalently, $2x + y = 4$) are solutions. For example, $x = 2$, $y = 0$; $x = 0$, $y = 4$; $x = -2$, $y = 8$; $x = 4$, $y = -4$; and so on, are solutions. There are, in fact, infinitely many values of x and y for which $2x + y = 4$, so the original system has infinitely many solutions. We will write the solutions of the original system either as

$$y = 4 - 2x$$

where x can be any real number, or as

$$x = 2 - \tfrac{1}{2}y$$

where y can be any real number.

FIGURE 5

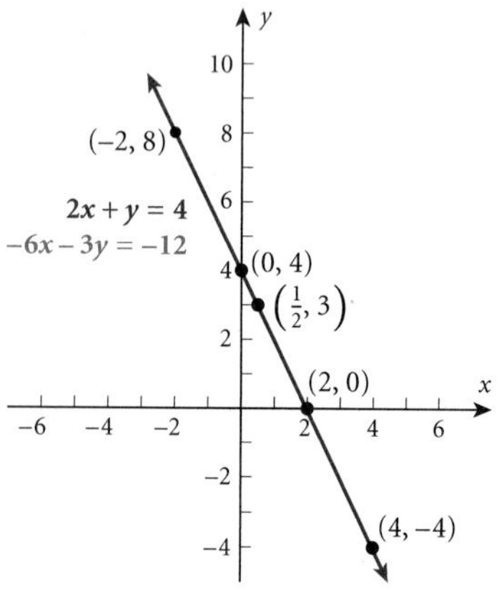

Figure 5 illustrates the situation presented in Example 9. Notice that the graphs of the two equations are lines, each with slope -2 and each with y-intercept $(0, 4)$. The lines are coincident. Notice also that Equation (2) in the original system is just -3 times Equation (1), indicating that the two equations are dependent.

For the system in Example 9 we can find some of the infinite number of solutions by assigning values to x and then finding $y = 4 - 2x$.

When we express the solution in this way, we call x a **parameter.** Thus:

If $x = 4$, then $y = -4$.	This is the point $(4, -4)$ on the graph.
If $x = 0$, then $y = 4$.	This is the point $(0, 4)$ on the graph.
If $x = \frac{1}{2}$, then $y = 3$.	This is the point $(\frac{1}{2}, 3)$ on the graph.

Alternatively, if we express the solution in the form $x = 2 - \frac{1}{2}y$, then y is the parameter and we can assign values to y in order to find x.

$$\text{If } y = -4, \text{ then } x = 2 - \tfrac{1}{2}(-4) = 4$$
$$\text{If } y = 0, \text{ then } x = 2 - \tfrac{1}{2}(0) = 2$$
$$\text{If } y = 8, \text{ then } x = 2 - \tfrac{1}{2}(8) = -2$$

NOW WORK PROBLEM 23.

Three Equations Containing Three Variables

Just as with a system of two linear equations containing two variables, a system of three linear equations containing three variables also has either (1) exactly one solution (a consistent system with independent equations), or (2) no solution (an inconsistent system), or (3) infinitely many solutions (a consistent system with dependent equations).

We can view the problem of solving a system of three linear equations containing three variables as a geometry problem. The graph of each equation in such a system is a plane in space. A system of three linear equations containing three variables represents three planes in space. Figure 6 illustrates some of the possibilities.

Recall that a **solution** to a system of equations consists of values for the variables that are solutions of each equation of the system. For example, $x = 3$, $y = -1$, $z = -5$ is a solution to the system of equations

$$\begin{cases} x + y + z = -3 & (1) \quad 3 + (-1) + (-5) = -3 \\ 2x - 3y + 6z = -21 & (2) \quad 2(3) - 3(-1) + 6(-5) = 6 + 3 - 30 = -21 \\ -3x + 5y = -14 & (3) \quad -3(3) + 5(-1) = -9 - 5 = -14 \end{cases}$$

because these values of the variables are solutions of each equation.

FIGURE 6

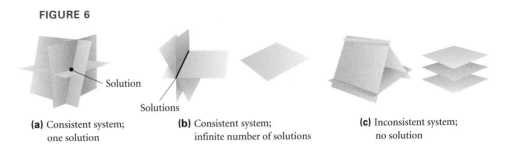

(a) Consistent system;   (b) Consistent system;   (c) Inconsistent system;
    one solution   infinite number of solutions   no solution

Typically, when solving a system of three linear equations containing three variables, we use the method of elimination. Recall that the idea behind the method of elimination is to form equivalent equations so that adding two of the equations eliminates a variable.

Solve systems of three   5   Let's see how elimination works on a system of three equations containing three
equations containing three   variables.
variables

EXAMPLE 10   Solving a System of Three Linear Equations with Three Variables

Use the method of elimination to solve the system of equations.

$$\begin{cases} x + y - z = -1 & (1) \\ 4x - 3y + 2z = 16 & (2) \\ 2x - 2y - 3z = 5 & (3) \end{cases}$$

SOLUTION   For a system of three equations, we attempt to eliminate one variable at a time, using pairs of equations, until an equation with a single variable remains. Our plan of attack on this system will be to use Equation (1) to eliminate the variable x from Equations (2) and (3).

We begin by multiplying each side of Equation (1) by -4 and adding the result to Equation (2). (Do you see why? The coefficients of x are now opposites of each other.) We also multiply Equation (1) by -2 and add the result to Equation (3). Notice that these two procedures result in the removal of the x-variable from Equations (2) and (3).

$$\begin{array}{ll} x + y - z = -1 & \text{(1) Multiply by } -4 \\ 4x - 3y + 2z = 16 & \text{(2)} \end{array} \qquad \begin{array}{ll} -4x - 4y + 4z = 4 & \text{(1)} \\ 4x - 3y + 2z = 16 & \text{(2)} \\ \hline -7y + 6z = 20 & \text{Add} \end{array}$$

$$\begin{array}{ll} x + y - z = -1 & \text{(1) Multiply by } -2 \\ 2x - 2y - 3z = 5 & \text{(3)} \end{array} \qquad \begin{array}{ll} -2x - 2y + 2z = 2 & \text{(1)} \\ 2x - 2y - 3z = 5 & \text{(3)} \\ \hline -4y - z = 7 & \text{Add} \end{array}$$

$$\begin{cases} x + y - z = -1 & (1) \\ -7y + 6z = 20 & (2) \\ -4y - z = 7 & (3) \end{cases}$$

We now concentrate on Equations (2) and (3), treating them as a system of two equations containing two variables. It is easier to eliminate z. We multiply each side of Equation (3) by 6 and add Equations (2) and (3). The result is the new Equation (3).

$$\begin{aligned} -7y + 6z &= 20 \quad (2) \\ -4y - z &= 7 \quad (3) \text{ Multiply by 6} \end{aligned} \qquad \begin{aligned} -7y + 6z &= 20 \quad (2) \\ -24y - 6z &= 42 \quad (3) \\ \hline -31y &= 62 \quad \text{Add} \end{aligned} \qquad \begin{cases} x + y - z = -1 & (1) \\ -7y + 6z = 20 & (2) \\ -31y = 62 & (3) \end{cases}$$

We now solve Equation (3) for y by dividing both sides of the equation by -31.

$$\begin{cases} x + y - z = -1 & (1) \\ -7y + 6z = 20 & (2) \\ y = -2 & (3) \end{cases}$$

Back-substitute $y = -2$ in Equation (2) and solve for z.

$$\begin{aligned} -7y + 6z &= 20 && (2) \\ -7(-2) + 6z &= 20 && \text{Substitute } y = -2 \text{ in (2).} \\ 6z &= 6 && \text{Subtract 14 from each side.} \\ z &= 1 && \text{Divide each side by 6.} \end{aligned}$$

Finally, we back-substitute $y = -2$ and $z = 1$ in Equation (1) and solve for x.

$$\begin{aligned} x + y - z &= -1 && (1) \\ x + (-2) - 1 &= -1 && \text{Substitute } y = -2 \text{ and } z = 1 \text{ in (1).} \\ x - 3 &= -1 && \text{Simplify.} \\ x &= 2 && \text{Add 3 to each side.} \end{aligned}$$

The solution of the original system is $x = 2$, $y = -2$, $z = 1$. You should verify this solution.

Look back over the solution given in Example 10. Note the pattern of removing one of the variables from two of the equations, followed by solving this system of two equations and two unknowns. Although which variables to remove is your choice, the methodology remains the same for all systems.

NOW WORK PROBLEM 35.

Identify inconsistent systems of equations containing three variables 6

The previous example was a consistent system that had a unique solution. The next two examples deal with the two other possibilities that may occur.

EXAMPLE 11 An Inconsistent System of Linear Equations

Solve: $\begin{cases} 2x + y - z = -2 & (1) \\ x + 2y - z = -9 & (2) \\ x - 4y + z = 1 & (3) \end{cases}$

SOLUTION Our plan of attack is the same as in Example 10. However, in this system, it seems easiest to eliminate the variable z first. Do you see why?

Multiply each side of Equation (1) by -1 and add the result to Equation (2). Add Equations (2) and (3).

$$
\begin{array}{rl}
-2x - y + z = 2 & \text{(1) Multiply by } -1. \\
x + 2y - z = -9 & \text{(2)} \\
\hline
-x + y = -7 & \text{Add} \\
\\
x + 2y - z = -9 & \text{(2)} \\
x - 4y + z = 1 & \text{(3)} \\
\hline
2x - 2y = -8 & \text{Add}
\end{array}
\longrightarrow
\begin{cases}
2x + y - z = -2 & (1) \\
-x + y = -7 & (2) \\
2x - 2y = -8 & (3)
\end{cases}
$$

We now concentrate on Equations (2) and (3), treating them as a system of two equations containing two variables. Multiply each side of Equation (2) by 2 and add the result to Equation (3).

$$
\begin{array}{rl}
-x + y = -7 & \text{(2) Multiply by 2.} \\
2x - 2y = -8 & \text{(3)}
\end{array}
\quad
\begin{array}{rl}
-2x + 2y = -14 & \text{(2)} \\
2x - 2y = -8 & \text{(3)} \\
\hline
0 = -22 & \text{Add}
\end{array}
\longrightarrow
\begin{cases}
2x + y - z = -2 & (1) \\
-x + y = -7 & (2) \\
0 = -22 & (3)
\end{cases}
$$

Equation (3) has no solution and the system is inconsistent.

NOW WORK PROBLEM 37.

Express the solutions of a system of dependent equations containing three variables **7**

Now let's look at a system of dependent equations.

EXAMPLE 12 Solving a System of Dependent Equations

$$
\text{Solve:} \quad
\begin{cases}
x - 2y - z = 8 & (1) \\
2x - 3y + z = 23 & (2) \\
4x - 5y + 5z = 53 & (3)
\end{cases}
$$

SOLUTION Multiply each side of Equation (1) by -2 and add the result to Equation (2). Also, multiply each side of Equation (1) by -4 and add the result to Equation (3).

$$
\begin{array}{rl}
x - 2y - z = 8 & \text{(1) Multiply by } -2. \\
2x - 3y + z = 23 & \text{(2)}
\end{array}
\quad
\begin{array}{rl}
-2x + 4y + 2z = -16 & \text{(1)} \\
2x - 3y + z = 23 & \text{(2)} \\
\hline
y + 3z = 7 & \text{Add}
\end{array}
$$

$$
\begin{array}{rl}
x - 2y - z = 8 & \text{(1) Multiply by } -4. \\
4x - 5y + 5z = 53 & \text{(3)}
\end{array}
\quad
\begin{array}{rl}
-4x + 8y + 4z = -32 & \text{(1)} \\
4x - 5y + 5z = 53 & \text{(2)} \\
\hline
3y + 9z = 21 & \text{Add}
\end{array}
\longrightarrow
\begin{cases}
x - 2y - z = 8 & (1) \\
y + 3z = 7 & (2) \\
3y + 9z = 21 & (3)
\end{cases}
$$

Treat Equations (2) and (3) as a system of two equations containing two variables, and eliminate the y-variable by multiplying each side of Equation (2) by -3 and adding the result to Equation (3).

$$
\begin{array}{rl}
y + 3z = 7 & \text{Multiply by } -3. \\
3y + 9z = 21 &
\end{array}
\quad
\begin{array}{rl}
-3y - 9z = -21 & \\
3y + 9z = 21 & \\
\hline
0 = 0 & \text{Add}
\end{array}
\longrightarrow
\begin{cases}
x - 2y - z = 8 & (1) \\
y + 3z = 7 & (2) \\
0 = 0 & (3)
\end{cases}
$$

The original system is equivalent to a system containing two equations, so the equations are dependent and the system has infinitely many solutions. If we let z represent any real number, then, solving Equation (2) for y, we determine that $y = -3z + 7$. Substitute this expression into Equation (1) to determine x in terms of z.

$$\begin{aligned} x - 2y - z &= 8 && \text{(1)} \\ x - 2(-3z + 7) - z &= 8 && \text{Substitute } y = -3z + 7 \text{ in (1).} \\ x + 6z - 14 - z &= 8 && \text{Remove parentheses.} \\ x + 5z &= 22 && \text{Combine like terms.} \\ x &= -5z + 22 && \text{Solve for } x. \end{aligned}$$

We will write the solution to the system as

$$\begin{cases} x = -5z + 22 \\ y = -3z + 7 \end{cases}$$

where z, the parameter, can be any real number.

To find specific solutions to the system, choose any value of z and use the equations $x = -5z + 22$ and $y = -3z + 7$ to determine x and y. For example, if $z = 0$, then $x = 22$ and $y = 7$, and if $z = 1$, then $x = 17$ and $y = 4$.

NOW WORK PROBLEM 39.

EXERCISE 2.1 Answers to Odd-Numbered Problems Begin on Page AN-9.

In Problems 1–10, decide whether the values of the variables listed are solutions of the system of equations.

1. $\begin{cases} 2x - y = 5 \\ 5x + 2y = 8 \end{cases}$ $x = 2, y = -1$

2. $\begin{cases} 3x + 2y = 2 \\ x - 7y = -30 \end{cases}$ $x = 2, y = 4$

3. $\begin{cases} 3x + 4y = 4 \\ \frac{1}{2}x - 3y = -\frac{1}{2} \end{cases}$ $x = 2, y = \frac{1}{2}$

4. $\begin{cases} 2x + \frac{1}{2}y = 0 \\ 3x - 4y = -\frac{19}{2} \end{cases}$ $x = -\frac{1}{2}, y = 2$

5. $\begin{cases} x - y = 3 \\ \frac{1}{2}x + y = 3 \end{cases}$ $x = 4, y = 1$

6. $\begin{cases} x - y = 3 \\ -3x + y = 1 \end{cases}$ $x = -2, y = -5$

7. $\begin{cases} 3x + 3y + 2z = 4 \\ x - y - z = 0 \\ 2y - 3z = -8 \end{cases}$ $x = 1, y = -1, z = 2$

8. $\begin{cases} 4x - z = 7 \\ 8x + 5y - z = 0 \\ -x - y + 5z = 6 \end{cases}$ $x = 2, y = -3, z = 1$

9. $\begin{cases} 3x + 3y + 2z = 4 \\ x - 3y + z = 10 \\ 5x - 2y - 3z = 8 \end{cases}$ $x = 2, y = -2, z = 2$

10. $\begin{cases} 4x - 5z = 6 \\ 5y - z = -17 \\ -x - 6y + 5z = 24 \end{cases}$ $x = 4, y = -3, z = 2$

In Problems 11–46, solve each system of equations. If the system has no solution, say that it is inconsistent.

11. $\begin{cases} x + y = 8 \\ x - y = 4 \end{cases}$

12. $\begin{cases} x + 2y = 5 \\ x + y = 3 \end{cases}$

13. $\begin{cases} 5x - y = 13 \\ 2x + 3y = 12 \end{cases}$

14. $\begin{cases} x + 3y = 5 \\ 2x - 3y = -8 \end{cases}$

15. $\begin{cases} 3x = 24 \\ x + 2y = 0 \end{cases}$

16. $\begin{cases} 4x + 5y = -2 \\ -2y = -4 \end{cases}$

17. $\begin{cases} 3x - 6y = 2 \\ 5x + 4y = 1 \end{cases}$

18. $\begin{cases} 2x + 4y = \frac{2}{3} \\ 3x - 5y = -10 \end{cases}$

19. $\begin{cases} 2x + y = 1 \\ 4x + 2y = 3 \end{cases}$

20. $\begin{cases} x - y = 5 \\ -3x + 3y = 2 \end{cases}$

21. $\begin{cases} 2x - y = 0 \\ 3x + 2y = 7 \end{cases}$

22. $\begin{cases} 3x + 3y = -1 \\ 4x + y = \dfrac{8}{3} \end{cases}$

23. $\begin{cases} x + 2y = 4 \\ 2x + 4y = 8 \end{cases}$

24. $\begin{cases} 3x - y = 7 \\ 9x - 3y = 21 \end{cases}$

25. $\begin{cases} 2x - 3y = -1 \\ 10x + y = 11 \end{cases}$

26. $\begin{cases} 3x - 2y = 0 \\ 5x + 10y = 4 \end{cases}$

27. $\begin{cases} 2x + 3y = 6 \\ x - y = \dfrac{1}{2} \end{cases}$

28. $\begin{cases} \dfrac{1}{2}x + y = -2 \\ x - 2y = 8 \end{cases}$

29. $\begin{cases} \dfrac{1}{2}x + \dfrac{1}{3}y = 3 \\ \dfrac{1}{4}x - \dfrac{2}{3}y = -1 \end{cases}$

30. $\begin{cases} \dfrac{1}{3}x - \dfrac{3}{2}y = -5 \\ \dfrac{3}{4}x + \dfrac{1}{3}y = 11 \end{cases}$

31. $\begin{cases} 3x - 5y = 3 \\ 15x + 5y = 21 \end{cases}$

32. $\begin{cases} 2x - y = -1 \\ x + \dfrac{1}{2}y = \dfrac{3}{2} \end{cases}$

33. $\begin{cases} x - y = 6 \\ 2x - 3z = 16 \\ 2y + z = 4 \end{cases}$

34. $\begin{cases} 2x + y = -4 \\ -2y + 4z = 0 \\ 3x - 2z = -11 \end{cases}$

35. $\begin{cases} x - 2y + 3z = 7 \\ 2x + y + z = 4 \\ -3x + 2y - 2z = -10 \end{cases}$

36. $\begin{cases} 2x + y - 3z = -2 \\ -2x + 2y + z = -9 \\ 3x - 4y - 3z = 15 \end{cases}$

37. $\begin{cases} x - y - z = 1 \\ 2x + 3y + z = 2 \\ 3x + 2y = 0 \end{cases}$

38. $\begin{cases} 2x - 3y - z = 0 \\ -x + 2y + z = 5 \\ 3x - 4y - z = 1 \end{cases}$

39. $\begin{cases} x - y - z = 1 \\ -x + 2y - 3z = -4 \\ 3x - 2y - 7z = 0 \end{cases}$

40. $\begin{cases} 2x - 3y - z = 0 \\ 3x + 2y + 2z = 2 \\ x + 5y + 3z = 2 \end{cases}$

41. $\begin{cases} 2x - 2y + 3z = 6 \\ 4x - 3y + 2z = 0 \\ -2x + 3y - 7z = 1 \end{cases}$

42. $\begin{cases} 3x - 2y + 2z = 6 \\ 7x - 3y + 2z = -1 \\ 2x - 3y + 4z = 0 \end{cases}$

43. $\begin{cases} x + y - z = 6 \\ 3x - 2y + z = -5 \\ x + 3y - 2z = 14 \end{cases}$

44. $\begin{cases} x - y + z = -4 \\ 2x - 3y + 4z = -15 \\ 5x + y - 2z = 12 \end{cases}$

45. $\begin{cases} x + 2y - z = -3 \\ 2x - 4y + z = -7 \\ -2x + 2y - 3z = 4 \end{cases}$

46. $\begin{cases} x + 4y - 3z = -8 \\ 3x - y + 3z = 12 \\ x + y + 6z = 1 \end{cases}$

---

47. **Dimensions of a Floor** The perimeter of a rectangular floor is 90 feet. Find the dimensions of the floor if the length is twice the width.

48. **Dimensions of a Field** The length of fence required to enclose a rectangular field is 3000 meters. What are the dimensions of the field if the difference between its length and width is 50 meters?

49. **Agriculture** According to the U.S. Department of Agriculture, in 1996–1997, the production cost for planting corn was \$246 per acre and the cost for planting soybeans was \$140 per acre. The average farm used 445 acres of land to raise corn and soybeans and budgeted \$85,600 for planting these crops. If all the land and all the money budgeted is used, how many acres of each crop should they plant?

Source: USDA, National Agricultural Statistics Service.

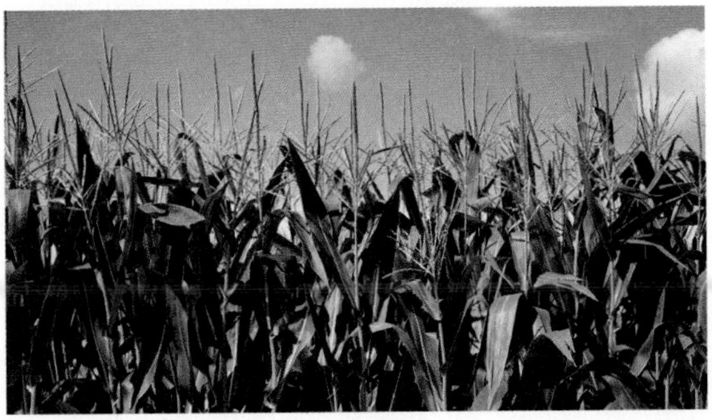

50. **Movie Theater Tickets** A movie theater charges \$9.00 for adults and \$7.00 for senior citizens. On a day when 325 people paid an admission, the total receipts were \$2495. How many who paid were adults? How many were seniors?

51. **Mixing Nuts** A store sells cashews for \$5.00 per pound and peanuts for \$1.50 per pound.The manager decides to mix 30 pounds of peanuts with some cashews and sell the mixture for \$3.00 per pound. How many pounds of cashews should be mixed with the peanuts so that the mixture will produce the same revenue as would selling the nuts separately?

52. **Financial Planning** A recently retired couple need \$12,000 per year to supplement their Social Security. They have \$150,000 to invest to obtain this income. They have decided on two investment options: AA bonds yielding 10% per annum and a Bank Certificate yielding 5%.

(a) How much should be invested in each to realize exactly \$12,000?

(b) If, after two years, the couple require \$14,000 per year in income, how should they reallocate their investment to achieve the new amount?

53. Cost of Food in Japan In Kyotoshi, Japan, the cost of three bowls of noodles and two cartons of fresh milk is 2153 yen. Three cartons of fresh milk cost 89 yen more than one bowl of noodles. What is the cost of a bowl of noodles? A carton of fresh milk?

Source: Statistics Bureau and Statistics Center, Ministry of Public Management, Home Affairs, Posts and Telecommunications, Japan, 2002.

54. Cost of Fast Food Four large cheeseburgers and two chocolate shakes cost a total of \$7.90. Two shakes cost 15¢ more than one cheeseburger. What is the cost of a cheeseburger? A shake?

55. Computing a Refund The grocery store we use does not mark prices on its goods. My wife went to this store, bought three 1-pound packages of bacon and two cartons of eggs, and paid a total of \$7.45. Not knowing that she went to the store, I also went to the same store, purchased two 1-pound packages of bacon and three cartons of eggs, and paid a total of \$6.45. Now we want to return two 1-pound packages of bacon and two cartons of eggs. How much will be refunded?

56. Blending Coffees A coffee manufacturer wants to market a new blend of coffee that will cost \$5 per pound by mixing \$3.75-per-pound coffee and \$8-per-pound coffee. What amounts of the \$3.75-per-pound coffee and \$8-per-pound coffee should be blended to obtain the desired mixture? [*Hint*: Assume the total weight of the desired blend is 100 pounds.]

57. Pharmacy A doctor's prescription calls for a daily intake of liquid containing 40 mg of vitamin C and 30 mg of vitamin D. Your pharmacy stocks two liquids that can be used: one contains 20% vitamin C and 30% vitamin D, the other 40% vitamin C and 20% vitamin D. How many milligrams of each liquid should be mixed to fill the prescription?

58. Pharmacy A doctor's prescription calls for the creation of pills that contain 12 units of vitamin B_{12} and 12 units of vitamin E. Your pharmacy stocks two powders that can be used to make these pills: one contains 20% vitamin B_{12} and 30% vitamin E, the other 40% vitamin B_{12} and 20% vitamin E. How many units of each powder should be mixed in each pill?

59. Diet Preparation A 600- to 700-pound yearling horse needs 33.0 grams of calcium and 21.0 grams of phosphorus per day for a healthy diet. A farmer provides a combination of rolled oats and molasses to provide those nutrients. Rolled oats provide 0.41 grams of calcium per pound and 1.95 grams of phosphorus per pound, while molasses provides 3.35 grams of calcium per pound and 0.36 grams of phosphorus per pound. How many pounds each of rolled oats and molasses should the farmer feed the yearling in order to meet the daily requirements?

Source: Balancing Rations for Horses, R. D. Setzler, Washington State University.

60. Restaurant Management A restaurant manager wants to purchase 200 sets of dishes. One design costs \$25 per set, while another costs \$45 per set. If she only has \$7400 to spend, how many of each design should be ordered?

61. Theater Revenues A Broadway theater has 500 seats, divided into orchestra, main, and balcony seating. Orchestra seats sell for \$50, main seats for \$35, and balcony seats for \$25. If all the seats are sold, the gross revenue to the theater is \$17,100. If all the main and balcony seats are sold, but only half the orchestra seats are sold, the gross revenue is \$14,600. How many are there of each kind of seat?

62. Theater Revenues A movie theater charges \$8.00 for adults, \$4.50 for children, and \$6.00 for senior citizens. One day the theater sold 405 tickets and collected \$2320 in receipts. There were twice as many children's tickets sold as adult tickets. How many adults, children, and senior citizens went to the theater that day?

63. Investments Kelly has \$20,000 to invest. As her financial planner, you recommend that she diversify into three investments: Treasury Bills that yield 5% simple interest, Treasury Bonds that yield 7% simple interest, and corporate bonds that yield 10% simple interest. Kelly wishes to earn \$1390 per year in income. Also, Kelly wants her investment in Treasury Bills to be \$3000 more than her investment in corporate bonds. How much money should Kelly place in each investment?

Q **64.** Make up a system of two linear equations containing two variables that has:

(a) No solution
(b) Exactly one solution
(c) Infinitely many solutions

Give the three systems to a friend to solve and critique.

Q **65.** Write a brief paragraph outlining your strategy for solving a system of two linear equations containing two variables.

Q **66.** Do you prefer the method of substitution or the method of elimination for solving a system of two linear equations containing two variables? Give reasons.

2.2 Systems of Linear Equations: Matrix Method

OBJECTIVES
1. Write the augmented matrix of a system of linear equations
2. Write the system from the augmented matrix
3. Perform row operations on a matrix
4. Solve systems of linear equations using matrices
5. Express the solution of a system with an infinite number of solutions

The systematic approach of the method of elimination for solving a system of linear equations provides another method of solution that involves a simplified notation using a *matrix*.

A **matrix** is defined as a rectangular array of numbers, enclosed by brackets. The numbers are referred to as the **entries** of the matrix. A matrix is further identified by naming its *rows* and *columns*. Some examples of matrices are

$$\begin{array}{c} \\ \text{Row 1} \\ \text{Row 2} \\ \text{Row 3} \end{array} \begin{array}{c} \begin{array}{cc} \text{Column 1} & \text{Column 2} \end{array} \\ \begin{bmatrix} 8 & 0 \\ 1 & 3 \\ -2 & 4 \end{bmatrix} \end{array} \qquad \begin{array}{c} \\ \text{Row 1} \\ \text{Row 2} \end{array} \begin{array}{c} \begin{array}{ccc} \text{Column 1} & \text{Column 2} & \text{Column 3} \end{array} \\ \begin{bmatrix} 4 & 1 & -3 \\ 2 & 1 & 2 \end{bmatrix} \end{array} \qquad \begin{array}{c} \\ \text{Row 1} \end{array} \begin{array}{c} \begin{array}{cc} \text{Column 1} & \text{Column 2} \end{array} \\ \begin{bmatrix} 4 & 3 \end{bmatrix} \end{array}$$

(a) (b) (c)

Matrix Representation of a System of Linear Equations

Consider the following two systems of two linear equations containing two variables

$$\begin{cases} x + 4y = 14 \\ 3x - 2y = 0 \end{cases} \quad \text{and} \quad \begin{cases} u + 4v = 14 \\ 3u - 2v = 0 \end{cases}$$

We observe that, except for the symbols used to represent the variables, these two systems are identical. As a result, we can dispense altogether with the letters used to symbolize the variables, provided we have some means of keeping track of them. A matrix serves us well in this regard.

When a matrix is used to represent a system of linear equations, it is called the **augmented matrix** of the system. For example, the system

$$\begin{cases} x + 4y = 14 & (1) \\ 3x - 2y = 0 & (2) \end{cases}$$

can be represented by the augmented matrix

$$\begin{array}{r} \\ \\ \text{Row 1 [Equation (1)]} \\ \text{Row 2 [Equation (2)]} \end{array} \begin{array}{c} \begin{array}{ccc} \text{Column 1} & \text{Column 2} & \text{Column 3} \\ x & y & \text{right-hand side} \end{array} \\ \left[\begin{array}{cc|c} 1 & 4 & 14 \\ 3 & -2 & 0 \end{array}\right] \end{array}$$

Here it is understood that column 1 contains the coefficients of the variable x, column 2 contains the coefficients of the variable y, and column 3 contains the numbers to the right of the equal sign. Each row of the matrix represents an equation of the system. Although not required, it has become customary to place a vertical bar in the matrix as a reminder of the equal sign.

In this book we shall follow the practice of using x and y to denote the variables for systems containing two variables. We will use x, y, and z for systems containing three variables; we will use subscripted variables (x_1, x_2, x_3, x_4, etc.) for systems containing four or more variables.

In writing the augmented matrix of a system, the variables of each equation must be on the left side of the equal sign and the constants on the right side. A variable that does not appear in an equation has a coefficient of 0.

1 **EXAMPLE 1 Writing the Augmented Matrix of a System of Linear Equations**

Write the augmented matrix of each system of equations.

(a) $\begin{cases} 3x - 4y = -6 & (1) \\ 2x - 3y = -5 & (2) \end{cases}$ **(b)** $\begin{cases} 2x - y + z = 0 & (1) \\ x + z - 1 = 0 & (2) \\ x + 2y - 8 = 0 & (3) \end{cases}$ **(c)** $\begin{cases} 3x_1 - x_2 + x_3 + x_4 = 5 & (1) \\ 2x_1 + 6x_3 = 2 & (2) \end{cases}$

SOLUTION **(a)** The augmented matrix is

$$\left[\begin{array}{cc|c} 3 & -4 & -6 \\ 2 & -3 & -5 \end{array}\right]$$

(b) Care must be taken that the system be written so that the coefficients of all variables are present (if any variable is missing, its coefficient is 0). Also, all constants must be to the right of the equal sign. We need to rearrange the given system as follows:

$$\begin{cases} 2x - y + z = 0 & (1) \\ x + z - 1 = 0 & (2) \\ x + 2y - 8 = 0 & (3) \end{cases}$$

$$\begin{cases} 2x - y + z = 0 & (1) \\ x + 0 \cdot y + z = 1 & (2) \\ x + 2y + 0 \cdot z = 8 & (3) \end{cases}$$

The augmented matrix is

$$\left[\begin{array}{ccc|c} 2 & -1 & 1 & 0 \\ 1 & 0 & 1 & 1 \\ 1 & 2 & 0 & 8 \end{array}\right]$$

(c) The augmented matrix is

$$\left[\begin{array}{cccc|c} 3 & -1 & 1 & 1 & 5 \\ 2 & 0 & 6 & 0 & 2 \end{array}\right]$$

NOW WORK PROBLEM 1.

Given an augmented matrix, we can write the corresponding system of equations.

2 **EXAMPLE 2**   **Writing the System of Linear Equations from the Augmented Matrix**

Write the system of linear equations corresponding to each augmented matrix.

(a) $\begin{bmatrix} 5 & 2 & | & 13 \\ -3 & 1 & | & -10 \end{bmatrix}$ (b) $\begin{bmatrix} 3 & -1 & -1 & | & 7 \\ 4 & 0 & 2 & | & 8 \\ 0 & 1 & 1 & | & 0 \end{bmatrix}$

SOLUTION   (a) The matrix has two rows and so represents a system of two equations. The two columns to the left of the vertical bar indicate that the system has two variables. If x and y are used to denote these variables, the system of equations is

$$\begin{cases} 5x + 2y = 13 & (1) \\ -3x + y = -10 & (2) \end{cases}$$

(b) Since the augmented matrix has three rows, it represents a system of three equations. Since there are three columns to the left of the vertical bar, the system contains three variables. If x, y, and z are the three variables, the system of equations is

$$\begin{cases} 3x - y - z = 7 & (1) \\ 4x + 2z = 8 & (2) \\ y + z = 0 & (3) \end{cases}$$

▶

Row Operations on a Matrix

Perform row operations on **3** a matrix

Row operations on a matrix are used to solve systems of equations when the system is written as an augmented matrix. There are three basic row operations.

> **Row Operations**
>
> 1. Interchange any two rows.
> 2. Replace a row by a nonzero multiple of that row.
> 3. Replace a row by the sum of that row and a constant nonzero multiple of some other row.

These three row operations correspond to the three rules given earlier for obtaining an equivalent system of equations. When a row operation is performed on a matrix, the resulting matrix represents a system of equations equivalent to the system represented by the original matrix.

For example, consider the augmented matrix

$$\begin{bmatrix} 1 & 2 & | & 3 \\ 4 & -1 & | & 2 \end{bmatrix}$$

Suppose that we want to apply a row operation to this matrix that results in a matrix whose entry in row 2, column 1 is a 0. The row operation to use is

Multiply each entry in row 1 by -4 and add the result to the corresponding entries in row 2.   **(1)**

If we use R_2 to represent the new entries in row 2 and we use r_1 and r_2 to represent the original entries in rows 1 and 2, respectively, then we can represent the row operation in statement (1) by

$$R_2 = -4r_1 + r_2$$

Then

$$\left[\begin{array}{cc|c} 1 & 2 & 3 \\ 4 & -1 & 2 \end{array}\right] \xrightarrow[R_2 = -4r_1 + r_2]{} \left[\begin{array}{cc|c} 1 & 2 & 3 \\ -4(1) + 4 & -4(2) + (-1) & -4(3) + 2 \end{array}\right] = \left[\begin{array}{cc|c} 1 & 2 & 3 \\ 0 & -9 & -10 \end{array}\right]$$

As desired, we now have the entry 0 in row 2, column 1.

EXAMPLE 3 Applying a Row Operation to an Augmented Matrix

Apply the row operation $R_2 = -3r_1 + r_2$ to the augmented matrix

$$\left[\begin{array}{cc|c} 1 & -2 & 2 \\ 3 & -5 & 9 \end{array}\right]$$

SOLUTION The row operation $R_2 = -3r_1 + r_2$ tells us that the entries in row 2 are to be replaced by the entries obtained after multiplying each entry in row 1 by -3 and adding the result to the corresponding entries in row 2. Thus,

$$\left[\begin{array}{cc|c} 1 & -2 & 2 \\ 3 & -5 & 9 \end{array}\right] \xrightarrow[R_2 = -3r_1 + r_2]{} \left[\begin{array}{cc|c} 1 & -2 & 2 \\ -3(1) + 3 & (-3)(-2) + (-5) & -3(2) + 9 \end{array}\right] = \left[\begin{array}{cc|c} 1 & -2 & 2 \\ 0 & 1 & 3 \end{array}\right]$$

NOW WORK PROBLEM 13.

EXAMPLE 4 Finding a Particular Row Operation

Using the augmented matrix

$$\left[\begin{array}{cc|c} 1 & -2 & 2 \\ 0 & 1 & 3 \end{array}\right]$$

find a row operation that will result in this augmented matrix having a 0 in row 1, column 2.

SOLUTION We want a 0 in row 1, column 2. This result can be accomplished by multiplying row 2 by 2 and adding the result to row 1. That is, we apply the row operation $R_1 = 2r_2 + r_1$.

$$\left[\begin{array}{cc|c} 1 & -2 & 2 \\ 0 & 1 & 3 \end{array}\right] \xrightarrow[R_1 = 2r_2 + r_1]{} \left[\begin{array}{cc|c} 2(0) + 1 & 2(1) + (-2) & 2(3) + 2 \\ 0 & 1 & 3 \end{array}\right] = \left[\begin{array}{cc|c} 1 & 0 & 8 \\ 0 & 1 & 3 \end{array}\right]$$

A word about the notation that we have introduced. A row operation such as $R_1 = 2r_2 + r_1$ changes the entries in row 1. Note also that for this type of row operation we change the entries in a given row by multiplying the entries in some other row by an appropriate nonzero number and adding the results to the original entries of the row to be changed.

Solving a System of Linear Equations Using Matrices

Solve systems of linear equations using matrices **4**

To solve a system of linear equations using matrices, we use row operations on the augmented matrix of the system to obtain a matrix that is in *row echelon form*.

A matrix is in **row echelon form** when

1. The entry in row 1, column 1 is a 1, and 0s appear below it.
2. The first nonzero entry in each row after the first row is a 1, 0s appear below it, and it appears to the right of the first nonzero entry in any row above.
3. Any rows that contain all 0s to the left of the vertical bar appear at the bottom.

For example, for a system of two linear equations containing two variables with a unique solution, the augmented matrix is in row echelon form if it is of the form

$$\left[\begin{array}{cc|c} 1 & a & b \\ 0 & 1 & c \end{array}\right]$$

where a, b, and c are real numbers. The second row tells us that $y = c$. We can then find the value of x by back-substituting $y = c$ into the equation given by the first row: $x + ay = b$ and solving for x.

For a system of three equations containing three variables with a unique solution, the augmented matrix is in row echelon form if it is of the form

$$\left[\begin{array}{ccc|c} 1 & a & b & d \\ 0 & 1 & c & e \\ 0 & 0 & 1 & f \end{array}\right]$$

where a, b, c, d, e, and f are real numbers. The last row of the augmented matrix states that $z = f$. We can then determine the value of y using back-substitution with $z = f$, since row 2 represents the equation $y + cz = e$. Finally, x is determined using back-substitution again.

Two advantages of solving a system of equations by writing the augmented matrix in row echelon form are the following:

1. The process is algorithmic; that is, it consists of repetitive steps that can be programmed on a computer.
2. The process works on any system of linear equations, no matter how many equations or variables are present.

Let's see how row operations are used to solve a system of linear equations. To see what is happening, we'll write the corresponding system of equations next to the matrix obtained after a row operation is performed.

EXAMPLE 5 **Solving a System of Linear Equations Using Matrices (Row Echelon Form)**

Solve: $\begin{cases} 4x + 3y = 11 \\ x - 3y = -1 \end{cases}$

SOLUTION We write the augmented matrix that represents this system:

$$\left[\begin{array}{rr|r} 4 & 3 & 11 \\ 1 & -3 & -1 \end{array}\right] \qquad \begin{cases} 4x + 3y = 11 \\ x - 3y = -1 \end{cases}$$

The next step is to place a 1 in row 1, column 1. An interchange of rows 1 and 2 is the easiest way to do this.

$$\left[\begin{array}{rr|r} 4 & 3 & 11 \\ 1 & -3 & -1 \end{array}\right] \xrightarrow[\substack{R_1 = r_2 \\ R_2 = r_1}]{} \left[\begin{array}{rr|r} 1 & -3 & -1 \\ 4 & 3 & 11 \end{array}\right] \qquad \begin{cases} x - 3y = -1 \\ 4x + 3y = 11 \end{cases}$$

Now we want a 0 under the entry 1 in column 1. (This eliminates the variable x from the second equation.) We use the row operation $R_2 = -4r_1 + r_2$.

$$\left[\begin{array}{rr|r} 1 & -3 & -1 \\ 4 & 3 & 11 \end{array}\right] \xrightarrow[R_2 = -4r_1 + r_2]{} \left[\begin{array}{rr|r} 1 & -3 & -1 \\ 0 & 15 & 15 \end{array}\right] \qquad \begin{cases} x - 3y = -1 \\ 15y = 15 \end{cases}$$

Now we want the entry 1 in row 2, column 2. (This makes it easy to solve for y.) We use $R_2 = \frac{1}{15}r_2$.

$$\left[\begin{array}{rr|r} 1 & -3 & -1 \\ 0 & 15 & 15 \end{array}\right] \xrightarrow[R_2 = \frac{1}{15}r_2]{} \left[\begin{array}{rr|r} 1 & -3 & -1 \\ 0 & 1 & 1 \end{array}\right] \qquad \begin{cases} x - 3y = -1 \\ y = 1 \end{cases}$$

This matrix is the row echelon form of the augmented matrix. The second row of the matrix on the right represents the equation $y = 1$. Using $y = 1$, we back-substitute into the equation $x - 3y = -1$ (from the first row) to get

$$\begin{aligned} x - 3y &= -1 \\ x - 3(1) &= -1 \qquad y = 1 \\ x &= 2 \end{aligned}$$

The solution of the system is $x = 2, y = 1$.

NOW WORK PROBLEM 33.

The steps we used to solve the system of linear equations in Example 5 can be summarized as follows:

Matrix Method for Solving a System of Linear Equations (Row Echelon Form)

STEP 1 Write the augmented matrix that represents the system.

STEP 2 Perform row operations that place the entry 1 in row 1, column 1.

STEP 3 Perform row operations that leave the entry 1 in row 1, column 1 unchanged, while causing 0s to appear below it in column 1.

STEP 4 Perform row operations that place the entry 1 in row 2, column 2, but leave the entries in columns to the left unchanged. If it is impossible to place a 1 in row 2, column 2, then proceed to place a 1 in row 2, column 3. Once a 1 is in place, perform row operations to place 0s below it.

(If any rows are obtained that contain only 0s on the left side of the vertical bar, place such rows at the bottom of the matrix.)

> **STEP 5** Now repeat Step 4, placing a 1 in the next row, but one column to the right. Continue until the bottom row or the vertical bar is reached.
>
> **STEP 6** The matrix that results is the row echelon form of the augmented matrix. Analyze the system of equations corresponding to it to solve the original system.

In the next example, we solve a system of three linear equations containing three variables using the steps of the matrix method.

EXAMPLE 6   **Solving a System of Linear Equations Using the Matrix Method (Row Echelon Form)**

Solve: $\begin{cases} x - y + z = 8 & (1) \\ 2x + 3y - z = -2 & (2) \\ 3x - 2y - 9z = 9 & (3) \end{cases}$

SOLUTION   **STEP 1** The augmented matrix of the system is

$$\begin{bmatrix} 1 & -1 & 1 & | & 8 \\ 2 & 3 & -1 & | & -2 \\ 3 & -2 & -9 & | & 9 \end{bmatrix}$$

STEP 2 Because the entry 1 is already present in row 1, column 1, we can go to Step 3.
STEP 3 Perform the row operations $R_2 = -2r_1 + r_2$ and $R_3 = -3r_1 + r_3$. Each of these leaves the entry 1 in row 1, column 1 unchanged, while causing 0's to appear under it.*

$$\begin{bmatrix} 1 & -1 & 1 & | & 8 \\ 2 & 3 & -1 & | & -2 \\ 3 & -2 & -9 & | & 9 \end{bmatrix} \xrightarrow[\substack{R_2 = -2r_1 + r_2 \\ R_3 = -3r_1 + r_3}]{} \begin{bmatrix} 1 & -1 & 1 & | & 8 \\ 0 & 5 & -3 & | & -18 \\ 0 & 1 & -12 & | & -15 \end{bmatrix}$$

STEP 4 The easiest way to obtain the entry 1 in row 2, column 2 without altering column 1 is to interchange rows 2 and 3 (another way would be to multiply row 2 by $\frac{1}{5}$, but this introduces fractions).

$$\begin{bmatrix} 1 & -1 & 1 & | & 8 \\ 0 & 1 & -12 & | & -15 \\ 0 & 5 & -3 & | & -18 \end{bmatrix}$$

To get a 0 under the 1 in row 2, column 2, perform the row operation $R_3 = -5r_2 + r_3$.

$$\begin{bmatrix} 1 & -1 & 1 & | & 8 \\ 0 & 1 & -12 & | & -15 \\ 0 & 5 & -3 & | & -18 \end{bmatrix} \xrightarrow{R_3 = -5r_2 + r_3} \begin{bmatrix} 1 & -1 & 1 & | & 8 \\ 0 & 1 & -12 & | & -15 \\ 0 & 0 & 57 & | & 57 \end{bmatrix}$$

You should convince yourself that doing both of these simultaneously is the same as doing the first followed by the second.

STEP 5 Continuing, we obtain a 1 in row 3, column 3 by using $R_3 = \frac{1}{57} r_3$.

$$\left[\begin{array}{rrr|r} 1 & -1 & 1 & 8 \\ 0 & 1 & -12 & -15 \\ 0 & 0 & 57 & 57 \end{array}\right] \xrightarrow[R_3 = \frac{1}{57} r_3]{} \left[\begin{array}{rrr|r} 1 & -1 & 1 & 8 \\ 0 & 1 & -12 & -15 \\ 0 & 0 & 1 & 1 \end{array}\right]$$

STEP 6 The matrix on the right is the row echelon form of the augmented matrix. The system of equations represented by the matrix in row echelon form is

$$\begin{cases} x - y + \quad z = \quad 8 & (1) \\ \quad y - 12z = -15 & (2) \\ \qquad\quad z = \quad 1 & (3) \end{cases}$$

Using $z = 1$, we back-substitute to get

$$\begin{cases} x - y + \quad 1 = \quad 8 & (1) \\ \quad y - 12(1) = -15 & (2) \end{cases} \xrightarrow{\text{Simplify.}} \begin{cases} x - y = \quad 7 & (1) \\ \quad y = -3 & (2) \end{cases}$$

We get $y = -3$, and back-substituting into $x - y = 7$, we find that $x = 4$. The solution of the system is $x = 4, y = -3, z = 1$.

GRAPHING UTILITY SOLUTION A graphing utility can be used to obtain the row echelon form of the augmented matrix. The augmented matrix of the system given in Example 6 is

$$\left[\begin{array}{rrr|r} 1 & -1 & 1 & 8 \\ 2 & 3 & -1 & -2 \\ 3 & -2 & -9 & 9 \end{array}\right]$$

We enter this matrix into our graphing utility and name it A. See Figure 7(a). Using the REF (row echelon form) command on matrix A, we obtain the results shown in Figure 7(b). Since the entire matrix does not fit on the screen, we need to scroll right to see the rest of it. See Figure 7(c).

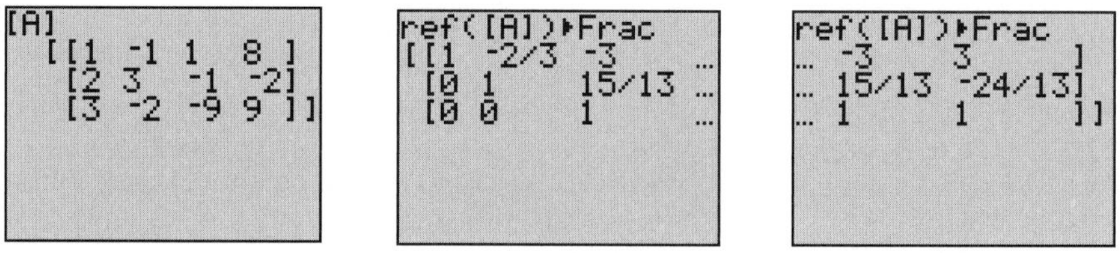

FIGURE 7

(a) (b) (c)

The system of equations represented by the matrix in row echelon form is

$$\begin{cases} x - \frac{2}{3}y - \quad 3z = \quad 3 & (1) \\ \quad y + \frac{15}{13}z = -\frac{24}{13} & (2) \\ \qquad\quad z = \quad 1 & (3) \end{cases}$$

Using $z = 1$, we back-substitute to get

$$\begin{cases} x - \frac{2}{3}y - \quad 3(1) = \quad 3 & (1) \\ \quad y + \frac{15}{13}(1) = -\frac{24}{13} & (2) \end{cases} \xrightarrow{\text{Simplify.}} \begin{cases} x - \frac{2}{3}y = 6 & (1) \\ \quad y = -\frac{39}{13} = -3 & (2) \end{cases}$$

Solving the second equation for y, we find that $y = -3$. Back-substituting $y = -3$ into $x - \frac{2}{3}y = 6$, we find that $x = 4$. The solution of the system is $x = 4, y = -3, z = 1$.

Notice that the row echelon form of the augmented matrix using the graphing utility differs from the row echelon form in our algebraic solution, yet both matrices provide the same solution! This is because the two solutions used different row operations to obtain the row echelon form. In all likelihood, the two solutions parted ways in Step 4 of the algebraic solution, where we avoided introducing fractions by interchanging rows 2 and 3.

Sometimes it is advantageous to write a matrix in **reduced row echelon form.** In this form, row operations are used to obtain entries that are 0 above (as well as below) the leading 1 in a row. For example, the row echelon form obtained in the algebraic solution to Example 6 is

$$\left[\begin{array}{rrr|r} 1 & -1 & 1 & 8 \\ 0 & 1 & -12 & -15 \\ 0 & 0 & 1 & 1 \end{array}\right]$$

To write this matrix in reduced row echelon form, we proceed as follows:

$$\left[\begin{array}{rrr|r} 1 & -1 & 1 & 8 \\ 0 & 1 & -12 & -15 \\ 0 & 0 & 1 & 1 \end{array}\right] \xrightarrow[R_1 = r_2 + r_1]{} \left[\begin{array}{rrr|r} 1 & 0 & -11 & -7 \\ 0 & 1 & -12 & -15 \\ 0 & 0 & 1 & 1 \end{array}\right] \xrightarrow[\substack{R_1 = 11r_3 + r_1 \\ R_2 = 12r_3 + r_2}]{} \left[\begin{array}{rrr|r} 1 & 0 & 0 & 4 \\ 0 & 1 & 0 & -3 \\ 0 & 0 & 1 & 1 \end{array}\right]$$

The matrix is now written in reduced row echelon form. The advantage of writing the matrix in this form is that the solution to the system, $x = 4, y = -3, z = 1$, is readily found, without the need to back-substitute. Another advantage will be seen in Section 2.6, where the inverse of a matrix is discussed.

FIGURE 8

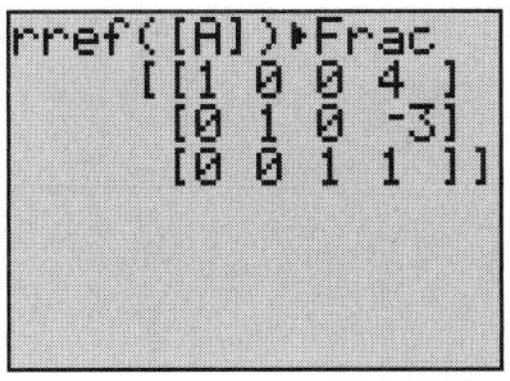

COMMENT: Most graphing utilities also have the ability to put a matrix in reduced row echelon form. Figure 8 shows the reduced row echelon form of the augmented matrix from Example 6 using the RREF command on a TI-83 Plus graphing calculator.

NOW WORK PROBLEMS 49 AND 59.

EXAMPLE 7 Calculating Production Output

FoodPerfect Corporation manufactures three models of the Perfect Foodprocessor. Each Model X processor requires 30 minutes of electrical assembly, 40 minutes of mechanical assembly, and 30 minutes of testing; each Model Y requires 20 minutes of electrical assembly, 50 minutes of mechanical assembly, and 30 minutes of testing; and each Model Z requires 30 minutes of electrical assembly, 30 minutes of mechanical assembly, and 20 minutes of testing. If 2500 minutes of electrical assembly, 3500 minutes of mechanical assembly, and 2400 minutes of testing are used in one day, how many of each model will be produced?

SOLUTION The table on page 74 summarizes the given information:

	Model			Time Used
	X	Y	Z	
Electrical Assembly	30	20	30	2500
Mechanical Assembly	40	50	30	3500
Testing	30	30	20	2400

We assign variables to represent the unknowns:

$$x = \text{Number of Model X produced}$$
$$y = \text{Number of Model Y produced}$$
$$z = \text{Number of Model Z produced}$$

Based on the table, we obtain the following system of equations:

$$\begin{cases} 30x + 20y + 30z = 2500 \\ 40x + 50y + 30z = 3500 \\ 30x + 30y + 20z = 2400 \end{cases} \xrightarrow[\text{equation by 10.}]{\text{Divide each}} \begin{cases} 3x + 2y + 3z = 250 & (1) \\ 4x + 5y + 3z = 350 & (2) \\ 3x + 3y + 2z = 240 & (3) \end{cases}$$

The augmented matrix of this system is

$$\left[\begin{array}{ccc|c} 3 & 2 & 3 & 250 \\ 4 & 5 & 3 & 350 \\ 3 & 3 & 2 & 240 \end{array}\right]$$

We could obtain a 1 in row 1, column 1 by using the row operation $R_1 = \frac{1}{3}r_1$, but the introduction of fractions is best avoided. Instead, we use

$$R_2 = -1r_1 + r_2$$

to place a 1 in row 2, column 1. The result is

$$\left[\begin{array}{ccc|c} 3 & 2 & 3 & 250 \\ 4 & 5 & 3 & 350 \\ 3 & 3 & 2 & 240 \end{array}\right] \xrightarrow[R_2 = -r_1 + r_2]{} \left[\begin{array}{ccc|c} 3 & 2 & 3 & 250 \\ 1 & 3 & 0 & 100 \\ 3 & 3 & 2 & 240 \end{array}\right]$$

Next, interchange row 1 and row 2:

$$\left[\begin{array}{ccc|c} 3 & 2 & 3 & 250 \\ 1 & 3 & 0 & 100 \\ 3 & 3 & 2 & 240 \end{array}\right] \xrightarrow[\substack{R_1 = r_2 \\ R_2 = r_1}]{} \left[\begin{array}{ccc|c} 1 & 3 & 0 & 100 \\ 3 & 2 & 3 & 250 \\ 3 & 3 & 2 & 240 \end{array}\right]$$

Use $R_2 = -3r_1 + r_2$ and $R_3 = -3r_1 + r_3$ to obtain

$$\left[\begin{array}{ccc|c} 1 & 3 & 0 & 100 \\ 3 & 2 & 3 & 250 \\ 3 & 3 & 2 & 240 \end{array}\right] \xrightarrow[\substack{R_2 = -3r_1 + r_2 \\ R_3 = -3r_1 + r_3}]{} \left[\begin{array}{ccc|c} 1 & 3 & 0 & 100 \\ 0 & -7 & 3 & -50 \\ 0 & -6 & 2 & -60 \end{array}\right]$$

We use $R_2 = -1r_2$ followed by $R_2 = r_3 + r_2$:

$$\left[\begin{array}{ccc|c} 1 & 3 & 0 & 100 \\ 0 & -7 & 3 & -50 \\ 0 & -6 & 2 & -60 \end{array}\right] \xrightarrow[R_2 = -1r_2]{} \left[\begin{array}{ccc|c} 1 & 3 & 0 & 100 \\ 0 & 7 & -3 & 50 \\ 0 & -6 & 2 & -60 \end{array}\right] \xrightarrow[R_2 = r_3 + r_2]{} \left[\begin{array}{ccc|c} 1 & 3 & 0 & 100 \\ 0 & 1 & -1 & -10 \\ 0 & -6 & 2 & -60 \end{array}\right]$$

Next, use $R_3 = 6r_2 + r_3$ to obtain

$$\left[\begin{array}{rrr|r} 1 & 3 & 0 & 100 \\ 0 & 1 & -1 & -10 \\ 0 & -6 & 2 & -60 \end{array}\right] \xrightarrow[R_3 = 6r_2 + r_3]{} \left[\begin{array}{rrr|r} 1 & 3 & 0 & 100 \\ 0 & 1 & -1 & -10 \\ 0 & 0 & -4 & -120 \end{array}\right]$$

Next, use $R_3 = -\frac{1}{4}r_3$. The result is

$$\left[\begin{array}{rrr|r} 1 & 3 & 0 & 100 \\ 0 & 1 & -1 & -10 \\ 0 & 0 & -4 & -120 \end{array}\right] \xrightarrow[R_3 = -\frac{1}{4}r_3]{} \left[\begin{array}{rrr|r} 1 & 3 & 0 & 100 \\ 0 & 1 & -1 & -10 \\ 0 & 0 & 1 & 30 \end{array}\right]$$

The matrix is now in row echelon form. We find $z = 30$. From row 2, we have $y - z = -10$ so that $y = z - 10 = 30 - 10 = 20$. Finally, from row 1, we have $x + 3y = 100$ so $x = -3y + 100 = -60 + 100 = 40$. The solution of the system is $x = 40$, $y = 20$, $z = 30$. In one day 40 Model X, 20 Model Y, and 30 Model Z processors were produced.

NOW WORK PROBLEM 73.

The matrix method for solving a system of linear equations also identifies systems that have infinitely many solutions and systems that are inconsistent. Let's see how.

5 **EXAMPLE 8**

Solving a System of Linear Equations Using Matrices (Infinitely Many Solutions)

Solve: $\begin{cases} 2x - 3y = 5 \\ 4x - 6y = 10 \end{cases}$

SOLUTION The augmented matrix representing this system is

$$\left[\begin{array}{rr|r} 2 & -3 & 5 \\ 4 & -6 & 10 \end{array}\right]$$

To place a 1 in row 1, column 1, we use $R_1 = \frac{1}{2}r_1$:

$$\left[\begin{array}{rr|r} 2 & -3 & 5 \\ 4 & -6 & 10 \end{array}\right] \xrightarrow[R_1 = \frac{1}{2}r_1]{} \left[\begin{array}{rr|r} 1 & -\dfrac{3}{2} & \dfrac{5}{2} \\ 4 & -6 & 10 \end{array}\right]$$

To place a 0 in column 1, row 2, we use $R_2 = -4r_1 + r_2$:

$$\left[\begin{array}{rr|r} 1 & -\dfrac{3}{2} & \dfrac{5}{2} \\ 4 & -6 & 10 \end{array}\right] \xrightarrow[R_2 = -4r_1 + r_2]{} \left[\begin{array}{rr|r} 1 & -\dfrac{3}{2} & \dfrac{5}{2} \\ 0 & 0 & 0 \end{array}\right]$$

The system of equations looks like

$$\begin{cases} x - \dfrac{3}{2}y = \dfrac{5}{2} \\ 0x + 0y = 0 \end{cases}$$

The second equation is true for any choice of x and y, so all numbers x and y that obey the first equation are solutions of the system. Since any point on the line $x - \frac{3}{2}y = \frac{5}{2}$ is a solution, there are an infinite number of solutions.

Using y as parameter, we can list some of these solutions by assigning values to y and then calculating x from the equation $x = \frac{3}{2}y + \frac{5}{2}$.

If $y = 0$, then $x = \frac{5}{2}$, so $x = \frac{5}{2}, y = 0$ is a solution.
If $y = 1$, then $x = 4$, so $x = 4, y = 1$ is a solution.
If $y = 5$, then $x = 10$, so $x = 10, y = 5$ is a solution.
If $y = -3$, then $x = -2$, so $x = -2, y = -3$ is a solution.

And so on.

We write the solution as

$$x = \frac{3}{2}y + \frac{5}{2}, \quad y \text{ is any real number}$$

NOW WORK PROBLEM 43.

Let's solve a system of three equations containing three variables.

EXAMPLE 9 Solving a System of Linear Equations Using Matrices (Infinitely Many Solutions)

Solve: $\begin{cases} 6x - y - z = 4 & (1) \\ -12x + 2y + 2z = -8 & (2) \\ 5x + y - z = 3 & (3) \end{cases}$

SOLUTION We start with the augmented matrix of the system. We then use row operations to obtain a 1 in row 1, column 1 and 0's in the remainder of column 1.

$$\left[\begin{array}{rrr|r} 6 & -1 & -1 & 4 \\ -12 & 2 & 2 & -8 \\ 5 & 1 & -1 & 3 \end{array}\right] \xrightarrow{R_1 = -r_3 + r_1} \left[\begin{array}{rrr|r} 1 & -2 & 0 & 1 \\ -12 & 2 & 2 & -8 \\ 5 & 1 & -1 & 3 \end{array}\right] \xrightarrow[R_3 = -5r_1 + r_3]{R_2 = 12r_1 + r_2} \left[\begin{array}{rrr|r} 1 & -2 & 0 & 1 \\ 0 & -22 & 2 & 4 \\ 0 & 11 & -1 & -2 \end{array}\right]$$

Obtaining a 1 in row 2, column 2 without altering column 1 can be accomplished by $R_2 = -\frac{1}{22}r_2$, or by $R_3 = \frac{1}{11}r_3$ and interchanging rows 2 and 3, or by $R_2 = \frac{23}{11}r_3 + r_2$. We shall use the first of these.

$$\left[\begin{array}{rrr|r} 1 & -2 & 0 & 1 \\ 0 & -22 & 2 & 4 \\ 0 & 11 & -1 & -2 \end{array}\right] \xrightarrow{R_2 = -\frac{1}{22}r_2} \left[\begin{array}{rrr|r} 1 & -2 & 0 & 1 \\ 0 & 1 & -\frac{1}{11} & -\frac{2}{11} \\ 0 & 11 & -1 & -2 \end{array}\right] \xrightarrow{R_3 = -11r_2 + r_3} \left[\begin{array}{rrr|r} 1 & -2 & 0 & 1 \\ 0 & 1 & -\frac{1}{11} & -\frac{2}{11} \\ 0 & 0 & 0 & 0 \end{array}\right]$$

This matrix is in row echelon form. Because the bottom row consists entirely of 0s, the system actually consists of only two equations.

$$\begin{cases} x - 2y = 1 & (1) \\ y - \frac{1}{11}z = -\frac{2}{11} & (2) \end{cases}$$

From the second equation we get $y = \frac{1}{11}z - \frac{2}{11}$, and then we back-substitute the solution for y from the second equation into the first equation to get

$$x = 2y + 1 = 2\left(\frac{1}{11}z - \frac{2}{11}\right) + 1 = \frac{2}{11}z + \frac{7}{11}$$

The original system is equivalent to the system

$$\begin{cases} x = \dfrac{2}{11}z + \dfrac{7}{11} & (1) \\ y = \dfrac{1}{11}z - \dfrac{2}{11} & (2) \end{cases}$$

where z, the parameter, can be any real number.

Let's look at the situation. The original system of three equations is equivalent to a system containing two equations. This means that any values of x, y, z that satisfy both

$$x = \frac{2}{11}z + \frac{7}{11} \quad \text{and} \quad y = \frac{1}{11}z - \frac{2}{11}$$

will be solutions. For example, $z = 0$, $x = \frac{7}{11}$, $y = -\frac{2}{11}$; $z = 1$, $x = \frac{9}{11}$, $y = -\frac{1}{11}$; and $z = -1$, $x = \frac{5}{11}$, $y = -\frac{3}{11}$ are some of the solutions of the original system. There are, in fact, infinitely many values of x, y, and z for which the two equations are satisfied. That is, the original system has infinitely many solutions. We will write the solution of the original system as

$$\begin{cases} x = \dfrac{2}{11}z + \dfrac{7}{11} \\ y = \dfrac{1}{11}z - \dfrac{2}{11} \end{cases}$$

where z, the parameter, can be any real number.

We can also find the solution by writing the augmented matrix in reduced row echelon form. Starting with the row echelon form, we have

$$\left[\begin{array}{ccc|c} 1 & -2 & 0 & 1 \\ 0 & 1 & -\dfrac{1}{11} & -\dfrac{2}{11} \\ 0 & 0 & 0 & 0 \end{array}\right] \xrightarrow[R_1 = 2r_2 + r_1]{} \left[\begin{array}{ccc|c} 1 & 0 & -\dfrac{2}{11} & \dfrac{7}{11} \\ 0 & 1 & -\dfrac{1}{11} & -\dfrac{2}{11} \\ 0 & 0 & 0 & 0 \end{array}\right]$$

The matrix on the right is in reduced row echelon form. The corresponding system of equations is

$$\begin{cases} x - \dfrac{2}{11}z = \dfrac{7}{11} & (1) \\ y - \dfrac{1}{11}z = -\dfrac{2}{11} & (2) \end{cases}$$

or, equivalently,

$$\begin{cases} x = \dfrac{2}{11}z + \dfrac{7}{11} & (1) \\ y = \dfrac{1}{11}z - \dfrac{2}{11} & (2) \end{cases}$$

where z, the parameter, can be any real number.

NOW WORK PROBLEM 51.

EXAMPLE 10 Solving a System of Linear Equations Using Matrices

$$\text{Solve: } \begin{cases} x + y + z = 6 \\ 2x - y - z = 3 \\ x + 2y + 2z = 0 \end{cases}$$

SOLUTION We proceed as follows, beginning with the augmented matrix.

$$\left[\begin{array}{ccc|c} 1 & 1 & 1 & 6 \\ 2 & -1 & -1 & 3 \\ 1 & 2 & 2 & 0 \end{array}\right] \xrightarrow[\substack{R_2 = -2r_1 + r_2 \\ R_3 = -r_1 + r_3}]{} \left[\begin{array}{ccc|c} 1 & 1 & 1 & 6 \\ 0 & -3 & -3 & -9 \\ 0 & 1 & 1 & -6 \end{array}\right] \xrightarrow[\substack{\text{Interchange} \\ \text{rows 2 and 3.}}]{} \left[\begin{array}{ccc|c} 1 & 1 & 1 & 6 \\ 0 & 1 & 1 & -6 \\ 0 & -3 & -3 & -9 \end{array}\right] \xrightarrow[R_3 = 3r_2 + r_3]{} \left[\begin{array}{ccc|c} 1 & 1 & 1 & 6 \\ 0 & 1 & 1 & -6 \\ 0 & 0 & 0 & -27 \end{array}\right]$$

This matrix is in row echelon form. The bottom row is equivalent to the equation

$$0x + 0y + 0z = -27$$

which has no solution. The original system is inconsistent.

NOW WORK PROBLEM 23.

SUMMARY After finding the row echelon form of the augmented matrix of a system of two linear equations containing two variables, one of the following matrices will result.

$\left[\begin{array}{cc|c} 1 & a & c \\ 0 & 1 & d \end{array}\right]$	**Unique solution:** $y = d$ Back-substitute to find x.
$\left[\begin{array}{cc|c} a & b & c \\ 0 & 0 & 0 \end{array}\right]$	**Infinite number of solutions:** $ax + by = c$ Either x or y can be used as parameter.
$\left[\begin{array}{cc|c} a & b & c \\ 0 & 0 & \text{nonzero number} \end{array}\right]$	**No solution**

EXERCISE 2.2 Answers to Odd-Numbered Problems Begin on Page AN-10.

In Problems 1–12, write the augmented matrix of each system of equations.

1. $\begin{cases} 2x - 3y = 5 \\ x - y = 3 \end{cases}$

2. $\begin{cases} 4x + y = 5 \\ 2x + y = 5 \end{cases}$

3. $\begin{cases} 2x + y + 6 = 0 \\ 3x + y = -1 \end{cases}$

4. $\begin{cases} -3x - y = -3 \\ 4x - y + 2 = 0 \end{cases}$

5. $\begin{cases} 2x - y - z = 0 \\ x - y + z = 1 \\ 3x - y = 2 \end{cases}$

6. $\begin{cases} x + y + z = 3 \\ 2x + z = 0 \\ 3x - y - z = 1 \end{cases}$

7. $\begin{cases} 2x - 3y + z - 7 = 0 \\ x + y - z = 1 \\ 2x + 2y - 3z + 4 = 0 \end{cases}$

8. $\begin{cases} 5x - 3y + 6z + 1 = 0 \\ -x - y + z = 1 \\ 2x + 3y + 5 = 0 \end{cases}$

9. $\begin{cases} 4x_1 - x_2 + 2x_3 - x_4 = 4 \\ x_1 + x_2 + 6 = 0 \\ 2x_2 - x_3 + x_4 = 5 \end{cases}$

10. $\begin{cases} 3x_1 - 5x_2 + x_3 = 2 \\ x_1 - x_2 + x_3 = 6 \\ 2x_1 + x_3 + 4 = 0 \end{cases}$

11. $\begin{cases} x_1 - x_2 + x_3 - x_4 = 0 \\ 2x_1 + 3x_2 - x_3 + 4x_4 = 5 \end{cases}$

12. $\begin{cases} x_1 + x_2 + x_3 + x_4 = 4 \\ x_1 - 2x_2 + 3x_3 - 4x_4 = 5 \end{cases}$

In Problems 13–20, perform each row operation on the given augmented matrix.

13. $\left[\begin{array}{rr|r} 1 & -3 & -2 \\ 2 & -5 & 5 \end{array}\right]$ (a) $R_2 = -2r_1 + r_2$

14. $\left[\begin{array}{rr|r} 1 & -3 & -3 \\ 2 & -5 & -4 \end{array}\right]$ (a) $R_2 = -2r_1 + r_2$

15. $\left[\begin{array}{rrr|r} 1 & -3 & 4 & 3 \\ 2 & -5 & 6 & 6 \\ -3 & 3 & 4 & 6 \end{array}\right]$ (a) $R_2 = -2r_1 + r_2$ (b) $R_3 = 3r_1 + r_3$

16. $\left[\begin{array}{rrr|r} 1 & -3 & 3 & -5 \\ 2 & -5 & -3 & -5 \\ -3 & -2 & 4 & 6 \end{array}\right]$ (a) $R_2 = -2r_1 + r_2$ (b) $R_3 = 3r_1 + r_3$

17. $\left[\begin{array}{rrr|r} 1 & -3 & 2 & -6 \\ 2 & -5 & 3 & -4 \\ -3 & -6 & 2 & 6 \end{array}\right]$ (a) $R_2 = -2r_1 + r_2$ (b) $R_3 = 3r_1 + r_3$

18. $\left[\begin{array}{rrr|r} 1 & -3 & -4 & -6 \\ 2 & -5 & 6 & -6 \\ -3 & 1 & 4 & 6 \end{array}\right]$ (a) $R_2 = -2r_1 + r_2$ (b) $R_3 = 3r_1 + r_3$

19. $\left[\begin{array}{rrr|r} 1 & -3 & 1 & -2 \\ 2 & -5 & 6 & -2 \\ -3 & 1 & 4 & 6 \end{array}\right]$ (a) $R_2 = -2r_1 + r_2$ (b) $R_3 = 3r_1 + r_3$

20. $\left[\begin{array}{rrr|r} 1 & -3 & -1 & 2 \\ 2 & -5 & 2 & 6 \\ -3 & -6 & 4 & 6 \end{array}\right]$ (a) $R_2 = -2r_1 + r_2$ (b) $R_3 = 3r_1 + r_3$

In Problems 21–32, the row echelon form of a system of linear equations is given. Write the system of equations corresponding to the given matrix. Use x, y; or x, y, z; or x_1, x_2, x_3, x_4 as variables. Determine whether the system is consistent or inconsistent. If it is consistent, give the solution.

21. $\left[\begin{array}{rr|r} 1 & 2 & 5 \\ 0 & 1 & -1 \end{array}\right]$

22. $\left[\begin{array}{rr|r} 1 & -3 & -4 \\ 0 & 1 & 0 \end{array}\right]$

23. $\left[\begin{array}{rrr|r} 1 & 2 & 3 & 1 \\ 0 & 1 & 4 & 2 \\ 0 & 0 & 0 & 3 \end{array}\right]$

24. $\left[\begin{array}{rrr|r} 1 & 2 & -1 & 0 \\ 0 & 1 & -1 & 1 \\ 0 & 0 & 0 & 2 \end{array}\right]$

25. $\left[\begin{array}{rrr|r} 1 & 0 & 2 & -1 \\ 0 & 1 & -4 & -2 \\ 0 & 0 & 0 & 0 \end{array}\right]$

26. $\left[\begin{array}{rrr|r} 1 & 0 & 4 & 4 \\ 0 & 1 & 3 & 2 \\ 0 & 0 & 0 & 0 \end{array}\right]$

27. $\left[\begin{array}{rrrr|r} 1 & 2 & -1 & 1 & 1 \\ 0 & 1 & 4 & 1 & 2 \\ 0 & 0 & 1 & 2 & 3 \end{array}\right]$

28. $\left[\begin{array}{rrrr|r} 1 & 2 & 4 & 0 & 1 \\ 0 & 1 & -1 & 2 & 2 \\ 0 & 0 & 1 & 3 & 0 \end{array}\right]$

29. $\left[\begin{array}{rrrr|r} 1 & 2 & 0 & 4 & 2 \\ 0 & 1 & 1 & 3 & 3 \\ 0 & 0 & 0 & 0 & 0 \end{array}\right]$

30. $\left[\begin{array}{rrrr|r} 1 & 0 & 3 & 0 & 1 \\ 0 & 1 & 4 & 3 & 2 \\ 0 & 0 & 1 & 2 & 3 \end{array}\right]$

31. $\left[\begin{array}{rrrr|r} 1 & -2 & 0 & 1 & -2 \\ 0 & 1 & -3 & 2 & 2 \\ 0 & 0 & 1 & -1 & 0 \\ 0 & 0 & 0 & 0 & 0 \end{array}\right]$

32. $\left[\begin{array}{rrrr|r} 1 & 3 & 0 & 4 & 1 \\ 0 & 1 & 2 & -1 & 2 \\ 0 & 0 & 1 & 2 & 3 \\ 0 & 0 & 0 & 1 & 0 \end{array}\right]$

In Problems 33–58, solve each system of equations using matrices. If the system has no solution, say it is inconsistent.

33. $\begin{cases} x + y = 6 \\ 2x - y = 0 \end{cases}$

34. $\begin{cases} x - y = 2 \\ 2x + y = 1 \end{cases}$

35. $\begin{cases} 2x + y = 5 \\ x - y = 1 \end{cases}$

36. $\begin{cases} 3x + 2y = 7 \\ x + y = 3 \end{cases}$

37. $\begin{cases} 2x + 3y = 7 \\ 3x - y = 5 \end{cases}$

38. $\begin{cases} 2x - 3y = 5 \\ 3x + y = 2 \end{cases}$

39. $\begin{cases} 2x - 3y = 6 \\ 6x - 9y = 10 \end{cases}$

40. $\begin{cases} 3x + 9y = 4 \\ 2x + 6y = 1 \end{cases}$

41. $\begin{cases} 2x - 3y = 0 \\ 4x + 9y = 5 \end{cases}$

42. $\begin{cases} 3x - 4y = 3 \\ 6x + 2y = 1 \end{cases}$

43. $\begin{cases} 2x + 6y = 4 \\ 5x + 15y = 10 \end{cases}$

44. $\begin{cases} 3x + 5y = 5 \\ 6x + 10y = 10 \end{cases}$

45. $\begin{cases} \frac{1}{2}x + \frac{1}{3}y = 2 \\ x + y = 5 \end{cases}$

46. $\begin{cases} x - \frac{1}{4}y = 0 \\ \frac{1}{2}x + \frac{1}{2}y = \frac{5}{2} \end{cases}$

47. $\begin{cases} x + y = 1 \\ 3x - 2y = \frac{4}{3} \end{cases}$

48. $\begin{cases} 4x - y = \frac{11}{4} \\ 3x + y = \frac{5}{2} \end{cases}$

49. $\begin{cases} 2x + y + z = 6 \\ x - y - z = -3 \\ 3x + y + 2z = 7 \end{cases}$

50. $\begin{cases} x + y + z = 5 \\ 2x - y + z = 2 \\ x + 2y - z = 3 \end{cases}$

51. $\begin{cases} 2x - 2y - z = 2 \\ 2x + 3y + z = 2 \\ 3x + 2y = 0 \end{cases}$

52. $\begin{cases} 2x - y - z = -5 \\ x + y + z = 2 \\ x + 2y + 2z = 5 \end{cases}$

53. $\begin{cases} 2x + y - z = 2 \\ x + 3y + 2z = 1 \\ x + y + z = 2 \end{cases}$

54. $\begin{cases} 2x + 2y + z = 6 \\ x - y - z = -2 \\ x - 2y - 2z = -5 \end{cases}$

55. $\begin{cases} x + y - z = 0 \\ 4x + 4y - 4z = -1 \\ 2x + y + z = 2 \end{cases}$

56. $\begin{cases} x + y - z = 0 \\ 4x + 2y - 4z = 0 \\ x + 2y + z = 0 \end{cases}$

57. $\begin{cases} 3x + y - z = \frac{2}{3} \\ 2x - y + z = 1 \\ 4x + 2y = \frac{8}{3} \end{cases}$

58. $\begin{cases} x + y = 1 \\ 2x - y + z = 1 \\ x + 2y + z = \frac{8}{3} \end{cases}$

In Problems 59–64, use a graphing utility to find the row echelon form (REF) and the reduced row echelon form (RREF) of the augmented matrix of each of the following systems. Solve each system. If the system has no solution, say it is inconsistent.

59. $\begin{cases} 2x - 2y + z = 2 \\ x - \frac{1}{2}y + 2z = 1 \\ 2x + \frac{1}{3}y - z = 0 \end{cases}$

60. $\begin{cases} x + y = -1 \\ x - z = 0 \\ y - z = 1 \end{cases}$

61. $\begin{cases} x + y + z = 4 \\ x - y - z = 0 \\ y - z = -4 \end{cases}$

62. $\begin{cases} 2x + y + z = 6 \\ x - y - z = -3 \\ 3x + y + 2z = 7 \end{cases}$

63. $\begin{cases} x_1 + x_2 + x_3 + x_4 = 20 \\ x_2 + x_3 + x_4 = 0 \\ x_3 + x_4 = 13 \\ x_2 - 2x_4 = -5 \end{cases}$

64. $\begin{cases} x_1 - 2x_2 + 3x_3 - 4x_4 = 40 \\ 4x_2 + 6x_4 = -10 \\ x_3 - x_4 = 12 \\ x_2 + 2x_4 = -10 \end{cases}$

65. **Theater Seating** The Fox Theatre in St. Louis offers three levels of seating. One group of patrons buys 4 mezzanine tickets and 6 lower balcony tickets for \$444. Another group spends \$614 for 2 mezzanine tickets, 7 lower balcony tickets, and 8 middle balcony tickets. A third group purchases 3 lower balcony tickets and 12 middle balcony tickets for \$474. What is the individual price of a mezzanine ticket, a lower balcony ticket, and a middle balcony ticket?

Sources: Fox Theatre, St. Louis, Missouri, 2002; MetroTix, 2003.

66. **Mixture** A store sells almonds for \$6 per pound, cashews for \$5 per pound, and peanuts for \$2 per pound. One week the manager decides to prepare 100 16-ounce packages of nuts by mixing 40 pounds of peanuts with some almonds and cashews. Each package will be sold for \$4. How many pounds of almonds and cashews should be mixed with the peanuts so that the mixture will produce the same revenue as selling the nuts separately?

67. **Laboratory Work Stations** A chemistry laboratory can be used by 38 students at one time. The laboratory has 16 work stations, some set up for 2 students each and the others set up for 3 students each. How many are there of each kind of work station?

68. **Cost of Fast Food** One group of people purchased 10 hot dogs and 5 soft drinks at a cost of \$12.50. A second group bought 7 hot dogs and 4 soft drinks at a cost of \$9. What is the cost of a single hot dog? A single soft drink?

69. **Financial Planning** Carletta has \$10,000 to invest. As her financial consultant, you recommend that she invest in Treasury Bills that yield 6%, Treasury Bonds that yield 7%, and corporate bonds that yield 8%. Carletta wants to have an annual income of \$680, and the amount invested in corporate bonds must be half that invested in Treasury Bills. Find the amount in each investment.

70. **Financial Planning** John has \$20,000 to invest. As his financial consultant, you recommend that he invest in Treasury Bills that yield 5%, Treasury bonds that yield 7%, and corporate Bonds that yield 9%. John wants to have an annual income of \$1280, and the amount invested in Treasury Bills must be two times the amount invested in corporate bonds. Find the amount in each investment.

71. **Diet Preparation** A hospital dietician is planning a meal consisting of three foods whose ingredients are summarized as follows:

	Chicken (2-oz. Serving)	Potatoes (1/2-cup serving)	Spinach (1-cup serving)
Grams of Protein	14	1	6
Grams of Carbohydrates	0	18	8
Grams of Fat	4.5	0	1

Determine the number of servings of each food needed to create a meal containing 30 grams of protein, 38 grams of carbohydrates, and 7 grams of fat.

Source: Food and Nutrition Service, United States Department of Agriculture, 2002.

72. **Mixture** Sally's Girl Scout troop is selling cookies for the Christmas season. There are three different kinds of cookies in three different containers: *bags* that hold 1 dozen chocolate chip and 1 dozen oatmeal; *gift boxes* that hold 2 dozen chocolate chip, 1 dozen mint, and 1 dozen oatmeal; and *cookie tins* that hold 3 dozen mint and 2 dozen chocolate chip. Sally's mother is having a Christmas party and wants 6 dozen oatmeal; 10 dozen mint, and 14 dozen chocolate chip cookies. How can Sally fill her mother's order?

73. **Production** A citrus company completes the preparation of its products by cleaning, filling, and labeling bottles. Each case of orange juice requires 10 minutes in the cleaning machine, 4 minutes in the filling machine, and 2 minutes in the labeling machine. For each case of tomato juice, the times are 12 minutes of cleaning, 4 minutes of filling, and 1 minute of labeling. Pineapple juice requires 9 minutes of cleaning, 6 minutes of filling, and 1 minute of labeling per case. If the company runs the cleaning machine for 398 minutes, the filling machine for 164 minutes, and the labeling machine for 58 minutes, how many cases of each type of juice are prepared?

74. **Production** The manufacture of an automobile requires painting, drying, and polishing. The Rome Motor Company produces three types of cars: the Centurion, the Tribune, and the Senator. Each Centurion requires 8 hours for painting, 2 hours for drying, and 1 hour for polishing. A Tribune needs 10 hours for painting, 3 hours for drying, and 2 hours for polishing. It takes 16 hours of painting, 5 hours of drying, and 3 hours of polishing to prepare a Senator. If the company uses 240 hours for painting, 69 hours for drying, and 41 hours for polishing in a given month, how many of each type of car are produced?

75. **Inventory Control** An art teacher finds that colored paper can be bought in three different packages. The first package has 20 sheets of white paper, 15 sheets of blue paper, and 1 sheet of red paper. The second package has 3 sheets of blue paper and 1 sheet of red paper. The last package has 40 sheets of white paper and 30 sheets of blue paper. Suppose he needs 200 sheets of white paper, 180 sheets of blue paper, and 12 sheets of red paper. How many of each type of package should he order?

76. **Inventory Control** An interior decorator has ordered 12 cans of sunset paint, 35 cans of brown, and 18 cans of fuchsia. The paint store has special pair packs, containing 1 can each of sunset and fuchsia; darkening packs, containing 2 cans of sunset, 5 cans of brown, and 2 cans of fuchsia; and economy packs, containing 3 cans of sunset, 15 cans of brown, and 6 cans of fuchsia. How many of each type of pack should the paint store send to the interior decorator?

77. **Packaging** A recreation center wants to purchase compact discs (CDs) to be used in the center. There is no requirement as to the artists. The only requirement is that they purchase 40 rock CDs, 32 western CDs, and 14 blues CDs. There are three different shipping packages offered by the company. They are an *assorted* carton, containing 2 rock CDs, 4 western CDs, and 1 blues CD; a *mixed* carton containing 4 rock and 2 western CDs; and a *single* carton containing 2 blues CDs. What combination of these packages is needed to fill the center's order?

78. **Production** A luggage manufacturer produces three types of luggage: economy, standard, and deluxe. The company produces 1000 pieces of luggage at a cost of $20, $25, and $30 for the economy, standard, and deluxe luggage, respectively. The manufacturer has a budget of $20,700. Each economy luggage requires 6 hours of labor, each standard luggage requires 10 hours of labor, and each deluxe model requires 20 hours of labor. The manufacturer has a maximum of 6800 hours of labor available. If the manufacturer sells all the luggage, consumes the entire budget, and uses all the available labor, how many of each type of luggage should be produced?

79. **Mixture** Suppose that a store has three sizes of cans of nuts. The *large* size contains 2 pounds of peanuts and 1 pound of cashews. The *mammoth* size contains 1 pound of walnuts, 6 pounds of peanuts, and 2 pounds of cashews. The *giant* size contains 1 pound of walnuts, 4 pounds of peanuts, and 2 pounds of cashews. Suppose that the store receives an order for 5 pounds of walnuts, 26 pounds of peanuts, and 12 pounds of cashews. How can it fill this order with the given sizes of cans?

80. **Mixture** Suppose that the store in Problem 79 receives a new order for 6 pounds of walnuts, 34 pounds of peanuts, and 15 pounds of cashews. How can this order be filled with the given cans?

81. Write a brief paragraph or two that outlines your strategy for solving a system of linear equations using matrices.

82. When solving a system of linear equations using matrices, do you prefer to place the augmented matrix in row echelon form or in reduced row echelon form? Give reasons for your choice.

83. Make up a system of three linear equations containing three variables that has:
(a) No solution
(b) Exactly one solution
(c) Infinitely many solutions

Give the three systems to a friend to solve and critique.

2.3 Systems of *m* Linear Equations Containing *n* Variables

OBJECTIVES
1 Analyze the reduced row echelon form of an augmented matrix
2 Solve a system of *m* linear equations containing *n* variables
3 Express the solution of a system with an infinite number of solutions

We saw in the previous two sections that systems of two linear equations containing two variables and systems of three linear equations containing three variables each have either one solution, no solution, or infinitely many solutions. As it turns out, no matter how many equations are in a system of linear equations and no matter how many variables a system has, only these three possibilities can arise.

For example, the system of three linear equations containing four variables

$$\begin{cases} x_1 + 3x_2 + 5x_3 + x_4 = 2 \\ 2x_1 + 3x_2 + 4x_3 + 2x_4 = 1 \\ x_1 + 2x_2 + 3x_3 + x_4 = 1 \end{cases}$$

will have either no solution, one solution, or infinitely many solutions.

A general definition of a system of *m* linear equations containing *n* variables is given next.

System of *m* Linear Equations Containing *n* Variables

A system of *m* linear equations containing *n* variables $x_1, x_2, \ldots, x_n$ is of the form

$$\begin{cases} a_{11}x_1 + a_{12}x_2 + \cdots + a_{1n}x_n = b_1 \\ a_{21}x_1 + a_{22}x_2 + \cdots + a_{2n}x_n = b_2 \\ a_{31}x_1 + a_{32}x_2 + \cdots + a_{3n}x_n = b_3 \\ \quad\vdots \qquad\quad \vdots \qquad\qquad\quad \vdots \qquad\quad \vdots \\ a_{i1}x_1 + a_{i2}x_2 + \cdots + a_{in}x_n = b_i \\ \quad\vdots \qquad\quad \vdots \qquad\qquad\quad \vdots \qquad\quad \vdots \\ a_{m1}x_1 + a_{m2}x_2 + \cdots + a_{mn}x_n = b_m \end{cases}$$

where a_{ij} and b_i are real numbers, $i = 1, 2 \ldots, m, j = 1, 2, \ldots, n$.

A **solution** of a system of *m* linear equations containing *n* variables $x_1, x_2, \ldots, x_n$ is any ordered set $(x_1, x_2, \ldots, x_n)$ of real numbers for which *each* of the *m* linear equations of the system is satisfied.

Reduced Row Echelon Form

A system of m linear equations containing n variables will have either no solution, one solution, or infinitely many solutions. We can determine which of these possibilities occurs and, if solutions exist, find them, by performing row operations on the augmented matrix of the system until we arrive at the reduced row echelon form of the augmented matrix.

Let's review the conditions required for the reduced row echelon form:

Conditions for the Reduced Row Echelon Form of a Matrix

1. The first nonzero entry in each row is 1 and it has 0s above it and below it.
2. The leftmost 1 in any row is to the right of the leftmost 1 in the row above.
3. Any rows that contain all 0s to the left of the vertical bar appear at the bottom.

The next two examples will help you recognize when an augmented matrix is in reduced row echelon form.

EXAMPLE 1 **Examples of Matrices That Are in Reduced Row Echelon Form**

(a) $\left[\begin{array}{cc|c} 1 & 0 & 2 \\ 0 & 1 & 3 \end{array}\right]$ (b) $\left[\begin{array}{ccc|c} 1 & 0 & -3 & 4 \\ 0 & 1 & -2 & 2 \end{array}\right]$

(c) $\left[\begin{array}{cccc|c} 1 & -2 & 0 & & 1 \\ 0 & 0 & 1 & & 3 \\ 0 & 0 & 0 & & 5 \end{array}\right]$ (d) $\left[\begin{array}{cc|c} 1 & 0 & 3 \\ 0 & 1 & 4 \\ 0 & 0 & 0 \end{array}\right]$

EXAMPLE 2 **Examples of Matrices That Are Not in Reduced Row Echelon Form**

(a) $\left[\begin{array}{cc|c} 1 & 0 & 0 \\ 0 & 0 & 0 \\ 0 & 1 & 0 \end{array}\right]$ The second row contains all 0s and the third does not—this violates the rule that states that any rows with all 0s are at the bottom.

(b) $\left[\begin{array}{ccc|c} 1 & 0 & 2 & 4 \\ 0 & 2 & 4 & 3 \\ 0 & 0 & 0 & 1 \end{array}\right]$ The first nonzero entry in row 2 is not a 1.

(c) $\left[\begin{array}{ccc|c} 1 & 0 & 0 & 0 \\ 0 & 0 & 1 & 1 \\ 0 & 1 & 0 & 0 \end{array}\right]$ The leftmost 1 in the third row is not to the right of the leftmost 1 in the row above it.

NOW WORK PROBLEM 1.

Analyze the reduced row echelon form of an augmented matrix 1

Now let's analyze the reduced row echelon form of an augmented matrix.

EXAMPLE 3 Analyzing the Reduced Row Echelon Form of an Augmented Matrix

The matrix

$$\left[\begin{array}{ccc|c} 1 & 0 & 0 & 3 \\ 0 & 1 & 0 & 8 \\ 0 & 0 & 1 & -4 \end{array}\right]$$

is the reduced row echelon form of the augmented matrix of a system of three linear equations containing three variables. If the variables are x, y, z, this matrix represents the system of equations

$$\begin{cases} x = 3 \\ y = 8 \\ z = -4 \end{cases}$$

The system has one solution: $x = 3, y = 8, z = -4$.

EXAMPLE 4 Analyzing the Reduced Row Echelon Form of an Augmented Matrix

The matrix

$$\left[\begin{array}{ccc|c} 1 & 0 & 3 & 0 \\ 0 & 1 & 2 & 0 \\ 0 & 0 & 0 & 1 \\ 0 & 0 & 0 & 0 \end{array}\right]$$

is the reduced row echelon form of the augmented matrix of a system of four equations containing three variables. If the variables are x, y, z, the equation represented by the third row is

$$0 \cdot x + 0 \cdot y + 0 \cdot z = 1 \qquad \text{or} \qquad 0 = 1$$

Since $0 = 1$ is a contradiction, we conclude the system is inconsistent.

EXAMPLE 5 Analyzing the Reduced Row Echelon Form of an Augmented Matrix

The matrix

$$\left[\begin{array}{cccc|c} 1 & 0 & 0 & 2 & 5 \\ 0 & 1 & 0 & 1 & 2 \\ 0 & 0 & 1 & 3 & 4 \end{array}\right]$$

is the reduced row echelon form of a system of three equations containing four variables. If x_1, x_2, x_3, x_4 are the variables, the system of equations is

$$\begin{cases} x_1 + 2x_4 = 5 \\ x_2 + x_4 = 2 \\ x_3 + 3x_4 = 4 \end{cases} \qquad \text{or} \qquad \begin{cases} x_1 = -2x_4 + 5 \\ x_2 = -x_4 + 2 \\ x_3 = -3x_4 + 4 \end{cases}$$

The system has infinitely many solutions. In this form, the variable x_4 is the parameter. We assign values to the parameter x_4 from which the variables x_1, x_2, x_3 can be calculated. Some of the possibilities are

$$\text{If } x_4 = 0, \text{ then } x_1 = 5, x_2 = 2, x_3 = 4.$$
$$\text{If } x_4 = 1, \text{ then } x_1 = 3, x_2 = 1, x_3 = 1.$$
$$\text{If } x_4 = 2, \text{ then } x_1 = 1, x_2 = 0, x_3 = -2.$$

And so on.

NOW WORK PROBLEM 17.

Let's review the procedure for obtaining the reduced row echelon form of a matrix. Then we will use this procedure to solve systems of linear equations.

EXAMPLE 6 Finding the Reduced Row Echelon Form of a Matrix

Find the reduced row echelon form of

$$A = \left[\begin{array}{cc|c} 1 & -1 & 2 \\ 2 & -3 & 2 \\ 3 & -5 & 2 \end{array}\right]$$

SOLUTION The entry in row 1, column 1 is 1. We proceed to obtain a matrix in which all the remaining entries in column 1 are 0s. We can obtain such a matrix by performing the row operations

$$R_2 = -2r_1 + r_2$$
$$R_3 = -3r_1 + r_3$$

The new matrix is

$$\left[\begin{array}{cc|c} 1 & -1 & 2 \\ 0 & -1 & -2 \\ 0 & -2 & -4 \end{array}\right]$$

We want the entry in row 2, column 2 (now -1), to be 1. By multiplying row 2 by -1, $R_2 = (-1)r_2$, we obtain

$$\left[\begin{array}{cc|c} 1 & -1 & 2 \\ 0 & 1 & 2 \\ 0 & -2 & -4 \end{array}\right]$$

Now we want the entry in row 1, column 2 and in row 3, column 2 to be 0. This can be accomplished by applying the row operations

$$R_1 = r_2 + r_1$$
$$R_3 = 2r_2 + r_3$$

The new matrix is

$$\left[\begin{array}{cc|c} 1 & 0 & 4 \\ 0 & 1 & 2 \\ 0 & 0 & 0 \end{array}\right]$$

This is the reduced row echelon form of A.

2 **EXAMPLE 7** **Solving a System of Three Linear Equations Containing Two Variables**

Solve: $\begin{cases} x - y = 2 \\ 2x - 3y = 2 \\ 3x - 5y = 2 \end{cases}$

SOLUTION The augmented matrix of this system is

$$\left[\begin{array}{cc|c} 1 & -1 & 2 \\ 2 & -3 & 2 \\ 3 & -5 & 2 \end{array}\right]$$

Using the solution to Example 6, the reduced row echelon form of this augmented matrix is

$$\left[\begin{array}{cc|c} 1 & 0 & 4 \\ 0 & 1 & 2 \\ 0 & 0 & 0 \end{array}\right]$$

We conclude that the system has the solution $x = 4, y = 2$.

COMMENT: A graphing utility can be used to solve systems of m linear equations containing n variables. Check the solution to Example 7 using the RREF feature of your graphing utility.

EXAMPLE 8 **Solving a System of Four Linear Equations Containing Three Variables**

Solve: $\begin{cases} x - y + 2z = 2 \\ 2x - 3y + 2z = 1 \\ 3x - 5y + 2z = -3 \\ -4x + 12y + 8z = 10 \end{cases}$

SOLUTION We need to find the reduced row echelon form of the augmented matrix of this system, namely,

$$\left[\begin{array}{ccc|c} 1 & -1 & 2 & 2 \\ 2 & -3 & 2 & 1 \\ 3 & -5 & 2 & -3 \\ -4 & 12 & 8 & 10 \end{array}\right]$$

The entry 1 is already present in row 1, column 1. To obtain 0s elsewhere in column 1, we use the row operations

$$R_2 = -2r_1 + r_2 \qquad R_3 = -3r_1 + r_3 \qquad R_4 = 4r_1 + r_4$$

The new matrix is

$$\left[\begin{array}{ccc|c} 1 & -1 & 2 & 2 \\ 0 & -1 & -2 & -3 \\ 0 & -2 & -4 & -9 \\ 0 & 8 & 16 & 18 \end{array}\right]$$

To obtain the entry 1 in row 2, column 2, we use $R_2 = -r_2$, obtaining

$$\left[\begin{array}{rrr|r} 1 & -1 & 2 & 2 \\ 0 & 1 & 2 & 3 \\ 0 & -2 & -4 & -9 \\ 0 & 8 & 16 & 18 \end{array}\right]$$

To obtain 0s elsewhere in column 2, we use

$$R_1 = r_2 + r_1 \qquad R_3 = 2r_2 + r_3 \qquad R_4 = -8r_2 + r_4$$

The new matrix is

$$\left[\begin{array}{rrr|r} 1 & 0 & 4 & 5 \\ 0 & 1 & 2 & 3 \\ 0 & 0 & 0 & -3 \\ 0 & 0 & 0 & -6 \end{array}\right]$$

We can stop here even though the matrix is not in reduced row echelon form. Because the third row yields the equation

$$0 \cdot x + 0 \cdot y + 0 \cdot z = -3$$

we conclude the system is inconsistent.

NOW WORK PROBLEM 31.

Express the solution of a system with an infinite number of solutions 3

Infinite Number of Solutions

We have seen several examples of systems of linear equations that have an infinite number of solutions. Let's look at a few more examples.

EXAMPLE 9 **Solving a System of Two Linear Equations Containing Three Variables**

Solve: $\begin{cases} x + y + z = 7 \\ x - y - 3z = 1 \end{cases}$

SOLUTION The augmented matrix of the system is

$$\left[\begin{array}{rrr|r} 1 & 1 & 1 & 7 \\ 1 & -1 & -3 & 1 \end{array}\right]$$

The reduced row echelon form (as you should verify) is

$$\left[\begin{array}{rrr|r} 1 & 0 & -1 & 4 \\ 0 & 1 & 2 & 3 \end{array}\right]$$

The system of equations represented by this matrix is

$$\begin{cases} x - z = 4 \\ y + 2z = 3 \end{cases} \quad \text{or} \quad \begin{cases} x = z + 4 \\ y = -2z + 3 \end{cases} \tag{1}$$

The system has infinitely many solutions. In the form (1), the variable z is the parameter. We can assign any value to z and use it to compute values of x and y.

✓ **CHECK:** We check the solution to Example 9 as follows:

$$x + y + z = (z + 4) + (-2z + 3) + z = 7 + z - 2z + z = 7$$
$$x - y - 3z = (z + 4) - (-2z + 3) - 3z = 1 + z + 2z - 3z = 1$$

The solution is verified.

The next example illustrates a system having an infinite number of solutions with two parameters.

EXAMPLE 10 **Solving a System of Three Linear Equations Containing Four Variables**

Solve: $$\begin{cases} x_1 + x_2 + 2x_3 + 2x_4 = 2 \\ x_1 + x_3 + x_4 = 0 \\ x_2 + x_3 + x_4 = 2 \end{cases}$$

SOLUTION The augmented matrix of the system is

$$\left[\begin{array}{cccc|c} 1 & 1 & 2 & 2 & 2 \\ 1 & 0 & 1 & 1 & 0 \\ 0 & 1 & 1 & 1 & 2 \end{array}\right]$$

The reduced row echelon form (as you should verify) is

$$\left[\begin{array}{cccc|c} 1 & 0 & 1 & 1 & 0 \\ 0 & 1 & 1 & 1 & 2 \\ 0 & 0 & 0 & 0 & 0 \end{array}\right]$$

The equations represented by this system are

$$\begin{cases} x_1 + x_3 + x_4 = 0 \\ x_2 + x_3 + x_4 = 2 \end{cases}$$

We can rewrite this system in the form

$$\begin{cases} x_1 = -x_3 - x_4 \\ x_2 = -x_3 - x_4 + 2 \end{cases} \quad (2)$$

The system has infinitely many solutions. In the form (2), the system has two parameters x_3 and x_4. Solutions are obtained by assigning the two parameters x_3 and x_4 arbitrary values. Some choices are shown in Table 1 below.

TABLE 1

x_3	x_4	x_1	x_2	Solution (x_1, x_2, x_3, x_4)
0	0	0	2	(0, 2, 0, 0)
1	0	−1	1	(−1, 1, 1, 0)
0	2	−2	0	(−2, 0, 0, 2)

The variables used as parameters are not unique. We could have chosen x_1 and x_4 as parameters by rewriting Equations (2) in the following manner.

From the first equation

$$x_3 = -x_1 - x_4$$

We can replace the parameter x_3 in the second equation by the above to produce

$$x_2 = 2 - x_3 - x_4 = 2 + x_1 + x_4 - x_4 = 2 + x_1$$

We then obtain the system

$$\begin{cases} x_2 = x_1 + 2 \\ x_3 = -x_1 - x_4 \end{cases}$$

showing the solution with x_1 and x_4 as parameters.

NOW WORK PROBLEM 27.

EXAMPLE 11 Investment Goals

A couple have \$25,000 available and want to invest in U.S. Savings Bonds. As of January 2003, the rates were 3.2% for EE/E bonds, 4% for I bonds, and 1.5% for HH/H bonds. Their goal is to invest in all three types of bonds in such a way that they obtain \$500 in interest per year. Prepare a table showing the various ways this couple can achieve their goal.

Source: US Department of the Treasury, 2003.

SOLUTION Let x represent the amount invested in EE/E bonds, y represent the amount invested in I bonds, and z represent the amount invested in HH/H bonds. Since the couple have \$25,000 to invest and want a \$500 return on their investment, we need to solve the system of equations

$$\begin{cases} x + y + z = 25000 & (1) \\ .032x + .04y + .015z = 500 & (2) \end{cases}$$

The augmented matrix of this system and its reduced row echelon form (which you should verify) are given by

$$\left[\begin{array}{ccc|c} 1 & 1 & 1 & 25{,}000 \\ .032 & .04 & .015 & 500 \end{array}\right] \longrightarrow \left[\begin{array}{ccc|c} 1 & 0 & 3.125 & 62{,}500 \\ 0 & 1 & -2.125 & -37{,}500 \end{array}\right]$$

The augmented matrix represents the system of equations

$$\begin{cases} x + 3.125z = 62{,}500 \\ y - 2.125z = -37{,}500 \end{cases} \xrightarrow[\text{Solve for } x \text{ and } y.]{} \begin{cases} x = 62{,}500 - 3.125z \\ y = -37{,}500 + 2.125z \end{cases} \qquad (3)$$

where z is the parameter. Since we want to invest in each of the three bond types, the conditions $x > 0$, $y > 0$, and $z > 0$ must hold. So we can determine from the system (3), that

$$62{,}500 - 3.125z > 0 \text{ so } z < 20{,}000$$
$$-37{,}500 + 2.125z > 0 \text{ so } z > 17{,}647$$

Several of the possible solutions are listed in Table 2 on page 90. The couple's final decision on asset allocation will usually reflect their attitude toward risk.

TABLE 2

Amount in EE/E	Amount in I	Amount in HH/H
\$6250	750	18,000
5469	1281	18,250
4688	1812	18,500
3906	2344	18,750
3125	2875	19,000
2344	3406	19,250
1563	3937	19,500
781	4469	19,750

NOW WORK PROBLEM 53.

SUMMARY

Steps for Solving a System of *m* Linear Equations Containing *n* Variables

STEP 1 Write the augmented matrix.
STEP 2 Find the reduced row echelon form of the augmented matrix.
STEP 3 Analyze this matrix to determine if the system has no solution, one solution, or infinitely many solutions.

EXERCISE 2.3 Answers to Odd-Numbered Problems Begin on Page AN-11.

In Problems 1–12, tell whether the given matrix is in reduced row echelon form. If it is not, tell why.

1. $\left[\begin{array}{cc|c} 1 & 2 & 3 \\ 0 & 0 & 0 \\ 0 & 0 & 1 \end{array}\right]$

2. $\left[\begin{array}{cc|c} 1 & 2 & 3 \\ 0 & 0 & 0 \\ 0 & 0 & 0 \end{array}\right]$

3. $\left[\begin{array}{cc|c} 1 & 1 & 0 \\ 0 & 1 & 0 \end{array}\right]$

4. $\left[\begin{array}{cc|c} 1 & 0 & 3 \\ 0 & 1 & 0 \end{array}\right]$

5. $\left[\begin{array}{c|c} 0 & 1 \\ 1 & 0 \end{array}\right]$

6. $\left[\begin{array}{cc|c} 0 & 1 & 0 \\ 0 & 0 & 1 \\ 0 & 0 & 0 \end{array}\right]$

7. $\left[\begin{array}{cc|c} 1 & 2 & 1 \\ 0 & 0 & 0 \end{array}\right]$

8. $\left[\begin{array}{cc|c} 1 & 0 & 8 \\ 0 & 2 & 9 \end{array}\right]$

9. $\left[\begin{array}{cccc|c} 1 & 0 & 0 & 0 & 0 \\ 0 & 0 & 1 & 2 & 0 \\ 0 & 0 & 0 & 0 & 1 \\ 0 & 0 & 0 & 0 & 0 \end{array}\right]$

10. $\left[\begin{array}{cc|c} 1 & 1 & 0 \\ 0 & 0 & 2 \end{array}\right]$

11. $\left[\begin{array}{cc|c} 1 & 0 & 1 \\ 0 & 1 & 2 \\ 0 & 0 & 0 \end{array}\right]$

12. $\left[\begin{array}{cc|c} 1 & 0 & 2 \\ 0 & 1 & 2 \end{array}\right]$

In Problems 13–28, the reduced row echelon form of the augmented matrix of a system of linear equations is given. Tell whether the system has one solution, no solution, or infinitely many solutions. Write the solutions or, if there is no solution, say the system is inconsistent.

13. $\left[\begin{array}{cc|c} 1 & 1 & 1 \\ 0 & 0 & 0 \end{array}\right]$

14. $\left[\begin{array}{cc|c} 1 & 0 & 0 \\ 0 & 0 & 1 \end{array}\right]$

15. $\left[\begin{array}{cc|c} 1 & 0 & 4 \\ 0 & 1 & 5 \end{array}\right]$

16. $\left[\begin{array}{ccc|c} 1 & 0 & 0 & 0 \\ 0 & 1 & 0 & 0 \\ 0 & 0 & 1 & 6 \end{array}\right]$

17. $\left[\begin{array}{ccc|c} 1 & 0 & -2 & 6 \\ 0 & 1 & 3 & 1 \end{array}\right]$

18. $\left[\begin{array}{ccc|c} 1 & 0 & 0 & 0 \\ 0 & 1 & 0 & 5 \\ 0 & 0 & 0 & 0 \end{array}\right]$

19. $\left[\begin{array}{ccc|c} 1 & 2 & 0 & 1 \\ 0 & 0 & 1 & 2 \\ 0 & 0 & 0 & 0 \end{array}\right]$

20. $\left[\begin{array}{ccc|c} 1 & 2 & 0 & 0 \\ 0 & 0 & 1 & 0 \\ 0 & 0 & 0 & 1 \end{array}\right]$

21. $\left[\begin{array}{ccc|c} 1 & 0 & 0 & 1 \\ 0 & 1 & 0 & 2 \\ 0 & 0 & 0 & 0 \end{array}\right]$

22. $\left[\begin{array}{cccc|c} 1 & 0 & 1 & -1 & 0 \\ 0 & 1 & 2 & 1 & 1 \\ 0 & 0 & 0 & 0 & 0 \end{array}\right]$

23. $\left[\begin{array}{ccc|c} 1 & 0 & 0 & -1 \\ 0 & 1 & 0 & 3 \\ 0 & 0 & 1 & 4 \\ 0 & 0 & 0 & 0 \end{array}\right]$

24. $\left[\begin{array}{cccc|c} 1 & 2 & 0 & 0 & -4 \\ 0 & 0 & 1 & 0 & -3 \\ 0 & 0 & 0 & 1 & 2 \\ 0 & 0 & 0 & 0 & 0 \end{array}\right]$

25. $\left[\begin{array}{ccc|c} 1 & 0 & -1 & 1 \\ 0 & 1 & 2 & 1 \end{array}\right]$

26. $\left[\begin{array}{cc|c} 1 & 0 & 1 \\ 0 & 1 & 1 \\ 0 & 0 & 0 \end{array}\right]$

27. $\left[\begin{array}{cccc|c} 1 & 0 & 0 & -1 & 4 \\ 0 & 1 & 2 & 3 & 0 \end{array}\right]$

28. $\left[\begin{array}{cccc|c} 1 & 0 & 2 & 4 & -1 \\ 0 & 1 & 3 & 5 & -2 \end{array}\right]$

In Problems 29–52, solve each system of equations by finding the reduced row echelon form of the augmented matrix. If there is no solution, say the system is inconsistent.

29. $\begin{cases} x + y = 3 \\ 2x - y = 3 \end{cases}$

30. $\begin{cases} x - y = 5 \\ 2x + 3y = 15 \end{cases}$

31. $\begin{cases} 3x - 3y = 12 \\ 3x + 2y = -3 \\ 2x + y = 4 \end{cases}$

32. $\begin{cases} 6x + y = 8 \\ x - 3y = -5 \\ 2x + y = 2 \end{cases}$

33. $\begin{cases} 2x - 4y = 8 \\ x - 2y = 4 \\ -x + 2y = -4 \end{cases}$

34. $\begin{cases} 3x + y = 8 \\ 6x + 2y = 16 \\ -9x - 3y = -24 \end{cases}$

35. $\begin{cases} 2x + y + 3z = -1 \\ -x + y + 3z = 8 \\ 2x - 2y - 6z = -16 \end{cases}$

36. $\begin{cases} x + 2y + 3z = 5 \\ -2x + 6y + 4z = 0 \\ 2x + 4y + 6z = 10 \end{cases}$

37. $\begin{cases} x - y = 1 \\ y - z = 6 \\ x + z = -1 \end{cases}$

38. $\begin{cases} 2x - y + 3z = 0 \\ x + 2y - z = 5 \\ 2y + z = 1 \end{cases}$

39. $\begin{cases} x_1 + x_2 = 7 \\ x_2 - x_3 + x_4 = 5 \\ x_1 - x_2 + x_3 + x_4 = 6 \\ x_2 - x_4 = 10 \end{cases}$

40. $\begin{cases} x_1 + x_2 + x_3 + x_4 = 0 \\ 2x_1 - x_2 - x_3 + x_4 = 0 \\ x_1 - x_2 - x_3 + x_4 = 0 \\ x_1 + x_2 - x_3 - x_4 = 0 \end{cases}$

41. $\begin{cases} x_1 + 2x_2 + 3x_3 - x_4 = 0 \\ 3x_1 - x_4 = 4 \\ x_2 - x_3 - x_4 = 2 \end{cases}$

42. $\begin{cases} 2x - 3y + 4z = 7 \\ x - 2y + 3z = 2 \end{cases}$

43. $\begin{cases} x - y + z = 5 \\ 2x - 2y + 2z = 8 \end{cases}$

44. $\begin{cases} x + y + z = 3 \\ x - y + z = 7 \\ x - y - z = 1 \end{cases}$

45. $\begin{cases} 3x - y + 2z = 3 \\ 3x + 3y + z = 3 \\ 3x - 5y + 3z = 12 \end{cases}$

46. $\begin{cases} x + y - z = 12 \\ 3x - y = 1 \\ 2x - 3y + 4z = 3 \end{cases}$

47. $\begin{cases} x_1 + x_2 + x_3 + x_4 = 4 \\ 2x_1 - x_2 + x_3 = 0 \\ 3x_1 + 2x_2 + x_3 - x_4 = 6 \\ x_1 - 2x_2 - 2x_3 + 2x_4 = -1 \end{cases}$

48. $\begin{cases} x_1 + x_2 + x_3 + x_4 = 4 \\ -x_1 + 2x_2 + x_3 = 0 \\ 2x_1 + 3x_2 + x_3 - x_4 = 6 \\ -2x_1 + x_2 - 2x_3 + 2x_4 = -1 \end{cases}$

49. $\begin{cases} 2x - y - z = 0 \\ x - y - z = 1 \\ 3x - y - z = 2 \end{cases}$

50. $\begin{cases} x + y + z = 3 \\ 2x + y + z = 0 \\ 3x + y + z = 1 \end{cases}$

51. $\begin{cases} 2x - y + z = 6 \\ 3x - y + z = 6 \\ 4x - 2y + 2z = 12 \end{cases}$

52. $\begin{cases} x - y + z = 2 \\ 2x - 3y + z = 0 \\ 3x - 3y + 3z = 6 \end{cases}$

53. **Investment Allocation** Look again at Example 11. Suppose the couple now require \$800 interest per year. Prepare a table that shows various ways this couple can achieve their goal.

54. **Investment Allocation** Look again at Example 11. Suppose the couple still require \$500 in interest per year, but the interest rate on HH/H bonds increases to 2%. Prepare a table showing various ways the couple can achieve their goal.

55. Investment Allocation Look again at Example 11. Suppose the interest rate on I bonds goes down to 3.5%. Can the couple maintain their goal to obtain $500 per year in interest? Prepare a table that shows various investment options that the couple can use to achieve their goal.

56. Inventory Control Three species of bacteria will be kept in one test tube and will feed on three resources. Each member of the first species consumes 3 units of the first resource and 1 unit of the third. Each bacterium of the second type consumes 1 unit of the first resource and 2 units each of the second and third. Each bacterium of the third type consumes 2 units of the first resource and 4 each of the second and third. If the test tube is supplied daily with 12,000 units of the first resource, 12,000 units of the second, and 14,000 units of the third, how many of each species can coexist in equilibrium in the test tube so that all of the supplied resources are consumed? Prepare a table that shows some of the possibilities.

57. Cost of Fast Food One group of customers bought 8 deluxe hamburgers, 6 orders of large fries, and 6 large colas for $26.10. A second group ordered 10 deluxe hamburgers, 6 large fries, and 8 large colas and paid $31.60. Is there sufficient information to determine the price of each food item? If not, construct a table showing the various possibilities. Assume the hamburgers cost between $1.75 and $2.25, the fries between $0.75 and $1.00, and the colas between $0.60 and $0.90.

58. Use the information given in Problem 57 and add a third group that purchased 3 deluxe hamburgers, 2 large fries, and 4 colas for $10.95. Is there now sufficient information to determine the price of each food item?

59. Financial Planning Three retired couples each require an additional annual income of $2000 per year. As their financial consultant, you recommend that they invest some money in Treasury Bills that yield 7%, some money in corporate bonds that yield 9%, and some money in junk bonds that yield 11%. Prepare a table for each couple showing the various ways that their goals can be achieved:

(a) If the first couple has $20,000 to invest
(b) If the second couple has $25,000 to invest
(c) If the third couple has $30,000 to invest
(d) What advice would you give each couple regarding the amount to invest and the choices available?
[*Hint*: Higher yields generally carry more risk.]

60. Financial Planning A young couple has $25,000 to invest. As their financial consultant, you recommend that they invest some money in Treasury Bills that yield 7%, some money in corporate bonds that yield 9%, and some money in junk bonds that yield 11%. Prepare a table showing the various ways this couple can achieve the following goals:

(a) The couple want $1500 per year in income.
(b) The couple want $2000 per year in income.
(c) The couple want $2500 per year in income.
(d) What advice would you give this couple regarding the income that they require and the choices available?
[*Hint*: Higher yields generally carry more risk.]

61. Pharmacy A doctor's prescription calls for a daily intake of liquid containing 40 mg of vitamin C and 30 mg of vitamin D. Your pharmacy stocks three liquids that can be used: one contains 20% vitamin C and 30% vitamin D; a second, 40% vitamin C and 20% vitamin D; and a third, 30% vitamin C and 50% vitamin D. Create a table showing the possible combinations that could be used to fill the prescription.

62. Pharmacy A doctor's prescription calls for the creation of pills that contain 12 units of vitamin B_{12} and 12 units of vitamin E. Your pharmacy stocks three powders that can be used to make these pills: one contains 20% vitamin B_{12} and 30% vitamin E; a second, 40% vitamin B_{12} and 20% vitamin E; and a third, 30% vitamin B_{12} and 40% vitamin E. Create a table showing the possible combinations of each powder that could be mixed in each pill.

63. Make up a system of three linear equations containing four variables that has infinitely many solutions. How many parameters will be in the solution to this system? Solve the system and create a table showing various solutions to the system of equations.

64. Make up a system of two linear equations containing four variables that has infinitely many solutions. How many parameters will be in the solution to this system? Solve the system and create a table showing various solutions to the system of equations.

2.4 Matrix Algebra

PREPARING FOR THIS SECTION *Before getting started, review the following:*

> Properties of Real Numbers (Appendix A, Section A.1, pp. 622–636)

OBJECTIVES
1. Find the dimension of a matrix
2. Find the sum of two matrices
3. Work with properties of matrices
4. Find the difference of two matrices
5. Find scalar multiples of a matrix

Matrices can be added, subtracted, and multiplied. They also possess many of the algebraic properties of real numbers. **Matrix algebra** is the study of these properties. Its importance lies in the fact that many situations in both pure and applied mathematics involve rectangular arrays of numbers. In fact, in many branches of business and the biological and social sciences, it is necessary to express and use data in a rectangular array. Let's look at an example.

Let's begin with an example that illustrates how matrices can be used to conveniently represent an array of information.

EXAMPLE 1 Arranging Data in a Matrix

In a survey of 900 people, the following information was obtained:

200	Males	Thought federal defense spending was too high
150	Males	Thought federal defense spending was too low
45	Males	Had no opinion
315	Females	Thought federal defense spending was too high
125	Females	Thought federal defense spending was too low
65	Females	Had no opinion

We can arrange the above data in a rectangular array as follows:

	Too High	Too Low	No Opinion
Male	200	150	45
Female	315	125	65

or as the matrix

$$\begin{bmatrix} 200 & 150 & 45 \\ 315 & 125 & 65 \end{bmatrix}$$

This matrix has two rows (representing males and females) and three columns (representing "too high," "too low," and "no opinion").

We now give a general definition for a matrix.

Definition of a Matrix

A **matrix** is defined as a rectangular array of the form:

$$
\begin{array}{c}
\\
\text{Row } 1\\
\text{Row } 2\\
\vdots\\
\text{Row } i\\
\vdots\\
\text{Row } m
\end{array}
\begin{array}{c}
\begin{array}{cccccc}
\text{Column } 1 & \text{Column } 2 & & \text{Column } j & & \text{Column } n
\end{array}\\
\begin{bmatrix}
a_{11} & a_{12} & \cdots & a_{1j} & \cdots & a_{1n}\\
a_{21} & a_{22} & \cdots & a_{2j} & \cdots & a_{2n}\\
\vdots & \vdots & & \vdots & & \vdots\\
a_{i1} & a_{i2} & \cdots & a_{ij} & \cdots & a_{in}\\
\vdots & \vdots & & \vdots & & \vdots\\
a_{m1} & a_{m2} & \cdots & a_{mj} & \cdots & a_{mn}
\end{bmatrix}
\end{array}
\qquad (1)
$$

The symbols $a_{11}, a_{12}, \ldots$ of a matrix are referred to as the **entries** (or **elements**) of the matrix. Each entry a_{ij} of the matrix has two indices: the **row index,** i, and the **column index,** j. The symbols $a_{i1}, a_{i2}, \ldots, a_{in}$ represent the entries in the ith row, and the symbols $a_{1j}, a_{2j}, \ldots, a_{mj}$ represent the entries in the jth column.

We shall use capital letters to denote matrices. If we denote the matrix in display (1) above by A, then we can abbreviate A by

$$A = [a_{ij}] \qquad i = 1, 2, \ldots, m \qquad j = 1, 2, \ldots, n.$$

The matrix A has m rows and n columns.

Dimension of a Matrix

The **dimension of a matrix** A is determined by the number of rows and the number of columns in the matrix. If a matrix A has m rows and n columns, we denote the dimension of A by $m \times n$, read as "m by n."

For a 2×3 matrix, remember that the first number 2 denotes the number of rows and the second number 3 is the number of columns. A matrix with 3 rows and 2 columns is of dimension 3×2.

Square Matrix

If a matrix A has the same number of rows as it has columns, it is called a **square matrix.**

In a square matrix $A = [a_{ij}]$ the entries for which $i = j$, namely a_{11}, a_{22}, a_{33}, a_{44}, and so on, are the **diagonal entries** of A.

EXAMPLE 2 Arranging Data as a Matrix

In the recent U.S. census the following figures were obtained with regard to the city of Oak Lawn. Each year 7% of city residents move to the suburbs and 1% of the people in the suburbs move to the city. This situation can be represented by the matrix

$$P = \begin{array}{r} \\ \text{City} \\ \text{Suburbs} \end{array} \begin{array}{c} \begin{array}{cc} \text{City} & \text{Suburbs} \end{array} \\ \begin{bmatrix} 0.93 & 0.07 \\ 0.01 & 0.99 \end{bmatrix} \end{array}$$

Here, the entry in row 1, column 2—0.07—indicates that 7% of city residents move to the suburbs. The matrix P is a square matrix and its dimension is 2×2. The diagonal entries are 0.93 and 0.99.

A **row matrix** is a matrix with 1 row of entries. A **column matrix** is a matrix with 1 column of entries. Row matrices and column matrices are also referred to as **row vectors** and **column vectors,** respectively.

1 EXAMPLE 3 Finding the Dimension of a Matrix

Find the dimension of each matrix. Say if the matrix is a square matrix, or a row matrix, or a column matrix.

(a) $\begin{bmatrix} 5 & -1 \\ 2 & 5 \end{bmatrix}$ **(b)** $\begin{bmatrix} 2 & 3 & 0 \\ 1 & -2 & 5 \end{bmatrix}$ **(c)** $[1 \quad 0 \quad 4]$ **(d)** $\begin{bmatrix} 2 \\ 1 \end{bmatrix}$ **(e)** $[9]$

SOLUTION **(a)** 2×2, a square matrix
(b) 2×3
(c) 1×3, a row matrix
(d) 2×1, a column matrix
(e) 1×1, a square matrix, a row matrix, and a column matrix

NOW WORK PROBLEMS 3 AND 63.

Equality of Matrices

In an algebra system, it is important to know when two quantities are equal. So, we ask, "When, if at all, are two matrices equal?"

Let's try to arrive at a sound definition for equality of matrices by requiring equal matrices to have certain desirable properties. First, it would seem necessary that two equal matrices have the same dimension—that is, that they both be $m \times n$ matrices. Next, it would seem necessary that their entries be identical numbers. With these two restrictions, we define equality of matrices.

Equality of Matrices

Two matrices A and B are **equal** if they are of the same dimension and if corresponding entries are equal. In this case we write $A = B$, read as "matrix A is equal to matrix B."

EXAMPLE 4 Determining Equality of Matrices

In order for the two matrices

$$\begin{bmatrix} p & q \\ 1 & 0 \end{bmatrix} \quad \text{and} \quad \begin{bmatrix} 2 & 4 \\ n & 0 \end{bmatrix}$$

to be equal, we must have $p = 2, q = 4$, and $n = 1$.

EXAMPLE 5 Determining Equality of Matrices

Let A and B be two matrices given by

$$A = \begin{bmatrix} x + y & 6 \\ 2x - 3 & 2 - y \end{bmatrix} \qquad B = \begin{bmatrix} 5 & 5x + 2 \\ y & x - y \end{bmatrix}$$

Determine if there are values of x and y so that A and B are equal.

SOLUTION Both A and B are 2×2 matrices so $A = B$ if

$$x + y = 5 \quad (1) \qquad 6 = 5x + 2 \quad (2)$$
$$2x - 3 = y \quad (3) \qquad 2 - y = x - y \quad (4)$$

Here we have four equations containing the two variables x and y. From Equation (4) we see that $x = 2$. Using this value in Equation (1), we obtain $y = 3$. But $x = 2, y = 3$ do not satisfy either Equation (2) or Equation (3). Hence, there are *no* values for x and y satisfying all four equations. This means A and B can never be equal.

NOW WORK PROBLEM 13.

Find the sum of two matrices 2

Addition of Matrices

In an algebra system there are operations, like adding and subtracting. Can two matrices be added? And, if so, what is the rule or law for addition of matrices?

EXAMPLE 6 Adding Two Matrices

Motors, Inc., produces three models of cars: a sedan, a convertible, and an SUV. If the company wishes to compare the units of raw material and the units of labor involved in 1 month's production of each of these models, the rectangular array displayed in Table 3 below may be used to present the data:

TABLE 3

	Sedan Model	Convertible Model	SUV Model
Units of Material	23	16	10
Units of Labor	7	9	11

The same information may be written concisely as the matrix

$$A = \begin{bmatrix} 23 & 16 & 10 \\ 7 & 9 & 11 \end{bmatrix}$$

Suppose the next month's production is

$$B = \begin{bmatrix} 18 & 12 & 9 \\ 14 & 6 & 8 \end{bmatrix}$$

in which the pattern of recording units and models remains the same.

The total production for the 2 months can be displayed by the matrix

$$C = \begin{bmatrix} 41 & 28 & 19 \\ 21 & 15 & 19 \end{bmatrix}$$

since the number of units of material for sedan models is $41 = 23 + 18$, the number of units of material for convertible models is $28 = 16 + 12$, and so on.

This leads to the following definition.

Addition of Matrices

We define the sum $A + B$ of two matrices A and B with the same dimension as the matrix consisting of the sum of corresponding entries from A and B. That is, if $A = [a_{ij}]$ and $B = [b_{ij}]$ are two $m \times n$ matrices, the **sum** $A + B$ is the $m \times n$ matrix $[a_{ij} + b_{ij}]$.

EXAMPLE 7 Adding Two Matrices

(a)
$$\begin{bmatrix} 23 & 16 & 10 \\ 7 & 9 & 11 \end{bmatrix} + \begin{bmatrix} 18 & 12 & 9 \\ 14 & 6 & 8 \end{bmatrix} = \begin{bmatrix} 23+18 & 16+12 & 10+9 \\ 7+14 & 9+6 & 11+8 \end{bmatrix} = \begin{bmatrix} 41 & 28 & 19 \\ 21 & 15 & 19 \end{bmatrix}$$

(b)
$$\begin{bmatrix} 0.6 & 0.4 \\ 0.1 & 0.9 \end{bmatrix} + \begin{bmatrix} 2.3 & 0.6 \\ 1.8 & 5.2 \end{bmatrix} = \begin{bmatrix} 0.6+2.3 & 0.4+0.6 \\ 0.1+1.8 & 0.9+5.2 \end{bmatrix} = \begin{bmatrix} 2.9 & 1.0 \\ 1.9 & 6.1 \end{bmatrix}$$

Notice that it is possible to add two matrices only if their dimensions are the same. Also, the dimension of the sum of two matrices is the same as that of the two original matrices.

The following pairs of matrices cannot be added since they are of different dimensions:

$$A = \begin{bmatrix} 1 & 2 \\ 7 & 2 \end{bmatrix} \quad \text{and} \quad B = \begin{bmatrix} 1 \\ -3 \end{bmatrix}$$

$$A = [2 \quad 3] \quad \text{and} \quad B = [1 \quad 1 \quad 1]$$

$$A = \begin{bmatrix} -1 & 7 & 0 \\ 2 & \frac{1}{2} & 0 \end{bmatrix} \quad \text{and} \quad B = \begin{bmatrix} -1 & 2 \\ 3 & 0 \\ 1 & 5 \end{bmatrix}$$

NOW WORK PROBLEM 25.

Properties of Matrices

Work with properties of 3 matrices

It turns out that the usual rules for the addition of real numbers (such as the commutative property and associative property) are also valid for matrix addition.

EXAMPLE 8 Demonstrating the Commutative Property

Let $A = \begin{bmatrix} 1 & 5 \\ 7 & -3 \end{bmatrix}$ and $B = \begin{bmatrix} 3 & -2 \\ 4 & 1 \end{bmatrix}$

Then

$$A + B = \begin{bmatrix} 1 & 5 \\ 7 & -3 \end{bmatrix} + \begin{bmatrix} 3 & -2 \\ 4 & 1 \end{bmatrix} = \begin{bmatrix} 1+3 & 5+(-2) \\ 7+4 & -3+1 \end{bmatrix}$$
$$= \begin{bmatrix} 4 & 3 \\ 11 & -2 \end{bmatrix}$$
$$B + A = \begin{bmatrix} 3 & -2 \\ 4 & 1 \end{bmatrix} + \begin{bmatrix} 1 & 5 \\ 7 & -3 \end{bmatrix} = \begin{bmatrix} 4 & 3 \\ 11 & -2 \end{bmatrix}$$

This leads us to formulate the following property for addition.

Commutative Property for Addition

If A and B are two matrices of the same dimension, then

$$A + B = B + A$$

The associative property for addition of matrices is also true.

Associative Property for Addition

If A, B, and C are three matrices of the same dimension, then

$$A + (B + C) = (A + B) + C$$

The fact that addition of matrices is associative means that the notation $A + B + C$ is *not* ambiguous, since $(A + B) + C = A + (B + C)$.

NOW WORK PROBLEM 45.

A matrix in which all entries are 0 is called a **zero matrix.** We use the symbol **0** to represent a zero matrix of any dimension.

For real numbers, 0 has the property that $x + 0 = x$ for any x. A property of a zero matrix is that $A + \mathbf{0} = A$, provided the dimension of $\mathbf{0}$ is the same as that of A.

EXAMPLE 9 Demonstrating a Property of the Zero Matrix

Let

$$A = \begin{bmatrix} 3 & 4 & -\dfrac{1}{2} \\ \sqrt{2} & 0 & 3 \end{bmatrix}$$

Then

$$A + \mathbf{0} = \begin{bmatrix} 3 & 4 & -\dfrac{1}{2} \\ \sqrt{2} & 0 & 3 \end{bmatrix} + \begin{bmatrix} 0 & 0 & 0 \\ 0 & 0 & 0 \end{bmatrix}$$

$$= \begin{bmatrix} 3+0 & 4+0 & -\dfrac{1}{2}+0 \\ \sqrt{2}+0 & 0+0 & 3+0 \end{bmatrix} = \begin{bmatrix} 3 & 4 & -\dfrac{1}{2} \\ \sqrt{2} & 0 & 3 \end{bmatrix} = A$$

If A is any matrix, the **additive inverse** of A, denoted by $-A$, is the matrix obtained by replacing each entry in A by its negative.

EXAMPLE 10 Finding the Additive Inverse of a Matrix

If

$$A = \begin{bmatrix} -3 & 0 \\ 5 & -2 \\ 1 & 3 \end{bmatrix} \quad \text{then} \quad -A = \begin{bmatrix} 3 & 0 \\ -5 & 2 \\ -1 & -3 \end{bmatrix}$$

Additive Inverse Property

For any matrix A, we have the property that

$$\boxed{A + (-A) = \mathbf{0}}$$

NOW WORK PROBLEM 47.

Subtraction of Matrices

Find the difference of two matrices 4

Now that we have defined the sum of two matrices and the additive inverse of a matrix, it is natural to ask about the *difference* of two matrices. As you will see, subtracting matrices and subtracting real numbers are much the same kind of process.

Subtraction of Matrices

We define the **difference** $A - B$ of two matrices A and B with the same dimension as the matrix consisting of the difference of corresponding entries from A and B. That is, if $A = [a_{ij}]$ and $B = [b_{ij}]$ are two $m \times n$ matrices, the **difference** $A - B$ is the $m \times n$ matrix $[a_{ij} - b_{ij}]$.

EXAMPLE 11 Subtracting Two Matrices

Let
$$A = \begin{bmatrix} 2 & 3 & 4 \\ 1 & 0 & 2 \end{bmatrix} \quad \text{and} \quad B = \begin{bmatrix} -2 & 1 & -1 \\ 3 & 0 & 3 \end{bmatrix}$$

Then
$$A - B = \begin{bmatrix} 2 & 3 & 4 \\ 1 & 0 & 2 \end{bmatrix} - \begin{bmatrix} -2 & 1 & -1 \\ 3 & 0 & 3 \end{bmatrix}$$
$$= \begin{bmatrix} 2-(-2) & 3-1 & 4-(-1) \\ 1-3 & 0-0 & 2-3 \end{bmatrix} = \begin{bmatrix} 4 & 2 & 5 \\ -2 & 0 & -1 \end{bmatrix}$$

Notice that the difference $A - B$ is nothing more than the matrix formed by subtracting the entries in B from the corresponding entries in A.

Using the matrices A and B from Example 11, we find that
$$B - A = \begin{bmatrix} -2 & 1 & -1 \\ 3 & 0 & 3 \end{bmatrix} - \begin{bmatrix} 2 & 3 & 4 \\ 1 & 0 & 2 \end{bmatrix}$$
$$= \begin{bmatrix} -2-2 & 1-3 & -1-4 \\ 3-1 & 0-0 & 3-2 \end{bmatrix} = \begin{bmatrix} -4 & -2 & -5 \\ 2 & 0 & 1 \end{bmatrix}$$

Observe that $A - B \neq B - A$, illustrating that matrix subtraction, like subtraction of real numbers, is not commutative.

EXAMPLE 12 Adding and Subtracting Matrices

Suppose that
$$A = \begin{bmatrix} 2 & 4 & 8 & -3 \\ 0 & 1 & 2 & 3 \end{bmatrix} \quad \text{and} \quad B = \begin{bmatrix} -3 & 4 & 0 & 1 \\ 6 & 8 & 2 & 0 \end{bmatrix}$$

Find: **(a)** $A + B$ **(b)** $A - B$

SOLUTION **(a)**
$$A + B = \begin{bmatrix} 2 & 4 & 8 & -3 \\ 0 & 1 & 2 & 3 \end{bmatrix} + \begin{bmatrix} -3 & 4 & 0 & 1 \\ 6 & 8 & 2 & 0 \end{bmatrix}$$
$$= \begin{bmatrix} 2+(-3) & 4+4 & 8+0 & -3+1 \\ 0+6 & 1+8 & 2+2 & 3+0 \end{bmatrix} \quad \text{Add corresponding entries.}$$
$$= \begin{bmatrix} -1 & 8 & 8 & -2 \\ 6 & 9 & 4 & 3 \end{bmatrix}$$

(b)
$$A - B = \begin{bmatrix} 2 & 4 & 8 & -3 \\ 0 & 1 & 2 & 3 \end{bmatrix} - \begin{bmatrix} -3 & 4 & 0 & 1 \\ 6 & 8 & 2 & 0 \end{bmatrix}$$
$$= \begin{bmatrix} 2-(-3) & 4-4 & 8-0 & -3-1 \\ 0-6 & 1-8 & 2-2 & 3-0 \end{bmatrix} \quad \text{Subtract corresponding entries.}$$
$$= \begin{bmatrix} 5 & 0 & 8 & -4 \\ -6 & -7 & 0 & 3 \end{bmatrix}$$

COMMENT: Graphing utilities make the sometimes tedious process of matrix algebra easy. Let's see how a graphing utility adds and subtracts matrices by solving Example 12 using a graphing utility.

GRAPHING UTILITY SOLUTION   Enter the matrices into a graphing utility. Name them $[A]$ and $[B]$. Figure 9 shows the results of adding and subtracting $[A]$ and $[B]$.

FIGURE 9

```
[A]+[B]
  [[-1 8 8 -2]
   [6  9 4 3 ]]
[A]-[B]
  [[5  0  8 -4]
   [-6 -7 0 3 ]]
```

▶ ▶

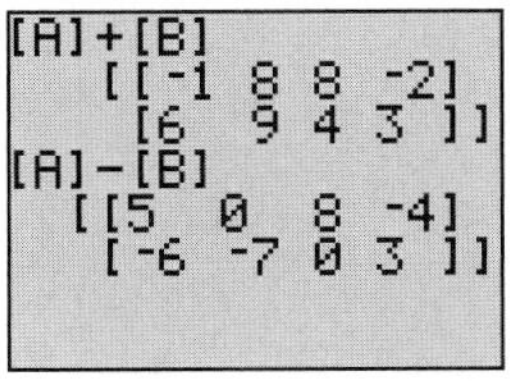

NOW WORK PROBLEM 33.

Multiplying a Matrix by a Number

Let's return to the production of Motors, Inc., during the month specified in Example 6. The matrix A describing this production is

$$A = \begin{bmatrix} 23 & 16 & 10 \\ 7 & 9 & 11 \end{bmatrix}$$

Let's assume that for 3 consecutive months, the monthly production remained the same. Then the total production for the 3 months is simply the sum of the matrix A taken 3 times. If we represent the total production by the matrix T, then

$$T = \begin{bmatrix} 23 & 16 & 10 \\ 7 & 9 & 11 \end{bmatrix} + \begin{bmatrix} 23 & 16 & 10 \\ 7 & 9 & 11 \end{bmatrix} + \begin{bmatrix} 23 & 16 & 10 \\ 7 & 9 & 11 \end{bmatrix}$$

$$= \begin{bmatrix} 23 + 23 + 23 & 16 + 16 + 16 & 10 + 10 + 10 \\ 7 + 7 + 7 & 9 + 9 + 9 & 11 + 11 + 11 \end{bmatrix}$$

$$= \begin{bmatrix} 3 \cdot 23 & 3 \cdot 16 & 3 \cdot 10 \\ 3 \cdot 7 & 3 \cdot 9 & 3 \cdot 11 \end{bmatrix} = \begin{bmatrix} 69 & 48 & 30 \\ 21 & 27 & 33 \end{bmatrix}$$

In other words, when we add the matrix A 3 times, we multiply each entry of A by 3. This leads to the following definition.

Scalar Multiplication

Let A be an $m \times n$ matrix and let c be a real number, called a **scalar**. The product of the matrix A by the scalar c, called **scalar multiplication**, is the $m \times n$ matrix cA, whose entries are the product of c and the corresponding entries of A. That is, if $A = [a_{ij}]$, then $cA = [ca_{ij}]$.

When multiplying a matrix by a real number, each entry of the matrix is multiplied by the number. Notice that the dimension of A and the dimension of the product cA are the same.

5

EXAMPLE 13 Finding Scalar Multiples of a Matrix

Suppose that

$$A = \begin{bmatrix} 3 & 1 & 5 \\ -2 & 0 & 6 \end{bmatrix}, \qquad B = \begin{bmatrix} 4 & 1 & 0 \\ 8 & 1 & -3 \end{bmatrix}, \qquad C = \begin{bmatrix} 9 & 0 \\ -3 & 6 \end{bmatrix}$$

Find: **(a)** $4A$ **(b)** $\frac{1}{3}C$ **(c)** $3A - 2B$

SOLUTION **(a)** $4A = 4\begin{bmatrix} 3 & 1 & 5 \\ -2 & 0 & 6 \end{bmatrix} = \begin{bmatrix} 4 \cdot 3 & 4 \cdot 1 & 4 \cdot 5 \\ 4(-2) & 4 \cdot 0 & 4 \cdot 6 \end{bmatrix} = \begin{bmatrix} 12 & 4 & 20 \\ -8 & 0 & 24 \end{bmatrix}$

(b) $\frac{1}{3}C = \frac{1}{3}\begin{bmatrix} 9 & 0 \\ -3 & 6 \end{bmatrix} = \begin{bmatrix} \frac{1}{3} \cdot 9 & \frac{1}{3} \cdot 0 \\ \frac{1}{3} \cdot (-3) & \frac{1}{3} \cdot 6 \end{bmatrix} = \begin{bmatrix} 3 & 0 \\ -1 & 2 \end{bmatrix}$

(c) $3A - 2B = 3\begin{bmatrix} 3 & 1 & 5 \\ -2 & 0 & 6 \end{bmatrix} - 2\begin{bmatrix} 4 & 1 & 0 \\ 8 & 1 & -3 \end{bmatrix}$

$$= \begin{bmatrix} 3 \cdot 3 & 3 \cdot 1 & 3 \cdot 5 \\ 3(-2) & 3 \cdot 0 & 3 \cdot 6 \end{bmatrix} - \begin{bmatrix} 2 \cdot 4 & 2 \cdot 1 & 2 \cdot 0 \\ 2 \cdot 8 & 2 \cdot 1 & 2(-3) \end{bmatrix}$$

$$= \begin{bmatrix} 9 & 3 & 15 \\ -6 & 0 & 18 \end{bmatrix} - \begin{bmatrix} 8 & 2 & 0 \\ 16 & 2 & -6 \end{bmatrix}$$

$$= \begin{bmatrix} 1 & 1 & 15 \\ -22 & -2 & 24 \end{bmatrix}$$

GRAPHING UTILITY SOLUTION Enter the matrices [A], and [B], and [C] into a graphing utility. Figure 10 shows the required computations.

FIGURE 10

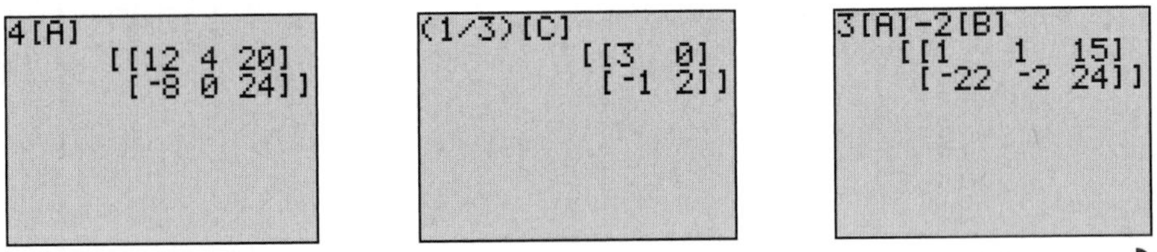

(a) (b) (c)

NOW WORK PROBLEM 35.

We list next some of the algebraic properties of scalar multiplication. Let h and k be real numbers, and let A and B be $m \times n$ matrices. Then

Properties of Scalar Multiplication

Let k and h be two real numbers and let A and B be two matrices of dimension $m \times n$. Then

$$k(hA) = (kh)A \qquad (2)$$

$$(k + h)A = kA + hA \qquad (3)$$

$$k(A + B) = kA + kB \qquad (4)$$

Properties (2), (3), and (4) are illustrated in the following example.

EXAMPLE 14 **Using Properties of Scalar Multiplication**

For $A = \begin{bmatrix} 2 & -3 & -1 \\ 5 & 6 & 4 \end{bmatrix}$ and $B = \begin{bmatrix} -3 & 0 & 4 \\ 2 & -1 & 5 \end{bmatrix}$

show that

(a) $5[2A] = 10A$ **(b)** $(4+3)A = 4A + 3A$ **(c)** $3[A+B] = 3A + 3B$

SOLUTION **(a)** $5[2A] = 5\begin{bmatrix} 4 & -6 & -2 \\ 10 & 12 & 8 \end{bmatrix} = \begin{bmatrix} 20 & -30 & -10 \\ 50 & 60 & 40 \end{bmatrix}$

$$10A = \begin{bmatrix} 20 & -30 & -10 \\ 50 & 60 & 40 \end{bmatrix}$$

(b) $(4+3)A = 7A = \begin{bmatrix} 14 & -21 & -7 \\ 35 & 42 & 28 \end{bmatrix}$

$$4A + 3A = \begin{bmatrix} 8 & -12 & -4 \\ 20 & 24 & 16 \end{bmatrix} + \begin{bmatrix} 6 & -9 & -3 \\ 15 & 18 & 12 \end{bmatrix} = \begin{bmatrix} 14 & -21 & -7 \\ 35 & 42 & 28 \end{bmatrix}$$

(c) $3[A+B] = 3\begin{bmatrix} -1 & -3 & 3 \\ 7 & 5 & 9 \end{bmatrix} = \begin{bmatrix} -3 & -9 & 9 \\ 21 & 15 & 27 \end{bmatrix}$

$$3A + 3B = \begin{bmatrix} 6 & -9 & -3 \\ 15 & 18 & 12 \end{bmatrix} + \begin{bmatrix} -9 & 0 & 12 \\ 6 & -3 & 15 \end{bmatrix} = \begin{bmatrix} -3 & -9 & 9 \\ 21 & 15 & 27 \end{bmatrix}$$

EXERCISE 2.4 Answers to Odd-Numbered Problems Begin on Page AN-12.

In Problems 1–12, find the dimension of each matrix. Say if the matrix is a square matrix, or a row matrix, or a column matrix.

1. $\begin{bmatrix} 3 & 2 \\ -1 & 3 \end{bmatrix}$

2. $\begin{bmatrix} -1 & 0 \\ 0 & 5 \end{bmatrix}$

3. $\begin{bmatrix} 2 & 1 & -3 \\ 1 & 0 & -1 \end{bmatrix}$

4. $\begin{bmatrix} 1 & 2 \\ 2 & 1 \\ 0 & -3 \end{bmatrix}$

5. $\begin{bmatrix} 4 & 0 \\ -1 & 2 \\ 5 & 8 \end{bmatrix}$

6. $\begin{bmatrix} 0 & -3 & 6 \\ 0 & 5 & 2 \\ 1 & 8 & 7 \end{bmatrix}$

7. $\begin{bmatrix} 1 & 4 \\ -2 & 8 \\ 0 & 0 \end{bmatrix}$

8. $\begin{bmatrix} 8 & 1 & 0 & 0 \\ 3 & -4 & 0 & 0 \end{bmatrix}$

9. $\begin{bmatrix} 4 \\ 1 \end{bmatrix}$

10. $\begin{bmatrix} 2 & 1 & -3 \end{bmatrix}$

11. $[2]$

12. $[0]$

In Problems 13–24, determine whether the given statements are true or false. If false, tell why.

13. $\begin{bmatrix} 0 \\ 1 \end{bmatrix} = \begin{bmatrix} 0 & 1 \end{bmatrix}$

14. $\begin{bmatrix} 3 & 2 \\ -1 & 0 \end{bmatrix} = \begin{bmatrix} 3 & 2 \\ -1 & 4 \end{bmatrix}$

15. $\begin{bmatrix} 5 & 0 \\ 0 & 1 \end{bmatrix}$ is square

16. $\begin{bmatrix} 3 & 2 & 1 \\ 4 & -1 & 0 \end{bmatrix}$ is 3×2

17. $\begin{bmatrix} x & 2 \\ 4 & 0 \end{bmatrix} = \begin{bmatrix} 3 & 2 \\ 4 & 0 \end{bmatrix}$ if $x = 3$

18. $\begin{bmatrix} x & y \\ 0 & 0 \end{bmatrix} = \begin{bmatrix} x & y \end{bmatrix}$

19. $\begin{bmatrix} 5 & 0 \\ 1 & 1 \end{bmatrix} = \begin{bmatrix} 2+3 & 0 \\ 1 & 1 \end{bmatrix}$

20. $\begin{bmatrix} 1 & 0 \\ 0 & 1 \end{bmatrix} = \begin{bmatrix} 3-2 & 3-3 \\ 3-3 & 3-2 \end{bmatrix}$

21. $2\begin{bmatrix} 1 & 0 \\ 0 & 2 \end{bmatrix} = \begin{bmatrix} 2 & 0 \\ 0 & 4 \end{bmatrix}$

22. $-\begin{bmatrix} 8 & 0 \\ 5 & -1 \end{bmatrix} = \begin{bmatrix} -8 & 0 \\ 5 & 1 \end{bmatrix}$

23. $\begin{bmatrix} 8 \\ 1 \end{bmatrix} + \begin{bmatrix} 2 \\ 9 \end{bmatrix} = [10]$

24. $[6 \quad 0] + [1] = [7 \quad 1]$

In Problems 25–32, perform the indicated operations. Express your answer as a single matrix.

25. $\begin{bmatrix} 3 & -1 \\ 4 & 2 \end{bmatrix} + \begin{bmatrix} -2 & 2 \\ 2 & 5 \end{bmatrix}$

26. $\begin{bmatrix} 2 & 4 \\ 3 & -4 \end{bmatrix} - \begin{bmatrix} 8 & 9 \\ 0 & 7 \end{bmatrix}$

27. $3\begin{bmatrix} 2 & 6 & 0 \\ 4 & -2 & 1 \end{bmatrix}$

28. $-3\begin{bmatrix} 2 & 1 \\ -2 & 1 \\ 0 & 3 \end{bmatrix}$

29. $2\begin{bmatrix} 1 & -1 & 8 \\ 2 & 4 & 1 \end{bmatrix} - 3\begin{bmatrix} 0 & -2 & 8 \\ 1 & 4 & 1 \end{bmatrix}$

30. $6\begin{bmatrix} 2 & 1 \\ 3 & 1 \\ -1 & 0 \end{bmatrix} + 4\begin{bmatrix} 6 & -4 \\ -2 & -3 \\ 0 & 1 \end{bmatrix}$

31. $3\begin{bmatrix} a & 8 \\ b & 1 \\ c & -2 \end{bmatrix} + 5\begin{bmatrix} 2a & 6 \\ -b & -2 \\ -c & 0 \end{bmatrix}$

32. $2\begin{bmatrix} 2x & y & z \\ 2 & -4 & 8 \end{bmatrix} - 3\begin{bmatrix} -3x & 4y & 2z \\ 6 & -1 & 4 \end{bmatrix}$

In Problems 33–50, use the matrices below. For Problems 33–44 perform the indicated operation(s); for Problems 45–50 verify the indicated property.

$$A = \begin{bmatrix} 2 & -3 & 4 \\ 0 & 2 & 1 \end{bmatrix} \quad B = \begin{bmatrix} 1 & -2 & 0 \\ 5 & 1 & 2 \end{bmatrix} \quad C = \begin{bmatrix} -3 & 0 & 5 \\ 2 & 1 & 3 \end{bmatrix}$$

33. $A - B$

34. $B + C$

35. $2A - 3C$

36. $3C - 4B$

37. $(A + B) - 2C$

38. $4C + (A - B)$

39. $3A + 4(B + C)$

40. $(A + B) + 3C$

41. $2(A - B) - C$

42. $2A - 5(B + C)$

43. $3A - B - 6C$

44. $3A + 2B - 4C$

45. Verify the commutative property for addition by finding $A + B$ and $B + A$.

46. Verify the associative property for addition by finding $(A + B) + C$ and $A + (B + C)$.

47. Verify the additive inverse property by showing that $A + (-A) = 0$.

48. Verify Property (2) of scalar multiplication by finding $2(3A)$ and $6A$.

49. Verify Property (3) of scalar multiplication by finding $2B + 3B$ and $5B$.

50. Verify Property (4) of scalar multiplication by finding $2(A + C)$ and $2A + 2C$.

51. Find x and z so that

$$\begin{bmatrix} x \\ 4 \end{bmatrix} = \begin{bmatrix} -4 \\ z \end{bmatrix}$$

52. Find x, y, and z so that

$$\begin{bmatrix} x+y & -2 \\ 4 & 10 \end{bmatrix} = \begin{bmatrix} 6 & x-y \\ 4 & z \end{bmatrix}$$

53. Find x and y so that

$$\begin{bmatrix} x-2y & 0 \\ -2 & 6 \end{bmatrix} = \begin{bmatrix} 3 & 0 \\ -2 & x+y \end{bmatrix}$$

54. Find x, y, and z so that

$$\begin{bmatrix} x-2 & 3 & 2z \\ 6y & x & 2y \end{bmatrix} = \begin{bmatrix} y & z & 6 \\ 18z & y+2 & 6z \end{bmatrix}$$

55. Find x, y, and z so that

$$[2 \quad 3 \quad -4] + [x \quad 2y \quad z] = [6 \quad -9 \quad 2]$$

56. Find x and y so that

$$\begin{bmatrix} 3 & -2 & 2 \\ 1 & 0 & -1 \end{bmatrix} + \begin{bmatrix} x-y & 2 & -2 \\ 4 & x & 6 \end{bmatrix} = \begin{bmatrix} 6 & 0 & 0 \\ 5 & 2x+y & 5 \end{bmatrix}$$

In Problems 57–62, use a graphing utility to perform the indicated operations on the matrices given below.

$$A = \begin{bmatrix} -1 & -1 & 3 & 0 \\ 2 & 6 & 2 & 2 \\ -4 & 2 & 3 & 2 \\ 7 & 0 & 5 & -1 \end{bmatrix} \quad B = \begin{bmatrix} -1 & 2 & 4 & 5 \\ 2 & 0 & 5 & 3 \\ 0.5 & 6 & -7 & 11 \\ 5 & -1 & 2 & 7 \end{bmatrix} \quad C = \begin{bmatrix} 13 & -8 & 7 & 0 \\ 0 & 5 & 0 & -2 \\ 5 & 0 & 7 & 0 \\ 7 & 7 & 7 & 7 \end{bmatrix}$$

57. $A + B$

58. $3C - 2B$

59. $C - 3(A + B)$

60. $2(A - B) + \frac{1}{2}C$

61. $3(B + C) - A$

62. $\frac{1}{3}(A + 2C) - B$

---

63. Prison Populations In 2000, local jails and state and federal prisons contained nearly 2 million people, as follows: 613,534 prisoners were in local jails, of which 11.5% were female; 1,236,476 prisoners were in state prisons, of which 93.4% were male; and 145,416 were in federal prisons, of which 7.0% were female. Express this information using a 2×3 matrix. Label the rows MALE and FEMALE and the columns LOCAL, STATE, FED.

Source: United States Department of Justice, 2001.

64. Nail Production XYZ Company produces steel and aluminum nails. One week 25 gross of $\frac{1}{2}$-inch steel nails and 45 gross of 1-inch steel nails were produced. Suppose 13 gross of $\frac{1}{2}$-inch aluminum nails, 20 gross of 1-inch aluminum nails, 35 gross of 2-inch steel nails, and 23 gross of 2-inch aluminum nails were also made. Write a 2×3 matrix depicting this. Could you also write a 3×2 matrix for this situation?

65. College Degrees by Gender Post-secondary degrees include associate, bachelor's, master's, and doctoral degrees. Projections for 2003–2004 are as follows: 582,000 associate degrees, of which 218,000 will be awarded to men; 1,251,000 bachelor's degrees, of which 714,000 will be awarded to women; 442,000 master's degrees, of which 261,000 will be awarded to women; and 47,100 doctoral degrees, of which 26,700 will be awarded to men. Write a 2×4 matrix depicting this. Could you also write a 4×2 matrix for this situation?

Source: Digest of Education Statistics, National Center for Education Statistics, 2001.

66. Katy, Mike, and Danny go to the candy store. Katy buys 5 sticks of gum, 2 ice cream cones, and 20 jelly beans. Mike buys 2 sticks of gum, 15 jelly beans, and 3 candy bars. Danny buys 1 stick of gum, 1 ice cream cone, and 4 candy bars. Write a matrix depicting this situation.

67. Use a matrix to display the information given below, which was obtained in a survey of voters. Label the rows UNDER \$25,000 and OVER \$25,000 and label the columns DEMOCRATS, REPUBLICANS, INDEPENDENTS.

351	Democrats earning under \$25,000
271	Republicans earning under \$25,000
73	Independents earning under \$25,000
203	Democrats earning \$25,000 or more
215	Republicans earning \$25,000 or more
55	Independents earning \$25,000 or more

68. Listing Stocks One day on the New York Stock Exchange, 800 issues went up and 600 went down. Of the 800 up issues, 200 went up more than \$1 per share. Of the 600 down issues, 50 went down more than \$1 per share. Express this information in a 2×2 matrix. Label the rows UP and DOWN and the columns MORE THAN \$1 and LESS THAN \$1.

69. Surveys In a survey of 1000 college students, the following information was obtained: 500 were liberal arts and sciences (LAS) majors, of which 50% were female; 300 were engineering (ENG) majors, of which 75% were male; and the remaining were education (EDUC) majors, of which 60% were female. Express this information using a 2×3 matrix. Label the rows MALE and FEMALE and the columns LAS, ENG, and EDUC.

70. Sales of Cars The sales figures for two car dealers during June showed that dealer A sold 100 compacts, 50 intermediates, and 40 full-size cars, while dealer B sold 120 compacts, 40 intermediates, and 35 full-size cars. During July, dealer A sold 80 compacts, 30 intermediates, and 10 full-size cars, while dealer B sold 70 compacts, 40 intermediates, and 20 full-size cars. Total sales over the 3-month period of June–August revealed that dealer A sold 300 compacts, 120 intermediates, and 65 full-size cars. In the same 3-month period, dealer B sold 250 compacts, 100 intermediates, and 80 full-size cars.

(a) Write 2×3 matrices summarizing sales data for June, July, and the 3-month period for each dealer.
(b) Use matrix addition to find the sales over the 2-month period for June and July for each dealer.
(c) Use matrix subtraction to find the sales in August for each dealer.

2.5 Multiplication of Matrices

OBJECTIVES 1 Find the product of two matrices
2 Work with properties of matrices

While addition and subtraction of matrices and the product of a scalar and a matrix are fairly straightforward, defining the *product* of two matrices requires a bit more detail.

We explain first what we mean by the product of a row matrix (a matrix with one row) with a column matrix (a matrix with one column).

Let's look at a simple example. In a given month suppose 23 units of material and 7 units of labor were required to manufacture 4-door sedans. We can represent this by the column matrix

$$\begin{bmatrix} 23 \\ 7 \end{bmatrix}$$

Also, suppose the cost per unit of material is \$450 and the cost per unit of labor is \$600. We represent these costs by the row matrix

$$\begin{bmatrix} 450 & 600 \end{bmatrix}$$

The total cost of producing the sedans is calculated as follows:

$$\begin{aligned} \text{Total cost} &= (\text{Cost per unit of material}) \times (\text{Units of material}) \\ &\quad + (\text{Cost per unit of labor}) \times (\text{Units of labor}) \\ &= 450 \times 23 + 600 \times 7 = 14{,}550 \end{aligned}$$

In terms of the matrix representations,

$$\text{Total cost} = \begin{bmatrix} 450 & 600 \end{bmatrix} \begin{bmatrix} 23 \\ 7 \end{bmatrix} = 450 \times 23 + 600 \times 7 = 14{,}550$$

This leads us to formulate the following definition for multiplying a row matrix times a column matrix:

If $R = [r_1\ r_2\ \ldots\ r_n]$ is a row matrix of dimension $1 \times n$ and $C = \begin{bmatrix} c_1 \\ c_2 \\ \vdots \\ c_n \end{bmatrix}$ is a column matrix of dimension $n \times 1$, then by the **product of R and C** we mean the number

$$RC = r_1c_1 + r_2c_2 + r_3c_3 + \ldots + r_nc_n$$

EXAMPLE 1 **Finding the Product of a 1 × 3 Row Matrix and a 3 × 1 Column Matrix**

If

$$R = [1 \quad 5 \quad 3] \quad \text{and} \quad C = \begin{bmatrix} 2 \\ -1 \\ 4 \end{bmatrix}$$

then the product of R and C is

$$RC = [1 \quad 5 \quad 3]\begin{bmatrix} 2 \\ -1 \\ 4 \end{bmatrix} = 1 \cdot 2 + 5 \cdot (-1) + 3 \cdot 4 = 9$$

Notice that for the product of a row matrix R and a column matrix C to be defined, if R is a $1 \times n$ row matrix, then C must have dimension $n \times 1$.

EXAMPLE 2 **Finding the Product of a 1 × 4 Row Matrix and a 4 × 1 Column Matrix**

Let

$$R = [1 \quad 0 \quad 1 \quad 5] \quad \text{and} \quad C = \begin{bmatrix} 0 \\ -11 \\ 0 \\ 8 \end{bmatrix}$$

Then the product of R and C is

$$RC = [1 \quad 0 \quad 1 \quad 5]\begin{bmatrix} 0 \\ -11 \\ 0 \\ 8 \end{bmatrix} = 1 \cdot 0 + 0 \cdot (-11) + 1 \cdot 0 + 5 \cdot 8 = 40$$

NOW WORK PROBLEM 1.

Given two matrices A and B, the rows of A can be thought of as row matrices, while the columns of B can be thought of as column matrices. This observation will be used in the following definition.

Multiplication of Matrices

Let A denote an $m \times r$ matrix, and let B denote an $r \times n$ matrix. The **product** AB is defined as the $m \times n$ matrix whose entry in row i, column j is the product of the ith row of A and the jth column of B.

An example will help to clarify the definition.

1 **EXAMPLE 3** **Finding the Product of Two Matrices**

Find the product AB if

$$A = \begin{bmatrix} 2 & 4 & -1 \\ 5 & 8 & 0 \end{bmatrix} \quad \text{and} \quad B = \begin{bmatrix} 2 & 5 & 1 & 4 \\ 4 & 8 & 0 & 6 \\ -3 & 1 & -2 & -1 \end{bmatrix}$$

SOLUTION First, we note that A is 2×3 and B is 3×4, so the product AB is defined and will be a 2×4 matrix.

Suppose that we want the entry in row 2, column 3 of AB. To find it, we find the product of the row vector from row 2 of A and the column vector from column 3 of B.

$$\underset{\text{Row 2 of } A}{[5 \quad 8 \quad 0]} \overset{\text{Column 3 of } B}{\begin{bmatrix} 1 \\ 0 \\ -2 \end{bmatrix}} = 5 \cdot 1 + 8 \cdot 0 + 0(-2) = 5$$

So far we have

$$AB = \begin{bmatrix} \rule{1em}{0.4pt} & \rule{1em}{0.4pt} & \overset{\text{Column 3}}{\rule{1em}{0.4pt}} & \rule{1em}{0.4pt} \\ \rule{1em}{0.4pt} & \rule{1em}{0.4pt} & 5 & \rule{1em}{0.4pt} \end{bmatrix} \text{Row 2}$$

Now, to find the entry in row 1, column 4 of AB, we find the product of row 1 of A and column 4 of B.

$$\underset{\text{Row 1 of } A}{[2 \quad 4 \quad -1]} \overset{\text{Column 4 of } B}{\begin{bmatrix} 4 \\ 6 \\ -1 \end{bmatrix}} = 2 \cdot 4 + 4 \cdot 6 + (-1)(-1) = 33$$

Continuing in this fashion, we find AB.

$$AB = \begin{bmatrix} 2 & 4 & -1 \\ 5 & 8 & 0 \end{bmatrix} \begin{bmatrix} 2 & 5 & 1 & 4 \\ 4 & 8 & 0 & 6 \\ -3 & 1 & -2 & -1 \end{bmatrix}$$

$$= \begin{bmatrix} \text{Row 1 of } A \\ \text{times} \\ \text{column 1 of } B & \text{Row 1 of } A \\ \text{times} \\ \text{column 2 of } B & \text{Row 1 of } A \\ \text{times} \\ \text{column 3 of } B & \text{Row 1 of } A \\ \text{times} \\ \text{column 4 of } B \\ \text{Row 2 of } A \\ \text{times} \\ \text{column 1 of } B & \text{Row 2 of } A \\ \text{times} \\ \text{column 2 of } B & \text{Row 2 of } A \\ \text{times} \\ \text{column 3 of } B & \text{Row 2 of } A \\ \text{times} \\ \text{column 4 of } B \end{bmatrix}$$

$$= \begin{bmatrix} 2 \cdot 2 + 4 \cdot 4 + (-1)(-3) & 2 \cdot 5 + 4 \cdot 8 + (-1)1 & 2 \cdot 1 + 4 \cdot 0 + (-1)(-2) & 33 \text{ (from earlier)} \\ 5 \cdot 2 + 8 \cdot 4 + 0(-3) & 5 \cdot 5 + 8 \cdot 8 + 0 \cdot 1 & 5 \text{ (from earlier)} & 5 \cdot 4 + 8 \cdot 6 + 0(-1) \end{bmatrix}$$

$$= \begin{bmatrix} 23 & 41 & 4 & 33 \\ 42 & 89 & 5 & 68 \end{bmatrix}$$

▶

A graphing utility can be used to multiply two matrices. Use a graphing utility to do Example 3.

GRAPHING UTILITY SOLUTION   Enter the matrices A and B into a graphing utility. Figure 11 shows the product AB.

FIGURE 11

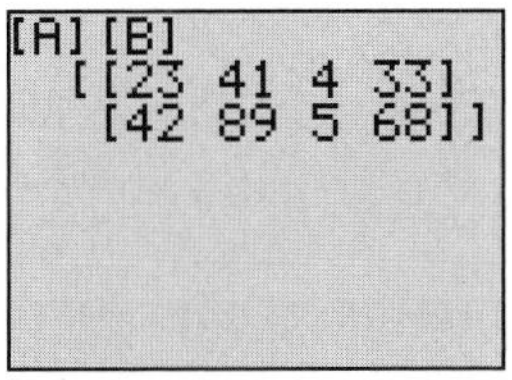

NOW WORK PROBLEM 27.

EXAMPLE 4   **Manufacturing Cost**

One month's production at Motors, Inc., may be given in matrix form as

$$
\begin{array}{c}
\text{Sedan Convertible SUV}\\
A = \begin{bmatrix} 23 & 16 & 10 \\ 7 & 9 & 11 \end{bmatrix} \begin{array}{l}\text{Units of material}\\ \text{Units of labor}\end{array}
\end{array}
$$

Suppose that in this month's production, the cost for each unit of material is \$450 and the cost for each unit of labor is \$600. What is the total cost to manufacture the sedans, the convertibles, and the SUVs?

SOLUTION   For sedans, the cost is 23 units of material at \$450 each, plus 7 units of labor at \$600 each, for a total cost of

$$23 \cdot \$450 + 7 \cdot \$600 = 10{,}350 + 4200 = \$14{,}550$$

Similarly, for convertibles, the total cost is

$$16 \cdot \$450 + 9 \cdot \$600 = 7200 + 5400 = \$12{,}600$$

Finally, for SUVs, the total cost is

$$10 \cdot \$450 + 11 \cdot \$600 = 4500 + 6600 = \$11{,}100$$

If we represent the cost of units of material and units of labor by the 1×2 matrix

$$
\begin{array}{c}
\begin{array}{cc}\text{Unit cost of} & \text{Unit cost of}\\ \text{material} & \text{labor}\end{array}\\
U = \begin{bmatrix} 450 & 600 \end{bmatrix}
\end{array}
$$

then the total cost is the product UA.

$$
UA = \begin{bmatrix} 450 & 600 \end{bmatrix} \begin{bmatrix} 23 & 16 & 10 \\ 7 & 9 & 11 \end{bmatrix}
$$

$$
= \begin{bmatrix} 450 \cdot 23 + 600 \cdot 7 & 450 \cdot 16 + 600 \cdot 9 & 450 \cdot 10 + 600 \cdot 11 \end{bmatrix}
$$

$$
= \begin{bmatrix} 14{,}550 & 12{,}600 & 11{,}100 \end{bmatrix}
$$

We conclude that the total cost to manufacture the sedans is \$14,550, to manufacture the convertibles is \$12,600, and to manufacture the SUVs is \$11,100.

  NOW WORK PROBLEM 59.

Spreadsheets, such as Excel, can also be used to multiply matrices. Such utilities are especially useful when the dimensions of the matrices are large. Use Excel to do Example 4.

SOLUTION **STEP 1** Enter the matrices for A and U into an Excel spreadsheet as follows:

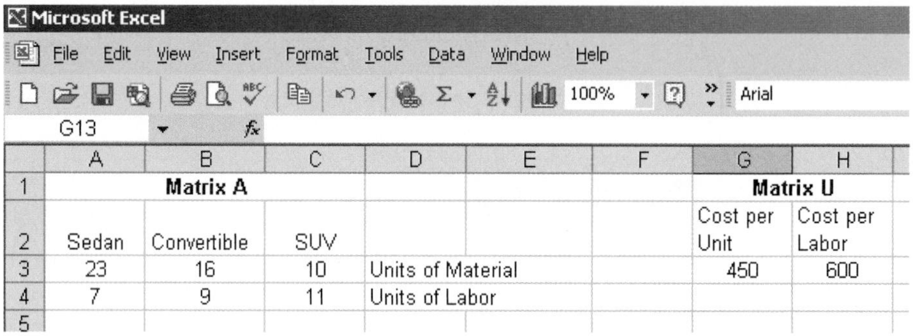

STEP 2 Highlight the cells that are to contain the product matrix. Since U is a 1×2 matrix and A is a 2×3 matrix, the product must be a 1×3 matrix.

STEP 3 Type: =MMULT(*highlight U, highlight A*)

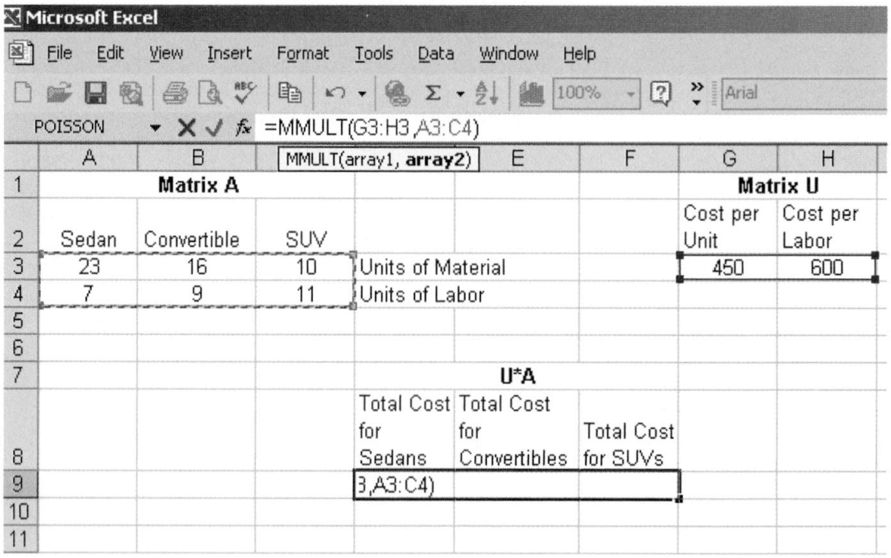

STEP 4 Press Ctrl-Shift-Enter at the same time.

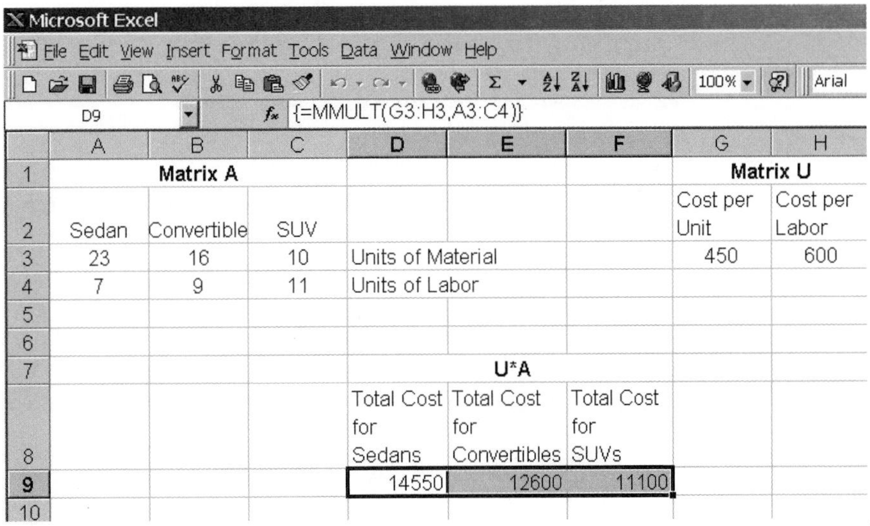

Microsoft Excel

File Edit View Insert Format Tools Data Window Help

D9 | {=MMULT(G3:H3,A3:C4)}

	A	B	C	D	E	F	G	H
1		Matrix A					Matrix U	
2	Sedan	Convertible	SUV				Cost per Unit	Cost per Labor
3	23	16	10	Units of Material			450	600
4	7	9	11	Units of Labor				
5								
6								
7					U*A			
8				Total Cost for Sedans	Total Cost for Convertibles	Total Cost for SUVs		
9				14550	12600	11100		
10								

NOW WORK PROBLEM 59 USING EXCEL.

Properties of Matrix Multiplication

Work with properties of **2** matrices

If A is a matrix of dimension $m \times r$ (which has r columns) and B is a matrix of dimension $p \times n$ (which has p rows) and if $r \neq p$, the product AB is not defined.

> Multiplication of matrices is possible only if the number of columns of the matrix on the left equals the number of rows of the matrix on the right.
>
> If A is of dimension $m \times r$ and B is of dimension $r \times n$, then the product AB is of dimension $m \times n$.

FIGURE 12

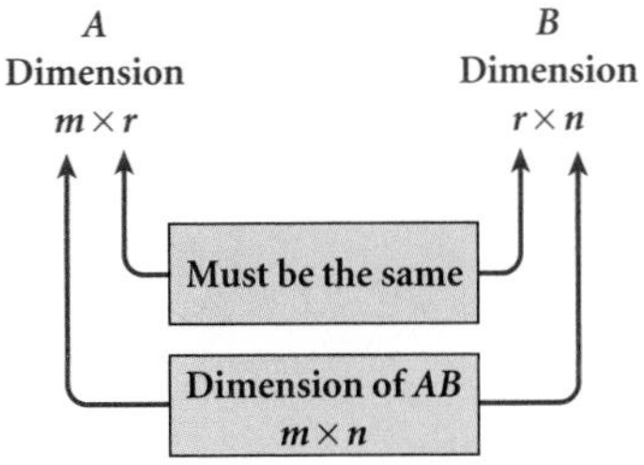

Figure 12 illustrates the result stated above.

For example, for the matrices given in Example 3 the product BA is not defined, because B is 3×4 and A is 2×3.

Another result of matrix multiplication is illustrated in the next example.

EXAMPLE 5 Multiplying Two Matrices

If

$$A = \begin{bmatrix} 2 & 1 & 3 \\ 1 & -1 & 0 \end{bmatrix} \quad \text{and} \quad B = \begin{bmatrix} 1 & 0 \\ 2 & 1 \\ 3 & 2 \end{bmatrix}$$

find: **(a)** AB **(b)** BA

SOLUTION **(a)** $AB = \underset{2 \times 3}{\begin{bmatrix} 2 & 1 & 3 \\ 1 & -1 & 0 \end{bmatrix}} \underset{3 \times 2}{\begin{bmatrix} 1 & 0 \\ 2 & 1 \\ 3 & 2 \end{bmatrix}} = \underset{2 \times 2}{\begin{bmatrix} 13 & 7 \\ -1 & -1 \end{bmatrix}}$

(b) $BA = \underset{3 \times 2}{\begin{bmatrix} 1 & 0 \\ 2 & 1 \\ 3 & 2 \end{bmatrix}} \underset{2 \times 3}{\begin{bmatrix} 2 & 1 & 3 \\ 1 & -1 & 0 \end{bmatrix}} = \underset{3 \times 3}{\begin{bmatrix} 2 & 1 & 3 \\ 5 & 1 & 6 \\ 8 & 1 & 9 \end{bmatrix}}$

Notice in Example 5 that AB is 2×2 and BA is 3×3. It is possible for both AB and BA to be defined yet be unequal. In fact, even if A and B are both $n \times n$ matrices, so that AB and BA are each defined and $n \times n$, AB and BA will usually be unequal.

EXAMPLE 6 Multiplying Two Square Matrices

If

$$A = \begin{bmatrix} 2 & 1 \\ 0 & 4 \end{bmatrix} \quad \text{and} \quad B = \begin{bmatrix} -3 & 1 \\ 1 & 2 \end{bmatrix}$$

find: **(a)** AB **(b)** BA

SOLUTION **(a)** $AB = \begin{bmatrix} 2 & 1 \\ 0 & 4 \end{bmatrix} \begin{bmatrix} -3 & 1 \\ 1 & 2 \end{bmatrix} = \begin{bmatrix} -5 & 4 \\ 4 & 8 \end{bmatrix}$

(b) $BA = \begin{bmatrix} -3 & 1 \\ 1 & 2 \end{bmatrix} \begin{bmatrix} 2 & 1 \\ 0 & 4 \end{bmatrix} = \begin{bmatrix} -6 & 1 \\ 2 & 9 \end{bmatrix}$

The preceding examples demonstrate that an important property of real numbers, the commutative property of multiplication, is not shared by matrices.

Matrix multiplication is not commutative. That is, in general,

AB is not equal to BA

Matrix multiplication is associative.

Associative Property of Multiplication

Let A be a matrix of dimension $m \times r$, let B be a matrix of dimension $r \times p$, and let C be a matrix of dimension $p \times n$. Then matrix multiplication is **associative.** That is,

$$A(BC) = (AB)C$$

The resulting matrix ABC is of dimension $m \times n$.

Notice the limitations that are placed on the dimensions of the matrices in order for multiplication to be associative.

Distributive Property

Let A be a matrix of dimension $m \times r$. Let B and C be matrices of dimension $r \times n$. Then the **distributive property** states that

$$A(B + C) = AB + AC$$

The resulting matrix $AB + AC$ is of dimension $m \times n$.

NOW WORK PROBLEM 41.

The Identity Matrix

For an $n \times n$ square matrix, the entries located in row i, column i, $1 \leq i \leq n$, are called the **diagonal entries.** An $n \times n$ square matrix whose diagonal entries are 1s, while all other entries are 0s, is called the **identity matrix I_n.** For example,

$$I_2 = \begin{bmatrix} 1 & 0 \\ 0 & 1 \end{bmatrix}, \quad I_3 = \begin{bmatrix} 1 & 0 & 0 \\ 0 & 1 & 0 \\ 0 & 0 & 1 \end{bmatrix}$$

and so on.

EXAMPLE 7 Finding the Product of a Matrix and the Identity Matrix I_2

For

$$A = \begin{bmatrix} 3 & 2 \\ -4 & 5 \end{bmatrix}$$

compute **(a)** AI_2 **(b)** I_2A

SOLUTION **(a)** $AI_2 = \begin{bmatrix} 3 & 2 \\ -4 & 5 \end{bmatrix}\begin{bmatrix} 1 & 0 \\ 0 & 1 \end{bmatrix} = \begin{bmatrix} 3 & 2 \\ -4 & 5 \end{bmatrix} = A$

(b) $I_2A = \begin{bmatrix} 1 & 0 \\ 0 & 1 \end{bmatrix}\begin{bmatrix} 3 & 2 \\ -4 & 5 \end{bmatrix} = \begin{bmatrix} 3 & 2 \\ -4 & 5 \end{bmatrix} = A$

Example 7 demonstrates a more general result.

If A is a square matrix of dimension $n \times n$, then $AI_n = I_nA = A$.

For square matrices the identity matrix plays the role that the number 1 plays for multiplication in the set of real numbers.

When the matrix A is not square, care must be taken when forming the products AI and IA. For example, if

$$A = \begin{bmatrix} 1 & 2 \\ 3 & 2 \\ 1 & 1 \end{bmatrix}$$

then A is of dimension 3×2 and

$$AI_2 = A\begin{bmatrix} 1 & 0 \\ 0 & 1 \end{bmatrix} = \begin{bmatrix} 1 & 2 \\ 3 & 2 \\ 1 & 1 \end{bmatrix}\begin{bmatrix} 1 & 0 \\ 0 & 1 \end{bmatrix} = \begin{bmatrix} 1 & 2 \\ 3 & 2 \\ 1 & 1 \end{bmatrix} = A$$

Although the product I_2A is not defined, we can calculate the product I_3A as follows:

$$I_3A = \begin{bmatrix} 1 & 0 & 0 \\ 0 & 1 & 0 \\ 0 & 0 & 1 \end{bmatrix}\begin{bmatrix} 1 & 2 \\ 3 & 2 \\ 1 & 1 \end{bmatrix} = \begin{bmatrix} 1 & 2 \\ 3 & 2 \\ 1 & 1 \end{bmatrix} = A$$

The above observations can be generalized as follows:

Identity Property

If A is a matrix of dimension $m \times n$ and if I_n denotes the identity matrix of dimension $n \times n$, and I_m denotes the identity matrix of dimension $m \times m$, then

$$I_mA = A \quad \text{and} \quad AI_n = A$$

EXERCISE 2.5 Answers to Odd-Numbered Problems Begin on Page AN-13.

In Problems 1–16, find the product.

1. $[1 \quad 3]\begin{bmatrix} 2 \\ 4 \end{bmatrix}$

2. $[-1 \quad 4]\begin{bmatrix} 5 \\ 2 \end{bmatrix}$

3. $[1 \quad -2 \quad 3]\begin{bmatrix} 0 \\ 1 \\ 2 \end{bmatrix}$

4. $[-1 \quad 1 \quad 0]\begin{bmatrix} 1 \\ -1 \\ 1 \end{bmatrix}$

5. $[1 \quad 4]\begin{bmatrix} 2 & 0 \\ 4 & -2 \end{bmatrix}$

6. $\begin{bmatrix} 2 & 0 \\ 4 & -2 \end{bmatrix}\begin{bmatrix} 2 \\ 4 \end{bmatrix}$

7. $\begin{bmatrix} 2 & 0 \\ 4 & -2 \end{bmatrix}\begin{bmatrix} 2 & 1 \\ 3 & -2 \end{bmatrix}$

8. $\begin{bmatrix} 1 & 4 \\ -1 & 2 \end{bmatrix}\begin{bmatrix} 3 & 0 \\ -2 & 2 \end{bmatrix}$

9. $[1 \quad -2 \quad 3]\begin{bmatrix} 0 & 1 \\ 1 & 2 \\ 2 & 3 \end{bmatrix}$

10. $\begin{bmatrix} 1 & -2 & 3 \\ 4 & 0 & 6 \end{bmatrix}\begin{bmatrix} 0 \\ 1 \\ 2 \end{bmatrix}$

11. $\begin{bmatrix} 1 & -2 & 3 \\ 4 & 0 & 6 \end{bmatrix}\begin{bmatrix} 0 & -2 \\ 1 & 0 \\ 2 & -4 \end{bmatrix}$

12. $\begin{bmatrix} 1 & -2 & 3 \\ 4 & 0 & 6 \end{bmatrix}\begin{bmatrix} -1 & 2 & 1 \\ 1 & 3 & 0 \\ 0 & 4 & -2 \end{bmatrix}$

13. $\begin{bmatrix} 2 & 0 \\ 4 & -2 \\ 6 & -1 \end{bmatrix}\begin{bmatrix} 2 & 1 \\ 3 & -2 \end{bmatrix}$ **14.** $\begin{bmatrix} 1 & 4 \\ -1 & 2 \end{bmatrix}\begin{bmatrix} 2 & 0 & 6 \\ -1 & 4 & 1 \end{bmatrix}$ **15.** $\begin{bmatrix} 1 & -1 & 6 \\ 2 & 0 & -1 \\ 3 & 1 & 2 \end{bmatrix}\begin{bmatrix} 3 & 2 \\ 0 & 1 \\ 1 & 0 \end{bmatrix}$ **16.** $\begin{bmatrix} 2 & 0 & 1 \\ -1 & 1 & 1 \end{bmatrix}\begin{bmatrix} 1 & 0 & 0 \\ 2 & 1 & 2 \\ 3 & 2 & 4 \end{bmatrix}$

In Problems 17–26, let A be of dimension 3×4, B be of dimension 3×3, C be of dimension 2×3, and D be of dimension 3×2. Determine which of the following expressions are defined and, for those that are, give the dimension.

17. BA **18.** CD **19.** AB **20.** DC **21.** $(BA)C$

22. $A(CD)$ **23.** $BA + A$ **24.** $CD + BA$ **25.** $DC + B$ **26.** $CB - A$

In Problems 27–42, use the matrices given below. For Problems 27–40 perform the indicated operation(s); for Problems 41 and 42 verify the indicated property.

$$A = \begin{bmatrix} 1 & 2 \\ 0 & 4 \end{bmatrix} \quad B = \begin{bmatrix} 1 & 2 & 3 \\ -1 & 4 & -2 \end{bmatrix} \quad C = \begin{bmatrix} 3 & 1 \\ 4 & -1 \\ 0 & 2 \end{bmatrix} \quad D = \begin{bmatrix} 1 & 0 & 4 \\ 0 & 1 & 2 \\ 0 & -1 & 1 \end{bmatrix} \quad E = \begin{bmatrix} 3 & -1 \\ 4 & 2 \end{bmatrix}$$

27. AB **28.** DC **29.** BC **30.** AA **31.** $(D + I_3)C$

32. $DC + C$ **33.** EI_2 **34.** I_3D **35.** $(2E)B$ **36.** $E(2B)$

37. $-5E + A$ **38.** $3A + 2E$ **39.** $3CB + 4D$ **40.** $2EA - 3BC$

41. Verify the associative property of matrix multiplication by finding $D(CB)$ and $(DC)B$.

42. Verify the distributive property by finding $(A + E)B$ and $AB + EB$.

In Problems 43–50, use a graphing utility or EXCEL and the matrices given below to perform the indicated operations.

$$A = \begin{bmatrix} -1 & -1 & 3 & 0 \\ 2 & 6 & 2 & 2 \\ -4 & 2 & 3 & 2 \\ 7 & 0 & 5 & -1 \end{bmatrix} \quad B = \begin{bmatrix} -1 & 2 & 4 & 5 \\ 2 & 0 & 5 & 3 \\ 0.5 & 6 & -7 & 11 \\ 5 & -1 & 2 & 7 \end{bmatrix} \quad C = \begin{bmatrix} 13 & -8 & 7 & 0 \\ 0 & 5 & 0 & -2 \\ 5 & 0 & 7 & 0 \\ 7 & 7 & 7 & 7 \end{bmatrix}$$

43. AB **44.** BA **45.** $(AB)C$ **46.** $A(BC)$

47. $B(A + C)$ **48.** $(A + C)B$ **49.** $A(2B - 3C)$ **50.** $(A + B)(A - B)$

51. For

$$A = \begin{bmatrix} 1 & -1 \\ 2 & 0 \end{bmatrix} \quad \text{and} \quad B = \begin{bmatrix} 3 & 2 \\ -2 & 4 \end{bmatrix}$$

find AB and BA. Notice that $AB \neq BA$.

52. Show that, for all values a, b, c, and d, the matrices

$$A = \begin{bmatrix} a & b \\ -b & a \end{bmatrix} \quad \text{and} \quad B = \begin{bmatrix} c & d \\ -d & c \end{bmatrix}$$

are commutative; that is, $AB = BA$.

53. For what numbers x will the following be true?

$$[x \quad 4 \quad 1]\begin{bmatrix} 2 & 1 & 0 \\ 1 & 0 & 2 \\ 0 & 2 & 4 \end{bmatrix}\begin{bmatrix} x \\ -7 \\ \frac{5}{4} \end{bmatrix} = 0$$

54. Let

$$A = \begin{bmatrix} 1 & 2 & 5 \\ 2 & 4 & 10 \\ -1 & -2 & -5 \end{bmatrix}$$

Show that $A \cdot A = A^2 = \mathbf{0}$. Thus the rule in the real number system that if $a^2 = 0$, then $a = 0$ does not hold for matrices.

55. What must be true about a, b, c, and d if we demand that $AB = BA$ for the following matrices?

$$A = \begin{bmatrix} a & b \\ c & d \end{bmatrix} \qquad B = \begin{bmatrix} 1 & 1 \\ -1 & 1 \end{bmatrix}$$

Assume that

$$\begin{bmatrix} a & b \\ c & d \end{bmatrix} \neq \begin{bmatrix} 1 & 0 \\ 0 & 1 \end{bmatrix}$$

56. Let

$$A = \begin{bmatrix} a & b \\ b & a \end{bmatrix}$$

Find a and b such that $A^2 + A = \mathbf{0}$, where $A^2 = A \cdot A$.

57. For the matrix

$$A = \begin{bmatrix} a & 1-a \\ 1+a & -a \end{bmatrix}$$

show that $A^2 = A \cdot A = I_2$.

58. Find the row vector $[x_1 \quad x_2]$ for which

$$[x_1 \quad x_2]\begin{bmatrix} \frac{1}{2} & \frac{1}{2} \\ \frac{1}{4} & \frac{3}{4} \end{bmatrix} = [x_1 \quad x_2]$$

under the condition that $x_1 + x_2 = 1$. Here, the row vector $[x_1 \quad x_2]$ is called a **fixed vector** of the matrix

$$\begin{bmatrix} \frac{1}{2} & \frac{1}{2} \\ \frac{1}{4} & \frac{3}{4} \end{bmatrix}$$

59. Department Store Purchases Lee went to a department store and purchased 6 pairs of pants, 8 shirts, and 2 jackets. Chan purchased 2 pairs of pants, 5 shirts, and 3 jackets. If pants are \$25 each, shirts are \$18 each, and jackets are \$39 each, use matrix multiplication to find the amounts spent by Lee and Chan.

60. Factory Production Suppose a factory is asked to produce three types of products, which we will call P_1, P_2, P_3. Suppose the following purchase order was received: $P_1 = 7$, $P_2 = 12$, $P_3 = 5$. Represent this order by a row vector and call it P:

$$P = [7 \quad 12 \quad 5]$$

To produce each of the products, raw material of four kinds is needed. Call the raw material M_1, M_2, M_3, and M_4. The matrix below gives the amount of material needed for each product:

$$Q = \begin{array}{c} \\ P_1 \\ P_2 \\ P_3 \end{array}\begin{array}{c} \begin{matrix} M_1 & M_2 & M_3 & M_4 \end{matrix} \\ \begin{bmatrix} 2 & 3 & 1 & 12 \\ 7 & 9 & 5 & 20 \\ 8 & 12 & 6 & 15 \end{bmatrix} \end{array}$$

Suppose the cost for each of the materials M_1, M_2, M_3, and M_4 is \$10, \$12, \$15, and \$20, respectively. The cost vector is

$$C = \begin{bmatrix} 10 \\ 12 \\ 15 \\ 20 \end{bmatrix}$$

Compute each of the following and interpret each one:

(a) PQ (b) QC (c) PQC

Problems 61–69 require the following definition:

Powers of Matrices For a square matrix A it is possible to find $A \cdot A = A^2$. We can also compute

$$A^n = \underbrace{A \cdot A \cdot \ldots \cdot A}_{n \text{ factors}}$$

In Problems 61–64, find A^2, A^3, and A^4.

61. $A = \begin{bmatrix} 1 & 0 \\ 3 & 2 \end{bmatrix}$

62. $A = \begin{bmatrix} 3 & 1 \\ -2 & -1 \end{bmatrix}$

63. $A = \begin{bmatrix} 1 & 0 \\ 0 & 1 \end{bmatrix}$

64. $A = \begin{bmatrix} \frac{1}{2} & \frac{1}{2} \\ \frac{1}{4} & \frac{3}{4} \end{bmatrix}$

65. Can you guess what A^n looks like for the matrix given in Problem 63?

66. Can you guess what A^n looks like for the matrix given in Problem 64?

In Problems, 67–69 use a graphing utility to compute A^2, A^{10}, and A^{15}.

67. $A = \begin{bmatrix} -0.5 & -1 & 0.3 & 0 & 0.3 \\ 2 & 1.6 & 1 & -1 & 0.4 \\ -4 & 2 & 0.7 & 2 & 0.2 \\ 1 & 0 & 0 & -1 & 0 \\ 0 & 0 & -0.9 & 0 & 0 \end{bmatrix}$

68. $A = \begin{bmatrix} -1 & 0.02 & 0.24 & 0 \\ 2 & 0 & 0 & 1.3 \\ 0.5 & 6 & -0.7 & 1.1 \\ 2.5 & -1 & 0.02 & 0.7 \end{bmatrix}$

69. $A = \begin{bmatrix} 0 & 1 & 0 \\ 1 & 0 & 1 \\ 1 & 1 & 1 \end{bmatrix}$

70. Make up two matrices A and B for which $AB = BA$.

71. Make up two square matrices A and B for which AB and BA are both defined, but $AB \neq BA$.

72. Make up two matrices A and B for which AB is defined but BA is not.

2.6 The Inverse of a Matrix

OBJECTIVES
1. Find the inverse of a matrix
2. Use the inverse of a matrix to solve a system of equations

The *inverse* of a matrix, if it exists, plays the role in matrix algebra that the reciprocal of a number plays in the set of real numbers. For example, the product of 2 and its reciprocal, $\frac{1}{2}$, equals 1, the multiplicative identity. The product of a matrix and its inverse equals the identity matrix.

Inverse of a Matrix

Let A be a matrix of dimension $n \times n$. A matrix B of dimension $n \times n$ is called the **inverse** of A if $AB = BA = I_n$. We denote the inverse of a matrix A, if it exists, by A^{-1}.

EXAMPLE 1 Verifying That One Matrix Is the Inverse of Another Matrix

Show that $\begin{bmatrix} \frac{1}{2} & -\frac{1}{2} \\ 0 & 1 \end{bmatrix}$ is the inverse of $\begin{bmatrix} 2 & 1 \\ 0 & 1 \end{bmatrix}$.

SOLUTION Since

$$\begin{bmatrix} 2 & 1 \\ 0 & 1 \end{bmatrix} \cdot \begin{bmatrix} \frac{1}{2} & -\frac{1}{2} \\ 0 & 1 \end{bmatrix} = \begin{bmatrix} 1 & 0 \\ 0 & 1 \end{bmatrix}$$

and

$$\begin{bmatrix} \frac{1}{2} & -\frac{1}{2} \\ 0 & 1 \end{bmatrix} \begin{bmatrix} 2 & 1 \\ 0 & 1 \end{bmatrix} = \begin{bmatrix} 1 & 0 \\ 0 & 1 \end{bmatrix}$$

the required condition is met.

NOW WORK PROBLEM 1.

The next example provides a technique for finding the inverse of a matrix. Although this technique is not the one we shall ultimately use, it is illustrative.

EXAMPLE 2 Finding the Inverse of a Matrix

Find the inverse of the matrix $A = \begin{bmatrix} 2 & 1 \\ 0 & 1 \end{bmatrix}$.

SOLUTION We begin by assuming that this matrix has an inverse of the form

$$A^{-1} = \begin{bmatrix} a & b \\ c & d \end{bmatrix}$$

Then the product of A and A^{-1} must be the identity matrix:

$$\begin{bmatrix} 2 & 1 \\ 0 & 1 \end{bmatrix} \begin{bmatrix} a & b \\ c & d \end{bmatrix} = \begin{bmatrix} 1 & 0 \\ 0 & 1 \end{bmatrix}$$

Multiplying the matrices on the left side, we get

$$\begin{bmatrix} 2a + c & 2b + d \\ c & d \end{bmatrix} = \begin{bmatrix} 1 & 0 \\ 0 & 1 \end{bmatrix}$$

The condition for equality requires that

$$2a + c = 1 \qquad 2b + d = 0 \qquad c = 0 \qquad d = 1$$

Using $c = 0$ and $d = 1$ in the first two equations, we find

$$a = \tfrac{1}{2} \qquad b = -\tfrac{1}{2} \qquad c = 0 \qquad d = 1$$

The inverse of

$$A = \begin{bmatrix} 2 & 1 \\ 0 & 1 \end{bmatrix} \quad \text{is} \quad A^{-1} = \begin{bmatrix} a & b \\ c & d \end{bmatrix} = \begin{bmatrix} \frac{1}{2} & -\frac{1}{2} \\ 0 & 1 \end{bmatrix}$$

Sometimes a square matrix does not have an inverse.

EXAMPLE 3 Showing a Matrix Has No Inverse

Show that the matrix below does not have an inverse.

$$A = \begin{bmatrix} 0 & 1 \\ 0 & 0 \end{bmatrix}$$

SOLUTION We proceed as in Example 2 by assuming that A does have an inverse. It will be of the form

$$A^{-1} = \begin{bmatrix} a & b \\ c & d \end{bmatrix}$$

The product of A and A^{-1} must be the identity matrix.

$$\begin{bmatrix} 0 & 1 \\ 0 & 0 \end{bmatrix} \begin{bmatrix} a & b \\ c & d \end{bmatrix} = \begin{bmatrix} 1 & 0 \\ 0 & 1 \end{bmatrix}$$

Performing the multiplication on the left side, we have

$$\begin{bmatrix} c & d \\ 0 & 0 \end{bmatrix} = \begin{bmatrix} 1 & 0 \\ 0 & 1 \end{bmatrix}$$

But these two matrices can never be equal (look at row 2, column 2: $0 \neq 1$). We conclude that our assumption that A has an inverse is false. That is, A does not have an inverse.

NOW WORK PROBLEM 21.

So far, we have shown that a square matrix may or may not have an inverse. What about nonsquare matrices? Can they have inverses? The answer is "No." By definition, whenever a matrix has an inverse, it will commute with its inverse under multiplication. So if the nonsquare matrix A had the alleged inverse B, then AB would have to be equal to BA. But the fact that A is not square causes AB and BA to have different dimensions and prevents them from being equal. So such a B could not exist.

A nonsquare matrix has no inverse.

The procedure used in Example 2 to find the inverse, if it exists, of a square matrix becomes quite involved as the dimension of the matrix gets larger. A more efficient method that uses the reduced row-echelon form of a matrix is provided next.

Reduced Row-Echelon Technique for Finding Inverses

We will introduce this technique by looking at an example.

EXAMPLE 4 Finding the Inverse of a Matrix

Find the inverse of the matrix

$$A = \begin{bmatrix} 4 & 2 \\ 3 & 1 \end{bmatrix}$$

SOLUTION Assuming A has an inverse, we will denote it by

$$X = \begin{bmatrix} x_1 & x_2 \\ x_3 & x_4 \end{bmatrix}$$

Then the product of A and X is the identity matrix of dimension 2×2. That is,

$$AX = I_2$$

$$\begin{bmatrix} 4 & 2 \\ 3 & 1 \end{bmatrix}\begin{bmatrix} x_1 & x_2 \\ x_3 & x_4 \end{bmatrix} = \begin{bmatrix} 1 & 0 \\ 0 & 1 \end{bmatrix}$$

Performing the multiplication on the left yields

$$\begin{bmatrix} 4x_1 + 2x_3 & 4x_2 + 2x_4 \\ 3x_1 + x_3 & 3x_2 + x_4 \end{bmatrix} = \begin{bmatrix} 1 & 0 \\ 0 & 1 \end{bmatrix}$$

This matrix equation can be written as the following system of four equations containing four variables:

$$\begin{cases} 4x_1 + 2x_3 = 1 \\ 3x_1 + x_3 = 0 \end{cases} \qquad \begin{matrix} 4x_2 + 2x_4 = 0 \\ 3x_2 + x_4 = 1 \end{matrix}$$

We find the solution to be

$$x_1 = -\tfrac{1}{2} \qquad x_2 = 1 \qquad x_3 = \tfrac{3}{2} \qquad x_4 = -2$$

The inverse of A is

$$A^{-1} = \begin{bmatrix} -\frac{1}{2} & 1 \\ \frac{3}{2} & -2 \end{bmatrix}$$

Let's look at this example more closely. The system of four equations containing four variables can be written as two systems:

$$\text{(a)} \begin{cases} 4x_1 + 2x_3 = 1 \\ 3x_1 + x_3 = 0 \end{cases} \qquad \text{(b)} \begin{cases} 4x_2 + 2x_4 = 0 \\ 3x_2 + x_4 = 1 \end{cases}$$

Their augmented matrices are

$$\text{(a)} \left[\begin{array}{cc|c} 4 & 2 & 1 \\ 3 & 1 & 0 \end{array}\right] \qquad \text{(b)} \left[\begin{array}{cc|c} 4 & 2 & 0 \\ 3 & 1 & 1 \end{array}\right]$$

Since the matrix A appears in both (a) and (b), any row operation we perform on (a) and (b) can be performed on the single augmented matrix that combines the two right-hand columns. We denote this matrix by $A|I_2$ and write

$$[A|I_2] = \left[\begin{array}{cc|cc} 4 & 2 & 1 & 0 \\ 3 & 1 & 0 & 1 \end{array}\right]$$

If we perform the row operations on $[A|I_2]$ needed to obtain the reduced row echelon form of A, we get

$$\left[\begin{array}{cc|cc} 1 & 0 & -\frac{1}{2} & 1 \\ 0 & 1 & \frac{3}{2} & -2 \end{array}\right]$$

The 2×2 matrix on the right-hand side of the vertical bar is A^{-1}.

This example illustrates the general procedure:

To find the inverse, if it exists, of a square matrix of dimension $n \times n$, follow these steps:

Steps for Finding the Inverse of a Matrix

STEP 1 Form the matrix $[A|I_n]$.

STEP 2 Using row operations, write $[A|I_n]$ in reduced row echelon form.

STEP 3 If the resulting matrix is of the form $[I_n|B]$, that is, if the identity matrix appears on the left side of the bar, then B is the inverse of A. Otherwise, A has no inverse.

Let's work another example.

1 **EXAMPLE 5** **Finding the Inverse of a Matrix**

Find the inverse of

$$A = \begin{bmatrix} 1 & 1 & 2 \\ 2 & 1 & 0 \\ 1 & 2 & 2 \end{bmatrix}$$

SOLUTION **STEP 1** Since A is of dimension 3×3, we use the identity matrix I_3. The matrix $[A|I_3]$ is

$$\left[\begin{array}{ccc|ccc} 1 & 1 & 2 & 1 & 0 & 0 \\ 2 & 1 & 0 & 0 & 1 & 0 \\ 1 & 2 & 2 & 0 & 0 & 1 \end{array}\right]$$

STEP 2 We proceed to obtain the reduced row echelon form of this matrix:

Use $\begin{matrix} R_2 = -2r_1 + r_2 \\ R_3 = -1r_1 + r_3 \end{matrix}$ to obtain $\left[\begin{array}{ccc|ccc} 1 & 1 & 2 & 1 & 0 & 0 \\ 0 & -1 & -4 & -2 & 1 & 0 \\ 0 & 1 & 0 & -1 & 0 & 1 \end{array}\right]$

Use $R_2 = -1r_2$ to obtain $\left[\begin{array}{ccc|ccc} 1 & 1 & 2 & 1 & 0 & 0 \\ 0 & 1 & 4 & 2 & -1 & 0 \\ 0 & 1 & 0 & -1 & 0 & 1 \end{array}\right]$

Use $\begin{matrix} R_1 = -1r_2 + r_1 \\ R_3 = -1r_2 + r_3 \end{matrix}$ to obtain $\left[\begin{array}{ccc|ccc} 1 & 0 & -2 & -1 & 1 & 0 \\ 0 & 1 & 4 & 2 & -1 & 0 \\ 0 & 0 & -4 & -3 & 1 & 1 \end{array}\right]$

Use $R_3 = -\frac{1}{4}r_3$ to obtain $\left[\begin{array}{ccc|ccc} 1 & 0 & -2 & -1 & 1 & 0 \\ 0 & 1 & 4 & 2 & -1 & 0 \\ 0 & 0 & 1 & \frac{3}{4} & -\frac{1}{4} & -\frac{1}{4} \end{array}\right]$

Use $\begin{matrix} R_1 = 2r_3 + r_1 \\ R_2 = -4r_3 + r_2 \end{matrix}$ to obtain $\left[\begin{array}{ccc|ccc} 1 & 0 & 0 & \frac{1}{2} & \frac{1}{2} & -\frac{1}{2} \\ 0 & 1 & 0 & -1 & 0 & 1 \\ 0 & 0 & 1 & \frac{3}{4} & -\frac{1}{4} & -\frac{1}{4} \end{array}\right]$

The matrix $[A|I_3]$ is in reduced row echelon form.

STEP 3 Since the identity matrix I_3 appears on the left side, the matrix appearing on the right is the inverse. That is,

$$A^{-1} = \begin{bmatrix} \frac{1}{2} & \frac{1}{2} & -\frac{1}{2} \\ -1 & 0 & 1 \\ \frac{3}{4} & -\frac{1}{4} & -\frac{1}{4} \end{bmatrix}$$

You should verify that $AA^{-1} = A^{-1}A = I_3$.

NOW WORK PROBLEM 15.

COMMENT: A graphing utility can be used to find the inverse of a matrix. Use a graphing utility to do Example 5.

GRAPHING UTILITY SOLUTION Enter the matrix A into a graphing utility and use the $\boxed{x^{-1}}$ key to obtain $[A]^{-1}$. Figure 13 shows A^{-1}.

FIGURE 13

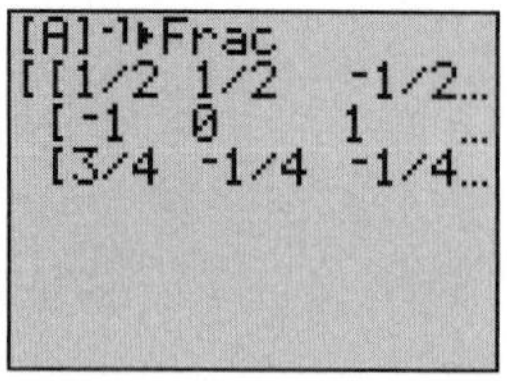

NOW WORK PROBLEM 51 USING A GRAPHING UTILITY.

Use EXCEL to do Example 5.

SOLUTION **STEP 1** Enter the matrix A.

STEP 2 Highlight the cells that will contain A^{-1}; A^{-1} is a 3×3 matrix.

STEP 3 Type: **=MINVERSE** (*highlight matrix A*). The Excel screen should look like the one below.

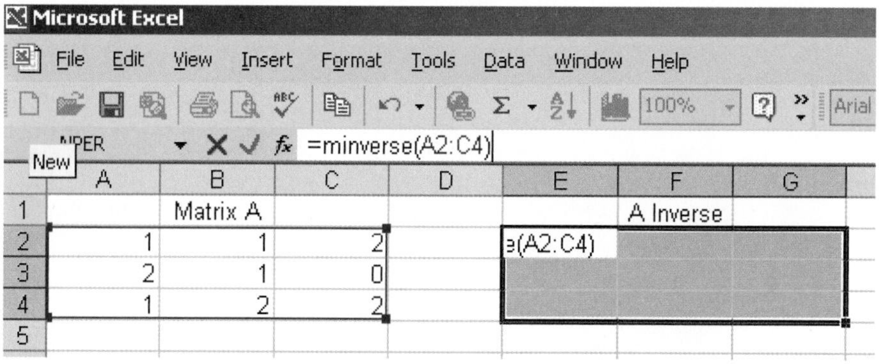

STEP 4 Press **Ctrl-Shift-Enter** all at the same time. The inverse will be displayed in the highlighted cells, as in the screen below.

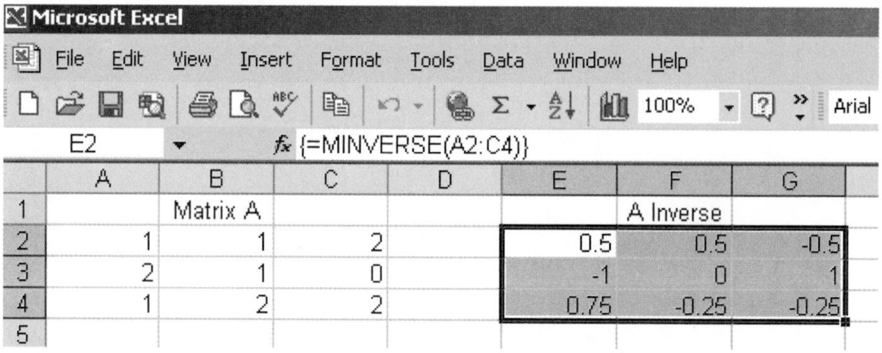

✓**CHECK:** Multiply $A*A^{-1}$ using the steps below. This product should be the identity matrix.

1. Highlight the cells that are to contain the product matrix.
2. Type: **=MMULT**(*highlight A*, *highlight* A^{-1})
3. Press **Ctrl-Shift-Enter** at the same time.

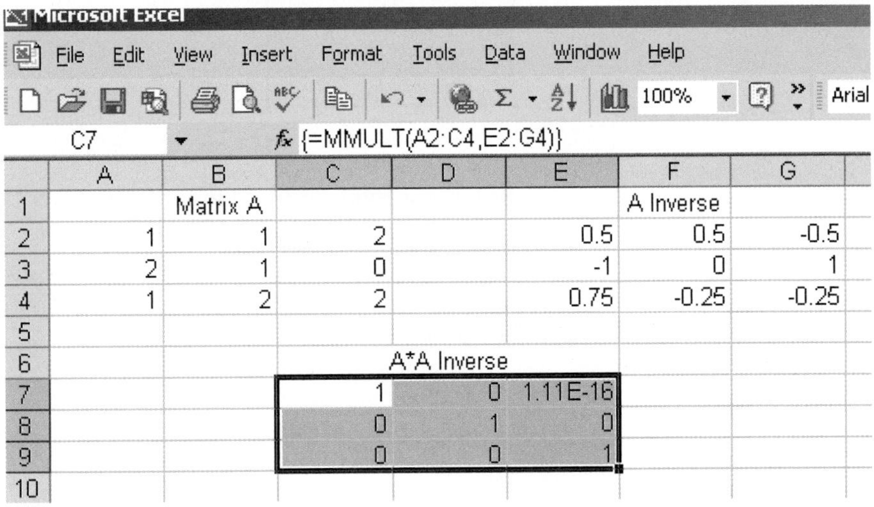

	A	B	C	D	E	F	G
1		Matrix A				A Inverse	
2	1	1	2		0.5	0.5	-0.5
3	2	1	0		-1	0	1
4	1	2	2		0.75	-0.25	-0.25
5							
6				A*A Inverse			
7			1	0	1.11E-16		
8			0	1	0		
9			0	0	1		
10							

Notice that $a_{13} = $ 1.11E-16 is in scientific notation. In decimal notation it is 0.000000000000000111, very close to zero. So, $A*A^{-1} = \begin{bmatrix} 1 & 0 & 0 \\ 0 & 1 & 0 \\ 0 & 0 & 1 \end{bmatrix}$, as required.

NOW WORK PROBLEM 51 USING EXCEL.

EXAMPLE 6 Showing that a Matrix Has No Inverse

Show that the matrix given below has no inverse.

$$\begin{bmatrix} 3 & 2 \\ 6 & 4 \end{bmatrix}$$

SOLUTION We set up the matrix

$$\left[\begin{array}{cc|cc} 3 & 2 & 1 & 0 \\ 6 & 4 & 0 & 1 \end{array}\right]$$

Use $R_1 = \frac{1}{3}r_1$ to obtain $\left[\begin{array}{cc|cc} 1 & \frac{2}{3} & \frac{1}{3} & 0 \\ 6 & 4 & 0 & 1 \end{array}\right]$

Use $R_2 = -6r_1 + r_2$ to obtain $\left[\begin{array}{cc|cc} 1 & \frac{2}{3} & \frac{1}{3} & 0 \\ 0 & 0 & -2 & 1 \end{array}\right]$

The 0s in row 2 tell us we cannot get the identity matrix. This, in turn, tells us the original matrix has no inverse.

COMMENT: If a matrix has no inverse, a graphing utility will display an Error message. See Figure 14 for the result of doing Example 6.

FIGURE 14

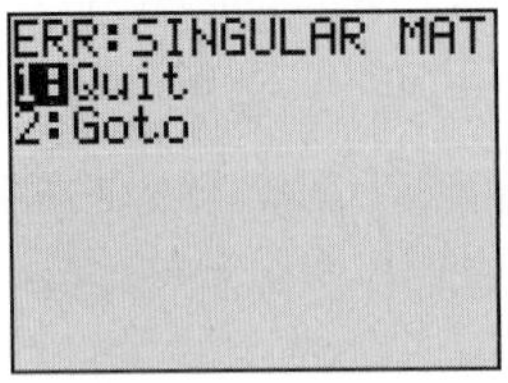

Solving a System of *n* Linear Equations Containing *n* Variables Using Inverses

Use the inverse of a matrix to solve a system of equations 2

The inverse of a matrix can be used to solve a system of n linear equations containing n variables. We begin with a system of three linear equations containing three variables:

$$\begin{cases} x + y + 2z = 1 \\ 2x + y = 2 \\ x + 2y + 2z = 3 \end{cases}$$

If we let

$$A = \begin{bmatrix} 1 & 1 & 2 \\ 2 & 1 & 0 \\ 1 & 2 & 2 \end{bmatrix} \quad X = \begin{bmatrix} x \\ y \\ z \end{bmatrix} \quad B = \begin{bmatrix} 1 \\ 2 \\ 3 \end{bmatrix}$$

the above system of equations can be written compactly as the matrix equation

$$AX = B$$

In general, any system of n linear equations containing n variables $x_1, x_2, \ldots, x_n$, can be written in the form

$$AX = B$$

where A is the $n \times n$ matrix of the coefficients of the unknowns, B is an $n \times 1$ column matrix whose entries are the numbers appearing to the right of each equal sign in the system, and X is an $n \times 1$ column matrix containing the n variables.

To find X, we start with the matrix equation $AX = B$ and use properties of matrices. Our assumption is that the $n \times n$ matrix A has an inverse A^{-1}.

$$\begin{aligned} AX &= B && \text{A has an inverse } A^{-1} \\ A^{-1}(AX) &= A^{-1}B && \text{Multiply both sides by } A^{-1} \\ (A^{-1}A)X &= A^{-1}B && \text{Apply the Associative Property on the left side} \\ I_nX &= A^{-1}B && \text{Apply the Inverse Property: } A^{-1}A = I_n \\ X &= A^{-1}B && \text{Apply the Identity Property: } I_nX = X \end{aligned}$$

This leads us to formulate the following result:

A system of n linear equations containing n variables

$$\boxed{AX = B}$$

for which A is a square matrix and A^{-1} exists, always has a unique solution that is given by

$$\boxed{X = A^{-1}B}$$

We use the above result in the next example.

EXAMPLE 7 Solving a System of Equations Using Inverses

Solve the system of equations: $\begin{cases} x + y + 2z = 1 \\ 2x + y = 2 \\ x + 2y + 2z = 3 \end{cases}$

SOLUTION Here

$$A = \begin{bmatrix} 1 & 1 & 2 \\ 2 & 1 & 0 \\ 1 & 2 & 2 \end{bmatrix} \quad X = \begin{bmatrix} x \\ y \\ z \end{bmatrix} \quad B = \begin{bmatrix} 1 \\ 2 \\ 3 \end{bmatrix}$$

From Example 5 we know A has the inverse.

$$A^{-1} = \begin{bmatrix} \frac{1}{2} & \frac{1}{2} & -\frac{1}{2} \\ -1 & 0 & 1 \\ \frac{3}{4} & -\frac{1}{4} & -\frac{1}{4} \end{bmatrix}$$

Based on the result just stated, the solution X of the system is

$$X = A^{-1}B$$

$$\begin{bmatrix} x \\ y \\ z \end{bmatrix} = \begin{bmatrix} \frac{1}{2} & \frac{1}{2} & -\frac{1}{2} \\ -1 & 0 & 1 \\ \frac{3}{4} & -\frac{1}{4} & -\frac{1}{4} \end{bmatrix} \begin{bmatrix} 1 \\ 2 \\ 3 \end{bmatrix} = \begin{bmatrix} 0 \\ 2 \\ -\frac{1}{2} \end{bmatrix}$$

The solution is $x = 0, y = 2, z = -\frac{1}{2}$.

NOW WORK PROBLEMS 39 AND 41.

The method used to solve the system in Example 7 requires that A have an inverse. If, for a system of equations $AX = B$, the matrix A has no inverse, then the system must be analyzed using the methods discussed in Section 2.3.

The method used in Example 7 for solving a system of equations is particularly useful for applications in which the constants appearing to the right of the equal sign change while the coefficients of the variables on the left side do not. See Problems 39–50 for an illustration. See also the discussion of Leontief models in Section 2.7.

EXCEL

Use EXCEL to do Example 7.

SOLUTION From Example 5, we know the inverse of A is $A^{-1} = \begin{bmatrix} 0.5 & 0.5 & -0.5 \\ -1 & 0 & 1 \\ 0.75 & -0.25 & -0.25 \end{bmatrix}$.

To find the solution to the system of equations, multiply $A^{-1} * B$, where $B = \begin{bmatrix} 1 \\ 2 \\ 3 \end{bmatrix}$.

STEP 1 Highlight the cells that are to contain the product matrix. This will be a 3×1 matrix.

STEP 2 Type: **=MMULT** (*highlight* A^{-1}, *highlight* B)

STEP 3 Press **Ctrl-Shift-Enter** at the same time, to get $A^{-1}B$.

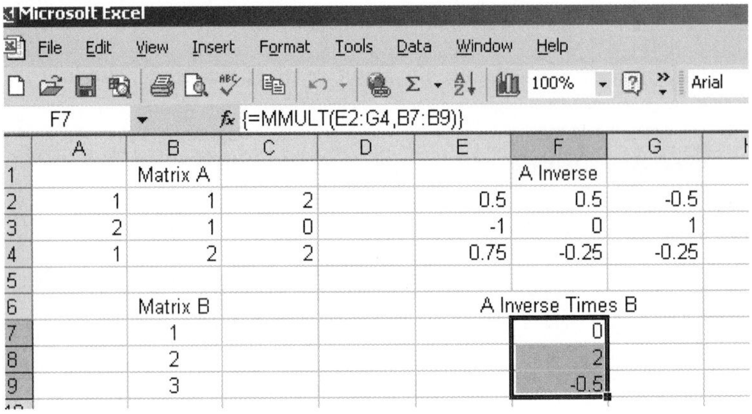

	A	B	C	D	E	F	G
1		Matrix A				A Inverse	
2	1	1	2		0.5	0.5	-0.5
3	2	1	0		-1	0	1
4	1	2	2		0.75	-0.25	-0.25
5							
6		Matrix B			A Inverse Times B		
7		1				0	
8		2				2	
9		3				-0.5	

So, $x = 0, y = 2$, and $z = -0.5$.

NOW WORK PROBLEMS 39 AND 41 USING EXCEL.

EXERCISE 2.6 Answers to Odd-Numbered Problems Begin on Page AN-14.

In Problems 1–6, show that the given matrices are inverses of each other.

1. $\begin{bmatrix} 1 & 2 \\ 2 & 3 \end{bmatrix} \begin{bmatrix} -3 & 2 \\ 2 & -1 \end{bmatrix}$

2. $\begin{bmatrix} 1 & 5 \\ 2 & 0 \end{bmatrix} \begin{bmatrix} 0 & \frac{1}{2} \\ \frac{1}{5} & -\frac{1}{10} \end{bmatrix}$

3. $\begin{bmatrix} -1 & -2 \\ 3 & 4 \end{bmatrix} \begin{bmatrix} 2 & 1 \\ -\frac{3}{2} & -\frac{1}{2} \end{bmatrix}$

4. $\begin{bmatrix} 1 & 3 \\ 2 & -1 \end{bmatrix} \begin{bmatrix} \frac{1}{7} & \frac{3}{7} \\ \frac{2}{7} & -\frac{1}{7} \end{bmatrix}$

5. $\begin{bmatrix} 1 & 2 & 3 \\ 2 & 3 & 4 \\ 1 & 2 & 1 \end{bmatrix} \begin{bmatrix} -\frac{5}{2} & 2 & -\frac{1}{2} \\ 1 & -1 & 1 \\ \frac{1}{2} & 0 & -\frac{1}{2} \end{bmatrix}$

6. $\begin{bmatrix} 1 & 3 & 3 \\ 1 & 4 & 3 \\ 1 & 3 & 4 \end{bmatrix} \begin{bmatrix} 7 & -3 & -3 \\ -1 & 1 & 0 \\ -1 & 0 & 1 \end{bmatrix}$

In Problems 7–20, find the inverse of each matrix using the reduced row-echelon technique.

7. $\begin{bmatrix} 3 & 7 \\ 2 & 5 \end{bmatrix}$

8. $\begin{bmatrix} 4 & 1 \\ 3 & 1 \end{bmatrix}$

9. $\begin{bmatrix} 1 & -1 \\ 3 & -4 \end{bmatrix}$

10. $\begin{bmatrix} 5 & 3 \\ 3 & 2 \end{bmatrix}$

11. $\begin{bmatrix} 2 & 1 \\ 4 & 3 \end{bmatrix}$

12. $\begin{bmatrix} 2 & 3 \\ 2 & -2 \end{bmatrix}$

13. $\begin{bmatrix} 0 & 0 & 1 \\ 0 & 1 & 0 \\ 1 & 0 & 0 \end{bmatrix}$

14. $\begin{bmatrix} -1 & 1 & 0 \\ 1 & 0 & 2 \\ 3 & 1 & 0 \end{bmatrix}$

15. $\begin{bmatrix} 1 & 1 & -1 \\ 3 & -1 & 0 \\ 2 & -3 & 4 \end{bmatrix}$

16. $\begin{bmatrix} 1 & 1 & 1 \\ 2 & 1 & 1 \\ 1 & 1 & 2 \end{bmatrix}$

17. $\begin{bmatrix} 1 & 1 & -1 \\ 2 & 1 & 1 \\ 1 & 0 & 1 \end{bmatrix}$

18. $\begin{bmatrix} 2 & 3 & -1 \\ 1 & 1 & 1 \\ 0 & 2 & -1 \end{bmatrix}$

19. $\begin{bmatrix} 1 & 1 & 0 & 0 \\ 0 & 1 & -1 & 1 \\ 1 & -1 & 1 & 1 \\ 0 & 1 & 0 & -1 \end{bmatrix}$

20. $\begin{bmatrix} 1 & 2 & -3 & -2 \\ 0 & 1 & 4 & -2 \\ 3 & -1 & 4 & 0 \\ 2 & 1 & 0 & 3 \end{bmatrix}$

In Problems 21–26, show that each matrix has no inverse.

21. $\begin{bmatrix} 4 & 6 \\ 2 & 3 \end{bmatrix}$

22. $\begin{bmatrix} -1 & 2 \\ 3 & -6 \end{bmatrix}$

23. $\begin{bmatrix} -8 & 4 \\ -4 & 2 \end{bmatrix}$

24. $\begin{bmatrix} 2 & 10 \\ 1 & 5 \end{bmatrix}$

25. $\begin{bmatrix} 1 & 1 & 1 \\ 3 & -4 & 2 \\ 0 & 0 & 0 \end{bmatrix}$

26. $\begin{bmatrix} -1 & 2 & 3 \\ 5 & 2 & 0 \\ 2 & -4 & -6 \end{bmatrix}$

In Problems 27–34, find the inverse, if it exists, of each matrix.

27. $\begin{bmatrix} 1 & 1 \\ 1 & 2 \end{bmatrix}$

28. $\begin{bmatrix} 2 & 1 \\ 1 & 1 \end{bmatrix}$

29. $\begin{bmatrix} 3 & -2 \\ 0 & 4 \end{bmatrix}$

30. $\begin{bmatrix} 4 & -1 \\ -2 & 0 \end{bmatrix}$

31. $\begin{bmatrix} 3 & 2 \\ 6 & 4 \end{bmatrix}$

32. $\begin{bmatrix} 4 & 2 \\ 2 & 1 \end{bmatrix}$

33. $\begin{bmatrix} 1 & -2 & -1 \\ -2 & 5 & 4 \\ 3 & -8 & -5 \end{bmatrix}$

34. $\begin{bmatrix} 1 & 1 & -1 \\ -2 & -1 & 4 \\ 3 & 2 & -8 \end{bmatrix}$

35. Find the inverse of both

$$A = \begin{bmatrix} 1 & 2 \\ 2 & -1 \end{bmatrix} \quad \text{and} \quad B = \begin{bmatrix} 1 & 3 \\ 2 & 1 \end{bmatrix}$$

to determine $A^{-1} - B^{-1}$.

36. Find the inverse of

$$A = \begin{bmatrix} 1 & -4 \\ 2 & -3 \end{bmatrix} \quad \text{and} \quad B = \begin{bmatrix} 2 & -2 \\ 3 & 2 \end{bmatrix}$$

Determine $A^{-1} - B^{-1}$.

37. Write the matrix product $A^{-1}B$ used to find the solution to the system $AX = B$.

$$\begin{cases} x + 3y + 2z = 2 \\ 2x + 7y + 3z = 1 \\ x \qquad\quad + 6z = 3 \end{cases}$$

38. Write the matrix product $A^{-1}B$ used to find the solution to the system $AX = B$.

$$\begin{cases} x + 2y + 2z = 3 \\ 2x + 5y + 7z = 2 \\ 2x + y - 4z = 4 \end{cases}$$

In Problems 39–50, solve each system of equations by the method of Example 7. For Problems 39–44, use the inverse found in Problem 7. For Problems 45–50, use the inverse found in Problem 15.

39. $\begin{cases} 3x + 7y = 10 \\ 2x + 5y = 2 \end{cases}$

40. $\begin{cases} 3x + 7y = -4 \\ 2x + 5y = -3 \end{cases}$

41. $\begin{cases} 3x + 7y = 13 \\ 2x + 5y = 9 \end{cases}$

42. $\begin{cases} 3x + 7y = 0 \\ 2x + 5y = 14 \end{cases}$

43. $\begin{cases} 3x + 7y = 12 \\ 2x + 5y = -4 \end{cases}$

44. $\begin{cases} 3x + 7y = -2 \\ 2x + 5y = 10 \end{cases}$

45. $\begin{cases} x + y - z = 3 \\ 3x - y = -4 \\ 2x - 3y + 4z = 6 \end{cases}$

46. $\begin{cases} x + y - z = 6 \\ 3x - y = 8 \\ 2x - 3y + 4z = -3 \end{cases}$

47. $\begin{cases} x + y - z = 12 \\ 3x - y = -4 \\ 2x - 3y + 4z = 16 \end{cases}$

48. $\begin{cases} x + y - z = -8 \\ 3x - y = 12 \\ 2x - 3y + 4z = -2 \end{cases}$

49. $\begin{cases} x + y - z = 0 \\ 3x - y = -8 \\ 2x - 3y + 4z = -6 \end{cases}$

50. $\begin{cases} x + y - z = 21 \\ 3x - y = 12 \\ 2x - 3y + 4z = 14 \end{cases}$

In Problems 51–56, use a graphing utility or EXCEL to find the inverse, if it exists, of each matrix.

51. $\begin{bmatrix} 25 & 61 & -12 \\ 18 & -2 & 4 \\ 8 & 35 & 21 \end{bmatrix}$

52. $\begin{bmatrix} 18 & -3 & 4 \\ 6 & -20 & 14 \\ 10 & 25 & -15 \end{bmatrix}$

53. $\begin{bmatrix} 44 & 21 & 18 & 6 \\ -2 & 10 & 15 & 5 \\ 21 & 12 & -12 & 4 \\ -8 & -16 & 4 & 9 \end{bmatrix}$

54. $\begin{bmatrix} 16 & 22 & -3 & 5 \\ 21 & -17 & 4 & 8 \\ 2 & 8 & 27 & 20 \\ 5 & 15 & -3 & -10 \end{bmatrix}$

55. $A = \begin{bmatrix} 3 & 0 & 2 & -1 & 3 \\ -2 & 1 & 2 & 3 & 0 \\ 2 & 2 & 1 & 1 & -1 \\ 1 & 2 & 0 & 2 & -3 \\ 4 & 0 & -1 & 1 & -1 \end{bmatrix}$

56. $A = \begin{bmatrix} 0 & 0 & 2 & -1 & 3 \\ -2 & 0 & 2 & 3 & 0 \\ 2 & 2 & 0 & 0 & -1 \\ 1 & 2 & 0 & 2 & -3 \\ 4 & 4 & 0 & 0 & -2 \end{bmatrix}$

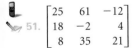

In Problems 57–60, use the idea behind Example 7 and either a graphing utility or EXCEL to solve each system of equations.

57. $\begin{cases} 25x + 61y - 12z = 10 \\ 18x - 12y + 7z = -9 \\ 3x + 4y - z = 12 \end{cases}$
58. $\begin{cases} 25x + 61y - 12z = 5 \\ 18x - 12y + 7z = -3 \\ 3x + 4y - z = 12 \end{cases}$
59. $\begin{cases} 25x + 61y - 12z = 21 \\ 18x - 12y + 7z = 7 \\ 3x + 4y - z = -2 \end{cases}$
60. $\begin{cases} 25x + 61y - 12z = 25 \\ 18x - 12y + 7z = 10 \\ 3x + 4y - z = -4 \end{cases}$

61. Show that the inverse of $A = \begin{bmatrix} a & b \\ c & d \end{bmatrix}$ is given by the formula $A^{-1} = \begin{bmatrix} \dfrac{d}{\Delta} & \dfrac{-b}{\Delta} \\ \dfrac{-c}{\Delta} & \dfrac{a}{\Delta} \end{bmatrix}$

where $\Delta = ad - bc \neq 0$. The number Δ is called the **determinant** of A.

In Problems 62–65, use the result of Problem 61 to find the inverse of each matrix.

62. $\begin{bmatrix} 1 & 2 \\ 2 & 3 \end{bmatrix}$
63. $\begin{bmatrix} 1 & 5 \\ 2 & 0 \end{bmatrix}$
64. $\begin{bmatrix} -1 & -2 \\ 3 & 4 \end{bmatrix}$
65. $\begin{bmatrix} 1 & 2 \\ 8 & 15 \end{bmatrix}$

2.7 Applications: Leontief Model; Cryptography; Accounting; The Method of Least Squares*

OBJECTIVES
1 Determine relative income using a closed Leontief model
2 Determine production levels necessary to meet forecasted demand using an open Leontief model
3 Use matrices in cryptography
4 Determine the full cost of manufactured products
5 Find the transpose of a matrix
6 Use the method of least squares to find the line of best fit

Application 1: Leontief Models

The Leontief models in economics are named after Wassily Leontief, who received the Nobel Prize in economics in 1973. These models can be characterized as a description of an economy in which input equals output or, in other words, consumption equals production. That is, the models assume that whatever is produced is always consumed.

Leontief models are of two types: closed, in which the entire production is consumed by those participating in the production; and open, in which some of the production is consumed by those who produce it and the rest of the production is consumed by external bodies.

In the *closed model* we seek the relative income of each participant in the system. In the *open model* we seek the amount of production needed to achieve a forecast demand, when the amount of production needed to achieve current demand is known.

THE CLOSED MODEL

We begin with an example to illustrate the idea.

**Each application is optional, and the applications given are independent of one another.*

1 **EXAMPLE 1** **Determining Relative Income Using a Closed Leontief Model**

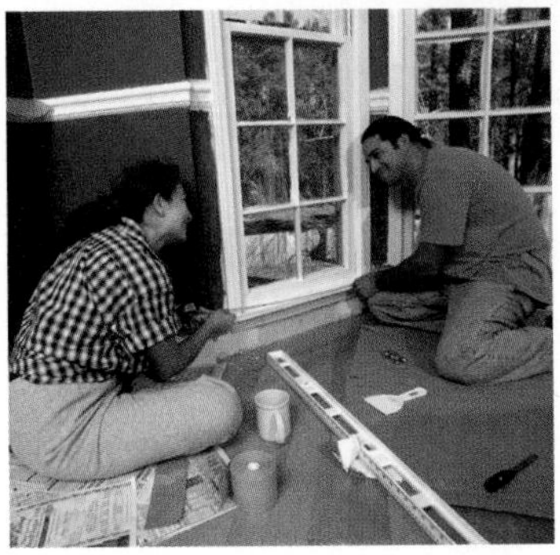

Three homeowners, Juan, Luis, and Carlos, each with certain skills, agreed to pool their talents to make repairs on their houses. As it turned out, Juan spent 20% of his time on his own house, 40% of his time on Luis' house, and 40% on Carlos' house. Luis spent 10% of his time on Juan's house, 50% of his time on his own house, and 40% on Carlos' house. Of Carlos' time, 60% was spent on Juan's house, 10% on Luis', and 30% on his own. Now that the projects are finished, they need to figure out how much money each should get for his work, including the work performed on his own house, so that the amount paid by each person equals the amount received by each one. They agreed in advance that the payment to each one should be approximately $3000.00.

SOLUTION We place the information given in the problem in a 3×3 matrix, as follows:

	Work done by Juan	Luis	Carlos
Proportion of work done on Juan's house	0.2	0.1	0.6
Proportion of work done on Luis' house	0.4	0.5	0.1
Proportion of work done on Carlos' house	0.4	0.4	0.3

Next, we define the variables:

$$x = \text{Juan's wages}$$
$$y = \text{Luis' wages}$$
$$z = \text{Carlos' wages}$$

We require that the amount paid out by each one equals the amount received by each one. Let's analyze this requirement, by looking just at the work done on Juan's house. Juan's wages are x. Juan's expenditures for work done on his house are $0.2x + 0.1y + 0.6z$. Juan's wages and expenditures are required to be equal, so

$$x = 0.2x + 0.1y + 0.6z$$

Similarly,

$$y = 0.4x + 0.5y + 0.1z$$
$$z = 0.4x + 0.4y + 0.3z$$

These three equations can be written compactly as

$$\begin{bmatrix} x \\ y \\ z \end{bmatrix} = \begin{bmatrix} 0.2 & 0.1 & 0.6 \\ 0.4 & 0.5 & 0.1 \\ 0.4 & 0.4 & 0.3 \end{bmatrix} \begin{bmatrix} x \\ y \\ z \end{bmatrix}$$

Some computation reduces the system to

$$\begin{cases} 0.8x - 0.1y - 0.6z = 0 \\ -0.4x + 0.5y - 0.1z = 0 \\ -0.4x - 0.4y + 0.7z = 0 \end{cases}$$

Solving for x, y, z, we find that

$$x = \frac{31}{36}z \qquad y = \frac{32}{36}z$$

where z is the parameter. To get solutions that fall close to \$3000, we set $z = 3600$.*
The wages to be paid out are therefore

$$x = \$3100 \qquad y = \$3200 \qquad z = \$3600$$

The matrix in Example 1, namely,

$$\begin{bmatrix} 0.2 & 0.1 & 0.6 \\ 0.4 & 0.5 & 0.1 \\ 0.4 & 0.4 & 0.3 \end{bmatrix}$$

is called an **input–output matrix.**

In the general closed model we have an economy consisting of n components. Each component produces an *output* of some goods or services, which, in turn, is completely used up by the n components. The proportionate use of each component's output by the economy makes up the input–output matrix of the economy. The problem is to find suitable pricing levels for each component so that income equals expenditure.

Closed Leontief Model

In general, an input–output matrix for a closed Leontief model is of the form

$$\boldsymbol{A = [a_{ij}] \quad i, j = 1, 2, \ldots, n}$$

where the a_{ij} represent the fractional amount of goods or services used by i and produced by j. For a closed model the sum of each column equals 1 (this is the condition that all production is consumed internally) and $0 \le a_{ij} \le 1$ for all entries (this is the restriction that each entry is a fraction).

If A is the input–output matrix of a closed system with n components and $\boldsymbol{X}$ is a column vector representing the price of each output of the system, then

$$\boldsymbol{X = AX}$$

represents the requirement that income equal expenditure.

For example, the first entry of the matrix equality $X = AX$ requires that

$$x_1 = a_{11}x_1 + a_{12}x_2 + \ldots + a_{1n}x_n$$

**Other choices for z are, of course, possible. The choice of which value to use is up to the homeowners. No matter what choice is made, the amount paid by each one equals the amount received by each one.*

The right side represents the price paid by component 1 for the goods it uses, while x_1 represents the income of component 1; we are requiring they be equal.

We can rewrite the equation $X = AX$ as

$$X - AX = \mathbf{0}$$
$$I_n X - AX = \mathbf{0}$$
$$(I_n - A)X = \mathbf{0}$$

This matrix equation, which represents a system of equations in which the right-hand side is always **0**, is called a **homogeneous system of equations.** It can be shown that if the entries in the input–output matrix A are positive and if the sum of each column of A equals 1, then this system has a one-parameter solution; that is, we can solve for $n - 1$ of the variables in terms of the remaining one, which serves as the parameter. This parameter serves as a "scale factor."

NOW WORK PROBLEM 1.

THE OPEN MODEL

Determine production levels necessary to meet forecasted demand using an open Leontief model 2

For the open model, in addition to internal consumption of goods produced, there is an outside demand for the goods produced. This outside demand may take the form of exportation of goods or may be the goods needed to support consumer demand. Again, however, we make the assumption that whatever is produced is also consumed.

For example, suppose an economy consists of three industries R, S, and T, and suppose each one produces a single product. We assume that a portion of R's production is used by each of the three industries, while the remainder is used by consumers.

The same is true of the production of S and T. To organize our thoughts, we construct a table that describes the interaction of the use of R, S, and T's production over some fixed period of time. See Table 4.

TABLE 4

		Amounts Consumed				
		R	*S*	*T*	Consumer	Total
Amounts Produced	*R*	50	20	40	70	180
	S	20	30	20	90	160
	T	30	20	20	50	120

All entries in the table are in appropriate units, say, in dollars. The first row (row R) represents the production in dollars of industry R (input). Out of the total of \$180 worth of goods produced, R, S, and T use \$50, \$20, and \$40, respectively, for the production of their goods, while consumers purchase the remaining \$70 for their consumption (output). Observe that input equals output since everything produced by R is used up by R, S, T, and consumers.

The second and third rows are interpreted in the same way.

An important observation is that the goal of R's production is to produce \$70 worth of goods, since this is the demand of consumers. In order to meet this demand, R must produce a total of \$180, since the difference, \$110, is required internally by R, S, and T.

Suppose, however, that consumer demand is expected to change. To effect this change, how much should each industry now produce? For example, in Table 4, current demand for R, S, and T can be represented by a **demand vector:**

$$D_0 = \begin{bmatrix} 70 \\ 90 \\ 50 \end{bmatrix}$$

But suppose marketing forecasts predict that in 3 years the demand vector will be

$$D_3 = \begin{bmatrix} 60 \\ 110 \\ 60 \end{bmatrix}$$

Here, the demand for item R has decreased; the demand for item S has significantly increased, and the demand for item T is higher. Given the current total output of R, S, and T at 180, 160, and 120, respectively, what must it be in 3 years to meet this projected demand?

The solution of this type of forecasting problem is derived from the *open Leontief model* in input–output analysis. In using input–output analysis to obtain a solution to such a forecasting problem, we take into account the fact that the output of any one of these industries is affected by changes in the other two, since the total demand for, say, R in 3 years depends not only on consumer demand for R, but also on consumer demand for S and T. That is, the industries are interrelated.

To obtain the solution, we need to determine how much of each of the three products R, S, and T is required to produce 1 unit of R. For example, to obtain 180 units of R requires the use of 50 units of R, 20 units of S, and 30 units of T (the entries in column 1). Forming the ratios, we find that to produce 1 unit of R requires $\frac{50}{180} = 0.278$ of R, $\frac{20}{180} = 0.111$ of S, and $\frac{30}{180} = 0.167$ of T. If we want, say, x units of R, we will require $0.278x$ units of R, $0.111x$ units of S, and $0.167x$ units of T.

Continuing in this way, we can construct the matrix

$$A = \begin{array}{c} \\ R \\ S \\ T \end{array} \begin{array}{c} \begin{array}{ccc} R & S & T \end{array} \\ \begin{bmatrix} 0.278 & 0.125 & 0.333 \\ 0.111 & 0.188 & 0.167 \\ 0.167 & 0.125 & 0.167 \end{bmatrix} \end{array}$$

Observe that column 1 represents the amounts of R, S, T required for 1 unit of R; column 2 represents the amounts of R, S, and T required for 1 unit of S; and column 3 represents the amounts of R, S, and T required for 1 unit of T. For example, the entry in row 3, column 2 (0.125), represents the amount of T needed to produce 1 unit of S.

As a result of placing the entries this way, if

$$X = \begin{bmatrix} x \\ y \\ z \end{bmatrix}$$

represents the total output required to obtain a given demand, the product AX represents the amount of R, S, and T required for internal consumption. The condition that production = consumption requires that

Internal consumption + Consumer demand = Total output

In terms of the matrix A, the total output X, and the demand vector D, this requirement is equivalent to the equation

$$AX + D = X$$

In this equation we seek to find X for a prescribed demand D. The matrix A is calculated as above for some initial production process.*

EXAMPLE 2 **Finding Production Levels to Meet Future Demand: The Open Leontief Model**

For the data given in Table 4, find the total output X required to achieve a future demand of

$$D_3 = \begin{bmatrix} 60 \\ 110 \\ 60 \end{bmatrix}$$

SOLUTION We need to solve for X in

$$AX + D_3 = X$$

Simplifying, we have

$$[I_3 - A]X = D_3$$

Solving for X, we have

$$\begin{aligned} X &= [I_3 - A]^{-1} \cdot D_3 \\ &= \begin{bmatrix} 0.722 & -0.125 & -0.333 \\ -0.111 & 0.812 & -0.167 \\ -0.167 & -0.125 & 0.833 \end{bmatrix}^{-1} \begin{bmatrix} 60 \\ 110 \\ 60 \end{bmatrix} \\ &= \begin{bmatrix} 1.6048 & 0.3568 & 0.7131 \\ 0.2946 & 1.3363 & 0.3857 \\ 0.3660 & 0.2721 & 1.4013 \end{bmatrix} \begin{bmatrix} 60 \\ 110 \\ 60 \end{bmatrix} \\ &= \begin{bmatrix} 178.33 \\ 187.81 \\ 135.96 \end{bmatrix} \end{aligned}$$

The total output of R, S, and T required for the forecast demand D_3 is

$$x = 178.322 \qquad y = 187.811 \qquad z = 135.969$$

NOW WORK PROBLEM 9.

**The entries in A can be checked by using the requirement that $AX + D_0 = X$, for D_0 = Initial demand and X = Total output. For our example, it must happen that*

$$\begin{matrix} \textit{Internal consumption} & + & \textit{Consumer demand} & = & \textit{Total output} \\ AX & + & D_0 & = & X \\ \begin{bmatrix} 0.278 & 0.125 & 0.333 \\ 0.111 & 0.188 & 0.167 \\ 0.167 & 0.125 & 0.167 \end{bmatrix} \begin{bmatrix} 180 \\ 160 \\ 120 \end{bmatrix} & + & \begin{bmatrix} 70 \\ 90 \\ 50 \end{bmatrix} & = & \begin{bmatrix} 180 \\ 160 \\ 120 \end{bmatrix} \end{matrix}$$

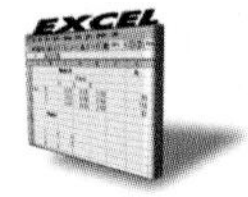

Use EXCEL to do Example 2.

SOLUTION   We need to solve $AX + D_3 = X$ for X. The solution to this equation is

$$X = (I_3 - A)^{-1} \cdot D_3$$

We begin by using EXCEL to compute $(I - A)^{-1} \cdot D$.

STEP 1   Enter matrices A, I, and D_3 into an EXCEL spreadsheet.

	File	Edit	View	Insert	Format	Tools	Data	Window	Help				
		A	B	C	D	E	F	G					
1				**Matrix A**					**D₃**				
2					Output								
3				R	S	T							
4		Input	R	0.278	0.125	0.333			60				
5			S	0.111	0.188	0.167			110				
6			T	0.167	0.125	0.167			60				
7													
8			**Matrix I**										
9													
10													
11		1	0	0									
12		0	1	0									
13		0	0	1									
14													

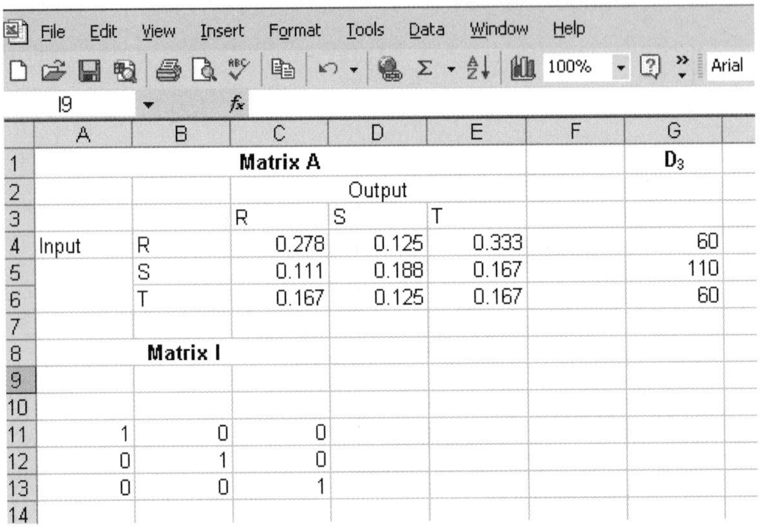

STEP 2   Compute $I - A$.

    (a)   Highlight the cells where the difference is to go. This must be a 3×3 matrix.

    (b)   Type $= (highlight\ I) - (highlight\ A)$ and press **Ctrl-Shift-Enter.**

Microsoft Excel

AVERAGE   =C4:E6-A9:C11

	A	B	C	D	E	F	G	H
1			**Matrix A**				**D₃**	
2				Output				
3			R	S	T			
4	Input	R	0.278	0.278	0.278		60	
5		S	0.278	0.278	0.278		110	
6		T	0.278	0.278	0.278		60	
7								
8		**Matrix I**			I - A			
9	1	0	0	A9:C11				
10	0	1	0					
11	0	0	1					
12								
13								

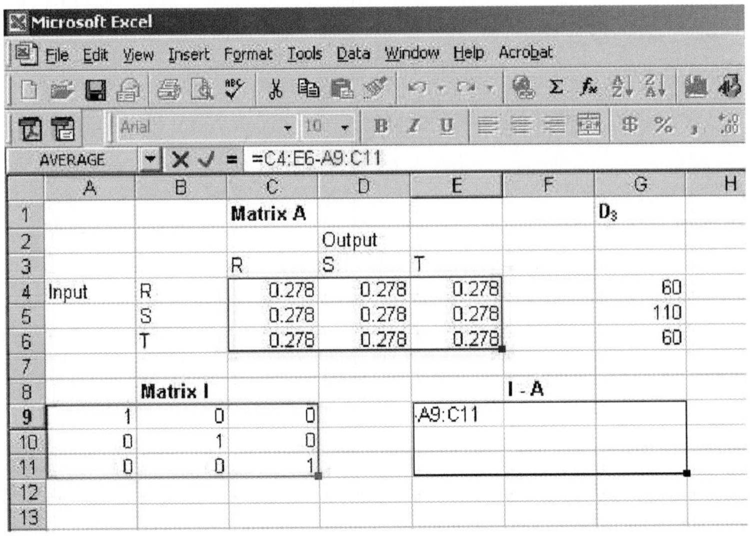

STEP 3 Compute $(I - A)^{-1}$

   (a) Highlight the cells that will contain $(I - A)^{-1}$; $(I - A)^{-1}$ is a 3×3 matrix.

   (b) Type: **=MINVERSE**(*highlight matrix $(I - A)$*) in the formula bar, and press **Ctrl-Shift-Enter.**

STEP 4 Compute $(I - A)^{-1} D$.

   (a) Highlight the cells that are to contain the product matrix. This will be a 3×1 matrix.

   (b) Type: **=MMULT**(*highlight$(I - A)^{-1}$, highlight D*), and press **Ctrl-Shift-Enter.**

The results are given below.

File   Edit   View   Insert   Format   Tools   Data   Window   Help

F16   ▼ f_x {=MMULT(A16:C18,G4:G6)}

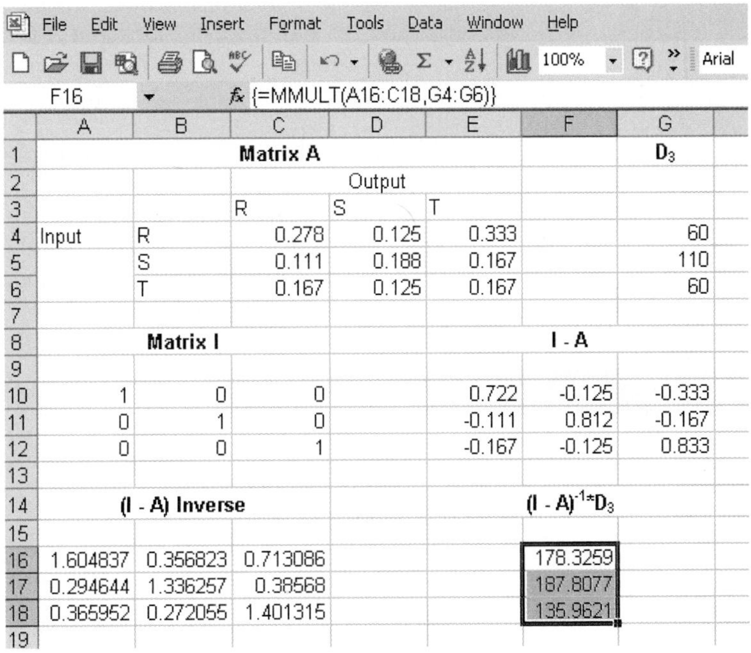

	A	B	C	D	E	F	G
1			Matrix A				D₃
2				Output			
3			R	S	T		
4	Input	R	0.278	0.125	0.333		60
5		S	0.111	0.188	0.167		110
6		T	0.167	0.125	0.167		60
7							
8		Matrix I				I - A	
9							
10	1	0	0			0.722	-0.125	-0.333
11	0	1	0			-0.111	0.812	-0.167
12	0	0	1			-0.167	-0.125	0.833
13							
14		(I - A) Inverse				(I - A)⁻¹*D₃	
15							
16	1.604837	0.356823	0.713086			178.3259	
17	0.294644	1.336257	0.38568			187.8077	
18	0.365952	0.272055	1.401315			135.9621	
19							

So $x = 178.33$, $y = 187.81$, $z = 135.96$.

NOW WORK PROBLEM 9 USING EXCEL.

The general open model can be described as follows: Suppose there are n industries in the economy. Each industry produces some goods or services, which are partially consumed by the n industries, while the rest are used to meet a prescribed current demand. Given the output required of each industry to meet current demand, what should the output of each industry be to meet some different future demand?

Open Leontief Model

The matrix $A = [a_{ij}], i, j = 1, \ldots, n$ of the open model is defined to consist of entries a_{ij}, where a_{ij} is the amount of output of industry j required for one unit of output of industry i. If X is a column vector representing the production of each industry in the system and D is a column vector representing future demand for goods produced in the system, then

$$\boxed{X = AX + D}$$

From the equation above we find

$$[I_n - A]X = D$$

It can be shown that the matrix $I_n - A$ has an inverse, provided each entry in A is positive and the sum of each column in A is less than 1. Under these conditions we may solve for X to get

$$X = [I_n - A]^{-1} \cdot D$$

This form of the solution is particularly useful since it allows us to find X for a variety of demands D by doing one calculation: $[I_n - A]^{-1}$.

We conclude by noting that the use of an input–output matrix to solve forecasting problems assumes that each industry produces a single commodity and that no technological advances take place in the period of time under investigation (in other words, the proportions found in the matrix A are fixed).

EXERCISE 2.7 APPLICATION 1 Answers to Odd-Numbered Problems Begin on Page AN-15.

In Problems 1–4, find the relative wages of each person for the given closed input–output matrix. In each case take the wages of C to be the parameter and use z = C's wages = \$30,000.

1.
$$\begin{array}{c} \\ A \\ B \\ C \end{array}\begin{array}{c} \begin{array}{ccc} A & B & C \end{array} \\ \begin{bmatrix} \frac{1}{2} & \frac{1}{3} & \frac{1}{4} \\ \frac{1}{4} & \frac{1}{3} & \frac{1}{4} \\ \frac{1}{4} & \frac{1}{3} & \frac{1}{2} \end{bmatrix} \end{array}$$

2.
$$\begin{array}{c} \\ A \\ B \\ C \end{array}\begin{array}{c} \begin{array}{ccc} A & B & C \end{array} \\ \begin{bmatrix} \frac{1}{4} & \frac{2}{3} & \frac{1}{2} \\ \frac{1}{2} & \frac{1}{6} & \frac{1}{4} \\ \frac{1}{4} & \frac{1}{6} & \frac{1}{4} \end{bmatrix} \end{array}$$

3.
$$\begin{array}{c} \\ A \\ B \\ C \end{array}\begin{array}{c} \begin{array}{ccc} A & B & C \end{array} \\ \begin{bmatrix} 0.2 & 0.3 & 0.1 \\ 0.6 & 0.4 & 0.2 \\ 0.2 & 0.3 & 0.7 \end{bmatrix} \end{array}$$

4.
$$\begin{array}{c} \\ A \\ B \\ C \end{array}\begin{array}{c} \begin{array}{ccc} A & B & C \end{array} \\ \begin{bmatrix} 0.4 & 0.3 & 0.2 \\ 0.2 & 0.3 & 0.3 \\ 0.4 & 0.4 & 0.5 \end{bmatrix} \end{array}$$

5. For the three industries R, S, and T in the open Leontief model of Table 4 on page 131, compute the total output vector X if the forecast demand vector is

$$D_2 = \begin{bmatrix} 80 \\ 90 \\ 60 \end{bmatrix}$$

6. Rework Problem 5 if the forecast demand vector is

$$D_4 = \begin{bmatrix} 100 \\ 80 \\ 60 \end{bmatrix}$$

7. Closed Leontief Model A society consists of four individuals: a farmer, a builder, a tailor, and a rancher (who produces meat products). Of the food produced by the farmer, $\frac{3}{10}$ is used by the farmer, $\frac{2}{10}$ by the builder, $\frac{2}{10}$ by the tailor, and $\frac{3}{10}$ by the rancher. The builder's production is utilized 30% by the farmer, 30% by the builder, 10% by the tailor, and 30% by the rancher. The tailor's production is used in the ratios $\frac{3}{10}$, $\frac{3}{10}$, $\frac{1}{10}$, and $\frac{3}{10}$ by the farmer, builder, tailor, and rancher, respectively. Finally, meat products are used 20% by each of the farmer, builder, and tailor, and 40% by the rancher. What is the relative income of each if the rancher's income is scaled at \$25,000?

8. If in Problem 7 the meat production utilization changes so that it is used equally by all four individuals, while everyone else's production utilization remains the same, what are the relative incomes?

9. **Open Leontief Model** Suppose the interrelationships between the production of two industries R and S in a given year are given in the table:

	R	S	Current Consumer Demand	Total Output
R	30	40	60	130
S	20	10	40	70

If the forecast demand in 2 years is

$$D_2 = \begin{bmatrix} 80 \\ 40 \end{bmatrix}$$

what should the total output X be?

10. Open Leontief Model Suppose the interrelationships between the production of five industries (manufacturing, electric power, petroleum, transportation, and textiles) in a given year are as listed in the table:

	Manu.	E.P.	Petr.	Tran.	Text.
Manu.	0.2	0.12	0.15	0.18	0.1
E.P.	0.17	0.11	0	0.19	0.28
Petr.	0.11	0.11	0.12	0.46	0.12
Tran.	0.1	0.14	0.18	0.17	0.19
Text.	0.16	0.18	0.02	0.1	0.3

If the demand vector is given by

$$D = \begin{bmatrix} 100 \\ 100 \\ 200 \\ 100 \\ 100 \end{bmatrix}$$

what should the production vector X be?

Application 2: Cryptography

Use matrices in cryptography 3

Our second application is to *cryptography*, the art of writing or deciphering secret codes. We begin by giving examples of elementary codes.

EXAMPLE 1 Encoding a Message

A message can be encoded by associating each letter of the alphabet with some other letter of the alphabet according to a prescribed pattern. For example, we might have

$$\begin{array}{cccccccccccccccccccccccccc} A & B & C & D & E & F & G & H & I & J & K & L & M & N & O & P & Q & R & S & T & U & V & W & X & Y & Z \\ \downarrow & \downarrow & \downarrow & \downarrow & \downarrow & \downarrow & \downarrow & \downarrow & \downarrow & \downarrow & \downarrow & \downarrow & \downarrow & \downarrow & \downarrow & \downarrow & \downarrow & \downarrow & \downarrow & \downarrow & \downarrow & \downarrow & \downarrow & \downarrow & \downarrow & \downarrow \\ C & D & E & F & G & H & I & J & K & L & M & N & O & P & Q & R & S & T & U & V & W & X & Y & Z & A & B \end{array}$$

With the above code the word MESSAGE would become OGUUCIG.

EXAMPLE 2 Encoding a Message

Another code may associate numbers with the letters of the alphabet. For example, we might have

A	B	C	D	E	F	G	H	I	J	K	L	M	N	O	P	Q	R	S	T	U	V	W	X	Y	Z
↓	↓	↓	↓	↓	↓	↓	↓	↓	↓	↓	↓	↓	↓	↓	↓	↓	↓	↓	↓	↓	↓	↓	↓	↓	↓
26	25	24	23	22	21	20	19	18	17	16	15	14	13	12	11	10	9	8	7	6	5	4	3	2	1

In this code the word PEACE looks like 11 22 26 24 22.

Both the above codes have one important feature in common. The association of letters with the coding symbols is made using a one-to-one correspondence so that no possible ambiguities can arise.

Suppose we want to encode the following message:

TOP SECURITY CLEARANCE

If we decide to divide the message into pairs of letters, the message becomes

TO PS EC UR IT YC LE AR AN CE

(If there is a letter left over, we arbitrarily assign *Z* to the last position.) Using the correspondence of letters to numbers given in Example 2, and writing each pair of letters as a column vector, we obtain

$$\begin{bmatrix} T \\ O \end{bmatrix} = \begin{bmatrix} 7 \\ 12 \end{bmatrix} \quad \begin{bmatrix} P \\ S \end{bmatrix} = \begin{bmatrix} 11 \\ 8 \end{bmatrix} \quad \begin{bmatrix} E \\ C \end{bmatrix} = \begin{bmatrix} 22 \\ 24 \end{bmatrix} \quad \begin{bmatrix} U \\ R \end{bmatrix} = \begin{bmatrix} 6 \\ 9 \end{bmatrix} \quad \begin{bmatrix} I \\ T \end{bmatrix} = \begin{bmatrix} 18 \\ 7 \end{bmatrix}$$

$$\begin{bmatrix} Y \\ C \end{bmatrix} = \begin{bmatrix} 2 \\ 24 \end{bmatrix} \quad \begin{bmatrix} L \\ E \end{bmatrix} = \begin{bmatrix} 15 \\ 22 \end{bmatrix} \quad \begin{bmatrix} A \\ R \end{bmatrix} = \begin{bmatrix} 26 \\ 9 \end{bmatrix} \quad \begin{bmatrix} A \\ N \end{bmatrix} = \begin{bmatrix} 26 \\ 13 \end{bmatrix} \quad \begin{bmatrix} C \\ E \end{bmatrix} = \begin{bmatrix} 24 \\ 22 \end{bmatrix}$$

Next, we arbitrarily choose a 2×2 matrix A, which we know has an inverse A^{-1} (we'll see why later). Let's choose

$$A = \begin{bmatrix} 2 & 3 \\ 1 & 2 \end{bmatrix}$$

The inverse A^{-1} of A is

$$A^{-1} = \begin{bmatrix} 2 & -3 \\ -1 & 2 \end{bmatrix}$$

Now we transform the column vectors representing the message by multiplying each of them on the left by matrix A:

$$A\begin{bmatrix} T \\ O \end{bmatrix} = \begin{bmatrix} 2 & 3 \\ 1 & 2 \end{bmatrix}\begin{bmatrix} 7 \\ 12 \end{bmatrix} = \begin{bmatrix} 50 \\ 31 \end{bmatrix}$$

$$A\begin{bmatrix} P \\ S \end{bmatrix} = \begin{bmatrix} 2 & 3 \\ 1 & 2 \end{bmatrix}\begin{bmatrix} 11 \\ 8 \end{bmatrix} = \begin{bmatrix} 46 \\ 27 \end{bmatrix}$$

$$A\begin{bmatrix} E \\ C \end{bmatrix} = \begin{bmatrix} 2 & 3 \\ 1 & 2 \end{bmatrix}\begin{bmatrix} 22 \\ 24 \end{bmatrix} = \begin{bmatrix} 116 \\ 70 \end{bmatrix}$$

$$A\begin{bmatrix} U \\ R \end{bmatrix} = \begin{bmatrix} 2 & 3 \\ 1 & 2 \end{bmatrix}\begin{bmatrix} 6 \\ 9 \end{bmatrix} = \begin{bmatrix} 39 \\ 24 \end{bmatrix}$$

$$A\begin{bmatrix} I \\ T \end{bmatrix} = \begin{bmatrix} 2 & 3 \\ 1 & 2 \end{bmatrix}\begin{bmatrix} 18 \\ 7 \end{bmatrix} = \begin{bmatrix} 57 \\ 32 \end{bmatrix}$$
$$A\begin{bmatrix} Y \\ C \end{bmatrix} = \begin{bmatrix} 2 & 3 \\ 1 & 2 \end{bmatrix}\begin{bmatrix} 2 \\ 24 \end{bmatrix} = \begin{bmatrix} 76 \\ 50 \end{bmatrix}$$
$$A\begin{bmatrix} L \\ E \end{bmatrix} = \begin{bmatrix} 2 & 3 \\ 1 & 2 \end{bmatrix}\begin{bmatrix} 15 \\ 22 \end{bmatrix} = \begin{bmatrix} 96 \\ 59 \end{bmatrix}$$
$$A\begin{bmatrix} A \\ R \end{bmatrix} = \begin{bmatrix} 2 & 3 \\ 1 & 2 \end{bmatrix}\begin{bmatrix} 26 \\ 9 \end{bmatrix} = \begin{bmatrix} 79 \\ 24 \end{bmatrix}$$
$$A\begin{bmatrix} A \\ N \end{bmatrix} = \begin{bmatrix} 2 & 3 \\ 1 & 2 \end{bmatrix}\begin{bmatrix} 26 \\ 13 \end{bmatrix} = \begin{bmatrix} 79 \\ 52 \end{bmatrix}$$
$$A\begin{bmatrix} C \\ E \end{bmatrix} = \begin{bmatrix} 2 & 3 \\ 1 & 2 \end{bmatrix}\begin{bmatrix} 24 \\ 22 \end{bmatrix} = \begin{bmatrix} 76 \\ 116 \end{bmatrix}$$

The coded message is

50 31 46 27 116 70 39 24 57 32 76 50 96 59 79 24 79 52 76 116

To decode or unscramble the above message, pair the numbers in 2×1 column vectors. Then on the left multiply each column vector by A^{-1}. For example, the first two column vectors then become

$$A^{-1}\begin{bmatrix} 50 \\ 31 \end{bmatrix} = \begin{bmatrix} 2 & -3 \\ -1 & 2 \end{bmatrix}\begin{bmatrix} 50 \\ 31 \end{bmatrix} = \begin{bmatrix} 7 \\ 12 \end{bmatrix} = \begin{bmatrix} T \\ O \end{bmatrix}$$
$$A^{-1}\begin{bmatrix} 46 \\ 17 \end{bmatrix} = \begin{bmatrix} 2 & -3 \\ -1 & 2 \end{bmatrix}\begin{bmatrix} 46 \\ 27 \end{bmatrix} = \begin{bmatrix} 11 \\ 8 \end{bmatrix} = \begin{bmatrix} P \\ S \end{bmatrix} \text{ etc.}$$

Continuing in this way, the original message is obtained.

NOW WORK PROBLEM 1.

EXAMPLE 3 Encoding a Message

The message to be encoded is

SEPTEMBER IS OKAY

The encoded message is to be formed using triplets of numbers and the 3×3 matrix

$$A = \begin{bmatrix} 1 & 0 & 0 \\ 3 & 1 & 5 \\ -2 & 0 & 1 \end{bmatrix} \text{ whose inverse is } A^{-1} = \begin{bmatrix} 1 & 0 & 0 \\ -13 & 1 & -5 \\ 2 & 0 & 1 \end{bmatrix}.$$

Use the following association of letters to numbers:

A	B	C	D	E	F	G	H	I	J	K	L	M	N	O	P	Q	R	S	T	U	V	W	X	Y	Z
↓	↓	↓	↓	↓	↓	↓	↓	↓	↓	↓	↓	↓	↓	↓	↓	↓	↓	↓	↓	↓	↓	↓	↓	↓	↓
1	2	3	4	5	6	7	8	9	10	11	12	13	14	15	16	17	18	19	20	21	22	23	24	25	26

SOLUTION When we divide the message into triplets of letters, we obtain

SEP TEM BER ISO KAY

Now multiply the matrix A times each column vector of the message.

[*Note*: If additional letters had been required to complete a triplet, we would have used *Z* or *YZ*.]

$$A\begin{bmatrix} S \\ E \\ P \end{bmatrix} = \begin{bmatrix} 1 & 0 & 0 \\ 3 & 1 & 5 \\ -2 & 0 & 1 \end{bmatrix}\begin{bmatrix} 19 \\ 5 \\ 16 \end{bmatrix} = \begin{bmatrix} 19 \\ 142 \\ -22 \end{bmatrix}$$

$$A\begin{bmatrix} T \\ E \\ M \end{bmatrix} = \begin{bmatrix} 1 & 0 & 0 \\ 3 & 1 & 5 \\ -2 & 0 & 1 \end{bmatrix}\begin{bmatrix} 20 \\ 5 \\ 13 \end{bmatrix} = \begin{bmatrix} 20 \\ 130 \\ -27 \end{bmatrix}$$

$$A\begin{bmatrix} B \\ E \\ R \end{bmatrix} = \begin{bmatrix} 1 & 0 & 0 \\ 3 & 1 & 5 \\ -2 & 0 & 1 \end{bmatrix}\begin{bmatrix} 2 \\ 5 \\ 18 \end{bmatrix} = \begin{bmatrix} 2 \\ 101 \\ 14 \end{bmatrix}$$

$$A\begin{bmatrix} I \\ S \\ O \end{bmatrix} = \begin{bmatrix} 1 & 0 & 0 \\ 3 & 1 & 5 \\ -2 & 0 & 1 \end{bmatrix}\begin{bmatrix} 9 \\ 19 \\ 15 \end{bmatrix} = \begin{bmatrix} 9 \\ 121 \\ -3 \end{bmatrix}$$

$$A\begin{bmatrix} K \\ A \\ Y \end{bmatrix} = \begin{bmatrix} 1 & 0 & 0 \\ 3 & 1 & 5 \\ -2 & 0 & 1 \end{bmatrix}\begin{bmatrix} 11 \\ 1 \\ 25 \end{bmatrix} = \begin{bmatrix} 11 \\ 109 \\ 3 \end{bmatrix}$$

The coded message is

19 142 −22 20 130 −27 2 101 14 9 121 −3 11 109 3

To decode the message in Example 3, form 3×1 column vectors of the numbers in the coded message and multiply on the left by A^{-1}.

The above are elementary examples of encoding and decoding. Modern-day cryptography uses sophisticated computer-implemented codes that depend on higher-level mathematics.

EXERCISE 2.7 APPLICATION 2 Answers to Odd-Numbered Problems Begin on Page AN-15.

In Problems 1–5, use the correspondence

A	B	C	D	E	F	G	H	I	J	K	L	M	N	O	P	Q	R	S	T	U	V	W	X	Y	Z
↓	↓	↓	↓	↓	↓	↓	↓	↓	↓	↓	↓	↓	↓	↓	↓	↓	↓	↓	↓	↓	↓	↓	↓	↓	↓
1	2	3	4	5	6	7	8	9	10	11	12	13	14	15	16	17	18	19	20	21	22	23	24	25	26

1. The matrix

$$A = \begin{bmatrix} 2 & 3 \\ 1 & 2 \end{bmatrix}$$

was used to code the following messages:

(a) 64 36 75 47 75 47 49 29 60 36 60 37 53 33 71 39 35 18

(b) 76 49 64 41 95 55 43 24 69 42 59 32 77 45 27 15 37 21 128 77

Use the inverse to decode each message.

2. Use the inverse of matrix

$$A = \begin{bmatrix} 1 & 0 & 0 \\ 3 & 1 & 5 \\ -2 & 0 & 1 \end{bmatrix}$$

to decode the message

4 152 17 15 106 −22 1 50 3 1 52 7 19 195 −12

3. The matrix A used to encode a message has as its inverse the matrix

$$A^{-1} = \begin{bmatrix} 2 & -3 \\ -1 & 2 \end{bmatrix}$$

Decode the message

70 39 62 41 113 69 93 57 51 28 29 15 54 33 14 9 62 40 67 43 116 71

4. Use the matrix

$$A = \begin{bmatrix} 1 & 0 & 0 \\ 3 & 1 & 5 \\ -2 & 0 & 1 \end{bmatrix}$$

to encode the message: I HAVE TICKETS TO THE FINAL FOUR.

5. Use the matrix $A = \begin{bmatrix} 2 & 3 \\ 1 & 2 \end{bmatrix}$ to encode each of the following messages:

(a) I AM GOING TO DISNEY
(b) WE SURF THE NET
(c) LETS GO CLUBBING

6. Use the matrix

$$A = \begin{bmatrix} 1 & 0 & 0 \\ 3 & 1 & 5 \\ -2 & 0 & 1 \end{bmatrix}$$

to encode the messages given in Problem 5.

Application 3: Accounting

Determine the full cost of manufactured products 4

Consider a firm that has two types of departments, production and service. The production departments produce goods that can be sold in the market, and the service departments provide services to the production departments. A major objective of the cost accounting process is the determination of the full cost of manufactured products on a per unit basis. This requires an allocation of indirect costs, first, from the service department (where they are incurred) to the producing department in which the goods are manufactured and, second, to the specific goods themselves. For example, an accounting department usually provides accounting services for service departments as well as for the production departments. The indirect costs of service rendered by a service department must be determined in order to correctly assess the production departments. The total costs of a service department consist of its direct costs (salaries, wages, and materials) and its indirect costs (charges for the services it receives from other service departments). The nature of the problem and its solution are illustrated by the following example.

EXAMPLE 1 Finding Total Monthly Costs

Consider a firm with two production departments, P_1 and P_2, and three service departments, S_1, S_2, and S_3. These five departments are listed in the leftmost column of Table 5.

TABLE 5

Department	Total Costs	Direct Costs, Dollars	Indirect Costs for Services from Departments		
			S_1	S_2	S_3
S_1	x_1	600	$0.25x_1$	$0.15x_2$	$0.15x_3$
S_2	x_2	1100	$0.35x_1$	$0.20x_2$	$0.25x_3$
S_3	x_3	600	$0.10x_1$	$0.10x_2$	$0.35x_3$
P_1	x_4	2100	$0.15x_1$	$0.25x_2$	$0.15x_3$
P_2	x_5	1500	$0.15x_1$	$0.30x_2$	$0.10x_3$
Totals		5900	x_1	x_2	x_3

The total monthly costs of these departments are unknown and are denoted by x_1, x_2, x_3, x_4, x_5. The direct monthly costs of the five departments are shown in the third column of the table. The fourth, fifth, and sixth columns show the allocation of charges for the services of S_1, S_2, and S_3 to the various departments. Since the total cost for each department is its direct costs plus its indirect costs, the first three rows of the table yield the total costs for the three service departments:

$$\begin{aligned} x_1 &= 600 + 0.25x_1 + 0.15x_2 + 0.15x_3 \\ x_2 &= 1100 + 0.35x_1 + 0.20x_2 + 0.25x_3 \\ x_3 &= 600 + 0.10x_1 + 0.10x_2 + 0.35x_3 \end{aligned}$$

Let X, C, and D denote the following matrices:

$$X = \begin{bmatrix} x_1 \\ x_2 \\ x_3 \end{bmatrix} \qquad C = \begin{bmatrix} 0.25 & 0.15 & 0.15 \\ 0.35 & 0.20 & 0.25 \\ 0.10 & 0.10 & 0.35 \end{bmatrix} \qquad D = \begin{bmatrix} 600 \\ 1100 \\ 600 \end{bmatrix}$$

Then the system of equations above can be written in matrix notation as

$$X = D + CX$$

which is equivalent to

$$[I_3 - C]X = D$$

The total costs of the three service departments can be obtained by solving this matrix equation for X:

$$X = [I_3 - C]^{-1}D$$

Now

$$[I_3 - C] = \begin{bmatrix} 0.75 & -0.15 & -0.15 \\ -0.35 & 0.80 & -0.25 \\ -0.10 & -0.10 & 0.65 \end{bmatrix}$$

from which it can be shown that

$$[I_3 - C]^{-1} = \begin{bmatrix} 1.57 & 0.36 & 0.50 \\ 0.79 & 1.49 & 0.76 \\ 0.36 & 0.28 & 1.73 \end{bmatrix}$$

It is significant that the inverse of $[I_3 - C]$ exists, and that all of its entries are nonnegative. Because of this and the fact that the matrix D contains only nonnegative entries, the matrix X will also have only nonnegative entries. This means there is a meaningful solution to the accounting problem:

$$X = [I_3 - C]^{-1}D = \begin{bmatrix} 1638.00 \\ 2569.00 \\ 1562.00 \end{bmatrix}$$

Thus, x_1 = \$1638.00, x_2 = \$2569.00, and x_3 = \$1562.00. All direct and indirect costs can now be determined by substituting these values in Table 5, as shown in Table 6.

From Table 6 we learn that department P_1 pays \$1122.25 for the services it receives from S_1, S_2, S_3, and P_2 pays \$1172.60 for the services it receives from these departments. The procedure we have followed charges the direct costs of the service departments to the production departments, and each production department is charged according to the services it utilizes. Furthermore, the total cost for P_1 and P_2 is \$5894.85, and this figure approximates the sum of the direct costs of the three service departments and the two production departments. The results are consistent with conventional accounting procedure. Discrepancies that occur are due to rounding.

TABLE 6

Department	Total Costs, Dollars	Direct Costs, Dollars	Indirect Costs for Services from Departments, Dollars		
			S_1	S_2	S_3
S_1	1629.15	600	409.50	385.35	234.30
S_2	2577.60	1100	573.30	513.80	390.50
S_3	1567.40	600	163.80	256.90	546.70
P_1	3222.25	2100	245.70	642.25	234.30
P_2	2672.60	1500	245.70	770.70	156.20

Finally, a comment should be made about the allocation of charges for services as shown in Table 5. How is it determined that 25% of the total cost x_1 of S_1 should be charged to S_1, 35% to S_2, 10% to S_3, 15% to P_1, and 15% to P_2? The services of each department can be measured in some suitable unit, and each department can be charged according to the number of these units of service it receives. If 20% of the accounting items concern a given department, that department is charged 20% of the total cost of the accounting department. When services are not readily measurable, the allocation basis is subjectively determined.

EXERCISE 2.7 APPLICATION 3 Answers to Odd-Numbered Problems Begin on Page AN-15.

1. Consider the accounting problem described by the data in the table:

Department	Total Costs	Direct Costs, Dollars	Indirect Costs for Services from Departments	
			S_1	S_2
S_1	x_1	2,000	$\frac{1}{9}x_1$	$\frac{3}{9}x_2$
S_2	x_2	1,000	$\frac{3}{9}x_1$	$\frac{1}{9}x_2$
P_1	x_3	2,500	$\frac{1}{9}x_1$	$\frac{2}{9}x_2$
P_2	x_4	1,500	$\frac{3}{9}x_1$	$\frac{1}{9}x_2$
P_3	x_5	3,000	$\frac{1}{9}x_1$	$\frac{2}{9}x_2$
Totals		10,000	x_1	x_2

Determine whether this accounting problem has a solution. If it does, find the total costs. Prepare a table similar to Table 6. Show that the total of the service charges allocated to P_1, P_2, and P_3 is equal to the sum of the direct costs of the service departments S_1 and S_2.

2. Follow the directions of Problem 1 for the accounting problem described by the following data:

Department	Total Costs	Direct Costs, Dollars	Indirect Costs for Services from Departments		
			S_1	S_2	S_3
S_1	x_1	500	$0.20x_1$	$0.10x_2$	$0.10x_3$
S_2	x_2	1000	$0.40x_1$	$0.15x_2$	$0.30x_3$
S_3	x_3	500	$0.10x_1$	$0.05x_2$	$0.30x_3$
P_1	x_4	2000	$0.20x_1$	$0.35x_2$	$0.20x_3$
P_2	x_5	1500	$0.10x_1$	$0.35x_2$	$0.10x_3$
Totals		5500	x_1	x_2	x_3

Application 4: The Method of Least Squares

In Section 1.4 we were given a set of data points (x, y) and we drew the scatter diagram representing the relation between x and y. We analyzed the scatter diagram to determine if the relation between the variables was linear. When it was linear, we selected two points and found an equation of the line containing the points. We then compared our line to the *line of best fit,* found using a graphing utility. In this application we define the *line of best fit* and learn how matrices are used to find the *line of best fit.*

The method of least squares refers to a technique that is often used in data analysis to find the "best" linear equation that fits a given collection of experimental data. (We shall see a little later just what we mean by the word *best.*) It is a technique employed by statisticians, economists, business forecasters, and those who try to interpret and analyze data. Before we begin our discussion, we will need to define the *transpose* of a matrix.

Transpose

Let A be a matrix of dimension $m \times n$. The **transpose** of A, written A^T, is the $n \times m$ matrix obtained from A by interchanging the rows and columns of A.

The first row of A^T is the first column of A; the second row of A^T is the second column of A; and so on.

5 **EXAMPLE 1 Finding the Transpose of a Matrix**

If

$$A = \begin{bmatrix} 1 & 2 & 3 \\ 0 & -1 & 2 \end{bmatrix} \qquad B = \begin{bmatrix} 1 & 1 \\ 0 & 1 \\ 2 & 3 \end{bmatrix} \qquad C = [1 \quad 0 \quad -1]$$

then

$$A^T = \begin{bmatrix} 1 & 0 \\ 2 & -1 \\ 3 & 2 \end{bmatrix} \qquad B^T = \begin{bmatrix} 1 & 0 & 2 \\ 1 & 1 & 3 \end{bmatrix} \qquad C^T = \begin{bmatrix} 1 \\ 0 \\ -1 \end{bmatrix}$$

Note how dimensions are reversed when computing A^T: A has the dimension 2×3, while A^T has the dimension 3×2, and so on.

We begin our discussion of the method of least squares with an example.

Suppose a product has been sold over time at various prices and that we have some data that show the demand for the product (in thousands of units) in terms of its price (in dollars). If we use x to represent price and y to represent demand, the data might look like the information in Table 7.

TABLE 7

Price, x	4	5	9	12
Demand, y	9	8	6	3

Suppose also that we have reason to assume that y and x are *linearly related.* That is, we assume we can write

$$y = mx + b$$

for some, as yet unknown, m and b. In other words, we are assuming that when demand is graphed against price, the resulting graph will be a line. Our belief that y and x are linearly related might be based on past experience or economic theory.

If we plot the data from Table 7 in a scatter diagram (see Figure 15), it seems pretty clear that no single line passes through the plotted points. Though no line will pass through all the points, we might still ask: "Is there a line that 'best fits' the points?" That is, we seek a line of the sort in Figure 16.

FIGURE 15

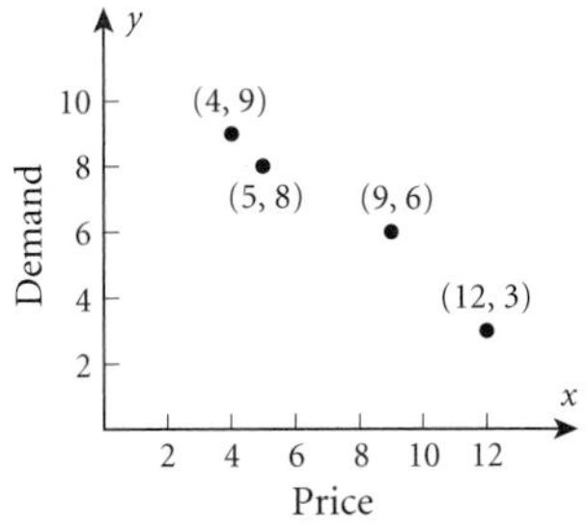

FIGURE 16

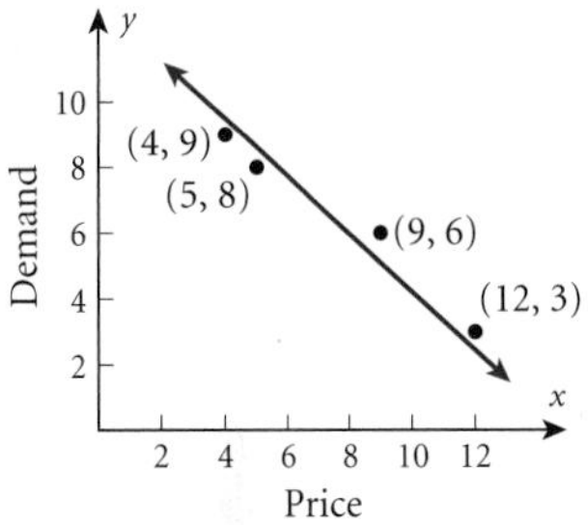

Statement of the Problem

Given a set of noncollinear points, find the line that "best fits" these points.

We still need to clarify our use of the phrase "best fits." First, though, suppose there are four data points, labeled from left to right as (x_1, y_1), (x_2, y_2), and so on. See Figure 17. Now suppose the equation of the line L in Figure 17 is

$$y = mx + b$$

where we do not know m and b. If (x_1, y_1) is actually on the line, it would be true that

$$y_1 = mx_1 + b$$

FIGURE 17

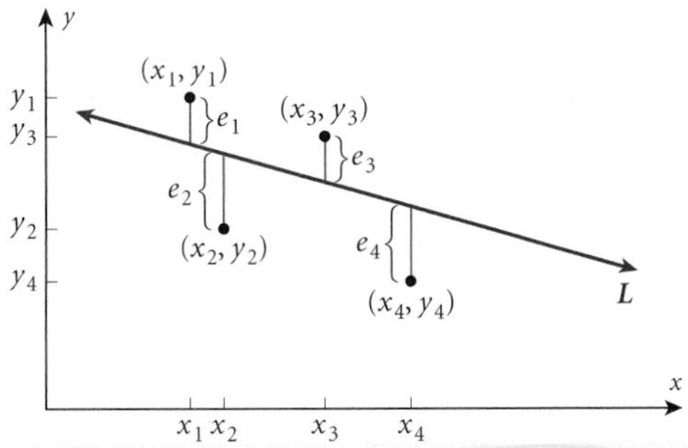

But, since we can't be sure (x_1, y_1) lies on L, the above equation may not be true. That is, there may be a nonzero difference between y_1 and $mx_1 + b$. We designate this difference by e_1, where e_1 represents the 'error'. Then

$$e_1 = y_1 - (mx_1 + b)$$

so, solving for y, we have

$$y_1 = mx_1 + b + e_1$$

What we have just said about (x_1, y_1) we can repeat for the remaining data points (x_i, y_i), $i = 2, 3$, and 4, obtaining

$$e_i = y_i - (mx_i + b)$$

or, equivalently,

$$y_i = mx_i + b + e_i \qquad i = 1, 2, 3, 4 \tag{1}$$

Geometrically, $|e_i|$ measures the vertical distance between the sought-after line L and the data point (x_i, y_i).

Recall that our problem is to find a line L that "best fits" the data points (x_i, y_i). Intuitively, we would seek a line L for which the e_i's are all small. Since some e_i's may be positive and some may be negative, it will not do to just add them up. To eliminate the signs, we square each e_i. The method of least squares consists of finding a line L for which the sum of the squares of the e_i's is as small as possible. Since finding the line $y = mx + b$ is the same as finding m and b, we restate the least squares problem for this example as follows:

The Least Squares Problem

Given the data points (x_i, y_i), $i = 1, 2, 3, 4$, find m and b satisfying

$$\boxed{y_i = mx_i + b + e_i \qquad i = 1, 2, 3, 4} \tag{2}$$

so that $e_1^2 + e_2^2 + e_3^2 + e_4^2$ is minimized.

A solution to the problem can be compactly expressed using matrix language. Since a derivation requires either calculus or more advanced linear algebra, we omit the proof and simply present the solution:

Solution to the Least Squares Problem

The line of best fit

$$\boxed{y = mx + b} \tag{3}$$

to a set of four points $(x_1, y_1), (x_2, y_2), (x_3, y_3), (x_4, y_4)$ is obtained by solving the system of two equations in two unknowns

$$\boxed{A^T AX = A^T Y} \tag{4}$$

for m and b, where

$$A = \begin{bmatrix} x_1 & 1 \\ x_2 & 1 \\ x_3 & 1 \\ x_4 & 1 \end{bmatrix} \qquad X = \begin{bmatrix} m \\ b \end{bmatrix} \qquad Y = \begin{bmatrix} y_1 \\ y_2 \\ y_3 \\ y_4 \end{bmatrix} \tag{5}$$

6 **EXAMPLE 2** **Finding the Line of Best Fit**

Use least squares to find the line of best fit for the points

$$(4,\ 9),\ (5,\ 8),\ (9,\ 6),\ (12,\ 3)$$

SOLUTION We use Equations (5) to set up the matrices A and Y:

$$A = \begin{bmatrix} 4 & 1 \\ 5 & 1 \\ 9 & 1 \\ 12 & 1 \end{bmatrix} \qquad Y = \begin{bmatrix} 9 \\ 8 \\ 6 \\ 3 \end{bmatrix}$$

Equation (4), $A^TAX = A^TY$, becomes

$$\begin{bmatrix} 4 & 5 & 9 & 12 \\ 1 & 1 & 1 & 1 \end{bmatrix} \begin{bmatrix} 4 & 1 \\ 5 & 1 \\ 9 & 1 \\ 12 & 1 \end{bmatrix} \begin{bmatrix} m \\ b \end{bmatrix} = \begin{bmatrix} 4 & 5 & 9 & 12 \\ 1 & 1 & 1 & 1 \end{bmatrix} \begin{bmatrix} 9 \\ 8 \\ 6 \\ 3 \end{bmatrix}$$

This reduces to a system of two equations in two unknowns:

$$\begin{cases} 266m + 30b = 166 \\ 30m + 4b = 26 \end{cases}$$

The solution, which you can verify, is given by

$$m = -\tfrac{29}{41} \qquad \text{and} \qquad b = \tfrac{484}{41}$$

The least squares solution to the problem is given by the line

$$y = -\tfrac{29}{41}x + \tfrac{484}{41}$$

Note, for example, that corresponding to the value $x = 5$, the above line of "best fit" has $y = -\frac{29}{41} \cdot 5 + \frac{484}{41} = 8.27$, while the experimentally observed demand had value 8. Were we to use our computed line to approximate the connection between price and demand, we would predict that a selling price of $x = 6$ would yield a demand of $y = -\frac{29}{41} \cdot 6 + \frac{484}{41} = 7.56$.

COMMENT: The line of best fit obtained in Example 2 is

$$y = -\tfrac{29}{41}x + \tfrac{484}{41} = -0.7073170732x + 11.80487805$$

In Figure 18, we compare the line of best fit obtained above to that generated by a graphing utility and show the graph of the line of best fit on the scatter diagram.

FIGURE 18

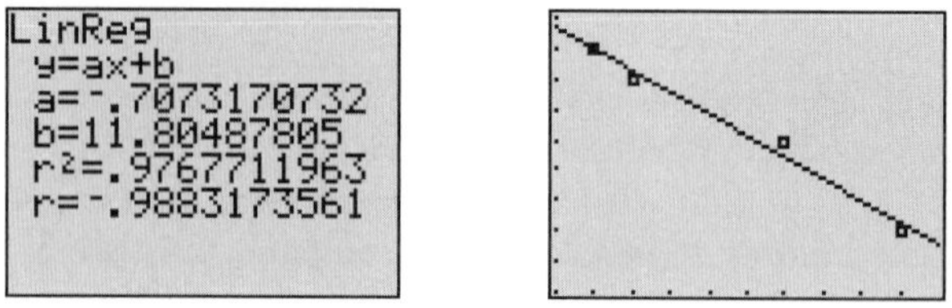

THE GENERAL LEAST SQUARES PROBLEM

We presented the method of least squares using a particular example with four data points. It is easy to extend this analysis.

The general least squares problem consists of finding the line of best fit to a set of n data points:

$$(x_1, y_1), (x_2, y_2), \ldots, (x_n, y_n) \tag{6}$$

If we let

$$A = \begin{bmatrix} x_1 & 1 \\ x_2 & 1 \\ \vdots & \vdots \\ x_n & 1 \end{bmatrix} \qquad X = \begin{bmatrix} m \\ b \end{bmatrix} \qquad Y = \begin{bmatrix} y_1 \\ y_2 \\ \vdots \\ y_n \end{bmatrix}$$

then the line of best fit to the data points in Equation (6) is

$$y = mx + b$$

where $X = \begin{bmatrix} m \\ b \end{bmatrix}$ is found by solving the system of two equations in two unknowns,

$$A^T AX = A^T Y \tag{7}$$

It can be shown that the system given by Equation (7) always has a unique solution provided the data points do not all lie on the same vertical line.

The least squares techniques presented here along with more elaborate variations are frequently used today by statisticians and researchers.

In Problems 1–6, compute A^T.

1. $A = \begin{bmatrix} 4 & 1 & 2 \\ 3 & 1 & 0 \end{bmatrix}$

2. $A = \begin{bmatrix} 5 & 2 & -1 \\ 1 & 3 & 6 \\ 1 & -1 & 2 \end{bmatrix}$

3. $A = \begin{bmatrix} 1 & 11 \\ 0 & 12 \\ 1 & 4 \end{bmatrix}$

4. $A = \begin{bmatrix} -1 & 6 & 4 \end{bmatrix}$

5. $A = \begin{bmatrix} 8 \\ 6 \\ 3 \end{bmatrix}$

6. $A = \begin{bmatrix} 5 & 3 \\ 3 & 7 \end{bmatrix}$

7. **Supply Equation** The following table shows the supply (in thousands of units) of a product at various prices (in dollars):

Price, x	3	5	6	7
Supply, y	10	13	15	16

   (a) Find the least squares line of best fit to the above data.
   (b) Use the equation of this line to estimate the supply of the product at a price of $8.

8. **Study Time vs. Performance** Data giving the number of hours a person studied compared to his or her performance on an exam are given in the table:

Hours Studied x	Exam Score y
0	50
2	74
4	85
6	90
8	92

   (a) Find a least squares line of best fit to the above data.
   (b) What prediction would this line make for a student who studied 9 hours?

9. **Advertising vs. Sales** A business would like to determine the relationship between the amount of money spent on advertising and its total weekly sales. Over a period of 5 weeks it gathers the following data:

Amount Spent on Advertising (in thousands) x	Weekly Sales Volume (in thousands) y
10	50
17	61
11	55
18	60
21	70

Find a least squares line of best fit to the above data.

10. **Drug Concentration vs. Time** The following data show the connection between the number of hours a drug has been in a person's body and its concentration in the body.

Number of Hours	Drug Concentration (parts per million)
2	2.1
4	1.6
6	1.4
8	1.0

   (a) Fit a least squares line to the above data.
   (b) Use the equation of the line to estimate the drug concentration after 5 hours.

11. A matrix is **symmetric** if $A^T = A$. Which of the following matrices are symmetric?

(a) $\begin{bmatrix} 1 & 1 & 2 \\ 1 & 0 & 1 \\ 3 & 2 & 3 \end{bmatrix}$ (b) $\begin{bmatrix} 0 & 1 & 3 \\ 1 & 4 & 7 \\ 3 & 7 & 5 \end{bmatrix}$ (c) $\begin{bmatrix} 1 & 2 & 3 & 0 \\ 2 & 4 & 5 & 0 \\ 3 & 5 & 1 & 0 \end{bmatrix}$

Need a symmetric matrix be square?

12. Show that the matrix $A^T A$ is always symmetric.

Chapter 2 Review

OBJECTIVES

Section		You should be able to	Review Exercises
2.1	1	Solve systems of equations by substitution	1–6
	2	Solve systems of equations by elimination	1–6
	3	Identify inconsistent systems of equations containing two variables	4, 5
	4	Express the solutions of a system of dependent equations containing two variables	6
	5	Solve systems of three equations containing three variables	7–10, 60–63
	6	Identify inconsistent systems of equations containing three variables	10
	7	Express the solutions of a system of dependent equations containing three variables	9, 61–63
2.2	1	Write the augmented matrix of a system of linear equations	15–32
	2	Write the system from the augmented matrix	11–14
	3	Perform row operations on a matrix	15–32
	4	Solve systems of linear equations using matrices	15–32
	5	Express the solutions of a system with an infinite number of solutions	22
2.3	1	Analyze the reduced row echelon form of an augmented matrix	13–14, 33–36
	2	Solve a system of m linear equations containing n variables	27–32
	3	Express the solutions of a system with an infinite number of solutions	27–30
2.4	1	Find the dimension of a matrix	37–48
	2	Find the sum of two matrices	37, 45
	3	Work with properties of matrices	49–50
	4	Find the difference of two matrices	38, 46
	5	Find scalar multiples of a matrix	39, 40
2.5	1	Find the product of two matrices	41–42
	2	Work with properties of matrices	49, 50
2.6	1	Find the inverse of a matrix	51–56
	2	Use the inverse of a matrix to solve a system of equations	15–26
2.7	1	Determine relative income using a closed Leontief model	66
	2	Determine production levels necessary to meet forecasted demand using an open Leontief model	67
	3	Use matrices in cryptography	68
	4	Determine the full cost of manufactured products	69
	5	Find the transpose of a matrix	47–48
	6	Use the method of least squares to find the line of best fit	64–65

THINGS TO KNOW

Systems of Linear Equations
Systems with a solution are consistent and either have a unique solution or have infinitely many solutions
Systems with no solution are inconsistent

Solving Systems of Linear Equations
Using substitution (p. 53)
Using elimination (p. 55)
Using matrices: row echelon method (pp. 70–71)
Using matrices: reduced row echelon method (p. 73)

Matrix (pp. 65 and 94)
Augmented matrix of a system of linear equations (p. 65)
Row operations (p. 67)
Dimension of a matrix (p. 94)
Square matrix (p. 94)
Zero matrix (p. 98)
Identity matrix (p. 113)
Inverse of a matrix (p. 117)

TRUE–FALSE ITEMS **Answers are on page AN-15.**

T F **1.** Matrices of the same dimension can always be added.

T F **2.** Matrices of the same dimension can always be multiplied.

T F **3.** A square matrix will always have an inverse.

T F **4.** The reduced row echelon form of a matrix A is unique.

T F **5.** If A and B are each of dimension 4×4, then $AB = BA$.

T F **6.** Matrix addition is always defined.

T F **7.** Matrix multiplication is commutative.

FILL IN THE BLANKS **Answers are on page AN-15.**

1. If matrix A is of dimension 3×4 and matrix B is of dimension 4×2, then AB is of dimension ________.

2. A system of three linear equations containing three variables has either ________ solution, or no solution, or ________ ________ solutions.

3. If A is a matrix of dimension 3×4, the 3 tells the number of ________ and the 4 tells the number of ________ .

4. If $AB = I$, the identity matrix, then B is called the ________ of A.

5. If B is a 2×3 matrix and BA^2 is defined, then A is a matrix of dimension ________.

6. If A is a 4×5 matrix and AB^3 is defined, then B is a matrix of dimension ________.

REVIEW EXERCISES **Answers to odd-numbered problems begin on page AN-15.**
Blue problem numbers represent the author's suggestions for a Practice Test.

In Problems 1–10, solve each system of equations algebraically using the method of substitution or the method of elimination. If the system has no solution, say it is inconsistent.

1. $\begin{cases} 2x - y = 5 \\ 5x + 2y = 8 \end{cases}$

2. $\begin{cases} 2x + 3y = 2 \\ 7x - y = 3 \end{cases}$

3. $\begin{cases} x - 2y - 4 = 0 \\ 3x + 2y - 4 = 0 \end{cases}$

4. $\begin{cases} x - 3y + 4 = 0 \\ \frac{1}{2}x - \frac{3}{2}y + \frac{4}{3} = 0 \end{cases}$

5. $\begin{cases} 3x - 2y = 8 \\ x - \frac{2}{3}y = 12 \end{cases}$

6. $\begin{cases} 2x + 5y = 10 \\ 4x + 10y = 20 \end{cases}$

7. $\begin{cases} x + 2y - z = 6 \\ 2x - y + 3z = -13 \\ 3x - 2y + 3z = -16 \end{cases}$

8. $\begin{cases} x + 5y - z = 2 \\ 2x + y + z = 7 \\ x - y + 2z = 11 \end{cases}$

9. $\begin{cases} 2x - 4y + z = -15 \\ x + 2y - 4z = 27 \\ 5x - 6y - 2z = -3 \end{cases}$

10. $\begin{cases} x - 4y + 3z = 15 \\ -3x + y - 5z = -5 \\ -7x - 5y - 9z = 10 \end{cases}$

In Problems 11–14, write the system of equations corresponding to the given augmented matrix. In Problems 13–14, analyze the solution.

11. $\left[\begin{array}{cc|c} 3 & 2 & 8 \\ 1 & 4 & -1 \end{array}\right]$

12. $\left[\begin{array}{ccc|c} 1 & 2 & 5 & -2 \\ 5 & 0 & -3 & 8 \\ 2 & -1 & 0 & 0 \end{array}\right]$

13. $\left[\begin{array}{ccc|c} 1 & 0 & 0 & 4 \\ 0 & 1 & 0 & 6 \\ 0 & 0 & 1 & -1 \end{array}\right]$

14. $\left[\begin{array}{ccc|c} 1 & 0 & 3 & 5 \\ 0 & 1 & 2 & 8 \end{array}\right]$

In Problems 15–32, use matrices to find the solution, if it exists, of each system of linear equations. If the system has infinitely many solutions, write the solution using parameters and then list at least three solutions. If the system has no solution, say it is inconsistent.

15. $\begin{cases} -5x + 2y = -2 \\ -3x + 3y = 4 \end{cases}$

16. $\begin{cases} -3x + 2y = 3 \\ -3x + 4y = 4 \end{cases}$

17. $\begin{cases} x + 2y + 5z = 6 \\ 3x + 7y + 12z = 23 \\ x + 4y = 25 \end{cases}$

18. $\begin{cases} x + 2y - z = -4 \\ 3x + 7y - z = -21 \\ x + 4y - 6z = -17 \end{cases}$

19. $\begin{cases} x + 2y + 7z = 2 \\ 3x + 7y + 18z = -1 \\ x + 4y + 2z = -13 \end{cases}$

20. $\begin{cases} x + 2y - 7z = -1 \\ 3x + 7y - 24z = 6 \\ x + 4y - 12z = 26 \end{cases}$

21. $\begin{cases} 2x - y + z = 1 \\ x + y - z = 2 \\ 3x - y + z = 0 \end{cases}$

22. $\begin{cases} 2x + 3y - z = 5 \\ x - y + z = 1 \\ 3x - 3y + 3z = 3 \end{cases}$

23. $\begin{cases} y - 2z = 6 \\ 3x + 2y - z = 2 \\ 4x + 3z = -1 \end{cases}$

24. $\begin{cases} 2x - y + 3z = 5 \\ x + 2z = 0 \\ 3x + 2y + z = -3 \end{cases}$

25. $\begin{cases} x - 3y = 5 \\ 3y + z = 0 \\ 2x - y + 2z = 2 \end{cases}$

26. $\begin{cases} x - z = 2 \\ 2x - y = 4 \\ x + y + z = 6 \end{cases}$

27. $\begin{cases} 3x + y - 2z = 3 \\ x - 2y + z = 4 \end{cases}$

28. $\begin{cases} 2x - y - 3z = 0 \\ x - 2y + z = 4 \end{cases}$

29. $\begin{cases} x + 2y - z = 5 \\ 2x - y + 2z = 0 \end{cases}$

30. $\begin{cases} x - y + 2z = 6 \\ 2x + 2y - z = -1 \end{cases}$

31. $\begin{cases} 2x - y = 6 \\ x - 2y = 0 \\ 3x - y = 6 \end{cases}$

32. $\begin{cases} x - 2y = 0 \\ 2x + y = 5 \\ x - 3y = 6 \end{cases}$

In Problems 33–36, analyze the solution, if any, of the system of equations having the given augmented matrix.

33. $\left[\begin{array}{ccc|c} 1 & 4 & 3 & 4 \\ 0 & 1 & 0 & -1 \\ 0 & 0 & 1 & 1 \end{array}\right]$

34. $\left[\begin{array}{ccc|c} 1 & 0 & -3 & 4 \\ 0 & 1 & 0 & 8 \\ 0 & 0 & 1 & 0 \end{array}\right]$

35. $\left[\begin{array}{cccc|c} 1 & 0 & 0 & 2 & 1 \\ 0 & 1 & 1 & 2 & 2 \\ 0 & 0 & 1 & 0 & 3 \end{array}\right]$

36. $\left[\begin{array}{cccc|c} 1 & 0 & 3 & 1 & 2 \\ 0 & 1 & 4 & 2 & 1 \end{array}\right]$

In Problems 37–50, use the following matrices to compute each expression. State the dimension of the resulting matrix.

$$A = \begin{bmatrix} 1 & 0 \\ 2 & 4 \\ -1 & 2 \end{bmatrix} \qquad B = \begin{bmatrix} 4 & -3 & 0 \\ 1 & 1 & -2 \end{bmatrix} \qquad C = \begin{bmatrix} 3 & -4 \\ 1 & 5 \\ 5 & -2 \end{bmatrix}$$

37. $A + C$

38. $A - C$

39. $6A$

40. $-4B$

41. AB

42. BA

43. CB

44. BC

45. $(A + C)B$

46. $B(2C - A)$

47. A^{T}

48. B^{T}

49. $C + 0$

50. $A + (-A)$

In Problems 51–56, find the inverse, if it exists, of each matrix.

51. $\begin{bmatrix} 3 & 0 \\ -2 & 1 \end{bmatrix}$

52. $\begin{bmatrix} 4 & 1 \\ 3 & 1 \end{bmatrix}$

53. $\begin{bmatrix} 4 & 2 \\ 6 & 3 \end{bmatrix}$

54. $\begin{bmatrix} 1 & 2 & 3 \\ 2 & 4 & 5 \\ 3 & 5 & 6 \end{bmatrix}$

55. $\begin{bmatrix} 4 & 3 & -1 \\ 0 & 2 & 2 \\ 3 & -1 & 0 \end{bmatrix}$

56. $\begin{bmatrix} 1 & 2 & -3 \\ 4 & 6 & 2 \\ -1 & -6 & 9 \end{bmatrix}$

57. What must be true about x, y, z, w, if we require $AB = BA$ for the matrices

$$A = \begin{bmatrix} x & y \\ z & w \end{bmatrix} \quad \text{and} \quad B = \begin{bmatrix} 1 & 1 \\ -1 & 1 \end{bmatrix}$$

58. Let $t = [t_1 \ t_2]$, with $t_1 + t_2 = 1$, and let $A = \begin{bmatrix} \frac{1}{4} & \frac{3}{4} \\ \frac{2}{3} & \frac{1}{3} \end{bmatrix}$.

Find t such that $tA = t$.

59. **Mixture** Sweet Delight Candies, Inc., sells boxes of candy consisting of creams and caramels. Each box sells for $4 and holds 50 pieces of candy (all pieces are the same size). If the caramels cost $0.05 to produce and the creams cost $0.10 to produce, how many caramels and creams should be in each box for no profit and no loss? Would you increase or decrease the number of caramels in order to obtain a profit?

60. Cookie Orders A cookie company makes three kinds of cookies—oatmeal raisin, chocolate chip, and shortbread—packaged in small, medium, and large boxes. The small box contains 1 dozen oatmeal raisin and 1 dozen chocolate chip; the medium box has 2 dozen oatmeal raisin, 1 dozen chocolate chip, and 1 dozen shortbread; the large box contains 2 dozen oatmeal raisin, 2 dozen chocolate chip, and 3 dozen shortbread. If you require exactly 15 dozen oatmeal raisin, 10 dozen chocolate chip, and 11 dozen shortbread cookies, how many of each size box should you buy?

61. Mixture Problem A store sells almonds for $6 per pound, cashews for $5 per pound, and peanuts for $2 per pound. One week the manager decides to prepare 100 16-ounce packages of nuts by mixing the peanuts, almonds, and cashews. Each package will be sold for $4. The mixture is to produce the same revenue as selling the nuts separately. Prepare a table that shows some of the possible ways the manager can prepare the mixture.

62. Financial Planning Three retired couples each require an additional annual income of $1800 per year. As their financial consultant, you recommend they invest some money in Treasury Bills that yield 6%, some money in corporate bonds that yield 8%, and some money in junk bonds that yield 10%. Prepare a table for each couple showing the various ways their goal can be achieved

(a) If the first couple has $20,000 to invest
(b) If the second couple has $25,000 to invest
(c) If the third couple has $30,000 to invest

63. Financial Planning A retired couple have $40,000 to invest. As their financial consultant, you recommend they invest some money in Treasury Bills that yield 6%, some money in corporate bonds that yield 8%, and some money in junk bonds that yield 10%. Prepare a table showing the various ways this couple can achieve the following goals:

(a) They want $2500 per year in income.
(b) They want $3000 per year in income.
(c) They want $3500 per year in income.

64. Cost of College

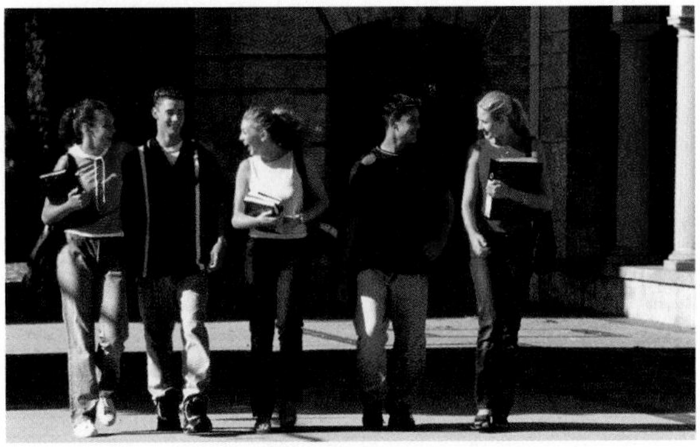

The National Center for Education Statistics collects data on the costs of college. The data in the table show the average cost of tuition, fees, room, and board paid by full-time undergraduate students at degree-granting institutions for select years.

Year	Average Cost
1976–77	$2,275
1981–82	3,489
1987–88	5,494
1991–92	7,077
1995–96	8,800
2000–01	10,876

(a) Draw a scatter diagram of the data, and comment on its apparent shape. [*Hint*: Represent 1976–77 by 1, 1981–82 by 6, 1987–88 by 12, and so on.]
(b) Find the least squares line of best fit relating years to average cost.
(c) Interpret the slope of the line found in part (b).
(d) Use the line of best fit to predict the cost of tuition, fees, room, and board for the school year 2003–04.
(e) Use the least squares line to predict the cost of college in 2005–06.
(f) Compute the line of best fit using a graphing utility and compare it to the line obtained in part (b).

Source: Digest of Education Statistics, National Center for Education Statistics, 2001.

65. The U.S. Bureau of the Census publishes estimates of emigration from the United States to other countries. The following table lists the estimated emigration rates for selected years from 1991–2001.

Year	Emigrates
1991	252,000
1993	258,000
1996	267,000
1998	278,000
2000	287,000
2001	293,000

Use these data in the following questions.

(a) Draw a scatter diagram, and comment on its apparent shape. [*Hint*: Use 1 for 1991, 3 for 1993, 6 for 1996 and so on.]
(b) Find the least squares line of best fit relating years to emigration numbers.
(c) Interpret the slope of the line found in part (b).
(d) Use the line of best fit to predict the emigration from the United States in 2005.
(e) Use the least squares line to estimate the number of people who emigrated from the United States in 1995.
(f) Compute the line of best fit using a graphing utility and compare it to the line obtained in part (b).

Source: U.S. Bureau of the Census, Internet release, 2003.

66. In a professional cooperative a physician, an attorney, and a financial planner trade services. They agree that the value of each one's work is approximately $20,000. At the end of the period the physician had spent 30% of his time on his own care, 40% on the attorney's care, and 30% on the financial planner's care. The attorney had spent 50% of her time on the physician's affairs, 20% on her own affairs, and 30% on the financial planner's affairs. Finally, the financial planner spent 20% of his time managing the physician's portfolio, 20% managing the attorney's portfolio, and 60% managing his own portfolio. How should each of these three professionals be paid for their work?

67. Suppose three corporations each manufacture a different product, although each needs the others' goods to produce its own. Moreover, suppose some of each manufacturer's production is consumed by individuals outside the system. Current internal and external consumptions are given in the table below, but it has been forecast that consumer demand will change in the next several years and that in 5 years consumers will demand 80 units of product A, 40 units of product B, and 80 units of product C. Find the total output needed to be made by each corporation to meet the predicted demand.

	A	B	C	Consumer	Total
A	100	50	40	60	250
B	20	10	30	40	100
C	30	40	30	100	200

68. Use the correspondence

A	B	C	D	E	F	G	H	I	J	K	L	M	N	O	P
26	25	24	23	22	21	20	19	18	17	16	15	14	13	12	11

Q	R	S	T	U	V	W	X	Y	Z
10	9	8	7	6	5	4	3	2	1

and the matrices (I) $A = \begin{bmatrix} 3 & 1 \\ 2 & 2 \end{bmatrix}$ and (II) $B = \begin{bmatrix} 1 & 4 & 2 \\ 2 & 0 & 2 \\ 0 & 0 & 4 \end{bmatrix}$

to encode the following messages:

(a) HEY DUDE WHATS UP
(b) CALL ME ON MY CELL

Then use (I) to decode the message: 75 78 45 42 72 64 35 42 17 26 67 62.

69. A local firm has two service departments, purchasing and legal, and three production departments, P_1, P_2, and P_3. The total costs of each department are unknown and are denoted x_1, x_2, x_3, x_4, and x_5. The direct costs of manufacture and the indirect costs of service are listed in the table on page 155.

Department	Total Costs	Direct Cost in Dollars	Indirect Costs for Services	
S_1	x_1	800	$0.20x_1$	$0.10x_2$
S_2	x_2	4000	$0.10x_1$	$0.30x_2$
P_1	x_3	1500	$0.20x_1$	$0.10x_2$
P_2	x_4	500	$0.30x_1$	$0.20x_2$
P_3	x_5	1200	$0.20x_1$	$0.30x_2$
Totals		8000	x_1	x_2

Find the total costs. Prepare a table similar to Table 6 on page 143. Show that the total of the service charges allocated to P_1, P_2, and P_3 is equal to the sum of the direct costs of the service departments S_1 and S_2.

Chapter 2 Project

THE OPEN LEONTIEF MODEL AND THE AMERICAN ECONOMY

The open Leontief model, introduced in Section 2.7, is used to model the economies of many countries throughout the world, as well as the global economy itself. The first example Leontief himself used was the American economy in 1947. He divided the economy into 42 industries (or **sectors**) for his original model. For purposes of this example, the data from the 42 sectors used by Leontief has been collected into just 3: agriculture, manufacturing, and services (shown in Table 1). Of course, the open sector (which consumes only goods and services) or demand vector is also present.

TABLE 1 EXCHANGE OF GOODS AND SERVICES IN THE U.S. FOR 1947 (IN BILLIONS OF 1947 DOLLARS)

	Agriculture	Manufacturing	Services	Open Sector	Total Gross Output
Agriculture	34.69	4.92	5.62	39.24	84.56
Manufacturing	5.28	61.82	22.99	60.02	163.43
Services	10.45	25.95	42.03	130.65	219.03

1. Create the matrix A and the demand vector D_0 for the open Leontief model from Table 1.

2. Calculate the production vector X_0 for the open Leontief model from Table 1.

3. Now suppose that the demand vector is changed to

$$D_1 = \begin{bmatrix} 40.24 \\ 60.02 \\ 130.65 \end{bmatrix}$$

Notice that the only change from the original demand vector D_0 to the new demand vector D_1 is the addition of 1 unit of consumer demand in the agricultural sector. Calculate a new production vector X_1 for this new demand vector D_1.

4. Compute $X_1 - X_0$.

You should find that the difference in the production vectors X_0 and X_1 is the first column of the matrix $(I - A)^{-1}$. This is not a coincidence. It can be shown that the difference in the old and new production vectors must be the first column of $(I - A)^{-1}$. This observation leads to an important interpretation of the entries of $(I - A)^{-1}$:

Interpretation: The (i, j) entry in the matrix $(I - A)^{-1}$ is the amount by which industry i must change its production level to satisfy an increase of one unit in the demand from industry j.

5. How much would the service production level need to increase if demand from the agriculture sector is increased by 1 unit?

6. How much would the manufacturing production level need to increase if demand from the agricultural sector is increased by 1 unit?

The most up-to-date available input–output table for the American economy is the 1998 table. The entire table (which is officially called an I-O Use table) may be found in the publication "Annual Input–Output Accounts of the U.S. Economy, 1998," which may be downloaded from the Bureau of Economic Analysis website: www.bea.gov/bea/an2.htm. This document provides a good overview of different types of tables, as well as some applications of the 1998 table. The table for 1998 divides the economy into nearly 500 sectors, which are then consolidated into over 90 sectors. There is also a table that further consolidates the economy into just 9 sectors: This table is given as Table 2.

TABLE 2 EXCHANGE OF GOODS AND SERVICES IN THE U.S. FOR 1998 (IN MILLIONS OF 1998 DOLLARS)

Industry	Agric.	Mining	Const.	Manuf.	T.C.U.	Trade	F.I.R.	Services	Other	Demand	Total Output
Agriculture	68682	78	5860	144622	154	1816	11476	12310	567	34940	283290
Mining	368	31478	7368	81722	52354	31	6	32	3061	−39241	147738
Construction	3369	4693	895	28756	47369	12694	66515	28785	25895	787208	1006179
Manufacturing	49395	14510	299429	1380590	70485	68005	19318	340944	17593	1611520	3953584
Transportation, Communication, Utilities	12625	12652	24847	179922	200933	68214	52626	120762	22872	586248	1250307
Trade	13948	3498	81671	230668	15091	32765	4925	68036	2646	1103110	1552311
Finance, Insurance, Real Estate	20647	33253	16485	71167	40283	108418	445679	243750	7945	1520718	2547553
Services	8998	5851	103708	240141	144495	219223	191363	530971	13585	2214382	3479631
Other	166	29	1076	13826	3306	11226	28196	24713	3034	1032052	1211707

7. Find the consumption matrix A and demand vector D_0 for this table.

8. Find the production vector X for the consumption matrix A and demand vector D_0.

9. If the demand for construction increases by 1 million, how much will the production of the transportation, communication, and utilities sector have to increase to compensate?

10. Which three sectors are most affected by an increase of $1 million of demand for construction?

11. If the entire jth column of the matrix $(I - A)^{-1}$ is added, the result is the amount of new production in the **entire economy** caused by an increase of $1 million in demand for the jth sector. How much new production is produced in the entire economy by an increase of $1 million in demand for construction? This amount is referred to in economic literature as a **backward linkage.**

12. What would it mean for the (i, j) entry in $(I - A)^{-1}$ to be zero?

MATHEMATICAL QUESTIONS FROM PROFESSIONAL EXAMS*

Use the following information to answer Problems 1–4:
Akron, Inc. owns 80% of the capital stock of Benson Company and 70% of the capital stock of Cashin, Inc. Benson Company owns 15% of the capital stock of Cashin, Inc. Cashin, Inc., in turn, owns 25% of the capital stock of Akron, Inc. These ownership interrelationships are illustrated in the diagram.

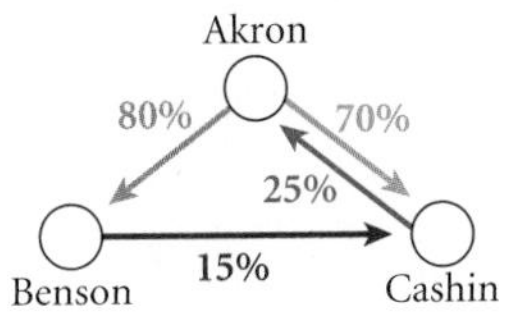

Net income before adjusting for interests in intercompany net income for each corporation follows:

Akron, Inc.	*$190,000*
Benson Co.	*$170,000*
Cashin, Inc.	*$230,000*

Ignore all income tax considerations.

A_e = *Akron's consolidated net income; that is, its net income plus its share of the consolidated net income of Benson and Cashin*

B_e = *Benson's consolidated net income; that is, its net income plus its share of the consolidated net income of Cashin*

C_e = *Cashin's consolidated net income; that is, its net income plus its share of the consolidated net income of Akron*

1. **CPA Exam** The equation, in a set of simultaneous equations, which computes A_e is

(a) $A_e = .75(190{,}000 + .8B_e + .7C_e)$
(b) $A_e = 190{,}000 + .8B_e + .7C_e$
(c) $A_e = .75(190{,}000) + .8(170{,}000) + .7(230{,}000)$
(d) $A_e = .75(190{,}000) + .8B_e + .7C_e$

2. **CPA Exam** The equation, in a set of simultaneous equations, which computes B_e is

(a) $B_e = 170{,}000 + .15C_e - .75A_e$
(b) $B_e = 170{,}000 + .15C_e$
(c) $B_e = .2(170{,}000) + .15(230{,}000)$
(d) $B_e = .2(170{,}000) + .15C_e$

3. **CPA Exam** Cashin's minority interest in consolidated net income is

(a) $.15(230{,}000)$ (b) $230{,}000 + .25A_e$
(c) $.15(230{,}000) + .25A_e$ (d) $.15C_e$

4. **CPA Exam** Benson's minority interest in consolidated net income is

(a) $34,316 (b) $25,500 (c) $45,755 (d) $30,675

CHAPTER 3

Linear Programming: Geometric Approach

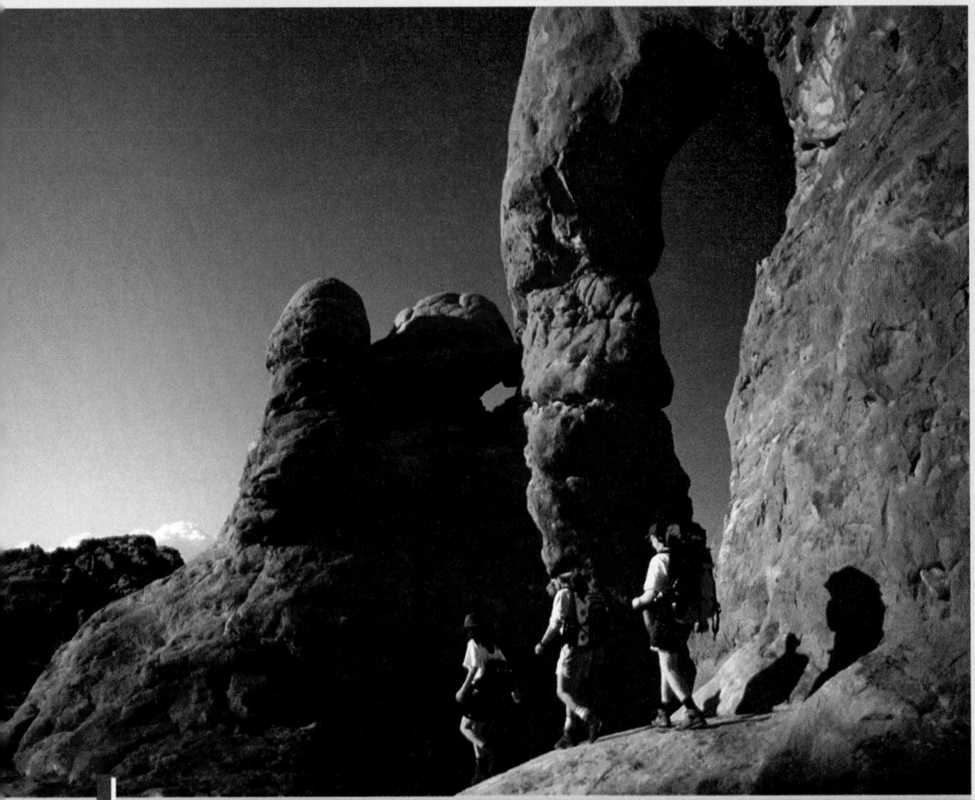

OUTLINE

It's the weekend after midterms, and a hiking trip to Pinnacles National Monument in California is on the agenda. Hiking in Pinnacles will require some advance planning. Some trails go through caves, so a flashlight is needed. Others go to the top of low mountain peaks. No matter where the hiking ends up, some food, such as a trail mix, will be required. Peanuts and raisins sound good. But in what proportions should they be mixed? And what about meeting some minimum calorie requirements? What about carbohydrates and protein? And, of course, fat should be minimized! Fortunately, this chapter was covered before midterms, so these questions can be answered. The Chapter Project at the end of the chapter will guide you.

A LOOK BACK, A LOOK FORWARD

In Chapter 1, we discussed linear equations and some applications that involve linear equations. In Chapter 2, we studied systems of linear equations. In this chapter, we discuss systems of linear inequalities and an extremely important application involving linear equations and systems of linear inequalities: *linear programming.*

Whenever the analysis of a problem leads to minimizing or maximizing a linear expression in which the variables must obey a collection of linear inequalities, a solution may be obtained using linear programming techniques.

Historically, linear programming problems evolved out of the need to solve problems involving resource allocation by the U.S. Army during World War II. Among those who worked on such problems was George Dantzig, who later gave a general formulation of the linear programming problem and offered a method for solving it, called the *simplex method.* This method is discussed in Chapter 4.

In this chapter we study ways to solve linear programming problems that involve only two variables. As a result, we can use a geometric approach utilizing the graph of a system of linear inequalities to solve the problem.

3.1 Systems of Linear Inequalities

PREPARING FOR THIS SECTION *Before getting started, review the following:*

- > Inequalities (Appendix A, Section A.2, pp. 636–642)
- > Lines (Section 1.1, pp. 2–14)
- > Pairs of Lines (Section 1.2, pp. 19–23)

OBJECTIVES
1. Graph linear inequalities
2. Graph systems of linear inequalities

In Section 1.1, we discussed linear equations (linear equalities) in two variables x and y. Recall that these are equations of the form

$$Ax + By = C \tag{1}$$

where A, B, and C are real numbers and A and B are not both zero. If in Equation (1) we replace the equal sign by an inequality symbol, namely, one of the symbols $<, >, \leq$, or $\geq$, we obtain a **linear inequality in two variables** x and y. For example, the expressions

$$3x + 2y \geq 4 \qquad 2x - 3y < 0 \qquad 3x + 5y > -8$$

are each linear inequalities in two variables. The first of these is called a **nonstrict inequality** since the expression is satisfied when $3x + 2y = 4$, as well as when $3x + 2y > 4$. The remaining two linear inequalities are **strict.**

The Graph of a Linear Inequality

Graph linear inequalities 1

The **graph of a linear inequality** in two variables x and y is the set of all points (x, y) for which the inequality is satisfied.

Let's look at an example.

EXAMPLE 1 Graphing a Linear Inequality

Graph the inequality: $2x + 3y \geq 6$

SOLUTION The inequality $2x + 3y \geq 6$ is equivalent to $2x + 3y > 6$ or $2x + 3y = 6$. So we begin by graphing the line L: $2x + 3y = 6$, noting that any point on the line must satisfy the inequality $2x + 3y \geq 6$. See Figure 1(a).

FIGURE 1

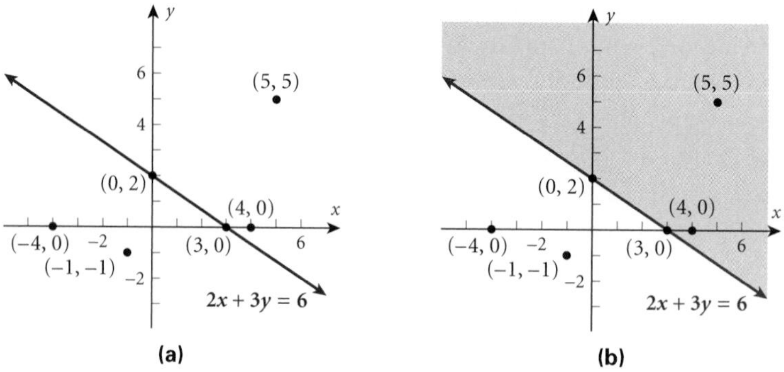

Now let's test a few points, such as $(-1, -1)$, $(5, 5)$, $(4, 0)$, $(-4, 0)$, to see if they satisfy the inequality. We do this by substituting the coordinates of each point into the inequality and determining whether the result is ≥ 6 or < 6.

	$2x \quad + 3y$	*Conclusion*
$(-1, -1)$:	$2(-1) + 3(-1) = -2 - 3 = -5 < 6$	Not part of graph
$(5, 5)$:	$2(5) \quad + 3(5) \quad = 25 > 6$	Part of graph
$(4, 0)$:	$2(4) \quad + 3(0) \quad = 8 > 6$	Part of graph
$(-4, 0)$:	$2(-4) + 3(0) \quad = -8 < 6$	Not part of graph

Notice that the two points $(4, 0)$ and $(5, 5)$ that are part of the graph both lie on one side of L, while the points $(-4, 0)$ and $(-1, -1)$ (not part of the graph) lie on the other side of L. This is not an accident. The graph of the inequality consists of all points on the same side of L as $(4, 0)$ and $(5, 5)$. The shaded region of Figure 1(b) illustrates the graph of the inequality.

The inequality in Example 1 was *nonstrict,* so the *corresponding line was part of the graph of the inequality*. If the inequality is *strict,* the *corresponding line is not part of the graph of the inequality*. We will indicate a strict inequality by using dashes to graph the line.

Let's outline the procedure for graphing a linear inequality:

Steps for Graphing a Linear Inequality

STEP 1 Graph the corresponding linear equation, a line L. If the inequality is nonstrict, graph L using a solid line; if the inequality is strict, graph L using dashes.

STEP 2 Select a test point P not on the line L.

STEP 3 Substitute the coordinates of the test point P into the given inequality. If the coordinates of this point P satisfy the linear inequality, then all points on the same side of L as the point P satisfy the inequality. If the coordinates of the point P do not satisfy the linear inequality, then all points on the opposite side of L from P satisfy the inequality.

EXAMPLE 2 Graphing a Linear Inequality

Graph the linear inequality: $2x - y < -4$

SOLUTION The corresponding linear equation is the line

$$L: \quad 2x - y = -4$$

Since the inequality is strict, points on L are not part of the graph of the linear inequality. When we graph L, we use a dashed line to indicate this fact. See Figure 2(a).

We select a point not on the line L to be tested, for example, (0, 0):

$$2x - y = 2(0) - 0 = 0 \qquad 2x - y < -4$$

Since 0 is not less then -4, the point (0, 0) does not satisfy the inequality. As a result, all points on the opposite side of L from (0, 0) satisfy the inequality. The graph of $2x - y < -4$ is the shaded region of Figure 2(b).

FIGURE 2

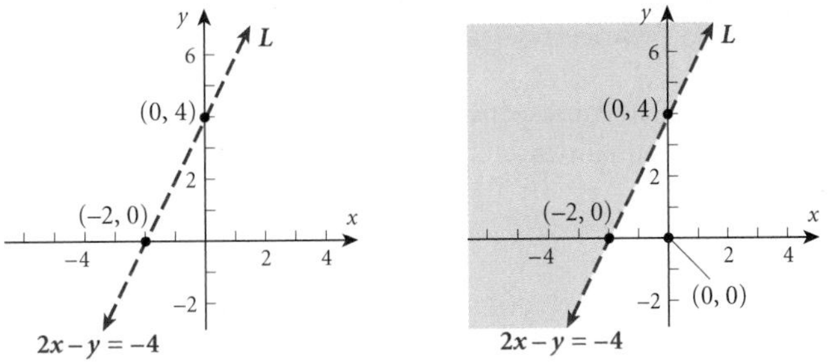

(a) (b)

NOW WORK PROBLEM 7.

EXAMPLE 3 Graphing Linear Inequalities

Graph: **(a)** $x \leq 3$ **(b)** $2x \leq y$

SOLUTION

(a) The corresponding linear equation is $x = 3$, a vertical line. If we choose (0, 0) as the test point, we find that it satisfies the inequality $[0 \leq 3]$, so all points to the left of, and on, the vertical line also satisfy the inequality. See Figure 3(a).

(b) The corresponding linear equation is $2x = y$. Since its graph passes through (0, 0), we choose the point (0, 2) as the test point. The inequality is satisfied by the point $(0, 2)[2(0) \leq 2]$, so all points on the same side of the line as (0, 2) also satisfy the inequality. See Figure 3(b).

FIGURE 3

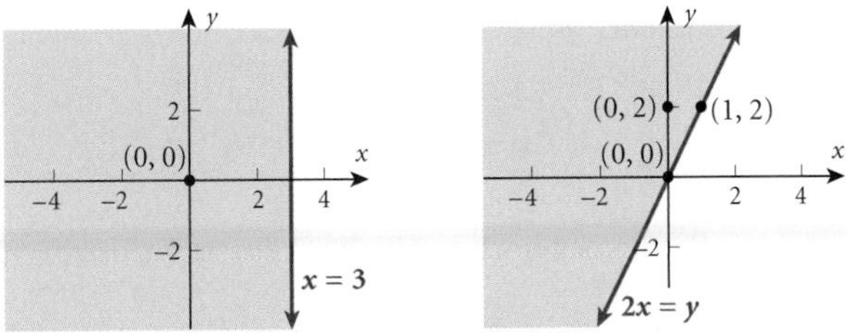

(a) (b)

The set of points belonging to the graph of a linear inequality [for example, the shaded region in Figure 3(b)] is called a **half-plane.**

NOW WORK PROBLEM 5.

COMMENT: A graphing utility can be used to obtain the graph of a linear inequality. See "Using a graphing utility to graph inequalities" in Appendix C, Section C.4, for a discussion.

Systems of Linear Inequalities

Graph systems of linear inequalities 2

A **system of linear inequalities** is a collection of two or more linear inequalities. To **graph** a system of two inequalities we locate all points whose coordinates satisfy each of the linear inequalities of the system.

EXAMPLE 4 Determining Whether a Point Belongs to the Graph of a System of Two Linear Inequalities

Determine which of the following points are part of the graph of the system of linear inequalities:

$$\begin{cases} 2x + y \leq 6 & (1) \\ x - y \geq 3 & (2) \end{cases}$$

(a) $P_1 = (6, 0)$ **(b)** $P_2 = (3, 5)$ **(c)** $P_3 = (0, 0)$ **(d)** $P_4 = (3, -2)$

SOLUTION We check to see if the given point satisfies each of the inequalities of the system.

(a) $P_1 = (6, 0)$

$$\underset{2x+y\leq 6}{2x + y = 2(6) + 0 = 12} \qquad \underset{x-y\geq 3}{x - y = 6 - 0 = 6}$$

P_1 satisfies inequality (2) but not inequality (1), so P_1 is not part of the graph of the system.

(b) $P_2 = (3, 5)$

$$\underset{2x+y\leq 6}{2x + y = 2(3) + 5 = 11} \qquad \underset{x-y\geq 3}{x - y = 3 - 5 = -2}$$

P_2 satisfies neither inequality (1) nor inequality (2), so P_2 is not part of the graph of the system.

(c) $P_3 = (0, 0)$

$$\underset{2x+y\leq 6}{2x + y = 2(0) + 0 = 0} \qquad \underset{x-y\geq 3}{x - y = 0 - 0 = 0}$$

P_3 satisfies inequality (1) but not inequality (2), so P_3 is not part of the graph of the system.

(d) $P_4 = (3, -2)$

$$\underset{2x+y\leq 6}{2x + y = 2(3) + (-2) = 4} \qquad \underset{x-y\geq 3}{x - y = 3 - (-2) = 5}$$

P_4 satisfies both inequality (1) and inequality (2), so P_4 is part of the graph of the system.

NOW WORK PROBLEM 13.

Let's graph the information from Example 4. Figure 4(a) shows the graph of each of the lines $2x + y = 6$ and $x - y = 3$ and the four points P_1, P_2, P_3, and P_4. Notice that because the two lines of the system intersect, the plane is divided into four regions. Since the graph of each linear inequality of the system is a half-plane, the graph of the system of linear inequalities is the intersection of these two half-planes. As a result, the region containing P_4 is the graph of the system. See Figure 4(b).

We use the method described above in the next example.

FIGURE 4

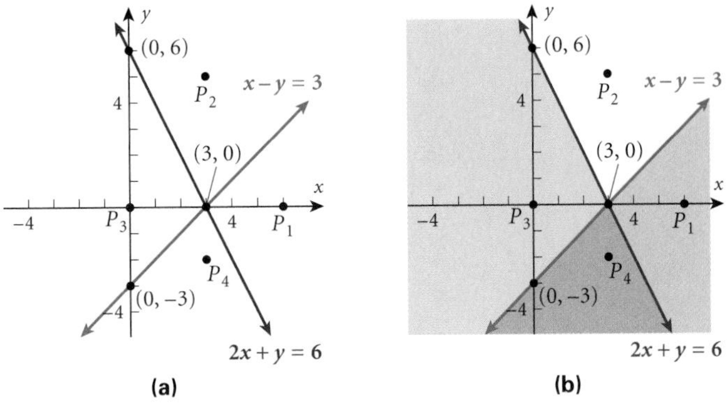

EXAMPLE 5 Graphing a System of Two Linear Inequalities

Graph the system: $\begin{cases} 2x - y \leq -4 \\ x + y \geq -1 \end{cases}$

SOLUTION First we graph each inequality separately. See Figures 5(a) and 5(b).

The solution of the system consists of all points common to these two half-planes. The dark blue shaded region in Figure 6 represents the solution of the system.

FIGURE 5

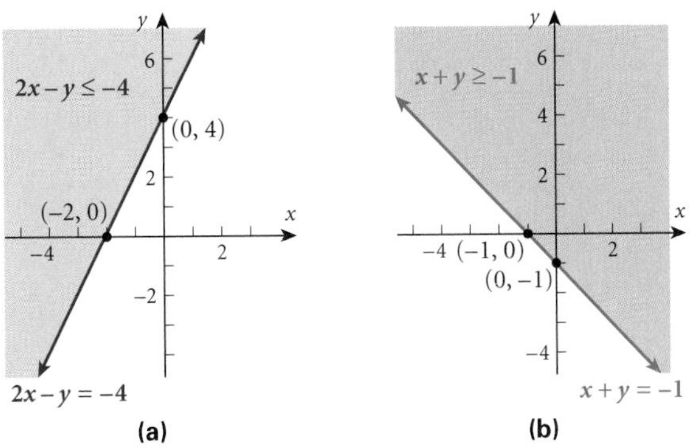

FIGURE 6

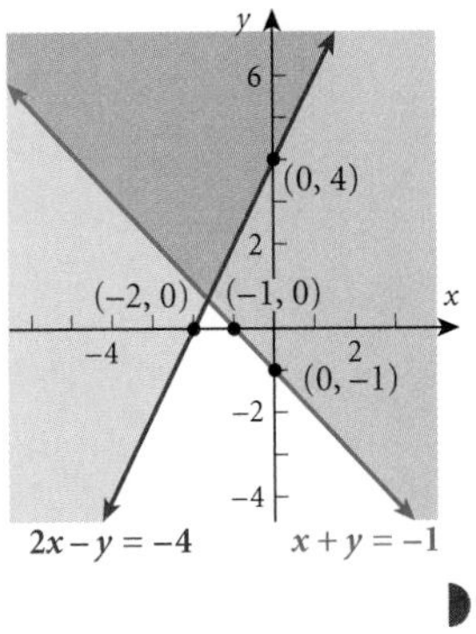

NOW WORK PROBLEM 17.

The lines in the system of linear inequalities given in Example 5 intersect. If the two lines of a system of two linear inequalities are parallel, the system of linear inequalities may or may not have a solution. Examples of such situations follow.

EXAMPLE 6 Graphing a System of Two Linear Inequalities

Graph the system: $\begin{cases} 2x - y \leq -4 \\ 2x - y \leq -2 \end{cases}$

SOLUTION First we graph each inequality separately. See Figures 7(a) and 7(b). The grey shaded region in Figure 8 represents the solution of the system.

FIGURE 7

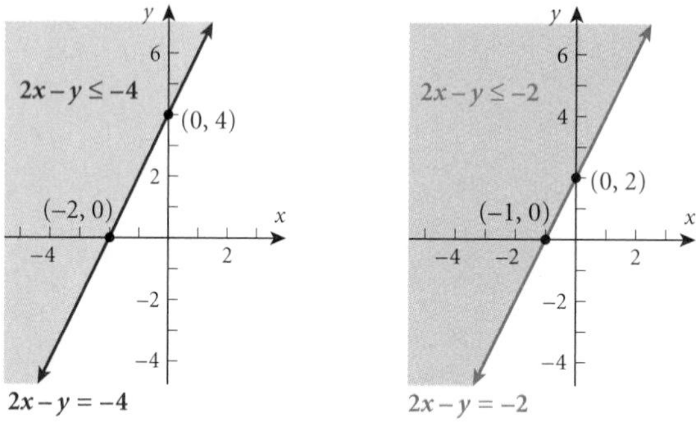

(a) **(b)**

FIGURE 8

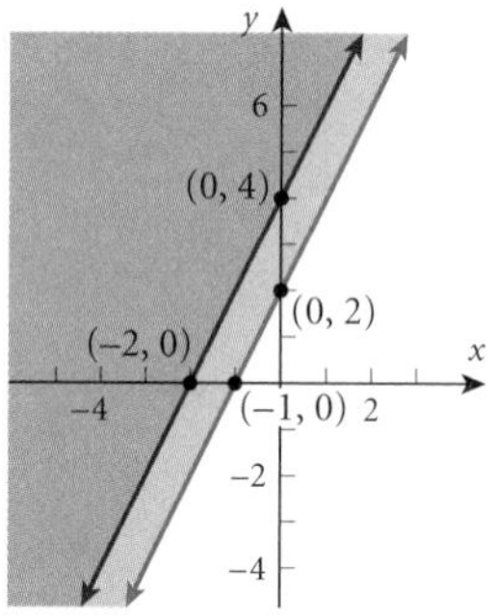

Notice that the solution of this system is the same as that of the single linear inequality $2x - y \leq -4$.

EXAMPLE 7 Graphing a System of Two Linear Inequalities

The solution of the system

$$\begin{cases} 2x - y \geq -4 \\ 2x - y \leq -2 \end{cases}$$

is the grey shaded region in Figure 9.

FIGURE 9

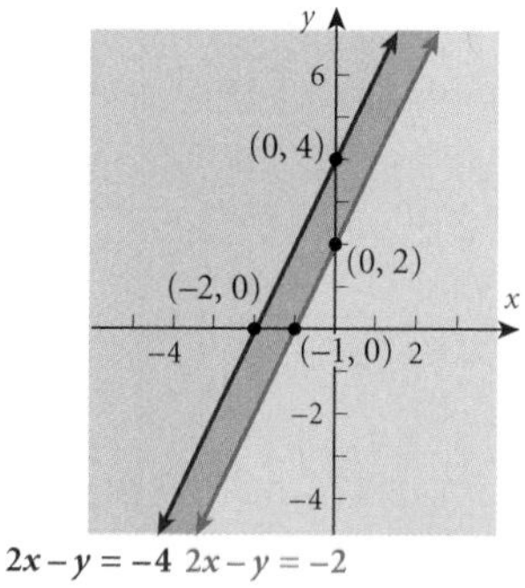

EXAMPLE 8 Graphing a System of Two Linear Inequalities

The system

$$\begin{cases} 2x - y \leq -4 \\ 2x - y \geq -2 \end{cases}$$

has no solution, as Figure 10 indicates, because the two half-planes have no points in common.

FIGURE 10

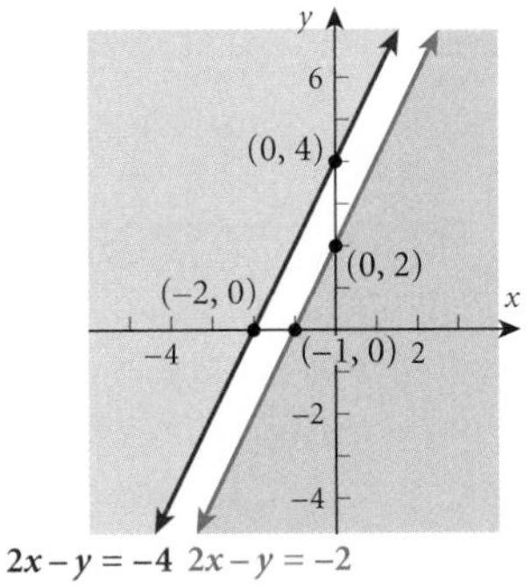

Until now, we have considered systems of only two linear inequalities. The next example is of a system of four linear inequalities. As we shall see, the technique for graphing such systems is the same as that used for graphing systems of two linear inequalities in two variables.

EXAMPLE 9 Graphing a System of Four Linear Inequalities

Graph the system: $\begin{cases} x + y \geq 2 \\ 2x + y \geq 3 \\ x \geq 0 \\ y \geq 0 \end{cases}$

SOLUTION Again we first graph the four lines:

$$L_1: \quad x + y = 2$$
$$L_2: \quad 2x + y = 3$$
$$L_3: \quad x = 0 \quad \text{(the } y\text{-axis)}$$
$$L_4: \quad y = 0 \quad \text{(the } x\text{-axis)}$$

The lines L_1 and L_2 intersect at the point $(1, 1)$. (Do you see why?) The inequalities $x \geq 0$ and $y \geq 0$ indicate that the graph of the system lies in quadrant I. The graph of the system consists of that part of the graphs of the inequalities $x + y \geq 2$ and $2x + y \geq 3$ that lies in quadrant I. See Figure 11.

FIGURE 11

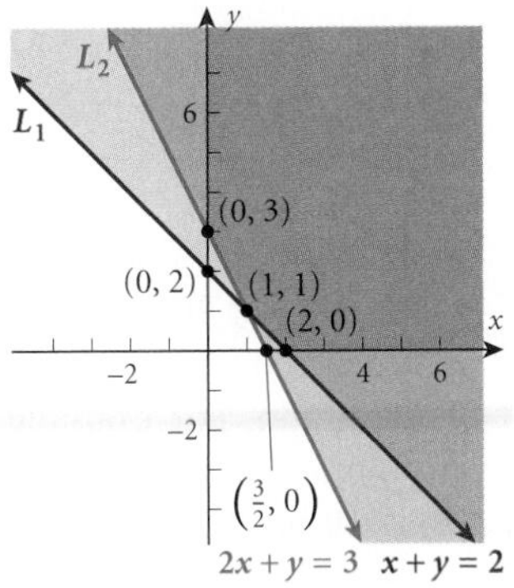

EXAMPLE 10   **Graphing a System of Four Linear Inequalities**

Graph the system:
$$\begin{cases} x + y \le 2 \\ 2x + y \le 3 \\ x \ge 0 \\ y \ge 0 \end{cases}$$

SOLUTION   The lines associated with these linear inequalities are the same as those of the previous example. Again the inequalities $x \ge 0$, $y \ge 0$ indicate that the graph of the system lies in quadrant I. The graph of the system consists of that part of the graphs of the inequalities $x + y \le 2$ and $2x + y \le 3$ that lie in quadrant I. See Figure 12.

FIGURE 12

$2x + y = 3 \quad x + y = 2$

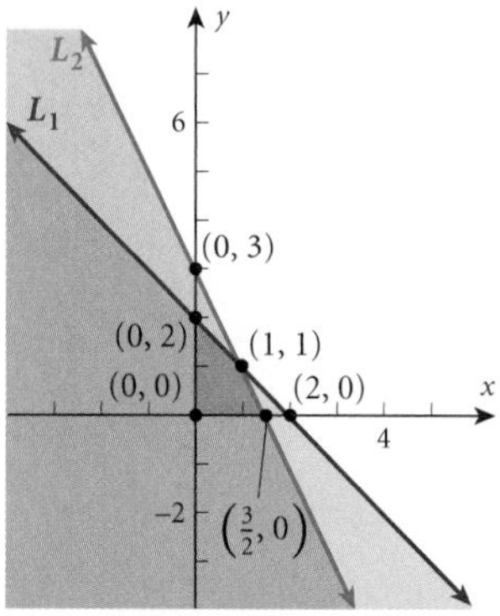

Some Terminology

Compare the graphs of the systems of linear inequalities given in Figures 11 and 12. The graph in Figure 11 is said to be **unbounded** in the sense that it extends infinitely far in some direction. The graph in Figure 12 is **bounded** in the sense that it can be enclosed by some circle of sufficiently large radius. See Figure 13.

FIGURE 13

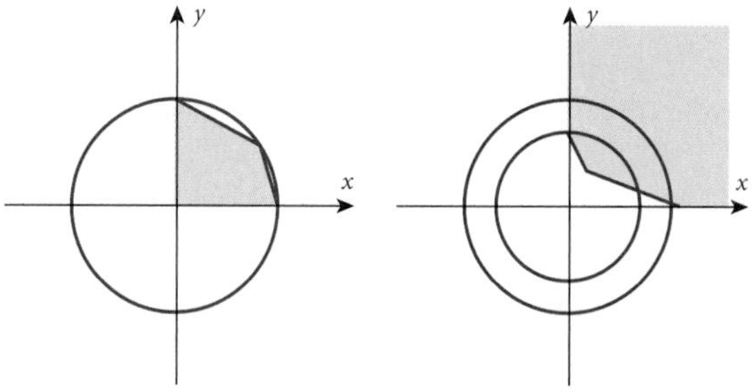

(a) Bounded graph   **(b)** Unbounded graph

The boundary of each of the graphs in Figures 11 and 12 consists of line segments. In fact, the graph of any system of linear inequalities will have line segments as boundaries. The point of intersection of two line segments that form the boundary is called a **corner point** of the graph. For example, the graph of the system given in Example 9 has the corner points $(0, 3)$, $(1, 1)$, and $(2, 0)$. See Figure 11. The graph of the system given in Example 10 has the corner points $(0, 2)$, $(0, 0)$, $\left(\frac{3}{2}, 0\right)$, $(1, 1)$. See Figure 12.

We shall soon see that the corner points of the graph of a system of linear inequalities play a major role in the procedure for solving linear programming problems.

NOW WORK PROBLEMS 25 AND 29.

Application

EXAMPLE 11 Analyzing a Mixture Problem

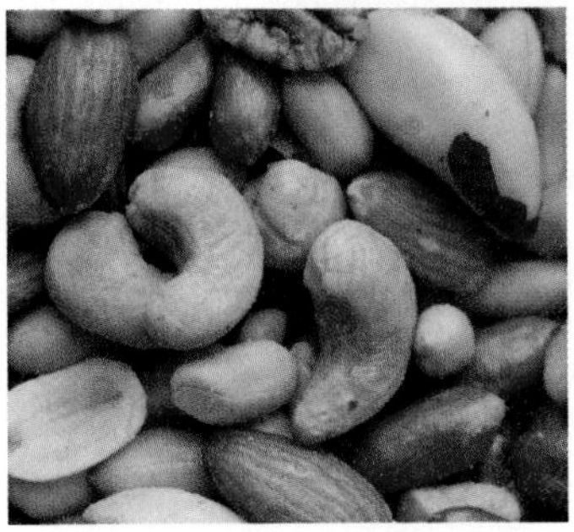

Nutt's Nuts has 75 pounds of cashews and 120 pounds of peanuts. These are to be mixed in 1-pound packages as follows: a low-grade mixture that contains 4 ounces of cashews and 12 ounces of peanuts and a high-grade mixture that contains 8 ounces of cashews and 8 ounces of peanuts.

(a) Use x to denote the number of packages of the low-grade mixture and use y to denote the number of packages of the high-grade mixture to be made and write a system of linear inequalities that describes the possible number of each kind of package.

(b) Graph the system and list its corner points.

SOLUTION **(a)** We begin by naming the variables:

$$x = \text{Number of packages of low-grade mixture}$$
$$y = \text{Number of packages of high-grade mixture}$$

First, we note that the only meaningful values for x and y are nonnegative values. We restrict x and y so that

$$x \geq 0 \qquad \text{and} \qquad y \geq 0$$

Next, we note that there is a limit to the number of pounds of cashews and peanuts available. That is, the total number of pounds of cashews cannot exceed 75 pounds (1200 ounces), and the number of pounds of peanuts cannot exceed 120 pounds (1920 ounces). This means that

$$\left(\begin{array}{c}\text{Ounces of}\\ \text{cashews}\\ \text{required}\\ \text{for low-grade}\\ \text{mixture}\end{array}\right)\left(\begin{array}{c}\text{Number of}\\ \text{packages of}\\ \text{low-grade}\\ \text{mixture}\end{array}\right)+\left(\begin{array}{c}\text{Ounces of}\\ \text{cashews}\\ \text{required for}\\ \text{high-grade}\\ \text{mixture}\end{array}\right)\left(\begin{array}{c}\text{Number of}\\ \text{packages}\\ \text{of high-}\\ \text{grade}\\ \text{mixture}\end{array}\right)\begin{array}{c}\text{cannot}\\ \text{exceed}\end{array} 1200$$

$$\left(\begin{array}{c}\text{Ounces of}\\ \text{peanuts}\\ \text{required}\\ \text{for low-grade}\\ \text{mixture}\end{array}\right)\left(\begin{array}{c}\text{Number of}\\ \text{packages of}\\ \text{low-grade}\\ \text{mixture}\end{array}\right)+\left(\begin{array}{c}\text{Ounces of}\\ \text{peanuts}\\ \text{for high-}\\ \text{grade}\\ \text{mixture}\end{array}\right)\left(\begin{array}{c}\text{Number of}\\ \text{packages}\\ \text{of high-}\\ \text{grade}\\ \text{mixture}\end{array}\right)\begin{array}{c}\text{cannot}\\ \text{exceed}\end{array} 1920$$

In terms of the data given and the variables introduced, we can write these statements compactly as

$$4x + 8y \leq 1200$$
$$12x + 8y \leq 1920$$

The system of linear inequalities that gives the possible values x and y can take on is

$$\begin{cases} 4x + 8y \leq 1200 \\ 12x + 8y \leq 1920 \\ x \geq 0 \\ y \geq 0 \end{cases}$$

(b) The system of linear inequalities given above can be simplified to the equivalent form

$$\begin{cases} x + 2y \leq 300 & (1) \\ 3x + 2y \leq 480 & (2) \\ x \geq 0 & (3) \\ y \geq 0 & (4) \end{cases}$$

The graph of the system is given in Figure 14. Notice that the corner points of the graph are labeled. Three are easy to identify by inspection: (0, 0), (0, 150), and (160, 0). We found the remaining one (90, 105) by solving the system of equations

$$\begin{cases} x + 2y = 300 & (1) \\ 3x + 2y = 480 & (2) \end{cases}$$

By subtracting the first equation from the second, we find $2x = 180$ or $x = 90$. Back-substituting in the first equation, we find $y = 105$.

FIGURE 14

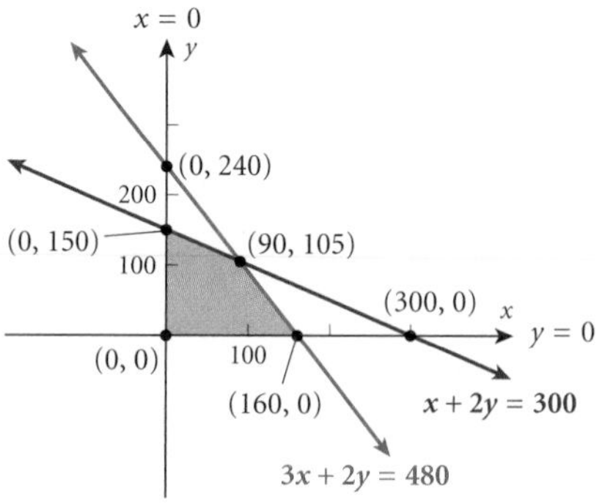

EXERCISE 3.1 Answers to Odd-Numbered Problems Begin on Page AN-17.

In Problems 1–12, graph each inequality.

1. $x \geq 0$ **2.** $y \geq 0$ **3.** $x < 4$ **4.** $y \leq 6$

5. $y \geq 1$ **6.** $x > 2$ 7. $2x + 3y \leq 6$ **8.** $3x + 2y \geq 6$

9. $5x + y \geq 10$ **10.** $x + 2y > 4$ **11.** $x + 5y < 5$ **12.** $3x + y \leq 3$

13. Without graphing, determine which of the points $P_1 = (3, 8)$, $P_2 = (12, 9)$, $P_3 = (5, 1)$ are part of the graph of the following system:

$$\begin{cases} x + 3y \geq 0 \\ -3x + 2y \geq 0 \end{cases}$$

14. Without graphing, determine which of the points $P_1 = (9, -5)$, $P_2 = (12, -4)$, $P_3 = (4, 1)$ are part of the graph of the following system:

$$\begin{cases} x + 4y \leq 0 \\ 5x + 2y \geq 0 \end{cases}$$

15. Without graphing, determine which of the points $P_1 = (2, 3)$, $P_2 = (10, 10)$, $P_3 = (5, 1)$ are part of the graph of the following system:

$$\begin{cases} 3x + 2y \geq 0 \\ x + y \leq 15 \end{cases}$$

16. Without graphing, determine which of the points $P_1 = (2, 6)$, $P_2 = (12, 4)$, $P_3 = (4, 2)$ are part of the graph of the following system:

$$\begin{cases} 2x - 5y \leq 0 \\ x + 3y \leq 15 \end{cases}$$

In Problems 17–24, determine which region a, b, c, or d represents the graph of the given system of linear inequalities. The regions a, b, c, and d are nonoverlapping regions bounded by the indicated lines.

17. $\begin{cases} 5x - 4y \leq 8 \\ 2x + 5y \leq 23 \end{cases}$

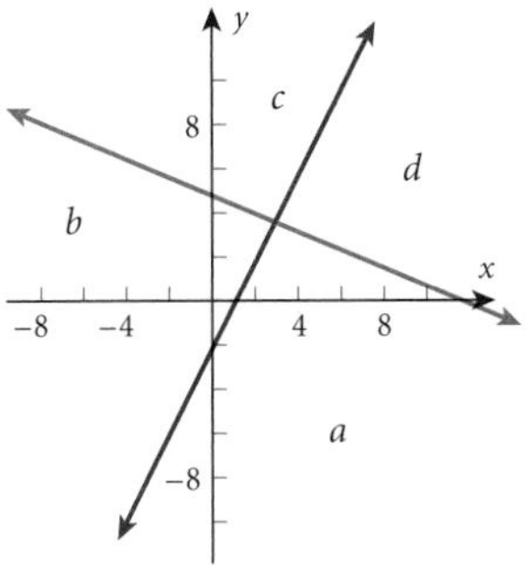

18. $\begin{cases} 4x - 5y \leq 0 \\ 4x + 2y \leq 28 \end{cases}$

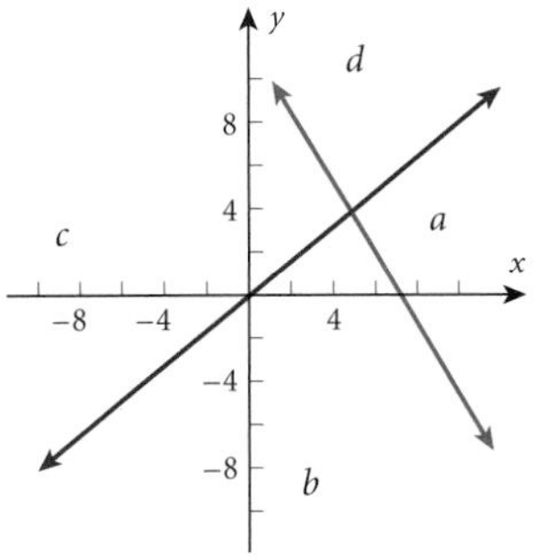

19. $\begin{cases} 2x - 3y \geq -3 \\ 2x + 3y \leq 16 \end{cases}$

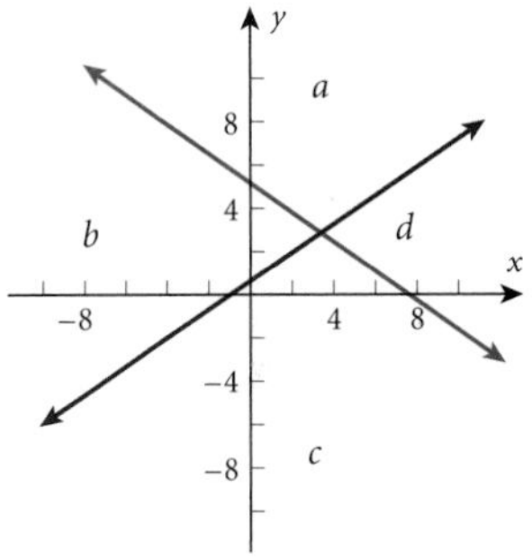

20. $\begin{cases} 6x - 5y \leq 5 \\ 2x + 4y \geq 30 \end{cases}$

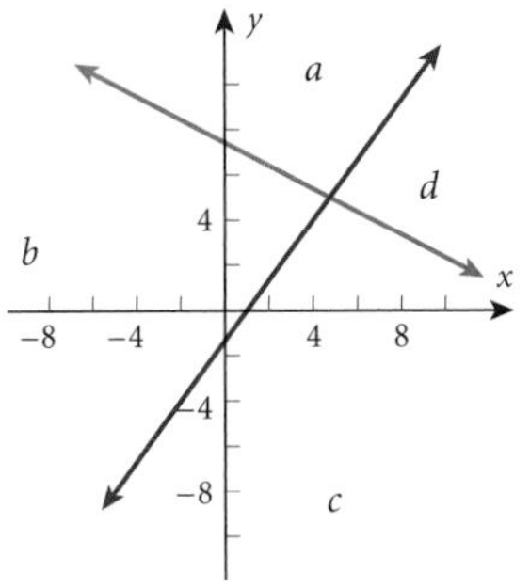

21. $\begin{cases} 5x - 3y \geq 3 \\ 2x + 6y \geq 30 \end{cases}$

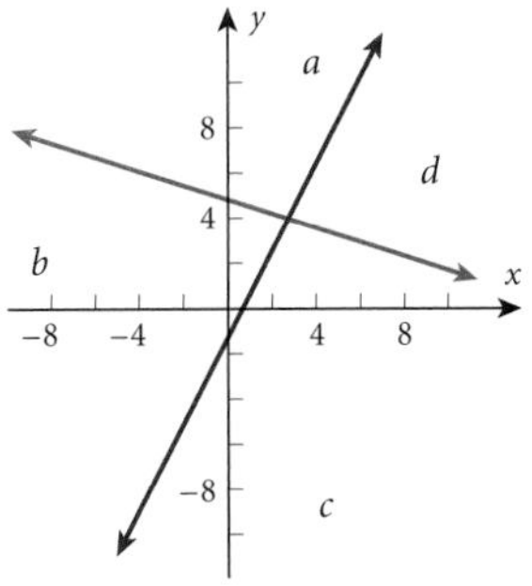

22. $\begin{cases} 5x - 5y \geq 10 \\ 6x + 4y \geq 48 \end{cases}$

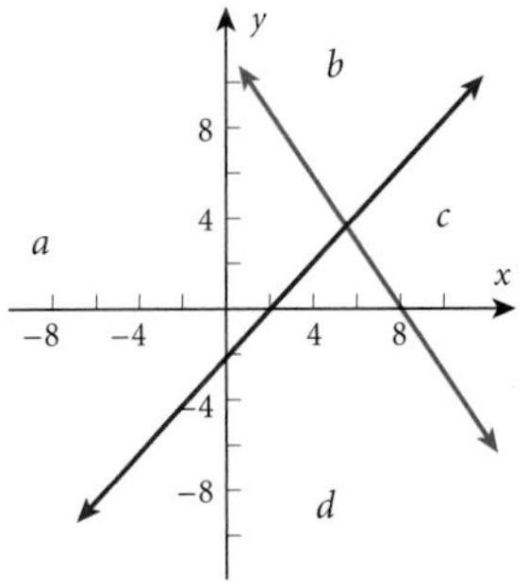

23. $\begin{cases} 5x - 4y \leq 0 \\ 2x + 4y \leq 28 \end{cases}$

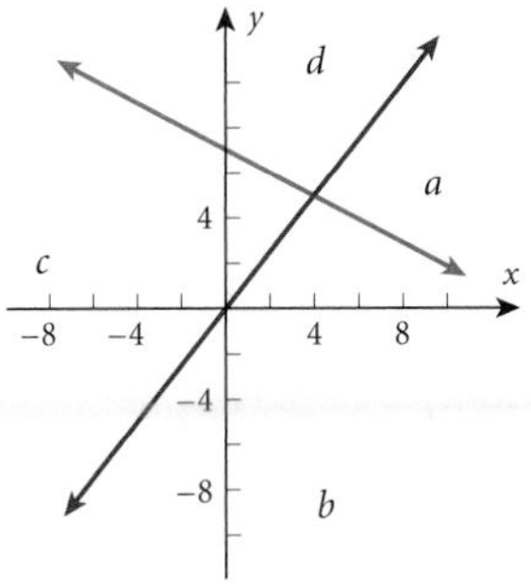

24. $\begin{cases} 2x - 5y \leq -5 \\ 3x + 5y \leq 30 \end{cases}$

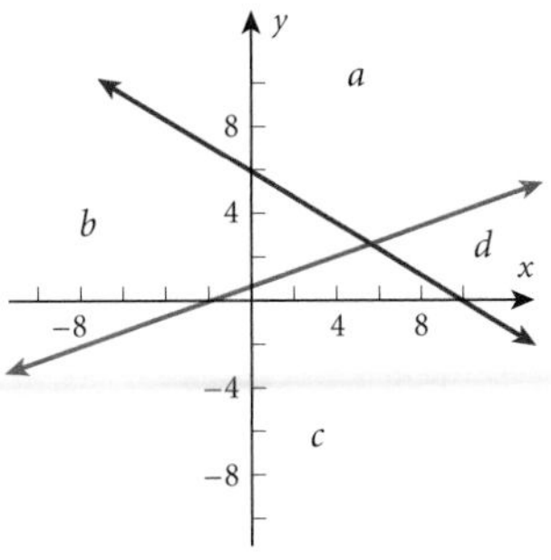

In Problems 25–36, graph each system of linear inequalities. Tell whether the graph is bounded or unbounded and list each corner point of the graph.

25. $\begin{cases} x + y \geq 2 \\ x \geq 0 \\ y \geq 0 \end{cases}$

26. $\begin{cases} 2x + 3y \leq 6 \\ x \geq 0 \\ y \geq 0 \end{cases}$

27. $\begin{cases} x + y \geq 2 \\ 2x + 3y \leq 6 \\ x \geq 0 \\ y \geq 0 \end{cases}$

28. $\begin{cases} x + y \geq 2 \\ 2x + 3y \leq 12 \\ 3x + 2y \leq 12 \\ x \geq 0 \\ y \geq 0 \end{cases}$

29. $\begin{cases} x + y \geq 2 \\ x + y \leq 8 \\ 2x + y \leq 10 \\ x \geq 0 \\ y \geq 0 \end{cases}$

30. $\begin{cases} x + y \geq 2 \\ x + y \leq 8 \\ x + 2y \geq 1 \\ x \geq 0 \\ y \geq 0 \end{cases}$

31. $\begin{cases} x + y \geq 2 \\ 2x + 3y \leq 12 \\ 3x + y \leq 12 \\ x \geq 0 \\ y \geq 0 \end{cases}$

32. $\begin{cases} x + y \geq 2 \\ 2x + y \geq 3 \\ x \geq 0 \\ y \geq 0 \end{cases}$

33. $\begin{cases} x + 2y \geq 1 \\ y \leq 4 \\ x \geq 0 \\ y \geq 0 \end{cases}$

34. $\begin{cases} x + 2y \geq 1 \\ x + 2y \leq 10 \\ x + y \geq 2 \\ x + y \leq 8 \\ x \geq 0 \\ y \geq 0 \end{cases}$

35. $\begin{cases} x + 2y \geq 2 \\ x + y \leq 4 \\ 3x + y \leq 3 \\ x \geq 0 \\ y \geq 0 \end{cases}$

36. $\begin{cases} 2x + y \geq 2 \\ 3x + 2y \leq 6 \\ x + y \geq 2 \\ x \geq 0 \\ y \geq 0 \end{cases}$

---

37. Rework Example 11 if 60 pounds of cashews and 90 pounds of peanuts are available.

38. Rework Example 11 if the high-grade mixture contains 10 ounces of cashews and 6 ounces of peanuts.

39. **Manufacturing** Mike's Famous Toy Trucks company manufactures two kinds of toy trucks—a dumpster and a tanker. In the manufacturing process, each dumpster requires 3 hours of grinding and 4 hours of finishing, while each tanker requires 2 hours of grinding and 3 hours of finishing. The company has two grinders and three finishers, each of whom works at most 40 hours per week.

(a) Using x to denote the number of dumpsters and y to denote the number of tankers, write a system of linear inequalities that describes the possible numbers of each truck that can be manufactured.
(b) Graph the system and list its corner points.

40. Manufacturing Repeat Problem 39 if one grinder and two finishers, each of whom works at most 40 hours per week, are available.

41. Financial Planning A retired couple have up to \$25,000 to invest. As their financial adviser, you recommend they place at least \$15,000 in Treasury Bills yielding 6% and at most \$10,000 in corporate bonds yielding 9%.

(a) Using x to denote the amount of money invested in Treasury Bills and y to denote the amount invested in corporate bonds, write a system of inequalities that describes this situation.
(b) Graph the system and list its corner points.
(c) Interpret the meaning of each corner point in relation to the investments it represents.

42. Financial Planning Use the information supplied in Problem 41, along with the fact that the couple will invest at least \$20,000, to answer parts (a), (b), and (c).

43. Nutrition A farmer prepares feed for livestock by combining two types of grain. Each unit of the first grain contains 1 unit of protein and 5 units of iron while each unit of the second grain contains 2 units of protein and 1 unit of iron. Each animal must receive at least 5 units of protein and 16 units of iron each day.

(a) Write a system of linear inequalities that describes the possible amounts of each grain the farmer needs to prepare.
(b) Graph the system and list the corner points.

44. Investment Strategy Kathleen wishes to invest up to a total of \$40,000 in class AA bonds and stocks. Furthermore, she believes that the amount invested in class AA bonds should be at most one-third of the amount invested in stocks.

(a) Write a system of linear inequalities that describes the possible amount of investments in each security.
(b) Graph the system and list the corner points.

45. Nutrition To maintain an adequate daily diet, nutritionists recommend the following: at least 85 g of carbohydrate, 70 g of fat, and 50 g of protein. An ounce of food A contains 5 g of carbohydrate, 3 g of fat, and 2 g of protein, while an ounce of food B contains 4 g of carbohydrate, 3 g of fat, and 3 g of protein.

(a) Write a system of linear inequalities that describes the possible quantities of each food.
(b) Graph the system and list the corner points.

46. Transportation A microwave company has two plants, one on the East Coast and one in the Midwest. It takes 25 hours (packing, transportation, and so on) to transport an order of microwaves from the eastern plant to its central warehouse and 20 hours from the Midwest plant to its central warehouse. It costs \$80 to transport an order from the eastern plant to the central warehouse and \$40 from the midwestern plant to its central warehouse. There are 1000 work-hours available for packing, transportation, and so on, and \$3000 for transportation cost.

(a) Write a system of linear inequalities that describes the transportation system.
(b) Graph the system and list the corner points.

Q 47. Make up a system of linear inequalities that has no solution.

Q 48. Make up a system of linear inequalities that has a single point as solution.

3.2 A Geometric Approach to Linear Programming Problems

OBJECTIVES
1 Identify a linear programming problem
2 Solve a linear programming problem

Identify a linear programming problem 1

To help see the characteristics of a linear programming problem, we look again at Example 11 of the previous section.

Nutt's Nuts has 75 pounds of cashews and 120 pounds of peanuts. These are to be mixed in 1-pound packages as follows: a low-grade mixture that contains 4 ounces of cashews and 12 ounces of peanuts and a high-grade mixture that contains 8 ounces of cashews and 8 ounces of peanuts.

Suppose that in addition to the information given above, we also know what the profit will be on each type of mixture. For example, suppose the profit is \$0.25 on each package of the low-grade mixture and is \$0.45 on each package of the high-grade mixture. The question of importance to the manager is "How many packages of each type of mixture should be prepared to maximize the profit?"

If P symbolizes the profit, x the number of packages of low-grade mixture, and y the number of high-grade packages, then the question can be restated as "What are the values of x and y so that the expression

$$P = \$0.25x + \$0.45y$$

is a maximum?"

This problem is typical of a **linear programming problem.** It requires that a certain linear expression, in this case the profit, be maximized. This linear expression is called the **objective function.** Furthermore, the problem requires that the maximum profit be achieved under certain restrictions or **constraints,** each of which are linear inequalities involving the variables. The linear programming problem may be restated as

Maximize

$$P = \$0.25x + \$0.45y \quad \text{Objective function}$$

subject to the conditions that

$$\begin{cases} x + 2y \leq 300 & \text{Cashew constraint} \\ 3x + 2y \leq 480 & \text{Peanut constraint} \\ x \geq 0 & \text{Nonnegativity constraint} \\ y \geq 0 & \text{Nonnegativity constraint} \end{cases}$$

In general, every linear programming problem has two components:

1. A linear objective function to be maximized or minimized.
2. A collection of linear inequalities that must be satisfied simultaneously.

Linear Programming Problem

A **linear programming problem** in two variables, x and y, consists of maximizing or minimizing an **objective function**

$$\boxed{z = Ax + By}$$

where A and B are given real numbers, not both zero, subject to certain conditions or **constraints** expressible as a system of linear inequalities in x and y.

Let's look at this definition more closely. To maximize (or minimize) the quantity $z = Ax + By$ means to locate the points (x, y) that result in the largest (or smallest) value of z. But not all points (x, y) are eligible. Only the points that obey *all* the constraints are potential solutions. We refer to such points as **feasible points.**

Solve a linear programming problem 2

In a linear programming problem we want to find the feasible point that maximizes (or minimizes) the objective function.

By a **solution to a linear programming problem** we mean a feasible point (x, y), together with the value of the objective function at that point, which maximizes (or minimizes) the objective function. If none of the feasible points maximizes (or minimizes) the objective function, or if there are no feasible points, then the linear programming problem has no solution.

EXAMPLE 1 Solving a Linear Programming Problem

Minimize the quantity

$$z = x + 2y$$

subject to the constraints

$$\begin{cases} x + y \geq 1 \\ x \geq 0 \\ y \geq 0 \end{cases}$$

SOLUTION The objective function to be minimized is $z = x + 2y$. The constraints are the linear inequalities

$$\begin{cases} x + y \geq 1 \\ x \geq 0 \\ y \geq 0 \end{cases}$$

FIGURE 15

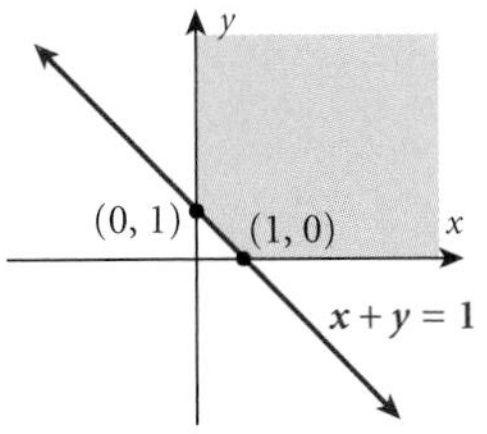

The shaded portion of Figure 15 illustrates the set of feasible points.

To see if there is a smallest z, we graph $z = x + 2y$ for some choice of z, say, $z = 3$. See Figure 16. By moving the line $x + 2y = 3$ parallel to itself, we can observe what happens for different values of z. Since we want a minimum value for z, we try to move $z = x + 2y$ down as far as possible while keeping some part of the line within the set of feasible points. The "best" solution is obtained when the line just touches a corner point of the set of feasible points. If you refer to Figure 16, you will see that the best solution is $x = 1, y = 0$, which yields $z = 1$. There is no other feasible point for which z is smaller.

FIGURE 16

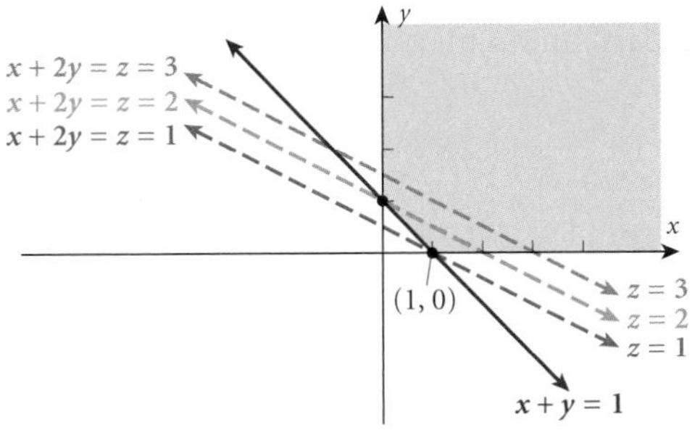

The next example illustrates a linear programming problem that has no solution.

EXAMPLE 2 A Linear Programming Problem without a Solution

Maximize the quantity

$$z = x + 2y$$

subject to the constraints

$$\begin{cases} x + y \geq 1 \\ x \geq 0 \\ y \geq 0 \end{cases}$$

SOLUTION First, we graph the constraints. The shaded portion of Figure 17 illustrates the set of feasible points.

The graphs of the objective function $z = x + 2y$ for $z = 2$, $z = 8$, and $z = 12$ are also shown in Figure 17. Observe that we continue to get larger values for z by moving the graph of the objective function upward. But there is no feasible point that will make z *largest*. No matter how large a value is assigned to z, there is a feasible point that

FIGURE 17

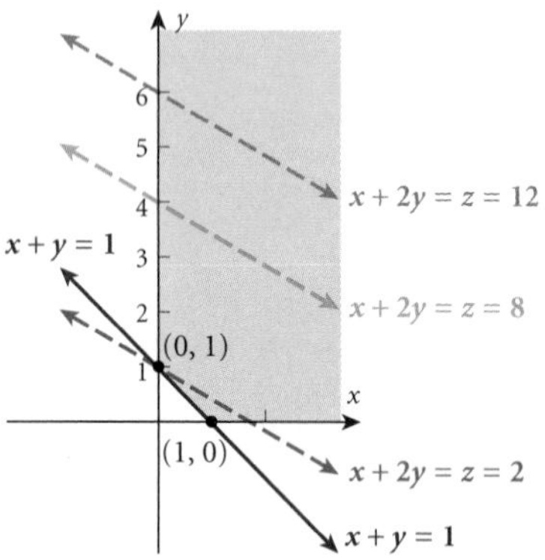

will give a larger value. Since there is no feasible point that makes z largest, we conclude that this linear programming problem has no solution.

Examples 1 and 2 demonstrate that sometimes a linear programming problem has a solution and sometimes it does not. The next result gives conditions on the set of feasible points that determine when a solution to a linear programming problem exists.

Existence of a Solution

Consider a linear programming problem with the set R of feasible points and objective function $z = Ax + By$.

1. If R is bounded, then z has both a maximum and a minimum value on R.
2. If R is unbounded and $A \geq 0$, $B \geq 0$, and the constraints include $x \geq 0$ and $y \geq 0$, then z has a minimum value on R but not a maximum value (see Example 2).
3. If R is the empty set, then the linear programming problem has no solution and z has neither a maximum nor a minimum value.

In Example 1 we found that the feasible point that minimizes z occurs at a corner point. This is not an unusual situation. If there are feasible points minimizing (or maximizing) the objective function, at least one will be at a corner point of the set of feasible points.

Fundamental Theorem of Linear Programming

If a linear programming problem has a solution, it is located at a corner point of the set of feasible points; if a linear programming problem has multiple solutions, at least one of them is located at a corner point of the set of feasible points. In either case the corresponding value of the objective function is unique.

The result just stated indicates that it is possible for a feasible point that is not a corner point to minimize (or maximize) the objective function. For example, if the slope of the objective function is the same as the slope of one of the boundaries of the set of feasible points and if the two adjacent corner points are solutions, then so are all the points on the line segment joining them. The following example illustrates this situation.

EXAMPLE 3 A Linear Programming Problem with Multiple Solutions

Minimize the quantity

$$z = x + 2y$$

subject to the constraints

$$\begin{cases} x + y \geq 1 \\ 2x + 4y \geq 3 \\ x \geq 0 \\ y \geq 0 \end{cases}$$

SOLUTION Again we first graph the constraints. The shaded portion of Figure 18 illustrates the set of feasible points.

FIGURE 18

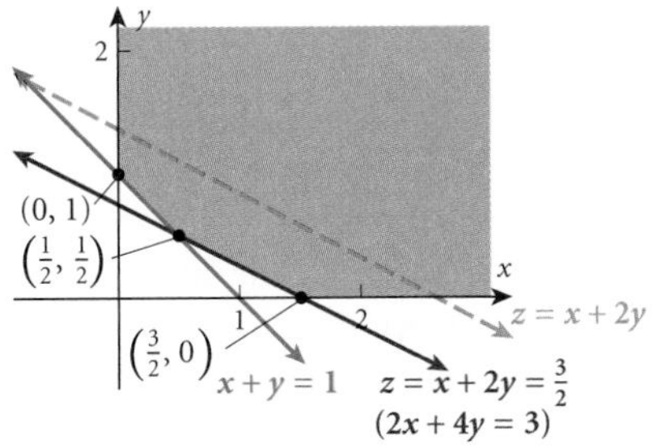

If we graph the objective equation $z = x + 2y$ for some choice of z and move it down, we find that a minimum is reached when $z = \frac{3}{2}$. In fact, any point on the line $2x + 4y = 3$ between the adjacent corner points $(\frac{1}{2}, \frac{1}{2})$ and $(\frac{3}{2}, 0)$ and including these corner points will minimize the objective function. Of course, the reason any feasible point on $2x + 4y = 3$ minimizes the objective equation $z = x + 2y$ is that these two lines each have slope $-\frac{1}{2}$. This linear programming problem has infinitely many solutions.

NOW WORK PROBLEM 1.

Since the objective function attains its maximum or minimum value at the corner points of the set of feasible points, we can outline a procedure for solving a linear programming problem provided that it has a solution.

Steps for Solving a Linear Programming Problem

If a linear programming problem has a solution, follow these steps to find it:

STEP 1 Write an expression for the quantity that is to be maximized or minimized (the objective function).
STEP 2 Determine all the constraints and graph the set of feasible points.
STEP 3 List the corner points of the set of feasible points.
STEP 4 Determine the value of the objective function at each corner point.
STEP 5 Select the maximum or minimum value of the objective function.

Let's look at some examples.

EXAMPLE 4 Solving a Linear Programming Problem

Maximize and minimize the objective function

$$z = x + 5y$$

subject to the constraints

$$\begin{cases} x + 4y \leq 12 & (1) \\ x \leq 8 & (2) \\ x + y \geq 2 & (3) \\ x \geq 0 & (4) \\ y \geq 0 & (5) \end{cases}$$

SOLUTION The objective function is $z = x + 5y$ and the constraints consist of a system of five linear inequalities. We proceed to graph the system of five linear inequalities. The shaded portion of Figure 19 illustrates the graph, the set of feasible points. Since this set is bounded, we know a solution to the linear programming problem exists. Notice in Figure 19 that we have labeled each line from the system of linear inequalities. We have also labeled the corner points.

FIGURE 19

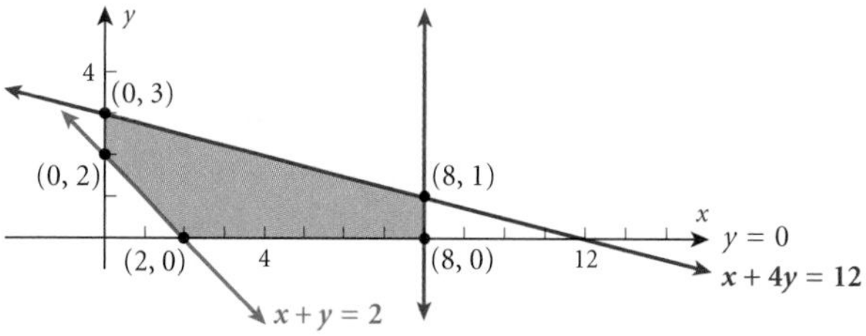

The corner points of the set of feasible points are

$$(0,\ 3) \qquad (8,\ 1) \qquad (8,\ 0) \qquad (2,\ 0) \qquad (0,\ 2)$$

To find the maximum and minimum value of the objective function $z = x + 5y$, we construct Table 1:

TABLE 1

Corner Point (x, y)	Value of Objective Function $z = x + 5y$
(0, 3)	$z = 0 + 5(3) = 15$
(8, 1)	$z = 8 + 5(1) = 13$
(8, 0)	$z = 8 + 5(0) = 8$
(2, 0)	$z = 2 + 5(0) = 2$
(0, 2)	$z = 0 + 5(2) = 10$

The maximum value of z is 15, and it occurs at the point (0, 3). The minimum value of z is 2, and it occurs at the point (2, 0).

NOW WORK PROBLEMS 17 AND 29.

Now let's solve the problem of the cashews and peanuts that we discussed at the start of this section.

EXAMPLE 5 Maximizing Profit

Maximize

$$P = 0.25x + 0.45y$$

subject to the constraints

$$\begin{cases} x + 2y \le 300 & (1) \\ 3x + 2y \le 480 & (2) \\ x \ge 0 & (3) \\ y \ge 0 & (4) \end{cases}$$

SOLUTION Before applying the method of this chapter to solve this problem, let's discuss a solution that might be suggested by intuition. Namely, since the profit is higher for the high-grade mixture, you might think that Nutt's Nuts should prepare as many packages of the high-grade mixture as possible. If this were done, then there would be a total of 150 packages (8 ounces divides into 75 pounds of cashews exactly 150 times) and the total profit would be

$$150(0.45) = \$67.50$$

As we shall see, this is not the best solution to the problem.

To obtain the maximum profit, we use linear programming. We reproduce here in Figure 20 the graph of the set of feasible points we obtained earlier (see Figure 14, page 168)

FIGURE 20

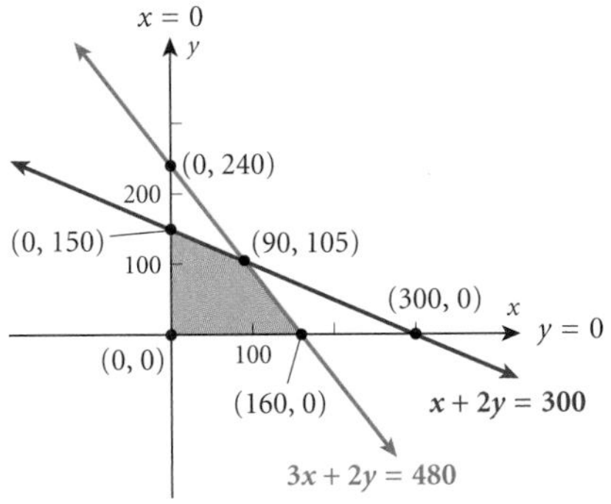

Notice that this set is bounded. The corner points of the set of feasible points are

$$(0,\ 0) \qquad (0,\ 150) \qquad (160,\ 0) \qquad (90,\ 105)$$

It remains only to evaluate the objective function at each corner point: see Table 2.

TABLE 2

Corner Point (x, y)	Value of Objective Function $P = (\$0.25)x + (\$0.45)y$
(0, 0)	$P = (0.25)(0) + (0.45)(0) = 0$
(0, 150)	$P = (0.25)(0) + (0.45)(150) = \$67.50$
(160, 0)	$P = (0.25)(160) + (0.45)(0) = \$40.00$
(90, 105)	$P = (0.25)(90) + (0.45)(105) = \$69.75$

A maximum profit is obtained if 90 packages of low-grade mixture and 105 packages of high-grade mixture are made. The maximum profit obtainable under the conditions described is \$69.75.

NOW WORK PROBLEM 49.

EXERCISE 3.2 Answers to Odd-Numbered Problems Begin on Page AN-19.

In Problems 1–10, the given figure illustrates the graph of the set of feasible points of a linear programming problem. Find the maximum and minimum values of each objective function.

1. $z = 2x + 3y$
2. $z = 3x + 2y$
3. $z = x + y$
4. $z = 3x + 3y$
5. $z = x + 6y$
6. $z = 6x + y$
7. $z = 3x + 4y$
8. $z = 4x + 3y$
9. $z = 10x + y$
10. $z = x + 10y$

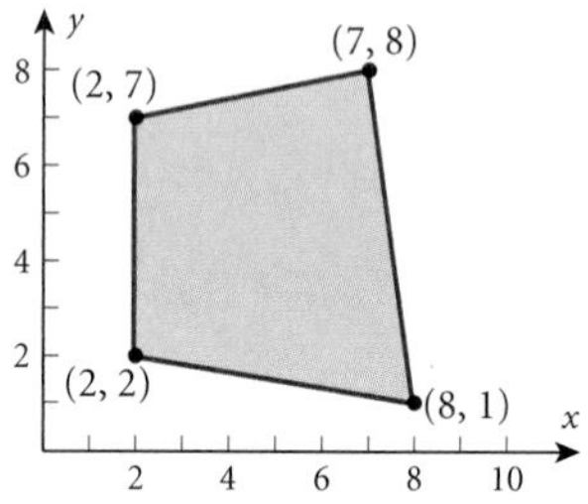

In Problems 11–16, list the corner points for each collection of constraints of a linear programming problem.

11. $\begin{cases} x \le 13 \\ 4x + 3y \ge 12 \\ x \ge 0 \\ y \ge 0 \end{cases}$

12. $\begin{cases} x \le 8 \\ 2x + 3y \ge 6 \\ x \ge 0 \\ y \ge 0 \end{cases}$

13. $\begin{cases} y \le 10 \\ x + y \le 15 \\ x \ge 0 \\ y \ge 0 \end{cases}$

14. $\begin{cases} y \le 8 \\ 2x + y \ge 10 \\ x \ge 0 \\ y \ge 0 \end{cases}$

15. $\begin{cases} x \le 10 \\ y \le 8 \\ 4x + 3y \ge 12 \\ x \ge 0 \\ y \ge 0 \end{cases}$

16. $\begin{cases} x \le 9 \\ y \le 12 \\ 2x + 3y \le 24 \\ x \ge 0 \\ y \ge 0 \end{cases}$

In Problems 17–24, maximize (if possible) the quantity $z = 5x + 7y$ subject to the given constraints.

17. $\begin{cases} x + y \le 2 \\ y \ge 1 \\ x \ge 0 \\ y \ge 0 \end{cases}$

18. $\begin{cases} 2x + 3y \le 6 \\ x \le 2 \\ x \ge 0 \\ y \ge 0 \end{cases}$

19. $\begin{cases} x + y \ge 2 \\ 2x + 3y \le 6 \\ x \ge 0 \\ y \ge 0 \end{cases}$

20. $\begin{cases} x + y \ge 2 \\ 2x + 3y \le 12 \\ 3x + 2y \le 12 \\ x \ge 0 \\ y \ge 0 \end{cases}$

21. $\begin{cases} x + y \ge 2 \\ x + y \le 8 \\ 2x + y \le 10 \\ x \ge 0 \\ y \ge 0 \end{cases}$

22. $\begin{cases} x + y \ge 2 \\ x + y \le 8 \\ x + 2y \ge 1 \\ x + 2y \le 10 \\ x \ge 0 \\ y \ge 0 \end{cases}$

23. $\begin{cases} x + y \le 10 \\ x \ge 6 \\ x \ge 0 \\ y \ge 0 \end{cases}$

24. $\begin{cases} x + y \le 8 \\ y \ge 2 \\ x \ge 0 \\ y \ge 0 \end{cases}$

In Problems 25–32, minimize (if possible) the quantity $z = 2x + 3y$ subject to the given constraints.

25. $\begin{cases} x + y \le 2 \\ y \le x \\ x \ge 0 \\ y \ge 0 \end{cases}$

26. $\begin{cases} 3x + y \le 3 \\ y \ge x \\ x \ge 0 \\ y \ge 0 \end{cases}$

27. $\begin{cases} x + y \ge 2 \\ x + 3y \le 12 \\ 3x + y \le 12 \\ x \ge 0 \\ y \ge 0 \end{cases}$

28. $\begin{cases} x + y \le 8 \\ 2x + 3y \ge 6 \\ x + y \ge 2 \\ x \ge 0 \\ y \ge 0 \end{cases}$

29. $\begin{cases} x + y \ge 2 \\ x + y \le 10 \\ 2x + 3y \le 6 \\ x \ge 0 \\ y \ge 0 \end{cases}$

30. $\begin{cases} 2y \le x \\ x + 2y \le 10 \\ x + 2y \ge 4 \\ x \ge 0 \\ y \ge 0 \end{cases}$

31. $\begin{cases} x + 2y \ge 1 \\ x + 2y \le 10 \\ y \ge 2x \\ x + y \le 8 \\ x \ge 0 \\ y \ge 0 \end{cases}$

32. $\begin{cases} 2x + y \ge 2 \\ x + y \le 6 \\ 2x \ge y \\ x \ge 0 \\ y \ge 0 \end{cases}$

In Problems 33–40, find the maximum and minimum values (if possible) of the given objective function subject to the constraints

$$\begin{cases} x + y \le 10 \\ 2x + y \ge 10 \\ x + 2y \ge 10 \\ x \ge 0 \\ y \ge 0 \end{cases}$$

33. $z = x + y$

34. $z = 2x + 3y$

35. $z = 5x + 2y$

36. $z = x + 2y$

37. $z = 3x + 4y$

38. $z = 3x + 6y$

39. $z = 10x + y$

40. $z = x + 10y$

---

41. Find the maximum and minimum values of $z = 18x + 30y$ subject to the constraints $3x + 3y \ge 9$, $-x + 4y \le 12$, and $4x - y \le 12$, where $x \ge 0$ and $y \ge 0$.

42. Find the maximum and minimum values of $z = 20x + 16y$ subject to the constraints $4x + 3y \ge 12$, $-2x + 4y \le 16$, and $6x - y \le 18$, where $x \ge 0$ and $y \ge 0$.

43. Find the maximum and minimum values of $z = 7x + 6y$ subject to the constraints $2x + 3y \ge 6$, $-3x + 4y \le 8$, and $5x - y \le 15$, where $x \ge 0$ and $y \ge 0$.

44. Find the maximum and minimum values of $z = 6x + 3y$ subject to the constraints $2x + 2y \ge 4$, $-x + 5y \le 10$, and $3x - 3y \le 6$, where $x \ge 0$ and $y \ge 0$.

45. Maximize $z = -20x + 30y$ subject to the constraints $0 \le x \le 15$, $0 \le y \le 10$, $5x + 3y \ge 15$, and $-3x + 3y \le 21$, where $x \ge 0$ and $y \ge 0$.

46. Maximize $z = -10x + 10y$ subject to the constraints $0 \le x \le 15$, $0 \le y \le 10$, $6x + y \ge 6$, and $-3x + y \le 7$, where $x \ge 0$ and $y \ge 0$.

47. Maximize $z = -12x + 24y$ subject to the constraints $0 \le x \le 15$, $0 \le y \le 10$, $3x + 3y \ge 9$, and $-3x + 2y \le 14$, where $x \ge 0$ and $y \ge 0$.

48. Maximize $z = -20x + 10y$ subject to the constraints $0 \le x \le 15$, $0 \le y \le 10$, $4x + 3y \ge 12$, and $-3x + y \le 7$, where $x \ge 0$ and $y \ge 0$.

49. In Example 5, if the profit on the low-grade mixture is \$0.30 per package and the profit on the high-grade mixture is \$0.40 per package, how many packages of each mixture should be made for a maximum profit?

3.3 Applications

OBJECTIVES **1** Solve applied problems

Solve applied problems 1 In this section, several situations that lead to linear programming problems are presented.

EXAMPLE 1 Maximizing Profit

Mike's Famous Toy Trucks manufactures two kinds of toy trucks—a standard model and a deluxe model. In the manufacturing process each standard model requires 2 hours of grinding and 2 hours of finishing, and each deluxe model needs 2 hours of grinding and 4 hours of finishing. The company has two grinders and three finishers, each of whom works at most 40 hours per week. Each standard model toy truck brings a profit of \$3 and each deluxe model a profit of \$4. Assuming that every truck made will be sold, how many of each should be made to maximize profits?

SOLUTION First, we name the variables:

$$x = \text{Number of standard models made}$$
$$y = \text{Number of deluxe models made}$$

The quantity to be maximized is the profit, which we denote by P:

$$P = \$3x + \$4y$$

This is the objective function. To manufacture one standard model requires 2 grinding hours and to make one deluxe model requires 2 grinding hours. The number of grinding hours needed to manufacture x standard and y deluxe models is

$$2x + 2y$$

But the total amount of grinding time available is only 80 hours per week. This means we have the constraint

$$2x + 2y \leq 80 \quad \text{Grinding time constraint}$$

Similarly, for the finishing time we have the constraint

$$2x + 4y \leq 120 \quad \text{Finishing time constraint}$$

By simplifying each of these constraints and adding the nonnegativity constraints $x \geq 0$ and $y \geq 0$, we may list all the constraints for this problem:

$$\begin{cases} x + y \leq 40 & (1) \\ x + 2y \leq 60 & (2) \\ x \geq 0 & (3) \\ y \geq 0 & (4) \end{cases}$$

Figure 21 illustrates the set of feasible points, which is bounded.

FIGURE 21

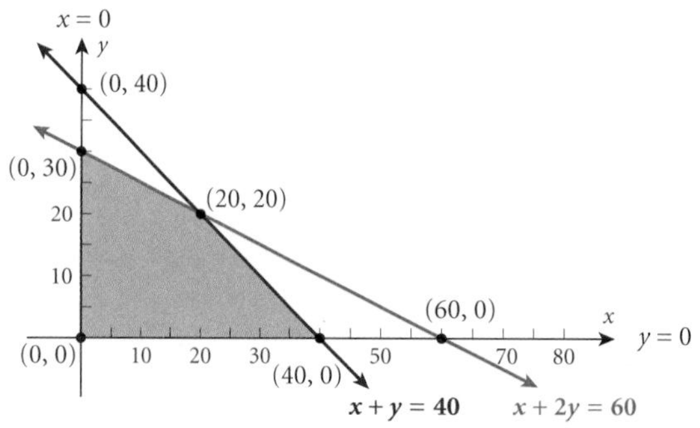

The corner points of the set of feasible points are

$$(0, 0) \qquad (0, 30) \qquad (40, 0) \qquad (20, 20)$$

Table 3 lists the corresponding values of the objective equation:

TABLE 3

Corner Point (x, y)	Value of Objective Function $P = \$3x + \$4y$
$(0, 0)$	$P = 0$
$(0, 30)$	$P = \$120$
$(40, 0)$	$P = \$120$
$(20, 20)$	$P = 3(20) + 4(20) = \$140$

A maximum profit is obtained if 20 standard trucks and 20 deluxe trucks are manufactured. The maximum profit is \$140.

NOW WORK PROBLEM 1.

EXAMPLE 2 Financial Planning

A retired couple have up to \$30,000 to invest in fixed-income securities. Their broker recommends investing in two bonds: one a AAA bond yielding 8%; the other a B^+ bond paying 12%. After some consideration, the couple decide to invest at most \$12,000 in the B^+-rated bond and at least \$6000 in the AAA bond. They also want the amount invested in the AAA bond to exceed or equal the amount invested in the B^+ bond. What should the broker recommend if the couple (quite naturally) want to maximize the return on their investment?

SOLUTION First, we name the variables:

$$x = \text{Amount invested in the AAA bond}$$
$$y = \text{Amount invested in the B}^+ \text{ bond}$$

The quantity to be maximized, the couple's return on investment, which we denote by P, is

$$P = 0.08x + 0.12y$$

This is the objective function. The conditions specified by the problem are

Up to \$30,000 available to invest	$x + y \le 30{,}000$
Invest at most \$12,000 in the B^+ bond	$y \le 12{,}000$
Invest at least \$6000 in the AAA bond	$x \ge 6000$
Amount in the AAA bond must exceed or equal the amount in the B^+ bond	$x \ge y$

In addition, we must have the conditions $x \ge 0$ and $y \ge 0$. The total list of constraints is

$$\begin{cases} x + y \le 30{,}000 & (1) \\ y \le 12{,}000 & (2) \\ x \ge 6000 & (3) \\ x \ge y & (4) \\ x \ge 0 & (5) \\ y \ge 0 & (6) \end{cases}$$

Figure 22 illustrates the set of feasible points, which is bounded. The corner points of the set of feasible points are

$$(6000, 0) \qquad (6000, 6000) \qquad (12000, 12000) \qquad (18000, 12000) \qquad (30000, 0)$$

FIGURE 22

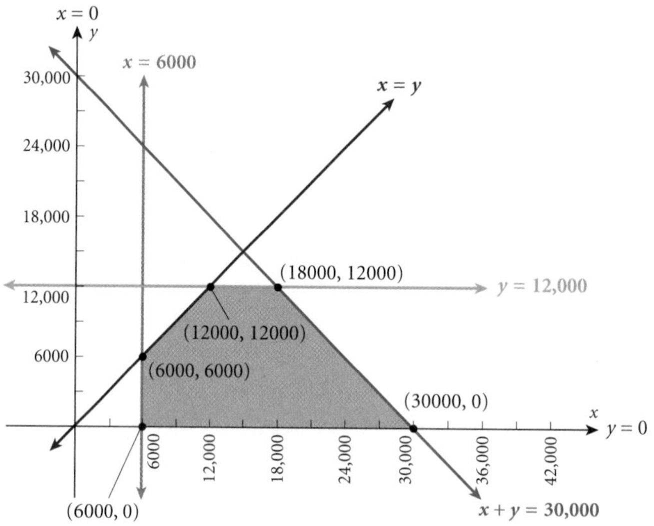

The corresponding return on investment at each corner point is

$$P = 0.08(6000) + 0.12(0) = \$480$$
$$P = 0.08(6000) + 0.12(6000) = 480 + 720 = \$1200$$
$$P = 0.08(12{,}000) + 0.12(12{,}000) = 960 + 1440 = \$2400$$
$$P = 0.08(18{,}000) + 0.12(12{,}000) = 1440 + 1440 = \$2880$$
$$P = 0.08(30{,}000) + 0.12(0) = \$2400$$

The maximum return on investment is \$2880, obtained by placing \$18,000 in the AAA bond and \$12,000 in the B^+ bond.

NOW WORK PROBLEM 3.

EXAMPLE 3 Manufacturing Vitamin Pills—Maximizing Profit

A pharmaceutical company makes two types of vitamins at its New Jersey plant—a high-potency, antioxidant vitamin and a vitamin enriched with added calcium. Each high-potency vitamin contains, among other things, 500 mg of vitamin C and 40 mg of calcium and generates a profit of \$0.10 per tablet. A calcium-enriched vitamin tablet contains 100 mg of vitamin C and 400 mg of calcium and generates a profit of \$0.05 per tablet. Each day the company has available 235 kg of vitamin C and 156 kg of calcium for use. Assuming all vitamins made are sold, how many of each type of vitamin should be manufactured to maximize profit?

Source: Centrum Vitamin Supplements.

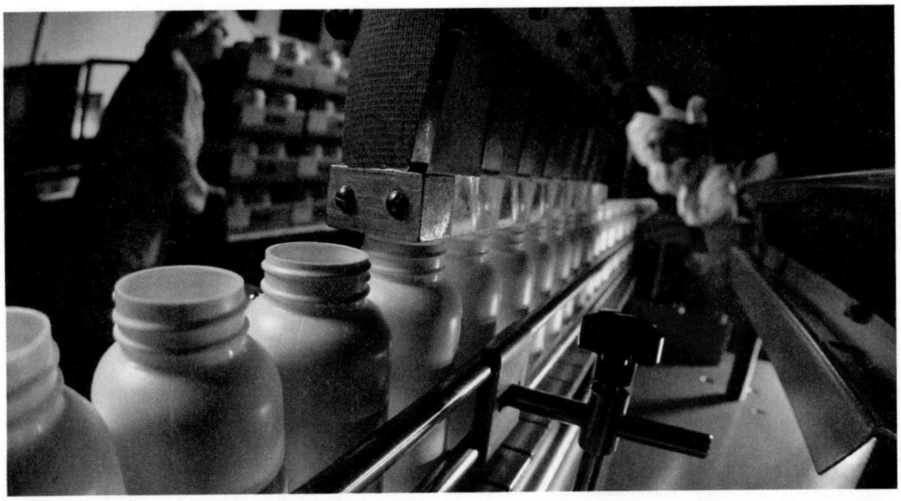

SOLUTION First we name the variables:

x = Number (in thousands) of high-potency vitamins to be produced

y = Number (in thousands) of calcium-enriched vitamins to be produced

We want to maximize the profit, P, which is given by:

$$P = 0.10x + 0.05y \quad x \text{ and } y \text{ in thousands}$$

Since 1 kg = 1,000,000 mg, the constraints, in mg, take the form

$$\begin{cases} 500x + 100y \leq 235{,}000 & \text{vitamin C constraint (in thousands of mg)} \\ 40x + 400y \leq 156{,}000 & \text{calcium constraint (in thousands of mg)} \\ x \geq 0 & \text{non-negativity constraints (in thousands)} \\ y \geq 0 & \end{cases}$$

Figure 23 illustrates the set of feasible points, which is bounded. The corner points of the set of feasible points are

$$(0, 0) \qquad (0, 390) \qquad (400, 350) \qquad (470, 0)$$

FIGURE 23

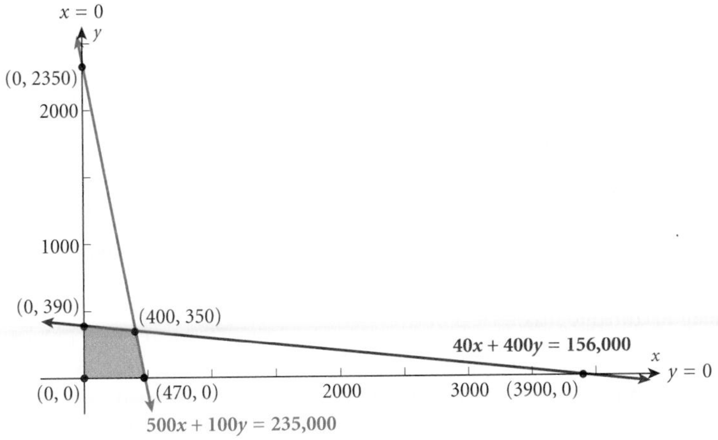

Since x and y are in thousands, the profit corresponding to each corner point is

$$P = 0.10(0) + 0.05(0) = 0 = \$0$$
$$P = 0.10(0) + 0.05(390) = 19.5 \text{ thousand} = \$19{,}500$$
$$P = 0.10(400) + 0.05(350) = 57.5 \text{ thousand} = \$57{,}500$$
$$P = 0.10(470) + 0.05(0) = 47 \text{ thousand} = \$47{,}000$$

The maximum profit is \$57,500, obtained when 400,000 high-potency vitamins are produced ($x = 400$ thousand units) and 350,000 calcium-enriched vitamins are produced ($y = 350$ thousand units).

NOW WORK PROBLEM 5.

EXERCISE 3.3 Answers to Odd-Numbered Problems Begin on Page AN-19.

1. **Optimal Land Use** A farmer has 70 acres of land available on which to grow some soybeans and some corn. The cost of cultivation per acre, the workdays needed per acre, and the profit per acre are indicated in the table:

	Soybeans	Corn	Total Available
Cultivation Cost per Acre	\$60	\$30	\$1800
Days of Work per Acre	3 days	4 days	120 days
Profit per Acre	\$300	\$150	

As indicated in the last column, the acreage to be cultivated is limited by the amount of money available for cultivation costs and by the number of working days that can be put into this part of the business. Find the number of acres of each crop that should be planted in order to maximize the profit.

2. **Manufacturing** A factory manufactures two products, each requiring the use of three machines. The first machine can be used at most 70 hours; the second machine at most 40 hours; and the third machine at most 90 hours. The first product requires 2 hours on machine 1, 1 hour on machine 2, and 1 hour on machine 3; the second product requires 1 hour each on machines 1 and 2, and 3 hours on machine 3. If the profit is \$40 per unit for the first product and \$60 per unit for the second product, how many units of each product should be manufactured to maximize profit?

3. **Investment Strategy** An investment broker wants to invest up to \$20,000. She can purchase a type A bond yielding a 10% return on the amount invested, and she can purchase a type B bond yielding a 15% return on the amount invested. She wants to invest at least as much in the type A bond as in the type B bond. She will also invest at least \$5000 in the type A bond and no more than \$8000 in the type B bond. How much should she invest in each type of bond to maximize her return?

4. **Diet** A diet is to contain at least 400 units of vitamins, 500 units of minerals, and 1400 calories. Two foods are available: F_1, which costs \$0.05 per unit, and F_2, which costs \$0.03 per unit. A unit of food F_1 contains 2 units of vitamins, 1 unit of minerals, and 4 calories; a unit of food F_2 contains 1 unit of vitamins, 2 units of minerals, and 4 calories. Find the minimum cost for a diet that consists of a mixture of these two foods and also meets the minimal nutrition requirements.

5. **Manufacturing Vitamin Pills** After changing suppliers, the pharmaceutical company in Example 3 has 300 kg of vitamin C and 220 kg of calcium available each day for the manufacture of the high-potency, antioxidant vitamins and vitamins enriched with added calcium. If each high-potency vitamin contains, among other things, 500 mg of vitamin C and 40 mg of calcium and generates a profit of \$0.10 per tablet, and each calcium-enriched vitamin tablet contains 100 mg of vitamin C and 400 mg of calcium and generates a profit of \$0.05 per tablet, how many of each type of vitamin should be manufactured to maximize profit?

Source: Centrum Vitamin Supplements.

6. **Investment Strategy** A financial consultant wishes to invest up to a total of \$30,000 in two types of securities, one that yields 10% per year and another that yields 8% per year. Furthermore, she believes that the amount invested in the first security should be at most one-third of the amount invested in the second security. What investment program should the consultant pursue in order to maximize income?

7. **Scheduling** Blink Appliances has a sale on microwaves and stoves. Each microwave requires 2 hours to unpack and set up, and each stove requires 1 hour. The storeroom space is limited to 50 items. The budget of the store allows only 80 hours of employee time for unpacking and setup. Microwaves sell for \$300 each, and stoves sell for \$200 each. How many of each should the store order to maximize revenue?

8. **Transportation** An appliance company has a warehouse and two terminals. To minimize shipping costs, the manager must decide how many appliances should be shipped to each terminal. There is a total supply of 1200 units in the warehouse and a demand for 400 units in terminal A and 500 units in terminal B. It costs \$12 to ship each unit to terminal A and \$16 to ship to terminal B. How many units should be shipped to each terminal in order to minimize cost?

9. **Pension Fund Investments** A pension fund has decided to invest \$45,000 in the two high-yield stocks listed in the table below.

	Price per Share 2/21/03	Yield
Duke Energy Corp.	\$14	8%
Eastman Kodak	\$30	6%

This pension fund has decided to invest at least 25% of the \$45,000 in each of the two stocks. Further, it has been decided that at most 63% of the \$45,000 can be invested in either one of the stocks. How many shares of each stock should be purchased in order to maximize the annual yield, while meeting the stipulated requirements? What is the annual yield in dollars for the optimal investment plan?

Source: Yahoo! Finance. Prices and yields have been rounded.

10. **Pollution Control** A chemical plant produces two items A and B. For each item A produced, 2 cubic feet of carbon monoxide and 6 cubic feet of sulfur dioxide are emitted into the atmosphere; to produce item B, 4 cubic feet of carbon monoxide and 3 cubic feet of sulfur dioxide are emitted into the atmosphere. Government pollution standards permit the manufacturer to emit a maximum of 3000 cubic feet of carbon monoxide and 5400 cubic feet of sulfur dioxide per week. The manufacturer can sell all of the items that it produces and makes a profit of \$1.50 per unit for item A and \$1.00 per unit for item B. Determine the number of units of each item to be produced each week to maximize profit without exceeding government standards.

11. **Baby Food Servings** Gerber Banana Plum Granola costs \$0.89 per 5.5-oz serving; each serving contains 140 calories, 31 g of carbohydrates, and 0% of the recommended daily allowance of vitamin C. Gerber Mixed Fruit Carrot Juice costs \$0.79 per 4-oz serving; each serving contains 60 calories, 13 g of carbohydrates, and 100% of the recommended daily allowance of vitamin C. Determine how many servings of each of the above foods would be needed to provide a child at least 160 calories, 40 g of carbohydrates, and 70% of the recommended daily allowance of vitamin C at minimum cost. Fractions of servings are permitted.

Source: Gerber website and Safeway Stores, Inc.

12. **Production Scheduling** A company produces two types of steel. Type 1 requires 2 hours of melting, 4 hours of cutting, and 10 hours of rolling per ton. Type 2 requires 5 hours of melting, 1 hour of cutting, and 5 hours of rolling per ton. Forty hours are available for melting, 20 for cutting, and 60 for rolling. Each ton of Type 1 produces \$240 profit, and each ton of Type 2 yields \$80 profit. Find the maximum profit and the production schedule that will produce this profit.

13. **Website Ads** Nielson's Net Ratings for the month of December 2002 indicated that in the United States, AOL had a unique audience of about 76.4 million people and Yahoo! had a unique audience of about 66.2 million people. An advertising company wants to purchase website ads to promote a new product. Suppose that the monthly cost of an ad on the AOL website is \$1200 and the monthly cost of an ad on the Yahoo! website is \$1100. Determine how many months an ad should run on each website to maximize the number of people who would be exposed to it. Assume that, for future months, the monthly website audience remains the same as given for December 2002. Also assume that the advertising budget is \$35,000 and that it has been decided to advertise on Yahoo! for at least ten months.

Source: Nielson's Net Ratings.

14. **Diet** Danny's Chicken Farm is a producer of frying chickens. In order to produce the best fryers possible, the regular chicken feed is supplemented by four vitamins. The minimum amount of each vitamin required per 100 ounces of feed is: vitamin 1, 50 units; vitamin 2, 100 units; vitamin 3, 60 units; vitamin 4, 180 units. Two supplements are available: supplement I costs \$0.03 per ounce and contains 5 units of vitamin 1 per ounce, 25 units of vitamin 2 per ounce, 10 units of vitamin 3 per ounce, and 35 units of vitamin 4 per ounce. Supplement II costs \$0.04 per ounce and contains 25 units of vitamin 1 per ounce, 10 units of vitamin 2 per ounce, 10 units of vitamin 3 per ounce, and 20 units of vitamin 4 per ounce. How much of each supplement should Danny buy to add to each 100 ounces of feed in order to minimize his cost, but still have the desired vitamin amounts present?

15. **Maximizing Income** J. B. Rug Manufacturers has available 1200 square yards of wool and 1000 square yards of nylon for

the manufacture of two grades of carpeting: high-grade, which sells for \$500 per roll, and low-grade, which sells for \$300 per roll. Twenty square yards of wool and 40 square yards of nylon are used in a roll of high-grade carpet, and 40 square yards of nylon are used in a roll of low-grade carpet. Forty work-hours are required to manufacture each roll of the high-grade carpet, and 20 work-hours are required for each roll of the low-grade carpet, at an average cost of \$6 per work-hour. A maximum of 800 work-hours are available. The cost of wool is \$5 per square yard and the cost of nylon is \$2 per square yard. How many rolls of each type of carpet should be manufactured to maximize income? [*Hint*: Income = Revenue from sale − (Production cost for material + labor)]

16. The rug manufacturer in Problem 15 finds that maximum income occurs when no high-grade carpet is produced. If the price of the low-grade carpet is kept at \$300 per roll, in what price range should the high-grade carpet be sold so that income is maximized by selling some rolls of each type of carpet? Assume all other data remain the same.

Chapter 3 Review

OBJECTIVES

Section	You should be able to	Review Exercises
3.1	1 Graph linear inequalities	1–4
	2 Graph systems of linear inequalities	9–14
3.2	1 Identify a linear programming problem	33–38
	2 Solve a linear programming problem	15–32
3.3	1 Solve applied problems	33–38

THINGS TO KNOW

Graphs of Inequalities (p. 160)
The graph of a strict inequality is represented by a dashed line and the half-plane satisfying the inequality.
The graph of a nonstrict inequality is represented by a solid line and the half-plane satisfying the inequality.

Graphs of Systems of Linear Inequalities (p. 162)
The graph of a system of linear inequalities is the set of all points that satisfy each inequality in the system.

Bounded (p. 166)
The graph is called bounded if some circle can be drawn around it.
The graph is called unbounded if it extends infinitely far in at least one direction.

Corner Point (p. 166)
A corner point is the intersection of two line segments that form the boundary of the graph of a system of linear inequalities.

Linear Programming (p. 172)
Maximize (or minimize) a linear objective function, $z = Ax + By$, subject to certain conditions, or constraints, expressible as linear inequalities in x and y. A feasible point (x, y) is a point that satisfies the constraints of a linear programming problem.

Solution to a Linear Programming Problem (p. 172)
A solution to a linear programming problem is a feasible point that maximizes (or minimizes) the objective function together with the value of the objective function at that point.

Location of Solution (p. 174)
If a linear programming problem has a solution, it is located at a corner point of the graph of the feasible points.
If a linear programming problem has multiple solutions, at least one of them is located at a corner point of the graph of the feasible points.
In either case, the corresponding value of the objective function is unique.

TRUE–FALSE ITEMS Answers are on page AN-20.

T F **1.** The graph of a system of linear inequalities may be bounded or unbounded.

T F **2.** The graph of the set of constraints of a linear programming problem, under certain conditions, could have a circle for a boundary.

T F **3.** The objective function of a linear programming problem is always a linear equation involving the variables.

T F **4.** In a linear programming problem, there may be more than one point that maximizes or minimizes the objective function.

T F **5.** Some linear programming problems will have no solution.

T F **6.** If a linear programming problem has a solution, it is located at the center of the set of feasible points.

FILL IN THE BLANKS Answers are on page AN-20.

1. The graph of a linear inequality in two variables is called a ____________.

2. In a linear programming problem the quantity to be maximized or minimized is referred to as the ____________ function.

3. The points that obey the collection of constraints of a linear programming problem are called ____________ points.

4. A linear programming problem will always have a solution if the set of feasible points is ____________.

5. If a linear programming problem has a solution, it is located at a ____________ of the set of feasible points.

REVIEW EXERCISES Answers to odd-numbered problems begin on page AN-20.

Blue problem numbers indicate the author's suggestions for use in a practice test.

In Problems 1–4, graph each linear inequality.

1. $x + 3y \le 0$

2. $4x + y \ge 0$

3. $5x + y > 10$

4. $2x + 3y < 6$

5. Without graphing, determine which of the points $P_1 = (4, -3), P_2 = (2, -6), P_3 = (8, -3)$ are part of the graph of the following system:

$$\begin{cases} x + 2y \le 8 \\ 2x - y \ge 4 \end{cases}$$

6. Without graphing, determine which of the points $P_1 = (8, 6), P_2 = (2, -5), P_3 = (4, 1)$ are part of the graph of the following system:

$$\begin{cases} 5x - y \ge 2 \\ x - 4y \le -2 \end{cases}$$

In Problems 7–8, determine which region—a, b, c, or d—represents the graph of the given system of linear inequalities.

7. $\begin{cases} 6x - 4y \le 12 \\ 3x + 2y \le 18 \end{cases}$

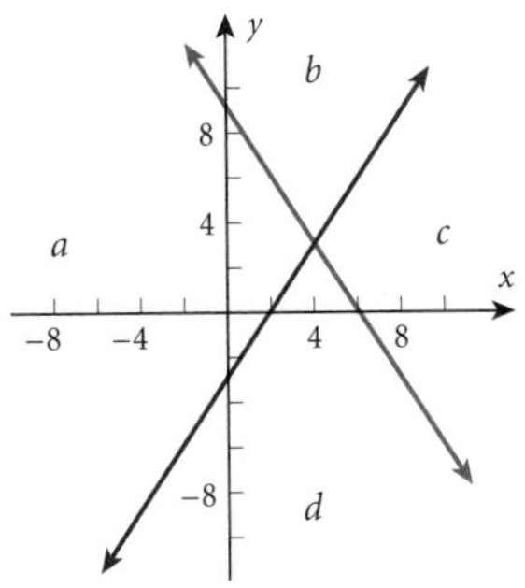

8. $\begin{cases} 6x - 5y \ge 5 \\ 6x + 6y \le 60 \end{cases}$

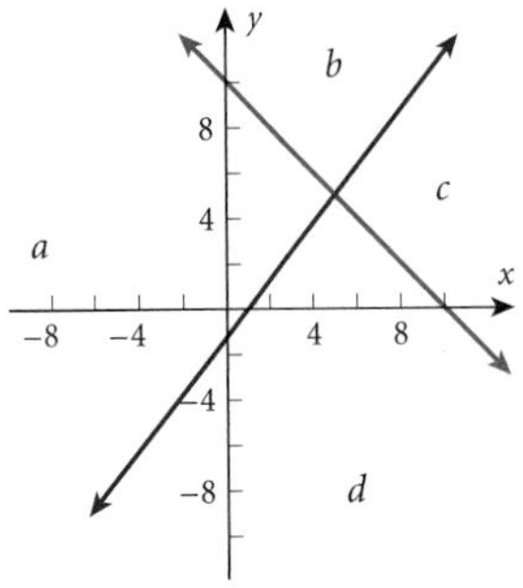

In Problems 9–14, graph each system of linear inequalities. Locate the corner points and tell whether the graph is bounded or unbounded.

9. $\begin{cases} 3x + 2y \le 12 \\ x + y \ge 4 \\ x \ge 0 \\ y \ge 0 \end{cases}$

10. $\begin{cases} x + y \le 8 \\ 2x + y \ge 4 \\ x \ge 0 \\ y \ge 0 \end{cases}$

11. $\begin{cases} x + 2y \ge 4 \\ 3x + y \le 6 \\ x \ge 0 \\ y \ge 0 \end{cases}$

12. $\begin{cases} 2x + y \geq 4 \\ 3x + 2y \geq 6 \\ x \geq 0 \\ y \geq 0 \end{cases}$

13. $\begin{cases} 3x + 2y \geq 6 \\ 3x + 2y \leq 12 \\ x + 2y \leq 8 \\ x \geq 0 \\ y \geq 0 \end{cases}$

14. $\begin{cases} x + y \geq 8 \\ x + 2y \geq 10 \\ y \leq 8 \\ x \geq 0 \\ y \geq 0 \end{cases}$

In Problems 15–22, use the constraints below to solve each linear programming problem.

$$\begin{cases} x + 2y \leq 40 \\ 2x + y \leq 40 \\ x + y \geq 10 \\ x \geq 0 \\ y \geq 0 \end{cases}$$

15. Maximize $z = x + y$
16. Maximize $z = 2x + 3y$
17. Minimize $z = 5x + 2y$
18. Minimize $z = 3x + 2y$
19. Maximize $z = 2x + y$
20. Maximize $z = x + 2y$
21. Minimize $z = 2x + 5y$
22. Minimize $z = x + y$

In Problems 23–26, maximize and minimize (if possible) the quantity $z = 15x + 20y$ subject to the given constraints.

23. $\begin{cases} x \leq 5 \\ y \leq 8 \\ 3x + 4y \geq 12 \\ x \geq 0 \\ y \geq 0 \end{cases}$

24. $\begin{cases} x + 2y \geq 4 \\ 3x + 2y \geq 6 \\ x \geq 0 \\ y \geq 0 \end{cases}$

25. $\begin{cases} 2x + 3y \leq 22 \\ x \leq 5 \\ y \leq 6 \\ x \geq 0 \\ y \geq 0 \end{cases}$

26. $\begin{cases} x + 2y \leq 20 \\ x + 10y \geq 36 \\ 5x + 2y \geq 36 \\ x \geq 0 \\ y \geq 0 \end{cases}$

In Problems 27–32, solve each linear programming problem.

27. Maximize

$$z = 2x + 3y$$

subject to the constraints

$$\begin{cases} x \leq 9 \\ y \leq 8 \\ x + y \geq 3 \\ x \geq 0 \\ y \geq 0 \end{cases}$$

28. Maximize

$$z = 4x + y$$

subject to the constraints

$$\begin{cases} x \leq 7 \\ y \leq 8 \\ x + y \geq 2 \\ x \geq 0 \\ y \geq 0 \end{cases}$$

29. Maximize

$$z = x + 2y$$

subject to the constraints

$$\begin{cases} x + y \geq 1 \\ y \leq 2x \\ x \leq 8 \\ y \leq 8 \\ x \geq 0 \\ y \geq 0 \end{cases}$$

30. Maximize

$$z = 3x + 4y$$

subject to the constraints

$$\begin{cases} x + 2y \geq 2 \\ 3x + 2y \leq 12 \\ y \leq 5 \\ x \geq 0 \\ y \geq 0 \end{cases}$$

31. Minimize

$$z = 3x + 2y$$

subject to the constraints

$$\begin{cases} x + 2y \geq 8 \\ 3x + y \geq 6 \\ x \leq 8 \\ x \geq 0 \\ y \geq 0 \end{cases}$$

32. Minimize

$$z = 2x + 5y$$

subject to the constraints

$$\begin{cases} x \leq 10 \\ y \leq 12 \\ 2x + y \geq 10 \\ x - y \geq 0 \\ x \geq 0 \\ y \geq 0 \end{cases}$$

33. Maximizing Profit A ski manufacturer makes two types of skis: downhill and cross-country. Using the information given in the table below, how many of each type of ski should be made for a maximum profit to be achieved? What is the maximum profit?

	Downhill	Cross-Country	Maximum Time Available
Manufacturing Time per Ski	2 hours	1 hour	40 hours
Finishing Time per Ski	1 hour	1 hour	32 hours
Profit per Ski	\$70	\$50	

34. Rework Problem 33 if the manufacturing unit has a maximum of 48 hours available.

35. Nutrition Katy needs at least 60 units of carbohydrates, 45 units of protein, and 30 units of fat each month. From each pound of food A, she receives 5 units of carbohydrates, 3 of protein, and 4 of fat. Food B contains 2 units of carbohydrates, 2 units of protein, and 1 unit of fat per pound. If food A costs \$1.30 per pound and food B costs \$0.80 per pound, how many pounds of each food should Katy buy each month to keep costs at a minimum?

36. Production Scheduling A company sells two types of shoes. The first uses 2 units of leather and 2 units of synthetic material and yields a profit of \$8 per pair. The second type requires 5 units of leather and 1 unit of synthetic material and gives a profit of \$10 per pair. If there are 40 units of leather and 16 units of synthetic material available, how many pairs of each type of shoe should be manufactured to maximize profit? What is the maximum profit?

37. Baby Food Servings Gerber Banana Oatmeal and Peach costs \$0.79 per 4-oz serving; each serving contains 90 calories, 19 g of carbohydrates, and 45% of the recommended daily allowance of vitamin C. Gerber Mixed Fruit Juice costs \$0.65 per 4-oz serving; each serving contains 60 calories, 15 g of carbohydrates, and 100% of the recommended daily allowance of vitamin C. Determine how many servings of each of the above foods would be needed to provide a child with at least 130 calories, 30 g of carbohydrates, and 60% of the recommended daily allowance of vitamin C at minimum cost. Fractions of servings are permitted.

Source: Gerber website and Safeway Stores, Inc.

38. Maximizing Profit A company makes two explosives: type I and type II. Due to storage problems, a maximum of 100 pounds of type I and 150 pounds of type II can be mixed and packaged each week. One pound of type I takes 60 hours to mix and 70 hours to package; 1 pound of type II takes 40 hours to mix and 40 hours to package. The mixing department has at most 7200 work-hours available each week, and packaging has at most 7800 work-hours available. If the profit for 1 pound of type I is \$60 and for 1 pound of type II is \$40, what is the maximum profit possible each week?

39. A pharmaceutical company makes two types of vitamins at its New Jersey plant—a high-potency, antioxidant vitamin and a vitamin enriched with added calcium. Each high-potency vitamin contains, among other things, 500 mg of vitamin C, 40 mg of calcium, and 100 mg of magnesium and generates a profit of \$0.10 per tablet. A calcium-enriched vitamin tablet contains 100 mg of vitamin C, 400 mg of calcium, and 40 mg of magnesium and generates a profit of \$0.05 per tablet. Each day the company has available 300 kg of vitamin C, 122 kg of calcium, and 65 kg of magnesium for use in the production of the vitamins. How many of each type of vitamin should be manufactured to maximize profit?

Source: Centrum Vitamin Supplements.

40. Mixture A company makes two kinds of animal food, A and B, which contain two food supplements. It takes 2 pounds of the first supplement and one pound of the second to make a dozen cans of food A, and 4 pounds of the first supplement and 5 pounds of the second to make a dozen cans of food B. On a certain day 80 pounds of the first supplement and 70 pounds of the second are available. How many cans of food A and food B should be made to maximize company profits, if the profit on a dozen cans of food A is \$3.00 and the profit on a dozen cans of food B is \$10.00.

Chapter 3 Project

BALANCING NUTRIENTS

In preparing a recipe you must decide what ingredients and how much of each ingredient you will use. In these health-conscious days, you may also want to consider the amount of certain nutrients in your recipe. You may even be interested in minimizing some quantities (like calories or fat) or maximizing others (like carbohydrates or protein). Linear programming techniques can help to do this.

For example, consider making a very simple trail mix from dry-roasted, unsalted peanuts and seedless raisins. Table 1 lists the amounts of various dietary quantitites for these ingredients. The amounts are given per serving of the ingredient.

TABLE 1 NUTRIENTS IN PEANUTS AND RAISINS

Nutrient	Peanuts Serving Size = 1 Cup	Raisins Serving Size = 1 Cup
Calories (kcal)	850	440
Protein (g)	34.57	4.67
Fat(g)	72.50	.67
Carbohydrates (g)	31.40	114.74

Source: USDA National Nutrient Database for Standard Reference. www.nal.gov/fnic/foodcomp.

Suppose that you want to make at most 6 cups of trail mix for a day hike. You don't want either ingredient to dominate the mixture, so you want the amount of raisins to be at least $\frac{1}{2}$ of the amount of peanuts and the amount of peanuts to be at least $\frac{1}{2}$ of the amount of raisins. You want the entire amount of trail mix you make to have fewer than 4000 calories, and you want to maximize the amount of carbohydrates in the mix.

1. Let x be the number of cups of peanuts you will use, let y be the number of cups of raisins you will use, and let c be the amount of carbohydrates in the mix. Find the objective function.
2. What constraints must be placed on the objective function?
3. Graph the set of feasible points for this problem.
4. Find the number of cups of peanuts and raisins that maximize the amount of carbohydrates in the mix.
5. How many grams of carbohydrates are in a cup of the final mix? How many calories?
6. Under all of the constraints given above, what recipe for trail mix will maximize the amount of protein in the mix? How many grams of protein are in a cup of this mix? How many calories?
7. Suppose you decide to eat at least 3 cups of the trail mix. Keeping the constraints given above, what recipe for trail mix will minimize the amount of fat in the mix?
8. How many grams of carbohydrates are in this mix?
9. How many grams of protein are in this mix?
10. Which of the three trail mixes would you use? Why?

MATHEMATICAL QUESTIONS FROM PROFESSIONAL EXAMS*

Use the following information to do Problems 1–3:
The Random Company manufactures two products, Zeta and Beta. Each product must pass through two processing operations. All materials are introduced at the start of process 1. There are no work-in-process inventories. Random may produce either one product exclusively or various combinations of both products subject to the following constraints:

	Process No. 1	Process No. 2	Contribution Margin per Unit
Hours required to produce one unit of			
Zeta	1 hour	1 hour	$4.00
Beta	2 hours	3 hours	5.25
Total capacity in hours per day	1000 hours	1275 hours	

A shortage of technical labor has limited Beta production to 400 units per day. There are no constraints on the production of Zeta other than the hour constraints in the above schedule. Assume that all relationships between capacity and production are linear, and that all of the above data and relationships are deterministic rather than probabilistic.

1. **CPA Exam** Given the objective to maximize total contribution margin, what is the production constraint for process 1?

   (a) Zeta + Beta $\leq$ 1000
   (b) Zeta + 2 Beta $\leq$ 1000
   (c) Zeta + Beta $\geq$ 1000
   (d) Zeta + 2 Beta $\geq$ 1000

2. **CPA Exam** Given the objective to maximize total contribution margin, what is the labor constraint for production of Beta?

   (a) Beta $\leq$ 400 (b) Beta $\geq$ 400
   (c) Beta $\leq$ 425 (d) Beta $\geq$ 425

3. **CPA Exam** What is the objective function of the data presented?

   (a) Zeta + 2 Beta = $9.25
   (b) ($4.00)Zeta + 3($5.25)Beta = Total contribution margin
   (c) ($4.00)Zeta + ($5.25)Beta = Total contribution margin
   (d) 2($4.00)Zeta + 3($5.25)Beta = Total contribution margin

4. **CPA Exam** Williamson Manufacturing intends to produce two products, X and Y. Product X requires 6 hours of time on machine 1 and 12 hours of time on machine 2. Product Y requires 4 hours of time on machine 1 and no time on machine 2. Both machines are available for 24 hours. Assuming that the objective function of the total contribution margin is $\$2X + \$1Y$, what product mix will produce the maximum profit?

   (a) No units of product X and 6 units of product Y.
   (b) 1 unit of product X and 4 units of product Y.
   (c) 2 units of product X and 3 units of product Y.
   (d) 4 units of product X and no units of product Y.

5. **CPA Exam** Quepea Company manufactures two products, Q and P, in a small building with limited capacity. The selling price, cost data, and production time are given below:

	Product *Q*	Product *P*
Selling price per unit	$20	$17
Variable costs of producing and selling a unit	$12	$13
Hours to produce a unit	3	1

Based on this information, the profit maximization objective function for a linear programming solution may be stated as

(a) Maximize $\$20Q + \$17P$.
(b) Maximize $\$12Q + \$13P$.
(c) Maximize $\$3Q + \$1P$.
(d) Maximize $\$8Q + \$4P$.

6. **CPA Exam** Patsy, Inc., manufactures two products, X and Y. Each product must be processed in each of three departments: machining, assembling, and finishing. The hours needed to produce one unit of product per department and the maximum possible hours per department follow:

	Production Hours per Unit		
Department	X	Y	Maximum Capacity in Hours
Machining	2	1	420
Assembling	2	2	500
Finishing	2	3	600

Other restrictions follow:

$$X \geq 50 \quad Y \geq 50$$

The objective function is to maximize profits where profit $= \$4X + \$2Y$. Given the objective and constraints, what is the most profitable number of units of X and Y, respectively, to manufacture?

(a) 150 and 100 (b) 165 and 90
(c) 170 and 80 (d) 200 and 50

7. **CPA Exam** Milford Company manufactures two models, medium and large. The contribution margin expected is \$12 for the medium model and \$20 for the large model. The medium model is processed 2 hours in the machining department and 4 hours in the polishing department. The large model is processed 3 hours in the machining department and 6 hours in the polishing department. How would the formula for determining the maximization of total contribution margin be expressed?

(a) $5X + 10Y$ (b) $6X + 9Y$
(c) $12X + 20Y$ (d) $12X(2 + 4) + 20Y(3 + 6)$

8. **CPA Exam** Hale Company manufactures products A and B, each of which requires two processes, polishing and grinding. The contribution margin is \$3 for product A and \$4 for product B. The illustration shows the maximum number of units of each product that may be processed in the two departments.

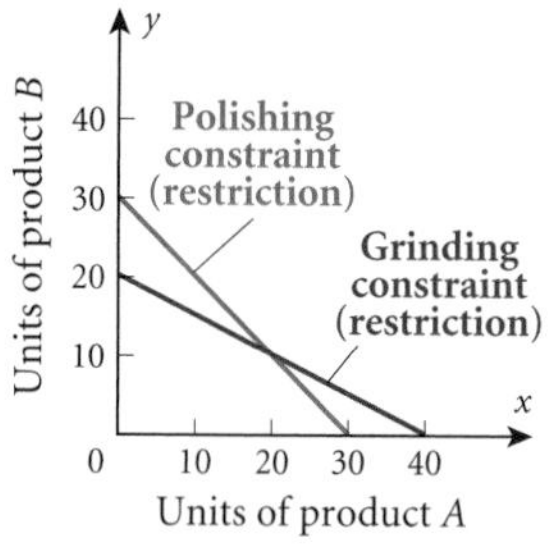

Considering the constraints (restrictions) on processing, which combination of products A and B maximizes the total contribution margin?

(a) 0 units of A and 20 units of B.
(b) 20 units of A and 10 units of B.
(c) 30 units of A and 0 units of B.
(d) 40 units of A and 0 units of B.

9. **CPA Exam** Johnson, Inc., manufactures product X and product Y, which are processed as follows:

	Machine A	Machine B
Product X	6 hours	4 hours
Product Y	9 hours	5 hours

The contribution margin is \$12 for product X and \$7 for product Y. The available time daily for processing the two products is 120 hours for machine A and 80 hours for machine B. How would the restriction (constraint) for machine B be expressed?

(a) $4X + 5Y$ (b) $4X + 5Y \leq 80$
(c) $6X + 9Y \leq 120$ (d) $12X + 7Y$

10. **CPA Exam** A small company makes only two products, with the following two production constraints representing two machines and their maximum availability:

$$2X + 3Y \leq 18$$
$$2X + Y \leq 10$$

where $X =$ Units of the first product
$Y =$ Units of the second product

If the profit equation is $Z = \$4X + \$2Y$, the maximum possible profit is

(a) \$20 (b) \$21 (c) \$18 (d) \$24
(e) Some profit other than those given above

Questions 11–13 are based on the Jarten Company, which manufactures and sells two products. Demand for the two products has grown to such a level that Jarten can no longer meet the demand with its facilities. The company can work a total of 600,000 direct labor-hours annually using three shifts. A total of 200,000 hours of machine time is available annually. The company plans to use linear programming to determine a production schedule that will maximize its net return.

The company spends $2,000,000 in advertising and promotion and incurs $1,000,000 for general and administrative costs. The unit sale price for model A is $27.50; model B sells for $75.00 each. The unit manufacturing requirements and unit cost data are as shown below. Overhead is assigned on a machine-hour (MH) basis.

	Model A		Model B	
Raw material		$ 3		$ 7
Direct labor	1 DLH @ $8	8	1.5 DLH @ $8	12
Variable overhead	0.5 MH @ $12	6	2.0 MH @ $12	24
Fixed overhead	0.5 MH @ $4	2	2.0 MH @ $4	8
		$19		$ 51

11. **CMA Exam** The objective function that would maximize Jarten's net income is

(a) $10.50A + 32.00B$ (b) $8.50A + 24.00B$
(c) $27.50A + 75.00B$ (d) $19.00A + 51.00B$
(e) $17.00A + 43.00B$

12. **CMA Exam** The constraint function for the direct labor is

(a) $1A + 1.5B \leq 200{,}000$ (b) $8A + 12B \leq 600{,}000$
(c) $8A + 12B \leq 200{,}000$ (d) $1A + 1.5B \leq 4{,}800{,}000$
(e) $1A + 1.5B \leq 600{,}000$

13. **CMA Exam** The constraint function for the machine capacity is

(a) $6A + 24B \leq 200{,}000$
(b) $(1/0.5)A + (1.5/2.0)B \leq 800{,}000$
(c) $0.5A + 2B \leq 200{,}000$
(d) $(0.5 + 0.5)A + (2 + 2)B \leq 200{,}000$
(e) $(0.5 \times 1) + (1.5 \times 2.00) \leq (200{,}000 \times 600{,}000)$

14. **CPA Exam** Boaz Company manufactures two models, medium (X) and large (Y). The contribution margin expected is $24 for the medium model and $40 for the large model. The medium model is processed 2 hours in the machining department and 4 hours in the polishing department. The large model is processed 3 hours in the machining department and 6 hours in the polishing department. If total contribution margin is to be maximized, using linear programming, how would the objective function be expressed?

(a) $24X(2 + 4) + 40Y(3 + 6)$ (b) $24X + 40Y$
(c) $6X + 9Y$ (d) $5X + 10Y$

CHAPTER 4

Linear Programming: Simplex Method

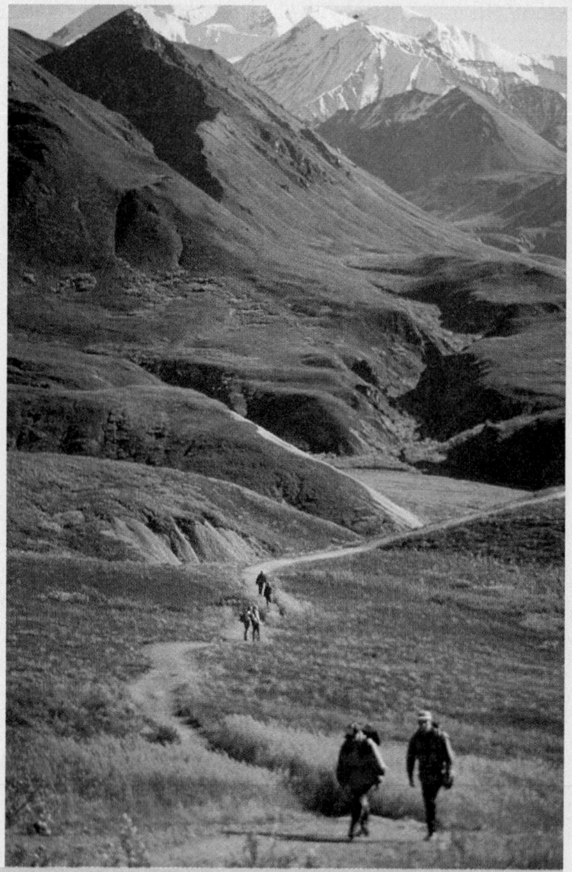

OUTLINE

That hiking trip to Pinnacles is still going to happen. But when you told the others who are going about your trail mix of unsalted peanuts and raisins, they all said "borrring." It was decided to use the unsalted peanuts and raisins mixed with M&Ms and salted mini-pretzels. Now it's more complicated to determine in what proportions these four ingredients should be mixed. And there remain the issues of carbohydrates, protein, calories, and fat. Fortunately, you also finished the simplex method before midterm, so you can handle the problem just fine. The Chapter Project at the end of the chapter will guide you.

A LOOK BACK, A LOOK FORWARD

In Chapter 3 we described a geometric method (using graphs) for solving linear programming problems. Unfortunately, this method is useful only when there are no more than two variables and the number of constraints is small.

Since most practical linear programming problems involve systems of several hundred linear inequalities containing several hundred variables, more sophisticated techniques need to be used. One of these methods is the *simplex method*, the subject of this chapter.

The **simplex method** is a way to solve linear programming problems involving many inequalities and variables. This method was developed by George Dantzig in 1946 and is particularly well suited for computerization. In 1984, Narendra Karmarkar of Bell Laboratories discovered a way of solving large linear programming problems that improves on the simplex method.

A discussion of LINDO, a software package that closely mimics the simplex method, may be found in Appendix B.

4.1 The Simplex Tableau; Pivoting

PREPARING FOR THIS SECTION *Before getting started, review the following:*

> Row Operations (Section 2.3, pp. 67–68)

> Linear Programming (Section 3.2, pp. 171–178)

OBJECTIVES
1 Determine a maximum problem is in standard form
2 Set up the initial simplex tableau
3 Perform pivot operations
4 Analyze a tableau

Introduction

If we have a large number of either variables or constraints, it is still true that if an optimal solution exists, it will be found at a corner point of the set of feasible points. In fact, we could find these corner points by writing all the equations corresponding to the inequalities of the problem and then proceeding to solve all possible combinations of these equations. We would, of course, have to discard any solutions that are not feasible (because they do not satisfy one or more of the constraints). Then we could evaluate the objective function at the remaining corner points.

Just how difficult is this procedure? Well, if there were just 4 variables and 7 constraints, we would have to solve all possible combinations of 4 equations chosen from a set of 7 equations—that would be 35 solutions in all. Each of these solutions would then have to be tested for feasibility. So even for this relatively small number of variables and constraints, the work would be quite tedious. In the real world of applications, it is fairly common to encounter problems with *hundreds*, even *thousands*, of variables and constraints. Of course, such problems must be solved by computer. Even so, choosing a more efficient problem-solving strategy might reduce the computer's running time from hours to seconds, or, for very large problems, from years to hours.

A more systematic approach would involve choosing a solution at one corner point of the feasible set, then moving from there to another corner point at which the objective function has a better value, and continuing in this way until the best possible value is found. The simplex method is a very efficient and popular way of doing this.

Standard Form of a Maximum Problem

Determine a maximum problem is in standard form 1

A linear programming problem in which the objective function is to be maximized is referred to as a **maximum problem**. Such problems are said to be in **standard form** provided two conditions are met:

Standard Form of a Maximum Problem

Condition 1 All the variables are nonnegative.

Condition 2 Every other constraint is written as a linear expression that is less than or equal to a positive constant.

EXAMPLE 1 Determining Whether a Maximum Problem Is in Standard Form

Determine which of the following maximum problems are in standard form.*

(a) Maximize

$$z = 5x_1 + 4x_2$$

subject to the constraints

$$\begin{aligned} 3x_1 + 4x_2 &\leq 120 \\ 4x_1 + 3x_2 &\leq 20 \\ x_1 \geq 0, \quad x_2 &\geq 0 \end{aligned}$$

(b) Maximize

$$z = 8x_1 + 2x_2 + 3x_3$$

subject to the constraints

$$\begin{aligned} 4x_1 + 8x_2 \qquad &\leq 120 \\ 3x_2 + 4x_3 &\leq 120 \\ x_1 \geq 0, \quad x_2 &\geq 0 \end{aligned}$$

(c) Maximize

$$z = 6x_1 - 8x_2 + x_3$$

subject to the constraints

$$\begin{aligned} 3x_1 + x_2 \qquad &\leq 10 \\ 4x_1 - x_2 \qquad &\leq 5 \\ x_1 - x_2 - x_3 &\geq -3 \\ x_1 \geq 0, \quad x_2 \geq 0, \quad x_3 &\geq 0 \end{aligned}$$

**Due to the nature of solving linear programming problems using the simplex method, we shall find it convenient to use subscripted variables throughout this chapter.*

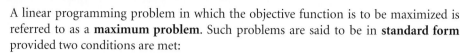

(d) Maximize

$$z = 8x_1 + x_2$$

subject to the constraints

$$3x_1 + 4x_2 \geq 2$$
$$x_1 + x_2 \leq 6$$
$$x_1 \geq 0, \qquad x_2 \geq 0$$

SOLUTION   **(a)** This is a maximum problem containing two variables x_1 and x_2. Since both variables are nonnegative and since the other constraints

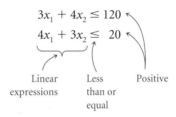

are each written as linear expressions less than or equal to a positive constant, we conclude the maximum problem is in standard form.

(b) This is a maximum problem containing three variables $x_1, x_2,$ and x_3. Since the variable x_3 is not given as nonnegative, the maximum problem is not in standard form.

(c) This is a maximum problem containing three variables $x_1, x_2,$ and x_3. Each variable is nonnegative. The set of constraints

$$3x_1 + x_2 \qquad \leq 10$$
$$4x_1 - x_2 \qquad \leq 5$$
$$x_1 - x_2 - x_3 \geq -3$$

contains $x_1 - x_2 - x_3 \geq -3$, which is not a linear expression that is less than or equal to a positive constant. The maximum problem is not in standard form. Notice, however, that by multiplying this constraint by -1, we get

$$-x_1 + x_2 + x_3 \leq 3$$

which is in the desired form. Although the maximum problem as stated is not in standard form, it can be modified to conform to the requirements of the standard form.

(d) The maximum problem contains two variables x_1 and x_2, each of which is nonnegative. Of the other constraints, the first one, $3x_1 + 4x_2 \geq 2$ does not conform. The maximum problem is not in standard form. Notice that we cannot modify this problem to place it in standard form. Even though multiplying by -1 will change the $\geq$ to $\leq$, in so doing the 2 will change to -2.   ▶

NOW WORK PROBLEMS 1 AND 11.

Slack Variables and the Simplex Tableau

Set up the initial simplex **2**
tableau

In order to apply the simplex method to a maximum problem, we need to first

1. Introduce *slack variables.*
2. Construct the *initial simplex tableau.*

We will show how these steps are done by working with a specific maximum problem in standard form. (This problem is the same as Example 1, Section 3.3, page 180.)

Maximize

$$P = 3x_1 + 4x_2$$

subject to the constraints

$$2x_1 + 4x_2 \leq 120$$
$$2x_1 + 2x_2 \leq 80$$
$$x_1 \geq 0, \qquad x_2 \geq 0$$

First, observe that this maximum problem is in standard form.

Next, recall that when we say that $2x_1 + 4x_2 \leq 120$, we mean that there is a number greater than or equal to 0, which we designate by s_1, such that

$$2x_1 + 4x_2 + s_1 = 120, \quad s_1 \geq 0$$

This number s_1 is a variable. It must be nonnegative since it is the difference between 120 and a number that is less than or equal to 120. We call it a **slack variable** since it "takes up the slack" between the left and right sides of the inequality.

Similarly, for the constraint $2x_1 + 2x_2 \leq 80$, we introduce the slack variable s_2:

$$2x_1 + 2x_2 + s_2 = 80, \quad s_2 \geq 0$$

Finally, we write the objective function $P = 3x_1 + 4x_2$ as

$$P - 3x_1 - 4x_2 = 0$$

In effect, we have now replaced our original system of constraints and the objective function by a system of three equations containing five variables, P, x_1, x_2, s_1, and s_2:

$$\left. \begin{array}{l} 2x_1 + 4x_2 + s_1 \qquad = 120 \\ 2x_1 + 2x_2 \qquad + s_2 = 80 \end{array} \right\} \quad \text{Constraints}$$

$$P - 3x_1 - 4x_2 \qquad\qquad = 0 \quad \text{Objective function}$$

where $\qquad x_1 \geq 0 \qquad x_2 \geq 0 \qquad s_1 \geq 0 \qquad s_2 \geq 0$

To solve the maximum problem is to find the particular solution (P, x_1, x_2, s_1, s_2) that gives the largest possible value for P. The augmented matrix for this system is given below:

$$\begin{array}{ccccc} P & x_1 & x_2 & s_1 & s_2 \end{array}$$
$$\left[\begin{array}{ccccc|c} 0 & 2 & 4 & 1 & 0 & 120 \\ 0 & 2 & 2 & 0 & 1 & 80 \\ 1 & -3 & -4 & 0 & 0 & 0 \end{array} \right]$$

If we write the augmented matrix in the form given next, we have the **initial simplex tableau** for the maximum problem:

BV	P	x_1	x_2	s_1	s_2	RHS	
s_1	0	2	4	1	0	120	
s_2	0	2	2	0	1	80	**(1)**
P	1	−3	−4	0	0	0	

The bottom row of the initial simplex tableau represents the objective function and is called the **objective row**. The rows above it represent the constraints. We separate the objective row from these rows with a horizontal rule. Notice that we have written the symbol for each variable above the column in which its coefficients appear. The notation **BV** stands for **basic variables.** These are the variables that have a coefficient of 1 and 0 elsewhere in their column. The notation **RHS** stands for **right-hand side**, that is, the numbers to the right of the equal sign in each equation.

So far, we have seen this much of the simplex method:

For a maximum problem in standard form:

1. The constraints are changed from inequalities to equations by the introduction of additional variables—one for each constraint and all nonnegative—called **slack variables.**
2. These equations, together with one that describes the objective function, are placed in the **initial simplex tableau.**

EXAMPLE 2 Setting Up the Initial Simplex Tableau

The following maximum problems are in standard form. For each one introduce slack variables and set up the initial simplex tableau.

(a) Maximize

$$P = 3x_1 + 2x_2 + x_3$$

subject to the constraints

$$\begin{aligned} 3x_1 + x_2 + x_3 &\le 30 \\ 5x_1 + 2x_2 + x_3 &\le 24 \\ x_1 + x_2 + 4x_3 &\le 20 \\ x_1 \ge 0 \quad x_2 \ge 0 \quad x_3 &\ge 0 \end{aligned}$$

(b) Maximize

$$P = x_1 + 4x_2 + 3x_3 + x_4$$

subject to the constraints

$$\begin{aligned} 2x_1 + x_2 \qquad\qquad &\le 10 \\ 3x_1 + x_2 + x_3 + 2x_4 &\le 18 \\ x_1 + x_2 + x_3 + x_4 &\le 14 \\ x_1 \ge 0 \quad x_2 \ge 0 \quad x_3 \ge 0 \quad x_4 &\ge 0 \end{aligned}$$

SOLUTION **(a)** We write the objective function in the form

$$P - 3x_1 - 2x_2 - x_3 = 0$$

For each constraint we introduce a nonnegative slack variable to obtain the following system of equations:

$$\begin{aligned} 3x_1 + x_2 + x_3 + s_1 \qquad\qquad &= 30 \\ 5x_1 + 2x_2 + x_3 \qquad + s_2 \qquad &= 24 \\ x_1 + x_2 + 4x_3 \qquad\qquad + s_3 &= 20 \end{aligned}$$

where

$$x_1 \geq 0 \quad x_2 \geq 0 \quad x_3 \geq 0$$
$$s_1 \geq 0 \quad s_2 \geq 0 \quad s_3 \geq 0$$

These equations, together with the objective function P, give the initial simplex tableau:

BV	P	x_1	x_2	x_3	s_1	s_2	s_3	RHS
s_1	0	3	1	1	1	0	0	30
s_2	0	5	2	1	0	1	0	24
s_3	0	1	1	4	0	0	1	20
P	1	−3	−2	−1	0	0	0	0

(b) We write the objective function in the form

$$P - x_1 - 4x_2 - 3x_3 - x_4 = 0$$

For each constraint we introduce a nonnegative slack variable to obtain the system of equations

$$\begin{aligned} 2x_1 + x_2 \qquad\qquad + s_1 \qquad\qquad &= 10 \\ 3x_1 + x_2 + x_3 + 2x_4 \qquad + s_2 \qquad &= 18 \\ x_1 + x_2 + x_3 + x_4 \qquad\qquad + s_3 &= 14 \end{aligned}$$

where

$$x_1 \geq 0 \quad x_2 \geq 0 \quad x_3 \geq 0 \quad x_4 \geq 0$$
$$s_1 \geq 0 \quad s_2 \geq 0 \quad s_3 \geq 0$$

These equations, together with the objective function P, give the initial simplex tableau:

BV	P	x_1	x_2	x_3	x_4	s_1	s_2	s_3	RHS
s_1	0	2	1	0	0	1	0	0	10
s_2	0	3	1	1	2	0	1	0	18
s_3	0	1	1	1	1	0	0	1	14
P	1	−1	−4	−3	−1	0	0	0	0

Notice that in each initial simplex tableaux an identity matrix appears under the columns headed by P and the slack variables. Notice too that the right-hand column (RHS) always contains nonnegative constants.

NOW WORK PROBLEM 17.

The Pivot Operation

Perform pivot operations 3

Before going any further in our discussion of the simplex method, we need to discuss the matrix operation known as *pivoting*. The first thing one does in a pivot operation is to choose a *pivot element*. However, for now the pivot element will be specified in advance. The method of selecting pivot elements in the simplex tableau will be shown in the next section.

Pivoting

To **pivot** a matrix about a given element, called the **pivot element**, is to apply certain row operations so that the pivot element is replaced by a 1 and all other entries in the same column, called the **pivot column,** become 0s.

Steps for Pivoting

STEP 1 In the pivot row where the pivot element appears, divide each entry by the pivot element, which we assume is not 0. This causes the pivot element to become 1.

STEP 2 Obtain 0s elsewhere in the pivot column by performing row operations using the revised pivot row.

The steps for pivoting utilize two variations of the three row operations for matrices, namely:

Row Operations Used in Pivoting

STEP 1 Replace the pivot row by a positive multiple of that same row.

STEP 2 Replace a row by the sum of that row and a multiple of the pivot row.

Warning! Step 2 requires row operations that must involve the pivot row.

We continue with the initial simplex tableau given in Display (1), page 198, to illustrate the pivot operation.

EXAMPLE 3 Performing a Pivot Operation

Perform a pivot operation on the initial simplex tableau given in Display (1) and repeated below in Display (2), where the pivot element is circled, and the pivot row and pivot column are marked by arrows:

$$\begin{array}{c} \\ \text{BV} \\ \rightarrow s_1 \\ s_2 \\ \hline P \end{array} \begin{array}{c} \\ \\ \\ \\ \\ \end{array} \begin{array}{cccccc|c} & & \downarrow & & & & \\ P & x_1 & x_2 & s_1 & s_2 & & \text{RHS} \\ 0 & 2 & \textcircled{4} & 1 & 0 & & 120 \\ 0 & 2 & 2 & 0 & 1 & & 80 \\ \hline 1 & -3 & -4 & 0 & 0 & & 0 \end{array} \qquad (2)$$

SOLUTION In this tableau the pivot column is column x_2 and the pivot row is row s_1. Step 1 of the pivoting procedure tells us to divide the pivot row by 4, so we use the row operation

$$R_1 = \frac{1}{4}r_1$$

$$
\begin{array}{c}
\begin{matrix} P & x_1 & x_2 & s_1 & s_2 & \text{RHS} \end{matrix}\\
\left[\begin{array}{ccccc|c}
0 & \frac{1}{2} & \textcircled{1} & \frac{1}{4} & 0 & 30 \\
0 & 2 & 2 & 0 & 1 & 80 \\
\hline
1 & -3 & -4 & 0 & 0 & 0
\end{array}\right]
\end{array}
$$

For Step 2 the pivot row is row 1. To obtain 0s elsewhere in the pivot column, we multiply row 1 by -2 and add it to row 2; then we multiply row 1 by 4 and add it to row 3. The row operations specified are

$$R_2 = -2r_1 + r_2 \qquad R_3 = 4r_1 + r_3$$

The new tableau looks like this:

$$
\begin{array}{c|c}
\text{BV} & \begin{matrix} P & x_1 & x_2 & s_1 & s_2 & \text{RHS} \end{matrix}\\
\begin{matrix} x_2 \\ s_2 \\ P \end{matrix} &
\left[\begin{array}{ccccc|c}
0 & \frac{1}{2} & \textcircled{1} & \frac{1}{4} & 0 & 30 \\
0 & 1 & 0 & -\frac{1}{2} & 1 & 20 \\
\hline
1 & -1 & 0 & 1 & 0 & 120
\end{array}\right]
\end{array}
\qquad (3)
$$

This completes the pivot operation since the pivot column (column x_2) has a 1 in the pivot row x_2 and has 0s everywhere else.

This process should look familiar. It is similar to the one used in Chapter 2 to obtain the reduced row echelon form of a matrix.

Analyzing a Tableau

Analyze a tableau **4**

Just what has the pivot operation done? To see, look again at the initial simplex tableau, in Display (2). Observe that the entries in columns P, s_1, and s_2 form an identity matrix (I_3, to be exact). This makes it easy to solve for P, s_1, and s_2 using the other variables as parameters:

$$
\begin{array}{lcl}
2x_1 + 4x_2 + s_1 \qquad = 120 & \text{or} & s_1 = -2x_1 - 4x_2 + 120 \\
2x_1 + 2x_2 \qquad + s_2 = 80 & \text{or} & s_2 = -2x_1 - 2x_2 + 80 \\
P - 3x_1 - 4x_2 \qquad = 0 & \text{or} & P = 3x_1 + 4x_2
\end{array}
$$

The variables s_1, s_2, and P are the original basic variables (BV) listed in the tableau.

After pivoting, we obtain the tableau given in Display (3). Notice that in this form, it is easy to solve for P, x_2, and s_2 in terms of x_1 and s_1.

$$
\begin{aligned}
x_2 &= -\frac{1}{2}x_1 - \frac{1}{4}s_1 + 30 \\
s_2 &= -x_1 + \frac{1}{2}s_1 + 20 \\
P &= x_1 - s_1 + 120
\end{aligned}
\qquad (4)
$$

The variables x_2, s_2, and P are the new basic variables of the tableau. The variables x_1 and s_1 are the **nonbasic variables.** The result of pivoting is that x_2 becomes a basic variable, while s_1 becomes a nonbasic variable.

Notice in Equations (4) that if we let the value of the nonbasic variables x_1 and s_1 equal 0, then the basic variables P, x_2, and s_2 equal the entries across from them in the right-hand side (RHS) of the tableau in Display (3). For the tableau in Display (3) the current value of the objective function is $P = 120$, obtained for $x_1 = 0$, $s_1 = 0$. The values of x_2 and s_2 are $x_2 = 30$, $s_2 = 20$. Because $P = x_1 - s_1 + 120$, $x_1 \geq 0$, and $s_1 \geq 0$, the value of P can be increased beyond 120 when $x_1 > 0$ and $s_1 = 0$. So we have not maximized P yet.

We summarize this discussion below.

Analyzing a Tableau

To obtain the current values of the objective function and the basic variables in a tableau, follow these steps:

STEP 1 From the tableau, write the equation corresponding to each row.
STEP 2 Solve the bottom equation for P and the remaining equations for the basic variables.
STEP 3 Set each nonbasic variable equal to zero to obtain the current values of P and the basic variables.

NOW WORK PROBLEM 29.

EXAMPLE 4 Performing a Pivot Operation; Analyzing a Tableau

Perform another pivot operation on the tableau given in Display (3). Use the circled pivot element in the tableau below. Then analyze the new tableau.

BV	P	x_1 ↓	x_2	s_1	s_2	RHS
x_2	0	$\frac{1}{2}$	1	$\frac{1}{4}$	0	30
→ s_2	0	①	0	$-\frac{1}{2}$	1	20
P	1	-1	0	1	0	120

SOLUTION Since the pivot element happens to be a 1 in this case, we skip Step 1. For Step 2 we perform the row operations

$$R_1 = -\frac{1}{2}r_2 + r_1 \qquad R_3 = r_2 + r_3$$

The result is

BV	P	x_1	x_2	s_1	s_2	RHS
x_2	0	0	1	$\frac{1}{2}$	$-\frac{1}{2}$	20
x_1	0	1	0	$-\frac{1}{2}$	1	20
P	1	0	0	$\frac{1}{2}$	1	140

(5)

In the tableau given in Display (5), the new basic variables are x_2, x_1, and P. The variables s_1 and s_2 are the nonbasic variables. The result of pivoting caused x_1 to become a basic variable and s_2 to become a nonbasic variable. Finally, the equations represented by Display (5) can be written as

$$\begin{aligned} x_2 &= -\frac{1}{2}s_1 + \frac{1}{2}s_2 + 20 \\ x_1 &= \frac{1}{2}s_1 - s_2 + 20 \\ P &= -\frac{1}{2}s_1 - s_2 + 140 \end{aligned}$$

If we let the nonbasic variables s_1 and s_2 equal 0, then the current values of the basic variables are $P = 140$, $x_2 = 20$, and $x_1 = 20$. The second pivot has improved the value of P from 120 to 140.

Because $P = -\frac{1}{2}s_1 - s_2 + 140$ and $s_1 \geq 0$ and $s_2 \geq 0$, the value of P cannot increase beyond 140 (any values of s_1 and s_2, other than 0, reduce the value of P). So we have maximized P.

COMMENT: The pivot operation can be done using a graphing utility such as the TI-83Plus. The augmented matrix contained in the initial tableau is entered into the calculator. Once the pivot element has been identified, the elementary row operations built into the graphing utility are used to obtain a new tableau with a pivot element of one and all other elements in the pivot column replaced with zeros. This is best seen using an example.

EXAMPLE 5 Performing a Pivot Operation Using a Graphing Utility

The following maximum problem is in standard form,

Maximize

$$P = 2x_1 + 3x_2$$

subject to the constraints

$$\begin{aligned} 3x_1 + 2x_2 &\leq 200 \\ x_1 + 3x_2 &\leq 150 \\ x_1 \geq 0 \qquad x_2 &\geq 0 \end{aligned}$$

SOLUTION By introducing nonnegative slack variables s_1 and s_2, and by rewriting the objective function in the form

$$P - 2x_1 - 3x_2 = 0$$

we obtain the system of equations

$$\begin{aligned} 3x_1 + 2x_2 + s_1 \qquad &= 200 \\ x_1 + 3x_2 + \qquad s_2 &= 150 \\ P - 2x_1 - 3x_2 \qquad &= \ \ 0 \end{aligned}$$

where $\quad x_1 \geq 0 \quad x_2 \geq 0 \quad s_1 \geq 0 \quad s_2 \geq 0$

The initial simplex tableau is

	BV	P	x_1	x_2 ↓	s_1	s_2	RHS
	s_1	0	3	2	1	0	200
→	s_2	0	1	(3)	0	1	150
	P	1	−2	−3	0	0	0

(6)

The pivot element 3 is circled and the pivot row and pivot column are marked with an arrow.

At this point we enter the initial tableau (6) into the graphing utility. See Figure 1. The matrix has 3 rows and 6 columns and has been named matrix $[A]$. Using the pivot element 3 as indicated above, we follow the steps below using the elementary row operations in the calculator to obtain the new tableau.

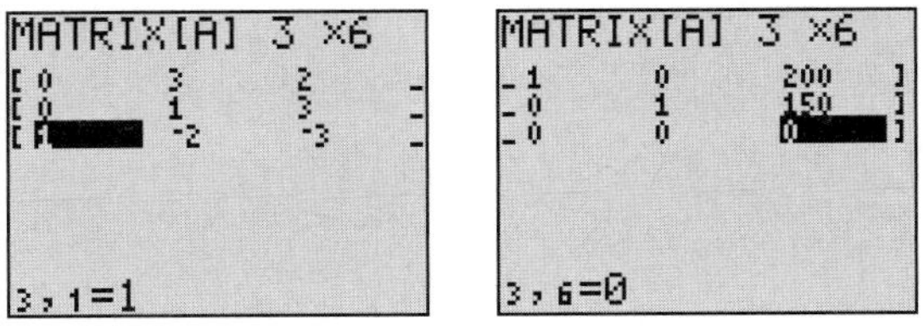

FIGURE 1 The initial tableau.

STEP 1 Make the pivot element one by dividing the *pivot row* by 3. On a TI 83Plus, the operation is:

matrix math E: *row((hit enter)

*row(1/3, [A], 2)

STEP 2 Obtain 0s elsewhere in the *pivot column* by performing row operations using the revised pivot row from Step 1. To obtain 0s elsewhere in the pivot column, we multiply row 2 by -2 and add it to row 1, then we multiply row 2 by 3 and add it to row 3. On a TI 83Plus, the operations are:

matrix math F: *row+((hit enter)

*row+(−2, ANS, 2, 1)

matrix math F: *row+((hit enter)

*row+(3, ANS, 2, 3)

Figure 2 shows the result. In Figure 2 the pivot column has a 1 in the pivot row and 0s everywhere else, as the pivot operation is complete.

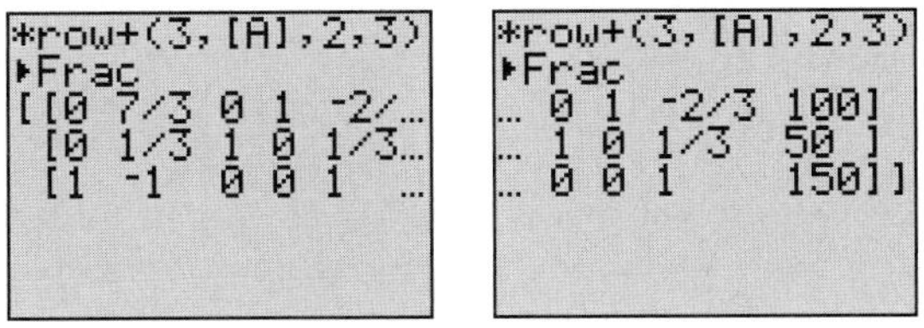

FIGURE 2 The simplex tableau after one complete iteration.

EXERCISE 4.1 Answers to Odd-Numbered Problems Begin on Page AN-21.

In Problems 1–10, determine which maximum problems are in standard form. Do not attempt to solve them!

1. Maximize

$$P = 2x_1 + x_2$$

subject to the constraints

$$\begin{aligned} x_1 + x_2 &\le 5 \\ 2x_1 + 3x_2 &\le 2 \\ x_1 \ge 0 \quad x_2 &\ge 0 \end{aligned}$$

2. Maximize

$$P = 3x_1 + 4x_2$$

subject to the constraints

$$\begin{aligned} 3x_1 + x_2 &\le 6 \\ x_1 + 4x_2 &\le 74 \\ x_1 \ge 0 \quad x_2 &\ge 0 \end{aligned}$$

3. Maximize

$$P = 3x_1 + x_2 + x_3$$

subject to the constraints

$$\begin{aligned} x_1 + x_2 + x_3 &\le 6 \\ 2x_1 + 3x_2 + 4x_3 &\le 10 \\ x_1 &\ge 0 \end{aligned}$$

4. Maximize

$$P = 2x_1 + x_2 + 4x_3$$

subject to the constraints

$$\begin{aligned} 2x_1 + x_2 + x_3 &\le 10 \\ x_2 &\ge 0 \end{aligned}$$

5. Maximize

$$P = 3x_1 + x_2 + x_3$$

subject to the constraints

$$\begin{aligned} x_1 + x_2 + x_3 &\le 8 \\ 2x_1 + x_2 + 4x_3 &\ge 6 \\ x_1 \ge 0 \quad x_2 &\ge 0 \end{aligned}$$

6. Maximize

$$P = 2x_1 + x_2 + 4x_3$$

subject to the constraints

$$\begin{aligned} 2x_1 + x_2 + x_3 &\le -1 \\ x_1 \ge 0 \quad x_2 &\ge 0 \end{aligned}$$

7. Maximize

$$P = 2x_1 + x_2$$

subject to the constraints

$$\begin{aligned} x_1 + x_2 &\ge -6 \\ 2x_1 + x_2 &\le 4 \\ x_1 \ge 0 \quad x_2 &\ge 0 \end{aligned}$$

8. Maximize

$$P = 3x_1 + x_2$$

subject to the constraints

$$\begin{aligned} x_1 + 3x_2 &\le 4 \\ 2x_1 - x_2 &\ge 1 \\ x_1 \ge 0 \quad x_2 &\ge 0 \end{aligned}$$

9. Maximize

$$P = 2x_1 + x_2 + 3x_3$$

subject to the constraints

$$\begin{aligned} x_1 + x_2 - x_3 &\leq 10 \\ x_2 + x_3 &\leq 4 \\ x_1 \geq 0 \quad x_2 \geq 0 \quad x_3 &\geq 0 \end{aligned}$$

10. Maximize

$$P = 2x_1 + 2x_2 + 3x_3$$

subject to the constraints

$$\begin{aligned} x_1 - x_2 + x_3 &\leq 6 \\ x_1 &\leq 4 \\ x_1 \geq 0 \quad x_2 \geq 0 \quad x_3 &\geq 0 \end{aligned}$$

In Problems 11–16, each maximum problem is not in standard form. Determine if the problem can be modified so as to be in standard form. If it can, write the modified version.

11. Maximize

$$P = x_1 + x_2$$

subject to the constraints

$$\begin{aligned} 3x_1 - 4x_2 &\leq -6 \\ x_1 + x_2 &\leq 4 \\ x_1 \geq 0 \quad x_2 &\geq 0 \end{aligned}$$

12. Maximize

$$P = 2x_1 + 3x_2$$

subject to the constraints

$$\begin{aligned} -4x_1 + 2x_2 &\geq -8 \\ x_1 - x_2 &\leq 6 \\ x_1 \geq 0 \quad x_2 &\geq 0 \end{aligned}$$

13. Maximize

$$P = x_1 + x_2 + x_3$$

subject to the constraints

$$\begin{aligned} x_1 + x_2 + x_3 &\leq 6 \\ 4x_1 + 3x_2 &\geq 12 \\ x_1 \geq 0 \quad x_2 \geq 0 \quad x_3 &\geq 0 \end{aligned}$$

14. Maximize

$$P = 2x_1 + x_2 + 3x_3$$

subject to the constraints

$$\begin{aligned} x_1 + x_2 + x_3 &\geq -8 \\ x_1 - x_2 &\leq -6 \\ x_1 \geq 0 \quad x_2 \geq 0 \quad x_3 &\geq 0 \end{aligned}$$

15. Maximize

$$P = 2x_1 + x_2 + 3x_3$$

subject to the constraints

$$\begin{aligned} -x_1 + x_2 + x_3 &\geq -6 \\ 2x_1 - 3x_2 &\geq -12 \\ x_3 &\leq 2 \\ x_1 \geq 0 \quad x_2 \geq 0 \quad x_3 &\geq 0 \end{aligned}$$

16. Maximize

$$P = x_1 + x_2 + x_3$$

subject to the constraints

$$\begin{aligned} 2x_1 - x_2 + 3x_3 &\leq 8 \\ x_1 - x_2 &\geq 6 \\ x_3 &\leq 4 \\ x_1 \geq 0 \quad x_2 \geq 0 \quad x_3 &\geq 0 \end{aligned}$$

In Problems 17–24, each maximum problem is in standard form. For each one introduce slack variables and set up the initial simplex tableau.

17. Maximize

$$P = 2x_1 + x_2 + 3x_3$$

subject to the constraints

$$\begin{aligned} 5x_1 + 2x_2 + x_3 &\leq 20 \\ 6x_1 - x_2 + 4x_3 &\geq 24 \\ x_1 + x_2 + 4x_3 &\leq 16 \\ x_1 \geq 0 \quad x_2 \geq 0 \quad x_3 &\geq 0 \end{aligned}$$

18. Maximize

$$P = 3x_1 + 2x_2 + x_3$$

subject to the constraints

$$\begin{aligned} 3x_1 + 2x_2 - x_3 &\leq 10 \\ x_1 - x_2 + 3x_3 &\leq 12 \\ 2x_1 + x_2 + x_3 &\leq 6 \\ x_1 \geq 0 \quad x_2 \geq 0 \quad x_3 &\geq 0 \end{aligned}$$

19. Maximize

$$P = 3x_1 + 5x_2$$

subject to the constraints

$$\begin{aligned} 2.2x_1 - 1.8x_2 &\leq 5 \\ 0.8x_1 + 1.2x_2 &\leq 2.5 \\ x_1 + x_2 &\leq 0.1 \\ x_1 \geq 0 \quad x_2 &\geq 0 \end{aligned}$$

20. Maximize

$$P = 2x_1 + 3x_2$$

subject to the constraints

$$\begin{aligned} 1.2x_1 - 2.1x_2 &\leq 0.5 \\ 0.3x_1 + 0.4x_2 &\leq 1.5 \\ x_1 + x_2 &\leq 0.7 \\ x_1 \geq 0 \quad x_2 &\geq 0 \end{aligned}$$

21. Maximize

$$P = 2x_1 + 3x_2 + x_3$$

subject to the constraints

$$\begin{aligned} x_1 + x_2 + x_3 &\leq 50 \\ 3x_1 + 2x_2 + x_3 &\leq 10 \\ x_1 \geq 0 \quad x_2 \geq 0 \quad x_3 &\geq 0 \end{aligned}$$

22. Maximize

$$P = x_1 + 4x_2 + 2x_3$$

subject to the constraints

$$\begin{aligned} 3x_1 + x_2 + x_3 &\leq 10 \\ x_1 + x_2 + 3x_3 &\leq 5 \\ x_1 \geq 0 \quad x_2 \geq 0 \quad x_3 &\geq 0 \end{aligned}$$

23. Maximize

$$P = 3x_1 + 4x_2 + 2x_3$$

subject to the constraints

$$\begin{aligned} 3x_1 + x_2 + 4x_3 &\le 5 \\ x_1 - x_2 \qquad\quad &\le 5 \\ 2x_1 - x_2 + x_3 &\le 6 \\ x_1 \ge 0 \quad x_2 \ge 0 \quad x_3 &\ge 0 \end{aligned}$$

24. Maximize

$$P = 2x_1 + x_2 + 3x_3$$

subject to the constraints

$$\begin{aligned} 2x_1 + x_2 + x_3 &\le 2 \\ x_1 - x_2 \qquad\quad &\le 4 \\ 2x_1 + x_2 - x_3 &\le 5 \\ x_1 \ge 0 \quad x_2 \ge 0 \quad x_3 &\ge 0 \end{aligned}$$

In Problems 25–28, each maximum problem can be modified so as to be in standard form. Write the modified version and, for each one, introduce slack variables and set up the initial simplex tableau.

25. Maximize

$$P = x_1 + 2x_2 + 5x_3$$

subject to the constraints

$$\begin{aligned} -x_1 + 2x_2 + 3x_3 &\ge -10 \\ 3x_1 + x_2 - x_3 &\ge -12 \\ x_1 \ge 0 \quad x_2 \ge 0 \quad x_3 &\ge 0 \end{aligned}$$

26. Maximize

$$P = 2x_1 + 4x_2 + x_3$$

subject to the constraints

$$\begin{aligned} -2x_1 - 3x_2 + x_3 &\ge -8 \\ -3x_1 + x_2 - 2x_3 &\ge -12 \\ 2x_1 + x_2 + x_3 &\le 10 \\ x_1 \ge 0 \quad x_2 \ge 0 \quad x_3 &\ge 0 \end{aligned}$$

27. Maximize

$$P = 2x_1 + 3x_2 + x_3 + 6x_4$$

subject to the constraints

$$\begin{aligned} -x_1 + x_2 + 2x_3 + x_4 &\le 10 \\ x_1 - x_2 + x_3 - x_4 &\ge -8 \\ x_1 + x_2 + x_3 + x_4 &\le 9 \\ x_1 \ge 0 \quad x_2 \ge 0 \quad x_3 \ge 0 \quad x_4 &\ge 0 \end{aligned}$$

28. Maximize

$$P = x_1 + 5x_2 + 3x_3 + 6x_4$$

subject to the constraints

$$\begin{aligned} x_1 - x_2 + 2x_3 - 2x_4 &\ge -8 \\ -x_1 + x_2 + x_3 - x_4 &\ge -10 \\ x_1 + x_2 + x_3 + x_4 &\le 12 \\ x_1 \ge 0 \quad x_2 \ge 0 \quad x_3 \ge 0 \quad x_4 &\ge 0 \end{aligned}$$

In Problems 29–33, perform a pivot operation on each tableau. The pivot element is circled. Using the new tableau obtained, write the corresponding system of equations. Indicate the current values of the objective function and the basic variables.

29.

BV	P	x_1	x_2	s_1	s_2	RHS
s_1	0	1	$\textcircled{2}$	1	0	300
s_2	0	3	2	0	1	480
P	1	−1	−2	0	0	0

30.

BV	P	x_1	x_2	s_1	s_2	RHS
s_1	0	1	4	1	0	100
s_2	0	$\textcircled{2}$	5	0	1	50
P	1	−2	−1	0	0	0

31.

BV	P	x_1	x_2	x_3	s_1	s_2	s_3	RHS
s_1	0	1	2	4	1	0	0	24
s_2	0	2	−1	1	0	1	0	32
s_3	0	3	2	$\textcircled{4}$	0	0	1	18
P	1	−1	−2	−3	0	0	0	0

32.

BV	P	x_1	x_2	x_3	s_1	s_2	s_3	RHS
s_1	0	1	2	1	1	0	0	6
s_2	0	2	3	1	0	1	0	12
s_3	0	1	−2	$\textcircled{3}$	0	0	1	0
P	1	−1	−2	−3	0	0	0	0

33.

BV	P	x_1	x_2	x_3	x_4	s_1	s_2	s_3	s_4	RHS
s_1	0	−3	0	1	0	1	0	0	0	20
s_2	0	2	0	0	$\textcircled{1}$	0	1	0	0	24
s_3	0	0	−3	1	0	0	0	1	0	28
s_4	0	0	−3	0	1	0	0	0	1	24
P	1	−1	−2	−3	−4	0	0	0	0	0

4.2 The Simplex Method: Solving Maximum Problems in Standard Form

OBJECTIVES
1. Solve maximum problems in standard form using the simplex method
2. Determine if a tableau is final, requires additional pivoting, or indicates no solution

Solve maximum problems in standard form using the simplex method 1

We are now ready to state the details of the simplex method for solving a maximum problem. This method requires that the problem be in standard form and that the problem be placed in an initial simplex tableau with slack variables.

The Simplex Method for Solving a Maximum Problem in Standard Form

STEP 1 Begin with the initial simplex tableau of a maximum problem in standard form.

STEP 2 Identify the smallest entry in the objective row.

STEP 3 If it is nonnegative, STOP. A solution has been found.

If it is negative, the pivot column is found. (In case of ties, choose any one of the tied entries.)

STEP 4 If all the entries in the pivot column are zero or negative, STOP. The problem is unbounded and has no solution.

For each positive entry above the objective row in the pivot column, compute the value of the entry in the far right column (RHS) divided by the corresponding positive entry of the pivot column. (Note that the objective row is ignored)

STEP 5 The pivot row is the one having the smallest nonnegative quotient. (In case of ties, choose any one of the tied entries.)

STEP 6 The pivot element is the entry in the pivot row and pivot column. (Note that the pivot element can never be in the objective row.) Pivot and repeat the process from Step 2 until a STOP is obtained.

Let's go through the process.

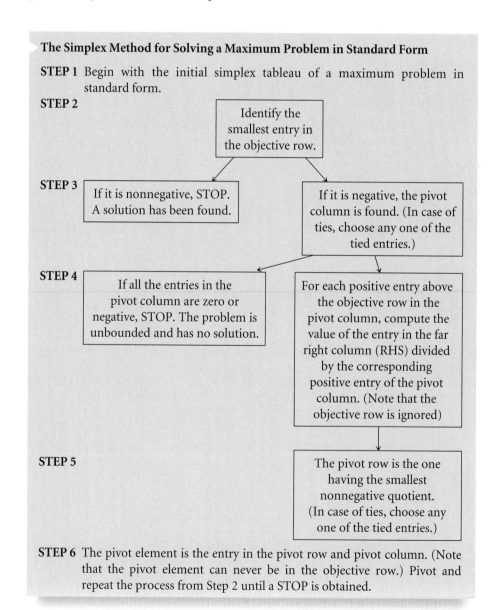

EXAMPLE 1 Using the Simplex Method

Maximize

$$P = 3x_1 + 4x_2$$

subject to the constraints

$$\begin{aligned} 2x_1 + 4x_2 &\le 120 \\ 2x_1 + 2x_2 &\le 80 \\ x_1 \ge 0 \quad & x_2 \ge 0 \end{aligned}$$

SOLUTION **STEP 1** This is a maximum problem in standard form. To obtain the initial simplex tableau, we proceed as follows: The objective function is written in the form

$$P - 3x_1 - 4x_2 = 0$$

After introducing slack variables s_1 and s_2, the constraints take the form

$$\begin{aligned} 2x_1 + 4x_2 + s_1 \qquad &= 120 \\ 2x_1 + 2x_2 \qquad + s_2 &= 80 \end{aligned}$$

where

$$x_1 \ge 0 \quad x_2 \ge 0$$
$$s_1 \ge 0 \quad s_2 \ge 0$$

The initial simplex tableau is

BV	P	x_1	x_2	s_1	s_2	RHS
s_1	0	2	4	1	0	120
s_2	0	2	2	0	1	80
P	1	−3	−4	0	0	0

STEP 2 The smallest entry in the objective row is −4.

STEP 3 Since −4 is negative, the pivot column is column x_2.

STEP 4 For each positive entry above the objective row in the pivot column, form the quotient of the corresponding RHS entry divided by the positive entry.

BV	***Positive entry, x_2***	***RHS***	***Quotient***
s_1	4	120	$120 \div 4 = 30$
s_2	2	80	$80 \div 2 = 40$

STEP 5 The smallest nonnegative value is 30, so the pivot row is row s_1. The tableau below shows the pivot element (circled) and the current values.

	BV	P	x_1	↓ x_2	s_1	s_2	RHS	Current values
→	s_1	0	2	(4)	1	0	120	$s_1 = 120$
	s_2	0	2	2	0	1	80	$s_2 = 80$
	P	1	−3	−4	0	0	0	$P = 0$

(1)

STEP 6 The pivot element is 4. Next, we pivot by using the row operations:

1. $R_1 = \frac{1}{4}r_1$
2. $R_2 = -2r_1 + r_2, \qquad R_3 = 4r_1 + r_3$

After pivoting, we have this tableau:

BV	P	x_1	x_2	s_1	s_2	RHS	Current values
x_2	0	$\frac{1}{2}$	1	$\frac{1}{4}$	0	30	$x_2 = 30$
s_2	0	1	0	$-\frac{1}{2}$	1	20	$s_2 = 20$
P	1	-1	0	1	0	120	$P = 120$

(2)

Notice that x_2 is now a basic variable, replacing s_1. The objective row is

$$P = x_1 - s_1 + 120$$

P has a current value of 120. Because $P = x_1 - s_1 + 120$ and $x_1 \geq 0$ and $s_1 \geq 0$, the value of P can be increased beyond 120, when $x_1 > 0$ and $s_1 = 0$. So we have not maximized P yet.

We continue with the simplex method at Step 2.

STEP 2 The smallest entry in the objective row is -1.

STEP 3 Since -1 is negative, the pivot column is column x_1.

STEP 4 For each positive entry above the objective row in the pivot column, form the quotient of the corresponding RHS entry by the positive entry.

BV	*Positive entry, x_2*	*RHS*	*Quotient*
x_2	$\frac{1}{2}$	30	$30 \div \frac{1}{2} = 60$
s_2	1	20	$20 \div 1 = 20$

STEP 5 The smallest nonnegative value is 20, so the pivot row is row s_2. The tableau below shows the pivot element (circled).

BV	P	↓ x_1	x_2	s_1	s_2	RHS	Current values
x_2	0	$\frac{1}{2}$	1	$\frac{1}{4}$	0	30	$x_2 = 30$
$\rightarrow s_2$	0	(1)	0	$-\frac{1}{2}$	1	20	$s_2 = 20$
P	1	-1	0	1	0	120	$P = 120$

STEP 6 The pivot element is 1. Next, we pivot by using the row operations:

$$R_1 = -\frac{1}{2}r_2 + r_1 \qquad R_3 = r_2 + r_3$$

The result is this tableau:

BV	P	x_1	x_2	s_1	s_2	RHS	Current values
x_2	0	0	1	$\frac{1}{2}$	$-\frac{1}{2}$	20	$x_2 = 20$
x_1	0	1	0	$-\frac{1}{2}$	1	20	$x_1 = 20$
P	1	0	0	$\frac{1}{2}$	1	140	$P = 140$

(3)

We continue with the simplex method at Step 2.

STEP 2 The smallest entry in the objective row is 0.

STEP 3 Since it is nonnegative, we STOP. To see why we STOP, write the equation from the objective row, namely,

$$P = -\frac{1}{2}s_1 - s_2 + 140$$

Since $s_1 \geq 0$ and $s_2 \geq 0$, any positive value of s_1 or s_2 would make the value of P smaller than 140. By choosing $s_1 = 0$ and $s_2 = 0$, we obtain the largest possible value for P, namely, 140. If we write the equations from the second and third rows, substituting 0 for s_1 and s_2, we have

$$x_2 = -\frac{1}{2}s_1 + \frac{1}{2}s_2 + 20 \qquad \text{so} \quad x_2 = 20 \text{ if } s_1 = 0, s_2 = 0$$

$$x_1 = \frac{1}{2}s_1 - s_2 + 20 \qquad \text{so} \quad x_1 = 20 \text{ if } s_1 = 0, s_2 = 0$$

The maximum value of P is

$$P = 140$$

and it occurs at

$$x_1 = 20 \qquad x_2 = 20 \qquad s_1 = 0 \qquad s_2 = 0$$

The tableau in Display (3) is called the **final tableau** because with this tableau the maximum value is found.

SUMMARY To summarize, each tableau obtained by applying the simplex method provides information about the current status of the solution:

1. The RHS entry in the objective row gives the current value of the objective function.
2. The remaining entries in the RHS column give the current values of the corresponding basic variables.

We can trace what happened each time we pivoted. The maximum problem just solved is the same as one we solved using geometrical methods in Section 3.3. Refer to Example 1, page 180, and look at Figure 21, the graph of the set of feasible solutions. Now look at Display (1), where the current values are $s_1 = 120$, $s_2 = 80$, and $P = 0$. The values of x_1 and x_2 are each 0. So, we are at the origin of the set of feasible solutions. After the first pivot, the current values are $x_2 = 30$, $s_2 = 20$, and $P = 120$ [see Display (2)]. The values of the remaining variables are $x_1 = 0$ and $s_1 = 0$. Now we are at the point $(x_1, x_2) = (0, 30)$ of the set of feasible solutions. After the next pivot, the current values are $x_2 = 20$, $x_1 = 20$, and $P = 140$. See Display (3). The remaining variables have the values $s_1 = 0$ and $s_2 = 0$. Now we are at the point (20, 20) of the set of feasible solutions. So the pivot process moves us from one corner point to another corner point, until we reach the corner point that maximizes P.

Next, we discuss the analysis of a simplex tableau in more detail.

2 **EXAMPLE 2** **Analyzing Tableaus**

Determine whether each tableau

(1) is a final tableau. [If it is, give the solution.]
(2) requires additional pivoting. [If so, identify the pivot element.]
(3) indicates no solution.

(a)

BV	P	x_1	x_2	s_1	s_2	RHS
s_1	0	0	0	1	1	40
x_1	0	1	−1	0	1	20
P	1	0	−2	0	1	20

(b)

BV	P	x_1	x_2	s_1	s_2	RHS
x_1	0	1	0	2	−1	40
x_2	0	0	1	−1	1	20
P	1	0	0	1	2	220

(c)

BV	P	x_1	x_2	s_1	s_2	RHS
x_1	0	1	0	1	0.5	60
x_2	0	3	1	0	0.25	125
P	1	−4	0	0	2	10

SOLUTION **(a)** The smallest entry in the objective row is -2, which is negative. The pivot column is column x_2. Since all the entries in the pivot column are zero or negative, the problem is unbounded and has no solution.

(b) The objective row contains no negative entries, so this is a final tableau. The solution is

$$P = 220 \qquad \text{when} \qquad x_1 = 40 \quad x_2 = 20 \quad s_1 = 0 \quad s_2 = 0$$

(c) The smallest entry in the objective row is -4, which is negative. The pivot column is column x_1. Since both entries in the pivot column are positive, compute the quotients of the RHS divided by the corresponding entries in the pivot column

$$60 \div 1 = 60 \qquad 125 \div 3 = 41\frac{2}{3}$$

Since $41\frac{2}{3} < 60$, x_2 is the pivot row and 3 is the pivot element.

NOW WORK PROBLEM 1.

EXAMPLE 3 Solving a Maximum Problem Using the Simplex Method

Maximize

$$P = 6x_1 + 8x_2 + x_3$$

subject to the constraints

$$\begin{aligned} 3x_1 + 5x_2 + 3x_3 &\leq 20 \\ x_1 + 3x_2 + 2x_3 &\leq 9 \\ 6x_1 + 2x_2 + 5x_3 &\leq 30 \\ x_1 \geq 0 \quad x_2 \geq 0 &\quad x_3 \geq 0 \end{aligned}$$

SOLUTION Note that the problem is in standard form. By introducing slack variables s_1, s_2, and s_3, the constraints take the form

$$
\begin{aligned}
3x_1 + 5x_2 + 3x_3 + s_1 \qquad\qquad &= 20 \\
x_1 + 3x_2 + 2x_3 \qquad + s_2 \qquad &= 9 \\
6x_1 + 2x_2 + 5x_3 \qquad\qquad + s_3 &= 30
\end{aligned}
$$

where

$$
\begin{matrix}
x_1 \geq 0 & x_2 \geq 0 & x_3 \geq 0 \\
s_1 \geq 0 & s_2 \geq 0 & s_3 \geq 0
\end{matrix}
$$

Since

$$P - 6x_1 - 8x_2 - x_3 = 0$$

the initial simplex tableau is

BV	P	x_1	x_2 ↓	x_3	s_1	s_2	s_3	RHS	Current values	RHS ÷ x_2
s_1	0	3	5	3	1	0	0	20	$s_1 = 20$	$20 \div 5 = 4$
→ s_2	0	1	$\boxed{3}$	2	0	1	0	9	$s_2 = 9$	$9 \div 3 = 3$
s_3	0	6	2	5	0	0	1	30	$s_3 = 30$	$30 \div 2 = 15$
P	1	-6	-8	-1	0	0	0	0	$P = 0$	

The pivot column is found by locating the column containing the smallest entry in the objective row (-8 in column x_2). The pivot row is obtained by dividing each entry in the RHS column by the corresponding entry in the pivot column and selecting the smallest nonnegative quotient. The pivot row is row s_2. The pivot element is 3, which is circled. After pivoting, the new tableau is

BV	P	x_1 ↓	x_2	x_3	s_1	s_2	s_3	RHS	Current values	RHS ÷ x_1
→ s_1	0	$\boxed{\frac{4}{3}}$	0	$-\frac{1}{3}$	1	$-\frac{5}{3}$	0	5	$s_1 = 5$	$5 \div \frac{4}{3} = 3.75$
x_2	0	$\frac{1}{3}$	1	$\frac{2}{3}$	0	$\frac{1}{3}$	0	3	$x_2 = 3$	$3 \div \frac{1}{3} = 9$
s_3	0	$\frac{16}{3}$	0	$\frac{11}{3}$	0	$-\frac{2}{3}$	1	24	$s_3 = 24$	$24 \div \frac{16}{3} = 4.5$
P	1	$-\frac{10}{3}$	0	$\frac{13}{3}$	0	$\frac{8}{3}$	0	24	$P = 24$	

The value of P has improved to 24, but the negative entry, $-\frac{10}{3}$, in the objective row indicates further improvement is possible. We determine the next pivot element to be $\frac{4}{3}$. After pivoting, we obtain the tableau

BV	P	x_1	x_2	x_3	s_1	s_2 ↓	s_3	RHS	Current values	RHS ÷ s_2
x_1	0	1	0	$-\frac{1}{4}$	$\frac{3}{4}$	$-\frac{5}{4}$	0	$\frac{15}{4}$	$s_1 = \frac{15}{4}$	
x_2	0	0	1	$\frac{3}{4}$	$-\frac{1}{4}$	$\frac{3}{4}$	0	$\frac{7}{4}$	$x_2 = \frac{7}{4}$	$\frac{7}{4} \div \frac{3}{4} = \frac{7}{3}$
→ s_3	0	0	0	5	-4	$\boxed{6}$	1	4	$s_3 = 4$	$4 \div 6 = \frac{2}{3}$
P	1	0	0	$\frac{7}{2}$	$\frac{5}{2}$	$-\frac{3}{2}$	0	$\frac{73}{2}$	$P = \frac{73}{2}$	

The value of P has improved to $\frac{73}{2}$, but, since we still observe a negative entry in the objective row, we pivot again. (Remember, in finding the pivot row, we ignore the objective row and any rows in which the pivot column contains a negative number or zero—in this case, row x_1, containing $-\frac{5}{4}$, is ignored.) The new tableau is

BV	P	x_1	x_2	x_3	s_1	s_2	s_3	RHS	Current values
x_1	0	1	0	$\frac{19}{24}$	$-\frac{1}{12}$	0	$\frac{5}{24}$	$\frac{55}{12}$	$s_1 = \frac{55}{12}$
x_2	0	0	1	$\frac{1}{8}$	$\frac{1}{4}$	0	$-\frac{1}{8}$	$\frac{5}{4}$	$x_2 = \frac{5}{4}$
s_2	0	0	0	$\frac{5}{6}$	$-\frac{2}{3}$	1	$\frac{1}{6}$	$\frac{2}{3}$	$s_3 = \frac{2}{3}$
P	1	0	0	$\frac{19}{4}$	$\frac{3}{2}$	0	$\frac{1}{4}$	$\frac{75}{2}$	$P = \frac{75}{2}$

This is a final tableau since all the entries in the objective row are nonnegative. The objective (bottom) row yields the equation

$$P = -\frac{19}{4}x_3 - \frac{3}{2}s_1 - \frac{1}{4}s_3 + \frac{75}{2}$$

The maximum value of P is $P = \frac{75}{2}$, obtained when $x_3 = 0$, $s_1 = 0$, and $s_3 = 0$. From the rows x_1, x_2, and s_2 of the final tableau we have the equations

$$\begin{aligned} x_1 &= -\frac{19}{24}x_3 + \frac{1}{12}s_1 - \frac{5}{24}s_3 + \frac{55}{12} \\ x_2 &= -\frac{1}{8}x_3 - \frac{1}{4}s_1 + \frac{1}{8}s_3 + \frac{5}{4} \\ s_2 &= -\frac{5}{6}x_3 + \frac{2}{3}s_1 - \frac{1}{6}s_3 + \frac{2}{3} \end{aligned}$$

Using $x_3 = 0$, $s_1 = 0$, and $s_3 = 0$ in these equations, we find

$$x_1 = \frac{55}{12} \qquad x_2 = \frac{5}{4} \qquad s_2 = \frac{2}{3} \tag{4}$$

The solution of the maximum problem is $P = \frac{75}{2}$, obtained when $x_1 = \frac{55}{12}$, $x_2 = \frac{5}{4}$, and $x_3 = 0$.

Note: this solution may also be found by looking at the current values of the final tableau.

NOW WORK PROBLEM 9.

EXAMPLE 4 Solving a Maximum Problem Using the Simplex Method

Maximize

$$P = 4x_1 + 4x_2 + 3x_3$$

subject to the constraints

$$\begin{aligned} 2x_1 + 3x_2 + x_3 &\le 6 \\ x_1 + 2x_2 + 3x_3 &\le 6 \\ x_1 + x_2 + x_3 &\le 5 \end{aligned}$$

$$x_1 \ge 0 \qquad x_2 \ge 0 \qquad x_3 \ge 0$$

SOLUTION The problem is in standard form. We introduce slack variables s_1, s_2, s_3, and write the constraints as

$$\begin{aligned} 2x_1 + 3x_2 + x_3 + s_1 &= 6 \\ x_1 + 2x_2 + 3x_3 + s_2 &= 6 \\ x_1 + x_2 + x_3 + s_3 &= 5 \end{aligned}$$

where

$$x_1 \ge 0 \qquad x_2 \ge 0 \qquad x_3 \ge 0$$
$$s_1 \ge 0 \qquad s_2 \ge 0 \qquad s_3 \ge 0$$

Since $P - 4x_1 - 4x_2 - 3x_3 = 0$, the initial simplex tableau is

BV	P	x_1	x_2	x_3	s_1	s_2	s_3	RHS	RHS $\div x_1$
s_1	0	(2)	3	1	1	0	0	6	$6 \div 2 = 3$
s_2	0	1	2	3	0	1	0	6	$6 \div 1 = 6$
s_3	0	1	1	1	0	0	1	5	$5 \div 1 = 5$
P	1	-4	-4	-3	0	0	0	0	

Since -4 is the smallest negative entry in the objective row, we have a tie for the pivot column between the columns x_1 and x_2. We choose (arbitrarily) as pivot column the column x_1. The pivot row is row s_1. (Do you see why?) The pivot element is 2, which is circled. After pivoting, we obtain the following tableau:

BV	P	x_1	x_2	x_3	s_1	s_2	s_3	RHS	Current values	RHS $\div x_3$
x_1	0	1	$\frac{3}{2}$	$\frac{1}{2}$	$\frac{1}{2}$	0	0	3	$x_1 = 3$	$3 \div \frac{1}{2} = 6$
s_2	0	0	$\frac{1}{2}$	($\frac{5}{2}$)	$-\frac{1}{2}$	1	0	3	$s_2 = 3$	$3 \div \frac{5}{2} = \frac{6}{5}$
s_3	0	0	$-\frac{1}{2}$	$\frac{1}{2}$	$-\frac{1}{2}$	0	1	2	$s_3 = 2$	$2 \div \frac{1}{2} = 4$
P	1	0	2	-1	2	0	0	12	$P = 12$	

The pivot element is $\frac{5}{2}$, which is circled. After pivoting, we obtain

BV	P	x_1	x_2	x_3	s_1	s_2	s_3	RHS	Current values
x_1	0	1	$\frac{7}{5}$	0	$\frac{3}{5}$	$-\frac{1}{5}$	0	$\frac{12}{5}$	$x_1 = \frac{12}{5}$
x_3	0	0	$\frac{1}{5}$	1	$-\frac{1}{5}$	$\frac{2}{5}$	0	$\frac{6}{5}$	$x_3 = \frac{6}{5}$
s_3	0	0	$-\frac{3}{5}$	0	$-\frac{2}{5}$	$-\frac{1}{5}$	1	$\frac{7}{5}$	$s_3 = \frac{7}{5}$
P	1	0	$\frac{11}{5}$	0	$\frac{9}{5}$	$\frac{2}{5}$	0	$\frac{66}{5}$	$P = \frac{66}{5}$

This is a final tableau. The solution is $P = \frac{66}{5}$, obtained when $x_1 = \frac{12}{5}$, $x_2 = 0$, $x_3 = \frac{6}{5}$.

NOW WORK EXAMPLE 4 AGAIN, THIS TIME CHOOSING COLUMN X_2 AS THE PIVOT COLUMN FOR THE FIRST PIVOT.

EXAMPLE 5 Maximizing Profit

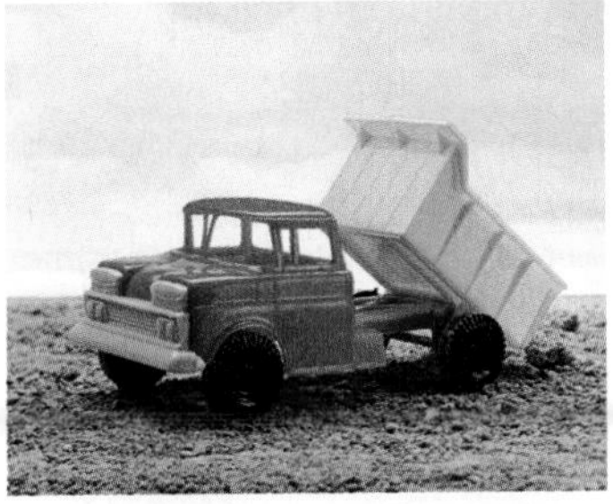

Mike's Famous Toy Trucks specializes in making four kinds of toy trucks: a delivery truck, a dump truck, a garbage truck, and a gasoline truck. Three machines—a metal casting machine, a paint spray machine, and a packaging machine—are used in the production of these trucks. The time, in hours, each machine works to make each type of truck and the profit for each truck are given in Table 1. The maximum time available per week for each machine is given as: metal casting, 4000 hours; paint spray, 1800 hours; and packaging, 1000 hours. How many of each type truck should be produced to maximize profit? Assume that every truck made is sold.

SOLUTION Let x_1, x_2, x_3, and x_4 denote the number of delivery trucks, dump trucks, garbage trucks, and gasoline trucks, respectively, to be made. If P denotes the profit to be maximized, we have this problem:

TABLE 1

	Delivery Truck	Dump Truck	Garbage Truck	Gasoline Truck	Maximum Time
Metal Casting	2 hours	2.5 hours	2 hours	2 hours	4000 hours
Paint Spray	1 hour	1.5 hours	1 hour	2 hours	1800 hours
Packaging	0.5 hour	0.5 hour	1 hour	1 hour	1000 hours
Profit	\$0.50	\$1.00	\$1.50	\$2.00	

Maximize

$$P = 0.5x_1 + x_2 + 1.5x_3 + 2x_4$$

subject to the constraints

$$\begin{aligned} 2x_1 + 2.5x_2 + 2x_3 + 2x_4 &\le 4000 \\ x_1 + 1.5x_2 + x_3 + 2x_4 &\le 1800 \\ 0.5x_1 + 0.5x_2 + x_3 + x_4 &\le 1000 \end{aligned}$$

$$x_1 \ge 0 \qquad x_2 \ge 0 \qquad x_3 \ge 0 \qquad x_4 \ge 0$$

Since this problem is in standard form, we introduce slack variables s_1, s_2, and s_3, write the initial simplex tableau, and solve:

BV	P	x_1	x_2	x_3	x_4	s_1	s_2	s_3	RHS	Current values
s_1	0	2	2.5	2	2	1	0	0	4000	$s_1 = 4000$
s_2	0	1	1.5	1	(2)	0	1	0	1800	$s_2 = 1800$
s_3	0	0.5	0.5	1	1	0	0	1	1000	$s_3 = 1000$
P	1	−0.5	−1	−1.5	−2	0	0	0	0	$P = 0$

BV	P	x_1	x_2	x_3	x_4	s_1	s_2	s_3	RHS	Current values
s_1	0	1	1	1	0	1	−1	0	2200	$s_1 = 2200$
x_4	0	0.5	0.75	0.5	1	0	0.5	0	900	$x_4 = 900$
s_3	0	0	−0.25	(0.5)	0	0	−0.5	1	100	$s_3 = 100$
P	1	0.5	0.5	−0.5	0	0	1	0	1800	$P = 1800$

BV	P	x_1	x_2	x_3	x_4	s_1	s_2	s_3	RHS	Current values
s_1	0	1	1.5	0	0	1	0	−2	2000	$s_1 = 2000$
x_4	0	0.5	1	0	1	0	1	−1	800	$x_4 = 800$
x_3	0	0	−0.5	1	0	0	−1	2	200	$x_3 = 200$
P	1	0.5	0.25	0	0	0	0.5	1	1900	$P = 1900$

This is a final tableau. The maximum profit is $P =$ \$1900, and it is attained for

$$x_1 = 0 \qquad x_2 = 0 \qquad x_3 = 200 \qquad x_4 = 800$$

The practical considerations of the situation described in Example 5 are that delivery trucks and dump trucks are too costly to produce or too little profit is being gained from their sale. Since the slack variable s_1 has a value of 2000 for maximum P and, since s_1 represents the number of hours the metal casting machine is idle, it may be possible to release this machine for other duties. Also note that both the paint spray and packaging machines are operating at full capacity. This means that to increase productivity, more paint spray and packaging capacity is required.

NOW WORK PROBLEM 25.

Use Excel to solve Example 5.

Maximize

$$P = 0.5x_1 + x_2 + 1.5x_3 + 2x_4$$

subject to the constraints

$$\begin{aligned} 2x_1 + 2.5x_2 + 2x_3 + 2x_4 &\le 4000 \\ x_1 + 1.5x_2 + x_3 + 2x_4 &\le 1800 \\ 0.5x_1 + 0.5x_2 + x_3 + x_4 &\le 1000 \end{aligned}$$

$$x_1 \ge 0 \qquad x_2 \ge 0 \qquad x_3 \ge 0 \qquad x_4 \ge 0$$

SOLUTION **STEP 1** Set up an Excel spreadsheet containing the variables, the objective function, and the constraints as follows:

A	B	C
1. Variables		
2. Delivery Truck, x_1	0	
3. Dump Truck, x_2	0	
4. Garbage Truck, x_3	0	
5. Gasoline Truck, x_4	0	
7. Objective		
9. Maximize Profit	=0.5*B2+1*B3+1.5*B4+2*B5	
11. Constraints		
12.	***Amount Used***	***Maximum***
13. Metal Casting	=2*B2+2.5*B3+2*B4+2*B5	4000
14. Paint Spray	=B2+1.5*B3+1*B4+2*B5	1800
15. Packaging	=0.5*B2+0.5*B3+B4+B5	1000

You should obtain:

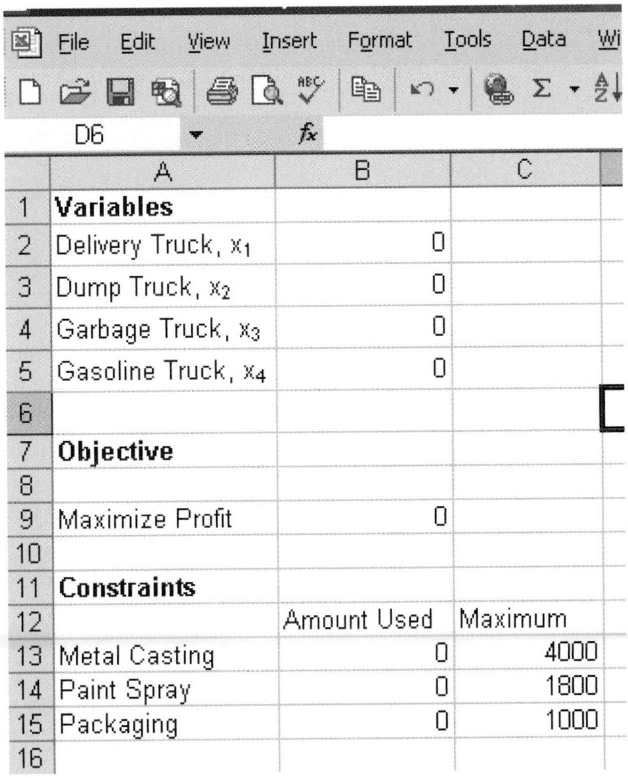
File Edit View Insert Format Tools Data Wi
D6
A B C
1 Variables
2 Delivery Truck, x1 0
3 Dump Truck, x2 0
4 Garbage Truck, x3 0
5 Gasoline Truck, x4 0
6
7 Objective
8
9 Maximize Profit 0
10
11 Constraints
12 Amount Used Maximum
13 Metal Casting 0 4000
14 Paint Spray 0 1800
15 Packaging 0 1000
16

STEP 2 Click on Tools and then Solver to get the screen below.

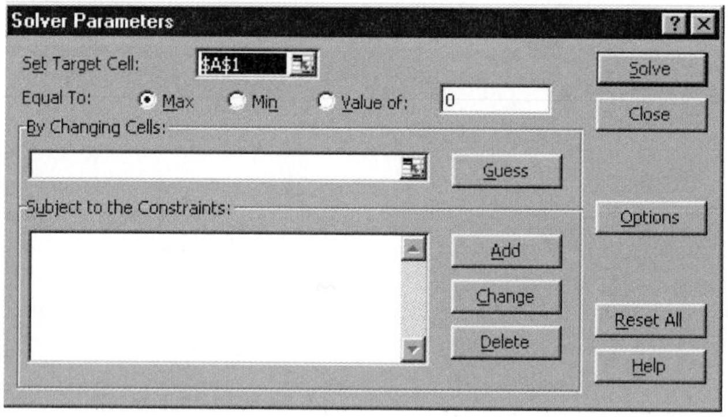

- **Set Target Cell:** is the cell of the optimization function.
- **Equal To:** Allows you to maximize or minimize the objective function.
- **By Changing Cells:** The variable cells.
- **Subject to the Constraints:** The system of inequalities.

STEP 3 Input Set Target Cell, B9, Equal To, Max, By Changing Cells, B2 to B9. Just click on the appropriate cells.

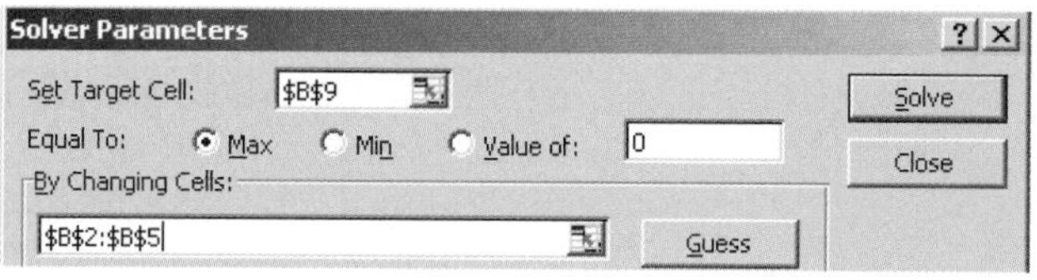

STEP 4 Add the constraints.

- Click cursor into Subject to the Constraints entry box.
- Press Add button. You should see:

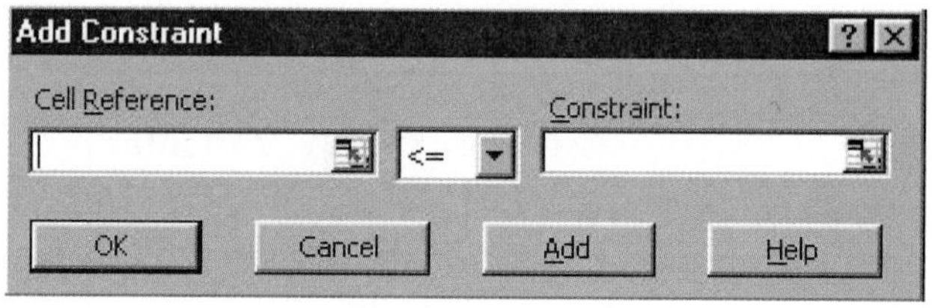

- Cell Reference is the cell containing the formula.
- Constraint is the cell containing the maximum.
- You must also enter the constraint that the variables must be greater than or equal to zero.
- To enter the Metal Casting constraint, put the cursor in the Cell Reference box and click on cell B13; then put the cursor in the Constraint box and click on C13.

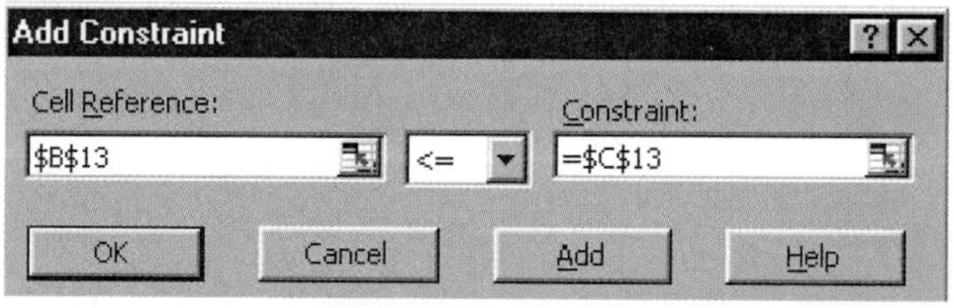

- Click on Add button and repeat this for Paint Spray and Packaging constraints.
- To add the constraint that the variables are greater than or equal to zero:
  - Click on the variable cells.
  - Select >= in the middle box.
  - Type in zero in the constraint box.

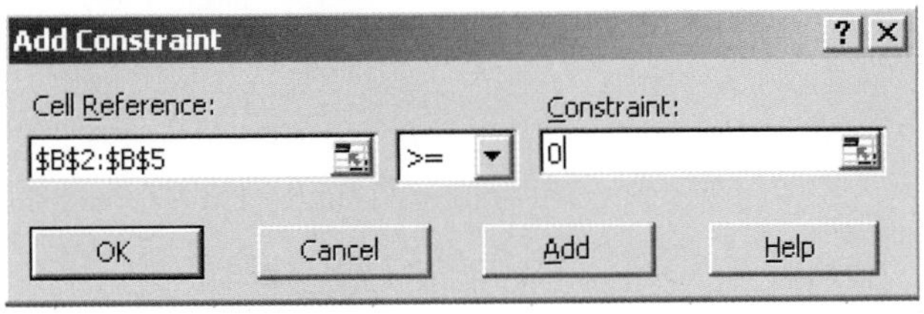

- When you are finished, click on OK.

STEP 5 Your Solver box should look like this:

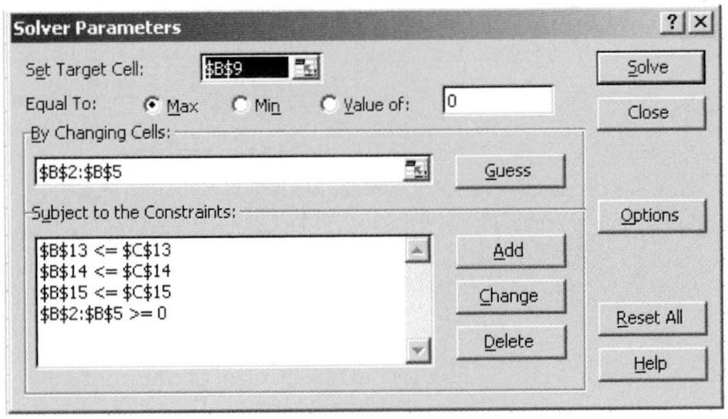

STEP 6 Click on Options box and make sure Assume Linear Model is checked.

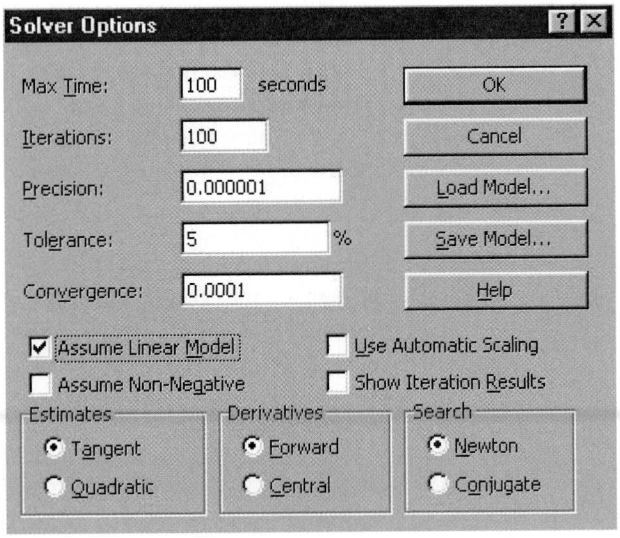

STEP 7 Find the solution: Click on Solve and highlight Answer.

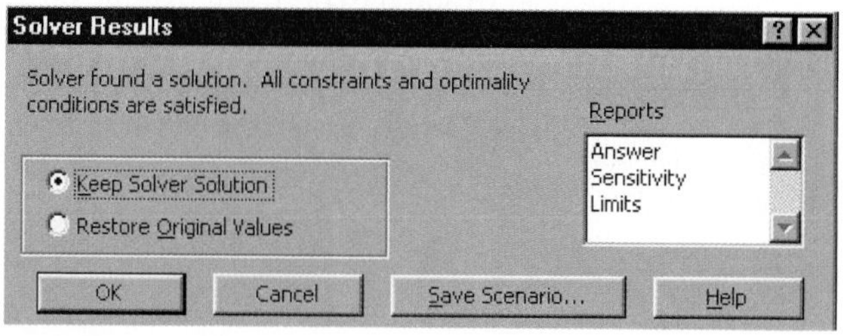
Solver Results
Solver found a solution. All constraints and optimality conditions are satisfied.
Reports
Answer
Sensitivity
Limits
Keep Solver Solution
Restore Original Values
OK
Cancel
Save Scenario...
Help

The solution is in the Excel spread sheet below.

Variables		
Delivery Truck, x_1	0	
Dump Truck, x_2	0	
Garbage Truck, x_3	200	
Gasoline Truck, x_4	800	
Objective		
Maximize Profit	1900	
Constraints		
	Amount Used	Maximum
Metal Casting	2000	4000
Paint Spray	1800	1800
Packaging	1000	1000

To get a maximum profit of \$1900, you should produce 0 delivery trucks, 0 dump trucks, 200 garbage trucks, and 800 gasoline trucks. Only 2000 hours of metal casting are used. The other constraints are used at full capacity.

NOW WORK PROBLEM 25 USING EXCEL.

Analyzing the Simplex Method

To justify some of the steps in the simplex method, we analyze more carefully what we did in Example 5.

The reason we choose the most negative entry in the objective row is that it is the negative of the *largest* coefficient in the objective function:

$$P = 0.5x_1 + x_2 + 1.5x_3 + 2x_4$$

In the initial tableau we set $x_1 = x_2 = x_3 = x_4 = 0$, which results in a profit $P = 0$. Of course, this profit can easily be improved by manufacturing some trucks. The profit per gasoline truck, x_4, is \$2.00, while the profit from the other trucks is \$0.50, \$1.00, and \$1.50, respectively. So it is more effective to increase x_4 than to increase x_1, x_2, or x_3. But by how much can x_4 be increased? That is determined by the constraints.

Referring to Table 1, we see that each gasoline truck takes 2 hours to cast, 2 hours to paint, and 1 hour to package. The limited time available for each task restricts the number of trucks that can be made. So we look at the limits for producing x_4:

metal casting constraint: 4000 hours available, 2 hours per truck; 2000 gasoline trucks can be manufactured.
paint spraying constraint: 1800 hours available, 2 hours per truck; 900 gasoline trucks can be manufactured.
packaging constraint: 1000 hours available, 1 hour per truck; 1000 gasoline trucks can be manufactured.

The paint spraying constraint indicates that we can make at most 900 gasoline trucks. So we choose row s_2 to be the pivot row. This will increase x_4 to as large an amount as possible. After pivoting, the tableau indicates that if we make 900 gasoline trucks, but no other trucks ($x_1 = x_2 = x_3 = 0$), we will make a profit P of \$1800. However, the negative value in the objective row indicates that the profit can be improved if we increase x_3.

Increasing production of x_3, while increasing profit, will also impact the amount of x_4 that can be produced, since some painting time will need to be freed up to produce x_3. This is illustrated in the final tableau, which shows the maximum profit is obtained by producing $x_3 = 200$ garbage trucks and $x_4 = 800$ gasoline trucks.

Each iteration of the simplex method consists in choosing an "entering" variable from the nonbasic variables and a "leaving" variable from the basic variables, using a selective criterion so that the value of the objective function is not decreased (sometimes the value may remain unchanged). The iterative procedure stops when no variable can enter from the nonbasic variables.

The reasoning behind the simplex method for standard maximum problems is not too complicated, and the process of "moving to a better solution" is made quite easy simply by following the rules. Briefly, the pivoting strategy works like this:

> The choice of the pivot column forces us to pivot the variable that apparently improves the value of the objective function most effectively.
>
> The choice of the pivot row prevents us from making this variable *too large* to be feasible.

Geometry of the Simplex Method

The maximum value (provided it exists) of the objective function will occur at one of the corner points of the set of feasible points. The simplex method is designed to move from corner point to corner point of the set of feasible points, at each stage improving the value of the objective function until a solution is found. More precisely, the geometry behind the simplex method is outlined below.

1. A given tableau corresponds to a corner point of the set of feasible points.
2. The operation of pivoting moves us to an adjacent corner point, where the objective function has a value at least as large as it did at the previous corner point.
3. The process continues until the final tableau is reached—which produces a corner point that maximizes the objective function.

Though drawings depicting this process can be rendered in only two or three dimensions (that is, when the objective function has only two or three variables in it), the same interpretation can be shown to hold regardless of the number of variables involved. Let's look at an example.

EXAMPLE 6 The Geometry of the Simplex Method

Maximize

$$P = 3x_1 + 5x_2$$

subject to the constraints

$$x_1 + x_2 \le 60$$
$$x_1 + 2x_2 \le 80$$
$$x_1 \ge 0 \qquad x_2 \ge 0$$

SOLUTION The feasible region is shown in Figure 3.

Below, we apply the simplex method, indicating the corner point corresponding to each tableau and the value of the objective function there. You should supply the details.

FIGURE 3

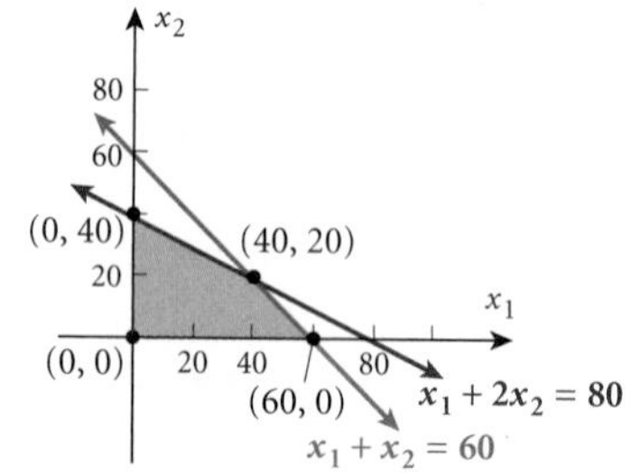

Tableau							*Corner Point* (x_1, x_2)	*Value of* $P = 3x_1 + 5x_2$ *at the Corner Point*
BV	P	x_1	x_2	s_1	s_2	RHS		
s_1	0	1	1	1	0	60		
s_2	0	1	⓶ (2, circled)	0	1	80	(0, 0)	0
P	1	−3	−5	0	0	0		
			(Pivot)					
BV	P	x_1	x_2	s_1	s_2	RHS		
s_1	0	$\frac{1}{2}$ (circled)	0	1	$-\frac{1}{2}$	20	(0, 40)	200
x_2	0	$\frac{1}{2}$	1	0	$\frac{1}{2}$	40		
P	1	$-\frac{1}{2}$	0	0	$\frac{5}{2}$	200		
BV	P	x_1	x_2	s_1	s_2	RHS		
x_1	0	1	0	2	−1	40		
x_2	0	0	1	−1	1	20	(40, 20)	220
P	1	0	0	1	2	220		
			(Final tableau)					(Maximum value)

The Unbounded Case

So far in our discussion, it has always been possible to continue to choose pivot elements until the problem has been solved. But it may turn out that all the entries in a column of a tableau are 0 or negative at some stage. If this happens, it means that the problem is *unbounded* and a maximum solution does not exist.

For example, consider the tableau

BV	P	x_1	x_2	s_1	s_2	RHS
s_1	0	−1	1	1	0	2
s_2	0	1	−1	0	1	2
P	1	−2	−1	0	0	0

FIGURE 4

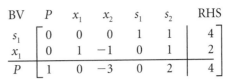

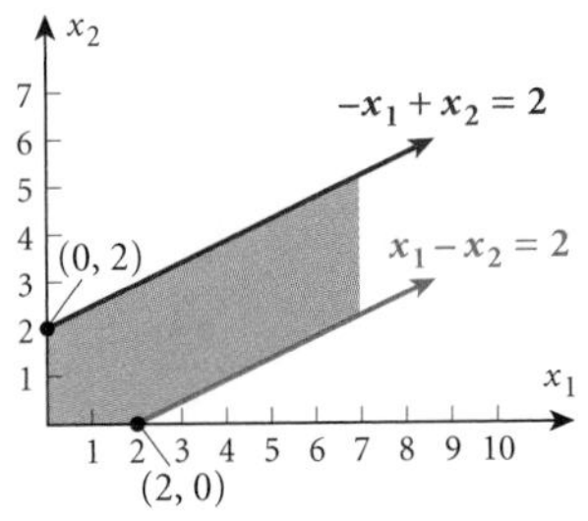

The pivot element is in column x_1 row s_2. After pivoting, the tableau becomes

BV	P	x_1	x_2	s_1	s_2	RHS
s_1	0	0	0	1	1	4
x_1	0	1	−1	0	1	2
P	1	0	−3	0	2	4

Now the only negative entry in the objective (bottom) row is in column x_2, and it is impossible to choose a pivot element in that column. To see why, notice that the constraints are $-x_1 + x_2 \le 2$, $x_1 - x_2 \le 2$, $x_1 \ge 0$, and $x_2 \ge 0$. The set of feasible points shown in Figure 4, is unbounded, so the maximum problem has no solution.

Summary of the Simplex Method

The general procedure for solving a maximum problem in standard form using the simplex method can be outlined as follows:

1. The maximum problem is stated in standard form as

Maximize

$$P = c_1x_1 + c_2x_2 + \cdots + c_nx_n$$

subject to the constraints

$$\begin{aligned} a_{11}x_1 + a_{12}x_2 + \cdots + a_{1n}x_n &\le b_1 \\ a_{21}x_1 + a_{22}x_2 + \cdots + a_{2n}x_n &\le b_2 \\ &\vdots \\ a_{m1}x_1 + a_{m2}x_2 + \cdots + a_{mn}x_n &\le b_m \end{aligned}$$

where

$$x_1 \ge 0, x_2 \ge 0, \ldots, x_n \ge 0$$

$$b_1 > 0, b_2 > 0, \ldots, b_m > 0.$$

2. Introduce slack variables $s_1, s_2, \ldots, s_m$ so that the constraints take the form of equalities:

$$\begin{aligned} a_{11}x_1 + a_{12}x_2 + \cdots + a_{1n}x_n + s_1 &= b_1 \\ a_{21}x_1 + a_{22}x_2 + \cdots + a_{2n}x_n + s_2 &= b_2 \\ &\vdots \\ a_{m1}x_1 + a_{m2}x_2 + \cdots + a_{mn}x_n + s_m &= b_m \end{aligned}$$

$$x_1 \ge 0, x_2 \ge 0, \ldots, x_n \ge 0$$

$$s_1 \ge 0, s_2 \ge 0, \ldots, s_m \ge 0$$

3. Write the objective function in the form

$$P - c_1x_1 - c_2x_2 - \cdots - c_nx_n = 0$$

4. Set up the initial simplex tableau

BV	P	x_1	$x_2 \cdots x_n$	s_1	s_2	s_m	RHS
s_1	0	a_{11}	$a_{12} \cdots a_{1n}$	1	$0 \cdots 0$		b_1
s_2	0	a_{21}	$a_{22} \cdots a_{2n}$	0	$1 \cdots 0$		b_2
$\vdots$	$\vdots$	$\vdots$	$\vdots \quad \vdots$	$\vdots$	$\vdots$	$\vdots$	$\vdots$
s_m	0	a_{m1}	$a_{m2} \cdots a_{mn}$	0	$0 \cdots 1$		b_m
P	1	$-c_1$	$-c_2 \cdots -c_n$	0	$0 \cdots 0$		0

5. Pivot until

(a) All the entries in the objective row are nonnegative. This is a final tableau from which a solution can be read.

Or until

(b) The pivot column is a column whose entries are negative or zero. In this case the problem is unbounded and there is no solution.

The flowchart in Figure 5 illustrates the steps to be used in solving standard maximum linear programming problems.

FIGURE 5

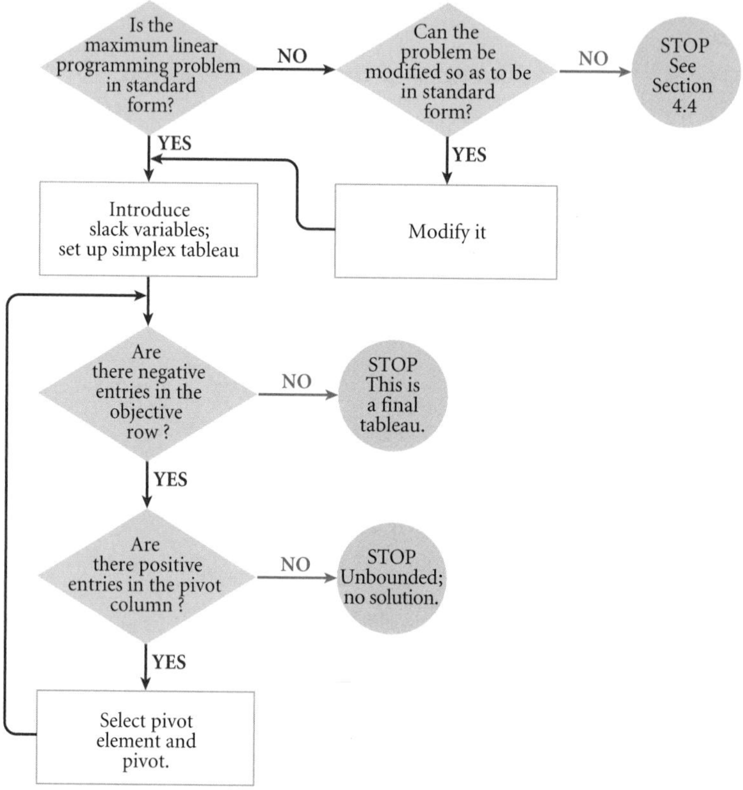

EXERCISE 4.2 Answers to Odd-Numbered Problems Begin on Page AN-22.

In Problems 1–8, determine which of the following statements is true about each tableau:
(a) It is the final tableau.
(b) It requires additional pivoting.
(c) It indicates no solution to the problem.
If the answer is (a), write down the solution; if the answer is (b), indicate the pivot element.

1.

BV	P	x_1	x_2	s_1	s_2	RHS
s_1	0	1	0	1	$-\frac{1}{2}$	20
x_2	0	$\frac{1}{2}$	1	0	$\frac{1}{4}$	30
P	1	-1	0	1	1	120

2.

BV	P	x_1	x_2	s_1	s_2	RHS
x_1	0	1	0	1	$-\frac{1}{2}$	20
x_2	0	0	1	$-\frac{1}{2}$	$\frac{1}{2}$	20
P	1	0	0	1	$\frac{1}{2}$	140

3.

BV	P	x_1	x_2	s_1	s_2	RHS
s_1	0	0	$\frac{1}{14}$	1	$-\frac{1}{7}$	$\frac{186}{21}$
x_1	0	1	$\frac{12}{7}$	0	$\frac{4}{7}$	$\frac{32}{7}$
P	1	0	$\frac{12}{7}$	0	$\frac{32}{7}$	$\frac{256}{7}$

4.

BV	P	x_1	x_2	s_1	s_2	RHS
s_1	0	$\frac{1}{4}$	$\frac{1}{2}$	1	0	10
s_2	0	$\frac{7}{4}$	3	0	1	8
P	1	−8	−12	0	0	0

5.

BV	P	x_1	x_2	s_1	s_2	RHS
x_1	0	1	−2	0	4	24
s_1	0	0	−2	1	4	36
P	1	5	−10	12	4	20

6.

BV	P	x_1	x_2	s_1	s_2	RHS
s_1	0	1	3	1	0	30
s_2	0	2	1	0	1	12
P	1	−2	−5	0	0	0

7.

BV	P	x_1	x_2	s_1	s_2	s_3	RHS
x_2	0	0	1	−2	0	1	6
s_2	0	0	0	1	1	4	6
x_1	0	1	0	1	0	−1	1
P	1	0	0	−10	0	−5	110

8.

BV	P	x_1	x_2	s_1	s_2	s_3	RHS
x_2	0	2	1	0	0	−1	8
s_2	0	−1	0	0	1	5	5
s_1	0	1	0	1	0	−1	1
P	1	10	0	0	0	−15	120

In Problems 9–24, use the simplex method to solve each maximum problem.

9. Maximize

$$P = 5x_1 + 7x_2$$

subject to

$$\begin{aligned} 2x_1 + 3x_2 &\leq 12 \\ 3x_1 + x_2 &\leq 12 \\ x_1 \geq 0 \quad x_2 &\geq 0 \end{aligned}$$

10. Maximize

$$P = x_1 + 5x_2$$

subject to

$$\begin{aligned} 2x_1 + x_2 &\leq 10 \\ x_1 + 2x_2 &\leq 10 \\ x_1 \geq 0 \quad x_2 &\geq 0 \end{aligned}$$

11. Maximize

$$P = 5x_1 + 7x_2$$

subject to

$$\begin{aligned} x_1 + 2x_2 &\leq 2 \\ 2x_1 + x_2 &\leq 2 \\ x_1 \geq 0 \quad x_2 &\geq 0 \end{aligned}$$

12. Maximize

$$P = 5x_1 + 4x_2$$

subject to

$$\begin{aligned} x_1 + x_2 &\leq 2 \\ 2x_1 + 3x_2 &\leq 6 \\ x_1 \geq 0 \quad x_2 &\geq 0 \end{aligned}$$

13. Maximize

$$P = 3x_1 + x_2$$

subject to

$$\begin{aligned} x_1 + x_2 &\leq 2 \\ 2x_1 + 3x_2 &\leq 12 \\ 3x_1 + x_2 &\leq 12 \\ x_1 \geq 0 \quad x_2 &\geq 0 \end{aligned}$$

14. Maximize

$$P = 3x_1 + 5x_2$$

subject to

$$\begin{aligned} 2x_1 + x_2 &\leq 4 \\ x_1 + 2x_2 &\leq 6 \\ x_1 \geq 0 \quad x_2 &\geq 0 \end{aligned}$$

15. Maximize

$$P = 2x_1 + x_2 + x_3$$

subject to

$$\begin{aligned} -2x_1 + x_2 - 2x_3 &\leq 4 \\ x_1 - 2x_2 + x_3 &\leq 2 \\ x_1 \geq 0 \quad x_2 \geq 0 \quad x_3 &\geq 0 \end{aligned}$$

16. Maximize

$$P = 4x_1 + 2x_2 + 5x_3$$

subject to

$$\begin{aligned} x_1 + 3x_2 + 2x_3 &\leq 30 \\ 2x_1 + x_2 + 3x_3 &\leq 12 \\ x_1 \geq 0 \quad x_2 \geq 0 \quad x_3 &\geq 0 \end{aligned}$$

17. Maximize

$$P = 2x_1 + x_2 + 3x_3$$

subject to

$$\begin{aligned} x_1 + 2x_2 + x_3 &\leq 25 \\ 3x_1 + 2x_2 + 3x_3 &\leq 30 \\ x_1 \geq 0 \quad x_2 \geq 0 \quad x_3 &\geq 0 \end{aligned}$$

18. Maximize

$$P = 6x_1 + 3x_2 + 2x_3$$

subject to

$$\begin{aligned} 2x_1 + 2x_2 + 3x_3 &\leq 30 \\ 2x_1 + 2x_2 + x_3 &\leq 12 \\ x_1 \geq 0 \quad x \geq 0 \quad x_3 &\geq 0 \end{aligned}$$

19. Maximize

$$P = 2x_1 + 4x_2 + x_3 + x_4$$

subject to

$$\begin{aligned} 2x_1 + x_2 + 2x_3 + 3x_4 &\leq 12 \\ 2x_2 + x_3 + 2x_4 &\leq 20 \\ 2x_1 + x_2 + 4x_3 \quad &\leq 16 \\ x_1 \geq 0 \quad x_2 \geq 0 \quad x_3 \geq 0 \quad x_4 &\geq 0 \end{aligned}$$

20. Maximize

$$P = 2x_1 + 4x_2 + x_3$$

subject to

$$\begin{aligned} -x_1 + 2x_2 + 3x_3 &\leq 6 \\ -x_1 + 4x_2 + 5x_3 &\leq 5 \\ x_1 + 5x_2 + 7x_3 &\leq 7 \\ x_1 \geq 0 \quad x_2 \geq 0 \quad x_3 &\geq 0 \end{aligned}$$

21. Maximize

$$P = 2x_1 + x_2 + x_3$$

subject to

$$\begin{aligned} x_1 + 2x_2 + 4x_3 &\le 20 \\ 2x_1 + 4x_2 + 4x_3 &\le 60 \\ 3x_1 + 4x_2 + x_3 &\le 90 \\ x_1 \ge 0 \quad x_2 \ge 0 \quad x_3 &\ge 0 \end{aligned}$$

22. Maximize

$$P = x_1 + 2x_2 + 4x_3$$

subject to

$$\begin{aligned} 8x_1 + 5x_2 - 4x_3 &\le 30 \\ -2x_1 + 6x_2 + x_3 &\le 5 \\ -2x_1 + 2x_2 + x_3 &\le 15 \\ x_1 \ge 0 \quad x_2 \ge 0 \quad x_3 &\ge 0 \end{aligned}$$

23. Maximize

$$P = x_1 + 2x_2 + 4x_3 - x_4$$

subject to

$$\begin{aligned} 5x_1 \qquad + 4x_3 + 6x_4 &\le 20 \\ 4x_1 + 2x_2 + 2x_3 + 8x_4 &\le 40 \\ x_1 \ge 0 \quad x_2 \ge 0 \quad x_3 \ge 0 \quad x_4 &\ge 0 \end{aligned}$$

24. Maximize

$$P = x_1 + 2x_2 - x_3 + 3x_4$$

subject to

$$\begin{aligned} 2x_1 + 4x_2 + 5x_3 + 6x_4 &\le 24 \\ 4x_1 + 4x_2 + 2x_3 + 2x_4 &\le 4 \\ x_1 \ge 0 \quad x_2 \ge 0 \quad x_3 \ge 0 \quad x_4 &\ge 0 \end{aligned}$$

25. Process Utilization A jean manufacturer makes three types of jeans, each of which goes through three manufacturing phases—cutting, sewing, and finishing. The number of minutes each type of product requires in each of the three phases is given below:

Jean	Cutting	Sewing	Finishing
I	8	12	4
II	12	18	8
III	18	24	12

There are 5200 minutes of cutting time, 6000 minutes of sewing time, and 2200 minutes of finishing time each day. The company can sell all the jeans it makes and make a profit of \$3 on each Jean I, \$4.50 on each Jean II, and \$6 on each Jean III. Determine the number of jeans in each category that should be made each day to maximize profits.

26. Process Utilization A company manufactures three types of toys *A*, *B*, and *C*. Each requires rubber, plastic, and aluminum as listed below:

Toy	Rubber	Plastic	Aluminum
A	2	2	4
B	1	2	2
C	1	2	4

The company has available 600 units of rubber, 800 units of plastic, and 1400 units of aluminum. The company makes a profit of \$4, \$3, and \$2 on toys *A*, *B*, and *C*, respectively. Assuming all toys manufactured can be sold, determine a production order so that profit is maximum.

27. Scheduling Products A, B, and C are sold door-to-door. Product A costs \$3 per unit, takes 10 minutes to sell (on the average), and costs \$0.50 to deliver to the customer. Product B costs \$5, takes 15 minutes to sell, and is left with the customer at the time of sale. Product C costs \$4, takes 12 minutes to sell, and costs \$1.00 to deliver. During any week a salesperson is allowed to draw up to \$500 worth of A, B, and C (at cost) and is allowed delivery expenses not to exceed \$75. If a salesperson's selling time is not expected to exceed 30 hours (1800 minutes) in a week, and if the salesperson's profit (net after all expenses) is \$1 each on a unit of A or B and \$2 on a unit of C, what combination of sales of A, B, and C will lead to maximum profit and what is this maximum profit?

28. Resource Allocations Suppose that a large hospital classifies its surgical operations into three categories according to their length and charges a fee of \$600, \$900, and \$1200, respectively, for each of the categories. The average time of the operations in the three categories is 30 minutes, 1 hour, and 2 hours, respectively; the hospital has four operating rooms, each of which can be used for 10 hours per day. If the total number of operations cannot exceed 60, how many of each type should the hospital schedule to maximize its revenues?

29. Mixture The Lee refinery blends high and low octane gasoline into three intermediate grades: regular, premium, and super premium. The regular grade consists of 60% high octane and 40% low octane, the premium consists of 70% high octane and 30% low octane, and the super premium consists of 80% high octane and 20% low octane. The company has available 140,000 gallons of high octane and 120,000 gallons of low octane, but can mix only 225,000 gallons. Regular gas sells for \$1.20 per gallon, premium sells for \$1.30 per gallon, and super premium sells for \$1.40 per gallon. How many gallons of each grade should the company mix in order to maximize revenues?

30. Mixture Repeat Problem 29 under the additional assumption that the combined total number of gallons produced by the refinery cannot exceed 200,000 gallons.

31. Investment A financial consultant has at most $90,000 to invest in stocks, corporate bonds, and municipal bonds. The average yields for stocks, corporate bonds, and municipal bonds is 10%, 8%, and 6%, respectively. Determine how much she should invest in each security to maximize the return on her investments, if she has decided that her investment in stocks should not exceed half her funds, and that the difference between her investment in corporate bonds and municipal bonds should not be more than 20% of her investment.

32. Investment Repeat Problem 31 under the assumption that no more than $25,000 can be invested in stocks and that the difference between her investment in corporate bonds and municipal bonds should not be more than 40% of her investment.

33. Crop Planning A farmer has at most 200 acres of farmland suitable for cultivating crops A, B, and C. The costs for cultivating crops A, B, and C are $40, $50, and $30 per acre, respectively. The farmer has a maximum of $18,000 available for land cultivation. Crops A, B, and C require 20, 30, and 15 hours per acre of labor, respectively, and there is a maximum of 4200 hours of labor available. If the farmer expects to make a profit of $70, $90, and $50 per acre on crops A, B, and C, respectively, how many acres of each crop should he plant in order to maximize his profit?

34. Crop Planning Repeat Problem 33 if the farmer modified his allocations as follows:

	Cost of Cultivating			Maximum
	Crop A	Crop B	Crop C	Available
Cost	$30	$40	$20	$12,000
Hours	10	20	18	3,600
Profit	$50	$60	$40	

35. Mixture Problem Nutt's Nut Company has 500 pounds of peanuts, 100 pounds of pecans, and 50 pounds of cashews on hand. They package three types of 5-pound cans of nuts: can I contains 3 pounds of peanuts, 1 pound of pecans, and 1 pound of cashews; can II contains 4 pounds of peanuts, $\frac{1}{2}$ pound of pecans, and $\frac{1}{2}$ pound of cashews; and can III contains 5 pounds of peanuts. The selling price is $28 for can I, $24 for can II, and $20 for can III. How many cans of each kind should be made to maximize revenue?

36. Maximizing Profit One of the methods used by the Alexander Company to separate copper, lead, and zinc from ores is the flotation separation process. This process consists of three steps: oiling, mixing, and separation. These steps must be applied for 2, 2, and 1 hour, respectively, to produce 1 unit of copper; 2, 3, and 1 hour, respectively, to produce 1 unit of lead; and 1, 1, and 3 hours, respectively, to produce 1 unit of zinc. The oiling and separation phases of the process can be in operation for a maximum of 10 hours a day, while the mixing phase can be in operation for a maximum of 11 hours a day. The Alexander Company makes a profit of $45 per unit of copper, $30 per unit of lead, and $35 per unit of zinc. The demand for these metals is unlimited. How many units of each metal should be produced daily by use of the flotation process to achieve the highest profit?

37. Maximizing Profit A wood cabinet manufacturer produces cabinets for television consoles, stereo systems, and radios, each of which must be assembled, decorated, and crated. Each television console requires 3 hours to assemble, 5 hours to decorate, and 0.1 hour to crate, and returns a profit of $10. Each stereo system requires 10 hours to assemble, 8 hours to decorate, and 0.6 hour to crate, and returns a profit of $25. Each radio requires 1 hour to assemble, 1 hour to decorate, and 0.1 hour to crate, and returns a profit of $3. The manufacturer has 30,000, 40,000, and 120 hours available weekly for assembling, decorating, and crating, respectively. How many units of each product should be manufactured to maximize profit?

38. Maximizing Profit The finishing process in the manufacture of cocktail tables and end tables requires sanding, staining, and varnishing. The time in minutes required for each finishing process is given below:

	Sanding	Staining	Varnishing
End table	8	10	4
Cocktail table	4	4	8

The equipment required for each process is used on one table at a time and is available for 6 hours each day. If the profit on each cocktail table is $20 and on each end table is $15, how many of each should be manufactured each day in order to maximize profit?

39. Maximizing Profit A large TV manufacturer has warehouse facilities for storing its 52-inch color TVs in Chicago, New York, and Denver. Each month the city of Atlanta is shipped at most four hundred 52-inch TVs. The cost of transporting each TV to Atlanta from Chicago, New York, and Denver averages $20, $20, and $40, respectively, while the cost of labor required for packing averages $6, $8, and $4, respectively. Suppose $10,000 is allocated each month for transportation costs and $3000 is allocated for labor costs. If the profit on each TV made in Chicago is $50, in New York is $80, and in Denver is $40, how should monthly shipping arrangements be scheduled to maximize profit?

4.3 Solving Minimum Problems in Standard Form Using the Duality Principle*

PREPARING FOR THIS SECTION *Before getting started, review the following:*

> Matrix Algebra (Section 2.4, pp. 93–103)

OBJECTIVES
1 Determine a minimum problem is in standard form
2 Obtain the dual problem of a minimum problem in standard form
3 Solve a minimum problem in standard form using the duality principle

Determine a minimum problem is in standard form 1

Standard Form of a Minimum Problem

A linear programming problem in which the objective function is to be minimized is referred to as a **minimum problem**. Such problems are said to be in **standard form** provided the following three conditions are met:

Standard Form of a Minimum Problem

CONDITION 1 All the variables are nonnegative.
CONDITION 2 All other constraints are written as linear expressions that are greater than or equal to a constant.
CONDITION 3 The objective function is expressed as a linear expression with nonnegative coefficients.

EXAMPLE 1 Determining a Minimum Problem Is in Standard Form

Determine which of the following minimum problems are in standard form.

(a) Minimize

$$C = 2x_1 + 3x_2$$

subject to the constraints

$$x_1 + 3x_2 \geq 24$$
$$2x_1 + x_2 \geq 18$$
$$x_1 \geq 0 \qquad x_2 \geq 0$$

(b) Minimize

$$C = 3x_1 - x_2 + 4x_3$$

subject to the constraints

$$3x_1 + x_2 + x_3 \geq 12$$
$$x_1 + x_2 + x_3 \geq 8$$
$$x_1 \geq 0 \qquad x_2 \geq 0 \qquad x_3 \geq 0$$

SOLUTION

(a) Since all three conditions are met, this minimum problem is in standard form.

(b) Conditions 1 and 2 are met, but Condition 3 is not, since the coefficient of x_2 in the objective function is negative. This minimum problem is not in standard form.

**The solution of general minimum problems is discussed in Section 4.4. If you plan to cover Section 4.4, this section may be omitted without loss of continuity.*

(c) Minimize

$$C = 2x_1 + x_2 + x_3$$

subject to the constraints

$$\begin{aligned} x_1 - 3x_2 + x_3 &\le 12 \\ x_1 + x_2 + x_3 &\ge 1 \end{aligned}$$

$$x_1 \ge 0 \qquad x_2 \ge 0 \qquad x_3 \ge 0$$

SOLUTION

(c) Conditions 1 and 3 are met, but Condition 2 is not, since the first constraint

$$x_1 - 3x_2 + x_3 \le 12$$

is not written with a $\ge$ sign. The minimum problem as stated is not in standard form. Notice, however, that by multiplying by -1, we can write this constraint as

$$-x_1 + 3x_2 - x_3 \ge -12$$

Written in this way, the minimum problem is in standard form.

(d) Minimize

$$C = 2x_1 + x_2 + 3x_3$$

subject to the constraints

$$\begin{aligned} -x_1 + 2x_2 + x_3 &\ge -2 \\ x_1 + x_2 + x_3 &\ge 6 \end{aligned}$$

$$x_1 \ge 0 \qquad x_2 \ge 0 \qquad x_3 \ge 0$$

(d) Conditions 1, 2, and 3 are each met, so this minimum problem is in standard form.

NOW WORK PROBLEM 1.

The Duality Principle

Obtain the dual problem of a minimum problem in standard form 2

One technique for solving a minimum problem in standard form was developed by John von Neumann and others. The solution (if it exists) is found by solving a related maximum problem, called the **dual problem**. The next example illustrates how to obtain the dual problem.

EXAMPLE 2 Obtaining the Dual Problem of a Minimum Problem in Standard Form

Obtain the dual problem of the following minimum problem:

Minimize

$$C = 300x_1 + 480x_2$$

subject to the constraints

$$\begin{aligned} x_1 + 3x_2 &\ge 0.25 \\ 2x_1 + 2x_2 &\ge 0.45 \end{aligned}$$

$$x_1 \ge 0 \qquad x_2 \ge 0$$

SOLUTION First notice that the minimum problem is in standard form. We begin by writing a matrix that represents the constraints and the objective function:

$$\begin{array}{c} \\ \\ \\ \end{array}\begin{array}{cc|c} x_1 & x_2 & \\ \end{array}$$

$$\left[\begin{array}{cc|c} 1 & 3 & 0.25 \\ 2 & 2 & 0.45 \\ 300 & 480 & 0 \end{array}\right] \quad \begin{array}{l} \text{Constraint: } x_1 + 3x_2 \ge 0.25 \\ \text{Constraint: } 2x_1 + 2x_2 \ge 0.45 \\ \text{Objective function: } C = 300x_1 + 480x_2 \end{array}$$

Now form the matrix that has as columns the rows of the above matrix by taking column 1 above and writing it as row 1 below, taking column 2 above and writing it as row 2 below, and taking column 3 above and writing it as row 3 below.

$$\left[\begin{array}{cc|c} 1 & 2 & 300 \\ 3 & 2 & 480 \\ 0.25 & 0.45 & 0 \end{array}\right]$$

This matrix is called the **transpose** of the first matrix.

From this matrix, create the following maximum problem:

Maximize

$$P = 0.25y_1 + 0.45y_2$$

subject to the conditions

$$y_1 + 2y_2 \leq 300$$
$$3y_1 + 2y_2 \leq 480$$
$$y_1 \geq 0 \qquad y_2 \geq 0$$

The maximum problem is the dual of the given minimum problem.

Notice that the dual of a minimum problem in standard form is a maximum problem in standard form so it can be solved by using techniques discussed in the previous section. The significance of this is expressed in the following principle:

Von Neumann Duality Principle

Suppose a minimum problem in standard form has a solution. The minimum value of the objective function of the minimum problem in standard form equals the maximum value of the objective function of the dual problem, a maximum problem in standard form.

So, one way to solve a minimum problem in standard form is to form the dual problem and solve it. Another way to solve minimum problems (even those not in standard form) is given in Section 4.4.

The steps to use for obtaining the dual problem are listed below.

Steps for Obtaining the Dual Problem

STEP 1 Write the minimum problem in standard form.
STEP 2 Construct a matrix that represents the constraints and the objective function.
STEP 3 Interchange the rows and columns to form the matrix of the dual problem.
STEP 4 Translate this matrix into a maximum problem in standard form.

EXAMPLE 3 Obtaining the Dual Problem of a Minimum Problem in Standard Form

Find the dual of the following minimum problem:

Minimize

$$C = 2x_1 + 3x_2$$

subject to

$$\begin{aligned} 2x_1 + x_2 &\geq 6 \\ x_1 + 2x_2 &\geq 4 \\ x_1 + x_2 &\geq 5 \\ x_1 \geq 0 \qquad x_2 &\geq 0 \end{aligned}$$

SOLUTION **STEP 1** The minimum problem is in standard form.

STEP 2 The matrix that represents the constraints and the objective function is

$$\left[\begin{array}{cc|c} 2 & 1 & 6 \\ 1 & 2 & 4 \\ 1 & 1 & 5 \\ 2 & 3 & 0 \end{array}\right] \quad \begin{array}{l} \text{Constraint: } 2x_1 + x_2 \geq 6 \\ \text{Constraint: } x_1 + 2x_2 \geq 4 \\ \text{Constraint: } x_1 + x_2 \geq 5 \\ \text{Objective function: } C = 2x_1 + 3x_2 \end{array}$$

STEP 3 Interchanging rows and columns, we obtain the matrix

$$\left[\begin{array}{ccc|c} 2 & 1 & 1 & 2 \\ 1 & 2 & 1 & 3 \\ 6 & 4 & 5 & 0 \end{array}\right] \quad \begin{array}{l} \text{Constraint: } 2y_1 + y_2 + y_3 \leq 2 \\ \text{Constraint: } y_1 + 2y_2 + y_3 \leq 3 \\ \text{Objective function: } P = 6y_1 + 4y_2 + 5y_3 \end{array}$$

STEP 4 This matrix represents the following maximum problem:

Maximize

$$P = 6y_1 + 4y_2 + 5y_3$$

subject to

$$\begin{aligned} 2y_1 + y_2 + y_3 &\leq 2 \\ y_1 + 2y_2 + y_3 &\leq 3 \\ y_1 \geq 0 \qquad y_2 \geq 0 \qquad y_3 &\geq 0 \end{aligned}$$

This maximum problem is in standard form and is the dual problem of the minimum problem.

Some observations about Example 3:

1. The variables (x_1, x_2) of the minimum problem are different from the variables of its dual problem (y_1, y_2, y_3).
2. The minimum problem has three constraints and two variables, while the dual problem has two constraints and three variables. (In general, if a minimum problem has m constraints and n variables, its dual problem will have n constraints and m variables.)

3. The inequalities defining the constraints are $\geq$ for the minimum problem and $\leq$ for the maximum problem.
4. Since the coefficients in the objective function to be minimized are positive, the dual problem has nonnegative numbers to the right of the $\leq$ signs.
5. We follow the custom of denoting an objective function by C (for *Cost*), if it is to be minimized; and P (for *Profit*), if it is to be maximized.

NOW WORK PROBLEM 7.

3 **EXAMPLE 4** **Solving a Minimum Problem in Standard Form by Using the Duality Principle**

Solve the maximum problem of Example 3 by the simplex method, and thereby obtain the solution for the minimum problem.

SOLUTION We introduce slack variables s_1 and s_2 to obtain

$$\begin{aligned} 2y_1 + y_2 + y_3 + s_1 \qquad &= 2, \quad s_1 \geq 0 \\ y_1 + 2y_2 + y_3 \qquad + s_2 &= 3, \quad s_2 \geq 0 \end{aligned}$$

The initial simplex tableau is

BV	P	y_1	y_2	y_3	s_1	s_2	RHS
s_1	0	(2)	1	1	1	0	2
s_2	0	1	2	1	0	1	3
P	1	−6	−4	−5	0	0	0

The pivot element, 2, is circled. After pivoting, we obtain this tableau:

BV	P	y_1	y_2	y_3	s_1	s_2	RHS	Current value
y_1	0	1	$\frac{1}{2}$	($\frac{1}{2}$)	$\frac{1}{2}$	0	1	$y_1 = 1$
s_2	0	0	$\frac{3}{2}$	$\frac{1}{2}$	$-\frac{1}{2}$	1	2	$s_2 = 2$
P	1	0	−1	−2	3	0	6	$P = 6$

The pivot element, $\frac{1}{2}$, is circled. After pivoting, we obtain this tableau:

BV	P	y_1	y_2	y_3	s_1	s_2	RHS	Current value
y_3	0	2	1	1	1	0	2	$y_3 = 2$
s_2	0	−1	1	0	−1	1	1	$s_2 = 1$
P	1	4	1	0	5	0	10	$P = 10$

This is a final tableau, so an optimal solution has been found. We read from it that the solution to the maximum problem is

$$P = 10 \qquad y_1 = 0 \qquad y_2 = 0 \qquad y_3 = 2$$

The duality principle states that the minimum value of the objective function in the original problem is the same as the maximum value in the dual; that is,

$$C = 10$$

But which values of x_1 and x_2 will yield this minimum value? As it turns out, the value of x_1 is found in the objective row in column $s_1 (x_1 = 5)$ and x_2 is found in the objective row in column $s_2 (x_2 = 0)$. As a consequence, the solution to the minimum problem can be read from the right end of the objective row of the final tableau of the maximum problem:

$$x_1 = 5 \qquad x_2 = 0 \qquad C = 10$$

We summarize how to solve a minimum linear programming problem below.

Solving a Minimum Problem in Standard Form Using the Duality Principle

STEP 1 Write the dual (maximum) problem.
STEP 2 Solve this maximum problem by the simplex method.
STEP 3 The minimum value of the objective function (C) will appear in the lower right corner of the final tableau; it is equal to the maximum value of the dual objective function (P). The values of the variables that give rise to the minimum value are located in the objective row in the slack variable columns.

NOW WORK PROBLEM 13.

EXAMPLE 5 Solving a Minimum Problem in Standard Form Using the Duality Principle

Minimize

$$C = 6x_1 + 8x_2 + x_3$$

subject to

$$\begin{aligned} 3x_1 + 5x_2 + 3x_3 &\geq 20 \\ x_1 + 3x_2 + 2x_3 &\geq 9 \\ 6x_1 + 2x_2 + 5x_3 &\geq 30 \end{aligned}$$
$$x_1 \geq 0 \qquad x_2 \geq 0 \qquad x_3 \geq 0$$

SOLUTION This minimum problem is in standard form. The matrix representing this problem is

$$\left[\begin{array}{ccc|c} 3 & 5 & 3 & 20 \\ 1 & 3 & 2 & 9 \\ 6 & 2 & 5 & 30 \\ 6 & 8 & 1 & 0 \end{array}\right] \begin{array}{l} \text{Constraint: } 3x_1 + 5x_2 + 3x_3 \geq 20 \\ \text{Constraint: } x_1 + 3x_2 + 2x_3 \geq 9 \\ \text{Constraint: } 6x_1 + 2x_2 + 5x_3 \geq 30 \\ \text{Objective function: } C = 6x_1 + 8x_2 + x_3 \end{array}$$

We interchange rows and columns to get

$$\left[\begin{array}{ccc|c} 3 & 1 & 6 & 6 \\ 5 & 3 & 2 & 8 \\ 3 & 2 & 5 & 1 \\ 20 & 9 & 30 & 0 \end{array}\right] \begin{array}{l} \text{Constraint: } 3y_1 + y_2 + 6y_3 \leq 6 \\ \text{Constraint: } 5y_1 + 3y_2 + 2y_3 \leq 8 \\ \text{Constraint: } 3y_1 + 2y_2 + 5y_3 \leq 1 \\ \text{Objective function: } P = 20y_1 + 9y_2 + 30y_3 \end{array}$$

The dual problem is:

Maximize

$$P = 20y_1 + 9y_2 + 30y_3$$

subject to

$$\begin{aligned} 3y_1 + y_2 + 6y_3 &\le 6 \\ 5y_1 + 3y_2 + 2y_3 &\le 8 \\ 3y_1 + 2y_2 + 5y_3 &\le 1 \end{aligned}$$

$$y_1 \ge 0 \qquad y_2 \ge 0 \qquad y_3 \ge 0$$

We introduce nonnegative slack variables s_1, s_2, and s_3. The initial tableau for this problem is

BV	P	y_1	y_2	y_3	s_1	s_2	s_3	RHS
s_1	0	3	1	6	1	0	0	6
s_2	0	5	3	2	0	1	0	8
s_3	0	3	2	5	0	0	1	1
P	1	−20	−9	−30	0	0	0	0

The final tableau (as you should verify) is

BV	P	y_1	y_2	y_3	s_1	s_2	s_3	RHS	Current value
s_1	0	0	−1	1	1	0	−1	5	$s_1 = 5$
s_2	0	0	$-\frac{1}{3}$	$-\frac{19}{3}$	0	1	$-\frac{5}{3}$	$\frac{19}{3}$	$s_2 = \frac{19}{3}$
y_1	0	1	$\frac{2}{3}$	$\frac{5}{3}$	0	0	$\frac{1}{3}$	$\frac{1}{3}$	$y_1 = \frac{1}{3}$
P	1	0	$\frac{13}{3}$	$\frac{10}{3}$	0	0	$\frac{20}{3}$	$\frac{20}{3}$	$P = \frac{20}{3}$

The solution to the maximum problem is

$$P = \frac{20}{3} \qquad y_1 = \frac{1}{3} \qquad y_2 = 0 \qquad y_3 = 0$$

For the minimum problem, the values of x_1, x_2, and x_3 are read as the entries in the objective row in the columns under s_1, s_2, and s_3, respectively. The solution to the minimum problem is

$$x_1 = 0 \qquad x_2 = 0 \qquad x_3 = \frac{20}{3}$$

and the minimum value is $C = \dfrac{20}{3}$.

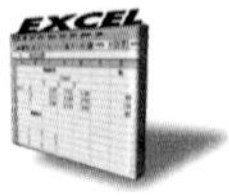

Use Excel to solve Example 5.

SOLUTION **STEP 1** Set up the Excel spreadsheet. The spreadsheet below is presented with the formulas revealed.

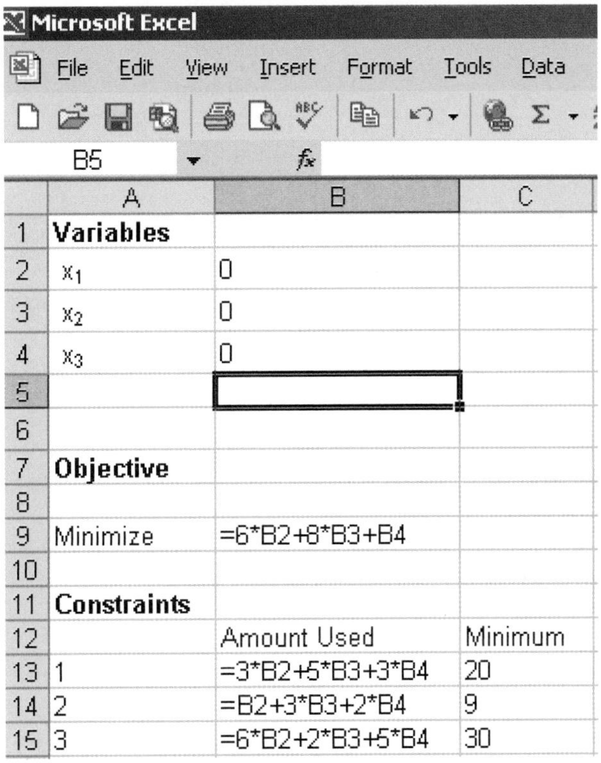

	A	B	C
1	**Variables**		
2	x_1	0	
3	x_2	0	
4	x_3	0	
5			
6			
7	**Objective**		
8			
9	Minimize	=6*B2+8*B3+B4	
10			
11	**Constraints**		
12		Amount Used	Minimum
13	1	=3*B2+5*B3+3*B4	20
14	2	=B2+3*B3+2*B4	9
15	3	=6*B2+2*B3+5*B4	30

STEP 2 Set up Solver. Solver can be found under the toolbar command Tools.

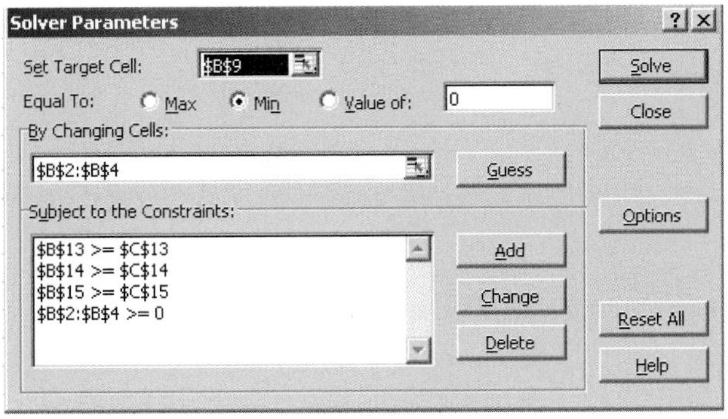

The only differences between maximize and minimize problems are

(a) Equal To: must be min.
(b) Subject to the Constraints: must be >=.

STEP 3 Click on Solve. The solution is given below.

Variables	
x_1	0
x_2	0
x_3	6.666666667

Objective

Minimize 6.666666667

Constraints

	Amount Used	Minimum
1	20	20
2	13.33333333	9
3	33.33333333	30

So $x_1 = 0, x_2 = 0, x_3 = 6.6667$ with a minimum value of $C = 6.6667$.

EXERCISE 4.3 Answers to Odd-Numbered Problems Begin on Page AN-22.

In Problems 1–6, determine which of the following minimum problems are in standard form.

1. Minimize
$$C = 2x_1 + 3x_2$$
subject to the constraints
$$\begin{aligned} 4x_1 - x_2 &\geq 2 \\ x_1 + x_2 &\geq 1 \\ x_1 \geq 0 \quad x_2 &\geq 0 \end{aligned}$$

2. Minimize
$$C = 3x_1 + 5x_2$$
subject to the constraints
$$\begin{aligned} 3x_1 - x_2 &\geq 4 \\ x_1 - 2x_2 &\geq 3 \\ x_1 \geq 0 \quad x_2 &\geq 0 \end{aligned}$$

3. Minimize
$$C = 2x_1 - x_2$$
subject to the constraints
$$\begin{aligned} 2x_1 - x_2 &\geq 1 \\ -2x_1 &\geq -3 \\ x_1 \geq 0 \quad x_2 &\geq 0 \end{aligned}$$

4. Minimize
$$C = 2x_1 + 3x_2$$
subject to the constraints
$$\begin{aligned} x_1 - x_2 &\leq 3 \\ 2x_1 + 3x_2 &\geq 4 \\ x_1 \geq 0 \quad x_2 &\geq 0 \end{aligned}$$

5. Minimize
$$C = 3x_1 + 7x_2 + x_3$$
subject to the constraints
$$\begin{aligned} x_1 + x_3 &\leq 6 \\ 2x_1 + x_2 &\geq 4 \\ x_1 \geq 0 \quad x_2 \geq 0 \quad x_3 &\geq 0 \end{aligned}$$

6. Minimize
$$C = x_1 - x_2 + x_3$$
subject to the constraints
$$\begin{aligned} x_1 + x_2 &\geq 6 \\ 2x_1 - x_3 &\geq 4 \\ x_1 \geq 0 \quad x_2 \geq 0 \quad x_3 &\geq 0 \end{aligned}$$

In Problems 7–12, write the dual problem of each minimum problem.

7. Minimize
$$C = 2x_1 + 3x_2$$
subject to
$$\begin{aligned} x_1 + x_2 &\geq 2 \\ 2x_1 + 3x_2 &\geq 6 \\ x_1 \geq 0 \quad x_2 &\geq 0 \end{aligned}$$

8. Minimize
$$C = 3x_1 + 4x_2$$
subject to
$$\begin{aligned} 2x_1 + x_2 &\geq 2 \\ 2x_1 + x_2 &\geq 6 \\ x_1 \geq 0 \quad x_2 &\geq 0 \end{aligned}$$

9. Minimize
$$C = 3x_1 + x_2 + x_3$$
subject to
$$\begin{aligned} x_1 + x_2 + x_3 &\geq 5 \\ 2x_1 + x_2 &\geq 4 \\ x_1 \geq 0 \quad x_2 \geq 0 \quad x_3 &\geq 0 \end{aligned}$$

10. Minimize
$$C = 2x_1 + x_2 + x_3$$
subject to
$$\begin{aligned} 2x_1 + x_2 + x_3 &\geq 4 \\ x_1 + 2x_2 + x_3 &\geq 6 \\ x_1 \geq 0 \quad x_2 \geq 0 \quad x_3 &\geq 0 \end{aligned}$$

11. Minimize
$$C = 3x_1 + 4x_2 + x_3 + 2x_4$$
subject to
$$\begin{aligned} x_1 + x_2 + x_3 + 2x_4 &\geq 60 \\ 3x_1 + 2x_2 + x_3 + 2x_4 &\geq 90 \\ x_1 \geq 0 \quad x_2 \geq 0 \quad x_3 \geq 0 \quad x_4 &\geq 0 \end{aligned}$$

12. Minimize
$$C = 2x_1 + x_2 + 4x_3 + x_4$$
subject to
$$\begin{aligned} 2x_1 + x_2 + x_3 + x_4 &\geq 80 \\ 2x_1 + 3x_2 + x_3 + 2x_4 &\geq 100 \\ x_1 \geq 0 \quad x_2 \geq 0 \quad x_3 \geq 0 \quad x_4 &\geq 0 \end{aligned}$$

In Problems 13–20, solve each minimum problem.

13. Minimize

$$C = 6x_1 + 3x_2$$

subject to

$$\begin{aligned} x_1 + x_2 &\geq 2 \\ 2x_1 + 6x_2 &\geq 6 \\ x_1 \geq 0 \quad x_2 &\geq 0 \end{aligned}$$

14. Minimize

$$C = 3x_1 + 4x_2$$

subject to

$$\begin{aligned} x_1 + x_2 &\geq 3 \\ 2x_1 + x_2 &\geq 4 \\ x_1 \geq 0 \quad x_2 &\geq 0 \end{aligned}$$

15. Minimize

$$C = 6x_1 + 3x_2$$

subject to

$$\begin{aligned} x_1 + x_2 &\geq 4 \\ 3x_1 + 4x_2 &\geq 12 \\ x_1 \geq 0 \quad x_2 &\geq 0 \end{aligned}$$

16. Minimize

$$C = 2x_1 + 3x_2 + 4x_3$$

subject to

$$\begin{aligned} x_1 - 2x_2 - 3x_3 &\geq -2 \\ x_1 + x_2 + x_3 &\geq 2 \\ 2x_1 + x_3 &\geq 3 \\ x_1 \geq 0 \quad x_2 \geq 0 \quad x_3 &\geq 0 \end{aligned}$$

17. Minimize

$$C = x_1 + 2x_2 + x_3$$

subject to

$$\begin{aligned} x_1 - 3x_2 + 4x_3 &\geq 12 \\ 3x_1 + x_2 + 2x_3 &\geq 10 \\ x_1 - x_2 - x_3 &\geq -8 \\ x_1 \geq 0 \quad x_2 \geq 0 \quad x_3 &\geq 0 \end{aligned}$$

18. Minimize

$$C = x_1 + 2x_2 + 4x_3$$

subject to

$$\begin{aligned} x_1 - x_2 + 3x_3 &\geq 4 \\ 2x_1 + 2x_2 - 3x_3 &\geq 6 \\ -x_1 + 2x_2 + 3x_3 &\geq 2 \\ x_1 \geq 0 \quad x_2 \geq 0 \quad x_3 &\geq 0 \end{aligned}$$

19. Minimize

$$C = x_1 + 4x_2 + 2x_3 + 4x_4$$

subject to

$$\begin{aligned} x_1 + x_3 &\geq 1 \\ x_2 + x_4 &\geq 1 \\ -x_1 - x_2 - x_3 - x_4 &\geq -3 \\ x_1 \geq 0 \quad x_2 \geq 0 \quad x_3 \geq 0 \quad x_4 &\geq 0 \end{aligned}$$

20. Minimize

$$C = x_1 + 2x_2 + 3x_3 + 4x_4$$

subject to

$$\begin{aligned} x_1 + x_3 &\geq 1 \\ x_2 + x_4 &\geq 1 \\ -x_1 - x_2 - x_3 - x_4 &\geq -3 \\ x_1 \geq 0 \quad x_2 \geq 0 \quad x_3 \geq 0 \quad x_4 &\geq 0 \end{aligned}$$

21. Diet Preparation Mr. Jones needs to supplement his diet with at least 50 mg of calcium and 8 mg of iron daily. The minerals are available in two types of vitamin pills, P and Q. Pill P contains 5 mg of calcium and 2 mg of iron, while Pill Q contains 10 mg of calcium and 1 mg of iron. If each P pill costs 3 cents and each Q pill costs 4 cents, how could Mr. Jones minimize the cost of adding the minerals to his diet? What would the daily minimum cost be?

22. Production Schedule A company owns two mines. Mine A produces 1 ton of high-grade ore, 3 tons of medium-grade ore, and 5 tons of low-grade ore each day. Mine B produces 2 tons of each grade of ore per day. The company needs at least 80 tons of high-grade ore, at least 160 tons of medium-grade ore, and at least 200 tons of low-grade ore. How many days should each mine be operated to minimize costs if it costs $2000 per day to operate each mine?

23. Production Schedule Argus Company makes three products: A, B, and C. Each unit of A costs $4, each unit of B costs $2, and each unit of C costs $1 to produce. Argus must produce at least 20 As, 30 Bs, and 40 Cs, and cannot produce fewer than 200 total units of As, Bs, and Cs combined. Minimize Argus's costs.

24. Diet Planning A health clinic dietician is planning a meal consisting of three foods whose ingredients are summarized as follows:

	One Unit of		
	Food I	Food II	Food III
Units of protein	10	15	20
Units of carbohydrates	1	2	1
Units of iron	4	8	1
Calories	80	120	100

The dietician wishes to determine the number of units of each food to use to create a balanced meal containing at least 40 units of protein, 6 units of carbohydrates, and 12 units of iron, with as few calories as possible. How many units of each food should be used in order to minimize calories?

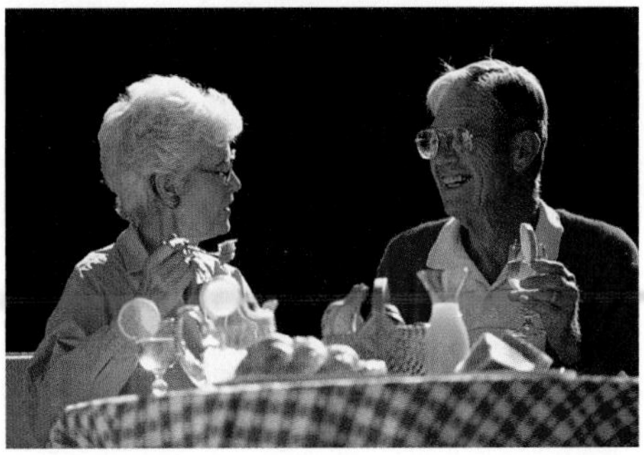

25. Menu Planning Fresh Starts Catering offers the following lunch menu:

Menu		
Lunch #1	Soup, salad, sandwich	\$6.20
Lunch #2	Salad, pasta	\$7.40
Lunch #3	Salad, sandwich, pasta	\$9.10

Mrs. Mintz and her friends would like to order at least 4 bowls of soup, 9 salads, 6 sandwiches, and 5 orders of pasta and keep the cost as low as possible. Compose their order.

26. Inventory Control A department store stocks three brands of toys: A, B, and C. Each unit of brand A occupies 1 square foot of shelf space, each unit of brand B occupies 2 square feet, and each unit of brand C occupies 3 square feet. The store has 120 square feet available for storage. Surveys show that the store should have on hand at least 12 units of brand A and at least 30 units of A and B combined. Each unit of brand A costs the store \$8, each unit of brand B \$6, and each unit of brand C \$10. Minimize the cost to the store.

4.4 The Simplex Method with Mixed Constraints

OBJECTIVES

1 Solve a maximum problem with mixed constraints

2 Solve a minimum problem with mixed constraints

3 Solve maximum/minimum problems with equality constraints

So far, we have developed the simplex method only for solving linear programming problems in standard form. In this section we develop the simplex method for linear programming problems that cannot be written in standard form.

The Simplex Method for a Maximum Problem with Mixed Constraints

Solve a maximum problem with mixed constraints 1

Recall that for a maximum problem in standard form each constraint must be of the form

$$a_1x_1 + a_2x_2 + \cdots + a_nx_n \le b_1 \qquad b_1 > 0$$

That is, each constraint is a linear expression *less than or equal to a positive constant.* When the constraints are of any other form (greater than or equal to, or equal to, or with a zero or negative constraint on the right), we have what are called **mixed constraints.** The following example illustrates the simplex method for solving problems with mixed constraints.

EXAMPLE 1 Solving a Maximum Problem with Mixed Constraints

Maximize

$$P = 20x_1 + 15x_2$$

subject to the constraints

$$\begin{aligned} x_1 + x_2 &\ge 7 \quad (1)\\ 9x_1 + 5x_2 &\le 45 \quad (2)\\ 2x_1 + x_2 &\ge 8 \quad (3)\\ x_1 \ge 0 \quad x_2 &\ge 0 \end{aligned}$$

SOLUTION We first observe this is a maximum problem that is not in standard form. Second, it cannot be modified so as to be in standard form.

STEP 1 Write each constraint, except the nonnegative constraints, as an inequality with the variables on the left side of a $\leq$ sign.

For Step 1, we merely multiply the first and third inequalities by -1. The result is that the constraints become

$$\begin{aligned} -x_1 - x_2 &\leq -7 \quad (1) \\ 9x_1 + 5x_2 &\leq 45 \quad (2) \\ -2x_1 - x_2 &\leq -8 \quad (3) \\ x_1 \geq 0 \qquad x_2 &\geq 0 \end{aligned}$$

STEP 2 Introduce nonnegative slack variables on the left side of each inequality to form an equality.

For Step 2, we introduce the slack variables s_1, s_2, s_3 to obtain

$$\begin{aligned} -x_1 - x_2 + s_1 \qquad\qquad &= -7 \\ 9x_1 + 5x_2 \qquad + s_2 \qquad &= 45 \\ -2x_1 - x_2 \qquad\qquad + s_3 &= -8 \end{aligned}$$

where $\quad x_1 \geq 0 \quad x_2 \geq 0 \quad s_1 \geq 0 \quad s_2 \geq 0 \quad s_3 \geq 0$

STEP 3 Set up the initial simplex tableau.

BV	P	x_1	x_2	s_1	s_2	s_3	RHS	Current value
s_1	0	-1	-1	1	0	0	-7	$s_1 = -7$
s_2	0	9	5	0	1	0	45	$s_2 = 45$
s_3	0	-2	-1	0	0	1	-8	$s_3 = -8$
P	1	-20	-15	0	0	0	0	$P = 0$

(1)

This initial tableau represents the solution $x_1 = 0$, $x_2 = 0$, $s_1 = -7$, $s_2 = 45$, $s_3 = -8$. This is not a feasible point. The two negative entries in the right-hand column violate the nonnegativity requirement. Whenever this occurs, the simplex algorithm requires an *alternative pivoting strategy.*

Alternative Pivoting Strategy

STEP 4 Whenever negative entries occur in the right-hand column RHS of the constraint equations, the pivot element is selected as follows:

Pivot row: Identify the negative entries on the RHS and their corresponding basic variables (BV). (Ignore the objective row.) If any of the basic variables is an x-variable, choose the one with the smallest subscript. Otherwise choose the slack variable with the smallest subscript. The basic variable BV chosen identifies the pivot row. Because the objective row is ignored, it can never be the pivot row.

Pivot column: Go from left to right along the pivot row until a negative entry is found. (Ignore the RHS.) This entry identifies the pivot column and is the pivot element. If there are no negative entries in the pivot row except the one in the RHS column, then the problem has no solution.

STEP 5 Pivot.

1. If, in the new tableau, negative entries appear on the RHS of the constraint equations, repeat Step 4.
2. If, in the new tableau, only nonnegative entries appear on the RHS of the constraint equations, then the tableau represents a maximum problem in standard form and the steps on page 208 of Section 4.2 are followed.*

To continue with the example, we notice there are negative entries in the RHS column of Display (1). We follow Step 4.

STEP 4 The RHS column has negative entries in the rows corresponding to the basic variables s_1 and s_3. Neither of these is an x-variable so row s_1, the row with slack variable having the smallest subscript, is the pivot row. Going across from left to right along row s_1, the first negative entry we find is -1 in column x_1. The pivot column is x_1 and -1 is the pivot element.

STEP 5 Pivot.

BV	P	x_1	x_2	s_1	s_2	s_3	RHS		BV	P	x_1	x_2	s_1	s_2	s_3	RHS
s_1	0	(−1)	−1	1	0	0	−7		x_1	0	1	1	−1	0	0	7
s_2	0	9	5	0	1	0	45	Pivot →	s_2	0	0	−4	9	1	0	−18
s_3	0	−2	−1	0	0	1	−8		s_3	0	0	1	−2	0	1	6
P	1	−20	−15	0	0	0	0		P	1	0	5	−20	0	0	140

The new tableau has a negative entry, -18, in the RHS column, so we repeat Step 4.

STEP 4 The pivot row corresponds to the slack variable s_2. Going across, the first negative entry is -4, so the pivot column is column x_2. The pivot element is -4.

It can be shown that the alternative pivoting strategy will always lead to a tableau that represents a maximum problem in standard form. See Alan Sultan, Linear Programming, *Academic Press, 1993.*

STEP 5 Pivot.

BV	P	x_1	x_2	s_1	s_2	s_3	RHS
x_1	0	1	1	-1	0	0	7
s_2	0	0	-4	9	1	0	-18
s_3	0	0	1	-2	0	1	6
P	1	0	5	-20	0	0	140

Pivot →

BV	P	x_1	x_2	s_1	s_2	s_3	RHS
x_1	0	1	0	$\frac{5}{4}$	$\frac{1}{4}$	0	$\frac{5}{2}$
x_2	0	0	1	$-\frac{9}{4}$	$-\frac{1}{4}$	0	$\frac{9}{2}$
s_3	0	0	0	$\frac{1}{4}$	$\frac{1}{4}$	1	$\frac{3}{2}$
P	1	0	0	$-\frac{35}{4}$	$\frac{5}{4}$	0	$\frac{235}{2}$

Since the new tableau has only nonnegative entries in the RHS column, it represents a maximum problem in standard form. Since the tableau is not a final tableau (the objective row contains a negative entry), we use the standard pivoting strategy of Section 4.2 found on page 208. The pivot column is column s_1. Form the quotients:

$$\frac{5}{2} \div \frac{5}{4} = 2 \qquad \frac{3}{2} \div \frac{1}{4} = 6$$

The smaller of these is 2, so the pivot row is row x_1. The pivot element is $\frac{5}{4}$.

BV	P	x_1	x_2	s_1	s_2	s_3	RHS
x_1	0	1	0	$\frac{5}{4}$	$\frac{1}{4}$	0	$\frac{5}{2}$
x_2	0	0	1	$-\frac{9}{4}$	$-\frac{1}{4}$	0	$\frac{9}{2}$
s_3	0	0	0	$\frac{1}{4}$	$\frac{1}{4}$	1	$\frac{3}{2}$
P	1	0	0	$-\frac{35}{4}$	$\frac{5}{4}$	0	$\frac{235}{2}$

Pivot →

BV	P	x_1	x_2	s_1	s_2	s_3	RHS
s_1	0	$\frac{4}{5}$	0	1	$\frac{1}{5}$	0	2
x_2	0	$\frac{9}{5}$	1	0	$\frac{1}{5}$	0	9
s_3	0	$-\frac{1}{5}$	0	0	$\frac{1}{5}$	1	1
P	1	7	0	0	3	0	135

This is a final tableau. The maximum value of P is 135, and it is achieved when $x_1 = 0, x_2 = 9, s_1 = 2, s_2 = 0, s_3 = 1$.

NOW WORK PROBLEM 1.

The Simplex Method for a Minimum Problem with Mixed Constraints

Solve a minimum problem with mixed constraints 2

In general, a minimum problem can be changed to a maximum problem by using the fact that minimizing z is the same as maximizing $P = -z$. The following example illustrates this.

EXAMPLE 2 Solving a Minimum Problem with Mixed Constraints

Minimize

$$z = 5x_1 + 6x_2$$

subject to the constraints

$$\begin{aligned} x_1 + x_2 &\le 10 \\ x_1 + 2x_2 &\ge 12 \\ 2x_1 + x_2 &\ge 12 \\ x_1 &\ge 3 \\ x_1 \ge 0 \qquad x_2 &\ge 0 \end{aligned}$$

SOLUTION We change the problem from minimizing $z = 5x_1 + 6x_2$ to maximizing $P = -z = -5x_1 - 6x_2$ and follow the steps for a mixed-constraint problem.

STEP 1 Write each constraint with $\leq$.

$$\begin{aligned} x_1 + x_2 &\leq 10 \\ -x_1 - 2x_2 &\leq -12 \\ -2x_1 - x_2 &\leq -12 \\ -x_1 &\leq -3 \end{aligned}$$

STEP 2 Introduce nonnegative slack variables to form equalities:

$$\begin{aligned} x_1 + x_2 + s_1 \qquad\qquad\qquad &= 10 \\ -x_1 - 2x_2 \qquad + s_2 \qquad\qquad &= -12 \\ -2x_1 - x_2 \qquad\qquad + s_3 \qquad &= -12 \\ -x_1 \qquad\qquad\qquad\qquad + s_4 &= -3 \end{aligned}$$

where $x_1 \geq 0 \quad x_2 \geq 0 \quad s_1 \geq 0 \quad s_2 \geq 0 \quad s_3 \geq 0 \quad s_4 \geq 0$

The objective function is: $P = -5x_1 - 6x_2$.

STEP 3 Set up the initial simplex tableau

BV	P	x_1	x_2	s_1	s_2	s_3	s_4	RHS
s_1	0	1	1	1	0	0	0	10
s_2	0	−1	−2	0	1	0	0	−12
s_3	0	−2	−1	0	0	1	0	−12
s_4	0	−1	0	0	0	0	1	−3
P	1	5	6	0	0	0	0	0

Because of the negative entries in the RHS column, we follow the alternative pivoting strategy given in Step 4.

STEP 4 The pivot row is s_2. The pivot column is x_1. The pivot element is -1.

STEP 5 Pivot.

BV	P	x_1	x_2	s_1	s_2	s_3	s_4	RHS
s_1	0	1	1	1	0	0	0	10
s_2	0	(−1)	−2	0	1	0	0	−12
s_3	0	−2	−1	0	0	1	0	−12
s_4	0	−1	0	0	0	0	1	−3
P	1	5	6	0	0	0	0	0

$\xrightarrow{\text{Pivot}}$

BV	P	x_1	x_2	s_1	s_2	s_3	s_4	RHS
s_1	0	0	−1	1	1	0	0	−2
x_1	0	1	2	0	−1	0	0	12
s_3	0	0	3	0	−2	1	0	12
s_4	0	0	2	0	−1	0	1	9
P	1	0	−4	0	5	0	0	−60

The new tableau has a negative entry, -2, in the RHS column. Repeat Step 4.

STEP 4 The pivot row is s_1; the pivot column is x_2; the pivot element is -1.

STEP 5 Pivot.

BV	P	x_1	x_2	s_1	s_2	s_3	s_4	RHS
s_1	0	0	(−1)	1	1	0	0	−2
x_1	0	1	2	0	−1	0	0	12
s_3	0	0	3	0	−2	1	0	12
s_4	0	0	2	0	−1	0	1	9
P	1	0	−4	0	5	0	0	−60

$\xrightarrow{\text{Pivot}}$

BV	P	x_1	x_2	s_1	s_2	s_3	s_4	RHS
x_2	0	0	1	−1	−1	0	0	2
x_1	0	1	0	2	1	0	0	8
s_3	0	0	0	3	1	1	0	6
s_4	0	0	0	2	1	0	1	5
P	1	0	0	−4	1	0	0	−52

Since the new tableau has only nonnegative entries in the RHS column (remember, the objective row is ignored), it represents a maximum problem in standard form. Since the tableau is not a final tableau (the objective row contains a negative entry), we use the standard pivoting strategy of Section 4.2. The pivot column is column s_1. Form the quotients:

$$8 \div 2 = 4, \qquad 6 \div 3 = 2, \qquad 5 \div 2 = 2.5$$

The smallest of these is 2. The pivot row is row s_3. The pivot element is 3.

BV	P	x_1	x_2	s_1	s_2	s_3	s_4	RHS
x_2	0	0	1	-1	-1	0	0	2
x_1	0	1	0	2	1	0	0	8
s_3	0	0	0	③	1	1	0	6
s_4	0	0	0	2	1	0	1	5
P	1	0	0	-4	1	0	0	-52

$\xrightarrow{\text{Pivot}}$

BV	P	x_1	x_2	x_3	s_1	s_2	s_3	RHS
x_2	0	0	1	0	$-\frac{2}{3}$	$\frac{1}{3}$	0	4
x_1	0	1	0	0	$\frac{1}{3}$	$-\frac{2}{3}$	0	4
s_1	0	0	0	1	$\frac{1}{3}$	$\frac{1}{3}$	0	2
s_4	0	0	0	0	$\frac{1}{3}$	$-\frac{2}{3}$	1	1
P	1	0	0	0	$\frac{7}{3}$	$\frac{4}{3}$	0	-44

This is a final tableau. The maximum value of P is -44, so the minimum value of z is 44. This occurs when $x_1 = 4, x_2 = 4, s_1 = 2, s_2 = 0, s_3 = 0, s_4 = 1$.

So, to solve a minimum linear programming problem, change it to a maximum linear programming problem as follows:

Steps for Solving a Minimum Problem

STEP 1 If z is to be minimized, let $P = -z$.

STEP 2 Solve the linear programming problem: Maximize P subject to the same constraints as the minimum problem.

STEP 3 Use the principle that

$$\text{Minimum of } z = -\text{Maximum of } P$$

NOW WORK PROBLEM 7.

Equality Constraints

3 Solve maximum/minimum problems with equality constraints

So far, all our constraints used $\leq$ or $\geq$. What can be done if one of the constraints is an equality? One method is to replace the $=$ constraint with the two constraints $\leq$ and $\geq$. The next example illustrates this method.

EXAMPLE 3 **Solving a Minimum Problem with Equality Constraints**

Minimize

$$z = 7x_1 + 5x_2 + 6x_3$$

subject to the constraints

$$\begin{aligned} x_1 + x_2 + x_3 &= 10 \\ x_1 + 2x_2 + 3x_3 &\leq 19 \\ 2x_1 + 3x_2 \quad\quad &\geq 21 \end{aligned}$$

$$x_1 \geq 0 \qquad x_2 \geq 0 \qquad x_3 \geq 0$$

SOLUTION We wish to maximize $P = -z = -7x_1 - 5x_2 - 6x_3$ subject to the constraints

$$\begin{aligned} x_1 + x_2 + x_3 &\le 10 \\ x_1 + x_2 + x_3 &\ge 10 \\ x_1 + 2x_2 + 3x_3 &\le 19 \\ 2x_1 + 3x_2 \quad &\ge 21 \\ x_1 \ge 0 \quad x_2 \ge 0 \quad x_3 &\ge 0 \end{aligned}$$

STEP 1 Rewrite the constraints with $\le$:

$$\begin{aligned} x_1 + x_2 + x_3 &\le 10 \\ -x_1 - x_2 - x_3 &\le -10 \\ x_1 + 2x_2 + 3x_3 &\le 19 \\ -2x_1 - 3x_2 \quad &\le -21 \end{aligned}$$

STEP 2 Introduce nonnegative slack variables:

$$\begin{aligned} x_1 + x_2 + x_3 + s_1 \qquad\qquad &= 10 \\ -x_1 - x_2 - x_3 \quad + s_2 \qquad &= -10 \\ x_1 + 2x_2 + 3x_3 \qquad + s_3 \quad &= 19 \\ -2x_1 - 3x_2 \qquad\qquad + s_4 &= -21 \end{aligned}$$

where $x_1 \ge 0 \quad x_2 \ge 0 \quad x_3 \ge 0 \quad s_1 \ge 0 \quad s_2 \ge 0 \quad s_3 \ge 0 \quad s_4 \ge 0$

STEP 3 Set up the initial simplex tableau

BV	P	x_1	x_2	x_3	s_1	s_2	s_3	s_4	RHS
s_1	0	1	1	1	1	0	0	0	10
s_2	0	−1	−1	−1	0	1	0	0	−10
s_3	0	1	2	3	0	0	1	0	19
s_4	0	−2	−3	0	0	0	0	1	−21
P	1	7	5	6	0	0	0	0	0

Because of the negative entries in the RHS column, we follow the alternative pivoting strategy.

STEP 4 The pivot row is row s_2; the pivot column is column x_1. The pivot element is -1.

STEP 5 Pivot.

BV	P	x_1	x_2	x_3	s_1	s_2	s_3	s_4	RHS
s_1	0	1	1	1	1	0	0	0	10
s_2	0	(−1)	−1	−1	0	1	0	0	−10
s_3	0	1	2	3	0	0	1	0	19
s_4	0	−2	−3	0	0	0	0	1	−21
P	1	7	5	6	0	0	0	0	0

Pivot →

BV	P	x_1	x_2	x_3	s_1	s_2	s_3	s_4	RHS
s_1	0	0	0	0	1	1	0	0	0
x_1	0	1	1	1	0	−1	0	0	10
s_3	0	0	1	2	0	1	1	0	9
s_4	0	0	−1	2	0	−2	0	1	−1
P	1	0	−2	−1	0	7	0	0	−70

The new tableau has a negative entry, -1, in the RHS column, so we repeat Step 4.

STEP 4 The pivot row is row s_4. Going across, the first negative entry is -1, so the pivot column is column x_2. The pivot element is -1.

BV	P	x_1	x_2	x_3	s_1	s_2	s_3	s_4	RHS
s_1	0	0	0	0	1	1	0	0	0
x_1	0	1	1	1	0	-1	0	0	10
s_3	0	0	1	2	0	1	1	0	9
s_4	0	0	(-1)	2	0	-2	0	1	-1
P	1	0	-2	-1	0	7	0	0	-70

Pivot →

BV	P	x_1	x_2	x_3	s_1	s_2	s_3	s_4	RHS
s_1	0	0	0	0	1	1	0	0	0
x_1	0	1	0	3	0	-3	0	1	9
s_3	0	0	0	4	0	-1	1	1	8
x_2	0	0	1	-2	0	2	0	-1	1
P	1	0	0	-5	0	11	0	-2	-68

Since the new tableau has only nonnegative entries in the RHS column (remember, the objective row is ignored), it represents a maximum problem in standard form. Since the tableau is not a final tableau (the objective row contains negative entries), we use the standard pivoting strategy of Section 4.2. The pivot column is column x_3. Form the nonnegative quotients:

$$9 \div 3 = 3, \qquad 8 \div 4 = 2$$

The smaller of these is 2, so the pivot row is row s_3. The pivot element is 4.

BV	P	x_1	x_2	x_3	s_1	s_2	s_3	s_4	RHS
s_1	0	0	0	0	1	1	0	0	0
x_1	0	1	0	3	0	-3	0	1	9
s_3	0	0	0	(4)	0	-1	1	1	8
x_2	0	0	1	-2	0	2	0	-1	1
P	1	0	0	-5	0	11	0	-2	-68

Pivot →

BV	P	x_1	x_2	x_3	s_1	s_2	s_3	s_4	RHS
s_1	0	0	0	0	1	1	0	0	0
x_1	0	1	0	0	0	$-\frac{9}{4}$	$-\frac{3}{4}$	$\frac{1}{4}$	3
x_3	0	0	0	1	0	$-\frac{1}{4}$	$\frac{1}{4}$	$\frac{1}{4}$	2
x_2	0	0	1	0	0	$\frac{3}{2}$	$\frac{1}{2}$	$-\frac{1}{2}$	5
P	1	0	0	0	0	$\frac{39}{4}$	$\frac{5}{4}$	$-\frac{3}{4}$	-58

This is not a final tableau. The pivot column is column s_4. Form the quotients:

$$3 \div \frac{1}{4} = 12, \qquad 2 \div \frac{1}{4} = 8$$

The smaller of these is 8. The pivot row is x_3; the pivot element is $\frac{1}{4}$.

BV	P	x_1	x_2	x_3	s_1	s_2	s_3	s_4	RHS
s_1	0	0	0	0	1	1	0	0	0
x_1	0	1	0	0	0	$-\frac{9}{4}$	$-\frac{3}{4}$	$\frac{1}{4}$	3
x_3	0	0	0	1	0	$-\frac{1}{4}$	$\frac{1}{4}$	$\boxed{\frac{1}{4}}$	2
x_2	0	0	1	0	0	$\frac{3}{2}$	$\frac{1}{2}$	$-\frac{1}{2}$	5
P	1	0	0	0	0	$\frac{39}{4}$	$\frac{5}{4}$	$-\frac{3}{4}$	-58

	BV	P	x_1	x_2	x_3	s_1	s_2	s_3	s_4	RHS
	s_1	0	0	0	0	1	1	0	0	0
Pivot →	x_1	0	1	0	-1	0	-2	-1	0	1
	s_4	0	0	0	4	0	-1	1	1	8
	x_2	0	0	1	2	0	1	1	0	9
	P	1	0	0	3	0	9	2	0	-52

This is a final tableau. The maximum value of P is -52, so the minimum value of z is 52. This occurs when $x_1 = 1, x_2 = 9, x_3 = 0, s_1 = 0, s_2 = 0, s_3 = 0, s_4 = 8$.

✓**CHECK:** Verify the results obtained in Example 3 using Excel.

NOW WORK PROBLEM 5.

EXAMPLE 4 Minimizing Cost

The Red Tomato Company operates two plants for canning its tomatoes and has two warehouses for storing the finished products until they are purchased by retailers. The schedule shown in the table represents the per case shipping costs from plant to warehouse.

		Warehouse	
		A	B
Plant	I	\$0.25	\$0.18
	II	\$0.25	\$0.14

Each week plant I can produce at most 450 cases and plant II can produce at most 350 cases of tomatoes. Also, each week warehouse A requires at least 300 cases and warehouse B requires at least 500 cases. If we represent the number of cases shipped from plant I to warehouse A by x_1, from plant I to warehouse B by x_2, and so on, the above data can be represented by the following table:

		Warehouse		Maximum Available
		A	B	
Plant	I	x_1	x_2	450
	II	x_3	x_4	350
Minimum Demand		300	500	

The company wants to arrange its shipments from the plants to the warehouses so that the requirements of the warehouses are met and shipping costs are kept at a minimum. How should the company proceed?

SOLUTION The linear programming problem is stated as follows:

Minimize the cost equation

$$C = 0.25x_1 + 0.18x_2 + 0.25x_3 + 0.14x_4$$

subject to

$$\begin{aligned} x_1 + x_2 &\le 450 \\ x_3 + x_4 &\le 350 \\ x_1 + x_3 &\ge 300 \\ x_2 + x_4 &\ge 500 \end{aligned}$$

$$x_1 \ge 0 \quad x_2 \ge 0 \quad x_3 \ge 0 \quad x_4 \ge 0$$

We shall maximize

$$P = -C = -0.25x_1 - 0.18x_2 - 0.25x_3 - 0.14x_4$$

subject to the same constraints.

STEP 1 Write each constraint with $\le$.

$$\begin{aligned} x_1 + x_2 &\le 450 \\ x_3 + x_4 &\le 350 \\ -x_1 - x_3 &\le -300 \\ -x_2 - x_4 &\le -500 \end{aligned}$$

STEP 2 Introduce nonnegative slack variables to form equalities:

$$\begin{aligned} x_1 + x_2 + s_1 &= 450 \\ x_3 + x_4 + s_2 &= 350 \\ -x_1 - x_3 + s_3 &= -300 \\ -x_2 - x_4 + s_4 &= -500 \end{aligned}$$

$$x_1 \ge 0 \quad x_2 \ge 0 \quad x_3 \ge 0 \quad x_4 \ge 0 \quad s_1 \ge 0 \quad s_2 \ge 0 \quad s_3 \ge 0 \quad s_4 \ge 0$$

STEP 3 Set up the initial simplex tableau:

BV	P	x_1	x_2	x_3	x_4	s_1	s_2	s_3	s_4	RHS
s_1	0	1	1	0	0	1	0	0	0	450
s_2	0	0	0	1	1	0	1	0	0	350
s_3	0	−1	0	−1	0	0	0	1	0	−300
s_4	0	0	−1	0	−1	0	0	0	1	−500
P	1	0.25	0.18	0.25	0.14	0	0	0	0	0

Because of the negative entries in the RHS column, we follow the alternative pivoting strategy.

STEP 4 The pivot row is row s_3; the pivot column is column x_1. The pivot element is -1.

STEP 5 Pivot.

BV	P	x_1	x_2	x_3	x_4	s_1	s_2	s_3	s_4	RHS
s_1	0	1	1	0	0	1	0	0	0	450
s_2	0	0	0	1	1	0	1	0	0	350
s_3	0	(−1)	0	−1	0	0	0	1	0	−300
s_4	0	0	−1	0	−1	0	0	0	1	−500
P	1	0.25	0.18	0.25	0.14	0	0	0	0	0

	BV	P	x_1	x_2	x_3	x_4	s_1	s_2	s_3	s_4	RHS
	s_1	0	0	1	−1	0	1	0	1	0	150
Pivot →	s_2	0	0	0	1	1	0	1	0	0	350
	x_1	0	1	0	1	0	0	0	−1	0	300
	s_4	0	0	−1	0	−1	0	0	0	1	−500
	P	1	0	0.18	0	0.14	0	0	0.25	0	−75

The new tableau has a negative entry, −500, in the RHS column. Repeat Step 4.

STEP 4 The pivot row is row s_4; the pivot column is column x_2; the pivot element is −1.

STEP 5 Pivot.

	BV	P	x_1	x_2	x_3	x_4	s_1	s_2	s_3	s_4	RHS
	s_1	0	0	1	−1	0	1	0	1	0	150
	s_2	0	0	0	1	1	0	1	0	0	350
	x_1	0	1	0	1	0	0	0	−1	0	300
	s_4	0	0	(−1)	0	−1	0	0	0	1	−500
	P	1	0	0.18	0	0.14	0	0	0.25	0	−75

	BV	P	x_1	x_2	x_3	x_4	s_1	s_2	s_3	s_4	RHS
	s_1	0	0	0	−1	−1	1	0	1	1	−350
Pivot →	s_2	0	0	0	1	1	0	1	0	0	350
	x_1	0	1	0	1	0	0	0	−1	0	300
	x_2	0	0	1	0	1	0	0	0	−1	500
	P	1	0	0	0	−0.04	0	0	0.25	0.18	−165

The new tableau has a negative entry, −350, in the RHS column, so we repeat Step 4.

STEP 4 The pivot row is row s_1. Going across, the first negative entry is −1, so the pivot column is column x_3. The pivot element is −1.

STEP 5 Pivot.

	BV	P	x_1	x_2	x_3	x_4	s_1	s_2	s_3	s_4	RHS
	s_1	0	0	0	(−1)	−1	1	0	1	1	−350
	s_2	0	0	0	1	1	0	1	0	0	350
	x_1	0	1	0	1	0	0	0	−1	0	300
	x_2	0	0	1	0	1	0	0	0	−1	500
	P	1	0	0	0	−0.04	0	0	0.25	0.18	−165

	BV	P	x_1	x_2	x_3	x_4	s_1	s_2	s_3	s_4	RHS
	x_3	0	0	0	1	1	−1	0	−1	−1	350
Pivot →	s_2	0	0	0	0	0	1	1	1	1	0
	x_1	0	1	0	0	−1	1	0	0	1	−50
	x_2	0	0	1	0	1	0	0	0	−1	500
	P	1	0	0	0	−0.04	0	0	0.25	0.18	−165

The new tableau has a negative entry, −50, in the RHS column. Repeat Step 4.

STEP 4 The pivot row is row x_1; the pivot column is column x_4; the pivot element is −1.

STEP 5 Pivot.

BV	P	x_1	x_2	x_3	x_4	s_1	s_2	s_3	s_4	RHS
x_3	0	0	0	1	1	−1	0	−1	−1	350
s_2	0	0	0	0	0	1	1	1	1	0
x_1	0	1	0	0	(−1)	1	0	0	1	−50
x_2	0	0	1	0	1	0	0	0	−1	500
P	1	0	0	0	−0.04	0	0	0.25	0.18	−165

Pivot →

BV	P	x_1	x_2	x_3	x_4	s_1	s_2	s_3	s_4	RHS
x_3	0	1	0	1	0	0	0	−1	0	300
s_2	0	0	0	0	0	1	1	1	1	0
x_4	0	−1	0	0	1	−1	0	0	−1	50
x_2	0	1	1	0	0	1	0	0	0	450
P	1	−0.04	0	0	0	−0.04	0	0.25	0.14	−163

Since the new tableau has only nonnegative entries in the RHS column (remember, the objective row is ignored), it represents a maximum problem in standard form. Since the tableau is not a final tableau (the objective row contains negative entries), we use the standard pivoting strategy of Section 4.2. The pivot column is column s_1 (column x_1 could also be chosen). Form the quotients:

$$0 \div 1 = 0 \qquad 450 \div 1 = 450$$

The smaller of these is 0. The pivot row is row s_2. The pivot element is 1.

BV	P	x_1	x_2	x_3	x_4	s_1	s_2	s_3	s_4	RHS
x_3	0	1	0	1	0	0	0	−1	0	300
s_2	0	0	0	0	0	(1)	1	1	1	0
x_4	0	−1	0	0	1	−1	0	0	−1	50
x_2	0	1	1	0	0	1	0	0	0	450
P	1	−0.04	0	0	0	−0.04	0	0.25	0.14	−163

Pivot →

BV	P	x_1	x_2	x_3	x_4	s_1	s_2	s_3	s_4	RHS
x_3	0	1	0	1	0	0	0	−1	0	300
s_1	0	0	0	0	0	1	1	1	1	0
x_4	0	−1	0	0	1	0	1	1	0	50
x_2	0	1	1	0	0	0	−1	−1	−1	450
P	1	−0.04	0	0	0	0	0.04	0.29	0.18	−163

The tableau is not a final tableau since the objective row contains a negative entry. We use the standard pivoting strategy of Section 4.2. The pivot column is column x_1. Form the quotients:

$$300 \div 1 = 300 \qquad 450 \div 1 = 450$$

The smaller of these is 300. The pivot row is row x_3, the pivot element is 1.

BV	P	x_1	x_2	x_3	x_4	s_1	s_2	s_3	s_4	RHS
x_3	0	(1)	0	1	0	0	0	−1	0	300
s_1	0	0	0	0	0	1	1	1	1	0
x_4	0	−1	0	0	1	0	1	1	0	50
x_2	0	1	1	0	0	0	−1	−1	−1	450
P	1	−0.04	0	0	0	0	0.04	0.29	0.18	−163

	BV	P	x_1	x_2	x_3	x_4	s_1	s_2	s_3	s_4	RHS
	x_1	0	1	0	1	0	0	0	−1	0	300
Pivot →	s_1	0	0	0	0	0	1	1	1	1	0
	x_4	0	0	0	1	1	0	1	0	0	350
	x_2	0	0	1	−1	0	0	−1	0	−1	150
	P	1	0	0	0.04	0	0	0.04	0.25	0.18	−151

This is a final tableau. The maximum value of P is -151, so the minimum cost C is \$151. This occurs when $x_1 = 300$, $x_2 = 150$, $x_3 = 0$, $x_4 = 350$. This means plant I should deliver 300 cases to warehouse A and 150 cases to warehouse B; and plant II should deliver 350 cases to warehouse B to keep costs at the minimum (\$151).

✓ **CHECK:** Verify the results obtained in Example 4 using Excel.

EXERCISE 4.4 Answers to Odd-Numbered Problems Begin on Page AN-23.

In Problems 1–6, use the mixed-constraint method to solve each maximum problem.

1. Maximize

$$P = 3x_1 + 4x_2$$

subject to the constraints

$$\begin{aligned} x_1 + x_2 &\le 12 \\ 5x_1 + 2x_2 &\ge 36 \\ 7x_1 + 4x_2 &\ge 14 \\ x_1 \ge 0 \quad x_2 &\ge 0 \end{aligned}$$

2. Maximize

$$P = 5x_1 + 2x_2$$

subject to the constraints

$$\begin{aligned} x_1 + x_2 &\ge 11 \\ 2x_1 + 3x_2 &\ge 24 \\ x_1 + 3x_2 &\le 18 \\ x_1 \ge 0 \quad x_2 &\ge 0 \end{aligned}$$

3. Maximize

$$P = 3x_1 + 2x_2 - x_3$$

subject to the constraints

$$\begin{aligned} x_1 + 3x_2 + x_3 &\le 9 \\ 2x_1 + 3x_2 - x_3 &\ge 2 \\ 3x_1 - 2x_2 + x_3 &\ge 5 \\ x_1 \le 0 \quad x_2 \ge 0 \quad x_3 &\ge 0 \end{aligned}$$

4. Maximize

$$P = 3x_1 + 2x_2 - x_3$$

subject to the constraints

$$\begin{aligned} 2x_1 - x_2 - x_3 &\le 2 \\ x_1 + 2x_2 + x_3 &\ge 2 \\ x_1 - 3x_2 - 2x_3 &\le -5 \\ x_1 \ge 0 \quad x_2 \ge 0 \quad x_3 &\ge 0 \end{aligned}$$

5. Maximize

$$P = 3x_1 + 2x_2$$

subject to the constraints

$$\begin{aligned} 2x_1 + x_2 &\le 4 \\ x_1 + x_2 &= 3 \\ x_1 \ge 0 \quad x_2 &\ge 0 \end{aligned}$$

6. Maximize

$$P = 45x_1 + 27x_2 + 18x_3 + 36x_4$$

subject to the constraints

$$\begin{aligned} 5x_1 + x_2 + x_3 + 8x_4 &= 30 \\ 2x_1 + 4x_2 + 3x_3 + 2x_4 &= 30 \\ x_1 \ge 0 \quad x_2 \ge 0 \quad x_3 &\ge 0 \\ x_4 &\ge 0 \end{aligned}$$

In Problems 7–8, use the mixed-constraint method to solve each minimum problem.

7. Minimize

$$z = 6x_1 + 8x_2 + x_3$$

subject to the constraints

$$\begin{aligned} 3x_1 + 5x_2 + 3x_3 &\ge 20 \\ x_1 + 3x_2 + 2x_3 &\ge 9 \\ 6x_1 + 2x_2 + 5x_3 &\ge 30 \\ x_1 + x_2 + x_3 &\le 10 \\ x_1 \ge 0 \quad x_2 \ge 0 \quad x_3 &\ge 0 \end{aligned}$$

8. Minimize

$$z = 2x_1 + x_2 + x_3$$

subject to the constraints

$$\begin{aligned} 3x_1 - x_2 - 4x_3 &\le -12 \\ x_1 + 3x_2 + 2x_3 &\ge 10 \\ x_1 - x_2 + x_3 &\le 8 \\ x_1 \ge 0 \quad x_2 \ge 0 \quad x_3 &\ge 0 \end{aligned}$$

9. Shipping Private Motors, Inc., has two plants, M1 and M2, which manufacture engines; the company also has two assembly plants, A1 and A2, which assemble the cars. M1 can produce at most 600 engines per week. M2 can produce at most 400 engines per week. A1 needs at least 500 engines per week and A2 needs at least 300 engines per week. Following is a table of charges to ship engines to assembly plants.

	A1	A2
M1	\$400	\$100
M2	\$200	\$300

How many engines should be shipped each week from each engine plant to each assembly plant? [*Hint*: Consider four variables: x_1 = number of units shipped from M1 to A1, x_2 = number of units shipped from M1 to A2, x_3 = number of units shipped from M2 to A1, and x_4 = number of units shipped from M2 to A2.]

10. Minimizing Materials Quality Oak Tables, Inc., has an individual who does all its finishing work, and it wishes to use him in this capacity at least 6 hours each day. The assembly area can be used at most 8 hours each day. The company has three models of oak tables, T1, T2, T3. T1 requires 1 hour for assembly, 2 hours for finishing, and 9 board feet of oak. T2 requires 1 hour for assembly, 1 hour for finishing, and 9 board feet of oak. T3 requires 2 hours for assembly, 1 hour for finishing, and 3 board feet of oak. If we wish to minimize the board feet of oak used, how many of each model should be made?

11. Mixture Minimize the cost of preparing the following mixture, which is made up of three foods, I, II, III. Food I costs \$2 per unit, food II costs \$1 per unit, and food III costs \$3 per unit. Each unit of food I contains 2 ounces of protein and 4 ounces of carbohydrate; each unit of food II has 3 ounces of protein and 2 ounces of carbohydrate; and each unit of food III has 4 ounces of protein and 2 ounces of carbohydrate. The mixture must contain at least 20 ounces of protein and 15 ounces of carbohydrate.

12. Advertising A local appliance store has decided on an advertising campaign utilizing newspaper and radio. Each dollar spent on newspaper advertising is expected to reach 50 people in the "Under \$25,000" and 40 in the "Over \$25,000" bracket. Each dollar spent on radio advertising is expected to reach 70 people in the "Under \$25,000" and 20 people in the "Over \$25,000" bracket. If the store wants to reach at least 100,000 people in the "Under \$25,000" and at least 120,000 in the "Over \$25,000" bracket, how should it proceed so that the cost of advertising is minimized?

13. Shipping Schedule A television manufacturer must fill orders from two retailers. The first retailer, R_1, has ordered 55 television sets, while the second retailer, R_2, has ordered 75 sets. The manufacturer has the television sets stored in two warehouses, W_1 and W_2. There are 100 sets in W_1 and 120 sets in W_2. The shipping costs per television set are: \$8 from W_1 to R_1; \$12 from W_1 to R_2; \$13 from W_2 to R_1; \$7 from W_2 to R_2. Find the number of television sets to be shipped from each warehouse to each retailer if the total shipping cost is to be a minimum. What is this minimum cost?

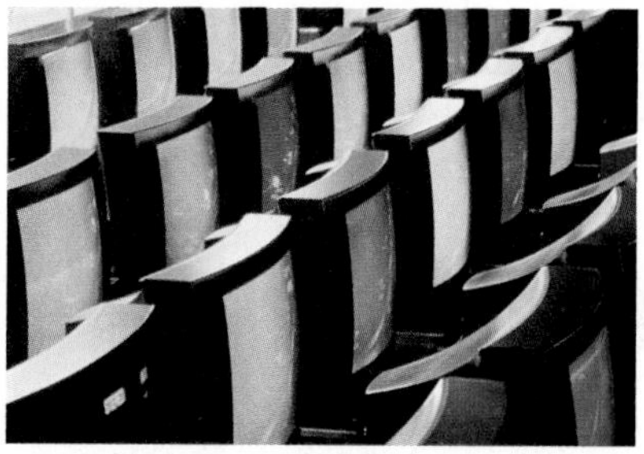

14. Shipping Schedule A motorcycle manufacturer must fill orders from two dealers. The first dealer, D_1, has ordered 20 motorcycles, while the second dealer, D_2, has ordered 30 motorcycles. The manufacturer has the motorcycles stored in two warehouses, W_1 and W_2. There are 40 motorcycles in W_1 and 15 in W_2. The shipping costs per motorcycle are as follows: \$15 from W_1 to D_1; \$13 from W_1 to D_2; \$14 from W_2 to D_1; \$16 from W_2 to D_2. Under these conditions, find the number of motorcycles to be shipped from each warehouse to each dealer if the total shipping cost is to be held to a minimum. What is this minimum cost?

15. Minimizing Airfares Suppose that a particular computer company has 12 sales representatives based in New York City and 18 sales representatives based in San Francisco. To staff an upcoming trade show in Dallas and another trade show in Chicago, this company wants to send at least 15 sales representatives to Dallas and at least 10 sales representatives to Chicago. Also, at least 5 of the sales representatives in San Francisco must be sent to Dallas. According to Travelocity, for March 2003, the lowest round-trip airfares (including taxes and fees) for a one-week stay, with tickets purchased two weeks in advance, are as follows: San Francisco to Dallas: \$340; San Francisco to Chicago: \$180; New York (LaGuardia) to Dallas: \$280; New York (LaGuardia) to Chicago: \$180. Formulate and solve a linear programming problem to determine how many sales representatives this company should send from New York City to each trade show and how many should be sent from San Francisco to each trade show in order to satisfy the staffing requirements while minimizing the total airfare required to send the representatives.

Source: Travelocity.

16. Maximizing Salespeople at a Trade Show Suppose that a particular computer company has 12 sales representatives based in New York City and 18 sales representatives based in San Francisco. To staff an upcoming trade show in Dallas and another trade show in Chicago, this company wants to send at least 5 of these representatives to Dallas and at least 5 of these sales representatives to Chicago. According to Travelocity, for March 2003, the lowest round-trip airfares (including taxes and fees) for a one-week stay, with tickets purchased two weeks in advance, are as follows: San Francisco to Dallas: $340; San Francisco to Chicago: $180; New York (LaGuardia) to Dallas: $280; New York (LaGuardia) to Chicago: $180. For these two trade shows, this computer company has a total airfare budget of $4830. Formulate and solve a linear programming problem to determine how many sales representatives this company should send from New York City to each trade show and how many sales representatives should be sent from San Francisco to each trade show in order to satisfy the staffing requirements and stay within the company's airfare budget while maximizing the total number of sales representatives that are sent.

Source: Travelocity.

Chapter 4 Review

OBJECTIVES

Section		You should be able to	Review Exercises
4.1	1	Determine a maximum problem is in standard form	1–8
	2	Set up the initial simplex tableau	9–14
	3	Perform pivot operations	15a–22a
	4	Analyze a tableau	15b–22b
4.2	1	Solve maximum problems in standard form using the simplex method	23–26
	2	Determine if a tableau is final, requires additional pivoting, or indicates no solution	15c–22c; 23–26, 47, 48, 51
4.3	1	Determine a minimum problem is in standard form	27–32
	2	Obtain the dual problem of a minimum problem in standard form	33–36
	3	Solve a minimum problem in standard form using the duality principle	37–40
4.4	1	Solve a maximum problem with mixed constraints	41, 42, 52, 53, 54
	2	Solve a minimum problem with mixed constraints	43, 44
	3	Solve maximum/minimum problems with equality constraints	45, 46, 49, 50

TRUE–FALSE ITEMS Answers are on page AN-23.

T F 1. For a maximum problem in standard form, each of the constraints, with the exception of the nonnegativity constraints, is written with a ≤ symbol.

T F 2. For a maximum problem in standard form, the slack variables are sometimes negative.

T F 3. Once the pivot element is identified in a tableau, the pivot operation causes the pivot element to become a 1 and causes the remaining entries in the pivot column to become 0s.

T F 4. The pivot element is sometimes in the objective row.

T F 5. One way to solve a minimum problem in standard form is to first solve its dual, which is a maximum problem.

T F 6. Another way to solve a minimum problem is to solve the maximum problem whose objective function is the negative of the minimum problem's objective function.

FILL IN THE BLANKS Answers are on page AN-23

1. The constraints of a maximum problem in standard form are changed from an inequality to an equation by introducing ____________ ____________.

2. For a maximum problem in standard form, the pivot ____________ is located by selecting the most negative entry in the objective row.

3. For a minimum problem to be in standard form, all the constraints must be written with ____________ signs.

4. The ____________ ____________ ____________ principle states that the optimal solution of a minimum linear programming problem, if it exists, has the same value as the optimal solution of the maximum problem, which is its dual.

REVIEW EXERCISES Answers to odd-numbered problems begin on page AN-23.

Blue Problem numbers indicate the author's suggestions for a practice test.

In Problems 1–8, determine which maximum problems are in standard form.

1. Maximize
$$P = 2x_1 + 3x_2$$
subject to the constraints
$$\begin{aligned} x_1 + 2x_2 &\le 4 \\ 3x_1 + x_2 &\le 8 \\ x_1 \ge 0 \quad x_2 &\ge 0 \end{aligned}$$

2. Maximize
$$P = x_1 + 5x_2$$
subject to the constraints
$$\begin{aligned} x_1 + x_2 &\le 5 \\ 2x_1 - x_2 &\le 8 \\ x_1 + 3x_2 &\le 15 \\ x_1 \ge 0 \quad x_2 &\ge 0 \end{aligned}$$

3. Maximize
$$P = 5x_1 + x_2 + x_3$$
subject to the constraints
$$\begin{aligned} x_1 + x_2 + x_3 &\le 30 \\ x_1 + 4x_3 &\le 10 \\ 2x_1 + x_2 + x_3 &\le 15 \\ x_1 \ge 0 \quad x_2 \ge 0 \quad x_3 &\ge 0 \end{aligned}$$

4. Maximize
$$P = 2x_1 + 3x_2 + x_3$$
subject to the constraints
$$\begin{aligned} x_1 + 2x_2 + x_3 &\le 12 \\ 2x_1 + x_2 + 4x_3 &\le 20 \\ x_1 + \quad + x_3 &\le 8 \\ x_1 \ge 0 \quad x_2 \ge 0 \quad x_3 &\ge 0 \end{aligned}$$

5. Maximize
$$P = x_1 + 2x_2 + x_3$$
subject to the constraints
$$\begin{aligned} x_1 + 3x_2 + 3x_3 &\ge 4 \\ 2x_1 + 5x_2 + x_3 &\le 8 \\ x_1 \ge 0 \quad x_2 \ge 0 \quad x_3 &\ge 0 \end{aligned}$$

6. Maximize
$$P = 5x_1 + 2x_2 + x_3$$
subject to the constraints
$$\begin{aligned} x_1 + 3x_2 + x_3 &\le 50 \\ x_1 + 2x_2 + 5x_3 &= 8 \\ x_1 \ge 0 \quad x_3 &\ge 0 \end{aligned}$$

7. Maximize
$$P = x_1 + 3x_2$$
subject to the constraints
$$\begin{aligned} x_1 + x_2 &\le 10 \\ 3x_1 + x_2 &\le 20 \\ x_1 &\ge 0 \end{aligned}$$

8. Maximize
$$P = 4x_1 + 5x_2$$
subject to the constraints
$$\begin{aligned} x_1 + 3x_2 &\ge 10 \\ 3x_1 + x_2 &\ge 20 \\ x_1 \ge 0 \quad x_2 &\ge 0 \end{aligned}$$

In Problems 9–14, set up the initial tableau for each maximum problem.

9. Maximize
$$P = 2x_1 + x_2 + 3x_3$$
subject to the constraints
$$\begin{aligned} 2x_1 + 5x_2 + x_3 &\le 100 \\ x_1 + 3x_2 + x_3 &\le 80 \\ 2x_1 + 3x_2 + 3x_3 &\le 120 \\ x_1 \ge 0 \quad x_2 \ge 0 \quad x_3 &\ge 0 \end{aligned}$$

10. Maximize
$$P = x_1 + 5x_2 + 3x_2$$
subject to the constraints
$$\begin{aligned} x_1 + x_2 + x_3 &\le 50 \\ 2x_1 + 3x_2 + x_3 &\le 80 \\ x_1 + 2x_2 + 5x_3 &\le 100 \\ x_1 \ge 0 \quad x_2 \ge 0 \quad x_3 &\ge 0 \end{aligned}$$

11. Maximize
$$P = 6x_1 + 3x_2$$
subject to the constraints
$$\begin{aligned} x_1 + 5x_2 &\le 200 \\ 5x_1 + 3x_2 &\le 450 \\ x_1 + x_2 &\le 120 \\ x_1 \ge 0 \quad x_2 &\ge 0 \end{aligned}$$

12. Maximize

$$P = 2x_1 + x_2$$

subject to the constraints

$$\begin{aligned} x_1 + x_2 &\leq 20 \\ 2x_1 + 3x_2 &\leq 50 \\ 3x_1 + x_2 &\leq 30 \\ x_1 \geq 0 \quad x_2 &\geq 0 \end{aligned}$$

13. Maximize

$$P = x_1 + 2x_2 + x_3 + 4x_4$$

subject to the constraints

$$\begin{aligned} x_1 + 3x_2 + x_3 + 2x_4 &\leq 20 \\ 4x_1 + x_2 + x_3 + 6x_4 &\leq 80 \\ x_1 \geq 0 \quad x_2 \geq 0 \quad x_3 \geq 0 \quad x_4 &\geq 0 \end{aligned}$$

14. Maximize

$$P = x_1 + x_2 + 3x_3 + 4x_4$$

subject to the constraints

$$\begin{aligned} 2x_1 + 3x_2 + 5x_3 + x_4 &\leq 200 \\ x_1 + x_2 + 6x_3 + x_4 &\leq 180 \\ x_1 \geq 0 \quad x_2 \geq 0 \quad x_3 \geq 0 \quad x_4 &\geq 0 \end{aligned}$$

In Problems 15–22, use each tableau to:
(a) Choose the pivot element and perform a pivot.
(b) Write the system of equations resulting from pivoting in part (a).
(c) Determine if the new tableau: (1) is the final tableau and, if so, write the solution; (2) requires additional pivoting and, if so, choose the new pivot element, or (3) indicates no solution exists for the problem.

15.

BV	P	x_1	x_2	s_1	s_2	RHS
x_1	0	1	5	1	0	40
s_2	0	0	2	2	1	10
P	1	0	−1	0	3	120

16.

BV	P	x_1	x_2	s_1	s_2	RHS
s_1	0	2	3	1	0	12
s_2	0	3	2	0	1	12
P	1	−3	−5	0	0	0

17.

BV	P	x_1	x_2	x_3	s_1	s_2	RHS
s_1	0	1	1	−1	1	0	10
s_2	0	0	1	1	0	1	4
P	1	−2	−1	−3	0	0	0

18.

BV	P	x_1	x_2	x_3	s_1	s_2	s_3	RHS
s_1	0	−1	1	3	1	0	0	12
s_2	0	1	2	1	0	1	0	6
s_3	0	2	3	−1	0	0	1	10
P	1	−2	−3	−1	0	0	0	0

19.

BV	P	x_1	x_2	s_1	s_2	RHS
s_1	0	0.5	0.5	1	0	1
s_2	0	1	1.5	0	1	3
P	1	−2.5	−2	0	0	0

20.

BV	P	x_1	x_2	s_1	s_2	RHS
x_2	0	1	1	1	0	2
s_2	0	3	0	−1	1	2
P	1	−3	0	7	0	14

21.

BV	P	x_1	x_2	x_3	s_1	s_2	s_3	RHS
s_1	0	−1	0	1	1	−1	0	7
x_2	0	−1	1	5	0	1	0	5
s_3	0	1	0	3	0	−5	1	3
P	1	−3	0	4	0	0	0	5

22.

BV	P	x_1	x_2	x_3	s_1	s_2	s_3	RHS
s_1	0	$\frac{8}{3}$	$\frac{5}{3}$	$-\frac{4}{3}$	1	0	0	10
s_2	0	$-\frac{2}{3}$	2	$\frac{1}{3}$	0	1	0	$\frac{5}{3}$
s_3	0	$-\frac{2}{3}$	$\frac{2}{3}$	$\frac{1}{3}$	0	0	1	5
P	1	$-\frac{1}{3}$	$-\frac{2}{3}$	$-\frac{4}{3}$	0	0	0	0

In Problems 23–26, use the simplex method to solve each maximum problem.

23. Maximize

$$P = 100x_1 + 200x_2 + 50x_3$$

subject to the constraints

$$\begin{aligned} 5x_1 + 5x_2 + 10x_3 &\leq 1000 \\ 10x_1 + 8x_2 + 5x_3 &\leq 2000 \\ 10x_1 + 5x_2 &\leq 500 \\ x_1 \geq 0 \quad x_2 \geq 0 \quad x_3 &\geq 0 \end{aligned}$$

24. Maximize

$$P = x_1 + 2x_2 + x_3$$

subject to the constraints

$$\begin{aligned} 3x_1 + x_2 + x_3 &\leq 3 \\ x_1 - 10x_2 - 4x_3 &\leq 20 \\ x_1 \geq 0 \quad x_2 \geq 0 \quad x_3 &\geq 0 \end{aligned}$$

25. Maximize

$$P = 40x_1 + 60x_2 + 50x_3$$

subject to the constraints

$$\begin{aligned} 2x_1 + 2x_2 + x_3 &\le 8 \\ x_1 - 4x_2 + 3x_3 &\le 12 \\ x_1 \ge 0 \quad x_2 \ge 0 \quad x_3 &\ge 0 \end{aligned}$$

26. Maximize

$$P = 2x_1 + 8x_2 + 10x_3 + x_4$$

subject to the constraints

$$\begin{aligned} x_1 + 2x_2 + x_3 + x_4 &\le 50 \\ 3x_1 + x_2 + 2x_3 + x_4 &\le 100 \\ x_1 \ge 0 \quad x_2 \ge 0 \quad x_3 \ge 0 \quad x_4 &\ge 0 \end{aligned}$$

In Problems 27–32, determine if the minimum problems are in standard form.

27. Minimize

$$C = 2x_1 + 3x_2$$

subject to the constraints

$$\begin{aligned} x_1 + 2x_2 &\ge 4 \\ 3x_1 + x_2 &\ge 12 \\ x_1 \ge 0 \quad x_2 &\ge 0 \end{aligned}$$

28. Minimize

$$C = x_1 + 3x_2$$

subject to the constraints

$$\begin{aligned} 3x_1 + 2x_2 &\ge 20 \\ x_1 + 5x_2 &\ge 40 \\ x_1 \ge 0 \quad x_2 &\ge 0 \end{aligned}$$

29. Minimize

$$C = 2x_1 + x_2 + 3x_3$$

subject to the constraints

$$\begin{aligned} x_1 + x_2 + x_3 &\le 15 \\ x_1 - x_2 + 3x_3 &\le 20 \\ 2x_1 + 2x_2 + 4x_3 &\le 25 \\ x_1 \ge 0 \quad x_2 \ge 0 \quad x_3 &\ge 0 \end{aligned}$$

30. Minimize

$$C = x_1 + 2x_2 + x_3$$

subject to the constraints

$$\begin{aligned} 2x_1 + 3x_2 + x_3 &\le 400 \\ x_1 + 6x_2 + 3x_3 &\le 250 \\ 3x_1 + x_2 + x_3 &\le 300 \\ x_1 \ge 0 \quad x_2 \ge 0 \quad x_3 &\ge 0 \end{aligned}$$

31. Minimize

$$C = x_1 + 2x_2 + x_3$$

subject to the constraints

$$\begin{aligned} 3x_1 + x_2 + x_3 &\ge 50 \\ x_1 + 2x_2 + x_3 &\le 60 \\ x_1 \ge 0 \quad x_2 \ge 0 \quad x_3 &\ge 0 \end{aligned}$$

32. Minimize

$$C = 2x_1 + 2x_2 + x_3$$

subject to the constraints

$$\begin{aligned} 2x_1 + 3x_2 + x_3 &\ge 100 \\ x_1 + 2x_2 + 2x_3 &= 160 \\ x_1 \ge 0 \quad x_3 &\ge 0 \end{aligned}$$

In Problems 33–36, write the dual of the minimum problem.

33. Minimize

$$C = 2x_1 + x_2$$

subject to the constraints

$$\begin{aligned} 2x_1 + 2x_2 &\ge 8 \\ x_1 - x_2 &\ge 2 \\ x_1 \ge 0 \quad x_2 &\ge 0 \end{aligned}$$

34. Minimize

$$C = 4x_1 + 2x_2$$

subject to the constraints

$$\begin{aligned} x_1 + 2x_2 &\ge 4 \\ x_1 + 4x_2 &\ge 6 \\ x_1 \ge 0 \quad x_2 &\ge 0 \end{aligned}$$

35. Minimize

$$C = 5x_1 + 4x_2 + 3x_3$$

subject to the constraints

$$\begin{aligned} x_1 + x_2 + x_3 &\ge 100 \\ 2x_1 + x_2 \qquad &\ge 50 \\ x_1 \ge 0 \quad x_2 \ge 0 \quad x_3 &\ge 0 \end{aligned}$$

36. Minimize

$$C = 2x_1 + x_2 + 3x_3 + x_4$$

subject to the constraints

$$\begin{aligned} x_1 + x_2 + x_3 + x_4 &\ge 50 \\ 3x_1 + x_2 + 2x_3 + x_4 &\ge 100 \\ x_1 \ge 0 \quad x_2 \ge 0 \quad x_3 \ge 0 \quad x_4 &\ge 0 \end{aligned}$$

37. Solve Problem 33 using the duality principle.

38. Solve Problem 34 using the duality principle.

39. Solve Problem 35 using the duality principle.

40. Solve Problem 36 using the duality principle.

In Problems 41–46, use the mixed constraints method to solve each linear programming problem.

41. Maximize

$$P = 3x_1 + 5x_2$$

subject to the constraints

$$\begin{aligned} x_1 + x_2 &\geq 2 \\ 2x_1 + 3x_2 &\leq 12 \\ 3x_1 + 2x_2 &\leq 12 \\ x_1 \geq 0 \quad & x_2 \geq 0 \end{aligned}$$

42. Maximize

$$P = 2x_1 + 4x_2$$

subject to the constraints

$$\begin{aligned} 2x_1 + x_2 &\geq 4 \\ x_1 + x_2 &\leq 9 \\ x_1 \geq 0 \quad & x_2 \geq 0 \end{aligned}$$

43. Minimize

$$C = 2x_1 + 3x_2$$

subject to the constraints

$$\begin{aligned} x_1 + x_2 &\geq 3 \\ x_1 + x_2 &\leq 9 \\ x_1 \geq 0 \quad & x_2 \geq 0 \end{aligned}$$

44. Minimize

$$C = 3x_1 + 6x_2 + 2x_3$$

subject to the constraints

$$\begin{aligned} x_2 + x_3 &\leq 4 \\ x_1 + 4x_2 + x_3 &\leq 3 \\ x_1 + x_2 + x_3 &\geq 1 \\ x_1 \geq 0 \quad x_2 \geq 0 \quad & x_3 \geq 0 \end{aligned}$$

45. Maximize

$$P = 300x_1 + 200x_2 + 450x_3$$

subject to the constraints

$$\begin{aligned} 4x_1 + 3x_2 + 5x_3 &\leq 140 \\ x_1 + x_2 + x_3 &= 30 \\ x_1 \geq 0 \quad x_2 \geq 0 \quad & x_3 \geq 0 \end{aligned}$$

46. Maximize

$$P = 2x_1 + x_2 + 3x_3$$

subject to the constraints

$$\begin{aligned} x_1 + x_2 + 2x_3 &\geq 6 \\ 3x_1 + 2x_2 + x_3 &\leq 24 \\ 5x_1 - 15x_2 + 5x_3 &= 10 \\ x_1 \geq 0 \quad x_2 \geq 0 \quad & x_3 \geq 0 \end{aligned}$$

---

47. Mixture A brewery manufactures three types of beer—lite, regular, and dark. Each vat of lite beer requires 6 bags of barley, 1 bag of sugar, and 1 bag of hops. Each vat of regular beer requires 4 bags of barley, 3 bags of sugar, and 1 bag of hops. Each vat of dark beer requires 2 bags of barley, 2 bags of sugar, and 4 bags of hops. Each day the brewery has 800 bags of barley, 600 bags of sugar, and 300 bags of hops available. The brewery realizes a profit of \$10 per vat of lite beer, \$20 per vat of regular beer, and \$30 per vat of dark beer. How many vats of lite, regular, and dark beer should be brewed in order to maximize profits? What is the maximum profit?

48. Management The manager of a supermarket meat department finds that there are 160 pounds of round steak, 600 pounds of chuck steak, and 300 pounds of pork in stock on Saturday morning. From experience, the manager knows that half of these quantities can be sold as straight cuts. The remaining meat will have to be ground into hamburger patties and picnic patties for which there is a large weekend demand. Each pound of hamburger patties contains 20% ground round and 60% ground chuck. Each pound of picnic patties contains 30% ground pork and 50% ground chuck. The remainder of each product consists of an inexpensive nonmeat filler that the store has in unlimited quantities. How many pounds of each product should be made if the objective is to maximize the amount of meat used to make the patties?

49. Scheduling An automobile manufacturer must fill orders from two dealers. The first dealer, D_1, has ordered 40 cars, while the second dealer, D_2, has ordered 25 cars. The manufacturer has the cars stored in two locations, W_1 and W_2. There are 30 cars in W_1 and 50 cars in W_2. The shipping costs per car are as follows: \$180 from W_1 to D_1; \$150 from W_1 to D_2; \$160 from W_2 to D_1; \$170 from W_2 to D_2. Under these conditions, how many cars should be shipped from each storage location to each dealer so as to minimize the total shipping costs? What is this minimum cost?

50. Minimizing Cost The ACE Meat Market makes up a combination package of ground beef and ground pork for meat loaf. The ground beef is 75% lean (75% beef, 25% fat) and costs the market 70¢ per pound. The ground pork is 60% lean (60% pork, 40% fat) and costs the market 50¢ per pound. If the meat loaf is to be at least 70% lean, how much ground beef and ground pork should be mixed to keep cost at a minimum?

51. Optimal Land Use A farmer has 1000 acres of land on which corn, wheat, or soybeans can be grown. Each acre of corn costs \$100 for preparation, requires 7 days of labor, and yields a profit of \$30. An acre of wheat costs \$120 to prepare, requires 10 days of labor, and yields \$40 profit. An acre of soybeans costs \$70 to prepare, requires 8 days of labor, and yields \$40 profit. If the farmer has \$10,000 for preparation and can count on enough workers to supply 8000 days of labor, how many acres should be devoted to each crop to maximize profits?

52. **Production Control** RCA manufacturing received an order for a machine. The machine is to weigh 150 pounds. The two raw materials used to produce the machine are A, with a cost of $4 per unit, and B, with a cost of $8 per unit. At least 14 units of B and no more than 20 units of A must be used. Each unit of A weighs 5 pounds; each unit of B weighs 10 pounds. How much of each type of raw material should be used for each machine if we wish to minimize cost?

53. **Pension Funds** A particular pension fund has decided to invest $50,000 in the four high-yield stocks listed in the table below.

Stock	Price per Share 2-21-03	Yield
Duke Energy Corp.	$14	8%
Eastman Kodak	$30.00	6.00%
General Motors	$34.00	6.00%
H. J. Heinz	$31.50	5.00%

The pension fund has decided to invest at least 10% of the $50,000 in H. J. Heinz stock and at least 20% of the $50,000 in each of the other three stocks. Further, it has been decided that at most 50% of the $50,000 can be invested in Duke Energy Corp. and General Motors combined. Formulate and solve a linear programming problem to determine how many shares of each stock should be purchased in order to maximize the annual yield for the investment, while conforming to the above requirements. Assume that it is possible to purchase a fraction of a share. What is the annual yield in dollars for the optimal investment plan?

Source: Yahoo! Finance.

54. **Pension Funds** A pension fund has decided to invest $50,000 in the four high-yield stocks listed in the table below.

Stock	Price per share 2-21-03	Yield	Price/ Earnings Ratio
Duke Energy Corp.	$14.00	8%	11.4
Eastman Kodak	$30.00	6%	11.2
General Motors	$34.00	6%	15.0
H. J. Heinz	$31.50	5%	13.6

Also, this pension fund has decided to invest at least 10% of the $50,000 in H. J. Heinz stock and at least 20% of the $50,000 in each of the other three stocks. Further, it has been decided that at most 50% of the $50,000 can be invested in Duke Energy Corp. and General Motors combined. Formulate and solve a linear programming problem to determine how many shares of each stock should be purchased in order to minimize the average price/earning ratio of the stocks purchased, while conforming to the above requirements. Note that to minimize the average price/earning ratio of the stocks purchased, it is sufficient to minimize the sum of the four products obtained by multiplying the price/earning ratio for each stock and the amount invested in that stock. Assume that it is possible to purchase a fraction of a share. What is the average price/earning ratio (rounded to the nearest hundredth) for the optimal investment plan? What is the annual yield in dollars for the optimal investment plan?

Source: Yahoo! Finance.

Chapter 4 Project

BALANCING NUTRIENTS—CONTINUED

In the Chapter 3 Project, we discussed making a very simple trail mix from dry-roasted unsalted peanuts and seedless raisins. We used graphing techniques to find which recipes maximized carbohydrates and protein and which minimized fat. The simplex method allows us to consider and to solve more complicated recipes.

Suppose we add two more ingredients to our trail mix: M&Ms Plain Chocolate Candies and hard, salted mini-pretzels. Table 1 lists the amounts of various dietary quantities for these ingredients, including serving sizes. The amounts are given per serving of the ingredient:

TABLE 1 NUTRIENTS AND OTHER QUANTITIES—TRAIL MIX INGREDIENTS

Nutrient	Peanuts (1 cup)	Raisins (1 cup)	M&Ms (1 cup)	Mini-Pretzels (1 cup)
Calories (kcal)	854.10	435.00	1023.96	162.02
Protein (g)	34.57	4.67	9.01	3.87
Fat (g)	72.50	.67	43.95	1.49
Carbohydrates (g)	31.40	114.74	148.12	33.68

Source: USDA National Nutrient Database for Standard Reference at www.nal.usda.gov/fnic/foodcomp.

Suppose that you want to make at most 10 cups of trail mix for a day hike. You don't want any of the ingredients to dominate the mixture, so you want each of the ingredients to contribute at least 10% of the total volume of mix made. You want the entire amount of trail mix you make to have fewer than 7000 calories, and you want to maximize the amount of carbohydrates in the mix.

1. Let x_1 be the number of cups of peanuts you will use in the mix, x_2 the number of cups of raisins, x_3 the number of cups of M&Ms, and x_4 the number of cups of mini-pretzels. Let C be the amount of carbohydrates in the mix. Find the objective function.
2. What constraints must be placed on the objective function?
3. Find the number of cups of peanuts, raisins, M&Ms, and mini-pretzels that maximize the amount of carbohydrates in the mix.
4. How many grams of carbohydrates are in a cup of the final mix? How many calories?
5. Under all the constraints given above, what recipe for trail mix will maximize the amount of protein in the mix? How many grams of protein are in a cup of this mix? How many calories?
6. Consider making a batch of trail mix under the following conditions: You still want to make at most 10 cups of trail mix, and you still want each of the ingredients to contribute at least 10% of the total volume of mix made. You want the entire amount of trail mix you make to have at least 1000 grams of carbohydrates, and you want to minimize the amount of fat in the mix. What recipe for trail mix will minimize the amount of fat in the mix?
7. How much fat, protein, and carbohydrates are in a cup of this mix?
8. How many calories are in a cup of this mix?

MATHEMATICAL QUESTIONS FROM PROFESSIONAL EXAMS*

Use the following information to answer Problems 1–4.

CPA Exam *The Ball Company manufactures three types of lamps, which are labeled A, B, and C. Each lamp is processed in two departments—I and II. Total available work-hours per day for departments I and II are 400 and 600, respectively. No additional labor is available. Time requirements and profit per unit for each lamp type are as follows:*

	A	B	C
Work-hours required in department I	2	3	1
Work-hours required in department II	4	2	3
Profit per unit (sales price less all variable costs)	\$5	\$4	\$3

The company has assigned you, as the accounting member of its profit planning committee, to determine the number of types of A, B, and C lamps that it should produce in order to maximize its total profit from the sale of lamps. The following questions relate to a linear programming model that your group has developed.

1. The coefficients of the objective function would be

   (a) 4, 2, 3 (b) 2, 3, 1 (c) 5, 4, 3 (d) 400, 600

2. The constraints in the model would be

   (a) 2, 3, 1 (b) 5, 4, 3 (c) 4, 2, 3 (d) 400, 600

3. The constraint imposed by the available work-hours in department I could be expressed as

   (a) $4X_1 + 2X_2 + 3X_3 \leq 400$
   (b) $4X_1 + 2X_2 + 3X_3 \geq 400$
   (c) $2X_1 + 3X_2 + 1X_3 \leq 400$
   (d) $2X_1 + 3X_2 + 1X_3 \geq 400$

4. The most types of lamps that would be included in the optimal solution would be

   (a) 2 (b) 1 (c) 3 (d) 0

5. **CPA Exam** In a system of equations for a linear programming model, what can be done to equalize an inequality such as $3X + 2Y \leq 15$?

   (a) Nothing.
   (b) Add a slack variable.
   (c) Add a tableau.
   (d) Multiply each element by -1.

Use the following information to answer Problems 6 and 7.

CPA Exam *The Golden Hawk Manufacturing Company wants to maximize the profits on product A, B, and C. The contribution margin for each product follows:*

Product	Contribution Margin
A	\$2
B	\$5
C	\$4

The production requirements and departmental capacities, by departments, are as follows:

Department	Production Requirements by Product (Hours)		
	A	B	C
Assembling	2	3	2
Painting	1	2	2
Finishing	2	3	1

Department	Departmental Capacity (Total Hours)
Assembling	30,000
Painting	38,000
Finishing	28,000

6. What is the profit maximization formula for the Golden Hawk Company?

(a) $\$2A + \$5B + \$4C = X$ (where X = Profit)
(b) $5A + 8B + 5C \leq 96{,}000$
(c) $\$2A + \$5B + \$4C \leq X$ (where X = Profit)
(d) $\$2A + \$5B + \$4C = 96{,}000$

7. What is the constraint for the painting department of the Golden Hawk Company?

(a) $1A + 2B + 2C \geq 38{,}000$
(b) $\$2A + \$5B + \$4C \geq 38{,}000$
(c) $1A + 2B + 2C \leq 38{,}000$
(d) $2A + 3B + 2C \leq 30{,}000$

8. **CPA Exam** Watch Corporation manufactures products *A*, *B*, and *C*. The daily production requirements are shown at the right.

What is Watch's objective function in determining the daily production of each unit?

(a) $A + B + C \leq \$60$
(b) $\$3A + \$6B + \$7C = \$60$
(c) $A + B + C \leq$ Profit
(d) $\$10A + \$20B + \$30C$ = Profit

Product	Profit per Unit	Hours Required per Unit per Department: Machining	Plating	Polishing
A	\$10	1	1	1
B	\$20	3	1	2
C	\$30	2	3	2
Total Hours per Day per Department		16	12	6

***CPA Exam** Problems 9–11 are based on a company that uses a linear programming model to schedule the production of three products. The per-unit selling prices, variable costs, and labor time required to produce these products are presented below. Total labor time available is 200 hours.*

Product	Selling Price	Variable Cost	Labor (Hours)
A	\$4.00	\$1.00	2
B	\$2.00	\$ .50	2
C	\$3.50	\$1.50	3

9. The objective function to maximize the company's gross profit (Z) is

(a) $4A + 2B + 3.5C = Z$
(b) $2A + 2B + 3C = Z$
(c) $5A + 2.5B + 5C = Z$
(d) $3A + 1.5B + 2C = Z$
(e) $A + B + C = Z$

10. The constraint of labor time available is represented by

(a) $2A + 2B + 3C \leq 200$
(b) $2A + 2B + 3C \geq 200$
(c) $A + B + C \geq 200$
(d) $4A + 2B + 3.5C = 200$
(e) $A/2 + B/2 + C/3 = 200$

11. A linear programming model produces an optimal solution by

(a) Ignoring resource constraints.
(b) Minimizing production costs.
(c) Minimizing both variable production costs and labor costs.
(d) Maximizing the objective function subject to resource constraints.
(e) Finding the point at which various resource constraints intersect.

CHAPTER 5

Finance

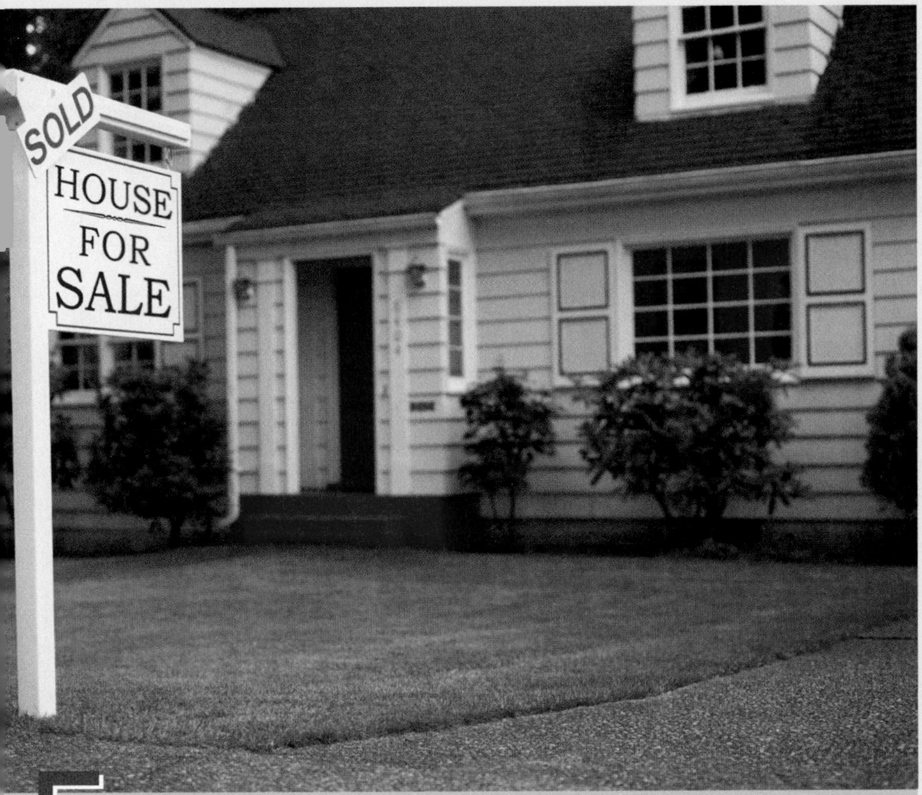

OUTLINE

Even though you're in college now, at some time, probably not too far in the future, you will be thinking of buying a house. And, unless you've won the lottery, you will need to borrow a substantial part of the purchase price in the form of a mortgage. You will quickly discover that mortgages are very complicated and come in many shapes and sizes. One choice you will need to make is whether "paying points" for a mortgage makes sense. The Chapter Project at the end of this chapter takes you through some of the possibilities that can occur when "points" are involved.

A LOOK BACK. A LOOK FORWARD

In earlier mathematics courses, you studied percents. In later courses, you studied exponential expressions and logarithmic expressions, which are reviewed in Appendix A, Section A.4. In this chapter we apply these topics to the field of finance, discussing simple interest, compound interest, simple annuities, sinking funds, and amortization.

5.1 Interest

OBJECTIVES
1. Solve problems involving percents
2. Solve problems involving simple interest
3. Solve problems involving discounted loans

Interest is money paid for the use of money. The total amount of money borrowed, whether by an individual from a bank in the form of a loan or by a bank from an individual in the form of a savings account, is called the **principal**.

The **rate of interest** is the amount charged for the use of the principal for a given length of time, usually on a yearly, or *per annum*, basis. Rates of interest are usually expressed as percents: 10% per annum, 14% per annum, $7\frac{1}{2}\%$ per annum, and so on.

Solve problems involving percents 1

The word **percent** means "per hundred." The familiar symbol for percent means to divide by one hundred. For example,

$$1\% = \frac{1}{100} = 0.01 \quad 12\% = \frac{12}{100} = 0.12 \quad 0.3\% = \frac{0.3}{100} = \frac{3}{1000} = 0.003$$

By reversing these ideas, we can write decimals as percents.

$$0.35 = \frac{35}{100} = 35\% \quad 1.25 = \frac{125}{100} = 125\% \quad 0.005 = \frac{5}{1000} = \frac{0.5}{100} = 0.5\% = \frac{1}{2}\%$$

NOW WORK PROBLEMS 1 AND 9.

EXAMPLE 1 Working with Percents

(a) Find 12% of 80.
(b) What percent of 40 is 18?
(c) 8 is 15% of what number?

SOLUTION **(a)** The English word "of" translates to "multiply" when percents are involved. For example,

$$12\% \text{ of } 80 = 12\% \text{ times } 80 = (0.12)\cdot(80) = 9.6$$

(b) Let x represent the unknown percent. Then

$$x\% \text{ of } 40 = 18 \qquad \text{What \% of 40 is 18}$$

$$\frac{x}{100}\cdot 40 = 18 \qquad x\% = \frac{x}{100}$$

$$40x = 1800 \qquad \text{Multiply both sides by 100}$$

$$x = \frac{1800}{40} \qquad \text{Divide both sides by 40}$$

$$x = 45$$

So, 45% of 40 is 18.

(c) Let x represent the number. Then

$$
\begin{aligned}
8 &= 15\% \text{ of } x \\
8 &= 0.15x && 15\% = 0.15 \\
x &= \frac{8}{0.15} && \text{Divide both sides by } 0.15 \\
x &= 53.33 && \text{Use a calculator}
\end{aligned}
$$

So, 8 is 15% of 53.33.

NOW WORK PROBLEMS 17, 23, AND 27.

EXAMPLE 2 Computing a State Income Tax

A resident of Illinois has base income, after adjustment for deductions, of $18,000. The state income tax on this base income is 3%. What tax is due?

SOLUTION We must find 3% of $18,000. We convert 3% to its decimal equivalent and then multiply by $18,000.

$$3\% \text{ of } \$18{,}000 = (0.03)(\$18{,}000) = \$540$$

The state income tax due is $540.00.

Simple Interest

Solve problems involving simple interest **2**

The easiest type of interest to deal with is called *simple interest*.

Simple Interest

Simple interest is interest computed on the principal for the entire period it is borrowed.

Simple Interest Formula

If a principal P is borrowed at a simple interest rate of $r\%$ per annum (where $r\%$ is expressed as a decimal) for a period of t years, the interest charge I is

$$\boxed{I = Prt} \qquad (1)$$

The **amount** A owed at the end of t years is the sum of the principal P borrowed and the interest I charged: That is,

$$\boxed{A = P + I = P + Prt = P(1 + rt)} \tag{2}$$

EXAMPLE 3 Computing Interest and the Amount Due

A loan of \$250 is made for 9 months at a simple interest rate of 10% per annum. What is the interest charge? What amount is due after 9 months?

SOLUTION The actual period the money is borrowed for is 9 months, which is $\frac{9}{12} = \frac{3}{4}$ of a year. The interest charge is the product of the amount borrowed, \$250; the annual rate of interest, 10%, expressed as a decimal, 0.10; and the length of time in years, $\frac{3}{4}$. We use Formula (1) to find the interest charge I.

$$I = Prt$$

$$\text{Interest charge} = (\$250)(0.10)\left(\frac{3}{4}\right) = \$18.75$$

Using Formula (2), the amount A due after 9 months is

$$A = P + I = 250 + 18.75 = \$268.75$$

NOW WORK PROBLEM 31.

EXAMPLE 4 Computing the Rate of Interest

A person borrows \$1000 for a period of 6 months. What simple interest rate is being charged if the amount A that must be repaid after 6 months is \$1045?

SOLUTION The principal P is \$1000, the period is $\frac{1}{2}$ year (6 months), and the amount A owed after 6 months is \$1045. We substitute the known values of A, P, and t in Formula (2), and then solve for r, the interest rate.

$$A = P + Prt \qquad \text{Formula (2)}$$

$$1045 = 1000 + 1000r\left(\frac{1}{2}\right) \qquad A = 1045;\ P = 1000;\ t = \frac{1}{2}$$

$$45 = 500r \qquad \text{Subtract 1000 from each side and simplify.}$$

$$r = \frac{45}{500} = 0.09 \qquad \text{Solve for } r.$$

The per annum rate of interest is 9%.

NOW WORK PROBLEM 37.

EXAMPLE 5 **Computing the Amount Due**

A company borrows $1,000,000 for 1 month at a simple interest rate of 9% per annum. How much must the company pay back at the end of 1 month?

SOLUTION The principal P is $1,000,000, the period t is $\frac{1}{12}$ year, and the interest rate r is 0.09. The amount A that must be paid back is

$$A = P(1 + rt) \qquad \text{Formula (2)}$$

$$\begin{aligned} A &= 1{,}000{,}000\left[1 + 0.09\left(\frac{1}{12}\right)\right] \\ &= 1{,}000{,}000(1.0075) \\ &= \$1{,}007{,}500 \end{aligned}$$

At the end of 1 month, the company must pay back $1,007,500.

Discounted Loans

Solve problems involving discounted loans 3

If a lender deducts the interest from the amount of the loan at the time the loan is made, the loan is said to be **discounted**. The interest deducted from the amount of the loan is the **discount**. The amount the borrower receives is called the **proceeds**.

Discounted Loans

Let r be the per annum rate of interest, t the time in years, and L the amount of the loan. Then the proceeds R is given by

$$\boxed{R = L - Lrt = L(1 - rt)} \qquad (3)$$

where Lrt is the discount, the interest deducted from the amount of the loan.

EXAMPLE 6 **Computing the Proceeds of a Discounted Loan**

A borrower signs a note for a discounted loan and agrees to pay the lender $1000 in 9 months at a 10% rate of interest. How much does this borrower receive?

SOLUTION The amount of the loan is $L = 1000$. The rate of interest is $r = 10\% = 0.10$. The time is $t = 9$ months $= \frac{9}{12}$ year. The discount is

$$Lrt = \$1000(0.10)\left(\frac{9}{12}\right) = \$75$$

The discount is deducted from the loan amount of $1000, so that the proceeds, the amount the borrower receives, is

$$R = L - Lrt = 1000 - 75 = \$925$$

NOW WORK PROBLEM 43.

EXAMPLE 7 Computing the Simple Interest on a Discounted Loan

What simple rate of interest is the borrower in Example 6 paying on the \$925 that was borrowed for 9 months and paid back in the amount of \$1000?

SOLUTION The principal P is \$925, t is $\frac{9}{12} = \frac{3}{4}$ of a year, and the amount A is \$1000. If r is the simple rate of interest, then, from Formula (2),

$$\begin{aligned} A &= P + Prt && \text{Formula (2)} \\ 1000 &= 925 + 925r\left(\frac{3}{4}\right) && A = 1000;\ P = 925;\ t = \frac{3}{4} \\ 75 &= 693.75r && \text{Simplify.} \\ r &= \frac{75}{693.75} = 0.108108 && \text{Solve for } r. \end{aligned}$$

The equivalent simple rate of interest for this loan is 10.81%.

EXAMPLE 8 Finding the Amount of a Discounted Loan

You wish to borrow \$10,000 for 3 months. If the person you are borrowing from offers a discounted loan at 8%, how much must you repay at the end of 3 months?

SOLUTION The amount you borrow, the proceeds, is $R =$ \$10,000; the interest rate is $r = 0.08$; and the time t is $\frac{3}{12} = \frac{1}{4}$ year. From Formula (3), the amount L you repay obeys the equation

$$\begin{aligned} R &= L(1 - rt) && \text{Formula (3)} \\ 10{,}000 &= L\left[1 - 0.08\left(\frac{1}{4}\right)\right] && R = 10{,}000;\ r = 0.08;\ t = \tfrac{1}{4} \\ 10{,}000 &= 0.98L && \text{Simplify.} \\ L &= \frac{10{,}000}{0.98} = \$10{,}204.08 && \text{Solve for } L. \end{aligned}$$

You will repay \$10,204.08 for this loan.

NOW WORK PROBLEM 47.

Treasury Bills

Treasury bills (T-bills) are short-term securities issued by the Federal Reserve. The bills do not specify a rate of interest. They are sold at public auction with financial institutions making competitive bids. For example, a financial institution may bid \$982,400 for a 3-month \$1 million treasury bill. At the end of 3 months the financial institution receives \$1 million, which includes the interest earned and the cost of the T-bill. This bidding process is an example of a discounted loan.

EXAMPLE 9 Bidding on Treasury Bills

How much should a bank bid if it wants to earn 1.2% simple interest on a 6-month \$1 million treasury bill?*

**On February 6, 2003, 6-month treasury bills were yielding 1.16%.*

SOLUTION The maturity value, the amount to be repaid to the bank by the government, is $L = \$1{,}000{,}000$. The rate of interest is $r = 1.2\% = 0.012$. The time is $t = 6$ months $= \frac{1}{2}$ year. The proceeds R to the government are

$$\begin{aligned} R = L(1 - rt) &= \$1{,}000{,}000\left[1 - 0.012\left(\frac{1}{2}\right)\right] \\ &= 1{,}000{,}000(0.994) \\ &= 994{,}000 \end{aligned}$$

The bank should bid \$994,000 if it wants to earn 1.2%.

NOW WORK PROBLEM 61.

EXERCISE 5.1 Answers to Odd-Numbered Problems Begin on Page AN-24.

In Problems 1–8, write each decimal as a percent.

1. 0.60
2. 0.40
3. 1.1
4. 1.2
5. 0.06
6. 0.07
7. 0.0025
8. 0.0015

In Problems 9–16, write each percent as a decimal.

9. 25%
10. 15%
11. 100%
12. 300%
13. 6.5%
14. 4.3%
15. 73.4%
16. 92%

In Problems 17–30, calculate the indicated quantity.

17. 15% of 1000
18. 20% of 500
19. 18% of 100
20. 10% of 50
21. 210% of 50
22. 135% of 1000
23. What percent of 80 is 4?
24. What percent of 60 is 5?
25. What percent of 5 is 8?
26. What percent of 25 is 45?
27. 20 is 8% of what number?
28. 25 is 12% of what number?
29. 50 is 15% of what number?
30. 40 is 18% of what number?

In Problems 31–36, find the interest due on each loan.

31. \$1000 is borrowed for 3 months at 4% simple interest.
32. \$100 is borrowed for 6 months at 8% simple interest.
33. \$500 is borrowed for 9 months at 12% simple interest.
34. \$800 is borrowed for 8 months at 5% simple interest.
35. \$1000 is borrowed for 18 months at 10% simple interest.
36. \$100 is borrowed for 24 months at 12% simple interest.

In Problems 37–42, find the simple interest rate for each loan.

37. \$1000 is borrowed; the amount owed after 6 months is \$1050.
38. \$500 is borrowed; the amount owed after 8 months is \$600.
39. \$300 is borrowed; the amount owed after 12 months is \$400.
40. \$600 is borrowed; the amount owed after 9 months is \$660.
41. \$900 is borrowed; the amount owed after 10 months is \$1000.
42. \$800 is borrowed; the amount owed after 3 months is \$900.

In Problems 43–46, find the proceeds for each discounted loan.

43. \$1200 repaid in 6 months at 10%.

44. \$500 repaid in 8 months at 9%.

45. \$2000 repaid in 24 months at 8%.

46. \$1500 repaid in 18 months at 10%.

In Problems 47–50, find the amount you must repay for each discounted loan. What is the equivalent simple interest for this loan?

47. The proceeds is \$1200 for 6 months at 10%.

48. The proceeds is \$500 for 8 months at 9%.

49. The proceeds is \$2000 for 24 months at 8%.

50. The proceeds is \$1500 for 18 months at 10%.

---

51. **Buying a Stereo** Madalyn wants to buy a \$500 stereo set in 9 months. How much should she invest at 3% simple interest to have the money then?

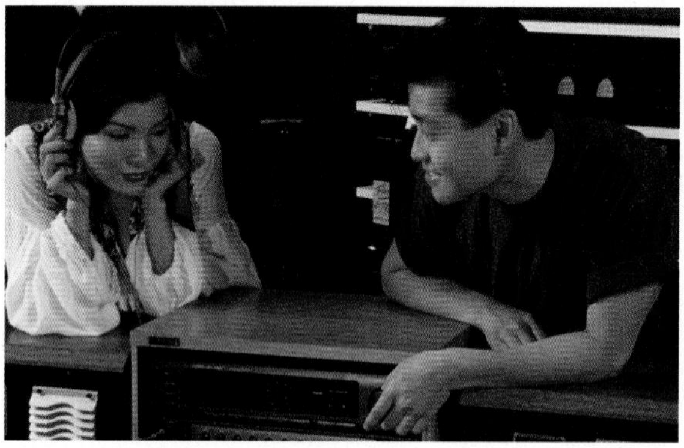

52. **Interest on a Loan** Mike borrows \$10,000 for a period of 3 years at a simple interest rate of 10%. Determine the interest due on the loan.

53. **Term of a Loan** Tami borrowed \$600 at 8% simple interest. The amount of interest paid was \$156. What was the length of the loan?

54. **Equipment Loan** The owner of a restaurant would like to borrow \$12,000 from a bank to buy some equipment. The bank will give the owner a discounted loan at an 11% rate of interest for 9 months. What maturity value should be used so that the owner will receive \$12,000?

55. **Comparing Loans** You need to borrow \$1000 right now, but can repay the loan in 6 months. Since you want to pay as little interest as possible, which type of loan should you take: a discounted loan at 9% per annum or a simple interest loan at 10% per annum?

56. **Comparing Loans** You need to borrow \$5000 right now, but can repay the loan in 9 months. Since you want to pay as little interest as possible, which type of loan should you take: a discounted loan at 8% per annum or a simple interest loan at 8.5% per annum?

57. **Comparing Loans** You need to borrow \$4000 right now, but can repay the loan in 1 year. Since you want to pay as little interest as possible, which type of loan should you take: a discounted loan at 6% per annum or a simple interest loan at 6.3% per annum?

58. **Comparing Loans** You need to borrow \$5000 right now, but can repay the loan in 18 months. Since you want to pay as little interest as possible, which type of loan should you take: a discounted loan at 5.3% per annum or a simple interest loan at 5.6% per annum?

59. **Comparing Loans** Ruth would like to borrow \$2000 for one year from a bank. She is given a choice of a simple interest loan at 12.3% or a discounted loan at 12.1%. What should she do?

60. Refer to Problem 59. If Ruth only needs to borrow the \$2000 for 3 months, what should she do?

61. **Bidding on Treasury Bills** A bank wants to earn 2% simple interest on a 3-month \$1 million treasury bill. How much should they bid?

62. **Bidding on Treasury Bills** How much should a bank bid on a 6-month \$3 million treasury bill to earn 3% simple interest?

63. **T-Bill Auctions** As the result of a treasury bill auction, 6-month T-bills were issued on January 16, 2003, at a price of \$993.78 per \$1000 face value. How much interest would be received on the maturity date, July 17, 2003, for an investment of \$10,000? What is the per annum simple interest rate for this investment, rounded to the nearest thousandth of a percent?

Source: United States Treasury.

64. T-Bill Auctions As the result of a treasury bill auction, 3-month T-bills were issued on January 30, 2003, at a price of \$997.12 per \$1000 face value. How much interest would be received on the maturity date, May 1, 2003, for an investment of \$100,000? What is the per annum simple interest rate for this investment, rounded to the nearest thousandth of a percent?

Source: United States Treasury.

65. T-Bill Auctions As the result of a treasury bill auction, 1-month T-bills were issued on January 23, 2003, at a price of \$999.12 per \$1000 face value. How much interest would be received on the maturity date, February 20, 2003, for an investment of \$5000? What is the per annum simple interest rate for this investment, rounded to the nearest thousandth of a percent?

Source: United States Treasury.

66. Price of Dell Computer Corporation Stock Year-end stock prices for one share of Dell stock are as follows:

1997:	\$84.00
1998:	\$73.19
1999:	\$51.00
2000:	\$17.44
2001:	\$27.18
2002:	\$26.74

Taking into account that there was a two-for-one stock split of Dell shares in March 1998, again in September 1998, and again in March 1999, for each year after 1997, determine the annual percent increase or decrease in the value of Dell stock beginning with one share of stock at the end of 1997. In which year was the percent increase the greatest? In which year was the percent decrease the greatest?

Source: Yahoo!; Finance; Historical Prices.

5.2 Compound Interest

PREPARING FOR THIS SECTION *Before getting started, review the following:*

> Exponents and Logarithms (Appendix A, Section A. 3 pp. 645–649)

OBJECTIVES
1. Solve problems involving compound interest
2. Find the effective rate of interest
3. Solve problems involving present value

Compound Interest Formula

In working with problems involving interest, we use the term **payment period** as follows:

Annually	Once per year
Semiannually	Twice per year
Quarterly	4 times per year
Monthly	12 times per year
Daily	365 times per year*

If the interest due at the end of each payment period is added to the principal, so that the interest computed for the next payment period is based on this new amount of the old principal plus interest, then the interest is said to have been **compounded.** That is, **compound interest** is interest paid on the initial principal and previously earned interest.

**Some banks use 360 times per year.*

EXAMPLE 1 Computing Compound Interest

A bank pays interest of 4% per annum compounded quarterly. If $200 is placed in a savings account and the quarterly interest is left in the account, how much money is in the account after 1 year?

SOLUTION At the end of the first quarter (3 months) the interest earned is

$$I = Prt = (\$200)(0.04)\left(\frac{1}{4}\right) = \$2.00$$

The new principal is $P + I = \$200 + \$2 = \$202$. The interest on this principal at the end of the second quarter is

$$I = (\$202)(0.04)\left(\frac{1}{4}\right) = \$2.02$$

The interest at the end of the third quarter on the principal of $\$202 + \$2.02 = \$204.02$ is

$$I = (\$204.02)(0.04)\left(\frac{1}{4}\right) = \$2.04$$

The interest at the end of the fourth quarter on the principal of $\$204.02 + \$2.04 = \$206.06$ is

$$I = (\$206.06)(0.04)\left(\frac{1}{4}\right) = \$2.06$$

After 1 year, the total in the savings account is $\$206.06 + \$2.06 = \$208.12$.
These results are shown in Figure 1.

FIGURE 1

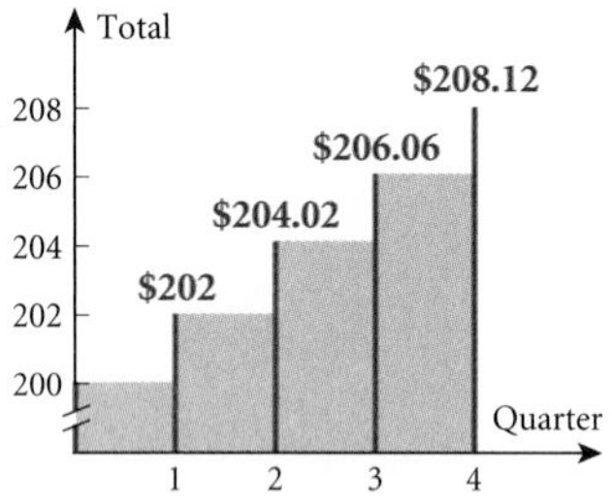

Let's develop a formula for computing the amount when interest is compounded. Suppose r is the per annum rate of interest compounded each payment period. Then the rate of interest per payment period is

$$i = \frac{\text{Per annum rate of interest}}{\text{Number of payment periods}}$$

For example, if the annual rate of interest is 10% and the compounding is monthly, then there are 12 payment periods per year and

$$i = \frac{0.10}{12} = 0.00833$$

If 18% is the annual rate compounded daily (365 payment periods), then

$$i = \frac{0.18}{365} = 0.000493$$

If P is the principal and i is the interest rate per payment period, then the amount A_1 at the end of the first payment period is

$$A_1 = P + Pi = P(1 + i)$$

At the end of the second payment period, and subsequent ones, the amounts are

$$A_2 = A_1 + A_1 i = A_1(1 + i) = P(1 + i)(1 + i) = P(1 + i)^2$$
$$A_3 = A_2 + A_2 i = A_2(1 + i) = P(1 + i)^2(1 + i) = P(1 + i)^3$$
$$\vdots$$
$$A_n = A_{n-1} + A_{n-1} i = A_{n-1}(1 + i) = P(1 + i)^{n-1}(1 + i) = P(1 + i)^n$$

Compound Interest Formula

The amount A_n accrued on a principal P after n payment periods at i interest per payment period is

$$\boxed{A_n = P(1 + i)^n} \tag{1}$$

In working with the Compound Interest Formula (1), we use a calculator with a $\boxed{y^x}$ key.* To use this key, enter the value of y, press $\boxed{y^x}$, enter the value of x, and press $\boxed{=}$ or $\boxed{\text{ENTER}}$.

1 **EXAMPLE 2 Working with the Compound Interest Formula**

If \$1000 is invested at an annual rate of interest of 10%, what is the amount after 5 years if the compounding takes place

(a) Annually? **(b)** Monthly? **(c)** Daily?

How much interest is earned in each case?

SOLUTION We use the Compound Interest Formula (1). The principal is $P = \$1000$.

(a) For annual compounding, $i = 0.10$ and $n = 5$. The amount A is

$$A = P(1 + i)^n = (\$1000)(1 + 0.10)^5 = (\$1000)(1.61051) = \$1610.51$$

The interest earned is

$$A - P = \$1610.51 - \$1000.00 = \$610.51$$

(b) For monthly compounding, there are $n = 5 \cdot 12 = 60$ payment periods over 5 years. The interest rate per payment period is $i = \frac{0.10}{12}$. The amount A is

$$A = P(1 + i)^n = (\$1000)\left(1 + \frac{0.10}{12}\right)^{60} = (\$1000)(1.64531) = \$1645.31$$

The interest earned is

$$A - P = \$1645.31 - \$1000.00 = \$645.31$$

*On a graphing calculator, the key used is $\boxed{\wedge}$.

(c) For daily compounding, there are $n = 5 \cdot 365 = 1825$ payment periods over 5 years. The interest rate per payment period is $i = \frac{0.10}{365}$. The amount A is

$$A = P(1+i)^n = (\$1000)\left(1 + \frac{0.10}{365}\right)^{1825} = (\$1000)(1.64861) = \$1648.61$$

The interest earned is

$$A - P = \$1648.61 - \$1000.00 = \$648.61$$

The results of Example 2 are summarized in Table 1.

TABLE 1

Per Annum Rate	Compounding Method	Interest Rate per Payment Period	Initial Principal	Amount after 5 Years	Interest Earned
10%	Annual	0.10	\$1000	\$1610.51	\$610.51
10%	Monthly	0.00833	\$1000	\$1645.31	\$645.31
10%	Daily	0.000274	\$1000	\$1648.61	\$648.61

NOW WORK PROBLEM 1.

EXAMPLE 3 Comparing Compound Rates of Interest with Simple Interest

(a) If \$100 is invested at an annual rate of interest of 10% compounded monthly, what is the interest earned after 1 year?

(b) What simple rate of interest is required to obtain this amount of interest?

SOLUTION **(a)** We begin with a principal of \$100 and proceed to compute the amount after 1 year at 10% compounded monthly. For monthly compounding at 10%, there are $n = 12$ payment periods, and the interest rate per period is $i = \frac{0.10}{12}$. The amount A is

$$A = P(1+i)^n = \$100\left(1 + \frac{0.10}{12}\right)^{12} = \$110.47$$

The interest earned is $\$110.47 - \$100 = \$10.47$.

(b) The simple interest rate r required to earn interest of \$10.47 on a principal of \$100 after one year obeys the equation

$$\begin{aligned} I &= Prt && \text{Simple Interest Formula} \\ 10.47 &= (100)r(1) && I = 10.47;\ P = 100;\ t = 1 \\ r &= .1047 && \text{Solve for } r \end{aligned}$$

So, a simple interest rate of 10.47% is required to obtain interest equal to that obtained using 10% compounded monthly.

Let's look at the effect of various compounding periods on a principal of \$100 after 1 year using a rate of interest of 8% per annum.

Annual compounding:	$A = \$100(1 + 0.08)^1$	$= \$108.00$
Semiannual compounding:	$A = \$100\left(1 + \dfrac{0.08}{2}\right)^2$	$= \$108.16$
Quarterly compounding(s)	$A = \$100\left(1 + \dfrac{0.08}{4}\right)^4$	$= \$108.24$
Monthly compounding:	$A = \$100\left(1 + \dfrac{0.08}{12}\right)^{12}$	$= \$108.30$
Daily compounding:	$A = \$100\left(1 + \dfrac{0.08}{365}\right)^{365}$	$= \$108.33$

With semiannual compounding, the amount \$108.16 could have been obtained with a simple interest of 8.16%. We describe this by saying that the *effective rate of interest* of 8% compounded semiannually is 8.16%.

Find the effective rate of interest 2

In general, the **effective rate of interest** is the equivalent annual simple rate of interest that yields the same amount as compounding does after one year.

Table 2 summarizes the calculations given above for 8%.

TABLE 2

Rate	Compounding Period	Effective Rate of Interest
8%	Compounded Semiannually	8.16%
8%	Compounded Quarterly	8.24%
8%	Compounded Monthly	8.3%
8%	Compounded Daily	8.33%

NOW WORK PROBLEM 17.

EXAMPLE 4 Comparing Certificates of Deposit

Three local banks offer the following 1-year certificates of deposit (CDs):

(a) Simple interest of 5.2% per annum
(b) 5% per annum compounded monthly
(c) $4\frac{3}{4}$% per annum compounded daily

Which CD results in the most interest?

SOLUTION To compare the three CDs, we compute the amount \$1000 (any other amount could also be used) will grow to in each case.

(a) At simple interest of 5.2%, \$1000 will grow to

$$A = P + Prt = \$1000 + \$1000(0.052)(1) = \$1052.00$$

(b) There are 12 payment periods and the rate of interest per payment period is $i = \frac{0.05}{12}$. The amount A that \$1000 will grow to is

$$A = P(1 + i)^n = \$1000\left(1 + \frac{0.05}{12}\right)^{12} = \$1051.16$$

(c) There are 365 payment periods and the rate of interest per payment period is $i = \frac{0.0475}{365}$. The amount A that \$1000 will grow to is

$$A = P(1+i)^n = \$1000\left(1 + \frac{0.0475}{365}\right)^{365} = \$1048.64$$

The CD offering 5.2% simple interest results in the most interest earned.

NOW WORK PROBLEM 31.

Present Value

Solve problems involving present value

3 The Compound Interest Formula states that a principal P earning an interest rate per payment period i will, after n payment periods, be worth the amount A, where

$$A = P(1+i)^n$$

If we solve for P, we obtain

$$\boxed{P = \frac{A}{(1+i)^n} = A(1+i)^{-n}} \tag{2}$$

In this formula P is called the **present value** of the amount A due at the end of n interest periods at i interest per payment period. In other words, P is the amount that must be invested for n payment periods at i interest per payment period in order to accumulate the amount A.

The Compound Interest Formula and the Present Value Formula can be used to solve many different kinds of problems. The examples below illustrate some of these applications.

EXAMPLE 5 Computing the Present Value of $10,000

How much money should be invested now at 8% per annum so that after 2 years the amount will be $10,000 when the interest is compounded

(a) Annually? **(b)** Monthly? **(c)** Daily?

SOLUTION In this problem we want to find the principal P needed now to get the amount $A =$ $10,000 after 2 years. That is, we want to find the present value of $10,000.

(a) Since compounding is once per year for 2 years, $n = 2$ and $i = 0.08$. The present value P of $10,000 is

$$P = A(1+i)^{-n} = 10{,}000(1+0.08)^{-2} = 10{,}000(0.8573388) = \$8573.39$$

(b) Since compounding is 12 times per year for 2 years, $n = 24$ and $i = \frac{0.08}{12}$. The present value P of $10,000 is

$$P = A(1+i)^{-n} = 10{,}000\left(1 + \frac{0.08}{12}\right)^{-24} = 10{,}000(0.852596) = \$8525.96$$

(c) Since compounding is 365 times per year for 2 years, $n = 730$ and $i = \frac{0.08}{365}$. The present value P of $10,000 is

$$P = A(1+i)^{-n} = 10{,}000\left(1 + \frac{0.08}{365}\right)^{-730} = 10{,}000(0.8521587) = \$8521.59$$

NOW WORK PROBLEM 7.

EXAMPLE 6 Finding the Rate of Interest to Double an Investment

What annual rate of interest compounded annually should you seek if you want to double your investment in 5 years?

SOLUTION If P is the principal and we want P to double, the amount A will be $2P$. We use the Compound Interest Formula (1) with $n = 5$ to find i:

$$A = P(1 + i)^n \qquad \text{Formula (1)}$$

$$2P = P(1 + i)^5 \qquad A = 2P, n = 5$$

$$2 = (1 + i)^5 \qquad \text{Cancel the } Ps.$$

$$\sqrt[5]{2} = 1 + i \qquad \text{Take the 5th root of each side.}$$

$$i = \sqrt[5]{2} - 1 = 1.148698 - 1 = 0.148698 \qquad \text{Solve for } i.$$

↑ $\sqrt[5]{2} = 2^{1/5} = 2^{0.2}$
Use the $\boxed{y^x}$ key on your calculator.

The annual rate of interest needed to double the principal in 5 years is 14.87%.

NOW WORK PROBLEM 19.

EXAMPLE 7 Finding the Time Required to Double/Triple an Investment*

(a) How long will it take for an investment to double in value if it earns 5% compounded monthly?
(b) How long will it take to triple at this rate?

SOLUTION **(a)** If P is the initial investment and we want P to double, the amount A will be $2P$. We use the Compound Interest Formula (1) with $i = \frac{0.05}{12}$. Then

$$A = P(1 + i)^n \qquad \text{Formula (1)}$$

$$2P = P\left(1 + \frac{0.05}{12}\right)^n \qquad A = 2P, i = \frac{0.05}{12}$$

$$2 = (1.0041667)^n \qquad \text{Simplify.}$$

$$n = \log_{1.0041667} 2 = \frac{\log_{10} 2}{\log_{10} 1.0041667} = 166.7 \text{ months}$$

↑ Apply the definition of a logarithm. ↑ Change-of-base formula ↑ Payment period measured in months

It will take about 13 years 11 months to double the investment.

(b) To triple the investment, the amount A is $3P$. Thus,

$$A = P(1 + i)^n \qquad \text{Formula (1)}$$

$$3P = P\left(1 + \frac{0.05}{12}\right)^n \qquad A = 3P, i = \frac{0.05}{12}$$

$$3 = (1.0041667)^n \qquad \text{Simplify.}$$

$$n = \log_{1.0041667} 3 = \frac{\log_{10} 3}{\log_{10} 1.0041667} = 264.2 \text{ months}$$

↑ Apply the definition of a logarithm

It will take about 22 years to triple the investment.

NOW WORK PROBLEM 21.

Requires a knowledge of logarithms, especially the Change-of-Base Formula. See Appendix A, Section A.3, for a review.

EXERCISE 5.2 Answers to Odd-Numbered Problems Begin on Page AN-24.

In Problems 1–6, find the amount if:

1. \$1000 is invested at 4% compounded monthly for 36 months.
2. \$100 is invested at 6% compounded monthly for 20 months.
3. \$500 is invested at 5% compounded annually for 3 years.
4. \$200 is invested at 10% compounded annually for 10 years.
5. \$800 is invested at 6% compounded daily for 200 days.
6. \$400 is invested at 7% compounded daily for 180 days.

In Problems 7–10, find the principal needed now to get each amount.

7. To get \$100 in 6 months at 4% compounded monthly
8. To get \$500 in 1 year at 6% compounded annually
9. To get \$500 in 1 year at 7% compounded daily
10. To get \$800 in 2 years at 5% compounded monthly.
11. If \$1000 is invested at 4% compounded

    (a) Annually (b) Semiannually
    (c) Quarterly (d) Monthly

    what is the amount after 3 years? How much interest is earned?
12. If \$2000 is invested at 5% compounded

    (a) Annually (b) Semiannually
    (c) Quarterly (d) Monthly

    what is the amount after 5 years? How much interest is earned?
13. If \$1000 is invested at 6% compounded quarterly, what is the amount after

    (a) 2 years? (b) 3 years? (c) 4 years?
14. If \$2000 is invested at 4% compounded quarterly, what is the amount after

    (a) 2 years? (b) 3 years? (c) 4 years?
15. If a bank pays 3% compounded semiannually, how much should be deposited now to have \$5000

    (a) 4 years later? (b) 8 years later?
16. If a bank pays 2% compounded quarterly, how much should be deposited now to have \$10,000

    (a) 5 years later? (b) 10 years later?
17. Find the effective rate of interest for

    (a) 8% compounded semiannually
    (b) 4% compounded monthly
18. Find the effective rate of interest for

    (a) 6% compounded monthly
    (b) 14% compounded semiannually
19. What annual rate of interest compounded annually is required to double an investment in 3 years?
20. What annual rate of interest compounded annually is required to double an investment in 10 years?
21. Approximately how long will it take to triple an investment at 10% compounded annually?
22. Approximately how long will it take to triple an investment at 9% compounded annually?
23. Mr. Nielsen wants to borrow \$1000 for 2 years. He is given the choice of (a) a simple interest loan of 12% or (b) a loan at 10% compounded monthly. Which loan results in less interest due?
24. Rework Problem 23 if the simple interest loan is 15% and the other loan is at 14% compounded daily.
25. What principal is needed now to get \$1000 in 1 year at 9% compounded annually? How much should be invested to get \$1000 in 2 years?
26. Repeat Problem 25 using 9% compounded daily.
27. Find the effective rate of interest for $5\frac{1}{4}$% compounded quarterly.
28. Repeat Problem 27 using 6% compounded quarterly.
29. What interest rate compounded quarterly will give an effective interest rate of 7%?
30. Repeat Problem 29 using 10%.

In Problems 31–34, which of the two rates would yield the larger amount in 1 year?
Hint: Start with a principal of \$10,000 in each instance.

31. 6% compounded quarterly or $6\frac{1}{4}$% compounded annually
32. 9% compounded quarterly or $9\frac{1}{4}$% compounded annually
33. 9% compounded monthly or 8.8% compounded daily
34. 8% compounded semiannually or 7.9% compounded daily

35. **Future Price of a Home** If the price of homes rises an average of 5% per year for the next 4 years, what will be the selling price of a home that is selling for $90,000 today 4 years from today? Express your answer rounded to the nearest hundred dollars.

36. **Amount Due on a Charge Card** A department store charges 1.25% per month on the unpaid balance for customers with charge accounts (interest is compounded monthly). A customer charges $200 and does not pay her bill for 6 months. What is the bill at that time?

37. **Amount Due on a Charge Card** A major credit card company has a finance charge of 1.5% per month on the outstanding indebtedness. Caryl charged $600 and did not pay her bill for 6 months. What is the bill at that time?

38. **Buying a Car** Laura wishes to have $8000 available to buy a car in 3 years. How much should she invest in a savings account now so that she will have enough if the bank pays 8% interest compounded quarterly?

39. **Cost of College—Public** The average annual undergraduate college costs (tuition, room and board) for 2000–2001 for public 4-year institutions were $8655 (in-state tuition: $3506; room and board: $5149). The Bureau of Labor Statistics indicated that the inflation rate for tuition in 2001 was 5.09%, while the general inflation rate for room and board (Consumer Price Index) was 2.85%. Assuming that these inflation rates remain constant for the next 5 years, determine the projected annual undergraduate college costs for public 4-year institutions for 2005–2006.

Source: United States Department of Education, National Center for Education Statistics, Integrated Postsecondary Education Data System (IPEDS), "Fall Enrollment" and "Institutional Characteristics" survey (August 2001); FinAid.org, "The Smart Student Guide to Financial Aid"; and U. S. Department of Labor, Bureau of Labor Statistics.

40. **Cost of College—Private** The average annual undergraduate college costs (tuition, room and board) for 2000–2001 for private 4-year institutions were $21,907 (tuition: $15,531; room and board: $6376). The Bureau of Labor Statistics indicated that the inflation rate for tuition in 2001 was 5.09%, while the general inflation rate for room and board (Consumer Price Index) was 2.85%. Assuming that these inflation rates are constant for the next 5 years, determine the projected annual undergraduate college costs for private 4-year institutions for 2005–2006.

Source: United States Department of Education, National Center for Education Statistics, Integrated Postsecondary Education Data System (IPEDS), "Fall Enrollment" and "Institutional Characteristics" surveys (August, 2001); FinAid.org. "The Smart Student Guide to Financial Aid"; and U.S. Department of Labor, Bureau of Labor Statistics.

41. **Down Payment on a House** Tami and Todd will need $40,000 for a down payment on a house in 4 years. How much should they invest in a savings account now so that they will be able to do this? The bank pays 8% compounded quarterly.

42. **Saving for College** A newborn child receives a $3000 gift toward a college education. How much will the $3000 be worth in 17 years if it is invested at 10% compounded quarterly?

43. **Gifting** A child's grandparents have opened a $6000 savings account for the child on the day of her birth. The account pays 8% compounded semiannually. The child will be allowed to withdraw the money when she reaches the age of 25. What will the account be worth at that time?

44. **Future Price of a House** What will a $90,000 house cost 5 years from now if the inflation rate over that period averages 5% compounded annually? Express your answer rounded to the nearest hundred dollars.

45. **Deciding on a Stock Purchase** Jack is considering buying 1000 shares of a stock that sells at $15 per share. The stock pays no dividends. From the history of the stock, Jack is certain that he will be able to sell it 4 years from now at $20 per share. Jack's goal is not to make any investment unless it returns at least 7% compounded quarterly. Should Jack buy the stock?

46. Repeat Problem 45 if Jack requires a return of at least 14% compounded quarterly.

47. **Value of an IRA** An Individual Retirement Account (IRA) has $2000 in it, and the owner decides not to add any more money to the account other than the interest earned at 9% compounded quarterly. How much will be in the account 25 years from the day the account was opened?

48. **Return on Investment** If Jack sold a stock for $35,281.50 (net) that cost him $22,485.75 three years ago, what annual compound rate of return did Jack make on his investment?

49. **National Debt** The national debt was approximately $4.8 trillion on January 1, 1995, and approximately $6.406 trillion on January 1, 2003. Find the annual rate of growth of the national debt in this 8-year period, assuming a constant annual rate of growth. If this rate of growth continues, what would be the projected national debt on January 1, 2010?

Source: United States Treasury Department.

50. Cost of Energy The average U.S. household spent $2868 on energy use in the year 2000. This included $1492 for gasoline, $910 for electricity, $383 for natural gas, and $83 for fuel oil and kerosene. Assuming a constant annual inflation rate of 2.85%, which was the general inflation rate in 2001, determine the projected amounts that the average U. S. household will spend on each of the forms of energy listed above in the year 2005.

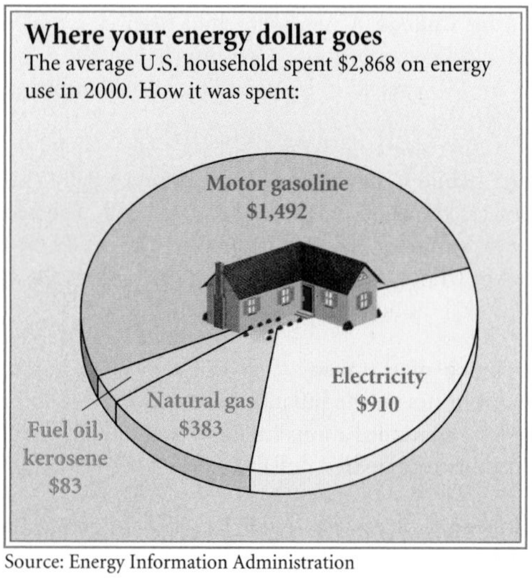
Where your energy dollar goes
The average U.S. household spent $2,868 on energy use in 2000. How it was spent:
Motor gasoline
$1,492
Electricity
$910
Natural gas
$383
Fuel oil,
kerosene
$83
Source: Energy Information Administration

Source: *USA Today*, 12/1/02, based on Energy Information Administration data; and the U.S. Department of Labor, Bureau of Labor Statistics.

For Problems 51–53, zero coupon bonds are used. A ***zero coupon bond*** *is a bond that is sold now at a discount and will pay its face value at some time in the future when it matures; no interest payments are made.*

51. Saving for College Tami's grandparents are considering buying a $40,000 face value zero coupon bond at birth so that she will have enough money for her college education 17 years later. If money is worth 8% compounded annually, what should they pay for the bond?

52. Price of a Bond How much should a $10,000 face value zero coupon bond, maturing in 10 years, be sold for now if its rate of return is to be 8% compounded annually?

53. Rate of Return of a Bond If you pay $12,485.52 for a $25,000 face value zero coupon bond that matures in 8 years, what is your annual compound rate of return?

54. Effective Rates of Interest A bank advertises that it pays interest on saving accounts at the rate of 6.25% compounded daily.

(a) Find the effective rate if the bank uses 360 days in determining the daily rate.
(b) What if 365 days is used?

Problems 55 and 56 require logarithms.

55. Length of Investment How many years will it take for an initial investment of $10,000 to grow to $25,000? Assume a rate of interest of 6% compounded daily.

56. Length of Investment How many years will it take for an initial investment of $25,000 to grow to $80,000? Assume a rate of interest of 7% compounded daily.

Use the following discussion for Problems 57–64. ***Inflation*** *erodes the purchasing power of money. For example, suppose there is an annual rate of inflation of 3%. Then $1000 worth of purchasing power now will be worth only $970 in one year. In general, for an annual rate of inflation of r%, the amount A that $P will purchase after n years is*

$$A = P(1 - r)^n$$

where r is a decimal.

57. Suppose the inflation rate is 3%. After 2 years, how much will $1000 purchase?

58. Suppose the inflation rate is 4%. After 2 years, how much will $1000 purchase?

59. Suppose the inflation rate is 3%. After 5 years, how much will $1000 purchase?

60. Suppose the inflation rate is 4%. After 5 years, how much will $1000 purchase?

Problems 61–64 require logarithms.

61. Suppose the inflation rate is 3%. How long is it until purchasing power is halved?

62. Suppose the inflation rate is 4%. How long is it until purchasing power is halved?

63. Suppose the inflation rate is 6%. How long is it until purchasing power is halved?

64. Suppose the inflation rate is 9%. How long is it until purchasing power is halved?

5.3 Annuities; Sinking Funds

PREPARING FOR THIS SECTION *Before getting started, review the following:*

> Exponents and Logarithms (Appendix A, Section A.3, pp. 645–649)

> Geometric Sequences (Appendix A, Section A.4, pp. 649–654)

OBJECTIVES
1 Solve problems involving annuities
2 Solve problems involving sinking funds

Annuities

Solve problems involving annuities 1

In the previous section we saw how to compute the future value of an investment when a fixed amount of money is deposited in an account that pays interest compounded periodically. Often, however, people and financial institutions do not deposit money and then sit back and watch it grow. Rather, money is invested in small amounts at periodic intervals. Examples of such investments are annual life insurance premiums, monthly deposits in a bank, installment loan payments, and dollar averaging in the stock market with 401(k) or 403(b) accounts.

An **annuity** is a sequence of equal periodic deposits. The periodic deposits can be annual, semiannual, quarterly, monthly, or any other fixed length of time. When the deposits are made at the same time the interest is credited, the annuity is termed **ordinary.** We shall concern ourselves only with ordinary annuities in this book.

The **amount of an annuity** is the sum of all deposits made plus all interest accumulated.

EXAMPLE 1 Finding the Amount of an Annuity

Find the amount of an annuity after 5 deposits if each deposit is equal to \$100 and is made on an annual basis at an interest rate of 10% per annum compounded annually.

SOLUTION After 5 deposits the first \$100 deposit will have accumulated interest compounded annually at 10% for 4 years. Using the Compound Interest Formula, the value A_1 of the first deposit of \$100 after 4 years is

$$A_1 = \$100(1 + 0.10)^4 = \$100(1.4641) = \$146.41$$

The second deposit of \$100, made 1 year after the first deposit, will accumulate interest compounded at 10% for 3 years. Its value A_2 after 3 years is

$$A_2 = \$100(1 + 0.10)^3 = \$100(1.331) = \$133.10$$

Similarly, the third, fourth, and fifth deposits will have the values

FIGURE 2

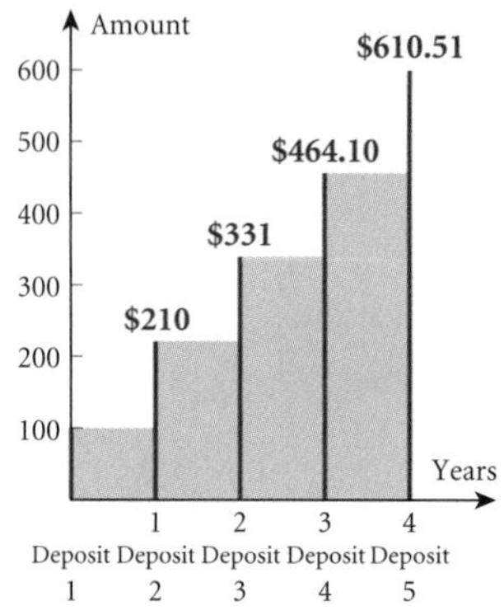

$$A_3 = \$100(1 + 0.10)^2 = \$100(1.21) = \$121.00$$
$$A_4 = \$100(1 + 0.10)^1 = \$100(1.10) = \$110.00$$
$$A_5 = \$100$$

The amount of the annuity after 5 deposits is

$$\begin{aligned} A_1 + A_2 + A_3 + A_4 + A_5 &= \$146.41 + \$133.10 + \$121.00 \\ &\quad + \$110.00 + \$100.00 \\ &= \$610.51 \end{aligned}$$

Figure 2 illustrates the growth of the annuity described above.

We seek a formula for finding the amount of an annuity. Suppose that the interest an annuity earns is i percent per payment period (expressed as a decimal). For example, if an account pays 12% compounded monthly (12 times a year) then $i = \frac{0.12}{12} = 0.01$. If an account pays 8% compounded quarterly (4 times a year) then $i = \frac{0.08}{4} = 0.02$.

Now, suppose that n deposits of $\$P$ each are made at the beginning of each payment period. When the last deposit is made, the first deposit of $\$P$ has earned interest compounded for $n - 1$ payment periods, the second deposit of $\$P$ has earned interest compounded for $n - 2$ payment periods, and so on. Table 3 shows the value of each deposit after n deposits have been made.

TABLE 3

Deposit	1	2	3	. . .	$n - 1$	n
Amount	$P(1 + i)^{n-1}$	$P(1 + i)^{n-2}$	$P(1 + i)^{n-3}$	. . .	$P(1 + i)$	P

The amount A of the annuity is the sum of the amounts shown in Table 3; that is,

$$\begin{aligned} A &= P(1 + i)^{n-1} + P(1 + i)^{n-2} + \cdots + P(1 + i) + P \\ &= P[1 + (1 + i) + \cdots + (1 + i)^{n-1}] \end{aligned}$$

The expression in brackets is the sum of a geometric sequence* with n terms and common ratio $1 + i$. As a result,

$$1 + (1 + i) + \cdots + (1 + i)^{n-1} = \frac{1 - (1 + i)^n}{1 - (1 + i)} = \frac{1 - (1 + i)^n}{-i} = \frac{(1 + i)^n - 1}{i}$$

We have established the following result.

Amount of an Annuity

Suppose P is the deposit in dollars made at each payment period for an annuity paying i percent interest per payment period. The amount A of the annuity after n deposits is

$$A = P\frac{(1 + i)^n - 1}{i} \qquad (1)$$

**The sum of the first n terms of the geometric sequence with common ratio r is*

$$1 + r + r^2 + \cdots + r^{n-1} = \frac{1 - r^n}{1 - r}$$

See Appendix A, Section A.4, for a more detailed discussion.

Be careful in using Formula (1). Formula (1) gives the amount A of an annuity after n deposits. The nth deposit is made after $n - 1$ compounding periods. Refer to Table 3.

Let's do some examples.

EXAMPLE 2 Finding the Amount of an Annuity

Find the amount of an annuity after 5 years if a deposit of \$100 is made each year, at 10% compounded annually. How much interest is earned?

SOLUTION The deposit is $P = \$100$. The number of deposits is $n = 5$ and the interest per payment period is $i = 0.10$. Using Formula (1), the amount A after 5 years is

$$A = P\left[\frac{(1+i)^n - 1}{i}\right] = 100\left[\frac{(1+0.10)^5 - 1}{0.10}\right] = \$100(6.1051) = \$610.51$$

The interest earned is the amount after 5 years less the 5 annual payments of \$100 each:

$$\text{Interest earned} = \text{A} - 500 = 610.51 - 500 = \$110.51$$

NOW WORK PROBLEM 1.

EXAMPLE 3 Finding the Amount of an Annuity

Mary decides to put aside \$100 every month in an insurance fund that pays 8% compounded monthly. After making 8 deposits, how much money does Mary have?

SOLUTION This is an annuity with $P = \$100$, $n = 8$ deposits, and interest $i = \frac{0.08}{12}$ per payment period. Using Formula (1), the amount A after 8 deposits is

$$A = P\left[\frac{(1+i)^n - 1}{i}\right] = 100\left[\frac{\left(1 + \dfrac{0.08}{12}\right)^8 - 1}{\dfrac{0.08}{12}}\right] = \$100(8.1892) = \$818.92$$

Mary has \$818.92 after making 8 deposits.

EXAMPLE 4 Saving for College

To save for his son's college education, Mr. Graff decides to put \$50 aside every month in a credit union account paying 10% interest compounded monthly. If he begins this savings program when his son is 3 years old, how much will he have saved by the time his son is 18 years old?

SOLUTION When his son is 18 years old, Mr. Graff will have made his 180th deposit (15 years $\times$ 12 deposits per year). This is an annuity with $P = \$50$, $n = 180$ deposits, and interest $i = \frac{0.10}{12}$ per payment period. The amount A saved is

$$A = 50\left[\frac{\left(1 + \frac{0.10}{12}\right)^{180} - 1}{\frac{0.10}{12}}\right] = \$50(414.4703) = \$20{,}723.52$$

When his son is 18 years old, Mr. Graff will have saved \$20,723.52.

NOW WORK PROBLEM 21.

EXAMPLE 5 Funding an IRA

Joe, at age 35, decides to invest in an IRA. He will put aside \$2000 per year for the next 30 years. How much will he have at age 65 if his rate of return is assumed to be 10% per annum?

SOLUTION This is an annuity with $P = \$2000$, $n = 30$ deposits, and interest $i = 0.10$ per payment period. The amount A in Joe's IRA after 30 years is

$$A = 2000\left[\frac{(1 + 0.10)^{30} - 1}{0.10}\right] = \$2000(164.49402) = \$328{,}988.05$$

Under this plan, Joe will have \$328,988.05 in his IRA when he is 65.

EXAMPLE 6 Funding an IRA

If, in Example 5, Joe had begun his IRA at age 25, instead of 35, what would his IRA be worth at age 65?

SOLUTION Now there are $n = 40$ deposits of \$2000 per year. The amount A of this annuity is

$$A = 2000\left[\frac{(1 + 0.10)^{40} - 1}{0.10}\right] = \$2000(442.59256) = \$885{,}185.11$$

Joe will have \$885,185.11 in his IRA when he is 65 if he begins the IRA at age 25.

NOW WORK PROBLEM 23.

EXAMPLE 7 Finding the Time it Takes to Reach a Certain Savings Goal*

How long does it take to save \$500,000 if you place \$500 per month in an account paying 6% compounded monthly?

SOLUTION This is an annuity in which the amount A is \$500,000, each deposit is $P = \$500$, and the interest is $i = \frac{0.06}{12}$ per payment period. We seek the number n of deposits needed to reach \$500,000.

**This example requires a knowledge of logarithms, especially the Change-of-Base Formula. A review of logarithms appears in Appendix A, Section A.3.*

$$A = P\frac{(1+i)^n - 1}{i} \qquad \text{Formula (1)}$$

$$500{,}000 = 500\,\frac{\left(1+\frac{0.06}{12}\right)^n - 1}{\frac{0.06}{12}} \qquad A = 500{,}000;\ P = 500;\ i = \frac{0.06}{12}$$

$$5 = \left(1+\frac{0.06}{12}\right)^n - 1 \qquad \text{Simplify}$$

$$6 = (1.005)^n \qquad \text{Simplify}$$

$$n = \log_{1.005} 6 = \frac{\log_{10} 6}{\log_{10} 1.005} = 359.25 \text{ months}$$

↑ Apply the definition of a logarithm. ↑ Change-of-base formula ↑ Payment period in months

It takes $359\frac{1}{4}$ months (almost 30 years) to save the \$500,000.

NOW WORK PROBLEM 35.

Sinking Funds

Solve problems involving sinking funds 2

A person with a debt may decide to accumulate sufficient funds to pay off the debt by agreeing to set aside enough money each month (or quarter, or year) so that when the debt becomes payable, the money set aside each month plus the interest earned will equal the debt. The fund created by such a plan is called a **sinking fund**.

In working with sinking funds, we generally seek the deposit or payment required to reach a certain goal. In other words, we seek the payment P required to obtain the amount A in Formula (1).

Let's look at an example.

EXAMPLE 8 Funding a Bond Obligation

The Board of Education received permission to issue \$4,000,000 in bonds to build a new high school. The board is required to make payments every 6 months into a sinking fund paying 8% compounded semiannually. At the end of 12 years the bond obligation will be retired. What should each payment be?

SOLUTION This is an example of a sinking fund. The payment P required twice a year to accumulate \$4,000,000 in 12 years (24 payments at a rate of interest of $i = \frac{0.08}{2}$ per payment period) obeys Formula (1).

$$A = P\frac{(1+i)^n - 1}{i} \qquad \text{Formula (1)}$$

$$4{,}000{,}000 = P\,\frac{\left(1+\frac{0.08}{2}\right)^{24} - 1}{\frac{0.08}{2}} \qquad A = 4{,}000{,}000;\ n = 24;\ i = \frac{0.08}{2}$$

$$4{,}000{,}000 = P(39.082604) \qquad \text{Simplify}$$

$$P = \$102{,}347.33 \qquad \text{Solve for } P.$$

The board will need to make a payment of \$102,347.33 every 6 months to redeem the bonds in 12 years.

NOW WORK PROBLEM 11.

EXAMPLE 9 Funding a Loan

A woman borrows \$3000, which will be paid back to the lender in one payment at the end of 5 years. She agrees to pay interest monthly at an annual rate of 12%. At the same time she sets up a sinking fund in order to repay the loan at the end of 5 years. She decides to make monthly payments into her sinking fund, which earns 8% interest compounded monthly.

(a) What is the monthly sinking fund payment?
(b) Construct a table that shows how the sinking fund grows over time.
(c) How much does she need each month to be able to pay the interest on the loan and make the sinking fund payment?

SOLUTION **(a)** The sinking fund payment is the value of P in Formula (1):

$$A = P\frac{(1+i)^n - 1}{i} \qquad \text{Formula (1)}$$

Here A equals the amount to be accumulated, namely, $A = \$3000$, $n = 60$ (5 years of monthly payments), and $i = \frac{0.08}{12}$. The sinking fund payment P obeys

$$3000 = P\left[\frac{\left(1+\frac{0.08}{12}\right)^{60} - 1}{\frac{0.08}{12}}\right]$$

$$300 = P(73.476856)$$

$$P = \$40.83$$

The monthly sinking fund payment is \$40.83

(b) Table 4 shows the growth of the sinking fund over time. For example, the total in the account after payment number 12 is obtained by using Formula (1). Since the monthly payment of \$40.83 has been made for 12 months at 8% compounded monthly, the total amount in the account at this point in time is

TABLE 4

Payment Number	Sinking Fund Deposit, \$	Cumulative Deposits	Accumulated Interest, \$	Total, \$
1	40.83	40.83	0	40.83
12	40.83	489.96	18.37	508.83
24	40.83	979.92	78.93	1058.85
36	40.83	1469.88	185.19	1655.07
48	40.83	1959.84	340.93	2300.77
60	40.77	2449.74	550.26	3000.00

$$\text{Total} = \$40.83\left[\frac{\left(1+\frac{0.08}{12}\right)^{12}-1}{\frac{0.08}{12}}\right] = \$40.83(12.449926) = \$508.33$$

The deposit for payment number 60, the final payment, is only \$40.77 because a deposit of \$40.83 results in a total of \$3000.06.

(c) The monthly interest payment due on the loan of \$3000 at 12% interest is found using the simple interest formula.

$$I = 3000(0.12)\left(\frac{1}{12}\right) = \$30$$

The woman needs to be able to pay \$40.83 + \$30 = \$70.83 each month.

NOW WORK PROBLEM 27.

Use Excel to solve Example 9, parts (a) and (b).

SOLUTION **(a)** Use the Payment Value function in Excel.
The syntax of the Payment Value function is:

PMT(rate, nper, pv, fv, type)

rate is the interest rate per period.
nper is the total number of payments.
pv is the principal value.
fv is the balance after the last payment is made.
type is the number 0 or 1 and indicates when payments are due. Use 0 for an ordinary annuity.

STEP 1 Set up the Excel spreadsheet as shown below:

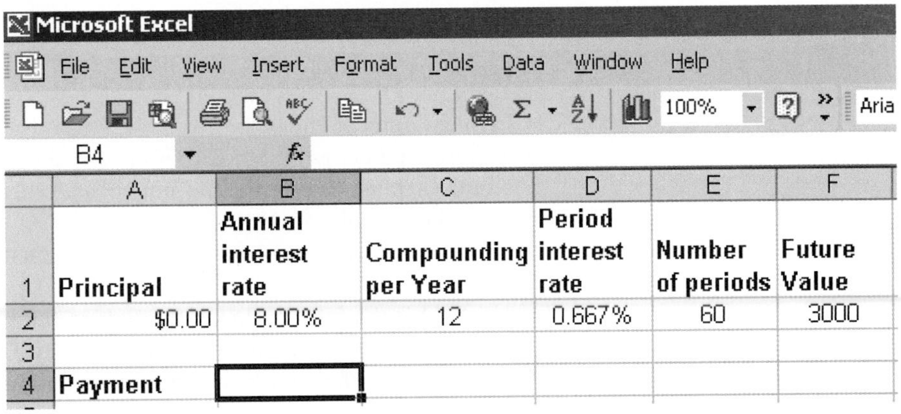

	A	B	C	D	E	F
1	Principal	Annual interest rate	Compounding per Year	Period interest rate	Number of periods	Future Value
2	\$0.00	8.00%	12	0.667%	60	3000
3						
4	Payment					

STEP 2 Insert the Payment function in B4 as shown below in the *fx* bar.

Microsoft Excel

File Edit View Insert Format Tools Data Window Help

100% Arial

B4 *fx* =PMT(D2,E2,A2,F2,0)

	A	B	C	D	E	F
1	**Principal**	**Annual interest rate**	**Compounding per Year**	**Period interest rate**	**Number of periods**	**Future Value**
2	$0.00	8.00%	12	0.667%	60	3000
3						
4	**Payment**	($40.83)				
5						

She has to deposit $40.83 each month into her sinking fund.

(b) Construct a table that shows how the sinking fund grows over time.

STEP 1 Set up the column headings in row 7.

Microsoft Excel

File Edit View Insert Format Tools Data Window Help

100% Arial

G14 *fx*

	A	B	C	D	E	F
1	**Principal**	**Annual interest rate**	**Compounding per Year**	**Period interest rate**	**Number of periods**	**Future Value**
2	$0.00	8.00%	12	0.667%	60	3000
3						
4	**Payment**	($40.83)				
5						
6						
7	**Payment Number**	**Sinking Fund Deposit**	**Cumulative Deposits**	**Accumulated Interest**	**Total**	
8						

STEP 2 Set up row 8.

- Payment number will be 1.
- Sinking fund deposit is $40.83. This is stored in cell B4. $40.83 is the deposit for each month; it must be positive and not change when the rest of the table is completed. Enter the formula =−1*B4 in B8.
- Cumulative deposits is number of payments times $40.83. Enter the formula =A8*B8 in C8.
- The total is the future value of the sinking fund. The inputs are fixed except for the payment number, A8. Enter the formula =FV(D2,A8,B4,A2,0).
- Accumulated interest is found by subtracting the cumulative deposits from the total. Enter the formula =E8 − C8 into D8.

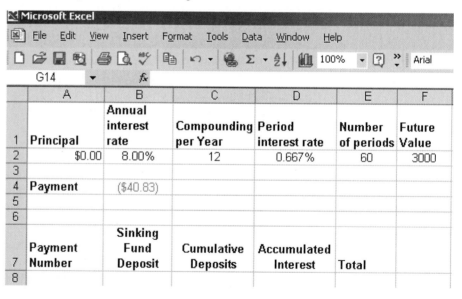

The formulas are given in the table below.

7	Payment Number	Sinking Fund Deposit	Cumulative Deposits	Accumulated Interest	Total
8	1	=-1*B4	=A8*B8	=E8-C8	=FV(D2,A8,B4,A2,0)
9					

STEP 3 To complete the table, capture row 8 and drag down two rows. Change the payment number in A9 to 6 and A10 to 12. Highlight rows 9 and 10 and click and drag through row 18.

The completed table is given below.

Microsoft Excel

File Edit View Insert Format Tools Data Window Help

Arial 10

J25 =

	A	B	C	D	E	F
1	Principal	Annual interest rate	Compounding per year	Period interest rate	Number of periods	Future value
2	$0.00	8.00%	12	0.667%	60	3000
3						
4	Payment	($40.83)				
5						
6						
7	Payment Number	Sinking Fund Deposit	Cumulative Deposits	Accumulated Interest	Total	
8	1	$40.83	$40.83	($0.00)	$40.83	
9	6	$40.83	$244.97	$4.12	$249.09	
10	12	$40.83	$489.95	$18.37	$508.32	
11	18	$40.83	$734.92	$43.17	$778.08	
12	24	$40.83	$979.89	$78.93	$1,058.82	
13	30	$40.83	$1,224.86	$126.12	$1,350.98	
14	36	$40.83	$1,469.84	$185.19	$1,655.03	
15	42	$40.83	$1,714.81	$256.63	$1,971.44	
16	48	$40.83	$1,959.78	$340.94	$2,300.72	
17	54	$40.83	$2,204.75	$438.64	$2,643.39	
18	60	$40.83	$2,449.73	$550.27	$3,000.00	
19						

NOW WORK PROBLEM 27 USING EXCEL.

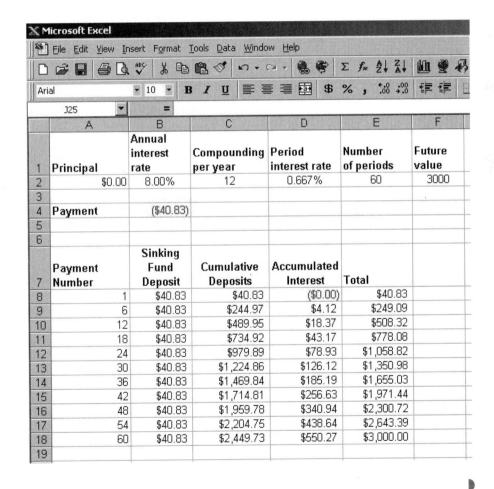

EXAMPLE 10 Depletion Investment

A gold mine is expected to yield an annual net income of \$200,000 for the next 10 years, after which it will be worthless. An investor wants an annual return on his investment of 18%. If he can establish a sinking fund earning 10% annually, how much should he be willing to pay for the mine?

SOLUTION Let x denote the price to be paid for the mine. Then $0.18x$ represents an 18% annual Return On Investment (ROI).

The annual Sinking Fund Contribution (SFC) needed to recover the purchase price x in 10 years obeys Formula (1), where $A = x$, $n = 10$, $i = 0.10$, and $P = \text{SFC}$. Then

$$A = P\frac{(1+i)^n - 1}{i} \qquad \text{Formula (1)}$$

$$x = \text{SFC}\,\frac{(1+0.10)^{10} - 1}{0.10} \qquad A = x,\ n = 10,\ i = 0.10,\ P = \text{SFC}$$

$$x = \text{SFC}\,(15.9374246)$$

$$\text{SFC} = 0.0627454\,x \qquad \text{Solve for SFC.}$$

The required annual Return On Investment (ROI) plus the annual Sinking Fund Contribution equals the annual net income of \$200,000.

$$\begin{pmatrix}\text{ROI}\\ \text{Annual return}\\ \text{on investment}\end{pmatrix} + \begin{pmatrix}\text{SFC}\\ \text{Annual sinking}\\ \text{fund contribution}\end{pmatrix} = \text{Annual income}$$

$$0.18x + (0.0627453)x = \$200{,}000$$

$$0.2427453x = \$200{,}000$$

$$x = \$823{,}909$$

A purchase price of \$823,909 will achieve the investor's goals.

NOW WORK PROBLEM 29.

EXERCISE 5.3 Answers to Odd-Numbered Problems Begin on Page AN-25.

In Problems 1–10, find the amount of each annuity.

1. After ten annual deposits of \$100 at 10% compounded annually.
2. After twelve monthly deposits of \$200 at 5% compounded monthly.
3. After twelve monthly deposits of \$400 at 12% compounded monthly.
4. After five annual deposits of \$1000 at 10% compounded annually.
5. After thirty-six monthly deposits of \$200 at 6% compounded monthly.
6. After forty semiannual deposits of \$2000 at 5% compounded semiannually.
7. After sixty monthly deposits of \$100 at 6% compounded monthly.
8. After eight quarterly deposits of \$1000 at 4% compounded quarterly.
9. After ten annual deposits of \$9000 at 5% compounded annually.
10. After twenty annual deposits of \$5000 at 4% compounded annually.

In Problems 11–20, find the payment required for each sinking fund.

11. The amount required is \$10,000 after 5 years at 5% compounded monthly. What is the monthly payment?
12. The amount required is \$5000 after 180 days at 4% compounded daily. What is the daily payment?
13. The amount required is \$20,000 after $2\frac{1}{2}$ years at 6% compounded quarterly. What is the quarterly payment?

14. The amount required is $50,000 after 10 years at 7% compounded semiannually. What is the semiannual payment?

15. The amount required is $25,000 after 6 months at $5\frac{1}{2}$% compounded monthly. What is the monthly payment?

16. The amount required is $100,000 after 25 years at $4\frac{1}{4}$% compounded annually. What is the annual payment?

17. The amount required is $5000 after 2 years at 4% compounded monthly. What is the monthly payment?

18. The amount required is $5000 after 2 years at 4% compounded semiannually. What is the semiannual payment?

19. The amount required is $9000 after 4 years at 5% compounded annually. What is the annual payment?

20. The amount required is $9000 after 2 years at 5% compounded quarterly. What is the quarterly payment?

21. **Market Value of a Mutual Fund** Al invests $2500 a year in a mutual fund for 15 years. If the market value of the fund increases 7% per year, what will be the value of the fund after 15 deposits?

22. **Saving for a Car** Sheila wants to invest an amount every 3 months so that she will have $12,000 in 3 years to buy a new car. The account pays 8% compounded quarterly. How much should she deposit each quarter to have $12,000 after 12 deposits?

23. **Value of an Annuity** Todd and Tami pay $300 every 3 months for 6 years into an ordinary annuity paying 8% compounded quarterly. What is the value of the annuity after the 24 deposits?

24. **Saving for a House** In 4 years Colleen and Bill would like to have $30,000 for a down payment on a house. How much should they deposit each month into an account paying 9% compounded monthly to have $30,000 after 48 deposits?

25. **Funding a Pension** Dan wishes to have $350,000 in a pension fund 20 years from now. How much should he deposit each month into an account paying 9% compounded monthly to have $350,000 after 240 deposits?

26. **Funding a Keogh Plan** Pat has a Keogh retirement plan (this type of plan is tax-deferred until money is withdrawn). If deposits of $7500 are made each year into an account paying 8% compounded annually, how much will be in the account after 25 years?

27. **Sinking Fund Payment** A company establishes a sinking fund to provide for the payment of a $100,000 debt maturing in 4 years. Contributions to the fund are to be made each year. Find the amount of each annual deposit if interest is 8% per annum. Prepare a table showing the annual growth of the sinking fund.

28. **Paying Off Bonds** A state has $5,000,000 worth of bonds that are due in 20 years. A sinking fund is established to pay off the debt. If the state can earn 10% annually on its money, what is the annual sinking fund deposit needed? Prepare a table showing the growth of the sinking fund every 5 years.

29. **Depletion Investment** An investor wants to know the amount she should pay for an oil well expected to yield an annual return of $30,000 for the next 30 years, after which the well will be dry. Find the amount she should pay to yield a 14% annual return if a sinking fund earns 10% annually.

30. **Time Needed for a Million Dollars** If you deposit $10,000 every year into an account paying 8% compounded annually, how long will it take to accumulate $1,000,000?

31. **Bond Payments** A city has issued bonds to finance a new library. The bonds have a total face value of $1,000,000 and are payable in 10 years. A sinking fund has been opened to retire the bonds. If the interest rate on the fund is 8% compounded quarterly, what are the quarterly payments?

32. **Value of an IRA**

(a) Tanya invested $2000 in an IRA each year for 10 years earning 8% compounded annually. At the end of 10 years she ceased the IRA payments, but continued to invest her accumulated amount at 8% compounded annually, for the next 30 years. What was the value of her IRA investment at the end of 10 years? What was the value of her investment at the end of the next 30 years?

(b) Carol started her IRA investment in the 11th year and invested $2000 per year for the next 30 years at 8% compounded annually. What was the value of her investment at the end of 30 years?

(c) Who had more money at the end of the period?

33. **Managing a Condo** The Crown Colony Condo Association is required by law to set aside funds to replace its roof. The current cost to replace the roof is $100,000 and it will need to be replaced in 20 years. The cost of a roof is expected to increase at the rate of 3% per year. The Condo can invest in Treasuries yielding 6% paid semiannually.

(a) What will the roof cost in 20 years?

(b) If the Condo invests in the Treasuries, what semiannual payment is required to have the funds to replace the roof in 20 years?

34. **Managing a Condo** The Crown Colony Condo Association is required by law to set aside funds to replace its common-area carpet. The current cost to replace the carpet is $20,000 and it will need to be replaced in 6 years. The cost of carpet is expected to increase at the rate of 2% per year. The Condo can invest in Treasuries yielding 5% paid semiannually.

(a) What will the carpet cost in 6 years?

(b) If the Condo invests in the Treasuries, what semiannual payment is required to have the funds to replace the carpet in 6 years?

Problems 35 and 36 require logarithms.

35. How many years will it take to save $1,000,000 if you place $600 per month in an account that earns 7% compounded monthly?

36. How many years will it take to save $1,000,000 if you place $1000 per month in an account that earns 6% compounded monthly?

37. Saving for a Car In January 2003, a new Honda Accord EX with manual transmission was listed at $21,600. Assuming a constant annual inflation rate of 1.58%, which was the annual percent increase in the U. S. Consumer Price Index for 2002, calculate a projection of what a new Honda Accord EX with manual transmission would cost in January 2007. Assume that the sales tax rate where you plan to purchase this vehicle in January 2007 will be 9.25%. Beginning in January 2003, suppose that you want to make a monthly payment into a sinking fund that earns interest at a 2.75% annual rate, compounded monthly, in order to accumulate funds to purchase a new Honda in 4 years. What monthly payment will produce a balance in this sinking fund in January 2007, which would be equal to the projected cost of a 2007 Honda Accord EX as calculated above plus sales tax?

Source: American Honda Motor Co., Inc., and U. S. Department of Labor, Bureau of Labor Statistics.

38. Saving for a Down Payment on a Home The median price of an existing single-family detached home in California was $270,210 in November 2001 and increased to $328,310 in November 2002. If the median price of an existing single-family detached home in California continues to increase at the annual rate experienced between November 2001 and November 2002, determine the projected median price of an existing single-family detached home in California in November 2006. Suppose that you wish to make monthly payments into a sinking fund between November 2002 and November 2006 to generate a down payment of 10% of this projected median price. Find the monthly payment required for this sinking fund, assuming that this sinking fund earns 2.9%, compounded monthly.

Source: San Francisco *Business Times* based on data from the California Association of Realtors.

39. Saving for College The average annual undergraduate college in-state tuition nationally for 2000–2001 for public 4-year institutions was $3506. The Bureau of Labor Statistics indicated that the inflation rate for tuition in 2001 was 5.09%. Assuming that this inflation rate is constant for the next 16 years, determine the projected annual undergraduate college tuition for public 4-year institutions for 2012–2013, 2013–2014, 2014–2015, and 2015–2016. Next, determine a quarterly sinking fund payment, beginning with the fourth quarter of 2002, so that at the end of 2012 the sinking fund will have the amount needed to fund 4 years of tuition, as calculated above. Assume that this sinking fund earns interest at a 4.2% annual rate, compounded quarterly. Assume that no additional quarterly payments to this sinking fund will be made after the fourth quarter of 2012, that tuition is due to be paid at the end of the fourth quarter of 2012 for the year 2012–2013, at the end of the fourth quarter of 2013 for the year 2013–2014, and so on and that funds which remain in this sinking fund after 2012 will continue to earn interest at a 4.2% annual rate, compounded quarterly.

Source: United States Department of Education, National Center for Education Statistics, Integrated Postsecondary Education Data System (IPEDS), "Fall Enrollment" and "Institutional Characteristics" surveys (August, 2001), FinAid.org, "The Smart Student Guide to Financial Aid"; and U. S. Department of Labor, Bureau of Labor Statistics.

40. Saving for College The average annual undergraduate college tuition nationally for 2000–2001 for private 4-year institutions was $15,531. The Bureau of Labor Statistics indicated that the inflation rate for tuition in 2001 was 5.09%. Assuming that this inflation rate is constant for the next 16 years, determine the projected annual undergraduate college tuition for private 4-year institutions for 2012–2013, 2013–2014, 2014–2015, and 2015–2016. Next, determine a quarterly sinking fund payment, beginning with the fourth quarter of 2002, so that at the end of 2012 the sinking fund will have the amount needed to fund 4 years of tuition, as calculated above. Assume that this sinking fund earns interest at a 4.2% annual rate, compounded quarterly. Assume that no additional quarterly payments to this sinking fund will be made after the fourth quarter of 2012, that tuition is due to be paid at the end of the fourth quarter of 2012 for the year 2012–2013, at the end of the fourth quarter of 2013 for the year 2013–2014, and so on and that funds which remain in this sinking fund after 2012 will continue to earn interest at a 4.2% annual rate, compounded quarterly.

Source: United States Department of Education, National Center for Education Statistics, Integrated Postsecondary Education Data System (IPEDS), "Fall Enrollment" and "Institutional Characteristics" surveys (August, 2001), FinAid.org, "The Smart Student Guide to Financial Aid"; and U. S. Department of Labor, Bureau of Labor Statistics.

5.4 Present Value of an Annuity; Amortization

PREPARING FOR THIS SECTION *Before getting started, review the following:*

> Exponents and Logarithms (Appendix A, Section A.3, pp. 645–649)

OBJECTIVES
1 Solve problems involving the present value of an annuity
2 Solve problems involving amortization

In Section 5.2 we defined present value (as it relates to the compound interest formula) as the amount of money needed now to obtain an amount A in the future. A similar idea is used for periodic withdrawals.

Suppose you want to withdraw \$10,000 per year each year for the next 5 years from a retirement account that earns 10% compounded annually. How much money is required initially in this account for this to happen? In fact, the amount is the *present value* of each of the \$10,000 withdrawals. This leads to the following definition.

The **present value** of an annuity is the sum of the present values of the withdrawals. In other words, the present value of an annuity is the amount of money needed now so that if it is invested at i percent, n equal dollar amounts can be withdrawn without any money left over.

EXAMPLE 1 Finding the Present Value of an Annuity

Compute the amount of money required to pay out \$10,000 per year for 5 years at 10% compounded annually.

SOLUTION For the first \$10,000 withdrawal, the present value V_1 (the dollars needed now to withdraw \$10,000 in one year) is

$$V_1 = \$10{,}000(1 + 0.10)^{-1} = \$10{,}000(0.9090909) = \$9090.91$$

For the second \$10,000 withdrawal, the present value V_2 (the dollars needed now to withdraw \$10,000 in two years) is

$$V_2 = \$10{,}000(1 + 0.10)^{-2} = \$10{,}000(0.826446) = \$8264.46$$

Similarly,

$$V_3 = \$10{,}000(1 + 0.10)^{-3} = \$10{,}000(0.7513148) = \$7513.15$$
$$V_4 = \$10{,}000(1 + 0.10)^{-4} = \$10{,}000(0.6830134) = \$6830.13$$
$$V_5 = \$10{,}000(1 + 0.10)^{-5} = \$10{,}000(0.6209213) = \$6209.21$$

The present value V for 5 withdrawals of \$10,000 each is

$$\begin{aligned} V &= V_1 + V_2 + V_3 + V_4 + V_5 \\ &= \$9090.91 + \$8264.46 + \$7513.15 + \$6830.13 + \$6209.21 \\ &= \$37{,}907.86 \end{aligned}$$

A person would need \$37,907.86 now, invested at 10% per annum in order to withdraw \$10,000 per year for the next 5 years.

Table 5 summarizes these results. Table 6 lists the amount at the beginning of each year.

TABLE 5

Withdrawal	Present Value
1st	$\$10{,}000(1.10)^{-1} = \$9{,}090.91$
2nd	$\$10{,}000(1.10)^{-2} = \$8{,}264.46$
3rd	$\$10{,}000(1.10)^{-3} = \$7{,}513.15$
4th	$\$10{,}000(1.10)^{-4} = \$6{,}830.13$
5th	$\$10{,}000(1.10)^{-5} = \$6{,}209.21$
Total	\$37,907.86

TABLE 6

Year	Amount at the Beginning of the Year	Add Interest	Subtract Withdrawal
1	37,907.86	3,790.79	10,000.00
2	31,698.65	3,169.87	10,000.00
3	24,868.52	2,486.85	10,000.00
4	17,355.37	1,735.54	10,000.00
5	9,090.91	909.09	10,000.00
6	0		

We seek a formula for the present value of an annuity. Suppose an annuity earns interest of i percent per payment period and suppose we wish to make n withdrawals of $\$P$ at each payment period. The amount V_1 required for the first withdrawal (the present value of $\$P$) is

$$V_1 = P(1+i)^{-1}$$

The amount V_2 required for the second withdrawal is

$$V_2 = P(1+i)^{-2}$$

The amount V_n required for the nth withdrawal is

$$V_n = P(1+i)^{-n}$$

The present value V of the annuity is the sum of the amounts $V_1, V_2, \ldots, V_n$

$$\begin{aligned} V = V_1 + \cdots + V_n &= P(1+i)^{-1} + \cdots + P(1+i)^{-n} \\ &= P(1+i)^{-n}\,[1 + (1+i) + \cdots + (1+i)^{n-1}] \end{aligned}$$

The expression in brackets is the sum of the first n terms of a geometric sequence, whose ratio is $1+i$. As a result,

$$V = P(1+i)^{-n}\frac{1-(1+i)^n}{1-(1+i)} = P\frac{(1+i)^{-n}-1}{-i} = P\frac{1-(1+i)^{-n}}{i}$$

Present Value of an Annuity

Suppose an annuity earns interest at the rate of i percent per payment period. If n withdrawals of $\$P$ are made at each payment period, the amount V required obeys

$$V = P\frac{1-(1+i)^{-n}}{i} \tag{1}$$

Here V is called the present value of the annuity.

1 **EXAMPLE 2** **Getting By in College**

A student requires \$200 each month for the next 10 months to cover miscellaneous expenses at school. A money market fund will pay her interest of 2% per annum each month. How much money should she ask for from her parents so that she can withdraw \$200 each month for 10 months?

SOLUTION We seek the present value of an annuity. The monthly withdrawal is \$200, the interest rate is $i = \frac{0.02}{12}$ per month, and the number of withdrawals is $n = 10$. The money required for this is given by Formula (1):

$$V = \$200\left[\frac{1-\left(1+\frac{0.02}{12}\right)^{-10}}{\frac{0.02}{12}}\right] = \$200\,(9.90894) = \$1981.79$$

She should ask her parents for \$1981.79

NOW WORK PROBLEM 1.

EXAMPLE 3 **Finding the Cost of a Car**

A man agrees to pay \$300 per month for 48 months to pay off a car loan. If interest of 12% per annum is charged monthly, how much did the car originally cost? How much interest was paid?

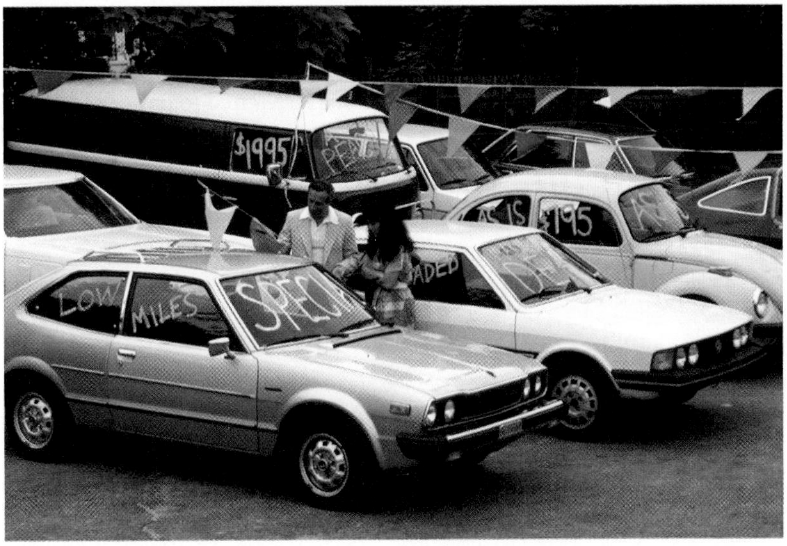

SOLUTION This is the same as asking for the present value V of an annuity of \$300 per month at 12% for 48 months. The original cost of the car is

$$V = 300\left[\frac{1-\left(1+\frac{0.12}{12}\right)^{-48}}{\frac{0.12}{12}}\right] = \$300(37.9739595) = \$11{,}392.19$$

The total payment is $(\$300)(48) = \$14{,}400$. The interest paid is

$$\$14{,}400 - \$11{,}392.19 = \$3007.81$$

NOW WORK PROBLEM 15.

Amortization

We can look at Example 3 differently. What it also says is that if the man pays \$300 per month for 48 months with an interest of 12% compounded monthly, then the car is his. In other words, he *amortized* the debt in 48 equal monthly payments. (The Latin word *mort* means "death." Paying off a loan is regarded as "killing" it.) A loan with a fixed rate of interest is said to be **amortized** if both principal and interest are paid by a sequence of equal payments made over equal periods of time.

When a loan of V dollars is amortized at a rate of interest i per payment period over n payment periods, the customary question is, "What is the payment P?" In other words, in amortization problems, we want to find the amount of payment P that, after n payment periods at the rate of interest i per payment period, gives us a present value equal to the amount of the loan. We need to find P in the formula

$$V = P\left[\frac{1-(1+i)^{-n}}{i}\right]$$

Solving for P, we find

$$P = V\left[\frac{1-(1+i)^{-n}}{i}\right]^{-1} = V\left[\frac{i}{1-(1+i)^{-n}}\right]$$

Amortization

The payment P required to pay off a loan of V dollars borrowed for n payment periods at a rate of interest i per payment period is

$$\boxed{P = V\left[\frac{i}{1-(1+i)^{-n}}\right]} \qquad (2)$$

2 **EXAMPLE 4** **Finding the Payment for an Amortized Loan**

What monthly payment is necessary to pay off a loan of \$800 at 10% per annum **(a)** In 2 years? **(b)** In 3 years? **(c)** What total amount is paid out for each loan?

SOLUTION **(a)** For the 2-year loan $V = \$800$, $n = 24$, and $i = \frac{0.10}{12}$. The monthly payment P is

$$P = 800\left[\frac{\frac{0.10}{12}}{1-\left(1+\frac{0.10}{12}\right)^{-24}}\right] = \$800(0.04614493) = \$36.92$$

(b) For the 3-year loan $V = \$800$, $n = 36$, and $i = \frac{0.10}{12}$. The monthly payment P is

$$P = 800\left[\frac{\frac{0.10}{12}}{1-\left(1+\frac{0.10}{12}\right)^{-36}}\right] = \$800(0.03226719) = \$25.81$$

(c) For the 2-year loan, the total amount paid out is (36.92)(24) = \$886.08; for the 3-year loan, the total amount paid out is (\$25.81)(36) = \$929.16.

NOW WORK PROBLEM 9.

EXAMPLE 5 Mortgage Payments

Mr. and Mrs. Corey have just purchased a \$300,000 house and have made a down payment of \$60,000. They can amortize the balance (\$300,000 − \$60,000 = \$240,000) at 6% for 30 years.

(a) What are the monthly payments?
(b) What is their total interest payment?
(c) After 20 years, what equity do they have in their house (that is, what is the sum of the down payment and the amount paid on the loan)?

SOLUTION **(a)** The monthly payment P needed to pay off the loan of \$240,000 at 6% for 30 years (360 months) is

$$P = \$240{,}000\left[\frac{\frac{0.06}{12}}{1-\left(1+\frac{0.06}{12}\right)^{-360}}\right] = \$240{,}000(0.0059955) = \$1438.92$$

(b) The total paid out for the loan is

$$(\$1438.92)(360) = \$518{,}011.20$$

The interest on this loan amounts to

$$\$518{,}011.20 - \$240{,}000 = \$278{,}011.20$$

(c) After 20 years (240 months) there remains 10 years (or 120 months) of payments. The present value of the loan is the present value of a monthly payment of \$1438.92 for 120 months at 6%, namely,

$$V = \$1438.92\left[\frac{1-\left(1+\frac{0.06}{12}\right)^{-120}}{\frac{0.06}{12}}\right]$$
$$= (\$1438.92)(90.073453) = \$129{,}608.49$$

The amount paid on the loan is

$$(\text{Original loan amount}) - (\text{Present value}) = \$240{,}000 - \$129{,}608.49 = \$110{,}391.51$$

The equity after 20 years is

$$(\text{Down payment}) + (\text{Amount paid on loan}) = \$60{,}000 + \$110{,}391.51 = \$170{,}391.51$$

This equity does not include any appreciation in the value of the house over the 20 year period.

Table 7 gives a partial schedule of payments for the loan in Example 5. It is interesting to observe how slowly the amount paid on the loan increases early in the payment schedule, with very little of the payment used to reduce principal, and how quickly the amount paid on the loan increases during the last 5 years.

TABLE 7

Payment Number	Monthly Payment	Principal	Interest	Amount Paid on Loan
1	$1,438.92	$ 238.92	$1,200.00	$ 238.92
60	$1,438.92	$ 320.67	$1,118.26	$ 16,669.54
120	$1,438.92	$ 432.53	$1,006.39	$ 39,154.26
180	$1,438.92	$ 538.42	$ 855.50	$ 69,482.77
240	$1,438.92	$ 786.94	$ 651.98	$110,391.39
300	$1,438.92	$1,061.47	$ 377.45	$165,570.99
360	$1,438.92	$1,431.76	$ 7.16	$240,000.00

NOW WORK PROBLEM 11.

Use EXCEL to solve Example 5, parts (a) and (c).

SOLUTION **(a)** Use the Excel function PMT to find the monthly payments. The function has the following syntax:

PMT(*rate, nper, pv, fv, type*)

rate is the interest rate per period.
nper is the total number of payment periods
pv is the initial deposit or present value
fv is the future value
type is the number 0 or 1: use 0 for an ordinary annuity

The values of the parameters of the PMT function are:

rate is .06/12
nper is 30*12 = 360
pv is $240,000
fv is 0
type is 0

The Excel spreadsheet is given below.

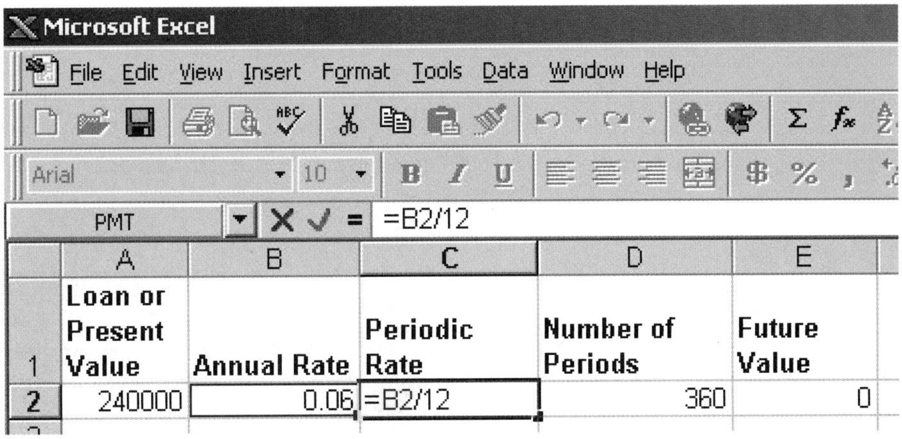

In cell B4 type =PMT(*C2,D2,A2,E2,0*)

The final Excel spreadsheet for part (a) is given below.

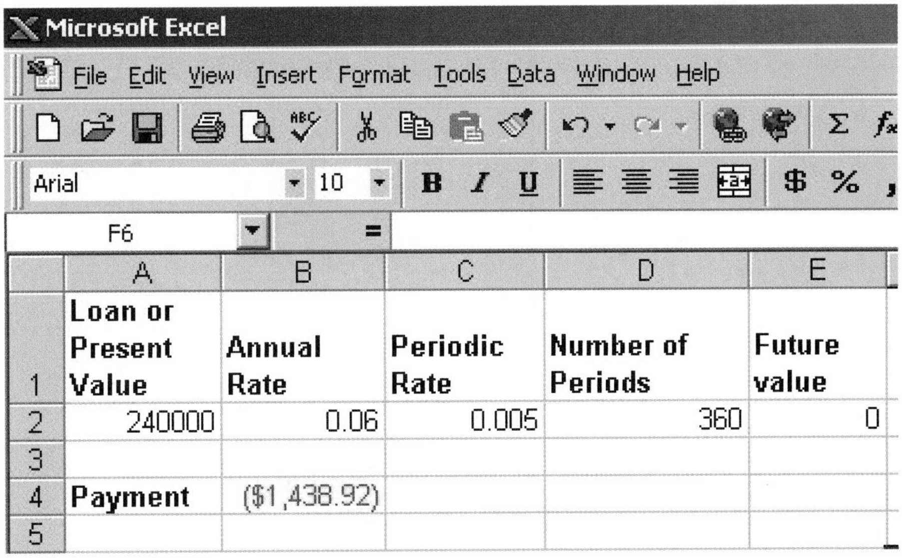

Mr. And Mrs. Corey's monthly payments are $1438.92.

(c) To create an amortization table, use the Excel function PPMT to compute the monthly principal.

- Set up the table headings in row 7.
- In A8 enter 1 for payment number 1.
- In B8 enter the formula = −1 *B4. (Recall the $ keeps the cell constant.)
- In C8 enter the formula = −1 *PPMT(C2,A8,D2,A2,E2).
- In D8 enter the formula =B8 − C8.
- In E8 enter the formula = 240,000 + FV(C2, A8,B4,A2, 0).
- Highlight row 8, click and drag 2 rows (filling rows 9 and 10).
- Change A9 to 60 and A10 to 120.
- Highlight rows 9 and 10; click and drag the two rows to row 14 to get the rest of the table.

The completed table is given below.

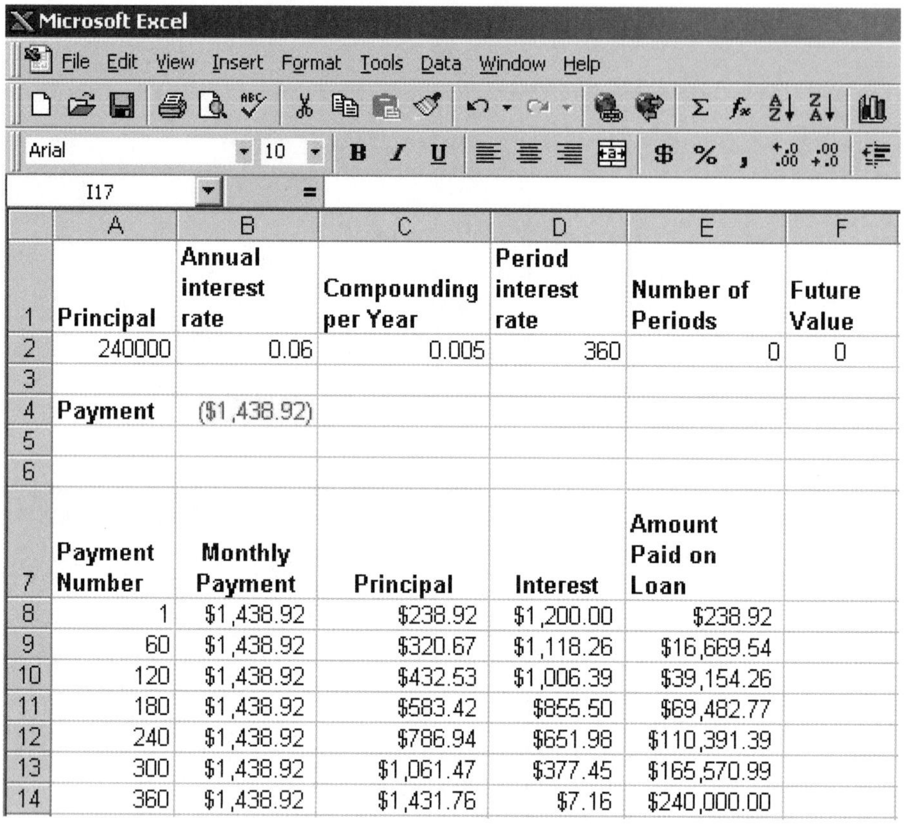

	A	B	C	D	E	F
1	Principal	Annual interest rate	Compounding per Year	Period interest rate	Number of Periods	Future Value
2	240000	0.06	0.005	360	0	0
3						
4	Payment	($1,438.92)				
5						
6						
7	Payment Number	Monthly Payment	Principal	Interest	Amount Paid on Loan	
8	1	$1,438.92	$238.92	$1,200.00	$238.92	
9	60	$1,438.92	$320.67	$1,118.26	$16,669.54	
10	120	$1,438.92	$432.53	$1,006.39	$39,154.26	
11	180	$1,438.92	$583.42	$855.50	$69,482.77	
12	240	$1,438.92	$786.94	$651.98	$110,391.39	
13	300	$1,438.92	$1,061.47	$377.45	$165,570.99	
14	360	$1,438.92	$1,431.76	$7.16	$240,000.00	

NOW WORK PROBLEM 11 USING EXCEL.

EXAMPLE 6 Inheritance

When Mr. Nicholson died, he left an inheritance of \$15,000 for his family to be paid to them over a 10-year period in equal amounts at the end of each year. If the \$15,000 is invested at 10% per annum, what is the annual payout to the family?

SOLUTION This example asks what annual payment is needed at 10% for 10 years to disperse \$15,000. That is, we can think of the \$15,000 as a loan amortized at 10% for 10 years. The payment needed to pay off the loan is the yearly amount Mr. Nicholson's family will receive. The yearly payout P is

$$P = \$15{,}000\left[\frac{0.10}{1 - (1 + 0.10)^{-10}}\right]$$
$$= \$15{,}000(0.16274539) = \$2441.18$$

EXAMPLE 7 Determining Retirement Income

Joan is 20 years away from retiring and starts saving \$100 a month in an account paying 6% compounded monthly. When she retires, she wishes to withdraw a fixed amount each month for 25 years. What will this fixed amount be?

SOLUTION After 20 years the amount A accumulated in her account is the amount of an annuity with a monthly payment of \$100 and an interest rate of $i = \frac{0.06}{12}$.

$$A = 100\left[\frac{\left(1 + \frac{0.06}{12}\right)^{240} - 1}{\frac{0.06}{12}}\right]$$

$$= 100 \cdot (462.0408951) = \$46{,}204.09$$

The amount P she can withdraw for 300 payments (25 years) at 6% compounded monthly is

$$P = 46{,}204.09\left[\frac{\frac{0.06}{12}}{1 - \left(1 + \frac{0.06}{12}\right)^{-300}}\right] = \$297.69$$

NOW WORK PROBLEM 7.

EXERCISE 5.4 Answers to Odd-Numbered Problems Begin on Page AN-25.

In Problems 1–6, find the present value of each annuity.

1. The withdrawal is to be \$500 per month for 36 months at 10% compounded monthly.

2. The withdrawal is to be \$1000 per year for 3 years at 8% compounded annually.

3. The withdrawal is to be \$100 per month for 9 months at 12% compounded monthly.

4. The withdrawal is to be \$400 per month for 18 months at 5% compounded monthly.

5. The withdrawal is to be \$10,000 per year for 20 years at 10% compounded annually.

6. The withdrawal is to be \$2000 per month for 3 years at 4% compounded monthly.

7. **Value of a IRA** A husband and wife contribute \$4000 per year to an IRA paying 10% compounded annually for 20 years. What is the value of their IRA? How much can they withdraw each year for 25 years at 10% compounded annually?

8. Rework Problem 7 if the interest rate is 8%.

9. **Loan Payments** What monthly payment is needed to pay off a loan of \$10,000 amortized at 12% compounded monthly for 2 years?

10. **Loan Payments** What monthly payment is needed to pay off a loan of \$500 amortized at 12% compounded monthly for 2 years?

11. In Example 5, if Mr. and Mrs. Corey amortize the \$240,000 loan at 6% for 20 years, what is their monthly payment?

12. In Example 5, if Mr. and Mrs. Corey amortize their \$240,000 loan at 5% for 15 years, what is their monthly payment?

13. In Example 6, if Mr. Nicholson left \$15,000 to be paid over 20 years in equal yearly payments and if this amount were invested at 12%, what would the annual payout be?

14. Joan has a sum of \$30,000 that she invests at 10% compounded monthly. What equal monthly payments can she receive over a 10-year period? Over a 20-year period?

15. **Planning Retirement** Mr. Doody, at age 65, can expect to live for 20 years. If he can invest at 10% per annum compounded monthly, how much does he need now to guarantee himself \$250 every month for the next 20 years?

16. **Planning Retirement** Sharon, at age 65, can expect to live for 25 years. If she can invest at 10% per annum compounded monthly, how much does she need now to guarantee herself \$300 every month for the next 25 years?

17. **House Mortgage** A couple wish to purchase a house for \$200,000 with a down payment of \$40,000. They can amortize the balance either at 8% for 20 years or at 9% for 25 years. Which monthly payment is greater? For which loan is the total interest paid greater? After 10 years, which loan provides the greater equity?

18. **House Mortgage** A couple have decided to purchase a \$250,000 house using a down payment of \$20,000. They can amortize the balance at 8% for 25 years. (a) What is their monthly payment? (b) What is the total interest paid? (c) What is their equity after 5 years? (d) What is the equity after 20 years?

19. **Planning Retirement** John is 45 years old and wants to retire at 65. He wishes to make monthly deposits in an account paying 9% compounded monthly so when he retires he can withdraw \$300 a month for 30 years. How much should John deposit each month?

20. **Cost of a Lottery** The grand prize in an Illinois lottery is \$6,000,000, which will be paid out in 20 equal annual payments of \$300,000 each. Assume the first payment of \$300,000 is made, leaving the state with the obligation to pay out \$5,700,000 in 19 equal yearly payments of \$300,000 each. How much does the state need to deposit in an account paying 6% compounded annually to achieve this?

21. **College Expenses** Dan works during the summer to help with expenses at school the following year. He is able to save \$100 each week for 12 weeks, and he invests it at 6% compounded weekly.

    (a) How much does he have after 12 weeks?
    (b) When school starts, Dan will begin to withdraw equal amounts from this account each week. What is the most Dan can withdraw each week for 34 weeks?

22. Repeat Problem 21 if the rate of interest is 8%. What if it is only 4%?

23. **Analyzing a Town House Purchase** Mike and Yola have just purchased a town house for \$200,000. They obtain financing with the following terms: a 20% down payment and the balance to be amortized over 30 years at 9%.

    (a) What is their down payment?
    (b) What is the loan amount?
    (c) How much is their monthly payment on the loan?
    (d) How much total interest do they pay over the life of the loan?
    (e) If they pay an additional \$100 each month toward the loan, when will the loan be paid?
    (f) With the \$100 additional monthly payment, how much total interest is paid over the life of the loan?

24. **House Mortgage** Mr. Smith obtained a 25-year mortgage on a house. The monthly payments are \$2247.57 (principal and interest) and are based on a 7% interest rate. How much did Mr. Smith borrow? How much interest will be paid?

25. **Car Payments** A car costs \$12,000. You put 20% down and amortize the rest with equal monthly payments over a 3-year period at 15% to be compounded monthly. What will the monthly payment be?

26. **Cost of Furniture** Jay pays \$320 per month for 36 months for furniture, making no down payment. If the interest charged is 0.5% per month on the unpaid balance, what was the original cost of the furniture? How much interest did he pay?

27. **Paying for Restaurant Equipment** A restaurant owner buys equipment costing \$20,000. If the owner pays 10% down and amortizes the rest with equal monthly payments over 4 years at 12% compounded monthly, what will be the monthly payments? How much interest is paid?

28. **Refinancing a Mortgage** A house that was bought 8 years ago for \$150,000 is now worth \$300,000. Originally, the house was financed by paying 20% down with the rest financed through a 25-year mortgage at 10.5% interest. The owner (after making 96 equal monthly payments) is in need of cash, and would like to refinance the house. The finance company is willing to loan 80% of the new value of the house amortized over 25 years with the same interest rate. How much cash will the owner receive after paying the balance of the original loan?

29. **Comparing Mortgages** A home buyer is purchasing a \$140,000 house. The down payment will be 20% of the price of the house, and the remainder will be financed by a 30-year mortgage at a rate of 9.8% interest compounded monthly. What will the monthly payment be? Compare the monthly payments and the total amounts of interest paid if a 15-year mortgage is chosen instead of a 30-year mortgage.

30. **Mortgage Payments** Mr. and Mrs. Hoch are interested in building a house that will cost \$180,000. They intend to use the \$60,000 equity in their present house as a down payment on the new one and will finance the rest with a 25-year mortgage at an interest rate of 10.2% compounded monthly. How large will their monthly payment be on the new house?

Problems 31 and 32 require logarithms.

31. **IRA Withdrawals** How long will it take to exhaust an IRA of \$100,000 if you withdraw \$2000 every month? Assume a rate of interest of 5% compounded monthly.

32. **IRA Withdrawals** How long will it take to exhaust an IRA of \$200,000 if you withdraw \$3000 every month? Assume a rate of interest of 4% compounded monthly.

33. **Refinancing a Home Loan** Suppose that a couple decide to refinance a 30-year home loan, after making monthly payments for 5 years. The original loan amount was \$312,000 with an annual interest rate of 6.825%, compounded

monthly. This couple accept an offer from a California bank to refinance their loan with no closing costs. The new loan will be a 25-year loan and will have a lower interest rate, 6.125%, compounded monthly. With this refinancing, by how much will the monthly payments be reduced? Over the full term of the new loan, how much total interest will be saved?

Source: http://www.fremontbank.com on January 6, 2003.

34. **Refinancing a Home Loan** Suppose that a couple decide to refinance a 30-year home loan, after making monthly payments for 3 years. The original loan amount was \$285,000 with an annual interest rate of 6.75%, compounded monthly. This couple accept an offer from a California bank to refinance their loan with no closing costs. The new loan will be a 15-year loan but will have a lower interest rate, 5.5%, compounded monthly. With this refinancing, by how much will the monthly payments be increased? Over the full term of the new loan, how much total interest will be saved?

Source: http://www.fremontbank.com on January 6, 2003.

35. **Prepaying a Home Loan** After making minimum payments for 4 years on a 30-year home loan, a couple decide to pay an additional \$150 per month toward the principal. The original loan amount was \$235,000 with an annual interest rate of 6.125%, compounded monthly. By how many months (rounded to the nearest tenth) will the term of this loan be reduced with this additional monthly payment? Over the life of this loan, how much interest will be saved?

Source: http://www.fremontbank.com on January 6, 2003.

36. **Adjustable Rate Mortgages** A couple decide to accept an adjustable rate, 30-year mortgage, in which the interest rate is fixed for the first 5 years and then may be adjusted annually beginning with year 6. Suppose that this couple borrow \$305,000 with the annual interest rate, compounded monthly, for the first 5 years set at 5.5%. Suppose that the interest rate is increased to 6.75% for the remaining term of the loan. Calculate the monthly payment during the first 5 years and after the first 5 years. Find the amount of interest that would be paid over the term of this loan.

Source: http://www.fremontbank.com on January 6, 2003.

5.5 Annuities and Amortization Using Recursive Sequences

PREPARING FOR THIS SECTION *Before getting started, review the following:*

> Recursive Sequences (Appendix A, Section A.4, pp. 649–654)

OBJECTIVES
1 Use sequences and a graphing utility to solve annuity problems
2 Use sequences and a graphing utility to solve amortization problems

Annuities

1 Use sequences and a graphing utility to solve annuity problems

In Section 5.2 we developed the compound interest formula, which gives the future value when a fixed amount of money is deposited in an account that pays interest compounded periodically. Often, though, money is invested in equal amounts at periodic intervals. An **annuity** is a sequence of equal periodic deposits. The periodic deposits may be made annually, quarterly, monthly, or daily.

When deposits are made at the same time that the interest is credited, the annuity is called **ordinary**. We will only deal with ordinary annuities here. The **amount of an annuity** is the sum of all deposits made plus all interest paid.

Suppose that the initial amount deposited in an annuity is \$$M$, the periodic deposit is \$$P$, and the per annum rate of interest is r% (expressed as a decimal) compounded N times per year*. The periodic deposit is made at the same time that the interest is credited,

We use N to represent the number of times interest is compounded per annum instead of n, since n is the traditional symbol used with sequences to denote the term of the sequence.

so N deposits are made per year. The amount A_n of the annuity after n deposits will equal the amount of the annuity after $n - 1$ deposits, A_{n-1}, plus the interest earned on this amount plus the periodic deposit P. That is,

$$\underset{\substack{\text{Amount}\\ \text{after}\\ n \text{ deposits}}}{\underset{\uparrow}{A_n}} = \underset{\substack{\text{Amount}\\ \text{in previous}\\ \text{period}}}{\underset{\uparrow}{A_{n-1}}} + \underset{\substack{\text{Interest}\\ \text{earned}}}{\underset{\uparrow}{\frac{r}{N}A_{n-1}}} + \underset{\substack{\text{Periodic}\\ \text{deposit}}}{\underset{\uparrow}{P}} = \left(1 + \frac{r}{N}\right)A_{n-1} + P$$

We have established the following result:

Annuity Formula

If $A_0 = M$ represents the initial amount deposited in an annuity that earns r% per annum compounded N times per year, and if P is the periodic deposit made at each payment period, then the amount A_n of the annuity after n deposits is given by the recursive sequence

$$A_0 = M, \qquad A_n = \left(1 + \frac{r}{N}\right)A_{n-1} + P, \qquad n \geq 1 \tag{1}$$

Formula (1) may be explained as follows: the money in the account initially, A_0, is $\$M$; the money in the account after $n - 1$ payments, A_{n-1}, earns interest $\frac{r}{n}$ during the nth period; so when the periodic payment of P dollars is added, the amount after n payments, A_n, is obtained.

EXAMPLE 1 Saving for Spring Break

A trip to Cancun during spring break will cost \$450 and full payment is due March 2. To have the money, a student, on September 1, deposits \$100 in a savings account that pays 4% per annum compounded monthly. On the first of each month, the student deposits \$50 in this account.

(a) Find a recursive sequence that explains how much is in the account after n months.
(b) Use the TABLE feature to list the amounts of the annuity for the first 6 months.
(c) After the deposit on March 1 is made, is there enough in the account to pay for the Cancun trip?
(d) If the student deposits \$60 each month, will there be enough for the trip?

SOLUTION **(a)** The initial amount deposited in the account is $A_0 = \$100$. The monthly deposit is $P = \$50$, and the per annum rate of interest is $r = 0.04$ compounded $N = 12$ times per year. The amount A_n in the account after n monthly deposits is given by the recursive sequence

$$A_0 = 100, \qquad A_n = \left(1 + \frac{r}{N}\right)A_{n-1} + P = \left(1 + \frac{0.04}{12}\right)A_{n-1} + 50$$

(b) In SEQuence mode on a TI-83, enter the sequence $\{A_n\}$ and create Table 8. On September 1 ($n = 0$), there is \$100 in the account. After the first payment on

October 1, the value of the account is \$150.33. After the second payment on November 1, the value of the account is \$200.83. After the third payment on December 1, the value of the account is \$251.50, and so on.

(c) On March 1 ($n = 6$), there is only \$404.53, not enough to pay for the trip to Cancun.

(d) If the periodic deposit, P, is \$60, then on March 1, there is \$465.03 in the account, enough for the trip. See Table 9.

TABLE 8

n	u(n)
0	100
1	150.33
2	200.83
3	251.5
4	302.34
5	353.35
6	404.53

u(n)■(1+.04/12)...

TABLE 9

n	u(n)
0	100
1	160.33
2	220.87
3	281.6
4	342.54
5	403.68
6	465.03

u(n)■(1+.04/12)...

NOW WORK PROBLEM 5.

Amortization

Use sequences and a graphing utility to solve amortization problems

2 Recursive sequences can also be used to compute information about loans. When equal periodic payments are made to pay off a loan, the loan is said to be **amortized**.

Amortization Formula

If \$$B$ is borrowed at an interest rate of r% (expressed as a decimal) per annum compounded monthly, the balance A_n due after n monthly payments of \$$P$ is given by the recursive sequence

$$A_0 = B, \qquad A_n = \left(1 + \frac{r}{12}\right)A_{n-1} - P, \qquad n \geq 1 \tag{2}$$

Formula (2) may be explained as follows: The initial loan balance is \$$B$. The balance due A_n after n payments will equal the balance due previously, A_{n-1}, plus the interest charged on that amount reduced by the periodic payment P.

EXAMPLE 2 Mortgage Payments

John and Wanda borrowed \$180,000 at 7% per annum compounded monthly for 30 years to purchase a home. Their monthly payment is determined to be \$1197.54.

(a) Find a recursive formula that represents their balance after each payment of \$1197.54 has been made.

(b) Determine their balance after the first payment is made.

(c) When will their balance be below \$170,000?

SOLUTION **(a)** We use Formula (2) with $A_0 = 180{,}000$, $r = 0.07$, and $P = \$1197.54$. Then

$$A_0 = 180{,}000 \qquad A_n = \left(1 + \frac{0.07}{12}\right)A_{n-1} - 1197.54$$

(b) In SEQuence mode on a TI-83, enter the sequence $\{A_n\}$ and create Table 10. After the first payment is made, the balance is $A_1 = \$179{,}852$.

(c) Scroll down until the balance is below \$170,000. See Table 11. After the 58th payment is made ($n = 58$), the balance is below \$170,000.

TABLE 10

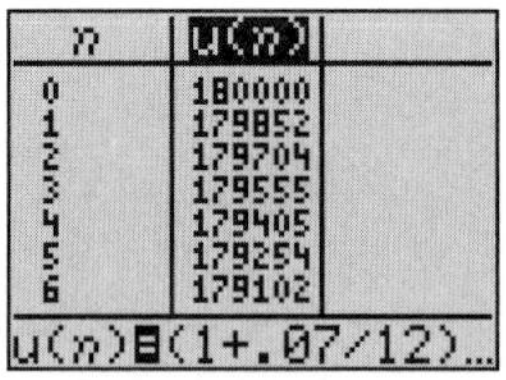

n	u(n)
0	180000
1	179852
2	179704
3	179555
4	179405
5	179254
6	179102

u(n)■(1+.07/12)...

TABLE 11

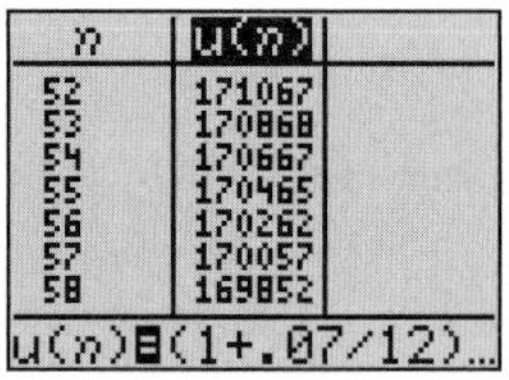

n	u(n)
52	171067
53	170868
54	170667
55	170465
56	170262
57	170057
58	169852

u(n)■(1+.07/12)...

NOW WORK PROBLEM 1.

EXERCISE 5.5 Answers to Odd-Numbered Problems Begin on Page AN-26.

1. **Credit Card** Debt John has a balance of \$3000 on his credit card that charges 1% interest per month on any unpaid balance. John can afford to pay \$100 toward the balance each month. His balance each month after making a \$100 payment is given by the recursively defined sequence

$$B_0 = \$3000, \quad B_n = 1.01B_{n-1} - 100$$

(a) Determine John's balance after making the first payment. That is, determine B_1.
(b) Using a graphing utility, determine when John's balance will be below \$2000. How many payments of \$100 have been made?
(c) Using a graphing utility, determine when John will pay off the balance. What is the total of all the payments?
(d) What was John's interest expense?

2. **Car Loans** Phil bought a car by taking out a loan for \$18,500 at 0.5% interest per month. Phil's normal monthly payment is \$434.47 per month, but he decides that he can afford to pay \$100 extra toward the balance each month. His balance each month is given by the recursively defined sequence

$$B_0 = \$18{,}500, \quad B_n = 1.005B_{n-1} - 534.47$$

(a) Determine Phil's balance after making the first payment. That is, determine B_1.
(b) Using a graphing utility, determine when Phil's balance will be below \$10,000. How many payments of \$534.47 have been made?
(c) Using a graphing utility, determine when Phil will pay off the balance. What is the total of all the payments?
(d) What was Phil's interest expense?

3. **Trout Population** A pond currently has 2000 trout in it. A fish hatchery decides to add an additional 20 trout each month. In addition, it is known that the trout population is growing 3% per month. The size of the population after n months is given by the recursively defined sequence

$$p_0 = 2000, \quad p_n = 1.03p_{n-1} + 20$$

(a) How many trout are in the pond at the end of the second month? That is, what is p_2?
(b) Using a graphing utility, determine how long it will be before the trout population reaches 5000.

4. **Environmental Control** The Environmental Protection Agency (EPA) determines that Maple Lake has 250 tons of pollutants as a result of industrial waste and that 10% of the pollutants present is neutralized by solar oxidation every year. The EPA imposes new pollution control laws that result in 15 tons of new pollutants entering the lake each year. The amount of pollutants in the lake at the end of each year is given by the recursively defined sequence

$$p_0 = 250, \quad p_n = 0.9p_{n-1} + 15$$

(a) Determine the amount of pollutants in the lake at the end of the second year. That is, determine p_2.
(b) Using a graphing utility, provide pollutant amounts for the next 20 years.
(c) What is the equilibrium level of pollution in Maple Lake? That is, what is $\lim_{n \to \infty} p_n$?

5. **Roth IRA** On January 1, 1999, Bob decides to place \$500 at the end of each quarter into a Roth Individual Retirement Account.

(a) Find a recursive formula that represents Bob's balance at the end of each quarter if the rate of return is assumed to be 8% per annum compounded quarterly.
(b) How long will it be before the value of the account exceeds \$100,000?
(c) What will be the value of the account in 25 years when Bob retires?

6. Education IRA On January 1, 1999, John's parents decide to place \$45 at the end of each month into an Education IRA.

(a) Find a recursive formula that represents the balance at the end of each month if the rate of return is assumed to be 6% per annum compounded monthly.
(b) How long will it be before the value of the account exceeds \$4000?
(c) What will be the value of the account in 16 years when John goes to college?

7. Home Loan Bill and Laura borrowed \$150,000 at 6% per annum compounded monthly for 30 years to purchase a home. Their monthly payment is determined to be \$899.33.

(a) Find a recursive formula for their balance after each monthly payment has been made.
(b) Determine Bill and Laura's balance after the first payment.
(c) Using a graphing utility, create a table showing Bill and Laura's balance after each monthly payment.
(d) Using a graphing utility, determine when Bill and Laura's balance will be below \$140,000.
(e) Using a graphing utility, determine when Bill and Laura will pay off the balance.
(f) Determine Bill and Laura's interest expense when the loan is paid.
(g) Suppose that Bill and Laura decide to pay an additional \$100 each month on their loan. Answer parts (a) to (f) under this scenario.
(h) Is it worthwhile for Bill and Laura to pay the additional \$100? Explain.

8. Home Loan Jodi and Jeff borrowed \$120,000 at 6.5% per annum compounded monthly for 30 years to purchase a home. Their monthly payment is determined to be \$758.48.

(a) Find a recursive formula for their balance after each monthly payment has been made.
(b) Determine Jodi and Jeff's balance after the first payment.
(c) Using a graphing utility, create a table showing Jodi and Jeff's balance after each monthly payment.
(d) Using a graphing utility, determine when Jodi and Jeff's balance will be below \$100,000.
(e) Using a graphing utility, determine when Jodi and Jeff will pay off the balance.
(f) Determine Jodi and Jeff's interest expense when the loan is paid.
(g) Suppose that Jodi and Jeff decide to pay an additional \$100 each month on their loan. Answer parts (a) to (f) under this scenario.
(h) Is it worthwhile for Jodi and Jeff to pay the additional \$100? Explain.

5.6 Applications: Leasing; Capital Expenditure; Bonds

OBJECTIVE 1 Solve applied problems

1 In this section we solve various applied problems.

Leasing

EXAMPLE 1 Lease or Purchase

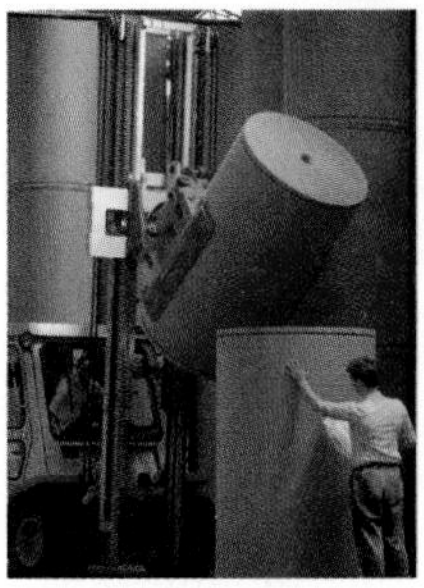

A corporation may obtain a particular machine either by leasing it for 4 years (the useful life) at an annual rent of \$1000 or by purchasing the machine for \$3000.

(a) Which alternative is preferable if the corporation can invest money at 10% per annum?
(b) What if it can invest at 14% per annum?

SOLUTION **(a)** Suppose the corporation may invest money at 10% per annum. The present value of an annuity of \$1000 for 4 years at 10% equals \$3169.87, which exceeds the purchase price. Therefore, purchase is preferable.
(b) Suppose the corporation may invest at 14% per annum. The present value of an annuity of \$1000 for 4 years at 14% equals \$2913.71, which is less than the purchase price. Leasing is preferable.

NOW WORK PROBLEM 1.

Capital Expenditure

EXAMPLE 2 Selecting Equipment

A corporation is faced with a choice between two machines, both of which are designed to improve operations by saving on labor costs. Machine A costs \$8000 and will generate an annual labor savings of \$2000. Machine B costs \$6000 and will save \$1800 annually. Machine A has a useful life of 7 years while machine B has a useful life of only 5 years. Assuming that the time value of money (the investment opportunity rate) of the corporation is 10% per annum, which machine is preferable? (Assume annual compounding and that the savings is realized at the end of each year.)

SOLUTION Machine A costs \$8000 and has a life of 7 years. Since an annuity of \$1 for 7 years at 10% interest has a present value of \$4.87, the cost of machine A may be considered the present value of an annuity:

$$\frac{\$8000}{4.87} = \$1642.71$$

The \$1642.71 may be termed the *equivalent annual cost* of machine A. Similarly, the equivalent annual cost of machine B may be calculated by reference to the present value of an annuity of 5 years, namely,

$$\frac{\$6000}{3.79} = \$1583.11$$

The net annual savings of each machine is given in Table 12:

TABLE 12

	A	*B*
Labor savings	\$2000.00	\$1800.00
Equivalent annual cost	1642.71	1583.11
Net savings	\$ 357.29	\$ 216.89

Machine A is preferable.

NOW WORK PROBLEM 3.

Bonds

We begin our third application with some definitions of terms concerning corporate bonds.

Face Amount (Face Value or Par Value)

The **face amount** or **denomination** of a bond (normally \$1000) is the amount paid to the bondholder at maturity. It is also the amount usually paid by the bondholder when the bond is originally issued.

Nominal Interest (Coupon Rate)

The contractual periodic interest payments on a bond.

Nominal interest is normally quoted as an annual percentage of the face amount. Nominal interest payments are conventionally made semiannually, so semiannual periods are used for compound interest calculations. For example, if a bond has a face amount of \$1000 and a coupon rate of 8%, then every 6 months the owner of the bond would receive $(\$1000)(0.08)(\frac{1}{2}) = \$40$.

But, because of market conditions, such as the current prime rate of interest, or the discount rate set by the Federal Reserve Board, or changes in the credit rating of the company issuing the bond, the price of a bond will fluctuate. When the bond price is higher than the face amount, it is trading at a **premium;** when it is lower, it is trading at a **discount.** For example, a bond with a face amount of \$1000 and a coupon rate of 8% may trade in the marketplace at a price of \$1100, which means the **true yield** is less than 8%.

To obtain the **true interest rate** of a bond, we view the bond as a combination of an annuity of semiannual interest payments plus a single future amount payable at maturity. The price of a bond is therefore the sum of the present value of the annuity of semiannual interest payments plus the present value of the single future payment at maturity. This present value is calculated by discounting at the true interest rate and assuming semiannual discounting periods.

EXAMPLE 3 Pricing Bonds

A bond has a face amount of \$1000 and matures in 10 years. The nominal interest rate is 8.5%. What is the price of the bond to yield a true interest rate of 8%?

SOLUTION

STEP 1 Calculate the amount of each semiannual interest payment:

$$(\$1000)(\tfrac{1}{2})(0.085) = \$42.50$$

STEP 2 Calculate the present value of the annuity of semiannual payments:

Amount of each payment from Step 1		\$ 42.50
Number of payments (2 × 10 years): 20		
True interest rate per period (half of stated true rate): 4%		
Factor from formula for V	$\dfrac{1-(1+.04)^{-20}}{.04} =$	13.5903
Present value of interest payments (\$42.50)(13.5903)		\$ 577.59

STEP 3 Calculate the present value of the amount payable at maturity:

Amount payable at maturity		\$ 1000
Number of semiannual compounding periods before maturity: 20		
True interest rate per period: 4%		
Factor from formula for $(1+i)^{-n}$	$(1+.04)^{-20} =$	0.45639
Present value of maturity value (\$1000)(0.45639)		\$ 456.39

STEP 4 Determine the price of the bond:

Present value of interest payments	$ 577.59
Present value of maturity payment	456.39
Price of bond (Add)	$1033.98

NOW WORK PROBLEM 7.

Use Excel to Solve Example 3.

SOLUTION   **STEP 1** Calculate the amount of each semiannual interest payment:

$$(\$1000)(.5)(0.085) = \$42.50$$

STEP 2 Calculate the present value of the annuity of semiannual payments. Use the Excel function PV(*rate, nper, pmt, fv, type*).

rate is the interest rate per period, 4%.
nper is the total number of payments, 20.
pmt is the payment made each period, $42.50.
fv is the future value of the annuity or loan, 0.
type is set to 0 for an ordinary annuity.

The result is given in the Excel spreadsheet below.

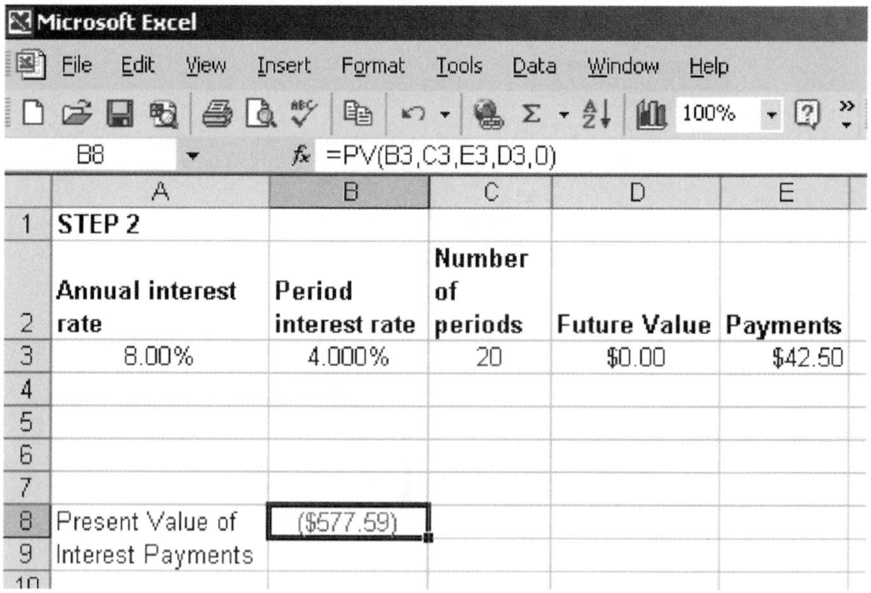

STEP 3 Calculate the present value of the amount payable at maturity. Use the Excel function PV(*rate, nper, pmt, fv, type*).

rate is the interest rate per period, 4%.
nper is the total number of payments, 20.
pmt is the payment made each period, $0.00.
fv is the future value of the annuity or loan, 1000.
type is set to 0 for an ordinary annuity.

STEP 4 Determine the price of the bond. Add the two present values. The entire problem's Excel spreadsheet is given below.

Microsoft Excel

File Edit View Insert Format Tools Data Window Help

100%

B18 f_x =B8+B13

	A	B	C	D	E
1	**STEP 2**				
2	**Annual interest rate**	**Period interest rate**	**Number of periods**	**Future Value**	**Payments**
3	8.00%	4.000%	20	$0.00	$42.50
4					
5					
6					
7					
8	Present Value of	($577.59)			
9	Interest Payments				
10					
11	**Step 3**			**Future Value**	**Payments**
12				$1,000.00	$0.00
13	Present Value of	($456.39)			
14	maturity Value				
15					
16	**Step 4**				
17					
18	Bond Price	-$1,033.98			

NOW WORK PROBLEM 7 USING EXCEL.

EXERCISE 5.6 Answers to Odd-Numbered Problems Begin on Page AN-26.

1. **Leasing Problem** A corporation may obtain a machine either by leasing it for 5 years (the useful life) at an annual rent of $2000 or by purchasing the machine for $8100. If the corporation can borrow money at 10% per annum, which alternative is preferable?

2. If the corporation in Problem 1 can borrow money at 14% per annum, which alternative is preferable?

3. **Capital Expenditure Analysis** Machine A costs $10,000 and has a useful life of 8 years, and machine B costs $8000 and has a useful life of 6 years. Suppose machine A generates an annual labor savings of $2000 while machine B generates an annual labor savings of $1800. Assuming the time value of money (investment opportunity rate) is 10% per annum, which machine is preferable?

4. In Problem 3, if the time value of money is 14% per annum, which machine is preferable?

5. **Corporate Bonds** A bond has a face amount of $1000 and matures in 15 years. The nominal interest rate is 9%. What is the price of the bond that will yield an effective interest rate of 8%?

6. For the bond in Problem 5 what is the price of the bond to yield an effective interest rate of 10%?

7. **Treasury Notes** Determine the selling price, per $1000 maturity value, of a 10-year treasury note with a nominal interest rate of 4.000% and a true interest rate of 4.095%.

Source: U.S. Treasury. (This is an actual U.S. Treasury note auctioned in November 2002.)

8. **Treasury Notes** Determine the selling price, per $1000 maturity value, of a 5-year treasury note with a nominal interest rate of 3.000% and a true interest rate of 3.030%.

Source: U.S. Treasury. (This is an actual U.S. Treasury note auctioned in November 2002.)

9. **Treasury Notes** Determine the true interest rate of a 10-year treasury note with a nominal interest rate of 4.375% if the selling price, per $1000 maturity value, is $998.80.

Source: U.S. Treasury. (This is an actual U.S. Treasury note auctioned in August 2002.)

10. **Treasury Notes** Determine the true interest rate of a 5-year treasury note with a nominal interest rate of 3.250% if the selling price, per $1000 maturity value, is $995.52.

Source: U.S. Treasury. (This is an actual U.S. Treasury note auctioned in August 2002.)

Chapter 5 Review

OBJECTIVES

Section		You should be able to	Review Exercises
5.1	1	Solve problems involving percents	1–11
	2	Solve problems involving simple interest	12, 13, 14, 19a
	3	Solve problems involving discounted loans	15, 16
5.2	1	Solve problems involving compound interest	17, 18, 19b, 22, 23, 46
	2	Find the effective rate of interest	24, 25, 43
	3	Solve problems involving present value	20, 21, 38, 44
5.3	1	Solve problems involving annuities	28, 36a, 37, 45, 47
	2	Solve problems involving sinking funds	32, 33, 35, 39, 42, 48
5.4	1	Solve problems involving the present value of an annuity	26, 27, 34, 36b, 40
	2	Solve problems involving amortization	29, 30, 31, 41, 49
5.5	1	Use sequences and a graphing utility to solve annuity problems	26, 27, 28, 34, 36, 37, 40, 45, 47
	2	Use sequences and a graphing utility to solve amortization problems	29, 30, 31, 41, 49
5.6	1	Solve applied problems	50–54

IMPORTANT FORMULAS

Simple Interest Formula $I = Prt$

Discounted Loans $R = L - Lrt$

Compound Interest Formula $A_n = P(1 + i)^n$

Amount of an Annuity $A = P\dfrac{(1 + i)^n - 1}{i}$

Present Value of an Annuity $V = P\dfrac{1 - (1 + i)^{-n}}{i}$

Amortization $P = V\dfrac{i}{1 - (1 + i)^{-n}}$

TRUE–FALSE ITEMS Answers are on page AN-26.

T F **1.** Simple interest is interest computed on the principal for the entire period it is borrowed.

T F **2.** The amount A accrued on a principal P after five payment periods at i interest per payment period is

$$A = P(1 + i)^5$$

T F **3.** The effective rate of interest for 10% compounded daily is about 9.58%.

T F **4.** The present value of an annuity is the sum of all the present values of the payment less any interest.

FILL IN THE BLANKS Answers are on page AN-26.

1. For a discounted loan, the amount the borrower receives is called the ____________.

2. In the formula $P = A(1 + i)^{-n}$, P is called the ____________ ____________ of the amount A due at the end of n interest periods at i interest per payment period.

3. The amount of an ____________ is the sum of all deposits made plus all interest accumulated.

4. A loan with a fixed rate of interest is ____________ if both principal and interest are paid by a sequence of equal payments over equal periods of time.

REVIEW EXERCISES Answers to odd-numbered problems begin on page AN-26.

Blue numbered problems indicate the author's suggestion for use in a practice test.

In Problems 1–9, calculate the indicated quantity.

1. 3% of 500
2. 20% of 1200
3. 140% of 250
4. What percent of 200 is 40?
5. What percent of 350 is 75?
6. What percent of 50 is 125?
7. 12 is 15% of what number?
8. 25 is 6% of what number?
9. 11 is 0.5% of what number?

10. Sales Tax The sales tax in New York City is 8.5%. If the total charge for a DVD player (including tax) is \$162.75, what does the DVD player itself cost?

11. Sales Tax Indiana has a sales tax of 6%. Dan purchased gifts for his family worth \$330.00. How much sales tax will Dan have to pay on his purchase?

12. Find the interest I charged and amount A due, if \$400 is borrowed for 9 months at 12% simple interest.

13. Dan borrows \$500 at 9% per annum simple interest for 1 year and 2 months. What is the interest charged, and what is the amount due?

14. Interest Due Jim borrows \$14,000 for a period of 4 years at 6% simple interest. Determine the interest due on the loan.

15. Loan Amount Warren needs \$15,000 for a new machine for his auto repair shop. He obtains a 2-year discounted loan at 12% interest. How much must he repay to settle his debt?

16. Treasury Bills How much should a bank bid for a 15-month, \$5,000 treasury bill in order to earn 2.5% simple interest?

17. Find the amount of an investment of \$100 after 2 years and 3 months at 10% compounded monthly.

18. Mike places \$200 in a savings account that pays 4% per annum compounded monthly. How much is in his account after 9 months?

19. **Choosing a Car Loan** A car dealer offers Mike the choice of two loans:

(a) \$3000 for 3 years at 12% per annum simple interest
(b) \$3000 for 3 years at 10% per annum compounded monthly
(c) Which loan costs Mike the least?

20. A mutual fund pays 9% per annum compounded monthly. How much should I invest now so that 2 years from now I will have \$100 in the account?

21. **Saving for a Bicycle** Katy wants to buy a bicycle that costs \$75 and will purchase it in 6 months. How much should she put in her savings account for this if she can get 10% per annum compounded monthly?

22. **Doubling Money** Marcia has \$220,000 saved for her retirement. How long will it take for the investment to double in value if it earns 6% compounded semiannually?

23. **Doubling Money** What annual rate of interest will allow an investment to double in 12 years?

24. **Effective Rate of Interest** A bank advertises that it pays $3\frac{3}{4}\%$ interest compounded monthly. What is the effective rate of interest?

25. **Effective Rate of Interest** What interest rate compounded quarterly has an effective interest rate of 6%?

26. **Saving for a Car** Mike decides he needs \$500 1 year from now to buy a used car. If he can invest at 8% compounded monthly, how much should he save each month to buy the car?

27. **Saving for a House** Mr. and Mrs. Corey are newlyweds and want to purchase a home, but they need a down payment of \$40,000. If they want to buy their home in 2 years, how much should they save each month in their savings account that pays 3% per annum compounded monthly?

28. **True Cost of a Car** Mike has just purchased a used car and will make equal payments of \$50 per month for 18 months at 12% per annum charged monthly. How much did the car actually cost? Assume no down payment.

29. **House Mortgage** Mr. and Mrs. Ostedt have just purchased an \$400,000 home and made a 25% down payment. The balance can be amortized at 10% for 25 years.

(a) What are the monthly payments?
(b) How much interest will be paid?
(c) What is their equity after 5 years?

30. **Inheritance Payouts** An inheritance of \$25,000 is to be paid in equal amounts over a 5-year period at the end of each year. If the \$25,000 can be invested at 10% per annum, what is the annual payment?

31. **House Mortgage** A mortgage of \$125,000 is to be amortized at 9% per annum for 25 years. What are the monthly payments? What is the equity after 10 years?

32. **Paying Off Construction Bonds** A state has \$8,000,000 worth of construction bonds that are due in 25 years. What annual sinking fund deposit is needed if the state can earn 10% per annum on its money?

33. **Depletion Problem** How much should Mr. Graff pay for a gold mine expected to yield an annual return of \$20,000 and to have a life expectancy of 20 years, if he wants to have a 15% annual return on his investment and he can set up a sinking fund that earns 10% a year?

34. **Retirement Income** Mr. Doody, at age 70, is expected to live for 15 years. If he can invest at 12% per annum compounded monthly, how much does he need now to guarantee himself \$300 every month for the next 15 years?

35. **Depletion Problem** An oil well is expected to yield an annual net return of \$25,000 for the next 15 years, after which it will run dry. An investor wants a return on his investment of 20%. He can establish a sinking fund earning 10% annually. How much should he pay for the oil well?

36. **Saving for the Future** Hal deposited \$100 in an account paying 9% per annum compounded monthly for 25 years. At the end of the 25 years Hal retires. What is the largest amount he may withdraw monthly for the next 35 years?

37. **Saving for College** Mr. Jones wants to save for his son's college education. If he deposits \$500 every 6 months at 6% compounded semiannually, how much will he have on hand at the end of 8 years?

38. How much money should be invested at 8% compounded quarterly in order to have \$20,000 in 6 years?

39. Bill borrows \$1000 at 5% compounded annually. He is able to establish a sinking fund to pay off the debt and interest in 7 years. The sinking fund will earn 8% compounded quarterly. What should be the size of the quarterly payments into the fund?

40. A man has \$50,000 invested at 12% compounded quarterly at the time he retires. If he wishes to withdraw money every 3 months for the next 7 years, what will be the size of his withdrawals?

41. How large should monthly payments be to amortize a loan of \$3000 borrowed at 12% compounded monthly for 2 years?

42. **Paying off School Bonds** A school board issues bonds in the amount of \$20,000,000 to be retired in 25 years. How much must be paid into a sinking fund at 6% compounded annually to pay off the total amount due?

43. What effective rate of interest corresponds to a nominal rate of 9% compounded monthly?

44. If an amount was borrowed 5 years ago at 6% compounded quarterly, and \$6000 is owed now, what was the original amount borrowed?

45. **Trust Fund Payouts** John is the beneficiary of a trust fund set up for him by his grandparents. If the trust fund amounts to \$20,000 earning 8% compounded semiannually and he is to receive the money in equal semiannual installments for the next 15 years, how much will he receive each 6 months?

46. **IRA** Mike has \$4000 in his IRA. At 6% annual interest compounded monthly, how much will he have at the end of 20 years?

47. **Retirement Funds** An employee gets paid at the end of each month and \$60 is withheld from her paycheck for a retirement fund. The fund pays 1% per month (equivalent to 12% annually compounded monthly). What amount will be in the fund at the end of 30 months?

48. \$6000 is borrowed at 10% compounded semiannually. The amount is to be paid back in 5 years. If a sinking fund is established to repay the loan and interest in 5 years, and the fund earns 8% compounded quarterly, how much will have to be paid into the fund every 3 months?

49. **Buying a Car** A student borrowed \$4000 from a credit union toward purchasing a car. The interest rate on such a loan is 14% compounded quarterly, with payments due every quarter. The student wants to pay off the loan in 4 years. Find the quarterly payment.

50. **Leasing Decision** A firm can lease office furniture for 5 years (its useful life) for \$20,000 a year, or it can purchase the furniture for \$80,000. Which is a better choice if the firm can invest its money at 10% per annum?

51. **Leasing Decision** A company can lease 10 trucks for 4 years (their useful life) for \$50,000 a year, or it can purchase them for \$175,000. If the company can earn 5% per annum on its money, which choice is preferable, leasing or purchasing?

52. **Capital Expenditures** In an effort to increase productivity, a corporation decides to purchase a new piece of equipment. Two models are available; both reduce labor costs. Model A costs \$50,000, saves \$12,000 per year in labor costs, and has a useful life of 10 years. Model B costs \$42,000, saves \$10,000 annually in labor costs, and has a useful life of 8 years. If the time value of money is 10% per annum, what piece of equipment provides a better investment?

53. **Corporate Bonds** A bond has a face amount of \$10,000 and matures in 8 years. The nominal rate of interest on the bond is 6.25%. At what price would the bond yield a true rate of interest of 6.5%?

54. **Treasury Notes** Determine the selling price, per \$1000 maturity value, of a 20 year treasury note with a nominal interest rate of 5.0% and a true interest rate of 5.1%.

Chapter 5 Project

MORTGAGE POINTS: SHOULD YOU BUY THEM?

When you are shopping for a house, you are usually also shopping for a mortgage. Banks and other lending institutions offer an amazing variety of mortgages. Mortgages may differ in their amount, their rates of interest, and their term (the time needed to pay off the mortgage). If you know the amount, the rate of interest, and the term, you can use the amortization formula from Section 5.4 to figure out your monthly payment, determine the total amount of interest you will pay, and even create a schedule of payments for the loan.

Another difference among mortgages concerns *points*.* A **point** is a fee equal to some percentage of the loan amount and is usually paid at the time of closing—when you pay your down payment and other fees and take possession of the house. Paying points sometimes lessens the total amount of a mortgage. Since paying points is considered to be prepaid interest on your mortgage, often the lender will offer a reduced rate for mortgages with points. The big question for the home buyer is whether to pay points. As we shall see, the

The discussion here pertains to a certain type of points called* ***discount points**. Other types of points behave differently.*

answer depends on the amount of the mortgage, how large a down payment you can make, and the amount of time you expect to spend in your house.

Suppose you are thinking of buying a house in Charlotte, North Carolina, that costs $150,000. You plan to make a down payment of $30,000, so you will need a mortgage of $120,000. A local lender, Hartland Mortgage Centers, gives you four quotes on 30-year, fixed rate mortgages. See Table 1.

TABLE 1 Mortgage Rates and Points—February 2003

Rate	Points	Fee Paid for Points
5.50	0.00	$ 0.00
5.38	0.50	$ 600.00
5.25	1.10	$1320.00
5.00	2.70	$3240.00

1. Compute your monthly payment for the loan with no points.

2. Now consider the loan with 0.5 points. Compute the monthly payment you would have in this case.

3. The loan with 0.5 points gives you a lower monthly payment. How much is the difference?

4. If you take the loan with 0.5 points, you will eventually make back the $600 you spent for the points in reduced payments. How many months will it take for you to make back your points fee?

5. If you plan to leave your house before this "break-even" time, then which of the two loans makes better sense?

6. Compare the other two loan options given in the table with the two you have already computed. What are the monthly payments for those loans? How many months does it take to regain the points fee in each of these cases?

7. Does it really make sense to compare the monthly payments as we have calculated them? Sometimes it does, but consider the following situation. If you were willing to pay $3240 for the points for the last loan, it is reasonable to expect that you would also be willing to add that $3240 to the down payment if you were going to take the loan with no points. Likewise, you should be willing to add $3240 − $600 = $2640 to the down payment for the loan with 0.5 points and to add $3240 − $1320 = $1920 to the down payment for the loan with 1.10 points. Recalculate the monthly payment for each loan by filling in Table 2.

Rate	Points	Fee Paid for Points	Loan Amount	Monthly Payment
5.50	0.00	$ 0.00		
5.38	0.50	$ 600.00		
5.25	1.10	$1320.00		
5.00	2.70	$3240.00		

8. Now which mortgage gives you the lowest payment?

MATHEMATICAL QUESTIONS FROM PROFESSIONAL EXAMS*

1. **CPA Exam** Which of the following should be used to calculate the amount of the equal periodic payments that could be equivalent to an outlay of $3000 at the time of the last payment?

   (a) Amount of 1
   (b) Amount of an annuity of 1
   (c) Present value of an annuity of 1
   (d) Present value of 1

2. **CPA Exam** A businessman wants to withdraw $3000 (including principal) from an investment fund at the end of each year for 5 years. How should he compute his required initial investment at the beginning of the first year if the fund earns 6% compounded annually?

   (a) $3000 times the amount of an annuity of $1 at 6% at the end of each year for 5 years
   (b) $3000 divided by the amount of an annuity of $1 at 6% at the end of each year for 5 years
   (c) $3000 times the present value of an annuity of $1 at 6% at the end of each year for 5 years
   (d) $3000 divided by the present value of an annuity of $1 at 6% at the end of each year for 5 years

3. **CPA Exam** A businesswoman wants to invest a certain sum of money at the end of each year for 5 years. The investment will earn 6% compounded annually. At the end of 5 years, she will need a total of $30,000 accumulated. How should she compute the required annual investment?

   (a) $30,000 times the amount of an annuity of $1 at 6% at the end of each year for 5 years
   (b) $30,000 divided by the amount of an annuity of $1 at 6% at the end of each year for 5 years

(c) \$30,000 times the present value of an annuity of \$1 at 6% at the end of each year for 5 years
(d) \$30,000 divided by the present value of an annuity of \$1 at 6% at the end of each year for 5 years

4. **CPA Exam** Shaid Corporation issued \$2,000,000 of 6%, 10-year convertible bonds on June 1, 1993, at 98 plus accrued interest. The bonds were dated April 1, 1993, with interest payable April 1 and October 1. Bond discount is amortized semiannually on a straight-line basis.

On April 1, 1994, \$500,000 of these bonds were converted into 500 shares of \$20 par value common stock. Accrued interest was paid in cash at the time of conversion.

What was the effective interest rate on the bonds when they were issued?

(a) 6% (b) Above 6% (c) Below 6%
(d) Cannot be determined from the information given.

Items 5–7 apply to the appropriate use of present value tables. Given below are the present value factors for \$1.00 discounted at 8% for 1 to 5 periods. Each of the following items is based on 8% interest compounded annually from day of deposit to day of withdrawal.

Periods	Present Value of \$1 Discounted at 8% per Period
1	0.926
2	0.857
3	0.794
4	0.735
5	0.681

5. **CPA Exam** What amount should be deposited in a bank today to grow to \$1000 3 years from today?

(a) $\frac{\$1000}{0.794}$ (b) $\$1000 \times 0.926 \times 3$
(c) $(\$1000 \times 0.926) + (\$1000 \times 0.857) + (\$1000 \times 0.794)$
(d) $\$1000 \times 0.794$

6. **CPA Exam** What amount should an individual have in her bank account today before withdrawal if she needs \$2000 each year for 4 years with the first withdrawal to be made today and each subsequent withdrawal at 1-year intervals? (She is to have exactly a zero balance in her bank account after the fourth withdrawal.)

(a) $\$2000 + (\$2000 \times 0.926) + (\$2000 \times 0.857) + (\$2000 \times 0.794)$
(b) $\frac{\$2000}{0.735} \times 4$
(c) $(\$2000 \times 0.926) + (\$2000 \times 0.857) + (\$2000 \times 0.794) + (\$2000 \times 0.735)$
(d) $\frac{\$2000}{0.926} \times 4$

7. **CPA Exam** If an individual put \$3000 in a savings account today, what amount of cash would be available 2 years from today?

(a) $\$3000 \times 0.857$ (b) $\$3000 \times 0.857 \times 2$
(c) $\frac{\$3000}{0.857}$ (d) $\frac{\$3000}{0.926} \times 2$

CHAPTER 6

Sets; Counting Techniques

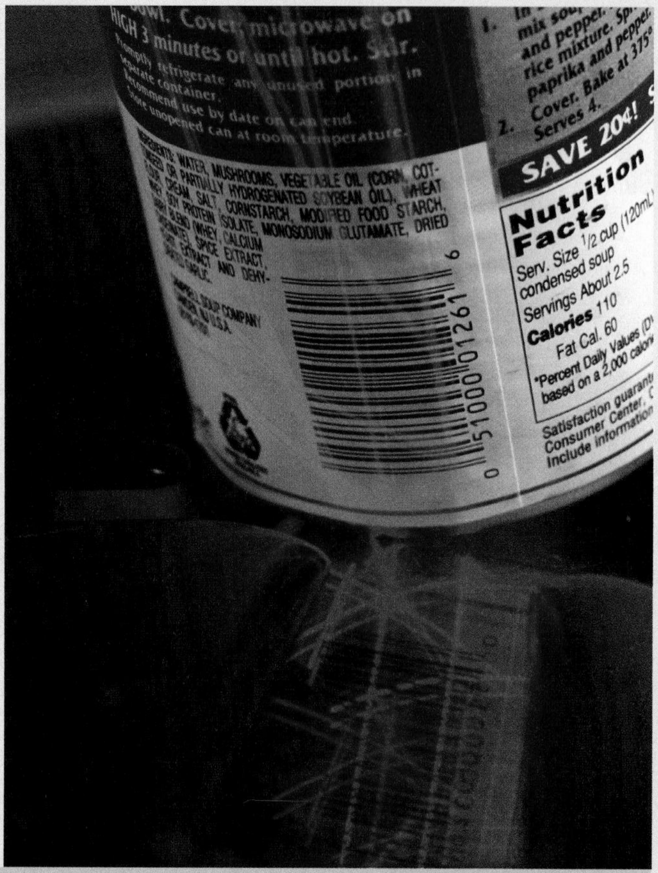

OUTLINE

Take a look at the back cover of your book. Do you see the bar code and the ISBN number? Have you ever wondered what all the numbers and hyphens mean? You know it identifies your book, but what else does it do? And what about the bar code on that bottle of water you just bought? You know, the UPC code. What do those numbers mean? Every product you buy, from food to drugs to clothing—just about everything—has a bar code that identifies the item purchased. How is this possible using just 12 digits? In this chapter we discuss counting techniques that, together with the Chapter Project, will explain these everyday codes.

A LOOK BACK, A LOOK FORWARD

In Chapters 1–4, we studied various applications of linear equations and linear inequalities. Then in Chapter 5, we investigated the mathematics of finance. In this chapter, we take up another new topic, *sets*. We will define an algebra for sets, much like we did for matrices in Chapter 2. Most importantly, we will discuss techniques for counting the number of elements in a set. We will use counting techniques to determine *probabilities*, the subject of Chapter 7. These techniques are also used in computer science to analyze algorithms and to study stacks and queues.

6.1 Sets

OBJECTIVES

1. Identify relations between pairs of sets
2. Find the union and intersection of two sets
3. Find the complement of a set
4. Use Venn diagrams

Set Notation

When we want to treat a collection of distinct objects as a whole, we use the idea of a **set**. For example, the set of **digits** consists of the collection of numbers 0, 1, 2, 3, 4, 5, 6, 7, 8, and 9. If we use the symbol D to denote the set of digits, then we can write

$$D = \{0, 1, 2, 3, 4, 5, 6, 7, 8, 9\}$$

In this notation the braces { } are used to enclose the objects, or **elements,** in the set. This method of denoting a set is called the **roster method.** A second way to denote a set is to use **set-builder notation,** where the set D of digits is written as

$$D = \{ \quad x \quad | \quad x \text{ is a digit}\}$$

$$\uparrow \; \uparrow \; \uparrow \quad \uparrow \quad \uparrow \quad \uparrow$$

Read as "D is the set of all x such that x is a digit."

EXAMPLE 1 Examples of Set Notation

(a) $E = \{x | x \text{ is an even digit}\} = \{0, 2, 4, 6, 8\}$
(b) $O = \{x | x \text{ is an odd digit}\} = \{1, 3, 5, 7, 9\}$

EXAMPLE 2 Using Set Notation

(a) Let A denote the set that consists of all possible outcomes resulting from tossing a coin two times. If we let H denote heads and T denote tails, then the set A can be written as

$$A = \{HH, HT, TH, TT\}$$

where, for instance, TH means the first toss resulted in tails and the second toss resulted in heads.

(b) Let B denote the set consisting of all possible arrangements of the digits without repetition. Some typical elements of B are

1478906532 4875326019 3214569870

The number of elements in B is very large, so listing all of them is impractical. Here we can write B using set-builder notation:

$$B = \{x \mid x \text{ is a ten-digit number in which no digit is repeated}\}$$

The elements of a set are never repeated. That is, we would never write $\{3, 2, 2\}$; the correct listing is $\{3, 2\}$. Finally, the order in which the elements of a set are listed does not make any difference. The three sets

$$\{3, 2, 4\} \qquad \{2, 3, 4\} \qquad \{4, 3, 2\}$$

are different listings of the same set. The *elements* of a set distinguish the set—not the *order* in which the elements are written.

A set that has no elements is called the **empty set** or **null set** and is denoted by the symbol $\varnothing$.

Identify relations between pairs of sets **1**

We begin to develop an algebra for sets by defining what we mean by two sets being *equal.*

Equality of Sets

Let A and B be two sets. We say that A **is equal to** B, written as

$$A = B$$

if and only if A and B have the same elements. If A and B are not equal, we write

$$A \neq B$$

For example, $\{1, 2, 3\} = \{1, 3, 2\}$, but $\{1, 2\} \neq \{1, 2, 3\}$.

Subset

Let A and B be two sets. We say that A **is a subset of** B or that A **is contained in** B, written as

$$A \subseteq B$$

if and only if every element of A is also an element of B. If A is not a subset of B, we write

$$A \nsubseteq B$$

For example, $\{1, 2\} \subseteq \{1, 2, 3\}$, but $\{1, 2, 3, 4\} \nsubseteq \{1, 2\}$.

NOW WORK PROBLEMS 1 AND 3.

We can state the definition of subset in another way: $A \subseteq B$ if and only if whenever x is in A, then x is in B for all x. This latter way of interpreting the meaning of $A \subseteq B$ is useful for obtaining various laws that sets obey. For example, it follows that for any set A, $A \subseteq A$; that is, every set is a subset of itself. Do you see why? Whenever x is in A, then x is in A!

When we say that A is a subset of B, it is equivalent to saying "there are no elements in set A that are not also elements in set B." In particular, suppose $A = \varnothing$. Since the empty set $\varnothing$ has no elements, there is no element of the set $\varnothing$ that is not also in B. That is,

$$\varnothing \subseteq B, \qquad \text{for any set } B$$

Proper Subset

Let A and B be two sets. We say that A **is a proper subset of** B or that A **is properly contained in** B, written as

$$\boxed{A \subset B}$$

if and only if every element in the set A is also in the set B, but there is at least one element in set B that is *not* in set A. If A is *not* a proper subset of B, we write

$$\boxed{A \not\subset B}$$

For example, $\{1, 2\} \subset \{1, 2, 3, 4\}$, but $\{1, 2, 3\} \not\subset \{1, 2\}$. Also, $\{1, 2\} \not\subset \{1, 2\}$. Do you see why? The set to the right of the $\subset$ symbol must have an element not found in the set to the left of the $\subset$.

Notice that "A is a proper subset of B" means that there are *no* elements of A that are not also elements of B, but there is at least one element of B that is not in A. For example, if B is any nonempty set, that is, any set having at least one element, then

$$\varnothing \subset B$$

Also, $A \not\subset A$; that is, a set is never a proper subset of itself.

The following example illustrates some uses of the three relationships, $=$, $\subseteq$, and $\subset$, just defined.

EXAMPLE 3 Identifying Relationships between Sets

Consider three sets A, B, and C given by

$$A = \{1, 2, 3\} \qquad B = \{1, 2, 3, 4, 5\} \qquad C = \{3, 2, 1\}$$

Some of the relationships between pairs of these sets are

(a) $A = C$ **(b)** $A \subseteq B$ **(c)** $A \subseteq C$ **(d)** $A \subset B$ **(e)** $C \subseteq A$ **(f)** $\varnothing \subseteq A$

In comparing the two definitions of *subset* and *proper subset,* you should notice that if a set A is a subset of a set B, then either A is a proper subset of B or else A equals B. That is,

$$A \subseteq B \quad \text{if and only if either} \quad A \subset B \text{ or } A = B$$

Also, if A is a proper subset of B, we can infer that A is a subset of B, but A does not equal B. That is,

$$A \subset B \qquad \text{if and only if} \qquad A \subseteq B \text{ and } A \neq B$$

We can think of the relationship $\subset$ as a refinement of $\subseteq$. On the other hand, the relationship $\subseteq$ is an extension of $\subset$, in the sense that $\subseteq$ may include equality whereas with $\subset$, equality cannot be included.

In applications the elements that may be considered are usually limited to some specific all-encompassing set. For example, in discussing students eligible to graduate from Midwestern University, the discussion would be limited to students enrolled at the university.

Universal Set

The **universal set** U is defined as the set consisting of all elements under consideration.

If A is any set and if U is the universal set, then every element in A must be in U (since U consists of all elements under consideration). As a result,

$$A \subseteq U$$

for *any* set A.

It is convenient to represent a set as the interior of a circle. Two or more sets may be depicted as circles enclosed in a rectangle, which represents the universal set. The circles may or may not overlap, depending on the situation. Such diagrams of sets are called **Venn diagrams**. See Figure 1.

FIGURE 1

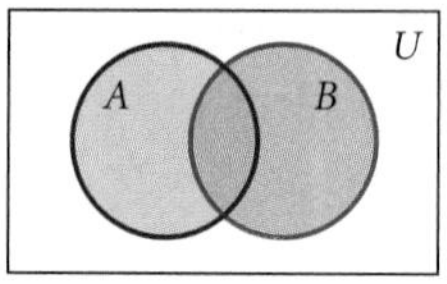

Operations on Sets

Continuing to develop an algebra of sets, we introduce operations that may be performed on sets.

Union of Two Sets

Let A and B be any two sets. The **union** of A with B, written as

$$\boxed{A \cup B}$$

and read as "A union B" or as "A or B," is defined to be the set consisting of those elements either in A or in B or in both A and B. That is,

$$\boxed{A \cup B = \{x \mid x \text{ is in } A \text{ or } x \text{ is in } B\}}$$

For example, if $A = \{1, 2\}$ and $B = \{2, 3\}$, then $A \cup B = \{1, 2, 3\}$.

FIGURE 2

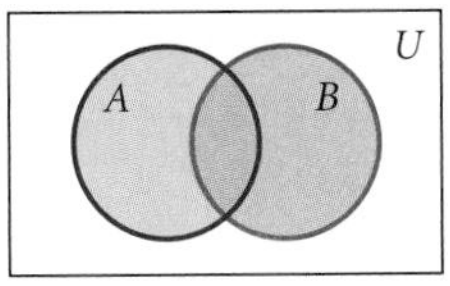

In the Venn diagram in Figure 2, the shaded area corresponds to $A \cup B$.

Warning! In English, the word *or* has two meanings: the *inclusive or* "A or B" means A or B or both. The *exclusive or* "A or B" means A or B but *not* both. In mathematics, A or B means the inclusive *or* from English.

Intersection of Two Sets

Let A and B be any two sets. The **intersection** of A with B, written as

$$\boxed{A \cap B}$$

and read as "A intersect B" or as "A and B," is defined as the set consisting of those elements that are in both A and B. That is,

$$\boxed{A \cap B = \{x \mid x \text{ is in } A \text{ and } x \text{ is in } B\}}$$

FIGURE 3

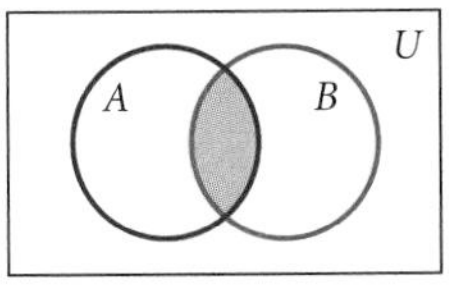

For example, if $A = \{1, 2\}$ and $B = \{2, 3\}$, then $A \cap B = \{2\}$.

In other words, to find the intersection of two sets A and B means to find the elements *common* to A *and* B. In the Venn diagram in Figure 3 the shaded region is $A \cap B$.

For any set A it follows that

$$A \cup \varnothing = A \qquad A \cap \varnothing = \varnothing$$

2 **EXAMPLE 4** **Finding the Union and Intersection of Two Sets**

Use the sets

$$A = \{1, 3, 5\} \qquad B = \{3, 4, 5, 6\} \qquad C = \{6, 7\}$$

to find

(a) $A \cup B$ **(b)** $A \cap B$ **(c)** $A \cap C$

SOLUTION **(a)** $A \cup B = \{1, 3, 5\} \cup \{3, 4, 5, 6\} = \{1, 3, 4, 5, 6\}$
(b) $A \cap B = \{1, 3, 5\} \cap \{3, 4, 5, 6\} = \{3, 5\}$
(c) $A \cap C = \{1, 3, 5\} \cap \{6, 7\} = \varnothing$

NOW WORK PROBLEMS 11 AND 13.

EXAMPLE 5 **Interpreting the Intersection of Two Sets**

Let T be the set of all taxpayers and let S be the set of all people over 65 years of age. Describe $T \cap S$.

SOLUTION $T \cap S$ is the set of all taxpayers who are also over 65 years of age.

Disjoint Sets

If two sets A and B have no elements in common, that is, if

$$A \cap B = \varnothing$$

then A and B are called **disjoint sets.**

FIGURE 4

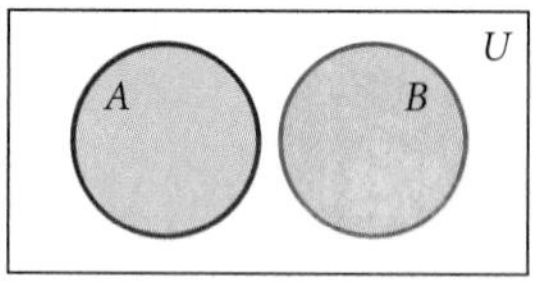

For example, if $A = \{1, 2\}$ and $B = \{3, 4, 5\}$, then $A \cap B = \varnothing$

Two disjoint sets A and B are illustrated in the Venn diagram in Figure 4. Notice that the circles corresponding to A and B do not overlap anywhere because $A \cap B$ is empty.

EXAMPLE 6 Tossing a Die*

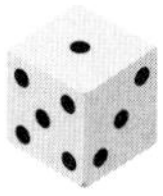

Suppose that a die is tossed. What is the universal set? Let A be the set of outcomes in which an even number turns up; let B be the set of outcomes in which an odd number shows. Find A and B. Find $A \cap B$.

SOLUTION Since only a 1, 2, 3, 4, 5, or 6 can result when a die is tossed, the universal set is $U = \{1, 2, 3, 4, 5, 6\}$. It follows that

$$A = \{2, 4, 6\} \qquad B = \{1, 3, 5\}$$

We note that A and B have no elements in common since an even number and an odd number cannot occur simultaneously. As a result, $A \cap B = \varnothing$ and the sets A and B are disjoint sets.

Suppose we consider all the employees of some company as our universal set U. Let A be the subset of employees who smoke. Then all the nonsmokers will make up the subset of U that is called the *complement* of the set of smokers.

Complement of a Set

Let A be any set. The **complement** of A, written as

$$\overline{A} \qquad (\text{or } A')$$

is defined as the set consisting of elements in the universe U that are not in A. Thus

$$\overline{A} = \{x \mid x \text{ is not in } A\}$$

FIGURE 5

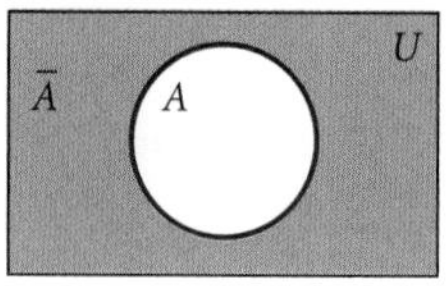

The shaded region in Figure 5 illustrates the complement, $\overline{A}$.

For any set A it follows that

$$A \cup \overline{A} = U \qquad A \cap \overline{A} = \varnothing \qquad \overline{\overline{A}} = A$$

*A **die** (plural **dice**) is a cube with 1, 2, 3, 4, 5, or 6 dots showing on the six faces.*

3 **EXAMPLE 7** **Finding the Complement of a Set**

Use the sets

$$U = \{a, b, c, d, e, f\} \qquad A = \{a, b, c\} \qquad B = \{a, c, f\}$$

to list the elements of the following sets:

(a) $\overline{A}$ **(b)** $\overline{B}$ **(c)** $\overline{A \cup B}$

(d) $\overline{A} \cap \overline{B}$ **(e)** $\overline{A \cap B}$ **(f)** $\overline{A} \cup \overline{B}$

SOLUTION **(a)** $\overline{A}$ consists of all the elements in U that are not in A:

$$\overline{A} = \{d, e, f\}.$$

(b) Similarly,

$$\overline{B} = \{b, d, e\}.$$

(c) To find $\overline{A \cup B}$, we first list the elements in $A \cup B$:

$$A \cup B = \{a, b, c, f\}$$

The complement of the set $A \cup B$ is then

$$\overline{A \cup B} = \{d, e\}$$

(d) From parts (a) and (b) we find that

$$\overline{A} \cap \overline{B} = \{d, e\}$$

(e) As in part (c), we first list the elements in $A \cap B$:

$$A \cap B = \{a, c\}$$

Then

$$\overline{A \cap B} = \{b, d, e, f\}$$

(f) From parts (a) and (b) we find that

$$\overline{A} \cup \overline{B} = \{b, d, e, f\}$$

NOW WORK PROBLEM 19.

The answers to parts (c) and (d) in Example 7 are the same, and so are the results from parts (e) and (f). This is no coincidence. There are two important properties involving intersections and unions of complements of sets. They are known as *De Morgan's properties*:

De Morgan's Properties

Let A and B be any two sets. Then

(a) $\overline{A \cup B} = \overline{A} \cap \overline{B}$ (b) $\overline{A \cap B} = \overline{A} \cup \overline{B}$.

De Morgan's properties state that all we need to do to form the complement of a union (or intersection) of sets is to form the complements of the individual sets and then change the union symbol to an intersection (or the intersection to a union). We can use a Venn diagram to verify De Morgan's Properties.

4 EXAMPLE 8 Using a Venn Diagram to Verify De Morgan's Properties

Use a Venn diagram to verify that $\overline{A \cup B} = \overline{A} \cap \overline{B}$.

SOLUTION First we draw two diagrams, as shown in Figure 6.

FIGURE 6

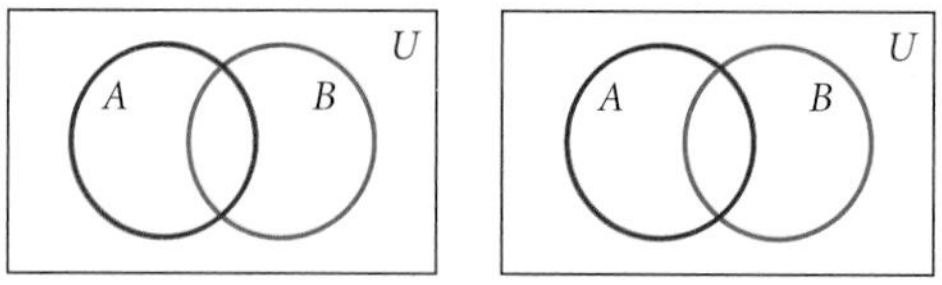

(a) (b)

We will use the diagram on the left for $\overline{A \cup B}$ and the one on the right for $\overline{A} \cap \overline{B}$.

Figure 7 illustrates the completed Venn diagrams of these sets.

FIGURE 7

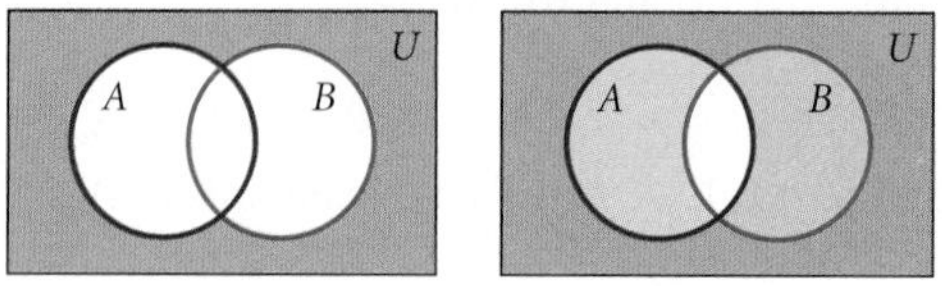

$\overline{A \cup B}$ $\overline{A} \cap \overline{B}$

(a) (b)

In Figure 7(a), $\overline{A \cup B}$ is represented by the shaded region, and in Figure 7(b), $\overline{A} \cap \overline{B}$ is represented by the same region. This illustrates that the two sets $\overline{A \cup B}$ and $\overline{A} \cap \overline{B}$ are equal.

EXAMPLE 9 Using Venn Diagrams to Illustrate a Set

Use a Venn diagram to illustrate the set $(A \cup B) \cap C$.

SOLUTION First we construct Figure 8(a). Then we shade $A \cup B$ in green and C in yellow as in Figure 8(b). The region where the colors overlap is the set $(A \cup B) \cap C$.

FIGURE 8

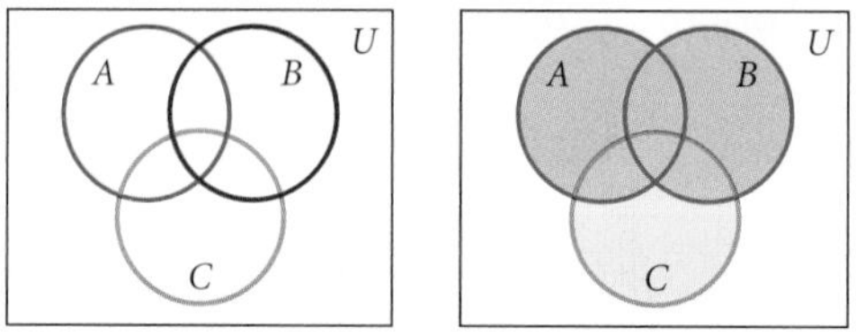

(a) (b)

EXAMPLE 10 Using a Venn Diagram to Illustrate the Equality of Two Sets

Use a Venn diagram to illustrate the following equality:

$$A \cup B = (A \cap \overline{B}) \cup (A \cap B) \cup (\overline{A} \cap B)$$

SOLUTION We begin with Figure 9. There $A \cap \overline{B}$ is the purple region. Now shade the regions $A \cap \overline{B}$, $A \cap B$, and $\overline{A} \cap B$, as shown in Figure 10. The union of the three regions in Figure 10 is the set $A \cup B$. See Figure 11.

FIGURE 9

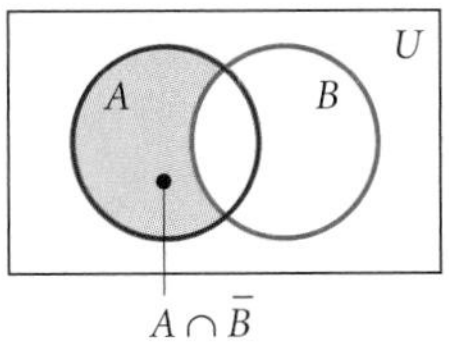

FIGURE 10

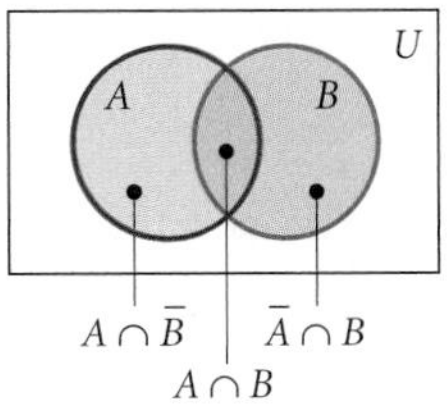

FIGURE 11

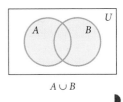

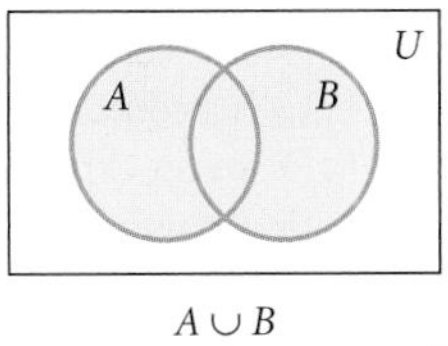

NOW WORK PROBLEMS 23(a) AND (g).

EXERCISE 6.1 Answers to Odd-Numbered Problems Begin on Page AN-27.

In Problems 1–10, tell whether the given statement is true or false.

1. $\{1, 3, 2\} = \{2, 1, 3\}$
2. $\{2, 3\} \subseteq \{1, 2, 3\}$
3. $\{6, 7, 9\} \subseteq \{1, 6, 9\}$
4. $\{4, 8, 2\} = \{2, 4, 6, 8\}$
5. $\{6, 7, 9\} \subset \{1, 6, 9\}$
6. $\{4, 8, 2\} \subset \{2, 4, 6, 8\}$
7. $\{1, 2\} \cap \{2, 3, 4\} = \{2\}$
8. $\{2, 3\} \cap \{2, 3, 4\} = \{2, 3, 4\}$
9. $\{4, 5\} \cap \{1, 2, 3, 4\} \subseteq \{4\}$
10. $\{1, 4\} \cup \{2, 3\} \subseteq \{1, 2, 3, 4\}$

In Problems 11–18, write each expression as a single set.

11. $\{1, 2, 3\} \cap \{2, 3, 4, 5\}$
12. $\{0, 1, 3\} \cup \{2, 3, 4\}$
13. $\{1, 2, 3\} \cup \{2, 3, 4, 5\}$
14. $\{0, 1, 2\} \cap \{2, 3, 4\}$
15. $\{2, 4, 6, 8\} \cap \{1, 3, 5, 7\}$
16. $\{2, 4, 6\} \cup \{1, 3, 5\}$
17. $\{a, b, e\} \cup \{d, e, f, q\}$
18. $\{a, e, m\} \cup \{p, o, m\}$

19. If U = universal set = $\{0, 1, 2, 3, 4, 5, 6, 7, 8, 9\}$ and if $A = \{0, 1, 5, 7\}$, $B = \{2, 3, 5, 8\}$, $C = \{5, 6, 9\}$, find
    (a) $A \cup B$ (b) $B \cap C$
    (c) $A \cap B$ (d) $\overline{A \cap B}$
    (e) $\overline{A} \cap \overline{B}$ (f) $A \cup (B \cap A)$
    (g) $(C \cap A) \cap (\overline{A})$ (h) $(A \cap B) \cup (B \cap C)$

20. If U = universal set = $\{1, 2, 3, 4, 5\}$ and if $A = \{3, 5\}$, $B = \{1, 2, 3\}$, $C = \{2, 3, 4\}$, find
    (a) $\overline{A} \cap \overline{B}$ (b) $(A \cup B) \cap C$
    (c) $A \cup (B \cap C)$ (d) $(A \cup B) \cap (A \cup C)$
    (e) $\overline{A \cap C}$ (f) $\overline{A \cup B}$
    (g) $\overline{A} \cup \overline{B}$ (h) $(A \cap B) \cup C$

21. Let U = {All letters of the alphabet}, $A = \{b, c, d\}$, and $B = \{c, e, f, g\}$. List the elements of the sets:
    (a) $A \cup B$ (b) $A \cap B$
    (c) $\overline{A} \cap \overline{B}$ (d) $\overline{A} \cup \overline{B}$

22. Let $U = \{a, b, c, d, e, f\}$, $A = \{b, c\}$, and $B = \{c, d, e\}$. List the elements of the sets:
    (a) $A \cup B$ (b) $A \cap B$
    (c) $\overline{A}$ (d) $\overline{B}$
    (e) $\overline{A \cap B}$ (f) $\overline{A \cup B}$

23. Use Venn diagrams to illustrate the following sets:

(a) $\overline{A} \cap B$ (b) $(\overline{A} \cap \overline{B}) \cup C$
(c) $A \cap (A \cup B)$ (d) $A \cup (A \cap B)$
(e) $(A \cup B) \cap (A \cup C)$ (f) $A \cup (B \cap C)$
(g) $A = (A \cap B) \cup (A \cap \overline{B})$
(h) $B = (A \cap B) \cup (\overline{A} \cap B)$

24. Use Venn diagrams to illustrate the following Properties:

(a) $A \cap (B \cup C) = (A \cap B) \cup (A \cap C)$ (Distributive property)
(b) $A \cap (A \cup B) = A$ (Absorption property)
(c) $\overline{A \cap B} = \overline{A} \cup \overline{B}$ (De Morgan's property)
(d) $(A \cup B) \cup C = A \cup (B \cup C)$ (Associative property)

In Problems 25–30, use

$A = \{x \mid x \text{ is a customer of IBM}\}$
$B = \{x \mid x \text{ is a secretary employed by IBM}\}$
$C = \{x \mid x \text{ is a computer operator at IBM}\}$
$D = \{x \mid x \text{ is a stockholder of IBM}\}$
$E = \{x \mid x \text{ is a member of the Board of Directors of IBM}\}$

to describe each set in words.

25. $A \cap E$

26. $B \cap D$

27. $A \cup D$

28. $C \cap E$

29. $\overline{A} \cap D$

30. $A \cup \overline{D}$

In Problems 31–36, use

$U = \{\text{All college students}\}$
$M = \{\text{All male students}\}$
$S = \{\text{All students who smoke}\}$
$F = \{\text{All Freshmen}\}$

to describe each set in words.

31. $M \cap S$

32. $M \cup S$

33. $\overline{M} \cup \overline{F}$

34. $\overline{M} \cap \overline{S}$

35. $F \cap S \cap M$

36. $F \cup S \cup M$

---

37. List all the subsets of $\{a, b, c\}$.

38. List all the subsets of $\{a, b, c, d\}$.

6.2 The Number of Elements in a Set

OBJECTIVES
1 Use the counting formula
2 Use Venn diagrams to analyze survey data

When we count objects, what we are actually doing is taking each object to be counted and matching each of these objects exactly once to the counting numbers 1, 2, 3, and so on, until *no* objects remain. Even before numbers had names and symbols assigned to them, this method of counting was used. Prehistoric peoples used rocks to determine how many cattle did not return from pasture. As each cow left, a rock was placed aside. As each cow returned, a rock was removed from the pile. If rocks remained after all the cows returned, it was then known that some cows were missing.

If A is any set, we will denote by $c(A)$ the number of elements in A. For example, for the set L of letters in the alphabet,

$$L = \{a, b, c, d, e, f, \ldots, x, y, z\}$$

we write $c(L) = 26$ and say "the number of elements in L is 26."

Also, for the set

$$N = \{1, 2, 3, 4, 5\}$$

we write $c(N) = 5$.

The empty set $\varnothing$ has no elements, and we write

$$c(\varnothing) = 0$$

If the number of elements in a set is zero or a positive integer, we say that the set is **finite.** Otherwise, the set is said to be **infinite.** The area of mathematics that deals with the study of finite sets is called **finite mathematics.**

EXAMPLE 1 Analyzing Survey Data

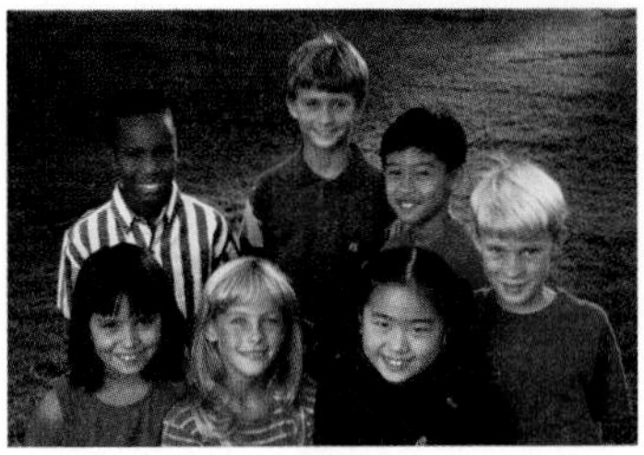

A survey of a group of people indicated there were 25 with brown eyes and 15 with black hair. If 10 people had both brown eyes and black hair and 23 people had neither, how many people interviewed have either brown eyes or black hair? How many people in all were interviewed?

SOLUTION Let A denote the set of people with brown eyes and B the set of people with black hair. Then the data given tell us

$$c(A) = 25 \qquad c(B) = 15 \qquad c(A \cap B) = 10$$

The number of people with either brown eyes or black hair is $c(A \cup B)$. But $c(A \cup B)$ cannot be $c(A) + c(B)$, since those with both characteristics would then be counted twice. To obtain $c(A \cup B)$, we need to subtract those with both characteristics from $c(A) + c(B)$ to avoid counting them twice. So,

$$c(A \cup B) = c(A) + c(B) - c(A \cap B) = 25 + 15 - 10 = 30$$

There are 30 people with either brown eyes or black hair.

The sum of people found either in A or in B and those found neither in A nor in B is the total interviewed. The total number of people interviewed is

$$30 + 23 = 53$$

FIGURE 12

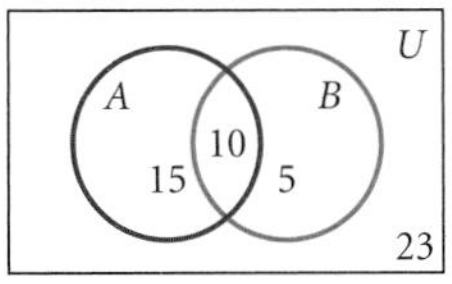

Figure 12 illustrates how a Venn diagram can be used to represent this situation.

In constructing Figure 12, we begin by placing 10 in $A \cap B$. Since $c(A) = 25$, A requires 15 more. Similarly, B requires 5 more. Now we can see that $c(A \cup B) = 15 + 10 + 5 = 30$.

In Example 1 we discovered the following important relationship:

Counting Formula

Let A and B be two finite sets. Then

$$c(A \cup B) = c(A) + c(B) - c(A \cap B) \tag{1}$$

1 **EXAMPLE 1 Using the Counting Formula**

Let $A = \{a, b, c, d, e\}$, $B = \{a, e, g, u, w, z\}$. Find $c(A)$, $c(B)$, $c(A \cap B)$, and $c(A \cup B)$.

SOLUTION $c(A) = 5$ and $c(B) = 6$. To find $c(A \cap B)$, we note that $A \cap B = \{a, e\}$ so that $c(A \cap B) = 2$. Since $A \cup B = \{a, b, c, d, e, g, u, w, z\}$, we have $c(A \cup B) = 9$. This checks with the formula

$$c(A \cup B) = c(A) + c(B) - c(A \cap B) = 5 + 6 - 2 = 9$$

NOW WORK PROBLEM 7.

Applications

2 **EXAMPLE 3 Analyzing a Consumer Survey Using Venn Diagrams**

In a survey of 75 consumers, 12 indicated they were going to buy a new car, 18 said they were going to buy a new refrigerator, and 24 said they were going to buy a new stove. Of these, 6 were going to buy both a car and a refrigerator, 4 were going to buy a car and a stove, and 10 were going to buy a stove and refrigerator. One person indicated he was going to buy all three items.

(a) How many were going to buy none of these items?
(b) How many were going to buy only a car?
(c) How many were going to buy only a stove?
(d) How many were going to buy only a refrigerator?
(e) How many were going to buy a stove and refrigerator, but not a car?

SOLUTION We denote the sets of people buying cars, refrigerators, and stoves by C, R, and S, respectively. Then we know from the data given that

$$c(C) = 12 \qquad c(R) = 18 \qquad c(S) = 24$$
$$c(C \cap R) = 6 \qquad c(C \cap S) = 4 \qquad c(S \cap R) = 10$$
$$c(C \cap R \cap S) = 1$$

We use the information given above in the reverse order and put it into a Venn diagram. Beginning with the fact that $c(C \cap R \cap S) = 1$, we place a 1 in that set, as shown in Figure 13(a).

FIGURE 13

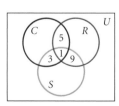

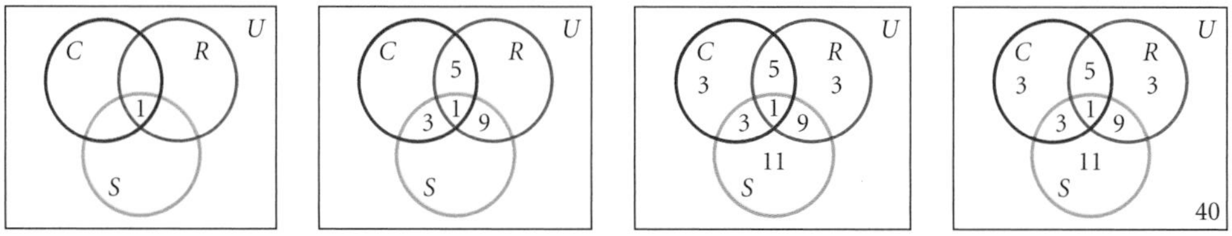

(a) (b) (c) (d)

Now $c(C \cap R) = 6$, $c(C \cap S) = 4$, and $c(S \cap R) = 10$. We place $6 - 1 = 5$ in the proper region (giving a total of 6 in the set $C \cap R$). Similarly, we place 3 and 9 in the proper regions for the sets $C \cap S$ and $S \cap R$. See Figure 13(b).

Now $c(C) = 12$ and 9 of these 12 are already accounted for. Also, $c(R) = 18$ with 15 accounted for and $c(S) = 24$ with 13 accounted for. See Figure 13(c).

Finally, the number in $\overline{C \cup R \cup S}$ is the total of 75 less those accounted for in C, R, and S, namely, $3 + 5 + 1 + 3 + 3 + 9 + 11 = 35$. Thus

$$c(\overline{C \cup R \cup S}) = 75 - 35 = 40$$

See Figure 13(d). From this figure, we can see that

(a) 40 were going to buy none of the items;
(b) 3 were going to buy only a car;
(c) 11 were going to buy only a stove;
(d) 3 were going to buy only a refrigerator;
(e) 9 were going to buy a stove and refrigerator, but not a car.

NOW WORK PROBLEM 17.

EXAMPLE 4 Analyzing Data

In a survey of 10,281 people restricted to those who were either black or male or over 18 years of age, the following data were obtained:

Black:	3490	Black males:	1745	Black male over 18:	239
Male:	5822	Over 18 and male:	859		
Over 18:	4722	Over 18 and black:	1341		

The data are inconsistent. Why?

SOLUTION We denote the set of people who were black by B, male by M, and over 18 by H. Then we know that

$$c(B) = 3490 \qquad c(M) = 5822 \qquad c(H) = 4722$$
$$c(B \cap M) = 1745 \qquad c(H \cap M) = 859 \qquad c(H \cap B) = 1341$$
$$c(H \cap M \cap B) = 239$$

FIGURE 14

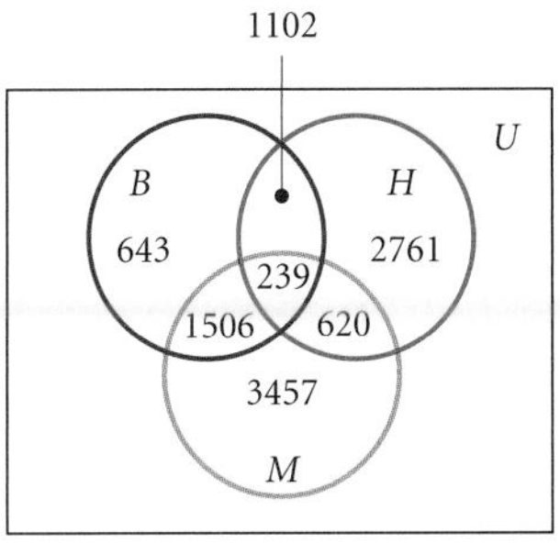

We construct the Venn diagram shown in Figure 14. This means that

$$239 + 1102 + 620 + 1506 + 3457 + 643 + 2761 = 10{,}328$$

people were interviewed. However, it is given that only 10,281 were interviewed. This means the data are inconsistent.

EXERCISE 6.2 Answers to Odd-Numbered Problems Begin on Page AN-27.

In Problems 1–6, use the sets $A = \{1, 2, 3, 4, 5, 6\}$ *and* $B = \{2, 4, 6, 8\}$ *to find the number of elements in each set.*

1. A **2.** B **3.** $A \cap B$ **4.** $A \cup B$ **5.** $(A \cap B) \cup A$ **6.** $(B \cap A) \cup B$

7. Find $c(A \cup B)$, given that $c(A) = 4$, $c(B) = 3$, and $c(A \cap B) = 2$.

8. Find $c(A \cup B)$, given that $c(A) = 14$, $c(B) = 11$, and $c(A \cap B) = 6$.

9. Find $c(A \cap B)$, given that $c(A) = 5$, $c(B) = 4$, and $c(A \cup B) = 7$.

10. Find $c(A \cap B)$, given that $c(A) = 8$, $c(B) = 9$, and $c(A \cup B) = 16$.

11. Find $c(A)$, given that $c(B) = 8$, $c(A \cap B) = 4$, and $c(A \cup B) = 14$.

12. Find $c(B)$, given that $c(A) = 10$, $c(A \cap B) = 5$, and $c(A \cup B) = 29$.

13. Motors, Inc., manufactured 325 cars with automatic transmissions, 216 with power steering, and 89 with both these options. How many cars were manufactured if every car has at least one option?

14. Suppose that out of 1500 first-year students at a certain college, 350 are taking history, 300 are taking mathematics, and 270 are taking both history and mathematics. How many first-year students are taking history or mathematics?

In Problems 15–24, use the data in the figure to answer each question.

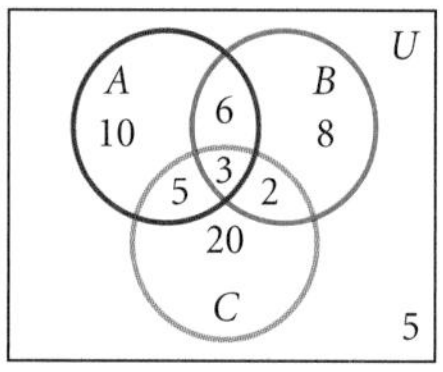

15. How many elements are in set A?

16. How many elements are in set B?

17. How many elements are in A or B?

18. How many elements are in B or C?

19. How many elements are in A but not B?

20. How many elements are in B but not C?

21. How many elements are in A or B or C?

22. How many elements are in neither A nor B nor C?

23. How many elements are in A and B and C?

24. How many elements are in U?

25. Voting Patterns Suppose the influence of religion and age on voting preference is given by the table.

	Age		
	Below 35	35–54	Over 54
Protestant Voting Republican	82	152	111
Protestant Voting Democratic	42	33	15
Catholic Voting Republican	27	33	7
Catholic Voting Democratic	44	47	33

Find

(a) The number of voters who are Catholic or Republican.
(b) The number of voters who are Catholic or over 54.
(c) The number voting Democratic below 35 or over 54.

26. The Venn diagram illustrates the number of seniors (S), female students (F), and students on the dean's list (D) at a small western college. Describe each number in terms of the sets S, F, or D.

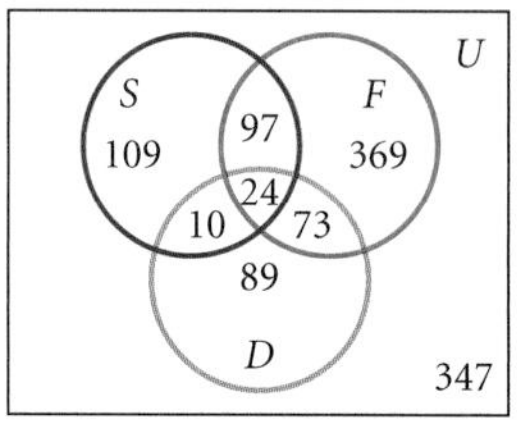

27. At a small midwestern college:

31	female seniors were on the dean's list
62	women were on the dean's list who were not seniors
45	male seniors were on the dean's list
87	female seniors were not on the dean's list
96	male seniors were not on the dean's list
275	women were not seniors and were not on the dean's list
89	men were on the dean's list who were not seniors
227	men were not seniors and were not on the dean's list

(a) How many were seniors?
(b) How many were women?
(c) How many were on the dean's list?
(d) How many were seniors on the dean's list?
(e) How many were female seniors?
(f) How many were women on the dean's list?
(g) How many were students at the college?

28. Survey Analysis In a survey of 75 college students, it was found that of the three weekly news magazines *Time*, *Newsweek*, and *U.S. News & World Report*:

23	read *Time*
18	read *Newsweek*
14	read *U.S. News & World Report*
10	read *Time* and *Newsweek*
9	read *Time* and *U.S. News & World Report*
8	read *Newsweek* and *U.S. News World & Report*
5	read all three

(a) How many read none of these three magazines?
(b) How many read *Time* alone?
(c) How many read *Newsweek* alone?
(d) How many read *U.S. News & World Report* alone?
(e) How many read neither *Time* nor *Newsweek*?
(f) How many read *Time* or *Newsweek* or both?

29. Car Sales Of the cars sold during the month of July, 90 had air conditioning, 100 had automatic transmissions, and 75 had power steering. Five cars had all three of these extras. Twenty cars had none of these extras. Twenty cars had only air conditioning; 60 cars had only automatic transmissions; and 30 cars had only power steering. Ten cars had both automatic transmission and power steering.

(a) How many cars had both power steering and air conditioning?
(b) How many had both automatic transmission and air conditioning?
(c) How many had neither power steering nor automatic transmission?
(d) How many cars were sold in July?
(e) How many had automatic transmission or air conditioning or both?

30. Incorrect Information A staff member at a large engineering school was presenting data to show that the students there received a liberal education as well as a scientific one. "Look at our record," she said. "Out of our senior class of 500 students, 281 are taking English, 196 are taking English and history, 87 are taking history and a foreign language, 143 are taking a foreign language and English, and 36 are taking all of these." She was fired. Why?

31. Blood Classification Blood is classified as being either Rh-positive or Rh-negative and according to type. If blood contains an A antigen, it is type A; if it has a B antigen, it is type B; if it has both A and B antigens, it is type AB; and if it has neither antigen, it is type O. Use a Venn diagram to illustrate these possibilities. How many different possibilities are there?

32. Survey Analysis A survey of 52 families from a suburb of Chicago indicated that there was a total of 241 children below the age of 18. Of these, 109 were male; 132 were below the age of 11; 143 had played Little League; 69 males were below the age of 11; 45 females under 11 had played Little League; and 30 males under 11 had played Little League.

(a) How many children over 11 and under 18 had played Little League?
(b) How many females under 11 did not play Little League?

33. Survey Analysis Of 100 personal computer users surveyed: 27 use IBM; 35 use Macs, 35 use Dell, 10 use both IBM and Macs, 10 use both IBM and Dell, 10 use both Macs and Dell, 3 use all three; and 30 use another computer brand. How many people exclusively use one of the three brands mentioned, that is, only IBM or only Macs or only Dell?

34. List all the subsets of $\{a, b, c\}$. How many are there?

35. List all the subsets of $\{a, b, c, d\}$. How many are there?

6.3 The Multiplication Principle

OBJECTIVE 1 Use the multiplication principle

In this section we introduce a general principle of counting, called the *Multiplication Principle.* We begin with an example.

EXAMPLE 1 Counting the Number of Ways a Certain Trip Can Be Taken

In traveling from New York to Los Angeles, Mr. Doody wishes to stop over in Chicago. If he has five different routes to choose from in driving from New York to Chicago and has three routes to choose from in driving from Chicago to Los Angeles, in how many ways can Mr. Doody travel from New York to Los Angeles?

SOLUTION Traveling from New York to Los Angeles through Chicago consists of making two choices:

Choose a route from New York to Chicago Task 1	Choose a route from Chicago to Los Angeles Task 2

In Figure 15 we see after the five routes from New York to Chicago there are three routes from Chicago to Los Angeles.

FIGURE 15

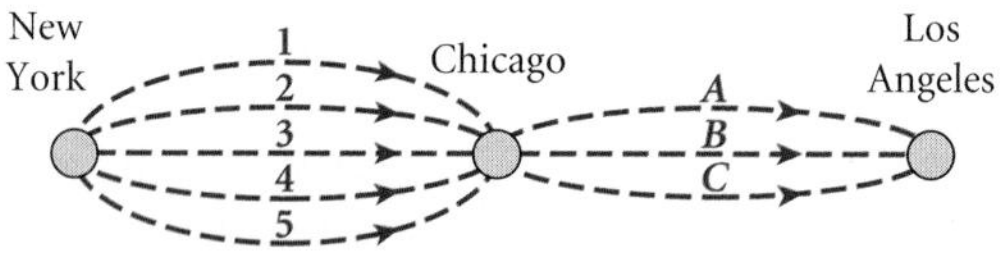

These different routes can be enumerated as

$$1A, 1B, 1C \quad 2A, 2B, 2C \quad 3A, 3B, 3C \quad 4A, 4B, 4C \quad 5A, 5B, 5C$$

In all, there are $5 \cdot 3 = 15$ different routes.

Notice that the total number of ways the trip can be taken is simply the product of the number of ways of doing Task 1 with the number of ways of doing Task 2.

The different routes in Example 1 can also be depicted in a **tree diagram.** See Figure 16.

FIGURE 16

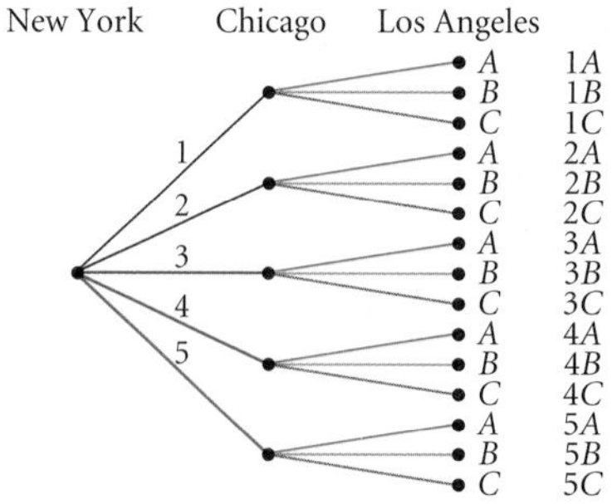

NOW WORK PROBLEM 1.

The example just solved demonstrates a general type of counting problem, which can be solved by the multiplication principle.

Multiplication Principle of Counting

If a task consists of a sequence of choices in which there are p selections for the first choice, q selections for the second choice, r selections for the third choice, and so on, then the task of making these selections can be done in

$$\boxed{p \cdot q \cdot r \cdot \ldots}$$

different ways.

1 **EXAMPLE 2 Forming Codes**

How many two-symbol codewords can be formed if the first symbol is a letter (uppercase) and the second symbol is a digit?

SOLUTION It sometimes helps to begin by listing some of the possibilities. The code consists of a letter (uppercase) followed by a digit, so some possibilities are A1, A2, B3, X0, and so on. The task consists of making two selections: the first selection requires choosing an uppercase letter (26 choices) and the second task requires choosing a digit (10 choices). By the Multiplication Principle, there are

$$26 \cdot 10 = 260$$

different codewords of the type described.

NOW WORK PROBLEM 7.

EXAMPLE 3 Combination Locks

A particular type of combination lock has 10 numbers on it.

(a) How many sequences of four numbers can be formed to open the lock?
(b) How many sequences can be formed if no number is repeated?

SOLUTION **(a)** Each of the four numbers can be chosen in 10 ways. By the Multiplication Principle, there are

$$10 \cdot 10 \cdot 10 \cdot 10 = 10{,}000$$

different sequences.

(b) If no number can be repeated, then there are 10 choices for the first number, only 9 for the second number, 8 for the third number, and 7 for the fourth number. By the Multiplication Principle, there are

$$10 \cdot 9 \cdot 8 \cdot 7 = 5040$$

different sequences.

EXAMPLE 4 Counting the Number of Ways Four Offices Can Be Filled

In a city election there are four candidates for mayor, three candidates for vice-mayor, six candidates for treasurer, and two for secretary. In how many ways can these four offices be filled?

SOLUTION The task of filling an office can be divided into four consecutive operations:

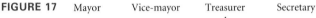

Select a mayor	Select a vice-mayor	Select a treasurer	Select a secretary

Corresponding to each of the four possible mayors, there are three vice-mayors. These two offices can be filled in $4 \cdot 3 = 12$ different ways. Also, corresponding to each of these 12 possibilities, we have six different choices for treasurer—giving $12 \cdot 6 = 72$ different possibilities. Finally, to each of these 72 possibilities there correspond two choices for secretary. In all, these offices can be filled in $4 \cdot 3 \cdot 6 \cdot 2 = 144$ different ways. A partial illustration is given by the tree diagram in Figure 17.

FIGURE 17

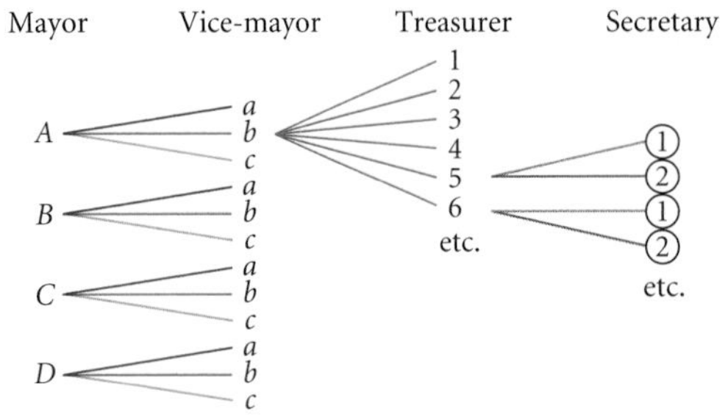

NOW WORK PROBLEM 9.

EXAMPLE 5 License Plates in Maryland

License plates in the state of Maryland consist of three letters of the alphabet followed by three digits.

(a) The Maryland system will allow how many possible license plates?
(b) Of these, how many will have all their digits distinct?
(c) How many will have distinct digits and distinct letters?

SOLUTION **(a)** There are six positions on the license plate to be filled, the first three by letters and the last three by digits. The first three positions can be filled in any one of 26 ways, while the remaining three positions can each be filled in any of 10 ways. The total number of license plates, by the Multiplication Principle, is then

$$26 \cdot 26 \cdot 26 \cdot 10 \cdot 10 \cdot 10 = 17{,}576{,}000$$

(b) Here, the tasks involved in filling the digit positions are slightly different. The first digit can be any one of 10, but the second digit can be only any one of 9 (we cannot duplicate the first digit); there are only 8 choices for the third digit (we cannot duplicate either the first or the second). By the Multiplication Principle there are

$$26 \cdot 26 \cdot 26 \cdot 10 \cdot 9 \cdot 8 = 12{,}654{,}720$$

plates with no repeated digit.

(c) If the letters and digits are each to be distinct, then the total number of possible, license plates is

$$26 \cdot 25 \cdot 24 \cdot 10 \cdot 9 \cdot 8 = 11{,}232{,}000$$

NOW WORK PROBLEM 15.

EXERCISE 6.3 Answers to Odd-Numbered Problems Begin on Page AN-28.

1. There are 2 roads between towns A and B. There are 4 roads between towns B and C. How many different routes may one travel from town A to town C through town B?

2. A woman has 4 blouses and 5 skirts. How many different outfits can she wear?

3. XYZ Company wants to build a complex consisting of a factory, office building, and warehouse. If the building contractor has 3 different kinds of factories, 2 different office buildings, and 4 different warehouses, how many models must be built to show all possibilities to XYZ Company?

4. Cars, Inc., has 3 different car models and 8 color schemes. If you are one of the dealers, how many cars must you display to show each possibility?

5. **Choosing an Outfit** A man has 3 pairs of shoes, 8 pairs of socks, 4 pairs of slacks, and 9 sweaters. How many outfits can he wear?

6. A house has 3 outside doors and 12 windows. In how many ways can a person enter the house through a window and exit through a door?

7. **License Plates** How many license plates consisting of 2 letters (uppercase) followed by 2 digits are possible?

8. **Lunch Selections** A restaurant offers 3 different salads, 8 different main courses, 10 different desserts, and 4 different drinks. How many different lunches—each consisting of a salad, a main course, a dessert, and a drink—are possible?

9. **Arranging Books** Five different mathematics books are to be arranged on a student's desk. How many arrangements are possible?

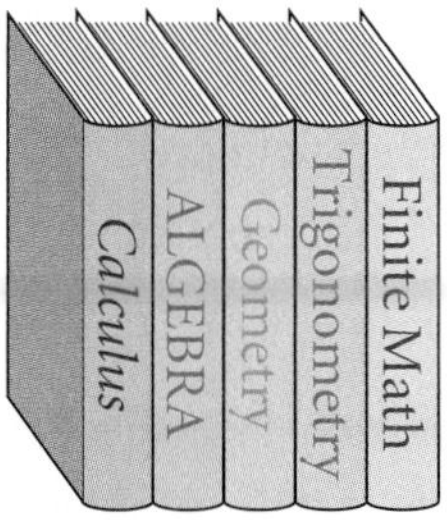

10. How may ways can 6 people be seated in a row of 6 seats?

In Problems 11–13, use the following discussion.
Students log on to the California Virtual In Campus with a user name consisting of eight characters: four uppercase letters of the alphabet followed by four digits.

Source: California Virtual Campus.

11. **User Names** How many user names are theoretically possible for this system?

12. **User Names** How many user names for this system have no repeated letters or digits?

13. **User Names** How many user names for this system have no matching adjacent letters or digits?

14. **Baseball** In the World Series, the National League champion plays the American League champion. There are 16 teams in the National League and 14 teams in the American League. How many different theoretical match-ups of two teams are possible in the World Series?

Source: Major League Baseball, 2003.

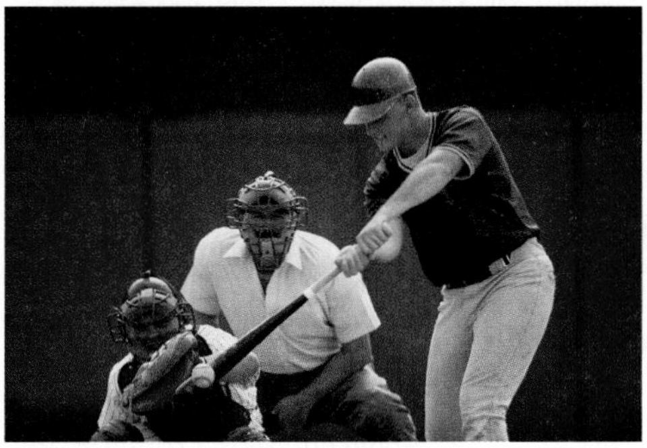

15. **Codes** How many 4-letter code words are possible using the first 6 letters of the alphabet with no letters repeated? How many codes are there when letters are allowed to repeat?

16. Telephone Numbers Find the number of 7-digit telephone numbers

(a) With no repeated digits (lead 0 is allowed)
(b) With no repeated digits (lead 0 not allowed)
(c) With repeated digits allowed including a lead 0

17. How many ways are there to rank 7 candidates who apply for a job?

18. (a) How many different ways are there to arrange the 7 letters in the word PROBLEM?
(b) If we insist that the letter P comes first, how many ways are there?
(c) If we insist that the letter P comes first and the letter M last, how many ways are there?

19. Testing On a math test there are 10 multiple-choice questions with 4 possible answers and 15 true–false questions. In how many possible ways can the 25 questions be answered?

20. Using the digits 1, 2, 3, and 4, how many different 4-digit numbers can be formed?

21. License Plate Possibilities How many different license plate numbers can be made using 2 letters followed by 4 digits, if

(a) Letters and digits may be repeated?
(b) Letters may be repeated, but digits are not repeated?
(c) Neither letters nor digits may be repeated?

22. Security A system has 7 switches, each of which may be either open or closed. The state of the system is described by indicating for each switch whether it is open or closed. How many different states of the system are there?

23. Product Choice An automobile manufacturer produces 3 different models. Models A and B can come in any of 3 body styles; model C can come in only 2 body styles. Each car also comes in either black or green. How many distinguishable car types are there?

24. Telephone Numbers How many 7-digit numbers can be formed if the first digit cannot be 0 or 9 and if the last digit is greater than or equal to 2 and less than or equal to 3? Repeated digits are allowed.

25. Home Choices A contractor constructs homes with 5 different choices of exterior finish, 3 different roof arrangements, and 4 different window designs. How many different types of homes can be built?

26. License Plate Possibilities A license plate consists of 1 letter, excluding O and I, followed by a 4-digit number that cannot have a 0 in the lead position. How many different plates are possible?

27. Bytes Using only the digits 0 and 1, how many different numbers consisting of 8 digits can be formed?

28. Stock Portfolios As a financial planner, you are asked to select one stock from each of the following groups: 8 DOW stocks, 15 NASDAQ stocks, and 4 global stocks. How many different portfolios are possible?

29. Combination Locks A combination lock has 50 numbers on it. To open it, you turn counterclockwise to a number, then rotate clockwise to a second number, and then counterclockwise to the third number. How many different lock combinations are there?

30. Opinion Polls An opinion poll is to be conducted among college students. Eight multiple-choice questions, each with 3 possible answers, will be asked. In how many different ways can a student complete the poll if exactly one response is given to each question?

31. Path Selection in a Maze The maze below is constructed so that a rat must pass through a series of one-way doors. How many different paths are there from start to finish?

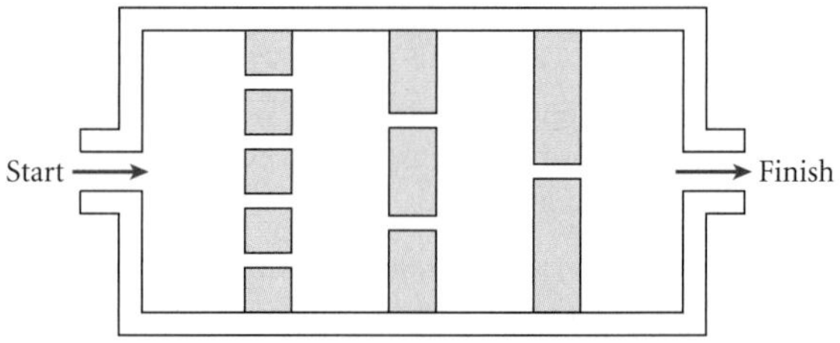

32. How many 3-letter code words are possible using the first 10 letters of the alphabet if

(a) No letter can be repeated?
(b) Letters can be repeated?
(c) Adjacent letters cannot be the same?

6.4 Permutations

OBJECTIVES
1 Evaluate factorials
2 Solve counting problems involving permutations [distinct, with repetition]
3 Solve counting problems involving permutations [distinct, without repetitions]

In the next two sections we use the Multiplication Principle to discuss two general types of counting problems, called *permutations* and *combinations*. These concepts arise often in applications, especially in probability.

Factorial

Before discussing permutations, we introduce a useful shorthand notation—the *factorial symbol.*

Factorial

The symbol $n!$, read as **"*n* factorial,"** is defined as

$$\begin{aligned} 0! &= 1 \\ 1! &= 1 \\ 2! &= 2 \cdot 1 = 2 \\ 3! &= 3 \cdot 2 \cdot 1 = 6 \\ 4! &= 4 \cdot 3 \cdot 2 \cdot 1 = 24 \end{aligned}$$

and, in general, for $n \geq 1$ an integer,

$$n! = n \cdot (n-1) \cdot (n-2) \cdot \ldots 3 \cdot 2 \cdot 1 \qquad \textbf{(1)}$$

To compute $n!$, we find the product of all consecutive integers from n down to 1, inclusive, or from 1 up to n, inclusive. Remember that, by definition, $0! = 1$.

Look at Equation (1). It follows that

$$n! = n\,(n-1)! \qquad (2)$$

We shall find Formula (2) useful in evaluating expressions containing factorials.

1 **EXAMPLE 1** **Evaluating Expressions Containing Factorials**

(a) $5! = 5 \cdot 4 \cdot 3 \cdot 2 \cdot 1 = 120$

(b) $\dfrac{5!}{4!} = \dfrac{5 \cdot \cancel{4!}}{\cancel{4!}} = 5$

(c) $\dfrac{52!}{5!47!} = \dfrac{52 \cdot 51 \cdot 50 \cdot 49 \cdot 48 \cdot \cancel{47!}}{5 \cdot 4 \cdot 3 \cdot 2 \cdot 1 \cdot \cancel{47!}} = 2{,}598{,}960$

(d) $\dfrac{7!}{(7-5)!5!} = \dfrac{7!}{2!5!} = \dfrac{7 \cdot 6 \cdot \cancel{5!}}{2!\cancel{5!}} = \dfrac{7 \cdot 6}{2} = 21$

(e) $\dfrac{50 \cdot 49 \cdot 48 \cdot 47 \cdot 46}{50!} = \dfrac{50 \cdot 49 \cdot 48 \cdot 47 \cdot 46}{50 \cdot 49 \cdot 48 \cdot 47 \cdot 46 \cdot 45!} = \dfrac{1}{45!}$

NOW WORK PROBLEM 1.

Factorials grow very quickly. Compare the following:

$$\begin{aligned} 5! &= 120 \\ 10! &= 3{,}628{,}800 \\ 15! &= 1{,}307{,}674{,}368{,}000 \end{aligned}$$

In fact, if your calculator has a factorial key, you will find that $69! = 1.71 \cdot 10^{98}$, while 70! produces an error message—indicating you have exceeded the range of the calculator. Because of this, it is important to "cancel out" factorials whenever possible so that "out of range" errors may be avoided. For example, to calculate 100!/95!, we write

$$\frac{100!}{95!} = \frac{100 \cdot 99 \cdot 98 \cdot 97 \cdot 96 \cdot \cancel{95!}}{\cancel{95!}} = 9.03 \cdot 10^9$$

Permutations

We begin with the definition

> A **permutation** is an ordered arrangement of r objects chosen from n objects.

We discuss three types of permutations:

1. The n objects are distinct (different), and repetition is allowed in the selection of r of them. [Distinct, with repetition]
2. The n objects are distinct (different), and repetition is not allowed in the selection of r of them, where $r \leq n$. [Distinct, without repetition]
3. The n objects are not distinct, and we use all of them in the arrangement. [Not distinct]

We take up the first two types here and deal with the third type in the next section. The first type of permutation is handled using the Multiplication Principle.

2 **EXAMPLE 2** **Counting Airport Codes [Permutation: Distinct, with Repetition]**

The International Airline Transportation Association (IATA) assigns three-letter codes to represent airport locations. For example, the airport code for Ft. Lauderdale, Florida is FLL. Notice that repetition is allowed in forming this code. How many airport codes are possible?

SOLUTION We are choosing 3 letters from 26 letters and arranging them in order. In the ordered arrangement a letter may be repeated. This is an example of a permutation with repetition in which 3 objects are chosen from 26 distinct objects.

The task of counting the number of such arrangements consists of making three selections. Each selection requires choosing a letter of the alphabet (26 choices). By the Multiplication Principle, there are

$$26 \cdot 26 \cdot 26 = 17,576$$

different airport codes.

The solution given to Example 2 can be generalized.

> **Permutations: Distinct Objects, with Repetition**
>
> The number of ordered arrangements of r objects chosen from n objects, in which the n objects are distinct and repetition is allowed, is n^r.

NOW WORK PROBLEM 27.

Next we discuss permutations in which the objects are distinct and repetition is not allowed. We begin with an example.

3 **EXAMPLE 3** **Forming Codes [Permutation: Distinct, without Repetition]**

Suppose that we wish to establish a three-letter code using any of the 26 uppercase letters of the alphabet, but we require that no letter be used more than once. How many different three-letter codes are there?

SOLUTION Some of the possibilities are: ABC, ABD, ABZ, ACB, CBA, and so on. The task consists of making three selections. The first selection requires choosing from 26 letters. Because no letter can be used more than once, the second selection requires choosing from 25 letters. The third selection requires choosing from 24 letters. (Do you see why?) By the Multiplication Principle, there are

$$26 \cdot 25 \cdot 24 = 15{,}600$$

different three-letter codes with no letter repeated.

For this second type of permutation, we introduce the following symbol.

> The symbol $P(n, r)$ represents the number of ordered arrangements of r objects chosen from n distinct objects, where $r \leq n$ and repetition is not allowed.

For example, the question posed in Example 3 asks for the number of ways that the 26 letters of the alphabet can be arranged in order using three nonrepeated letters. The answer is

$$P(26, 3) = 26 \cdot 25 \cdot 24 = 15{,}600$$

To arrive at a formula for $P(n, r)$, we note that the task of obtaining an ordered arrangement of n objects in which only $r \leq n$ of them are used, without repeating any of them, requires making r selections. For the first selection, there are n choices; for the second selection, there are $n - 1$ choices; for the third selection, there are $n - 2$ choices; . . .; for the rth selection, there are $n - (r - 1)$ choices. By the Multiplication Principle, we have

$$\overset{\text{1st}}{} \quad \overset{\text{2nd}}{} \quad \overset{\text{3rd}}{} \qquad \overset{r\text{th}}{}$$
$$P(n, r) = n \cdot (n - 1) \cdot (n - 2) \cdot \ldots \cdot [n - (r - 1)]$$
$$= n \cdot (n - 1) \cdot (n - 2) \cdot \ldots \cdot (n - r + 1)$$

This formula for $P(n, r)$ can be compactly written using factorial notation.

$$P(n, r) = n \cdot (n - 1) \cdot (n - 2) \cdot \ldots \cdot (n - r + 1)$$
$$= n \cdot (n - 1) \cdot (n - 2) \cdot \ldots \cdot (n - r + 1) \cdot \frac{(n - r) \cdot \ldots \cdot 3 \cdot 2 \cdot 1}{(n - r) \cdot \ldots \cdot 3 \cdot 2 \cdot 1} = \frac{n!}{(n - r)!}$$

Permutations of r Objects Chosen from n Distinct Objects without Repetition

The number of arrangements of n objects using $r \leq n$ of them, in which

1. the n objects are distinct,
2. once an object is used it cannot be repeated, and
3. order is important,

is given by the formula

$$\boxed{P(n, r) = \frac{n!}{(n-r)!}} \qquad (3)$$

EXAMPLE 4 Computing Permutations

Evaluate: **(a)** $P(7, 3)$ **(b)** $P(6, 1)$ **(c)** $P(52, 5)$

SOLUTION We shall work parts (a) and (b) in two ways.

(a)

$$P(7, 3) = \underbrace{7 \cdot 6 \cdot 5}_{3 \text{ factors}} = 210$$

or

$$P(7, 3) = \frac{7!}{(7-3)!} = \frac{7!}{4!} = \frac{7 \cdot 6 \cdot 5 \cdot \cancel{4!}}{\cancel{4!}} = 210$$

(b)

$$P(6, 1) = \underbrace{6}_{1 \text{ factor}} = 6$$

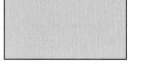

or

$$P(6, 1) = \frac{6!}{(6-1)!} = \frac{6!}{5!} = \frac{6 \cdot \cancel{5!}}{\cancel{5!}} = 6$$

FIGURE 18

(c) Figure 18 shows the solution using a TI-83 plus graphing calculator:

$$P(52, 5) = 311{,}875{,}200.$$

NOW WORK PROBLEM 13.

Here is a list of short problems with their solutions given in $P(n, r)$ notation.

Problem	***Solution***
Find the number of ways of choosing five people from a group of 10 and arranging them in a line.	$P(10, 5)$
Find the number of six-letter "words" that can be formed with no letter repeated.	$P(26, 6)$
Find the number of seven-digit telephone numbers, with no repeated digit (allow 0 for a first digit).	$P(10, 7)$
Find the number of ways of arranging eight people in a line.	$P(8, 8)$

In each of the previous examples, notice that the objects are distinct, no object is repeated, and order is important.

NOW WORK PROBLEM 21.

EXAMPLE 5 The Birthday Problem

All we know about Shannon, Patrick, and Ryan is that they have different birthdays. If we listed all the possible ways this could occur, how many would there be? Assume that there are 365 days in a year.

SOLUTION This is an example of a permutation in which 3 birthdays are selected from a possible 365 days, and no birthday may repeat itself. The number of ways that this can occur is

$$P(365, 3) = \frac{365!}{(365-3)!} = \frac{365 \cdot 364 \cdot 363 \cdot \cancel{362!}}{\cancel{362!}} = 365 \cdot 364 \cdot 363 = 48{,}228{,}180$$

There are 48,228,180 ways in a group of three people that each can have a different birthday.

NOW WORK PROBLEM 37.

EXAMPLE 6 Answering Test Questions

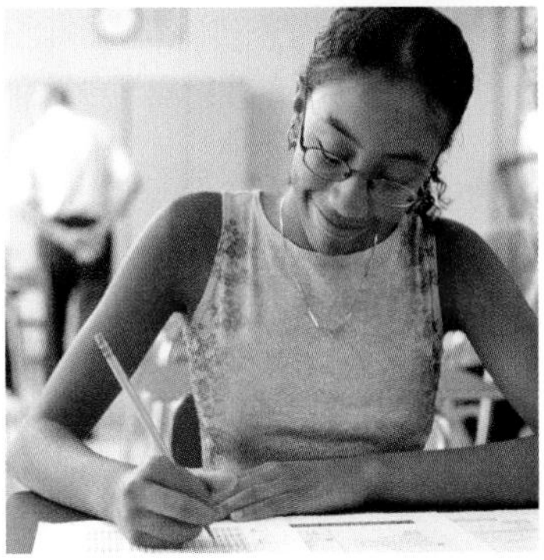

A student has six questions on an examination and is allowed to answer the questions in any order. In how many different orders could the student answer these questions?

SOLUTION We seek the number of ordered arrangements of the six questions using all six of them. The number is given by

$$P(6, 6) = \frac{6!}{(6-6)!} = \frac{6!}{0!} = \frac{6!}{1} = 720$$

Example 6 leads us to formulate the next result.

> The number of permutations (arrangements) of n distinct objects using all n of them is given by
>
> $$\boxed{P(n, n) = n!}$$

For example, in a class of n students, there are $n!$ ways of positioning all the students in a line.

NOW WORK PROBLEM 29.

EXAMPLE 7 **Arranging Books on a Shelf**

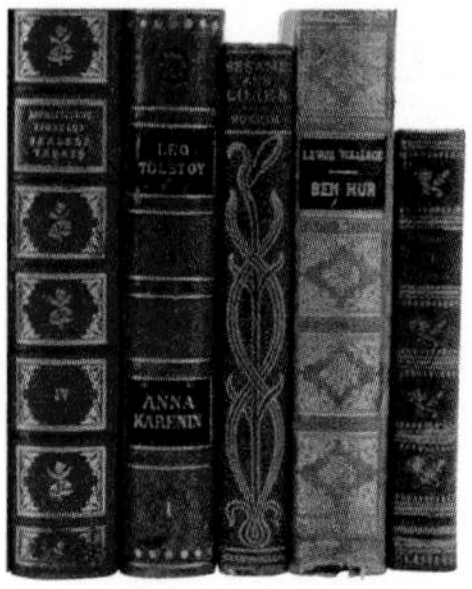

You own eight mathematics books and six computer science books and wish to fill seven positions on a shelf. If the first four positions are to be occupied by math books and the last three by computer science books, in how many ways can this be done?

SOLUTION We think of the problem as consisting of two tasks. Task 1 is to fill the first four positions with four of the eight mathematics books. This can be done in $P(8, 4)$ ways. Task 2 is to fill the remaining three positions with three of six computer books. This can be done in $P(6, 3)$ ways. By the Multiplication Principle, the seven positions can be filled in

$$P(8, 4) \cdot P(6, 3) = \frac{8!}{4!} \cdot \frac{6!}{3!} = 8 \cdot 7 \cdot 6 \cdot 5 \cdot 6 \cdot 5 \cdot 4 = 201{,}600 \text{ ways}$$

EXERCISE 6.4 Answers to Odd-Numbered Problems Begin on Page AN-28.

In Problems 1–20, evaluate each expression.

1. $\dfrac{5!}{2!}$
2. $\dfrac{8!}{2!}$
3. $\dfrac{6!}{3!}$
4. $\dfrac{9!}{3!}$
5. $\dfrac{10!}{8!}$
6. $\dfrac{11!}{9!}$
7. $\dfrac{9!}{8!}$
8. $\dfrac{10!}{9!}$
9. $\dfrac{8!}{2!6!}$
10. $\dfrac{9!}{3!6!}$
11. $P(7, 2)$
12. $P(8, 1)$
13. $P(8, 7)$
14. $P(6, 6)$
15. $P(6, 0)$
16. $P(6, 4)$
17. $\dfrac{8!}{(8-3)!3!}$
18. $\dfrac{9!}{(9-5)!5!}$
19. $\dfrac{6!}{(6-6)!6!}$
20. $\dfrac{7!}{(0-0)!7!}$

21. List all the ordered arrangements of 5 objects a, b, c, d, and e, choosing 3 at a time without repetition. What is $P(5, 3)$?
22. List all the ordered arrangements of 5 objects a, b, c, d, and e, choosing 2 at a time without repetition. What is $P(5, 2)$?
23. List all the ordered arrangements of 4 objects 1, 2, 3, and 4, choosing 3 at a time without repetition. What is $P(4, 3)$?
24. List all the ordered arrangements of 6 objects 1, 2, 3, 4, 5, and 6, choosing 2 at a time without repetition. What is $P(6, 2)$?
25. **Forming Codes** How many two-letter codes can be formed using the letters A, B, C, and D? Repeated letters are allowed.
26. **Forming Codes** How many two-letter codes can be formed using the letters A, B, C, D, and E? Repeated letters are allowed.
27. **Forming Numbers** How many three-digit numbers can be formed using the digits 0 and 1? Repeated digits are allowed.
28. **Forming Numbers** How many three-digit numbers can be formed using the digits 0, 1, 2, 3, 4, 5, 6, 7, 8, and 9? Repeated digits are allowed.
29. **Lining People Up** In how many ways can 4 people be lined up?
30. **Stacking Boxes** In how many ways can 5 different boxes be stacked?
31. **Forming Codes** How many different three-letter codes are there if only the letters A, B, C, D, and E can be used and no letter can be used more than once?
32. **Forming Codes** How many different four-letter codes are there if only the letters A, B, C, D, E, and F can be used and no letter can be used more than once?
33. How many ways are there to seat 5 people in 8 chairs?

34. How many ways are there to seat 4 people in a 6-passenger automobile?

35. Stocks on the NYSE Companies whose stocks are listed on the New York Stock Exchange (NYSE) have their company name represented by either 1, 2, or 3 letters (repetition of letters is allowed). What is the maximum number of companies that can be listed on the NYSE?

36. Stocks on the NASDAQ Companies whose stocks are listed on the NASDAQ stock exchange have their company name represented by either 4 or 5 letters (repetition of letters is allowed). What is the maximum number of companies that can be listed on the NASDAQ?

37. Birthday Problem In how many ways can 2 people each have different birthdays? Assume that there are 365 days in a year.

38. Birthday Problem In how many ways can 5 people each have different birthdays? Assume that there are 365 days in a year.

39. Arranging Letters (a) How many different ways are there to arrange the 6 letters in the word SUNDAY?
(b) If we insist that the letter *S* come first, how many ways are there?
(c) If we insist that the letter *S* come first and the letter *Y* be last, how many ways are there?

40. Arranging Books There are 5 different French books and 5 different Spanish books. How many ways are there to arrange them on a shelf if
(a) Books of the same language must be grouped together, French on the left, Spanish on the right?
(b) French and Spanish books must alternate in the grouping, beginning with a French book?

41. Distributing Books In how many ways can 8 different books be distributed to 12 children if no child gets more than one book?

42. A computer must assign each of 4 outputs to one of 8 different printers. In how many ways can it do this provided no printer gets more than one output?

43. Lottery Tickets From 1500 lottery tickets that are sold, 3 tickets are to be selected for first, second, and third prizes. How many possible outcomes are there?

44. Personnel Assignment A salesperson is needed in each of 7 different sales territories. If 10 equally qualified persons apply for the jobs, in how many ways can the jobs be filled?

45. Choosing Officers A club has 15 members. In how many ways can 4 officers consisting of a president, vice-president, secretary, and treasurer be chosen?

46. Psychology Testing In an ESP experiment a person is asked to select and arrange 3 cards from a set of 6 cards labeled *A*, *B*, *C*, *D*, *E*, and *F*. Without seeing the card, a second person is asked to give the arrangement. Determine the number of possible responses by the second person if he simply guesses.

6.5 Combinations

OBJECTIVES
1 Solve counting problems involving combinations
2 Solve counting problems involving permutations [*n* objects, not all distinct]

Permutations focus on the order in which objects are arranged. However, in many cases, order is not important. For example, in a draw poker hand, the order in which you receive the cards is not relevant—all that matters is what cards are received. That is, with poker hands, we are concerned only with what *combination* of cards we have—not the particular order of the cards.

Combinations

Solve counting problems involving combinations 1

The following examples illustrate the distinction between selections in which order is important and those for which order is not important.

EXAMPLE 1 Arranging Letters

From the four letters *a*, *b*, *c*, *d*, choose two without repeating any letter

(a) If order is important **(b)** If order is not important

SOLUTION **(a)** If order is important, there are $P(4, 2) = 4 \cdot 3 = 12$ possible selections, namely,

ab *ac* *ad* *ba* *bc* *bd* *ca* *cb* *cd* *da* *db* *dc*

See Figure 19.

FIGURE 19

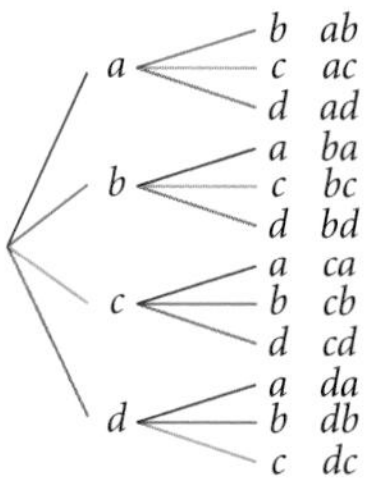

$P(4, 2) = 12$

(b) If order is not important, only 6 of the 12 selections found in part (a) are listed, namely,

ab *ac* *ad* *bc* *bd* *cd*

Notice that the number of ordered selections, 12, is $2! = 2$ times the number of unordered selections, 6. The reason is that each unordered selection consists of two letters that allow for 2! rearrangements. For example, the selection *ab* in the unordered list gives rise to *ab* and *ba* in the ordered list.

EXAMPLE 2 Arranging Letters

From the four letters *a*, *b*, *c*, *d*, choose three without repeating any letter

(a) If order is important **(b)** If order is not important

SOLUTION **(a)** If order is important, there are $P(4, 3) = 4 \cdot 3 \cdot 2 = 24$ possible selections, namely,

abc *abd* *acb* *acd* *adb* *adc* *bac* *bad* *bca* *bcd* *bda* *bdc*
cab *cad* *cba* *cbd* *cda* *cdb* *dab* *dac* *dba* *dbc* *dca* *dcb*

See Figure 20.

FIGURE 20

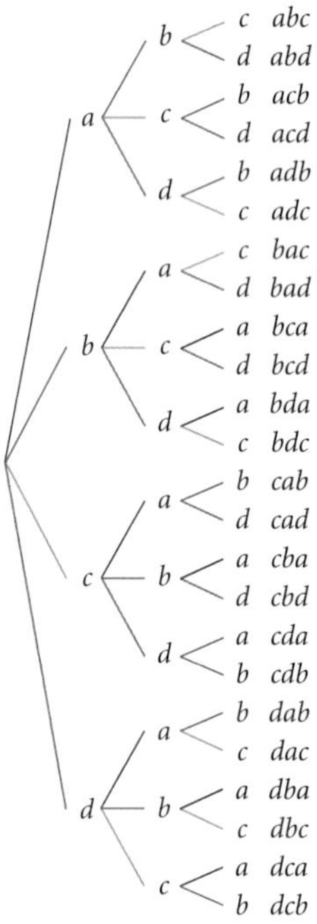

$P(4, 3) = 24$

(b) If order is not important, only 4 of the 24 selections found in part (a) are listed, namely,

abc *abd* *acd* *bcd*

Notice that the number of ordered selections, 24, is $3! = 6$ times the number of unordered selections, 4. The reason is that each unordered selection consists of three letters that allow for 3! rearrangements. For example, the selection *abc* in the unordered list gives rise to *abc*, *acb*, *bac*, *bca*, *cab*, *cba* in the ordered list.

Unordered selections are called *combinations.*

A **combination** is an arrangement, without regard to order, of r objects selected from n distinct objects without repetition, where $r \leq n$. The symbol $C(n, r)$ represents the number of combinations of n distinct objects using r of them.

$C(n, r)$ is also referred to as the number of **combinations of n objects taken r at a time.**

To obtain a formula for $C(n, r)$, we first observe that each unordered selection of r objects will give rise to $r!$ ordered selections. Thus the number of ordered selections, $P(n, r)$, is $r!$ times the number of unordered selections, $C(n, r)$. That is,

$$P(n, r) = r!C(n, r)$$

Using this formula for $P(n, r)$, we find that

$$C(n, r) = \frac{P(n, r)}{r!} = \frac{n!}{(n-r)!\,r!}$$

We have proved the following result:

Number of Combinations of n Distinct Objects Taken r at a Time

The number of arrangements of n objects using $r \leq n$ of them, in which

1. the n objects are distinct,
2. once an object is used, it cannot be repeated, and
3. order is not important,

is given by the formula

$$C(n, r) = \frac{n!}{(n-r)!r!} \quad (1)$$

EXAMPLE 3 Using Formula (1)

Use Formula (1) to find the value of each expression.

(a) $C(3, 1)$ **(b)** $C(6, 3)$ **(c)** $C(n, n)$ **(d)** $C(n, 0)$ **(e)** $C(52, 5)$

SOLUTION **(a)** $C(3, 1) = \frac{3!}{(3-1)!1!} = \frac{3!}{2!1!} = \frac{3 \cdot \cancel{2} \cdot 1}{\cancel{2} \cdot 1 \cdot 1} = 3$

(b) $C(6, 3) = \frac{6!}{(6-3)!3!} = \frac{6 \cdot 5 \cdot 4 \cdot \cancel{3!}}{3! \cdot \cancel{3!}} = \frac{\cancel{6} \cdot 5 \cdot 4}{\cancel{6}} = 20$

(c) $C(n, n) = \frac{n!}{(n-n)!n!} = \frac{\cancel{n!}}{0!\cancel{n!}} = \frac{1}{1} = 1$

(d) $C(n, 0) = \frac{n!}{(n-0)!0!} = \frac{\cancel{n!}}{\cancel{n!}0!} = \frac{1}{1} = 1$

(e) Figure 21 shows the solution using a TI-83 graphing calculator:

$$C(52, 5) = 2{,}598{,}960.$$

FIGURE 21

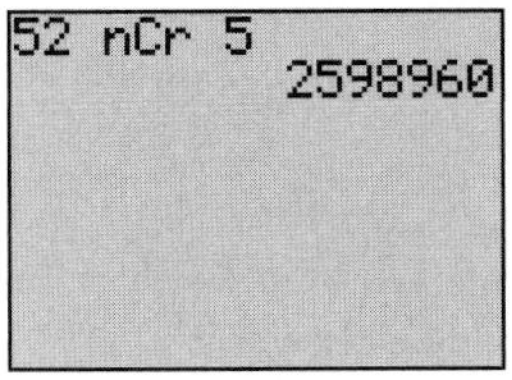

NOW WORK PROBLEM 1.

EXAMPLE 4 Forming Committees

From 5 faculty members, a committee of 2 is to be formed. In how many ways can this be done?

SOLUTION The formation of committees is an example of a combination. The 5 faculty members are all different. The members of the committee are distinct. Order is not important.

(On committees it is membership, not the order of selection, that is important.) For the situation described, we can form

$$C(5, 2) = \frac{5!}{3!2!} = 10$$

different committees.

Here are some other problems that are examples of combinations. The solutions are given in $C(n, r)$ notation.

Problem	*Solution*
Find the number of ways of selecting 4 people from a group of 6	$C(6, 4)$
Find the number of committees of 6 that can be formed from the U.S. Senate (100 members)	$C(100, 6)$
Find the number of ways of selecting 5 courses from a catalog containing 200	$C(200, 5)$

NOW WORK PROBLEM 9.

EXAMPLE 5 Playing Cards

From a deck of 52 cards a hand of 5 cards is dealt. How many different hands are possible?

SOLUTION Such a hand is an unordered selection of 5 cards from a deck of 52. So the number of different hands is

$$C(52, 5) = \frac{52!}{47!5!} = \frac{52 \cdot 51 \cdot 50 \cdot 49 \cdot 48 \cdot \cancel{47!}}{5 \cdot 4 \cdot 3 \cdot 2 \cdot 1 \cdot \cancel{47!}} = 2{,}598{,}960$$

EXAMPLE 6 Six-Bit Strings

A bit is a 0 or a 1. A six-bit string is a sequence of length six consisting of 0s and 1s. How many six-bit strings contain

(a) Exactly one 1? **(b)** Exactly two 1s?

SOLUTION **(a)** To form a six-bit string having one 1, we only need to specify where the single 1 is located (the other positions are 0s). The location for the 1 can be chosen in

$$C(6, 1) = 6 \text{ ways}$$

(b) Here, we must choose two of the six positions to contain 1s. Hence, there are

$$C(6, 2) = 15 \text{ such strings}$$

NOW WORK PROBLEM 19.

EXAMPLE 7 Forming Committees

In how many ways can a committee consisting of 2 faculty members and 3 students be formed if 6 faculty members and 10 students are eligible to serve on the committee?

SOLUTION The problem can be separated into two parts: the number of ways that the faculty members can be chosen, $C(6, 2)$, and the number of ways that the student members can be chosen, $C(10, 3)$. By the Multiplication Principle, the committee can be formed in

$$C(6, 2) \cdot C(10, 3) = \frac{6!}{4!2!} \cdot \frac{10!}{7!3!} = \frac{6 \cdot 5 \cdot \cancel{4!}}{\cancel{4!}2!} \cdot \frac{10 \cdot 9 \cdot 8 \cdot \cancel{7!}}{\cancel{7!}3!}$$

$$= \frac{30}{2} \cdot \frac{720}{6} = 1800 \text{ ways}$$

NOW WORK PROBLEM 27.

EXAMPLE 8 Forming Committees

From 6 women and 4 men a committee of 3 is to be formed. The committee must include at least 2 women. In how many ways can this be done?

SOLUTION A committee of 3 that includes at least 2 women will contain either exactly 2 women and 1 man, or exactly 3 women and 0 men. Since no committee can contain exactly 2 women and simultaneously exactly 3 women, once we have counted the number of ways a committee with exactly 2 women and the number of ways a committee with exactly 3 women can be formed, their sum will give the number of ways exactly 2 women or exactly 3 women can be on the committee. [Refer to the Counting Formula (1) on page 328, noting that the sets are disjoint.]

Following the solution to Example 7, a committee of exactly 2 women and 1 man can be formed from 6 women, 4 men in $C(6, 2) \cdot C(4, 1)$ ways, while a committee of exactly 3 women, 0 men can be formed in $C(6, 3) \cdot C(4, 0)$ ways. A committee of 3 consisting of at least 2 women can be formed in

$$C(6, 2) \cdot C(4, 1) + C(6, 3) \cdot C(4, 0) = \frac{6!}{4!2!} \cdot \frac{4!}{3!1!} + \frac{6!}{3!3!} \cdot \frac{4!}{4!0!}$$

$$= 15 \cdot 4 + 20 \cdot 1 = 60 + 20 = 80 \text{ ways}$$

Permutations Involving *n* Objects That Are Not All Distinct

Solve counting problems involving permutations [*n* objects, not all distinct] **2**

Our previous discussion of permutations required that the objects we were arranging be distinct. We now examine what happens when some of the objects are the same.

EXAMPLE 9 Forming Words

How many three-letter words (real or imaginary) can be formed from the letters in the word

(a) MAD? **(b)** DAD?

SOLUTION **(a)** The three distinct letters in MAD can be rearranged in

$$P(3, 3) = 3! = 6 \text{ ways}$$

They are MAD, MDA, AMD, ADM, DAM, DMA

(b) Straightforward listing shows that there are only three ways of rearranging the letters in the word DAD:

DAD, DDA, and ADD

The word DAD in Example 9 contains 2 Ds, and it is this duplication that results in fewer rearrangements for DAD than for MAD. In the next example we describe a way of dealing with the problem of duplication.

EXAMPLE 10 Forming Words

How many distinct "words" can be formed using all the letters of the six-letter word

M A M M A L?

SOLUTION Any such word will have 6 letters formed from 3 Ms, 2 As, and 1 L. To form a word, think of six blank positions that will have to be filled by the above letters.

$$\underset{1}{\_}\quad\underset{2}{\_}\quad\underset{3}{\_}\quad\underset{4}{\_}\quad\underset{5}{\_}\quad\underset{6}{\_}$$

We separate the construction of a word into three tasks.

Task 1 Choose 3 of the positions for the Ms.
Task 2 Choose 2 of the remaining positions for the As.
Task 3 Choose the remaining position for the L.

Doing this sequence of tasks will result in a word and, conversely, every rearrangement of MAMMAL can be interpreted as resulting from this sequence of tasks.

Task 1 can be done in $C(6, 3)$ ways. There are now three positions left for the 2 As, so Task 2 can be done in $C(3, 2)$ ways. Five blanks have been filled, so that the L must go in the remaining blank. That is, Task 3 can be done in $C(1, 1)$ way. The Multiplication Principle says that the number of rearrangements is

$$C(6,3)\cdot C(3,2)\cdot C(1,1) = \frac{6!}{3!3!}\cdot\frac{3!}{1!2!}\cdot\frac{1!}{0!1!} = \frac{6!}{3!2!1!}$$

The form of the answer in Example 10 is suggestive of a general result. Had the letters in MAMMAL been distinct, there would have been $P(6, 6) = 6!$ rearrangements possible. This is the numerator of the answer. The presence of 3 Ms, 2 As, and 1 L reduces the number of different words as shown in the denominator above. This reasoning can be used to derive the following general result.

Permutation Involving n Objects That Are Not Distinct

The number of permutations of n objects, of which n_1 are of one kind, n_2 are of a second kind, . . . , and n_k are of a kth kind, is given by

$$\boxed{\frac{n!}{n_1!\cdot n_2!\cdot\ldots\cdot n_k!}} \qquad (2)$$

where $n_1 + n_2 + \cdots + n_k = n$.

NOW WORK PROBLEM 23.

EXAMPLE 11 Arranging Flags

How many different vertical arrangements are possible for 10 flags if 2 are white, 3 are red, and 5 are blue?

SOLUTION Here we want the different arrangements of 10 objects, which are not all different. Using formula (2), we have

$$\frac{10!}{2!3!5!} = \frac{10 \cdot 9 \cdot 8 \cdot 7 \cdot \cancel{6} \cdot \cancel{5!}}{2 \cdot \cancel{3} \cdot \cancel{2} \cdot \cancel{5!}} = 2520 \text{ different arrangements}$$

EXERCISE 6.5 Answers to Odd-Numbered Problems Begin on Page AN-28.

In Problems 1–8, find the value of each expression.

1. $C(6, 4)$ 2. $C(5, 4)$ 3. $C(7, 2)$ 4. $C(8, 7)$ 5. $C(5, 1)$ 6. $C(8, 1)$ 7. $C(8, 6)$ 8. $C(8, 4)$

9. List all the combinations of 5 objects a, b, c, d, and e taken 3 at a time. What is $C(5, 3)$?

10. List all the combinations of 5 objects a, b, c, d, and e taken 2 at a time. What is $C(5, 2)$?

11. List all the combinations of 4 objects 1, 2, 3, and 4 taken 3 at a time. What is $C(4, 3)$?

12. List all the combinations of 6 objects 1, 2, 3, 4, 5, and 6 taken 3 at a time. What is $C(6, 3)$?

13. **Establishing Committees** In how many ways can a committee of 4 students be formed from a pool of 7 students?

14. **Establishing Committees** In how many ways can a committee of 3 professors be formed from a department having 8 professors?

15. **Tenure Selection** A math department is allowed to tenure 4 of 17 eligible teachers. In how many ways can the selection for tenure be made?

16. **Bridge Hands** How many different hands are possible in a bridge game? (A bridge hand consists of 13 cards dealt from a deck of 52 cards.)

17. **Forming a Committee** There are 20 students in the Math Club. In how many ways can a subcommittee of 3 members be formed?

18. **Relay Teams** How many different relay teams of 4 persons can be chosen from a group of 10 runners?

19. **Eight-Bit Strings** How many eight-bit strings contain exactly three 1s?

20. **Eight-Bit Strings** How many eight-bit strings contain exactly two 1s?

21. **Choosing Presidents** In January 2003, George W. Bush was president of the United States; he is the 43rd president of the United States. Suppose that a group of historians want to select the five presidencies that had the greatest influence on the history of the United States. In how many ways could a collection of five presidencies be selected?

22. **Lottery Tickets** A state of Maryland million dollar lottery ticket consists of 6 distinct numbers chosen from the range 00 through 99. The order in which the numbers appear on the ticket is irrelevant. How many distinct lottery tickets can be issued?

23. **Forming Words** How many different 9-letter words (real or imaginary) can be formed from the letters in the word ECONOMICS?

24. **Forming Words** How many different 11-letter words (real or imaginary) can be formed from the letters in the word MATHEMATICS?

25. **Arranging Lights** How many different ways can 3 red, 4 yellow, and 5 blue bulbs be arranged in a string of Christmas tree lights with 12 sockets?

26. **Arranging Trees** In how many ways can 3 apple trees, 4 peach trees, and 2 plum trees be arranged along a fence line if one does not distinguish between trees of the same kind?

27. **Forming a Committee** A student dance committee is to be formed consisting of 2 boys and 3 girls. If the membership is to be chosen from 4 boys and 8 girls, how many different committees are possible?

28. **Forming a Committee** The student relations committee of a college consists of 2 administrators, 3 faculty members, and 5 students. Four administrators, 8 faculty members, and

20 students are eligible to serve. How many different committees are possible?

29. Forming Teams In how many ways can 12 children be placed on 3 distinct teams of 3, 5, and 4 members?

30. Forming Committees A group of 9 people is going to be formed into committees of 4, 3, and 2 people. How many committees can be formed if

(a) A person can serve on any number of committees?
(b) No person can serve on more than one committee?

31. Forming Committees The U.S. Senate has 100 members. Suppose it is desired to place each senator on exactly 1 of 7 possible committees. The first committee has 22 members, the second has 13, the third has 10, the fourth has 5, the fifth has 16, and the sixth and seventh have 17 apiece. In how many ways can these committees be formed?

32. Hockey There are 15 teams in the Eastern Conference of the National Hockey League for the 2002–2003 season, divided into 3 divisions of 5 teams each. Eight of these teams will participate in the playoffs. The division winner from each division goes to the playoffs; in addition, the next 5 ranked teams in the conference, based on regular season records, will participate in the playoffs. How many different collections of eight teams from the Eastern Conference could go to the playoffs?

Source: National Hockey League.

33. Baseball For the 2002 baseball season, there were 16 teams in the National League. These teams were divided into 3 divisions as follows: 5 teams in the East Division, 6 teams in the Central Division, and 5 teams in the West Division. Four of these teams participate in the playoffs. The division winner from each division goes to the playoffs; in addition, the remaining team with the best record participates in the playoffs as the wild card team. How many different collections of four teams from the National League could go to the playoffs?

Source: Major League Baseball.

34. Baseball For the 2002 baseball season, there were 14 teams in the American League. These teams were divided into 3 divisions as follows: 5 teams in the East Division, 5 teams in the Central Division, and 4 teams in the West Division. Four of these teams participate in the playoffs. The division winner from each division goes to the playoffs; in addition, the remaining team with the best record participates in the playoffs as the wild card team. How many different collections of four teams from the American League could go to the playoffs?

Source: Major League Baseball.

In Problems 35–43, use the multiplication principle, permutations, or combinations, as appropriate.

35. Test Panel Selection A sample of 8 persons is selected for a test from a group containing 40 smokers and 15 nonsmokers. In how many ways can the 8 persons be selected?

36. Resource Allocation A trucking company has 8 trucks and 6 drivers available when requests for 4 trucks are received. How many different ways are there of selecting the trucks and the drivers to meet these requests?

37. Group Selection From a group of 5 people we are required to select a different person to participate in each of 3 different tests. In how many ways can the selections be made?

38. Congressional Committees In the U.S. Congress a conference committee is to be composed of 5 senators and 4 representatives. In how many ways can this be done? (There are 435 representatives and 100 senators.)

39. Quality Control A box contains 24 light bulbs. The quality control engineer will pick a sample of 4 light bulbs for inspection. How many different samples are there?

40. Investment Selection An investor is going to invest \$21,000 in 3 stocks from a list of 12 prepared by his broker. How many different investments are possible if

(a) \$7000 is to be invested in each stock?
(b) \$10,000 is to be invested in one stock, \$6000 in another, and \$5000 in the third?
(c) \$8000 is to be invested in each of 2 stocks and \$5000 in a third stock?

41. Rating A sportswriter makes a preseason guess of the top 15 university basketball teams (in order) from among 50 major university teams. How many different possibilities are there?

42. Packaging A manufacturer produces 8 different items. He packages assortments of equal parts of 3 different items. How many different assortments can be packaged?

43. The digits 0 through 9 are written on 10 cards. Four different cards are drawn, and a 4-digit number is formed. How many different 4-digit numbers can be formed in this way?

6.6 The Binomial Theorem

OBJECTIVE 1 Expand a binomial

We begin with the famous Pascal triangle.

Pascal's Triangle

Sometimes the notation $\binom{n}{r}$ read as "from n choose r," is used in place of $C(n, r)$. $\binom{n}{r}$ is called a **binomial coefficient** because of its connection with the binomial theorem. A triangular display of $\binom{n}{r}$ for $n = 0$ to $n = 6$ is given in Figure 22. This triangular display is called **Pascal's triangle.**

For example, $\binom{5}{2} = 10$ is found in the row marked $n = 5$ and on the diagonal marked $r = 2$.

In the Pascal triangle, successive entries can be obtained by adding the two nearest entries in the row above it. The shaded triangles in Figure 22 illustrate this. For example, $1 + 3 = 4$, $10 + 5 = 15$, and $1 + 5 = 6$.

FIGURE 22

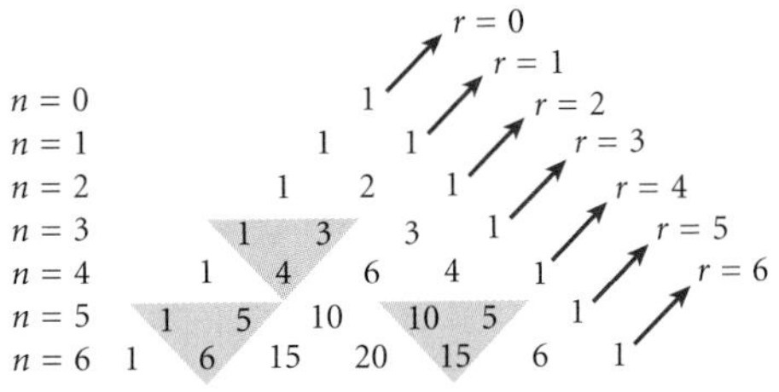

The Pascal triangle, as the figure indicates, is symmetric. When n is even, the largest entry occurs in the middle, and corresponding entries on either side are equal. When n is odd, there are two equal middle entries with corresponding equal entries on either side.

The reasons behind these properties of Pascal's triangle as well as other properties of binomial coefficients are discussed later in this section.

The Binomial Theorem

The *binomial theorem* deals with the problem of expanding an expression of the form $(x + y)^n$, where n is a positive integer.

Expressions such as $(x + y)^2$ and $(x + y)^3$ are not too difficult to expand. For example,

$$(x + y)^2 = x^2 + 2xy + y^2$$
$$(x + y)^3 = (x + y)^2(x + y) = (x^2 + 2xy + y^2)(x + y) = x^3 + 3x^2y + 3xy^2 + y^3$$

However, expanding expressions such as $(x + y)^6$ or $(x + y)^8$ by the normal process of multiplication would be tedious and time consuming. It is here that the binomial theorem is especially useful.

Recall that

$$C(n, r) = \binom{n}{r} = \frac{n!}{(n - r)!\, r!}$$

Since $\binom{2}{0} = 1$, $\binom{2}{1} = 2$, and $\binom{2}{2} = 1$, we can write the expression

$$(x + y)^2 = x^2 + 2xy + y^2$$

in the form

$$(x + y)^2 = \binom{2}{0}x^2 + \binom{2}{1}xy + \binom{2}{2}y^2$$

Since $\binom{3}{0} = 1$, $\binom{3}{1} = 3$, $\binom{3}{2} = 3$, and $\binom{3}{3} = 1$, the expansion of $(x + y)^3$ can be written as

$$(x + y)^3 = x^3 + 3x^2y + 3xy^2 + y^3 = \binom{3}{0}x^3 + \binom{3}{1}x^2y + \binom{3}{2}xy^2 + \binom{3}{3}y^3$$

Similarly, the expansion of $(x + y)^4$ can be written as

$$(x + y)^4 = x^4 + 4x^3y + 6x^2y^2 + 4xy^3 + y^4$$
$$= \binom{4}{0}x^4 + \binom{4}{1}x^3y + \binom{4}{2}x^2y^2 + \binom{4}{3}xy^3 + \binom{4}{4}y^4$$

The binomial theorem generalizes this pattern.

Binomial Theorem

If n is a positive integer,

$$(x + y)^n = \binom{n}{0}x^n + \binom{n}{1}x^{n-1}y + \binom{n}{2}x^{n-2}y^2 + \cdots + \binom{n}{k}x^{n-k}y^k + \cdots + \binom{n}{n}y^n \quad (1)$$

Observe that the powers of x begin at n and decrease by 1, while the powers of y begin with 0 and increase by 1. Also, the coefficient of the term involving y^k is always $\binom{n}{k}$.

Let's get some practice using the binomial theorem.

1 **EXAMPLE 1 Expanding a Binomial**

Expand $(x + y)^6$ using the binomial theorem.

SOLUTION

$$(x + y)^6 = \binom{6}{0}x^6 + \binom{6}{1}x^5y + \binom{6}{2}x^4y^2 + \binom{6}{3}x^3y^3$$
$$+ \binom{6}{4}x^2y^4 + \binom{6}{5}xy^5 + \binom{6}{6}y^6$$
$$= x^6 + 6x^5y + 15x^4y^2 + 20x^3y^3 + 15x^2y^4 + 6xy^5 + y^6$$

Note that the coefficients in the expansion of $(x + y)^6$ are the entries in the Pascal triangle for $n = 6$. (See Figure 22, page 351.)

NOW WORK PROBLEM 1.

EXAMPLE 2 Finding a Particular Coefficient

Find the coefficient of x^3y^4 in the expansion of $(x + y)^7$.

SOLUTION The expansion of $(x + y)^7$ is

$$(x + y)^7 = \binom{7}{0}x^7 + \binom{7}{1}x^6y + \binom{7}{2}x^5y^2 + \binom{7}{3}x^4y^3 + \binom{7}{4}x^3y^4 + \binom{7}{5}x^2y^5 + \binom{7}{6}xy^6 + \binom{7}{7}y^7$$

The coefficient of x^3y^4 is

$$\binom{7}{4} = \frac{7 \cdot 6 \cdot 5}{3 \cdot 2 \cdot 1} = 35$$

NOW WORK PROBLEM 7.

EXAMPLE 3 Expanding a Binomial

Expand $(x + 2y)^4$ using the binomial theorem.

SOLUTION Here, we let "$2y$" play the role of "y" in the binomial theorem. We then get

$$(x + 2y)^4 = \binom{4}{0}x^4 + \binom{4}{1}x^3(2y) + \binom{4}{2}x^2(2y)^2 + \binom{4}{3}x(2y)^3 + \binom{4}{4}(2y)^4$$
$$= x^4 + 8x^3y + 24x^2y^2 + 32xy^3 + 16y^4$$

NOW WORK PROBLEM 3.

To explain why the binomial theorem is true, we take a close look at what happens when we compute $(x + y)^3$. Think of $(x + y)^3$ as the product of three factors, namely,

$$(x + y)^3 = (x + y)(x + y)(x + y)$$

Were we to multiply these three factors together without any attempt at simplification or collecting of terms, we would get

$$\underset{\text{Factor 1}}{(x + y)}\underset{\text{Factor 2}}{(x + y)}\underset{\text{Factor 3}}{(x + y)} = xxx + xyx + yxx + yyx + xxy + xyy + yxy + yyy$$

Notice that the terms on the right yield all possible products that can be formed by picking either an x or y from each of the three factors on the left. For example,

xyx results from choosing an x from factor 1, a y from factor 2, and an x from factor 3

Now, the number of terms on the right that will simplify to, say, xy^2, will be those terms that resulted from choosing y's from two of the factors and an x from the remaining factor. How many such terms are there? There are as many as there are ways of choosing two of the three factors to contribute y's—that is, there are $C(3, 2) = \binom{3}{2}$ such terms. This is why the coefficient of xy^2 in the expansion of $(x + y)^3$ is $\binom{3}{2}$.

In general, if we think of $(x + y)^n$ as the product of n factors,

$$(x + y)^n = \underbrace{(x + y) \cdot (x + y) \cdot \, . \; . \; . \, \cdot (x + y)}_{n \text{ factors}}$$

then, upon multiplying out and simplifying, there will be as many terms of the form $x^{n-k} y^k$ as there are ways of choosing k of the n factors to contribute y's (and the remaining $n - k$ to contribute x's). There are $C(n, k)$ ways of making this choice. So the coefficient of $x^{n-k} y^k$ is $\binom{n}{k}$, and this is the assertion of the binomial theorem.

Binomial Identities

Binomial coefficients have some interesting properties. For example, it turns out that $\binom{n}{k}$ and $\binom{n}{n-k}$ are equal. We can explain this equality as follows: Suppose we wanted to pick a team of k players from n people. Then choosing those k who will play is the same as choosing those $n - k$ who will not. So the number of ways of choosing the players equals the number of ways of choosing the nonplayers. The players can be chosen in $C(n, k) = \binom{n}{k}$ ways, while those to be left out can be chosen in $\binom{n}{n-k}$ ways, and the equality follows.

EXAMPLE 4 Proving a Property of Binomial Coefficients

Show that

$$\binom{n}{k} = \binom{n}{n - k}$$

SOLUTION By definition,

$$\binom{n}{k} = \frac{n!}{(n - k)!k!}$$

while

$$\binom{n}{n - k} = \frac{n!}{(n - k)![n - (n - k)]!}$$

Since $n - (n - k) = k$, a comparison of the expressions shows that they are equal. ◗

Based on Example 4, we have

$$\binom{5}{3} = \binom{5}{2} \qquad \binom{10}{2} = \binom{10}{8}$$

and so on. This identity accounts for the symmetry in the rows of Pascal's triangle.

EXAMPLE 5 Proving a Property of Binomial Coefficients

Show that

$$\binom{n}{k} = \binom{n - 1}{k} + \binom{n - 1}{k - 1}$$

SOLUTION We could expand both sides of the above identity using the definition of binomial coefficients and, after some algebra, demonstrate the equality. But we choose the route of posing a problem that we solve two different ways. Equating the two solutions will prove the identity.

A committee of k is to be chosen from n people. The total number of ways this can be done is $C(n, k) = \binom{n}{k}$.

We now count the total number of committees a different way. Assume that one of the n people is Jennifer. We compute

(1) those committees not containing Jennifer

and

(2) those committees containing Jennifer

The number of committees of type 1 is $\binom{n-1}{k}$ since the k people must be chosen from the $n - 1$ people who are not Jennifer. The number of committees of type 2 will correspond to the number of ways we can choose the $k - 1$ people other than Jennifer to be on the committee. So the number of committees of type 2 is given by $\binom{n-1}{k-1}$. Since the number of committees of type 1 plus the number of committees of type 2 equals the total number of committees, our identity follows.

For example, $\binom{8}{5} = \binom{7}{5} + \binom{7}{4}$. It is precisely this identity that explains the reason why an entry in Pascal's triangle can be obtained by adding the nearest two entries in the row above it. Due to its recursive character, the identity in Example 5 is sometimes used in computer programs that evaluate binomial coefficients.

NOW WORK PROBLEM 19.

EXAMPLE 6 Establishing a Relationship of Specific Binomial Coefficients

Show that

$$\binom{6}{3} = \binom{2}{2} + \binom{3}{2} + \binom{4}{2} + \binom{5}{2}$$

SOLUTION We make repeated use of the identity in Example 5. So

$$\begin{aligned}
\binom{6}{3} &= \binom{5}{3} + \binom{5}{2} \\
&= \left[\binom{4}{3} + \binom{4}{2}\right] + \binom{5}{2} && \text{Apply the identity to } \binom{5}{3}. \\
&= \left[\binom{3}{3} + \binom{3}{2}\right] + \binom{4}{2} + \binom{5}{2} && \text{Apply the identity to } \binom{4}{3}. \\
&= \binom{2}{2} + \binom{3}{2} + \binom{4}{2} + \binom{5}{2} && \text{Since } \binom{2}{2} = \binom{3}{3} = 1
\end{aligned}$$

EXAMPLE 7 Proving a Property of Binomial Coefficients

Show that

$$\binom{n}{0} + \binom{n}{1} + \binom{n}{2} + \cdots + \binom{n}{n} = 2^n$$

SOLUTION We make use of the binomial theorem. Since the binomial theorem is valid for all x and y, we may set $x = y = 1$ in Equation (1). This gives

$$2^n = (1+1)^n = \binom{n}{0} + \binom{n}{1} + \binom{n}{2} + \cdots + \binom{n}{n}$$

This shows, for example, that the sum of the elements in the row marked $n = 6$ of Pascal's triangle is $2^6 = 64$. The result in Example 7 can be used to find the number of subsets of a set with n elements. $\binom{n}{0}$ gives the number of subsets with 0 elements; $\binom{n}{1}$ the number of subsets with 1 element; $\binom{n}{2}$ the number of subsets with 2 elements; and so on. The sum $\binom{n}{0} + \binom{n}{1} + \cdots + \binom{n}{n}$ is the total number of subsets of a set with n elements. The result in Example 7 can then be rephrased as follows:

A set with n elements has 2^n subsets.

For example, a set with 5 elements has $2^5 = 32$ subsets.

NOW WORK PROBLEM 11.

EXAMPLE 8 Proving a Property of Binomial Coefficients

Show that

$$\binom{n}{0} - \binom{n}{1} + \binom{n}{2} - \cdots + (-1)^n\binom{n}{n} = 0$$

(The last term will be preceded by a plus or minus sign depending on whether n is even or odd.)

SOLUTION We again make use of the binomial theorem. This time we let $x = 1$ and $y = -1$ in Equation (1). This produces

$$\begin{aligned}(1-1)^n &= \binom{n}{0} + \binom{n}{1}(-1) + \binom{n}{2}(-1)^2 + \binom{n}{3}(-1)^3 \\ &\quad + \cdots + \binom{n}{n}(-1)^n \\ 0 &= \binom{n}{0} - \binom{n}{1} + \binom{n}{2} - \cdots + (-1)^n\binom{n}{n}\end{aligned}$$

We mention an interpretation of this identity by examining the instance where $n = 5$. The identity gives

$$\binom{5}{0} - \binom{5}{1} + \binom{5}{2} - \binom{5}{3} + \binom{5}{4} - \binom{5}{5} = 0$$

Rearranging some terms yields

$$\binom{5}{0} + \binom{5}{2} + \binom{5}{4} = \binom{5}{1} + \binom{5}{3} + \binom{5}{5}$$

This says that a set with 5 elements has as many subsets containing an even number of elements as it has subsets containing an odd number of elements.

EXERCISE 6.6 Answers to Odd-Numbered Problems Begin on Page AN-28.

In Problems 1–6, use the binomial theorem to expand each expression.

1. $(x + y)^5$
2. $(x + y)^4$
3. $(x + 3y)^3$
4. $(2x + y)^3$
5. $(2x - y)^4$
6. $(x - y)^4$

7. What is the coefficient of x^2y^3 in the expansion of $(x + y)^5$?
8. What is the coefficient of x^2y^6 in the expansion of $(x + y)^8$?
9. What is the coefficient of x^8 in the expansion of $(x + 3)^{10}$?
10. What is the coefficient of x^3 in the expansion of $(x + 2)^5$?
11. How many different subsets can be chosen from a set with 8 elements?
12. How many different subsets can be chosen from a set with 50 elements?
13. How many nonempty subsets does a set with 10 elements have?
14. How many subsets with an even number of elements does a set with 10 elements have?
15. How many subsets with an odd number of elements does a set with 10 elements have?
16. Show that
$$\binom{8}{5} = \binom{4}{4} + \binom{5}{4} + \binom{6}{4} + \binom{7}{4}$$
17. Show that
$$\binom{10}{7} = \binom{6}{6} + \binom{7}{6} + \binom{8}{6} + \binom{9}{6}$$
18. Show that
$$\binom{7}{1} + \binom{7}{3} + \binom{7}{5} + \binom{7}{7} = 2^6$$
19. Replace $\binom{11}{6} + \binom{11}{5}$ by a single binomial coefficient.
20. Replace $\binom{8}{8} + \binom{9}{8} + \binom{10}{8}$ by a single binomial coefficient.
21. Show that
$$k\binom{n}{k} = n\binom{n-1}{k-1}$$

Chapter 6 Review

OBJECTIVES

Section		You should be able to	Review Exercises
6.1	1	Identify relations between pairs of sets	1–16
	2	Find the union and intersection of two sets	3, 4, 6, 9, 11–13, 15–18, 23–28
	3	Find the complement of a set	18 a, c–f, 24, 26, 28
	4	Use Venn diagrams	19–22
6.2	1	Use the counting formula	29–34
	2	Use Venn diagrams to analyze survey data	35, 36
6.3	1	Use the Multiplication Principle	63, 64, 68b, 71–75, 80
6.4	1	Evaluate factorials	39–58
	2	Solve counting problems involving permutations [distinct, with repetition]	65, 70a, 80, 82
	3	Solve counting problems involving permutations [distinct, without repetition]	59–61, 60–62, 66a, 67a, 68, 70b, 71, 72, 77–79
6.5	1	Solve counting problems involving combinations	58, 66b, 67b, 69, 70c, 73–76, 81
	2	Solve counting problems involving permutations [*n* objects, not all distinct]	83–86
6.6	1	Expand a binomial	87–90

THINGS TO KNOW

Set (p. 317)		Well-defined collection of distinct objects, called elements
Empty set (p. 318)	$\varnothing$	Set that has no elements
Equality (p. 318)	$A = B$	A and B have the same elements.
Subset (p. 318)	$A \subseteq B$	Each element of A is an element of B.
Proper Subset (p. 319)	$A \subset B$	Each element of A is an element of B. but there is at least one element in B not in A.
Universal set (p. 320)	U	Set consisting of all the elements that we wish to consider
Union (p. 320)	$A \cup B$	Set consisting of elements that belong to either A or B, or both
Intersection (p. 321)	$A \cap B$	Set consisting of elements that belong to both A and B
Complement (p. 322)	$\overline{A}$	Set consisting of elements of the universal set that are not in A
Finite set (p. 327)		The number of elements in the set is a nonnegative integer.
Infinite set (p. 327)		A set that is not finite
Counting Formula (p. 327)		$c(A \cup B) = c(A) + c(B) - c(A \cap B)$
Multiplication Principle (p. 333)		If a task consists of a sequence of choices in which there are p selections for the first choice, q selections for the second choice, and so on, then the task of making these selections can be done in $p \cdot q \cdot \ldots$ different ways.
Factorial (p. 337)	$n! = n(n-1)\ldots(3)(2)(1)$	The product of the first n positive integers
Permutation (p. 338)		An ordered arrangement of r objects chosen from n objects.
(p. 338)	n^r	The n objects are distinct (different) and repetition is allowed in the selection of r of them.
(p. 340)	$P(n, r) = n(n-1) \cdot \ldots \cdot [n-(r-1)] = \dfrac{n!}{(n-r)!}$	The n objects are distinct (different) and repetition is not allowed in the selection of r of them, where $r \leq n$.
(p. 348)	$\dfrac{n!}{n_1!n_2!\cdots n_k!}$	The number of permutations of n objects of which n_1 are of one kind, n_2 are of a second kind, ..., and n_k are of a kth kind, where $n = n_1 + n_2 + \ldots + n_k$
Combination (pp. 344–345)	$C(n, r) = \dfrac{P(n, r)}{r!} = \dfrac{n!}{(n-r)!r!}$	An arrangement, without regard to order, of r objects selected from n distinct objects without repetition, where $r \leq n$.

Binomial Theorem (p. 352)

$$(x+y)^n = \binom{n}{0}x^n + 1\binom{n}{1}x^{n-1}y + \binom{n}{2}x^{n-2}y^2 + \cdots + \binom{n}{k}x^{n-k}y^k + \cdots + \binom{n}{n}y^n$$

TRUE–FALSE ITEMS **Answers are on page AN-29.**

T F **1.** If $A \cup B = A \cap B$, then $A = B$.

T F **2.** If A and B are disjoint sets, then $c(A \cup B) = c(A) + c(B)$.

T F **3.** The number of permutations of 4 different objects taken 4 at a time is 12.

T F **4.** $C(5, 3) = 20$

T F **5.** $5! = 120$

T F **6.** $\frac{7!}{6!} = \frac{7}{6}$

T F **7.** In the binomial expansion of $(x+1)^7$, the coefficient of x^4 is 4.

FILL IN THE BLANKS Answers are on page AN-29.

1. Two sets that have no elements in common are called _____________.

2. The number of different selections of r objects from n objects in which (a) the n objects are different, (b) no object is repeated more than once in a selection, and (c) order is important is called a _____________.

3. If in 2 above, condition (c) is replaced by "order is not important," we have a _____________.

4. A triangular display of the number of combinations is called the _____________ triangle.

5. The numbers $\binom{n}{r}$ are sometimes called _____________ _____________.

6. To expand an expression such as $(x + y)^n$, n a positive integer, we can use the _____________ _____________.

7. The coefficient of x^3 in the expansion of $(x + 2)^5$ is _____________.

REVIEW EXERCISES Answers to odd-numbered problems begin on page AN-29.

Blue problem numbers indicate the author's suggestions for a practice test.

In Problems 1–16, replace the blank by any of the symbol(s) $\subset$, $\subseteq$, $=$ that result in a true statement. If none result in a true statement, write "None of these." More than one answer may be possible.

1. $\varnothing$ ____ $\{0\}$
2. $\{0\}$ ____ $\{1, 0, 3\}$
3. $\{5, 6\} \cap \{2, 6\}$ ____ $\{8\}$
4. $\{3\}$ ____ $\{2, 3\} \cup \{3, 4\}$
5. $\{8, 9\}$ ____ $\{9, 10, 11\}$
6. $\{1\}$ ____ $\{1, 3, 5\} \cup \{3, 4\}$
7. $\{5\}$ ____ $\{0, 5\}$
8. $\varnothing$ ____ $\{1, 2, 3\}$
9. $\varnothing$ ____ $\{1, 2\} \cap \{3, 4, 5\}$
10. $\{2, 3\}$ ____ $\{3, 4\}$
11. $\{1, 2\}$ ____ $\{1\} \cup \{2\}$
12. $\{5\}$ ____ $\{1\} \cup \{2, 3\}$
13. $\{4, 5\} \cap \{5, 6\}$ ____ $\{4, 5\}$
14. $\{6, 8\}$ ____ $\{8, 9, 10\}$
15. $\{6, 7, 8\} \cap \{6\}$ ____ $\{6\}$
16. $\{4\}$ ____ $\{6, 8\} \cap \{4, 8\}$

---

17. For the sets

$$A = \{1, 3, 5, 6, 8\} \quad B = \{2, 3, 6, 7\} \quad C = \{6, 8, 9\}$$

find

(a) $(A \cap B) \cup C$ (b) $(A \cap B) \cap C$
(c) $(A \cup B) \cap B$ (d) $B \cup \varnothing$
(e) $A \cap \varnothing$ (f) $(A \cup B) \cup C$

18. For the sets U = universal set $= \{1, 2, 3, 4, 5, 6, 7\}$ and

$$A = \{1, 3, 5, 6\} \quad B = \{2, 3, 6, 7\} \quad C = \{4, 6, 7\}$$

find:

(a) $\overline{A \cap B}$ (b) $(B \cap C) \cap A$
(c) $\overline{B} \cup \overline{A}$ (d) $C \cup \overline{C}$
(e) $\overline{\overline{B}}$ (f) $A \cap \overline{B}$

19. Use a Venn diagram to illustrate the following sets:

(a) $A \cup \overline{B}$ (b) $(A \cap B) \cup \overline{B}$
(c) $B \cap \overline{A}$ (d) $(A \cup B) \cap C$
(e) $(A \cap B) \cap C$ (f) $(\overline{\overline{B} \cup C})$

20. Use Venn diagrams to illustrate the following properties:

(a) $(A \cap B) \cap C = A \cap (B \cap C)$
(b) $\overline{A \cap B} = \overline{A} \cup \overline{B}$
(c) $A \cup (B \cap C) = (A \cup B) \cap (A \cup C)$
(d) $\overline{\overline{A}} = A$

21. Draw a Venn diagram that illustrates $A \cap B = \varnothing$.

22. Draw a Venn diagram that illustrates $A \subset B$.

In Problems 23–28, use

U = Universal set $= \{x | x$ is a state in the United States$\}$
$A = \{a | a$ is a state whose name begins with the letter $A\}$
$V = \{v | v$ is a state whose name ends with a vowel$\}$
$E = \{e | e$ is a state that lies east of the Mississippi River$\}$

to describe each set in words.

23. $A \cup V$
24. $A \cap \overline{V}$
25. $V \cap E$
26. $\overline{E}$
27. $(A \cup V) \cap E$
28. $\overline{A \cup V}$

29. If A and B are sets and if $c(A) = 24$, $c(A \cup B) = 33$, and $c(B) = 12$, find $c(A \cap B)$.

30. If $c(A) = 10$, $c(B) = 8$, and $c(A \cap B) = 2$, find $c(A \cup B)$.

31. (a) If $c(A \cap B) = 0$, $c(A) = 3$, and $c(B) = 17$, find $c(A \cup B)$.
(b) What can you conclude about the relation between A and B?

32. (a) If $c(A) = 14$, $c(B) = 8$, and $c(A \cap B) = 8$, find $c(A \cup B)$.
(b) What can you conclude about the relation between A and B?

33. Suppose a die is tossed. Let A denote that the outcome is an even number; let B denote that the outcome is a number less than or equal to 3. How many elements are in $A \cap B$? How many elements are in $A \cup B$?

34. Suppose a die is tossed. Let A denote the outcome is an even number; let B denote that the outcome is divisible by 3. How many elements are in $A \cap B$? How many elements are in $A \cup B$?

35. **Car Options** During June, Colleen's Motors sold 75 cars with air conditioning, 95 with power steering, and 100 with automatic transmissions. Twenty cars had all three options. 10 cars had none of these options, and 10 cars were sold that had only air conditioning. In addition, 50 cars had both automatic transmissions and power steering, and 60 cars had both automatic transmissions and air conditioning.

(a) How many cars were sold in June?
(b) How many cars had only power steering?

36. **Student Survey** In a survey of 125 college students, it was found that of three newspapers, the *Wall Street Journal, New York Times,* and *Chicago Tribune:*

60	read the *Chicago Tribune*
40	read the *New York Times*
15	read the *Wall Street Journal*
25	read the *Chicago Tribune* and *New York Times*
8	read the *New York Times* and *Wall Street Journal*
3	read the *Chicago Tribune* and *Wall Street Journal*
1	read all three

(a) How many read none of these papers?
(b) How many read only the *Chicago Tribune*?
(c) How many read neither the *Chicago Tribune* nor the *New York Times*?

37. If U = universal set = $\{1, 2, 3, 4, 5\}$ and $B = \{1, 4, 5\}$, find all sets A for which $A \cap B = \{1\}$.

38. If U = universal set = $\{1, 2, 3, 4, 5\}$ and $B = \{1, 4, 5\}$, find all sets A for which $A \cup B = \{1, 2, 4, 5\}$.

In Problem 39–58, evaluate each expression.

39. $0!$

40. $(8-3)!$

41. $\frac{7!}{4!}$

42. $\frac{10!}{2!\,8!}$

43. $\frac{12!}{11!}$

44. $\frac{6!}{(6-3)!\,3!}$

45. $P(5, 2)$

46. $P(6, 6)$

47. $P(12, 1)$

48. $P(7, 6)$

49. $P(100, 2)$

50. $P(75, 3)$

51. $C(10, 2)$

52. $C(10, 8)$

53. $C(6, 6)$

54. $C(8, 0)$

55. $\binom{7}{4}$

56. $\binom{5}{5}$

57. $\binom{9}{1}$

58. $\binom{9}{8}$

59. In how many different ways can a committee of 3 people be formed from a group of 5 people?

60. In how many different ways can 4 people line up?

61. In how many different ways can 3 books be placed on a shelf?

62. In how many different ways can 3 people be seated in 4 chairs?

63. How many house styles are possible if a contractor offers 3 choices of roof designs, 4 choices of window designs, and 6 choices of brick?

64. In a cafeteria the \$6.95 lunch menu lets you choose 1 salad, 1 entree, 1 dessert, and 1 beverage. If today's menu features 3 different salads, 5 different entrees, 6 different desserts, and 10 different beverages, how many distinct meals could one order?

65. How many different answers are possible in a true-false test consisting of 10 questions?

66. You are to set up a code of 2-digit words using the digits 1, 2, 3, 4 without using any digit more than once.

(a) What is the maximum number of words in such a language?
(b) If all words of the form *ab* and *ba* are the same, how many words are possible?

67. You are to set up a code of 3-digit words using the digits 1, 2, 3, 4, 5, 6 without using any digit more than once in the same word.

(a) What is the maximum number of words in such a language?
(b) If the words 124, 142, etc., designate the same word, how many different words are possible?

68. **Program Selection** A ceremony is to include 7 speeches and 6 musical selections.

(a) How many programs are possible?
(b) How many programs are possible if speeches and musical selections are to be alternated?

69. **Forming Committees** There are 7 boys and 6 girls willing to serve on a committee. How many 7-member committees are possible if a committee is to contain:

(a) 3 boys and 4 girls?
(b) At least one member of each sex?

70. **Choosing Double-Dip Cones** Juan's Ice Cream Parlor offers 31 different flavors to choose from and specializes in double-dip cones.

(a) How many different cones are there to choose from if you may select the same flavor for each dip?
(b) How many different cones are there to choose from if you cannot repeat any flavor? Assume that a cone with vanilla on top of chocolate is different from a cone with chocolate on top of vanilla.
(c) How many different cones are there if you consider any cone having chocolate on top and vanilla on the bottom the same as having vanilla on top and chocolate on the bottom?

71. **Arranging Books** A person has 4 history, 5 English, and 6 mathematics books. How many ways can the books be arranged on a shelf if books on the same subject must be together?

72. **Choosing Names** A newborn child can be given 1, 2, or 3 names. In how many ways can a child be named if we can choose from 100 names?

73. **Forming Committees** In how many ways can a committee of 8 boys and 5 girls be formed if there are 10 boys and 11 girls eligible to serve on the committee?

74. **Football Teams** A football squad has 7 linemen, 11 linebackers, and 9 safeties. How many different teams composed of 5 linemen, 3 linebackers, and 3 safeties can be formed?

75. In how many ways can we choose three words, one each from five 3-letter words, six 4-letter words, and eight 5-letter words?

76. In how many ways can 5 girls and 3 boys be divided into 2 teams of 4 if each team is to include at least 1 boy?

77. **Program Selection** A meeting is to be addressed by 5 speakers, A, B, C, D, and E. In how many ways can the speakers be ordered if B must come first?

78. **Program Selection** A meeting is to be addressed by 5 speakers, A, B, C, D, E. In how many ways can the speakers be ordered if B must not precede A?

79. **Program Selection** A meeting is to be addressed by 5 speakers, A, B, C, D, E. In how many ways can the speakers be ordered if B is to speak immediately after A?

80. **License Plate Numbers** An automobile license number contains 1 or 2 letters followed by a 4-digit number. Compute the maximum number of different licenses.

81. There are 5 rotten plums in a crate of 25 plums. How many samples of 4 of the 25 plums contain

(a) Only good plums?
(b) Three good plums and 1 rotten plum?
(c) One or more rotten plums?

82. An admissions test given by a university contains 10 true–false questions. Eight or more of the questions must be answered correctly in order to be admitted.

(a) How many different ways can the answer sheet be filled out?
(b) How many different ways can the answer sheet be filled out so that 8 or more questions are answered correctly?

83. How many 6-letter words (real or imaginary) can be made from the word FINITE?

84. How many 7-letter words (real or imaginary) can be made from the word MESSAGE?

85. **Arranging Books** Jessica has 10 books to arrange on a shelf. How many different arrangements are possible if she has 3 identical copies of *Harry Potter and the Sorcerer's Stone* and 2 identical copies of *Finite Mathematics*?

86. **Arranging Pennants** Mike has 9 pennants to arrange on a pole. How many different arrangements are possible if 4 pennants are blue, 2 are yellow, and 3 are green?

87. Expand $(x + 2)^4$.

88. Expand $(x - 1)^5$.

89. What is the coefficient of x^3 in the expansion of $(x + 2)^7$?

90. What is the coefficient of x^4 in the expansion of $(2x + 1)^6$?

Chapter 6 Project

IDENTIFICATION NUMBERS: HOW MANY ARE ENOUGH?

If you look carefully around you, you will see batches of numbers and letters on almost everything: on products in every store, on books, and certainly on the credit cards in your wallet. Even though they may contain letters of the alphabet, these numbers are called **identification numbers:** They serve to identify the product, the book, or the person who owns the credit card.

Producing identification numbers is a lot more difficult than just deciding to print numbers on products or books. Every system of identification numbers has a plan that tells exactly how to assign those numbers. These plans must confront several issues. First, there must be enough numbers for all of the things that you expect to number. Since giving the same identification number to different items would defeat the purpose of having an identification number, it is important that the plan have a way to give a unique number to each item. It is also necessary to be able to check for errors in the identification number. This check helps to prevent misidentifying items. As a result, most identification number plans include a number, called a **check digit,** for this purpose. We consider two examples of identification numbers that you have probably already seen many times.

The Universal Product Code: The Universal Product Code, or UPC, appears on just about every item you buy from a major store. It is designed to help the store track inventory and to make price changes easier. Figure 1 shows three examples: the first is from a box of Bigelow tea, the second is from a box of Xerox toner cartridges, and the third is from the front page of a newspaper.

There are two parts to the code: a bar code and a UPC number. Notice that the UPC number has 12 digits. The first 6 digits identify the producer and the next 5 digits identify the exact product. The final digit is the check digit.

FIGURE 1 Universal Product Codes

0 72310 00192 3

0 95205 60881 6
UPC

7 68663 30909 9

Of the 6 digits identifying the producer, the first digit is particularly important. This digit, which appears at the far left of the code and is called the **number system character,** tells the type of product. Table 1 lists the number system characters currently in use. The numbers 1, 8, and 9 have been reserved for future use. The 5 digits that identify the product can be any digits whatsoever. Once the first 11 digits are known, the check digit is calculated from them.

TABLE 1 UPC NUMBER SYSTEM CHARACTERS

0	Standard UPC number (for any type of product)
2	Random-weight items (produce, meats, etc.)
3	Pharmaceuticals
4	In-store marketing (gift cards, store-specific coupons)
5	Manufacturer's coupons
6	Standard UPC number (for any type of product)
7	Standard UPC number (for any type of product)

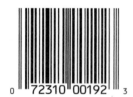

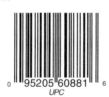

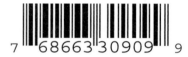

1. Under the current system, how many different producer numbers are possible?
2. If all digits were allowed for the number system character, how many more producer numbers would be created?
3. How many possible correct UPC codes are there? Assume all possible producer numbers are allowed.

The International Standard Book Number: The International Standard Book Number, or ISBN, is used to identify works published anywhere in the world. The number is used to track and identify books quickly. Table 2 shows some examples.

TABLE 2 ISBN NUMBERS FOR VARIOUS BOOKS

A Beautiful Mind by Sylvia Nasar	0-684-81906-6
Theoretical Probability for Applications by Sidney C. Port	0-471-63216-3
Biblia Hebraica Stuttgartensia	3-438-05218-0
Gesammelte Werke in Einzelausgaben by Arthur Schnitzler	3-596-21961-2
Le Véritable Amour by Jacqueline Harpman	2-930311-10-X
Understanding Search Engines by Michael Berry and Murray Browne	0-89871-437-0
Authorizing an End by Donald C. Polaski	90-04-11607-9
Solstices by Ananda Devi	99903-23-34-8
MYSTERY BOOK by MYSTERY AUTHOR	0-471-32899-5

Notice that each ISBN number has 10 digits, which are divided into 4 blocks with 3 hyphens. The final block always contains just 1 digit: the check digit. You might notice in passing that the check digit can be an X—that is a story for another day. (See Problem 8.) The other three blocks can contain anywhere from 1 to 7 numbers, but the total number of digits in the three blocks must always be 9. Only digits are used for these groups; no letters are allowed.

The first block tells the language or country in which the book was published. Books from English-speaking areas are assigned the numbers 0 or 1, those from French-speaking areas are assigned 2, those from German-speaking areas are assigned 3, and so on. Since a smaller country will likely produce fewer books, these countries get longer country codes. For example, the code for the Netherlands is 90, and the code for Mauritius (an island nation in the Indian Ocean) is 99903. The second block identifies the publisher. For example, John Wiley and Sons has publisher number 471, so we see that this company published *Theoretical Probability for Applications* as well as the *MYSTERY BOOK*. The third block identifies the individual book. Notice that the number of digits in this block limits the number of different titles a publisher can publish without needing to get another publisher number. Thus Ancrage Press, which published *Le Véritable Amour*, only has two digits (or 100 titles) for its use. Publishers that produce large numbers of titles are given shorter publisher numbers to accommodate more titles.

4. How many different titles may John Wiley and Sons publish using the publisher number 471?
5. How many different titles may be published in Mauritius? [*Hint:* In this case, four digits define the publisher and the title. First consider how many different four-digit blocks are possible. Then consider how many different ways a hyphen may be inserted into the four-digit block.]
6. How many correct ISBN codes are possible for a book from the English-speaking area?
7. In practice, the number of correct ISBN codes is limited by restrictions on both the format of the country code and the format of the publisher number. For the English-speaking area, the publisher number must be in the following ranges: 00–19, 200–699, 7000–8499, 85000–89999, 900000–949999, and 9500000–9999999. With these restrictions on the publisher number, how many correct ISBN codes are possible for a book from the English-speaking area?
8. Research the use of X as the check digit. Write a brief report on what you find.

CHAPTER 7

Probability

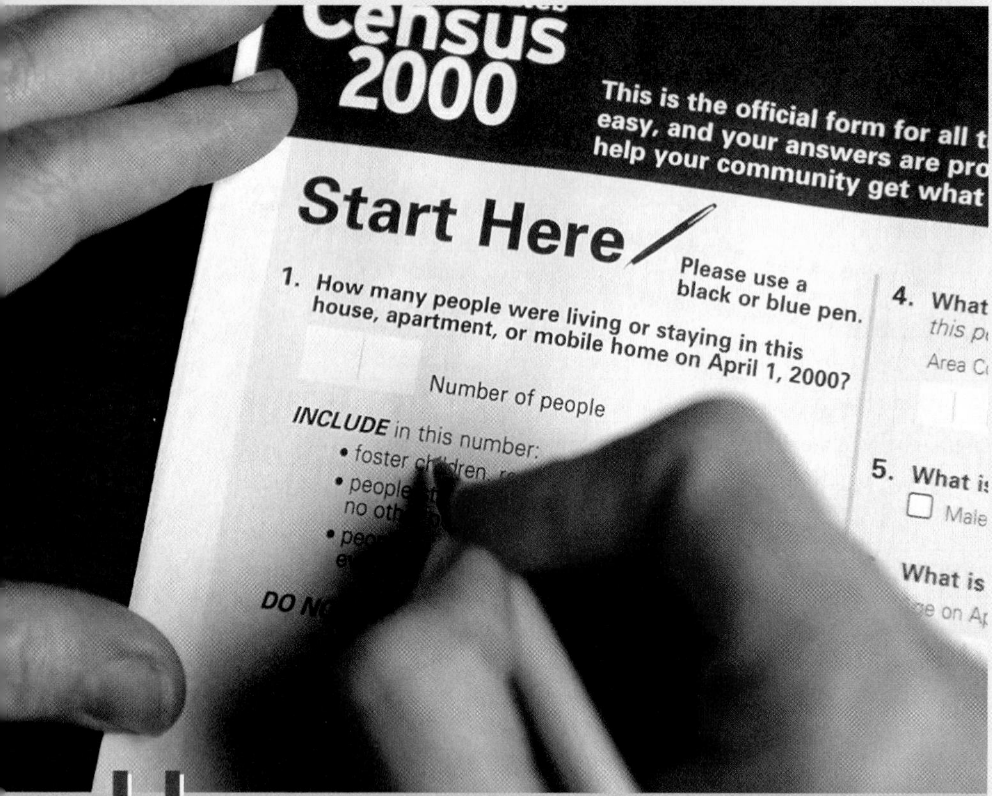

OUTLINE

Have you ever participated in a survey? You know, someone asks you a question and you give an answer. But what if the question is one you'd rather not answer. Suppose you were asked to answer "Yes" or "No" to the question "Do you feel you are overweight?" If you're not overweight, you'd probably answer honestly. But what would you do if you *were* overweight? Would you lie? If you could be assured of the confidentiality of your response, would that make a difference? For example, suppose the person asking the question said to you, "Flip a coin and don't tell me the face. If it's heads, check "Yes"; if it's tails, answer the question honestly. Then drop the paper in the box." This process guarantees the confidentiality of your response. But how does the surveyor get the information he wants? As it turns out, conditional probability can be used to complete the survey. After you study Section 7.5, look at the Chapter Project for details.

A LOOK BACK, A LOOK FORWARD

In Chapter 6 we defined sets, including the empty set and the universal set. We developed an algebra for sets, defining the operations of union and intersection of sets and the complement of a set. Then we solved many problems involving counting. In this chapter we use these ideas to develop *probability,* a part of mathematics that deals with the investigation of regular features of random events.

7.1 Sample Spaces and the Assignment of Probabilities

PREPARING FOR THIS SECTION *Before getting started, review the following:*

> Sets (Section 6.1, pp. 317–325)

> The Multiplication Principle (Section 6.3, pp. 332–335)

OBJECTIVES

1. Find a sample space
2. Assign probabilities
3. Construct a probability model
4. Find probabilities involving equally likely outcomes

Introduction

Certain phenomena in the real world may be considered *chance phenomena.* These phenomena do not always produce the same observed outcome, and the outcome of any given observation of the phenomena may not be predictable. But they have a long-range behavior known as *statistical regularity.* Some examples of such cases, called *random events,* follow.

Tossing a fair coin gives a result that is either heads or tails. For any one throw, we cannot predict the result, although it is obvious that it is determined by definite causes (such as the initial velocity of the coin, the initial angle of throw, and the surface on which the coin rests). Even though some of these causes can be controlled, we cannot predetermine the result of any particular toss. The result of tossing a coin is a *random event.*

Although we cannot predict the result of any particular toss of the coin, if we perform a long sequence of tosses, we expect that the number of heads is approximately equal to the number of tails. That is, it seems *reasonable* to say that in any toss of this fair coin, heads or tails is *equally likely* to occur. As a result, we might *assign a probability* of $\frac{1}{2}$ for obtaining a head (or tail) on a particular toss.

In throwing an ordinary die, we cannot predict the result with certainty. The result of throwing a die is a *random event.* We do know that one of the faces 1, 2, 3, 4, 5, or 6 will occur.

The appearance of any particular face of the die is an *outcome.* If we perform a long series of tosses, any face is as *likely* to occur as any other, provided the die is fair. Here we might *assign a probability* of $\frac{1}{6}$ for obtaining a particular face.

The sex of a newborn baby is either male or female. This, too, is an example of a *random event.*

Our intuition tells us that a boy baby and a girl baby are *equally likely* to occur. If we follow this reasoning, we might *assign a probability* of $\frac{1}{2}$ to having a girl baby. However, if we consult the data about births in the United States found in Table 1, we see that it might be more accurate to assign a probability of .488 to having a girl baby.

TABLE 1

Year of Birth	Number of Births (in thousands) Boys b	Girls g	Total Number of Births (in thousands) $b+g$	Ratio of Births $\frac{b}{b+g}$	$\frac{g}{b+g}$
2001	2,058	1,968	4,026	.511	.489
2000	2,077	1,982	4,059	.512	.488
1999	2,027	1,933	3,960	.512	.488
1998	2,041	1,940	3,981	.513	.487
1997	2,035	1,936	3,971	.512	.488
1996	2,010	1,910	3,920	.513	.487
1995	1,996	1,903	3,899	.512	.488
1994	2,023	1,930	3,953	.512	.488
1993	2,049	1,951	4,000	.512	.488
1992	2,082	1,983	4,065	.512	.488
1991	2,102	2,009	4,111	.511	.489
1990	2,129	2,029	4,158	.512	.488

Source: U.S. Department of Health and Human Services, *Monthly Vital Statistics Report,* National Center for Health Statistics June 2003.

These examples demonstrate that in studying a sequence of random experiments it is not possible to forecast individual results. These are subject to irregular, random fluctuations that cannot be exactly predicted. However, if the number of observations is large—that is, if we deal with a *mass phenomenon*—some regularity appears. This leads to the study of probability.

In studying probability we are concerned with experiments, real or conceptual, and their outcomes. In this study we try to formulate in a precise manner a mathematical theory that closely resembles the experiment in question. The first stage of development of such a mathematical theory is the building of what is termed a *probability model.* This model is then used to analyze and predict outcomes of the experiment. The purpose of this section is to learn how a probability model can be constructed.

Sample Spaces

We begin by writing the associated *sample space* of an experiment; that is, we write all outcomes that can occur as a result of the experiment as a set.

1 **EXAMPLE 1** **Finding a Sample Space**

Flip a coin. Find a sample space for this experiment.

SOLUTION The experiment consists of flipping a coin. The only possible outcomes are heads (H) and tails (T). Therefore, a sample space for the experiment is the set $\{H, T\}$.

EXAMPLE 2 Finding a Sample Space

Consider an experiment in which, for the sake of simplicity, one die is green and the other is red. When the dice are rolled, the set of outcomes consists of all the different ways that the dice may come to rest, that is, the set of all *possibilities* that can occur as a result of the experiment. Find a sample space for this experiment.

SOLUTION An application of the Multiplication Principle reveals that the number of outcomes of this experiment is 36 since there are 6 possible ways for the green die to come up and 6 ways for the red die to come up.

We can use a tree diagram to obtain a list of the 36 outcomes. See Figure 1. Figure 2 gives a graphical representation of the 36 outcomes. The sample space consists of the 36 ordered pairs listed.

FIGURE 1

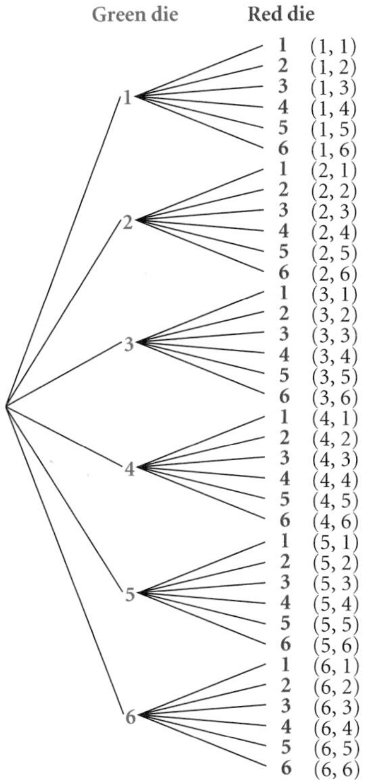

FIGURE 2

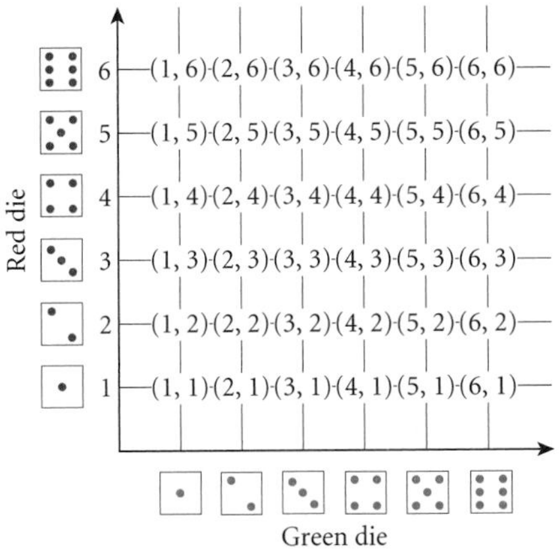

Sample Space; Outcome

A **sample space** S, associated with a real or conceptual experiment, is the set of all possibilities that can occur as a result of the experiment. Each element of a sample space S is called an **outcome.**

We list below some experiments and their sample spaces.

Experiment	*Sample Space*
(a) A spinner is marked from 1 to 8. An experiment consists of spinning the dial once.	$\{1, 2, 3, 4, 5, 6, 7, 8\}$
(b) An experiment consists of tossing two coins, a penny and a nickel, and observing whether the coins match (M) or do not match (D).	$\{M, D\}$
(c) An experiment consists of tossing a coin twice and observing the number of heads that appear.	$\{0, 1, 2\}$

(d) An experiment consists of tossing a coin twice and observing whether the coin falls heads (H) or tails (T). — $\{HH, HT, TH, TT\}$

(e) An experiment consists of tossing three coins and observing whether the coins fall heads (H) or tails (T). — $\{HHH, HHT, HTH, HTT, THH, THT, TTH, TTT\}$

(f) An experiment consists of selecting three manufactured parts from the production process and observing whether they are acceptable (A) or defective (D). — $\{AAA, AAD, ADA, ADD, DAA, DAD, DDA, DDD\}$

NOW WORK PROBLEM 1.

The sample space of an experiment plays the same role in probability as the universal set does in set theory.

In this chapter we confine our attention to those cases for which the sample space is finite, that is, to those situations in which it is possible to have only a finite number of outcomes.

Notice that in our definition we say *a* sample space, rather than *the* sample space, since an experiment can be described in many different ways. In general, it is a safe guide to include as much detail as possible in the description of the outcomes of the experiment in order to answer all pertinent questions concerning the result of the experiment.

EXAMPLE 3 Finding a Sample Space

Consider the set of all different types of families with three children. Describe a sample space for the experiment of selecting one family from the set of all possible three-child families.

SOLUTION One way of describing a sample space is by denoting the number of girls in the family. The only possibilities are members of the set

$$\{0, 1, 2, 3\}$$

That is, a three-child family can have 0, 1, 2, or 3 girls. This sample space has four outcomes.

Another way of describing a sample space for this experiment is by first defining B and G as "boy" and "girl," respectively. Then the sample space would be given as

$$\{BBB, BBG, BGB, BGG, GBB, GBG, GGB, GGG\}$$

where BBB means first child is a boy, second child is a boy, third child is a boy, and so on. This experiment can be depicted by the tree diagram in Figure 3. The experiment has $2 \cdot 2 \cdot 2 = 8$ possible outcomes, as the Multiplication Principle indicates.

FIGURE 3

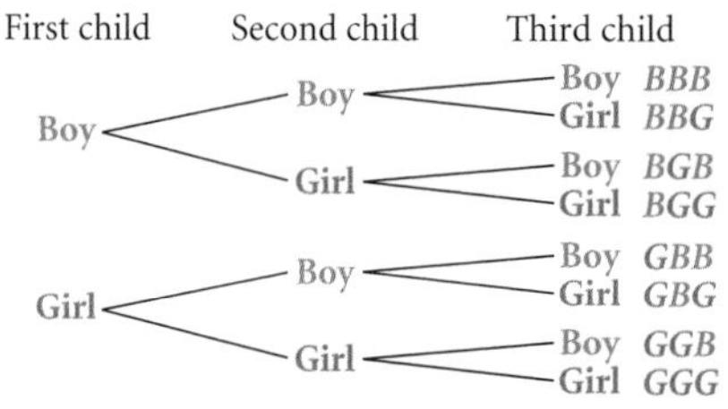
First child
Second child
Third child
Boy
Girl
Boy
Girl
Boy
Girl
Boy BBB
Girl BBG
Boy BGB
Girl BGG
Boy GBB
Girl GBG
Boy GGB
Girl GGG

NOW WORK PROBLEM 7.

Assignment of Probabilities

We are now ready to give a definition of the probability of an outcome of a sample space.

Probability of an Outcome

Suppose the sample space S of an experiment has n outcomes given by

$$S = \{e_1, e_2, \dots, e_n\}$$

To each outcome of S, we assign a real number, $P(e)$, called the **probability of the outcome** e, so that

$$P(e_1) \geq 0, P(e_2) \geq 0, \dots, P(e_n) \geq 0 \quad (1)$$

and

$$P(e_1) + P(e_2) + \dots + P(e_n) = 1 \quad (2)$$

Condition (1) states that each probability assignment must be nonnegative. Condition (2) states that the sum of all the probability assignments must equal one. Only assignments of probabilities satisfying both conditions (1) and (2) are valid.

2 **EXAMPLE 4** **Valid Probability Assignments**

Let a die be thrown. A sample space S is then

$$S = \{1, 2, 3, 4, 5, 6\}$$

There are six outcomes in S: 1, 2, 3, 4, 5, 6.

One valid assignment of probabilities is

$$P(1) = \frac{1}{6} \quad P(2) = \frac{1}{6} \quad P(3) = \frac{1}{6} \quad P(4) = \frac{1}{6} \quad P(5) = \frac{1}{6} \quad P(6) = \frac{1}{6}$$

This choice is consistent with the definition since the probability assigned each outcome is nonnegative and their sum is 1. This assignment is made when the die is **fair**.

Another assignment that is valid is

$$P(1) = 0 \quad P(2) = 0 \quad P(3) = \frac{1}{3} \quad P(4) = \frac{2}{3} \quad P(5) = 0 \quad P(6) = 0$$

This assignment, although unnatural, is made when the die is "loaded" in such a way that only a 3 or a 4 can occur and the 4 is twice as likely as the 3 to occur.

Many other assignments can also be made that are valid.

NOW WORK PROBLEM 23.

The sample space and the assignment of probabilities to each outcome of an experiment constitutes a *probability model* for the experiment.

Constructing a Probability Model

To construct a probability model requires two steps:

STEP 1 Find a sample space. List all the possible outcomes of the experiment, or, if this is not easy to do, determine the number of outcomes of the experiment.

STEP 2 Assign to each outcome e a probability $P(e)$ so that

(a) $P(e) \geq 0$
(b) The sum of all the probabilities assigned to the outcomes equals 1.

Sometimes a probability model is referred to as a **stochastic model.**

3 **EXAMPLE 5 Constructing a Probability Model**

A coin is tossed. The coin is weighted so that heads (H) is 5 times more likely to occur than tails (T). Construct a probability model for this experiment.

SOLUTION **STEP 1** A sample space S for this experiment is

$$S = \{T, H\}$$

STEP 2 Let x denote the probability that tails occurs. Then

$$P(T) = x \qquad \text{and} \qquad P(H) = 5x$$

Since the sum of all the probability assignments must equal 1, we have

$$\begin{aligned} P(H) + P(T) = 5x + x &= 1 \\ 6x &= 1 \\ x &= \frac{1}{6} \end{aligned}$$

As a result, we assign the probabilities

$$P(H) = \frac{5}{6} \qquad P(T) = \frac{1}{6}$$

NOW WORK PROBLEM 27.

The above discussion constitutes the construction of a probability model for the experiment.

Equally Likely Outcomes

When the same probability is assigned to each outcome of a sample space, the outcomes are termed **equally likely outcomes.**

Equally likely outcomes often occur when items are selected randomly. For example, in randomly selecting 1 person from a group of 10, the probability of selecting a particular individual is $\frac{1}{10}$. If a card is chosen randomly from a deck of 52 cards, the probability of drawing a particular card is $\frac{1}{52}$. In general, if a sample space S has n equally likely outcomes, the probability assigned to each outcome is $\frac{1}{n}$.

Probability of an Event E in a Sample Space with Equally Likely Outcomes

Let a sample space S be given by

$$S = \{e_1, e_2, \ldots, e_n\}$$

Suppose each of the outcomes $e_1, \ldots, e_n$ is equally likely to occur. An event E in S is any subset of S. If E contains m of the n outcomes in S, then

$$P(E) = \frac{m}{n} \tag{3}$$

4 **EXAMPLE 6** **Finding Probabilities Involving Equally Likely Outcomes**

If a person is chosen randomly from a group of 100 people, 60 female and 40 male, the probability a male is chosen is $\frac{40}{100}$.

This leads to the following formulation:

Probability of an Event E in a Sample Space with Equally Likely Outcomes

If the sample space S of an experiment has n equally likely outcomes, and the event E in S occurs in m ways, then the probability of event E, written as $P(E)$, is $\frac{m}{n}$. That is,

$$P(E) = \frac{\text{Number of possible ways the event } E \text{ can take place}}{\text{Number of outcomes in } S} = \frac{c(E)}{c(S)} \tag{4}$$

To compute the probability of an event E in which the outcomes are equally likely, count the number $c(E)$ of outcomes in E, and divide by the total number $c(S)$ of outcomes in the sample space.

EXAMPLE 7 **Finding Probabilities Involving Equally Likely Outcomes**

A jar contains 10 marbles; 5 are solid color, 4 are speckled, and 1 is clear.

(a) If one marble is picked at random, what is the probability it is speckled?
(b) If one marble is picked at random, what is the probability it is clear?

SOLUTION The experiment is an example of one in which the outcomes are equally likely. That is, no one marble is more likely to be picked than another. If S is the sample space, then there are 10 possible outcomes in S, so $c(S) = 10$.

(a) Define the event E: A speckled marble is picked. There are 4 ways E can occur. Using Formula (4), we find

$$P(E) = \frac{c(E)}{c(S)} = \frac{4}{10} = \frac{2}{5}$$

(b) Define the event F: A clear marble is picked. Then there is 1 way for F to occur. Using Formula (4), we find

$$P(F) = \frac{c(F)}{c(S)} = \frac{1}{10}$$

NOW WORK PROBLEM 31.

EXAMPLE 8 Finding Probabilities Involving Equally Likely Outcomes

In the experiment of tossing two fair dice, find the probability of each of the following events:

(a) The sum of the faces is 3.
(b) The sum of the faces is 7.
(c) The sum of the faces is 7 or 3.
(d) The sum of the faces is 7 and 3.

SOLUTION The tree diagram shown in Figure 4 illustrates the experiment. Since the dice are fair, the 36 outcomes in the sample space S are equally likely. For this experiment, $c(S) = 36$. We assign a probability of $\frac{1}{36}$ to each one. The probability model is constructed.

FIGURE 4

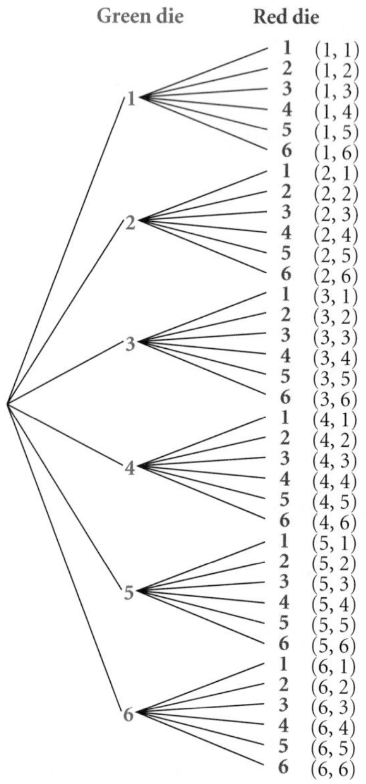
Green die
Red die
1
2
3
4
5
6
(1, 1)
(1, 2)
(1, 3)
(1, 4)
(1, 5)
(1, 6)
(2, 1)
(2, 2)
(2, 3)
(2, 4)
(2, 5)
(2, 6)
(3, 1)
(3, 2)
(3, 3)
(3, 4)
(3, 5)
(3, 6)
(4, 1)
(4, 2)
(4, 3)
(4, 4)
(4, 5)
(4, 6)
(5, 1)
(5, 2)
(5, 3)
(5, 4)
(5, 5)
(5, 6)
(6, 1)
(6, 2)
(6, 3)
(6, 4)
(6, 5)
(6, 6)

(a) The sum of the faces is 3 when the event $A = \{(1, 2), (2, 1)\}$ occurs. Since $c(A) = 2$, we have

$$P(A) = \frac{c(A)}{c(S)} = \frac{2}{36} = \frac{1}{18}$$

(b) The sum of the faces is 7 when the event $B = \{(1, 6), (2, 5), (3, 4), (4, 3), (5, 2), (6, 1)\}$ occurs. Since $c(B) = 6$, we have

$$P(B) = \frac{c(B)}{c(S)} = \frac{6}{36} = \frac{1}{6}$$

(c) The sum of the faces is 7 or 3 if and only if the outcome is an element of $A \cup B$, where A and B are the sets given in parts (a) and (b). Then

$$A \cup B = \{(2, 1), (1, 2), (1, 6), (2, 5), (3, 4), (4, 3), (5, 2), (6, 1)\} \text{ and } c(A \cup B) = 8.$$

Then

$$P(A \cup B) = \frac{c(A \cup B)}{c(S)} = \frac{8}{36} = \frac{2}{9}$$

(d) The sum of the faces is 3 and 7 if and only if the outcome is an element of $A \cap B$. Since $A \cap B = \varnothing$. We have $c(A \cap B) = 0$. Then

$$P(A \cap B) = \frac{c(A \cap B)}{c(S)} = \frac{0}{36} = 0$$

NOW WORK PROBLEM 41.

EXERCISE 7.1 Answers to Odd-Numbered Problems Begin on Page AN-30.

In Problems 1–6, describe a sample space associated with each experiment. List the outcomes of each sample space. In each experiment we are interested in whether the coin falls heads (H) or tails (T).

1. Tossing 2 coins
2. Tossing 1 coin twice
3. Tossing a coin 3 times
4. Tossing 3 coins
5. Tossing a coin 2 times and then a die
6. Tossing two coins and then a die

In Problems 7–14, use the pictured spinners to list the outcomes of a sample space associated with each experiment.

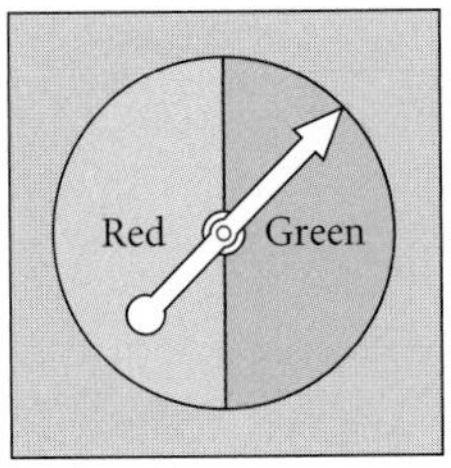

Spinner 1

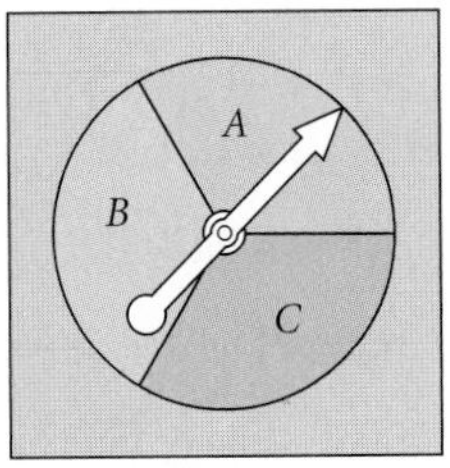

Spinner 2

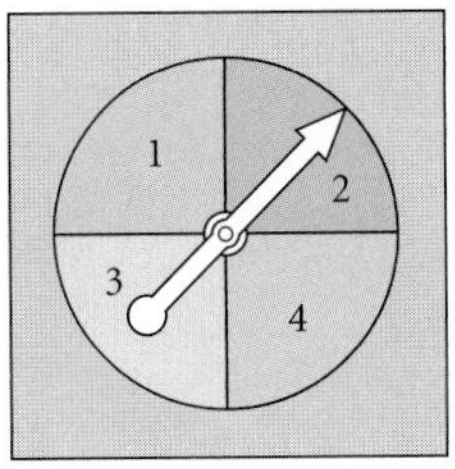

Spinner 3

7. First spinner 1 is spun and then spinner 2 is spun.
8. First spinner 2 is spun and then spinner 3 is spun.
9. Spinner 2 is spun twice.
10. Spinner 3 is spun twice.
11. Spinner 2 is spun twice and then spinner 3 is spun.
12. Spinner 3 is spun once and then spinner 2 is spun twice.
13. Spinners 1, 2, and 3 are each spun once in this order.
14. Spinners 3, 2, and 1 are each spun once in this order.

In Problems 15–22, find the number of outcomes of a sample space associated with each experiment.

15. Tossing a coin 4 times
16. Tossing a coin 5 times
17. Tossing 3 dice
18. Tossing 2 dice and then a coin
19. Selecting 2 cards (without replacement) from a regular deck of cards (Assume order is not important.)
20. Selecting 3 cards (without replacement) from a regular deck of 52 cards (Assume order is not important.)
21. Picking two letters from the alphabet (repetitions allowed; assume order is important)
22. Picking two letters from the alphabet (no repetitions allowed; assume order is important)

In Problems 23–26, consider the experiment of tossing a coin twice. The table below lists six possible assignments of probabilities for this experiment:

Assignments	Sample Space			
	HH	*HT*	*TH*	*TT*
1	$\frac{1}{4}$	$\frac{1}{4}$	$\frac{1}{4}$	$\frac{1}{4}$
2	0	0	0	1
3	$\frac{3}{16}$	$\frac{5}{16}$	$\frac{5}{16}$	$\frac{3}{16}$
4	$\frac{1}{2}$	$\frac{1}{2}$	$-\frac{1}{2}$	$\frac{1}{2}$
5	$\frac{1}{8}$	$\frac{1}{4}$	$\frac{1}{4}$	$\frac{1}{8}$
6	$\frac{1}{9}$	$\frac{2}{9}$	$\frac{2}{9}$	$\frac{4}{9}$

23. Which of the assignments of probabilities are valid?

24. Which of the assignments of probabilities should be used if the coin is known to be fair, that is, the outcomes are equally likely?

25. Which of the assignments of probabilities should be used if the coin is known to always come up tails?

26. Which of the assignments of probabilities should be used if tails is twice as likely as heads to occur?

27. **Constructing a Probability Model** A coin is tossed once. The coin is weighted so that heads is three times as likely as tails to occur. Construct a probability model for this experiment.

28. Constructing a Probability Model A coin is tossed once. The coin is weighted so that tails is twice as likely as heads to occur. Construct a probability model for the experiment.

29. Constructing a Probability Model A die is tossed. The die is weighted so that an odd-numbered face is twice as likely as an even-numbered face to appear. Construct a probability model for this experiment.

30. Constructing a Probability Model A die is tossed. The die is weighted so that a 6 appears half the time. The other faces each have the same probability assignment. Construct a probability model for this experiment.

In Problems 31–38, a ball is picked at random from a box containing 3 white, 5 red, 8 blue, and 7 green balls. Find the probability of each event.

31. White ball is picked.

32. Blue ball is picked.

33. Green ball is picked.

34. Red ball is picked.

35. White or red ball is picked.

36. Green or blue ball is picked.

37. Neither red nor green ball is picked.

38. Red or white or blue ball is picked.

In Problems 39–44, the experiment consists of tossing 2 fair dice. Find the probability of each event.

39. $A = \{(1, 4), (4, 1)\}$

40. $B = \{(1, 5), (2, 4), (3, 3), (4, 2), (5, 1)\}$

41. $C = \{(1, 4), (2, 4), (3, 4), (4, 4)\}$

42. $D = \{(1, 2), (2, 1), (2, 4), (4, 2), (3, 6), (6, 3)\}$

43. $E = \{(1, 1), (2, 2), (3, 3), (4, 4), (5, 5), (6, 6)\}$

44. $F = \{(1, 3), (2, 2), (3, 1)\}$

In Problems 45–54, a card is drawn at random from a regular deck of 52 cards. Find the probability of each event.*

45. The ace of hearts is drawn.

46. An ace is drawn.

**A regular deck of cards has 52 cards. There are four suits of 13 cards each. The suits are called clubs (black), diamonds (red), hearts (red), and spades (black). In each suit the 13 cards are labeled A (ace), 2, 3, 4, 5, 6, 7, 8, 9, 10, J (jack), Q (queen), and K (king).*

47. A spade is drawn.

48. A red card is drawn.

49. A picture card (J, Q, K) is drawn.

50. A number card (A, 2, 3, 4, 5, 6, 7, 8, 9, 10) is drawn.

51. A card with a number less than 6 is drawn (count A as 1).

52. A card with a value of 10 or higher is drawn.

53. A card that is not an ace is drawn.

54. A card that is either a queen or king of any suit is drawn.

55. Health Insurance Coverage The U.S. Census Bureau reported that 231,533,000 Americans were covered by health insurance in the year 2000 and 42,554,000 Americans were not covered by health insurance in that year. What is the probability that a randomly selected American in the year 2000 was covered by health insurance? Write the answer as a decimal, rounded to the nearest thousandth.

Source: U.S. Census Bureau, Current Population Survey, September 28, 2000.

56. Referring to Problem 55, what is the probability that a randomly selected American in the year 2000 was not covered by health insurance?

Sometimes experiments are simulated using a random number function instead of actually performing the experiment. In Problems 57–62, use a graphing utility to simulate each experiment.*

57. Tossing a Fair Coin Consider the experiment of tossing a fair coin. Simulate the experiment using a random number function, considering a toss to be tails (T) if the result is less than 0.5, and considering a toss to be heads (H) if the result is greater than or equal to 0.5. [*Note*: Most utilities repeat the action of the last entry if you simply press the ENTER, or EXE, key again.] Repeat the experiment 10 times. Using these 10 outcomes of the experiment you can estimate the probabilities $P(H)$ and $P(T)$ by the ratios

$$P(H) \approx \frac{\text{Number of times } H \text{ occurred}}{10}$$

$$P(T) \approx \frac{\text{Number of times } T \text{ occurred}}{10}$$

What are the actual probabilities? How close are the results of the experiment to the actual values?

58. Urn and Balls Consider the experiment of choosing a ball from an urn containing 18 red and 12 white balls. Simulate the experiment using a random number function, considering a selection to be a red ball (R) if the result is less than 0.6, and considering a toss to be a white ball (W) if the result is greater than or equal to 0.6. [*Note*: Most utilities repeat the action of the last entry if you simply press the ENTER, or EXE, key again.] Repeat the experiment 10 times. Using these 10 outcomes of the experiment you can estimate the probabilities $P(R)$ and $P(W)$ by the ratios

$$P(R) \approx \frac{\text{Number of times } R \text{ occurred}}{10}$$

$$P(W) \approx \frac{\text{Number of times } W \text{ occurred}}{10}$$

What are the actual probabilities? How close are the results of the experiment to the actual values?

59. Tossing a Loaded Coin Consider the experiment of tossing a loaded coin. Simulate the experiment using a random number function, considering a toss to be tails (T) if the result is less than 0.25, and considering a toss to be heads (H) if the result is greater than or equal to 0.25. [*Note*: Most utilities repeat the action of the last entry if you simply press the ENTER, or EXE, key again.] Repeat the experiment 20 times. Using these 20 outcomes of the experiment you can estimate the probabilities $P(H)$ and $P(T)$ by the ratios

$$P(H) \approx \frac{\text{Number of times } H \text{ occurred}}{20}$$

$$P(T) \approx \frac{\text{Number of times } T \text{ occurred}}{20}$$

What are the actual probabilities? How close are the results of the experiment to the actual values?

**Most graphing utilities have a random number function (usually RAND or RND) for generating numbers between 0 and 1. Check your user's manual to see how to use this function.*

60. Urn and Balls Consider the experiment of choosing a ball from an urn containing 15 red and 35 white balls. Simulate the experiment using a random number function, considering a selection to be a red ball (R) if the result is less than 0.3, and considering a toss to be a white ball (W) if the result is greater than or equal to 0.3. [*Note*: Most utilities repeat the action of the last entry if you simply press the ENTER, or EXE, key again.] Repeat the experiment 10 times. Using these 10 outcomes of the experiment you can estimate the probabilities $P(R)$ and $P(W)$ by the ratios

$$P(R) \approx \frac{\text{Number of times } R \text{ occurred}}{10}$$

$$P(W) \approx \frac{\text{Number of times } W \text{ occurred}}{10}$$

What are the actual probabilities? How close are the results of the experiment to the actual values?

61. Jar and Marbles Consider an experiment of choosing a marble from a jar containing 5 red, 2 yellow, and 8 white marbles. Simulate the experiment using a random number function on your calculator, considering a selection to be a red marble (R) if the result is less than or equal to 0.33, a yellow marble (Y) if the result is between 0.33 and 0.47, and a white marble (W) if the result is greater than or equal to 0.47. [*Note*: Most calculators repeat the action of the last entry if you simply press the ENTER, or EXE, key again.] Repeat the experiment 10 times. Using these 10 outcomes of the experiment, you can estimate the probabilities $P(R)$, $P(Y)$, and $P(W)$ by the ratios

$$P(R) \approx \frac{\text{Number of times } R \text{ occurred}}{10}$$

$$P(Y) \approx \frac{\text{Number of times } Y \text{ occurred}}{10}$$

$$P(W) \approx \frac{\text{Number of times } W \text{ occurred}}{10}$$

What are the actual probabilities? How close are the results of the experiment to the actual values?

62. Poker Game The probabilities in a game of poker that Adam, Beatrice, or Cathy win are .22, .60, and .18, respectively. Simulate the game using a random number function on your calculator, considering that Adam won (A) if the result is less than or equal to 0.22, that Beatrice won (B) if the result is between 0.22 and 0.82, and that Cathy won (C) if the result is greater than or equal to 0.82. [*Note*: Most calculators repeat the action of the last entry if you simply press the ENTER, or EXE, key again.] Repeat the experiment 20 times; that is, simulate 20 games. Using these 20 outcomes of the experiment you can estimate the probabilities $P(A)$, $P(B)$, and $P(C)$ by the ratios

$$P(A) \approx \frac{\text{Number of times } A \text{ won}}{20}$$

$$P(B) \approx \frac{\text{Number of times } B \text{ won}}{20}$$

$$P(C) \approx \frac{\text{Number of times } C \text{ won}}{20}$$

What are the actual values? How close are the results of the experiment to the actual values?

7.2 Properties of the Probability of an Event

PREPARING FOR THIS SECTION *Before getting started, review the following:*

> Sets (Section 6.1, pp. 317–325)

OBJECTIVES
1 Find the probability of an event
2 Find the probability of E or F, when E and F are mutually exclusive
3 Use the Additive Rule
4 Use a Venn diagram to find probabilities
5 Find the probability of the complement of an event
6 Compute odds

In Section 7.1 we defined the probability of an event E in a sample space S with equally likely outcomes. We begin this section with a definition for the probability of an event E in a sample space S whose outcomes have been assigned valid probabilities, not necessarily all the same.

Events and Simple Events

Event; Simple Event

An **event** is any subset of the sample space. If an event has exactly one element, that is, consists of only one outcome, it is called a **simple event.**

For example, consider an experiment discussed in Section 7.1: three-child families. A sample space S for this experiment is

$$S = \{BBB, BBG, BGB, BGG, GBB, GBG, GGB, GGG\}$$

The event E that the family consists of exactly two boys is

$$E = \{BBG, BGB, GBB\}$$

The event E is the union of the three simple events $\{BBG\}$, $\{BGB\}$, $\{GBB\}$. That is,

$$E = \{BBG\} \cup \{BGB\} \cup \{GBB\}$$

In fact, every event can be written as the union of simple events.

Since a sample space S is also an event, we can express a sample space as the union of simple events. If the sample space S consists of n outcomes,

$$S = \{e_1, e_2, \ldots, e_n\}$$

then

$$S = \{e_1\} \cup \{e_2\} \cup \cdots \cup \{e_n\}$$

Suppose valid probabilities have been assigned to each outcome of S. What do we mean by the probability of an event E of S?

Let S be a sample space and let E be any event of S. It is clear that either $E = \varnothing$ or E is a simple event or E is the union of two or more simple events. We give the following definition,

Probability of an Event

If $E = \varnothing$, the event E is **impossible.** We define the **probability of $\varnothing$** as

$$\boxed{P(\varnothing) = 0}$$

If $E = \{e\}$ is a simple event, then $P(E) = P(e)$; that is, $P(E)$ equals the probability assigned to the outcome e.

$$\boxed{P(E) = P(e)}$$

If E is the union of r simple events $\{e_1\}, \{e_2\}, \ldots, \{e_r\}$, we define the **probability of E** to be the sum of the probabilities assigned to each simple event in E. That is,

$$\boxed{P(E) = P(e_1) + P(e_2) + \cdots + P(e_r)} \qquad (1)$$

If the sample space S is given by

$$S = \{e_1, e_2, \ldots, e_n\}$$

then the probability of S is

$$P(S) = P(e_1) + \cdots + P(e_n) = 1$$

The probability of S, the sample space, is 1.

1 **EXAMPLE 1 Finding the Probability of an Event**

Each of the eight possible blood types is listed in Table 2 along with the percent of the U.S. population having that type. What is the probability that a randomly selected person in the United States has a blood type that is Rh-negative?

Source: AABB Facts about Blood, 7/1/98.

TABLE 2

O Positive—38%
A Positive—34%
B Positive—9%
O Negative—7%
A Negative—6%
AB Positive—3%
B Negative—2%
AB Negative—1%

SOLUTION The sample space S consists of the eight blood types, and the percents, when given as decimals, are the probability assignments for each element of the sample space. For example, the simple event "O positive" is assigned the probability .38. Notice that the assignments given are valid. Now, let E be the event that a randomly selected person in the United States has a blood type that is Rh-negative. We seek $P(E)$. Since E is the union of the simple events {O negative}, {A negative}, {AB negative}, and {B negative}, by Equation (1) we have

$$\begin{aligned} P(E) &= P\{\text{O negative}\} + P\{\text{A negative}\} + P\{\text{AB negative}\} + P\{\text{B negative}\} \\ &= .07 + .06 + .01 + .02 = .16 \end{aligned}$$

The probability is .16, or 16%, that a randomly selected person in the United States has Rh-negative blood.

NOW WORK PROBLEM 1.

Mutually Exclusive Events

Mutually Exclusive Events

Two or more events of a sample space S are said to be **mutually exclusive** if and only if they have no outcomes in common.

If we treat mutually exclusive events as sets, they are disjoint.

The following result gives us a way of computing probabilities for mutually exclusive events.

Probability of E or F if E and F Are Mutually Exclusive

Let E and F be two events of a sample space S. If E and F are mutually exclusive, that is, if $E \cap F = \varnothing$, then the probability of E or F is the sum of their probabilities. That is,

$$P(E \cup F) = P(E) + P(F) \quad \text{if } E \cap F = \varnothing \tag{2}$$

Since E and F can be written as a union of simple events in which no simple event of E appears in F and no simple event of F appears in E, the result follows.

2 **EXAMPLE 2** **Finding the Probability of *E* or *F*, when *E*, *F* Are Mutually Exclusive**

In the experiment of tossing two fair dice, what is the probability of obtaining either a sum of 7 or a sum of 11?

SOLUTION Refer to Figure 4, p. 372, for a tree diagram depicting this experiment. The sample spaces has 36 equally likely outcomes. Each simple event of the sample space is assigned the probability $\frac{1}{36}$. Let E and F be the events

$$E: \quad \text{Sum is } 7 \qquad F: \quad \text{Sum is } 11$$

Since a sum of 7 can occur in six different ways, the probability of E is

$$P(E) = \frac{c(E)}{c(S)} = \frac{6}{36} = \frac{1}{6}$$

Similarly, since a sum of 11 can occur in two different ways, the probability of F is

$$P(F) = \frac{c(F)}{c(S)} = \frac{2}{36} = \frac{1}{18}$$

The two events E and F are mutually exclusive. By Equation (2), the probability that the sum is 7 or 11, that is, the probability of the event E or F, is

$$P(E \cup F) = P(E) + P(F) = \frac{6}{36} + \frac{2}{36} = \frac{8}{36} = \frac{2}{9}$$

NOW WORK PROBLEM 15.

Additive Rule

The following result, called the **Additive Rule,** provides a technique for finding the probability of the union of two events whether they are mutually exclusive or not.

Additive Rule

For any two events E and F of a sample space S,

$$\mathbf{P(E \cup F) = P(E) + P(F) - P(E \cap F)} \tag{3}$$

This result concerning probabilities is closely related to the Counting Formula discussed in Chapter 6 (page 328). A proof is outlined in Problem 43.

3 **EXAMPLE 3** **Using the Additive Rule**

If $P(E) = .30$, $P(F) = .20$, and $P(E \cup F) = .40$, find $P(E \cap F)$.

SOLUTION Using the Additive Rule Equation (3), we know that

$$P(E \cup F) = P(E) + P(F) - P(E \cap F)$$

Since $P(E) = .30$, $P(F) = .20$, and $P(E \cup F) = .40$, we have

$$.40 = .30 + .20 - P(E \cap F)$$

Solving for $P(E \cap F)$, we find

$$P(E \cap F) = .30 + .20 - .40 = .10$$

NOW WORK PROBLEM 13.

EXAMPLE 4 Using the Additive Rule

Consider the two events

E: A shopper spends at least \$40 for food

F: A shopper spends at least \$15 for meat

Based on recent studies, we assign the following probabilities to the events E and F:

$$P(E) = .56 \qquad P(F) = .63$$

Now suppose the probability that a shopper spends at least \$40 for food and \$15 for meat is .33. What is the probability that a shopper spends at least \$40 for food or at least \$15 for meat?

SOLUTION From the information given, we conclude that $P(E \cap F) = .33$. Since we are looking for the probability of $E \cup F$, we use the Additive Rule and find that

$$\begin{aligned} P(E \cup F) &= P(E) + P(F) - P(E \cap F) \\ &= .56 + .63 - .33 = .86 \end{aligned}$$

The probability is .86 that a shopper will spend at least \$40 for food or at least \$15 for meat.

Use a Venn diagram to find probabilities **4**

Using Venn Diagrams

FIGURE 5

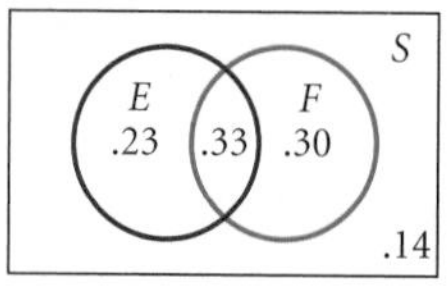

Often a Venn diagram is helpful in solving probability problems. A Venn diagram depicting the information of Example 4 is given in Figure 5. To obtain the diagram, we begin with the fact that $P(E \cap F) = .33$. Since $P(E) = .56$ and $P(F) = .63$, we fill in E with $.56 - .33 = .23$ and we fill in F with $.63 - .33 = .30$. Since $P(S) = 1$, we complete Figure 5 by entering $1 - (.23 + .33 + .30) = .14$. Now, for example, it is easy to see that the probability of E, but not F, is .23. The probability of neither E nor F is .14.

EXAMPLE 5 Using a Venn Diagram to Find Probabilities

In an experiment with two fair dice, consider the events

E: The sum of the faces is 8

F: Doubles are thrown

What is the probability of obtaining E, but not F?

SOLUTION We write down the elements of E and F.

$$E = \{(2,6), (3,5), (4,4), (5,3), (6,2)\}$$
$$F = \{(1,1), (2,2), (3,3), (4,4), (5,5), (6,6)\}$$
$$E \cap F = \{(4,4)\}$$

The probabilities of each of these events is

$$P(E) = \frac{5}{36} \qquad P(F) = \frac{6}{36} \qquad P(E \cap F) = \frac{1}{36}$$

FIGURE 6

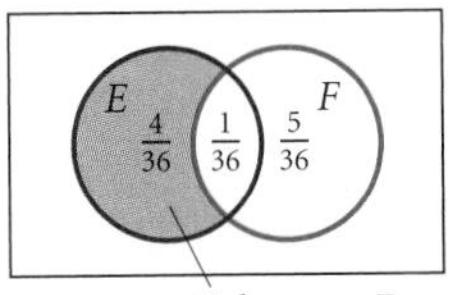

E, but not F

Now construct a Venn diagram. See Figure 6. The probability of E, but not F, is $\frac{4}{36} = \frac{1}{9}$.

We could also have solved Example 5 by actually finding the set E, but not F, namely:

$$\{(2,6), (3,5), (5,3), (6,2)\}$$

Since this set has four simple events, the probability is

$$P(E, \text{but not } F) = \frac{4}{36} = \frac{1}{9}$$

as before.

NOW WORK PROBLEM 21.

Properties of the Probability of an Event

The probability of any outcome of a sample space S is nonnegative. Furthermore, since any event E of S is the union of outcomes in S, and since $P(S) = 1$, it follows that

$$0 \le P(E) \le 1$$

Properties of the Probability of an Event

The probability of an event E of a sample space S has the following three properties:

(I) $0 \le P(E) \le 1$ for every event E of S

(II) $P(\varnothing) = 0$ **and** $P(S) = 1$

(III) $P(E \cup F) = P(E) + P(F) - P(E \cap F)$ for any two events E and F of S

Complement of an Event

5 Find the probability of the complement of an event

Let E be an event of a sample space S. The complement of E is the event "Not E" in S. The next result gives a relationship between their probabilities.

Probability of the Complement of an Event

Let E be an event of a sample space S. Then

$$\boxed{P(\overline{E}) = 1 - P(E)} \tag{4}$$

where $\overline{E}$ is the complement of E.

Proof We know that

$$S = E \cup \overline{E} \qquad E \cap \overline{E} = \varnothing$$

Since E and $\overline{E}$ are mutually exclusive,

$$P(S) = P(E) + P(\overline{E})$$

Since $P(S) = 1$ (Property II), it follows that

$$1 = P(E) + P(\overline{E})$$
$$P(\overline{E}) = 1 - P(E)$$

This result gives us a tool for finding the probability that an event does not occur if we know the probability that it does occur. The probability $P(\overline{E})$ that E does not occur is obtained by subtracting the probability $P(E)$ that E does occur from 1. We will see shortly that it is sometimes easier to find $P(E)$ by finding $P(\overline{E})$ and using Formula (4), than it is to proceed directly.

EXAMPLE 6 Using Formula (4)

A study of people over 40 with an MBA degree shows that it is reasonable to assign a probability of .756 that such a person will have annual earnings in excess of \$80,000. The probability that such a person will earn \$80,000 or less is then

$$1 - .756 = .244$$

NOW WORK PROBLEM 9.

EXAMPLE 7 Using Formula (4)

In an experiment of tossing two fair dice, find:

(a) The probability that the sum of the faces is less than or equal to 3
(b) The probability that the sum of the faces is greater than 3

SOLUTION The sample space S consists of 36 equally likely outcomes.

(a) The number of outcomes in the event E, "the sum of the faces is less than or equal to 3," is 3.

$$P(E) = \frac{3}{36} = \frac{1}{12}$$

(b) The event "the sum of the faces is greater than 3" is the complement of the event E defined in part (a). Since we seek $P(\overline{E})$, we can use the result of part (a) and Formula (4):

$$P(\overline{E}) = 1 - P(E) = 1 - \frac{1}{12} = \frac{11}{12}$$

That is, the probability that the sum of the faces is greater than 3 is $\frac{11}{12}$.

Compare the work required to do part (a) first and then apply Formula (4) (as we did in Example 7) to solving part (b) by listing the elements of the set and then computing the probability. We'll see that in many cases the approach of finding $P(E)$ first to get $P(\overline{E})$ is not only easier, but sometimes is the only choice available.

Odds

Compute odds 6

In many instances the probability of an event may be expressed as *odds*—either *odds for* an event or *odds against* an event.

If E is an event:

The **odds for E** are $\dfrac{P(E)}{P(\overline{E})}$ or $P(E)$ to $P(\overline{E})$

The **odds against E** are $\dfrac{P(\overline{E})}{P(E)}$ or $P(\overline{E})$ to $P(E)$

EXAMPLE 8 Computing Odds Given a Probability

Suppose the probability of the event

$$E: \quad \text{It will rain}$$

is .3. The odds for rain are

$$\frac{.3}{.7} \quad \text{or} \quad 3 \text{ to } 7$$

The odds against rain are

$$\frac{.7}{.3} \quad \text{or} \quad 7 \text{ to } 3$$

NOW WORK PROBLEM 33.

To obtain the probability of the event E when either the odds for E or the odds against E are known, we use the following formulas:

If the odds for E are a to b, then

$$P(E) = \frac{a}{a+b} \tag{5}$$

If the odds against E are a to b, then

$$P(E) = \frac{b}{a+b}$$

The proof of Equation (5) is outlined in Problem 44.

EXAMPLE 9 Computing a Probability from Odds

(a) The odds for a Republican victory in the next presidential election are 7 to 5. What is the probability that a Republican victory occurs?

(b) The odds against the Chicago Cubs winning the league pennant are 200 to 1. What is the probability that the Cubs win the pennant?

SOLUTION **(a)** The event E is "A Republican victory occurs." The odds for E are 7 to 5. Thus,

$$P(E) = \frac{7}{7+5} = \frac{7}{12} \approx .583$$

(b) The event F is "The Cubs win the pennant." The odds against F are 200 to 1. Thus

$$P(F) = \frac{1}{200+1} = \frac{1}{201} \approx .00498$$

NOW WORK PROBLEM 41.

EXERCISE 7.2 Answers to Odd-Numbered Problems Begin on Page AN-30.

Problems 1–4, require the following discussion:
Blood Types *Each of the eight possible blood types is listed in the table to the right along with the percent of the U.S. population having that type.*

O Positive—38%	A Negative—6%
A Positive—34%	AB Positive—3%
B Positive—9%	B Negative—2%
O Negative—7%	AB Negative—1%

Source: AABB Facts about Blood, 7/1/98.

1. What is the probability that a randomly selected person in the United States has a blood type that is Rh-positive?

2. What is the probability that a randomly selected person in the United States has a blood type that is type O?

3. What is the probability that a randomly selected person in the United States has a blood type that contains the A antigen? [*Hint*: AB blood contain both the A antigen and the B antigen.]

4. What is the probability that a randomly selected person in the United States has a blood type that contains both the A and the B antigens?

Problems 5–8 use the discussion below. The table below lists the top U.S. Internet service providers (ISPs) listed in order of decreasing number of subscribers, as of the third quarter of 2002:

ISP	Number of Subscribers (in millions)
America Online	26.7
MSN	9.0
EarthLink	4.8
United Online (Net Zero and Juno Online)	4.8
SBC/Prodigy	3.7
CompuServe (AOL owned)	3.0
Road Runner (AOL owned)	2.3
AT&T Broadband	1.9
Verizon	1.6
Bell South	1.6
Other US ISPs	89.6

Source: ISP-planet: Top U.S. ISPs by Subscriber Q3 2002, November 15, 2002.

Compute the probability of each of the following events. Round answers to the nearest tenth.

5. A randomly selected ISP subscriber subscribes to America Online or MSN.

6. A randomly selected ISP subscriber subscribes to one of the top three ISPs.

7. A randomly selected ISP subscriber does not subscribe to any of the top ten ISPs.

8. A randomly selected ISP subscriber subscribes to EarthLink, CompuServe, or AT&T Broadband.

In Problems 9–14, find the probability of the indicated event if $P(A) = .25$ and $P(B) = .40$.

9. $P(\overline{A})$

10. $P(\overline{B})$

11. $P(A \cup B)$ if A, B are mutually exclusive

12. $P(A \cap B)$ if A, B are mutually exclusive

13. $P(A \cup B)$ if $P(A \cap B) = .15$

14. $P(A \cap B)$ if $P(A \cup B) = .55$

---

15. In tossing 2 fair dice, are the events "Sum is 2" and "Sum is 12" mutually exclusive? What is the probability of obtaining either a 2 or a 12?

16. In tossing 2 fair dice, are the events "Sum is 6" and "Sum is 8" mutually exclusive? What is the probability of obtaining either a 6 or an 8?

17. **Chicago Bears** The Chicago Bears football team has a probability of winning of .65 and of tying of .05. What is their probability of losing?

18. **Chicago Black Hawks** The Chicago Black Hawks hockey team has a probability of winning of .6 and a probability of losing of .25. What is the probability of a tie?

19. **Likelihood of Passing** Jenny is taking courses in both mathematics and English. She estimates her probability of passing mathematics at .4 and English at .6, and she estimates her probability of passing at least one of them at .8. What is her probability of passing both courses?

20. **Dropping a Course** After midterm exams, Jenny (see Problem 19) reassesses her probability of passing mathematics to .7. She feels her probability of passing at least one of these courses is still .8, but she has a probability of only .1 of passing both courses. If her probability of passing English is less than .4, she will drop English. Should she drop English? Why?

21. Let A and B be events of a sample space S and let $P(A) = .5$, $P(B) = .4$, and $P(A \cap B) = .2$. Find the probabilities of each of the following events:

    (a) A or B (b) A but not B
    (c) B but not A (d) Neither A nor B

22. If A and B represent two mutually exclusive events such that $P(A) = .35$ and $P(B) = .60$, find each of the following probabilities:

    (a) $P(A \cup B)$ (b) $P(\overline{A \cup B})$ (c) $P(\overline{B})$
    (d) $P(\overline{A})$ (e) $P(A \cap B)$

23. **Car Repair** At the Milex tune-up and brake repair shop, the manager has found that a car will require a tune-up with a probability of .6, a brake job with a probability of .1, and both with a probability of .02.

    (a) What is the probability that a car requires either a tune-up or a brake job?
    (b) What is the probability that a car requires a tune-up but not a brake job?
    (c) What is the probability that a car requires neither type of repair?

24. **Factory Shortages** A factory needs two raw materials, say, E and F. The probability of not having an adequate supply of material E is .06, whereas the probability of not having an adequate supply of material F is .04. A study shows that the probability of a shortage of both E and F is .02. What is the probability of the factory being short of either material E or F?

25. **TV Sets** In a survey of the number of TV sets in a house, the following probability table was constructed:

Number of TV sets	0	1	2	3	4 or more
Probability	.05	.24	.33	.21	.17

Find the probability of a house having

(a) 1 or 2 TV sets (b) 1 or more TV sets
(c) 3 or fewer TV sets (d) 3 or more TV sets
(e) Fewer than 2 TV sets (f) Not even 1 TV set
(g) 1, 2, or 3 TV sets (h) 2 or more TV sets

26. **Supermarket Lines** Through observation it has been determined that the probability for a given number of people waiting in line at a particular checkout register of a supermarket is as shown in the table:

Number waiting in line	0	1	2	3	4 or more
Probability	.10	.15	.20	.24	.31

Find the probability of

(a) At most 2 people in line
(b) At least 2 people in line
(c) At least 1 person in line

In Problems 27–32, determine the probability of E for the given odds.

27. 3 to 1 for E

28. 4 to 1 against E

29. 7 to 5 against E

30. 2 to 9 for E

31. 1 to 1 for E (even)

32. 50 to 1 for E

In Problems 33–36, determine the odds for and against each event for the given probability.

33. $P(E) = .6$

34. $P(H) = \frac{1}{4}$

35. $P(F) = \frac{3}{4}$

36. $P(G) = .1$

37. If two fair dice are thrown, what are the odds of obtaining a sum of 7? a sum of 11? a sum of 7 or 11?

38. If the odds for event A are 1 to 5 and the odds for event B are 1 to 3, what are the odds for the event A or B, assuming the event A and B is impossible?

39. The probability of a person getting a job interview is .54; what are the odds against getting the interview?

40. If the probability of war is .6, what are the odds against war?

41. **Track** In a track contest the odds that A will win are 1 to 2, and the odds that B will win are 2 to 3. Find the probability and the odds that A or B wins the race, assuming a tie is impossible.

42. DUI It has been estimated that in 70% of the fatal accidents involving two cars, at least one of the drivers is drunk. If you hear of a two-car fatal accident, what odds should you give a friend for the event that at least one of the drivers was drunk?

43. Prove the Additive Rule, Equation (3) on page 379.
[*Hint*: From Example 9 in Section 6.1 (page 324) we have]

$$E \cup F = (E \cap \overline{F}) \cup (E \cap F) \cup (\overline{E} \cap F)$$

Since $E \cap \overline{F}$, $E \cap F$, and $\overline{E} \cap F$ are pairwise disjoint, we can write

$$P(E \cup F) = P(E \cap \overline{F}) + P(E \cap F) + P(\overline{E} \cap F) \qquad \text{(a)}$$

We may write the sets E and F in the form

$$E = (E \cap F) \cup (E \cap \overline{F})$$
$$F = (E \cap F) \cup (\overline{E} \cap F)$$

Since $E \cap F$ and $E \cap \overline{F}$ are disjoint and $E \cap F$ and $\overline{E} \cap F$ are disjoint, we have

$$P(E) = P(E \cap F) + P(E \cap \overline{F})$$
$$P(F) = P(E \cap F) + P(\overline{E} \cap F) \qquad \text{(b)}$$

Now combine (a) and (b).]

44. Prove Equation (5) on page 383.
[*Hint*: If the odds for E are a to b, then, by the definition of odds,

$$\frac{P(E)}{P(\overline{E})} = \frac{a}{b}$$

But $P(\overline{E}) = 1 - P(E)$. So

$$\frac{P(E)}{1 - P(E)} = \frac{a}{b}$$

Now solve for $P(E)$.]

45. Generalize the Additive Rule by showing the probability of the occurrence of at least one of the three events A, B, C is given by

$$\begin{aligned} P(A \cup B \cup C) = {} & P(A) + P(B) + P(C) \\ & - P(A \cap B) - P(A \cap C) \\ & - P(B \cap C) + P(A \cap B \cap C) \end{aligned}$$

7.3 Probability Problems Using Counting Techniques

PREPARING FOR THIS SECTION *Before getting started, review the following:*

- > The Multiplication Principle (Section 6.3, pp. 332–335)
- > Permutations (Section 6.4, pp. 336–342)
- > Combinations (Section 6.5, pp. 343–349)

OBJECTIVES 1 Find the probability of events using counting techniques

Find the probability of events using counting techniques 1

In this section we see how the counting techniques studied in Chapter 6 can be used to find probabilities.

EXAMPLE 1 Finding Probabilities Using Counting Techniques

From a box containing four white, three yellow, and one green ball, two balls are drawn one at a time without replacing the first before the second is drawn. Find the probability that one white and one yellow ball are drawn.

SOLUTION The experiment consists of selecting 2 balls from 8 balls. The number of ways that 2 balls can be drawn from 8 balls is $C(8, 2) = 28$. We define the events E and F as

E: A white ball is drawn

F: A yellow ball is drawn

The number of ways a white ball can be drawn is $C(4, 1) = 4$ and the number of ways a yellow ball can be drawn is $C(3, 1) = 3$. By the Multiplication Principle, the number of ways a white and yellow ball can be drawn is

$$C(4, 1) \cdot C(3, 1) = 4 \cdot 3 = 12$$

Since the outcomes are equally likely, the probability of drawing a white ball and a yellow ball is

$$P(E \cap F) = \frac{C(4, 1) \cdot C(3, 1)}{C(8, 2)} = \frac{12}{28} = \frac{3}{7}$$

EXAMPLE 2 Finding Probabilities Using Counting Techniques

A box contains 12 light bulbs, of which 5 are defective. All bulbs look alike and have equal probability of being chosen. Three lightbulbs are selected and placed in a box.

(a) What is the probability that all 3 are defective?
(b) What is the probability that exactly 2 are defective?
(c) What is the probability that at least 2 are defective?

SOLUTION The number of elements in the sample space S is equal to the number of combinations of 12 light bulbs taken 3 at a time, namely,

$$C(12, 3) = \frac{12!}{3!9!} = 220$$

(a) Define E as the event, "3 bulbs are defective." Then E can occur in $C(5, 3)$ ways, that is, the number of ways in which 3 defective bulbs can be chosen from 5 defective ones. The probability $P(E)$ is

$$P(E) = \frac{C(5, 3)}{C(12, 3)} = \frac{\frac{5!}{3!2!}}{220} = \frac{10}{220} = .04545$$

(b) Define F as the event, "2 bulbs are defective." To obtain 2 defective bulbs when 3 are chosen requires that we select 2 defective bulbs from the 5 available defective ones and 1 good bulb from the 7 good ones. By the Multiplication Principle this can be done in the following number of ways:

$$C(5, 2) \cdot C(7, 1) = \frac{5!}{2!3!} \cdot \frac{7!}{1!6!} = 10 \cdot 7 = 70$$

↗ Number of ways to select 2 defectives from 5 defectives

↖ Number of ways to select 1 good bulb from 7 good ones

The probability of selecting exactly 2 defective bulbs is therefore

$$P(F) = P(\text{exactly 2 defectives}) = \frac{C(5, 2) \cdot C(7, 1)}{C(12, 3)} = \frac{70}{220} = .31818$$

(c) Define G as the event, "At least 2 are defective." The event G is equivalent to asking for the probability of selecting either exactly 2 or exactly 3 defective bulbs. Since these events are mutually exclusive, the sum of their probabilities will give the probability of G. Using the results found in parts (a) and (b), we find

$$\begin{aligned} P(G) &= P(\text{Exactly 2 defectives}) + P(\text{Exactly 3 defectives}) \\ &= P(F) + P(E) = .31818 + .04545 = .36363 \end{aligned}$$

NOW WORK PROBLEM 1.

EXAMPLE 3 Finding Probabilities Using Counting Techniques

A fair coin is tossed 10 times.

(a) What is the probability of obtaining exactly 5 heads and 5 tails?
(b) What is the probability of obtaining between 4 and 6 heads, inclusive?

SOLUTION The number of elements in the sample space S is found by using the Multiplication Principle. Each toss results in a head (H) or a tail (T). Since the coin is tossed 10 times, we have

$$c(S) = \underbrace{2 \cdot 2 \cdot \ldots \cdot 2}_{\text{10 twos}} = 2^{10}$$

The outcomes are equally likely since the coin is fair.

(a) Any sequence that contains 5 heads and 5 tails is determined once the position of the 5 heads (or 5 tails) is known. The number of ways we can position 5 heads in a sequence of 10 slots is $C(10, 5)$. The probability of the event E: Exactly 5 heads, 5 tails is

$$P(E) = \frac{c(E)}{c(S)} = \frac{C(10, 5)}{2^{10}} = .2461$$

(b) Let F be the event: Between 4 and 6 heads, inclusive. To obtain between 4 and 6 heads is equivalent to the event: Exactly 4 heads or exactly 5 heads or exactly 6 heads. Since these are mutually exclusive (for example, it is impossible to obtain exactly 4 heads and exactly 5 heads when tossing a coin 10 times), we have

$$\begin{aligned} P(F) &= P(\text{4 heads or 5 heads or 6 heads}) \\ &= P(\text{4 heads}) + P(\text{5 heads}) + P(\text{6 heads}) \end{aligned}$$

Proceeding as we did in part (a), we find

$$\begin{aligned} P(F) &= \frac{C(10, 4)}{2^{10}} + \frac{C(10, 5)}{2^{10}} + \frac{C(10, 6)}{2^{10}} = .2051 + .2461 + .2051 \\ &= .6563 \end{aligned}$$

NOW WORK PROBLEM 3.

EXAMPLE 4 Finding Probabilities Using Counting Techniques

What is the probability that a four-digit telephone extension has one or more repeated digits? Assume no one digit is more likely than another to be used.

SOLUTION By the Multiplication Principle, there are $10^4 = 10{,}000$ distinct four-digit telephone extensions, so the number of outcomes in the sample space is

$$c(S) = 10{,}000$$

We wish to find the probability that a four-digit telephone extension has one or more repeated digits. Define the event E as follows:

E: The extension has one or more repeated digits

Since it is difficult to count the outcomes of this event, we first compute the probability of the complement, namely,

$\overline{E}$: No repeated digits in four-digit extension

To find $c(\overline{E})$, we use the Multiplication Principle. The number of four-digit extensions that have *no* repeated digits is

$$c(\overline{E}) = 10 \cdot 9 \cdot 8 \cdot 7 = 5040$$

Hence,

$$P(\overline{E}) = \frac{c(\overline{E})}{c(S)} = \frac{5040}{10{,}000} = .504$$

Therefore, the probability of one or more repeated digits is

$$P(E) = 1 - P(\overline{E}) = 1 - .504 = .496$$

NOW WORK PROBLEM 7.

The next example is very similar to the one just completed.

EXAMPLE 5 Birthday Problem

An interesting problem, called the **birthday problem,** is to find the probability that in a group of r people there are at least two people who have the same birthday (the same month and day of the year).

SOLUTION We assume a person is as likely to be born on one day as another.

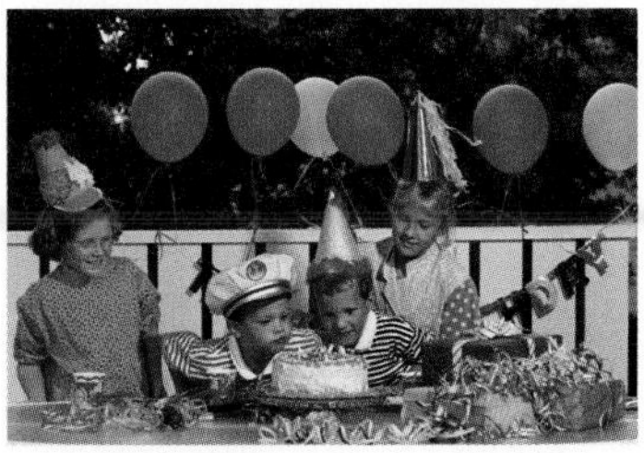

We first determine the number of outcomes in the sample space S. There are 365 possibilities for each person's birthday (we exclude February 29 for simplicity). Since there are r people in the group, there are 365^r possibilities for the birthdays. [For one person in the group, there are 365 days on which his or her birthday can fall; for two people, there are $(365)(365) = 365^2$ pairs of days; and, in general, using the Multiplication Principle, for r people there are 365^r possibilities.] That is,

$$c(S) = 365^r$$

These outcomes are equally likely.

We wish to find the probability of the event E:

$$E: \quad \text{at least two people have the same birthday.}$$

It is difficult to count the elements in this set; it is much easier to count the elements of the complement:

$$\overline{E}: \quad \text{No two people have the same birthday}$$

We proceed to find $c(\overline{E})$ as follows:

Choose one person at random. There are 365 possibilities for his or her birthday. Choose a second person. There are 364 possibilities for this birthday, if no two people are to have the same birthday. Choose a third person. There are 363 possibilities left for this birthday. Finally, we arrive at the rth person. There are $365 - (r - 1)$ possibilities left for this birthday. By the Multiplication Principle, the total number of possibilities is

$$c(\overline{E}) = 365 \cdot 364 \cdot 363 \cdot \ldots \cdot (365 - r + 1).$$

The probability of event $\overline{E}$ is

$$P(\overline{E}) = \frac{c(\overline{E})}{c(S)} = \frac{365 \cdot 364 \cdot 363 \cdot \ldots \cdot (365 - r + 1)}{365^r}$$

The probability that two or more people have the same birthday is then

$$P(E) = 1 - P(\overline{E})$$

For example, in a group of eight people, the probability that two or more will have the same birthday is

$$\begin{aligned} P(E) &= 1 - \frac{365 \cdot 364 \cdot 363 \cdot 362 \cdot 361 \cdot 360 \cdot 359 \cdot 358}{365^8} \\ &= 1 - .93 \\ &= .07 \end{aligned}$$

Table 3 gives the probabilities for two or more people having the same birthday. Notice that the probability is better than $\frac{1}{2}$ for any group of 23 or more people.

TABLE 3

	Number of People															
	5	10	15	20	21	22	23	24	25	30	40	50	60	70	80	90
Probability That 2 or More Have Same Birthday	.027	.117	.253	.411	.444	.476	.507	.538	.569	.706	.891	.970	.994	.99916	.99991	.99999

NOW WORK PROBLEM 13.

EXAMPLE 6 Finding Probabilities Using Counting Techniques

If the letters in the word MISTER are randomly scrambled, what is the probability that the resulting rearrangement has the I preceding the E?

SOLUTION The 6 letters in the word MISTER can be rearranged in $P(6, 6) = 6!$ ways. We need to count the number of arrangements in which the letter I precedes the letter E. We can view the problem as consisting of two tasks.

Task 1 is to choose two of the six positions for the I and E, with I preceding E. There are $P(6, 2)$ ways of arranging I and E in six positions, $\frac{1}{2}$ of which have I before E. Task 1 can be done in $\dfrac{P(6, 2)}{2}$ ways.

Task 2 would be to arrange the letters MSTR in the remaining four positions. Task 2 can be done in 4! ways.

By the Multiplication Principle, the number of arrangements with the I preceding the E is

$$\frac{P(6, 2)}{2} \cdot 4! = \frac{6 \cdot 5}{2} \cdot 4! = 15 \cdot 4!$$

So the desired probability is

$$\frac{15 \cdot 4!}{6!} = \frac{15}{6 \cdot 5} = \frac{15}{30} = \frac{1}{2}$$

The answer should be intuitively plausible—we would expect one half of the arrangements to have I before E and the other half to have E before I.

NOW WORK PROBLEM 17.

EXERCISE 7.3 Answers to Odd-Numbered Problems Begin on Page AN-31.

1. **Defective Refrigerator** Through a mix-up on the production line, 6 defective refrigerators were shipped out with 44 good ones. If 5 are selected at random, what is the probability that all 5 are defective? What is the probability that at least 2 of them are defective?

2. **Defective Transformers** In a shipment of 50 transformers, 10 are known to be defective. If 30 transformers are picked at random, what is the probability that all 30 are nondefective? Assume that all transformers look alike and have an equal probability of being chosen.

3. A fair coin is tossed 5 times.
   (a) Find the probability that exactly 3 heads appear.
   (b) Find the probability that no heads appear.

4. A fair coin is tossed 6 times.
   (a) Find the probability that exactly 1 tail appears.
   (b) Find the probability that no more than 1 tail appears.

5. A pair of fair dice are tossed 3 times.
   (a) Find the probability that the sum of seven appears 3 times.
   (b) Find the probability that a sum of 7 or 11 appears at least twice.

6. A pair of fair dice are tossed 5 times.
   (a) Find the probability that the sum is never 2.
   (b) Find the probability that the sum is never 7.

7. What is the probability that a seven-digit phone number has one or more repeated digits?

8. What is the probability that a seven-digit phone number contains the number 7?

9. Five letters, with repetition allowed, are selected from the alphabet. What is the probability that none is repeated?

10. Four letters, with repetition allowed, are selected from the alphabet. What is the probability that none of them is a vowel (a, e, i, o, u)?

11. **Best Movies** Based on revenue (including re-releases) adjusted to account for inflation, the top four movies of all time, in alphabetical order, are *E. T. The Extra-Terrestrial*, *Gone with the Wind*, *Star Wars*, and *The Sound of Music*. What is the probability that a random ordering of these four movies would match a listing of these movies, based on revenue, from highest revenue to lowest revenue?

    Source: Canadian Broadcasting Corporation (CBC), CBC News Online, May 16, 2002.

12. **Birthday Problem** What is the probability that, in a group of 3 people, at least 2 were born in the same month (disregard day and year)?

13. **Birthday Problem** What is the probability that, in a group of 6 people, at least 2 were born in the same month (disregard day and year)?

14. A box contains 100 slips of paper numbered from 1 to 100. If 3 slips are drawn in succession with replacement, what is the probability that at least 2 of them have the same number?

15. **Birthday Problem** Find the probability that 2 or more U.S. senators have the same birthday. (There are 100 Senators.)

16. **Birthday Problem** Find the probability that 2 or more members of the House of Representatives have the same birthday. (There are 435 representatives.)

17. **Rearranging Letters** If the five letters in the word VOWEL are rearranged, what is the probability the L will precede the E?

18. **Rearranging Letters** If the four letters in the word MATH are rearranged, what is the probability the word begins with the letters TH?

19. **Rearranging Letters** If the five letters in the word VOWEL are rearranged, what is the probability the word will begin with L?

20. **Rearranging Letters** If the seven letters in the word DEFAULT are rearranged, what is the probability the word will end in E and begin with T?

Problems 21–23 use the following discussion.

National League Baseball Divisional Playoffs *For the 2002 baseball season, there were 16 teams in the National League. These teams were divided into 3 divisions as follows: 5 teams in the East Division, 6 teams in the Central Division, and 5 teams in the West Division. Four of these teams participate in the playoffs. The division winner from each division goes to the playoffs; in addition, from the 13 teams that were not division winners, the team with the best record participates in the playoffs as the wild card team.*

Source: Major League Baseball.

21. What is the probability that two particular teams from the West Division, the San Francisco Giants and the Los Angeles, Dodgers, both go to the playoffs, assuming that each possible collection of 4 playoff teams from the National League is equally likely to occur?

22. What is the probability that one particular team from the West Division, the San Francisco Giants, goes to the playoffs and another team from the West Division, the Los Angeles Dodgers, does not go to the playoffs, assuming that each possible collection of 4 playoff teams from the National League is equally likely to occur?

23. What is the probability that the wild card team is a team in the Central Division, assuming that each possible collection of 4 playoff teams from the National League is equally likely to occur?

24. Five cards are dealt at random from a regular deck of 52 playing cards. Find the probability that

    (a) All are hearts.
    (b) Exactly 4 are spades.
    (c) Exactly 2 are clubs.

25. In a game of bridge find the probability that a hand of 13 cards consists of 5 spades, 4 hearts, 3 diamonds, and 1 club.

26. Find the probability of obtaining each of the following poker hands:

    (a) Royal flush (10, J, Q, K, A all of the same suit)
    (b) Straight flush (5 cards in sequence in a single suit, but not a royal flush)
    (c) Four of a kind (4 cards of the same face value)
    (d) Full house (one pair and one triple of the same face values)
    (e) Flush (5 nonconsecutive cards each of the same suit)
    (f) Straight (5 consecutive cards, not all of the same suit)

27. **Elevator Problem** An elevator starts with 5 passengers and stops at 8 floors. Find the probability that no 2 passengers leave at the same floor. Assume that all arrangements of discharging the passengers have the same probability.

7.4 Conditional Probability

OBJECTIVES
1. Find conditional probabilities
2. Find probabilities using the Product Rule
3. Find probabilities using a tree diagram

Whenever we compute the probability of an event, we do it relative to the entire sample space. When we ask for the probability $P(E)$ of the event E, this probability $P(E)$ represents an appraisal of the likelihood that a chance experiment will produce an outcome in the set E relative to the sample space S.

However, sometimes we want to compute the probability of an event E of a sample space relative to another event F of the same sample space. That is, if we have *prior* information that the outcome must be in a set F, this information should be used to reappraise the likelihood that the outcome will also be in E. This reappraised probability is denoted by $P(E|F)$, and is read as the *probability of E given F*. It represents the answer to the question, "How probable is E, given that F has occurred?"

Let's work some examples to illustrate conditional probability before giving the definition.

EXAMPLE 1 Example of Conditional Probability

FIGURE 7

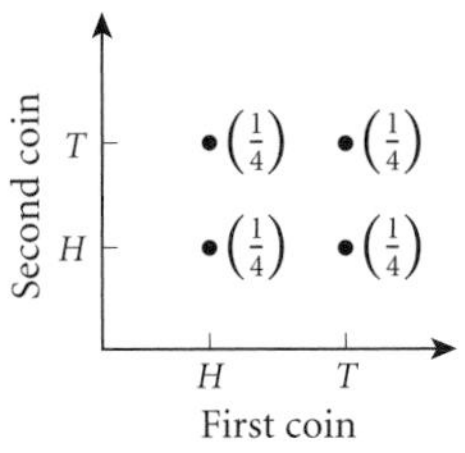

Consider the experiment of flipping two fair coins. As we have previously seen, a sample space S is

$$S = \{HH, HT, TH, TT\}$$

Figure 7 illustrates the sample space and, for convenience, the probability of each outcome.

Suppose the experiment is performed by another person and we have no knowledge of the result, but we are informed that at least one tail was tossed. This information means the outcome HH could not have occurred. But the remaining outcomes HT, TH, TT are still possible. See Figure 8. How does this alter the probabilities of the remaining outcomes?

FIGURE 8

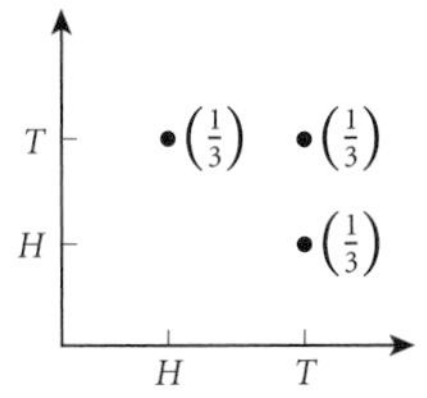

For instance, we might be interested in calculating the probability of the event $\{TT\}$. The three outcomes TH, HT, TT were each assigned the probability $\frac{1}{4}$ *before* we knew the information that at least one tail occurred, so it is not reasonable to assign them this same probability now. Since only three equally likely outcomes are now possible, we assign to each of them the probability $\frac{1}{3}$.

EXAMPLE 2 Example of Conditional Probability

Suppose a population of 1000 people includes 70 accountants and 520 females. There are 40 females who are accountants. A person is chosen at random, and we are told the person is female. The probability the person is an accountant, given that the person is female, is $\frac{40}{520}$. The ratio $\frac{40}{520} = \frac{1}{13}$ represents the *conditional probability* of the event E (accountant) assuming the event F (the person chosen is female).

We shall use the symbol $P(E|F)$, read "the probability of E given F," to denote conditional probability. For Example 2, if we let E be the event "A person chosen at random is an accountant" and we let F be the event "A person chosen at random is female," we would write

$$P(E|F) = \frac{40}{520} = \frac{1}{13}$$

FIGURE 9

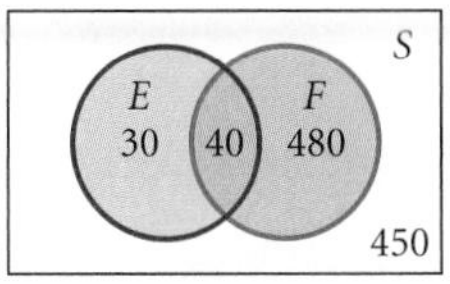

Figure 9 illustrates that in computing $P(E|F)$ in Example 2, we form the ratio of the numbers of those entries in E and in F with the numbers that are in F. Since $c(E \cap F) = 40$ and $c(F) = 520$, then

$$P(E|F) = \frac{c(E \cap F)}{c(F)} = \frac{40}{520} = \frac{1}{13}$$

Notice that

$$P(E|F) = \frac{\dfrac{c(E \cap F)}{c(S)}}{\dfrac{c(F)}{c(S)}} = \frac{P(E \cap F)}{P(F)}$$

With this result in mind we define conditional probability as follows:

Conditional Probability

Let E and F be events of a sample space S and suppose $P(F) > 0$. The **conditional probability of the event E, assuming the event F,** denoted by $P(E|F)$, is defined as

$$P(E|F) = \frac{P(E \cap F)}{P(F)} \quad \textbf{(1)}$$

We give a proof of Formula (1) for sample spaces involving equally likely outcomes.

PROOF Suppose E and F are two events for a particular experiment. Assume that the sample space S for this experiment has n equally likely outcomes. Suppose event F has m outcomes, while $E \cap F$ has k outcomes ($k \leq m$). Since the outcomes are equally likely, we have

$$P(F) = \frac{c(F)}{c(S)} = \frac{m}{n} \quad \textbf{(2)}$$

and

$$P(E \cap F) = \frac{c(E \cap F)}{c(S)} = \frac{k}{n} \quad \textbf{(3)}$$

We wish to compute $P(E|F)$, the probability that E occurs given that F has occurred. Since we assume F has occurred, look only at the m outcomes in F. Of these m outcomes, there are k outcomes where E also occurs (since $E \cap F$ has k outcomes). See Figure 10. This means that

FIGURE 10

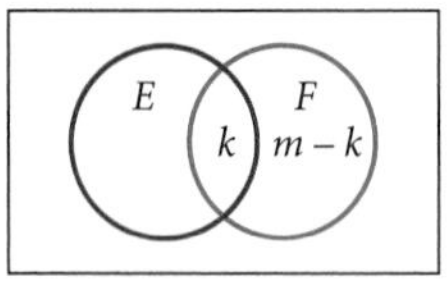

$$P(E|F) = \frac{k}{m} \quad \textbf{(4)}$$

Now, divide the numerator and the denominator of Equation (4) by n and use Equations (2) and (3) to get

$$P(E|F) = \frac{\dfrac{k}{n}}{\dfrac{m}{n}} = \frac{P(E \cap F)}{P(F)}$$

1 **EXAMPLE 3** **Finding a Conditional Probability**

Consider a three-child family for which the sample space S is

$$S = \{BBB, BBG, BGB, BGG, GBB, GBG, GGB, GGG\}$$

We assume that each outcome is equally likely, so that each outcome is assigned a probability of $\frac{1}{8}$. Let E be the event, "The family has exactly two boys" and let F be the event

"The first child is a boy." What is the probability that the family has two boys, given that the first child is a boy?

SOLUTION We want to find $P(E|F)$. The events E and F are

$$E = \{BBG, BGB, GBB\} \qquad F = \{BBB, BBG, BGB, BGG\}$$

Since $E \cap F = \{BBG, BGB\}$, we have

$$P(E \cap F) = \frac{1}{4} \qquad P(F) = \frac{1}{2}$$

Now we use Equation (1) to get

$$P(E|F) = \frac{P(E \cap F)}{P(F)} = \frac{\frac{1}{4}}{\frac{1}{2}} = \frac{1}{2}$$

FIGURE 11

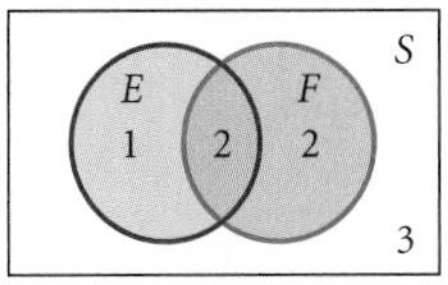

Since the sample space S of Example 3 consists of equally likely outcomes, we can also compute $P(E|F)$ using the Venn diagram in Figure 11. Then

$$P(E|F) = \frac{c(E \cap F)}{c(F)} = \frac{2}{4} = \frac{1}{2}$$

Let's analyze the situation in Example 3 more carefully. See Figure 11. The event E is "the family has exactly 2 boys." We have computed the probability of event E, knowing event F has occurred. This means we are computing a probability relative to a *new sample space F*. That is, F is treated as the universal set and we should consider only that part of E that is included in F, namely, $E \cap F$. So if E is any subset of the sample space S, then $P(E|F)$ provides the reappraisal of the likelihood that an outcome of the experiment will be in the set E if we have prior information that it must be in the set F.

NOW WORK PROBLEM 3.

EXAMPLE 4 Finding Probabilities

In a certain experiment the events E and F have the characteristics

$$P(E) = .7 \qquad P(F) = .8 \qquad P(E \cap F) = .6$$

Find

(a) $P(E \cup F)$ **(b)** $P(E|F)$ **(c)** $P(F|E)$ **(d)** $P(\bar{E}|\bar{F})$

SOLUTION We can visualize the situation by using a Venn diagram. See Figure 12.

FIGURE 12

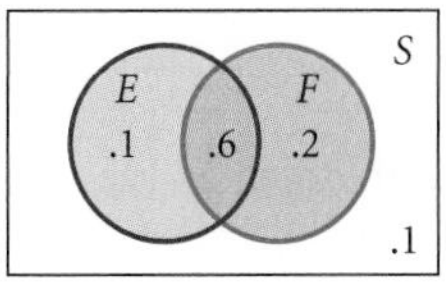

(a) To find $P(E \cup F)$, we use the Additive Rule.

$$P(E \cup F) = P(E) + P(F) - P(E \cap F) = .7 + .8 - .6 = .9$$

(b) To find the conditional probability $P(E|F)$, we use Equation (1).

$$P(E|F) = \frac{P(E \cap F)}{P(F)} = \frac{.6}{.8} = .75$$

(c) To find the conditional probability $P(F|E)$, we use a variation of Equation (1), replacing E by F and F by E. The result is

$$P(F|E) = \frac{P(F \cap E)}{P(E)} = \frac{.6}{.7} = .857$$

(d) To find the conditional probability $P(\overline{E} \mid \overline{F})$, we use a variation of Equation (1), replacing E by $\overline{E}$ and F by $\overline{F}$. The result is

$$P(\overline{E} \mid \overline{F}) = \frac{P(\overline{E} \cap \overline{F})}{P(\overline{F})}$$

Since $\overline{E} \cap \overline{F} = \overline{E \cup F}$ (De Morgan's property), we have

$$P(\overline{E} \mid \overline{F}) = \frac{P(\overline{E \cup F})}{P(\overline{F})} \underset{\substack{\uparrow \\ \text{Use the probability of a} \\ \text{complement}}}{=} \frac{1 - P(E \cup F)}{1 - P(F)} = \frac{1 - .9}{1 - .8} = \frac{.1}{.2} = .50$$

NOW WORK PROBLEMS 23 AND 25.

Product Rule

If in Formula (1) we multiply both sides of the equation by $P(F)$, we obtain the following useful relationship, which is referred to as the **Product Rule:**

Product Rule

For two events E and F, the probability of the event E and F is given by

$$P(E \cap F) = P(F) \cdot P(E \mid F) \tag{5}$$

The next example illustrates how the Product Rule is used to compute the probability of an event that is itself a sequence of two events.

2 **EXAMPLE 5** **Finding Probabilities Using the Product Rule**

Two cards are drawn at random (without replacement) from a regular deck of 52 cards. What is the probability that the first card is a diamond and the second is red?

SOLUTION Define the events

A: The first card is a diamond

B: The second card is red

We seek the probability of A and B, namely $P(A \cap B)$.
Since there are 52 cards in the deck, of which 13 are diamonds, it follows that

$$P(A) = \frac{13}{52} = \frac{1}{4}$$

If A occurred, it means that there are only 51 cards left in the deck, of which 25 are red, so

$$P(B \mid A) = \frac{25}{51}$$

By the Product Rule,

$$P(A \cap B) = P(B \cap A) = P(A) \cdot P(B \mid A) = \frac{1}{4} \cdot \frac{25}{51} = \frac{25}{204}$$

Find probabilities using a tree diagram **3**

A tree diagram is helpful for problems like that of Example 5. See Figure 13. The branch leading to A: Diamond has probability $\frac{13}{52} = \frac{1}{4}$; the branch leading to Heart also has probability $\frac{13}{52} = \frac{1}{4}$; and the branch leading to Club (or Spade) has probability $\frac{13}{52} = \frac{1}{4}$. These must add up to 1 since no other possibilities (branches) are possible. The branch from A to Red is the conditional probability of drawing a red card on the second draw after a diamond on the first draw, $P(\text{Red}|\text{Diamond})$, which is $\frac{25}{51}$. The branch from A to Black is the conditional probability $P(\text{Black}|\text{Diamond})$, which is $\frac{26}{51}$ (26 black cards and 51 total cards remain after a diamond on the first draw). The remaining entries are obtained similarly. Notice that the probability the first card is a diamond and the second a red corresponds to tracing the top branch of the tree. The Product Rule then tells us to multiply the branch probabilities.

FIGURE 13

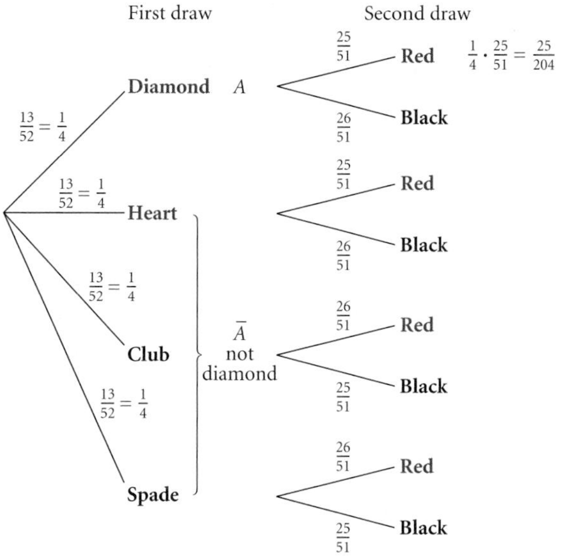

A further advantage of using a tree diagram is that it enables us to easily answer other questions about the experiment. For example, based on Figure 13, we see that

(a) Probability the first card is a heart and the second is black equals $\dfrac{1}{4} \cdot \dfrac{26}{51}$

(b) Probability the first card is a club and the second is red equals $\dfrac{1}{4} \cdot \dfrac{26}{51}$

(c) Probability the first card is a spade and the second is black equals $\dfrac{1}{4} \cdot \dfrac{25}{51}$

NOW WORK PROBLEM 33.

EXAMPLE 6 Using a Tree Diagram and the Product Rule

From a box containing four white, three yellow, and one green ball, two balls are drawn one at a time without replacing the first before the second is drawn. Use a tree diagram to find the probability that one white and one yellow ball are drawn.

SOLUTION To fix our ideas, we define the events

W: White ball drawn $\quad Y$: Yellow ball drawn $\quad G$: Green ball drawn

The tree diagram for this experiment is given in Figure 14.

FIGURE 14

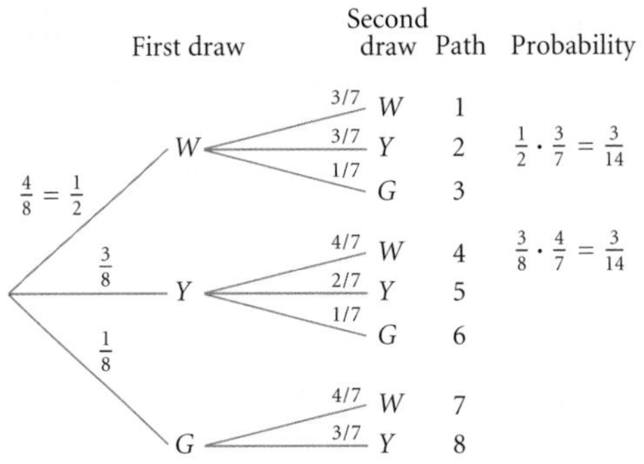

The event of drawing one white ball and one yellow ball can occur in two ways: drawing a white ball first and then a yellow ball (path 2 of the tree diagram in Figure 14), or drawing a yellow ball first and then a white ball (path 4).

Consider path 2. Since four of the eight balls are white, on the first draw we have

$$P(W) = P(W \text{ on 1st}) = \frac{4}{8} = \frac{1}{2}$$

Since one white ball has been removed, leaving seven balls in the box, of which three are yellow, on the second draw we have

$$P(Y|W) = P(Y \text{ on 2nd} | W \text{ on 1st}) = \frac{3}{7}$$

For path 2 we have

$$P(W) \cdot P(Y|W) = P(W \text{ on 1st}) \cdot P(Y \text{ on 2nd} | W \text{ on 1st}) = \frac{1}{2} \cdot \frac{3}{7} = \frac{3}{14}$$

Similarly, for path 4 we have

$$P(Y) \cdot P(W|Y) = P(Y \text{ on 1st}) \cdot P(W \text{ on 2nd} | Y \text{ on 1st}) = \frac{3}{8} \cdot \frac{4}{7} = \frac{3}{14}$$

Since the two events are mutually exclusive, the probability of drawing one white ball and one yellow ball is the sum of these two probabilities:

$$\frac{3}{14} + \frac{3}{14} = \frac{6}{14} = \frac{3}{7}$$

Compare the solution to Example 6 given here with the solution to Example 1 of Section 7.3 (page 387). Which of the two solutions do you prefer? Why?

NOW WORK PROBLEM 35.

EXAMPLE 7 Using a Tree Diagram to Find Conditional Probabilities

Motors, Inc., has two plants to manufacture cars. Plant I manufactures 80% of the cars and plant II manufactures 20%. At plant I, 85 out of every 100 cars are rated standard quality or better. At plant II, only 65 out of every 100 cars are rated standard quality or better. We would like to answer the following questions:

(a) What is the probability that a customer obtains a standard quality car if he buys a car from Motors, Inc.?

(b) What is the probability that the car came from plant I if it is known that the car is of standard quality?

SOLUTION   We begin with a tree diagram. See Figure 15.

FIGURE 15

.80 Plant I $\begin{cases} \text{.85 Standard or better} & (.80)(.85) = .68 \\ \text{.15 Not} \end{cases}$

.20 Plant II $\begin{cases} \text{.65 Standard or better} & (.20)(.65) = .13 \\ \text{.35 Not} \end{cases}$

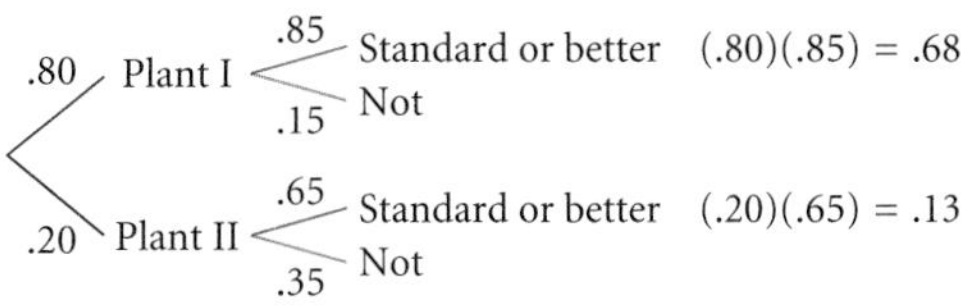

We define the following events:

I: Car came from plant I   II: Car came from plant II

A: Car is of standard quality

(a) There are two ways a standard car can be obtained: either it is standard and came from plant I, or else it is standard and came from plant II. By the Product Rule,

$$P(A \cap I) = P(I) \cdot P(A|I) = (.8)(.85) = .68$$
$$P(A \cap II) = P(II) \cdot P(A|II) = (.2)(.65) = .13$$

Since $A \cap I$ and $A \cap II$ are mutually exclusive, we have

$$P(A) = P(A \cap I) + P(A \cap II) = .68 + .13 = .81$$

(b) To compute $P(I|A)$, we use the definition of conditional probability.

$$P(I|A) = \frac{P(I \cap A)}{P(A)} = \frac{.68}{.81} = .8395$$

NOW WORK PROBLEM 59.

EXERCISE 7.4   **Answers to Odd-Numbered Problems Begin on Page AN-31.**

In Problems 1–8, use the Venn diagram below to find each probability.

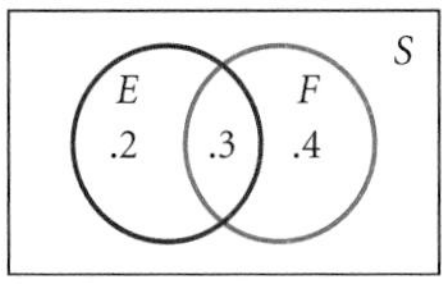

1. $P(E)$

2. $P(F)$

3. $P(E|F)$

4. $P(F|E)$

5. $P(E \cap F)$

6. $P(E \cup F)$

7. $P(\overline{E})$

8. $P(\overline{F})$

9. If E and F are events with $P(E) = .2$, $P(F) = .4$, and $P(E \cap F) = .1$, find the probability of E given F. Find $P(F|E)$.

10. If E and F are events with $P(E) = .5$, $P(F) = .6$, and $P(E \cap F) = .3$, find the probability of E given F. Find $P(F|E)$.

11. If E and F are events with $P(E \cap F) = .2$ and $P(E|F) = .4$, find $P(F)$.

12. If E and F are events with $P(E \cap F) = .2$ and $P(E|F) = .6$, find $P(F)$.

13. If E and F are events with $P(F) = \frac{5}{13}$ and $P(E|F) = \frac{4}{5}$, find $P(E \cap F)$.

14. If E and F are events with $P(F) = .38$ and $P(E|F) = .46$, find $P(E \cap F)$.

15. If E and F are events with $P(E \cap F) = \frac{1}{3}$, $P(E|F) = \frac{1}{2}$, and $P(F|E) = \frac{2}{3}$, find
    (a) $P(E)$ (b) $P(F)$

16. If E and F are events with $P(E \cap F) = .1$, $P(E|F) = .25$, and $P(F|E) = .125$, find
    (a) $P(E)$ (b) $P(F)$

In Problems 17–22, find each probability by referring to the following tree diagram:

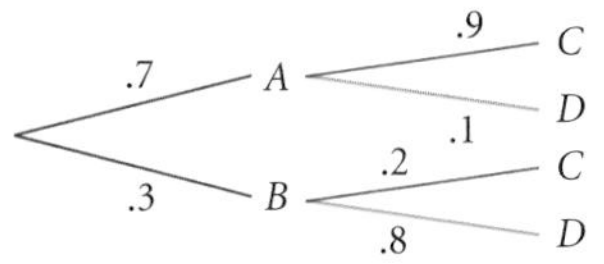

17. $P(C)$

18. $P(D)$

19. $P(C|A)$

20. $P(D|A)$

21. $P(C|B)$

22. $P(D|B)$

In Problems 23–28, E and F are events in a sample space S for which

$$P(E) = .5 \qquad P(F) = .4 \qquad P(E \cup F) = .8$$

Find each probability

23. $P(E \cap F)$

24. $P(E|F)$

25. $P(F|E)$

26. $P(\overline{E}|F)$

27. $P(E|\overline{F})$

28. $P(\overline{E}|\overline{F})$

29. For a 3-child family, find the probability of exactly 2 girls, given that the first child is a girl.

30. For a 3-child family, find the probability of exactly 1 girl, given that the first child is a boy.

31. A fair coin is tossed 4 successive times. Find the probability of obtaining 4 heads. Does the probability change if we are told that the second throw resulted in a head?

32. A pair of fair dice is thrown and we are told that at least one of them shows a 2. If we know this, what is the probability that the total is 7?

33. Two cards are drawn at random (without replacement) from a regular deck of 52 cards. What is the probability that the first card is a heart and the second is red? What is the probability that the first card is red and the second is a heart?

34. If 2 cards are drawn from a regular deck of 52 cards without replacement, what is the probability that the second card is a queen? What is the probability that both cards are queens?

35. From a box containing 3 white, 2 green, and 1 yellow ball, 2 balls are drawn at a time without replacing the first before the second is drawn. Find the probability that 1 white and 1 yellow ball are drawn.

36. A box contains 2 red, 4 green, 1 black, and 8 yellow marbles. If 2 marbles are selected without replacement, what is the probability that one is red and one is green?

37. A card is drawn at random from a regular deck of 52 cards. What is the probability that

(a) The card is a red ace?
(b) The card is a red ace if it is known an ace was picked?
(c) The card is a red ace if it is known a red card was picked?

38. A card is drawn at random from a regular deck of 52 cards. What is the probability that

(a) The card is a black jack?
(b) The card is a black jack if it is known a jack was picked?
(c) The card is a black jack if it is known a black card was picked?

39. In a small town it is known that 20% of the families have no children, 30% have 1 child, 20% have 2 children, 16% have 3 children, 8% have 4 children, and 6% have 5 or more children. Find the probability that a family has more than 2 children if it is known that it has at least 1 child.

40. A sequence of 2 cards is drawn from a regular deck of 52 cards (without replacement). What is the probability that the first card is red and the second is black?

In Problems 41–52, use the table to obtain probabilities for events in a sample space S.

	E	F	G	Totals
H	.10	.06	.08	.24
I	.30	.14	.32	.76
Totals	.40	.20	.40	1.00

For Problems 41–48, read each probability directly from the table:

41. $P(E)$

42. $P(G)$

43. $P(H)$

44. $P(I)$

45. $P(E \cap H)$

46. $P(E \cap I)$

47. $P(G \cap H)$

48. $P(G \cap I)$

For Problems 49–52, use Equation (1) and the appropriate values from the previous table to compute the conditional probability.

49. $P(E|H)$

50. $P(E|I)$

51. $P(G|H)$

52. $P(G|I)$

53. Blood Types Each of the eight possible blood types is listed below with the percent of the U.S. population having that blood type:

O Positive—38%
A Positive—34%
B Positive—9%
O Negative—7%
A Negative—6%
AB Positive—3%
B Negative—2%
AB Negative—1%

Source: AABB Facts about Blood, 7/1/98.

Let E be the event that a randomly selected person in the United States has type B blood. Let F be the event that a randomly selected person in the United States has Rh-positive blood. Find $P(E|F)$ and $P(F|E)$.

54. Voting Preferences A recent poll of residents in a certain community revealed the following information about voting preferences:

	Democrat	Republican	Independent
Male	50	40	30
Female	60	30	25

Events M, F, D, R, and I are defined as follows:

M: Resident is male.
F: Resident is female.
D: Resident is a Democrat.
R: Resident is a Republican.
I: Resident is an Independent.

Find

(a) $P(F|I)$ (b) $P(R|F)$ (c) $P(M|D)$
(d) $P(D|M)$ (e) $P(M|R \cup I)$ (f) $P(I|M)$

55. Graduate Profiles The following table summarizes the graduating class of a midwestern university:

	Arts and Sciences A	Education E	Business B	Total
Male, M	342	424	682	1448
Female, F	324	102	144	570
Total	666	526	826	2018

A student is selected at random from the graduating class. Find the probability that the student

(a) is male.
(b) is receiving an arts and sciences degree.
(c) is a female receiving a business degree.
(d) is a female, given that the student is receiving an education degree.
(e) is receiving an arts and sciences degree, given that the student is a male.
(f) is a female, given that the student is receiving an arts and sciences degree or an education degree.
(g) is not receiving a business degree and is male.
(h) is female, given that an education degree is not received.

56. Beer Preferences The following data are the result of a survey conducted by a marketing company to determine beer preferences.

	Do Not Drink Beer N	Light Beer L	Regular Beer R	Total
Female, F	224	420	622	1266
Male, M	196	512	484	1192
Total	420	932	1106	2458

A respondent is selected at random. Find the probability that

(a) the respondent does not drink beer.
(b) the respondent is a female.
(c) the respondent is a female who prefers regular beer.
(d) the respondent prefers regular beer, given that the respondent is male.
(e) the respondent is male, given that the respondent prefers regular beer.
(f) the respondent is female, given that the respondent prefers regular beer or does not drink beer.

57. 12-Sided Dice Consider a pair of dice such that each die has the shape of a regular dodecahedron, a 12-sided polyhedron in which each face is a regular pentagon. The faces of each of these dice are numbered with the integers from 1 through 12. An experiment consists of rolling this pair of dice and observing the top faces. Let F be the event that at least one of the top faces shows the number 5. Let E be the event that the total of the numbers on the top faces is equal to 14. Find $P(E|F)$.

Source: The Bone Rollers' Guide to Polyhedral Solids.

58. Of the first-year students in a certain college, it is known that 40% attended private secondary schools and 60% attended public schools. The registrar reports that 30% of all students who attended private schools maintain an A average in their first year at college and that 24% of all first-year students had an A average. At the end of the year, one student is chosen at random from the class. If the student has an A average, what is the conditional probability that the student attended a private school? [*Hint*: Use a tree diagram.]

59. In a rural area in the north, registered Republicans outnumber registered Democrats by 3 to 1. In a recent election all Democrats voted for the Democratic candidate and enough Republicans also voted for the Democratic candidate so that the Democrat won by a ratio of 5 to 4. If a voter is selected at random, what is the probability he or she is Republican? What is the probability a voter is Republican, if it is known that he or she voted for the Democratic candidate?

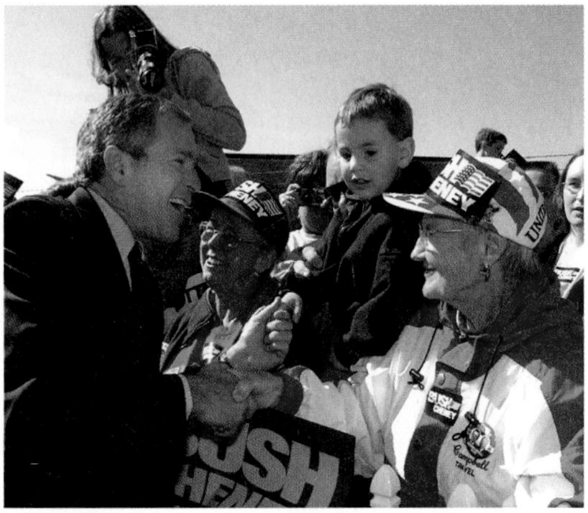

60. "Temp Help" uses a preemployment test to screen applicants for the job of programmer. The test is passed by 70% of the applicants. Among those who pass the test, 85% complete training successfully. In an experiment a random sample of applicants who do not pass the test is also employed. Training is successfully completed by only 40% of this group. If no preemployment test is used, what percentage of applicants would you expect to complete the training successfully?

61. Cigarette Smoking in the US The American Heart Association reported that, for people 18 years of age or older in the United States, an estimated 26.0 million men (25.7%) and 22.7 million women (21.0%) are cigarette smokers. Find the probability that a randomly selected person 18 years of age or older in the United States is a cigarette smoker. Write the answer as a decimal, rounded to the nearest thousandth.

Source: American Heart Association, January 30, 2003.

62. Cigarette Smoking in Britain The British government reported that, in Great Britain in the year 2001, of the 19.913 million men 16 years of age or older, 28% smoked cigarettes; of the 21.987 million women 16 years of age or older, 26% smoked cigarettes. Find the probability that a randomly selected person 16 years of age or older in Great Britain in the year 2001 was a cigarette smoker. Write the answer as a decimal, rounded to the nearest thousandth.

Source: Living in Britain—2001 (published December 17, 2002), which contains information from the General Household Survey (GHS) conducted by the Social Survey Division of the Office for National Statistics, Great Britain.

Problems 63–64 require the following discussion.

Craps *In a popular dice game, the player rolls a pair of fair dice. If the total number of spots is 7 or 11, the player wins. If the total number of spots is 2, 3, or 12, the player loses. If the player rolls any other total, the total rolled becomes the player's "point" and the player continues to roll the pair of dice. To win the game after establishing the player's "point," the player must re-roll a total equal to the "point" that has been established before rolling a total of 7. The player loses at this stage if a total of 7 is rolled while attempting to re-roll the player's "point."*

Source: The International Bone Rollers Guild, Games with Ordinary Dice, from *According to Hoyle,* by Richard L. Frey.

63. Let E be the event that the player rolls a total of 8 on the first roll. (Therefore, the player's "point" is 8.) Let W be the event that the player wins the game. Find $P(W|E)$.
[*Hint*: Use the formula for the sum of the first n terms of a geometric series given in Appendix A.4 and let n approach infinity.]

64. Let E be the event that the player rolls a total of 5 on the first roll. (Therefore, the player's "point" is 5.) Let W be the event that the player wins the game. Find $P(W|E)$.
[*Hint*: Use the formula for the sum of the first n terms of a geometric series given in Appendix A.4 and let n approach infinity.]

65. If E and F are two events with $P(E) > 0$ and $P(F) > 0$, show that

$$P(F) \cdot P(E|F) = P(E) \cdot P(F|E)$$

66. Show that $P(E|E) = 1$ when $P(E) \neq 0$.

67. Show that $P(E|F) + P(\overline{E}|F) = 1$.

68. If S is the sample space, show that $P(E|S) = P(E)$.

69. If $P(E) > 0$ and $P(E|F) = P(E)$, show that $P(F|E) = P(F)$.

7.5 Independent Events

OBJECTIVES
1. **Show two events are independent**
2. **Find $P(E \cap F)$ for independent events E and F**
3. **Find probabilities for several independent events**

One of the most important concepts in probability is that of independence. In this section we define what is meant by two events being *independent*. First, however, we provide an intuitive idea of the meaning of independent events.

EXAMPLE 1 Example of Independent Events

Consider a group of 36 students. Define the events E and F as

E: Student has blue eyes $\quad$ F: Student is female

With regard to these two characteristics, suppose it is found that the 36 students are distributed as shown in Table 4.

TABLE 4

	Blue Eyes E	Not Blue Eyes $\overline{E}$	Totals
Female, F	12	12	24
Male, $\overline{F}$	6	6	12
Totals	18	18	36

What is $P(E|F)$?

SOLUTION If we choose a student at random, the following probabilities can be obtained from the table:

$$P(E) = \frac{18}{36} = \frac{1}{2} \qquad P(F) = \frac{24}{36} = \frac{2}{3}$$

$$P(E \cap F) = \frac{12}{36} = \frac{1}{3}$$

Then we find that

$$P(E|F) = \frac{P(E \cap F)}{P(F)} = \frac{\frac{1}{3}}{\frac{2}{3}} = \frac{1}{2} = P(E)$$

In Example 1 the probability of E given F equals the probability of E. This situation can be described by saying that the information that the event F has occurred does not affect the probability of the event E. If this is the case, we say that *E is independent of F.*

E Is Independent of *F*

Let E and F be two events of a sample space S with $P(F) > 0$. **The event E is independent of the event F** if and only if

$$\boxed{P(E|F) = P(E)}$$

The result that follows is intuitively obvious. The proof is outlined in Problem 49.

Let E, F be events for which $P(E) > 0$ and $P(F) > 0$. If E is independent of F, then F is independent of E.

This result forms the basis for the following definition.

Independent Events

If two events E and F have positive probabilities and if the event E is independent of F, then F is also independent of E. In this case E and F are called **independent events.**

The following result will be used frequently. Its proof is outlined in Problem 50.

Criterion for Independent Events

Two events E and F of a sample space S are independent events if and only if

$$\boxed{P(E \cap F) = P(E) \cdot P(F)} \qquad (1)$$

That is, the probability of E *and* F is equal to the product of the probability of E and the probability of F.

The Criterion for Independent Events actually consists of two statements:

1. If $P(E \cap F) = P(E) \cdot P(F)$ for two events E and F, then the events E and F are independent.
2. If two events E and F are independent, then $P(E \cap F) = P(E) \cdot P(F)$.

We use statement 1 to show that two events are independent.
We use statement 2 when we know or are told two events are independent.

The next example shows how statement 1 is used to show the independence of two events.

1 **EXAMPLE 2** **Showing Two Events Are Independent**

Suppose $P(E) = \frac{1}{4}$, $P(F) = \frac{2}{3}$, and $P(E \cap F) = \frac{1}{6}$. Show that E and F are independent events.

SOLUTION We see if Equation (1) holds.

$$P(E) \cdot P(F) = \frac{1}{4} \cdot \frac{2}{3} = \frac{1}{6}$$

$$P(E \cap F) = \frac{1}{6}$$

Since Equation (1) holds, by the Criterion for Independence, E and F are independent events.

NOW WORK PROBLEM 5.

EXAMPLE 3 **Showing Two Events Are Independent**

Suppose a red die and a green die are thrown. Let event E be "Throw a 5 with the red die," and let event F be "Throw a 6 with the green die." Show that E and F are independent events.

SOLUTION In this experiment the events E and F are

$$E = \{(5, 1), (5, 2), (5, 3), (5, 4), (5, 5), (5, 6)\}$$
$$F = \{(1, 6), (2, 6), (3, 6), (4, 6), (5, 6), (6, 6)\}$$

Then

$$P(E) = \frac{6}{36} = \frac{1}{6} \qquad P(F) = \frac{6}{36} = \frac{1}{6}$$

Also, the event E and F is

$$E \cap F = \{(5, 6)\}$$

so that

$$P(E \cap F) = \frac{1}{36}$$

Since $P(E) \cdot P(F) = \frac{1}{6} \cdot \frac{1}{6} = \frac{1}{36} = P(E \cap F)$, E and F are independent events.

EXAMPLE 4 Testing for Independent Events

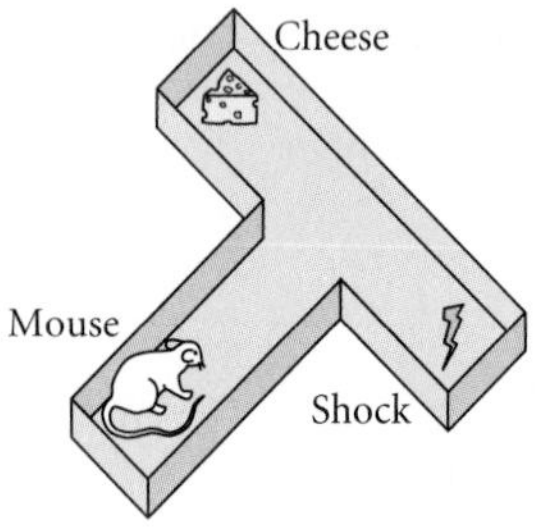

In a T-maze a mouse may run to the right, R, or to the left, L. Suppose its behavior in making such "choices" is random so that R and L are equally likely outcomes. Suppose a mouse is put through the T-maze three times. Define events E, G, and H as

E: Run to the right 2 or more consecutive times
G: Run to the left on first trial
H: Run to the right on second trial

(a) Show that E and G are not independent.
(b) Show that G and H are independent.

SOLUTION **(a)** The events E and G are

$$E = \{\text{RRL, LRR, RRR}\}$$
$$G = \{\text{LLL, LLR, LRL, LRR}\}$$

The sample space S has eight elements, so that

$$P(E) = \frac{3}{8} \qquad P(G) = \frac{1}{2}$$

Also, the event E and G is

$$E \cap G = \{\text{LRR}\}$$

so that

$$P(E \cap G) = \frac{1}{8}$$

Since $P(E \cap G) \neq P(E) \cdot P(G)$, the events E and G are not independent.

(b) The event H is

$$H = \{\text{RRL, RRR, LRL, LRR}\}$$

so that

$$P(H) = \frac{1}{2}$$

The event G and H and its probability are

$$G \cap H = \{\text{LRL, LRR}\} \qquad P(G \cap H) = \frac{1}{4}$$

Since $P(G \cap H) = P(G) \cdot P(H)$, the events G and H are independent.

Example 4 illustrates that the question of whether two events are independent can be answered simply by determining whether Equation (1) is satisfied. Although we may often suspect two events E and F as being independent, our intuition must be checked by computing $P(E)$, $P(F)$, and $P(E \cap F)$ and determining whether $P(E \cap F) = P(E) \cdot P(F)$.

NOW WORK PROBLEM 13.

If E and F are two independent events and $P(E)$ and $P(F)$ are known, then, as statement 2 asserts, we can find $P(E \cap F)$, the probability of the event E and F.

Warning! The two events E and F must be independent in order to use Equation (1) to find $P(E \cap F)$.

2

EXAMPLE 5 Finding $P(E \cap F)$ for Independent Events E and F

Suppose E and F are independent events with $P(E) = .4$ and $P(F) = .3$. Find $P(E \cap F)$.

SOLUTION Since E and F are independent, we use Equation (1) to find $P(E \cap F)$.

$$\begin{aligned} P(E \cap F) &= P(E) \cdot P(F) \quad \text{Equation (1)} \\ &= (.4) \cdot (.3) \\ &= .12 \end{aligned}$$

NOW WORK PROBLEM 1.

The next example shows how we use the idea of independent events in the construction of a probability model.

EXAMPLE 6 Constructing a Probability Model

A fair coin is tossed. If it comes up heads (H), then a fair die is rolled. If it comes up tails (T), then the coin is tossed once more. Construct a probability model for this experiment.

SOLUTION This experiment consists of two stages: the coin toss and, depending on what occurs, the toss of a die or the coin again. See Figure 16 for a tree diagram depicting the experiment. As the tree diagram illustrates, a sample space S for this experiment is

$$S = \{H1, H2, H3, H4, H5, H6, TH, TT\}$$

It remains to assign valid probabilities to each of these simple events. Since the coin is fair, we have $P(H) = P(T) = \frac{1}{2}$. Since the die is also fair, we have $P(1) = P(2) = P(3) = P(4) = P(5) = P(6) = \frac{1}{6}$, as shown in Figure 16.

Now, think of the branches of the tree as events. Then the branches that lead to $H1$ consist of the events "the coin shows heads" and "the die shows 1". These are independent events, so the probability of H followed by 1 is the product $P(H)P(1) = (\frac{1}{2})(\frac{1}{6}) = \frac{1}{12}$.

Similarly, the probability of H followed by 2 is also $\frac{1}{12}$. The probability of T followed by H is $P(T)P(H) = (\frac{1}{2})(\frac{1}{2}) = \frac{1}{4}$.

FIGURE 16

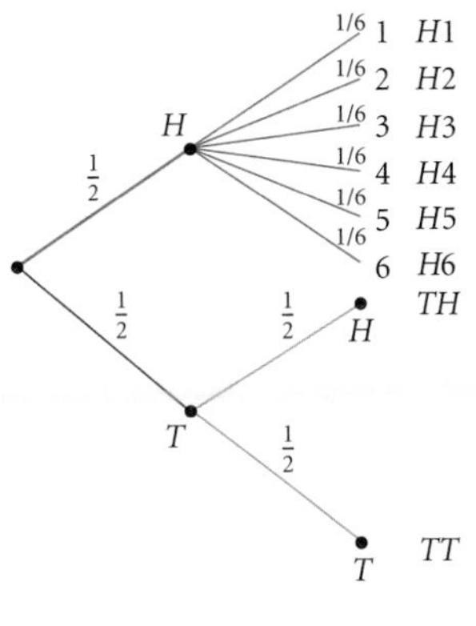

Based on all this, we assign the probabilities to each outcome in S follows:

$$P(H1) = P(H2) = P(H3) = P(H4) = P(H5) = P(H6) = \frac{1}{12}$$
$$P(TH) = P(TT) = \frac{1}{4}$$

This assignment of probabilities has the desired characteristics:

(a) Each probability assignment is nonnegative
(b) The sum of the probability assignments is 1

$$\left[\frac{1}{12} + \frac{1}{12} + \frac{1}{12} + \frac{1}{12} + \frac{1}{12} + \frac{1}{12} + \frac{1}{4} + \frac{1}{4} = 1\right]$$

The above discussion constitutes the construction of a probability model for the experiment.

NOW WORK PROBLEM 17.

For some probability models, an assumption of independence is made. In such instances when two events are independent, Equation (1) may be used to compute the probability that both events occur. The following example illustrates such a situation.

EXAMPLE 7 Planting Seeds

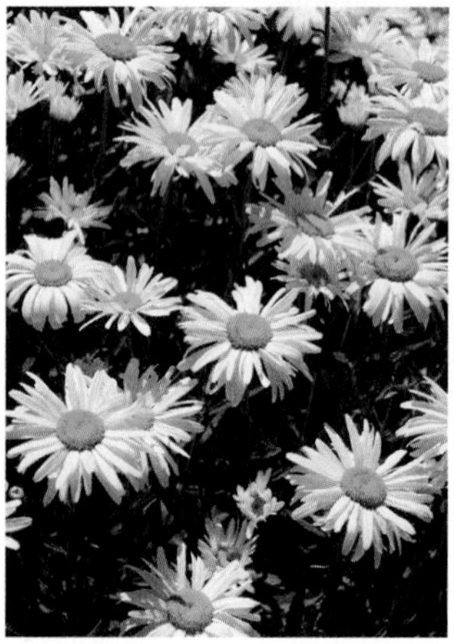

In a group of seeds, $\frac{1}{4}$ of which should produce white plants, the best germination that can be obtained is 75%. If one seed is planted, what is the probability that it will grow into a white plant?

SOLUTION Let G and W be the events

G: The plant will grow
W: The seed will produce a white plant

Assume that W and G are independent events.

Then the probability that the plant grows and is white, namely, $P(W \cap G)$ is

$$P(W \cap G) = P(W) \cdot P(G) = \frac{1}{4} \cdot \frac{3}{4} = \frac{3}{16}$$

A white plant will grow 3 out of 16 times.

NOW WORK PROBLEM 31.

There is a danger that mutually exclusive events and independent events may be confused. A source of this confusion is the common expression, "Events are independent if they have nothing to do with each other." This expression provides a description of independence when applied to everyday events; but when it is applied to sets, it suggests nonoverlapping. Nonoverlapping sets are mutually exclusive but in general are not independent. See Problem 44.

Find probabilities for several independent events

3 Independence for More than Two Events

The concept of independence can be applied to more than two events:

Independent Events

A set $\{E_1, E_2, \ldots, E_n\}$ of n events is called **independent** if the occurrence of one or more of them does not change the probability of any of the others. It can be shown that, for such events,

$$P(E_1 \cap E_2 \cap \ldots \cap E_n) = P(E_1) \cdot P(E_2) \cdot \ldots \cdot P(E_n) \tag{2}$$

EXAMPLE 8 Using Equation (2)

A new skin cream can cure skin infection 90% of the time. If five randomly selected people with skin infections use this cream, assuming independence, what is the probability that

(a) All five are cured? **(b)** All five still have the infection?

SOLUTION **(a)** Let

E_1: First person does not have infection
E_2: Second person does not have infection
E_3: Third person does not have infection
E_4: Fourth person does not have infection
E_5: Fifth person does not have infection

Then, since events E_1, E_2, E_3, E_4, E_5, are given to be independent,

$$\begin{aligned} P(\text{all 5 are cured}) &= P(E_1 \cap E_2 \cap E_3 \cap E_4 \cap E_5) \\ &= P(E_1) \cdot P(E_2) \cdot P(E_3) \cdot P(E_4) \cdot P(E_5) \\ &= (.9)^5 \\ &= .59 \end{aligned}$$

(b) Let

$\overline{E}_i$: ith person has an infection, $i = 1, 2, 3, 4, 5$

Then

$$\begin{aligned} P(\overline{E}_i) &= 1 - .9 = .1 \\ P(\text{all 5 are infected}) &= P(\overline{E}_1 \cap \overline{E}_2 \cap \overline{E}_3 \cap \overline{E}_4 \cap \overline{E}_5) \\ &= P(\overline{E}_1) \cdot P(\overline{E}_2) \cdot P(\overline{E}_3) \cdot P(\overline{E}_4) \cdot P(\overline{E}_5) \\ &= (.1)^5 \\ &= .00001 \end{aligned}$$

EXERCISE 7.5 Answers to Odd-Numbered Problems Begin on Page AN-31.

1. If E and F are independent events and if $P(E) = .4$ and $P(F) = .6$, find $P(E \cap F)$.

2. If E and F are independent events and if $P(E) = .6$ and $P(E \cap F) = .2$, find $P(F)$.

3. If E and F are independent events, find $P(F)$ if $P(E) = .2$ and $P(E \cup F) = .3$.

4. If E and F are independent events, find $P(E)$ if $P(F) = .3$ and $P(E \cup F) = .6$.

5. Suppose E and F are two events such that $P(E) = \frac{4}{21}$, $P(F) = \frac{7}{12}$, and $P(E \cap F) = \frac{2}{9}$. Are E and F independent?

6. If E and F are two events such that $P(E) = .25$, $P(F) = .36$, and $P(E \cap F) = .09$, then are E and F independent?

7. If E and F are two independent events with $P(E) = .2$, and $P(F) = .4$, find

(a) $P(E|F)$ (b) $P(F|E)$
(c) $P(E \cap F)$ (d) $P(E \cup F)$

8. If E and F are independent events with $P(E) = .3$ and $P(F) = .5$, find

(a) $P(E|F)$ (b) $P(F|E)$
(c) $P(E \cap F)$ (d) $P(E \cup F)$

9. If E, F, and G are three independent events with $P(E) = \frac{2}{3}$, $P(F) = \frac{3}{7}$, and $P(G) = \frac{2}{21}$, then find $P(E \cap F \cap G)$.

10. If E_1, E_2, E_3, and E_4 are four independent events with $P(E_1) = .6$, $P(E_2) = .3$, $P(E_3) = .5$, and $P(E_4) = .4$, find $P(E_1 \cap E_2 \cap E_3 \cap E_4)$.

11. If $P(E) = .3$, $P(F) = .2$, and $P(E \cup F) = .4$, what is $P(E|F)$? Are E and F independent?

12. If $P(E) = .4$, $P(F) = .6$, and $P(E \cup F) = .7$, what is $P(E|F)$? Are E and F independent?

13. A fair die is rolled. Let E be the event "1, 2, or 3 is rolled" and let F be the event "3, 4, or 5 is rolled." Are E and F independent?

14. A loaded die is rolled. The probabilities for this die are $P(1) = P(2) = P(4) = P(5) = \frac{1}{8}$ and $P(3) = P(6) = \frac{1}{4}$. Are the events defined in Problem 13 independent in this case?

15. A 12-Sided Die Consider a fair die that has the shape of a regular dodecahedron, a 12-sided polyhedron in which each face is a regular pentagon. The faces of this die are numbered with the integers from 1 through 12. An experiment consists of rolling this dodecahedral die and observing the number on the top face.

(a) Let E be the event that the number on the top face is a 6 or less. Let F be the event that the number on the top face is an odd number. Are E and F independent events? Justify your answer.
(b) Let E be the event that the number on the top face is an 8 or more. Let F be the event that the number on the top face is an even number. Are E and F independent events? Justify your answer.

Source: The Bone Rollers' Guide to Polyhedral Solids.

16. A 20-Sided Die Consider a fair die that has the shape of a regular icosahedron, a 20-sided polyhedron in which each face is an equilateral triangle. The faces of this die are numbered with the integers from 1 through 20. An experiment consists of rolling this icosahedral die and observing the number on the top face.

(a) Let E be the event that the number on the top face is an 11 or more. Let F be the event that the number on the top face is a 9, 10, 11, or 12. Are E and F independent events? Justify your answer.
(b) Let E be the event that the number on the top face is a 10 or less. Let F be the event that the number on the top face is an odd number. Are E and F independent events? Justify your answer.

Source: The Bone Rollers' Guide to Polyhedral Solids.

In Problems 17–26, the experiment consists of tossing a fair die and then a fair coin. Construct a probability model for this experiment. Then find the probability of each event.

17. The coin comes up heads.

18. The coin comes up tails.

19. The die comes up 4.

20. The die comes up 1.

21. The die does not come up 4.

22. The die does not come up 1.

23. The die comes up 5 or 6.

24. The die comes up 1 or 2.

25. The die comes up 3, 4, or 5 and the coin comes up heads.

26. The coin comes up tails and the die comes up a number less than 4.

27. In a T-maze a mouse may turn to the right (R) and receive a mild shock, or to the left (L) and get a piece of cheese. Its behavior in making such "choices" is studied by psychologists. Suppose a mouse runs a T-maze 3 times. List the set of all possible outcomes and assign valid probabilities to each outcome under the assumption that the first two times the

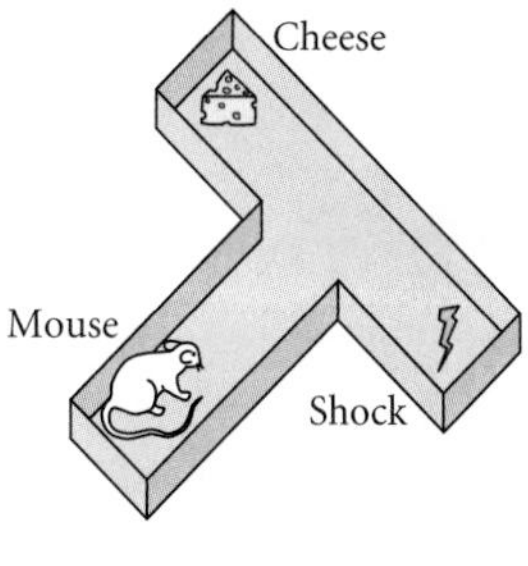

maze is run the mouse chooses equally between left and right, but on the third run, the mouse is twice as likely to choose cheese. Find the probability of each of the events listed.

(a) E: Run to the right 2 consecutive times.
(b) F: Never run to the right.
(c) G: Run to the left on the first trial.
(d) H: Run to the right on the second trial.

Assume the mouse has no memory so the trials are independent.

28. A first card is drawn at random from a regular deck of 52 cards and is then put back into the deck. A second card is drawn. What is the probability that

(a) The first card is a club?
(b) The second card is a heart, given that the first is a club?
(c) The first card is a club and the second is a heart?
(d) The first card is an ace?
(e) The second card is a king, given that the first card is an ace?
(f) The first card is an ace and the second is a king?

29. A fair coin is tossed twice. Define the events E and F to be

E: A head turns up on the first throw of a fair coin.
F: A tail turns up on the second throw of a fair coin.

Show that E and F are independent events.

30. A die is loaded so that

$$P(1) = P(2) = P(3) = \frac{1}{4} \qquad P(4) = P(5) = P(6) = \frac{1}{12}$$

If $A = \{1, 2\}$, $B = \{2, 3\}$, $C = \{1, 3\}$, show that any pair of these events is independent.

31. Cardiovascular Disease Records show that a child of parents with heart disease has a probability of $\frac{3}{4}$ of inheriting the disease. Assuming independence, what is the probability that, for a couple with heart disease that have two children:

(a) Both children have heart disease.
(b) Neither child has heart disease.
(c) Exactly one child has heart disease.

32. Hitting a Target A marksman hits a target with probability $\frac{4}{5}$. Assuming independence for successive firings, find the probabilities of getting

(a) One miss followed by two hits
(b) Two misses and one hit (in any order)

33. A coin is loaded so that tails is three times as likely as heads. If the coin is tossed three times, find the probability of getting

(a) All tails
(b) Two heads and one tail (in any order)

34. Sex of Newborns The probability of a newborn baby being a girl is .49. Assuming that the sex of one baby is independent of the sex of all other babies—that is, the events are independent—what is the probability that all four babies born in a certain hospital on one day are girls?

35. Recovery Rate The recovery rate from a flu is .9. If 4 people have this flu, what is the probability that

(a) All will recover? (b) Exactly 2 will recover?
(c) At least 2 will recover?

Assume independence.

36. Germination In a group of seeds, $\frac{1}{3}$ of which should produce violets, the best germination that can be obtained is 60%. If one seed is planted, what is the probability that it will grow into a violet? Assume independence.

37. A box has 10 marbles in it, 6 red and 4 white. Suppose we draw a marble from the box, replace it, and then draw another. Find the probability that

(a) Both marbles are red
(b) Just one of the two marbles is red

38. Survey In a survey of 100 people, categorized as drinkers or nondrinkers, with or without a liver ailment, the following data were obtained:

	Liver Ailment F	No Liver Ailment $\bar{F}$
Drinkers, E	52	18
Nondrinkers, $\bar{E}$	8	22

(a) Are the events E and F independent?
(b) Are the events $\bar{E}$ and $\bar{F}$ independent?
(c) Are the events E and $\bar{F}$ independent?

39. Insurance By examining the past driving records of 840 randomly selected drivers over a period of 1 year, the following data were obtained.

	Under 25 U	Over 25 $\bar{U}$	Totals
Accident, A	40	5	45
No Accident, $\bar{A}$	285	510	795
Totals	325	515	840

(a) What is the probability of a driver having an accident, given that the person is under 25?
(b) What is the probability of a driver having an accident, given that the person is over 25?
(c) Are events U and A independent?
(d) Are events U and $\bar{A}$ independent?
(e) Are events $\bar{U}$ and A independent?
(f) Are events $\bar{U}$ and $\bar{A}$ independent?

40. **Voting Patterns** The following data show the number of voters in a sample of 1000 from a large city, categorized by religion and their voting preference.

	Demo-crat D	Repub-lican R	Indepen-dent	Totals
Catholic, C	160	150	90	400
Protestant, P	220	220	60	500
Jewish, J	20	30	50	100
Totals	400	400	200	1000

(a) Find the probability a person is a Democrat.
(b) Find the probability a person is a Catholic.
(c) Find the probability a person is Catholic, knowing the person is a Democrat.
(d) Are the events R and D independent?
(e) Are the events P and R independent?

41. **Election** A candidate for office believes that $\frac{2}{3}$ of registered voters in her district will vote for her in the next election. If two registered voters are independently selected at random, what is the probability that

(a) Both of them will vote for her in the next election?
(b) Neither will vote for her in the next election?
(c) Exactly one of them will vote for her in the next election?

42. A woman has 10 keys but only 1 fits her door. She tries them successively (without replacement). Find the probability that a key fits in exactly 5 tries.

43. **Chevalier de Mere's Problem** Which of the following random events do you think is more likely to occur?

(a) To obtain a 1 on at least one die in a simultaneous throw of four fair dice
(b) To obtain at least one pair of 1s in a series of 24 throws of a pair of fair dice
[*Hint*: Part (a): $P(\text{No 1s are obtained}) = 5^4/6^4 = \frac{625}{1296} = .4823$. Part (b): The probability of not obtaining a double 1 on any given toss is $\frac{35}{36}$. So $P(\text{no double 1s are obtained}) = (\frac{35}{36})^{24} = .509$.]

44. Give an example of two events that are

(a) Independent, but not mutually exclusive
(b) Not independent, but mutually exclusive
(c) Not independent and not mutually exclusive

45. Show that whenever two events are both independent and mutually exclusive, then at least one of them is impossible.

46. Let E be any event. If F is an impossible event, show that E and F are independent.

47. Show that if E and F are independent events, so are $\overline{E}$ and $\overline{F}$. [*Hint*: Use De Morgan's properties.]

48. Show that if E and F are independent events and if $P(E) \neq 0$, $P(F) \neq 0$, then E and F are not mutually exclusive.

49. Suppose $P(E) > 0$, $P(F) > 0$ and E is independent of F. Show that F is independent of E.
[*Hint*. First we note that

$$P(F) \cdot P(E|F) = P(F) \cdot \frac{P(E \cap F)}{P(F)} = P(E \cap F)$$

and

$$P(E) \cdot P(F|E) = P(E) \cdot \frac{P(F \cap E)}{P(E)} = P(E \cap F)$$

Then $P(F) \cdot P(E|F) = P(E) \cdot P(F|E)$. Now use the fact that E is independent of F—that is, $P(E|F) = P(E)$—to show that F is independent of E—that is, $P(F|E) = P(F)$.]

50. Prove Equation (1) on page 404.
[*Hint*: If E and F are independent events, then

$$P(E|F) = \frac{P(E \cap F)}{P(F)} \quad \text{and} \quad P(E|F) = P(E)$$

Now show that $P(E \cap F) = P(E) \cdot P(F)$. Conversely, suppose $P(E \cap F) = P(E) \cdot P(F)$. Use the fact that

$$P(E|F) = \frac{P(E \cap F)}{P(F)}$$

to show that $P(E|F) = P(E)$.]

Chapter 7 Review

OBJECTIVES

Section		You should be able to	Review Exercises
7.1	1	Find a sample space	1–4
	2	Assign probabilities	5–8, 27, 28
	3	Construct a probability model	9a, 10a, 11a, 12a
	4	Find probabilities involving equally likely outcomes	9b, 10b, 11b, 12b, 13–14

7.2	1	Find the probability of an event	17, 18, 20, 27, 28
	2	Find the probability of *E* or *F*, when *E* and *F* are mutually exclusive	23, 24, 25, 26, 29, 30
	3	Use the Additive Rule	15, 16, 19b, 20b, 21b, 22a
	4	Use a Venn diagram to find probabilities	37, 38
	5	Find the probability of the complement of an event	15b, 16b, 19a, 20a, 21a, 21c, 22b, 23a, 23b, 24a, 24b
	6	Compute odds	31–34
7.3	1	Find the probability of events using counting techniques	39, 40, 45–50
7.4	1	Find conditional probabilities	37a,b, 38a, 38b, 41c, 42c, 43, 44
	2	Find probabilities using the Product Rule	41a, 42a, 43, 44
	3	Find probabilities using a tree diagram	41, 42
7.5	1	Show two events are independent	35, 36, 43c, 43d, 44f
	2	Find $P(E \cap F)$ for independent events *E* and *F*	40, 51, 52
	3	Find probabilities for several independent events	39, 53, 54

IMPORTANT FORMULAS

Equally Likely Outcomes $P(E) = \dfrac{c(E)}{c(S)}$

Odds for *E* are *a* to *b* $P(E) = \dfrac{a}{a+b}$

Additive Rule For any two events *E* and *F*, $P(E \cup F) = P(E) + P(F) - P(E \cap F)$

Mutually Exclusive Events If *E* and *F* are mutually exclusive events, $P(E \cup F) = P(E) + P(F)$

Complementary Events $P(\overline{E}) = 1 - P(E)$

Conditional Probability $P(E|F) = \dfrac{P(E \cap F)}{P(F)}$

Product Rule $P(E \cap F) = P(F) \cdot P(E|F)$

E* is independent of *F $P(E|F) = P(E)$

Criterion for Independent Events *E*, *F* are independent if and only if $P(E \cap F) = P(E) \cdot P(F)$

TRUE–FALSE ITEMS Answers are on page AN-32.

T F **1.** If the odds for an event *E* are 2 to 1, then $P(E) = \frac{2}{3}$.

T F **2.** The conditional probability of *E* given *F* is

$$P(E|F) = \frac{P(E \cap F)}{P(E)}$$

T F **3.** If two events in a sample space have no outcomes in common, they are said to be independent.

T F **4.** $P(E|F) = P(F|E)$ for any events *E* and *F*.

T F **5.** $P(E) + P(\overline{E}) = 1$

T F **6.** If $P(E) = .4$ and $P(F) = .3$, then $P(E \cup F)$ must be .7.

T F **7.** If $P(E \cap F) = 0$, then *E* and *F* are said to be mutually exclusive.

T F **8.** If $P(E \cup F) = .7$, $P(E) = .4$, and $P(F) = .3$, then $P(E \cap F) = 0$.

FILL IN THE BLANKS Answers are on page AN-32.

1. If $S = \{a, b, c, d\}$ is a sample space, the outcomes are equally likely, and $E = \{a, b\}$, then $P(\overline{E}) =$ ________.

2. If a coin is tossed five times, the number of outcomes in the sample space of this experiment is ________.

3. If an event is certain to occur, then its probability is ________. If an event is impossible, its probability is ________.

4. If $P(\overline{E}) = .2$, then $P(E) =$ ________.

5. If $P(E) = .6$, the odds ____________ E are 3 to 2.

6. When each outcome in a sample space is assigned the same probability, the outcomes are termed ____________ ____________.

7. If two events in a sample space have no outcomes in common, they are said to be ____________.

REVIEW EXERCISES

Answers to odd-numbered problems begin on page AN-32.
Blue problem numbers indicate the author's suggestions for use in a practice test.

1. **Tossing a Coin** A coin is tossed 5 times. We are interested in the number of heads that shows. List the outcomes of the sample space.

2. **Rolling a Die** A die is rolled 10 times. We are interested in the number of times an even number shows. List the outcomes of the sample space.

3. A survey of families with 2 children is made, and the gender of the children is recorded. Describe the sample space and draw a tree diagram of this experiment.

4. The spinner pictured is spun 3 times. Each time the color is noted. List the outcomes of the sample space. Draw a tree diagram of the experiment

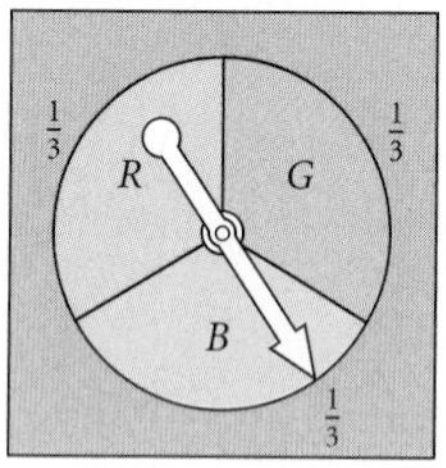

5. **Jars and Coins** A jar has 15 coins: 4 pennies, 5 dimes, and 6 quarters. A coin is selected from the jar. Assign valid probabilities to the outcomes of this experiment.

6. **Choosing a Ball** A bag has 12 golf balls, 5 of which are cracked. A golf ball is chosen from the bag at random. Assign valid probabilities to this experiment.

7. **Tossing a Weighted Die** A die is weighted so that 2 and 5 appear twice as often as 1, 3, 4, and 6. The die is tossed. Assign valid probabilities to this experiment.

8. **Tossing an Unfair Coin** A coin is not fair. When tossed, heads shows up 4 times more often than tails. Assign valid probabilities to this experiment.

9. **Genders of Children** Families with 4 children were surveyed.
   (a) Construct a probability model describing the possible number of girls in the family. (Assume the probability a child is a girl is 0.5.)
   (b) Find the probability
      (i) No children are girls
      (ii) Exactly 2 children are girls
      (iii) Only 1 child is a girl
      (iv) At least one child is a boy.

10. A fair coin is tossed three times.
   (a) Construct a probability model corresponding to this experiment.
   (b) Find the probabilities of the following events:
      (i) The first toss is tails.
      (ii) The first toss is heads.
      (iii) Either the first toss is tails or the third toss is heads.
      (iv) At least one of the tosses is heads.
      (v) There are at least 2 tails.
      (vi) No tosses are heads.

11. **Tossing a Coin** A fair coin is tossed three times and the number of tails is counted.
   (a) Construct a probability model for this experiment.
   (b) Find the probability that
      (i) all 3 tosses are tails.
      (ii) no tosses are tails.
      (iii) exactly 2 tosses are tails.
      (iv) at least 2 tosses are tails.

12. **Gender of Children** Families with 4 children are surveyed. Gender of the children according to birth order is determined.
   (a) Construct a probability model describing the experiment.
   (b) Find the probability that
      (i) the first 2 children are girls.
      (ii) all children are boys.
      (iii) at least one child is a girl.
      (iv) the first child and the last child are girls.

13. A jar contains 3 white marbles, 2 yellow marbles, 4 red marbles, and 5 blue marbles. Two marbles are picked at random. What is the probability that
   (a) Both are blue? (b) Exactly 1 is blue?
   (c) At least 1 is blue?

14. Drawing Cards Two cards are drawn from a 52-card deck. What is the probability that

(a) 1 card is black and 1 card is red?
(b) Both cards are the same color?
(c) Both cards are red?

15. Let A and B be events with $P(A) = .3$, $P(B) = .5$, and $P(A \cap B) = .2$. Find the probability that

(a) A or B happens.
(b) A does not happen.
(c) Neither A nor B happens.
(d) Either A does not happen or B does not happen.

16. Let A and B be events with $P(A) = .6$, $P(B) = .8$, and $P(A \cap B) = 4$. Find the probability that

(a) Either A or B happens.
(b) B does not happen.
(c) Neither A nor B happen.

17. A loaded die is rolled 400 times, and the following outcomes are recorded:

Face	1	2	3	4	5	6
No. Times Showing	32	45	84	74	92	73

Estimate the probability of rolling a

(a) 3 (b) 5 (c) 6

18. Tossing an Unfair Coin An unfair coin is tossed 300 times and outcomes are recorded

Outcomes	H	T
No. Times Occurring	243	57

Estimate the probability of tossing a

(a) head (b) tail

19. A survey of a group of criminals shows that 65% came from low-income families, 40% from broken homes, and 30% came from low-income families and broken homes. Define

E: Criminal came from low-income family

F: Criminal came from broken home

A criminal is selected at random.

(a) Find the probability that the criminal is not from a low-income family.
(b) Find the probability that the criminal comes from a broken home or a low-income family.
(c) Are E and F mutually exclusive?

20. Working Students A survey of a group of 18- to 22-year-olds revealed that 63% were college students, 50% held jobs, and 35% did both. Define

E: The respondent is a college student.

F: The respondent has a job.

A respondent is selected at random.

(a) Find the probability that the respondent does not have a job.
(b) Find the probability that the respondent is a student or holds a job.
(c) Are E and F mutually exclusive?

21. If E and F are events with $P(E \cup F) = \frac{5}{8}$, $P(E \cap F) = \frac{1}{3}$, and $P(E) = \frac{1}{2}$, find

(a) $P(\overline{E})$ (b) $P(F)$ (c) $P(\overline{F})$

22. If $P(E) = .2$, $P(F) = .6$, and $P(E \cap F) = .1$, find

(a) $P(E \cup F)$ (b) $P(\overline{E})$ (c) $P(\overline{E \cap F})$

23. If E and F represent mutually exclusive events, $P(E) = .30$, and $P(F) = .45$, find each of the probabilities:

(a) $P(\overline{E})$ (b) $P(\overline{F})$
(c) $P(E \cap F)$ (d) $P(E \cup F)$
(e) $P(\overline{E \cap F})$ (f) $P(\overline{E \cup F})$
(g) $P(\overline{E} \cup \overline{F})$ (h) $P(\overline{E} \cap \overline{F})$

24. If $P(E) = .25$, $P(F) = .3$, and $P(E \cup F) = .55$, find each of the probabilities:

(a) $P(\overline{E})$ (b) $P(\overline{F})$
(c) $P(E \cap F)$ (d) $P(\overline{E \cap F})$
(e) $P(\overline{E} \cap \overline{F})$ (f) $P(\overline{E} \cup \overline{F})$

25. Throwing Dice A pair of dice are thrown. Define

E: The sum of the faces is 8.

F: The sum of the faces is even.

Find the probability of throwing either an 8 or an even sum.

26. Choosing a Card A card is drawn from a 52-card deck. Define

E: The card is red.

F: The card is a face card (J, Q, or K).

Find the probability that it is either red or a face card.

27. Consider the experiment of spinning the spinner shown in the figure 3 times. (Assume the spinner cannot fall on a line.)

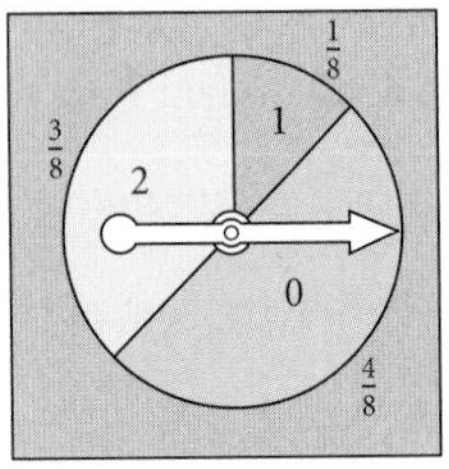

(a) Are all outcomes equally likely?
(b) If not, which of the outcomes has the highest probability?
(c) Let F be the event, "Each digit will occur exactly once." Find $P(F)$.

28. M&M Candies A bowl contains 22 M&M candies; 4 are red, 6 are green, and 12 are blue. A piece of candy is chosen, and its color is noted.

(a) Is each color equally likely to be picked?
(b) Which outcome has the highest probability of occurring?
(c) Suppose you took 3 pieces of candy without looking. Define the event

E: Each color is represented.

What is $P(E)$?

29. Tossing Dice Two dice are tossed and the sum of the spots is observed. What is the probability of obtaining a sum of 5, 7, or 9?

30. Tossing Dodecahedra Two dodecahedra (12-sided dice) are tossed. See Section 7.4, Problem 15. The sum of the top faces is observed. What is the probability of observing a sum of 15 or 20?

31. What are the odds in favor of a 5 when a fair die is thrown?

32. Gender of Children A family chosen at random has 4 children. What are the odds that all four are boys?

33. A bettor is willing to give 7 to 6 odds that the Bears will win the NFL title. What is the probability of the Bears winning?

34. Football Pool You want to enter a football pool. The daily paper states the odds the Giants will win Sunday's game are 5:3. What is the probability the Giants will win Sunday's game? What is the probability they will lose?

35. A biased coin is such that the probability of heads (H) is $\frac{1}{4}$ and the probability of tails (T) is $\frac{3}{4}$. Show that in flipping this coin twice, the events E and F defined below are independent.

E: A head turns up on the first throw.

F: A tail turns up on the second throw.

36. Rolling a Loaded Die A die is loaded so that when it is rolled an outcome of 6 is 3 times more likely to occur than any other number. The die is rolled twice. Define

E: A 3 appears on the first roll.

F: A 6 appears on the second roll.

Show that the events E and F are independent.

37. The records of Midwestern University show that in one semester, 38% of the students failed mathematics, 27% of the students failed physics, and 9% of the students failed mathematics and physics. A student is selected at random.

(a) If a student failed physics, what is the probability that he or she failed mathematics?
(b) If a student failed mathematics, what is the probability that he or she failed physics?
(c) What is the probability that he or she failed mathematics or physics?

38. Keeping in Shape A health insurer surveyed its clients and learned that 42% of them exercised regularly, 51% of them ate healthy diets, and 30% of those surveyed did both. A client is selected at random.

(a) If the client exercises regularly, what is the probability he or she eats a healthy diet?
(b) What is the probability the client exercises regularly given it is known he or she eats a healthy diet?
(c) What is the probability that the client either eats a healthy diet or exercises regularly?

39. A pair of fair dice is thrown 3 times. What is the probability that on the first toss the sum of the 2 dice is even, on the second toss the sum is less than 6, and on the third toss the sum is 7?

40. Coins and Dice A fair coin is flipped and then a fair die is thrown. What is the probability of getting heads on the coin and an even number on the die?

41. In a certain population of people, 25% are blue-eyed and 75% are brown-eyed. Also, 10% of the blue-eyed people are left-handed and 5% of the brown-eyed people are left-handed.

(a) What is the probability that a person chosen at random is blue-eyed and left-handed?
(b) What is the probability that a person chosen at random is left-handed?
(c) What is the probability that a person is blue-eyed, given that the person is left-handed?

42. **College Majors** At a local college 55% of the students are female and 45% are male. Also 40% of the female students are education majors, and 15% of the males are education majors.

(a) What is the probability a student selected at random is male and an education major?
(b) What is the probability a student selected is an education major?
(c) What is the probability a student is female given the person is an education major?

43. **Score Distribution** Two forms of a standardized math exam were given to 100 students. The following are the results.

Score	Form A	Form B	Total
Over 80%	8	12	20
Under 80%	32	48	80
Totals	40	60	100

(a) What is the probability that a student who scored over 80% took form A?
(b) What is the probability that a student who took form A scored over 80%?
(c) Show that the events "scored over 80%" and "took form A" are independent.
(d) Are the events "scored over 80%" and "took form B" independent?

44. **ACT Scores** The following data compare ACT scores of students with their performance in the classroom [based on a maximum 4.0 grade point average (GPA)].

GPA	Below 21	22–27	Above 28	Total
3.6–4	8	56	104	168
3.0–3.5	47	70	30	147
Below 3	47	34	4	85
Totals	102	160	138	400

A graduating student is selected at random. Find the probability that

(a) The student scored above 28.
(b) The student's GPA is 3.6–4.
(c) The student scored above 28 with GPA of 3.6–4.
(d) The student's ACT score was in the 22–27 range.
(e) The student had a 3.0–3.5 GPA.
(f) Show that "ACT above 28" and "GPA below 3" are not independent.

45. Three envelopes are addressed for 3 secret letters written in invisible ink. A secretary randomly places each of the letters in an envelope and mails them. What is the probability that at least 1 person receives the correct letter?

46. Jones lives at O (see the figure). He owns 5 gas stations located 4 blocks away (dots). Each afternoon he checks on one of his gas stations. He starts at O. At each intersection he flips a fair coin. If it shows heads, he will head north (N); otherwise, he will head toward the east (E). What is the probability that he will end up at gas station G before coming to one of the other stations?

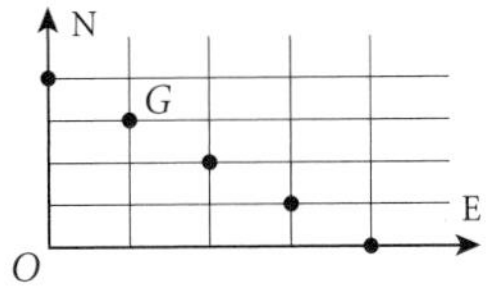

47. **Defective Machine** The calibration of a filling machine broke, causing it to underfill 10 jars of jam. The jars were accidentally mixed in with 62 properly filled jars. All the jars look the same. Four jars are chosen from the batch and weighed. What is the probability

(a) All 4 jars are underweight?
(b) Exactly 2 jars are underweight?
(c) At most 1 jar is underweight?

48. **Broken Calculators** Three broken calculators were inadvertently packed in a case of 12 calculators. Two were chosen from the case.

(a) What is the probability both are broken?
(b) What is the probability neither is broken?
(c) What is the probability at least 1 is broken?

49. **Birthday Problem** Fifteen people are in a room. Each is asked the day of the month (e.g., 1–31) he or she was born. What is the probability each person was born on a different day of the month?

50. **Birthday Problem** There are 5 people in the Smith family. What is the probability at least 2 of them have birthdays in the same month?

51. E and F are independent events. Find $P(E|F)$ if $P(F) = .4$ and $P(E \cap F) = .2$.

52. E and F are independent events. Find $P(E)$ if $P(E \cup F) = .6$ and $P(F) = .1$.

53. **Shooting Free Throws** A basketball player hits 70% of his free throws. Assuming independence on successive throws, what is the probability of

(a) Missing the first throw and then getting 3 in a row?
(b) Making 10 free throws in a row?

54. **Choosing Marbles** A box contains 10 marbles (identical except for color); 3 are blue and 7 are clear. A marble is drawn, checked for color, and returned to the box. This is done 5 times.

(a) What is the probability of drawing a blue marble each time?
(b) What is the probability of drawing a blue marble 2 times and a clear marble 3 times?
(c) What is the probability of choosing a clear marble at least once?

55. **Buying a Car** A car dealer has 20 cars in stock of a given model; 8 are black and 12 are red. Two of the black cars have a tan interior, and 6 of the red cars have tan interiors. You choose a car with a tan interior. What is the probability the car is black?

56. **Majoring in Business** At the College at Old Westbury, 20% of the students are business majors and the rest major in something else. Although 70% of the business majors take Finite Mathematics, only 10% of the other majors take Finite Mathematics. A student is chosen at random. She is taking Finite Mathematics. What is the probability she is a business major?

Chapter 7 Project

ASKING SENSITIVE QUESTIONS USING CONDITIONAL PROBABILITY

In your journalism class, a group project places you in charge of determining what proportion of the students at your school are overweight. You decide to conduct a survey of a sample of the student body, and (of course) you will need to ask the survey respondents the question "Are you overweight?" There is a problem in being so direct, however. Even though the survey is anonymous, some respondents might not want to answer this question truthfully. They might be afraid that you could somehow figure out who they were by their responses, or they just might not care to share that information with you. You could call a question such as "Are you overweight?" **a sensitive question.** Nevertheless, you want to estimate the proportion of overweight students at your school. How can you get an idea of that proportion without driving some of your respondents to lie? It turns out that conditional probability provides a way to handle this problem.

Consider conducting the survey in the following manner: Before answering the sensitive question, the respondent flips a coin. We assume that it is a fair coin; that is, heads appears with probability $\frac{1}{2}$. If the flip is tails, the respondent is instructed to answer "Yes" to the sensitive question; if the flip is heads, the respondent is instructed to answer the question truthfully. Since no one except the respondent has any idea whether the flip was a head or a tail, no one except the respondent knows whether a "Yes" response means "the coin came up tails" or "I am overweight." In this way, the respondent might be more inclined to answer the question truthfully if required to do so.

Define the events:

E: The respondent answer "Yes"

H: The flip is a head

T: The flip is a tail

After you tabulate the results to your survey, you will know $P(E)$, the proportion of your respondents that answered "Yes".

1. Why are you interested in finding $P(E|H)$?
2. Find an equation for $P(E)$ that involves $P(E|H)$ and $P(E|T)$.
3. Explain why $P(E|T) = 1$, then solve the equation found above for $P(E|H)$.
4. If 82 of 100 respondents answered "Yes," what is your best estimate for the proportion of overweight students at your school?

Suppose that you want to get a bit more information: You want respondents to check one of the three boxes labeled "underweight," "less than 20 pounds overweight," or "more than 20 pounds overweight." Once again, use a coin-flipping strategy. If the flip is tails, the respondent is instructed to check the "more than 20 pounds overweight" box; if the flip is heads, the respondent is instructed to answer the question truthfully.

Define the events:

A: The respondent answers "underweight."

B: The respondent answers "less than 20 pounds overweight."

C: The respondent answers "more than 20 pounds overweight."

H: The flip is heads.

T: The flip is tails.

5. Express $P(A|H)$ as an equation.
6. Express $P(B|H)$ as an equation.
7. Express $P(C|H)$ as an equation.

Suppose you survey 300 students and get the following results:

Number checking "Underweight"	30
Number checking "Less than 20 pounds overweight"	50
Number checking "More than 20 pounds overweight"	220

8. What is your best estimate for the proportion of students at your school who are underweight, less than 20 pounds overweight, and more than 20 pounds overweight?
9. Write a one-page report that summarizes your findings and the methodology you used.

MATHEMATICAL QUESTIONS FROM PROFESSIONAL EXAMS

1. **Actuary Exam—Part II** If P and Q are events having positive probability in the same sample space S such that $P \cap Q = \varnothing$, then all of the following pairs are independent EXCEPT

   (a) $\varnothing$ and P (b) P and Q
   (c) P and S (d) P and $P \cap Q$
   (e) $\varnothing$ and the complement of P

2. **Actuary Exam—Part II** A box contains 12 varieties of candy and exactly 2 pieces of each variety. If 12 pieces of candy are selected at random, what is the probability that a given variety is represented?

   (a) $\frac{2^{12}}{(12!)^2}$ (b) $\frac{2^{12}}{24!}$ (c) $\frac{2^{12}}{\binom{24}{12}}$ (d) $\frac{11}{46}$ (e) $\frac{35}{46}$

3. **Actuary Exam—Part II** What is the probability that a 3-card hand drawn at random and without replacement from a regular deck consists entirely of black cards?

   (a) $\frac{1}{17}$ (b) $\frac{2}{17}$ (c) $\frac{1}{8}$ (d) $\frac{3}{17}$ (e) $\frac{4}{17}$

4. **Actuary Exam—Part II** Events S and T are independent with $\Pr(S) < \Pr(T)$, $\Pr(S \cap T) = \frac{6}{25}$, and $\Pr(S|T) + \Pr(T|S) = 1$. What is $\Pr(s)$?

   (a) $\frac{1}{25}$ (b) $\frac{1}{5}$ (c) $\frac{5}{25}$ (d) $\frac{2}{5}$ (e) $\frac{3}{5}$

5. **Actuary Exam—Part II** What is the least number of independent times that an unbiased die must be thrown to make the probability that all throws do not give the same result greater than .999?

   (a) 3 (b) 4 (c) 5 (d) 6 (e) 7

6. **Actuary Exam—Part II** In a group of 20,000 men and 10,000 women, 6% of the men and 3% of the women have a certain affliction. What is the probability that an afflicted member of the group is a man?

   (a) $\frac{3}{5}$ (b) $\frac{2}{3}$ (c) $\frac{3}{4}$ (d) $\frac{4}{5}$ (e) $\frac{8}{9}$

7. **Actuary Exam—Part II** An unbiased die is thrown 2 independent times. Given that the first throw resulted in an even number, what is the probability that the sum obtained is 8?

   (a) $\frac{5}{36}$ (b) $\frac{1}{6}$ (c) $\frac{4}{21}$ (d) $\frac{7}{36}$ (e) $\frac{1}{3}$

8. **Actuary Exam—Part II** If the events S and T have equal probability and are independent with $\Pr(S \cap T) = p > 0$, then $\Pr(S) =$

   (a) $\sqrt{p}$ (b) p^2 (c) $\frac{p}{2}$ (d) p (e) $2p$

9. **Actuary Exam—Part II** The probability that both S and T occur, the probability that S occurs and T does not, and the probability that T occurs and S does not are all equal to p. What is the probability that either S or T occurs?

   (a) p (b) $2p$ (c) $3p$ (d) $3p^2$ (e) p^3

10. **Actuary Exam—Part II** What is the probability that a bridge hand contains 1 card of each denomination (i.e., 1 ace, 1 king, 1 queen, . . . , 1 three, 1 two)?

    (a) $\frac{13!}{13^{13}}$ (b) $\frac{4^{13}}{\binom{52}{13}}$ (c) $\frac{\binom{52}{4}}{\binom{52}{13}}$

    (d) $\left(\frac{1}{13}\right)^{13}$ (e) $\frac{13^4}{\binom{52}{13}}$

CHAPTER 8

Additional Probability Topics

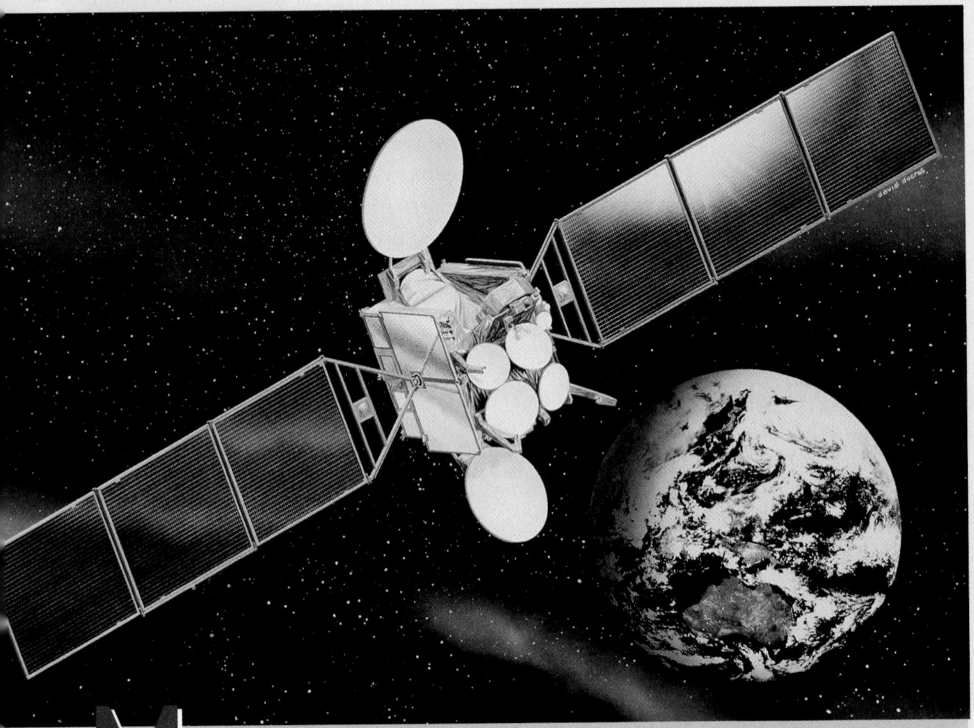

OUTLINE

Most manufactured products go through a series of tests before they are put on the market for sale. Cars, for example, have various parts tested and, if found defective, they are replaced. The car you buy is a version that has had its parts tested for endurance and reliability. But what if each time you tested a product for reliable parts, you weakened the parts. That's okay if you can replace the weakened parts by reliable substitutes. But what if the replacement parts also needed to be tested? Then the number of tests you need to ensure reliability is in conflict with the fact that with each test the item becomes less reliable! This happens in the manufacturing of satellites, for example. The material in this chapter provides a means for resolving the conflict of testing versus reliability, and the Chapter Project will guide you through the process.

A LOOK BACK, A LOOK FORWARD

In Chapter 6 we discussed sets and counting techniques, and in Chapter 7 we studied probability. We used facts from Chapter 6 in our studies in Chapter 7. In this chapter we continue our study of probability and discuss *Bayes' Formula*, which involves sample spaces that can be divided into nonintersecting sets. We also discuss the *binomial probability model*, which investigates experiments that are repetitive, so counting techniques will be employed. Both of these topics are very rich in applications. *Expected value* uses probability to help make decisions. Then, we take up applications in the field of operations research and transmission of data. The chapter closes with a brief introduction to *random variables*.

8.1 Bayes' Formula

PREPARING FOR THIS SECTION *Before getting started, review the following:*

> Sets (Section 6.1, pp 317–325)

> Conditional Probability (Section 7.4, pp. 392–399)

OBJECTIVES
1. Solve probability problems: sample space partitioned into two sets
2. Solve probability problems: sample space partitioned into three sets
3. Use Bayes' Formula to solve probability problems
4. Find *a priori* and *a posteriori* probabilities

In this section we consider experiments with sample spaces that will be divided or partitioned into two (or more) mutually exclusive events. This study involves a further application of conditional probabilities and leads us to the famous Bayes' Formula, named after Thomas Bayes, who first published it in 1763.

We begin by considering the following example.

EXAMPLE 1 Introduction to Bayes' Formula

Given two urns, suppose urn I contains four black and seven white balls. Urn II contains three black, one white, and four yellow balls. We select an urn at random and then draw a ball. What is the probability that we obtain a black ball?

SOLUTION Let U_{I} and U_{II} stand for the events "Urn I is chosen" and "Urn II is chosen," respectively. Similarly, let B, W, Y stand for the event that "a black," "a white," or "a yellow" ball is chosen, respectively.

SOLUTION A Example 1 can be solved using the tree diagram shown in Figure 1.

FIGURE 1

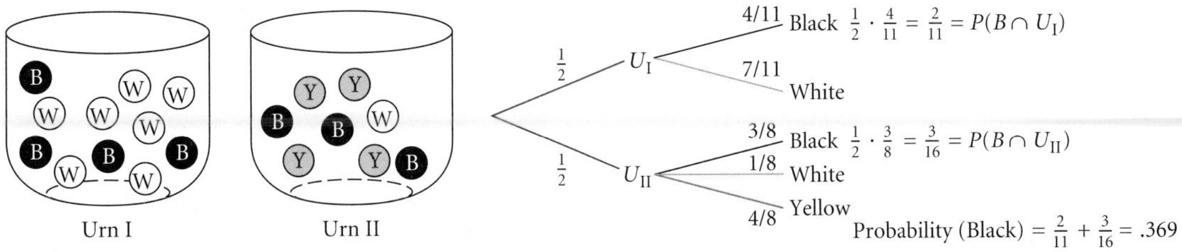

Then

$$P(B) = \frac{2}{11} + \frac{3}{16} = \frac{65}{176} = .369$$

SOLUTION B Let's see how we can solve Example 1 without using a tree diagram. First we observe that

$$P(U_{\text{I}}) = P(U_{\text{II}}) = \frac{1}{2}$$

$$P(B|U_{\text{I}}) = \frac{4}{11} \qquad P(B|U_{\text{II}}) = \frac{3}{8}$$

The event B can be written as

$$B = (B \cap U_{\text{I}}) \cup (B \cap U_{\text{II}})$$

Since $B \cap U_{\text{I}}$ and $B \cap U_{\text{II}}$ are disjoint, we add their probabilities. Then

$$P(B) = P(B \cap U_{\text{I}}) + P(B \cap U_{\text{II}}) \tag{1}$$

Using the Product Rule, we have

$$P(B \cap U_{\text{I}}) = P(U_{\text{I}}) \cdot P(B|U_{\text{I}}) \qquad P(B \cap U_{\text{II}}) = P(U_{\text{II}}) \cdot P(B|U_{\text{II}}) \tag{2}$$

Combining (1) and (2), we have

$$\begin{aligned} P(B) &= P(U_{\text{I}}) \cdot P(B|U_{\text{I}}) + P(U_{\text{II}}) \cdot P(B|U_{\text{II}}) \\ &= \frac{1}{2} \cdot \frac{4}{11} + \frac{1}{2} \cdot \frac{3}{8} = \frac{2}{11} + \frac{3}{16} = .369 \end{aligned}$$

Partitions

The preceding discussion leads to the following generalization. Suppose A_1 and A_2 are two nonempty, mutually exclusive events of a sample space S and the union of A_1 and A_2 is S; that is,

FIGURE 2

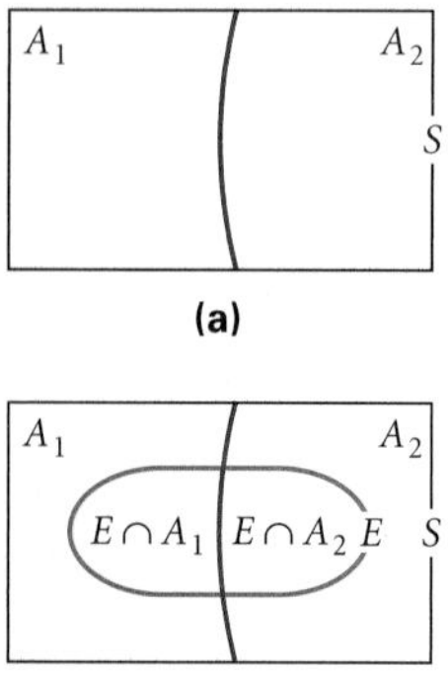

$$A_1 \neq \varnothing \qquad A_2 \neq \varnothing \qquad A_1 \cap A_2 = \varnothing \qquad S = A_1 \cup A_2$$

In this case we say that A_1 and A_2 form a **partition** of S. See Figure 2(a). Now if we let E be any event in S, we may write the set E in the form

$$E = (E \cap A_1) \cup (E \cap A_2)$$

See Figure 2(b).

The sets $E \cap A_1$ and $E \cap A_2$ are disjoint since

$$(E \cap A_1) \cap (E \cap A_2) = (E \cap E) \cap (A_1 \cap A_2) = E \cap \varnothing = \varnothing$$

Using the Product Rule, the probability of E is

$$\begin{aligned} P(E) &= P(E \cap A_1) + P(E \cap A_2) \\ &= P(A_1) \cdot P(E|A_1) + P(A_2) \cdot P(E|A_2) \end{aligned} \tag{3}$$

Solve probability problems: sample space partitioned into two sets 1

Formula (3) is used to find the probability of an event E of a sample space when the sample space is partitioned into two sets A_1 and A_2. See Figure 3 for a tree diagram depicting Formula (3). You may find it easier to construct Figure 3 to obtain Formula (3) than to memorize it.

FIGURE 3

$$P(E) = P(A_1) \cdot P(E|A_1) + P(A_2) \cdot P(E|A_2)$$

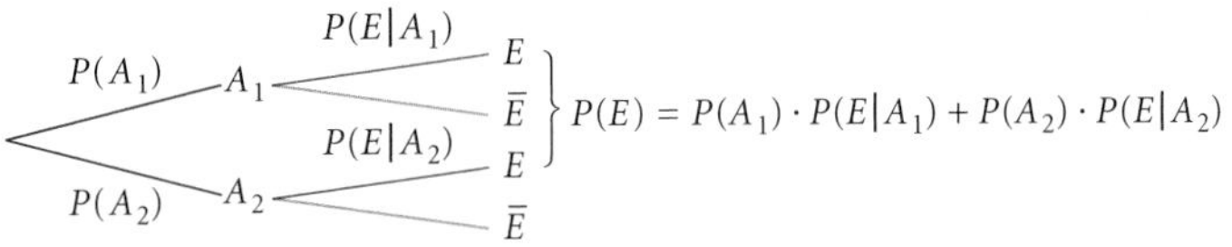

EXAMPLE 2   Admissions Tests for Medical School

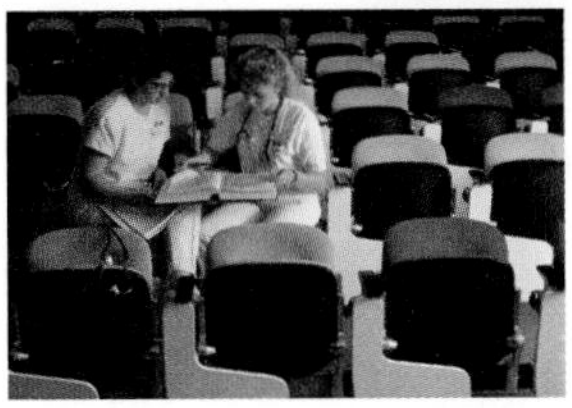

Of the applicants to a medical school, 80% are eligible to enter and 20% are not. To aid in the selection process, an admissions test is administered that is designed so that an eligible candidate will pass 90% of the time, while an ineligible candidate will pass only 30% of the time. What is the probability that an applicant for admission will pass the admissions test?

SOLUTION A   Figure 4 provides a tree diagram solution.

FIGURE 4

.8 / Eligible .9 Pass (.8)(.9) = .72
                .1 Fail
.2 \ Ineligible .3 Pass (.2)(.3) = .06
                  .7 Fail

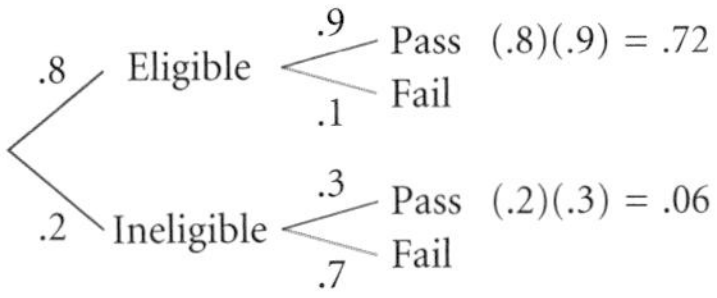

Probability of passing = .72 + .06 = .78

SOLUTION B   The sample space S consists of the applicants for admission, and S can be partitioned into the following two events:

A_1: Eligible applicant A_2: Ineligible applicant

FIGURE 5

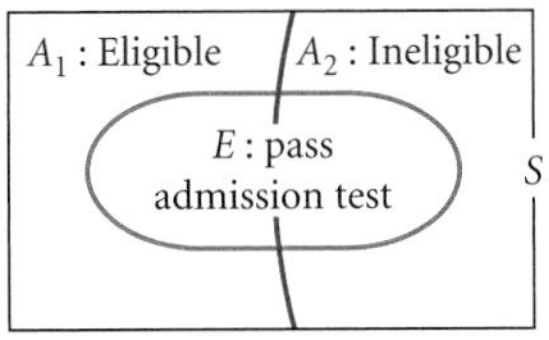

These two events are disjoint, and their union is S. See Figure 5.
   The event E is

A_1 : Eligible A_2 : Ineligible

E : pass admission test

S

E: Applicant passes admissions test

Now

$$P(A_1) = .8 \qquad P(A_2) = .2$$
$$P(E|A_1) = .9 \qquad P(E|A_2) = .3$$

Using Formula (3), we have

$$P(E) = P(A_1) \cdot P(E|A_1) + P(A_2) \cdot P(E|A_2) = (.8)(.9) + (.2)(.3) = .78$$

The probability that an applicant will pass the admissions test is .78.

NOW WORK PROBLEM 15.

If we partition a sample space S into three sets A_1, A_2, and A_3 so that

$$S = A_1 \cup A_2 \cup A_3$$
$$A_1 \cap A_2 = \varnothing \qquad A_2 \cap A_3 = \varnothing \qquad A_1 \cap A_3 = \varnothing$$
$$A_1 \neq \varnothing \qquad A_2 \neq \varnothing \qquad A_3 \neq \varnothing$$

FIGURE 6

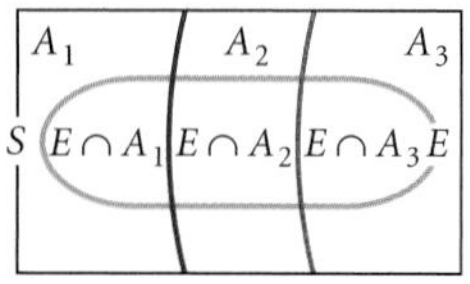

we may write any set E in S in the form

$$E = (E \cap A_1) \cup (E \cap A_2) \cup (E \cap A_3)$$

See Figure 6.

Since $E \cap A_1$, $E \cap A_2$, and $E \cap A_3$ are disjoint, the probability of event E is

$$\begin{aligned} P(E) &= P(E \cap A_1) + P(E \cap A_2) + P(E \cap A_3) \\ &= P(A_1) \cdot P(E|A_1) + P(A_2) \cdot P(E|A_2) + P(A_3) \cdot P(E|A_3) \end{aligned} \qquad \textbf{(4)}$$

Solve probability problems: sample space partitioned into three sets 2

Formula (4) is used to find the probability of an event E of a sample space when the sample space is partitioned into three sets A_1, A_2, and A_3. See Figure 7 for a tree diagram depicting Formula (4). Again, you may find it easier to construct Figure 7 to obtain Formula (4) than to memorize it.

FIGURE 7

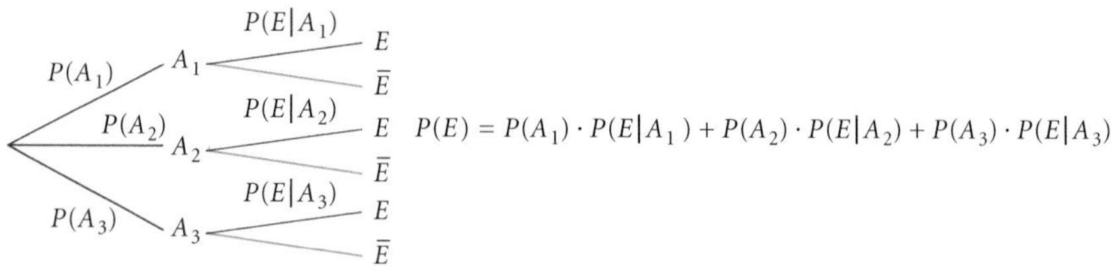

EXAMPLE 3 Determining Quality Control

Three machines, I, II, and III, manufacture .4, .5, and .1 of the total production in a plant, respectively. The percentage of defective items produced by I, II, and III is 2%, 4%, and 1%, respectively. For an item chosen at random, what is the probability that it is defective?

SOLUTION A Figure 8 gives a tree diagram solution.

FIGURE 8

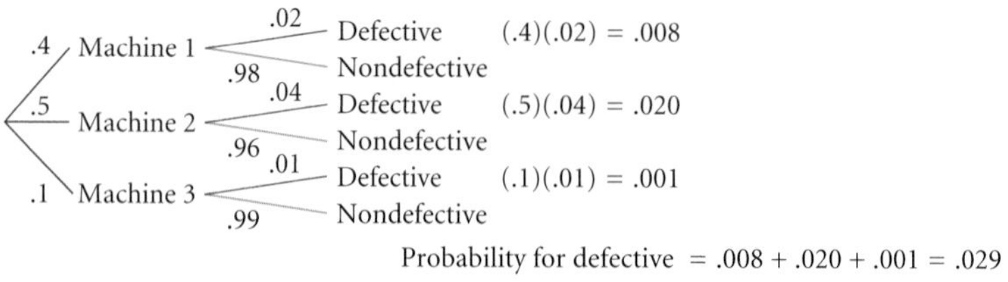

SOLUTION B The sample space S is partitioned into three events A_1, A_2, and A_3 defined as follows:

A_1: Item produced by machine I

A_2: Item produced by machine II

A_3: Item produced by machine III

The events A_1, A_2, and A_3 are mutually exclusive, and their union is S. Define the event E in S to be

$$E: \text{ Item is defective}$$

Now

$$P(A_1) = .4 \qquad P(A_2) = .5 \qquad P(A_3) = .1$$
$$P(E|A_1) = .02 \qquad P(E|A_2) = .04 \qquad P(E|A_3) = .01$$

Using Formula (4), we see that

$$\begin{aligned} P(E) &= (.4)(.02) + (.5)(.04) + (.1)(.01) \\ &= .008 + .020 + .001 = .029 \end{aligned}$$

NOW WORK PROBLEM 17.

To generalize Formulas (3) and (4) to a sample space S partitioned into n subsets, we require the following definition:

Partition

A sample space S is **partitioned** into n subsets $A_1, A_2 \ldots, A_n$, provided:

(a) Each subset is nonempty.
(b) The intersection of any two of the subsets is empty.
(c) $A_1 \cup A_2 \cup \cdots \cup A_n = S$

Let S be a sample space and let $A_1, A_2, A_3, \ldots, A_n$ *be n* events that form a partition of the set S. If E is any event in S, then

$$E = (E \cap A_1) \cup (E \cap A_2) \cup \cdots \cup (E \cap A_n)$$

Since $E \cap A_1, E \cap A_2, \ldots, E \cap A_n$ are mutually exclusive events, we have

$$P(E) = P(E \cap A_1) + P(E \cap A_2) + \cdots + P(E \cap A_n) \qquad (5)$$

In (5) replace $P(E \cap A_1)$, $P(E \cap A_2)$, . . . , $P(E \cap A_n)$ using the Product Rule. Then we obtain the formula

$$\boxed{P(E) = P(A_1) \cdot P(E|A_1) + P(A_2) \cdot P(E|A_2) + \cdots + P(A_n) \cdot P(E|A_n)} \qquad (6)$$

EXAMPLE 4 Admissions Tests for Medical School

In Example 2 suppose an applicant passes the admissions test. What is the probability that he or she was among those eligible; that is, what is the probability $P(A_1|E)$?

SOLUTION By the definition of conditional probability,

$$P(A_1|E) = \frac{P(A_1 \cap E)}{P(E)} = \frac{P(A_1) \cdot P(E|A_1)}{P(E)}$$

But $P(E)$ is given by (6) when $n = 2$. Then

$$P(A_1|E) = \frac{P(A_1) \cdot P(E|A_1)}{P(A_1) \cdot P(E|A_1) + P(A_2) \cdot P(E|A_2)} \tag{7}$$

Using the information supplied in Example 2, we find

$$P(A_1|E) = \frac{(.8)(.9)}{.78} = \frac{.72}{.78} = .923$$

The admissions test is a reasonably effective device. Less than 8% of the students passing the test are ineligible.

NOW WORK PROBLEM 19.

Bayes' Formula

Equation (7) is a special case of Bayes' Formula when the sample space is partitioned into two sets A_1 and A_2. The general formula is given below.

Bayes' Formula

Let S be a sample space partitioned into n events, $A_1, \ldots, A_n$. Let E be any event of S for which $P(E) > 0$. The probability of the event $A_j (j = 1, 2, \ldots, n)$, given the event E, is

$$\begin{aligned} P(A_j|E) &= \frac{P(A_j) \cdot P(E|A_j)}{P(E)} \\ &= \frac{P(A_j) \cdot P(E|A_j)}{P(A_1) \cdot P(E|A_1) + P(A_2) \cdot P(E|A_2) + \cdots + P(A_n) \cdot P(E|A_n)} \end{aligned} \tag{8}$$

The proof is left as an exercise (see Problem 42).

Use Bayes' Formula to solve probability problems 3

The following example illustrates a use for Bayes' formula when the sample space is partitioned into three events.

EXAMPLE 5 Quality Control: Source of Defective Cars

Motors, Inc., has three plants. Plant I produces 35% of the car output, plant II produces 20%, and plant III produces the remaining 45%. One percent of the output of plant I is defective, as is 1.8% of the output of plant II and 2% of the output of plant III. The annual total output of Motors, Inc., is 1,000,000 cars. A car is chosen at random from the annual output and it is found to be defective. What is the probability that it came from plant I? Plant II? Plant III?

SOLUTION We first define the following events:

E: Car is defective
A_1: Car produced by plant I
A_2: Car produced by plant II
A_3: Car produced by plant III

Then, the probabilities we seek are:

$P(A_1|E)$: the probability a car was produced by plant I, given that it was defective.
$P(A_2|E)$: the probability a car was produced by plant II, given that it was defective.
$P(A_3|E)$: the probability a car was produced by plant III, given that it was defective.

To find these probabilities, we first need to find $P(E)$.
From the data given in the problem we know the following probabilities:

$$\begin{aligned} P(A_1) &= .35 \qquad & P(E|A_1) &= .010 \\ P(A_2) &= .20 \qquad & P(E|A_2) &= .018 \\ P(A_3) &= .45 \qquad & P(E|A_3) &= .020 \end{aligned} \tag{9}$$

Now,

$A_1 \cap E$ is the event "Produced by plant I and is defective"
$A_2 \cap E$ is the event "Produced by plant II and is defective"
$A_3 \cap E$ is the event "Produced by plant III and is defective"

From the Product Rule we find

$$\begin{aligned} P(A_1 \cap E) &= P(A_1) \cdot P(E|A_1) = (.35)(.010) = .0035 \\ P(A_2 \cap E) &= P(A_2) \cdot P(E|A_2) = (.20)(.018) = .0036 \\ P(A_3 \cap E) &= P(A_3) \cdot P(E|A_3) = (.45)(.020) = .0090 \end{aligned}$$

Since $E = (A_1 \cap E) \cup (A_2 \cap E) \cup (A_3 \cap E)$, we have

$$\begin{aligned} P(E) &= P(A_1 \cap E) + P(A_2 \cap E) + P(A_3 \cap E) \\ &= .0035 + .0036 + .0090 \\ &= .0161 \end{aligned}$$

The probability that a defective car is chosen is .0161. See Figure 9.

FIGURE 9

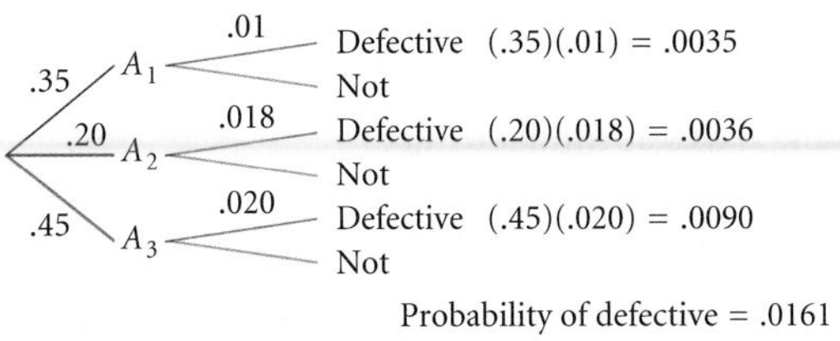

Now, we use Bayes' Formula to find $P(A_1|E)$, $P(A_2|E)$, and $P(A_3|E)$.

$$P(A_1|E) = \frac{P(A_1)\cdot P(E|A_1)}{P(A_1)\cdot P(E|A_1) + P(A_2)\cdot P(E|A_2) + P(A_3)\cdot P(E|A_3)}$$

$$= \frac{P(A_1)\cdot P(E|A_1)}{P(E)} = \frac{(.35)(.01)}{.0161} = .217 \qquad (10)$$

$$P(A_2|E) = \frac{P(A_2)\cdot P(E|A_2)}{P(E)} = \frac{.0036}{.0161} = .224$$

$$P(A_3|E) = \frac{P(A_3)\cdot P(E|A_3)}{P(E)} = \frac{.0090}{.0161} = .559$$

Given that a defective car is chosen, the probability it came from plant A_1 is .217, from plant A_2 is .224, and from plant A_3 is .559.

NOW WORK PROBLEM 27.

A Priori, A Posteriori Probabilities

Find *a priori* and *a posteriori* probabilities

4 In Bayes' Formula the probabilities $P(A_j)$ are referred to as *a priori* probabilities, while the $P(A_j|E)$ are called *a posteriori* probabilities. We use Example 5 to explain the reason for this terminology. Knowing nothing else about a car, the probability that it was produced by plant I is given by $P(A_1)$, so $P(A_1)$ can be regarded as a "before the fact," or *a priori,* probability. With the additional information that the car is defective, we reassess the likelihood of whether it came from plant I and compute $P(A_1|E)$. Then $P(A_1|E)$ can be viewed as an "after the fact," or *a posteriori,* probability.

Note that $P(A_1) = .35$, while $P(A_1|E) = .217$. So the knowledge that a car is defective decreases the chance that it came from plant I.

EXAMPLE 6 Testing for Cancer

The residents of a community are examined for cancer. The examination results are classified as positive (+) if a malignancy is suspected, and as negative (−) if there are no indications of a malignancy. If a person has cancer, the probability of a positive result from the examination is .98. If a person does not have cancer, the probability of a positive result is .15. If 5% of the community has cancer, what is the probability of a person not having cancer if the examination is positive?

SOLUTION Define the following events:

A_1: Person has cancer

A_2: Person does not have cancer

E: Examination is positive

We want to know the probability of a person not having cancer if it is known that the examination is positive; that is, we wish to find $P(A_2|E)$. We are given the probabilities:

$$P(A_1) = .05 \qquad P(A_2) = .95$$
$$P(E|A_1) = .98 \qquad P(E|A_2) = .15$$

Using Bayes' formula, we get

$$P(A_2|E) = \frac{P(A_2)\cdot P(E|A_2)}{P(A_1)\cdot P(E|A_1) + P(A_2)\cdot P(E|A_2)}$$

$$= \frac{(.95)(.15)}{(.05)(.98) + (.95)(.15)} = .744$$

So, even if the examination is positive, the person examined is more likely not to have cancer than to have cancer. The reason the test is designed this way is that it is better for a healthy person to be examined more thoroughly than for someone with cancer to go undetected. Simply stated, the test is useful because of the high probability (.98) that a person with cancer will not go undetected.

NOW WORK PROBLEM 29.

EXAMPLE 7 Car Repair Diagnosis

The manager of a car repair shop knows from past experience that when a call is received from a person whose car will not start, the probabilities for various troubles (assuming no two can occur simultaneously) are as given in Table 1.

TABLE 1

Event	Trouble	Probability
A_1	Flooded	.3
A_2	Battery cable loose	.2
A_3	Points bad	.1
A_4	Out of gas	.3
A_5	Something else	.1

The manager also knows that if the person will hold the gas pedal down and try to start the car, the probability that it will start (E) is

$$P(E|A_1) = .9 \qquad P(E|A_2) = 0 \qquad P(E|A_3) = .2$$
$$P(E|A_4) = 0 \qquad P(E|A_5) = .2$$

(a) If a person has called and is instructed to "hold the pedal down . . . ," what is the probability that the car will start?

(b) If the car does start after holding the pedal down, what is the probability that the car was flooded?

SOLUTION **(a)** We need to compute $P(E)$. Using Formula (6) for $n = 5$ (the sample space is partitioned into five disjoint sets), we have

$$\begin{aligned} P(E) &= P(A_1) \cdot P(E|A_1) + P(A_2) \cdot P(E|A_2) + P(A_3) \cdot P(E|A_3) \\ &\quad + P(A_4) \cdot P(E|A_4) + P(A_5) \cdot P(E|A_5) \\ &= (.3)(.9) + (.2)(0) + (.1)(.2) + (.3)(0) + (.1)(.2) \\ &= .27 + .02 + .02 = .31 \end{aligned}$$

See Figure 10.

FIGURE 10

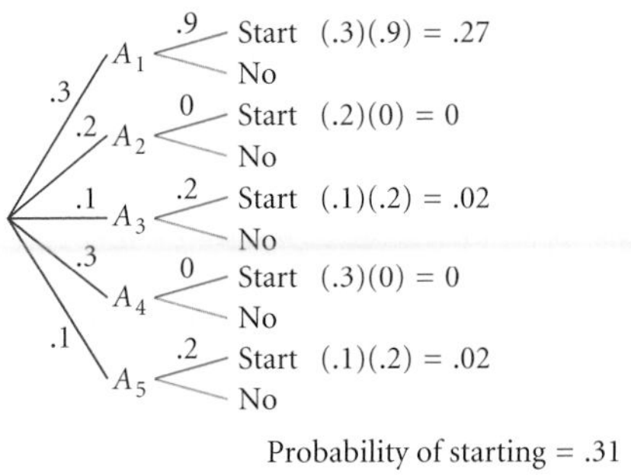

(b) We use Bayes' formula to compute the *a posteriori* probability $P(A_1|E)$:

$$P(A_1|E) = \frac{P(A_1) \cdot P(E|A_1)}{P(E)} = \frac{(.3)(.9)}{.31} = .87$$

The probability that the car was flooded, after it is known that holding down the pedal started the car, is .87.

EXERCISE 8.1 Answers to Odd-Numbered Problems Begin on Page AN-33.

In Problems 1–14, find the indicated probabilities by referring to the tree diagram below.

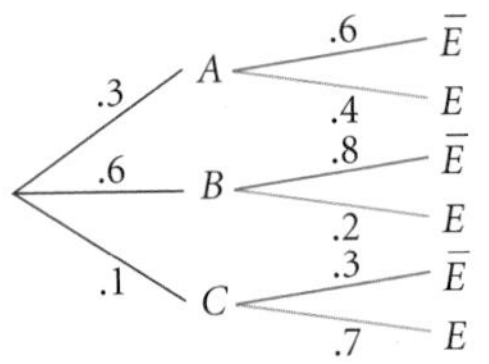

1. $P(E|A)$ **2.** $P(\bar{E}|A)$ **3.** $P(E|B)$

4. $P(\bar{E}|B)$ **5.** $P(E|C)$ **6.** $P(\bar{E}|C)$

7. $P(E)$ **8.** $P(\bar{E})$ **9.** $P(A|E)$

10. $P(B|\bar{E})$ **11.** $P(C|E)$ **12.** $P(A|\bar{E})$

13. $P(B|E)$ **14.** $P(C|\bar{E})$

15. Events A_1 and A_2 form a partition of a sample space S with $P(A_1) = .4$ and $P(A_2) = .6$. If E is an event in S with $P(E|A_1) = .03$ and $P(E|A_2) = .02$, compute $P(E)$.

16. Events A_1 and A_2 form a partition of a sample space S with $P(A_1) = .3$ and $P(A_2) = .7$. If E is an event in S with $P(E|A_1) = .04$ and $P(E|A_2) = .01$, compute $P(E)$.

17. Events A_1, A_2, and A_3 form a partition of a sample space S with $P(A_1) = .6$, $P(A_2) = .2$, and $P(A_3) = .2$. If E is an event in S with $P(E|A_1) = .01$, $P(E|A_2) = .03$, and $P(E|A_3) = .02$, compute $P(E)$.

18. Events A_1, A_2, and A_3 form a partition of a sample space S with $P(A_1) = .3$, $P(A_2) = .2$, and $P(A_3) = .5$. If E is an event in S with $P(E|A_1) = .01$, $P(E|A_2) = .02$, and $P(E|A_3) = .02$, compute $P(E)$.

19. Use the information in Problem 15 to find $P(A_1|E)$ and $P(A_2|E)$.

20. Use the information in Problem 16 to find $P(A_1|E)$ and $P(A_2|E)$.

21. Use the information in Problem 17 to find $P(A_1|E)$, $P(A_2|E)$, and $P(A_3|E)$.

22. Use the information in Problem 18 to find $P(A_1|E)$, $P(A_2|E)$, and $P(A_3|E)$.

23. In Example 3 (page 424) suppose it is known that a defective item was produced. Find the probability that it came from machine I. From machine II. From machine III.

24. In Example 5 (page 426) suppose $P(A_1) = P(A_2) = P(A_3) = \frac{1}{3}$. Find $P(A_2|E)$ and $P(A_3|E)$.

25. In Example 7 (page 429), compute the *a posteriori* probabilities $P(A_2|E)$, $P(A_3|E)$, $P(A_4|E)$, and $P(A_5|E)$.

26. In Example 6 (page 428) compute $P(A_1|E)$.

27. Three jars contain colored balls as follows:

Jar	Red, *R*	White, *W*	Blue, *B*
I	5	6	5
II	3	4	9
III	7	5	4

One jar is chosen at random and a ball is withdrawn. The ball is red. What is the probability that it came from jar I? From jar II? From jar III? [*Hint*: Define the events E: Ball selected is red; U_I: jar I selected; U_{II}: jar II selected; and U_{III}: jar III selected.] Determine $P(U_I|E)$, $P(U_{II}|E)$, and $P(U_{III}|E)$ by using Bayes' Formula.

28. Car Production Cars are being produced by two factories, but factory I produces twice as many cars as factory II in a given time. Factory I is known to produce 2% defectives and factory II produces 1% defectives. A car is examined and found to be defective. What are the *a priori* and *a posteriori* probabilities that the car was produced by factory I?

29. Color Blindness According to the 2000 U.S. Census, 50.9% of the U.S. population is female. NBC news reported that 1 out of 12 males, but only 1 out of 250 females is color-blind. Given that a person randomly chosen from the U.S. population is color-blind, what is the probability that the person is a male? Round your answer to the nearest thousandth.

Source: U.S. Census Bureau and NBC News, December 15, 1999.

30. **Medical Diagnosis** In a certain small town, 16% of the population developed lung cancer. If 45% of the population are smokers, and 85% of those developing lung cancer are smokers, what is the probability that a smoker in this population will develop lung cancer?

31. **Voting Pattern** In Cook County, 55% of the registered voters are Democrats, 30% are Republicans, and 15% are independents. During a recent election, 35% of the Democrats voted, 65% of the Republicans voted, and 75% of the independents voted. What is the probability that someone who voted is a Democrat? Republican? Independent?

32. **Quality Control** A computer manufacturer has three assembly plants. Records show that 2% of the computers shipped from plant A turn out to be defective, as compared to 3% of those that come from plant B and 4% of those that come from plant C. In all, 30% of the manufacturer's total production comes from plant A, 50% from plant B, and 20% from plant C. If a customer finds that his computer is defective, what is the probability it came from plant B?

33. **Oil Drilling** An oil well is to be drilled in a certain location. The soil there is either rock (probability .53), clay (probability .21), or sand. If it is rock, a geological test gives a positive result with 35% accuracy; if it is clay, this test gives a positive result with 48% accuracy; and if it is sand, the test gives a positive result with 75% accuracy. Given that the test is positive, what is the probability that the soil is rock? What is the probability that the soil is clay? What is the probability that the soil is sand?

34. **Oil Drilling** A geologist is using seismographs to test for oil. It is found that if oil is present, the test gives a positive result 95% of the time, and if oil is not present, the test gives a positive result 2% of the time. Oil is actually present in 1% of the cases tested. If the test shows positive, what is the probability that oil is present?

35. **Political Polls** In conducting a political poll, a pollster divides the United States into four sections: Northeast (N), containing 40% of the population; South (S), containing 10% of the population; Midwest (M), containing 25% of the population; and West (W), containing 25% of the population. From the poll it is found that in the next election 40% of the people in the Northeast say they will vote for Republicans, in the South 56% will vote Republican, in the Midwest 48% will vote Republican, and in the West 52% will vote Republican. What is the probability that a person chosen at random will vote Republican? Assuming a person votes Republican, what is the probability that he or she is from the Northeast?

36. **TB Screening** Suppose that if a person with tuberculosis is given a TB screening, the probability that his or her condition will be detected is .90. If a person without tuberculosis is given a TB screening, the probability that he or she will be diagnosed incorrectly as having tuberculosis is .3. Suppose, further, that 11% of the adult residents of a certain city have tuberculosis. If one of these adults is diagnosed as having tuberculosis based on the screening, what is the probability that he or she actually has tuberculosis? Interpret your result.

37. **Detective Columbo** An absent-minded nurse is to give Mr. Brown a pill each day. The probability that the nurse forgets to administer the pill is $\frac{2}{3}$. If he receives the pill, the probability that Mr. Brown will die is $\frac{1}{3}$. If he does not get his pill, the probability that he will die is $\frac{3}{4}$. Mr. Brown died. What is the probability that the nurse forgot to give Mr. Brown the pill?

38. **Marketing** To introduce a new beer, a company conducted a survey. It divided the United States into four regions: eastern, northern, southern, and western. The company estimates that 35% of the potential customers for the beer are in the eastern region, 30% are in the northern region, 20% are in the southern region, and 15% are in the western region. The survey indicates that 50% of the potential customers in the eastern region, 40% of the potential customers in the northern region, 36% of the potential customers in the southern region, and 42% of those in the western region will buy the beer. If a potential customer chosen at random indicates that he or she will buy the beer, what is the probability that the customer is from the southern region?

39. **College Majors** Data collected by the Office of Admissions of a large midwestern university indicate the following choices made by the members of the freshman class regarding their majors:

Major	Percentage of Freshmen Choosing Major	Female (in percent)	Male (in percent)
Engineering	26	40	60
Business	30	35	65
Education	9	80	20
Social science	12	52	48
Natural science	12	56	44
Humanities	9	65	35
Other	2	51	49

What is the probability that a female student selected at random from the freshman class is majoring in engineering?

40. **Testing for HIV** An article in the *New York Times* some time ago reported that college students are beginning to routinely ask to be tested for the AIDS virus. The standard test for the HIV virus is the Elias test, which tests for the presence of HIV antibodies. It is estimated that this test has a 99.8% sensitivity and a 99.8% specificity. A 99.8% sensitivity means that, in a large-scale screening test, for every 1000 people tested who have the virus we can expect 998 people to test positive and 2 to have a false negative test. A 99.8% specificity means that, in a large-scale screening test, for every 1000 people tested who do not have the virus we can expect 998 people to have a negative test and 2 to have a false positive test.

    (a) The *New York Times* article remarks that it is estimated that about 2 in every 1000 college students have the HIV virus. Assume that a large group of randomly chosen college students, say 100,000, are tested by the Elias test. If a student tests positive, what is the chance that this student has the HIV virus?
    (b) What would this probability be for a population at high risk, where 5% of the population has the HIV virus?
    (c) Suppose Jack tested positive on an Elias test. Another Elias test* is performed and the results are positive again. Assuming that the tests are independent, what is the probability that Jack has the HIV virus?

41. **Medical Test** A scientist designed a medical test for a certain disease. Among 100 patients who have the disease, the test will show the presence of the disease in 97 cases out of 100, and will fail to show the presence of the disease in the remaining 3 cases out of 100. Among those who do not have the disease, the test will erroneously show the presence of the disease in 4 cases out of 100, and will show that there is no disease in the remaining 96 cases out of 100.

    (a) What is the probability that a patient who tested positive on this test actually has the disease, if it is estimated that 20% of the population has the disease?
    (b) What is the probability that a patient who tested positive on this test actually has the disease, if it is estimated that 4% of the population has the disease?
    (c) What is the probability that a patient who took the test twice and tested positive both times actually has the disease, if it is estimated that 4% of the population has the disease?

42. Prove Bayes' Formula (8).

43. Show that $P(E|F) = 1$ if F is a subset of E and $P(F) \neq 0$.

**Actually, in practice, if a person tests positive on an Elias test, then two more Elias tests are carried out. If either is positive, then one more confirmatory test, called the Western blot test, is carried out. If this is positive, the person is assumed to have the HIV virus.*

8.2 The Binomial Probability Model

PREPARING FOR THIS SECTION *Before getting started, review the following:*

- > Combinations (Section 6.5, pp. 343–349)
- > Independent Events (Section 7.5, pp. 403–409)

OBJECTIVE 1 Find binomial probabilities

Bernoulli Trials

In this section we study experiments that can be analyzed by using a probability model called the *binomial probability model.* The model was first studied by J. Bernoulli about 1700, and for this reason the model is sometimes referred to as a *Bernoulli trial.*

The **binomial probability model** is a sequence of trials, each of which consists of repetition of a single experiment. We assume the outcome of one experiment does not affect the outcome of any other one; that is, we assume the trials to be independent. Furthermore, we assume that there are only two possible outcomes for each trial and label them A, for *success,* and F, for *failure.* We denote the probability of success by $p = P(A)$; $p = P(A)$ remains the same from trial to trial. In addition, since there are only two outcomes in each trial, the probability of failure, denoted by q, must be $1 - p$, so

$$q = 1 - p = P(F)$$

Notice that $p + q = 1$.

Any random experiment for which the binomial probability model is appropriate is called a *Bernoulli trial.*

Bernoulli Trial

Random experiments are called **Bernoulli trials** if

(a) The same experiment is repeated several times.
(b) There are only two possible outcomes, success and failure, on each trial.
(c) The repeated trials are independent.
(d) The probability of each outcome remains the same for each trial.

Many real-world situations have the characteristics of the binomial probability model. For example, in repeatedly running a subject through a T-maze, we may label a turn to the left by A and a turn to the right by F. The assumption of independence of each trial is equivalent to presuming the subject has no memory.

In opinion polls one person's response is independent of any other person's response, and we may designate the answer "Yes" by an A and any other answer ("No" or "Don't know") by an F.

In testing TVs, we have a sequence of independent trials (each test of a particular TV is a trial), and we label a nondefective TV with an A and a defective one with an F.

In determining whether 9 out of 12 persons will recover from a tropical disease, we assume that each of the 12 persons has the same chance of recovery from the disease and that their recoveries are independent (they are not treated by the same doctor or in the same hospital). We may designate "recovery" by A and "nonrecovery" by F.

Next we consider an experiment that will lead us to formulate a general expression for the probability of obtaining k successes in a sequence of n Bernoulli trials ($k \leq n$).

EXAMPLE 1 Tossing a Coin Three Times

Discuss the experiment that consists of tossing a coin three times.

SOLUTION We define a success as heads (H) and a failure as tails (T). The coin may be fair or loaded, so we let p be the probability of heads and q be the probability of tails.

The sample space for this experiment is

$$\{HHH, \quad HHT, \quad HTH, \quad HTT, \quad THH, \quad THT, \quad TTH, \quad TTT\}$$

Here, as before, HHT means that the first two tosses are heads and the third is tails.

The tree diagram in Figure 11 lists all the possible outcomes and their respective probabilities.

FIGURE 11

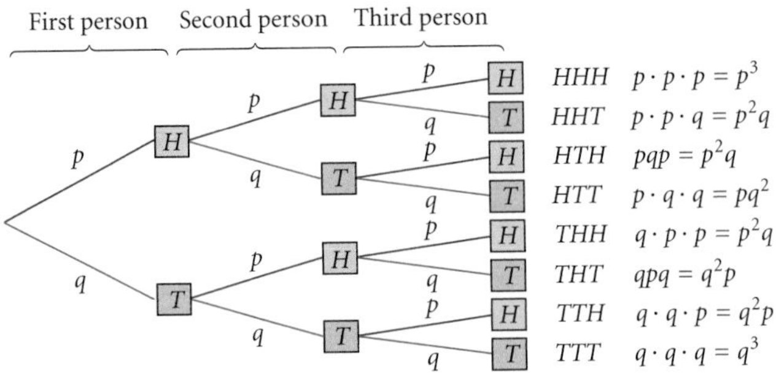

The outcome *HHH*, in which all three tosses are heads, has a probability $ppp = p^3$ since each Bernoulli trial has a probability p of resulting in heads.

If we wish to calculate the probability that exactly two heads appear, we consider only the outcomes

$$HHT, HTH, THH$$

The outcome *HHT* has probability $ppq = p^2q$ since the two heads each have probability p and the tail has probability q. In the same way the other two outcomes, *HTH* and *THH*, in which there are two heads and one tail, also have probabilities p^2q. Therefore, the probability that exactly two heads appear is equal to the sum of the probabilities of the three outcomes and, hence, is given by $3p^2q$.

In a similar way we can compute the following probabilities:

$$P\text{ (Exactly 1 head and 2 tails)} = 3pq^2$$
$$P\text{ (All 3 heads)} = p^3$$
$$P\text{ (All 3 tails)} = q^3$$

Binomial Probabilities

When considering tosses in excess of three, it would be extremely tedious to solve problems of this type using a tree diagram. This is why it is desirable to find a general formula.

Suppose the probability of a success in a Bernoulli trial is p and suppose we wish to find the probability of exactly k successes in n repeated trials. One possible outcome is

$$\underbrace{AAA \cdots A}_{k \text{ successes}} \cdot \underbrace{FF \cdots F}_{n-k \text{ failures}} \tag{1}$$

where k successes come first, followed by $n - k$ failures. The probability of this outcome is

$$\underbrace{ppp \cdots p}_{k \text{ factors}} \cdot \underbrace{qq \cdots q}_{n-k \text{ factors}} = p^k q^{n-k}$$

The k successes could also be obtained by rearranging the letters A and F in Display (1) above. Then the number of such sequences must equal the number of ways of choosing k of the n trials to contain successes—namely, $C(n, k) = \binom{n}{k}$. If we multiply this number by the probability of obtaining any one such sequence, we arrive at the following general result:

Formula for $b(n, k; p)$

In a Bernoulli trial the probability of exactly k successes in n trials is given by

$$b(n, k; p) = \binom{n}{k} p^k \cdot q^{n-k} = \frac{n!}{k!(n-k)!} p^k \cdot q^{n-k} \qquad (2)$$

where p is the probability of success and $q = 1 - p$ is the probability of failure.

The symbol $b(n, k; p)$ represents the probability of exactly k successes in n trials and is called a **binomial probability.**

NOW WORK PROBLEM 1.

1 **EXAMPLE 2** **Example of a Bernoulli Trial; Finding a Binomial Probability**

A common example of a Bernoulli trial is the experiment of tossing a fair coin.

1. There are exactly two possible mutually exclusive outcomes on each trial or toss (heads or tails).
2. The probability of a particular outcome (say, heads) remains constant from trial to trial (toss to toss).
3. The outcome on any trial (toss) is independent of the outcome on any other trial (toss).

Find the probability of obtaining exactly one tail in six tosses of a fair coin.

SOLUTION Let T denote the outcome "Tail shows" and let H denote the outcome "Head shows." Using Formula (2), in which $k = 1$ (one success), $n = 6$ (the number of trials), and $p = \frac{1}{2} = P(T)$ (the probability of success), we obtain

$$P(\text{Exactly 1 success}) = b\left(6, 1; \frac{1}{2}\right) = \binom{6}{1}\left(\frac{1}{2}\right)^1\left(\frac{1}{2}\right)^{6-1} = \frac{6}{64} = .0938$$

NOW WORK PROBLEM 15.

COMMENT: A graphing utility can be used to compute binomial probabilities. Figure 12 shows $b(6, 1; .5)$ using a TI-83 Plus calculator.

FIGURE 12

Use EXCEL to solve Example 2. Make a table to find the probabilities of getting exactly 0, 1, 2, 3, 4, 5, or 6 tails.

SOLUTION

STEP 1 Use the Excel function BINOMDIST (). The function is found under the category *Statistical.*

STEP 2 The syntax for the function is

BINOMDIST(*# of successes, # of trials, probability of success, cumulative*)

Cumulative is a logical variable. Set *cumulative* equal to FALSE to find the probability.

STEP 3 Set up the Excel spreadsheet.

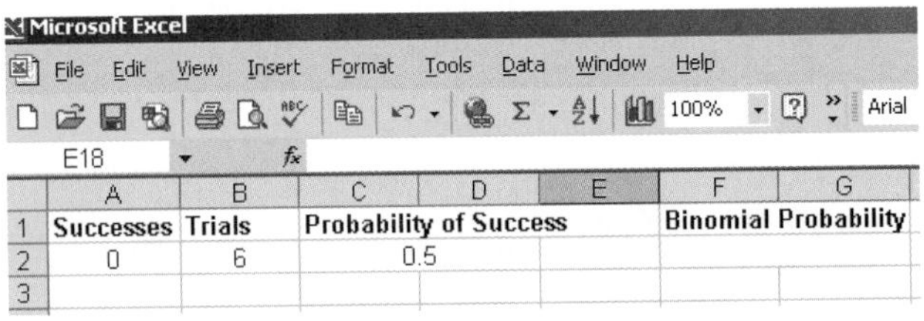

STEP 4 Enter the function in F2. Keep number of trials and probability of success constant by using $.

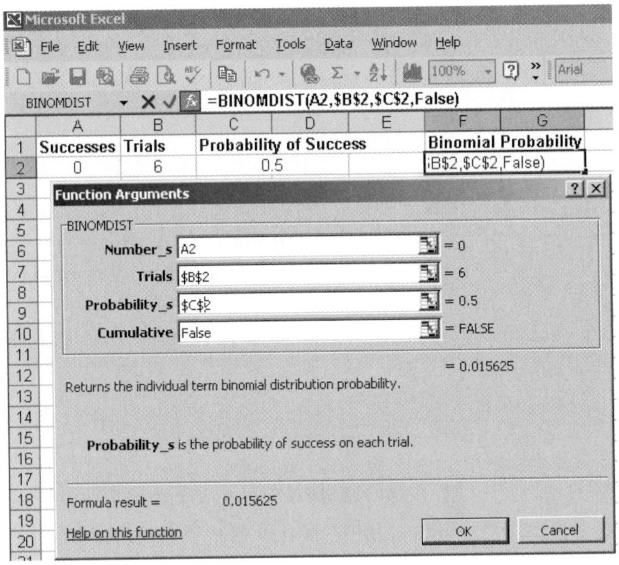

STEP 5 The probability of getting zero tails when tossing a coin 6 times is given below.

Successes	*Trials*	*Probability of Success*	*Binomial Probability*
0	6	0.5	0.015625

STEP 6 To find the probability of getting exactly 1, 2, 3, 4, 5, or 6 tails, highlight row 2 and click and drag to row 8.

Successes	*Trials*	*Probability of Success*	*Binomial Probability*
0	6	0.5	0.015625
1	6	0.5	0.09375
2	6	0.5	0.234375
3	6	0.5	0.3125
4	6	0.5	0.234375
5	6	0.5	0.09375
6	6	0.5	0.015625

NOW WORK PROBLEM 15 USING EXCEL.

EXAMPLE 3 Baseball

A baseball pitcher gives up a hit on the average of once every fifth pitch. If nine pitches are thrown, what is the probability that

(a) Exactly three pitches result in hits?
(b) No pitch results in a hit?
(c) Eight or more pitches result in hits?
(d) No more than seven pitches result in hits?

SOLUTION In this example $p = P(\text{Success}) = P(\text{Allowing a hit}) = \frac{1}{5} = .2$. The number of trials is the number of pitches, so $n = 9$. Finally, k is the number of pitches that result in hits.

(a) If exactly three pitches result in hits, then $k = 3$. Using $n = 9$, $p = .2$, and $q = 1 - p = .8$, we find

$$P(\text{Exactly 3 hits}) = b(9, 3; .2) = \binom{9}{3}(.2)^3\,(.8)^6 = .1762$$

(b) If none of the nine pitches resulted in a hit, $k = 0$.

$$P(\text{Exactly 0 hits}) = b(9, 0; .2) = \binom{9}{0}(.2)^0\,(.8)^9 = .1342$$

(c) Nine pitches are thrown, so the event "eight or more pitches result in hits" is equivalent to the events: "exactly 8 result in hits" or "exactly 9 result in hits." Since these events are mutually exclusive, we have

$$\begin{aligned} P(\text{At least 8 hits}) &= P(\text{Exactly 8 hits}) + P(\text{Exactly 9 hits}) \\ &= b(9, 8; .2) + b(9, 9; .2) = .0000189 \end{aligned}$$

(d) "No more than seven pitches result in hits" means 0, 1, 2, 3, 4, 5, 6, or 7 pitches result in a hit. We could add $b(9, 0; .2)$, $b(9, 1; .2)$, and so on, but it is easier to use the formula $P(E) = 1 - P(\overline{E})$. The complement of "No more than 7" is "at least 8." We use the result of part (c) to find:

$$\begin{aligned} P(\text{No more than 7 hits}) &= 1 - P\,(\text{At least 8 hits}) \\ &= 1 - [b(9, 8; .2) + b(9, 9; .2)] = .999981 \end{aligned}$$

Look again at part (c) in Example 3. When we seek the probability of *at least* k successes (or *at most* k successes), it is necessary to find the individual probabilities and, because the events are mutually exclusive, add them. The next example looks once more at this type of situation.

NOW WORK PROBLEM 35.

COMMENT: A graphing utility can be used to compute cumulative probabilities. Figure 13 shows the probability of 0 through 7 successes in 9 trials where the probability of success is 0.2 using a TI-83 Plus. This is the same result obtained in Example 3(d).

FIGURE 13

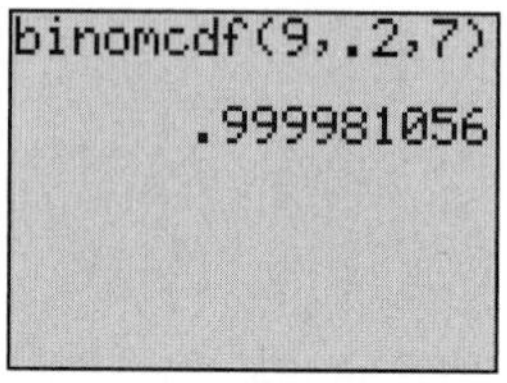

EXAMPLE 4 Quality Control

A machine produces light bulbs to meet certain specifications, and 80% of the bulbs produced meet these specifications. A sample of six bulbs is taken from the machine's production and placed in a box. What is the probability that at least three of them fail to meet the specifications?

SOLUTION In this example the number of trials is $n = 6$. We are looking for the probability of the event

$$E: \quad \text{At least 3 fail to meet specifications}$$

Since the experiment consists of choosing 6 bulbs, the event E is the union of the mutually exclusive events: "Exactly 3 fail," "Exactly 4 fail," "Exactly 5 fail," and "Exactly 6 fail." We use Formula (2) for $n = 6$ and $k = 3$, 4, 5, and 6. Since the probability of a bulb failing to meet specifications is .20, we have

$$P(\text{Exactly 3 fail}) = b(6, 3; .20) = .0819$$
$$P(\text{Exactly 4 fail}) = b(6, 4; .20) = .0154$$
$$P(\text{Exactly 5 fail}) = b(6, 5; .20) = .0015$$
$$P(\text{Exactly 6 fail}) = b(6, 6; .20) = .0001$$

Therefore,

$$P(\text{At least 3 fail}) = P(E) = .0819 + .0154 + .0015 + .0001 = .0989$$

Another way of getting this answer is to compute the probability of the complementary event

$$\overline{E}: \quad \text{Fewer than 3 fail}$$

Then

$$\begin{aligned} P(\overline{E}) &= P(\text{Exactly 2 fail}) + P(\text{Exactly 1 fail}) + P(\text{Exactly 0 fail}) \\ &= b(6, 2; .20) + b(6, 1; .20) + b(6, 0; .20) \\ &= .2458 + .3932 + .2621 = .9011 \end{aligned}$$

As a result,

$$P(\text{At least 3 fail}) = 1 - P(\overline{E}) = 1 - .9011 = .0989$$

as before.

NOW WORK PROBLEM 29.

EXAMPLE 5 **Product Testing**

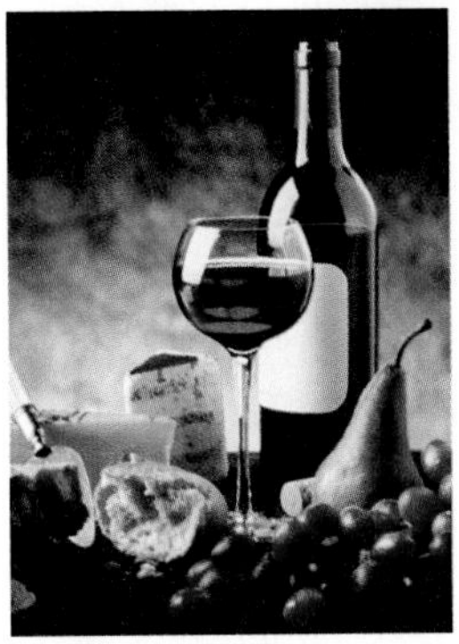

A man claims to be able to distinguish between two kinds of wine with 90% accuracy and presents his claim to an agency interested in promoting the consumption of one of the two kinds of wine. The following experiment is conducted to check his claim. The man is to taste the two types of wine and distinguish between them. This is to be done nine times with a 3-minute break after each taste. It is agreed that if the man is correct at least six out of the nine times, he will be hired.

The main questions to be asked are, on the one hand, whether the above procedure gives sufficient protection to the hiring agency against a person guessing and, on the other hand, whether the man is given a sufficient chance to be hired if he is really a wine connoisseur.

SOLUTION The number of trials is $n = 9$. To answer the first question, let's assume that the man is guessing. Then in each trial he has a probability of $\frac{1}{2}$ of identifying the wine correctly. Let k be the number of correct identifications. Let's compute the binomial probability for $k = 6, 7, 8, 9$ to find the likelihood of the man being hired (at least 6 correct) while guessing ($p = \frac{1}{2}$):

$$\begin{aligned} b(9, 6; \tfrac{1}{2}) + b(9, 7; \tfrac{1}{2}) + b(9, 8; \tfrac{1}{2}) + b(9, 9; \tfrac{1}{2}) &= .1641 + .0703 + .0176 + .0020 \\ &= .2540 \end{aligned}$$

There is a likelihood of .254 that he will pass if he is just guessing.

To answer the second question in the case where the claim is true ($p = .90$), we need to find the sum of the probabilities $b(9, k; .90)$ for $k = 6, 7, 8, 9$:

$$\begin{aligned} b(9, 6; .90) + b(9, 7; .90) + b(9, 8; .90) + b(9, 9; .90) &= .0446 + .1722 + .3874 + .3874 \\ &= .9916 \end{aligned}$$

Notice that the test in Example 5 is fair to the man since it practically assures him the position if his claim is true. However, the company may not like the test because 25% of the time a person who guesses will pass the test.

NOW WORK PROBLEM 43.

Application

EXAMPLE 6 **Testing a Serum or Vaccine**

Suppose that the normal rate of infection of a certain disease in cattle is 25%. To test a newly discovered serum, healthy animals are injected with the serum. How can we evaluate the result of the experiment?

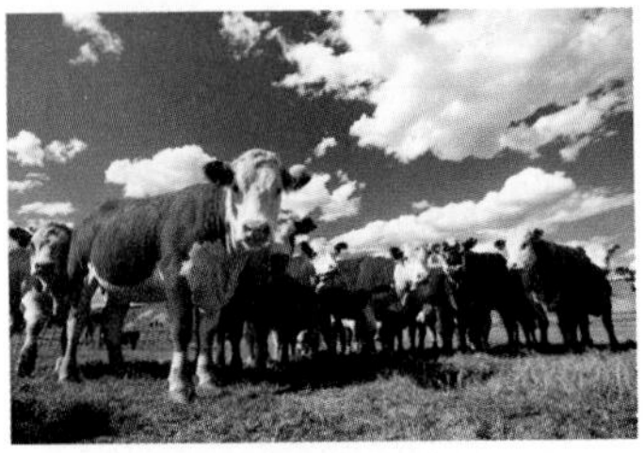

SOLUTION For an absolutely worthless serum, the probability that exactly k of n test animals remain free from infection equals $b(n, k; .75)$. For $k = n = 10$, this probability is about $b(10, 10; .75) = .056$. Thus, if out of 10 test animals none is infected, this may be taken as an indication that the serum has had an effect, although it is not conclusive proof. Notice that, without serum, the probability that out of 17 animals at most 1 catches the infection is $b(17, 0; .25) + b(17, 1; .25) = .0501$. Therefore, there is *stronger evidence* in favor of the serum if out of 17 test animals at most 1 gets infected than if out of 10 all remain healthy. For $n = 23$, the probability of at most 2 animals catching the infection is about .0492 and, thus, at most 2 failures out of 23 is again better evidence for the serum than at most 1 out of 17 or 0 out of 10.

EXERCISE 8.2 Answers to Odd-Numbered Problems Begin on Page AN-33.

In Problems 1–14, use Formula (2), page 435, to compute each binomial probability.

1. $b(7, 4; .20)$
2. $b(8, 5; .30)$
3. $b(15, 8; .80)$
4. $b(8, 5; .70)$
5. $b(15, 10; \frac{1}{2})$
6. $b(12, 6; .90)$
7. $b(15, 3; .3) + b(15, 2; .3) + b(15, 1; .3) + b(15, 0; .3)$
8. $b(8, 6; .4) + b(8, 7; .4) + b(8, 8; .4)$
9. $n = 3, \quad k = 2, \quad p = \dfrac{1}{3}$
10. $n = 3, \quad k = 1, \quad p = \dfrac{1}{3}$
11. $n = 3, \quad k = 0, \quad p = \dfrac{1}{6}$
12. $n = 3, \quad k = 3, \quad p = \dfrac{1}{6}$
13. $n = 5, \quad k = 3, \quad p = \dfrac{2}{3}$
14. $n = 5, \quad k = 0, \quad p = \dfrac{2}{3}$

15. Find the probability of obtaining exactly 6 successes in 10 trials when the probability of success is .3.

16. Find the probability of obtaining exactly 5 successes in 9 trials when the probability of success is .2.

17. Find the probability of obtaining exactly 9 successes in 12 trials when the probability of success is .8.

18. Find the probability of obtaining exactly 8 successes in 15 trials when the probability of success is .75.

19. Find the probability of obtaining at least 5 successes in 8 trials when the probability of success is .30.

20. Find the probability of obtaining at most 3 successes in 7 trials when the probability of success is .20.

In Problems 21–26, a fair coin is tossed 8 times.

21. What is the probability of obtaining exactly 1 head?

22. What is the probability of obtaining exactly 2 heads?

23. What is the probability of obtaining at least 5 tails?

24. What is the probability of obtaining at most 2 tails?

25. What is the probability of obtaining exactly 2 heads if it is known that at least 1 head appeared?

26. What is the probability of obtaining exactly 3 heads if it is known that at least 1 head appeared?

27. In five rolls of two fair dice, what is the probability of obtaining a sum of 7 exactly twice?

28. In seven rolls of two fair dice, what is the probability of obtaining a sum of 11 exactly three times?

29. **Quality Control** Suppose that 5% of the items produced by a factory are defective. If 8 items are chosen at random, what is the probability that

    (a) Exactly 1 is defective? (b) Exactly 2 are defective?
    (c) At least 1 is defective? (d) Fewer than 3 are defective?

30. **Opinion Poll** Suppose that 60% of the voters intend to vote for a conservative candidate. What is the probability that a survey polling 8 people reveals that 3 or fewer intend to vote for a conservative candidate?

31. **Family Structure** What is the probability that a family with exactly 6 children will have 3 boys and 3 girls?

32. **Family Structure** What is the probability that in a family of 7 children:

    (a) 4 are girls?
    (b) At least 2 are girls?
    (c) At least 2 and not more than 4 are girls?

33. An experiment is performed 4 times, with 2 possible outcomes, F (failure) and S (success), with probabilities $\frac{1}{4}$ and $\frac{3}{4}$, respectively.

    (a) Draw a tree diagram describing the experiment.
    (b) Calculate the probability of exactly 2 successes and 2 failures by using the tree diagram from Part (a).
    (c) Verify your answer to Part (b) by using Formula (2).

34. **Batting Averages** For a baseball player with a .250 batting average, what is the probability that the player will have at least 2 hits in 4 times at bat? What is the probability of at least 1 hit in 4 times at bat?

35. **Target Shooting** If the probability of hitting a target is $\frac{2}{3}$ and 10 shots are fired independently, what is the probability of the target being hit at least twice?

36. **Quality Control** A television manufacturer tests a random sample of 15 picture tubes to determine whether any are defective. The probability that a picture tube is defective has been found from past experience to be .03.

(a) What is the probability that there are no defective tubes in the sample?
(b) What is the probability that more than 2 of the tubes are defective?

37. **True–False Tests** In a 15-item true–false examination, what is the probability that a student who guesses on each question will get at least 10 correct answers? If another student has .8 probability of correctly answering each question, what is the probability that this student will answer at least 12 questions correctly?

38. **Screening Employees** To screen prospective employees, a company gives a 10-question multiple-choice test. Each question has 4 possible answers, of which 1 is correct. The chance of answering the questions correctly by just guessing is $\frac{1}{4}$ or 25%. Find the probability of answering, by chance:

(a) Exactly 3 questions correctly.
(b) No questions correctly.
(c) At least 8 questions correctly.
(d) No more than 7 questions correctly.

39. **Opinion Poll** Mr. Austin and Ms. Moran are running for public office. A survey conducted just before the day of election indicates that 60% of the voters prefer Mr. Austin and 40% prefer Ms. Moran. If 8 people are chosen at random and asked their preference, find the probability that all 8 people will express a preference for Ms. Moran.

40. **Working Habits** If 30% of the workers at a large factory bring their lunch each day, what is the probability that in a randomly selected sample of 8 workers

(a) Exactly 2 bring their lunch each day?
(b) At least 2 bring their lunch each day?
(c) No one brings lunch?
(d) No more than 3 bring lunch each day?

41. **Heart Attack** Approximately 23% of North American unexpected deaths are due to heart attacks. What is the probability that 4 of the next 10 unexpected deaths reported in a certain community will be due to heart attacks?

42. **Support for the President** A Fox News/Opinion Dynamics Poll conducted with 900 registered voters in the United States on February 25–26, 2003, found that 55% approved of the job George W. Bush was doing as president. Assuming that this poll reflects national opinion, if 30 voters were randomly selected, what is the probability that a majority of these voters would approve of the job George W. Bush is doing as president?

Source: Fox News Opinion/Dynamics Poll.

43. **Product Testing** A supposed coffee connoisseur claims she can distinguish between a cup of instant coffee and a cup of drip coffee 80% of the time. You give her 6 cups of coffee and tell her that you will grant her claim if she correctly identifies at least 5 of the 6 cups.

(a) What are her chances of having her claim granted if she is in fact only guessing?
(b) What are her chances of having her claim rejected when in fact she really does have the ability she claims?

44. **Opinion Poll** Opinion polls based on small samples often yield misleading results. Suppose 65% of the people in a city are opposed to a bond issue and the others favor it. If 7 people are asked for their opinion, what is the probability that a majority of them will favor the bond issue?

45. **The Aging Population** The 2000 United States Census showed that 12.4% of the United States population was 65 years of age or older. Suppose that in the year 2000, 10 people were randomly selected from the United States population.

(a) What is the probability that exactly four of these people would be 65 years of age or older?
(b) What is the probability that none of these people would be 65 years of age or older?
(c) What is the probability that at most five of these people would be 65 years of age or older?

Source: United States Census Bureau.

46. **Age Distribution** The 2000 United States Census showed that 87.6% of the United States population was younger than 65 years old. Suppose that in the year 2000, 10 people were randomly selected from the United States population.

(a) What is the probability that exactly 8 of these people would be younger than 65?
(b) What is the probability that all of these people would be younger than 65?
(c) What is the probability that at most five of these people would be 65 years of age or older?

Source: United States Census Bureau, 2000.

Sometimes experiments are simulated using a random number function instead of actually performing the experiment.*

In Problems 47–50, use a graphing utility to simulate each experiment.

47. Tossing a Fair Coin Consider the experiment of tossing a coin 4 times, counting the number of heads occurring in these 4 tosses. Simulate the experiment using a random number function on your calculator, considering a toss to be tails (T) if the result is less than 0.5, and considering a toss to be heads (H) if the result is greater than or equal to 0.5. Record the number of heads in 4 tosses. [*Note*: Most calculators repeat the action of the last entry if you simply press the ENTER, or EXE, key again.] Repeat the experiment 10 times, obtaining a sequence of 10 numbers. Using these 10 numbers you can estimate the probability of k heads, $P(k)$, for each $k = 0, 1, 2, 3, 4$, by the ratio

$$\frac{\text{Number of times } k \text{ appears in your sequence}}{10}$$

Enter your estimates in the table below. Calculate the actual probabilities using the binomial probability formula, and enter these numbers in the table. How close are your numbers to the actual values?

k	Your Estimate of $P(k)$	Actual Value of $P(k)$
0		
1		
2		
3		
4		

48. Tossing a Loaded Coin Consider the experiment of tossing a loaded coin 4 times, counting the number of heads occurring in these 4 tosses. Simulate the experiment using a random number function on your calculator, considering a toss to be tails (T) if the result is less than 0.80, and considering a toss to be heads (H) if the result is greater than or equal to 0.80. Record the number of heads in 4 tosses. Repeat the experiment 10 times, obtaining a sequence of 10 numbers. Using these 10 numbers you can estimate the probability of k heads, $P(k)$, for each $k = 0, 1, 2, 3, 4$, by the ratio

$$\frac{\text{Number of times } k \text{ appears in your sequence}}{10}$$

Enter your estimates in the table below. Calculate the actual probabilities using the binomial probability formula, and enter these numbers in the table. How close are your numbers to the actual values? Why is this coin loaded?

k	Your Estimate of $P(k)$	Actual Value of $P(k)$
0		
1		
2		
3		
4		

49. Tossing a Fair Coin Consider the experiment of tossing a fair coin 8 times, counting the number of heads occurring in these 8 tosses. Simulate the experiment using a random number function on your calculator, considering a toss to be tails (T) if the result is less than 0.50, and considering a toss to be heads (H) if the result is greater than or equal to 0.50. Record the number of heads in 8 tosses. Repeat the experiment 10 times, obtaining a sequence of 10 numbers. Using these 10 numbers you can estimate the probability of 3 heads, $P(3)$, by the ratio

$$\frac{\text{Number of times 3 appears in your sequence}}{10}$$

Calculate the actual probability using the binomial probability formula. How close is your estimate to the actual value?

50. Tossing a Loaded Coin Consider the experiment of tossing a loaded coin 8 times, counting the number of heads occurring in these 8 tosses. Simulate the experiment using a random number function on your calculator, considering a toss to be tails (T) if the result is less than 0.80, and considering a toss to be heads (H) if the result is greater than or equal to 0.80. Record the number of heads in 8 tosses. Repeat the experiment 10 times, obtaining a sequence of 10 numbers. Using these 10 numbers you can estimate the probability of 3 heads, $P(3)$, by the ratio

$$\frac{\text{Number of times 3 appears in your sequence}}{10}$$

Calculate the actual probability using the binomial probability formula. How close is your estimate to the actual value?

**Most graphing utilities have a random number function (usually RAND or RND) for generating numbers between 0 and 1. Check your user's manual to see how to use this function on your graphing calculator.*

8.3 Expected Value

PREPARING FOR THIS SECTION *Before getting started, review the following:*

> Sample Spaces and Assignment of Probabilities (Section 7.1, pp. 365–373)

OBJECTIVES
1 Compute expected values
2 Solve applied problems involving expected value
3 Find expected value in a Bernoulli trial

An important concept, which uses probability, is *expected value.*

EXAMPLE 1 Examples of Expected Value

(a) Suppose that 1000 tickets are sold for a raffle that has the following prizes: one \$300 prize, two \$100 prizes, and one hundred \$1 prizes. Then, of the 1000 tickets, 1 ticket has a cash value of \$300, 2 are worth \$100, and 100 are worth \$1, while the remaining are worth \$0. The *expected* (average) *value* of a ticket is then

$$\begin{aligned} E &= \frac{\$300 + (\$100 + \$100) + \overbrace{(\$1 + \$1 + \cdots + \$1)}^{100 \text{ times}} + \overbrace{(\$0 + \$0 + \cdots + \$0)}^{897 \text{ times}}}{1000} \\ &= \$300 \cdot \frac{1}{1000} + \$100 \cdot \frac{2}{1000} + \$1 \cdot \frac{100}{1000} + \$0 \cdot \frac{897}{1000} \\ &= \frac{\$600}{1000} = \$0.60 \end{aligned}$$

If the raffle is to be nonprofit to all, the charge for each ticket should be \$0.60.

The situation above can also be viewed as follows: if we entered such a raffle many times, $\frac{1}{1000}$ of the time we would win \$300, $\frac{2}{1000}$ of the time we would win \$100, and so on, with our winnings in the long run averaging \$0.60 per ticket.

(b) Suppose that you are to receive \$3.00 each time you obtain two heads when flipping a coin two times and \$0 otherwise. Then the *expected value* is

$$E = \$3.00 \cdot \frac{1}{4} + \$0 \cdot \frac{3}{4} = \$0.75$$

This means, if the game is to be fair, that you should be willing to pay \$0.75 each time you toss the coins.

(c) A game consists of flipping a single coin. If a head shows, the player loses \$1; but if a tail shows, the player wins \$2. Thus half the time the player loses \$1 and the other half the player wins \$2. The expected value E of the game is

$$E = \$2 \cdot \frac{1}{2} + (-\$1) \cdot \frac{1}{2} = \$0.50$$

The player is expected to win an average of \$0.50 on each play.

In each of the above examples, we arrive at the expected value E by multiplying the amount earned for a given outcome times the probability for that outcome to occur, and adding all the products.

Look back at Example 1(a). In the expression for the expected value in the raffle problem, the term $\$300 \cdot \frac{1}{1000}$ is the pairing of the value \$300 with its corresponding probability $\frac{1}{1000}$, namely, the probability of having a \$300 ticket. Likewise, the probability of getting a \$1 ticket is $\frac{100}{1000}$ and this produces the term $\$1 \cdot \frac{100}{1000}$ in the expression for the expected value, E. So the expression for E is obtained by multiplying each ticket value by the probability of its occurrence and adding the results.

In Example 1(b), the term $\$3 \cdot \frac{1}{4}$ is the pairing of the value \$3.00 with its corresponding probability $\frac{1}{4}$, namely, the probability of obtaining *HH*. Likewise, the probability of getting *HT*, *TH*, and *TT* is $\frac{3}{4}$ with payoff 0, and this produces $\$0 \cdot \frac{3}{4}$ in the expression for the expected value.

Finally, in Example 1(c) the term $\$2 \cdot \frac{1}{2}$ is the pairing of the value \$2.00 with its corresponding probability $\frac{1}{2}$, namely, the probability of obtaining T. Likewise, the probability of getting H is $\frac{1}{2}$, with payoff $-\$1$.

This leads to the following definition.

Expected Value

Let S be a sample space and let $A_1, A_2, \ldots, A_n$ be n events of S that form a partition of S. Let $p_1, p_2, \ldots, p_n$ be the probabilities of the events $A_1, A_2, \ldots, A_n$, respectively. If each event $A_1, A_2, \ldots, A_n$ is assigned the payoff $m_1, m_2, \ldots, m_n$, respectively, the **expected value** E corresponding to these payoffs is

$$E = m_1 \cdot p_1 + m_2 \cdot p_2 + \cdots + m_n \cdot p_n \tag{1}$$

The term *expected value* should not be interpreted as a value that actually occurs in the experiment. In Example 1(a) there was no raffle ticket costing \$.60. Rather, this number represents the average value of a raffle ticket.

In gambling, E is interpreted as the average winnings expected for the player in the long run. If E is positive, we say that the game is **favorable** to the player; if $E = 0$, we say the game is **fair;** and if E is negative, we say the game is **unfavorable** to the player.

When the payoff assigned to an outcome of an experiment is positive, it can be interpreted as a profit, winnings, gain, etc. When it is negative, it represents losses, penalties, deficits, etc.

The following steps outline the general procedure involved in determining expected value.

Steps for Computing Expected Value

STEP 1 Partition S into n events $A_1, A_2, \ldots, A_n$.

STEP 2 Determine the probability $p_1, p_2, \ldots, p_n$ of each event $A_1, A_2, \ldots, A_n$. Since these events constitute a partition of S, it follows that

$$p_1 + p_2 + \cdots + p_n = 1.$$

STEP 3 Assign payoff values $m_1, m_2, \ldots, m_n$ to each event $A_1, A_2, \ldots, A_n$.

STEP 4 Calculate $E = m_1 \cdot p_1 + m_2 \cdot p_2 + \cdots + m_n \cdot p_n$.

NOW WORK PROBLEM 1.

1 **EXAMPLE 2** **Computing Expected Value**

Consider the experiment of rolling a fair die. The player recovers an amount of dollars equal to the number of dots on the face that turns up, except when face 5 or 6 turns up, in which case the player will lose \$5 or \$6, respectively. What is the expected value of the game?

SOLUTION **STEP 1** The sample space is $S = \{1, 2, 3, 4, 5, 6\}$. Each of these 6 outcomes constitutes an event that forms a partition of S.

STEP 2 Since each outcome (event) is equally likely, the probability of each one is $\frac{1}{6}$.

STEP 3 Since the player wins \$1 for the outcome 1, \$2 for the outcome 2, \$3 for a 3, and \$4 for a 4 and the player loses \$5 for a 5 or equivalently wins −\$5 for a 5 and loses \$6 for a 6 or equivalently wins −\$6 for a 6, we assign payoffs (the winnings of the player) as follows:

event	1	2	3	4	5	6
payoff (winnings)	\$1	\$2	\$3	\$4	−\$5	−\$6

STEP 4 The expected value of the game is

$$E = \$1 \cdot \frac{1}{6} + \$2 \cdot \frac{1}{6} + \$3 \cdot \frac{1}{6} + \$4 \cdot \frac{1}{6} + (-\$5) \cdot \frac{1}{6} + (-\$6) \cdot \frac{1}{6}$$
$$= -\$\frac{1}{6} = -\$0.167$$

The player would expect to lose an average of 16.7 cents on each throw.

EXAMPLE 3 **Computing Expected Value**

What is the expected number of heads in tossing a fair coin three times?

SOLUTION **STEP 1** The sample space is

$$S = \{HHH, HHT, HTH, HTT, THH, THT, TTH, TTT\}$$

Since we are interested in the number of heads and since a coin tossed three times will result in 0, 1, 2, or 3 heads, we partition S as follows:

$$\underbrace{A_1 = \{TTT\}}_{\text{0 heads}} \quad \underbrace{A_2 = \{HTT, THT, TTH\}}_{\text{1 head}} \quad \underbrace{A_3 = \{HHT, HTH, THH\}}_{\text{2 heads}} \quad \underbrace{A_4 = \{HHH\}}_{\text{3 heads}}$$

STEP 2 The probability of each of these events is

$$P(A_1) = \frac{1}{8}, \qquad P(A_2) = \frac{3}{8} \qquad P(A_3) = \frac{3}{8} \qquad P(A_4) = \frac{1}{8}$$

STEP 3 Since we are interested in the expected number of heads, we assign payoffs of 0, 1, 2, and 3, respectively, to A_1, A_2, A_3, and A_4.

STEP 4 The expected number of heads can now be found by multiplying each payoff by its corresponding probability and finding the sum of these values.

$$\text{Expected number of heads} = E = 0 \cdot \frac{1}{8} + 1 \cdot \frac{3}{8} + 2 \cdot \frac{3}{8} + 3 \cdot \frac{1}{8} = \frac{3}{2} = 1.5$$

On average, tossing a coin three times will result in 1.5 heads.

NOW WORK PROBLEM 11.

Applications

2 **EXAMPLE 4** **Bidding for Oil Wells**

An oil company may bid for only one of two contracts for oil drilling in two different locations. If oil is discovered at location I, the profit to the company will be \$3,000,000. If no oil is found, the company's loss will be \$250,000. If oil is discovered at location II, the profit will be \$4,000,000. If no oil is found, the loss will be \$500,000. The probability of discovering oil at location I is .7, and at location II it is .6. Which field should the company bid on; that is, for which location is expected profit highest?

SOLUTION In the first field the company expects to discover oil .7 of the time at a profit of \$3,000,000. It would not discover oil .3 of the time at a loss of \$250,000. The expected profit E_{I} is therefore

$$E_{\text{I}} = (\$3{,}000{,}000)(.7) + (-\$250{,}000)(.3) = \$2{,}025{,}000$$

Similarly, for the second field, the expected profit E_{II} is

$$E_{\text{II}} = (\$4{,}000{,}000)(.6) + (-\ \$500{,}000)(.4) = \$2{,}200{,}000$$

Since the expected profit for the second field exceeds that for the first, the oil company should bid on the second field.

NOW WORK PROBLEM 19.

EXAMPLE 5 **Evaluating Insurance**

Mr. Richmond is producing an outdoor concert. He estimates that he will make \$300,000 if it does not rain and make \$60,000 if it does rain. The weather bureau predicts that the chance of rain is .34 for the day of the concert.

(a) What are Mr. Richmond's expected earnings from the concert?
(b) An insurance company is willing to insure the concert for \$150,000 against rain for a premium of \$30,000. If he buys this policy, what are his expected earnings from the concert?
(c) Based on the expected earnings, should Mr. Richmond buy an insurance policy?

SOLUTION The sample space consists of two outcomes: R: Rain or N: No rain. The probability of rain is $P(R) = .34$; the probability of no rain is $P(N) = .66$.

(a) If it rains, the earnings are \$60,000; if there is no rain, the earnings are \$300,000. The expected earnings E are

$$\begin{aligned} E &= 60{,}000 \cdot P(R) + 300{,}000 \cdot P(N) \\ &= 60{,}000\,(.34) + 300{,}000\,(.66) \\ &= \$218{,}400 \end{aligned}$$

(b) With insurance, if it rains, the earnings are

$$\underset{\text{Insurance payout}}{150{,}000} + \underset{\text{Concert profit}}{60{,}000} - \underset{\text{Cost of insurance}}{30{,}000} = 180{,}000$$

With insurance, if it does not rain, the earnings are

$$\underset{\text{No insurance payout}}{0} + \underset{\text{Concert profit}}{300{,}000} - \underset{\text{Cost of insurance}}{30{,}000} = 270{,}000$$

With insurance, the expected earnings E are

$$\begin{aligned} E &= 180{,}000 \cdot P(R) + 270{,}000 \cdot P(N) \\ &= 180{,}000\,(.34) + 270{,}000\,(.66) \\ &= \$239{,}400 \end{aligned}$$

(c) Since the expected earnings are higher with insurance, it would be better to obtain the insurance.

EXAMPLE 6 Quality Control

A laboratory contains 10 electron microscopes, of which 2 are defective. If all microscopes are equally likely to be chosen and if 4 are chosen, what is the expected number of defective microscopes?

SOLUTION The sample of 4 microscopes can contain 0, 1, or 2 defective microscopes. The probability p_0 that none in the sample is defective is

$$p_0 = \frac{C(2,0)\cdot C(8,4)}{C(10,4)} = \frac{1}{3}$$

Similarly, the probabilities p_1 and p_2 for 1 and 2 defective microscopes are

$$p_1 = \frac{C(2,1)\cdot C(8,3)}{C(10,4)} = \frac{8}{15} \quad \text{and} \quad p_2 = \frac{C(2,2)\cdot C(8,2)}{C(10,4)} = \frac{2}{15}$$

Since we are interested in determining the expected number of defective microscopes, we assign a payoff of 0 to the outcome "0 defectives are selected," a payoff of 1 to the outcome "1 defective is chosen," and a payoff of 2 to the outcome "2 defectives are chosen." The expected value E is then

$$E = 0 \cdot p_0 + 1 \cdot p_1 + 2 \cdot p_2 = \frac{8}{15} + \frac{4}{15} = \frac{4}{5}$$

Of course, we cannot have $\frac{4}{5}$ of a defective microscope. However, we can interpret this to mean that in the long run, such a sample will average just under 1 defective microscope.

We point out that $\frac{4}{5}$ is a reasonable answer for the expected number of defective microscopes since $\frac{1}{5}$ of the microscopes in the laboratory are defective and we are selecting a random sample consisting of 4 of these microscopes.

Expected Value for Bernoulli Trials

In 100 tosses of a coin, what is the expected number of heads? If a student guesses at random on a true–false exam with 50 questions, what is her expected grade? These are specific instances of the following more general question:

In n trials of a Bernoulli process, what is the expected number of successes?

We now compute this expected value. As before, p denotes the probability of success on any individual trial.

If $n = 1$ (one trial), then the expected number of successes is

$$E = 1 \cdot p + 0 \cdot (1 - p) = p$$

If $n = 2$ (two trials), then either 0, 1, or 2 successes can occur. The expected number E of successes is

$$E = 2 \cdot p^2 + 1 \cdot 2p(1 - p) + 0 \cdot (1 - p)^2 = 2p$$

If $n = 3$ (three trials), then either 0, 1, 2, or 3 successes can occur. The expected number E of successes is

$$\begin{aligned} E &= 3 \cdot p^3 + 2 \cdot 3p^2(1 - p) + 1 \cdot 3p(1 - p)^2 + 0 \cdot (1 - p)^3 \\ &= 3p^3 + 6p^2 - 6p^3 + 3p - 6p^2 + 3p^3 \\ &= 3p \end{aligned}$$

This would seem to suggest that with n trials the expected value E would be given by $E = np$. This is indeed the case and we have the following result:

Expected Value for Bernoulli Trials

In a Bernoulli process with n trials the expected number of successes is

$$\boxed{E = np}$$

where p is the probability of success on any single trial.

A derivation of this result is included a little later in this section. The intuitive idea behind the result is fairly simple. Thinking of probabilities as percentages, if success results p percent of the time, then out of n attempts, p percent of them, namely, np, should be successful.

3 **EXAMPLE 7 Expected Value in a Bernoulli Trial**

In flipping a fair coin five times, what is the expected number of tails?

SOLUTION The six events 0 tails, 1 tail, 2 tails, 3 tails, 4 tails, 5 tails are events that constitute a partition of the sample space. The respective probability of each of these events is

$$\binom{5}{0}\left(\frac{1}{2}\right)^5 \quad \binom{5}{1}\left(\frac{1}{2}\right)^5 \quad \binom{5}{2}\left(\frac{1}{2}\right)^5 \quad \binom{5}{3}\left(\frac{1}{2}\right)^5 \quad \binom{5}{4}\left(\frac{1}{2}\right)^5 \quad \binom{5}{5}\left(\frac{1}{2}\right)^5$$

If we assign the payoffs 0, 1, 2, 3, 4, 5 respectively to each event, then the expected number of tails is

$$\begin{aligned} E = 0 &\cdot \binom{5}{0}\left(\frac{1}{2}\right)^5 + 1 \cdot \binom{5}{1}\left(\frac{1}{2}\right)^5 + 2 \cdot \binom{5}{2}\left(\frac{1}{2}\right)^5 \\ &+ 3 \cdot \binom{5}{3}\left(\frac{1}{2}\right)^5 + 4 \cdot \binom{5}{4}\left(\frac{1}{2}\right)^5 + 5 \cdot \binom{5}{5}\left(\frac{1}{2}\right)^5 = \frac{5}{2} \end{aligned}$$

Using the result $E = np$ is much easier. For $n = 5$ and $p = \frac{1}{2}$, we obtain $E = (5)(\frac{1}{2}) = \frac{5}{2}$.

NOW WORK PROBLEM 21.

EXAMPLE 8 Expected Value in a Bernoulli Trial

A multiple-choice test contains 100 questions, each with four choices. If a person guesses, what is the expected number of correct answers?

SOLUTION This is an example of a Bernoulli trial. The probability for success (a correct answer) when guessing is $p = \frac{1}{4}$. Since there are $n = 100$ questions, the expected number of correct answers is

$$E = np = (100)\left(\frac{1}{4}\right) = 25$$

Derivation of $E = np$

The derivation is an exercise in handling binomial coefficients and using the binomial theorem. We will make use of the following identity:

$$k\binom{n}{k} = n\binom{n-1}{k-1} \tag{1}$$

Equation (1) can be established by expanding both sides using the definition of a binomial coefficient. (See Problem 21, Exercise 6.6, page 357.)

Recall that the probability of obtaining exactly k successes in n trials is given by $b(n, k; p) = \binom{n}{k}p^k q^{n-k}$. Then the expected number of successes is

$$E = 0 \cdot \binom{n}{0} p^0 q^n + 1 \cdot \binom{n}{1} p^1 q^{n-1} + 2 \cdot \binom{n}{2} p^2 q^{n-2} + \cdots + \underbrace{k}_{\text{No. of successes}}\underbrace{\binom{n}{k} p^k q^{n-k}}_{\text{Corresponding probability}} + \cdots + n\binom{n}{n} p^n q^0$$

Using Equation (1) above,

$$E = n\binom{n-1}{0} pq^{n-1} + n\binom{n-1}{1} p^2 q^{n-2} + \cdots + n\binom{n-1}{k-1} p^k q^{n-k} + \cdots + n\binom{n-1}{n-1} p^n$$

Now factor out an n and a p from each of the terms on the right to get

$$E = np\left[\binom{n-1}{0} q^{n-1} + \binom{n-1}{1} pq^{n-2} + \cdots + \binom{n-1}{k-1} p^{k-1} q^{(n-1)-(k-1)} + \cdots + \binom{n-1}{n-1} p^{n-1}\right]$$

The expression in brackets is $(q + p)^{n-1}$. To see why, use the binomial theorem. As a result,

$$E = np(q + p)^{n-1}$$

Since $p + q = 1$, the result $E = np$ follows.

EXERCISE 8.3 Answers to Odd-Numbered Problems Begin on Page AN-34.

1. For the data given below, compute the expected value.

Outcome	e_1	e_2	e_3	e_4
Probability	.4	.2	.1	.3
Payoff	2	3	−2	0

2. For the data below, compute the expected value.

Outcome	e_1	e_2	e_3	e_4
Probability	$\frac{1}{3}$	$\frac{1}{6}$	$\frac{1}{4}$	$\frac{1}{4}$
Payoff	1	0	4	−2

3. **Weather and Attendance** Attendance at a football game in a certain city results in the following pattern. If it is extremely cold, the attendance will be 30,000; if it is cold, it will be 40,000; if it is moderate, 60,000; and if it is warm, 80,000. If the probabilities for extremely cold, cold, moderate, and warm are .08, .42, .42, and .08, respectively, how many fans are expected to attend the game?

4. **Analyzing a Game** A player rolls a fair die and receives a number of dollars equal to the number of dots appearing on the face of the die. What is the least the player should expect to pay in order to play the game?

5. **Analyzing a Game** Mary will win $8 if she draws an ace from a set of 10 different cards from ace to 10. How much should she pay for one draw?

6. **Analyzing a Game** Thirteen playing cards, ace through king, are placed randomly with faces down on a table. The prize for guessing correctly the face of any given card is $1. What would be a fair price to pay for a guess?

7. **Analyzing a Game** David gets $10 if he throws a double on a single throw of a pair of dice. How much should he pay for a throw?

8. **Analyzing a Game** You pay $1 to toss 2 coins. If you toss 2 heads, you get $2 (including your $1); if you toss only 1 head, you get back your $1; and if you toss no heads, you lose your $1. Is this a fair game to play?

9. **Raffles** In a raffle 1000 tickets are being sold at $1.00 each. The first prize is $100, and there are 3 second prizes of $50 each. By how much does the price of a ticket exceed its expected value?

10. **Raffles** In a raffle 1000 tickets are being sold at $1.00 each. The first prize is $100. There are 2 second prizes of $50 each, and 5 third prizes of $10 each. Jenny buys 1 ticket. How much more than the expected value of the ticket does she pay?

11. **Analyzing a Game** A fair coin is tossed 3 times, and a player wins $3 if 3 tails occur, wins $2 if 2 tails occur, and loses $3 if no tails occur. If 1 tail occurs, no one wins.
    (a) What is the expected value of the game?
    (b) Is the game fair?
    (c) If the answer to part (b) is "No," how much should the player win or lose for a toss of exactly 1 tail to make the game fair?

12. **Analyzing a Game** Colleen bets $1 on a 2-digit number. She wins $75 if she draws her number from the set of all 2-digit numbers, $\{00, 01, 02, \ldots, 99\}$; otherwise, she loses her $1.
    (a) Is this game fair to the player?
    (b) How much is Colleen expected to lose in a game?

13. Two teams have played each other 14 times. Team A won 9 games, and team B won 5 games. They will play again next week. Bob offers to bet $6 on team A while you bet $4 on team B. The winner gets the $10. Is the bet fair to you in view of the past records of the two teams? Explain your answer.

14. A department store wants to sell 11 purses that cost the store $40 each and 32 purses that cost the store $10 each. If all purses are wrapped in 43 identical boxes and if each customer picks a box randomly, find
    (a) Each customer's expectation.
    (b) The department store's expected profit if it charges $15 for each box.

15. Sarah draws a card from a deck of 52 cards. She receives 40¢ for a heart, 50¢ for an ace, and 90¢ for the ace of hearts. If the cost of a draw is 15¢, should she play the game? Explain.

16. **Family Size** The following data give information about family size in the United States for a household containing a husband and wife where the husband is in the 30–34 age bracket:

Number of Children	0	1	2	3
Proportion of Families	10.2%	15.9%	31.8%	42.1%

A family is chosen at random. Find the expected number of children in the family.

17. Assume that the odds for a certain race horse to win are 7 to 5. If a bettor wins $5 when the horse wins, how much should he bet to make the game fair?

18. **Roulette** In roulette there are 38 equally likely possibilities: the numbers 1–36, 0, and 00 (double zero). See the figure on page 451. What is the expected value for a gambler who bets $1 on number 15 if she wins $35 each time the number

15 turns up and loses \$1 if any other number turns up? If the gambler plays the number 15 for 200 consecutive times, what is the total expected gain?

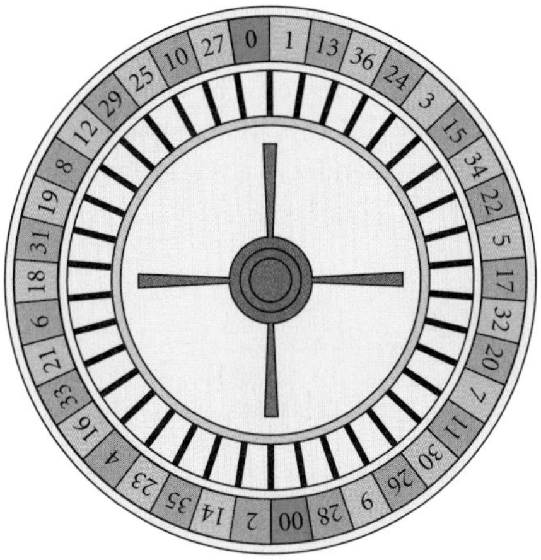

19. **Site Selection** A company operating a chain of supermarkets plans to open a new store in 1 of 2 locations. They conducted a survey of the 2 locations and estimated that the first location will show an annual profit of \$15,000 if it is successful and a \$3000 loss otherwise. For the second location, the estimated annual profit is \$20,000 if successful and a \$6000 loss otherwise. The probability of success at each location is $\frac{1}{2}$. What location should the management decide on in order to maximize its expected profit?

20. For Problem 19 assume the probability of success at the first location is $\frac{2}{3}$ and at the second location is $\frac{1}{3}$. What location should be chosen?

21. Find the number of times the face 5 is expected to occur in a sequence of 2000 throws of a fair die.

22. What is the expected number of tails that will turn up if a fair coin is tossed 582 times?

23. A certain kind of light bulb has been found to have a .02 probability of being defective. A shop owner receives 500 light bulbs of this kind. How many of these bulbs are expected to be defective?

24. A student enrolled in a math course has a .9 probability of passing the course. In a class of 20 students, how many would you expect to fail the math course?

25. **Drug Reaction** A doctor has found that the probability that a patient who is given a certain drug will have unfavorable reactions to the drug is .002. If a group of 500 patients is going to be given the drug, how many of them does the doctor expect to have unfavorable reactions?

26. A true–false test consisting of 30 questions is scored by subtracting the number of wrong answers from the number of right ones. Find the expected number of correct answers of a student who just guesses on each question. What will the test score be?

27. A coin is weighted so that $P(H) = \frac{1}{4}$ and $P(T) = \frac{3}{4}$. Find the expected number of tosses of the coin required in order to obtain either a head or 4 tails.

28. A box contains 3 defective bulbs and 9 good bulbs. If 5 bulbs are drawn from the box without replacement, what is the expected number of defective bulbs?

29. **Aircraft Use** An airline must decide which of two aircraft it will use on a flight from New York to Los Angeles. Aircraft A has a seating capacity of 200, while aircraft B has a capacity of 300. Previous experience has allowed the airline to estimate the number of passengers on the flight as follows:

Number of Passengers	150	180	200	250	300
Probability	.2	.3	.2	.2	.1

Regardless of aircraft used, the average cost of a ticket is \$500, but there are different operating costs attached to each aircraft. There is a fixed cost (fuel, crew, etc.) of \$16,000 attached to using aircraft A, while aircraft B has a fixed cost of \$18,000. There is also a per passenger cost (meals, luggage, added fuel) of \$200 for aircraft A and \$230 for aircraft B. Which aircraft should the airline schedule so that it maximizes its profit on the flight?

30. Prove that if the numerical values assigned to the outcomes of an experiment that has expected value E are all multiplied by the constant k, then the expected value of the new experiment is $k \cdot E$. Similarly, if to all the numerical values we add the same constant k, prove that the expected value of the new experiment is $E + k$.

8.4 Applications

OBJECTIVE 1 Solve applied problems

1 The field of **operations research**, the science of making optimal or best decisions, has experienced remarkable growth and development since the 1940s. Let's work some examples from operations research that utilize expected value.

EXAMPLE 1 Car Rentals

A national car rental agency rents cars for \$16 per day (gasoline and mileage are additional expenses to the customer). The daily cost per car (for example, lease costs and overhead) is \$6 per day. The daily profit to the company is \$10 per car if the car is rented, and the company incurs a daily loss of \$6 per car if the car is not rented. The daily profit depends on two factors: the demand for cars and the number of cars the company has available to rent. Previous rental records show that the daily demand is as given in Table 2:

TABLE 2

Number of Customers	8	9	10	11	12
Probability	.10	.10	.30	.30	.20

Find the expected number of customers and determine the optimal number of cars the company should have available for rental. (This is the number that yields the largest expected profit.)

SOLUTION The expected number of customers is

$$8(.1) + 9(.1) + 10(.3) + 11(.3) + 12(.2) = 10.4$$

If 10.4 customers are expected, how many cars should be on hand? Surely, the number should probably not exceed 11 since fewer than 11 customers are expected. However, the number may not be the integer closest to 10.4 since costs play a major role in the determination of profit. We need to compute the expected profit for each possible number of cars. The largest expected profit will tell us how many cars to have on hand.

For example, if there are 10 cars available, the expected profit for 8, 9, or 10 customers is

$$68(.1) + 84(.1) + 100(.8) = \$95.20$$

We obtain the entry 68(.1) by noting that the 10 cars cost the company \$60, and 8 cars rented with probability .10 bring in \$128, for a profit of \$68. Similarly, we obtain the entry 84(.1) by noting that the 10 cars cost the company \$60, and 9 cars rented with probability .10 bring in \$144, for a profit of \$84. The entry 100(.8) is obtained since for 10 or more customers (probability .3 + .3 + .2 = .8) the profit is

$$10 \times \$16 - \$60 = \$100.$$

Table 3 lists the expected profit for 8 to 12 available cars. The optimal stock size is 11 cars, since this number of cars maximizes expected profit.

TABLE 3

Number of Available Cars	8	9	10	11	12
Expected Profit	\$80.00	\$88.40	\$95.20	\$97.20	\$94.40

NOW WORK PROBLEM 1.

EXAMPLE 2 Quality Control

A factory produces electronic components, and each component must be tested. If the component is good, it will allow the passage of current; if the component is defective, it will block the passage of current. Let p denote the probability that a component is good. See Figure 14. With this system of testing, a large number of components requires an equal number of tests. This increases the production cost of the electronic components since it requires one test per component. To reduce the number of tests, a quality control engineer proposes, instead, a new testing procedure: Connect the components pairwise in series, as shown in Figure 15.

FIGURE 14

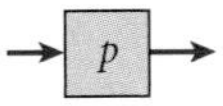

FIGURE 15

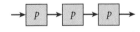

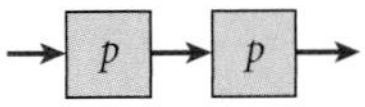

If the current passes two components in series, then both components are good and only one test is required. The probability that two components are good is p^2. If the current does not pass, the components must be sent individually to the quality control department, where each component is tested separately. In this case three tests are required. The probability that three tests are needed is $1 - p^2$ (1 minus probability of success p^2). The expected number of tests for a pair of components is

$$E = 1 \cdot p^2 + 3 \cdot (1 - p^2) = p^2 + 3 - 3p^2 = 3 - 2p^2$$

The number of tests saved for a pair is

$$2 - (3 - 2p^2) = 2p^2 - 1$$

The number of tests saved per component is

$$\frac{2p^2 - 1}{2} = p^2 - \frac{1}{2} \text{ tests}$$

The greater the probability p that the component is good, the greater the saving. For example, if p is almost 1, we have a saving of almost $1 - \frac{1}{2}$ or $\frac{1}{2}$, which is 50% of the original number of tests needed. Of course, if p is small, say, less than .7, we do not save anything since $(.7)^2 - \frac{1}{2}$ is less than 0, and we are wasting tests.

If the reliability of the components manufactured in Example 2 is very high, it might even be advisable to make larger groups. Suppose three components are connected in series. See Figure 16.

FIGURE 16

For individual testing we need three tests. For group testing we have

1 test needed with probability p^3

4 tests needed with probability $1 - p^3$

The expected number of tests is

$$E = 1 \cdot p^3 + 4 \cdot (1 - p^3) = 4 - 3p^3 \text{ tests}$$

The number of tests saved per component is

$$\frac{3p^3 - 1}{3} = p^3 - \frac{1}{3} \text{ tests}$$

In a similar way, we can show that if the components are arranged in groups of four connected in series, then the number of tests saved per component is

$$p^4 - \frac{1}{4} \text{ tests}$$

In general, for groups of n, the number of tests saved per component is

$$p^n - \frac{1}{n} \text{ tests}$$

Notice from the above formula that as n, the group size, gets very large, the number of tests saved per component gets very, very small.

To determine the optimal group size for $p = .9$, we refer to Table 4. From the table we can see that the optimal group size is four, resulting in a substantial saving of approximately 41%.

TABLE 4

Group Size	Expected Tests Saved per Component $p = .9$	Percent Saving
2	$p^2 - \frac{1}{2} = .81 - .50 = .31$	31
3	$p^3 - \frac{1}{3} = .729 - .333 = .396$	39.6
4	$p^4 - \frac{1}{4} = .6561 - .25 = .4061$	40.61
5	$p^5 - \frac{1}{5} = .59049 - .2 = .39049$	39.05
6	$p^6 - \frac{1}{6} = .531 - .167 = .364$	36.4
7	$p^7 - \frac{1}{7} = .478 - .143 = .335$	33.5
8	$p^8 - \frac{1}{8} = .430 - .125 = .305$	30.5

We also note that larger group sizes do not increase savings.

NOW WORK PROBLEM 3.

EXAMPLE 3 Optimizing Hiring

A \$75,000 oil detector is lowered under the sea to detect oil fields, and it becomes detached from the ship. If the instrument is not found within 24 hours, it will crack under the pressure of the sea. It is assumed that a scuba diver will find it with probability .85, but it costs \$500 to hire each diver. How many scuba divers should be hired?

SOLUTION Let's assume that x scuba divers are hired. The probability that they will fail to discover the instrument is $.15^x$, so the instrument will be found with probability $1 - .15^x$.

The expected gain from hiring the scuba divers is

$$\$75{,}000(1 - .15^x) = \$75{,}000 - \$75{,}000(.15^x)$$

while the cost for hiring them is

$$\$500 \cdot x$$

The expected net gain, denoted by $E(x)$, is

$$E(x) = \$75{,}000 - \$75{,}000(.15^x) - \$500x$$

The problem is then to choose x so that $E(x)$ is maximum.

We begin by evaluating $E(x)$ for various values of x:

$$E(1) = \$75{,}000 - \$75{,}000(.15^1) - \$500(1) = \$63{,}250.00$$
$$E(2) = \$75{,}000 - \$75{,}000(.15^2) - \$500(2) = \$72{,}312.50$$
$$E(3) = \$75{,}000 - \$75{,}000(.15^3) - \$500(3) = \$73{,}246.88$$
$$E(4) = \$75{,}000 - \$75{,}000(.15^4) - \$500(4) = \$72{,}962.03$$

$$E(5) = \$75{,}000 - \$75{,}000(.15^5) - \$500(5) = \$72{,}494.30$$
$$E(6) = \$75{,}000 - \$75{,}000(.15^6) - \$500(6) = \$71{,}999.15$$

The expected net gain is optimal when three divers are hired. Note that hiring additional divers does not necessarily increase expected net gain. In fact, the expected net gain declines if more than three divers are hired.

NOW WORK PROBLEM 5.

Error Correction in the Digital Electronic Transmission of Data

Electronically transmitted data, be it from computer to computer or from a satellite to a ground station, is normally in the form of strings of 0s and 1s—that is, in binary form. Bursts of noise or faults in relays, for example, may at times garble the transmission and produce errors so that the message received is not the same as the one originally sent. For example,

001	⟼	101
Message sent	Noisy channel	Message received

is a transmission where the message received is in error since the initial 0 has been changed to a 1.

A naive way of trying to protect against such error would be to repeat the message. So instead of transmitting 001, we would send 001001. Then, were the same error to creep in as it did before, the received message would be

101001

The receiver would certainly know an error had occurred since the last half of the message is not a duplicate of the first half. But she would have no way of recovering the original message since she would not know where the error happened. For example, she would not be able to distinguish between the two messages

001001 and 101101

There are more sophisticated ways of coding binary data with redundancy that not only allow the detection of errors but simultaneously permit their location and correction so that the original message can be recovered. These are referred to as *error-correcting codes* and are commonly used today in computer-implemented transmissions. One such is the (7, 4) Hamming code named after Richard Hamming, a former researcher at AT&T Laboratories. It is a code of length seven, meaning that an individual message is a string consisting of seven items, each of which is either a 0 or 1. (The 4 refers to the fact that the first four elements in the string can be freely chosen by the sender, while the remaining three are determined by a fixed rule and constitute the redundancy that gives the code its error correction capability.) The (7, 4) Hamming code is capable of locating and correcting a single error. That is, if during transmission a 1 has been changed to a 0 or vice versa in one of the seven locations, the Hamming code is capable of detecting and correcting this. While we will not explain how or why the Hamming code works, we will analyze the benefit obtained by its use.

By an error we mean that an individual 1 has been changed to a 0 or that a 0 has been changed to a 1. We assume that the probability of an error happening remains constant during transmission, and we designate this probability by q. So $p = 1 - q$ is the probability that an individual symbol remains unchanged. This is summarized in the diagram. (In practice, values of q are normally small and values of p close to 1 since we would normally be using a relatively reliable channel.)

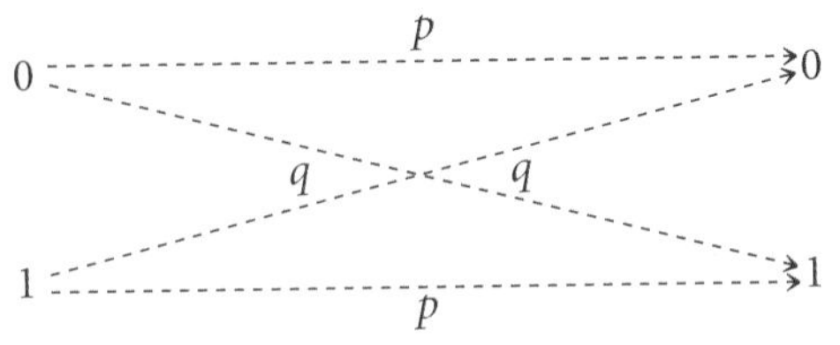

We also assume that errors occur randomly and independently. In short, we can think of the transmission of a binary string of length seven as a Bernoulli trial, with failure corresponding to a symbol being received in error.

For a message of length seven, if *no* coding were used, the probability that the receiver would get the correct message would be

$$b(7, 7; p) = p^7$$

since none of the seven symbols could have been altered. For $p = .98$, this gives $(.98)^7 = .8681$.

Using the Hamming code, the receiver will get the correct message even if one error has occurred. Hence, the corresponding probability of correct reception would be

$$\underbrace{b(7, 7; p)}_{\text{No errors}} + \underbrace{b(7, 6; p)}_{\text{1 error}} = p^7 + 7p^6q$$

For $p = .98$ this now gives $.8681 + .1240 = .9921$, which shows a considerable improvement.

There are codes in use that correct more than a single error. One such code, known by the initials of its originators as a BCH code, is a code of length 15 (messages are binary strings of length 15) that corrects up to two errors. Sending a message of length 15 with no attempt at coding would result in a probability of $b(15, 15; p) = p^{15}$ of the correct message being received. For $p = .98$ this gives $b(15, 15; .98) = .7386$. Using the BCH code that can correct two or fewer errors, the probability that a message will be correctly received becomes

$$\underbrace{b(15, 15; p)}_{\text{No errors}} + \underbrace{b(15, 14; p)}_{\text{1 error}} + \underbrace{b(15, 13; p)}_{\text{2 errors}}$$

Evaluating this for $p = .98$ we get

$$.7386 + .2261 + .0323 = .9970$$

Codes that can correct a high number of errors are clearly very desirable. Yet a basic result in the theory of codes states that as the error-correcting capability of a code increases, so, of necessity, must its length. But lengthier codes require more time for transmission and are clumsier to use. Here, speed and correctness are at odds.

EXERCISE 8.4 Answers to Odd-Numbered Problems Begin on Page AN-34.

1. **Market Assessment** A car agency has fixed costs of \$10 per car per day and the revenue for each car rented is \$30 per day. The daily demand is given in the table:

Number of Customers	7	8	9	10	11
Probability	.10	.20	.40	.20	.10

Find the expected number of customers. Determine the optimal number of cars the company should have on hand each day. What is the expected profit in this case?

2. In Example 2 suppose $p = .8$. Show that the optimal group size is 3.

3. In Example 2 suppose $p = .95$. Show that the optimal group size is 5.

4. In Example 2 suppose $p = .99$. Compute savings for group sizes 10, 11, and 12, and show that 11 is the optimal group size. Determine the percent saving.

5. In Example 3 suppose the probability of the scuba divers discovering the instrument is .95. Find
   (a) An equation expressing the net expected gain.
   (b) The number x of scuba divers that maximizes the net gain.

6. In Example 2 compute the expected number of tests saved per component if on the first test the current does not pass through 2 components in series, but on the second test the current does pass through 1 of them. A third test is not made (since the other component is obviously defective).

7. **Hamming Code** There is a Hamming code of length 15 that corrects a single error. Assuming $p = .98$, find the probability that a message transmitted using this code will be correctly received.

8. **Golay Code** There is a binary code of length 23 (called the Golay code) that can correct up to 3 errors. With $p = .98$, find the probability that a message transmitted using the Golay code will be correctly received.

8.5 Random Variables

OBJECTIVE 1 Use random variables

When we perform an experiment, we are often interested not in a particular outcome, but rather in some number associated with that outcome. For example, in tossing a coin three times we may be interested in the number of heads obtained. Similarly, the gambler throwing a pair of dice in a crap game is interested in the sum of the faces rather than the particular number on each face.

In each of these examples we are interested in numbers that are associated with experimental outcomes. This process of assigning a number to each outcome is called *random variable* assignment.

Random Variable

A **random variable** is a rule that assigns a number to each outcome of an experiment.

Use random variables 1

Consider the experiment of tossing a fair coin three times. Table 5 shows the outcomes in the sample space and the number of heads associated with each outcome. Table 6 shows the probability of the events 0 heads, 1 head, 2 heads, and 3 heads for this experiment.

TABLE 5

Sample Space	Number of Heads
e_1:*HHH*	3
e_2: *HHT*	2
e_3: *HTH*	2
e_4: *THH*	2
e_5: *HTT*	1
e_6: *THT*	1
e_7: *TTH*	1
e_8: *TTT*	0

TABLE 6

Number of Heads Obtained in Three Flips of a Coin	Probability
0	$\frac{1}{8}$
1	$\frac{3}{8}$
2	$\frac{3}{8}$
3	$\frac{1}{8}$

The role of the random variable is to transform the original sample space {*HHH*, *HHT*, *HTH*, *HTT*, *THH*, *THT*, *TTH*, *TTT*} into a new sample space that consists of the number of heads that occur: {0, 1, 2, 3}. If X denotes the random variable, then

$$X(e_1) = X(HHH) = 3 \qquad X(e_2) = X(HHT) = 2 \qquad X(e_3) = X(HTH) = 2$$
$$X(e_4) = X(THH) = 2 \qquad X(e_5) = X(HTT) = 1 \qquad X(e_6) = X(THT) = 1$$
$$X(e_7) = X(TTH) = 1 \qquad X(e_8) = X(TTT) = 0$$

Based on Table 6, we may write

$$\text{Probability}(X = 0) = \frac{1}{8} \qquad \text{Probability}(X = 1) = \frac{3}{8}$$
$$\text{Probability}(X = 2) = \frac{3}{8} \qquad \text{Probability}(X = 3) = \frac{1}{8}$$

Probability Distribution

The discussion so far illustrates two features of a random variable:

1. The values the random variable can assume.
2. The probabilities associated with each of these values.

Because values assumed by a random variable can be used to symbolize all outcomes associated with a given experiment, the probability assigned to a simple event can now be assigned as the likelihood that the random variable takes on the corresponding value.

If a random variable X has the values

$$x_1, x_2, x_3, \ldots, x_n \tag{1}$$

then the rule given by

$$p(x) = P(X = x)$$

where x assumes the values in (1) is called the **probability distribution** of X, or simply, the distribution of X.

It follows that $p(x)$ is a probability distribution of X if it satisfies the following two conditions:

1. $0 \le p(x) \le 1$ for $x = x_1, x = x_2, \ldots, x = x_n$
2. $p(x_1) + p(x_2) + \cdots + p(x_n) = 1$

where $x_1, x_2, \ldots, x_n$ are the values of X.

TABLE 7

x_i	x_1	x_2	$\cdots$	x_n
p_i	p_1	p_2	$\cdots$	p_n

Suppose the probability distribution for the random variable X is given in Table 7, where $p_i = p(x_i)$. Then the **expected value of X,** denoted by $E(X)$, is

$$E(X) = x_1p_1 + x_2p_2 + \cdots + x_np_n$$

Consider again the experiment of tossing a fair coin three times and denote the probability distribution by

$$p(x) \qquad \text{where } x = 0, 1, 2, \text{ or } 3$$

For instance, $p(3)$ is the probability of getting exactly three heads, that is,

$$p(3) = P(X = 3) = \frac{1}{8}$$

In a similar way, we define $p(0)$, $p(1)$, and $p(2)$.

Probability distributions may also be represented graphically, as shown in Figure 17. The graph of a probability distribution is often called a *histogram*.

FIGURE 17

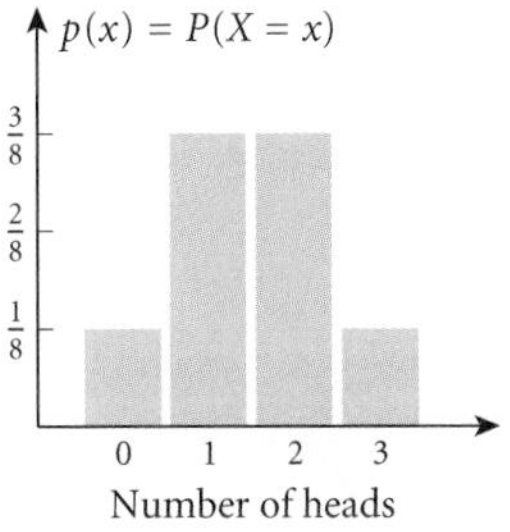

Binomial Probability Using a Probability Distribution

We can now state the formula developed for the binomial probability model in terms of its probability distribution.

The binomial probability that assigns probabilities to the number of successes in n trials is an example of a probability distribution. For this distribution X is the random variable whose value for any outcome of the experiment is the number of successes obtained. For this distribution we may write

$$P(X = k) = b(n, k; p) = \binom{n}{k} p^k q^{n-k}$$

where $P(X = k)$ denotes the probability that the random variable equals k, that is, that exactly k successes are obtained.

EXERCISE 8.5 Answers to Odd-Numbered Problems Begin on Page AN-34.

In Problems 1–6, list the values of the given random variable X together with the probability distributions.

1. A fair coin is tossed two times and X is the random variable whose value for an element in the sample space is the number of heads obtained.

2. A fair die is tossed once. The random variable X is the number showing on the top face.

3. The random variable X is the number of female children in a family with 3 children. (Assume the probability of a female birth is $\frac{1}{2}$.)

4. A job applicant takes a 3-question true–false examination and guesses on each question. Let X be the number of right answers minus the number of wrong answers.

5. An urn contains 4 red balls and 6 white balls. Three balls are drawn with replacement. The random variable X is the number of red balls that are drawn.

6. A couple getting married will have 3 children, and the random variable X denotes the number of boys they will have.

7. For the data given below, compute the expected value.

Outcome	e_1	e_2	e_3	e_4
Probability	.4	.2	.1	.3
x_i	2	3	−2	0

8. For the data below, compute the expected value.

Outcome	e_1	e_2	e_3	e_4
Probability	$\frac{1}{3}$	$\frac{1}{6}$	$\frac{1}{4}$	$\frac{1}{4}$
x_i	1	0	4	−2

Sometimes experiments are simulated using a random number function instead of actually performing the experiment. In Problems 9–14, use a graphing utility to simulate each experiment.*

9. **Rolling a Fair Die** Consider the experiment of rolling a die, and let the random variable X denote the number showing on the top face. Simulate the experiment using a random number function on your calculator, considering a roll to have the outcome k if the value of the random number function is between $(k-1)\cdot 0.167$ and $k\cdot 0.167$. Record the outcome. Repeat the experiment 50 times, obtaining a sequence of 50 numbers. [*Note*: Most calculators repeat the action of the last entry if you simply press the ENTER, or EXE, key again.] Using these 50 numbers you can estimate the probability $P(X=k)$, for $k=1, 2, 3, 4, 5, 6$, by the ratio

$$\frac{\text{Number of times } k \text{ appears in your sequence}}{50}$$

Enter your estimates in the table. Calculate the actual probabilities, and enter these numbers in the table. How close are your numbers to the actual values?

k	Your Estimate of $P(X=k)$	Actual Value of $P(X=k)$
1		
2		
3		
4		
5		
6		

10. Use a random number function to select a value for the random variable X. Repeat this experiment 50 times. Count the number of times the random variable X is between 0.6 and 0.9. Calculate the ratio

$$R=\frac{\text{Number of times the random variable } X \text{ is between 0.6 and 0.9}}{50}$$

What value of R did you obtain? Calculate the actual probability $P(0.6 \le X < 0.9)$.

11. Use a random number function to select a value for the random variable X. Repeat this experiment 50 times. Count the number of times the random variable X is between 0.1 and 0.3. Calculate the ratio

$$R=\frac{\text{Number of times the random variable } X \text{ is between 0.1 and 0.3}}{50}$$

What value of R did you obtain? Calculate the actual probability $P(0.1 \le X < 0.3)$.

12. **Rolling an Octahedron** Consider the experiment of rolling an octahedron (a regular polyhedron with all 8 faces congruent), and let the random variable X denote the number showing on the top face. Simulate the experiment using a random number function on your calculator, considering a roll to have the outcome k if the value of the random number function is between $(k-1)/8$ and $k/8$, for $k=1, 2, 3, 4, 5, \ldots, 8$. Record the outcome. Repeat the experiment 50 times, obtaining a sequence of 50 numbers. Using these 50 numbers you can estimate the probability $P(X=k)$, for $k=1, 2, 3, 4, 5, 6, 7, 8$, by the ratio

$$\frac{\text{Number of times } k \text{ appears in your sequence}}{50}$$

Enter your estimates in the table shown. Calculate the actual probabilities and enter these numbers in the table. How close are your numbers to the actual values?

k	Your Estimate of $P(X=k)$	Actual Value of $P(X=k)$
1		
2		
3		
4		
5		
6		
7		
8		

13. **Rolling a Dodecahedron** Consider the experiment of rolling a dodecahedron (a regular polyhedron with all 12 faces congruent), and let the random variable X denote the number showing on the top face. Simulate the experiment using a random number function on your calculator, considering a roll to have the outcome k if the value of the random number function is between $(k-1)/12$ and $k/12$, for $k=1, 2, 3, 4, 5, \ldots, 12$. Record the outcome. Repeat the experiment 50 times, obtaining a sequence of 50 numbers. Using these 50 numbers you can estimate the probability $P(X=2)$ by the ratio

$$\frac{\text{Number of times 2 appears in your sequence}}{50}$$

**Most graphing utilities have a random number function (usually RAND or RND) for generating numbers between 0 and 1. Check your user's manual to see how to use this function on your graphing calculator.*

Calculate the actual probability, $P(X = 2)$, and compare these values. How close is your estimate to the actual value?

14. Rolling an Icosahedron Consider the experiment of rolling an icosahedron (a regular polyhedron with all 20 faces congruent), and let the random variable X denote the number showing on the top face. Simulate the experiment using a random number function on your calculator, considering a roll to have the outcome k if the value of the random number function is between $(k - 1)/20$ and $k/20$, for $k = 1, 2, 3, 4, 5, \ldots, 20$. Record the outcome. Repeat the experiment 50 times, obtaining a sequence of 50 numbers. Using these 50 numbers you can estimate the probability $P(X = 5)$ by the ratio

$$\frac{\text{Number of times 5 appears in your sequence}}{50}$$

Calculate the actual probability, $P(X = 5)$, and compare these values. How close is your estimate to the actual value?

Chapter 8 Review

OBJECTIVES

Section	You should be able to	Review Exercises
8.1	1 Solve probability problems: Sample space partitioned into two sets	1–4
	2 Solve probability problems: Sample space partitioned into three sets	9–16
	3 Use Bayes' Formula to solve probability problems	5–8, 17–22, 23–26
	4 Find *a priori* and *a posteriori* probabilities	23–26
8.2	1 Find binomial probabilities	27–32
8.3	1 Compute expected values	33–40, 49c, 50c, 51c, 52c, 53, 54
	2 Solve applied problems involving expected value	41, 42
	3 Find expected value in a Bernoulli trial	43, 44
8.4	1 Solve applied problems	45–48
8.5	1 Use random variables	49–54

IMPORTANT FORMULAS

Probability of an Event E in a Partitioned Sample Space

$$P(E) = P(A_1) \cdot P(E|A_1) + P(A_2) \cdot P(E|A_2) + P(A_3) \cdot P(E|A_3) + \cdots + P(A_n) \cdot P(E|A_n)$$

Bayes' Formula

$$P(A_j|E) = \frac{P(A_j) \cdot P(E|A_j)}{P(E)}$$

Binomial Probability Formula

$$b(n, k; p) = \binom{n}{k} p^k q^{n-k}, \quad q = 1 - p$$

Expected Value for Bernoulli Trials

$$E = np$$

TRUE–FALSE ITEMS Answers are on page AN-34.

T F **1.** $b(3, 2; \frac{1}{2}) = 3 \cdot (\frac{1}{2})^3$

T F **2.** $P(A_1|E) = P(A_1)P(E)$

T F **3.** The expected value of an experiment is never negative.

T F **4.** Bayes' formula is useful for computing *a posteriori* probability.

T F **5.** $b(n, k; p)$ gives the probability of exactly n successes in k trials.

T F **6.** In flipping a fair coin 10 times, the expected number of heads is 5.

FILL IN THE BLANKS Answers are on page AN-34.

1. The formula

$$P(A_1|E) = \frac{P(A_1) \cdot P(E|A_1)}{P(E)}$$

is called __________ __________ .

2. Random experiments are called Bernoulli trials if
(a) The same experiment is repeated several times.
(b) There are only two possible outcomes, success and failure.
(c) The repeated trials are __________ .
(d) The probability of each outcome remains __________ for each trial.

3. If an experiment has n outcomes that are assigned the payoffs $m_1, m_2, \ldots, m_n$, occurring with probabilities $p_1, p_2, \ldots, p_n$, then $E = m_1p_1 + m_2p_2 + \cdots + m_np_n$ is the __________ __________ .

4. A random variable on a sample space S is a rule that assigns __________ to each element in S.

5. If X is a random variable assuming the values $x_1, x_2, \ldots x_n$, then $E(X) = x_1p(x_1) + x_2p(x_2) + \cdots + x_np(x_n)$ is the __________ of X.

REVIEW EXERCISES Answers to odd-numbered problems begin on page AN-34.

Blue-numbered problems indicate the author's suggestions for a practice test.

In Problems 1–8, use the tree diagram below to find the indicated probability:

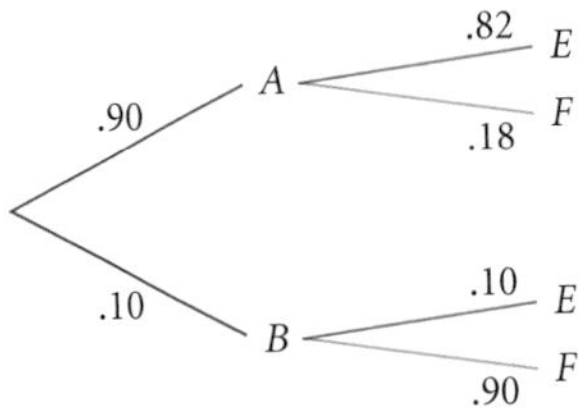

1. $P(E|A)$ **2.** $P(F|A)$ **3.** $P(E|B)$ **4.** $P(F|B)$ **5.** $P(A|E)$ **6.** $P(A|F)$ **7.** $P(B|E)$ **8.** $P(B|F)$

In Problems 9–22, use the tree diagram below to find the indicated probability:

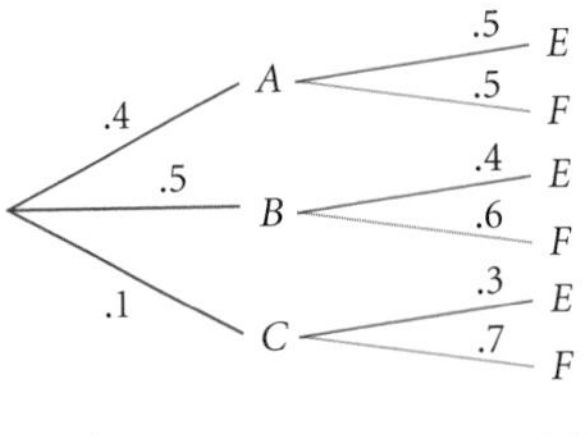

9. $P(E|A)$ **10.** $P(F|A)$ **11.** $P(E|B)$ **12.** $P(F|B)$ **13.** $P(E|C)$

14. $P(F|C)$ **15.** $P(E)$ **16.** $P(F)$ **17.** $P(A|E)$ **18.** $P(A|F)$

19. $P(B|E)$ **20.** $P(B|F)$ **21.** $P(C|E)$ **22.** $P(C|F)$

23. The table below indicates a survey conducted by a deodorant producer:

	Like the Deodorant	Did Not Like the Deodorant	No Opinion
Group I	180	60	20
Group II	110	85	12
Group III	55	65	7

Let the events E, F, G, H, and K be defined as follows:

E: Customer likes the deodorant
F: Customer does not like the deodorant
G: Customer is from group I
H: Customer is from group II
K: Customer is from group III

Find

(a) $P(E|G)$ (b) $P(G|E)$ (c) $P(H|E)$
(d) $P(K|E)$ (e) $P(F|G)$ (f) $P(G|F)$
(g) $P(H|F)$ (h) $P(K|F)$

24. **Quality Control** In a factory three machines A_1, A_2, A_3, produce 55%, 30%, and 15% of total production, respectively. The percentage of defective output of these machines is 1%, 2%, and 3%, respectively. An item is chosen at random and it is defective. What is the probability that it came from machine A_1? From A_2? From A_3?

25. **Test for Cancer** A lung cancer test has been found to have the following reliability. The test detects 85% of the people who have cancer and does not detect 15% of these people. Among the noncancerous group it detects 92% of the people not having cancer, whereas 8% of this group are detected erroneously as having lung cancer. Statistics show that about 1.8% of the population has lung cancer. Suppose an individual is given the test for lung cancer and it detects the disease. What is the probability that the person actually has lung cancer?

26. **Test for Tuberculosis** A faster and simpler diagnostic test for the presence of tuberculosis bacilli called "FASTPlaqueTB" was evaluated in Cape Town, South Africa, and found to have the following reliability. For those subjects known to have the tuberculosis bacilli, the FASTPlaqueTB test was positive 70% of the time. For those subjects known not to have the tuberculosis bacilli, the FASTPlaqueTB test was negative 99% of the time. Statistics indicate that about 14.2% of the people in Cape Town have the tuberculosis bacilli.

(a) What is the probability that a randomly selected person from Cape Town will have a positive FASTPlaqueTB test result?
(b) What is the probability that a person who tests positive actually has tuberculosis bacilli?

27. **Advertising** Management believes that 1 out of 5 people watching a television advertisement about their new product will purchase the product. Five people who watched the advertisement are picked at random. What is the probability that none of these people will purchase the product? That exactly 3 will purchase the product?

28. **Baseball** Suppose that the probability of a player hitting a home run is $\frac{1}{20}$. In 5 tries what is the probability that the player hits at least 1 home run?

29. **Guessing on a True–False Exam** In a 12-item true–false examination, a student guesses on each question.

(a) What is the probability that the student will obtain all correct answers?
(b) If 7 correct answers constitute a passing grade, what is the probability that the student will pass?
(c) What are the odds in favor of passing?

30. **Guessing on a True–False Exam** In a 20-item true–false examination, a student guesses on each question.

(a) What is the probability that the student will obtain all correct answers?
(b) If 12 correct answers constitute a passing grade, what is the probability that the student will pass?
(c) What are the odds in favor of passing?

31. **Throwing Dice** Find the probability of throwing a sum of 11 at least 3 times in 5 throws of a pair of fair dice.

32. **Throwing Dice** Find the probability of throwing a sum of 7 at least twice in 5 throws of a pair of fair dice.

33. What is the expected number of girls in families having exactly 3 children?

34. **Tossing Coins** What is the expected number of heads when a coin is tossed 7 times?

35. **Evaluating a Game** In a certain game a player has the probability $\frac{1}{7}$ of winning a prize worth \$90 and the probability $\frac{1}{3}$ of winning another prize worth \$50. What is the expected return of the game for the player?

36. **Evaluating a Game** Frank pays \$0.70 to play a certain game. He draws 2 balls (together) from a bag containing 2 red balls and 4 green balls. He receives \$1 for each red ball that he draws. Has he paid too much? By how much?

37. **Playing the Lottery** In a lottery 1000 tickets are sold at \$0.25 each. There are 3 cash prizes: \$100, \$50, and \$30. Alice buys 8 tickets.

(a) What would have been a fair price for a ticket?
(b) How much extra did Alice pay?

38. **Evaluating a Game** The figure shows a spinning game for which a person pays \$0.30 to purchase an opportunity to spin the dial. The numbers in the figure indicate the amount of payoff and its corresponding probability in parentheses. Find the expected value of this game. Is the game fair?

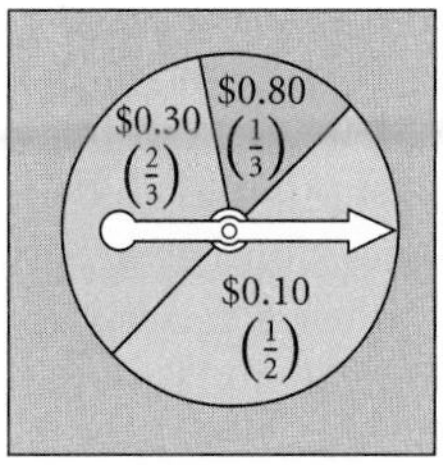

39. Evaluating a Game Consider the 3 boxes in the figure. The game is played in 2 stages. The first stage is to choose a ball from box A. If the result is a ball marked I, then we go to box I and select a ball from there. If the ball is marked II, then we select a ball from box II. The number drawn on the second stage is the gain. Find the expected value of this game.

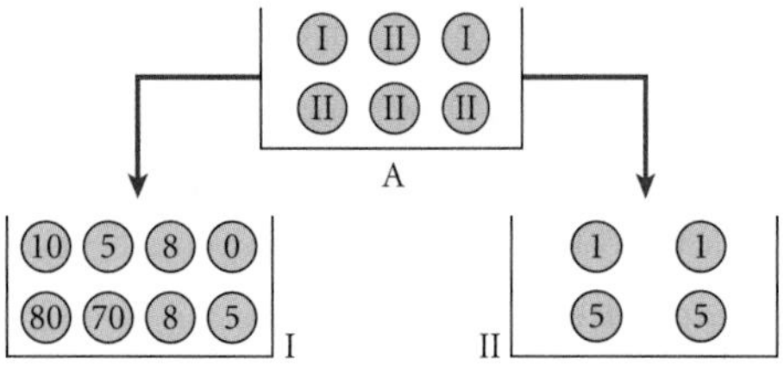

40. European Roulette A European roulette wheel has only 37 compartments, 18 red, 18 black, and 1 green. A player will be paid $2 (including his $1 bet) if he picks correctly the color of the compartment in which the ball finally rests. Otherwise, he loses $1. Is the game fair to the player?

41. Insuring against Weather The profits from an outdoor sporting event are affected by inclement weather. The promoter of the event expects it to net a profit of $750,000 if the weather is dry but only $20,000 if the weather is wet. The historical probability of rain is 5% for the day of the event.

(a) What is the expected profit from the event?
(b) The promoter can buy $500,000 weather insurance for a premium of $50,000. Based on expected profit, would it be wise to buy insurance? Explain your reasoning.

42. Insuring against Weather Consider the previous problem, but suppose the chance of rain is 20% and the insurance costs $70,000. Based on the new information,

(a) What is the expected profit from the event?
(b) Based on expected profit, would it be wise to buy insurance? Explain your reasoning.

43. Rolling Dice The probability of rolling a double when 2 fair dice are rolled is $\frac{1}{6}$. How many doubles should you expect if you roll two dice 500 times?

44. Guessing on a Test A multiple choice test has 100 questions, and each question has 5 possible answer choices, only 1 of which is correct. If a student guesses on every question, how many should he expect to get wrong?

45. Error Correction There is a binary code of length 31 [called the binary (31, 31, p) BCH code] used for some pagers that can correct up to two errors. If the probability that an individual character is transmitted correctly is .97, find the probability that a message transmitted using this code will be correctly received.

Source: The Error Correcting Codes (ECC) Site.

46. Ski Rentals Swen's Ski Rental rents skis, boots, and poles for $20 a day. The daily cost per set of skis is $6. It includes maintenance, storage, and overhead. Daily profits depend on daily demand for skis and the number of sets available. Swen knows that on a typical weekend the daily demand for skis is given in the table.

Number of Customers	90	91	92	93	94	95	96	97
Probability	.01	.10	.20	.20	.30	.10	.05	.04

How many sets of skis should Swen have ready for rental to maximize his profit?

47. Blood Testing A group of 1000 people is subjected to a blood test that can be administered in 2 ways: (1) each person can be tested separately (in this case 1000 tests are required) or (2) the blood samples of 30 people can be pooled and analyzed together. If we use the second way and the test is negative, then 1 test suffices for 30 people. If the test is positive, each of the 30 people can then be tested separately, and, in all, 30 + 1 tests are required for the 30 people. Assume the probability p that the test is positive is the same for all people and that the people to be tested are independent.

(a) What is the probability that the test for a pooled sample of 30 people will be positive?
(b) What is the expected number of tests necessary under plan 2?

48. Allergy Testing A person's blood needs to be tested for an allergic reaction to 200 known allergens. The tests can either be done individually, resulting in 200 tests, or they can be done by combining 20 allergens and testing the blood. In the second method, if the test is negative then one test suffices for the 20 allergens, but if it is positive then each allergen is tested separately, resulting in 20 + 1 tests for 20 allergens. Assume the probability, p, that the test is positive is the same for each allergen and that each allergen is independent.

(a) What is the probability that a test is positive using the second method?
(b) What is the expected number of tests needed if the second method is utilized?
(c) What is the number of tests saved per individual when the second method is used?

49. Defective Pens A box contains 100 pens, 10 of which do not work. Five pens are taken from the box without replacement. The random variable X represents the number of defective pens in the sample.

(a) List the values of X.
(b) Find the probability distribution of X.
(c) Find the expected value of X.

50. Quality Control A production process manufactures 50 items per minute. The probability of manufacturing a defective item is .04. The quality control team selects 3 items and tests them for defects. The random variable X represents the number of defective items in the sample.

(a) List the values of X.
(b) Find the probability distribution of X.
(c) Find the expected value of X.

51. Flowers A bag of 10 flower bulbs has 7 bulbs that produce white flowers and 3 bulbs that produce yellow flowers. Three bulbs are chosen from the bag and are planted. The random variable X is the number of yellow flowering plants that grow.

(a) List the values of X.
(b) Find the probability distribution of X.
(c) Find the expected value of X.

52. Flowers Refer to Problem 51. Suppose you plant 5 bulbs. The random variable X is the number of yellow flowering plants that grow.

(a) List the values of X.
(b) Find the probability distribution of X.
(c) Find the expected value of X.

53. Number of Senior Citizens The 2000 U.S. Census showed that 12.4% of the U.S. population was 65 years of age or older. Suppose that 5 people were randomly selected from the U.S. population. The random variable X is the number of people selected who are 65 years of age or older. Find the probability distribution of X, and find the expected value of X.

Source: United States Census Bureau.

54. The 2000 U.S. Census showed that 54.1% of persons 5 years of age or older were living in the same house in the year 2000 as they were in the year 1995. Suppose 350 people, 5 years of age or older, were selected from the U.S. population. The random variable X is the number of people selected who were living in the same house in the year 2000 as they were in the year 1995. Find $E(X)$, the expected value of X.

Source: United States Census Bureau.

Chapter 8 Project

SEQUENTIAL TESTING

Bayesian techniques are extremely powerful tools for the analysis of test data. For example, a satellite manufacturer must have a rigorous test program to ensure that a spacecraft will perform at the desired level when in orbit. However, because spacecraft are very expensive and there is a chance that the tests themselves will cause a component to wear out, making it more prone to failure, the goal is to run as few tests as possible. How are these conflicting goals resolved? Let's examine a typical test and see.

One common test is a vibration and thermal vacuum test or a "shake and bake." In this test, a spacecraft is placed into a vessel that simulates the harsh conditions of space and then it is shaken violently, after which system functions are verified. Each "shake and bake" has a probability of detecting whether there is a problem with the spacecraft and also a probability of wearing out a component. Suppose that each "shake and bake" has a 0.9 probability of detecting a defect in the spacecraft: In other words, if the spacecraft has a defect, the test will detect it with a probability of 0.9; if the spacecraft does not have a defect, it will always pass the test. Let's investigate how each subsequent test affects the system reliability. We assume that each test is independent. Define the following events:

E: Spacecraft passes the test
F: Spacecraft fails the test

1. If the spacecraft is defective, then $P(E) = .1$ and $P(F) = .9$. What is the probability a defective spacecraft is detected using four or fewer tests?

2. If the spacecraft is not defective, find $P(E)$ and $P(F)$.

3. Spacecraft are extremely complicated, and many of the parts are made by hand. As a result, there is a high likelihood that a given spacecraft will be defective in some way during initial testing. An engineer estimates that the probability is .05 that a newly manufactured spacecraft is without defects. Suppose this spacecraft passes one test.

(a) What is the probability the spacecraft is good?
(b) What is the probability the spacecraft is defective?

[*Hint*: Define these events G: The spacecraft is good; and D: The spacecraft is defective.]

4. Suppose the spacecraft is tested a second time and passes. How does this change the probability that the spacecraft is good?

5. Suppose the spacecraft is tested a third time and passes. What is the probability it is good?

6. Suppose the spacecraft is tested a fourth time and passes. What is the probability it is good?

7. Fill in the table below using the results of Problems 3, 4, 5, and 6.

Number of Tests	0	1	2	3	4
P(G)					

8. Suppose the probability of premature wear-out increases depending on the number of tests that are performed, according to the table below.

Number of Tests	Probability of Premature Wear-out While in Orbit
0	.01
1	.02
2	.05
3	.10
4	.20

The mission director wants you to analyze the tradeoff between the number of tests used and the possibility of premature wear-out while in orbit. She asks you to create a table that shows the probability the spacecraft will fail while in orbit if it has been given (a) 1 test; (b) 2 tests; (c) 3 tests; (d) 4 tests. A spacecraft will fail while in orbit if (1) it is defective or (2) it is good and a part wears out prematurely while in orbit.

[*Hint*: Define the events: W—A part wears out prematurely while in orbit; and A—The spacecraft fails while in orbit.]

Q **9.** Based on the results of Problem 8, how many tests would you recommend? Be sure to give reasons.

MATHEMATICAL QUESTIONS FROM PROFESSIONAL EXAMS*

1. **Actuary Exam—Part II** What is the probability that 10 independent tosses of an unbiased coin result in no fewer than 1 head and no more than 9 heads?

(a) $(\frac{1}{2})^9$ (b) $1 - 11(\frac{1}{2})^9$ (c) $1 - 11(\frac{1}{2})^{10}$
(d) $1 - (\frac{1}{2})^9$ (e) $1 - (\frac{1}{2})^{10}$

2. **CPA Exam** The Stat Company wants more information on the demand for its products. The following data are relevant:

Units Demanded	Probability of Unit Demand	Total Cost of Units Demanded
0	.10	$0
1	.15	1.00
2	.20	2.00
3	.40	3.00
4	.10	4.00
5	.05	5.00

What is the total expected value or payoff with perfect information?

(a) $2.40 (b) $7.40 (c) $9.00 (d) $9.15

3. **CPA Exam** Your client wants your advice on which of 2 alternatives he should choose. One alternative is to sell an investment now for $10,000. Another alternative is to hold the investment 3 days, after which he can sell it for a certain selling price based on the following probabilities:

Selling Price	Probability
$ 5,000	.4
$ 8,000	.2
$12,000	.3
$30,000	.1

Using probability theory, which of the following is the most reasonable statement?

(a) Hold the investment 3 days because the expected value of holding exceeds the current selling price.
(b) Hold the investment 3 days because of the chance of getting $30,000 for it.
(c) Sell the investment now because the current selling price exceeds the expected value of holding.
(d) Sell the investment now because there is a 60% chance that the selling price will fall in 3 days.

4. **CPA Exam** The Polly Company wishes to determine the amount of safety stock that it should maintain for product D that will result in the lowest cost.

The following information is available:

Stockout cost	$80 per occurrence
Carrying cost of safety stock	$2 per unit
Number of purchase orders	5 per year

The available options open to Polly are as follows:

Units of Safety Stock	10	20	30	40	50	55
Probability	5%	30%	30%	20%	10%	5%

The number of units of safety stock that will result in the lowest cost is

(a) 20 (b) 40 (c) 50 (d) 55

5. **CPA Exam** The ARC Radio Company is trying to decide whether to introduce as a new product a wrist "radiowatch" designed for shortwave reception of exact time as broadcast by the National Bureau of Standards. The "radiowatch" would be priced at $60, which is exactly twice the variable cost per unit to manufacture and sell it. The incremental fixed costs necessitated by introducing this new product would amount to $240,000 per year. Subjective estimates of the probable demand for the product are shown in the following probability distribution:

Annual Demand	6,000	8,000	10,000	12,000	14,000	16,000
Probability	.2	.2	.2	.2	.1	.1

The expected value of demand for the new product is

(a) 11,000 units (b) 10,200 units (c) 9000 units
(d) 10,600 units (e) 9800 units

6. **CPA Exam** In planning its budget for the coming year, King Company prepared the following payoff probability distribution describing the relative likelihood of monthly sales volume levels and related contribution margins for product A:

Monthly Sales Volume	Contribution Margin	Probability
4,000	$ 80,000	.20
6,000	120,000	.25
8,000	160,000	.30
10,000	200,000	.15
12,000	240,000	.10

What is the expected value of the monthly contribution margin for product A?

(a) $140,000 (b) $148,000
(c) $160,000 (d) $180,000

7. **CPA Exam** A decision tree has been formulated for the possible outcomes of introducing a new product line. Branches related to alternative 1 reflect the possible payoffs from introducing the product without an advertising campaign. The branches for alternative 2 reflect the possible payoffs with an advertising campaign costing $40,000.

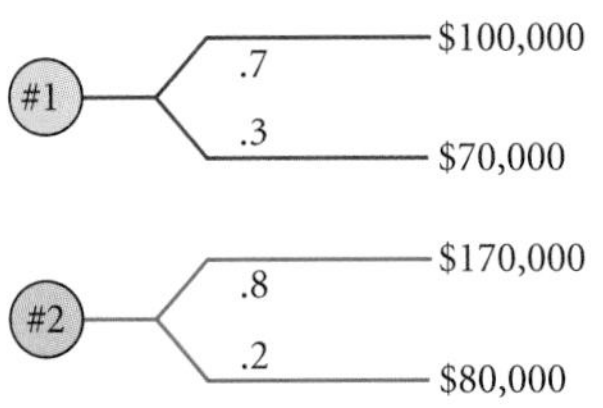

The expected values of alternatives 1 and 2, respectively, are

(a) #1: $(.7 \times \$100{,}000) + (.3 \times \$70{,}000)$
#2: $(.8 \times \$170{,}000) + (.2 \times \$80{,}000)$
(b) #1: $(.7 \times \$100{,}000) + (.3 \times \$70{,}000)$
#2: $(.8 \times \$130{,}000) + (.2 \times \$40{,}000)$
(c) #1: $(.7 \times \$100{,}000) + (.3 \times \$70{,}000)$
#2: $(.8 \times \$170{,}000) + (.2 \times \$80{,}000) - \$40{,}000$
(d) #1: $(.7 \times \$100{,}000) + (.3 \times \$70{,}000) - \$40{,}000$
#2: $(.8 \times \$170{,}000) + (.2 \times \$80{,}000) - \$40{,}000$

8. **CPA Exam** A battery manufacturer warrants its automobile batteries to perform satisfactorily for as long as the owner keeps the car. Auto industry data show that only 20% of car buyers retain their cars for 3 years or more. Historical data suggest the following:

Number of Years Owned	Probability of Battery Failure	Battery Exchange Costs	Percentage of Failed Batteries Returned
Less than 3 years	.4	\$50	75%
3 years or more	.6	\$20	50%

If 50,000 batteries were sold this year, what is the estimated warranty cost?

(a) \$375,000 (b) \$435,000
(c) \$500,000 (d) \$660,000

CHAPTER 9

Statistics

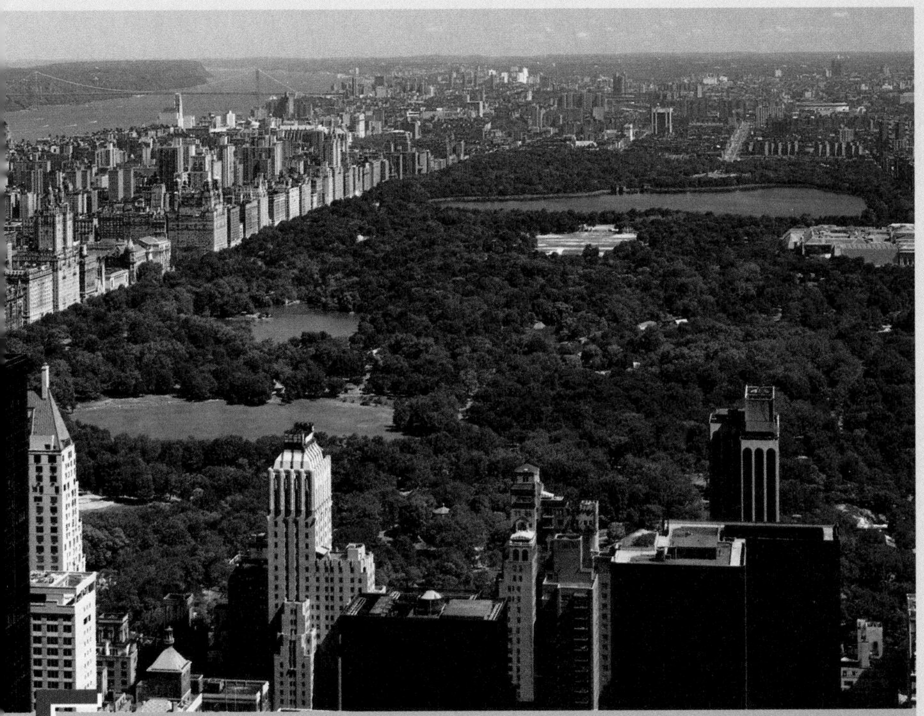

OUTLINE

Final exams are finally over and a weekend trip to New York City next month in June is the plan. Weather can be a problem wherever one travels, but New York in June seems a safe choice. But is it? The highest temperature on record is 101°F and the lowest a brisk 44°F. Wow, what a difference! Can we rely on average temperatures to feel more secure? How do we use information about average temperatures to assess the likelihood the trip will not be affected by extreme weather? This chapter on statistics will give you the background to do an analysis, and the Chapter 9 Project will guide you. Have a great trip!

A LOOK BACK, A LOOK FORWARD

This chapter, like many of the chapters in this book, can be covered right away, but a better sense of the subject is obtained if probability is studied first. Rectangular coordinates, studied earlier in Chapter 1, are also used here.

This chapter provides only an introduction to statistics, an area of mathematics that is as rich and as varied as algebra and geometry are. When you finish this chapter, you will have been exposed to many of the concepts studied in more detail in a full course, giving you a head start. If you don't intend to take a full course in statistics, then this chapter will give you an overview of the terminology and applications studied in statistics.

9.1 Introduction to Statistics: Data and Sampling

PREPARING FOR THIS SECTION *Before getting started, review the following:*

> Equally Likely Outcomes (Section 7.1, pp. 370–373)

OBJECTIVES
1 Identify continuous and discrete variables
2 Obtain a simple random sample
3 Identify sources of bias in a sample

Statistics is the science of collecting, arranging, analyzing, and interpreting information. By making observations, statisticians collect **data** in the form of measurements or counts. A measurable characteristic is called a **variable.** If a variable can assume any real value between certain limits, it is called a **continuous variable.** It is called a **discrete variable** if it can assume only a finite set of values or as many values as there are whole numbers. Examples of continuous variables are weight, height, length, and time. Examples of discrete variables are the number of votes a candidate gets or the number of cars sold.

Identify continuous and discrete variables 1

It is important to be able to identify whether a variable is continuous or discrete because it determines the kind of statistical analysis that can be used.

EXAMPLE 1 Identifying Continuous and Discrete Variables

State the variable in the following experiments and determine whether it is continuous or discrete.

(a) Roll a die and observe the number of dots.
(b) Measure the length of time a smoker can blow into a tube before taking a breath.
(c) Count the number of defective parts produced by a certain machine in an hour.

SOLUTION

(a) The variable is the number of dots. It is discrete because it can have only the values 1, 2, 3, 4, 5, and 6.
(b) The variable is time. It is continuous because it can take on any value within certain limits, say from 0 to 90 seconds.
(c) The variable is the number of defective parts. It is discrete; it is a whole number.

NOW WORK PROBLEMS 1 AND 3.

The set of observations made of the variable in an experiment is called **data.** In collecting data it is often impossible or impractical to observe an entire group, called the **population.** So instead of examining the entire population, a small segment, called a **sample,** is chosen to be observed. It would be difficult, for example, to question all cigarette smokers in order to study the effects of smoking. Therefore, a sample of smokers is usually selected for sampling.

The method of selecting the sample is important if we want the results to be reliable. It is essential that the individuals in the sample have characteristics similar to those of the population they are intended to represent.

Obtain a simple random sample 2

There are several ways to select a representative sample. The simplest and most reliable method of sampling is called **simple random sampling.** In a simple random sample, every member of the population has an equal probability of being selected into the sample.

EXAMPLE 2 Obtaining a Simple Random Sample

Janet is the president of the Marketing Society. She wants to know the opinion of the members concerning a proposed guest speaker. There are 150 members in the club, and rather than ask each one's opinion, Janet decides to ask a simple random sample of 10 members. How can Janet choose the sample?

SOLUTION Janet needs to choose 10 members of the Marketing Society, and she wants every member of the society to have an equal opportunity to be chosen. She writes each member's name on a separate sheet of paper and puts the papers in a box. She mixes the papers and selects 10 sheets. She asks the 10 members whose names she chose their opinions.

NOW WORK PROBLEM 13.

A more efficient way to select a simple random sample would be to assign a number to each member of the population. Then a random number generator or a list of random numbers can be used to identify the members that will make up the simple random sample.

COMMENT: Graphing utilities have random generators built into them. Check your user's manual to learn how to generate random numbers on your graphing utility.

Identify sources of bias in a sample 3

If the random sample is not chosen in such a way that every member of the population has an equal probability of being selected, a **biased sample** could result. Bias occurs when a segment of the population is either overrepresented or underrepresented in the sample. The data collected from a biased sample often do not reflect the population accurately. For example, if we want to study the relationship between smoking cigarettes and lung cancer, we cannot choose a sample of smokers who all live in the same location. The individuals chosen might have characteristics peculiar to their environment that give a false impression with respect to all smokers.

EXAMPLE 3 Identify Sources of Bias in a Sample

Janet decides that it is too difficult to obtain a simple random sample of the members of the Marketing Society, so she asks 10 freshmen from her Introduction to Marketing class their opinions about the proposed speaker. Identify a possible source of bias in this sample.

SOLUTION Janet pooled only freshmen. As a result, sophomore, junior, and senior members of the Marketing Society were underrepresented in the sample. If the speaker had appeal to a particular level of student, the results would not represent the opinion of the population.

NOW WORK PROBLEM 19.

When a sample is representative of a population, important conclusions about the population can often be inferred from analysis of the data collected. The phase of statistics dealing with conditions under which such inference is valid is called **inductive statistics** or **statistical inference.** Since such inference cannot be absolutely certain, the language of probability is often used in stating conclusions. When a meteorologist makes a weather forecast, weather data collected over a large region are studied; based on the study, the forecast is given in terms of chances. A typical forecast might be, "There is a 20% chance of rain tomorrow."

EXERCISE 9.1 Answers to Odd-Numbered Problems Begin on Page AN-35.

In Problems 1–12, identify, the variable in each experiment and determine whether it is continuous or discrete.

1. A statistician observes the number of heads that occur when a coin is tossed 1000 times.
2. A statistician counts the number of red pieces of candy in each of 100 one ounce bags of M&Ms.
3. The Environmental Protection Agency (EPA) determines the average gas mileage for each of 500 randomly chosen Honda Accords.
4. A General Electric (GE) quality control team measures how long it takes each of 500 light bulbs to burn out.
5. A statistician measures the time a person who arrives at a neighborhood McDonald's between noon and 1:00 P.M. waits in line.
6. A pharmaceutical firm measures the time it takes for a drug to take effect.
7. A marketing analyst surveys 500 people and asks the number of airplane flights they took in the last 12 months.
8. A sales manager counts the number of calls made in a day before a sale is made.
9. An urban planner counts the number of people who cross a particular intersection between 1:00 P.M. and 3:00 P.M.
10. A carpet manufacturer counts the number of missed stitches in a 100-yard roll of carpeting.
11. A quality control team measures how long a phone battery lasts.
12. A psychologist observes how many times a subject blinks when being told a scary story.

In Problems 13–18, list some possible ways to choose random samples for each study.

13. A study to determine opinion about a certain television program.
14. A study to detect defective radio resistors.
15. A study of the opinions of people toward Medicare.
16. A study to determine opinions about an election of a U.S. president.
17. A study of the number of savings accounts per family in the United States.
18. A national study of the monthly budget for a family of four.
19. The following is an example of a biased sample: In a study of political party preferences, poorly dressed interviewers obtained a significantly greater proportion of answers favoring Democratic party candidates in their samples than did their well-dressed and wealthier-looking counterparts. Give two more examples of biased samples.
20. In a study of the number of savings accounts per family, a sample of accounts totaling less than $10,000 was taken and, from the owners of these accounts, information about the total number of accounts owned by all family members was obtained. Criticize this sample.

Q 21. It is customary for news reporters to sample the opinions of a few people to find out how the population at large feels about the events of the day. A reporter questions people on a downtown street corner. Is there anything wrong with such an approach?

Q 22. In 1936 the *Literary Digest* conducted a poll to predict the presidential election. Based on its poll, it predicted the election of Landon over Roosevelt. In the actual election, Roosevelt won. The sample was taken by drawing the mailing list from telephone directories and lists of car owners. What was wrong with the sample?

9.2 Representing Data Graphically: Bar Graphs; Pie Charts

OBJECTIVES
1 Construct a bar graph
2 Construct a pie chart
3 Analyze a graph

The data as collected are called **raw data.** Raw data are often organized to make them more useful. The organization of data involves the presentation of the collected measurements or counts in a form suitable for determining logical conclusions. Usually tables or graphs are used to represent the collected data. There are two popular methods for graphically displaying data: **bar graphs** and **pie charts.** Both are appropriate when the data can be separated into categories. A bar graph will show the number or the percent of data that are in each category, while a pie chart will show only the percent of data in each category.

Construct a bar graph 1

We begin with an example showing how to construct a bar graph from data.

EXAMPLE 1 Constructing a Bar Graph

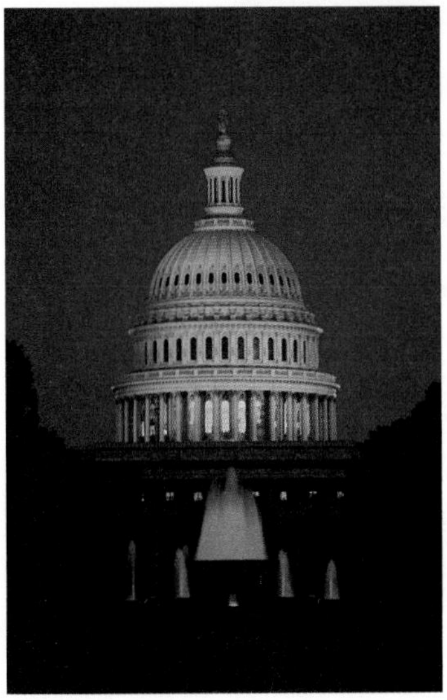

For the fiscal year 2001 (October 2000–September 2001), the federal government spent a total of $1.9 trillion. The breakdown of expenditures (in billions of dollars) is given in Table 1. Construct a bar graph of the data.

TABLE 1

Social Security, Medicare, and other retirement	$684
National defense, veterans, and foreign affairs	$342
Net interest (interest on the public debt)	$190
Physical, human, and community development	$190
Social programs	$342
Law enforcement and general government	$ 38
Surplus to pay down debt	$114

Source: Department of the Treasury

SOLUTION A horizontal axis is used to indicate the category of spending and a vertical axis is used to represent the amount spent in each category. For each category of spending we draw rectangles of equal width whose height represents the amount spent in the category. The rectangles will not touch each other. See Figure 1 on page 474.

FIGURE 1

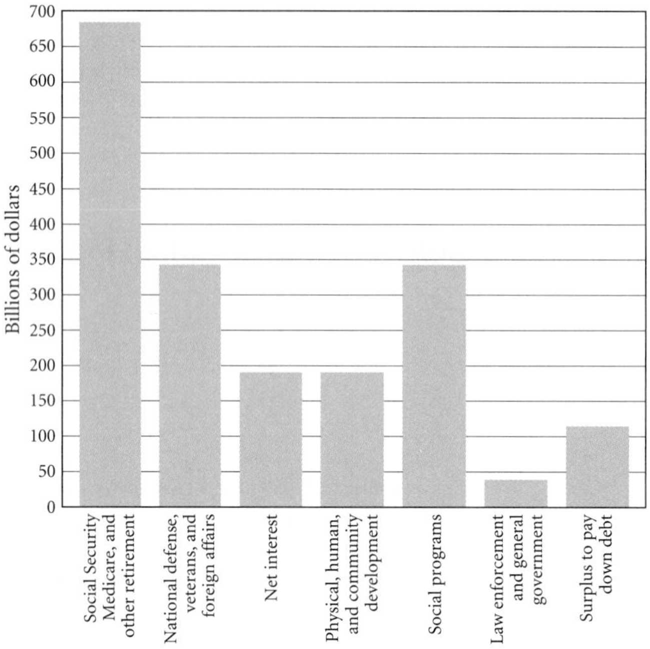

NOW WORK PROBLEM 1(a)–(c).

Construct a pie chart ☐2 Now we show how to construct a pie chart using the data given in Table 1.

EXAMPLE 2   Constructing a Pie Chart

Use the data given in Table 1 to construct a pie chart.

SOLUTION   To construct a pie chart a circle is divided into sectors, one sector for each category of data. The size of each sector is proportional to the total amount spent. Since Social Security, Medicare, and other retirement is $684 billion and total spending is $1.9 trillion = $1900 billion, the percent of data in this category is $\frac{684}{1900} = 0.36 = 36\%$. Therefore, Social Security, Medicare, and other retirement will make up 36% of the pie chart. Since a circle has 360°, the degree measure of the sector for this category of spending is $0.36(360°) \approx$ 130°. Following this procedure for the remaining categories of spending, we obtain Table 2.

TABLE 2

Category	Spending (in billions)	Percent of Total Spending	Degree Measure of Sector*
Social Security, Medicare, and other retirement	$684	0.36 = 36%	130°
National defense, veterans, and foreign affairs	$342	0.18 = 18%	65°
Net interest (interest on the public debt)	$190	0.10 = 10%	36°
Physical, human, and community development	$190	0.10 = 10%	36°
Social programs	$342	0.18 = 18%	65°
Law enforcement and general government	$ 38	0.02 = 2%	7°
Surplus used to pay down debt	$114	0.06 = 6%	22°

*The data in column 4 does not add up to 360° due to rounding.

To construct a pie chart by hand, we use a protractor to approximate the angles for each sector. See Figure 2.

FIGURE 2

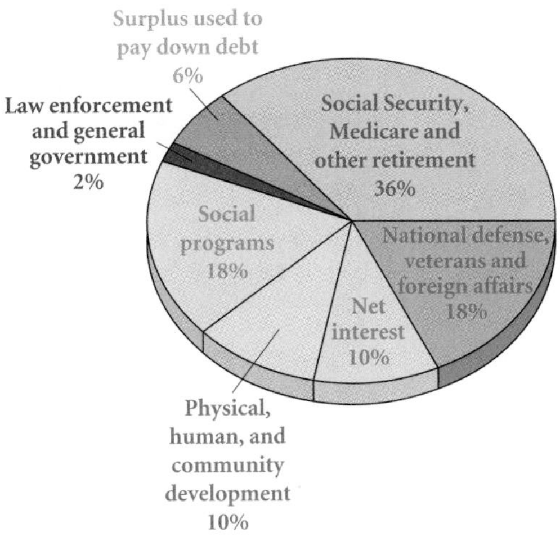
Surplus used to pay down debt 6%
Law enforcement and general government 2%
Social Security, Medicare and other retirement 36%
Social programs 18%
National defense, veterans and foreign affairs 18%
Net interest 10%
Physical, human, and community development 10%

NOW WORK PROBLEM 3.

Use Excel to construct a pie chart for the data in Table 1.

STEP 1 Enter the data into Excel, highlight it, and select the Chart Wizard.

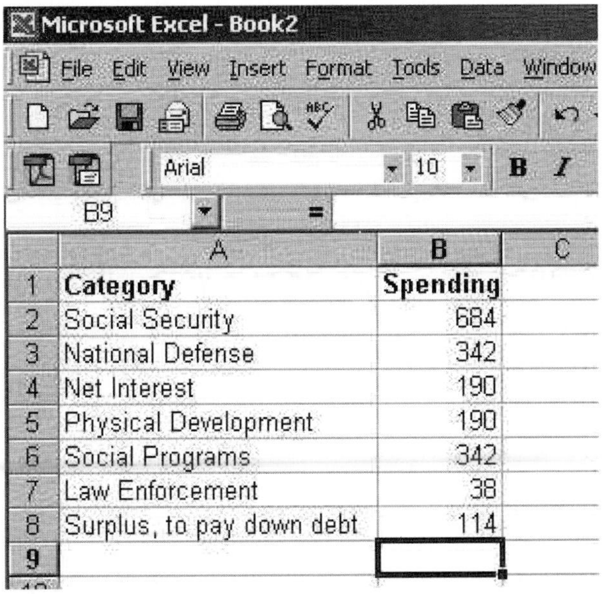
Microsoft Excel - Book2
File Edit View Insert Format Tools Data Window
Arial 10 B I
B9 =

	A	B	C
1	**Category**	**Spending**	
2	Social Security	684	
3	National Defense	342	
4	Net Interest	190	
5	Physical Development	190	
6	Social Programs	342	
7	Law Enforcement	38	
8	Surplus, to pay down debt	114	
9			

STEP 2 For the chart type, select Pie.

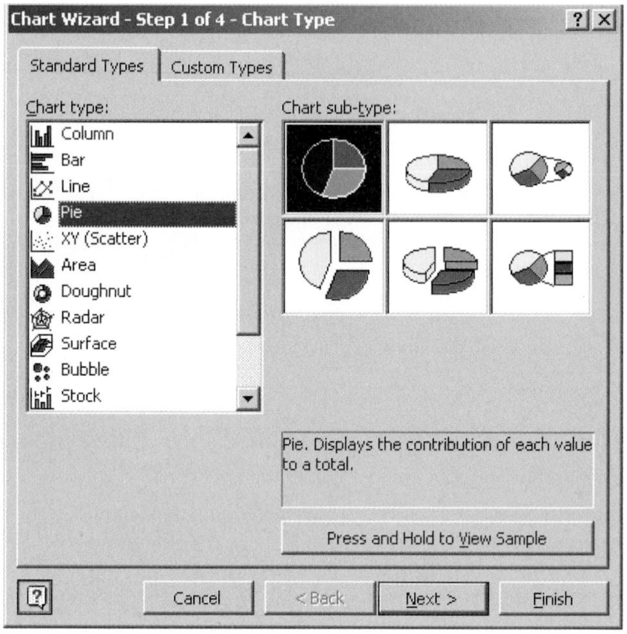

STEP 3 Click Next. The basic pie chart will appear.
Click Next again. This screen provides the options for the pie chart.
Click on Legend. Turn the legend off by clicking in the box Show Legend. Then highlight the Data Label tab.

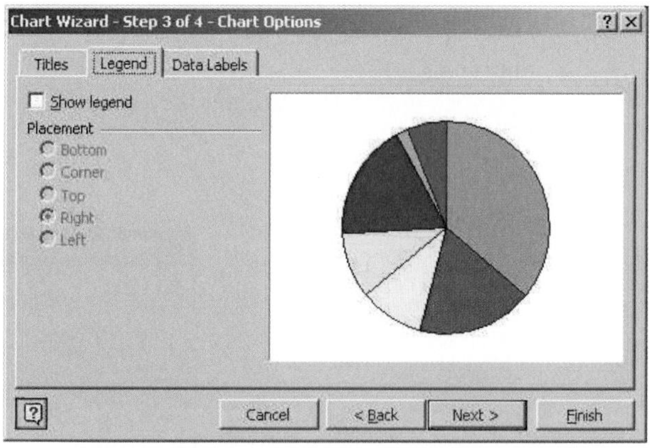

STEP 4 Click on Category name and Percentage.

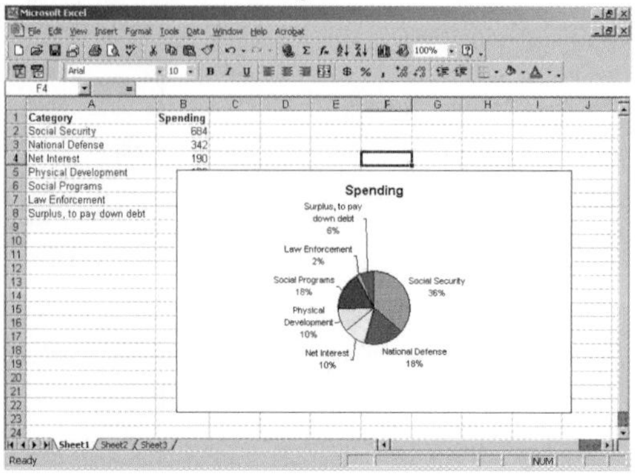

NOW WORK PROBLEM 3(b) USING EXCEL.

Analyze a graph 3

One reason for graphing data by drawing bar graphs or pie charts is to quickly determine certain information about the data.

EXAMPLE 3 Analyzing a Pie Chart

The pie chart in Figure 3 represents the sources of revenue for the United States federal government in its fiscal year 2001. Answer the questions below using the pie chart. Total revenue for the fiscal year 2001 was $2.0 trillion.

FIGURE 3

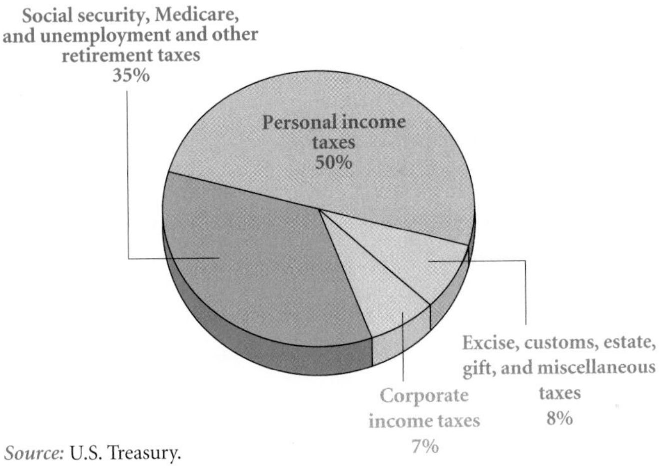

Source: U.S. Treasury.

(a) What is the largest source of revenue for the federal govenment? What is the amount?

(b) What is the smallest source of revenue for the federal government? What is the amount?

(c) How much revenue does the government collect from Social Security, Medicare, etc.

SOLUTION

(a) The largest source of revenue for the federal government is personal income taxes. The government collects about $1.0 trillion from this source.

(b) The smallest source of revenue for the federal government is Corporate Income Taxes. The government collects

$$(0.08)\ (2.0 \text{ trillion}) = (0.08)(2000 \text{ billion}) = \$160 \text{ billion}$$

from this source.

(c) The revenue collected from Social Security, Medicare, etc. is

$$(0.35)(2.0 \text{ trillion}) = (0.35)(2000 \text{ billion}) = \$700 \text{ billion}$$

NOW WORK PROBLEM 11.

EXERCISE 9.2 Answers to Odd-Numbered Problems Begin an Page AN-35.

1. **Household Income** The data below represent the median income (in dollars) of households by region of the country in 2001.

Region	Median Income
Northeast	$57,000
Midwest	54,096
South	46,688
West	51,966

Source: US Bureau of the Census, 2003.

(a) Draw a bar graph of the data.
(b) Which region has the highest median income?
(c) Which region has the lowest median income?
(d) Discuss the use of a pie chart to represent this situation.

2. **Household Income** The data below represent the median income (in dollars) of households by region of the country in 1995.

Region	Median Income
Northeast	$36,111
Midwest	35,839
South	30,942
West	35,979

Source: U.S. Bureau of the Census, 1996.

(a) Draw a bar graph of the data.
(b) Which region has the highest median income?
(c) Which region has the lowest median income?
(d) Comparing the results to those from Problem 1, which region had the largest increase in median income between 1995 and 2001?
(e) Discuss the use of a pie chart to represent this situation.

3. **Populatlon Distribution** The data below represent the number of families (in thousands) by region of the country in 2001.

Region	Number of Families
Northeast	14,131
Midwest	17,448
South	29,021
West	15,740

Source: US Bureau of the Census, 2003.

(a) Draw a bar graph of the data.
(b) Draw a pie chart of the data.
(c) Which chart seems to summarize the data better? Why?
(d) Which region has the most families?
(e) Which region has the fewest families?
(f) Do you think that these are misleading statistics? Explain.

4. **Family Size** The data below represent the number (in thousands) of families by size of the family in 2001.

Family Size	Number of Families
2	32,847
3	16,574
4	14,978
5	6,549
6	2,194
7 or more	1,197

Source: US Bureau of the Census, 2003.

(a) Draw a bar graph of the data.
(b) Draw a pie chart of the data.
(c) Which chart seems to summarize the data better? Why?
(d) What is the most common size of a family in the United States?
(e) What percent of families have 5 or more members?
(f) If you were a marketing consultant planning to market a product to families, what size families would you target? Why?

5. **Household Income** The data that follow represent the median income (in dollars) of families by type of household in 2001.

Family Household	Median Income
Married—couple families	$56,735
Female householder—no spouse	28,142
Male householder—no spouse	36,590

Source: U.S. Census Bureau 2003, Current Population Survey.

(a) Draw a bar graph of the data.
(b) Which household type has the highest median income?
(c) Which household type has the lowest median income?
(d) Why do you think that there is such a large discrepancy?

6. **Household Income** The data that follow represent the median income (in dollars) of families by type of household in 1995.

Family Household	Median Income
Married—couple families	$47,129
Female householder—no spouse	21,348
Male householder—no spouse	33,534

Source: U.S. Census Bureau 1996, Current Population Survey.

(a) Draw a bar graph of the data.
(b) Which household type has the highest median income?
(c) Which household type has the lowest median income?
(d) Why do you think that there is such a large discrepancy?

(e) Comparing the data here with that in Problem 5, which group had the largest percent increase in median income from 1995 to 2001?

7. Causes of Death The data below represent the causes of death for Americans in 2000.

Cause of Death	Number
Heart disease	710,760
Cancer	553,091
Stroke	167,661
Respiratory diseases	187,322
Accidents	97,900
Diabetes	69,301
Alzheimer's	49,558
Kidney failure	37,251
Septicemia	31,224

Source: National Center for Health Statistics, 2003.

(a) Draw a bar graph of the data.
(b) Draw a pie chart of the data.
(c) Which graph seems to summarize the data better?
(d) What was the leading cause of death among Americans in 2000?

8. Automobile Costs The following data represent the total amount spent (in billions of dollars) in the United States on automobiles and their maintenance in 1999.

Category	Expenditures
New, used cars	200.2
Tires, accessories	49.3
Gas and oil	141.1
Insurance less claims	43.0
Interest on debt	40.6
Registration and fees	9.1
Repairs	178.3

Source: U.S. Census, 2002.

(a) Draw a bar grap of the data.
(b) Draw a pie chart of the data.
(c) Which graph seems to summarize the data better?
(d) After purchasing the car, what is the largest expense?
(e) What portion of the expenditures is spent on debt?

9. On-Time Performance The bar graph below represents the overall percentage of reported flight operations arriving on time in February 2003.

(a) Which airline has the highest percentage of on-time flights?
(b) Which airline has the lowest percentage of on-time flights?
(c) What percentage of United Airlines' flights are on time?

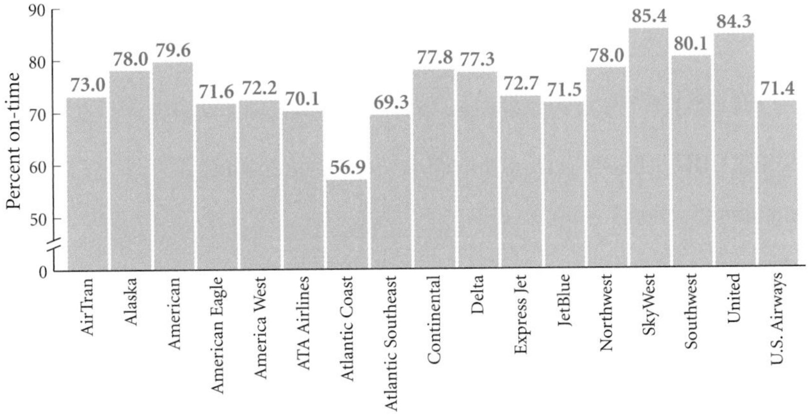

Source: United States Department of Transportation.

10. Income Required for a Loan The bar graph to the right shows the minimum annual income required for a $100,000 loan using interest rates available on 12/1/96. Taxes and insurance are assumed to be $230 monthly.

(a) What minimum annual income is needed to qualify for a 5/1 year ARM (Adjustable Rate Mortgage)?
(b) Which loan type requires the most annual income? What minimum annual income is required for this loan type?

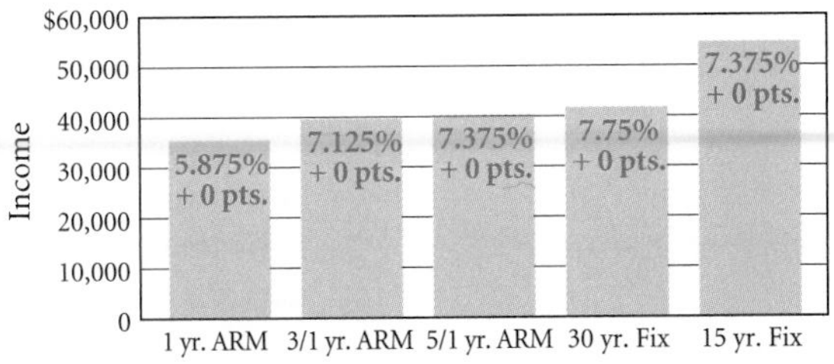

11. **Consumer Price Index** The Consumer Price Index (CPI) is an index that measures inflation. It is calculated by obtaining the prices of a market basket of goods each month. The market basket, along with the percentages of each product, is given in the following pie chart:

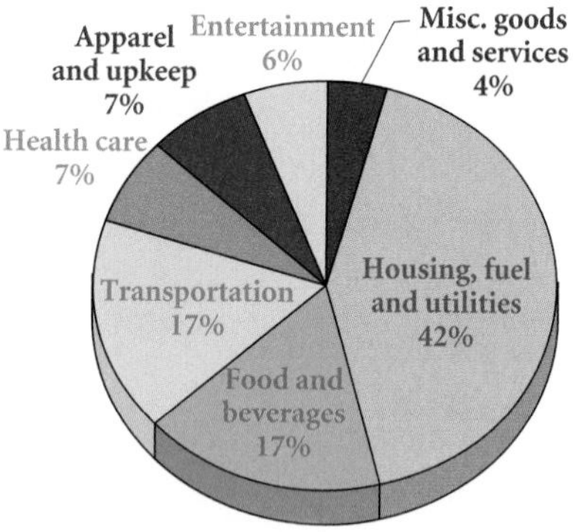

Source: Bureau of Labor Statistics.

(a) What is the largest component of the CPI?
(b) What is the smallest component of the CPI?
(c) Senior citizens spend about 14% of their income on health care. Why do you think they feel the CPI weight for health care is too low?

12. **Asset Allocation** According to financial planners, an individual's investment mix should change over a person's lifetime. The longer an individual's time horizon, the more the individual should invest in stocks. A financial planner suggested that Jim's retirement portfolio be diversified according to the mix provided in the pie chart below, since Jim has 40 years to retirement.

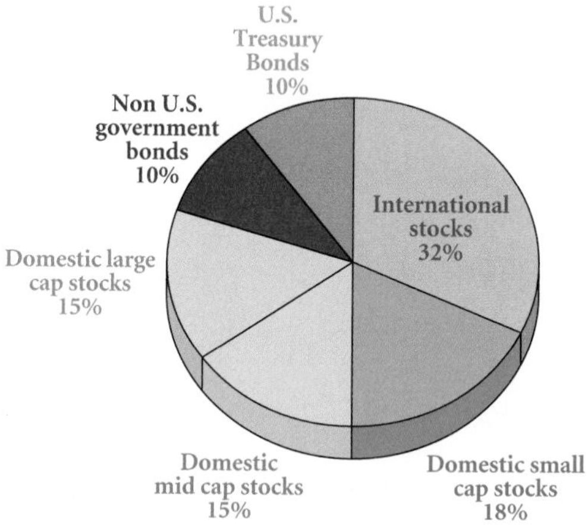

(a) How much should Jim invest in stocks?
(b) How much should Jim invest in bonds?
(c) How much should Jim invest in domestic (U.S.) stocks?
(d) The return on bonds over long periods of time is less than that of stocks. Explain why you think the financial planner recommended what she did for bonds.

9.3 Organization of Data

PREPARING FOR THIS SECTION *Before getting started, review the following:*

> Rectangular Coordinates (Section 1.1, pp. 2–3)

OBJECTIVES
1 Form a frequency table
2 Construct a line chart
3 Group data into class intervals
4 Build a histogram
5 Draw a frequency polygon
6 Draw a cumulative frequency distribution

Often studies result in data represented by a large collection of numbers. If the data are to be interpreted, they must be organized. In this section we discuss the organization of data. One way to organize data is by using a **frequency table** and **line chart.**

Frequency Tables; Line Charts

We begin with an example in which the weights of 71 children have been recorded.

1 **EXAMPLE 1** **Forming a Frequency Table**

Table 3 lists the weights of a random sample of 71 children selected from a group of 10,000.

(a) List the data from smallest to highest in a table.
(b) Form a frequency table of the data.

TABLE 3 WEIGHTS OF 71 STUDENTS, IN POUNDS

69	71	71	55	52	55	58	58	58	62	67	94
82	94	95	89	89	104	93	93	58	62	67	62
94	85	92	75	75	79	75	82	94	105	115	104
105	109	94	92	89	85	85	89	95	92	105	71
72	72	79	79	85	72	79	119	89	72	72	69
79	79	69	93	85	93	79	85	85	69	79	

SOLUTION **(a)** Certain information available from the sample becomes more evident once the data are ordered according to some scheme. If the 71 measurements are written from smallest to highest, we obtain Table 4.

TABLE 4

52	55	55	58	58	58	58	62	62	62	67	67
69	69	69	69	71	71	71	72	72	72	72	72
75	75	75	79	79	79	79	79	79	79	79	82
82	85	85	85	85	85	85	85	89	89	89	89
89	92	92	92	93	93	93	93	94	94	94	94
94	95	95	104	104	105	105	105	109	115	119	

(b) Once Table 4 has been constructed, it becomes easy to present the data in a **frequency table.** This is done as follows: Tally marks are used to record the occurrence of weights. Then the **frequency** f with which each weight occurs is listed. See Table 5.

TABLE 5

Score	Tally	Frequency, f	Score	Tally	Frequency, f
52	/	1	85	𝍸 //	7
55	//	2	89	𝍸	5
58	////	4	92	///	3
62	///	3	93	////	4
67	//	2	94	𝍸	5
69	////	4	95	//	2
71	///	3	104	//	2
72	𝍸	5	105	///	3
75	///	3	109	/	1
79	𝍸 ///	8	115	/	1
82	//	2	119	/	1

NOW WORK PROBLEM 1(a).

2 **EXAMPLE 2** **Constructing a Line Chart**

Use Table 5 to construct a line chart for the data given in Table 3.

SOLUTION A **line chart** is obtained using rectangular coordinates. The vertical axis (y-axis) denotes the frequency f and the horizontal axis (x-axis) denotes the weight data. Then points are plotted according to the information in Table 5 and a vertical line is drawn from each point to the horizontal axis. See Figure 4.

FIGURE 4

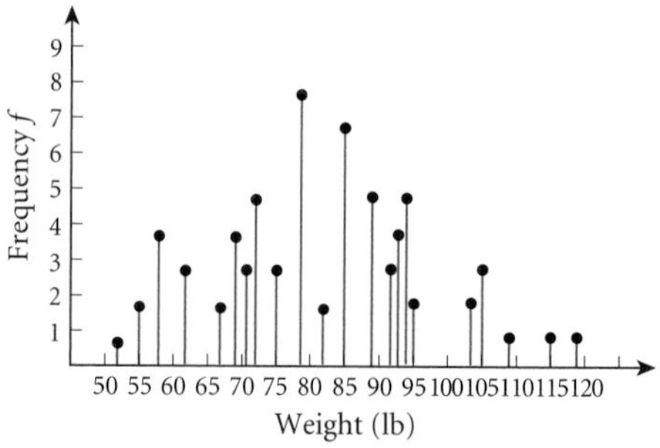

NOW WORK PROBLEM 1(b).

Grouping Data; Histograms

When data are collected and few repeated entries are obtained, the data are usually easier to organize by grouping them and constructing a *histogram*.

Table 6 lists the monthly electric bills of a sample of 71 residential customers, beginning with the smallest bill.

TABLE 6 MONTHLY ELECTRIC BILLS

52.30	55.61	55.71	58.01	58.41	58.51	58.91	62.33	62.50	62.71
67.13	67.23	69.51	69.67	69.80	69.82	71.34	71.65	71.83	72.15
72.22	72.41	72.59	72.67	75.11	75.71	75.82	79.03	79.06	79.09
79.15	79.28	79.32	79.51	79.62	82.32	82.61	85.09	85.13	85.25
85.31	85.41	85.51	85.58	89.21	89.32	89.49	89.61	89.78	92.41
92.63	92.89	93.05	93.19	93.28	93.91	94.17	94.28	94.31	94.52
94.71	95.32	95.51	104.31	104.71	105.21	105.37	105.71	109.34	115.71
119.38									

The first step in grouping data is to calculate the *range.*

> The **range** of a set of numbers is the difference between the largest and the smallest number in the set. That is
>
> **Range = (Largest value) − (Smallest value)**

For the data in Table 6 the range is

$$\text{Range} = 119.38 - 52.30 = 67.08$$

To group this data, we divide the range into intervals of equal size, called **class intervals.** Table 7 shows the data using 14 class intervals, each of size 5, beginning with the interval 50–54.99.

TABLE 7

	Class Interval	Tally	Frequency
1	50–54.99	/	1
2	55–59.99	𝍸 /	6
3	60–64.99	///	3
4	65–69.99	𝍸 /	6
5	70–74.99	𝍸 ///	8
6	75–79.99	𝍸 𝍸 /	11
7	80–84.99	//	2
8	85–89.99	𝍸 𝍸 //	12
9	90–94.99	𝍸 𝍸 //	12
10	95–99.99	//	2
11	100–104.99	//	2
12	105–109.99	////	4
13	110–114.99		0
14	115–119.99	//	2

Group data into class intervals **3**

Example 3 below shows how to group the data of Table 6 into class intervals of size \$10.

EXAMPLE 3 Grouping Data into Class Intervals

For the monthly electric bills listed in Table 6, group the data into class intervals each of size \$10.

SOLUTION First we determine the first class interval. Since the smallest bill is \$52.30 and we want a class interval of size \$10, we choose \$50 to \$59.99 as the first interval. The intervals will be from \$50.00 to \$59.99, from \$60.00 to \$69.99, up to the interval \$110 to \$119.99. We can stop here since the largest bill, \$119.38, is in this interval. Table 8 shows the result of using class intervals of size \$10.

TABLE 8

	Class Interval	Tally	Frequency
1	50–59.99	~~////~~ //	7
2	60–69.99	~~////~~ ////	9
3	70–79.99	~~////~~ ~~////~~ ~~////~~ ////	19
4	80–89.99	~~////~~ ~~////~~ ////	14
5	90–99.99	~~////~~ ~~////~~ ////	14
6	100–109.99	~~////~~ /	6
7	110–119.99	//	2

NOW WORK PROBLEM 1(c).

We use Tables 7 and 8 to introduce some vocabulary.

The class intervals shown in Tables 7 and 8 each begin at 50 and end at 119.99, so as to include all the data from Table 6. The first number in a class interval is called the **lower class limit;** the second number is called the **upper class limit.** We choose these limits so that each item in Table 6 can be assigned to one and only one class interval. The **midpoint** of a class interval is defined as

$$\text{Midpoint} = \frac{\text{Upper class limit} + \text{Lower class limit}}{2}$$

The **class width** is the difference between consecutive lower class limits.

When the data are represented in the form of Table 7 (or Table 8), they are said to be **grouped data.** Notice that once raw data are converted to grouped data, it is impossible to retrieve or recover the original data. The best we can do is to choose the midpoint of each class interval as a representative for each class. In Table 7, for example, the actual electric bills of \$105.21, \$105.37, \$105.71, and \$109.34 are viewed as being represented by the midpoint of the class interval from 105 to 109.99, namely $(105 + 109.99)/2 = 107.50$.

Next we present the grouped data of Table 8 in a graph, called a **histogram.**

4

EXAMPLE 4 Building a Histogram

Build a histogram for the grouped data of Table 8.

SOLUTION To build a histogram for the data in Table 8, we construct a set of adjoining rectangles having as base the size of the class interval and as height the frequency of occurrence of data in that particular interval. The center of the base is the midpoint of each class interval. Figure 5 shows the histogram for the data in Table 8.

FIGURE 5

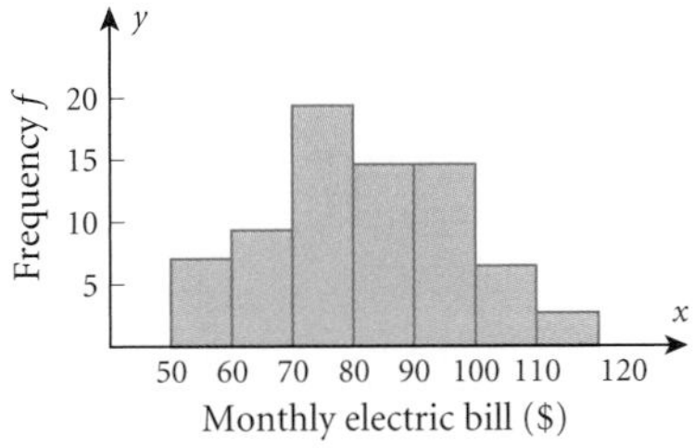

NOW WORK PROBLEM 1(d).

EXCEL

Use Excel to work Examples 3 and 4.

SOLUTION **STEP 1** Enter the data given in Table 6 in column B. [This is the tedious part. Usually the data will be entered electronically.]

In column D, enter the maximum value of the class intervals of the frequency table. These are called the **Bin values** in Excel.

Electric Bills	*Bin*
52.3	49.99
55.61	59.99
55.71	69.99
58.01	79.99
58.41	89.99
58.51	99.99
58.91	109.99
62.33	119.99

STEP 2 Click on Tools and then Data Analysis. (If Data Analysis isn't available, click on Add-ins.)

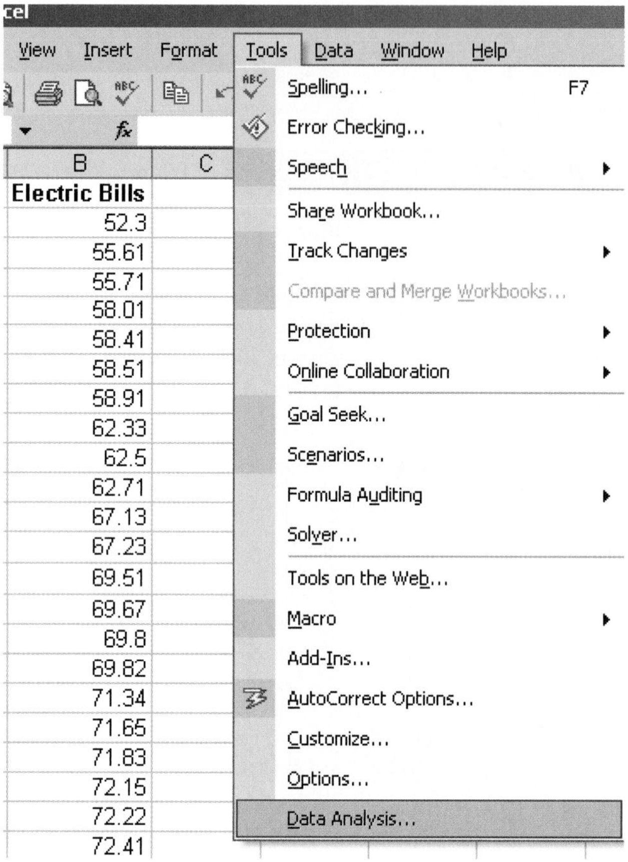

STEP 3 Highlight Histogram and click on OK.

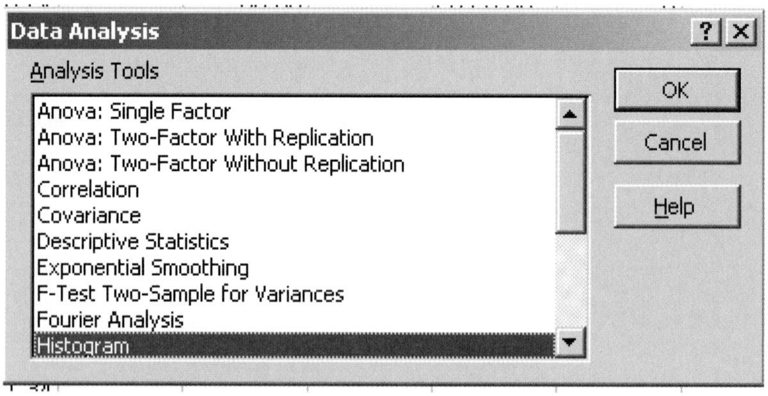

STEP 4 **Input Range** is the cells containing the data.
Bin Range is the cells containing the class intervals.
Output Range is the upper left cell where the frequency table will go.
New Worksheet Ply is the name of the histogram.

Cumulative Percentage gives the percentage of the data less than or equal to the bin value.
Chart Output gives the histogram.

To enter the data, click and drag on the appropriate cells.

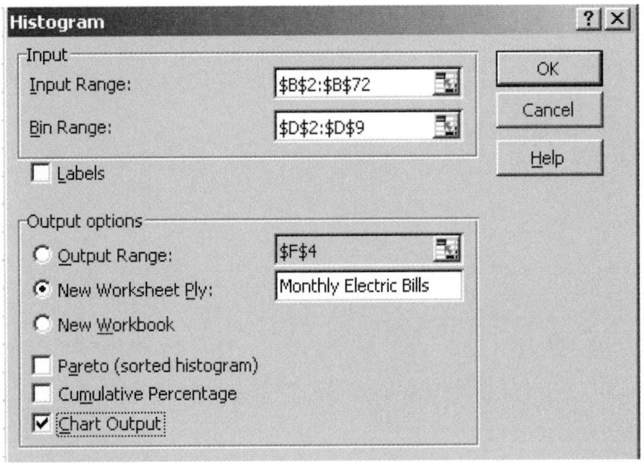

The frequency table and histogram are produced after clicking on OK.

Bin	*Frequency*
49.99	0
59.99	7
69.99	9
79.99	19
89.99	14
99.99	14
109.99	6
119.99	2
More	0

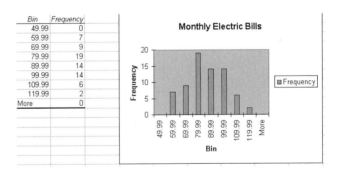

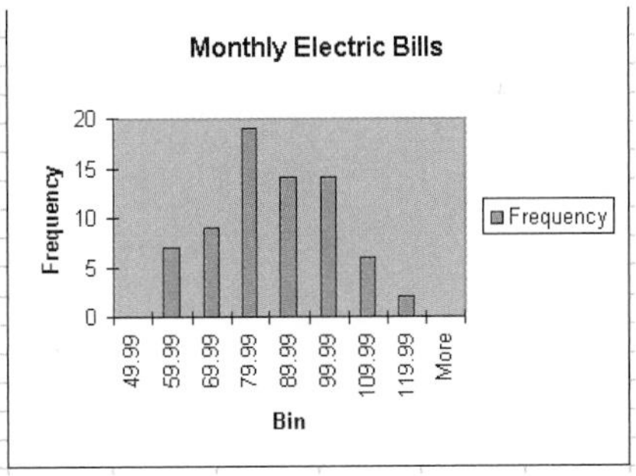

To solve Example 4, merely change the bin numbers to the class intervals in Table 8.

Bin	*Frequency*
Less than	
49.99	0
50–59.99	7
60–69.99	9
70–79.99	19
80–89.99	14
90–99.99	14
100–109.99	6
110–119.99	2
More	0

STEP 5 To eliminate the spaces between the bars, double click on one of the bars to get the screen below.

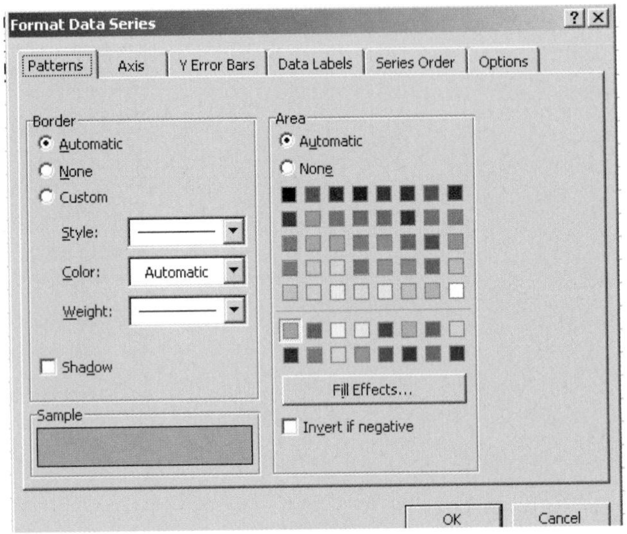

Select the Options tab and make the gap with zero.

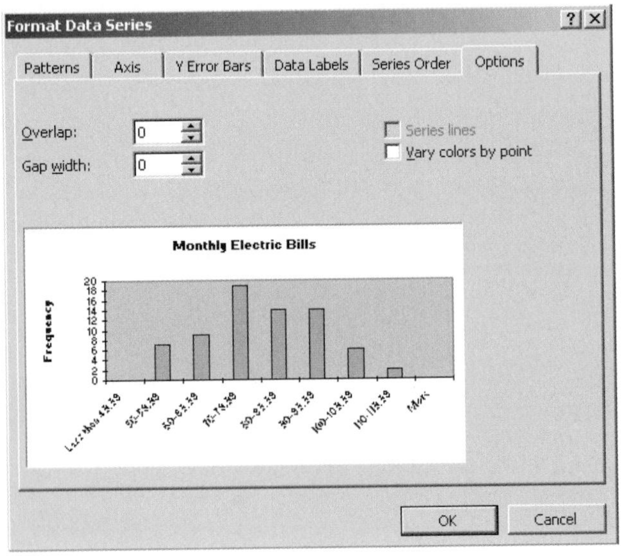

The final histogram is given below.

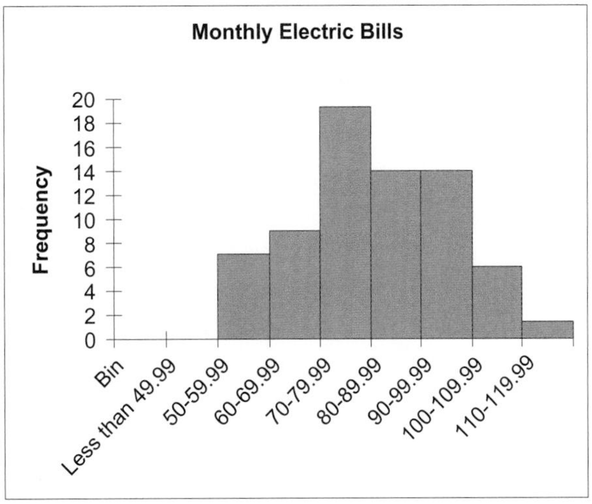

NOW WORK PROBLEMS 1(a), (b), AND (d) USING EXCEL.

Frequency Polygons

Draw a frequency polygon 5

Look again at Figure 5. If we connect all the midpoints of the tops of the rectangles in Figure 5, we obtain a line graph called a **frequency polygon.** (In order not to leave the graph hanging, we will connect it to the horizontal axis on each side.) Figure 6 shows the frequency polygon for the histogram obtained in Example 4.

FIGURE 6

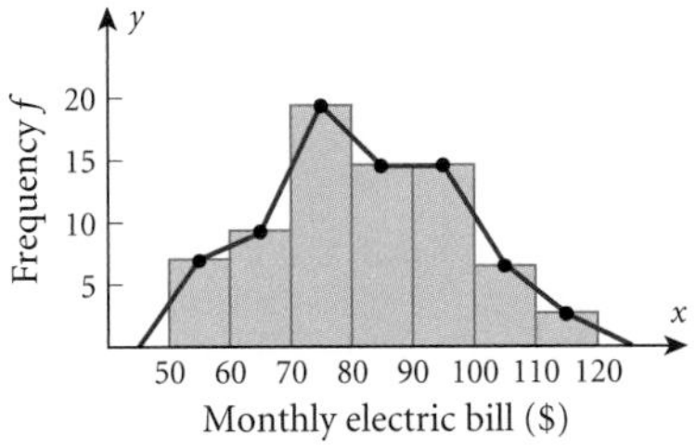

NOW WORK PROBLEM 1(e).

Sometimes it is useful to determine how many data points fall below a certain value. This is called *calculating cumulative frequencies,* and it is done by adding the frequencies from all class intervals less than or equal to the class interval being considered.

EXAMPLE 5 Creating a Cumulative Frequency Table

Create a cumulative frequency table for the data in Table 8.

SOLUTION We compute the cumulative frequency by determining the frequency of measurements below a given point. For the data in Table 8 we start in the lowest class interval ($50–$59.99) and note that there are 7 bills in this interval. So we put 7 in the column labeled *cf* (cumulative frequency) of Table 9 in the row for $50–$59.99. Next we add the number of bills in the second class interval ($60–$69.99), 9, to the cumulative total from the first interval $7 + 9 = 16$, and place 16 in column *cf* of the second interval. We then continue with the third interval, adding 19 to 16, and place the sum, 35, in the *cf* column of the third interval, and so on. The last entry in the *cf* column should equal the total number of bills in the sample. The numbers in the *cf* column are called the **cumulative frequencies.**

TABLE 9

Class Interval	Tally	*f*	*cf*
50–59.99	𝍸 //	7	7
60–69.99	𝍸 ////	9	16
70–79.99	𝍸 𝍸 𝍸 ////	19	35
80–89.99	𝍸 𝍸 ////	14	49
90–99.99	𝍸 𝍸 ////	14	63
100–109.99	𝍸 /	6	69
110–119.99	//	2	71

NOW WORK PROBLEM 1(f).

Draw a cumulative frequency distribution 6

The graph in which the horizontal axis represents class intervals and the vertical axis represents cumulative frequencies is called a **cumulative frequency distribution.**

EXAMPLE 6 Draw a Cumulative Frequency Distribution

Use the cumulative frequency table in Table 9 to draw a cumulative frequency distribution.

SOLUTION We draw two axes and put the class intervals on the horizontal axis and the cumulative frequencies on the vertical axis. We then plot each cumulative frequency at the upper class limit of each class interval. Finally we connect the points with straight line segments. See Figure 7.

FIGURE 7

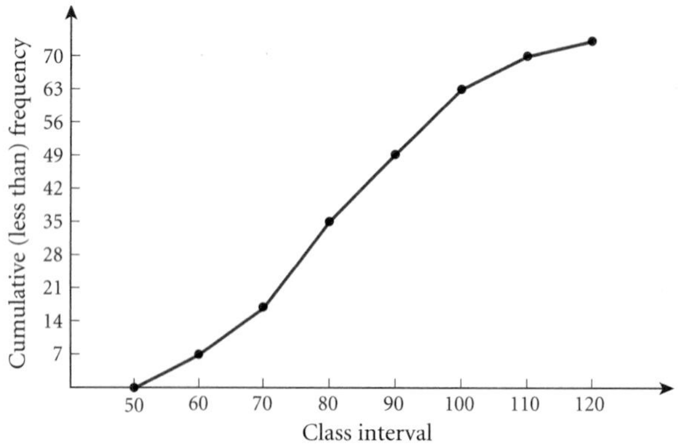

NOW WORK PROBLEM 1(g).

EXERCISE 9.3 Answers to Odd-Numbered Problems Begin on Page AN-36.

1. The following scores were made on a 60-item test:

25	30	34	37	41	42	46	49	53
26	31	34	37	41	42	46	50	53
28	31	35	37	41	43	47	51	54
29	32	36	38	41	44	48	52	54
30	33	36	39	41	44	48	52	55
30	33	37	40	42	45	48	52	

(a) Set up a frequency table for the above data.
(b) Draw a line chart for the data.
(c) Group the data into class intervals of size 2, beginning with the interval 24–25.9.
(d) Build the histogram for the data.
(e) Draw the frequency polygon for this histogram.
(f) Find the cumulative frequencies.
(g) Draw the cumulative frequency distribution.

2. For the test scores given in Problem 1,

(a) Group the data into class intervals of size 5, beginning with the interval 24–28.9.
(b) Build the histogram for the data.
(c) Draw the frequency polygon for the histogram.
(d) Find the cumulative frequencies.
(e) Draw the cumulative frequency distribution.

3. For Table 5 (p. 482) in the text:

(a) Group the data into class intervals of size 5 beginning with the interval 50–54.9.
(b) Build the histogram for the data.
(c) Draw the frequency polygon for this histogram.
(d) Find the cumulative frequencies.
(e) Draw the cumulative frequency distribution.

4. For Table 5 (p. 482) in the text:

(a) Group the data into class intervals of size 10, beginning with the interval 50–59.9. How many class intervals now exist?
(b) Build the histogram for the data.
(c) Draw the frequency polygon for this histogram.
(d) Find the cumulative frequencies.
(e) Draw the cumulative frequency distribution.
(f) Discuss the advantages/disadvantages of using a class interval size of 10 compared to a smaller interval size such as 5.

5. **Licensed Drivers in Florida** The histogram that follows represents the number of licensed drivers between the ages of 20 and 84 in the state of Florida.

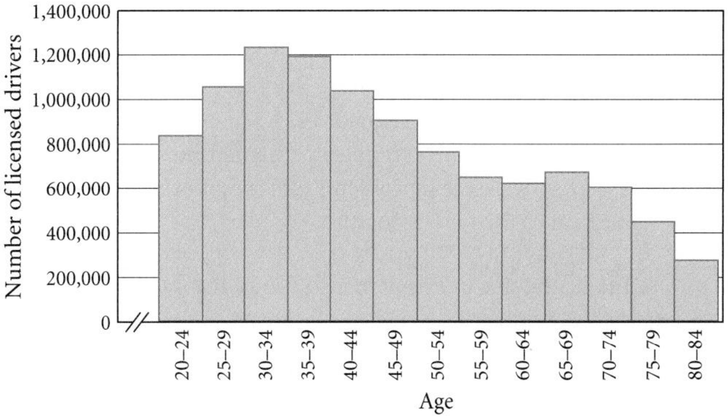

(a) Determine the number of class intervals.
(b) What is the lower class limit of the first class interval? What is the upper class limit of the first class interval?
(c) Determine the class width.
(d) How many licensed drivers are 70 to 84 years old?
(e) Which class interval has the most licensed drivers?
(f) Which class interval has the fewest licensed drivers?
(g) Draw a frequency polygon for the given data.

Source: Federal Highway Administration.

6. **IQ Scores** The histogram below represents the IQ scores of students enrolled in College Algebra at a local university.

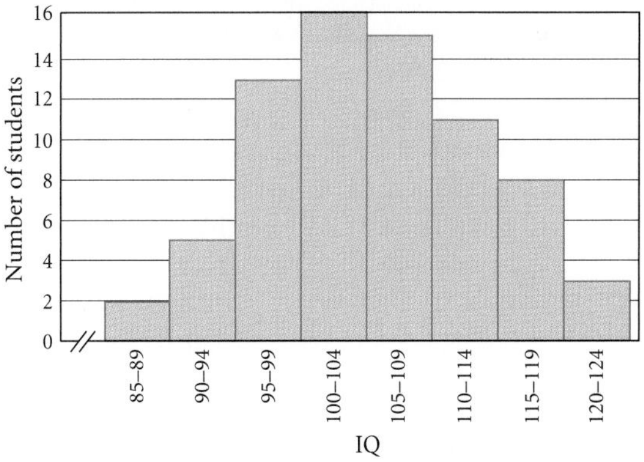

(a) Determine the number of class intervals.
(b) What is the lower class limit of the first class interval? What is the upper class limit of the first class interval?
(c) Determine the class width.
(d) How many students have an IQ between 100 and 104?
(e) How many students have an IQ above 110?
(f) How many students are enrolled in College Algebra?
(g) Draw a frequency polygon for the given data.

7. Use the histogram in Problem 5 to group the data into class intervals if width 10.

(a) Determine the number of class intervals.
(b) What is the lower class limit for the last interval? What is the upper class limit of the last interval?
(c) Build a histogram of the data using the new class intervals of size 10.
(d) Draw a frequency polygon for this histogram.
(e) Find the cumulative frequencies.
(f) Draw the cumulative frequency distribution.
(g) Discuss the possible problems that might arise in the last interval of the new histogram.

8. Use the histogram in Problem 6 to group the data into class intervals of width 10.

(a) Determine the number of class intervals.
(b) What is the lower class limit for the last interval? What is the upper class limit of the last interval?
(c) Build a histogram of the data using the new class intervals of size 10.
(d) Draw a frequency polygon for this histogram.
(e) Find the cumulative frequencies.
(f) Draw the cumulative frequency distribution.
(g) Discuss the possible problems that might arise in the last interval in this new histogram.

9. Licensed Drivers in Tennessee The frequency table below provides the number of licensed drivers between the ages of 20 and 84 in the state of Tennessee in 1994.

Age	Number of Licensed Drivers
20–24	345,941
25–29	374,629
30–34	428,748
35–39	439,137
40–44	414,344
45–49	372,814
50–54	292,460
55–59	233,615
60–64	204,235
65–69	181,977
70–74	150,347
75–79	100,068
80–84	50,190

Source: Federal Highway Administration.

(a) Determine the number of class intervals.
(b) What is the lower class limit of the first class interval? What is the upper class limit of the first class interval?
(c) Determine the class width.
(d) Build the histogram for the data.
(e) Draw a frequency polygon for the data.
(f) Which age group has the most licensed drivers?
(g) Which age group has the fewest licensed drivers?

10. Licensed Drivers in Hawaii The frequency table below provides the number of licensed drivers between the ages of 20 and 84 in the state of Hawaii in 1994.

Age	Number of Licensed Drivers
20–24	65,951
25–29	78,119
30–34	91,976
35–39	92,557
40–44	87,430
45–49	75,978
50–54	55,199
55–59	39,678
60–64	35,650
65–69	33,885
70–74	26,125
75–79	14,990
80–84	6,952

Source: Federal Highway Administration.

(a) Determine the number of class intervals.
(b) What is the lower class limit of the first class interval? What is the upper class limit of the first class interval?
(c) Determine the class width.
(d) Build the histogram for the data.
(e) Draw a frequency polygon of the data.
(f) Which age group has the most licensed drivers?
(g) Which age group has the fewest licensed drivers?

11. Undergraduate Tuition The data below represent the cost of undergraduate tuition at four-year colleges for

Tuition (Dollars)	Number of 4-Year Colleges
0–999	10
1000–1999	7
2000–2999	45
3000–3999	66
4000–4999	84
5000–5999	84
6000–6999	97
7000–7999	118
8000–8999	138
9000–9999	110
10,000–10,999	104
11,000–11,999	82
12,000–12,999	61
13,000–13,999	34
14,000–14,999	29

Source: The College Board, New York, NY, Annual Survey of Colleges 1992 and 1993.

1992–1993 having tuition amounts ranging from $0 through $14,999.

(a) Determine the number of class intervals.
(b) What is the lower class limit of the first class interval? What is the upper class limit of the first class interval?
(c) Determine the class width.
(d) Build the histogram for the data.
(e) Draw a frequency polygon of the data.
(f) What range of tuition occurs most frequently?

12. Undergraduate Tuition The data below represent the cost of undergraduate tuition at four-year colleges for 1993–1994 having tuition amounts ranging from $0 through $14,999.

Tuition (Dollars)	Number of 4-Year Colleges	Tuition (Dollars)	Number of 4-Year Colleges
0–999	8	8000–8999	112
1000–1999	5	9000–9999	118
2000–2999	28	10,000–10,999	106
3000–3999	48	11,000–11,999	90
4000–4999	76	12,000–12,999	70
5000–5999	65	13,000–13,999	59
6000–6999	81	14,000–14,999	23
7000–7999	96		

Source: The College Board, New York, NY, Annual Survey of Colleges 1993 and 1994.

(a) Determine the number of class intervals.
(b) What is the lower class limit of the first class interval? What is the upper limit of the first class interval?
(c) Determine the class width.
(d) Build the histogram for the data.
(e) Draw a frequency polygon of the data.
(f) What range of tuition occurs most frequently?

13. Birth Rates The following data give the 2001 birth rates per 1000 people for 20 states.

State	Birth Rate	State	Birth Rate	State	Birth Rate
CT	12.9	AL	13.9	AK	16.0
AZ	17.1	CA	15.5	DE	13.9
FL	13.2	GA	16.5	IL	15.0
KS	14.5	LA	14.9	MI	13.4
NV	16.1	NJ	14.0	OR	13.5
PA	12.0	SC	14.1	TN	14.0
VA	14.0	WI	12.9		

Source: National Center of Health Statistics, 2003.

(a) Group the data into 8 class intervals of equal width beginning with the interval 11.3–11.9.
(b) Build the histogram for the data.
(c) Draw the frequency polygon for this histogram.

14. Hourly Earnings The following table gives the average hourly earnings of production workers in manufacturing during 1993 in 20 states.

State	Hourly Earnings	State	Hourly Earnings
AK	11.14	MA	12.36
AL	10.36	ME	11.40
AR	9.36	MI	15.35
CA	12.37	NH	11.61
CT	13.01	OH	14.05
DC	13.18	PA	12.09
HI	11.98	RI	10.22
IL	12.04	TN	10.33
IN	13.17	VT	11.81
KY	11.48	WI	12.17

Source: U.S. Census Bureau 1994.

(a) Group the data into 7 class intervals of equal width beginning with the interval 9.00–9.99.
(b) Build the histogram for the data.
(c) Draw the frequency polygon for this histogram.

15. HIV Death Rates The following data give the 2001 death rates (number of deaths per 100,000 people) from HIV-related illnesses for 20 states.

State	Death Rate	State	Death Rate	State	Death Rate
AL	4.6	AZ	3.2	CA	4.4
CT	6.1	FL	11.7	HI	2.3
IA	0.8	KY	1.6	MD	10.6
MI	2.5	MO	3.0	NV	4.7
NJ	10.1	NY	12.2	ND	0.1
OR	1.1	VT	0.1	VA	4.1
WI	1.4	IL	3.9		

Source: National Center of Health Statistics, 2003.

(a) Group the data into 7 class intervals of equal width beginning with the interval 0.0–1.9.
(b) Build the histogram for the data.
(c) Draw the frequency polygon for this histogram.

16. Cancer Death Rates The table lists the 2001 death rates (number of deaths per 100,000 people) attributed to cancer for 20 states.

(a) Group the data into 5 class intervals of equal width beginning with the interval 150–174.9.
(b) Build the histogram for these data.
(c) Draw the frequency polygon for this histogram.

State	Death Rate	State	Death Rate	State	Death Rate
AL	223.5	AZ	185.8	CA	158.1
CT	214.3	FL	255.6	HI	164.8
IA	224.1	KY	231.0	MD	197.2
MI	190.9	MO	220.7	NV	200.1
NJ	220.3	NY	203.5	ND	213.9
OR	208.0	VT	207.4	VA	194.1
WI	200.9	IL	224.1		

Source: National Center of Health Statistics, 2003.

9.4 Measures of Central Tendency

OBJECTIVES
1 Compute the mean of a set of data
2 Compute the mean for grouped data
3 Find the median of a set of data
4 Compute the median for grouped data
5 Identify the mode of a set of data

The idea of taking an *average* is familiar to practically everyone. Often we hear people talk about average salary, average height, average grade, and so on. The idea of average is so commonly used it should not surprise you to learn that several kinds of averages have been introduced in statistics.

Averages are called *measures of central tendency* because they estimate the "center" of the data collected. The three most common measures of central tendency are the *(arithmetic) mean, median,* and *mode.* Of these, the mean is the one most commonly used.

Mean

The **arithmetic mean,** or **mean,** of a set of n real numbers $x_1, x_2, \dots x_n$ is defined as the number

$$\overline{X} = \frac{x_1 + x_2 + \cdots + x_n}{n} \tag{1}$$

where n is the number of items in the set.

1 **EXAMPLE 1** **Computing the Mean of a Set of Data**

Professor Murphy tested 8 students in her Finite Mathematics class. The scores on the test were 100, 94, 85, 79, 70, 69, 65, and 62. What is the mean?

SOLUTION To compute the mean $\overline{X}$, we add up the 8 scores and divide by 8.

$$\overline{X} = \frac{100 + 94 + 85 + 79 + 70 + 69 + 65 + 62}{8} = \frac{624}{8} = 78$$

NOW WORK PROBLEM 1(a).

An interesting fact about the mean is that the sum of deviations of each item from the mean is zero. In Example 1 the deviation of each score from the mean $\overline{X} = 78$ is (100 − 78), (94 − 78), (85 − 78), (79 − 78), (70 − 78), (69 − 78), (65 − 78), and (62 − 78). Table 10 lists each score, the mean, and the deviation from the mean. If we add the deviations from the mean, we obtain a sum of zero.

TABLE 10

Score	Mean	Deviation from Mean
62	78	−16
65	78	−13
69	78	−9
70	78	−8
79	78	1
85	78	7
94	78	16
100	78	22
		Sum of Deviations: 0

For any set of data the following result is true:

The sum of the deviations from the mean is zero.

As a matter of fact, we could have defined the mean as that real number for which the sum of the deviations is zero.

Another interesting fact about the mean is given below.

If Y is any guessed or assumed mean (which may be any real number) and if d_j denotes the deviation of each item of the data from the assumed mean ($d_j = x_j - Y$), then the actual mean is

$$\overline{X} = Y + \frac{d_1 + d_2 + \cdots + d_n}{n} \tag{2}$$

Look again at Example 1. We know that the actual mean is 78. Suppose we had guessed the mean to be 52. Then, using Formula (2), we obtain

$$\overline{X} = 52 + \frac{(100-52)+(94-52)+(85-52)+(79-52)+(70-52)+(69-52)+(65-52)+(62-52)}{8}$$

$$= 52 + 26 = 78$$

which agrees with the mean computed in Example 1.

Compute the mean of grouped data **2**

A method for computing the mean for grouped data given in a frequency table is illustrated in the following example.

EXAMPLE 2 Finding the Mean for Grouped Data

Find the mean for the grouped data given in Table 8 (repeated in the first two columns of Table 11 below).

SOLUTION We follow the steps below

STEP 1 Take the midpoint (m_i) of each of the class intervals as a reference point and enter the result in column 3 of Table 11. For example, the midpoint of the class interval 80–89.99 is $\frac{80 + 89.99}{2} = 85$.

STEP 2 Multiply the entry m_i in column 3 by the frequency f_i for that class interval and enter the product in column 4, which is labeled $f_i m_i$.

STEP 3 Add the entries in column 4; sum of $f_i m_i = 5775$.

TABLE 11

Class Interval	f_i	m_i	$f_i m_i$
50– 59.99	7	55	385
60– 69.99	9	65	585
70– 79.99	19	75	1425
80– 89.99	14	85	1190
90– 99.99	14	95	1330
100–109.99	6	105	630
110–119.99	2	115	230
	$n = 71$		5775 = Sum of $f_i m_i$

STEP 4 The mean $\overline{X}$ is then computed by dividing the sum of $f_i m_i$ by the number n of entries. That is,

$$\overline{X} = \frac{5775}{71} = 81.34$$

Mean For Grouped Data

For grouped data, the mean $\overline{X}$ is given as

$$\overline{X} = \frac{\Sigma f_i m_i}{n} \tag{3}$$

where

Σ = Means add the entries $f_i m_i$

f_i = Number of entries in the ith class interval

m_i = Midpoint of ith class interval

n = Number of items

COMMENT: Graphing utilities can be used to find measures of central tendency for both grouped and ungrouped data. Figure 8 illustrates the calculation of the mean for the grouped data in Example 2 on a TI-83Plus. The midpoint for each interval was entered into L_1 and the frequency into L_2. Then one-variable statistics were computed using L_1 and L_2. The mean is $\overline{X} = 81.34$ (rounded to two decimal places). Consult your user's manual for the commands needed to compute measures of central tendency on your graphing utility.

FIGURE 8

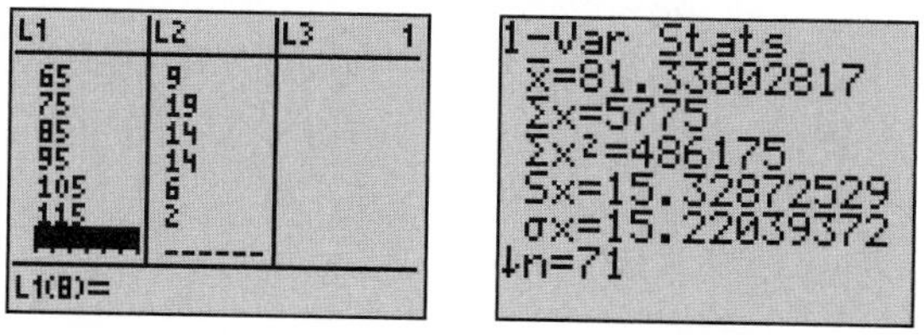

When data are grouped, the original data are lost due to grouping. As a result, the mean obtained by using formula (3) is only an approximation to the actual mean. The reason for this is that using formula (3) amounts to computing the weighted average midpoint of a class interval (weighted by the frequency of scores in that interval) and does not compute $\overline{X}$ exactly.

NOW WORK PROBLEM 13(a).

The Median of a Set of Data

Median

The **median** of a set of real numbers arranged in order of magnitude is the middle value if the number of items is odd, and it is the mean of the two middle values if the number of items is even.

3 **EXAMPLE 3** **Finding the Median of a Set of Data**

(a) The set of data 2, 2, 3, 4, 5, 7, 7, 7, 11 has median 5.
(b) The set of data 2, 2, 3, 3, 4, 5, 7, 7, 7, 11 has median 4.5 since

$$\frac{4 + 5}{2} = 4.5$$

NOW WORK PROBLEM 1(b).

Compute the median for **4**
grouped data

Finding the median for grouped data requires more work. As with the mean, the median for grouped data only approximates the actual median that would have been obtained prior to grouping the data. We use Example 4 below to show the steps.

EXAMPLE 4 **Computing the Median for Grouped Data**

Find the median for the grouped data in Table 12. (Based on Table 8, p. 484)

TABLE 12

	Class Interval	Tally	Frequency
1	50– 59.99	〢〢 //	7
2	60– 69.99	〢〢 ////	9
3	70– 79.99	〢〢 〢〢 〢〢 ////	19
4	80– 89.99	〢〢 〢〢 ////	14
5	90– 99.99	〢〢 〢〢 ////	14
6	100–109.99	〢〢 /	6
7	110–119.99	//	2

SOLUTION We follow the steps below.

STEP 1 Find the interval containing the median.
   The median is the middle value when the 71 items are in ascending order. So the median in this sample is the 36th entry. We start counting the tallies, beginning with the interval 50–59.99, until we come as close as we can to 36. This brings us through the interval 70–79.99, since through this interval there are 35 data items. The median will lie in the interval 80–89.99.

STEP 2 In the interval containing the median, count the number p of items remaining to reach the median.
   Since we have accounted for 35 data items, there is left $36 - 35 = 1$ item, so $p = 1$.

STEP 3 If q is the frequency for the interval containing the median and i is its size, then the interpolation factor is

$$\textbf{interpolation factor} = \frac{p}{q} \cdot i$$

For the grouped data in Table 12, we have $q = 14$ and $i = 10$. Then

$$\text{interpolation factor} = \frac{1}{14} \cdot (10) = 0.71$$

STEP 4 The median M is

$$\mathrm{M} = \begin{bmatrix} \textbf{lower limit of interval} \\ \textbf{containing the median} \end{bmatrix} + [\textbf{interpolation factor}]$$

For the grouped data in Table 12, the median M is

$$\mathrm{M} = 80 + 0.71 = 80.71$$

To summarize,

Median for Grouped Data

The median M for grouped data is given by

$$M = \begin{bmatrix} \textbf{lower limit of interval} \\ \textbf{containing the median} \end{bmatrix} + \left[\frac{p}{q} \cdot i \right] \tag{4}$$

where

p = Number required to reach the median from the lower limit of the interval containing the median
q = Number of data entries in this interval
i = Size of this interval

NOW WORK PROBLEM 13(b).

As with the mean, the median of a set of grouped data is an approximation to the actual median since it is obtained from grouped data. For Example 4, if we go back to the original data listed in Table 6, we obtain the true median $M = 82.32$.

The median of a set of data or grouped data is sometimes called the **fiftieth percentile** and is denoted by C_{50} to indicate that 50% of the data are less than or equal to it. Similarly, we can define C_{25}, or the first quartile, as the point that separates the lowest 25% of the data from the upper 75%, and C_{75}, or the third quartile, as the point that separates the lower 75% of the data from the upper 25% of the data.

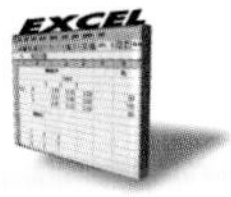

Use Excel to find the mean and the median of the data given in Table 6. (Refer to the Excel example given on page 485).

SOLUTION Excel has the statistical functions Average( ) and Median( ).

STEP 1 Insert a function.

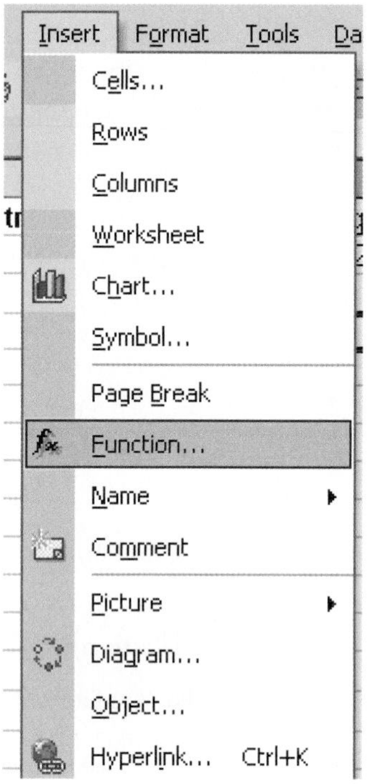

STEP 2 Select Statistical, then Average.

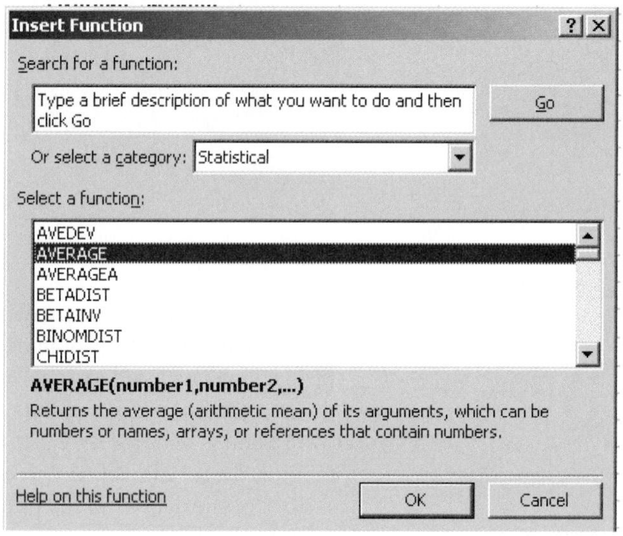

STEP 3 There are two ways to insert the cells that are used to find the average and the median

(a) Highlight the cells.

or

(b) Type the location of the first cell followed by a colon and the location of the last cell.

Average	*Median*
=AVERAGE(B2:B72)	=MEDIAN(B2:B72)

The average and median are given below.

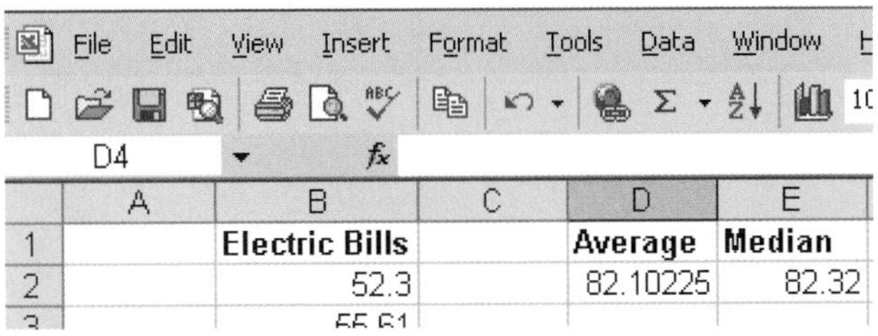

NOW WORK PROBLEMS 1(a) AND 1(b) USING EXCEL.

The Mode of a Set of Data

Mode

The **mode** of a set of real numbers is the value that occurs with the greatest frequency exceeding a frequency of 1.

Identify the mode of a set of data 5

The mode does not necessarily exist, and if it does, it is not always unique.

EXAMPLE 5 Set of Data with No Mode

The set of data 2, 3, 4, 5, 7, 15 has no mode.

NOW WORK PROBLEM 1(c).

EXAMPLE 6 A Bimodal Set of Data

The set of data 2, 2, 2, 3, 3, 7, 7, 7, 11, 15 has two modes, 2 and 7, and is called **bimodal.**

When data have been listed in a grouped frequency table, the mode is defined as the midpoint of the interval consisting of the largest number of cases. For example, the mode for the data in Table 8, page 484, is 75 (the midpoint of the interval 70–79.99).

EXAMPLE 7 Finding the Mode

For the data listed in Table 5, page 482, the mode is 79 (8 is the highest frequency).

Sample Mean and Population Mean

Depending on the nature of the data, there are two types of means: the *sample mean* and the *population mean.*

Sample Mean

If the data used in computing the mean is a sample $x_1, x_2, x_3, \ldots x_n$, of n items taken from the population of N items, $n < N$, the **sample mean** $\bar{X}$ is defined as the number

$$\bar{X} = \frac{x_1 + x_2 + x_3 + \cdots + x_n}{n} \quad (5)$$

Population Mean

If the data used in computing the mean are from the entire population $x_1, x_2, x_3, \ldots x_N$, of N items, the **population mean** $\boldsymbol{\mu}$ (the Greek letter mu) is defined as

$$\mu = \frac{x_1 + x_2 + x_3 + \cdots + x_N}{N} \quad (6)$$

Notice that Formula (5) and Formula (6) for the calculation of the mean are identical: In both cases the items are added up and the result is divided by the number of items.

The reason for making a distinction between a sample mean $\bar{X}$ and a population mean μ will be made clear in the next section.

EXERCISE 9.4 Answers to Odd-Numbered Problems Begin on Page AN-41.

In Problems 1–8, compute (a) the mean, (b) the median, and (c) the mode of each set of data.

1. 21, 25, 43, 36
2. 16, 18, 24, 30
3. 55, 55, 80, 92, 70
4. 90, 80, 82, 82, 70
5. 65, 82, 82, 95, 70
6. 62, 71, 83, 90, 75
7. 48, 65, 80, 92, 80, 75
8. 95, 90, 91, 82, 80, 80

9. **Baseball Players** At the end of 2003 spring training, the ages of the 31 men on the New York Yankees roster were:

24	28	41	32	37	32	31	33
35	30	31	34	32	40	32	27
32	24	32	29	24	24	36	30
38	30	29	32	24	32	35	

Source: Major League Baseball, March 2003.

(a) Find the mean age of the players.
(b) Find the median age of the players.
(c) Identify the modal age of the players if one exists.

10. **Baseball Players** At the end of 2003 spring training, the ages of the 34 men on the 2002 World Series Champion Anaheim Angels roster were:

35	27	32	22	24	30	33	30
33	21	29	22	27	28	33	28
33	30	29	25	30	25	27	25
30	26	27	30	30	29	28	32
34	28						

Source: Major League Baseball, March 2003.

(a) Find the mean age of the players.
(b) Find the median age of the players.
(c) Identify the modal age of the players if one exists.

11. Investments If an investor purchased 50 shares of IBM stock at $85 per share, 90 shares at $105 per share, 120 shares at $110 per share, and another 75 shares at $130 per share, what is the mean cost per share?

12. Revenue If a farmer sells 120 bushels of corn at $4 per bushel, 80 bushels at $4.10 per bushel, 150 bushels at $3.90 per bushel, and 120 bushels at $4.20 per bushel, what is the mean income per bushel?

13. **Mother's Age** The following data give the number of births in the United States in the year 2000 by the age of the mother. The numbers of children are given in thousands.

Age of Mother	No. of Births
10–14	9
15–19	469
20–24	1018
25–29	1088
30–34	929
35–39	452
40–44	90
45–49	4

Source: National Center for Health Statistics, 2003.

(a) Find the mean age of a woman who became a mother in 2000.
(b) Find the median age of a woman who became a mother in 2000.

14. Multiple Births The following data give the number of multiple births (two or more children) in the United States in the year 2000 by age of the mother.

Age of Mother	No. of Births
10–14	118
15–19	7,560
20–24	22,813
25–29	32,402
30–34	36,729
35–39	21,314
40–44	4,445
45–49	838

Source: National Center for Health Statistics, 2003.

(a) Find the mean age of a mother whose pregnancy resulted in multiple births in 2000.
(b) Find the median age of a mother whose pregnancy resulted in multiple births in 2000.

15. Yearly Sales According to an article in the *Wall Street Journal,* for companies having fewer than 5000 employees the average sales per employee were as follows:

Size of Company (Number of Employees)	Sales per Employee (Thousands of Dollars)
1–4	112
5–19	128
20–99	127
100–499	118
500–4999	120

(a) Compute the mean sales per employee for all firms having fewer than 5000 employees.
(b) Compute the median sales per employee for all firms having fewer than 5000 employees.

16. The distribution of the monthly earnings of 1155 secretaries in May 1994 in the Chicago metropolitan area is summarized in the table below. Find (a) the mean salary and (b) the median salary.

Monthly Earnings, $	Number of Secretaries
950–1199.99	25
1200–1449.99	55
1450–1699.99	325
1700–1949.99	410
1950–2199.99	215
2200–2449.99	75
2450–2699.99	50

17. Licensed Drivers in Tennessee Refer to the data provided in Problem 9, Exercise 9.3 and find

(a) The mean age of a licensed driver in Tennessee.
(b) The median age of a licensed driver.

18. Licensed Drivers in Hawaii Refer to the data provided in Problem 10, Exercise 9.3 and find

(a) The mean age of a licensed driver in Hawaii.
(b) The median age of a licensed driver.

19. Undergraduate Tuition 1992–1993 Refer to the data provided in Problem 11, Exercise 9.3 and find the mean tuition at a four-year college in 1992–1993.

20. Undergraduate Tuition 1993–1994 Refer to the data provided in Problem 12, Exercise 9.3 and find the mean tuition at a four-year college in 1993–1994.

21. Faculty Salary The annual salaries of five faculty members in the mathematics department at a large university are \$34,000, \$35,000, \$36,000, \$36,500, and \$65,000.

(a) Compute the mean and the median.
(b) Which measure describes the situation more realistically?
(c) If you were among the four lower-paid members, which measure would you use to describe the situation? What if you were the one making \$65,000?

22. Family Wealth Use the data in the table to estimate

Net Worth (\$)	Percentage of Families 1985	1990	1995
0–4,999	33	24	11
5,000–9,999	5	7	5
10,000–24,999	12	10	12
25,000–49,999	16	19	22
50,000–99,999	17	19	22
100,000–249,999	12	14	17
250,000–499,999	3	4	6
500,000 or more	2	3	5

(a) The mean net worth of U.S. families in each year.
(b) The median net worth in each year.
Use a class midpoint of \$1 million for the class 500,000 or more.
(c) Discuss whether you think the mean or the median is a better measure of central tendency in this problem.
(d) Why is there such a discrepancy between the two measures of central tendency?

9.5 Measures of Dispersion

OBJECTIVES
1. Find the standard deviation for sample data
2. Find the standard deviation for grouped data
3. Use Chebychev's theorem

EXAMPLE 1 Comparing Mean and Median for a Set of Sample Data

Find the mean and median for each of the following sets of sample data:

$$S_1: \quad 4, 6, 8, 10, 12, 14, 16$$
$$S_2: \quad 4, 7, 9, 10, 11, 13, 16$$

SOLUTION For S_1, the mean $\overline{X}_1$ and median M_1 are

$$\overline{X}_1 = \frac{4 + 6 + 8 + 10 + 12 + 14 + 16}{7} = \frac{70}{7} = 10 \qquad M_1 = 10$$

For S_2, the mean $\overline{X}_2$ and median M_2 are

$$\overline{X}_2 = \frac{4 + 7 + 9 + 10 + 11 + 13 + 16}{7} = \frac{70}{7} = 10 \qquad M_2 = 10$$

Notice that each set of scores has the same mean and the same median. Now look at Figure 9. Do you see that the data in S_1 seem to be more spread out from the mean 10 than those in S_2?

FIGURE 9

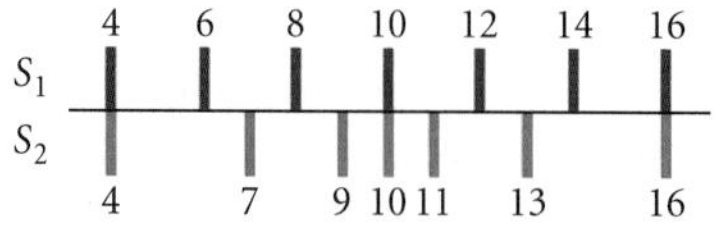

We seek a way to measure the extent to which scores are spread out. Such measures are called *measures of dispersion.*

Range

The simplest measure of dispersion is the **range,** which we have already defined as the difference between the largest value and the smallest value. For each of the sets S_1 and S_2 in Example 1, the range is $16 - 4 = 12$. We conclude that the range is a poor measure of dispersion since it depends on only two data items and tells us nothing about how the rest of the data are spread out.

Variance

Another measure of dispersion is the *variance,* a measure of deviation from the mean. Recall that the sum of the deviations from the mean add up to zero, so just using deviations from the mean will not be of much use. Since some of the deviations are positive and some are negative, by squaring the deviations and adding them up, we obtain a positive number, called the **variance.** Like the mean, we distinguish between the *population variance* and the *sample variance.*

Population Variance

If the data used are from the entire population $x_1, x_2, x_3, \ldots x_N$, the **population variance** σ^2, the Greek letter sigma squared, is defined as

$$\sigma^2 = \frac{(x_1 - \mu)^2 + (x_2 - \mu)^2 + \cdots + (x_N - \mu)^2}{N} \tag{1}$$

where

$$\mu = \frac{x_1 + x_2 + x_3 + \cdots + x_N}{N}$$

is the population mean and N is the number of items in the population.

Sample Variance

If the data used are a sample $x_1, x_2, x_3, \ldots x_n$ of the entire population, the **sample variance** S^2 is defined as

$$S^2 = \frac{(x_1 - \overline{X})^2 + (x_2 - \overline{X})^2 + \cdots + (x_n - \overline{X})^2}{n - 1} \tag{2}$$

where

$$\overline{X} = \frac{x_1 + x_2 + x_3 + \cdots + x_n}{n}$$

is the sample mean and n is the number of items in the sample.

Notice that we divide by $n - 1$ in the formula for the sample variance [Formula (2)]. The reason for this is that statisticians have found that when using the sample variance to estimate the population variance, a better agreement is obtained by dividing by $n - 1$.

EXAMPLE 2 Finding the Sample Variance for a Set of Sample Data

Calculate the sample variance for sets S_1 and S_2 of Example 1.

SOLUTION For S_1, $\overline{X} = 10$ and $n = 7$, so that

$$S_1^2 = \frac{(4-10)^2 + (6-10)^2 + (8-10)^2 + (10-10)^2 + (12-10)^2 + (14-10)^2 + (16-10)^2}{7-1} = 18.67$$

For S_2, $\overline{X} = 10$ and $n = 7$, so that

$$S_2^2 = \frac{(4-10)^2 + (7-10)^2 + (9-10)^2 + (10-10)^2 + (11-10)^2 + (13-10)^2 + (16-10)^2}{7-1} = 15.33$$

Since the sample variance for set S_1 is larger than the sample variance for set S_2, we conclude that the data in set S_1 are more widely dispersed than the data in set S_2, confirming what we saw in Figure 9.

Standard Deviation

In computing the variance, we square the deviations from the mean. This means, for example, that if our data represent dollars, then the variance has the units "dollars squared." To remedy this, we use the square root of the variance, called the *standard deviation.*

Standard Deviation of Sample Data*

The **standard deviation** of a set $x_1, x_2 \ldots, x_n$ of n data items taken from the population is defined as

$$S = \sqrt{\frac{(x_1 - \overline{X})^2 + (x_2 - \overline{X})^2 + \cdots + (x_n - \overline{X})^2}{n-1}} = \sqrt{\frac{\Sigma(x_i - \overline{X})^2}{n-1}} \quad \textbf{(3)}$$

where $\overline{X}$ is the sample mean.

**The standard deviation σ of population data is defined as*

$$\sigma = \sqrt{\frac{(x_1 - \mu)^2 + (x_2 - \mu)^2 + \cdots + (x_N - \mu)^2}{N}} = \sqrt{\frac{\Sigma(x_i - \mu)^2}{N}} \quad (4)$$

where μ is the mean and $x_1, x_2, x_3, \ldots, x_N$ is the population.

For the sample data in Example 1 the standard deviation for S_1 is

$$S_1 = \sqrt{\frac{36 + 16 + 4 + 0 + 4 + 16 + 36}{7 - 1}} = \sqrt{\frac{112}{6}} = \sqrt{18.67} = 4.32$$

and the standard deviation for S_2 is

$$S_2 = \sqrt{\frac{36 + 9 + 1 + 0 + 1 + 9 + 36}{7 - 1}} = \sqrt{\frac{92}{6}} = \sqrt{15.33} = 3.92$$

Again, the fact that the standard deviation of the set S_2 is less than the standard deviation of the set S_1 indicates that the data of S_2 are more clustered around the mean than those of S_1.

1 **EXAMPLE 3** **Finding the Standard Deviation of Sample Data**

Find the standard deviation for the sample data

100, 90, 90, 85, 80, 75, 75, 75, 70, 70, 65, 65, 60, 40, 40, 40

SOLUTION The sample mean is

$$\overline{X} = \frac{100 + 2 \cdot 90 + 85 + 80 + 3 \cdot 75 + 2 \cdot 70 + 2 \cdot 65 + 60 + 3 \cdot 40}{15} = 70$$

The deviations from the mean and their squares are computed in Table 13 on page 508. The standard deviation is

$$S = \sqrt{\frac{4950}{15}} = 18.17$$

NOW WORK PROBLEM 1.

EXAMPLE 4 **Finding the Standard Deviation of Sample Data**

Find the standard deviation for the sample data

80, 80, 80, 80, 75, 75, 75, 75, 70, 70, 65, 65, 60, 60, 55, 55

SOLUTION Here the mean is $\overline{X} = 70$ for the 16 scores. Table 14 on page 508 gives the deviations from the mean and their squares. The standard deviation is

$$S = \sqrt{\frac{1200}{15}} = 8.94$$

These two examples show that although the samples have the same mean, 70, and the same sample size, 16, the data in Example 3 deviate further from the mean than do the data in Example 4.

In general, a relatively small standard deviation indicates that the measures tend to cluster close to the mean, and a relatively large standard deviation shows that the measures are widely scattered from the mean.

TABLE 13

Scores, x	Deviation from the Mean, $x - \overline{X}$	Deviation Squared, $(x - \overline{X})^2$
40	−30	900
40	−30	900
40	−30	900
60	−10	100
65	−5	25
65	−5	25
70	0	0
70	0	0
75	5	25
75	5	25
75	5	25
80	10	100
85	15	225
90	20	400
90	20	400
100	30	900
Mean = 70 $n = 16$	Sum = 0	Sum = 4950

TABLE 14

Scores, x	Deviation from the Mean, $x - \overline{X}$	Deviation Squared, $(x - \overline{X})^2$
55	−15	225
55	−15	225
60	−10	100
60	−10	100
65	−5	25
65	−5	25
70	0	0
70	0	0
75	5	25
75	5	25
75	5	25
75	5	25
80	10	100
80	10	100
80	10	100
80	10	100
Mean = 70 $n = 16$	Sum = 0	Sum = 1200

Standard Deviation for Grouped Data

To find the standard deviation for grouped data, we use the formula

$$s = \sqrt{\frac{(m_1 - \overline{X})^2 \cdot f_1 + (m_2 - \overline{X})^2 \cdot f_2 + \cdots + (m_k - \overline{X})^2 \cdot f_k}{n - 1}} = \sqrt{\frac{\Sigma[(m_i - \overline{X})^2 \cdot f_i]}{n - 1}} \tag{5}$$

where $m_1, m_2, \ldots, m_k$ are the class midpoints; $f_1, f_2, \ldots, f_k$ are the respective frequencies; n is the number of data items, that is, $n = f_1 + f_2 + \ldots + f_k$; and $\overline{X}$ is the mean.

2 **EXAMPLE 5** **Finding Standard Deviation for Grouped Data**

Find the standard deviation for the grouped data given in Table 15.

SOLUTION We have already found (see Example 2, p. 496) that the mean for this grouped data is

$$\overline{X} = 81.34$$

TABLE 15

	Class Interval	Frequency
1	50– 59.99	7
2	60– 69.99	9
3	70– 79.99	19
4	80– 89.99	14
5	90– 99.99	14
6	100–109.99	6
7	110–119.99	2

The class midpoints are 55, 65, 75, 85, 95, 105, and 115. The deviations of the mean from the class midpoints, their squares, and the products of the squares by the respective frequencies are listed in Table 16.

TABLE 16

Class Midpoint	f_i	$m_i - \overline{X}$	$(m_i - \overline{X})^2$	$(m_i - \overline{X})^2 \cdot f_i$
55	7	−26.3	691.69	4841.83
65	9	−16.3	265.69	2391.21
75	19	−6.3	39.69	754.11
85	14	3.7	13.69	191.66
95	14	13.7	187.69	2627.66
105	6	23.7	561.69	3370.14
115	2	33.7	1,135.69	2271.38
Sum	71			16,447.99

Using Formula (5), the standard deviation is

$$S = \sqrt{\frac{16,447.99}{70}} = 15.33$$

NOW WORK PROBLEM 7.

A little computation shows that the sum of the deviations of the approximate mean from the class midpoints is not exactly zero. This is due to the fact that we are using an approximation to the mean. Remember, we cannot compute the exact mean for grouped data.

Use Excel to find the standard deviation of the grouped data given in Table 15. (Refer to the Excel example given on page 499).

SOLUTION Use the Excel function STDEV*() and highlight the cells used to compute the standard deviation.

Electric Bills	*Standard Deviation*
52.3	
55.61	15.33

COMMENT: Finding standard deviations by hand is extremely tedious. We have just seen that Excel calculates the standard deviation quickly and accurately. Graphing utilities are also designed to compute the standard deviation of a set of data points. Figure 10 illustrates the standard deviation of the grouped data from Table 15 on the TI-83 Plus. Here the class midpoints are put in L_1 and the class frequencies are put in L_2, as in Figure 10(a). The standard deviation is given by Sx, seen in Figure 10(b). Note that the population standard deviation is also calculated and is given by σx. Consult your user's manual to find the method of calculating standard deviations on your graphing utility.

FIGURE 10

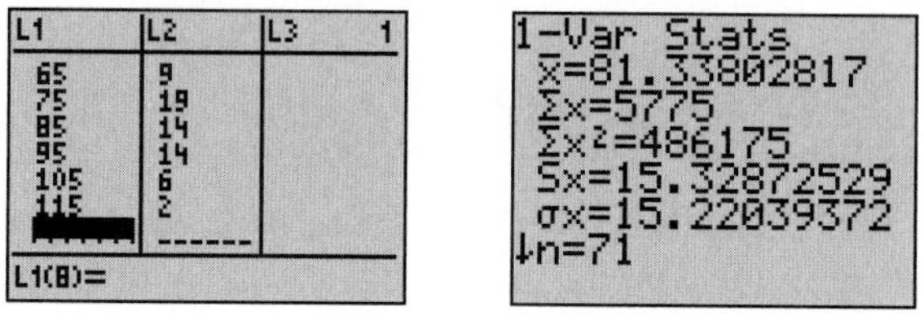

(a) (b)

Chebychev's Theorem

Suppose we are observing an experiment with numerical outcomes and that the experiment has mean population μ and population standard deviation σ. We wish to estimate the probability that a randomly chosen outcome lies within k units of the mean.

Chebychev's Theorem*

For any distribution of numbers with population mean μ and population standard deviation σ, the probability that a randomly chosen outcome lies between $\mu - k$ and $\mu + k$ is at least $1 - \dfrac{\sigma^2}{k^2}$.

3 **EXAMPLE 6** **Using Chebychev's Theorem**

Suppose that an experiment with numerical outcomes has population mean 4 and population standard deviation 1. Use Chebychev's theorem to estimate the probability that an outcome lies between 2 and 6.

SOLUTION Here, $\mu = 4$, $\sigma = 1$. Since we wish to estimate the probability that an outcome lies between 2 and 6, the value of k is $k = 6 - \mu = 6 - 4 = 2$ (or $k = \mu - 2 = 4 - 2 = 2$). Then by Chebychev's theorem, the desired probability is at least

$$1 - \frac{\sigma^2}{k^2} = 1 - \frac{1}{2^2} = 1 - \frac{1}{4} = .75$$

That is, we expect at least 75% of the outcomes of this experiment to lie between 2 and 6.

NOW WORK PROBLEM 21.

*Named after the nineteenth-century Russian mathematician P. L. Chebychev.

EXAMPLE 7 Using Chebychev's Theorem

An office supply company sells boxes containing 100 paper clips. Because of the packaging procedure, not every box contains exactly 100 clips. From previous data it is known that the average number of clips in a box is indeed 100 and the standard deviation is 2.8. If the company ships 10,000 boxes, estimate the number of boxes having between 94 and 106 clips, inclusive.

SOLUTION Our experiment involves counting the number of clips in the box. For this experiment we have $\mu = 100$ and $\sigma = 2.8$. Therefore, by Chebychev's theorem the fraction of boxes having between $100 - 6$ and $100 + 6$ clips ($k = 6$) should be at least

$$1 - \frac{(2.8)^2}{6^2} = 1 - .22 = .78$$

That is, we expect at least 78% of 10,000 boxes, or about 7800 boxes to have between 94 and 106 clips.

The importance of Chebychev's theorem stems from the fact that it applies to *any* data—only the population mean and population standard deviation must be known. However, the estimate is a crude one. Other results (such as the *normal distribution* given later) produce more accurate estimates about the probability of falling within k units of the mean.

EXERCISE 9.5 Answers to Odd-Numbered Problems Begin on Page AN-42.

In Problems 1–6, compute the standard deviation for each set of sample data.

1. 4, 5, 9, 9, 10, 14, 25
2. 6, 8, 10, 10, 11, 12, 18
3. 62, 58, 70, 70
4. 55, 65, 80, 80, 90
5. 85, 75, 62, 78, 100
6. 92, 82, 75, 75, 82

In Problems 7 and 8, calculate the mean and the standard deviation of the sample data below.

7.

Class	Frequency
10–16	1
17–23	3
24–30	10
31–37	12
38–44	5
45–51	2

8.

Class	Frequency
0–3	2
4–7	5
8–11	8
12–15	6
16–19	3

9. **Light Bulb Life** A sample of 6 light bulbs was chosen and their lifetimes measured. The bulbs lasted 968, 893, 769, 845, 922, and 915 hours. Calculate the mean and the standard deviation of the lifetimes of the light bulbs.

10. **Aptitude Scores** A sample of 25 applicants for admission to Midwestern University had the following scores on the quantitative part of an aptitude test:

591	570	425	472	555
490	415	479	517	570
606	614	542	607	441
502	506	603	488	460
550	551	420	590	482

Find the mean and standard deviation of these scores.

11. Baseball Players At the end of 2003 spring training, the ages of the 31 men on the New York Yankees roster were:

24	28	41	32	37	32	31	33
35	30	31	34	32	40	32	27
32	24	32	29	24	24	36	30
38	30	29	32	24	32	35	

Source: Major League Baseball, March 2003.

(a) Find the range of the players ages.
(b) Find the standard deviation, assuming sample data.
(c) Find the standard deviation assuming population data.
(d) Decide whether the data are sample or population. Give reasons.

12. Baseball Player At the end of 2003 spring training, the ages of the 34 men on the 2002 World Series champion Anaheim Angel roster were:

35	27	32	22	24	30	33	30
33	21	29	22	27	28	33	28
33	30	29	25	30	25	27	25
30	26	27	30	30	29	28	32
34	28						

Source: Major League Baseball, March 2003.

(a) Find the range of the players ages.
(b) Find the standard deviation assuming sample data.
(c) Find the standard deviation assuming population data.
(d) Decide whether the data are sample or population. Give reasons.

13. Mother's Age The following data give the number of births in the United States in 2000 by the age of the mother. (The numbers of births are given in thousands.)

Age of Mother	No. of Births
< 15	9
15–19	469
20–24	1018
25–29	1088
30–34	929
35–39	452
40–44	90
45–49	4

Source: U.S. Census Bureau 2003.

(a) Are these sample data or population data? Justify your reasoning.
(b) Find the standard deviation of the mothers' ages.

14. Charge Accounts A department store takes a sample of its customer charge accounts and finds the following:

Outstanding Balance	Number of Accounts
0–49	15
50–99	41
100–149	80
150–199	60
200–249	8

Find the mean and the standard deviation of the outstanding balances.

15. Earthquakes The data below list the number of earthquakes recorded worldwide during 1998 that measured below 8 on the Richter scale.

Magnitude	Earthquakes
0–0.9	2,389
1.0–1.9	752
2.0–2.9	3,851
3.0–3.9	5,639
4.0–4.9	6,943
5.0–5.9	832
6.0–6.9	113
7.0–7.9	10

Source: National Earthquake Information Center.

(a) Are these sample data or population data?
(b) Find the mean magnitude of the earthquakes worldwide in 1998.
(c) Find the standard deviation of the magnitude of the earthquakes recorded in 1998.

16. Earthquakes The data below list the number of earthquakes recorded in the United States in 2001 that measured below 8 on the Richter scale.

Magnitude	Earthquakes
0–0.9	374
1.0–1.9	2
2.0–2.9	604
3.0–3.9	760
4.0–4.9	268
5.0–5.9	37
6.0–6.9	6
7.0–7.9	1

Source: National Earthquake Information Center.

(a) Are these sample data or population data?
(b) Find the mean magnitude of the earthquakes in the United States in 2001.
(c) Find the standard deviation of the magnitude of the earthquakes recorded in the United States in 2001.

17. Licensed Drivers in Tennessee Refer to the data in Problem 9, Exercise 9.3.

(a) Find the standard deviation assuming sample data.
(b) Find the standard deviation assuming population data.
(c) Decide whether the data are sample or population. Give reasons.

18. Licensed Drivers in Hawaii Refer to the data provide in Problem 10, Exercise 9.3.

(a) Find the standard deviation assuming sample data.
(b) Find the standard deviation assuming population data.
(c) Decide whether the data are sample or population. Give reasons.

19. Undergraduate Tuition 1992–1993 Refer to the data provided in Problem 11, Exercise 9.3.

(a) Are these data from a population or from a sample? Justify your answer.
(b) Find the standard deviation of tuition.

20. Undergraduate Tuition 1993–1994 Refer to the data provided in Problem 12, Exercise 9.3.

(a) Are these data from a population or from a sample? Justify your answer.
(b) Find the standard deviation of tuition.

21. Suppose that an experiment with numerical outcomes has mean 25 and standard deviation 3. Use Chebychev's theorem to tell what percent of outcomes lie

(a) Between 19 and 31.
(b) Between 20 and 30.
(c) Between 16 and 34.
(d) Less than 19 or more than 31.
(e) Less than 16 or more than 34.

22. Cost of Meat A survey reveals that the mean price for a pound of beef is \$3.20 with a standard deviation of \$.40. Use Chebychev's theorem to determine the probability that a randomly selected pound of beef costs

(a) Between \$2.80 and \$3.60.
(b) Between \$2.50 and \$3.90.
(c) Between \$2.20 and \$4.20.
(d) Between \$2.40 and \$4.00.
(e) Less than \$2.40 or more than \$4.00.

23. Quality Control A watch company determines that the number of defective watches in each box averages 6 with standard deviation 2. Suppose that 1000 boxes are produced. Estimate the number of boxes having between 0 and 12 defective watches.

24. Sales The average sale at a department store is \$51.25, with a standard deviation of \$8.50. Find the smallest interval such that by Chebychev's Theorem at least 90% of the store's sales fall within it.

25. Annual Births The table gives the number of live births in the United States for 1995 through 2000.

Year	Births
2000	4,051,814
1999	3,959,417
1998	3,941,553
1997	3,880,894
1996	3,891,494
1995	3,899,689

Source: National Center for Health Statistics, 2003.

(a) Are these population or sample data? Justify your reasoning.
(b) Calculate the mean number of births over the six-year period.
(c) Calculate the standard deviation of births over the six-year period.
(d) Are the mean and the standard deviations you calculated exact or are they approximations? Explain your reasoning.
(e) Do the number of births differ greatly over the six-year period? Justify your reasoning.

26. Fishing The number of salmon caught in each of two rivers over the past 15 years is as follows:

River I		River II	
Number Caught	Years	Number Caught	Years
500–1499	4	750–1249	2
1500–2499	8	1350–1799	3
2500–3499	2	1800–2249	4
3500–4499	1	2250–2699	4
		2700–3149	2

(a) Are these population or sample data?
(b) Find the mean and the standard deviation number of fish caught in each river.
(c) Are the mean and the standard deviation calculated in part (b) approximate or exact? Explain.
(d) Using the statistics you calculated, discuss which river should be preferred for fishing.

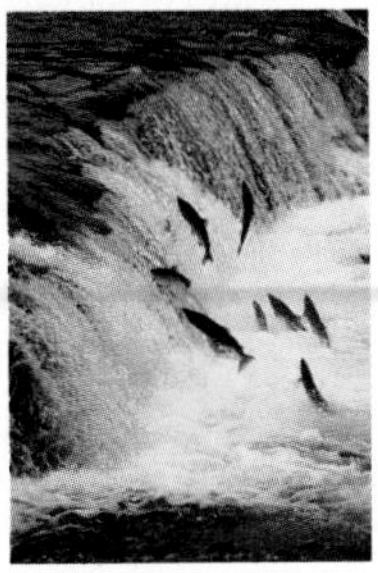

9.6 The Normal Distribution

PREPARING FOR THIS SECTION *Before getting started, review the following:*

> Equally Likely Outcomes (Section 7.1, pp. 370–373)

> Binomial Probabilities (Section 8.2, pp. 432–435)

OBJECTIVES
1. Compute a Z-score
2. Use the standard normal curve
3. Approximate a binomial distribution by the standard normal distribution

Frequency polygons or frequency distributions can assume almost any shape or form, depending on the data. However, the data obtained from many experiments often follow a common pattern. For example, heights of adults, weights of adults, and test scores all lead to data that have the same kind of frequency distribution. This distribution is referred to as the **normal distribution** or the **Gaussian distribution.** Because it occurs so often in practical situations, it is generally regarded as the most important distribution, and much statistical theory is based on it. The graph of the normal distribution, called the **normal curve,** is the bell-shaped curve shown in Figure 11.

FIGURE 11

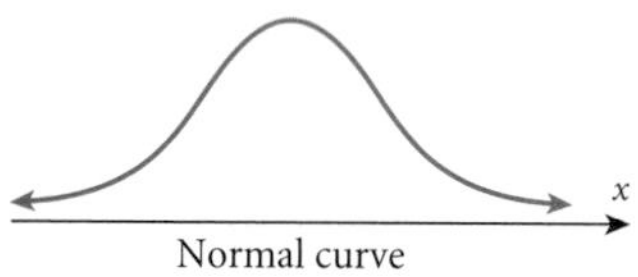

Normal curve

Some properties of the normal distribution are listed below.

Properties of the Normal Distribution

1. Normal curves are bell-shaped and are symmetric with respect to a vertical line at the mean μ. See Figure 12.
2. The mean, median, and mode of a normal distribution are equal.
3. Regardless of the shape, the area enclosed by the normal curve and the x-axis is always equal to 1 square unit. The shaded region in Figure 12 has an area of 1 square unit.
4. The probability that an outcome of a normally distributed experiment is between a and b equals the area under the associated normal curve from $x = a$ to $x = b$. See the shaded region in Figure 13.
5. The standard deviation of a normal distribution plays a major role in describing the area under the normal curve. As shown in Figure 14, the standard deviation is related to the area under the normal curve as follows:
   (a) About 68.27% of the total area under the curve is within 1 standard deviation of the mean (from $\mu - \sigma$ to $\mu + \sigma$).
   (b) About 95.45% of the total area under the curve is within 2 standard deviations of the mean (from $\mu - 2\sigma$ to $\mu + 2\sigma$).
   (c) About 99.73% of the total area under the curve is within 3 standard deviations of the mean (from $\mu - 3\sigma$ to $\mu + 3\sigma$).

FIGURE 12

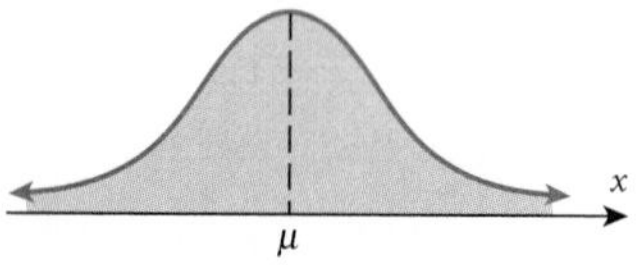

FIGURE 13

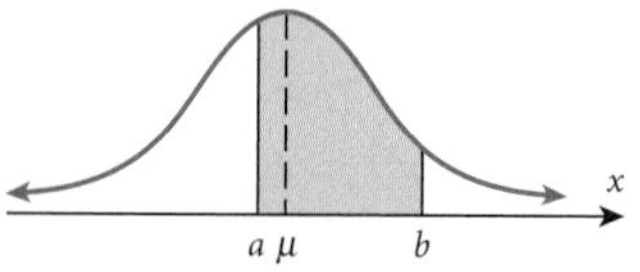

Probability between a and b = area of the shaded region.

It is also worth noting that, in theory, the normal curve will never touch the x-axis but will extend to infinity in either direction.

FIGURE 14

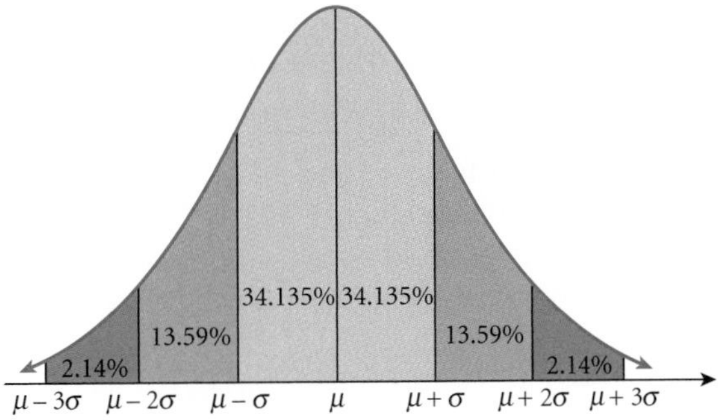

NOW WORK PROBLEM 1.

EXAMPLE 1 Using the Normal Curve to Analyze IQ Scores

At Jefferson High School the average IQ score of the 1200 students is 100, with a standard deviation of 15. The IQ scores have a normal distribution.

(a) How many students have an IQ between 85 and 115?
(b) How many students have an IQ between 70 and 130?
(c) How many students have an IQ between 55 and 145?
(d) How many students have an IQ under 55 or over 145?
(e) How many students have an IQ over 145?

SOLUTION Figure 15 shows a normal distribution with mean = 100 and standard deviation = 15.

FIGURE 15

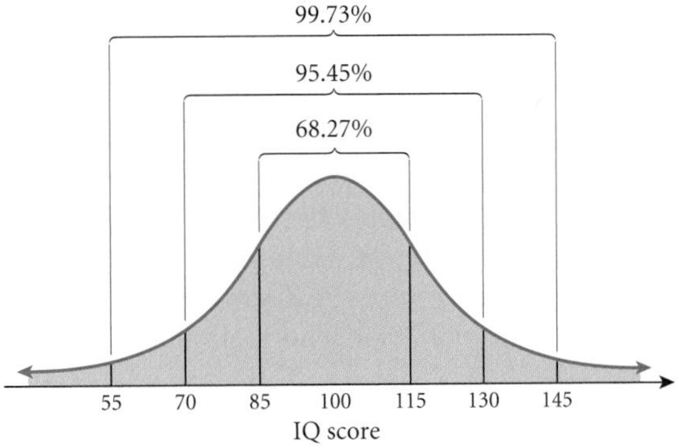

(a) The IQ scores have a normal distribution and the mean is 100. Since the standard deviation σ is 15, then 1σ either side of the mean is from 85 to 115. By Property 5(a) we know that 68.27% of 1200, or

$$(0.6827)(1200) = 819 \text{ students}$$

have IQs between 85 and 115.

(b) The scores from 70 to 130 extend 2σ $(=30)$ either side of the mean. By Property 5(b) we know that 95.45% of 1200, or

$$(0.9545)(1200) = 1145 \text{ students}$$

have IQs between 70 and 130.

(c) The scores from 55 to 145 extend 3σ ($= 45$) either side of the mean. By Property 5(c) we know that 99.73% of 1200, or

$$(0.9973)(1200) = 1197 \text{ students}$$

have IQs between 55 and 145.

(d) There are three students ($1200 - 1197$) who have scores that are not between 55 and 145.

(e) One or two students have IQs above 145.

NOW WORK PROBLEM 15.

A normal distribution is completely determined by the mean μ and the standard deviation σ. Normal distributions of data with different means or different standard deviations give rise to different normal curves.

Figure 16 indicates how the normal curve changes when the standard deviation changes. Each normal curve has the same mean 0.

As the standard deviation increases, the normal curve spreads out [Figure 16(c)], indicating a greater likelihood for the outcomes to be far from the mean. A compressed curve [Figure 16(a)] indicates that the outcomes are more likely to be close to the mean.

FIGURE 16

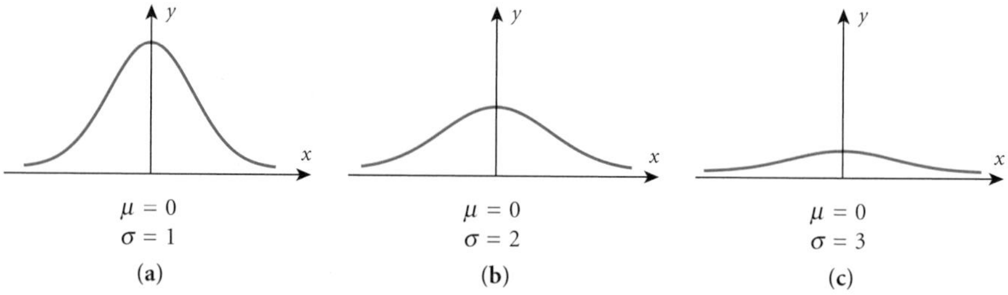

Standard Normal Curve

It would be a hopeless task to attempt to tabulate areas under a normal curve for every conceivable value of μ and σ. Fortunately, we are able to transform all the observations to one table—the table corresponding to the so-called **standard normal curve,** which is the normal curve for which $\mu = 0$ and $\sigma = 1$. This is accomplished by introducing new data, called *Z-scores.* A *Z*-score transforms any normal value with a mean, μ, and a standard deviation, σ, to a normal value with mean 0 and standard deviation 1.

Z-Score

The ***Z*-score** of a data item x is defined as

$$Z = \frac{\textbf{Difference between } x \textbf{ and } \mu}{\textbf{Standard deviation}} = \frac{x - \mu}{\sigma} \qquad (1)$$

where

x = Original data point

μ = Mean of the original data

σ = Standard deviation of the original data

The transformed data obtained using Equation (1) will always have a *zero mean* and a *standard deviation* of one. Such data are said to be expressed in **standard units** or **standard scores.**

A Z-score can be interpreted as the number of standard deviations that the original score is away from its mean. So, by expressing data in terms of standard units, it becomes possible to make a comparison of distributions.

1

EXAMPLE 2 Computing *Z*-Scores

On a test, 80 is the mean and 7 is the standard deviation. What is the Z-score of a score of

(a) 88? **(b)** 62?

Interpret your results. Graph the normal curve.

SOLUTION **(a)** Here, 88 is the original score. Using Equation (1) with $x = 88$, $\mu = 80$, $\sigma = 7$, we get

$$Z = \frac{x - \mu}{\sigma} = \frac{88 - 80}{7} = \frac{8}{7} = 1.1429$$

(b) Here, 62 is the original score. Using Equation (1) with $x = 62$, $\mu = 80$, and $\sigma = 7$, we get

$$Z = \frac{62 - 80}{7} = \frac{-18}{7} = -2.5714$$

The Z-score of 1.1429 tells us that the original score of 88 is 1.1429 standard deviations *above* the mean. The Z-score of -2.5714 tells us that the original score of 62 is 2.5714 standard deviations *below* the mean. A negative Z-score always means that the score is below the mean. See Figure 17 for a graph of the normal curve.

FIGURE 17

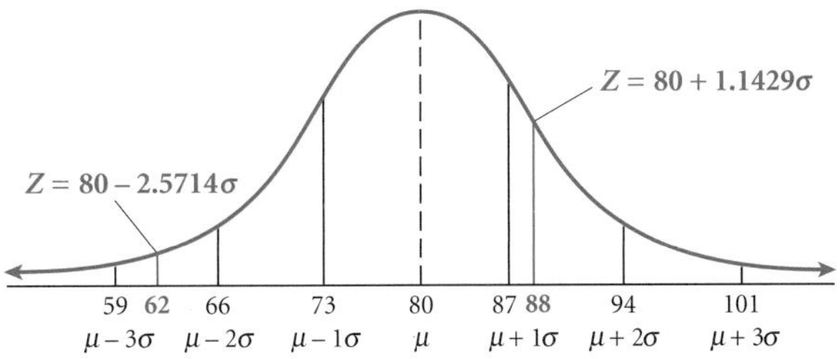

FIGURE 18

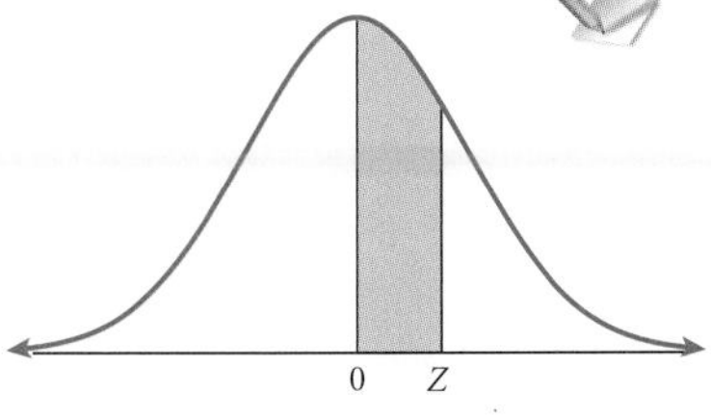

NOW WORK PROBLEM 5.

The curve in Figure 18 with mean $\mu = 0$ and standard deviation $\sigma = 1$ is the standard normal curve. For this curve the areas between $Z = -1$ and 1, $Z = -2$ and 2, $Z = -3$ and 3 are equal, respectively, to 68.27%, 95.45%, and 99.73% of the total area under the curve, which is 1. To find the areas cut off between other points, we proceed as in the following example.

2

EXAMPLE 3 Using the Standard Normal Curve

(a) Find the area, that is, find the proportion of cases, included between 0 and 0.6 on a standard normal curve. Refer to the shaded area in Figure 19(a).

(b) Find the area, that is, find the proportion of cases, included between 0.6 and 1.86 on a standard normal curve. Refer to the shaded area in Figure 19(b).

FIGURE 19

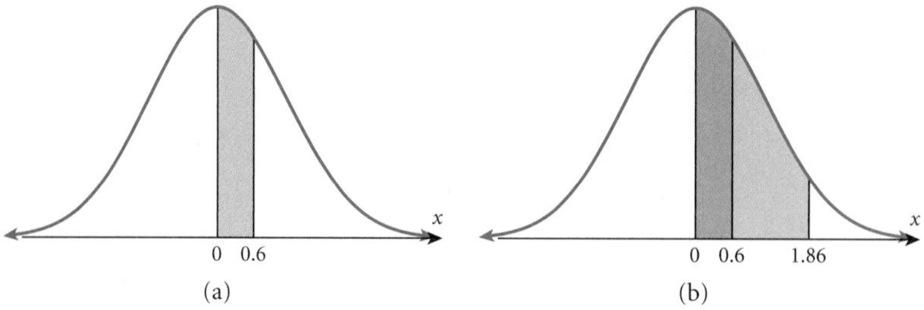

SOLUTION We use the **standard normal curve table,** provided in Table I, which is given on the inside back cover.

(a) To find the area between 0 and 0.6, we find $Z = 0.6$ in the table. Corresponding to $Z = 0.6$ is the value 0.2257, which is the area between the mean 0 and 0.6. In other words, 22.57% of the cases will lie between 0 and 0.6.

(b) We begin by checking the table to find the area of the curve cut off between the mean and a point equivalent to a standard score of 0.6 from the mean. This value is 0.2257, as we found in part (a). Next, we continue down the table in the left-hand column until we come to a standard score of 1.8. By looking across the row to the column below 0.06, we find that 0.4686 of the area is included between the mean and 1.86. Then the area of the curve between these two points is the difference between the two areas, $0.4686 - 0.2257$, which is 0.2429. We can then state that approximately 24.29% of the cases fall between 0.6 and 1.86, or that *the probability of a score falling between these two points is about .2429.*

NOW WORK PROBLEM 7(a).

In the next example we take two points that are on different sides of the mean.

EXAMPLE 4 Using the Standard Normal Curve

We want to determine the area of the standard normal curve that falls between a standard score of -0.39 and one of 1.86.

SOLUTION See Figure 20.

FIGURE 20

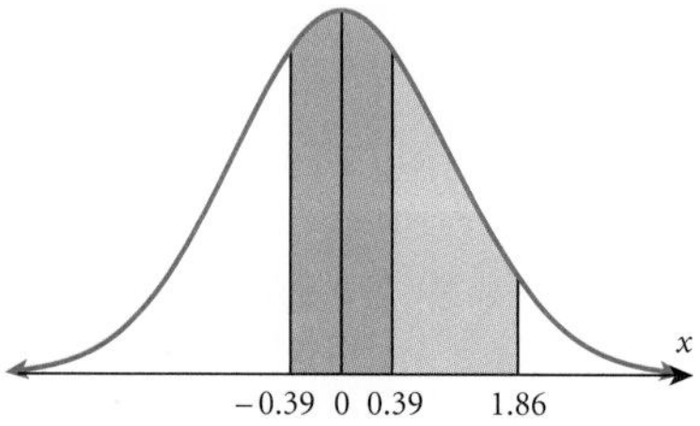

There are no values for negative standard scores in Table I. Because of the symmetry of normal curves, standard scores equal in absolute value, give equal areas when taken from the mean. From Table I we find that a standard score for 0.39 cuts off an area of 0.1517 between it and the mean. A standard score of 1.86 includes 0.4686 of the area of the curve between it and the mean. The area included between -0.39 and 1.86 is then equal to the sum of these two areas, $0.1517 + 0.4686$, which is 0.6203. Approximately 62.03% of the area is between -0.39 and 1.86. In other words, the probability of a score falling between these two points is about .6203.

EXAMPLE 5 Using the Standard Normal Curve

The scores on a test are normally distributed with a mean of 78 and a standard deviation of 7. What is the probability that a test chosen at random has a score between 80 and 90?

SOLUTION To find the probability of obtaining a score between 80 and 90 on the test, we need to find the area under a normal curve from $x_1 = 80$ to $x_2 = 90$.

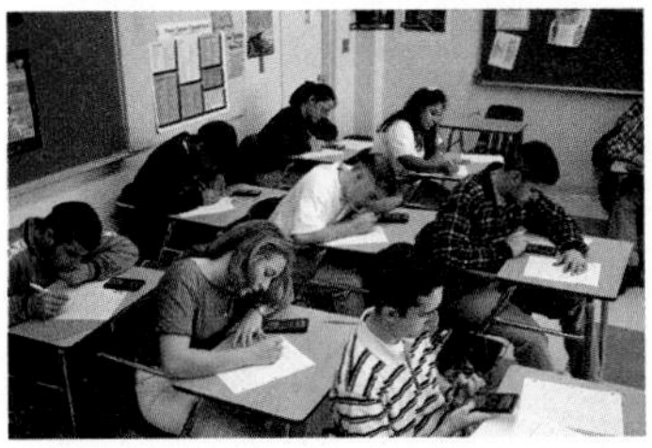

We begin by finding the Z-scores Z_1 of $x_1 = 80$ and Z_2 of $x_2 = 90$.

$$Z_1 = \frac{x_1 - \mu}{\sigma} = \frac{80 - 78}{7} = 0.28$$

$$Z_2 = \frac{x_2 - \mu}{\sigma} = \frac{90 - 78}{7} = 1.71$$

From Table I, the area A_1 from the mean to Z_1 and the area A_2 from the mean to Z_2 are $A_1 = 0.1103$ and $A_2 = 0.4564$.

See Figure 21. The area between Z_1 and Z_2 is the difference between A_2 and A_1.

$$A_2 - A_1 = 0.4564 - 0.1103 = 0.3461$$

So, the probability of obtaining a test score between 80 and 90 is .3461, or 34.61%.

FIGURE 21

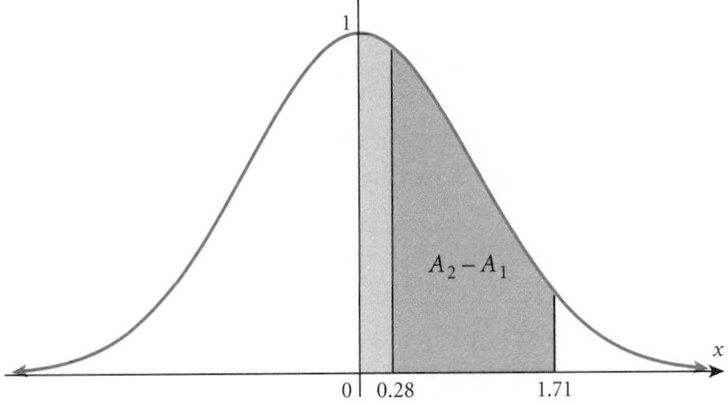

COMMENT: A graphing utility can be used to find the area under a portion of a normal curve. The graphing utility uses the actual distribution mean and standard deviation and computes the area beneath the normal curve over the interval of interest all in

FIGURE 22

one step. See Figure 22. The information needed for computing the solution is $x_1 = 80$, $x_2 = 90$, $\mu = 78$, and $\sigma = 7$. The result, normal cdf(80, 90, 78, 7) = .344 indicates that for the normal curve with mean 78 and standard deviation 7, 34.4% of the area beneath the curve is between $x_1 = 80$, and $x_2 = 90$, so the probability of obtaining a test score between 80 and 90 on the test is .344 or 34.4%.

EXAMPLE 6 Comparing Exam Scores

A student receives a grade of 82 on a final examination in biology for which the mean is 73 and the standard deviation is 9. In his final examination in sociology, for which the mean grade is 81 and the standard deviation is 15, he receives an 89. In which examination is his relative standing higher?

SOLUTION In their present forms these distributions are not comparable since they have different means and, more important, different standard deviations. In order to compare the data, we transform the data to standard scores. For the biology test data the Z-score for the student's examination score of 82 is

$$Z = \frac{82 - 73}{9} = \frac{9}{9} = 1$$

For the sociology test data, the Z-score for the student's examination score of 89 is

$$Z = \frac{89 - 81}{15} = \frac{8}{15} = 0.533$$

This means the student's score in the biology exam is 1 standard unit above the mean, while his score in the sociology exam is 0.533 standard unit above the mean. So the student's **relative standing** is higher in biology.

NOW WORK PROBLEM 23.

The Normal Curve as an Approximation to the Binomial Distribution

Approximate a binomial distribution by the standard normal distribution 3

We start with an example.

EXAMPLE 7 Finding the Frequency Distribution for a Binomial Probability

Consider an experiment in which a fair coin is tossed 10 times. Find the frequency distribution for the probability of tossing a head.

SOLUTION The probability for obtaining exactly k heads is given by a binomial distribution $b(10, k; \frac{1}{2})$. The distribution is given in Table 17. If we graph this frequency distribution, we obtain the line chart shown in Figure 23. When we connect the tops of the lines of the line chart, we obtain a *normal curve,* as shown.

This particular distribution for $n = 10$ and $p = \frac{1}{2}$ is not a result of the choice of n or p. As a matter of fact, the line chart for any binomial probability $b(n, k; p)$ will give an approximation to a normal curve. You should verify this for the cases in which $n = 15$, $p = .3$, and $n = 8, p = \frac{3}{4}$.

TABLE 17

No. of Heads	Probability $b(10, k; \frac{1}{2})$
0	.0010
1	.0098
2	.0439
3	.1172
4	.2051
5	.2461
6	.2051
7	.1172
8	.0439
9	.0098
10	.0010

FIGURE 22

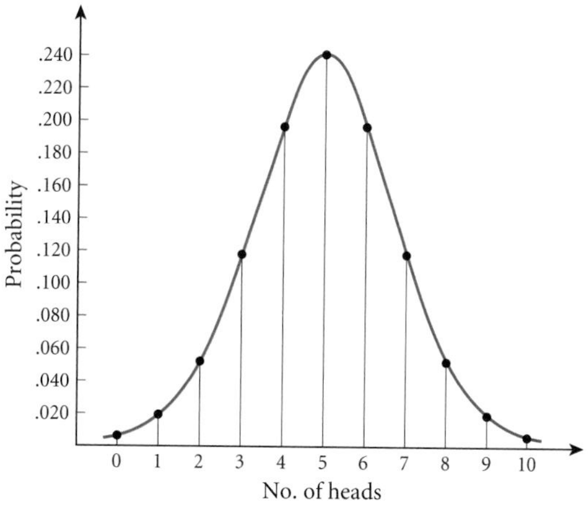

Probabilities associated with binomial experiments are readily obtainable from the formula $b(n, k; p)$ when n is small. If n is large, we can compute the binomial probabilities by an approximating procedure using a normal curve. It turns out that the normal distribution provides a very good approximation to the binomial distribution when n is large or p is close to $\frac{1}{2}$.

The mean μ for the binomial distribution is given by $\mu = np$ (see Expected Value for Bernoulli Trials). It can be shown that the standard deviation is $\sigma = \sqrt{npq}$, where $q = 1 - p$.

EXAMPLE 8 Quality Control

A company manufactures 60,000 pencils each day. Quality control studies have shown that, on the average, 4% of the pencils are defective. A random sample of 500 pencils is selected from each day's production and tested. What is the probability that in the sample there are

(a) At least 12 and no more than 24 defective pencils?
(b) 32 or more defective pencils?

SOLUTION **(a)** Since $n = 500$ is very large, it is appropriate to use a normal curve approximation for the binomial distribution. With $n = 500$ and $p = .04$, we have

$$\mu = np = 500(.04) = 20 \qquad \sigma = \sqrt{npq} = \sqrt{500(.04)(.96)} = 4.38$$

To find the approximate probability that the number of defective pencils in a sample is at least 12 and no more than 24, we find the area under a normal curve from $x = 12$ to $x = 24$. See Figure 24.

FIGURE 24

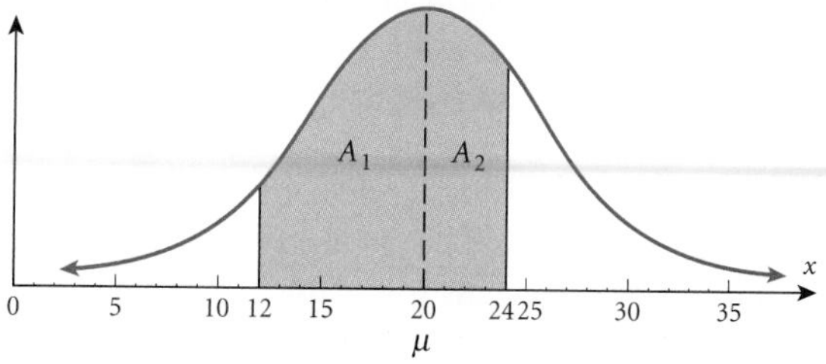

Areas A_1 and A_2 are found by converting to Z-scores and using the standard normal curve table, Table I.

$$x = 12: \quad Z_1 = \frac{x - \mu}{\sigma} = \frac{12 - 20}{4.38} = -1.83 \qquad A_1 = 0.4664$$

$$x = 24: \quad Z_2 = \frac{x - \mu}{\sigma} = \frac{24 - 20}{4.38} = 0.91 \qquad A_2 = 0.3186$$

$$\text{Total area} = A_1 + A_2 = 0.4664 + 0.3186 = 0.785$$

The approximate probability of the number of defective pencils in the sample being at least 12 and no more than 24 is .785.

(b) We want to find the area A_2 indicated in Figure 25. We know that the area to the right of the mean is 0.5, and if we subtract the area A_1 from 0.5, we will obtain A_2. First, we find the area A_1:

$$Z = \frac{x - \mu}{\sigma} = \frac{32 - 20}{4.38} = 2.74 \qquad A_1 = 0.4969$$

Then

$$A_2 = 0.5 - A_1 = 0.5 - 0.4969 = 0.0031$$

The approximate probability of finding 32 or more defective pencils in the sample is 0.0031.

FIGURE 25

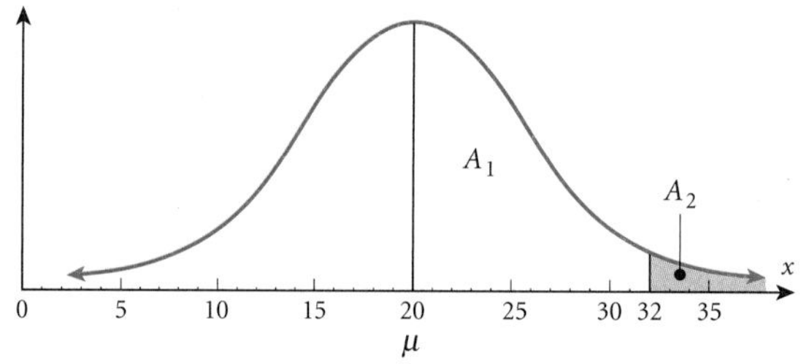

NOW WORK PROBLEM 35.

COMMENT: Graphing utilities can calculate exact values for the binomial distribution even if the number of Bernoulli trials, n, is large. This makes the normal approximation less important than it had been in the past. Check the user's manual for your graphing utility to learn how to compute the cumulative binomial distribution. Then redo Example 8 with exact values, and compare the exact solutions to the approximations obtained in Example 8.

The results of Example 8, using a TI-83 Plus graphing calculator, are shown in Figures 26 and 27. When calculating the probability that the number of defective pencils is between 12 and 24, we find the difference between the probability that the number of defective pencils is less than or equal to 24 and the probability that the number of defective pencils is less than or equal to 11. See Figure 26.

FIGURE 26

FIGURE 27

To find the probability that there are at least 32 defective pencils, we use one minus the probability of the complement. That is, we calculate one minus the probability there are 31 or fewer defective pencils, as illustrated in Figure 27.

The solutions obtained with the graphing utility differ from the normal approximations in Example 8 because the approximation uses the continuous variables of a normal distribution in place of the discrete variables of the binomial distribution.

Some statisticians use a correction factor in the normal approximation to improve the approximated results.

EXERCISE 9.6 Answers to Odd-Numbered Problems Begin on Page AN-42.

In Problems 1–4, determine μ and σ by inspection for each normal curve.

1.
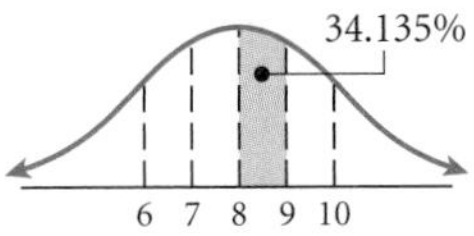

2.
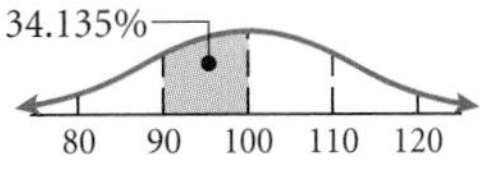

3.
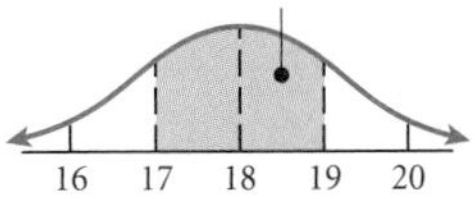

4.
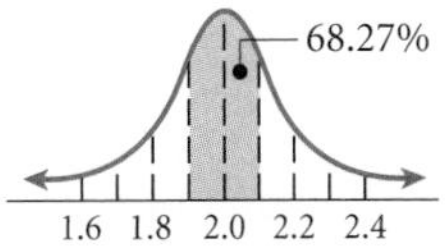

5. Given a normal distribution with a mean of 13.1 and a standard deviation of 9.3, find the Z-score equivalent of the following scores in this distribution:

(a) $x = 7$ (b) $x = 9$ (c) $x = 13$
(d) $x = 29$ (e) $x = 37$ (f) $x = 41$

6. Given a normal distribution with a mean of 15.2 and a standard deviation of 5.1, find the Z-score equivalent of the following scores in this distribution:

(a) $x = 8$ (b) $x = 9$ (c) $x = 16$
(d) $x = 22$ (e) $x = 23$ (f) $x = 25$

7. Given the following Z-scores on a standard normal distribution, find the area under the standard normal curve between each score and the mean.

(a) $Z = 0.89$ (b) $Z = 1.10$ (c) $Z = 3.06$
(d) $Z = -1.22$ (e) $Z = 2.30$ (f) $Z = -0.75$

8. Given the following Z-scores on a standard normal distribution, find the area under the standard normal curve between each score and the mean.

(a) $Z = 1.85$ (b) $Z = 1.15$ (c) $Z = -1.60$
(d) $Z = 2.50$ (e) $Z = -2.31$ (f) $Z = 0.25$

9. **Assigning Grades** An instructor assigns grades in an examination according to the following procedure:

A if score exceeds $\mu + 1.6\sigma$
B if score is between $\mu + 0.6\sigma$ and $\mu + 1.6\sigma$
C if score is between $\mu - 0.3\sigma$ and $\mu + 0.6\sigma$
D if score is between $\mu - 1.4\sigma$ and $\mu - 0.3\sigma$
F if score is below $\mu - 1.4\sigma$

What percent of the class receives each grade, assuming that the scores are normally distributed?

10. **Assigning Grades** Professor Morgan uses a normal distribution to assign grades in his Finite Mathematics class. He assigns an A to students scoring more than 2 standard deviations above the mean and an F to students scoring more than two standard deviations below the mean. He assigns a B to students who score between 1.2 and 2 standard deviations above the mean and a D to students who score between 1.6 and 2 standard deviations below the mean. All other students get a C. What percent of the class receive each grade assuming the scores are normally distributed?

In Problems 11–14, use Table I, the Standard Normal Curve Table, to find the area of each shaded region under the standard normal curve.

11.
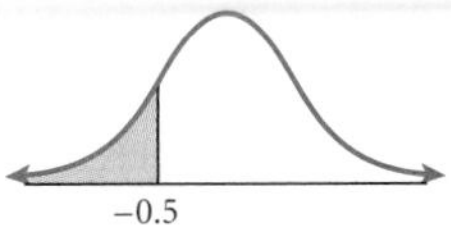

12.
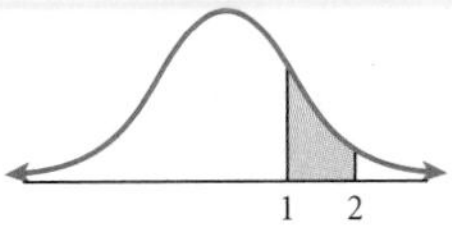

13.
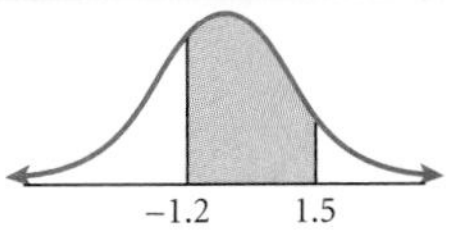

14.
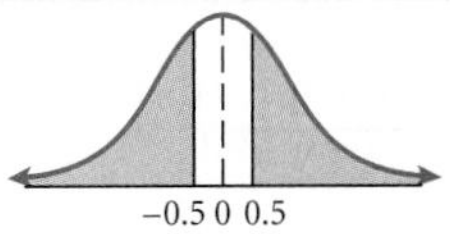

15. **Women's Heights** The average height of 2000 women in a random sample is 64 inches. The standard deviation is 2 inches. The heights have a normal distribution.

(a) How many women in the sample are between 62 and 66 inches tall?
(b) How many women in the sample are between 60 and 68 inches tall?
(c) How many women in the sample are between 58 and 70 inches tall?
(d) How many women in the sample are more than 70 inches tall?
(e) How many women in the sample are shorter than 58 inches?

16. **Weight of Corn Flakes in a Box** Corn flakes come in a box that says it holds a mean weight of 16 ounces of cereal. The standard deviation is 0.1 ounce. Suppose that the manufacturer packages 600,000 boxes with weights that have a normal distribution.

(a) How many boxes weigh between 15.9 and 16.1 ounces?
(b) How many boxes weigh between 15.8 and 16.2 ounces?
(c) How many boxes weigh between 15.7 and 16.3 ounces?
(d) How many boxes weigh under 15.7 or over 16.3 ounces?
(e) How many boxes weigh under 15.7 ounces?

17. **Student Weights** The weight of 100 college students closely follows a normal distribution with a mean of 130 pounds and a standard deviation of 5.2 pounds.

(a) How many of these students would you expect to weigh at least 142 pounds?
(b) What range of weights would you expect to include the middle 70% of the students in this group?

18. **Time to Do Taxes** The Internal Revenue Service claims it takes an average of 6 hours to complete a 1040 tax form. Assuming the time to complete the form is normally distributed with a standard deviation of 30 minutes:

(a) What percent of people would you expect to complete the form in less than 5 hours?
(b) What time interval would you expect to include the middle 50% of the tax filers?

Source: 2002 Form 1040 Instructions.

19. **Life Expectancy of Clothing** If the average life of a certain make of clothing is 40 months with a standard deviation of 7 months, what percentage of these clothes can be expected to last from 28 months to 42 months? Assume that clothing lifetime follows a normal distribution.

20. **Life Expectancy of Shoes** Records show that the average life expectancy of a pair of shoes is 2.2 years with a standard deviation of 1.7 years. A manufacturer guarantees that shoes lasting less than a year are replaced free. For every 1000 pairs sold, how many pairs should the manufacturer expect to replace free? Assume a normal distribution.

21. **Movie Theater Attendance** The attendance over a weekly period of time at a movie theater is normally distributed with a mean of 10,000 and a standard deviation of 1000 persons. Find

(a) The number in the lowest 70% of the attendance figures.
(b) The percent of attendance figures that falls between 8500 and 11,000 persons.
(c) The percent of attendance figures that differs from the mean by 1500 persons or more.

22. **Test Scores** Scores on an aptitude test are normally distributed with a mean of 980 and a standard deviation of 110. If 10,000 students take the test,

(a) How many would you expect to score between 900 and 1200?
(b) How many would you expect to score above 1400?
(c) How many would you expect to score below 750?

23. **Comparing Test Scores** Colleen, Mary, and Kathleen are vying for a position as editor. Colleen, who is tested with group I, gets a score of 76 on her test; Mary, who is tested with group II, gets a score of 89; and Kathleen, who is tested with group III, gets a score of 21. If the average score for group I is 82, for group II is 93, and for group III is 24, and if the standard deviation for each group is 7, 2, and 9, respectively, which person has the highest relative standing?

24. In Mathematics 135 the average final grade is 75.0 and the standard deviation is 10.0. The professor's grade distribution shows that 15 students with grades from 68.0 to 82.0 received Cs. Assuming the grades follow a normal distribution, how many students are in Mathematics 135?

25. (a) Draw the line chart and frequency curve for the probability of a head in an experiment in which a biased coin is tossed 15 times and the probability that a head occurs is .3. [*Hint:* Find $b(15, k; .30)$ for $k = 0, 1, \ldots, 15$.]
(b) Discuss the shape of the curve.
(c) What are the mean and the standard deviation of the distribution?

26. Follow the same directions as in Problem 25 for an experiment in which a biased coin is tossed 25 times and the probability that a head appears is $\frac{3}{4}$.

In Problems 27–32, suppose a binomial experiment consists of 750 trials and the probability of success for each trial is .4. Then

$$\mu = np = 300 \qquad and \qquad \sigma = \sqrt{npq} = \sqrt{(750)(.4)(.6)} = 13.4$$

Approximate the probability of obtaining the number of successes indicated by using a normal curve.

27. 285–315

28. 280–320

29. 300 or more

30. 300 or less

31. 325 or more

32. 275 or less

33. Lifetime Batting Averages A baseball player has a lifetime batting average of .250. If, in a season, this player comes to bat 300 times, what is the probability that at least 80 and no more than 90 hits occur? What is the probability that 85 or more hits occur?

34. Hitting a Target A skeet shooter has a long-established probability of hitting a target of .75. If, in a particular session, 200 attempts are made, what is the probability that at least 135 and no more than 160 are successful? What is the probability that 160 or more are successful?

35. Quality Control A company manufactures 100,000 packages of jelly beans each week. On average, 1% of the packages do not seal properly. A random sample of 500 packages is selected at the end of the week. What is the probability that in this sample at least 10 are not properly sealed?

36. Quality Control A company manufactures 200,000 pairs of pantyhose per week. On average 1% of the pairs are defective. A random sample of 300 pairs are selected each week. What is the probability that more than 5 pairs in the sample are defective?

37. Graph the standard normal curve using a graphing utility. For what value of x does the function assume its maximum? The equation is given by

$$y = \frac{1}{\sqrt{2\pi}} e^{-(1/2)x^2}$$

38. Graph the normal curve with $\mu = 10$ and $\sigma = 2$ using a graphing utility. For what value of x does the function assume its maximum? The equation is given by

$$y = \frac{1}{2\sqrt{2\pi}} e^{-(1/8)(x-10)^2}$$

39. Quality Control Refer to Problem 35. Use a graphing utility to compute the exact probability that the sample contains at least 10 bags of improperly sealed jelly beans.

40. Quality Control Refer to Problem 36. Use a graphing utility to compute the exact probability that the sample contains at least 5 pairs that are defective?

41. Quality Control Refer to Problem 35. Suppose that each week a random sample of 500 packages of jelly beans are selected and every week there are at least 12 improperly sealed bags of jelly beans. What would you conclude about the statistics given in Problem 35?

42. Quality Control Refer to Problem 36. Suppose that each week a random sample of 300 pairs of pantyhose are selected and every week there are at least 8 defective pairs in the sample. What would you conclude about the statistics given in Problem 36?

Chapter 9 | Review

OBJECTIVES

IMPORTANT FORMULAS

Mean for Sample Data (p. 494)

$$\overline{X} = \frac{x_1 + x_2 + \cdots + x_n}{n}$$

Mean for Population Data (p. 502)

$$\mu = \frac{x_1 + x_2 + \cdots + x_N}{N}$$

Mean for Grouped Data (p. 497)

$$\overline{X} = \frac{\Sigma f_i m_i}{n}$$

Standard Deviation for Sample Data (p. 506)

$$S = \sqrt{\frac{(x_1 - \overline{X})^2 + (x_2 - \overline{X})^2 + \cdots + (x_n - \overline{X})^2}{n - 1}}$$

$$= \sqrt{\frac{\Sigma(x_i - \overline{X})^2}{n - 1}}$$

Standard Deviation for Population Data (p. 506)

$$\sigma = \sqrt{\frac{(x_1 - \mu)^2 + (x_2 - \mu)^2 + \cdots + (x_N - \mu)^2}{N}}$$

$$= \sqrt{\frac{\Sigma(x_i - \mu)^2}{N}}$$

Standard Deviation for Grouped Scores (p. 508)

$$S = \sqrt{\frac{(m_1 - \overline{X})^2 \cdot f_1 + (m_2 - \overline{X})^2 \cdot f_2 + \cdots + (m_k - \overline{X})^2 \cdot f_k}{n - 1}} = \sqrt{\frac{\Sigma[(m_i - \overline{X})^2 \cdot f_i]}{n - 1}}$$

Z-Score (p. 516) $Z = \dfrac{x - \mu}{\sigma}$

TRUE–FALSE ITEMS **Answers are on page AN-43.**

T F 1. The range of a set of numbers is the difference between the standard deviation and the mean.

T F 2. Two sets of scores can have the same mean and median, yet be different.

T F 3. A relatively small standard deviation indicates that measures are widely scattered from the mean.

T F 4. The sum of the deviations from the mean is zero for ungrouped data.

T F 5. For the normal distribution, approximately 68.27% of the total area under the curve is within 2 standard deviations of the mean.

FILL IN THE BLANKS **Answers are on page AN-43.**

1. The three most common measures of central tendency are (a) ________ (b) ________ (c) ________.

2. The square root of the variance is called ________.

3. The graph of the normal distribution has a ________ shape.

4. The formula $\frac{x - \mu}{\sigma}$ is called the ________ of x.

5. The formula $1 - \frac{\sigma^2}{k^2}$ measures the probability that a randomly chosen variable lies between ________ and ________.

REVIEW EXERCISES **Answers to odd-numbered problems begin on page AN-43.**
Blue problem numbers indicate the author's suggestions for a practice test.

In Problems 1–6, determine the variable in the experiment and state whether it is continuous or discrete.

1. A municipal hospital epidemiologist measures the circumference of each newborn baby's head.

2. An immigration officer asks each alien how long he/she expects to stay in the country.

3. A pollster counts how many people are in favor of a new law.

4. A statistician counts the number of hits a major league baseball player gets in a season.

5. A quality control manager lists the number of defective products manufactured each day.

6. A human resources clerk records the days an employee calls in sick.

7. **Opinion Poll** The student activities president wants to poll students for their opinions regarding a band to play on campus. Describe how she can choose a simple random sample of 100 students to poll.

8. **Retirement Plans** An employee benefits officer wants to poll employees of a bank regarding their retirement plans. How can he choose a simple random sample of 50 employees?

9. **Opinion Poll** An urban planner wants to poll the citizens of East Norwich regarding their opinions about a proposed road construction project that would interfere with rush hour traffic for the next 18 months. He asks 200 people as they leave the neighborhood supermarket between 10:00 A.M. and 3:00 P.M. on Thursday. What is wrong with the urban planner's sample?

10. **Theater Attendance** A researcher wants to determine how often the citizens of Muttontown attend the theater. She interviews 50 people at the local senior citizens' center. What is wrong with this researcher's sample?

11. The following data show how office workers in Chicago get to work:

Means of Transportation	Percentage
Ride alone	64
Car pool	5
Ride bus	30
Other	1

(a) Construct a bar graph of the data.
(b) Construct a pie chart of the data.
(c) Which of the graphs seems more informative to you? Why?

12. Getting to Work According to the 1990 Census, Americans traveled to work in the following ways. The numbers are given in thousands.

Means of Transportation	Number
Drove	99,592
Public transportation	5,852
Walked	4,489
Other	1,730

Source: U.S. Census Bureau, 2002.

(a) Construct a bar graph of the data.
(b) Construct a pie chart of the data.
(c) Which of the graphs seems more informative to you? Why?

13. To study their attitudes toward a new product, 1000 people were interviewed. Their response is given in the following table:

Attitude	No. of Responses
Do not like	420
Like	360
Like very much	220

(a) Construct a bar graph of the data.
(b) Construct a pie chart of the data.
(c) Which of the graphs seems more informative to you? Why?

14. Vacation Travel An Internet travel agency surveyed 1000 of its clients to determine their mode of transportation for their last vacation. The agency found that 635 people drove, 280 flew, and 85 cruised to their destination.

(a) Construct a bar graph of the data.
(b) Construct a pie chart of the data.
(c) Which of the graphs seems more informative to you? Why?

15. College Enrollment The pie chart represents the race or ethnicity of students enrolled in four-year colleges in the United States in the fall of 1997. A total of 8,897,000 students were enrolled.

Enrollment in Four-Year Colleges by Race/Ethnicity 1997

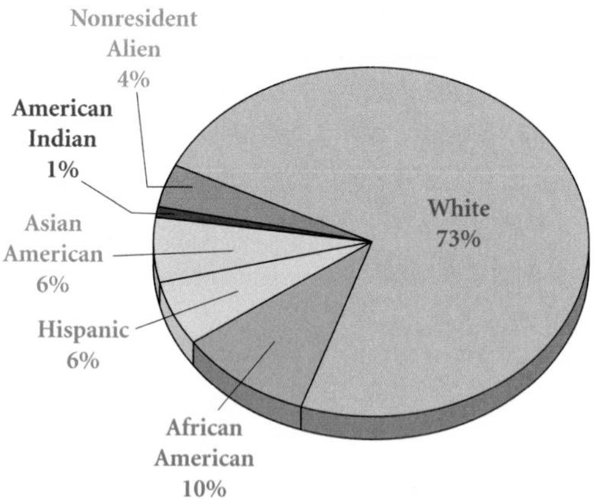

(a) What group made up the smallest percentage of four-year college enrollment?
(b) If Asian-Americans represent less than 4% of the total population in the United States, were they underrepresented or overrepresented in four-year colleges in 1997?
(c) Approximately how many Hispanics were enrolled in four-year colleges in 1997?

Source: U.S. Department of Education, National Center for Education Statistics, 1999.

16. In 1998 Mark McGwire hit 62 home runs, breaking Roger Maris' major league record. The bar graph shows where he hit each home run ball.

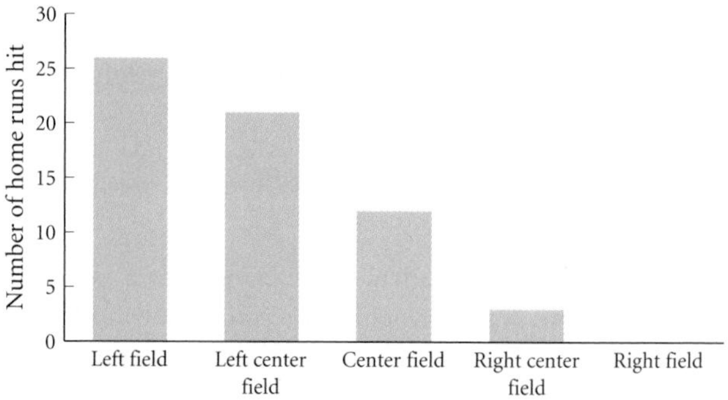

(a) In what direction were the fewest of his home runs hit?
(b) What percent of the home runs were hit to left field?
(c) Where were almost 20% of the home runs hit?
(d) If you were a fielding coach for the opposing team and Mark McGwire was at bat, what advice would you give to your outfielders?

17. **Educational Attainment** The bar graph depicts the highest level of education attained by people in the United States in 2000.

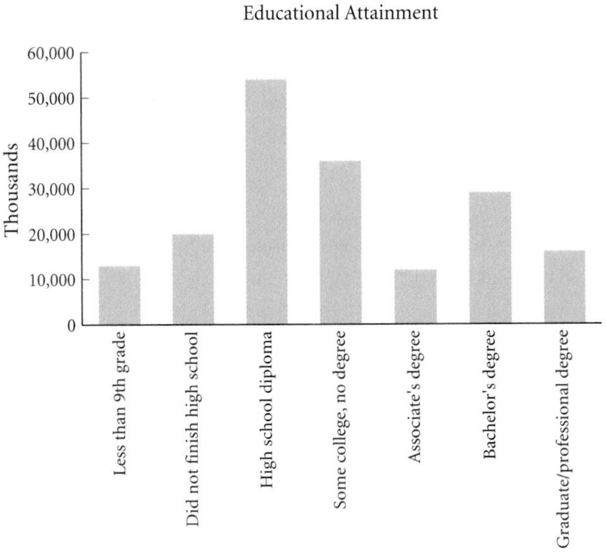

(a) What is the highest level of education that most Americans have obtained?
(b) How many Americans have at least a bachelor's degree?
(c) How many people in the United States do not have a high school diploma?
(d) How many people have gone to college but do not have a bachelor's degree?

Source: U.S. Census Bureau, 2003.

18. **College Enrollment** The pie chart represents the race or ethnicity of students enrolled in two-year colleges in the United States in the fall of 1997. A total of 5,606,000 students were enrolled.

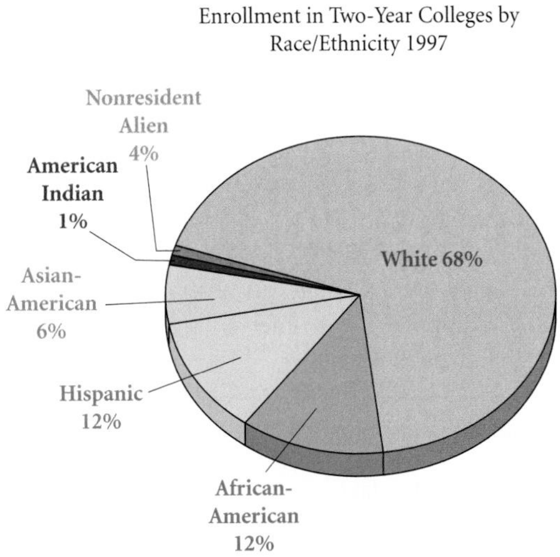

(a) What group represents the largest portion of the student population?
(b) Approximately how many American Indians were enrolled in a two-year college in 1997?
(c) Approximately how many African-Americans were enrolled in a two-year college in 1997?

Source: U.S. Department of Education, National Center for Education Statistics, 1999.

19. **Test Scores** The following scores were made on a math exam:

80	99	82	21	100	55	80	62	78	52
21	73	80	44	72	63	91	85	33	66
78	42	87	90	90	70	48	75	83	77
63	85	69	80	41	87	66	52	71	60
74	70	73	95	89	41	92	68	100	72

(a) Set up a frequency table for the above data. What is the range?
(b) Draw a line chart for the data.
(c) Build a histogram for the data using a class interval of size 10 beginning with the interval 20.0–29.9.
(d) Draw a frequency polygon on the histogram.
(e) Find the cumulative frequencies.
(f) Draw the cumulative frequency distribution.

20. **Test Scores** Use the data from Problem 19:

(a) Group the data into class intervals of size 5, starting with the interval 6.0–10.9.
(b) Build a histogram for the data using a class interval of size 5.
(c) Draw a frequency polygon on the histogram.
(d) Find the cumulative frequencies.
(e) Draw the cumulative frequency distribution.

21. **Track Championship** At the high school track championships, 42 boys participated in the 1600 meter race. Their times (rounded to the nearest second) were:

4 min 30 sec		4 min 12 sec		4 min 46 sec		4 min 15 sec		4 min 46 sec		5 min 50 sec	
4	30	4	22	4	39	4	56	4	50	5	01
5	08	5	12	5	20	5	20	5	31	5	18
5	06	4	40	4	52	4	36	5	02	5	06
5	12	5	31	5	37	5	40	5	55	5	43
5	48	5	40	6	01	6	32	6	12	6	10
6	40	6	02	7	05	7	15	6	30	5	20

(a) Set up a frequency table for the above data. What is the range?
(b) Draw a line chart for the data.
(c) Group the data into class intervals of size 30 seconds, starting with the interval 4 minutes 0 seconds–4 minutes 29 seconds.
(d) Build the histogram for the data.
(e) Draw a frequency polygon on the histogram.
(f) Find the cumulative frequencies.
(g) Draw the cumulative frequency distribution.

22. Track Championship Use the data in Problem 21.

(a) Group the data into class intervals of size 1 minute, starting with the interval 4 minutes 0 seconds–4 minutes 59 seconds.
(b) Build the histogram for the data.
(c) Draw a frequency polygon on the histogram.
(d) Find the cumulative frequencies.
(e) Draw the cumulative frequency distribution.

23. Baseball Players At the end of 2003 spring training, the age of the 31 men on the New York Yankees roster were:

24 28 41 32 37 32 31 33 35 30 31 34
32 40 32 27 32 24 32 29 24 24 36 30
38 30 29 32 24 32 35

(a) Make a frequency table of the players' ages.
(b) Construct a line chart of the data.
(c) Group the data into class intervals of size 5, beginning with the interval 20.0–24.9. How many class intervals are there?
(d) Build the histogram for the data.
(e) Draw the frequency polygon on the histogram.
(f) Find the cumulative frequencies.
(g) Draw the cumulative frequency distribution.

Source: Major League Baseball, March 2003.

24. Baseball Players At the end of 2003 spring training, the ages of the 34 men on the 2002 World Series Champion Anaheim Angels roster were:

35 27 32 22 24 30 33 30 33 21 29 22
27 28 33 28 33 30 29 25 30 25 27 25
30 26 27 30 30 29 28 32 34 28

(a) Make a frequency table of the players ages.
(b) Construct a line chart of the data.
(c) Group the data into class intervals of size 5, beginning with the interval 20.0–24.9. How many class intervals are there?
(d) Build the histogram for the data.
(e) Draw the frequency polygon on the histogram.
(f) Find the cumulative frequencies.
(g) Draw the cumulative frequency distribution.

Source: Major League Baseball, March 2003.

25. **Tax Rates** The following table gives the percentage of marginal tax rates for married couples filing jointly for taxable income for 2002.

Income	Marginal Rates in Percent
0–12,000	10
12,000–46,700	15
46,700–112,850	27
112,850–171,950	30
171,950–307,050	35
Over 307,050	38.6

(a) Graph the data using a histogram.
(b) Draw a frequency polygon on the histogram.

Source: Internal Revenue Service, 2002.

26. **Tax Rates** The following table gives the percentage of marginal tax rates for single individuals for their taxable income in 2002.
(a) Graph the data using a histogram.
(b) Draw a frequency polygon on the histogram.

Source: Internal Revenue Service, 2002.

Income	Marginal Rates in Percent
0–6,000	10
6,000–27,950	15
27,950–67,500	27
67,500–141,250	30
141,250–307,050	35
Over 307,050	38.6

In Problems 27–32, find (a) the mean, (b) the median, (c) the mode, if it exists, (d) the range, and (e) the standard deviation of each sample of data.

27. 12, 10, 8, 2, 0, 4, 10, 5, 4, 8, 0

28. 195, 5, 2, 2, 2, 2, 1, 0

29. 2, 5, 5, 7, 7, 7, 9, 9, 11, 100

30. −5, 2, 3, 8, −6, 9, 11, 10, −2, 8, −4, 0

31. 5, 7, 7, 9, 10, 11, 1, 6, 2, 12

32. 12, 10, 8, 10, 6, 10, 14

33. In which of the sets of data in Problems 27–32 is the mean a poor measure of central tendency? Why?

34. Give an example of two sets of scores for which the means are the same and the standard deviations are different.

35. Give one advantage of the mean over the median. Give an example.

36. Give an advantage of using the standard deviation over the variance.

37. In seven different rounds of golf, Joe scores 74, 72, 76, 81, 77, 76, and 73.
(a) Do you think these are sample data or population data? Explain your answer.
(b) What is Joe's mean golf score?
(c) Find the standard deviation of Joe's golf scores.

38. The 10 fish caught on Tuesday weighed 16.2 pounds, 15 pounds, 12.3 pounds, 20 pounds, 8 pounds, 6.5 pounds, 8 pounds, 10.8 pounds, 12 pounds, and 9 pounds.
(a) Do you think these are sample data or population data? Explain your answer.
(b) What is the mean weight of the fish?
(c) Find the standard deviation of the fishes' weights.

39. **Population Distribution** The U.S. Census Bureau gives information about the age distribution of persons living in the United States. The following data give the number (in thousands) of females for various ages in 2000.

Age	Population	Age	Population
1–4	9,263	50–54	9,049
5–9	9,611	55–59	6,992
10–14	9,765	60–64	5,670
15–19	9,668	65–69	5,087
20–24	9,162	70–74	4,972
25–29	8,855	75–79	4,316
30–34	9,890	80–84	3,072
35–39	11,087	85–89	1,834
40–44	11,473	90–94	871
45–49	10,202	95–99	286
		100+	56

(a) Approximate the mean age of a female in 2000.
(b) Approximate the median age of a female in 2000.
(c) Approximate the standard deviation of the age of the females.

Source: U.S. Bureau of the Census, 2003.

40. Population Distribution The U.S. Census Bureau gives information about the age distribution of persons living in the United States. The following data give the number (in thousands) of males for various ages in the year 2000.

Age	Population
1–4	9,682
5–9	10,072
10–14	10,252
15–19	10,226
20–24	9,531
25–29	8,769
30–34	9,674
35–39	10,956
40–44	11,296
45–49	9,856

Age	Population
50–54	8,577
55–59	6,461
60–64	5,087
65–69	4,330
70–74	3,886
75–79	3,109
80–84	1,896
85–89	900
90–94	326
95–99	83
100+	12

(a) Approximate the mean age of a male in 2000.
(b) Approximate the median age of a male in 2000.
(c) Approximate the standard deviation of males' ages.

Source: U.S Bureau of the Census, 2003.

In Problems 41–44, use Chebychev's theorem to find the probability.

41. **Jelly Jars** A machine fills jars with 12 ounces of jam. The machine is not exact and jars have a standard deviation of 0.05 ounce. In a lot of 1000 jars, how many jars would you expect to have between 11.9 and 12.1 ounces of jam?

42. **Machining** A die cast stamps out metal washers with a mean outer diameter of 2.5 cm. and a variance of 0.01 cm. In a shipment of 10,000 washers, how many would you expect to have an outer diameter between 2.3 and 2.7 cm.?

43. **Sacks of Potatoes** Potatoes are packed in 10 pound bags. Each bag has a mean weight of 10 pounds with a standard deviation of 0.25 pound. What is the probability that a bag picked at random weighs less than 9.5 pounds or more than 10.5 pounds?

44. **Shrimp by the Pound** Shrimp are priced by size; the larger the shrimp, the more expensive the cost. Jumbo shrimp have 9 shrimp to the pound with a standard deviation of 0.75 shrimp. What is the probability that a pound of jumbo shrimp contains fewer than 8 or more than 10 shrimp?

In Problems 45–50, compute the Z-score for x, using the given population mean and standard deviation.

45. $\mu = 10, \sigma = 3, x = 8$

46. $\mu = 14, \sigma = 3, x = 20$

47. $\mu = 1, \sigma = 5, x = 8$

48. $\mu = 140, \sigma = 15, x = 125$

49. $\mu = 55, \sigma = 3, x = 60$

50. $\mu = 32, \sigma = 13, x = 10$

In Problems 51–54, calculate the area under the normal curve.

51. Between $Z = -1.35$ and $Z = -2.75$

52. Between $Z = 1.2$ and $Z = 1.75$

53. Between $Z = -0.75$ and $Z = 2.1$

54. Between $Z = -1.5$ and $Z = 0.34$

55. A normal distribution has a mean of 25 and a standard deviation of 5.

(a) What proportion of the scores fall between 20 and 30?
(b) What proportion of the scores will lie above 35?

56. A set of 600 scores is normally distributed. How many scores would you expect to find:

(a) Between $\pm\ 1\sigma$ of the mean?
(b) Between 1σ and 3σ above the mean?
(c) Between $\pm\ \frac{2}{3}\sigma$ of the mean?

57. **Average Life of a Dog** The average life expectancy of a dog is 14 years, with a standard deviation of about 1.25 years. Assuming that the life spans of dogs are normally distributed, approximately how many dogs will die before reaching the age of 10 years, 4 months?

58. **Life of a Battery** The average life span of a phone battery is 20 months with a standard deviation of 1.5 months. Assuming the battery life is normally distributed, what is the probability that a battery chosen at random will last more than 24 months.

59. **Comparing Test Scores** Bob got an 89 on the final exam in mathematics and a 79 on the sociology exam. In the mathematics class the average grade was 79 with a standard deviation of 5, and in the sociology class the average grade was 72 with a standard deviation of 3.5. Assuming that the grades in both subjects were normally distributed, in which class did Bob rank higher?

60. **Comparing Test Scores** Marty took 3 tests in science this semester. His grades were 58, 79, and 83. If the mean score and standard deviation of the tests were 50 and 7, 73 and 4, and 75 and 8, respectively, on which test did Marty have the highest standing?

61. Suppose it is known that the number of items produced in a factory has a mean of 40. If the variance of a week's production is known to equal 25, then what can be said about the probability that this week's production will be between 30 and 50?

62. From past experience a teacher knows that the test scores of students taking an examination have a mean of 75 and a variance of 25. What can be said about the probability that a student will score between 65 and 85?

63. **Blood Tests** A blood test for an antigen gives a positive result with a probability of 0.7. In testing 200 blood samples, what is the probability of obtaining more than 160 positive results?

64. **Color-Blindness** The probability a male is color-blind is 0.30. In a sample of 1000 males, what is the probability that fewer than 200 are color-blind?

65. **Toll Gates** An automatic gate malfunctions, failing to open, with a probability 0.05. What is the probability that in 500 operations, the gate fails to work between 20 and 30 times?

66. **Licensing Tests** A professional licensing test has a 60% pass rate. What is the probability that, in a group of 200 randomly selected test takers, between 110 and 125 pass the test?

Chapter 9 Project

"AVERAGE" WEATHER

The daily weather forecast that appears in a local newspaper or on television often contains information regarding the average high and low temperatures, air quality, and other information. This information helps a reader or viewer to make a comparison between historical and actual temperatures, rainfall, humidity level, pollen, air pollutants, and so on. These figures are the result of the application of statistics to a large collection of data. In this example, we compare the actual weather to the reported averages.

Part of what makes travel comfortable and enjoyable is good weather. Suppose that you want to visit New York City, which is notorious for extreme weather. To avoid the cold of winter and the extreme heat of summer, you decide to visit in June. According to *USA Today*, the average high temperature is 80°F, the average low temperature is 63°F, and the average rainfall is 3.6 inches. The highest temperature recorded was 101°F and the lowest temperature recorded was 44°F. However, you are curious about the deviation from these averages and discover the data shown in Table 1 for June 2002.

TABLE 1

Day	High (°F)	Low (°F)	Precipitation (inches)	Day	High (°F)	Low (°F)	Precipitation (inches)
1	85	64	0	16	76	59	0.24
2	78	62	0	17	77	59	0.02
3	74	56	0	18	78	58	0
4	72	60	0	19	79	62	0.04
5	85	65	0	20	81	63	0
6	78	58	1.69	21	81	65	0
7	68	56	0.72	22	87	67	0
8	71	53	0	23	87	70	0
9	83	56	0	24	88	75	0
10	80	65	0	25	83	74	0
11	89	64	0	26	92	75	0.16
12	85	60	0.21	27	89	70	0.7
13	66	57	0	28	84	70	0
14	63	56	0.63	29	85	69	0
15	63	55	0.09	30	87	70	0

Sources: USA Today; National Oceanic and Atmospheric Administration.

We shall investigate how June 2002 compared to an average month.

1. Calculate the mean high temperature, the mean low temperature, and the total rainfall.

2. The *mean temperature* for the month is defined as the mean of the daily temperatures (the midpoint of the high and low temperatures). What is the mean temperature for this month?

3. Show that we can calculate the mean monthly temperature by calculating the mean of the mean high temperature and the mean low temperature.

4. What are the median high and the median low temperatures for the month? How do they compare with the corresponding mean temperatures? Interpret the differences.

5. Is there a modal high temperature? If so, what is it? Is there a modal low temperature? If so, what is it?

6. Compare the mean, median, and modal high temperatures. Do the same for the low temperatures. What do you conclude?

7. Calculate the standard deviation of the high and low temperatures. A normal distribution is symmetric around its mean, so the mean, the median, and the mode are all equal. What kind of errors might one expect if one were to use the normal distribution to estimate the probability of obtaining a high temperature between 90°F and 92°F degrees on a day in June based on the 2002 data? [*Hint*: Try a geometric argument. That is, use a graph.]

8. Make a frequency table for the daily high temperatures and one for the daily low temperatures. Then group each set of data into class intervals of equal degree width. Use 51°–55°F as the first interval.

9. Build a histogram for each set of grouped data and draw a frequency polygon on each histogram.

10. Examine the histograms and determine the modes of the high and low temperature frequency distributions. If the earlier calculations indicated that the month was close to the average, what do the modes tell you?

11. Find the approximate mean high temperature for June 2002 and the approximate standard deviation using the grouped data. Are they good approximations of the actual mean high temperature and its standard deviation? Explain your reasoning.

12. Repeat Problem 11 for the grouped low temperatures.

MATHEMATICAL QUESTIONS FROM PROFESSIONAL EXAMS*

1. **Actuary Exam—Part II** Under the hypothesis that a pair of dice are fair, the probability is approximately .95 that the number of 7s appearing in 180 throws of the dice will lie within $30 \pm K$. What is the value of K?

   (a) 2 (b) 4 (c) 6 (d) 8 (e) 10

2. **Actuary Exam—Part II** If X is normally distributed with mean μ and variance μ^2 and if $P(-4 < X < 8) = .9974$, then $\mu =$

   (a) 1 (b) 2 (c) 4 (d) 6 (e) 8

3. **Actuary Exam—Part II** A manufacturer makes golf balls whose weights average 1.62 ounces, with a standard deviation of 0.05 ounce. What is the probability that the weight of a group of 100 balls will lie in the interval 162 ± 0.5 ounces?

   (a) .18 (b) .34 (c) .68 (d) .84 (e) .96

CHAPTER 10

Functions and Their Graphs

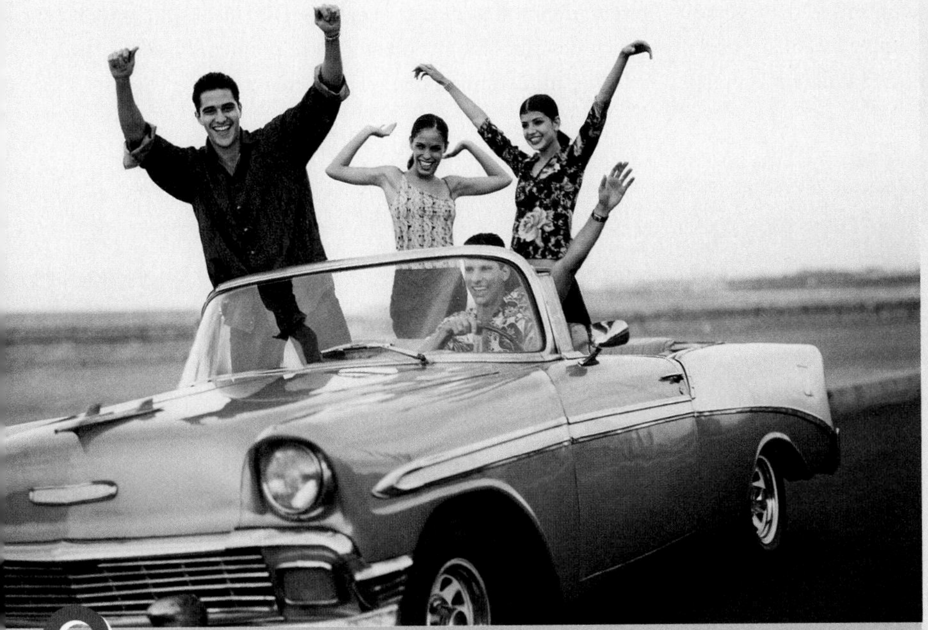

OUTLINE

On the way home from college, you and a friend decide to stop off in Charlotte, North Carolina. Because you have only one full day to see the sights, you decide that renting a car is the best way to see the most. But which car rental company should you use? Naturally, the cheapest! But what is the cheapest? Is it the one with unlimited mileage or the one with a better daily rate and a mileage charge? The mathematics of this chapter provides the background for solving this problem. The Chapter Project at the end of the chapter will help you understand how to make the best decision.

A LOOK BACK, A LOOK FORWARD

In Chapter 1 we studied rectangular coordinates and used them to graph linear equations of the form $Ax + By = C$, where either $A \neq 0$ or $B \neq 0$. We continue the study of graphing equations in two variables here. In particular, we look at a special type of equation involving two variables, called a *function*. We will define what a function is, how to graph functions, what properties functions have, and develop a "library" of functions.

The word function apparently was introduced by René Descartes in 1637. For him, a function simply meant any positive integral power of a variable x. Gottfried Wilhelm Leibniz (1646–1716), who always emphasized the geometric side of mathematics, used the word function to denote any quantity associated with a curve, such as the coordinates of a point on the curve. Leonhard Euler (1707–1783) employed the word to mean any equation or formula involving variables and constants. His idea of a function is similar to the one most often seen in courses that precede calculus. Later, the use of functions in investigating heat flow equations led to a very broad definition, due to Lejeune Dirichlet (1805–1859), which describes a function as a correspondence between two sets. It is his definition that we use here.

10.1 Graphs of Equations

PREPARING FOR THIS SECTION *Before getting started, review the following:*

- Evaluating Algebraic Expressions (Appendix A, Section A.2, pp. A-18–A-19)
- Rectangular coordinates (Chapter 1, Section 1.1, pp. 2–3)
- Solving equations (Appendix A, Section A.6, pp. A-49–A-58)
- Lines (Chapter 1, Section 1.1, pp. 3–15)

OBJECTIVES
1. Graph equations by plotting points
2. Find intercepts from a graph
3. Find intercepts from an equation
4. Test an equation for symmetry with respect to the (a) x-axis, (b) y-axis, and (c) origin

An **equation in two variables,** say x and y, is a statement in which two expressions involving x and y are equal. The expressions are called the **sides** of the equation. Since an equation is a statement, it may be true or false, depending on the value of the variables. Any values of x and y that result in a true statement are said to **satisfy** the equation.

For example, the following are all equations in two variables x and y:

$$x^2 + y^2 = 5 \qquad 2x^2 - y = 6 \qquad y = 2x + 5 \qquad x^2 = y^3$$

The first of these, $x^2 + y^2 = 5$, is satisfied for $x = 1$, $y = 2$, since $1^2 + 2^2 = 1 + 4 = 5$. Other choices of x and y also satisfy this equation. It is not satisfied for $x = 2$ and $y = 3$, since $2^2 + 3^2 = 4 + 9 = 13 \neq 5$.

The **graph of an equation** in two variables x and y consists of the set of points in the xy-plane whose coordinates (x, y) satisfy the equation.

For example, as we learned in Chapter 1, the graph of any equation of the form $Ax + By = C$, where either $A \neq 0$ or $B \neq 0$, is a line.

Graphs play an important role in helping us to visualize the relationships that exist between two variable quantities. Figure 1 shows the monthly closing prices of Intel stock from August 31, 2002, through August 31, 2003. For example, the closing price on May 31, 2003, was about \$21 per share.

FIGURE 1 Monthly closing prices of Intel stock, 8/31/02 to 8/31/03.

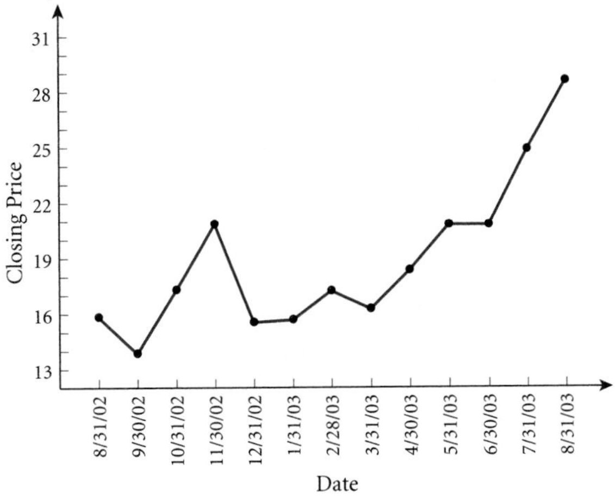

EXAMPLE 1 Determining Whether a Point Is on the Graph of an Equation

Determine if the following points are on the graph of the equation $2x - y = 6$.

(a) $(2, 3)$ **(b)** $(2, -2)$

SOLUTION **(a)** For the point $(2, 3)$, we check to see if $x = 2, y = 3$ satisfy the equation $2x - y = 6$.

$$2x - y = 2(2) - 3 = 4 - 3 = 1 \neq 6$$

The equation is not satisfied, so the point $(2, 3)$ is not on the graph.

(b) For the point $(2, -2)$, we have

$$2x - y = 2(2) - (-2) = 4 + 2 = 6$$

The equation is satisfied, so the point $(2, -2)$ is on the graph.

NOW WORK PROBLEM 23.

EXAMPLE 2 Graphing an Equation by Plotting Points

Graph the equation: $y = x^2$

SOLUTION Table 1 on page 540 provides several points on the graph. In Figure 2 we plot these points and connect them with a smooth curve to obtain the graph (a *parabola*).

The graph of the equation shown in Figure 2 does not show all points. For example, the point $(5, 25)$ is a part of the graph of $y = x^2$, but it is not shown. Since the graph of $y = x^2$ could be extended out as far as we please, we use arrows to indicate that the pattern shown continues. It is important when illustrating a graph to present enough of the graph so that any viewer of the illustration will "see" the rest of it as an obvious continuation of what is actually there. This is referred to as a **complete graph.**

One way to obtain a complete graph of an equation is to plot a sufficient number of points on the graph until a pattern becomes evident. Then these points are connected

TABLE 1

x	$y = x^2$	(x, y)
−4	16	(−4, 16)
−3	9	(−3, 9)
−2	4	(−2, 4)
−1	1	(−1, 1)
0	0	(0, 0)
1	1	(1, 1)
2	4	(2, 4)
3	9	(3, 9)
4	16	(4, 16)

FIGURE 2

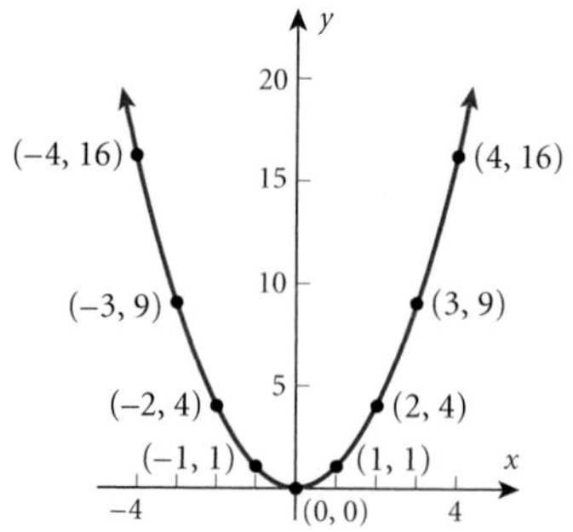

with a smooth curve following the suggested pattern. But how many points are sufficient? Sometimes knowledge about the equation tells us. For example, we learned in Chapter 1 that if an equation is of the form $y = mx + b$, then its graph is a line. In this case, only two points are needed to obtain the graph.

One purpose of this book is to investigate the properties of equations in order to decide whether a graph is complete. At first we shall graph equations by plotting a sufficient number of points. Shortly, we shall investigate various techniques that will enable us to graph an equation without plotting so many points. Other times we shall graph equations based solely on properties of the equation.

COMMENT: Another way to obtain the graph of an equation is to use a graphing utility. Read Section C.2, *Using a Graphing Utility to Graph Equations,* in Appendix C.

EXAMPLE 3 Graphing an Equation by Plotting Points

Graph the equation: $y = x^3$

SOLUTION We set up Table 2, listing several points on the graph. Figure 3 illustrates some of these points and the graph of $y = x^3$.

TABLE 2

x	$y = x^3$	(x, y)
−3	−27	(−3, −27)
−2	−8	(−2, −8)
−1	−1	(−1, −1)
0	0	(0, 0)
1	1	(1, 1)
2	8	(2, 8)
3	27	(3, 27)

FIGURE 3

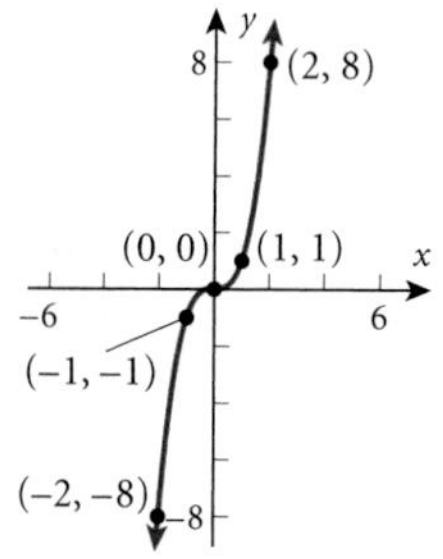

EXAMPLE 4 Graphing an Equation by Plotting Points

Graph the equation: $x = y^2$

SOLUTION We set up Table 3, listing several points on the graph. In this case, because of the form of the equation, we assign some numbers to y and find corresponding values of x. Figure 4 illustrates some of these points and the graph of $x = y^2$.

TABLE 3

y	$x = y^2$	(x, y)
−3	9	(9, −3)
−2	4	(4, −2)
−1	1	(1, −1)
0	0	(0, 0)
1	1	(1, 1)
2	4	(4, 2)
3	9	(9, 3)
4	16	(16, 4)

FIGURE 4

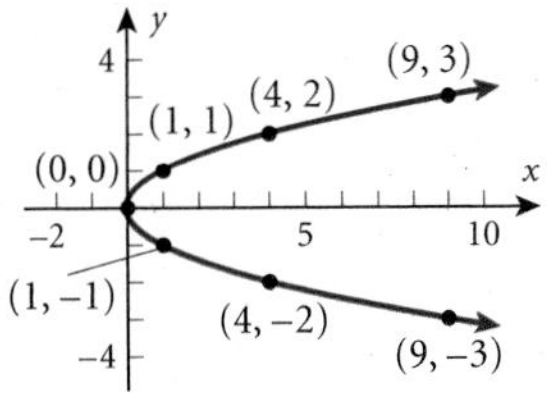

If we restrict y so that $y \geq 0$, the equation $x = y^2, y \geq 0$, may be written equivalently as $y = \sqrt{x}$. The portion of the graph of $x = y^2$ in quadrant I is therefore the graph of $y = \sqrt{x}$. See Figure 5.

FIGURE 5

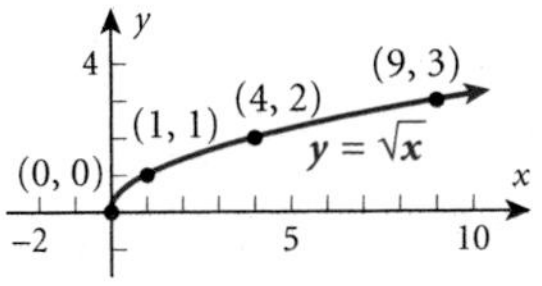

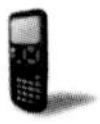

COMMENT: To see the graph of the equation $x = y^2$ on a graphing calculator, you will need to graph two equations: $Y_1 = \sqrt{x}$ and $Y_2 = -\sqrt{x}$. See Figure 6. We discuss why a little later in this chapter.

FIGURE 6

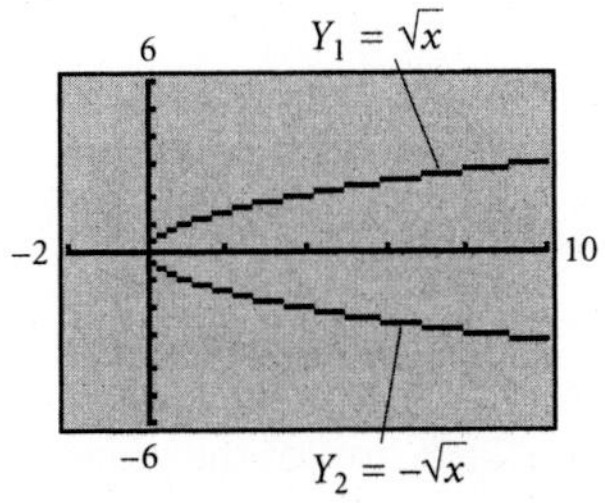

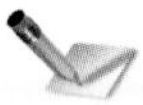

NOW WORK PROBLEM 45.

We said earlier that we would discuss techniques that reduce the number of points required to graph an equation. Two such techniques involve finding *intercepts* and checking for *symmetry*.

Find intercepts from a graph **2**

Intercepts

The points, if any, at which a graph crosses or touches the coordinate axes are called the **intercepts.** See Figure 7. The point at which the graph crosses or touches the x-axis is an **x-intercept,** and the point at which the graph crosses or touches the y-axis is a **y-intercept.**

FIGURE 7

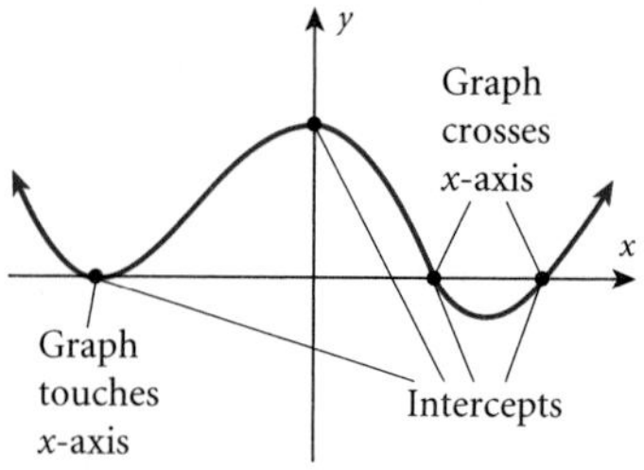

EXAMPLE 5 Finding Intercepts from a Graph

FIGURE 8

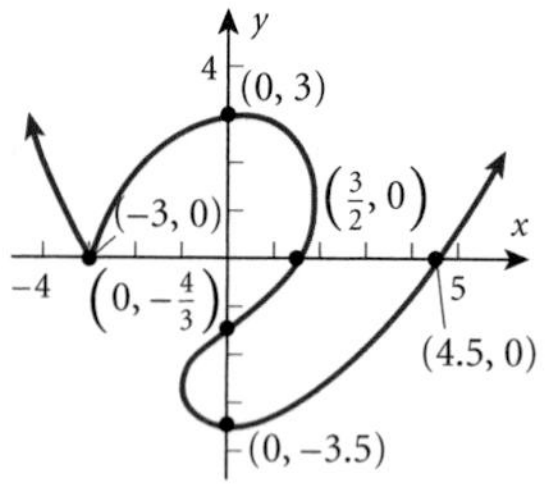

Find the intercepts of the graph in Figure 8. What are its x-intercepts? What are its y-intercepts?

SOLUTION The intercepts of the graph are the points

$$(-3, 0), \quad (0, 3), \quad \left(\frac{3}{2}, 0\right), \quad \left(0, -\frac{4}{3}\right), \quad (0, -3.5), \quad (4.5, 0)$$

The x-intercepts are $(-3, 0)$, $\left(\frac{3}{2}, 0\right)$, and $(4.5, 0)$. The y-intercepts are $(0, 3)$, $\left(0, -\frac{4}{3}\right)$, and $(0, -3.5)$.

NOW WORK PROBLEM 11(a).

Find intercepts from an equation **3**

The intercepts of the graph of an equation can be found from the equation by using the fact that points on the x-axis have y-coordinates equal to 0, and points on the y-axis have x-coordinates equal to 0.

Procedure for Finding Intercepts

1. To find the x-intercept(s), if any, of the graph of an equation, let $y = 0$ in the equation and solve for x.
2. To find the y-intercept(s), if any, of the graph of an equation, let $x = 0$ in the equation and solve for y.

Because the x-intercepts of the graph of an equation are those x-values for which $y = 0$, they are also called the **zeros** (or **roots**) of the equation.

EXAMPLE 6 Finding Intercepts from an Equation

Find the x-intercept(s) and the y-intercept(s), if any, of the graph of $y = x^2 - 4$. Graph $y = x^2 - 4$.

SOLUTION To find the x-intercept(s), we let $y = 0$ and obtain the equation

$$x^2 - 4 = 0$$

$$(x + 2)(x - 2) = 0 \quad \text{Factor.}$$

$$x + 2 = 0 \quad \text{or} \quad x - 2 = 0 \quad \text{Zero-Product Property.}$$

$$x = -2 \quad \text{or} \quad x = 2$$

The equation has the solution set $\{-2, 2\}$. The x-intercepts are $(-2, 0)$ and $(2, 0)$.

To find the y-intercept(s), we let $x = 0$ and obtain the equation

$$y = -4$$

The y-intercept is $(0, -4)$.

Since $x^2 \geq 0$ for all x, we deduce from the equation $y = x^2 - 4$ that $y \geq -4$ for all x. This information, the intercepts, and the points from Table 4, enable us to graph $y = x^2 - 4$. See Figure 9.

TABLE 4

x	$y = x^2 - 4$	(x, y)
−3	5	(−3, 5)
−1	−3	(−1, −3)
1	−3	(1, −3)
3	5	(3, 5)

FIGURE 9

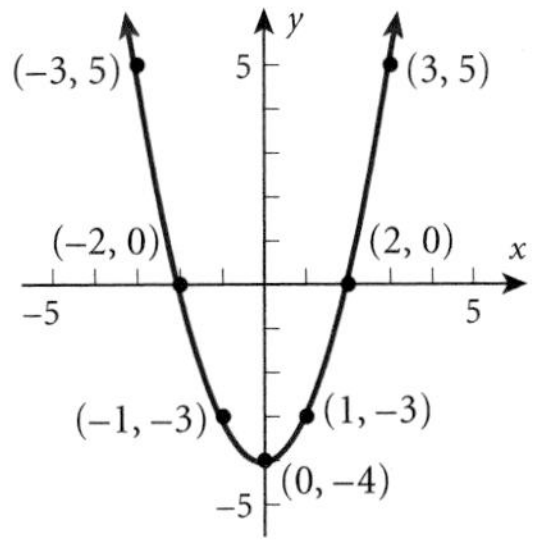

NOW WORK PROBLEM 33 (List the Intercepts).

COMMENT: For many equations, finding intercepts may not be so easy. In such cases, a graphing utility can be used. Read Section C.5, *Using a Graphing Utility to Locate Intercepts and Check for Symmetry* in Appendix C, to find out how a graphing utility locates intercepts.

Symmetry

We have just seen the role that intercepts play in obtaining key points on the graph of an equation. Another helpful tool for graphing equations involves *symmetry,* particularly symmetry with respect to the x-axis, the y-axis, and the origin.

> A graph is said to be **symmetric with respect to the x-axis** if, for every point (x, y) on the graph, the point $(x, -y)$ is also on the graph.

FIGURE 10 Symmetry with respect to the x-axis

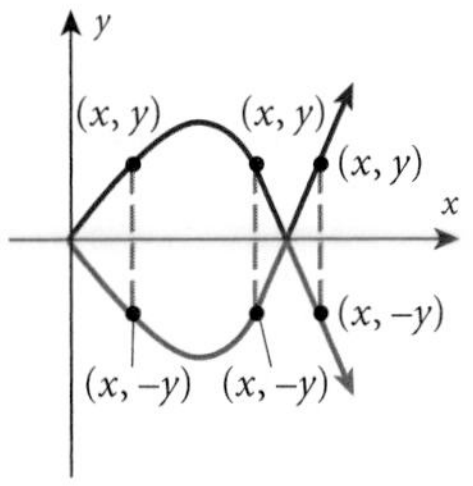

Figure 10 illustrates the definition. Notice that, when a graph is symmetric with respect to the x-axis, the part of the graph above the x-axis is a reflection or mirror image of the part below it, and vice versa.

EXAMPLE 7 **Points Symmetric with Respect to the *x*-Axis**

If a graph is symmetric with respect to the *x*-axis and the point (3, 2) is on the graph, then the point (3, −2) is also on the graph.

A graph is said to be **symmetric with respect to the *y*-axis** if, for every point (x, y) on the graph, the point $(-x, y)$ is also on the graph.

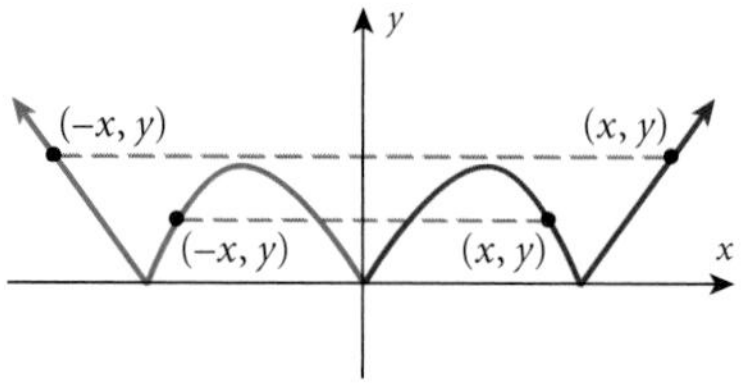

FIGURE 11 Symmetry with respect to the *y*-axis

Figure 11 illustrates the definition. Notice that, when a graph is symmetric with respect to the *y*-axis, the part of the graph to the right of the *y*-axis is a reflection of the part to the left of it, and vice versa.

EXAMPLE 8 **Points Symmetric with Respect to the *y*-Axis**

If a graph is symmetric with respect to the *y*-axis and the point (5, 8) is on the graph, then the point (−5, 8) is also on the graph.

A graph is said to be **symmetric with respect to the origin** if, for every point (x, y) on the graph, the point $(-x, -y)$ is also on the graph.

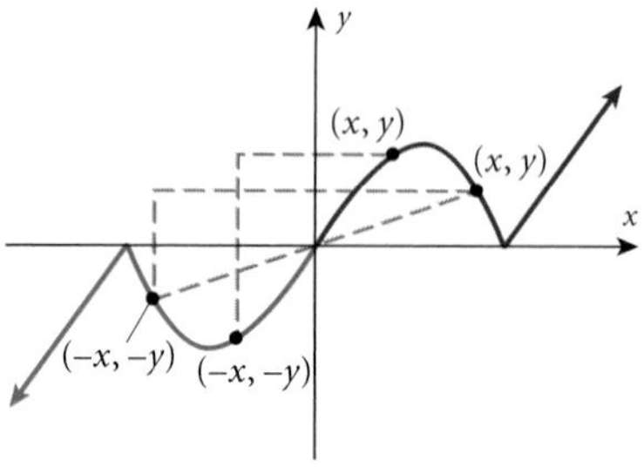

FIGURE 12 Symmetry with respect to the origin

Figure 12 illustrates the definition. Notice that symmetry with respect to the origin may be viewed in two ways:

1. As a reflection about the *y*-axis, followed by a reflection about the *x*-axis.
2. As a projection along a line through the origin so that the distances from the origin are equal.

EXAMPLE 9 **Points Symmetric with Respect to the Origin**

If a graph is symmetric with respect to the origin and the point (4, 2) is on the graph, then the point (−4, −2) is also on the graph.

NOW WORK PROBLEMS 1 AND 11(b).

Test an equation for symmetry with respect to the (a) *x*-axis, (b) *y*-axis, and (c) origin **4**

When the graph of an equation is symmetric with respect to a coordinate axis or the origin, the number of points that you need to plot in order to see the pattern is reduced. For example, if the graph of an equation is symmetric with respect to the *y*-axis, then, once points to the right of the *y*-axis are plotted, an equal number of points on the graph can be obtained by reflecting them about the *y*-axis. Because of this, before we graph an equation, we first want to determine whether it has any symmetry. The following tests are used for this purpose.

Tests for Symmetry

To test the graph of an equation for symmetry with respect to the

x-Axis Replace y by $-y$ in the equation. If an equivalent equation results, the graph of the equation is symmetric with respect to the x-axis.

y-Axis Replace x by $-x$ in the equation. If an equivalent equation results, the graph of the equation is symmetric with respect to the y-axis.

Origin Replace x by $-x$ and y by $-y$ in the equation. If an equivalent equation results, the graph of the equation is symmetric with respect to the origin.

Let's look at an equation that we have already graphed to see how these tests are used.

EXAMPLE 10 Testing an Equation for Symmetry ($x = y^2$)

(a) To test the graph of the equation $x = y^2$ for symmetry with respect to the x-axis, we replace y by $-y$ in the equation, as follows:

$$x = y^2 \quad \text{Original equation.}$$
$$x = (-y)^2 \quad \text{Replace } y \text{ by } -y.$$
$$x = y^2 \quad \text{Simplify.}$$

When we replace y by $-y$, the result is the same equation. The graph is symmetric with respect to the x-axis.

(b) To test the graph of the equation $x = y^2$ for symmetry with respect to the y-axis, we replace x by $-x$ in the equation:

$$x = y^2 \quad \text{Original equation.}$$
$$-x = y^2 \quad \text{Replace } x \text{ by } -x.$$

Because we arrive at the equation $-x = y^2$, which is not equivalent to the original equation, we conclude that the graph is not symmetric with respect to the y-axis.

(c) To test for symmetry with respect to the origin, we replace x by $-x$ and y by $-y$:

$$x = y^2 \quad \text{Original equation.}$$
$$-x = (-y)^2 \quad \text{Replace } x \text{ by } -x \text{ and } y \text{ by } -y.$$
$$-x = y^2 \quad \text{Simplify.}$$

The resulting equation, $-x = y^2$, is not equivalent to the original equation. We conclude that the graph is not symmetric with respect to the origin.

Figure 13(a) on page 546 illustrates the graph of $x = y^2$. In forming a table of points on the graph of $x = y^2$, we can restrict ourselves to points whose y-coordinates are positive. Once these are plotted and connected, a reflection about the x-axis (because of the symmetry) provides the rest of the graph.

Figures 13(b) and (c) illustrate two other equations, $y = x^2$ and $y = x^3$, that we graphed earlier. Test each of these equations for symmetry to verify the conclusions stated in Figures 13(b) and (c). Notice how the existence of symmetry reduces the number of points that we need to plot.

FIGURE 13

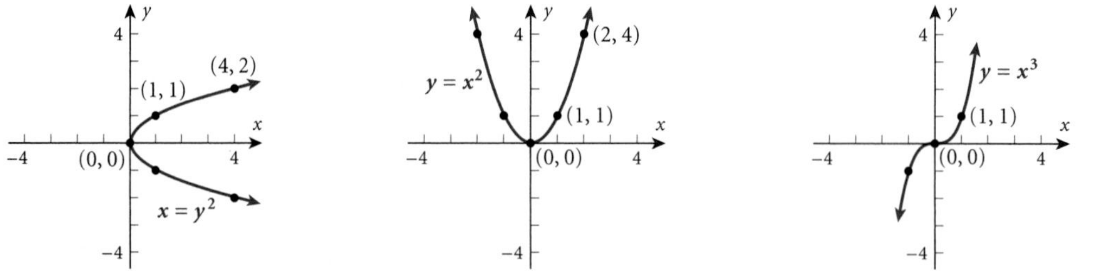

(a) Symmetry with respect to the x-axis

(b) Symmetry with respect to the y-axis

(c) Symmetry with respect to the origin

 NOW WORK PROBLEM 33 (Test for Symmetry).

EXAMPLE 11   **Graphing the Equation** $y = \dfrac{1}{x}$

Graph the equation: $y = \dfrac{1}{x}$

Find any intercepts and check for symmetry first.

SOLUTION   We check for intercepts first. If we let $x = 0$, we obtain a 0 denominator, which is not defined. We conclude that there is no y-intercept. If we let $y = 0$, we get the equation $\dfrac{1}{x} = 0$, which has no solution. We conclude that there is no x-intercept. The graph of $y = \dfrac{1}{x}$ does not cross or touch the coordinate axes.

Next we check for symmetry:

TABLE 5

x	$y = \dfrac{1}{x}$	(x, y)
$\dfrac{1}{10}$	10	$\left(\dfrac{1}{10}, 10\right)$
$\dfrac{1}{3}$	3	$\left(\dfrac{1}{3}, 3\right)$
$\dfrac{1}{2}$	2	$\left(\dfrac{1}{2}, 2\right)$
1	1	$(1, 1)$
2	$\dfrac{1}{2}$	$\left(2, \dfrac{1}{2}\right)$
3	$\dfrac{1}{3}$	$\left(3, \dfrac{1}{3}\right)$
10	$\dfrac{1}{10}$	$\left(10, \dfrac{1}{10}\right)$

x-Axis   Replacing y by $-y$ yields $-y = \dfrac{1}{x}$, which is not equivalent to $y = \dfrac{1}{x}$.

y-Axis   Replacing x by $-x$ yields $y = \dfrac{1}{-x}$, which is not equivalent to $y = \dfrac{1}{x}$.

Origin   Replacing x by $-x$ and y by $-y$ yields $-y = \dfrac{1}{-x} = -\dfrac{1}{x}$, which is equivalent to $y = \dfrac{1}{x}$.

The graph is symmetric with respect to the origin.

Finally, we set up Table 5, listing several points on the graph. Because of the symmetry with respect to the origin, we use only positive values of x.

From Table 5 we infer that if x is a large and positive number then $y = \dfrac{1}{x}$ is a positive number close to 0. We also infer that if x is a positive number close to 0 then $y = \dfrac{1}{x}$ is a large and positive number. Armed with this information, we can graph the equation. Figure 14 illustrates some of these points and the graph of $y = \dfrac{1}{x}$. Observe how the absence of intercepts and the existence of symmetry with respect to the origin were utilized.

FIGURE 14

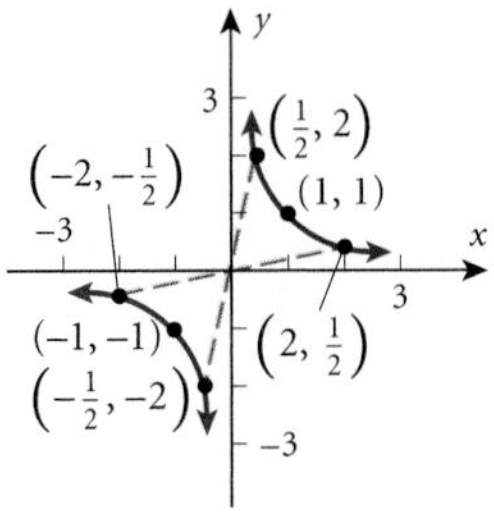

COMMENT: Look at Figure 14. The line $y = 0$ (the x-axis) is called a horizontal *asymptote* of the graph. The line $x = 0$ (the y-axis) is called a vertical *asymptote* of the graph. We will discuss asymptotes in more detail in Chapter 12.

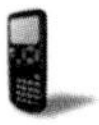

COMMENT: Refer to Example 3 in Section C.5 of Appendix C for the graph of $y = \dfrac{1}{x}$ using a graphing utility.

EXERCISE 10.1 Answers to Odd-Numbered Problems Begin on Page AN-48.

In Problems 1–10, plot each point. Then plot the point that is symmetric to it with respect to (a) the x-axis; (b) the y-axis; (c) the origin.

1. $(3, 4)$
2. $(5, 3)$
3. $(-2, 1)$
4. $(4, -2)$
5. $(1, 1)$
6. $(-1, -1)$
7. $(-3, -4)$
8. $(4, 0)$
9. $(0, -3)$
10. $(-3, 0)$

In Problems 11–22, the graph of an equation is given.
(a) List the intercepts of the graph.
(b) Based on the graph, tell whether the graph is symmetric with respect to the x-axis, the y-axis, and/or the origin.

11.

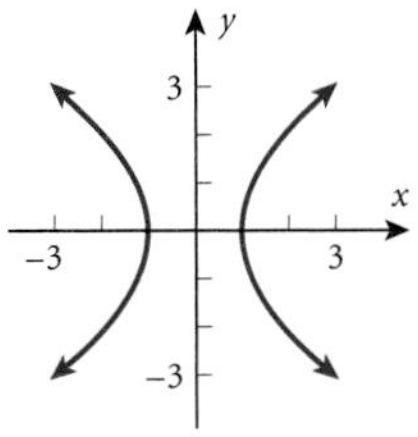

12.

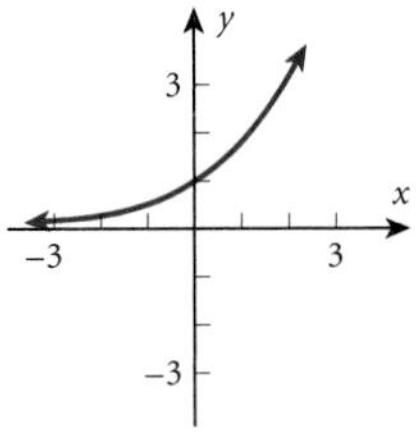

13.

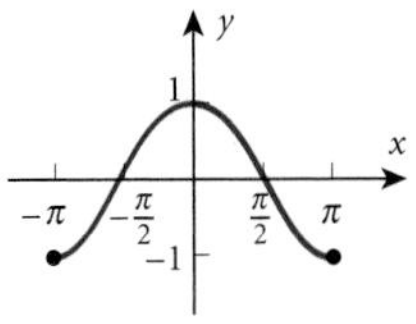

14.

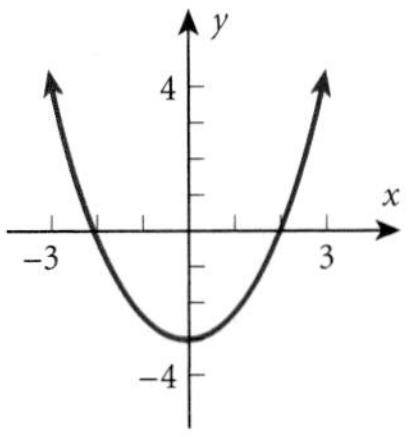

15.

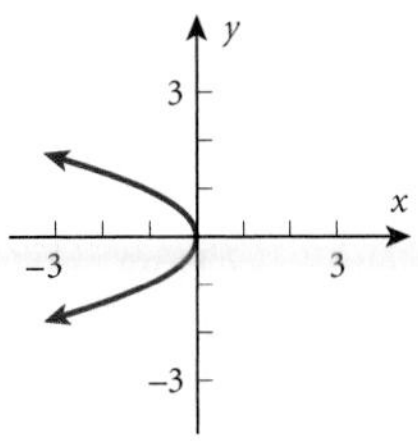

16.

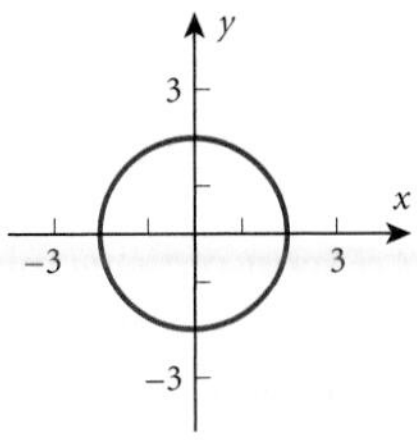

17.

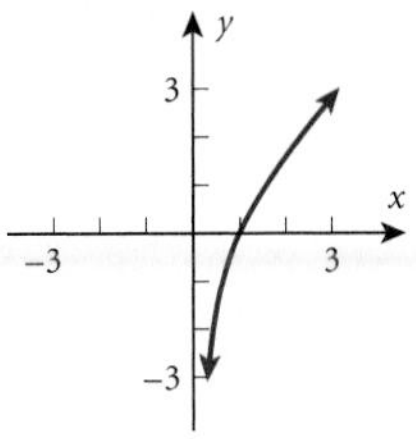

18.

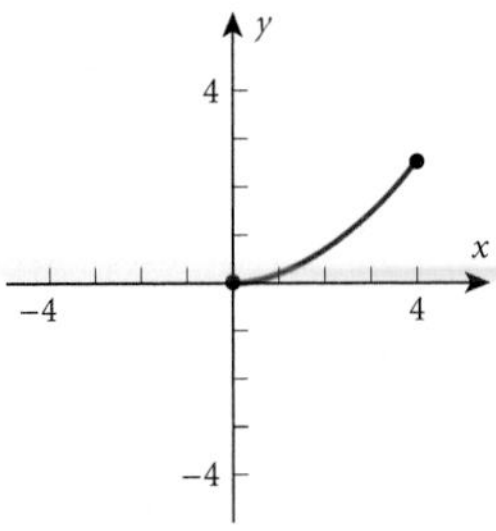

19.

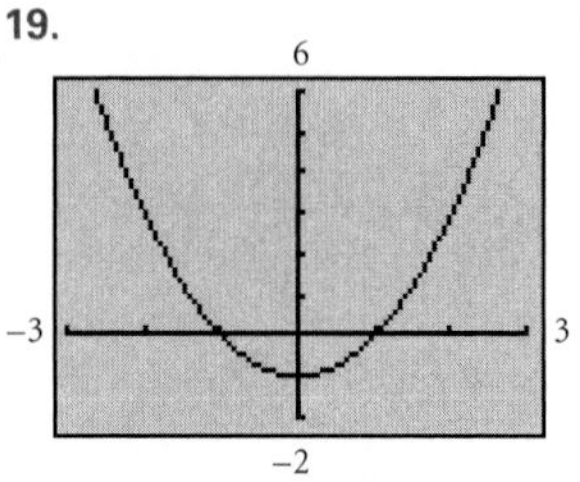

20.

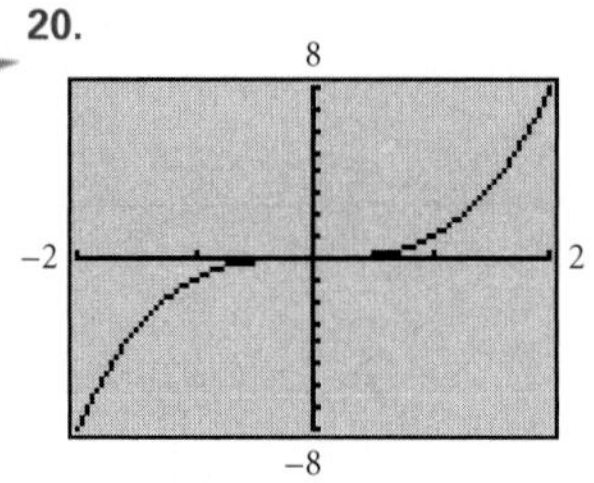

21.

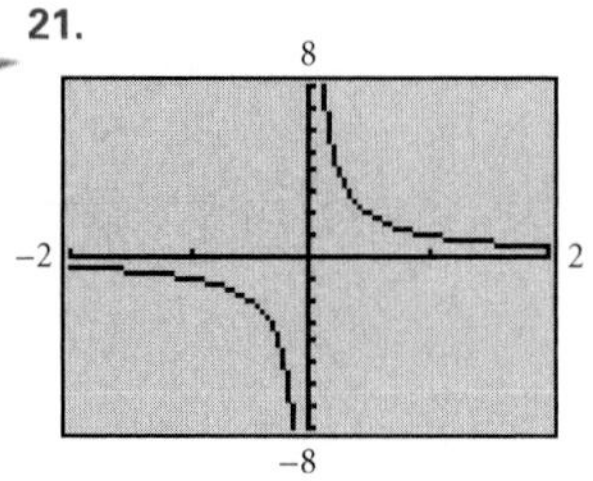

22.

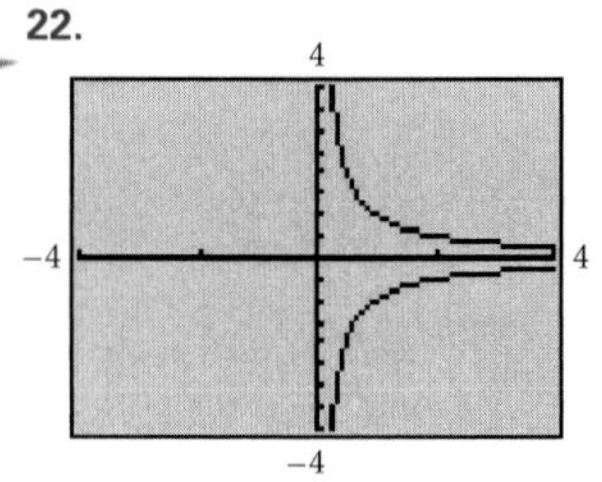

In Problems 23–28, determine whether the given points are on the graph of the equation.

23. Equation: $y = x^4 - \sqrt{x}$
Points: $(0, 0); (1, 1); (-1, 0)$

24. Equation: $y = x^3 - 2\sqrt{x}$
Points: $(0, 0); (1, 1); (1, -1)$

25. Equation: $y^2 = x^2 + 9$
Points: $(0, 3); (3, 0); (-3, 0)$

26. Equation: $y^3 = x + 1$
Points: $(1, 2); (0, 1); (-1, 0)$

27. Equation: $x^2 + y^2 = 4$
Points: $(0, 2); (-2, 2); (\sqrt{2}, \sqrt{2})$

28. Equation: $x^2 + 4y^2 = 4$
Points: $(0, 1); (2, 0); \left(2, \frac{1}{2}\right)$

In Problems 29–44, list the intercepts and test for symmetry.

29. $x^2 = y$
30. $y^2 = x$
31. $y = 3x$
32. $y = -5x$
33. $x^2 + y - 9 = 0$
34. $y^2 - x - 4 = 0$
35. $9x^2 + 4y^2 = 36$
36. $4x^2 + y^2 = 4$
37. $y = x^3 - 27$
38. $y = x^4 - 1$
39. $y = x^2 - 3x - 4$
40. $y = x^2 + 4$
41. $y = \dfrac{3x}{x^2 + 9}$
42. $y = \dfrac{x^2 - 4}{2x^4}$
43. $y = |x|$
44. $y = \sqrt{x}$

In Problems 45–48, graph each equation by plotting points.

45. $y = x^3 - 1$
46. $x = y^2 + 1$
47. $y = 2\sqrt{x}$
48. $y = x^2 + 2$

49. If $(a, 2)$ is a point on the graph of $y = 3x + 5$, what is a?

50. If $(2, b)$ is a point on the graph of $y = x^2 + 4x$, what is b?

51. If (a, b) is a point on the graph of $2x + 3y = 6$, write an equation that relates a to b.

52. If $(2, 0)$ and $(0, 5)$ are points on the graph of $y = mx + b$, what are m and b?

In Problem 53, you may use a graphing utility, but it is not required.

53. (a) Graph $y = \sqrt{x^2}$, $y = x$, $y = |x|$, and $y = (\sqrt{x})^2$, noting which graphs are the same.
(b) Explain why the graphs of $y = \sqrt{x^2}$ and $y = |x|$ are the same.
(c) Explain why the graphs of $y = x$ and $y = (\sqrt{x})^2$ are not the same.
(d) Explain why the graphs of $y = \sqrt{x^2}$ and $y = x$ are not the same.

54. Make up an equation with the intercepts $(2, 0)$, $(4, 0)$, and $(0, 1)$. Compare your equation with a friend's equation. Comment on any similarities.

55. An equation is being tested for symmetry with respect to the x-axis, the y-axis, and the origin. Explain why, if two of these symmetries are present, the remaining one must also be present.

56. Draw a graph that contains the points $(-2, -1)$, $(0, 1)$, $(1, 3)$, and $(3, 5)$. Compare your graph with those of other students. Are most of the graphs almost straight lines? How many are "curved"? Discuss the various ways that these points might be connected.

10.2 Functions

PREPARING FOR THIS SECTION *Before getting started, review the following:*

- > Intervals (Appendix A, Section A.7, pp. A-61–A-62)
- > Evaluating Algebraic Expressions; Domain of a Variable (Appendix A, Section A.2, pp. A-18–A-19)
- > Square Root Method (Appendix A, Section A.6, pp. A-53–A-54)
- > Solving Inequalities (Appendix A, Section A.7, pp. A-63–A-65)

OBJECTIVES
1 Find the value of a function
2 Find the difference quotient of a function
3 Find the domain of a function
4 Solve applied problems involving functions

In many applications a correspondence (such as an equation) exists between two variables. For example, the relation between the revenue R resulting from the sale of x items selling for \$10 each may be expressed by the equation $R = 10x$. If we know how many items have been sold, then we can calculate the revenue by using the equation $R = 10x$. This equation is an example of a *function*.

As another example, suppose that an icicle falls off a building from a height of 64 feet above the ground. According to a law of physics, the distance s (in feet) of the icicle from the ground after t seconds is given (approximately) by the formula $s = 64 - 16t^2$. When $t = 0$ seconds, the icicle is $s = 64$ feet above the ground. After 1 second, the icicle is $s = 64 - 16(1)^2 = 48$ feet above the ground. After 2 seconds, the icicle strikes the ground. The formula $s = 64 - 16t^2$ provides a way of finding the distance s for any time t $(0 \le t \le 2)$. There is a correspondence between each time t in the interval $0 \le t \le 2$ and the distance s. We say that the distance s is a function of the time t because:

1. There is a correspondence between the set of times and the set of distances.
2. There is exactly one distance s obtained for any time t in the interval $0 \le t \le 2$.

Let's now look at the definition of a function.

Definition of Function

Let X and Y be two nonempty sets of real numbers. A **function** from X into Y is a correspondence that associates with each number in X exactly one number in Y.

FIGURE 15

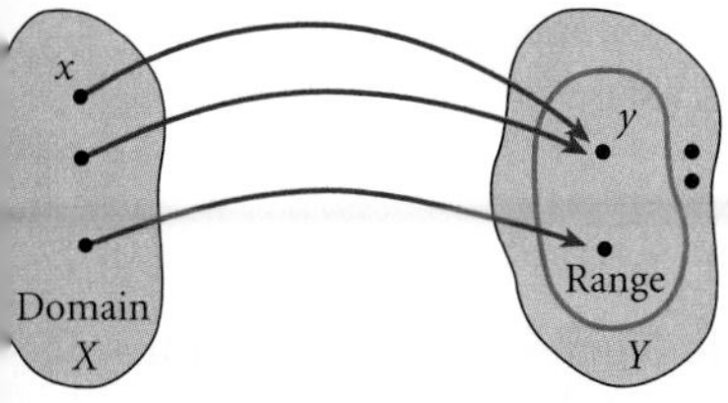

The set X is called the **domain** of the function. For each number x in X, the corresponding number y in Y is called the **value** of the function at x, or the **image** of x. The set of all images of the numbers in the domain is called the **range** of the function. See Figure 15.

Since there may be some numbers in Y that are not the image of any x in X, it follows that the range of a function is a subset of Y, as shown in Figure 15.

EXAMPLE 1   Example of a Function

Consider the function defined by the equation

$$y = 2x - 5, \qquad 1 \le x \le 6$$

Notice that for each number x there corresponds exactly one number y. For example, if $x = 1$, then $y = 2(1) - 5 = -3$. If $x = 3$, then $y = 2(3) - 5 = 1$. For this reason, the equation is a function. Since we restrict the numbers x to the real numbers between 1 and 6, inclusive, the domain of the function is $\{x \mid 1 \le x \le 6\}$. The function specifies that in order to get the image of x we multiply x by 2 and then subtract 5 from this product. ▶

Find the value of a function ☐1 **Function Notation**

Functions are often denoted by letters such as f, F, g, G, and others. If f is a function, then for each number x in its domain the corresponding image in the range is designated by the symbol $f(x)$, read as "f of x" or as "f at x." We refer to $f(x)$ as the **value of f at the number x;** $f(x)$ is the number that results when x is given and the function f is applied; $f(x)$ does *not* mean "f times x." For example, the function given in Example 1 may be written as $y = f(x) = 2x - 5$, $1 \le x \le 6$. Then $f(1) = -3$.

Figure 16 illustrates some other functions. Notice that in every function illustrated, for each x in the domain there is one value in the range.

FIGURE 16

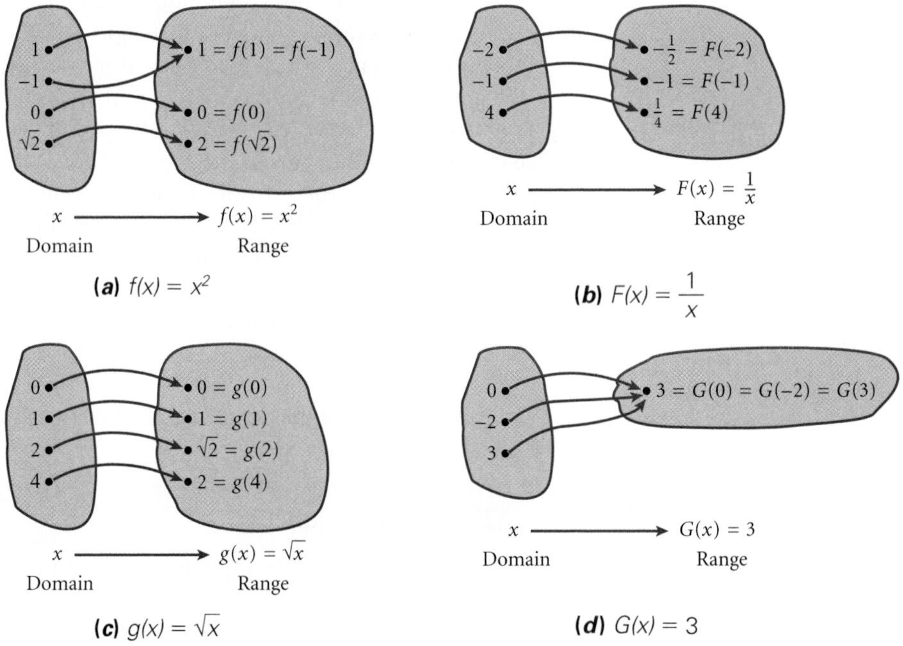

(a) $f(x) = x^2$

(b) $F(x) = \dfrac{1}{x}$

(c) $g(x) = \sqrt{x}$

(d) $G(x) = 3$

FIGURE 17

Input x

Output $y = f(x)$

f

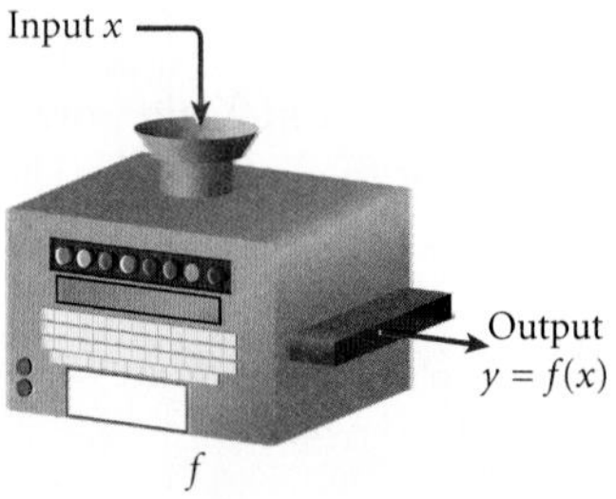

Sometimes it is helpful to think of a function f as a machine that receives as input a number from the domain, manipulates it, and outputs the value. See Figure 17.

The restrictions on this input/output machine are as follows:

1. It only accepts numbers from the domain of the function.
2. For each input, there is exactly one output (which may be repeated for different inputs).

For a function $y = f(x)$, the variable x is called the **independent variable,** because it can be assigned any of the permissible numbers from the domain. The variable y is called the **dependent variable,** because its value depends on x.

Any symbol can be used to represent the independent and dependent variables. For example, if f is the *cube function*, then f can be given by $f(x) = x^3$ or $f(t) = t^3$ or $f(z) = z^3$. All three functions are the same. Each tells us to cube the independent variable. In practice, the symbols used for the independent and dependent variables are based on common usage, such as using C for cost in business.

The independent variable is also called the **argument** of the function. Thinking of the independent variable as an argument can sometimes make it easier to find the value of a function. For example, if f is the function defined by $f(x) = x^3$, then f tells us to cube the argument. For example, $f(2)$ means to cube 2, $f(a)$ means to cube the number a, and $f(x + h)$ means to cube the quantity $x + h$.

EXAMPLE 2 Finding Values of a Function

For the function f defined by $f(x) = 2x^2 - 3x$, evaluate

(a) $f(3)$ **(b)** $f(x) + f(3)$ **(c)** $f(-x)$

(d) $-f(x)$ **(e)** $f(x + 3)$ **(f)** $\dfrac{f(x+h) - f(x)}{h}, h \neq 0$

SOLUTION **(a)** We substitute 3 for x in the equation for f to get

$$f(3) = 2(3)^2 - 3(3) = 18 - 9 = 9$$

(b) $f(x) + f(3) = (2x^2 - 3x) + 9 = 2x^2 - 3x + 9$

(c) We substitute $-x$ for x in the equation for f.

$$f(-x) = 2(-x)^2 - 3(-x) = 2x^2 + 3x$$

(d) $-f(x) = -(2x^2 - 3x) = -2x^2 + 3x$

(e) We substitute $x + 3$ for x in the equation for f.

$$\begin{aligned} f(x+3) &= 2(x+3)^2 - 3(x+3) \quad \text{Notice the use of parentheses here.} \\ &= 2(x^2 + 6x + 9) - 3x - 9 \\ &= 2x^2 + 12x + 18 - 3x - 9 \\ &= 2x^2 + 9x + 9 \end{aligned}$$

(f)
$$\begin{aligned} \frac{f(x+h) - f(x)}{h} &\underset{\substack{\uparrow \\ f(x+h) = 2(x+h)^2 - 3(x+h)}}{=} \frac{[2(x+h)^2 - 3(x+h)] - [2x^2 - 3x]}{h} \\ &= \frac{2(x^2 + 2xh + h^2) - 3x - 3h - 2x^2 + 3x}{h} \quad \text{Simplify.} \\ &= \frac{2x^2 + 4xh + 2h^2 - 3h - 2x^2}{h} \\ &= \frac{4xh + 2h^2 - 3h}{h} \\ &= \frac{h(4x + 2h - 3)}{h} \quad \text{Factor out } h. \\ &= 4x + 2h - 3 \quad \text{Cancel } h. \end{aligned}$$

Notice in this example that $f(x + 3) \neq f(x) + f(3)$ and $f(-x) \neq -f(x)$.

NOW WORK PROBLEM 1.

The expression in part (f) of Example 2 is called the *difference quotient* of f, an important expression in calculus.

Difference Quotient

The **difference quotient** of a function $y = f(x)$ is :

$$\frac{f(x + h) - f(x)}{h} \qquad h \neq 0$$

2 **EXAMPLE 3 Finding Difference Quotients**

Find the difference quotient of

(a) $f(x) = b$ **(b)** $f(x) = mx + b$ **(c)** $f(x) = x^2$

where m and b are constants.

SOLUTION **(a)** If $f(x) = b$, then $f(x + h) = b$ and the difference quotient of f is

$$\frac{f(x + h) - f(x)}{h} = \frac{b - b}{h} = \frac{0}{h} = 0$$

(b) If $f(x) = mx + b$, then $f(x + h) = m(x + h) + b = mx + mh + b$ and the difference quotient of f is

$$\frac{f(x + h) - f(x)}{h} = \frac{mx + mh + b - (mx + b)}{h} = \frac{mh}{h} = m$$

(c) If $f(x) = x^2$, then $f(x + h) = (x + h)^2 = x^2 + 2xh + h^2$ and the difference quotient of f is

$$\frac{f(x + h) - f(x)}{h} = \frac{x^2 + 2xh + h^2 - x^2}{h} = \frac{(2x + h)h}{h} = 2x + h$$

NOW WORK PROBLEM 9.

Most calculators have special keys that enable you to find the value of certain commonly used functions. For example, you should be able to find the square function $f(x) = x^2$, the square root function $f(x) = \sqrt{x}$, the reciprocal function $f(x) = \dfrac{1}{x} = x^{-1}$, and many others that will be discussed later in the book (such as $\ln x$ and $\log x$). Verify the results of Example 6, which follows, on your calculator.

EXAMPLE 4 Finding Values of a Function on a Calculator

(a) $f(x) = x^2$; $f(1.234) = 1.522756$

(b) $f(x) = \dfrac{1}{x}$; $F(1.234) = 0.8103727715$

(c) $g(x) = \sqrt{x}$; $g(1.234) = 1.110855526$

COMMENT: Graphing calculators can be used to evaluate any function that you wish. Figure 18 shows the result obtained in Example 2(a) on a TI-83 graphing calculator with the function to be evaluated, $f(x) = 2x^2 - 3x$, in Y_1.*

FIGURE 18

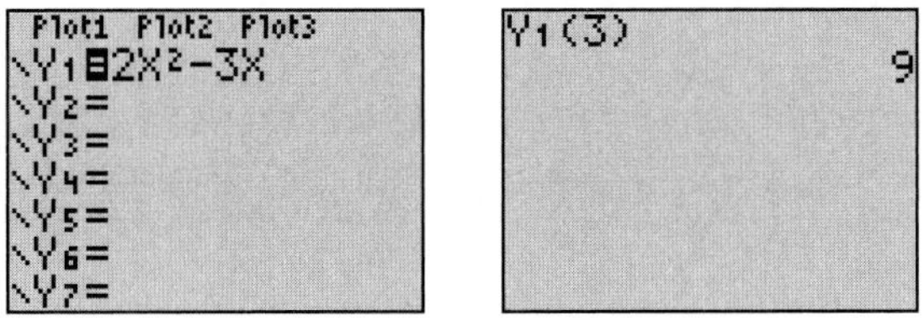

Implicit Form of a Function

In general, when a function f is defined by an equation in x and y, we say that the function f is given **implicitly.** If it is possible to solve the equation for y in terms of x, then we write $y = f(x)$ and say that the function is given **explicitly.** For example,

Implicit Form	*Explicit Form*
$3x + y = 5$	$y = f(x) = -3x + 5$
$x^2 - y = 6$	$y = f(x) = x^2 - 6$
$xy = 4$	$y = f(x) = \dfrac{4}{x}$

Not all equations in x and y define a function $y = f(x)$. If an equation is solved for y and two or more values of y can be obtained for a given x, then the equation does not define a function.

EXAMPLE 5 Determining Whether an Equation Is a Function

Determine if the equation $x^2 + y^2 = 1$ is a function.

SOLUTION To determine whether the equation $x^2 + y^2 = 1$ is a function, we need to solve the equation for y.

$$x^2 + y^2 = 1$$
$$y^2 = 1 - x^2$$
$$y = \pm\sqrt{1 - x^2} \quad \text{Square Root Method.}$$

For values of x between -1 and 1, two values of y result. This means that the equation $x^2 + y^2 = 1$ does not define a function.

NOW WORK PROBLEM 23.

COMMENT: The explicit form of a function is the form required by a graphing calculator. Now do you see why it is necessary to graph some equations in two "pieces"?

We list next a summary of some important facts to remember about a function f.

**Consult your owner's manual for the required keystrokes.*

SUMMARY IMPORTANT FACTS ABOUT FUNCTIONS

(a) To each x in the domain of f, there is exactly one image $f(x)$ in the range; however, a number in the range can result from more than one x in the domain.

(b) f is the symbol that we use to denote the function. It is symbolic of the equation that we use to get from an x in the domain to $f(x)$ in the range.

(c) If $y = f(x)$, then the function f is given explicitly; x is called the independent variable, or argument, of f and y is called the dependent variable or the value of f at x.

Find the domain of a function

3 Domain of a Function

Often the domain of a function f is not specified; instead, only the equation defining the function is given. In such cases, we agree that the domain of f is the largest set of real numbers for which the value $f(x)$ is a real number. The domain of a function f is the same as the domain of the variable x in the expression $f(x)$.

EXAMPLE 6 Finding the Domain of a Function

Find the domain of each of the following functions:

(a) $f(x) = x^2 + 5x$ **(b)** $g(x) = \dfrac{3x}{x^2 - 4}$ **(c)** $h(t) = \sqrt{4 - 3t}$

SOLUTION **(a)** The function f tells us to square a number and then add five times the number. Since these operations can be performed on any real number, we conclude that the domain of f is all real numbers.

(b) The function g tells us to divide $3x$ by $x^2 - 4$. Since division by 0 is not defined, the denominator $x^2 - 4$ can never be 0, so x can never equal -2 or 2. The domain of the function g is $\{x \mid x \neq -2, x \neq 2\}$.

(c) The function h tells us to take the square root of $4 - 3t$. But only nonnegative numbers have real square roots, so the expression under the square root must be nonnegative. This requires that

$$\begin{aligned} 4 - 3t &\geq 0 \\ -3t &\geq -4 \\ t &\leq \frac{4}{3} \end{aligned}$$

The domain of h is $\left\{t \mid t \leq \dfrac{4}{3}\right\}$ or the interval $\left(-\infty, \dfrac{4}{3}\right]$.

NOW WORK PROBLEM 33.

If x is in the domain of a function f, we shall say that **f is defined at x,** or **$f(x)$ exists.** If x is not in the domain of f, we say that **f is not defined at x,** or **$f(x)$ does not exist.** For example, if $f(x) = \dfrac{x}{x^2 - 1}$, then $f(0)$ exists, but $f(1)$ and $f(-1)$ do not exist. (Do you see why?)

We have not said much about finding the range of a function. The reason is that when a function is defined by an equation it is often difficult to find the range. Therefore, we shall usually be content to find just the domain of a function when only the rule for the function is given. We shall express the domain of a function using inequalities, interval notation, set notation, or words, whichever is most convenient.

Applications

Solve applied problems involving functions 4

When we use functions in applications, the domain may be restricted by physical or geometric considerations. For example, the domain of the function f defined by $f(x) = x^2$ is the set of all real numbers. However, if f is used to obtain the area of a square when the length x of a side is known, then we must restrict the domain of f to the positive real numbers, since the length of a side can never be 0 or negative.

EXAMPLE 7 Constructing a Cost Function

The cost per square foot to build a house is \$110.

(a) Express the cost C as a function of x, the number of square feet.
(b) What is the cost to build a 2000-square-foot house?

SOLUTION **(a)** The cost C of building a house containing x square feet is $110x$ dollars. A function expressing this relationship is

$$C(x) = 110x$$

where x is the independent variable and C is the dependent variable. In this setting the domain is $\{x|x > 0\}$ since a house cannot have 0 or negative square feet.

(b) The cost to build a 2000-square-foot house is

$$C(2000) = 110(2000) = \$220{,}000$$

NOW WORK PROBLEM 49.

Observe in the solution to Example 7 that we used the symbol C in two ways: it is used to name the function, and it is used to symbolize the dependent variable. This double use is common in applications and should not cause any difficulty.

EXAMPLE 8 Determining the Cost of Removing Pollutants

The cost of eliminating a large part of the pollutants from the atmosphere (or from water) is relatively cheap. However, removing the last traces of pollutants results in a significant increase in cost. A typical relationship between the cost C, in millions of dollars, for removal and the percent x of pollutant removed is given by the function

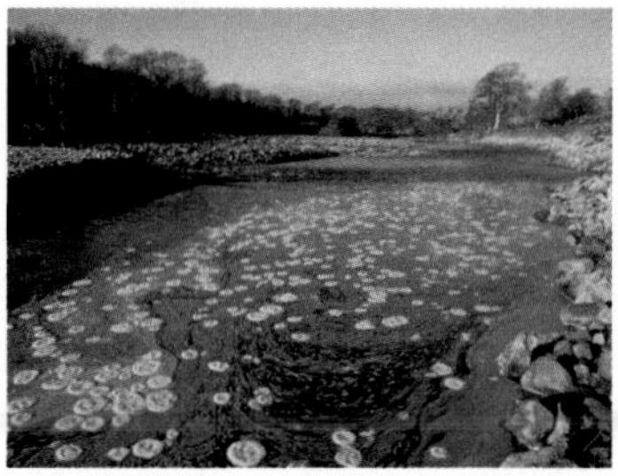

$$C(x) = \frac{3x}{105 - x}$$

Since x is a percentage, the domain of C consists of all real numbers x for which $0 \le x \le 100$. The cost of removing 0% of the pollutant is

$$C(0) = 0$$

The cost of removing 50% of the pollutant is

$$C(50) = \frac{150}{55} = 2.727 \text{ million dollars}$$

The costs of removing 60% and 70% are

$$C(60) = \frac{180}{45} = 4 \text{ million dollars}$$

and

$$C(70) = \frac{210}{35} = 6 \text{ million dollars}$$

Observe that the cost of removing an additional 10% of the pollutant after 50% had been removed is \$1,273,000, while the cost of removing an additional 10% after 60% is removed is \$2,000,000.

Revenue, Cost, and Profit Functions **Revenue** is the amount of money derived from the sale of a product and equals the price of the product times the quantity of the product that is actually sold. But the price and the quantity sold are not independent. As the price falls, the demand for the product increases; and when the price rises, the demand decreases.

The equation that relates the price p of a quantity bought and the amount x of a quantity demanded is called the **demand equation.** If in this equation we solve for p, we have

$$p = d(x)$$

The function d is called the **price function** and $d(x)$ is the price per unit when x units are demanded. If x is the number of units sold and $d(x)$ is the price for each unit, the **revenue function** $R(x)$ is defined as

$$R(x) = xp = x\,d(x)$$

If we denote the **cost function** by $C(x)$, then the **profit function** $P(x)$ is defined as

$$P(x) = R(x) - C(x)$$

EXAMPLE 9 Constructing a Revenue Function

No matter how much wheat a farmer can grow, it can be sold at \$4 per bushel. Find the price function. What is the revenue function?

SOLUTION Since the price per bushel is fixed at \$4 per bushel, the price function is

$$p = \$4$$

The revenue function is

$$R(x) = xp = 4x$$

EXAMPLE 10 Constructing a Revenue Function

The manager of a toy store has observed that each week 1000 toy trucks are sold at a price of \$5 per truck. When there is a special sale, the trucks sell for \$4 each and 1200 per week are sold. Assuming a linear price function, construct the price function. What is the revenue function?

SOLUTION Let p be the price of each truck and let x be the number sold. If the price function $p = d(x)$ is linear, then we know that $(x_1, p_1) = (1000, 5)$ and $(x_2, p_2) = (1200, 4)$ are two points on the line $p = d(x)$. The slope of the line is

$$\frac{p_2 - p_1}{x_2 - x_1} = \frac{4 - 5}{1200 - 1000} = \frac{-1}{200} = -\frac{1}{200}$$

Use the point–slope form of the equation of a line:

$$p - p_1 = m(x - x_1) \qquad \text{Point–slope form.}$$

$$p - 5 = -\frac{1}{200}(x - 1000) \qquad m = -\frac{1}{200};\ p_1 = 5;\ x_1 = 1000$$

$$p = -\frac{1}{200}x + 10$$

The price function is

$$p = d(x) = -\frac{1}{200}x + 10$$

The revenue function is

$$R(x) = xp = x\left(-\frac{1}{200}x + 10\right) = -\frac{1}{200}x^2 + 10x$$

The price function obtained in Example 10 is not meant to reflect extreme situations. For example, we do not expect to sell $x = 0$ trucks nor do we expect to sell too many trucks in excess of 1500, since even during a special sale only 1200 are sold. The price function does represent the relationship between price and quantity in a certain range—in this case, perhaps $500 < x < 1500$.

NOW WORK PROBLEM 51.

SUMMARY

We list here some of the important vocabulary introduced in this section, with a brief description of each term.

Function	A relation between two sets of real numbers so that each number x in the first set, the domain, has corresponding to it exactly one number y in the second set. The range is the set of y values of the function for the x values in the domain. A function f may be defined implicitly by an equation involving x and y or explicitly by writing $y = f(x)$.
Unspecified domain	If a function f is defined by an equation and no domain is specified, then the domain will be taken to be the largest set of real numbers for which the equation defines a real number.
Function notation	$y = f(x)$ f is a symbol for the function. x is the independent variable or argument. y is the dependent variable. $f(x)$ is the value of the function at x, or the image of x.

EXERCISE 10.2 Answers to Odd-Numbered Problems Begin on Page AN-49.

In Problems 1–8, find the following values for each function:

(a) $f(0)$ (b) $f(1)$ (c) $f(-1)$ (d) $f(-x)$ (e) $-f(x)$ (f) $f(x+1)$ (g) $f(2x)$ (h) $f(x+h)$

1. $f(x) = 3x^2 + 2x - 4$
2. $f(x) = -2x^2 + x - 1$
3. $f(x) = \dfrac{x}{x^2 + 1}$
4. $f(x) = \dfrac{x^2 - 1}{x + 4}$
5. $f(x) = |x| + 4$
6. $f(x) = \sqrt{x^2 + x}$
7. $f(x) = \dfrac{2x + 1}{3x - 5}$
8. $f(x) = 1 - \dfrac{1}{(x + 2)^2}$

In Problems 9–16, find the difference quotient of f, that is, find $\dfrac{f(x+h) - f(x)}{h}$, $h \neq 0$, *for each function. Be sure to simplify.*

9. $f(x) = 4x + 3$
10. $f(x) = -3x + 1$
11. $f(x) = x^2 - x + 4$
12. $f(x) = x^2 + 5x - 1$
13. $f(x) = x^3$
14. $f(x) = x^3 - 2x$
15. $f(x) = x^4$
16. $f(x) = \dfrac{1}{x}$

In Problems 17–28, determine whether the equation is a function.

17. $y = x^2$
18. $y = x^3$
19. $y = \dfrac{1}{x}$
20. $y = |x|$
21. $y^2 = 4 - x^2$
22. $y = \pm\sqrt{1 - 2x}$
23. $x = y^2$
24. $x + y^2 = 1$
25. $y = 2x^2 - 3x + 4$
26. $y = \dfrac{3x - 1}{x + 2}$
27. $2x^2 + 3y^2 = 1$
28. $x^2 - 4y^2 = 1$

In Problems 29–42, find the domain of each function.

29. $f(x) = -5x + 4$
30. $f(x) = x^2 + 2$
31. $f(x) = \dfrac{x}{x^2 + 1}$
32. $f(x) = \dfrac{x^2}{x^2 + 1}$
33. $g(x) = \dfrac{x}{x^2 - 16}$
34. $h(x) = \dfrac{2x}{x^2 - 4}$
35. $F(x) = \dfrac{x - 2}{x^3 + x}$
36. $G(x) = \dfrac{x + 4}{x^3 - 4x}$
37. $h(x) = \sqrt{3x - 12}$
38. $G(x) = \sqrt{1 - x}$
39. $f(x) = \dfrac{4}{\sqrt{x - 9}}$
40. $f(x) = \dfrac{x}{\sqrt{x - 4}}$
41. $p(x) = \sqrt{\dfrac{2}{x - 1}}$
42. $q(x) = \sqrt{-x - 2}$

---

43. If $f(x) = 2x^3 + Ax^2 + 4x - 5$ and $f(2) = 5$, what is the value of A?
44. If $f(x) = 3x^2 - Bx + 4$ and $f(-1) = 12$, what is the value of B?
45. If $f(x) = \dfrac{3x + 8}{2x - A}$ and $f(0) = 2$, what is the value of A? Where is f undefined?
46. If $f(x) = \dfrac{2x - B}{3x + 4}$ and $f(2) = \dfrac{1}{2}$, what is the value of B? Where is f undefined?
47. If $f(x) = \dfrac{2x - A}{x - 3}$ and $f(4) = 0$, what is the value of A? Where is f not defined?
48. If $f(x) = \dfrac{x - B}{x - A}$, $f(2) = 0$, and $f(1)$ is undefined, what are the values of A and B?
49. **Constructing Functions** Express the gross salary G of a person who earns \$10 per hour as a function of the number x of hours worked.

50. Constructing Functions Tiffany, a commissioned salesperson, earns \$100 base pay plus \$10 per item sold. Express her gross salary G as a function of the number x of items sold.

51. Demand Equation The price p and the quantity x sold of a certain product obey the demand equation

$$p = -\frac{1}{5}x + 100 \qquad 0 \le x \le 500$$

Express the revenue $R = xp$ as a function of x.

52. Demand Equation The price p and the quantity x sold of a certain product obey the demand equation

$$p = -\frac{1}{4}x + 100 \qquad 0 \le x \le 400$$

Express the revenue $R = xp$ as a function of x.

53. Demand Equation The price p and the quantity x sold of a certain product obey the demand equation

$$x = -20p + 100 \qquad 0 \le p \le 5$$

Express the revenue $R = xp$ as a function of x.

54. Demand Equation The price p and the quantity x sold of a certain product obey the demand equation

$$x = -5p + 500 \qquad 0 \le p \le 100$$

Express the revenue $R = xp$ as a function of x.

55. Wheat Production The amount of wheat planted annually in the United States is given in the table.

Year	1990	1991	1992	1993	1994	1995
Thousand Acres	77,041	69,881	72,219	72,168	70,349	69,031

Year	1996	1997	1998	1999	2000
Thousand Acres	75,105	70,412	65,821	62,714	62,629

If 1990 is taken as year 0, the number (in thousands) of acres of wheat planted in the United States can be approximated by $A(t) = -119t^2 + 113t + 73{,}367$. If this function remains valid, project the number (in thousands) of acres of wheat that will be planted in 2010.

Source: U.S. Department of Agriculture.

56. Wheat Production Suppose the number (in thousands) of acres of wheat in 2010 is 28,027, which is consistent with the current trend. If, at that time, there is a movement to increase the number of acres of wheat planted annually, and the equation selected to achieve this is $A(t) = 28{,}027 + 200\sqrt{t}$, how much wheat would be planted in 2020? Assume $t = 0$ for the year 2010.

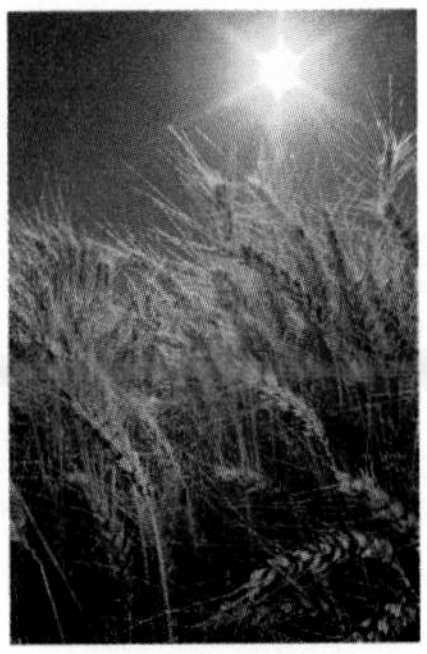

57. SAT Scores The data for the mathematics scores on the SAT can be approximated by $S(t) = -0.04t^3 + 0.43t^2 + 0.24t + 506$, where t is the number of years since 1994. If the trend continues, what would be the expected score in 2010?

58. Page Design A page with dimensions of 11 inches by 7 inches has a border of uniform width x surrounding the printed matter of the page, as shown in the figure. Write a formula for the area A of the printed part of the page as a function of the width x of the border. Give the domain and range of A.

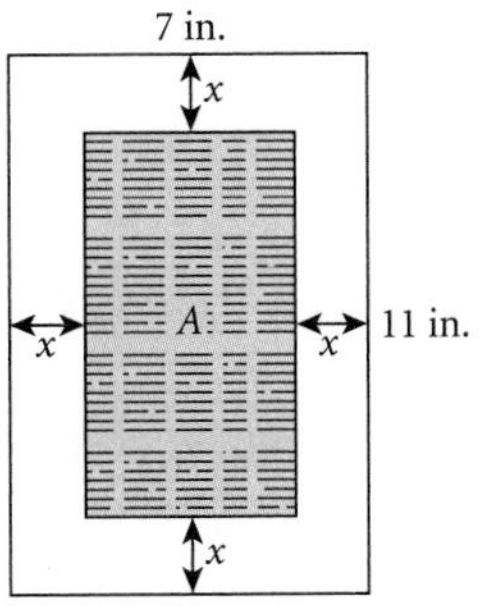

59. Cost of Flying An airplane crosses the Atlantic Ocean (3000 miles) with an airspeed of 500 miles per hour. The cost C (in dollars) per passenger is

$$C(x) = 100 + \frac{x}{10} + \frac{36{,}000}{x}$$

where x is the ground speed (airspeed $\pm$ wind).

(a) What is the cost per passenger for quiescent (no-wind) conditions?

(b) What is the cost per passenger with a head wind of 50 miles per hour?

(c) What is the cost per passenger with a tail wind of 100 miles per hour?

(d) What is the cost per passenger with a head wind of 100 miles per hour?

60. Cable Installation A cable TV company is asked to provide service to a customer whose house is located 2 miles from the road along which the cable is buried. The nearest connection box for the cable is located 5 miles down the road.

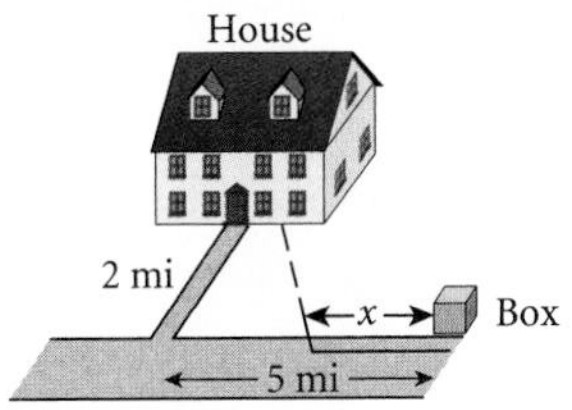

(a) If the installation cost is \$100 per mile along the road and \$140 per mile off the road, express the total cost C of installation as a function of the distance x (in miles) from the connection box to the point where the cable installation turns off the road.

(b) What is the domain of C?

(c) Compute the cost for $x = 1, x = 2, x = 3$, and $x = 4$.

61. Some functions f have the property that $f(a + b) = f(a) + f(b)$ for all real numbers a and b. Which of the following functions have this property?

(a) $h(x) = 2x$ **(b)** $g(x) = x^2$

(c) $F(x) = 5x - 2$ **(d)** $G(x) = \frac{1}{x}$

62. Are the functions $f(x) = x - 1$ and $g(x) = \frac{x^2 - 1}{x + 1}$ the same? Explain.

63. Investigate when, historically, the use of the function notation $y = f(x)$ first appeared.

10.3 Graphs of Functions; Properties of Functions

PREPARING FOR THIS SECTION *Before getting started, review the following:*

- Rectangular Coordinates (Chapter 1, Section 1.1, pp. 2–3)
- Intervals (Appendix A, Section A.7, pp. A-61–A-62)
- Slope of a Line (Chapter 1, Section 1.1, pp. 7–10)
- Point–Slope Equation of a Line (Chapter 1, Section 1.1, p. 11)

OBJECTIVES
1. Identify the graph of a function
2. Obtain information from or about the graph of a function
3. Determine even and odd functions from a graph
4. Identify even and odd functions from the equation
5. Use a graph to determine where a function is increasing, is decreasing, or is constant
6. Use a graph to locate local maxima and minima
7. Use a graphing utility to approximate local maxima and minima and to determine where a function is increasing or decreasing
8. Find the average rate of change of a function

In applications, a graph often demonstrates more clearly the relationship between two variables than, say, an equation or table would. For example, Table 6 shows the price per share of Intel stock at the end of each month from 8/31/02 through 8/31/03. If we plot these data using the date as the x-coordinate and the price as the y-coordinate and then connect the points, we obtain Figure 19.

TABLE 6

Date	Closing Price ($)
8/31/02	15.86
9/30/02	13.89
10/31/02	17.30
11/30/02	20.88
12/31/02	15.57
1/31/03	15.70
2/28/03	17.26
3/31/03	16.28
4/30/03	18.37
5/31/03	20.82
6/30/03	20.81
7/31/03	24.89
8/31/03	28.59

FIGURE 19 Monthly closing prices of Intel stock 8/31/02 through 8/31/03.

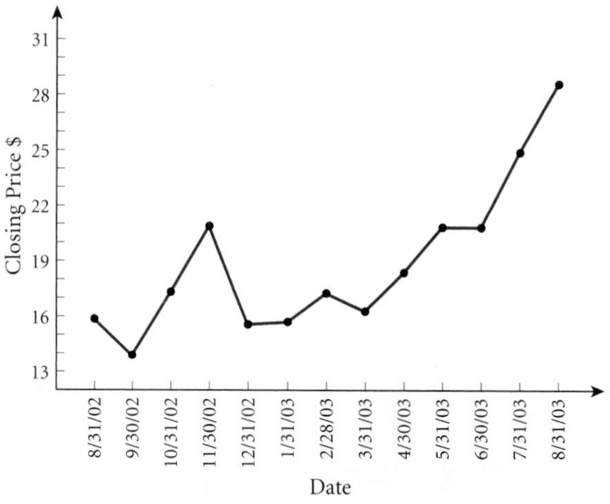

We can see from the graph that the price of the stock was rising rapidly from 9/30/02 through 11/30/02 and was falling slightly from 8/31/02 through 9/30/02. The graph also shows that the lowest price occurred at the end of September, 2002 whereas the highest occurred at the end of August, 2003. Equations and tables, on the other hand, usually require some calculations and interpretation before this kind of information can be "seen."

Look again at Figure 19. The graph shows that for each date on the horizontal axis there is only one price on the vertical axis. Thus, the graph represents a function, although the exact rule for getting from date to price is not given.

When a function is defined by an equation in x and y, the **graph of the function** is the graph of the equation, that is, the set of points (x, y) in the xy-plane that satisfies the equation.

For example, the graph of the function $f(x) = mx + b$ is a line with slope m and y-intercept $(0, b)$. Because of this, functions of the form $f(x) = mx + b$ are called **linear functions.**

COMMENT: When we select a viewing rectangle to graph a function, the values of Xmin, Xmax give the domain that we wish to view, while Ymin, Ymax give the range that we wish to view. These settings usually do not represent the actual domain and range of the function.

Identify the graph of a function 1

Not every collection of points in the xy-plane represents the graph of a function. Remember, for a function, each number x in the domain has exactly one image y in the range. This means that the graph of a function cannot contain two points with the same x-coordinate and different y-coordinates. Therefore, the graph of a function must satisfy the following **vertical-line test.**

Vertical-Line Test

A set of points in the xy-plane is the graph of a function if and only if every vertical line intersects the graph in at most one point.

In other words, if any vertical line intersects a graph at more than one point, the graph is not the graph of a function.

EXAMPLE 1 Identifying the Graph of a Function

Which of the graphs in Figure 20 are graphs of functions?

FIGURE 20

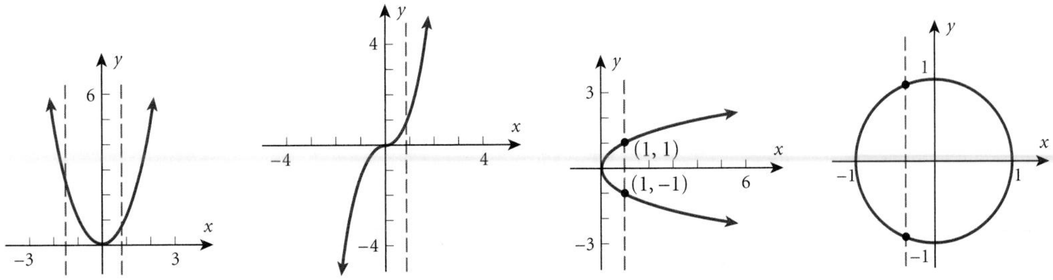

(a) $y = x^2$ (b) $y = x^3$ (c) $x = y^2$ (d) $x^2 + y^2 = 1$

SOLUTION The graphs in Figures 20(a) and 20(b) are graphs of functions, because every vertical line intersects each graph in at most one point. The graphs in Figures 20(c) and 20(d) are not graphs of functions, because there is a vertical line that intersects each graph in more than one point.

NOW WORK PROBLEM 5.

Obtain information from or about the graph of a function **2**

If (x, y) is a point on the graph of a function f, then y is the value of f at x; that is, $y = f(x)$. The next example illustrates how to obtain information about a function if its graph is given.

FIGURE 21

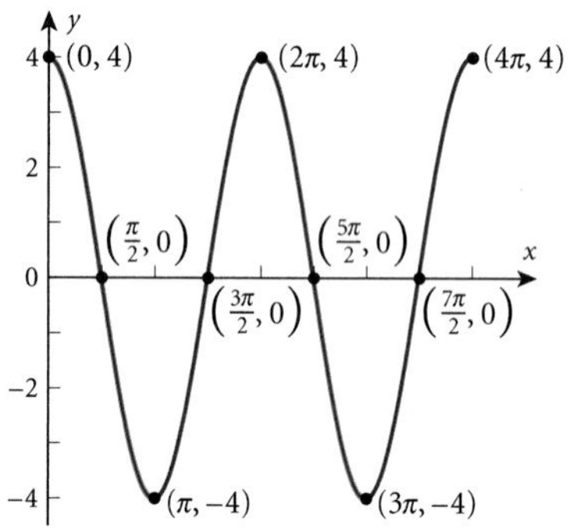

EXAMPLE 2 Obtaining Information from the Graph of a Function

Let f be the function whose graph is given in Figure 21.

(a) What is $f(0), f\left(\frac{3\pi}{2}\right)$, and $f(3\pi)$?

(b) What is the domain of f?

(c) What is the range of f?

(d) List the intercepts. (Recall that these are the points, if any, where the graph crosses or touches the coordinate axes.)

(e) How often does the line $y = 2$ intersect the graph?

(f) For what values of x does $f(x) = -4$?

(g) For what values of x is $f(x) > 0$?

SOLUTION **(a)** Since $(0, 4)$ is on the graph of f, the y-coordinate 4 is the value of f at the x-coordinate 0; that is, $f(0) = 4$. In a similar way, we find that when $x = \frac{3\pi}{2}$ then $y = 0$, so $f\left(\frac{3\pi}{2}\right) = 0$. When $x = 3\pi$, then $y = -4$, so $f(3\pi) = -4$.

(b) To determine the domain of f, we notice that the points on the graph of f will have x-coordinates between 0 and 4π, inclusive; and for each number x between 0 and 4π there is a point $(x, f(x))$ on the graph. The domain of f is $\{x \mid 0 \le x \le 4\pi\}$ or the interval $[0, 4\pi]$.

(c) The points on the graph all have y-coordinates between -4 and 4, inclusive; and for each such number y there is at least one number x in the domain. The range of f is $\{y \mid -4 \le y \le 4\}$ or the interval $[-4, 4]$.

(d) The intercepts are $(0, 4)$, $\left(\frac{\pi}{2}, 0\right)$, $\left(\frac{3\pi}{2}, 0\right)$, $\left(\frac{5\pi}{2}, 0\right)$, and $\left(\frac{7\pi}{2}, 0\right)$.

(e) Draw the horizontal line $y = 2$ on the graph in Figure 21. Then we find that it intersects the graph four times.

(f) Since $(\pi, -4)$ and $(3\pi, -4)$ are the only points on the graph for which $y = f(x) = -4$, we have $f(x) = -4$ when $x = \pi$ and $x = 3\pi$.

(g) To determine where $f(x) > 0$, we look at Figure 21 and determine the x-values for which the y-coordinate is positive. This occurs on the intervals $\left[0, \frac{\pi}{2}\right)$, $\left(\frac{3\pi}{2}, \frac{5\pi}{2}\right)$, and $\left(\frac{7\pi}{2}, 4\pi\right]$. Using inequality notation, $f(x) > 0$ for $0 \le x < \frac{\pi}{2}$, $\frac{3\pi}{2} < x < \frac{5\pi}{2}$, and $\frac{7\pi}{2} < x \le 4\pi$.

When the graph of a function is given, its domain may be viewed as the shadow created by the graph on the x-axis by vertical beams of light. Its range can be viewed as the shadow created by the graph on the y-axis by horizontal beams of light. Try this technique with the graph given in Figure 21.

NOW WORK PROBLEMS 3, 13, and 31(a) and (b).

EXAMPLE 3 Obtaining Information about the Graph of a Function

Consider the function: $f(x) = \dfrac{x}{x+2}$

(a) Is the point $\left(1, \dfrac{1}{2}\right)$ on the graph of f?

(b) If $x = -1$, what is $f(-1)$? What point is on the graph of f?

(c) If $f(x) = 2$, what is x? What point is on the graph of f?

SOLUTION **(a)** When $x = 1$, then

$$f(x) = \frac{x}{x+2}$$
$$f(1) = \frac{1}{1+2} = \frac{1}{3}$$

The point $\left(1, \dfrac{1}{3}\right)$ is on the graph of f; the point $\left(1, \dfrac{1}{2}\right)$ is not.

(b) If $x = -1$, then

$$f(x) = \frac{x}{x+2}$$
$$f(-1) = \frac{-1}{-1+2} = -1$$

The point $(-1, -1)$ is on the graph of f.

(c) If $f(x) = 2$, then

$$f(x) = 2$$
$$\frac{x}{x+2} = 2$$
$$x = 2(x+2) \quad \text{Multiply both sides by } x+2.$$
$$x = 2x + 4 \quad \text{Remove parentheses.}$$
$$x = -4 \quad \text{Solve for } x.$$

If $f(x) = 2$, then $x = -4$. The point $(-4, 2)$ is on the graph of f.

NOW WORK PROBLEM 17.

EXAMPLE 4 Average Cost Function

The average cost $\overline{C}$ of manufacturing x computers per day is given by the function

$$\overline{C}(x) = 0.56x^2 - 34.39x + 1212.57 + \frac{20{,}000}{x}$$

Determine the average cost of manufacturing the following:

(a) 30 computers in a day.
(b) 40 computers in a day.
(c) 50 computers in a day.
(d) Graph the function $\overline{C} = \overline{C}(x), 0 < x \leq 80$.
(e) Create a TABLE with TblStart = 1 and ΔTbl = 1.* Which value of x minimizes the average cost?

SOLUTION **(a)** The average cost of manufacturing $x = 30$ computers is

$$\overline{C}(30) = 0.56(30)^2 - 34.39(30) + 1212.57 + \frac{20{,}000}{30} = \$1351.54$$

(b) The average cost of manufacturing $x = 40$ computers is

$$\overline{C}(40) = 0.56(40)^2 - 34.39(40) + 1212.57 + \frac{20{,}000}{40} = \$1232.97$$

(c) The average cost of manufacturing $x = 50$ computers is

$$\overline{C}(50) = 0.56(50)^2 - 34.39(50) + 1212.57 + \frac{20{,}000}{50} = \$1293.07$$

(d) See Figure 22 for the graph of $\overline{C} = \overline{C}(x)$.

FIGURE 22

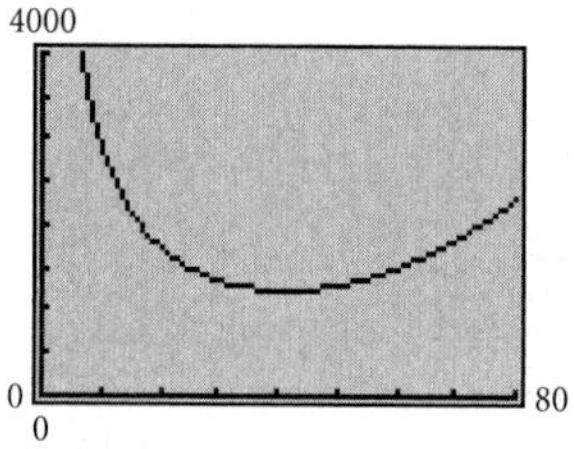

(e) With the function $\overline{C} = \overline{C}(x)$ in Y_1, we create Table 7. We scroll down until we find a value of x for which Y_1 is smallest. Table 8 shows that manufacturing $x = 41$ computers minimizes the average cost at \$1231.75 per computer.

TABLE 7

X	Y1	
1	21179	
2	11146	
3	7781.1	
4	6084	
5	5054.6	
6	4359.7	
7	3856.4	

Y1■.56X²-34.39X...

TABLE 8

X	Y1	
38	1240.7	
39	1235.9	
40	1233	
41	1231.7	
42	1232.2	
43	1234.4	
44	1238.1	

Y1=1231.74487805

NOW WORK PROBLEM 79.

It is easiest to obtain the graph of a function $y = f(x)$ by knowing certain properties that the function has and the impact of these properties on the way that the graph will look. We describe next some properties of functions that we will use in subsequent chapters.

We begin with intercepts and symmetry.

*Consult your user's manual for the keystrokes to use.

Intercepts

If $x = 0$ is in the domain of a function $y = f(x)$, then the y-intercept of the graph of f is obtained by finding the value of f at 0, which is $f(0)$. The x-intercepts of the graph of f, if there are any, are obtained by finding the solutions of the equation $f(x) = 0$.

The x-intercepts of the graph of a function f are called the **zeros of f.**

Even and Odd Functions

Determine even and odd functions from a graph 3

The words *even* and *odd*, when applied to a function f, describe the symmetry that exists for the graph of the function.

A function f is even if and only if whenever the point (x, y) is on the graph of f then the point $(-x, y)$ is also on the graph. Using function notation, we define an even function as follows:

A function f is **even** if, for every number x in its domain, the number $-x$ is also in the domain and

$$f(-x) = f(x)$$

A function f is odd if and only if whenever the point (x, y) is on the graph of f then the point $(-x, -y)$ is also on the graph. Using function notation, we define an odd function as follows:

A function f is **odd** if, for every number x in its domain, the number $-x$ is also in the domain and

$$f(-x) = -f(x)$$

Refer to Section 10.1, where the tests for symmetry are explained. The following results are then evident.

Theorem

A function is even if and only if its graph is symmetric with respect to the y-axis.
A function is odd if and only if its graph is symmetric with respect to the origin.

EXAMPLE 5 Determining Even and Odd Functions from the Graph

Determine whether each graph given in Figure 23 is the graph of an even function, an odd function, or a function that is neither even nor odd.

FIGURE 23

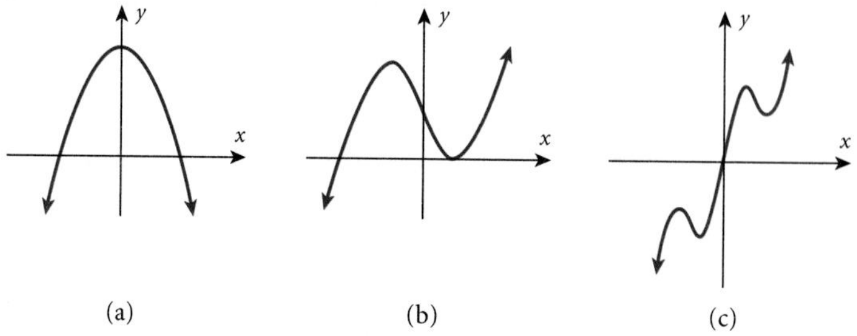

SOLUTION The graph in Figure 23(a) is that of an even function, because the graph is symmetric with respect to the y-axis. The function whose graph is given in Figure 23(b) is neither even nor odd, because the graph is neither symmetric with respect to the y-axis nor symmetric with respect to the origin. The function whose graph is given in Figure 23(c) is odd, because its graph is symmetric with respect to the origin.

NOW WORK PROBLEM 31(d).

4 Identify even and odd functions from the equation

In the next example, we use algebraic techniques to verify whether a given function is even, odd, or neither.

EXAMPLE 6 Identifying Even and Odd Functions Algebraically

Determine whether each of the following functions is even, odd, or neither. Then determine whether the graph is symmetric with respect to the y-axis or with respect to the origin.

(a) $f(x) = x^2 - 5$
(b) $g(x) = x^3 - 1$
(c) $h(x) = 5x^3 - x$
(d) $F(x) = |x|$

SOLUTION **(a)** To determine whether f is even, odd, or neither, we replace x by $-x$ in $f(x) = x^2 - 5$. Then

$$f(-x) = (-x)^2 - 5 = x^2 - 5 = f(x)$$

Since $f(-x) = f(x)$, we conclude that f is an even function, and the graph is symmetric with respect to the y-axis.

(b) We replace x by $-x$ in $g(x) = x^3 - 1$. Then

$$g(-x) = (-x)^3 - 1 = -x^3 - 1$$

Since $g(-x) \neq g(x)$ and $g(-x) \neq -g(x) = -(x^3 - 1) = -x^3 + 1$, we conclude that g is neither even nor odd. The graph is not symmetric with respect to the y-axis nor is it symmetric with respect to the origin.

(c) We replace x by $-x$ in $h(x) = 5x^3 - x$. Then

$$h(-x) = 5(-x)^3 - (-x) = -5x^3 + x = -(5x^3 - x) = -h(x)$$

Since $h(-x) = -h(x)$, h is an odd function, and the graph of h is symmetric with respect to the origin.

(d) We replace x by $-x$ in $F(x) = |x|$. Then

$$F(-x) = |-x| = |-1| \cdot |x| = |x| = F(x)$$

Since $F(-x) = F(x)$, F is an even function, and the graph of F is symmetric with respect to the y-axis.

NOW WORK PROBLEM 57.

Increasing and Decreasing Functions

5 Use a graph to determine where a function is increasing, is decreasing, or is constant

Consider the graph given in Figure 24. If you look from left to right along the graph of the function, you will notice that parts of the graph are rising, parts are falling, and parts are horizontal. In such cases, the function is described as *increasing, decreasing,* or *constant,* respectively.

FIGURE 24

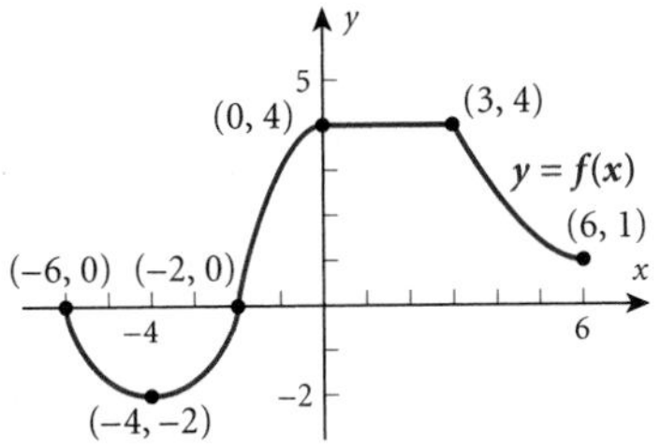

EXAMPLE 7 Determining Where a Function Is Increasing, Decreasing, or Constant from Its Graph

Where is the function in Figure 24 increasing? Where is it decreasing? Where is it constant?

SOLUTION To answer the question of where a function is increasing, where is it decreasing, and where it is constant, we use strict inequalities involving the independent variable x, or we use open intervals of x-coordinates. The graph in Figure 24 is rising (increasing) from the point $(-4, -2)$ to the point $(0, 4)$, so we conclude that it is increasing on the open interval $(-4, 0)$ or for $-4 < x < 0$. The graph is falling (decreasing) from the point $(-6, 0)$ to the point $(-4, -2)$ and from the point $(3, 4)$ to the point $(6, 1)$. We conclude that the graph is decreasing on the open intervals $(-6, -4)$ and $(3, 6)$ or for $-6 < x < -4$ and $3 < x < 6$. The graph is constant on the open interval $(0, 3)$ or for $0 < x < 3$.

More precise definitions follow:

A function f is **increasing** on an open interval I if, for any choice of x_1 and x_2 in I, with $x_1 < x_2$, we have $f(x_1) < f(x_2)$.

A function f is **decreasing** on an open interval I if, for any choice of x_1 and x_2 in I, with $x_1 < x_2$, we have $f(x_1) > f(x_2)$.

A function f is **constant** on an open interval I if, for all choices of x in I, the values $f(x)$ are equal.

Figure 25 illustrates the definitions. The graph of an increasing function goes up from left to right, the graph of a decreasing function goes down from left to right, and the graph of a constant function remains at a fixed height.

FIGURE 25

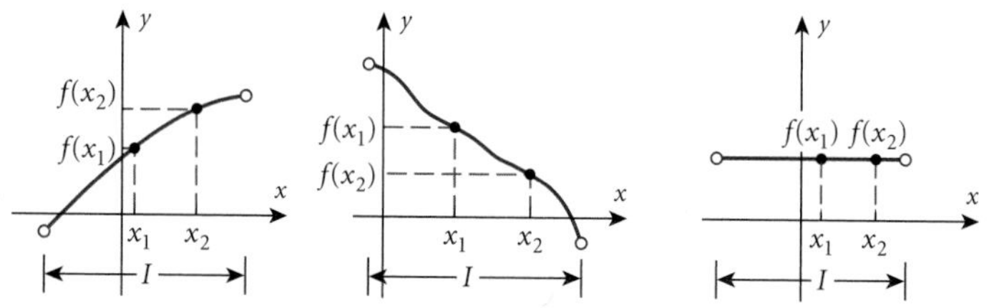

(a) For $x_1 < x_2$ in I, $f(x_1) < f(x_2)$; f is increasing on I

(b) For $x_1 < x_2$ in I, $f(x_1) > f(x_2)$; f is decreasing on I

(c) For all x in I, the values of f are equal; f is constant on I

In Chapter 14 we develop a method for determining where a function is increasing or decreasing using calculus.

NOW WORK PROBLEMS 21, 23, 25, and 31(c).

Local Maximum; Local Minimum

6 Use a graph to locate local maxima and minima

When the graph of a function is increasing to the left of $x = c$ and decreasing to the right of $x = c$, then at c the value of f is largest. This value is called a *local maximum* of f. See Figure 26(a).

FIGURE 26

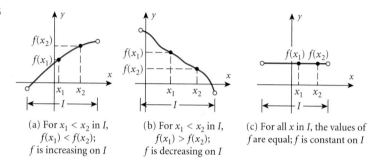

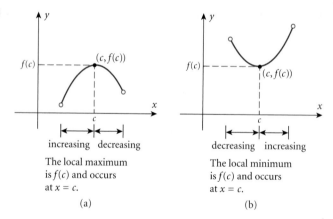

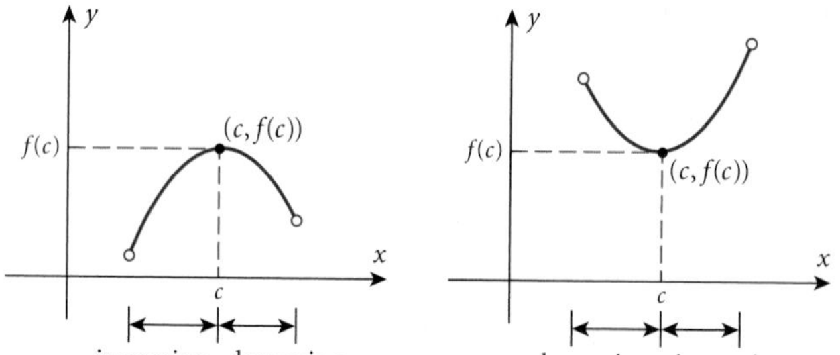

(a) The local maximum is $f(c)$ and occurs at $x = c$.

(b) The local minimum is $f(c)$ and occurs at $x = c$.

When the graph of a function is decreasing to the left of $x = c$ and is increasing to the right of $x = c$, then at c the value of f is the smallest. This value is called a *local minimum* of f. See Figure 26(b).

A function f has a **local maximum at c** if there is an open interval I containing c so that, for all $x \neq c$ in I, $f(x) < f(c)$. We call $f(c)$ a **local maximum of** f.

A function f has a **local minimum at c** if there is an open interval I containing c so that, for all $x \neq c$ in I, $f(x) > f(c)$. We call $f(c)$ a **local minimum of** f.

If f has a local maximum at c, then the value of f at c is greater than the values of f near c. If f has a local minimum at c, then the value of f at c is less than the values of f near c. The word *local* is used to suggest that it is only near c that the value $f(c)$ is largest or smallest.

EXAMPLE 8 Finding Local Maxima and Local Minima from the Graph of a Function and Determining Where the Function Is Increasing, Decreasing, or Constant

FIGURE 27

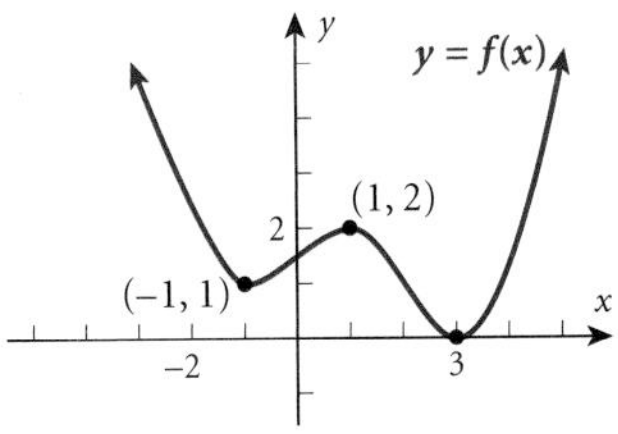

Figure 27 shows the graph of a function f.

(a) At what number(s), if any, does f have a local maximum?
(b) What are the local maxima?
(c) At what number(s), if any, does f have a local minimum?
(d) What are the local minima?
(e) List the intervals on which f is increasing.
(f) List the intervals on which f is decreasing.

SOLUTION

(a) f has a local maximum at 1, since for all x close to 1, $x \neq 1$, we have $f(x) < f(1)$.
(b) The local maximum is $f(1) = 2$.
(c) f has local minima at -1 and at 3.
(d) The local minima are $f(-1) = 1$ and $f(3) = 0$.
(e) The function whose graph is given in Figure 27 is increasing on the interval $(-1, 1)$. It is also increasing for all values of x greater than 3. That is, the function is increasing on the intervals $(-1, 1)$ and $(3, \infty)$ or for $-1 < x < 1$ and $x > 3$.
(f) The function is decreasing for all values of x less than -1. It is also decreasing on the interval $(1, 3)$. That is, the function is decreasing on the intervals $(-\infty, -1)$ and $(1, 3)$ or for $x < -1$ and $1 < x < 3$.

NOW WORK PROBLEMS 27 AND 29.

In Chapter 14, we use calculus to determine the local maxima and the local minima of a function.

A graphing utility may be used to approximate these values by using the MAXIMUM and MINIMUM features.*

7

EXAMPLE 9 Using a Graphing Utility to Approximate Local Maxima and Minima and to Determine Where a Function Is Increasing or Decreasing

(a) Use a graphing utility to graph $f(x) = 6x^3 - 12x + 5$ for $-2 < x < 2$. Approximate where f has a local maximum and where f has a local minimum.
(b) Determine where f is increasing and where it is decreasing.

SOLUTION

(a) Graphing utilities have a feature that finds the maximum or minimum point of a graph within a given interval. Graph the function f for $-2 < x < 2$. Using MAXIMUM, we find that the local maximum is 11.53 and it occurs at $x = -0.82$, rounded to two decimal places. See Figure 28(a). Using MINIMUM, we find that the local minimum is -1.53 and it occurs at $x = 0.82$, rounded to two decimal places. See Figure 28(b).

Consult your owner's manual for the appropriate keystrokes.

FIGURE 28

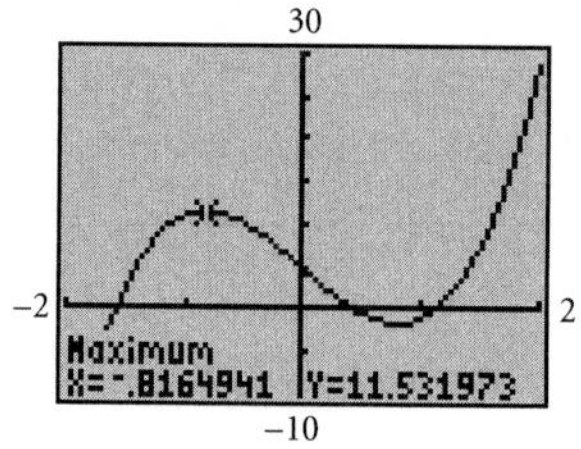

(a)

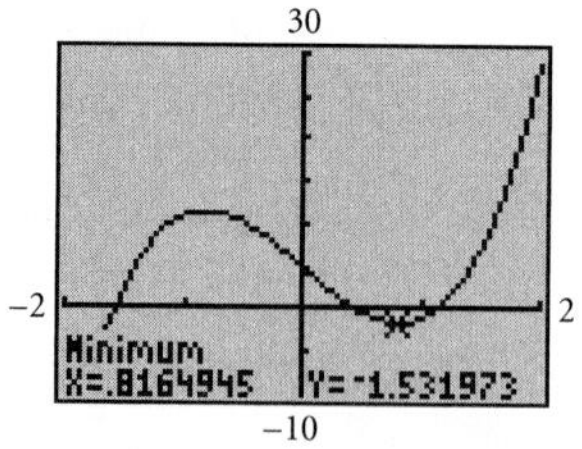

(b)

(b) Looking at Figures 28(a) and (b), we see that the graph of f is increasing from $x = -2$ to $x = -0.82$ and from $x = 0.82$ to $x = 2$, so f is increasing on the intervals $(-2, -0.82)$ and $(0.82, 2)$ or for $-2 < x < -0.82$ and $0.82 < x < 2$. The graph is decreasing from $x = -0.82$ to $x = 0.82$, so f is decreasing on the interval $(-0.82, 0.82)$ or for $-0.82 < x < 0.82$.

NOW WORK PROBLEM 69.

Average Rate of Change

Find the average rate of change of a function 8

Often we are interested in the rate at which functions change. To find the average rate of change of a function between any two points on its graph, we calculate the slope of the line containing the two points.

If c is in the domain of a function $y = f(x)$, the **average rate of change of f** from c to x is defined as

$$\text{Average rate of change} = \frac{\Delta y}{\Delta x} = \frac{f(x) - f(c)}{x - c}, \quad x \neq c \qquad (1)$$

Recall that the symbol Δy in (1) is the "change in y," and Δx is the "change in x." The average rate of change of f is the change in y divided by the change in x.

EXAMPLE 10 Finding the Average Rate of Change

Find the average rate of change of $f(x) = 3x^2$:

(a) From 1 to 3 **(b)** From 1 to 5 **(c)** From 1 to 7

SOLUTION **(a)** The average rate of change of $f(x) = 3x^2$ from 1 to 3 is

$$\frac{\Delta y}{\Delta x} = \frac{f(3) - f(1)}{3 - 1} = \frac{27 - 3}{3 - 1} = \frac{24}{2} = 12$$

(b) The average rate of change of $f(x) = 3x^2$ from 1 to 5 is

$$\frac{\Delta y}{\Delta x} = \frac{f(5) - f(1)}{5 - 1} = \frac{75 - 3}{5 - 1} = \frac{72}{4} = 18$$

(c) The average rate of change of $f(x) = 3x^2$ from 1 to 7 is

$$\frac{\Delta y}{\Delta x} = \frac{f(7) - f(1)}{7 - 1} = \frac{147 - 3}{7 - 1} = \frac{144}{6} = 24$$

The average rate of change of a function has an important geometric interpretation. Look at the graph of $y = f(x)$ in Figure 29. We have labeled two points on the graph: $(c, f(c))$ and $(x, f(x))$. The line containing these two points is called a **secant line;** its slope is

$$m_{\text{sec}} = \frac{f(x) - f(c)}{x - c}$$

FIGURE 29

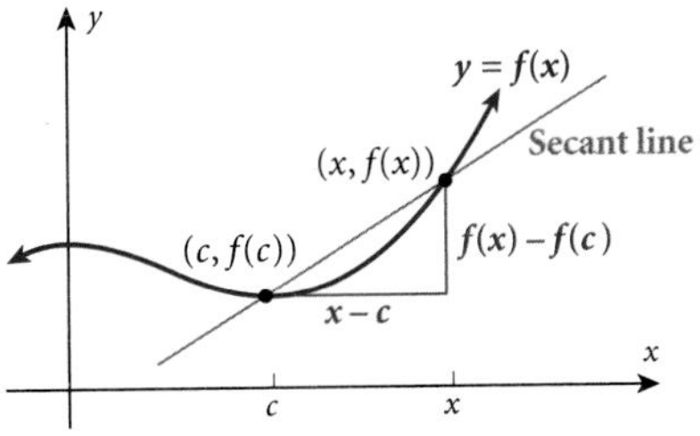

Slope of the Secant Line

The average rate of change of a function equals the slope of the secant line containing two points on its graph.

EXAMPLE 11 Finding the Average Rate of Change of a Function

(a) Find the average rate of change of $f(x) = 2x^2 - 3x$ from 1 to x.
(b) Use this result to find the slope of the secant line containing $(1, f(1))$ and $(2, f(2))$.
(c) Find an equation of this secant line.

SOLUTION **(a)** The average rate of change of f from 1 to x is

$$\frac{\Delta y}{\Delta x} = \frac{f(x) - f(1)}{x - 1} \qquad x \neq 1$$

$$= \frac{2x^2 - 3x - (-1)}{x - 1} \qquad f(x) = 2x^2 - 3x; f(1) = 2 \cdot 1^2 - 3(1) = -1$$

$$= \frac{2x^2 - 3x + 1}{x - 1} \qquad \text{Simplify.}$$

$$= \frac{(2x - 1)(x - 1)}{x - 1} \qquad \text{Factor the numerator.}$$

$$= 2x - 1 \qquad x \neq 1; \text{cancel } x - 1.$$

(b) The slope of the secant line containing $(1, f(1))$ and $(2, f(2))$ is the average rate of change of f from 1 to 2. Using $x = 2$ in part (a), we obtain $m_{\text{sec}} = 2(2) - 1 = 3$.

(c) Use the point–slope form to find the equation of the secant line.

$$y - y_1 = m_{\text{sec}}(x - x_1) \qquad \text{Point–slope form of the secant line.}$$

$$y + 1 = 3(x - 1) \qquad x_1 = 1, y_1 = f(1) = -1; m_{\text{sec}} = 3$$

$$y + 1 = 3x - 3$$

$$y = 3x - 4 \qquad \text{Slope–intercept form of the secant line.}$$

NOW WORK PROBLEM 49.

EXERCISE 10.3 Answers to Odd-Numbered Problems Begin on Page AN-50.

In Problems 1–12, determine whether the graph is that of a function by using the vertical-line test. If it is, use the graph to find:

(a) Its domain and range; (b) The intercepts, if any; (c) Any symmetry with respect to the x-axis, the y-axis, or the origin.

1.

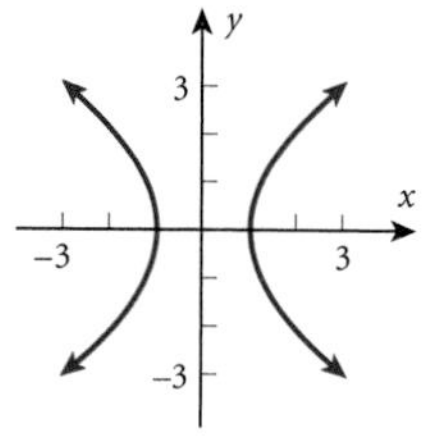

2.

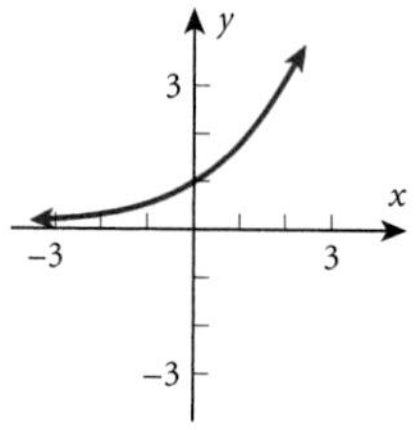

3.

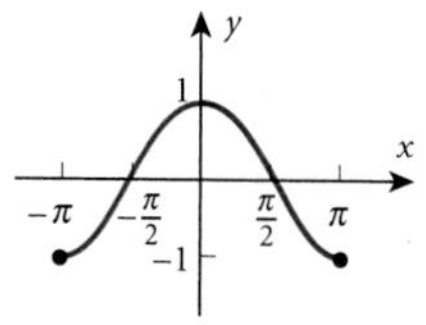

4.

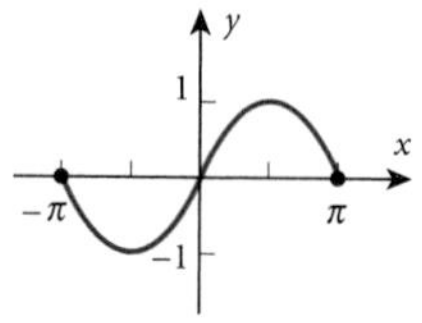

5.

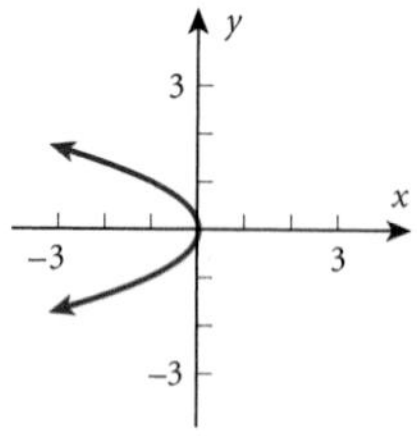

6.

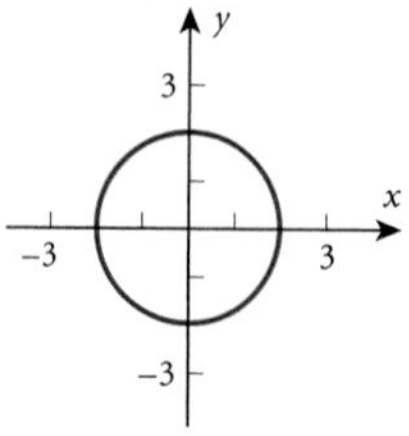

7.

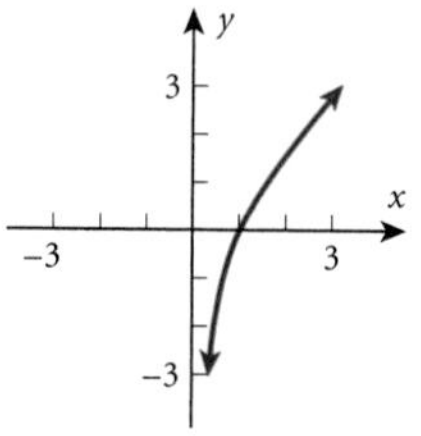

8.

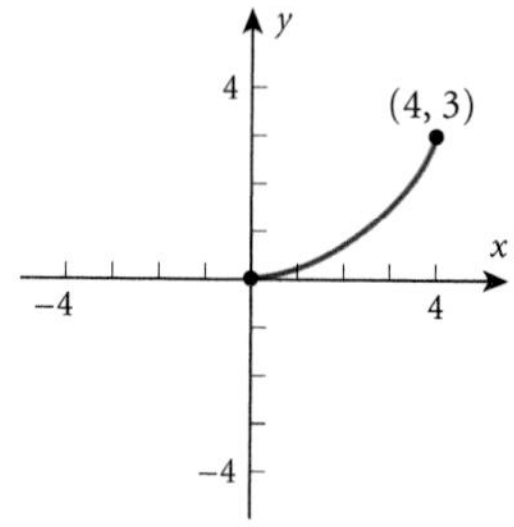

9.

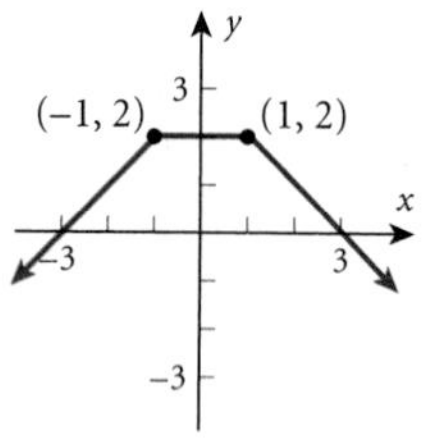

10.

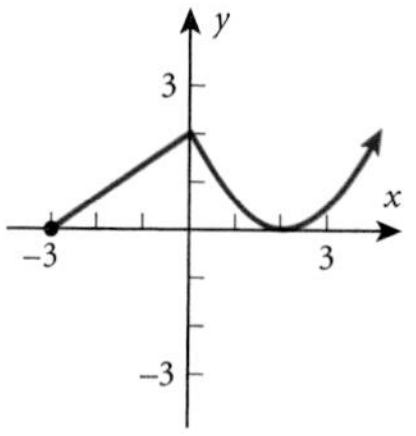

11.

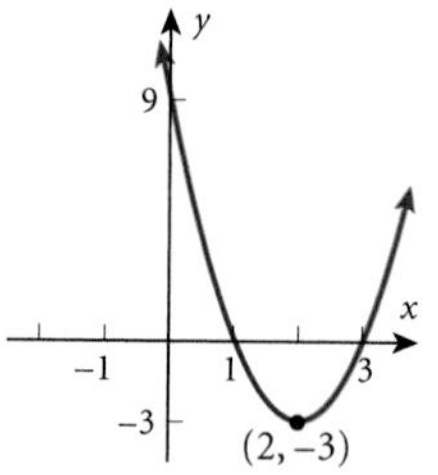

12.

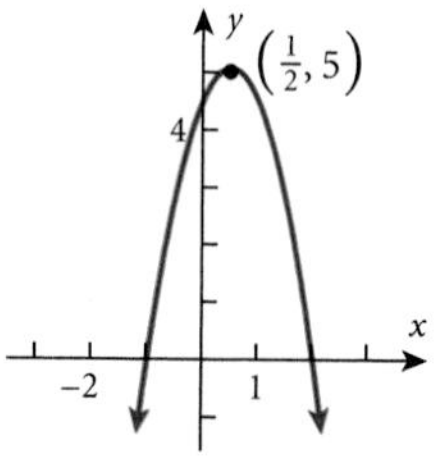

13. Use the graph of the function f given below to answer parts (a)–(n).

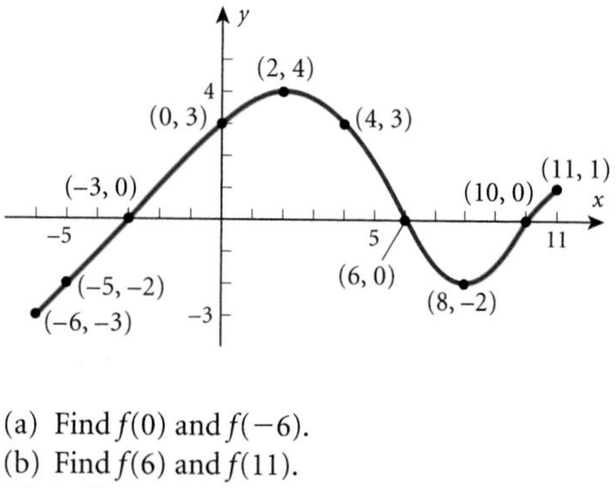

(a) Find $f(0)$ and $f(-6)$.
(b) Find $f(6)$ and $f(11)$.
(c) Is $f(3)$ positive or negative?
(d) Is $f(-4)$ positive or negative?
(e) For what numbers x is $f(x) = 0$?
(f) For what numbers x is $f(x) > 0$?
(g) What is the domain of f?
(h) What is the range of f?
(i) What are the x-intercepts?
(j) What is the y-intercept?
(k) How often does the line $y = \frac{1}{2}$ intersect the graph?
(l) How often does the line $x = 5$ intersect the graph?
(m) For what values of x does $f(x) = 3$?
(n) For what values of x does $f(x) = -2$?

14. Use the graph of the function f given below to answer parts (a)–(n).

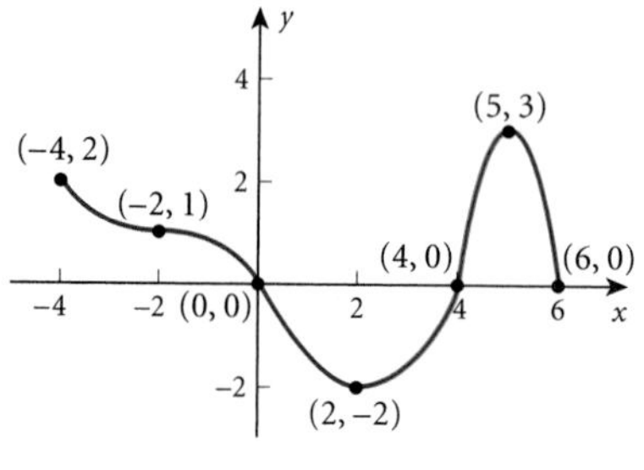

(a) Find $f(0)$ and $f(6)$.
(b) Find $f(2)$ and $f(-2)$.
(c) Is $f(3)$ positive or negative?
(d) Is $f(-1)$ positive or negative?

(e) For what numbers is $f(x) = 0$?
(f) For what numbers is $f(x) < 0$?
(g) What is the domain of f?
(h) What is the range of f?
(i) What are the x-intercepts?
(j) What is the y-intercept?
(k) How often does the line $y = -1$ intersect the graph?
(l) How often does the line $x = 1$ intersect the graph?
(m) For what value of x does $f(x) = 3$?
(n) For what value of x does $f(x) = -2$?

In Problems 15–20, answer the questions about the given function.

15. $f(x) = 2x^2 - x - 1$

(a) Is the point $(-1, 2)$ on the graph of f?
(b) If $x = -2$, what is $f(x)$? What point is on the graph of f?
(c) If $f(x) = -1$, what is x? What points(s) are on the graph of f?
(d) What is the domain of f?
(e) List the x-intercepts, if any, of the graph of f.
(f) List the y-intercept, if there is one, of the graph of f.

16. $f(x) = -3x^2 + 5x$

(a) Is the point $(-1, 2)$ on the graph of f?
(b) If $x = -2$, what is $f(x)$? What point is on the graph of f?
(c) If $f(x) = -2$, what is x? What point(s) are on the graph of f?
(d) What is the domain of f?
(e) List the x-intercepts, if any, of the graph of f.
(f) List the y-intercept, if there is one, of the graph of f.

17. $f(x) = \dfrac{x + 2}{x - 6}$

(a) Is the point $(3, 14)$ on the graph of f?
(b) If $x = 4$, what is $f(x)$? What point is on the graph of f?
(c) If $f(x) = 2$, what is x? What point(s) are on the graph of f?
(d) What is the domain of f?
(e) List the x-intercepts, if any, of the graph of f.
(f) List the y-intercept, if there is one, of the graph of f.

18. $f(x) = \dfrac{x^2 + 2}{x + 4}$

(a) Is the point $\left(1, \dfrac{3}{5}\right)$ on the graph of f?
(b) If $x = 0$, what is $f(x)$? What point is on the graph of f?
(c) If $f(x) = \dfrac{1}{2}$, what is x? What point(s) are on the graph of f?
(d) What is the domain of f?
(e) List the x-intercepts, if any, of the graph of f.
(f) List the y-intercept, if there is one, of the graph of f.

19. $f(x) = \dfrac{2x^2}{x^4 + 1}$

(a) Is the point $(-1, 1)$ on the graph of f?
(b) If $x = 2$, what is $f(x)$? What point is on the graph of f?
(c) If $f(x) = 1$, what is x? What points(s) are on the graph of f?
(d) What is the domain of f?
(e) List the x-intercepts, if any, of the graph of f.
(f) List the y-intercept, if there is one, of the graph of f.

20. $f(x) = \dfrac{2x}{x - 2}$

(a) Is the point $\left(\dfrac{1}{2}, -\dfrac{2}{3}\right)$ on the graph of f?
(b) If $x = 4$, what is $f(x)$? What point is on the graph of f?
(c) If $f(x) = 1$, what is x? What point(s) are on the graph of f?
(d) What is the domain of f?
(e) List the x-intercepts, if any, of the graph of f.
(f) List the y-intercept, if there is one, of the graph of f.

In Problems 21–30, use the graph of the function f given below.

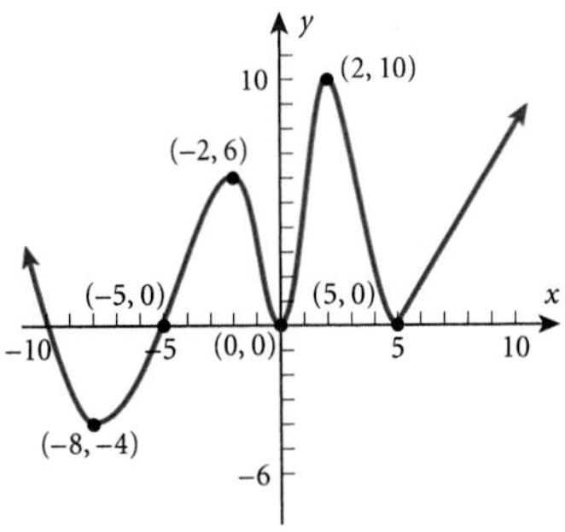

21. Is f increasing on the interval $(-8, -2)$?

22. Is f decreasing on the interval $(-8, -4)$?

23. Is f increasing on the interval $(2, 10)$?

24. Is f decreasing on the interval $(2, 5)$?

25. List the interval(s) on which f is increasing.

26. List the interval(s) on which f is decreasing.

27. Is there a local maximum at 2? If yes, what is it?

28. Is there a local maximum at 5? If yes, what is it?

29. List the numbers at which f has a local maximum. What are these local maxima?

30. List the numbers at which f has a local minimum. What are these local minima?

In Problems 31–38, the graph of a function is given. Use the graph to find:

(a) The intercepts, if any
(b) Its domain and range
(c) The intervals on which it is increasing, decreasing, or constant
(d) Whether it is even, odd, or neither

31.

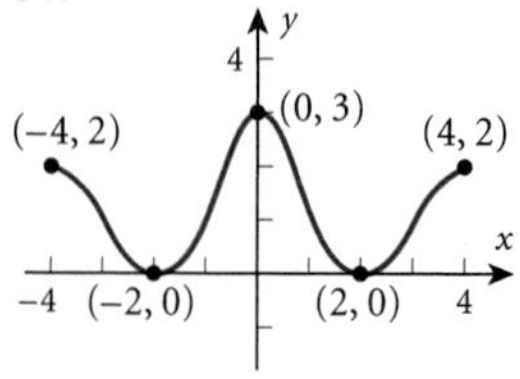

32.

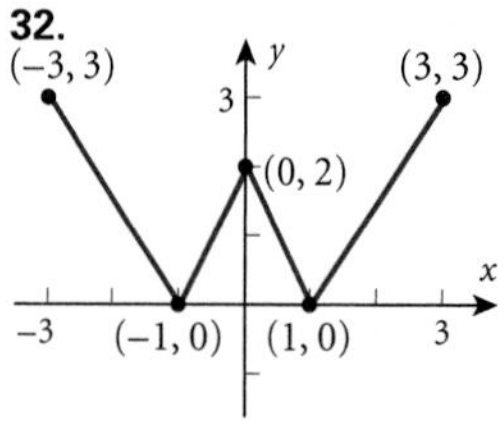

33.

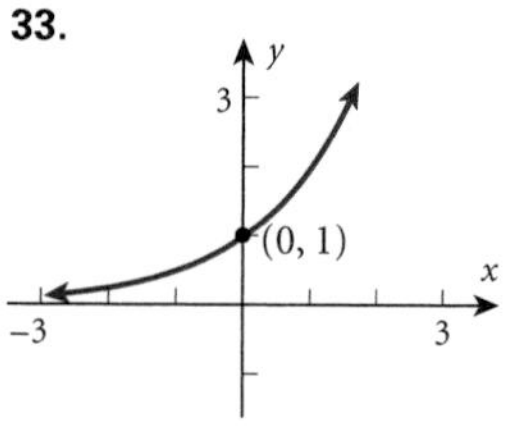

34.

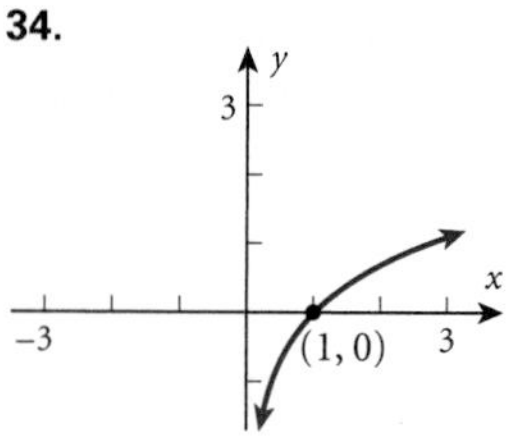

35.

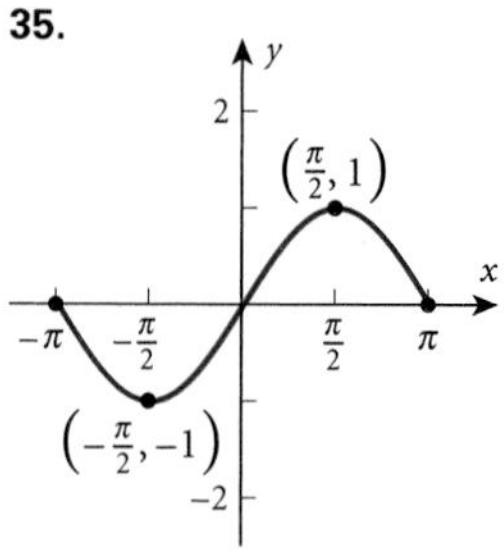

36.

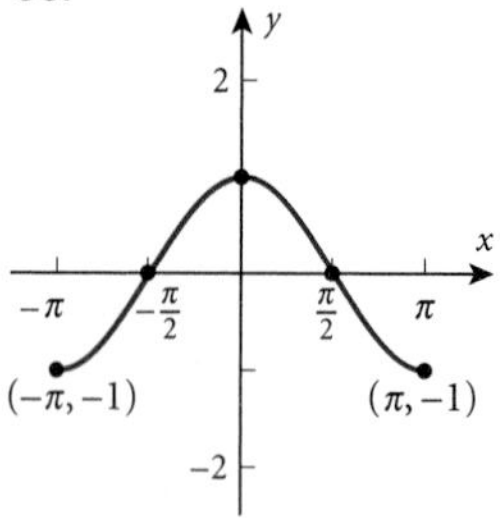

37.

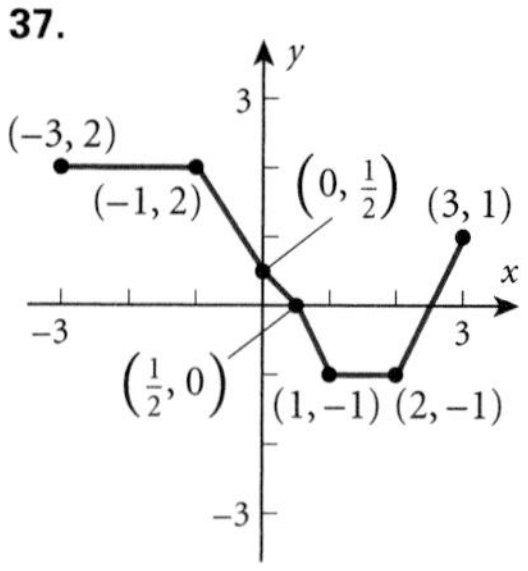

38.

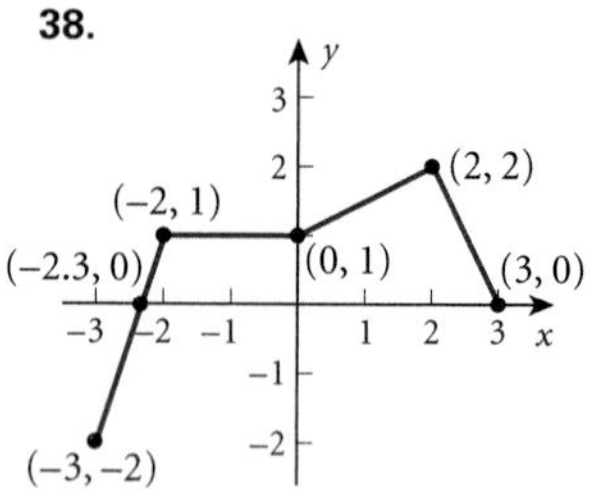

In Problems 39–42, the graph of a function f is given. Use the graph to find:

(a) *The numbers, if any, at which f has a local maximum. What are these local maxima?*
(b) *The numbers, if any, at which f has a local minimum. What are these local minima?*

39.

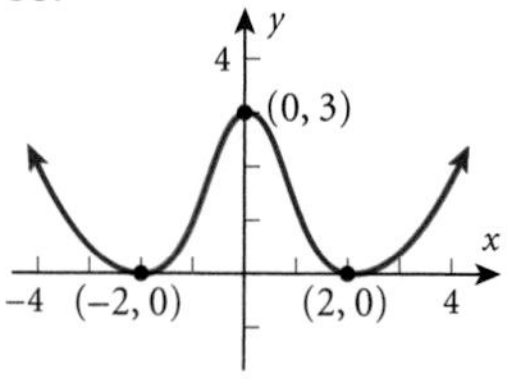

40.

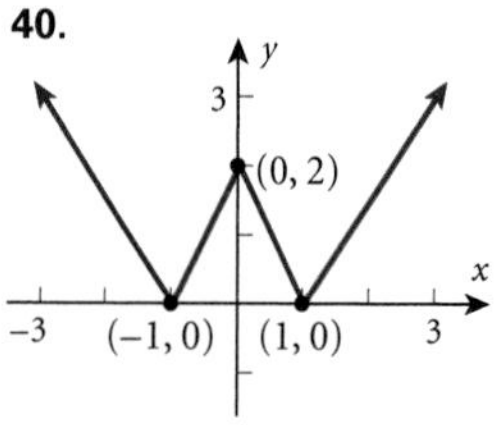

41.

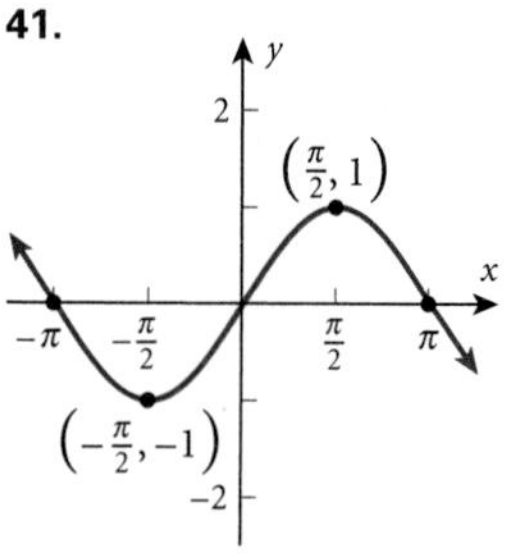

42.

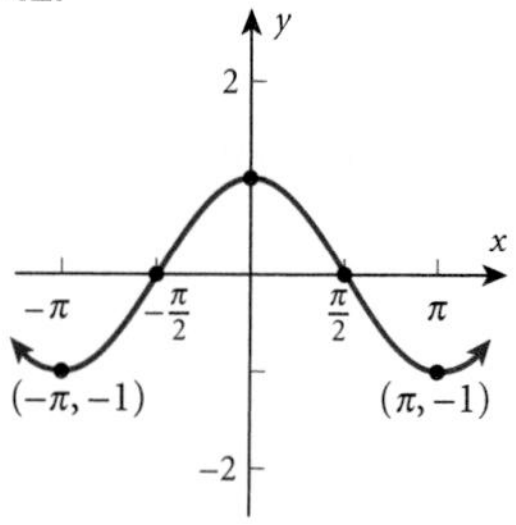

43. Find the average rate of change of $f(x) = -2x^2 + 4$
(a) from 0 to 2 (b) from 1 to 3 (c) from 1 to 4

44. Find the average rate of change of $f(x) = -x^3 + 1$
(a) from 0 to 2 (b) from 1 to 3 (c) from −1 to 1

In Problems 45–56, (a) for each function find the average rate of change of f from 1 to x:

$$\frac{f(x) - f(1)}{x - 1}, \quad x \neq 1$$

(b) *Use the result from part (a) to compute the average rate of change from $x = 1$ to $x = 2$. Be sure to simplify.*
(c) *Find an equation of the secant line containing $(1, f(1))$ and $(2, f(2))$.*

45. $f(x) = 5x$
46. $f(x) = -4x$
47. $f(x) = 1 - 3x$
48. $f(x) = x^2 + 1$
49. $f(x) = x^2 - 2x$
50. $f(x) = x - 2x^2$
51. $f(x) = x^3 - x$
52. $f(x) = x^3 + x$
53. $f(x) = \dfrac{2}{x + 1}$
54. $f(x) = \dfrac{1}{x^2}$
55. $f(x) = \sqrt{x}$
56. $f(x) = \sqrt{x + 3}$

Problems 57–68, determine algebraically whether each function is even, odd, or neither.

57. $f(x) = 4x^3$
58. $f(x) = 2x^4 - x^2$
59. $g(x) = -3x^2 - 5$
60. $h(x) = 3x^3 + 5$
61. $F(x) = \sqrt[3]{x}$
62. $G(x) = \sqrt{x}$
63. $f(x) = x + |x|$
64. $f(x) = \sqrt[3]{2x^2 + 1}$
65. $g(x) = \dfrac{1}{x^2}$
66. $h(x) = \dfrac{x}{x^2 - 1}$
67. $h(x) = \dfrac{-x^3}{3x^2 - 9}$
68. $F(x) = \dfrac{2x^2}{x^3 + 1}$

In Problems 69–76, use a graphing utility to graph each function over the indicated interval and approximate any local maxima and local minima. Determine where the function is increasing and where it is decreasing. Round answers to two decimal places.

69. $f(x) = x^3 - 3x + 2;\ (-2, 2)$

70. $f(x) = x^3 - 3x^2 + 5;\ (-1, 3)$

71. $f(x) = x^5 - x^3;\ (-2, 2)$

72. $f(x) = x^4 - x^2;\ (-2, 2)$

73. $f(x) = -0.2x^3 - 0.6x^2 + 4x - 6;\ (-6, 4)$

74. $f(x) = -0.4x^3 + 0.6x^2 + 3x - 2;\ (-4, 5)$

75. $f(x) = 0.25x^4 + 0.3x^3 - 0.9x^2 + 3;\ (-3, 2)$

76. $f(x) = -0.4x^4 - 0.5x^3 + 0.8x^2 - 2;\ (-3, 2)$

---

77. For the function $f(x) = x^2$, compute each average rate of change:

(a) from 0 to 1
(b) from 0 to 0.5
(c) from 0 to 0.1
(d) from 0 to 0.01
(e) from 0 to 0.001
(f) Graph each of the secant lines. Set the viewing rectangle to: Xmin = −0.2, Xmax = 1.2, Xscl = 0.1, Ymin = −0.2, Ymax = 1.2, Yscl = 0.1.
(g) What do you think is happening to the secant lines?
(h) What is happening to the slopes of the secant lines? Is there some number they are getting closer to? What is that number?

78. For the function $f(x) = x^2$, compute each average rate of change:

(a) from 1 to 2
(b) from 1 to 1.5
(c) from 1 to 1.1
(d) from 1 to 1.01
(e) from 1 to 1.001
(f) Graph each of the secant lines. Set the viewing rectangle to: Xmin = −0.5, Xmax = 2.5, Xscl = 0.1, Ymin = −1, Ymax = 4, Yscl = 0.1.
(g) What do you think is happening to the secant lines?
(h) What is happening to the slopes of the secant lines? Is there some number they are getting closer to? What is that number?

79. Motion of a Golf Ball A golf ball is hit with an initial velocity of 130 feet per second at an inclination of 45° to the horizontal. In physics, it is established that the height h of the golf ball is given by the function

$$h(x) = \frac{-32x^2}{130^2} + x$$

where x is the horizontal distance that the golf ball has traveled.

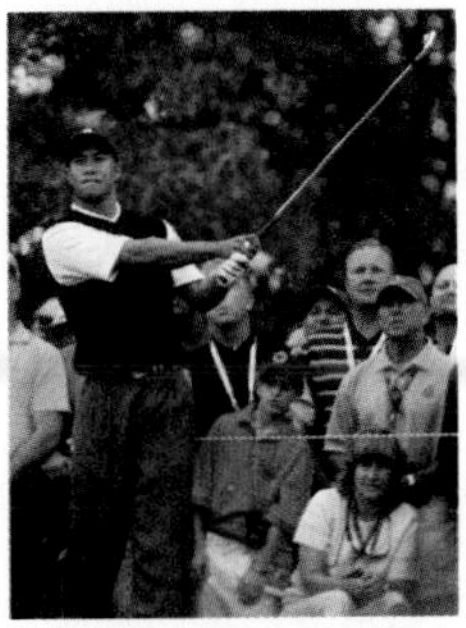

(a) Determine the height of the golf ball after it has traveled 100 feet.
(b) What is the height after it has traveled 300 feet?
(c) What is the height after it has traveled 500 feet?
(d) How far was the golf ball hit?
(e) Use a graphing utility to graph the function $h = h(x)$.
(f) Use a graphing utility to determine the distance that the ball has traveled when the height of the ball is 90 feet.
(g) Create a TABLE with TblStart = 0 and ΔTbl = 25.
(h) To the nearest 25 feet, how far does the ball travel before it reaches a maximum height? What is the maximum height?
(i) Adjust the value of ΔTbl until you determine the distance, to within 1 foot, that the ball travels before it reaches a maximum height.

80. Effect of Elevation on Weight If an object weighs m pounds at sea level, then its weight W (in pounds) at a height of h miles above sea level is given approximately by

$$W(h) = m\left(\frac{4000}{4000 + h}\right)^2$$

(a) If Amy weighs 120 pounds at sea level, how much will she weigh on Pike's Peak, which is 14,110 feet above sea level? (1 mile = 5280 feet)
(b) Use a graphing utility to graph the function $W = W(h)$. Use $m = 120$ pounds.
(c) Create a Table with TblStart = 0 and ΔTbl = 0.5 to see how the weight W varies as h changes from 0 to 5 miles.
(d) At what height will Amy weigh 119.95 pounds?
(e) Does your answer to part (d) seem reasonable?

81. Constructing an Open Box An open box with a square base is to be made from a square piece of cardboard 24 inches on a side by cutting out a square from each corner and turning up the sides (see the figure).

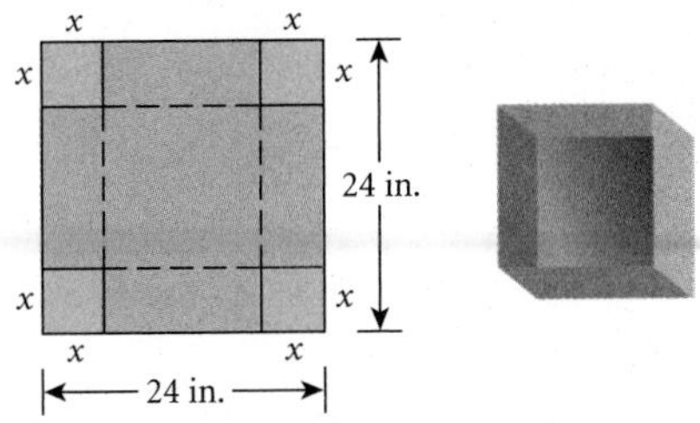

(a) Express the volume V of the box as a function of the length x of the side of the square cut from each corner.

(b) What is the volume if a 3-inch square is cut out?

(c) What is the volume if a 10-inch square is cut out?

(d) Graph $V = V(x)$. For what value of x is V largest?

82. **Constructing an Open Box** A open box with a square base is required to have a volume of 10 cubic feet.

(a) Express the amount A of material used to make such a box as a function of the length x of a side of the square base.

(b) How much material is required for a base 1 foot by 1 foot?

(c) How much material is required for a base 2 feet by 2 feet?

(d) Graph $A = A(x)$. For what value of x is A smallest?

83. **Minimum Average Cost** The average cost of producing x riding lawn mowers is given by

$$\overline{C}(x) = 0.3x^2 + 21x - 251 + \frac{2500}{x}$$

(a) Use a graphing utility to graph $\overline{C}$.

(b) Determine the number of riding lawn mowers to produce in order to minimize average cost.

(c) What is the minimum average cost?

84. Match each function with the graph that best describes the situation. Discuss the reason for your choice.

(a) The cost of building a house as a function of its square footage

(b) The height of an egg dropped from a 300-foot building as a function of time

(c) The height of a human as a function of time

(d) The demand for Big Macs as a function of price

(e) The height of a child on a swing as a function of time

$$(\text{I}) \qquad (\text{II}) \qquad (\text{III}) \qquad (\text{IV}) \qquad (\text{V})$$

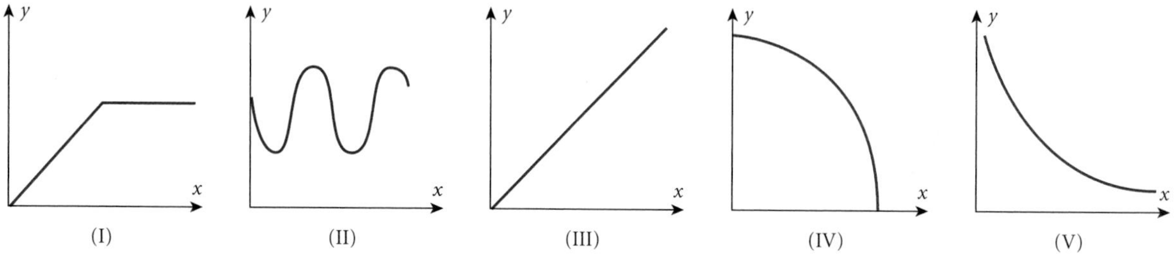

85. Match each function with the graph that best describes the situation. Discuss the reason for your choice.

(a) The temperature of a bowl of soup as a function of time

(b) The number of hours of daylight per day over a two-year period

(c) The population of Florida as a function of time

(d) The distance of a car traveling at a constant velocity as a function of time

(e) The height of a golf ball hit with a 7-iron as a function of time

$$(\text{I}) \qquad (\text{II}) \qquad (\text{III}) \qquad (\text{IV}) \qquad (\text{V})$$

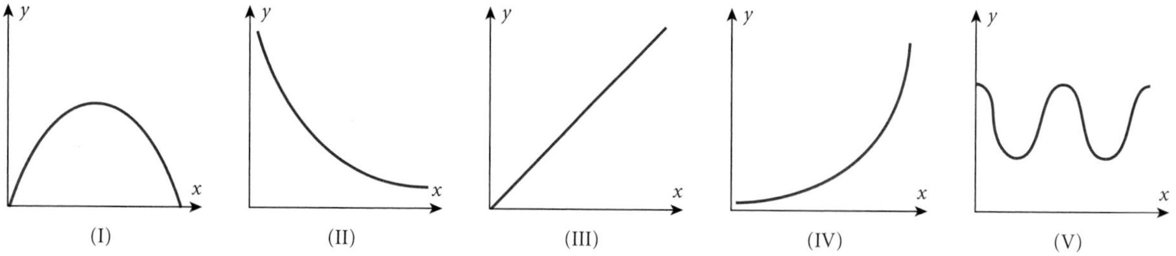

86. Draw the graph of a function that has the following characteristics:

Domain: all real numbers

Range: all real numbers

Intercepts: $(0, -3)$ and $(2, 0)$

A local maximum of -2 is at -1; a local minimum of -6 is at 2. Compare your graph with others. Comment on any differences.

87. Redo Problem 86 with the following additional information:

Increasing on $(-\infty, -1), (2, \infty)$

Decreasing on $(-1, 2)$

Again compare your graph with others and comment on any differences.

88. How many x-intercepts can a function defined on an interval have if it is increasing on that interval? Explain.

89. Suppose that a friend of yours does not understand the idea of increasing and decreasing functions. Provide an explanation complete with graphs that clarifies the idea.

90. Can a function be both even and odd? Explain.

91. Describe how you would proceed to find the domain and range of a function if you were given its graph. How would your strategy change if, instead, you were given the equation defining the function?

92. Is a graph that consists of a single point the graph of a function? If so, can you write the equation of such a function?

93. Define some functions that pass through (0, 0) and (1, 1) and are increasing for $x \geq 0$. Begin your list with $y = \sqrt{x}$, $y = x$, and $y = x^2$. Can you propose a general result about such functions?

10.4 Library of Functions; Piecewise-defined Functions

PREPARING FOR THIS SECTION *Before getting started, review the following:*

- Square Roots (Appendix A, Section A.2, pp. A-22–A-23)
- Absolute Value (Appendix A, Section A.2, p. A-17–A-18)
- *n*th Roots (Appendix A, Section A.8, pp. A-73–A-74)

OBJECTIVES
1 Graph the functions listed in the library of functions
2 Graph piecewise-defined functions

FIGURE 30

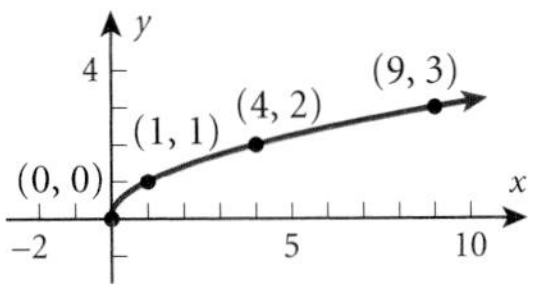

We now introduce a few more functions to add to our list of important functions. We begin with the *square root function.*

In Section 10.1 we graphed the equation $x = y^2$. If we solve the equation for y and restrict y so that $y \geq 0$, the equation $x = y^2$, $y \geq 0$, can be written as $y = f(x) = \sqrt{x}$. Figure 30 shows a graph of $f(x) = \sqrt{x}$.

Based on the graph of $f(x) = \sqrt{x}$, we have the following properties:

Properties of $f(x) = \sqrt{x}$

1. The domain and range of $f(x) = \sqrt{x}$ are the set of nonnegative real numbers.
2. The x-intercept of the graph of $f(x) = \sqrt{x}$ is $(0, 0)$. The y-intercept of the graph of $f(x) = \sqrt{x}$ is also $(0, 0)$.
3. The function is neither even nor odd.
4. It is increasing on the interval $(0, \infty)$.

EXAMPLE 1 Graphing the Cube Root Function

(a) Determine whether $f(x) = \sqrt[3]{x}$ is even, odd, or neither. State whether the graph of f is symmetric with respect to the y-axis or symmetric with respect to the origin.
(b) Determine the intercepts, if any, of the graph of $f(x) = \sqrt[3]{x}$.
(c) Graph $f(x) = \sqrt[3]{x}$.

SOLUTION (a) Because

$$f(-x) = \sqrt[3]{-x} = -\sqrt[3]{x} = -f(x)$$

the function is odd. The graph of f is symmetric with respect to the origin.

(b) Since $f(0) = \sqrt[3]{0} = 0$, the y-intercept is $(0, 0)$. The x-intercept is found by solving the equation $f(x) = 0$.

$$f(x) = 0$$
$$\sqrt[3]{x} = 0 \qquad f(x) = \sqrt[3]{x}$$
$$x = 0 \qquad \text{Cube both sides of the equation.}$$

The x-intercept is also $(0, 0)$.

(c) We use the function to form Table 9 and obtain some points on the graph. Because of the symmetry with respect to the origin, we only need to find points (x, y) for which $x \geq 0$. Figure 31 shows the graph of $f(x) = \sqrt[3]{x}$.

TABLE 9

x	$y = \sqrt[3]{x}$	(x, y)
0	0	$(0, 0)$
$\frac{1}{8}$	$\frac{1}{2}$	$\left(\frac{1}{8}, \frac{1}{2}\right)$
1	1	$(1, 1)$
2	$\sqrt[3]{2} \approx 1.26$	$(2, \sqrt[3]{2})$
8	2	$(8, 2)$

FIGURE 31

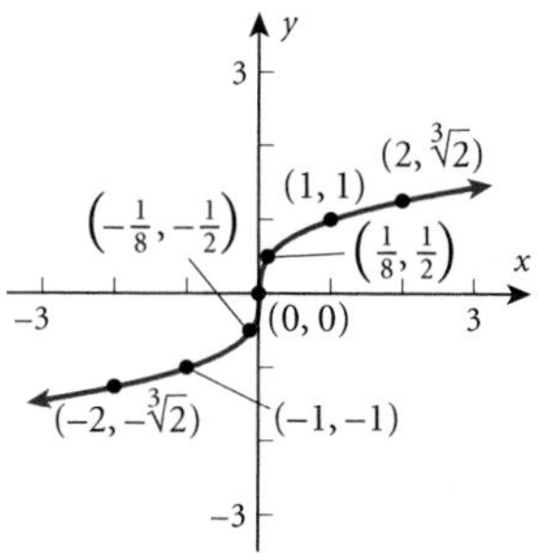

From the results of Example 1 and Figure 31, we have the following properties of the cube root function.

> **Properties of $f(x) = \sqrt[3]{x}$**
>
> **1.** The domain and range of $f(x) = \sqrt[3]{x}$ are the set of real numbers.
> **2.** The x-intercept of the graph of $f(x) = \sqrt[3]{x}$ is $(0, 0)$. The y-intercept of the graph of $f(x) = \sqrt[3]{x}$ is also $(0, 0)$.
> **3.** The function is odd.
> **4.** It is increasing on the interval $(-\infty, \infty)$.

EXAMPLE 2 Graphing the Absolute Value Function

(a) Determine whether $f(x) = |x|$ is even, odd, or neither. State whether the graph of f is symmetric with respect to the y-axis or symmetric with respect to the origin.
(b) Determine the intercepts, if any, of the graph of $f(x) = |x|$.
(c) Graph $f(x) = |x|$.

SOLUTION **(a)** Because

$$f(-x) = |-x| = |x| = f(x)$$

the function is even. The graph of f is symmetric with respect to the y-axis.

(b) Since $f(0) = |0| = 0$, the y-intercept is $(0, 0)$. The x-intercept is found by solving the equation $f(x) = |x| = 0$. So the x-intercept is also $(0, 0)$.

(c) We use the function to form Table 10 and obtain some points on the graph. Because of the symmetry with respect to the y-axis, we only need to find points (x, y) for which $x \geq 0$. Figure 32 shows the graph of $f(x) = |x|$.

TABLE 10

x	$y = |x|$	(x, y)
0	0	$(0, 0)$
1	1	$(1, 1)$
2	2	$(2, 2)$
3	3	$(3, 3)$

FIGURE 32

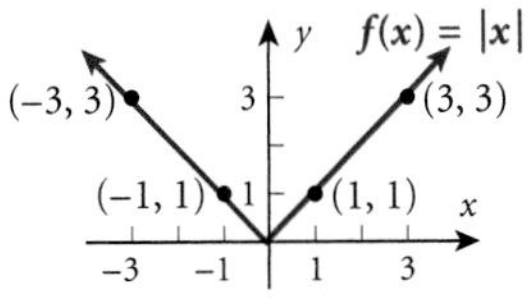

From the results of Example 2 and Figure 32, we have the following properties of the absolute value function.

Properties of $f(x) = |x|$

1. The domain of $f(x) = |x|$ is the set of real numbers; the range is the set of nonnegative real numbers.

2. The x-intercept of the graph of $f(x) = |x|$ is $(0, 0)$. The y-intercept of the graph of $f(x) = |x|$ is also $(0, 0)$.

3. The function is even.

4. It is decreasing on the interval $(-\infty, 0)$. It is increasing on the interval $(0, \infty)$.

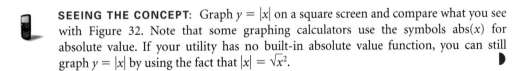

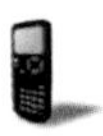

SEEING THE CONCEPT: Graph $y = |x|$ on a square screen and compare what you see with Figure 32. Note that some graphing calculators use the symbols abs(x) for absolute value. If your utility has no built-in absolute value function, you can still graph $y = |x|$ by using the fact that $|x| = \sqrt{x^2}$.

Library of Functions

Graph the functions listed $\boxed{1}$ **in the library of functions**

We now provide a summary of the key functions that we have encountered. In going through this list, pay special attention to the properties of each function, particularly to the shape of each graph. Knowing these graphs will lay the foundation for later graphing techniques.

Linear Function

$$f(x) = mx + b \qquad m \text{ and } b \text{ are real numbers}$$

FIGURE 33 Linear function

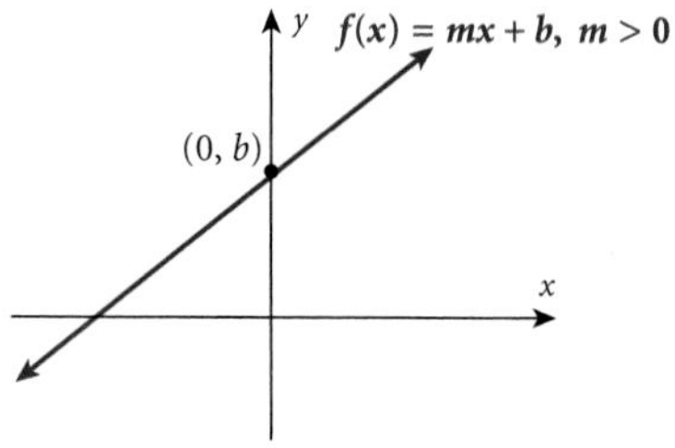

See Figure 33.

The domain of a **linear function** is the set of all real numbers. The graph of this function is a nonvertical line with slope m and y-intercept $(0, b)$. A linear function is increasing if $m > 0$, decreasing if $m < 0$, and constant if $m = 0$.

Constant Function

$$f(x) = b \qquad b \text{ is a real number}$$

FIGURE 34 Constant function

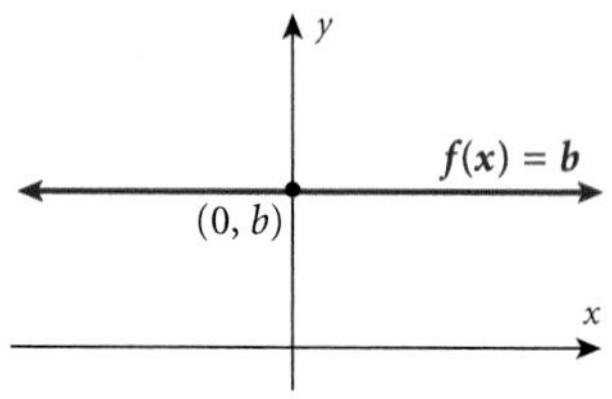

See Figure 34.

A **constant function** is a special linear function ($m = 0$). Its domain is the set of all real numbers; its range is the set consisting of a single number b. Its graph is a horizontal line whose y-intercept is $(0, b)$. The constant function is an even function whose graph is constant over its domain.

Identity Function

$$f(x) = x$$

FIGURE 35 Identity function

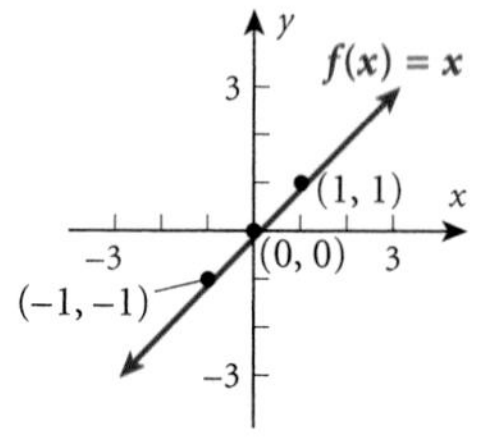

See Figure 35.

The **identity function** is also a special linear function. Its domain and range are the set of all real numbers. Its graph is a line whose slope is $m = 1$ and whose only intercept is (0,0). The line consists of all points for which the x-coordinate equals the y-coordinate. The identity function is an odd function that is increasing over its domain. Note that the graph bisects quadrants I and III.

Square Function

$$f(x) = x^2$$

FIGURE 36 Square function

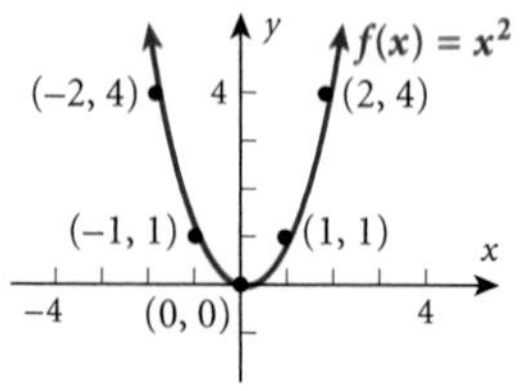

See Figure 36.

The domain of the **square function** f is the set of all real numbers; its range is the set of nonnegative real numbers. The graph of this function is a parabola whose only intercept is $(0, 0)$. The square function is an even function that is decreasing on the interval $(-\infty, 0)$ and increasing on the interval $(0, \infty)$.

FIGURE 37 Cube function

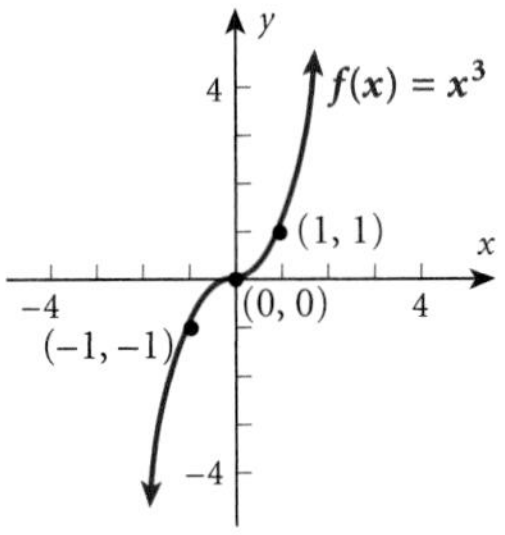

Cube Function

$$f(x) = x^3$$

See Figure 37.

The domain and range of the **cube function** are the set of all real numbers. The only intercept of the graph is $(0, 0)$. The cube function is odd and is increasing on the interval $(-\infty, \infty)$.

Square Root Function

$$f(x) = \sqrt{x}$$

FIGURE 38
Square root function

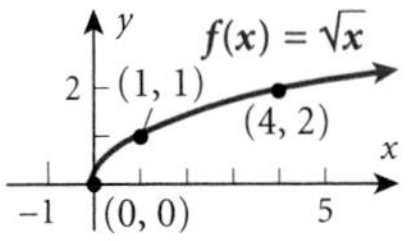

See Figure 38.

The domain and range of the **square root function** are the set of nonnegative real numbers. The only intercept of the graph is $(0, 0)$. The square root function is neither even nor odd and is increasing on the interval $(0, \infty)$.

FIGURE 39 Cube root function

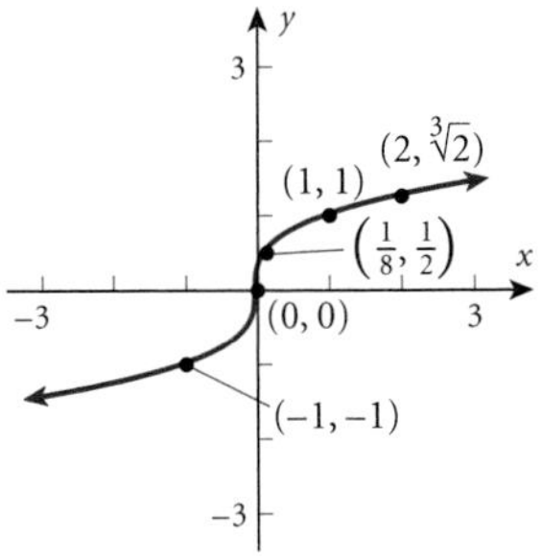

Cube Root Function

$$f(x) = \sqrt[3]{x}$$

See Figure 39.

The domain and the range of the **cube root function** is the set of all real numbers. The intercept of the graph is at $(0, 0)$. The cube root function is an odd function that is increasing on the interval $(-\infty, \infty)$.

Reciprocal Function

$$f(x) = \frac{1}{x}$$

FIGURE 40
Reciprocal function

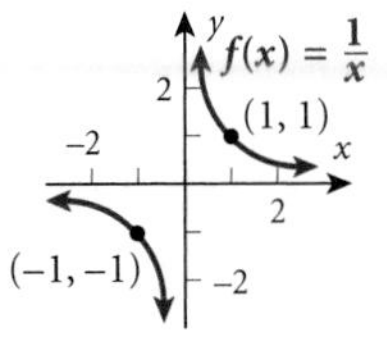

Refer to Example 11, p. 546 for a discussion of the equation $y = \dfrac{1}{x}$. See Figure 40.

The domain and range of the **reciprocal function** are the set of all nonzero real numbers. The graph has no intercepts. The reciprocal function is decreasing on the intervals $(-\infty, 0)$ and $(0, \infty)$ and is an odd function.

Absolute Value Function

$$f(x) = |x|$$

See Figure 41.

The domain of the **absolute value function** is the set of all real numbers; its range is the set of nonnegative real numbers. The intercept of the graph is at (0, 0). If $x \geq 0$, then $f(x) = x$, and the graph of f is part of the line $y = x$; if $x < 0$, then $f(x) = -x$, and the graph of f is part of the line $y = -x$. The absolute value function is an even function; it is decreasing on the interval $(-\infty, 0)$ and is increasing on the interval $(0, \infty)$.

FIGURE 41
Absolute value function

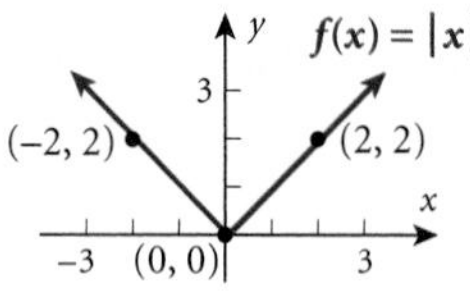

The notation int(x) stands for the largest integer less than or equal to x. For example,

$$\text{int}(1) = 1 \quad \text{int}(2.5) = 2 \quad \text{int}(\tfrac{1}{2}) = 0 \quad \text{int}(-\tfrac{3}{4}) = -1 \quad \text{int}(\pi) = 3$$

This type of correspondence occurs frequently enough in mathematics that we give it a name.

Greatest Integer Function

$$f(x) = \text{int}(x) = \text{Greatest integer less than or equal to } x$$

NOTE: Some books use the notation $f(x) = [x]$ instead of int(x).

We obtain the graph of $f(x) = \text{int}(x)$ by plotting several points. See Table 11. For values of x, $-1 \leq x < 0$, the value of $f(x) = \text{int}(x)$ is -1; for values of x, $0 \leq x < 1$, the value of f is 0. See Figure 42 for the graph.

TABLE 11

x	$y = \text{int}(x)$	(x, y)
-1	-1	$(-1, -1)$
$-\frac{1}{2}$	-1	$\left(-\frac{1}{2}, -1\right)$
$-\frac{1}{4}$	-1	$\left(-\frac{1}{4}, -1\right)$
0	0	(0, 0)
$\frac{1}{4}$	0	$\left(\frac{1}{4}, 0\right)$
$\frac{1}{2}$	0	$\left(\frac{1}{2}, 0\right)$
$\frac{3}{4}$	0	$\left(\frac{3}{4}, 0\right)$

FIGURE 42
Greatest integer function

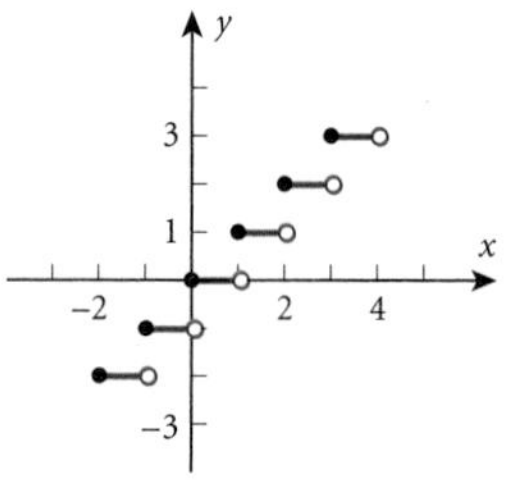

The domain of the **greatest integer function** is the set of all real numbers; its range is the set of integers. The y-intercept of the graph is (0, 0). The x-intercepts lie in the interval [0, 1). The greatest integer function is neither even nor odd. It is constant on every interval of the form $[k, k + 1)$, for k an integer. In Figure 42, we use a solid dot to indicate, for example, that at $x = 1$ the value of f is $f(1) = 1$; we use an open circle to illustrate that the function does not assume the value of 0 at $x = 1$.

From the graph of the greatest integer function, we can see why it is also called a **step function.** At $x = 0$, $x = \pm 1$, $x = \pm 2$, and so on, this function exhibits what is called a *discontinuity;* that is, at integer values, the graph suddenly "steps" from one value to another without taking on any of the intermediate values. For example, to the immediate left of $x = 3$, the y-coordinates are 2, and to the immediate right of $x = 3$, the y-coordinates are 3.

COMMENT: When graphing a function, you can choose either the **connected mode,** in which points plotted on the screen are connected, making the graph appear without any breaks, or the **dot mode,** in which only the points plotted appear. When graphing the greatest integer function with a graphing utility, it is necessary to be in the **dot mode.** This is to prevent the utility from "connecting the dots" when $f(x)$ changes from one integer value to the next. See Figure 43.

FIGURE 43

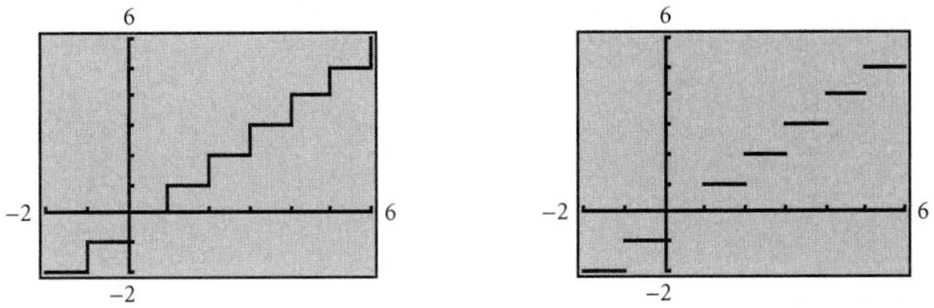

(a) Connected mode (b) Dot mode

The functions that we have discussed so far are basic. Whenever you encounter one of them, you should see a mental picture of its graph. For example, if you encounter the function $f(x) = x^2$, you should see in your mind's eye a picture like Figure 36.

NOW WORK PROBLEMS 1–8.

Piecewise-defined Functions

Graph piecewise-defined functions 2

Sometimes a function is defined differently on different parts of its domain. For example, the absolute value function $f(x) = |x|$ is actually defined by two equations: $f(x) = x$ if $x \geq 0$ and $f(x) = -x$ if $x < 0$. For convenience, we generally combine these equations into one expression as

$$f(x) = |x| = \begin{cases} x & \text{if } x \geq 0 \\ -x & \text{if } x < 0 \end{cases}$$

When functions are defined by more than one equation, they are called **piecewise-defined** functions.

Let's look at another example of a piecewise-defined function.

EXAMPLE 3 Analyzing a Piecewise-Defined Function

The function f is defined as

$$f(x) = \begin{cases} -x + 1 & \text{if } -1 \leq x < 1 \\ 2 & \text{if } x = 1 \\ x^2 & \text{if } x > 1 \end{cases}$$

(a) Find $f(0)$, $f(1)$, and $f(2)$.
(b) Determine the domain of f.
(c) Graph f.
(d) Use the graph to find the range of f.

SOLUTION **(a)** To find $f(0)$, we observe that when $x = 0$ the equation for f is given by $f(x) = -x + 1$. So we have

$$f(0) = -0 + 1 = 1$$

When $x = 1$, the equation for f is $f(x) = 2$. So

$$f(1) = 2$$

When $x = 2$, the equation for f is $f(x) = x^2$. So

$$f(2) = 2^2 = 4$$

FIGURE 44

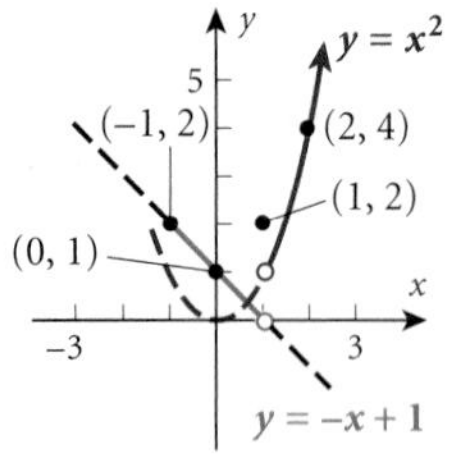

(b) To find the domain of f, we look at its definition. We conclude that the domain of f is $\{x|x \geq -1\}$, or the interval $[-1, \infty)$.

(c) To graph f, we graph "each piece." First we graph the line $y = -x + 1$ and keep only the part for which $-1 \leq x < 1$. Then we plot the point $(1, 2)$ because, when $x = 1$, $f(x) = 2$. Finally, we graph the parabola $y = x^2$ and keep only the part for which $x > 1$. See Figure 44.

(d) From the graph, we conclude that the range of f is $\{y|y > 0\}$, or the interval $(0, \infty)$.

NOW WORK PROBLEM 21.

EXAMPLE 4 Cost of Electricity

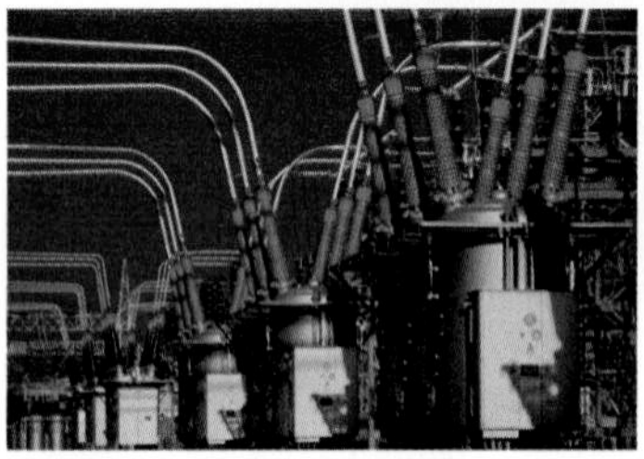

In May 2003, Commonwealth Edison Company supplied electricity to residences for a monthly customer charge of \$7.58 plus 8.275¢ per kilowatt-hour (kWhr) for the first 400 kWhr supplied in the month and 6.574¢ per kWhr for all usage over 400 kWhr in the month.

(a) What is the charge for using 300 kWhr in a month?
(b) What is the charge for using 700 kWhr in a month?
(c) If C is the monthly charge for x kWhr, express C as a function of x.

Source: Commonwealth Edison Co., Chicago, Illinois, 2003.

SOLUTION **(a)** For 300 kWhr, the charge is \$7.58 plus 8.275¢ = \$0.08275 per kWhr. That is,

$$\text{Charge} = \$7.58 + \$0.08275(300) = \$32.41$$

(b) For 700 kWhr, the charge is \$7.58 plus 8.275¢ per kWhr for the first 400 kWhr plus 6.574¢ per kWhr for the 300 kWhr in excess of 400. That is,

$$\text{Charge} = \$7.58 + \$0.08275(400) + \$0.06574(300) = \$60.40$$

(c) If $0 \leq x \leq 400$, the monthly charge C (in dollars) can be found by multiplying x times \$0.08275 and adding the monthly customer charge of \$7.58. So, if $0 \leq x \leq 400$, then $C(x) = 0.08275x + 7.58$. For $x > 400$, the charge is $0.08275(400) + 7.58 + 0.06574(x - 400)$, since $x - 400$ equals the usage in excess of 400 kWhr, which costs \$0.06574 per kWhr. That is, if $x > 400$, then

$$\begin{aligned} C(x) &= 0.08275(400) + 7.58 + 0.06574(x - 400) \\ &= 40.68 + 0.06574(x - 400) \\ &= 0.06574x + 14.38 \end{aligned}$$

FIGURE 45

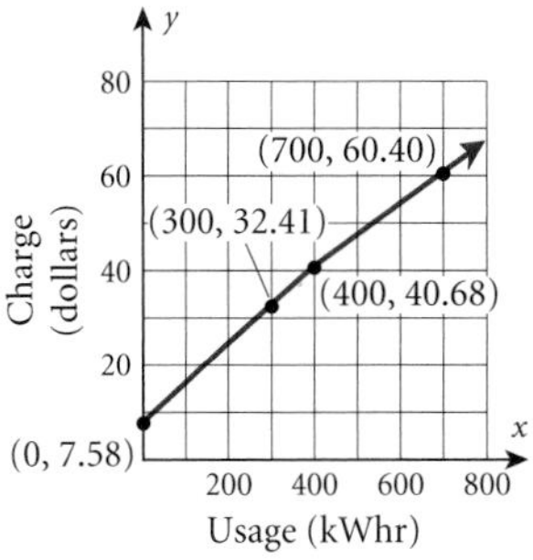

The rule for computing C follows two equations:

$$C(x) = \begin{cases} 0.08275x + 7.58 & \text{if } 0 \leq x \leq 400 \\ 0.06574x + 14.38 & \text{if } x > 400 \end{cases}$$

See Figure 45 for the graph.

EXERCISE 10.4 Answers to Odd-Numbered Problems Begin on Page AN-53.

In Problems 1–8, match each graph to the function whose graph most resembles the one given.

A. Constant function *B. Linear function* *C. Square function*

D. Cube function *E. Square root function* *F. Reciprocal function*

G. Absolute value function *H. Cube root function*

1. 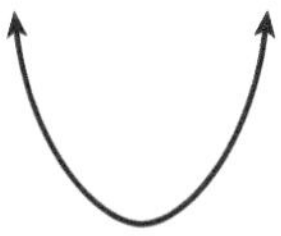

2. 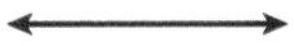

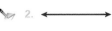

3. 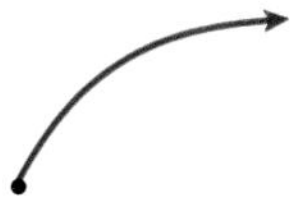

4. 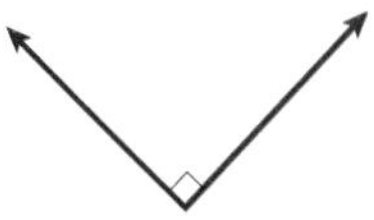

5. 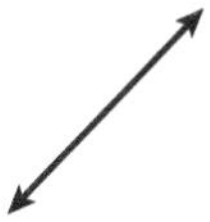

6. 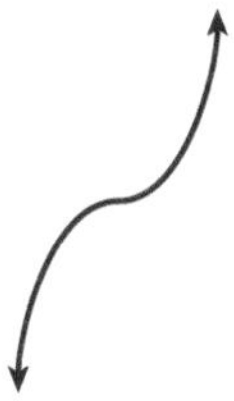

7. 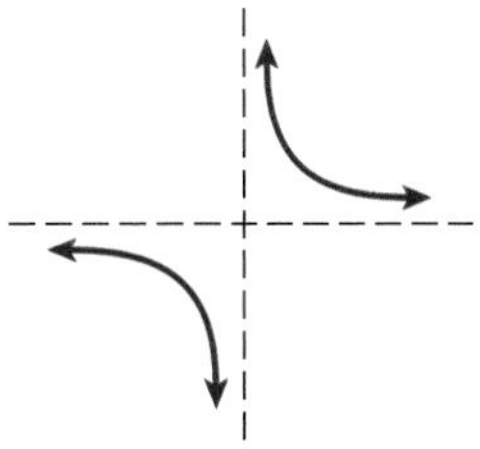

8. 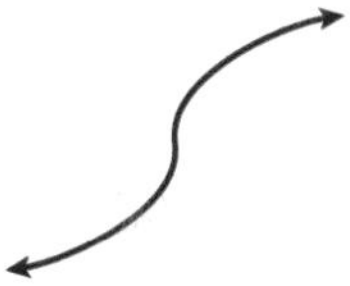

In Problems 9–16, sketch the graph of each function. Be sure to label at least three points on the graph.

9. $f(x) = x$ **10.** $f(x) = x^2$ **11.** $f(x) = x^3$ **12.** $f(x) = \sqrt{x}$

13. $f(x) = \dfrac{1}{x}$ **14.** $f(x) = |x|$ **15.** $f(x) = \sqrt[3]{x}$ **16.** $f(x) = 3$

17. If $f(x) = \begin{cases} x^2 & \text{if } x < 0 \\ 2 & \text{if } x = 0 \\ 2x + 1 & \text{if } x > 0 \end{cases}$

find: (a) $f(-2)$ (b) $f(0)$ (c) $f(2)$

18. If $f(x) = \begin{cases} x^3 & \text{if } x < 0 \\ 3x + 2 & \text{if } x \geq 0 \end{cases}$

find: (a) $f(-1)$ (b) $f(0)$ (c) $f(1)$

19. If $f(x) = \text{int}\,(2x)$, find: (a) $f(1.2)$ (b) $f(1.6)$ (c) $f(-1.8)$

20. If $f(x) = \text{int}\left(\dfrac{x}{2}\right)$, find: (a) $f(1.2)$ (b) $f(1.6)$ (c) $f(-1.8)$

In Problems 21–32:

(a) Find the domain of each function. *(b) Locate any intercepts.*
(c) Graph each function *(d) Based on the graph, find the range.*

21. $f(x) = \begin{cases} 2x & \text{if } x \neq 0 \\ 1 & \text{if } x = 0 \end{cases}$

22. $f(x) = \begin{cases} 3x & \text{if } x \neq 0 \\ 4 & \text{if } x = 0 \end{cases}$

23. $f(x) = \begin{cases} -2x + 3 & x < 1 \\ 3x - 2 & x \geq 1 \end{cases}$

24. $f(x) = \begin{cases} x + 3 & x < -2 \\ -2x - 3 & x \geq -2 \end{cases}$

25. $f(x) = \begin{cases} x + 3 & -2 \leq x < 1 \\ 5 & x = 1 \\ -x + 2 & x > 1 \end{cases}$

26. $f(x) = \begin{cases} 2x + 5 & -3 \leq x < 0 \\ -3 & x = 0 \\ -5x & x > 0 \end{cases}$

27. $f(x) = \begin{cases} 1 + x & \text{if } x < 0 \\ x^2 & \text{if } x \geq 0 \end{cases}$

28. $f(x) = \begin{cases} \dfrac{1}{x} & \text{if } x < 0 \\ \sqrt[3]{x} & \text{if } x \geq 0 \end{cases}$

29. $f(x) = \begin{cases} |x| & \text{if } -2 \leq x < 0 \\ 1 & \text{if } x = 0 \\ x^3 & \text{if } x > 0 \end{cases}$

30. $f(x) = \begin{cases} 3 + x & \text{if } -3 \leq x < 0 \\ 3 & \text{if } x = 0 \\ \sqrt{x} & \text{if } x > 0 \end{cases}$

31. $f(x) = 2 \text{ int }(x)$

32. $f(x) = \text{int }(2x)$

In Problems 33–36, the graph of a piecewise-defined function is given. Write a definition for each function.

33.

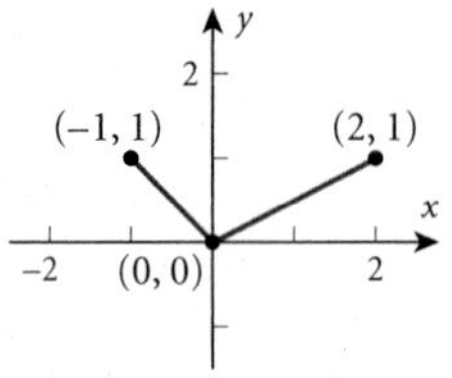

34.

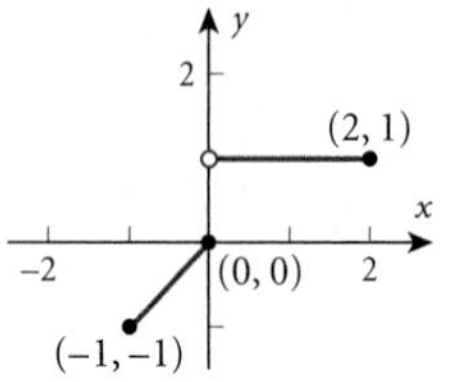

35.

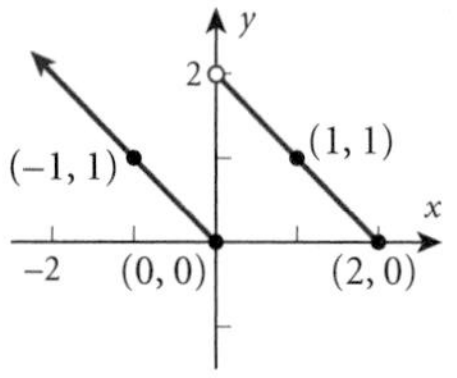

36.

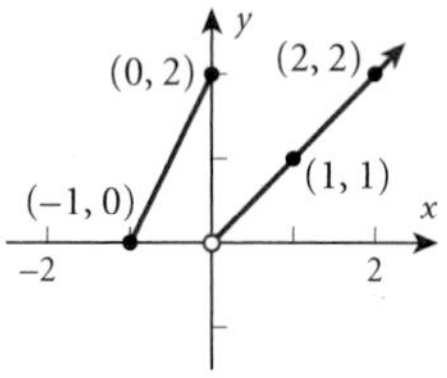

37. Cell Phone Service Sprint PCS offers a monthly cellular phone plan for $39.99. It includes 350 anytime minutes plus $0.25 per minute for additional minutes. The following function is used to compute the monthly cost for a subscriber

$$C(x) = \begin{cases} 39.99 & \text{if } 0 < x \leq 350 \\ 0.25x - 47.51 & \text{if } x > 350 \end{cases}$$

where x is the number of anytime minutes used. Compute the monthly cost of the cellular phone for the following anytime minutes:

(a) 200 (b) 365 (c) 351

Source: Sprint PCS.

38. First-class Letter According to the U.S. Postal Service, first-class mail is used for personal and business correspondence. Any mailable item may be sent as first-class mail. It includes postcards, letters, large envelopes, and small packages. The maximum weight is 13 ounces. The following function is used to compute the cost of mailing a first-class item.

$$C(x) = \begin{cases} 0.37 & \text{if } 0 < x \leq 1 \\ 0.23 \text{ int}(x) + 0.37 & \text{if } 1 < x \leq 13 \end{cases}$$

where x is the weight of the item in ounces. Compute the cost of mailing the following items first-class:

(a) A letter weighing 4.3 ounces
(b) A postcard weighing 0.4 ounces
(c) A package weighing 12.2 ounces

Source: United States Postal Service.

39. Cost of Natural Gas In May 2003, the People Gas Company had the following rate schedule for natural gas usage in single-family residences:

Monthly service charge	$9.45
Per therm service charge	
1st 50 therms	$0.36375/therm
Over 50 therms	$0.11445/therm
Gas charge	0.6338/therm

(a) What is the charge for using 50 therms in a month?
(b) What is the charge for using 500 therms in a month?
(c) Construct a function that relates the monthly charge C for x therms of gas.
(d) Graph this function.

Source: The Peoples Gas Company, Chicago, Illinois, 2003.

40. Cost of Natural Gas In May 2003, Nicor Gas had the following rate schedule for natural gas usage in single-family residences:

Monthly customer charge	$6.45
Distribution charge	
1st 20 therms	$0.2012/therm
Next 30 therms	$0.1117/therm
Over 50 therms	$0.0374/therm
Gas supply charge	$0.7268/therm

(a) What is the charge for using 40 therms in a month?
(b) What is the charge for using 202 therms in a month?
(c) Construct a function that gives the monthly charge C for x therms of gas.
(d) Graph this function.

Source: Nicor Gas, Aurora, Illinois, 2003.

41. Wind Chill The wind chill factor represents the equivalent air temperature at a standard wind speed that would produce the same heat loss as the given temperature and wind speed. One formula for computing the equivalent temperature is

$$W = \begin{cases} t & 0 \le v < 1.79 \\ 33 - \dfrac{(10.45 + 10\sqrt{v} - v)(33 - t)}{22.04} & 1.79 \le v \le 20 \\ 33 - 1.5958(33 - t) & v > 20 \end{cases}$$

where v represents the wind speed (in meters per second) and t represents the air temperature (°C). Compute the wind chill for the following:

(a) An air temperature of 10°C and a wind speed of 1 meter per second (m/sec).
(b) An air temperature of 10°C and a wind speed of 5 m/sec.
(c) An air temperature of 10°C and a wind speed of 15 m/sec.
(d) An air temperature of 10°C and a wind speed of 25 m/sec.
(e) Explain the physical meaning of the equation corresponding to $0 \le v < 1.79$.
(f) Explain the physical meaning of the equation corresponding to $v > 20$.

42. Wind Chill Redo Problem 41(a)–(d) for an air temperature of −10°C

43. Federal Income Tax Two 2003 Tax Rate Schedules are given in the accompanying tables. If x equals taxable income and y equals the tax due, construct a function $y = f(x)$ for Schedule X.

Revised 2003 Tax Rate Schedules

	If TAXABLE INCOME Is Over	But Not Over	Then The TAX is This Amount	Plus This %	Of the Excess Over
SCHEDULE X—					
Single	$0	$ 7,000	$ 0.00	10%	$ 0.00
	$ 7,000	$ 28,400	$ 700.00	15%	$ 7,000
	$ 28,400	$ 68,800	$ 3,910.00	25%	$ 28,400
	$ 68,800	$143,500	$14,010.00	28%	$ 68,800
	$143,500	$311,950	$34,926.00	33%	$143,500
	$311,950	—	$90,514.50	35%	$311,950

	If TAXABLE INCOME Is Over	But Not Over	Then The TAX is This Amount	Plus This %	Of the Excess Over
SCHEDULE Y-1—					
Married Filing Jointly or Qualifying Widow(er)	$0	$ 14,000	$ 0.00	10%	$ 0.00
	$ 14,000	$ 56,800	$ 1,400.00	15%	$ 14,000
	$ 56,800	$114,650	$ 7,820.00	25%	$ 56,800
	$114,650	$174,700	$22,282.50	28%	$114,650
	$174,700	$311,950	$39,096.50	33%	$174,700
	$311,950	—	$84,389.00	35%	$311,950

Source: Internal Revenue Service.

44. Federal Income Tax Refer to the revised 2003 tax rate schedules. If x equals the taxable income and y equals the tax due, construct a function $y = f(x)$ for Schedule Y-1.

45. Exploration Graph $y = x^2$. Then on the same screen graph $y = x^2 + 2$, followed by $y = x^2 + 4$, followed by $y = x^2 - 2$. What pattern do you observe? Can you predict the graph of $y = x^2 - 4$? Of $y = x^2 + 5$?

46. Exploration Graph $y = x^2$. Then on the same screen graph $y = (x - 2)^2$, followed by $y = (x - 4)^2$, followed by $y = (x + 2)^2$. What pattern do you observe? Can you predict the graph of $y = (x + 4)^2$? Of $y = (x - 5)^2$?

47. Exploration Graph $y = x^2$. Then on the same screen graph $y = -x^2$. What pattern do you observe? Now try $y = |x|$ and $y = -|x|$. What do you conclude?

48. Exploration Graph $y = \sqrt{x}$. Then on the same screen graph $y = \sqrt{-x}$. What pattern do you observe? Now try $y = 2x + 1$ and $y = 2(-x) + 1$. What do you conclude?

49. Exploration Graph $y = x^3$. Then on the same screen graph $y = (x - 1)^3 + 2$. Could you have predicted the result?

50. Exploration Graph $y = x^2$, $y = x^4$, and $y = x^6$ on the same screen. What do you notice is the same about each graph? What do you notice that is different?

51. Exploration Graph $y = x^3$, $y = x^5$, and $y = x^7$ on the same screen. What do you notice is the same about each graph? What do you notice that is different?

52. Consider the equation $y = \begin{cases} 1 & \text{if } x \text{ is rational} \\ 0 & \text{if } x \text{ is irrational} \end{cases}$

Is this a function? What is its domain? What is its range? What is its y-intercept, if any? What are its x-intercepts, if any? Is it even, odd, or neither? How would you describe its graph?

10.5 Graphing Techniques: Shifts and Reflections

OBJECTIVES
1. Graph functions using horizontal and vertical shifts
2. Graph functions using reflections about the x-axis or y-axis

At this stage, if you were asked to graph any of the functions defined by $y = x$, $y = x^2$, $y = x^3$, $y = \sqrt{x}$, $y = \sqrt[3]{x}$, $y = |x|$, or $y = \dfrac{1}{x}$, your response should be, "Yes, I recognize these functions and know the general shapes of their graphs." (If this is not your answer, review the previous section, Figures 35 through 41.)

Sometimes we are asked to graph a function that is "almost" like one that we already know how to graph. In this section, we look at some of these functions and develop techniques for graphing them. Collectively, these techniques are referred to as **transformations.**

Graph functions using horizontal and vertical shifts

1 Vertical Shifts

EXAMPLE 1 Vertical Shift Up

Use the graph of $f(x) = x^2$ to obtain the graph of $g(x) = x^2 + 3$.

SOLUTION We begin by obtaining some points on the graphs of f and g. For example, when $x = 0$, then $y = f(0) = 0$ and $y = g(0) = 3$. When $x = 1$, then $y = f(1) = 1$ and $y = g(1) = 4$. Table 12 lists these and a few other points on each graph. We conclude that the graph of g is identical to that of f, except that it is shifted vertically up 3 units. See Figure 46.

TABLE 12

x	$y = f(x) = x^2$	$y = g(x) = x^2 + 3$
−2	4	7
−1	1	4
0	0	3
1	1	4
2	4	7

FIGURE 46

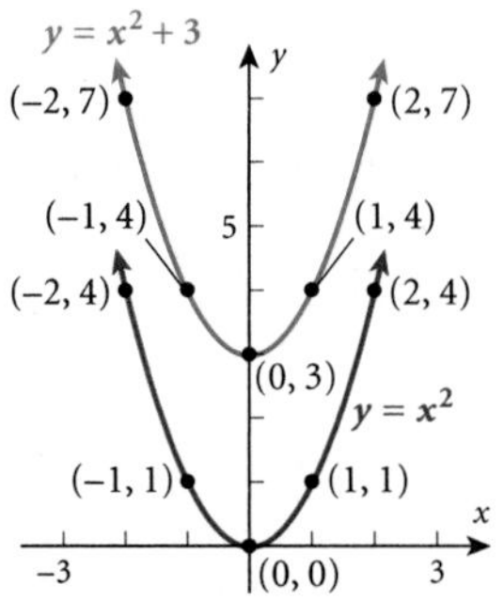

SEEING THE CONCEPT: On the same screen, graph each of the following functions:

$$Y_1 = x^2$$
$$Y_2 = x^2 + 1$$
$$Y_3 = x^2 + 2$$
$$Y_4 = x^2 - 1$$
$$Y_5 = x^2 - 2$$

FIGURE 47

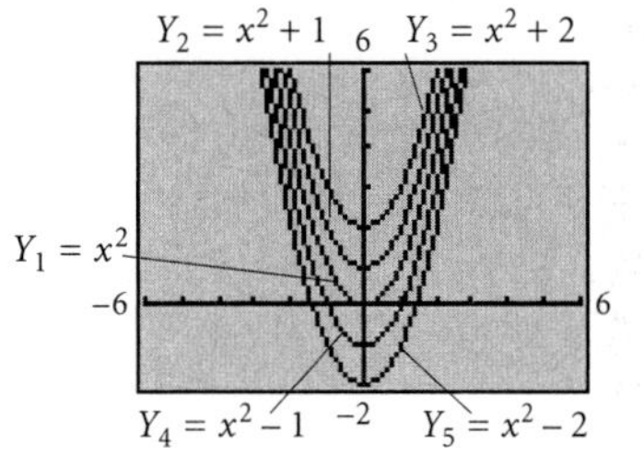

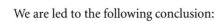

Figure 47 illustrates the graphs. You should have observed a general pattern. With $Y_1 = x^2$ on the screen, the graph of $Y_2 = x^2 + 1$ is identical to that of $Y_1 = x^2$, except that it is shifted vertically up 1 unit. Similarly, $Y_3 = x^2 + 2$ is identical to that of $Y_1 = x^2$, except that it is shifted vertically up 2 units. The graph of $Y_4 = x^2 - 1$ is identical to that of $Y_1 = x^2$, except that it is shifted vertically down 1 unit.

We are led to the following conclusion:

> If a real number k is added to the right side of a function $y = f(x)$, the graph of the new function $y = f(x) + k$ is the graph of f **shifted vertically up** (if $k > 0$) or **down** (if $k < 0$).

Let's look at another example.

EXAMPLE 2 Vertical Shift Down

Use the graph of $f(x) = x^2$ to obtain the graph of $g(x) = x^2 - 4$.

SOLUTION Table 13 lists some points on the graphs of f and g. Notice that each y-coordinate of g is 4 units less than the corresponding y-coordinate of f. The graph of g is identical to that of f, except that it is shifted down 4 units. See Figure 48.

TABLE 13

x	$y = f(x) = x^2$	$y = g(x) = x^2 - 4$
−2	4	0
−1	1	−3
0	0	−4
1	1	−3
2	4	0

FIGURE 48

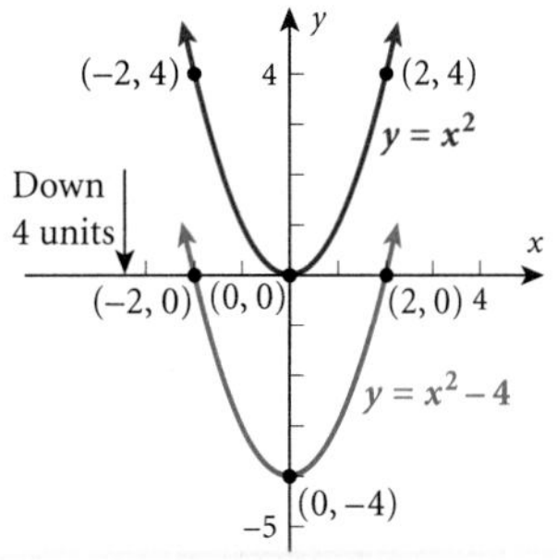

NOW WORK PROBLEMS 11 AND 21.

Horizontal Shifts

EXAMPLE 3 Horizontal Shift to the Right

Use the graph of $f(x) = x^2$ to obtain the graph of $g(x) = (x - 2)^2$.

SOLUTION The function $g(x) = (x - 2)^2$ is basically a square function. Table 14 lists some points on the graphs of f and g. Note that when $f(x) = 0$ then $x = 0$, and when $g(x) = 0$, then $x = 2$. Also, when $f(x) = 4$, then $x = -2$ or 2, and when $g(x) = 4$, then $x = 0$ or 4. We conclude that the graph of g is identical to that of f, except that it is shifted 2 units to the right. See Figure 49.

TABLE 14

x	$y = f(x) = x^2$	$y = g(x) = (x - 2)^2$
−2	4	16
0	0	4
2	4	0
4	16	4

FIGURE 49

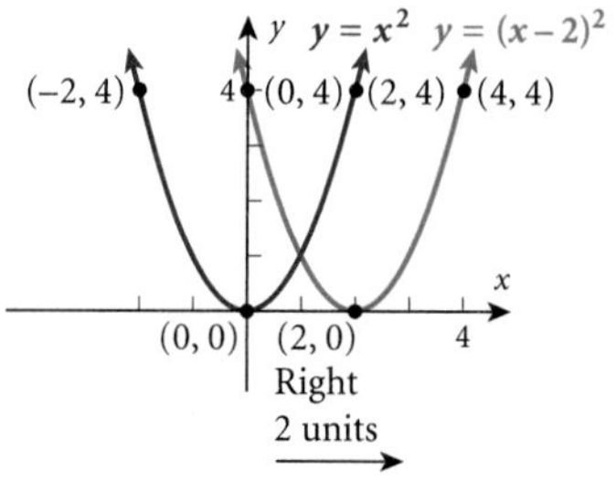

SEEING THE CONCEPT: On the same screen, graph each of the following functions:

$$Y_1 = x^2$$
$$Y_2 = (x - 1)^2$$
$$Y_3 = (x - 3)^2$$
$$Y_4 = (x + 2)^2$$

FIGURE 50

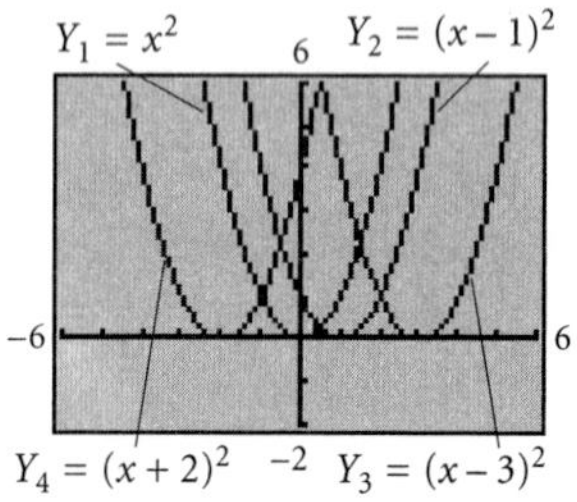

Figure 50 illustrates the graphs.

You should have observed the following pattern. With the graph of $Y_1 = x^2$ on the screen, the graph of $Y_2 = (x - 1)^2$ is identical to that of $Y = x^2$, except that it is shifted horizontally to the right 1 unit. Similarly, the graph of $Y_3 = (x - 3)^2$ is identical to that of $Y_1 = x^2$, except that it is shifted horizontally to the right 3 units. Finally, the graph of $Y_4 = (x + 2)^2$ is identical to that of $Y_1 = x^2$, except that it is shifted horizontally to the left 2 units.

We are led to the following conclusion.

> If the argument x of a function f is replaced by $x - h$, h a real number, the graph of the new function $y = f(x - h)$ is the graph of f **shifted horizontally left** (if $h < 0$) or **right** (if $h > 0$).

EXAMPLE 4 Horizontal Shift to the Left

Use the graph of $f(x) = x^2$ to obtain the graph of $g(x) = (x + 4)^2$.

SOLUTION The function $g(x) = (x + 4)^2$ is basically a square function. Its graph is the same as that of f, except that it is shifted horizontally 4 units to the left. (Do you see why? $(x + 4)^2 = [x - (-4)]^2$) See Figure 51.

FIGURE 51

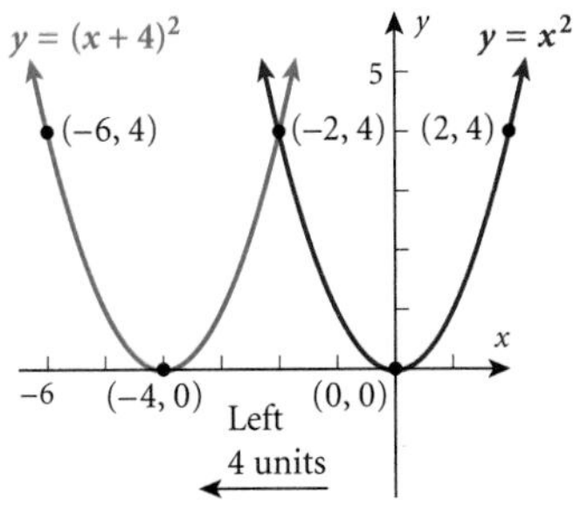

y = (x + 4)^2
y = x^2
y
x
5
(−6, 4)
(−2, 4)
(2, 4)
−6
(−4, 0)
(0, 0)
Left
4 units

NOW WORK PROBLEMS 9 AND 25.

Vertical and horizontal shifts are sometimes combined.

EXAMPLE 5 Combining Vertical and Horizontal Shifts

Graph the function: $f(x) = (x + 3)^2 - 5$

SOLUTION We graph f in steps. First, we note that the rule for f is basically a square function, so we begin with the graph of $y = x^2$ as shown in Figure 52(a). Next, to get the graph of $y = (x + 3)^2$, we shift the graph of $y = x^2$ horizontally 3 units to the left. See Figure 52(b). Finally, to get the graph of $y = (x + 3)^2 - 5$, we shift the graph of $y = (x + 3)^2$ vertically down 5 units. See Figure 52(c). Note the points plotted on each graph. Using key points can be helpful in keeping track of the transformation that has taken place.

FIGURE 52

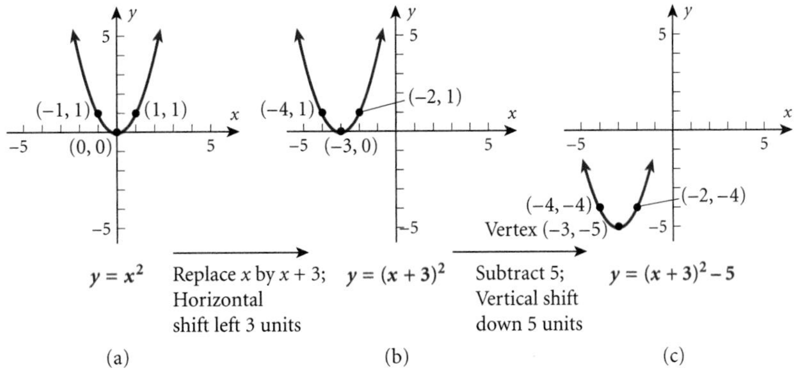

y
x
5
−5
(−1, 1)
(1, 1)
(0, 0)
(−4, 1)
(−2, 1)
(−3, 0)
(−4, −4)
(−2, −4)
Vertex (−3, −5)
y = x^2
Replace x by x + 3; Horizontal shift left 3 units
y = (x + 3)^2
Subtract 5; Vertical shift down 5 units
y = (x + 3)^2 − 5
(a)
(b)
(c)

CHECK: Graph $Y_1 = f(x) = (x + 3)^2 - 5$ and compare the graph to Figure 52(c).

In Example 5, if the vertical shift had been done first, followed by the horizontal shift, the final graph would have been the same. Try it for yourself.

NOW WORK PROBLEM 27.

Graph functions using reflections about the x-axis or y-axis

2 Reflections about the x-Axis and the y-Axis

EXAMPLE 6 Reflection about the x-Axis

Graph the function: $f(x) = -x^2$

SOLUTION We begin with the graph of $y = x^2$, as shown in Figure 53. For each point (x, y) on the graph of $y = x^2$, the point $(x, -y)$ is on the graph of $y = -x^2$, as indicated in Table 15. We can draw the graph of $y = -x^2$ by reflecting the graph of $y = x^2$ about the x-axis. See Figure 53.

TABLE 15

x	$y = x^2$	$y = -x^2$
-2	4	-4
-1	1	-1
0	0	0
1	1	-1
2	4	-4

FIGURE 53

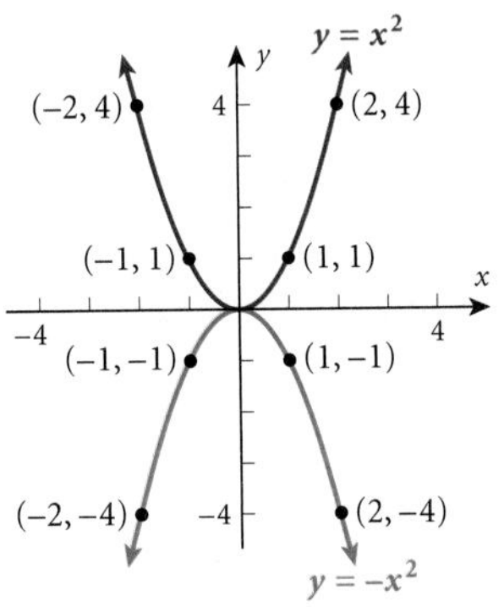

When the right side of the function $y = f(x)$ is multiplied by -1, the graph of the new function $y = -f(x)$ is the **reflection about the x-axis** of the graph of the function $y = f(x)$.

NOW WORK PROBLEM 29.

EXAMPLE 7 Reflection about the y-Axis

Graph the function: $f(x) = \sqrt{-x}$

SOLUTION First, notice that the domain of f consists of all real numbers x for which $-x \geq 0$ or, equivalently, $x \leq 0$. To get the graph of $f(x) = \sqrt{-x}$, we begin with the graph of $y = \sqrt{x}$. For each point (x, y) on the graph of $y = \sqrt{x}$, the point $(-x, y)$ is on the graph of $y = \sqrt{-x}$. We obtain the graph of $y = \sqrt{-x}$ by reflecting the graph of $y = \sqrt{x}$ about the y-axis. See Figure 54.

FIGURE 54

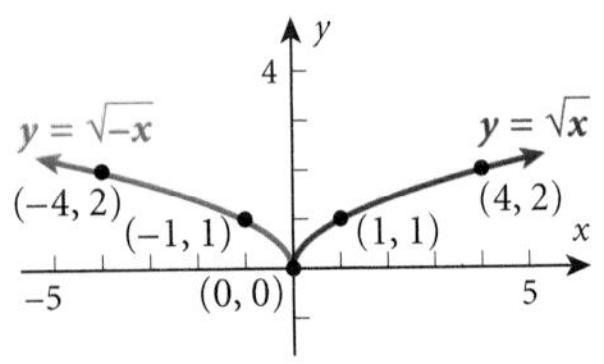

When the graph of the function $y = f(x)$ is known, the graph of the new function $y = f(-x)$ is the **reflection about the y-axis** of the graph of the function $y = f(x)$.

NOW WORK PROBLEMS 13 AND 37.

SUMMARY: GRAPHING TECHNIQUES

Table 16 summarizes the graphing procedures that we have just discussed.

TABLE 16

To Graph:	Draw the Graph of *f* and:	Functional Change to *f* (*x*)
Vertical shifts		
$y = f(x) + k, \quad k > 0$	Raise the graph of f by k units.	Add k to $f(x)$.
$y = f(x) - k, \quad k > 0$	Lower the graph of f by k units.	Subtract k from $f(x)$.
Horizontal shifts		
$y = f(x + h), \quad h > 0$	Shift the graph of f to the left h units.	Replace x by $x + h$.
$y = f(x - h), \quad h > 0$	Shift the graph of f to the right h units.	Replace x by $x - h$.
Reflection about the x-axis		
$y = -f(x)$	Reflect the graph of f about the x-axis.	Multiply $f(x)$ by -1.
Reflection about the y-axis		
$y = f(-x)$	Reflect the graph of f about the y-axis.	Replace x by $-x$.

The examples that follow combine some of the procedures outlined in this section to get the required graph.

EXAMPLE 8 Determining the Function Obtained from a Series of Transformations

Find the function that is finally graphed after the following three transformations are applied to the graph of $y = |x|$.

1. Shift left 2 units. **2.** Shift up 3 units. **3.** Reflect about the y-axis.

SOLUTION

1. Shift left 2 units: Replace x by $x + 2$. $y = |x + 2|$

2. Shift up 3 units: Add 3. $y = |x + 2| + 3$

3. Reflect about the y-axis: Replace x by $-x$. $y = |-x + 2| + 3$

NOW WORK PROBLEM 15.

EXAMPLE 9 Combining Graphing Procedures

Graph the function: $f(x) = \sqrt{1 - x} + 2$

SOLUTION We use the following steps to get the graph of $y = \sqrt{1 - x} + 2$:

STEP 1: $y = \sqrt{x}$ Square root function.

STEP 2: $y = \sqrt{x + 1}$ Replace x by $x + 1$; horizontal shift left 1 unit.

STEP 3: $y = \sqrt{-x + 1} = \sqrt{1 - x}$ Replace x by $-x$; reflect about y-axis.

STEP 4: $y = \sqrt{1 - x} + 2$ Add 2; vertical shift up 2 units.

See Figure 55.

FIGURE 55

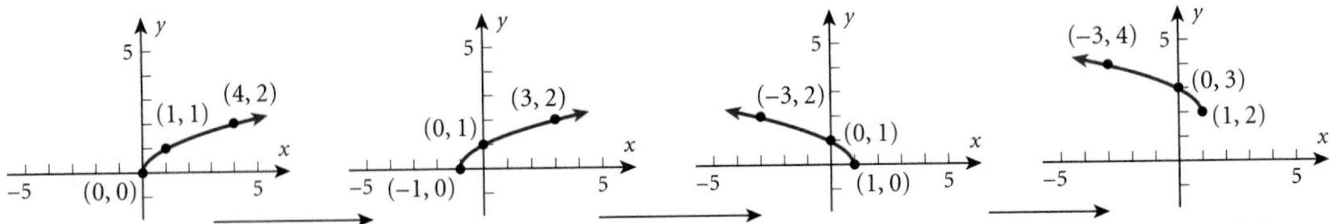

(a) $y = \sqrt{x}$ Replace x by $x + 1$;   (b) $y = \sqrt{x + 1}$ Replace x by $-x$;   (c) $y = \sqrt{-x + 1}$ Add 2;   (d) $y = \sqrt{1 - x} + 2$
Horizontal shift   Reflect $= \sqrt{1 - x}$ Vertical shift
left 1 unit   about y-axis   up 2 units

EXERCISE 10.5   **Answers to Odd-Numbered Problems Begin on Page AN-55.**

In Problems 1–8, match each graph to one of the following functions.

A. $y = x^2 + 2$ B. $y = -x^2 + 2$ C. $y = |x| + 2$ D. $y = -|x| + 2$

E. $y = (x - 2)^2$ F. $y = -(x + 2)^2$ G. $y = |x - 2|$ H. $y = -|x + 2|$

1.
2.
3.
4.
5.
6.
7.
8.
In Problems 9–14, write the function whose graph is the graph of $y = x^3$, but is:

9. Shifted to the right 4 units

10. Shifted to the left 4 units

11. Shifted up 4 units

12. Shifted down 4 units

13. Reflected about the y-axis

14. Reflected about the x-axis

In Problems 15–18, find the function that is finally graphed after the following transformations are applied to the graph of $y = \sqrt{x}$.

15. (1) Shift up 2 units
(2) Reflect about the x-axis
(3) Reflect about the y-axis

16. (1) Reflect about the x-axis
(2) Shift right 3 units
(3) Shift down 2 units

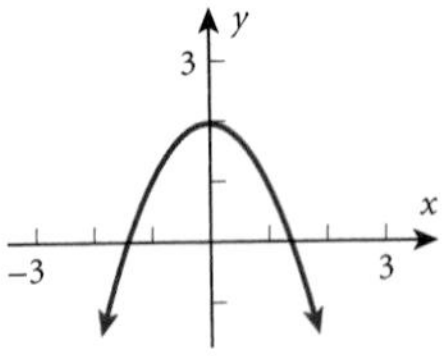

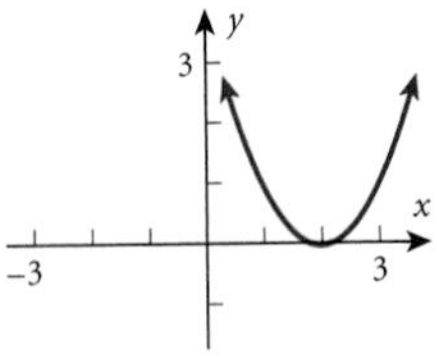

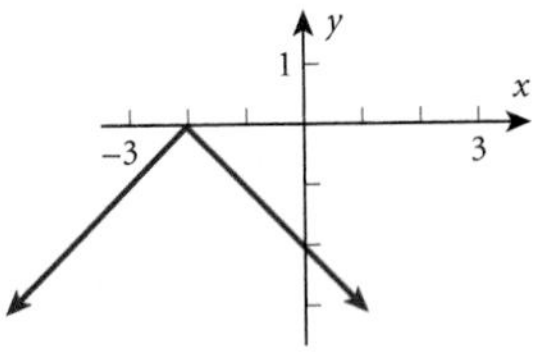

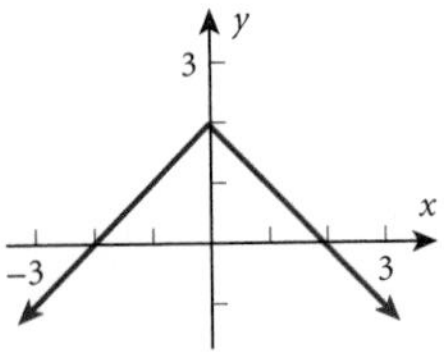

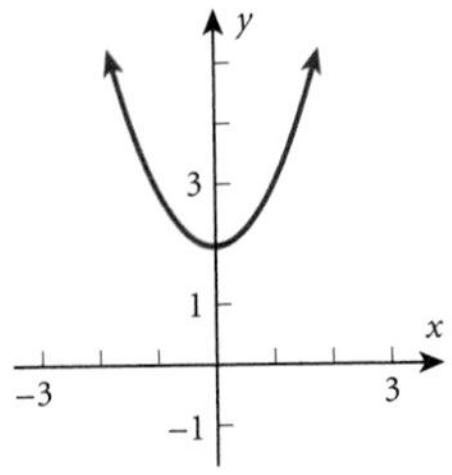

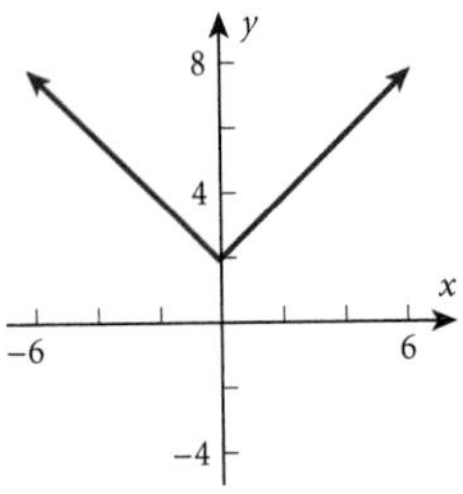

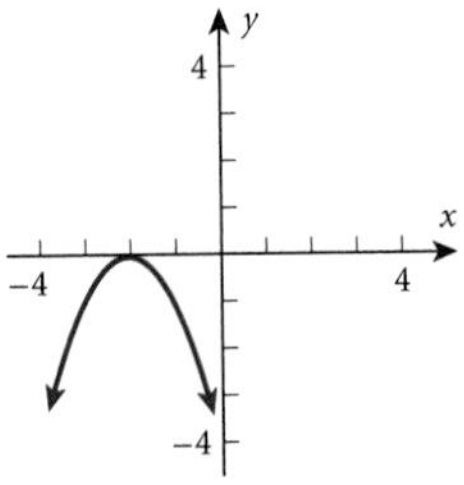

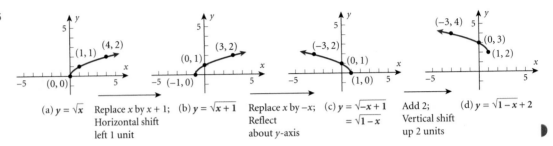

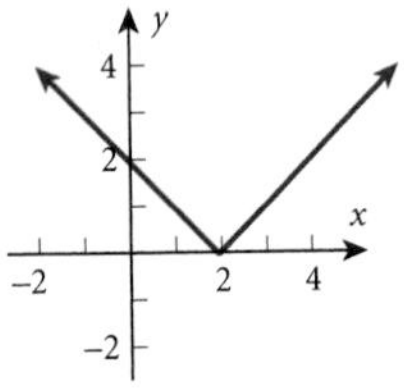

17. (1) Reflect about the x-axis
(2) Shift up 2 units
(3) Shift left 3 units

18. (1) Shift up 2 units
(2) Reflect about the y-axis
(3) Shift left 3 units

19. If (3, 0) is a point on the graph of $y = f(x)$, which of the following must be on the graph of $y = -f(x)$?
(a) $(0, 3)$ (b) $(0, -3)$ (c) $(3, 0)$ (d) $(-3, 0)$

20. If (3, 0) is a point on the graph of $y = f(x)$, which of the following must be on the graph of $y = f(-x)$?
(a) $(0, 3)$ (b) $(0, -3)$ (c) $(3, 0)$ (d) $(-3, 0)$

In Problems 21–38, graph each function using the techniques of shifting and/or reflecting. Start with the graph of the basic function (for example, $y = x^2$) and show all stages.

21. $f(x) = x^2 - 1$
22. $f(x) = x^2 + 4$
23. $g(x) = x^3 + 1$
24. $g(x) = x^3 - 1$
25. $h(x) = \sqrt{x - 2}$
26. $h(x) = \sqrt{x + 1}$
27. $f(x) = (x - 1)^3 + 2$
28. $f(x) = (x + 2)^3 - 3$
29. $f(x) = -\sqrt[3]{x}$
30. $f(x) = -\sqrt{x}$
31. $g(x) = |-x|$
32. $g(x) = \sqrt[3]{-x}$
33. $h(x) = -x^3 + 2$
34. $h(x) = \dfrac{1}{-x} + 2$
35. $g(x) = \sqrt{x - 2} + 1$
36. $g(x) = |x + 1| - 3$
37. $h(x) = \sqrt{-x} - 2$
38. $f(x) = -(x + 1)^3 - 1$

In Problems 39–40, the graph of a function f is illustrated. Use the graph of f as the first step toward graphing each of the following functions:

(a) $F(x) = f(x) + 3$ (b) $G(x) = f(x + 2)$ (c) $P(x) = -f(x)$ (d) $H(x) = f(x + 1) - 2$ (e) $g(x) = f(-x)$

39.

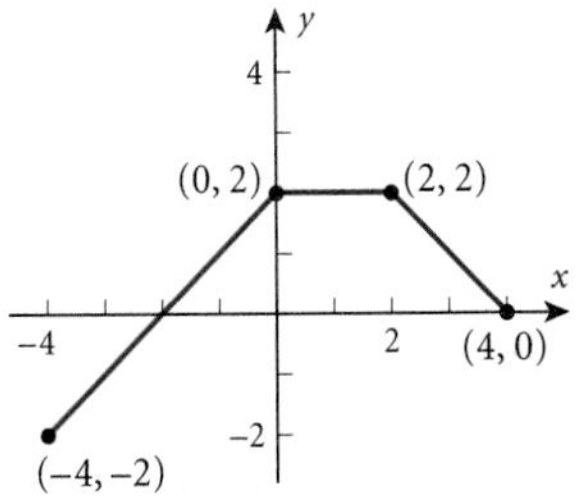

40.

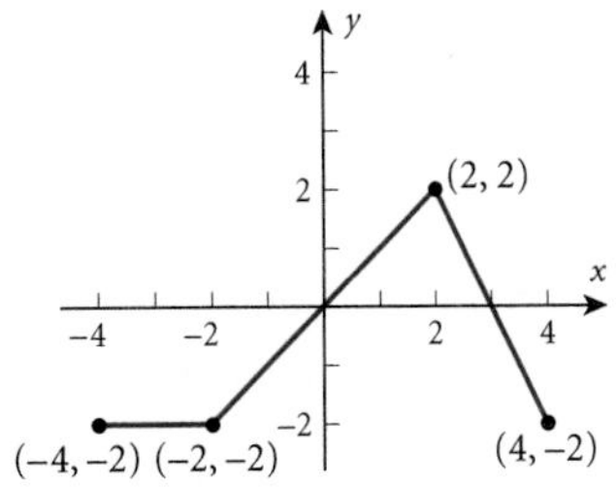

41. Exploration

(a) Use a graphing utility to graph $y = x + 1$ and $y = |x + 1|$.
(b) Graph $y = 4 - x^2$ and $y = |4 - x^2|$.
(c) Graph $y = x^3 + x$ and $y = |x^3 + x|$.
(d) What do you conclude about the relationship between the graphs of $y = f(x)$ and $y = |f(x)|$?

42. Exploration

(a) Use a graphing utility to graph $y = x + 1$ and $y = |x| + 1$.
(b) Graph $y = 4 - x^2$ and $y = 4 - |x|^2$.
(c) Graph $y = x^3 + x$ and $y = |x|^3 + |x|$.
(d) What do you conclude about the relationship between the graphs of $y = f(x)$ and $y = f(|x|)$?

43. The graph of a function f is illustrated in the figure.
(a) Draw the graph of $y = |f(x)|$.
(b) Draw the graph of $y = f(|x|)$.

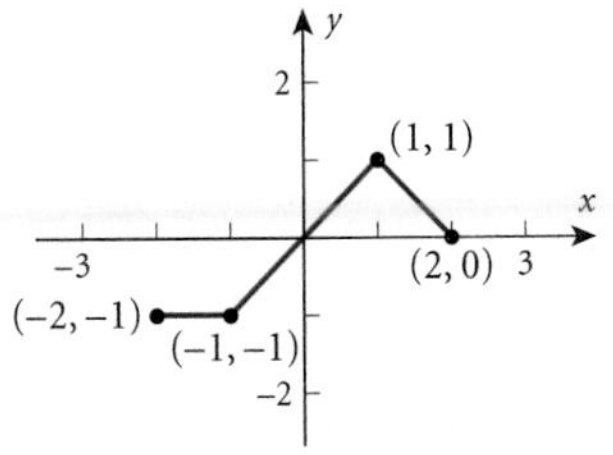

44. The graph of a function f is illustrated in the figure.
(a) Draw the graph of $y = |f(x)|$.
(b) Draw the graph of $y = f(|x|)$.

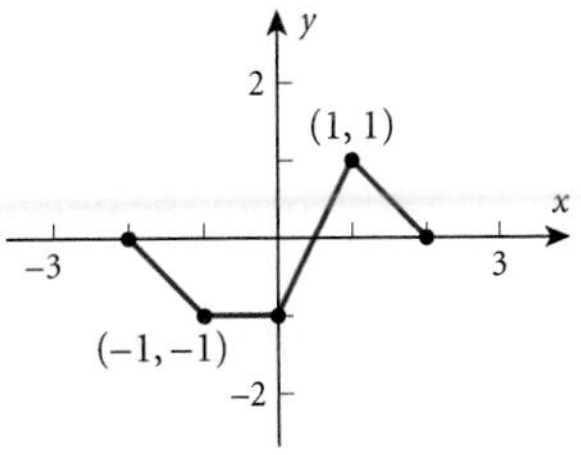

Chapter 10 Review

OBJECTIVES

Section		You should be able to	Review Exercises
10.1	1	Graph equations by plotting points	1
	2	Find intercepts from a graph	2, 27(b), 28(c), 29(f), 30(f)
	3	Find intercepts from an equation	3–10, 53(b)–56(b)
	4	Test an equation for symmetry with respect to the (a) x-axis, (b) y-axis, and (c) origin	3–10
10.2	1	Find the value of a function	11–16, 27(c), 28(b), 69, 70
	2	Find the difference quotient of a function	25, 26
	3	Find the domain of a function	17–24, 53(a)–56(a)
	4	Solve applied problems involving functions	71–78
10.3	1	Identify the graph of a function	49
	2	Obtain information from or about the graph of a function	27–30
	3	Determine even and odd functions from a graph	29(e), 30(e)
	4	Determine even and odd functions from the equation	31–38
	5	Use a graph to determine where a function is increasing, is decreasing, or is constant	29(b), 30(b)
	6	Use a graph to locate local maxima and minima	29(c), 30(c)
	7	Use a graphing utility to approximate local maxima and minima and to determine where a function is increasing or decreasing	39–42
	8	Find the average rate of change of a function	43–48
10.4	1	Graph the functions listed in the library of functions	50–52
	2	Graph piecewise-defined functions	53(c)–56(c)
10.5	1	Graph functions using horizontal and vertical shifts	57–60, 63, 64, 65(c), (d), 66(c), (d)
	2	Graph functions using reflections about the x-axis or y-axis	61, 62, 65(a), (b), 66(a), (b)

THINGS TO KNOW

Function (p. 549)
A relation between two sets of real numbers so that each number x in the first set, the domain, has corresponding to it exactly one number y in the second set. The range is the set of y values of the function for the x values in the domain.
x is the independent variable; y is the dependent variable.
A function f may be defined implicitly by an equation involving x and y or explicitly by writing $y = f(x)$.

Function notation (p. 550)
$y = f(x)$
f is a symbol for the function.
x is the argument, or independent variable.
y is the dependent variable.

Domain (p. 554)
If unspecified, the domain of a function f is the largest set of real numbers for which $f(x)$ is a real number.

Difference quotient of f (p. 552) $\dfrac{f(x+h)-f(x)}{h}, \quad h \neq 0$

Vertical-line test (p. 561) A set of points in the plane is the graph of a function if and only if every vertical line intersects the graph in at most one point.

Even function f (p. 565) $f(-x) = f(x)$ for every x in the domain ($-x$ must also be in the domain).

Odd function f (p. 565) $f(-x) = -f(x)$ for every x in the domain ($-x$ must also be in the domain).

Increasing function (p. 567) A function f is increasing on an open interval I if, for any choice of x_1 and x_2 in I, with $x_1 < x_2$, we have $f(x_1) < f(x_2)$.

Decreasing function (p. 567) A function f is decreasing on an open interval I if, for any choice of x_1 and x_2 in I, with $x_1 < x_2$, we have $f(x_1) > f(x_2)$.

Constant function (p. 567) A function f is constant on an interval I if, for all choices of x in I, the values of $f(x)$ are equal.

Local maximum (p. 568) A function f has a local maximum at c if there is an open interval I containing c so that, for all $x \neq c$ in I, $f(x) < f(c)$.

Local minimum (p. 568) A function f has a local minimum at c if there is an open interval I containing c so that, for all $x \neq c$ in I, $f(x) > f(c)$.

Average rate of change of a function (p. 570) The average rate of change of f from c to x is

$$\frac{\Delta y}{\Delta x} = \frac{f(x)-f(c)}{x-c}, \quad x \neq c$$

LIBRARY OF FUNCTIONS

Constant function (p. 580)
$f(x) = b$
Graph is a horizontal line with y-intercept $(0, b)$.

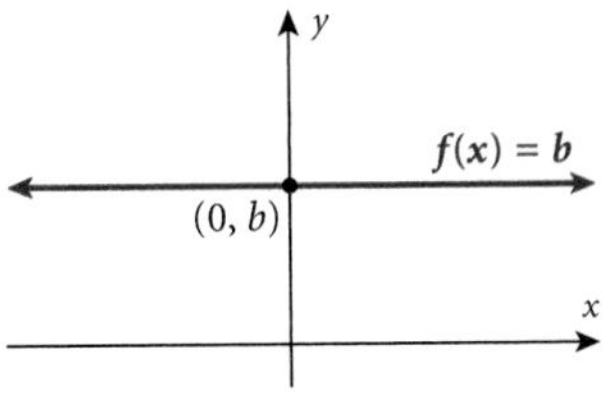

Linear function (p. 579)
$f(x) = mx + b$
Graph is a line with slope m and y-intercept $(0, b)$.

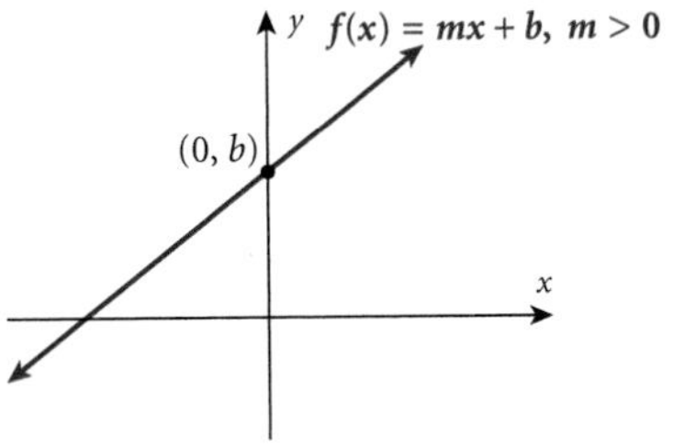

Identity function (p. 580)
$f(x) = x$
Graph is a line with slope 1 and y-intercept $(0, 0)$.

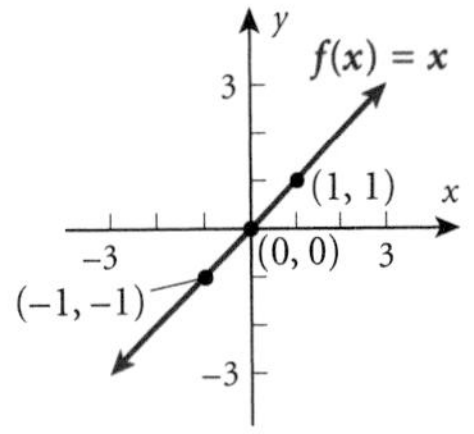

Square function (p. 580)
$f(x) = x^2$
Graph is a parabola with intercept at $(0, 0)$.

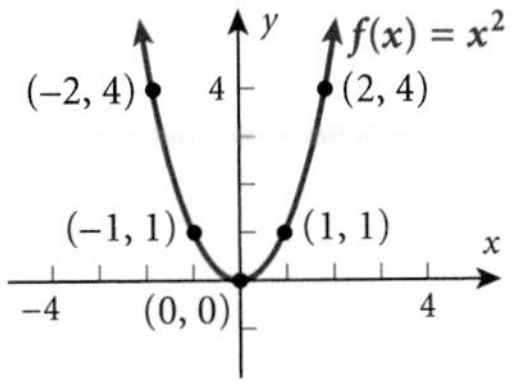

Cube function (p. 581)
$f(x) = x^3$

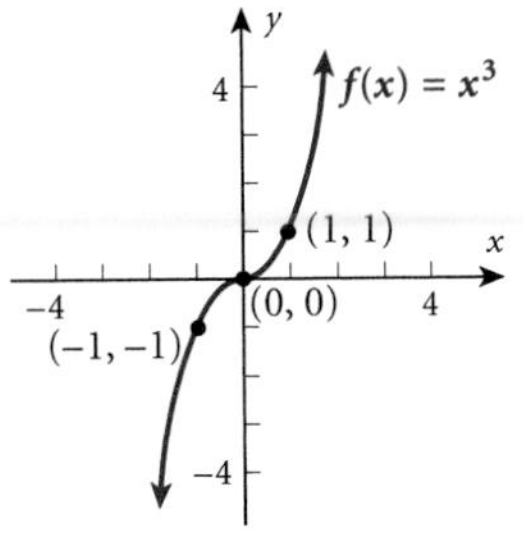

Square root function (p. 581)
$f(x) = \sqrt{x}$

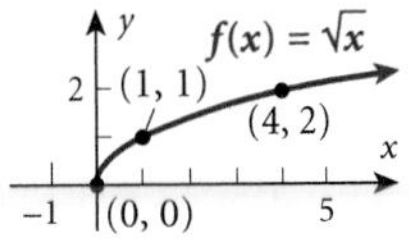

Cube root function (p. 581)

$f(x) = \sqrt[3]{x}$

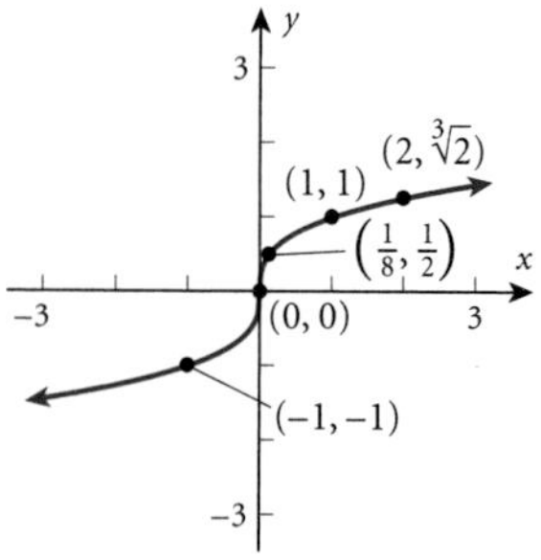

Absolute value function (p. 582)

$f(x) = |x|$

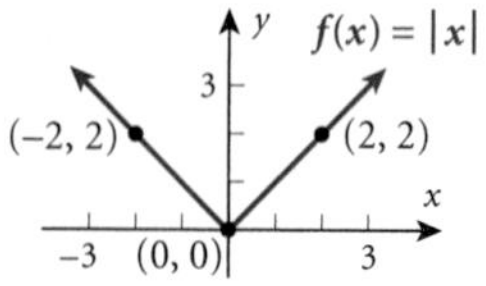

Reciprocal function (p. 581)

$f(x) = \dfrac{1}{x}$

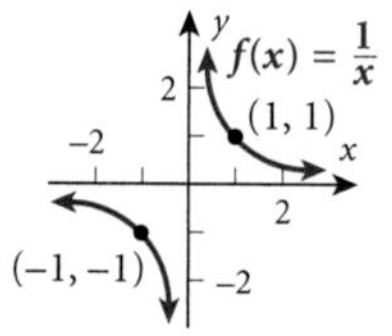

Greatest integer function (p. 582)

$f(x) = \text{int}(x)$

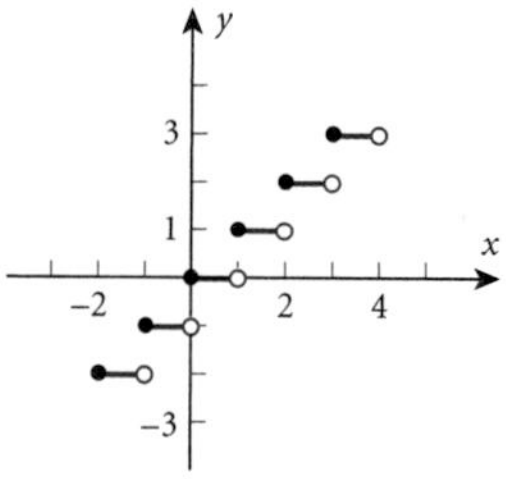

TRUE–FALSE ITEMS **Answers are on page AN-57.**

T F **1.** The domain of the function $f(x) = \sqrt{x}$ is the set of all real numbers.

T F **2.** For any function f, it follows that

$$f(x + h) = f(x) + f(h).$$

T F **3.** A function can have more than one y-intercept.

T F **4.** The graph of a function $y = f(x)$ always crosses the y-axis.

T F **5.** The y-intercept of the graph of the function $y = f(x)$, whose domain is all real numbers, is $(0, f(0))$.

T F **6.** To obtain the graph of $y = f(x + 2) - 3$, shift the graph of $y = f(x)$ horizontally to the right 2 units and vertically down 3 units.

FILL IN THE BLANKS **Answers are on page AN-57.**

1. If f is a function defined by the equation $y = f(x)$, then x is called the __________ variable and y is the __________ variable.

2. A set of points in the xy-plane is the graph of a function if and only if every __________ line intersects the graph in at most one point.

3. If the point $(5, -3)$ is a point on the graph of f, then $f($__________$) =$ __________.

4. If the point $(-1, 2)$ is on the graph of $f(x) = ax^2 + 4$, then $a =$ __________.

5. Suppose that the x-intercepts of the graph of $y = f(x)$ are $(-2, 0)$, $(1, 0)$, and $(5, 0)$. The x-intercepts of $y = f(x + 3)$ are __________, __________, and __________.

REVIEW EXERCISES **Answers to odd-numbered problems begin on page AN-57. Blue problem numbers indicate the author's suggestion for a practice test.**

1. Graph $y = x^2 + 4$ by plotting points.
2. List the intercepts of the graph shown.

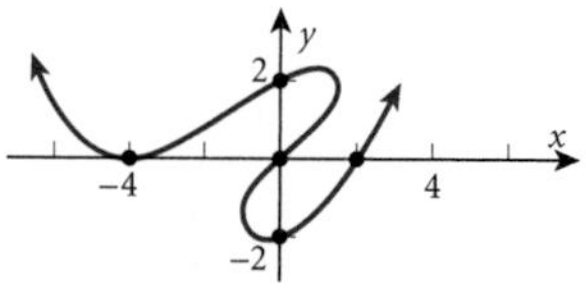

In Problems 3–10, list the intercepts and test for symmetry with respect to the x-axis, the y-axis, and the origin.

3. $2x = 3y^2$
4. $y = 5x$
5. $x^2 + 4y^2 = 16$
6. $9x^2 - y^2 = 9$
7. $y = x^4 + 2x^2 + 1$
8. $y = x^3 - x$
9. $x^2 + x + y^2 + 2y = 0$
10. $x^2 + 4x + y^2 - 2y = 0$

In Problems 11–16, find the following for each function:

(a) $f(2)$ (b) $f(-2)$ (c) $f(-x)$ (d) $-f(x)$ (e) $f(x-2)$ (f) $f(2x)$

11. $f(x) = \dfrac{3x}{x^2 - 1}$
12. $f(x) = \dfrac{x^2}{x + 1}$
13. $f(x) = \sqrt{x^2 - 4}$
14. $f(x) = |x^2 - 4|$
15. $f(x) = \dfrac{x^2 - 4}{x^2}$
16. $f(x) = \dfrac{x^3}{x^2 - 9}$

In Problems 17–24, find the domain of each function.

17. $f(x) = \dfrac{x}{x^2 - 9}$
18. $f(x) = \dfrac{3x^2}{x - 2}$
19. $f(x) = \sqrt{2 - x}$
20. $f(x) = \sqrt{x + 2}$
21. $h(x) = \dfrac{\sqrt{x}}{|x|}$
22. $g(x) = \dfrac{|x|}{x}$
23. $f(x) = \dfrac{x}{x^2 + 2x - 3}$
24. $F(x) = \dfrac{1}{x^2 - 3x - 4}$

In Problems 25–26, find the difference quotient of each function f, that is, find $\dfrac{f(x+h) - f(x)}{h}$, $h \neq 0$.

25. $f(x) = -2x^2 + x + 1$
26. $f(x) = 3x^2 - 2x + 4$

27. Using the graph of the function f shown below,
(a) Find the domain and range of f.
(b) List the intercepts.
(c) Find $f(-2)$.
(d) For what values of x does $f(x) = -3$?
(e) Solve $f(x) > 0$.

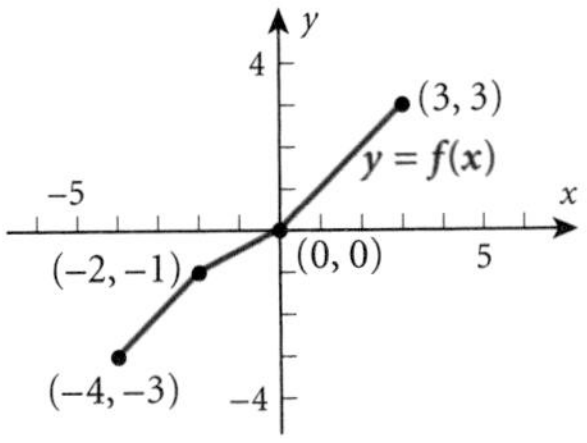

28. Using the graph of the function g shown below,
(a) Find the domain and range of g.
(b) Find $g(-1)$.
(c) List the intercepts of g.
(d) For what value of x does $g(x) = -3$?
(e) Solve $g(x) > 0$.

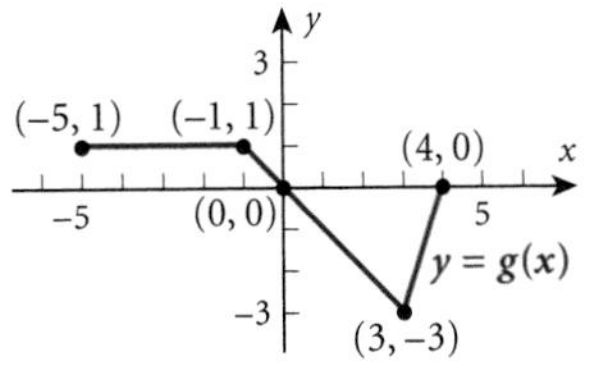

In Problems 29 and 30, use the graph of the function f to find:

(a) The domain and the range of f.
(b) The intervals on which f is increasing, decreasing, or constant.
(c) The local minima and local maxima.
(d) Whether the graph is symmetric with respect to the x axis, the y-axis, or the origin.
(e) Whether the function is even, odd, or neither.
(f) The intercepts, if any.

29.

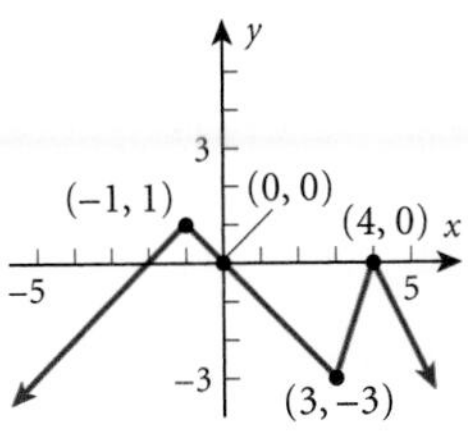

30.

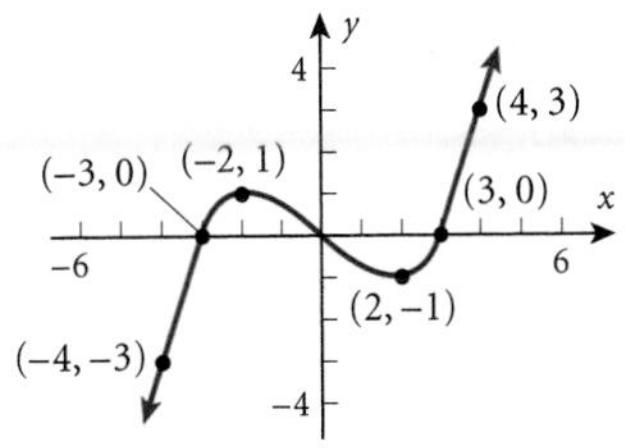

In Problems 31–38, determine (algebraically) whether the given function is even, odd, or neither.

31. $f(x) = x^3 - 4x$ **32.** $g(x) = \dfrac{4 + x^2}{1 + x^4}$ **33.** $h(x) = \dfrac{1}{x^4} + \dfrac{1}{x^2} + 1$ **34.** $F(x) = \sqrt{1 - x^3}$

35. $G(x) = 1 - x + x^3$ **36.** $H(x) = 1 + x + x^2$ **37.** $f(x) = \dfrac{x}{1 + x^2}$ **38.** $g(x) = \dfrac{1 + x^2}{x^3}$

In Problems 39–42, use a graphing utility to graph each function over the indicated interval. Approximate any local maxima and local minima. Determine where the function is increasing and where it is decreasing.

39. $f(x) = 2x^3 - 5x + 1 \quad (-3, 3)$ **40.** $f(x) = -x^3 + 3x - 5 \quad (-3, 3)$

41. $f(x) = 2x^4 - 5x^3 + 2x + 1 \quad (-2, 3)$ **42.** $f(x) = -x^4 + 3x^3 - 4x + 3 \quad (-2, 3)$

In Problems 43–44, find the average rate of change of f.

(*a*) from *1 to 2* (*b*) *from 0 to 1* (*c*) *from 2 to 4*

43. $f(x) = 8x^2 - x$ **44.** $f(x) = 2x^3 + x$

In Problems 45–48, find the average rate of change from 2 to x for each function f. Be sure to simplify.

45. $f(x) = 2 - 5x$ **46.** $f(x) = 2x^2 + 7$ **47.** $f(x) = 3x - 4x^2$ **48.** $f(x) = x^2 - 3x + 2$

49. Which of the following are graphs of functions.

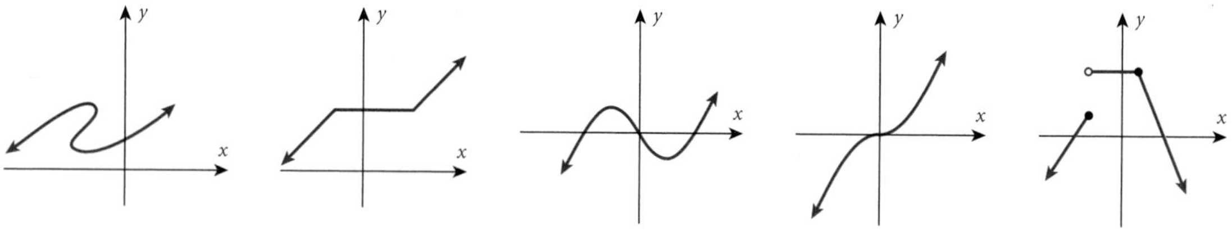

(a) (b) (c) (d) (e)

In Problems 50–52, sketch the graph of each function. Be sure to label at least three points.

50. $f(x) = |x|$ **51.** $f(x) = \sqrt[3]{x}$ **52.** $f(x) = \sqrt{x}$

In Problems 53–56:

(a) *Find the domain of each function.*
(b) *Locate any intercepts.*
(c) *Graph each function.*
(d) *Based on the graph, find the range.*

53. $f(x) = \begin{cases} 3x & -2 < x \le 1 \\ x + 1 & x > 1 \end{cases}$ **54.** $f(x) = \begin{cases} x - 1 & -3 < x < 0 \\ 3x - 1 & x \ge 0 \end{cases}$

55. $f(x) = \begin{cases} x & -4 \le x < 0 \\ 1 & x = 0 \\ 3x & x > 0 \end{cases}$ **56.** $f(x) = \begin{cases} x^2 & -2 \le x \le 2 \\ 2x - 1 & x > 2 \end{cases}$

In Problems 57–64, graph each function using shifting and/or reflections. Identify any intercepts of the graph. State the domain and, based on the graph, find the range.

57. $F(x) = |x| - 4$ **58.** $f(x) = |x| + 4$ **59.** $h(x) = \sqrt{x - 1}$ **60.** $h(x) = \sqrt{x} - 1$

61. $f(x) = \sqrt{1 - x}$ **62.** $f(x) = -\sqrt{x + 3}$ **63.** $h(x) = (x - 1)^2 + 2$ **64.** $h(x) = (x + 2)^2 - 3$

65. For the graph of the function f shown below, draw the graph of:

(a) $y = f(-x)$ (b) $y = -f(x)$
(c) $y = f(x + 2)$ (d) $y = f(x) + 2$

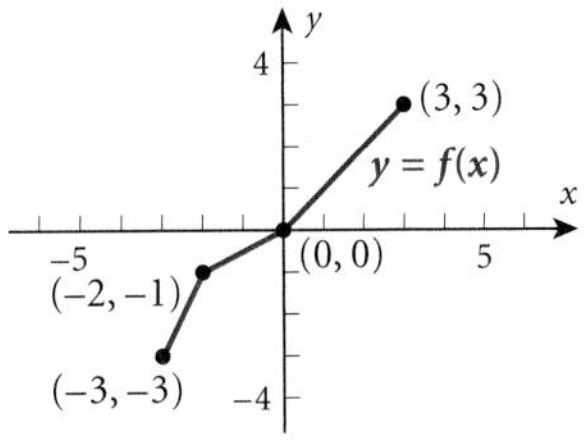

66. For the graph of the function g shown below, draw the graph of:

(a) $y = g(-x)$ (b) $y = -g(x)$
(c) $y = g(x + 2)$ (d) $y = g(x) + 2$

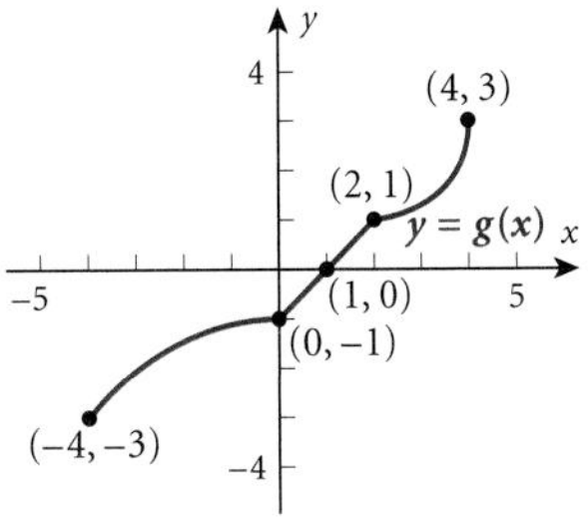

67. Given that f is a linear function, $f(4) = -5$ and $f(0) = 3$, write the equation that defines f.

68. Given that g is a linear function with slope $= -4$ and $g(-2) = 2$, write the equation that defines g.

69. A function f is defined by

$$f(x) = \frac{Ax + 5}{6x - 2}$$

If $f(1) = 4$, find A.

70. A function g is defined by

$$g(x) = \frac{A}{x} + \frac{8}{x^2}$$

If $g(-1) = 0$, find A.

71. **Volume of a Cylinder** The volume V of a right circular cylinder of height h and radius r is $V = \pi r^2 h$. If the height is twice the radius, express the volume V as a function of r.

72. **Volume of a Cone** The volume V of a right circular cone is $V = \frac{1}{3}\pi r^2 h$. If the height is twice the radius, express the volume V as a function of r.

73. **Demand Equation** The price p and the quantity x sold of a certain product obey the demand equation

$$p = -\frac{1}{6}x + 100, \quad 0 \le x \le 600$$

(a) Express the revenue R as a function of x. (Remember, $R = xp$.)
(b) What is the revenue if 200 units are sold?

74. **Demand Equation** The price p and the quantity x sold of a certain product obey the demand equation

$$p = -\frac{1}{3}x + 100, \quad 0 \le x \le 300$$

(a) Express the revenue R as a function of x.
(b) What is the revenue if 100 units are sold?

75. **Demand Equation** The price p and the quantity x sold of a certain product obey the demand equation

$$x = -5p + 100, \quad 0 \le p \le 20$$

(a) Express the revenue R as a function of x.
(b) What is the revenue if 15 units are sold?

76. **Demand Equation** The price p and the quantity x sold of a certain product obey the demand equation

$$x = -20p + 500, \quad 0 \le p \le 25$$

(a) Express the revenue R as a function of x.
(b) What is the revenue if 20 units are sold?

77. **Cost of a Drum** A drum in the shape of a right circular cylinder is required to have a volume of 500 cubic centimeters. The top and bottom are made of material that costs 6¢ per square centimeter; the sides are made of material that costs 4¢ per square centimeter.

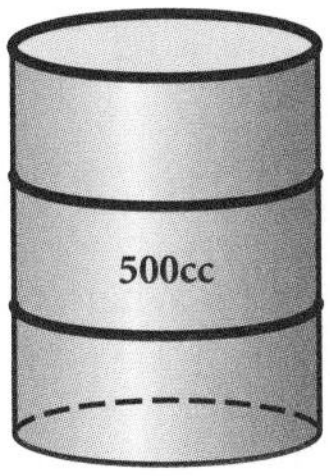

(a) Express the total cost C of the material as a function of the radius r of the cylinder. [**Hint:** The volume V of a right circular cylinder of height h and radius r is $V = \pi r^2 h$.]
(b) What is the cost if the radius is 4 cm?
(c) What is the cost if the radius is 8 cm?
(d) Graph $C = C(r)$. For what value of r is the cost C least?

78. Material Needed to Make a Drum A steel drum in the shape of a right circular cylinder is required to have a volume of 100 cubic feet.

(a) Express the amount A of material required to make the drum as a function of the radius r of the cylinder.

[**Hint:** The surface area S of a right circular cylinder of height h and radius r is $S = 2\pi r^2 + 2\pi rh$.]

(b) How much material is required if the drum is of radius 3 feet?

(c) Of radius 4 feet?

(d) Of radius 5 feet?

(e) Graph $A = A(r)$. For what value of r is A smallest?

Chapter 10 Project

For the one-day sightseeing trip to Charlotte, North Carolina, you and your friend decide to rent a mid-size car and, naturally, you want to do this as cheaply as possible.

You begin by contacting two well-known car rental companies: Avis and Enterprise. Avis offers a mid-size car for \$64.99 per day with unlimited mileage, so the number of miles you actually drive the car will not matter. Enterprise, on the other hand, offers a mid-size car for \$45.87 per day with 150 free miles, but will charge \$0.25 per mile for each mile in excess of 150 miles. Enterprise is the better deal as long as you drive less than 150 miles, but at what point, if any, will Avis be better? We'll use piecewise-defined functions to arrive at the answer.

1. Let x denote the number of miles the rental car is driven. Find the function $A = A(x)$ that gives the cost of driving the Avis car x miles. What kind of function will this be?
2. Find the function $E = E(x)$ that gives the cost of driving the Enterprise car x miles. Remember that the rule for computing this cost changes when x exceeds 150 miles, so a piecewise function is required.
3. Graph the functions $A = A(x)$ and $E = E(x)$ on the same set of axes. At what number of miles does the Avis rental car become a better choice?
4. In an effort to find an even better deal, you contact SaveALot Car Rental. They offer a mid-size car for \$36.99 per day with 100 free miles, but each mile in excess of 100 will cost \$0.30. Find the function $S = S(x)$ that gives the cost of driving the SaveALot car x miles. Graph this function along with $A = A(x)$ and $E = E(x)$.
5. Determine the number of miles driven and the companies that minimize the cost of the car rental.
6. In one last attempt to save money, you contact USave Car Rental and are offered a mid-size car for \$35.99 per day with 50 free miles, but each mile in excess of 50 will cost \$0.35. Find the function $U = U(x)$ that gives the cost of driving the USave car x miles. Graph this function along with the other three.
7. Determine the mileage and companies that minimize the cost of renting.
8. Comment on which car rental company you would use. Be sure to provide reasons.

MATHEMATICAL QUESTIONS FROM PROFESSIONAL EXAMS*

1. **Actuary Exam—Part I** What is the range of the function $f(x) = x^2$ with domain $(-1, 2]$?

(a) $1 < y \le 4$ (b) $1 < y < 4$
(c) $0 < y \le 4$ (d) $0 \le y < 4$
(e) $1 \le y \le 4$

2. **Actuary Exam—Part I** If $f(x + 1) = x^3 + 6x^2 + x + 3$ for all real x, then $f(4) = ?$

(a) 22 (b) 42 (c) 57 (d) 87 (e) 116

3. **Actuary Exam—Part I** What is the largest possible subset of $\mathbb{R}$ (real numbers) that can be a domain for the function $f(x) = \sqrt{x^3 - x}$?

(a) $[1, \infty)$
(b) $\{0\} \cup [1, \infty)$
(c) $[-1, 0] \cup [1, \infty)$
(d) $(-\infty, -1] \cup [1, \infty)$
(e) $(-\infty, -1] \cup \{0\} \cup [1, \infty)$

CHAPTER 11

Classes of Functions

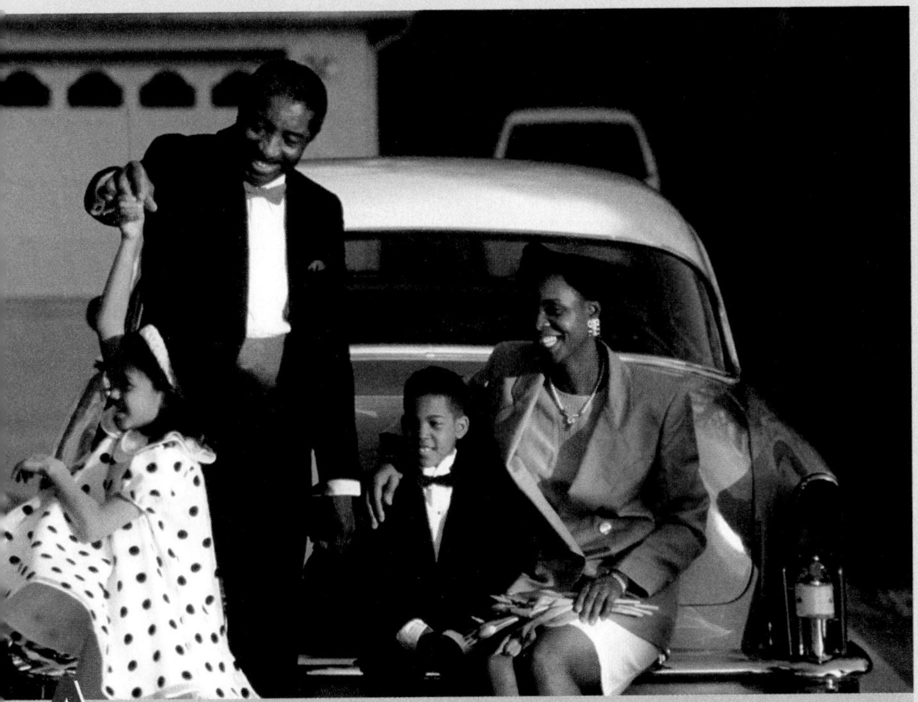

OUTLINE

Almost everyday you can pick up a newspaper and read about some prediction. Headlines appear like 'World population to increase 46% by 2050'. Or 'Projected growth in population of North America will be 41.8% over the next 50 years. Of course, predictions like these are important for proper planning and allocation of future resources. But where did these numbers come from? The Chapter Project provides one method of making predictions about future populations.

A LOOK BACK, A LOOK FORWARD

In Chapter 10, we began our discussion of functions. We defined domain and range and independent and dependent variables; we found the value of a function and graphed functions. We continued our study of functions by listing the properties that a function might have, like being even or odd, and we created a library of functions, naming key functions and listing their properties, including their graphs.

In this chapter, we look at four general classes of functions—polynomial functions, rational functions, exponential functions, and logarithmic functions—and examine their properties. Polynomial functions are arguably the simplest expressions in algebra. For this reason, they are often used to approximate other, more complicated functions. Rational functions are simply ratios of polynomial functions.

We begin with a discussion of quadratic functions, a type of polynomial function.

11.1 Quadratic Functions

PREPARING FOR THIS SECTION ***Before getting started, review the following:***

> Completing the Square (Appendix A, Section A.6, p. A-54)
> Square Root Method (Appendix A, Section A.6, pp. A-53–A-54)
> Quadratic Equations (Appendix A, Section A.6, pp. A-52–A-58)
> Geometry Review (Appendix A, Section A.9, pp. A-82–A-83)

OBJECTIVES
1. Locate the vertex and axis of symmetry of a quadratic function
2. Graph quadratic functions
3. Find the maximum or the minimum value of a quadratic function
4. Use the maximum or the minimum value of a quadratic function to solve applied problems

A *quadratic function* is a function that is defined by a second-degree polynomial in one variable.

A **quadratic function** is a function of the form

$$f(x) = ax^2 + bx + c \qquad (1)$$

where a, b, and c are real numbers and $a \neq 0$. The domain of a quadratic function is the set of all real numbers.

Many applications require a knowledge of the properties of the graph of a quadratic function. For example, suppose that Texas Instruments collects the data shown in Table 1 that relate the number of calculators sold at the price p per calculator. Since the price of a product determines the quantity that will be purchased, we treat price as the independent variable.

TABLE 1

Price per Calculator, p (Dollars)	Number of Calculators, x
60	11,100
65	10,115
70	9,652
75	8,731
80	8,087
85	7,205
90	6,439

A linear relationship between the number of calculators and the price p per calculator may be given by the equation

$$x = 21{,}000 - 150p$$

Then the revenue R derived from selling x calculators at the price p per calculator is

$$R = xp$$
$$R(p) = (21{,}000 - 150p)p$$
$$= -150p^2 + 21{,}000p$$

So the revenue R is a quadratic function of the price p. Figure 1 illustrates the graph of this revenue function, whose domain is $0 \le p \le 140$, since both x and p must be nonnegative. Later in this section we shall determine the price p that maximizes revenue.

A second situation in which a quadratic function appears involves the motion of a projectile. Based on Newton's second law of motion (force equals mass times acceleration, $F = ma$), it can be shown that, ignoring air resistance, the path of a projectile propelled upward at an inclination to the horizontal is the graph of a quadratic function. See Figure 2 for an illustration. Later in this section we shall analyze the path of a projectile.

FIGURE 1 Graph of a revenue function: $R = -150p^2 + 21{,}000p$

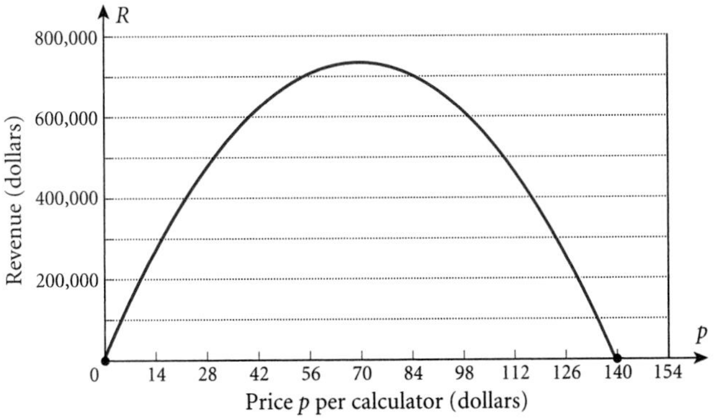

FIGURE 2 Path of a cannonball

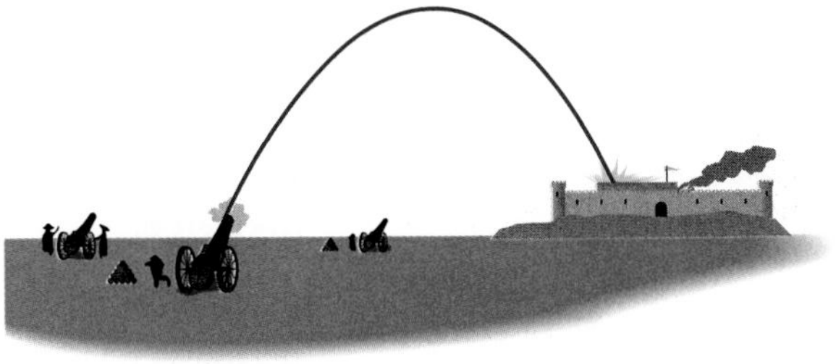

Graphing Quadratic Functions

We know how to graph the quadratic function $f(x) = x^2$. Figure 3 on page 606 shows the graph of three functions of the form $f(x) = ax^2$, $a > 0$, for $a = 1$, $a = \frac{1}{2}$, and $a = 3$. Notice that the larger the value of a, the "narrower" the graph, and the smaller the value of a, the "wider" the graph.

Figure 4 shows the graphs of $f(x) = ax^2$ for $a < 0$. Notice that these graphs are reflections about the x-axis of the graphs in Figure 3. Based on the results of these two figures, we can draw some general conclusions about the graph of $f(x) = ax^2$. First, as $|a|$ increases, the graph becomes *narrower*, and as $|a|$ gets closer to zero, the graph gets *wider*. Second, if a is positive, then the graph opens *up*, and if a is negative, the graph opens *down*.

The graphs in Figures 3 and 4 are typical of the graphs of all quadratic functions, which we call **parabolas**. Refer to Figure 5, where two parabolas are pictured. The one

FIGURE 3

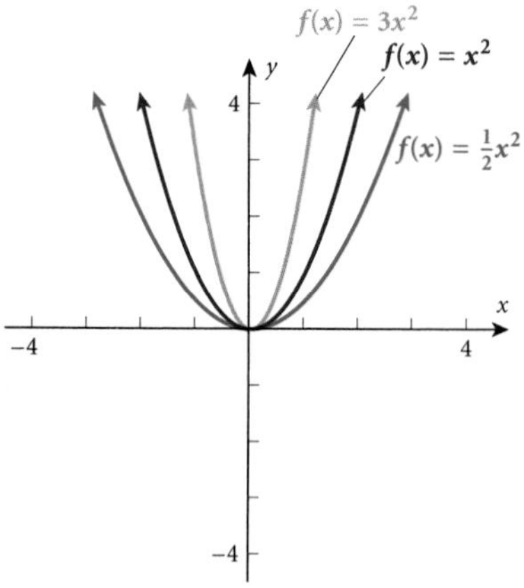

FIGURE 4

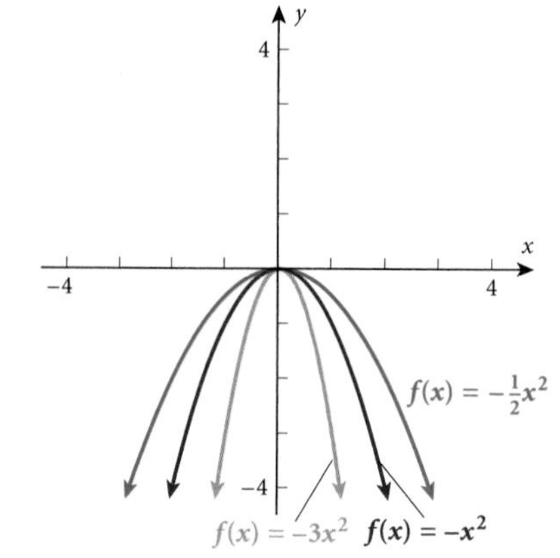

FIGURE 5 Graphs of a quadratic function, $f(x) = ax^2 + bx + c,\ a \neq 0$

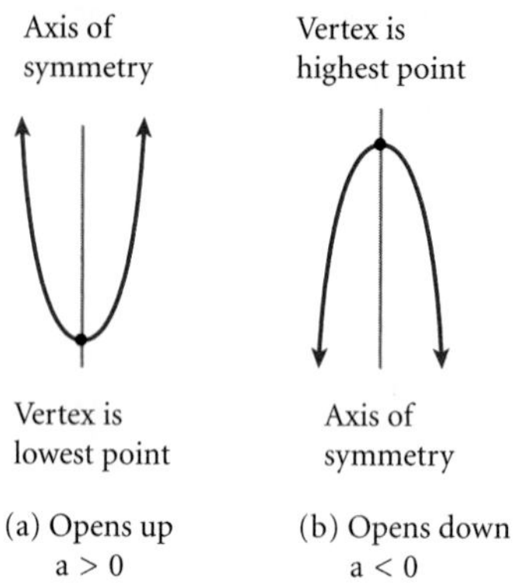

on the left **opens up** and has a lowest point; the one on the right **opens down** and has a highest point. The lowest or highest point of a parabola is called the **vertex.** The vertical line passing through the vertex in each parabola in Figure 5 is called the **axis of symmetry** (sometimes abbreviated to **axis**) of the parabola. Because the parabola is symmetric about its axis, the axis of symmetry of a parabola can be used to find additional points on the parabola.

The parabolas shown in Figure 5 are the graphs of a quadratic function $f(x) = ax^2 + bx + c,\ a \neq 0$. Notice that the coordinate axes are not included in the figure. Depending on the values of a, b, and c, the axes could be anywhere. The important fact is that the shape of the graph of a quadratic function will look like one of the parabolas in Figure 5.

A key element in graphing a quadratic function is locating the vertex. To find a formula, we begin with a quadratic function $f(x) = ax^2 + bx + c,\ a \neq 0$, and complete the square in x.

$$\begin{aligned}
f(x) &= ax^2 + bx + c && a \neq 0\\
&= a\left(x^2 + \frac{b}{a}x\right) + c && \text{Factor out } a \text{ from } ax^2 + bx\\
&= a\left(x^2 + \frac{b}{a}x + \frac{b^2}{4a^2}\right) + c - a\left(\frac{b^2}{4a^2}\right) && \text{Complete the square by adding and subtracting } a\frac{b^2}{4a^2}\text{. Look closely at this step!}\\
&= a\left(x + \frac{b}{2a}\right)^2 + c - \frac{b^2}{4a} && \text{Factor the perfect square.}\\
f(x) &= a\left(x + \frac{b}{2a}\right)^2 + \frac{4ac - b^2}{4a} && c - \frac{b^2}{4a} = c \cdot \frac{4a}{4a} - \frac{b^2}{4a} = \frac{4ac - b^2}{4a} \qquad (2)
\end{aligned}$$

Suppose that $a > 0$. If $x = -\dfrac{b}{2a}$, then the term $a\left(x + \dfrac{b}{2a}\right)^2 = a\left(-\dfrac{b}{2a} + \dfrac{b}{2a}\right)^2 = a \cdot 0 = 0$. For any other value of x, the term $a\left(x + \dfrac{b}{2a}\right)^2$ will be positive. [Do you see why? $\left(x + \dfrac{b}{2a}\right)^2$ is positive for $x \neq -\dfrac{b}{2a}$ because it is a nonzero quantity squared, a is positive, and the product of two positive quantities is positive.] Because $a\left(x + \dfrac{b}{2a}\right)^2$ is zero if $x = -\frac{b}{2a}$ and is positive if $x \neq -\frac{b}{2a}$, the value of the function f given in (2) will be smallest when $x = -\frac{b}{2a}$. That is, if $a > 0$, the parabola opens up, the vertex is at $\left(-\frac{b}{2a}, f\left(-\frac{b}{2a}\right)\right)$, and the vertex is a minimum point.

Similarly, if $a < 0$, the term $a\left(x + \frac{b}{2a}\right)^2$ is zero for $x = -\frac{b}{2a}$ and is negative if $x \neq -\frac{b}{2a}$. In this case the largest value of $f(x)$ occurs when $x = -\frac{b}{2a}$. That is, if $a < 0$, the parabola opens down, the vertex is at $\left(-\frac{b}{2a}, f\left(-\frac{b}{2a}\right)\right)$, and the vertex is a maximum point.

We summarize these remarks as follows:

Properties of the Graph of a Quadratic Function

$$f(x) = ax^2 + bx + c, \quad a \neq 0$$

Vertex $= \left(-\frac{b}{2a}, f\left(-\frac{b}{2a}\right)\right)$ Axis of symmetry: the line $x = -\frac{b}{2a}$

Parabola opens up if $a > 0$; the vertex is a minimum point.
Parabola opens down if $a < 0$; the vertex is a maximum point.

1 **EXAMPLE 1** **Locating the Vertex and Axis of Symmetry**

Locate the vertex and axis of symmetry of the parabola defined by $f(x) = -3x^2 + 6x + 1$. Does it open up or down?

SOLUTION For this quadratic function, $a = -3$, $b = 6$, and $c = 1$. The x-coordinate of the vertex is

$$-\frac{b}{2a} = -\frac{6}{-6} = 1$$

The y-coordinate of the vertex is

$$f\left(-\frac{b}{2a}\right) = f(1) = -3 + 6 + 1 = 4$$

The vertex is located at the point $(1, 4)$. The axis of symmetry is the line $x = 1$. Because $a = -3 < 0$, the parabola opens down.

Graph quadratic functions 2

The facts that we gathered in Example 1, together with the location of the intercepts, usually provide enough information to graph $f(x) = ax^2 + bx + c, a \neq 0$.

The y-intercept is found by finding the value of f at $x = 0$; that is, by finding $f(0) = c$.

The x-intercepts, if there are any, are found by solving the equation

$$f(x) = ax^2 + bx + c = 0$$

This equation has two, one, or no real solutions, depending on whether the discriminant $b^2 - 4ac$ is positive, 0, or negative. Depending on the value of the discriminant, the graph of f has x-intercepts as follows:

The x-Intercepts of a Quadratic Function

1. If the discriminant $b^2 - 4ac > 0$, the graph of $f(x) = ax^2 + bx + c$ has two distinct x-intercepts and so will cross the x-axis in two places.
2. If the discriminant $b^2 - 4ac = 0$, the graph of $f(x) = ax^2 + bx + c$ has one x-intercept and touches the x-axis at its vertex.
3. If the discriminant $b^2 - 4ac < 0$, the graph of $f(x) = ax^2 + bx + c$ has no x-intercept and so will not cross or touch the x-axis.

Figure 6 illustrates these possibilities for parabolas that open up.

FIGURE 6
$f(x) = ax^2 + bx + c,\ a > 0$

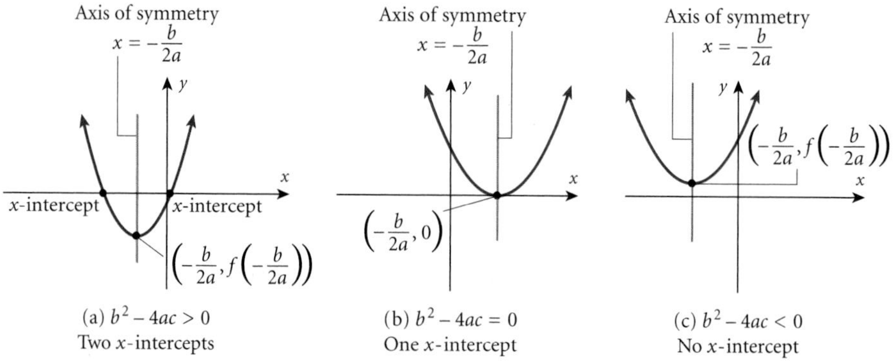

EXAMPLE 2 Graphing a Quadratic Function Using Its Vertex, Axis, and Intercepts

Use the information from Example 1 and the locations of the intercepts to graph $f(x) = -3x^2 + 6x + 1$.

SOLUTION In Example 1, we found the vertex to be at $(1, 4)$ and the axis of symmetry to be $x = 1$. The y-intercept is found by letting $x = 0$. Since $f(0) = 1$, the y-intercept is $(0, 1)$. The x-intercepts are found by solving the equation $f(x) = 0$. This results in the equation

$$-3x^2 + 6x + 1 = 0 \qquad a = -3, b = 6, c = 1$$

The discriminant $b^2 - 4ac = (6)^2 - 4(-3)(1) = 36 + 12 = 48 > 0$, so the equation has two real solutions and the graph has two x-intercepts. Using the quadratic formula, we find that

$$x = \frac{-b + \sqrt{b^2 - 4ac}}{2a} = \frac{-6 + \sqrt{48}}{-6} = \frac{-6 + 4\sqrt{3}}{-6} \approx -0.15$$

and

$$x = \frac{-b - \sqrt{b^2 - 4ac}}{2a} = \frac{-6 - \sqrt{48}}{-6} = \frac{-6 - 4\sqrt{3}}{-6} \approx 2.15$$

FIGURE 7

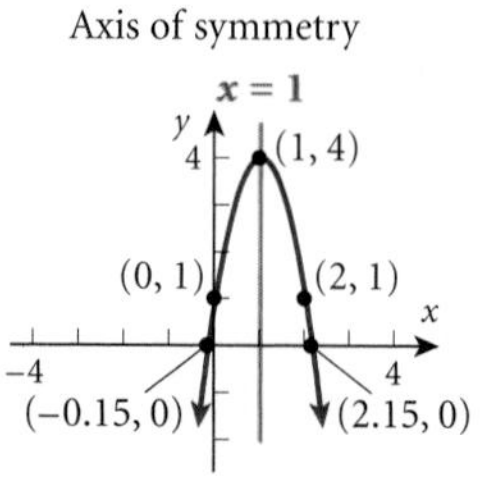

The x-intercepts are approximately $(-0.15, 0)$ and $(2.15, 0)$.

The graph is illustrated in Figure 7. Notice how we used the y-intercept $(0, 1)$ and the axis of symmetry, $x = 1$, to obtain the additional point $(2, 1)$ on the graph.

CHECK: Graph $f(x) = -3x^2 + 6x + 1$. Use ROOT or ZERO to locate the two x-intercepts and use MAXIMUM to locate the vertex.

NOW WORK PROBLEM 9.

If the graph of a quadratic function has only one x-intercept or none, it is usually necessary to plot an additional point to obtain the graph.

EXAMPLE 3 Graphing a Quadratic Function Using Its Vertex, Axis, and Intercepts

Graph $f(x) = x^2 - 6x + 9$ by determining whether the graph opens up or down. Find its vertex, axis of symmetry, y-intercept, and x-intercepts, if any.

FIGURE 8

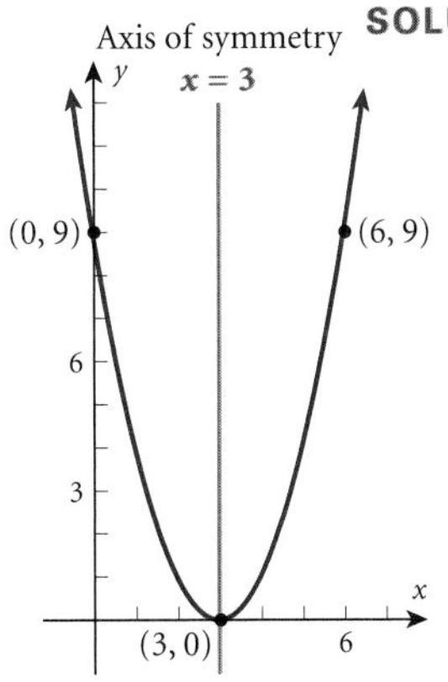

SOLUTION For $f(x) = x^2 - 6x + 9$, we have $a = 1$, $b = -6$, and $c = 9$. Since $a = 1 > 0$, the parabola opens up. The x-coordinate of the vertex is

$$-\frac{b}{2a} = -\frac{-6}{2(1)} = 3$$

The y-coordinate of the vertex is

$$f(3) = (3)^2 - 6(3) + 9 = 0$$

So the vertex is at $(3, 0)$. The axis of symmetry is the line $x = 3$. Since $f(0) = 9$, the y-intercept is $(0, 9)$. Since the vertex $(3, 0)$ lies on the x-axis, the graph touches the x-axis at the x-intercept. By using the axis of symmetry and the y-intercept $(0, 9)$, we can locate the additional point $(6, 9)$ on the graph. See Figure 8.

NOW WORK PROBLEM 17.

EXAMPLE 4 Graphing a Quadratic Function Using Its Vertex, Axis, and Intercepts

Graph $f(x) = 2x^2 + x + 1$ by determining whether the graph opens up or down. Find its vertex, axis of symmetry, y-intercept, and x-intercepts, if any.

SOLUTION For $f(x) = 2x^2 + x + 1$, we have $a = 2$, $b = 1$, and $c = 1$. Since $a = 2 > 0$, the parabola opens up. The x-coordinate of the vertex is

FIGURE 9

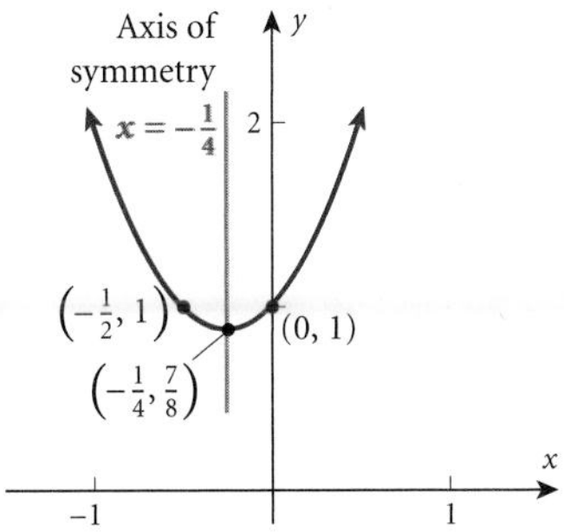

$$-\frac{b}{2a} = -\frac{1}{4}$$

The y-coordinate of the vertex is

$$k = f\left(-\frac{1}{4}\right) = 2\left(\frac{1}{16}\right) + \left(-\frac{1}{4}\right) + 1 = \frac{7}{8}$$

So the vertex is at $\left(-\frac{1}{4}, \frac{7}{8}\right)$. The axis of symmetry is the line $x = -\frac{1}{4}$. Since $f(0) = 1$, the y-intercept is $(0, 1)$. The x-intercept(s), if any, obey the equation $2x^2 + x + 1 = 0$. Since the discriminant $b^2 - 4ac = (1)^2 - 4(2)(1) = -7 < 0$, this equation has no real solutions, and therefore the graph has no x-intercepts. We use the point (0, 1) and the axis of symmetry $x = -\frac{1}{4}$ to locate the additional point $\left(-\frac{1}{2}, 1\right)$ on the graph. See Figure 9.

NOW WORK PROBLEM 21.

SUMMARY

Steps for Graphing a Quadratic Function $f(x) = ax^2 + bx + c, \quad a \neq 0$

STEP 1: Determine the vertex, $\left(-\frac{b}{2a}, f\left(-\frac{b}{2a}\right)\right)$.

STEP 2: Determine the axis of symmetry, $x = -\frac{b}{2a}$.

STEP 3: Determine the y-intercept by finding $f(0) = c$.

STEP 4: Evaluate the discriminant $b^2 - 4ac$.

(a) If $b^2 - 4ac > 0$, then the graph of the quadratic function has two x-intercepts, which are found by solving the equation $ax^2 + bx + c = 0$.

(b) If $b^2 - 4ac = 0$, the vertex is the x-intercept.

(c) If $b^2 - 4ac < 0$, there are no x-intercepts.

STEP 5: Determine an additional point if $b^2 - 4ac \leq 0$ by using the y-intercept and the axis of symmetry.

STEP 6: Plot the points and draw the graph.

Quadratic Models

When a mathematical model leads to a quadratic function, the properties of this quadratic function can provide important information about the model. For example, for a quadratic revenue function, we can find the maximum revenue; for a quadratic cost function, we can find the minimum cost.

To see why, recall that the graph of a quadratic function $f(x) = ax^2 + bx + c$ is a parabola with vertex at $\left(-\frac{b}{2a}, f\left(-\frac{b}{2a}\right)\right)$. This vertex is the highest point on the graph if $a < 0$ and the lowest point on the graph if $a > 0$. If the vertex is the highest point, ($a < 0$), then $f\left(-\frac{b}{2a}\right)$ is the **maximum value** of f. If the vertex is the lowest point, ($a > 0$), then $f\left(-\frac{b}{2a}\right)$ is the **minimum value** of f.

This property of the graph of a quadratic function enables us to answer questions involving optimization (finding maximum or minimum values) in models involving quadratic functions.

3 **EXAMPLE 5** **Finding the Maximum or Minimum Value of a Quadratic Function**

Determine whether the quadratic function

$$f(x) = x^2 - 4x + 7$$

has a maximum or minimum value. Then find the maximum or minimum value.

SOLUTION We compare $f(x) = x^2 - 4x + 7$ to $f(x) = ax^2 + bx + c$. We conclude that $a = 1$, $b = -4$, and $c = 7$. Since $a > 0$, the graph of f opens up, so the vertex is a minimum point. The minimum value occurs at

$$x = -\frac{b}{2a} = -\frac{-4}{2(1)} = \frac{4}{2} = 2$$

$\uparrow$ $a = 1, b = -4$

The minimum value is

$$f\left(-\frac{b}{2a}\right) = f(2) = 2^2 - 4(2) + 7 = 4 - 8 + 7 = 3$$

NOW WORK PROBLEM 29.

4 **EXAMPLE 6** **Maximizing Revenue**

The marketing department at Texas Instruments has found that, when certain calculators are sold at a price of p dollars per unit, the revenue R (in dollars) as a function of the price p is

$$R(p) = -150p^2 + 21{,}000p$$

What unit price should be established in order to maximize revenue? If this price is charged, what is the maximum revenue?

SOLUTION The revenue R is

$$R(p) = -150p^2 + 21{,}000p \quad R(p) = ap^2 + bp + c$$

The function R is a quadratic function with $a = -150$, $b = 21{,}000$, and $c = 0$. Because $a < 0$, the vertex is the highest point of the parabola. The revenue R is therefore a maximum when the price p is

$$p = -\frac{b}{2a} = -\frac{21{,}000}{2(-150)} = \frac{-21{,}000}{-300} = \$70.00$$

The maximum revenue R is

$$R(70) = -150(70)^2 + 21{,}000(70) = \$735{,}000$$

See Figure 10 for an illustration.

FIGURE 10

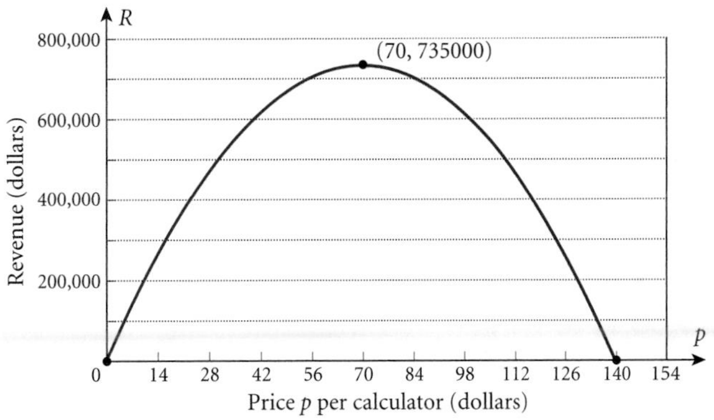

NOW WORK PROBLEM 37.

EXAMPLE 7 Maximizing the Area Enclosed by a Fence

A farmer has 2000 yards of fence to enclose a rectangular field. What is the largest area that can be enclosed?

FIGURE 11

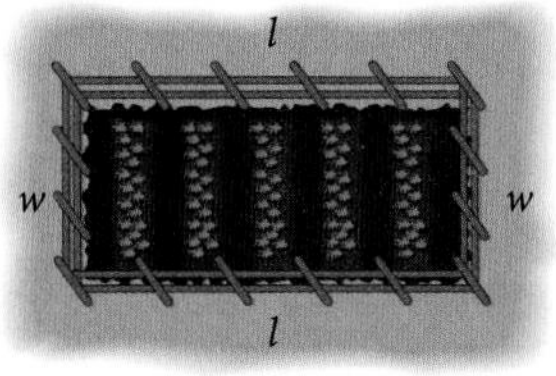

SOLUTION Figure 11 illustrates the situation. The available fence represents the perimeter of the rectangle. If ℓ is the length and w is the width, then

$$\text{perimeter} = 2\ell + 2w$$

$$2\ell + 2w = 2000 \tag{3}$$

The area A of the rectangle is

$$A = \ell w$$

To express A in terms of a single variable, we solve equation (3) for w and substitute the result in $A = \ell w$. Then A involves only the variable ℓ. [You could also solve equation (3) for ℓ and express A in terms of w alone. Try it!]

$$2\ell + 2w = 2000 \qquad \text{Equation (3).}$$

$$2w = 2000 - 2\ell \qquad \text{Solve for } w.$$

$$w = \frac{2000 - 2\ell}{2} = 1000 - \ell$$

Then the area A is

$$A = \ell w = \ell(1000 - \ell) = -\ell^2 + 1000\ell$$

Now, A is a quadratic function of ℓ.

$$A(\ell) = -\ell^2 + 1000\ell \qquad a = -1, b = 1000, c = 0$$

Since $a < 0$, the vertex is a maximum point on the graph of A. The maximim value occurs at

$$\ell = -\frac{b}{2a} = -\frac{1000}{2(-1)} = 500$$

The maximum value of A is

$$A\left(-\frac{b}{2a}\right) = A(500) = -500^2 + 1000(500)$$

$$= -250{,}000 + 500{,}000 = 250{,}000$$

The largest area that can be enclosed by 2000 yards of fence in the shape of a rectangle is 250,000 square yards.

FIGURE 12

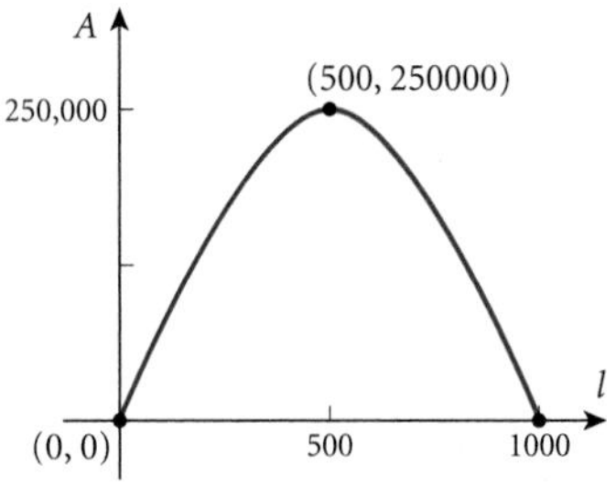

Figure 12 shows the graph of $A(\ell) = -\ell^2 + 1000\ell$.

NOW WORK PROBLEM 43.

EXAMPLE 8 Analyzing the Motion of a Projectile

A projectile is fired from a cliff 500 feet above the water at an inclination of 45° to the horizontal, with a muzzle velocity of 400 feet per second. In physics, it is established

FIGURE 13

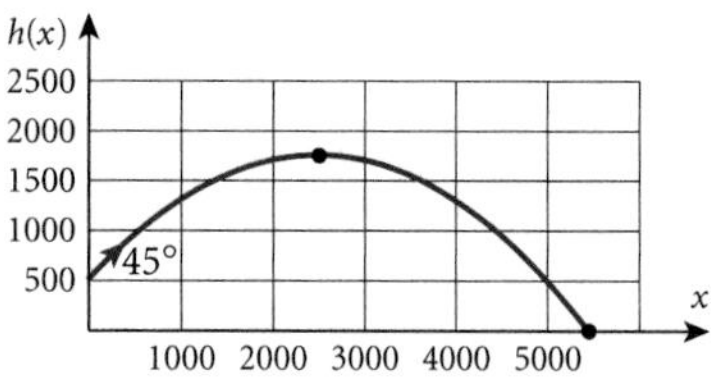

that the height h of the projectile above the water is given by

$$h(x) = \frac{-32x^2}{(400)^2} + x + 500$$

where x is the horizontal distance of the projectile from the base of the cliff. See Figure 13.

(a) Find the maximum height of the projectile.
(b) How far from the base of the cliff will the projectile strike the water?

SOLUTION **(a)** The height of the projectile is given by a quadratic function.

$$h(x) = \frac{-32x^2}{(400)^2} + x + 500 = \frac{-1}{5000}x^2 + x + 500 \qquad a = \frac{-1}{5000}, \quad b = 1, \quad c = 500$$

We are looking for the maximum value of h. Since the maximum value is obtained at the vertex, we compute

$$x = -\frac{b}{2a} = -\frac{1}{2\left(\frac{-1}{5000}\right)} = \frac{5000}{2} = 2500$$

The maximum height of the projectile is

$$\begin{aligned} h(2500) &= \frac{-1}{5000}(2500)^2 + 2500 + 500 \\ &= -1250 + 2500 + 500 = 1750 \text{ ft} \end{aligned}$$

(b) The projectile will strike the water when the height is zero. To find the distance x traveled, we need to solve the equation

$$h(x) = \frac{-1}{5000}x^2 + x + 500 = 0$$

We use the quadratic formula with

$$b^2 - 4ac = 1 - 4\left(\frac{-1}{5000}\right)(500) = 1.4$$

$$x = \frac{-1 \pm \sqrt{1.4}}{2\left(\frac{-1}{5000}\right)} \approx \begin{cases} -458 \\ 5458 \end{cases}$$

We discard the negative solution and find that the projectile will strike the water at a distance of about 5458 feet from the base of the cliff.

SEEING THE CONCEPT: Graph

$$h(x) = \frac{-1}{5000}x^2 + x + 500, \qquad 0 \le x \le 5500$$

Use MAXIMUM to find the maximum height of the projectile, and use ROOT or ZERO to find the distance from the base of the cliff to where the projectile strikes the water. Compare your results with those obtained in Example 8. TRACE the path of the projectile. How far from the base of the cliff is the projectile when its height is 1000 ft? 1500 ft?

NOW WORK PROBLEM 47.

EXERCISE 11.1 Answers to Odd-Numbered Problems Begin on Page AN-60.

In Problems 1–8, match each graph to one the following functions without using a graphing utility.

1. $f(x) = x^2 - 1$ **2.** $f(x) = -x^2 - 1$ **3.** $f(x) = x^2 - 2x + 1$ **4.** $f(x) = x^2 + 2x + 1$

5. $f(x) = x^2 - 2x + 2$ **6.** $f(x) = x^2 + 2x$ **7.** $f(x) = x^2 - 2x$ **8.** $f(x) = x^2 + 2x + 2$

(A)
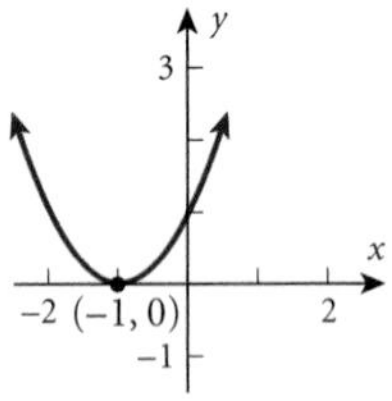

(B)
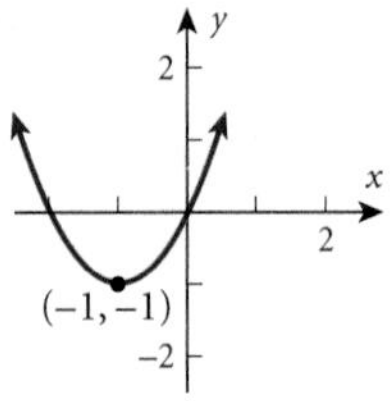

(C)
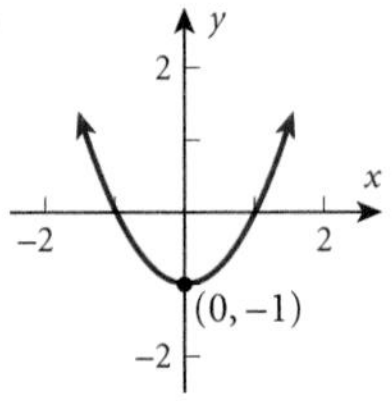

(D)
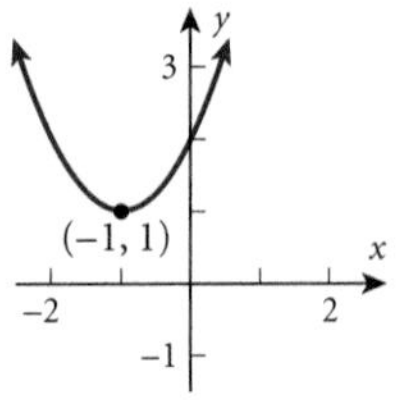

(E)
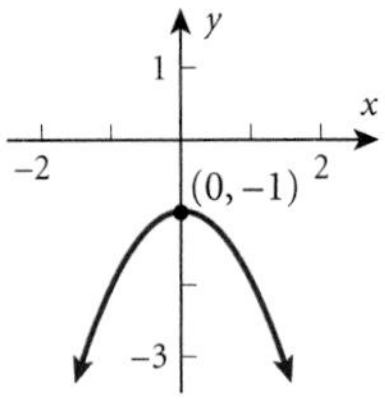

(F)
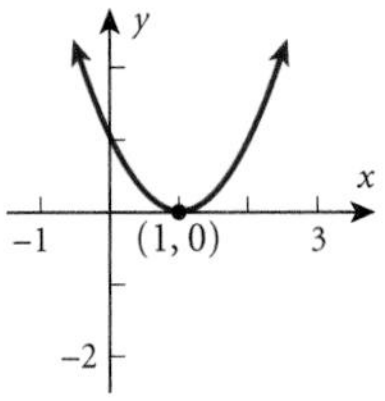

(G)
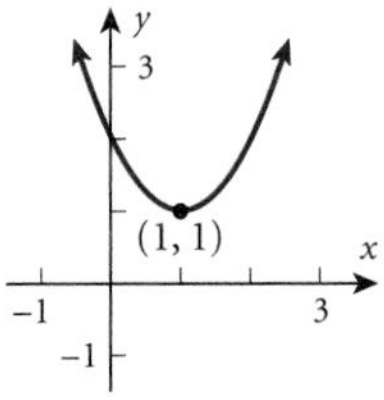

(H)
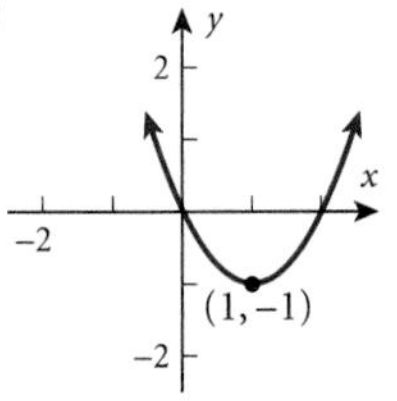

In Problems 9–26, graph each quadratic function by determining whether its graph opens up or down and by finding its vertex, axis of symmetry, y-intercept, and x-intercepts, if any. Determine the domain and the range of the function. Determine where the function is increasing and where it is decreasing.

9. $f(x) = x^2 + 2x$ **10.** $f(x) = x^2 - 4x$ **11.** $f(x) = -x^2 - 6x$ **12.** $f(x) = -x^2 + 4x$

13. $f(x) = 2x^2 - 8x$ **14.** $f(x) = 3x^2 + 18x$ **15.** $f(x) = x^2 + 2x - 8$ **16.** $f(x) = x^2 - 2x - 3$

17. $f(x) = x^2 + 2x + 1$ **18.** $f(x) = x^2 + 6x + 9$ **19.** $f(x) = 2x^2 - x + 2$ **20.** $f(x) = 4x^2 - 2x + 1$

21. $f(x) = -2x^2 + 2x - 3$ **22.** $f(x) = -3x^2 + 3x - 2$ **23.** $f(x) = 3x^2 + 6x + 2$ **24.** $f(x) = 2x^2 + 5x + 3$

25. $f(x) = -4x^2 - 6x + 2$ **26.** $f(x) = 3x^2 - 8x + 2$

In Problems 27–34, determine, without graphing, whether the given quadratic function has a maximum value or a minimum value and then find the value.

27. $f(x) = 2x^2 + 12x$ **28.** $f(x) = -2x^2 + 12x$ **29.** $f(x) = 2x^2 + 12x - 3$ **30.** $f(x) = 4x^2 - 8x + 3$

31. $f(x) = -x^2 + 10x - 4$ **32.** $f(x) = -2x^2 + 8x + 3$ **33.** $f(x) = -3x^2 + 12x + 1$ **34.** $f(x) = 4x^2 - 4x$

Answer Problems 35 and 36 using the following discussion: A quadratic function of the form $f(x) = ax^2 + bx + c$ with $b^2 - 4ac > 0$ may also be written in the form $f(x) = a(x - r_1)(x - r_2)$, where r_1 and r_2 are the x-intercepts of the graph of the quadratic function.

35. (a) Find a quadratic function whose x-intercepts are -3 and 1 with $a = 1$; $a = 2$; $a = -2$; $a = 5$.
(b) How does the value of a affect the intercepts?
(c) How does the value of a affect the axis of symmetry?
(d) How does the value of a affect the vertex?
(e) Compare the x-coordinate of the vertex with the midpoint of the x-intercepts. What might you conclude?

36. (a) Find a quadratic function whose x-intercepts are -5 and 3 with $a = 1$; $a = 2$; $a = -2$; $a = 5$.
(b) How does the value of a affect the intercepts?
(c) How does the value of a affect the axis of symmetry?
(d) How does the value of a affect the vertex?
(e) Compare the x-coordinate of the vertex with the midpoint of the x-intercepts. What might you conclude?

37. **Maximizing Revenue** Suppose that the manufacturer of a gas clothes dryer has found that, when the unit price is p dollars, the revenue R (in dollars) is

$$R(p) = -4p^2 + 4000p$$

What unit price should be established for the dryer to maximize revenue? What is the maximum revenue?

38. **Maximizing Revenue** The John Deere company has found that the revenue from sales of heavy-duty tractors is a function of the unit price p that it charges. If the revenue R is

$$R(p) = -\frac{1}{2}p^2 + 1900p$$

what unit price p should be charged to maximize revenue? What is the maximum revenue?

39. **Demand Equation** The price p and the quantity x sold of a certain product obey the demand equation

$$p = -\frac{1}{6}x + 100, \qquad 0 \le x \le 600$$

(a) Express the revenue R as a function of x. (Remember, $R = xp$.)
(b) What is the revenue if 200 units are sold?
(c) What quantity x maximizes revenue? What is the maximum revenue?
(d) What price should the company charge to maximize revenue?

40. **Demand Equation** The price p and the quantity x sold of a certain product obey the demand equation

$$p = -\frac{1}{3}x + 100, \qquad 0 \le x \le 300$$

(a) Express the revenue R as a function of x.
(b) What is the revenue if 100 units are sold?
(c) What quantity x maximizes revenue? What is the maximum revenue?
(d) What price should the company charge to maximize revenue?

41. **Demand Equation** The price p and the quantity x sold of a certain product obey the demand equation

$$x = -5p + 100, \qquad 0 \le p \le 20$$

(a) Express the revenue R as a function of x.
(b) What is the revenue if 15 units are sold?
(c) What quantity x maximizes revenue? What is the maximum revenue?
(d) What price should the company charge to maximize revenue?

42. **Demand Equation** The price p and the quantity x sold of a certain product obey the demand equation

$$x = -20p + 500, \quad 0 \le p \le 25$$

(a) Express the revenue R as a function of x.
(b) What is the revenue if 20 units are sold?
(c) What quantity x maximizes revenue? What is the maximum revenue?
(d) What price should the company charge to maximize revenue?

43. **Enclosing a Rectangular Field** David has available 400 yards of fencing and wishes to enclose a rectangular area.

(a) Express the area A of the rectangle as a function of the width x of the rectangle.
(b) For what value of x is the area largest?
(c) What is the maximum area?

44. **Enclosing a Rectangular Field** Beth has 3000 feet of fencing available to enclose a rectangular field.

(a) Express the area A of the rectangle as a function of x, where x is the length of the rectangle.
(b) For what value of x is the area largest?
(c) What is the maximum area?

45. **Enclosing the Most Area with a Fence** A farmer with 4000 meters of fencing wants to enclose a rectangular plot that borders on a river. If the farmer does not fence the side along the river, what is the largest area that can be enclosed? (See the figure.)

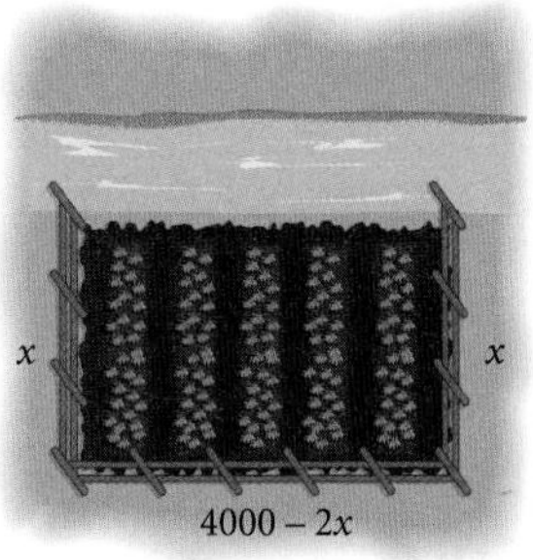

46. **Enclosing the Most Area with a Fence** A farmer with 2000 meters of fencing wants to enclose a rectangular plot that borders on a straight highway. If the farmer does not fence the side along the highway, what is the largest area that can be enclosed?

47. **Analyzing the Motion of a Projectile** A projectile is fired from a cliff 200 feet above the water at an inclination of 45° to the horizontal, with a muzzle velocity of 50 feet per second. The height h of the projectile above the water is given by

$$h(x) = \frac{-32x^2}{(50)^2} + x + 200$$

where x is the horizontal distance of the projectile from the base of the cliff.

(a) How far from the base of the cliff is the height of the projectile a maximum?
(b) Find the maximum height of the projectile.

(c) How far from the base of the cliff will the projectile strike the water?

(d) Using a graphing utility, graph the function h, $0 \leq x \leq 200$.

(e) When the height of the projectile is 100 feet above the water, how far is it from the cliff?

48. Analyzing the Motion of a Projectile A projectile is fired at an inclination of 45° to the horizontal, with a muzzle velocity of 100 feet per second. The height h of the projectile is given by

$$h(x) = \frac{-32x^2}{(100)^2} + x$$

where x is the horizontal distance of the projectile from the firing point.

(a) How far from the firing point is the height of the projectile a maximum?

(b) Find the maximum height of the projectile.

(c) How far from the firing point will the projectile strike the ground?

(d) Using a graphing utility, graph the function h, $0 \leq x \leq 350$.

(e) When the height of the projectile is 50 feet above the ground, how far has it traveled horizontally?

49. Constructing Rain Gutters A rain gutter is to be made of aluminum sheets that are 12 inches wide by turning up the edges 90°. What depth will provide maximum cross-sectional area, allowing the most water to flow?

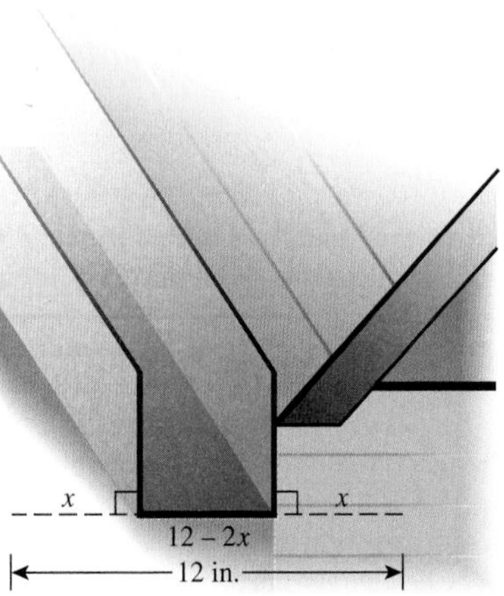

50. Norman Windows A Norman window has the shape of a rectangle surmounted by a semicircle of diameter equal to the width of the rectangle (see the figure). If the perimeter of the window is 20 feet, what dimensions will admit the most light (maximize the area)?
[**Hint:** Circumference of a circle = $2\pi r$; area of a circle = πr^2, where r is the radius of the circle.]

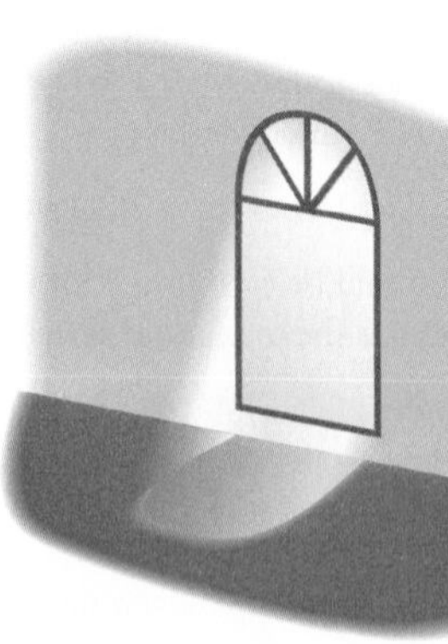

51. Constructing a Stadium A track and field playing area is in the shape of a rectangle with semicircles at each end (see the figure). The inside perimeter of the track is to be 400 meters. What should the dimensions of the rectangle be so that the area of the rectangle is a maximum?

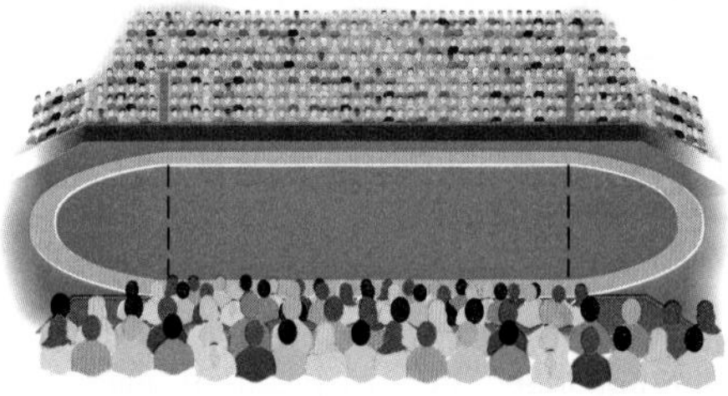

52. Architecture A special window has the shape of a rectangle surmounted by an equilateral triangle (see the figure). If the perimeter of the window is 16 feet, what dimensions will admit the most light?
[**Hint:** Area of an equilateral triangle = $\frac{\sqrt{3}}{4}x^2$, where x is the length of a side of the triangle.]

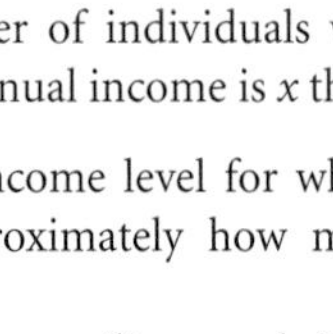

53. Hunting The function $H(x) = -1.01x^2 + 114.3x + 451.0$ models the number of individuals who engage in hunting activities whose annual income is x thousand dollars.

(a) What is the income level for which there are the most hunters? Approximately how many hunters earn this amount?

(b) Using a graphing utility, graph $H = H(x)$. Are the number of hunters increasing or decreasing for individuals earning between \$20,000 and \$40,000?

Source: National Sporting Goods Association.

54. Advanced Degrees The function

$$P(x) = -0.008x^2 + 0.868x - 11.884$$

models the percentage of the U.S. population whose age is given by x that have earned an advanced degree (more than a bachelor's degree) in March 2000.

(a) What is the age for which the highest percentage of Americans have earned an advanced degree? What is the highest percentage?

(b) Using a graphing utility, graph $P = P(x)$. Is the percentage of Americans that have earned an advanced degree increasing or decreasing for individuals between the ages of 40 and 50?

Source: U.S. Census Bureau.

55. Male Murder Victims The function

$$M(x) = 0.76x^2 - 107.00x + 3854.18$$

models the number of male murder victims who are x years of age $(20 \le x < 90)$.

(a) Use the model to approximate the number of male murder victims who are $x = 23$ years of age.

(b) At what age is the number of male murder victims 1456?

(c) Using a graphing utility, graph $M = M(x)$.

(d) Based on the graph obtained in part (c), describe what happens to the number of male murder victims as age increases.

Source: Federal Bureau of Investigation.

56. Health Care Expenditures The function

$$H(x) = 0.004x^2 - 0.197x + 5.406$$

models the percentage of total income that an individual that is x years of age spends on health care.

(a) Use the model to approximate the percentage of total income an individual 45 years of age spends on health care.

(b) At what age is the percentage of income spent on health care 10%?

(c) Using a graphing utility, graph $H = H(x)$.

(d) Based on the graph obtained in part (c), describe what happens to the percentage of income spent on health care as individuals age.

Source: Bureau of Labor Statistics.

57. Chemical Reactions A self-catalytic chemical reaction results in the formation of a compound that causes the formation ratio to increase. If the reaction rate V is given by

$$V(x) = kx(a - x), \qquad 0 \le x \le a$$

where k is a positive constant, a is the initial amount of the compound, and x is the variable amount of the compound, for what value of x is the reaction rate a maximum?

58. Calculus: Simpson's Rule The figure shows the graph of $y = ax^2 + bx + c$. Suppose that the points $(-h, y_0)$, $(0, y_1)$, and (h, y_2) are on the graph. It can be shown that the area enclosed by the parabola, the x-axis, and the lines $x = -h$ and $x = h$ is

$$\text{Area} = \frac{h}{3}(2ah^2 + 6c)$$

Show that this area may also be given by

$$\text{Area} = \frac{h}{3}(y_0 + 4y_1 + y_2)$$

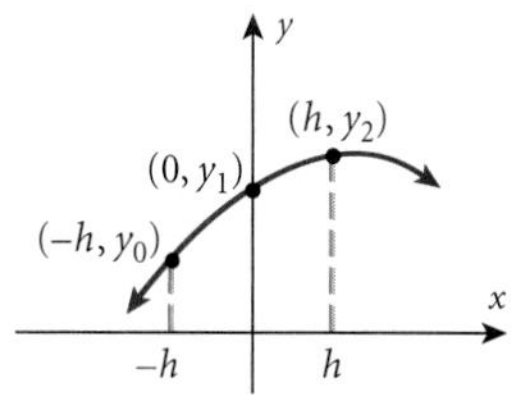

59. Use the result obtained in Problem 58 to find the area enclosed by $f(x) = -5x^2 + 8$, the x-axis, and the lines $x = -1$ and $x = 1$.

60. Use the result obtained in Problem 58 to find the area enclosed by $f(x) = 2x^2 + 8$, the x-axis, and the lines $x = -2$ and $x = 2$.

61. Use the result obtained in Problem 58 to find the area enclosed by $f(x) = x^2 + 3x + 5$, the x-axis, and the lines $x = -4$ and $x = 4$.

62. Use the result obtained in Problem 58 to find the area enclosed by $f(x) = -x^2 + x + 4$, the x-axis, and the lines $x = -1$ and $x = 1$.

63. A rectangle has one vertex on the line $y = 10 - x$, $x > 0$, another at the origin, one on the positive x-axis, and one on the positive y-axis. Find the largest area A that can be enclosed by the rectangle.

64. Let $f(x) = ax^2 + bx + c$, where a, b, and c are odd integers. If x is an integer, show that $f(x)$ must be an odd integer. [**Hint:** x is either an even integer or an odd integer.]

65. Make up a quadratic function that opens down and has only one x-intercept. Compare yours with others in the class. What are the similarities? What are the differences?

66. On one set of coordinate axes, graph the family of parabolas $f(x) = x^2 + 2x + c$ for $c = -3$, $c = 0$, and $c = 1$. Describe the characteristics of a member of this family.

67. On one set of coordinate axes, graph the family of parabolas $f(x) = x^2 + bx + 1$ for $b = -4$, $b = 0$, and $b = 4$. Describe the general characteristics of this family.

68. State the circumstances under which the graph of a quadratic function $f(x) = ax^2 + bx + c$ has no x-intercepts.

69. Why does the graph of a quadratic function open up if $a > 0$ and down if $a < 0$?

70. Refer to Example 6 on page 611. Notice that if the price charged for the calculators is \$0 or \$140 the revenue is \$0. It is easy to explain why revenue would be \$0 if the priced charged is \$0, but how can revenue be \$0 if the price charged is \$140?

11.2 Power Functions; Polynomial Functions; Rational Functions

PREPARING FOR THIS SECTION *Before getting started, review the following:*

- Polynomials (Appendix A, Section A.5, pp. A-38–A-41)
- Graphing Techniques (Chapter 10, Section 10.5, pp. 588–594)
- Domain of a Function (Chapter 10, Section 10.2, pp. 554–555)

OBJECTIVES
1 Know the properties of power functions
2 Graph functions using shifts and/or reflections
3 Identify polynomial functions and their degree
4 Find the end behavior of a polynomial function
5 Find the domain of a rational function

We begin by discussing *power functions,* a special kind of polynomial.

Power Functions

A **power function of degree *n*** is a function of the form

$$f(x) = ax^n \tag{1}$$

where a is a real number, $a \neq 0$, and $n > 0$ is an integer.

In other words, a power function is a function that is defined by a single monomial.

The graph of a power function of degree 1, $f(x) = ax$, is a straight line, with slope a, that passes through the origin. The graph of a power function of degree 2, $f(x) = ax^2$, is a parabola, with vertex at the origin, that opens up if $a > 0$ and down if $a < 0$.

Know the properties of power functions 1

We begin with power functions of even degree of the form $f(x) = x^n$, $n \geq 2$ and n even. The domain of f is the set of all real numbers, and the range is the set of nonnegative real numbers. Such a power function is an even function (do you see why?), so its graph is symmetric with respect to the y-axis. Its graph always contains the origin and the points $(-1, 1)$ and $(1, 1)$.

If $n = 2$, the graph is the familiar parabola $y = x^2$ that opens up, with vertex at the origin. If $n \geq 4$, the graph of $f(x) = x^n$, n even, will be closer to the x-axis than the parabola $y = x^2$, if $-1 < x < 1$, and farther from the x-axis than the parabola $y = x^2$, if $x < -1$ or if $x > 1$. Figure 14(a) illustrates this conclusion. Figure 14(b) shows the graphs of $y = x^4$ and $y = x^8$ for comparison.

From Figure 14, we can see that as n increases the graph of $f(x) = x^n$, $n \geq 2$ and n even, tends to flatten out near the origin and to increase very rapidly when x is far from 0. For large n, it may appear that the graph coincides with the x-axis near the origin, but it does not; the graph actually touches the x-axis only at the origin (see Table 2). Also, for large n, it may appear that for $x < -1$ or for $x > 1$ the graph is vertical, but it is not; it is only increasing very rapidly in these intervals. If the graphs were enlarged many times, these distinctions would be clear.

FIGURE 14

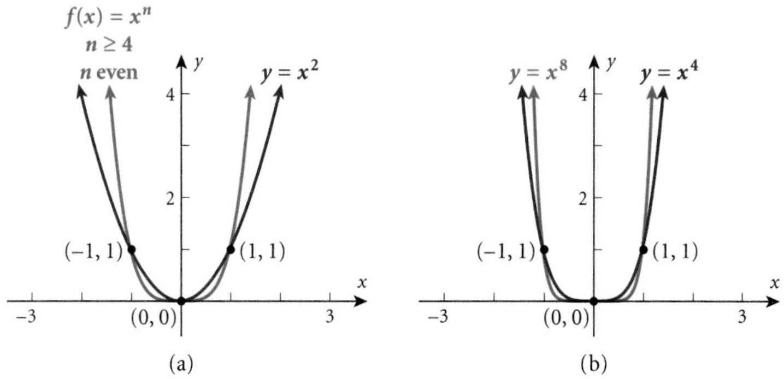

TABLE 2

	$x = 0.1$	$x = 0.3$	$x = 0.5$
$f(x) = x^8$	10^{-8}	0.0000656	0.0039063
$f(x) = x^{20}$	10^{-20}	$3.487 \cdot 10^{-11}$	0.000001
$f(x) = x^{40}$	10^{-40}	$1.216 \cdot 10^{-21}$	$9.095 \cdot 10^{-13}$

SEEING THE CONCEPT: Graph $Y_1 = x^4$, $Y_2 = x^8$, and $Y_3 = x^{12}$ using the viewing rectangle $-2 \leq x \leq 2$, $-4 \leq y \leq 16$. Then graph each again using the viewing rectangle $-1 \leq x \leq 1$, $0 \leq y \leq 1$. See Figure 15. TRACE along one of the graphs to confirm that for x close to 0 the graph is above the x-axis and that for $x > 0$ the graph is increasing.

FIGURE 15

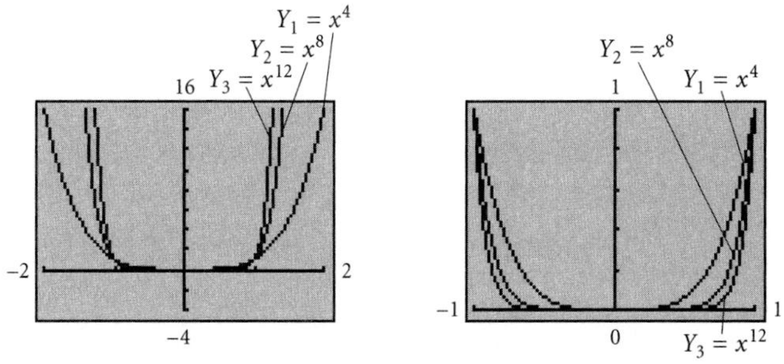

Properties of Power Functions, $f(x) = x^n$, n Is an Even Integer

1. The domain is the set of all real numbers. The range is the set of nonnegative real numbers.
2. The graph always contains the points $(0, 0)$, $(1, 1)$, and $(-1, 1)$.
3. The graph is symmetric with respect to the y-axis; the function is even.
4. As the exponent n increases in magnitude, the graph becomes more vertical when $x < -1$ or $x > 1$; but for x near the origin, the graph tends to flatten out and lie closer to the x-axis.

FIGURE 16

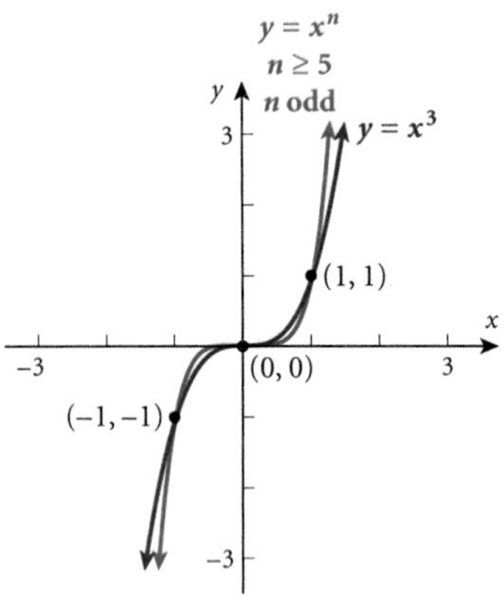

Now we consider power functions of odd degree of the form $f(x) = x^n$, $n \geq 3$ and n odd. The domain and range of f are the set of real numbers. Such a power function is an odd function (do you see why?), so its graph is symmetric with respect to the origin. Its graph always contains the origin and the points $(-1, -1)$ and $(1, 1)$.

The graph of $f(x) = x^n$ when $n = 3$ has been shown several times and is repeated in Figure 16. If $n \geq 5$, the graph of $f(x) = x^n$, n odd, will be closer to the x-axis than that of $y = x^3$, if $-1 < x < 1$, and farther from the x-axis than that of $y = x^3$, if $x < -1$ or if $x > 1$. Figure 16 also illustrates this conclusion.

Figure 17 shows the graph of $y = x^5$ and the graph of $y = x^9$ for further comparison.

It appears that each graph coincides with the x-axis near the origin, but it does not; each graph actually touches the x-axis only at the origin. Also, it appears that as x increases the graph becomes vertical, but it does not; each graph is increasing very rapidly.

SEEING THE CONCEPT: Graph $Y_1 = x^3$, $Y_2 = x^7$, and $Y_3 = x^{11}$ using the viewing rectangle $-2 \leq x \leq 2$, $-16 \leq y \leq 16$. Then graph each again using the viewing rectangle $-1 \leq x \leq 1$, $-1 \leq y \leq 1$. See Figure 18. TRACE along one of the graphs to confirm that the graph is increasing and only touches the x-axis at the origin.

FIGURE 17

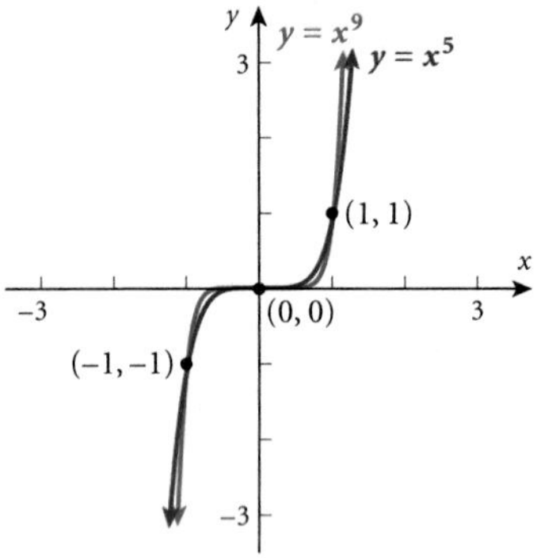

FIGURE 18

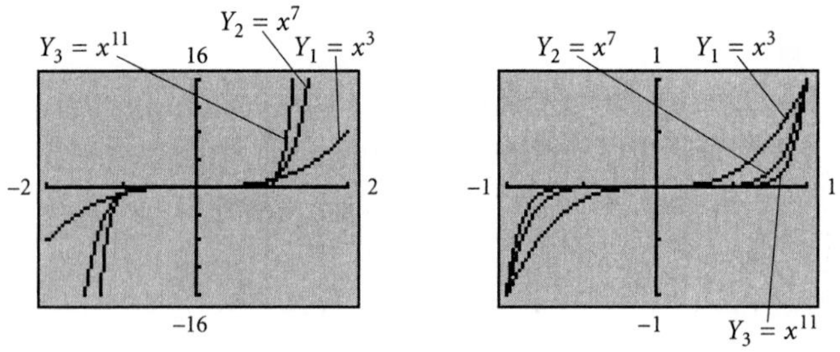

To summarize:

Properties of Power Functions, $f(x) = x^n$, n Is an Odd Integer

1. The domain and range are the set of all real numbers.
2. The graph always contains the points $(0, 0)$, $(1, 1)$, and $(-1, -1)$.
3. The graph is symmetric with respect to the origin; the function is odd.
4. As the exponent n increases in magnitude, the graph becomes more vertical when $x < -1$ or $x > 1$; but for x near the origin, the graph tends to flatten out and lie closer to the x-axis.

2 **EXAMPLE 1** **Graphing Functions Using Shifts and/or Reflections**

Graph the function $f(x) = -x^3 + 1$

SOLUTION We begin with the power function $y = x^3$. See Figure 19 for the steps.

FIGURE 19

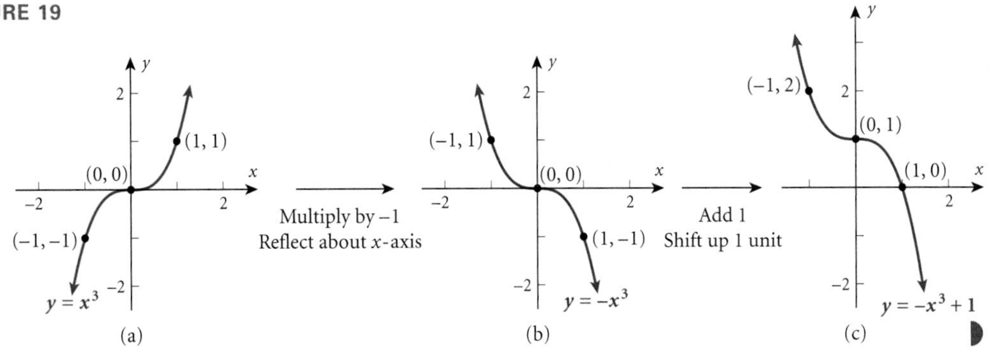

NOW WORK PROBLEMS 1 AND 5.

Polynomial Functions

Polynomial functions are among the simplest expressions in algebra. They are easy to evaluate: only addition and repeated multiplication are required. Because of this, they are often used to approximate other, more complicated functions. In this section, we investigate characteristics of this important class of function.

A **polynomial function** is a function of the form

$$f(x) = a_n x^n + a_{n-1}x^{n-1} + \cdots + a_1 x + a_0 \qquad (2)$$

where $a_n, a_{n-1}, \ldots, a_1, a_0$ are real numbers and n is a nonnegative integer. The domain of a polynomial function consists of all real numbers.

A polynomial function is a function whose rule is given by a polynomial in one variable. The **degree** of a polynomial function is the degree of the polynomial in one variable, that is, the largest power of x that appears.

3 **EXAMPLE 2** **Identifying Polynomial Functions and Their Degree**

Determine which of the following are polynomial functions. For those that are, state the degree; for those that are not, tell why not.

(a) $f(x) = 2 - 3x^4$ **(b)** $g(x) = \sqrt{x}$ **(c)** $h(x) = \dfrac{x^2 - 2}{x^3 - 1}$

(d) $F(x) = 0$ **(e)** $G(x) = 8$ **(f)** $H(x) = -2x^3(x - 1)^2$

SOLUTION **(a)** f is a polynomial function of degree 4.

(b) g is not a polynomial function. The variable x is raised to the $\frac{1}{2}$ power, which is not a nonnegative integer.

(c) h is not a polynomial function. It is the ratio of two polynomials, and the polynomial in the denominator is of positive degree.

(d) F is the zero polynomial function; it is not assigned a degree.

(e) G is a nonzero constant function, a polynomial function of degree 0 since $G(x) = 8 = 8x^0$.

(f) $H(x) = -2x^3(x-1)^2 = -2x^3(x^2 - 2x + 1) = -2x^5 + 4x^4 - 2x^3$. So H is a polynomial function of degree 5. Do you see how to find the degree of H without multiplying out?

NOW WORK PROBLEMS 11 AND 15.

We have already discussed in detail polynomial functions of degrees 0, 1, and 2. See Table 3 for a summary of the properties of the graphs of these polynomial functions.

TABLE 3

Degree	Form	Name	Graph
No degree	$f(x) = 0$	Zero function	The x-axis
0	$f(x) = a_0, \quad a_0 \neq 0$	Constant function	Horizontal line with y-intercept $(0, a_0)$
1	$f(x) = a_1x + a_0, \quad a_1 \neq 0$	Linear function	Nonvertical, nonhorizontal line with slope a_1 and y-intercept $(0, a_0)$
2	$f(x) = a_2x^2 + a_1x + a_0, \quad a_2 \neq 0$	Quadratic function	Parabola: Graph opens up if $a_2 > 0$; graph opens down if $a_2 < 0$. The y intercept is $(0, a_0)$.

One of the objectives of this book is to analyze the graph of a polynomial function. You will learn that the graph of every polynomial function is both smooth and continuous. By *smooth*, we mean that the graph contains no sharp corners or cusps; by *continuous*, we mean that the graph has no gaps or holes and can be drawn without lifting pencil from paper. Later we use calculus to define these concepts more carefully. See Figures 20(a) and (b).

FIGURE 20

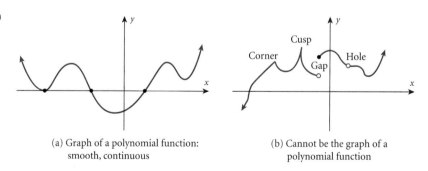

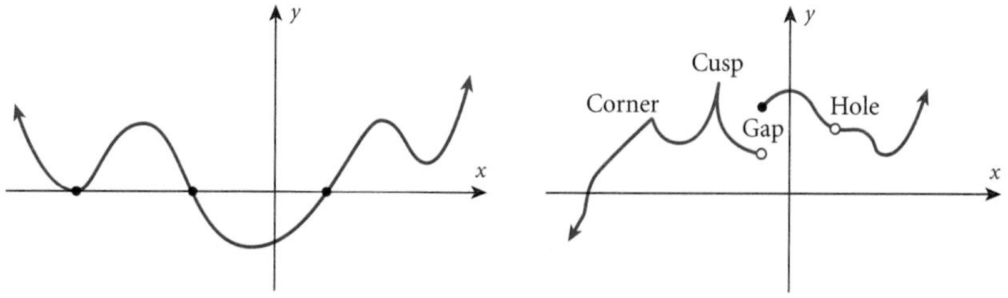

(a) Graph of a polynomial function: smooth, continuous

(b) Cannot be the graph of a polynomial function

Figure 21 shows the graph of a polynomial function with four x-intercepts. Notice that at the x-intercepts the graph must either cross the x-axis or touch the x-axis. Consequently, between consecutive x-intercepts the graph is either above the x-axis or below the x-axis. Notice also that the graph has two local maxima and two local minima. In Chapter 14 we will use calculus to locate the local maxima and minima of polynomial functions so that we can draw a complete graph.

FIGURE 21

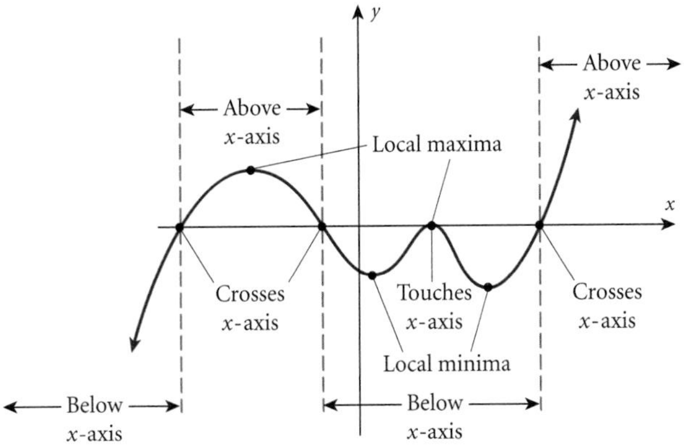

The behavior of the graph of a function for large values of x, either positive or negative, is referred to as its **end behavior.**

For polynomial functions, we have this important result.

End Behavior

For large values of x, either positive or negative, the graph of the polynomial

$$f(x) = a_n x^n + a_{n-1}x^{n-1} + \cdots + a_1 x + a_0$$

resembles the graph of the power function

$$y = a_n x^n$$

4 **EXAMPLE 3** **Finding the End Behavior of a Polynomial Function**

(a) For large values of x, the graph of the polynomial function

$$f(x) = 3x^5 - 4x^4 + 8x - 4$$

resembles that of the power function $y = 3x^5$.

(b) For large values of x, the graph of the polynomial function

$$f(x) = -4x^7 + 2x^5 - 4x^2 + 2x - 10$$

resembles that of the power function $y = -4x^7$.

NOW WORK PROBLEM 25.

Look back at Figures 14 and 16. Based on the above theorem and the previous discussion on power functions, the end behavior of a polynomial can only be of four types. See Figure 22.

FIGURE 22   End behavior

$$f(x) = a_n x^n + a_{n-1} x^{n-1} + \cdots + a_1 x + a_0$$

| $n \geq 2$ even | | $n \geq 3$ odd | |
| $a_n > 0$ | $a_n < 0$ | $a_n > 0$ | $a_n < 0$ |

(a)
$n \geq 2$ even; $a_n > 0$

(b)
$n \geq 2$ even; $a_n < 0$

(c)
$n \geq 3$ odd; $a_n > 0$

(d)
$n \geq 3$ odd; $a_n < 0$

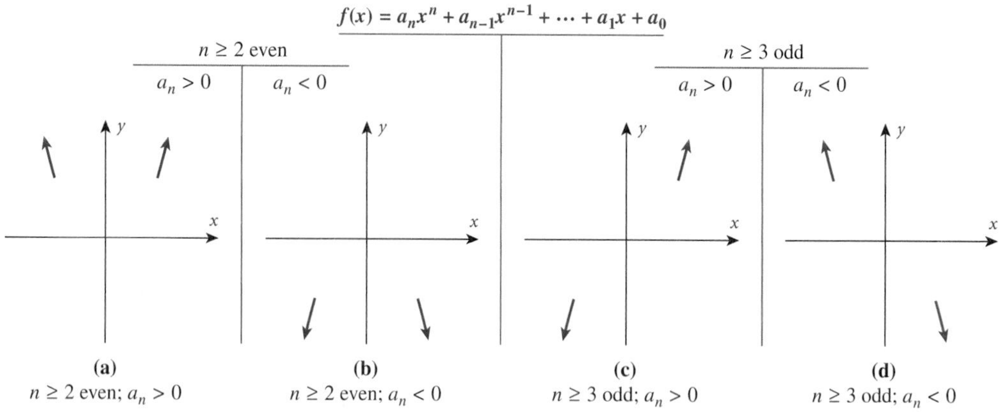

For example, consider the polynomial function $f(x) = -2x^4 + x^3 + 4x^2 - 7x + 1$. The graph of f will resemble the graph of the power function $y = -2x^4$ for large values of x, either positive or negative. The graph of f will look like Figure 22(b) for large values of x.

Rational Functions

Ratios of integers are called *rational numbers*. Similarly, ratios of polynomial functions are called *rational functions*.

A **rational function** is a function of the form

$$R(x) = \frac{p(x)}{q(x)}$$

where p and q are polynomial functions and q is not the zero polynomial. The domain of a rational function consists of all real numbers except those for which the denominator q is 0.

5   EXAMPLE 4   Finding the Domain of a Rational Function

(a) The domain of $R(x) = \dfrac{2x^2 - 4}{x + 5}$ consists of all real numbers x except -5, that is, $\{x \mid x \neq -5\}$.

(b) The domain of $R(x) = \dfrac{1}{x^2 - 4}$ consists of all real numbers x except -2 and 2, that is, $\{x \mid x \neq -2, x \neq 2\}$.

(c) The domain of $R(x) = \dfrac{x^3}{x^2 + 1}$ consists of all real numbers.

(d) The domain of $R(x) = \dfrac{-x^2 + 2}{3}$ consists of all real numbers.

(e) The domain of $R(x) = \dfrac{x^2 - 1}{x - 1}$ consists of all real numbers x except 1, that is, $\{x \mid x \neq 1\}$.

It is important to observe that the functions

$$R(x) = \frac{x^2 - 1}{x - 1} \quad \text{and} \quad f(x) = x + 1$$

are not equal, since the domain of R is $\{x \mid x \neq 1\}$ and the domain of f is all real numbers.

If $R(x) = \dfrac{p(x)}{q(x)}$ is a rational function and if p and q have no common factors, then the rational function R is said to be in **lowest terms.** For a rational function $R(x) = \dfrac{p(x)}{q(x)}$ in lowest terms, the zeros, if any, of the numerator are the x-intercepts of the graph of R and so will play a major role in the graph of R. The zeros of the denominator of R [that is, the numbers x, if any, for which $q(x) = 0$], although not in the domain of R, also play a major role in the graph of R. We will discuss this role in Chapter 12.

NOW WORK PROBLEM 31.

EXERCISE 11.2 Answers to Odd-Numbered Problems Begin on Page AN-62.

1. Name three points on the graph of the power function $f(x) = x^5$.

2. Name three points on the graph of the power function $f(x) = x^6$.

3. The graph of the power function $f(x) = x^5$ is symmetric with respect to the ________.

4. The graph of the power function $f(x) = x^6$ is symmetric with respect to the ________.

In Problems 5–10, graph each function using shifts and/or reflections. Be sure to label at least three points.

5. $f(x) = x^6 + 2$
6. $f(x) = x^5 - 3$
7. $f(x) = -x^5 + 2$
8. $f(x) = -x^4 + 9$
9. $f(x) = (x - 2)^4$
10. $f(x) = (x + 3)^5$

In Problems 11–22, determine which functions are polynomial functions. For those that are, state the degree. For those that are not, tell why not.

11. $f(x) = 4x + x^3$
12. $f(x) = 5x^2 + 4x^4$
13. $g(x) = \dfrac{1 - x^2}{2}$
14. $h(x) = 3 - \dfrac{1}{2}x$
15. $f(x) = 1 - \dfrac{1}{x}$
16. $f(x) = x(x - 1)$
17. $g(x) = x^{3/2} - x^2 + 2$
18. $h(x) = \sqrt{x}(\sqrt{x} - 1)$
19. $F(x) = 5x^4 - \pi x^3 + \dfrac{1}{2}$
20. $F(x) = \dfrac{x^2 - 5}{x^3}$
21. $G(x) = 2(x - 1)^2(x^2 + 1)$
22. $G(x) = -3x^2(x + 2)^3$

In Problems 23–28, find the power function that the graph of f resembles for large values of x. That is, find the end behavior of each polynomial function.

23. $f(x) = 3x^4 - 2x^2 + 1$
24. $f(x) = 4x^5 - 6x^3 - x$
25. $f(x) = -2x^5 + 8x^4$
26. $f(x) = -3x^4 - 5x + 1$
27. $f(x) = 5(x + 1)^2(x - 2)$
28. $f(x) = 6x(x^2 + 4)^2$

In Problems 29–40, find the domain of each rational function.

29. $R(x) = \dfrac{4x}{x - 3}$
30. $R(x) = \dfrac{5x^2}{3 + x}$
31. $H(x) = \dfrac{-4x^2}{(x - 2)(x + 4)}$
32. $G(x) = \dfrac{6}{(x + 3)(4 - x)}$
33. $F(x) = \dfrac{3x(x - 1)}{2x^2 - 5x - 3}$
34. $Q(x) = \dfrac{-x(1 - x)}{3x^2 + 5x - 2}$
35. $R(x) = \dfrac{x}{x^3 - 8}$
36. $R(x) = \dfrac{x}{x^4 - 1}$
37. $H(x) = \dfrac{3x^2 + x}{x^2 + 4}$
38. $G(x) = \dfrac{x - 3}{x^4 + 1}$
39. $R(x) = \dfrac{3(x^2 - x - 6)}{4(x^2 - 9)}$
40. $F(x) = \dfrac{-2(x^2 - 4)}{3(x^2 + 4x + 4)}$

41. Union Membership The percentage of the labor force who are union members is given below for 1930–2000.

Year	1930	1940	1950	1960	1970	1980	1990	2000
Percentage	11.6	26.9	31.5	31.4	27.3	21.9	16.1	13.2

This data can be modeled by the polynomial function

$$u(t) = 11.93 + 1.9t - 0.052t^2 + 0.00037t^3$$

where t is the number of years since 1930.

(a) Use $u(t)$ to find the percentage of union membership in 2000.

(b) Find $u(75)$. Write a sentence explaining what it means.

Source: Bureau of Labor Statistics, U.S. Department of Labor.

11.3 Exponential Functions

PREPARING FOR THIS SECTION ***Before getting started, review the following:***

> Exponents (Appendix A, Section A.2, pp. A19–A-21 and Section A.8, pp. A73–A-78)

> Slope of a Line (Chapter 1, Section 1.1, pp. 7–10)

> Graphing Techniques (Chapter 10, Section 10.5, pp. 588–594)

> Solving Equations (Appendix A, Section A.6, pp. A-49–A-58)

OBJECTIVES

1 Evaluate exponents

2 Graph exponential functions

3 Define the number *e*

4 Solve exponential equations

Evaluate exponents 1

In Appendix A, Section A.8, we give a definition for raising a real number a to a rational power. Based on that discussion, we gave meaning to expressions of the form

$$a^r$$

where the base a is a positive real number and the exponent r is a rational number.

But what is the meaning of a^x, where the base a is a positive real number and the exponent x is an irrational number? Although a rigorous definition requires advanced methods, the basis for the definition is easy to follow: Select a rational number r that is formed by truncating (removing) all but a finite number of digits from the irrational number x. Then it is reasonable to expect that

$$a^x \approx a^r$$

For example, take the irrational number $\pi = 3.14159\ldots$. Then, an approximation to a^π is

$$a^\pi \approx a^{3.14}$$

where the digits after the hundredths position have been removed from the value for π. A better approximation would be

$$a^\pi \approx a^{3.14159}$$

where the digits after the hundred-thousandths position have been removed. Continuing in this way, we can obtain approximations to a^π to any desired degree of accuracy.

Most calculators have an $\boxed{x^y}$ key or a caret key $\boxed{\wedge}$ for working with exponents. To evaluate expressions of the form a^x, enter the base a, then press the $\boxed{x^y}$ key (or the $\boxed{\wedge}$ key), enter the exponent x, and press $\boxed{=}$ (or $\boxed{\text{enter}}$).

EXAMPLE 1 Using a Calculator to Evaluate Powers of 2

Using a calculator, evaluate:

(a) $2^{1.4}$ **(b)** $2^{1.41}$ **(c)** $2^{1.414}$ **(d)** $2^{1.4142}$ **(e)** $2^{\sqrt{2}}$

SOLUTION **(a)** $2^{1.4} \approx 2.639015822$ **(b)** $2^{1.41} \approx 2.657371628$
(c) $2^{1.414} \approx 2.66474965$ **(d)** $2^{1.4142} \approx 2.665119089$
(e) $2^{\sqrt{2}} \approx 2.665144143$

NOW WORK PROBLEM 1.

It can be shown that the Laws of Exponents hold for real exponents.

Laws of Exponents

If s, t, a, and b are real numbers, with $a > 0$ and $b > 0$, then

$$a^s \cdot a^t = a^{s+t} \qquad (a^s)^t = a^{st} \qquad (ab)^s = a^s \cdot b^s$$
$$1^s = 1 \qquad a^{-s} = \frac{1}{a^s} = \left(\frac{1}{a}\right)^s \qquad a^0 = 1 \tag{1}$$

We are now ready for the following definition:

An **exponential function** is a function of the form

$$f(x) = a^x$$

where a is a positive real number ($a > 0$) and $a \neq 1$. The domain of f is the set of all real numbers.

We exclude the base $a = 1$ because this function is simply the constant function $f(x) = 1^x = 1$. We also need to exclude the bases that are negative, because, otherwise, we would have to exclude many values of x from the domain, such as $x = \frac{1}{2}$ and $x = \frac{3}{4}$. [Recall that $(-2)^{1/2} = \sqrt{-2}$, $(-3)^{3/4} = \sqrt[4]{(-3)^3} = \sqrt[4]{-27}$, and so on, are not defined in the system of real numbers.]

COMMENT: It is important to distinguish a power function $g(x) = x^n$, $n \geq 2$ an integer, from an exponential function $f(x) = a^x$, $a > 0$, $a \neq 1$, a real. In a power function, the base is a variable and the exponent is a constant. In an exponential function, the base is a constant and the exponent is a variable.

Some examples of exponential functions are

$$f(x) = 2^x, \quad F(x) = \left(\frac{1}{3}\right)^x$$

Notice that in each example, the base is a constant and the exponent is a variable.

You may wonder what role the base a plays in the exponential function $f(x) = a^x$. We use the following Exploration to find out.

EXPLORATION

(a) Evaluate $f(x) = 2^x$ at $x = -2, -1, 0, 1, 2$, and 3.
(b) Evaluate $g(x) = 3x + 2$ at $x = -2, -1, 0, 1, 2$, and 3.
(c) Comment on the pattern that exists in the values of f and g.

Result

(a) Table 4 shows the values of $f(x) = 2^x$ for $x = -2, -1, 0, 1, 2$, and 3.
(b) Table 5 shows the values of $g(x) = 3x + 2$ for $x = -2, -1, 0, 1, 2$, and 3.

TABLE 4

x	$f(x) = 2^x$
-2	$f(-2) = 2^{-2} = \frac{1}{2^2} = \frac{1}{4}$
-1	$\frac{1}{2}$
0	1
1	2
2	4
3	8

TABLE 5

x	$g(x) = 3x + 2$
-2	$g(-2) = 3(-2) + 2 = -4$
-1	-1
0	2
1	5
2	8
3	11

(c) In Table 4 we notice that each value of the exponential function $f(x) = a^x = 2^x$ could be found by multiplying the previous value of the function by the base, $a = 2$. For example,

$$f(-1) = 2 \cdot f(-2) = 2 \cdot \frac{1}{4} = \frac{1}{2}, \quad f(0) = 2 \cdot f(-1) = 2 \cdot \frac{1}{2} = 1, \quad f(1) = 2 \cdot f(0) = 2 \cdot 1 = 2$$

and so on.

Put another way, we see that the ratio of consecutive outputs is constant for unit increases in the inputs. The constant equals the value of the base of the exponential function a. For example, for the function $f(x) = 2^x$, we notice that

$$\frac{f(-1)}{f(-2)} = \frac{\frac{1}{2}}{\frac{1}{4}} = 2, \qquad \frac{f(1)}{f(0)} = \frac{2}{1} = 2, \qquad \frac{f(x+1)}{f(x)} = \frac{2^{x+1}}{2^x} = 2$$

and so on.

From Table 5 we see that ratios of consecutive outputs of $g(x) = 3x + 2$ are not constant. For example,

$$\frac{g(-1)}{g(-2)} = \frac{-1}{-4} = \frac{1}{4} \neq \frac{g(1)}{g(0)} = \frac{5}{2}$$

Instead, because $g(x) = 3x + 2$ is a linear function, for unit increases in the input, the outputs increase by a fixed amount equal to the value of the slope, 3. ▶

The conclusions reached in the Exploration lead to the following theorem.

For an exponential function $f(x) = a^x, a > 0, a \neq 1$, if x is any real number, then

$$\boxed{\frac{f(x + 1)}{f(x)} = a}$$

Proof

$$\frac{f(x + 1)}{f(x)} = \frac{a^{x+1}}{a^x} = a^{x+1-x} = a^1 = a \qquad ■$$

 NOW WORK PROBLEM 11.

Graphs of Exponential Functions

First, we graph the exponential function $f(x) = 2^x$.

2 **EXAMPLE 2**   **Graphing an Exponential Function**

Graph the exponential function: $f(x) = 2^x$

SOLUTION   The domain of $f(x) = 2^x$ consists of all real numbers. We begin by locating some points on the graph of $f(x) = 2^x$, as listed in Table 6 on page 630.

Since $2^x > 0$ for all x, the range of f is the interval $(0, \infty)$. From this, we conclude that the graph has no x-intercepts, and, in fact, the graph will lie above the x-axis. As Table 6 indicates, the y-intercept is 1. Table 6 also indicates that as x becomes unbounded in the negative direction, the value of $f(x) = 2^x$ get closer and closer to 0. This means that the line $y = 0$ (the x-axis) is a horizontal asymptote to the graph as x becomes unbounded in the negative direction.* This gives us the end behavior of the graph for x large and negative.

---

* A horizontal asymptote is a line that the graph of a function gets closer and closer to, as x becomes unbounded in the positive (or negative) direction. We discuss asymptotes in detail in Chapter 12.

TABLE 6

x	$f(x) = 2^x$
-10	$2^{-10} \approx 0.00098$
-3	$2^{-3} = \frac{1}{8}$
-2	$2^{-2} = \frac{1}{4}$
-1	$2^{-1} = \frac{1}{2}$
0	$2^0 = 1$
1	$2^1 = 2$
2	$2^2 = 4$
3	$2^3 = 8$
10	$2^{10} = 1024$

To determine the end behavior for x large and positive, look again at Table 6. As x becomes unbounded in the positive direction, $f(x) = 2^x$ grows very quickly, causing the graph of $f(x) = 2^x$ to rise very rapidly. It is apparent that f is an increasing function.

Using all this information, we plot some of the points from Table 6 and connect them with a smooth, continuous curve, as shown in Figure 23.

FIGURE 23

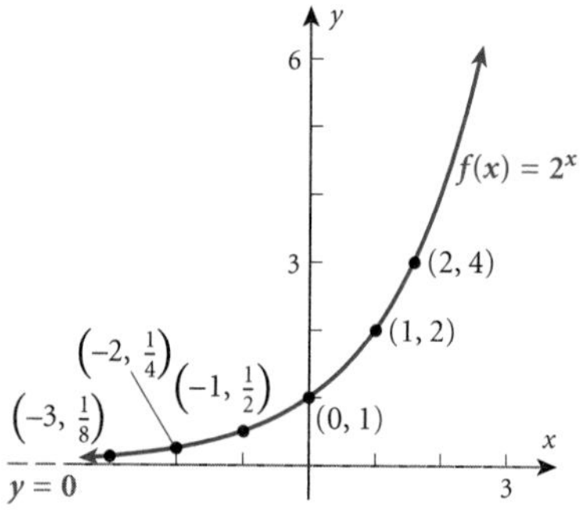

As we shall see, graphs that look like the one in Figure 23 occur very frequently in a variety of situations. For example, look at the graph in Figure 24, which illustrates the closing price of a share of Harley Davidson stock. Investors might conclude from this graph that the price of Harley Davidson is *behaving exponentially*; that is, the graph exhibits rapid, or exponential, growth.

FIGURE 24

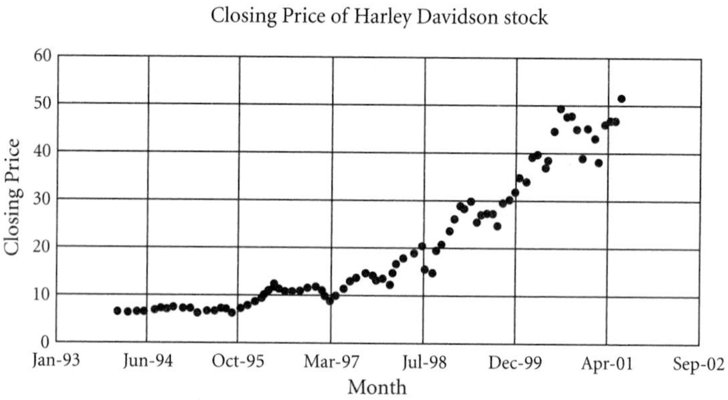

FIGURE 25

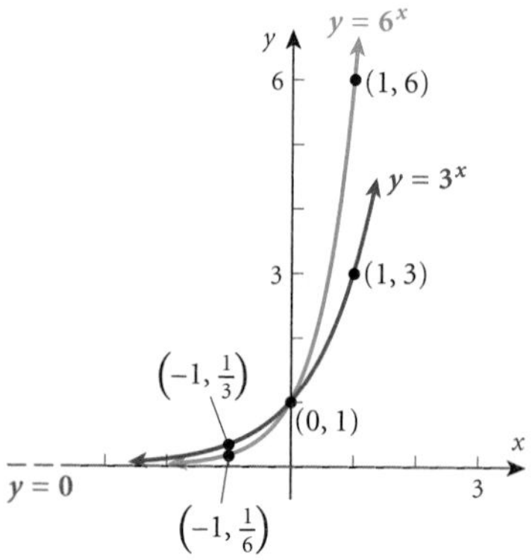

We shall have more to say about situations that lead to exponential growth later. For now, we continue to seek properties of exponential functions.

The graph of $f(x) = 2^x$ in Figure 23 is typical of all exponential functions that have a base larger than 1. Such functions are increasing functions. Their graphs lie above the x-axis, pass through the point $(0, 1)$, and thereafter rise rapidly as x becomes unbounded in the positive direction. As x becomes unbounded in the negative direction, the line $y = 0$ (the x-axis) is a horizontal asymptote. Finally, the graphs are smooth and continuous, with no corners or gaps.

Figure 25 illustrates the graphs of two more exponential functions whose bases are larger than 1. Notice that for the larger base the graph is steeper when $x > 0$ and is closer to the x-axis when $x < 0$.

SEEING THE CONCEPT: Graph $y = 2^x$ and compare what you see to Figure 23. Clear the screen and graph $y = 3^x$ and $y = 6^x$ and compare what you see to Figure 25. Clear the screen and graph $y = 10^x$ and $y = 100^x$. What viewing rectangle seems to work best?

The following list summarizes the information that we have about $f(x) = a^x$, $a > 1$.

FIGURE 26

Properties of the Exponential Function $f(x) = a^x, a > 1$

1. The domain is the set of all real numbers; the range is the set of positive real numbers.
2. There is no x-intercept; the y-intercept is $(0, 1)$.
3. The line $y = 0$ (the x-axis) is a horizontal asymptote as x becomes unbounded in the negative direction.
4. $f(x) = a^x$, $a > 1$, is an increasing function.
5. The graph of f contains the points $(0, 1)$, $(1, a)$, and $\left(-1, \dfrac{1}{a}\right)$.
6. The graph of f is smooth and continuous, with no corners or gaps. See Figure 26.

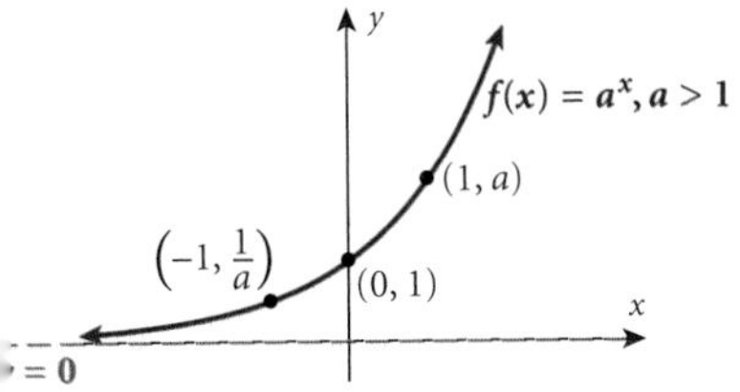

Now we consider $f(x) = a^x$ when $0 < a < 1$.

EXAMPLE 3   Graphing an Exponential Function

Graph the exponential function: $f(x) = \left(\dfrac{1}{2}\right)^x$

TABLE 7

x	$f(x) = \left(\dfrac{1}{2}\right)^x$
-10	$\left(\dfrac{1}{2}\right)^{-10} = 1024$
-3	$\left(\dfrac{1}{2}\right)^{-3} = 8$
-2	$\left(\dfrac{1}{2}\right)^{-2} = 4$
-1	$\left(\dfrac{1}{2}\right)^{-1} = 2$
0	$\left(\dfrac{1}{2}\right)^{0} = 1$
1	$\left(\dfrac{1}{2}\right)^{1} = \dfrac{1}{2}$
2	$\left(\dfrac{1}{2}\right)^{2} = \dfrac{1}{4}$
3	$\left(\dfrac{1}{2}\right)^{3} = \dfrac{1}{8}$
10	$\left(\dfrac{1}{2}\right)^{10} \approx 0.00098$

SOLUTION The domain of $f(x) = \left(\dfrac{1}{2}\right)^x$ consists of all real numbers. As before, we locate some points on the graph by creating Table 7. Since $\left(\dfrac{1}{2}\right)^x > 0$ for all x, the range of f is the interval $(0, \infty)$. The graph lies above the x-axis and so has no x-intercepts. The y-intercept is $(0, 1)$. As x becomes unbounded in the negative direction, $f(x) = \left(\dfrac{1}{2}\right)^x$ grows very quickly. As x becomes unbounded in the positive direction, the values of $f(x)$ approach 0. The line $y = 0$ (the x-axis) is a horizontal asymptote as x becomes unbounded in the positive direction. It is apparent that f is a decreasing function. Figure 27 illustrates the graph.

FIGURE 27

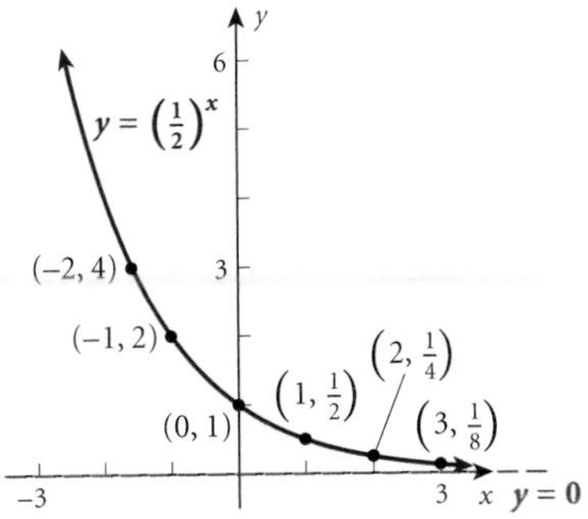

We could have obtained the graph of $y = \left(\frac{1}{2}\right)^x$ from the graph of $y = 2^x$ using a reflection. If $f(x) = 2^x$, then $f(-x) = 2^{-x} = \frac{1}{2^x} = \left(\frac{1}{2}\right)^x$. The graph of $y = \left(\frac{1}{2}\right)^x = 2^{-x}$ is a reflection about the y-axis of the graph of $y = 2^x$. See Figures 28(a) and (b).

FIGURE 28

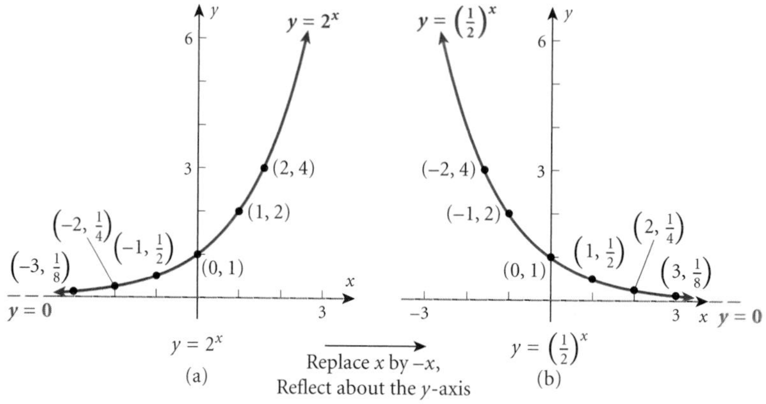

SEEING THE CONCEPT: Using a graphing utility, simultaneously graph

(a) $Y_1 = 3^x, \quad Y_2 = \left(\frac{1}{3}\right)^x$ **(b)** $Y_1 = 6^x, \quad Y_2 = \left(\frac{1}{6}\right)^x$

Conclude that the graph of $Y_2 = \left(\frac{1}{a}\right)^x$, for $a > 0$, is the reflection about the y-axis of the graph of $Y_1 = a^x$.

FIGURE 29

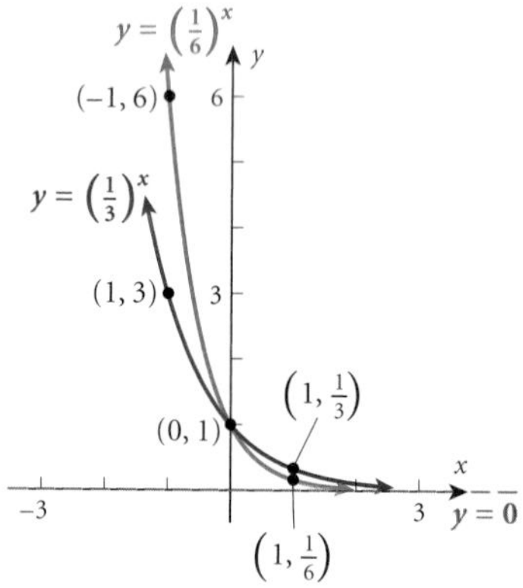

The graph of $f(x) = \left(\frac{1}{2}\right)^x$ in Figure 28(b) is typical of all exponential functions that have a base between 0 and 1. Such functions are decreasing. Their graphs lie above the x-axis and pass through the point (0,1). The graphs rise rapidly as x becomes unbounded in the negative direction. As x becomes unbounded in the positive direction, the x-axis is a horizontal asymptote. Finally, the graphs are smooth and continuous with no corners or gaps.

Figure 29 illustrates the graphs of two more exponential functions whose bases are between 0 and 1. Notice that the choice of a base closer to 0 results in a graph that is steeper when $x < 0$ and closer to the x-axis when $x > 0$.

SEEING THE CONCEPT: Graph $Y = \left(\frac{1}{2}\right)^x$ and compare what you see to Figure 28(b). Clear the screen and graph $Y_1 = \left(\frac{1}{3}\right)^x$ and $Y_2 = \left(\frac{1}{6}\right)^x$ and compare what you see to Figure 29. Clear the screen and graph $Y_1 = \left(\frac{1}{10}\right)^x$ and $Y_2 = \left(\frac{1}{100}\right)^x$. What viewing rectangle seems to work best?

The following list summarizes the information that we have about the function $f(x) = a^x, 0 < a < 1$.

Properties of the Graph of an Exponential Function $f(x) = a^x, 0 < a < 1$

1. The domain is the set of all real numbers; the range is the set of positive real numbers.
2. There is no x-intercept; the y-intercept is (0,1).
3. The line $y = 0$ (the x-axis) is a horizontal asymptote as x become unbounded in the positive direction.
4. $f(x) = a^x, 0 < a < 1$, is a decreasing function.
5. The graph of f contains the points (0, 1), (1, a), and $\left(-1, \frac{1}{a}\right)$.
6. The graph of f is smooth and continuous, with no corners or gaps. See Figure 30.

FIGURE 30

$f(x) = a^x, 0 < a < 1$

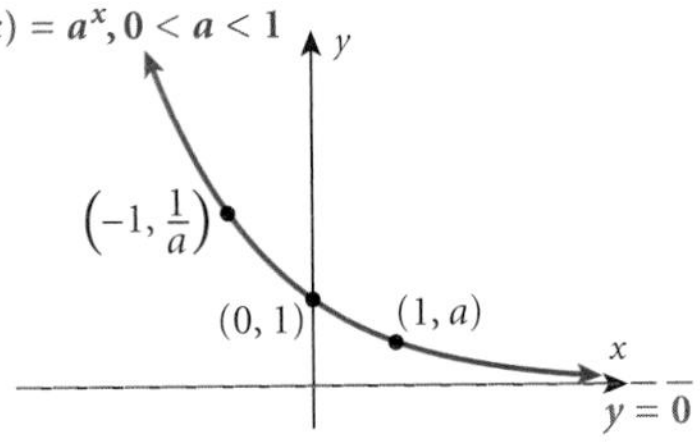

EXAMPLE 4 **Graphing Exponential Functions Using Shifts and/or Reflections**

Graph $f(x) = 2^{-x} - 3$ and determine the domain, range, and horizontal asymptote of f.

SOLUTION We begin with the graph of $y = 2^x$. Figure 31 shows the various steps.

FIGURE 31

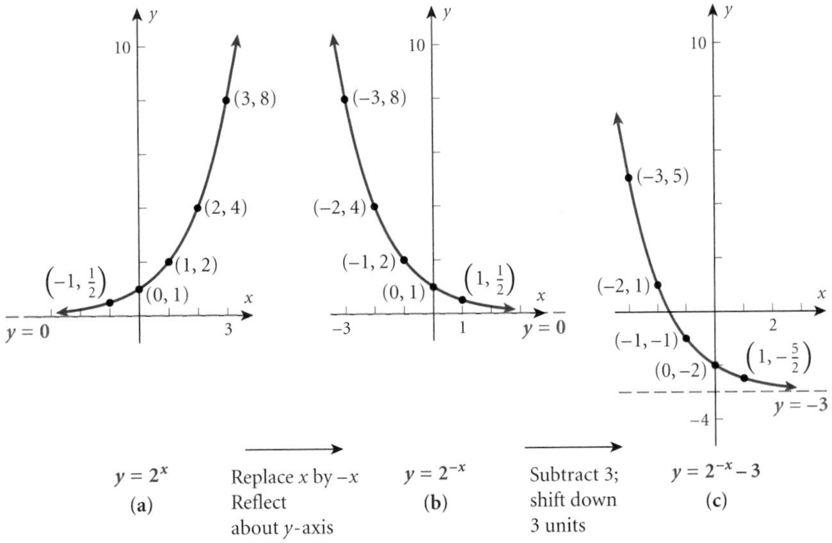

As Figure 31(c) illustrates, the domain of $f(x) = 2^{-x} - 3$ is the interval $(-\infty, \infty)$ and the range is the interval $(-3, \infty)$. The horizontal asymptote of f is the line $y = -3$.

NOW WORK PROBLEM 27.

The Base e

Define the number e

3 Many applied problems require the use of an exponential function whose base is a certain irrational number, symbolized by the letter e.

Let's look now at one way of arriving at this important number e.

The **number *e*** is defined as the number that the expression

$$\left(1 + \frac{1}{n}\right)^n \tag{2}$$

approaches as n becomes unbounded in the positive direction. In calculus, this is expressed using limit notation as

$$e = \lim_{n \to \infty}\left(1 + \frac{1}{n}\right)^n$$

Table 8 illustrates what happens to the defining expression (2) as n becomes unbounded in the positive direction. The last number in the last column in the table gives e correct to nine decimal places and is the same as the entry given for e on your calculator (if expressed correctly to nine decimal places).

TABLE 8

n	$\frac{1}{n}$	$1 + \frac{1}{n}$	$\left(1 + \frac{1}{n}\right)^n$
1	1	2	2
2	0.5	1.5	2.25
5	0.2	1.2	2.48832
10	0.1	1.1	2.59374246
100	0.01	1.01	2.704813829
1,000	0.001	1.001	2.716923932
10,000	0.0001	1.0001	2.718145927
100,000	0.00001	1.00001	2.718268237
1,000,000	0.000001	1.000001	2.718280469
1,000,000,000	10^{-9}	$1 + 10^{-9}$	2.718281827

TABLE 9

x	e^x
−2	0.14
−1	0.37
0	1
1	2.72
2	7.39

FIGURE 32
$y = e^x$

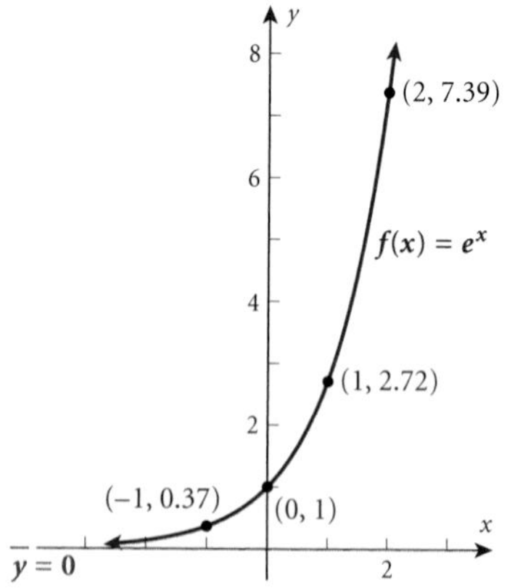

The exponential function $f(x) = e^x$, whose base is the number e, occurs with such frequency in applications that it is usually referred to as *the* exponential function. Indeed, most calculators have the key $\boxed{e^x}$ or $\boxed{\exp(x)}$, which may be used to evaluate the exponential function for a given value of x. (Consult your owner's manual if there is no such key.)

Now use your calculator to approximate e^x for $x = -2$, $x = -1$, $x = 0$, $x = 1$, and $x = 2$, as we have done to create Table 9.

The graph of the exponential function $f(x) = e^x$ is given in Figure 32. Since $2 < e < 3$, the graph of $y = e^x$ lies between the graphs of $y = 2^x$ and $y = 3^x$. Do you see why? (Refer to Figures 23 and 25.)

SEEING THE CONCEPT: Graph $Y_1 = e^x$ and compare what you see to Figure 32. Use eVALUEate or TABLE to verify the entries in Table 9. Now graph $Y_2 = 2^x$ and $Y_3 = 3^x$

on the same screen as $Y_1 = e^x$. Notice that the graph of $Y_1 = e^x$ lies between these two graphs.

EXAMPLE 5 Graphing Exponential Functions Using Shifts and/or Reflections

Graph $f(x) = -e^x + 1$ and determine the domain, range, and horizontal asymptote of f.

SOLUTION We begin with the graph of $y = e^x$. Figure 33 shows the various steps.

FIGURE 33

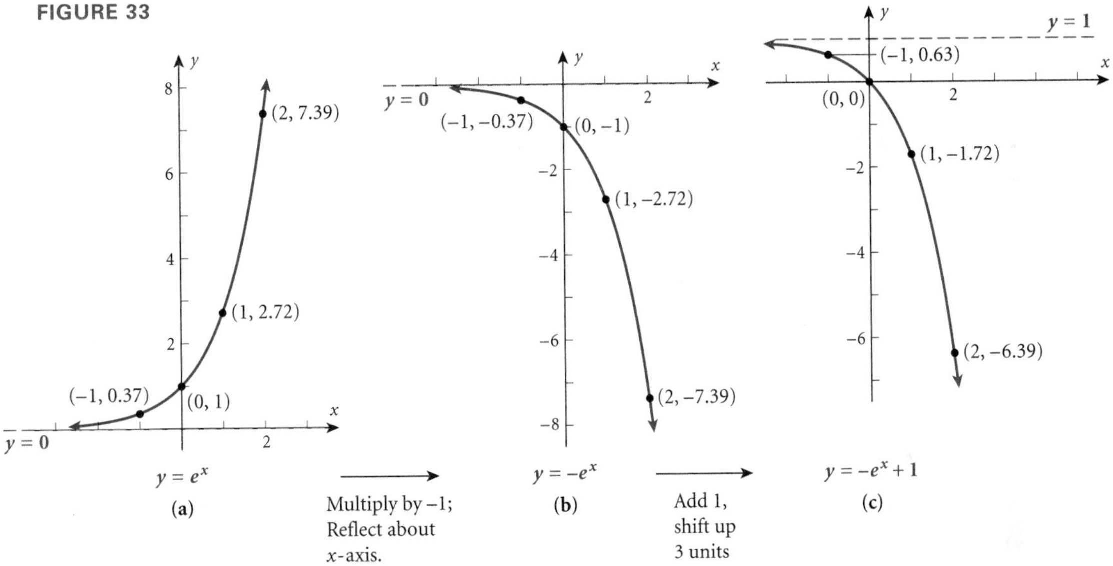

As Figure 33(c) illustrates, the domain of $f(x) = -e^x + 1$ is the interval $(-\infty, \infty)$ and the range is the interval $(-\infty, 1)$. The horizontal asymptote is the line $y = 1$.

NOW WORK PROBLEM 31.

Exponential Equations

Equations that involve terms of the form a^x, $a > 0$, $a \neq 1$, are often referred to as **exponential equations.** Such equations can sometimes be solved by appropriately applying the Laws of Exponents and statement (3) below.

$$\text{If } a^u = a^v, \quad \text{then } u = v \tag{3}$$

To use property (3), each side of the equality must be written with the same base.

4 EXAMPLE 6 Solving an Exponential Equation

Solve: $3^{x+1} = 81$

SOLUTION Since $81 = 3^4$, we can write the equation as

$$3^{x+1} = 81$$
$$3^{x+1} = 3^4$$

Now we have the same base, 3, on each side, so we can apply property (3) to obtain

$$x + 1 = 4$$
$$x = 3$$

NOW WORK PROBLEM 35.

EXAMPLE 7 Solving an Exponential Equation

Solve: $e^{-x^2} = (e^x)^2 \cdot \dfrac{1}{e^3}$

SOLUTION We use Laws of Exponents first to get the base e on the right side.

$$(e^x)^2 \cdot \frac{1}{e^3} = e^{2x} \cdot e^{-3} = e^{2x-3}$$

As a result,

$$e^{-x^2} = e^{2x-3}$$
$$-x^2 = 2x - 3 \quad \text{Apply Property (3).}$$
$$x^2 + 2x - 3 = 0 \quad \text{Place the quadratic equation in standard form.}$$
$$(x + 3)(x - 1) = 0 \quad \text{Factor.}$$
$$x = -3 \quad \text{or} \quad x = 1 \quad \text{Use the Zero-Product Property.}$$

The solution set is $\{-3, 1\}$.

Application

Many applications involve the exponential function. Let's look at one.

EXAMPLE 8 Exponential Probability

Between 9:00 PM and 10:00 PM cars arrive at Burger King's drive-thru at the rate of 12 cars per hour (0.2 car per minute). The following formula from probability can be used to determine the probability that a car will arrive within t minutes of 9:00 PM.

$$F(t) = 1 - e^{-0.2t}$$

(a) Determine the probability that a car will arrive within 5 minutes of 9 PM (that is, before 9:05 PM).

(b) Determine the probability that a car will arrive within 30 minutes of 9 PM (before 9:30 PM).

(c) What value does F approach as t becomes unbounded in the positive direction?

(d) Graph $F(t) = 1 - e^{-0.2t}$, $t > 0$. Use eVALUEate or TABLE to compare the values of F at $t = 5$ [part (a)] and at $t = 30$ [part (b)].

(e) Within how many minutes of 9 PM will the probability of a car arriving equal 50%? [**Hint:** Use TRACE or TABLE].

SOLUTION **(a)** The probability that a car will arrive within 5 minutes is found by evaluating $F(t)$ at $t = 5$.

$$F(5) = 1 - e^{-0.2(5)} \approx 0.63212$$

↑ Use a calculator

We conclude that there is a 63% probability that a car will arrive within 5 minutes.

(b) The probability that a car will arrive within 30 minutes is found by evaluating $F(t)$ at $t = 30$.

$$F(30) = 1 - e^{-0.2(30)} \approx 0.9975$$

↑ Use a calculator

There is a 99.75% probability that a car will arrive within 30 minutes.

(c) As time passes, the probability that a car will arrive increases. The value that F approaches can be found by letting t become unbounded in the positive direction. Since $e^{-0.2t} = \frac{1}{e^{0.2t}}$, it follows that $e^{-0.2t}$ approaches 0 as t becomes unbounded in the positive direction. Thus, F approaches 1 as t becomes unbounded in the positive direction.

FIGURE 34

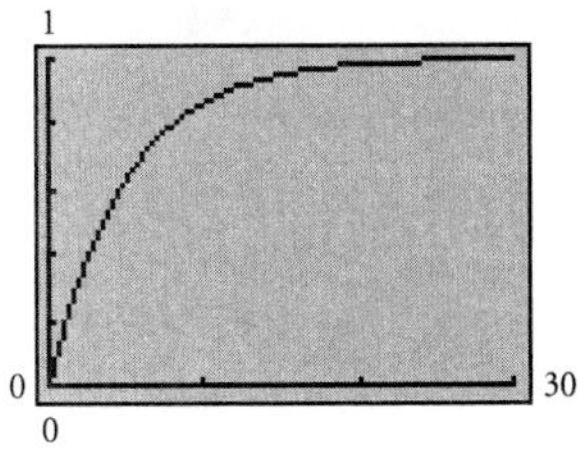

(d) See Figure 34 for the graph of F.

(e) Within 3.5 minutes of 9 PM, the probability of a car arriving equals 50%.

NOW WORK PROBLEM 63.

SUMMARY Properties of the Exponential Function

$f(x) = a^x$, $a > 1$ Domain: the interval $(-\infty, \infty)$; Range: the interval $(0, \infty)$; x-intercept: none; y-intercept: $(0, 1)$ horizontal asymptote: the line $y = 0$ (the x-axis), as x becomes unbounded in the negative direction increasing; smooth; continuous See Figure 26 for a typical graph.

$f(x) = a^x$, $0 < a < 1$ Domain: the interval $(-\infty, \infty)$; Range: the interval $(0, \infty)$; x-intercept: none; y-intercept: $(0, 1)$ horizontal asymptote: the line $y = 0$ (the x-axis), as x becomes unbounded in the positive direction decreasing; smooth; continuous See Figure 30 for a typical graph.

If $a^u = a^v$, then $u = v$.

EXERCISE 11.3 Answers to Odd-Numbered Problems Begin on Page AN-62.

In Problems 1–10, approximate each number using a calculator. Express your answer rounded to four decimal places.

1. (a) $3^{2.2}$ (b) $3^{2.23}$ (c) $3^{2.236}$ (d) $3^{\sqrt{5}}$

2. (a) $5^{1.7}$ (b) $5^{1.73}$ (c) $5^{1.732}$ (d) $5^{\sqrt{3}}$

3. (a) $2^{3.14}$ (b) $2^{3.141}$ (c) $2^{3.1415}$ (d) 2^{π}

4. (a) $2^{2.7}$ (b) $2^{2.71}$ (c) $2^{2.718}$ (d) 2^{e}

5. (a) $3.1^{2.7}$ (b) $3.14^{2.71}$ (c) $3.141^{2.718}$ (d) π^e

6. (a) $2.7^{3.1}$ (b) $2.71^{3.14}$ (c) $2.718^{3.141}$ (d) e^π

7. $e^{1.2}$

8. $e^{-1.3}$

9. $e^{-0.85}$

10. $e^{2.1}$

In Problems 11–18, determine whether the given function is exponential or not. For those that are exponential functions, identify the value of a. [**Hint:** *Look at the ratio of consecutive values.*]

11.

x	$f(x)$
−1	3
0	6
1	12
2	18
3	30

12.

x	$g(x)$
−1	2
0	5
1	8
2	11
3	14

13.

x	$H(x)$
−1	$\frac{1}{4}$
0	1
1	4
2	16
3	64

14.

x	$F(x)$
−1	$\frac{2}{3}$
0	1
1	$\frac{3}{2}$
2	$\frac{9}{4}$
3	$\frac{27}{8}$

15.

x	$f(x)$
−1	$\frac{3}{2}$
0	3
1	6
2	12
3	24

16.

x	$g(x)$
−1	6
0	1
1	0
2	3
3	10

17.

x	$H(x)$
−1	2
0	4
1	6
2	8
3	10

18.

x	$F(x)$
−1	$\frac{1}{2}$
0	$\frac{1}{4}$
1	$\frac{1}{8}$
2	$\frac{1}{16}$
3	$\frac{1}{32}$

In Problems 19–26, the graph of an exponential function is given. Match each graph to one of the following functions.

A. $y = 3^x$ B. $y = 3^{-x}$ C. $y = -3^x$ D. $y = -3^{-x}$

E. $y = 3^x - 1$ F. $y = 3^{x-1}$ G. $y = 3^{1-x}$ H. $y = 1 - 3^x$

19.

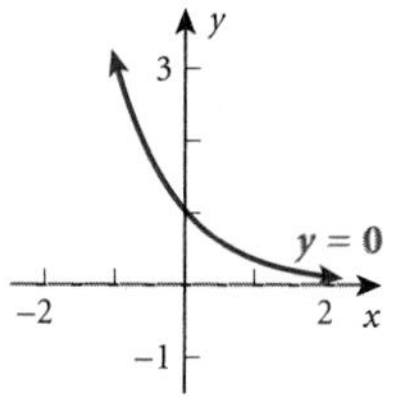

20.

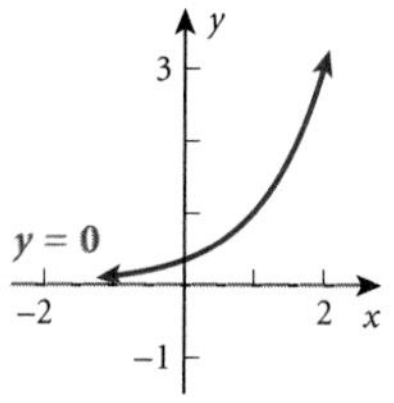

21.

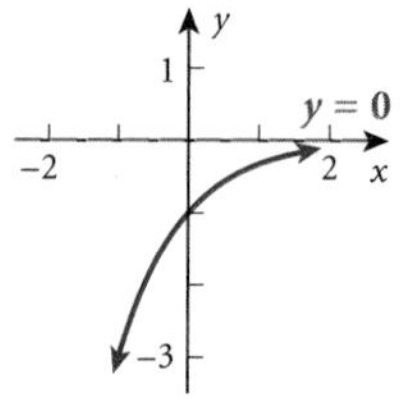

22.

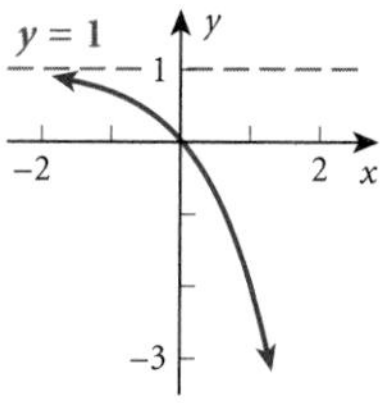

23.

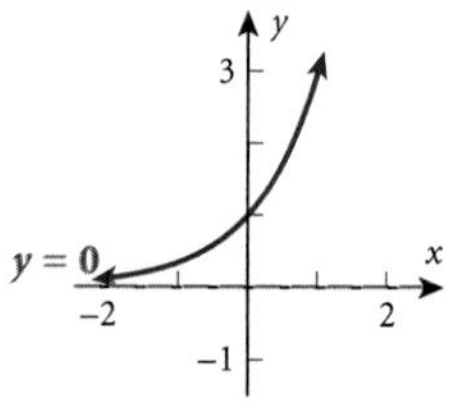

24.

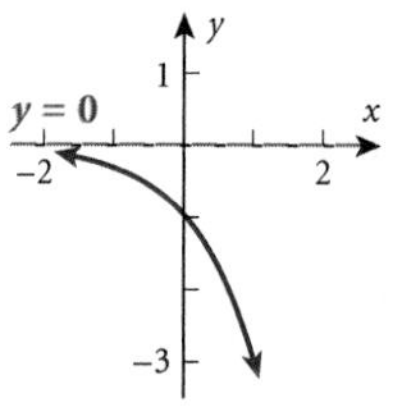

25.

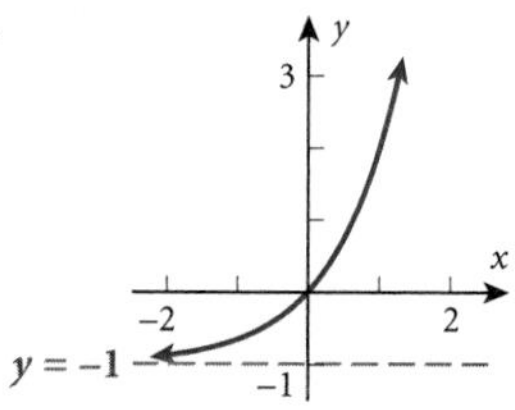

26.

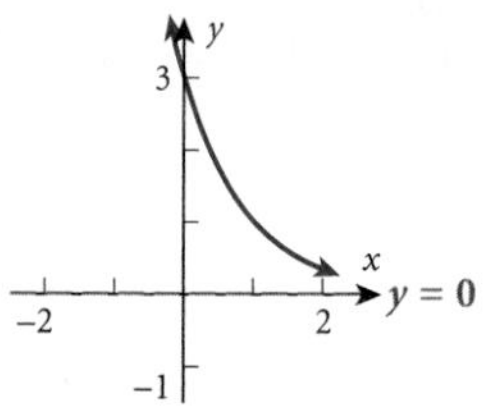

In Problems 27–34, use shifts and/or reflections to graph each function. Determine the domain, range, and horizontal asymptote of each function.

27. $f(x) = 2^x + 1$ **28.** $f(x) = 2^{x+2}$ **29.** $f(x) = 3^{-x} - 2$ **30.** $f(x) = -3^x + 1$

31. $f(x) = e^{-x}$ **32.** $f(x) = -e^x$ **33.** $f(x) = e^{x-2} - 1$ **34.** $f(x) = -e^x - 1$

In Problems 35–48, solve each equation.

35. $2^{2x+1} = 4$ **36.** $5^{1-2x} = \frac{1}{5}$ **37.** $3^{x^3} = 9^x$ **38.** $4^{x^2} = 2^x$

39. $8^{x^2-2x} = \frac{1}{2}$ **40.** $9^{-x} = \frac{1}{3}$ **41.** $2^x \cdot 8^{-x} = 4^x$ **42.** $\left(\frac{1}{2}\right)^{1-x} = 4$

43. $\left(\frac{1}{5}\right)^{2-x} = 25$ **44.** $4^x - 2^x = 0$ **45.** $4^x = 8$ **46.** $9^{2x} = 27$

47. $e^{x^2} = (e^{3x}) \cdot \frac{1}{e^2}$ **48.** $(e^4)^x \cdot e^{x^2} = e^{12}$

49. If $4^x = 7$, what does 4^{-2x} equal? **50.** If $2^x = 3$, what does 4^{-x} equal?

51. If $3^{-x} = 2$, what does 3^{2x} equal? **52.** If $5^{-x} = 3$, what does 5^{3x} equal?

In Problems 53–56, determine the exponential function whose graph is given.

53.

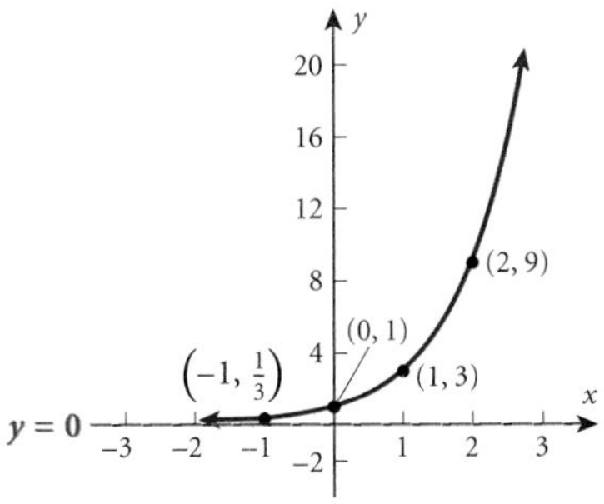

54.

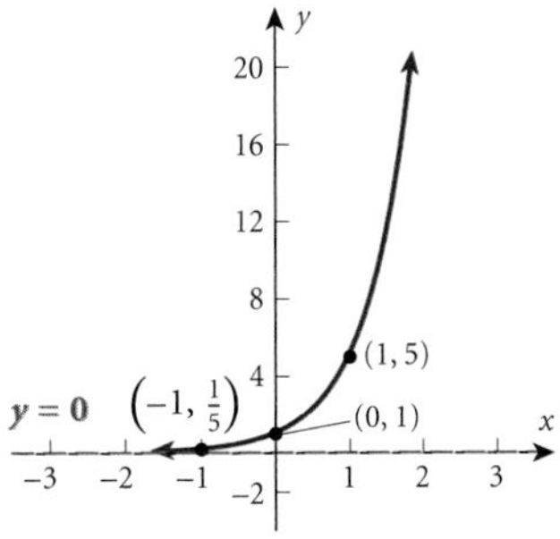

55.

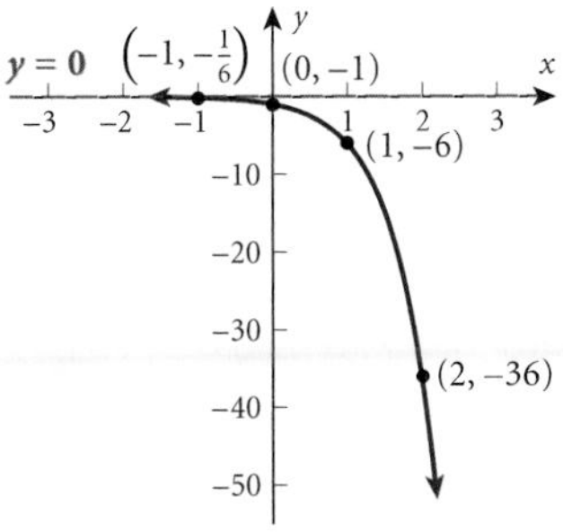

56.

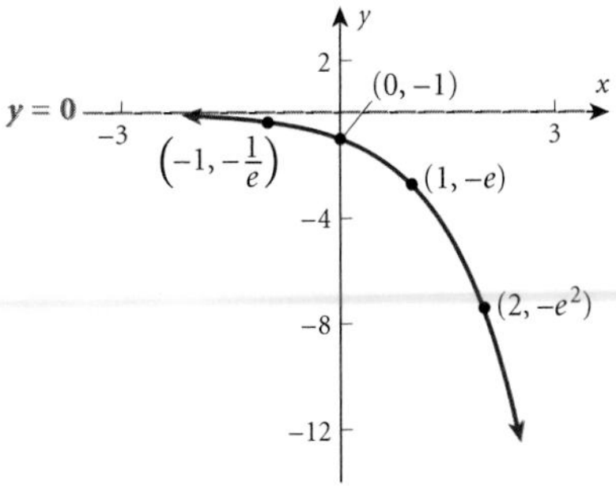

57. Optics If a single pane of glass obliterates 3% of the light passing through it, then the percent p of light that passes through n successive panes is given approximately by the function

$$p(n) = 100e^{-0.03n}$$

(a) What percent of light will pass through 10 panes?
(b) What percent of light will pass through 25 panes?

58. Atmospheric Pressure The atmospheric pressure p on a balloon or airplane decreases with increasing height. This pressure, measured in millimeters of mercury, is related to the number of kilometers h above sea level by the function

$$p(h) = 760e^{-0.145h}$$

(a) Find the atmospheric pressure at a height of 2 kilometers (over 1 mile).
(b) What is it at a height of 10 kilometers (over 30,000 feet)?

59. Space Satellites The number of watts w provided by a space satellite's power supply after a period of d days is given by the function

$$w(d) = 50e^{-0.004d}$$

(a) How much power will be available after 30 days?
(b) How much power will be available after 1 year (365 days)?

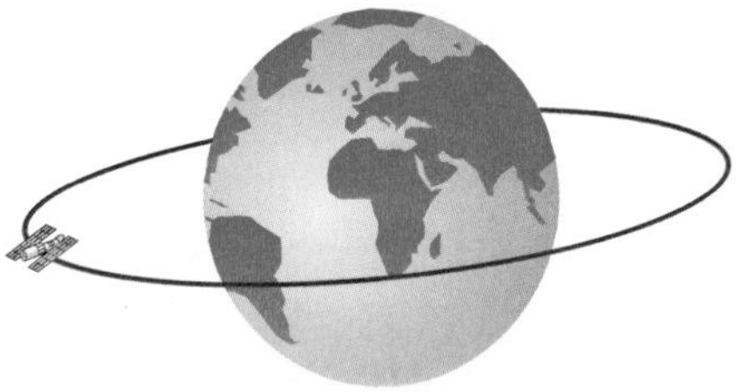

60. Healing of Wounds The normal healing of wounds can be modeled by an exponential function. If A_0 represents the original area of the wound and if A equals the area of the wound after n days, then the function

$$A(n) = A_0e^{-0.35n}$$

describes the area of the wound on the nth day following an injury when no infection is present to retard the healing. Suppose that a wound initially had an area of 100 square millimeters.

(a) If healing is taking place, how large will the area of the wound be after 3 days?
(b) How large will it be after 10 days?

61. Drug Medication The function

$$D(h) = 5e^{-0.4h}$$

can be used to find the number of milligrams D of a certain drug that is in a patient's bloodstream h hours after the drug has been administered. How many milligrams will be present after 1 hour? After 6 hours?

62. Spreading of Rumors A model for the number of people N in a college community who have heard a certain rumor is

$$N = P(1 - e^{-0.15d})$$

where P is the total population of the community and d is the number of days that have elapsed since the rumor began. In a community of 1000 students, how many students will have heard the rumor after 3 days?

63. **Exponential Probability** Between 12:00 PM and 1:00 PM, cars arrive at Citibank's drive-thru at the rate of 6 cars per hour (0.1 car per minute). The following formula from probability can be used to determine the probability that a car will arrive within t minutes of 12:00 PM:

$$F(t) = 1 - e^{-0.1t}$$

(a) Determine the probability that a car will arrive within 10 minutes of 12:00 PM (that is, before 12:10 PM).
(b) Determine the probability that a car will arrive within 40 minutes of 12:00 PM (before 12:40 PM).
(c) What value does F approach as t becomes unbounded in the positive direction?
(d) Graph F using your graphing utility.
(e) Using TRACE, determine how many minutes are needed for the probability to reach 50%.

64. Exponential Probability Between 5:00 PM and 6:00 PM, cars arrive at Jiffy Lube at the rate of 9 cars per hour (0.15 car per minute). The following formula from probability can be used to determine the probability that a car will arrive within t minutes of 5:00 PM:

$$F(t) = 1 - e^{-0.15t}$$

(a) Determine the probability that a car will arrive within 15 minutes of 5:00 PM (that is, before 5:15 PM).
(b) Determine the probability that a car will arrive within 30 minutes of 5:00 PM (before 5:30 PM).
(c) What value does F approach as t becomes unbounded in the positive direction?
(d) Graph F using your graphing utility.
(e) Using TRACE, determine how many minutes are needed for the probability to reach 60%.

65. Poisson Probability Between 5:00 PM and 6:00 PM, cars arrive at McDonald's drive-thru at the rate of 20 cars per hour. The following formula from probability can be used to determine the probability that x cars will arrive between 5:00 PM and 6:00 PM.

$$P(x) = \frac{20^x e^{-20}}{x!}$$

where

$$x! = x \cdot (x - 1) \cdot (x - 2) \cdot \cdots \cdot 3 \cdot 2 \cdot 1$$

(a) Determine the probability that $x = 15$ cars will arrive between 5:00 PM and 6:00 PM.
(b) Determine the probability that $x = 20$ cars will arrive between 5:00 PM and 6:00 PM.

66. Poisson Probability People enter a line for the *Demon Roller Coaster* at the rate of 4 per minute. The following formula from probability can be used to determine the probability that x people will arrive within the next minute.

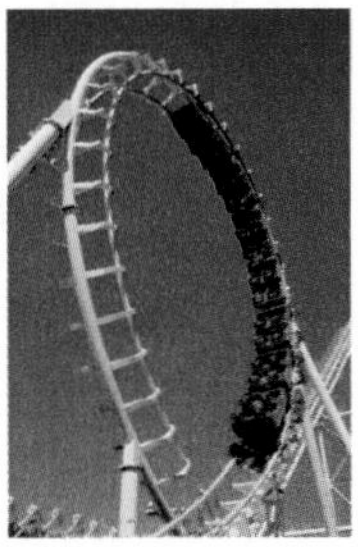

$$P(x) = \frac{4^x e^{-4}}{x!}$$

where

$$x! = x \cdot (x-1) \cdot (x-2) \cdot \cdots \cdot 3 \cdot 2 \cdot 1$$

(a) Determine the probability that $x = 5$ people will arrive within the next minute.
(b) Determine the probability that $x = 8$ people will arrive within the next minute.

67. Depreciation The price p of a Honda Civic DX Sedan that is x years old is given by

$$p(x) = 16{,}630(0.90)^x$$

(a) How much does a 3-year-old Civic DX Sedan cost?
(b) How much does a 9-year-old Civic DX Sedan cost?

68. Learning Curve Suppose that a student has 500 vocabulary words to learn. If the student learns 15 words after 5 minutes, the function

$$L(t) = 500(1 - e^{-0.0061t})$$

approximates the number of words L that the student will learn after t minutes.

(a) How many words will the student learn after 30 minutes?
(b) How many words will the student learn after 60 minutes?

69. Alternating Current in a *RL* Circuit The equation governing the amount of current I (in amperes) after time t (in seconds) in a single RL circuit consisting of a resistance R (in ohms), an inductance L (in henrys), and an electromotive force E (in volts) is

$$I = \frac{E}{R}[1 - e^{-(R/L)t}]$$

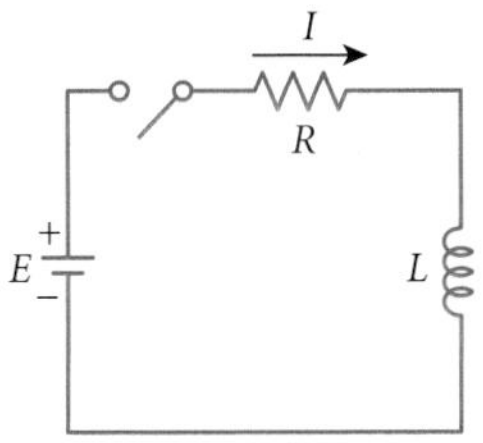

(a) If $E = 120$ volts, $R = 10$ ohms, and $L = 5$ henrys, how much current I_1 is flowing after 0.3 second? After 0.5 second? After 1 second?
(b) Graph the function $I = I_1(t)$, measuring I along the y-axis and t along the x-axis.
(c) What is the maximum current?
(d) If $E = 120$ volts, $R = 5$ ohms, and $L = 10$ henrys, how much current I_2 is flowing after 0.3 second? After 0.5 second? After 1 second?
(e) Graph the function $I = I_2(t)$ on the same screen as $I_1(t)$.
(f) What is the maximum current?

70. Alternating Current in a *RC* Circuit The equation governing the amount of current I (in amperes) after time t (in microseconds) in a single RC circuit consisting of a resistance R (in ohms), a capacitance C (in microfarads), and an electromotive force E (in volts) is

$$I = \frac{E}{R}e^{-t/(RC)}$$

(a) If $E = 120$ volts, $R = 2000$ ohms, and $C = 1.0$ microfarad, how much current I_1 is flowing initially ($t = 0$)? After 1000 microseconds? After 3000 microseconds?
(b) Graph the function $I = I_1(t)$, measuring I along the y-axis and t along the x-axis.
(c) What is the maximum current?
(d) If $E = 120$ volts, $R = 1000$ ohms, and $C = 2.0$ microfarads, how much current I_2 is flowing initially? After 1000 microseconds? After 3000 microseconds?
(e) Graph the function $I = I_2(t)$ on the same screen as $I_1(t)$.
(f) What is the maximum current?

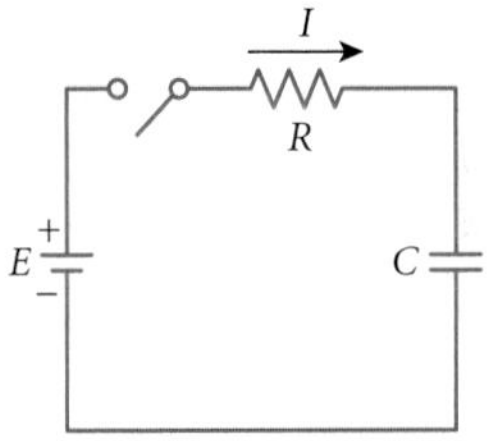

71. Another Formula for *e* Use a calculator to compute the values of

$$2 + \frac{1}{2!} + \frac{1}{3!} + \cdots + \frac{1}{n!}$$

for $n = 4, 6, 8$, and 10. Compare each result with e.
[**Hint:** $1! = 1$, $2! = 2 \cdot 1$, $3! = 3 \cdot 2 \cdot 1$, $n! = n(n-1) \cdot \cdots \cdot (3)(2)(1)$]

72. Another Formula for *e* Use a calculator to compute the first five values of the expression. The first one is $2 + 1 = 3$; the second one is $2 + \dfrac{1}{1+1} = 2.5$. Compare the values to e.

$$2 + \cfrac{1}{1 + \cfrac{1}{2 + \cfrac{2}{3 + \cfrac{3}{4 + \cfrac{4}{\text{etc.}}}}}}$$

73. Difference Quotient If $f(x) = a^x$, show that

$$\frac{f(x+h) - f(x)}{h} = a^x\left(\frac{a^h - 1}{h}\right)$$

74. If $f(x) = a^x$, show that $f(A + B) = f(A) \cdot f(B)$.

75. If $f(x) = a^x$, show that $f(-x) = \frac{1}{f(x)}$.

76. If $f(x) = a^x$, show that $f(\alpha x) = [f(x)]^\alpha$.

77. Relative Humidity The relative humidity is the ratio (expressed as a percent) of the amount of water vapor in the air to the maximum amount that it can hold at a specific temperature. The relative humidity, R, is found using the following formula:

$$R = 10^{\frac{4221}{T+459.4} - \frac{4221}{D+459.4} + 2}$$

where T is the air temperature (in °F) and D is the dew point temperature (in °F).

(a) Determine the relative humidity if the air temperature is 50° Fahrenheit and the dew point temperature is 41° Fahrenheit.
(b) Determine the relative humidity if the air temperature is 68° Fahrenheit and the dew point temperature is 59° Fahrenheit.
(c) What is the relative humidity if the air temperature and the dew point temperature are the same?

78. Historical Problem Pierre de Fermat (1601–1665) conjectured that the function

$$f(x) = 2^{(2^x)} + 1$$

for $x = 1, 2, 3, \ldots$, would always have a value equal to a prime number. But Leonhard Euler (1707–1783) showed that this formula fails for $x = 5$. Use a calculator to determine the prime numbers produced by f for $x = 1, 2, 3, 4$. Then show that $f(5) = 641 \times 6{,}700{,}417$, which is not prime.

Problems 79 and 80 provide definitions for two other functions.

79. The **hyperbolic sine function,** designated by $\sinh x$, is defined as

$$\sinh x = \frac{1}{2}(e^x - e^{-x})$$

(a) Show that $f(x) = \sinh x$ is an odd function.
(b) Graph $f(x) = \sinh x$ using a graphing utility.

80. The **hyperbolic cosine function,** designated by $\cosh x$, is defined as

$$\cosh x = \frac{1}{2}(e^x + e^{-x})$$

(a) Show that $f(x) = \cosh x$ is an even function.
(b) Graph $f(x) = \cosh x$ using a graphing utility.
(c) Refer to Problem 79. Show that, for every x,

$$(\cosh x)^2 - (\sinh x)^2 = 1.$$

81. The bacteria in a 4-liter container double every minute. After 60 minutes the container is full. How long did it take to fill half the container? Explain your reasoning.

82. Explain in your own words what the number e is. Provide at least two applications that require the use of this number.

83. Do you think that there is a power function that increases more rapidly than an exponential function whose base is greater than 1? Explain.

84. As the base a of an exponential function $f(x) = a^x$, $a > 1$, increases, what happens to the behavior of its graph for $x > 0$? What happens to the behavior of the graph for $x < 0$?

85. The graphs of $y = a^{-x}$ and $y = \left(\frac{1}{a}\right)^x$ are identical. Why?

11.4 Logarithmic Functions

PREPARING FOR THIS SECTION *Before getting started, review the following:*

> Solving Inequalities (Appendix A, Section A.7, pp. A-63–A-65)
> Vertical Line Test (Chapter 10, Section 10.3, p. 561)
> Graphing Techniques (Chapter 10, Section 10.5, pp. 588–594)

OBJECTIVES
1 Change exponential expressions to logarithmic expressions
2 Change logarithmic expressions to exponential expressions
3 Evaluate logarithmic functions
4 Find the domain of a logarithmic function
5 Graph logarithmic functions
6 Solve logarithmic equations

We begin with the exponential function

$$y = 3^x$$

If we interchange the variables x and y, we obtain the equation

$$x = 3^y$$

FIGURE 35

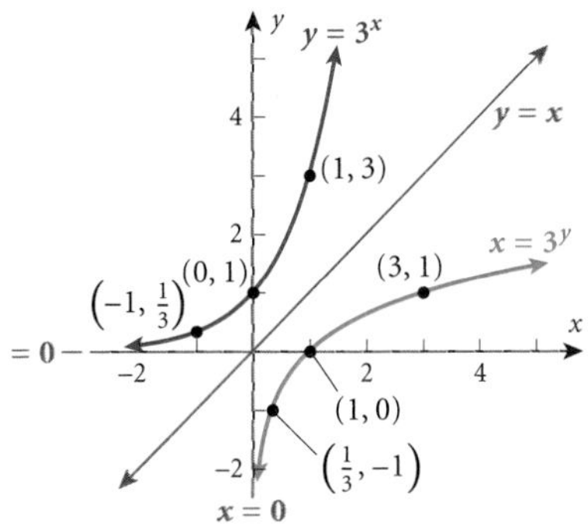

Let's compare the graphs of these two equations. For example, the point $(1, 3)$ is on the graph of $y = 3^x$ and the point $(3, 1)$ is on the graph of $x = 3^y$. Also, the point $(0,1)$ is on the graph of $y = 3^x$ and the point $(1,0)$ is on the graph of $x = 3^y$. In general, if the point (a, b) is on the graph of $y = 3^x$, then the point (b, a) will be on the graph of $x = 3^y$. See Figure 35.

Notice in Figure 35 that we show the line $y = x$. You should see that the graphs of $y = 3^x$ and $x = 3^y$ are symmetric with respect to the line $y = x$, a fact we shall not prove. This means we could have obtained the graph of $x = a^y$, $a > 0$, $a \neq 1$, by reflecting the graph of $y = a^x$ about the line $y = x$.

Look again at Figure 35. We see from the graph of the equation $x = 3^y$ that it is the graph of a function. (Do you see why? Apply the Vertical Line Test.) We call this function a *logarithmic function*. The general definition is given next.

The **logarithmic function to the base *a*,** where $a > 0$ and $a \neq 1$, is denoted by $y = \log_a x$ (read as "y is the logarithm to the base a of x") and is defined by

$$y = \log_a x \quad \text{if and only if} \quad x = a^y$$

The domain of the logarithmic function $y = \log_a x$ is $x > 0$.

A *logarithm* is merely a name for a certain exponent.

EXAMPLE 1 Relating Logarithms to Exponents

(a) If $y = \log_3 x$, then $x = 3^y$. For example, $2 = \log_3 9$ is equivalent to $9 = 3^2$.

(b) If $y = \log_5 x$, then $x = 5^y$. For example, $-1 = \log_5\left(\frac{1}{5}\right)$ is equivalent to $\frac{1}{5} = 5^{-1}$.

1 EXAMPLE 2 Changing Exponential Expressions to Logarithmic Expressions

Change each exponential expression to an equivalent expression involving a logarithm.

(a) $1.2^3 = m$ **(b)** $e^b = 9$ **(c)** $a^4 = 24$

SOLUTION We use the fact that $y = \log_a x$ and $x = a^y$, $a > 0$, $a \neq 1$, are equivalent.

(a) If $1.2^3 = m$, then $3 = \log_{1.2} m$. **(b)** If $e^b = 9$, then $b = \log_e 9$.

(c) If $a^4 = 24$, then $4 = \log_a 24$.

NOW WORK PROBLEM 1.

2 **EXAMPLE 3** **Changing Logarithmic Expressions to Exponential Expressions**

Change each logarithmic expression to an equivalent expression involving an exponent.

(a) $\log_a 4 = 5$ **(b)** $\log_e b = -3$ **(c)** $\log_3 5 = c$

SOLUTION **(a)** If $\log_a 4 = 5$, then $a^5 = 4$. **(b)** If $\log_e b = -3$, then $e^{-3} = b$.

(c) If $\log_3 5 = c$, then $3^c = 5$.

NOW WORK PROBLEM 13.

To find the exact value of a logarithm, we write the logarithm in exponential notation and use the fact that if $a^u = a^v$ then $u = v$.

3 **EXAMPLE 4** **Finding the Exact Value of a Logarithmic Expression**

Find the exact value of

(a) $\log_2 16$ **(b)** $\log_3 \frac{1}{27}$

SOLUTION **(a)** $y = \log_2 16$

$2^y = 16$ Change to exponential form.

$2^y = 2^4$ $16 = 2^4$

$y = 4$ Equate exponents.

Therefore, $\log_2 16 = 4$.

(b) $y = \log_3 \frac{1}{27}$

$3^y = \frac{1}{27}$ Change to exponential form.

$3^y = 3^{-3}$ $\frac{1}{27} = \frac{1}{3^3} = 3^{-3}$

$y = -3$ Equate exponents.

Therefore, $\log_3 \frac{1}{27} = -3$.

NOW WORK PROBLEM 25.

The domain of a logarithmic function consists of the *positive* real numbers, so the argument of a logarithmic function must be greater than zero.

4 **EXAMPLE 5** **Finding the Domain of a Logarithmic Function**

Find the domain of each logarithmic function

(a) $F(x) = \log_2(x - 5)$ **(b)** $g(x) = \log_5\left(\frac{1 + x}{1 - x}\right)$ **(c)** $h(x) = \log_{1/2}|x|$

SOLUTION **(a)** The domain of F consists of all x for which $x - 5 > 0$, that is, all $x > 5$, or using interval notation, $(5, \infty)$.

(b) The domain of g is restricted to

$$\frac{1 + x}{1 - x} > 0$$

Solving this inequality, we find that the domain of g consists of all x between -1 and 1, that is, $-1 < x < 1$, or using interval notation, $(-1, 1)$.

(c) Since $|x| > 0$, provided that $x \neq 0$, the domain of h consists of all nonzero real numbers, or using interval notation, $(-\infty, 0)$ or $(0, \infty)$.

NOW WORK PROBLEMS 37 AND 39.

Graph logarithmic functions **5**

Graphs of Logarithmic Functions

As we said earlier, we can obtain the graph of $x = a^y$, $a > 0$, $a \neq 1$, or equivalently the graph of $y = \log_a x$, by reflecting the graph of $y = a^x$ about the line $y = x$. For example, to graph $y = \log_2 x$, graph $y = 2^x$ and reflect it about the line $y = x$. See Figure 36. To graph $y = \log_{1/3} x$, graph $y = \left(\frac{1}{3}\right)^x$ and reflect it about the line $y = x$. See Figure 37.

FIGURE 36

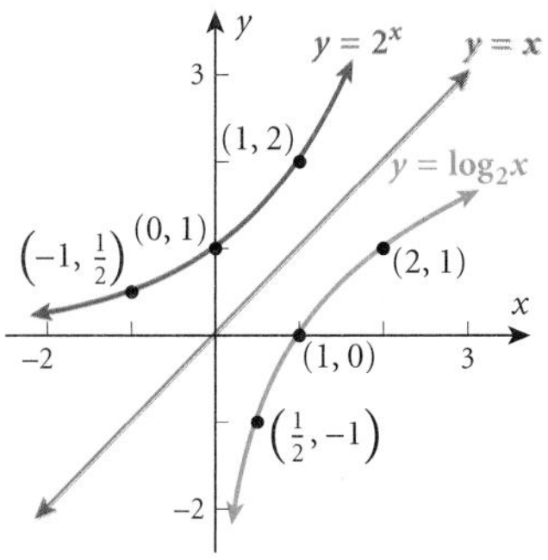

FIGURE 37

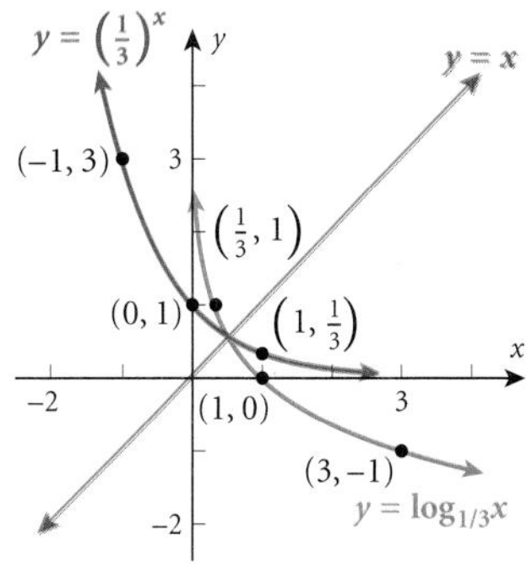

The graph of the logarithmic function $y = \log_a x$ is the reflection about the line $y = x$ of the graph of the exponential function $y = a^x$, as shown in Figures 38 and 39.

FIGURE 38

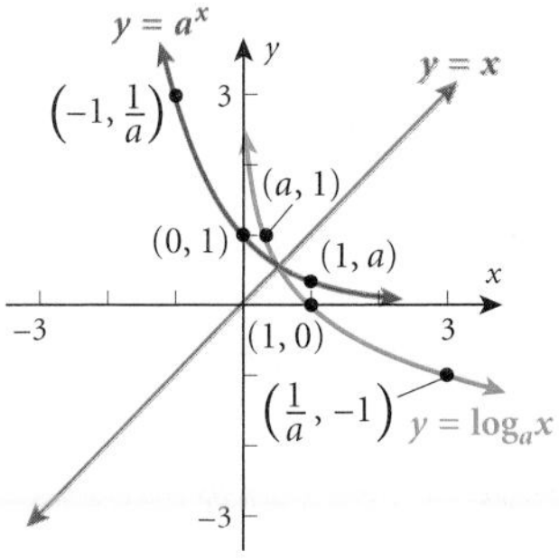

$0 < a < 1$

FIGURE 39

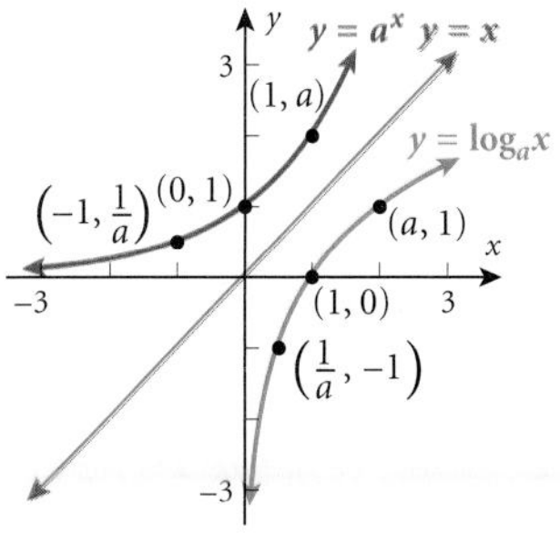

$a > 1$

NOW WORK PROBLEM 51.

The graphs in Figures 38 and 39 lead to the following result.

Properties of the Graph of a Logarithmic Function $f(x) = \log_a x$

1. The domain is the set of positive real numbers; the range is all real numbers.
2. The x-intercept of the graph is $(1, 0)$ There is no y-intercept.
3. The line $x = 0$ (the y-axis) is a vertical asymptote of the graph.
4. A logarithmic function is decreasing if $0 < a < 1$ and is increasing if $a > 1$.
5. The graph of f contains the points $(1, 0)$, $(a, 1)$, and $\left(\frac{1}{a}, -1\right)$.
6. The graph is smooth and continuous, with no corners or gaps.

If the base of a logarithmic function is the number e, then we have the **natural logarithm function.** This function occurs so frequently in applications that it is given a special symbol, **ln** (from the Latin, *logarithmus naturalis*). That is,

$$y = \log_e x = \ln x \quad \text{if and only if} \quad x = e^y \tag{1}$$

We can obtain the graph of $y = \ln x$ by reflecting the graph of $y = e^x$ about the line $y = x$. See Figure 40.

FIGURE 40

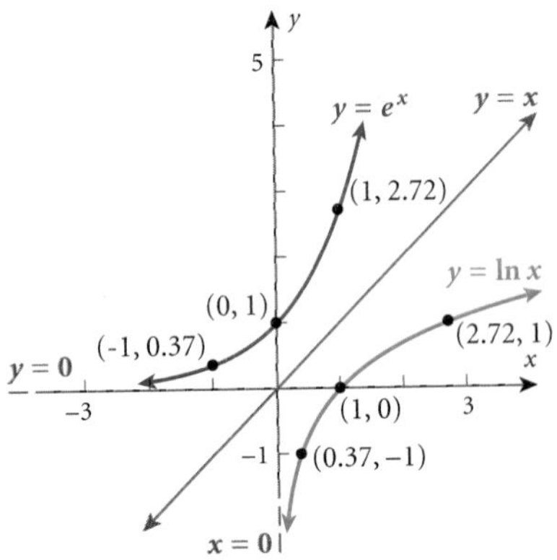

TABLE 10

x	$\ln x$
0.5	-0.69
2	0.69
3	1.10

Using a calculator with an $\boxed{\ln}$ key, we can obtain other points on the graph of $f(x) = \ln x$. See Table 10.

SEEING THE CONCEPT: Graph $Y_1 = e^x$ and $Y_2 = \ln x$ on the same square screen. Use eVALUEate to verify the points on the graph given in Figure 41. Do you see the symmetry of the two graphs with respect to the line $y = x$?

EXAMPLE 6 Graphing Logarithmic Functions Using Shifts and/or Reflections

Graph $f(x) = -\ln(x + 2)$ by starting with the graph of $y = \ln x$. Determine the domain, range, and vertical asymptote of f.

SOLUTION The domain of f consists of all x for which

$$x + 2 > 0 \quad \text{or} \quad x > -2$$

so the domain is $\{x \mid x > -2\}$.

To obtain the graph of $y = -\ln(x + 2)$, we use the steps illustrated in Figure 41.

FIGURE 41

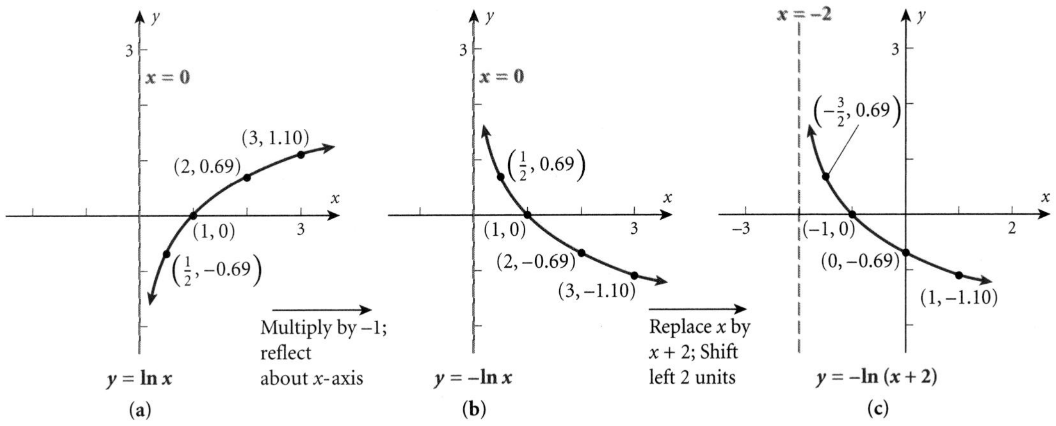

The range of $f(x) = -\ln(x + 2)$ is the interval $(-\infty, \infty)$, and the vertical asymptote is $x = -2$. [Do you see why? The original asymptote ($x = 0$) is shifted to the left 2 units.]

NOW WORK PROBLEM 63.

If the base of a logarithmic function is the number 10, then we have the **common logarithm function.** If the base a of the logarithmic function is not indicated, it is understood to be 10. Thus,

$$y = \log x \quad \text{if and only if} \quad x = 10^y$$

We can obtain the graph of $y = \log x$ by reflecting the graph of $y = 10^x$ about the line $y = x$. See Figure 42.

FIGURE 42

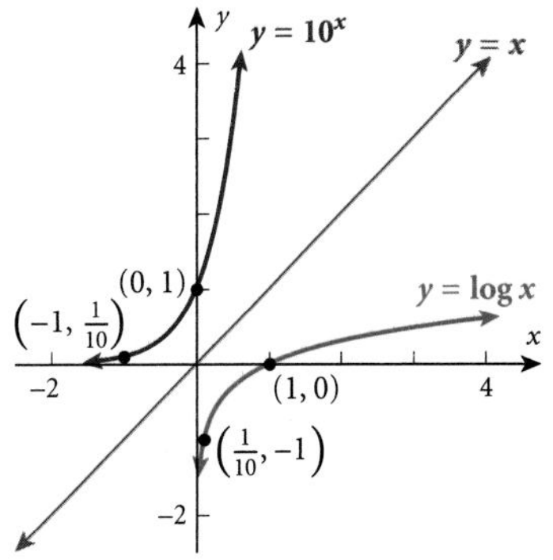

Logarithmic Equations

Equations that contain logarithms are called **logarithmic equations.** Care must be taken when solving logarithmic equations. Be sure to check each apparent solution in the original equation and discard any that are extraneous. In the expression $\log_a M$, remember that a and M are positive and $a \neq 1$.

Some logarithmic equations can be solved by changing from a logarithmic expression to an exponential expression.

6 **EXAMPLE 7 Solving a Logarithmic Equation**

Solve: **(a)** $\log_3(4x - 7) = 2$ **(b)** $\log_x 64 = 2$

SOLUTION **(a)** We can obtain an exact solution by changing the logarithm to exponential form

$$\begin{aligned} \log_3(4x - 7) &= 2 \\ 4x - 7 &= 3^2 && \text{Change to exponential form.} \\ 4x - 7 &= 9 \\ 4x &= 16 \\ x &= 4 \end{aligned}$$

CHECK: $\log_3(4x - 7) = \log_3(16 - 7) = \log_3 9 = 2 \quad 3^2 = 9$

(b) We can obtain an exact solution by changing the logarithm to exponential form.

$$\begin{aligned} \log_x 64 &= 2 \\ x^2 &= 64 && \text{Change to exponential form.} \\ x &= \pm\sqrt{64} = \pm 8 \end{aligned}$$

The base of a logarithm is always positive. As a result, we discard -8; the only solution is 8.

CHECK: $\log_8 64 = 2 \quad 8^2 = 64.$

EXAMPLE 8 Using Logarithms to Solve Exponential Equations

Solve: $e^{2x} = 5$

SOLUTION We can obtain an exact solution by changing the exponential equation to logarithmic form.

$$\begin{aligned} e^{2x} &= 5 \\ \ln 5 &= 2x && \text{Change to a logarithmic expression.} \\ x &= \frac{\ln 5}{2} && \text{Exact solution.} \\ &\approx 0.805 && \text{Approximate solution.} \end{aligned}$$

NOW WORK PROBLEMS 71 AND 83.

EXAMPLE 9 Alcohol and Driving

The concentration of alcohol in a person's blood is measurable. Recent medical research suggests that the risk R (given as a percent) of having an accident while driving a car can be modeled by the equation

$$R = 6e^{kx}$$

where x is the variable concentration of alcohol in the blood and k is a constant.

(a) Suppose that a concentration of alcohol in the blood of 0.04 results in a 10% risk ($R = 10$) of an accident. Find the constant k in the equation.

(b) Using this value of k, what is the risk if the concentration is 0.17?

(c) Using the same value of k, what concentration of alcohol corresponds to a risk of 100%?

(d) If the law asserts that anyone with a risk of having an accident of 20% or more should not have driving privileges, at what concentration of alcohol in the blood should a driver be arrested and charged with a DUI (Driving Under the Influence)?

SOLUTION **(a)** For a concentration of alcohol in the blood of 0.04 and a risk of 10%, we let $x = 0.04$ and $R = 10$ in the equation and solve for k.

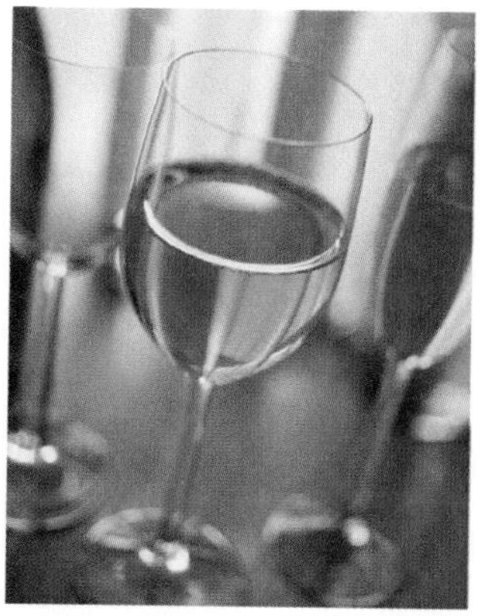

$$R = 6e^{kx}$$

$$10 = 6e^{k(0.04)} \qquad R = 10; x = 0.04.$$

$$\frac{10}{6} = e^{0.04k} \qquad \text{Divide both sides by 6.}$$

$$0.04k = \ln\frac{10}{6} = 0.5108256 \qquad \text{Change to a logarithmic expression.}$$

$$k = 12.77 \qquad \text{Solve for } k.$$

(b) Using $k = 12.77$ and $x = 0.17$ in the equation, we find the risk R to be

$$R = 6e^{kx} = 6e^{(12.77)(0.17)} = 52.6$$

For a concentration of alcohol in the blood of 0.17, the risk of an accident is about 52.6%.

(c) Using $k = 12.77$ and $R = 100$ in the equation, we find the concentration x of alcohol in the blood to be

$$R = 6e^{kx}$$

$$100 = 6e^{12.77x} \qquad R = 100; k = 12.77.$$

$$\frac{100}{6} = e^{12.77x} \qquad \text{Divide both sides by 6.}$$

$$12.77x = \ln\frac{100}{6} = 2.8134 \qquad \text{Change to a logarithmic expression.}$$

$$x = 0.22 \qquad \text{Solve for } x.$$

For a concentration of alcohol in the blood of 0.22, the risk of an accident is 100%.

(d) Using $k = 12.77$ and $R = 20$ in the equation, we find the concentration x of alcohol in the blood to be

$$R = 6e^{kx}$$

$$20 = 6e^{12.77x} \qquad R = 20; k = 12.77$$

$$\frac{20}{6} = e^{12.77x}$$

$$12.77x = \ln\frac{20}{6} = 1.204$$

$$x = 0.094$$

A driver with a concentration of alcohol in the blood of 0.094 or more (9.4%) should be arrested and charged with DUI.

[**NOTE:** Most states use 0.08 or 0.10 as the blood alcohol content at which a DUI citation is given.]

SUMMARY Properties of the Logarithmic Function

$f(x) = \log_a x,\ a > 1$ ($y = \log_a x$ means $x = a^y$): Domain: the interval $(0, \infty)$; Range: the interval $(-\infty, \infty)$; x-intercept: $(1, 0)$; y-intercept: none; vertical asymptote: $x = 0$ (y-axis); increasing.

See Figure 43 for a typical graph.

$f(x) = \log_a x,\ 0 < a < 1$ ($y = \log_a x$ means $x = a^y$): Domain: the interval $(0, \infty)$; Range: the interval $(-\infty, \infty)$; x-intercept: $(1, 0)$; y-intercept: none; vertical asymptote: $x = 0$ (y-axis); decreasing.

See Figure 44 for a typical graph.

FIGURE 43

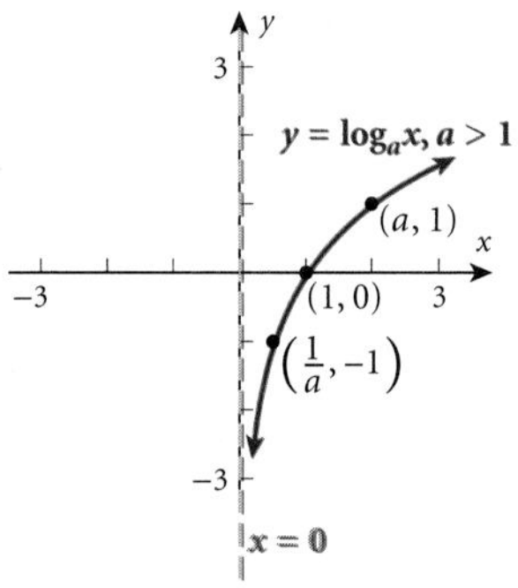

FIGURE 44

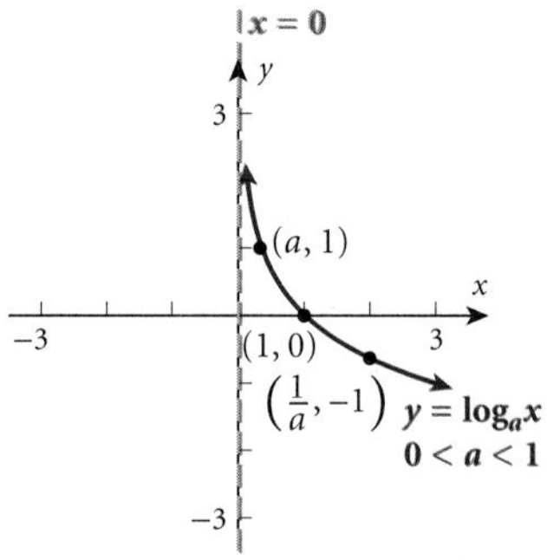

EXERCISE 11.4 Answers to Odd-Numbered Problems Begin on Page AN-64.

In Problems 1–12, change each exponential expression to an equivalent expression involving a logarithm.

1. $9 = 3^2$
2. $16 = 4^2$
3. $a^2 = 1.6$
4. $a^3 = 2.1$
5. $1.1^2 = M$
6. $2.2^3 = N$
7. $2^x = 7.2$
8. $3^x = 4.6$
9. $x^{\sqrt{2}} = \pi$
10. $x^{\pi} = e$
11. $e^x = 8$
12. $e^{2.2} = M$

In Problems 13–24, change each logarithmic expression to an equivalent expression involving an exponent.

13. $\log_2 8 = 3$
14. $\log_3\left(\frac{1}{9}\right) = -2$
15. $\log_a 3 = 6$
16. $\log_b 4 = 2$
17. $\log_3 2 = x$
18. $\log_2 6 = x$
19. $\log_2 M = 1.3$
20. $\log_3 N = 2.1$
21. $\log_{\sqrt{2}} \pi = x$
22. $\log_{\pi} x = \frac{1}{2}$
23. $\ln 4 = x$
24. $\ln x = 4$

In Problems 25–36, find the exact value of each logarithm without using a calculator.

25. $\log_2 1$
26. $\log_8 8$
27. $\log_5 25$
28. $\log_3\left(\frac{1}{9}\right)$

29. $\log_{1/2} 16$ **30.** $\log_{1/3} 9$ **31.** $\log_{10} \sqrt{10}$ **32.** $\log_5 \sqrt[3]{25}$

33. $\log_{\sqrt{2}} 4$ **34.** $\log_{\sqrt{3}} 9$ **35.** $\ln \sqrt{e}$ **36.** $\ln e^3$

In Problems 37–44, find the domain of each function.

37. $f(x) = \ln(x - 3)$ **38.** $g(x) = \ln(x - 1)$ 39. $F(x) = \log_2 x^2$ **40.** $H(x) = \log_5 x^3$

41. $f(x) = 3 - 2\log_4 \frac{x}{2}$ **42.** $g(x) = 8 + 5\ln(2x)$ **43.** $f(x) = \sqrt{\ln x}$ **44.** $g(x) = \frac{1}{\ln x}$

In Problems 45–48, use a calculator to evaluate each expression. Round your answer to three decimal places.

45. $\ln \frac{5}{3}$ **46.** $\frac{\ln 5}{3}$ **47.** $\frac{\ln \frac{10}{3}}{0.04}$ **48.** $\frac{\ln \frac{2}{3}}{-0.1}$

49. Find a so that the graph of $f(x) = \log_a x$ contains the point $(2, 2)$.

50. Find a so that the graph of $f(x) = \log_a x$ contains the point $\left(\frac{1}{2}, -4\right)$.

In Problems 51–54, graph each logarithmic function.

51. $y = \log_3 x$ **52.** $y = \log_{1/3} x$ **53.** $y = \log_{1/5} x$ **54.** $y = \log_5 x$

In Problems 55–62, the graph of a logarithmic function is given. Match each graph to one of the following functions:

A. $y = \log_3 x$ B. $y = \log_3(-x)$ C. $y = -\log_3 x$ D. $y = -\log_3(-x)$

E. $y = \log_3 x - 1$ F. $y = \log_3(x - 1)$ G. $y = \log_3(1 - x)$ H. $y = 1 - \log_3 x$

55.

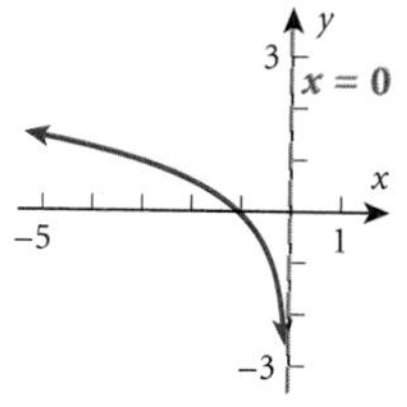

56.

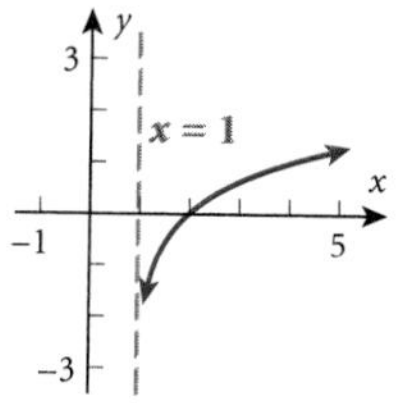

57.

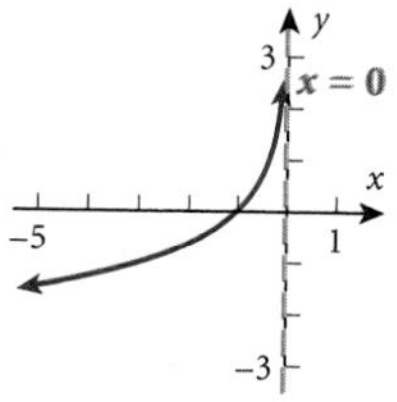

58.

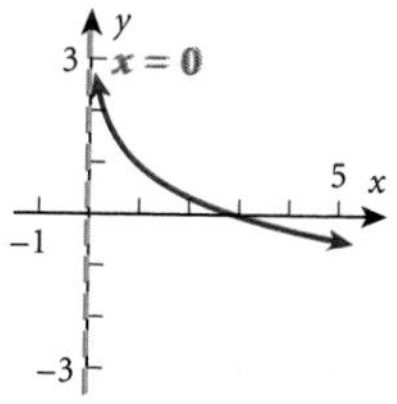

59.

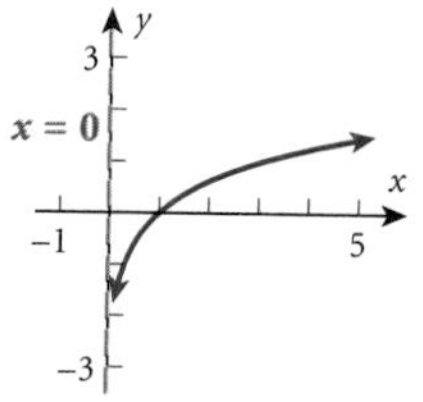

60.

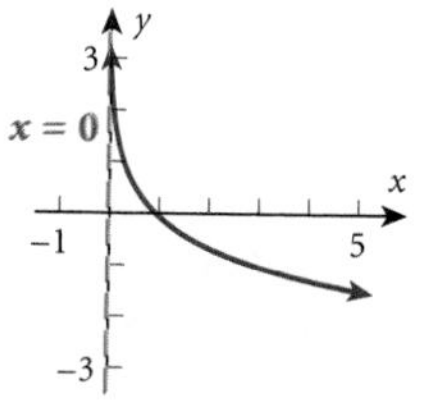

61.

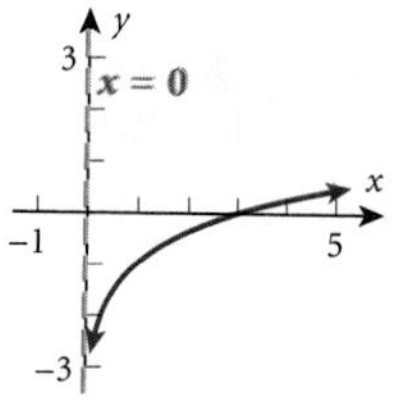

62.

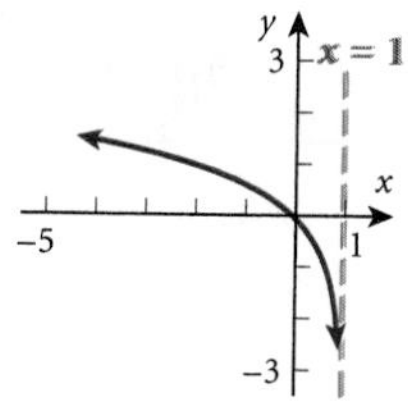

In Problems 63–70, use shifts and/or reflections to graph each function. Determine the domain, range, and vertical asymptote of each function.

63. $f(x) = \ln(x + 4)$ **64.** $f(x) = \ln(x - 3)$ **65.** $f(x) = 2 + \ln x$ **66.** $f(x) = -\ln(-x)$

67. $f(x) = \log(x - 4)$ **68.** $f(x) = \log(x + 5)$ 69. $h(x) = \log x + 2$ **70.** $g(x) = \log x - 5$

In Problems 71–90, solve each equation.

71. $\log_3 x = 2$
72. $\log_5 x = 3$
73. $\log_2(2x + 1) = 3$
74. $\log_3(3x - 2) = 2$
75. $\log_x 4 = 2$
76. $\log_x\left(\frac{1}{8}\right) = 3$
77. $\ln e^x = 5$
78. $\ln e^{-2x} = 8$
79. $\log_4 64 = x$
80. $\log_5 625 = x$
81. $\log_3 243 = 2x + 1$
82. $\log_6 36 = 5x + 3$
83. $e^{3x} = 10$
84. $e^{-2x} = \frac{1}{3}$
85. $e^{2x+5} = 8$
86. $e^{-2x+1} = 13$
87. $\log_3(x^2 + 1) = 2$
88. $\log_5(x^2 + x + 4) = 2$
89. $\log_2 8^x = -3$
90. $\log_3 3^x = -1$

91. **Chemistry** The pH of a chemical solution is given by the formula

$$\text{pH} = -\log_{10}[\text{H}^+]$$

where $[\text{H}^+]$ is the concentration of hydrogen ions in moles per liter. Values of pH range from 0 (acidic) to 14 (alkaline). Distilled water has a pH of 7.

(a) What is the pH of a solution for which $[\text{H}^+]$ is 0.1?
(b) What is the pH of a solution for which $[\text{H}^+]$ is 0.01?
(c) What is the pH of a solution for which $[\text{H}^+]$ is 0.001?
(d) What happens to pH as the hydrogen ion concentration decreases?
(e) Determine the hydrogen ion concentration of an orange (pH = 3.5).
(f) Determine the hydrogen ion concentration of human blood (pH = 7.4).

92. **Diversity Index** Shannon's diversity index is a measure of the diversity of a population. The diversity index is given by the formula

$$H = -(p_1 \log p_1 + p_2 \log p_2 + \cdots + p_n \log p_n)$$

where p_1 is the proportion of the population that is species 1, p_2 is the proportion of the population that is species 2, and so on.

The distribution of race in the United States in 2000 was as follows:

Race	Proportion
American Indian or Native Alaskan	0.014
Asian	0.041
Black or African American	0.128
Hispanic	0.124
Native Hawaiian or Pacific Islander	0.003
White	0.690

Source: U.S. Census Bureau.

(a) Compute the diversity index of the United States in 2000.
(b) The largest value of the diversity index is given by $H_{max} = \log(S)$, where S is the number of categories of race. Compute H_{max}.
(c) The evenness ratio is given by $E_H = \frac{H}{H_{max}}$, where $0 \le E_H \le 1$. If $E_H = 1$, there is complete evenness. Compute the evenness ratio for the United States.
(d) Obtain the distribution of race for the United States in 1990 from the Census Bureau. Compute Shannon's diversity index. Is the United States becoming more diverse? Why?

93. **Atmospheric Pressure** The atmospheric pressure p on a balloon or an aircraft decreases with increasing height. This pressure, measured in millimeters of mercury, is related to the height h (in kilometers) above sea level by the formula.

$$p = 760e^{-0.145h}$$

(a) Find the height of an aircraft if the atmospheric pressure is 320 millimeters of mercury.
(b) Find the height of a mountain if the atmospheric pressure is 667 millimeters of mercury.

94. **Healing of Wounds** The normal healing of wounds can be modeled by an exponential function. If A_0 represents the original area of the wound and if A equals the area of the wound after n days, then the formula

$$A = A_0e^{-0.35n}$$

describes the area of the wound on the nth day following an injury when no infection is present to retard the healing. Suppose that a wound initially had an area of 100 square millimeters.

(a) If healing is taking place, how many days should pass before the wound is one-half its original size?
(b) How long before the wound is 10% of its original size?

95. **Exponential Probability** Between 12:00 PM and 1:00 PM, cars arrive at Citibank's drive-thru at the rate of 6 cars per hour (0.1 car per minute). The following formula from

statistics can be used to determine the probability that a car will arrive within t minutes of 12:00 PM.

$$F(t) = 1 - e^{-0.1t}$$

(a) Determine how many minutes are needed for the probability to reach 50%.
(b) Determine how many minutes are needed for the probability to reach 80%.
(c) Is it possible for the probability to equal 100%? Explain.

96. Exponential Probability Between 5:00 PM and 6:00 PM, cars arrive at Jiffy Lube at the rate of 9 cars per hour (0.15 car per minute). The following formula from statistics can be used to determine the probability that a car will arrive within t minutes of 5:00 PM.

$$F(t) = 1 - e^{-0.15t}$$

(a) Determine how many minutes are needed for the probability to reach 50%.
(b) Determine how many minutes are needed for the probability to reach 80%.

97. Drug Medication The formula

$$D = 5e^{-0.4h}$$

can be used to find the number of milligrams D of a certain drug that is in a patient's bloodstream h hours after the drug has been administered. When the number of milligrams reaches 2, the drug is to be administered again. What is the time between injections?

98. Spreading of Rumors A model for the number of people N in a college community who have heard a certain rumor is

$$N = P(1 - e^{-0.15d})$$

where P is the total population of the community and d is the number of days that have elapsed since the rumor began. In a community of 1000 students, how many days will elapse before 450 students have heard the rumor?

99. Current in an *RL* Circuit The equation governing the amount of current I (in amperes) after time t (in seconds) in a simple RL circuit consisting of a resistance R (in ohms), an inductance L (in henrys), and an electromotive force E (in volts) is

$$I = \frac{E}{R}[1 - e^{-(R/L)t}]$$

If $E = 12$ volts, $R = 10$ ohms, and $L = 5$ henrys, how long does it take to obtain a current of 0.5 ampere? Of 1.0 ampere?

100. Learning Curve Psychologists sometimes use the function

$$L(t) = A(1 - e^{-kt})$$

to measure the amount L learned at time t. The number A represents the amount to be learned, and the number k measures the rate of learning. Suppose that a student has an amount A of 200 vocabulary words to learn. A psychologist determines that the student learned 20 vocabulary words after 5 minutes.

(a) Determine the rate of learning k.
(b) Approximately how many words will the student have learned after 10 minutes?
(c) After 15 minutes?
(d) How long does it take for the student to learn 180 words?

101. U.S. Population According to the U.S. Census Bureau, the population of the United States is projected to be 298,710,000 on January 1, 2010. Suppose this projection is correct, but after January 1, 2010, the population grows according to $P(t) = 298{,}710{,}000 + 10{,}000{,}000 \log t$. Project the population, to the nearest thousand, on January 1, 2020.

[**Hint:** $t = 1$ corresponds to the year 2010.]

102. U.S. Population A reasonable projection for the population of the United States on January 1, 2025, is 336,566,000.

(a) If 2025 is taken as year 1, and the formula for year 2025 onward that represents a new trend in the population growth is $P(t) = 336{,}566{,}000 + 8{,}000{,}000 \log t$, what would be the population in 2045?
(b) Is this higher or lower than the current U.S. Census Bureau's estimate of 363,077,000?

Loudness of Sound *Problems 103–106 use the following discussion: The* **loudness** *$L(x)$, measured in decibels, of a sound of intensity x, measured in watts per square meter, is defined as $L(x) = 10 \log \frac{x}{I_0}$, where $I_0 = 10^{-12}$ watt per square meter is the least intense sound that a human ear can detect.*

Determine the loudness, in decibels, of each of the following sounds.

103. Normal conversation: intensity of $x = 10^{-7}$ watt per square meter.

104. Heavy city traffic: intensity of $x = 10^{-3}$ watt per square meter.

105. Amplified rock music: intensity of 10^{-1} watt per square meter.

106. Diesel truck traveling 40 miles per hour 50 feet away: intensity 10 times that of a passenger car traveling 50 miles per hour 50 feet away whose loudness is 70 decibels.

Problems 107 and 108 use the following discussion: The ***Richter scale*** *is one way of converting seismographic readings into numbers that provide an easy reference for measuring the magnitude M of an earthquake. All earthquakes are compared to a* ***zero-level earthquake*** *whose seismographic reading measures 0.001 millimeter at a distance of 100 kilometers from the epicenter. An earthquake whose seismographic reading measures x millimeters has* ***magnitude*** *M(x) given by*

$$M(x) = \log\left(\frac{x}{x_0}\right)$$

where $x_0 = 10^{-3}$ is the reading of a zero-level earthquake the same distance from its epicenter. Determine the magnitude of the following earthquakes.

107. Magnitude of an Earthquake Mexico City in 1985: seismographic reading of 125,892 millimeters 100 kilometers from the center.

108. Magnitude of an Earthquake San Francisco in 1906: seismographic reading of 7943 millimeters 100 kilometers from the center.

109. Alcohol and Driving The concentration of alcohol in a person's blood is measurable. Suppose that the risk R (given as a percent) of having an accident while driving a car can be modeled by the equation

$$R = 3e^{kx}$$

where x is the variable concentration of alcohol in the blood and k is a constant.

(a) Suppose that a concentration of alcohol in the blood of 0.06 results in a 10% risk ($R = 10$) of an accident. Find the constant k in the equation.
(b) Using this value of k, what is the risk if the concentration is 0.17?
(c) Using the same value of k, what concentration of alcohol corresponds to a risk of 100%?
(d) If the law asserts that anyone with a risk of having an accident of 15% or more should not have driving privileges, at what concentration of alcohol in the blood should a driver be arrested and charged with a DUI?
(e) Compare this situation with that of Example 9. If you were a lawmaker, which situation would you support? Give your reasons.

110. Is there any function of the form $y = x^{\alpha}$, $0 < \alpha < 1$, that increases more slowly than a logarithmic function whose base is greater than 1? Explain.

111. In the definition of the logarithmic function, the base a is not allowed to equal 1. Why?

112. Critical Thinking In buying a new car, one consideration might be how well the price of the car holds up over time. Different makes of cars have different depreciation rates. One way to compute a depreciation rate for a car is given here. Suppose that the current prices of a certain Mercedes automobile are as follows:

Age in Years					
New	1	2	3	4	5
\$38,000	\$36,600	\$32,400	\$28,750	\$25,400	\$21,200

(a) Use the formula New = Old(e^{Rt}) to find R, the annual depreciation rate, for a specific time t.
(b) When might be the best time to trade in the car?
(c) Consult the NADA ("blue") book and compare two like models that you are interested in. Which has the better depreciation rate?

11.5 Properties of Logarithms

OBJECTIVES
1. Work with the properties of logarithms
2. Write a logarithmic expression as a sum or difference of logarithms
3. Write a logarithmic expression as a single logarithm
4. Evaluate logarithms whose base is neither 10 nor e

Work with the properties of logarithms 1

Logarithms have some very useful properties that can be derived directly from the definition and the laws of exponents.

EXAMPLE 1 Establishing Properties of Logarithms

(a) Show that $\log_a 1 = 0$. **(b)** Show that $\log_a a = 1$.

SOLUTION **(a)** This fact was established when we graphed $y = \log_a x$ (see Figure 25). To show the result algebraically, let $y = \log_a 1$. Then

$$\begin{aligned} y &= \log_a 1 & & \\ a^y &= 1 & & \text{Change to an exponent.} \\ a^y &= a^0 & & a^0 = 1 \\ y &= 0 & & \text{Solve for } y. \\ \log_a 1 &= 0 & & y = \log_a 1 \end{aligned}$$

(b) Let $y = \log_a a$. Then

$$\begin{aligned} y &= \log_a a & & \\ a^y &= a & & \text{Change to an exponent.} \\ a^y &= a^1 & & a^1 = a \\ y &= 1 & & \text{Solve for } y. \\ \log_a a &= 1 & & y = \log_a a \end{aligned}$$

To summarize:

$$\boxed{\log_a 1 = 0 \quad \log_a a = 1}$$

Properties of Logarithms

In the properties given next, M and a are positive real numbers, with $a \neq 1$, and r is any real number.

The number $\log_a M$ is the exponent to which a must be raised to obtain M. That is,

$$\boxed{a^{\log_a M} = M} \qquad (1)$$

The logarithm to the base a of a raised to a power equals that power. That is,

$$\boxed{\log_a a^r = r} \qquad (2)$$

The proof uses the fact that $x = a^y$ and $y = \log_a x$ are equivalent.

PROOF: Since $x = a^y$ and $y = \log_a x$ are equivalent, we have

$$a^{\log_a x} = a^y = x$$

Now let $x = M$ to obtain equation (1).

To prove (2), we use the fact that $x = a^y$ and $y = \log_a x$ are equivalent. Then,

$$\log_a a^y = \log_a x = y$$

Now let $y = r$ to obtain equation (2).

EXAMPLE 2 Using Properties (1) and (2)

(a) $2^{\log_2 \pi} = \pi$ **(b)** $\log_{0.2} 0.2^{-\sqrt{2}} = -\sqrt{2}$ **(c)** $\ln e^{kt} = kt$

NOW WORK PROBLEM 3.

Other useful properties of logarithms are given below.

Properties of Logarithms

In the following properties M, N, and a are positive real numbers, with $a \neq 1$, and r is any real number.

The Log of a Product Equals the Sum of the Logs

$$\log_a(MN) = \log_a M + \log_a N \tag{3}$$

The Log of a Quotient Equals the Difference of the Logs

$$\log_a\left(\frac{M}{N}\right) = \log_a M - \log_a N \tag{4}$$

The Log of a Power Equals the Product of the Power and the Log

$$\log_a M^r = r \log_a M \tag{5}$$

We shall derive properties (3) and (5) and will leave the derivation of property (4) as an exercise (see Problem 95).

PROOF OF PROPERTY (3) Let $A = \log_a M$ and let $B = \log_a N$. These expressions are equivalent to the exponential expressions

$$a^A = M \quad \text{and} \quad a^B = N$$

Now

$$\begin{aligned} \log_a(MN) &= \log_a(a^A a^B) = \log_a a^{A+B} && \text{Law of Exponents.} \\ &= A + B && \text{Property (2) of logarithms.} \\ &= \log_a M + \log_a N \end{aligned}$$

PROOF OF PROPERTY (5) Let $A = \log_a M$. This expression is equivalent to

$$a^A = M$$

Now

$$\begin{aligned} \log_a M^r &= \log_a (a^A)^r = \log_a a^{rA} && \text{Law of Exponents.} \\ &= rA && \text{Property (2) of logarithms.} \\ &= r \log_a M \end{aligned}$$

NOW WORK PROBLEM 7.

Logarithms can be used to transform products into sums, quotients into differences, and powers into factors. Such transformations prove useful in certain types of calculus problems.

2 **EXAMPLE 3** **Writing a Logarithmic Expression as a Sum of Logarithms**

Write $\log_a(x\sqrt{x^2 + 1})$, $x > 0$, as a sum of logarithms. Express all powers as factors.

SOLUTION

$$\begin{aligned} \log_a (x\sqrt{x^2 + 1}) &= \log_a x + \log_a \sqrt{x^2 + 1} && \text{Property (3).} \\ &= \log_a x + \log_a (x^2 + 1)^{1/2} \\ &= \log_a x + \frac{1}{2}\log_a (x^2 + 1) && \text{Property (5).} \end{aligned}$$

EXAMPLE 4 **Writing a Logarithmic Expression as a Difference of Logarithms**

Write

$$\ln \frac{x^2}{(x - 1)^3}, \quad x > 1$$

as a difference of logarithms. Express all powers as factors.

SOLUTION

$$\ln \frac{x^2}{(x - 1)^3} = \underset{\uparrow\ \text{Property (4)}}{} \ln x^2 - \ln(x - 1)^3 = \underset{\uparrow\ \text{Property (5)}}{} 2 \ln x - 3 \ln(x - 1)$$

EXAMPLE 5 **Writing a Logarithmic Expression as a Sum and Difference of Logarithms**

Write

$$\log_a \frac{\sqrt{x^2 + 1}}{x^3(x + 1)^4}, \quad x > 0$$

as a sum and difference of logarithms. Express all powers as factors.

SOLUTION

$$\begin{aligned} \log_a \frac{\sqrt{x^2 + 1}}{x^3(x + 1)^4} &= \log_a \sqrt{x^2 + 1} - \log_a [x^3(x + 1)^4] && \text{Property (4).} \\ &= \log_a \sqrt{x^2 + 1} - [\log_a x^3 + \log_a (x + 1)^4] && \text{Property (3).} \\ &= \log_a (x^2 + 1)^{1/2} - \log_a x^3 - \log_a (x + 1)^4 \\ &= \frac{1}{2}\log_a (x^2 + 1) - 3 \log_a x - 4 \log_a (x + 1) && \text{Property (5).} \end{aligned}$$

CAUTION: In using properties (3) through (5), be careful about the values that the variable may assume. For example, the domain of the variable for $\log_a x$ is $x > 0$ and for $\log_a(x - 1)$ it is $x > 1$. If we add these functions, the domain of the sum function is $x > 1$. That is, the equality

$$\log_a x + \log_a(x - 1) = \log_a[x(x - 1)]$$

is true only for $x > 1$.

NOW WORK PROBLEM 39.

Another use of properties (3) through (5) is to write sums and/or differences of logarithms with the same base as a single logarithm.

3 **EXAMPLE 6 Writing Expressions as a Single Logarithm**

Write each of the following as a single logarithm.

(a) $\log_a 7 + 4\log_a 3$ **(b)** $\frac{2}{3}\ln 8 - \ln(3^4 - 8)$

(c) $\log_a x + \log_a 9 + \log_a(x^2 + 1) - \log_a 5$

SOLUTION **(a)**

$$\begin{aligned} \log_a 7 + 4\log_a 3 &= \log_a 7 + \log_a 3^4 && \text{Property (5).} \\ &= \log_a 7 + \log_a 81 \\ &= \log_a(7 \cdot 81) && \text{Property (3).} \\ &= \log_a 567 \end{aligned}$$

(b)

$$\begin{aligned} \frac{2}{3}\ln 8 - \ln(3^4 - 8) &= \ln 8^{2/3} - \ln(81 - 8) && \text{Property (5).} \\ &= \ln 4 - \ln 73 \\ &= \ln\left(\frac{4}{73}\right) && \text{Property (4).} \end{aligned}$$

(c)

$$\begin{aligned} \log_a x + \log_a 9 + \log_a(x^2 + 1) - \log_a 5 &= \log_a(9x) + \log_a(x^2 + 1) - \log_a 5 \\ &= \log_a[9x(x^2 + 1)] - \log_a 5 \\ &= \log_a\left[\frac{9x(x^2 + 1)}{5}\right] \end{aligned}$$

WARNING: A common error made by some students is to express the logarithm of a sum as the sum of logarithms.

$$\log_a(M + N) \quad \text{is not equal to} \quad \log_a M + \log_a N$$

Correct statement $\log_a(MN) = \log_a M + \log_a N$ Property (3).

Another common error is to express the difference of logarithms as the quotient of logarithms.

$$\log_a M - \log_a N \quad \text{is not equal to} \quad \frac{\log_a M}{\log_a N}$$

Correct statement $\log_a M - \log_a N = \log_a\left(\frac{M}{N}\right)$ Property (4).

A third common error is to express a logarithm raised to a power as the product of the power times the logarithm.

	$(\log_a M)^r$	is not equal to	$r \log_a M$
Correct statement	$\log_a M^r = r \log_a M$		Property (5).

NOW WORK PROBLEM 45.

Two other properties of logarithms that we need to know are given next.

Properties of Logarithms

In the following properties, M, N, and a are positive real numbers, with $a \neq 1$.

$$\text{If } M = N, \text{ then } \log_a M = \log_a N. \tag{6}$$

$$\text{If } \log_a M = \log_a N, \text{ then } M = N. \tag{7}$$

When property (6) is used, we start with the equation $M = N$ and say "take the logarithm of both sides" to obtain $\log_a M = \log_a N$.

Using a Calculator to Evaluate Logarithms with Bases Other Than 10 or *e*

Logarithms to the base 10, common logarithms, were used to facilitate arithmetic computations before the widespread use of calculators. Natural logarithms, that is, logarithms whose base is the number e, remain very important because they arise frequently in the study of natural phenomena.

Common logarithms are usually abbreviated by writing **log,** with the base understood to be 10, just as natural logarithms are abbreviated by **ln,** with the base understood to be e.

Most calculators have both [log] and [ln] keys to calculate the common logarithm and natural logarithm of a number. Let's look at an example to see how to approximate logarithms having a base other than 10 or e.

4 **EXAMPLE 7 Approximating Logarithms Whose Base Is Neither 10 Nor *e***

Approximate $\log_2 7$. Round the answer to four decimal places.

SOLUTION Let $y = \log_2 7$. Then

$2^y = 7$	Change to an exponential expression.
$\ln 2^y = \ln 7$	Property (6).
$y \ln 2 = \ln 7$	Property (5).
$y = \dfrac{\ln 7}{\ln 2}$	Exact solution.
$y \approx 2.8074$	Approximate solution rounded to four decimal places.

Example 7 shows how to approximate a logarithm whose base is 2 by changing to logarithms involving the base e. In general, we use the **Change-of-Base Formula.**

Change-of-Base Formula

If $a \neq 1, b \neq 1$, and M are positive real numbers, then

$$\log_a M = \frac{\log_b M}{\log_b a} \tag{8}$$

PROOF We derive this formula as follows: Let $y = \log_a M$. Then

$$a^y = M \quad \text{Change to an exponential expression.}$$
$$\log_b a^y = \log_b M \quad \text{Property (6).}$$
$$y \log_b a = \log_b M \quad \text{Property (5).}$$
$$y = \frac{\log_b M}{\log_b a} \quad \text{Solve for } y.$$
$$\log_a M = \frac{\log_b M}{\log_b a} \quad y = \log_a M$$

Since calculators have keys only for [log] and [ln], in practice the Change-of-Base Formula uses either $b = 10$ or $b = e$. That is,

$$\log_a M = \frac{\log M}{\log a} \quad \text{and} \quad \log_a M = \frac{\ln M}{\ln a} \tag{9}$$

EXAMPLE 8 Using the Change-of-Base Formula

Approximate: **(a)** $\log_5 89$ **(b)** $\log_{\sqrt{2}} \sqrt{5}$

Round answers to four decimal places.

SOLUTION **(a)** $\log_5 89 = \dfrac{\log 89}{\log 5} \approx \dfrac{1.949390007}{0.6989700043} \approx 2.7889$

or

$$\log_5 89 = \frac{\ln 89}{\ln 5} \approx \frac{4.48863637}{1.609437912} \approx 2.7889$$

(b) $\log_{\sqrt{2}} \sqrt{5} = \dfrac{\log \sqrt{5}}{\log \sqrt{2}} \approx 2.3219$

or

$$\log_{\sqrt{2}} \sqrt{5} = \frac{\ln \sqrt{5}}{\ln \sqrt{2}} \approx 2.3219$$

COMMENT: To graph logarithmic functions when the base is different from e or 10 requires the Change-of-Base Formula. For example, to graph $y = \log_2 x$, we would instead graph $y = \dfrac{\ln x}{\ln 2}$. Try it.

NOW WORK PROBLEMS 11 AND 59.

SUMMARY **Properties of Logarithms**

In the summary that follows, $a > 0, a \neq 1$, and $b > 0, b \neq 1$; also, $M > 0$ and $N > 0$.

Definition	$y = \log_a x$ means $x = a^y$
Properties of logarithms	$\log_a 1 = 0 \quad \log_a a = 1 \quad a^{\log_a M} = M \quad \log_a a^r = r$
	$\log_a(MN) = \log_a M + \log_a N$
	$\log_a\left(\dfrac{M}{N}\right) = \log_a M - \log_a N$
	$\log_a M^r = r \log_a M$
	If $M = N$, then $\log_a M = \log_a N$.
	If $\log_a M = \log_a N$, then $M = N$.
Change-of-Base Formula	$\log_a M = \dfrac{\log_b M}{\log_b a}$

EXERCISE 11.5 Answers to Odd-Numbered Problems Begin on Page AN-65.

In Problems 1–16, use properties of logarithms to find the exact value of each expression. Do not use a calculator.

1. $\log_3 3^{71}$ **2.** $\log_2 2^{-13}$ **3.** $\ln e^{-4}$ **4.** $\ln e^{\sqrt{2}}$

5. $2^{\log_2 7}$ **6.** $e^{\ln 8}$ **7.** $\log_8 2 + \log_8 4$ **8.** $\log_6 9 + \log_6 4$

9. $\log_6 18 - \log_6 3$ **10.** $\log_8 16 - \log_8 2$ **11.** $\log_2 6 \cdot \log_6 4$ **12.** $\log_3 8 \cdot \log_8 9$

13. $3^{\log_3 5 - \log_3 4}$ **14.** $5^{\log_5 6 + \log_5 7}$ **15.** $e^{\log_{e^2} 16}$ **16.** $e^{\log_{e^2} 9}$

In Problems 17–24, suppose that $\ln 2 = a$ *and* $\ln 3 = b$. *Use properties of logarithms to write each logarithm in terms of a and b.*

17. $\ln 6$ **18.** $\ln \dfrac{2}{3}$ **19.** $\ln 1.5$ **20.** $\ln 0.5$

21. $\ln 8$ **22.** $\ln 27$ **23.** $\ln \sqrt[5]{6}$ **24.** $\ln \sqrt[4]{\dfrac{2}{3}}$

In Problems 25–44, write each expression as a sum and/or difference of logarithms. Express powers as factors.

25. $\log_5(25x)$ **26.** $\log_3 \dfrac{x}{9}$ **27.** $\log_2 z^3$ **28.** $\log_7(x^5)$

29. $\ln(ex)$ **30.** $\ln \dfrac{e}{x}$ **31.** $\ln(xe^x)$ **32.** $\ln \dfrac{x}{e^x}$

33. $\log_a(u^2v^3)$, $u > 0, v > 0$

34. $\log_2\left(\frac{a}{b^2}\right)$, $a > 0, b > 0$

35. $\ln(x^2\sqrt{1-x})$, $0 < x < 1$

36. $\ln(x\sqrt{1+x^2})$, $x > 0$

37. $\log_2\left(\frac{x^3}{x-3}\right)$, $x > 3$

38. $\log_5\left(\frac{\sqrt[3]{x^2+1}}{x^2-1}\right)$, $x > 1$

39. $\log\left[\frac{x(x+2)}{(x+3)^2}\right]$, $x > 0$

40. $\log\left[\frac{x^3\sqrt{x+1}}{(x-2)^2}\right]$, $x > 2$

41. $\ln\left[\frac{x^2-x-2}{(x+4)^2}\right]^{1/3}$, $x > 2$

42. $\ln\left[\frac{(x-4)^2}{x^2-1}\right]^{2/3}$, $x > 4$

43. $\ln\frac{5x\sqrt{1+3x}}{(x-4)^3}$, $x > 4$

44. $\ln\left[\frac{5x^2\sqrt[3]{1-x}}{4(x+1)^2}\right]$, $0 < x < 1$

In Problems 45–58, write each expression as a single logarithm.

45. $3\log_5 u + 4\log_5 v$

46. $2\log_3 u - \log_3 v$

47. $\log_3\sqrt{x} - \log_3 x^3$

48. $\log_2\left(\frac{1}{x}\right) + \log_2\left(\frac{1}{x^2}\right)$

49. $\log_4(x^2-1) - 5\log_4(x+1)$

50. $\log(x^2+3x+2) - 2\log(x+1)$

51. $\ln\left(\frac{x}{x-1}\right) + \ln\left(\frac{x+1}{x}\right) - \ln(x^2-1)$

52. $\log\left(\frac{x^2+2x-3}{x^2-4}\right) - \log\left(\frac{x^2+7x+6}{x+2}\right)$

53. $8\log_2\sqrt{3x-2} - \log_2\left(\frac{4}{x}\right) + \log_2 4$

54. $21\log_3\sqrt[3]{x} + \log_3(9x^2) - \log_3 9$

55. $2\log_a(5x^3) - \frac{1}{2}\log_a(2x+3)$

56. $\frac{1}{3}\log(x^3+1) + \frac{1}{2}\log(x^2+1)$

57. $2\log_2(x+1) - \log_2(x+3) - \log_2(x-1)$

58. $3\log_5(3x+1) - 2\log_5(2x-1) - \log_5 x$

In Problems 59–66, use the Change-of-Base Formula and a calculator to evaluate each logarithm. Round your answer to three decimal places.

59. $\log_3 21$

60. $\log_5 18$

61. $\log_{1/3} 71$

62. $\log_{1/2} 15$

63. $\log_{\sqrt{2}} 7$

64. $\log_{\sqrt{5}} 8$

65. $\log_\pi e$

66. $\log_\pi \sqrt{2}$

In Problems 67–72, graph each function using a graphing utility and the Change-of-Base Formula.

67. $y = \log_4 x$

68. $y = \log_5 x$

69. $y = \log_2(x+2)$

70. $y = \log_4(x-3)$

71. $y = \log_{x-1}(x+1)$

72. $y = \log_{x+2}(x-2)$

In Problems 73–82, express y as a function of x. The constant C is a positive number.

73. $\ln y = \ln x + \ln C$

74. $\ln y = \ln(x + C)$

75. $\ln y = \ln x + \ln(x+1) + \ln C$

76. $\ln y = 2\ln x - \ln(x+1) + \ln C$

77. $\ln y = 3x + \ln C$

78. $\ln y = -2x + \ln C$

79. $\ln(y-3) = -4x + \ln C$

80. $\ln(y+4) = 5x + \ln C$

81. $3\ln y = \frac{1}{2}\ln(2x+1) - \frac{1}{3}\ln(x+4) + \ln C$

82. $2\ln y = -\frac{1}{2}\ln x + \frac{1}{3}\ln(x^2+1) + \ln C$

83. Find the value of $\log_2 3 \cdot \log_3 4 \cdot \log_4 5 \cdot \log_5 6 \cdot \log_6 7 \cdot \log_7 8$.

84. Find the value of $\log_2 4 \cdot \log_4 6 \cdot \log_6 8$.

85. Find the value of $\log_2 3 \cdot \log_3 4 \cdot \cdots \cdot \log_n(n+1) \cdot \log_{n+1} 2$.

86. Find the value of $\log_2 2 \cdot \log_2 4 \cdot \cdots \cdot \log_2 2^n$.

87. Show that $\log_a(x + \sqrt{x^2 - 1}) + \log_a(x - \sqrt{x^2 - 1}) = 0$.

88. Show that $\log_a(\sqrt{x} + \sqrt{x - 1}) + \log_a(\sqrt{x} - \sqrt{x - 1}) = 0$.

89. Show that $\ln(1 + e^{2x}) = 2x + \ln(1 + e^{-2x})$.

90. Difference Quotient If $f(x) = \log_a x$, show that

$$\frac{f(x+h) - f(x)}{h} = \log_a\left(1 + \frac{h}{x}\right)^{1/h}, \quad h \neq 0.$$

91. If $f(x) = \log_a x$, show that $-f(x) = \log_{1/a} x$.

92. If $f(x) = \log_a x$, show that $f(AB) = f(A) + f(B)$.

93. If $f(x) = \log_a x$, show that $f\left(\frac{1}{x}\right) = -f(x)$.

94. If $f(x) = \log_a x$, show that $f(x^\alpha) = \alpha f(x)$.

95. Show that $\log_a\left(\frac{M}{N}\right) = \log_a M - \log_a N$, where a, M, and N are positive real numbers, with $a \neq 1$.

96. Show that $\log_a\left(\frac{1}{N}\right) = -\log_a N$, where a and N are positive real numbers, with $a \neq 1$.

97. Graph $Y_1 = \log(x^2)$ and $Y_2 = 2\log(x)$ on your graphing utility. Are they equivalent? What might account for any differences in the two functions?

11.6 Continuously Compounded Interest

OBJECTIVES
1. Use the compound interest formula
2. Find the present value of a dollar amount
3. Find the time required to double an investment
4. Find the rate of interest needed to double an investment

Suppose a principal P is to be invested at an annual rate of interest r, which is compounded n times per year. The interest earned on a principal P at each compounding period is then $P \cdot \frac{r}{n}$. The amount A after 1 year with

1 compoundings per year (annually)

$$A = P + P \cdot \left(\frac{r}{1}\right) = P \cdot \left(1 + \frac{r}{1}\right)$$

2 compoundings per year (semiannually)

$$A = P \cdot \left(1 + \frac{r}{2}\right) + P \cdot \left(1 + \frac{r}{2}\right)\left(\frac{r}{2}\right)$$

$$= P \cdot \left(1 + \frac{r}{2}\right)\left(1 + \frac{r}{2}\right) = P \cdot \left(1 + \frac{r}{2}\right)^2$$

4 compoundings per year (quarterly)

$$A = P \cdot \left(1 + \frac{r}{4}\right)^3 + P \cdot \left(1 + \frac{r}{4}\right)^3\left(\frac{r}{4}\right)$$

$$= P \cdot \left(1 + \frac{r}{4}\right)^3\left(1 + \frac{r}{4}\right) = P \cdot \left(1 + \frac{r}{4}\right)^4$$

⋮

n compoundings per year per year

$$A = P \cdot \left(1 + \frac{r}{n}\right)^n$$

What happens to the amount A after 1 year if the number of times, n, that the interest is compounded per year gets larger and larger? The answer turns out to involve the number e.

Rewrite the expression for A as follows:

$$A = P\cdot\left(1 + \frac{r}{n}\right)^n = P\cdot\left[\left(1 + \frac{r}{n}\right)^{n/r}\right]^r$$

To simplify the calculation, let

$$k = \frac{n}{r} \quad \text{so} \quad \frac{1}{k} = \frac{r}{n}$$

We substitute to get

$$A = P\cdot\left[\left(1 + \frac{r}{n}\right)^{n/r}\right]^r = P\cdot\left[\left(1 + \frac{1}{k}\right)^k\right]^r$$

As n gets larger and larger, so does k and, since $\left(1 + \frac{1}{k}\right)^k$ approaches e as k becomes unbounded in the positive direction, it follows that

$$P\cdot\left[\left(1 + \frac{1}{k}\right)^k\right]^r \to P\cdot[(e)]^r = Pe^r$$

That is, no matter how often the interest is compounded during the year, the amount after 1 year has the definite ceiling Pe^r. When interest is compounded so that the amount after 1 year is Pe^r, we say that the interest is **compounded continuously.**

For example, the amount A due to investing \$1000 for 1 year at an annual rate of 10% compounded continuously is

$$A = 1000e^{0.1} = \$1105.17$$

The formula $A = Pe^r$ gives the amount A after 1 year resulting from investing a principal P at the annual rate of interest r compounded continuously.

Compound Interest Formula–Continuous Compounding

The amount A due to investing a principal P for a period of t years at the annual rate of interest r compounded continuously is

$$\boxed{A = Pe^{rt}} \qquad (1)$$

1 **EXAMPLE 1 Using the Compound Interest Formula**

If \$1000 is invested at 10% compounded continuously, how much is in the account

(a) after 3 years **(b)** after 5 years

SOLUTION **(a)** If \$1000 is invested at 10% compounded continuously, the amount A after 3 years is

$$A = Pe^{rt} = 1000e^{(0.1)(3)} = 1000e^{0.3} = \$1349.86$$

(b) After 5 years the amount A is

$$A = 1000e^{(0.1)(5)} = \$1648.72$$

NOW WORK PROBLEM 1.

The Compound Interest Formula states that a principal P earning an annual rate of interest r compounded continuously will, after t years, be worth the amount A, where

$$A = Pe^{rt}$$

If we solve for P, we obtain

$$P = \frac{A}{e^{rt}} = Ae^{-rt} \qquad (2)$$

In this formula P is called the **present value** of the amount A. In other words, P is the amount that must be invested now in order to accumulate the amount A in t years.

2 **EXAMPLE 2 Computing the Present Value of $10,000**

How much money should be invested now at 8% per annum compounded continuously, so that after 2 years the amount will be $10,000?

SOLUTION In this problem we want to find the principal P needed now to get the amount $A =$ \$10,000 after $t = 2$ years. That is, we want to find the present value of $10,000. We use formula (2) with $r = .08$:

$$P = Ae^{-rt} = \$10{,}000e^{-.08(2)} = \$8521.44$$

If you invest $8521.44 now at 8% per annum compounded continuously, you will have $10,000 after 2 years.

NOW WORK PROBLEM 5.

3 **EXAMPLE 3 Finding the Time Required to Double an Investment**

Find the time required to double an investment if the rate of interest is 5% compounded continuously.

SOLUTION If P is the principal invested, it will double when the amount $A = 2P$. We use the Compound Interest Formula (1).

$$\begin{aligned} A &= Pe^{rt} && \text{Formula (1).} \\ 2P &= Pe^{0.05t} && A = 2P;\ r = 0.05 \\ 2 &= e^{0.05t} && \text{Cancel the } P\text{'s.} \\ 0.05t &= \ln 2 && \text{Change to a logarithmic expression.} \\ t &= 13.863 && \text{Solve for } t. \end{aligned}$$

It will take almost 14 years to double an investment at 5% compounded continuously.

NOW WORK PROBLEM 15.

EXAMPLE 4 **Finding the Rate of Interest to Double an Investment**

Find the rate of interest required to double an investment in 8 years if the interest is compounded continuously.

SOLUTION If P is the principle invested, it will double when the amount $A = 2P$. We use the Compound Interest Formula (1).

$$A = Pe^{rt} \quad \text{Formula (1).}$$
$$2P = Pe^{8r} \quad A = 2P;\ t = 8.$$
$$2 = e^{8r} \quad \text{Cancel the } P\text{'s.}$$
$$8r = \ln 2 \quad \text{Change to a logarithmic expression.}$$
$$r = 0.0866 \quad \text{Solve for } r.$$

It will require an interest rate of about 8.7% compounded continuously to double an investment in 5 years.

NOW WORK PROBLEM 13.

EXERCISE 11.6 **Answers to Odd-Numbered Problems Begin on Page AN-66.**

In Problems 1–4, find the amount if:

1. \$1000 is invested at 4% compounded continuously for 3 years.
2. \$100 is invested at 6% compounded continuously for $1\frac{1}{2}$ years.
3. \$500 is invested at 5% compounded continuously for 3 years.
4. \$200 is invested at 10% compounded continuously for 10 years.

In Problems 5–8, find the principal needed now to get each amount.

5. To get \$100 in 6 months at 4% compounded continuously.
6. To get \$500 in 1 year at 6% compounded continuously.
7. To get \$500 in 1 year at 7% compounded continuously.
8. To get \$800 in 2 years at 5% compounded continuously.
9. If \$1000 is invested at 2% compounded continuously, what is the amount after 1 year? How much interest is earned?
10. If \$2000 is invested at 5% compounded continuously, what is the amount after 5 years? How much interest is earned?
11. If a bank pays 3% compounded continuously, how much should be deposited now to have \$5000
    (a) 4 years later? (b) 8 years later?
12. If a bank pays 2% compounded continuously, how much should be deposited now to have \$10,000
    (a) 5 years later? (b) 10 years later?
13. What annual rate of interest compounded continuously is required to double an investment in 3 years?
14. What annual rate of interest compounded continuously is required to double an investment in 10 years?
15. Approximately how long will it take to triple an investment at 10% compounded continuously?
16. Approximately how long will it take to triple an investment at 9% compounded continuously?
17. What principal is needed now to get \$1000 in 1 year at 9% compounded continuously? How much should be invested to get \$1000 in 2 years?
18. **Buying a Car** Laura wishes to have \$8000 available to buy a car in 3 years. How much should she invest in a savings account now so that she will have enough if the bank pays 8% interest compounded continuously?

19. Down Payment on a House Tami and Todd will need \$40,000 for a down payment on a house in 4 years. How much should they invest in a savings account now so that they will be able to do this? The bank pays 3% compounded continuously.

20. Saving for College A newborn child receives a \$3000 gift toward a college education. How much will the \$3000 be worth in 17 years if it is invested at 10% compounded continuously?

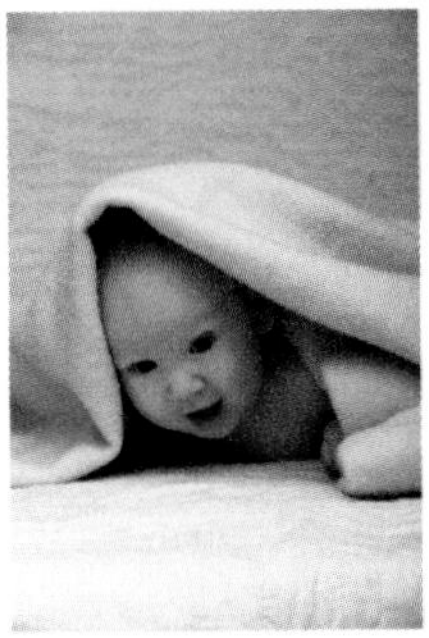

21. What annual rate of interest compounded continuously is required to triple an investment in 5 years?

22. What annual rate of interest compounded continuously is required to triple an investment in 7 years?

Problems 23 and 24 require the following discussion: In asking for the time required to double an investment, we use formula (1) with $A = 2P$.

$A = Pe^{rt}$	Formula (1).
$2P = Pe^{rt}$	$A = 2P$
$2 = e^{rt}$	Cancel the P's.
$\ln 2 = rt$	Change to a logarithm.
$t = \dfrac{\ln 2}{r} = \dfrac{0.6931471806}{r}$	Solve for t.

23. The Rule of 70 This rule uses the approximation $\ln 2 = 0.70$ to find t. Compare the approximation given by the rule of 70 to the actual solution if

(a) $r = 1\%$ (b) $r = 5\%$ (c) $r = 10\%$

24. The Rule of 72 This rule uses the approximation $\ln 2 = 0.72$. Compare the approximation given by the rule of 72 to the actual solution if

(a) $r = 1\%$ (b) $r = 8\%$ (c) $r = 12\%$

Chapter 11 Review

OBJECTIVES

Section	You should be able to	Review Exercises
11.1	1 Locate the vertex and axis of symmetry of a quadratic function	1–10
	2 Graph quadratic functions	1–10
	3 Find the maximum or the minimum value of a quadratic function	11–16
	4 Use the maximum or the minimum value of a quadratic function to solve applied problems	91–93, 99, 100
11.2	1 Know the properties of power functions	17, 18
	2 Graph functions using shifts and/or reflections	19–22
	3 Identify polynomial functions and their degree	23–26
	4 Find the end behavior of a polynomial function	27, 28
	5 Find the domain of a rational function	29–32

11.3	1	Evaluate exponential functions	33 (a), (c), 34 (a), (c), 94(a)
	2	Graph exponential functions	67–70
	3	Define the number *e*	44, 47, 69, 70
	4	Solve exponential equations	73–76, 79, 80, 82
11.4	1	Change exponential expressions to logarithmic expressions	35, 36
	2	Change logarithmic expressions to exponential expressions	37, 38
	3	Evaluate logarithmic functions	33 (b), (d), 34 (b), (d), 43–48, 89, 90, 95 (a), (b), 96
	4	Find the domain of a logarithmic function	39–42
	5	Graph logarithmic functions	71, 72
	6	Solve logarithmic equations	77, 78, 81
11.5	1	Work with the properties of logarithms	43–52
	2	Write a logarithmic expression as a sum or difference of logarithms	53-58
	3	Write a logarithmic expression as a single logarithm	59–64
	4	Evaluate logarithms whose base is neither 10 nor *e*	65–66
11.6	1	Use the Compound Interest Formula	83, 84, 97
	2	Find the present value of a dollar amount	85, 86
	3	Find the time required to double an investment	87
	4	Find the rate of interest needed to double an investment	88

THINGS TO KNOW

Quadratic function (pp. 607 and 608)

$f(x) = ax^2 + bx + c, \quad a \neq 0$

Graph is a parabola that opens up if $a > 0$ and opens down if $a < 0$.

Vertex: $\left(-\dfrac{b}{2a}, f\left(-\dfrac{b}{2a}\right)\right)$

Axis of symmetry: $x = -\dfrac{b}{2a}$

y-intercept: Found by evaluating $f(0) = c$.

x-intercept (s): Found by finding the real solutions, if any, of the equation $ax^2 + bx + c = 0$.

Polynomial function (p. 621)

$f(x) = a_n x^n + a_{n-1}x^{n-1} + \cdots + a_1 x + a_0, \quad a_n \neq 0, n \geq 0$ an integer

Domain: all real numbers

Rational function (p. 624)

$R(x) = \dfrac{p(x)}{q(x)}$

p, q are polynomial functions. Domain: $\{x \mid q(x) \neq 0\}$

Exponential Function (pp. 631 and 633)

$f(x) = a^x, \quad a > 1$

Domain: the interval $(-\infty, \infty)$; Range: the interval $(0, \infty)$; x-intercept: none; y-intercept: $(0, 1)$ horizontal asymptote: the line $y = 0$ (the x-axis), as x becomes unbounded in the negative direction; increasing; smooth; continuous

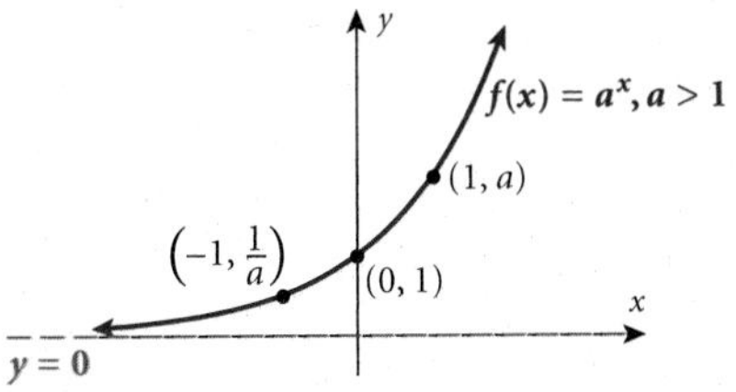

$f(x) = a^x, 0 < a < 1$

Domain: the interval $(-\infty, \infty)$; Range: the interval $(0, \infty)$; x-intercept: none; y-intercept: $(0, 1)$ horizontal asymptote: the line $y = 0$ (the x-axis), as x becomes unbounded in the positive direction; decreasing; smooth; continuous

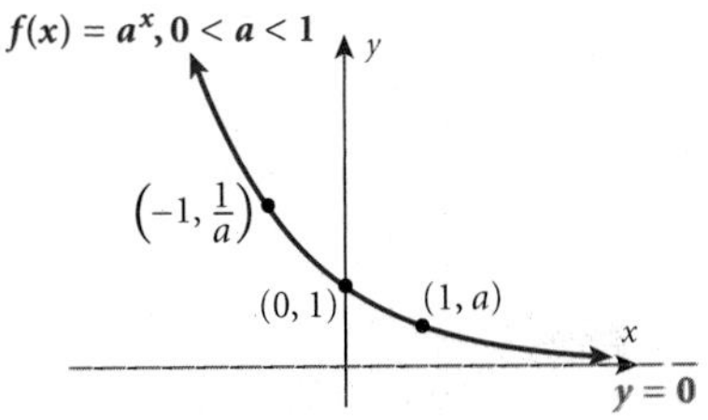

Number *e* (p. 634)

Value approached by the expression $\left(1 + \frac{1}{n}\right)^n$ as n becomes unbounded in the positive direction

That is, $\lim_{n \to \infty} \left(1 + \frac{1}{n}\right)^n = e.$

Property of exponents (p. 635)

If $a^u = a^v$, then $u = v$.

Logarithmic Functions (p. 646)

$f(x) = \log_a x, \quad a > 1$
($y = \log_a x$ means $x = a^y$)

Domain: the interval $(0, \infty)$; Range: the interval $(-\infty, \infty)$; x-intercept: $(1, 0)$; y-intercept: none; vertical asymptote: the line $x = 0$ (the y-axis); increasing; smooth; continuous

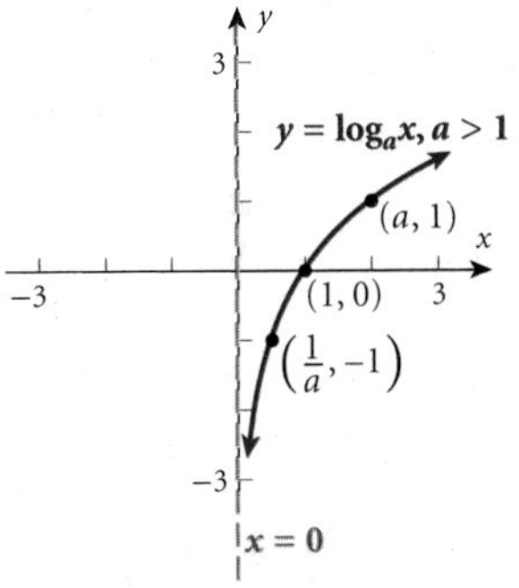

$f(x) = \log_a x, \quad 0 < a < 1$
($y = \log_a x$ means $x = a^y$)

Domain: the interval $(0, \infty)$; Range: the interval $(-\infty, \infty)$; x-intercept: $(1, 0)$; y-intercept: none; vertical asymptote: the line $x = 0$ (the y-axis); decreasing; smooth; continuous

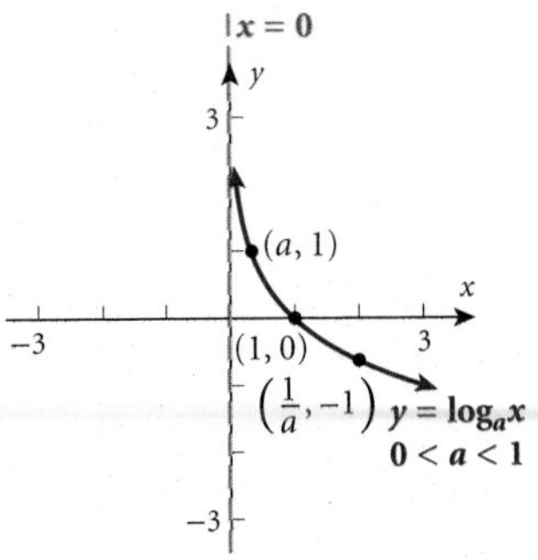

Natural Logarithm (p. 646)

$y = \ln x$ means $x = e^y$.

Properties of logarithms (pp. 655, 656, and 659)	$\log_a 1 = 0 \quad \log_a a = 1 \quad a^{\log_a M} = M \quad \log_a a^r = r$ $\log_a(MN) = \log_a M + \log_a N \quad \log_a\left(\frac{M}{N}\right) = \log_a M - \log_a N$ $\log_a M^r = r \log_a M$ If $M = N$, then $\log_a M = \log_a N$. If $\log_a M = \log_a N$, then $M = N$.
Change-of-Base Formula (p. 660)	$\log_a M = \dfrac{\log_b M}{\log_b a}$
Compound Interest Formula (continuously compounded interest) (p. 664)	$A = Pe^{rt}$ A = amount; P = principal; r = rate of interest t = time (in years)

TRUE–FALSE ITEMS **Answers are on page AN-66.**

T F **1.** The graph of $f(x) = 2x^2 + 3x - 4$ opens up.

T F **2.** The x-coordinate of the vertex of $f(x) = -x^2 + 4x + 5$ is $f(2)$.

T F **3.** If the discriminant $b^2 - 4ac = 0$, the graph of $f(x) = ax^2 + bx + c$, $a \neq 0$, will touch the x-axis at its vertex.

T F **4.** The graph of an exponential function $f(x) = a^x$, $0 < a < 1$, is decreasing.

T F **5.** The graphs of $y = 3^x$ and $y = \left(\frac{1}{3}\right)^x$ are identical.

T F **6.** The range of the exponential function $f(x) = a^x$, $a > 0$, $a \neq 1$, is the set of all real numbers.

T F **7.** If $y = \log_a x$, then $y = a^x$.

T F **8.** The graph of every logarithmic function $f(x) = \log_a x$, $a > 0$, $a \neq 1$, will contain the points $(1, 0)$, $(a, 1)$ and $\left(\frac{1}{a}, -1\right)$.

T F **9.** $\log_a (MN) = \log_a (M + N)$, $M > 0$, $N > 0$, $a > 0$

FILL IN THE BLANKS **Answers are on page AN-66.**

1. The graph of a quadratic function is called a __________.

2. The vertical line passing through the vertex of a parabola is called the __________.

3. The x-coordinate of the vertex of $f(x) = ax^2 + bx + c$, $a \neq 0$, is __________.

4. The graph of every exponential function $f(x) = a^x$, $a > 0$, $a \neq 1$, passes through three points: __________, __________, and __________.

5. If the graph of the exponential function $f(x) = a^x$, $a > 0$, $a \neq 1$, is decreasing, then a must be less than __________.

6. If $3^x = 3^4$, then $x =$ __________.

7. The domain of the logarithmic function $f(x) = \log_a x$ is __________.

8. The graph of every logarithmic function $f(x) = \log_a x$, $a > 0$, $a \neq 1$, passes through three points: __________, __________, and __________.

9. If the graph of a logarithmic function $f(x) = \log_a x$, $a > 0$, $a \neq 1$ is increasing, then its base must be larger than __________.

10. $\log_a M^r =$ __________, $a > 0$, $M > 0$, $r > 0$.

REVIEW EXERCISES **Answers to odd-numbered problems begin on page AN-66. Blue problem numbers indicate the author's suggestions for a practice test.**

In Problems 1–10, graph each quadratic function by determining whether its graph opens up or down and by finding its vertex, axis of symmetry, y-intercept, and x-intercepts, if any.

1. $f(x) = (x - 2)^2 + 2$ **2.** $f(x) = (x + 1)^2 - 4$ **3.** $f(x) = \frac{1}{4}x^2 - 16$ **4.** $f(x) = -\frac{1}{2}x^2 + 2$

5. $f(x) = -4x^2 + 4x$

6. $f(x) = 9x^2 - 6x + 3$

7. $f(x) = \frac{9}{2}x^2 + 3x + 1$

8. $f(x) = -x^2 + x + \frac{1}{2}$

9. $f(x) = 3x^2 + 4x - 1$

10. $f(x) = -2x^2 - x + 4$

In Problems 11–16, determine whether the given quadratic function has maximum value or a minimum value, and then find the value.

11. $f(x) = 3x^2 - 6x + 4$

12. $f(x) = 2x^2 + 8x + 5$

13. $f(x) = -x^2 + 8x - 4$

14. $f(x) = -x^2 - 10x - 3$

15. $f(x) = -3x^2 + 12x + 4$

16. $f(x) = -2x^2 + 4$

17. Name three points the graph of the power function $f(x) = x^6$ contains.

18. The graph of the power function $f(x) = x^5$ is symmetric with respect to the ___________.

In Problems 19–22, graph each function using shifts and/or reflections. Be sure to label at least three points.

19. $f(x) = x^4 + 2$

20. $f(x) = (x - 3)^3$

21. $f(x) = -x^5 + 1$

22. $f(x) = -x^4 - 2$

In Problems 23–26, determine which functions are polynomial functions. For those that are, state the degree. For those that are not, tell why not.

23. $f(x) = 4x^5 - 3x^2 + 5x - 2$

24. $f(x) = \frac{3x^5}{2x + 1}$

25. $f(x) = 3x^2 + 5x^{1/2} - 1$

26. $f(x) = 3$

In Problems 27–28, find the power function that the graph of f resembles for large values of x. That is, find the end behavior of each polynomial function.

27. $f(x) = -2x^4 + 3x^3 - 6x + 4$

28. $f(x) = 5x^6 - 7x^4 + 8x - 10$

In Problems 29–32, find the domain of each rational function.

29. $R(x) = \frac{x + 2}{x^2 - 9}$

30. $R(x) = \frac{x^2 + 4}{x - 2}$

31. $R(x) = \frac{x^2 + 3x + 2}{(x + 2)^2}$

32. $R(x) = \frac{x^3}{x^3 - 1}$

In Problems 33 and 34, suppose that $f(x) = 3^x$ and $g(x) = \log_3 x$.

33. Evaluate the following: (a) $f(4)$ (b) $g(9)$ (c) $f(-2)$ (d) $g\left(\frac{1}{27}\right)$

34. Evaluate the following: (a) $f(1)$ (b) $g(81)$ (c) $f(-4)$ (d) $g\left(\frac{1}{243}\right)$

In Problems 35 and 36, convert each exponential expression to an equivalent expression involving a logarithm. In Problems 37 and 38, convert each logarithmic expression to an equivalent expression involving an exponent.

35. $5^2 = z$

36. $a^5 = m$

37. $\log_5 u = 13$

38. $\log_a 4 = 3$

In Problems 39–42, find the domain of each logarithmic function.

39. $f(x) = \log(3x - 2)$

40. $F(x) = \log_5(2x + 1)$

41. $H(x) = \log_2(-3x + 2)$

42. $F(x) = \ln(-2x - 9)$

In Problems 43–52, evaluate each expression. Do not use a calculator

43. $\log_2\left(\frac{1}{8}\right)$ **44.** $\ln e$ **45.** $\log_3 81$ **46.** $\log 10$ **47.** $\ln e^2$

48. $\log_3\left(\frac{1}{3}\right)$ **49.** $\ln e^{\sqrt{2}}$ **50.** $e^{\ln 0.1}$ **51.** $2^{\log_2 0.4}$ **52.** $\log_2 2^{\sqrt{3}}$

In Problems 53–58, write each expression as the sum and/or difference of logarithms. Express powers as factors.

53. $\log_3\left(\frac{uv^2}{w}\right)$, $u > 0, v > 0, w > 0$ **54.** $\log_2(a^2\sqrt{b})^4$, $a > 0, b > 0$ **55.** $\log(x^2\sqrt{x^3+1})$, $x > 0$

56. $\log_5\left(\frac{x^2+2x+1}{x^2}\right)$, $x > 0$ **57.** $\ln\left(\frac{x\sqrt[3]{x^2+1}}{x-3}\right)$, $x > 3$ **58.** $\ln\left(\frac{2x+3}{x^2-3x+2}\right)^2$, $x > 2$

In Problems 59–64, write each expression as a single logarithm.

59. $3\log_4 x^2 + \frac{1}{2}\log_4\sqrt{x}$

60. $-2\log_3\left(\frac{1}{3}\right) + \frac{1}{3}\log_3\sqrt{x}$

61. $\ln\left(\frac{x-1}{x}\right) + \ln\left(\frac{x}{x+1}\right) - \ln(x^2-1)$

62. $\log(x^2-9) - \log(x^2+7x+12)$

63. $2\log 2 + 3\log x - \frac{1}{2}[\log(x+3) + \log(x-2)]$

64. $\frac{1}{2}\ln(x^2+1) - 4\ln\frac{1}{2} - \frac{1}{2}[\ln(x-4) + \ln x]$

In Problems 65 and 66, use the Change-of-Base Formula and a calculator to evaluate each logarithm. Round your answer to three decimal places.

65. $\log_4 19$ **66.** $\log_2 21$

In Problems 67–72, use shifts and/or reflections to graph each function. Determine the domain, range, and any asymptotes.

67. $f(x) = 2^{x-3}$ **68.** $f(x) = -2^x + 3$ **69.** $f(x) = 1 - e^x$

70. $f(x) = 3 - e^{-x}$ **71.** $f(x) = 3 + \ln x$ **72.** $f(x) = 4 - \ln(-x)$

In Problems 73–82, solve each equation.

73. $4^{1-2x} = 2$ **74.** $8^{6+3x} = 4$ **75.** $3^{x^2+x} = \sqrt{3}$ **76.** $4^{x-x^2} = \frac{1}{2}$ **77.** $\log_x 64 = -3$

78. $\log_{\sqrt{2}} x = -6$ **79.** $9^{2x} = 27^{3x-4}$ **80.** $25^{2x} = 5^{x^2-12}$ **81.** $\log_3(x-2) = 2$ **82.** $2^{x+1} \cdot 8^{-x} = 4$

83. Find the amount of an investment of \$100 after 2 years and 3 months at 10% compounded continuously.

84. Mike places \$200 in a savings account that pays 4% per annum compounded continuously. How much is in his account after 9 months?

85. A bank pays 4% per annum compounded continuously. How much should I invest now so that 2 years from now I will have \$1000 in the account?

86. Saving for a Bicycle Katy wants to buy a bicycle that costs \$75 and will purchase it in 6 months. How much should she put in her savings account now if she can get 10% per annum compounded continuously?

87. **Doubling Money** Marcia has \$220,000 saved for her retirement. How long will it take for the investment to double in value if it earns 6% compounded continuously?

88. **Doubling Money** What annual rate of interest is required to double an investment in 4 years?

In Problems 89 and 90, use the following result: If x is the atmospheric pressure (measured in millimeters of mercury), then the formula for the altitude h(x) (measured in meters above sea level) is

$$h(x) = (30T + 8000)\log\left(\frac{P_0}{x}\right)$$

where T is the temperature (in degrees Celsius) and P_0 is the atmospheric pressure at sea level, which is approximately 760 millimeters of mercury.

89. **Finding the Altitude of an Airplane** At what height is a Piper Cub whose instruments record an outside temperature of 0°C and a barometric pressure of 300 millimeters of mercury?

90. **Finding the Height of a Mountain** How high is a mountain if instruments placed on its peak record a temperature of 5°C and a barometric pressure of 500 millimeters of mercury?

91. **Landscaping** A landscape engineer has 200 feet of border to enclose a rectangular pond. What dimensions will result in the largest pond?

92. **Enclosing the Most Area with a Fence** A farmer with 10,000 meters of fencing wants to enclose a rectangular field and then divide it into two plots with a fence parallel to one of the sides (see the figure). What is the largest area that can be enclosed?

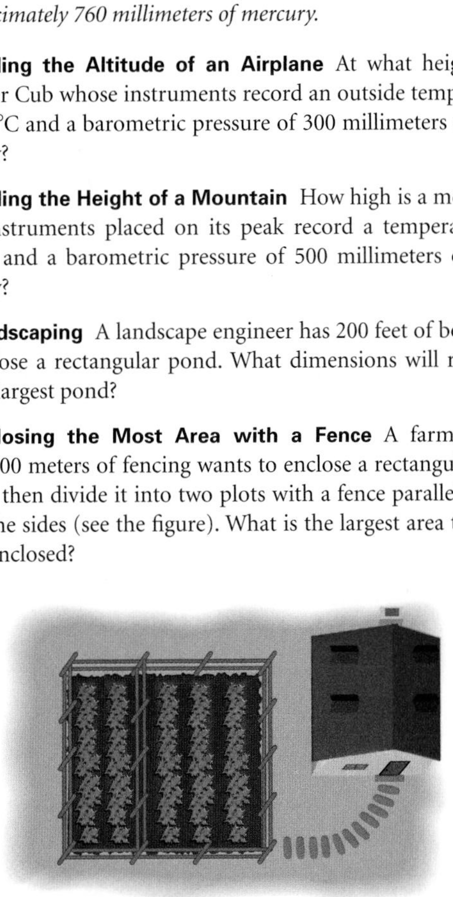

93. **Architecture** A special window in the shape of a rectangle with semicircles at each end is to be constructed so that the outside dimensions are 100 feet in length. See the illustration. Find the dimensions of the rectangle that maximizes its area.

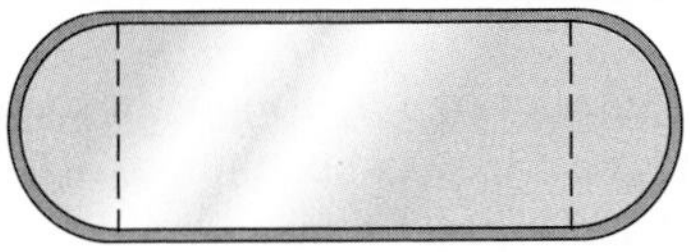

94. **Amplifying Sound** An amplifier's power output P (in watts) is related to its decibel voltage gain d by the formula $P = 25e^{0.1d}$.

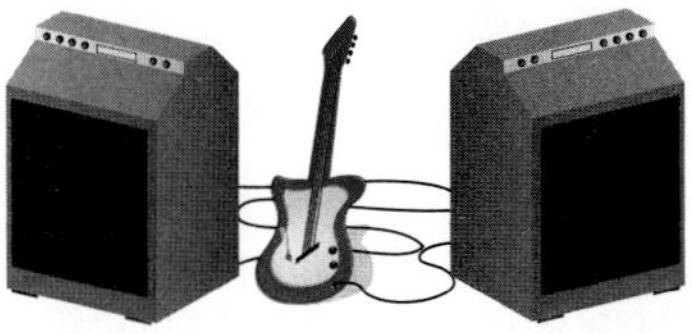

(a) Find the power output for a decibel voltage gain of 4 decibels.

(b) For a power output of 50 watts, what is the decibel voltage gain?

95. **Limiting Magnitude of a Telescope** A telescope is limited in its usefulness by the brightness of the star it is aimed at and by the diameter of its lens. One measure of a star's brightness is its *magnitude;* the dimmer the star, the larger its magnitude. A formula for the limiting magnitude L of a telescope, that is, the magnitude of the dimmest star that it can be used to view, is given by

$$L = 9 + 5.1\log d$$

where d is the diameter (in inches) of the lens.

(a) What is the limiting magnitude of a 3.5-inch telescope?
(b) What diameter is required to view a star of magnitude 14?

96. **Salvage Value** The number of years n for a piece of machinery to depreciate to a known salvage value can be found using the formula

$$n = \frac{\log s - \log i}{\log(1 - d)}$$

where s is the salvage value of the machinery, i is its initial value, and d is the annual rate of depreciation.

(a) How many years will it take for a piece of machinery to decline in value from \$90,000 to \$10,000 if the annual rate of depreciation is 0.20 (20%)?

(b) How many years will it take for a piece of machinery to lose half of its value if the annual rate of depreciation is 15%?

97. Funding an IRA First Colonial Bankshares Corporation advertised the following IRA investment plans.

TARGET IRA PLANS For each \$5000 Maturity Value Desired	
Deposit	For a Term of:
\$620.17	20 Years
\$1045.02	15 Years
\$1760.92	10 Years
\$2967.26	5 Years

(a) Assuming continuous compounding, what was the annual rate of interest that they offered?
(b) First Colonial Bankshares claims that \$4000 invested today will have a value of over \$32,000 in 20 years. Use the answer found in part (a) to find the actual value of \$4000 in 20 years. Assume continuous compounding.

98. Find the point on the line $y = x$ that is closest to the point $(3, 1)$.

[**Hint:** Find the minimum value of the function $f(x) = d^2$, where d is the distance from $(3, 1)$ to a point on the line.]

99. Minimizing Marginal Cost The marginal cost of a product can be thought of as the cost of producing one additional unit of output. For example, if the marginal cost of producing the 50th product is \$6.20, then it cost \$6.20 to increase production from 49 to 50 units of output. Callaway Golf Company has determined that the marginal cost C of manufacturing x Big Bertha golf clubs may be expressed by the quadratic function

$$C(x) = 4.9x^2 - 617.4x + 19{,}600$$

(a) How many clubs should be manufactured to minimize the marginal cost?
(b) At this level of production, what is the marginal cost?

100. Violent Crimes The function $V(t) = -10.0t^2 + 39.2t + 1862.6$ models the number V (in thousands) of violent crimes committed in the United States t years after 1990 based on data obtained from the Federal Bureau of Investigation. So $t = 0$ represents 1990, $t = 1$ represents 1991, and so on.

(a) Determine the year in which the most violent crimes were committed.
(b) Approximately how many violent crimes were committed during this year?
(c) Using a graphing utility, graph $V = V(t)$. Were the number of violent crimes increasing or decreasing during the years 1994 to 1998?

Chapter 11 Project

The table below lists historical data on the population of Houston, Texas from 1850 through 2000. Use the information provided to do the following problems.

Year	Population (in thousands)
2000	1953.6
1990	1630.6
1980	1595.1
1970	1232.8
1960	938.2
1950	596.2
1940	384.5
1930	292.4
1920	138.3
1910	78.8
1900	44.6
1890	27.6
1880	16.5
1870	9.4
1860	4.8
1850	2.4

Source: U.S. Census.

1. Plot the data measuring population (in thousands) on the y-axis and time (in years) on the x-axis. Use $t = 0$ for 1850, $t = 10$ for 1860, and so on.

2. Assume the relationship between population P and time t is exponential. That is, assume $P = P_0 a^{t \div 10}$, where $P_0 = P(0)$. Then find a as follows:
Form a table that lists the ratios of the populations for consecutive years. For example, from 1850 to 1860 the ratio is $\frac{4.8}{2.4} = 2$. From 1860 to 1870 the ratio is $\frac{9.4}{4.8} = 1.96$. Now find the average of all these ratios. Use this number for the base a of the function $P = P_0 a^t$. (Refer to the theorem on page 629 to review the basis for this).
3. (a) According to the result found above, what is the population of Houston in 2000?
(b) How close is this to the actual population as given in the table?
(c) Can you provide an explanation for any differences?
(d) Use the exponential growth function to predict the population in 2010.
(e) What is the predicted population of Houston in 2050?
4. Write the exponential function found in Problem 2 in the form $P = P_o e^{kt}$. Here k is the growth rate of the population. [**Hint:** $a^t = e^{kt} = (e^k)^t$. Now solve for k.]
5. What is the growth rate of the population of Houston?

6.2–6.5 Repeat Problems 2–5 using the data from the table, but beginning with 1900 instead of 1850.

7. Compare the results of Problems 3 and 6.3. What differences do you get in the exponential growth function and the prediction of future population of Houston? Explain any differences. Which predictions seem more reasonable?

8.2–8.5 Repeat Problems 2–5 using the data from the table, but beginning with 1950 instead of 1850.

9. Compare the results of Problems 3, 6.3, and 8.3. What differences do you get in the exponential growth function and the prediction of future population of Houston? Explain any differences. Which predictions seem more reasonable?
10. Use a graphing utility to find the exponential curve of best fit using the data from the table beginning with 1950 and ending with 2000.
11. Work Problems 3, 4, and 5 using the result found in Problem 10. Compare the results with those of Problems 8.3, 8.4, and 8.5. Explain any differences.

MATHEMATICAL QUESTIONS FROM PROFESSIONAL EXAMS*

1. **Actuary Exam—Part I** If $\log_6 2 = b$, which of the following is equal to $\log_6(4 \cdot 27)$?

(a) $3 - b$ (b) $2 + b$ (c) $5b$ (d) $3b$
(e) $2b + \log_6 27$

2. **Actuary Exam—Part II** If b and c are real numbers and there are two different real numbers x such that $(e^x)^2 + be^x + c = 0$, which of the following must be true?

I. $b^2 - 4c > 0$ II. $b < 0$ III. $c > 0$

(a) None (b) I only (c) I and II only
(d) I and III only (e) I, II, and III

3. **Actuary Exam—Part I** If $y = \frac{e^x - e^{-x}}{2}$, then find x in terms of y.

(a) $x = \frac{e^y - e^{-y}}{2}$ (b) $x = \ln y$
(c) $x = \ln(y + \sqrt{y^2 + 1})$ (d) $x = \ln 2y$
(e) $x = \ln(y + y^2)$

4. **Actuary Exam—Part I** If $(\log_x x)(\log_5 x) = 3$ then $x = ?$

(a) 3 (b) 5 (c) 25 (d) 75 (e) 125

5. **Actuary Exam—Part I** $(\log_a b)(\log_b a) = ?$

(a) 0 (b) 1 (c) $\log_a b$ (d) $(\log_a b)^2$ (e) $\frac{1}{\log_b a}$

6. **Actuary Exam—Part I** The product $(\log_3 2)(\log_2 9)$ is equal to:

(a) 2 (b) 3 (c) 9 (d) $\sqrt[3]{9}$ (e) $\sqrt{2}$

7. **Actuary Exam—Part I** For what real value of $x > 1$ is $e^{2\ln(x-1)} = 4$?

(a) 3 (b) 9 (c) 17 (d) $1 + 4e^{-2}$ (e) $\frac{2 + e^4}{2}$

CHAPTER 12

The Limit of a Function

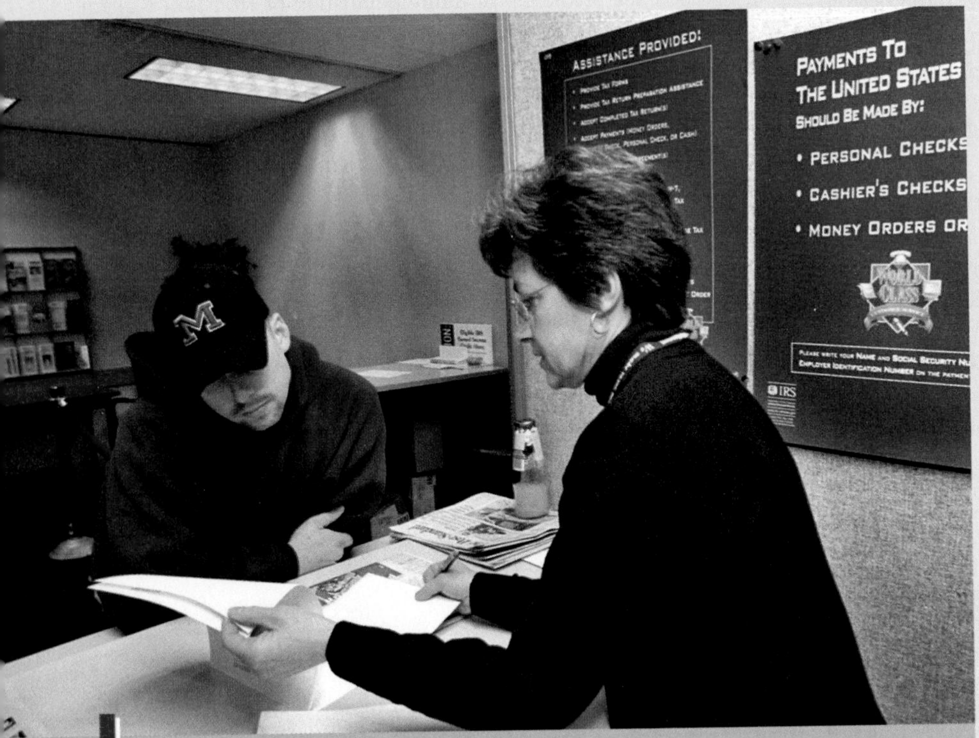

OUTLINE

It's almost April 15th, time to start thinking about taxes. As usual, a headache starts, especially when you look at the tax rate schedules the IRS supplies (see page 712). You begin to wonder what tax bracket you are in. You see that the tax rate can be as low as 10% and as high as 35%. If your taxable income is less than \$28,400 the tax rate is 15%, but if your taxable income is just over \$28,400, the tax rate jumps from 15% to 25%. That's a big difference. Does that mean the amount you pay in taxes will also have a big jump? No, that wouldn't make any sense. But how can you have a jump in the tax rate, but not in the amount you pay? The Chapter Project will provide an answer and the discussion in this chapter will help explain.

A LOOK BACK, A LOOK FORWARD

In Chapter 10 we defined a function and many of the properties that functions can have. In Chapter 11 we discussed classes of functions and properties that the classes have. With this as background we are ready to study the *limit of a function.* This concept is the bridge that takes us from the mathematics of algebra and geometry to the mathematics of calculus.

Calculus actually consists of two parts: the *differential calculus,* which we discuss in Chapters 13 and 14 and the *integral calculus,* discussed in Chapters 15 and 16. In Chapter 17 we study the calculus of functions of two or more variables.

In differential calculus we introduce another property of functions, namely the *derivative of a function.* We shall find that the derivative opens up a way for doing many applied problems in business, economics, and social sciences. Many of these applications involve an analysis of the graph of a function.

12.1 Finding Limits Using Tables and Graphs

PREPARING FOR THIS SECTION *Before getting started, review the following:*

- > Evaluating Functions (Chapter 10, Section 10.2, pp. 550–551)
- > Piecewise-defined Functions (Chapter 10, Section 10.4, pp. 583–585)
- > Library of Functions (Chapter 10, Section 10.4, pp. 579–583)

OBJECTIVES
1. Find a limit using a table
2. Find a limit using a graph

The idea of the limit of a function is what connects algebra and geometry to calculus. In working with the limit of a function, we encounter notation of the form

$$\lim_{x \to c} f(x) = N$$

This is read as "the limit of $f(x)$ as x approaches c equals the number N." Here f is a function defined on some open interval containing the number c; f need not be defined at c, however.

We may describe the meaning of $\lim_{x \to c} f(x) = N$ as follows:

> For all values of x approximately equal to c, with $x \neq c$, the corresponding value $f(x)$ is approximately equal to N.

Another description of $\lim_{x \to c} f(x) = N$ is

> As x gets closer to c, but remains unequal to c, the corresponding value of $f(x)$ gets closer to N.

Tables generated with the help of a calculator are useful for finding limits.

1 EXAMPLE 1 Finding a Limit Using a Table

Find: $\lim_{x \to 3}(5x^2)$

SOLUTION Here $f(x) = 5x^2$ and $c = 3$. We choose values of x close to 3, arbitrarily starting with 2.99. Then we select additional numbers that get closer to 3, but remain less than 3. Next we choose values of x greater than 3, starting with 3.01, that get closer to 3. Finally, we evaluate f at each choice to obtain Table 1.

TABLE 1

x	2.99	2.999	2.9999 →	← 3.0001	3.001	3.01
$f(x) = 5x^2$	44.701	44.97	44.997 →	← 45.003	45.030	45.301

From Table 1, we infer that as x gets closer to 3 the value of $f(x) = 5x^2$ gets closer to 45. That is,

$$\lim_{x \to 3}(5x^2) = 45$$

TABLE 2

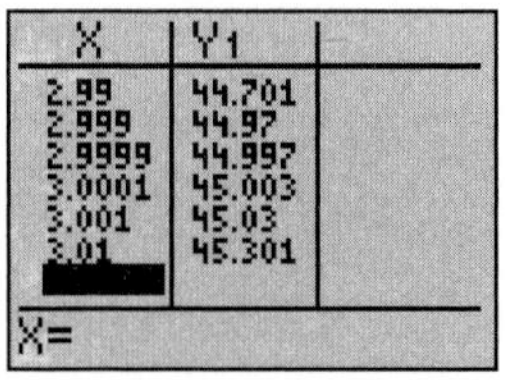

X	Y1
2.99	44.701
2.999	44.97
2.9999	44.997
3.0001	45.003
3.001	45.03
3.01	45.301

X=

When choosing the values of x in a table, the number to start with and the subsequent entries are arbitrary. However, the entries should be chosen so that the table makes it clear what the corresponding values of f are getting close to.

COMMENT: A graphing utility with a TABLE feature can be used to generate the entries. Table 2 shows the result using a TI-83 Plus.

NOW WORK PROBLEMS 1 AND 9.

EXAMPLE 2 Finding a Limit Using a Table

Find: **(a)** $\lim_{x \to 2} \dfrac{x^2 - 4}{x - 2}$ **(b)** $\lim_{x \to 2}(x + 2)$

SOLUTION **(a)** Here $f(x) = \dfrac{x^2 - 4}{x - 2}$ and $c = 2$. Notice that the domain of f is $\{x \mid x \neq 2\}$, so f is not defined at 2. We proceed to choose values of x close to 2 and evaluate f at each choice, as shown in Table 3.

TABLE 3

x	1.99	1.999	1.9999 →	← 2.0001	2.001	2.01
$f(x) = \dfrac{x^2 - 4}{x - 2}$	3.99	3.999	3.9999 →	← 4.0001	4.001	4.01

We infer that as x gets closer to 2 the value of $f(x) = \dfrac{x^2 - 4}{x - 2}$ gets closer to 4. That is,

$$\lim_{x \to 2} \frac{x^2 - 4}{x - 2} = 4$$

(b) Here $g(x) = x + 2$ and $c = 2$. The domain of g is all real numbers. As before, we choose values of x close to 2, and evaluate the function f at each choice. See Table 4.

TABLE 4

x	1.99	1.999	1.9999 →	← 2.0001	2.001	2.01
$f(x) = x + 2$	3.99	3.999	3.9999 →	← 4.0001	4.001	4.01

We infer that as x gets closer to 2 the value of $g(x)$ gets closer to 4. That is,

$$\lim_{x \to 2} (x + 2) = 4$$

CHECK: Use a graphing utility with a TABLE feature to verify the results obtained in Example 2.

The conclusion that $\lim_{x \to 2}(x + 2) = 4$ could have been obtained without the use of Table 4; as x gets closer to 2, it follows that $x + 2$ will get closer to $2 + 2 = 4$.

Also, for part (a), you are right if you make the observation that, since $x \neq 2$, then

$$f(x) = \frac{x^2 - 4}{x - 2} = \frac{\cancel{(x - 2)}(x + 2)}{\cancel{x - 2}} = x + 2, \qquad x \neq 2$$

Now it is easy to conclude that

$$\lim_{x \to 2} \frac{x^2 - 4}{x - 2} = \lim_{x \to 2}(x + 2) = 4$$

The graph of a function f can also be of help in finding limits. See Figure 1.

FIGURE 1

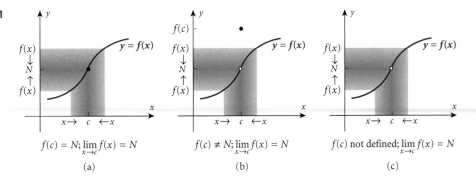

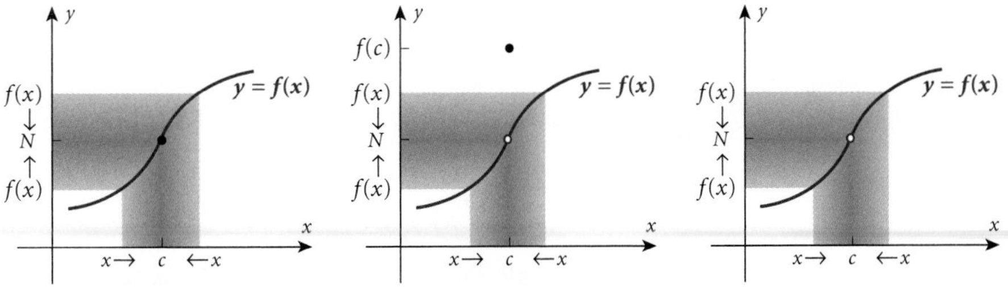

$f(c) = N; \lim_{x \to c} f(x) = N$ (a)

$f(c) \neq N; \lim_{x \to c} f(x) = N$ (b)

$f(c)$ not defined; $\lim_{x \to c} f(x) = N$ (c)

In each graph, notice that as x gets closer to c, the value of f gets closer to the number N. We conclude that

$$\lim_{x \to c} f(x) = N$$

This is the conclusion regardless of the value of f at c. In Figure 1(a), $f(c) = N$, and in Figure 1(b), $f(c) \neq N$. Figure 1(c) illustrates that $\lim_{x \to c} f(x) = N$, even if f is not defined at c.

2

EXAMPLE 3 Finding a Limit by Graphing

Find: $\lim_{x \to 2} f(x)$ if $f(x) = \begin{cases} 3x - 2 & \text{if } x \neq 2 \\ 3 & \text{if } x = 2 \end{cases}$

SOLUTION The function f is a piecewise-defined function. Its graph is shown in Figure 2. We conclude from the graph that $\lim_{x \to 2} f(x) = 4$.

FIGURE 2

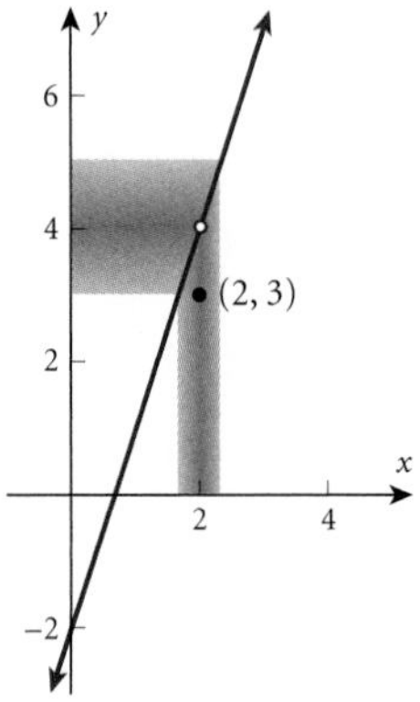

Notice in Example 3 that the value of f at 2, that is, $f(2) = 3$, plays no role in the conclusion that $\lim_{x \to 2} f(x) = 4$. In fact, even if f were undefined at 2, it would still happen that $\lim_{x \to 2} f(x) = 4$.

NOW WORK PROBLEMS 17 AND 23.

Sometimes there is no *single* number that the value of f gets closer to as x gets closer to c. In this case, we say that **f has no limit as x approaches c** or that $\lim_{x \to c} f(x)$ **does not exist.**

EXAMPLE 4 A Function That Has No Limit at 0

FIGURE 3

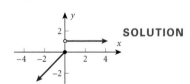

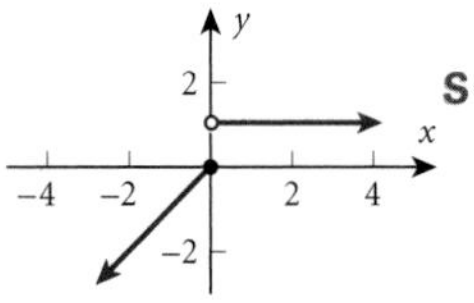

Find: $\lim_{x \to 0} f(x)$ if $f(x) = \begin{cases} x & \text{if } x \leq 0 \\ 1 & \text{if } x > 0 \end{cases}$

SOLUTION See Figure 3. As x gets closer to 0, but remains negative, the value of f also gets closer to 0. As x gets closer to 0, but remains positive, the value of f always equals 1. Since there is no single number that the values of f are close to when x is close to 0, we conclude that $\lim_{x \to 0} f(x)$ does not exist.

NOW WORK PROBLEM 35.

EXAMPLE 5 Using a Graphing Utility to Find a Limit

Find: $\lim_{x \to 2} \dfrac{x^3 - 2x^2 + 4x - 8}{x^4 - 2x^3 + x - 2}$

SOLUTION We create Table 5. From the table we conclude that

$$\lim_{x \to 2} \frac{x^3 - 2x^2 + 4x - 8}{x^4 - 2x^3 + x - 2} = 0.889$$

rounded to three decimal places.

TABLE 5

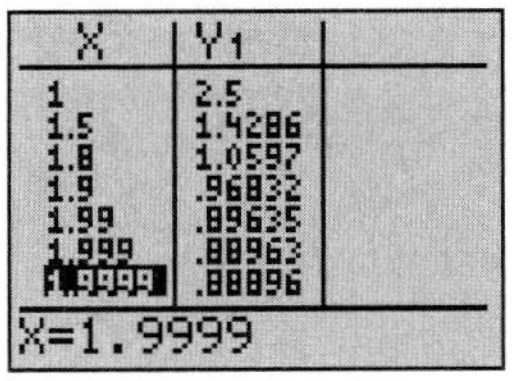

X	Y_1
1	2.5
1.5	1.4286
1.8	1.0597
1.9	.96832
1.99	.89635
1.999	.88963
1.9999	.88896

X=1.9999

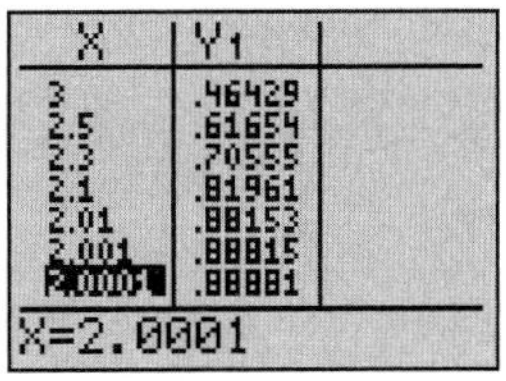

X	Y_1
3	.46429
2.5	.61654
2.3	.70555
2.1	.81961
2.01	.88153
2.001	.88815
2.0001	.88881

X=2.0001

NOW WORK PROBLEM 41.

In the next section, we will see how to obtain exact solutions to limits like the one in Example 5.

EXERCISE 12.1 Answers to Odd-Numbered Problems Begin on Page AN-68.

In Problems 1–8, complete each table and evaluate the indicated limit.

1.

x	0.9	0.99	0.999
$f(x) = 2x$			
x	1.1	1.01	1.001
$f(x) = 2x$			

$\lim_{x \to 1} f(x) =$ ______

2.

x	1.9	1.99	1.999
$f(x) = x + 3$			
x	2.1	2.01	2.001
$f(x) = x + 3$			

$\lim_{x \to 2} f(x) =$ ______

3.

x	−0.1	−0.01	−0.001
$f(x) = x^2 + 2$			
x	0.1	0.01	0.001
$f(x) = x^2 + 2$			

$\lim_{x \to 0} f(x) =$ ______

4.

x	−1.1	−1.01	−1.001
$f(x) = x + 2$			
x	−0.9	−0.99	−0.999
$f(x) = x^2 - 2$			

$\lim_{x \to -1} f(x) =$ ______

5.

x	−1.9	−1.99	−1.999
$f(x) = \dfrac{x^2 - 4}{x + 2}$			
x	−2.1	−2.01	−2.001
$f(x) = \dfrac{x^2 - 4}{x + 2}$			

$\lim_{x \to -2} f(x) =$ ______

6.

x	−1.1	−1.01	−1.001
$f(x) = \dfrac{x^2 - 1}{x + 1}$			
x	−0.9	−0.99	−0.999
$f(x) = \dfrac{x^2 - 1}{x + 1}$			

$\lim_{x \to -1} f(x) =$ ______

7.

x	-1.1	-1.01	-1.001
$f(x) = \dfrac{x^3+1}{x+1}$			
x	-0.9	-0.99	-0.999
$f(x) = \dfrac{x^3+1}{x+1}$			

$\lim_{x \to -1} f(x) =$ ______

8.

x	2.9	2.99	2.999
$f(x) = \dfrac{x^3-27}{x-3}$			
x	3.1	3.01	3.001
$f(x) = \dfrac{x^3-27}{x-3}$			

$\lim_{x \to 3} f(x) =$ ______

In Problems 9–16, use a table to find the indicated limit.

9. $\lim_{x\to 2}(4x^3)$
10. $\lim_{x\to 3}(2x^2+1)$
11. $\lim_{x\to 0}\dfrac{x+1}{x^2+1}$
12. $\lim_{x\to 0}\dfrac{2-x}{x^2+4}$
13. $\lim_{x\to 4}\dfrac{x^2-4x}{x-4}$
14. $\lim_{x\to 3}\dfrac{x^2-9}{x^2-3x}$
15. $\lim_{x\to 0}(e^x+1)$
16. $\lim_{x\to 0}\dfrac{e^x-e^{-x}}{2}$

In Problems 17–22, use the graph shown to determine if the limit exists. If it does, find it.

17.
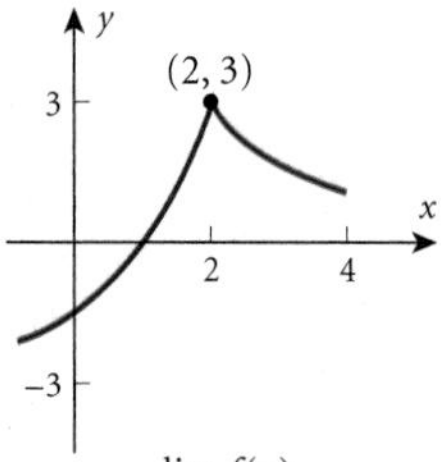

$\lim_{x\to 2} f(x)$

18.
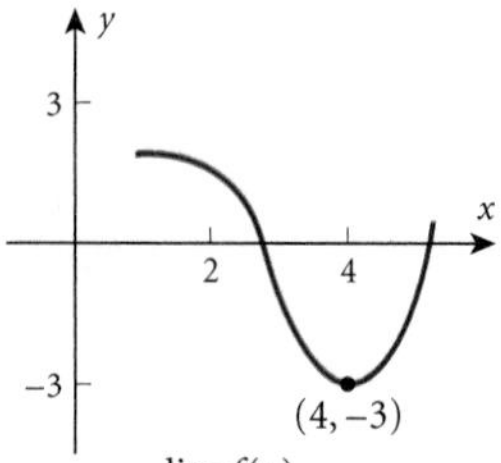

$\lim_{x\to 4} f(x)$

19.
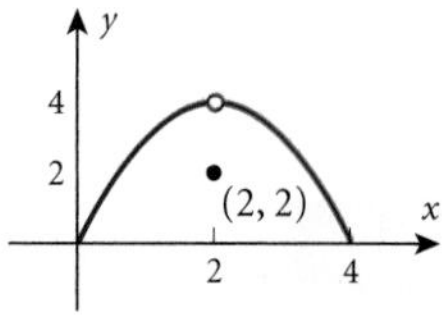

$\lim_{x\to 2} f(x)$

20.
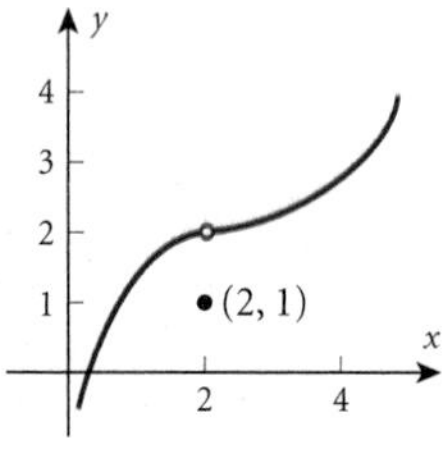

$\lim_{x\to 2} f(x)$

21.
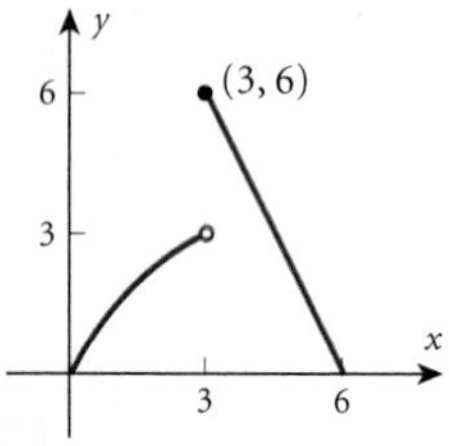

$\lim_{x\to 3} f(x)$

22.
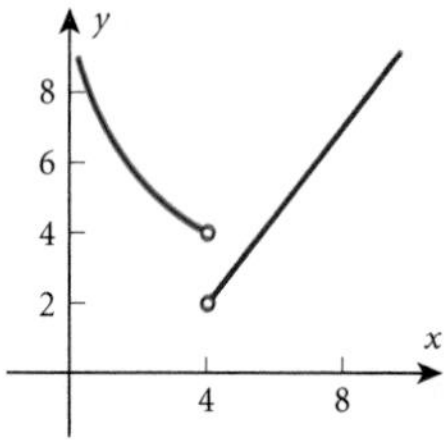

$\lim_{x\to 4} f(x)$

In Problems 23–40, graph each function. Use the graph to find the indicated limit, if it exists.

23. $\lim_{x\to 4} f(x),\quad f(x) = 3x+1$
24. $\lim_{x\to -1} f(x),\quad f(x) = 2x-1$
25. $\lim_{x\to 2} f(x),\quad f(x) = 1-x^2$
26. $\lim_{x\to -1} f(x),\quad f(x) = x^3-1$
27. $\lim_{x\to -3} f(x),\quad f(x) = |x|-2$
28. $\lim_{x\to 4} f(x),\quad f(x) = \sqrt{x+5}$
29. $\lim_{x\to 0} f(x),\quad f(x) = e^x$
30. $\lim_{x\to 1} f(x),\quad f(x) = \ln x$
31. $\lim_{x\to -1} f(x),\quad f(x) = \dfrac{1}{x}$
32. $\lim_{x\to 8} f(x),\quad f(x) = \sqrt[3]{x}$
33. $\lim_{x\to 0} f(x),\quad f(x) = \begin{cases} x^2 & x<0 \\ 2x & x\ge 0\end{cases}$
34. $\lim_{x\to 0} f(x),\quad f(x) = \begin{cases} x-1 & x<0 \\ 3x-1 & x\ge 0\end{cases}$
35. $\lim_{x\to 1} f(x),\quad f(x) = \begin{cases} 3x & x\le 1 \\ x+1 & x>1\end{cases}$
36. $\lim_{x\to 2} f(x),\quad f(x) = \begin{cases} x^2 & x\le 2 \\ 2x-1 & x>2\end{cases}$
37. $\lim_{x\to 0} f(x),\quad f(x) = \begin{cases} x & x<0 \\ 1 & x=0 \\ 3x & x>0\end{cases}$

38. $\lim_{x\to 0} f(x), \quad f(x) = \begin{cases} 1 & x < 0 \\ -1 & x > 0 \end{cases}$

39. $\lim_{x\to 0} f(x), \quad f(x) = \begin{cases} e^x - 1 & x \le 0 \\ x^2 & x > 0 \end{cases}$

40. $\lim_{x\to 0} f(x), \quad f(x) = \begin{cases} e^x & x \le 0 \\ 1 - x & x > 0 \end{cases}$

In Problems 41–46, use a graphing utility to find the indicated limit. Round answers to two decimal places.

41. $\lim_{x\to 1} \dfrac{x^3 - x^2 + x - 1}{x^4 - x^3 + 2x - 2}$

42. $\lim_{x\to -1} \dfrac{x^3 + x^2 + 3x + 3}{x^4 + x^3 + 2x + 2}$

43. $\lim_{x\to 2} \dfrac{x^3 - 2x^2 + 4x - 8}{x^2 + x - 6}$

44. $\lim_{x\to 1} \dfrac{x^3 - x^2 + 3x - 3}{x^2 + 3x - 4}$

45. $\lim_{x\to -1} \dfrac{x^3 + 2x^2 + x}{x^4 + x^3 + 2x + 2}$

46. $\lim_{x\to 3} \dfrac{x^3 - 3x^2 + 4x - 12}{x^4 - 3x^3 + x - 3}$

12.2 Techniques for Finding Limits of Functions

PREPARING FOR THIS SECTION *Before getting started, review the following:*

- Power Functions (Chapter 11, Section 11.2, pp. 618–621)
- Polynomial Functions (Chapter 11, Section 11.2, pp. 621–624)
- Rational Functions (Chapter 11, Section 11.2, pp. 624–625)
- Library of Functions (Chapter 10, Section 10.4, pp. 579–583)
- Average Rate of Change (Chapter 10, Section 10.3, pp. 570–571)

OBJECTIVES
1. Find the limit of a sum, a difference, and a product
2. Find the limit of a polynomial function
3. Find the limit of a function involving a power or a root
4. Find the limit of a quotient
5. Find the limit of an average rate of change

We can find the limit of most functions by developing two formulas involving limits and by using properties of limits.

Two Formulas: $\lim_{x\to c} b$ **and** $\lim_{x\to c} x$

Limit of the Constant Function

For the constant function $f(x) = b$,

$$\lim_{x\to c} f(x) = \lim_{x\to c} b = b \tag{1}$$

where c is any number.

Limit of the Identity Function

For the identity function $f(x) = x$,

$$\lim_{x\to c} f(x) = \lim_{x\to c} x = c \tag{2}$$

where c is any number.

We use graphs to establish formulas (1) and (2). Since the graph of a constant function is a horizontal line, it follows that, no matter how close x is to c, the corresponding value of $f(x)$ equals b. That is, $\lim_{x\to c} b = b$. See Figure 4.

The identity function is $f(x) = x$. For any choice of c, as x gets closer to c, the corresponding value of $f(x)$ is just as close to c. That is, $\lim_{x\to c} x = c$. See Figure 5.

FIGURE 4

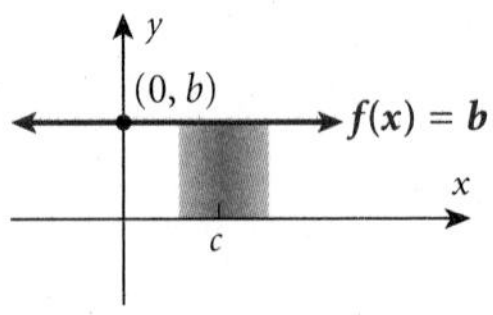

FIGURE 5

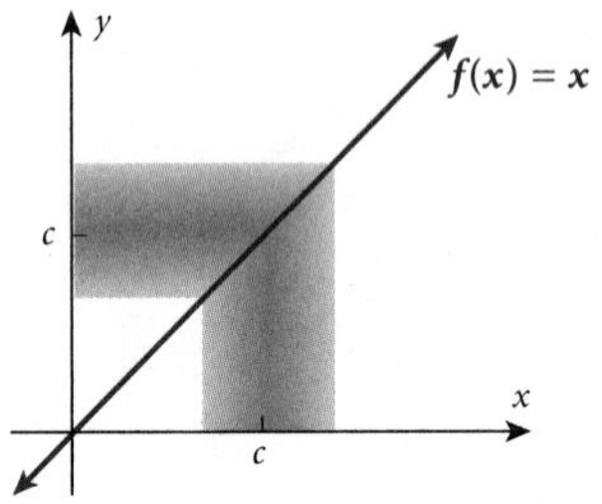

EXAMPLE 1 **Using Formulas (1) and (2)**

(a) $\lim_{x\to 3} 5 = 5$ **(b)** $\lim_{x\to 3} x = 3$ **(c)** $\lim_{x\to 0} (-8) = -8$ **(d)** $\lim_{x\to -1/2} x = -\frac{1}{2}$

NOW WORK PROBLEM 1.

Formulas (1) and (2), when used with the algebraic properties that follow, enable us to evaluate limits of more complicated functions.

Algebraic Properties of Limits

Find the limit of a sum, a difference, and a product

1 In the following properties, we assume that f and g are two functions for which both $\lim_{x\to c} f(x)$ and $\lim_{x\to c} g(x)$ exist.

Limit of a Sum

$$\lim_{x\to c}[f(x) + g(x)] = \lim_{x\to c} f(x) + \lim_{x\to c} g(x) \tag{3}$$

In words, the limit of the sum of two functions equals the sum of their limits.

EXAMPLE 2 **Finding the Limit of a Sum**

Find: $\lim_{x\to -3} (x + 4)$

SOLUTION The limit we seek is the sum of the two functions $f(x) = x$ and $g(x) = 4$. From Formulas (1) and (2), we know that

$$\lim_{x \to -3} f(x) = \lim_{x \to -3} x = -3 \quad \text{and} \quad \lim_{x \to -3} g(x) = \lim_{x \to -3} 4 = 4$$

From Formula (3), it follows that

$$\lim_{x \to -3} (x + 4) = \lim_{x \to -3} x + \lim_{x \to -3} 4 = -3 + 4 = 1$$

Limit of a Difference

$$\lim_{x \to c}[f(x) - g(x)] = \lim_{x \to c} f(x) - \lim_{x \to c} g(x) \quad \textbf{(4)}$$

In words, the limit of the difference of two functions equals the difference of their limits.

EXAMPLE 3 Finding the Limit of a Difference

Find: $\lim_{x \to 4}(6 - x)$

SOLUTION The limit we seek is the difference of the two functions $f(x) = 6$ and $g(x) = x$. From Formulas (1) and (2), we know that

$$\lim_{x \to 4} f(x) = \lim_{x \to 4} 6 = 6 \quad \text{and} \quad \lim_{x \to 4} g(x) = \lim_{x \to 4} x = 4$$

From Formula (4), it follows that

$$\lim_{x \to 4}(6 - x) = \lim_{x \to 4} 6 - \lim_{x \to 4} x = 6 - 4 = 2$$

Limit of a Product

$$\lim_{x \to c}[f(x) \cdot g(x)] = \left[\lim_{x \to c} f(x)\right]\left[\lim_{x \to c} g(x)\right] \quad \textbf{(5)}$$

In words, the limit of the product of two functions equals the product of their limits.

EXAMPLE 4 Finding the Limit of a Product

Find: $\lim_{x \to -5} (-4x)$

SOLUTION The limit we seek is the product of the two functions $f(x) = -4$ and $g(x) = x$. From Formulas (1) and (2), we know that

$$\lim_{x \to -5} f(x) = \lim_{x \to -5} (-4) = -4 \quad \text{and} \quad \lim_{x \to -5} g(x) = \lim_{x \to -5} x = -5$$

From Formula (5), it follows that

$$\lim_{x \to -5} (-4x) = \left[\lim_{x \to -5} (-4)\right]\left[\lim_{x \to -5} x\right] = (-4)(-5) = 20$$

EXAMPLE 5 Finding Limits Using Algebraic Properties

Find: **(a)** $\lim_{x \to -2} (3x - 5)$ **(b)** $\lim_{x \to 2}(5x^2)$

SOLUTION **(a)** $\lim_{x \to -2} (3x - 5) = \lim_{x \to -2} (3x) - \lim_{x \to -2} 5 = \left[\lim_{x \to -2} 3\right]\left[\lim_{x \to -2} x\right] - \lim_{x \to -2} 5$

$$= (3)(-2) - 5 = -6 - 5 = -11$$

(b) $\lim_{x \to 2}(5x^2) = \left[\lim_{x \to 2} 5\right]\left[\lim_{x \to 2} x^2\right] = 5 \cdot \lim_{x \to 2}(x \cdot x) = 5 \cdot \left[\lim_{x \to 2} x\right]\left[\lim_{x \to 2} x\right]$

$$= 5 \cdot 2 \cdot 2 = 20$$

NOW WORK PROBLEM 5.

Notice in the solution to part (b) of Example 5 that $\lim_{x \to 2} (5x^2) = 5 \cdot 2^2$.

Limit of a Power Function

If $n \geq 1$ is a positive integer and a is a constant, then

$$\lim_{x \to c}(ax^n) = ac^n \quad (6)$$

for any number c. That is, if $f(x) = ax^n$, then

$$\lim_{x \to c} f(x) = f(c) \quad (7)$$

Proof

$$\lim_{x \to c}(ax^n) = \left[\lim_{x \to c} a\right]\left[\lim_{x \to c} x^n\right] = a\left[\lim_{x \to c} (\underbrace{x \cdot x \cdot x \cdot \ldots \cdot x}_{n \text{ factors}})\right]$$

$$= a\underbrace{\left[\lim_{x \to c} x\right]\left[\lim_{x \to c} x\right]\left[\lim_{x \to c} x\right]\ldots\left[\lim_{x \to c} x\right]}_{n \text{ factors}}$$

$$= a \cdot \underbrace{c \cdot c \cdot c \cdot \ldots \cdot c}_{n \text{ factors}} = ac^n$$

■

EXAMPLE 6 Finding the Limit of a Power Function

Find: $\lim_{x \to 2} (-4x^3)$

SOLUTION $\lim_{x \to 2}(-4x^3) = -4 \cdot 2^3 = -4 \cdot 8 = -32$

Find the limit of a polynomial function

2 Since a polynomial is a sum of power functions, we can use Formula (6) and repeated use of Formula (3) to obtain the following result:

Limit of a Polynomial Function

If P is a polynomial function, then

$$\lim_{x \to c} P(x) = P(c) \tag{8}$$

for any number c.

Proof If P is a polynomial function, that is, if

$$P(x) = a_n x^n + a_{n-1} x^{n-1} + \cdots + a_1 x + a_0$$

then

$$\begin{aligned} \lim_{x \to c} P(x) &= \lim_{x \to c}[a_n x^n + a_{n-1} x^{n-1} + \cdots + a_1 x + a_0] \\ &= \lim_{x \to c}(a_n x^n) + \lim_{x \to c}(a_{n-1} x^{n-1}) + \cdots + \lim_{x \to c}(a_1 x) + \lim_{x \to c} a_0 && \text{Formula (3)} \\ &= a_n c^n + a_{n-1} c^{n-1} + \cdots + a_1 c + a_0 && \text{Formula (6)} \\ &= P(c) \end{aligned}$$

■

Formula (8) states that to find the limit of a polynomial as x approaches c, all we need to do is to evaluate the polynomial at c.

EXAMPLE 7 Finding the Limit of a Polynomial Function

Find: $\lim_{x \to 2}[5x^4 - 6x^3 + 3x^2 + 4x - 2]$

SOLUTION

$$\begin{aligned} \lim_{x \to 2}[5x^4 - 6x^3 + 3x^2 + 4x - 2] &= 5 \cdot 2^4 - 6 \cdot 2^3 + 3 \cdot 2^2 + 4 \cdot 2 - 2 \\ &= 5 \cdot 16 - 6 \cdot 8 + 3 \cdot 4 + 8 - 2 \\ &= 80 - 48 + 12 + 6 = 50 \end{aligned}$$

NOW WORK PROBLEM 7.

Limit of a Power or Root

If $\lim_{x \to c} f(x)$ exists and if $n \geq 2$ is a positive integer, then

$$\lim_{x \to c}[f(x)]^n = \left[\lim_{x \to c} f(x)\right]^n \tag{9}$$

and

$$\lim_{x \to c} \sqrt[n]{f(x)} = \sqrt[n]{\lim_{x \to c} f(x)} \tag{10}$$

In Formula (10), we require that both $\sqrt[n]{f(x)}$ and $\sqrt[n]{\lim_{x \to c} f(x)}$ be defined.

3 **EXAMPLE 8** **Finding the Limit of a Function Involving a Power or a Root**

Find: **(a)** $\lim_{x\to 1}(3x-5)^4$ **(b)** $\lim_{x\to 0}\sqrt{5x^2+8}$ **(c)** $\lim_{x\to -1}(5x^3-x+3)^{4/3}$

SOLUTION **(a)** $\lim_{x\to 1}(3x-5)^4 = \left[\lim_{x\to 1}(3x-5)\right]^4 = (-2)^4 = 16$

(b) $\lim_{x\to 0}\sqrt{5x^2+8} = \sqrt{\lim_{x\to 0}(5x^2+8)} = \sqrt{8} = 2\sqrt{2}$

(c) $$\lim_{x\to -1}(5x^3-x+3)^{4/3} = \sqrt[3]{\lim_{x\to -1}(5x^3-x+3)^4}$$
$$= \sqrt[3]{\left[\lim_{x\to -1}(5x^3-x+3)\right]^4} = \sqrt[3]{(-1)^4} = \sqrt[3]{1} = 1$$

NOW WORK PROBLEM 17.

Limit of a Quotient

$$\lim_{x\to c}\left[\frac{f(x)}{g(x)}\right] = \frac{\lim_{x\to c} f(x)}{\lim_{x\to c} g(x)} \qquad (11)$$

provided that $\lim_{x\to c} g(x) \neq 0$. In words, the limit of the quotient of two functions equals the quotient of their limits, provided that the limit of the denominator is not zero.

Since a rational function is a quotient of polynomials, we can use Formulas (8) and (11) to establish the following result.

Limit of a Rational Function

If R is a rational function and if c is in the domain of R, then

$$\lim_{x\to c} R(x) = R(c) \qquad (12)$$

Proof Suppose $R(x) = \frac{p(x)}{q(x)}$, where p and q are polynomial functions. If c is in the domain of R, then $q(c) \neq 0$. By Formula (8), $\lim_{x\to c} p(x) = p(c)$ and $\lim_{x\to c} q(x) = q(c)$. Since $q(c) \neq 0$, by Formula (11) we have

$$\lim_{x\to 0} R(x) = \lim_{x\to c}\frac{p(x)}{q(x)} \underset{\uparrow}{=} \frac{\lim_{x\to c} p(x)}{\lim_{x\to c} q(x)} \underset{\uparrow}{=} \frac{p(c)}{q(c)} = R(c)$$

Formula (11) Formula (8)

4 **EXAMPLE 9** **Finding the Limit of a Rational Function**

Find: $\lim_{x \to 1} \dfrac{5x^3 - x + 2}{3x + 4}$

SOLUTION The limit we seek is the limit of a rational function whose domain is $\left\{x \middle| x \neq -\dfrac{4}{3}\right\}$. Since 1 is in the domain, we use Formula (12).

$$\lim_{x \to 1} \frac{5x^3 - x + 2}{3x + 4} = \frac{5 \cdot 1^3 - 1 + 2}{3 \cdot 1 + 4} = \frac{6}{7}$$

NOW WORK PROBLEM 15.

When the limit of the denominator of a quotient is zero, Formula (11) cannot be used. In such cases, other strategies need to be used. Let's look at an example.

EXAMPLE 10 **Finding the Limit of a Quotient**

Find: $\lim_{x \to 3} \dfrac{x^2 - x - 6}{x^2 - 9}$

SOLUTION The domain of the rational function $R(x) = \dfrac{x^2 - x - 6}{x^2 - 9}$ is $\{x | x \neq -3, x \neq 3\}$. Since 3 is not in the domain, we cannot use Formula (12). Also, the limit of the denominator equals zero, so Formula (11) cannot be used. Instead, we notice that the expression can be factored as

$$\frac{x^2 - x - 6}{x^2 - 9} = \frac{(x - 3)(x + 2)}{(x - 3)(x + 3)}$$

When we compute a limit as x approaches 3, we are interested in the values of the function when x is close to 3, but unequal to 3. Since $x \neq 3$, we can cancel the $(x - 3)$'s. Formula (11) can then be used.

$$\lim_{x \to 3} \frac{x^2 - x - 6}{x^2 - 9} = \lim_{x \to 3} \frac{\cancel{(x - 3)}(x + 2)}{\cancel{(x - 3)}(x + 3)} = \frac{\lim_{x \to 3}(x + 2)}{\lim_{x \to 3}(x + 3)} = \frac{5}{6}$$

Now let's work Example 5 of Section 12.1.

EXAMPLE 11 **Finding Limits Using Algebraic Properties**

Find: $\lim_{x \to 2} \dfrac{x^3 - 2x^2 + 4x - 8}{x^4 - 2x^3 + x - 2}$

SOLUTION The limit of the denominator is zero, so Formula (11) cannot be used. We factor the expression.

$$\frac{x^3 - 2x^2 + 4x - 8}{x^4 - 2x^3 + x - 2} \underset{\substack{\uparrow \\ \text{Factor by grouping}}}{=} \frac{x^2(x - 2) + 4(x - 2)}{x^3(x - 2) + 1(x - 2)} \underset{\substack{\uparrow \\ \text{Factor}}}{=} \frac{(x^2 + 4)(x - 2)}{(x^3 + 1)(x - 2)}$$

Then,

$$\lim_{x\to 2}\frac{x^3-2x^2+4x-8}{x^4-2x^3+x-2}=\lim_{x\to 2}\frac{(x^2+4)\cancel{(x-2)}}{(x^3+1)\cancel{(x-2)}}=\frac{8}{9}$$

which is exact.

Compare the exact solution above with the approximate solution found in Example 5 of Section 12.1.

5 EXAMPLE 12 Finding the Limit of an Average Rate of Change

Find the limit as x approaches 2 of the average rate of change of the function

$$f(x)=x^2+3x$$

from 2 to x.

SOLUTION The average rate of change of f from 2 to x is

$$\frac{\Delta y}{\Delta x}=\frac{f(x)-f(2)}{x-2}=\frac{x^2+3x-10}{x-2}\underset{\text{Factor the numerator}}{\overset{\uparrow}{=}}\frac{(x+5)(x-2)}{x-2}$$

The limit as x approaches 2 of the average rate of change is

$$\lim_{x\to 2}\frac{f(x)-f(2)}{x-2}=\lim_{x\to 2}\frac{x^2+3x-10}{x-2}=\lim_{x\to 2}\frac{(x+5)\cancel{(x-2)}}{\cancel{x-2}}=7$$

NOW WORK PROBLEM 35.

SUMMARY To find exact values for $\lim_{x\to c} f(x)$, try the following:

1. If f is a polynomial Function or if f is a rational function and c is in the domain, then $\lim_{x\to c} f(x)=f(c)$ [Formula (8) or Formula (12)].
2. If f is a polynomial raised to a power or is the root of a polynomial, use Formulas (8) and (9) with Formula (7).
3. If f is a quotient and the limit of the denominator is not zero, use the fact that the limit of a quotient is the quotient of the limits.
4. If f is a quotient and the limit of the denominator is zero, use other techniques, such as factoring.

EXERCISE 12.2 Answers to Odd-Numbered Problems Begin on Page AN-69.

In Problems 1–32, find each limit.

1. $\lim_{x\to 1} 5$
2. $\lim_{x\to 1}(-3)$
3. $\lim_{x\to 4} x$
4. $\lim_{x\to -3} x$
5. $\lim_{x\to 2}(3x+2)$
6. $\lim_{x\to 3}(2-5x)$
7. $\lim_{x\to -1}(3x^2-5x)$
8. $\lim_{x\to 2}(8x^2-4)$
9. $\lim_{x\to 1}(5x^4-3x^2+6x-9)$
10. $\lim_{x\to -1}(8x^5-7x^3+8x^2+x-4)$
11. $\lim_{x\to 1}(x^2+1)^3$
12. $\lim_{x\to 2}(3x-4)^2$
13. $\lim_{x\to 1}\sqrt{5x+4}$
14. $\lim_{x\to 0}\sqrt{1-2x}$
15. $\lim_{x\to 0}\frac{x^2-4}{x^2+4}$
16. $\lim_{x\to 2}\frac{3x+4}{x^2+x}$
17. $\lim_{x\to 2}(3x-2)^{5/2}$
18. $\lim_{x\to -1}(2x+1)^{5/3}$

19. $\lim\limits_{x \to 2} \dfrac{x^2 - 4}{x^2 - 2x}$

20. $\lim\limits_{x \to -1} \dfrac{x^2 + x}{x^2 - 1}$

21. $\lim\limits_{x \to -3} \dfrac{x^2 - x - 12}{x^2 - 9}$

22. $\lim\limits_{x \to -3} \dfrac{x^2 + x - 6}{x^2 + 2x - 3}$

23. $\lim\limits_{x \to 1} \dfrac{x^3 - 1}{x - 1}$

24. $\lim\limits_{x \to 1} \dfrac{x^4 - 1}{x - 1}$

25. $\lim\limits_{x \to -1} \dfrac{(x + 1)^2}{x^2 - 1}$

26. $\lim\limits_{x \to 2} \dfrac{x^3 - 8}{x^2 - 4}$

27. $\lim\limits_{x \to 1} \dfrac{x^3 - x^2 + x - 1}{x^4 - x^3 + 2x - 2}$

28. $\lim\limits_{x \to -1} \dfrac{x^3 + x^2 + 3x + 3}{x^4 + x^3 + 2x + 2}$

29. $\lim\limits_{x \to 2} \dfrac{x^3 - 2x^2 + 4x - 8}{x^2 + x - 6}$

30. $\lim\limits_{x \to 1} \dfrac{x^3 - x^2 + 3x - 3}{x^2 + 3x - 4}$

31. $\lim\limits_{x \to -1} \dfrac{x^3 + 2x^2 + x}{x^4 + x^3 + 2x + 2}$

32. $\lim\limits_{x \to 3} \dfrac{x^3 - 3x^2 + 4x - 12}{x^4 - 3x^3 + x - 3}$

In Problems 33–44, find the limit as x approaches c of the average rate of change of each function from c to x.

33. $c = 2$; $f(x) = 5x - 3$

34. $c = -2$; $f(x) = 4 - 3x$

35. $c = 3$; $f(x) = x^2$

36. $c = 3$; $f(x) = x^3$

37. $c = -1$; $f(x) = x^2 + 2x$

38. $c = -1$; $f(x) = 2x^2 - 3x$

39. $c = 0$; $f(x) = 3x^3 - 2x^2 + 4$

40. $c = 0$; $f(x) = 4x^3 - 5x + 8$

41. $c = 1$; $f(x) = \dfrac{1}{x}$

42. $c = 1$; $f(x) = \dfrac{1}{x^2}$

43. $c = 4$; $f(x) = \sqrt{x}$

44. $c = 1$; $f(x) = \sqrt{x}$

In Problems 45–52, assume that $\lim\limits_{x \to c} f(x) = 5$ and $\lim\limits_{x \to c} g(x) = 2$ to find each limit.

45. $\lim\limits_{x \to c} [2f(x)]$

46. $\lim\limits_{x \to c} [f(x) - g(x)]$

47. $\lim\limits_{x \to c} [g(x)^3]$

48. $\lim\limits_{x \to c} \dfrac{f(x)}{g(x)}$

49. $\lim\limits_{x \to c} \dfrac{4}{f(x)}$

50. $\lim\limits_{x \to c} \dfrac{3}{g(x)}$

51. $\lim\limits_{x \to c} [4f(x) - 5g(x)]$

52. $\lim\limits_{x \to c} [8f(x) \cdot g(x)]$

12.3 One-Sided Limits; Continuous Functions

PREPARING FOR THIS SECTION ***Before getting started, review the following:***

- > Piecewise-defined Functions (Chapter 10, Section 10.4, pp. 583–585)
- > Library of Functions (Chapter 10, Section 10.4, pp. 579–583)
- > Properties of the Logarithmic Function (Chapter 11, Section 11.4, p. 646)
- > Domain of Rational Functions (Chapter 11, Section 11.2, p. 624)
- > Properties of the Exponential Function (Chapter 11, Section 11.3, pp. 631–633)

OBJECTIVES
1 Find the one-sided limits of a function
2 Determine whether a function is continuous

1 Find the one-sided limits of a function

Earlier we described $\lim\limits_{x \to c} f(x) = N$ by saying that as x gets closer to c, but remains unequal to c, the corresponding value of $f(x)$ gets closer to N. Whether we use a numerical argument or the graph of the function f, the variable x can get closer to c in only two ways: either by approaching c from the left, through numbers less than c, or by approaching c from the right, through numbers greater than c.

If we only approach c from one side, we have a **one-sided limit.** The notation

$$\lim_{x \to c^-} f(x) = L$$

sometimes called the **left limit,** read as "the limit of $f(x)$ as x approaches c from the left equals L," may be described by the following statement:

> As x gets closer to c, but remains less than c, the corresponding value of $f(x)$ gets closer to L.

The notation $x \to c^-$ is used to remind us that x is less than c.

The notation

$$\lim_{x \to c^+} f(x) = R$$

sometimes called the **right limit,** read as "the limit of $f(x)$ as x approaches c from the right equals R," may be described by the following statement:

> As x gets closer to c, but remains greater than c, the corresponding value of $f(x)$ gets closer to R.

The notation $x \to c^+$ is used to remind us that x is greater than c.

Figure 6 illustrates left and right limits.

FIGURE 6

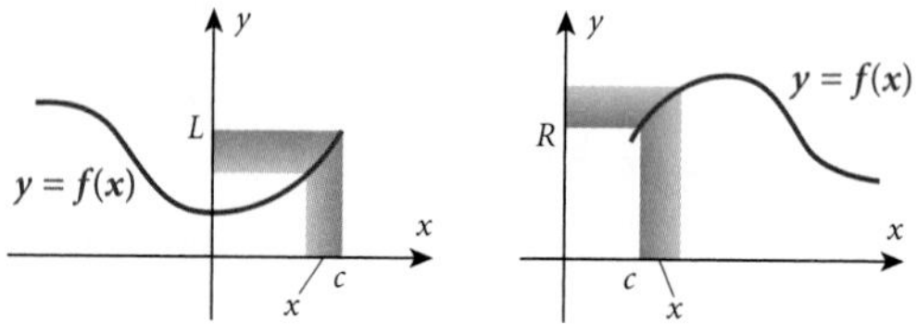

$x < c$
$\lim_{x \to c^-} f(x) = L$
(a)

$x > c$
$\lim_{x \to c^+} f(x) = R$
(b)

The left and right limits can be used to determine whether $\lim_{x \to c} f(x)$ exists. See Figure 7.

FIGURE 7

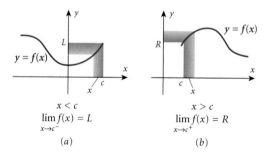

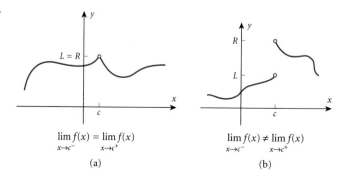

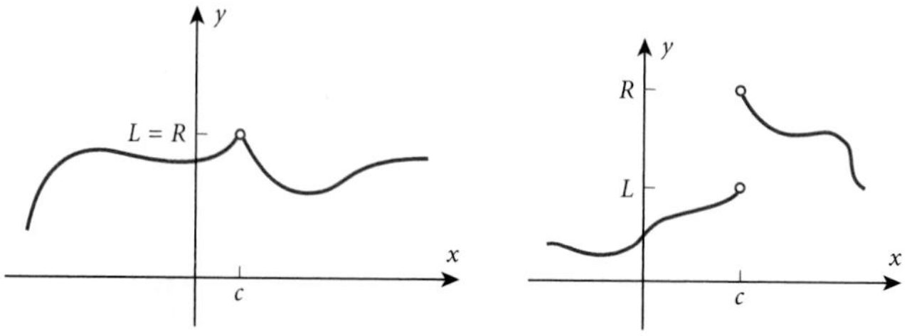

$\lim_{x \to c^-} f(x) = \lim_{x \to c^+} f(x)$
(a)

$\lim_{x \to c^-} f(x) \neq \lim_{x \to c^+} f(x)$
(b)

As Figure 7(a) illustrates, $\lim_{x \to c} f(x)$ exists and equals the common value of the left limit and the right limit ($L = R$). In Figure 7(b), we see that $\lim_{x \to c} f(x)$ does not exist and that $L \neq R$. This leads us to the following result:

> Suppose that $\lim_{x \to c^-} f(x) = L$ and $\lim_{x \to c^+} f(x) = R$. Then $\lim_{x \to c} f(x)$ exists if and only if $L = R$. Furthermore, if $L = R$, then $\lim_{x \to c} f(x) = L(=R)$.

Collectively, the left and right limits of a function are called **one-sided limits** of the function.

EXAMPLE 1 Finding One-Sided Limits of a Function

For the function

$$f(x) = \begin{cases} 2x - 1 & \text{if } x < 2 \\ 1 & \text{if } x = 2 \\ x - 2 & \text{if } x > 2 \end{cases}$$

find: **(a)** $\lim_{x \to 2^-} f(x)$ **(b)** $\lim_{x \to 2^+} f(x)$ **(c)** $\lim_{x \to 2} f(x)$

SOLUTION Figure 8 shows the graph of f.

FIGURE 8

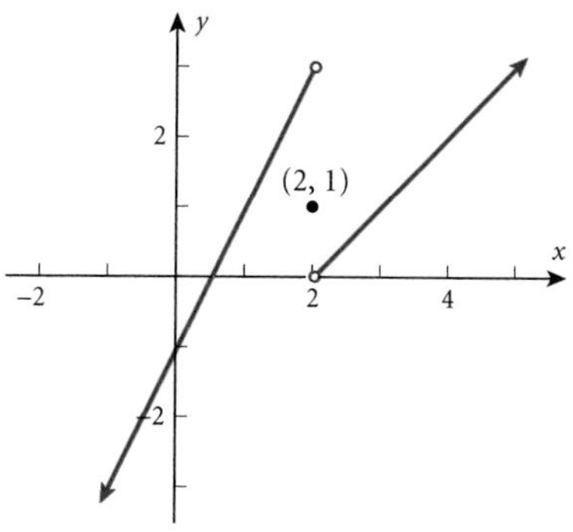

(a) To find $\lim_{x \to 2^-} f(x)$, we look at the values of f when x is close to 2, but less than 2. Since $f(x) = 2x - 1$ for such numbers, we conclude that

$$\lim_{x \to 2^-} f(x) = \lim_{x \to 2^-} (2x - 1) = 3$$

(b) To find $\lim_{x \to 2^+} f(x)$, we look at the values of f when x is close to 2, but greater than 2. Since $f(x) = x - 2$ for such numbers, we conclude that

$$\lim_{x \to 2^+} f(x) = \lim_{x \to 2^+} (x - 2) = 0$$

(c) Since the left and right limits are unequal, $\lim_{x \to 2} f(x)$ does not exist.

NOW WORK PROBLEMS 9 AND 23.

Continuous Functions

2 Determine whether a function is continuous

We have observed that the value of a function f at c, namely $f(c)$, plays no role in determining the one-sided limits of f at c. What is the role of the value of a function at c and its one-sided limits at c? Let's look at some of the possibilities. See Figure 9.

FIGURE 9

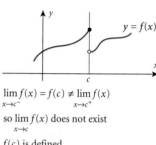

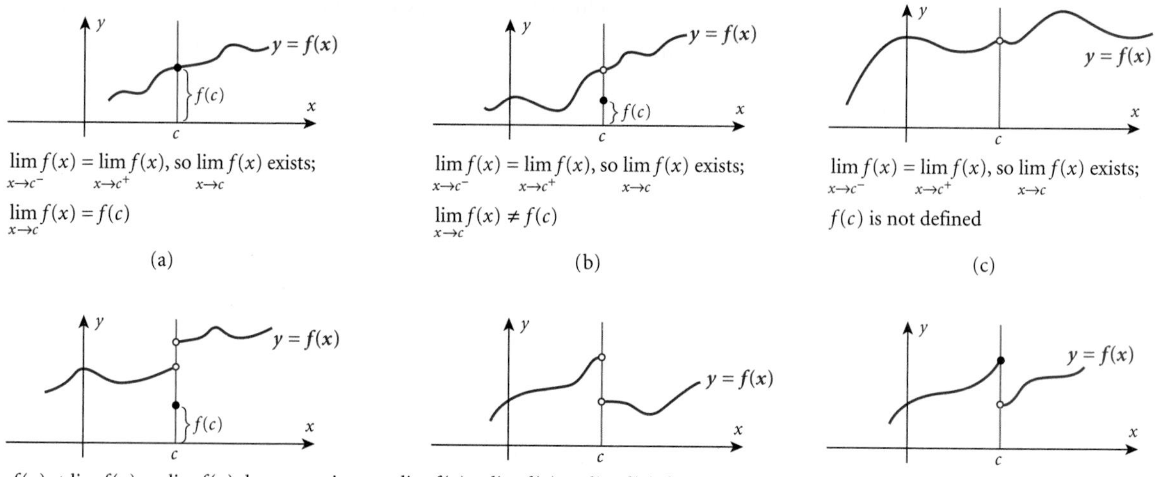

(a) $\lim_{x \to c^-} f(x) = \lim_{x \to c^+} f(x)$, so $\lim_{x \to c} f(x)$ exists; $\lim_{x \to c} f(x) = f(c)$

(b) $\lim_{x \to c^-} f(x) = \lim_{x \to c^+} f(x)$, so $\lim_{x \to c} f(x)$ exists; $\lim_{x \to c} f(x) \neq f(c)$

(c) $\lim_{x \to c^-} f(x) = \lim_{x \to c^+} f(x)$, so $\lim_{x \to c} f(x)$ exists; $f(c)$ is not defined

(d) $\lim_{x \to c^-} f(x) \neq \lim_{x \to c^+} f(x)$, so $\lim_{x \to c} f(x)$ does not exist; $f(c)$ is defined

(e) $\lim_{x \to c^-} f(x) \neq \lim_{x \to c^+} f(x)$, so $\lim_{x \to c} f(x)$ does not exist; $f(c)$ is not defined

(f) $\lim_{x \to c^-} f(x) = f(c) \neq \lim_{x \to c^+} f(x)$ so $\lim_{x \to c} f(x)$ does not exist; $f(c)$ is defined

Much earlier in this book, we said that a function f was *continuous* if its graph could be drawn without lifting pencil from paper. In looking at Figure 9, the only graph that has this characteristic is the graph in Figure 9(a), for which the one-sided limits at c each exist and are equal to the value of f at c. This leads us to the following definition:

> A function f is **continuous** at c if:
>
> **1.** f is defined at c; that is, c is in the domain of f so that $f(c)$ equals a number.
>
> **2.** $\lim_{x \to c^-} f(x) = f(c)$
>
> **3.** $\lim_{x \to c^+} f(x) = f(c)$
>
> In other words, a function f is continuous at c if
>
> $$\lim_{x \to c} f(x) = f(c)$$

If f is not continuous at c, we say that f is **discontinuous at c.** Each of the functions whose graphs appear in Figures 9(b) to 9(f) is discontinuous at c.

NOW WORK PROBLEM 15.

Look again at Formula (8) on page 687. Based on (8), we conclude that a polynomial function is continuous at every number. Look at Formula (12). We conclude that a rational function is continuous at every number, except numbers at which it is not defined.

As we mentioned in Chapter 11, pp. 631, 632, 633, and 646, the exponential and logarithmic functions are continuous at every number in their domain. Look at the graphs of the square root function, the absolute value function, and the greatest integer function on pages 581–583. We see that the square root function and absolute value function are continuous at every number in their domain. The function $f(x) = \text{int}(x)$ is continuous except for $x =$ an integer, where a jump occurs in the graph.

Piecewise-defined functions require special attention.

EXAMPLE 2 Determining Where a Piecewise-Defined Function Is Continuous

Determine the numbers at which the following function is continuous.

$$f(x) = \begin{cases} x^2 & \text{if } x \leq 0 \\ x + 1 & \text{if } 0 < x < 2 \\ 5 - x & \text{if } 2 \leq x \leq 5 \end{cases}$$

SOLUTION The "pieces" of f, that is, $y = x^2$, $y = x + 1$, and $y = 5 - x$, are each continuous for every number since they are polynomials. In other words, when we graph the pieces, we will not lift our pencil. When we graph the function f, however, we have to be careful, because the pieces change at $x = 0$ and at $x = 2$. So the numbers we need to investigate further are $x = 0$ and $x = 2$.

For $x = 0$: $\quad f(0) = 0^2 = 0$

$$\lim_{x \to 0^-} f(x) = \lim_{x \to 0^-} x^2 = 0$$

$$\lim_{x \to 0^+} f(x) = \lim_{x \to 0^+} (x + 1) = 1$$

FIGURE 10

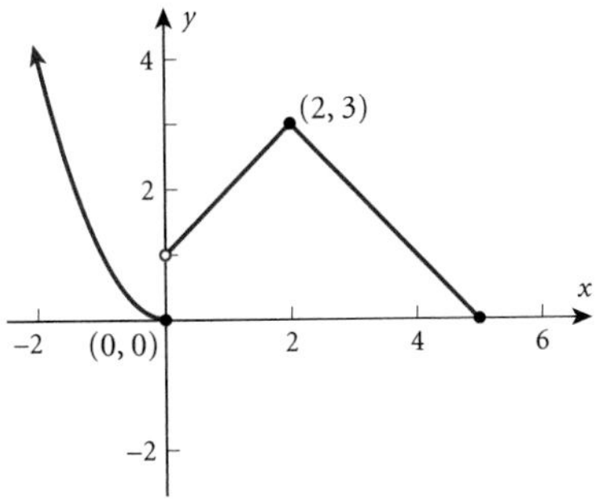

Since $\lim_{x \to 0^+} f(x) \neq f(0)$, we conclude that f is not continuous at $x = 0$.

For $x = 2$: $\quad f(2) = 5 - 2 = 3$

$$\lim_{x \to 2^-} f(x) = \lim_{x \to 2^-} (x + 1) = 3$$

$$\lim_{x \to 2^+} f(x) = \lim_{x \to 2^+} (5 - x) = 3$$

We conclude that f is continuous at $x = 2$.

The function f is continuous for all x, except $x = 0$. The graph of f, given in Figure 10, demonstrates this conclusion.

NOW WORK PROBLEMS 41 AND 49.

SUMMARY Continuity Properties

Function	*Domain*	*Property*
Polynomial function	All real numbers	Continuous at every number in the domain
Rational function $R(x) = \dfrac{p(x)}{q(x)}$ p, q are polynomials	$\{x \mid q(x) \neq 0\}$	Continuous at every number in the domain
Exponential function	All real numbers	Continuous at every number in the domain
Logarithmic function	Positive real numbers	Continuous at every number in the domain

EXERCISE 12.3 Answers to Odd-Numbered Problems Begin on Page AN-69.

In Problems 1–20, use the accompanying graph of $y = f(x)$.

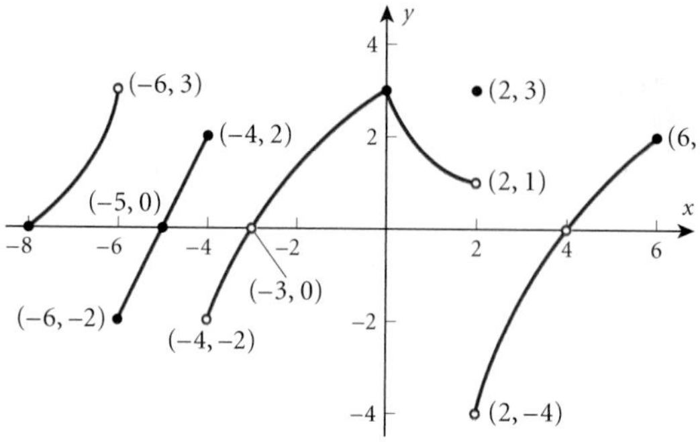

1. What is the domain of f?

2. What is the range of f?

3. Find the x-intercept(s), if any, of f.

4. Find the y-intercept(s), if any, of f.

5. Find $f(-8)$ and $f(-4)$.

6. Find $f(2)$ and $f(6)$.

7. Find $\lim_{x \to -6^-} f(x)$.

8. Find $\lim_{x \to -6^+} f(x)$.

9. Find $\lim_{x \to -4^-} f(x)$.

10. Find $\lim_{x \to -4^+} f(x)$.

11. Find $\lim_{x \to 2^-} f(x)$.

12. Find $\lim_{x \to 2^+} f(x)$.

13. Does $\lim_{x \to 4} f(x)$ exist? If it does, what is it?

14. Does $\lim_{x \to 0} f(x)$ exist? If it does, what is it?

15. Is f continuous at -6?

16. Is f continuous at -4?

17. Is f continuous at 0?

18. Is f continuous at 2?

19. Is f continuous at 4?

20. Is f continuous at 5?

In Problems 21–32, find the one-sided limit.

21. $\lim_{x \to 1^+} (2x + 3)$

22. $\lim_{x \to 2^-} (4 - 2x)$

23. $\lim_{x \to 1^-} (2x^3 + 5x)$

24. $\lim_{x \to -2^+} (3x^2 - 8)$

25. $\lim_{x \to 0^-} e^x$

26. $\lim_{x \to 0^+} e^x$

27. $\lim_{x \to 2^+} \dfrac{x^2 - 4}{x - 2}$

28. $\lim_{x \to 1^-} \dfrac{x^3 - x}{x - 1}$

29. $\lim_{x \to -1^-} \dfrac{x^2 - 1}{x^3 + 1}$

30. $\lim_{x \to 0^+} \dfrac{x^3 - x^2}{x^4 + x^2}$

31. $\lim_{x \to -2^+} \dfrac{x^2 + x - 2}{x^2 + 2x}$

32. $\lim_{x \to -4^-} \dfrac{x^2 + x - 12}{x^2 + 4x}$

In Problems 33–48, determine whether f is continuous at c. Justify your answer.

33. $f(x) = x^3 - 3x^2 + 2x - 6 \quad c = 2$

34. $f(x) = 3x^2 - 6x + 5 \quad c = -3$

35. $f(x) = \dfrac{x^2 + 5}{x - 6} \quad c = 3$

36. $f(x) = \dfrac{x^3 - 8}{x^2 + 4} \quad c = 2$

37. $f(x) = \dfrac{x + 3}{x - 3} \quad c = 3$

38. $f(x) = \dfrac{x - 6}{x + 6} \quad c = -6$

39. $f(x) = \dfrac{x^3 + 3x}{x^2 - 3x} \quad c = 0$

40. $f(x) = \dfrac{x^2 - 6x}{x^2 + 6x} \quad c = 0$

41. $f(x) = \begin{cases} \dfrac{x^3 + 3x}{x^2 - 3x} & \text{if } x \neq 0 \\ 1 & \text{if } x = 0 \end{cases} \quad c = 0$

42. $f(x) = \begin{cases} \dfrac{x^2 - 6x}{x^2 + 6x} & \text{if } x \neq 0 \\ -2 & \text{if } x = 0 \end{cases} \quad c = 0$

43. $f(x) = \begin{cases} \dfrac{x^3 + 3x}{x^2 - 3x} & \text{if } x \neq 0 \\ -1 & \text{if } x = 0 \end{cases} \quad c = 0$

44. $f(x) = \begin{cases} \dfrac{x^2 - 6x}{x^2 + 6x} & \text{if } x \neq 0 \\ -1 & \text{if } x = 0 \end{cases} \quad c = 0$

45. $f(x) = \begin{cases} \dfrac{x^3 - 1}{x^2 - 1} & \text{if } x < 1 \\ 2 & \text{if } x = 1 \\ \dfrac{3}{x + 1} & \text{if } x > 1 \end{cases} \quad c = 1$

46. $f(x) = \begin{cases} \dfrac{x^2 - 2x}{x - 2} & \text{if } x < 2 \\ 2 & \text{if } x = 2 \\ \dfrac{x - 4}{x - 1} & \text{if } x > 2 \end{cases} \quad c = 2$

47. $f(x) = \begin{cases} 2e^x & \text{if } x < 0 \\ 2 & \text{if } x = 0 \\ \dfrac{x^3 + 2x^2}{x^2} & \text{if } x > 0 \end{cases} \quad c = 0$

48. $f(x) = \begin{cases} 3e^{-x} & \text{if } x < 0 \\ 3 & \text{if } x = 0 \\ \dfrac{x^3 + 3x^2}{x^2} & \text{if } x > 0 \end{cases} \quad c = 0$

In Problems 49–62, find the numbers at which f is continuous. At which numbers is f discontinuous?

49. $f(x) = 2x + 3$

50. $f(x) = 4 - 3x$

51. $f(x) = 3x^2 + x$

52. $f(x) = -3x^3 + 7$

53. $f(x) = 4 \ln x$

54. $f(x) = -2 \ln(x - 3)$

55. $f(x) = 3e^x$

56. $f(x) = 4e^{-x}$

57. $f(x) = \dfrac{2x + 5}{x^2 - 4}$

58. $f(x) = \dfrac{x^2 - 4}{x^2 - 9}$

59. $f(x) = \dfrac{x - 3}{\ln x}$

60. $f(x) = \dfrac{\ln x}{x - 3}$

61. $f(x) = \begin{cases} 3x + 1 & \text{if } x \leq 0 \\ -x^2 & \text{if } 0 < x \leq 2 \\ \dfrac{1}{2}x - 5 & \text{if } x > 2 \end{cases}$

62. $f(x) = \begin{cases} -1 & \text{if } x < -2 \\ x + 1 & \text{if } -2 \leq x \leq 1 \\ x^2 + x + 1 & \text{if } x > 1 \end{cases}$

63. Cell Phone Service Sprint PCS offers a monthly cellular phone plan for \$39.99. It includes 350 anytime minutes plus \$0.25 per minute for additional minutes. The following function is used to compute the monthly cost for a subscriber

$$C(x) = \begin{cases} 39.99 & \text{if } 0 < x \leq 350 \\ 0.25x - 47.51 & \text{if } x > 350 \end{cases}$$

where x is the number of anytime minutes used.

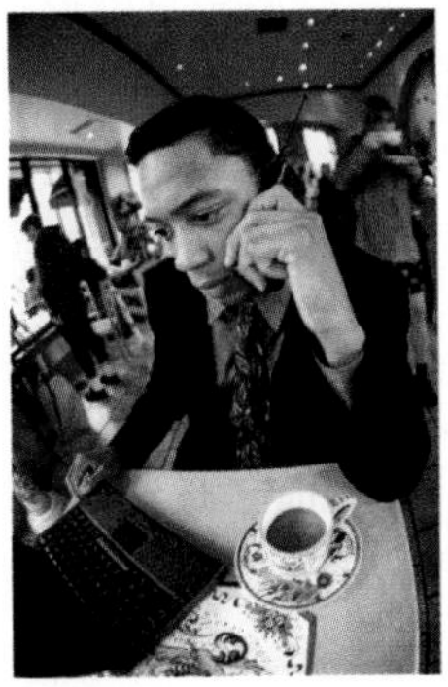

(a) Find $\lim_{x \to 350^-} C(x)$

(b) Find $\lim_{x \to 350^+} C(x)$

(c) Is C continuous at 350?

(d) Give an explanation for your answer in part (c).

64. First-class Letter According to the U.S. Postal Service, first-class mail is used for personal and business correspondence. Any mailable item may be sent as first-class mail. It includes postcards, letters, large envelopes, and small packages. The maximum weight is 13 ounces. The following function is used to compute the cost of mailing an item first-class.

$$C(x) = \begin{cases} 0.37 & \text{if } 0 < x \leq 1 \\ 0.23 \text{ int}(x) + 0.37 & \text{if } 1 < x \leq 13 \end{cases}$$

where x is the weight of the package in ounces.

(a) Find $\lim_{x \to 1^-} C(x)$

(b) Find $\lim_{x \to 1^+} C(x)$

(c) Is C continuous at 1?

(d) Give an explanation for your answer in part (c).

65. Wind Chill The wind chill factor represents the equivalent air temperature at a standard wind speed that would produce the same heat loss as the given temperature and wind speed. One formula for computing the equivalent temperature is

$$W = \begin{cases} t & 0 \leq v < 1.79 \\ 33 - \dfrac{(10.45 + 10\sqrt{v} - v)(33 - t)}{22.04} & 1.79 \leq v \leq 20 \\ 33 - 1.5958(33 - t) & v > 20 \end{cases}$$

where v represents the wind speed (in meters per second) and t represents the air temperature (in °C). Suppose $t = 10$°C.

(a) Write the function $W = W(v)$

(b) Find $\lim_{v \to 0^+} W(v)$

(c) Find $\lim_{v \to 1.79^-} W(v)$

(d) Find $\lim_{v \to 1.79^+} W(v)$

(e) Find $W(1.79)$.

(f) Is W continuous at $v = 1.79$?

(g) Round the answers obtained in parts (c), (d), and (e) to two decimal places. Now is W continuous at $v = 1.79$?

(h) Comment on your answers to parts (f) and (g).

(i) Find $\lim_{v \to 20^-} W(v)$

(j) Find $\lim_{v \to 20^+} W(v)$

(k) Find $W(20)$.

(l) Is W continuous at $v = 20$?

(m) Round the answers obtained in parts (i), (j) and (k) to two decimal places. Now is W continuous at $v = 20$?

(n) Comment on your answers to parts (l) and (m).

66. Rework Problem 65 where $t = 0°$ C.

12.4 Limits at Infinity; Infinite Limits; End Behavior; Asymptotes

PREPARING FOR THIS SECTION ***Before getting started, review the following:***

- Rational Functions (Chapter 11, Section 11.2, pp. 624–625)
- End Behavior of a Polynomial Function (Chapter 11, Section 11.2, pp. 623–624)
- Graph of $f(x) = \frac{1}{x}$ (Chapter 10, Section 10.1, p. 546)

OBJECTIVES

1. Find limits at infinity
2. Find infinite limits
3. Find horizontal asymptotes
4. Find vertical asymptotes
5. Analyze the graph of a rational function at points of discontinuity

In Section 12.2 we described $\lim_{x \to c} f(x) = N$ by saying that the value of f can be made as close as we please to N by choosing numbers x sufficiently close to c. It was understood that N and c were real numbers. In this section we extend the language of limits to allow c to be ∞ or $-\infty$ (*limits at infinity*) and to allow N to be ∞ or $-\infty$ (*infinite limits*).* These limits, it turns out, are useful for finding the end behavior and locating *asymptotes* which help in obtaining the graph of certain functions.

We begin with limits at infinity.

Limits at Infinity

Let's look again at the graph of the function $f(x) = \frac{1}{x}$, whose domain is $\{x \mid x \neq 0\}$, that was discussed in Section 10.1, p. 546. See Figure 11.

FIGURE 11

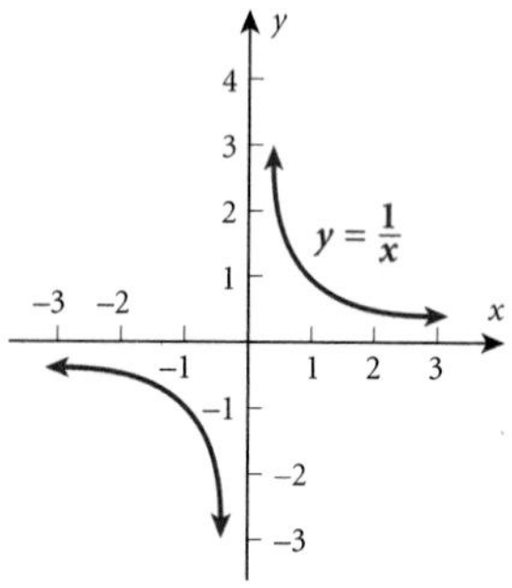

Table 6 illustrates that the values of f can be made as close as we please to 0 as x becomes unbounded in the positive direction.

TABLE 6

x	1	10	100	1000	10,000	100,000
$f(x) = \frac{1}{x}$	1	0.1	0.01	0.001	0.0001	0.00001

**Remember that the symbols ∞ (infinity) and $-\infty$ (minus infinity) are not numbers. Infinity expresses the idea of unboundedness in the positive direction; minus infinity expresses the idea of unboundedness in the negative direction.*

This conclusion is expressed by saying that $f(x) = \frac{1}{x}$ has the limit 0 as x approaches ∞ and is symbolized by writing

$$\lim_{x \to \infty} \frac{1}{x} = 0 \tag{1}$$

In the same way, we can write

$$\lim_{x \to -\infty} \frac{1}{x} = 0 \tag{2}$$

to indicate that $\frac{1}{x}$ can be made as close as we please to 0 as x becomes unbounded in the negative direction. We summarize statements (1) and (2) by saying that $f(x) = \frac{1}{x}$ has **limits at infinity.**

Recall that as x becomes unbounded in the positive direction or unbounded in the negative direction, the graph of a polynomial function

$$f(x) = a_n x^n + a_{n-1} x^{n-1} + \ldots + a_1 x + a_0 \qquad a_n \neq 0,$$

behaves the same as the graph of $y = a_n x^n$. In other words, as $x \to -\infty$ or as $x \to \infty$, we can replace $a_n x^n + a_{n-1} x^{n-1} + \ldots + a_1 x + a_0$ by $a_n x^n$. We use this fact to find limits of rational functions at infinity.

1 **EXAMPLE 1 Finding Limits at Infinity**

Find: **(a)** $\lim_{x \to \infty} \frac{3x - 2}{4x - 1}$ **(b)** $\lim_{x \to \infty} \frac{5x^2 - 3x + 2}{x^3 + 5}$

SOLUTION **(a)** $\lim_{x \to \infty} \frac{3x - 2}{4x - 1} = \lim_{x \to \infty} \frac{3x}{4x}$ As $x \to \infty$, $3x - 2 = 3x$ and $4x - 1 = 4x$

$$= \lim_{x \to \infty} \frac{3}{4} = \frac{3}{4}$$

(b) We follow the same procedure as in part (a):

$$\lim_{x \to \infty} \frac{5x^2 - 3x + 2}{x^3 + 5} = \lim_{x \to \infty} \frac{5x^2}{x^3} = 5 \lim_{x \to \infty} \frac{1}{x} = 0$$

$\uparrow$ $\lim_{x \to \infty} \frac{1}{x} = 0$

NOW WORK PROBLEM 1.

Infinite Limits

Again we use the function $f(x) = \frac{1}{x}$, whose graph is given in Figure 11, to introduce the idea of **infinite limits.** Table 7 gives values of f for selected numbers x that are close to 0 and positive:

TABLE 7

x	1	0.1	0.01	0.001	0.0001	0.00001
$f(x) = \frac{1}{x}$	1	10	100	1000	10,000	100,000

We see that as x gets closer to 0 from the right, the value of $f(x) = \frac{1}{x}$ is becoming unbounded in the positive direction. We express this fact by writing

$$\lim_{x \to 0^+} \frac{1}{x} = \infty \tag{3}$$

Similarly, we use the notation

$$\lim_{x\to 0^-} \frac{1}{x} = -\infty \tag{4}$$

to indicate that as x gets closer to 0, and is negative, the values of $\frac{1}{x}$ are becoming unbounded in the negative direction. We summarize (3) and (4) by saying that $f(x) = \frac{1}{x}$ has **one-sided infinite limits** at 0.

2 EXAMPLE 2 Finding Infinite Limits

Find: $\lim_{x\to 4^+} \frac{2-x}{x-4}$

SOLUTION As x gets closer to 4, $x > 4$, then $2 - x$ gets closer to -2 and $\frac{1}{x-4}$ is positive and unbounded. As a result, as $x \to 4^+$, the expression $\frac{2-x}{x-4}$ is negative and unbounded. That is,

$$\lim_{x\to 4^+} \frac{2-x}{x-4} = -\infty$$

NOW WORK PROBLEM 13.

We now apply the ideas of limits at infinity and infinite limits to the problem of finding end behavior and locating horizontal asymptotes.

End Behavior; Horizontal Asymptotes

The limit at infinity of a function provides information about the end behavior of the graph. This limit can be infinite, indicating that the graph is becoming unbounded as $x \to -\infty$ or as $x \to \infty$. When this limit is a number, a horizontal asymptote describes the end behavior of the graph.

For example, if $\lim_{x\to\infty} f(x) = N$, it means that as x becomes unbounded in the positive direction, the value of f can be made as close as we please to N. That is, the graph of $y = f(x)$ for x sufficiently positive is as close as we please to the horizontal line $y = N$. Similarly, $\lim_{x\to-\infty} f(x) = M$ means that the graph of $y = f(x)$ for x sufficiently negative is as close as we please to the horizontal line $y = M$. These lines are called **horizontal asymptotes** of the graph of f. See Figure 12.

FIGURE 12

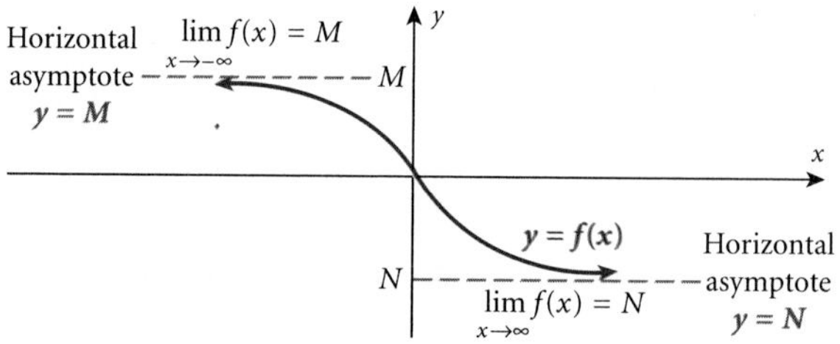

3 EXAMPLE 3 Finding Horizontal Asymptotes

Find the horizontal asymptotes, if any, of the graph of

$$f(x) = \frac{4x^2}{x^2+2}$$

SOLUTION To find any horizontal asymptotes, we need to examine two limits: $\lim_{x\to\infty} f(x)$ and $\lim_{x\to-\infty} f(x)$.

$$\lim_{x\to\infty} f(x) = \lim_{x\to\infty} \frac{4x^2}{x^2+2} = \lim_{x\to\infty} \frac{4x^2}{x^2} = 4$$

We conclude that the line $y = 4$ is a horizontal asymptote of the graph when x is sufficiently positive.

$$\lim_{x \to -\infty} f(x) = \lim_{x \to -\infty} \frac{4x^2}{x^2 + 2} = \lim_{x \to -\infty} \frac{4x^2}{x^2} = 4$$

We conclude that the line $y = 4$ is a horizontal asymptote of the graph when x is sufficiently negative.

These conclusions also explain the end behavior of the graph.

Vertical Asymptotes

Infinite limits are used to find vertical asymptotes. Figure 13 illustrates some of the possibilities that can occur when a function has an infinite limit.

FIGURE 13

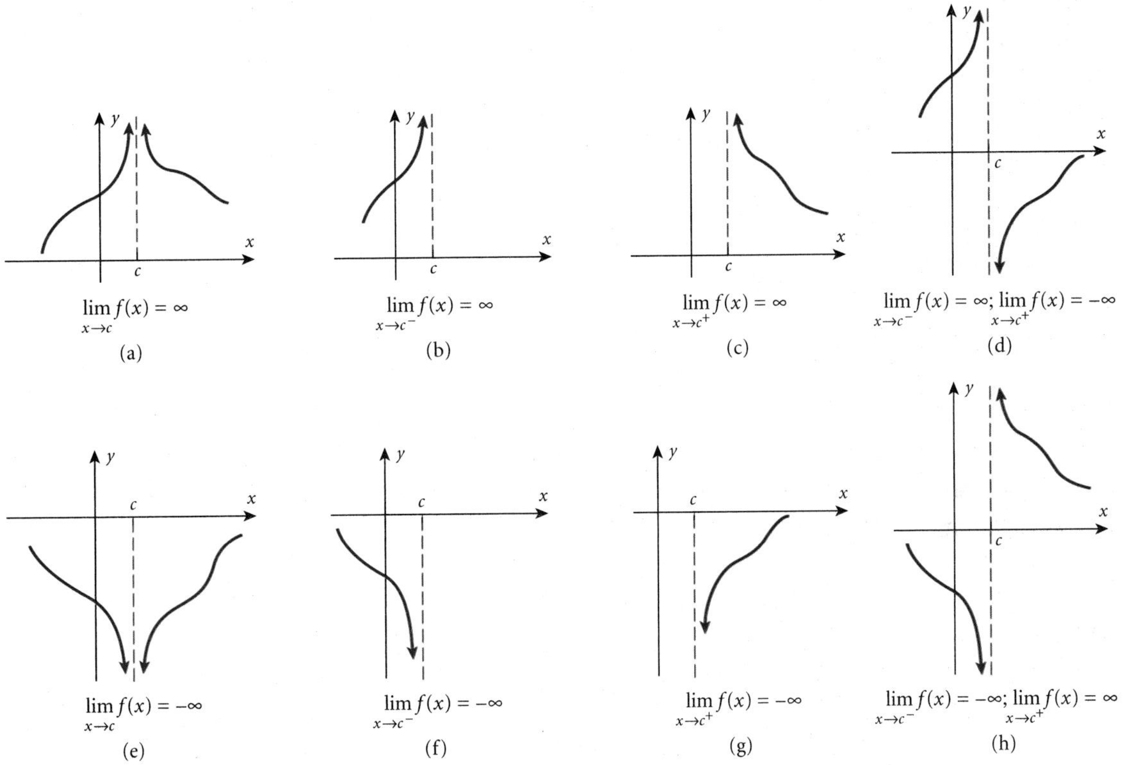

$\lim_{x \to c} f(x) = \infty$ (a)

$\lim_{x \to c^-} f(x) = \infty$ (b)

$\lim_{x \to c^+} f(x) = \infty$ (c)

$\lim_{x \to c^-} f(x) = \infty$; $\lim_{x \to c^+} f(x) = -\infty$ (d)

$\lim_{x \to c} f(x) = -\infty$ (e)

$\lim_{x \to c^-} f(x) = -\infty$ (f)

$\lim_{x \to c^+} f(x) = -\infty$ (g)

$\lim_{x \to c^-} f(x) = -\infty$; $\lim_{x \to c^+} f(x) = \infty$ (h)

Whenever

$$\lim_{x \to c^-} f(x) = \infty \text{ (or } -\infty) \quad \text{or} \quad \lim_{x \to c^+} f(x) = \infty \text{ (or } -\infty)$$

we call the line $x = c$ a **vertical asymptote** of the graph of f.

For rational functions, vertical asymptotes can occur at numbers at which the rational function is not defined.

4 **EXAMPLE 4** **Finding Vertical Asymptotes**

Find the vertical asymptotes, if any, of the rational function

$$R(x) = \frac{x^2}{x - 4}$$

SOLUTION The domain of the rational function R is $\{x \mid x \neq 4\}$. To examine the behavior of the graph of R near 4, where R is not defined, we look at

$$\lim_{x \to 4} R(x) = \lim_{x \to 4} \frac{x^2}{x - 4}$$

This will require that we examine the one-sided limits of R at 4.

$\lim_{x \to 4^-} R(x)$: Since $x \to 4^-$, we know $x < 4$, so $x - 4 < 0$. Since $x^2 \geq 0$, it follows that the expression $\frac{x^2}{x-4}$ is negative and becomes unbounded as $x \to 4^-$. That is,

$$\lim_{x \to 4^-} R(x) = \lim_{x \to 4^-} \frac{x^2}{x - 4} = -\infty$$

$\lim_{x \to 4^+} R(x)$: Since $x \to 4^+$, we know $x > 4$, so $x - 4 > 0$. Since $x^2 \geq 0$, it follows that the expression $\frac{x^2}{x-4}$ is positive and becomes unbounded as $x \to 4^+$. That is,

$$\lim_{x \to 4^+} R(x) = \lim_{x \to 4^+} \frac{x^2}{x - 4} = \infty$$

We conclude that the graph of R has a vertical asymptote at $x = 4$.

The graph of $R(x) = \frac{x^2}{x-4}$, based on the information obtained in Example 3, will exhibit the behavior shown in Figure 14 near $x = 4$.

FIGURE 14

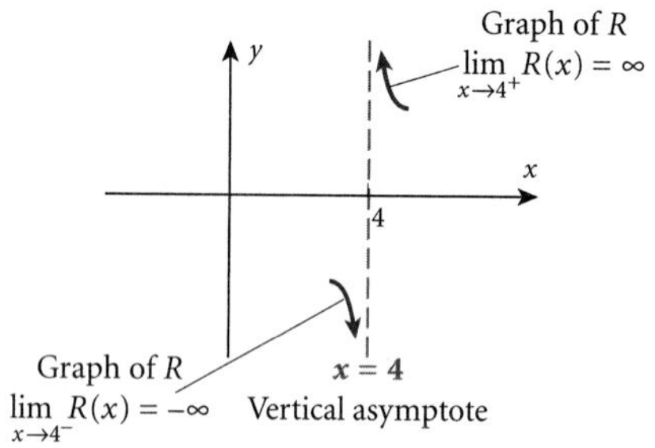

NOW WORK PROBLEM 25.

Numbers at which a rational function is not defined are referred to as **points of discontinuity.** Example 4 demonstrates that a rational function can have a vertical asymptote at a point of discontinuity. Sometimes a rational function has a hole at a point of discontinuity.

5 **EXAMPLE 5** **Analyzing the Graph of a Rational Function at Points of Discontinuity**

(a) Determine the numbers at which the rational function

$$R(x) = \frac{x - 2}{x^2 - 6x + 8}$$

is continuous.

(b) Use limits to analyze the graph of R at any points of discontinuity.

(c) Graph R.

SOLUTION **(a)** Since $R(x) = \dfrac{x-2}{(x-2)(x-4)}$, the domain of R is $\{x \mid x \neq 2, x \neq 4\}$.

We conclude that R is discontinuous at both 2 and 4. (Condition 1 of the definition is violated.) Since R is a rational function, R is continuous at every number, except 2 and 4.

(b) To determine the behavior of the graph at the points of discontinuity, 2 and 4, we look at $\lim_{x\to 2} R(x)$ and $\lim_{x\to 4} R(x)$.

For $\lim_{x\to 2} R(x)$, we have

$$\lim_{x\to 2} R(x) = \lim_{x\to 2} \frac{\cancel{x-2}}{\cancel{(x-2)}(x-4)} = \lim_{x\to 2} \frac{1}{x-4} = -\frac{1}{2}$$

As x gets closer to 2, the graph of R gets closer to $-\dfrac{1}{2}$. Since R is not defined at 2, the graph will have a hole at $\left(2, -\dfrac{1}{2}\right)$.

For $\lim_{x\to 4} R(x)$, we have

$$\lim_{x\to 4} R(x) = \lim_{x\to 4} \frac{\cancel{x-2}}{\cancel{(x-2)}(x-4)} = \lim_{x\to 4} \frac{1}{x-4}$$

Since the limit of the denominator is 0, we use one-sided limits to investigate $\lim_{x\to 4} \dfrac{1}{x-4}$.

If $x < 4$ and x is getting closer to 4, the value of $\dfrac{1}{x-4} < 0$ and is becoming unbounded; that is, $\lim_{x\to 4^-} R(x) = -\infty$.

If $x > 4$ and x is getting closer to 4, the value of $\dfrac{1}{x-4} > 0$ and is becoming unbounded; that is, $\lim_{x\to 4^+} R(x) = \infty$.

FIGURE 15

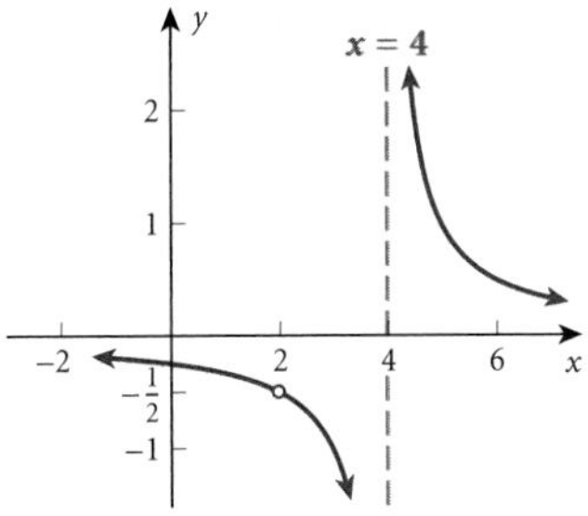

The graph of R will have a vertical asymptote at $x = 4$.

(c) It is easiest to graph R by observing that

$$\text{if } x \neq 2, \quad \text{then } R(x) = \frac{\cancel{x-2}}{\cancel{(x-2)}(x-4)} = \frac{1}{x-4}$$

So the graph of R is the graph of $y = \dfrac{1}{x}$ shifted to the right 4 units with a hole at $\left(2, -\dfrac{1}{2}\right)$ See Figure 15.

Example 4 illustrates the following general result.

> The graph of a rational function will have either a vertical asymptote or a hole at numbers at which it is not defined.

NOW WORK PROBLEM 29.

Next we list certain properties of the exponential function and the logarithmic function that involve limits at infinity and infinite limits.

Based on Figure 16, we conclude that:

$\lim_{x \to -\infty} e^x = 0$: The line $y = 0$ (the x-axis) is a horizontal asymptote as $x \to -\infty$.

$\lim_{x \to \infty} e^x = \infty$: The graph of $y = e^x$ becomes unbounded as $x \to \infty$.

See Figure 17. We conclude that:

$\lim_{x \to 0^-} \ln x = -\infty$: The graph of $y = \ln x$ has a vertical asymptote as $x \to 0^+$.

$\lim_{x \to \infty} \ln x = \infty$: The graph of $y = \ln x$ becomes unbounded as $x \to \infty$.

FIGURE 16

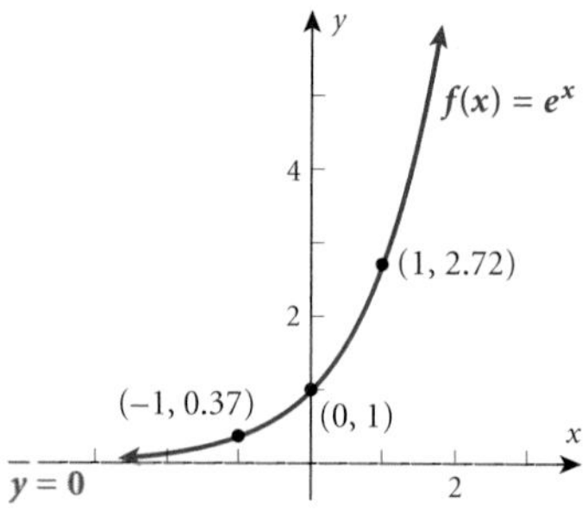

FIGURE 17

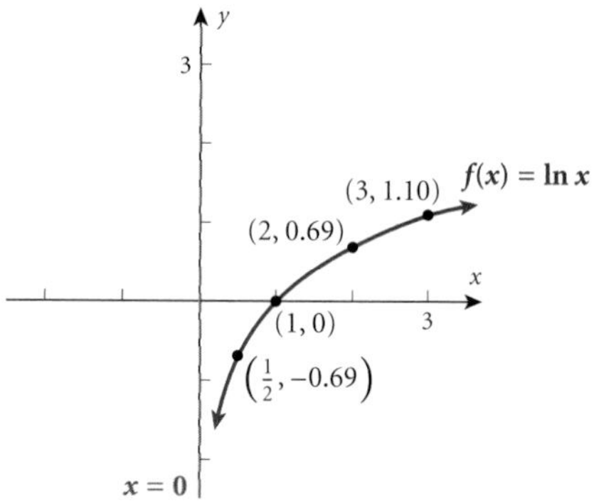

Application

EXAMPLE 6 Analyzing an Average Cost Function

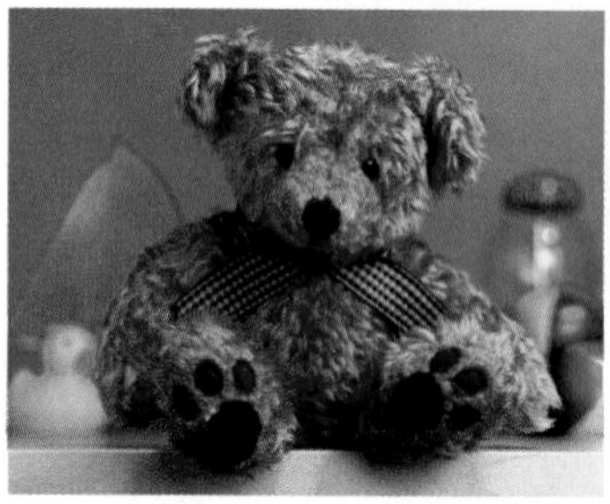

A company estimates that the fixed costs for producing a new toy are \$50,000 and the variable costs are \$3 per toy.

(a) Express the cost C of producing x toys as a function $C = C(x)$.
(b) Find the domain of C.
(c) Find the average cost of producing x toys; that is, find $\overline{C}(x) = \dfrac{C(x)}{x}$.
(d) Find the domain of $\overline{C}$.
(e) Find $\lim_{x \to 0^+} \overline{C}(x)$ and interpret the answer.
(f) Find $\lim_{x \to \infty} \overline{C}(x)$ and interpret the answer.

SOLUTION **(a)** The cost C of producing x toys is

$$C(x) = 50{,}000 + 3x$$

(b) The domain of C is $\{x \mid x \geq 0\}$.

(c) The average cost function $\overline{C}(x)$ is

$$\overline{C}(x) = \frac{C(x)}{x} = \frac{50{,}000 + 3x}{x} = \frac{50{,}000}{x} + 3$$

(d) The domain of $\overline{C}$ is $\{x \mid x > 0\}$.

(e) $\lim_{x \to 0^+} \overline{C}(x) = \lim_{x \to 0^+} \left(\dfrac{50{,}000}{x} + 3 \right) = \infty$

The average cost of producing close to zero toys will be unbounded. Notice that

$$C(1) = \$50{,}003 \quad \text{and} \quad C(2) = \$25{,}003$$

As you would expect, the average cost of producing a few toys is very high due to fixed costs.

FIGURE 18

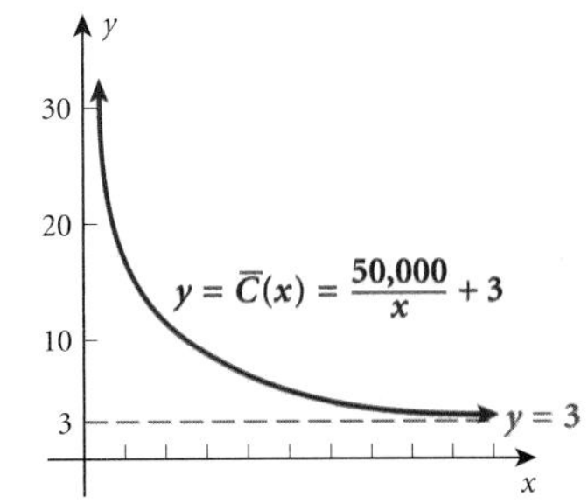

$y = \overline{C}(x) = \dfrac{50,000}{x} + 3$

$y = 3$

(f) $\lim\limits_{x\to\infty} \overline{C}(x) = \lim\limits_{x\to\infty}\left(\dfrac{50,000}{x} + 3\right) = \lim\limits_{x\to\infty}\dfrac{50,000}{x} + \lim\limits_{x\to\infty} 3 = 0 + 3 = 3$

The line $y = 3$ is a horizontal asymptote to the graph of $\overline{C} = \overline{C}(x)$. This means the average cost of producing x toys will never go lower than \$3 per toy. This is expected since the cost of producing each toy is \$3. In other words, the more toys produced, the closer the average cost gets to the unit cost.

Figure 18 shows the graph of $\overline{C} = \overline{C}(x)$. As the graph illustrates, the more units that are produced, the closer the average cost will get to the unit cost, and the less important the fixed cost becomes.

 NOW WORK PROBLEM 39.

SUMMARY

Function Name	End Behavior	Asymptotes
Polynomial Function $P(x) = a_n x^n + a_{n-1}x^{n-1} + \ldots + a_1 x + a_0,\ a \ne 0$ Domain: all real numbers	Behaves like the graph of $y = a_n x^n$	None
Rational Function $R(x) = \dfrac{p(x)}{q(x)},\ p, q$ polynomials Domain: $\{x \mid q(x) \ne 0\}$	Either unbounded or has a horizontal asymptote	If $\lim\limits_{x\to\infty} R(x) = N$, then $y = N$ is a horizontal asymptote. At numbers where $q(x) = 0$ either a hole or a vertical asymptote occurs.
Exponential Function $f(x) = e^x$ Domain: all real numbers	Unbounded as $x \to \infty$; $\lim\limits_{x\to\infty} e^x = \infty$	The line $y = 0$ (the x-axis) is a horizontal asymptote as $x \to -\infty$; $\lim\limits_{x\to-\infty} e^x = 0$
Logarithmic Function $f(x) = \ln x$ Domain: Positive real numbers	Unbounded as $x \to \infty$; $\lim\limits_{x\to\infty} \ln x = \infty$	The line $x = 0$ (the y-axis) is a vertical asymptote as $x \to 0^+$; $\lim\limits_{x\to 0^+} \ln x = -\infty$

EXERCISE 12.4   Answers to Odd-Numbered Problems Begin on Page AN-70.

In Problems 1–12, find each limit at infinity.

1. $\lim\limits_{x\to\infty} \dfrac{x^3 + x^2 + 2x - 1}{x^3 + x + 1}$

2. $\lim\limits_{x\to\infty} \dfrac{2x^2 - 5x + 2}{5x^2 + 7x - 1}$

3. $\lim\limits_{x\to\infty} \dfrac{2x + 4}{x - 1}$

4. $\lim\limits_{x\to\infty} \dfrac{x + 1}{x}$

5. $\lim\limits_{x\to\infty} \dfrac{3x^2 - 1}{x^2 + 4}$

6. $\lim\limits_{x\to-\infty} \dfrac{x^3 - 2x^2 + 1}{4x^3 + 5x + 4}$

7. $\lim\limits_{x\to-\infty} \dfrac{5x^3 - 1}{x^4 + 1}$

8. $\lim\limits_{x\to-\infty} \dfrac{x^2 + 1}{x^3 - 1}$

9. $\lim\limits_{x\to\infty} \dfrac{5x^3 + 3}{x^2 + 1}$

10. $\lim\limits_{x\to\infty} \dfrac{6x^2 + x}{x - 3}$

11. $\lim\limits_{x\to-\infty} \dfrac{4x^5}{x^2 + 1}$

12. $\lim\limits_{x\to-\infty} \dfrac{3x^6}{4x^3 - 1}$

In Problems 13–20, find each limit.

13. $\lim\limits_{x\to 2^+} \dfrac{1}{x - 2}$

14. $\lim\limits_{x\to-1^+} \dfrac{4}{x + 1}$

15. $\lim\limits_{x\to 1^-} \dfrac{x}{(x - 1)^2}$

16. $\lim\limits_{x\to-1^+} \dfrac{x^2}{(x + 1)^2}$

17. $\lim\limits_{x\to 1^+} \dfrac{x^2 + 1}{x^3 - 1}$

18. $\lim\limits_{x\to 3^-} \dfrac{6x^2 + x}{x - 3}$

19. $\lim\limits_{x\to 2^-} \dfrac{1 - x}{3x - 6}$

20. $\lim\limits_{x\to 5^+} \dfrac{2 - x}{5 - x}$

In Problems 21–26, locate all horizontal and vertical asymptotes, if any, of the function f.

21. $f(x) = 3 + \frac{1}{x^2}$

22. $f(x) = 2 - \frac{1}{x^2}$

23. $f(x) = \frac{2x^2}{(x-1)^2}$

24. $f(x) = \frac{3x-1}{x+1}$

25. $f(x) = \frac{x^2}{x^2-4}$

26. $f(x) = \frac{x}{x^2-1}$

27. Use the graph of f below for parts (a)–(p).

(a) What is the domain of f?
(b) What is the range of f?
(c) What are the intercepts, if any, of f?
(d) What is $f(-2)$?
(e) What is x if $f(x) = 4$?
(f) Where is f discontinuous?
(g) List the vertical asymptotes, if any.
(h) List the horizontal asymptotes, if any.
(i) List any local maxima.
(j) List any local minima.
(k) Where is f increasing?
(l) Where is f decreasing?
(m) What is $\lim_{x\to-\infty} f(x)$?
(n) What is $\lim_{x\to\infty} f(x)$?
(o) What is $\lim_{x\to 6^-} f(x)$?
(p) What is $\lim_{x\to 6^+} f(x)$?

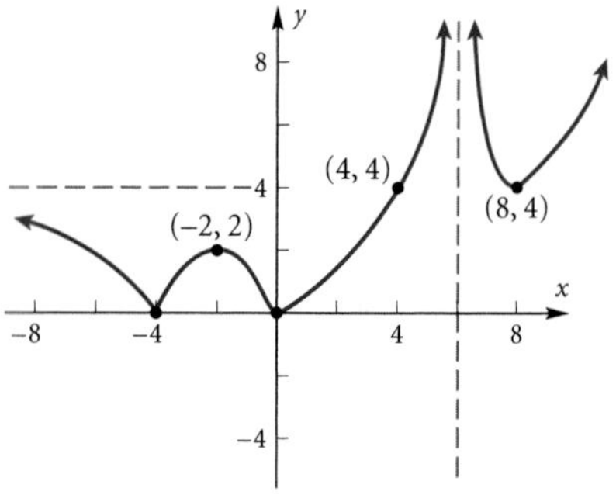

28. Use the graph of f below for parts (a)–(p).

(a) What is the domain of f?
(b) What is the range of f?
(c) What are the intercepts, if any, of f?
(d) What is $f(6)$?
(e) What is x if $f(x) = -2$?
(f) Where is f discontinuous?
(g) List the vertical asymptotes, if any.
(h) List the horizontal asymptotes, if any.
(i) List any local maxima.
(j) List any local minima.
(k) Where is f increasing?
(l) Where is f decreasing?
(m) What is $\lim_{x\to-\infty} f(x)$?
(n) What is $\lim_{x\to\infty} f(x)$?
(o) What is $\lim_{x\to -6^-} f(x)$?
(p) What is $\lim_{x\to -6^+} f(x) = ?$

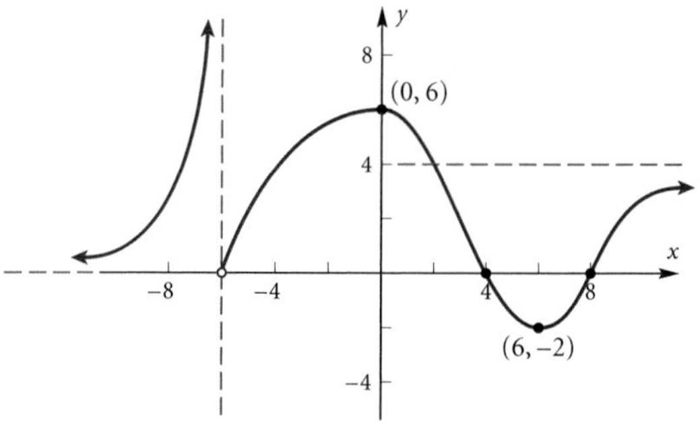

In Problems 29–32, R is discontinuous at c. Use limits to analyze the graph of R at c.

29. $R(x) = \frac{x-1}{x^2-1}$ $c = -1$ and $c = 1$

30. $R(x) = \frac{3x+6}{x^2-4}$ $c = -2$ and $c = 2$

31. $R(x) = \frac{x^2+x}{x^2-1}$ $c = -1$ and $c = 1$

32. $R(x) = \frac{x^2+4x}{x^2-16}$ $c = -4$ and $c = 4$

In Problems 33–38, determine where each rational function is undefined. Determine whether an asymptote or a hole appears at such numbers.

33. $R(x) = \frac{x^3-x^2+x-1}{x^4-x^3+8x-8}$

34. $R(x) = \frac{x^3+x^2+3x+3}{x^4+x^3+8x+8}$

35. $R(x) = \frac{x^3-2x^2+4x-8}{x^2+x-6}$

36. $R(x) = \frac{x^3-x^2+3x-3}{x^2+3x-4}$

37. $R(x) = \frac{x^3+2x^2+x}{x^4+x^3+x+1}$

38. $R(x) = \frac{x^3-3x^2+4x-12}{x^4-3x^3+x-3}$

39. A company has fixed costs of \$79,000 to produce x calculators at a cost of \$10 per calculator.

(a) Express the cost C of producing x calculators as a function $C = C(x)$.
(b) Find the domain of C.
(c) Find the average cost of producing x calculators; that is, find $\overline{C}(x) = \frac{C(x)}{x}$.
(d) Find the domain of $\overline{C}$.
(e) Find $\lim_{x \to 0^+} \overline{C}(x)$ and interpret the answer.
(f) Find $\lim_{x \to \infty} \overline{C}(x)$ and interpret the answer.

40. A company has fixed costs of \$85,000 to produce x cell phones at a cost of \$4 per cell phone.

(a) Express the cost C of producing x cell phones as a function $C = C(x)$.
(b) Find the domain of C.
(c) Find the average cost of producing x cell phones; that is, find $\overline{C}(x) = \frac{C(x)}{x}$.
(d) Find the domain of $\overline{C}$.
(e) Find $\lim_{x \to 0^+} \overline{C}(x)$ and interpret the answer.
(f) Find $\lim_{x \to \infty} \overline{C}(x)$ and interpret the answer.

41. Pollution Control The cost C, in thousands of dollars, for removal of a pollutant from a certain lake is

$$C(x) = \frac{5x}{100 - x} \qquad 0 \le x < 100$$

where x is the percent of pollutant removed.

(a) Find $\lim_{x \to 100^-} C(x)$.
(b) Is it possible to remove 100% of the pollutant? Explain.

42. Drug Concentration The concentration C of a certain drug in a patient's bloodstream t hours after injection is given by

$$C(t) = \frac{0.3t}{t^2 + 2}$$

milligrams per cubic centimeter.

(a) Find the horizontal asymptote of $C(t)$.
(b) Interpret your answer.

43. Draw the graph of a function $y = f(x)$ that has the following characteristics:
Domain: all real numbers, except 2

$\lim_{x \to -\infty} f(x) = 0 \quad \lim_{x \to \infty} f(x) = \infty \quad \lim_{x \to 2^-} f(x) = -\infty$
$\lim_{x \to 2^+} f(x) = 5 \quad f(0) = 0$
local maximum of 5 at $x = -2$
local minimum of 3 at $x = 4$
increasing on $(-\infty, -2)$ and on $(4, \infty)$
decreasing on $(-2, 2)$ and on $(2, 4)$

44. Draw the graph of a function $y = f(x)$ that has the following characteristics:
Domain: all real numbers, except -3 and 4

$\lim_{x \to -\infty} f(x) = 3 \quad \lim_{x \to \infty} f(x) = 4$
$\lim_{x \to -3^-} f(x) = -\infty \quad \lim_{x \to -3^+} f(x) = 0$
$f(-8) = 0 \quad f(0) = 0 \quad f(10) = 0$
local maximum of 5 at $x = -2$
increasing on $(-3, -2)$ and on $(4, \infty)$
decreasing on $(-\infty, -3)$ and on $(-2, 4)$

Chapter 12 Review

OBJECTIVES

Section		You should be able to	Review Exercises
12.1	1	Find a limit using a table	1, 2
	2	Find a limit using a graph	3, 4, 47(j), (k)
12.2	1	Find the limit of a sum, a difference, and a product	5, 6 15, 16
	2	Find the limit of a polynomial	5, 6
	3	Find the limit of a power or a root	7–10, 13, 14
	4	Find the limit of a quotient	17–20, 23–26
	5	Find the limit of an average rate of change	48–52
12.3	1	Find the one-sided limits of a function	11, 12, 21, 22, 47(g), (h), (i)
	2	Determine whether a function is continuous	35–42, 47(l), (m), (n), (o), (p), (q)

12.4	**1** Find limits at infinity	27, 28, 31, 32, 34, 47(t)
	2 Find infinite limits	29, 30, 33, 47(k)
	3 Find horizontal asymptotes	43–46, 47(v)
	4 Find vertical asymptotes	43–46, 47(v)
	5 Analyze the graph of a rational function at points of discontinuity	53–56

THINGS TO KNOW

Limit (p. 677)

$$\lim_{x \to c} f(x) = N$$

As x gets closer to c, $x \neq c$, the value of f gets closer to N.

Limit Formulas (p. 683)

$$\lim_{x \to c} b = b$$

The limit of a constant is the constant.

$$\lim_{x \to c} x = c$$

The limit of x as x approaches c is c.

Limit Properties

$$\lim_{x \to c} [f(x) + g(x)] = \lim_{x \to c} f(x) + \lim_{x \to c} g(x) \quad \textbf{(p. 684)}$$

The limit of a sum equals the sum of the limits.

$$\lim_{x \to c} [f(x) - g(x)] = \lim_{x \to c} f(x) - \lim_{x \to c} g(x) \quad \textbf{(p. 685)}$$

The limit of a difference equals the difference of the limits.

$$\lim_{x \to c} [f(x) \cdot g(x)] = \left[\lim_{x \to c} f(x)\right] \cdot \left[\lim_{x \to c} g(x)\right] \quad \textbf{(p. 685)}$$

The limit of a product equals the product of the limits.

$$\lim_{x \to c} \left[\frac{f(x)}{g(x)}\right] = \frac{\lim_{x \to c} f(x)}{\lim_{x \to c} g(x)} \quad \textbf{(p. 688)}$$

provided that $\lim_{x \to c} g(x) \neq 0$

The limit of a quotient equals the quotient of the limits, provided that the limit of the denominator is not zero.

Limit of a Polynomial Function (p. 687)

$$\lim_{x \to c} P(x) = P(c), \text{ where } P \text{ is a polynomial}$$

Limit of a Power (p. 687)

$$\lim_{x \to c} [f(x)]^n = \left[\lim_{x \to c} f(x)\right]^n$$

Limit of a Root (p. 687)

$$\lim_{x \to c} \sqrt[n]{f(x)} = \sqrt[n]{\lim_{x \to c} f(x)}$$

Continuous Function (p. 694)

$$\lim_{x \to c} f(x) = f(c)$$

Horizontal Asymptote (p. 700)

If $\lim_{x \to -\infty} f(x) = M$, then $y = M$ is a horizontal asymptote for the graph of f as $x \to -\infty$

If $\lim_{x \to \infty} f(x) = N$, then $y = N$ is a horizontal asymptote for the graph of f as $x \to \infty$

Vertical Asymptote (p. 701)

If $\lim_{x \to c^-} f(x) = -\infty$, then $x = c$ is a vertical asymptote for the graph of f as $x \to c^-$

If $\lim_{x \to c^+} f(x) = -\infty$, then $x = c$ is a vertical asymptote for the graph of f as $x \to c^+$

If $\lim_{x \to c^-} f(x) = \infty$, then $x = c$ is a vertical asymptote for the graph of f as $x \to c^-$

If $\lim_{x \to c^+} f(x) = \infty$, then $x = c$ is a vertical asymptote for the graph of f as $x \to c^+$

TRUE–FALSE ITEMS **Answers are on page AN-71.**

T F 1. The limit of the sum of two functions equals the sum of their limits, provided that each limit exists.

T F 2. The limit of a function f as x approaches c always equals $f(c)$.

T F 3. $\lim_{x\to 4} \frac{x^2 - 16}{x - 4} = 8$

T F 4. The function $f(x) = \frac{5x^2}{x^2 + 4}$ is continuous at $x = -2$.

T F 5. The limit of a quotient of two functions equals the quotient of their limits, provided that each limit exists and the limit of the denominator is not zero.

T F 6. The graph of a rational function might have both asymptotes and holes.

T F 7. The graph of an exponential function has an asymptote.

FILL-IN-THE-BLANKS **Answers are on page AN-71.**

1. The notation ________ may be described by saying, "For x approximately equal to c, but $x \neq c$, the value $f(x)$ is approximately equal to N."

2. If $\lim_{x\to c} f(x) = N$ and f is continuous at c, then $f(c)$ ________ N.

3. If there is no single number that the value of f approaches when x is close to c, then $\lim_{x\to c} f(x)$ does ________ .

4. When $\lim_{x\to c} f(x) = f(c)$, we say that f is ________ at c.

5. $\lim_{x\to c} \frac{f(x)}{g(x)} = \frac{\lim_{x\to c} f(x)}{\lim_{x\to c} g(x)}$, provided that $\lim_{x\to c} f(x)$ and $\lim_{x\to c} g(x)$ each exist and $\lim_{x\to c} g(x)$ ________ 0.

6. If $\lim_{x\to c^-} f(x) = L$ and $\lim_{x\to c^+} f(x) = R$, then $\lim_{x\to c} f(x)$ exists provided that L ________ R.

7. If, for a function $y = f(x)$, $\lim_{x\to\infty} f(x) = 2$, then ________ is a ________ asymptote of the graph of f.

REVIEW EXERCISES **Answers to odd-numbered problems begin on page AN-71.**
Blue problem numbers indicate the author's suggestions for use in a practice test.

1. Use a table to find $\lim_{x\to 2} \frac{x^3 - 8}{x - 2}$

2. Use a table to find $\lim_{x\to 2} \left(\frac{1}{3}\right)^x$

3. Use a graph to find $\lim_{x\to 0} f(x)$, where

$$f(x) = \begin{cases} x^2 & \text{if } x < 0 \\ 2 & \text{if } x = 0 \\ e^x - 1 & \text{if } x > 0 \end{cases}$$

4. Use a graph to find $\lim_{x\to 2} g(x)$, where

$$g(x) = \begin{cases} x^2 + 1 & \text{if } x < 2 \\ 5 & \text{if } x = 2 \\ 3x - 2 & \text{if } x > 2 \end{cases}$$

In Problems 5–34, find the limit.

5. $\lim_{x\to 2} (3x^2 - 2x + 1)$

6. $\lim_{x\to 1} (-2x^3 + x + 4)$

7. $\lim_{x\to -2} (x^2 + 1)^2$

8. $\lim_{x\to -2} (x^3 + 1)^2$

9. $\lim_{x\to 3} \sqrt{x^2 + 7}$

10. $\lim_{x\to -2} \sqrt[3]{x + 10}$

11. $\lim_{x\to 1^-} \sqrt{1 - x^2}$

12. $\lim_{x\to 2^+} \sqrt{3x - 2}$

13. $\lim_{x\to 2} (5x + 6)^{3/2}$

14. $\lim_{x\to -3} (15 - 3x)^{-3/2}$

15. $\lim_{x\to -1} (x^2 + x + 2)(x^2 - 9)$

16. $\lim_{x\to 3} (3x + 4)(x^2 + 1)$

17. $\lim_{x\to 1} \frac{x - 1}{x^3 - 1}$

18. $\lim_{x\to -1} \frac{x^2 - 1}{x^2 + x}$

19. $\lim_{x\to -3} \frac{x^2 - 9}{x^2 - x - 12}$

20. $\lim_{x\to -3} \frac{x^2 + 2x - 3}{x^2 - 9}$

21. $\lim_{x\to -1^-} \frac{x^2 - 1}{x^3 - 1}$

22. $\lim_{x\to 2^+} \frac{x^2 - 4}{x^3 - 8}$

23. $\lim_{x\to 2} \frac{x^3 - 8}{x^3 - 2x^2 + 4x - 8}$

24. $\lim_{x\to 1} \frac{x^3 - 1}{x^3 - x^2 + 3x - 3}$

25. $\lim_{x\to 3} \frac{x^4 - 3x^3 + x - 3}{x^3 - 3x^2 + 2x - 6}$

26. $\lim_{x \to -1} \frac{x^4 + x^3 + 2x + 2}{x^3 + x^2}$

27. $\lim_{x \to \infty} \frac{5x^4 - 8x^3 + x}{3x^4 + x^2 + 5}$

28. $\lim_{x \to \infty} \frac{8x^3 - x^2 - 5}{2x^3 - 10x + 1}$

29. $\lim_{x \to 3^-} \frac{x^2}{x - 3}$

30. $\lim_{x \to 2^+} \frac{5x}{x - 2}$

31. $\lim_{x \to \infty} \frac{8x^4 - x^2 + 2}{-4x^3 + 1}$

32. $\lim_{x \to \infty} \frac{8x^2 + 2}{4x^3 + x}$

33. $\lim_{x \to -3^+} \frac{1 - 9x^2}{x^2 - 9}$

34. $\lim_{x \to -\infty} \frac{1 - 9x^2}{1 - 4x^2}$

In Problems 35–42, determine whether f is continuous at c.

35. $f(x) = 3x^4 - x^2 + 2 \quad c = 5$

36. $f(x) = \frac{x^2 - 9}{x + 10} \quad c = 2$

37. $f(x) = \frac{x^2 - 4}{x + 2} \quad c = -2$

38. $f(x) = \frac{x^2 + 6x}{x^2 - 6x} \quad c = 0$

39. $f(x) = \begin{cases} \frac{x^2 - 4}{x + 2} & \text{if } x \neq -2 \\ 4 & \text{if } x = -2 \end{cases} \quad c = -2$

40. $f(x) = \begin{cases} \frac{x^2 + 6x}{x^2 - 6x} & \text{if } x \neq 0 \\ 1 & \text{if } x = 0 \end{cases} \quad c = 0$

41. $f(x) = \begin{cases} \frac{x^2 - 4}{x + 2} & \text{if } x \neq -2 \\ -4 & \text{if } x = -2 \end{cases} \quad c = -2$

42. $f(x) = \begin{cases} \frac{x^2 + 6x}{x^2 - 6x} & \text{if } x \neq 0 \\ -1 & \text{if } x = 0 \end{cases} \quad c = 0$

In Problems 43–46, locate all horizontal asymptotes and all vertical asymptotes of each function.

43. $f(x) = \frac{3x}{x^2 - 1}$

44. $f(x) = \frac{4x^2 - 2x + 1}{x^2 - 4}$

45. $f(x) = \frac{5x}{x + 2}$

46. $f(x) = \frac{x^3}{x - 3}$

47. *In Problem 47, use the accompanying graph of $y = f(x)$.*

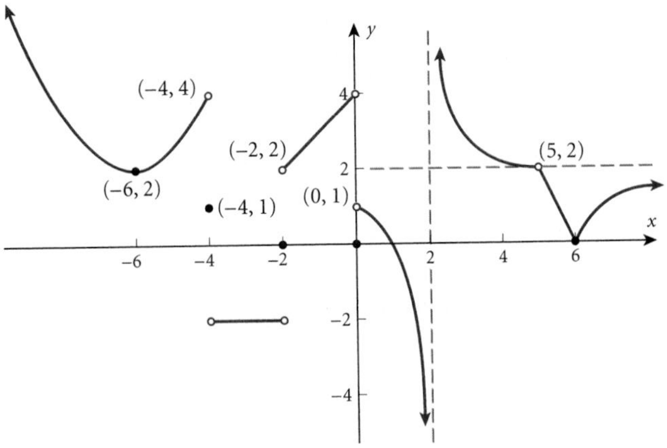

(a) What is the domain of f?
(b) What is the range of f?
(c) Find the x-intercept(s), if any, of f.
(d) Find the y-intercept(s), if any, of f.
(e) Find $f(-6)$ and $f(-4)$.
(f) Find $f(-2)$ and $f(6)$.
(g) Find $\lim_{x \to -4^-} f(x)$ and $\lim_{x \to -4^+} f(x)$.
(h) Find $\lim_{x \to -2^-} f(x)$ and $\lim_{x \to -2^+} f(x)$.
(i) Find $\lim_{x \to 5^-} f(x)$ and $\lim_{x \to 5^+} f(x)$.
(j) Does $\lim_{x \to 0} f(x)$ exist? If it does, what is it?
(k) Does $\lim_{x \to 2} f(x)$ exist? If it does, what is it?
(l) Is f continuous at -2?
(m) Is f continuous at -4?
(n) Is f continuous at 0?
(o) Is f continuous at 2?
(p) Is f continuous at 4?
(q) Is f continuous at 5?
(r) Where is f increasing?
(s) Where is f decreasing?
(t) Find $\lim_{x \to -\infty} f(x)$ and $\lim_{x \to \infty} f(x)$.
(u) List any local maxima and local minima.
(v) List any horizontal or vertical asymptotes.

In Problems 48–52, find the limit as $x \to c$ of the average rate of change of $f(x)$ from c to x.

48. $c = 5;\ f(x) = 1 - x^2$

49. $c = -2;\ f(x) = 2x^2 - 3x$

50. $c = 2;\ f(x) = 4 - 3x + x^2$

51. $c = 3;\ f(x) = \frac{x}{x - 1}$

52. $c = 2;\ f(x) = \frac{x - 1}{x}$

In Problems 53 and 54, R is discontinuous at c. Use limits to analyze the graph of R at c.

53. $R(x) = \dfrac{x + 4}{x^2 - 16}$ at $c = -4$ and $c = 4$

54. $R(x) = \dfrac{3x^2 + 6x}{x^2 - 4}$ at $c = -2$ and $c = 2$

In Problems 55 and 56, determine where each rational function is undefined. Determine whether an asymptote or a hole appears at such numbers.

55. $R(x) = \dfrac{x^3 - 2x^2 + 4x - 8}{x^2 - 11x + 18}$

56. $R(x) = \dfrac{x^3 + 3x^2 - 2x - 6}{x^2 + x - 6}$

---

57. Draw the graph of a function $y = f(x)$ that has the following characteristics:

Domain: all real numbers, except -2 and 4

$\lim_{x \to -\infty} f(x) = \infty$

$\lim_{x \to \infty} f(x) = 5$

$\lim_{x \to -2^-} f(x) = \infty$

$\lim_{x \to -2^+} f(x) = -\infty$

$\lim_{x \to 4^-} f(x) = 0$

$\lim_{x \to 4^+} f(x) = 0$

$f(0) = 1$

local maximum of 5 at $x = 2$

local minimum of 3 at $x = -4$

increasing on $(-4, -2)$, on $(-4, 2)$, and on $(4, \infty)$

decreasing on $(-\infty, -4)$ and on $(2, 4)$

58. A company has fixed costs of \$158,000 to produce x scooters at a cost of \$115 per scooter.

(a) Express the cost C of producing x scooters as a function $C = C(x)$.

(b) Find the domain of C.

(c) Find the average cost of producing x scooters; that is, find $\overline{C}(x) = \dfrac{C(x)}{x}$.

(d) Find the domain of $\overline{C}$.

(e) Find $\lim_{x \to 0^+} \overline{C}(x)$ and interpret the answer.

(f) Find $\lim_{x \to \infty} \overline{C}(x)$ and interpret the answer.

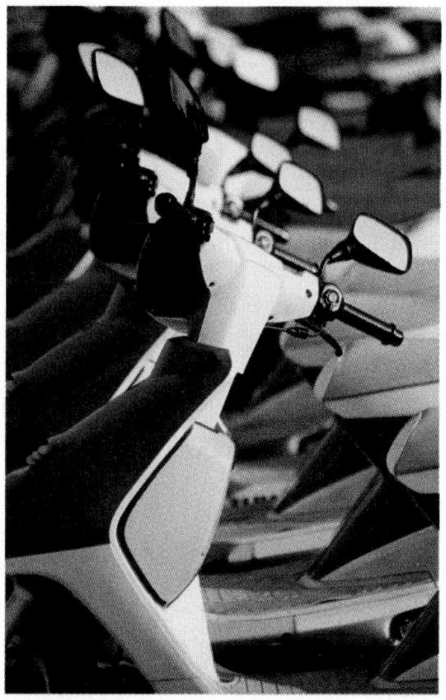

59. Advertising The sale of a new product over a period of time is expected to follow the relationship

$$S(x) = \frac{2000x^2}{3.5x^2 + 1000}$$

where x is the amount of money spent on advertising.

(a) Evaluate $\lim_{x \to \infty} S(x)$.

(b) Interpret your answer.

Chapter 12 Project

TAX RATES AND CONTINUOUS FUNCTIONS

Even though it may seem that the notion of continuity is too abstract for any real application, in some cases we really need a function to be continuous. In other cases, discontinuities may be either unavoidable or even desirable. In this Chapter Project, we look at tax rates.

We consider two functions: the function that gives the tax rate and the function that gives the amount of tax paid.

On page 712 is the 2003 tax rate schedule for single taxpayers.

If Taxable Income Is		Then The Tax Is		
Over	But Not Over	This Amount	Plus This %	Of The Excess Over
$0	$7,000	$0.00	10%	$0.00
$7,000	$28,400	$700.00	15%	$7,000.00
$28,400	$68,800	$3,910.00	25%	$28,400.00
$ 68,800	$143,500	$14,010.00	28%	$68,800.00
$143,500	$311,950	$34,926.00	33%	$143,500.00
$311,950	—	$90,514.00	35%	$311,950.00

Source: Internal Revenue Service.

1. Construct a function $R = R(x)$ that gives the tax rate R as a function of taxable income x.

Hint: Use a piecewise defined function

2. Graph $R = R(x)$.

3. Is R continuous? If not, where does it fail to be continuous?

4. Explain why R has to be discontinuous if 'tax brackets' are required.

5. Construct a function $A = A(x)$ that gives the amount A of tax due as a function of taxable income x.

6. Graph the function $A = A(x)$.

7. Is A continuous? If not, where does it fail to be continuous?

8. What if the third row of the tax table were changed slightly to:

If Taxable Income Is		Then The Tax Is		
Over	But Not Over	This Amount	Plus This %	Of The Excess Over
$28,400	$68,800	**$4,000.00**	25%	$28,400.00

Show that $A = A(x)$ now has a discontinuity at $x =$ \$28,400. Would a tax system using this method of taxation be fair? Why or why not? Why is continuity a desirable property for $A = A(x)$ to have?

9. Here is a partial 2003 tax rate schedule for married taxpayers filing jointly. Fill in the blanks with appropriate values to make the amount of tax paid a continuous function of taxable income.

If Taxable Income Is		Then The Tax Is		
Over	But Not Over	This Amount	Plus This %	Of The Excess Over
$0	$14,000	$0.00	10%	$0.00
$14,000	$56,800		15%	$14,000.00
$56,800	$114,650		25%	$56,800.00
$114,650	$174,700		28%	$114,650.00
$174,700	$311,950		33%	$174,700.00
$311,950	—		35%	$311,950.00

MATHEMATICAL QUESTIONS FROM PROFESSIONAL EXAMS*

1. Actuary Exam—Part I $\lim_{x \to 3} \dfrac{x^2 - x - 6}{x^2 - 9}$

(a) 0 (b) $\dfrac{5}{6}$ (c) 1 (d) $\dfrac{5}{3}$ (e) Undefined

2. Actuary Exam—Part I For which of the following functions does $\lim_{x \to 0} f(x)$ exist?

I. $f(x) = \begin{cases} -1 & \text{if } x < 0 \\ 0 & \text{if } x = 0 \\ 1 & \text{if } x > 0 \end{cases}$

II. $f(x) = \begin{cases} |x| & \text{if } x \neq 0 \\ 1 & \text{if } x = 0 \end{cases}$

III. $f(x) = \begin{cases} \dfrac{x^2 + x}{x} & \text{if } x \neq 0 \\ 1 & \text{if } x = 0 \end{cases}$

(a) I (b) II (c) III (d) I, III (e) II, III

3. Actuary Exam—Part I What is the value of $\lim_{h \to 0} \dfrac{\sqrt{2 + h} - \sqrt{2}}{h}$?

(a) Undefined (b) 0 (c) $\dfrac{\sqrt{2}}{2}$ (d) $\dfrac{\sqrt{2}}{4}$ (e) $+\infty$

4. Actuary Exam—Part I $f(x) = \dfrac{x^3 + 1}{x + 1}$. For what values of x is f continuous at x?

(a) $x \leq -1$ (b) $x \geq -1$ (c) $x > -1$ (d) $x < -1$ (e) $x \neq -1$

CHAPTER 13

The Derivative of a Function

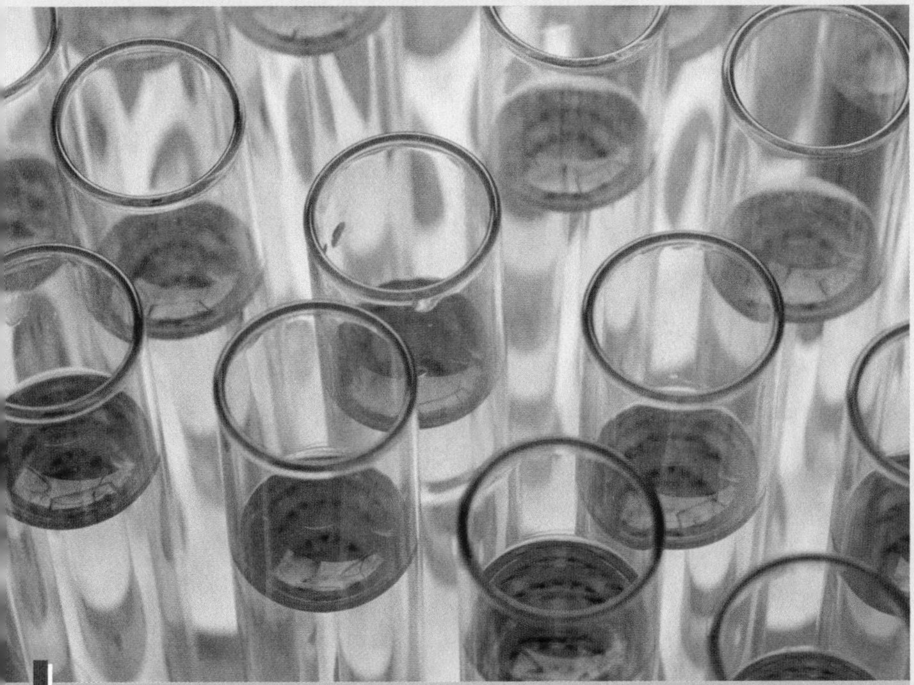

OUTLINE

If you had to give a blood test to 3000 people to test for the presence of Hepatitis C, how would you do it? Would you test each person individually, which would require 3000 tests? That could be expensive, especially if each test were to cost \$100. How can you use fewer tests? One way is to pool the samples. What sample size would you use? If you pooled the blood of 100 people and the test came out negative, then one test was used instead of 100. But if the test came out positive, then you would need to test each one separately, which now requires $1 + 100 = 101$ tests. Maybe the sample size to use is 200. Could it be 300? And the likelihood of a positive test must play a role as well. Sounds like a very hard problem. But with the discussion in this chapter and the Chapter Project to guide you, you can find the best sample size to use so cost is least.

A LOOK BACK, A LOOK FORWARD

In Chapter 10, we discussed various properties that functions have, such as intercepts, even/odd, increasing/decreasing, local maxima and minima, and average rate of change. In Chapter 11 we discussed classes of functions and listed some properties that these classes possess. In Chapter 12 we began our study of the calculus by discussing limits of functions and continuity of functions. Now we are ready to define another property of functions: the *derivative of a function.*

The cofounders of calculus are generally recognized to be Gottfried Wilhelm von Leibniz (1646–1716) and Sir Isaac Newton (1642–1727). Newton approached calculus by solving a physics problem involving falling objects, while Leibniz approached calculus by solving a geometry problem. Surprisingly, the solution of these two problems led to the same mathematical concept: the derivative. We shall discuss the physics problem later in this chapter. We shall address the geometry problem, referred to as *The Tangent Problem,* now.

13.1 The Definition of a Derivative

PREPARING FOR THIS SECTION ***Before getting started, review the following:***

- Average Rate of Change (Chapter 10, Section 10.3, p. 570)
- Secant Line (Chapter 10, Section 10.3, p. 571)
- Factoring (Appendix A, Section A.5, pp. A-40–A-43)
- Point–slope Form of a Line (Chapter 1, Section 1.1, p. 7)
- Difference Quotient (Chapter 10, Section 10.2, p. 552)

OBJECTIVES
1. Find an equation of the tangent line to the graph of a function
2. Find the derivative of a function at a number c
3. Find the derivative of a function using the difference quotient
4. Find the instantaneous rate of change of a function
5. Find marginal cost and marginal revenue

The Tangent Problem

The geometry question that motivated the development of calculus was "What is the slope of the tangent line to the graph of a function $y = f(x)$ at a point P on its graph?" See Figure 1.

FIGURE 1

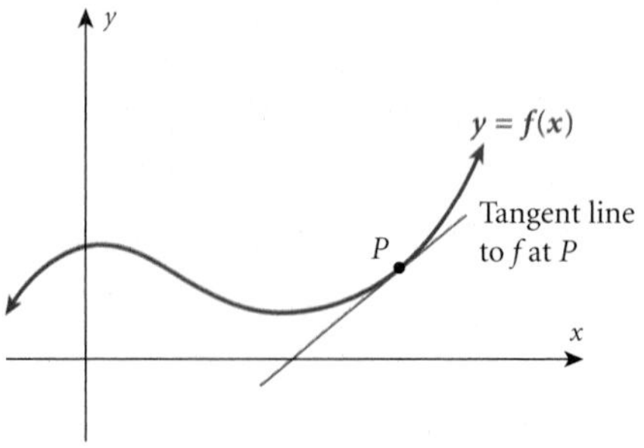

FIGURE 2

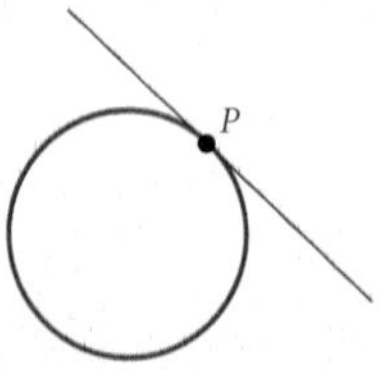

We first need to define what we mean by a *tangent* line. In high school geometry, the tangent line to a circle is defined as the line that intersects the graph in exactly one point. Look at Figure 2. Notice that the tangent line just touches the graph of the circle.

This definition, however, does not work in general. Look at Figure 3. The lines L_1 and L_2 only intersect the graph in one point P, but neither touches the graph at P. Additionally, the tangent line L_T shown in Figure 4 touches the graph of f at P, but also intersects the graph elsewhere. So how should we define the tangent line to the graph of f at a point P?

FIGURE 3

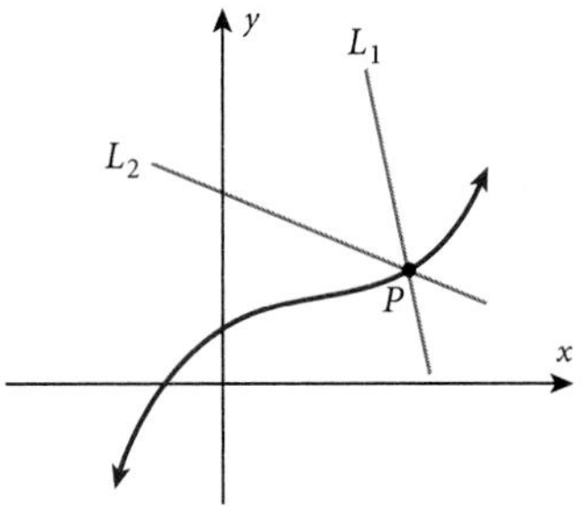

FIGURE 4

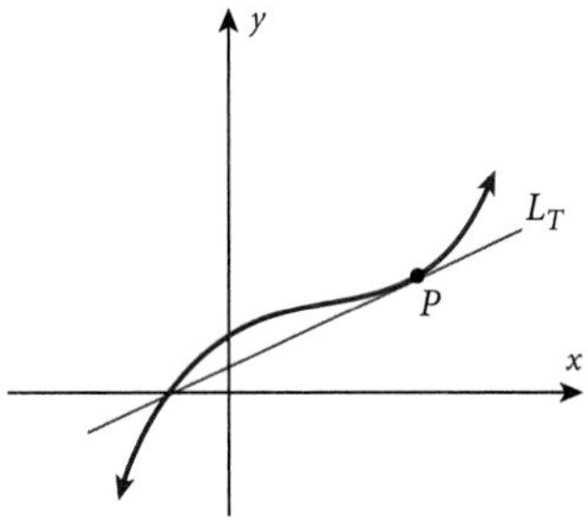

The tangent line L_T to the graph of a function $y = f(x)$ at a point P necessarily contains the point P. To find an equation for L_T using the point–slope form of the equation of a line, it remains to find the slope $m_{\tan}$ of the tangent line.

Suppose that the coordinates of the point P are $(c, f(c))$. Locate another point $Q = (x, f(x))$ on the graph of f. The line containing P and Q is a secant line. (Refer to Section 10.3.) The slope $m_{\sec}$ of the secant line is

$$m_{\sec} = \frac{f(x) - f(c)}{x - c}$$

FIGURE 5

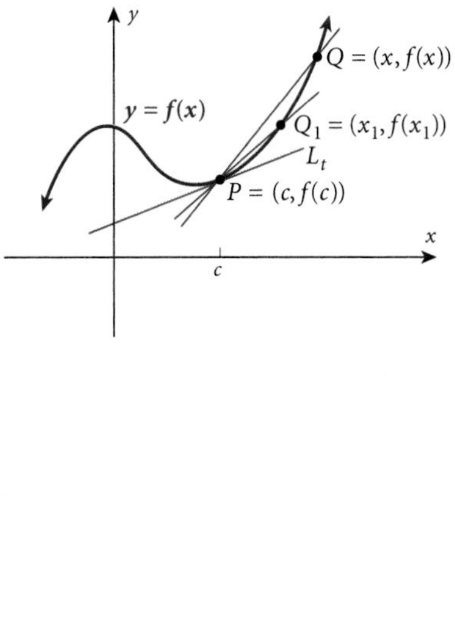

Now look at Figure 5.

As we move along the graph of f from Q toward P, we obtain a succession of secant lines. The closer we get to P, the closer the secant line is to the tangent line. The limiting position of these secant lines is the tangent line. Therefore, the limiting value of the slopes of these secant lines equals the slope of the tangent line. But, as we move from Q toward P, the values of x get closer to c. Therefore,

$$m_{\tan} = \lim_{x \to c} m_{\sec} = \lim_{x \to c} \frac{f(x) - f(c)}{x - c}$$

The **tangent line** to the graph of a function $y = f(x)$ at a point $P = (c, f(c))$ on its graph is defined as the line containing the point P whose slope is

$$m_{\tan} = \lim_{x \to c} \frac{f(x) - f(c)}{x - c} \qquad \textbf{(1)}$$

provided that this limit exists.

If $m_{\tan}$ exists, an equation of the tangent line is

$$y - f(c) = m_{\tan}(x - c) \qquad \textbf{(2)}$$

1 **EXAMPLE 1 Finding an Equation of the Tangent Line**

Find an equation of the tangent line to the graph of $f(x) = \dfrac{x^2}{4}$ at the point $\left(1, \dfrac{1}{4}\right)$. Graph the function and the tangent line.

SOLUTION The tangent line contains the point $\left(1, \dfrac{1}{4}\right)$. The slope of the tangent line to the graph of $f(x) = \dfrac{x^2}{4}$ at $\left(1, \dfrac{1}{4}\right)$ is

$$m_{\tan} = \lim_{x\to 1}\frac{f(x) - f(1)}{x - 1} = \lim_{x\to 1}\frac{\dfrac{x^2}{4} - \dfrac{1}{4}}{x - 1} = \lim_{x\to 1}\frac{\dfrac{x^2 - 1}{4}}{x - 1} = \lim_{x\to 1}\frac{\cancel{(x-1)}(x+1)}{4\cancel{(x-1)}}$$

$$= \lim_{x\to 1}\frac{x+1}{4} = \frac{1}{2}$$

An equation of the tangent line is

$$y - \frac{1}{4} = \frac{1}{2}(x - 1) \qquad y - f(c) = m_{\tan}(x - c)$$

$$y = \frac{1}{2}x - \frac{1}{4}$$

Figure 6 shows the graph of $y = \dfrac{x^2}{4}$ and the tangent line at $\left(1, \dfrac{1}{4}\right)$.

FIGURE 6

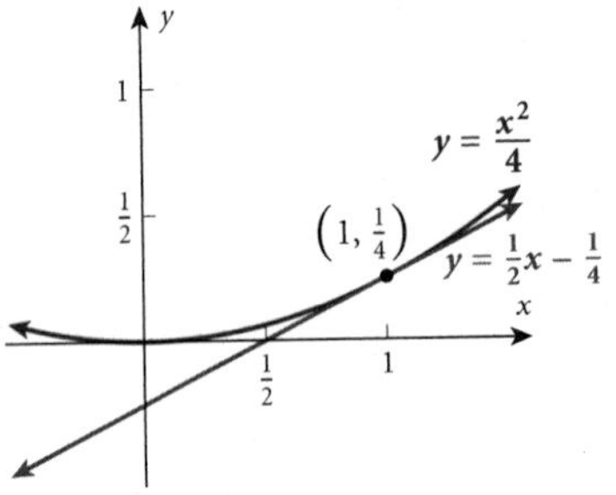

NOW WORK PROBLEM 3.

The limit in Formula (1) has an important generalization: it is called the *derivative of f at c.*

The Derivative of a Function at a Number c.

Let $y = f(x)$ denote a function f. If c is a number in the domain of f, the **derivative of f at c,** denoted by $f'(c)$, read "f prime of c," is defined as

$$f'(c) = \lim_{x\to c}\frac{f(x) - f(c)}{x - c} \tag{3}$$

provided that this limit exists.

The steps for finding the derivative of a function are listed below:

Steps for Finding the Derivative of a Function at c

STEP 1 Find $f(c)$.

STEP 2 Subtract $f(c)$ from $f(x)$ to get $f(x) - f(c)$ and form the quotient

$$\frac{f(x) - f(c)}{x - c}$$

STEP 3 Find the limit (if it exists) of the quotient found in Step 2 as $x \to c$:

$$f'(c) = \lim_{x \to c} \frac{f(x) - f(c)}{x - c}$$

2 **EXAMPLE 2** **Finding the Derivative of a Function at a Number c**

Find the derivative of $f(x) = 2x^2 - 5x$ at 2. That is, find $f'(2)$.

SOLUTION **Step 1:** $f(2) = 2(4) - 5(2) = -2$

Step 2: $$\frac{f(x) - f(2)}{x - 2} = \frac{(2x^2 - 5x) - (-2)}{x - 2} = \frac{2x^2 - 5x + 2}{x - 2} = \frac{(2x - 1)(x - 2)}{x - 2}$$

Step 3: The derivative of f at 2 is

$$f'(2) = \lim_{x \to 2} \frac{f(x) - f(2)}{x - 2} = \lim_{x \to 2} \frac{(2x - 1)\cancel{(x - 2)}}{\cancel{x - 2}} = 3$$

NOW WORK PROBLEM 13.

Example 2 provides a way of finding the derivative at 2 analytically. Graphing utilities have built-in procedures to approximate the derivative of a function at any number c. Consult your owner's manual for the appropriate keystrokes.

EXAMPLE 3 **Finding the Derivative of a Function Using a Graphing Utility**

Use a graphing utility to find the derivative of $f(x) = 2x^2 - 5x$ at 2. That is, find $f'(2)$.

SOLUTION Figure 7 shows the solution using a TI-83 Plus graphing calculator.

FIGURE 7

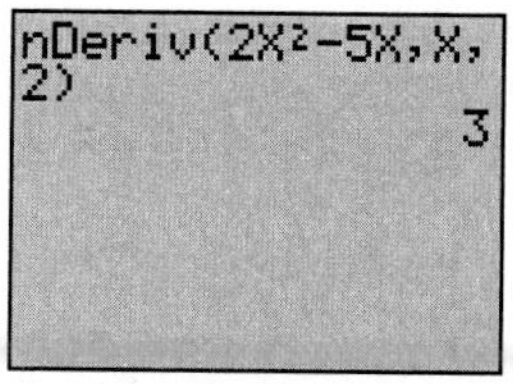

So $f'(2) = 3$.

NOW WORK PROBLEM 45.

EXAMPLE 4 Finding the Derivative of a Function at c

Find the derivative of $f(x) = x^2$ at c. That is, find $f'(c)$.

SOLUTION Since $f(c) = c^2$, we have

$$\frac{f(x) - f(c)}{x - c} = \frac{x^2 - c^2}{x - c} = \frac{(x + c)(x - c)}{x - c}$$

The derivative of f at c is

$$f'(c) = \lim_{x \to c} \frac{f(x) - f(c)}{x - c} = \lim_{x \to c} \frac{(x + c)\cancel{(x - c)}}{\cancel{x - c}} = 2c$$

As Example 4 illustrates, the derivative of $f(x) = x^2$ exists and equals $2c$ for any number c. In other words, the derivative is itself a function and, using x for the independent variable, we can write $f'(x) = 2x$. The function f' is called the **derivative function of f** or the **derivative of f.** We also say that f is **differentiable.** The instruction "differentiate f" means "find the derivative of f".

It is usually easier to find the derivative function by using another form. We derive this alternate form as follows:

Formula (3) for the derivative of f at c is

$$f'(c) = \lim_{x \to c} \frac{f(x) - f(c)}{x - c} \qquad \text{Formula (3).}$$

Let $h = x - c$. Then $x = c + h$ and

$$\frac{f(x) - f(c)}{x - c} = \frac{f(c + h) - f(c)}{h}$$

Since $h = x - c$, then, as $x \to c$, it follows that $h \to 0$. As a result,

$$f'(c) = \lim_{x \to c} \frac{f(x) - f(c)}{x - c} = \lim_{h \to 0} \frac{f(c + h) - f(c)}{h} \tag{4}$$

Now replace c by x in (4). This gives us the following formula for finding the derivative of f at any number x.

Formula for the Derivative of a Function $y = f(x)$ at x

$$f'(x) = \lim_{h \to 0} \frac{f(x + h) - f(x)}{h} \tag{5}$$

That is, the derivative of the function f is the limit as $h \to 0$ of its difference quotient.

3

EXAMPLE 5 Using the Difference Quotient to Find a Derivative

(a) Use Formula (5) to find the derivative of $f(x) = x^2 + 2x$.
(b) Find $f'(0)$, $f'(-1)$, $f'(3)$.

SOLUTION **(a)** First, we find the difference quotient of $f(x) = x^2 + 2x$.

$$\begin{aligned}\frac{f(x+h)-f(x)}{h} &= \frac{[(x+h)^2 + 2(x+h)] - [x^2+2x]}{h} \\ &= \frac{x^2 + 2xh + h^2 + 2x + 2h - x^2 - 2x}{h} \\ &= \frac{2xh + h^2 + 2h}{h} && \text{Simplify.} \\ &= \frac{h(2x+h+2)}{h} && \text{Factor out } h. \\ &= 2x + h + 2 && \text{Cancel the } h\text{'s.}\end{aligned}$$

The derivative of f is the limit of the difference quotient as $h \to 0$. That is,

$$f'(x) = \lim_{h\to 0} \frac{f(x+h)-f(x)}{h} = \lim_{h\to 0}(2x+h+2) = 2x+2$$

(b) Since

$$f'(x) = 2x + 2$$

we have

$$\begin{aligned} f'(0) &= 2\cdot 0 + 2 = 2 \\ f'(-1) &= 2(-1) + 2 = 0 \\ f'(3) &= 2(3) + 2 = 8 \end{aligned}$$

NOW WORK PROBLEM 59.

Instantaneous Rate of Change

In Chapter 10 we defined the average rate of change of a function f from c to x as

$$\frac{\Delta y}{\Delta x} = \frac{f(x)-f(c)}{x-c}$$

The limit as x approaches c of the average rate of change of f, based on Formula (3), is the derivative of f at c. As a result, we call the derivative of f at c the **instantaneous rate of change of f with respect to x at c.** That is,

$$\left(\begin{matrix}\text{Instantaneous rate of} \\ \text{change of } f \text{ with respect to } x \text{ at } c\end{matrix}\right) = f'(c) = \lim_{x\to c}\frac{f(x)-f(c)}{x-c} \qquad \textbf{(6)}$$

4 **EXAMPLE 6** **Finding the Instantaneous Rate of Change**

During a month-long advertising campaign, the total sales S of a magazine were given by the function

$$S(x) = 5x^2 + 100x + 10{,}000$$

where x represents the number of days of the campaign, $0 \le x \le 30$.

(a) What is the average rate of change of sales from $x = 10$ to $x = 20$ days?
(b) What is the instantaneous rate of change of sales when $x = 10$ days?

SOLUTION **(a)** Since $S(10) = 11{,}500$ and $S(20) = 14{,}000$, the average rate of change of sales from $x = 10$ to $x = 20$ is

$$\frac{\Delta S}{\Delta x} = \frac{S(20) - S(10)}{20 - 10} = \frac{14{,}000 - 11{,}500}{10} = 250 \text{ magazines per day}$$

(b) The instantaneous rate of change of sales when $x = 10$ is the derivative of S at 10.

$$S'(10) = \lim_{x \to 10} \frac{S(x) - S(10)}{x - 10} = \lim_{x \to 10} \frac{[(5x^2 + 100x + 10{,}000) - 11{,}500]}{x - 10}$$

$$= \lim_{x \to 10} \frac{5(x^2 + 20x - 300)}{x - 10}$$

$$= 5 \lim_{x \to 10} \frac{(x + 30)(x - 10)}{x - 10} = 5 \lim_{x \to 10} (x + 30) = 5 \cdot 40 = 200$$

The instantaneous rate of change of S at 10 is 200 magazines per day.

We interpret the results of Example 6 as follows: The fact that the average rate of sales from $x = 10$ to $x = 20$ is $\frac{\Delta S}{\Delta x} = 250$ magazines per day indicates that on the 10th day of the campaign, we can expect to average 250 magazines per day of additional sales if we continue the campaign for 10 more days. The fact that $S'(10) = 200$ magazines per day indicates that on the 10th day of the campaign, one more day of advertising will result in additional sales of 200 magazines per day.

NOW WORK PROBLEM 59.

Find marginal cost and marginal revenue 5

Application to Economics: Marginal Analysis

Economics is one of the many fields in which calculus has been used to great advantage. Economists have a special name for the application of derivatives to problems in economics—it is **marginal analysis.** Whenever the term *marginal* appears in a discussion involving cost functions or revenue functions, it signals the presence of derivatives in the background.

Marginal Cost

Suppose $C = C(x)$ is the cost of producing x units. Then the derivative $C'(x)$ is called the **marginal cost.**

We interpret the marginal cost as follows. Since

$$C'(x) = \lim_{h \to 0} \frac{C(x + h) - C(x)}{h}$$

it follows, for small values of h, that

$$C'(x) \approx \frac{C(x + h) - C(x)}{h}$$

That is to say,

$$C'(x) \approx \frac{\text{cost of increasing production from } x \text{ to } x + h}{h}$$

In most practical situations x is very large. Because of this, many economists let $h = 1$, which is small compared to large x. Then, marginal cost may be interpreted as

$$C'(x) = C(x+1) - C(x) = \text{cost of increasing production by one unit}$$

EXAMPLE 7 Finding Marginal Cost

Suppose that the cost in dollars for a weekly production of x tons of steel is given by the function:

$$C(x) = \frac{1}{10}x^2 + 5x + 1000$$

(a) Find the marginal cost.
(b) Find the cost and marginal cost when $x = 1000$ tons.
(c) Interpret $C'(1000)$.

SOLUTION **(a)** The marginal cost is the derivative $C'(x)$. We use the difference quotient of $C(x)$ to find $C'(x)$.

$$C'(x) = \lim_{h\to 0}\frac{C(x+h) - C(x)}{h} = \lim_{h\to 0}\frac{\left[\frac{1}{10}(x+h)^2 + 5(x+h) + 1000\right] - \left[\frac{1}{10}x^2 + 5x + 1000\right]}{h}$$

$$= \lim_{h\to 0}\frac{\frac{1}{10}(x^2 + 2xh + h^2) + 5x + 5h - \frac{1}{10}x^2 - 5x}{h}$$

$$= \lim_{h\to 0}\frac{\frac{1}{5}xh + \frac{1}{10}h^2 + 5h}{h} = \lim_{h\to 0}\left(\frac{1}{5}x + \frac{1}{10}h + 5\right) = \frac{1}{5}x + 5$$

(b) We evaluate $C(x)$ and $C'(x)$ at $x = 1000$. The cost when $x = 1000$ tons is

$$C(1000) = \frac{1}{10}(1000)^2 + 5 \cdot 1000 + 1000 = \$106{,}000$$

The marginal cost when $x = 1000$ tons is

$$C'(1000) = \frac{1}{5} \cdot 1000 + 5 = \$205 \text{ per ton}$$

(c) $C'(1000) = \$205$ per ton means that the cost of producing one additional ton of steel after 1000 tons have been produced is \$205.

The average cost of producing one more ton of steel after the 1000th ton is

$$\frac{\Delta C}{\Delta x} = \frac{C(1001) - C(1000)}{1001 - 1000}$$

$$= \left(\frac{1}{10} \cdot 1001^2 + 5 \cdot 1001 + 1000\right) - \left(\frac{1}{10} \cdot 1000^2 + 5 \cdot 1000 + 1000\right)$$

$$= \$205.10/\text{ton}$$

Observe that the average cost differs from the marginal cost by only 0.1 dollar/ton, which is less than $\frac{1}{20}$th of 1%. Since the marginal cost is usually easier to compute than the average cost, the marginal cost is often used to approximate the cost of producing one additional unit.

The money received by our hypothetical steel producer when he sells his product is the revenue. Specifically, let $R = R(x)$ be the total revenue received from selling x tons. Then the derivative $R'(x)$ is called the **marginal revenue.** For this example, marginal revenue, like marginal cost, is measured in dollars per ton. An approximate value for $R'(x)$ is obtained by noting again that

$$R'(x) \approx \frac{R(x + h) - R(x)}{h}$$

When x is large, then $h = 1$ is small by comparison, so that

$$R'(x) = R(x + 1) - R(x) = \text{revenue resulting from the sale of one additional unit}$$

This is the interpretation many economists give to marginal revenue.

EXAMPLE 8 Finding Marginal Revenue

Suppose that the weekly revenue R for the sale of x tons of steel is given by the formula

$$R = x^2 + 5x$$

(a) Find the marginal revenue.
(b) Find the revenue and marginal revenue when $x = 1000$ tons.
(c) Interpret $R'(1000)$.

SOLUTION **(a)** The marginal revenue is the derivative $R'(x)$. We use the difference quotient of $R(x)$ to find $R'(x)$.

$$\begin{aligned} R'(x) = \lim_{h\to 0}\frac{R(x + h) - R(x)}{h} &= \lim_{h\to 0}\frac{[(x + h)^2 + 5(x + h)] - [x^2 + 5x]}{h} \\ &= \lim_{h\to 0}\frac{x^2 + 2xh + h^2 + 5x + 5h - x^2 - 5x}{h} \\ &= \lim_{h\to 0}\frac{2xh + h^2 + 5h}{h} = \lim_{h\to 0}(2x + h + 5) = 2x + 5 \end{aligned}$$

(b) The revenue when $x = 1000$ tons is

$$R(1000) = (1000)^2 + 5(1000) = \$1{,}005{,}000$$

The marginal revenue when $x = 1000$ tons is

$$R'(1000) = 2(1000) + 5 = \$2005 \text{ per ton}$$

(c) $R'(1000) = \$2005/\text{ton}$ means that the revenue obtained from selling one additional ton of steel after 1000 tons have been sold is \$2005.

The average revenue derived from selling one additional ton after 1000 tons have been sold is

$$\frac{\Delta R}{\Delta x} = \frac{R(1001) - R(1000)}{1001 - 1000} = 1{,}007{,}006 - 1{,}005{,}000 = \$2006/\text{ton}$$

Observe that the actual average revenue differs from the marginal revenue by only \$1/ton, or 0.05%. Since the marginal revenue is usually easier to compute than the average revenue, the marginal revenue is often used to approximate the revenue from selling one additional unit.

NOW WORK PROBLEM 63.

SUMMARY The derivative of a function $y = f(x)$ at c is defined as

$$f'(c) = \lim_{x \to c} \frac{f(x) - f(c)}{x - c}$$

The derivative $f'(x)$ of a function $y = f(x)$ is

$$f'(x) = \lim_{h \to 0} \frac{f(x+h) - f(x)}{h}$$

In geometry, $f'(c)$ equals the slope of the tangent line to the graph of f at the point $(c, f(c))$.

In applications, if two variables are related by the function $y = f(x)$, then $f'(c)$ equals the instantaneous rate of change of f with respect to x at c.

In economics, the derivative of a cost function is the marginal cost and the derivative of a revenue function is the marginal revenue.

EXERCISE 13.1 Answers to Odd-Numbered Problems Begin on Page AN-72.

In Problems 1–12, find the slope of the tangent line to the graph of f at the given point. What is an equation of the tangent line? Graph f and the tangent line.

1. $f(x) = 3x + 5$ at $(1, 8)$

2. $f(x) = -2x + 1$ at $(-1, 3)$

3. $f(x) = x^2 + 2$ at $(-1, 3)$

4. $f(x) = 3 - x^2$ at $(1, 2)$

5. $f(x) = 3x^2$ at $(2, 12)$

6. $f(x) = -4x^2$ at $(-2, -16)$

7. $f(x) = 2x^2 + x$ at $(1, 3)$

8. $f(x) = 3x^2 - x$ at $(0, 0)$

9. $f(x) = x^2 - 2x + 3$ at $(-1, 6)$

10. $f(x) = -2x^2 + x - 3$ at $(1, -4)$

11. $f(x) = x^3 + x^2$ at $(-1, 0)$

12. $f(x) = x^3 - x^2$ at $(1, 0)$

In Problems 13–24, find the derivative of each function at the given number.

13. $f(x) = -4x + 5$ at 3

14. $f(x) = -4 + 3x$ at 1

15. $f(x) = x^2 - 3$ at 0

16. $f(x) = 2x^2 + 1$ at -1

17. $f(x) = 2x^2 + 3x$ at 1

18. $f(x) = 3x^2 - 4x$ at 2

19. $f(x) = x^3 + 4x$ at 0

20. $f(x) = 2x^3 - x^2$ at 0

21. $f(x) = x^3 + x^2 - 2x$ at 1

22. $f(x) = x^3 - 2x^2 + x$ at 1

23. $f(x) = \frac{1}{x}$ at 1

24. $f(x) = \frac{1}{x^2}$ at 1

In Problems 25–36, find the derivative of f using the difference quotient.

25. $f(x) = 2x$

26. $f(x) = 3x$

27. $f(x) = 1 - 2x$

28. $f(x) = 5 - 3x$

29. $f(x) = x^2 + 2$

30. $f(x) = 2x^2 - 3$

31. $f(x) = 3x^2 - 2x + 1$

32. $f(x) = 2x^2 + x + 1$

33. $f(x) = x^3$

34. $f(x) = \frac{1}{x}$

35. $f(x) = mx + b$

36. $f(x) = ax^2 + bx + c$

In Problems 37–44, find

(a) The average rate of change as x changes from 1 to 3.
(b) The instantaneous rate of change at 1.

37. $f(x) = 3x + 4$ **38.** $f(x) = 2x - 6$ **39.** $f(x) = 3x^2 + 1$ **40.** $f(x) = 2x^2 + 1$

41. $f(x) = x^2 + 2x$ **42.** $f(x) = x^2 - 4x$ **43.** $f(x) = 2x^2 - x + 1$ **44.** $f(x) = 2x^2 + 3x - 2$

In Problems 45–54, find the derivative of each function at the given number using a graphing utility.

45. $f(x) = 3x^3 - 6x^2 + 2$ at -2

46. $f(x) = -5x^4 + 6x^2 - 10$ at 5

47. $f(x) = \dfrac{-x^3 + 1}{x^2 + 5x + 7}$ at 8

48. $f(x) = \dfrac{-5x^4 + 9x + 3}{x^3 + 5x^2 - 6}$ at -3

49. $f(x) = xe^x$ at 0 **50.** $f(x) = xe^x$ at 1 **51.** $f(x) = x^2 e^x$ at 1

52. $f(x) = x^2 e^x$ at 0 **53.** $f(x) = xe^{-x}$ at 1 **54.** $f(x) = x^2 e^{-x}$ at 2

---

55. Does the tangent line to the graph of $y = x^2$ at (1, 1) pass through the point (2, 5)?

56. Does the tangent line to the graph of $y = x^3$ at (1, 1) pass through the point (2, 5)?

57. A dive bomber is flying from right to left along the graph of $y = x^2$. When a rocket bomb is released, it follows a path that approximately follows the tangent line. Where should the pilot release the bomb if the target is at (1, 0)?

58. Answer the question in Problem 57 if the plane is flying from right to left along the graph of $y = x^3$.

59. **Ticket Sales** The cumulative ticket sales for the 12 days preceding a popular concert is given by

$$S = 4x^2 + 50x + 5000$$

where x, $1 \le x \le 12$, represents the number of days before the concert.

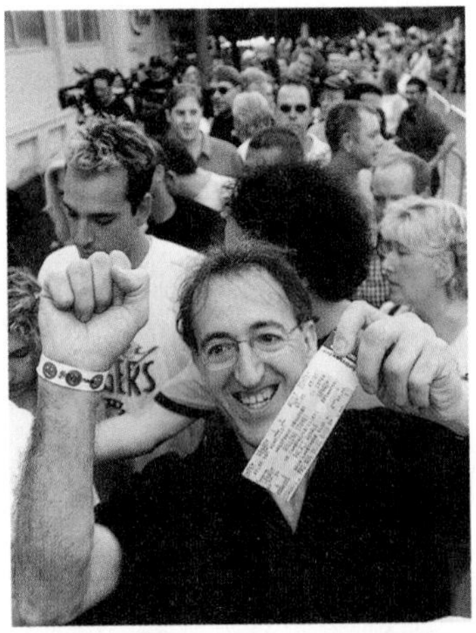

(a) What is the average rate of change in sales from day 1 to day 5?
(b) What is the average rate of change in sales from day 1 to day 10?
(c) What is the average rate of change in sales from day 5 to day 10?
(d) What is the instantaneous rate of change in sales on day 5?
(e) What is the instantaneous rate of change in sales on day 10?

60. Computer Sales The weekly revenue R, in dollars, due to selling x computers is

$$R(x) = -20x^2 + 1000x$$

(a) Find the average rate of change in revenue obtained from selling 5 additional computers after the 20th has been sold.
(b) Find the marginal revenue.
(c) Find the marginal revenue at $x = 20$.
(d) Interpret the answers found in (a) and (c).
(e) For what value of x is $R'(x) = 0$?

61. Supply and Demand Suppose $S(x) = 50x^2 - 50x$ is the supply function describing the number of crates of grapefruit a farmer is willing to supply to the market for x dollars per crate.

(a) How many crates is the farmer willing to supply for \$10 per crate?
(b) How many crates is the farmer willing to supply for \$13 per crate?
(c) Find the average rate of change in supply from \$10 per crate to \$13 per crate.
(d) Find the instantaneous rate of change in supply at $x = 10$.
(e) Interpret the answers found in (c) and (d).

62. Glucose Conversion In a metabolic experiment, the mass M of glucose decreases over time t according to the formula

$$M = 4.5 - 0.03t^2$$

(a) Find the average rate of change of the mass from $t = 0$ to $t = 2$.
(b) Find the instantaneous rate of change of mass at $t = 0$.
(c) Interpret the answers found in (a) and (b).

63. Cost and Revenue Functions For a certain production facility, the cost function is

$$C(x) = 2x + 5$$

and the revenue function is

$$R(x) = 8x - x^2$$

where x is the number of units (in thousands) produced and sold and R and C are measured in millions of dollars. Find:

(a) The marginal revenue.
(b) The marginal cost.
(c) The break-even point(s) [the number(s) x for which $R(x) = C(x)$].
(d) The number x for which marginal revenue equals marginal cost.
(e) Graph $C(x)$ and $R(x)$ on the same set of axes.

64. Cost and Revenue Functions For a certain production facility, the cost function is

$$C(x) = x + 5$$

and the revenue function is

$$R(x) = 12x - 2x^2$$

where x is the number of units (in thousands) produced and sold and R and C are measured in millions of dollars. Find:

(a) The marginal revenue.
(b) The marginal cost.
(c) The break-even point(s) [the number(s) x for which $R(x) = C(x)$].
(d) The number x for which marginal revenue equals marginal cost.
(e) Graph $C(x)$ and $R(x)$ on the same set of axes.

65. Demand Equation The price p per ton of cement when x tons of cement are demanded is given by the equation

$$p = -10x + 2000$$

dollars. Find:

(a) The revenue function $R = R(x)$
Hint: $R = xp$, where p is the unit price.
(b) The marginal revenue.
(c) The marginal revenue at $x = 100$ tons.
(d) The average rate of change in revenue from $x = 100$ to $x = 101$ tons.
(e) Interpret the answers found in (c) and (d).

66. Demand Equation The cost function and demand equation for a certain product are

$$C(x) = 50x + 40{,}000 \quad \text{and} \quad p = 100 - 0.01x$$

Find:

(a) The revenue function.
(b) The marginal revenue.
(c) The marginal cost.
(d) The break-even point(s).
(e) The number x for which marginal revenue equals marginal cost.

67. Demand Equation A certain item can be produced at a cost of \$10 per unit. The demand equation for this item is

$$p = 90 - 0.02x$$

where p is the price in dollars and x is the number of units. Find:

(a) The revenue function.
(b) The marginal revenue.
(c) The marginal cost.
(d) The break-even point(s).
(e) The number x for which marginal revenue equals marginal cost.

68. Instantaneous Rate of Change A circle of radius r has area $A = \pi r^2$ and circumference $C = 2\pi r$. If the radius changes from r to $(r + h)$, find the:

(a) Change in area.
(b) Change in circumference.
(c) Average rate of change of area with respect to the radius.
(d) Average rate of change of the circumference with respect to the radius.
(e) Instantaneous rate of change of area with respect to the radius.
(f) Instantaneous rate of change of the circumference with respect to the radius.

69. Instantaneous Rate of Change The volume V of a right circular cylinder of height 3 feet and radius r feet is $V = V(r) = 3\pi r^2$. Find the instantaneous rate of change of the volume with respect to the radius r at $r = 3$.

70. Instantaneous Rate of Change The surface area S of a sphere of radius r feet is $S = S(r) = 4\pi r^2$. Find the instantaneous rate of change of the surface area with respect to the radius r at $r = 2$ feet.

13.2 The Derivative of a Power Function; Sum and Difference Formulas

OBJECTIVES
1. Find the derivative of a power function
2. Find the derivative of a constant times a function
3. Find the derivative of a polynomial function

In the previous section, we found the derivative $f'(x)$ of a function $y = f(x)$ by using the difference quotient:

$$f'(x) = \lim_{h \to 0} \frac{f(x+h) - f(x)}{h} \qquad (1)$$

We use this form for the derivative to derive formulas for finding derivatives.

We begin by considering the constant function $f(x) = b$, where b is a real number. Since the graph of the constant function f is a horizontal line (see Figure 8), the tangent line to f at any point is also a horizontal line. Since the derivative equals the slope of the tangent line to the graph of a function f at a point, then the derivative of f should be 0.

FIGURE 8

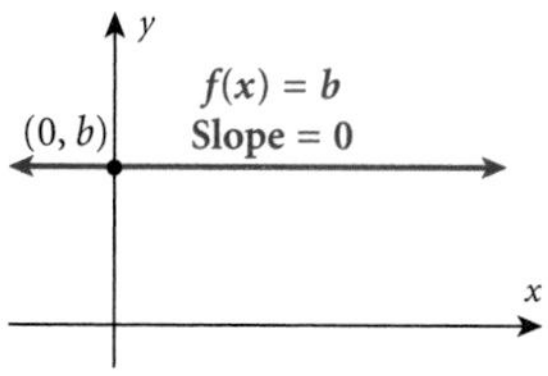

Analytically, the derivative is obtained by using Formula (1). The difference quotient of $f(x) = b$ is

$$\frac{f(x+h) - f(x)}{h} = \frac{b - b}{h} = \frac{0}{h} = 0$$

The derivative of $f(x) = b$ is

$$f'(x) = \lim_{h \to 0} \frac{f(x+h) - f(x)}{h} = \lim_{h \to 0} 0 = 0$$

Derivative of the Constant Function

For the constant function $f(x) = b$, the derivative is $f'(x) = 0$. In other words, the derivative of a constant is 0.

Besides the **prime notation** f', there are several other ways to denote the derivative of a function $y = f(x)$. The most common ones are

$$y' \quad \text{and} \quad \frac{dy}{dx}$$

The notation $\frac{dy}{dx}$, often referred to as the **Leibniz notation,** may also be written as

$$\frac{dy}{dx} = \frac{d}{dx}y = \frac{d}{dx}f(x)$$

where $\frac{d}{dx}f(x)$ is an instruction to compute the derivative of the function f with respect to its independent variable x. A change in the symbol used for the independent variable does not affect the meaning. If $s = f(t)$ is a function of t, then $\frac{ds}{dt}$ is an instruction to differentiate f with respect to t.

In terms of the Leibniz notation, if b is a constant, then

$$\boxed{\frac{d}{dx}b = 0} \qquad (2)$$

EXAMPLE 1 **Finding the Derivative of a Constant Function**

(a) If $f(x) = 5$, then $f'(x) = 0$.

(b) If $y = -1.7$, then $y' = 0$.

(c) If $y = \frac{2}{3}$, then $\frac{dy}{dx} = 0$.

(d) If $s = f(t) = \sqrt{5}$, then $\frac{ds}{dt} = f'(t) = 0$.

In subsequent work with derivatives we shall use the prime notation or the Leibniz notation, or sometimes a mixture of the two, depending on which is more convenient.

NOW WORK PROBLEM 1.

Derivative of a Power Function

We now investigate the derivative of the power function $f(x) = x^n$, where n is a positive integer, to see if a pattern appears.

For $f(x) = x$, $n = 1$, we have

$$f'(x) = \lim_{h\to 0}\frac{f(x+h) - f(x)}{h} = \lim_{h\to 0}\frac{(x+h) - x}{h} = \lim_{h\to 0}\frac{h}{h} = \lim_{h\to 0} 1 = 1$$

For $f(x) = x^2$, $n = 2$, we have

$$\begin{aligned} f'(x) = \lim_{h\to 0}\frac{f(x+h) - f(x)}{h} &= \lim_{h\to 0}\frac{(x+h)^2 - x^2}{h} = \lim_{h\to 0}\frac{x^2 + 2xh + h^2 - x^2}{h} \\ &= \lim_{h\to 0}\frac{2xh + h^2}{h} = \lim_{h\to 0}\frac{h(2x+h)}{h} \\ &= \lim_{h\to 0}(2x + h) = 2x \end{aligned}$$

For $f(x) = x^3$, $n = 3$, we have

$$\begin{aligned} f'(x) = \lim_{h\to 0}\frac{f(x+h) - f(x)}{h} &= \lim_{h\to 0}\frac{(x+h)^3 - x^3}{h} = \lim_{h\to 0}\frac{x^3 + 3x^2h + 3xh^2 + h^3 - x^3}{h} \\ &= \lim_{h\to 0}\frac{3x^2h + 3xh^2 + h^3}{h} = \lim_{h\to 0}\frac{h(3x^2 + 3xh + h^2)}{h} = \lim_{h\to 0}(3x^2 + 3xh + h^2) = 3x^2 \end{aligned}$$

In the Leibniz notation, these results take the form

$$\frac{d}{dx}x = 1 \qquad \frac{d}{dx}x^2 = 2x \qquad \frac{d}{dx}x^3 = 3x^2$$

This pattern suggests the following formula:

Derivative of $f(x) = x^n$

For the power function $f(x) = x^n$, n a positive integer, the derivative is $f'(x) = nx^{n-1}$. That is,

$$\frac{d}{dx}x^n = nx^{n-1} \tag{3}$$

Formula (3) may be stated in words as follows:

The derivative with respect to x of x raised to the power n, where n is a positive integer, is n times x raised to the power $n - 1$.

Problems 73 and 74 outline proofs of Formula (3).

1 **EXAMPLE 2 Finding the Derivative of a Power Function**

(a) If $f(x) = x^6$, then $f'(x) = 6x^{6-1} = 6x^5$

(b) $\frac{d}{dt}t^5 = 5t^4$

(c) $\frac{d}{dx}x = 1$ $\quad \frac{d}{dx}x^1 = 1\cdot x^{1-1} = 1\cdot x^0 = 1\cdot 1 = 1$

NOW WORK PROBLEM 3.

EXAMPLE 3 Finding the Derivative of a Power Function at a Number

Find $f'(4)$ if $f(x) = x^3$

SOLUTION We use Formula (3).

$$f'(x) = 3x^2 \quad \text{Use Formula (3).}$$
$$f'(4) = 3(4)^2 = 48 \quad \text{Substitute 4 for } x.$$

Formula (3) allows us to compute some derivatives with ease. However, do not forget that a derivative is, in actuality, the limit of a difference quotient.

The next formula is used often.

Derivative of a Constant Times a Function

The derivative of a constant times a function equals the constant times the derivative of the function. That is, if c is a constant and f is a differentiable function, then

$$\frac{d}{dx}[cf(x)] = c\frac{d}{dx}f(x) \tag{4}$$

Proof We prove Formula (4) as follows.

$$\begin{aligned}
\frac{d}{dx}[cf(x)] &= \lim_{h\to 0}\frac{cf(x+h)-cf(x)}{h} && \text{Use the difference quotient of } cf(x).\\
&= \lim_{h\to 0} c\,\frac{f(x+h)-f(x)}{h} && \text{Factor out } c.\\
&= \lim_{h\to 0} c\cdot \lim_{h\to 0}\frac{f(x+h)-f(x)}{h} && \text{The limit of a product is the product of the limits.}\\
&= c\frac{d}{dx}f(x) && \text{The limit of a constant is the constant; } \frac{d}{dx}f(x)=\lim_{h\to 0}\frac{f(x+h)-f(x)}{h}
\end{aligned}$$

The usefulness and versatility of this formula are often overlooked, especially when the constant appears in the denominator. Note that

$$\frac{d}{dx}\left[\frac{f(x)}{c}\right] = \frac{d}{dx}\left[\frac{1}{c}f(x)\right] = \frac{1}{c}\frac{d}{dx}[f(x)]$$

Always be on the lookout for constant factors *before* differentiating.

2 **EXAMPLE 4 Finding the Derivative of a Constant Times a Function**

(a) If $f(x) = 10x^3$, then

$$f'(x) = \frac{d}{dx}(10x^3) = 10\frac{d}{dx}x^3 = 10\cdot 3x^2 = 30x^2$$

(b) $\frac{d}{dx}\left(\frac{x^5}{10}\right) = \frac{1}{10}\frac{d}{dx}x^5 = \frac{1}{10}\cdot 5x^4 = \frac{1}{2}x^4$

(c) $\frac{d}{dt}(6t) = 6\frac{d}{dt}t = 6\cdot 1 = 6$

(d) $\frac{d}{dx}\left(\frac{2\sqrt{3}}{3}x^3\right) = \frac{2\sqrt{3}}{3}\frac{d}{dx}x^3 = \frac{2\sqrt{3}}{3}\cdot 3x^2 = 2\sqrt{3}\,x^2$

NOW WORK PROBLEM 7.

Sum and Difference Formulas

Derivative of a Sum

The derivative of the sum of two differentiable functions equals the sum of their derivatives. That is,

$$\frac{d}{dx}[f(x)+g(x)] = \frac{d}{dx}f(x) + \frac{d}{dx}g(x) \qquad (5)$$

A proof is given at the end of this section.

This formula states that functions that are sums can be differentiated "term by term."

EXAMPLE 5 Finding the Derivative of a Function

Find the derivative of: $f(x) = x^2 + 4x$

SOLUTION The function f is the sum of the two functions x^2 and $4x$. We can differentiate term by term.

$$\frac{d}{dx}f(x) = \frac{d}{dx}(x^2 + 4x) = \frac{d}{dx}x^2 + \frac{d}{dx}(4x) = 2x + 4\frac{d}{dx}x = 2x + 4$$

Formula (5) — Formulas (3) and (4) — $\frac{d}{dx}x = 1$

Derivative of a Difference

The derivative of the difference of two differentiable functions equals the difference of their derivatives. That is,

$$\frac{d}{dx}[f(x) - g(x)] = \frac{d}{dx}f(x) - \frac{d}{dx}g(x) \qquad \textbf{(6)}$$

Formulas (5) and (6) extend to sums and differences of more than two functions. Since a polynomial function is a sum (or difference) of power functions, we can find the derivative of any polynomial function by using a combination of Formulas (2), (3), (4), (5), and (6).

3 **EXAMPLE 6 Finding the Derivative of a Polynomial Function**

Find the derivative of: $f(x) = 6x^4 - 3x^2 + 10x - 8$

SOLUTION

$$\begin{aligned} f'(x) &= \frac{d}{dx}(6x^4 - 3x^2 + 10x - 8) \\ &= \frac{d}{dx}(6x^4) - \frac{d}{dx}(3x^2) + \frac{d}{dx}(10x) - \frac{d}{dx}8 && \text{Use Formulas (5) and (6).} \\ &= 6\frac{d}{dx}x^4 - 3\frac{d}{dx}x^2 + 10\frac{d}{dx}x - 0 && \text{Use Formulas (2) and (4).} \\ &= 24x^3 - 6x + 10 && \text{Use Formula (3); Simplify.} \end{aligned}$$

NOW WORK PROBLEM 21.

EXAMPLE 7 Finding the Derivative of a Polynomial Function

If $f(x) = \frac{x^4}{2} - 2x + 3$, find

(a) $f'(x)$ **(b)** $f'(-1)$

SOLUTION **(a)** $f'(x) = \frac{4x^3}{2} - 2 + 0 = 2x^3 - 2$

(b) $f'(-1) = 2(-1)^3 - 2 = -4$

NOW WORK PROBLEM 33.

EXAMPLE 8 **Analyzing a Cost Function**

The daily cost C, in dollars, of producing dishwashers is

$$C(x) = 1000 + 72x - 0.06x^2 \qquad 0 \leq x \leq 60$$

where x represents the number of dishwashers produced.

(a) Find the daily cost of producing 50 dishwashers.
(b) Find the marginal cost function.
(c) Find $C'(50)$ and interpret its meaning.
(d) Use the marginal cost to estimate the cost of producing 51 dishwashers.
(e) Find the actual cost of producing 51 dishwashers. Compare the actual cost of making 51 dishwashers to the estimated cost of producing 51 dishwashers found in part (d).
(f) Find the actual cost of producing the 51st dishwasher.
(g) The average cost function is defined as $\overline{C}(x) = \dfrac{C(x)}{x}$, $0 < x \leq 60$. Find the average cost function for producing x dishwashers.
(h) Find the average cost of producing 51 dishwashers.

SOLUTION **(a)** The daily cost of producing 50 dishwashers is

$$C(50) = 1000 + 72(50) - 0.06(50)^2 = \$4450$$

(b) The marginal cost function is

$$C'(x) = \frac{d}{dx}(1000 + 72x - 0.06x^2) = 72 - 0.12x$$

(c) $C'(50) = 72 - 0.12(50) = \$66.$
The marginal cost of producing 50 dishwashers may be interpreted as the cost to produce the 51st dishwasher.

(d) From part (a) the cost to produce 50 dishwashers is \$4450. If the 51st costs \$66, then the cost to produce 51 dishwashers will be

$$\$4450 + \$66 = \$4516$$

(e) The actual cost to produce 51 dishwashers is

$$C(51) = \$1000 + 72(51) - 0.06(51)^2 = \$4515.94$$

There is a difference of \$0.06 between the actual cost and the cost obtained using the marginal cost.

(f) The actual cost of producing the 51st dishwasher is

$$C(51) - C(50) = \$4515.94 - \$4450 = \$65.94$$

(g) The average cost function is

$$\overline{C}(x) = \frac{C(x)}{x} = \frac{1000 + 72x - 0.06x^2}{x} = \frac{1000}{x} + 72 - 0.06x$$

(h) The average cost of producing 51 dishwashers is

$$\overline{C}(51) = \frac{1000}{51} + 72 - 0.06(51) = \$88.55$$

NOW WORK PROBLEM 65.

Proof of the Sum Formula We prove Formula (5) as follows. To compute

$$\frac{d}{dx}[f(x) + g(x)]$$

we need to find the limit of the difference quotient of $f(x) + g(x)$.

$$\begin{aligned}\frac{d}{dx}[f(x) + g(x)] &= \lim_{h\to 0}\frac{[f(x+h) + g(x+h)] - [f(x) + g(x)]}{h}\\ &= \lim_{h\to 0}\frac{[f(x+h) - f(x)] + [g(x+h) - g(x)]}{h}\\ &= \lim_{h\to 0}\left[\frac{f(x+h) - f(x)}{h} + \frac{g(x+h) - g(x)}{h}\right]\\ &= \lim_{h\to 0}\left[\frac{f(x+h) - f(x)}{h}\right] + \lim_{h\to 0}\left[\frac{g(x+h) - g(x)}{h}\right]\\ &= \frac{d}{dx}f(x) + \frac{d}{dx}g(x)\end{aligned}$$

■

Proof of the Difference Formula The proof uses Formulas (4) and (5).

$$\begin{aligned}\frac{d}{dx}[f(x) - g(x)] &= \frac{d}{dx}[f(x) + (-1)g(x)]\\ &= \frac{d}{dx}f(x) + \frac{d}{dx}[(-1)g(x)]\\ &= \frac{d}{dx}f(x) + (-1)\frac{d}{dx}g(x)\\ &= \frac{d}{dx}f(x) - \frac{d}{dx}g(x)\end{aligned}$$

■

EXERCISE 13.2 Answers to Odd-Numbered Problems Begin on Page AN-73.

In Problems 1–20, find the derivative of each function.

1. $f(x) = 4$
2. $f(x) = -2$
3. $f(x) = x^5$
4. $f(x) = x^4$
5. $f(x) = 6x^2$
6. $f(x) = -8x^3$
7. $f(t) = \dfrac{t^4}{4}$
8. $f(t) = \dfrac{t^3}{6}$
9. $f(x) = x^2 + x$
10. $f(x) = x^2 - x$
11. $f(x) = x^3 - x^2 + 1$
12. $f(x) = x^4 - x^3 + x$
13. $f(t) = 2t^2 - t + 4$
14. $f(t) = 3t^3 - t^2 + t$
15. $f(x) = \dfrac{1}{2}x^8 + 3x + \dfrac{2}{3}$
16. $f(x) = \dfrac{2}{3}x^6 - \dfrac{1}{2}x^4 + 2$
17. $f(x) = \dfrac{1}{3}(x^5 - 8)$
18. $f(x) = \dfrac{x^3 + 2}{5}$
19. $f(x) = ax^2 + bx + c$ a, b, c are constants
20. $f(x) = ax^3 + bx^2 + cx + d$ a, b, c, d are constants

In Problems 21–28, find the indicated derivative.

21. $\dfrac{d}{dx}(-6x^2 + x + 4)$
22. $\dfrac{d}{dx}(8x^3 - 6x^2 + 2x)$
23. $\dfrac{d}{dt}(-16t^2 + 80t)$
24. $\dfrac{d}{dt}(-16t^2 + 64t)$
25. $\dfrac{dA}{dr}$ if $A = \pi r^2$
26. $\dfrac{dC}{dr}$ if $C = 2\pi r$
27. $\dfrac{dV}{dr}$ if $V = \dfrac{4}{3}\pi r^3$
28. $\dfrac{dP}{dt}$ if $P = 0.2t$

In Problems 29–38, find the value of the derivative at the indicated number.

29. $f(x) = 4x^2$ at $x = -3$

30. $f(x) = -10x^3$ at $x = -2$

31. $f(x) = 2x^2 - x$ at $x = 4$

32. $f(x) = x^4 - 2x^2$ at $x = 2$

33. $f(t) = -\frac{1}{3}t^3 + 5t$ at $t = 3$

34. $f(t) = -\frac{1}{4}t^4 + \frac{1}{2}t^2 + 4$ at $t = 1$

35. $f(x) = \frac{1}{2}(x^6 - x^4)$ at $x = 1$

36. $f(x) = \frac{1}{3}(x^6 + x^3 + 1)$ at $x = -1$

37. $f(x) = ax^2 + bx + c$ at $x = -\frac{b}{2a}$
a, b, c are constants

38. $f(x) = ax^3 + bx^2 + cx + d$ at $x = 0$
a, b, c, d are constants

In Problems 39–48, find the value of $\frac{dy}{dx}$ *at the indicated point.*

39. $y = x^4$ at $(1, 1)$

40. $y = x^4$ at $(2, 16)$

41. $y = x^2 - 14$ at $(4, 2)$

42. $y = x^3 + 1$ at $(3, 28)$

43. $y = 3x^2 - x$ at $(-1, 4)$

44. $y = x^2 - 3x$ at $(-1, 4)$

45. $y = \frac{1}{2}x^2$ at $\left(1, \frac{1}{2}\right)$

46. $y = x^3 - x^2$ at $(1, 0)$

47. $y = 2 - 2x + x^3$ at $(2, 6)$

48. $y = 2x^2 - \frac{1}{2}x + 3$ at $(0, 3)$

In Problems 49–50, find the slope of the tangent line to the graph of the function at the indicated point. What is an equation of the tangent line?

49. $f(x) = x^3 + 3x - 1$ at $(0, -1)$

50. $f(x) = x^4 + 2x - 1$ at $(1, 2)$

In Problems 51–56, find those x, if any, at which $f'(x) = 0$.

51. $f(x) = 3x^2 - 12x + 4$

52. $f(x) = x^2 + 4x - 3$

53. $f(x) = x^3 - 3x + 2$

54. $f(x) = x^4 - 4x^3$

55. $f(x) = x^3 + x$

56. $f(x) = x^5 - 5x^4 + 1$

---

57. Find the point(s), if any, on the graph of the function $y = 9x^3$ at which the tangent line is parallel to the line $3x - y + 2 = 0$.

58. Find the points(s), if any, on the graph of the function $y = 4x^2$ at which the tangent line is parallel to the line $2x - y - 6 = 0$.

59. Two lines through the point $(1, -3)$ are tangent to the graph of the function $y = 2x^2 - 4x + 1$. Find the equations of these two lines.

60. Two lines through the point $(0, 2)$ are tangent to the graph of the function $y = 1 - x^2$. Find the equations of these two lines.

61. Marginal Cost The cost per day, $C(x)$, in dollars, of producing x pairs of eyeglasses is

$$C(x) = 0.2x^2 + 3x + 1000$$

(a) Find the average cost due to producing 10 additional pairs of eyeglasses after 100 have been produced.
(b) Find the marginal cost.
(c) Find the marginal cost at $x = 100$.
Q (d) Interpret $C'(100)$.

62. Toy Truck Sales At Dan's Toy Store, the revenue R, in dollars, derived from selling x electric trucks is

$$R(x) = -0.005x^2 + 20x$$

(a) What is the average rate of change in revenue obtained from selling 10 additional trucks after 1000 have been sold?
(b) What is the marginal revenue?
(c) What is the marginal revenue at $x = 1000$?
Q (d) Interpret $R'(1000)$.
(e) For what value of x is $R'(x) = 0$?

63. Medicine The French physician Poiseville discovered that the volume V of blood (in cubic centimeters) flowing through a clogged artery with radius R (in centimeters) can be modeled by

$$V(R) = kR^4$$

where k is a positive constant.

(a) Find the derivative $V'(R)$.
(b) Find the rate of change of volume for a radius of 0.3 cm.
(c) Find the rate of change of volume for a radius of 0.4 cm.
Q (d) If the radius of a clogged artery is increased from 0.3 cm to 0.4 cm, estimate the effect on the volume of blood flowing through the enlarged artery.

64. Respiration Rate A human being's respiration rate R (in breaths per minute) is given by

$$R = -10.35p + 0.59p^2$$

where p is the partial pressure of carbon dioxide in the lungs. Find the rate of change in respiration rate when $p = 50$.

65. **Analyzing a Cost Function** The daily cost C of producing microwave ovens is

$$C(x) = 2000 + 50x - 0.05x^2, \quad 0 \le x \le 50$$

where x represents the number of microwave ovens produced.

(a) Find the daily cost of producing 40 microwave ovens.
(b) Find the marginal cost function.
(c) Find $C'(40)$ and interpret its meaning.
(d) Use the marginal cost to estimate the cost of producing 41 microwave ovens.
(e) Find the actual cost of producing 41 microwave ovens. Compare the actual cost of making 41 microwave ovens to the estimated cost of producing 41 microwave ovens.
(f) Find the actual cost of producing the 41st microwave oven.
(g) The average cost function is defined as $\overline{C}(x) = \dfrac{C(x)}{x}$, $0 < x \le 50$. Find the average cost function for producing x microwave ovens.
(h) Find the average cost of producing 41 microwave ovens.
(i) Compare your answers from parts (c) and (f). Give explanations for the differences.

66. Analyzing a Cost Function The daily cost C of producing small televisions is

$$C(x) = 1500 + 25x - 0.05x^2, \; 0 \le x \le 100$$

where x represents the number of televisions produced.

(a) Find the daily cost of producing 70 televisions.
(b) Find the marginal cost function.
(c) Find $C'(70)$ and interpret its meaning.
(d) Use the marginal cost to estimate the cost of producing 71 televisions.
(e) Find the actual cost of producing 71 televisions. Compare the actual cost of making 71 televisions to the estimated cost of producing 71 televisions.
(f) Find the actual cost of manufacturing the 71st television.
(g) The average cost function is defined as $\overline{C}(x) = \dfrac{C(x)}{x}$, $0 < x \le 100$. Find the average cost function for producing x television.
(h) Find the average cost of producing 71 televisions.
(i) Compare your answers from parts (c) and (f). Give explanations for the differences.

67. Price of Beans The price in dollars per cwt for beans from 1993 through 2002 can be modeled by the polynomial function $p(t) = 0.007t^3 - 0.63t^2 + 0.005t + 6.123$, where t is in years, and $t = 0$ corresponds to 1993.

(a) Find the marginal price of beans for the year 1995.
(b) Find the marginal price for beans for the year 2002.
(c) How do you interpret the two marginal prices? What is the trend?

68. Price of Beans The price in dollars per cwt for beans from 1993 through 2002 can also be modeled by the polynomial function, $p(t) = -0.002t^4 + 0.044t^3 - 0.335t^2 + 0.750t + 5.543$, where t is in years and $t = 0$ corresponds to 1993.

(a) Find the marginal price for beans for the year 1995.
(b) Find the marginal price for beans for the year 2002.
(c) How do you interpret the two marginal prices? What is the trend?
(d) Explain why there might be two different functions that can model the price of beans.

69. Instantaneous Rate of Change The volume V of a sphere of radius r feet is $V = V(r) = \frac{4}{3}\pi r^3$. Find the instantaneous rate of change of the volume with respect to the radius r at $r = 2$ feet.

70. Instantaneous Rate of Change The volume V of a cube of side x meters is $V = V(x) = x^3$. Find the instantaneous rate of change of the volume with respect to the side x at $x = 3$ meters.

71. Work Output The relationship between the amount $A(t)$ of work output and the elapsed time t, $t \ge 0$, was found through empirical means to be

$$A(t) = a_3t^3 + a_2t^2 + a_1t + a_0$$

where a_0, a_1, a_2, a_3 are constants. Find the instantaneous rate of change of work output at time t.

72. Consumer Price Index The consumer price index (CPI) of an economy is described by the function

$$I(t) = -0.2t^2 + 3t + 200 \qquad 0 \le t \le 10$$

where $t = 0$ corresponds to the year 2000.

(a) What was the average rate of increase in the CPI over the period from 2000 to 2003?
(b) At what rate was the CPI of the economy changing in 2003? in 2006?

73. Use the binomial theorem to prove Formula (3), p. 728.

[**Hint:** $(x + h)^n - x^n = x^n + nx^{n-1}h + \dfrac{n(n-1)}{2}x^{n-2}h^2 + \cdots + h^n - x^n = nx^{n-1}h + h^2 \cdot$ (terms involving x and h). Now apply Formula (1), page 286.]

74. Use the following factoring rule to prove Formula (3), p. 728.

$$\begin{aligned} f(x) &= x^n - c^n \\ &= (x - c)(x^{n-1} + x^{n-2}c + x^{n-3}c^2 + \cdots + c^{n-1}) \end{aligned}$$

Now apply Formula (3), page 716, to find $f'(c)$.

13.3 Product and Quotient Formulas

OBJECTIVES
1 Find the derivative of a product
2 Find the derivative of a quotient
3 Find the derivative of $f(x) = x^n$, n is a negative integer

The Derivative of a Product

In the previous section we learned that the derivative of the sum or the difference of two functions is simply the sum or the difference of their derivatives. The natural inclination at this point may be to assume that differentiating a product or quotient of two functions is as simple. But this is not the case, as illustrated for the case of a product of two functions. Consider

$$F(x) = f(x) \cdot g(x) = (3x^2 - 3)(2x^3 - x) \tag{1}$$

where $f(x) = 3x^2 - 3$, and $g(x) = 2x^3 - x$. The derivative of $f(x)$ is $f'(x) = 6x$ and the derivative of $g(x)$ is $g'(x) = 6x^2 - 1$. The product of these derivatives is

$$f'(x) \cdot g'(x) = 6x(6x^2 - 1) = 36x^3 - 6x \tag{2}$$

To see if this is equal to the derivative of the product, we first multiply out the right side of equation (1) and then differentiate. Then

$$F(x) = (3x^2 - 3)(2x^3 - x) = 6x^5 - 9x^3 + 3x$$

so that

$$F'(x) = 30x^4 - 27x^2 + 3 \tag{3}$$

Since equations (2) and (3) are not equal, we conclude that the derivative of a product *is not* equal to the product of the derivatives.

The formula for finding the derivative of the product of two functions is given below:

Derivative of a Product

The derivative of the product of two differentiable functions equals the first function times the derivative of the second plus the second function times the derivative of the first. That is,

$$\frac{d}{dx}[f(x)g(x)] = f(x)\frac{d}{dx}g(x) + g(x)\frac{d}{dx}f(x) \tag{4}$$

The following version of Formula (4) may help you remember it.

$$\frac{d}{dx}(\text{first} \cdot \text{second}) = \text{first} \cdot \frac{d}{dx}\text{second} + \text{second} \cdot \frac{d}{dx}\text{first}$$

1 **EXAMPLE 1 Finding the Derivative of a Product**

Find the derivative of: $F(x) = (x^2 + 2x - 5)(x^3 - 1)$

SOLUTION The function F is the product of the two functions $f(x) = x^2 + 2x - 5$ and $g(x) = x^3 - 1$ so that, by Formula (4), we have

$$\begin{aligned} F'(x) &= (x^2 + 2x - 5)\left[\frac{d}{dx}(x^3 - 1)\right] + (x^3 - 1)\left[\frac{d}{dx}(x^2 + 2x - 5)\right] && \text{Use Formula (4).} \\ &= (x^2 + 2x - 5)(3x^2) + (x^3 - 1)(2x + 2) && \text{Differentiate.} \\ &= 3x^4 + 6x^3 - 15x^2 + 2x^4 + 2x^3 - 2x - 2 && \text{Simplify.} \\ &= 5x^4 + 8x^3 - 15x^2 - 2x - 2 && \text{Simplify.} \end{aligned}$$

Now that you know the formula for the derivative of a product, be careful not to use it unnecessarily. When one of the factors is a constant, you should use the formula for the derivative of a constant times a function. For example, it is easier to work

$$\frac{d}{dx}[5(x^2 + 1)] = 5\frac{d}{dx}(x^2 + 1) = (5)(2x) = 10x$$

than it is to work

$$\begin{aligned} \frac{d}{dx}[5(x^2 + 1)] &= 5\left[\frac{d}{dx}(x^2 + 1)\right] + (x^2 + 1)\left(\frac{d}{dx}5\right) \\ &= (5)(2x) + (x^2 + 1)(0) = 10x \end{aligned}$$

NOW WORK PROBLEM 1.

The Derivative of a Quotient

As in the case of a product, the derivative of a quotient is *not* the quotient of the derivatives.

Derivative of a Quotient

The derivative of the quotient of two differentiable functions is equal to the denominator times the derivative of the numerator minus the numerator times the derivative of the denominator, all divided by the square of the denominator.

$$\frac{d}{dx}\left[\frac{f(x)}{g(x)}\right] = \frac{g(x)\frac{d}{dx}f(x) - f(x)\frac{d}{dx}g(x)}{[g(x)]^2} \quad \text{where } g(x) \neq 0 \tag{5}$$

You may want to memorize the following version of Formula (5):

$$\frac{d}{dx}\frac{\text{numerator}}{\text{denominator}} = \frac{(\text{denominator})\frac{d}{dx}(\text{numerator}) - (\text{numerator})\frac{d}{dx}(\text{denominator})}{(\text{denominator})^2}$$

2 **EXAMPLE 2** **Finding the Derivative of a Quotient**

Find the derivative of: $F(x) = \dfrac{x^2 + 1}{x - 3}$

SOLUTION Here, the function F is the quotient of $f(x) = x^2 + 1$ and $g(x) = x - 3$. We use Formula (5) to get

$$\frac{d}{dx}\left(\frac{x^2 + 1}{x - 3}\right) = \frac{(x - 3)\frac{d}{dx}(x^2 + 1) - (x^2 + 1)\frac{d}{dx}(x - 3)}{(x - 3)^2}$$ Use Formula (5).

$$= \frac{(x - 3)(2x) - (x^2 + 1)(1)}{(x - 3)^2}$$ Differentiate.

$$= \frac{2x^2 - 6x - x^2 - 1}{(x - 3)^2}$$ Simplify.

$$= \frac{x^2 - 6x - 1}{(x - 3)^2}$$ Simplify.

NOW WORK PROBLEM 9.

We shall follow the practice of leaving our answers in factored form as shown in Example 2.

EXAMPLE 3 **Finding the Derivative of a Quotient**

Find the derivative of: $y = \dfrac{(1 - 3x)(2x + 1)}{3x - 2}$

SOLUTION We shall solve the problem in two ways.

Method 1 Use the formula for the derivative of a quotient right away.

$$y' = \frac{d}{dx}\frac{(1 - 3x)(2x + 1)}{3x - 2}$$

$$= \frac{(3x - 2)\frac{d}{dx}[(1 - 3x)(2x + 1)] - (1 - 3x)(2x + 1)\frac{d}{dx}(3x - 2)}{(3x - 2)^2}$$ Use Formula (5).

$$= \frac{(3x - 2)\left[(1 - 3x)\frac{d}{dx}(2x + 1) + (2x + 1)\frac{d}{dx}(1 - 3x)\right] - (1 - 3x)(2x + 1)\cdot 3}{(3x - 2)^2}$$ Use Formula (4); Differentiate.

$$= \frac{(3x-2)[(1-3x)(2)+(2x+1)(-3)]-(-6x^2-x+1)(3)}{(3x-2)^2}$$ Differentiate, Simplify.

$$= \frac{(3x-2)[2-6x-6x-3]-(-18x^2-3x+3)}{(3x-2)^2}$$ Simplify.

$$= \frac{(3x-2)(-12x-1)-(-18x^2-3x+3)}{(3x-2)^2}$$ Simplify.

$$= \frac{-36x^2+21x+2+18x^2+3x-3}{(3x-2)^2}$$ Simplify.

$$= \frac{-18x^2+24x-1}{(3x-2)^2}$$ Simplify.

Method 2 First, multiply the factors in the numerator and then apply the formula for the derivative of a quotient.

$$y = \frac{(1-3x)(2x+1)}{3x-2} = \frac{-6x^2-x+1}{3x-2}$$

Now use Formula (5):

$$y' = \frac{d}{dx}\frac{-6x^2-x+1}{3x-2}$$

$$= \frac{(3x-2)\frac{d}{dx}(-6x^2-x+1)-(-6x^2-x+1)\frac{d}{dx}(3x-2)}{(3x-2)^2}$$ Formula (5).

$$= \frac{(3x-2)(-12x-1)-(-6x^2-x+1)(3)}{(3x-2)^2}$$ Differentiate.

$$= \frac{-36x^2+21x+2+18x^2+3x-3}{(3x-2)^2}$$ Simplify.

$$= \frac{-18x^2+24x-1}{(3x-2)^2}$$ Simplify.

As you can see from this example, looking at alternative methods may make the differentiation easier. Which method did you find easier?

The Derivative of $f(x) = x^n$, n a Negative Integer

Find the derivative of $f(x) = x^n$, n a negative integer **3**

In the previous section, we learned that the derivative of a power function $f(x) = x^n$, $n \geq 1$ an integer, is $f'(x) = nx^{n-1}$.

The formula for the derivative of x raised to a negative integer exponent follows the same form.

The derivative of $f(x) = x^n$, where n is *any* integer, is n times x to the $n-1$ power. That is,

$$\frac{d}{dx}x^n = nx^{n-1} \quad \text{for any integer } n \qquad (6)$$

The proof is left as an exercise. See Problem 54.

EXAMPLE 4 Using Formula (6)

(a) $\frac{d}{dx}x^{-3} = -3x^{-4} = \frac{-3}{x^4}$

(b) $\frac{d}{dx}\frac{4}{x^2} = \frac{d}{dx}(4x^{-2}) = 4\frac{d}{dx}x^{-2} = 4(-2x^{-3}) = \frac{-8}{x^3}$

(c) $\frac{d}{dx}\left(x + \frac{2}{x}\right) = \frac{d}{dx}(x + 2x^{-1}) = \frac{d}{dx}x + 2\frac{d}{dx}x^{-1} = 1 + 2(-1)x^{-2} = 1 - \frac{2}{x^2}$

NOW WORK PROBLEM 17.

EXAMPLE 5 Finding the Derivative of a Function

Find the derivative of: $g(x) = \left(1 - \frac{1}{x^2}\right)(x + 1)$

SOLUTION Since $g(x)$ is the product of two simpler functions, we begin by applying the formula for the derivative of a product:

$$g'(x) = \left(1 - \frac{1}{x^2}\right)\frac{d}{dx}(x+1) + (x+1)\frac{d}{dx}\left(1 - \frac{1}{x^2}\right) \quad \text{Derivative of a product.}$$

$$= \left(1 - \frac{1}{x^2}\right)(1) + (x+1)\frac{d}{dx}(1 - x^{-2}) \quad \text{Differentiate; } \frac{1}{x^2} = x^{-2}.$$

$$= 1 - \frac{1}{x^2} + (x+1)(2x^{-3}) \quad \text{Differentiate.}$$

$$= 1 - \frac{1}{x^2} + \frac{2(x+1)}{x^3} \quad \text{Simplify.}$$

$$= 1 - \frac{1}{x^2} + \frac{2x}{x^3} + \frac{2}{x^3} \quad \text{Simplify.}$$

$$= 1 + \frac{1}{x^2} + \frac{2}{x^3} \quad \text{Simplify.}$$

Alternatively, we could have solved Example 5 by multiplying the factors first. Then

$$g(x) = \left(1 - \frac{1}{x^2}\right)(x+1) = x + 1 - \frac{1}{x} - \frac{1}{x^2}$$

so

$$g'(x) = \frac{d}{dx}\left(x + 1 - \frac{1}{x} - \frac{1}{x^2}\right) = \frac{d}{dx}x + \frac{d}{dx}1 - \frac{d}{dx}\frac{1}{x} - \frac{d}{dx}\frac{1}{x^2}$$

$$= 1 + 0 - \frac{d}{dx}x^{-1} - \frac{d}{dx}x^{-2} = 1 - (-1)x^{-2} - (-2)x^{-3} = 1 + \frac{1}{x^2} + \frac{2}{x^3}$$

EXAMPLE 6 Application

The value $V(t)$, in dollars, of a car t years after its purchase is given by the equation

$$V(t) = \frac{8000}{t} + 5000 \qquad 1 \le t \le 5$$

Graph the function $V = V(t)$.
Then find:

(a) The average rate of change in value from $t = 1$ to $t = 4$.
(b) The instantaneous rate of change in value.
(c) The instantaneous rate of change in value after 1 year.
(d) The instantaneous rate of change in value after 3 years.
Q **(e)** Interpret the answers to (c) and (d).

SOLUTION The graph of $V = V(t)$ is given in Figure 9.

FIGURE 9

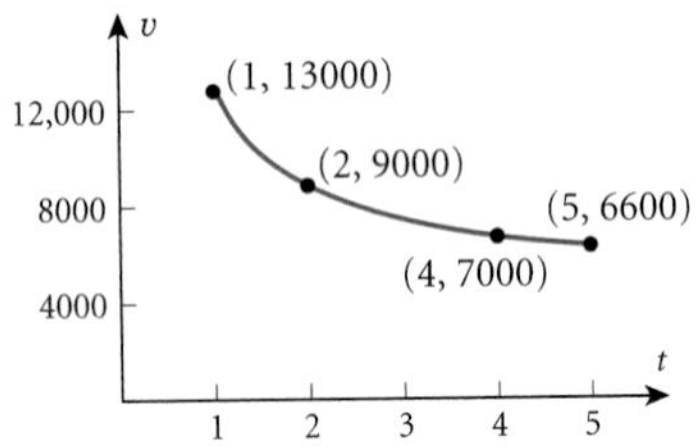

(a) The average rate of change in value from $t = 1$ to $t = 4$ is given by

$$\frac{V(4) - V(1)}{4 - 1} = \frac{7000 - 13{,}000}{3} = -2000$$

So the average rate of change in value from $t = 1$ to $t = 4$ is $-\$2000$ per year. That is, the value of the car is decreasing at the rate of \$2000 per year.

(b) The derivative $V'(t)$ of $V(t)$ equals the instantaneous rate of change in the value of the car.

$$V'(t) = \frac{d}{dt}\left(\frac{8000}{t} + 5000\right) = \frac{d}{dt}\frac{8000}{t} + \frac{d}{dt}(5000)$$
$$= \frac{d}{dt}8000t^{-1} + 0 = 8000(-1)t^{-2} = -\frac{8000}{t^2}$$

Notice that $V'(t) < 0$; we interpret this to mean that the value of the car is decreasing over time.

(c) After 1 year, $V'(1) = -\frac{8000}{1} = -\$8000/\text{year}$

(d) After 3 years, $V'(3) = -\frac{8000}{9} = -\$888.89/\text{year}$

Q **(e)** $V'(1) = -\$8000$ means that the value of the car after 1 year will decline by approximately \$8000 over the next year; $V'(3) = -\$888.89$ means that the value of the car after 3 years will decline by approximately \$888.89 over the next year.

NOW WORK PROBLEM 41.

SUMMARY Each of the derivative formulas given so far can be written without reference to the independent variable of the function. If f and g are differentiable functions, we have the following formulas:

Derivative of a constant times a function	$(cf)' = cf'$
Derivative of a sum	$(f + g)' = f' + g'$
Derivative of a difference	$(f - g)' = f' - g'$
Derivative of a product	$(f \cdot g)' = f \cdot g' + g \cdot f'$
Derivative of a quotient	$\left(\frac{f}{g}\right)' = \frac{g \cdot f' - f \cdot g'}{g^2}$

EXERCISE 13.3 Answers to Odd-Numbered Problems Begin on Page AN-74.

In Problems 1–8, find the derivative of each function by using the formula for the derivative of a product.

1. $f(x) = (2x + 1)(4x - 3)$
2. $f(x) = (3x - 4)(2x + 5)$
3. $f(t) = (t^2 + 1)(t^2 - 4)$
4. $f(t) = (t^2 - 3)(t^2 + 4)$
5. $f(x) = (3x - 5)(2x^2 + 1)$
6. $f(x) = (3x^2 - 1)(4x + 1)$
7. $f(x) = (x^5 + 1)(3x^3 + 8)$
8. $f(x) = (x^6 - 2)(4x^2 + 1)$

In Problems 9–20, find the derivative of each function.

9. $f(x) = \dfrac{x}{x + 1}$
10. $f(x) = \dfrac{x + 4}{x^2}$
11. $f(x) = \dfrac{3x + 4}{2x - 1}$
12. $f(x) = \dfrac{3x - 5}{4x + 1}$
13. $f(x) = \dfrac{x^2}{x - 4}$
14. $f(x) = \dfrac{x}{x^2 - 4}$
15. $f(x) = \dfrac{2x + 1}{3x^2 + 4}$
16. $f(x) = \dfrac{2x^2 - 1}{5x + 2}$
17. $f(t) = \dfrac{-2}{t^2}$
18. $f(t) = \dfrac{4}{t^3}$
19. $f(x) = 1 + \dfrac{1}{x} + \dfrac{1}{x^2}$
20. $f(x) = 1 - \dfrac{1}{x} + \dfrac{1}{x^2}$

In Problems 21–24, find the slope of the tangent line to the graph of the function f at the indicated point. What is an equation of the tangent line?

21. $f(x) = (x^3 - 2x + 2)(x + 1)$ at $(1, 2)$
22. $f(x) = (2x^2 - 5x + 1)(x - 3)$ at $(1, 4)$
23. $f(x) = \dfrac{x^3}{x + 1}$ at $\left(1, \dfrac{1}{2}\right)$
24. $f(x) = \dfrac{x^2}{x - 1}$ at $\left(-1, -\dfrac{1}{2}\right)$

In Problems 25–28, find those x, if any, at which $f'(x) = 0$.

25. $f(x) = (x^2 - 2)(2x - 1)$
26. $f(x) = (3x^2 - 3)(2x^3 - x)$
27. $f(x) = \dfrac{x^2}{x + 1}$
28. $f(x) = \dfrac{x^2 + 1}{x}$

In Problems 29–40, find y'.

29. $y = x^2(3x - 2)$
30. $y = (x^2 + 2)(x - 1)$
31. $y = (x^2 + 4)(4x^2 + 3)$
32. $y = (2x + 3)(x^3 + x^2)$
33. $y = \dfrac{2x + 3}{3x + 5}$
34. $y = \dfrac{3x - 2}{4x - 3}$
35. $y = \dfrac{x^2}{x^2 - 4}$
36. $y = \dfrac{x^3}{x - 1}$
37. $y = \dfrac{(3x + 4)(2x - 3)}{2x + 1}$
38. $y = \dfrac{(2 - 3x)(1 - x)}{x + 2}$
39. $y = \dfrac{4x^3}{x^2 + 4}$
40. $y = \dfrac{3x^4}{x^3 + 1}$

41. **Value of a Car** The value V of a car after t years is

$$V(t) = \frac{10{,}000}{t} + 6000 \qquad 1 \le t \le 6$$

Graph $V = V(t)$.

(a) What is the average rate of change in value from $t = 2$ to $t = 5$?
(b) What is the instantaneous rate of change in value?
(c) What is the instantaneous rate of change after 2 years?
(d) What is the instantaneous rate of change after 5 years?
(e) Interpret the answers found in (c) and (d).

42. **Value of a Painting** The value V of a painting t years after it is purchased is

$$V(t) = \frac{100t^2 + 50}{t} + 400 \qquad 1 \le t \le 5$$

(a) What is the average rate of change in value from $t = 1$ to $t = 3$?
(b) What is the instantaneous rate of change in value?
(c) What is the instantaneous rate of change after 1 year?
(d) What is the instantaneous rate of change after 3 years?
(e) Interpret the answers found in (c) and (d).

43. Demand Equation The demand equation for a certain commodity is

$$p = 10 + \frac{40}{x} \qquad 1 \le x \le 10$$

where p is the price in dollars when x units are demanded. Find:

(a) The revenue function.
(b) The marginal revenue.
(c) The marginal revenue for $x = 4$.
(d) The marginal revenue for $x = 6$.

44. Cost Function The cost of fuel in operating a luxury yacht is given by the equation

$$C(s) = \frac{-3s^2 + 1200}{s}$$

where s is the speed of the yacht. Find the rate at which the cost is changing when $s = 10$.

45. Price–Demand Function The price–demand function for calculators is given by

$$D(p) = \frac{100{,}000}{p^2 + 10p + 50} \qquad 5 \le p \le 20$$

where D is the quantity demanded and p is the unit price in dollars.

(a) Find $D'(p)$, the rate of change of demand with respect to price.
(b) Find $D'(5)$, $D'(10)$, and $D'(15)$.
(c) Interpret the results found in part (b).

46. Height of a Balloon The height, in kilometers, that a balloon will rise in t hours is given by the formula

$$s = s(t) = \frac{t^2}{2 + t}$$

Find the rate at which the balloon is rising after
(a) 10 minutes
(b) 20 minutes.

47. Population Growth A population of 1000 bacteria is introduced into a culture and grows in number according to the formula

$$P(t) = 1000\left(1 + \frac{4t}{100 + t^2}\right)$$

where t is measured in hours. Find the rate at which the population is growing when

(a) $t = 1$ (b) $t = 2$ (c) $t = 3$ (d) $t = 4$

48. Drug Concentration The concentration of a certain drug in a patient's bloodstream t hours after injection is given by

$$C(t) = \frac{0.4t}{2t^2 + 1}$$

Find the rate at which the concentration of the drug is changing with respect to time. At what rate is the concentration changing

(a) 10 minutes after the injection?
(b) 30 minutes after the injection?
(c) 1 hour after the injection?
(d) 3 hours after the injection?

49. Intensity of Illumination The intensity of illumination I on a surface is inversely proportional to the square of the distance r from the surface to the source of light. If the intensity is 1000 units when the distance is 1 meter, find the rate of change of the intensity with respect to the distance when the distance is 10 meters.

50. Cost Function The cost C, in thousands of dollars, for removal of pollution from a certain lake is

$$C(x) = \frac{5x}{110 - x}$$

where x is the percent of pollutant removed. Find:

(a) $C'(x)$, the rate of change of cost with respect to the amount of pollutant removed.
(b) Compute $C'(10)$, $C'(20)$, $C'(70)$, $C'(90)$.
(c) Interpret the answers found in part (b).

51. Cost Function An airplane crosses the Atlantic Ocean (3000 miles) with an airspeed of 500 miles per hour. The cost C (in dollars) per person is

$$C(x) = 100 + \frac{x}{10} + \frac{36{,}000}{x}$$

where x is the ground speed (airspeed $\pm$ wind). Find:

(a) The marginal cost.
(b) The marginal cost at a ground speed of 500 mph.
(c) The marginal cost at a ground speed of 550 mph.
(d) The marginal cost at a ground speed of 450 mph.

52. Average Cost Function If C is the total cost function then $\overline{C}(x) = \frac{C(x)}{x}$ is defined as the **average cost function,** that is, the cost per unit produced. Suppose a company estimates that the total cost of producing x units of a certain product is given by

$$C(x) = 400 + 0.02x + 0.0001x^2$$

Then the average cost is given by

$$\overline{C}(x) = \frac{C(x)}{x} = \frac{400}{x} + 0.02 + 0.0001x$$

(a) Find the marginal average cost $\overline{C}'(x)$.
(b) Find the marginal average cost at $x = 200$, 300, and 400.
(c) Interpret your results.

53. Satisfaction and Reward The relationship between satisfaction S and total reward r has been found to be

$$S(r) = \frac{ar}{g - r}$$

where $g \geq 0$ is the predetermined goal level and $a > 0$ is the perceived justice per unit of reward.

(a) Show that the instantaneous rate of change of satisfaction with respect to reward is inversely proportional to the square of the difference between the personal goal of the individual and the amount of reward received.

(b) Interpret the equation obtained in part (a).

54. Prove Formula (6).

Hint: If $n < 0$, then $-n > 0$. Now use the fact that

$$\frac{d}{dx}x^n = \frac{d}{dx}\frac{1}{x^{-n}}$$

and use the quotient formula.

13.4 The Power Rule

OBJECTIVES
1. Find derivatives using the Power Rule
2. Find derivatives using the Power Rule and other derivative formulas

When a function is of the form $y = [g(x)]^n$, n an integer, the formula used to find the derivative y' is called the *Power Rule.* Let's see if we can guess this formula by finding the derivative of $y = [g(x)]^n$ when $n = 2$, $n = 3$, and $n = 4$.

If $n = 2$,

$$\frac{d}{dx}[g(x)]^2 = \frac{d}{dx}[g(x)g(x)] \underset{\uparrow}{=} g(x)\,g'(x) + g(x)g'(x) = 2g(x)g'(x)$$

Derivative of a product

If $n = 3$,

$$\frac{d}{dx}[g(x)]^3 = \frac{d}{dx}\{[g(x)]^2g(x)\} = [g(x)]^2g'(x) + g(x)\left\{\frac{d}{dx}[g(x)]^2\right\}$$
$$= [g(x)]^2g'(x) + g(x)[2g(x)g'(x)] = 3[g(x)]^2g'(x)$$

If $n = 4$,

$$\frac{d}{dx}[g(x)]^4 = \frac{d}{dx}\{[g(x)]^3g(x)\} = [g(x)]^3g'(x) + g(x)\left\{\frac{d}{dx}[g(x)]^3\right\}$$
$$= [g(x)]^3g'(x) + g(x)\{3[g(x)]^2g'(x)\} = 4[g(x)]^3g'(x)$$

Let's summarize what we've found:

$$\frac{d}{dx}[g(x)]^2 = 2g(x)g'(x)$$

$$\frac{d}{dx}[g(x)]^3 = 3[g(x)]^2g'(x)$$

$$\frac{d}{dx}[g(x)]^4 = 4[g(x)]^3g'(x)$$

These results suggest the following formula:

The Power Rule

If g is a differentiable function and n is any integer, then

$$\frac{d}{dx}[g(x)]^n = n[g(x)]^{n-1}g'(x) \tag{1}$$

Note the similarity between the Power Rule and the formula for the derivative of a power function:

$$\frac{d}{dx}x^n = nx^{n-1}$$

The main difference between these formulas is the factor $g'(x)$. Be sure to remember to include $g'(x)$ when using Formula (1).

1 **EXAMPLE 1** **Using the Power Rule to Find a Derivative**

Find the derivative of the function: $f(x) = (x^2 + 1)^3$

SOLUTION We could, of course, expand the right-hand side and proceed according to techniques discussed earlier. However, the usefulness of the Power Rule is that it enables us to find derivatives of functions like this without resorting to tedious (and sometimes impossible) computation.

The function $f(x) = (x^2 + 1)^3$ is the function $g(x) = x^2 + 1$ raised to the power 3. Using the Power Rule,

$$\frac{d}{dx}f(x) = \frac{d}{dx}(x^2+1)^3 \underset{\uparrow}{=} 3(x^2+1)^2\frac{d}{dx}(x^2+1)$$

Use the Power Rule: $g(x) = x^2 + 1$

$$= 3(x^2+1)^2(2x) = 6x(x^2+1)^2$$

NOW WORK PROBLEM 1.

EXAMPLE 2 **Using the Power Rule**

Find the derivative $f'(x)$.

(a) $f(x) = \dfrac{1}{(x^3+4)^5}$ **(b)** $f(x) = \dfrac{1}{(x^2+4)^3}$

SOLUTION **(a)** We write $f(x)$ as $f(x) = (x^3 + 4)^{-5}$. Then we use the Power Rule:

$$f'(x) = \frac{d}{dx}(x^3+4)^{-5} \underset{\uparrow}{=} -5(x^3+4)^{-6}\frac{d}{dx}(x^3+4)$$

Use the Power Rule.

$$= -5(x^3+4)^{-6}(3x^2) = \frac{-15x^2}{(x^3+4)^6}$$

(b) $$f'(x) = \frac{d}{dx}\frac{1}{(x^2+4)^3} = \frac{d}{dx}(x^2+4)^{-3} \underset{\substack{\uparrow \\ \text{Use the Power Rule}}}{=} -3(x^2+4)^{-4}\frac{d}{dx}(x^2+4)$$

$$= -3(x^2+4)^{-4} \cdot 2x = \frac{-6x}{(x^2+4)^4}$$

Often, we must use at least one other derivative formula along with the Power Rule to differentiate a function. Here are two examples.

2 **EXAMPLE 3** **Using the Power Rule with Other Derivative Formulas**

Find the derivative of the function: $f(x) = x(x^2+1)^3$

SOLUTION The function f is the product of x and $(x^2+1)^3$. We begin by using the formula for the derivative of a product. That is,

$$f'(x) = x\frac{d}{dx}(x^2+1)^3 + (x^2+1)^3\frac{d}{dx}x \quad \text{Formula for the derivative of a product.}$$

We continue by using the Power Rule:

$$\begin{aligned} f'(x) &= x\left[3(x^2+1)^2\frac{d}{dx}(x^2+1)\right] + (x^2+1)^3 \cdot 1 && \text{Power Rule; } \frac{d}{dx}x = 1. \\ &= x\,[3(x^2+1)^2(2x)] + (x^2+1)^3 && \text{Differentiate.} \\ &= (x^2+1)^2(6x^2) + (x^2+1)^2(x^2+1) && \text{Simplify.} \\ &= (x^2+1)^2[6x^2 + (x^2+1)] && \text{Factor.} \\ &= (x^2+1)^2(7x^2+1) && \text{Simplify.} \end{aligned}$$

NOW WORK PROBLEM 7.

EXAMPLE 4 **Using the Power Rule with Other Derivative Formulas**

Find the derivative of the function: $f(x) = \left(\frac{3x+2}{4x^2-5}\right)^5$

SOLUTION Here, f is the quotient $\frac{3x+2}{4x^2-5}$ raised to the power 5. We begin by using the Power Rule and then use the formula for the derivative of a quotient:

$$\begin{aligned} f'(x) &= 5\left(\frac{3x+2}{4x^2-5}\right)^4\left[\frac{d}{dx}\left(\frac{3x+2}{4x^2-5}\right)\right] && \text{Power Rule.} \\ &= 5\left(\frac{3x+2}{4x^2-5}\right)^4\left[\frac{(4x^2-5)\frac{d}{dx}(3x+2) - (3x+2)\frac{d}{dx}(4x^2-5)}{(4x^2-5)^2}\right] && \text{Formula for the derivative of a quotient.} \\ &= 5\left(\frac{3x+2}{4x^2-5}\right)^4\left[\frac{(4x^2-5)(3) - (3x+2)(8x)}{(4x^2-5)^2}\right] && \text{Differentiate.} \\ &= \frac{5(3x+2)^4(-12x^2-16x-15)}{(4x^2-5)^6} && \text{Simplify.} \end{aligned}$$

NOW WORK PROBLEM 19.

Application

The revenue $R = R(x)$ derived from selling x units of a product at a price p per unit is

$$R = xp$$

where $p = d(x)$ is the demand equation, namely, the equation that gives the price p when the number x of units demanded is known. The marginal revenue is then the derivative of R with respect to x:

$$R'(x) = \frac{d}{dx}(xp) = p + x\frac{dp}{dx} \qquad (2)$$

It is sometimes easier to find the marginal revenue by using Formula (2) instead of differentiating the revenue function directly.

EXAMPLE 5 Finding the Marginal Revenue

Suppose the price p in dollars per ton when x tons of polished aluminum are demanded is given by the equation

$$p = \frac{2000}{x + 20} - 10, \quad 0 < x < 60$$

Find:

(a) The rate of change of price with respect to x.
(b) The revenue function.
(c) The marginal revenue.
(d) The marginal revenue at $x = 20$ and $x = 40$.
(e) Interpret the answers found in part (d).

SOLUTION **(a)** The rate of change of price with respect to x is the derivative $\frac{dp}{dx}$.

$$\frac{dp}{dx} = \frac{d}{dx}\left(\frac{2000}{x+20} - 10\right) = \frac{d}{dx}2000(x+20)^{-1} - \frac{d}{dx}10$$

$$= -2000\,(x+20)^{-2}\frac{d}{dx}(x+20) - 0 = \frac{-2000}{(x+20)^2}$$

↑ Power Rule

(b) The revenue function is

$$R(x) = xp = x\left[\frac{2000}{x+20} - 10\right] = \frac{2000x}{x+20} - 10x$$

(c) Using Formula (2), the marginal revenue is

$$R'(x) = p + x\frac{dp}{dx} \qquad \text{Formula (2).}$$

$$= \left[\frac{2000}{x+20} - 10\right] + x\left(\frac{-2000}{(x+20)^2}\right) \qquad \text{Use the result from (a).}$$

$$= \frac{2000}{x+20} - 10 - \frac{2000x}{(x+20)^2} \qquad \text{Simplify.}$$

(d) Using the result from part (c), we find

$$R'(20) = \frac{2000}{40} - 10 - \frac{2000(20)}{(40)^2} = \$15.00/\text{ton}$$

$$R'(40) = \frac{2000}{60} - 10 - \frac{2000(40)}{(60)^2} = \$1.11/\text{ton}$$

(e) $R'(20) = \$15.00/\text{ton}$ means that the next ton of aluminum sold will generate \$15.00 in revenue.
$R'(40) = \$1.11/\text{ton}$ means that the next ton of aluminum sold will generate \$1.11 in revenue.

NOW WORK PROBLEM 31.

EXERCISE 13.4 Answers to Odd-Numbered Problems Begin on Page AN-74.

In Problems 1–28, find the derivative of each function using the Power Rule.

1. $f(x) = (2x - 3)^4$
2. $f(x) = (5x + 4)^3$
3. $f(x) = (x^2 + 4)^3$
4. $f(x) = (x^2 - 1)^4$
5. $f(x) = (3x^2 + 4)^2$
6. $f(x) = (9x^2 + 1)^2$
7. $f(x) = x(x + 1)^3$
8. $f(x) = x(x - 4)^2$
9. $f(x) = 4x^2(2x + 1)^4$
10. $f(x) = 3x^2(x^2 + 1)^3$
11. $f(x) = [x(x - 1)]^3$
12. $f(x) = [x(x + 4)]^4$
13. $f(x) = (3x - 1)^{-2}$
14. $f(x) = (2x + 3)^{-3}$
15. $f(x) = \dfrac{4}{x^2 + 4}$
16. $f(x) = \dfrac{3}{x^2 - 9}$
17. $f(x) = \dfrac{-4}{(x^2 - 9)^3}$
18. $f(x) = \dfrac{-2}{(x^2 + 2)^4}$
19. $f(x) = \left(\dfrac{x}{x + 1}\right)^3$
20. $f(x) = \left(\dfrac{x^2}{x + 5}\right)^4$
21. $f(x) = \dfrac{(2x + 1)^4}{3x^2}$
22. $f(x) = \dfrac{(3x + 4)^3}{9x}$
23. $f(x) = \dfrac{(x^2 + 1)^3}{x}$
24. $f(x) = \dfrac{(3x^2 + 4)^2}{2x}$
25. $f(x) = \left(x + \dfrac{1}{x}\right)^3$
26. $f(x) = \left(x - \dfrac{1}{x}\right)^4$
27. $f(x) = \dfrac{3x^2}{(x^2 + 1)^2}$
28. $g(x) = \dfrac{2x^3}{(x^2 - 4)^2}$

---

29. Car Depreciation A certain car depreciates according to the formula

$$V(t) = \frac{29{,}000}{1 + 0.4t + 0.1t^2}$$

where V is the value of the car in dollars at time t in years. Find the rate at which the car is depreciating:

(a) 1 year after purchase. (b) 2 years after purchase.
(c) 3 years after purchase. (d) 4 years after purchase.

30. Demand Function The demand function for a certain calculator is given by

$$d(x) = \frac{100}{0.02x^2 + 1} \qquad 0 \le x \le 20$$

where x (in thousands of units) is the quantity demanded per week and $d(x)$ is the unit price in dollars.

(a) Find $d'(x)$.
(b) Find $d'(10)$, $d'(15)$, and $d'(20)$ and interpret your results.
(c) Find the revenue function.
(d) Find the marginal revenue.

31. Demand Equation The price p in dollars per pound when x pounds of a certain commodity are demanded is

$$p = \frac{10{,}000}{5x + 100} - 5 \qquad 0 < x < 90$$

Find:

(a) The rate of change of price with respect to x.
(b) The revenue function.
(c) The marginal revenue.
(d) The marginal revenue at $x = 10$ and at $x = 40$.
(e) Interpret the answers to (d).

32. Revenue Function The weekly revenue R in dollars resulting from the sale of x DVD players is

$$R(x) = \frac{100x^5}{(x^2 + 1)^2} \qquad 0 \le x \le 100$$

Find:

(a) The marginal revenue.
(b) The marginal revenue at $x = 40$.
(c) The marginal revenue at $x = 60$.
(d) Interpret the answers to (b) and (c).

33. Amino Acids A protein disintegrates into amino acids according to the formula

$$M = \frac{28}{t + 2}$$

where M, the mass of the protein, is measured in grams and t is time measured in hours.

(a) Find the average rate of change in mass from $t = 0$ to $t = 2$ hours.

(b) Find $M'(0)$.

(c) Interpret the answers to (a) and (b).

13.5 The Derivatives of the Exponential and Logarithmic Functions; the Chain Rule

PREPARING FOR THIS SECTION *Before getting started, review the following:*

> Exponential Functions (Chapter 11, Section 11.3, pp. 626–637)

> Change-of-Base Formula (Chapter 11, Section 11.5, p. 660)

> Logarithmic Functions (Chapter 11, Section 11.4, pp. 642–650)

OBJECTIVES

1 Find the derivative of functions involving e^x

2 Find a derivative using the Chain Rule

3 Find the derivative of functions involving $\ln x$

4 Find the derivative of functions involving $\log_a x$ and a^x

Up to now, our discussion of finding derivatives has been focused on polynomial functions (derivative of a sum or difference of power functions), rational functions (derivative of a quotient of two polynomials), and these functions raised to an integer power (the Power Rule). In this section we present formulas for finding the derivative of the exponential and logarithmic functions.

The Derivative of $f(x) = e^x$

We begin the discussion of the derivative of $f(x) = e^x$ by considering the function

$$f(x) = a^x \qquad a > 0, \qquad a \neq 1$$

To find the derivative of $f(x) = a^x$, we use the formula for finding the derivative of f at x using the difference quotient, namely:

$$f'(x) = \lim_{h \to 0} \frac{f(x + h) - f(x)}{h}$$

For $f(x) = a^x$, we have

$$f'(x) = \frac{d}{dx} a^x = \lim_{h \to 0} \frac{a^{x+h} - a^x}{h} \underset{\uparrow}{=} \lim_{h \to 0} \left[a^x \left(\frac{a^h - 1}{h} \right) \right] = a^x \lim_{h \to 0} \frac{a^h - 1}{h}$$

Factor out a^x

Suppose we seek $f'(0)$. Assuming the limit on the right exists and equals some number, it follows (since $a^0 = 1$) that the derivative of $f(x) = a^x$ at 0 is

$$f'(0) = \lim_{h \to 0} \frac{a^h - 1}{h}$$

This limit equals the slope of the tangent line to the graph of $f(x) = a^x$ at the point (0, 1). The value of this limit depends upon the choice of a. Observe in Figure 10 that the slope of the tangent line to the graph of $f(x) = 2^x$ at (0, 1) is less than 1, and that the slope of the tangent line to the graph of $f(x) = 3^x$ at (0, 1) is greater than 1.

FIGURE 10

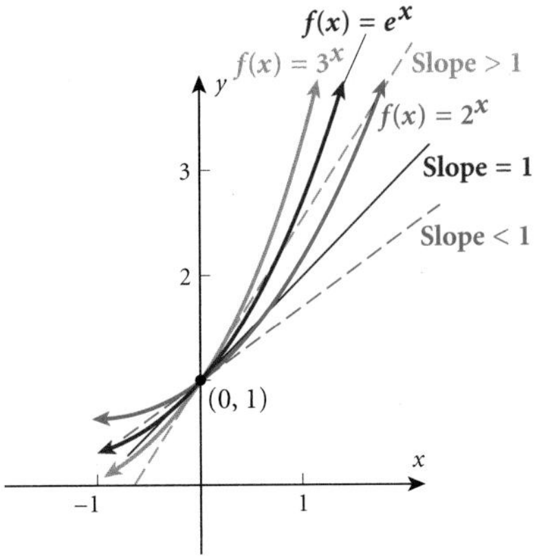

From this, we conclude there is a number a, $2 < a < 3$, for which the slope of the tangent line to the graph of $f(x) = a^x$ at (0, 1) is exactly 1. The function $f(x) = a^x$ for which $f'(0) = 1$ is the function $f(x) = e^x$, whose base is the number e, that we introduced in Chapter 11. A further property of the number e is that

$$\lim_{h \to 0} \frac{e^h - 1}{h} = 1$$

Using this result, we find that

$$\frac{d}{dx} e^x = \lim_{h \to 0} \frac{e^{x+h} - e^x}{h} = \lim_{h \to 0} \frac{e^x(e^h - 1)}{h} = e^x \lim_{h \to 0} \frac{e^h - 1}{h} = e^x \cdot 1 = e^x$$

Derivative of $f(x) = e^x$

The derivative of the exponential function $f(x) = e^x$ is e^x. That is,

$$\frac{d}{dx} e^x = e^x \qquad (1)$$

The simple nature of Formula (1) is one of the reasons the exponential function $f(x) = e^x$ appears so frequently in applications.

1 **EXAMPLE 1 Finding the Derivative of Functions Involving e^x**

Find the derivative of each function:

(a) $f(x) = x^2 + e^x$ **(b)** $f(x) = xe^x$ **(c)** $f(x) = \dfrac{e^x}{x}$

SOLUTION **(a)** Use the formula for the derivative of a sum. Then

$$f'(x) = \frac{d}{dx}(x^2 + e^x) = \frac{d}{dx}x^2 + \frac{d}{dx}e^x = 2x + e^x$$

(b) Use the formula for the derivative of a product. Then

$$f'(x) = \frac{d}{dx}(xe^x) = x\frac{d}{dx}e^x + e^x\frac{d}{dx}x = xe^x + e^x \cdot 1 = e^x(x + 1)$$

(c) Use the formula for the derivative of a quotient. Then

$$f'(x) = \frac{d}{dx}\frac{e^x}{x} \underset{\text{Derivative of a quotient}}{\uparrow =} \frac{x\frac{d}{dx}e^x - e^x\frac{d}{dx}x}{x^2} \underset{\text{Differentiate}}{\uparrow =} \frac{xe^x - e^x \cdot 1}{x^2} \underset{\text{Factor}}{\uparrow =} \frac{(x-1)e^x}{x^2}$$

NOW WORK PROBLEM 3.

To find the derivative of other functions involving e^x and to find the derivative of the logarithmic function requires a formula called *the Chain Rule.*

The Chain Rule

The Power Rule is a special case of a more general, and more powerful formula, called the *Chain Rule.* This formula enables us to find the derivative of a *composite function.*

Consider the function $y = (2x + 3)^2$. If we write $y = f(u) = u^2$ and $u = g(x) = 2x + 3$, then, by a substitution process, we can obtain the original function, namely, $y = f(u) = f(g(x)) = (2x + 3)^2$. This process is called **composition** and the function $y = (2x + 3)^2$ is called the **composite function** of $y = f(u) = u^2$ and $u = g(x) = 2x + 3$.

EXAMPLE 2 Finding a Composite Function

Find the composite function of

$$y = f(u) = \sqrt{u} \qquad \text{and} \qquad u = g(x) = x^2 + 4$$

SOLUTION The composite function is

$$y = f(u) = \sqrt{u} = \sqrt{g(x)} = \sqrt{x^2 + 4}$$

The Chain Rule will require that we find the components of a composite function.

EXAMPLE 3 Decomposing a Composite Function

(a) If $y = (5x + 1)^3$, then $y = u^3$ and $u = 5x + 1$.
(b) If $y = (x^2 + 1)^{-2}$, then $y = u^{-2}$ and $u = x^2 + 1$.
(c) If $y = \dfrac{5}{(2x + 3)^3}$, then $y = \dfrac{5}{u^3}$ and $u = 2x + 3$.

In the above examples, the composite function was "broken up" into simpler functions. The Chain Rule provides a way to use these simpler functions to find the derivative of the composite function.

The Chain Rule

Suppose f and g are differentiable functions. If $y = f(u)$ and $u = g(x)$, then, after substitution, y is a function of x. The Chain Rule states that the derivative of y with respect to x is the derivative of y with respect to u times the derivative of u with respect to x. That is,

$$\boxed{\frac{dy}{dx} = \frac{dy}{du} \cdot \frac{du}{dx}} \qquad (2)$$

2 **EXAMPLE 4** **Finding a Derivative Using the Chain Rule**

Use the Chain Rule to find the derivative of $y = (5x + 1)^3$

SOLUTION We break up y into simpler functions: If $y = (5x + 1)^3$, then $y = u^3$ and $u = 5x + 1$. To find $\frac{dy}{dx}$, we first find $\frac{dy}{du}$ and $\frac{du}{dx}$:

$$\frac{dy}{du} = \frac{d}{du} u^3 = 3u^2 \quad \text{and} \quad \frac{du}{dx} = \frac{d}{dx}(5x + 1) = 5$$

By the Chain Rule,

$$\frac{dy}{dx} = \frac{dy}{du} \cdot \frac{du}{dx} = 3u^2 \cdot 5 = 15u^2 \underset{\substack{\uparrow \\ u = 5x + 1}}{=} 15(5x + 1)^2$$

Notice that when using the Chain Rule, we must substitute for u in the expression for $\frac{dy}{du}$ so that we obtain a function of x.

NOW WORK PROBLEM 9.

EXAMPLE 5 **Finding a Derivative Using the Chain Rule**

Find the derivative of $y = e^{x^2}$

SOLUTION We break up y into simpler functions. If $y = e^{x^2}$, then $y = e^u$ and $u = x^2$. Now use the Chain Rule to find $y' = \frac{dy}{dx}$. Since $\frac{dy}{du} = e^u$ and $\frac{du}{dx} = 2x$, we have

$$\frac{dy}{dx} = \frac{dy}{du} \cdot \frac{du}{dx} = e^u \cdot 2x \underset{\substack{\uparrow \\ u = x^2}}{=} 2xe^{x^2}$$

The result of Example 5 can be generalized.

Derivative of $y = e^{g(x)}$

The derivative of a composite function $y = e^{g(x)}$, where g is a differentiable function, is

$$\boxed{\frac{d}{dx} e^{g(x)} = e^{g(x)} \frac{d}{dx} g(x) = e^{g(x)} g'(x)} \qquad (3)$$

The proof is left as an exercise. See Problem 76.

EXAMPLE 6 Finding the Derivative of Functions of the Form $e^{g(x)}$

Find the derivative of each function:

(a) $f(x) = 4e^{2x}$ **(b)** $f(x) = e^{x^2+1}$

SOLUTION **(a)** Use Formula (3) with $g(x) = 2x$. Then

$$f'(x) = \frac{d}{dx}(4e^{2x}) = 4\frac{d}{dx}e^{2x} \underset{\substack{\uparrow \\ \text{Formula (3)}}}{=} 4 \cdot e^{2x}\frac{d}{dx}(2x) = 4e^{2x}(2) = 8e^{2x}$$

(b) Use Formula (3) with $g(x) = x^2 + 1$. Then

$$f'(x) = \frac{d}{dx}e^{x^2+1} \underset{\substack{\uparrow \\ \text{Formula (3)}}}{=} e^{x^2+1}\frac{d}{dx}(x^2+1) = e^{x^2+1}(2x) = 2xe^{x^2+1}$$

NOW WORK PROBLEM 23.

EXAMPLE 7 Finding the Derivative of Functions Involving e^x

Find the derivative of each function:

(a) $f(x) = xe^{x^2}$ **(b)** $f(x) = \frac{x}{e^x}$ **(c)** $f(x) = (e^x)^2$

SOLUTION **(a)** The function f is the product of two simpler functions, so we start with the formula for the derivative of a product.

$$f'(x) = \frac{d}{dx}(xe^{x^2}) \underset{\substack{\uparrow \\ \text{Derivative of a Product}}}{=} x\frac{d}{dx}e^{x^2} + e^{x^2}\frac{d}{dx}x \underset{\substack{\uparrow \\ \text{Formula (3); } \frac{d}{dx}x = 1}}{=} x \cdot e^{x^2} \cdot \frac{d}{dx}x^2 + e^{x^2} \cdot 1 = xe^{x^2} \cdot 2x + e^{x^2} \underset{\substack{\uparrow \\ \text{Factor}}}{=} e^{x^2}(2x^2 + 1)$$

(b) We could use the formula for the derivative of a quotient, but it is easier to rewrite f in the form $f(x) = xe^{-x}$ and use the formula for the derivative of a product.

$$f'(x) = \frac{d}{dx}(xe^{-x}) \underset{\substack{\uparrow \\ \text{Derivative of a Product}}}{=} x\frac{d}{dx}e^{-x} + e^{-x}\frac{d}{dx}x \underset{\substack{\uparrow \\ \text{Formula (3)}}}{=} x \cdot e^{-x}\frac{d}{dx}(-x) + e^{-x} \cdot 1 \underset{\substack{\uparrow \\ \frac{d}{dx}(-x) = -1}}{=} xe^{-x}(-1) + e^{-x} \underset{\substack{\uparrow \\ \text{Factor}}}{=} e^{-x}(1 - x)$$

(c) Here the function f is e^x raised to the power 2. We first apply a Law of Exponents and write $f(x) = (e^x)^2 = e^{2x}$. Then we can use Formula (3).

$$f'(x) = \frac{d}{dx}e^{2x} = e^{2x}\frac{d}{dx}(2x) = e^{2x} \cdot 2 = 2e^{2x}$$

CAUTION: Notice the difference between e^{x^2} and $(e^x)^2$. In e^{x^2}, e is raised to the power x^2; in $(e^x)^2$, the parentheses tell us e^x is raised to the power 2.

NOW WORK PROBLEM 29.

The Derivative of $f(x) = \ln x$

To find the derivative of $f(x) = \ln x$, we observe that if $y = \ln x$, then $e^y = x$. That is,

$$e^{\ln x} = x$$

If we differentiate both sides with respect to x, we obtain

$$\frac{d}{dx}e^{\ln x} = \frac{d}{dx}x$$

$$e^{\ln x}\frac{d}{dx}\ln x = 1 \qquad \text{Apply Formula (3) on the left.}$$

$$\frac{d}{dx}\ln x = \frac{1}{e^{\ln x}} \qquad \text{Solve for } \frac{d}{dx}\ln x.$$

$$\frac{d}{dx}\ln x = \frac{1}{x} \qquad e^{\ln x} = x.$$

We have proved the following formula:

Derivative of $f(x) = \ln x$

If $f(x) = \ln x$, then $f'(x) = \frac{1}{x}$. That is,

$$\boxed{\frac{d}{dx}\ln x = \frac{1}{x}} \tag{4}$$

3 **EXAMPLE 8 Finding the Derivative of Functions Involving ln x**

Find the derivative of each function.

(a) $f(x) = x^2 + \ln x$ **(b)** $f(x) = x\ln x$

SOLUTION **(a)** Use the formula for the derivative of a sum. Then

$$f'(x) = \frac{d}{dx}(x^2 + \ln x) = \frac{d}{dx}x^2 + \frac{d}{dx}\ln x = 2x + \frac{1}{x}$$

(b) Use the formula for the derivative of a product. Then

$$f'(x) = \frac{d}{dx}(x\ln x) = x\frac{d}{dx}\ln x + \ln x\frac{d}{dx}x$$

$$= x\left(\frac{1}{x}\right) + (\ln x)(1) = 1 + \ln x$$

NOW WORK PROBLEM 35.

To differentiate the natural logarithm of a function $g(x)$, namely, $\ln g(x)$, use the following formula.

Derivative of $\ln g(x)$

The formula for finding the derivative of the composite function $f(x) = \ln g(x)$, where g is a differentiable function, is

$$\boxed{\frac{d}{dx}\ln g(x) = \frac{\frac{d}{dx}g(x)}{g(x)} = \frac{g'(x)}{g(x)}} \tag{5}$$

The proof uses the Chain Rule and is left as an exercise. See Problem 77.

EXAMPLE 9 Finding the Derivative of Functions Involving ln x

Find the derivative of each function.

(a) $f(x) = \ln(x^2 + 1)$ **(b)** $f(x) = (\ln x)^2$

SOLUTION **(a)** The function $f(x) = \ln(x^2 + 1)$ is of the form $f(x) = \ln g(x)$. We use Formula (5) with $g(x) = x^2 + 1$. Then,

$$f'(x) = \frac{d}{dx}\ln(x^2+1) \underset{\substack{\uparrow \\ \text{Formula (5)}}}{=} \frac{\frac{d}{dx}(x^2+1)}{x^2+1} = \frac{2x}{x^2+1}$$

(b) The function $f(x)$ is $\ln x$ raised to the power 2. We use the Power Rule. Then

$$f'(x) = \frac{d}{dx}(\ln x)^2 \underset{\substack{\uparrow \\ \text{Power Rule}}}{=} 2\ln x\left(\frac{d}{dx}\ln x\right) = (2\ln x)\cdot\frac{1}{x} = \frac{2\ln x}{x}$$

NOW WORK PROBLEM 45.

Find the derivative of functions involving $\log_a x$ and a^x

4 The Derivative of $f(x) = \log_a x$ and $f(x) = a^x$

To find the derivative of the logarithm function $f(x) = \log_a x$ for any base a, we use the Change-of-Base Formula. Then

$$f(x) = \log_a x = \frac{\log_e x}{\log_e a} = \frac{\ln x}{\ln a}$$

Since $\ln a$ is a constant, we have

$$f'(x) = \frac{d}{dx}\log_a x = \frac{d}{dx}\frac{\ln x}{\ln a} = \frac{1}{\ln a}\cdot\frac{d}{dx}\ln x = \frac{1}{\ln a}\cdot\frac{1}{x} = \frac{1}{x\ln a}$$

We have the formula

Derivative of $f(x) = \log_a x$

If $f(x) = \log_a x$, then $f'(x) = \dfrac{1}{x\ln a}$. That is,

$$\frac{d}{dx}\log_a x = \frac{1}{x\ln a} \qquad (6)$$

EXAMPLE 10 Finding the Derivative of $\log_2 x$

Find the derivative of: $f(x) = \log_2 x$

SOLUTION Using Formula (6), we have

$$f'(x) = \frac{d}{dx}\log_2 x \underset{\substack{\uparrow \\ \text{Formula (6)}}}{=} \frac{1}{x\ln 2}$$

NOW WORK PROBLEM 47.

To find the derivative of $f(x) = a^x$, where $a > 0$, $a \neq 1$, is any real constant, we use the definition of a logarithm and the Change-of-Base Formula. If $y = a^x$, we have

$$x = \log_a y \quad \text{Definition of a logarithm.}$$

$$x = \frac{\ln y}{\ln a} \quad \text{Apply the Change-of-Base Formula.}$$

$$x = \frac{\ln a^x}{\ln a} \quad \text{Substitute } y = a^x.$$

Now, we differentiate both sides with respect to x:

$$\frac{d}{dx} x = \frac{d}{dx} \frac{\ln a^x}{\ln a}$$

$$1 = \frac{1}{\ln a} \cdot \frac{d}{dx} \ln a^x \quad \ln a \text{ is a constant.}$$

$$1 = \frac{1}{\ln a} \cdot \frac{\frac{d}{dx} a^x}{a^x} \quad \text{Use Formula (5).}$$

$$1 = \frac{\frac{d}{dx} a^x}{a^x \ln a} \quad \text{Simplify.}$$

$$\frac{d}{dx} a^x = a^x \ln a \quad \text{Solve for } \frac{d}{dx} a^x.$$

We have derived the formula:

Derivative of $f(x) = a^x$

The derivative of $f(x) = a^x$, $a > 0$, $a \neq 1$, is $f'(x) = a^x \ln a$. That is,

$$\boxed{\frac{d}{dx} a^x = a^x \ln a} \qquad (7)$$

EXAMPLE 11 Finding the Derivative of 2^x

Find the derivative of: $f(x) = 2^x$

SOLUTION Using Formula (7), we have

$$f'(x) = \frac{d}{dx} 2^x \underset{\uparrow}{=} 2^x \ln 2$$

Formula (7)

NOW WORK PROBLEM 51.

EXAMPLE 12 Maximizing Profit

At a Notre Dame football weekend, the demand for game-day t-shirts is given by

$$p = 30 - 5 \ln\left(\frac{x}{100} + 1\right)$$

where p is the price of the shirt in dollars and x is the number of shirts demanded.

(a) At what price can 1000 t-shirts be sold?
(b) At what price can 5000 t-shirts be sold?
(c) Find the marginal demand for 1000 t-shirts and interpret the answer.
(d) Find the marginal demand for 5000 t-shirts and interpret the answer.
(e) Find the revenue function $R = R(x)$.
(f) Find the marginal revenue from selling 1000 t-shirts and interpret the answer.
(g) Find the marginal revenue from selling 5000 t-shirts and interpret the answer.
(h) If each t-shirt costs \$4, find the profit function $P = P(x)$.
(i) What is the profit if 1000 t-shirts are sold?
(j) What is the profit if 5000 t-shirts are sold?
(k) Use the TABLE feature of a graphing utility to find the quantity x (to the nearest hundred) that maximizes profit.
(l) What price should be charged for a t-shirt to maximize profit?

SOLUTION **(a)** For $x = 1000$, the price p is

$$p = 30 - 5\ln\left(\frac{1000}{100} + 1\right) = \$18.01$$

(b) For $x = 5000$, the price p is

$$p = 30 - 5\ln\left(\frac{5000}{100} + 1\right) = \$10.34$$

(c) The marginal demand for x shirts is

$$p'(x) = \frac{dp}{dx} = \frac{d}{dx}\left[30 - 5\ln\left(\frac{x}{100} + 1\right)\right] \underset{\text{Use Formula (5)}}{=} -5 \cdot \frac{\frac{1}{100}}{\frac{x}{100} + 1} \underset{\text{Multiply by } \frac{100}{100}}{=} \frac{-5}{x + 100}$$

For $x = 1000$,

$$p'(1000) = \frac{-5}{1000 + 100} = -\$0.0045$$

This means that another t-shirt will be demanded if the price is reduced by \$0.0045.

(d) Use the result for $p'(x)$ found in part (c). Then for $x = 5000$, we have

$$p'(5000) = \frac{-5}{5000 + 100} = -\$0.00098$$

This means that another t-shirt will be demanded if the price is reduced by \$0.00098.

(e) The revenue function $R = R(x)$ is

$$R = xp = x\left[30 - 5\ln\left(\frac{x}{100} + 1\right)\right]$$

(f) The marginal revenue is

$$R'(x) = \frac{d}{dx}[xp(x)] = xp'(x) + p(x) \qquad \text{Derivative of a product.}$$

$$= x \cdot \frac{-5}{x + 100} + 30 - 5\ln\left(\frac{x}{100} + 1\right) \qquad \text{Use the result of part (c).}$$

$$= \frac{-5x}{x + 100} + 30 - 5\ln\left(\frac{x}{100} + 1\right) \qquad \text{Simplify.}$$

If $x = 1000$,

$$R'(1000) = \frac{-5000}{1100} + 30 - 5 \ln 11 = \$13.47$$

The revenue received for selling the 1001st t-shirt is $13.47.

(g) If $x = 5000$

$$R'(5000) = \frac{-25,000}{5100} + 30 - 5 \ln 51 = \$5.44$$

The revenue received for selling the 5001st t-shirt is $5.44.

(h) The cost C for x t-shirts is $C = 4x$, so the profit function P is

$$P = P(x) = R(x) - C(x) = x \left[30 - 5 \ln \left(\frac{x}{100} + 1 \right) \right] - 4x$$

$$= 26x - 5x \ln \left(\frac{x}{100} + 1 \right)$$

(i) If $x = 1000$, the profit is

$$P(1000) = 26(1000) - 5(1000) \ln \left(\frac{1000}{100} + 1 \right) = \$14,010.52$$

(j) If $x = 5000$, the profit is

FIGURE 11

$$P(5000) = 26(5000) - 5(5000) \ln \left(\frac{5000}{100} + 1 \right) = \$31,704.36$$

X	Y1
6500	32836
6600	32845
6700	32846
6800	32840
6900	32827
7000	32806
7100	32778
X=6700	

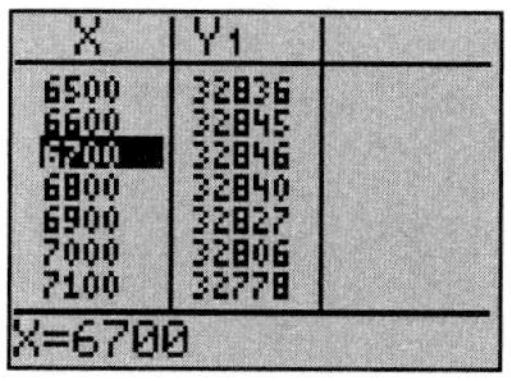

(k) See Figure 11. For $x = 6700$ t-shirts, the profit is largest. ($32,846).
(l) If $x = 6700$, the price p is

$$p(6700) = 30 - 5 \ln \left(\frac{6700}{100} + 1 \right) = \$8.90$$

SUMMARY

$$\frac{d}{dx} e^x = e^x \qquad \frac{d}{dx} e^{g(x)} = e^{g(x)} g'(x) \qquad \frac{d}{dx} a^x = a^x \ln a$$

$$\frac{d}{dx} \ln x = \frac{1}{x} \qquad \frac{d}{dx} \ln g(x) = \frac{g'(x)}{g(x)} \qquad \frac{d}{dx} \log_a x = \frac{1}{x \ln a}$$

EXERCISE 13.5 **Answers to Odd-Numbered Problems Begin on Page AN-75.**

In Problems 1–8, find the derivative of each function.

1. $f(x) = x^3 - e^x$

2. $f(x) = 2e^x - x$

3. $f(x) = x^2 e^x$

4. $f(x) = x^3 e^x$

5. $f(x) = \dfrac{e^x}{x^2}$

6. $f(x) = \dfrac{5x}{e^x}$

7. $f(x) = \dfrac{4x^2}{e^x}$

8. $f(x) = \dfrac{3x^3}{e^x}$

In Problems 9–20, form the composite function $y = f(x)$. Then find $\dfrac{dy}{dx}$ using the Chain Rule.

9. $y = u^5, \quad u = x^3 + 1$

10. $y = u^3, \quad u = 2x + 5$

11. $y = \dfrac{u}{u + 1}, \quad u = x^2 + 1$

12. $y = \frac{u-1}{u}, \quad u = x^2 - 1$

13. $y = (u+1)^2, \quad u = \frac{1}{x}$

14. $y = (u^2 - 1)^3, \quad u = \frac{1}{x+2}$

15. $y = (u^3 - 1)^5, \quad u = x^{-2}$

16. $y = (u^2 + 4)^4, \quad u = x^{-2}$

17. $y = u^3, \quad u = e^x$

18. $y = 4u^2, \quad u = e^x$

19. $y = e^u, \quad u = x^3$

20. $y = e^u, \quad u = \frac{1}{x}$

21. Find the derivative y' of $y = (x^3 + 1)^2$ by:

(a) Using the Chain Rule.
(b) Using the Power Rule.
(c) Expanding and then differentiating.

22. Follow the directions in Problem 21 for the function $y = (x^2 - 2)^3$.

In Problems 23–54, find the derivative of each function.

23. $f(x) = e^{5x}$

24. $f(x) = e^{-3x}$

25. $f(x) = 8e^{-x^2}$

26. $f(x) = -e^{3x^2}$

27. $f(x) = x^2 e^{x^2}$

28. $f(x) = x^3 e^{x^2}$

29. $f(x) = 5(e^x)^3$

30. $f(x) = 4(e^x)^4$

31. $f(x) = \frac{x^2}{e^x}$

32. $f(x) = \frac{8x}{e^{-x}}$

33. $f(x) = \frac{(e^x)^2}{x}$

34. $f(x) = \frac{e^{-2x}}{x^2}$

35. $f(x) = x^2 - 3\ln x$

36. $f(x) = 5\ln x - 2x$

37. $f(x) = x^2 \ln x$

38. $f(x) = x^3 \ln x$

39. $f(x) = 3\ln(5x)$

40. $f(x) = -2\ln(3x)$

41. $f(x) = x\ln(x^2 + 1)$

42. $f(x) = x^2 \ln(x^2 + 1)$

43. $f(x) = x + 8\ln(3x)$

44. $f(x) = 3\ln(2x) - 5x$

45. $f(x) = 8(\ln x)^3$

46. $f(x) = 2(\ln x)^4$

47. $f(x) = \log_3 x$

48. $f(x) = x + \log_4 x$

49. $f(x) = x^2 \log_2 x$

50. $f(x) = x^3 \log_3 x$

51. $f(x) = 3^x$

52. $f(x) = x + 4^x$

53. $f(x) = x^2 \cdot 2^x$

54. $f(x) = x^3 \cdot 3^x$

In Problems 55–62, find an equation of the tangent line to the graph of each function at the given point.

55. $f(x) = e^{3x}$ at $(0, 1)$

56. $f(x) = e^{4x}$ at $(0, 1)$

57. $f(x) = \ln x$ at $(1, 0)$

58. $f(x) = \ln(3x)$ at $(1, 0)$

59. $f(x) = e^{3x-2}$ at $\left(\frac{2}{3}, 1\right)$

60. $f(x) = e^{-x^2}$ at $\left(1, \frac{1}{e}\right)$

61. $f(x) = x\ln x$ at $(1, 0)$

62. $f(x) = \ln x^2$ at $(1, 0)$

---

63. Find the equation of the tangent line to $y = e^x$ that is parallel to the line $y = x$.

64. Find the equation of the tangent line to $y = e^{3x}$ that is parallel to the line $y = -\frac{1}{2}x$.

65. Weber–Fechner Law When a certain drug is administered, the reaction R to the dose x is given by the **Weber–Fechner law:**

$$R = 5.5\ln x + 10$$

(a) Find the reaction rate for a dose of 5 units.
(b) Find the reaction rate for a dose of 10 units.
(c) Interpret the results of parts (a) and (b).

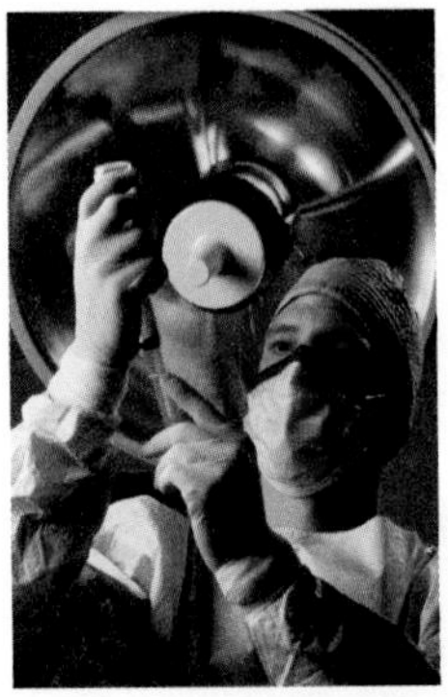

66. Marginal Cost The cost (in dollars) of producing x units (measured in thousands) of a certain product is found to be

$$C(x) = 20 + \ln(x + 1)$$

Find the marginal cost.

67. Atmospheric Pressure The atmospheric pressure at a height of x meters above sea level is $P(x) = 10^4 e^{-0.00012x}$ kilograms per square meter. What is the rate of change of the pressure with respect to the height at $x = 500$ meters? At $x = 700$ meters?

68. Revenue Revenue sales analysis of a new toy by Toys Inc. indicates that the relationship between the unit price p and the monthly sales x of its new toy is given by the equation

$$p = 10e^{-0.04x}$$

Find

(a) The revenue function $R = R(x)$.
(b) The marginal revenue when $x = 200$.

69. Market Penetration The function

$$A(t) = 102 - 90e^{-0.21t}$$

expresses the relationship between A, the percentage of the market penetrated by DVD players, and t, the time in years, where $t = 0$ corresponds to the year 2000.

(a) Find the rate of change of A with respect to time.
(b) Evaluate $A'(5)$ and interpret your result.
(c) Evaluate $A'(10)$ and interpret your result.
(d) Evaluate $A'(30)$ and interpret your result.

70. Sales Because of lack of promotion, the yearly sales S of a product decline according to the equation

$$S(t) = 3000e^{-0.80t}$$

where t is the time. Find

(a) The rate of change of sales with respect to time.
(b) The rate of change of sales at $t = 0.5$.
(c) The rate of change of sales at $t = 2$.
(d) Interpret the results of (b) and (c). Explain the difference.

71. Advertising The function

$$S(x) = 100{,}000 + 400{,}000 \ln x$$

expresses the relation between sales (in dollars) of a product and the advertising for the product, where x is in thousands of dollars. Find

(a) The rate of change of S with respect to x.
(b) $S'(10)$.
(c) $S'(20)$.
(d) Interpret $S'(10)$ and $S'(20)$. Explain the difference.

72. Depreciation of a Car A car depreciates according to the function

$$V(t) = 35{,}000\, e^{-0.25t}$$

where t is measured in years and V represents the value of the car in dollars.

(a) What is the value of the car after 1 year?
(b) What is the value of the car after 5 years?
(c) Find $V'(t)$, the rate of depreciation.
(d) Interpret the result. What do you think the sign of $V'(t)$ represents?
(e) What is the depreciation rate after 1 year?
(f) What is the depreciation rate after 5 years?
(g) Interpret your answers to parts (e) and (f).

73. Maximizing Profit At the Super Bowl, the demand for game-day t-shirts is given by

$$p = 50 - 4 \ln\left(\frac{x}{100} + 1\right)$$

where p is the price of the shirt in dollars and x is the number of shirts demanded.

(a) At what price can 1000 t-shirts be sold?
(b) At what price can 5000 t-shirts be sold?
(c) Find the marginal demand for 1000 t-shirts and interpret the answer.
(d) Find the marginal demand for 5000 t-shirts and interpret the answer.
(e) Find the revenue function $R = R(x)$.
(f) Find the marginal revenue from selling 1000 t-shirts and interpret the answer.
(g) Find the marginal revenue from selling 5000 t-shirts and interpret the answer.
(h) If each t-shirt costs \$4, find the profit function $P = P(x)$.
(i) What is the profit if 1000 t-shirts are sold?
(j) What is the profit if 5000 t-shirts are sold?
(k) Use the TABLE feature of a graphing utility to find the quantity x that maximizes profit.
(l) What price should be charged for a t-shirt to maximize profit?

74. Mean Earnings The mean earnings of workers 18 years old and over are given in the table below.

Year	1975	1980	1985	1990	1995	2000
Mean Earnings	8,552	12,665	17,181	21,793	26,792	32,604

Source: U.S. Bureau of the Census, Current Population Survey.

The data can be modeled by the function

$$E(t) = 8550 + 280t \ln t$$

where t is the number of years since 1974. Find

(a) The rate of change of E with respect to t.
(b) The rate of change at $t = 21$ (year 1995).
(c) The rate of change at $t = 26$ (year 2000).
(d) The rate of change at $t = 31$ (year 2005).
(e) Compare the answer to parts (b), (c), and (d). Explain the differences.

75. Price of Tomatoes The price of one pound of tomatoes from 1998 through 2003 are given in the table:

Year	1998	1999	2000	2001	2002	2003
Price	\$0.473	\$0.489	\$0.490	\$0.500	\$0.509	\$0.526

Source: The Bureau of Labor Statistics.

The price of tomatoes can be modeled by the function

$$p(t) = 0.470 + 0.026 \ln t$$

where t is the number of years since 1997. Find

(a) The rate of change of p with respect to t.
(b) The rate of change at $t = 5$ (year 2002).
(c) The rate of change at $t = 10$ (year 2007).
(d) Interpret the answers to parts (b) and (c). Explain the difference.

76 Prove Formula (3).
Hint: Use the Chain Rule with $y = e^u$, $u = g(x)$.

77. Prove Formula (5).
Hint: Use the Chain Rule with $y = \ln u$, $u = g(x)$.

13.6 Higher-Order Derivatives

OBJECTIVES
1 Find the first derivative and the second derivative of a function
2 Solve applied problems involving velocity and acceleration

The derivative of a function $y = f(x)$ is also a function. For example, if

$$f(x) = 6x^3 - 3x^2 + 2x - 5$$

(a polynomial function of degree 3), then

$$f'(x) = 18x^2 - 6x + 2$$

(a polynomial function of degree 2).

The derivative of the function $f'(x)$ is called the **second derivative of f** and is denoted by $f''(x)$. For the function f above,

$$f''(x) = \frac{d}{dx} f'(x) = \frac{d}{dx}(18x^2 - 6x + 2) = 36x - 6$$

By continuing in this fashion, we can find the third derivative $f'''(x)$, the fourth derivative $f^{(4)}(x)$, and so on, provided that these derivatives exist.*

NOW WORK PROBLEM 3.

The first, second, and third derivatives of the function

$$f(x) = 3x^3 - 2x^2 + 5x - 6$$

*The symbols $f'(x), f''(x)$, and so on for higher-order derivatives have several parallel notations. If $y = f(x)$, we may write

$$y' = f'(x) = \frac{dy}{dx} = \frac{d}{dx} f(x)$$

$$y'' = f''(x) = \frac{d^2y}{dx^2} = \frac{d^2}{dx^2} f(x)$$

$$y''' = f'''(x) = \frac{d^3y}{dx^3} = \frac{d^3}{dx^3} f(x)$$

$$\vdots$$

$$y^{(n)} = f^{(n)}(x) = \frac{d^ny}{dx^n} = \frac{d^n}{dx^n} f(x)$$

are

$$f'(x) = 9x^2 - 4x + 5$$
$$f''(x) = \frac{d}{dx} f'(x) = 18x - 4$$
$$f'''(x) = \frac{d}{dx} f''(x) = 18$$

For this function f, observe that $f^{(4)}(x) = 0$ and that all derivatives of order 5 or more also equal 0.

The result obtained in this example can be generalized:

For a polynomial function f of degree n, we have

$$f(x) = a_n x^n + a_{n-1}x^{n-1} + \cdots + a_1 x + a_0, \quad a_n \neq 0$$
$$f'(x) = na_n x^{n-1} + (n-1)a_{n-1}x^{n-2} + \cdots + a_1$$

The first derivative of a polynomial function of degree n is a polynomial function of degree $n - 1$. By continuing the differentiation process, it follows that the nth-order derivative of f

$$f^{(n)}(x) = n(n-1)(n-2) \cdot \ldots \cdot (3)(2)(1)a_n = n!a_n$$

is a polynomial of degree 0, a constant, so all derivatives of order greater than n will equal 0.

NOW WORK PROBLEM 35.

In some applications it is important to find both the first and second derivatives of a function and to solve for those numbers x that make these derivatives equal 0.

1 **EXAMPLE 1** **Finding the First and Second Derivatives of a Function**

For $f(x) = 4x^3 - 12x^2 + 2$, find those numbers x, if any, at which the derivative $f'(x) = 0$. For what numbers x will $f''(x) = 0$?

SOLUTION

$$f'(x) = 12x^2 - 24x = 12x(x-2) = 0 \quad \text{when} \quad x = 0 \text{ or } x = 2$$
$$f''(x) = 24x - 24 = 24(x-1) = 0 \quad \text{when} \quad x = 1$$

NOW WORK PROBLEM 27.

Velocity and Acceleration

Solve applied problems involving velocity and acceleration 2

We mentioned at the beginning of this chapter that Sir Isaac Newton discovered calculus by solving a physics problem involving falling objects. We take up the problem of analyzing falling objects next.

We begin with the definition of *average velocity:*

Average Velocity

The **average velocity** is the ratio of the change in distance to the change in time. If s denotes distance and t denotes time, we have

$$\text{Average velocity} = \frac{\text{total distance}}{\text{elapsed time}} = \frac{\Delta s}{\Delta t}$$

EXAMPLE 2 Finding the Average Velocity

Mr. Doody and his family left on a car trip Saturday morning at 5 A.M. and arrived at their destination at 11 A.M. When they began the trip, the car's odometer read 26,700 kilometers, and when they arrived it read 27,000 kilometers. What was the average velocity for the trip?

SOLUTION

$$\text{Average velocity} = \frac{\Delta s}{\Delta t} = \frac{\text{total distance}}{\text{elapsed time}}$$

The total distance is $27{,}000 - 26{,}700 = 300$ kilometers and the elapsed time is $11 - 5 = 6$ hours. The average velocity is

$$\text{Average velocity} = \frac{300}{6} = 50 \text{ kilometers per hour}$$

Sometimes Mr. Doody will be traveling faster and sometimes slower, but the *average* velocity is 50 kilometers per hour.

Average velocity provides information about velocity over an interval of time, but provides little information about the velocity at a particular instant of time. To get such information we require the *instantaneous velocity.*

Instantaneous Velocity

The rate of change of distance with respect to time is called **(instantaneous) velocity.** If $s = s(t)$ is a function that describes the position s of a particle at time t, the velocity of the particle at time t is

$$v = \frac{ds}{dt} = s'(t)$$

EXAMPLE 3 Finding the Instantaneous Velocity

In physics it is shown that the height s of a ball thrown straight up with an initial speed of 80 feet per second (ft/sec) from a rooftop 96 feet high is

$$s = s(t) = -16t^2 + 80t + 96$$

where t is the elapsed time that the ball is in the air. The ball misses the rooftop on its way down and eventually strikes the ground. See Figure 12.

FIGURE 12

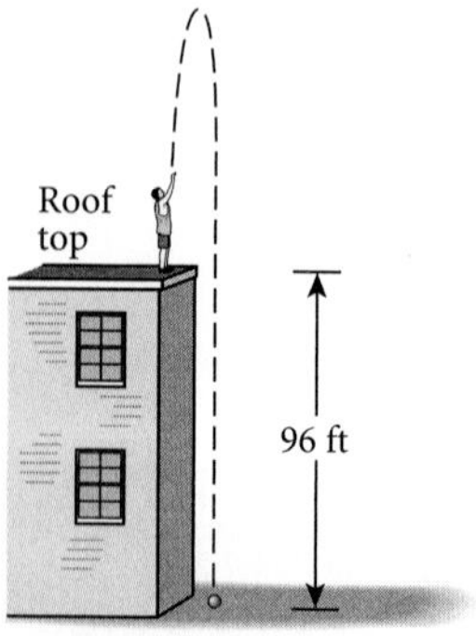

(a) When does the ball strike the ground? That is, how long is the ball in the air?
(b) At what time t will the ball pass the rooftop on its way down?
(c) What is the average velocity of the ball from $t = 0$ to $t = 2$?
(d) What is the instantaneous velocity of the ball at time t?
(e) What is the instantaneous velocity of the ball at $t = 2$?
(f) When is the instantaneous velocity of the ball equal to zero?
(g) What is the instantaneous velocity of the ball as it passes the rooftop on the way down?
(h) What is the instantaneous velocity of the ball when it strikes the ground?

SOLUTION **(a)** The ball strikes the ground when $s = s(t) = 0$.

$$
\begin{aligned}
-16t^2 + 80t + 96 &= 0 && s(t) = 0 \\
t^2 - 5t - 6 &= 0 && \text{Divide each side by } -16. \\
(t - 6)(t + 1) &= 0 && \text{Factor.} \\
t = 6 \quad \text{or} \quad t &= -1 && \text{Apply the Zero-Product Property; solve for } t.
\end{aligned}
$$

We discard the solution $t = -1$. The ball strikes the ground after 6 seconds.

(b) The ball passes the rooftop when $s = s(t) = 96$.

$$
\begin{aligned}
-16t^2 + 80t + 96 &= 96 && s(t) = 96 \\
t^2 - 5t &= 0 && \text{Simplify.} \\
t(t - 5) &= 0 && \text{Factor.} \\
t = 0 \quad \text{or} \quad t &= 5 && \text{Apply the Zero-Product Property; solve for } t.
\end{aligned}
$$

We discard the solution $t = 0$. The ball passes the rooftop on the way down after 5 seconds.

(c) The average velocity of the ball from $t = 0$ to $t = 2$ is

$$\frac{\Delta s}{\Delta t} = \frac{s(2) - s(0)}{2 - 0} = \frac{192 - 96}{2} = 48 \text{ ft/sec}$$

(d) The instantaneous velocity of the ball at time t is the derivative $s'(t)$; that is,

$$s'(t) = \frac{d}{dt}(-16t^2 + 80t + 96) = -32t + 80 = -16(2t - 5) \text{ ft/sec}$$

(e) At $t = 2$ sec, the instantaneous velocity of the ball is

$$s'(2) = -16(4 - 5) = 16 \text{ ft/sec}$$

(f) The instantaneous velocity of the ball is zero when

$$
\begin{aligned}
s'(t) &= 0 \\
-16(2t - 5) &= 0 \\
t &= \frac{5}{2} = 2.5 \text{ seconds}
\end{aligned}
$$

(g) From part (b), the ball passes the rooftop on the way down when $t = 5$ seconds. The instantaneous velocity at $t = 5$ is

$$s'(5) = -16(10 - 5) = -80 \text{ ft/sec}$$

At $t = 5$ seconds, the ball is traveling -80 ft/sec. When the instantaneous velocity is negative, it means that the direction of the object is downward. The ball is traveling 80 ft/sec in the downward direction when $t = 5$ seconds.

(h) The ball strikes the ground when $t = 6$. The instantaneous velocity when $t = 6$ is

$$s'(6) = -16(12 - 5) = -112 \text{ ft/sec}$$

The velocity of the ball at $t = 6$ sec is -112 ft/sec. Again, the negative value implies that the ball is traveling downward.

EXPLORATION: Determine the vertex of the quadratic function $s = s(t)$ given in Example 3. What do you conclude about instantaneous velocity when $s(t)$ is a maximum?

Acceleration

The **acceleration *a*** of a particle is defined as the instantaneous rate of change of velocity with respect to time. That is,

$$a = \frac{dv}{dt} = \frac{d}{dt}v = \frac{d}{dt}\left(\frac{ds}{dt}\right) = \frac{d^2s}{dt^2} = s''(t)$$

In other words, acceleration is the second derivative of the function $s = s(t)$ with respect to time.

EXAMPLE 4 Analyzing the Motion of an Object

A ball is thrown vertically upward from ground level with an initial velocity of 19.6 meters per second. The distance s (in meters) of the ball above the ground is

$$s = -4.9t^2 + 19.6t$$

where t is the number of seconds elapsed from the moment the ball is thrown.

(a) What is the velocity of the ball at the end of 1 second?
(b) When will the ball reach its highest point?
(c) What is the maximum height the ball reaches?
(d) What is the acceleration of the ball at any time t?
(e) How long is the ball in the air?
(f) What is the velocity of the ball upon impact?
(g) What is the total distance traveled by the ball?

SOLUTION **(a)** The velocity is

$$v = s'(t) = \frac{d}{dt}(-4.9t^2 + 19.6t) = -9.8t + 19.6$$

At $t = 1$, $v = s'(1) = 9.8$ meters per second.

(b) The ball will reach its highest point when it is stationary; that is, when $v = 0$.

$$v = -9.8t + 19.6 = 0 \quad \text{when } t = \frac{19.6}{9.8} = 2 \text{ seconds}$$

The ball reaches its maximum height 2 seconds after it is thrown.

(c) At $t = 2$, $s = s(2) = -4.9(4) + 19.6(2) = 19.6$ meters.

(d) $a = s''(t) = \dfrac{dv}{dt} = -9.8$ meters per second per second.

(e) We can answer this question in two ways. First, since the ball starts at ground level and it takes 2 seconds for the ball to reach its maximum height, it follows that it will take another 2 seconds to reach the ground, for a total time of 4 seconds in the air. The second way is to set $s = 0$ and solve for t:

$$-4.9t^2 + 19.6t = 0 \quad s(t) = 0$$
$$-4.9t(t - 4) = 0 \quad \text{Factor.}$$
$$t = 0 \quad \text{or} \quad t = 4 \quad \text{Apply the Zero-Product Property; solve for } t.$$

The ball is at ground level when $t = 0$ and when $t = 4$.

FIGURE 13

$t = 2; v(2) = 0; s(2) = 19.6$

$v(1) = 9.8$ meters per second

19.6 meters

$v(0) = 19.6$ meters per second

$t = 0$ $t = 4$

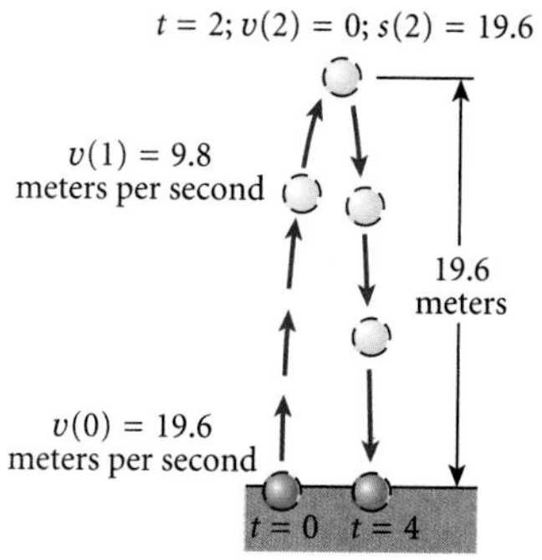

(f) Upon impact, $t = 4$. When $t = 4$,

$$v = s'(4) = (-9.8)(4) + 19.6 = -19.6 \text{ meters per second}$$

The minus sign here indicates that the direction of velocity is downward.

(g) The total distance traveled is

$$\text{Distance up} + \text{distance down} = 19.6 + 19.6 = 39.2 \text{ meters}$$

See Figure 13 for an illustration.

 NOW WORK PROBLEM 59.

In Example 4, the acceleration of the ball is constant, -9.8 meters/second/second. This is approximately true for all falling bodies provided air resistance is ignored. In fact, the constant is the same for all falling bodies, as Galileo (1564–1642) discovered in the sixteenth century. We can use calculus to see this. Galileo found by experimentation that all falling bodies obey the law that the distance they fall when they are dropped is proportional to the square of the time t it takes to fall that distance. Of importance is the fact that the constant of proportionality c is the same for all bodies. Thus, Galileo's law states that the distance s a body falls in time t is given by

$$s = -ct^2$$

The reason for the minus sign is that the body is falling and we have chosen our coordinate system so that the positive direction is up, along the vertical axis.

The velocity v of this freely falling body is

$$v = \frac{ds}{dt} = -2ct$$

and its acceleration a is

$$a = \frac{dv}{dt} = \frac{d^2s}{dt^2} = -2c$$

That is, the acceleration of a freely falling body is a constant. Usually, we denote this constant by $-g$ so that

$$a = -g$$

The number g is called the **acceleration of gravity**. For our planet, g may be approximated by 32 feet per second per second or 9.8 meters per second per second.* On the planet Jupiter, $g \approx 26$ meters per second per second, and on our moon, $g \approx 1.6$ meters per second per second.

EXAMPLE 5   **How High Can a Pitcher Throw a Ball?**

Safeco Field, home of the Seattle Mariners, has a 215-foot high retractable roof. Could a major league pitcher throw a ball up to the roof?

SOLUTION   To answer this question, we make the following assumptions:

(i) We assume the pitcher can throw the ball with an initial velocity v_0 of around 95 miles per hour ≈ 140 feet per second.

---

* The Earth is not perfectly round; it bulges slightly at the equator. But neither is it perfectly oval, and its mass is not distributed uniformly. As a result, the acceleration of any freely falling body varies slightly from these constants.

(ii) The ball is thrown upward from an initial height of 6 feet (h_0) (average height of a pitcher).

The equation

$$s(t) = -16t^2 + v_0 t + h_0$$

is a formula describing the path of the ball, where $s(t)$ is the height (in feet) at time t, h_0 is the height of the object at time $t = 0$, and v_0 is the initial velocity of the ball ($t = 0$). Based on the assumptions, we have

$$s(t) = -16t^2 + 140t + 6 \qquad v_0 = 140;\ h_0 = 6$$

The velocity $v = v(t)$ of the ball at time t is

$$v(t) = s'(t) = -32t + 140$$

The ball is at its highest when $v(t) = 0$

$$-32t + 140 = 0. \qquad v(t) = 0$$

$$t = \frac{140}{32} = 4.375 \text{ seconds} \qquad \text{Solve for } t.$$

The height of the ball when $t = 4.375$ seconds is

$$s(4.375) = -16(4.375)^2 + 140(4.375) + 6 = 312.25 \text{ feet.}$$

So, a baseball can be thrown up to the roof of Safeco Field.

EXERCISE 13.6 Answers to Odd-Numbered Problems Begin on Page AN-75.

In Problems 1–24, find the first derivative and the second derivative of each function.

1. $f(x) = 2x + 5$

2. $f(x) = 3x + 2$

3. $f(x) = 3x^2 + x - 2$

4. $f(x) = 5x^2 + 1$

5. $f(x) = -3x^4 + 2x^2$

6. $f(x) = -4x^3 + x^2 - 1$

7. $f(x) = \dfrac{1}{x}$

8. $f(x) = \dfrac{1}{x^2}$

9. $f(x) = x + \dfrac{1}{x}$

10. $f(x) = x - \dfrac{1}{x}$

11. $f(x) = \dfrac{x}{x+1}$

12. $f(x) = \dfrac{x+1}{x}$

13. $f(x) = e^x$

14. $f(x) = e^{-x}$

15. $f(x) = (x^2 + 4)^3$

16. $f(x) = (x^2 - 1)^4$

17. $f(x) = \ln x$

18. $f(x) = \ln(2x)$

19. $f(x) = xe^x$

20. $f(x) = x \ln x$

21. $f(x) = (e^x)^2$

22. $f(x) = (e^{-x})^2$

23. $f(x) = \dfrac{1}{\ln x}$

24. $f(x) = \dfrac{1}{e^{2x}}$

In Problems 25–32, for each function f, find:

(a) The domain of f.
(b) The derivative $f'(x)$.
(c) The domain of f'.
(d) Any numbers x for which $f'(x) = 0$.
(e) Any numbers x in the domain of f for which $f'(x)$ does not exist.
(f) The second derivative $f''(x)$.
(g) The domain of f''.

25. $f(x) = x^2 - 4$

26. $f(x) = x^2 + 2x$

27. $f(x) = x^3 - 9x^2 + 27x - 27$

28. $f(x) = x^3 - 6x^2 + 12x - 8$

29. $f(x) = 3x^4 - 12x^3 + 2$

30. $f(x) = x^4 - 4x + 2$

31. $f(x) = \frac{x}{x^2 - 4}$

32. $f(x) = \frac{x^2}{x^2 - 1}$

In Problems 33–38, find the indicated derivative.

33. $f^{(4)}(x)$ if $f(x) = x^3 - 3x^2 + 2x - 5$

34. $f^{(5)}(x)$ if $f(x) = 4x^3 + x^2 - 1$

35. $\frac{d^{20}}{dx^{20}}(8x^{19} - 2x^{14} + 2x^5)$

36. $\frac{d^{14}}{dx^{14}}(x^{13} - 2x^{10} + 5x^3 - 1)$

37. $\frac{d^8}{dx^8}\left(\frac{1}{8}x^8 - \frac{1}{7}x^7 + x^5 - x^3\right)$

38. $\frac{d^6}{dx^6}(x^6 + 5x^5 - 2x + 4)$

In Problems 39–42, find the velocity v and acceleration a of an object whose position s at time t is given.

39. $s = 16t^2 + 20t$

40. $s = 16t^2 + 10t + 1$

41. $s = 4.9t^2 + 4t + 4$

42. $s = 4.9t^2 + 5t$

In Problems 43–54, find a formula for the nth derivative of each function.

43. $f(x) = e^x$

44. $f(x) = e^{2x}$

45. $f(x) = \ln x$

46. $f(x) = \ln(2x)$

47. $f(x) = x \ln x$

48. $f(x) = x^2 \ln x$

49. $f(x) = (2x + 3)^n$

50. $f(x) = (4 - 3x)^n$

51. $f(x) = e^{ax}$

52. $f(x) = e^{-ax}$

53. $f(x) = \ln(ax)$

54. $f(x) = x^n \ln x$

---

55. If $y = e^{2x}$, find $y'' - 4y$

56. If $y = e^{-2x}$, find $y'' - 4y$

57. Find the second derivative of: $f(x) = x^2 g(x)$, where g' and g'' exist.

58. Find the second derivative of: $f(x) = \frac{g(x)}{x}$, where g' and g'' exist.

59. **Falling Body** A ball is thrown vertically upward with an initial velocity of 80 feet per second. The distance s (in feet) of the ball from the ground after t seconds is given by $s = s(t) = 6 + 80t - 16t^2$.

(a) What is the velocity of the ball after 2 seconds?
(b) When will the ball reach its highest point?
(c) What is the maximum height the ball reaches?
(d) What is the acceleration of the ball at any time t?
(e) How long is the ball in the air?
(f) What is the velocity of the ball upon impact?
(g) What is the total distance traveled by the ball?

60. **Falling Body** An object is propelled vertically upward with an initial velocity of 39.2 meters per second. The distance s (in meters) of the object from the ground after t seconds is $s = s(t) = -4.9t^2 + 39.2t$.

(a) What is the velocity of the object at any time t?
(b) When will the object reach its highest point?
(c) What is the maximum height?
(d) What is the acceleration of the object at any time t?
(e) How long is the object in the air?
(f) What is the velocity of the object upon impact?
(g) What is the total distance traveled by the object?

61. **Ballistics** A bullet is fired horizontally into a bale of paper. The distance s (in meters) the bullet travels in the bale of paper in t seconds is given by $s = 8 - (2 - t)^3$ for $0 \le t \le 2$. Find the velocity of the bullet after 1 second. Find the acceleration of the bullet at any time t.

62. **Falling Rocks on Jupiter** If a rock falls from a height of 20 meters on the planet Jupiter, then its height H after t seconds is approximately.

$$H(t) = 20 - 13t^2$$

(a) What is the average velocity of the rock from $t = 0$ to $t = 1$?
(b) What is the instantaneous velocity at time $t = 1$?
(c) What is the acceleration of the rock?
(d) When does the rock hit the ground?

63. **Falling Body** A rock is dropped from a height of 88.2 meters. In t seconds the rock falls $4.9t^2$ meters.

(a) How long does it take for the rock to hit the ground?
(b) What is the average velocity of the rock during the time it is falling?
(c) What is the average velocity of the rock for the first 3 seconds?
(d) What is the velocity of the rock when it hits the ground?

64. **Falling Body** A ball is thrown upward. The heights in feet of the ball is given by $s(t) = 100t - 16t^2$, where t is the time elapsed in seconds.

(a) What is the velocity of the ball when $t = 0$, $t = 1$, and $t = 4$ seconds?
(b) At what time does the ball strike the ground?
(c) At what time does the ball each its highest point?

13.7 Implicit Differentiation

PREPARING FOR THIS SECTION ***Before getting started, review the following:***

>> Implicit Form of a Function (Chapter 10, Section 10.2, p. 553)

OBJECTIVES 1 Find the derivative of a function defined implicitly

So far we have only discussed the derivative of a function that is given explicitly in the form $y = f(x)$. This expression of the relationship between x and y is said to be in *explicit form* because we have solved for the dependent variable y. For example, the equations

$$y = 7x - 2, \qquad s = -16t^2 + 10t + 100, \qquad v = 4h^2 - h$$

are all written in explicit form.

If the functional relationship between the independent variable x and the dependent variable y is not of this form, we say that x and y are related *implicitly*. For example, x and y are related implicitly in the expression

$$x^3 - y^4 - 3y + x = 6$$

In this equation, it is very difficult to find y as a function of x. How, then, do we go about finding the derivative $\frac{dy}{dx}$ in such a case?

The procedure for finding the derivative of y with respect to x when the functional relationship between x and y is given implicitly is called **implicit differentiation.** The procedure is to think of y as a function f of x, without actually expressing y in terms of x. If this requires differentiating terms like y^4, which we think of as $(f(x))^4$, then we use the Power Rule. The derivative of y^4 or $(f(x))^4$ is then

$$4(f(x))^3 f'(x) \qquad \text{or} \qquad 4y^3\frac{dy}{dx}$$

Let's look at an example.

1 **EXAMPLE 1 Differentiating Implicitly**

Find $\frac{dy}{dx}$ if

$$3x + 4y - 5 = 0$$

SOLUTION We begin by assuming that there is a differentiable function $y = f(x)$ implied by the above relationship. Then,

$$3x + 4f(x) - 5 = 0$$

We differentiate both sides of the equality with respect to x. Then

$$\frac{d}{dx}[3x + 4f(x) - 5] = \frac{d}{dx}0$$

$$\frac{d}{dx}(3x) + \frac{d}{dx}[4f(x)] - \frac{d}{dx}5 = \frac{d}{dx}0 \qquad \text{Derivative of a sum.}$$

$$3 + 4\frac{d}{dx}f(x) = 0 \qquad \text{Differentiate.}$$

Solving for $\frac{d}{dx}f(x)$, we find

$$\frac{d}{dx}f(x) = -\frac{3}{4}$$

Replacing $\frac{d}{dx}f(x)$ by $\frac{dy}{dx}$, we have

$$\frac{dy}{dx} = -\frac{3}{4}$$

For the function in Example 1, it is possible to solve for y as a function of x by algebraically solving for y.

$$3x + 4y - 5 = 0$$
$$4y = 5 - 3x$$
$$y = \frac{1}{4}(5 - 3x) = \frac{5}{4} - \frac{3}{4}x$$

Then

$$\frac{dy}{dx} = -\frac{3}{4}$$

which agrees with the result obtained using implicit differentiation. Often, though, it is very difficult, or even impossible, to actually solve for y in terms of x.

EXAMPLE 2 Differentiating Implicitly

Find $\frac{dy}{dx}$ if

$$3x^2 + 4y^2 = 2x$$

SOLUTION We again assume there is a differentiable function $y = f(x)$ implied by the above equation. We proceed to differentiate both sides of this equation with respect to x:

$$\frac{d}{dx}(3x^2 + 4y^2) = \frac{d}{dx}(2x)$$
$$\frac{d}{dx}(3x^2) + \frac{d}{dx}(4y^2) = 2$$
$$6x + 4\left[\frac{d}{dx}y^2\right] = 2$$

Using the Power Rule, $\frac{d}{dx}y^2 = 2y\frac{dy}{dx}$. Then

$$6x + 4\left(2y\frac{dy}{dx}\right) = 2$$
$$6x + 8y\frac{dy}{dx} = 2$$

This is a linear equation in $\frac{dy}{dx}$. Solving for $\frac{dy}{dx}$, we have

$$8y\frac{dy}{dx} = 2 - 6x$$
$$\frac{dy}{dx} = \frac{2 - 6x}{8y} = \frac{1 - 3x}{4y} \quad \text{provided} \quad y \neq 0.$$

Steps for Differentiating Implicitly

STEP 1: To find $\frac{dy}{dx}$ when x and y are related implicitly, assume that y is a differentiable function of x.

STEP 2: Differentiate both sides of the equation with respect to x by employing the Power Rule or the Chain Rule or other differentiation formulas.

STEP 3: Solve the resulting equation, which is linear in $\frac{dy}{dx}$, for $\frac{dy}{dx}$.

NOW WORK PROBLEM 1.

EXAMPLE 3 Differentiating Implicitly

Find $\frac{dy}{dx}$ if

$$x^2 + y^2 = e^x + e^y$$

SOLUTION

$$\frac{d}{dx}(x^2 + y^2) = \frac{d}{dx}(e^x + e^y) \quad \text{Differentiate both sides with respect to } x.$$

$$2x + \frac{d}{dx}y^2 = e^x + \frac{d}{dx}e^y \quad \text{Apply the sum formula; } \frac{d}{dx}x^2 = 2x; \frac{d}{dx}e^x = e^x.$$

$$2x + 2y\frac{dy}{dx} = e^x + e^y\frac{dy}{dx} \quad \text{Apply the Chain Rule on the right and the Power Rule on the left.}$$

We proceed to solve for $\frac{dy}{dx}$. First bring the terms involving $\frac{dy}{dx}$ to the left side and bring any other terms to the right side.

$$2y\frac{dy}{dx} - e^y\frac{dy}{dx} = e^x - 2x$$

$$(2y - e^y)\frac{dy}{dx} = e^x - 2x \quad \text{Factor.}$$

$$\frac{dy}{dx} = \frac{e^x - 2x}{2y - e^y} \quad \text{Divide both sides by } 2y - e^y.$$

NOW WORK PROBLEM 21.

EXAMPLE 4 Finding the Equation of a Tangent Line

Find the equation of the tangent line to the graph of $x^3 + xy + y^3 = 5$ at the point $(-1, 2)$.

SOLUTION The slope of the tangent line is $\frac{dy}{dx}$, which can be found by differentiating implicitly. We differentiate both sides with respect to x, obtaining

$$\frac{d}{dx}(x^3 + xy + y^3) = \frac{d}{dx}5$$

$$\frac{d}{dx}x^3 + \frac{d}{dx}(xy) + \frac{d}{dx}y^3 = 0$$

$$3x^2 + \left(x\frac{dy}{dx} + y\right) + 3y^2\frac{dy}{dx} = 0$$

$$(3x^2 + y) + (x + 3y^2)\frac{dy}{dx} = 0$$

$$(x + 3y^2)\frac{dy}{dx} = -(3x^2 + y)$$

Solving for $\frac{dy}{dx}$, we find

$$\frac{dy}{dx} = \frac{-(3x^2 + y)}{x + 3y^2} \qquad \text{provided} \qquad x + 3y^2 \neq 0$$

The derivative $\frac{dy}{dx}$ equals the slope of the tangent line to the graph at any point (x, y) for which $x + 3y^2 \neq 0$. In particular, for $x = -1$ and $y = 2$, we find the slope of the tangent line to the graph at $(-1, 2)$ to be

$$\frac{dy}{dx} = \frac{-(3 + 2)}{-1 + 12} = -\frac{5}{11} \qquad \frac{dy}{dx} = \frac{-(3x^2 + y)}{x + 3y^2};\ x = -1,\ y = 2$$

The equation of the tangent line at the point $(-1, 2)$ is

$$y - y_1 = m(x - x_1) \qquad \text{Point-slope form of a line.}$$

$$y - 2 = -\frac{5}{11}(x + 1) \qquad m = -\frac{5}{11},\ x_1 = -1,\ y_1 = 2$$

$$y - 2 = -\frac{5}{11}x - \frac{5}{11}$$

$$y = -\frac{5}{11}x + \frac{17}{11}$$

NOW WORK PROBLEM 35.

The prime notation y', y'', and so on, is usually used in finding higher-order derivatives for implicitly defined functions.

EXAMPLE 5 Finding First and Second Derivatives Implicitly

Using implicit differentiation, find y' and y'' in terms of x and y if

$$xy + y^2 - x^2 = 5$$

SOLUTION

$$\frac{d}{dx}(xy + y^2 - x^2) = \frac{d}{dx}5 \qquad \text{Differentiate both sides with respect to } x.$$

$$\frac{d}{dx}(xy) + \frac{d}{dx}y^2 - \frac{d}{dx}x^2 = 0$$

$$(xy' + y) + 2yy' - 2x = 0 \qquad (1)$$

$$(x + 2y)y' = 2x - y$$

$$y' = \frac{2x - y}{x + 2y} \qquad \text{provided} \qquad x + 2y \neq 0 \qquad (2)$$

It is easier to find y'' by differentiating (1) than by using (2):

$$\frac{d}{dx}(xy' + y + 2yy' - 2x) = \frac{d}{dx}0$$

$$\frac{d}{dx}(xy') + \frac{d}{dx}y + \frac{d}{dx}(2yy') - \frac{d}{dx}(2x) = 0$$

$$xy'' + y' + y' + 2y'(y') + 2yy'' - 2 = 0$$

$$y''(x + 2y) = 2 - 2y' - 2(y')^2$$

$$y'' = \frac{2 - 2y' - 2(y')^2}{x + 2y}$$

provided $x + 2y \neq 0$. To express y'' in terms of x and y, use (2). Then

$$y'' = \frac{2 - 2\left(\frac{2x - y}{x + 2y}\right) - 2\left(\frac{2x - y}{x + 2y}\right)^2}{x + 2y} \qquad y' = \frac{2x - y}{x + 2y}$$

$$= \frac{2(x + 2y)^2 - 2(2x - y)(x + 2y) - 2(2x - y)^2}{(x + 2y)^3} \qquad \text{Multiply by } \frac{(x + 2y)^2}{(x + 2y)^2}.$$

$$= \frac{2x^2 + 8xy + 8y^2 - 4x^2 - 6xy + 4y^2 - 8x^2 + 8xy - 2y^2}{(x + 2y)^3} \qquad \text{Simplify.}$$

$$= \frac{-10x^2 + 10xy + 10y^2}{(x + 2y)^3} \qquad \text{Simplify.}$$

$$= \frac{-10(x^2 - xy - y^2)}{(x + 2y)^3} \underset{\substack{\uparrow \\ x^2 - xy - y^2 = -5}}{=} \frac{50}{(x + 2y)^3}$$

NOW WORK PROBLEM 31.

Application

EXAMPLE 6 Finding Marginal Revenue

For a particular commodity, the demand equation is

$$3x^2 + 4p^2 = 1200 \qquad 0 < x < 20 \qquad 0 < p < 10\sqrt{3}$$

where x is the amount demanded and p is the price (in dollars). Find the marginal revenue when $x = 8$.

SOLUTION The revenue function is

$$R = xp$$

We could solve for p in the price demand function and then compute $\frac{dR}{dx}$. However, the technique introduced earlier is easier.

We differentiate $R = xp$ with respect to x remembering that p is a function of x. Then, by the rule for differentiating a product, we obtain the marginal revenue.

$$R'(x) = p + x\frac{dp}{dx} \qquad \frac{d}{dx}(xp) = x\frac{dp}{dx} + p\frac{d}{dx}x \tag{1}$$

To find $\frac{dp}{dx}$, we differentiate the demand equation implicitly:

$$3x^2 + 4p^2 = 1200$$

$$\frac{d}{dx}(3x^2 + 4p^2) = \frac{d}{dx}1200$$

$$6x + 8p\frac{dp}{dx} = 0$$

Solving for $\frac{dp}{dx}$, we have

$$\frac{dp}{dx} = \frac{-6x}{8p} = -\frac{3x}{4p}$$

Now substitute $\frac{dp}{dx}$ into equation (1). Then $R'(x)$ is

$$R'(x) = p + x\left(-\frac{3x}{4p}\right) = \frac{4p^2 - 3x^2}{4p}$$

When $x = 8$,

$$4p^2 = 1200 - 3x^2 = 1200 - 3(64) = 1200 - 192 = 1008$$

So $p^2 = \frac{1008}{4} = 252.$

Then $p = \sqrt{252} = 15.87$, so that the marginal revenue at $x = 8$ is

$$R'(8) = \frac{1008 - 192}{4(15.87)} = \$12.85 \text{ per unit}$$

EXERCISE 13.7 Answers to Odd-Numbered Problems Begin on Page AN-76.

In Problems 1–30, find $\frac{dy}{dx}$ by using implicit differentiation.

1. $x^2 + y^2 = 4$
2. $3x^2 - 2y^2 = 6$
3. $x^2y = 8$
4. $x^3y = 5$
5. $x^2 + y^2 - xy = 2$
6. $x^2y + xy^2 = x + 1$
7. $x^2 + 4xy + y^2 = y$
8. $x^2 + 2xy + y^2 = x$
9. $3x^2 + y^3 = 1$
10. $y^4 - 4x^2 = 5$
11. $4x^3 + 2y^3 = x^2$
12. $5x^2 + xy - y^2 = 0$
13. $\frac{1}{x^2} - \frac{1}{y^2} = 4$
14. $\frac{1}{x^2} + \frac{1}{y^2} = 6$
15. $\frac{1}{x} + \frac{1}{y} = 2$
16. $\frac{1}{x} - \frac{1}{y} = 4$
17. $x^2 + y^2 = ye^x$
18. $x^2 + y^2 = xe^y$
19. $\frac{x}{y} + \frac{y}{x} = 6e^x$
20. $x^2 + y^2 = 2ye^x$
21. $x^2 = y^2 \ln x$
22. $x^2 + y^2 = 2y^2 \ln x$
23. $(2x + 3y)^2 = x^2 + y^2$
24. $x^2 + y^2 = (3x - 4y)^2$
25. $(x^2 + y^2)^2 = (x - y)^3$
26. $(x^2 - y^2)^2 = (x + y)^3$
27. $(x^3 + y^3)^2 = x^2y^2$
28. $(x^3 - y^3)^2 = xy^2$
29. $y = e^{x^2 + y^2}$
30. $x = \ln(x^2 + y^2)$

In Problems 31–34, find y' and y'' in terms of x and y.

31. $x^2 + y^2 = 4$
32. $x^2 - y^2 = 1$
33. $xy + yx^2 = 2$
34. $4xy = x^2 + y^2$

In Problems 35–38, find the slope of the tangent line at the indicated point. Write an equation for this tangent line.

35. $x^2 + y^2 = 5$ at $(1, 2)$
36. $x^2 - y^2 = 8$ at $(3, 1)$
37. $e^{xy} = x$ at $(1, 0)$
38. $\ln(x^2 + y^2) = 1$ at $(0, \sqrt{e})$

In Problems 39–42, find those points (x, y), if there are any, where the tangent line is horizontal $\left(\dfrac{dy}{dx} = 0\right)$.

39. $x^2 + y^2 = 4$

40. $xy + y^2 - x^2 = 4x$

41. $y^2 + 4x^2 = 16$

42. $y^2 = 4x^2 + 4$

43. Given the equation $x + xy + 2y^2 = 6$:

(a) Find an expression for the slope of the tangent line at any point (x, y) on the graph.
(b) Write an equation for the tangent line to the graph at the point $(2, 1)$.
(c) Find the coordinates of all points (x, y) on the graph at which the slope of the tangent line equals the slope of the tangent line at $(2, 1)$.

44. The graph of the function $(x^2 + y^2)^2 = x^2 - y^2$ contains exactly four points at which the tangent line is horizontal. Find them.

45. Gas Pressure For ideal gases, **Boyle's law** states that pressure is inversely proportional to volume. A more realistic relationship between pressure P and volume V is given by **van der Waals equation**

$$P + \frac{a}{V^2} = \frac{C}{V - b}$$

where C is the constant of proportionality, a is a constant that depends on molecular attraction, and b is a constant that depends on the size of the molecules. Find the compressibility of the gas, which is measured by $\dfrac{dV}{dP}$.

46. Cost Function If the relationship between the cost C (in dollars) and the number x (in thousands) of units produced is

$$\frac{x^2}{9} - C^2 = 1 \qquad x > 0, \qquad C > 0$$

find the marginal cost by using implicit differentiation. For what number of units produced (approximately) does $C'(x) = 1$.

47. Master's Degrees in the U.S. In the 2000–01 academic year there were 430,164 Master's degrees granted in the United States. Assume the future relationship between the number N of Master's degrees granted and the years t since the 1999–2000 academic year is given by

$$e^{N(t)} = 430{,}163t + \frac{3t}{t^2 + 2}$$

(a) Find $\dfrac{dN}{dt}$.
(b) Evaluate $N'(t)$ for $t = 2$ and for $t = 4$.
(c) Interpret the answers to (b) and explain the difference.

48. Farm Income In 2002 the income from farming in the United States was 32.4 billion dollars. Suppose the relationship between farm income I in the United States and time t, where $t = 1$ corresponds to the year 2002, is given by

$$I(t) = \sqrt{32.4\,t}$$

(a) Find $I'(t)$.
[Hint: Square both sides and differentiate implicitly].
(b) Find $I'(3)$ and $I'(5)$.
(c) Interpret the answers found in part (b). Explain the difference.

Source: Bureau of Economic Analysis, 2002.

13.8 The Derivative of $f(x) = x^{p/q}$

PREPARING FOR THIS SECTION *Before getting started, review the following:*

>> n^{th} Roots; Rational Exponents (Appendix A, Section A.8, pp. A-73–A-78)

OBJECTIVES
1 Differentiate functions involving fractional exponents or radicals
2 Use the Chain Rule and the Power Rule with fractional exponents or radicals

So far we have developed formulas for finding the derivative of polynomials, rational functions, exponential functions, and logarithmic functions. In addition, with the Chain Rule and the Power Rule we can differentiate each of these functions when they are raised to an integer power. In this section we develop a formula that will handle powers that are rational numbers.

If p and q, $q > 2$, are integers, then $x^{p/q} = \sqrt[q]{x^p}$. As a result, once we know how to find derivatives involving rational exponents, we will also know how to handle derivatives involving radicals.

We begin by restating the formula for finding the derivative of x^n, where n is any integer, namely,

$$\frac{d}{dx}x^n = nx^{n-1}$$

In this section we show this result is true even when n is a rational number. That is, if $n = \frac{p}{q}$, where p and $q \neq 0$ are integers, then

$$\frac{d}{dx}x^{p/q} = \frac{p}{q}x^{p/q-1}$$

Let's look at an example.

EXAMPLE 1 Differentiating $f(x) = x^{3/2}$

Find $\frac{d}{dx}x^{3/2}$.

SOLUTION If we let $y = x^{3/2}$, then by squaring both sides, we find

$$y^2 = x^3$$

Differentiate implicitly to obtain

$$2yy' = 3x^2$$

Solving for y', we find that

$$y' = \frac{3x^2}{2y}$$

But $y = x^{3/2}$ so

$$y' = \frac{3x^2}{2x^{3/2}} = \frac{3}{2}x^{1/2}$$

That is,

$$\frac{d}{dx}x^{3/2} = \frac{3}{2}x^{1/2}$$

Example 1 illustrated that

$$\frac{d}{dx}x^{3/2} = \frac{3}{2}x^{(3/2)-1} = \frac{3}{2}x^{1/2}$$

To show that this formula is valid for any rational exponent, we proceed as follows.
Start with the function

$$y = x^{p/q}$$

If we raise both sides to the power q, we obtain

$$y^q = (x^{p/q})^q = x^p$$

We now differentiate implicitly to find

$$\begin{aligned}
\frac{d}{dx}y^q &= \frac{d}{dx}x^p \\
qy^{q-1}y' &= px^{p-1} \\
y' &= \frac{px^{p-1}}{qy^{q-1}} \underset{\uparrow\; y = x^{p/q}}{=} \frac{px^{p-1}}{q(x^{p/q})^{(q-1)}} = \frac{px^{p-1}}{qx^{p-(p/q)}} \\
&= \frac{p}{q}x^{p-1-p+(p/q)} \\
&= \frac{p}{q}x^{(p/q)-1}
\end{aligned}$$

Derivative of $f(x) = x^{p/q}$

The derivative of $f(x) = x^{p/q}$, where p and $q \neq 0$ are integers, is $f'(x) = \frac{p}{q} x^{p/q-1}$. That is,

$$\frac{d}{dx} x^{p/q} = \frac{p}{q} x^{p/q-1} \tag{1}$$

1 **EXAMPLE 2 Finding the Derivative of Functions Involving Fractional Exponents or Radicals**

(a) $\frac{d}{dx} x^{3/2} = \frac{3}{2} x^{(3/2)-1} = \frac{3}{2} x^{1/2}$

(b) $\frac{d}{dx} \sqrt{x} = \frac{d}{dx} x^{1/2} = \frac{1}{2} x^{(1/2)-1} = \frac{1}{2} x^{-1/2} = \frac{1}{2x^{1/2}} = \frac{1}{2\sqrt{x}}$

(c) $\frac{d}{dx} x^{-2/3} = -\frac{2}{3} x^{(-2/3)-1} = -\frac{2}{3} x^{-5/3} = \frac{-2}{3x^{5/3}}$

NOW WORK PROBLEM 1.

EXAMPLE 3 Finding the Derivative of a Function Involving Fractional Exponents

Find the derivative of: $f(x) = \frac{x^{1/2} - 2}{x^{1/2}}$

SOLUTION We shall solve the problem in two ways.

Method 1 Use the formula for the derivative of a quotient:

$$f'(x) = \frac{d}{dx} \frac{x^{1/2} - 2}{x^{1/2}} = \frac{x^{1/2} \frac{d}{dx}(x^{1/2} - 2) - (x^{1/2} - 2)\frac{d}{dx} x^{1/2}}{(x^{1/2})^2}$$

$$= \frac{x^{1/2} \cdot \frac{1}{2} x^{-1/2} - (x^{1/2} - 2) \cdot \frac{1}{2} x^{-1/2}}{x} = \frac{\frac{1}{2} - \frac{1}{2} + x^{-1/2}}{x} = \frac{x^{-1/2}}{x} = \frac{1}{x^{3/2}}$$

Method 2 Simplify first. Then the problem becomes that of finding the derivative of:

$$f(x) = \frac{x^{1/2} - 2}{x^{1/2}} = \frac{x^{1/2}}{x^{1/2}} - \frac{2}{x^{1/2}} = 1 - \frac{2}{x^{1/2}} = 1 - 2x^{-1/2}$$

Then,

$$f'(x) = \frac{d}{dx}(1 - 2x^{-1/2}) = \frac{d}{dx} 1 - 2\frac{d}{dx} x^{-1/2} = 0 - 2\left(-\frac{1}{2}\right) x^{-3/2} = x^{-3/2} = \frac{1}{x^{3/2}}$$

EXAMPLE 4 Finding the Derivative of a Function Involving Radicals

Find the derivative of: $f(x) = (2\sqrt{x} + 1)(\sqrt[3]{x} - 2)$

SOLUTION First we change each radical to its fractional exponent equivalent:

$$f(x) = (2x^{1/2} + 1)(x^{1/3} - 2)$$

Using the formula for the derivative of a product, we find

$$f'(x) = \frac{d}{dx}[(2x^{1/2} + 1)(x^{1/3} - 2)]$$

$$= (2x^{1/2} + 1)\frac{d}{dx}(x^{1/3} - 2) + (x^{1/3} - 2)\frac{d}{dx}(2x^{1/2} + 1) \qquad \text{Derivative of a product.}$$

$$= (2x^{1/2} + 1)\left(\frac{1}{3}x^{-2/3}\right) + (x^{1/3} - 2)(x^{-1/2}) \qquad \frac{d}{dx}(2x^{1/2} + 1) = 2 \cdot \frac{1}{2} \cdot x^{1/2-1} = x^{-1/2}$$

$$= \frac{2x^{1/2} + 1}{3x^{2/3}} + \frac{x^{1/3} - 2}{x^{1/2}} \qquad \text{Simplify.}$$

$$= \frac{x^{1/2}(2x^{1/2} + 1) + 3x^{2/3}(x^{1/3} - 2)}{3x^{2/3}x^{1/2}} \qquad \text{Add the quotients.}$$

$$= \frac{2x + x^{1/2} + 3x - 6x^{2/3}}{3x^{7/6}} \qquad \text{Simplify.}$$

$$= \frac{5x + x^{1/2} - 6x^{2/3}}{3x^{7/6}} \qquad \text{Simplify.}$$

NOW WORK PROBLEM 29.

The Power Rule can be extended as follows:

The Power Rule

If p and $q \neq 0$ are integers and g is a differentiable function, then

$$\frac{d}{dx}[g(x)]^{p/q} = \frac{p}{q}[g(x)]^{p/q-1} \cdot g'(x)$$

2 **EXAMPLE 5** **Using the Power Rule with Fractional Exponents and Radicals**

(a) $$\frac{d}{dx}(x^2 + 4)^{5/2} = \frac{5}{2}(x^2 + 4)^{3/2}\frac{d}{dx}(x^2 + 4)$$

$$= \frac{5}{2}(x^2 + 4)^{3/2} \cdot 2x = 5x(x^2 + 4)^{3/2}$$

(b) $$\frac{d}{dx}\sqrt{x^2 + 4} = \frac{d}{dx}(x^2 + 4)^{1/2} = \frac{1}{2}(x^2 + 4)^{-1/2}\frac{d}{dx}(x^2 + 4)$$

$$= \frac{1}{2}(x^2 + 4)^{-1/2} \cdot 2x = \frac{x}{\sqrt{x^2 + 4}}$$

(c) $$\frac{d}{dx}\frac{1}{\sqrt{4x + 3}} = \frac{d}{dx}(4x + 3)^{-1/2} = -\frac{1}{2}(4x + 3)^{-3/2}\frac{d}{dx}(4x + 3)$$

$$= -\frac{1}{2}(4x + 3)^{-3/2} \cdot 4 = \frac{-2}{(4x + 3)^{3/2}}$$

NOW WORK PROBLEM 7.

EXAMPLE 6 Using Implicit Differentiation

Find $\frac{dy}{dx}$ if $xy^2 - x + y^3\sqrt{x} + 5y = 10$

SOLUTION We use implicit differentiation to find $\frac{dy}{dx}$.

$$\frac{d}{dx}(xy^2 - x + y^3\sqrt{x} + 5y) = \frac{d}{dx}10$$

$$\frac{d}{dx}(xy^2) - \frac{d}{dx}x + \frac{d}{dx}(y^3\sqrt{x}) + \frac{d}{dx}(5y) = 0$$

Now use the formula for the derivative of a product to obtain

$$x\frac{d}{dx}y^2 + y^2\frac{d}{dx}x - 1 + y^3\frac{d}{dx}\sqrt{x} + \sqrt{x}\frac{d}{dx}y^3 + 5\frac{dy}{dx} = 0$$

$$x \cdot 2y\frac{dy}{dx} + y^2 - 1 + y^3\frac{1}{2\sqrt{x}} + \sqrt{x} \cdot 3y^2\frac{dy}{dx} + 5\frac{dy}{dx} = 0 \qquad \frac{d}{dx}\sqrt{x} = \frac{d}{dx}x^{1/2} = \frac{1}{2}x^{-1/2} = \frac{1}{2\sqrt{x}}$$

$$(2xy + 3\sqrt{x}\,y^2 + 5)\frac{dy}{dx} = 1 - y^2 - \frac{y^3}{2\sqrt{x}}$$

Then, if $2xy + 3\sqrt{x}\,y^2 + 5 \neq 0$,

$$\frac{dy}{dx} = \frac{1 - y^2 - \frac{y^3}{2\sqrt{x}}}{2xy + 3\sqrt{x}\,y^2 + 5} = \frac{2\sqrt{x}(1 - y^2) - y^3}{2\sqrt{x}\,(2xy + 3\sqrt{x}\,y^2 + 5)}$$

NOW WORK PROBLEM 33.

EXERCISE 13.8 Answers to Odd-Numbered Problems Begin on Page AN-77.

In Problems 1–32, find the derivative of each function.

1. $f(x) = x^{4/3}$
2. $f(x) = x^{5/2}$
3. $f(x) = x^{2/3}$
4. $f(x) = x^{3/4}$
5. $f(x) = \frac{1}{x^{1/2}}$
6. $f(x) = \frac{1}{x^{1/3}}$
7. $f(x) = (2x + 3)^{3/2}$
8. $f(x) = (3x + 4)^{4/3}$
9. $f(x) = (x^2 + 4)^{3/2}$
10. $f(x) = (x^3 + 1)^{4/3}$
11. $f(x) = \sqrt{2x + 3}$
12. $f(x) = \sqrt{4x - 5}$
13. $f(x) = \sqrt{9x^2 + 1}$
14. $f(x) = \sqrt{4x^2 + 1}$
15. $f(x) = 3x^{5/3} - 6x^{1/3}$
16. $f(x) = 4x^{5/4} - 8x^{1/4}$
17. $f(x) = x^{1/3}(x^2 - 4)$
18. $f(x) = x^{2/3}(x - 8)$
19. $f(x) = \frac{x}{\sqrt{x^2 - 4}}$
20. $f(x) = \frac{x^2}{\sqrt{x - 1}}$
21. $f(x) = \sqrt{e^x}$
22. $f(x) = e^{\sqrt{x}}$
23. $f(x) = \sqrt{\ln x}$
24. $f(x) = \ln\sqrt{x}$
25. $f(x) = e^{\sqrt[3]{x}}$
26. $f(x) = \sqrt[3]{e^x}$
27. $f(x) = \sqrt[3]{\ln x}$
28. $f(x) = \ln\sqrt[3]{x}$
29. $f(x) = \sqrt{x}\,e^x$
30. $f(x) = \sqrt{x}\ln x$
31. $f(x) = e^{2x}\sqrt{x^2 + 1}$
32. $f(x) = \ln\left[x\sqrt{x^2 + 1}\right]$

In Problems 33–40, use implicit differentiation to find y′.

33. $\sqrt{x} + \sqrt{y} = 4$
34. $\sqrt{x} - \sqrt{y} = 1$
35. $\sqrt{x^2 + y^2} = x$
36. $\sqrt{x^2 - y^2} = y$
37. $x^{1/3} + y^{1/3} = 1$
38. $x^{2/3} + y^{2/3} = 1$
39. $e^{\sqrt{x}} + e^{\sqrt{y}} = 4$
40. $\ln\sqrt{x^2 + y^2} = 4x$

In Problems 41–50, for each function f, find:

(a) *The domain of f.*
(b) *The derivative $f'(x)$.*
(c) *The domain of f'.*
(d) *Any numbers x for which $f'(x) = 0$.*
(e) *Any numbers x in the domain of f for which $f'(x)$ does not exist.*
(f) *The second derivative $f''(x)$.*
(g) *The domain of f''.*

41. $f(x) = \sqrt{x}$

42. $f(x) = \sqrt[3]{x}$

43. $f(x) = x^{2/3}$

44. $f(x) = x^{4/3}$

45. $f(x) = x^{2/3} + 2x^{1/3}$

46. $f(x) = x^{2/3} - 2x^{1/3}$

47. $f(x) = (x^2 - 1)^{2/3}$

48. $f(x) = (x^2 - 1)^{4/3}$

49. $f(x) = x\sqrt{1 - x^2}$

50. $f(x) = x^2\sqrt{4 - x}$

---

51. Enrollment Projection The Office of Admissions estimates that the total student enrollment in the University Division will be given by:

$$N(t) = -\frac{10{,}000}{\sqrt{1 + 0.1t}} + 11{,}000$$

where $N(t)$ denotes the number of students enrolled in the division t years from now.

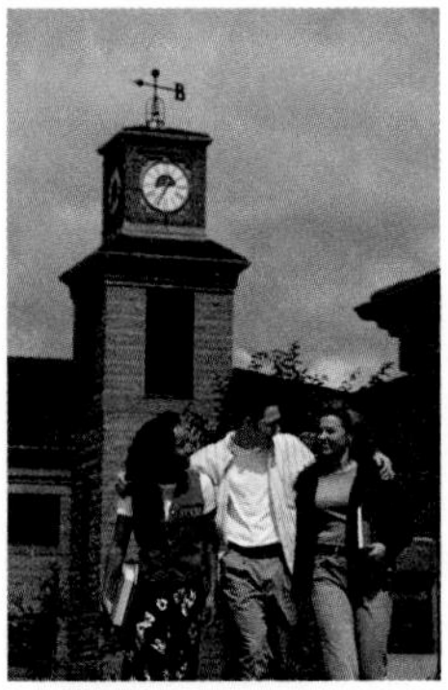

(a) Find an expression for $N'(t)$.
(b) How fast will student enrollment be increasing 10 years from now?

52. Learning Curve The psychologist L. L. Thurstone suggested the following equation for the time T it takes to memorize a list of n words:

$$T = f(n) = Cn\sqrt{n - b}$$

where C and b are constants depending upon the person and the task.

(a) Compute $\dfrac{dT}{dn}$ and interpret the result.
(b) Suppose that for a certain person and a certain task, $C = 2$ and $b = 2$. Compute $f'(10)$ and $f'(30)$.
(c) Interpret the results found in part (b).

53. Production Function The production of commodities sometimes requires several resources, such as land, labor, machinery, and the like. If there are two inputs that require the amounts x and y, then the output z is given by a function of two variables: $z = f(x, y)$. Here z is called a **production function.** For example, if we use x to represent land and y to represent capital, and z to be the amount of a particular commodity produced, a possible production function is

$$z = x^{0.5}y^{0.4}$$

Set z equal to a fixed amount produced and show that $\dfrac{dy}{dx} = -\dfrac{5y}{4x}$. This shows that the rate of change of capital with respect to land is always negative when the amount produced is fixed.

54. Price Function It is estimated that t months from now, the average price (in dollars) of a personal computer will be given by

$$P(t) = \frac{300}{1 + \frac{1}{6}\sqrt{t}} + 100 \qquad 0 \le t \le 60$$

(a) Find an expression for $P'(t)$.
(b) Compute $P'(0)$, $P'(16)$, and $P'(49)$.
(c) Interpret the answers found in part (b).

57. Pollution The amount of pollution in a certain lake is found to be

$$A(t) = (t^{1/4} + 3)^3$$

where t is measured in years and $A(t)$ is measured in appropriate units.

(a) What is the instantaneous rate of change of the amount of pollution?
(b) At what rate is the amount of pollution changing after 16 years?

59. A large container is being filled with water. After t hours there are $8t - 4t^{1/2}$ liters of water in the container. At what rate is the water filling the container (in liters per hour) when $t = 4$?

60. A young child travels s feet down a slide in t seconds, where $s = t^{3/2}$.

(a) What is the child's velocity after 1 second?
(b) If the slide is 8 feet long, with what velocity does the child strike the ground?

Chapter 13   Review

THINGS TO KNOW

Slope of the Tangent Line to $y = f(x)$ at the Point $(c, f(c))$ (p. 715)

$$m_{\tan} = \lim_{x \to c} \frac{f(x) - f(c)}{x - c}, \quad \text{provided the limit exists}$$

Equation of the Tangent Line at a Point $(c, f(c))$ (p. 715)

$$y - f(c) = f'(c)(x - c) \quad \text{provided } f'(c) \text{ exists}$$

Derivative of a Function at a Number c (p. 716)

$f'(c) = \lim\limits_{x \to c} \dfrac{f(x) - f(c)}{x - c}$,   provided the limit exists

Derivative of a Constant (pp. 726–727)

$\dfrac{d}{dx} b = 0$, b a constant

Derivative of a Constant Times a Function (p. 728)

$\dfrac{d}{dx} [cf(x)] = c \dfrac{d}{dx} f(x)$, c a constant

Derivative of a Difference (p. 730)

$\dfrac{d}{dx} [f(x) - g(x)] = \dfrac{d}{dx} f(x) - \dfrac{d}{dx} g(x)$

Derivative of a Quotient (p. 736)

$\dfrac{d}{dx} \left[\dfrac{f(x)}{g(x)} \right] = \dfrac{g(x) \dfrac{d}{dx} f(x) - f(x) \dfrac{d}{dx} g(x)}{[g(x)]^2}$

Derivative of $f(x) = e^x$ (p. 749)

$\dfrac{d}{dx} e^x = e^x$

Derivative of $f(x) = \ln x$ (p. 753)

$\dfrac{d}{dx} \ln x = \dfrac{1}{x}$

Derivative of $f(x) = \ln g(x)$ (p. 753)

$\dfrac{d}{dx} \ln g(x) = \dfrac{\dfrac{d}{dx} g(x)}{g(x)} = \dfrac{g'(x)}{g(x)}$

Derivative of $f(x) = a^x$ (p. 755)

$\dfrac{d}{dx} a^x = a^x \ln a$

Derivative of f at x (p. 718)

$f'(x) = \lim\limits_{h \to 0} \dfrac{f(x + h) - f(x)}{h}$ provided the limit exists

Derivative of $f(x) = x^n$ (pp. 728, 738, 776)

$\dfrac{d}{dx} x^n = nx^{n-1}$, n any rational number

Derivative of a Sum (p. 729)

$\dfrac{d}{dx} [f(x) + g(x)] = \dfrac{d}{dx} f(x) + \dfrac{d}{dx} g(x)$

Derivative of a Product (p. 735)

$\dfrac{d}{dx} [f(x) \cdot g(x)] = f(x) \dfrac{d}{dx} g(x) + g(x) \dfrac{d}{dx} f(x)$

Power Rule (p. 744)

$\dfrac{d}{dx} [g(x)]^n = n[g(x)]^{n-1} g'(x)$, n any rational number

Chain Rule (pp. 750–751)

If $y = f(u)$ and $u = g(x)$, then

$\dfrac{dy}{dx} = \dfrac{dy}{du} \cdot \dfrac{du}{dx}$

Derivative of $f(x) = e^{g(x)}$ (p. 751)

$\dfrac{d}{dx} e^{g(x)} = e^{g(x)} \dfrac{d}{dx} g(x) = e^{g(x)} g'(x)$

Derivative of $f(x) = \log_a x$ (p. 754)

$\dfrac{d}{dx} \log_a x = \dfrac{1}{x \ln a}$

TRUE–FALSE ITEMS Answers are on page AN-78.

T F **1.** The derivative of a function is the limit of a difference quotient.

T F **2.** The derivative of a product equals the product of the derivatives.

T F **3.** If $f(x) = \dfrac{1}{x}$, then $f'(x) = -\dfrac{1}{x^2}$.

T F **4.** The expression "rate of change of a function" means the derivative of the function.

T F **5.** Every function has a derivative at each number in its domain.

T F **6.** If $x^3 - y^3 = 1$, then $\dfrac{dy}{dx} = 3x^2 - 3y^2$.

T F **7.** The derivative of a function is the limit of an average rate of change.

FILL-IN-THE-BLANKS **Answers are on page AN-78.**

1. The derivative of f at c equals the slope of the ________ line to f at c.
2. If $C = C(x)$ denotes the cost C of producing x items, then $C'(x)$ is called the ________ ________.
3. The derivative of $f(x) = (x^2 + 1)^{3/2}$ may be obtained using either the ________ ________ or the ________ ________.
4. The acceleration of an object equals the rate of change of ________ with respect to time.
5. The fifth-order derivative of a polynomial of degree 4 equals ________.
6. The derivative of $x^3 - y^4x + 3y = 5$ is obtained using ________ differentiation.

REVIEW EXERCISES **Answers to odd-numbered problems begin on page AN-78.**

Blue problem numbers indicate the author's suggestions for use in a practice test.

In Problems 1–8, find the derivative of each function at the given number.

1. $f(x) = 2x + 15$ at 2
2. $f(x) = 4x - 6$ at -5
3. $f(x) = x^2 - 5$ at 2
4. $f(x) = 4x^2 + 1$ at -1
5. $f(x) = x^2 - 2x$ at 1
6. $f(x) = 3x^2 + 7x$ at 0
7. $f(x) = e^{3x}$ at 0
8. $f(x) = \ln x$ at 4

In Problems 9–12, find the derivative of f using the difference quotient.

9. $f(x) = 4x + 3$
10. $f(x) = x - 7$
11. $f(x) = 2x^2 + 1$
12. $f(x) = 7 - 3x^2$

In Problems 13–74, find the derivative of each function.

13. $f(x) = x^5$
14. $f(x) = x^3$
15. $f(x) = \dfrac{x^4}{4}$
16. $f(x) = -6x^2$
17. $f(x) = 2x^2 - 3x$
18. $f(x) = 3x^3 + \dfrac{2}{3}x^2 - 5x + 7$
19. $f(x) = 7(x^2 - 4)$
20. $f(x) = \dfrac{5(x + 6)}{7}$
21. $f(x) = 5(x^2 - 3x)(x - 6)$
22. $f(x) = (2x^3 + x)(x^2 - 5)$
23. $f(x) = 12x(8x^3 + 2x^2 - 5x + 2)$
24. $f(x) = \dfrac{6x^4 - 9x^2}{3x^2}$
25. $f(x) = \dfrac{2x + 2}{5x - 3}$
26. $f(x) = \dfrac{7x}{x - 5}$
27. $f(x) = 2x^{-12}$
28. $f(x) = 2x^3 + 5x^{-2}$
29. $f(x) = 2 + \dfrac{3}{x} + \dfrac{4}{x^2}$
30. $f(x) = \dfrac{1}{x} - \dfrac{1}{x^3}$
31. $f(x) = \dfrac{3x - 2}{x + 5}$
32. $f(x) = \dfrac{2x + 3}{x + 2}$
33. $f(x) = (3x^2 - 2x)^5$
34. $f(x) = (x^3 - 1)^3$
35. $f(x) = 7x(x^2 + 2x + 1)^2$
36. $f(x) = x(2x + 5)^2$
37. $f(x) = \left(\dfrac{x + 1}{3x + 2}\right)^2$
38. $f(x) = \left(\dfrac{5x}{x + 1}\right)^3$
39. $f(x) = \dfrac{7}{(x^3 + 4)^2}$
40. $f(x) = \dfrac{3}{(x^2 - 3x)^2}$
41. $f(x) = \left(3x + \dfrac{4}{x}\right)^3$
42. $f(x) = \left(\dfrac{2x^2 + 1}{x}\right)^4$
43. $f(x) = 3e^x + x^2$
44. $f(x) = 1 - e^x$
45. $f(x) = e^{3x+1}$
46. $f(x) = 2e^{x^2}$
47. $f(x) = e^x(2x^2 + 7x)$
48. $f(x) = (x + 1)e^{2x}$
49. $f(x) = \dfrac{1 + x}{e^x}$
50. $f(x) = \dfrac{e^{3x}}{x}$
51. $f(x) = \left(\dfrac{e^x}{3x}\right)^2$

52. $f(x) = (2x\,e^x)^3$

53. $f(x) = \ln(4x)$

54. $f(x) = 3\ln(3x) - 15x$

55. $f(x) = x^2 \ln x$

56. $f(x) = e^x \ln x$

57. $f(x) = \ln(2x^3 + 1)$

58. $f(x) = 7[e^{3x} + \ln(x + 2)]$

59. $f(x) = 2^x + x^2$

60. $f(x) = 10^x(x + 5)$

61. $f(x) = x + \log x$

62. $f(x) = x^4 \log_2 x$

63. $f(x) = \sqrt{x}$

64. $f(x) = \sqrt[3]{x} + 2x$

65. $f(x) = 3x^{5/3} + 5$

66. $f(x) = 2x^{2/3} + x^2$

67. $f(x) = \sqrt{x^2 - 3x}$

68. $f(x) = \sqrt{2x^3 + x^2}$

69. $f(x) = \dfrac{x + 1}{\sqrt{x + 5}}$

70. $f(x) = \dfrac{3x^2}{\sqrt{x^2 - 1}}$

71. $f(x) = (1 + x)\sqrt{e^x}$

72. $f(x) = \ln\sqrt{1 + x}$

73. $f(x) = \sqrt{x}\ln x$

74. $f(x) = \sqrt{e^x} \cdot \ln x^{1/2}$

In Problems 75–80, find the first derivative and the second derivative of each function.

75. $f(x) = x^3 - 8$

76. $f(x) = (5x + 3)^2$

77. $f(x) = e^{-3x}$

78. $f(x) = \ln(x^2)$

79. $f(x) = \dfrac{x}{2x + 1}$

80. $f(x) = e^x \ln x$

In Problems 81–84, find $\dfrac{dy}{dx}$ *by using implicit differentiation.*

81. $xy + 3y^2 = 10x$

82. $y^3 + y = x^2$

83. $xe^y = 4x^2$

84. $\ln(x + y) = 8x$

In Problems 85–87, find the slope of the tangent line to the graph of f at the given point. What is an equation of the tangent line?

85. $f(x) = 2x^2 + 3x - 7$ at $(-1, -8)$

86. $f(x) = \dfrac{x^2 + 1}{x - 1}$ at $(2, 5)$

87. $f(x) = x^2 + e^x$ at $(0, 1)$

88. Slope of a Tangent Line Find the slope of the tangent line at the indicated point:

$$4xy - y^2 = 3 \quad \text{at } (1, 3)$$

Write an equation of the tangent line.

89. For $f(x) = x^3 + 3x$, find:

(a) The average rate of change as x changes from 0 to 2.
(b) The instantaneous rate of change at $x = 2$.

90. For $f(x) = (4x - 5)^2$, find:

(a) The average rate of change as x changes from 2 to 5.
(b) The instantaneous rate of change at $x = 5$.

91. **Falling Rocks** A stone is dropped from a bridge that is 100 feet above the water. Its height h after t seconds is

$$h(t) = -16t^2 + 100$$

(a) How much time elapses before the stone hits the water?
(b) What is the average velocity of the stone during its fall?
(c) What is the instantaneous velocity of the stone as it hits the water?

92. Population Growth A small population is growing at the rate of

$$P(t) = 500\,e^{\frac{\ln 2}{3}t}$$

creatures per year.

(a) What is the average rate of growth of the population from year 0 to year 3?
(b) What is the instantaneous rate of growth of the population at year 3?

93. Throwing a Ball A major league pitcher throws a ball vertically upward with an initial velocity of 128 feet per second (87.3 miles per hour) from a height of 6 feet. The distance between the ball and the ground t seconds after the ball is thrown is given by

$$s(t) = -16t^2 + 128t + 6$$

(a) When does the ball reach its highest point?
(b) What is the maximum height the ball reaches?
(c) What is the total distance the ball travels?
(d) What is the velocity of the ball at any time t?

(e) At what time (if ever) is the velocity of the ball zero? Explain your answer.
(f) For how long is the ball in the air? (Round your answer to the nearest tenth of a second.)
(g) What is the velocity of the ball (rounded to the nearest tenth) when it hits the ground?
(h) What is the acceleration of the ball at any time t?
(i) What is the velocity of the ball when it has been in the air for 2 seconds? What is its velocity when it has been in the air for 6 seconds?
(j) Interpret the meaning of the different signs in the answers to part (i).

94. Throwing Balls on the Moon The same pitcher as in Problem 93 stands on the moon and again throws the ball vertically upward. This time the measurements are made in meters. The distance between the ball and the ground t seconds after the ball is thrown is given by

$$s(t) = -0.8t^2 + 39t + 2$$

(a) When does the ball reach its highest point?
(b) What is the maximum height the ball reaches?
(c) What is the total distance the ball travels?
(d) What is the velocity of the ball at any time t?
(e) At what time (if ever) is the velocity of the ball zero? Explain your answer.
(f) For how long is the ball in the air? (Round your answer to the nearest tenth of a second.)
(g) What is the velocity of the ball (rounded to the nearest tenth) when it hits the ground?
(h) What is the acceleration of the ball at any time t?
(i) What is the velocity of the ball when it has been in the air for 2 seconds? What is its velocity when it has been in the air for 6 seconds?
(j) Interpret the meaning of the different signs in the answers to part (i).

95. Business Analysis The cost function and the demand equation for a certain product are

$$C(x) = 15x + 550 \quad \text{and} \quad p = -0.50x + 75$$

Find:

(a) The revenue function.
(b) The marginal revenue.
(c) The marginal cost.
(d) The break even point(s).
Hint: the numbers for which $R(x) = C(x)$.
(e) The number x for which marginal revenue equals marginal cost.

96. Analyzing Revenue The cost function and the demand equation for a certain product are

$$C(x) = 60x + 7200 \qquad p = -2x + 300$$

Find:

(a) The revenue function.
(b) The marginal revenue.
(c) The marginal cost.
(d) The break even point(s).
Hint: the numbers for which $R(x) = C(x)$.
(e) The number x for which marginal revenue equals marginal cost.

Chapter 13 Project

TESTING BLOOD EFFICIENTLY

Suppose you are in charge of blood testing for a large organization (for example, drug testing for a corporation or disease testing for the armed forces). You have N people you need to test. One way to do this is to test each person individually. Then you must perform N tests, which could be costly.

You hit upon a clever idea: Take a sample of blood from each of x people, mix the samples, and test the mixture. This technique is called **pooling the sample.** If this "pooled" test comes out negative, then you're done with those x people. If the "pooled" test is positive, you would then test each of the x people individually. With this plan, to test x people, you would use either 1 test (potentially saving money in the process) or $x + 1$ tests (with a slight increase in cost from the earlier scheme). Is it worth changing to the new method? Let's try to find out by examining the cost of the new method.

1. Let p be the probability that the test of a single person is positive. Then $q = 1 - p$ is the probability that the test is negative. We assume that these probabilities are the same for all N people in the group to be tested. Let p_- be the probability that the test of the mixed sample of x people is negative and let p_+ be the probability that the test of the mixed sample is positive. Explain why $p_- = q^x$ and $p_+ = 1 - q^x$.

2. The number of tests we will have to do depends on the makeup of our group of x people. However, we can compute the long-term average of the number of tests we will have to do to get a result for each of the x people. Explain why the long-term average number of tests we will do to get a result for each of the x people is

$$1 \cdot p_- + (x+1) \cdot p_+ = x(1-q^x) + 1$$

3. For simplicity assume that x evenly divides N. Then to find a result for all N people, we will have to do $\frac{N}{x}$ groupings. Explain why the long-term average number of tests we will do to get a result for all N people is

$$\frac{N}{x}[x(1-q^x)+1] = N\left(1 - q^x + \frac{1}{x}\right)$$

4. Assume that each test costs K dollars to analyze. Explain why the cost function $C = C(x)$ for using the pooling method is

$$C(x) = KN\left(1 - q^x + \frac{1}{x}\right)$$

5. Suppose that you are testing for the presence of HIV in an adult population in North America. According to the Centers for Disease Control, 0.56% of this population carries HIV so the chance of a positive individual test is 0.0056. Then $q = 1 - 0.0056 = 0.9944$. The cost function is

$$C(x) = KN\left(1 - 0.9944)^x + \frac{1}{x}\right)$$

Show that the marginal cost function is

$$C'(x) = KN\left(-(0.9944)^x \ln 0.9944 - \frac{1}{x^2}\right)$$
$$= KN\left(0.0056(0.9944)^x - \frac{1}{x^2}\right)$$

---

6. We will want to determine whether $C'(x)$ is positive or negative: if $C'(x)$ is negative, then increasing x by 1 person should cause a decrease in the cost, while if $C'(x)$ is positive, then increasing x by 1 person should cause an increase in the cost. Unfortunately, it is difficult to solve these inequalities using algebra. Since K and N will not effect whether $C'(x)$ is positive or negative, we need only consider the values of the expression $.0056(.9944)^x - \frac{1}{x^2}$. The following table gives values for this expression.

x	1	2	3	4	5	6	7	8	9	10
$0.0056(0.9944)^x - \frac{1}{x^2}$	−0.99442	−0.24445	−0.10559	−0.05701	−0.03454	−0.02235	−0.01501	−0.01026	−0.00701	−0.00469

x	11	12	13	14	15	16	17	18	19	20
$0.0056(0.9944)^x - \frac{1}{x^2}$	−0.00299	−0.00169	−0.00070	0.00009	0.00072	0.00123	0.00164	0.00199	0.00228	0.00252

Find the most likely pair of x values for which the cost might be a minimum, and compare the cost at each of these x values. Which value of x makes the cost smaller?

7. How will the value of N change your answer?

8. How many tests do you expect to run?

9. If each test costs five dollars to administer and you are testing 1000 people, how much money do you expect to save by using the pooling procedure instead of testing all 1000 people?

10. Approximately 8.3% of the adult population of sub-Saharan Africa carries HIV. Use an analysis similar to that above to find the pool size x which minimizes the cost of testing N people.

MATHEMATICAL QUESTIONS FROM PROFESSIONAL EXAMS*

1. **Actuary Exam—Part I** $\frac{d}{dx}(x^2e^{x^2}) = ?$

   (a) $2xe^{x^2}$ (b) $\frac{x^3}{3}e^{x^2}$
   (c) $4x^2e^{x^2}$ (d) $2xe^{x^2} + x^4e^{x^2-1}$
   (e) $2x^3e^{x^2} + 2xe^{x^2}$

2. **Actuary Exam—Part I** If $f(x) = |x + 3|$, then $f'(x)$ is continuous for what values of x?

   (a) $x < 3$ (b) $x > 3$ (c) $x \neq 3$
   (d) $x \neq -3$ (e) All x

3. **Actuary Exam—Part I** If $y = be^{c^2+x^2}$, $\frac{dy}{dx} = ?$

   (a) $be^{c^2}2x$ (b) $2xbe^{x^2}$ (c) $b(c^2 + x^2)e^{c^2+x^2-1}$
   (d) $2bxe^{c^2+x^2}$ (e) $be^{c^2+x^2}$

4. **Actuary Exam—Part I** $\frac{d}{dx}(e^{x^2}) = ?$

   (a) $2xe^{x^2}$ (b) $\frac{x^3e^{x^2}}{3}$ (c) $4x^2e^{x^2}$
   (d) $3xe^{x^2} + x^4e^{x^2-1}$ (e) $2xe^{x^2} + 2x^3e^{x^2}$

5. **Actuary Exam—Part I** If $f'(x) = x\ f(x)$ for all real x and $f(-2) = 3$, then $f''(-2) =$

   (a) -18 (b) -1 (c) 1
   (d) 12 (e) 15

6. **Actuary Exam—Part I** A particle moves along the x-axis so that at any time $t \geq 0$ the position of the particle is given by $f(t) = \frac{t^3}{3} + t^2 - 2t - 2$. At what time t will the acceleration and the velocity of the particle be the same?

   (a) $t = 0$ (b) $t = 1$ (c) $t = 2$
   (d) $t = 2\sqrt{3}$ (e) $t = 6$

CHAPTER 14

Applications: Graphing Functions; Optimization

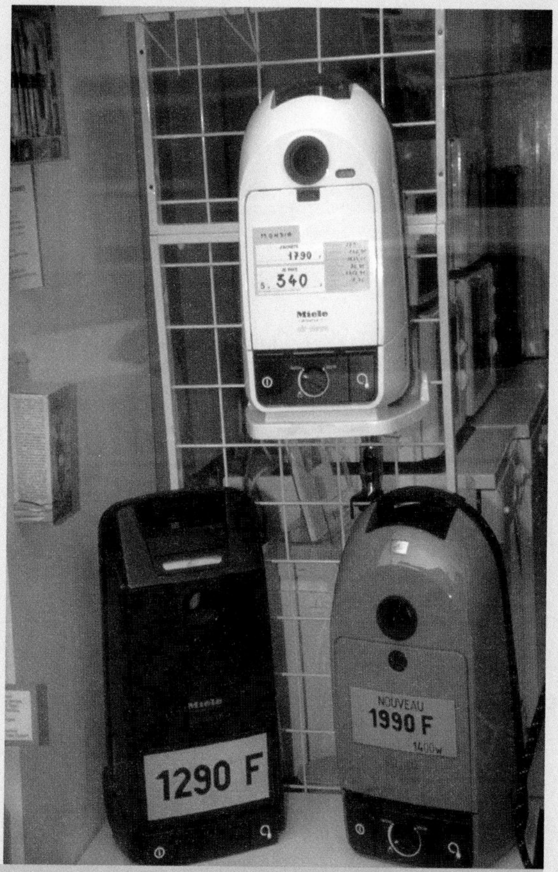

OUTLINE

Most businesses stock items that are either required for the business or are to be sold. This is usually referred to as *inventory*. There is always the issue of how much to have in inventory. The choices range from keeping inventory levels high and not reordering very often to keeping inventory levels low and ordering frequently. There are costs associated with maintaining inventory and there are costs associated with ordering, such as shipping charges. The Chapter Project at the end of this chapter will walk you through an analysis of inventory control and ways of minimizing costs.

A LOOK BACK, A LOOK FORWARD

In Chapter 13 we derived various formulas for finding the derivative of a function. In particular, we now know how to find the derivative of a polynomial function, a rational function, an exponential function, and a logarithm function. We also learned to differentiate implicitly, find the derivative of functions raised to a rational power, and use the Chain Rule. Now we are ready for applications involving the derivative of a function.

We shall use the derivative as a tool for obtaining the graph of a function as well as for finding the maximum and minimum values of a function. Then we shall see how the derivative can be used to solve certain problems in which related variables vary over time. Finally, we shall use the derivative as an approximation tool.

14.1 Horizontal and Vertical Tangent Lines; Continuity and Differentiability

PREPARING FOR THIS SECTION *Before getting started, review the following:*

- > Solving Equations (Appendix A, Section A.6, pp. A-49–A-58)
- > Library of Functions (Chapter 10, Section 10.4, pp. 579–583)
- > Tangent Lines; Derivative of a function at a Number *c* (Chapter 13, Section 13.1, pp. 714–716)
- > Lines (Chapter 1, Section 1.1, pp. 3–15)
- > Continuous Functions (Chapter 12, Section 12.3, pp. 693–695)

OBJECTIVES
1 Find horizontal tangent lines
2 Find vertical tangent lines
3 Discuss the graph of a function *f* where the derivative of *f* does not exist

We have seen that the slope of the tangent line to the graph of a function f equals the derivative. In particular, if the tangent line is horizontal, then its slope is zero, so the derivative of f will also be zero. See Figure 1. The following result is a consequence of this fact.

FIGURE 1

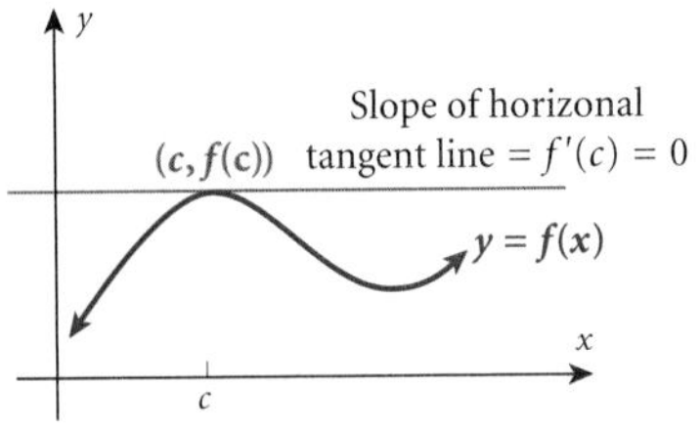

Criterion for a Horizontal Tangent Line

The tangent line to the graph of a differentiable function f at a point $(c, f(c))$ on the graph of f is horizontal if and only if $f'(c) = 0$.

To determine where the tangent line to the graph of $y = f(x)$ is horizontal, we need to solve the equation $f'(x) = 0$.

1 **EXAMPLE 1 Finding Horizontal Tangent Lines**

At what point(s) is the tangent line to the graph of

$$f(x) = x^3 + 3x^2 - 24x$$

horizontal?

SOLUTION We first find the derivative $f'(x)$:

$$f'(x) = 3x^2 + 6x - 24$$

Horizontal tangent lines occur where $f'(x) = 0$, so we solve the equation

$$\begin{aligned} 3x^2 + 6x - 24 &= 0 && f'(x) = 0 \\ 3(x^2 + 2x - 8) &= 0 && \text{Factor.} \\ 3(x + 4)(x - 2) &= 0 && \text{Factor.} \\ x + 4 = 0 \quad \text{or} \quad x - 2 &= 0 && \text{Zero-Product Property.} \\ x = -4 \quad \text{or} \quad x &= 2 && \text{Solve for } x. \end{aligned}$$

Now evaluate the function f at each of these numbers.

$$f(-4) = (-4)^3 + 3(-4)^2 - 24(-4) = 80 \qquad f(x) = x^3 + 3x^2 - 24x$$

and

$$f(2) = 2^3 + 3(2)^2 - 24(2) = -28$$

The tangent line to the graph of f is horizontal at the points $(-4, f(-4)) = (-4, 80)$ and $(2, f(2)) = (2, -28)$ on the graph of f.

Notice that we substitute $x = -4$ and $x = 2$ into the function $y = f(x)$ to find the y-coordinate of the point on the graph of f at which the tangent line is horizontal. Be careful not to substitute these values of x into the derivative $f'(x)$ because this will give the value of the derivative rather than the value of the function.

NOW WORK PROBLEM 3.

Vertical Tangent Lines

Recall that the slope of a vertical line is undefined. Look at Figure 2.

FIGURE 2

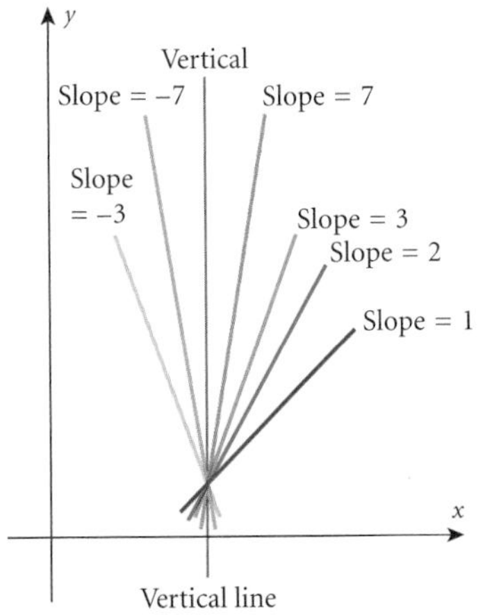

As a line moves toward being vertical, the absolute value of its slope becomes larger and larger, that is, it increases without bound. For vertical lines, the slope is unbounded. Since the slope of the tangent line to the graph of a function equals the derivative of the function, and since vertical lines have unbounded slope, we can formulate the following result about vertical tangent lines:

Conditions for a Vertical Tangent Line

The tangent line to the graph of a continuous function f at a point $(c, f(c))$ on the graph of f is vertical if $f'(x)$ is unbounded at $x = c$.

Two conditions must be present for a vertical tangent line at a point $(c, f(c))$ on the graph of a continuous function f:

1. There must be a point on the graph of f corresponding to $x = c$. That is, $x = c$ must be in the domain of the function f.
2. The derivative $f'(x)$ must be unbounded at $x = c$.

2

EXAMPLE 2 Finding Vertical Tangent Lines

At what point(s) is the tangent line to the graph of $f(x) = \sqrt[3]{x}$ vertical?

SOLUTION The cube root function f can be written as $f(x) = \sqrt[3]{x} = x^{1/3}$. Then, the derivative is

$$f'(x) = \frac{d}{dx}x^{1/3} = \frac{1}{3}x^{-2/3} = \frac{1}{3x^{2/3}} \qquad \frac{d}{dx}x^n = nx^{n-1}$$

The derivative $f'(x)$ is unbounded at $x = 0$. At $x = 0$, the value of f is $f(0) = 0^{1/3} = 0$. Since the conditions for a vertical tangent line have been met, there is a vertical tangent line to the graph of f at the point $(0, 0)$. See Figure 3 for the graph.

FIGURE 3

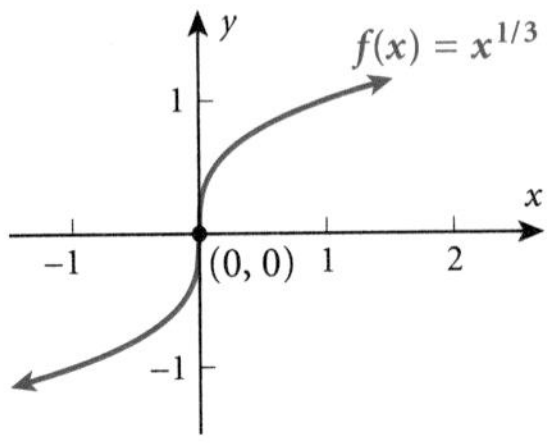

NOW WORK PROBLEM 7.

EXAMPLE 3 Finding the Horizontal and the Vertical Tangent Lines

At what point(s) is the tangent line to the graph of $f(x) = \dfrac{x^{2/3}}{x - 1}$ horizontal? At what point(s) is it vertical?

SOLUTION First, we determine that the domain of f is $\{x \mid x \neq 1\}$. Next we use the formula for the derivative of a quotient to find the derivative $f'(x)$:

$$f(x) = \frac{x^{2/3}}{x - 1}$$

$$f'(x) = \frac{(x-1)\dfrac{d}{dx}x^{2/3} - x^{2/3}\cdot\dfrac{d}{dx}(x-1)}{(x-1)^2} \qquad \text{Formula for the derivative of a quotient.}$$

$$= \frac{(x-1)\cdot\dfrac{2}{3}x^{-1/3} - x^{2/3}\cdot 1}{(x-1)^2} \qquad \text{Differentiate.}$$

$$= \frac{\dfrac{2(x-1)}{3x^{1/3}} - x^{2/3}}{(x-1)^2} \qquad \text{Simplify.}$$

$$= \frac{\dfrac{2(x-1) - 3x}{3x^{1/3}}}{(x-1)^2} \qquad \text{Write the numerator as a single quotient.}$$

$$= \frac{-x - 2}{3x^{1/3}(x-1)^2} \qquad \text{Simplify.}$$

$$= \frac{-(x+2)}{3x^{1/3}(x-1)^2} \qquad \text{Factor out } -1 \text{ in the numerator.}$$

We see that $f'(x) = 0$ if $x = -2$. We also see that $f'(x)$ is unbounded if $x = 0$ or if $x = 1$. However, since $x = 1$ is not in the domain of f, we disregard it.

Since

$$f(-2) = \frac{(-2)^{2/3}}{-2-1} = \frac{1.59}{-3} \approx -0.53 \qquad f(x) = \frac{x^{2/3}}{x-1}$$

and

$$f(0) = \frac{0^{2/3}}{0-1} = 0$$

we conclude that f has a horizontal tangent line at $(-2, -0.53)$ and a vertical tangent line at $(0, 0)$.

CHECK: Graph $y = \dfrac{x^{2/3}}{x-1}$. Can you see where the graph has a horizontal and a vertical tangent line?

Continuity and Differentiability

3 Discuss the graph of a function *f* where the derivative of *f* does not exist

In Chapter 12 we discussed an important property of functions, the property of being continuous, and in Chapter 13 we discussed another important property of functions, the derivative. As it turns out, there is an important result that relates these properties.

> **Continuity and Differentiability**
>
> Suppose c is a number in the domain of f. If f has a derivative at the number c, then f is continuous at c.

To put it another way, if a function f is not continuous at a number c, then it will have no derivative at c.

However, a function can be continuous at c, but not have a derivative at c. Let's look at an example.

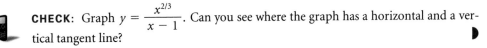

EXAMPLE 4 A Function Continuous at 0, For Which $f'(0)$ Does Not Exist

Discuss the derivative of the absolute value function $f(x) = |x|$ at 0.

SOLUTION The absolute value function $f(x) = |x|$ is in our library of functions. Its graph is given in Figure 4. Notice that the absolute value function is continuous at 0.

FIGURE 4

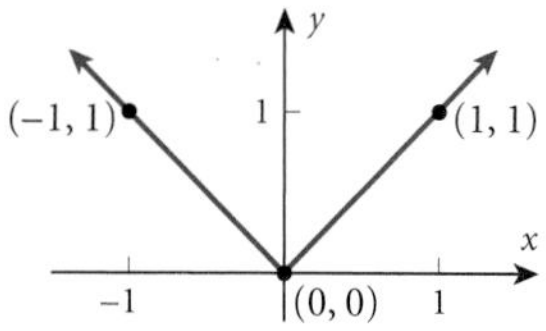

To find the derivative of $f(x) = |x|$ at 0, we use the definition given in Chapter 13, page 716.

$$f'(0) = \lim_{x \to 0} \frac{f(x) - f(0)}{x - 0} = \lim_{x \to 0} \frac{|x| - |0|}{x - 0} = \lim_{x \to 0} \frac{|x|}{x}$$

We examine the one-sided limits. Remember if $x < 0$, then $|x| = -x$ and if $x > 0$, then $|x| = x$.

$$\lim_{x \to 0^-} \frac{|x|}{x} = \lim_{x \to 0^-} \frac{-x}{x} = -1 \quad \text{and} \quad \lim_{x \to 0^+} \frac{|x|}{x} = \lim_{x \to 0^+} \frac{x}{x} = 1$$

Since the one-sided limits are unequal, we conclude that $\lim_{x \to 0} \frac{|x|}{x}$ does not exist and so $f'(0)$ does not exist.

Notice in Figure 4 the sharp point at (0, 0). To the left of (0, 0) the slope of the line is −1 and to the right of (0, 0) it is 1. At (0, 0), there is no tangent line since the derivative $f'(0)$ does not exist.

If f is continuous at c, the derivative at c may or may not exist. Figure 5 shows how the graph of a continuous function might look at points where the derivative does not exist.

FIGURE 5

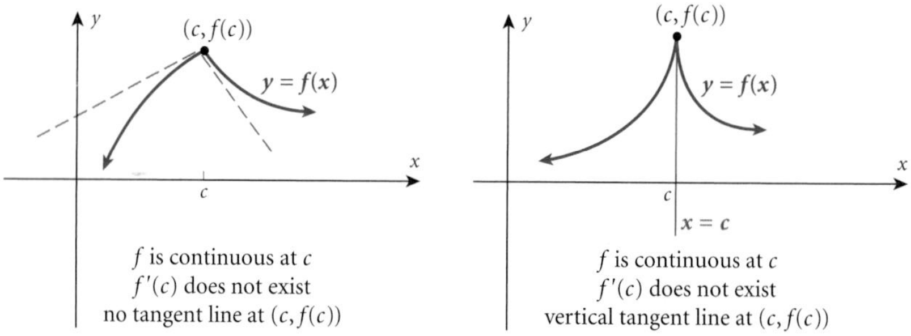
y
(c, f(c))
y = f(x)
x
c
f is continuous at c
f'(c) does not exist
no tangent line at (c, f(c))
y
(c, f(c))
y = f(x)
x
c
x = c
f is continuous at c
f'(c) does not exist
vertical tangent line at (c, f(c))

EXAMPLE 5 Discussing Continuity and Differentiability

For the function $f(x) = \begin{cases} 2x^2 & \text{if } x \le 2 \\ x^3 & \text{if } x > 2 \end{cases}$

(a) Is f continuous at 2?

(b) Does $f'(2)$ exist? If it does, what is its value?

(c) If f is continuous at 2 but $f'(2)$ does not exist, is there a vertical tangent line or no tangent line at 2?

(d) Graph f.

SOLUTION **(a)** $f(2) = 2 \cdot 2^2 = 8$. The one-sided limits are

$$\lim_{x \to 2^-} f(x) = \lim_{x \to 2^-} (2x^2) = 8, \qquad \lim_{x \to 2^+} f(x) = \lim_{x \to 2^+} x^3 = 8$$

Since $\lim_{x \to 2} f(x) = 8$ and $f(2) = 8$, the function f is continuous at 2.

(b) The derivative of f at 2 is $f'(2) = \lim_{x \to 2} \frac{f(x) - f(2)}{x - 2} = \lim_{x \to 2} \frac{f(x) - 8}{x - 2}$.

We look at the one-sided limits:

$$\lim_{x \to 2^-} \frac{f(x) - 8}{x - 2} = \lim_{x \to 2^-} \frac{2x^2 - 8}{x - 2} = \lim_{x \to 2^-} \left[2\frac{x^2 - 4}{x - 2}\right] = 2 \lim_{x \to 2^-} \frac{(x - 2)(x + 2)}{x - 2} = 2 \lim_{x \to 2^-} (x + 2) = 2 \cdot 4 = 8$$

$$\lim_{x \to 2^+} \frac{f(x) - 8}{x - 2} = \lim_{x \to 2^+} \frac{x^3 - 8}{x - 2} = \lim_{x \to 2^+} \frac{(x - 2)(x^2 + 2x + 4)}{x - 2} = \lim_{x \to 2^+} (x^2 + 2x + 4) = 12$$

We conclude that $f'(2)$ does not exist.

(c) Since the one-sided limits in (b) are unequal, there is no tangent line at the point (2, 8).

(d) See Figure 6 for the graph of f.

FIGURE 6

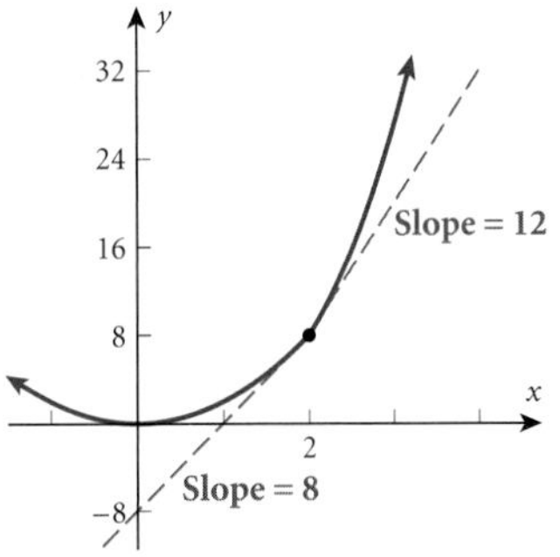

NOW WORK PROBLEM 31.

EXERCISE 14.1 Answers to Odd-Numbered Problems Begin on Page AN-79.

In Problems 1–26, find any points at which the graph of f has either a horizontal or a vertical tangent line.

1. $f(x) = x^2 - 4x$
2. $f(x) = x^2 + 2x$
3. $f(x) = -x^2 + 8x$
4. $f(x) = -x^2 - 12x + 1$
5. $f(x) = -2x^2 + 8x + 1$
6. $f(x) = -3x^2 + 12x$
7. $f(x) = 3x^{2/3} + 1$
8. $f(x) = 3x^{1/3} - 1$
9. $f(x) = -x^3 + 3x + 1$
10. $f(x) = -2x^3 + 6x^2 + 1$
11. $f(x) = 4x^{3/4} - 2$
12. $f(x) = 8x^{1/4} - 1$
13. $f(x) = x^5 - 10x^4$
14. $f(x) = x^5 + 5x$
15. $f(x) = 3x^5 + 20x^3 - 1$
16. $f(x) = 3x^5 - 5x^3 - 1$
17. $f(x) = x^{2/3} + 2x^{1/3}$
18 $f(x) = x^{2/3} - 2x^{1/3}$
19. $f(x) = x^{2/3}(x - 10)$
20. $f(x) = x^{2/3}(x - 15)$
21. $f(x) = x^{2/3}(x^2 - 16)$
22. $f(x) = x^{2/3}(x^2 - 4)$
23. $f(x) = \frac{x^{2/3}}{x - 2}$
24. $f(x) = \frac{x^{2/3}}{x + 1}$
25. $f(x) = \frac{x^{1/3}}{x - 1}$
26. $f(x) = \frac{x^{1/3}}{x + 1}$

In Problems 27–36, answer the following questions about the function f at c.

(a) *Is f continuous at c?*
(b) *Does f′(c) exist? If it does, what is its value?*
(c) *If f is continuous at c but f′(c) does not exist, is there a vertical tangent line or no tangent line at c?*
(d) *For Problems 31–36, graph f.*

27. $f(x) = x^{2/3}$ at 0
28. $f(x) = x^{2/5}$ at 0
29. $f(x) = \frac{x}{x - 1}$ at 1
30. $f(x) = \frac{x^2}{x - 4}$ at 4
31. $f(x) = \begin{cases} 3x & \text{if } x < 0 \\ x^2 & \text{if } x \geq 0 \end{cases}$ at 0
32. $f(x) = \begin{cases} 6x & \text{if } x < 1 \\ x^2 + 5 & \text{if } x \geq 1 \end{cases}$ at 1
33. $f(x) = \begin{cases} 4x & \text{if } x \leq 2 \\ x^2 & \text{if } x > 2 \end{cases}$ at 2
34. $f(x) = \begin{cases} -2x^2 & \text{if } x < 0 \\ x + 1 & \text{if } x \geq 0 \end{cases}$ at 0
35. $f(x) = \begin{cases} x^2 & \text{if } x < 0 \\ x^3 & \text{if } x \geq 0 \end{cases}$ at 0
36. $f(x) = \begin{cases} x^2 & \text{if } x < 0 \\ x^4 & \text{if } x \geq 0 \end{cases}$ at 0

14.2 Increasing and Decreasing Functions; the First Derivative Test

PREPARING FOR THIS SECTION *Before getting started, review the following:*

- Increasing and Decreasing Functions (Chapter 10, Section 10.3, pp. 567–568)
- Solving Inequalities (Appendix A, Section A.7, pp. A-60–A-70)
- Local Maxima and Local Minima (Chapter 10, Section 10.3, pp. 568–570)
- End Behavior; Asymptotes (Chapter 12, Section 12.4, pp. 700–705)

OBJECTIVES
1 Determine where a function is increasing and where it is decreasing
2 Use the first derivative test
3 Graph functions

Consider the graph of the function $y = f(x)$ given in Figure 7. The function is increasing on the intervals (a, b) and (c, d) while the function is decreasing on the intervals (b, c) and (d, e). Notice also that f has a local maximum at b and at d and a local minimum at c.

FIGURE 7

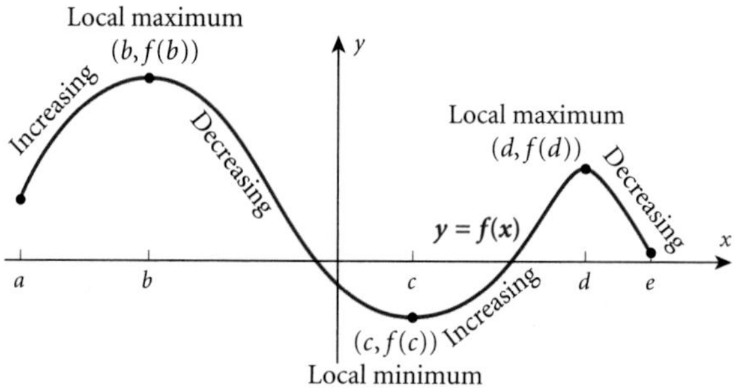

NOW WORK PROBLEMS 3 AND 7.

In this section we will give a test for differentiable functions that provides a straightforward way to determine the intervals on which the function is increasing or decreasing. This, in turn, will give us a way to locate the local maxima and the local minima.

Look at the graph of the function f given in Figure 8(a). On the interval (a, b), where f is increasing, we have drawn several tangent lines. Notice that each tangent line has a positive slope. This is characteristic of tangent lines of increasing functions. Since the derivative of f equals the slope of the tangent line, it follows that wherever the derivative is positive, the function f will be increasing.

Similarly, as Figure 8(b) illustrates, the tangent lines of a decreasing function have negative slope. Whenever the derivative of f is negative, the function f is decreasing. This leads us to the following test for an increasing or decreasing function.

FIGURE 8

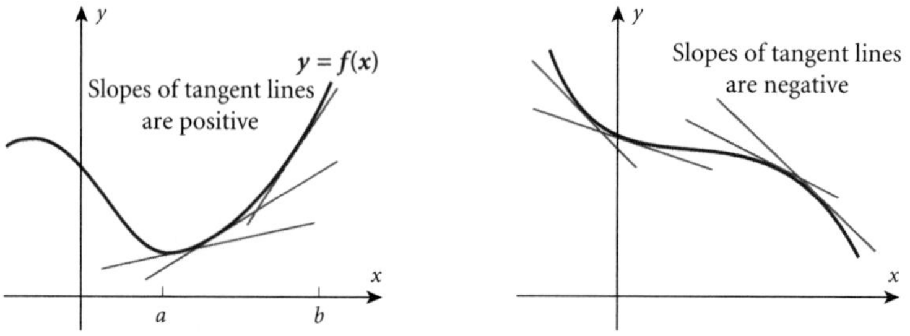

(a) (b)

Test for an Increasing or Decreasing Function

A differentiable function $y = f(x)$ is:

(a) Increasing on the interval (a, b) if $f'(x) > 0$ for all x in the interval (a, b).
(b) Decreasing on the interval (c, d) if $f'(x) < 0$ for all x in the interval (c, d).

To determine where the graph of a function is increasing or where it is decreasing, follow these steps:

Steps for Finding Where a Function Is Increasing and Where It Is Decreasing

STEP 1 Find the derivative $f'(x)$ of the function $y = f(x)$.
STEP 2 Set up a table that solves the two inequalities:

$$f'(x) > 0 \quad \text{and} \quad f'(x) < 0$$

1 **EXAMPLE 1**

Determining Where a Function Is Increasing and Where It Is Decreasing

Determine where the function

$$f(x) = x^3 - 6x^2 + 9x - 2$$

is increasing and where it is decreasing.

SOLUTION We follow the steps given above.

STEP 1 $f'(x) = 3x^2 - 12x + 9$
STEP 2 To set up the table, we first solve the equation $f'(x) = 0$.

$$f'(x) = 3x^2 - 12x + 9 = 3(x^2 - 4x + 3) = 3(x - 1)(x - 3)$$

The solutions of the equation $f'(x) = 3(x - 1)(x - 3) = 0$ are 1 and 3. These numbers separate the real number line into three parts: $-\infty < x < 1$, $1 < x < 3$, and $3 < x < \infty$. We construct Figure 9, using 0, 2, and 4 as test numbers to find the sign of $f'(x)$ on each interval.

FIGURE 9

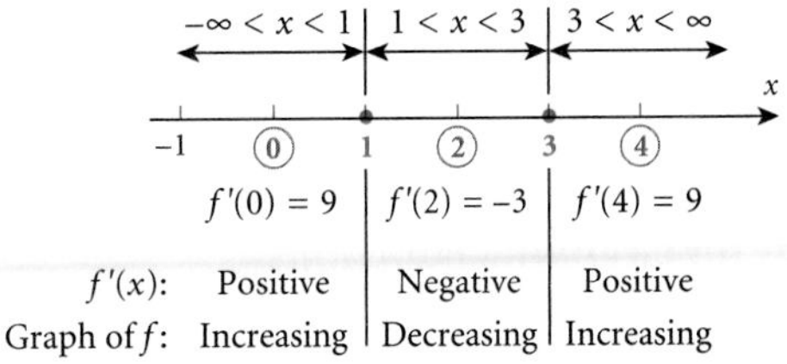

We conclude that the function f is increasing on the intervals $(-\infty, 1)$ and $(3, \infty)$ and is decreasing on the interval $(1, 3)$.

SEEING THE CONCEPT: Graph $y = x^3 - 6x^2 + 9x - 2$ and use TRACE to determine where f is increasing and where f is decreasing.

Local Maximum and Local Minimum

We have already observed that if a function f is increasing to the left of a point A on the graph of f and is decreasing to the right of A, then at A there is a local maximum since the graph of f is higher at A than at nearby points. Similarly, if f is decreasing to the left of a point B on the graph of f and is increasing to the right of B, then at B there is a local minimum. Since the first derivative of a function supplies information about where the function is increasing or decreasing, we call the test for locating local maxima and local minima the **First Derivative Test.**

First Derivative Test

Let f denote a differentiable function. Find the derivative of f and set up a table to determine where f is increasing and where it is decreasing.

1. If f is increasing to the left of a point A on the graph of f and is decreasing to the right of A, then at the point A there is a local maximum.
2. If f is decreasing to the left of a point B on the graph of f and is increasing to the right of B, then at the point B there is a local minimum.

2 **EXAMPLE 2** **Using the First Derivative Test**

Use the First Derivative Test to locate the local maxima and local minima, if any, of

$$f(x) = x^3 - 6x^2 + 9x - 2$$

SOLUTION This is the same function discussed in Example 1. If we refer back to Figure 9, we see that f is increasing for $-\infty < x < 1$ and is decreasing for $1 < x < 3$. When $x = 1$, we have $y = f(1) = 2$. By the First Derivative Test, f has a local maximum at the point $(1, 2)$.

Similarly, f is decreasing for $1 < x < 3$ and is increasing for $3 < x < \infty$. When $x = 3$, we have $y = f(3) = -2$. By the First Derivative Test, f has a local minimum at the point $(3, -2)$.

SEEING THE CONCEPT: Graph $y = x^3 - 6x^2 + 9x - 2$ and use MAXIMUM/MINIMUM to find the local maximum and the local minimum.

Graphing Functions

Graph functions 3

We can use the First Derivative Test to graph functions. In graphing a function $y = f(x)$, we follow these steps:

Steps for Graphing Functions

STEP 1 Find the domain of f.

STEP 2 Locate the intercepts of f (skip the x-intercepts if they are too hard to find).

STEP 3 Determine where the graph of f is increasing and where it is decreasing.
STEP 4 Find any local maxima or local minima of f by using the First Derivative Test.
STEP 5 Locate all points on the graph of f at which the tangent line is either horizontal or vertical.
STEP 6 Determine the end behavior and locate any asymptotes.

EXAMPLE 3 Graphing a Function

Graph the function

$$f(x) = x^3 - 12x$$

SOLUTION **STEP 1** The domain of f is all real numbers.
STEP 2 Let $x = 0$. Then $y = f(0) = 0$. The y-intercept is $(0, 0)$.
To find the x-intercepts, if any, let $y = 0$. Then

$$x^3 - 12x = 0 \qquad y = f(x) = 0$$
$$x(x^2 - 12) = 0 \qquad \text{Factor.}$$
$$x(x + 2\sqrt{3})(x - 2\sqrt{3}) = 0 \qquad \text{Factor.}$$
$$x = 0 \text{ or } x + 2\sqrt{3} = 0 \text{ or } x - 2\sqrt{3} = 0 \qquad \text{Apply the Zero-Product Property.}$$
$$x = 0 \text{ or } \quad x = -2\sqrt{3} \text{ or } \quad x = 2\sqrt{3} \qquad \text{Solve for } x.$$

The x-intercepts are $(0, 0)$, $(-2\sqrt{3}, 0)$, and $(2\sqrt{3}, 0)$.
STEP 3 To determine where the graph of f is increasing and where it is decreasing, we find $f'(x)$.

$$f(x) = x^3 - 12x$$
$$f'(x) = 3x^2 - 12 \underset{\uparrow \atop \text{Factor}}{=} 3(x^2 - 4) \underset{\uparrow \atop \text{Factor}}{=} 3(x + 2)(x - 2)$$

The solutions of the equation $f'(x) = 3(x + 2)(x - 2) = 0$ are -2 and 2. These numbers separate the number line into three parts:

$$-\infty < x < -2 \qquad -2 < x < 2 \qquad 2 < x < \infty$$

We construct Figure 10, using -3, 0, and 3 as test numbers for $f'(x)$.

FIGURE 10

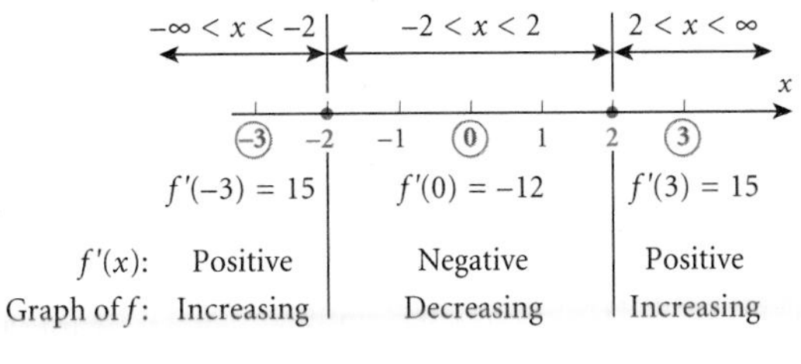

We conclude that the graph of f is increasing on the intervals $(-\infty, -2)$ and $(2, \infty)$. It is decreasing on the interval $(-2, 2)$.

STEP 4 From Figure 10, the graph of f is increasing for $-\infty < x < -2$ and is decreasing for $-2 < x < 2$. At $x = -2$, the graph changes from increasing to decreasing. Consequently, at the point $(-2, f(-2)) = (-2, 16)$ there is a local maximum. Similarly at $x = 2$, the graph changes from decreasing to increasing. At the point $(2, f(2)) = (2, -16)$ there is a local minimum.

STEP 5 The derivative of f is $f'(x) = 3x^2 - 12 = 3(x + 2)(x - 2)$. We see that $f'(x) = 0$ if $x = -2$ or if $x = 2$. The graph of f has a horizontal tangent line at the points $(-2, f(-2)) = (-2, 16)$ and $(2, f(2)) = (2, -16)$. There are no vertical tangent lines.

STEP 6 Since f is a polynomial function, its end behavior is like that of the power function $y = x^3$. Polynomial functions have no asymptotes.

To graph f we plot the intercepts, the local maximum, the local minimum, the points at which the tangent line is horizontal, and connect these points with a smooth curve. See Figure 11.

FIGURE 11

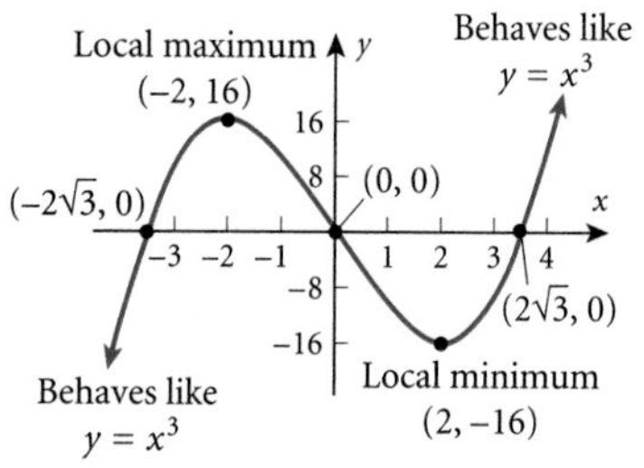

CHECK: Use a graphing utility to graph $f(x) = x^3 - 12x$. Be sure to choose a viewing rectangle that will show a complete graph. [Use Figure 11 as a guide.] Use TRACE to confirm where f is increasing and where f is decreasing. Use MAXIMUM and MINIMUM to confirm the local maximum and the local minimum. Use ZERO (or ROOT) to locate the x-intercepts. Use a TABLE to confirm the end behavior.

NOW WORK PROBLEM 9.

EXAMPLE 4 Graphing a Function

Graph the function

$$f(x) = x^4 - 4x^3$$

SOLUTION **STEP 1** f is a polynomial function so the domain of f is all real numbers.

STEP 2 Let $x = 0$. Then $y = f(0) = 0$. The y-intercept is $(0, 0)$.

To find the x-intercepts, if any, let $y = 0$. Then

$$x^4 - 4x^3 = 0 \qquad y = f(x) = 0$$
$$x^3(x - 4) = 0 \qquad \text{Factor.}$$
$$x^3 = 0 \quad \text{or} \quad x - 4 = 0 \qquad \text{Apply the Zero-Product Property.}$$
$$x = 0 \quad \text{or} \quad x = 4 \qquad \text{Solve for } x.$$

The x-intercepts are $(0, 0)$ and $(4, 0)$.

STEP 3 To determine where the graph of f is increasing and where it is decreasing, we find $f'(x)$:

$$f(x) = x^4 - 4x^3$$
$$f'(x) = 4x^3 - 12x^2 = 4x^2(x - 3)$$

The solutions of the equation $f'(x) = 4x^2(x-3) = 0$ are $x = 0$ and $x = 3$. These numbers separate the number line into three parts:

$$-\infty < x < 0 \qquad 0 < x < 3 \qquad 3 < x < \infty$$

We construct Figure 12, using -1, 1 and 4 as test numbers for $f'(x)$.

FIGURE 12

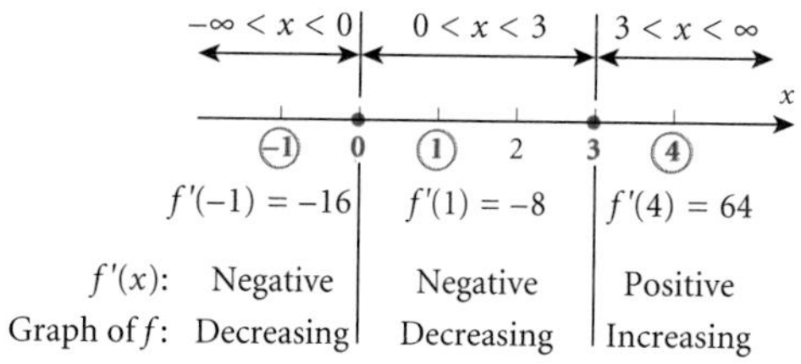

We conclude that the graph of f is decreasing on the intervals $(-\infty, 0)$ and $(0, 3)$. It is increasing on the interval $(3, \infty)$.

STEP 4 From Figure 12, the graph of f is decreasing for $-\infty < x < 3$ and is increasing for $3 < x < \infty$. At the point $(3, f(3)) = (3, -27)$, the graph changes from decreasing to increasing. Consequently, at the point $(3, -27)$ there is a local minimum.

STEP 5 The derivative of f is $f'(x) = 4x^3 - 12x^2 = 4x^2(x-3)$. We see that $f'(x) = 0$ if $x = 0$ or if $x = 3$. The graph of f has a horizontal tangent line at the points $(0, f(0)) = (0, 0)$ and $(3, f(3)) = (3, -27)$. There are no vertical tangent lines.

STEP 6 Since f is a polynomial function, its end behavior is like that of the power function $y = x^4$. Polynomial functions have no asymptotes.

To graph f, we plot the intercepts, the local minimum, the point $(0, 0)$, and the points at which the tangent line is horizontal, and connect these points with a smooth curve. See Figure 13.

FIGURE 13

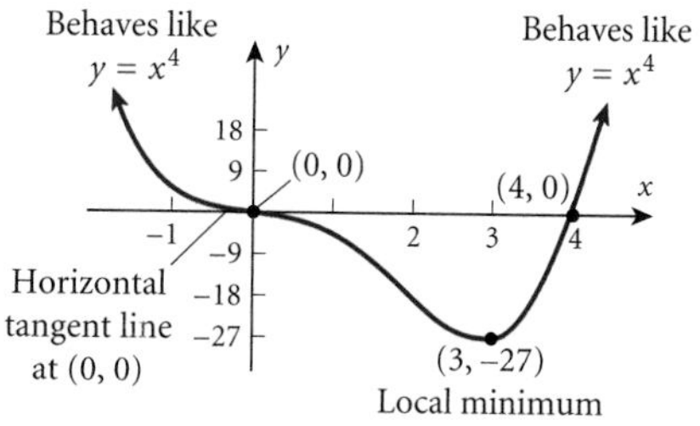

CHECK: Use a graphing utility to graph $f(x) = x^4 - 4x^3$. Be sure to choose a viewing rectangle that will show a complete graph. [Use Figure 13 as a guide.] Use TRACE to confirm where f is increasing and where f is decreasing. Use MINIMUM to confirm the local minimum. Use ZERO (or ROOT) to locate the x-intercepts. Use a TABLE to confirm the end behavior.

NOW WORK PROBLEM 19.

EXAMPLE 5 Graphing a Function

Graph the function

$$f(x) = 2x^{5/3} - 5x^{2/3}$$

SOLUTION **STEP 1** The domain of f is all real numbers.

STEP 2 Let $x = 0$. Then $y = f(0) = 0$. The y-intercept is $(0, 0)$.

To find the x-intercepts, if any, let $y = 0$. Then

$$2x^{5/3} - 5x^{2/3} = 0 \qquad y = f(x) = 0$$

$$x^{2/3}(2x - 5) = 0 \qquad \text{Factor out } x^{2/3}.$$

$$x^{2/3} = 0 \quad \text{or} \quad 2x - 5 = 0 \qquad \text{Apply the Zero-Product Property.}$$

$$x = 0 \quad \text{or} \quad x = \frac{5}{2} \qquad \text{Solve for } x.$$

The x-intercepts are $(0, 0)$ and $\left(\frac{5}{2}, 0\right)$.

STEP 3 To determine where the graph of f is increasing and where it is decreasing, we find $f'(x)$:

$$f(x) = 2x^{5/3} - 5x^{2/3}$$

$$f'(x) = 2 \cdot \frac{5}{3}x^{2/3} - 5 \cdot \frac{2}{3}x^{-1/3} = \frac{10x^{2/3}}{3} - \frac{10}{3x^{1/3}}$$

$$= \frac{10x - 10}{3x^{1/3}} = \frac{10(x - 1)}{3x^{1/3}} \qquad \text{Write with a common denominator.}$$

The factors that appear in the numerator and the denominator of $f'(x)$, $x - 1$ and $x^{1/3}$, equal 0 when $x = 1$ or $x = 0$. We use the numbers 0 and 1 to separate the number line into three parts:

$$-\infty < x < 0 \qquad 0 < x < 1 \qquad 1 < x < \infty$$

We construct Figure 14, using $-1, \frac{1}{2}$, and 2 as test numbers for $f'(x)$.

FIGURE 14

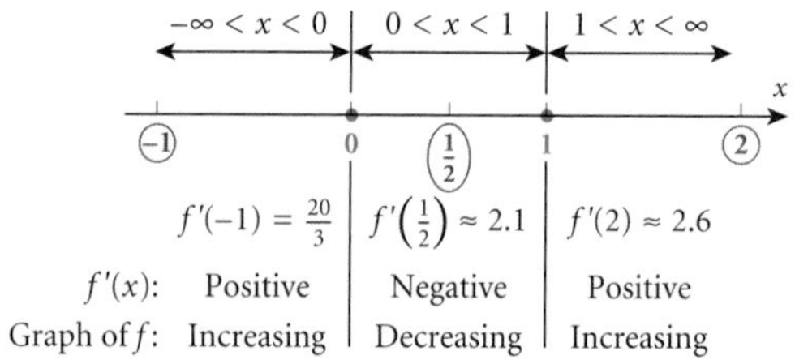

We conclude that the graph of f is increasing on the intervals $(-\infty, 0)$ and $(1, \infty)$; f is decreasing on the interval $(0, 1)$.

STEP 4 From Figure 14, we conclude that the graph of f is increasing for $-\infty < x < 0$ and is decreasing for $0 < x < 1$. At the point $(0, f(0)) = (0, 0)$, the graph changes from increasing to decreasing. Consequently, at the point $(0, 0)$ there is a local maximum.

Similarly, for $0 < x < 1$, the graph of f is decreasing and for $1 < x < \infty$, the graph of f is increasing. At the point $(1, f(1)) = (1, -3)$ there is a local minimum.

STEP 5 The derivative of f is $f'(x) = \dfrac{10(x-1)}{3x^{1/3}}$. We see that $f'(x) = 0$ if $x = 1$. The graph of f has a horizontal tangent line at the point $(1, f(1)) = (1, -3)$. Also, $f'(x)$ is unbounded if $x = 0$. So there is a vertical tangent line at the point $(0, f(0)) = (0, 0)$.

STEP 6 For the end behavior of f, we look at the two limits at infinity:

$$\lim_{x \to -\infty} f(x) = \lim_{x \to -\infty} (2x^{5/3} - 5x^{2/3}) = \lim_{x \to -\infty} (2x^{5/3}) = -\infty$$

$$\lim_{x \to \infty} f(x) = \lim_{x \to \infty} (2x^{5/3} - 5x^{2/3}) = \lim_{x \to \infty} (2x^{5/3}) = \infty$$

The graph of f becomes unbounded in the negative direction as $x \to -\infty$ and unbounded in the positive direction as $x \to \infty$. There are no asymptotes.

To graph f, we plot the intercepts, the local maximum, the local minimum, and the points at which the tangent line is horizontal and vertical, and connect these points. See Figure 15. Notice how the vertical tangent line and local maximum at (0, 0) are shown in the graph.

FIGURE 15

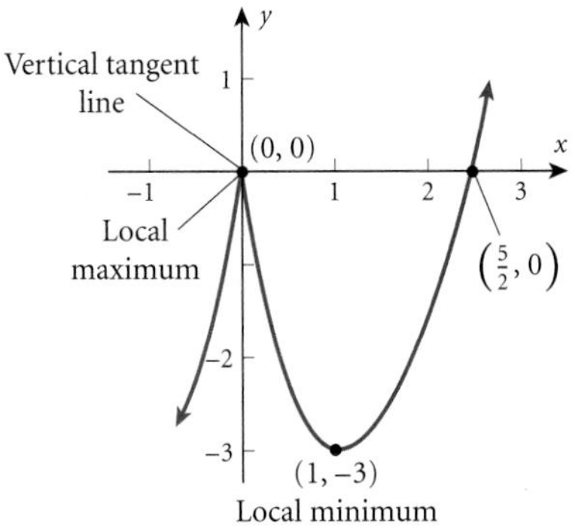

CHECK: Use a graphing utility to graph $f(x) = 2x^{5/3} - 5x^{2/3}$. Be sure to choose a viewing rectangle that will show a complete graph. [Use Figure 15 as a guide.] Use TRACE to confirm where f is increasing and where f is decreasing. Use MAXIMUM and MINIMUM to confirm the local maximum and the local minimum. Use ZERO (or ROOT) to locate the x-intercepts. Use a TABLE to confirm the end behavior.

NOW WORK PROBLEM 23.

EXAMPLE 6 Graphing a Function

Graph: $f(x) = \dfrac{x^2}{x^2 - 1}$

SOLUTION **STEP 1** The domain of f is $\{x | x \neq -1, x \neq 1\}$.

STEP 2 Let $x = 0$. Then $y = f(0) = 0$. The y-intercept is $(0, 0)$. Now let $y = 0$. Then $\dfrac{x^2}{(x^2 - 1)} = 0$, so $x = 0$. The x-intercept is also $(0, 0)$.

STEP 3 To find where the graph is increasing or decreasing, we find $f'(x)$:

$$f(x) = \frac{x^2}{x^2 - 1}$$

$$f'(x) = \frac{(x^2 - 1)\dfrac{d}{dx}x^2 - x^2\dfrac{d}{dx}(x^2 - 1)}{(x^2 - 1)^2} \qquad \text{Formula for the derivative of a quotient.}$$

$$= \frac{(x^2 - 1)(2x) - x^2(2x)}{(x^2 - 1)^2} \qquad \text{Differentiate.}$$

$$= \frac{-2x}{(x^2 - 1)^2} \qquad \text{Simplify.}$$

The numerator is zero for $x = 0$ and the denominator is zero for $x = -1$ and $x = 1$ so we use the numbers $-1, 0, 1$ to separate the number line into four parts:

$$-\infty < x < -1 \quad -1 < x < 0 \quad 0 < x < 1 \quad 1 < x < \infty$$

We construct Figure 16 using $-2, -\dfrac{1}{2}, \dfrac{1}{2}$, and 2 as test numbers for $f'(x)$.

FIGURE 16

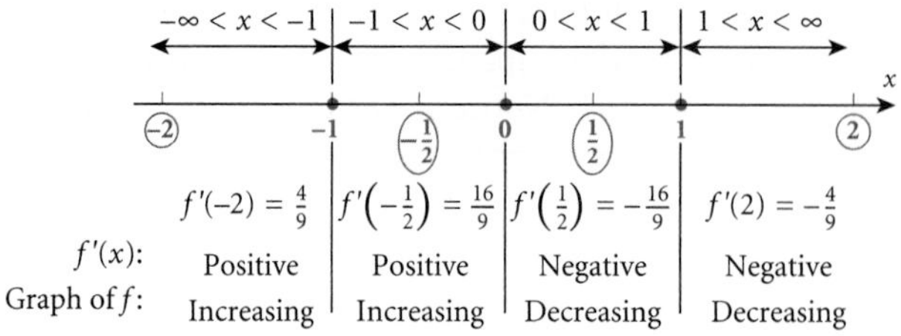

STEP 4 From Figure 16, we conclude the graph of f is increasing for $-\infty < x < -1$ and for $-1 < x < 0$. It is decreasing for $0 < x < 1$ and for $1 < x < \infty$. At the point $(0, f(0)) = (0, 0)$ there is a local maximum.

STEP 5 Since $f'(x) = \dfrac{-2x}{(x^2-1)}$, we conclude that $f'(0) = 0$ so the graph of f has a horizontal tangent line at $(0, 0)$. Since $x = -1$ and $x = 1$ are not in the domain of f, there are no vertical tangent lines at $x = -1$ or $x = 1$, even though $f'(x)$ is unbounded at $x = -1$ and at $x = 1$.

STEP 6 For the end behavior of f, we look at the limits at infinity. Since f is an even function $[f(-x) = f(x)]$, the graph is symmetric with respect to the y-axis. So we will only look at the limit as $x \to \infty$.

$$\lim_{x\to\infty} f(x) = \lim_{x\to\infty} \frac{x^2}{x^2-1} = \lim_{x\to\infty} \frac{x^2}{x^2} = 1$$

The line $y = 1$ is a horizontal asymptote to the graph of f as $x \to -\infty$ and as $x \to \infty$.

Since f is a rational function and f becomes unbounded for $x = -1$ and $x = 1$, the graph of f will have vertical asymptotes at $x = -1$ and at $x = 1$.

Further,

$$\lim_{x\to 1^-} f(x) = \lim_{x\to 1^-} \frac{x^2}{x^2-1} = \lim_{x\to 1^-} \frac{x^2}{x+1} \cdot \lim_{x\to 1^-} \frac{1}{x-1} = \frac{1}{2} \cdot \lim_{x\to 1^-} \frac{1}{x-1} = -\infty$$

$$\lim_{x\to 1^+} f(x) = \lim_{x\to 1^+} \frac{x^2}{x^2-1} = \lim_{x\to 1^+} \frac{x^2}{x+1} \cdot \lim_{x\to 1^+} \frac{1}{x-1} = \frac{1}{2} \cdot \lim_{x\to 1^+} \frac{1}{x-1} = \infty$$

Figure 17 shows the graph of f. The graph to the left of $(0, 0)$ is obtained using symmetry.

FIGURE 17

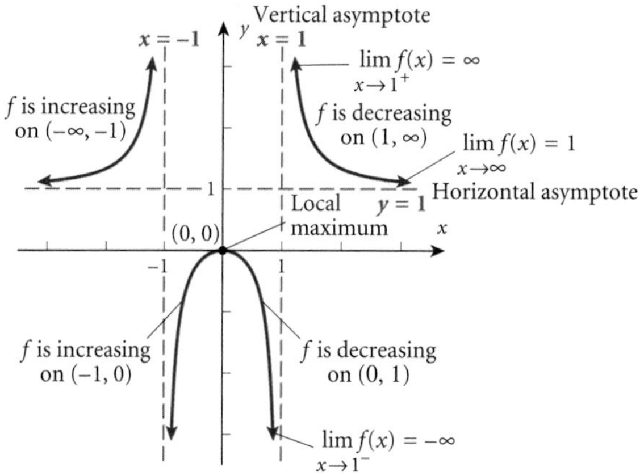

CHECK: Use a graphing utility to graph $f(x) = \dfrac{x^2}{x^2-1}$. Be sure to choose a viewing rectangle that will show a complete graph. [Use Figure 17 as a guide.] Use TRACE to confirm where f is increasing and where f is decreasing. Use MAXIMUM and MINIMUM to confirm the local maximum and the local minimum. Use ZERO (or ROOT) to locate the x-intercept. Use a TABLE to confirm the end behavior.

SUMMARY The box below summarizes what the first derivative tells us about the graph of a function.

Derivative of f	Graph of f
$f'(c) = 0$	Horizontal tangent line at the point $(c, f(c))$
$f'(c)$ is unbounded	Vertical tangent line at the point $(c, f(c))$
$f'(x) > 0$ for $a < x < b$	Increasing on the interval (a, b)
$f'(x) < 0$ for $a < x < b$	Decreasing on the interval (a, b)

EXERCISE 14.2 Answers to Odd-Numbered Problems Begin on Page AN-79.

In Problems 1–8, use the graph of $y = f(x)$ given on the right.

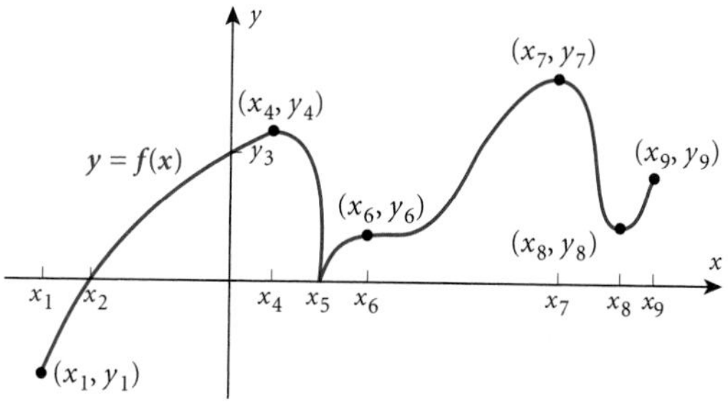

1. What is the domain of f?
2. List the intercepts of f.
3. On what intervals, if any, is the graph of f increasing?
4. On what intervals, if any, is the graph of f decreasing?
5. For what values of x does $f'(x) = 0$?
6. For what values of x is $f'(x)$ not defined?
7. List the point(s) at which f has a local maximum.
8. List the point(s) at which f has a local minimum.

In Problems 9–34, follow the six steps on pages 796 and 797 to graph f.

9. $f(x) = -2x^2 + 4x - 2$
10. $f(x) = -3x^2 + 12x$
11. $f(x) = x^3 - 9x^2 + 27x - 27$
12. $f(x) = x^3 - 6x^2 + 12x - 8$
13. $f(x) = 2x^3 - 15x^2 + 36x$
14. $f(x) = 2x^3 + 6x^2 + 6x$
15. $f(x) = -x^3 + 3x - 1$
16. $f(x) = -2x^3 + 6x^2 + 1$
17. $f(x) = 3x^4 - 12x^3 + 2$
18. $f(x) = x^4 - 4x + 2$
19. $f(x) = x^5 - 5x + 1$
20. $f(x) = x^5 + 5x^4 + 1$
21. $f(x) = 3x^5 - 20x^3 + 1$
22. $f(x) = 3x^5 - 5x^3 - 1$
23. $f(x) = x^{2/3} + 2x^{1/3}$
24. $f(x) = x^{2/3} - 2x^{1/3}$
25. $f(x) = (x^2 - 1)^{2/3}$
26. $f(x) = (x^2 - 1)^{4/3}$
27. $f(x) = \dfrac{8}{x^2 - 16}$
28. $f(x) = \dfrac{2}{x^2 - 4}$
29. $f(x) = \dfrac{x}{x^2 - 9}$
30. $f(x) = \dfrac{x - 1}{x^2}$
31. $f(x) = \dfrac{x^2}{x^2 - 4}$
32. $f(x) = \dfrac{x^2}{x^2 - 1}$
33. $f(x) = x \ln x$
34. $f(x) = \dfrac{\ln x}{x}$

35. **Ticket Sales** The cumulative ticket sales for the 10 days preceding a popular concert is given by

$$S(x) = 4x^2 + 50x + 5000$$

where x represents the 10 days leading up to the concert and $1 \le x \le 10$. Show that S is an increasing function.

36. **Production Cost** The cost C per day, in dollars, of producing x pairs of eyeglasses is

$$C(x) = 0.2x^2 + 3x + 1000$$

Show that C is an increasing function.

37. Sales of Toy Trucks At Dan's Toy Store, the revenue R, in dollars, derived from selling x electric trucks is

$$R(x) = -0.005x^2 + 20x$$

(a) Determine where the graph of R is increasing and where it is decreasing.
(b) How many trucks need to be sold to maximize revenue?
(c) What is the maximum revenue?
(d) Graph the function R.

38. Sales of Calculators. The weekly revenue R, in dollars, from selling x calculators is

$$R(x) = -20x^2 + 1000x$$

(a) Determine where the graph of R is increasing and where it is decreasing.
(b) How many calculators need to be sold to maximize revenue?
(c) What is the maximum revenue?
(d) Graph the function R.

39. Wheat Acreage in the US The amount of wheat planted annually in the United States is given in the table.

Year	1990	1991	1992	1993	1994	1995	1996
Thousand Acres	77,041	69,881	72,219	72,168	70,349	69,031	75,105

Year	1997	1998	1999	2000
Thousand Acres	70,412	65,821	62,714	62,629

Source: United States Department of Agriculture.

If 1990 is taken as year 0, the number of thousand acres of wheat planted in the United States can be approximated by

$$A(t) = -119.2t^2 + 113.4t + 73{,}367$$

(a) On $[0, 10]$, where is this function increasing?
(b) According to this model, will the acreage of wheat planted from 2004 to 2008 be increasing or decreasing?

40. SAT Scores in Math The data for the mathematics scores on the SAT can be approximated by

$$S(t) = -0.04t^3 + 0.43t^2 + 0.24t + 506,$$

where t is the number of years since 1994.

(a) On $[0, 10]$, where is this function increasing?
(b) According to this model, will SAT scores in math be increasing or decreasing from 2004 to 2008?

41. Corn Production The von Liebig model states that the yield of a plant, $f(x)$, measured in bushels, will respond to the amount of nitrogen in a fertilizer in the following fashion:

$$f(x) = -0.057 - 0.417x + 0.852\sqrt{x}$$

where x is the amount of the nitrogen in the fertilizer.

(a) For what amounts of nitrogen will the yield be increasing?
(b) For what amounts x of nitrogen will the yield be decreasing?

Source: Amer. Agr. Econ. **74**:1019–1028.

42. Cost Function Suppose the cost C of producing x items is given by the function $C(x) = \sqrt{x}$.

(a) Find the marginal cost function.
(b) Show that the marginal cost is a decreasing function.
(c) Find the average cost function $\overline{C}(x) = \dfrac{C(x)}{x}$.
(d) Show that the average cost function is a decreasing function.

In Problems 43–46, use the discussion that follows.

Rolle's Theorem If a function $y = f(x)$ has the following three properties:

1. It is continuous on the closed interval $[a, b]$
2. It is differentiable on the open interval (a, b)
3. $f(a) = f(b)$

then there is at least one number c, $a < c < b$, at which $f'(c) = 0$.

This result is called **Rolle's theorem.** See the figure.

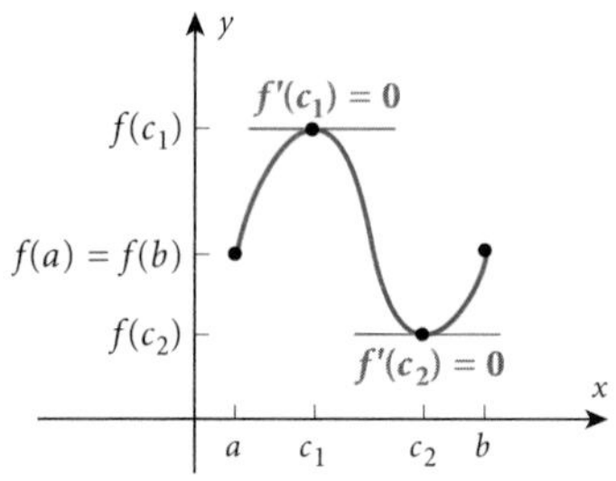

Verify Rolle's theorem by finding the number(s) c for each function on the interval indicated.

43. $f(x) = 2x^2 - 2x$ on $[0, 1]$

44. $f(x) = x^2 + 2x$ on $[-2, 0]$

45. $f(x) = x^4 - 1$ on $[-1, 1]$

46. $f(x) = x^4 - 2x^2 - 8$ on $[-2, 2]$

In Problems 47–50, use the discussion that follows.

Mean Value Theorem If $y = f(x)$ is a continuous function on the closed interval $[a, b]$ and is differentiable on the open interval (a, b), there is at least one number in the interval (a, b) at which the slope of the tangent line equals the slope of the line joining the points $(a, f(a))$ and $(b, f(b))$. That is, there is a number c, $a < c < b$, at which

$$f'(c) = \frac{f(b) - f(a)}{b - a}$$

This result is called the **Mean Value Theorem**. See the figure.

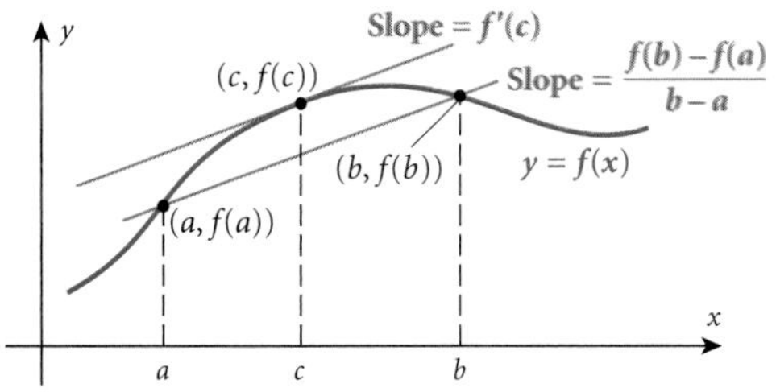

Verify the Mean Value Theorem by finding the number(s) c for each function on the interval indicated.

47. $f(x) = x^2$ on $[0, 3]$

48. $f(x) = x^3$ on $[0, 2]$

49. $f(x) = \dfrac{1}{x^2}$ on $[1, 2]$

50. $f(x) = x^{3/2}$ on $[0, 1]$

14.3 Concavity; the Second Derivative Test

OBJECTIVES
1. Determine the concavity of a graph
2. Find inflection points
3. Graph functions
4. Use the second derivative test
5. Solve applied problems

Consider the graphs of $y = x^2$, $x \geq 0$, and $y = \sqrt{x}$ as shown in Figure 18. Each graph starts at $(0, 0)$, is increasing, and passes through the point $(1, 1)$. However, the graph of $y = x^2$ increases rapidly while the graph of $y = \sqrt{x}$ increases slowly. We use the terms *concave up* or *concave down* to describe these characteristics of graphs.

FIGURE 18

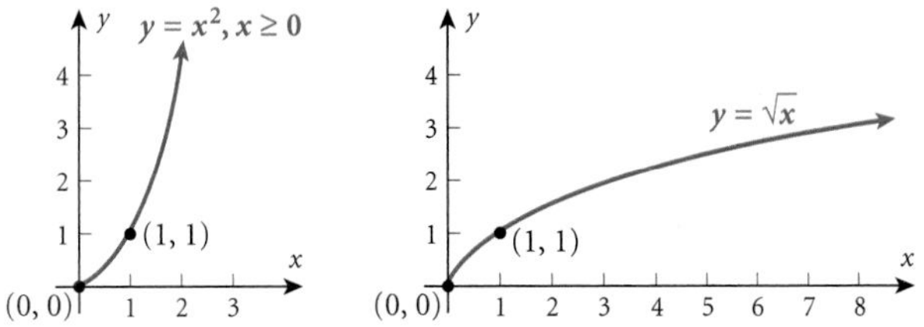

Now look at Figure 19.

FIGURE 19

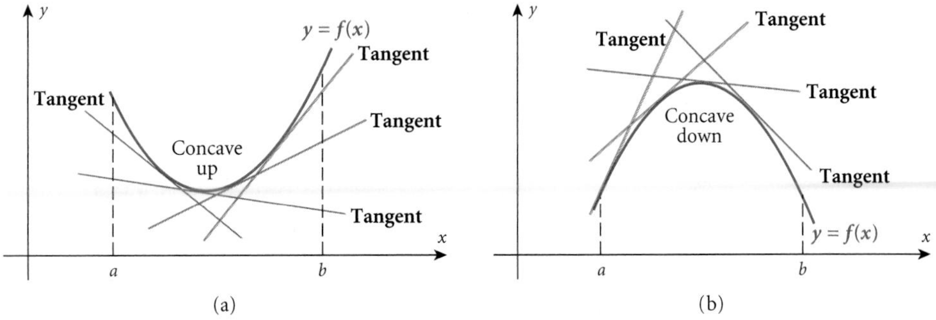

The graph of the function in Figure 19(a) is concave up on the interval (a, b), while the graph in Figure 19(b) is concave down on the interval (a, b). Notice that the tangent lines to the graph in Figure 19(a) lie below the graph, while the tangent lines to the graph in Figure 19(b) lie above the graph. This observation is the basis of the following definition.

Concave Up; Concave Down

Let f denote a function that is differentiable on the interval (a, b).

1. The graph of f is **concave up** on (a, b) if, throughout (a, b), the tangent lines to the graph of f lie below the graph.
2. The graph of f is **concave down** on (a, b), if throughout (a, b), the tangent lines to the graph of f lie above the graph.

NOW WORK PROBLEM 9.

Figure 19 also provides a test for determining whether a graph is concave up or concave down. Notice that in Figure 19(a), as one proceeds from left to right along the graph of f, the slopes of the tangent lines are increasing, starting off very negative and ending up very positive. Since the derivative $f'(x)$ equals the slope of a tangent line and these slopes are increasing, it follows that $f'(x)$ is an increasing function. So, its derivative $f''(x)$ must be positive on (a, b). Similarly, in Figure 19(b), the slopes of the tangent lines to the graph of f are decreasing. It follows that $f'(x)$ is a decreasing function and so $f''(x)$ must be negative on (a, b). As the preceding discussion shows, the second derivative provides information about the concavity of a function.

Test for Concavity

Let $y = f(x)$ be a function and let $f''(x)$ be its second derivative.

1. If $f''(x) > 0$ for all x in the interval (a, b), then the graph of f is concave up on (a, b).
2. If $f''(x) < 0$ for all x in the interval (a, b), then the graph of f is concave down on (a, b).

1 **EXAMPLE 1 Determining the Concavity of a Graph**

Determine where the graph of $f(x) = x^3 - 12x^2$ is concave up or concave down.

SOLUTION We proceed to find $f''(x)$.

$$\begin{aligned} f(x) &= x^3 - 12x^2 \\ f'(x) &= 3x^2 - 24x \\ f''(x) &= 6x - 24 \\ &= 6(x - 4) \end{aligned}$$

We need to solve the inequalities $f''(x) > 0$ and $f''(x) < 0$. We see that if $x < 4$, then $f''(x) < 0$. The graph of f is concave down on the interval $(-\infty, 4)$. If $x > 4$, then $f''(x) > 0$. The graph of f is concave up on the interval $(4, \infty)$.

EXAMPLE 2 Graphing Functions

Graph the function

$$f(x) = x^3 - 12x^2$$

SOLUTION We follow the 6 steps for graphing a function given on pp. 796–797.

STEP 1 The domain of f is all real numbers.
STEP 2 Let $x = 0$. Then $y = f(0) = 0$. The y-intercept is $(0, 0)$.
To find the x-intercepts, if any, let $y = 0$. Then

$$\begin{aligned} x^3 - 12x^2 &= 0 && y = f(x) = 0. \\ x^2(x - 12) &= 0 && \text{Factor.} \\ x^2 = 0 \quad \text{or} \quad x - 12 &= 0 && \text{Apply the Zero-Product Property.} \\ x = 0 \quad \text{or} \quad x &= 12 && \text{Solve for } x. \end{aligned}$$

The x-intercepts are $(0, 0)$ and $(12, 0)$.
STEP 3 To determine where the graph of f is increasing and where it is decreasing, we find $f'(x)$:

$$\begin{aligned} f(x) &= x^3 - 12x^2 \\ f'(x) &= 3x^2 - 24x && \text{Differentiate.} \\ &= 3x(x - 8) && \text{Factor.} \end{aligned}$$

The solutions of the equation $f'(x) = 3x(x - 8) = 0$ are 0 and 8. These numbers separate the number line into three parts:

$$-\infty < x < 0 \qquad 0 < x < 8 \qquad 8 < x < \infty$$

We construct Figure 20, using -1, 1, and 9 as test numbers for $f'(x)$. We conclude that the graph of f is increasing on the intervals $(-\infty, 0)$ and $(8, \infty)$; f is decreasing on the interval $(0, 8)$.

FIGURE 20

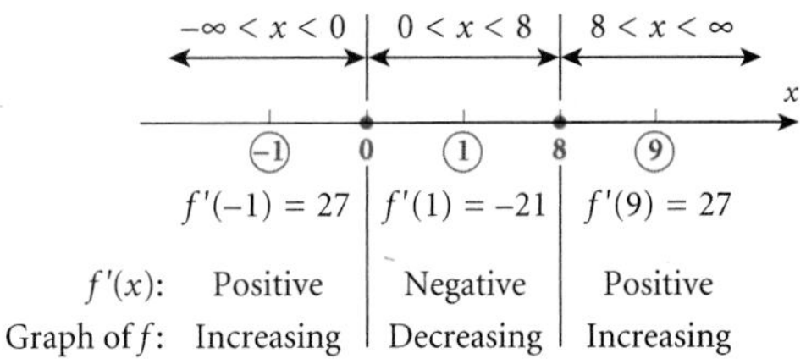

STEP 4 From Figure 20, the graph of f is increasing for $-\infty < x < 0$ and is decreasing for $0 < x < 8$. At $x = 0$, the graph changes from increasing to decreasing. Consequently, at the point $(0, f(0)) = (0, 0)$ there is a local maximum.

Similarly, at $x = 8$, the graph changes from decreasing to increasing. At the point $(8, f(8)) = (8, -256)$ there is a local minimum.
STEP 5 The derivative of f is $f'(x) = 3x(x - 8)$. We see that $f'(x) = 0$ if $x = 0$ or $x = 8$. The graph of f has a horizontal tangent line at the points $(0, 0)$ and $(8, -256)$. There are no vertical tangent lines.
STEP 6 Since f is a polynomial function, the graph of f behaves like that of $y = x^3$ for x unbounded. Polynomial functions have no asymptotes.

Finally, we use the result of Example 1, namely, that the graph of f is concave down on $(-\infty, 4)$ and is concave up on $(4, \infty)$.

To graph f, we plot the intercepts, the local maximum, the local minimum, the points at which the graph has a horizontal tangent line, and connect these points with a smooth curve, keeping the concavity of the graph in mind. See Figure 21.

FIGURE 21

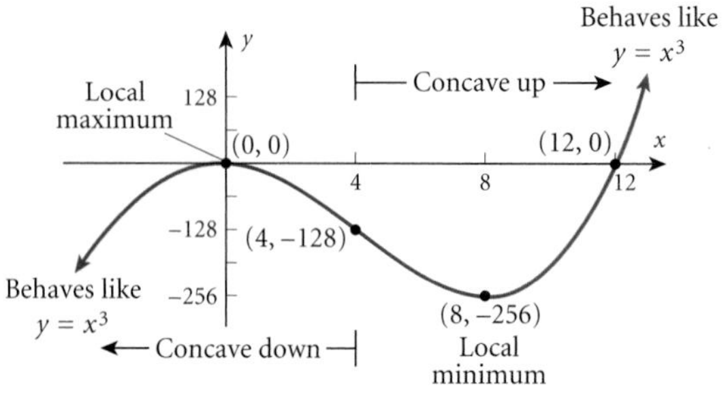

CHECK: Use a graphing utility to graph $f(x) = x^3 - 12x^2$. Be sure to choose a viewing rectangle that will show a complete graph. [Use Figure 22 as a guide.] Use TRACE to confirm where f is increasing and where f is decreasing. Use MAXIMUM and MINIMUM to confirm the local maximum and the local minimum. Use ZERO (or ROOT) to locate the x-intercepts. Use a TABLE to confirm the end behavior. Can you see where the graph is concave up and where it is concave down?

Look again at Figure 21. The point $(4, -128)$ on the graph of $f(x) = x^3 - 12x^2$ is the point at which the concavity of f changed from down to up. This point is called an *inflection point.*

Inflection Point

An **inflection point** of a function f is a point on the graph of f at which the concavity of f changes.

See Figure 22. The point $(b, f(b))$ is an inflection point of f.

FIGURE 22

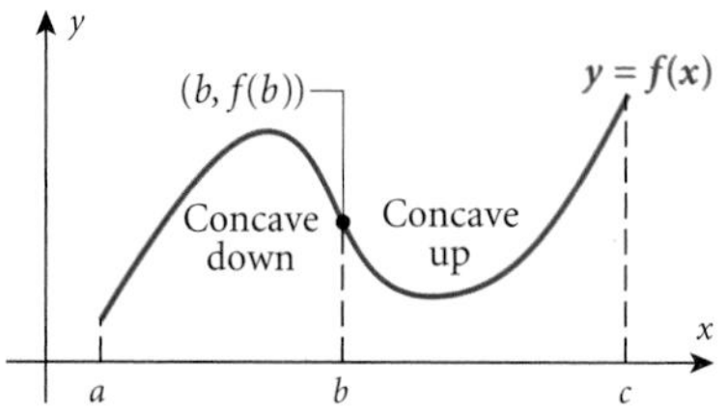

2 **EXAMPLE 3 Finding Inflection Points**

Find all the inflection points of

$$f(x) = x^2 - \frac{1}{x}$$

SOLUTION The domain of f is $\{x \mid x \neq 0\}$. To find the inflection points, we need to determine where the graph of f is concave up and where it is concave down. That is, we need to find $f''(x)$:

$$f(x) = x^2 - \frac{1}{x}$$

$$f'(x) = 2x + \frac{1}{x^2} \qquad \frac{d}{dx}\frac{1}{x} = \frac{d}{dx}x^{-1} = -1x^{-2} = -\frac{1}{x^2}$$

$$f''(x) = 2 - \frac{2}{x^3} \qquad \frac{d}{dx}\frac{1}{x^2} = \frac{d}{dx}x^{-2} = -2x^{-3} = \frac{-2}{x^3}$$

$$= \frac{2x^3 - 2}{x^3} \qquad \text{Write as a single quotient.}$$

$$= \frac{2(x^3 - 1)}{x^3} \qquad \text{Factor.}$$

$$= \frac{2(x - 1)(x^2 + x + 1)}{x^3} \qquad \text{Factor.}$$

The factors in the numerator and the denominator of $f''(x)$, $x - 1$, $x^2 + x + 1$, x^3, equal 0 if $x = 1$ or $x = 0$. The equation $x^2 + x + 1 = 0$ has no real solution. (Do you see why?) We use the numbers 0 and 1 to separate the number line into three parts:

$$-\infty < x < 0 \qquad 0 < x < 1 \qquad 1 < x < \infty$$

We construct Figure 23, using $-1, \frac{1}{2}$, and 2 as test numbers for $f''(x)$.

FIGURE 23

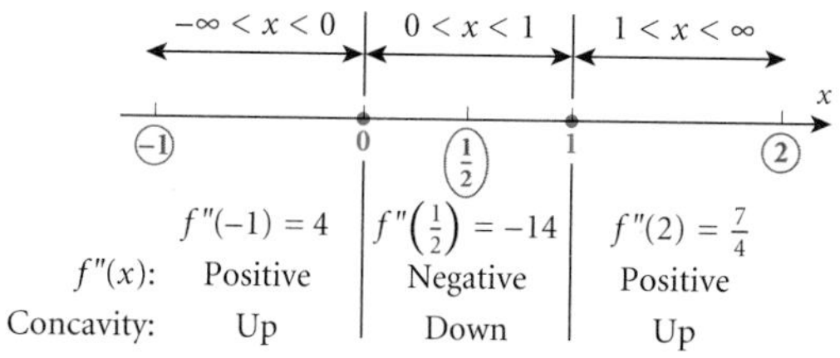

We conclude that the graph of f is concave up on the intervals $(-\infty, 0)$ and $(1, \infty)$. It is concave down on the interval $(0, 1)$. The only inflection point occurs at the point $(1, f(1)) = (1, 0)$, where the concavity changes from down to up. Note that even though the concavity changes from up to down at $x = 0$, there is no inflection point since 0 is not in the domain of f.

NOW WORK PROBLEM 13.

To the six steps we used earlier to graph a function, we now add Step 7:

Steps for Graphing Functions

STEP 1 Find the domain of f.

STEP 2 Locate the intercepts of f. (Skip the x-intercepts if they are too hard to find.)

STEP 3 Determine where the graph of f is increasing and where it is decreasing.

STEP 4 Find any local maxima or local minima of f by using the First Derivative Test.

STEP 5 Locate all points on the graph of f at which the tangent line is either horizontal or vertical.

STEP 6 Determine the end behavior and locate any asymptotes.

STEP 7 Locate the inflection points, if any, of the graph by determining the concavity of the graph.

3 **EXAMPLE 4** **Graphing Functions**

Graph the function

$$f(x) = 3x^5 - 5x^4$$

SOLUTION **STEP 1** The domain of f is all real numbers.

STEP 2 Let $x = 0$. Then $y = f(0) = 0$. The y-intercept is $(0, 0)$.

To find the x-intercepts, if any, let $y = 0$. Then

$$3x^5 - 5x^4 = 0 \qquad y = f(x) = 0.$$

$$x^4(3x - 5) = 0 \qquad \text{Factor.}$$

$$x^4 = 0 \quad \text{or} \quad 3x - 5 = 0 \qquad \text{Apply the Zero-Product Property.}$$

$$x = 0 \quad \text{or} \quad x = \frac{5}{3} \qquad \text{Solve for } x.$$

The x-intercepts are $(0, 0)$ and $\left(\frac{5}{3}, 0\right)$.

STEP 3 To determine where the graph of f is increasing and where it is decreasing, we find $f'(x)$:

$$f(x) = 3x^5 - 5x^4$$

$$f'(x) = 15x^4 - 20x^3 \qquad \text{Differentiate.}$$

$$= 5x^3(3x - 4) \qquad \text{Factor.}$$

The solutions of the equation $f'(x) = 5x^3(3x - 4) = 0$ are 0 and $\frac{4}{3}$. These numbers separate the number line into three parts:

$$-\infty < x < 0 \qquad 0 < x < \frac{4}{3} \qquad \frac{4}{3} < x < \infty$$

We construct Figure 24, using -1, 1, and 2 as test numbers for $f'(x)$.

FIGURE 24

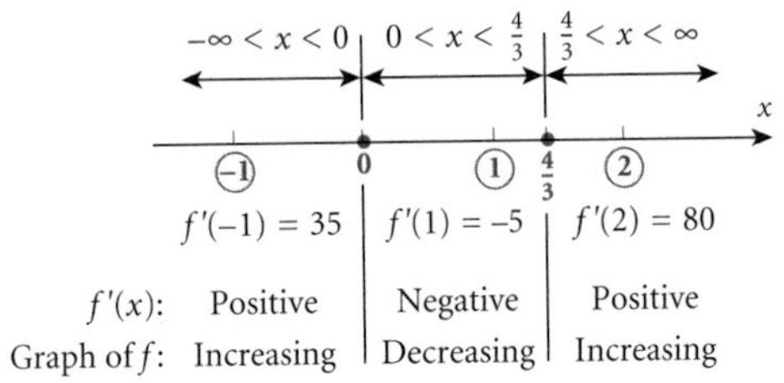

We conclude that the graph of f is increasing on the intervals $(-\infty, 0)$ and $\left(\frac{4}{3}, \infty\right)$. It is decreasing on the interval $\left(0, \frac{4}{3}\right)$.

STEP 4 From Figure 24, the graph of f is increasing for $-\infty < x < 0$ and is decreasing for $0 < x < \frac{4}{3}$. At $x = 0$, the graph changes from increasing to decreasing. Consequently, at the point $(0, f(0)) = (0, 0)$ there is a local maximum.

Similarly, at $x = \frac{4}{3}$, the graph changes from decreasing to increasing. At the point $\left(\frac{4}{3}, f\left(\frac{4}{3}\right)\right) = \left(\frac{4}{3}, -\frac{256}{81}\right)$ there is a local minimum.

STEP 5 The derivative of f is $f'(x) = 5x^3(3x - 4)$. We see that $f'(x) = 0$ if $x = 0$ or $x = \dfrac{4}{3}$. The graph of f has a horizontal tangent line at the points $(0, 0)$ and $\left(\dfrac{4}{3}, -\dfrac{256}{81}\right)$. There are no vertical tangent lines.

STEP 6 The function f is a polynomial function. Its graph will behave like $y = 3x^5$ for x unbounded. Polynomial functions have no asymptotes.

STEP 7 To locate the inflection points, if any, of f, we find $f''(x)$:

$$f(x) = 3x^5 - 5x^4$$
$$f'(x) = 15x^4 - 20x^3 \quad \text{Differentiate.}$$
$$f''(x) = 60x^3 - 60x^2 \quad \text{Differentiate.}$$
$$= 60x^2(x - 1) \quad \text{Factor.}$$

We use the numbers 0 and 1 to separate the number line into three parts:

$$-\infty < x < 0 \qquad 0 < x < 1 \qquad 1 < x < \infty$$

We construct Figure 25. Using $-1, \dfrac{1}{2}$, and 2 as test numbers for $f''(x)$.

FIGURE 25

$f''(-1) = -120 \quad f''\left(\dfrac{1}{2}\right) = -\dfrac{15}{2} \quad f''(2) = 240$

$f''(x)$:   Negative   Negative   Positive

Concavity: Down   Down   Up

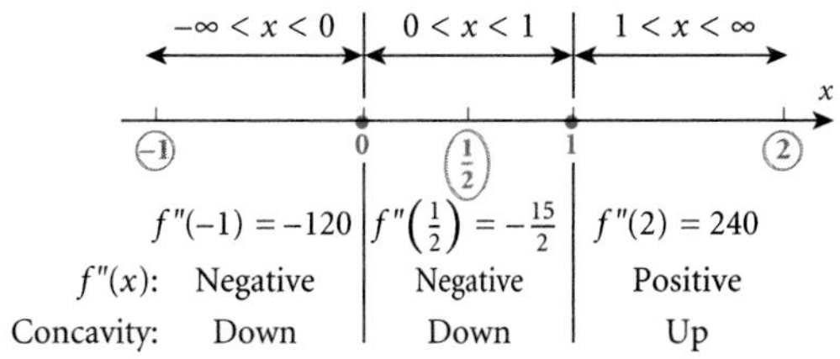

We conclude that the graph of f is concave down on $(-\infty, 1)$ and is concave up on $(1, \infty)$. Further, since the concavity changes at the point $(1, f(1)) = (1, -2)$, we conclude that $(1, -2)$ is an inflection point.

To graph f, we plot the intercepts, the local maximum, the local minimum, and the inflection point, and connect these points with a smooth curve. See Figure 26.

FIGURE 26

Local maximum

$(0, 0)$

Behaves like $y = 3x^5$

$\left(\dfrac{5}{3}, 0\right)$

Inflection point

$(1, -2)$

Behaves like $y = 3x^5$

$\left(\dfrac{4}{3}, -\dfrac{256}{81}\right)$

Local minimum

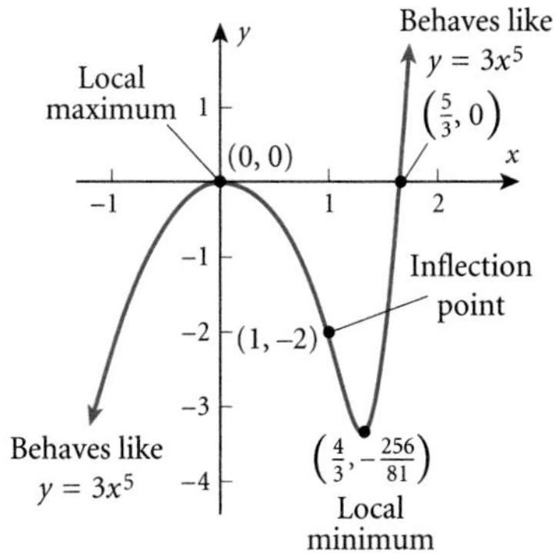

CHECK: Use a graphing utility to graph $f(x) = 3x^5 - 5x^4$. Be sure to choose a viewing rectangle that will show a complete graph. [Use Figure 26 as a guide.] Use TRACE to confirm where f is increasing and where f is decreasing. Use MAXIMUM and MINIMUM to confirm the local maximum and the local minimum. Use ZERO (or ROOT) to locate the x-intercepts. Use a TABLE to confirm the end behavior. Can you see where the graph is concave down and where it is concave up?

NOW WORK PROBLEM 29.

EXAMPLE 5 Graphing a Function

Graph the function: $f(x) = xe^{-x}$

SOLUTION **STEP 1** The domain of f is all real numbers.

STEP 2 Let $x = 0$. Then $y = f(0) = 0$, so $(0, 0)$ is the y-intercept.

Let $y = 0$. Then $xe^{-x} = 0$. Since $e^{-x} > 0$ for all x, the only solution is $x = 0$. The x-intercept is also $(0, 0)$.

STEP 3 To find where the graph of f is increasing and where it is decreasing, we find $f'(x)$:

$$f'(x) = \frac{d}{dx}(xe^{-x}) \underset{\uparrow}{=} x\frac{d}{dx}e^{-x} + e^{-x}\frac{d}{dx}x \underset{\uparrow}{=} x \cdot e^{-x}(-1) + e^{-x} \cdot 1$$

Derivative of a product; $\frac{d}{dx}e^{-x} = e^{-x}\frac{d}{dx}(-x) = e^{-x}(-1)$

$$= e^{-x}(-x + 1) = (1 - x)e^{-x}$$

Since $e^{-x} > 0$ for all x, it follows that $f'(x) > 0$ if $-\infty < x < 1$, and $f'(x) < 0$ if $1 < x < \infty$. The graph of f is increasing on the interval $(-\infty, 1)$ and is decreasing on the interval $(1, \infty)$.

STEP 4 Using the First Derivative Test, we conclude that at the point $(1, e^{-1}) = (1, 0.368)$ there is a local maximum.

STEP 5 Since $f'(1) = 0$, the tangent line is horizontal at the point $(1, 0.368)$.

STEP 6 To find the end behavior, we look at the limits at infinity and use a table. See Table 1. We conclude that

TABLE 1

x	$f(x) = xe^{-x}$
10	4.5 E^{-4}
100	3.7 E^{-42}
1000	0
$x \to \infty$	$f(x) \to 0$
-10	-2.2 E^5
-100	-2.7 E^{45}
-1000	-1.4 E^{89}
$x \to -\infty$	$f(x) \to -\infty$

$$\lim_{x\to-\infty} f(x) = \lim_{x\to-\infty} (xe^{-x}) = -\infty \quad \text{and} \quad \lim_{x\to\infty} f(x) = \lim_{x\to\infty} (xe^{-x}) = 0$$

This meas the graph of f becomes unbounded in the negative direction as $x \to -\infty$ and that $y = 0$ is a horizontal asymptotes as $x \to \infty$.

STEP 7 To locate any inflection points, we find $f''(x)$.

$$f''(x) = \frac{d}{dx}f'(x) = \frac{d}{dx}[(1 - x)e^{-x}] \underset{\uparrow}{=} (1 - x)\frac{d}{dx}e^{-x} + e^{-x}\frac{d}{dx}(1 - x)$$

Derivative of a product

$$= (1 - x) \cdot e^{-x}(-1) + e^{-x}(-1) \underset{\uparrow}{=} e^{-x}[(1 - x)(-1) + (-1)]$$

$$= (x - 2)e^{-x}$$

Factor out e^{-x}

It follows that $f''(x) < 0$ if $x < 2$, and $f''(x) > 0$ if $x > 2$. The graph of f is concave down on the interval $(-\infty, 2)$ and is concave up on the interval $(2, \infty)$. The point $(2, 2e^{-2}) = (2, 0.27)$ is an inflection point.

See Figure 27 for the graph.

FIGURE 27

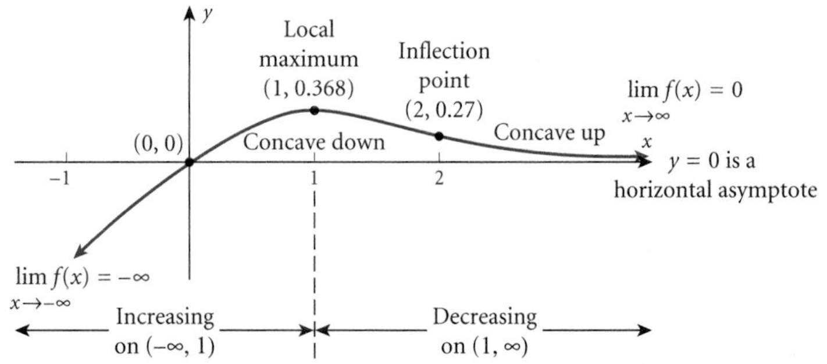

CHECK: Use a graphing utility to graph $f(x) = xe^{-x}$. Be sure to choose a viewing rectangle that will show a complete graph. [Use Figure 27 as a guide.] Use TRACE to confirm where f is increasing and where f is decreasing. Use MAXIMUM to confirm the local maximum. Use a TABLE to confirm the end behavior. Can you see where the graph is concave down and where it is concave up?

NOW WORK PROBLEM 45.

The Second Derivative Test

There is another test for finding the local maximum and local minimum that is sometimes easier to use than the First Derivative Test. This test is based on the geometric observation that the graph of a function has a local maximum at a point where the tangent line is horizontal and the graph is concave down. See Figure 28(a).

Similarly, the graph of a function has a local minimum at a point where the tangent line is horizontal and the graph is concave up. See Figure 28(b). This leads us to formulate the Second Derivative Test.

FIGURE 28

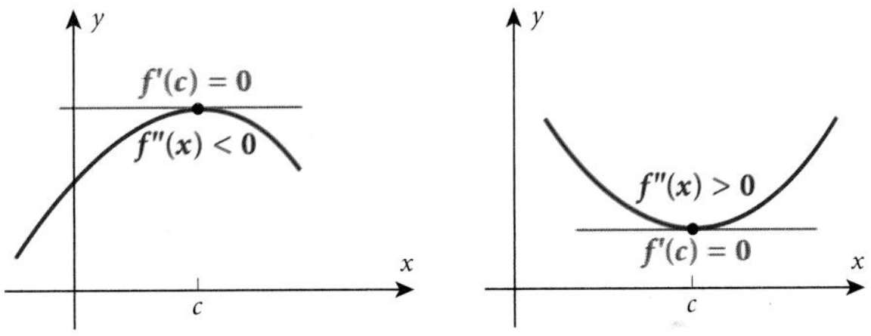

Local maximum

(a)

Local minimum

(b)

Second Derivative Test

Let $y = f(x)$ be a function that is differentiable on an open interval I and suppose that the second derivative $f''(x)$ exists on I. Also suppose c is a number in I for which $f'(c) = 0$.

1. If $f''(c) < 0$, then at the point $(c, f(c))$ there is a local maximum.
2. If $f''(c) > 0$, then at the point $(c, f(c))$ there is a local minimum.
3. If $f''(c) = 0$, the test is inconclusive and the First Derivative Test must be used.

Notice that the Second Derivative Test is used to locate those local maxima and local minima that occur where the tangent line is horizontal. Since local maxima and local minima can also occur at points on the graph at which the derivative does not exist, this test is usually only used for functions that are differentiable. Also note that the test provides no information if the second derivative is zero at c. Whenever either of these situations arise, you must use the First Derivative Test.

4 **EXAMPLE 6** **Using the Second Derivative Test**

Use the Second Derivative Test to find the local maxima and local minima of

$$f(x) = x^3 - 12x^2$$

SOLUTION First we find those numbers at which $f'(x) = 0$:

$$f(x) = x^3 - 12x^2$$
$$f'(x) = 3x^2 - 24x$$
$$= 3x(x - 8)$$

$f'(x) = 0$ at $x = 0$ and $x = 8$.

Next we evaluate $f''(x)$ at these numbers:

$$f''(x) = 6x - 24 \qquad \frac{d}{dx}f'(x) = \frac{d}{dx}(3x^2 - 24x) = 6x - 24$$

At $x = 0$, $f''(0) = -24 < 0$. By the Second Derivative Test, f has a local maximum at $(0, 0)$.

At $x = 8$, $f''(8) = 24 > 0$. By the Second Derivative Test, f has a local minimum at $(8, -256)$.

[The graph of $f(x) = x^3 - 12x^2$ was given earlier in Figure 21.]

NOW WORK PROBLEM 47.

EXAMPLE 7 **Finding a Graph**

Sketch the graph of a differentiable function $y = f(x)$, $0 \le x \le 6$, which has the following properties:

1. The points $(0, 4)$, $(1, 3)$, $(3, 5)$, $(5, 7)$, and $(6, 6)$ are on the graph.
2. $f'(1) = 0$ and $f'(5) = 0$; $f'(x)$ is not 0 anywhere else.
3. $f''(x) > 0$ for $x < 3$ and $f''(x) < 0$ for $x > 3$.

SOLUTION First we plot the points $(0, 4)$, $(1, 3)$, $(3, 5)$, $(5, 7)$, and $(6, 6)$. Since we are told that $f'(1) = 0$ and $f'(5) = 0$, we know the tangent lines to the graph at $(1, 3)$ and at $(5, 7)$ are horizontal. See Figure 29(a).

Since $f''(x) > 0$ if $x < 3$, it follows that $f''(1) > 0$. By the Second Derivative Test, at the point $(1, 3)$ there is a local minimum. Similarly, since $f''(x) < 0$ if $x > 3$, it follows that $f''(5) < 0$, so that at the point $(5, 7)$ there is a local maximum. See Figure 29(b).

Finally, since the graph is concave up for $x < 3$ and concave down for $x > 3$, it follows that the point $(3, 5)$ is an inflection point. See Figure 29(c).

FIGURE 29

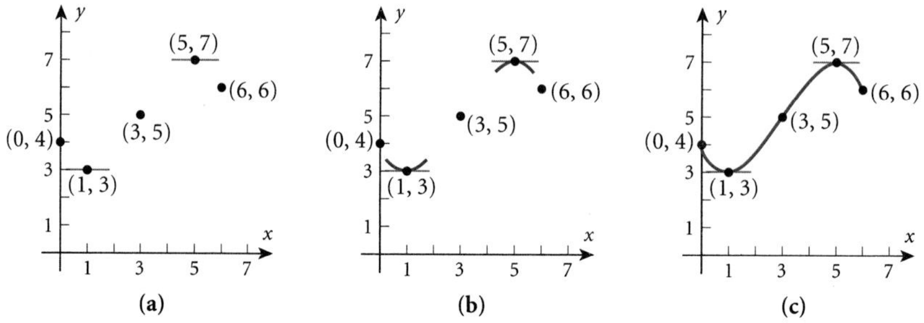

NOW WORK PROBLEM 55.

Applications

Interpreting an Inflection Point

EXAMPLE 8 **Interpreting an Inflection Point**

The sales $S(x)$, in thousands of dollars, of the Big Apple, a manufacturer of computers, is related to the amount x, in thousands of dollars, that the company spends on advertising by the function

$$S(x) = -0.02x^3 + 1.2x^2 + 1000 \qquad 0 \le x \le 60$$

Find the inflection point of the function S and discuss its significance.

SOLUTION The first and second derivatives of S are

$$S'(x) = -0.06x^2 + 2.4x = -0.06x(x - 40)$$
$$S''(x) = -0.12x + 2.4 = -0.12(x - 20)$$

To determine the concavity, we solve $S''(x) > 0$ and $S''(x) < 0$. Then,

$$S''(x) > 0 \quad \text{for} \quad x < 20 \qquad \text{and} \qquad S''(x) < 0 \quad \text{for} \quad x > 20$$

FIGURE 30

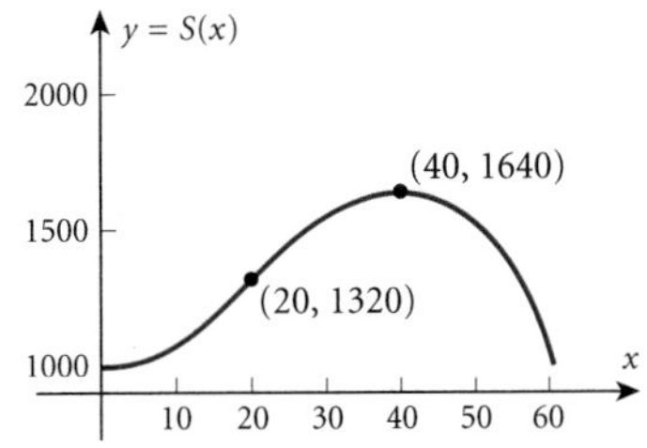

The point (20, 1320) is an inflection point of the function S. Figure 30 illustrates the graph of S.

Notice that the sales of the company increase very slowly at the beginning, but as advertising money increases, sales increase rapidly. This rapid sales growth indicates that consumers are responding to the advertisement. However, there is a point on the graph at which the *rate of growth* of sales changes from positive to negative—where the rate of growth changes from increasing to decreasing—resulting in a slower rate of increased sales. This point, commonly known as the point of *diminishing returns,* is the point of inflection of S, the point (20,1320).

Average Cost If the cost C of producing x units is given by the cost function $C = C(x)$, the **average cost $\overline{C}(x)$** of producing x units is defined as

$$\overline{C}(x) = \frac{C(x)}{x}$$

EXAMPLE 9 **Analyzing Marginal Cost and Average Cost**

Consider the cost function $C(x) = 1000 + \dfrac{1}{10}x^2$, where x is the number of units manufactured.

(a) What is the average cost?
(b) What is the minimum average cost?
(c) Find the marginal cost.
(d) Where does average cost equal marginal cost?
(e) Graph the average cost function and the marginal cost function using the same coordinate system.

SOLUTION   **(a)** The average cost $\overline{C}(x)$ is

$$\overline{C}(x) = \frac{C(x)}{x} = \frac{1000 + \frac{1}{10}x^2}{x} = \frac{1000}{x} + \frac{1}{10}x$$

(b) To find the minimum average cost, we use the Second Derivative Test. We begin by finding $\overline{C}'(x)$:

$$\overline{C}'(x) = \frac{-1000}{x^2} + \frac{1}{10} = \frac{-10{,}000 + x^2}{10x^2}$$

$$\overline{C}'(x) = 0 \quad \text{when } x^2 - 10{,}000 = 0 \quad \text{or} \quad x = \pm 100$$

We disregard $x = -100$, since the number of units manufactured must be positive. Next, we find the second derivative of $\overline{C}(x)$. Since

$$\overline{C}'(x) = \frac{-1000}{x^2} + \frac{1}{10}$$

we have

$$\overline{C}''(x) = \frac{d}{dx}\left(\frac{-1000}{x^2} + \frac{1}{10}\right) = \frac{d}{dx}\frac{-1000}{x^2} + 0 = -1000\frac{d}{dx}x^{-2} = -1000 \cdot (-2)x^{-3} = \frac{2000}{x^3}$$

Since $\overline{C}''(100) = \dfrac{2000}{1{,}000{,}000} > 0$, it follows by the Second Derivative Test that $\overline{C}(x)$ has a local minimum at $x = 100$. The minimum average cost is therefore

$$\overline{C}(100) = \frac{C(100)}{100} = \frac{2000}{100} = 20$$

(c) The cost function is $C(x) = 1000 + \dfrac{1}{10}x^2$. The marginal cost is

$$C'(x) = \frac{d}{dx}\left(1000 + \frac{1}{10}x^2\right) = \frac{1}{10}(2x) = \frac{1}{5}x$$

(d) Average cost equals marginal cost if

$$\overline{C}(x) = C'(x)$$

$$\frac{1000}{x} + \frac{1}{10}x = \frac{1}{5}x$$

$$\frac{1000}{x} = \frac{1}{10}x$$

$$x^2 = 10{,}000$$

$$x = 100$$

When 100 items are produced, average cost equals marginal cost.

(e) See Figure 31 for the graph.

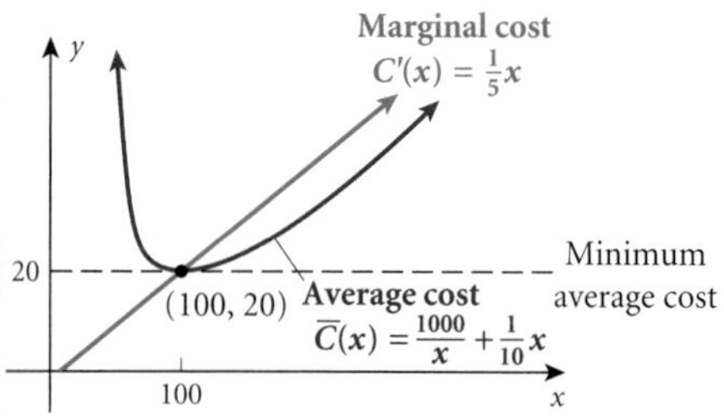

Notice that the minimum average cost occurs at the intersection of the average cost and the marginal cost. In fact, this is always the case.

> The minimum average cost occurs where the average cost and marginal cost are equal.

FIGURE 31

Marginal cost
$C'(x) = \frac{1}{5}x$

Minimum
average cost

$(100, 20)$ **Average cost**
$\overline{C}(x) = \frac{1000}{x} + \frac{1}{10}x$

NOW WORK PROBLEM 61.

Logistic Curves Many models in business and economics require the use of a *logistic curve*. Curves that describe a situation in which the rate of growth is slow at first, increases to a maximum rate, and then decreases are called **logistic curves,** or **saturation curves.** These curves are best characterized by their "S" shape. Figure 32 illustrates a typical general logistic curve.

FIGURE 32

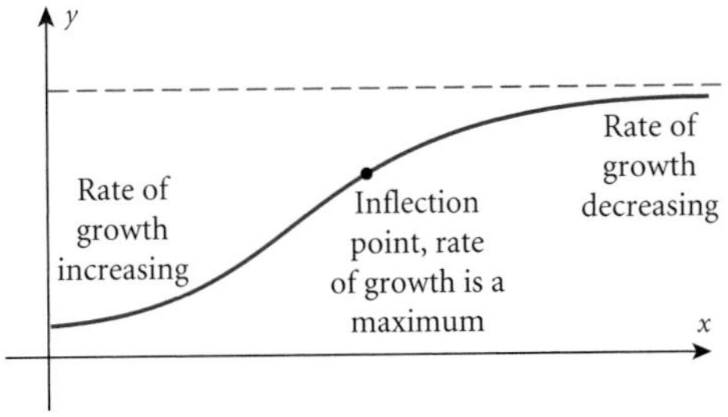

Logistic curve

EXAMPLE 10 Analyzing a Logistic Curve

The annual sales of a 42″ plasma TV are expected to follow the logistic curve

$$f(x) = \frac{10{,}000}{1 + 100e^{-x}}$$

where x is measured in years and $x = 0$ corresponds to 2000, the year production begins. Analyze the graph of this function and determine the year in which a maximum sales rate is achieved.

SOLUTION We follow the 7 steps for graphing a function. (p. 809).

STEP 1 The domain of this function is $\{x | x \geq 0\}$
STEP 2 If $x = 0$, then

$$y = f(0) = \frac{10{,}000}{1 + 100} = 99.01$$

The y-intercept is (0, 99.01). There is no x-intercept. The y-intercept represents the predicted number of plasma TV's sold when production begins.
STEP 3 The derivative of the function f is

$$f'(x) = \frac{d}{dx}[10{,}000(1 + 100e^{-x})^{-1}] = 10{,}000 \cdot (-1)(1 + 100e^{-x})^{-2} \cdot 100(-1)e^{-x} = \frac{1{,}000{,}000e^{-x}}{(1 + 100e^{-x})^2}$$

Since $e^{-x} > 0$ for all x, it follows that $f'(x) > 0$ for $x \geq 0$. The function f is increasing, which means that sales are increasing each year.
STEP 4 There are no local maxima since the graph is increasing.
STEP 5 There are no vertical or horizontal tangent lines.
STEP 6 As $x \to \infty$, we have $e^{-x} \to 0$. As a result,

$$\lim_{x\to\infty} f(x) = \lim_{x\to\infty} \frac{10{,}000}{1 + 100e^{-x}} = 10{,}000$$

This means $y = 10{,}000$ is a horizontal asymptote as x becomes unbounded in the positive direction. This number represents the upper estimate for sales.

STEP 7 To locate any inflection points, we find $f''(x)$. You should verify that

$$f''(x) = 1{,}000{,}000e^{-x}\left[\frac{100e^{-x} - 1}{(1 + 100e^{-x})^3}\right]$$

The sign of $f''(x)$ is controlled by the numerator since $1 + 100e^{-x} > 0$ for all x. Now $100e^{-x} - 1 = 0$ if $e^x = 100$, which happens if $x = \ln 100 = 4.6$. If $x < 4.6$, we have $100e^{-x} - 1 > 0$, so $f''(x) > 0$, and the graph of f is concave up. If $x > 4.6$, we have $100e^{-x} - 1 < 0$, so $f''(x) < 0$, and the graph of f is concave down. At (4.6, 5000) there is an inflection point. At this point, the first derivative, f', achieves its maximum value. That is, at 4.6 years, around the middle of 2004, the rate of growth in sales is a maximum.

The graph is given in Figure 33.

FIGURE 33

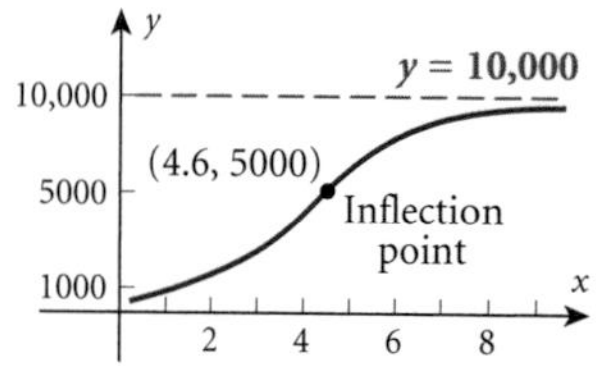

CHECK: Graph $y = \dfrac{10{,}000}{1 + 100e^{-x}}$ and compare the result with Figure 33.

EXERCISE 14.3 Answers to Odd-Numbered Problems Begin on Page AN-82.

In Problems 1–12, use the graph of $y = f(x)$ given on the right.

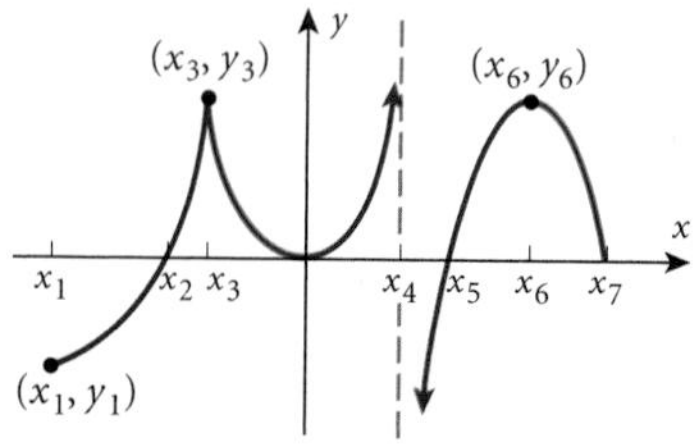

1. What is the domain of f?

2. List the intercepts of f.

3. On what intervals, if any, is the graph of f increasing?

4. On what intervals, if any, is the graph of f decreasing?

5. For what values of x does $f'(x) = 0$?

6. For what values of x does $f'(x)$ not exist?

7. List the point(s) at which f has a local maximum.

8. List the point(s) at which f has a local minimum.

9. On what intervals, if any, is the graph of f concave up?

10. On what intervals, if any, is the graph of f concave down?

11. List any asymptotes.

12. List any vertical tangent lines.

In Problems 13–28, determine the intervals on which the graph of f is concave up and concave down. List any inflection points.

13. $f(x) = x^3 - 6x^2 + 1$

14. $f(x) = x^3 + 3x^2 + 2$

15. $f(x) = x^4 - 2x^3 + 6x - 1$

16. $f(x) = x^4 + 2x^3 - 8x + 8$

17. $f(x) = 3x^5 - 5x^4 + 60x + 10$

18. $f(x) = 3x^5 + 5x^4 + 20x - 4$

19. $f(x) = 3x^5 - 10x^3 + 10x + 10$

20. $f(x) = 3x^5 - 10x^3 - 8x + 8$

21. $f(x) = x^5 - 10x^2 + 4$

22. $f(x) = x^5 + 10x^2 - 8x + 4$

23. $f(x) = 3x^{1/3} + 9x + 2$

24. $f(x) = 3x^{2/3} - 8x + 4$

25. $f(x) = x^{2/3}(x - 10)$

26. $f(x) = x^{2/3}(x - 15)$

27. $f(x) = x^{2/3}(x^2 - 16)$

28. $f(x) = x^{2/3}(x^2 - 4)$

In Problems 29–46, follow the seven steps on page 809 to graph f.

29. $f(x) = x^3 - 6x^2 + 1$

30. $f(x) = x^3 + 6x^2 + 2$

31. $f(x) = x^4 - 2x^2 + 1$

32. $f(x) = 2x^4 - 4x^2 + 2$

33. $f(x) = x^5 - 10x^4$

34. $f(x) = x^5 + 5x$

35. $f(x) = x^6 - 3x^5$

36. $f(x) = x^6 + 3x^5$

37. $f(x) = 3x^4 - 12x^3$

38. $f(x) = 3x^4 + 12x^3$

39. $f(x) = x^5 - 10x^2 + 4$

40. $f(x) = x^5 + 10x^2 + 2$

41. $f(x) = x^{2/3}(x - 10)$

42. $f(x) = x^{2/3}(x - 15)$

43. $f(x) = x^{2/3}(x^2 - 16)$

44. $f(x) = x^{2/3}(x - 4)$

45. $f(x) = xe^x$

46. $f(x) = x^2e^x$

In Problems 47–54, determine where $f'(x) = 0$. Use the Second Derivative Test to determine the local maxima and local minima of each function.

47. $f(x) = x^3 - 3x + 2$

48. $f(x) = x^3 - 12x - 4$

49. $f(x) = 3x^4 + 4x^3 - 3$

50. $f(x) = 3x^4 - 6x^2 + 4$

51. $f(x) = x^5 - 5x^4 + 2$

52. $f(x) = 3x^5 - 20x^3$

53. $f(x) = x + \dfrac{1}{x}$

54. $f(x) = 2x + \dfrac{1}{x^2}$

---

55. **Finding a Graph** Sketch the graph of a differentiable function $y = f(x)$ that has the following properties:

1. $(0, 10)$, $(6, 15)$, and $(10, 0)$ are on the graph.
2. $f'(6) = 0$ and $f'(10) = 0$; $f'(x)$ is not 0 anywhere else.
3. $f''(x) < 0$ for $x < 9$, $f''(9) = 0$, and $f''(x) > 0$ for $x > 9$.

56. **Finding a Graph** Sketch the graph of a differentiable function $y = f(x)$ that has the following properties:

1. $(-1, 3)$, $(1, 5)$, and $(3, 7)$ are on the graph.
2. $f'(3) = 0$ and $f'(-1) = 0$; $f'(x)$ is not 0 anywhere else.
3. $f''(x) > 0$ for $x < 1$, $f''(1) = 0$, and $f''(x) > 0$ for $x > 1$.

57. **Finding a Graph** Sketch the graph of a differentiable function $y = f(x)$ that has the following properties:

1. $(1, 5)$, $(2, 3)$ and $(3, 1)$ are on the graph.
2. $f'(1) = 0$ and $f'(3) = 0$; $f'(x)$ is not 0 anywhere else.
3. $f''(x) < 0$ for $x < 2$, $f''(2) = 0$, and $f''(x) > 0$ for $x > 2$.

58. **Finding a Graph** Sketch the graph of a differentiable function $y = f(x)$ that has the following properties:

1. Domain of $f(x)$ is $x \geq 0$.
2. $(0, 0)$ and $(6, 7)$ are on the graph.
3. $f'(x) > 0$ for $x > 0$.
4. $f''(x) < 0$ for $x < 6$, $f''(6) = 0$, and $f''(x) > 0$ for $x > 6$.

59. For the function $f(x) = ax^3 + bx^2$, determine a and b so that the point $(1, 6)$ is a point of inflection of $f(x)$.

60. Let $f(x) = ax^2 + bx + c$, where $a \neq 0$, b, and c are real numbers. Is it possible for $f(x)$ to have an inflection point? Explain your answer.

61. **Average Cost** The cost function for producing x items is

$$C(x) = 2x^2 + 50$$

(a) Find the average cost function.
(b) What is the minimum average cost?
(c) Find the marginal cost function.
(d) Graph the average cost function and the marginal cost function on the same set of axes. Label their point of intersection.
(e) Interpret the point of intersection.

62. **Average Cost** The cost function for producing x items is

$$C(x) = x^2 - 3x + 625$$

(a) Find the average cost function.
(b) What is the minimum average cost?
(c) Find the marginal cost function.
(d) Graph the average cost function and the marginal cost function on the same set of axes. Label their point of intersection.
(e) Interpret the point of intersection.

63. Average Cost The cost function for producing x items is

$$C(x) = 500 + 10x + \frac{x^2}{500}$$

(a) Find the average cost function.
(b) What is the minimum average cost?
(c) Find the marginal cost function.
(d) Graph the average cost function and the marginal cost function on the same set of axes. Label their point of intersection.
(e) Interpret the point of intersection.

64. Average Cost The cost function for producing x items is

$$C(x) = 800 + 0.04x + 0.0002x^2$$

(a) Find the average cost function.
(b) What is the minimum average cost?
(c) Find the marginal cost function.
(d) Graph the average cost function and the marginal cost function on the same set of axes. Label their point of intersection.
(e) Interpret the point of intersection.

65. Spread of Rumor In a city of 50,000 people, the number of people who have heard a certain rumor after t days obeys

$$N(t) = \frac{50{,}000}{1 + 49{,}999e^{-t}}$$

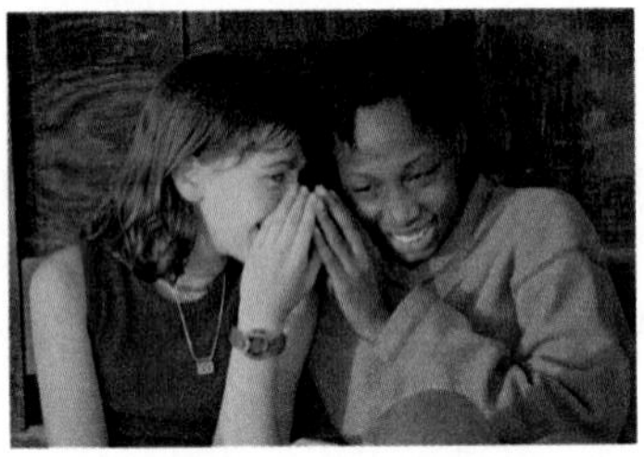

(a) Find the domain of N.
(b) Locate the y-intercept and the t-intercepts, if any.
(c) Determine where N is increasing and where it is decreasing.
(d) Determine where N is concave up and where it is concave down.
(e) Locate any inflection points.
(f) Graph N.
(g) At what time is the rumor spreading at the greatest rate?

66. Analyze the Function:

$$f(x) = \frac{2000}{1 + 4e^{-x}}$$

(a) Find the domain of f.
(b) Locate the y-intercept and the x-intercepts, if any.
(c) Determine where f is increasing and where it is decreasing.
(d) Determine where f is concave up and where it is concave down.
(e) Locate any inflection points.
(f) Graph f.

67. Bacteria Growth Rate A bacteria population grows from an initial population of 800 to a population $p(t)$ at time t (measured in days) according to the logistic curve

$$p(t) = \frac{800e^t}{1 + \frac{1}{10}(e^t - 1)}$$

(a) Determine the growth rate $p'(t)$.
(b) When is the growth rate a maximum?
(c) Determine $\lim_{t\to\infty} p(t)$, the equilibrium population.
(d) Graph the function.

68. Logistic Curve The sales of a new stereo system over a period of time are expected to follow the logistic curve

$$f(x) = \frac{5000}{1 + 5e^{-x}}$$

where x is measured in years.

(a) Determine the year in which the sales rate is a maximum.
(b) Graph the function.

69. Logistic Curve The sales of a new car model over a period of time are expected to follow the relationship

$$f(x) = \frac{20{,}000}{1 + 50e^{-x}}$$

where x is measured in months.

(a) Determine the month in which the sales rate is a maximum.
(b) Graph the function.

70. Spread of Disease In a town of 50,000 people, the number of people at time t who have influenza is

$$N(t) = \frac{10{,}000}{1 + 9999e^{-t}}$$

where t is measured in days. [Note that the flu is spread by the one person who has it at $t = 0$.]

(a) At what time t is the rate of spreading the greatest?
(b) Graph the function.

13.4 Optimization

PREPARING FOR THIS SECTION *Before getting started, review the following:*

>> Geometry Review (Appendix A, Section A.9, pp. A-82–A-83)

OBJECTIVES
1 Find the absolute maximum and the absolute minimum of a function
2 Solve applied problems

Absolute Maximum and Absolute Minimum

If you look again at the examples of the previous two sections, you will notice that the local maximum and local minimum, when they existed, always occurred at a number c at which $f'(c) = 0$ or at which $f'(c)$ did not exist.

Necessary Condition for a Local Maximum or Local Minimum

If a function f has a local maximum or a local minimum at c, then either $f'(c) = 0$ or $f'(c)$ does not exist.

Because of this result, we call the numbers c for which $f'(c) = 0$ or $f'(c)$ does not exist **critical numbers of f.** The corresponding points $(c, f(c))$ on the graph of f are called **critical points.**

The local maximum (and local minimum) of a function is a point on the graph that is higher (or lower) than nearby points on the graph. However, the value of the function at such points may not be the largest (or smallest) value of the function on its domain.

The largest value, if one exists, of a function on its domain is called the **absolute maximum** of the function. The smallest value, if it exists, of a function on its domain is called the **absolute minimum** of the function.

Let's look at some of the possibilities. See Figure 34.

FIGURE 34

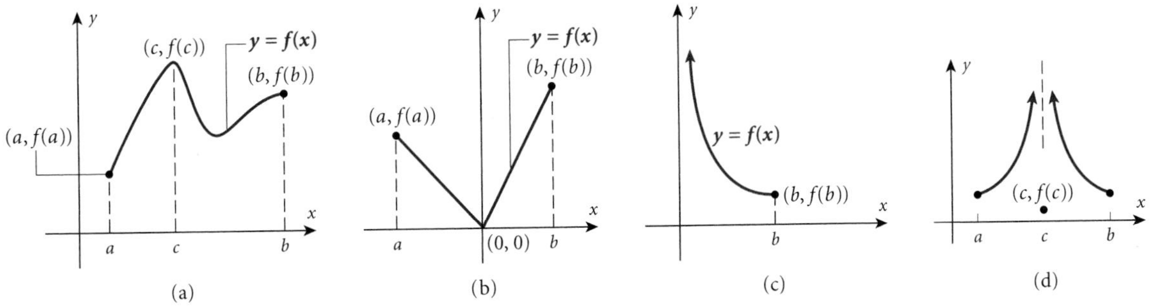

In Figure 34(a), the function f is continuous on the closed interval $[a, b]$. The absolute maximum of f is $f(c)$. Notice that c is a critical number since $f'(c) = 0$. The absolute minimum of f is $f(a)$.

In Figure 34(b), the function f is continuous on the closed interval $[a, b]$. The absolute maximum of f is $f(b)$. The absolute minimum of f is $f(0) = 0$. Notice that 0 is a critical number since $f'(0)$ does not exist.

In Figure 34(c), the function f is continuous on the interval $(0, b]$. The function f has no absolute maximum since there is no largest value of f on $(0, b]$. The absolute minimum is $f(b)$.

In Figure 34(d), the function f, whose domain is the closed interval $[a, b]$, is discontinuous at c. The function f has no absolute maximum since there is no largest value of f on $[a, b]$. The absolute minimum is $f(c)$.

The following result, which we state without proof, gives a condition under which a function will have an absolute maximum and an absolute minimum.

Condition for a Function to Have an Absolute Maximum and an Absolute Minimum

If a continuous function has as its domain a closed interval, the absolute maximum and absolute minimum exist.

Continuous functions defined on a closed interval will have an absolute maximum and an absolute minimum. Look again at Figures 34(a) and (b). We see that each function f is continuous on a closed interval $[a, b]$. Note that the absolute maximum and the absolute minimum occur either at a critical number or at an endpoint. This leads us to formulate the following test for finding the absolute maximum and the absolute minimum.

Test for Absolute Maximum and Absolute Minimum

If a continuous function $y = f(x)$ has a closed interval $[a, b]$ as its domain, we can find the absolute maximum (minimum) by choosing the largest (smallest) value from among the following:

1. Values of f at the critical numbers in the open interval (a, b)
2. $f(a)$
3. $f(b)$

If critical numbers of $y = f(x)$ are found that are not in the interval $[a, b]$, these critical numbers should be ignored since we are concerned only with the function on the interval $[a, b]$.

1 **EXAMPLE 1 Finding the Absolute Maximum and the Absolute Minimum**

Consider the function $f(x) = x^3 - 3x$. If the domain of f is $[0, 2]$, find the absolute maximum and the absolute minimum of f.

SOLUTION The function $f(x) = x^3 - 3x$, is a polynomial function so it is continuous on $[0, 2]$. To find the absolute maximum and absolute minimum, we first find $f'(x)$ so we can find any critical numbers of f,

$$f'(x) = 3x^2 - 3 = 3(x^2 - 1) = 3(x + 1)(x - 1)$$

The critical numbers of f are those values of x for which $f'(x)$ does not exist or for which $f'(x) = 0$. There are no values of x at which $f'(x)$ does not exist. The values of x for which $f'(x) = 0$ are found as follows;

$$f'(x) = 3(x + 1)(x - 1) = 0$$
$$x = -1 \quad \text{and} \quad x = 1$$

We ignore the critical number $x = -1$ since it is not in the domain, $0 \le x \le 2$. For the critical number $x = 1$, we have

$$f(1) = -2 \qquad f(x) = x^3 - 3x$$

The values of f at the endpoints 0 and 2 of the interval $[0, 2]$ are

$$f(0) = 0 \quad \text{and} \quad f(2) = 2$$

We choose the largest and smallest of -2, 0, and 2. The absolute maximum of f on $[0, 2]$ is 2 and the absolute minimum is -2.

NOW WORK PROBLEM 1.

EXAMPLE 2 Finding the Absolute Maximum and the Absolute Minimum

Find the absolute maximum and absolute minimum of the function $f(x) = x^{2/3}$ on the interval $[-1, 8]$. Graph f.

SOLUTION The function $f(x) = x^{2/3}$ is continuous on $[-1, 8]$. First, we locate the critical numbers, if any, by finding $f'(x)$.

$$f'(x) = \frac{d}{dx}x^{2/3} = \frac{2}{3}x^{-1/3} = \frac{2}{3x^{1/3}}$$

The only critical numbers are those for which $f'(x)$ does not exist, so the only critical number is $x = 0$, for which $f(0) = 0$.

Next, we find the values of f at the endpoints:

$$f(-1) = 1 \qquad \text{and} \qquad f(8) = 4$$

We choose the largest and smallest of 0, 1, and 4. The absolute maximum of $f(x) = x^{2/3}$ on $[-1, 8]$ is 4 and the absolute minimum is 0.

The graph of f is given in Figure 35.

FIGURE 35

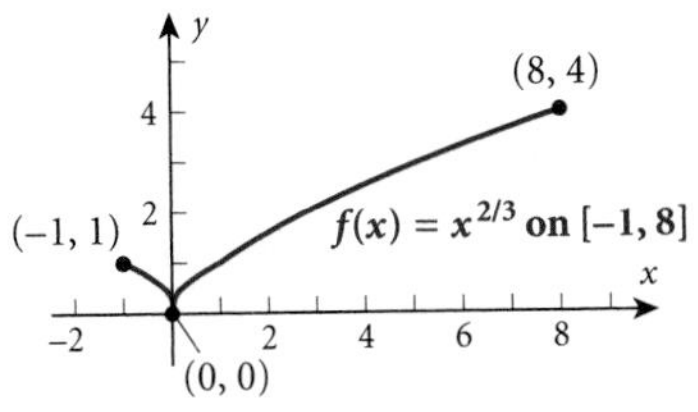

Applications

Solve applied problems 2

In general, each type of problem we will discuss requires some quantity to be minimized or maximized. We assume that the quantity we want to optimize can be represented by a function. Once this function is determined, the problem can be reduced to the question of determining at what number the function assumes its absolute maximum or absolute minimum.

Even though each applied problem has its unique features, it is possible to outline in a rough way a procedure for obtaining a solution. This five-step procedure is:

Steps for Solving Applied Problems

STEP 1 Identify the quantity for which a maximum or a minimum value is to be found.

STEP 2 Assign symbols to represent other variables in the problem. If possible, use an illustration to assist you.

STEP 3 Determine the relationships among these variables.

STEP 4 Express the quantity to be optimized as a function of one of these variables. Be sure to state the domain.

STEP 5 Apply the test for absolute maximum and absolute minimum to this function.

The following examples illustrate this procedure.

EXAMPLE 3 Maximizing Volume

From each corner of a square piece of sheet metal 18 centimeters on a side, remove a small square of side x centimeters and turn up the edges to form an open box. What should be the dimensions of the box so as to maximize the volume?

SOLUTION **STEP 1** The quantity to be maximized is the volume. Denote it by V.

STEP 2 Denote the dimensions of the side of the small square by x, as shown in Figure 36. Although the area of the sheet metal is fixed, the sides of the square can be changed and so are variables. Let y denote the portion left after cutting the x's to make the square.

FIGURE 36

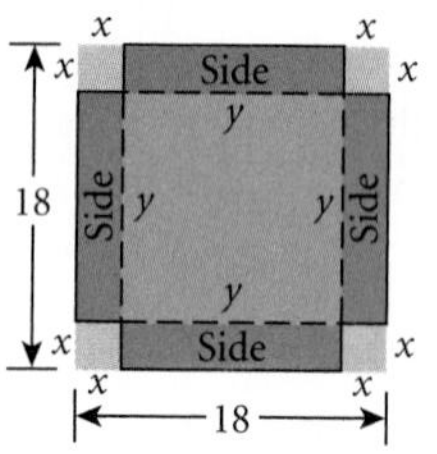

STEP 3 Then

$$y = 18 - 2x$$

STEP 4 The height of the box is x, while the area of the base of the box is y^2. The volume V is therefore

$$V = xy^2$$

To express V as a function of one variable, we substitute $y = 18 - 2x$ in the formula for V. This gives

$$V = V(x) = x(18 - 2x)^2 \qquad V = xy^2,\ y = 18 - 2x$$

This is the function to be maximized. Its domain is the set of real numbers. However, physically, the only numbers x that make sense are those between 0 and 9. We want to find the absolute maximum of

$$V(x) = x(18 - 2x)^2 \qquad 0 \le x \le 9$$

STEP 5 To find the number x that maximizes V, we differentiate and find the critical numbers, if any:

$$V'(x) = x\frac{d}{dx}(18 - 2x)^2 + (18 - 2x)^2\frac{d}{dx}x \qquad \text{Derivative of a product.}$$

$$\begin{aligned} V'(x) &= 2x(18 - 2x)(-2) + (18 - 2x)^2 && \text{Differentiate.} \\ &= (18 - 2x)[-4x + (18 - 2x)] && \text{Factor.} \\ &= (18 - 2x)(18 - 6x) && \text{Factor.} \end{aligned}$$

Now we set $V'(x) = 0$ and solve for x:

$$(18 - 2x)(18 - 6x) = 0 \qquad V'(x) = 0$$

$$18 - 2x = 0 \quad \text{or} \quad 18 - 6x = 0 \qquad \text{Apply the Zero-Product Property.}$$

$$x = 9 \quad \text{or} \quad x = 3 \qquad \text{Solve for } x.$$

The only critical number in the interval $(0, 9)$ is $x = 3$. We calculate the values of $V(x)$ at this critical number and at the endpoints of the interval $[0, 9]$:

$$V(0) = 0 \qquad V(3) = 3(18 - 6)^2 = 432 \qquad V(9) = 0$$

The maximum volume is 432 cubic centimeters and the dimensions of the box that yield the maximum volume are $x = 3$ centimeters deep by $y = 18 - 2(3) = 12$ centimeters on each side.

NOW WORK PROBLEM 25.

EXAMPLE 4 Maximizing Area

PLAYPEN PROBLEM* A manufacturer of playpens makes a model that can be opened at one corner and attached at right angles to a wall or the side of a house. If each side is 3 feet in length, the open configuration doubles the available area in which a child can play from 9 square feet to 18 square feet. See Figure 37(a).

Suppose hinges are placed at adjoining sides of the playpen to allow for a configuration like the one shown in Figure 37(b). If x is the distance between the two parallel sides of the playpen, then what value of x will maximize the area A? What is the maximum area?

FIGURE 37

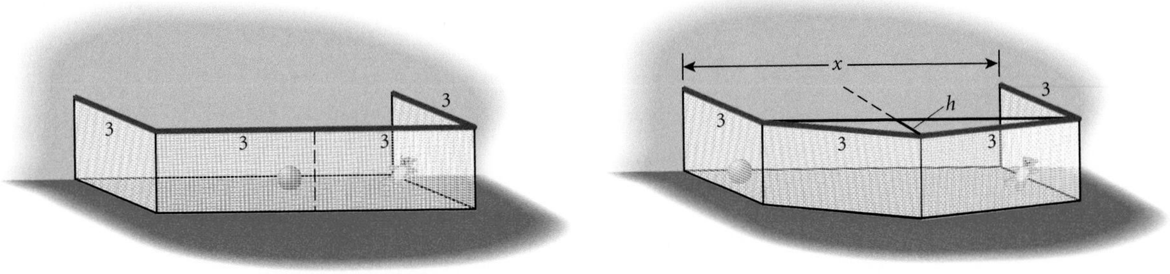

SOLUTION The variable x has domain $0 \leq x \leq 6$. Do you see why? The area A consists of the area of a rectangle (with width 3 and length x) and the area of an isosceles triangles (with base x and two equal sides of length 3). The height h of the triangle is obtained using the Pythagorean theorem:

$$h^2 = 3^2 - \left(\frac{x}{2}\right)^2 \qquad h^2 + \left(\frac{x}{2}\right)^2 = 3^2$$

$$h = \sqrt{9 - \frac{x^2}{4}} = \sqrt{\frac{36 - x^2}{4}} = \frac{1}{2}\sqrt{36 - x^2}$$

The area A enclosed by the playpen is

$$A = \text{area of rectangle} + \text{area of triangle}$$

$$= 3x + \frac{1}{2}xh = 3x + \frac{1}{2}x\left(\frac{1}{2}\sqrt{36 - x^2}\right)$$

$$A(x) = 3x + \frac{1}{4}x\sqrt{36 - x^2}$$

*Adapted from Proceedings, Summer Conference for College Teachers on Applied Mathematics, *University of Missouri–Rolla, 1971.*

To find the absolute maximum of $A = A(x), 0 \le x \le b$, we first find the critical numbers by examining $A'(x)$.

$$A'(x) = \frac{d}{dx}\left[3x + \frac{1}{4}x\sqrt{36 - x^2}\right]$$

$$= \frac{d}{dx}(3x) + \frac{1}{4}\frac{d}{dx}[x(36 - x^2)^{1/2}] \qquad \text{Derivative of a sum.}$$

$$= 3 + \frac{1}{4}\left[x\frac{d}{dx}(36 - x^2)^{1/2} + (36 - x^2)^{1/2}\frac{d}{dx}x\right] \qquad \text{Derivative of a product.}$$

$$= 3 + \frac{1}{4}\left[x \cdot \frac{1}{2} \cdot (36 - x^2)^{-1/2}(-2x) + (36 - x^2)^{1/2} \cdot 1\right] \qquad \text{Use the Power Rule.}$$

$$= 3 - \frac{1}{4} \cdot \frac{x^2}{\sqrt{36 - x^2}} + \frac{1}{4} \cdot \sqrt{36 - x^2} \qquad \text{Simplify.}$$

$$= \frac{12\sqrt{36 - x^2} - x^2 + (36 - x^2)}{4\sqrt{36 - x^2}} \qquad \text{Write as a single quotient.}$$

$$= \frac{12\sqrt{36 - x^2} + 36 - 2x^2}{4\sqrt{36 - x^2}} \qquad \text{Simplify.}$$

Set $A'(x) = 0$. Then

$$\frac{12\sqrt{36 - x^2} + 36 - 2x^2}{4\sqrt{36 - x^2}} = 0$$

$$12\sqrt{36 - x^2} + 36 - 2x^2 = 0 \qquad \text{Set the numerator equal to 0.}$$

$$12\sqrt{36 - x^2} = 2x^2 - 36 \qquad \text{Isolate the square root.}$$

$$144(36 - x^2) = 4x^4 - 144x^2 + 1296 \qquad \text{Square both sides.}$$

$$4x^4 = 3888 \qquad \text{Simplify.}$$

$$x = \sqrt[4]{\frac{3888}{4}} \approx 5.58 \qquad \text{Solve for } x.$$

$A'(x)$ does not exist at $x = -6$ and $x = 6$, but these values are not in the interval $(0, 6)$. The only critical number in $(0, 6)$ is 5.58.

Evaluate $A(x)$ at the endpoints $x = 0$ and $x = 6$, and at the critical number $x = 5.58$. The results are

$$A(0) = 0 \qquad A(6) = 18 \qquad A(5.58) = 19.82$$

A wall of length $2x = 2(5.58) = 11.16$ will maximize the area, and a configuration like the one in Figure 37(b), increases the play area by about 10% (from 18 square feet to 19.82 square feet).

The next two examples illustrate how to solve optimization problems in which the function to be optimized has a domain that is not a closed interval.

EXAMPLE 5 Minimizing Cost

A can company wants to produce a cylindrical container with a capacity of 1000 cubic centimeters. The top and bottom of the container must be made of material that costs \$0.05 per square centimeter, while the sides of the container can be made of material costing \$0.03 per square centimeter. Find the dimensions that will minimize the total cost of the container.

SOLUTION Figure 38 shows a cylindrical container and the area of its top, bottom, and lateral surfaces.

FIGURE 38

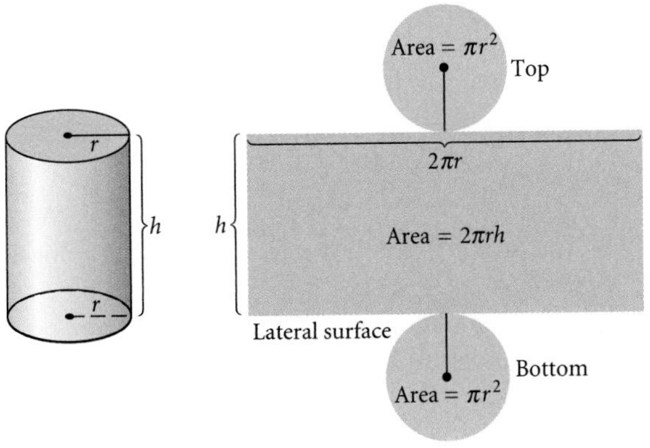

As indicated in the figure, if we let h stand for the height of the can and let r stand for the radius of the top and bottom, then the total area of the bottom and top is $2\pi r^2$ and the area of the lateral surface of the can is $2\pi rh$. The total cost C of manufacturing the can is

$$C = (\$0.05)(2\pi r^2) + (\$0.03)(2\pi rh) = 0.1\pi r^2 + 0.06\pi rh$$

This is the function we want to minimize.

The cost function is a function of two variables, h and r. But there is a relationship between h and r since the volume of the cylinder, $V = \pi r^2 h$, is fixed at 1000 cubic centimeters. That is,

$$\pi r^2 h = 1000$$

$$h = \frac{1000}{\pi r^2} \quad \text{Divide both sides by } \pi r^2.$$

Substituting this expression for h into the cost function C, we obtain

$$C = C(r) = 0.1\pi r^2 + 0.06\pi r\left(\frac{1000}{\pi r^2}\right) \quad C = 0.1\pi r^2 + 0.06\pi rh;\ h = \frac{1000}{\pi r^2}$$

$$= 0.1\pi r^2 + \frac{60}{r} \quad \text{Simplify.}$$

The domain of C is $\{r \mid r > 0\}$. The derivative of C with respect to r is

$$C'(r) = \frac{d}{dr}\left(0.1\pi r^2 + \frac{60}{r}\right) = 0.2\pi r - \frac{60}{r^2} \quad \frac{d}{dr}\frac{60}{r} = 60\frac{d}{dr}r^{-1} = -60r^{-2} = \frac{-60}{r^2}$$

$$= \frac{0.2\pi r^3 - 60}{r^2}$$

Set $C'(r) = 0$. Then

$$0.2\pi r^3 - 60 = 0 \quad \text{Set the numerator equal to 0.}$$

$$r^3 = \frac{300}{\pi} \quad \text{Solve for } r^3.$$

$$r = \sqrt[3]{\frac{300}{\pi}} \approx 4.57 \quad \text{Solve for } r.$$

Since $r > 0$, the only critical number is $r \approx 4.57$.

To use the Second Derivative Test, we need to evaluate C'' at this critical number.

$$C''(r) = \frac{d}{dr}\left(0.2\pi r - \frac{60}{r^2}\right) = 0.2\pi + \frac{120}{r^3}$$

and

$$C''\left(\sqrt[3]{\frac{300}{\pi}}\right) = 0.2\pi + \frac{120\pi}{300} > 0$$

By the Second Derivative Test, the cost is a local minimum for $r = \sqrt[3]{\dfrac{300}{\pi}} \approx 4.57$ centimeters.

Since the only physical constraint is that r be positive, this local minimum value is the absolute minimum. The corresponding height of this can is

$$h = \frac{1000}{\pi r^2} \approx 15.24 \text{ centimeters}$$

These are the dimensions that will minimize the cost of the material.

If the cost of the material is the same for the top, bottom, and lateral surfaces of a cylindrical container, then the minimum cost occurs when the surface area is minimum. It can be shown that for any fixed volume, the minimum surface area is obtained when the height equals twice the radius. See Problem 39.

NOW WORK PROBLEM 29.

SEEING THE CONCEPT: Graph the function $y = 0.1\pi x^2 + \dfrac{60}{x}$. Use MINIMUM to verify the result of Example 5.

Maximizing Tax Revenue In determining the tax rate on cars, telephones, etc., the government is always faced with the following problem: How large should the tax be so that the tax revenue will be as large as possible? Let's examine this situation. When the government places a tax on a product, the price of this product for the consumer may increase and the quantity demanded may decrease accordingly. A very large tax may cause the quantity demanded to diminish to zero with the result that no tax revenue is collected. On the other hand, if no tax is levied, there will be no tax revenue at all. The problem is to find the tax rate that optimizes tax revenue. (Tax revenue is the product of the tax rate times the actual quantity consumed.)

Let's assume that because of long-time experience in levying taxes, the government is able to determine that the relationship between the quantity q consumed of a certain product and the related tax rate t is

$$t = \sqrt{27 - 3q^2}$$

See Figure 39.

FIGURE 39

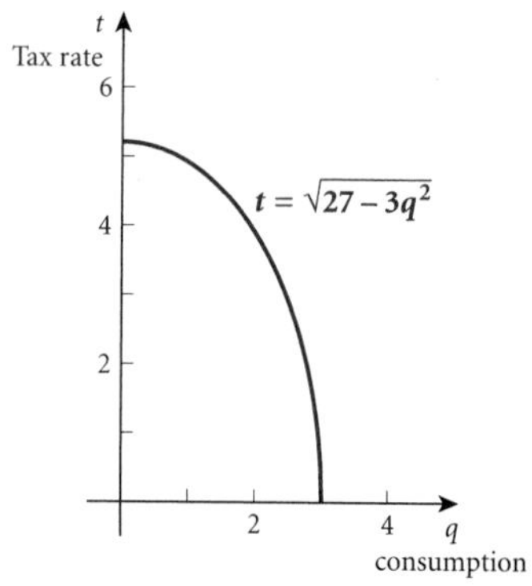

Notice that the relationship between tax rate and quantity consumed conforms to the restrictions discussed earlier. For example, when the tax rate $t = 0$, the quantity consumed is $q = 3$; when the tax is at a maximum ($t = 5.2\%$), the quantity consumed is zero.

The revenue R due to the tax rate t is the product of tax rate and the quantity consumed:

$$R = qt = q(27 - 3q^2)^{1/2}$$

where R is measured in millions of dollars. Since both q and t are assumed to be nonnegative, the domain of R is $0 \le q \le 3$.

To find the absolute maximum, we differentiate R with respect to q using the formula for the derivative of a product and the Power Rule:

$$R'(q) = \frac{d}{dq}\left[q(27 - 3q^2)^{1/2}\right]$$

$$= q\frac{d}{dq}(27 - 3q^2)^{1/2} + (27 - 3q^2)^{1/2}\frac{d}{dq}q \qquad \text{Derivative of a product.}$$

$$= q \cdot \frac{1}{2}(27 - 3q^2)^{-1/2}(-6q) + (27 - 3q^2)^{1/2} \qquad \text{Apply the Power Rule.}$$

$$= \frac{-3q^2}{(27 - 3q)^{1/2}} + (27 - 3q^2)^{1/2} \qquad \text{Simplify.}$$

$$= \frac{-3q^2 + 27 - 3q^2}{(27 - 3q^2)^{1/2}} = \frac{27 - 6q^2}{(27 - 3q^2)^{1/2}}$$

The critical numbers obey

$$27 - 6q^2 = 0 \qquad \text{and} \qquad 27 - 3q^2 = 0$$
$$q^2 = 4.5 \qquad\qquad q^2 = 9$$
$$q = \pm\sqrt{4.5} \qquad\qquad q = \pm 3$$

The only critical number in the interval $(0, 3)$ is $q = \sqrt{4.5} \approx 2.12$.

To find the absolute maximum, we compare the values of R at the endpoints $q = 0$ and $q = 3$ with its value at $q = 2.12$:

$$R(0) = 0 \quad R(\sqrt{4.5}) = \sqrt{4.5}\,\sqrt{13.5} = (2.12)(3.67) = 7.79 \quad R(3) = 0$$

The revenue is maximized at $q = 2.12$. The tax rate corresponding to maximum revenue is

$$t = \sqrt{27 - 3q^2} = \sqrt{13.5} = 3.67$$

This means that, for a tax rate of 3.67%, a maximum revenue $R = 7.79$ million dollars is generated.

NOW WORK PROBLEM 37.

Contraction of Windpipe While Coughing Coughing is caused by an increase in pressure in the lungs and is accompanied by a decrease in the diameter of the windpipe. From physics, the amount V of air flowing through the windpipe is related to the radius r of the windpipe and pressure difference p at each end by the equation $V = kpr^4$, where k is a constant. The radius r will decrease with increased pressure p according to the formula $r_0 - r = cp$, where r_0 is the radius of the windpipe when there is no difference in pressure and c is a positive constant. We wish to find the radius r that allows the most air to flow through the windpipe.

We shall restrict r so that

$$0 < \frac{r_0}{2} \le r \le r_0$$

The amount V of air flowing through the windpipe is given by

$$V = kpr^4$$

where $cp = r_0 - r$ from which $p = \dfrac{r - r_0}{c}$.

We wish to maximize

$$V(r) = k\left(\frac{r_0 - r}{c}\right)r^4 = \frac{k}{c}r_0r^4 - \frac{k}{c}r^5 \qquad \frac{r_0}{2} \le r \le r_0$$

The derivative is

$$V'(r) = 4\frac{k}{c}r_0r^3 - 5\frac{k}{c}r^4 = \frac{k}{c}r^3(4r_0 - 5r)$$

The only critical number is $r = \frac{4r_0}{5}$. (We exclude $r = 0$ because $0 < \frac{r_0}{2} \le r \le r_0$.) Using the test for an absolute maximum, we find

$$V\left(\frac{r_0}{2}\right) = \frac{k}{c}r_0 \cdot \frac{r_0^4}{16} - \frac{k}{c} \cdot \frac{r_0^5}{32} = \frac{kr_0^5}{32c} \qquad V(r_0) = 0 \qquad V\left(\frac{4r_0}{5}\right) = \frac{k}{c}r_0 \cdot \frac{256r_0^4}{625} - \frac{k}{c} \cdot \frac{1024}{3125}r_0^5 = \frac{256\,kr_0^5}{3125c}$$

Since $\frac{256kr_0^5}{3125c} > \frac{kr_0^5}{32c}$, the maximum air flow is obtained when the radius of the windpipe is $\frac{4}{5}r_0$; that is, the windpipe contracts by a factor of $\frac{1}{5}$.

Marginal Analysis The revenue derived from selling x units is $R(x) = x\,d(x)$, where $p = d(x)$ is the price function. If $C(x)$ is the cost of producing x units, the **profit function P**, assuming whatever is produced can be sold, is

$$P(x) = R(x) - C(x)$$

What quantity x will maximize profit?

To maximize $P(x)$, we find the critical numbers of P:

$$\frac{d}{dx}P(x) = \frac{d}{dx}[R(x) - C(x)] = \frac{d}{dx}R(x) - \frac{d}{dx}C(x) = R'(x) - C'(x) = 0$$

$$R'(x) = C'(x)$$

We apply the Second Derivative Test to the function P:

$$P''(x) = R''(x) - C''(x)$$

The profit P has a local maximum at a number x if $P''(x) < 0$. This will occur at a number x for which the marginal revenue function equals the marginal cost and $R''(x) < C''(x)$. The numbers x are restricted to a closed interval in which the endpoints should be tested separately.

Maximizing Profit

The equality

$$\boxed{R'(x) = C'(x)}$$

is the basis for the classical economic criterion for maximum profit—that marginal revenue and marginal cost be equal.

EXERCISE 14.4 Answers to Odd-Numbered Problems Begin on Page AN-85.

In Problems 1–24, find the absolute maximum and absolute minimum of each function f on the indicated interval.

1. $f(x) = x^2 + 2x$ on $[-3, 3]$
2. $f(x) = x^2 - 8x$ on $[-1, 10]$
3. $f(x) = 1 - 6x - x^2$ on $[0, 4]$
4. $f(x) = 4 - 2x - x^2$ on $[-2, 2]$
5. $f(x) = x^3 - 3x^2$ on $[1, 4]$
6. $f(x) = x^3 - 6x$ on $[-2, 2]$
7. $f(x) = x^4 - 2x^2 + 1$ on $[0, 1]$
8. $f(x) = 3x^4 - 4x^3$ on $[-2, 0]$
9. $f(x) = x^{2/3}$ on $[-1, 1]$
10. $f(x) = x^{1/3}$ on $[-1, 1]$
11. $f(x) = 2\sqrt{x}$ on $[1, 4]$
12. $f(x) = 4 - \sqrt{x}$ on $[0, 4]$
13. $f(x) = x\sqrt{1 - x^2}$ on $[-1, 1]$
14. $f(x) = x^2\sqrt{2 - x}$ on $[0, 2]$
15. $f(x) = \dfrac{x^2}{x - 1}$ on $\left[-1, \dfrac{1}{2}\right]$
16. $f(x) = \dfrac{x}{x^2 - 1}$ on $\left[-\dfrac{1}{2}, \dfrac{1}{2}\right]$
17. $f(x) = (x + 2)^2(x - 1)^{2/3}$ on $[-4, 5]$
18. $f(x) = (x - 1)^2(x + 1)^3$ on $[-2, 7]$
19. $f(x) = \dfrac{(x - 4)^{1/3}}{x - 1}$ on $[2, 12]$
20. $f(x) = \dfrac{(x + 3)^{2/3}}{x + 1}$ on $[-4, -2]$
21. $f(x) = xe^x$ on $[-10, 10]$
22. $f(x) = xe^{-x}$ on $[-10, 10]$
23. $f(x) = \dfrac{\ln x}{x}$ on $[1, 3]$
24. $f(x) = x \ln x$ on $[1, 5]$

25. **Best Dimensions for a Box** An open box with a square base is to be made from a square piece of cardboard 12 centimeters on a side by cutting out a square from each corner and turning up the sides. Find the dimensions of the box that yield the maximum volume.

26. **Best Dimensions for a Box** An open box with a square base is to be made from a square piece of cardboard 24 centimeters on a side by cutting out a square from each corner and turning up the sides. Find the dimensions of the box that yield the maximum volume.

27. **Best Dimensions for a Box** A box, open at the top with a square base, is to have a volume of 8000 cubic centimeters. What should the dimensions of the box be if the amount of material used is to be a minimum?

28. **Best Dimensions for a Closed Box** If the box in Problem 27 is to be closed on top, what should the dimensions of the box be if the amount of material used is to be a minimum?

29. **Best Dimensions for a Can** A cylindrical container is to be produced that will have a capacity of 4000 cubic centimeters. The top and bottom of the container are to be made of material that costs \$0.50 per square centimeter, while the side of the container is to be made of material costing \$0.40 per square centimeter. Find the dimensions that will minimize the total cost of the container.

30. **Best Dimensions for a Can** A cylindrical container is to be produced that will have a capacity of 10 cubic meters. The top and bottom of the container are to be made of a material that costs \$2 per square meter, while the side of the container is to be made of material costing \$1.50 per square meter. Find the dimensions that will minimize the total cost of the container.

31. **Placing Telephone Boxes** A telephone company is asked to provide telephone service to a customer whose house is located 2 kilometers away from the road along which the telephone lines run. The nearest telephone box is located 5 kilometers down this road. If the cost to connect the telephone line is \$50 per kilometer along the road and \$60 per kilometer away from the road, where along the road from the box should the company connect the telephone line so as to minimize construction cost?

Hint: Let x denote the distance from the box to the connection so that $5 - x$ is the distance from this point to the point on the road closest to the house.

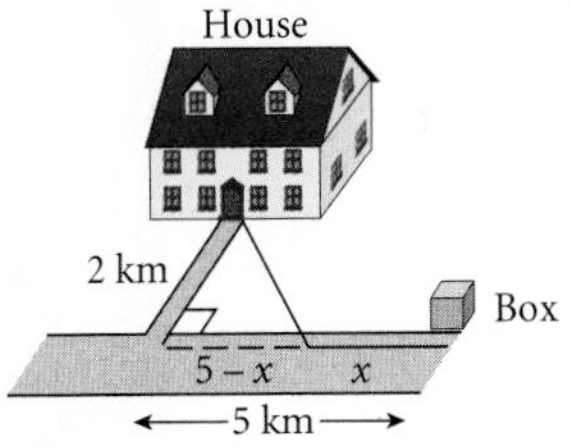

32. **Minimizing Travel Time** A small island is 3 kilometers from the nearest point P on the straight shoreline of a large lake. If a woman on the island can row her boat 2.5 kilometers per hour and can walk 4 kilometers per hour, where should she land her boat in order to arrive in the shortest time at a town 12 kilometers down the shore from P?

Hint: Let x be the distance from P to the landing point.

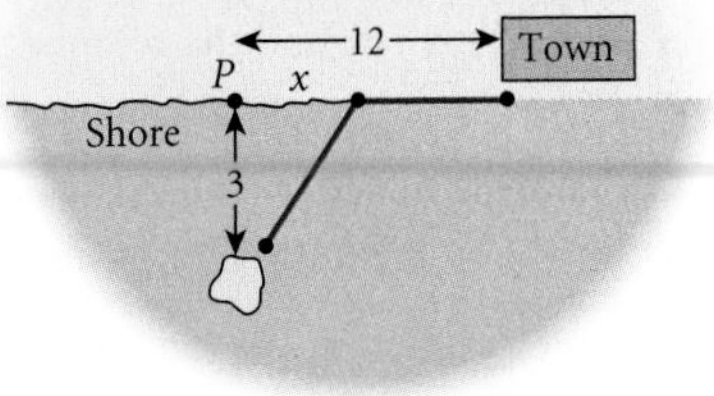

33. Most Economical Speed A truck has a top speed of 75 miles per hour and, when traveling at the rate of x miles per hour, consumes gasoline at the rate of $\frac{1}{200}\left[\left(\frac{1600}{x}\right)+x\right]$ gallon per mile. If the length of the trip is 200 miles and the price of gasoline is \$1.60 per gallon, the cost is

$$C(x) = 1.60\left(\frac{1600}{x} + x\right)$$

where $C(x)$ is measured in dollars. What is the most economical speed for the truck? Use the interval $[10, 75]$.

34. If the driver of the truck in Problem 33 is paid \$8 per hour, what is the most economical speed for the truck?

35. Page Layout A printer plans on having 50 square inches of printed matter per page and is required to allow for margins of 1 inch on each side and 2 inches on the top and bottom. What are the most economical dimensions for each page if the cost per page depends on the area of the page?

36. Dimensions for a Window A window is to be made in the shape of a rectangle surmounted by a semicircle with diameter equal to the width of the rectangle. See the figure. If the perimeter of the window is 22 feet, what dimensions will let in the most light?

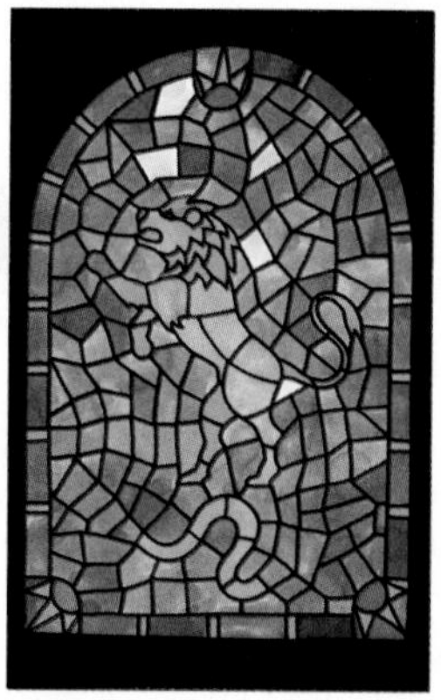

37. Tax Revenue On a particular product, government economists determine that the relationship between tax rate t and the quantity q consumed is

$$t + 3q^2 = 18$$

Graph this relationship and explain how it could be justified. Find the optimal tax rate and the revenue generated by this tax rate.

38. Most Economical Speed A truck has a top speed of 75 miles per hour and, when traveling at the rate of x miles per hour, consumes gasoline at the rate of $\frac{1}{200}\left[\left(\frac{1600}{x}\right)+x\right]$ gallon per mile. This truck is to be taken on a 200 mile trip by a driver who is to be paid at the rate of b dollars per hour plus a commission of c dollars. Since the time required for this trip at x miles per hour is $\frac{200}{x}$, the total cost, if gasoline costs a dollars per gallon, is

$$C(x) = \left(\frac{1600}{x} + x\right)a + \frac{200}{x}b + c$$

Find the most economical possible speed under each of the following sets of conditions:

(a) $b = 0$, $c = 0$ (b) $a = 1.50$, $b = 8.00$, $c = 500$
(c) $a = 1.60$, $b = 10.00$, $c = 0$

39. Prove that a cylindrical container of fixed volume V requires the least material (minimum surface area) when its height is twice its radius.

40. Transatlantic Crossing An airplane crosses the Atlantic Ocean (3000 miles) with an airspeed of 500 mph.

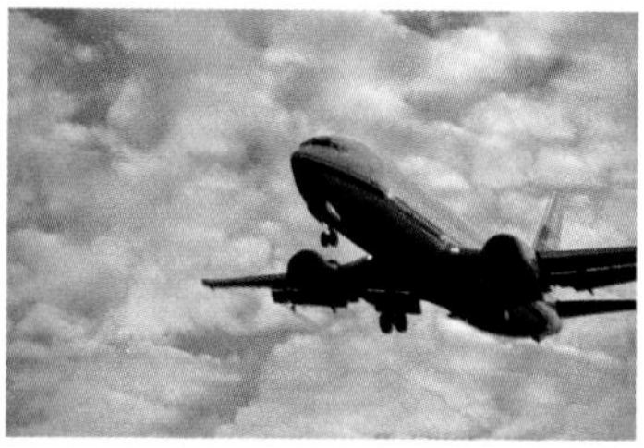

(a) Find the time saved with a 25 mile per hour tail wind.
(b) Find the time lost with a 50 mile per hour head wind.
(c) If the cost per passenger is

$$C(x) = 100 + \frac{x}{10} + \frac{36{,}000}{x}$$

where x is the ground speed and $C(x)$ is the cost in dollars, what is the cost per passenger when there is no wind?

(d) What is the cost with a tail wind of 25 miles per hour?
(e) What is the cost with a head wind of 50 miles per hour?
(f) What ground speed minimizes the cost?
(g) What is the minimum cost per passenger?

41. Drug Concentration The concentration C of a certain drug in the bloodstream t hours after injection into muscle tissue is given by

$$C(t) = \frac{2t}{16 + t^3}$$

When is the concentration greatest?

42. Profit Function The cost C in dollars of producing n machines is $C(n) = 10n + 0.05n^2 + 150{,}000$. If a machine is priced at \$$m$, the estimated number that would be sold is $n = 2000 - \frac{m}{5}$. At what price would the profit be maximum?

14.5 Elasticity of Demand

OBJECTIVE 1 Determine the elasticity of demand

In this section we will study how economists describe the effect that changes in price have on demand and revenue. Recall that a demand equation expresses the market price p which will generate a demand for exactly x units. Suppose the price p and the quantity x demanded for a certain product are related by the following demand equation:

$$p = 200 - 0.02x \qquad (1)$$

The equation states that in order to sell x units, the price must be set at $200 - 0.02x$ dollars. For example, to sell 5000 units, the price must be set at $200 - 0.02(5000) = \$100$ per unit.

In problems involving revenue, sales, and profit, it is customary to use the demand equation to express price as a function of demand. Since we are now interested in the effects that changes in price have on demand, it is more practical to express demand as a function of price. Equation (1) may be solved for x in terms of p to yield

$$x = \frac{1}{0.02}(200 - p) \qquad \text{or} \qquad x = f(p) = 50(200 - p)$$

This equation expresses quantity x as a function of the price. Since x and p must be nonnegative, we restrict p so that $0 \le p \le 200$.

FIGURE 40

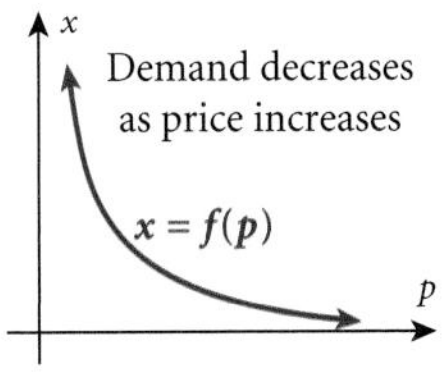

Usually, increasing the price of a commodity lowers the quantity demanded, while decreasing the price results in higher demand. Therefore, the typical demand function $x = f(p)$ is decreasing. See Figure 40. If $x = f(p)$ is a differentiable function, then $f'(p) < 0$ and $f'(p)$ equals the rate of change in quantity demanded with respect to price.

Elasticity measures the ratio of the relative change of the quantity demanded to the relative change of price. For the demand function $x = f(p)$ and the price p we have:

Elasticity

$$\text{The relative change in quantity demanded} = \frac{\Delta x}{x} = \frac{f(p + \Delta p) - f(p)}{f(p)}$$

$$\text{The relative change in price} = \frac{\Delta p}{p} \qquad (2)$$

$$\text{Elasticity} = \frac{\text{Relative change in quantity demanded}}{\text{Relative change in price}} = \frac{\dfrac{\Delta x}{x}}{\dfrac{\Delta p}{p}} \qquad (3)$$

Economists use elasticity to study the effect of price change on quantity demanded. Since elasticity depends on p and Δp, if we let $\Delta p \to 0$, we obtain an expression for the elasticity of demand at price p, denoted by $E(p)$:

$$\begin{aligned}
E(p) &= \lim_{\Delta p \to 0} \frac{\dfrac{\Delta x}{x}}{\dfrac{\Delta p}{p}} \\
&= \lim_{\Delta p \to 0} \frac{\dfrac{f(p+\Delta p)-f(p)}{f(p)}}{\dfrac{\Delta p}{p}} && x = f(p);\ \Delta x = f(p+\Delta p) - f(p) \\
&= \lim_{\Delta p \to 0} \left[\frac{f(p+\Delta p)-f(p)}{f(p)} \cdot \frac{p}{\Delta p}\right] \\
&= \lim_{\Delta p \to 0} \left[\frac{p}{f(p)} \cdot \frac{f(p+\Delta p)-f(p)}{\Delta p}\right] && \text{Rearrange terms.} \\
&= \lim_{\Delta p \to 0} \frac{p}{f(p)} \cdot \lim_{\Delta p \to 0} \frac{f(p+\Delta p)-f(p)}{\Delta p} && \text{Limit of a product equals the product of the limits.} \\
&= \frac{p}{f(p)} \lim_{\Delta p \to 0} \frac{f(p+\Delta p)-f(p)}{\Delta p} && \lim_{\Delta p \to 0} \frac{p}{f(p)} = \frac{p}{f(p)} \\
&= \frac{p}{f(p)} \cdot f'(p)
\end{aligned}$$

Elasticity of Demand

The **elasticity of demand $E(p)$ at a price p** for the demand function $x = f(p)$ is

$$E(p) = \frac{pf'(p)}{f(p)}$$

Since $f'(p)$ is always negative for a typical demand function, the quantity $\dfrac{pf'(p)}{f(p)}$ will be negative for all values of p. For convenience, economists prefer to work with positive numbers. Therefore the **price elasticity of demand** is taken to be $|E(p)|$. For a given price p, if $|E(p)| > 1$, the demand is said to be **elastic.** If $|E(p)| < 1$, the demand is said to be **inelastic.** If $|E(p)| = 1$, that is, if $E(p) = -1$, the demand is said to be **unit elastic.**

1 **EXAMPLE 1 Determine the Elasticity of Demand**

Suppose $x = f(p) = 5000 - 30p^2$ is the demand function for a certain commodity, where p is the price per pound and x is the quantity demanded in pounds.

(a) What quantity can be sold at \$10 per pound?
(b) Determine the elasticity of demand function $E(p)$.
(c) Interpret the elasticity of demand at $p = \$5$.
(d) Interpret the elasticity of demand at $p = \$10$.
(e) At what price is the elasticity of demand equal to -1? That is, at what price is demand unit elastic? Interpret unit elasticity.

SOLUTION **(a)** At $p = \$10$, $f(10) = 5000 - 30(10)^2 = 2000$. Therefore, 2000 pounds of the commodity can be sold at a price of \$10.

(b) $E(p) = \dfrac{p\,f'(p)}{f(p)} = \dfrac{p(-60p)}{5000 - 30p^2} = \dfrac{-60p^2}{5000 - 30p^2}$ $\quad f'(p) = \dfrac{d}{dp}(5000 - 30p^2) = -60p$

(c) The elasticity of demand at the price $p = \$5$ is $E(5)$.

$$E(5) = \frac{-60(5)^2}{5000 - 30(5)^2} = -\frac{1500}{4250} = -0.353$$

$$|E(5)| = 0.353$$

When the price is set at \$5 per pound, a small increase in price will result in a relative decrease in quantity demanded of about 0.353 times the relative increase in price. For instance, if the price is increased from \$5 by 10%, then the quantity demanded will decrease by $(0.353)(10\%) = 0.0353 = 3.53\%$.

(d) When $p = \$10$, we have

$$E(10) = \frac{-60(10)^2}{5000 - 30(10)^2} = -3$$

$$\text{or} \quad |E(10)| = 3$$

When the price is set at \$10, a small increase in price will result in a relative decrease in quantity demanded of 3 times the relative increase of price. For instance, a 10% price increase will result in a decrease in quantity demanded of approximately $3(10\%) = 30\%$.

(e) The demand is unit elastic if $E(p) = -1$.

$$\frac{-60p^2}{5000 - 30p^2} = -1 \qquad E(p) = \frac{-60p^2}{5000 - 30p^2} \text{ from part (b).}$$

$$-60p^2 = -5000 + 30p^2 \qquad \text{Clear fractions.}$$

$$-90p^2 = -5000 \qquad \text{Simplify.}$$

$$p^2 = 55.5556 \qquad \text{Divide by } -90.$$

$$p = 7.45 \qquad \text{Solve for } p.$$

If the price is set at \$7.45, a small increase in price will result in the same decrease in quantity demanded. For example, at this price, a 2% increase in price results in a 2% decrease in quantity demanded.

NOW WORK PROBLEM 1.

Revenue and Elasticity of Demand

The concept of elasticity has an interesting relationship to the total revenue $R(p)$:

(a) If the demand is elastic, then an increase in the price per unit will result in a decrease in total revenue.

(b) If the demand is inelastic, then an increase in the price per unit will result in an increase in total revenue.

(c) If the demand is unit elastic, then an increase in price will not change the total revenue.

We establish these relationships as follows:

Since total revenue is given by $R(p) = xp = f(p) \cdot p$, we calculate the marginal revenue to be

$$R'(p) = f(p) + pf'(p) \qquad \frac{d}{dp}[f(p) \cdot p] = f(p) \cdot \frac{d}{dp}p + p\frac{d}{dp}f(p) = f(p) + pf'(p)$$

$$= f(p)\left[1 + \frac{pf'(p)}{f(p)}\right] \qquad \text{Factor out } f(p);\ pf'(p) = f(p) \cdot \frac{pf'(p)}{f(p)}$$

$$= f(p)[1 + E(p)] \qquad E(p) = \frac{pf'(p)}{f(p)}$$

We know that $f(p) > 0$.

If the demand is elastic, then $|E(p)| > 1$. Since $E(p) < 0$, this means that $E(p) < -1$, so that $1 + E(p) < 0$. As a result, $R'(p) < 0$. In other words, $R(p)$ is decreasing. This means an increase in price will result in a decrease in total revenue when the demand is elastic.

If the demand is inelastic, then $|E(p)| < 1$, so that $E(p) > -1$ or $1 + E(p) > 0$. In this case, $R'(p) > 0$ so $R(p)$ is increasing. This implies that an increase in price will result in an increase in total revenue when the demand is inelastic.

Finally, if demand is unit elastic, then $|E(p)| = 1$. Then $E(p) = -1$ and $1 + E(p) = 0$. Then $R'(p) = 0$. When the first derivative of a function equals 0, the function is neither increasing nor decreasing. So this implies that when demand has unit elasticity, a small increase (or decrease) in price results in no change in total revenue.

NOW WORK PROBLEM 27.

EXERCISE 14.5 Answers to Odd-Numbered Problems Begin on Page AN-86.

1. Given the demand equation $p + \frac{1}{100}x = 40$
   (a) Express the demand x as a function of p.
   (b) Find the elasticity of demand $E(p)$.
   (c) What is the elasticity of demand when $p = \$5$? If the price is increased by 10%, what is the approximate change in demand?
   (d) What is the elasticity of demand when $p = \$15$? If the price is increased by 10%, what is the approximate change in demand?
   (e) What is the elasticity of demand when $p = \$20$? If the price is increased by 10%, what is the approximate change in demand?

2. Repeat Problem 1 for the demand equation

$$p + \frac{1}{200}x = 80$$

3. Given the demand equation $p + \frac{1}{200}x = 50$
   (a) Express the demand x as a function of p.
   (b) Find the elasticity of demand $E(p)$.
   (c) What is the elasticity of demand when $p = \$10$? If the price is increased by 5%, what is the approximate change in demand?
   (d) What is the elasticity of demand when $p = \$25$? If the price is increased by 5%, what is the approximate change in demand?
   (e) What is the elasticity of demand when $p = \$35$? If the price is increased by 5%, what is the approximate change in demand?

4. Repeat Problem 3 for the demand function

$$p + \frac{1}{200}x = 100$$

In Problems 5–16, a demand function is given. Find E(p) and determine if demand is elastic, inelastic, or unit elastic at the indicated price.

5. $x = f(p) = 600 - 3p \quad \text{at } p = 50$

6. $x = f(p) = 700 - 4p \quad \text{at } p = 40$

7. $x = f(p) = \dfrac{600}{p+4} \quad \text{at } p = 10$

8. $x = f(p) = \dfrac{500}{p+6} \quad \text{at } p = 10$

9. $x = f(p) = 10{,}000 - 10p^2 \quad \text{at } p = 10$

10. $x = f(p) = 2250 - p^2 \quad \text{at } p = 15$

11. $x = f(p) = \sqrt{100 - p} \quad \text{at } p = 10$

12. $x = f(p) = \sqrt{2500 - 2p^2} \quad \text{at } p = 25$

13. $x = f(p) = 40(4 - p)^3 \quad \text{at } p = 2$

14. $x = f(p) = 20(8 - p)^3 \quad \text{at } p = 4$

15. $x = f(p) = 20 - 3\sqrt{p} \quad \text{at } p = 4$

16. $x = f(p) = 30 - 4\sqrt{p} \quad \text{at } p = 20$

In Problems 17–20, use implicit differentiation to find the elasticity of demand at the indicated values of x and p.

17. $x^{1/2} + 2px + p^2 = 148; \quad x = 16, \quad p = 4$

18. $x^{3/2} + 2px + p^3 = 1088; \quad x = 4, \quad p = 10$

19. $2x^2 + 3px + 10p^2 = 600; \quad x = 10, \quad p = 5$

20. $3x^3 + x^2p^2 + 10p^3 = 3480; \quad x = 10; \quad p = 2$

The discussion about elasticity assumed that the demand x is a function of price p. However, in economics it is common to use x as the independent variable. Thus, if p = F(x) is the demand equation, then it can be shown the elasticity of demand is given by

$$E(x) = \frac{F(x)}{xF'(x)}$$

In Problems 21–26, use this formula to find the elasticity of demand at the given value of x.

21. $p = F(x) = 10 - \dfrac{1}{20}x \quad \text{at } x = 5$

22. $p = F(x) = 40 - \dfrac{1}{10}x \quad \text{at } x = 4$

23. $p = F(x) = 10 - 2x^2 \quad \text{at } x = 2$

24. $p = F(x) = 20 - 4x^2 \quad \text{at } x = 4$

25. $p = F(x) = 50 - 2\sqrt{x} \quad \text{at } x = 100$

26. $p = F(x) = 20 - 4\sqrt{x} \quad \text{at } x = 400$

---

27. Revenue and Elasticity A movie theater has a capacity of 1000 people. The number of people attending the show at a price of \$$p$ per ticket is $x = f(p) = \dfrac{6000}{p} - 500$. Currently the price is \$4 per ticket.

(a) Determine whether the demand is elastic or inelastic at \$4.
(b) If the price is increased, will revenue increase or decrease or remain the same?

28. Revenue and Elasticity The demand function for a rechargeable hand-held vacuum cleaner is given by

$$x = \frac{1}{5}(300 - p^2)$$

where x (measured in units of 100) is the quantity demanded per week and p is the unit price in dollars. The manufacturer would like to increase revenue.

(a) Is demand elastic, inelastic or unit elastic at $p = \$15$?
(b) Should the price of the unit be raised or lowered to increase revenue?

29. Revenue and Elasticity The demand function for digital watches is

$$x = \sqrt{300 - 6p}$$

where x is measured in hundreds of units.

(a) Is demand elastic or inelastic at $p = \$10$?
(b) If the price is lowered slightly, will revenue increase or decrease or remain the same.

30. Revenue and Elasticity A company wishes to increase its revenue by lowering the price of its product. The demand function for this product is

$$x = \frac{10{,}000}{p^2}$$

(a) Compute $E(p)$.
(b) Will the company succeed in raising its revenue?

31. Revenue and Elasticity When a wholesaler sold a certain product at \$15 per unit, sales were 2000 units each week. However, after a price rise of \$3, the average number of units sold decreased to 1800 per week. Assume that the demand function is linear.

(a) Determine the demand function.
(b) Find the elasticity of demand at the new price of \$18.
(c) Approximate the change in demand if the new price is increased by 5%.
(d) Will the price increase cause the revenue to increase or decrease or remain the same?

14.6 Related Rates

PREPARING FOR THIS SECTION *Before getting started, review the following:*

> Geometry Review (Appenidx A, Section A.9, pp. A-82–A-83)
> Chain Rule (Chapter 13, Section 13.5, pp. 750–751)
> Implicit Differentiation (Chapter 13, Section 13.7, pp. 768–773)

OBJECTIVE 1 Solve related rate problems

In all of the natural sciences and many of the social and behavioral sciences, quantities that are related, but vary with time, are encountered. For example, the pressure of an ideal gas of fixed volume is proportional to temperature, yet each of these quantities may change over a period of time. Problems involving rates of related variables are referred to as **related rate problems.** In such problems we normally want to find the rate at which one of the variables is changing at a certain time, when the rates at which the other variables are changing are known.

The usual procedure in such problems is to write an equation that relates all the time-dependent variables involved. Such a relationship is often obtained by investigating the geometric and/or physical conditions imposed by the problem. When this relationship is differentiated with respect to the time t, a new equation that involves the variables and their rates of change with respect to time is obtained.

For example, suppose x and y are two differentiable functions of time t; that is, $x = x(t)$, $y = y(t)$. And suppose they obey the equation

$$x^3 - y^3 + 2y - x - 199 = 0$$

Differentiate both sides of this equation with respect to the time t. Since x and y are functions of time t, we use the Chain Rule to obtain

$$3x^2\frac{dx}{dt} - 3y^2\frac{dy}{dt} + 2\frac{dy}{dt} - \frac{dx}{dt} = 0$$

This equation is valid for all times t under consideration and involves the derivatives of x and y with respect to t, as well as the variables themselves. Because the derivatives are related by this equation, we call them **related rates.** We can solve for one of these rates once the value of the other rate and the values of the variables are known. For example, if in the above equation at a specific time t, we know that $x = 5$, $y = 3$, and $\frac{dx}{dt} = 2$, then by direct substitution we can find that $\frac{dy}{dt} = \frac{148}{25}$. The following examples illustrate how related rates can be used to solve certain types of practical problems.

NOW WORK PROBLEM 1.

1 **EXAMPLE 1 Solving a Related Rate Problem**

A child throws a stone into a still millpond causing a circular ripple to spread. If the radius of the circle increases at the constant rate of 0.5 feet per second, how fast is the area of the ripple increasing when the radius of the ripple is 30 feet? See Figure 41.

FIGURE 41

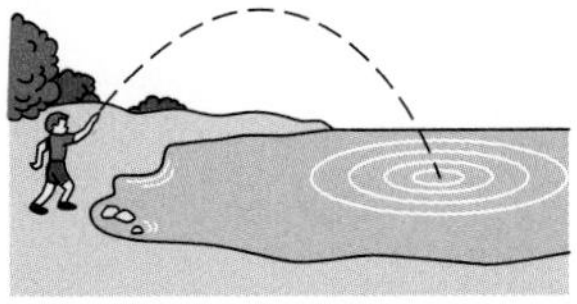

SOLUTION The variables involved are:

$$t = \text{time (in seconds) elapsed from the time the stone hits the water}$$
$$r = \text{radius of the ripple (in feet) after } t \text{ seconds}$$
$$A = \text{area of the ripple (in square feet) after } t \text{ seconds}$$

The rates involved are:

$$\frac{dr}{dt} = \text{the rate at which the radius is increasing with time}$$

$$\frac{dA}{dt} = \text{the rate at which the area is increasing with time}$$

We wish to find $\frac{dA}{dt}$ when $r = 30$; that is, we seek the rate at which the area of the ripple is increasing at the instant when $r = 30$. The relationship between A and r is given by the formula for the area of a circle:

$$A = \pi r^2 \tag{1}$$

Since A and r are functions of t, we differentiate both sides of (1) with respect to t, using the Chain Rule, to obtain

$$\frac{dA}{dt} = 2\pi r \frac{dr}{dt} \qquad \frac{dA}{dt} = \frac{dA}{dr} \cdot \frac{dr}{dt}; \ \frac{dA}{dr} = \frac{d}{dr}(\pi r^2) = 2\pi r \tag{2}$$

Since the radius increases at the rate of 0.5 feet per second, we know that

$$\frac{dr}{dt} = 0.5 \tag{3}$$

By substituting (3) into (2), we get

$$\frac{dA}{dt} = 2\pi r(0.5) = \pi r$$

When $r = 30$, the area of the ripple is increasing at the rate

$$\frac{dA}{dt} = \pi(30) = 30\pi \approx 94.25 \text{ square feet per second}$$

Example 1 illustrates some general guidelines that will prove helpful for solving related rate problems:

Steps for Solving Related Rate Problems

STEP 1 If possible, draw a picture illustrating the problem.
STEP 2 Identify the variables and assign symbols to them.
STEP 3 Identify and interpret rates of change as derivatives.
STEP 4 Write down what is known. Express all relationships among the variables by equations.
STEP 5 Obtain additional relationships among the variables and their derivatives by differentiating.
STEP 6 Substitute numerical values for the variables and the derivatives. Solve for the unknown rate.

Note: It is important to remember that the substitution of numerical values must occur after the differentiation process (Step 5). Also rates must be represented in appropriate units. Finally, remember a positive rate means the quantity is increasing; a negative rate means it is decreasing.

EXAMPLE 2 Solving a Related Rate Problem

A balloon in the form of a sphere is being inflated at the rate of 20 cubic feet per minute. Find the rate at which the surface area of the sphere is increasing at the instant when the radius of the sphere is 6 feet.

SOLUTION **STEP 1** See Figure 42.

FIGURE 42

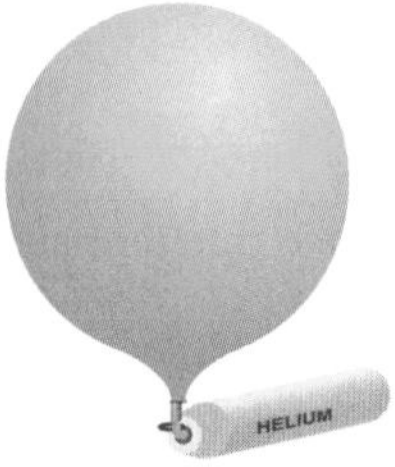

STEP 2 The variables of the problem are:

t = time (in minutes) measured from the moment inflation of the balloon begins
r = length (in feet) of the radius of the balloon at time t
V = volume (in cubic feet) of the balloon at time t
S = surface area (in square feet) of the balloon at time t

STEP 3 The rates of change are:

$$\frac{dr}{dt} = \text{the rate of change of radius with respect to time}$$
$$\frac{dV}{dt} = \text{the rate of change of volume with respect to time}$$
$$\frac{dS}{dt} = \text{the rate of change of surface area with respect to time}$$

STEP 4 We are given that $\frac{dV}{dt} = 20$ cubic feet per minute, and we seek $\frac{dS}{dt}$ when $r = 6$ feet. At any time t, the volume V of the balloon (a sphere) is $V = \frac{4}{3}\pi r^3$ and the surface area S of the balloon is $S = 4\pi r^2$.

STEP 5 Differentiate each of these equations with respect to the time t. Using the Chain Rule, we have

$$\frac{dV}{dt} = 4\pi r^2 \frac{dr}{dt} \qquad \frac{dV}{dt} = \frac{dV}{dr}\cdot\frac{dr}{dt};\ \frac{dV}{dr} = \frac{d}{dr}\left(\frac{4}{3}\pi r^3\right) = 4\pi r^2$$

$$\frac{dS}{dt} = 8\pi r \frac{dr}{dt} \qquad \frac{dS}{dt} = \frac{dS}{dr}\cdot\frac{dr}{dt};\ \frac{dS}{dr} = \frac{d}{dr}(4\pi r^2) = 8\pi r$$

In the equation for $\frac{dV}{dt}$, we solve for $\frac{dr}{dt}$ and substitute this quantity into the equation for $\frac{dS}{dt}$. Then

$$\frac{dV}{dt} = 4\pi r^2 \frac{dr}{dt}$$

$$\frac{dr}{dt} = \frac{1}{4\pi r^2}\frac{dV}{dt} \qquad \text{Solve for } \frac{dr}{dt}.$$

Then

$$\frac{dS}{dt} = 8\pi r \frac{dr}{dt} = 8\pi r \cdot \frac{1}{4\pi r^2}\frac{dV}{dt} = \frac{2}{r}\frac{dV}{dt}$$

STEP 6 At $r = 6$ and $\frac{dV}{dt} = 20$, we have

$$\frac{dS}{dt} = \left(\frac{2}{6}\right)(20) = 6.67 \text{ square feet per minute}$$

The surface area is increasing at the rate of 6.67 square feet per minute when the radius is 6 feet.

NOW WORK PROBLEM 13.

EXAMPLE 3 Solving a Related Rate Problem

A rectangular swimming pool 50 feet long and 25 feet wide is 15 feet deep at one end and 5 feet deep at the other. If water is pumped into the pool at the rate of 300 cubic feet per minute, at what rate is the water level rising when it is 7.5 feet deep at the deep end?

SOLUTION **STEP 1** A cross-sectional view of the pool is illustrated in Figure 43.

FIGURE 43

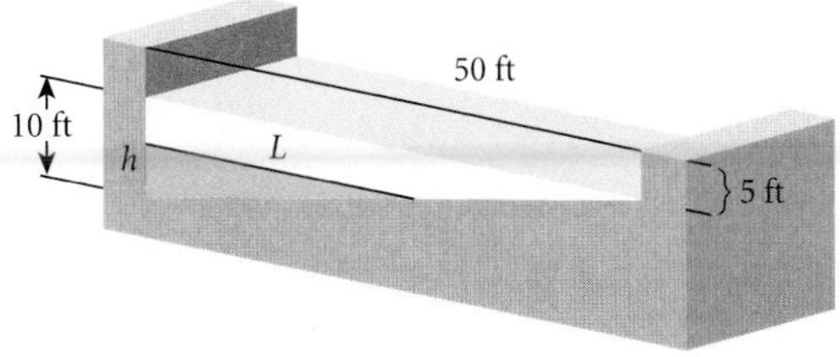

STEP 2 The variables involved are:

t = the time (in minutes) measured from the moment water begins to flow into the pool

h = the water level (in feet) measured at the deep end

L = the distance (in feet) from the deep end toward the short end measured at water level

V = the volume (in cubic feet) of water in the pool

STEP 3 The rates of change are:

$\frac{dV}{dt}$ = the rate of increase in volume at a given instant

$\frac{dh}{dt}$ = the rate of increase in height at a given instant

STEP 4 The volume V is related to L and h by the formula

$$V = (\text{Cross-sectional triangular area})(\text{width}) = \left(\frac{1}{2}Lh\right)(25) \text{ cubic feet} \qquad (4)$$

Using similar triangles, we see from Figure 43 that L and h are related by the equation

$$\frac{L}{h} = \frac{50}{10}$$

$$L = 5h$$

By replacing L by $5h$ in equation (4), we have

$$V = \frac{1}{2}(5h)(h)(25) = \frac{125}{2}h^2 \text{ cubic feet} \qquad (5)$$

Here, V and h are each functions of time t.

STEP 5 By differentiating (5) with respect to t, we obtain

$$\frac{dV}{dt} = 125h\frac{dh}{dt} \text{ cubic feet per minute} \qquad \frac{dV}{dt} = \frac{dV}{dh}\cdot\frac{dh}{dt};\ \frac{dV}{dh} = \frac{d}{dh}\frac{125}{2}h^2 = 125h \qquad (6)$$

STEP 6 We seek the rate at which the water level is rising, $\frac{dh}{dt}$, when $h = 7.5$ and the rate of water pumped into the pool is $\frac{dV}{dt} = 300$ cubic feet per minute.

Using equation (6), we find

$$300 = 125(7.5)\frac{dh}{dt}$$

$$\frac{dh}{dt} = \frac{300}{125(7.5)} = 0.32 \text{ feet per minute}$$

The water level is rising at the rate of 0.32 feet per minute when the height is 7.5 feet.

NOW WORK PROBLEM 15.

EXAMPLE 4 Solving a Related Rate Problem

Suppose that for a company manufacturing digital watches, the cost, revenue, and profit functions are given by

$$\begin{aligned} &\text{Cost function} && C(x) = 10{,}000 + 3x \\ &\text{Revenue function} && R(x) = 5x - \frac{x^2}{2000} \\ &\text{Profit function} && P(x) = R(x) - C(x) \end{aligned}$$

where x is the daily production of digital watches. Production is increasing at the rate of 50 watches per day when production is 1000 watches.

(a) Find the rate of increase in cost when production is 1000 watches.
(b) Find the rate of increase in revenue when production is 1000 watches.
(c) Find the rate of increase in profit when production is 1000 watches.

SOLUTION **STEP 1** No illustration here.

STEP 2 The variables involved are

$$\begin{aligned} t &= \text{the time in days} \\ x(t) &= \text{the production } x \text{ as a function of time} \\ C(t) &= \text{the cost as a function of time} \\ R(t) &= \text{the revenue as a function of time} \\ P(t) &= \text{the profit as a function of time} \end{aligned}$$

STEP 3 The rates involved are

$$\frac{dx}{dt} = 50, \text{ the rate at which production is increasing when } x = 1000$$

$$\frac{dC}{dt} = \text{the rate at which cost is increasing}$$

$$\frac{dR}{dt} = \text{the rate at which revenue is increasing}$$

$$\frac{dP}{dt} = \text{the rate at which profit is increasing}$$

(a) STEP 4 $C(x) = 10{,}000 + 3x$

STEP 5 $\dfrac{dC}{dt} = \dfrac{d}{dt}(10{,}000 + 3x) = \dfrac{d}{dt}(10{,}000) + \dfrac{d}{dt}(3x) = 3\dfrac{dx}{dt}$

STEP 6 Since

$$\frac{dx}{dt} = 50 \qquad \text{when} \qquad x = 1000$$

then

$$\frac{dC}{dt} = 3(50) = \$150 \text{ per day}$$

The cost is increasing at the rate of \$150 per day when production is 1000 watches.

(b) STEP 4 $R(x) = 5x - \dfrac{x^2}{2000}$

STEP 5 $\dfrac{dR}{dt} = \dfrac{d}{dt}\left(5x - \dfrac{x^2}{2000}\right) = \dfrac{d}{dt}(5x) - \dfrac{d}{dt}\left(\dfrac{x^2}{2000}\right) = 5\dfrac{dx}{dt} - \dfrac{x}{1000}\dfrac{dx}{dt}$

STEP 6 Since

$$\frac{dx}{dt} = 50 \quad \text{when} \quad x = 1000$$

then

$$\frac{dR}{dt} = 5(50) - \frac{1000}{1000}(50) = \$200 \text{ per day}$$

Revenue is increasing at the rate of \$200 per day when production is 1000 watches.

(c) STEP 4 $P = R - C$

STEP 5 $\frac{dP}{dt} = \frac{dR}{dt} - \frac{dC}{dt} = \$200 - \$150 = \$50$ per day

Profit is increasing at the rate of \$50 per day when production is 1000 watches.

NOW WORK PROBLEM 19.

EXERCISE 14.6 Answers to Odd-Numbered Problems Begin on Page AN-86.

In Problems 1–4, assume x and y are differentiable functions of t. Find $\frac{dx}{dt}$ *when* $x = 2$, $y = 3$, *and* $\frac{dy}{dt} = 2$.

1. $x^2 + y^2 = 13$
2. $x^2 - y^2 = -5$
3. $x^3y^2 = 72$
4. $x^2y^3 = 108$

5. Suppose h is a differentiable function of t and suppose that when $h = 3$, then $\frac{dh}{dt} = \frac{1}{12}$. Find $\frac{dV}{dt}$ if $V = 80h^2$.

6. Suppose x is a differentiable function of t and suppose that when $x = 15$, then $\frac{dx}{dt} = 3$. Find $\frac{dy}{dt}$ if $y^2 = 625 - x^2$.

7. Suppose h is a differentiable function of t and suppose that $\frac{dh}{dt} = \frac{5}{16}\pi$ when $h = 8$. Find $\frac{dV}{dt}$ if $V = \frac{1}{12}\pi h^3$.

8. Suppose x and y are differentiable functions of t and suppose that when $t = 20$, then $\frac{dx}{dt} = 5$, $\frac{dy}{dt} = 4$, $x = 150$, and $y = 80$. Find $\frac{ds}{dt}$ if $s^2 = x^2 + y^2$.

9. **Changing Volume** If each edge of a cube is increasing at the constant rate of 3 centimeters per second, how fast is the volume increasing when x, the length of an edge, is 10 centimeters long?

10. **Changing Volume** If the radius of a sphere is increasing at 1 centimeter per second, find the rate of change of its volume when the radius is 6 centimeters.

11. Consider a right triangle with hypotenuse of (fixed) length 45 centimeters and variable legs of lengths x and y, respectively. If the leg of length x increases at the rate of 2 centimeters per minute, how fast is y changing when x is 4 centimeters long?

12. **Increasing Surface Area** Air is pumped into a balloon with a spherical shape at the rate of 80 cubic centimeters per second. How fast is the surface area of the balloon increasing when the radius is 10 centimeters?

13. **Decreasing Surface Area** A spherical balloon filled with gas has a leak that permits the gas to escape at a rate of 1.5 cubic meters per minute. How fast is the surface area of the balloon shrinking when the radius is 4 meters?

14. **Heating a Plate** When a metal plate is heated, it expands. If the shape of the metal is circular and if its radius, as a result of expansion, increases at the rate of 0.02 centimeter per second, at what rate is the area of the surface increasing when the radius is 3 centimeters?

15. **Filling a Pool** A public swimming pool has a rectangular shape with the following dimensions: length 30 meters, width 15 meters, depth 3 meters at the adult side and 1 meter at the children's side. If water is pumped into the pool at the rate of 15 cubic meters per minute, how fast is the water level rising when it is 2 meters deep at the adult side?

16. **Cost, Revenue, Profit Functions** Suppose that for a company manufacturing computers, the cost, revenue, and profit functions are given by

$$C(x) = 85{,}000 + 300x$$

$$R(x) = 400x - \frac{x^2}{20}$$

$$P(x) = R(x) - C(x)$$

where x is the weekly production of computers. If production is increasing at the rate of 400 computers per week

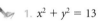

when production output is 5000 computers, find the rate of increase in:

(a) Cost (b) Revenue (c) Profit

17. Pollution Assume that oil spilled from a ruptured tanker forms a circular oil slick whose radius increases at a constant rate of 0.42 feet per minute $\left(\frac{dr}{dt} = 0.42\right)$. Estimate the rate $\frac{dA}{dt}$ (in square feet per minute) at which the area of the spill is increasing when the radius of the spill is 120 feet ($r = 120$).

18. Demand, Revenue Functions The marketing department of a computer manufacturer estimates that the demand q (in thousands of units per year) for a laptop is related to price by the demand equation $q = 200 - 0.9p$. Because of efficiency and technological advances, the prices are falling at a rate of \$30 per year $\left(\frac{dp}{dt} = -30\right)$. The current price of a laptop is \$650. At what rate $\frac{dR}{dt}$ are revenues changing?

19. Cost, Revenue, Profit The cost C and revenue R of a company are given by

$$C(x) = 5x + 5000 \qquad R(x) = 15x - \frac{x^2}{10{,}000}$$

where x is the daily production. Production is increasing at the rate of 100 units per day at a production level of 1000 units. Find:

(a) The rate of change in daily cost when production is 1000 units.

(b) The rate of change in daily revenue when production is 1000 units.

(c) Whether revenue is increasing or decreasing when production is 1000 units.

(d) The profit function.

(e) The rate of change in profit when production is 1000 units.

20. Rework Problem 19 if production is decreasing at the rate of 40 units per day at a production level of 2000 units.

21. Demand, Revenue Functions The marketing department of a manufacturing company estimates that the demand q (in thousands of units per year) for a plasma television is related to price by the demand equation $q = 10{,}000 - 0.9p$. Because of efficiency and technological advances, the prices are falling at a rate of \$100 per year $\left(\frac{dp}{dt} = -100\right)$. The current price of a television is \$7000. At what rate $\frac{dR}{dt}$ is revenue changing?

14.7 The Differential; Linear Approximations

PREPARING FOR THIS SECTION ***Before getting started, review the following:***

>> Definition of a Derivative (Chapter 13, Section 13.1, p. 718)

OBJECTIVES
1 Find differentials
2 Find linear approximations
3 Solve applied problems involving linear approximations

The Differential

In studying the derivative of a function $y = f(x)$, we use the notation $\frac{dy}{dx}$ to represent the derivative. The symbols dy and dx, called *differentials*, which appear in this notation may also be given their own meanings. To pursue this, recall that for a differentiable function f, the derivative is defined as

$$\frac{dy}{dx} = f'(x) = \lim_{\Delta x \to 0} \frac{\Delta y}{\Delta x} = \lim_{\Delta x \to 0} \frac{f(x + \Delta x) - f(x)}{\Delta x}$$

That is, the derivative f' is the limit of the ratio of the change in y to the change in x as Δx approaches 0, but $\Delta x \neq 0$. In other words, for Δx sufficiently close to 0, we can make $\frac{\Delta y}{\Delta x}$ as close as we please to $f'(x)$. We express this fact by writing

$$\frac{\Delta y}{\Delta x} \approx f'(x) \qquad \text{when} \qquad \Delta x \approx 0 \; (\Delta x \neq 0) \tag{1}$$

Another way of writing (1) is to write

$$\Delta y \approx f'(x)\Delta x \qquad \text{when} \qquad \Delta x \approx 0 \ (\Delta x \neq 0)$$

The quantity $f'(x)\,\Delta x$ is given a special name, the *differential of y*.

Differential

Let f denote a differentiable function and let Δx denote a change in x.

(a) The **differential of y**, denoted by dy, is defined as $dy = f'(x)\,\Delta x$.
(b) The **differential of x**, denoted by dx, is defined as $dx = \Delta x \neq 0$.

Using the notation of differentials, we can write

$$\boxed{dy = f'(x)\,dx} \tag{2}$$

Since $dx \neq 0$, (2) can be written as

$$\frac{dy}{dx} = f'(x) \tag{3}$$

The expression in (3) should look very familiar. Interestingly enough, we have given an independent meaning to the symbols dy and dx in such a way that, when dy is divided by dx, their quotient will be equal to the derivative. That is, the differential of y divided by the differential of x is equal to the derivative $f'(x)$. For this reason, *we may formally regard the derivative as a quotient of differentials.*

Note that the differential dy is a function of both x and dx. For example, the differential dy of the function $y = x^3$ is

$$dy = 3x^2\,dx$$

so that

if $x = 1$	and	$dx = 0.2$,	then	$dy = 3(1)^2(0.2) = 0.6$
if $x = 0.5$	and	$dx = 0.1$,	then	$dy = 3(0.5)^2(0.1) = 0.075$
if $x = 2$	and	$dx = -0.5$,	then	$dy = 3(2)^2(-0.5) = -6$

1 **EXAMPLE 1 Finding Differentials**

(a) If $y = x^2 + 3x - 5$ then $dy = (2x + 3)\,dx$. $\quad \frac{d}{dx}(x^2 + 3x - 5) = 2x + 3$

(b) If $y = \sqrt{x^2 + 4}$ then $dy = \dfrac{x}{\sqrt{x^2 + 4}}\,dx$. $\quad \frac{d}{dx}\sqrt{x^2 + 4} = \frac{d}{dx}(x^2 + 4)^{1/2} = \frac{1}{2}(x^2 + 4)^{-1/2} \cdot 2x = \frac{x}{\sqrt{x^2 + 4}}$

NOW WORK PROBLEM 1.

Differential Formulas

All the formulas derived earlier for finding derivatives carry over to differentials. The list below gives formulas for differentials next to the corresponding derivative formulas.

Derivative	Differential	
1. $\frac{d}{dx}c = 0$	1′. $dc = 0$	if c is constant
2. $\frac{d}{dx}(kx) = k$	2′. $d(kx) = k\,dx$	if k is constant
3. $\frac{d}{dx}(u+v) = \frac{du}{dx} + \frac{dv}{dx}$	3′. $d(u+v) = du + dv$	
4. $\frac{d}{dx}(uv) = u\frac{dv}{dx} + v\frac{du}{dx}$	4′. $d(uv) = u\,dv + v\,du$	
5. $\frac{d}{dx}\left(\frac{u}{v}\right) = \frac{v\frac{du}{dx} - u\frac{dv}{dx}}{v^2}$	5′. $d\left(\frac{u}{v}\right) = \frac{v\,du - u\,dv}{v^2}$	
6. $\frac{d}{dx}x^r = rx^{r-1}$	6′. $d(x^r) = rx^{r-1}\,dx$	r is a rational number
7. $\frac{d}{dx}e^x = e^x$	7′. $d(e^x) = e^x\,dx$	
8. $\frac{d}{dx}\ln x = \frac{1}{x}$	8′. $d(\ln x) = \frac{1}{x}dx$	

From now on, to find the differential of a function $y = f(x)$, either find the derivative $\frac{dy}{dx}$ and then multiply by dx or use formulas (1′) through (8′).

For example, if $y = x^3 + 2x + 1$, then the derivative is

$$\frac{dy}{dx} = 3x^2 + 2 \qquad \text{so that} \qquad dy = (3x^2 + 2)\,dx$$

Using the differential formulas,

$$\begin{aligned} dy &= d(x^3 + 2x + 1) = d(x^3) + d(2x) + d(1) \\ &= 3x^2\,dx + 2dx + 0 = (3x^2 + 2)\,dx \end{aligned}$$

Caution: The use of dy on the left side of an equation requires dx on the right side. That is, $dy = 3x^2 + 2$ is incorrect. The symbol d is an instruction to take the differential!

EXAMPLE 2 Finding Differentials

(a) $d(x^2 - 3x) = (2x - 3)dx \qquad \frac{d}{dx}(x^2 - 3x) = 2x - 3$

(b) $d(3y^4 - 2y + 4) = (12y^3 - 2)\,dy \qquad \frac{d}{dy}(3y^4 - 2y + 4) = 12y^3 - 2$

(c) $d(\sqrt{z^2 + 1}) = \frac{z}{\sqrt{z^2 + 1}}dz \qquad \frac{d}{dz}\sqrt{z^2 + 1} = \frac{d}{dz}(z^2 + 1)^{1/2} = \frac{1}{2}(z^2 + 1)^{-1/2}\cdot 2z = \frac{z}{\sqrt{z^2 + 1}}$

(d) $d(xe^x) = xe^x dx + e^x dx = (x + 1)e^x dx \qquad \frac{d}{dx}(xe^x) = x\frac{d}{dx}e^x + e^x\frac{d}{dx}x = xe^x + e^x = (x + 1)e^x$

The differential can be used to find the derivative of a function that is defined implicitly.

EXAMPLE 3 Using Differentials to Find Derivatives

Find $\dfrac{dy}{dx}$ and $\dfrac{dx}{dy}$ if $x^2 + y^2 = 2xy^2$.

SOLUTION We take the differential of each side:

$$d(x^2 + y^2) = d(2xy^2)$$

$$2x\,dx + 2y\,dy = 2(y^2\,dx + 2xy\,dy) \quad \text{Differential of a sum on the left; differential of a product on the right.}$$

$$x\,dx + y\,dy = y^2\,dx + 2xy\,dy \quad \text{Cancel the 2's.}$$

$$(y - 2xy)dy = (y^2 - x)\,dx \quad \text{Rearrange terms.}$$

$$\frac{dy}{dx} = \frac{y^2 - x}{y - 2xy} \quad \text{provided} \quad y - 2xy \neq 0 \quad \text{Solve for } \frac{dy}{dx}.$$

$$\frac{dx}{dy} = \frac{y - 2xy}{y^2 - x} \quad \text{provided} \quad y^2 - x \neq 0 \quad \text{Solve for } \frac{dx}{dy}.$$

NOW WORK PROBLEM 5.

Geometric Interpretation

We use Figure 44 to arrive at a geometric interpretation of the differentials dx and dy and their relationship to Δx and Δy. From the definition, the differential dx and the change Δx are equal. Therefore, we concentrate on the relationship between dy and Δy.

In Figure 44(a), $P = (x, y)$ is a point on the graph of $y = f(x)$ and $Q = (x + \Delta x, y + \Delta y)$ is a nearby point that is also on the graph of f. The slope of the tangent line to the graph of f at P is $f'(x)$. From the figure, it follows that

$$f'(x) = \frac{dy}{\Delta x} = \frac{dy}{dx} \quad \text{or} \quad dy = f'(x)\,dx$$

Figure 44(a) illustrates the case for which $dy < \Delta y$ and $\Delta x > 0$. The case for which $dy > \Delta y$ and $\Delta x > 0$ is illustrated in Figure 44(b). The remaining cases, in which $\Delta x = dx < 0$, have similar graphical representations.

FIGURE 44

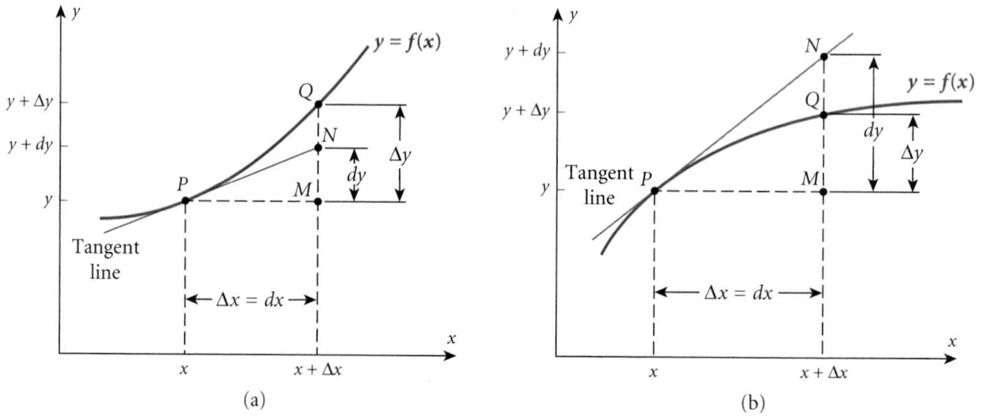

Linear Approximations

We now examine the relationship between Δy and dy. In Figure 44(a), the increment Δy is represented by the length of the line segment $|MQ|$. That is, $\Delta y - dy$ is the length of the line segment $|NQ|$. The size of $|NQ|$ equals the amount by which the graph departs from its tangent line. In fact, for $dx = \Delta x$ sufficiently small, the graph does not depart very much from its tangent line. As a result, the function whose graph is this tangent line is referred to as the *linear approximation to f near P.*

Approximating Δy

For $dx = \Delta x$ sufficiently small, the differential dy is a good approximation to Δy. That is,

$$\boxed{\Delta y \approx dy \quad \text{if} \quad \Delta x \approx 0} \tag{4}$$

We can use (4) to obtain the linear approximation to a function f near a point $P = (x_0, y_0)$ on f. Since

$$dy = f'(x_0)\,dx \underset{\substack{\uparrow \\ dx = \Delta x}}{=} f'(x_0)\,\Delta x \underset{\substack{\uparrow \\ \Delta x = x - x_0}}{=} f'(x_0)(x - x_0)$$

we find from (4) that

$$\Delta y \approx dy$$
$$f(x) - f(x_0) \approx f'(x_0)(x - x_0)$$
$$f(x) \approx f(x_0) + f'(x_0)(x - x_0) \qquad \text{Add } f(x_0) \text{ to both sides.}$$

Linear Approximation to f Near x_0

For $dx = \Delta x$ sufficiently small, that is, for x close to x_0,

$$\boxed{f(x) \approx f(x_0) + f'(x_0)(x - x_0)} \tag{5}$$

The line $y = f(x_0) + f'(x_0)(x - x_0)$ is called the **linear approximation to f near x_0.**

2 **EXAMPLE 4 Finding a Linear Approximation**

FIGURE 45

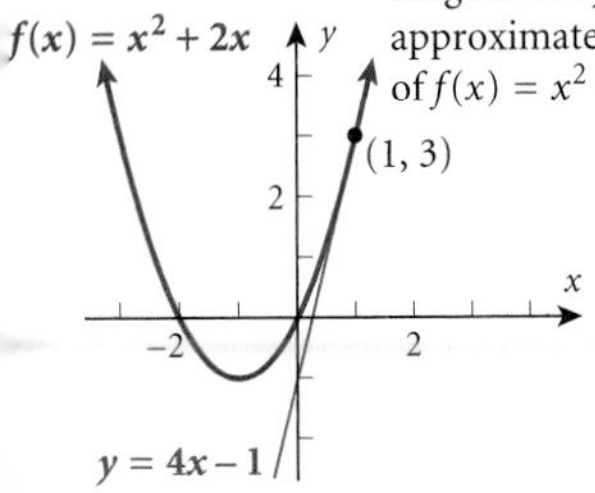

Find the linear approximation to $f(x) = x^2 + 2x$ near $x = 1$. Graph f and the linear approximation.

SOLUTION First, $f(1) = 3$. Next, $f'(x) = 2x + 2$ so that $f'(1) = 2(1) + 2 = 4$. By (5), the linear approximation to f near $x = 1$ is

$$f(x) \approx f(1) + f'(1)(x - 1) = 3 + 4(x - 1) = 4x - 1$$

Figure 45 illustrates the graph of f and the linear approximation $y = 4x - 1$ to f near 1.

NOW WORK PROBLEM 15.

The next two examples use (4).

3 **EXAMPLE 5** **Using Linear Approximations**

A bearing with a spherical shape has a radius of 3 centimeters when it is new. Find the approximate volume of the metal lost after it wears down to a radius of 2.971 centimeters.

SOLUTION   The exact volume of metal lost equals the change, ΔV, in the volume V of the sphere, where $V = \dfrac{4}{3}\pi r^3$. The change in the radius r is $\Delta r = 2.971 - 3 = -0.029$ centimeter. Since the change Δr is small, we can use the differential dV of volume to approximate the change ΔV in volume. Therefore,

$$\Delta V \approx dV = 4\pi r^2 dr = (4\pi)(9)(-0.029) \approx -3.28$$

$$\uparrow \qquad \uparrow \qquad \uparrow$$

Formula (4) $dV = V'(r)dr$ $r = 3$
$$dr = \Delta r = -0.029$$

The approximate loss in volume is 3.28 cubic centimeters.   ▸

NOW WORK PROBLEM 23.

The use of dy to approximate Δy when dx is small may also be helpful in approximating errors.

If Q is the quantity to be measured and if ΔQ is the change in Q, we define

$$\text{Relative error in } Q = \frac{|\Delta Q|}{Q} \qquad \text{Percentage error in } Q = \frac{|\Delta Q|}{Q}(100\%)$$

For example, if $Q = 50$ units and the change ΔQ in Q is measured to be 5 units, then

$$\text{Relative error in } Q = \frac{5}{50} = 0.10$$

$$\text{Percentage error in } Q = 10\%$$

EXAMPLE 6 **Using Linear Approximations**

Suppose a company manufactures spherical ball bearings with radius 3 centimeters, and the percentage error in the radius must be no more than 1%. What is the approximate percentage error for the surface area of the ball bearing?

SOLUTION   If S is the surface area of a sphere of radius r, then $S = 4\pi r^2$ and $\dfrac{\Delta r}{r} = 0.01$. The relative error $\dfrac{\Delta S}{S}$ we seek may be approximated by the use of differentials. That is,

$$\frac{\Delta S}{S} \underset{\uparrow}{\approx} \frac{dS}{S} \underset{\uparrow}{=} \frac{8\pi r\, dr}{4\pi r^2} \underset{\uparrow}{=} 2\frac{dr}{r} \underset{\uparrow}{=} 2\frac{\Delta r}{r} \underset{\uparrow}{=} 2(0.01) = 0.02$$

Formula (4) $dS = S'(r)dr$ Simplify $dr = \Delta r$ $\dfrac{\Delta r}{r} = 0.01$

The percentage error in the surface area is 2%.

In Example 6 the percentage error of 1% in the radius of the sphere means the radius will lie somewhere between 2.97 and 3.03 centimeters. But the percentage error of 2% in the surface area means the surface area lies within a factor of $\pm(0.02)$ of $S = 4\pi r^2 = 36\pi$; that is, it lies between $(0.98)(36\pi) = 35.28\pi = 110.84$ and $(1.02)(36\pi) = 36.72\pi = 115.36$ square centimeters. A rather small error in the radius results in a more significant range of possibilities for the surface area!

 NOW WORK PROBLEM 29.

EXERCISE 14.7 **Answers to Odd-Numbered Problems Begin on Page AN-87.**

In Problems 1–4, find the differential dy.

1. $y = x^3 - 2x + 1$

2. $y = 4(x^2 + 1)^{3/2}$

3. $y = \dfrac{x - 1}{x^2 + 2x - 8}$

4. $y = \sqrt{x^2 - 1}$

In Problems 5–10, find $\dfrac{dy}{dx}$ and $\dfrac{dx}{dy}$ using differentials.

5. $xy = 6$

6. $3x^2y + 2x - 10 = 0$

7. $x^2 + y^2 = 16$

8. $4xy^2 + yx^2 + 6 = 0$

9. $x^3 + y^3 = 3x^2y$

10. $2x^2 + y^3 = xy^2$

In Problems 11–14, find the indicated differential.

11. $d(\sqrt{x - 2})$

12. $d\left(\dfrac{1 - x}{1 + x}\right)$

13. $d(x^3 - x - 4)$

14. $d(x^2 + 5)^{2/3}$

In Problems 15–20, find the linear approximation to f near x_0. Graph f and the linear approximation.

15. $f(x) = x^2 - 2x + 1;\quad x_0 = 2$

16. $f(x) = x^3 - 1;\quad x_0 = 0$

17. $f(x) = \sqrt{x};\quad x_0 = 4$

18. $f(x) = x^{2/3};\quad x_0 = 1$

19. $f(x) = e^x;\quad x_0 = 0$

20. $f(x) = \ln x;\quad x_0 = 1$

---

21. Use equation (5) to find the approximate change in:

(a) $y = f(x) = x^2$ as x changes from 3 to 3.001

(b) $y = f(x) = \dfrac{1}{x + 2}$ as x changes from 2 to 1.98

22. Use equation (5) to find the approximate change in:

(a) $y = x^3$ as x changes from 3 to 3.01

(b) $y = \dfrac{1}{x - 1}$ as x changes from 2 to 1.98

23. A circular plate is heated and expands. If the radius of the plate increases from $r = 10$ centimeters to $r = 10.1$ centimeters, find the approximate increase in area of the top surface.

24. In a wooden block 3 centimeters thick, an existing circular hole with a radius of 2 centimeters is enlarged to a hole with a radius of 2.2 centimeters. Approximately what volume of wood is removed?

25. Find the approximate change in volume of a spherical balloon of radius 3 meters as the balloon swells to a radius of 3.1 meters.

26. A bee flies around the circumference of a circle traced on a ball with a radius of 7 centimeters at a constant distance of 2 centimeters from the ball. An ant travels along the circumference of the same circle on the ball. Approximately how many more centimeters does the bee travel in one trip around than does the ant?

27. If the percentage error in measuring the edge of a cube is 2%, what is the percentage error in computing its volume?

28. The radius of a spherical ball is computed by measuring the volume of the sphere (by finding how much water it displaces). The volume is found to be 40 cubic centimeters, with a percentage error of 1%. Compute the corresponding percentage error in the radius (due to the error in measuring the volume).

29. A manufacturer produces paper cups in the shape of a right circular cone with radius equal to one-fourth its height. Specifications call for the cups to have a diameter of 4 centimeters. After production, it is discovered that the diameters measure 3.9 centimeters. Assuming that the radius is still one-fourth of the height, what is the approximate loss in the capacity of the cup?

Hint: The volume V of a right circular cone of height h and radius r is $V = \frac{1}{3}\pi r^2 h$.

30. The oil pan of a car is shaped in the form of a hemisphere with a radius of 8 centimeters. The depth h of the oil is found to be 3 centimeters, with a percentage error of 10%. Approximate the percentage error in the volume. [*Hint:* The volume V for a spherical segment is $V = \frac{1}{3}\pi h^2(3r - h)$, where r is the radius.]

31. To find the height of a building, the length of the shadow of a 3 meter pole placed 9 meters from the building is measured. This measurement is found to be 1 meter, with a percentage error of 1%. What is the estimated height of the building? What is the percentage error in the estimate? See the figure.

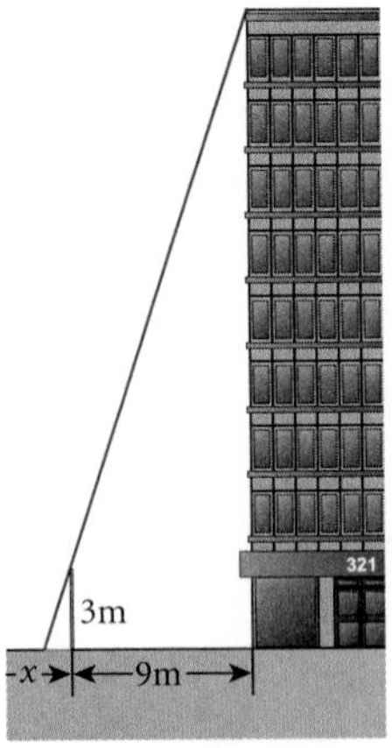

32. The period of the pendulum of a grandfather clock is

$$T = 2\pi\sqrt{\frac{l}{g}},$$

where l is the length (in meters) of the pendulum, T is the period (in seconds), and g is the acceleration due to gravity (9.8 meters per second per second). Suppose the length of the pendulum, a thin wire, increases by 1% due to an increase in temperature. What is the corresponding percentage error in the period? How much time will the clock lose each day?

33. Refer to Problem 32. If the pendulum of a grandfather clock is normally 1 meter long and the length is increased by 10 centimeters, how many minutes will the clock lose each day?

34. What is the approximate volume enclosed by a hollow sphere if its inner radius is 2 meters and its outer radius is 2.1 meters?

Chapter 14 Review

OBJECTIVES

Section	You should be able to	Review Exercises
14.1	1 Find horizontal tangent lines	1–4, 15(e)–22(e)
	2 Find vertical tangent lines	1–4, 15(e)–22(e)
	3 Discuss the graph of a function f where the derivative of f does not exist	5–8
14.2	1 Determine where a function is increasing and where it is decreasing	9(a)–14(a), 15(c)–22(c)
	2 Use the first derivative test	9(b)–14(b), 15(d)–22(d)
	3 Graph functions	15–22
14.3	1 Determine the concavity of a graph	15(g)–22(g)
	2 Find inflection points	15(g)–22(g), 66
	3 Graph functions	15–22
	4 Use the second derivative test	23–28
	5 Solve applied problems	53, 54

14.4	1 Find the absolute maximum and the absolute minimum of a function	29–34
	2 Solve applied problems	55–58
14.5	1 Determine the elasticity of demand	35–40
14.6	1 Solve related rate problems	47–52, 64
14.7	1 Find differentials	41–44
	2 Find linear approximations	45, 46
	3 Solve applied problems involving linear approximations	59–63

THINGS TO KNOW

Tangent Line	f is continuous; $(c, f(c))$ is a point on the graph of c.
Horizontal (p. 788)	$f'(c) = 0$.
Vertical (p. 790)	$f'(x)$ is unbounded at $x = c$.
Theorem on Continuity and Differentiability (p. 791)	If c is in the domain of f and if $f'(c)$ exists, then f is continuous at c.
Increasing Function on (a, b) (p. 795)	f is differentiable and $f'(x) > 0$ for all x in (a, b).
Decreasing Function on (c, d) (p. 795)	f is differentiable and $f'(x) < 0$ for all x in (c, d).
First Derivative Test (p. 796)	f is differentiable. If f is increasing to the left of a point A and decreasing on the right of A, then at A there is a local maximum. If f is decreasing to the left of a point B and decreasing on the right of B, then at B there is a local minimum.
Test for Concavity (p. 806)	If $f''(x) > 0$ for all x in (a, b), then f is concave up on (a, b). If $f''(x) < 0$ for all x in (c, d), then f is concave down on (c, d).
Inflection Point (p. 808)	A point on the graph at which the concavity changes.
Second Derivative Test (p. 814)	f is differentiable on an open interval I. $f''(x)$ exists on I. c is a number in I and $f'(c) = 0$. If $f''(c) < 0$, then at $(c, f(c))$ there is a local maximum. If $f''(c) > 0$, then at $(c, f(c))$ there is a local minimum. If $f''(c) = 0$ the test is inconclusive.
Average Cost Function (p. 815)	$\overline{C}(x) = \dfrac{C(x)}{x}$, where C is the cost function.
Test for Absolute Maximum and Absolute Minimum (p. 822)	f is continuous on a closed interval $[a, b]$. List the values of f at each critical number in the open interval (a, b). List $f(a)$ and $f(b)$. The largest of these is the absolute maximum; the smallest of these is the absolute minimum.
Elasticity of Demand (p. 834)	$E(p) = \dfrac{pf'(p)}{f(p)}$ where p is the price and $x = f(p)$ is the demand function.
Differential of $y = f(x)$ (p. 846)	f is differentiable; Δx = change in x. $dx = \Delta x$; $dy = f'(x)\Delta x = f'(x)dx$.
Linear Approximation to f near x_0 (p. 848)	For x close to x_0, $f(x) \approx f(x_0) + f'(x_0)(x - x_0)$.

TRUE–FALSE ITEMS **Answers are on page AN-87.**

T F **1.** If the derivative of a function f does not exist at c, then f has a vertical tangent line at c.

T F **2.** A differentiable function f is increasing on (a, b) if $f'(x) > 0$ throughout (a, b).

T F **3.** The absolute maximum of a function equals the value of the function at a critical number.

T F **4.** If $f''(x) > 0$ for all x in (a, b), then the graph of f is concave down on (a, b).

T F **5.** If x and y are two differentiable functions of t, then after differentiating $x^2 - y^3 + 4y - x = 100$, we obtain

$$2x\left(\frac{dx}{dt}\right) - 3y^2\left(\frac{dy}{dt}\right) + 4\left(\frac{dy}{dt}\right) - \left(\frac{dx}{dt}\right) = 0$$

T F **6.** If $y = x^2 + 2x$, then $dy = 2x + 2$.

FILL IN THE BLANKS **Answers are on page AN-87.**

1. The function $f(x)$ is __________ on (a, b) if $f'(x) < 0$ for all x in (a, b).

2. At a point $(c, f(c))$ there is a local minimum if the graph of the function is __________ to the left of the point and __________ to the right of the point.

3. A differentiable function is __________ __________ on (a, b) if the tangent lines to the graph at every point lie below its graph.

4. At a point $(c, f(c))$ on the graph of a differentiable function f for which $f'(c) = 0$ there is a __________ tangent line.

5. At an inflection point, the graph exhibits a change in __________.

6. If $y = f(x)$ is a differentiable function, then the differential of y is __________.

7. The function $y = f(x_0) + f'(x_0)(x - x_0)$ is called the __________ __________ to f at x_0.

REVIEW EXERCISES **Answers to odd-numbered problems begin on page AN-87.**

Blue problem numbers indicate the author's suggestions for a practice test.

In Problems 1–4, locate any points at which the graph of f has a horizontal or a vertical tangent line.

1. $f(x) = x^3 - x^2 + x + 15$ **2.** $f(x) = x^5 - 15x^3 + 5$ 3. $f(x) = \dfrac{x^{1/3}}{x + 4}$ **4.** $f(x) = x^{4/5}(x^2 - 14)$

In Problems 5–8, answer the following questions about the function f at c.

(a) *Is f continuous at c?*
(b) *Does $f'(c)$ exist? If it does, what is its value?*
(c) *If f is continuous at c, but $f'(c)$ does not exist, is there a vertical tangent line or no tangent line at c?*

5. $f(x) = 3x^{1/5}$ at $c = 0$

6. $f(x) = 5x^{4/5} - 2x$ at $c = 0$

7. $f(x) = \begin{cases} 3x + 1 & x < 3 \\ x^2 + 1 & x \geq 3 \end{cases}$ at $c = 3$

8. $f(x) = \begin{cases} 4x^2 & x \leq 1 \\ 5x^2 - 1 & x > 1 \end{cases}$ at $c = 1$

In Problems 9–14,
(a) *Determine the intervals on which the graph of f is increasing and the intervals on which it is decreasing.*
(b) *Use the First Derivative Test to identify any local maxima and minima (if they exist).*

9. $f(x) = \dfrac{1}{5}x^5 - x^3 - 4x$ **10.** $f(x) = 2x^4 - 9x^2$ **11.** $f(x) = \dfrac{x^2}{x^2 - 8}$

12. $f(x) = \dfrac{6x + 1}{x^2 + 5}$ **13.** $f(x) = 1 + 3e^{-x}$ **14.** $f(x) = xe^x$

In Problems 15–22, graph f by following the 7 steps listed.

(a) *STEP 1 Find the domain of f.*
(b) *STEP 2 Locate the intercepts of f.*
(c) *STEP 3 Determine where the graph is increasing and where it is decreasing.*
(d) *STEP 4 Find any local maxima or minima using the First Derivative Test.*
(e) *STEP 5 Locate all points on the graph of f at which the tangent line is either horizontal or vertical.*
(f) *STEP 6 Determine the end behavior and locate any asymptotes.*
(g) *STEP 7 Locate the inflection points, if any, of the graph by determining the concavity of the graph.*

15. $f(x) = x^3 - 3x^2 + 3x - 1$
16. $f(x) = 2x^3 - x^2 + 2$
17. $f(x) = x^5 - 5x$
18. $f(x) = x^5 + 5x^4$
19. $f(x) = x^{4/3} + 4x^{1/3}$
20. $f(x) = x^{4/3} - 4x^{1/3}$
21. $f(x) = \dfrac{2x}{x^2 + 1}$
22. $f(x) = \dfrac{4x}{x^2 + 4}$

In Problems 23–28, use the Second Derivative Test to determine the local maxima and local minima of each function.

23. $f(x) = 4x^3 - 3x$
24. $f(x) = 5x^4 + 7x^3 + 20$
25. $f(x) = x^4 - 2x^2$
26. $f(x) = x^4 + 2x^2$
27. $f(x) = xe^x$
28. $f(x) = xe^{-x}$

In Problems 29–34, find the absolute maximum and absolute minimum of each function on the given interval.

29. $f(x) = x^3 - 3x^2 + 3x - 1$ on $[0, 3]$
30. $f(x) = 2x^3 - x^2 + 2$ on $[0, 1]$
31. $f(x) = x^4 - 4x^3 + 4x^2$ on $[1, 3]$
32. $f(x) = x^4 - 2x^2$ on $[-1, 1]$
33. $f(x) = x^{4/3} - 4x^{1/3}$ on $[-1, 8]$
34. $f(x) = x^{4/3} + 4x^{1/3}$ on $[-1, 1]$

In Problems 35–38, a demand function is given. Find E(p) and determine if demand is elastic or inelastic or unit elastic at the indicated price.

35. $x = 1000 - 2p^2$ at $p = 20$
36. $x = \dfrac{1000}{p + 10}$ at $p = 12$
37. $x = \sqrt{500 - p^2}$ at $p = 10$
38. $x = \sqrt{2200 - 2p^2}$ at $p = 30$

---

39. **Pricing DVD Recorders** The demand function for a DVD player/recorder is $x = 40 - 2\sqrt{p}$ where p is the price of a unit.
(a) Is the demand function elastic, inelastic, or unit elastic at $p = \$300$?
(b) If the price is raised to \$310 will revenue increase, decrease, or remain the same?

40. **Cost of Cellular Service** The demand equation for a certain cellular phone package is given by $x = 60{,}000 - 1200p$, where p is the price of the package in dollars.
(a) Is demand elastic, inelastic, or unit elastic at $p = \$35$?
(b) If the price is raised to \$37, will revenue increase, decrease, or remain the same?

In Problems 41–44, find the differential dy.

41. $y = 3x^4 - 2x^3 + x$
42. $y = 3(x^2 - 1)^5$
43. $y = \dfrac{3 - 2x}{1 + x}$
44. $y = \sqrt[3]{2 + x^4}$

In Problems 45 and 46, find the linear approximation of f near x_0. Graph f and the linear approximation.

45. $f(x) = x^2 - 9$ $x_0 = 3$
46. $f(x) = \dfrac{1}{x}$ $x_0 = 1$

47. Suppose that x and y are both differentiable functions of t and $x^2 + y^2 = 8$. Find $\frac{dx}{dt}$ when $x = 2$, $y = 2$, and $\frac{dy}{dt} = 3$.

48. Suppose that x and y are both differentiable functions of t and $x^2 - y^2 = 5$. Find $\frac{dy}{dt}$ when $x = 4$, $y = 3$, and $\frac{dx}{dt} = 2$.

49. Suppose that x and y are both differentiable functions of t and $xy + 6x + y^3 = -2$. Find $\frac{dy}{dt}$ when $x = 2$, $y = -3$ and $\frac{dx}{dt} = 3$.

50. Suppose that x and y are both differentiable functions of t and $y^2 - 6x^2 = 3$. Find $\frac{dx}{dt}$ when $x = 2$, $y = -3$, and $\frac{dy}{dt} = 3$.

51. A balloon in the form of a sphere is being inflated at the rate of 10 cubic meters per minute. Find the rate at which the surface area of the sphere is increasing at the instant when the radius of the sphere is 3 meters.

Hint: The volume of a sphere is $V = \frac{4}{3}\pi r^3$; the surface area is $S = 4\pi r^2$.

52. A child throws a stone into a still millpond, causing a circular ripple to spread. If the radius of the circle increases at the constant rate of 0.5 meter per second, how fast is the area of the ripple increasing when the radius of the ripple is 20 meters?

53. **Average Cost** The cost function for producing x items is

$$C(x) = 5x^2 + 1125$$

(a) Find the average cost function.
(b) What is the minimum average cost?
(c) Find the marginal cost function.
(d) Graph the average cost function and the marginal cost function on the same set of axes. Label their point of intersection.
(e) Interpret the point of intersection.

54. **Average Cost** The cost function for producing x items is

$$C(x) = 1000 + x + x^2$$

(a) Find the average cost function.
(b) What is the minimum average cost?
(c) Find the marginal cost function.
(d) Graph the average cost function and the marginal cost function on the same set of axes. Label their point of intersection.
(e) Interpret the point of intersection.

55. **Maximizing Profit** The price function of a certain mobile home producer is

$$p(x) = 62{,}402.50 - 0.5x^2$$

where p is the price (in dollars) and x is the number of units sold. The cost of production for x units is

$$C(x) = 48{,}002.50 + 1500$$

How many units need to be sold to maximize profit?

56. **Maximizing Profit** A company's history shows that profit increases, as a result of advertising, according to

$$P(x) = 150 + 120x - 3x^2$$

where x is the number of dollars, in thousands, spent, on advertising. How much should be spent on advertising to maximize profit?

57. **Best Dimensions of a Can** A beer can is cylindrical and holds 500 cubic centimeters of beer. If the cost of the material used to make the sides, top, and bottom is the same, what dimensions should the can have to minimize cost?

58. **Setting Refrigerator Prices** A distributor of refrigerators has average monthly sales of 1500 refrigerators, each selling for \$300. From past experience, the distributor knows that a special month-long promotion will enable them to sell 200 additional refrigerators for each \$15 decrease in price. What should be charged for each refrigerator during the month of promotion in order to maximize revenue?

59. **Size of a Burn** A burn on a person's skin is in the shape of a circle, so that if r is the radius of the burn and A is the area of the burn, then $A = \pi r^2$. Use the differential to approximate the decrease in the area of the burn when the radius decreases from 10 to 8 millimeters.

60. **Related Rate** A spherical ball is being inflated. Find the approximate change in volume if the radius increases from 3 to 3.1 cm.

61. **Demand Function** The demand for peanuts (in hundreds of pounds) at a price of x dollars is

$$D(x) = -4x^3 - 3x^2 + 2000$$

Approximate the change in demand as the price changes from

(a) \$1.50 to \$2.00. (b) \$2.50 to \$3.50.

62. **Growth of a Tumor** A tumor is approximately spherical in shape. If the radius of the tumor changes from 11 to 13 millimeters, find the approximate change in volume.

63. Drug Concentration The concentration of a certain drug in the bloodstream x hours after being administered is

$$c(x) = \frac{3x}{4 + 2x^2}$$

Approximate the change in concentration as x changes from

(a) 1.2 to 1.3. (b) 2 to 2.25.

64. Wound Healing A wound on a person's skin is in the shape of a circle and is healing at the rate of 30 square millimeters per day. How fast is the radius r of the wound decreasing when $r = 10$ millimeters?

65. If a function f is differentiable for all x except 3, and if f has a local maximum at $(-1, 4)$ and a local minimum at $(3, -2)$, which of the following statements must be true?

(a) The graph of f has a point of inflection somewhere between $x = -1$ and $x = 3$.
(b) $f'(-1) = 0$
(c) The graph of f has a horizontal asymptote.
(d) The graph of f has a horizontal tangent line at $(3, -2)$.
(e) The graph of f intersects both axes.

66. Let $f(x)$ be a function with derivative

$$f'(x) = \frac{1}{1 + x^2}$$

Show that the graph of $f(x)$ has an inflection point at $x = 0$. Note that $f'(x) > 0$ for all x.

Chapter 14 Project

INVENTORY CONTROL

Suppose that you are employed by a local department store, and you are placed in charge of ordering vacuum cleaners. Based on past experience, you know that the store will sell 500 vacuum cleaners per year. You must decide how many times a year to order vacuum cleaners and how many to get with each order. You could order all 500 at the beginning of the year, but there will be a cost (the holding cost) for storing the unsold vacuum cleaners. You could order 5 at a time and place 100 orders over the course of the year, but there are costs for paperwork and shipping for each order you place. Perhaps there is an amount to order somewhere between 5 and 500 that minimizes the total cost: the holding cost plus the reorder cost. This number is called the lot size.

To simplify the mathematics, we make two assumptions:

1. Demand for vacuum cleaners remains constant through the year.
2. Stock is immediately replenished exactly when the inventory level of vacuum cleaners reaches zero.

With these assumptions, the graph of the number of vacuum cleaners in stock at any time will look like the graph shown. Here x is the lot size, the amount ordered each time inventory reaches zero. The graph signifies that the inventory is decreasing at a constant rate (lines with negative slope). The average number of items in stock is $\frac{x}{2}$.

Let D be the annual demand for the item. [For the vacuum cleaners, $D = 500$.] Our goal is to find a function $C = C(x)$,

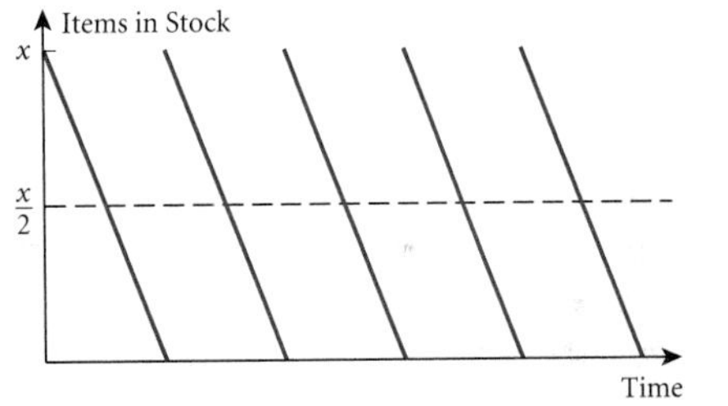

where C represents the holding costs plus the reorder costs and x is the lot size. We seek the value of x that minimizes C.

(a) Let H be the holding cost per year for each item you have in stock. Explain why your average holding costs for the year will be $\frac{Hx}{2}$.

(b) Let R be the cost for each reorder. Explain why you will make $\frac{D}{x}$ reorders during the year, and why your reorder cost will be $R \cdot \frac{D}{x} = \frac{DR}{x}$.

(c) What is the cost function $C = C(x)$?

(d) Show that $C(x)$ is minimized when $x = \sqrt{\frac{2DR}{H}}$. This value of x is called the "Wilson-Harris lot size."

(e) Apply the Wilson-Harris lot size formula to the vacuum cleaner problem. Assume that annual demand is 500 vacuum cleaners, that the holding cost is \$10 per vacuum cleaner, and that the reorder cost \$40. Find the lot size that will minimize the total cost. How many orders will you place over the course of a year?

(f) Suppose that in part (e) the holding cost is reduced to \$3 per vacuum cleaner. What lot size will now minimize the total cost? How many orders will you place?

(g) In many cases, the shipping charges (which are part of the reorder cost) vary with how many items are ordered. Suppose that the reorder cost is now $R + Sx$, where R is the cost of reordering exclusive of shipping costs, and S is the shipping charge per item. The cost function is now

$$C(x) = \frac{Hx}{2} + \frac{D(R + SX)}{x}.$$

Find the lot size that will now minimize the total cost.

MATHEMATICAL QUESTIONS FROM PROFESSIONAL EXAMS*†

1. **CPA Exam** The mathematical notation for the total cost for a business is $2X^3 + 4X^2 + 3X + 5$, where X equals production volume. Which of the following is the mathematical notation for the marginal cost function for this business?

(a) $2(X^3 + 2X^2 + 1.5X + 2.5)$ (b) $6X^2 + 8X + 3$
(c) $2X^3 + 4X^2 + 3X$ (d) $3X + 5$

2. **CPA Exam** The mathematical notation for the total cost function for a business is $4X^3 + 6X^2 + 2X + 10$ where X equals production volume. Which of the following is the mathematical notation for the average cost function for that business?

(a) $2(2X^2 + 3X + 2)$
(b) $2X^3 + 3X^2 + X + 5$
(c) $0.4X^3 + 0.6X^2 + 0.2X + 1$
(d) $4X^2 + 6X + 2 + \dfrac{10}{X}$

3. **CPA Exam** The mathematical notation for the average cost function for a business is $6X^3 + 4X^2 + 2X + 8 + 2/X$, where X equals production volume. What would be the mathematical notation for the total cost function for the business?

(a) The average cost function multiplied by X.
(b) The average cost function divided by X.
(c) The average cost function divided by $X/2$.
(d) The first derivative of the average cost function.

4. **CPA—Review** To find a minimum cost point given a total cost equation, the initial steps are to find the first derivative, set this derivative equal to zero, and solve the equation. Using the solution(s) so derived, what additional steps must be taken, and what result indicates a minimum?

(a) Substitute the solution(s) in the first derivative equation and a positive solution indicates a minimum.
(b) Substitute the solution(s) in the first derivative equation and a negative solution indicates a minimum.
(c) Substitute the solution(s) in the second derivative equation and a positive solution indicates a minimum.
(d) Substitute the solution(s) in the second derivative equation and a negative solution indicates a minimum.

5. **Actuary Exam—Part I** Figure A could represent the graph of which of the following?

(a) $y = x^3e^{x^2}$ (b) $y = xe^x$
(c) $y = xe^{-x}$ (d) $y = x^2e^x$
(e) $y = x^2e^{x^2}$

FIGURE A

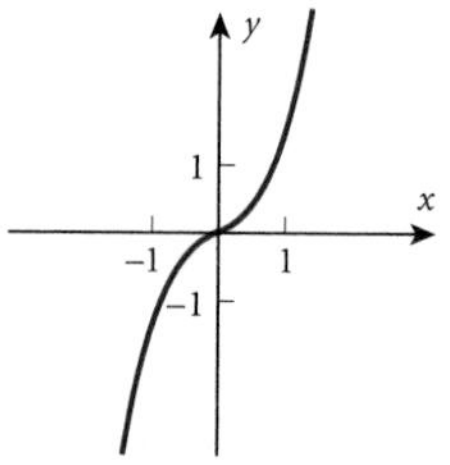

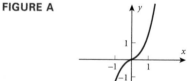

6. **Actuary Exam—Part I** According to classical economic theory, the business cycle peaks when employment reaches a maximum, relative to adjacent time periods. Employment can be approximated as a function of time, t, by a differentiable function $E(t)$. The graph of $E'(t)$ is pictured below.

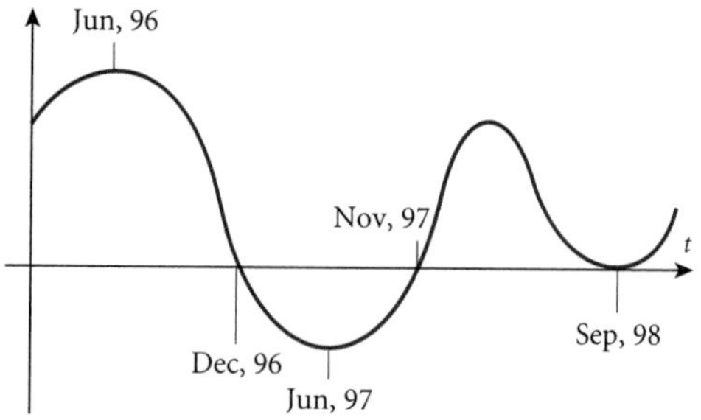

Which of the following points represents a peak in the business cycle?

(a) Jun, 96 (b) Dec, 96 (c) Jun, 97
(d) Nov, 97 (e) Sep, 98

7. An economist defines an "index of economic health," D, as follows:

$$D = E^2(100 - I)$$

where

E is the percent of the working-age population that is employed and I is the rate of inflation (expressed as a percent).

On June 30, 1996, employment is at 95% and is increasing at a rate of 2% per year, and the rate of inflation is at 6% and is increasing at a rate of 3% per year. Calculate the rate of change of D on June 30, 1996.

(a) $-9{,}503$ per year (b) $-9{,}500$ per year
(c) 0 per year (d) 8,645 per year
(e) 17,860 per year

8. **Actuary Exam—Part I** What values of x produce a relative minimum and relative maximum, respectively, for $f(x) = 2x^3 + 3x^2 - 12x - 5$?

(a) $-5, 0$ (b) $-2, 1$
(c) $1, -3$ (d) $1, -2$
(e) $2, -1$

9. **Actuary Exam—Part I** On which of the following intervals is the function $f(x) = \frac{1}{6}x^3 - 2x$ decreasing and concave upward?

(a) $(-2, 2)$ (b) $(-2, 0)$
(c) $(0, 2)$ (d) $(2, 3)$
(e) $(-2, 4)$

10. **Actuary Exam—Part I** What is the maximum value of $f(x) = \dfrac{x}{1 + x + x^2}$ on the closed interval $[-2, 2]$?

(a) $\frac{1}{4}$ (b) $\frac{2}{7}$ (c) $\frac{1}{3}$ (d) $\frac{1}{2}$ (e) 1

11. **Actuary Exam—Part I** If $f(x) = \dfrac{xe^{3x}}{1 + x}$ for $x \neq -1$, then $f'(1) =$

(a) $\frac{3e^3}{4}$ (b) $\frac{5e^3}{4}$ (c) $\frac{7e^3}{4}$ (d) $\frac{9e^3}{4}$
(e) $4e^3$

12. **Actuary Exam—Part I** Let $f(x) = 5 + 6x + 12x^2 - 2x^3 - x^4$, and let $g(x) = f'(x)$ for $-\infty < x < \infty$. At what value of x is $g(x)$ increasing most rapidly?

(a) -2 (b) $-\frac{1}{2}$ (c) $\frac{1}{2}$ (d) 1 (e) 2

13. **Actuary Exam—Part I** A water tank in the shape of a right circular cone has a height of 10 feet. The top rim of the tank is a circle with a radius of 4 feet. If water is being pumped into the tank at the rate of 2 cubic feet per minute, what is the rate of change of the water depth, in feet per minute, when the depth is 5 feet? (The volume V of a right circular cone is $V = \frac{1}{3}\pi r^2 h$, where r is its radius and h is its height.)

(a) $\frac{1}{2\pi}$ (b) $\frac{1}{\pi}$ (c) $\frac{3}{2\pi}$ (d) $\frac{2}{\pi}$ (e) $\frac{5}{2\pi}$

14. **Actuary Exam—Part I** What is the y-coordinate of the point on the curve $y = 2x^2 - 3x$ at which the slope of the tangent line is the same as that of the secant line between $x = 1$ and $x = 2$?

(a) -1 (b) 0 (c) 1 (d) 3 (e) 9

15. **Actuary Exam—Part I** Two vehicles start at time $t = 0$ from the same point A and travel along a straight line so that at each moment t the distance between the second vehicle and A equals the square of the distance between the first vehicle and A. After $t = 3$ hours, the first vehicle is moving at 3 kilometers per hour and is 9 kilometers away from A. What is the speed, in kilometers per hour, of the second vehicle at $t = 3$ hours?

(a) 9 (b) 18 (c) 27 (d) 54 (e) 81

16. **Actuary Exam—Part I** A cube of ice melts without changing shape at the uniform rate of 4 cm^3/min. What is the rate of change of surface area of the cube, in cm^2/min, when the volume of the cube is 125 cm^3?

(a) -4 (b) $\frac{-16}{5}$ (c) $\frac{-16}{6}$ (d) $\frac{60}{19}$
(e) $\frac{16}{5}$

CHAPTER 15

The Integral of a Function and Applications

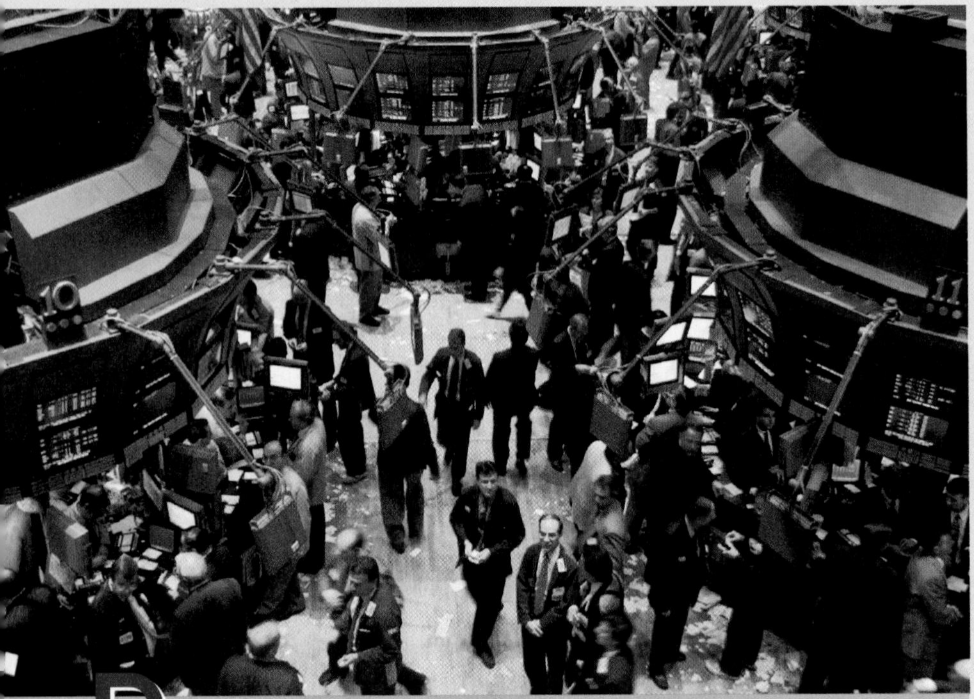

OUTLINE

Different families have different incomes. We all know that! But how is that income distributed? If we rank families in order of income they earn, will the lowest 20% have earnings equal to 20% of the total income earned by all families? Will the first 80% of these families have incomes that equal 80% of the total income earned by families? The answer is no. If we rank families by income earned, then the lowest 20% will have incomes that are less than 20% of the total income earned by all families.

We can determine what percent of total income of all families is earned by the lowest 20%. For example, in 2001, the percent of income earned by the lowest 20% was 3.5% of the total. The percent of income earned by the first 80% was 49.8%. (*Source:* U.S. Census Bureau.) An important question to ask is, What is the trend over time in these percents? And how can we measure the trend? The Chapter Project at the end of this chapter provides a methodology for measuring the distribution of income.

A LOOK BACK, A LOOK FORWARD

In Chapter 10 we discussed various properties that functions have, like increasing or decreasing, the average rate of change, the difference quotient, and so on. Chapter 11 identified classes of functions and properties they possessed. With Chapter 12 we began our study of the calculus and found limits of functions. Then in Chapter 13 we introduced the derivative of a functions, developing formulas for finding derivatives. We used these formulas in Chapter 14 to do applications.

Now in Chapter 15 we introduce the second important idea of the calculus, *the integral*. There are two types of integral: the *indefinite integral* and the *definite integral*. The indefinite integral involves the inverse process of differentiation, usually referred to as *finding the antiderivative* of a function. The definite integral plays a major role in applications to geometry (finding areas), to business and economics (marginal analysis, consumer's surplus, maximization of profit), and to solving differential equations.

15.1 Antiderivatives; the Indefinite Integral; Marginal Analysis

PREPARING FOR THIS SECTION *Before getting started, review the following:*

- > Derivative of a Power Function (Chapter 13, Section 13.2, p. 728)
- > Derivative of $f(x) = x^{p/q}$ (Chapter 13, Section 13.8, p. 776)
- > Derivative of $f(x) = e^x$ (Chapter 13, Section 13.5, p. 749)
- > Derivative of $f(x) = \ln x$ (Chapter 13, Section 13.5, p. 753)

OBJECTIVES
1 Find an antiderivative
2 Use integration formulas
3 Evaluate indefinite integrals
4 Find cost and revenue functions

Antiderivatives

Find an antiderivative 1

We have already learned that to each differentiable function f there corresponds a derivative function f'. It is also possible to ask the following question: If a function f is given, can we find a function F whose derivative is f? That is, is it possible to find a function F so that $F'(x) = f(x)$? If such a function F can be found, it is called *an antiderivative of f*.

Antiderivative

A function F is called an **antiderivative** of the function f if

$$F'(x) = f(x)$$

EXAMPLE 1 **Finding an Antiderivative**

An antiderivative of $f(x) = 2x$ is x^2, since

$$\frac{d}{dx}x^2 = 2x$$

Another antiderivative of $2x$ is $x^2 + 3$ since

$$\frac{d}{dx}(x^2 + 3) = 2x$$

This example leads us to suspect that the function $f(x) = 2x$ has an unlimited number of antiderivatives. Indeed, any of the functions x^2, $x^2 + \frac{1}{2}$, $x^2 + 2$, $x^2 + \sqrt{5}$, $x^2 - \pi$, $x^2 - 1$, $x^2 + K$, where K is a constant, has the property that its derivative is $2x$. So all functions of the form $F(x) = x^2 + K$, where K is a constant, are antiderivatives of $f(x) = 2x$. Are there others? The answer is no! As the following result tells us, *all* the antiderivatives of $f(x) = 2x$ are of the form $F(x) = x^2 + K$, where K is a constant.

Antiderivatives of *f*

If F is an antiderivative of f, then all the antiderivatives of f are of the form

$$\boxed{F(x) + K}$$

where K is a constant.

EXAMPLE 2 Finding All the Antiderivatives of a Function

All the antiderivatives of $f(x) = x^5$ are of the form

$$\frac{x^6}{6} + K$$

where K is a constant.

We check the answer by finding its derivative. Since

$$\frac{d}{dx}\left(\frac{x^6}{6} + K\right) = x^5$$

the answer checks.

EXAMPLE 3 Finding All the Antiderivatives of a Function

Find all the antiderivatives of: $f(x) = x^{1/2}$

SOLUTION The derivative of the function $\frac{2}{3}x^{3/2}$ is

$$\frac{d}{dx}\left(\frac{2}{3}x^{3/2}\right) = \frac{2}{3}\frac{d}{dx}x^{3/2} = \frac{2}{3}\left(\frac{3}{2}x^{1/2}\right) = x^{1/2}$$

So, $\frac{2}{3}x^{3/2}$ is an antiderivative of $f(x) = x^{1/2}$. All the antiderivatives of $f(x) = x^{1/2}$ are of the form

$$\frac{2}{3}x^{3/2} + K$$

where K is a constant.

In Example 3, you may ask how we knew to choose the function $\frac{2}{3}x^{3/2}$. First, we know that, for n a rational number,

$$\frac{d}{dx}x^n = nx^{n-1}$$

That is, differentiation of a power of x reduces the exponent by 1. Antidifferentiation is the inverse process, so it should increase the exponent by 1. This is how we obtained the $x^{3/2}$ part of $\frac{2}{3}x^{3/2}$. Second, the $\frac{2}{3}$ factor is needed so that, when we differentiate, we get $x^{1/2}$ and not $\frac{3}{2}x^{1/2}$.

Because

$$\frac{d}{dx}x^{n+1} = (n + 1)x^n$$

for n a rational number, if $n \neq -1$, it follows that:

Antiderivatives of $f(x) = x^n$

All the antiderivatives of $f(x) = x^n$ are of the form

$$F(x) = \frac{x^{n+1}}{n + 1} + K$$

where n is any rational number except -1, and K is a constant.

Notice that $n = -1$ is excluded from the formula. That is, we cannot use this formula to find the antiderative of $f(x) = x^{-1} = \frac{1}{x}$. The antiderivative of this function requires special attention and is considered a little later.

NOW WORK PROBLEM 1.

Indefinite Integrals

We use a special symbol to represent all the antiderivatives of a function—the **integral sign,** $\int$.

Indefinite Integral

Let F be an antiderivative of the function f. The **indefinite integral of *f*,** denoted by $\int f(x)\,dx$, is defined as

$$\int f(x)\,dx = F(x) + K$$

where K is a constant.

In other words, the indefinite integral of f equals all the antiderivatives of f.

In the expression $\int f(x)\,dx$, the integral sign $\int$ indicates that the operation of antidifferentiation is to be performed on the function f, and the dx reinforces the fact that the operation is to be performed with respect to the variable x. The function f is called the **integrand,** and the process of antidifferentiation is called **integration.**

Basic Integration Formulas

Based on the relationship between the process of differentiation and that of integration, or antidifferentiation, we can construct a list of formulas. These formulas may be verified by differentiating the right-hand side.

Basic Integration Formulas

If c is a real number and K is a constant,

$$\int c\,dx = cx + K \tag{1}$$

$$\int x^n\,dx = \frac{x^{n+1}}{n+1} + K \qquad n \neq -1 \tag{2}$$

$$\int [f(x) + g(x)]\,dx = \int f(x)\,dx + \int g(x)\,dx \tag{3}$$

$$\int [f(x) - g(x)]\,dx = \int f(x)\,dx - \int g(x)\,dx \tag{4}$$

$$\int cf(x)\,dx = c\int f(x)\,dx \tag{5}$$

As a special case of Formula (1), let $c = 1$. Then we find that

$$\int dx = \int 1\cdot dx = 1\cdot x + K = x + K$$

where K is a constant.

Formulas (3) and (4) state that the integral of a sum or a difference equals the sum or difference of the integrals.

Formula (5) states that a constant factor can be moved across an integral sign. Be careful! A variable factor cannot be moved across an integral sign.

2 **EXAMPLE 4** **Using Integration Formulas**

(a) $\int 5\,dx \underset{\uparrow}{=} 5x + K$

Formula (1)

(b) $\int x^5\,dx \underset{\uparrow}{=} \frac{x^{5+1}}{5+1} + K = \frac{x^6}{6} + K$

Formula (2)

(c) $$\int 3x^4\,dx \underset{\substack{\uparrow\\ \text{Formula (5)}}}{=} 3\int x^4\,dx \underset{\substack{\uparrow\\ \text{Formula (2)}}}{=} 3\,\frac{x^5}{5} + K = \frac{3}{5}x^5 + K$$

(d) $$\int (x^2 + x^3)\,dx \underset{\substack{\uparrow\\ \text{Formula (3)}}}{=} \int x^2\,dx + \int x^3\,dx \underset{\substack{\uparrow\\ \text{Formula (2)}}}{=} \frac{x^3}{3} + K_1 + \frac{x^4}{4} + K_2$$

$$= \frac{x^3}{3} + \frac{x^4}{4} + K$$

where $K = K_1 + K_2$.

NOW WORK PROBLEMS 9, 13, AND 19.

Formulas (3) and (4) can be combined and used for sums and differences of three or more functions.

EXAMPLE 5 Using Integration Formulas

$$\int\left(7x^5 + \frac{1}{2}x^2 - x\right)dx = \int 7x^5\,dx + \int \frac{1}{2}x^2\,dx - \int x\,dx \qquad \text{Formulas (3) and (4).}$$

$$= 7\int x^5\,dx + \frac{1}{2}\int x^2\,dx - \int x\,dx \qquad \text{Formula (5).}$$

$$= \left(7\cdot\frac{x^6}{6} + K_1\right) + \left(\frac{1}{2}\cdot\frac{x^3}{3} + K_2\right) - \left(\frac{x^2}{2} + K_3\right) \qquad \text{Formula (2).}$$

$$= \frac{7}{6}x^6 + \frac{1}{6}x^3 - \frac{1}{2}x^2 + K \qquad \text{Simplify.}$$

where $K = K_1 + K_2 - K_3$.

As Example 5 illustrates, we can now integrate any polynomial function.

NOW WORK PROBLEM 23.

Sometimes, it is necessary to use algebra to put the integrand in a form that matches one of the Basic Integration Formulas.

EXAMPLE 6 Using Integration Formulas

(a) $$\int \frac{1}{\sqrt{x}}\,dx \underset{\substack{\uparrow\\ \text{Algebra}}}{=} \int x^{-1/2}\,dx \underset{\substack{\uparrow\\ \text{Formula(2)}}}{=} \frac{x^{(-1/2)+1}}{-\frac{1}{2}+1} + K = \frac{x^{1/2}}{\frac{1}{2}} + K = 2x^{1/2} + K$$

(b) $$\int 4\sqrt[3]{x^5}\,dx \underset{\substack{\uparrow\\ \text{Formula (5)}}}{=} 4\int \sqrt[3]{x^5}\,dx \underset{\substack{\uparrow\\ \text{Algebra}}}{=} 4\int x^{5/3}\,dx \underset{\substack{\uparrow\\ \text{Formula (2)}}}{=} 4\,\frac{x^{5/3+1}}{\frac{5}{3}+1} + K = \frac{3}{2}x^{8/3} + K$$

(c) $$\int \frac{15\,dx}{x^5} \underset{\substack{\uparrow\\ \text{Formula (5)}}}{=} 15\int \frac{1}{x^5}\,dx \underset{\substack{\uparrow\\ \text{Algebra}}}{=} 15\int x^{-5}\,dx \underset{\substack{\uparrow\\ \text{Formula (2)}}}{=} 15\cdot\frac{x^{-4}}{-4} + K = \frac{-15}{4x^4} + K$$

(d) $$\int \frac{x^{3/2} - 2x}{\sqrt{x}}\,dx \underset{\text{Algebra}}{=} \int \frac{x^{3/2} - 2x}{x^{1/2}}\,dx \underset{\text{Algebra}}{=} \int \left(\frac{x^{3/2}}{x^{1/2}} - \frac{2x}{x^{1/2}}\right) dx$$

$$\underset{\text{Algebra}}{=} \int (x - 2x^{1/2})\,dx \underset{\text{Formulas (4) and (5)}}{=} \int x\,dx - 2\int x^{1/2}\,dx$$

$$\underset{\text{Formula (2)}}{=} \frac{x^2}{2} - \frac{2x^{3/2}}{\frac{3}{2}} + K = \frac{x^2}{2} - \frac{4}{3}x^{3/2} + K$$

NOW WORK PROBLEM 29.

Indefinite Integrals Involving Exponential and Logarithmic Functions

The next three integration formulas involve the exponential and logarithmic functions. Each one is a direct result of formulas developed in Chapter 13.

$$\int e^x\,dx = e^x + K \tag{6}$$

$$\int e^{ax}\,dx = \frac{1}{a}e^{ax} + K \qquad a \neq 0 \tag{7}$$

$$\int \frac{1}{x}\,dx = \ln|x| + K \tag{8}$$

where K is a constant.

Proof Formulas (6) and (7) follow from the facts that

$$\frac{d}{dx}e^x = e^x \qquad \text{and} \qquad \frac{d}{dx}\left[\frac{1}{a}e^{ax}\right] = \frac{1}{a}\cdot ae^{ax} = e^{ax}$$

Formula (8) requires more attention. To prove it, we need to show that

$$\frac{d}{dx}\ln|x| = \frac{1}{x}$$

Since $x \neq 0$ (do you see why?), we consider two cases: $x > 0$ and $x < 0$.

Case 1 $x > 0$:

$$\frac{d}{dx}\ln|x| \underset{\substack{|x| = x \\ \text{since } x > 0}}{=} \frac{d}{dx}\ln x = \frac{1}{x}$$

Case 2 $x < 0$:

$$\frac{d}{dx}\ln|x| \underset{\substack{|x| = -x \\ \text{since } x < 0}}{=} \frac{d}{dx}\ln(-x) \underset{\frac{d}{dx}\ln g(x) = \frac{\frac{d}{dx}g(x)}{g(x)}}{=} \frac{1}{-x}\frac{d}{dx}(-x) = \frac{1}{-x}(-1) = \frac{1}{x}$$

In each case,

$$\frac{d}{dx} \ln |x| = \frac{1}{x}$$

so that

$$\int \frac{1}{x}\, dx = \ln |x| + K$$

Formula (8) takes care of finding $\int \frac{1}{x}\, dx = \int x^{-1}\, dx$. Now we can find $\int x^n\, dx$ for any rational number n.

We have still not discussed the indefinite integral of $\ln x$. We postpone a discussion of $\int \ln x\, dx$ until Section 15.3, where we introduce *integration by parts.*

3 **EXAMPLE 7 Evaluating Indefinite Integrals**

Evaluate each of the following indefinite integrals

(a) $\int \frac{e^x + e^{-x}}{2}\, dx$ **(b)** $\int \frac{3x^7 - 4x}{x^2}\, dx$

SOLUTION **(a)**

$$\begin{aligned}
\int \frac{e^x + e^{-x}}{2}\, dx &= \frac{1}{2}\int (e^x + e^{-x})\, dx && \int c f(x)\, dx = c\int f(x)\, dx \\
&= \frac{1}{2}\left[\int e^x\, dx + \int e^{-x}\, dx\right] && \int [f(x) + g(x)]\, dx = \int f(x)\, dx + \int g(x)\, dx \\
&= \frac{1}{2}[e^x + (-e^{-x})] + K && \text{Formula (7).} \\
&= \frac{e^x - e^{-x}}{2} + K && \text{Simplify.}
\end{aligned}$$

(b)

$$\begin{aligned}
\int \frac{3x^7 - 4x}{x^2}\, dx &= \int\left(\frac{3x^7}{x^2} - \frac{4x}{x^2}\right) dx && \text{Algebra.} \\
&= \int\left(3x^5 - \frac{4}{x}\right) dx && \text{Algebra.} \\
&= \int 3x^5\, dx - \int \frac{4}{x}\, dx && \int \left[f(x) - g(x)\right] dx = \int f(x)\, dx - \int g(x)\, dx \\
&= 3\int x^5\, dx - 4\int \frac{1}{x}\, dx && \int c f(x)\, dx = c\int f(x)\, dx \\
&= 3 \cdot \frac{x^6}{6} - 4 \ln |x| + K && \text{Formulas (2) and (8).} \\
&= \frac{x^6}{2} - 4 \ln |x| + K && \text{Simplify.}
\end{aligned}$$

NOW WORK PROBLEM 35.

Application to Marginal Analysis

Find cost and revenue functions 4

As we discussed in Chapter 13, marginal revenue and marginal cost are defined as the derivative of the revenue function and cost function, respectively. As a result, if the marginal revenue (or marginal cost) is a known function, the revenue function R (or cost function C) may be found by using the process of integration. That is,

Finding Revenue and Cost from Marginal Revenue and Marginal Cost

$$\int R'(x)\,dx = R(x) + K \quad \text{and} \quad \int C'(x)\,dx = C(x) + K$$

where K is a constant.

The presence of a constant K in the integral of a function plays an important role in applications and will equal the particular value that the situation demands.

EXAMPLE 8 Finding a Cost Function

By experimenting with various production techniques, a manufacturer finds that the marginal cost of production is given by the function

$$C'(x) = 2x + 6$$

where x is the number of units produced and C' is the marginal cost in dollars. The fixed cost of production is known to be \$9. Find the cost of production.

SOLUTION The antiderivative of the marginal cost C' of production is the cost C of production. That is,

$$C(x) = \int C'(x)\,dx = \int (2x + 6)\,dx = x^2 + 6x + K$$

FIGURE 1

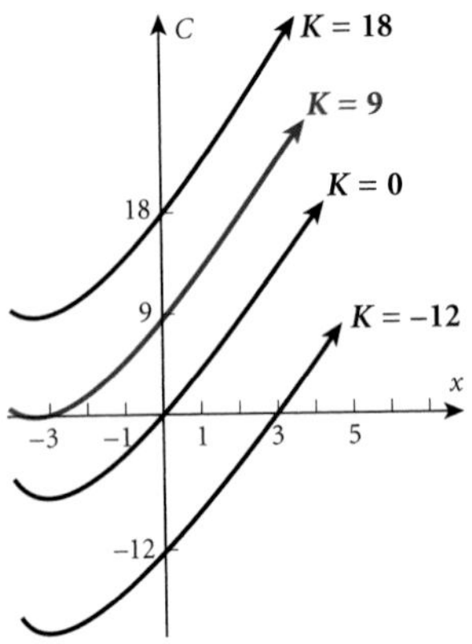

where K is a constant. We can find the value of the constant K by noting that of all the cost functions with derivative $2x + 6$, only one has a fixed cost of production of \$9, namely, the one whose cost is 9 when $x = 0$. We use this requirement to find the constant K.

$$C(x) = x^2 + 6x + K$$

$$C(0) = 0^2 + 6 \cdot 0 + K = 9 \qquad C(0) = 9$$

$$K = 9 \qquad \text{Solve for } K.$$

As a result, the cost function C is

$$C(x) = x^2 + 6x + 9$$

Figure 1 illustrates various cost functions whose marginal cost is $2x + 6$. The one having a fixed cost of \$9 is shown in color.

NOW WORK PROBLEM 43.

EXAMPLE 9 Finding a Profit Function

Suppose the manufacturer in Example 8 receives a price of \$60 per unit. This means the marginal revenue is \$60. That is,

$$R'(x) = 60$$

(a) Find the revenue function R.
(b) Find the profit function P.
(c) Find the sales volume that yields maximum profit.
(d) What is the profit at this sales volume?

SOLUTION **(a)** The revenue function R is the antiderivative of the marginal revenue function R'. That is,

$$R(x) = \int R'(x)\,dx = \int 60\,dx = 60x + K$$

Now, of all these revenue functions, there is only one for which revenue equals zero for $x = 0$ units sold. To find it, we need to find K. We find K as follows:

$$\begin{aligned} R(0) &= 60(0) + K = 0 && R(0) = 0 \\ K &= 0 && \text{Solve for } K. \end{aligned}$$

This means the revenue function is

$$R(x) = 60x$$

See Figure 2.

FIGURE 2

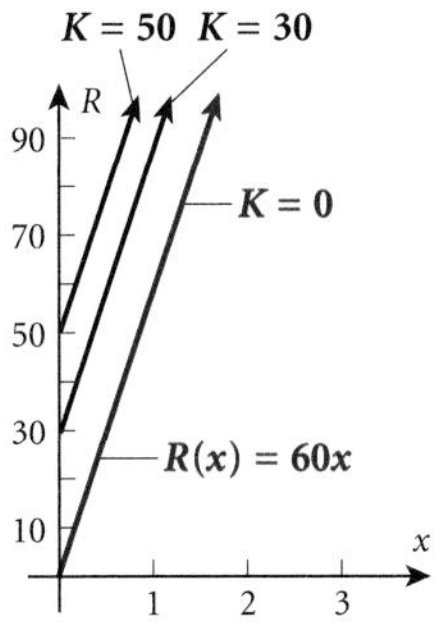

(b) The profit function P is the difference between revenue and cost. Since $R(x) = 60x$ and $C(x) = x^2 + 6x + 9$ (from Example 8), we have

$$P(x) = R(x) - C(x) = 60x - (x^2 + 6x + 9) = -x^2 + 54x - 9$$

(c) The maximum profit is obtained when marginal revenue equals marginal cost:

$$\begin{aligned} R'(x) &= C'(x) \\ 60 &= 2x + 6 \\ 2x &= 54 \\ x &= 27 \end{aligned}$$

When sales total 27 units, a maximum profit is obtained.

(d) The profit for sales of 27 units is

$$P(27) = -(27)^2 + 54(27) - 9 = \$720$$

NOW WORK PROBLEM 47.

EXAMPLE 10 Predicting the Size of a Mosquito Population

The size of a mosquito population is changing at the rate of $432t^2 - 5t^4$ per month where t is the time in months. If the current population is 40, what will the population size be 5 months from now?

SOLUTION If $P(t)$ is the population of mosquitoes at time t, then $P'(t)$ represents the rate of change of population with respect to time. That is,

$$P'(t) = 432t^2 - 5t^4$$

It follows that $P(t)$ is the antiderivative of $432t^2 - 5t^4$. That is,

$$P(t) = \int P'(t)\,dt = \int (432t^2 - 5t^4)\,dt = 144t^3 - t^5 + K$$

The constant K is determined from the condition that at $t = 0$ the population is 40. That is, $P(0) = 40$.

$$40 = 144(0^3) - 0^5 + K \quad P(0) = 40.$$
$$K = 40 \quad \text{Solve for } K.$$

As a result,

$$P(t) = 144t^3 - t^5 + 40$$

The population 5 months from now will be

$$P(5) = 144(5^3) - 5^5 + 40 = 14{,}915$$

mosquitoes.

EXERCISE 15.1 Answers to Odd-Numbered Problems Begin on Page AN-89.

In Problems 1-8, find all the antiderivatives of f.

1. $f(x) = x^3$
2. $f(x) = 6x^2$
3. $f(x) = 2x + 3$
4. $f(x) = 10 - 3x^2$
5. $f(x) = \frac{4}{x}$
6. $f(x) = e^x$
7. $f(x) = \sqrt[3]{x}$
8. $f(x) = e^{3x}$

In Problems 9–38, evaluate each indefinite integral.

9. $\int 3\,dx$
10. $\int -4\,dx$
11. $\int x\,dx$
12. $\int x^2\,dx$
13. $\int x^{1/3}\,dx$
14. $\int x^{4/3}\,dx$
15. $\int x^{-2}\,dx$
16. $\int x^{-3}\,dx$
17. $\int x^{-1/2}\,dx$
18. $\int x^{-2/3}\,dx$
19. $\int (2x^3 + 5x)\,dx$
20. $\int (3x^2 - 4x)\,dx$
21. $\int (x^2 + 2e^x)\,dx$
22. $\int (3x + 5e^x)\,dx$
23. $\int (x^3 - 2x^2 + x - 1)\,dx$
24. $\int (2x^4 + x^2 - 5)\,dx$
25. $\int \left(\frac{x - 1}{x}\right) dx$
26. $\int \left(\frac{x + 1}{x}\right) dx$
27. $\int \left(2e^x - \frac{3}{x}\right) dx$
28. $\int \left(\frac{8}{x} - e^{-x}\right) dx$
29. $\int \left(\frac{3\sqrt{x} + 1}{\sqrt{x}}\right) dx$
30. $\int \left(\frac{2\sqrt{x} - 4}{\sqrt{x}}\right) dx$
31. $\int \frac{x^2 - 4}{x + 2}\,dx$
32. $\int \frac{x^2 - 1}{x - 1}\,dx$
33. $\int x(x - 1)\,dx$
34. $\int x(x + 2)\,dx$
35. $\int \frac{3x^5 + 2}{x}\,dx$
36. $\int \frac{x^6 + x^2 + 1}{x^3}\,dx$
37. $\int \frac{4e^x + e^{2x}}{e^x}\,dx$
38. $\int \frac{3e^x + xe^{2x}}{xe^x}\,dx$

In Problems 39–42, find the revenue function R. Assume that revenue is zero when zero units are sold.

39. $R'(x) = 600$
40. $R'(x) = 350$
41. $R'(x) = 20x + 5$
42. $R'(x) = 50x - x^2$

In Problems 43–46, find the cost function C. Determine where the cost is a minimum.

43. $C'(x) = 14x - 2800$
Fixed cost = \$4300

44. $C'(x) = 6x - 2400$
Fixed cost = \$800

45. $C'(x) = 20x - 8000$
Fixed cost = \$500

46. $C'(x) = 15x - 3000$
Fixed cost = \$1000

47. **Profit Function** The marginal cost of production is found to be

$$C'(x) = 1000 - 20x + x^2$$

where x is the number of units produced. The fixed cost of production is \$9000. The manufacturer sets the price per unit at \$3400.

(a) Find the cost function.
(b) Find the revenue function.
(c) Find the profit function.
(d) Find the sales volume that yields maximum profit.
(e) What is the profit at this sales volume?
(f) Graph the revenue, cost, and profit functions.

48. **Profit** A company determines that the marginal cost of producing x units of a particular commodity during 1 day of operation is $C'(x) = 6x - 141$, where the production cost is in dollars. The selling price of the commodity is fixed at \$9 per unit and the fixed cost is \$1800 per day.

(a) Find the cost function.
(b) Find the revenue function.
(c) Find the profit function.
(d) What is the maximum profit that can be obtained in 1 day of operation?
(e) Graph the revenue, cost, and profit functions.

49. **Prison Population** In 1998 the total number of inmates in United States prisons was 592,462. Research indicates that this number will change at a rate of $7000t + 20{,}000$ where t represents the number of years elapsed since the year 1998. Predict the total number of inmates in United States prisons in the year 2008.

Source: Office of Justice Programs, United States Department of Justice.

50. **Undergraduate Tuition** Average annual undergraduate tuition paid by students at institutions of higher learning in the United States changes at the rate of $-0.14t + 225$ where t is the number of years elapsed since 1996. What will tuition be in 2007 if the cost of tuition in the year 2001 was \$5442?

Source: National Council of Education Statistics, United States Department of Education.

51. **Population Growth** It is estimated that the population of a certain town changes at the rate of $2 + t^{4/5}$ people per month. If the current population is 20,000, what will the population be in 10 months?

52. **Resource Depletion** The water currently used from a lake is estimated to amount to 150 million gallons a month. Water usage (in millions of gallons) is expected to increase at the rate of $3 + 0.01x$ after x months. What will the water usage be a year from now?

53. **Population Growth** There are currently 20,000 citizens of voting age in a small town. Demographics indicate that the voting population will change at the rate of $2.2t - 0.8t^2$ (in thousands of voting citizens), where t denotes time in years. How many citizens of voting age will there be 3 years from now?

54. **Air Pollution** An environmental study of a certain town suggests that t years from now the level of carbon monoxide in the air will be changing at the rate of $0.2t + 0.2$ parts per million per year. If the current level of carbon monoxide in the air is 3.8 parts per million, what will the level be 5 years from now?

55. **Chemical Reaction** The end product of a chemical reaction is produced at the rate of $\dfrac{\sqrt{t} - 1}{t}$ milligrams per minute. If the reaction started at time $t = 1$, determine the amount produced during the first 4 minutes.

56. **Free Fall** An object dropped from an airplane 3200 feet above the ground falls at the rate of

$$v(t) = 32t \text{ feet per second}$$

where t is given in seconds. If the object has fallen 576 feet after 6 seconds, how long after it is dropped will the object hit the ground?

57. **Water Depletion** A water reservoir is being filled at the rate of 15,000 gallons per hour. Due to increased consumption, the water in the reservoir is decreasing at the rate of $\frac{5}{2}t$ gallons per hour at time t. When will the reservoir be empty if the initial water volume was 100,000 gallons?

58. Verify the following statements:

(a) $\displaystyle\int (x \cdot \sqrt{x})\, dx \neq \int x\, dx \cdot \int \sqrt{x}\, dx$

(b) $\displaystyle\int x(x^2 + 1)\, dx \neq x\int (x^2 + 1)\, dx$

(c) $\displaystyle\int \frac{x^2 - 1}{x - 1}\, dx \neq \frac{\int (x^2 - 1)\, dx}{\int (x - 1)\, dx}$

15.2 Integration Using Substitution

PREPARING FOR THIS SECTION *Before getting started, review the following:*

> The Power Rule (Chapter 13, Section 13.4, p. 744)

> Differentials (Chapter 14, Section 14.7, pp. 845–847)

OBJECTIVE 1 Integrate using substitution

Indefinite integrals that cannot be evaluated by using Formulas (1)–(8) on pages 864 and 866 of Section 15.1 may sometimes be evaluated by the *substitution method.* This method involves the introduction of a function that changes the integrand into a form to which the formulas of Section 15.1 apply.

The basic idea behind integration by substitution is the Power Rule. To see how integration by substitution works, consider the following example.

EXAMPLE 1 Evaluating an Indefinite Integral

Evaluate: $\int (x^2 + 5)^3\, 2x\, dx$

SOLUTION By the Power Rule, we know that

$$\frac{d}{dx}[g(x)]^n = n[g(x)]^{n-1}g'(x)$$

With $g(x) = x^2 + 5$ and $n = 4$, we have

$$\frac{d}{dx}(x^2 + 5)^4 = 4(x^2 + 5)^3\, 2x$$

Then

$$\int 4(x^2 + 5)^3\, 2x\; dx = (x^2 + 5)^4 + K$$

where K is a constant. We apply Formula (5) on page 864 and move the 4 outside the integral.

$$4\int (x^2 + 5)^3\, 2x\; dx = (x^2 + 5)^4 + K$$

Now divide both sides by 4. Then

$$\int (x^2 + 5)^3\, 2x\; dx = \frac{1}{4}(x^2 + 5)^4 + K_1$$

where $K_1 = \dfrac{K}{4}$.

We may simplify the procedure used in Example 1 by *changing the variables.* Introduce the variable u, defined as

$$u = x^2 + 5$$

The differential of $u = u(x)$ is $du = u'(x)\, dx$. Then

$$du = 2x\, dx \qquad \frac{d}{dx}(x^2 + 5) = 2x$$

Now we express the integrand $(x^2 + 5)^3\, 2x$ in terms of u and the differential dx in terms of u and du.

$$\int (x^2+5)^3\, 2x\, dx \underset{\substack{u = x^2+5\\ du = 2x\,dx}}{\uparrow=} \int u^3\, du \underset{\text{Integrate}}{\uparrow=} \frac{1}{4}u^4 + K \underset{\substack{\text{Express in terms of } x;\\ u = x^2+5}}{\uparrow=} \frac{1}{4}(x^2+5)^4 + K$$

1 **EXAMPLE 2** **Integration Using Substitution**

Evaluate: $\displaystyle\int \frac{dx}{2x+1}$

SOLUTION We try the substitution $u = 2x + 1$ to see if it simplifies the integral. Then

$$du = 2\,dx \qquad \text{so} \qquad dx = \frac{du}{2}$$

$$\int \frac{dx}{2x+1} \underset{\substack{u = 2x+1\\ dx = \frac{du}{2}}}{\uparrow=} \int \frac{\frac{du}{2}}{u} \underset{\text{Simplify}}{\uparrow=} \int \frac{du}{2u} \underset{\substack{\text{Move } \frac{1}{2} \text{ outside}\\ \text{the integral}}}{\uparrow=} \frac{1}{2}\int \frac{du}{u} \underset{\text{Integrate}}{\uparrow=} \frac{1}{2}\ln|u| + K \underset{u = 2x+1}{\uparrow=} \frac{1}{2}\ln|2x+1| + K$$

NOW WORK PROBLEM 1.

EXAMPLE 3 **Integration Using Substitution**

Evaluate: $\displaystyle\int x\sqrt{x^2+1}\, dx$

SOLUTION We try the substitution $u = x^2 + 1$ to see if it simplifies the integral. Then

$$du = 2x\,dx \qquad \text{so} \qquad x\,dx = \frac{du}{2}$$

Now

$$\begin{aligned}
\int x\sqrt{x^2+1}\, dx &= \int \sqrt{x^2+1}\; x\, dx \\
&= \int \sqrt{u}\, \frac{du}{2} && u = x^2+1,\ x\,dx = \frac{du}{2} \\
&= \frac{1}{2}\int \sqrt{u}\, du && \text{Move } \tfrac{1}{2} \text{ outside the integral.} \\
&= \frac{1}{2}\int u^{1/2}\, du && \text{Change to a rational exponent.} \\
&= \frac{1}{2}\,\frac{u^{3/2}}{\frac{3}{2}} + K && \text{Integrate.} \\
&= \frac{(x^2+1)^{3/2}}{3} + K && u = x^2+1
\end{aligned}$$

Notice that the substitution $u = x^2 + 1$ in Example 3 worked because the x in the integrand along with dx gives du, except for the constant 2. If we try this same substitution to evaluate $\int \sqrt{x^2+1}\,dx$, we obtain

$$\int \sqrt{x^2+1}\,dx = \int \sqrt{u}\frac{du}{2\sqrt{u-1}} = \int \frac{\sqrt{u}}{2\sqrt{u-1}}\,du$$

In this case, the substitution results in an integrand that is *more complicated* than the original one.

EXAMPLE 4 Integration Using Substitution

Evaluate: $\int x^2 e^{x^3}\,dx$

SOLUTION Let $u = x^3$. Then $du = 3x^2\,dx$.

$$\int x^2 e^{x^3}\,dx = \int e^{x^3}\cdot x^2\,dx \underset{\substack{u=x^3 \\ \frac{du}{3}=x^2\,dx}}{=} \int e^u \frac{du}{3} \underset{\substack{\text{Move } \frac{1}{3} \text{ outside} \\ \text{the integral}}}{=} \frac{1}{3}\int e^u\,du \underset{\text{Integrate}}{=} \frac{1}{3}e^u + K \underset{u=x^3}{=} \frac{1}{3}e^{x^3} + K$$

EXAMPLE 5 Integration Using Substitution

Evaluate: $\int \frac{dx}{x\ln x}$

SOLUTION Let $u = \ln x$. Then $du = \frac{1}{x}\,dx = \frac{dx}{x}$. Then

$$\int \frac{dx}{x\ln x} = \int \frac{1}{\ln x}\frac{dx}{x} \underset{\text{Substitute}}{=} \int \frac{1}{u}\,du \underset{\text{Integrate}}{=} \ln|u| + K \underset{u=\ln x}{=} \ln|\ln x| + K$$

EXAMPLE 6 Integration Using Substitution

Evaluate: $\int \frac{dx}{2\sqrt{x}(1+\sqrt{x})^3}$

SOLUTION We try the substitution

$$u = 1 + \sqrt{x} = 1 + x^{1/2}$$

Then

$$du = \frac{1}{2}x^{-1/2}\,dx = \frac{1}{2\sqrt{x}}\,dx = \frac{dx}{2\sqrt{x}}$$

Now we substitute.

$$\int \frac{dx}{2\sqrt{x}(1+\sqrt{x})^3} \underset{\text{Rearrange}}{=} \int \frac{1}{(1+\sqrt{x})^3}\frac{dx}{2\sqrt{x}} \underset{\text{Substitute}}{=} \int \frac{1}{u^3}\,du = \int u^{-3}\,du \underset{\text{Integrate}}{=} \frac{u^{-2}}{-2} + K$$

$$\underset{\text{Simplify}}{=} -\frac{1}{2u^2} + K \underset{u=1+\sqrt{x}}{=} -\frac{1}{2(1+\sqrt{x})^2} + K$$

NOW WORK PROBLEM 27.

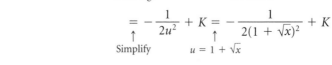

Sometimes more than one substitution will work. In the next example we use two different substitutions to evaluate the same integral.

EXAMPLE 7 Integrating Using Substitution

Evaluate: $\int x\sqrt{4+x}\,dx$

SOLUTION **Substitution I:** Let $u = 4 + x$. Then $du = dx$ and $x = u - 4$, so that

$$\begin{aligned}
\int x\sqrt{4+x}\,dx &= \int (u-4)\sqrt{u}\,du && x = u-4;\ u = 4+x;\ dx = du \\
&= \int \left(u^{3/2} - 4u^{1/2}\right) du && \sqrt{u} = u^{1/2};\ \text{multiply out.} \\
&= \frac{u^{5/2}}{\frac{5}{2}} - \frac{4u^{3/2}}{\frac{3}{2}} + K && \text{Integrate.} \\
&= \frac{2(4+x)^{5/2}}{5} - \frac{8(4+x)^{3/2}}{3} + K && u = 4+x
\end{aligned}$$

Substitution II: Let $u^2 = 4 + x$. Then $x = u^2 - 4$, and $dx = 2u\,dx$. Now substitute into the integral.

$$\begin{aligned}
\int x\sqrt{4+x}\,dx &= \int (u^2-4)u\,2u\,du && x = u^2-4;\ 4+x = u^2;\ dx = 2u\,du \\
&= 2\int (u^2-4)u^2\,du && \text{Move 2 outside the integral.} \\
&= 2\int (u^4-4u^2)\,du && \text{Multiply out.} \\
&= 2\int u^4\,du - 2\int 4u^2\,du && \text{The integral of a difference is the difference of the integrals.} \\
&= 2\int u^4\,du - 8\int u^2\,du && \text{Move 4 outside the second integral.} \\
&= 2\cdot\frac{u^5}{5} - 8\cdot\frac{u^3}{3} + K && \text{Integrate.} \\
&= \frac{2(4+x)^{5/2}}{5} - \frac{8(4+x)^{3/2}}{3} + K && u^2 = 4+x.
\end{aligned}$$

NOW WORK PROBLEM 17.

The idea behind the substitution method is to obtain an integral $\int h(u)\,du$ that is simpler than the original integral $\int f(x)\,dx$. When a substitution does not simplify the integral, other substitutions should be tried. If these do not work, other integration methods must be applied. Since integration, unlike differentiation, has no prescribed method, a lot of practice is required.

A summary of the method of integration by substitution is given in the flowchart in Figure 3.

FIGURE 3

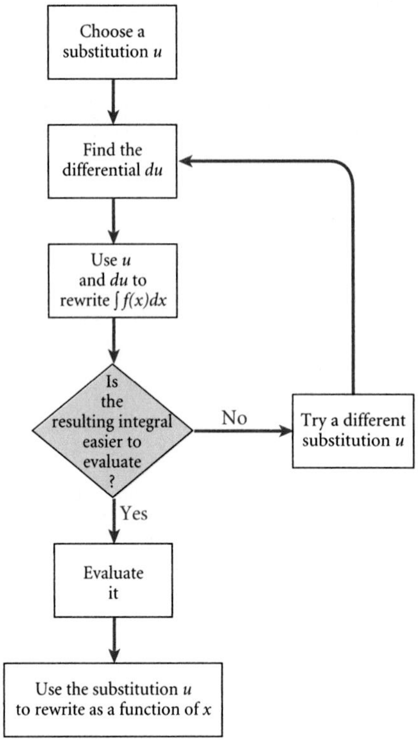

EXERCISE 15.2 Answers to Odd-Numbered Problems Begin on Page AN-90.

In Problems 1–34, evaluate each indefinite integral. Use the substitution method.

1. $\int (2x + 1)^5\,dx$
2. $\int (3x - 5)^4\,dx$
3. $\int e^{2x-3}\,dx$
4. $\int e^{3x+4}\,dx$
5. $\int (-2x + 3)^{-2}\,dx$
6. $\int (5 - 2x)^{-3}\,dx$
7. $\int x(x^2 + 4)^2\,dx$
8. $\int x(x^2 - 2)^3\,dx$
9. $\int e^{x^3+1}x^2\,dx$
10. $\int e^{2x^2+1}x\,dx$
11. $\int (e^{x^2} + e^{-x^2})\,x\,dx$
12. $\int (e^{x^2} - e^{-x^2})\,x\,dx$
13. $\int x^2(x^3 + 2)^6\,dx$
14. $\int x^2(x^3 - 1)^4\,dx$
15. $\int \frac{x}{\sqrt[3]{1 + x^2}}\,dx$
16. $\int \frac{x}{\sqrt[5]{1 - x^2}}\,dx$
17. $\int x\sqrt{x + 3}\,dx$
18. $\int x\sqrt{x - 3}\,dx$
19. $\int \frac{e^x}{e^x + 1}\,dx$
20. $\int \frac{e^{-x}}{e^{-x} + 4}\,dx$
21. $\int \frac{e^{\sqrt{x}}\,dx}{\sqrt{x}}$
22. $\int \frac{e^{\sqrt[3]{x}}\,dx}{x^{2/3}}$
23. $\int \frac{(x^{1/3} - 1)^6\,dx}{x^{2/3}}$
24. $\int \frac{(x^{1/3} + 2)^3}{x^{2/3}}\,dx$
25. $\int \frac{(x + 1)\,dx}{(x^2 + 2x + 3)^2}$
26. $\int \frac{(x + 4)\,dx}{(x^2 + 8x + 2)^3}$
27. $\int \frac{dx}{\sqrt{x}(1 + \sqrt{x})^4}$
28. $\int \frac{(3 - 2\sqrt{x})^2}{\sqrt{x}}\,dx$
29. $\int \frac{dx}{2x + 3}$
30. $\int \frac{dx}{3x - 5}$
31. $\int \frac{x\,dx}{4x^2 + 1}$
32. $\int \frac{x\,dx}{5x^2 - 2}$
33. $\int \frac{x + 1}{x^2 + 2x + 2}\,dx$
34. $\int \frac{2x - 1}{x^2 - x + 4}\,dx$

35. Depreciation of a Car The depreciation rate of a new car obeys the equation

$$V'(t) = -6000\, e^{-0.5t}$$

where V is the value of the car and t, in years, is the time from the date of purchase. If the car cost \$27,000 new, what is its value after 2 years? What is the value after 4 years?

36. Depreciation of a Car The depreciation rate of a new car obeys the equation

$$V'(t) = \frac{-16{,}000}{(2t+1)^2}, \quad 0 \le t \le 4$$

where V is the value of the car and t, in years, is the time from the date of purchase. If the car cost \$15,000 new, what is its value after 2 years? What is the value after 4 years?

37. Department of Agriculture Budget The annual budget outlay of the United States Department of Agriculture is growing at a rate of $B'(t) = 1.715e^{0.025t}$ where B is in billions of dollars per year and t is the number of years elapsed since 2001. The budget for 2001 was 68.6 billion dollars.

(a) Find an equation for $B(t)$.
(b) If the current rate of growth continues, predict when the budget will exceed \$100 billion.

Source: United States Department of Agriculture.

38. Revenue The marginal revenue (in thousands of dollars) from the sale of x tractors is

$$R'(x) = \frac{2x(x^2+10)^2}{1000}$$

Find the revenue function R if the revenue from the sale of 4 tractors is \$198,000.

39. Work Force Growth The number N of employees at the Ajex Steel Company is growing at a rate given by the equation

$$N'(t) = 20e^{0.01t}$$

where N is the number of people employed at time t, in years. The number of employees currently is 400.

(a) Find an equation for $N(t)$.
(b) How long will it take the work force to reach 800 employees?

40. Pollution An oil tanker is leaking oil and producing a circular oil slick that is growing at the rate of

$$R'(t) = \frac{30}{\sqrt{t+4}} \qquad t \ge 0$$

where R is the radius in meters of the slick after t minutes. Find the radius of the slick after 21 minutes if the radius is 0 when $t = 0$.

41. Derive the formula

$$\int (ax+b)^n\,dx = \frac{(ax+b)^{n+1}}{a(n+1)} + K \quad a \ne 0, n \ne -1$$

[**Hint:** Let $u = ax + b$.]

15.3 Integration by Parts

PREPARING FOR THIS SECTION *Before getting started, review the following:*

>> Formula for the Derivative of a Product (Chapter 13, Section 13.3, p. 735)

OBJECTIVE 1 Integrate by parts

Integrate by parts 1 Next, we discuss a method for evaluating indefinite integrals such as

$$\int xe^x\,dx \quad \text{and} \quad \int \ln x\,dx$$

for which the substitution technique does not work.

This method, called *integration by parts,* is based on the formula for the derivative of a product. Recall that if u and v are differentiable functions of x, then

$$\frac{d}{dx}(uv) = u\frac{dv}{dx} + v\frac{du}{dx}$$

By integrating both sides, we obtain

$$\int \frac{d}{dx}(uv)\,dx = \int\left(u\frac{dv}{dx} + v\frac{du}{dx}\right)dx$$

The integral on the left is just uv. On the right, use the fact that the integral of a sum is the sum of the integrals. Then,

$$uv = \int u\frac{dv}{dx}\,dx + \int v\frac{du}{dx}\,dx$$

Rearranging terms, we obtain

$$\boxed{\int u\frac{dv}{dx}\,dx = uv - \int v\frac{du}{dx}\,dx}$$

In abbreviated form, this formula may be written in the following way.

Integration by Parts Formula

$$\boxed{\int u\,dv = uv - \int v\,du} \tag{1}$$

To use the integration by parts formula, we choose u and dv so $\int u\,dv$ is equal to the integral we seek. We find the differential of u to obtain du and integrate dv to obtain v. Then we substitute into Formula (1). If we can integrate $\int v\,du$, the problem is solved.

The goal of this procedure, then, is to choose u and dv so that the term $\int v\,du$ is easier to integrate than the original problem, $\int u\,dv$. As the examples will illustrate, this usually happens when u is simplified by differentiation.

EXAMPLE 1 Using the Integration by Parts Formula

Evaluate: $\int xe^x\,dx$

SOLUTION To use the integration by parts formula, we choose u and dv so that

$$\int u\,dv = \int xe^x\,dx$$

and $\int v\,du$ is easier to evaluate than $\int u\,dv$. In this example, we choose

$$u = x \qquad \text{and} \qquad dv = e^x\,dx \tag{2}$$

With this choice, $\int u\,dv = \int xe^x\,dx$, as required. Now use equation (2) to find du and $v = \int dv$.

$$du = dx \quad \text{and} \quad v = \int dv \underset{\substack{\uparrow \\ dv = e^x dx}}{=} \int e^x\,dx = e^x \tag{3}$$

Notice in evaluating $\int dv$ that we only require a particular antiderivative of dv; we will add the constant of integration later.

Now substitute the results of (2) and (3) into (1). The result is

$$\int \overbrace{x}^{u}\ \overbrace{e^x\,dx}^{dv} = \overbrace{x}^{u}\ \overbrace{e^x}^{v} - \int \overbrace{e^x}^{v}\ \overbrace{dx}^{du} \qquad \text{Substitute into } \int u\,dv = uv - \int v\,du$$

$$= xe^x - e^x + K \qquad \text{Integrate } \int e^x\,dx\text{; add the constant of integration.}$$

NOW WORK PROBLEM 1.

Let's look once more at Example 1. Suppose we had chosen u and dv differently as

$$u = e^x \qquad \text{and} \qquad dv = x\,dx$$

This choice would have resulted in $\int u\,dv = \int xe^x\,dx$, as required, but now

$$du = e^x\,dx \qquad \text{and} \qquad v = \int dv = \int x\,dx = \frac{x^2}{2}$$

so equation (1) would have yielded

$$\int \overbrace{xe^x\,dx}^{u\,dv} = \overbrace{e^x\frac{x^2}{2}}^{uv} - \int \overbrace{\frac{x^2}{2}e^x\,dx}^{v\,du}$$

As you can see, instead of obtaining an integral that is easier to evaluate, we obtain one that is more complicated than the original. This means that an unwise choice of u and dv has been made.

Unfortunately, there are no general directions for choosing u and dv. Some hints you can use are given next.

Hints for Using Integration by Parts

1. Choose u and dv so that $\int u\,dv$ is the integral you seek. In choosing dv, dx is always a part of dv.
2. You must be able to find du and $v = \int dv$.
3. u and dv are chosen so that $\int v\,du$ is easier to evaluate than the original integral $\int u\,dv$. This often happens when u is simplified by differentiation.

In making an initial choice for u and dv, a certain amount of trial and error is used. If a selection appears to hold little promise, abandon it and try some other choice. If no choices work, it may be that some other technique of integration, such as substitution, should be tried.

Let's look at some more examples.

EXAMPLE 2 Using Integration by Parts

Evaluate: $\int x \ln x\,dx$

SOLUTION Choose

$$u = \ln x \quad \text{and} \quad dv = x\,dx \tag{4}$$

Then

$$\int u\,dv = \int (\ln x)(x\,dx) = \int x \ln x\,dx$$

as required.

Use (4) to find du and $\int v\,du$.

$$u = \ln x \qquad du = \frac{1}{x}\,dx \qquad \frac{d}{dx}\ln x = \frac{1}{x}$$

$$dv = x\,dx \qquad v = \frac{x^2}{2} \qquad v = \int dv = \int x\,dx = \frac{x^2}{2}$$

Integrate by parts.

$$\int x \ln x\,dx = \ln x \cdot \frac{x^2}{2} - \int \frac{x^2}{2}\,\frac{1}{x}\,dx \qquad \int u\,dv = uv - \int v\,du$$

$$= \frac{1}{2}x^2 \ln x - \frac{1}{2}\int x\,dx \qquad \text{Simplify. Move } \frac{1}{2} \text{ outside the integral.}$$

$$= \frac{1}{2}x^2 \ln x - \frac{1}{2}\cdot\frac{x^2}{2} + K \qquad \text{Integrate.}$$

$$= \frac{1}{2}x^2 \ln x - \frac{1}{4}x^2 + K$$

NOW WORK PROBLEM 3 AND 7.

Sometimes it is necessary to integrate by parts more than once to solve a particular problem, as illustrated by the next example.

EXAMPLE 3 Using Integration by Parts

Evaluate: $\int x^2 e^x\,dx$

SOLUTION Choose

$$u = x^2 \quad \text{and} \quad dv = e^x\,dx \qquad \int u\,dv = \int x^2 e^x\,dx$$

Then

$$du = 2x\,dx \quad \text{and} \quad v = e^x \qquad v = \int dv = \int e^x\,dx = e^x$$

Integrate by parts.

$$\int x^2 e^x\,dx = x^2 e^x - \int e^x \cdot 2x\,dx \qquad \int u\,dv = uv - \int v\,du$$

$$= x^2 e^x - 2\int x e^x\,dx \qquad \text{Move 2 outside the integral.}$$

Although we must still evaluate $\int x e^x\,dx$, we can see that the original integral has been replaced by a simpler one. In fact, in Example 1 we found (using integration by parts) that

$$\int x e^x\,dx = x e^x - e^x + K \qquad \text{Example 1.}$$

Using this result, we have

$$\int x^2 e^x\,dx = x^2 e^x - 2(x e^x - e^x) + K$$

$$= x^2 e^x - 2x e^x + 2e^x + K$$

EXAMPLE 4 **Formula for $\int \ln x\, dx$**

Find a formula for $\int \ln x\, dx$.

SOLUTION To find $\int \ln x\, dx$, we use the integration by parts formula. Choose $u = \ln x$ and $dv = dx$. Then $du = \frac{1}{x}\, dx$ and $v = x$, so

$$\int \ln x\, dx = x \ln x - \int x \frac{1}{x}\, dx \qquad \int u\, dv = uv - \int v\, du$$

$$= x \ln x - \int dx \qquad \text{Simplify.}$$

$$= x \ln x - x + K \qquad \text{Integrate.}$$

The integration by parts formula is useful for the evaluation of indefinite integrals that have integrands composed of e^x times a polynomial function of x or $\ln x$ times a polynomial function of x. It can also be used for other types of indefinite integrals that will not be discussed in this book.

EXERCISE 15.3 Answers to Odd-Numbered Problems Begin on Page AN-90.

1. Evaluate $\int x\, e^{4x}\, dx$. Choose $u = x$ and $dv = e^{4x}\, dx$.

2. Evaluate $\int x\, e^{-x}\, dx$. Choose $u = x$ and $dv = e^{-x}\, dx$.

In Problems 3–16, evaluate each indefinite integral. Use integration by parts.

3. $\int xe^{2x}\, dx$

4. $\int xe^{-3x}\, dx$

5. $\int x^2 e^{-x}\, dx$

6. $\int x^2 e^{2x}\, dx$

7. $\int \sqrt{x} \ln x\, dx$

8. $\int x(\ln x)^2\, dx$

9. $\int (\ln x)^2\, dx$

10. $\int \frac{\ln x}{x^2}\, dx$

11. $\int x^2 \ln 3x\, dx$

12. $\int x^2 \ln 5x\, dx$

13. $\int x^2(\ln x)^2\, dx$

14. $\int x^3(\ln x)^2\, dx$

15. $\int \frac{\ln x}{x^3}\, dx$

16. $\int \sqrt{x}\, (\ln \sqrt{x})^2\, dx$

17. Population Growth The growth rate of a colony of leaf-eater ants follows the equation

$$P'(t) = 90\sqrt{t} - 100\, te^{-t}$$

where P is the population at time t measured in days. If the population of the colony at time $t = 0$ is 5000, what is the population after 4 days? What is it after one week?

18. Population Growth The growth rate of a colony of slugs follows the equation

$$P'(t) = 180t + 300te^{-2t}$$

where P is the population at time t measured in days. If the population of the colony at time $t = 0$ is 5000, what is the population after 4 days? What is it after one week?

19. Depreciation of a Car The depreciation rate of a new car obeys the equation

$$V'(t) = -8000te^{-0.8t}$$

where V is the value of the car and t, in years, is the time from the date of purchase. If the car cost \$20,000 new, what is its value after 2 years? What is the value after 4 years?

15.4 The Definite Integral; Learning Curves; Total Sales over Time

OBJECTIVES
1. Evaluate a definite integral
2. Use properties of a definite integral
3. Solve applied problems involving definite integrals

We begin with an example illustrating the general idea of a *definite integral.*

EXAMPLE 1 Finding the Change in a Cost Function

The marginal cost of a certain firm is given by the equation

$$C'(x) = 4 - 0.2x \qquad 0 \le x \le 10$$

where C' is in thousands of dollars and the quantity x produced is in hundreds of units per day. If the number of units produced in a given day changes from two hundred to five hundred units, what is the change in cost?

SOLUTION If C is the cost function, the change in cost from $x = 2$ to $x = 5$ is

$$C(5) - C(2)$$

This is the number we seek. The cost C is an antiderivative of $C'(x) = 4 - 0.2x$ so

$$C(x) = \int C'(x)\,dx = \int (4 - 0.2x)\,dx = 4x - 0.1x^2 + K$$

We use this to compute $C(5) - C(2)$:

$$C(5) - C(2) = [4(5) - (0.1)(25) + K] - [4(2) - (0.1)(4) + K] = 9.9 \qquad (1)$$

The change in cost is 9.9 thousand dollars.

In this example, the change in C was computed by using an antiderivative of C', which is symbolized by $\int C'(x)\,dx$. To indicate that the change is from $x = 2$ to $x = 5$, we add to this notation as follows:

$$\text{Change in } C \text{ from 2 to 5} = \int_2^5 C'(x)\,dx$$

This form is called a *definite integral.*

Definite Integral

Let f denote a function that is continuous on a closed interval $[a, b]$. The **definite integral** from a to b of f equals the change from a to b in an antiderivative of f. If F is an antiderivative of f, the definite integral of f from a to b is

$$\int_a^b f(x)\,dx = F(b) - F(a) \qquad (2)$$

In $\int_a^b f(x)\,dx$, the numbers a and b are called the **lower limit of integration** and the **upper limit of integration,** respectively.

1 **EXAMPLE 2** **Evaluating a Definite Integral**

Evaluate: $\int_2^3 x^2\,dx$

SOLUTION First we find an antiderivative of $f(x) = x^2$.

One such antiderivative is $F(x) = \dfrac{x^3}{3}$. Then, based on (2), we have

$$\int_2^3 x^2\,dx = F(3) - F(2) = \frac{3^3}{3} - \frac{2^3}{3} = \frac{27}{3} - \frac{8}{3} = \frac{19}{3} \qquad F(x) = \frac{x^3}{3}$$

NOW WORK PROBLEM 1.

In computing $\int_a^b f(x)\,dx$, we find that the choice of an antiderivative of f does not matter. Look back at equation (1) in the solution to Example 1. The constant K drops out. Now look at Example 2. If we had used $F(x) = \dfrac{x^3}{3} + K$ as the antiderivative of x^2, we would have found that

$$\int_2^3 x^2\,dx = F(3) - F(2) = \left(\frac{27}{3} + K\right) - \left(\frac{8}{3} + K\right) = \frac{19}{3}$$

Again, the constant K drops out. This will always be the case.

Any antiderivative of f can be used to evaluate $\int_a^b f(x)\,dx$.

For convenience, we introduce new notation.

If F is an antiderivative of f, then

$$\int_a^b f(x)\,dx = F(x)\Big|_a^b = F(b) - F(a)$$

In terms of this new notation, to calculate $F(x)\big|_a^b$, first replace x by the upper limit b to obtain $F(b)$, and from this subtract $F(a)$, obtained by letting $x = a$.

EXAMPLE 3 **Evaluating Definite Integrals**

(a) $\displaystyle\int_{-1}^5 6x\,dx = (3x^2)\Big|_{-1}^5 = 3(5)^2 - 3(-1)^2 = 75 - 3 = 72$

(b) $\displaystyle\int_1^2 x^3\,dx = \frac{x^4}{4}\Big|_1^2 = \frac{2^4}{4} - \frac{1^4}{4} = \frac{16}{4} - \frac{1}{4} = \frac{15}{4}$

It is important to distinguish between the indefinite integral and the definite integral. The indefinite integral, a symbol for all the antiderivatives of a function, is a function. On the other hand, the definite integral is a number.

EXAMPLE 4 Evaluating a Definite Integral

Evaluate: $\int_1^4 \sqrt{x}\,dx$

SOLUTION An antiderivative of $\sqrt{x} = x^{1/2}$ is

$$\frac{x^{3/2}}{\frac{3}{2}} = \frac{2}{3}x^{3/2}$$

Then

$$\int_1^4 \sqrt{x}\,dx = \left(\frac{2}{3}x^{3/2}\right)\Big|_1^4 = \frac{2}{3}(4)^{3/2} - \frac{2}{3}(1)^{3/2} = \frac{16}{3} - \frac{2}{3} = \frac{14}{3}$$

NOW WORK PROBLEM 5.

Properties of the Definite Integral

We list some properties of the definite integral below.

Properties of the Definite Integral

If f is a continuous function that has an antiderivative on the interval $[a, b]$, then

$$\int_a^b f(x)\,dx = -\int_b^a f(x)\,dx \quad (3)$$

$$\int_a^a f(x)\,dx = 0 \quad (4)$$

In Equation (3), notice that the limits of integration have been interchanged. In Equation (4), notice that the limits of integration are the same.

2 **EXAMPLE 5 Using Properties of the Definite Integral**

(a) $\int_4^1 \sqrt{x}\,dx = -\int_1^4 \sqrt{x}\,dx = -\frac{14}{3}$ (first equality: Property (3); second: Example 4)

(b) $\int_1^1 x\,dx = 0$ (Property (4))

Properties (3) and (4) are an immediate consequence of the definition of a definite integral. Specifically, if F is an antiderivative of f, then

$$\int_a^b f(x)\,dx = F(b) - F(a) = -[F(a) - F(b)] = -\int_b^a f(x)\,dx$$

and

$$\int_a^a f(x)\,dx = F(a) - F(a) = 0$$

NOW WORK PROBLEM 35.

Properties of the Definite Integral

If f is a continuous function that has an antiderivative on the interval $[a, b]$, and if c is between a and b, then

$$\int_a^b f(x)\,dx = \int_a^c f(x)\,dx + \int_c^b f(x)\,dx \qquad (5)$$

If f is a continuous function that has an antiderivative on the interval $[a, b]$ and if c is a real number, then

$$\int_a^b cf(x)\,dx = c\int_a^b f(x)\,dx \qquad (6)$$

EXAMPLE 6 Using Properties of the Definite Integral

Suppose it is known that

$$\int_1^3 f(x)\,dx = 5 \quad \text{and} \quad \int_3^6 f(x) = 7$$

Find:

(a) $\int_1^6 f(x)\,dx$ **(b)** $\int_1^3 16f(x)\,dx$

SOLUTION **(a)** $\int_1^6 f(x)\,dx \underset{\substack{\uparrow \\ \text{Property (5)}}}{=} \int_1^3 f(x)\,dx + \int_3^6 f(x)\,dx = 5 + 7 = 12$

(b) $\int_1^3 16\,f(x)\,dx \underset{\substack{\uparrow \\ \text{Property (6)}}}{=} 16\int_1^3 f(x)\,dx = 16 \cdot 5 = 80$

Properties of the Definite Integral

If f and g are continuous functions that have antiderivatives on the interval $[a, b]$, then

$$\int_a^b [f(x) \pm g(x)]\,dx = \int_a^b f(x)\,dx \pm \int_a^b g(x)\,dx \qquad (7)$$

EXAMPLE 7   **Using Properties of the Definite Integral**

Suppose it is known that

$$\int_1^4 f(x)\,dx = 5 \qquad \int_1^4 g(x)\,dx = -3$$

Find:

(a) $\displaystyle\int_1^4 [f(x) + g(x)]\,dx$ **(b)** $\displaystyle\int_1^4 [3f(x) + 4g(x)]\,dx$

SOLUTION   **(a)** $\displaystyle\int_1^4 [f(x) + g(x)]\,dx = \underset{\underset{\text{Property (7)}}{\uparrow}}{} \int_1^4 f(x)\,dx + \int_1^4 g(x)\,dx = 5 + (-3) = 2$

(b) $\displaystyle\int_1^4 [3f(x) + 4g(x)]\,dx = \underset{\underset{\text{Property (3)}}{\uparrow}}{} \int_1^4 3f(x)\,dx + \int_1^4 4g(x)\,dx = \underset{\underset{\text{Property (6)}}{\uparrow}}{} 3\int_1^4 f(x)\,dx + 4\int_1^4 g(x)\,dx$

$$= 3\cdot 5 + 4\cdot(-3) = 3 \qquad\blacktriangleright$$

   NOW WORK PROBLEM 43.

Applications

3   **EXAMPLE 8**   **Finding the Cost Due To An Increase in Production**

The marginal cost function for producing x units is $3x^2 - 200x + 1500$ dollars. Find the increase in cost if production is increased from 90 to 100 units.

SOLUTION   If C equals the cost of producing x units, then

$$C'(x) = 3x^2 - 200x + 1500$$

The increase in cost due to a production increase from 90 to 100 units is

$$\begin{aligned}
C(100) - C(90) &= \int_{90}^{100} C'(x)\,dx \\
&= \int_{90}^{100} (3x^2 - 200x + 1500)\,dx \\
&= \left(x^3 - 200\frac{x^2}{2} + 1500x\right)\Big|_{90}^{100} \\
&= [1{,}000{,}000 - 100(10{,}000) + 1500(100)] - [(90)^3 - 100(8100) + 1500(90)] \\
&= 96{,}000 \text{ dollars} \qquad\blacktriangleright
\end{aligned}$$

We discuss below two additional applications involving definite integrals. The applications are independent of each other.

Learning Curves   Quite often, the managerial planning and control component of a production industry is faced with the problem of predicting labor time requirements and cost per unit of product. The tool used to achieve such predictions is the *learning curve.* The basic assumption made here is that, in certain production industries such as the assembling of televisions and cars, the worker learns from experience. As a result, the more often a worker repeats an operation, the more efficiently the job is performed and direct labor input per unit of product declines. If the *rate* of improvement is regular enough, the learning curve can be used to predict future reductions in labor requirements.

One function that might be used to model such a situation is

$$f(x) = cx^k$$

where $f(x)$ is the number of hours of direct labor required to produce the xth unit, $-1 \le k < 0$, and $c > 0$. The choice of x^k, with $-1 \le k < 0$, guarantees that, as the number of units x produced increases, the direct labor input decreases. See Figure 4.

FIGURE 4

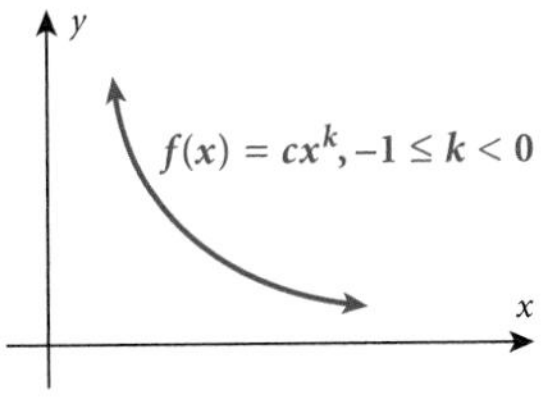

The function $f(x) = cx^k$ describes a rate of learning per unit produced. This rate is measured in terms of labor-hours per unit. As Figure 4 illustrates, the number of direct labor-hours declines as more items are produced.

Once a learning curve has been determined for a gross production process, it can be used as a predictor to determine the number of production hours for future work.

Learning Curves

For a learning curve $f(x) = cx^k$, the total number N of labor-hours required to produce units numbered a through b is

$$N = \int_a^b f(x)\,dx = \int_a^b cx^k\,dx$$

EXAMPLE 9 Applying Learning Curves

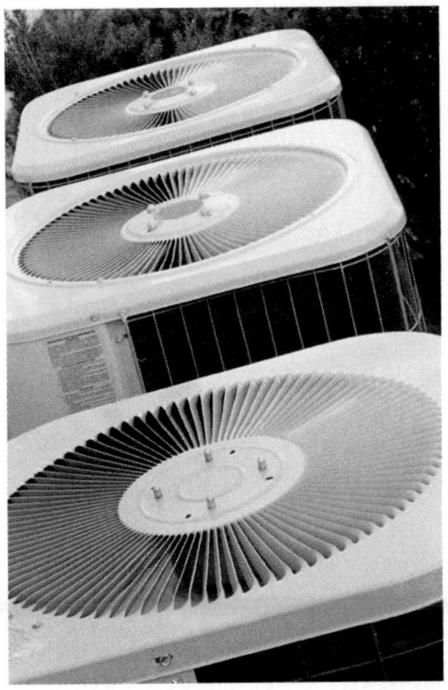

The Ace Air Conditioning Company manufactures air conditioners on an assembly line. From experience, it was determined that the first 100 air conditioners required 1272 labor-hours. For each 100 subsequent air conditioners (1 unit), fewer labor-hours were required according to the learning curve

$$f(x) = 1272x^{-0.25}$$

where $f(x)$ is the rate of labor-hours required to assemble the xth unit (each unit being 100 air conditioners). This curve was determined after 30 units (3000 air conditioners) had been manufactured.

The company is in the process of bidding for a large contract involving 5000 additional air conditioners, or 50 additional units. The company can estimate the labor-hours required to assemble these units by evaluating

$$N = \int_{30}^{80} 1272x^{-0.25}\,dx = \left.\frac{1272x^{0.75}}{0.75}\right|_{30}^{80} = 1696(80^{0.75} - 30^{0.75})$$

$$= 1696(26.75 - 12.82) = 23{,}627$$

The company can bid estimating the total labor-hours needed as 23,627.

NOW WORK PROBLEM 49.

Total Sales over Time

Total Sales over Time

When the rate of sales of a product is a known function, say, $f(t)$, where t is the time, the total sales of this product over a time period T are

$$\text{Total sales over time } T = \int_0^T f(t)\,dt$$

For example, suppose the rate of sales per day of a new product is given by

$$f(t) = 100 - 90e^{-t}$$

where t is the number of days the product is on the market. The total sales during the first 4 days are

$$\int_0^4 f(t)\,dt = \int_0^4 (100 - 90e^{-t})\,dt = (100t + 90e^{-t})\Big|_0^4 = 400 + 90e^{-4} - 90 = 311.6 \text{ units}$$

EXAMPLE 10 Finding Total Sales

A company has current sales of \$1,000,000 per month, and profit to the company averages 10% of sales. The company's past experience with a certain advertising strategy is that sales will increase by 2% per month over the length of the advertising campaign (12 months). The company now needs to decide whether to embark on a similar campaign that will have a total cost of \$130,000. The decision will be yes, provided the increase in sales due to the campaign results in profits that exceed \$13,000. (This is a 10% return on the advertising investment of \$130,000.)

SOLUTION Without advertising, the company has sales of \$12,000,000 over the 12 months. The monthly rate of sales during the advertising campaign obeys a growth curve of the form

$$\$1{,}000{,}000e^{0.02t}$$

where t is measured in months. The total sales after 12 months (the length of the campaign) are

$$\text{Total sales} = \int_0^{12} 1{,}000{,}000e^{0.02t}\,dt = \frac{1{,}000{,}000e^{0.02t}}{0.02}\Big|_0^{12}$$
$$= 50{,}000{,}000(e^{0.24} - 1) = \$13{,}562{,}458$$

The result is an increase in sales of $13{,}562{,}458 - 12{,}000{,}000 = 1{,}562{,}458$. The profit to the company is 10% of sales, so that the profit due to the increase in sales is

$$0.10(1{,}562{,}458) = \$156{,}246$$

This \$156,246 profit was achieved through the expenditure of \$130,000 in advertising. The advertising yielded a true profit of

$$\$156{,}246 - \$130{,}000 = \$26{,}246$$

Since this represents more than a 10% return on the cost of the advertising, the company should proceed with the advertising campaign.

NOW WORK PROBLEM 53.

EXERCISE 15.4 Answers to Odd-Numbered Problems Begin on Page AN-90.

In Problems 1–34, evaluate each definite integral.

1. $\int_1^2 (3x - 1)\, dx$
2. $\int_1^2 (2x + 1)\, dx$
3. $\int_0^1 (3x^2 + e^x)\, dx$
4. $\int_{-2}^0 (e^x + x^2)\, dx$
5. $\int_0^1 \sqrt{u}\, du$
6. $\int_1^4 \sqrt{u}\, du$
7. $\int_0^1 (t^2 - t^{3/2})\, dt$
8. $\int_1^4 (\sqrt{x} - 4x)\, dx$
9. $\int_{-2}^3 (x - 1)(x + 3)\, dx$
10. $\int_0^1 (z^2 + 1)^2\, dz$
11. $\int_1^2 \frac{x^2 - 1}{x^4}\, dx$
12. $\int_1^3 \frac{2 - x^2}{x^4}\, dx$
13. $\int_1^8 \left(\sqrt[3]{t^2} + \frac{1}{t}\right) dt$
14. $\int_1^4 \left(\sqrt{u} + \frac{1}{u}\right) du$
15. $\int_1^4 \frac{x + 1}{\sqrt{x}}\, dx$
16. $\int_1^9 \frac{\sqrt{x} + 1}{x^2}\, dx$
17. $\int_3^3 (5x^4 + 1)^{3/2}\, dx$
18. $\int_{-1}^1 (x + 1)^3\, dx$
19. $\int_{-1}^1 (x + 1)^2\, dx$
20. $\int_{-1}^{-1} \sqrt[3]{x^2 + 4}\, dx$
21. $\int_1^e \left(x - \frac{1}{x}\right) dx$
22. $\int_1^e \left(x + \frac{1}{x}\right) dx$
23. $\int_0^1 e^{-x}\, dx$
24. $\int_0^1 x^2 e^{x^3}\, dx$
25. $\int_1^3 \frac{dx}{x + 1}$
26. $\int_{-2}^2 e^{-7x/2}\, dx$
27. $\int_0^1 \frac{\sqrt{x}}{x^{3/2} + 1}\, dx$
28. $\int_2^3 \frac{dx}{x \ln x}$
29. $\int_1^3 xe^{2x}\, dx$
30. $\int_0^4 (1 + xe^{-x})\, dx$
31. $\int_1^2 xe^{-3x}\, dx$
32. $\int_1^3 x^2 \ln x\, dx$
33. $\int_1^5 \ln x\, dx$
34. $\int_1^2 x \ln x\, dx$

In Problems 35–38, evaluate each expression by applying properties of the definite integral.

35. $\int_2^2 e^{x^2}\, dx$
36. $\int_1^1 e^{-x^2}\, dx$
37. $\int_0^1 e^{-x^2}\, dx + \int_1^0 e^{-x^2}\, dx$
38. $\int_1^2 e^{x^2}\, dx + \int_2^1 e^{x^2}\, dx$

In Problems 39–46, evaluate each definite integral if it is known that

$$\int_1^3 f(x)dx = 4 \qquad \int_1^3 g(x)\, dx = -2 \qquad \int_3^6 f(x)\, dx = 8 \qquad \int_3^6 g(x)\, dx = 3$$

39. $\int_1^3 [f(x) + g(x)]\, dx$
40. $\int_1^3 [f(x) - g(x)]\, dx$
41. $\int_3^6 8f(x)\, dx$
42. $\int_3^6 -2f(x)\, dx$
43. $\int_3^6 [3f(x) + 4g(x)]\, dx$
44. $\int_3^6 [2f(x) - 3g(x)]\, dx$
45. $\int_1^6 f(x)\, dx$
46. $\int_1^6 g(x)\, dx$

---

47. **Finding a Cost Function** The marginal cost function for producing x units is $6x^2 - 100x + 1000$ dollars. Find the increase in cost if production is increased from 100 to 110 units.

48. **Finding a Revenue Function** The marginal revenue function for selling x units is $10 - 4x$. Find the increase in revenue if selling is increased from 10 to 12 units.

49. **Learning Curve** (a) Rework Example 9 for the learning curve $f(x) = 1272\, x^{-0.35}$.
   (b) Rework Example 9 for the learning curve $f(x) = 1272\, x^{-0.15}$.
   (c) Based on the answers to (a) and (b), explain the role of k in the learning curve $f(x) = cx^k$.

50. **Learning Curve** (a) Rework Example 9 for the learning curve $f(x) = 1500\, x^{-0.25}$. [This means it was determined that

the first 100 air conditioners (1 unit) required 1500 labor-hours.]

(b) Rework Example 9 for the learning curve $f(x) = 1000\,x^{-0.25}$.

(c) Based on the answers to (a) and (b), explain the role of c in the learning curve $f(x) = cx^k$.

51. US Budget Deficit The monthly budget deficit of the United States government between October 2002 and March 2003 can be modeled by $D(x) = -8.93x + 70$, where $D(x)$ is in billions of dollars and x is the number of months elapsed since September 2002. If this rate continues, project the total budget deficit for the 2002–2003 fiscal year (October 2002–September 2003).

Source: Bureau of Public Debt, United States Department of Treasury.

52. Interest Expense The Fiscal Year Interest Expense measures the total interest that accumulates on debt outstanding for a given fiscal year. From 1997–2002 the increase on this amount was modeled by $f(x) = -2.64x^2 + 15x + 342$, where x is the number of years since 1997 and $f(x)$ is measured in billions of dollars. If this rate continues, find the total Fiscal Year Interest Expense from 2003 to 2010.

53. Total Sales The rate of sales of a new product is given by

$$f(x) = 1200 - 950e^{-x}$$

where x is the number of months the product is on the market. Find the total sales during the first year.

54. Total Sales In Example 10, what decision should the company make if sales due to advertising increase by only 1.5% per month?

55. Learning Curve After producing 35 units, a company determines that its production facility is following a learning curve of the form $f(x) = 1000x^{-0.5}$, where $f(x)$ is the rate of labor-hours required to assemble the xth unit. How many total labor-hours should the company estimate are required to produce an additional 25 units?

56. Learning Curve Danny's Auto Shop has found that, after tuning up 50 cars, a learning curve of the form $f(x) = 1000x^{-1}$ is being followed. How many total labor-hours should the shop estimate are required to tune up an additional 50 cars?

57. Even Functions It can be shown that if f is an even function, then

$$\int_{-a}^{a} f(x)\,dx = 2\int_{0}^{a} f(x)\,dx \qquad a > 0$$

Verify the above formula by evaluating the following definite integrals:

(a) $\int_{-1}^{1} x^2\,dx$ (b) $\int_{-1}^{1} (x^4 + x^2)\,dx$

58. Odd Functions It can be shown that if f is an odd function, then

$$\int_{-a}^{a} f(x)\,dx = 0 \qquad a > 0$$

Verify the above formula by evaluating the following definite integrals:

(a) $\int_{-1}^{1} x\,dx$ (b) $\int_{-1}^{1} x^3\,dx$

15.5 Finding Areas; Consumer's Surplus, Producer's Surplus; Maximizing Profit over Time

PREPARING FOR THIS SECTION *Before getting started, review the following:*

> Geometry Formulas (Appendix A, Section A.9, pp. A-82–A-83)

> Interval Notation (Appendix A, Section A.7, pp. A-61–A-62)

OBJECTIVES
1. Find the area under a graph
2. Find the area enclosed by two graphs
3. Solve applied problems involving definite integrals

The development of the integral, like that of the derivative, was originally motivated to a large extent by attempts to solve a basic problem in geometry—namely, the *area problem.* The question is: Given a nonnegative function f, whose domain is the closed interval $[a, b]$, what is the area enclosed by the graph of f, the x-axis, and the vertical lines $x = a$ and $x = b$? Figure 5 illustrates the area to be found. We will refer to this area as the **area under the graph of f from a to b.**

FIGURE 5

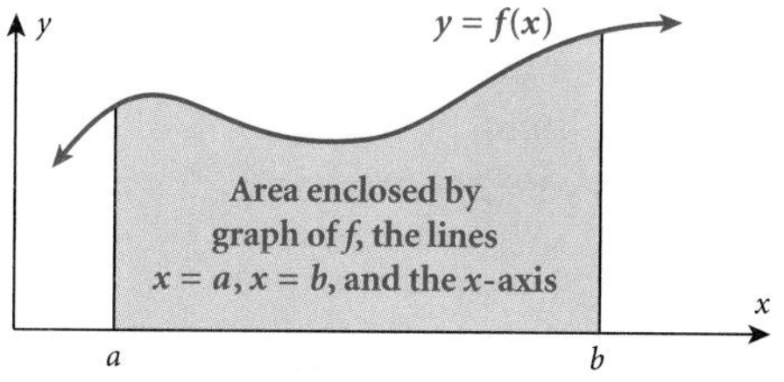

In plane geometry we learn how to find the area of certain geometric figures, such as squares, rectangles, and circles. For example, the area of a square with a side of length 3 feet is 9 square feet.

We also know that the area of a rectangle with length a units and width b units is ab square units.

All area problems have certain features in common. For example, whenever an area is computed, it is expressed as a number of square units; this number is never negative. One property of area is that it is nonnegative.

FIGURE 6

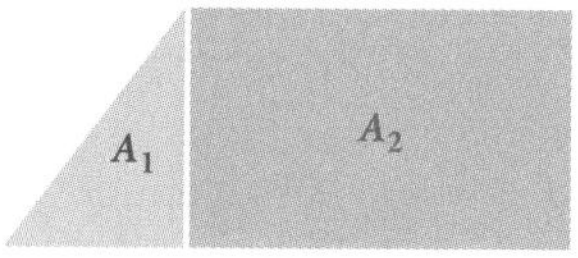

Consider the trapezoid shown in Figure 6. This trapezoid has been decomposed into two nonoverlapping geometric figures, a triangle (with area A_1) and a rectangle (with area A_2). The area of the trapezoid is the sum $A_1 + A_2$ of the two component areas. So, as long as two regions do not overlap (except perhaps for a common boundary), the total area can be found by adding the component areas. We sometimes call this the **additive property of area.**

Properties of Area

Two properties of area are:

I. Nonnegative Property Area ≥ 0

II. Additive Property If A and B are two nonoverlapping regions with areas that are known, then

$$\text{Total area of } A \text{ and } B = \text{area of } A + \text{area of } B$$

The next result gives a technique for evaluating areas such as the shaded region shown in Figure 5.

Area under a Graph

Suppose $y = f(x)$ is a continuous function defined on a closed interval I and $f(x) \geq 0$ for all points x in I. Then, for $a < b$ in I,

$$\text{Area under the graph of } f \text{ from } a \text{ to } b = \int_a^b f(x)\, dx \qquad (1)$$

FIGURE 7

$$\text{Area} = \int_a^b f(x)\,dx$$

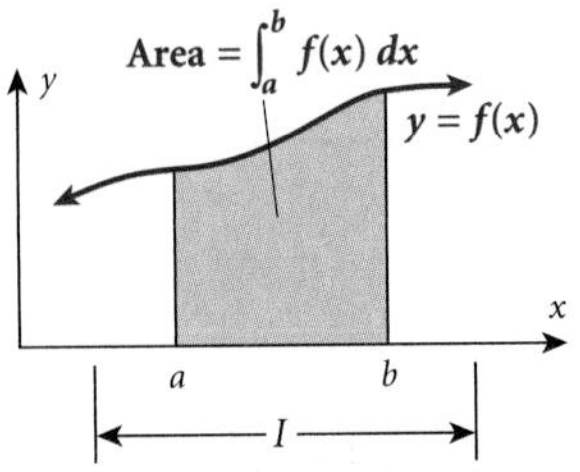

Figure 7 illustrates the above statement. A proof is given at the end of this section. We are now able to find the area under the graph of $y = f(x)$, provided three conditions are met:

1. f is continuous on $[a, b]$.
2. f is nonnegative on $[a, b]$, that is, $f(x) \ge 0$ for $a \le x \le b$.
3. An antiderivative for f can be found.

1 **EXAMPLE 1** **Finding the Area under a Graph**

Find the area under the graph $f(x) = x^2$, from $x = 0$ to $x = 1$.

SOLUTION First we graph f and show the area to be found. See Figure 8. Using Formula (1), the area we seek is given by the definite integral

FIGURE 8

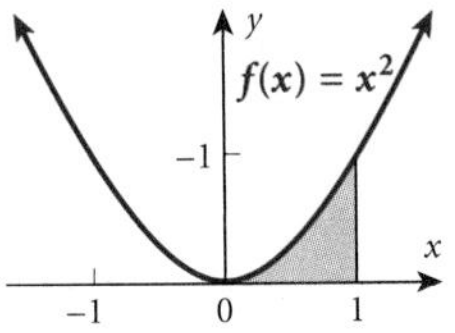

$$\int_0^1 x^2\,dx = \left.\frac{x^3}{3}\right|_0^1 = \frac{1}{3}$$

The area illustrated in Figure 8 is $\dfrac{1}{3}$ square unit.

NOW WORK PROBLEM 1.

Suppose a function f is continuous on the interval I, $a \le x \le b$ and has an antiderivative on I. Suppose $f(x) \le 0$ for $a \le x \le c$ and $f(x) \ge 0$ for $c \le x \le b$. How do we compute the area enclosed by $y = f(x)$, the x-axis, $x = a$, and $x = b$? See Figure 9.

FIGURE 9

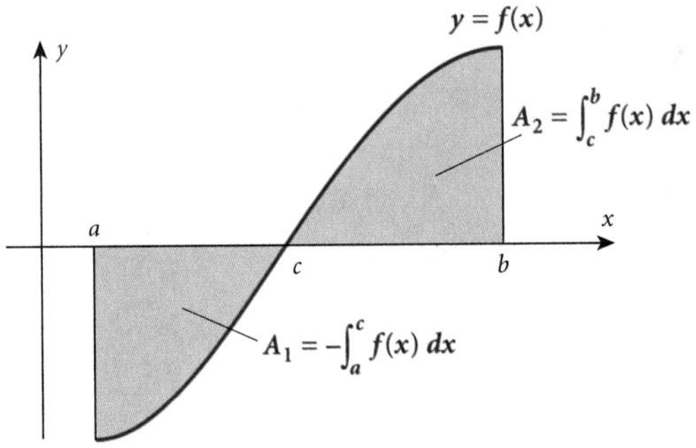

Notice in Figure 9 that the area A in question is composed of two nonoverlapping areas, A_1 and A_2, so that by the additive property of area,

$$A = A_1 + A_2$$

Also, we know that on the interval $[c, b]$, the function is nonnegative, so that

$$A_2 = \int_c^b f(x)\,dx$$

To find the area A_1, we note that, since $f(x) \le 0$ on $a \le x \le c$, then $-f(x) \ge 0$, and, by symmetry, the area A_1 equals

$$A_1 = \int_a^c [-f(x)]\,dx = -\int_a^c f(x)\,dx$$

The total area A we seek is therefore

$$A = A_1 + A_2 = -\int_a^c f(x)\,dx + \int_c^b f(x)\,dx$$

The next example illustrates this procedure for calculating area.

EXAMPLE 2 Finding an Area

Find the area A enclosed by $f(x) = x^3$, the x-axis, $x = -1$, and $x = \dfrac{1}{2}$.

SOLUTION The desired area is indicated by the shaded region in Figure 10. Notice that it is composed of two nonoverlapping regions: A_1, in which $f(x) \le 0$ over the interval $[-1, 0]$; and A_2, in which $f(x) \ge 0$ over the interval $[0, \frac{1}{2}]$.

FIGURE 10

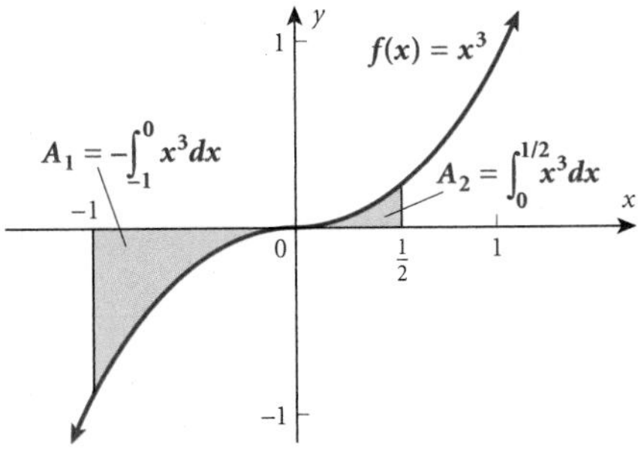

We use the additive property of area. Since $f(x) \le 0$ for $-1 \le x \le 0$,

$$A_1 = -\int_{-1}^{0} x^3\,dx = -\frac{x^4}{4}\Bigg|_{-1}^{0} = -\left[\frac{0^4}{4} - \frac{(-1)^4}{4}\right] = \frac{1}{4}$$

For the area A_2, we have

$$A_2 = \int_0^{1/2} x^3\,dx = \frac{x^4}{4}\Bigg|_0^{1/2} = \frac{\left(\frac{1}{2}\right)^4}{4} - \frac{0^4}{4} = \frac{1}{64}$$

The total area A is

$$A = A_1 + A_2 = \frac{1}{4} + \frac{1}{64} = \frac{17}{64} \text{ square units}$$

Example 2 illustrates the necessity of graphing the function before any attempt is made to compute the area. In subsequent examples we shall always graph the function before doing anything else.

EXAMPLE 3 Finding an Area

Find the area A enclosed by $f(x) = x^2 - 4$ and the x-axis from $x = 0$ to $x = 4$.

FIGURE 11

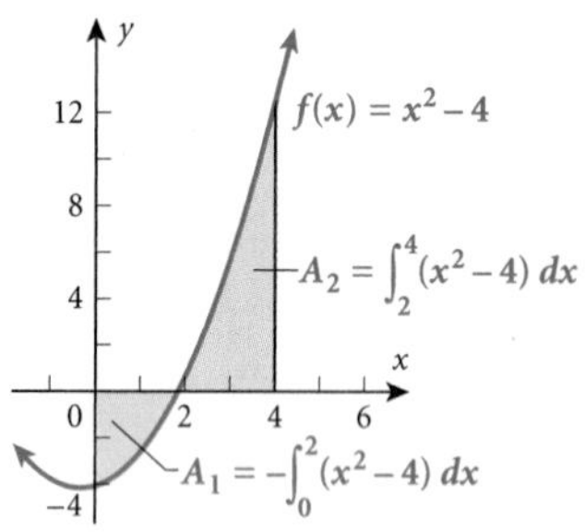

SOLUTION We begin by graphing f. See Figure 11. On the interval $[0, 4]$, the graph crosses the x-axis at $x = 2$ since $f(2) = 0$. Also, $f(x) \leq 0$ from $x = 0$ to $x = 2$, and $f(x) \geq 0$ from $x = 2$ to $x = 4$. The areas A_1 and A_2, as shown in Figure 11, are

$$A_1 = -\int_0^2 (x^2 - 4)\,dx = -\left(\frac{x^3}{3} - 4x\right)\Bigg|_0^2 = -\left(\frac{8}{3} - 8\right) = \frac{16}{3}$$

$$A_2 = \int_2^4 (x^2 - 4)\,dx = \left(\frac{x^3}{3} - 4x\right)\Bigg|_2^4 = \left(\frac{64}{3} - 16\right) - \left(\frac{8}{3} - 8\right)$$

$$= \frac{56}{3} - 8 = \frac{32}{3}$$

The area A is

$$A = A_1 + A_2 = \frac{16}{3} + \frac{32}{3} = \frac{48}{3} = 16 \text{ square units}$$

NOW WORK PROBLEM 5.

Area Enclosed by Two Graphs

The next example illustrates how to find the area enclosed by the graphs of two functions.

2 **EXAMPLE 4 Finding an Area Enclosed by Two Graphs**

Find the area A enclosed by the graphs of the functions

$$f(x) = 2x^2 \qquad \text{and} \qquad g(x) = 2x + 4$$

FIGURE 12

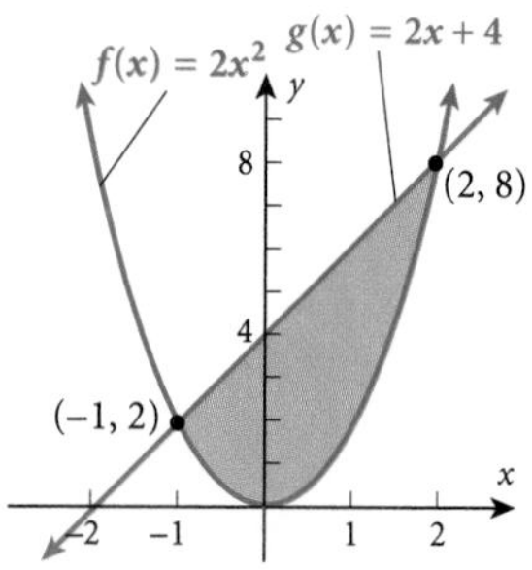

SOLUTION First, we graph each of the functions, as shown in Figure 12.

The area to be calculated (the shaded portion of Figure 12) lies under the graph of the line $g(x) = 2x + 4$ and above the graph of $f(x) = 2x^2$. To find this area, we first need to find the numbers x at which the graphs intersect, that is, all numbers x for which $f(x) = g(x)$. The solutions of this equation are obtained as follows:

$2x^2 = 2x + 4$	$f(x) = g(x)$
$x^2 - x - 2 = 0$	Simplify.
$(x + 1)(x - 2) = 0$	Factor.
$x + 1 = 0$ or $x - 2 = 0$	Apply the Zero-Product Property.
$x = -1$ or $x = 2$	Solve for x.

The points of intersection of the two graphs are $(-1, 2)$ and $(2, 8)$, as shown in Figure 12.

From Figure 12 we can see that if we subtract the area under $f(x) = 2x^2$, between $x = -1$ and $x = 2$, from the area under $g(x) = 2x + 4$, between $x = -1$ and $x = 2$, we will have the area A we seek. That is,

$$A = \int_{-1}^{2} g(x)\,dx - \int_{-1}^{2} f(x)\,dx \qquad A = \text{Area under } g - \text{Area under } f$$

$$= \int_{-1}^{2} (2x + 4)\,dx - \int_{-1}^{2} 2x^2\,dx \qquad g(x) = 2x + 4;\, f(x) = 2x^2$$

$$= \left(x^2 + 4x\right)\Big|_{-1}^{2} - 2 \cdot \frac{x^3}{3}\Big|_{-1}^{2} \qquad \text{Integrate.}$$

$$= [(4 + 8) - (1 - 4)] - 2 \cdot \left(\frac{8}{3} - \frac{-1}{3}\right)$$

$$= 9$$

NOW WORK PROBLEM 11.

The technique used in Example 4 can be used whenever we are asked to determine the area enclosed by the graphs of two continuous nonnegative functions f and g from $x = a$ to $x = b$.

FIGURE 13

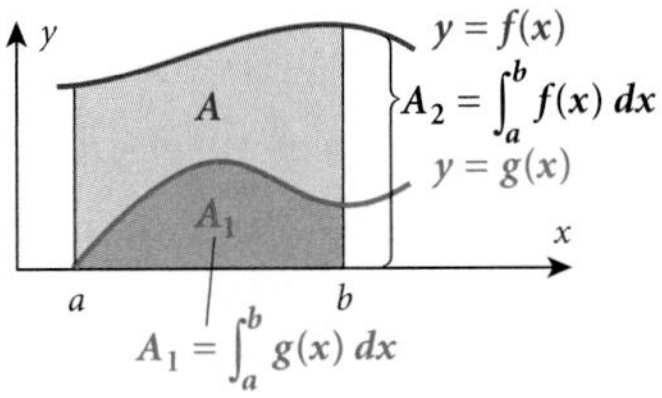

Suppose, as depicted in Figure 13, $f(x) \geq g(x) \geq 0$ for x in $[a, b]$, and we wish to determine the area enclosed by the graphs of f and g and the lines $x = a$, and $x = b$. If we denote this area by A, the area under f by A_2, and the area under g by A_1, then

$$A = A_2 - A_1$$

$$= \int_a^b f(x)\,dx - \int_a^b g(x)\,dx \qquad A_2 = \int_a^b f(x)dx;\, A_1 = \int_a^b g(x)dx$$

$$= \int_a^b [f(x) - g(x)]\,dx$$

The next example illustrates this formula.

EXAMPLE 5 Finding the Area Enclosed By Two Graphs

FIGURE 14

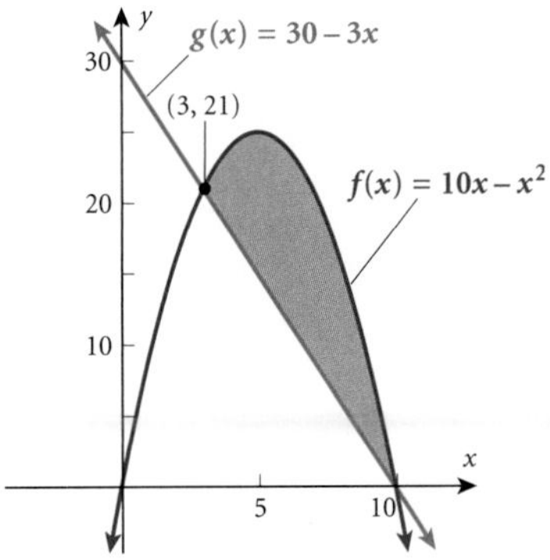

Find the area A enclosed by the graphs of the functions

$$f(x) = 10x - x^2 \quad \text{and} \quad g(x) = 30 - 3x$$

SOLUTION First, we graph the two functions. See Figure 14. The points of intersection of the two graphs were obtained by finding all numbers x for which $f(x) = g(x)$.

$$10x - x^2 = 30 - 3x \qquad f(x) = g(x)$$

$$x^2 - 13x + 30 = 0 \qquad \text{Place in standard form.}$$

$$(x - 3)(x - 10) = 0 \qquad \text{Factor.}$$

$$x - 3 = 0 \quad \text{or} \quad x - 10 = 0 \qquad \text{Apply the Zero-Product Property.}$$

$$x = 3 \quad \text{or} \quad x = 10 \qquad \text{Solve for } x.$$

The points where the two graphs intersect are (3, 21) and (10, 0). We also see that for $3 \leq x \leq 10$,

$$f(x) \geq g(x) \geq 0$$

The area A we seek, indicated by the shaded portion in Figure 14, is

$$A = \int_3^{10} [(10x - x^2) - (30 - 3x)]\,dx \qquad A = \int_3^{10} [f(x) - g(x)]\,dx$$

$$= \int_3^{10} [-x^2 + 13x - 30]\,dx \qquad \text{Simplify the integrand.}$$

$$= \left(-\frac{x^3}{3} + 13\frac{x^2}{2} - 30x\right)\Bigg|_3^{10} \qquad \text{Integrate.}$$

$$= \left[-\frac{10^3}{3} + 13\frac{10^2}{2} - 30(10)\right] - \left[-\frac{3^3}{3} + 13\frac{3^2}{2} - 30(3)\right] \qquad \text{Evaluate at the upper and lower limits.}$$

$$= -\frac{1000}{3} + \frac{1300}{2} - 300 + 9 - \frac{117}{2} + 90$$

$$= \frac{343}{6}$$

Remember, when computing area using the formula $\int_a^b [f(x) - g(x)]\,dx$, it must be true that $f(x) \geq g(x)$ on $[a, b]$. If this condition is not met, we break the area up into pieces on which the inequality does hold, compute each one separately, and use the Additive Property of area.

EXAMPLE 6 Finding the Area Enclosed by Two Graphs

Find the area A enclosed by the graphs of the functions

$$f(x) = x^3 \qquad \text{and} \qquad g(x) = -x^2 + 2x$$

SOLUTION First, we graph the two functions. See Figure 15.

FIGURE 15

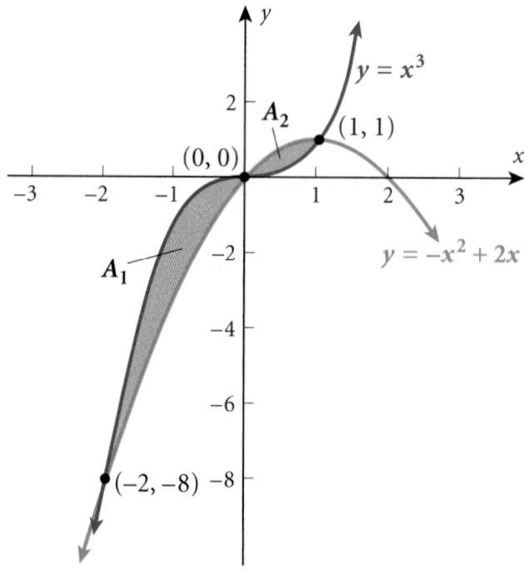

The points of intersection of the two graphs are found by solving the equation $f(x) = g(x)$.

$$
\begin{aligned}
x^3 &= -x^2 + 2x && f(x) = g(x) \\
x^3 + x^2 - 2x &= 0 && \text{Place in standard form.} \\
x(x^2 + x - 2) &= 0 && \text{Factor.} \\
x(x + 2)(x - 1) &= 0 && \text{Factor.} \\
x = 0 \quad \text{or} \quad x + 2 = 0 \quad &\text{or} \quad x - 1 = 0 && \text{Apply the Zero-Product Property.} \\
x = 0 \quad \text{or} \quad x = -2 \quad &\text{or} \quad x = 1 && \text{Solve for } x.
\end{aligned}
$$

The points of intersection are $(0, 0)$, $(-2, -8)$, and $(1, 1)$. Notice that $f(x) \geq g(x)$ on the interval $[-2, 0]$, while $g(x) \geq f(x)$ on the interval $[0, 1]$. To find the area A enclosed by the graphs of f and g, we compute the areas A_1 and A_2. Then,

$$
\begin{aligned}
A &= A_1 + A_2 \\
&= \int_{-2}^{0} [x^3 - (-x^2 + 2x)]dx + \int_{0}^{1} [(-x^2 + 2x) - x^3]\, dx && A_1 = \int_{-2}^{0} [f(x) - g(x)]\, dx;\ A_2 = \int_{0}^{1} [g(x) - f(x)]\, dx \\
&= \int_{-2}^{0} (x^3 + x^2 - 2x)dx + \int_{0}^{1} (-x^3 - x^2 + 2x)\, dx && \text{Simplify each integrand.} \\
&= \left(\frac{x^4}{4} + \frac{x^3}{3} - x^2\right)\Bigg|_{-2}^{0} + \left(\frac{-x^4}{4} - \frac{x^3}{3} + x^2\right)\Bigg|_{0}^{1} && \text{Integrate.} \\
&= 0 - \left(4 - \frac{8}{3} - 4\right) + \left(-\frac{1}{4} - \frac{1}{3} + 1\right) - 0 && \text{Evaluate each expression at the upper limit and the lower limit.} \\
&= \frac{8}{3} + \frac{5}{12} = \frac{37}{12} && \text{Simplify.}
\end{aligned}
$$

NOW WORK PROBLEM 25.

Applications

Solve applied problems involving definite integrals

3 We discuss two applications in business that involve area. They are independent of each other.

Consumer's Surplus; Producer's Surplus Suppose the price p a consumer is willing to pay for a quantity x of a particular commodity is governed by the demand curve

$$p = D(x)$$

In general, the demand function D is a decreasing function, indicating that as the price of the commodity increases, the quantity the consumer is willing to buy declines.

Suppose the price p that a producer is willing to charge for a quantity x of a particular commodity is governed by the supply curve

$$p = S(x)$$

In general, the supply function S is an increasing function since, as the price p of a commodity increases, the more the producer is willing to supply the commodity.

The point of intersection of the demand curve and the supply curve is called the **equilibrium point** E. If the coordinates of E are (x^*, p^*), then p^*, the **market price,** is the price a consumer is willing to pay for the commodity and p^* is also the price at which a producer is willing to sell; x^*, the **demand level,** is the quantity of the commodity purchased by the consumer and sold by the producer. See Figure 16.

FIGURE 16

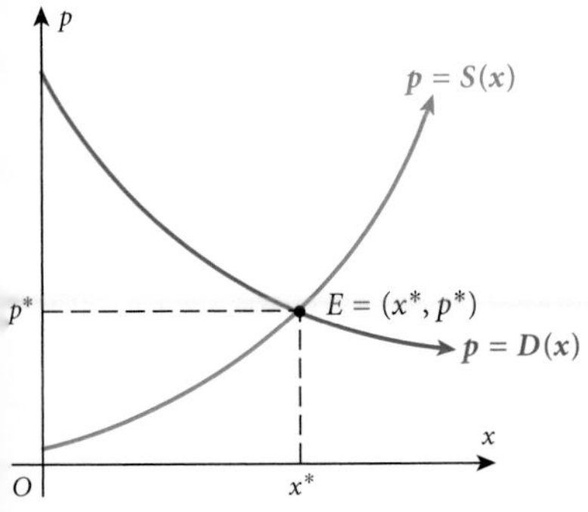

FIGURE 17

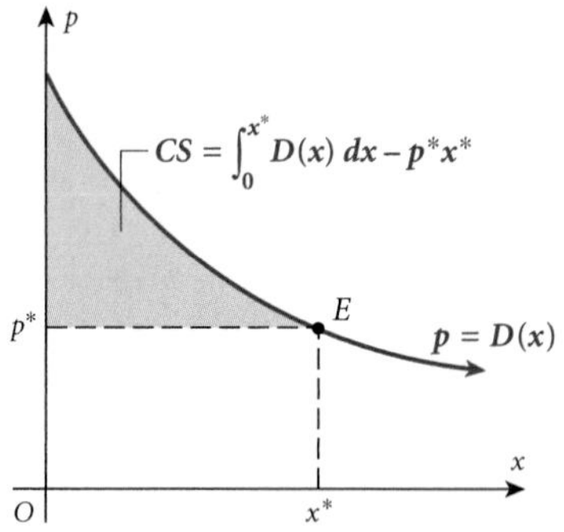

The revenue to the producer at a market price p^* and a demand level x^* is p^*x^* (the price per unit times the number of units). This revenue can be interpreted geometrically as the area of the rectangle Op^*Ex^* in Figure 16.

In a free market economy, there are times when some consumers would be willing to pay more for a commodity than the market price p^* that they actually do pay. The benefit of this to consumers—that is, the difference between what consumers *actually* paid and what they were *willing* to pay—is called the **consumer's surplus *CS*.** To obtain a formula for consumer's surplus *CS*, we use Figure 17 as a guide.

The quantity $\int_0^{x^*} D(x)\,dx$ is the area under the demand curve $D(x)$ from $x = 0$ to $x = x^*$ and represents the total revenue that would have been generated by the willingness of some consumers to pay more. By subtracting p^*x^* (the revenue actually achieved), the result is a surplus *CS* to the consumer. The formula for consumer's surplus is

$$CS = \int_0^{x^*} D(x)\,dx - p^*x^* \tag{2}$$

FIGURE 18

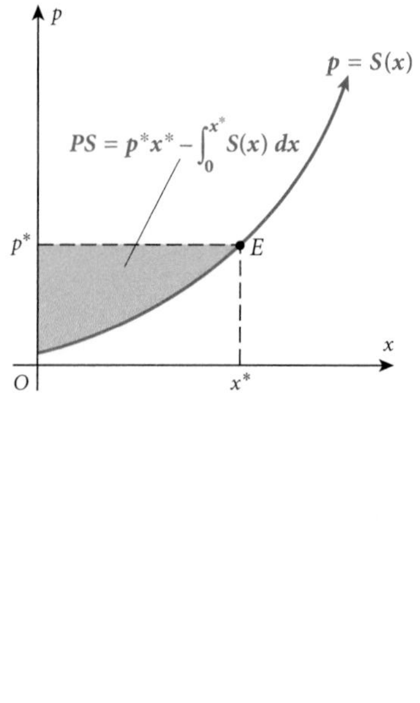

In a free market economy, there are also times when some producers would be willing to sell at a price below the market price p^* that the consumer actually pays. The benefit of this to the producer—that is, the difference between the revenue producers *actually* receive and what they would have been willing to receive—is called the **producer's surplus *PS*.** To obtain a formula for *PS*, we use Figure 18 as a guide.

The quantity $\int_0^{x^*} S(x)\,dx$ is the area under the supply curve $S(x)$ from $x = 0$ to $x = x^*$ and represents the total revenue that would have been generated by some producer's willingness to sell at a lower price. If we subtract this amount from p^*x^* (the revenue actually achieved), the result is a surplus to the producer, *PS*. The formula for producer's surplus is

$$PS = p^*x^* - \int_0^{x^*} S(x)\,dx \tag{3}$$

Example 7 illustrates a situation in which both the supply and demand curves are linear.

EXAMPLE 7 Finding Consumer's Surplus and Producer's Surplus

Find consumer's surplus *CS* and producer's surplus *PS* for the demand curve

$$D(x) = 18 - 3x$$

and the supply curve

$$S(x) = 3x + 6$$

where

$$p^* = D(x^*) = S(x^*)$$

SOLUTION   The equilibrium point is $E = (x^\star, p^\star)$, where $p^\star = D(x^\star) = S(x^\star)$ is the market price and $x^\star$ is the demand level. To find $x^\star$, we solve the equation

$$D(x^\star) = S(x^\star)$$
$$18 - 3x^\star = 3x^\star + 6$$
$$6x^\star = 12$$
$$x^\star = 2$$

To find $p^\star$, we evaluate $D(x^\star)$ or $S(x^\star)$:

$$p^\star = D(x^\star) = D(2) = 18 - 6 = 12$$

To find CS and PS, we use Formulas (2) and (3) with

$$D(x) = 18 - 3x, \qquad S(x) = 3x + 6, \qquad x^\star = 2, \qquad p^\star = 12$$

Then

$$CS = \int_0^{x^\star} D(x)\,dx - x^\star p^\star = \int_0^2 (18 - 3x)\,dx - (2)(12)$$

$$= \left(18x - 3 \cdot \frac{x^2}{2}\right)\Big|_0^2 - 24$$

$$= 36 - 6 - 24 = 6$$

$$PS = x^\star p^\star - \int_0^{x^\star} S(x)\,dx = (2)(12) - \int_0^2 (3x + 6)\,dx$$

$$= 24 - \left(3 \cdot \frac{x^2}{2} + 6x\right)\Big|_0^2$$

$$= 24 - (6 + 12) = 6$$

In this example, the consumer's surplus and the producer's surplus each equal $6. In general, PS and CS are unequal.   ▶

NOW WORK PROBLEM 35.

Maximizing Profit Over Time   The model introduced here concerns business operations of a special character. In oil drilling, mining, and other depletion operations, the initial revenue rate is generally higher than the revenue rate after a period of time has passed. That is, the revenue rate, as a function of time, is a decreasing function, because depletion is occurring.

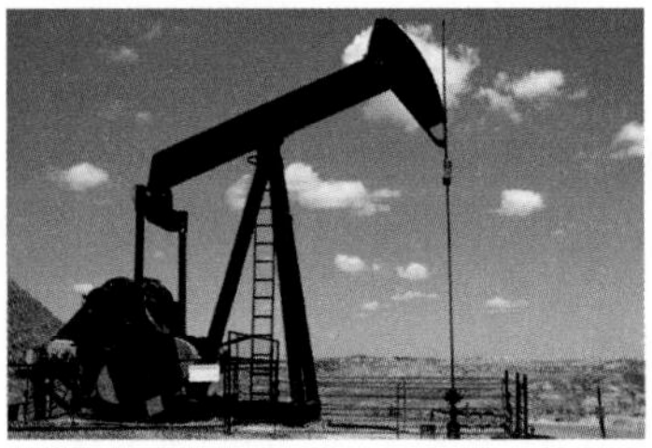

The cost rate of such operations generally increases with time because of inflation and other reasons. That is, the cost rate, as a function of time, is an increasing function. The problem that management faces is to determine the time t_{max} that maximizes the profit function $P = P(t)$.

To construct a model, we denote the cost function by $C = C(t)$ and the revenue function by $R = R(t)$, where t denotes time. This representation of cost and revenue deviates from the usual economic definitions of cost per unit times number of units, and price per unit times number of units. The derivatives C' and R', taken with respect to time, represent cost and revenue as time rates. Furthermore, we make the assumption that the revenue rate, say, dollars per week, is greater than the cost rate, also in dollars per week, at the beginning of the business operation under consideration. Also, as time goes on, we assume that the cost rate increases to the revenue rate and thereafter exceeds it. The optimum time at which the business operation should terminate is that point in time where the rates are equal. That is, the optimum time t_{max} obeys

$$C'(t_{max}) = R'(t_{max})$$

The profit rate P' is the difference between the revenue rate and the cost rate. That is,

$$P'(t) = R'(t) - C'(t)$$

Integrate each side with respect to t from 0 to t. Then

$$\int_0^t P'(t)\,dt = \int_0^t [R'(t) - C'(t)]\,dt$$

$$P(t) - P(0) = \int_0^t [R'(t) - C'(t)]\,dt$$

FIGURE 19

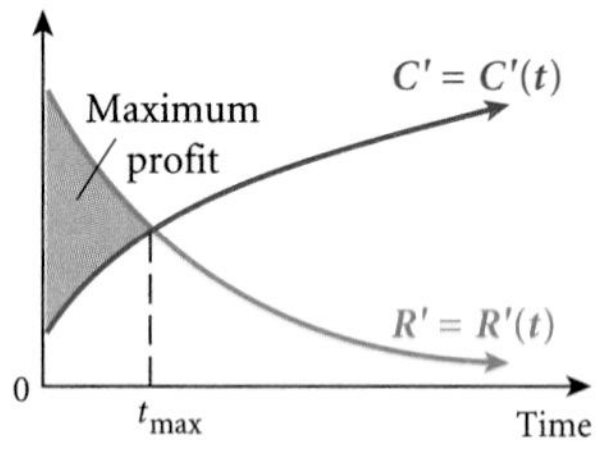

The maximum profit $P = P(t_{max})$ is obtained when $t = t_{max}$. Geometrically, the maximum profit $P(t_{max})$ is the area enclosed by the graphs of C' and R' from $t = 0$ to $t = t_{max}$. See Figure 19.

Notice that in Figure 19 the revenue rate function obeys the assumptions made in constructing the model: it is decreasing and it is high initially. Also, the cost rate function is increasing and is concave down, indicating that the cost rate eventually levels off.

EXAMPLE 8 Maximizing Profit over Time

The G-B Oil Company's revenue rate (in millions of dollars per year) at time t years is

$$R'(t) = 9 - t^{1/3}$$

and the corresponding cost rate function (also in millions of dollars) is

$$C'(t) = 1 + 3t^{1/3}$$

Determine how long the oil company should continue to operate and what the total profit will be at the end of the operation.

SOLUTION The time t_{max} of optimal termination is found when

$$\begin{aligned} R'(t) &= C'(t) \\ 9 - t^{1/3} &= 1 + 3t^{1/3} \\ 8 &= 4t^{1/3} \\ 2 &= t^{1/3} \\ t_{max} &= 8 \text{ years} \end{aligned}$$

See Figure 20.

FIGURE 20

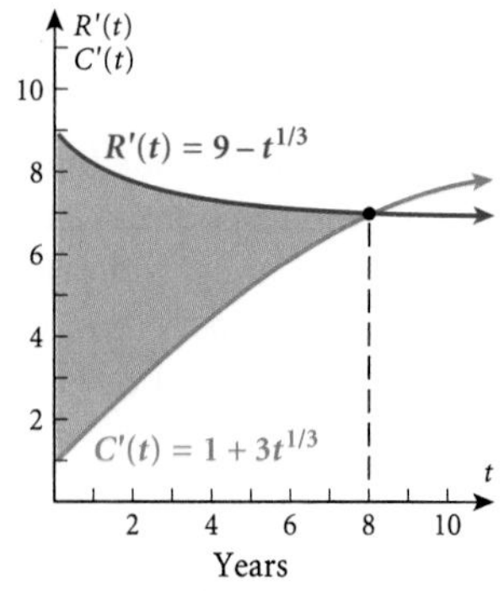

At $t_{max} = 8$, both the revenue and the cost rates are 7 million dollars per year. The profit $P(t_{max})$ is

$$\begin{aligned} P(t_{max}) &= \int_0^8 [R'(t) - C'(t)]\,dt \\ &= \int_0^8 [(9 - t^{1/3}) - (1 + 3t^{1/3})]\,dt \\ &= \int_0^8 [8 - 4t^{1/3}]\,dt \\ &= (8t - 3t^{4/3})\Big|_0^8 = 16 \text{ million dollars} \end{aligned}$$

In Example 8 we were forced to overlook the *fixed* cost for the cost function at time $t = 0$. This is because if $C = C(x)$ contains a constant (the fixed cost), then it becomes zero when we take the derivative C'. In the final analysis of the problem, total profit should be reduced by the amount corresponding to the fixed cost.

Justification of Area As a Definite Integral

Look at Figure 21. Choose a number c in I so that $c < a$. Suppose x in I is an arbitrary number for which $x > c$. Let $A(x)$ denote the area enclosed by $y = f(x)$ and the x-axis from c to x. We want to show that $A'(x) = f(x)$ for all x in $I, x > c$.

Now, choose $h > 0$ so that $x + h$ is in I. Then $A(x + h)$ is the area under the graph of $y = f(x)$ from c to $x + h$. See Figure 22. The difference $A(x + h) - A(x)$ is just the area under the graph of $y = f(x)$ from x to $x + h$. See Figure 23.

Next, we construct a rectangle with base h and area $A(x + h) - A(x)$. The height of the rectangle is then

$$\frac{A(x + h) - A(x)}{h}$$

Now, we superimpose this rectangle on Figure 23 to obtain Figure 24.

FIGURE 21

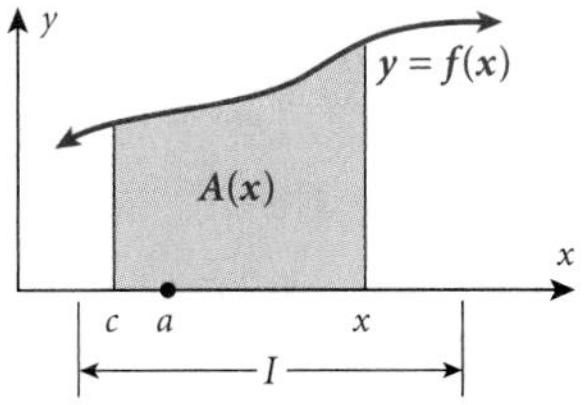

FIGURE 22

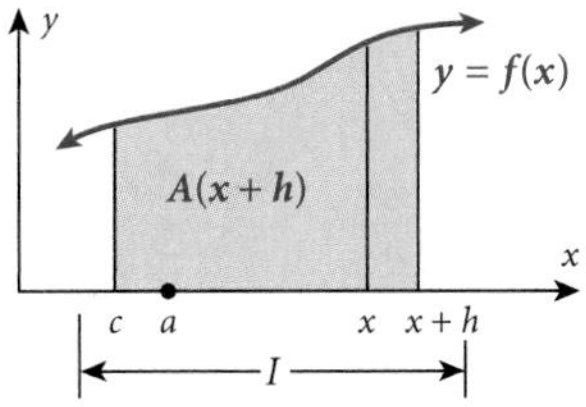

FIGURE 23

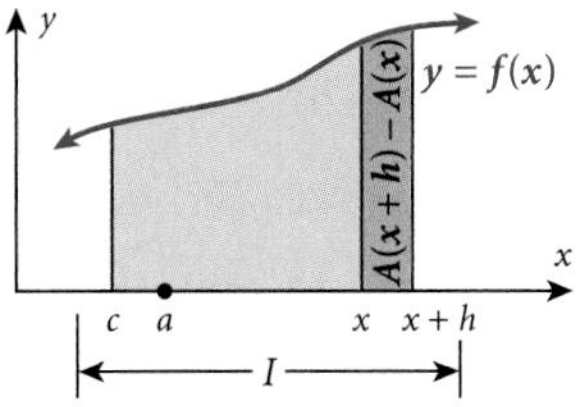

FIGURE 24

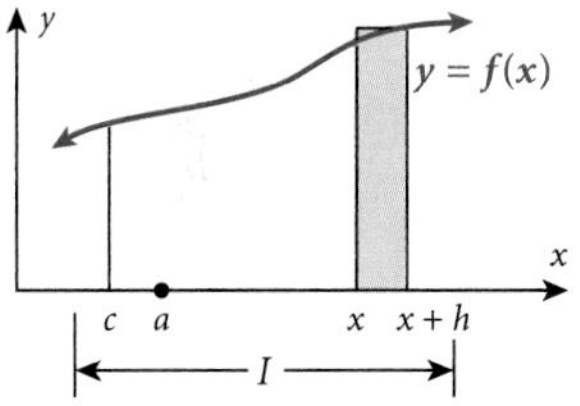

Since $y = f(x)$ is assumed to be a continuous function, and since both the rectangle and the shaded area have the same base and the same area, the upper edge of the rectangle must cross the graph of $y = f(x)$.

As we let $h \to 0^+$, the height of the rectangle tends to $f(x)$; that is,

$$\frac{A(x + h) - A(x)}{h} \to f(x) \quad \text{as} \quad h \to 0^+$$

A similar argument applies if we choose $h < 0$ and let $h \to 0^-$. As a result,

$$\lim_{h \to 0} \frac{A(x + h) - A(x)}{h} = f(x)$$

The limit on the left is the derivative of A. That is,

$$A'(x) = f(x)$$

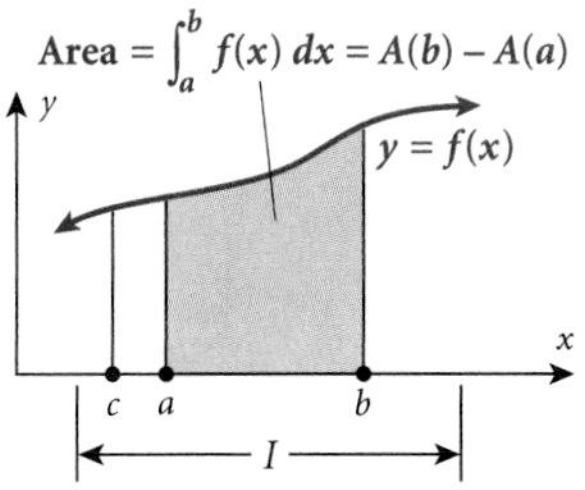

Since the choice of x is arbitrary (except for the condition that $x > c$), it follows that

$$A'(x) = f(x) \qquad \text{for all } x \text{ in } I, x > c$$

In other words, we have shown that the area A is an antiderivative of f on I. As a result,

$$\int_a^b f(x)\, dx = A(x)\Big|_a^b = A(b) - A(a)$$

But the area we want to find is the area under the graph of $y = f(x)$ from a to b. Since a and b are in I, this is the quantity $A(b) - A(a)$. See Figure 25. That is, we have

FIGURE 25

$$\text{Area} = \int_a^b f(x)\, dx = A(b) - A(a)$$

Area under the graph of $y = f(x)$ from a to b is $\displaystyle\int_a^b f(x)\, dx$

EXERCISE 15.5 Answers to Odd-Numbered Problems Begin on Page AN-91.

In Problems 1–10, find the area described. Be sure to sketch the graph first.

1. Under the graph of $f(x) = 3x + 2$ from 2 to 6
2. Under the graph of $f(x) = 3 - x$ from 0 to 3
3. Under the graph of $f(x) = x^2$ from 0 to 2
4. Under the graph of $f(x) = x^2$, from -2 to 1
5. Under the graph of $f(x) = x^2 - 1$, from -2 to 1
6. Under the graph of $f(x) = x^2 - 9$, from 0 to 6
7. Under the graph of $f(x) = \sqrt[3]{x}$, from -1 to 8
8. Under the graph of $f(x) = \sqrt[3]{x}$, from -8 to 1
9. Under the graph of $f(x) = e^x$, from 0 to 1
10. Under the graph of $f(x) = x^3$, from 0 to 1

In Problems 11–28, find the area enclosed by the graphs of the given functions. Draw a graph first.

11. $f(x) = x, \quad g(x) = 2x, \quad x = 0, \quad x = 1$
12. $f(x) = x, \quad g(x) = 3x, \quad x = 0, \quad x = 3$
13. $f(x) = x^2, \quad g(x) = x$
14. $f(x) = x^2, \quad g(x) = 4x$
15. $f(x) = x^2 + 1, \quad g(x) = x + 1$
16. $f(x) = x^2 + 1, \quad g(x) = 4x + 1$
17. $f(x) = \sqrt{x}, \quad g(x) = x^3$
18. $f(x) = x^2, \quad g(x) = x^3$
19. $f(x) = x^2, \quad g(x) = x^4$
20. $f(x) = \sqrt{x}, \quad g(x) = x^2$
21. $f(x) = x^2 - 4x, \quad g(x) = -x^2$
22. $f(x) = x^2 - 8x, \quad g(x) = -x^2$
23. $f(x) = 4 - x^2, \quad g(x) = x + 2$
24. $f(x) = 2 + x - x^2, \quad g(x) = -x - 1$
25. $f(x) = x^3, \quad g(x) = 4x$
26. $f(x) = x^3, \quad g(x) = 16x$
27. $y = x^2, \quad y = x, \quad y = -x$
28. $y = x^2 - 1, \quad y = x - 1, \quad y = -x - 1$

In Problems 29–34, an integral is given.
(a) *What area does the integral represent?*
(b) *Provide a graph that illustrates this area.*
(c) *Find the area.*

29. $\int_0^4 (3x + 1)\,dx$
30. $\int_1^3 (-2x + 7)\,dx$
31. $\int_2^5 (x^2 - 1)\,dx$
32. $\int_0^4 (16 - x^2)\,dx$
33. $\int_0^2 e^x\,dx$
34. $\int_e^{2e} \ln x\,dx$

35. **Consumer's Surplus; Producer's Surplus** Find the consumer's surplus and the producer's surplus for the demand curve

$$D(x) = -5x + 20$$

and the supply curve

$$S(x) = 4x + 8$$

Sketch the graphs and show the equilibrium point.

36. **Consumer's Surplus; Producer's Surplus** Follow the same directions as in Problem 35 if

$$D(x) = -0.4x + 15 \quad \text{and} \quad S(x) = 0.8x + 0.5$$

37. **Maximizing Profit Over Time** The revenue and the cost rate of Gold Star mining operation are, respectively,

$$R'(t) = 19 - t^{1/2} \quad \text{and} \quad C'(t) = 3 + 3t^{1/2}$$

where t is measured in years and R and C are measured in millions of dollars. Determine how long the operation should continue to maximize profit. Find the profit that can be generated during this period. Ignore any fixed costs.

38. **Consumer's Surplus** Find the consumer's surplus for the demand curve

$$D(x) = 50 - 0.025x^2$$

if it is known that the demand level x^* is 20 units.

39. **Mean Value Theorem for Integrals** If $y = f(x)$ is continuous on a closed interval $a \le x \le b$, then there is a number c, $a < c < b$, so that

$$\int_a^b f(x)\,dx = f(c)(b - a)$$

The interpretation of this result is that there is a rectangle with base $b - a$ and height $f(c)$, whose area is numerically equal to the area $\int_a^b f(x)\,dx$. See the illustration.

Verify this result by finding c for the functions below. Graph each function.

(a) $f(x) = x^2, \quad a = 0, \quad b = 1$

(b) $f(x) = \dfrac{1}{x^2}, \quad a = 1, \quad b = 4$

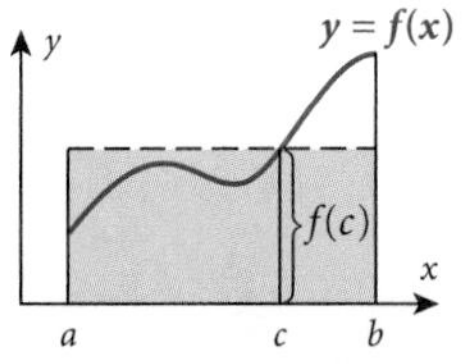

40. Show that the shaded area in the figure is $\frac{2}{3}$ of the area of the parallelogram $ABCD$. (This illustrates a result due to Archimedes concerning sectors of parabolas.)

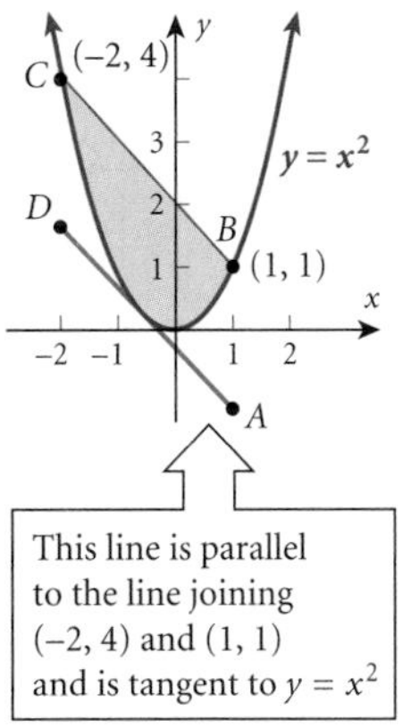

41. If $y = f(x)$ is continuous on the interval I and if it has an antiderivative on I, then for some a in I,

$$\frac{d}{dx}\int_a^x f(t)\,dt = f(x) \qquad \text{for } x > a \text{ in } I$$

This result gives us a technique for finding the derivative of a definite integral in which the lower limit is fixed and the upper limit is variable. Use this result to find:

(a) $\frac{d}{dx}\int_1^x t^2\,dt$

(b) $\frac{d}{dx}\int_2^x \sqrt{t^2 - 2}\,dt$

(c) $\frac{d}{dx}\int_5^x \sqrt{t^t + 2t}\,dt$

15.6 Approximating Definite Integrals

PREPARING FOR THIS SECTION ***Before getting started, review the following:***

>> Interval Notation (Appendix A, Section A.7, pp. A-61 – A-62)

OBJECTIVES
1 Approximate definite integrals using rectangles
2 Approximate definite integrals using a graphing utility

Up to now the evaluation of a definite integral

$$\int_a^b f(x)\ dx$$

has required that we find an antiderivative F of f so that

$$\int_a^b f(x)\ dx = F(x)\Big|_a^b = F(b) - F(a) \qquad \text{where} \qquad F'(x) = f(x)$$

But what if we can't find an antiderivative? In fact, sometimes it is impossible to find an antiderivative. In such situations, it is necessary to *approximate* the definite integral. One way is to use rectangles.

Approximate definite integrals using rectangles 1

We have already discussed the fact that when f is a continuous nonnegative function defined on the closed interval $[a, b]$, then the definite integral $\int_a^b f(x)\,dx$ equals the area under the graph of f from a to b. We will use this idea to obtain an approximation to $\int_a^b f(x)\,dx$.

Consider the graph of the function f in Figure 26(a) on page 904. The area under the graph of f from a to b is $\int_a^b f(x)\,dx$. Pick a number u in the interval $[a, b]$ and form the rectangle whose height is $f(u)$ and whose base is $b - a$. See Figure 26(b). The area of this rectangle provides a rough approximation to $\int_a^b f(x)\ dx$. That is,

$$\int_a^b f(x)\ dx \approx f(u)(b - a)$$

FIGURE 26

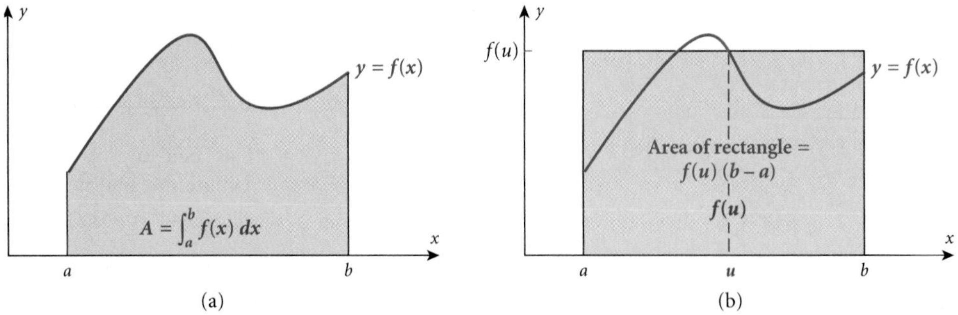

A better approximation to $\int_a^b f(x)\,dx$ can be obtained by dividing the interval $[a, b]$ into two subintervals of the same length. See Figure 27(a) where x_1 is the midpoint of $[a, b]$. Now pick a number u_1 between a and x_1, and a number u_2 between x_1 and b, and form two rectangles: One whose height is $f(u_1)$ and whose base is $x_1 - a$ and the other whose height is $f(u_2)$ and whose base is $b - x_1$. See Figure 27(b). The sum of the areas of these two rectangles provides an approximation to $\int_a^b f(x)\,dx$. That is,

$$\int_a^b f(x)\,dx \approx f(u_1)(x_1 - a) + f(u_2)(b - x_1)$$

FIGURE 27

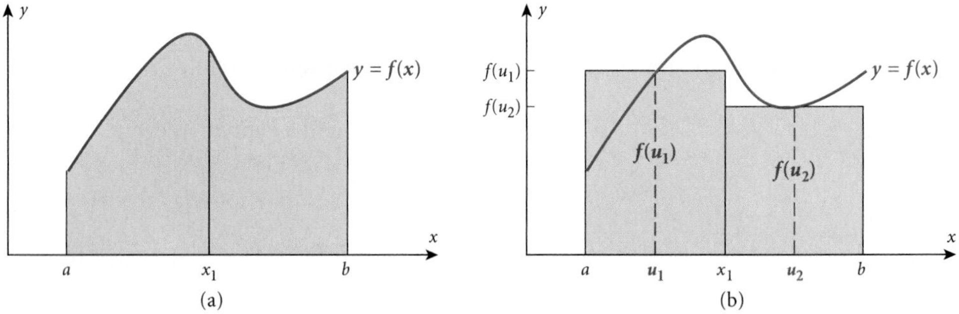

To get an even better approximation, divide the interval $[a, b]$ into three subintervals, each of the same length Δx. See Figure 28(a) where x_1 and x_2 are chosen so that

$$\Delta x = x_1 - a = x_2 - x_1 = b - x_2$$

Now pick numbers u_1, u_2, u_3 in each subinterval and form three rectangles of heights $f(u_1)$, $f(u_2)$, $f(u_3)$, each with the same base Δx. See Figure 28(b). The sum of the areas of these rectangles approximates $\int_a^b f(x)\,dx$. That is,

FIGURE 28

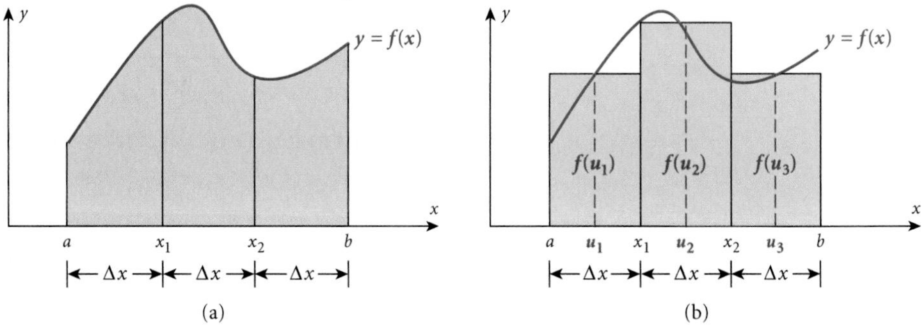

$$\int_a^b f(x)\,dx \approx f(u_1)\,\Delta x + f(u_2)\,\Delta x + f(u_3)\,\Delta x$$

The more subintervals we divide $[a, b]$ into, the better the approximation. Suppose we divide $[a, b]$ into n subintervals

$$[a, x_1], [x_1, x_2], \ldots, [x_{k-1}, x_k], \ldots, [x_{n-1}, b]$$

each of length $\Delta x = \dfrac{b-a}{n}$. See Figure 29(a). Now pick numbers $u_1, u_2, \ldots, u_n$ in each subinterval and form n rectangles of base Δx and heights $f(u_1), f(u_2), \ldots, f(u_n)$. See Figure 29(b). The sum of the areas of these rectangles approximates $\int_a^b f(x)\,dx$. That is,

$$\int_a^b f(x)\,dx \approx f(u_1)\,\Delta x + f(u_2)\,\Delta x + \cdots + f(u_n)\,\Delta x$$

FIGURE 29

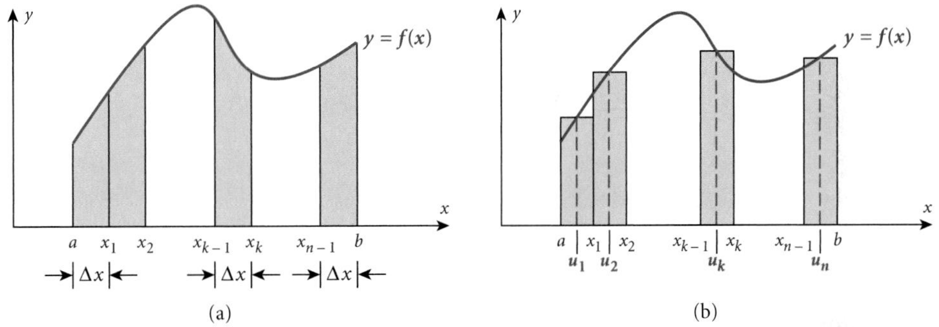

Let's summarize this process.

Steps for Approximating a Definite Integral

To approximate $\int_a^b f(x)\,dx$ follow these steps:

STEP 1 Divide the interval $[a, b]$ into n subintervals of equal length $\Delta x = \dfrac{b-a}{n}$.

The larger n is, the better your approximation will be.

STEP 2 Pick a number u in each subinterval and evaluate $f(u)$.

STEP 3 The sum

$$f(u_1)\Delta x + f(u_2)\Delta x + \cdots + f(u_n)\Delta x$$

approximates

$$\int_a^b f(x)\,dx$$

EXAMPLE 1 Approximating a Definite Integral

Approximate $\int_0^4 (3 + 2x)\,dx$ by dividing the interval $[0, 4]$ into four subintervals of equal length. Pick u_i as the left endpoint of each subinterval. Compare the approximation to the exact value.

SOLUTION Figure 30 illustrates the graph of $f(x) = 3 + 2x$ on $[0, 4]$.

FIGURE 30

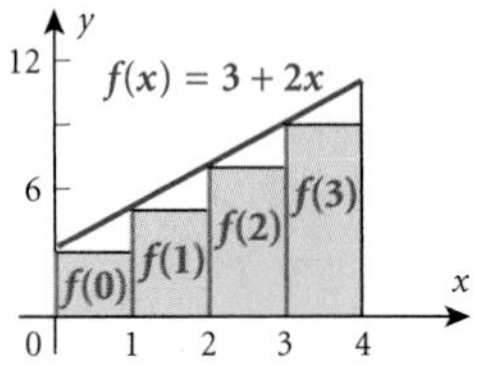

STEP 1 Divide $[0, 4]$ into four subintervals of equal length:

$$[0, 1], [1, 2], [2, 3], [3, 4]$$

STEP 2 Pick the left endpoint of each of these subintervals and evaluate f there:

$$f(0) = 3,\ f(1) = 5,\ f(2) = 7,\ f(3) = 9. \qquad f(x) = 3 + 2x$$

STEP 3
$$\int_0^4 (3 + 2x)\,dx \approx f(0)\cdot 1 + f(1)\cdot 1 + f(2)\cdot 1 + f(3)\cdot 1$$
$$= 3 + 5 + 7 + 9 = 24$$

This is an approximation of the integal. The exact value is

$$\int_0^4 (3 + 2x)\,dx = (3x + x^2)\Big|_0^4 = 12 + 16 = 28$$

NOW WORK PROBLEM 1.

EXAMPLE 2 Approximating a Definite Integral

Approximate: $\int_0^8 x^2\,dx$

(a) By dividing the interval $[0, 8]$ into four subintervals of equal length and picking u_i as the left end point of each subinterval.

(b) By dividing the interval $[0, 8]$ into eight subintervals of equal length and picking u_i as the right endpoint of each subinterval.

(c) Find the exact value.

SOLUTION **(a)** We divide $[0, 8]$ into four subintervals of equal length:

$$[0, 2], [2, 4], [4, 6], [6, 8]$$

Now we evaluate f at the left end point of each subinterval:

$$f(0) = 0,\ f(2) = 4,\ f(4) = 16,\ f(6) = 36$$

Since each subinterval is of length $\Delta x = 2$, we find

$$\int_0^8 x^2\,dx \approx f(0)\cdot 2 + f(2)\cdot 2 + f(4)\cdot 2 + f(6)\cdot 2$$
$$= 0 + 8 + 32 + 72 = 112$$

(b) We divide $[0, 8]$ into eight subintervals of equal length:

$$[0, 1], [1, 2], [2, 3], [3, 4], [4, 5], [5, 6], [6, 7], [7, 8]$$

Now we evaluate f at the right end point of each of these subintervals.

$$f(1) = 1,\ f(2) = 4,\ f(3) = 9,\ f(4) = 16,$$
$$f(5) = 25,\ f(6) = 36,\ f(7) = 49,\ f(8) = 64$$

Since each subinterval is of length $\Delta x = 1$, we find

$$\int_0^8 x^2 dx \approx f(1) + f(2) + f(3) + f(4) + f(5) + f(6) + f(7) + f(8) = 204$$

(c) The exact value of $\int_0^8 x^2\,dx$, is

$$\int_0^8 x^2\,dx = \frac{1}{3}x^3\Big|_0^8 = \frac{512}{3} = 170.7$$

NOW WORK PROBLEM 7.

Approximate definite integrals using a graphing utility

2 We can use a graphing utility to approximate a definite integral.

EXAMPLE 3 Using a Graphing Utility to Approximate a Definite Integral

FIGURE 31

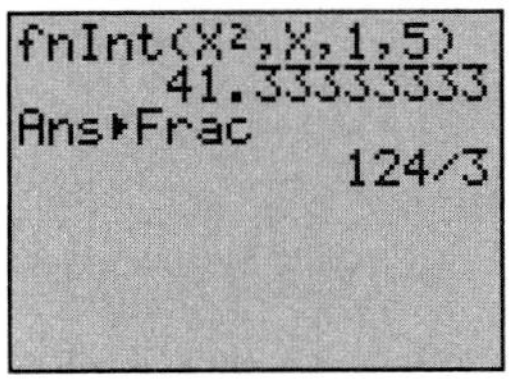

Use a graphing utility to approximate the area under the graph of $f(x) = x^2$ from 1 to 5. That is, evaluate the integral

$$\int_1^5 x^2\,dx$$

SOLUTION Figure 31 shows the result using a TI–83 Plus calculator. Consult your owner's manual for the proper keystrokes.

NOW WORK PROBLEM 19.

Riemann Sums

In the procedure for approximating a definite integral, suppose we subdivide each interval but don't require that each length be the same.

We begin with a continuous function f defined on a closed interval $[a, b]$. We partition, or divide, this interval into n subintervals, not necessarily of the same length. The point $a = x_0$ is the initial point, the first point of the subdivision is x_1, the second is $x_2, \ldots$, and the nth point is $b = x_n$. See Figure 32.

FIGURE 32

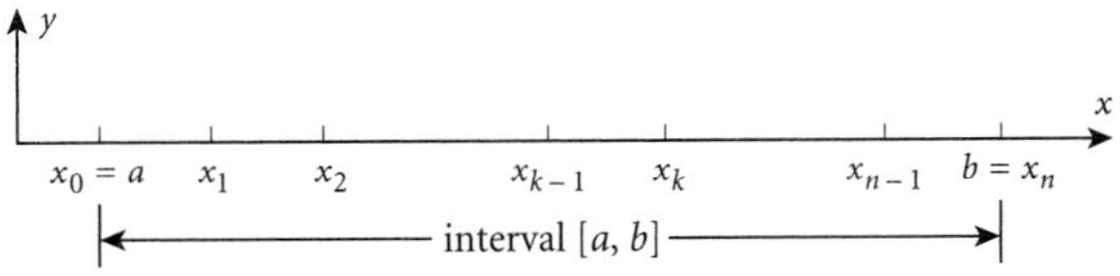

The original interval $[a, b]$ now consists of n subintervals; the length of each one is

First,	Second,	Third, . . . ,	kth, . . . ,	nth
$\Delta x_1 = x_1 - x_0,$	$\Delta x_2 = x_2 - x_1,$	$\Delta x_3 = x_3 - x_2, \ldots,$	$\Delta x_k = x_k - x_{k-1}, \ldots,$	$\Delta x_n = x_n - x_{n-1}$

We use the symbol Δ to denote the largest such length, which depends on how the partition itself has been chosen. We call Δ the **norm of the partition.**

Next, we concentrate on the function f. Pick a number in each subinterval (you may select a number in the interval or either endpoint, if you wish) and evaluate f at this number. To fix our ideas, let u_k denote the number chosen from the kth subinterval. The corresponding value of the function is $f(u_k)$. This represents the height of the function at u_k.

Multiply $f(u_1)$ times $\Delta x_1 = x_1 - x_0$, $f(u_2)$ times $\Delta x_2 = x_2 - x_1, \ldots, f(u_n)$ times $\Delta x_n = x_n - x_{n-1}$, and add these products. The result is the sum

$$f(u_1)\,\Delta x_1 + f(u_2)\,\Delta x_2 + \cdots + f(u_n)\,\Delta x_n$$

This is called a **Riemann sum** for the function f on $[a, b]$.

Now take the limit of this sum as the norm $\Delta \to 0$. If this limit exists, it is the definite integral of $f(x)$ from a to b. That is,

$$\int_a^b f(x)\,dx = \lim_{\Delta \to 0} [f(u_1)\,\Delta x_1 + f(u_2)\,\Delta x_2 + \cdots + f(u_n)\,\Delta x_n] \quad (1)$$

where $\Delta = \max(\Delta x_1, \Delta x_2, \ldots, \Delta x_n)$.

The above formula, in a more formal course in calculus, is taken as the definition of a definite integral. Then it can be proven as a theorem, called the **Fundamental Theorem of Calculus,** that

$$\int_a^b f(x)\,dx = F(b) - F(a)$$

where F is an antiderivative of f.

EXERCISE 15.6 Answers to Odd-Numbered Problems Begin on Page AN-94.

In Problems 1 and 2, refer to the illustration. The interval $[1, 3]$ is partitioned into two subintervals $[1, 2]$ and $[2, 3]$.

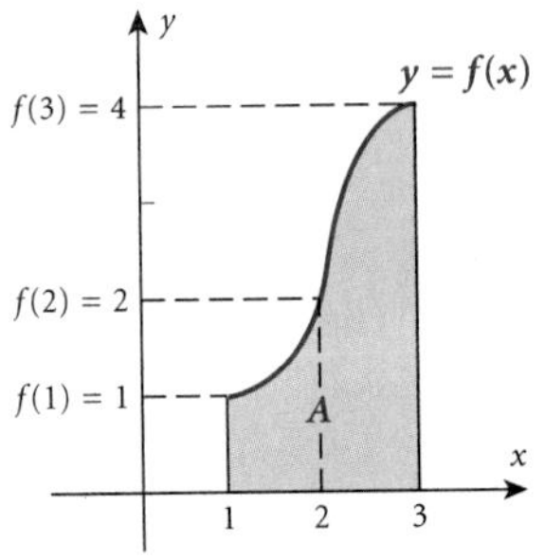

1. Approximate the area A, choosing u as the left endpoint of each subinterval.

2. Approximate the area A, choosing u as the right endpoint of each subinterval.

In Problems 3 and 4, refer to the illustration. The interval $[0, 8]$ *is partitioned into four subintervals* $[0, 2]$, $[2, 4]$, $[4, 6]$, *and* $[6, 8]$.

3. Approximate the area A, choosing u as the left endpoint of each subinterval.

4. Approximate the area A, choosing u as the right endpoint of each subinterval.

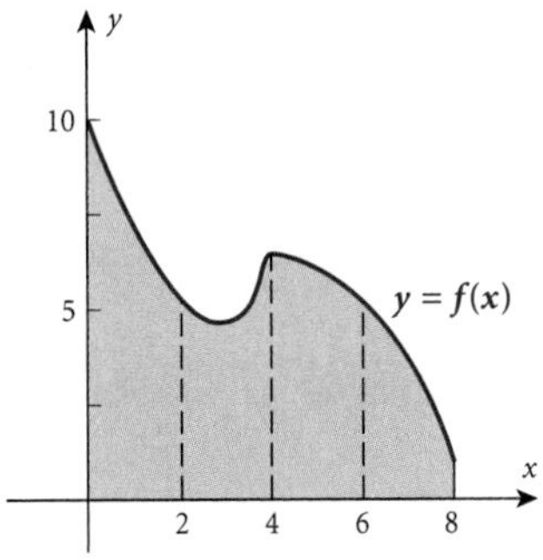

$f(0) = 10, f(2) = 6, f(4) = 7, f(6) = 5, f(8) = 1$

---

5. The function $f(x) = 3x$ is defined on the interval $[0, 6]$.

(a) Graph f.
In (b)–(e), approximate the area A under f from 0 to 6 as follows:
(b) By partitioning $[0, 6]$ into three subintervals of equal length and choosing u as the left endpoint of each subinterval.
(c) By partitioning $[0, 6]$ into three subintervals of equal length and choosing u as the right endpoint of each subinterval.
(d) By partitioning $[0, 6]$ into six subintervals of equal length and choosing u as the left endpoint of each subinterval.
(e) By partitioning $[0, 6]$ into six subintervals of equal length and choosing u as the right endpoint of each subinterval.
(f) What is the actual area A?

6. Repeat Problem 5 for $f(x) = 4x$.

7. The function $f(x) = -3x + 9$ is defined on the interval $[0, 3]$.

(a) Graph f.
In (b)–(e), approximate the area A under f from 0 to 3 as follows:
(b) By partitioning $[0, 3]$ into three subintervals of equal length and choosing u as the left endpoint of each subinterval.
(c) By partitioning $[0, 3]$ into three subintervals of equal length and choosing u as the right endpoint of each subinterval.
(d) By partitioning $[0, 3]$ into six subintervals of equal length and choosing u as the left endpoint of each subinterval.
(e) By partitioning $[0, 3]$ into six subintervals of equal length and choosing u as the right endpoint of each subinterval.
(f) What is the actual area A?

8. Repeat Problem 7 for $f(x) = -2x + 8$.

In Problems 9–16, a function f is defined over an interval $[a, b]$.
(a) *Graph f, indicating the area A under f from a to b.*
(b) *Approximate the area A by partitioning* $[a, b]$ *into four subintervals of equal length and choosing u as the left endpoint of each subinterval.*
(c) *Approximate the area A by partitioning* $[a, b]$ *into eight subintervals of equal length and choosing u as the left endpoint of each subinterval.*
(d) *Express the area A as an integral.*
(e) *Evaluate the integral.*

9. $f(x) = x^2 + 2, \quad [0, 4]$

10. $f(x) = x^2 - 4, \quad [2, 6]$

11. $f(x) = x^3, \quad [0, 4]$

12. $f(x) = x^3, \quad [1, 5]$

13. $f(x) = \dfrac{1}{x}, \quad [1, 5]$

14. $f(x) = \sqrt{x}, \quad [0, 4]$

15. $f(x) = e^x, \quad [-1, 3]$

16. $f(x) = e^{-x}, \quad [3, 7]$

In Problems 17–20, use a graphing utility to approximate each integral. Round your answer to two decimal places.

17. $\displaystyle\int_0^1 e^{x^2}\,dx$

18. $\displaystyle\int_{-2}^4 e^{x^2}\,dx$

19. $\displaystyle\int_1^5 \frac{e^x}{x}\,dx$

20. $\displaystyle\int_1^5 \frac{\ln x}{x}\,dx$

21. Consider the function $f(x) = \sqrt{1 - x^2}$ whose domain is the closed interval $[-1, 1]$.

(a) Graph f.
(b) Approximate the area under the graph of f from -1 to 1 by dividing $[-1, 1]$ into five subintervals, each of equal length. Choose u as the left end point.
(c) Approximate the area under the graph of f from -1 to 1 by dividing $[-1, 1]$ into ten subintervals each of equal length. Choose u as the left end point.
(d) Express the area as an integral.
(e) Evaluate the integral using a graphing utility.
(f) What is the actual area?
[**Hint:** The graph of f is a semi-circle.]

15.7 Differential Equations

PREPARING FOR THIS SECTION *Before getting started, review the following:*

- > Properties of Logarithms (Chapter 11, Section 11.5, pp. 654–661)
- > Logarithmic Equations (Chapter 11, Section 11.4, pp. 647–648)
- > Exponential Equations (Chapter 11, Section 11.3, pp. 635–636)
- > Differentials (Chapter 14, Section 14.7, pp. 845–847)

OBJECTIVES
1 Find a particular solution of a differential equation
2 Solve applied problems involving population growth
3 Solve applied problems involving radioactive decay
4 Solve applied problems involving price-demand equations

In studies of physical, chemical, biological, and other natural phenomena, scientists attempt, on the basis of long observation, to deduce mathematical laws that will describe and predict nature's behavior. Such laws often involve the derivatives of some unknown function F, and it is required to find this unknown function F.

For example, for a given function f it may be required to find all functions $y = F(x)$ so that

$$\boxed{\frac{dy}{dx} = f(x)} \qquad (1)$$

This equation is an example of what is called a **differential equation.** A function $y = F(x)$ for which $\frac{dy}{dx} = f(x)$ is a **solution** of the differential equation. The **general solution** of $\frac{dy}{dx} = f(x)$ consists of all the antiderivatives of f.

EXAMPLE 1 Finding the General Solution of a Differential Equation

The general solution of the differential equation

$$\frac{dy}{dx} = 5x^2 + 2 \qquad (2)$$

is

$$y = \frac{5x^3}{3} + 2x + K$$

where K is a constant.

A **particular solution** of the differential equation $\frac{dy}{dx} = f(x)$ occurs when K is assigned a particular value. When a particular solution is required, we use a **boundary condition.**

1 EXAMPLE 2 Finding a Particular Solution of a Differential Equation

In the differential equation (2) we require the general solution to obey the boundary condition that $y = 5$ when $x = 3$. Find the particular solution.

SOLUTION In the general solution,

$$y = \frac{5x^3}{3} + 2x + K$$

let $x = 3$ and $y = 5$. Then

$$5 = \frac{5(27)}{3} + (2)(3) + K \qquad x = 3;\ y = 1$$

$$5 = 45 + 6 + K \qquad \text{Simplify.}$$

$$K = -46 \qquad \text{Solve for } K.$$

The particular solution of the differential equation (2) with the boundary condition that $y = 5$ when $x = 3$ is

$$y = \frac{5x^3}{3} + 2x - 46$$

EXAMPLE 3 Finding a Particular Solution of a Differential Equation

Solve the differential equation below with the boundary condition that $y = -1$ when $x = 3$.

$$\frac{dy}{dx} = x^2 + 2x + 1$$

SOLUTION The general solution of the differential equation is

$$y = \frac{x^3}{3} + x^2 + x + K$$

where K is a constant. To determine the number K, we use the boundary condition. Then

$$-1 = \frac{3^3}{3} + 3^2 + 3 + K \qquad x = 3;\ y = -1$$

$$-1 = 9 + 9 + 3 + K \qquad \text{Simplify.}$$

$$K = -22 \qquad \text{Solve for } K.$$

The particular solution of the differential equation with the boundary condition that $y = -1$ when $x = 3$ is

$$y = \frac{x^3}{3} + x^2 + x - 22$$

NOW WORK PROBLEM 1.

Applications

The statement below describes many situations in business, biology, and the natural sciences.

> The amount A of a substance varies with time t in such a way that the time rate of change of A is proportional to A itself.

The mathematical formulation of this statement is the differential equation

$$\boxed{\frac{dA}{dt} = kA} \tag{3}$$

where $k \neq 0$ is a real number.

If $k > 0$, then equation (3) asserts that the time rate of change of A is positive, so that the amount A of the substance is increasing over time. If $k < 0$, then equation (3) asserts that the time rate of change of A is negative, so that the amount A of the substance is decreasing over time.

We seek a solution $A = A(t)$ to the differential equation (3). We begin by rewriting the equation in terms of differentials as

$$dA = kA\,dt$$

$$\frac{dA}{A} = k\,dt \qquad \text{Divide both sides by } A.$$

$$\int \frac{dA}{A} = \int k\,dt \qquad \text{Take the integral of each side.}$$

$$\ln A = kt + K \qquad \text{Integrate.}$$

To determine the constant K, we use the boundary condition that when $t = 0$, the initial amount present is $A_0 = A(0)$. When $t = 0$, we have

$$\ln A_0 = k(0) + K$$

$$K = \ln A_0$$

Replace K by $\ln A_0$ and proceed to solve for A.

$$\ln A = kt + \ln A_0 \qquad K = \ln A_0$$

$$\ln A - \ln A_0 = kt \qquad \text{Subtract } \ln A_0 \text{ from each side.}$$

$$\ln \frac{A}{A_0} = kt \qquad \log_a \frac{M}{N} = \log_a M - \log_a N$$

$$\frac{A}{A_0} = e^{kt} \qquad \text{Change to an exponential expression.}$$

The solution of the differential equation (3) is therefore

$$\boxed{A = A_0 e^{kt}} \tag{4}$$

When a function $A = A(t)$ varies according to the law given by the differential equation (3), or its solution, equation (4), it is said to follow the **exponential law,** or the **law of uninhibited growth or decay,** or the **law of continuously compounded interest.**

As previously noted, the sign of k determines whether A is increasing ($k > 0$) or decreasing ($k < 0$). Figure 33 illustrates the graphs of $A = A(t) = A_0 e^{kt}$ for $k > 0$ and for $k < 0$.

FIGURE 33

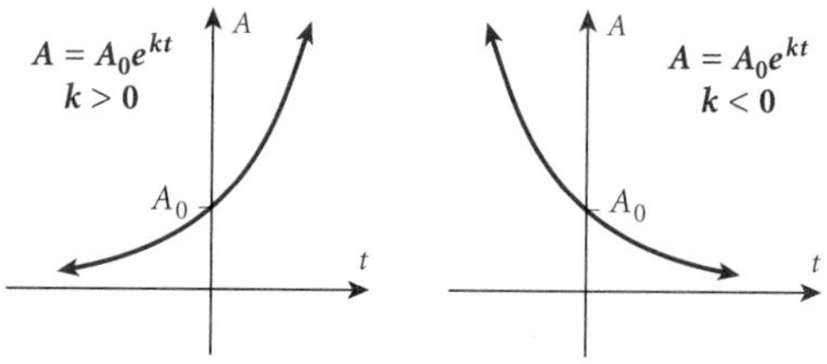

We present now three applications: to biology, radioactive decay, and business. They are independent of each other and may be covered in any order or omitted without loss of continuity.

Bacterial Growth Our first application is to bacterial growth.

2 **EXAMPLE 4 Growth of Bacteria**

Assume that the population of a colony of bacteria *increases at a rate proportional to the number present.** If the number of bacteria doubles in 5 hours, how long will it take for the bacteria to triple?

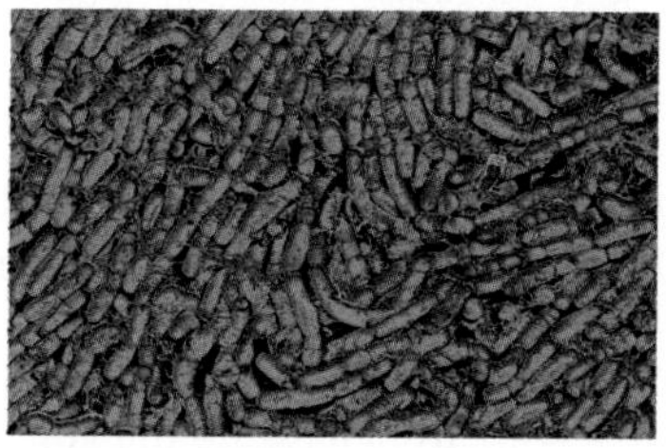

SOLUTION Let $N = N(t)$ be the number of bacteria present at time t. Then the assumption that this colony of bacteria increases at a rate proportional to the number present can be written as

$$\frac{dN}{dt} = kN \tag{5}$$

where k is a positive constant of proportionality. The solution of the differential equation (5) is

$$N(t) = N_0 e^{kt}$$

where N_0 is the initial number of bacteria in this colony. Since the number of bacteria doubles in 5 hours, we have

$$N(5) = 2N_0$$

But $N(5) = N_0 e^{k(5)}$, so that

$$\begin{aligned} N_0 e^{5k} &= 2N_0 \\ e^{5k} &= 2 && \text{Cancel the } N_0\text{'s.} \\ 5k &= \ln 2 && \text{Write as a logarithm.} \\ k &= \left(\frac{1}{5}\right)\ln 2 = 0.1386 && \text{Solve for } k. \end{aligned}$$

The function that gives the number N of bacteria present at time t is

$$N(t) = N_0 e^{0.1386t}$$

The time t that is required for this colony to triple must satisfy the equation

$$\begin{aligned} N(t) &= 3N_0 \\ N_0 e^{0.1386t} &= 3N_0 \\ e^{0.1386t} &= 3 && \text{Cancel the } N_0\text{'s.} \\ 0.1386t &= \ln 3 && \text{Write as a logarithm.} \\ t &= \frac{\ln 3}{0.1386} = 7.925 \text{ hours} && \text{Solve for } t. \end{aligned}$$

The bacteria will triple in size is just under 8 hours.

NOW WORK PROBLEM 11.

* *This is a model of uninhibited growth. However, after enough time has passed, growth will not continue at a rate proportional to the number present. Other factors, such as lack of living space, dwindling food supply, and so on, will start to affect the rate of growth. The model presented accurately reflects the way growth occurs in the early stages.*

Radioactive Decay Our second application is to radioactive decay, and, in particular, its use in carbon dating. For a radioactive substance, *the rate of decay is proportional to the amount present at a given time t*. That is, if $A = A(t)$ represents the amount of a radioactive substance at time t, we have

$$\frac{dA}{dt} = kA$$

where the constant k is negative and depends on the radioactive substance. The **half-life** of a radioactive substance is the time required for half of the substance to decay.

In carbon dating, we use the fact that all living organisms contain two kinds of carbon, carbon-12 (a stable carbon) and carbon-14 (a radioactive carbon). As a result, when an organism dies, the amount of carbon-12 present remains unchanged, while the amount of carbon-14 begins to decrease. This change in the amount of carbon-14 present relative to the amount of carbon-12 present makes it possible to calculate the time at which the organism lived.

3 **EXAMPLE 5** **Using Carbon Dating in Archaeology**

In the skull of an animal found in an archaeological dig, it was determined that about 20% of the original amount of carbon-14 was still present. If the half-life of carbon-14 is 5600 years, find the approximate age of the animal.

SOLUTION Let $A = A(t)$ be the amount of carbon-14 present in the skull at time t. Then A satisfies the differential equation $\frac{dA}{dt} = kA$, whose solution is

$$A(t) = A_0e^{kt}$$

where A_0 is the amount of carbon-14 present at time $t = 0$. To determine the constant k, we use the fact that when $t = 5600$, half of the original amount A_0 will remain. That is,

$$A(t) = A_0e^{kt}$$

$$\frac{1}{2}A_0 = A_0e^{5600k} \qquad A = \frac{1}{2}A_0 \text{ when } t = 5600.$$

$$\frac{1}{2} = e^{5600k} \qquad \text{Cancel the } A_0\text{'s.}$$

$$5600k = \ln\frac{1}{2} \qquad \text{Write as a logarithm.}$$

$$k = -0.000124 \qquad \text{Solve for } k.$$

The relationship between the amount A of carbon-14 and time t is therefore

$$A(t) = A_0e^{-(0.000124)t}$$

If the amount A of carbon-14 is 20% of the original amount A_0, we have

$$0.2A_0 = A_0e^{-(0.000124)t}$$

$$0.2 = e^{-(0.000124)t} \qquad \text{Cancel the } A_0\text{'s.}$$

$$-(0.000124)t = \ln 0.2 \qquad \text{Change to a logarithm.}$$

$$t = \frac{1.6094}{0.000124} = 12{,}979 \text{ years} \qquad \text{Solve for } t.$$

The animal found in the dig lived approximately 13,000 years ago.

NOW WORK PROBLEM 15.

Price-Demand Equations Our third application is to business: Suppose the marginal price of a product, when the demand is x units, is proportional to its price, p. This situation can be represented by the differential equation

$$\frac{dp}{dx} = kp$$

Since demand usually decreases as price increases, the constant k will be negative.

4 **EXAMPLE 6** **Pricing a Product**

A manufacturer finds that the marginal price of one of its products is proportional to the price. Past experience has shown that at a price of \$125 per unit, zero units are sold; but at a price of \$100 per unit, 100 units are sold.

(a) Find the price-demand equation.
(b) Find the approximate price that should be charged if the manufacturer wants to sell 50 units.

SOLUTION **(a)** If $p = p(x)$ represents the price when x units are demanded, then

$$\frac{dp}{dx} = kp, \quad k < 0$$

Rewriting the equation in terms of differentials we get

$$\frac{dp}{p} = k\,dx$$

We integrate each side to obtain

$$\ln p = kx + K$$

where K is a constant.

To determine the constant K we use the boundary condition that when $x = 0$, $p = 125$. Then

$$\ln 125 = k(0) + K$$
$$k = \ln 125$$

Substitute into $\ln p = \ln x + K$ and proceed to solve for p.

$$\ln p = kx + \ln 125$$
$$\ln p - \ln 125 = kx \qquad \text{Subtract ln 125 from each side.}$$
$$\ln \frac{p}{125} = kx \qquad \ln\left(\frac{M}{N}\right) = \ln M - \ln N$$
$$\frac{p}{125} = e^{kx} \qquad \text{Change to exponential form.}$$
$$p = 125e^{kx} \qquad \text{Solve for } p. \qquad (6)$$

We use the second boundary condition that when 100 units are demanded the price is \$100 to find k.

$$100 = 125e^{100k} \qquad p = 125e^{kx};\ p = \$100;\ x = 100$$
$$\frac{100}{125} = e^{100k} \qquad \text{Divide by 125.}$$
$$\ln\left(\frac{100}{125}\right) = 100\text{k} \qquad \text{Write as a logarithm.}$$
$$k = \frac{1}{100}\ln\left(\frac{100}{125}\right) \approx -0.00223 \qquad \text{Solve for } k.$$

From (6), the demand equation is

$$p(x) = 125e^{-0.00223x}$$

(b) To sell 50 units, the manufacturer should set the price at

$$p = 125e^{(-0.00223)50} = \$111.80$$

EXERCISE 15.7 Answers to Odd-Numbered Problems Begin on Page AN-95.

In Problems 1–10, solve each differential equation using the indicated boundary condition.

1. $\frac{dy}{dx} = x^2 - 1$, $y = 0$ when $x = 0$
2. $\frac{dy}{dx} = x^2 + 4$, $y = 1$ when $x = 0$
3. $\frac{dy}{dx} = x^2 - x$, $y = 3$ when $x = 3$
4. $\frac{dy}{dx} = x^2 + x$, $y = 5$ when $x = 3$
5. $\frac{dy}{dx} = x^3 - x + 2$, $y = 1$ when $x = -2$
6. $\frac{dy}{dx} = x^3 + x - 5$, $y = 1$ when $x = -2$
7. $\frac{dy}{dx} = e^x$, $y = 4$ when $x = 0$
8. $\frac{dy}{dx} = \frac{1}{x}$, $y = 0$ when $x = 1$
9. $\frac{dy}{dx} = \frac{x^2 + x + 1}{x}$, $y = 0$ when $x = 1$
10. $\frac{dy}{dx} = x + e^x$, $y = 4$ when $x = 0$

11. **Bacterial Growth** The rate of growth of bacteria is proportional to the amount present. If initially there are 100 bacteria and 5 minutes later there are 150 bacteria, how many bacteria will be present after 1 hour? How many are present after 90 minutes? How long will it take for the number of bacteria to reach 1,000,000?

12. Answer the questions given in Problem 11 if after 8 minutes the number of bacteria present grows from 100 to 150.

13. **Radioactive Decay** The half-life of radium is 1690 years. If 8 grams of radium are present now, how many grams will be present in 100 years?

14. **Radioactive Decay** If 25% of a radioactive substance disappears in 10 years, what is the half-life of the substance?

15. **Age of a Tree** A piece of charcoal is found to contain 30% of the carbon-14 it originally had. When did the tree from which the charcoal came die? Use 5600 years as the half-life of carbon-14.

16. **Age of a Fossil** A fossilized leaf contains 70% of a normal amount of carbon-14. How old is the fossil?

17. **Population Growth** The population growth of a colony of mosquitoes obeys the uninhibited growth equation. If there are 1500 mosquitoes initially, and there are 2500 mosquitoes after 24 hours, what is the size of the mosquito population after 3 days?

18. **Population Growth** The population of a suburb doubled in size in an 18-month period. If this growth continues and the current population is 8000, what will the population be in 4 years?

19. **Bacterial Growth** The number of bacteria in a culture is growing at a rate of $3000e^{2t/5}$. At $t = 0$, the number of bacteria present was 7500. Find the number present at $t = 5$.

20. **Bacterial Growth** At any time t, the rate of increase in the area of a culture of bacteria is twice the area of the culture. If the initial area of the culture is 10, then what is the area at time t?

21. **Bacterial Growth** The rate of change in the number of bacteria in a culture is proportional to the number present. In a certain laboratory experiment, a culture had 10,000 bacteria initially, 20,000 bacteria at time t_1 minutes, and 100,000 bacteria at $t_1 + 10$ minutes.
    (a) In terms of t only, find the number of bacteria in the culture at any time t minutes ($t \geq 0$).
    (b) How many bacteria were there after 20 minutes?
    (c) At what time were 20,000 bacteria observed? That is, find the value of t_1.

22. **Chemistry** Salt (NaCl) decomposes in water into sodium (Na^+) and chloride (Cl^-) ions at a rate proportional to its mass. If the initial amount of salt is 25 kilograms, and after 10 hours, 15 kilograms are left:
    (a) How much salt would be left after 1 day?
    (b) After how many hours would there be less than $\frac{1}{2}$ kilogram of salt left?

23. **Age of a Fossil** Radioactive beryllium is sometimes used to date fossils found in deep-sea sediment. The decay of radioactive beryllium satisfies the equation $\frac{dA}{dt} = -\alpha A$,

where $\alpha = 1.5 \times 10^{-7}$, and t is measured in years. What is the half-life of radioactive beryllium?

24. Pressure Atmospheric pressure P is a function of the altitude a above sea level and is given by the equation $\frac{dP}{da} = \beta P$, where β is a constant. The pressure is measured in millibars (mb). At sea level ($a = 0$), $P(0)$ is 1013.25 mb, which means that the atmosphere at sea level will support a column of mercury 1013.25 millimeters high at a standard temperature of 15 °C. At an altitude of $a = 1500$ meters, the pressure is 845.6 mb.

(a) What is the pressure at $a = 4000$ meters?
(b) What is the pressure at 10 kilometers?
(c) In California, the highest and lowest points are Mount Whitney (4418 meters) and Death Valley (86 meters below sea level). What is the difference in their atmospheric pressures?
(d) What is the atmospheric pressure at Mount Everest (elevation 8848 meters)?
(e) At what elevation is the atmospheric pressure equal to 1 mb?

25. Pricing a Product A manufacturer finds that the marginal price of one of its products is proportional to the price. The manufacturer also knows that at a price of \$300 per unit, no units are sold. At a price of \$150, a total of 200 units are sold.

(a) Find the price-demand equation.
(b) Find the price p that should be charged if the manufacturer wants to sell 300 units.
(c) What price should be charged if the manufacturer wants to sell 350 units?

26. Pricing a Product A manufacturer finds that the marginal price of one of its products is proportional to the price. The manufacturer also knows that at a price of \$500 per unit, no units are sold. At a price of \$300, a total of 150 units are sold.

(a) Find the price-demand equation.
(b) Find the price p that should be charged if the manufacturer wants to sell 200 units.
(c) What price should be charged if the manufacturer wants to sell 250 units?

Chapter 15 Review

OBJECTIVES

Section	You should be able to	Review Exercises
15.1	1 Find an antiderivative	1–6
	2 Use integration formulas	7–14, 17, 21, 22
	3 Evaluate indefinite integrals	7–30
	4 Find cost and revenue functions	31–36
15.2	1 Integrate using substitution	15, 16, 18-20, 23–26
15.3	1 Integration by parts	27–30
15.4	1 Evaluate a definite integral	37–48
	2 Use properties of a definite integral	49–52
	3 Solve applied problems involving definite integrals	63, 64, 66, 81–84
15.5	1 Find the area under a graph	53–56
	2 Find the area enclosed by two graphs	57–62
	3 Solve applied problems involving definite integrals	65, 85, 86
15.6	1 Approximate definite integrals using rectangles	67, 68
	2 Approximate definite integrals using a graphing utility	69, 70
15.7	1 Find a particular solution of a differential equation	71-76
	2 Solve applied problems involving population growth	77, 78
	3 Solve applied problems involving radioactive decay	79, 80
	4 Solve applied problems involving price-demand equations	87

THINGS TO KNOW

Indefinite Integral (p. 863)

$$\int f(x)\,dx = F(x) + K, \quad F'(x) = f(x) \text{ and } K \text{ is a constant}$$

Integration Formulas (pp. 864 and 866)
(c is a real number; K is a constant)

$$\int c\,dx = cx + K$$

$$\int x^n\,dx = \frac{x^{n+1}}{n+1} + K \quad n \neq -1 \text{ a rational number}$$

$$\int [f(x) \pm g(x)]\,dx = \int f(x)\,dx \pm \int g(x)\,dx$$

$$\int cf(x)\,dx = c\int f(x)\,dx$$

$$\int e^x\,dx = e^x + K$$

$$\int e^{ax}\,dx = \frac{1}{a}e^{ax} + K \quad a \neq 0$$

$$\int \frac{1}{x}\,dx = \ln|x| + K$$

Integration by Parts (p. 878)

$$\int u\,dv = uv - \int v\,du$$

Definite Integral (p. 882)

$$\int_a^b f(x)\,dx = F(b) - F(a), \quad F'(x) = f(x)$$

Properties of the Definite Integral (pp. 884–885)
(f and g are continuous functions on the interval $[a, b]$)

$$\int_a^b f(x)\,dx = -\int_b^a f(x)\,dx$$

$$\int_a^a f(x)\,dx = 0$$

$$\int_a^b f(x)\,dx = \int_a^c f(x)\,dx + \int_c^b f(x)\,dx \quad c \text{ is between } a \text{ and } b$$

$$\int_a^b cf(x)\,dx = c\int_a^b f(x)\,dx \quad c \text{ is a real number}$$

$$\int_a^b [f(x) \pm g(x)]\,dx = \int_c^b f(x)\,dx \pm \int_a^b g(x)\,dx$$

Area A under the graph of $y = f(x)$ from a to b. (p. 891)

$$A = \int_a^b f(x)\,dx \quad f(x) \geq 0 \text{ on } [a, b].$$

Approximating a Definite Integral (p. 895)

$$\int_a^b f(x)\,dx \approx f(u_1)\,\Delta x + f(u_2)\,\Delta x + \cdots + f(u_n)\,\Delta x, \; \Delta x = \frac{b-a}{n}$$

TRUE–FALSE ITEMS **Answers are on page AN-96.**

T F **1.** The integral of the sum of two functions equals the sum of their integrals.

T F **2.** $\displaystyle\int\left(\frac{1}{2}x^3 + x^{3/2} - 1\right)dx = \frac{x^4}{6} + \frac{2x^{7/2}}{7} - x + K$

T F **3.** $\displaystyle\int \frac{x^2+4}{x}\,dx = \frac{\frac{x^3}{3} + 4x}{\frac{x^2}{2}} + K$

T F **4.** $\displaystyle\int \frac{\sqrt{x^2+1}}{x}\,dx = \frac{1}{x}\int \sqrt{x^2+1}\,dx$

T F **5.** $\displaystyle\int \ln x\,dx = \frac{1}{x} + K$

T F 6. Any antiderivative of $f(x)$ can be used to evaluate $\int_a^b f(x)\,dx$.

T F 7. The definite integral $\int_a^b f(x)\,dx$, if it exists, is a number.

T F 8. $\int_a^b f(x)\,dx + \int_b^a f(x)\,dx = 0$

T F 9. $\int_0^1 x^2\,dx = 3$

T F 10. The area under the graph of $f(x) = x^4$ from 0 to 2 equals $\int_0^2 x^4\,dx$.

FILL-IN-THE-BLANKS **Answers are on page AN-96.**

1. A function F is called an antiderivative of the function f if ________.

2. The symbol ________ represents all the antiderivatives of a function f.

3. The formula $\int u\,dv = uv - \int v\,du$ is referred to as the ________ ________ ________ formula.

4. In $\int_a^b f(x)\,dx$, the numbers a and b are called the ________ and ________ ________ of ________, respectively.

5. $\int_a^a f(x)\,dx =$ ________.

6. If f is continuous on $[a, b]$ and $F' = f$, then $\int_a^b f(x)\,dx =$ ________.

7. The area under the graph of $f(x) = \sqrt{x^2+1}$ from 0 to 2 may be symbolized by the integral ________.

REVIEW EXERCISES **Answers to odd-numbered problems begin on page AN-96.**

Blue problem numbers indicate the author's suggestions for use in a practice test.

In Problems 1–6, find all the antiderivatives of the given function.

1. $f(x) = 6x^5$
2. $f(x) = 3x^2$
3. $f(x) = x^3 + x$
4. $f(x) = x + 7$
5. $f(x) = \dfrac{1}{\sqrt{x}}$
6. $f(x) = \dfrac{1}{2\sqrt{x}}$

In Problems 7–30, evaluate each indefinite integral.

7. $\int 7\,dx$
8. $\int \frac{1}{2}\,dx$
9. $\int (5x^3 + 2)\,dx$
10. $\int (x^2 + 3x)\,dx$
11. $\int (x^4 - 3x^2 + 6)\,dx$
12. $\int 12(x^3 + 6x^2 - 2x - 1)\,dx$
13. $\int \frac{3}{x}\,dx$
14. $\int \left(x^2 + \frac{4}{x}\right)dx$
15. $\int \frac{2x}{x^2-1}\,dx$
16. $\int \frac{3x^2+5}{x^3+5x}\,dx$
17. $\int e^{3x}\,dx$
18. $\int 5xe^{2x^2}\,dx$
19. $\int (x^3 + 3x)^5(x^2+1)\,dx$
20. $\int \frac{x}{\sqrt{1-x^2}}\,dx$
21. $\int 2x(x-3)\,dx$
22. $\int (x+1)(x-1)\,dx$
23. $\int e^{3x^2+x}(6x+1)\,dx$
24. $\int \frac{1}{x^4}e^{x^{-3}}\,dx$
25. $\int x\sqrt{x-5}\,dx$
26. $\int x(x+2)^{2/3}\,dx$
27. $\int xe^{4x}\,dx$

28. $\int 3x^2e^{x+4}\,dx$

29. $\int x^{-2}\ln 2x\;dx$

30. $\int (x^5+7)\ln x\;dx$

In Problems 31 and 32, find the revenue function R. Assume that the revenue is zero if there are no sales.

31. $R'(x) = 5x + 2$

32. $R'(x) = 12x - 5\sqrt{x}$

In Problems 33 and 34, find the cost function C, and determine where the cost is minimum.

33. $C'(x) = 5x + 120{,}000$; fixed costs are \$7500

34. $C'(x) = 1.3x - 2600$; fixed costs are \$2735

35. Revenue analysis (a) If sales of televisions have a marginal revenue given by $R'(x) = 500 - 0.01x$, where x is the number of televisions sold, find the revenue obtained by selling x televisions. Assume that the revenue is 0 if there are no sales.

(b) Find the number of televisions that need to be sold to maximize revenue.

(c) What is the maximum revenue that can be obtained from the sales of these televisions?

(d) If currently 35,000 televisions are sold, what is the total increase in revenue if sales increase to 40,000 units?

36. Cost analysis If the daily marginal cost of production of t-shirts has been measured to be $C'(x) = 0.06x - 6$ and fixed costs are known to be \$1000 per day,

(a) Find the cost function.

(b) Find the daily production that minimizes cost. What is the minimum daily cost of production?

(c) If the t-shirts are sold for \$10 each, find the revenue function.

(d) Find the daily break-even point.

(e) Find the daily sales volume that maximizes profit. What is the daily maximum profit?

In Problems 37–48, evaluate each definite integral.

37. $\int_{-2}^{1}(x^2+3x-1)\,dx$

38. $\int_{1}^{2}(x^3-1)\,dx$

39. $\int_{4}^{9}8\sqrt{x}\;dx$

40. $\int_{1}^{8}\sqrt[3]{x^2}\;dx$

41. $\int_{0}^{1}(e^x-e^{-x})\,dx$

42. $\int_{0}^{2}\frac{1}{x+2}\,dx$

43. $\int_{0}^{4}\frac{dx}{(3x+2)^2}$

44. $\int_{0}^{\sqrt{15}}\frac{x}{\sqrt{x^2+1}}\,dx$

45. $\int_{-2}^{2}e^{3x}\,dx$

46. $\int_{2}^{3}\frac{e^{1/x}}{x^2}\,dx$

47. $\int_{0}^{1}(x+2)e^{-x}\,dx$

48. $\int_{1}^{4}x\ln 3x\,dx$

In Problems 49–52, use the properties of the definite integral to evaluate each integral if it is known that

$$\int_0^5 f(x)\,dx = 3 \qquad \int_0^5 g(x)\,dx = 8 \qquad \int_5^9 f(x)\,dx = -2 \qquad \int_5^9 g(x)\,dx = 10$$

49. $\int_{0}^{9}f(x)\,dx$

50. $\int_{0}^{5}[2f(x)+g(x)]\,dx$

51. $\int_{9}^{5}g(x)\,dx$

52. $\int_{5}^{9}-3f(x)\,dx$

In Problems 53–62, find the area described. Be sure to sketch the graph first.

53. Under the graph of $f(x) = x^2 + 4$ from -1 to 2.

54. Under the graph of $f(x) = x^2 + 3x + 2$ from 0 to 3.

55. Under the graph of $f(x) = e^x + x$ from 0 to 1.

56. Under the graph of $f(x) = x^3 + 1$ from -1 to 2.

57. Enclosed by the graph of $f(x) = x^2 - x - 2$ and the x-axis from 0 to 3.

58. Enclosed by the graph of $f(x) = x^3 - 8$ and the x-axis from 0 to 3.

59. Enclosed by the graphs of $f(x) = x^2 - 4$ and $x + y = 2$.

60. Enclosed by the graphs of $f(x) = \sqrt{x}$ and $g(x) = \frac{1}{2}x$.

61. Enclosed by the graphs of $f(x) = x^3$ and $g(x) = 4x$.

62. Enclosed by the graphs of $f(x) = x^3$ and $g(x) = \sqrt[3]{x}$.

63. **Profit from Jeans** A clothing manufacturer produces x pairs of jeans per month. The company's marginal profit from the sale of the jeans is given by $P'(x) = 9 - 0.004x$. Currently the company is manufacturing 2000 pairs of jeans per month. Find the increased monthly profit if production increases to 2500 pairs.

64. **Total Maintenance Cost** Maintenance costs for a piece of machinery usually increase with the age of the machine. From experience, the production manager at a factory found that the rate of change of expense to maintain a particular type of machine is given by $f'(t) = 12t^2 + 2500$, where t is the age of the machine in years, and $f(t)$ is the total cost of maintenance.
(a) Find the total cost of maintenance for the years 0 through 5.
(b) Find the total cost of maintenance for years 5 through 10.
(c) Find the total cost of maintenance for the first 10 years of operation.
(d) If company policy dictates replacing the machine when total maintenance costs equal $100,000, what is the useful life of the machine?

65. **Useful Life of Vending Machines** It has been found that owning and operating a vending machine has an increasing cost rate and a decreasing revenue rate. At time t a particular machine generates a marginal revenue $R'(t) = -10t$, and has a marginal maintenance cost $C'(t) = 2t - 12$.
(a) Determine how long the owner should keep the machine to maximize profit.
(b) What is the total profit that the machine will generate in this time?

66. **Total Sales** The rate of sales of a certain product obeys
$$f(t) = 1340 - 850e^{-t}$$
where t is the number of years the product is on the market. Find the total sales during the first 5 years.

67. **Approximating Area** The function $f(x) = -2x + 10$ is defined on $[0, 4]$.
(a) Graph f.
Approximate the area A under f from 0 to 4 as follows:
(b) By partitioning $[0, 4]$ into 4 subintervals of equal length and choosing u as the left endpoint.
(c) By partitioning $[0, 4]$ into 4 subintervals of equal length and choosing u as the right endpoint.
(d) By partitioning $[0, 4]$ into 8 subintervals of equal length and choosing u as the left endpoint.
(e) By partitioning $[0, 4]$ into 8 subintervals of equal length and choosing u as the right endpoint.
(f) Express the area A as an integral.
(g) Evaluate the integral.

68. **Approximating Area** The function $f(x) = x + 3$ is defined on $[-1, 2]$.
(a) Graph f.
Approximate the area A under f from -1 to 2 as follows:
(b) By partitioning $[-1, 2]$ into 3 subintervals of equal length and choosing u as the left endpoint.
(c) By partitioning $[-1, 2]$ into 3 subintervals of equal length and choosing u as the right endpoint.
(d) By partitioning $[-1, 2]$ into 6 subintervals of equal length and choosing u as the left endpoint.
(e) By partitioning $[-1, 2]$ into 6 subintervals of equal length and choosing u as the right endpoint.
(f) Express the area A as an integral.
(g) Evaluate the integral.

69. Approximate $\int_1^{10} x \ln x \, dx$ with a graphing utility.

70. Approximate $\int_{-1}^{1} \sqrt{1 - x^2} \, dx$ with a graphing utility.

In Problems 71–76, solve each differential equation using the indicated boundary condition.

71. $\frac{dy}{dx} = x^2 + 5x - 10$; $y = 1$ when $x = 0$

72. $\frac{dy}{dx} = x^4 - 2x^2 + 1$; $y = 10$ when $x = 0$

73. $\frac{dy}{dx} = e^{2x} - x$; $y = 3$ when $x = 0$

74. $\frac{dy}{dx} = \frac{2}{2x + 5}$; $y = 0$ when $x = -2$

75. $\frac{dy}{dx} = 10y$; $y = 1$ when $x = 0$

76. $\frac{dy}{dx} = 15\sqrt{5x}$; $y = 100$ when $x = 5$

77. **Bacterial Growth** Bacteria grown in a certain culture increase at a rate proportional to the amount present. If there are 2000 bacteria present initially and the amount triples in 2 hours, how many bacteria will there be in $4\frac{1}{2}$ hours?

78. **Bugs in the House** A house becomes infected with fleas, whose population growth obeys the law of uninhibited growth. If initially a nest of 100 fleas hatched and after 2 days there are 500, how many fleas will be present in one week (7 days)?

79. **Prehistoric Civilization** An ancient burial ground is discovered in a countryside field. Carbon dating is done on the bones in the graves. It is found that 40% of the carbon-14 is remaining. How old is the burial ground? (Use 5600 years as the half-life of carbon-14.)

80. **Age of an Animal** The skeleton of an animal is found to contain 35% of the original amount of carbon-14. What is the approximate age of this animal? (Use 5600 years as the half-life of carbon-14).

81. **Expanding Economy** It is known that the economy of a certain developing nation is growing at a rate of $E'(t) = 0.02t^2 + t$, where t is time measured in years. This year the nation's economy totals 5 million dollars.
    (a) Find the function that measures the economy at time t.
    (b) Assuming the economy continues to expand at the current rate, what will it equal in 5 years?

82. **Total Cost** The marginal cost function for producing x units of a product is

$$C'(x) = 200 - \frac{1}{5}x$$

Find the increase in total cost if production increases from 100 to 150 units.

83. **Learning Curve** Margo Manufacturing introduced a new line of tennis ball servers equipped with electronic timing and delivery devices. It took the production crew 700 labor-hours to produce the first 50 servers, but production time is decreasing according to the learning curve

$$f(x) = 700\,x^{-0.1}$$

where $f(x)$ is the labor-hours needed to produce the xth production run of 50 servers and was determined after 8 production runs were completed. At that time a large order for 500 servers was received. How much time should Margo allow for labor to produce the additional 10 production runs of the tennis ball servers?

84. **Total Revenue** The marginal revenue function for selling x units in a week is

$$R'(x) = 50 - x$$

Find the increase in total revenue attained if sales increase from 30 to 40 units per week.

85. **Consumer's Surplus; Producer's Surplus** The demand curve is $D(x) = 12 - \dfrac{x}{50}$ and the supply curve is $S(x) = \dfrac{x}{20} + 5$.
    (a) Find the market price and the demand level of market equilibrium.
    (b) Find the consumer's surplus and the producer's surplus.
    (c) Sketch the graphs.

86. **Consumer's Surplus; Producer's Surplus** The demand curve is $D(x) = 20\,e^{-0.01x}$ and the supply curve is $S(x) = 0.005x^2$.
    (a) Find the market price and the demand level of market equilibrium (rounded to the nearest dollar).
    (b) Find the consumer's surplus and the producer's surplus.
    (c) Sketch the graphs.

87. **Pricing a Product** A manufacturer finds that the marginal price of one of its products is proportional to the price. The manufacturer also knows that at a price of \$800 per unit, no units are sold. At a price of \$600, a total of 80 units are sold.
    (a) Find the price-demand equation.
    (b) Find the price p that should be charged if the manufacturer wants to sell 200 units.
    (c) What price should be charged if the manufacturer wants to sell 250 units?

Chapter 15 Project

INEQUALITY OF INCOME DISTRIBUTION

The fact that income is distributed unequally in our economy is pretty obvious. Yet how can this inequality be measured? Certainly it would be useful to track some measure of inequality over time to see whether the inequality was getting better or worse. One way of measuring inequality of income distribution is to construct a function known as a **Lorenz curve.** To find this curve, we first rank every household in order of increasing income. The point $\left(\dfrac{x}{100}, \dfrac{x}{100}\right)$ is on the Lorenz curve if the bottom x % of the households receives y% of the income. For example, since the bottom 0% of the households receives 0% of the income and the bottom 100%

of the households receive 100% of the income, the points (0, 0) and (1, 1) must lie on every Lorenz curve. If income distribution were perfectly equal, then the bottom x % of the households would receive x% of the income, and the point (x, x) would be on the the Lorenz curve for any x. That is, if income distribution were perfectly equal, the Lorenz curve would be the line $y = x$.

But the income distribution in the American economy is not equal. According to the U.S. Census Bureau, household income in the United States in 2001 was distributed according to the following table:

2001 Percent of Income Earned by Bottom . . .				
20%	40%	60%	80%	95%
3.5%	12.2%	26.8%	49.8%	77.6%

Source: U.S. Census Bureau.

The Lorenz curve for 2001 contains the points (0, 0), (0.2, 0.035), (0.4, 0.122), (0.6, 0.268), (0.8, 0.498), (0.95, 0.776), and (1,1). We do not know other points on the Lorenz curve, but we can play "connect the dots" to get a realistic Lorenz curve:

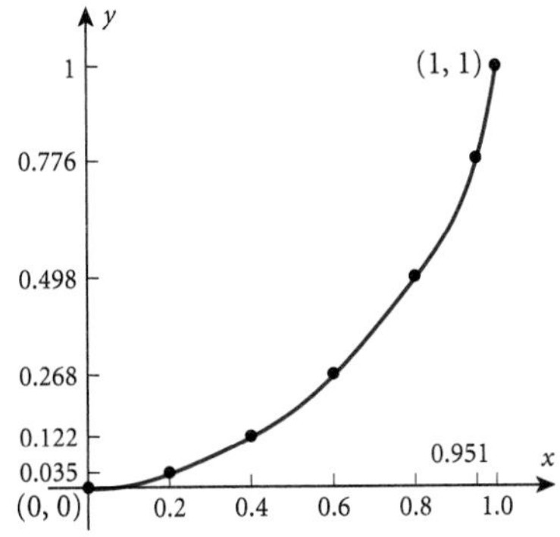

In fact, we can approximate this Lorenz curve by the polynomial function

$$L(x) = 0.297x^2 + 7.053x^3 - 27.988x^4 + 36.115x^5 + 6.87x^6 - 45.13x^7 + 23.783x^8.$$

Once we have the Lorenz curve, we can produce a number, called the **Gini coefficient,** which measures the inequality in the distribution of income. The Gini coefficient is defined to be the area between the line $y = x$ and the Lorenz curve $y = L(x)$ divided by the area under the line $y = x$. Using the figure, the Gini coefficient is

$$G = \frac{\text{Area of the green region}}{\text{Area of the green region} + \text{Area of the blue region}}$$

1. Express the area of the green region plus the area of the blue region as an integral.
2. Express the area of the blue region as an integral.

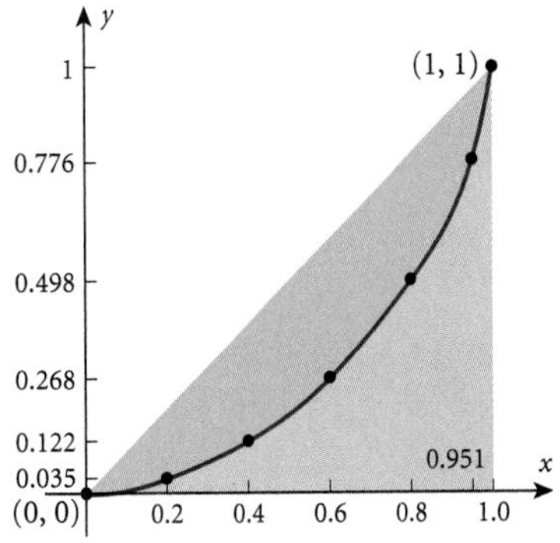

3. Show that the Gini coefficient is $G = 1 - 2\int_0^1 L(x)\,dx$.
4. Use the formula found in Problem 3 to find the Gini coefficient for the U.S. economy in 2001.
5. In 1997, income in the United States was distributed according to the following table.

1977 Percent of Income Earned by Bottom . . .				
20%	40%	60%	80%	95%
3.6%	12.5%	27.5%	50.7%	78.3%

In 1993, income in the United States was distributed according to the following table.

1993 Percent of Income Earned by Bottom . . .				
20%	40%	60%	80%	95%
3.6%	12.6%	27.7%	51.2%	79.0%

Just by looking at the distributions in the tables for 1993, 1997, and 2001, does it look like incomes are becoming less equally distributed or more equally distributed over this time span?

6. The Lorenz curve for 1997 may be approximated by

$$L(x) = 0.423x^2 + 6.083x^3 - 25.006x^4 + 32.938x^5 + 5.920x^6 - 41.316x^7 + 21.958x^8.$$

Find the Gini coefficient for the U.S. economy in 1997.

7. The Lorenz curve for 1993 may be approximated by

$$L(x) = .442x^2 + 5.8x^3 - 23.71x^4 + 31.036x^5 + 5.71x^6 - 38.842x^7 + 20.564x^8.$$

Find the Gini coefficients for the U.S. economy in 1993.

8. Compare the Gini coefficients you found in Problems 4, 6, and 7. Are incomes becoming less equally distributed or more equally distributed? Explain.

MATHEMATICAL QUESTIONS FROM PROFESSIONAL EXAMS*

1. Actuary Exam—Part I $\int_1^e \frac{1}{x} \ln x \, dx = ?$

(a) $\frac{1}{e}$ (b) $\frac{1}{2}$ (c) 1
(d) e (e) e^2

2. Actuary Exam—Part I $\int_0^1 x \ln x \, dx = ?$

(a) $-\infty$ (b) -2 (c) -1
(d) $-\frac{1}{4}$ (e) $-\frac{2}{9}$

3. Actuary Exam—Part I If $\int_1^b f(x)\, dx = b^2e^b - e$ for all $b > 0$, then for all $x > 0$, $f(x) = ?$

(a) x^2e^x (b) $\frac{x^3}{3}e^x$ (c) $x^2e^x + 2xe^x$
(d) $2xe^x$ (e) $x^2e^x - e^{x-1}$

4. Actuary Exam—Part I If the area of the region bounded by $y = f(x)$, the x-axis, and the lines $x = a$ and $x = b$ is given by $\int_a^b f(x)\, dx$, which of the following must be true?

(a) $a < b$ and $f(x) > 0$ (b) $a < b$ and $f(x) < 0$
(c) $a > b$ and $f(x) > 0$ (d) $a > b$ and $f(x) < 0$
(e) None of the above

5. Actuary Exam—Part I The rate of change of the population of a town in Pennsylvania at any time t is proportional to the population at time t. Four years ago, the population was 25,000. Now, the population is 36,000. Calculate what the population will be six years from now.

(a) 43,200 (b) 52,500 (c) 62,208
(d) 77,760 (e) 89,580

6. Actuary Exam—Part I For $x > 0$, $\int \frac{\ln (x^2)}{x} dx = ?$

(a) $\ln x^2 - \ln x + C$ (b) $\frac{1}{2}(\ln x)^2 + C$
(c) $(\ln x)^2 + C$ (d) $\ln(\ln x) + C$
(e) $\frac{1}{2}\ln(\ln x) + C$

CHAPTER 16

Other Applications and Extensions of the Integral

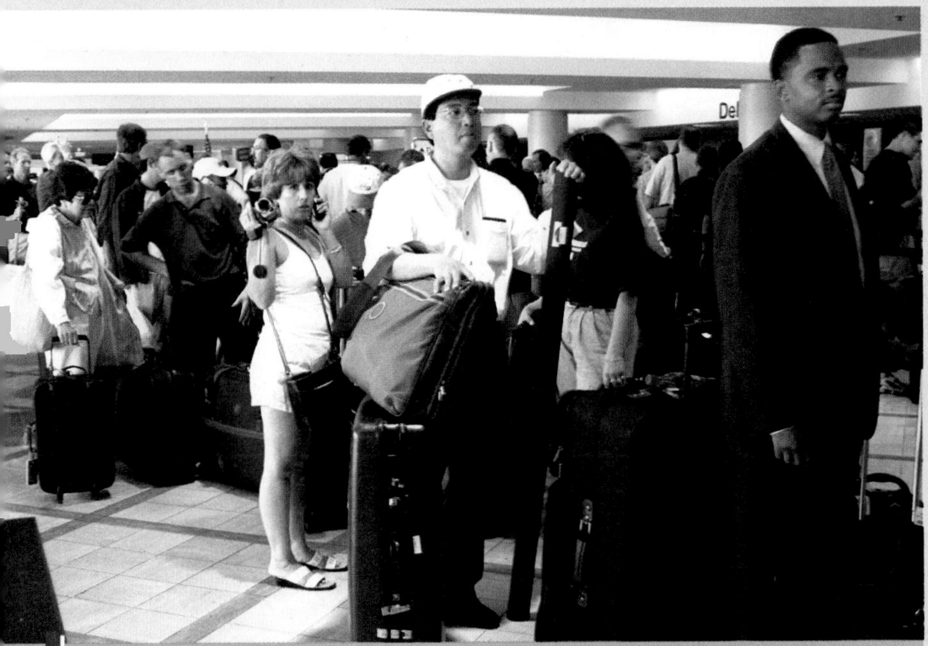

OUTLINE

It's Thanksgiving break and you are headed for the airport to fly home for turkey dinner. When you get to the airport, the check-in lines are backed up. As you wait in line, you wonder whether the number of people arriving equals the number that have been waited on. Or are the arrivals coming faster than they can be processed, making the lines even longer? If the airline knew how many people to expect in a given period of time, then additional employees could be used to handle high demand and fewer employees would be used when demand was low. How could the airline find out? The Chapter Project gives some insight into arrivals and how to predict waiting time.

A LOOK BACK, A LOOK FORWARD

In Chapter 15 we discussed the integral of a function, defining the indefinite integral and the definite integral. We gave applications in geometry (finding the area under the graph of a function), in business (marginal analysis, sales over time, learning curves, and maximizing profit over time), and in economics (consumer's surplus and producer's surplus). We also used integration to solve differential equations.

In this chapter we give additional applications of the integral, beginning with a discussion of improper integrals, an extension of the definite integral.

16.1 Improper Integrals

PREPARING FOR THIS SECTION *Before getting started, review the following:*

- > The Definite Integral (Chapter 15, Section 15.4, pp. 882–886)
- > Infinite Limits; Limits at Infinity (Chapter 12, Section 12.4, pp. 698–700)
- > Continuous Functions (Chapter 12, Section 12.3, pp. 691–695)

OBJECTIVE 1 Evaluate improper integrals

Recall that in defining $\int_a^b f(x)\,dx$ two basic assumptions are made:

1. The limits of integration a and b are both finite.
2. The function is continuous on $[a, b]$.

In many situations, one or both of these assumptions are not met. For example, one of the limits of integration might be infinity; or the function $y = f(x)$ might be discontinuous at some number in $[a, b]$. If either of the conditions (1) and (2) are not satisfied, then $\int_a^b f(x)\,dx$ is called an **improper integral.**

One Limit of Integration Is Infinite

We begin with an example.

EXAMPLE 1 Finding an Area

Find the area under the graph of $f(x) = \dfrac{1}{x^2}$ to the right of $x = 1$.

SOLUTION First we graph $f(x) = \dfrac{1}{x^2}$. See Figure 1. The area to the right of $x = 1$ is shaded. To find this area, we pick a number b to the right of $x = 1$. The area under the graph of $f(x) = \dfrac{1}{x^2}$ from $x = 1$ to $x = b$ is

FIGURE 1

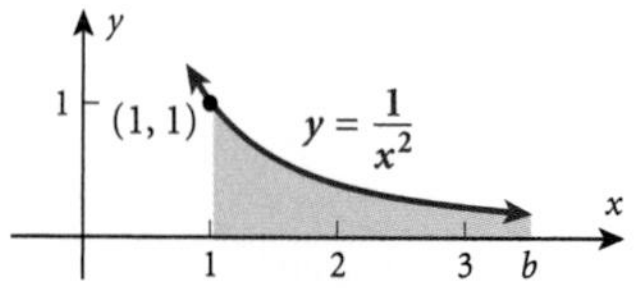

$$\int_1^b \frac{1}{x^2}\,dx = \left(-\frac{1}{x}\right)\Big|_1^b = -\frac{1}{b} + 1$$

This area depends on the choice of b. Now the area we seek is obtained by letting $b \to \infty$. Since

$$\lim_{b\to\infty}\left(-\frac{1}{b} + 1\right) = 1$$

we conclude that the area under the graph of $f(x) = x^2$ to the right of $x = 1$ is 1.

The area we found in Example 1 can be represented symbolically by the *improper integral*

$$\int_1^\infty \frac{1}{x^2}\,dx$$

and it can be evaluated by finding

$$\lim_{b\to\infty} \int_1^b \frac{1}{x^2}\,dx$$

This leads us to formulate the following definition.

Improper Integral

Suppose a function f is continuous on the interval $[a, \infty)$. The **improper integral,** $\int_a^\infty f(x)\,dx$, is defined as

$$\int_a^\infty f(x)\,dx = \lim_{b\to\infty} \int_a^b f(x)\,dx$$

provided this limit exists and is a real number. If this limit does not exist or if it is infinite, the improper integral has no value.

Suppose a function f is continuous on the interval $(-\infty, b]$. The **improper integral,** $\int_{-\infty}^b f(x)\,dx$, is defined as

$$\int_{-\infty}^b f(x)\,dx = \lim_{a\to-\infty} \int_a^b f(x)\,dx$$

provided that this limit exists and is a real number. If this limit does not exist or if it is infinite, the improper integral has no value.

1 **EXAMPLE 2** **Evaluating Improper Integrals**

Find the value, if there is one, of

(a) $\int_{-\infty}^0 e^x\,dx$ **(b)** $\int_1^\infty \frac{1}{x}\,dx$

SOLUTION **(a)**

$$\int_{-\infty}^0 e^x\,dx = \lim_{a\to-\infty} \int_a^0 e^x\,dx = \lim_{a\to-\infty} e^x\Big|_a^0$$

$$= \lim_{a\to-\infty} [1 - e^a] = 1 - \lim_{a\to-\infty} e^a = 1$$

$\uparrow$ $\lim_{a\to-\infty} e^a = 0$

(b)

$$\int_1^\infty \frac{1}{x}\,dx = \lim_{b\to\infty} \int_1^b \frac{1}{x}\,dx = \lim_{b\to\infty} \ln x\Big|_1^b = \lim_{b\to\infty} [\ln b - \ln 1] = \lim_{b\to\infty} \ln b = \infty$$

Since the limit is infinite, $\int_1^{\infty} \frac{1}{x}\, dx$ has no value.

NOW WORK PROBLEM 7.

The Integrand Is Discontinuous

The improper integrals studied in the above examples had infinity either as a lower limit of integration or as an upper limit of integration. A second type of improper integral occurs when the integrand f in $\int_a^b f(x)\, dx$ is discontinuous at either a or b, where a and b are both real numbers. Again, we illustrate the technique for evaluating this type of improper integral by an example.

EXAMPLE 3 Evaluating an Improper Integral

Evaluate: $\int_0^4 \frac{1}{\sqrt{x}}\, dx$

SOLUTION The function $f(x) = \frac{1}{\sqrt{x}}$ is not continuous at $x = 0$, but is continuous on the interval $(0, \infty)$. To evaluate the improper integral $\int_0^4 \frac{1}{\sqrt{x}}\, dx$ we proceed as follows:

$$\int_0^4 \frac{1}{\sqrt{x}}\, dx = \lim_{t \to 0^+} \int_t^4 x^{-1/2}\, dx = \lim_{t \to 0^+} \left. \frac{x^{1/2}}{\frac{1}{2}} \right|_t^4$$

$$= \lim_{t \to 0^+} (2\sqrt{4} - 2\sqrt{t}) = 4 - \lim_{t \to 0^+} 2\sqrt{t} = 4$$

This leads to the following definition.

Improper Integral

If f is continuous on $[a, b)$, but is discontinuous at b, we define the **improper integral,** $\int_a^b f(x)\, dx$, to be

$$\int_a^b f(x)\, dx = \lim_{t \to b^-} \int_a^t f(x)\, dx$$

provided that this limit exists and is a real number. If this limit does not exist or if it is infinite, then $\int_a^b f(x)\, dx$ has no value.

If f is continuous on $(a, b]$, but is discontinuous at a, we define the **improper integral,** $\int_a^b f(x)\, dx$, to be

$$\int_a^b f(x)\, dx = \lim_{t \to a^+} \int_t^b f(x)\, dx$$

provided that this limit exists as a real number. If this limit does not exist or if it is infinite, then $\int_a^b f(x)\,dx$ has no value.

Observe that when the upper limit is the point of discontinuity, the limit is left-handed and when the lower limit is the point of discontinuity, the limit is right-handed.

NOW WORK PROBLEM 13.

EXAMPLE 4 Finding an Area

Find the area A between the graph of $f(x) = \dfrac{1}{\sqrt[3]{x}}$ and the x-axis from $x = -1$ to $x = 0$.

SOLUTION We begin by graphing $f(x) = \dfrac{1}{\sqrt[3]{x}}$. See Figure 2. Since the area lies below the x-axis, the area is $A = -\int_{-1}^{0} \dfrac{1}{\sqrt[3]{x}}\,dx$.

FIGURE 2

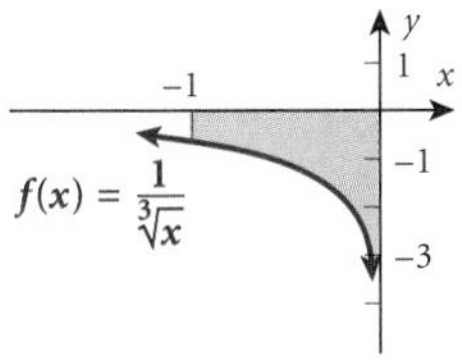

We observe that the integrand $f(x) = \dfrac{1}{\sqrt[3]{x}}$ is discontinuous at $x = 0$ (the upper limit of integration). So,

$$A = -\int_{-1}^{0} \frac{1}{\sqrt[3]{x}}\,dx = -\lim_{t\to 0^-} \int_{-1}^{t} \frac{1}{\sqrt[3]{x}}\,dx = -\lim_{t\to 0^-} \int_{-1}^{t} x^{-1/3}\,dx$$

$$= -\lim_{t\to 0^-} \left. \frac{x^{2/3}}{\frac{2}{3}} \right|_{-1}^{t} = -\lim_{t\to 0^-} \frac{3}{2}(t^{2/3} - 1) = \frac{3}{2}$$

The area is $\frac{3}{2}$ square units.

There are other types of improper integrals besides those discussed here. See Problems 20 and 21.

EXERCISE 16.1 Answers to Odd-Numbered Problems Begin on Page AN-98.

In Problems 1–6, determine which of the integrals are improper. Explain why they are, or are not, improper.

1. $\int_0^{\infty} x^2\,dx$ **2.** $\int_2^3 \dfrac{dx}{x-1}$ **3.** $\int_0^1 \dfrac{1}{x}\,dx$ **4.** $\int_{-1}^{1} \dfrac{x}{x^2+1}\,dx$ **5.** $\int_1^2 \dfrac{dx}{x-1}$ **6.** $\int_0^1 \dfrac{x}{x^2-1}\,dx$

In Problems 7–14, find the value, if any, of each improper integral.

7. $\int_1^{\infty} e^{-4x}\,dx$ **8.** $\int_{-\infty}^{-1} \dfrac{1}{x^3}\,dx$ **9.** $\int_0^{\infty} \sqrt{x}\,dx$ **10.** $\int_0^{\infty} xe^{-x}\,dx$

11. $\int_{-1}^{0} \dfrac{1}{\sqrt[5]{x}}\,dx$ **12.** $\int_2^4 \dfrac{x\,dx}{\sqrt{x^2-4}}$ **13.** $\int_0^1 \dfrac{1}{x}\,dx$ **14.** $\int_0^1 \dfrac{\ln x}{x}\,dx$

15. Find the area, if it exists, under the graph of $f(x) = \dfrac{1}{\sqrt{x}}$ from $x = 0$ to $x = 1$.

16. Find the area, if it exists, under the graph of $f(x) = \sqrt{x}$ to the right of $x = 0$.

17. Capital Value of Rental Property The capital value of a rental property, assuming it will last indefinitely, is given by the integral $\int_0^{\infty} Re^{-kt}\, dt$, where R is the annual rent and k is the current annual rate of interest on investments. Find the capital value of a typical apartment in the South Center Township area of Indianapolis, where annual rent is \$5124 and the current rate of interest on investments is 5%.

Source: CB Richard Ellis-Indianapolis.

18. Waiting Time The probability of waiting at least x minutes for Amtrak trains to arrive in Washington, D.C. between the hours of 7:30 A.M. and 1:30 A.M. is given by the integral $\int_x^{\infty} \frac{7}{3} e^{-7/3t}\, dt$, where x is measured in hours. Determine the probability that one will have to wait at least one hour for an Amtrak train to arrive.

Source: Amtrak.

19. Reaction Rate of a Drug The rate of reaction to a given dose of a drug at time t hours after administration is given by $r(t) = te^{-t^2}$ (measured in appropriate units).

(a) Why is it reasonable to define the **total reaction** as the area under the curve $y = r(t)$ from $t = 0$ to $t = \infty$?

(b) Evaluate the total reaction to the given dose of the drug.

20. If $y = f(x)$ is continuous, the improper integral $\int_{-\infty}^{\infty} f(x)\, dx$ is defined as

$$\int_{-\infty}^{\infty} f(x)\, dx = \int_{-\infty}^{0} f(x)\, dx + \int_{0}^{\infty} f(x)\, dx$$

provided that each of the improper integrals on the right has a value. Use this definition to find the value, if it exists, of

(a) $\int_{-\infty}^{\infty} e^x\, dx$ (b) $\int_{-\infty}^{\infty} \frac{x\, dx}{(x^2+1)^2}$

21. If $y = f(x)$ is continuous on $[a, b]$, except at a point c, $a < c < b$, the integral $\int_a^b f(x)\, dx$ is improper and is defined by

$$\int_a^b f(x)\, dx = \int_a^c f(x)\, dx + \int_c^b f(x)\, dx$$

provided that each of the improper integrals on the right has a value. Use this definition to evaluate

(a) $\int_{-1}^{1} \frac{1}{x^2}\, dx$ (b) $\int_0^4 \frac{x\, dx}{\sqrt[3]{x^2 - 4}}$

16.2 Average Value of a Function

PREPARING FOR THIS SECTION ***Before getting started, review the following:***

>> Approximating Definite Integrals (Chapter 15, Section 15.6, pp. 903–908)

OBJECTIVES
1. Find the average value of a function
2. Use average value in applications

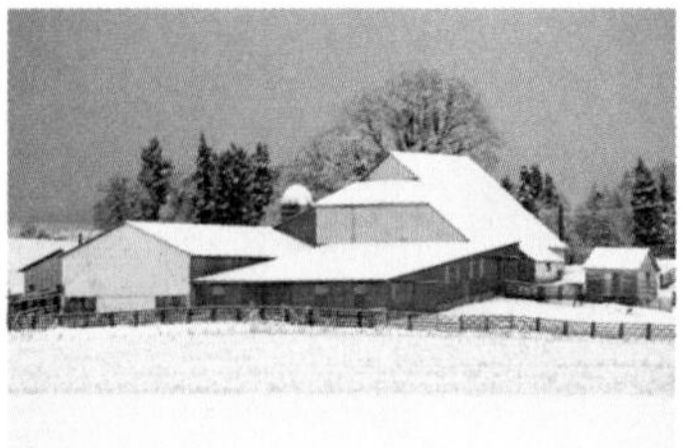

At the U.S. Weather Bureau, a continuous reading of the temperature over a 24-hour period is taken daily. To obtain the average daily temperature, 12 readings may be taken at 2-hour intervals beginning at midnight (0) and ending at 10 p.m. (22): $f(0), f(2), f(4), \ldots, f(20), f(22)$. The average temperature is then calculated as

$$\frac{f(0) + f(2) + f(4) + \cdots + f(20) + f(22)}{12}$$

This number represents a good approximation to the true average as long as there were no drastic temperature changes over any of the 2-hour intervals.

To improve the approximation, readings could be taken every hour. The average in this case would be

$$\frac{f(0) + f(1) + \cdots + f(22) + f(23)}{24}$$

An even better approximation would be obtained if readings were recorded every half hour.

In general, if $y = f(x)$ is a continuous function defined on the closed interval $[a, b]$, we can obtain the *average of f on* $[a, b]$ as follows:

Partition the interval $[a, b]$ into n subintervals

$$[a, x_1], \quad [x_1, x_2], \ldots, [x_{k-1}, x_k], \ldots, [x_{n-1}, b]$$

each of equal length $\Delta x = \dfrac{b - a}{n}$. This is the norm Δ of the partition. Pick a number in each subinterval and let these numbers be $u_1, u_2, \ldots, u_n$. An approximation of the average value of f over the interval $[a, b]$ is then the sum

$$\frac{f(u_1) + f(u_2) + \cdots + f(u_n)}{n} \tag{1}$$

If we multiply and divide the expression in (1) by $b - a$, we get

$$\frac{f(u_1) + f(u_2) + \cdots + f(u_n)}{n} = \frac{1}{b - a}\left[f(u_1)\frac{b - a}{n} + f(u_2)\frac{b - a}{n} + \cdots + f(u_n)\frac{b - a}{n}\right]$$

$$= \frac{1}{b - a}[f(u_1)\,\Delta x + f(u_2)\,\Delta x + \cdots + f(u_n)\,\Delta x]$$

This sum gives an approximation to the average value. As the norm $\Delta \to 0$, it provides a better and better approximation to the average value of f on $[a, b]$. Since this sum is a Riemann sum, its limit is a definite integral. This suggests the following definition:

Average Value of a Function over an Interval

The **average value** AV of a continuous function f over the interval $[a, b]$ is

$$AV = \frac{1}{b - a}\int_a^b f(x)\,dx \tag{2}$$

1 **EXAMPLE 1 Finding the Average Value of a Function**

The average value of $f(x) = x^3$ over the interval $[0, 2]$ is

$$AV = \frac{1}{2 - 0}\int_0^2 x^3\,dx = \frac{1}{2}\,\frac{x^4}{4}\bigg|_0^2 = \frac{1}{2}(4) = 2$$

NOW WORK PROBLEM 1.

Geometric Interpretation

The average value AV of a function f, as defined in (2), has an interesting geometric interpretation. If we rearrange the formula for AV, we obtain

$$(AV)(b - a) = \int_a^b f(x)\,dx \tag{3}$$

If $f(x) \geq 0$ on $[a, b]$, the right side of (3) represents the area under the graph of $y = f(x)$, from $x = a$ to $x = b$. The left side of the equation can be interpreted as the

FIGURE 3

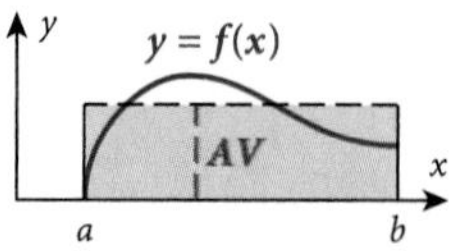

area of a rectangle of height AV and base $b - a$. As a result, equation (3) asserts that the average value of the function is the height of a rectangle with base $b - a$ and area equal to the area under the graph of f. See Figure 3.

Application

2 **EXAMPLE 2 Using the Average Value of a Function**

Suppose the current world population is $6 \cdot 10^9$ and the population in t years is assumed to grow exponentially at a 2% growth rate according to the law

$$P(t) = (6 \cdot 10^9)e^{0.02t}$$

What will be the average world population during the next 20 years?

SOLUTION The average value of the population during the next 20 years is

$$\begin{aligned} AV &= \frac{1}{20 - 0}\int_0^{20} P(t)\, dt \\ &= \frac{1}{20}\int_0^{20} (6 \cdot 10^9)e^{0.02t}\, dt \\ &= \frac{6 \cdot 10^9}{20}\int_0^{20} e^{0.02t}\, dt \\ &= 3 \cdot 10^8 \frac{e^{0.02t}}{0.02}\Bigg|_0^{20} \\ &= 3 \cdot 10^8 \frac{(e^{0.4} - 1)}{0.02} \\ &\approx 7.38 \cdot 10^9 \end{aligned}$$

NOW WORK PROBLEM 11.

EXERCISE 16.2 Answers to Odd-Numbered Problems Begin on Page AN-98.

In Problems 1–10, find the average value of the function f over the given interval.

1. $f(x) = x^2$, over $[0, 1]$
2. $f(x) = 2x^2$, over $[-4, 2]$
3. $f(x) = 1 - x^2$, over $[-1, 1]$
4. $f(x) = 16 - x^2$, over $[-4, 4]$
5. $f(x) = 3x$, over $[1, 5]$
6. $f(x) = 4x$, over $[-5, 5]$
7. $f(x) = -5x^4 + 4x - 10$, over $[-2, 2]$
8. $f(x) = 10x^4 - 2x + 7$, over $[-1, 2]$
9. $f(x) = e^x$, over $[0, 1]$
10. $f(x) = e^{-x}$, over $[0, 1]$

11. **Population Prediction** Rework Example 2 if the population function is given by $P(t) = (6 \cdot 10^9)e^{0.03t}$. (This is a 3% growth rate.)

12. **Population Prediction** Rework Example 2 if the growth rate is 1%.

13. **Average Temperature of a Rod** A rod 3 meters long is heated to $25x$ degrees Celsius, where x is the distance (in meters) from one end of the rod. Calculate the average temperature of the rod.

14. **Average Rainfall** The rainfall per day, measured in centimeters, x days after the beginning of the year is $0.00002(6511 + 366x - x^2)$. Estimate the average daily rainfall for the first 180 days of the year.

15. **Average Speed** A car starting from rest accelerates at the rate of 3 meters per second per second. Find its average speed over the first 8 seconds.

16. **Average Area** What is the average area of all circles with radii between 1 and 3 meters?

17. Average Annual Revenue The annual revenue for the Exxon-Mobil Corporation between 1997 and 2002 is given by $R(x) = -4.43x^3 + 46.17x^2 - 132.5x + 290$, where x is the number of years since 1996 and $R(x)$ is in billions of dollars. Find the average annual revenue during that time.

Source: Exxon-Mobil Corporation Financial Report 2002.

18. Deaths from Automobile Accidents Find the average number of annual deaths from automobile accidents in the United States from the period 1990–1999 if the annual number of deaths is given by $y = 38929 + 1443 \ln x$, where x is the number of years since 1989.

Source: National Highway Traffic Safety Administration's National Center for Statistical Analysis.

19. Average Rainfall in Baton Rouge For the first 90 days of the year the average daily rainfall in Baton Rouge, Louisiana is given by $r(x) = -0.000414x + 0.206748$, with $r(x)$ in inches and x the number of days since the beginning of the year. Find the average rainfall per day for the first 90 days of the year.

Source: National Weather Service.

16.3 Continuous Probability Functions

PREPARING FOR THIS SECTION *Before getting started, review the following:*

>> Area Under a Graph (Chapter 15, Section 15.5, pp. 890–897)

OBJECTIVES
1. Verify probability density functions
2. Find probabilities using a probability density function
3. Use the uniform density function
4. Use the exponential density function
5. Find expected value

TABLE 1

	Number of Heads	Probability
HHH	3	$\frac{1}{8}$
HHT	2	$\frac{1}{8}$
HTH	2	$\frac{1}{8}$
THH	2	$\frac{1}{8}$
HTT	1	$\frac{1}{8}$
THT	1	$\frac{1}{8}$
TTH	1	$\frac{1}{8}$
TTT	0	$\frac{1}{8}$

Random Variables

Intuitively, a *random variable* is a quantity that is measured in connection with a random experiment. For example, if the random experiment involves weighing individuals, then the weights of the individuals would be random variables. As another example, if the random experiment is to determine the time between arrivals of customers at a gas station, then the time between arrivals of customers at the gas station would be a random variable.

Let's consider some examples that demonstrate how to obtain random variables from random experiments.

When we perform a simple experiment, we are often interested not in a particular outcome, but rather in some number associated with that outcome. For example, in tossing a coin three times, we may be interested in the number of heads obtained, regardless of the particular sequence in which the heads appear. Similarly, the gambler throwing a pair of dice is generally more interested in the sum of the faces than in the particular number on each face.

Table 1 summarizes the results of the simple experiment of flipping a fair coin three times. The first column in the table gives a sample space of this experiment. The second column shows the number of heads for each outcome, and the third column shows the probability associated with each outcome.

Suppose that in this experiment we are interested only in the total number of heads. This information is given in Table 2.

The role of the random variable is to transform the original sample space {*HHH, HHT, HTH, HTT, THH, THT, TTH, TTT*} into a new sample space that consists of the

TABLE 2

Number of Heads Obtained in Three Flips of a Coin	Probability
0	$\frac{1}{8}$
1	$\frac{3}{8}$
2	$\frac{3}{8}$
3	$\frac{1}{8}$

number of heads that occur: {0, 1, 2, 3}. If X denotes the random variable, then X may take on any of the values 0, 1, 2, 3. From Table 2, the probability that the random variable X assumes the value 2 is

$$\text{Probability } (X = 2) = \frac{3}{8}$$

Also,

$$\text{Probability } (X = 5) = 0$$

We see that a random variable indicates the rule of correspondence between any member of a sample space and a number assigned to it. Because of this, a random variable is a function.

Random Variable

A **random variable** is a function that assigns a numerical value to each outcome of a sample space S.

We shall use the capital letter X to represent a random variable. In the coin-flipping example, the random variable X is

$$X(HHH) = 3 \quad X(HHT) = 2 \quad X(HTH) = 2 \quad X(THH) = 2$$
$$X(HTT) = 1 \quad X(THT) = 1 \quad X(TTH) = 1 \quad X(TTT) = 0$$

The random variable X indicates a relationship between the first two columns of Table 1 and pairs each outcome of the experiment with the real numbers 0, 1, 2, or 3.

Random variables fall into two classes: those related to *discrete sample spaces* and those associated with *continuous sample spaces.*

Discrete Random Variable

A sample space is discrete if it contains a finite number of outcomes or as many outcomes as there are counting numbers. A random variable is said to be **discrete** if it is defined over a discrete sample space.

Continuous Random Variable

Whenever a random variable has values that consist of an entire interval of real numbers, it is called a **continuous random variable.** In such cases, we also say the sample space is **continuous.**

Any practical problem that measures variables such as height, weight, time, and age will utilize a continuous random variable. As a result, the sample space associated with such experiments is continuous.

Probability Functions

The function that has as its domain the value of a random variable and has as its range the corresponding probability, is called a **probability function.**

For example, look again at Table 2. The left column contains values of the random variable and the right column contains the corresponding probability. The probability function f for this experiment has domain $\{0, 1, 2, 3\}$ and range $\left\{\frac{1}{8}, \frac{3}{8}\right\}$. Moreover,

$$f(0) = Pr(X = 0) = \frac{1}{8} \qquad f(1) = Pr(X = 1) = \frac{3}{8}$$

$$f(2) = Pr(X = 2) = \frac{3}{8} \qquad f(3) = Pr(X = 3) = \frac{1}{8}$$

EXAMPLE 1 Finding the Value of a Probability Function

In the experiment of one toss of two fair dice, compute the value of the probability function f at $x = 7$ where x is the sum of the dots shown on the faces of the dice. Construct a table that shows every value of f.

SOLUTION The domain of the probability function f for this experiment is $\{2, 3, 4, 5, 6, 7, 8, 9, 10, 11, 12\}$. The value of f at 7 equals the probability that the sum of the faces equals 7. That is,

$$f(7) = Pr(X = 7) = Pr\{(1, 6), (2, 5), (3, 4), (4, 3), (5, 2), (6, 1)\} = \frac{6}{36} = \frac{1}{6}$$

We compute the values $f(2), f(3), \ldots, f(12)$ of f in a similar way. These values are given in the following table:

Values of x	2	3	4	5	6	7	8	9	10	11	12
Probability Function of x, $f(x)$	$\frac{1}{36}$	$\frac{2}{36}$	$\frac{3}{36}$	$\frac{4}{36}$	$\frac{5}{36}$	$\frac{6}{36}$	$\frac{5}{36}$	$\frac{4}{36}$	$\frac{3}{36}$	$\frac{2}{36}$	$\frac{1}{36}$

We distinguish between two types of probability function. A **discrete probability function** is one for which the random variable is discrete. A **continuous probability function** is one for which the random variable is continuous. We shall discuss only continuous probability functions here.

Probability Density Functions

FIGURE 4

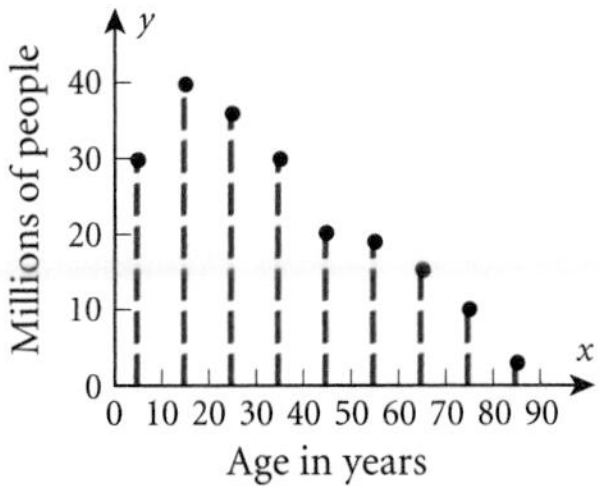

Continuous probability functions have as their domain the values that a continuous random variable can assume. For example, if a random experiment involves weighing individuals, the weight of each individual is a continuous random variable. Likewise, in the random experiment of determining the time between arrival of customers at a gas station, the time between arrivals is a continuous random variable.

Suppose the distribution of a population by age is given by data grouped in 10-year intervals. See Figure 4.

In this illustration there are 30 million people in the age group between 0 and 10, 40 million between 10 and 20, and so on. We also know that the total population is 200 million.

FIGURE 5

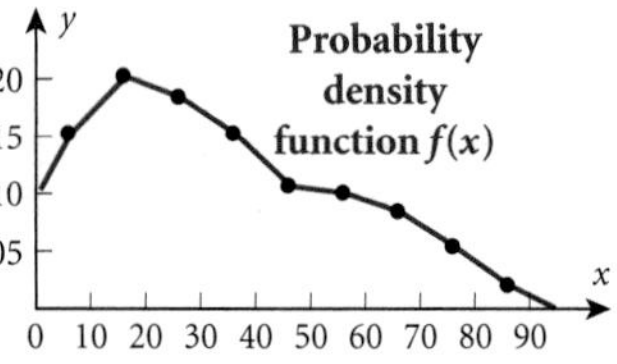

FIGURE 6

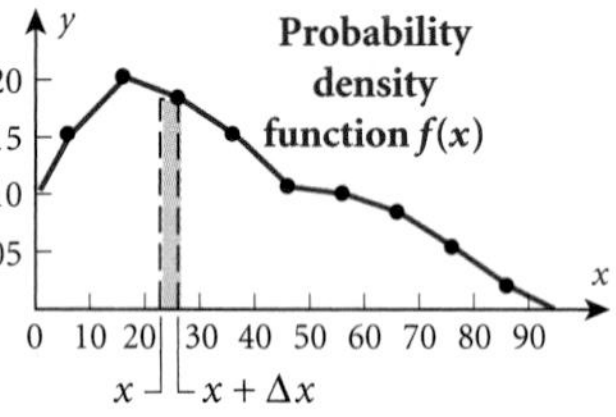

Since there are 40 million people in the age group 10–20, the probability that a person chosen at random is in this age group is $\frac{40}{200} = .20$. Figure 5 illustrates the distribution of probabilities for each age group. The function constructed by connecting the probability values by a smooth curve is an example of a *probability density function.*

When probabilities are associated with intervals, it is reasonable to assume that their values depend not only on the lengths of the intervals but also on their locations. For instance, there is no reason why the probability that a person is in the age group 10–20 should equal the probability that a person is in the age group 60–70, even though the two intervals have the same length. If we assume that there exists a function f with the values $f(x)$, then the probability that a person will be in a certain age group on a small interval from x to $x + \Delta x$ is approximately $f(x)\,\Delta x$. This approximation is given by the area of the shaded rectangle shown in Figure 6, where the height of the rectangle is $f(x)$ and the base is Δx.

In a similar manner, we obtain the probabilities of other age groups by computing the areas corresponding to different subintervals. The desired probability for the whole interval is approximately the sum of the areas

$$f(x_1)\,\Delta x + f(x_2)\,\Delta x + \cdots + f(x_n)\,\Delta x$$

where x_1 is in the first subinterval, x_2 is in the second, and so on. This leads us to the following definition.

Probability Density Function

A function f is a **probability density function** on the interval $[a, b]$ if two conditions are met:

$$\boxed{f(x) \geq 0 \quad \text{on } [a, b]} \tag{1}$$

and

$$\boxed{\int_a^b f(x)\,dx = 1} \tag{2}$$

where the interval $[a, b]$, possibly the interval $(-\infty, \infty)$, contains all values that the random variable X can assume.

With condition (1) in mind, the rationale behind condition (2) becomes apparent. Since the interval $[a, b]$ contains all the values the random variable can assume, the probability that a random variable lies between a and b must equal 1.

Using the fact that the area under the graph of f is given by a definite integral, the probability that a person is between the ages c and d, denoted by $Pr(c \leq X \leq d)$, is given by

$$Pr(c \leq X \leq d) = \int_c^d f(x)\,dx$$

For example, for the density function illustrated in Figure 5, the probability that a person is between 22 and 24 years of age is

$$\int_{22}^{24} f(x)\,dx$$

The discussion presented here leads to the following result.

The probability that the outcome of an experiment results in a value of a random variable X between c and d is given by

$$\boxed{Pr(c \le X \le d) = \int_c^d f(x)\,dx} \qquad (3)$$

where $f(x)$ is the probability density function of the random variable.

1 **EXAMPLE 1** **Verifying a Probability Density Function**

Show that the function $f(x) = \frac{3}{56}(5x - x^2)$ is a probability density function over the interval $[0, 4]$.

SOLUTION If f is indeed a probability density function, it has to satisfy conditions (1) and (2).

Condition (1) is satisfied since

$$f(x) = \frac{3}{56}(5x - x^2) = \frac{3}{56}x(5 - x) \ge 0$$

for all x in the interval $[0, 4]$.

To verify condition (2), we evaluate

$$\int_0^4 \frac{3}{56}(5x - x^2)\,dx = \frac{3}{56}\int_0^4 (5x - x^2)\,dx = \frac{3}{56}\left(\frac{5x^2}{2} - \frac{x^3}{3}\right)\Bigg|_0^4 = \frac{3}{56}\left[\frac{(5)16}{2} - \frac{64}{3}\right] = \frac{3}{56}\left(\frac{56}{3}\right) = 1$$

Condition (2) is also satisfied.

As a result, $f(x) = \frac{3}{56}(5x - x^2)$, for x in $0 \le x \le 4$, is a probability density function.

NOW WORK PROBLEM 1.

2 **EXAMPLE 2** **Find Probabilities Using a Probability Density Function**

Compute the probability that the random variable X with probability density function $f(x) = \frac{3}{56}(5x - x^2)$ assumes a value between 1 and 2.

SOLUTION To compute $Pr(1 \le X \le 2)$, we use equation (3):

$$Pr(1 \le X \le 2) = \int_1^2 \frac{3}{56}(5x - x^2)\,dx = \frac{3}{56}\left(\frac{5x^2}{2} - \frac{x^3}{3}\right)\Bigg|_1^2$$

$$= \frac{3}{56}\left[\left(\frac{(5)(4)}{2} - \frac{8}{3}\right) - \left(\frac{5}{2} - \frac{1}{3}\right)\right] = \frac{3}{56}\cdot\frac{31}{6} = \frac{31}{112}$$

NOW WORK PROBLEM 25.

How do we obtain a probability density function f? For individual random experiments, it is possible to construct them as we indicated in the example on age probabilities. However, the construction of a probability density function is usually difficult and depends on the nature of the experiment. Fortunately, several relatively simple probability density functions are available that can be used to fit most random experiments. In every example we discuss, the probability density function is given.

Uniform Density Function

The **uniform density function,** or the **uniform distribution,** the simplest of probability density functions, is one in which the random variable assumes all its values with equal probability. The probability density function for a uniformly distributed random variable is

$$f(x) = \begin{cases} \dfrac{1}{b-a} & \text{if } a \le x \le b \\ 0 & \text{if } x < a \text{ or } x > b \end{cases}$$

The graph of f is given in Figure 7.

FIGURE 7

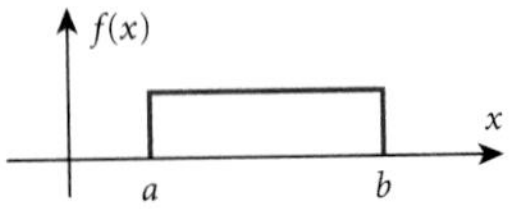

Notice that the function has the value 0 outside the interval $a \le x \le b$.

To verify that the uniform density function is a probability density function, we need to show that conditions (1) and (2) are satisfied. Since $b > a$, $f(x) = \dfrac{1}{b-a} > 0$ for $a \le x \le b$. Next,

$$\int_a^b f(x)\,dx = \int_a^b \frac{1}{b-a}\,dx = \frac{1}{b-a}\int_c^b dx = \frac{1}{b-a}\cdot x\Big|_a^b = \frac{1}{b-a}\cdot(b-a) = 1$$

The uniform density function satisfies conditions (1) and (2), and so is a probability density function.

3 **EXAMPLE 3** **Using the Uniform Density Function**

Trains leave a terminal every 40 minutes. What is the probability that a passenger arriving at a random time to catch a train will have to wait more than 10 minutes? Use a uniform density function.

SOLUTION Let T (time) be a random variable and assume it is uniformly distributed for $0 \le T \le 40$. The probability that the passenger must wait more than 10 minutes is

$$Pr(T \ge 10) = \int_{10}^{40} \frac{1}{40}\,dt = \frac{1}{40}\int_{10}^{40} dt = \frac{1}{40}\cdot t\Big|_{10}^{40} = \frac{1}{40}(40-10) = \frac{3}{4}$$

The probability is .75 that the passenger must wait more than 10 minutes.

Exponential Density Function

Exponential Density Function

Let X be a continuous random variable. Then X is said to be **exponentially distributed** if X has the probability density function

$$f(x) = \begin{cases} \lambda e^{-\lambda x} & \text{if } x \ge 0 \\ 0 & \text{if } x < 0 \end{cases}$$

where λ* is a positive constant. The function f is called the **exponential density function.**

*The Greek letter lambda.

FIGURE 8

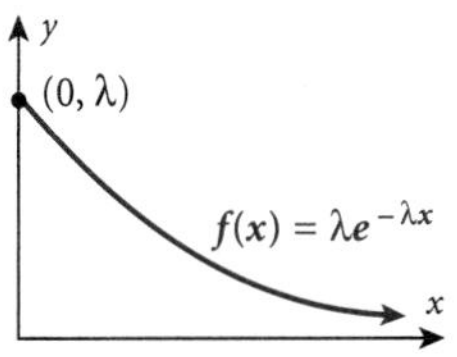

The graph of the exponential density function is given in Figure 8. The exponential density function is nonnegative for all x. Also,

$$\int_0^{\infty} f(x)\,dx = \underset{\substack{\uparrow \\ \text{Integral is improper}}}{\int_0^{\infty} \lambda e^{-\lambda x}\,dx} = \underset{\substack{\uparrow \\ \text{Move } \lambda \text{ outside the integral.}}}{\lim_{b\to\infty} \int_0^b \lambda e^{-\lambda x}\,dx} = \underset{\substack{\uparrow \\ \text{Integrate.}}}{\lim_{b\to\infty}\left[\lambda \int_0^b e^{-\lambda x}\,dx\right]} = \lim_{b\to\infty}\left[\lambda \cdot \frac{1}{-\lambda} \cdot e^{-\lambda x}\Big|_0^b\right]$$

$$= \lim_{b\to\infty} [(-1)(e^{-\lambda b} - 1)] \underset{\substack{\uparrow \\ \lim_{b\to\infty} e^{-\lambda b} = 0,\ \lambda > 0}}{=} 1$$

The exponential density function satisfies conditions (1) and (2) and so is a probability density function.

In general, any situation that deals with *waiting time* between successive events will lead to an exponential density function. In the exponential density function, the constant λ is the average number of arrivals per unit time and $\frac{1}{\lambda}$ is the average waiting time.

4

EXAMPLE 4 Using the Exponential Density Function

Airplanes arriving at an airport follow a pattern similar to the exponential density function with an average of $\lambda = 15$ arrivals per hour. Determine the probability of an arrival within 6 minutes (0.1 hour). Use an exponential density function.

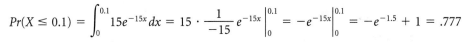

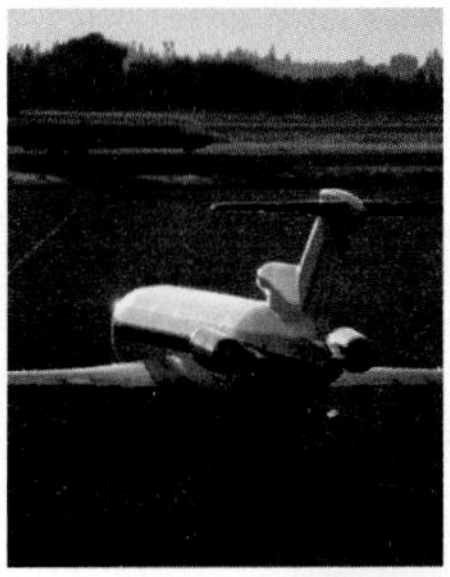

SOLUTION If X is the random variable associated with this random experiment, the probability that $X \le 0.1$ (6 minutes) is

$$Pr(X \le 0.1) = \int_0^{0.1} 15e^{-15x}\,dx = 15 \cdot \frac{1}{-15} e^{-15x}\Big|_0^{0.1} = -e^{-15x}\Big|_0^{0.1} = -e^{-1.5} + 1 = .777$$

The probability of an arrival within 6 minutes is .777.

NOW WORK PROBLEM 27.

EXAMPLE 5 Waiting Time to Find a Defective Product

From past data it is known that a certain machine normally produces 1 defective product every hour (60 minutes). To detect defective products, an inspector walks up to the machine and tests a continuous stream of products until she detects a defective one. Use an exponential density function to find the probability that

(a) the inspector waits less than 30 minutes before finding a defective product.
(b) the inspector must wait more than an hour and 15 minutes (75 minutes) before finding a defective product.

SOLUTION The average defective rate is 1 every 60 minutes, so we let $\lambda = \frac{1}{60}$ in the exponential density function

$$f(x) = \lambda e^{-\lambda x}$$

Then $f(x) = \frac{1}{60}e^{-x/60}$, $x \ge 0$. Suppose X is the random variable associated with this random experiment.

(a) The inspector waits less than 30 minutes before finding a defective product with probability

$$Pr(X \le 30) = \int_0^{30} \frac{1}{60} e^{-x/60}\, dx = -e^{-x/60}\Big|_0^{30} = -e^{-1/2} + 1 = .393$$

The probability the inspector waits less than 30 minutes is .393.

(b) First we compute the probability the inspector waits less than 75 minutes before finding a defective product.

$$Pr(X \le 75) = \int_0^{75} \frac{1}{60} e^{-x/60}\, dx = -e^{-x/60}\Big|_0^{75} = -e^{-5/4} + 1 = .713$$

The probability the inspector waits more than 75 minutes is $1 - .713 = .287$.

An alternative solution to part (b) of Example 5 is to evaluate

$$Pr(X \ge 75) = \int_{75}^{\infty} \frac{1}{60} e^{-x/60}\, dx.$$

Try it and compare to the solution given above.

Expected Value

Expected Value for a Continuous Random Variable

If X is a continuous random variable with the probability density function $f(x)$, $a \le x \le b$, the **expected value of X** is

$$E(X) = \int_a^b x f(x)\, dx$$

The expected value of the random variable X is the average (mean) value of the random variable. For example, if X is a random variable measuring heights, then $E(X)$ is the average (mean) height of the population.

5 **EXAMPLE 6 Finding the Expected Value**

A passenger arrives at a train terminal where trains arrive every 40 minutes. Determine the expected waiting time using a uniform density function.

SOLUTION Let the random variable T measure waiting time with uniform density function $f(t) = \frac{1}{40}$, where $0 \le T \le 40$. The expected value $E(T)$ is then

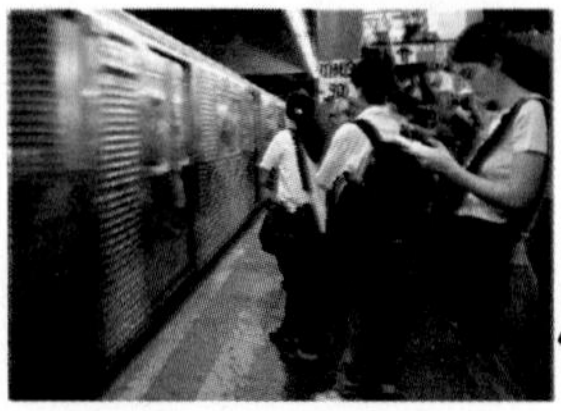

$$E(T) = \int_0^{40} t \frac{1}{40}\, dt = \frac{1}{40}\int_0^{40} t\, dt = \frac{1}{40} \cdot \frac{t^2}{2}\Big|_0^{40} = \frac{1}{40} \cdot \frac{1600}{2} = 20 \text{ minutes}$$

NOW WORK PROBLEM 17.

EXAMPLE 7 The Expected Value of the Uniform Density Function

Show that the expected value of the random variable with a uniform density function $f(x) = \frac{1}{(b-a)}$, $a \le x \le b$, is $\frac{a+b}{2}$.

SOLUTION Let X be a random variable with probability density function $f(x) = \frac{1}{b-a}$, $a \le x \le b$. Then

$$E(X) = \int_a^b x \frac{1}{b-a}\,dx = \frac{1}{b-a}\int_a^b x\,dx = \frac{1}{b-a}\cdot \frac{x^2}{2}\bigg|_a^b$$

$$= \frac{1}{b-a}\cdot\left(\frac{b^2}{2} - \frac{a^2}{2}\right) = \frac{1}{2}\cdot\frac{b^2-a^2}{b-a} = \frac{1}{2}\cdot\frac{(b-a)(b+a)}{b-a} = \frac{a+b}{2}$$

EXERCISE 16.3 Answers to Odd-Numbered Problems Begin on Page AN-98.

In Problems 1–8, verify that each function is a probability density function over the indicated interval.

1. $f(x) = \frac{1}{2}$ over $[0, 2]$
2. $f(x) = \frac{1}{5}$ over $[0, 5]$
3. $f(x) = 2x$ over $[0, 1]$
4. $f(x) = \frac{1}{8}x$ over $[0, 4]$
5. $f(x) = \frac{3}{250}(10x - x^2)$ over $[0, 5]$
6. $f(x) = \frac{6}{27}(3x - x^2)$ over $[0, 3]$
7. $f(x) = \frac{1}{x}$ over $[1, e]$
8. $f(x) = \frac{4}{3(x+1)^2}$ over $[0, 3]$

If $f(x) \ge 0$ is not a probability density function, we can find a constant k such that $kf(x)$ satisfies the condition $\int_a^b kf(x)\,dx = 1$. For the functions in Problems 9–16, determine the constant k that will make each one a probability density function over the interval indicated.

9. $f(x) = 1$ over $[0, 3]$
10. $f(x) = 1$ over $[0, 4]$
11. $f(x) = x$ over $[0, 2]$
12. $f(x) = x$ over $\left[0, \frac{1}{2}\right]$
13. $f(x) = 10x - x^2$ over $[0, 5]$
14. $f(x) = 10x - x^2$ over $[0, 8]$
15. $f(x) = \frac{1}{x}$ over $[1, 2]$
16. $f(x) = \frac{1}{(x+1)^3}$ over $[3, 7]$

In Problems 17–24, compute the expected value for each probability density function.

17. $f(x) = \frac{1}{2}$ over $[0, 2]$
18. $f(x) = \frac{1}{5}$ over $[0, 5]$
19. $f(x) = 2x$ over $[0, 1]$
20. $f(x) = \frac{1}{8}x$ over $[0, 4]$
21. $f(x) = \frac{3}{250}(10x - x^2)$ over $[0, 5]$
22. $f(x) = \frac{6}{27}(3x - x^2)$ over $[0, 3]$
23. $f(x) = \frac{1}{x}$ over $[1, e]$
24. $f(x) = \frac{4}{3(x+1)^2}$ over $[0, 3]$

25. A number x is selected at random from the interval $[0, 5]$. The probability density function for x is

$$f(x) = \frac{1}{5} \qquad \text{for } 0 \le x \le 5$$

Find the probability that a number is selected in the subinterval $[1, 3]$.

26. A number x is selected at random from the interval $[0, 10]$. The probability density function for x is

$$f(x) = \frac{1}{10} \qquad \text{for } 0 \le x \le 10$$

Find the probability that a number is selected in the subinterval $[6, 9]$.

27. **Time Between Telephone Calls** The time between incoming telephone calls at a hotel switchboard has an exponential density function with $\lambda = 0.5$ minute. What is the probability that there is an interval of at least 6 minutes between incoming calls?

28. **Waiting Time on a Phone Call** The length of the wait to speak to a customer representative at an airline is a random variable X. The average time a person is on hold is 6 minutes and follows an exponential distribution.

(a) Write the exponential density function which describes a caller's wait time.
(b) What is the probability that a caller is on hold between 5 and 10 minutes?
(c) What is the probability that a caller waits less than 1 minute for a customer representative?

29. **Time to Make a Choice** Let T be the random variable that a subject in a psychological testing program will make a certain choice after t seconds. If the probability density function is

$$f(t) = 0.4e^{-0.4t}$$

what is the probability that the subject will make the choice in less than 5 seconds?

30. **Waiting for a Bus** Buses on a certain route run every 50 minutes. What is the probability that a person arriving at a random stop along the route will have to wait at least 30 minutes? Assume that the random variable T is the time the person will have to wait and assume that T is uniformly distributed.

31. **Life of a Light Bulb** The length of time X a light bulb lasts is a random variable with an exponential probability distribution. Philips' 120 watt indoor spot lights have an average life of 2000 hours.

(a) What is the probability that Philips' 120 watt indoor spot light lasts between 1800 and 2200 hours?
(b) What is the probability that Philips' 120 watt indoor spot light lasts more than 2500 hours?

Source: Philips Lighting Company, Somerset, NJ.

32. **Life of a Light Bulb** The length of time X a light bulb lasts is a random variable with an exponential probability distribution. GE 100 watt Soft White light bulbs have an average life of 750 hours.

(a) What is the probability that a GE 100 watt Soft White light bulb lasts fewer than 500 hours?
(b) What is the probability that a GE 100 watt Soft White light bulb lasts between 750 and 1000 hours?
(c) What is the probability that a GE 100 watt Soft White light bulb lasts more than 1200 hours?
(d) Suppose you worked as manager of quality control at the GE production plant, and found that the 100 watt Soft White light bulbs consistently lasted more than 1200 hours. What might you conclude? Write a paragraph outlining a proposal you will make at next week's manager's meeting concerning the bulbs.

Source: General Electric Company, Cleveland, Ohio.

33. **Life of a Light Bulb** The length of time X a light bulb lasts is a random variable with an exponential probability distribution. GE 65 watt indoor spot lights have an average life of 2000 hours.

(a) What is the probability that a GE 65 watt indoor spot light lasts fewer than 1500 hours?
(b) What is the probability that a GE 65 watt indoor spot light lasts between 1750 and 2000 hours?
(c) What is the probability that a GE 65 watt indoor spot light lasts more than 1900 hours?
(d) Suppose you were the vice president of building maintenance, ordered a shipment of 1000 GE 65 watt bulbs, and found that after using half the bulbs, they consistently burned out before burning 1500 hours. What might you conclude? Explain.

Source: General Electric Company, Cleveland, Ohio.

34. **Waiting for Lunch** At a local Wendy's the lunch-time wait for service at the drive-in counter averages 4.5 minutes (270 seconds). If the wait is an exponentially distributed random variable X, what is the probability that a customer waits more than 5 minutes for an order?

35. **Express Lunch Special** A Pizza Hut manager guarantees that the lunch-time express pizza will be served within 10 minutes or it is free. He has found that the time to prepare, bake, and serve the pizza is an exponentially distributed

random variable with $\lambda = \frac{1}{8}$ minutes. What is the probability that a customer ordering the lunch-time express pizza will not have to pay for the pizza? That is, what is the probability that the customer waits longer than 10 minutes to be served after ordering a pizza?

36. Pizza Delivery At one time, Domino's Pizza guaranteed that their pizzas would be delivered within 30 minutes after ordering or the pizza was free. (The offer has been discontinued because it was found that delivery persons were involved in too many car accidents while rushing to deliver.) Time management experts had learned that the time it took to prepare, bake, and deliver the pizza is a random variable X and followed an exponential distribution with $\lambda = \frac{1}{24}$.

(a) What is the probability that a customer who calls and orders a pizza will get it free?
(b) Suppose that it is found that between 5 and 7 p.m. on Friday evenings it takes an extra 2 minutes to deliver the pizza. What is the probability that a customer calling during this time will not have to pay for the pizza?

37. Supermarket Lines A local supermarket has received complaints that the checkout lines are too slow. Management hires an industrial engineer who determines that it takes an average of 10 minutes for a customer who enters the line to complete checking out.

(a) Write the exponential density function which describes a customer's wait time.
(b) What is the probability that a customer entering the line takes between 7 and 12 minutes to complete the checkout process.
(c) What is the probability that a customer entering the line takes more than 15 minutes to checkout.
(d) If a mother entering the line has to pick up her child in 20 minutes, and it takes her 5 minutes to get to the bus stop, what is the probability she can get there on time?

38. Flight Arrivals On May 2, 2003, New Orleans International Airport had 14 flights arriving between 5 P.M. and 6 P.M. Determine the probability that a flight arrived within 3 minutes (.05 hour). Use an exponential density function.

Source: New Orleans International Airport.

39. Waiting Time Pairings at the 2003 HP Classic Golf Championship were scheduled to tee off every nine minutes. Determine the expected waiting time between pairings for spectators gathered on the 18th green.

Source: Professional Golfers Association.

40. Learning to Play a Game A manufacturer of educational games for children finds through extensive psychological research that the average time it takes for a child in a certain age group to learn the rules of the game is predicted by a **beta probability density function.**

$$f(x) = \begin{cases} \frac{1}{4500}(30x - x^2) & \text{if } 0 \le x \le 30 \\ 0 & \text{if } x < 0 \text{ or } x > 30 \end{cases}$$

where x is the time in minutes.

(a) Show that f is a probability density function.
(b) What is the probability a child will learn how to play the game within 10 minutes?
(c) What is the probability a child will learn the game after 20 minutes?
(d) What is the probability the game is learned in at least 10 minutes, but no more than 20 minutes?

41. Cost Estimates The probability density function that gives the probability that an electrical contractor's cost estimate is off by x percent is

$$f(x) = \frac{3}{56}(5x - x^2)$$

for x in the interval $[0, 4]$. On average, by what percent can the contractor be expected to be off?

42. Radioactive Decay Plutonium 239 decays continuously at a rate of .002845% per year (based on a half-life of 24,360 years). If X is the time that a randomly chosen plutonium atom will decay, a probability density function for X is the following:

$$f(x) = \begin{cases} .00002845e^{-.00002845x} & \text{if } x \ge 0 \\ 0 & \text{if } x < 0 \end{cases}$$

Use this probability density function to compute the probability that a plutonium atom will decay between 100 and 1000 years from now.

Source: Microsoft Encarta 97 Encyclopedia.

43. Gestation Time Suppose that X is the length of gestation in healthy humans. Then X is approximately normally distributed with a mean of 280 days and a standard deviation of 10 days. A probability density function for X is given by

$$f(x) = \frac{1}{10\sqrt{2\pi}} e^{-(x-280)^2/200}$$

(a) Use this probability density function to determine the probability that a healthy pregnant woman will have a pregnancy that lasts more than one week beyond the mean for the length of gestation in healthy humans.
(b) Determine the probability that a healthy pregnant woman will have a pregnancy such that the length of the

pregnancy is within one week of the mean for the length of gestation in healthy humans.

Source: School of Public Health and Health Sciences, University of Massachusetts.

44. Birth Weights Suppose that the random variable X is the weight of a baby at birth. The random variable X has a normal probability density function with mean approximately 114 ounces and standard deviation approximately 18 ounces. A probability density function for X is

$$f(x) = \frac{1}{18\sqrt{2\pi}} e^{-(x-114)^2/648}$$

Use this probability density function to determine the probability that a baby's weight at birth will be between 100 ounces and 125 ounces.

Source: Bao-Feng Feng, Department of Mathematics, University of Kansas.

Problems 45–46, require the following discussion:

Triangular probability distributions are often used to model business situations. The graph of a possible triangular probability function, f(x), is shown below. Such a distribution is used when we know only two pieces of information: the interval of possible values of the random variable X, that is, $a \le X \le b$, and the most likely value for the random variable X, that is, the value $X = c$ with the largest probability of occurring.

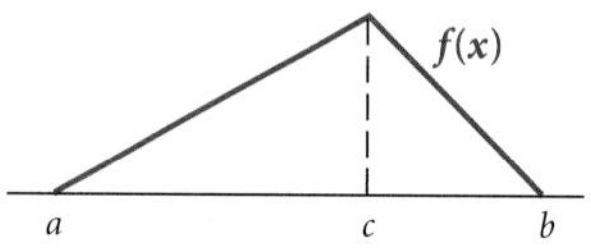

The probability density at $X = c$ can be calculated geometrically using the fact that

$$\int_a^b f(x)\,dx = 1$$

That is, the area of the triangle must be 1. Since $f(c)$ is the height of the triangle and $b - a$ is its base, we have

$$\frac{1}{2}(b - a)\cdot f(c) = 1$$

so

$$f(c) = \frac{2}{b - a}$$

Using this information, we can write the probability density function as a piecewise-defined function consisting of two line segments:

$$f(x) = \begin{cases} m_1x + b_1 & \text{if } a \le x \le c \\ m_2x + b_2 & \text{if } c < x \le b \end{cases}$$

Use this model in Problems 45 and 46.

45. Price of a New Car A business analyst predicts that a new car will cost between \$10,000 and \$20,000, with the most likely price being \$17,000.

(a) Using the triangular model above, find the probability density function $f(x)$.

(b) What is the probability that the car will cost less than \$15,000?

(c) Graph the function $f(x)$.

(d) Estimate the expected price of the car from the graph.

(e) Evaluate the expected price of the car.

(f) Why is expected price different from \$17,000?

46. Stock Market Analysis A stock market analyst predicts that a stock will be worth between \$15 and \$19, with the most likely value being \$16.

(a) Using the triangular model above, find the probability density function $f(x)$.

(b) What is the probability that the stock will be worth more than \$18?

(c) Graph the function $f(x)$.

(d) Estimate the expected value of the stock from the graph.

(e) Evaluate the expected value of the stock.

(f) Why is the expected value of the stock different from \$16?

Most graphing utilities have a random number function (usually RAND or RND) generating numbers between 0 and 1. Every time you use a random number function, a different number is selected. Check your user's manual to see how to use this feature of your graphing utility.

Sometimes probabilities are found by experimentation, that is, by performing an experiment. In Problems 47 and 48 use a random number function to perform an experiment.

47. Use a random number function to select a value for the random variable X. Repeat this experiment 50 times. [*Note:* Most calculators repeat the action of the last entry if you simply press the ENTER, or EXE, key again.] Count the number of times the random variable X is between 0.6 and 0.9.

(a) Calculate the ratio

$$R = \frac{\text{Number of times the random variable } X \text{ is between 0.6 and 0.9}}{50}$$

(b) Calculate the actual probability $Pr(0.6 \leq X < 0.9)$ using a uniform density function.

48. Use a random number function to select a value for the random variable X. Repeat this experiment 50 times. [*Note:* Most calculators repeat the action of the last entry if you simply press the ENTER, or EXE, key again.] Count the number of times the random variable X is between 0.1 and 0.3.

(a) Calculate the ratio

$$R = \frac{\text{Number of times the random variable } X \text{ is between 0.1 and 0.3}}{50}$$

(b) Calculate the actual probability $Pr(0.1 \leq X < 0.3)$ using a uniform density function.

49. The **variance** σ^2 associated with the probability density function f on $[a, b]$ is defined as

$$\sigma^2 = \int_a^b x^2 f(x)\, dx - [E(x)]^2$$

Verify that

$$\sigma^2 = \frac{(b-a)^2}{12}$$

for the uniform density function.

Chapter 16 Review

OBJECTIVES

Section	You should be able to	Review Exercises
16.1	1 Evaluate improper integrals	1–8
16.2	1 Find the average value of a function	9–14
	2 Use average value in applications	19–22
16.3	1 Verify probability density functions	15(a)–18(a), 25(a)– 27(a)
	2 Find probabilities using a probability density function	25(b)–27(b), 28(a) (b), 30, 34
	3 Use the uniform density function	23, 29, 30, 34
	4 Use the exponential density function	24, 31–33, 35, 36
	5 Find expected value	15(b)–18(b), 23(c), 24(c), 25(d)–27(d), 28(c)

THINGS TO KNOW

Improper Integrals (pp. 927, 928)	f is continuous on $[a, \infty)$	$\int_a^\infty f(x)\,dx = \lim_{b\to\infty}\int_a^b f(x)\,dx$, provided the limit exists
	f is continuous on $(-\infty, b]$	$\int_{-\infty}^b f(x)\,dx = \lim_{a\to-\infty}\int_a^b f(x)\,dx$, provided the limit exists
	f is continuous on $[a, b)$, and is not continuous at b	$\int_a^b f(x)\,dx = \lim_{t\to b^-}\int_a^t f(x)\,dx$, provided the limit exists
	f is continuous on $(a, b]$, and is not continuous at a	$\int_a^b f(x)\,dx = \lim_{t\to a^+}\int_t^b f(x)\,dx$, provided the limit exists
Average Value (p. 931)	f is continuous on $[a, b]$	$AV = \frac{1}{b-a}\int_a^b f(x)\,dx$
Probability Density Function (pp. 936–937)	$f(x) \geq 0$ and $\int_a^b f(x)\,dx = 1$	$Pr(c \leq X \leq d) = \int_c^d f(x)\,dx$
Expected Value for a Continuous Random Variable X (p. 940)	$E(x) = \int_a^b x\,f(x)\,dx$	

TRUE-FALSE ITEMS **Answers are on page AN-99.**

T F **1.** The average value of a function f over the interval $[a, b]$ equals $\dfrac{f(b) - f(a)}{b - a}$.

T F **2.** The integral $\int_0^1 \frac{1}{x-1}\,dx$ is improper.

T F **3.** A function f is a probability density function if $\int_a^b f(x)\,dx = 1$.

FILL-IN-THE-BLANKS **Answers are on page AN-99.**

1. For a continuous function f defined on the interval $[a, b]$, the number $\frac{1}{b-a}\int_a^b f(x)\,dx$ is called the ____________ ____________ of f.

2. A function that assigns a numerical value to each outcome of a sample space is called a ____________ ____________.

3. The function constructed by connecting probability values by a smooth curve is called a ____________ ____________ function.

4. If f is continuous on $[2, \infty)$, then $\int_2^\infty f(x)\,dx =$ ____________, provided this limit exists.

5. If f is discontinuous at 2, but continuous elsewhere, then $\int_0^2 f(x)\,dx =$ ____________, provided this limit exists.

REVIEW EXERCISES **Answers to odd-numbered problems begin on page AN-99. Blue problem numbers indicate the author's suggestions for use in a practice test.**

In Problems 1–6, find the value, if any, of each improper integral.

1. $\int_0^\infty 20e^{-20x}\,dx$

2. $\int_0^\infty \frac{1}{3}e^{-1/3x}\,dx$

3. $\int_0^8 \frac{1}{\sqrt[3]{x}}\,dx$

4. $\displaystyle\int_{1}^{10} \frac{4}{\sqrt{x-1}}\,dx$

5. $\displaystyle\int_{0}^{1} \frac{x+1}{x}\,dx$

6. $\displaystyle\int_{0}^{1} \frac{x^4+1}{x^2}\,dx$

7. Find the area, if it exists, between the graph of $f(x) = -e^{-x}$ and the x-axis to the right of $x = 0$.

8. Find the area, if it exists, under the graph of $f(x) = \frac{1}{x^2}$ to the right of $x = 1$.

In Problems 9–14, find the average value of each function f over the given interval.

9. $f(x) = x^3$ over $[-1, 3]$

10. $f(x) = \frac{1}{x}$ over $[1, 2]$

11. $f(x) = x^2 + x$ over $[2, 6]$

12. $f(x) = e^{4x}$ over $[0, 1]$

13. $f(x) = 3x^2$ over $[-2, 2]$

14. $f(x) = \frac{1}{2}x$ over $[2, 10]$

In Problems 15–18, (a) verify that each function is a probability density function; (b) compute the expected value of each one.

15. $f(x) = \frac{8}{9}x$ over $\left[0, \frac{3}{2}\right]$

16. $f(x) = \frac{1}{4}x^3$ over $[0, 2]$

17. $f(x) = 12x^3(1 - x^2)$ over $[0, 1]$

18. $f(x) = \frac{15}{7}(2x^2 - x^4)$ over $[0, 1]$

---

19. **Average Sales** The rate of sales of a certain product obeys

$$f(t) = 1340 - 850e^{-t}$$

where t is the number of years the product is on the market. Find the average yearly sales for the first 10 years.

20. **Profits from CD Sales** The weekly profit from the sales of a new CD is described by the function,

$$P(t) = 1000\,t\,e^{-t}$$

in dollars, where t is the number of weeks the CD is on the market. Find the average profit for the first 10 weeks of sales.

21. **Average Price of Sandals** The demand curve for pairs of sandals is given by

$$D(x) = 50\,e^{-0.01x}$$

Find the average price over the demand interval from 100 to 150 pairs of sandals.

22. **Average Price** A producer's supply curve for a product is

$$S(x) = 0.005x^2$$

Find the average price if the producer supplies between 50 and 80 units.

23. A random variable X obeys the uniform probability distribution

$$f(x) = \frac{1}{12} \text{ for } -2 \le x \le 10$$

(a) What is the probability that $X \le 1$?
(b) What is the probability that $X \ge 5$?
(c) What is the expected value of X?

24. A random variable X has the exponential probability distribution

$$f(x) = \frac{1}{20}e^{-1/20x} \quad x \ge 0$$

(a) Find $Pr(10 < X < 30)$
(b) Find $Pr(X > 30)$
(c) What is the expected value of X?

25. **Life Expectancy** A man who is currently 20 years old wants to purchase life insurance. The insurance company is interested in determining at what age X (in years) he is likely to die. The likelihood of his dying at age X is given by

$$f(x) = \frac{3}{635{,}840}(x^2 - 28x + 196)$$

over $[20, 100]$.
(a) Show that f is a probability density function.
(b) Find the probability that the man is likely to die at or before the age of 40.
(c) What is the probability that he will die at or before the age of 60?
(d) What is the man's expected age of death?

26. An experiment follows the function $f(x) = 6(x - x^2)$ for outcomes between 0 and 1.

(a) Show that f is a probability density function.
(b) Determine the probability that an outcome lies between $\frac{1}{3}$ and $\frac{1}{2}$.
(c) Determine the probability that an outcome lies between 0 and $\frac{3}{4}$.
(d) Find the expected value of the outcome.

27. The outcome X of an experiment lies between 0 and 2, and follows the function

$$f(x) = \frac{1}{2}x$$

(a) Show that f is a probability density function.
(b) What is the probability that $X < 1$?
(c) What is the probability that X is between 1 and 1.5?
(d) What is the probability that X is greater than 1.5?
(e) What is the expected value of X?

28. X is a random variable with a probability density function

$$f(x) = 3x^2 \qquad -1 \leq x \leq 0$$

(a) What is the probability that $X > -0.1$?
(b) What is the probability that $-0.5 < X < -0.75$?
(c) What is the expected value of X?

29. Waiting for Clock Chimes The Gastown Steam Clock in Vancouver, Canada, chimes every 15 minutes. A tourist walking around the city randomly arrives at the corner where the clock is located.

(a) What is the probability she will wait fewer than 3 minutes to hear the clock?
(b) What is the probability the tourist waits longer than 10 minutes?
(c) What is the expected wait time of a person arriving at the corner to hear the clock?

Use a uniform density function.

30. Uniform Probability Use a uniform density function. A number is randomly chosen from the interval [0, 1].

(a) What is the probability that the first decimal digit will be a 1?
(b) What is the probability that the first decimal digit will be greater than 4?

31. Fire Alarms A fire department in a medium-sized city receives an average of 2.5 calls each minute. Use an exponential distribution to find the probability that the switchboard is idle (no calls are received) for more than one minute.

32. Waiting for Lunch At a fast-food counter it takes an average of 3 minutes to get served. The service time X for a customer has an exponential probability density function.

(a) What fraction of the customers is served within 2 minutes?
(b) What is the probability that a customer has to wait at least 3 minutes to be served?

33. Life of a Light Bulb The length of time X a light bulb lasts is a random variable with an exponential probability distribution. Philips' *Dura Max* 3-way long life light bulbs have an average life of 1750 hours.

(a) What is the probability that a Philips' *Dura Max* bulb lasts between 1500 and 2000 hours?
(b) What is the probability that Philips' *Dura Max* bulb lasts more than 2000 hours?

Source: Philips Lighting Company, Somerset, NJ.

34. Quality Control A toy manufacturing machine produces a toy every 2 minutes. An inspector arrives at a random time and must wait X minutes for a toy.

(a) Find the probability density function for X.
(b) Find the probability that the inspector has to wait at least 1 minute for a toy to be produced.
(c) Find the probability that the inspector has to wait no more than 1 minute for a toy.

Use a uniform density function.

35. Collecting Tolls Cars approach a remote toll station at a rate of 40 cars per hour. The time between arrivals is a random variable X, which obeys the exponential probability density function

$$f(t) = \frac{2}{3}e^{-2t/3} \qquad t \geq 0$$

What is the probability that the attendant waits more than 1 minute for the first car?

36. Length of a Phone Call The length t of a telephone call to technical service at a computer software company follows an exponential distribution with $\lambda = 0.10$.

(a) What is the probability that a call lasts between 5 and 10 minutes?
(b) What is the probability that a call lasts more than 30 minutes?
(c) What is the expected length of a phone call?

Chapter 16 Project

A MATHEMATICAL MODEL FOR ARRIVALS

In this Chapter Project we try to see if the waiting times between arrivals really do have an exponential distribution. The following data set tabulates the dates from 1997 to 2001 on which a member of a large church in Charlotte, North Carolina died. Day 1 is January 1, 1997, while day 1461 is December 31, 2001.

{6, 47, 109, 113, 131, 152, 158, 178, 201, 206, 215, 250, 251, 252, 274, 276, 306, 311, 338, 382, 414, 429, 432, 440, 468, 478, 479, 498, 510, 530, 542, 554, 561, 597, 601, 602, 603, 605, 606, 629, 650, 653, 660, 666, 667, 669, 686, 732, 745, 760, 764, 782, 790, 803, 820, 828, 863, 870, 878, 881, 893, 896, 913, 960, 974, 998, 1004, 1005, 1050, 1068, 1088, 1094, 1095, 1146, 1151, 1171, 1185, 1273, 1274, 1282, 1288, 1334, 1348, 1364, 1388, 1412, 1423, 1437, 1441}

The first member of the church died on January 6, 1997, the second on February 16, 1997, and so on. Although it may seem a bit odd, this is a list of "arrivals." Notice that there were 89 deaths in 1461 days, which gives a rate of $\lambda = \dfrac{89}{1461} = 0.061$ arrivals per day. One can calculate that the waiting times between arrivals are

{6, 41, 62, 4, 18, 21, 6, 20, 23, 5, 9, 35, 1, 1, 22, 2, 30, 5, 27, 44, 32, 15, 3, 8, 28, 10, 1, 19, 12, 20, 12, 12, 7, 36, 4, 1, 1, 2, 1, 23, 21, 3, 7, 6, 1, 2, 17, 46, 13, 15, 4, 18, 8, 13, 17, 8, 35, 7, 8, 3, 12, 3, 17, 47, 14, 24, 6, 1, 45, 18, 20, 6, 1, 51, 5, 20, 14, 88, 1, 8, 6, 46, 14, 16, 24, 24, 11, 14, 4}

Notice how uneven these data are—in some cases a long time passes between deaths, while in other cases several deaths are bunched closely together.

1. Group the times t between deaths into 18 intervals: $0 \le t < 5$, $5 \le t < 10, \ldots, 85 \le t < 90$, and tally the number of deaths in each interval.

2. Build a histogram for the data tabulated in Problem 1. To build a histogram, we construct a set of adjoining rectangles, each having as base the size of the interval, 5, and as height the number of deaths in that particular interval.

3. Comment on the shape of the histogram built in Problem 2.

4. Refer to Problem 1. Find the relative frequency of the number of deaths by dividing the number of deaths in each time interval by 89, the total number of deaths that occurred in the four-year period.

5. Construct a relative frequency histogram. That is, a histogram, for which the height of each rectangle is the relative frequency of the interval.

6. Using $\lambda = 0.061$, and assuming that the waiting times are exponentially distributed, write the probability density function that will describe the time between deaths. Show that the distribution is a probability density function.

7. Using the probability density function from Problem 6, calculate the probability of the waiting time between deaths for each interval: $0 \le t < 5$ days, $5 \le t < 10$ days, $\ldots$, $85 \le t < 90$ days.

8. Compare the results from Problems 4 and 6. Does it seem that the waiting times from the data are exponentially distributed with $\lambda = 0.061$?

9. Graph $f(t) = 0.061e^{-0.061t}$ on the same set of axes as the relative frequency histogram. Comment on the graphs.

MATHEMATICAL QUESTIONS FROM PROFESSIONAL EXAMS*

1. **Actuary Exam—Part I** $\displaystyle\int_0^1 x \ln x \, dx =$

   (a) $-\infty$ (b) -2 (c) -1 (d) $-\dfrac{1}{4}$ (e) $-\dfrac{2}{9}$

2. **Actuary Exam—Part I** $\displaystyle\int_{-1}^{2} \frac{x}{\sqrt{x+1}} \, dx =$

   (a) $\dfrac{1}{2}$ (b) $\dfrac{2}{3}$ (c) 1 (d) $\dfrac{3}{2}$ (e) 0

3. **Actuary Exam—Part I** $\displaystyle\int_0^\infty \frac{x+1}{(x^2+2x+2)^2}\,dx =$

(a) 0 (b) $\frac{1}{4}$ (c) $\frac{1}{2}$ (d) 1 (e) ∞

4. **Actuary Exam—Part II** Two men patronize the same barber shop. If they both arrive independently between 3 P.M. and 4 P.M. on the same day and both stay for 15 minutes, what is the probability that they are in the barber shop for part or all of the same time?

(a) $\frac{3}{8}$ (b) $\frac{7}{16}$ (c) $\frac{1}{2}$ (d) $\frac{9}{16}$ (e) $\frac{5}{8}$

5. **Actuary Exam—Part II** Each day, X arrives at a point A between 8:00 and 9:00 A.M., his times of arrival being uniformly distributed. Y arrives independently at A between 8:30 and 9:00 A.M., his times of arrival also being uniformly distributed. What is the probability that Y arrives before X?

(a) $\frac{1}{8}$ (b) $\frac{1}{6}$ (c) $\frac{2}{9}$ (d) $\frac{1}{4}$ (e) $\frac{1}{2}$

CHAPTER 17

Calculus of Functions of Two or More Variables

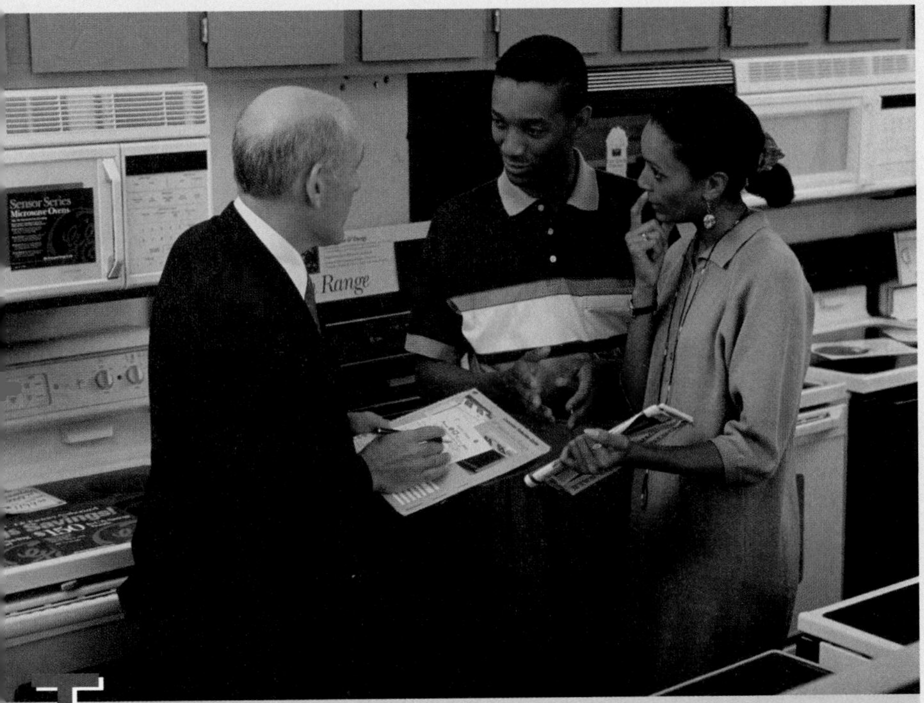

OUTLINE

The inventory control project given in Chapter 14 dealt with controlling the costs associated with the ordering and storage of vacuum cleaners. But stores would typically order and stock a variety of items, not just vacuum cleaners. How do we minimize costs when more than one item is involved? The Chapter Project at the end of this chapter will guide you to a solution when two items are involved.

A LOOK BACK, A LOOK FORWARD

Up to now, every discussion about functions dealt with a function $y = f(x)$, a function of one variable x. In this chapter we define functions that have more than one independent variable. We will present many of the same concepts introduced for functions of one variable. As you will see, often the discussions in this chapter will parallel earlier presentations.

17.1 Rectangular Coordinates in Space

PREPARING FOR THIS SECTION ***Before getting started, review the following:***

- Rectangular Coordinates (in the plane) (Chapter 1, Section 1.1, pp. 2–3)
- Completing the Square (Appendix A, Section A.6, p. A-54)
- Distance Formula (Appendix A, Section A.10, pp. A-85–A-88)
- Pythagorean Theorem (Appendix A, Section A.9, pp. A-80–A-82)

OBJECTIVES
1. Use the distance formula
2. Find the standard equation of a sphere

FIGURE 1

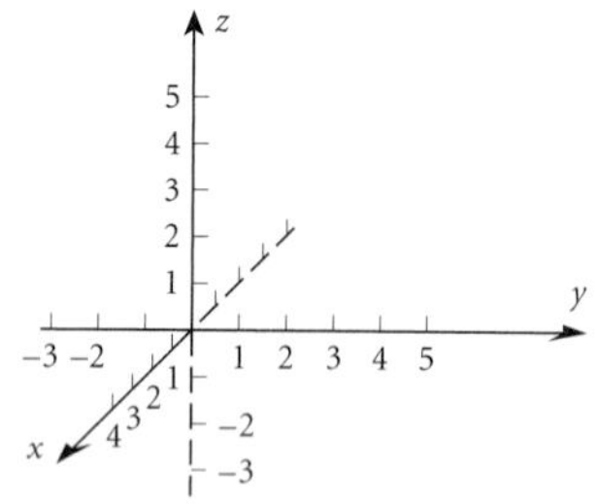

FIGURE 2

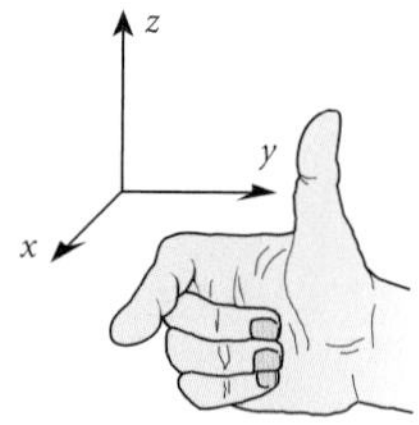

FIGURE 3

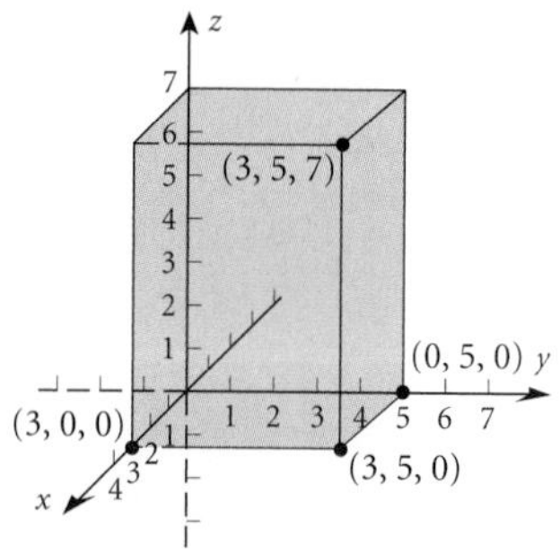

In Chapter 1 we established a correspondence between points on a line and real numbers. Then we showed that each point in a plane can be associated with an ordered pair of real numbers. Here we show that each point in (three-dimensional) space can be associated with an **ordered triple** of real numbers.

First we select a fixed point called the **origin.** Through the origin, we draw three mutually perpendicular lines. These are called the **coordinate axes,** and are labeled the ***x*-axis, *y*-axis,** and ***z*-axis.** On each of the three lines we choose one direction as positive and select an appropriate scale on each axis.

As indicated in Figure 1, we position the positive z-axis so that the system is **right-handed.** This conforms to the so-called **right-hand rule,** which asserts that if the index finger of the right hand points in the direction of the positive x-axis and the middle finger points in the direction of the positive y-axis, then the thumb will point in the direction of the positive z-axis.* See Figure 2.

Just as we did in one and two dimensions, we assign coordinates to each point P in space. Specifically, we identify each point P with an ordered triple of real numbers (x, y, z), and we refer to it as "the point (x, y, z)." So, "the point $(3, 5, 7)$" is the point for which $x = 3$, $y = 5$, $z = 7$, and, starting at the origin, we reach P by moving 3 units along the positive x-axis, then 5 units in the direction of the positive y-axis, and, finally, 7 units in the direction of the positive z-axis. Figure 3 illustrates the location of the point $(3, 5, 7)$ as well as the points $(3, 0, 0)$, $(3, 5, 0)$, and $(0, 5, 0)$. Observe that any point on the x-axis will have the form $(x, 0, 0)$. Similarly, $(0, y, 0)$ and $(0, 0, z)$ represent points on the y-axis and the z-axis, respectively.

In addition, all points of the form $(x, y, 0)$ constitute a plane called the ***xy*-plane.** This plane is perpendicular to the z-axis. Similarly, the points $(0, y, z)$ form the ***yz*-plane,** which is perpendicular to the x-axis; and the points $(x, 0, z)$ form the ***xz*-plane,**

**Although there are left-handed systems and left-handed rules, we shall adopt the usual convention and only use a right-handed system.*

FIGURE 4

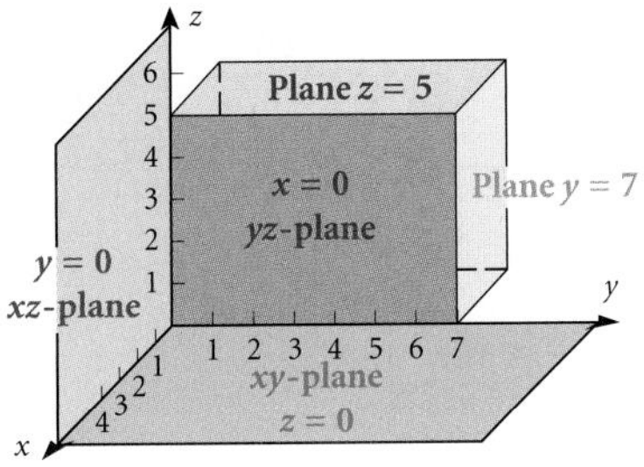

which is perpendicular to the y-axis (see Figure 4). Figure 4 also illustrates that points of the form (x, y, z), where $z = 5$, lie in a plane parallel to the xy-plane. Similarly, points (x, y, z), where $y = 7$, lie in a plane parallel to the xz-plane.

NOW WORK PROBLEMS 1 AND 13.

FIGURE 5

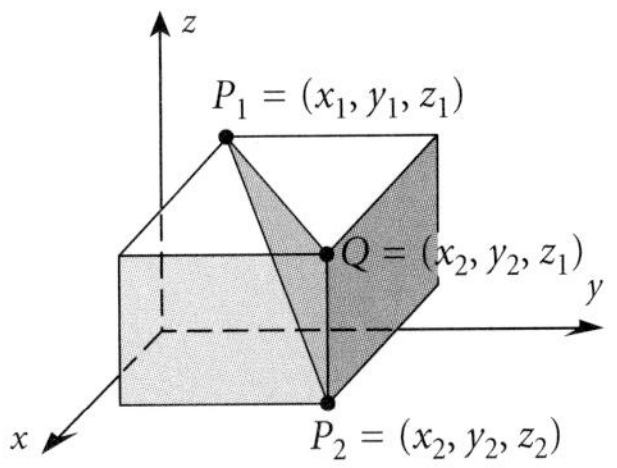

Distance in Space

To derive a formula for the distance $d = d(P_1, P_2)$ between two points $P_1 = (x_1, y_1, z_1)$ and $P_2 = (x_2, y_2, z_2)$ in space, we apply the Pythagorean theorem twice. As Figure 5 illustrates, we utilize the point $Q = (x_2, y_2, z_1)$. The first application of the Pythagorean theorem involves observing that the triangle QP_1P_2 is a right triangle in which the side P_1P_2 is the hypotenuse. As a result,

$$[d(P_1, P_2)]^2 = [d(P_1, Q)]^2 + [d(Q, P_2)]^2 \qquad (1)$$

The points P_1 and Q lie in a plane parallel to the xy-plane (the plane $z = z_1$). So, we can use the formula for distance in two dimensions and obtain

$$[d(P_1, Q)]^2 = (x_2 - x_1)^2 + (y_2 - y_1)^2 \qquad (2)$$

The points P_2 and Q lie along a line parallel to the z-axis so that

$$d(Q, P_2) = |z_2 - z_1| \quad \text{and} \quad [d(Q, P_2)]^2 = (z_2 - z_1)^2 \qquad (3)$$

Using the results of equations (2) and (3) in equation (1), we arrive at a formula for the distance between two points in space.

Distance Formula

The distance $d = d(P_1, P_2)$ between two points $P_1 = (x_1, y_1, z_1)$ and $P_2 = (x_2, y_2, z_2)$ in space is given by the formula

$$\boxed{d = \sqrt{(x_2 - x_1)^2 + (y_2 - y_1)^2 + (z_2 - z_1)^2}} \qquad \textbf{(4)}$$

1 **EXAMPLE 1** **Using the Distance Formula**

If $P_1 = (-1, 4, 2)$ and $P_2 = (6, -2, 3)$, find the distance d between P_1 and P_2.

SOLUTION We use Formula (4), with $x_1 = -1, y_1 = 4, z_1 = 2$ and $x_2 = 6, y_2 = -2, z_2 = 3$. Then

$$\begin{aligned} d &= \sqrt{[6-(-1)]^2 + [-2-4]^2 + [3-2]^2} \\ &= \sqrt{49 + 36 + 1} \\ &= \sqrt{86} \end{aligned}$$

NOW WORK PROBLEM 19.

The Sphere

FIGURE 6

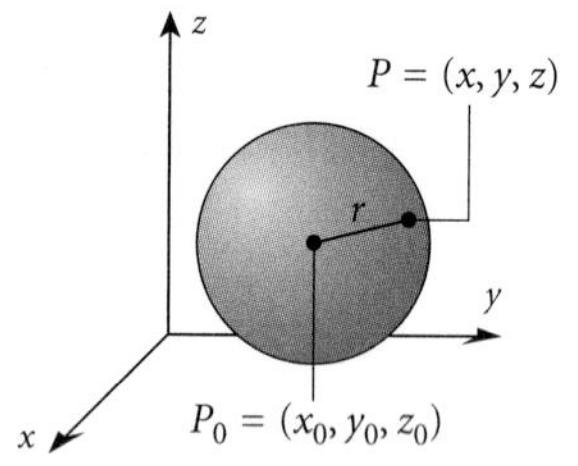

In space the collection of all points that are the same distance from some fixed point is called a **sphere.** The constant distance is called the **radius** and the fixed point is the **center** of the sphere. See Figure 6.

Any point $P = (x, y, z)$ on a sphere of radius r and center at the point $P_0 = (x_0, y_0, z_0)$ obeys

$$d(P, P_0) = r$$

By the Distance Formula (4), an equation of this sphere is

$$\sqrt{(x-x_0)^2 + (y-y_0)^2 + (z-z_0)^2} = r$$

Squaring both sides gives

$$(x-x_0)^2 + (y-y_0)^2 + (z-z_0)^2 = r^2$$

Standard Equation of a Sphere

The **standard equation of a sphere** with center at the point (x_0, y_0, z_0) and radius r is

$$\boxed{(x-x_0)^2 + (y-y_0)^2 + (z-z_0)^2 = r^2}$$

2 **EXAMPLE 2 Finding the Standard Equation of a Sphere**

The standard equation of a sphere with radius 2 and center at $(-1, 2, 0)$, is

$$(x-x_0)^2 + (y-y_0)^2 + (z-z_0)^2 = r^2$$

$$(x+1)^2 + (y-2)^2 + z^2 = 4 \qquad x_0 = -1, y_0 = 2, z_0 = 0, r = 2$$

NOW WORK PROBLEM 25.

When an equation of a sphere is given, it is sometimes necessary to complete the squares involving the variables in order to find the center and radius.

EXAMPLE 3 Finding the Center and Radius of a Sphere

Complete the squares in the expression

$$x^2 + y^2 + z^2 + 2x + 4y - 2z = 10$$

and show that it is the equation of a sphere. Find its center and radius.

SOLUTION Rewrite the given expression as

$$(x^2 + 2x) + (y^2 + 4y) + (z^2 - 2z) = 10$$

and complete the squares. The result is

$$(x^2 + 2x + 1) + (y^2 + 4y + 4) + (z^2 - 2z + 1) = 10 + 1 + 4 + 1$$
$$(x + 1)^2 + (y + 2)^2 + (z - 1)^2 = 16 = 4^2$$

This is the standard equation of a sphere with radius 4 and center at $(-1, -2, 1)$.

NOW WORK PROBLEM 31.

EXERCISE 17.1 Answers to Odd-Numbered Problems Begin on Page AN-100.

In Problems 1–6, plot each point.

1. $(1, 1, 1)$
2. $(0, 0, 1)$
3. $(0, 2, 5)$
4. $(-1, 5, 0)$
5. $(-3, 1, 0)$
6. $(4, -1, -3)$

In Problems 7–12, opposite vertices of a rectangular box whose edges are parallel to the coordinate axes are given. List the coordinates of the other six vertices of the box.

7. $(0, 0, 0)$; $(2, 1, 3)$
8. $(0, 0, 0)$; $(4, 2, 2)$
9. $(1, 2, 3)$; $(3, 4, 5)$
10. $(5, 6, 1)$; $(3, 8, 2)$
11. $(-1, 0, 2)$; $(4, 2, 5)$
12. $(-2, -3, 0)$; $(-6, 7, 1)$

In Problems 13–18, describe in words the set of all points (x, y, z) that satisfy the given conditions.

13. $y = 3$
14. $z = -3$
15. $x = 0$
16. $x = 1$ and $y = 0$
17. $z = 5$
18. $x = y$ and $z = 0$

In Problems 19–24, find the distance between each pair of points.

19. $(1, 3, 0)$ and $(4, 1, 2)$
20. $(3, 2, 1)$ and $(1, 2, 3)$
21. $(-1, 2, -3)$ and $(4, -2, 1)$
22. $(-2, 1, 3)$ and $(4, 0, -3)$
23. $(4, -2, -2)$ and $(3, 2, 1)$
24. $(2, -3, -3)$ and $(4, 1, -1)$

In Problems 25–28, find the standard equation of a sphere with radius r and center P_0.

25. $r = 1$; $P_0 = (3, 1, 1)$
26. $r = 2$; $P_0 = (1, 2, 2)$
27. $r = 3$; $P_0 = (-1, 1, 2)$
28. $r = 1$; $P_0 = (-3, 1, -1)$

In Problems 29–34, find the radius and center of each sphere.

29. $x^2 + y^2 + z^2 + 2x - 2y = 2$
30. $x^2 + y^2 + z^2 + 2x - 2z = -1$
31. $x^2 + y^2 + z^2 + 4x + 4y + 2z = 0$
32. $x^2 + y^2 + z^2 + 4x = 0$
33. $2x^2 + 2y^2 + 2z^2 - 8x + 4z = -2$
34. $3x^2 + 3y^2 + 3z^2 + 6x - 6y = 3$

In Problems 35–36, write the standard equation of the sphere described.

35. The endpoints of a diameter are $(-2, 0, 4)$ and $(2, 6, 8)$.
36. The endpoints of a diameter are $(1, 3, 6)$ and $(-3, 1, 4)$.

17.2 Functions and Their Graphs

PREPARING FOR THIS SECTION *Before getting started, review the following:*

- > Definition of a Function (Chapter 10, Section 10.2, pp. 549–551)
- > Completing the Square (Appendix A, Section A.6, p. A-54)

OBJECTIVES
1 Evaluate a function of two variables
2 Find the domain of a function of two variables

So far we have considered only functions of one independent variable, usually expressed explicitly by an equation $y = f(x)$. Often, models require functions of more than one variable.

For example, the cost of producing a certain item may depend on variables such as labor and material. In economic theory, supply and demand of a commodity often depend not only on the commodity's own price, but also on the prices of related commodities and other factors (such as income level, time of year, etc.).

Definition of a Function of Two Variables

A **function f of two variables** x and y is a relation $z = f(x, y)$ that assigns a unique real number z to each ordered pair (x, y) of real numbers in a subset D of the xy-plane. The set D is called the **domain** of the function f.

In the equation $z = f(x, y)$ we refer to z as the **dependent variable** and to x and y as the **independent variables.** The **range** of f consists of all real numbers $f(x, y)$ where (x, y) is in D. Figure 7 illustrates one way of depicting $z = f(x, y)$.

FIGURE 7

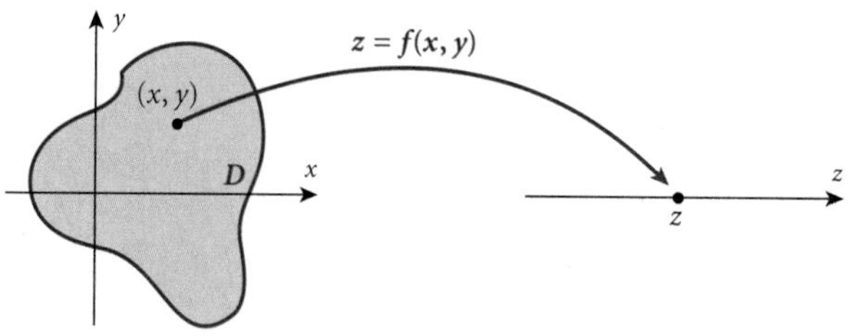

D is the domain of f

1 **EXAMPLE 1** **Evaluating a Function of Two Variables**

If $z = f(x, y) = x\sqrt{y} + xy^2$, find:

(a) $f(0, 0)$ **(b)** $f(2, 4)$ **(c)** $f(1, 9)$ **(d)** $f(x + \Delta x, y)$ **(e)** $f(x, y + \Delta y)$

SOLUTION **(a)** $f(0, 0) = 0\sqrt{0} + 0 \cdot 0^2 = 0$

(b) $f(2, 4) = 2\sqrt{4} + 2 \cdot 4^2 = 36$

(c) $f(1, 9) = 1\sqrt{9} + 1 \cdot 9^2 = 84$

(d) $f(x + \Delta x, y) = (x + \Delta x)\sqrt{y} + (x + \Delta x) \cdot y^2$

(e) $f(x, y + \Delta y) = x\sqrt{y + \Delta y} + x(y + \Delta y)^2$

NOW WORK PROBLEM 1.

As with functions of one variable, a function of two variables is usually given by an equation and, unless otherwise stated, the domain is taken to be the largest set of points in the plane for which this equation is defined in the real number system.

2 **EXAMPLE 2** **Finding the Domain of a Function of Two Variables**

Find the domain of each function. Graph each domain.

(a) $z = f(x, y) = \sqrt{x^2 + y^2 - 1}$
(b) $z = g(x, y) = \ln(y - x^2)$

FIGURE 8

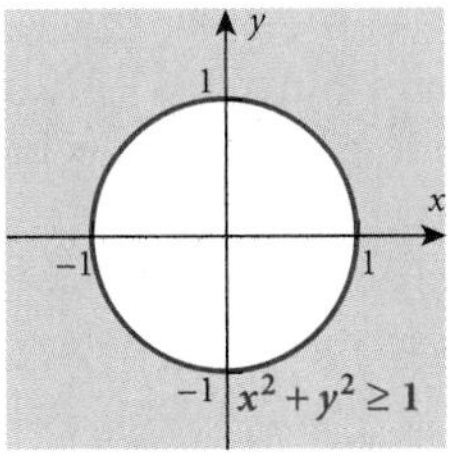

SOLUTION **(a)** The function f equals the square root of the expression $x^2 + y^2 - 1$. Since only square roots of nonnegative numbers are allowed in the real number system, we must have

$$x^2 + y^2 - 1 \geq 0$$
$$x^2 + y^2 \geq 1$$

This inequality describes the domain of f, namely the set of points (x, y) that are either on the circle $x^2 + y^2 = 1$ or outside of it. See Figure 8. Notice that we use a solid rule to show that the points on the circle are part of the domain.

(b) The function g is the natural logarithm of the expression $y - x^2$. Since only logarithms of positive numbers are allowed, we must have

FIGURE 9

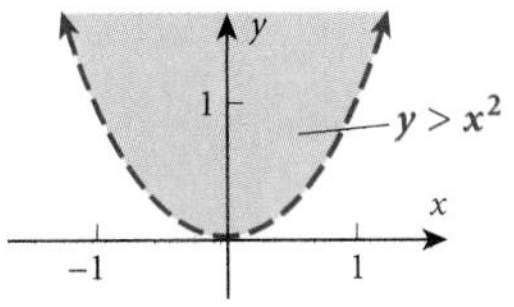

$$y - x^2 > 0$$

from which

$$y > x^2$$

This inequality describes the domain of g, namely, the set of points (x, y) "inside" the parabola $y = x^2$. See Figure 9. Notice that we use a dashed rule to show that the points on the parabola $y = x^2$ are not part of the domain.

FIGURE 10

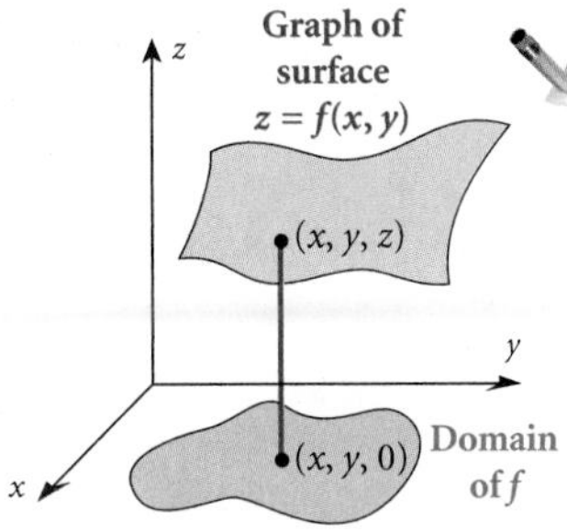

NOW WORK PROBLEM 17.

Graphing Functions of Two Variables

The graph of a function $z = f(x, y)$ of two variables, called a **surface,** consists of all points (x, y, z) for which $z = f(x, y)$ and (x, y) is in the domain of f. See Figure 10 for an illustration of a surface for which $f(x, y) \geq 0$.

EXAMPLE 3 Describing the Graph of a Function of Two Variables

Describe the graph of the function

$$z = f(x, y) = \sqrt{4 - x^2 - y^2}$$

SOLUTION First, notice that the domain of f consists of all points (x, y) for which $x^2 + y^2 \leq 4$. Since $z = f(x, y)$ is defined as a square root, it follows that $z \geq 0$, so the graph of f will lie above the xy-plane. By squaring both sides of $z = \sqrt{4-x^2-y^2}$, we find that

$$z^2 = 4 - x^2 - y^2, \quad z \geq 0$$
$$x^2 + y^2 + z^2 = 4, \quad z \geq 0$$

This is the equation of a hemisphere, with center at (0, 0, 0) and radius 2. Figure 11 illustrates the graph.

FIGURE 11

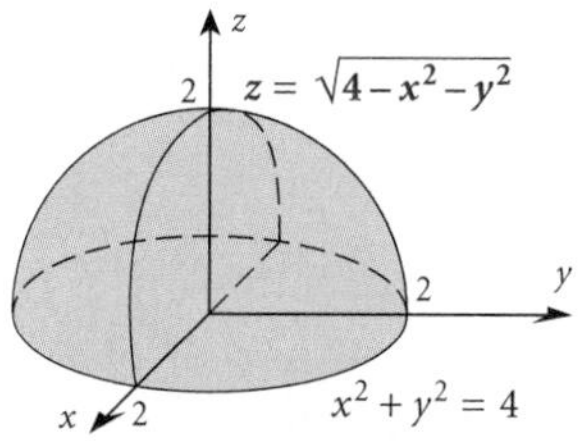

Obtaining the shape, or even a rough sketch, of the graph of most functions of two variables is a difficult task and is not taken up in this book. Graphing utilities may be used to generate *computer graphics* of surfaces in space.

For example, a portion of a surface $z = f(x, y)$ can be viewed in perspective and delineated by a rectangular mesh over a rectangular portion of the xy-plane. By changing the computer program, the point of view and other characteristics of the computer graphic may be altered. See Figure 12 for computer-generated graphs of several surfaces.

FIGURE 12

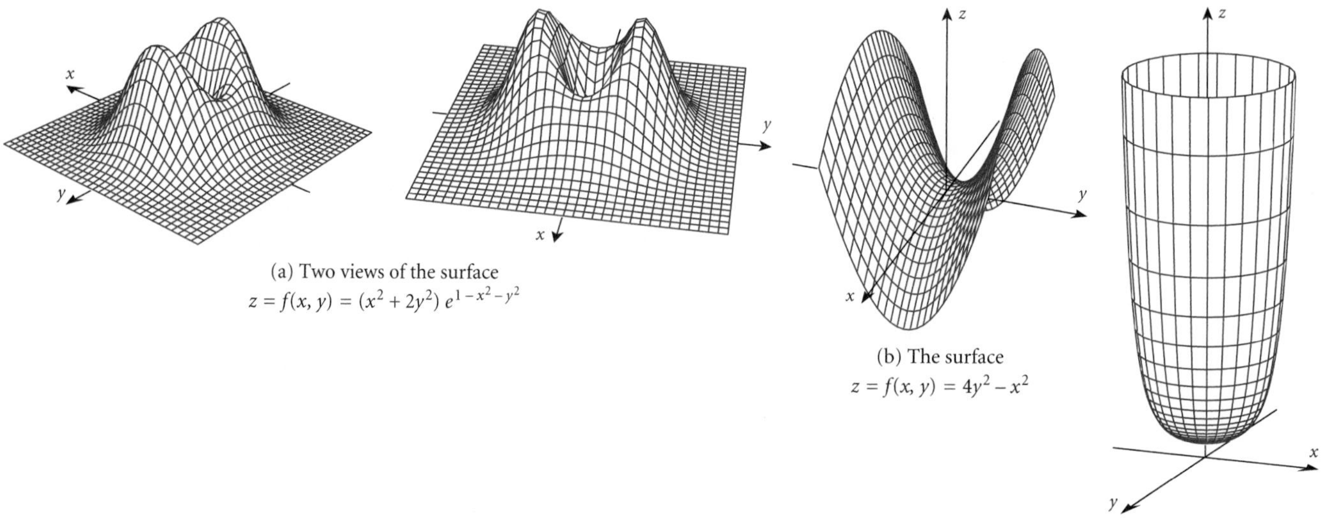

(a) Two views of the surface $z = f(x, y) = (x^2 + 2y^2)\, e^{1-x^2-y^2}$

(b) The surface $z = f(x, y) = 4y^2 - x^2$

(c) The surface $z = f(x, y) = e^{x^2+y^2}$

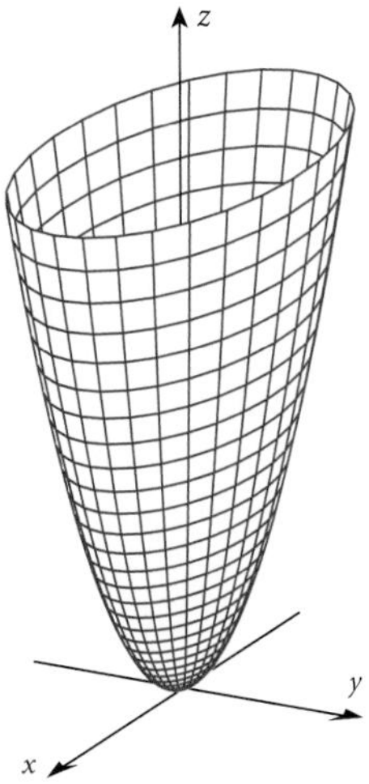

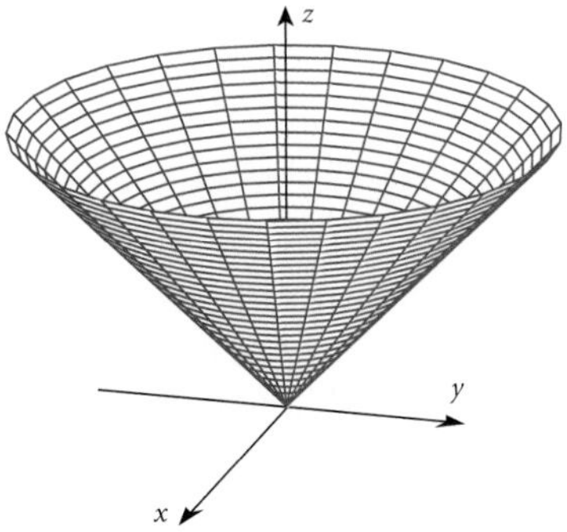

(e) The surface
$z = f(x, y) = \sqrt{x^2 + y^2}$

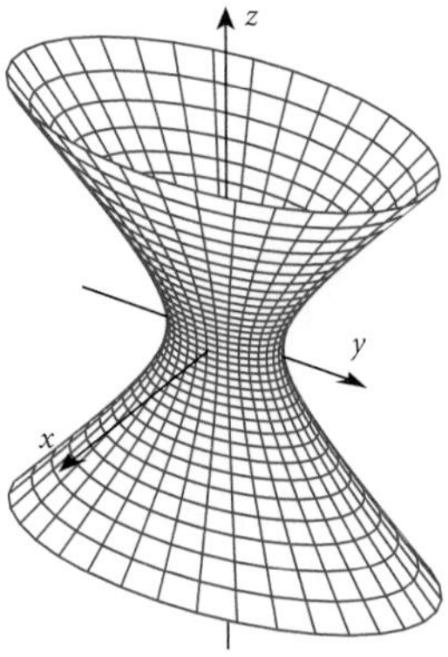

(f) The surface
$z^2 = 2x^2 + y^2 + 1$

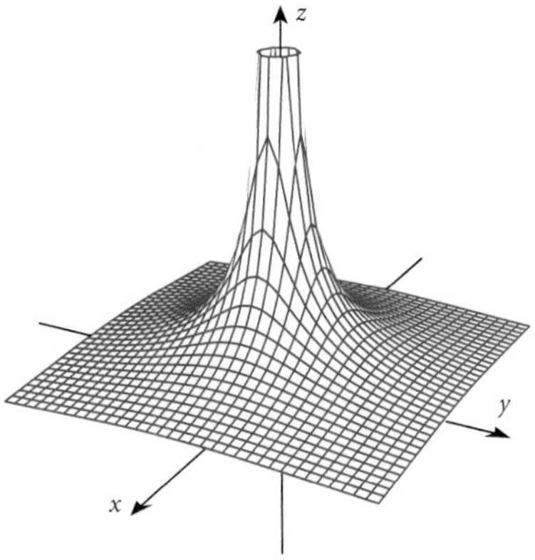

(g) The surface
$z = f(x, y) = \dfrac{4}{\sqrt{x^2 + y^2}}$

(d) The surface
$z = f(x, y) = x^2 + 4y^2$

Functions of Three Variables

The function f defined by $w = f(x, y, z)$ is a function of the three independent variables x, y, and z. For each ordered triple (x, y, z) in the domain, the function f assigns a value to w, the dependent variable. In this case, the domain is a collection of points in space. Figure 13 illustrates a way of depicting the function $w = f(x, y, z)$.

FIGURE 13

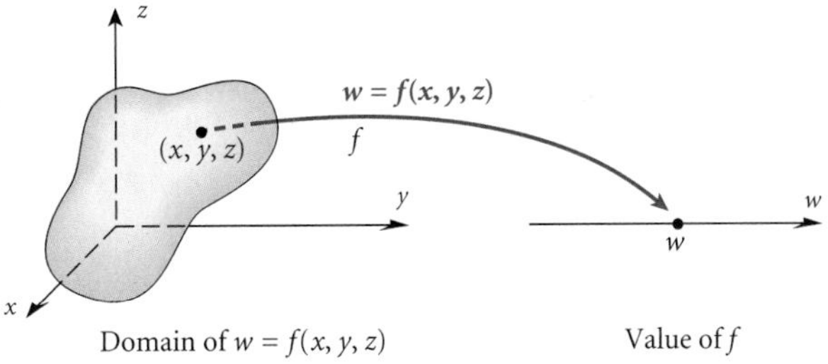

As with functions of two variables, a function of three variables is usually given by an equation, and, unless otherwise stated, the domain is taken to be the largest set of points in space for which this equation is defined in the real number system.

EXAMPLE 4 **Finding the Domain of a Function of Three Variables**

Find the domain of the function $w = f(x, y, z) = \sqrt{9 - x^2 - y^2 - z^2}$.

SOLUTION Since square roots of negative numbers are not defined, the domain of this function consists of all points for which $x^2 + y^2 + z^2 - 9 \leq 0$. Therefore, the domain consists of all points inside and on the sphere with center at $(0, 0, 0)$ and radius 3. See Figure 14 for the graph of the domain.

FIGURE 14

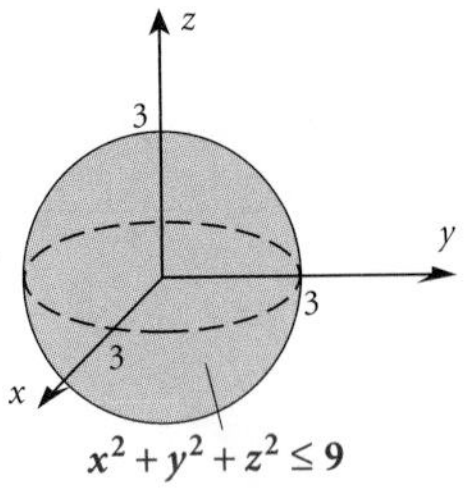

The graph of a function $w = f(x, y, z)$ of three variables consists of all points (x, y, z, w) for which (x, y, z) is in the domain of f and $w = f(x, y, z)$. Because this requires locating points in a four-dimensional space, it is impossible for us to draw such a graph.

NOW WORK PROBLEM 27.

EXERCISE 17.2 Answers to Odd-Numbered Problems Begin on Page AN-100.

In Problems 1–10, evaluate $f(2, 1)$.

1. $f(x, y) = x^2 + y$
2. $f(x, y) = x - y^2$
3. $f(x, y) = \sqrt{xy}$
4. $f(x, y) = x\sqrt{y}$
5. $f(x, y) = \dfrac{1}{2x + y}$
6. $f(x, y) = \dfrac{x}{x - 3y}$
7. $f(x, y) = \dfrac{x^2 - y}{x - y}$
8. $f(x, y) = \dfrac{x + y^2}{x^2 - y^2}$
9. $f(x, y) = \sqrt{4 - x^2y^2}$
10. $f(x, y) = \sqrt{9 - x^2y^2}$

11. Let $f(x, y) = 3x + 2y + xy$. Find:

    (a) $f(1, 0)$ (b) $f(0, 1)$ (c) $f(2, 1)$
    (d) $f(x + \Delta x, y)$ (e) $f(x, y + \Delta y)$

12. Let $f(x, y) = x^2y + x + 1$. Find:

    (a) $f(0, 0)$ (b) $f(0, 1)$ (c) $f(2, 1)$
    (d) $f(x + \Delta x, y)$ (e) $f(x, y + \Delta y)$

13. Let $f(x, y) = \sqrt{xy} + x$. Find:

    (a) $f(0, 0)$ (b) $f(0, 1)$ (c) $f(a^2, t^2)$; $a > 0, t > 0$
    (d) $f(x + \Delta x, y)$ (e) $f(x, y + \Delta y)$

14. Let $f(x, y) = e^{x+y}$. Find:

    (a) $f(0, 0)$ (b) $f(1, -1)$ (c) $f(x + \Delta x, y)$
    (d) $f(x, y + \Delta y)$

15. Let $f(x, y, z) = x^2y + y^2z$. Find:

    (a) $f(1, 2, 3)$ (b) $f(0, 1, 2)$ (c) $f(-1, -2, -3)$

16. Let $f(x, y, z) = 3x^2 + y^2 - 2z^2$. Find:

    (a) $f(1, 2, 3)$ (b) $f(0, 1, 2)$ (c) $f(-1, -2, -3)$

In Problems 17–30, find the domain of each function and give a description of the domain. For Problems 17–26, graph the domain.

17. $z = f(x, y) = \sqrt{x}\sqrt{y}$
18. $z = f(x, y) = \sqrt{xy}$
19. $z = f(x, y) = \sqrt{9 - x^2 - y^2}$
20. $z = f(x, y) = \sqrt{x^2 + y^2 - 16}$
21. $z = f(x, y) = \dfrac{\ln x}{\ln y}$
22. $z = f(x, y) = \ln \dfrac{x}{y}$
23. $z = f(x, y) = \dfrac{3}{x^2 + y^2 - 4}$
24. $z = f(x, y) = \dfrac{4}{9 - x^2 - y^2}$
25. $z = f(x, y) = \ln (x^2 + y^2)$
26. $z = f(x, y) = \ln (4x - y^2)$
27. $w = f(x, y, z) = \sqrt{x^2 + y^2 + z^2 - 16}$
28. $w = f(x, y, z) = \sqrt{9 - (x^2 + y^2 + z^2)}$
29. $w = f(x, y, z) = \dfrac{4}{x^2 + y^2 + z^2}$
30. $w = f(x, y, z) = \ln(x^2 + y^2 + z^2)$

31. For the function $z = f(x, y) = 3x + 4y$, find:

(a) $f(x + \Delta x, y)$
(b) $f(x + \Delta x, y) - f(x, y)$
(c) $\dfrac{f(x + \Delta x, y) - f(x, y)}{\Delta x}, \quad \Delta x \neq 0$
(d) $\lim_{\Delta x \to 0} \dfrac{f(x + \Delta x, y) - f(x, y)}{\Delta x}, \quad \Delta x \neq 0$

32. For the function $z = f(x, y) = 4x + 5y$, find:

(a) $f(x, y + \Delta y)$
(b) $f(x, y + \Delta y) - f(x, y)$
(c) $\dfrac{f(x, y + \Delta y) - f(x, y)}{\Delta y}, \quad \Delta y \neq 0$
(d) $\lim_{\Delta y \to 0} \dfrac{f(x, y + \Delta y) - f(x, y)}{\Delta y}, \quad \Delta y \neq 0$

33. Cost of Construction The cost of the bottom and top of a cylindrical tank is \$300 per square meter and the cost of the sides is \$500 per square meter. Write the total cost of constructing such a tank as a function of the radius r and height h (both is meters).

34. Cost of Construction The cost per square centimeter of the material to be used for an open rectangular box is \$4 for the bottom and \$2 for the other sides. Write the total cost of constructing such a box as a function of its bottom and side dimensions.

35. Baseball A pitcher's earned run average is given by

$$A(N, I) = 9\left(\frac{N}{I}\right)$$

where N is the total number of earned runs given up in I innings of pitching. Find

(a) $A(3, 4)$ (b) $A(6, 3)$ (c) $A(2, 9)$ (d) $A(3, 18)$

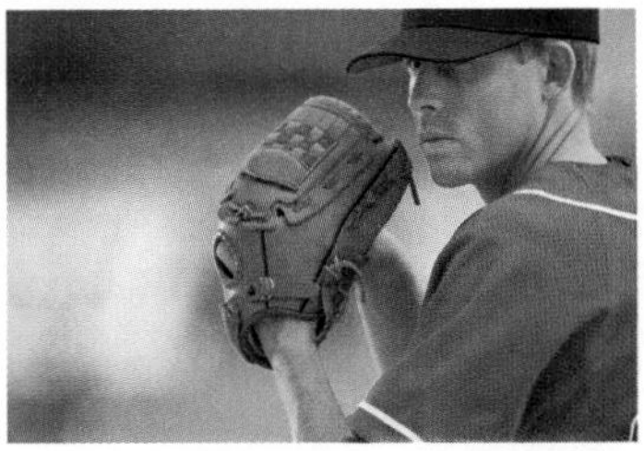

36. Intelligence Quotient In psychology, intelligence quotient (IQ) is measured by

$$\text{IQ} = f(M, C) = 100\frac{M}{C}$$

where M is a person's mental age and C is the person's chronological or actual age, $0 < C \leq 16$.

(a) Find the IQ of a 12-year-old child whose mental age is 10.
(b) Find the IQ of a 10-year-old child whose mental age is 12.
(c) If a 10-year-old girl has an IQ of 120, what is her mental age?

37. Cell Phone Usage One version of the Cingular Family Plan for cellular phone usage has a monthly cost of $B(x, y) = 79.99 + 0.4x + 0.02y$, where x is the number of minutes in excess of 500 during weekdays between 7:00 A.M. and 9:00 P.M. and y is the number of minutes in excess of 1500 at all other times. Find the total monthly bill for using 650 minutes weekdays between 7:00 A.M. and 9:00 P.M. and 1600 minutes at all other times.

Source: Cingular Wireless Phone Company.

38. Field Goal Percentages in the NBA In the National Basketball Association (NBA), the Adjusted Field Goal Percentage is determined by $f(x, y, s) = \dfrac{x + 1.5y}{s}$, where x is the number of two-point field goals made, y is the number of three-point field goals made, and s is the sum of two-point and three-point field goal attempts. Find the Adjusted Field Goal Percentages for the following NBA players for the 2002–2003 season.

(a) Stromile Swift of the Memphis Grizzlies, who made 235 out of 489 two-point attempts and 0 of 2 three-point attempts.
(b) Kobe Bryant of the Los Angeles Lakers, who made 744 out of 1600 two point field goal attempts, and 124 out of 324 three-point field goal attempts.

Source: National Basketball Association.

39. Heat Index At the 2000 Olympic Games in Sydney, Australia, the following formula was used to calculate the heat index (apparent temperature):

$$\begin{aligned} H = &-42.379 + 2.04901523\,t + 10.14333127\,r \\ &- 0.22475541\,tr - 0.00683783\,t^2 - 0.05481717\,r^2 \\ &+ 0.00122874\,t^2r + 0.00085282\,tr^2 - 0.00000199\,t^2r^2 \end{aligned}$$

where H = the heat index (in °F)
t = the air temperature (in °F)
r = the % relative humidity (e.g., $r = 75$ when the relative humidity is 75%)

To alert athletes and others, the Australian Bureau of Meteorology published the following effects for various apparent temperatures, H:

Heat Index H	Physical Effects
90°F – 104°F	Heat cramps or heat exhaustion possible
105°F – 130°F	Heat cramps or heat exhaustion likely, heatstroke possible
above 130°F	Heatstroke highly likely

(a) What is the heat index when the air temperature is 95°F and the relative humidity is 50%?
(b) If the air temperature is 97°F, what is the smallest relative humidity (to the nearest whole percent) that will result in a heat index of 105°F?
(c) If the air temperature is 102°F, what is the smallest relative humidity (to the nearest whole percent) that will result in a heat index of 130°F?

Source: Commonwealth Bureau of Meteorology, Melbourne, Australia.

40. Cobb-Douglas Model The production function for a toy manufacturer is given by the equation

$$Q(L, M) = 400L^{0.3}M^{0.7}$$

where Q is the output in units, L is the labor in hours, and M is the number of machine hours. Find

(a) $Q(19, 21)$ (b) $Q(21, 20)$

17.3 Partial Derivatives

PREPARING FOR THIS SECTION ***Before getting started, review the following:***

>> The Definition of the Derivative of a Function (Chapter 13, Section 13.1, p. 718)

OBJECTIVES
1. Find partial derivatives
2. Find the slope of the tangent line to a surface
3. Interpret partial derivatives
4. Find higher-order partial derivatives

For a function $y = f(x)$ of one independent variable, we have defined the derivative f'. For a function $z = f(x, y)$ of two independent variables, we introduce the idea of a *partial derivative.* A function $z = f(x, y)$ of two variables x and y will have two partial derivatives: f_x, the partial derivative of f with respect to x; and f_y, the partial derivative of f with respect to y.

The partial derivative of f with respect to x is found by differentiating f with respect to x while treating y as if it were a constant.
The partial derivative of f with respect to y is found by differentiating f with respect to y while treating x as if it were a constant.

For example, if $z = f(x, y) = 2xy + 3xy^2$, then we can find f_x by differentiating $z = 2xy + 3xy^2$ with respect to x, while treating y as if it were a constant. The result is

$$f_x = 2y + 3y^2 \quad \frac{dz}{dx} = \frac{d}{dx}(2xy + 3xy^2) = 2y + 3y^2 \text{ if } y \text{ is a constant}$$

Similarly, by treating x as if it were constant and differentiating with respect to y, we obtain

$$f_y = 2x + 6xy \qquad \frac{dz}{dy} = \frac{d}{dy}(2xy + 3xy^2) = 2x + 6xy \text{ if } x \text{ is a constant}$$

We define these partial derivatives as limits of certain difference quotients.

Partial Derivatives of $z = f(x, y)$

Let $z = f(x, y)$ be a function of two variables. Then the **partial derivative of f with respect to x and the partial derivative of f with respect to y** are functions f_x and f_y defined as follows:

$$f_x(x, y) = \lim_{\Delta x \to 0} \frac{f(x + \Delta x, y) - f(x, y)}{\Delta x} \tag{1}$$

$$f_y(x, y) = \lim_{\Delta y \to 0} \frac{f(x, y + \Delta y) - f(x, y)}{\Delta y} \tag{2}$$

provided these limits exist.

Observe the similarity between the above definitions and the definition of a derivative given in Chapter 13. Observe also that in $f_x(x, y)$, an increment Δx is given to x, while y is fixed; in $f_y(x, y)$, an increment Δy is given to y, while x is fixed.

Finding Partial Derivatives of $z = f(x, y)$

To find $f_x(x, y)$: Differentiate f with respect to x while treating y as a constant.

To find $f_y(x, y)$: Differentiate f with respect to y while treating x as a constant.

1 **EXAMPLE 1 Finding Partial Derivatives**

Find f_x and f_y for

$$z = f(x, y) = 4x^3 + 2x^2y + y^2$$

SOLUTION To find f_x, we treat y as a constant and differentiate $z = 4x^3 + 2x^2y + y^2$ with respect to x. The result is

$$f_x(x, y) = 12x^2 + 4xy$$

To find f_y, we treat x as a constant and differentiate $z = 4x^3 + 2x^2y + y^2$ with respect to y. The result is

$$f_y(x, y) = 2x^2 + 2y$$

EXAMPLE 2 Finding Partial Derivatives

(a) Find f_x and f_y for $z = f(x, y) = x^2 \ln y + ye^x$.
(b) Evaluate $f_x(2, 1)$ and $f_y(2, 1)$.

SOLUTION **(a)** $f_x(x, y) = 2x \ln y + ye^x \qquad f_y(x, y) = x^2 \cdot \dfrac{1}{y} + e^x = \dfrac{x^2}{y} + e^x$

(b) $f_x(2, 1) = 2 \cdot 2 \cdot \ln 1 + 1 \cdot e^2 = e^2 \qquad f_y(2, 1) = \dfrac{2^2}{1} + e^2 = 4 + e^2$

NOW WORK PROBLEM 1.

Another Notation

There is another notation used for the partial derivatives f_x and f_y of a function $z = f(x, y)$, which we introduce here:

$$f_x(x, y) = \frac{\partial f}{\partial x} = \frac{\partial z}{\partial x} \qquad f_y(x, y) = \frac{\partial f}{\partial y} = \frac{\partial z}{\partial y}$$

The symbols $\dfrac{\partial}{\partial x}$ and $\dfrac{\partial}{\partial y}$ read "the partial with respect to x" and "the partial with respect to y" respectively, denote operations performed on a function to obtain the partial derivatives with respect to x in the case of $\dfrac{\partial}{\partial x}$ and with respect to y in the case of $\dfrac{\partial}{\partial y}$. Using these notations sometimes makes it a little easier to find partial derivatives.

EXAMPLE 3 Finding Partial Derivatives

(a) $\dfrac{\partial}{\partial x}(3e^xy^2 - xy) = \dfrac{\partial}{\partial x}(3e^xy^2) - \dfrac{\partial}{\partial x}(xy) = 3y^2 \dfrac{\partial}{\partial x} e^x - y\dfrac{\partial}{\partial x}x = 3y^2e^x - y$

Treat y as a constant

(b) $\dfrac{\partial}{\partial y}(3e^xy^2 - xy) = \dfrac{\partial}{\partial y}(3e^xy^2) - \dfrac{\partial}{\partial y}(xy) = 3e^x \dfrac{\partial}{\partial y} y^2 - x\dfrac{\partial}{\partial y}y = 6e^xy - x$

Treat x as a constant

Geometric Interpretation

For a geometric interpretation of the partial derivatives of $z = f(x, y)$, we look at the graph of the surface $z = f(x, y)$. See Figure 15. In computing f_x, we hold y fixed, say, at $y = y_0$,

FIGURE 15

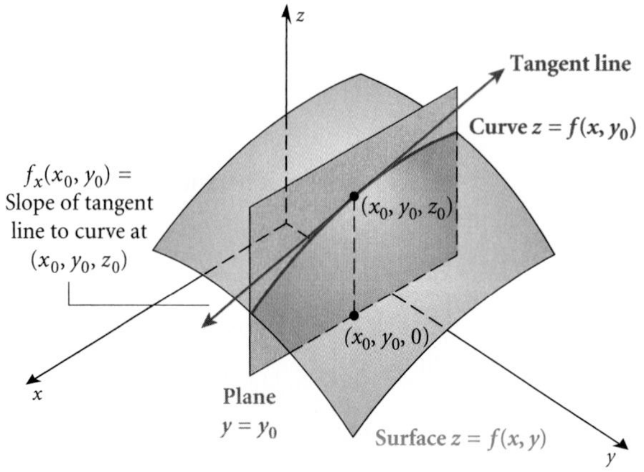

and then differentiate with respect to x. But holding y fixed at y_0 is equivalent to intersecting the surface $z = f(x, y)$ with the plane $y = y_0$, the result being the curve $z = f(x, y_0)$. The partial derivative f_x is the slope of the tangent line to this curve.

Geometric Interpretation of $f_x(x_0, y_0)$

The slope of the tangent line to the curve of intersection of the surface $z = f(x, y)$ and the plane $y = y_0$ at the point (x_0, y_0, z_0) on the surface equals $f_x(x_0, y_0)$.

Geometric Interpretation of $f_y(x_0, y_0)$

The slope of the tangent line to the curve of intersection of the surface $z = f(x, y)$ and the plane $x = x_0$ at the point (x_0, y_0, z_0) on the surface equals $f_y(x_0, y_0)$.

See Figure 16.

FIGURE 16

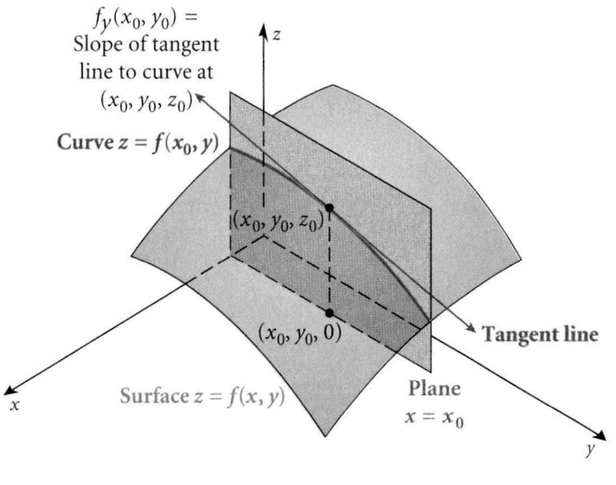

2 **EXAMPLE 4 Finding the Slope of the Tangent Line**

Find the slope of the tangent line to the curve of intersection of the surface $z = f(x, y) = 16 - x^2 - y^2$:

(a) With the plane $y = 2$ at the point $(1, 2, 11)$.
(b) With the plane $x = 1$ at the point $(1, 2, 11)$.

SOLUTION **(a)** The slope of the tangent line to the curve of intersection of the surface $z = 16 - x^2 - y^2$ and the plane $y = 2$ at any point is $f_x(x, y) = -2x$. At the point $(1, 2, 11)$, the slope is $f_x(1, 2) = -2(1) = -2$. See Figure 17(a) on page 526.

(b) The slope of the tangent line to the curve of intersection of the surface $z = 16 - x^2 - y^2$ and the plane $x = 1$ at any point is $f_y(x, y) = -2y$. At the point $(1, 2, 11)$, the slope is $f_y(1, 2) = -2(2) = -4$. See Figure 17(b).

FIGURE 17

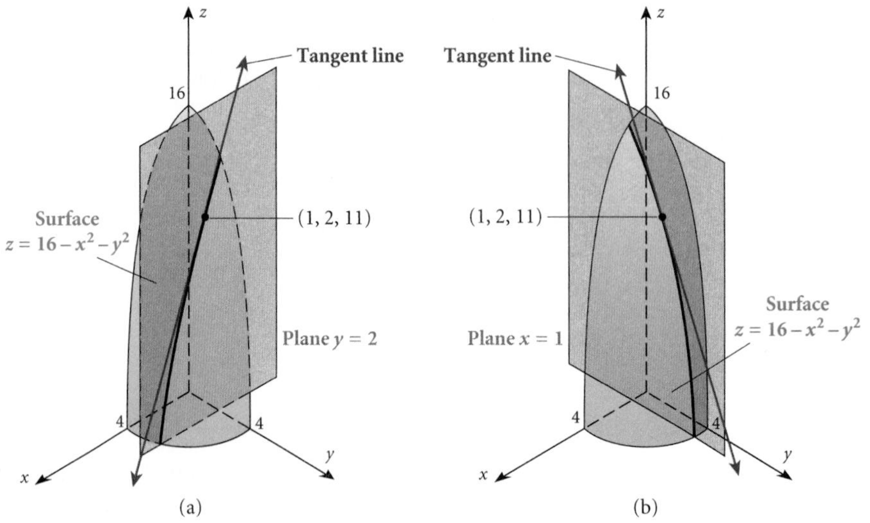

NOW WORK PROBLEM 31.

Applications

Marginal Analysis We have already noted the similarity of the definition of the partial derivatives of $z = f(x, y)$ and the definition of the derivative of a function of one variable. Let's look at how partial derivatives are interpreted in a business environment.

3 EXAMPLE 5 Interpreting Partial Derivatives

In many production processes, the total cost of manufacturing consists of a fixed cost and two variable costs: the cost of raw materials and the cost of labor. If the total cost is given by

$$C(x, y) = 180 + 18x + 40y$$

where x is the cost (in dollars) of raw materials and y is the cost (in dollars) of labor, find C_x and C_y and give an interpretation to your answer.

SOLUTION

$$C_x(x, y) = \frac{\partial C}{\partial x} = \frac{\partial}{\partial x}(180 + 18x + 40y) = 18$$

We interpret $C_x(x, y) = 18$ to mean that when the cost of labor y is held fixed, an increase of \$1 in the cost of raw materials causes an increase of \$18 in the total cost of the product. The partial derivative C_x measures the incremental cost due to an increase of \$1 in the cost of raw material, while labor costs are held fixed.

$$C_y(x, y) = \frac{\partial C}{\partial y} = \frac{\partial}{\partial y}(180 + 18x + 40y) = 40$$

The partial derivative $C_y(x, y) = 40$ means that when the cost of raw materials x is held fixed, an increase of \$1 in the cost of labor y causes a \$40 increase in the total cost of the product.

The partial derivative C_x is called the **marginal cost of raw material** and C_y is the **marginal cost of labor.**

NOW WORK PROBLEM 43.

Production Function The production of most goods requires the use of more than one component. If the quantity z of a good is produced by using the components x and

y, respectively, then the **production function**

$$z = f(x, y)$$

gives the amount of **output** z when the amounts x and y, the **inputs,** are used simultaneously.

Marginal Productivity

If a production function is given by $z = f(x, y)$, then the partial derivative $\frac{\partial z}{\partial x}$ of z with respect to x (with y held constant) is the **marginal productivity** of x or the **marginal product** of x. The partial derivative $\frac{\partial z}{\partial y}$ of z with respect to y (with x held constant) is the **marginal productivity** of y or the **marginal product** of y.

Notice that the marginal productivity of either input is the rate of increase of the total product as that input is increased, assuming that the amount of the other input remains constant.

Marginal productivity is usually positive—that is, as the amount of one input increases (with the amount of the other input held constant), the output also increases. However, as the input of one component increases, the output usually increases at a decreasing rate until the point is reached at which there is no further increase in output. Then, a decrease in total output occurs. This characteristic behavior of production functions is known as the **law of eventually diminishing marginal productivity.**

EXAMPLE 6 Finding Marginal Productivity

Let $z = 2x^{1/2}y^{1/2}$ be a production function. Find the marginal productivity of x and the marginal productivity of y.

SOLUTION The marginal productivity of x is

$$\frac{\partial z}{\partial x} = \frac{\partial}{\partial x}(2x^{1/2}y^{1/2}) = 2y^{1/2}\frac{\partial}{\partial x}x^{1/2} = 2y^{1/2}\cdot\frac{1}{2}x^{-1/2} = \frac{y^{1/2}}{x^{1/2}}$$

The marginal productivity of y is

$$\frac{\partial z}{\partial y} = \frac{\partial}{\partial y}(2x^{1/2}y^{1/2}) = 2x^{1/2}\frac{\partial}{\partial x}y^{1/2} = 2x^{1/2}\cdot\frac{1}{2}y^{-1/2} = \frac{x^{1/2}}{y^{1/2}}$$

Observe that $\frac{\partial z}{\partial x}$ is always positive, but decreases as x increases since y is held fixed. Similarly, $\frac{\partial z}{\partial y}$ is always positive, but decreases as y increases since x is held fixed.

Higher-Order Partial Derivatives

For a function $z = f(x, y)$ of two variables for which the limits (1) and (2) exist, there are two **first-order partial derivatives:** f_x and f_y. If it is possible to differentiate each of these partially with respect to x or partially with respect to y, there will result four **second-order partial derivatives,** namely,

$$f_{xx}(x, y) = \frac{\partial}{\partial x}f_x(x, y) = \frac{\partial}{\partial x}\frac{\partial z}{\partial x} = \frac{\partial^2 z}{\partial x^2} \qquad f_{xy}(x, y) = \frac{\partial}{\partial y}f_x(x, y) = \frac{\partial}{\partial y}\frac{\partial z}{\partial x} = \frac{\partial^2 z}{\partial y\,\partial x}$$

$$f_{yx}(x, y) = \frac{\partial}{\partial x}f_y(x, y) = \frac{\partial}{\partial x}\frac{\partial z}{\partial y} = \frac{\partial^2 z}{\partial x\,\partial y} \qquad f_{yy}(x, y) = \frac{\partial}{\partial y}f_y(x, y) = \frac{\partial}{\partial y}\frac{\partial z}{\partial y} = \frac{\partial^2 z}{\partial y^2}$$

The two second-order partial derivatives

$$\frac{\partial^2 z}{\partial x\,\partial y} = f_{yx}(x, y) \qquad \text{and} \qquad \frac{\partial^2 z}{\partial y\,\partial x} = f_{xy}(x, y)$$

are called **mixed partials.**

Observe the difference in the mixed partials. The notation f_{yx} means that first we should differentiate f partially with respect to y and then differentiate the result partially with respect to x—in that order! On the other hand, f_{xy} means we should differentiate with respect to x and then with respect to y.

> For most functions the two mixed partials are equal.

Although there are functions for which the mixed partials are unequal, they are rare and will not be encountered in this book.

4 **EXAMPLE 7 Finding Higher-Order Partial Derivatives**

Find all second-order partial derivatives of: $z = f(x, y) = x \ln y + ye^x$

SOLUTION First, we find the first-order partial derivatives $f_x(x, y)$ and $f_y(x, y)$.

$$f_x(x, y) = \frac{\partial}{\partial x}(x \ln y + ye^x) = \ln y + ye^x \qquad f_y(x, y) = \frac{\partial}{\partial y}(x \ln y + ye^x) = \frac{x}{y} + e^x$$

Then,

$$f_{xx} = \frac{\partial}{\partial x} f_x = \frac{\partial}{\partial x}(\ln y + ye^x) = ye^x \qquad f_{xy} = \frac{\partial}{\partial y} f_x = \frac{\partial}{\partial y}(\ln y + ye^x) = \frac{1}{y} + e^x$$

$$f_{yx} = \frac{\partial}{\partial x} f_y = \frac{\partial}{\partial x}\left(\frac{x}{y} + e^x\right) = \frac{1}{y} + e^x \qquad f_{yy} = \frac{\partial}{\partial y} f_y = \frac{\partial}{\partial y}\left(\frac{x}{y} + e^x\right) = \frac{-x}{y^2}$$

NOW WORK PROBLEM 7.

Functions of Three Variables

The idea of partial differentiation may be extended to a function of three variables. If $w = f(x, y, z)$ is a function of three variables, there will be three first-order partial derivatives: the partial derivative with respect to x is f_x; the partial derivative with respect to y is f_y; and the partial derivative with respect to z is f_z. Each of these is calculated by differentiating with respect to the indicated variable, while treating the other two as constants.

EXAMPLE 8 Finding Partial Derivatives

Find f_x, f_y, f_z, if $f(x, y, z) = 10x^2 y^3 z^4$.

SOLUTION

$$f_x(x, y, z) = \frac{\partial}{\partial x}(10x^2 y^3 z^4) = 10y^3 z^4 \frac{\partial}{\partial x} x^2 = 10y^3 z^4 \cdot 2x = 20xy^3z^4$$

$$f_y(x, y, z) = \frac{\partial}{\partial y}(10x^2 y^3 z^4) = 10x^2 z^4 \frac{\partial}{\partial y} y^3 = 10x^2 z^4 \cdot 3y^2 = 30x^2y^2z^4$$

$$f_z(x, y, z) = \frac{\partial}{\partial z}(10x^2 y^3 z^4) = 10x^2 y^3 \frac{\partial}{\partial z} z^4 = 10x^2 y^3 \cdot 4z^3 = 40x^2y^3z^3$$

NOW WORK PROBLEM 23.

EXERCISE 17.3 Answers to Odd-Numbered Problems Begin on Page AN-101.

In Problems 1–6, find f_x, f_y, $f_x(2, -1)$, and $f_y(-2, 3)$.

1. $f(x, y) = 3x - 2y + 3y^3$
2. $f(x, y) = 2x^3 - 3y + x^2$
3. $f(x, y) = (x - y)^2$
4. $f(x, y) = (x - y)^3$
5. $f(x, y) = \sqrt{x^2 + y^2}$
6. $f(x, y) = \sqrt{x^2 - y^2}$

In Problems 7–16, find f_x, f_y, f_{xx}, f_{yy}, f_{yx}, and f_{xy}.

7. $f(x, y) = y^3 - 2xy + y^2 - 12x^2$
8. $f(x, y) = x^3 - xy + 10y^2x$
9. $f(x, y) = xe^y + ye^x + x$
10. $f(x, y) = xe^x + xe^y + y$
11. $f(x, y) = \dfrac{x}{y}$
12. $f(x, y) = \dfrac{y}{x}$
13. $f(x, y) = \ln(x^2 + y^2)$
14. $f(x, y) = \ln(x^2 - y^2)$
15. $f(x, y) = \dfrac{10 - x + 2y}{xy}$
16. $f(x, y) = \dfrac{5 + 3x - 2y}{xy}$

In Problems 17–22, verify that $f_{xy} = f_{yx}$.

17. $f(x, y) = x^3 + y^2$
18. $f(x, y) = x^2 - y^3$
19. $f(x, y) = 3x^4y^2 + 7x^2y$
20. $f(x, y) = 5x^3y - 8xy^2$
21. $f(x, y) = \dfrac{y}{x^2}$
22. $f(x, y) = \dfrac{x}{y^2}$

In Problems 23–30, find f_x, f_y, f_z.

23. $f(x, y, z) = x^2y - 3xyz + z^3$
24. $f(x, y, z) = 3xy + 4yz + 8z^2$
25. $f(x, y, z) = xe^y + ye^z$
26. $f(x, y, z) = x \ln y + y \ln z$
27. $f(x, y, z) = x \ln(yz) + y \ln(xz)$
28. $f(x, y, z) = e^{(3x+4y+5z)}$
29. $f(x, y, z) = \ln(x^2 + y^2 + z^2)$
30. $f(x, y, z) = e^{(x^2+y^2+z^2)}$

In Problems 31–38, find the slope of the tangent line to the curve of intersection of the surface $z = f(x, y)$ with the given plane at the indicated point.

31. $z = f(x, y) = 5x^2 + 3y^2$; plane: $y = 3$; point: $(2, 3, 47)$
32. $z = f(x, y) = 2x^2 - 4y^2$; plane: $x = 2$; point: $(2, 3, -28)$
33. $z = f(x, y) = \sqrt{16 - x^2 - y^2}$; plane: $x = 1$; point: $(1, 2, \sqrt{11})$
34. $z = f(x, y) = \sqrt{x^2 - y^2}$; plane: $y = 0$; point: $(4, 0, 4)$
35. $z = f(x, y) = e^x \ln y$; plane: $x = 0$; point: $(0, 1, 0)$
36. $z = f(x, y) = e^{2x+3y}$; plane: $y = 0$; point: $(0, 0, 1)$
37. $z = f(x, y) = 2 \ln \sqrt{x^2 + y^2}$; plane: $x = 1$; point: $(1, 1, 2 \ln 2)$
38. $z = f(x, y) = e^{x^2+y^2}$; plane: $y = 0$; point: $(1, 0, e)$
39. If $z = x^2 + 4y^2$, show that $x\dfrac{\partial z}{\partial x} + y\dfrac{\partial z}{\partial y} = 2z$
40. If $z = xy^2$, show that $x\dfrac{\partial z}{\partial x} + y\dfrac{\partial z}{\partial y} = 3z$
41. If $z = \ln \sqrt{x^2 + y^2}$, show that $\dfrac{\partial^2 z}{\partial x^2} + \dfrac{\partial^2 z}{\partial y^2} = 0$
42. If $z = e^{x \ln y}$, find: $\dfrac{\partial^2 z}{\partial x^2}$, $\dfrac{\partial^2 z}{\partial y^2}$, and $\dfrac{\partial^2 z}{\partial x\, \partial y}$
43. **Demand for Butter** In a large town, the demand for butter (measured in pounds) is given by the formula

$$z = 1000 - 20x - 50y$$

where x (in dollars) is the average price per pound of butter and y (in dollars) is the average price per pound of margarine.

(a) Find the two first-order partial derivatives of z.
(b) Interpret each partial derivative.

44. **Marginal Productivity** The production function of a certain commodity is given by

$$P = 8I - I^2 + 3Ik + 50k - k^2$$

where I and k are the labor and capital inputs, respectively.

(a) Find the marginal productivities of I and k at $I = 2$ and $k = 5$.
(b) Interpret each marginal productivity.

45. Baseball A pitcher's earned run average is given by

$$A = 9\left(\frac{N}{I}\right)$$

where N is the total number of earned runs given up in I inning of pitching,

(a) Find both first-order partial derivatives.
(b) Evaluate the two-first order partial derivatives for the earned run average of Kevin Millwood of the Philadelphia Phillies who in the 2002 season gave up 78 earned runs in 217 innings.
(c) Interpret each partial derivative found in part (b).

46. Field Goal Percentage in the NBA The Adjusted Field Goal Percentage is determined by

$$f(x, y, z) = \frac{x + 1.5y}{s},$$

where x is the number of two-point field goals made, y is the number of three-point field goals made, and s is the total number of field goals attempted.

(a) Evaluate the three first-order partial derivatives for the Adjusted Field Goal Percentage function for Tracy McGrady, who made 656 of 1365 two point attempts and 173 out of 448 three-point attempts.
(b) Interpret each one of the three partial derivatives.

47. Heat Index At the 2000 Olympic Games in Sydney, Australia, the following formula was used to calculate the heat index (apparent temperature):

$$\begin{aligned} H = &-42.379 + 2.04901523t + 10.14333127r \\ &- 0.22475541tr - 0.00683783t^2 \\ &- 0.05481717r^2 + 0.00122874t^2r \\ &+ 0.00085282tr^2 - 0.00000199t^2r^2 \end{aligned}$$

where H = the heat index (in °F)
t = the air temperature (in °F)
r = the % relative humidity (e.g., $r = 75$ when the relative humidity is 75%)

(a) Find $\frac{\partial H}{\partial t}$.
(b) Give an interpretation to $\frac{\partial H}{\partial t}$.
(c) Find $\frac{\partial H}{\partial r}$.
(d) Give an interpretation to $\frac{\partial H}{\partial r}$.

Source: Commonwealth Bureau of Meteorology, Melbourne, Australia.

48. The Cobb-Douglas Model In 1928 the mathematician Charles W. Cobb and the economist Paul H. Douglas empirically derived a production model for the manufacturing sector of the United States. The table on the right shows the data used by Cobb and Douglas to construct their original model. In the table, P represents the output, K the capital input, and L the labor input of the manufacturing sector of the United States for the period 1899–1922. The data are scaled so that for the year 1899, $P = K = L = 100$. Using the model $P = aK^bL^{1-b}$ and multiple regression techniques, Cobb and Douglas determined the following relationship between P, K, and L:

$$P = 1.014651K^{.254134}L^{.745866} \approx 1.01K^{.25}L^{.75}$$

(a) For the Cobb-Douglas production model, find the partial derivatives: $\frac{\partial P}{\partial K}$ and $\frac{\partial P}{\partial L}$.
(b) Evaluate each partial derivative for $K = 226$ and $L = 152$. (These are the values for K and L in 1912.)
(c) Interpret these partial derivatives in terms of the Cobb-Douglas production model.

Year	P	K	L
1899	100	100	100
1900	101	107	105
1901	112	114	110
1902	122	122	118
1903	124	131	123
1904	122	138	116
1905	143	149	125
1906	152	163	133
1907	151	176	138
1908	126	185	121
1909	155	198	140
1910	159	208	144
1911	153	216	145
1912	177	226	152
1913	184	236	154
1914	169	244	149
1915	189	266	154
1916	225	298	182
1917	227	335	196
1918	223	366	200
1919	218	387	193
1920	231	407	193
1921	179	417	147
1922	240	431	161

Source: Cobb and Douglas, 1928.

49. If you are told that a function $z = f(x, y)$ has the two partial derivatives $f_x(x, y) = 3x - y$ and $f_y(x, y) = x - 3y$, should you believe it? Explain.

17.4 Local Maxima and Local Minima

PREPARING FOR THIS SECTION *Before getting started, review the following:*

- > First Derivative Test (Chapter 14, Section 14.2, pp. 795–796)
- > Second Derivative Test (Chapter 14, Section 14.3, p. 813)

OBJECTIVES

1. Find critical points
2. Determine the character of a critical point
3. Solve applied problems

We saw in Chapter 5 that an important application of the derivative is to find the local maxima and the local minima of a function of one variable. In this section we learn that the partial derivatives of a function of two variables are used in a similar way to find the local maxima and the local minima of a function of two variables.

Local Maximum; Local Minimum

Let $z = f(x, y)$ denote a function of two variables. We say that f has a **local maximum** at a point (x_0, y_0) if $f(x_0, y_0) \geq f(x, y)$ for all points (x, y) close to the point (x_0, y_0). Similarly, f has a **local minimum** at (x_1, y_1) if $f(x_1, y_1) \leq f(x, y)$ for all points (x, y) close to (x_1, y_1).

Figure 18 illustrates the graph of a function $z = f(x, y)$ that has several local maxima and minima.

FIGURE 18

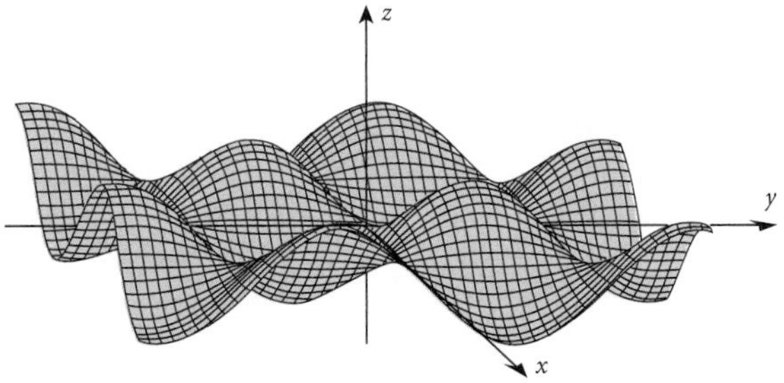

FIGURE 19

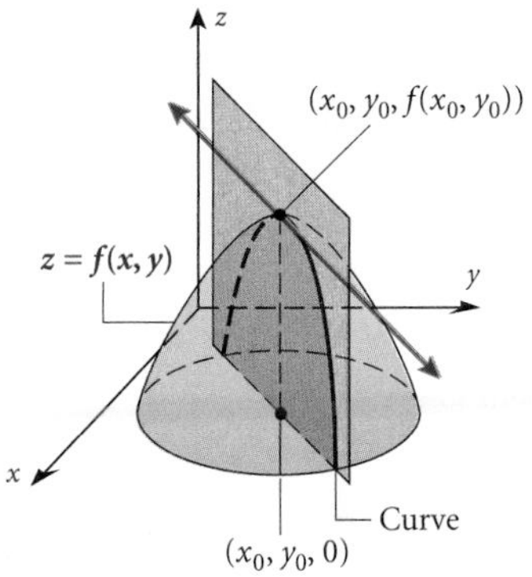

Suppose a function $z = f(x, y)$ has a local maximum at the point (x_0, y_0). Then the curve that results from the intersection of the surface $z = f(x, y)$ and any plane through the point $(x_0, y_0, 0)$ that is perpendicular to the xy-plane will have a local maximum at (x_0, y_0). See Figure 19.

In particular, the curve that results from intersecting $z = f(x, y)$ with the plane $x = x_0$ has this property. This means that if f_y exists at (x_0, y_0), then $f_y(x_0, y_0) = 0$. By a similar argument, we also have $f_x(x_0, y_0) = 0$. This leads us to formulate the following necessary condition for a local maximum and a local minimum:

A Necessary Condition for Local Maxima and Local Minima

Let $z = f(x, y)$ denote a function of two variables and let (x_0, y_0) be a point in its domain. Suppose f_x and f_y each exist at (x_0, y_0). If f has a local maximum at (x_0, y_0) or a local minimum at (x_0, y_0), then

$$f_x(x_0, y_0) = 0 \quad \text{and} \quad f_y(x_0, y_0) = 0 \tag{1}$$

From this theorem we see the importance of those points at which the partial derivatives exist and are zero simultaneously. Such points are called **critical points.**

We say that f has a **critical point** at (x_0, y_0) if

$$f_x(x_0, y_0) = 0 \quad \text{and} \quad f_y(x_0, y_0) = 0$$

FIGURE 20

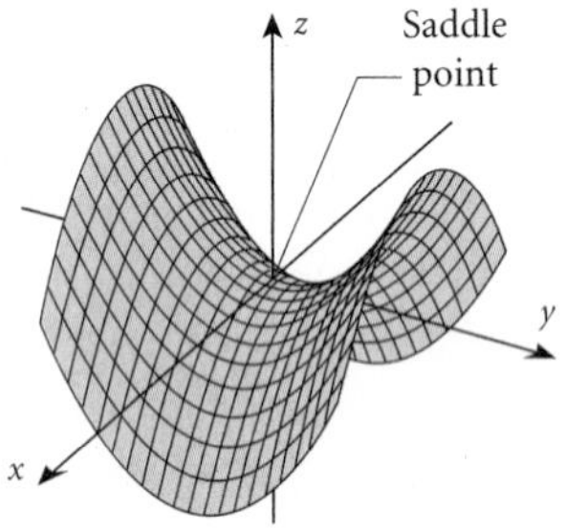

It can happen that f has a critical point at (x_0, y_0) but that the point is neither a local maximum nor a local minimum. Such points are called **saddle points.** Figure 20 illustrates a saddle point—and suggests justification for the name. Note that this saddle point is a maximum when viewed relative to x and a minimum when viewed relative to y.

1 **EXAMPLE 1 Finding Critical Points**

Find all the critical points of

$$z = f(x, y) = x^2 + y^4 - 2y^2 + 6$$

SOLUTION We compute the partial derivatives f_x and f_y, set each equal to zero, and solve the resulting system of equations.

$$f_x(x, y) = \frac{\partial}{\partial x}(x^2 + y^4 - 2y^2 + 6) = 2x = 0$$

$$f_y(x, y) = \frac{\partial}{\partial y}(x^2 + y^4 - 2y^2 + 6) = 4y^3 - 4y = 0$$

From the first equation, we obtain $x = 0$; from the second, we obtain

$$
\begin{aligned}
4y^3 - 4y &= 0 \\
4y(y^2 - 1) &= 0 && \text{Factor.} \\
4y(y + 1)(y - 1) &= 0 && \text{Factor.} \\
y = 0 \text{ or } y = -1 \text{ or } y &= 1 && \text{Solve for } y.
\end{aligned}
$$

We find that f has three critical points: $(0, 0)$, $(0, -1)$, and $(0, 1)$.

NOW WORK PROBLEM 1.

Tests for Local Maxima, Local Minima, Saddle Points

We still don't know, though, the character of these critical points. What we need is a test, or tests, to tell us whether these critical points are in fact local maxima, local minima, or saddle points. For functions that possess both first- and second-order partial derivatives (this is true of all the ones we will encounter in this book), the following tests may be used.

Test for a Local Maximum

Suppose (x_0, y_0) is a critical point of $z = f(x, y)$. If

1. $f_{xx}(x_0, y_0) < 0$
2. $D = f_{xx}(x_0, y_0) \cdot f_{yy}(x_0, y_0) - [f_{xy}(x_0, y_0)]^2 > 0$

then the function has a local maximum at (x_0, y_0).

Test for a Local Minimum

Suppose (x_0, y_0) is a critical point of $z = f(x, y)$. If

1. $f_{xx}(x_0, y_0) > 0$
2. $D = f_{xx}(x_0, y_0) \cdot f_{yy}(x_0, y_0) - [f_{xy}(x_0, y_0)]^2 > 0$

then the function f has a local minimum at (x_0, y_0).

Test for Saddle Point

Suppose (x_0, y_0) is a critical point of $z = f(x, y)$. If

$$D = f_{xx}(x_0, y_0) \cdot f_{yy}(x_0, y_0) - [f_{xy}(x_0, y_0)]^2 < 0$$

then the point (x_0, y_0) is a saddle point of f.

Two comments about using these tests:

1. If $D > 0$, then $f_{xx}(x_0, y_0)$ and $f_{yy}(x_0, y_0)$ are each of the same sign.
2. If $D = 0$, no information results.

2 **EXAMPLE 2 Determining the Character of a Critical Point**

Return to the function discussed in Example 1, $f(x, y) = x^2 + y^4 - 2y^2 + 6$, and determine the character of each critical point.

SOLUTION We need to compute the second-order partial derivatives. From Example 1, we have

$$f_x(x, y) = 2x \qquad \text{and} \qquad f_y(x, y) = 4y^3 - 4y$$

Then,

$$f_{xx}(x, y) = \frac{\partial}{\partial x} f_x(x, y) = \frac{\partial}{\partial x}(2x) = 2$$

$$f_{yy}(x, y) = \frac{\partial}{\partial y} f_y(x, y) = \frac{\partial}{\partial y}(4y^3 - 4y) = 12y^2 - 4$$

$$f_{xy}(x, y) = \frac{\partial}{\partial y} f_x(x, y) = \frac{\partial}{\partial y}(2x) = 0$$

Next, we evaluate these partial derivatives at each critical point to find the value of D.

(0, 0): $\quad f_{xx}(0, 0) = 2 \qquad f_{yy}(0, 0) = -4 \qquad f_{xy}(0, 0) = 0$

$$D = f_{xx}(0, 0) \cdot f_{yy}(0, 0) - [f_{xy}(0, 0)]^2 = 2 \cdot (-4) - 0^2 = -8 < 0$$

Since $D < 0$, the critical point (0, 0) is a saddle point of f. That is, the point (0, 0, 6) on the graph is a saddle point.

(0, −1): $\quad f_{xx}(0, -1) = 2 \qquad f_{yy}(0, -1) = 8 \qquad f_{xy}(0, -1) = 0$

$$D = f_{xx}(0, -1) \cdot f_{yy}(0, -1) - [f_{xy}(0, -1)]^2 = 2 \cdot 8 - 0^2 = 16 > 0$$

Since $f_{xx}(0, -1) = 2 > 0$, and $D > 0$, f has a local minimum at the critical point (0, −1). That is, at the point (0, −1, 5) on the graph, there is a local minimum.

(0, 1): $\quad f_{xx}(0, 1) = 2 \qquad f_{yy}(0, 1) = 8 \qquad f_{xy}(0, 1) = 0$

$$D = f_{xx}(0, 1) \cdot f_{yy}(0, 1) - [f_{xy}(0, 1)]^2 = 2 \cdot 8 - 0^2 = 16 > 0$$

Since $f_{xx}(0, 1) = 2 > 0$ and $D > 0$, f has a local minimum at the critical point (0, 1). That is, at the point (0, 1, 5) on the graph of f, there is a local minimum.

NOW WORK PROBLEM 7.

EXAMPLE 3 Determining the Character of a Critical Point

For the function

$$z = f(x, y) = x^2 + xy + y^2 - 6x + 6$$

find all critical points. Determine the character of each one.

SOLUTION First, we compute the first-order partial derivatives of $z = f(x, y)$.

$$f_x(x, y) = \frac{\partial}{\partial x}(x^2 + xy + y^2 - 6x + 6) = 2x + y - 6$$

$$f_y(x, y) = \frac{\partial}{\partial y}(x^2 + xy + y^2 - 6x + 6) = x + 2y$$

The critical points, if there are any, obey the system of equations

$$f_x(x, y) = 2x + y - 6 = 0 \qquad \text{and} \qquad f_y(x, y) = x + 2y = 0$$

Now, we solve these equations simultaneously, using substitution. From the equation $2x + y - 6 = 0$ we find $y = 6 - 2x$. Substituting into the equation $x + 2y = 0$, we find

$$\begin{aligned} x + 2(6 - 2x) &= 0 \\ x + 12 - 4x &= 0 \\ -3x &= -12 \\ x &= 4 \end{aligned}$$

Then $y = 6 - 2x = 6 - 2(4) = -2$. The only critical point of f is $(4, -2)$.

$$f_{xx}(x, y) = \frac{\partial}{\partial x} f_x(x, y) = \frac{\partial}{\partial x}(2x + y - 6) = 2$$

$$f_{xy}(x, y) = \frac{\partial}{\partial y} f_x(x, y) = \frac{\partial}{\partial y}(2x + y - 6) = 1$$

$$f_{yy}(x, y) = \frac{\partial}{dy} f_y(x, y) = \frac{\partial}{\partial y}(x + 2y) = 2$$

At the critical point $(4, -2)$, we have

$$f_{xx}(4, -2) = 2 \qquad f_{xy}(4, -2) = 1 \qquad f_{yy}(4, -2) = 2$$

and,

$$D = f_{xx}(4, -2) \cdot f_{yy}(4, -2) - [f_{xy}(4, -2)]^2 = 2 \cdot 2 - 1^2 = 3$$

Since $f_{xx}(4, -2) = 2 > 0$ and $D > 0$, f has a local minimum at the critical point $(4, -2)$. That is, at the point $(4, -2, -6)$ on the graph of f, there is a local minimum. ◗

Application

3 **EXAMPLE 4 Maximizing Profit**

The demand functions for two products are

$$p = 12 - 2x \qquad \text{and} \qquad q = 20 - y$$

where p and q are the respective prices (in thousands of dollars) for each product, and x and y are the respective amounts (in thousands of units) of each sold. Suppose the joint cost function is

$$C(x, y) = x^2 + 2xy + 2y^2$$

Find the revenue function and the profit function. Determine the prices and amounts that will maximize profit. What is the maximum profit?

SOLUTION The revenue function R is the sum of the revenues due to each product. That is,

$$R = R(x, y) = xp + yq = x(12 - 2x) + y(20 - y)$$

The profit function P is

$$\begin{aligned} P = P(x, y) &= R(x, y) - C(x, y) \\ &= x(12 - 2x) + y(20 - y) - (x^2 + 2xy + 2y^2) \\ &= 12x - 2x^2 + 20y - y^2 - x^2 - 2xy - 2y^2 \\ &= -3x^2 - 3y^2 - 2xy + 12x + 20y \end{aligned}$$

The first-order partial derivatives of P are

$$P_x(x, y) = \frac{\partial}{\partial x}(-3x^2 - 3y^2 - 2xy + 12x + 20y) = -6x - 2y + 12$$

$$P_y(x, y) = \frac{\partial}{\partial y}(-3x^2 - 3y^2 - 2xy + 12x + 20y) = -6y - 2x + 20$$

The critical points obey

$$-6x - 2y + 12 = 0 \qquad \text{and} \qquad -6y - 2x + 20 = 0$$

We solve the left equation $-6x - 2y + 12 = 0$ for y.

$$-2y = 6x - 12$$
$$y = -3x + 6$$

Substitute into the right equation and solve for x.

$$-6(-3x + 6) - 2x + 20 = 0 \qquad -6y - 2x + 20 = 0$$
$$18x - 36 - 2x + 20 = 0$$
$$16x = 16$$
$$x = 1$$

Then $y = -3x + 6 = -3(1) + 6 = 3$ so $(1, 3)$ is the critical point.

The second-order partial derivatives of P are

$$P_{xx}(x, y) = \frac{\partial}{\partial x}(-6x - 2y + 12) = -6$$

$$P_{xy}(x, y) = \frac{\partial}{\partial y}(-6x - 2y + 12) = -2$$

$$P_{yy}(x, y) = \frac{\partial}{\partial y}(-6y - 2x + 20) = -6$$

At the critical point $(1, 3)$, we see that

$$P_{xx}(1, 3) = -6 < 0$$
$$D = P_{xx}(1, 3) \cdot P_{yy}(1, 3) - [P_{xy}(1, 3)]^2 = (-6)(-6) - (-2)^2 = 32 > 0$$

We conclude that P has a local maximum at $(1, 3)$. For these quantities sold, namely, $x = 1000$ units and $y = 3000$ units, the corresponding prices p and q are $p = \$10{,}000$ and $q = \$17{,}000$. The maximum profit is $P(1, 3) = \$36{,}000$.

NOW WORK PROBLEM 25.

EXERCISE 17.4 Answers to Odd-Numbered Problems Begin on Page AN-102.

In Problems 1–6, find all the critical points of each function.

1. $f(x, y) = x^4 - 2x^2 + y^2 + 15$
2. $f(x, y) = x^2 - y^2 + 6x - 2y + 14$
3. $f(x, y) = 4xy - x^4 - y^4 + 12$
4. $f(x, y) = x^3 + 6xy + 3y^2 + 8$
5. $f(x, y) = x^4 + y^4$
6. $f(x, y) = xy + \dfrac{2}{x} + \dfrac{4}{y}$

In Problems 7–24, find all the critical points of each function f and determine at each one whether there is a local maximum, a local minimum, or a saddle point of f.

7. $f(x, y) = 3x^2 - 2xy + y^2$
8. $f(x, y) = x^2 - 2xy + 3y^2$
9. $f(x, y) = x^2 + y^2 - 3x + 12$
10. $f(x, y) = x^2 + y^2 - 6y + 10$
11. $f(x, y) = x^2 - y^2 + 4x + 8y$
12. $f(x, y) = x^2 - y^2 - 2x + 4y$

13. $f(x, y) = x^2 + 4y^2 - 4x + 8y - 1$

14. $f(x, y) = x^2 + y^2 - 4x + 2y - 4$

15. $f(x, y) = x^2 + y^2 + xy - 6x + 6$

16. $f(x, y) = x^2 + y^2 + xy - 8y$

17. $f(x, y) = 2 + x^2 - y^2 + xy$

18. $f(x, y) = x^2 - y^2 + 2xy$

19. $f(x, y) = x^3 - 6xy + y^3$

20. $f(x, y) = x^3 - 3xy - y^3$

21. $f(x, y) = x^3 + x^2y + y^2$

22. $f(x, y) = 3y^3 - x^2y + x$

23. $f(x, y) = \dfrac{y}{x + y}$

24. $f(x, y) = \dfrac{x}{x + y}$

25. **Economics** The demand functions for two products are $p = 12 - x$ and $q = 8 - y$, where p and q are the respective prices (in thousands of dollars), and x and y are the respective amounts (in thousands of units) of each product sold. If the joint cost function is $C(x, y) = x^2 + 2xy + 3y^2$, determine the quantities x, y and prices p, q that maximize profit. What is the maximum profit?

26. **Economics** The labor cost of a firm is given by the function

$$Q(x, y) = x^2 + y^3 - 6xy + 3x + 6y - 5$$

where x is the number of days required by a skilled worker and y is the number of days required by a semiskilled worker. Find the values of x and y for which the labor cost is a minimum.

27. **Maximizing Profit** A steel manufacturer produces two grades of steel, x tons of grade A and y tons of grade B. The cost C and revenue R are given in dollars by the formulas

$$C = \frac{1}{20}x^2 + 700x + y^2 - 150y - \frac{1}{2}xy$$

$$R = 2700x - \frac{3}{20}x^2 + 1000y - y^2 + \frac{1}{2}xy + 10{,}000$$

If $P = \text{Profit} = R - C$, find the production (in tons) of grades A and B that maximizes the manufacturer's profit.

28. A certain mountain is in the shape of the surface

$$z = 2xy - 2x^2 - y^2 - 8x + 6y + 4$$

(The unit of distance is 1000 feet.) If sea level is the xy-plane, how high is the mountain?

29. **Reaction to Drugs** Two drugs are used simultaneously as a treatment for a certain disease. The reaction R (measured in appropriate units) to x units of the first drug and y units of the second drug is

$$R(x, y) = x^2y^2(a - x)(b - y) \qquad 0 \le x \le a, \quad 0 \le y \le b$$

(a) For a fixed amount x of the first drug, what amount y of the second drug produces the maximum reaction?
(b) For a fixed amount y of the second drug, what amount x of the first drug produces the maximum reaction?
(c) If x and y are both variable, what amount of each maximizes the reaction?

30. **Reaction to Drugs** The reaction R to x units of a drug t hours after the drug has been administered is given by

$$R(x, t) = x^2(a - x)t^2e^{-t} \qquad 0 \le x \le a$$

For what amount x is the reaction as large as possible? When does the maximum reaction occur?

31. **Reaction to Drugs** The reaction y to an injection of x units of a certain drug, t hours after the injection, is given by

$$y = x^2(a - x)t \qquad 0 \le x \le a$$

Find the values of x and t, if any, that will maximize y.

32. **Metal Detector** A metal detector is used to locate an underground pipe. After several readings of the detector are taken, it is determined that the reading D at an arbitrary point (x, y), $x \ge 0$, $y \ge 0$, is given by

$$D = y(x - x^2) - x^2 \qquad \text{volts}$$

Determine the point (x, y) where the reading is largest.

33. **Parcel Post Regulations** United Parcel Service regulations state that individual packages can be up to 108 inches in length and 130 inches in combined length and girth (perimeter of a cross section) before additional charges apply.

(a) Find the dimensions of maximum volume of a rectangular box which can be sent without additional charges.
(b) What are the dimensions of maximum volume if the package is cylindrical?

Source: United Parcel Service.

Hint: The volume of a cylinder is $V = \pi r^2h$, where r is the radius and h is the height.

34. **Parcel Post Regulations** The U.S. Post Office regulations state that the combined (sum) length and girth of a parcel post package being sent first-class in the United States may not exceed 84 inches. If this combined length and girth exceeds 84 inches, extra postage will be charged according to weight. Find:

(a) The length, width, and height of a rectangular box of maximum volume that can be mailed first class, subject to the 84 inch restriction.
(b) The dimensions of a circular tube of maximum volume that can be mailed first class, subject to the 84 inch restriction.

Source: United States Postal Service.

35. **Waste Management** A car manufacturer uses x tons of steel at the rate of y tons per week. It is found that the waste

W due to storage and interplant distribution amounts to

$$W = \frac{1}{100}\left[\frac{1}{20}x^2 + 25y^2 - x(y + 4)\right] \text{ tons}$$

Determine the value of x and y for which waste is minimum.

36. Expansion of Gas The volume of a fixed amount of gas varies directly with the temperature and inversely with the pressure. That is, $V = k\left(\frac{T}{P}\right)$, where $k > 0$ is a constant and V, T, and P are the volume, temperature, and pressure, respectively.

(a) Calculate $\frac{\partial V}{\partial T}$ and $\frac{\partial V}{\partial P}$.

(b) Prove that

$$P \cdot \frac{\partial V}{\partial P} + T \cdot \frac{\partial V}{\partial T} = 0$$

17.5 Lagrange Multipliers

OBJECTIVE 1 Use the method of Lagrange multipliers

In the previous section we introduced a method to find the local maximum and the local minimum of a function of two variables without any constraints or conditions on the function or the variables. However, in many practical problems we are faced with maximizing or minimizing a function subject to conditions or constraints on the variables involved.

For example, a manufacturer may want to produce a box with a fixed volume so that the least amount of material is used.

Let's look again at an example we solved earlier.

EXAMPLE 1 Maximizing Area

A farmer wants to enclose a rectangular plot that borders on a straight river with a fence. He will not fence in the side along the river. If the farmer has 4000 meters of fencing, what is the largest area that can be enclosed?

SOLUTION Refer to Figure 21. If A is the area to be enclosed, then the problem is to find the maximum value of $A = xy$ subject to the condition that $2x + y = 4000$, that is, subject to the 4000 meters of fence that are available. To express the problem in terms of a single variable, we solve for y in the equation $2x + y = 4000$. Then $y = 4000 - 2x$ and the area A can be expressed in terms of x alone as

FIGURE 21

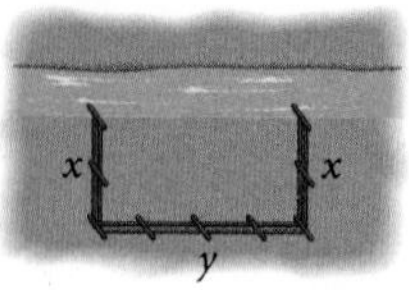

$$A = xy = x(4000 - 2x) = 4000x - 2x^2$$

This equation for A is easy to differentiate. We proceed to find the critical numbers of A:

$$A'(x) = \frac{d}{dx}(4000x - 2x^2) = 4000 - 4x = 0$$

$$x = 1000$$

Since $A''(x) = -4 < 0$, this critical number yields a maximum value for the area A, namely,

$$A = 4000(1000) - 2(1000)^2 = 2{,}000{,}000 \text{ square meters}$$

In Example 1, the problem required that we maximize $A = A(x, y) = xy$ subject to the side condition or *constraint* involving x and y, namely, that $2x + y = 4000$. We were able to the solve this problem using earlier techniques for two reasons:

1. In the equation $2x + y = 4000$ it was easy to solve for y in terms of x.
2. After substituting into $A = xy$, the area A became a function of the single variable x, which was easy to differentiate.

Suppose we want to maximize or minimize a function $z = f(x, y)$ subject to a constraint $g(x, y) = 0$ in which

1. It is *not* easy to solve the equation $g(x, y) = 0$ for x or for y, or
2. After substitution, the resulting function z of a single variable is not easy to differentiate.

In such cases, we can instead use the *method of Lagrange multipliers.* We describe the method below:

Consider a function $z = f(x, y)$ of two variables x and y, subject to a single constraint $g(x, y) = 0$. We introduce a new variable λ,* called the **Lagrange multiplier,** and construct the function

$$F(x, y, \lambda) = f(x, y) + \lambda g(x, y)$$

This new function F is a function of three variables x, y, and λ. The following result establishes the connection between the function F and the local maxima and the local minima of $z = f(x, y)$.

Method of Lagrange Multipliers

Suppose that, subject to the constraint $g(x, y) = 0$, the function $z = f(x, y)$ has a local maximum or a local minimum at the point (x_0, y_0). Form the function

$$\boxed{F(x, y, \lambda) = f(x, y) + \lambda g(x, y)}$$

Then there is a value of λ such that (x_0, y_0, λ) is a solution of the system of equations

$$\boxed{\begin{aligned} \frac{\partial F}{\partial x} &= \frac{\partial f}{\partial x} + \lambda\frac{\partial g}{\partial x} = 0 \\ \frac{\partial F}{\partial y} &= \frac{\partial f}{\partial y} + \lambda\frac{\partial g}{\partial y} = 0 \\ \frac{\partial F}{\partial \lambda} &= g(x, y) = 0 \end{aligned}} \tag{1}$$

provided all the partial derivatives exist.

In other words, the above result tells us that if we find all the solutions of the system of equations (1), then among the solutions we find the points at which $z = f(x, y)$ may have a local maximum or a local minimum subject to the condition $g(x, y) = 0$.

The Greek letter lambda.

1 **EXAMPLE 2** **Using the Method of Lagrange Multipliers**

Find the maximum value of

$$z = f(x, y) = xy$$

subject to the constraint

$$g(x, y) = x + y - 16 = 0$$

SOLUTION First, we construct the function F:

$$F(x, y, \lambda) = f(x, y) + \lambda g(x, y) = xy + \lambda(x + y - 16)$$

The system of equations (1) is

$$\frac{\partial F}{\partial x} = 0 \qquad \frac{\partial F}{\partial y} = 0 \qquad \frac{\partial F}{\partial \lambda} = 0$$

$$y + \lambda = 0 \qquad x + \lambda = 0 \qquad x + y - 16 = 0$$

Using the solutions of the first two equations, namely, $y = -\lambda$, $x = -\lambda$, in the third equation, we get

$$-\lambda - \lambda - 16 = 0 \quad \text{or} \quad \lambda = -8$$

Since $x = -\lambda$ and $y = -\lambda$, the only solution of the system is

$$x = 8 \qquad y = 8 \qquad \lambda = -8$$

We find that $z = f(x, y) = xy$ has a local maximum at (8, 8); the maximum value is $z = f(8, 8) = 64$.

In Example 2, we used the Method of Lagrange Multipliers to find the maximum value of z subject to a constraint. The steps we followed are outlined below.

Steps for Using the Method of Langrange Multipliers

STEP 1 Write the function to be maximized (or minimized) and the constraint in the form:

Find the maximum (or minimum) value of

$$z = f(x, y)$$

subject to the constraint

$$g(x, y) = 0$$

STEP 2 Construct the function F:

$$F(x, y, \lambda) = f(x, y) + \lambda g(x, y)$$

STEP 3 Set up the system of equations

$$\frac{\partial F}{\partial x} = 0$$

$$\frac{\partial F}{\partial y} = 0$$

$$\frac{\partial F}{\partial \lambda} = g(x, y) = 0$$

STEP 4 Solve the system of equations for x, y, and λ.

STEP 5 Evaluate $z = f(x, y)$ at each solution (x_0, y_0, λ) found in Step 4. Choose the maximum (or minimum) value of z.

NOW WORK PROBLEM 1.

EXAMPLE 3 Using the Method of Lagrange Multipliers

Find the minimum value of

$$z = f(x, y) = xy$$

subject to the constraint

$$g(x, y) = x^2 + y^2 - 4 = 0$$

SOLUTION **STEP 1:** The problem is already in the desired form.
STEP 2: We construct the function F:

$$F(x, y, \lambda) = f(x, y) + \lambda g(x, y) = xy + \lambda(x^2 + y^2 - 4)$$

STEP 3: The system of equations (1) is

$$\begin{aligned} \frac{\partial F}{\partial x} &= y + \lambda \cdot 2x = 0 && (1) \\ \frac{\partial F}{\partial y} &= x + \lambda \cdot 2y = 0 && (2) \\ \frac{\partial F}{\partial \lambda} &= x^2 + y^2 - 4 = 0 && (3) \end{aligned} \qquad (2)$$

STEP 4: We proceed to solve the system. From the first equation, we find that

$$y = -2x\lambda \qquad (3)$$

Substituting into the second equation, we obtain

$$\begin{aligned} x + 2y\lambda &= 0 && (2) \\ x + 2(-2x\lambda)\lambda &= 0 && y = -2x\lambda \\ x - 4x\lambda^2 &= 0 && \text{Simplify.} \\ x(1 - 4\lambda^2) &= 0 && \text{Factor out } x. \\ x = 0 \quad \text{or} \quad 1 - 4\lambda^2 &= 0 && \text{Apply the Zero-Product Property.} \end{aligned}$$

If $x = 0$, then from (3) we find $y = 0$. But $x = 0$ and $y = 0$ do not satisfy the constraint $g(x, y) = x^2 + y^2 - 4 = 0$. We discard $x = 0$. Then $1 - 4\lambda^2 = 0$.

$$1 - 4\lambda^2 = 0$$
$$\lambda^2 = \frac{1}{4}$$
$$\lambda = \pm\frac{1}{2}$$

Substituting these values for λ into equation (3), $y = 2x\lambda$, we find

$$y = x \qquad \text{or} \qquad y = -x$$

Since x and y are subject to the constraint $g(x, y) = x^2 + y^2 - 4 = 0$, we must have

$$x^2 + y^2 - 4 = x^2 + x^2 - 4 = 0$$
$$2x^2 = 4$$
$$x = \pm\sqrt{2}$$

Since $y = x$ or $y = -x$, the solutions of the system are

$$(\sqrt{2}, \sqrt{2}), \quad (\sqrt{2}, -\sqrt{2}), \quad (-\sqrt{2}, \sqrt{2}), \quad (-\sqrt{2}, -\sqrt{2})$$

STEP 5: We find the value of $z = f(x, y) = xy$ at each of these points.

$$f(\sqrt{2}, \sqrt{2}) = \sqrt{2}\sqrt{2} = 2$$
$$f(\sqrt{2}, -\sqrt{2}) = \sqrt{2}(-\sqrt{2}) = -2$$
$$f(-\sqrt{2}, \sqrt{2}) = -\sqrt{2}\sqrt{2} = -2$$
$$f(-\sqrt{2}, -\sqrt{2}) = (-\sqrt{2})(-\sqrt{2}) = 2$$

We see that $z = f(x, y)$ attains its minimum value at the two points $(-\sqrt{2}, \sqrt{2})$ and $(\sqrt{2}, -\sqrt{2})$. The minimum value is -2.

NOW WORK PROBLEM 3.

Application

EXAMPLE 4 Maximizing Profit

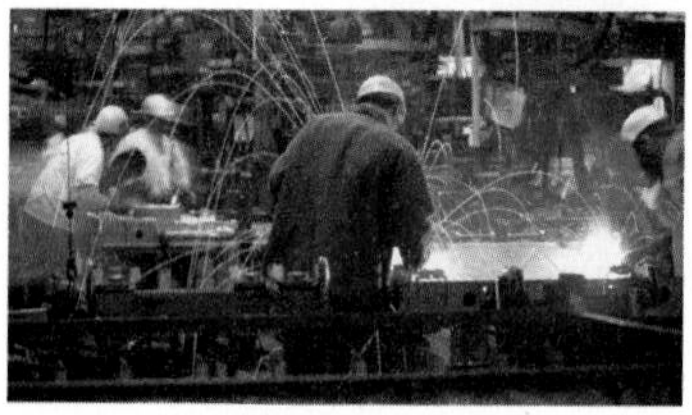

A manufacturer produces two types of engines, x units of type I and y units of type II. The joint profit function is given by

$$P(x, y) = x^2 + 3xy - 6y$$

To maximize profit, how many engines of each type should be produced if there must be a total of 42 engines produced?

SOLUTION **STEP 1:** The condition of a total of 42 engines constitutes the constraint of the problem. The constraint is

$$g(x, y) = x + y - 42 = 0$$

The problem is to

$$\text{Maximize} \quad P(x, y) = x^2 + 3xy - 6y$$
$$\text{subject to the constraint} \quad g(x, y) = x + y - 42 = 0$$

STEP 2: The function F is

$$F(x, y, \lambda) = P(x, y) + \lambda g(x, y) = x^2 + 3xy - 6y + \lambda(x + y - 42)$$

STEP 3: The system of equations (1) is

$$\frac{\partial F}{\partial x} = 2x + 3y + \lambda = 0 \quad (1)$$

$$\frac{\partial F}{\partial y} = 3x - 6 + \lambda = 0 \quad (2)$$

$$\frac{\partial F}{\partial \lambda} = x + y - 42 = 0 \quad (3)$$

STEP 4: From the middle equation $3x - 6 + \lambda = 0$, we have $3x = 6 - \lambda$ so $x = \frac{1}{3}(6 - \lambda)$. We subsitute into the first equation $2x + 3y + \lambda = 0$ and proceed to solve for y.

$$2x + 3y + \lambda = 0 \qquad (1)$$

$$2 \cdot \frac{1}{3}(6 - \lambda) + 3y + \lambda = 0 \qquad x = \frac{1}{3}(6 - \lambda)$$

$$4 - \frac{2}{3}\lambda + 3y + \lambda = 0 \qquad \text{Simplify.}$$

$$3y = -\frac{1}{3}\lambda - 4$$

$$y = -\frac{1}{9}\lambda - \frac{4}{3}$$

Now use the third equation.

$$x + y - 42 = 0 \qquad (3)$$

$$\frac{1}{3}(6 - \lambda) + \left(-\frac{1}{9}\lambda - \frac{4}{3}\right) = 42 \qquad x = \frac{1}{3}(6 - \lambda);\ y = -\frac{1}{9}\lambda - \frac{4}{3}$$

$$2 - \frac{1}{3}\lambda - \frac{1}{9}\lambda - \frac{4}{3} = 42 \qquad \text{Simplify.}$$

$$-\frac{4}{9}\lambda = 42 - \frac{2}{3}$$

$$-\frac{4}{9}\lambda = \frac{124}{3}$$

$$\lambda = -93$$

Then $x = \frac{1}{3}(6 - \lambda) = \frac{1}{3}(99) = 33$ and $y = -\frac{1}{9}\lambda - \frac{4}{3} = \frac{93}{9} - \frac{4}{3} = 9$.

The solution of the system is

$$x = 33 \qquad y = 9 \qquad \lambda = -93$$

STEP 5: The maximum profit is achieved for a production of $x = 33$ type I engines and $y = 9$ type II engines.

NOW WORK PROBLEM 17.

Function of Three Variables

One of the advantages of the method of Lagrange multipliers is that it extends easily to functions of three variables.

Method of Lagrange Multipliers

Suppose that, subject to the constraint $g(x, y, z) = 0$, the function $w = f(x, y, z)$ has a local maximum or a local minimum at the point (x_0, y_0, z_0). Form the function

$$\boxed{F(x, y, z) = f(x, y, z) + \lambda g(x, y, z)}$$

Then there is a value of λ so that (x_0, y_0, z_0, λ) is a solution of the system of equations:

$$\boxed{\begin{aligned} F_x(x, y, z, \lambda) &= f_x(x, y, z) + \lambda g_x(x, y, z) = 0 \\ F_y(x, y, z, \lambda) &= f_y(x, y, z) + \lambda g_y(x, y, z) = 0 \\ F_z(x, y, z, \lambda) &= f_z(x, y, z) + \lambda g_z(x, y, z) = 0 \\ F_\lambda(x, y, z, \lambda) &= g(x, y, z) = 0 \end{aligned}}$$

provided each of the partial derivatives exist.

EXAMPLE 5 Minimizing Cost

The material for a rectangular container costs \$3 per square foot for the bottom and \$2 per square foot for the sides and top. Find the dimensions of the container so that its volume is 12 cubic feet and the cost is minimum.

FIGURE 22

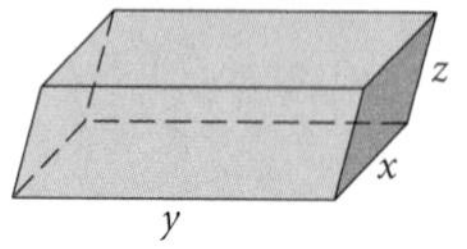

SOLUTION **STEP 1:** Refer to Figure 22. Let x and y (in feet) equal the length and width of the container and z (in feet) equal its height. The cost of the bottom is then $\$3xy$; the cost of the top is $\$2xy$; the cost of the sides is $\$2(2xz) + \$2(2yz)$. The volume is constrained to be $xyz = 12$ cubic feet. The problem is:

Minimize

$$C(x, y, z) = \underbrace{3xy}_{\text{Bottom}} + \underbrace{2xy}_{\text{Top}} + \underbrace{4yz + 4xz}_{\text{Sides}} = 5xy + 4yz + 4xz$$

subject to the constraint

$$g(x, y, z) = xyz - 12 = 0$$

STEP 2: Form the function

$$\begin{aligned} F(x, y, z, \lambda) &= C(x, y, z) + \lambda g(x, y, z) \\ &= 5xy + 4yz + 4xz + \lambda(xyz - 12) \end{aligned}$$

STEP 3: The system of equations to be solved is

$$\begin{aligned} F_x(x, y, z, \lambda) &= 5y + 4z + \lambda yz = 0 && (1) \\ F_y(x, y, z, \lambda) &= 5x + 4z + \lambda xz = 0 && (2) \\ F_z(x, y, z, \lambda) &= 4y + 4x + \lambda xy = 0 && (3) \\ F_\lambda(x, y, z, \lambda) &= xyz - 12 = 0 && (4) \end{aligned}$$

STEP 4: Since $x > 0, y > 0, z > 0$, we can solve for λ in the first three equations to get

$$\lambda = \frac{-(5y + 4z)}{yz} \qquad \lambda = \frac{-(5x + 4z)}{xz} \qquad \lambda = \frac{-4(y + x)}{xy}$$

From the first two of these, we find that

$$\frac{5y + 4z}{yz} = \frac{5x + 4z}{xz}$$

$$x(5y + 4z) = y(5x + 4z) \quad \text{Multiply both sides by } xyz \text{ and simplify.}$$

$$5xy + 4xz = 5xy + 4yz$$

$$4xz = 4yz \quad \text{Cancel the } z\text{'s.}$$

$$x = y$$

Now use the second two:

$$\frac{5x + 4z}{xz} = \frac{4y + 4x}{xy}$$

$$5xy + 4yz = 4yz + 4xz \quad \text{Multiply both sides by } xyz \text{ and simplify.}$$

$$5xy = 4xz$$

$$5y = 4z \quad \text{Cancel the } x\text{'s.}$$

$$y = \frac{4}{5}z \quad \text{Solve for } y.$$

Using these results in equation (4), we get

$$xyz - 12 = 0 \quad (4)$$

$$\frac{4}{5}z \cdot \frac{4}{5}z \cdot z = 12 \quad x = y = \frac{4}{5}z$$

$$z^3 = \frac{75}{4}$$

$$z = 2.657$$

STEP 5: The only solution is $x = y = \frac{4}{5}z = 2.125, z = 2.657$ feet. These are the dimensions of the container that minimize the cost of the container.

NOW WORK PROBLEM 23.

EXERCISE 17.5 Answers to Odd-Numbered Problems Begin on Page AN-103.

In Problems 1–12, use the Method of Lagrange Multipliers.

1. Find the maximum value of $z = f(x, y) = 3x + 4y$ subject to the constraint $g(x, y) = x^2 + y^2 - 9 = 0$.

2. Find the maximum value of $z = f(x, y) = 3xy$ subject to the constraint $g(x, y) = x^2 + y^2 - 4 = 0$.

3. Find the minimum value of $z = f(x, y) = x^2 + y^2$ subject to the constraint $g(x, y) = x + y - 1 = 0$.

4. Find the minimum value of $z = f(x, y) = 3x + 4y$ subject to the constraint $g(x, y) = x^2 + y^2 - 9 = 0$.

5. Find the maximum value of $z = f(x, y) = 12xy - 3y^2 - x^2$ subject to the constraint $g(x, y) = x + y - 16 = 0$.

6. Find the maximum value of $z = f(x, y) = xy$ subject to the constraint $g(x, y) = x + y - 8 = 0$.

7. Find the minimum value of $z = f(x, y) = 5x^2 + 6y^2 - xy$ subject to the constraint $g(x, y) = x + 2y - 24 = 0$.

8. Find the minimum value of $z = f(x, y) = x^2 + y^2$ subject to the constraint $g(x, y) = 2x + 3y - 4 = 0$.

9. Find the maximum value of $w = f(x, y, z) = xyz$ subject to the constraint $g(x, y, z) = x + 2y + 2z - 120 = 0$.

10. Find the maximum value of $w = f(x, y, z) = x + y + z$ subject to the constraint $g(x, y, z) = x^2 + y^2 + z^2 - 12 = 0$.

11. Find the minimum value of

$$w = f(x, y, z) = x^2 + y^2 + z^2 - x - 3y - 5z$$

subject to the constraint $g(x, y, z) = x + y + 2z - 20 = 0$.

12. Find the minimum value of $w = f(x, y, z) = 4x + 4y + 2z$ subject to the constraint $g(x, y, z) = x^2 + y^2 + z^2 - 9 = 0$.

13. Find two numbers x and y so that their product is a maximum while their sum is 100.

14. Find two numbers x and y so that the sum of their squares is a minimum while their sum is 100.

15. Find three numbers x, y, and z so that their sum is a maximum while the sum of their squares is 25.

16. Find three numbers x, y, and z so that their sum is a minimum while the sum of their squares is 25.

17. **Joint Cost Function** Let x and y be two types of items produced by a factory, and let

$$C = 18x^2 + 9y^2$$

be the joint cost of production of x and y. If $x + y = 54$, find x and y that minimize cost.

18. **Production Function** The production function of ABC Manufacturing is

$$P(x, y) = x^2 + 3xy - 6x$$

where x and y represent two different types of input. Find the amounts of x and y that maximize production if $x + y = 40$.

19. **Checked Luggage Requirements** The linear measurements (length + width + height) for baggage checked onto Delta Airlines flights must not exceed 62 inches. Find the dimensions of the rectangular box of greatest volume that meets this requirement.

Source: Delta Airlines.

20. **Fencing in an Area** The A Vinyl Fence Company prices its Cape Cod Concave fence, which is three feet tall, at \$14.11/ft. If a home builder has \$3000 available to spend on fencing a rectangular region, determine the largest area which can be enclosed.

Source: A Vinyl Fence Co. Inc, San Jose, CA.

21. **The Cobb-Douglas Model** (Refer to Problem 48 of Section 17.3 on page 970.) Apply the Cobb-Douglas production model $P = 1.01K^{.25} L^{.75}$ as follows: Suppose that each unit of capital (K) has a value of \$175 and each unit of labor (L) has a value of \$125.

(a) If there is a total of \$125,000 to invest in the economy, use the method of Lagrange multipliers to find the number of units of capital and the number of units of labor that will maximize the total production in the manufacturing sector of the economy.
(b) What is the maximum number of units of production that the manufacturing sector of the economy could generate under the given conditions?

22. **The Cobb-Douglas Model** Apply the Cobb-Douglas production model $P = 1.01K^{.25}L^{.75}$ as follows: Suppose that each unit of capital (K) has a value of \$125 and each unit of labor (L) has a value of \$175.

(a) If there is a total of \$125,000 to invest in the economy, use the method of Lagrange multipliers to find the number of units of capital and the number of units of labor that will maximize the total production in the manufacturing sector of the economy.
(b) What is the maximum number of units of production that the manufacturing sector of the economy could generate under the given conditions?
(c) Is this maximum value greater than, smaller than, or equal to the maximum value found in Problem 21? Explain.

23. **Minimizing Materials** A container producer wants to build a closed rectangular box with a volume of 175 cubic feet. Determine what the dimensions of the container should be so as to use the least amount of material in construction.

24. **Cost of a Box** A rectangular box, open at the top, is to be made from material costing \$2 per square foot. If the volume is to be 12 cubic feet, what dimensions will minimize the cost?

25. **Cost of a Box** A rectangular box is to have a bottom made from material costing \$2 per square foot while the top and sides are made from material costing \$1 per square foot. If the volume of the box is to be 18 cubic feet, what dimensions will minimize the cost?

17.6 The Double Integral

PREPARING FOR THIS SECTION *Before getting started, review the following:*

>> The Definite Integral (Chapter 15, Section 15.4, pp. 881–885)

OBJECTIVES
1 Evaluate partial integrals
2 Evaluate iterated integrals
3 Evaluate double integrals
4 Find the volume of a solid

The definite integral of a function of a single variable can be extended to functions of two variables. Integrals of a function of two variables are called **double integrals.** Recall that the definite integral of a function of one variable is defined over an interval. Double integrals, on the other hand, involve integration over a region of the plane.

An example of a double integral is

$$\iint_R 2x^2y\,dx\,dy$$

where the integrand is the function $f(x, y) = 2x^2y$ and R is some region of the x-y plane.

For example, R might be the rectangular region $1 \le x \le 2, 0 \le y \le 4$. See Figure 23.

FIGURE 23

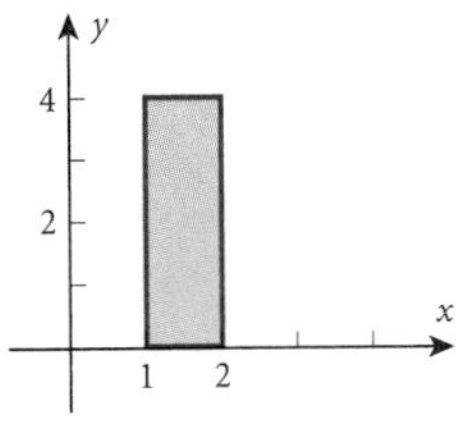

The evaluation of a double integral of a function f of two variables over a rectangular region is equivalent to the evaluation of a pair of definite integrals in which one of the integrations is performed *partially.* Partial integration is merely the reverse of partial differentiation. The symbol $\int_a^b f(x, y)\,dx$ is an instruction to hold y fixed and integrate with respect to x. The result will be a function of y alone. Similarly, $\int_c^d f(x, y)\,dy$ is an instruction to hold x fixed and integrate with respect to y. The result here is a function of x alone.

1 EXAMPLE 1 Evaluating Partial Integrals

Evaluate: **(a)** $\int_1^2 2x^2y\,dx$ **(b)** $\int_0^4 2x^2y\,dy$

SOLUTION **(a)** The dx tells us to integrate with respect to x, holding y as a constant. Then

$$\int_1^2 2x^2y\,dx \underset{\uparrow}{=} 2y\int_1^2 x^2\,dx \underset{\uparrow}{=} 2y \cdot \frac{x^3}{3}\Big|_1^2 = 2y \cdot \left(\frac{8}{3} - \frac{1}{3}\right) = 2y \cdot \frac{7}{3} = \frac{14y}{3}$$

Treat $2y$ as a constant Integrate

(b) The dy tells us to integrate with respect to y, holding x as a constant. Then

$$\int_0^4 2x^2y\,dy \underset{\uparrow}{=} 2x^2 \cdot \int_0^4 y\,dy \underset{\uparrow}{=} 2x^2 \cdot \frac{y^2}{2}\Big|_0^4 = 2x^2(8 - 0) = 16x^2$$

Treat $2x^2$ as a constant Integrate

EXAMPLE 2 Evaluating Partial Integrals

Evaluate: **(a)** $\int_1^2 (6x^2y + 3y^2)\,dy$ **(b)** $\int_1^2 (6x^2y + 3y^2)\,dx$

SOLUTION **(a)** The dy tells us to integrate with respect to y, holding x as a constant. Then

$$\int_1^2 (6x^2y + 3y^2)\,dy \underset{\substack{\uparrow \\ \text{Integral of a sum} \\ \text{is sum of the integrals}}}{=} \int_1^2 6x^2y\,dy + \int_1^2 3y^2\,dy \underset{\substack{\uparrow \\ \text{Treat } x \\ \text{as a constant}}}{=} 6x^2\int_1^2 y\,dy + 3\int_1^2 y^2\,dy \underset{\substack{\uparrow \\ \text{Integrate}}}{=} 6x^2 \cdot \frac{y^2}{2}\Big|_1^2 + 3 \cdot \frac{y^3}{3}\Big|_1^2$$

$$= 6x^2\left(2 - \frac{1}{2}\right) + 3\left(\frac{8}{3} - \frac{1}{3}\right) = 9x^2 + 7$$

(b) The dx tells us to integrate with respect to x, holding y as a constant. Then

$$\int_1^2 (6x^2y + 3y^2)\,dx \underset{\substack{\uparrow \\ \text{Integral of a sum} \\ \text{is sum of the integrals}}}{=} \int_1^2 6x^2y\,dx + \int_1^2 3y^2\,dx \underset{\substack{\uparrow \\ \text{Treat } y \\ \text{as a constant}}}{=} 6y\int_1^2 x^2\,dx + 3y^2\int_1^2 dx \underset{\substack{\uparrow \\ \text{Integrate}}}{=} 6y \cdot \frac{x^3}{3}\Big|_1^2 + 3y^2 \cdot x\Big|_1^2$$

$$= 6y\left(\frac{8}{3} - \frac{1}{3}\right) + 3y^2(2 - 1) = 14y + 3y^2$$

NOW WORK PROBLEM 3.

Integrals of the form

$$\int_a^b \left[\int_c^d f(x, y)\,dy\right] dx \quad \text{and} \quad \int_c^d \left[\int_a^b f(x, y)\,dx\right] dy$$

are called **iterated integrals.** In the iterated integral on the left, the function f is integrated partially with respect to y from c to d, resulting in a function of x that is then integrated from a to b. In the iterated integral on the right, the function f is integrated partially with respect to x from a to b, resulting in a function of y that is then integrated from c to d.

2

EXAMPLE 3 Evaluating Iterated Integrals

Evaluate: **(a)** $\int_0^4 \left[\int_1^2 2x^2y\,dx\right] dy$ **(b)** $\int_1^2 \left[\int_0^4 2x^2y\,dy\right] dx$

SOLUTION **(a)**

$$\int_0^4 \left[\int_1^2 2x^2y\,dx\right] dy \underset{\substack{\uparrow \\ \text{From Example 1, Part (a)}}}{=} \int_0^4 \frac{14y}{3}\,dy = \frac{7y^2}{3}\Big|_0^4 = \frac{112}{3}$$

(b)

$$\int_1^2 \left[\int_0^4 2x^2y\,dy\right] dx \underset{\substack{\uparrow \\ \text{From Example 1, Part (b)}}}{=} \int_1^2 16x^2\,dx = \frac{16x^3}{3}\Big|_1^2 = \frac{112}{3}$$

EXAMPLE 4 Evaluating Iterated Integrals

Evaluate:

(a) $\int_1^2 \left[\int_0^1 (6x^2y + 8y^3)\, dy \right] dx$ **(b)** $\int_0^1 \left[\int_1^2 (6x^2y + 8y^3)\, dx \right] dy$

SOLUTION **(a)** We evaluate the partial integral inside the brackets first. Then

$$\int_0^1 (6x^2y + 8y^3)\, dy = \int_0^1 6x^2y\, dy + \int_0^1 8y^3\, dy = 6x^2 \int_0^1 y\, dy + 8 \int_0^1 y^3\, dy$$

$$= 6x^2 \cdot \frac{y^2}{2}\Big|_0^1 + 8 \cdot \frac{y^4}{4}\Big|_0^1 = 6x^2\left(\frac{1}{2}\right) + 8\left(\frac{1}{4}\right) = 3x^2 + 2$$

Then

$$\int_1^2 \left[\int_0^1 (6x^2y + 8y^3) dy \right] dx = \int_1^2 (3x^2 + 2)\, dx = (x^3 + 2x)\Big|_1^2 = 12 - 3 = 9$$

(b) We evaluate the partial integral inside the brackets first. Then,

$$\int_1^2 (6x^2y + 8y^3)\, dx = \int_1^2 6x^2y\, dx + \int_1^2 8y^3\, dx = 6y \int_1^2 x^2\, dx + 8y^3 \int_1^2 dx$$

$$= 6y \cdot \frac{x^3}{3}\Big|_1^2 + 8y^3 \cdot x\Big|_1^2 = 6y\left(\frac{8}{3} - \frac{1}{3}\right) + 8y^3(2 - 1)$$

$$= 14y + 8y^3$$

Then

$$\int_0^1 \left[\int_1^2 (6x^2y + 8y^3)\, dx \right] dy = \int_0^1 (14y + 8y^3)\, dy = (7y^2 + 2y^4)\Big|_0^1 = 9 - 0 = 9$$

NOW WORK PROBLEM 19.

Notice in Example 4 that the integrand of the iterated integral in part (a) is the same as the one in part (b). Also notice that the limits of integration for x, $1 \le x \le 2$, and the limits of integration for y, $0 \le y \le 1$, are also the same. The difference between part (a) and part (b) is the order of the integration. Yet the answer obtained is the same.

Now look at Example 3. The same circumstances there led to equal answers as well. Examples 3 and 4 are special cases of the following result.

If $f(x, y)$ is a function that is continuous over a rectangular region R, $a \le x \le b$, $c \le y \le d$, then

$$\int_a^b \left[\int_c^d f(x, y)\, dy \right] dx = \int_c^d \left[\int_a^b f(x, y)\, dx \right] dy$$

Furthermore, the double integral of f over R has the value

$$\iint_R f(x, y)\, dx\, dy = \int_a^b \left[\int_c^d f(x, y)\, dy \right] dx = \int_c^d \left[\int_a^b f(xy)\, dx \right] dy$$

3 **EXAMPLE 5 Evaluating Double Integrals**

Evaluate $\iint_R 2xy\,dx\,dy$ if R is the rectangular region $1 \le x \le 2, 0 \le y \le 1$.

SOLUTION We choose to evaluate the double integral as follows:

$$\iint_R 2xy\,dx\,dy = \underbrace{\int_0^1 \left[\int_1^2 2xy\,dx\right] dy}_{\substack{1 \le x \le 2 \\ 0 \le y \le 1}} = \int_0^1 2y\left[\int_1^2 x\,dx\right] dy = \int_0^1 2y \cdot \left[\frac{x^2}{2}\Big|_1^2\right] dy = \int_0^1 2y \cdot \frac{3}{2}\,dy = \int_0^1 3y\,dy = \left(3\frac{y^2}{2}\right)\Big|_0^1 = \frac{3}{2}$$

NOW WORK EXAMPLE 5 using the iterated integral $\int_1^2 \left[\int_0^1 2xy\,dy\right] dx$.

NOW WORK PROBLEM 27.

Finding Volume by Using Double Integrals

Find the volume of a solid

4 One application of the definite integral $\int_a^b f(x)\,dx$ is to find the area under a curve. In a similar manner, the double integral is used to find the volume of a solid bounded above by the surface $z = f(x, y)$, below by the xy-plane, and on the sides by the vertical walls defined by the rectangular region R. See Figure 24.

FIGURE 24

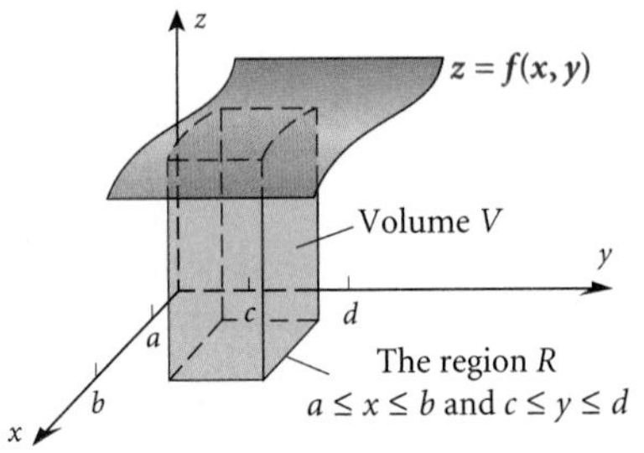

Volume

If $f(x, y) \ge 0$ over a rectangular region R: $a \le x \le b, c \le y \le d$, then the volume V of the solid under the graph of f and over the region R is

$$V = \iint_R f(x, y)\,dx\,dy$$

EXAMPLE 6 Finding the Volume of a Solid

Find the volume V of the solid under $f(x, y) = x^2 + y^2$ and over the rectangular region $0 \leq x \leq 2, 0 \leq y \leq 1$.

FIGURE 25

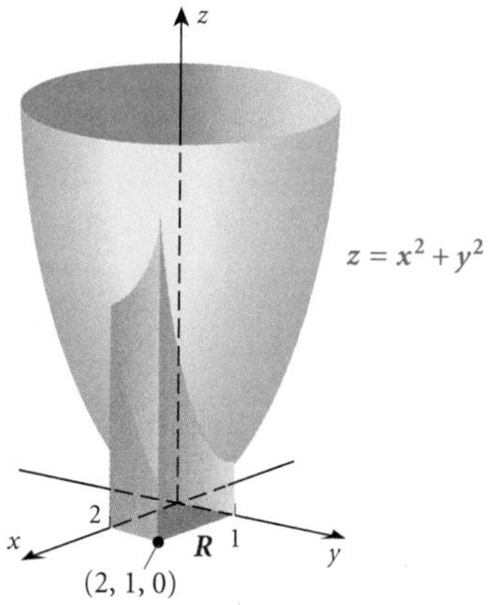

SOLUTION Figure 25 illustrates the volume we seek. The volume V is

$$V = \iint_R (x^2 + y^2)\, dx\, dy = \int_0^2 \left[\int_0^1 (x^2 + y^2)\, dy\right] dx$$

$$= \int_0^2 \left(x^2 y + \frac{y^3}{3}\right)\Bigg|_0^1 dx$$

$$= \int_0^2 \left(x^2 + \frac{1}{3}\right) dx$$

$$= \left(\frac{x^3}{3} + \frac{x}{3}\right)\Bigg|_0^2$$

$$= \frac{8}{3} + \frac{2}{3} = \frac{10}{3} \text{ cubic units}$$

NOW WORK PROBLEM 31.

EXERCISE 17.6 Answers to Odd-Numbered Problems Begin on Page AN-103.

In Problems 1–16, evaluate each partial integral.

1. $\int_0^2 (xy^3 + x^2)\, dx$

2. $\int_1^3 (xy^3 - x^2)\, dx$

3. $\int_2^4 (3x^2 y + 2x)\, dy$

4. $\int_0^1 (6xy^2 - 2xy + 3)\, dy$

5. $\int_2^3 (x + 3y)\, dx$

6. $\int_1^3 (6xy + 12x^2y^3)\, dy$

7. $\int_2^4 (4x - 6y + 7)\, dy$

8. $\int_1^3 (4x - 6y + 7)\, dx$

9. $\int_0^1 \frac{x^2}{\sqrt{1 + y^2}}\, dx$

10. $\int_0^2 \frac{y^3}{\sqrt{1 + x^2}}\, dy$

11. $\int_0^2 e^{x+y}\, dx$

12. $\int_0^2 e^{x+6y}\, dx$

13. $\int_0^4 e^{x-4y}\, dx$

14. $\int_2^4 e^{x+y}\, dy$

15. $\int_0^2 \frac{x}{\sqrt{y + 6}}\, dx$

16. $\int_1^4 \frac{y}{\sqrt{x^2 + 9}}\, dy$

In Problems 17–26, evaluate each iterated integral.

17. $\int_0^2 \left[\int_0^4 y\, dx\right] dy$

18. $\int_1^2 \left[\int_3^4 x\, dy\right] dx$

19. $\int_1^2 \left[\int_1^3 (x^2 + y)\, dx\right] dy$

20. $\int_0^1 \left[\int_2^3 (x + y)\, dy\right] dx$

21. $\int_0^1 \left[\int_1^2 (x^2 + y)\, dx\right] dy$

22. $\int_0^3 \left[\int_1^2 (x - y^2)\, dy\right] dx$

23. $\int_1^2 \left[\int_3^4 (4x + 2y + 5)\, dx\right] dy$

24. $\int_1^2 \left[\int_3^4 (6x + 4y + 7)\, dy\right] dx$

25. $\int_2^4 \left[\int_0^1 (6xy^2 - 2xy + 3)\, dy\right] dx$

26. $\int_0^2 \left[\int_1^3 (6xy + 12x^2y^3)\, dy\right] dx$

In Problems 27–30, evaluate each double integral over the indicated rectangular region R.

27. $\iint_R (y + 3x^2)\,dx\,dy \quad R: 0 \le x \le 2, \quad 1 \le y \le 3$

28. $\iint_R (x + 3y^2)\,dx\,dy \quad R: 0 \le x \le 3, \quad 0 \le y \le 4$

29. $\iint_R (x + y)\,dy\,dx \quad R: 0 \le x \le 2, \quad 1 \le y \le 4$

30. $\iint_R (x^2 - 2xy)\,dy\,dx \quad R: 0 \le x \le 2, \quad 1 \le y \le 4$

In Problems 31 and 32, find the volume under the surface of $f(x, y)$ and above the indicated rectangle.

31. $f(x, y) = 2x + 3y + 4 \quad 1 \le x \le 2, \quad 3 \le y \le 4$

32. $f(x, y) = x + y - 1 \quad 0 \le x \le 1, \quad 0 \le y \le 1$

Chapter 17 Review

OBJECTIVES

Section	You should be able to	Review Exercises
17.1	1 Use the distance formula	1–8
	2 Find the standard equation of a sphere	9–14
17.2	1 Evaluate a function of two variables	15–18
	2 Find the domain of a function of two variables	19–24
17.3	1 Find partial derivatives	25–32, 33, 34
	2 Find the slope of the tangent line to a surface	35–38
	3 Interpret partial derivatives	63–70
	4 Find higher-order partial derivatives	25–32
17.4	1 Find critical points	39(a)–44(a)
	2 Determine the character of a critical point	39(b)–44(b)
	3 Solve applied problems	71–74
17.5	1 Use the method of Lagrange multipliers	45–48, 73, 74
17.6	1 Evaluate partial integrals	49–52
	2 Evaluate iterated integrals	53–56
	3 Evaluate double integrals	57–60
	4 Find the volume of a solid	61, 62

THINGS TO KNOW

Distance Formula (p. 953)
The distance d from $P_1 = (x_1, y_1, z_1)$ to $P_2 = (x_2, y_2, z_2)$ is

$$d = \sqrt{(x_2 - x_1)^2 + (y_2 - y_1)^2 + (z_2 - z_1)^2}$$

Sphere (p. 954)
The equation of a sphere of radius r with center at (x_0, y_0, z_0) is

$$(x - x_0)^2 + (y - y_0)^2 + (z - z_0)^2 = r^2$$

Partial Derivatives (p. 963)

$$\frac{\partial f}{\partial x} = f_x(x, y) = \lim_{\Delta x \to 0} \frac{f(x + \Delta x, y) - f(x, y)}{\Delta x}$$

$$\frac{\partial f}{\partial y} = f_y(x, y) = \lim_{\Delta y \to 0} \frac{f(x, y + \Delta y) - f(x, y)}{\Delta y}$$

Tests for Local Maxima, Local Minima, and Saddle Points (p. 973)
(x_0, y_0) is a critical point if $f_x(x_0, y_0) = f_y(x_0, y_0) = 0$. Let

$$D = f_{xx}(x_0, y_0) \cdot f_{yy}(x_0, y_0) - [f_{xy}(x_0, y_0)]^2$$

If $D > 0$ and $f_{xx}(x_0, y_0) > 0$, f has a local minimum at (x_0, y_0).
If $D > 0$ and $f_{xx}(x_0, y_0) < 0$, f has a local maximum at (x_0, y_0).
If $D < 0$, f has a saddle point at (x_0, y_0).

Method of Lagrange Multipliers (pp. 979 and 984)
To find a local maximum or a local minimum for $z = f(x, y)$, subject to $g(x, y) = 0$, solve the system of equations

$$\frac{\partial f}{\partial x} + \lambda\frac{\partial g}{\partial x} = 0 \quad \frac{\partial f}{\partial y} + \lambda\frac{\partial g}{\partial y} = 0 \quad g(x, y) = 0$$

Double Integrals (pp. 989 and 990)
R: $a \le x \le b$, $c \le y \le d$

$$\iint_R f(x, y)\,dx\,dy = \int_a^b\left[\int_c^d f(x, y)\,dy\right]dx = \int_c^d\left[\int_a^b f(x, y)\,dx\right]dy$$

If $f(x, y) \ge 0$ on R, Volume $= V = \iint_R f(x, y)\,dx\,dy$

TRUE–FALSE ITEMS **Answers are on page AN-103.**

T F 1. The domain of a function of two variables is a set of points in the xy-plane.

T F 2. The partial derivative $f_x(x, y)$ of $z = f(x, y)$ is

$$f_x(x, y) = \lim_{\Delta x \to 0} \frac{f(x + \Delta x, y + \Delta y) - f(x, y)}{\Delta x}$$

provided the limit exists.

T F 3. For most functions in this book, $f_{xy} \ne f_{yx}$.

T F 4. If (x_0, y_0) is a critical point of $z = f(x, y)$ and if $f_{xx}(x_0, y_0) > 0$ and

$$D = f_{xx}(x_0, y_0) \cdot f_{yy}(x_0, y_0) - [f_{xy}(x_0, y_0)]^2 < 0$$

then f has a local minimum at (x_0, y_0).

FILL-IN-THE-BLANKS **Answers are on page AN-103.**

1. The graph of a function of two variables is called a __________.
2. If $f(x, y) = x^2y - \sqrt{xy}$, then $f(1, 2) =$ __________.
3. The partial derivative $f_y(x_0, y_0)$ equals the slope of the tangent line to the curve of intersection of the surface $z = f(x, y)$ and the plane __________ at the point (x_0, y_0, z_0) on the surface.
4. A critical point that is neither a local maximum nor a local minimum is a __________ __________.

REVIEW EXERCISES **Answers to odd-numbered problems begin on page AN-103. Blue problem numbers indicate the author's suggestions for a practice test.**

In Problems 1–6, find the distance between each pair of points.

1. $(2, 4, 0)$ and $(1, 6, -2)$
2. $(7, 2, 1)$ and $(1, 6, -2)$
3. $(6, 2, 1)$ and $(4, 6, 8)$
4. $(5, 8, 3)$ and $(7, 6, 2)$
5. $(0, 3, -1)$ and $(-3, 7, -1)$
6. $(6, 2, 3)$ and $(6, -10, -2)$
7. $(2, 2, 2)$ is the center of a sphere and $(3, 4, 0)$ is a point on its surface. Find the radius of the sphere.
8. The endpoints of the diameter of a sphere are $(3, 0, 2)$ and $(9, 0, -6)$. Find the radius of the sphere.
9. Find the standard equation of a sphere which has its center at $(-6, 3, 1)$ and has a radius of 2.
10. Find the standard equation of a sphere which has its center at $(0, 2, -1)$ and has a radius of 3.
11. What is the center and radius of the sphere described by $(x - 1)^2 + (y + 3)^2 + (z + 8)^2 = 25$?
12. What is the center and radius of the sphere described by $(x + 5)^2 + y^2 + z^2 = 16$?

In Problems 13 and 14, (a) find the standard equation of the sphere; (b) list its center and radius.

13. $x^2 + y^2 + z^2 - 2x + 8y - 6z = 10$

14. $x^2 + y^2 + z^2 - 6y + 2z = 6$

In Problems 15–18, evaluate each function (a) at the point $(1, -3)$; and (b) at the point $(4, -2)$.

15. $f(x, y) = 2x^2 + 6xy - y^3$

16. $f(x, y) = 3x^2y - x^2 + y^2$

17. $f(x, y) = \dfrac{x + 2y}{x - 3y}$

18. $f(x, y) = \dfrac{2x + y}{x^2 - y}$

In Problems 19–24, find the domain of each function. Give a description of its domain.

19. $z = f(x, y) = x^2 + 3y + 5$

20. $z = f(x, y) = 2xy - 5x + 10$

21. $z = f(x, y) = \ln(y - x^2 - 4)$

22. $z = f(x, y) = \ln(x - y^2)$

23. $z = f(x, y) = \sqrt{x^2 + y^2 + 4x - 5}$

24. $z = f(x, y) = \dfrac{25}{x^2 + y^2}$

In Problems 25–32, find $f_x(x, y)$, $f_y(x, y)$, $f_{xx}(x, y)$, $f_{xy}(x, y)$, $f_{yx}(x, y)$, and $f_{yy}(x, y)$ for each function.

25. $z = f(x, y) = x^2y + 4x$

26. $z = f(x, y) = x^2 + y^2 + 2xy$

27. $z = f(x, y) = y^2e^x + x \ln y$

28. $z = f(x, y) = \ln(x^2 + 3y)$

29. $z = f(x, y) = \sqrt{x^2 + y^2}$

30. $z = f(x, y) = \sqrt{x - 2y^2}$

31. $z = f(x, y) = e^x \ln(5x + 2y)$

32. $z = f(x, y) = (x + y^2)e^{3x}$

In Problems 33 and 34, find $f_x(x, y, z)$, $f_y(x, y, z)$, and $f_z(x, y, z)$.

33. $f(x, y, z) = 3x\,e^y + xy\,e^z - 12x^2y$

34. $f(x, y, z) = \ln|2xy + z|$

In Problems 35–38, find the slope of the tangent line to the curve of intersection of the surface $z = f(x, y)$ with the given plane at the indicated point.

35. $z = f(x, y) = 3xy^2$; plane: $y = 2$; point: $(1, 2, 12)$

36. $z = f(x, y) = 2x^2y + y \ln x$; plane: $y = 1$; point: $(1, 1, 2)$

37. $z = f(x, y) = xe^{xy}$; plane: $x = 1$; point: $(1, 0, 1)$

38. $z = f(x, y) = x \ln(xy)$; plane: $x = 1$; point: $(1, 1, 0)$

In Problems 39–44, (a) find all the critical points of each function; (b) determine the character of each critical point found in part (a).

39. $z = f(x, y) = xy - 6x - x^2 - y^2$

40. $z = f(x, y) = x^2 + 2x + y^2 + 4y + 10$

41. $z = f(x, y) = 2x - x^2 + 4y - y^2 + 10$

42. $z = f(x, y) = xy$

43. $z = f(x, y) = x^2 - 9y + y^2$

44. $z = f(x, y) = xy + 2y - 3x - 2$

In Problems 45–48, use the Method of Lagrange Multipliers.

45. Find the maximum value of $f(x, y) = 5x^2 + 3y^2 + xy$ subject to the constraint $g(x, y) = 2x - y - 20 = 0$.

46. Find the maximum value of $f(x, y) = x\sqrt{y}$ subject to the constraint $g(x, y) = 2x + y - 3000 = 0$.

47. Find the minimum value of $f(x, y) = x^2 + y^2$ subject to the constraint $g(x, y) = 2x + y - 4 = 0$.

48. Find the minimum value of $f(x, y) = x\,y^2$ subject to the constraint $g(x, y) = x^2 + y^2 - 1 = 0$.

In Problems 49–52, evaluate each partial integral.

49. $\int_0^2 (4x^2y - 12y)\,dx$

50. $\int_0^2 (4x^2y - 12y)\,dy$

51. $\int_{-1}^3 (6x^2y + 2y)\,dy$

52. $\int_{-1}^3 (6x^2y + 2y)\,dx$

In Problems 53–56, evaluate each iterated integral.

53. $\int_1^2 \left[\int_0^3 (6x^2 + 2x)\,dy\right] dx$

54. $\int_0^3 \left[\int_1^2 (6x^2 + 2x)\,dx\right] dy$

55. $\int_0^2 \left[\int_1^8 (x^2 + 2xy - y^2)dx\right] dy$

56. $\int_1^8 \left[\int_0^2 (x^2 + 2xy - y^2)\,dy\right] dx$

In Problems 57–60, evaluate each double integral over the indicated rectangular region R.

57. $\iint_R (2x + 4y)\,dy\,dx$

$R: -1 \le x \le 1, 1 \le y \le 3$

58. $\iint_R (3x + 2)\,dy\,dx$

$R: 0 \le x \le 2, 1 \le y \le 3$

59. $\iint_R (2xy)\,dy\,dx$

$R: 0 \le x \le 3, 1 \le y \le 2$

60. $\iint_R (x + y)^3\,dy\,dx$

$R: 0 \le x \le 4, 1 \le y \le 3$

In Problems 61 and 62, find the volume under the surface of $z = f(x, y)$ and above the indicated rectangle.

61. $f(x, y) = 2x + 2y + 1;\quad 1 \le x \le 8,\quad 0 \le y \le 6$

62. $f(x, y) = x^2 + y^2 - 4;\quad 0 \le x \le 2,\quad 0 \le y \le 2$

63. Production Function The Cobb-Douglas production function for a certain factory is

$$z = f(K, L) = 80\,K^{\frac{1}{4}}\,L^{\frac{3}{4}}$$

(a) Find: $\dfrac{\partial z}{\partial K}$ and $\dfrac{\partial z}{\partial L}$.

(b) Evaluate $\dfrac{\partial z}{\partial K}$ and $\dfrac{\partial z}{\partial L}$ when $K = \$800{,}000$ and $L = 20{,}000$ worker hours.

(c) To best improve productivity, should the factory increase the use of capital or labor? Explain your answer.

64. Production Function The productivity of a manufacturer of car parts approximately follows the Cobb-Douglas production function

$$z = f(K, L) = 50\,K^{\frac{2}{5}}\,L^{\frac{3}{5}}$$

(a) Find: $\dfrac{\partial z}{\partial K}$ and $\dfrac{\partial z}{\partial L}$.

(b) Evaluate $\dfrac{\partial z}{\partial K}$ and $\dfrac{\partial z}{\partial L}$ when $K = \$128{,}000$ and $L = 4000$ worker hours.

(c) To best improve productivity, should the manufacturer increase the use of capital or labor? Explain your answer.

65. Marginal Cost of Vacuum Cleaners The Vacitup Company manufactures two types of vacuum cleaners, the standard Vacu-Clean and deluxe Vacu-Clean Plus. The company's weekly cost function for manufacturing x standard and y deluxe vacuum cleaners is

$$C(x, y) = 1050 + 40x + 45y$$

Find $C_x(x, y)$ and $C_y(x, y)$ and interpret your answers.

66. Marginal Cost of Lawn Mowers A company manufactures two models, the standard gasoline powered lawn mower and a combination mower-mulcher. The monthly cost of producing x mowers and y mower-mulchers is given by the function

$$C(x, y) = 15{,}000 + 120x + 150y$$

Find $C_x(x, y)$ and $C_y(x, y)$ and interpret your answers.

67. Analyzing Revenue The demand functions for the vacuum cleaners manufactured in Problem 65 are given by

$$p = 350 - 6x + y$$
$$q = 400 + 2x - 8y$$

where p and q are the prices of the standard and deluxe vacuum cleaners respectively.

(a) Find the revenue function for the Vacitup Company.
(b) Find $R_x(x, y)$ and $R_y(x, y)$ and interpret your answers.

68. Analyzing Revenue The demand functions for the products manufactured in Problem 66, are given by

$$p = 1000 - 7x + y$$
$$q = 2500 + 2x - 50y$$

where p and q are the prices of the mower and the combination mower-mulcher respectively.

(a) Find the company's revenue function.
(b) Find $R_x(x, y)$ and $R_y(x, y)$ and interpret your answers.

69. Analyzing Profit Refer to Problems 65 and 67. If currently the Vacitup Company is manufacturing 50 standard and 30 deluxe vacuum cleaners each week,

(a) Find the profit function for the Vacitup Company.
(b) Evaluate $P_x(50, 30)$ and $P_y(50, 30)$ and interpret your answers.

70. Analyzing Profit Refer to Problems 66 and 68. If currently the company is manufacturing 100 lawn mowers and 40 combination mower-mulchers each month,

(a) Find the profit function.
(b) Evaluate $P_x(100, 40)$ and $P_y(100, 40)$ and interpret your answers.

71. Maximum Profit A supermarket sells two brands of refrigerated orange juice. The demand functions for the two products are

$$p = 9 - x \quad \text{and} \quad q = 21 - 2y$$

where p and q are the respective prices (in thousands of dollars), and x and y are the respective amounts (in thousands of units) of each brand. The joint cost to the supermarket is

$$C(x, y) = x + y + 225$$

(a) Determine the quantities x, y and the prices p, q that maximize profit.
(b) What is the maximum profit?

72. Maximum Profit A company produces two products at a total cost

$$C(x, y) = x^2 + 200x + y^2 + 100y - xy$$

where x and y represent the units of each product. The revenue function is

$$R(x, y) = 2000x - 2x^2 + 100y - y^2 + xy$$

(a) Find the number of units of each product that will maximize profit.
(b) What is the maximum profit?

73. Maximizing Productivity A manufacturer introduces a new product with a Cobb-Douglas production function of

$$P(x, y) = 10K^{0.3} L^{0.7}$$

where K represents the units of capital and L the units of labor needed to produce P units of the product. A total of \$51,000 has been budgeted for production, and each unit of labor costs the manufacturer \$100 and each unit of capital costs \$50.

(a) How should the \$51,000 be allocated between labor and capital to maximize production?
(b) What is the maximum number of units that can be produced?

74. Minimizing Material A rectangular cardboard box with an open top is to have a volume of 96 cubic feet. Find the dimensions of the box so that the amount of cardboard used is minimized.

Chapter 17 Project

MORE INVENTORY CONTROL

Due to your excellent work with the vacuum cleaner inventory in Chapter 14, you've been promoted to inventory specialist in the medium-sized household appliances department. You now need to order not only vacuum cleaners but also microwave ovens. Based on past experience, you know that the store will sell 500 vacuum cleaners and 800 microwave ovens per year, and you again must decide how many times a year to order each product, and how many to get with each order. There are two types of costs to consider. First, items must be stored when they arrive at the store, and this costs money. These costs are called the *holding costs associated with each product.* Second, each time you place an order for a product you incur costs for the paperwork, handling, and shipping. These costs we will call *reorder costs.* You will want to find the number of each product to order at a time (called the *lot size*) that minimizes the total of the holding costs and the reorder costs.

We assume that demand for each product is constant through the year, and that your stock is immediately replenished exactly when you run out of each product. If x is the lot size for vacuum cleaners and y is the lot size for microwave ovens, then the average number of vacuum cleaners you have in stock is $\frac{x}{2}$ and the average number of microwave ovens is $\frac{y}{2}$.

1. Assume that the annual holding costs are \$30 for each vacuum cleaner and \$15 for each microwave oven, and that the reorder costs are \$40 for vacuum cleaners and \$60 for microwave ovens. What is the total cost function?

2. Find the lot sizes x and y, give a minimum total cost, and confirm that this is indeed a local minimum.

3. To make your job slightly more difficult, you have discovered that your department only has access to 1000 cubic feet of storage space. You know that each vacuum cleaner uses 20 cubic feet of storage space and each microwave oven uses 10 cubic feet of storage space. How much space does your solution from Problem 2 require?

4. Since the solution in Problem 2 cannot be used, you will need to introduce a constraint function $g(x, y)$ to solve the problem. Such a function will equal 0 when all of the storage space is in use. Find $g(x, y)$.

5. Produce a system of equations that uses the Method of Lagrange Multipliers to solve the minimization problem.

6. Solve the system to find lot sizes x and y, which minimize the total cost function subject to the storage space constraint.

7. The solution will probably have noninteger solutions. Find integers x and y, which minimize the total cost function subject to the storage space constraint. How many orders will you make in the coming year for each item? How much of your storage space will you use on the average?

APPENDIX A

Review*

OUTLINE

A.1 Real Numbers

OBJECTIVES
1 Classify numbers
2 Evaluate numerical expressions
3 Work with properties of real numbers

Sets

When we want to treat a collection of similar but distinct objects as a whole, we use the idea of a **set**. For example, the set of *digits* consists of the collection of numbers 0, 1, 2, 3, 4, 5, 6, 7, 8, and 9. If we use the symbol D to denote the set of digits, then we can write

$$D = \{0, 1, 2, 3, 4, 5, 6, 7, 8, 9\}$$

In this notation, the braces { } are used to enclose the objects, or **elements,** in the set. This method of denoting a set is called the **roster method.** A second way to denote a set is to use **set-builder notation,** where the set D of digits is written as

$$D = \{x \mid x \text{ is a digit}\}$$

read as "D is the set of all x such that x is a digit."

EXAMPLE 1 Using Set-Builder Notation and the Roster Method

(a) $E = \{x \mid x \text{ is an even digit}\} = \{0, 2, 4, 6, 8\}$
(b) $O = \{x \mid x \text{ is an odd digit}\} = \{1, 3, 5, 7, 9\}$

*Based on material from *College Algebra, 7th ed.*, by Michael Sullivan. Used here with the permission of the author and Prentice-Hall, Inc.

In listing the elements of a set, we do not list an element more than once because the elements of a set are distinct. Also, the order in which the elements are listed is not relevant. For example, {2, 3} and {3, 2} both represent the same set.

If every element of a set A is also an element of a set B, then we say that A **is a subset of** B. If two sets A and B have the same elements, then we say that A **equals** B. For example, {1, 2, 3} is a subset of {1, 2, 3, 4, 5}, and {1, 2, 3} equals {2, 3, 1}.

Finally, if a set has no elements, it is called the **empty set,** or the **null set,** and is denoted by the symbol $\varnothing$.

Classification of Numbers

It is helpful to classify the various kinds of numbers that we deal with as sets. The **counting numbers,** or **natural numbers,** are the numbers in the set {1, 2, 3, 4, . . . }. (The three dots, called an **ellipsis,** indicate that the pattern continues indefinitely.) As their name implies, these numbers are often used to count things. For example, there are 26 letters in our alphabet; there are 100 cents in a dollar. The **whole numbers** are the numbers in the set {0, 1, 2, 3, . . . }, that is, the counting numbers together with 0.

The **integers** are the numbers in the set { . . . , −3, −2, −1, 0, 1, 2, 3, . . . }.

These numbers are useful in many situations. For example, if your checking account has \$10 in it and you write a check for \$15, you can represent the current balance as −\$5.

Notice that the set of counting numbers is a subset of the set of whole numbers. Each time we expand a number system, such as from the whole numbers to the integers, we do so in order to be able to handle new, and usually more complicated, problems. The integers allow us to solve problems requiring both positive and negative counting numbers, such as profit/loss, height above/below sea level, temperature above/below 0°F, and so on.

But integers alone are not sufficient for *all* problems. For example, they do not answer the question "What part of a dollar is 38 cents?" To answer such a question, we enlarge our number system to include *rational numbers.* For example, $\frac{38}{100}$ answers the question "What part of a dollar is 38 cents?"

A **rational number** is a number that can be expressed as a quotient $\dfrac{a}{b}$ of two integers. The integer a is called the **numerator,** and the integer b, which cannot be 0, is called the **denominator.** The rational numbers are the numbers in the set $\{x \mid x = \dfrac{a}{b}$, where a and b, $b \neq 0$, are integers$\}$.

Examples of rational numbers are $\frac{3}{4}, \frac{5}{2}, \frac{0}{4}, -\frac{2}{3}$, and $\frac{100}{3}$. Since $\frac{a}{1} = a$ for any integer a, it follows that the set of integers is a subset of the set of rational numbers.

Rational numbers may be represented as **decimals.** For example, the rational numbers $\frac{3}{4}, \frac{5}{2}, -\frac{2}{3}$, and $\frac{7}{66}$ may be represented as decimals by merely carrying out the indicated division:

$$\frac{3}{4} = 0.75 \qquad \frac{5}{2} = 2.5 \qquad -\frac{2}{3} = -0.666\ldots \qquad \frac{7}{66} = 0.1060606\ldots$$

Notice that the decimal representations of $\frac{3}{4}$ and $\frac{5}{2}$ terminate, or end. The decimal representations of $-\frac{2}{3}$ and $\frac{7}{66}$ do not terminate, but they do exhibit a pattern of repetition. For $-\frac{2}{3}$, the 6 repeats indefinitely; for $\frac{7}{66}$, the block 06 repeats indefinitely. It can be shown that every rational number may be represented by a decimal that either terminates or is nonterminating with a repeating block of digits, and vice versa.

On the other hand, there are decimals that do not fit into either of these categories. Such decimals represent **irrational numbers.** Every irrational number may be represented by a decimal that neither repeats nor terminates. In other words, irrational numbers cannot be written in the form $\frac{a}{b}$, where a and b, $b \neq 0$, are integers.

FIGURE 1

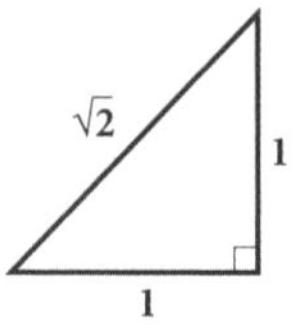

Irrational numbers occur naturally. For example, consider the isosceles right triangle whose legs are each of length 1. See Figure 1. The length of the hypotenuse is $\sqrt{2}$, an irrational number.

FIGURE 2 $\pi = \frac{C}{d}$

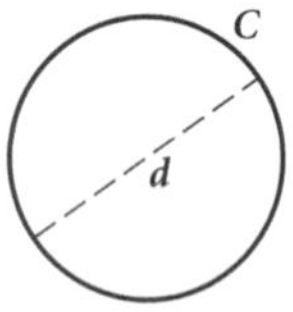

Also, the number that equals the ratio of the circumference C to the diameter d of any circle, denoted by the symbol π (the Greek letter pi), is an irrational number. See Figure 2.

> Together, the rational numbers and irrational numbers form the set of **real numbers.**

Figure 3 shows the relationship of various types of numbers.

FIGURE 3

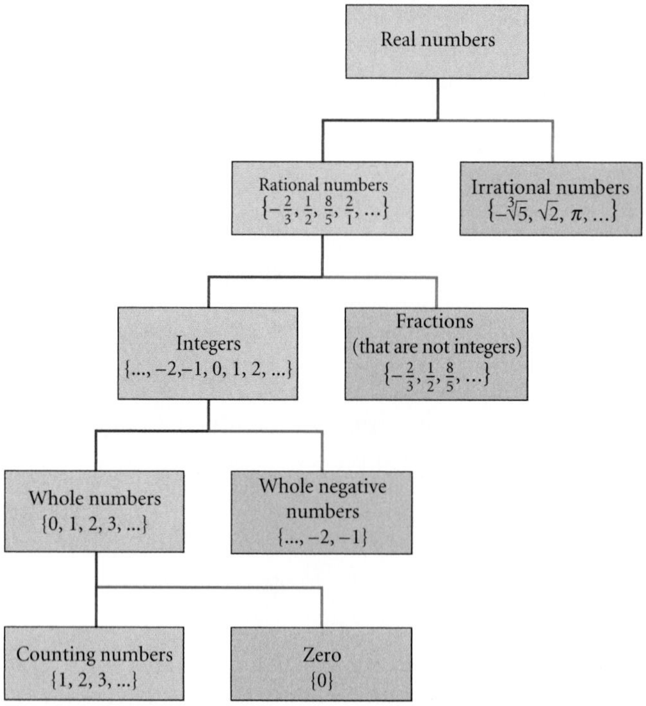

1 EXAMPLE 2 Classifying the Numbers in a Set

List the numbers in the set

$$\left\{-3, \frac{4}{3}, 0.12, \sqrt{2}, \pi, 2.151515\ \ldots\ \text{(where the block 15 repeats)}, 10\right\}$$

that are

(a) Natural numbers **(b)** Integers **(c)** Rational numbers
(d) Irrational numbers **(e)** Real numbers

SOLUTION **(a)** 10 is the only natural number.
(b) -3 and 10 are integers.
(c) $-3, \frac{4}{3}, 0.12, 2.151515 \ldots$, and 10 are rational numbers.
(d) $\sqrt{2}$ and π are irrational numbers.
(e) All the numbers listed are real numbers.

NOW WORK PROBLEM 3.

Approximations

Every decimal may be represented by a real number (either rational or irrational), and every real number may be represented by a decimal.

The irrational numbers $\sqrt{2}$ and π have decimal representations that begin as follows:

$$\sqrt{2} = 1.414213 \ldots \quad \pi = 3.14159 \ldots$$

In practice, decimals are generally represented by approximations. For example, using the symbol $\approx$ (read as "approximately equal to"), we can write

$$\sqrt{2} \approx 1.4142 \qquad \pi \approx 3.1416$$

In approximating decimals, we either *round off* or *truncate* to a given number of decimal places. The number of places establishes the location of the *final digit* in the decimal approximation.

Truncation: Drop all the digits that follow the specified final digit in the decimal.

Rounding: Identify the specified final digit in the decimal. If the next digit is 5 or more, add 1 to the final digit; if the next digit is 4 or less, leave the final digit as it is. Now truncate following the final digit.

EXAMPLE 3 Approximating a Decimal to Two Places

Approximate 20.98752 to two decimal places by

(a) Truncating **(b)** Rounding

SOLUTION For 20.98752, the final digit is 8, since it is two decimal places from the decimal point.

(a) To truncate, we remove all digits following the final digit 8. The truncation of 20.98752 to two decimal places is 20.98.
(b) To round, we examine the digit following the final digit 8, which is 7. Since 7 is 5 or more, we add 1 to the final digit 8 and truncate. The rounded form of 20.98752 to two decimal places is 20.99.

EXAMPLE 4 Approximating a Decimal to Two and Four Places

Number	*Rounded to Two Decimal Places*	*Rounded to Four Decimal Places*	*Truncated to Two Decimal Places*	*Truncated to Four Decimal Places*
(a) 3.14159	3.14	3.1416	3.14	3.1415
(b) 0.056128	0.06	0.0561	0.05	0.0561
(c) 893.46125	893.46	893.4613	893.46	893.4612

NOW WORK PROBLEM 7.

Calculators

Calculators are finite machines. As a result, they are incapable of displaying decimals that contain a large number of digits. For example, some calculators are capable of displaying only eight digits. When a number requires more than eight digits, the calculator either truncates or rounds. To see how your calculator handles decimals, divide 2 by 3. How many digits do you see? Is the last digit a 6 or a 7? If it is a 6, your calculator truncates; if it is a 7, your calculator rounds.

There are different kinds of calculators. An **arithmetic** calculator can only add, subtract, multiply, and divide numbers; therefore, this type is not adequate for this course. **Scientific** calculators have all the capabilities of arithmetic calculators and also contain **function keys** labelled ln, log, sin, cos, tan, x^y, inv, and so on. **Graphing** calculators have all the capabilities of scientific calculators and contain a screen on which graphs can be displayed.

For those who have access to a graphing calculator, we have included comments, examples, and exercises marked with a , indicating that a graphing calculator is required. We have also included Appendix C that explains some of the capabilities of a graphing calculator. The comments, examples, and exercises may be omitted without loss of continuity, if so desired.

Operations

In algebra, we use letters such as x, y, a, b, and c to represent numbers. The symbols used in algebra for the operations of addition, subtraction, multiplication, and division are $+$, $-$, $\cdot$, and /. The words used to describe the results of these operations are **sum, difference, product,** and **quotient.** Table 1 summarizes these ideas.

TABLE 1

Operation	Symbol	Words
Addition	$a + b$	Sum: a plus b
Subtraction	$a - b$	Difference: a minus b
Multiplication	$a\cdot b,\ (a)\cdot b,\ a\cdot(b),\ (a)\cdot(b),$ $ab,\ (a)b,\ a(b),\ (a)(b)$	Product: a times b
Division	a/b or $\dfrac{a}{b}$	Quotient: a divided by b

We generally avoid using the multiplication sign $\times$ and the division sign $\div$ so familiar in arithmetic. Notice also that when two expressions are placed next to each other without an operation symbol, as in ab, or in parentheses, as in $(a)(b)$, it is understood that the expressions, called **factors,** are to be multiplied.

We also prefer not to use mixed numbers. When mixed numbers are used, addition is understood; for example, $2\frac{3}{4}$ means $2 + \frac{3}{4}$. The use of a mixed number may be confusing because the absence of an operation symbol between two terms is generally taken to mean multiplication. The expression $2\frac{3}{4}$ is therefore written instead as 2.75 or as $\frac{11}{4}$.

The symbol =, called an **equal sign** and read as "equals" or "is," is used to express the idea that the number or expression on the left of the equal sign is equivalent to the number or expression on the right.

EXAMPLE 5 Writing Statements Using Symbols

(a) The sum of 2 and 7 equals 9. In symbols, this statement is written as $2 + 7 = 9$.

(b) The product of 3 and 5 is 15. In symbols, this statement is written as $3 \cdot 5 = 15$.

NOW WORK PROBLEM 19.

Order of Operations

2 Evalute numerical expressions

Consider the expression $2 + 3 \cdot 6$. It is not clear whether we should add 2 and 3 to get 5, and then multiply by 6 to get 30; or first multiply 3 and 6 to get 18, and then add 2 to get 20. To avoid this ambiguity, we have the following agreement.

> We agree that whenever the two operations of addition and multiplication separate three numbers, the multiplication operation will be performed first, followed by the addition operation.

For example, we find $2 + 3 \cdot 6$ as follows:

$$2 + 3 \cdot 6 = 2 + 18 = 20$$

EXAMPLE 6 Finding the Value of an Expression

Evaluate each expression.

(a) $3 + 4 \cdot 5$ **(b)** $8 \cdot 2 + 1$ **(c)** $2 + 2 \cdot 2$

SOLUTION

(a) $3 + 4 \cdot 5 = 3 + 20 = 23$ (Multiply first)

(b) $8 \cdot 2 + 1 = 16 + 1 = 17$ (Multiply first)

(c) $2 + 2 \cdot 2 = 2 + 4 = 6$

NOW WORK PROBLEM 31.

Look at Example 6, part (a). To first add 3 and 4 and then multiply the result by 5, we use parentheses and write $(3 + 4) \cdot 5$. Whenever parentheses appear in an expression, it means "perform the operations within the parentheses first!"

EXAMPLE 7 **Finding the Value of an Expression**

(a) $(5 + 3) \cdot 4 = 8 \cdot 4 = 32$
(b) $(4 + 5) \cdot (8 - 2) = 9 \cdot 6 = 54$

When we divide two expressions, as in

$$\frac{2 + 3}{4 + 8}$$

it is understood that the division bar acts like parentheses; that is,

$$\frac{2 + 3}{4 + 8} = \frac{(2 + 3)}{(4 + 8)}$$

The following list gives the rules for the order of operations.

Rules for the Order of Operations

1. Begin with the innermost parentheses and work outward. Remember that in dividing two expressions the numerator and denominator are treated as if they were enclosed in parentheses.
2. Perform multiplications and divisions, working from left to right.
3. Perform additions and subtractions, working from left to right.

EXAMPLE 8 **Finding the Value of an Expression**

Evaluate each expression.

(a) $8 \cdot 2 + 3$ **(b)** $5 \cdot (3 + 4) + 2$

(c) $\dfrac{2 + 5}{2 + 4 \cdot 7}$ **(d)** $2 + [4 + 2 \cdot (10 + 6)]$

SOLUTION **(a)** $8 \cdot 2 + 3 = 16 + 3 = 19$
↑ Multiply first

(b) $5 \cdot (3 + 4) + 2 = 5 \cdot 7 + 2 = 35 + 2 = 37$
↑ Parentheses first ↑ Multiply before adding

(c) $\dfrac{2 + 5}{2 + 4 \cdot 7} = \dfrac{2 + 5}{2 + 28} = \dfrac{7}{30}$

(d) $2 + [4 + 2 \cdot (10 + 6)] = 2 + [4 + 2 \cdot (16)]$
$= 2 + [4 + 32] = 2 + 36 = 38$

NOW WORK PROBLEMS 37 AND 45.

Properties of Real Numbers

Work with properties of real numbers 3

We have used the equal sign to mean that one expression is equivalent to another. Four important properties of equality are listed next. In this list, a, b, and c represent numbers.

1. The **reflexive property** states that a number always equals itself; that is, $a = a$.
2. The **symmetric property** states that if $a = b$ then $b = a$.
3. The **transitive property** states that if $a = b$ and $b = c$ then $a = c$.
4. The **principle of substitution** states that if $a = b$ then we may substitute b for a in any expression containing a.

Now, let's consider some other properties of real numbers. We begin with an example.

EXAMPLE 9 Commutative Properties

(a) $3 + 5 = 8$
$5 + 3 = 8$
$3 + 5 = 5 + 3$

(b) $2 \cdot 3 = 6$
$3 \cdot 2 = 6$
$2 \cdot 3 = 3 \cdot 2$

This example illustrates the **commutative property** of real numbers, which states that the order in which addition or multiplication takes place will not affect the final result.

Commutative Properties

$$a + b = b + a \tag{1a}$$

$$a \cdot b = b \cdot a \tag{1b}$$

Here, and in the properties listed next and on pages A-9–A-12, a, b, and c represent real numbers.

EXAMPLE 10 Associative Properties

(a) $2 + (3 + 4) = 2 + 7 = 9$
$(2 + 3) + 4 = 5 + 4 = 9$
$2 + (3 + 4) = (2 + 3) + 4$

(b) $2 \cdot (3 \cdot 4) = 2 \cdot 12 = 24$
$(2 \cdot 3) \cdot 4 = 6 \cdot 4 = 24$
$2 \cdot (3 \cdot 4) = (2 \cdot 3) \cdot 4$

The way we add or multiply three real numbers will not affect the final result. So, expressions such as $2 + 3 + 4$ and $3 \cdot 4 \cdot 5$ present no ambiguity, even though addition and multiplication are performed on one pair of numbers at a time. This property is called the **associative property.**

Associative Properties

$$a + (b + c) = (a + b) + c = a + b + c \tag{2a}$$

$$a \cdot (b \cdot c) = (a \cdot b) \cdot c = a \cdot b \cdot c \tag{2b}$$

The next property is perhaps the most important.

Distributive Property

$$a \cdot (b + c) = a \cdot b + a \cdot c \tag{3a}$$

The distributive property may be used in two different ways.

EXAMPLE 11 Distributive Property

(a) $2 \cdot (x + 3) = 2 \cdot x + 2 \cdot 3 = 2x + 6$ Use to remove parentheses.
(b) $3x + 5x = (3 + 5)x = 8x$ Use to combine two expressions.

NOW WORK PROBLEM 63.

The real numbers 0 and 1 have unique properties.

EXAMPLE 12 Identity Properties

(a) $4 + 0 = 0 + 4 = 4$ **(b)** $3 \cdot 1 = 1 \cdot 3 = 3$

The properties of 0 and 1 illustrated in Example 12 are called the **identity properties.**

Identity Properties

$$0 + a = a + 0 = a \tag{4a}$$
$$a \cdot 1 = 1 \cdot a = a \tag{4b}$$

We call 0 the **additive identity** and 1 the **multiplicative identity.**

For each real number a, there is a real number $-a$, called the **additive inverse** of a, having the following property:

Additive Inverse Property

$$a + (-a) = -a + a = 0 \tag{5a}$$

EXAMPLE 13 Finding an Additive Inverse

(a) The additive inverse of 6 is -6, because $6 + (-6) = 0$.
(b) The additive inverse of -8 is $-(-8) = 8$, because $-8 + 8 = 0$.

The additive inverse of a, that is, $-a$, is often called the *negative* of a or the *opposite* of a. The use of such terms can be dangerous, because they suggest that the additive inverse is a negative number, which it may not be. For example, the additive inverse of -3, namely $-(-3)$, equals 3, a positive number.

For each *nonzero* real number a, there is a real number $\frac{1}{a}$, called the **multiplicative inverse** of a, having the following property:

Multiplicative Inverse Property

$$a \cdot \frac{1}{a} = \frac{1}{a} \cdot a = 1 \quad \text{if } a \neq 0 \tag{5b}$$

The multiplicative inverse $\frac{1}{a}$ of a nonzero real number a is also referred to as the **reciprocal** of a.

EXAMPLE 14 Finding a Reciprocal

(a) The reciprocal of 6 is $\frac{1}{6}$, because $6 \cdot \frac{1}{6} = 1$.

(b) The reciprocal of -3 is $\frac{1}{-3}$, because $-3 \cdot \frac{1}{-3} = 1$.

(c) The reciprocal of $\frac{2}{3}$ is $\frac{3}{2}$, because $\frac{2}{3} \cdot \frac{3}{2} = 1$.

With these properties for adding and multiplying real numbers, we can now define the operations of subtraction and division as follows:

The **difference** $a - b$, also read "a less b" or "a minus b," is defined as

$$a - b = a + (-b) \tag{6}$$

To subtract b from a, add the opposite of b to a.

If b is a nonzero real number, the **quotient** $\frac{a}{b}$, also read as "a divided by b" or "the ratio of a to b," is defined as

$$\frac{a}{b} = a \cdot \frac{1}{b} \quad \text{if } b \neq 0 \tag{7}$$

EXAMPLE 15 Working with Differences and Quotients

(a) $8 - 5 = 8 + (-5) = 3$ **(b)** $4 - 9 = 4 + (-9) = -5$ **(c)** $\frac{5}{8} = 5 \cdot \frac{1}{8}$

For any number a, the product of a times 0 is always 0.

Multiplication by Zero

$$a \cdot 0 = 0 \tag{8}$$

For a nonzero number a, we have the following division properties.

Division Properties

$$\frac{0}{a} = 0 \qquad \frac{a}{a} = 1 \qquad \text{if } a \neq 0 \tag{9}$$

NOTE: Division by 0 is *not defined.* One reason is to avoid the following difficulty: $\frac{2}{0} = x$ means to find x such that $0 \cdot x = 2$. But $0 \cdot x$ equals 0 for all x, so there is *no* number x such that $\frac{2}{0} = x$.

Rules of Signs

$$\begin{array}{lll} a(-b) = -(ab) & (-a)b = -(ab) & (-a)(-b) = ab \\ -(-a) = a & \dfrac{a}{-b} = \dfrac{-a}{b} = -\dfrac{a}{b} & \dfrac{-a}{-b} = \dfrac{a}{b} \end{array} \tag{10}$$

EXAMPLE 16 **Applying the Rules of Signs**

(a) $2(-3) = -(2 \cdot 3) = -6$

(b) $(-3)(-5) = 3 \cdot 5 = 15$

(c) $\dfrac{3}{-2} = \dfrac{-3}{2} = -\dfrac{3}{2}$

(d) $\dfrac{-4}{-9} = \dfrac{4}{9}$

(e) $\dfrac{x}{-2} = \dfrac{1}{-2} \cdot x = -\dfrac{1}{2}x$

Cancellation Properties

$$\begin{array}{ll} ac = bc \text{ implies } a = b & \text{if } c \neq 0 \\ \dfrac{ac}{bc} = \dfrac{a}{b} & \text{if } b \neq 0, c \neq 0 \end{array} \tag{11}$$

EXAMPLE 17 Using the Cancellation Properties

(a) If $2x = 6$, then

$$2x = 6$$
$$2x = 2 \cdot 3 \quad \text{Factor 6.}$$
$$x = 3 \quad \text{Cancel the 2s.}$$

(b) $\dfrac{18}{12} = \dfrac{3 \cdot \not{6}}{2 \cdot \not{6}} = \dfrac{3}{2}$

Cancel the 6s.

NOTE: We follow the common practice of using slash marks to indicate cancellations.

Zero-Product Property

If $ab = 0$, then $a = 0$ or $b = 0$, or both. (12)

EXAMPLE 18 Using the Zero-Product Property

If $2x = 0$, then either $2 = 0$ or $x = 0$. Since $2 \neq 0$, it follows that $x = 0$.

Arithmetic of Quotients

$$\frac{a}{b} + \frac{c}{d} = \frac{ad}{bd} + \frac{bc}{bd} = \frac{ad + bc}{bd} \quad \text{if } b \neq 0, d \neq 0 \quad (13)$$

$$\frac{a}{b} \cdot \frac{c}{d} = \frac{ac}{bd} \quad \text{if } b \neq 0, d \neq 0 \quad (14)$$

$$\frac{\frac{a}{b}}{\frac{c}{d}} = \frac{a}{b} \cdot \frac{d}{c} = \frac{ad}{bc} \quad \text{if } b \neq 0, c \neq 0, d \neq 0 \quad (15)$$

EXAMPLE 19 Adding, Subtracting, Multiplying, and Dividing Quotients

(a) $\dfrac{2}{3} + \dfrac{5}{2} = \dfrac{2 \cdot 2}{3 \cdot 2} + \dfrac{3 \cdot 5}{3 \cdot 2} = \dfrac{2 \cdot 2 + 3 \cdot 5}{3 \cdot 2} = \dfrac{4 + 15}{6} = \dfrac{19}{6}$

By Equation (13)

(b) $$\frac{3}{5} - \frac{2}{3} \underset{\substack{\uparrow \\ \text{By Equation (6)}}}{=} \frac{3}{5} + \left(-\frac{2}{3}\right) \underset{\substack{\uparrow \\ \text{By Equation (10)}}}{=} \frac{3}{5} + \frac{-2}{3}$$

$$\underset{\substack{\uparrow \\ \text{By Equation (13)}}}{=} \frac{3 \cdot 3 + 5 \cdot (-2)}{5 \cdot 3} \underset{\substack{\uparrow \\ \text{By Equation (10)}}}{=} \frac{9 + (-10)}{15} = \frac{-1}{15} = -\frac{1}{15}$$

(c) $$\frac{8}{3} \cdot \frac{15}{4} \underset{\substack{\uparrow \\ \text{By Equation (14)}}}{=} \frac{8 \cdot 15}{3 \cdot 4} \underset{\substack{\uparrow \\ \text{Factor}}}{=} \frac{2 \cdot \cancel{4} \cdot \cancel{3} \cdot 5}{\cancel{3} \cdot \cancel{4} \cdot 1} \underset{\substack{\uparrow \\ \text{By Equation (11)}}}{=} \frac{2 \cdot 5}{1} = 10$$

NOTE: Slanting the cancellation marks in different directions for different factors, as shown here, is a good practice to follow, since it will help in checking for errors.

(d) $$\frac{\frac{3}{5}}{\frac{7}{9}} \underset{\substack{\uparrow \\ \text{By Equation (15)}}}{=} \frac{3}{5} \cdot \frac{9}{7} \underset{\substack{\uparrow \\ \text{By Equation (14)}}}{=} \frac{3 \cdot 9}{5 \cdot 7} = \frac{27}{35}$$

NOTE: In writing quotients, we shall follow the usual convention and write the quotient in lowest terms; that is, we write it so that any common factors of the numerator and the denominator have been removed using the cancellation properties, Equation (11). For example,

$$\frac{90}{24} = \frac{15 \cdot \cancel{6}}{4 \cdot \cancel{6}} = \frac{15}{4}$$

$$\frac{24x^2}{18x} = \frac{4 \cdot \cancel{6} \cdot x \cdot \cancel{x}}{3 \cdot \cancel{6} \cdot \cancel{x}} = \frac{4x}{3}, \quad x \neq 0$$

NOW WORK PROBLEMS 47, 51, AND 61.

Sometimes it is easier to add two fractions using *least common multiples* (LCM). The LCM of two numbers is the smallest number that each has as a common multiple.

EXAMPLE 20 Finding the Least Common Multiple of Two Numbers

Find the least common multiple of 15 and 12.

SOLUTION To find the LCM of 15 and 12, we look at multiples of 15 and 12.

15, 30, 45, **60**, 75, 90, 105, **120**, . . .

12, 24, 36, 48, **60**, 72, 84, 96, 108, **120**, . . .

The *common* multiples are in blue. The *least* common multiple is 60.

EXAMPLE 21 Using the Least Common Multiple to Add Two Fractions

Find: $\frac{8}{15} + \frac{5}{12}$

SOLUTION We use the LCM of the denominators of the fractions and rewrite each fraction using the LCM as a common denominator. The LCM of the denominators (12 and 15) is 60.

Rewrite each fraction using 60 as the denominator.

$$\frac{8}{15} + \frac{5}{12} = \frac{8}{15} \cdot \frac{4}{4} + \frac{5}{12} \cdot \frac{5}{5} = \frac{32}{60} + \frac{25}{60} = \frac{32 + 25}{60} = \frac{57}{60}$$

NOW WORK PROBLEM 55.

EXERCISE A.1 Answers to Odd-Numbered Problems Begin on Page AN-104.

In Problems 1–6, list the numbers in each set that are (a) natural numbers, (b) integers, (c) rational numbers, (d) irrational numbers, (e) real numbers.

1. $A = \left\{-6, \frac{1}{2}, -1.333\ldots \text{(the 3s repeat)}, \pi, 2, 5\right\}$

2. $B = \left\{-\frac{5}{3}, 2.060606\ldots \text{(the block 06 repeats)}, 1.25, 0, 1, \sqrt{5}\right\}$

3. $C = \left\{0, 1, \frac{1}{2}, \frac{1}{3}, \frac{1}{4}\right\}$

4. $D = \{-1, -1.1, -1.2, -1.3\}$

5. $E = \left\{\sqrt{2}, \pi, \sqrt{2} + 1, \pi + \frac{1}{2}\right\}$

6. $F = \left\{-\sqrt{2}, \pi + \sqrt{2}, \frac{1}{2} + 10.3\right\}$

In Problems 7–18, approximate each number (a) rounded and (b) truncated to three decimal places.

7. 18.9526 **8.** 25.86134 **9.** 28.65319 **10.** 99.05249 **11.** 0.06291 **12.** 0.05388

13. 9.9985 **14.** 1.0006 **15.** $\frac{3}{7}$ **16.** $\frac{5}{9}$ **17.** $\frac{521}{15}$ **18.** $\frac{81}{5}$

In Problems 19–28, write each statement using symbols.

19. The sum of 3 and 2 equals 5.

20. The product of 5 and 2 equals 10.

21. The sum of x and 2 is the product of 3 and 4.

22. The sum of 3 and y is the sum of 2 and 2.

23. 3 times y is 1 plus 2.

24. 2 times x is 4 times 6.

25. x minus 2 equals 6.

26. 2 minus y equals 6.

27. x divided by 2 is 6.

28. 2 divided by x is 6.

In Problems 29–62, evaluate each expression.

29. $9 - 4 + 2$ **30.** $6 - 4 + 3$ **31.** $-6 + 4 \cdot 3$ **32.** $8 - 4 \cdot 2$

33. $4 + 5 - 8$ **34.** $8 - 3 - 4$ **35.** $4 + \frac{1}{3}$ **36.** $2 - \frac{1}{2}$

37. $6 - [3 \cdot 5 + 2 \cdot (3 - 2)]$ **38.** $2 \cdot [8 - 3(4 + 2)] - 3$ **39.** $2 \cdot (3 - 5) + 8 \cdot 2 - 1$ **40.** $1 - (4 \cdot 3 - 2 + 2)$

41. $10 - [6 - 2 \cdot 2 + (8 - 3)] \cdot 2$ **42.** $2 - 5 \cdot 4 - [6 \cdot (3 - 4)]$

43. $(5 - 3) \cdot \frac{1}{2}$ **44.** $(5 + 4) \cdot \frac{1}{3}$ **45.** $\frac{4 + 8}{5 - 3}$ **46.** $\frac{2 - 4}{5 - 3}$

47. $\frac{3}{5} \cdot \frac{10}{21}$ **48.** $\frac{5}{9} \cdot \frac{3}{10}$ **49.** $\frac{6}{25} \cdot \frac{10}{27}$ **50.** $\frac{21}{25} \cdot \frac{100}{3}$

51. $\frac{3}{4} + \frac{2}{5}$ **52.** $\frac{4}{3} + \frac{1}{2}$ **53.** $\frac{5}{6} + \frac{9}{5}$ **54.** $\frac{8}{9} + \frac{15}{2}$

55. $\frac{5}{18} + \frac{1}{12}$ **56.** $\frac{2}{15} + \frac{8}{9}$ **57.** $\frac{1}{30} - \frac{7}{18}$ **58.** $\frac{3}{14} - \frac{2}{21}$

59. $\frac{3}{20} - \frac{2}{15}$ **60.** $\frac{6}{35} - \frac{3}{14}$ **61.** $\dfrac{\frac{5}{18}}{\frac{11}{27}}$ **62.** $\dfrac{\frac{5}{21}}{\frac{2}{35}}$

In Problems 63–74, use the Distributive Property to remove the parentheses.

63. $6(x + 4)$	64. $4(2x - 1)$	65. $x(x - 4)$	66. $4x(x + 3)$
67. $(x + 2)(x + 4)$	68. $(x + 5)(x + 1)$	69. $(x - 2)(x + 1)$	70. $(x - 4)(x + 1)$
71. $(x - 8)(x - 2)$	72. $(x - 4)(x - 2)$	73. $(x + 2)(x - 2)$	74. $(x - 3)(x + 3)$

75. Explain to a friend how the Distributive Property is used to justify the fact that $2x + 3x = 5x$.

76. Explain to a friend why $2 + 3 \cdot 4 = 14$, whereas $(2 + 3) \cdot 4 = 20$.

77. Explain why $2(3 \cdot 4)$ is not equal to $(2 \cdot 3) \cdot (2 \cdot 4)$.

78. Explain why $\frac{4 + 3}{2 + 5}$ is not equal to $\frac{4}{2} + \frac{3}{5}$.

79. Is subtraction commutative? Support your conclusion with an example.

80. Is subtraction associative? Support your conclusion with an example.

81. Is division commutative? Support your conclusion with an example.

82. Is division associative? Support your conclusion with an example.

83. If $2 = x$, why does $x = 2$?

84. If $x = 5$, why does $x^2 + x = 30$?

85. Are there any real numbers that are both rational and irrational? Are there any real numbers that are neither? Explain your reasoning.

86. Explain why the sum of a rational number and an irrational number must he irrational.

87. What rational number does the repeating decimal 0.9999 . . . equal?

A.2 Algebra Review

OBJECTIVES

1. Graph inequalities
2. Find distance on the real number line
3. Evaluate algebraic expressions
4. Determine the domain of a variable
5. Use the laws of exponents
6. Evaluate square roots
7. Use a calculator to evaluate exponents
8. Use scientific notation

The Real Number Line

The real numbers can be represented by points on a line called the **real number line.** There is a one-to-one correspondence between real numbers and points on a line. That is, every real number corresponds to a point on the line, and each point on the line has a unique real number associated with it.

Pick a point on the line somewhere in the center, and label it O. This point, called the **origin,** corresponds to the real number 0. See Figure 4. The point 1 unit to the right of O corresponds to the number 1. The distance between 0 and 1 determines the **scale** of the number line. For example, the point associated with the number 2 is twice as far from O as 1 is. Notice that an arrowhead on the right end of the line indicates the direction in which the numbers increase. Figure 4 also shows the points associated with the irrational numbers $\sqrt{2}$ and π. Points to the left of the origin correspond to the real numbers -1, -2, and so on.

FIGURE 4

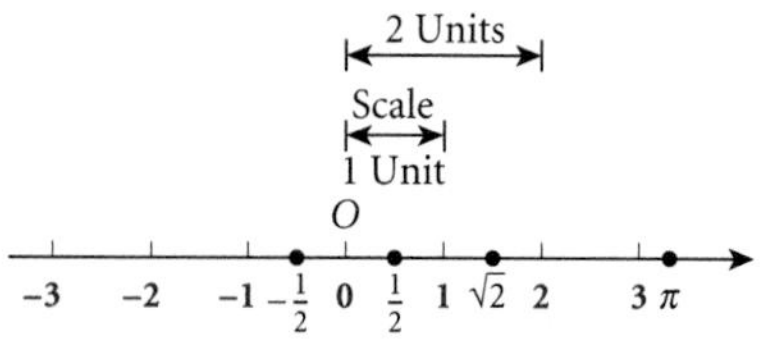

The real number associated with a point P is called the **coordinate** of P, and the line whose points have been assigned coordinates is called the **real number line.**

NOW WORK PROBLEM 1.

FIGURE 5

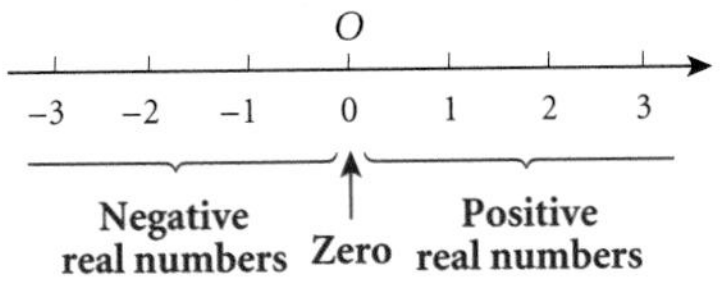

The real number line consists of three classes of real numbers, as shown in Figure 5:

1. The **negative real numbers** are the coordinates of points to the left of the origin O.
2. The real number **zero** is the coordinate of the origin O.
3. The **positive real numbers** are the coordinates of points to the right of the origin O.

Inequalities

FIGURE 6

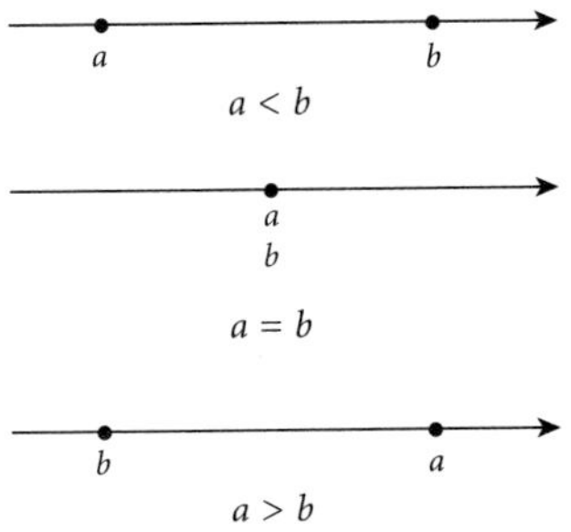

An important property of the real number line follows from the fact that, given two numbers (points) a and b, either a is to the left of b, a is at the same location as b, or a is to the right of b. See Figure 6.

If a is to the left of b, we say that "a is less than b" and write $a < b$. If a is to the right of b, we say that "a is greater than b" and write $a > b$. If a is at the same location as b, then $a = b$. If a is either less than or equal to b, we write $a \leq b$. Similarly, $a \geq b$ means that a is either greater than or equal to b. Collectively, the symbols $<$, $>$, $\leq$, and $\geq$ are called **inequality symbols.**

Note that $a < b$ and $b > a$ mean the same thing. It does not matter whether we write $2 < 3$ or $3 > 2$.

Furthermore, if $a < b$ or if $b > a$, then the difference $b - a$ is positive. Do you see why?

EXAMPLE 1 Using Inequality Symbols

(a) $3 < 7$ **(b)** $-8 > -16$ **(c)** $-6 < 0$

(d) $-8 < -4$ **(e)** $4 > -1$ **(f)** $8 > 0$

In Example 1(a), we conclude that $3 < 7$ either because 3 is to the left of 7 on the real number line or because the difference $7 - 3 = 4$ is a positive real number.

Similarly, we conclude in Example 1(b) that $-8 > -16$ either because -8 lies to the right of -16 on the real number line or because the difference $-8 - (-16) = -8 + 16 = 8$, is a positive real number.

Look again at Example 1. Note that the inequality symbol always points in the direction of the smaller number.

An **inequality** is a statement in which two expressions are related by an inequality symbol. The expressions are referred to as the **sides** of the inequality. Statements of the form $a < b$ or $b > a$ are called **strict inequalities,** whereas statements of the form $a \leq b$ or $b \geq a$ are called **nonstrict inequalities.**

Based on the discussion thus far, we conclude that

$a > 0$	is equivalent to	a is positive
$a < 0$	is equivalent to	a is negative

We sometimes read $a > 0$ by saying that "a is positive." If $a \geq 0$, then either $a > 0$ or $a = 0$, and we may read this as "a is nonnegative."

NOW WORK PROBLEMS 5 AND 15.

Graph inequalities **1**

We shall find it useful in later work to graph inequalities on the real number line.

EXAMPLE 2 Graphing Inequalities

(a) On the real number line, graph all numbers x for which $x > 4$.
(b) On the real number line, graph all numbers x for which $x \leq 5$.

SOLUTION **(a)** See Figure 7. Notice that we use a left parenthesis to indicate that the number 4 is *not* part of the graph.
(b) See Figure 8. Notice that we use a right bracket to indicate that the number 5 is part of the graph.

FIGURE 7

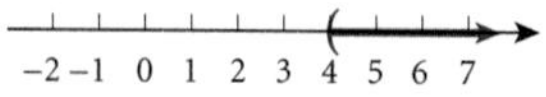

FIGURE 8

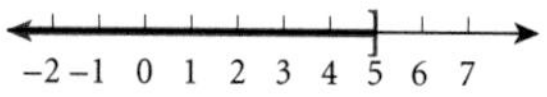

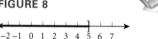

NOW WORK PROBLEM 21.

Absolute Value

The *absolute value* of a number a is the distance from 0 to a on the number line. For example, -4 is 4 units from 0; and 3 is 3 units from 0. See Figure 9. Thus, the absolute value of -4 is 4, and the absolute value of 3 is 3.

A more formal definition of absolute value is given next.

FIGURE 9

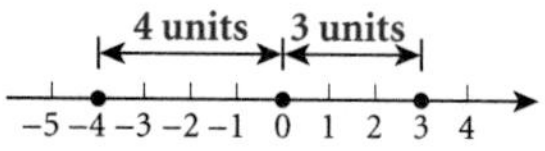

The **absolute value** of a real number a, denoted by the symbol $|a|$, is defined by the rules

$$|a| = a \quad \text{if } a \geq 0 \qquad \text{and} \qquad |a| = -a \quad \text{if } a < 0$$

For example, since $-4 < 0$, the second rule must be used to get $|-4| = -(-4) = 4$.

EXAMPLE 3 Computing Absolute Value

(a) $|8| = 8$ **(b)** $|0| = 0$ **(c)** $|-15| = -(-15) = 15$

Find distance on the real number line **2**

Look again at Figure 9. The distance from -4 to 3 is 7 units. This distance is the difference $3 - (-4)$, obtained by subtracting the smaller coordinate from the larger. However, since $|3 - (-4)| = |7| = 7$ and $|-4 - 3| = |-7| = 7$, we can use absolute value to calculate the distance between two points without being concerned about which is smaller.

If P and Q are two points on a real number line with coordinates a and b, respectively, the **distance between P and Q,** denoted by $d(P, Q)$, is

$$d(P, Q) = |b - a|$$

Since $|b - a| = |a - b|$, it follows that $d(P, Q) = d(Q, P)$.

EXAMPLE 4 Finding Distance on a Number Line

Let P, Q, and R be points on a real number line with coordinates -5, 7, and -3, respectively. Find the distance

(a) between P and Q **(b)** between Q and R

SOLUTION We begin with Figure 10.

FIGURE 10

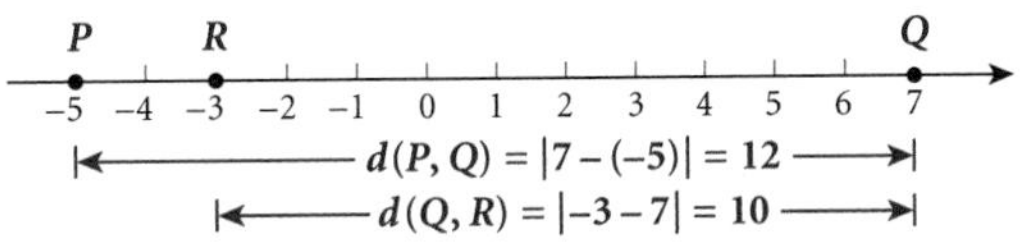

(a) $d(P, Q) = |7 - (-5)| = |12| = 12$

(b) $d(Q, R) = |-3 - 7| = |-10| = 10$

NOW WORK PROBLEM 27.

Constants and Variables

As we said earlier, in algebra we use letters such as x, y, a, b, and c to represent numbers. If the letter used is to represent *any* number from a given set of numbers, it is called a **variable.** A **constant** is either a fixed number, such as 5 or $\sqrt{3}$, or a letter that represents a fixed (possibly unspecified) number.

Constants and variables are combined using the operations of addition, subtraction, multiplication, and division to form *algebraic expressions.* Examples of algebraic expressions include

$$x + 3 \qquad \frac{3}{1 - t} \qquad 7x - 2y$$

Evaluate algebraic expressions 3

To evaluate an algebraic expression, substitute for each variable its numerical value.

EXAMPLE 5 Evaluating an Algebraic Expression

Evaluate each expression if $x = 3$ and $y = -1$.

(a) $x + 3y$ **(b)** $5xy$ **(c)** $\dfrac{3y}{2 - 2x}$ **(d)** $|-4x + y|$

SOLUTION **(a)** Substitute 3 for x and -1 for y in the expression $x + 3y$.

$$x + 3y \underset{\substack{\uparrow \\ x = 3,\, y = -1}}{=} 3 + 3(-1) = 3 + (-3) = 0$$

(b) If $x = 3$ and $y = -1$, then

$$5xy = 5(3)(-1) = -15$$

(c) If $x = 3$ and $y = -1$, then

$$\frac{3y}{2 - 2x} = \frac{3(-1)}{2 - 2(3)} = \frac{-3}{2 - 6} = \frac{-3}{-4} = \frac{3}{4}$$

(d) If $x = 3$ and $y = -1$, then

$$|-4x + y| = |-4(3) + (-1)| = |-12 + (-1)| = |-13| = 13$$

NOW WORK PROBLEMS 29 AND 37.

Determine the domain of a variable **4**

In working with expressions or formulas involving variables, the variables may be allowed to take on values from only a certain set of numbers. For example, in the formula for the area A of a circle of radius r, $A = \pi r^2$, the variable r is necessarily restricted to the positive real numbers. In the expression $\frac{1}{x}$, the variable x cannot take on the value 0, since division by 0 is not defined.

The set of values that a variable may assume is called the **domain of the variable.**

EXAMPLE 6 Finding the Domain of a Variable

The domain of the variable x in the expression

$$\frac{5}{x - 2}$$

is $\{x | x \neq 2\}$, since, if $x = 2$, the denominator becomes 0, which is not defined.

EXAMPLE 7 Circumference of a Circle

In the formula for the circumference C of a circle of radius r,

$$C = 2\pi r$$

the domain of the variable r, representing the radius of the circle, is the set of positive real numbers. The domain of the variable C, representing the circumference of the circle, is also the set of positive real numbers.

In describing the domain of a variable, we may use either set notation or words, whichever is more convenient.

NOW WORK PROBLEM 47.

Exponents

Use the laws of exponents **5**

Integer exponents provide a shorthand device for representing repeated multiplications of a real number. For example,

$$3^4 = 3 \cdot 3 \cdot 3 \cdot 3 = 81$$

Additionally, many formulas have exponents. For example,

- The formula for the horsepower rating H of an engine is

$$H = \frac{D^2N}{2.5}$$

  where D is the diameter of a cylinder and N is the number of cylinders.
- A formula for the resistance R of blood flowing in a blood vessel is

$$R = C\frac{L}{r^4}$$

  where L is the length of the blood vessel, r is the radius, and C is a positive constant.

If a is a real number and n is a positive integer, then the symbol $\boldsymbol{a^n}$ represents the product of n factors of a. That is,

$$a^n = \underbrace{a \cdot a \cdot \ldots \cdot a}_{n \text{ factors}} \tag{1}$$

Here it is understood that $a^1 = a$.

Then $a^2 = a \cdot a$, $a^3 = a \cdot a \cdot a$, and so on. In the expression a^n, a is called the **base** and n is called the **exponent,** or **power.** We read a^n as "a raised to the power n" or as "a to the nth power." We usually read a^2 as "a squared" and a^3 as "a cubed."

In working with exponents, the operation of *raising to a power* is performed before any other operation. As examples,

$$4 \cdot 3^2 = 4 \cdot 9 = 36 \qquad 2^2 + 3^2 = 4 + 9 = 13$$

$$-2^4 = -16 \qquad 5 \cdot 3^2 + 2 \cdot 4 = 5 \cdot 9 + 2 \cdot 4 = 45 + 8 = 53$$

Parentheses are used to indicate operations to be performed first. For example,

$$(-2)^4 = (-2)(-2)(-2)(-2) = 16 \qquad (2 + 3)^2 = 5^2 = 25$$

If $a \neq 0$, we define

$$a^0 = 1 \quad \text{if } a \neq 0$$

If $a \neq 0$ and if n is a positive integer, then we define

$$a^{-n} = \frac{1}{a^n} \quad \text{if } a \neq 0$$

Whenever you encounter a negative exponent, think "reciprocal."

EXAMPLE 8 Evaluating Expressions Containing Negative Exponents

(a) $2^{-3} = \frac{1}{2^3} = \frac{1}{8}$ **(b)** $x^{-4} = \frac{1}{x^4}$ **(c)** $\left(\frac{1}{5}\right)^{-2} = \frac{1}{\left(\frac{1}{5}\right)^2} = \frac{1}{\frac{1}{25}} = 25$

NOW WORK PROBLEMS 65 AND 85.

The following properties, called the **Laws of Exponents,** can be proved using the preceding definitions. In the list, a and b are real numbers, and m and n are integers.

Laws of Exponents

$$a^m a^n = a^{m+n} \qquad (a^m)^n = a^{mn} \qquad (ab)^n = a^n b^n$$

$$\frac{a^m}{a^n} = a^{m-n} = \frac{1}{a^{n-m}}, \quad \text{if } a \neq 0 \qquad \left(\frac{a}{b}\right)^n = \frac{a^n}{b^n}, \quad \text{if } b \neq 0$$

EXAMPLE 9 Using the Laws of Exponents

(a) $x^{-3} \cdot x^5 = x^{-3+5} = x^2, \quad x \neq 0$

(b) $(x^{-3})^2 = x^{-3\cdot 2} = x^{-6} = \frac{1}{x^6}, \quad x \neq 0$

(c) $(2x)^3 = 2^3 \cdot x^3 = 8x^3$

(d) $\left(\frac{2}{3}\right)^4 = \frac{2^4}{3^4} = \frac{16}{81}$

(e) $\frac{x^{-2}}{x^{-5}} = x^{-2-(-5)} = x^3, \quad x \neq 0$

NOW WORK PROBLEM 67.

EXAMPLE 10 Using the Laws of Exponents

Write each expression so that all exponents are positive.

(a) $\frac{x^5y^{-2}}{x^3y}, \quad x \neq 0, \quad y \neq 0$ **(b)** $\left(\frac{x^{-3}}{3y^{-1}}\right)^{-2}, \quad x \neq 0, \quad y \neq 0$

SOLUTION **(a)** $\frac{x^5y^{-2}}{x^3y} = \frac{x^5}{x^3} \cdot \frac{y^{-2}}{y} = x^{5-3} \cdot y^{-2-1} = x^2y^{-3} = x^2 \cdot \frac{1}{y^3} = \frac{x^2}{y^3}$

(b) $\left(\frac{x^{-3}}{3y^{-1}}\right)^{-2} = \frac{(x^{-3})^{-2}}{(3y^{-1})^{-2}} = \frac{x^6}{3^{-2}(y^{-1})^{-2}} = \frac{x^6}{\frac{1}{9}y^2} = \frac{9x^6}{y^2}$

NOW WORK PROBLEM 77.

Square Roots

Evaluate square roots 6

A real number is squared when it is raised to the power 2. The inverse of squaring is finding a **square root.** For example, since $6^2 = 36$ and $(-6)^2 = 36$, the numbers 6 and -6 are square roots of 36.

The symbol $\sqrt{\ }$, called a **radical sign,** is used to denote the **principal,** or nonnegative, square root. For example, $\sqrt{36} = 6$.

In general, if a is a nonnegative real number, the nonnegative number b, such that $b^2 = a$ is the **principal square root** of a, and is denoted by $b = \sqrt{a}$.

The following comments are noteworthy:

1. Negative numbers do not have square roots (in the real number system), because the square of any real number is *nonnegative.* For example, $\sqrt{-4}$ is not a real number, because there is no real number whose square is -4.
2. The principal square root of 0 is 0, since $0^2 = 0$. That is, $\sqrt{0} = 0$.
3. The principal square root of a positive number is positive.
4. If $c \geq 0$, then $(\sqrt{c})^2 = c$. For example, $(\sqrt{2})^2 = 2$ and $(\sqrt{3})^2 = 3$.

EXAMPLE 11 Evaluating Square Roots

(a) $\sqrt{64} = 8$ **(b)** $\sqrt{\dfrac{1}{16}} = \dfrac{1}{4}$ **(c)** $(\sqrt{1.4})^2 = 1.4$

(d) $\sqrt{(-3)^2} = |-3| = 3$

Examples 11(a) and (b) are examples of square roots of perfect squares, since $64 = 8^2$ and $\dfrac{1}{16} = \left(\dfrac{1}{4}\right)^2$.

Notice the need for the absolute value in Example 11(d). Since $a^2 \geq 0$, the principal square root of a^2 is defined whether $a > 0$ or $a < 0$. However, since the principal square root is nonnegative, we need the absolute value to ensure the nonnegative result.

In general, we have

$$\sqrt{a^2} = |a| \qquad (2)$$

EXAMPLE 12 Using Equation (2)

(a) $\sqrt{(2.3)^2} = |2.3| = 2.3$ **(b)** $\sqrt{(-2.3)^2} = |-2.3| = 2.3$ **(c)** $\sqrt{x^2} = |x|$

NOW WORK PROBLEM 73.

Calculator Use

Your calculator has either the caret key, [^], or the [x^y] key, which is used for computations involving exponents.

7 EXAMPLE 13 Exponents on a Graphing Calculator

Evaluate: $(2.3)^5$

SOLUTION Figure 11 shows the result using a TI-83 graphing calculator.

FIGURE 11

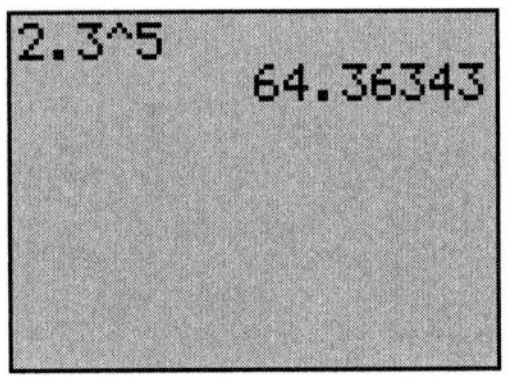

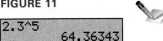

NOW WORK PROBLEM 103.

Scientific Notation

Measurements of physical quantities can range from very small to very large. For example, the mass of a proton is approximately 0.00000000000000000000000000167 kilogram and the mass of Earth is about 5,980,000,000,000,000,000,000,000 kilograms. These numbers obviously are tedious to write down and difficult to read, so we use exponents to rewrite each.

> When a number has been written as the product of a number x, where $1 \leq x < 10$, times a power of 10, it is said to be written in **scientific notation.**

In scientific notation,

$$\text{Mass of a proton} = 1.67 \times 10^{-27} \text{ kilogram}$$
$$\text{Mass of Earth} = 5.98 \times 10^{24} \text{ kilograms}$$

> **Converting a Decimal to Scientific Notation**
>
> To change a positive number into scientific notation:
>
> 1. Count the number N of places that the decimal point must be moved in order to arrive at a number x, where $1 \leq x < 10$.
> 2. If the original number is greater than or equal to 1, the scientific notation is $x \times 10^N$. If the original number is between 0 and 1, the scientific notation is $x \times 10^{-N}$.

8 **EXAMPLE 14**   **Using Scientific Notation**

Write each number in scientific notation.

(a) 9582   (b) 1.245   (c) 0.285   (d) 0.000561

SOLUTION   (a) The decimal point in 9582 follows the 2. Thus, we count

$$9 \quad 5 \quad 8 \quad 2 \quad .$$
$$\quad 3 \quad 2 \quad 1$$

stopping after three moves, because 9.582 is a number between 1 and 10. Since 9582 is greater than 1, we write

$$9582 = 9.582 \times 10^3$$

(b) The decimal point in 1.245 is between the 1 and 2. Since the number is already between 1 and 10, the scientific notation for it is $1.245 \times 10^0 = 1.245$.

(c) The decimal point in 0.285 is between the 0 and the 2. We count

$$0 \quad . \quad 2 \quad 8 \quad 5$$
$$\quad 1$$

stopping after one move, because 2.85 is a number between 1 and 10. Since 0.285 is between 0 and 1, we write

$$0.285 = 2.85 \times 10^{-1}$$

(d) The decimal point in 0.000561 is moved as follows:

$$0 \quad . \quad 0 \quad 0 \quad 0 \quad 5 \quad 6 \quad 1$$
$$\quad 1 \quad 2 \quad 3 \quad 4$$

As a result,

$$0.000561 = 5.61 \times 10^{-4}$$

  NOW WORK PROBLEM 109.

EXAMPLE 15   **Changing from Scientific Notation to a Decimal**

Write each number as a decimal.

(a) 2.1×10^4 (b) 3.26×10^{-5} (c) 1×10^{-2}

SOLUTION   (a) $2.1 \times 10^4 = 2 \quad . \quad 1 \quad 0 \quad 0 \quad 0 \quad \times 10^4 = 21{,}000$
$$1 \quad 2 \quad 3 \quad 4$$

(b) $3.26 \times 10^{-5} = 0 \quad 0 \quad 0 \quad 0 \quad 0 \quad 3 \quad . \quad 2 \quad 6 \times 10^{-5} = 0.0000326$
$$\phantom{3.26 \times 10^{-5} = 0 \quad}5 \quad 4 \quad 3 \quad 2 \quad 1$$

(c) $1 \times 10^{-2} = 0\ 0\ 1\ .\ \times 10^{-2} = 0.01$

2 1

On a calculator, a number such as 3.615×10^{12} is usually displayed as 3.615E12.

NOW WORK PROBLEM 117.

EXAMPLE 16 Using Scientific Notation

(a) The diameter of the smallest living cell is only about 0.00001 centimeter (cm). Express this number in scientific notation.

Source: Powers of Ten, Philip and Phylis Morrison.

(b) The surface area of Earth is about 1.97×10^8 square miles. Express the surface area as a whole number.

Source: 1998 Information Please Almanac.

SOLUTION **(a)** 0.00001 cm $= 1 \times 10^{-5}$ cm because the decimal point is moved five places and the number is less than 1.

(b) 1.97×10^8 square miles $=$ 197,000,000 square miles.

EXERCISE A.2 Answers to Odd-Numbered Problems Begin on Page AN-105.

1. On the real number line, label the points with coordinates $0, 1, -1, \frac{5}{2}, -2.5, \frac{3}{4}$, and 0.25.

2. Repeat Problem 1 for the coordinates $0, -2, 2, -1.5, \frac{3}{2}, \frac{1}{3}$, and $\frac{2}{3}$.

In Problems 3–12, replace the question mark by <, >, or =, whichever is correct.

3. $\frac{1}{2} \,?\, 0$
4. $5 \,?\, 6$
5. $-1 \,?\, -2$
6. $-3 \,?\, -\frac{5}{2}$
7. $\pi \,?\, 3.14$
8. $\sqrt{2} \,?\, 1.41$
9. $\frac{1}{2} \,?\, 0.5$
10. $\frac{1}{3} \,?\, 0.33$
11. $\frac{2}{3} \,?\, 0.67$
12. $\frac{1}{4} \,?\, 0.25$

In Problems 13–18, write each statement as an inequality.

13. x is positive
14. z is negative
15. x is less than 2
16. y is greater than -5
17. x is less than or equal to 1
18. x is greater than or equal to 2

In Problems 19–22, graph the numbers x on the real number line.

19. $x \geq -2$
20. $x < 4$
21. $x > -1$
22. $x \leq 7$

In Problems 23–28, use the real number line below to compute each distance.

$$A \quad B \quad C \quad D \quad \quad E$$

$$-4 \quad -3 \quad -2 \quad -1 \quad 0 \quad 1 \quad 2 \quad 3 \quad 4 \quad 5 \quad 6$$

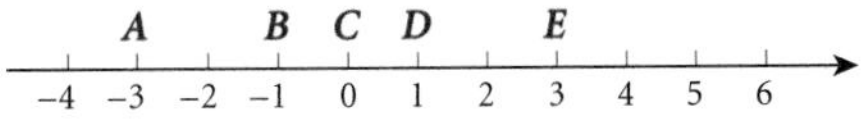

23. $d(C, D)$ **24.** $d(C, A)$ **25.** $d(D, E)$ **26.** $d(C, E)$ 27. $d(A, E)$ **28.** $d(D, B)$

In Problems 29–36, evaluate each expression if $x = -2$ and $y = 3$.

29. $x + 2y$ **30.** $3x + y$ **31.** $5xy + 2$ **32.** $-2x + xy$

33. $\dfrac{2x}{x - y}$ **34.** $\dfrac{x + y}{x - y}$ **35.** $\dfrac{3x + 2y}{2 + y}$ **36.** $\dfrac{2x - 3}{y}$

In Problems 37–46, find the value of each expression if $x = 3$ and $y = -2$.

37. $|x + y|$ **38.** $|x - y|$ **39.** $|x| + |y|$ **40.** $|x| - |y|$ **41.** $\dfrac{|x|}{x}$

42. $\dfrac{|y|}{y}$ **43.** $|4x - 5y|$ **44.** $|3x + 2y|$ **45.** $||4x| - |5y||$ **46.** $3|x| + 2|y|$

In Problems 47–54, determine which of the value(s) given below, if any, must be excluded from the domain of the variable in each expression:

(a) $x = 3$ (b) $x = 1$ (c) $x = 0$ (d) $x = -1$

47. $\dfrac{x^2 - 1}{x}$ **48.** $\dfrac{x^2 + 1}{x}$ **49.** $\dfrac{x}{x^2 - 9}$ **50.** $\dfrac{x}{x^2 + 9}$

51. $\dfrac{x^2}{x^2 + 1}$ **52.** $\dfrac{x^3}{x^2 - 1}$ **53.** $\dfrac{x^2 + 5x - 10}{x^3 - x}$ **54.** $\dfrac{-9x^2 - x + 1}{x^3 + x}$

In Problems 55–58, determine the domain of the variable x in each expression.

55. $\dfrac{4}{x - 5}$ **56.** $\dfrac{-6}{x + 4}$ **57.** $\dfrac{x}{x + 4}$ **58.** $\dfrac{x - 2}{x - 6}$

In Problems 59–62, use the formula $C = \frac{5}{9}(F - 32)$ for converting degrees Fahrenheit into degrees Celsius to find the Celsius measure of each Fahrenheit temperature.

59. $F = 32°$ **60.** $F = 212°$ **61.** $F = 77°$ **62.** $F = -4°$

In Problems 63–74, simplify each expression.

63. $(-4)^2$ **64.** -4^2 65. 4^{-2} **66.** -4^{-2} 67. $3^{-6} \cdot 3^4$ **68.** $4^{-2} \cdot 4^3$

69. $(3^{-2})^{-1}$ **70.** $(2^{-1})^{-3}$ **71.** $\sqrt{25}$ **72.** $\sqrt{36}$ 73. $\sqrt{(-4)^2}$ **74.** $\sqrt{(-3)^2}$

In Problems 75–84, simplify each expression. Express the answer so that all exponents are positive. Whenever an exponent is 0 or negative, we assume that the base is not 0.

75. $(8x^3)^2$ **76.** $(-4x^2)^{-1}$ 77. $(x^2 y^{-1})^2$ **78.** $(x^{-1}y)^3$ **79.** $\dfrac{x^2 y^3}{xy^4}$

80. $\dfrac{x^{-2}y}{xy^2}$ **81.** $\dfrac{(-2)^3 x^4 (yz)^2}{3^2 xy^3 z}$ **82.** $\dfrac{4x^{-2}(yz)^{-1}}{2^3 x^4 y}$ **83.** $\left(\dfrac{3x^{-1}}{4y^{-1}}\right)^{-2}$ **84.** $\left(\dfrac{5x^{-2}}{6y^{-2}}\right)^{-3}$

In Problems 85–96, find the value of each expression if $x = 2$ and $y = -1$.

85. $2xy^{-1}$
86. $-3x^{-1}y$
87. $x^2 + y^2$
88. x^2y^2
89. $(xy)^2$
90. $(x + y)^2$
91. $\sqrt{x^2}$
92. $(\sqrt{x})^2$
93. $\sqrt{x^2 + y^2}$
94. $\sqrt{x^2} + \sqrt{y^2}$
95. x^y
96. y^x

97. Find the value of the expression $2x^3 - 3x^2 + 5x - 4$ if $x = 2$. What is the value if $x = 1$?

98. Find the value of the expression $4x^3 + 3x^2 - x + 2$ if $x = 1$. What is the value if $x = 2$?

99. What is the value of $\dfrac{(666)^4}{(222)^4}$?

100. What is the value of $(0.1)^3(20)^3$?

In Problems 101–108, use a calculator to evaluate each expression. Round your answer to three decimal places.

101. $(8.2)^6$
102. $(3.7)^5$
103. $(6.1)^{-3}$
104. $(2.2)^{-5}$
105. $(-2.8)^6$
106. $-(2.8)^6$
107. $(-8.11)^{-4}$
108. $-(8.11)^{-4}$

In Problems 109–116, write each number in scientific notation.

109. 454.2
110. 32.14
111. 0.013
112. 0.00421
113. 32,155
114. 21,210
115. 0.000423
116. 0.0514

In Problems 117–124, write each number as a decimal.

117. 6.15×10^4
118. 9.7×10^3
119. 1.214×10^{-3}
120. 9.88×10^{-4}
121. 1.1×10^8
122. 4.112×10^2
123. 8.1×10^{-2}
124. 6.453×10^{-1}

In Problems 125–134, express each statement as an equation involving the indicated variables. State the domain of each variable.

125. **Area of a Rectangle** The area A of a rectangle is its length l times its width w.

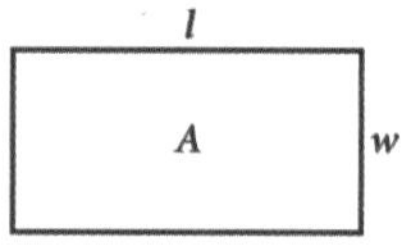

126. **Perimeter of a Rectangle** The perimeter P of a rectangle is twice the sum of its length l and its width w.

127. **Circumference of a Circle** The circumference C of a circle is π times its diameter d.

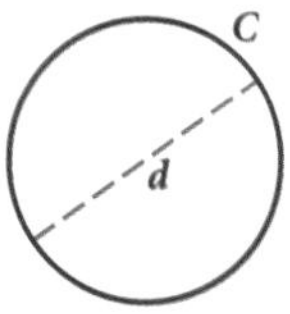

128. **Area of a Triangle** The area A of a triangle is one-half the base b times its height h.

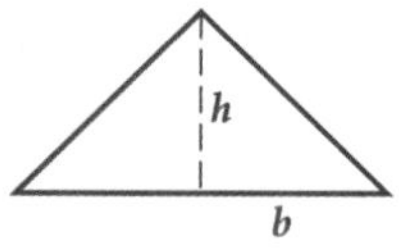

129. **Area of an Equilateral Triangle** The area A of an equilateral triangle is $\dfrac{\sqrt{3}}{4}$ times the square of the length x of one side.

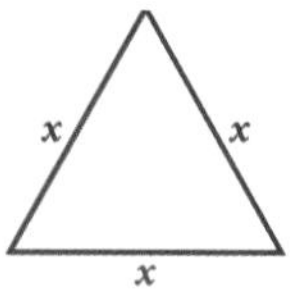

130. **Perimeter of an Equilateral Triangle** The perimeter P of an equilateral triangle is 3 times the length x of one side.

131. Volume of a Sphere The volume V of a sphere is $\frac{4}{3}$ times π times the cube of the radius r.

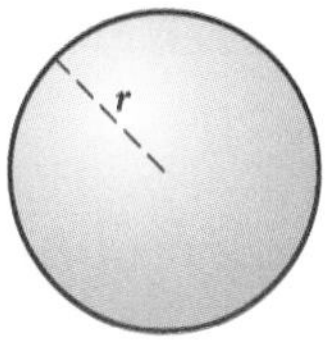

132. Surface Area of a Sphere The surface area S of a sphere is 4 times π times the square of the radius r.

133. Volume of a Cube The volume V of a cube is the cube of the length x of a side.

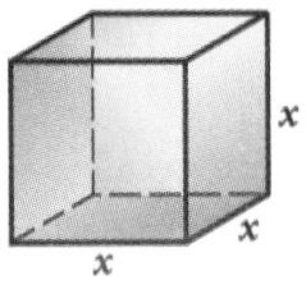

134. Surface Area of a Cube The surface area S of a cube is 6 times the square of the length x of a side.

135. Manufacturing Cost The weekly production cost C of manufacturing x watches is given by the formula $C = 4000 + 2x$, where the variable C is in dollars.

(a) What is the cost of producing 1000 watches?
(b) What is the cost of producing 2000 watches?

136. Balancing a Checkbook At the beginning of the month, Mike had a balance of \$210 in his checking account. During the next month, he deposited \$80, wrote a check for \$120, made another deposit of \$25, wrote two checks for \$60 and \$32, and was assessed a monthly service charge of \$5. What was his balance at the end of the month?

137. U.S. Voltage In the United States, normal household voltage is 115 volts. It is acceptable for the actual voltage x to differ from normal by at most 5 volts. A formula that describes this is

$$|x - 115| \leq 5$$

(a) Show that a voltage of 113 volts is acceptable.
(b) Show that a voltage of 109 volts is not acceptable.

138. Foreign Voltage In other countries, normal household voltage is 220 volts. It is acceptable for the actual voltage x to differ from normal by at most 8 volts. A formula that describes this is

$$|x - 220| \leq 8$$

(a) Show that a voltage of 214 volts is acceptable.
(b) Show that a voltage of 209 volts is not acceptable.

139. Making Precision Ball Bearings The FireBall Company manufactures ball bearings for precision equipment. One of their products is a ball bearing with a stated radius of 3 centimeters (cm). Only ball bearings with a radius within 0.01 cm of this stated radius are acceptable. If x is the radius of a ball bearing, a formula describing this situation is

$$|x - 3| \leq 0.01$$

(a) Is a ball bearing of radius $x = 2.999$ acceptable?
(b) Is a ball bearing of radius $x = 2.89$ acceptable?

140. Body Temperature Normal human body temperature is 98.6°F. A temperature x that differs from normal by at least 1.5°F is considered unhealthy. A formula that describes this is

$$|x - 98.6| \geq 1.5$$

(a) Show that a temperature of 97°F is unhealthy.
(b) Show that a temperature of 100°F is not unhealthy.

141. Does $\frac{1}{3}$ equal 0.333? If not, which is larger? By how much?

142. Does $\frac{2}{3}$ equal 0.666? If not, which is larger? By how much?

143. Is there a positive real number "closest" to 0? Explain.

144. I'm thinking of a number! It lies between 1 and 10; its square is rational and lies between 1 and 10. The number is larger than π. Correct to two decimal places, name the number. Now think of your own number, describe it, and challenge a fellow student to name it.

145. Write a brief paragraph that illustrates the similarities and differences between "less than" ($<$) and "less than or equal to" ($\leq$).

A.3 Exponents and Logarithms

OBJECTIVES
1 Evaluate expressions containing exponents
2 Change logarithms to exponents
3 Change exponents to logarithms
4 Evaluate logarithms using the change-of-base formula

Exponents

Evaluate expressions containing exponents 1

Integer exponents provide a convenient notation for repeated multiplication of a real number.

If a is a real number and n is a positive integer, then the **symbol a^n** represents the product of n factors of a. That is,

$$a^n = \underbrace{a \cdot a \cdot \ . \ . \ . \ \cdot a}_{n \text{ factors}}$$

where it is understood that $a^1 = a$.

In the expression a^n, read "a raised to the power n," a is called the **base** and n is called the **exponent** or **power.**

If n is a positive integer and $a \neq 0$, then we define

$$a^{-n} = \frac{1}{a^n} \qquad a \neq 0$$

Also, we define

$$a^0 = 1 \qquad a \neq 0$$

EXAMPLE 1 **Evaluating Exponents**

(a) $2^3 = 2 \cdot 2 \cdot 2 = 8$ **(b)** $3^{-2} = \frac{1}{3^2} = \frac{1}{9}$ **(c)** $5^0 = 1$

NOW WORK PROBLEM 1.

The **principal nth root of a real number a**, written $\sqrt[n]{a}$, is defined as follows:

(a) If $a > 0$ and n is even, then $\sqrt[n]{a}$ is the positive number x for which $x^n = a$.
(b) If $a < 0$ and n is even, then $\sqrt[n]{a}$ does not exist.
(c) If n is odd, then $\sqrt[n]{a}$ is the number x for which $x^n = a$.
(d) $\sqrt[n]{0} = 0$ for any n.

If $n = 2$, we write $\sqrt{a}$ in place of $\sqrt[2]{a}$.

EXAMPLE 2 **Evaluating Principal nth Roots**

(a) $\sqrt[3]{8} = 2$ because $2^3 = 8$

(b) $\sqrt{64} = 8$ because $8^2 = 64$ and $8 > 0$

(c) $\sqrt[3]{-27} = -3$ because $(-3)^3 = -27$

NOW WORK PROBLEM 7.

We use principal roots to define exponents that are rational. If a is a real number and $n \geq 2$ is an integer, then

$$a^{1/n} = \sqrt[n]{a}$$

provided $\sqrt[n]{a}$ exists.

If a is a real number and m, n are integers containing no common factors and $n \geq 2$, then

$$a^{m/n} = \sqrt[n]{a^m} = (\sqrt[n]{a})^m$$

provided $\sqrt[n]{a}$ exists.

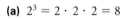

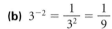

EXAMPLE 3 **Evaluating Rational Exponents**

(a) $4^{3/2} = (\sqrt{4})^3 = 2^3 = 8$

(b) $(-8)^{4/3} = (\sqrt[3]{-8})^4 = (-2)^4 = 16$

(c) $(32)^{-2/5} = (\sqrt[5]{32})^{-2} = 2^{-2} = \dfrac{1}{4}$

NOW WORK PROBLEM 13.

The Laws of Exponents establish some rules for working with exponents.

Laws of Exponents

If a and b are positive real numbers and r and s are rational numbers, then

$$a^r \cdot a^s = a^{r+s} \quad (a^r)^s = a^{rs} \quad (ab)^r = a^r b^r \quad a^{-r} = \frac{1}{a^r},$$

But what is the meaning of a^x, where the base a is a positive real number and the exponent x is an irrational number? Although a rigorous definition requires methods discussed in calculus, the basis for the definition is easy to follow: Select a rational number r that is formed by truncating (removing) all but a finite number of digits from the irrational number x. Then it is reasonable to expect that

$$a^x \approx a^r$$

For example, take the irrational number $\pi \approx 3.14159 \ldots$ An approximation to a^π is

$$a^\pi \approx a^{3.14}$$

where the digits after the hundredths position have been removed from the value for π. A better approximation would be

$$a^\pi \approx a^{3.14159}$$

where the digits after the hundred-thousandths position have been removed. Continuing in this way, we can obtain approximations to a^π to any desired degree of accuracy.

Most scientific calculators and all graphing utilities have an $\boxed{x^y}$ key (or a $\boxed{y^x}$ key) for working with exponents. To use this key, first enter the base x, then press the $\boxed{x^y}$ key, enter y, and press the $\boxed{=}$ or $\boxed{\text{ENTER}}$ key.

EXAMPLE 4 **Evaluating Exponents with a Calculator**

Evaluate

(a) $2^{1.4}$ **(b)** $2^{1.41}$ **(c)** $2.^{1.414}$ **(d)** $2^{1.4142}$ **(e)** $2^{\sqrt{2}}$

Round answers to four decimal places.

SOLUTION **(a)** $2^{1.4} \approx 2.6390$ **(b)** $2^{1.41} \approx 2.6574$ **(c)** $2^{1.414} \approx 2.6647$
(d) $2.^{1.4142} \approx 2.6651$ **(e)** $2^{\sqrt{2}} \approx 2.6651$

NOW WORK PROBLEM 19.

Logarithms

We have given meaning to a number a raised to the power x. Suppose $N = a^x$, $a > 0$, $a \neq 1$, x real. The definition of a logarithm is based on this exponential relationship.

The **logarithm to the base *a* of *N*** is symbolized by $\log_a N$ and is defined by

$$\log_a N = x \quad \text{if and only if} \quad N = a^x$$

2 **EXAMPLE 5 Changing Logarithms to Exponents**

(a) $\log_2 8 = x$ means $2^x = 8 = 2^3$, so $x = 3$
(b) $\log_3 9 = x$ means $3^x = 9 = 3^2$, so $x = 2$
(c) $\log_{10} N = 3$ means $N = 10^3 = 1000$
(d) $\log_4 N = 2$ means $N = 4^2 = 16$
(e) $\log_a 16 = 2$ means $16 = a^2$, so $a = 4$
(f) $\log_a 3 = -1$ means $3 = a^{-1}$, so $a = \frac{1}{3}$

NOW WORK PROBLEMS 25, 29, AND 33.

Change exponents to logarithms

3 As Example 5 illustrates, logarithms are another name for exponents. Let's see how we change exponents to logarithms.

EXAMPLE 6 Changing Exponents to Logarithms

(a) If $100 = (1.1)^n$, then $n = \log_{1.1} 100$
(b) If $350 = (1.25)^n$, then $n = \log_{1.25} 350$
(c) If $1000 = (1.005)^n$, then $n = \log_{1.005} 1000$

NOW WORK PROBLEM 37.

Change-of-Base Formula

Evaluate logarithms using the change-of-base formula

4 To evaluate the expressions shown in Example 6, we use a calculator and the following formula involving changing the base.

Change-of-Base Formula

$$\log_a M = \frac{\log_{10} M}{\log_{10} a} \qquad (1)$$

Consult your calculator manual to see how to evaluate $\log_{10} M$.

EXAMPLE 7 Using the Change-of-Base Formula

(a) $\log_{1.1} 100 = \dfrac{\log_{10} 100}{\log_{10} 1.1} = \dfrac{2}{0.0414} = 48.3177$

(b) $\log_{1.25} 350 = \dfrac{\log_{10} 350}{\log_{10} 1.25} = \dfrac{2.5441}{0.0969} = 26.2519$

(c) $\log_{1.005} 1000 = \dfrac{\log_{10} 1000}{\log_{10} 1.005} = \dfrac{3}{0.002166} = 1385.0021$

Comparing Examples 6 and 7, we find that:

(a) $100 = (1.1)^n$ means $n = \log_{1.1} 100 = 48.3177$
(b) $350 = (1.25)^n$ means $n = \log_{1.25} 350 = 26.2519$
(c) $1000 = (1.005)^n$ means $n = \log_{1.005} 1000 = 1385.0021$

NOW WORK PROBLEM 41.

EXERCISE A.3 Answers to Odd-Numbered Problems Begin on Page AN-105.

In Problems 1–18, evaluate each expression.

1. 4^3 2. 8^2 3. 2^{-3} 4. 4^{-2} 5. 8^0 6. $(-3)^0$

7. $\sqrt{16}$ 8. $\sqrt{9}$ 9. $\sqrt[3]{27}$ 10. $\sqrt[3]{-1}$ 11. $\sqrt[4]{16}$ 12. $\sqrt{0}$

13. $8^{2/3}$ 14. $9^{3/2}$ 15. $16^{-3/2}$ 16. $8^{-2/3}$ 17. $(-8)^{-2/3}$ 18. $(-27)^{-4/3}$

In Problems 19–24, approximate each number using a calculator. Round answers to four decimal places.

19. (a) $3^{2.2}$ (b) $3^{2.23}$ (c) $3^{2.236}$ (d) $3^{\sqrt{5}}$

20. (a) $5^{1.7}$ (b) $5^{1.73}$ (c) $5^{1.732}$ (d) $5^{\sqrt{3}}$

21. (a) $2^{3.14}$ (b) $2^{3.141}$ (c) $2^{3.1415}$ (d) 2^{π}

22. (a) $2^{2.7}$ (b) $2^{2.71}$ (c) $2^{2.718}$ (d) 2^{e}

23. (a) $3.1^{2.7}$ (b) $3.14^{2.71}$ (c) $3.141^{2.718}$ (d) π^{e}

24. (a) $2.7^{3.1}$ (b) $2.71^{3.14}$ (c) $2.718^{3.141}$ (d) e^{π}

In Problems 25–28, evaluate each expression by changing the logarithm to an exponent.

25. $\log_3 27$ 26. $\log_2 16$ 27. $\log_2 \dfrac{1}{2}$ 28. $\log_3 \dfrac{1}{9}$

In Problems 29–32, find N by changing the logarithm to an exponent.

29. $\log_2 N = 3$ 30. $\log_2 N = -2$ 31. $\log_3 N = -1$ 32. $\log_8 N = \dfrac{1}{3}$

In Problems 33–36, find a by changing the logarithm to an exponent.

33. $\log_a 8 = 3$ 34. $\log_a 4 = -1$ 35. $\log_a 9 = 2$ 36. $\log_a 16 = -2$

In Problems 37–40, change each exponent to a logarithm.

37. $5 = 2^x$ 38. $8 = 3^y$ 39. $10 = 1.1^t$ 40. $100 = 1.05^r$

In Problems 41–50, evaluate each logarithm using the Change-of-Base formula.

41. $\log_{1.1} 200$ **42.** $\log_{1.05} 200$ **43.** $\log_{1.005} 1000$ **44.** $\log_{1.001} 10$ **45.** $\log_{1.002} 20$

46. $\log_{1.005} 30$ **47.** $\log_{1.0005} 500$ **48.** $\log_{1.0001} 1000$ **49.** $\log_{1.003} 500$ **50.** $\log_{1.006} 500$

A.4 Recursively Defined Sequences; Geometric Sequences

OBJECTIVES
1. Write the terms of a sequence
2. Find the first term and common ratio of a geometric sequence
3. Add the terms of a geometric sequence

> A **sequence** is a rule that assigns a real number to each positive integer.

Write the terms of a sequence 1

A sequence is **often represented** by listing its values in order. For example, the sequence whose rule is to assign to each positive integer its reciprocal may be represented as

$$s_1 = 1, \quad s_2 = \frac{1}{2}, \quad s_3 = \frac{1}{3}, \quad s_4 = \frac{1}{4}, \quad \ldots, \quad s_n = \frac{1}{n}, \quad \ldots$$

or merely as the list

$$1, \quad \frac{1}{2}, \quad \frac{1}{3}, \quad \frac{1}{4}, \quad \ldots, \quad \frac{1}{n}, \quad \ldots$$

The list never ends, as the dots indicate. The real numbers in this ordered list are called the **terms** of the sequence. For the sequence above we have a rule for the nth term, namely, $s = 1/n$, so it is easy to find any term of the sequence. In this case we represent the sequence by placing braces around the formula for the nth term, writing $\{s_n\} = \{1/n\}$.

EXAMPLE 1 Writing the Terms of a Sequence

Write the first six terms of the sequence below.

$$\{a_n\} = \left\{\frac{n-1}{n}\right\}$$

SOLUTION

$$a_1 = \frac{1-1}{1} = 0$$

$$a_2 = \frac{2-1}{2} = \frac{1}{2}$$

$$a_3 = \frac{2}{3}$$

$$a_4 = \frac{3}{4}$$

$$a_5 = \frac{4}{5}$$

$$a_6 = \frac{5}{6}$$

NOW WORK PROBLEM 3.

EXAMPLE 2 Writing the Terms of a Sequence

Write the first six terms of the sequence below.

$$\{b_n\} = \left\{(-1)^{n-1}\left(\frac{2}{n}\right)\right\}$$

SOLUTION

$$b_1 = 2$$
$$b_2 = -1$$
$$b_3 = \frac{2}{3}$$
$$b_4 = -\frac{1}{2}$$
$$b_5 = \frac{2}{5}$$
$$b_6 = -\frac{1}{3}$$

NOW WORK PROBLEM 5.

A second way of defining a sequence is to assign a value to the first term (or the first few terms) and specify the nth term by a formula or equation that involves one or more of the terms preceding it. Sequences defined this way are said to be defined **recursively,** and the rule or formula is called a **recursive formula.**

EXAMPLE 3 Writing the Terms of a Recursively Defined Sequence

Write the first five terms of the recursively defined sequence given below.

$$s_1 = 1, \qquad s_n = 4s_{n-1}$$

SOLUTION The first term is given as $s_1 = 1$. To get the second term, we use $n = 2$ in the formula to get $s_2 = 4s_1 = 4 \cdot 1 = 4$. To get the third term, we use $n = 3$ in the formula to get $s_3 = 4s_2 = 4 \cdot 4 = 16$. To get the new term requires that we know the value of the preceding term. The first five terms are

$$s_1 = 1$$
$$s_2 = 4 \cdot 1 = 4$$
$$s_3 = 4 \cdot 4 = 16$$
$$s_4 = 4 \cdot 16 = 64$$
$$s_5 = 4 \cdot 64 = 256$$

NOW WORK PROBLEM 13.

EXAMPLE 4 Writing the Terms of a Recursively Defined Sequence

Write the first five terms of the recursively defined sequence given below.

$$u_1 = 1, \qquad u_2 = 1, \qquad u_{n+2} = u_n + u_{n+1}$$

SOLUTION We are given the first two terms. To get the third term requires that we know each of the previous two terms.

$$u_1 = 1$$
$$u_2 = 1$$
$$u_3 = u_1 + u_2 = 2$$
$$u_4 = u_2 + u_3 = 1 + 2 = 3$$
$$u_5 = u_3 + u_4 = 2 + 3 = 5$$

The sequence defined in Example 4 is called a **Fibonacci sequence,** and the terms of this sequence are called **Fibonacci numbers.** These numbers appear in a wide variety of applications.

EXAMPLE 5 Writing the Terms of a Recursively Defined Sequence

Write the first five terms of the recursively defined sequence given below.

$$f_1 = 1, \qquad f_{n+1} = (n + 1)f_n$$

SOLUTION Here,

$$f_1 = 1$$
$$f_2 = 2f_1 = 2 \cdot 1 = 2$$
$$f_3 = 3f_2 = 3 \cdot 2 = 6$$
$$f_4 = 4f_3 = 4 \cdot 6 = 24$$
$$f_5 = 5f_4 = 5 \cdot 24 = 120$$

NOW WORK PROBLEM 21.

Find the first term and common ratio of a geometric sequence

2 Geometric Sequences

When the ratio of successive terms of a sequence is always the same nonzero number, the sequence is called **geometric.** A **geometric sequence*** may be defined recursively as $a_1 = a, \quad \dfrac{a_{n+1}}{a_n} = r$, or as:

Geometric Sequence

$$a_1 = a, \qquad a_{n+1} = ra_n$$

where $a = a_1$ and $r \neq 0$ are real numbers. The number a is the **first term,** and the nonzero number r is called the **common ratio.**

The terms of a geometric sequence with first term a and common ratio r follow the pattern

$$a, \quad ar, \quad ar^2, \quad ar^3, \ \ldots$$

*Sometimes called a ***geometric progression.***

EXAMPLE 6 Identifying a Geometric Sequence

The sequence

$$2, \quad 6, \quad 18, \quad 54, \quad 162, \ldots$$

is geometric since the ratio of successive terms is 3. The first term is 2, and the common ratio is 3.

EXAMPLE 7 Finding the First Term and Common Ratio of a Geometric Sequence

Show that the sequence below is geometric. Find the first term and the common ratio.

$$\{t_n\} = \{4^n\}$$

SOLUTION The first term is $t_1 = 4^1 = 4$. The $(n + 1)$st and nth terms are

$$t_{n+1} = 4^{n+1} \qquad \text{and} \qquad t_n = 4^n$$

Their ratio is

$$\frac{t_{n+1}}{t_n} = \frac{4^{n+1}}{4^n} = 4$$

Because the ratio of successive terms is a nonzero number, the sequence $\{t_n\}$ is a geometric sequence with common ratio 4.

NOW WORK PROBLEM 27(a) AND (b).

EXAMPLE 8 Finding the First Term and Common Ratio of a Geometric Sequence

Show that the sequence below is geometric. Find the first term and the common ratio.

$$\{s_n\} = \{2^{-n}\}$$

SOLUTION The first term is $s_1 = 2^{-1} = \frac{1}{2}$. The $(n + 1)$st and nth terms of the sequence $\{s_n\}$ are

$$s_{n+1} = 2^{-(n+1)} \qquad \text{and} \qquad s_n = 2^{-n}$$

Their ratio is

$$\frac{s_{n+1}}{s_n} = \frac{2^{-(n+1)}}{2^{-n}} = 2^{-n-1+n} = 2^{-1} = \frac{1}{2}$$

Because the ratio of successive terms is a nonzero number, the sequence $\{s_n\}$ is geometric with common ratio $\frac{1}{2}$.

Adding the First *n* Terms of a Geometric Sequence

Add the terms of a geometric sequence **3**

The next result gives a formula for finding the sum of the first n terms of a geometric sequence.

Sum of the First n Terms of a Geometric Sequence

Let $\{a_n\}$ be a geometric sequence with first term a and common ratio r. The sum S_n of the first n terms of $\{a_n\}$ is

$$S_n = a\left(\frac{1 - r^n}{1 - r}\right) \qquad r \neq 0, 1 \tag{1}$$

Formula (1) may be derived as follows:

$$S_n = a + ar + \cdots + ar^{n-1} \tag{2}$$

Multiply each side by r to obtain

$$rS_n = ar + ar^2 + \cdots + ar^n \tag{3}$$

Now subtract (3) from (2). The result is

$$S_n - rS_n = a - ar^n$$
$$(1 - r)S_n = a(1 - r^n)$$

Since $r \neq 1$, we can solve for S_n:

$$S_n = a\left(\frac{1 - r^n}{1 - r}\right)$$

EXAMPLE 9 Adding the Terms of a Geometric Sequence

Find the sum S_n of the first n terms of the sequence $\{(\frac{1}{2})^n\}$; that is, find

$$\frac{1}{2} + \frac{1}{4} + \frac{1}{8} + \cdots + \left(\frac{1}{2}\right)^n$$

SOLUTION The sequence $\{(\frac{1}{2})^n\}$ is a geometric sequence with $a = \frac{1}{2}$ and $r = \frac{1}{2}$. The sum S_n we seek is the sum of the first n terms of the sequence, so we use Formula (1) to get

$$\begin{aligned} S_n &= \frac{1}{2} + \frac{1}{4} + \frac{1}{8} + \cdots + \left(\frac{1}{2}\right)^n \\ &= \frac{1}{2}\left[\frac{1 - \left(\frac{1}{2}\right)^n}{1 - \frac{1}{2}}\right] \\ &= \frac{1}{2}\left[\frac{1 - \left(\frac{1}{2}\right)^n}{\frac{1}{2}}\right] \\ &= 1 - \left(\frac{1}{2}\right)^n \end{aligned}$$

NOW WORK PROBLEM 27(c).

EXERCISE A.4 Answers to Odd-Numbered Problems Begin on Page AN-106.

In Problems 1–12, write the first five terms of each sequence.

1. $\{n\}$
2. $\{n^2 + 1\}$
3. $\left\{\frac{n}{n+1}\right\}$
4. $\left\{\frac{2n+3}{2n-1}\right\}$
5. $\{(-1)^{n+1}n^2\}$
6. $\left\{(-1)^{n-1}\left(\frac{n}{2n-1}\right)\right\}$
7. $\left\{\frac{2^n}{3^n+1}\right\}$
8. $\left\{\left(\frac{4}{3}\right)^n\right\}$
9. $\left\{\frac{(-1)^n}{(n+1)(n+2)}\right\}$
10. $\left\{\frac{3^n}{n}\right\}$
11. $\left\{\frac{n}{e^n}\right\}$
12. $\left\{\frac{n^2}{2^n}\right\}$

In Problems 13–26, sequence is defined recursively. Write the first five terms.

13. $a_1 = 1; a_{n+1} = 2 + a_n$
14. $a_1 = 3; a_{n+1} = 5 - a_n$
15. $a_1 = -2; a_{n+1} = n + a_n$
16. $a_1 = 1; a_{n+1} = n - a_n$
17. $a_1 = 5; a_{n+1} = 2a_n$
18. $a_1 = 2; a_{n+1} = -a_n$
19. $a_1 = 3; a_{n+1} = \frac{a_n}{n}$
20. $a_1 = -2; a_{n+1} = n + 3a_n$
21. $a_1 = 1; a_2 = 2; a_{n+2} = a_n a_{n+1}$
22. $a_1 = -1; a_2 = 1; a_{n+2} = a_{n+1} + na_n$
23. $a_1 = A; a_{n+1} = a_n + d$
24. $a_1 = A; a_{n+1} = ra_n, r \neq 0$
25. $a_1 = \sqrt{2}; a_{n+1} = \sqrt{2 + a_n}$
26. $a_1 = \sqrt{2}; a_{n+1} = \sqrt{\frac{a_n}{2}}$

In Problems 27–36, geometric sequence is given.
(a) Find the first term and the common ratio.
(b) Write out the first four terms.
(c) Find the sum of the first n *terms.*

27. $\{2^n\}$
28. $\{(-4)^n\}$
29. $\left\{-3\left(\frac{1}{2}\right)^n\right\}$
30. $\left\{\left(\frac{5}{2}\right)^n\right\}$
31. $\left\{\frac{2^{n-1}}{4}\right\}$
32. $\left\{\frac{3^n}{9}\right\}$
33. $\{2^{n/3}\}$
34. $\{3^{2n}\}$
35. $\left\{\frac{3^{n-1}}{2^n}\right\}$
36. $\left\{\frac{2^n}{3^{n-1}}\right\}$

A.5 Polynomials and Rational Expressions

OBJECTIVES
1. Recognize special products
2. Factor polynomials
3. Simplify rational expressions
4. Use the LCM to add rational expressions

As we said earlier, in algebra we use letters to represent real numbers. We shall use the letters at the end of the alphabet, such as x, y, and z, to represent variables and the letters at the beginning of the alphabet, such as a, b, and c, to represent constants. In the expressions $3x + 5$ and $ax + b$, it is understood that x is a variable and that a and b are constants, even though the constants a and b are unspecified. As you will find out, the context usually makes the intended meaning clear.

Now we introduce some basic vocabulary.

A **monomial** in one variable is the product of a constant and a variable raised to a nonnegative integer power; that is, a monomial is of the form

$$\boxed{ax^k}$$

where a is a constant, x is a variable, and $k \geq 0$ is an integer. The constant a is called the **coefficient** of the monomial. If $a \neq 0$, then k is called the **degree** of the monomial.

Examples of monomials follow:

Monomial	Coefficient	Degree	
$6x^2$	6	2	
$-\sqrt{2}x^3$	$-\sqrt{2}$	3	
3	3	0	Since $3 = 3 \cdot 1 = 3x^0$
$-5x$	-5	1	Since $-5x = -5x^1$
x^4	1	4	Since $x^4 = 1 \cdot x^4$

Two monomials ax^k and bx^k with the same degree and the same variable are called **like terms.** Such monomials when added or subtracted can be combined into a single monomial by using the distributive property. For example,

$$2x^2 + 5x^2 = (2 + 5)x^2 = 7x^2 \quad \text{and} \quad 8x^3 - 5x^3 = (8 - 5)x^3 = 3x^3$$

The sum or difference of two monomials having different degrees is called a **binomial.** The sum or difference of three monomials with three different degrees is called a **trinomial.** For example,

$x^2 - 2$ is a binomial.
$x^3 - 3x + 5$ is a trinomial.
$2x^2 + 5x^2 + 2 = 7x^2 + 2$ is a binomial.

A **polynomial** in one variable is an algebraic expression of the form

$$\boxed{a_nx^n + a_{n-1}x^{n-1} + \cdots + a_1x + a_0} \tag{1}$$

where $a_n, a_{n-1}, \ldots, a_1, a_0$ are constants,* called the **coefficients** of the polynomial, $n \geq 0$ is an integer, and x is a variable. If $a_n \neq 0$, it is called the **leading coefficient,** and n is called the **degree** of the polynomial.

The notation a_n is read as "a sub n." The number n is called a **subscript and should not be confused with an exponent. We use subscripts in order to distinguish one constant from another when a large or undetermined number of constants is required.*

The monomials that make up a polynomial are called its **terms.** If all the coefficients are 0, the polynomial is called the **zero polynomial,** which has no degree.

Polynomials are usually written in **standard form,** beginning with the nonzero term of highest degree and continuing with terms in descending order according to degree. If a power of x is missing, it is because its coefficient is zero. Examples of polynomials follow:

Polynomial	Coefficients	Degree
$3x^2 - 5 = 3x^2 + 0 \cdot x + (-5)$	$3, 0, -5$	2
$8 - 2x + x^2 = 1 \cdot x^2 - 2x + 8$	$1, -2, 8$	2
$5x + \sqrt{2} = 5x^1 + \sqrt{2}$	$5, \sqrt{2}$	1
$3 = 3 \cdot 1 = 3 \cdot x^0$	3	0
0	0	No degree

Although we have been using x to represent the variable, letters such as y or z are also commonly used.

$3x^4 - x^2 + 2$ is a polynomial (in x) of degree 4.
$9y^3 - 2y^2 + y - 3$ is a polynomial (in y) of degree 3.
$z^5 + \pi$ is a polynomial (in z) of degree 5.

Algebraic expressions such as

$$\frac{1}{x} \quad \text{and} \quad \frac{x^2 + 1}{x + 5}$$

are not polynomials. The first is not a polynomial because $\frac{1}{x} = x^{-1}$ has an exponent that is not a nonnegative integer. Although the second expression is the quotient of two polynomials, the polynomial in the denominator has degree greater than 0, so the expression cannot be a polynomial.

Recognize special products 1

Certain products, which we call **special products,** occur frequently in algebra. In the list that follows, x, a, b, c, and d are real numbers.

Difference of Two Squares

$$(x - a)(x + a) = x^2 - a^2 \qquad \textbf{(2)}$$

Squares of Binomials, or Perfect Squares

$$(x + a)^2 = x^2 + 2ax + a^2 \qquad \textbf{(3a)}$$
$$(x - a)^2 = x^2 - 2ax + a^2 \qquad \textbf{(3b)}$$

Miscellaneous Trinomials

$$(x + a)(x + b) = x^2 + (a + b)x + ab \quad \text{(4a)}$$

$$(ax + b)(cx + d) = acx^2 + (ad + bc)x + bd \quad \text{(4b)}$$

Cubes of Binomials, or Perfect Cubes

$$(x + a)^3 = x^3 + 3ax^2 + 3a^2x + a^3 \quad \text{(5a)}$$

$$(x - a)^3 = x^3 - 3ax^2 + 3a^2x - a^3 \quad \text{(5b)}$$

Difference of Two Cubes

$$(x - a)(x^2 + ax + a^2) = x^3 - a^3 \quad \text{(6)}$$

Sum of Two Cubes

$$(x + a)(x^2 - ax + a^2) = x^3 + a^3 \quad \text{(7)}$$

The special product formulas in equations (2) through (7) are used often, and their patterns should be committed to memory. But if you forget one or are unsure of its form, you should be able to derive it as needed.

EXAMPLE 1 Using Special Formulas

(a) $(x - 4)(x + 4) = x^2 - 4^2 = x^2 - 16$
(b) $(2x + 5)(3x - 1) = 6x^2 - 2x + 15x - 5 = 6x^2 + 13x - 5$
(c) $(x - 2)^3 = x^3 - 3(2)x^2 + 3(2)^2x - (2)^3 = x^3 - 6x^2 + 12x - 8$

NOW WORK PROBLEM 3.

Factor polynomials **2**

Factoring

Consider the following product:

$$(2x + 3)(x - 4) = 2x^2 - 5x - 12$$

The two polynomials on the left are called **factors** of the polynomial on the right. Expressing a given polynomial as a product of other polynomials, that is, finding the factors of a polynomial, is called **factoring.**

We shall restrict our discussion here to factoring polynomials in one variable into products of polynomials in one variable, where all coefficients are integers. We call this **factoring over the integers.**

Any polynomial can be written as the product of 1 times itself or as -1 times its additive inverse. If a polynomial cannot be written as the product of two other polynomials (excluding 1 and -1), then the polynomial is said to be **prime.** When a polynomial has been written as a product consisting only of prime factors, it is said to be **factored completely.** Examples of prime polynomials are

$$2, \quad 3, \quad 5, \quad x, \quad x+1, \quad x-1, \quad 3x+4$$

The first factor to look for in a factoring problem is a common monomial factor present in each term of the polynomial. If one is present, use the distributive property to factor it out. For example,

Polynomial	Common Monomial Factor	Remaining Factor	Factored Form
$2x+4$	2	$x+2$	$2x+4=2(x+2)$
$3x-6$	3	$x-2$	$3x-6=3(x-2)$
$2x^2-4x+8$	2	x^2-2x+4	$2x^2-4x+8=2(x^2-2x+4)$
$8x-12$	4	$2x-3$	$8x-12=4(2x-3)$
x^2+x	x	$x+1$	$x^2+x=x(x+1)$
x^3-3x^2	x^2	$x-3$	$x^3-3x^2=x^2(x-3)$
$6x^2+9x$	$3x$	$2x+3$	$6x^2+9x=3x(2x+3)$

The list of special products (2) through (7) given earlier provides a list of factoring formulas when the equations are read from right to left. For example, equation (2) states that if the polynomial is the difference of two squares, x^2-a^2, it can be factored into $(x-a)(x+a)$. The following example illustrates several factoring techniques.

EXAMPLE 2 Factoring Polynomials

Factor completely each polynomial.

(a) x^4-16 **(b)** x^3-1 **(c)** $9x^2-6x+1$

(d) $x^2+4x-12$ **(e)** $3x^2+10x-8$ **(f)** x^3-4x^2+2x-8

SOLUTION **(a)** $x^4-16=(x^2-4)(x^2+4)=(x-2)(x+2)(x^2+4)$

↑ Difference of squares ↑ Difference of squares

(b) $x^3 - 1 = (x - 1)(x^2 + x + 1)$

↑ Difference of cubes

(c) $9x^2 - 6x + 1 = (3x - 1)^2$

↑ Perfect square

(d) $x^2 + 4x - 12 = (x + 6)(x - 2)$

↑ The product of 6 and -2 is -12, and the sum of 6 and -2 is 4.

(e) $3x^2 + 10x - 8 = (3x - 2)(x + 4)$

$12x - 2x = 10x$; $3x^2$; -8

(f) $x^3 - 4x^2 + 2x - 8 = (x^3 - 4x^2) + (2x - 8)$

↑ Regroup

$= x^2(x - 4) + 2(x - 4) = (x^2 + 2)(x - 4)$

↑ Distributive property ↑ Distributive property

The technique used in Example 2(f) is called **factoring by grouping.**

NOW WORK PROBLEMS 17, 33, AND 69.

Rational Expressions

If we form the quotient of two polynomials, the result is called a **rational expression.** Some examples of rational expressions are

$$\text{(a)}\ \frac{x^3 + 1}{x} \qquad \text{(b)}\ \frac{3x^3 + x - 2}{x^5 + 5} \qquad \text{(c)}\ \frac{x}{x^2 - 1} \qquad \text{(d)}\ \frac{xy^2}{(x - y)^2}$$

Expressions (a), (b), and (c) are rational expressions in one variable, x, whereas (d) is a rational expression in two variables, x and y.

Rational expressions are described in the same manner as rational numbers. Thus, in expression (a), the polynomial $x^3 + 1$ is called the **numerator,** and x is called the **denominator.** When the numerator and denominator of a rational expression contain no common factors (except 1 and -1), we say that the rational expression is **reduced to lowest terms,** or **simplified.**

3 Simplify rational expressions

A rational expression is reduced to lowest terms by completely factoring the numerator and the denominator and canceling any common factors by using the cancellation property.

$$\frac{ac}{bc} = \frac{a}{b}, \qquad b \neq 0, \quad c \neq 0$$

For example,

$$\frac{x^2 - 1}{x^2 - 2x - 3} = \frac{(x - 1)\cancel{(x + 1)}}{(x - 3)\cancel{(x + 1)}} = \frac{x - 1}{x - 3}$$

EXAMPLE 3 Simplifying Rational Expressions

Reduce each rational expression to lowest terms.

(a) $\dfrac{x^2 + 4x + 4}{x^2 + 3x + 2}$ **(b)** $\dfrac{x^3 - 8}{x^3 - 2x}$ **(c)** $\dfrac{8 - 2x}{x^2 - x - 12}$

SOLUTION **(a)** $$\frac{x^2 + 4x + 4}{x^2 + 3x + 2} = \frac{\cancel{(x + 2)}(x + 2)}{\cancel{(x + 2)}(x + 1)} = \frac{x + 2}{x + 1}, \quad x \neq -2, -1$$

(b) $$\frac{x^3 - 8}{x^3 - 2x^2} = \frac{\cancel{(x - 2)}(x^2 + 2x + 4)}{x^2\cancel{(x - 2)}} = \frac{x^2 + 2x + 4}{x^2}, \quad x \neq 0, 2$$

(c) $$\frac{8 - 2x}{x^2 - x - 12} = \frac{2(4 - x)}{(x - 4)(x + 3)} = \frac{2(-1)\cancel{(x - 4)}}{\cancel{(x - 4)}(x + 3)} = \frac{-2}{x + 3}, \quad x \neq -3, 4$$

NOW WORK PROBLEM 75.

The rules for multiplying and dividing rational expressions are the same as the rules for multiplying and dividing rational numbers.

$$\frac{a}{b} \cdot \frac{c}{d} = \frac{ac}{bd}, \quad \text{if } b \neq 0, d \neq 0 \tag{8}$$

$$\frac{\frac{a}{b}}{\frac{c}{d}} = \frac{a}{b} \cdot \frac{d}{c} = \frac{ad}{bc}, \quad \text{if } b \neq 0, c \neq 0, d \neq 0 \tag{9}$$

In using equations (8) and (9) with rational expressions, be sure first to factor each polynomial completely so that common factors can be canceled. We shall follow the practice of leaving our answers in factored form.

EXAMPLE 4 Finding Products and Quotients of Rational Expressions

Perform the indicated operation and simplify the result. Leave your answer in factored form.

(a) $\dfrac{x^2 - 2x + 1}{x^3 + x} \cdot \dfrac{4x^2 + 4}{x^2 + x - 2}$ **(b)** $\dfrac{\dfrac{x + 3}{x^2 - 4}}{\dfrac{x^2 - x - 12}{x^3 - 8}}$

SOLUTION **(a)** $$\frac{x^2 - 2x + 1}{x^3 + x} \cdot \frac{4x^2 + 4}{x^2 + x - 2} = \frac{(x - 1)^2}{x(x^2 + 1)} \cdot \frac{4(x^2 + 1)}{(x + 2)(x - 1)}$$

$$= \frac{(x - 1)^{\cancel{2}}(4)\cancel{(x^2 + 1)}}{x\cancel{(x^2 + 1)}(x + 2)\cancel{(x - 1)}}$$

$$= \frac{4(x - 1)}{x(x + 2)}, \quad x \neq -2, 0, 1$$

(b)
$$\frac{\dfrac{x+3}{x^2-4}}{\dfrac{x^2-x-12}{x^3-8}} = \frac{x+3}{x^2-4} \cdot \frac{x^3-8}{x^2-x-12}$$
$$= \frac{x+3}{(x-2)(x+2)} \cdot \frac{(x-2)(x^2+2x+4)}{(x-4)(x+3)}$$
$$= \frac{\cancel{(x+3)}\cancel{(x-2)}(x^2+2x+4)}{\cancel{(x-2)}(x+2)(x-4)\cancel{(x+3)}}$$
$$= \frac{x^2+2x+4}{(x+2)(x-4)}, \quad x \neq -3, -2, 2, 4$$

NOW WORK PROBLEM 53.

If the denominators of two rational expressions to be added (or subtracted) are equal, we add (or subtract) the numerators and keep the common denominator. That is, if $\frac{a}{b}$ and $\frac{c}{b}$ are two rational expressions, then

$$\frac{a}{b} + \frac{c}{b} = \frac{a+c}{b} \qquad \frac{a}{b} - \frac{c}{b} = \frac{a-c}{b}, \qquad \text{if } b \neq 0 \tag{10}$$

EXAMPLE 5 Finding the Sum of Two Rational Expressions

Perform the indicated operation and simplify the result. Leave your answer in factored form.

$$\frac{2x^2-4}{2x+5} + \frac{x+3}{2x+5}, \quad x \neq -\frac{5}{2}$$

SOLUTION

$$\frac{2x^2-4}{2x+5} + \frac{x+3}{2x+5} = \frac{(2x^2-4)+(x+3)}{2x+5}$$
$$= \frac{2x^2+x-1}{2x+5} = \frac{(2x-1)(x+1)}{2x+5}$$

If the denominators of two rational expressions to be added or subtracted are not equal, we can use the general formulas for adding and subtracting quotients.

$$\frac{a}{b} + \frac{c}{d} = \frac{a \cdot d}{b \cdot d} + \frac{b \cdot c}{b \cdot d} = \frac{ad+bc}{bd}, \qquad \text{if } b \neq 0, d \neq 0$$
$$\frac{a}{b} - \frac{c}{d} = \frac{a \cdot d}{b \cdot d} - \frac{b \cdot c}{b \cdot d} = \frac{ad-bc}{bd}, \qquad \text{if } b \neq 0, d \neq 0 \tag{11}$$

EXAMPLE 6 **Finding the Difference of Two Rational Expressions**

Perform the indicated operation and simplify the result. Leave your answer in factored form.

$$\frac{x^2}{x^2 - 4} - \frac{1}{x}, \qquad x \neq -2, 0, 2$$

SOLUTION

$$\frac{x^2}{x^2 - 4} - \frac{1}{x} = \frac{x \cdot x^2}{x \cdot (x^2 - 4)} - \frac{1 \cdot (x^2 - 4)}{x \cdot (x^2 - 4)} = \frac{x^3 - (x^2 - 4)}{x \cdot (x^2 - 4)} = \frac{x^3 - x^2 + 4}{x(x - 2)(x + 2)}$$

Least Common Multiple (LCM)

If the denominators of two rational expressions to be added (or subtracted) have common factors, we usually do not use the general rules given by equation (11), since, in doing so, we make the problem more complicated than it needs to be. Instead, just as with fractions, we apply the **least common multiple (LCM) method** by using the polynomial of least degree that contains each denominator polynomial as a factor. Then we rewrite each rational expression using the LCM as the common denominator and use equation (10) to do the addition (or subtraction).

To find the least common multiple of two or more polynomials, first factor completely each polynomial. The LCM is the product of the different prime factors of each polynomial, each factor appearing the greatest number of times it occurs in each polynomial. The next example will give you the idea.

EXAMPLE 7 **Finding the Least Common Multiple**

Find the least common multiple of the following pair of polynomials:

$$x(x - 1)^2(x + 1) \quad \text{and} \quad 4(x - 1)(x + 1)^3$$

SOLUTION The polynomials are already factored completely as

$$x(x - 1)^2(x + 1) \quad \text{and} \quad 4(x - 1)(x + 1)^3$$

Start by writing the factors of the left-hand polynomial. (Alternatively, you could start with the one on the right.)

$$x(x - 1)^2(x + 1)$$

Now look at the right-hand polynomial. Its first factor, 4, does not appear in our list, so we insert it:

$$4x(x - 1)^2(x + 1)$$

The next factor, $x - 1$, is already in our list, so no change is necessary. The final factor is $(x + 1)^3$. Since our list has $x + 1$ to the first power only, we replace $x + 1$ in the list by $(x + 1)^3$. The LCM is

$$4x(x - 1)^2(x + 1)^3$$

Notice that the LCM is, in fact, the polynomial of least degree that contains $x(x - 1)^2(x + 1)$ and $4(x - 1)(x + 1)^3$ as factors.

The next example illustrates how the LCM is used for adding and subtracting rational expressions.

4 **EXAMPLE 8** **Using the LCM to Add Rational Expressions**

Perform the indicated operation and simplify the result. Leave your answer in factored form.

$$\frac{x}{x^2 + 3x + 2} + \frac{2x - 3}{x^2 - 1}, \qquad x \neq -2, -1, 1$$

SOLUTION First, we find the LCM of the denominators.

$$x^2 + 3x + 2 = (x + 2)(x + 1)$$
$$x^2 - 1 = (x - 1)(x + 1)$$

The LCM is $(x + 2)(x + 1)(x - 1)$. Next, we rewrite each rational expression using the LCM as the denominator.

$$\frac{x}{x^2 + 3x + 2} = \frac{x}{(x + 2)(x + 1)} = \frac{x(x - 1)}{(x + 2)(x + 1)(x - 1)}$$

↑ Multiply numerator and denominator by $x - 1$ to get the LCM in the denominator.

$$\frac{2x - 3}{x^2 - 1} = \frac{2x - 3}{(x - 1)(x + 1)} = \frac{(2x - 3)(x + 2)}{(x - 1)(x + 1)(x + 2)}$$

↑ Multiply numerator and denominator by $x + 2$ to get the LCM in the denominator.

Now we can add using equation (10).

$$\frac{x}{x^2 + 3x + 2} + \frac{2x - 3}{x^2 - 1} = \frac{x(x - 1)}{(x + 2)(x + 1)(x - 1)} + \frac{(2x - 3)(x + 2)}{(x + 2)(x + 1)(x - 1)}$$
$$= \frac{(x^2 - x) + (2x^2 + x - 6)}{(x + 2)(x + 1)(x - 1)}$$
$$= \frac{3x^2 - 6}{(x + 2)(x + 1)(x - 1)} = \frac{3(x^2 - 2)}{(x + 2)(x + 1)(x - 1)}$$

If we had not used the LCM technique to add the quotients in Example 8, but decided instead to use the general rule of equation (11), we would have obtained a more complicated expression, as follows:

$$\frac{x}{x^2 + 3x + 2} + \frac{2x - 3}{x^2 - 1} = \frac{x(x^2 - 1) + (x^2 + 3x + 2)(2x - 3)}{(x^2 + 3x + 2)(x^2 - 1)}$$
$$= \frac{3x^3 + 3x^2 - 6x - 6}{(x^2 + 3x + 2)(x^2 - 1)} = \frac{3(x^3 + x^2 - 2x - 2)}{(x^2 + 3x + 2)(x^2 - 1)}$$

Now we are faced with a more complicated problem of expressing this quotient in lowest terms. It is always best to first look for common factors in the denominators of expressions to be added or subtracted and to use the LCM if any common factors are found.

NOW WORK PROBLEM 57.

EXERCISE A.5 Answers to Odd-Numbered Problems Begin on Page AN-106.

In Problems 1–10, perform the indicated operations. Express each answer as a polynomial written in standard form.

1. $(10x^5 - 8x^2) + (3x^3 - 2x^2 + 6)$

2. $3(x^2 - 3x + 1) + 2(3x^2 + x - 4)$

3. $(x + a)^2 - x^2$

4. $(x - a)^2 - x^2$

5. $(x + 8)(2x + 1)$

6. $(2x - 1)(x + 2)$

7. $(x^2 + x - 1)(x^2 - x + 1)$

8. $(x^2 + 2x + 1)(x^2 - 3x + 4)$

9. $(x + 1)^3 - (x - 1)^3$

10. $(x + 1)^3 - (x + 2)^3$

In Problems 11–52, factor completely each polynomial. If the polynomial cannot be factored, say it is prime.

11. $x^2 - 36$

12. $x^2 - 9$

13. $1 - 4x^2$

14. $1 - 9x^2$

15. $x^2 + 7x + 10$

16. $x^2 + 5x + 4$

17. $x^2 - 2x + 8$

18. $x^2 - 4x + 5$

19. $x^2 + 4x + 16$

20. $x^2 + 12x + 36$

21. $15 + 2x - x^2$

22. $14 + 6x - x^2$

23. $3x^2 - 12x - 36$

24. $x^3 + 8x^2 - 20x$

25. $y^4 + 11y^3 + 30y^2$

26. $3y^3 - 18y^2 - 48y$

27. $4x^2 + 12x + 9$

28. $9x^2 - 12x + 4$

29. $3x^2 + 4x + 1$

30. $4x^2 + 3x - 1$

31. $x^4 - 81$

32. $x^4 - 1$

33. $x^6 - 2x^3 + 1$

34. $x^6 + 2x^3 + 1$

35. $x^7 - x^5$

36. $x^8 - x^5$

37. $5 + 16x - 16x^2$

38. $5 + 11x - 16x^2$

39. $4y^2 - 16y + 15$

40. $9y^2 + 9y - 4$

41. $1 - 8x^2 - 9x^4$

42. $4 - 14x^2 - 8x^4$

43. $x(x + 3) - 6(x + 3)$

44. $5(3x - 7) + x(3x - 7)$

45. $(x + 2)^2 - 5(x + 2)$

46. $(x - 1)^2 - 2(x - 1)$

47. $6x(2 - x)^4 - 9x^2(2 - x)^3$

48. $6x(1 - x^2)^4 - 24x^3(1 - x^2)^3$

49. $x^3 + 2x^2 - x - 2$

50. $x^3 - 3x^2 - x + 3$

51. $x^4 - x^3 + x - 1$

52. $x^4 + x^3 + x + 1$

In Problems 53–64, perform the indicated operation and simplify the result. Leave your answer in factored form.

53. $\dfrac{3x - 6}{5x} \cdot \dfrac{x^2 - x - 6}{x^2 - 4}$

54. $\dfrac{9x^2 - 25}{2x - 2} \cdot \dfrac{1 - x^2}{6x - 10}$

55. $\dfrac{4x^2 - 1}{x^2 - 16} \cdot \dfrac{x^2 - 4x}{2x + 1}$

56. $\dfrac{12}{x^2 - x} \cdot \dfrac{x^2 - 1}{4x - 2}$

57. $\dfrac{x}{x^2 - 7x + 6} - \dfrac{x}{x^2 - 2x - 24}$

58. $\dfrac{x}{x - 3} - \dfrac{x + 1}{x^2 + 5x - 24}$

59. $\dfrac{4}{x^2 - 4} - \dfrac{2}{x^2 + x - 6}$

60. $\dfrac{3}{x - 1} - \dfrac{x - 4}{x^2 - 2x + 1}$

61. $\dfrac{1}{x} - \dfrac{2}{x^2 + x} + \dfrac{3}{x^3 - x^2}$

62. $\dfrac{x}{(x - 1)^2} + \dfrac{2}{x} - \dfrac{x + 1}{x^3 - x^2}$

63. $\dfrac{1}{h}\left(\dfrac{1}{x + h} - \dfrac{1}{x}\right)$

64. $\dfrac{1}{h}\left[\dfrac{1}{(x + h)^2} - \dfrac{1}{x^2}\right]$

In Problems 65–74, expressions that occur in calculus are given. Factor completely each expression.

65. $2(3x + 4)^2 + (2x + 3) \cdot 2(3x + 4) \cdot 3$

66. $5(2x + 1)^2 + (5x - 6) \cdot 2(2x + 1) \cdot 2$

67. $2x(2x + 5) + x^2 \cdot 2$

68. $3x^2(8x - 3) + x^3 \cdot 8$

69. $2(x + 3)(x - 2)^3 + (x + 3)^2 \cdot 3(x - 2)^2$

70. $4(x + 5)^3(x - 1)^2 + (x + 5)^4 \cdot 2(x - 1)$

71. $(4x - 3)^2 + x \cdot 2(4x - 3) \cdot 4$

72. $3x^2(3x + 4)^2 + x^3 \cdot 2(3x + 4) \cdot 3$

73. $2(3x - 5) \cdot 3(2x + 1)^3 + (3x - 5)^2 \cdot 3(2x + 1)^2 \cdot 2$

74. $3(4x + 5)^2 \cdot 4(5x + 1)^2 + (4x + 5)^3 \cdot 2(5x + 1) \cdot 5$

In Problems 75–82, expressions that occur in calculus are given. Reduce each expression to lowest terms.

75. $\dfrac{(2x+3)\cdot 3-(3x-5)\cdot 2}{(3x-5)^2}$

76. $\dfrac{(4x+1)\cdot 5-(5x-2)\cdot 4}{(5x-2)^2}$

77. $\dfrac{x\cdot 2x-(x^2+1)\cdot 1}{(x^2+1)^2}$

78. $\dfrac{x\cdot 2x-(x^2-4)\cdot 1}{(x^2-4)^2}$

79. $\dfrac{(3x+1)\cdot 2x-x^2\cdot 3}{(3x+1)^2}$

80. $\dfrac{(2x-5)\cdot 3x^2-x^3\cdot 2}{(2x-5)^2}$

81. $\dfrac{(x^2+1)\cdot 3-(3x+4)\cdot 2x}{(x^2+1)^2}$

82. $\dfrac{(x^2+9)\cdot 2-(2x-5)\cdot 2x}{(x^2+9)^2}$

A.6 Solving Equations

PREPARING FOR THIS SECTION ***Before getting started, review the following:***

- > Factoring Polynomials (Appendix A, Section A.5, pp. A-40–A-43)
- > Zero-Product Property (Appendix A, Section A.1, p. A-12)
- > Square Roots (Appendix A, Section A.2, pp. A-22–A-23)
- > Absolute Value (Appendix A, Section A.2, pp. A-17–A-18)

OBJECTIVES
1. Solve equations
2. Solve quadratic equations by factoring
3. Know how to complete the square
4. Solve a quadratic equation by completing the square
5. Solve a quadratic equation using the quadratic formula

Solve equations **1**

An **equation in one variable** is a statement in which two expressions, at least one containing the variable, are equal. The expressions are called the **sides** of the equation. Since an equation is a statement, it may be true or false, depending on the value of the variable. Unless otherwise restricted, the admissible values of the variable are those in the domain of the variable. Those admissible values of the variable, if any, that result in a true statement are called **solutions,** or **roots,** of the equation. To **solve an equation** means to find all the solutions of the equation.

For example, the following are all equations in one variable, x:

$$x+5=9 \qquad x^2+5x=2x-2 \qquad \frac{x^2-4}{x+1}=0 \qquad \sqrt{x^2+9}=5$$

The first of these statements, $x+5=9$, is true when $x=4$ and false for any other choice of x. So, 4 is a solution of the equation $x+5=9$. We also say that 4 **satisfies** the equation $x+5=9$, because, when we substitute 4 for x, a true statement results.

Sometimes an equation will have more than one solution. For example, the equation

$$\frac{x^2-4}{x+1}=0$$

has $x=-2$ and $x=2$ as solutions.

Usually, we will write the solution of an equation in set notation. This set is called the **solution set** of the equation. For example, the solution set of the equation $x^2-9=0$ is $\{-3, 3\}$.

Some equations have no real solution. For example, $x^2 + 9 = 5$ has no real solution, because there is no real number whose square when added to 9 equals 5.

An equation that is satisfied for every choice of the variable for which both sides are defined is called an **identity.** For example, the equation

$$3x + 5 = x + 3 + 2x + 2$$

is an identity, because this statement is true for any real number x.

Two or more equations that have precisely the same solution set are called **equivalent equations.**

For example, all the following equations are equivalent, because each has only the solution $x = 5$:

$$2x + 3 = 13$$
$$2x = 10$$
$$x = 5$$

These three equations illustrate one method for solving many types of equations: Replace the original equation by an equivalent equation, and continue until an equation with an obvious solution, such as $x = 5$, is reached. The question, though, is "How do I obtain an equivalent equation?" In general, there are five ways to do so.

Procedures That Result in Equivalent Equations

1. Interchange the two sides of the equation:

   Replace $3 = x$ by $x = 3$

2. Simplify the sides of the equation by combining like terms, eliminating parentheses, and so on:

   Replace $(x + 2) + 6 = 2x + (x + 1)$

   by $x + 8 = 3x + 1$

3. Add or subtract the same expression on both sides of the equation:

   Replace $3x - 5 = 4$

   by $(3x - 5) + 5 = 4 + 5$

4. Multiply or divide both sides of the equation by the same nonzero expression:

   Replace $\dfrac{3x}{x - 1} = \dfrac{6}{x - 1}, \quad x \neq 1$

   by $\dfrac{3x}{x - 1} \cdot (x - 1) = \dfrac{6}{x - 1} \cdot (x - 1)$

5. If one side of the equation is 0 and the other side can be factored, then we may use the Zero-Product Property* and set each factor equal to 0:

   Replace $x(x - 3) = 0$

   by $x = 0$ or $x - 3 = 0$

WARNING: Squaring both sides of an equation does not necessarily lead to an equivalent equation.

The Zero-Product Property says that if $ab = 0$ then $a = 0$ or $b = 0$ or both equal 0.

Whenever it is possible to solve an equation in your head, do so. For example:

The solution of $2x = 8$ is $x = 4$.
The solution of $3x - 15 = 0$ is $x = 5$.

Often, though, some rearrangement is necessary.

EXAMPLE 1 Solving an Equation

Solve the equation: $3x - 5 = 4$

SOLUTION We replace the original equation by a succession of equivalent equations.

$$\begin{aligned} 3x - 5 &= 4 \\ (3x - 5) + 5 &= 4 + 5 && \text{Add 5 to both sides.} \\ 3x &= 9 && \text{Simplify.} \\ \frac{3x}{3} &= \frac{9}{3} && \text{Divide both sides by 3.} \\ x &= 3 && \text{Simplify.} \end{aligned}$$

The last equation, $x = 3$, has the single solution 3. All these equations are equivalent, so 3 is the only solution of the original equation, $3x - 5 = 4$.

Check: It is a good practice to check the solution by substituting 3 for x in the original equation.

$$\begin{aligned} 3x - 5 &= 4 \\ 3(3) - 5 &\stackrel{?}{=} 4 \\ 9 - 5 &\stackrel{?}{=} 4 \\ 4 &= 4 \end{aligned}$$

The solution checks.

NOW WORK PROBLEMS 15 AND 21.

In the next examples, we use the Zero-Product Property.

EXAMPLE 2 Solving Equations by Factoring

Solve the equations: **(a)** $x^2 = 4x$ **(b)** $x^3 - x^2 - 4x + 4 = 0$

SOLUTION **(a)** We begin by collecting all terms on one side. This results in 0 on one side and an expression to be factored on the other.

$$\begin{aligned} x^2 &= 4x \\ x^2 - 4x &= 0 \\ x(x - 4) &= 0 && \text{Factor.} \\ x = 0 \quad \text{or} \quad x - 4 &= 0 && \text{Apply the Zero-Product Property.} \\ x &= 4 \end{aligned}$$

The solution set is $\{0, 4\}$.

Check:

$x = 0$: $0^2 = 4 \cdot 0$ So 0 is a solution.

$x = 4$: $4^2 = 4 \cdot 4$ So 4 is a solution.

(b) We group the terms of $x^3 - x^2 - 4x + 4 = 0$ as follows:

$$(x^3 - x^2) - (4x - 4) = 0$$

Factor x^2 from the first grouping and 4 from the second.

$$x^2(x - 1) - 4(x - 1) = 0$$

This reveals the common factor $(x - 1)$, so we have

$$(x^2 - 4)(x - 1) = 0$$

$$(x - 2)(x + 2)(x - 1) = 0 \qquad \text{Factor again.}$$

$$x - 2 = 0 \quad \text{or} \quad x + 2 = 0 \quad \text{or} \quad x - 1 = 0 \qquad \text{Set each factor equal to 0.}$$

$$x = 2 \qquad x = -2 \qquad x = 1 \qquad \text{Solve.}$$

The solution set is $\{-2, 1, 2\}$.

Check: $x = -2$: $(-2)^3 - (-2)^2 - 4(-2) + 4 = -8 - 4 + 8 + 4 = 0$ -2 is a solution.

$x = 1$: $1^3 - 1^2 - 4(1) + 4 = 1 - 1 - 4 + 4 = 0$ 1 is a solution.

$x = 2$: $2^3 - 2^2 - 4(2) + 4 = 8 - 4 - 8 + 4 = 0$ 2 is a solution.

NOW WORK PROBLEM 25.

There are two points whose distance from the origin is 5 units, -5 and 5, so the equation $|x| = 5$ will have the solution set $\{-5, 5\}$.

EXAMPLE 3 Solving an Equation Involving Absolute Value

Solve the equation: $|x + 4| = 13$

SOLUTION There are two possibilities:

$$x + 4 = 13 \quad \text{or} \quad x + 4 = -13$$

$$x = 9 \qquad\qquad x = -17$$

The solution set is $\{-17, 9\}$.

WARNING: Since the absolute value of any real number is nonnegative, equations such as $|x| = -2$ have no solution.

NOW WORK PROBLEM 37.

Quadratic Equations

A **quadratic equation** is an equation equivalent to one written in the **standard form** $ax^2 + bx + c = 0$, where a, b, and c are real numbers and $a \neq 0$.

When a quadratic equation is written in the standard form, $ax^2 + bx + c = 0$, it may be possible to factor the expression on the left side as the product of two first-degree polynomials.

2

EXAMPLE 4 Solving a Quadratic Equation by Factoring

Solve the equation: $2x^2 = x + 3$

SOLUTION We put the equation in standard form by adding $-x - 3$ to both sides.

$$2x^2 = x + 3$$

$$2x^2 - x - 3 = 0 \qquad \text{Add } -x - 3 \text{ to both sides.}$$

The left side may now be factored as

$$(2x - 3)(x + 1) = 0$$

so that

$$\begin{aligned} 2x - 3 &= 0 \quad \text{or} \quad x + 1 = 0 \\ x &= \frac{3}{2} \quad \text{or} \quad x = -1 \end{aligned}$$

The solution set is $\left\{-1, \frac{3}{2}\right\}$.

When the left side factors into two linear equations with the same solution, the quadratic equation is said to have a **repeated solution.** We also call this solution a **root of multiplicity 2,** or a **double root.**

EXAMPLE 5 Solving a Quadratic Equation by Factoring

Solve the equation: $9x^2 - 6x + 1 = 0$

SOLUTION This equation is already in standard form, and the left side can be factored.

$$\begin{aligned} 9x^2 - 6x + 1 &= 0 \\ (3x - 1)(3x - 1) &= 0 \end{aligned}$$

so

$$x = \frac{1}{3} \quad \text{or} \quad x = \frac{1}{3}$$

This equation has only the repeated solution $\frac{1}{3}$.

NOW WORK PROBLEM 55.

The Square Root Method

Suppose that we wish to solve the quadratic equation

$$x^2 = p \tag{1}$$

where p is a nonnegative number. We proceed as in the earlier examples.

$$\begin{aligned} x^2 - p &= 0 && \text{Put in standard form.} \\ (x - \sqrt{p})(x + \sqrt{p}) &= 0 && \text{Factor (over the real numbers).} \\ x = \sqrt{p} \quad \text{or} \quad x &= -\sqrt{p} && \text{Solve.} \end{aligned}$$

We have the following result:

If $x^2 = p$ and $p \geq 0$, then $x = \sqrt{p}$ or $x = -\sqrt{p}$. **(2)**

When statement (2) is used, it is called the **Square Root Method.** In statement (2), note that if $p > 0$ the equation $x^2 = p$ has two solutions, $x = \sqrt{p}$ and $x = -\sqrt{p}$. We usually abbreviate these solutions as $x = \pm\sqrt{p}$, read as "x equals plus or minus the square root of p."

For example, the two solutions of the equation

$$x^2 = 4$$

are

$$x = \pm\sqrt{4} \quad \text{Use the Square Root Method.}$$

and, since $\sqrt{4} = 2$, we have

$$x = \pm 2$$

The solution set is $\{-2, 2\}$.

NOW WORK PROBLEM 69.

Completing The Square

Know how to complete the square

3 We now introduce the method of **completing the square.** The idea behind this method is to *adjust* the left side of a quadratic equation, $ax^2 + bx + c = 0$, so that it becomes a perfect square, that is, the square of a first-degree polynomial. For example, $x^2 + 6x + 9$ and $x^2 - 4x + 4$ are perfect squares because

$$x^2 + 6x + 9 = (x + 3)^2 \quad \text{and} \quad x^2 - 4x + 4 = (x - 2)^2$$

How do we adjust the left side? We do it by adding the appropriate number to the left side to create a perfect square. For example, to make $x^2 + 6x$ a perfect square, we add 9.

Let's look at several examples of completing the square when the coefficient of x^2 is 1:

Start	Add	Result
$x^2 + 4x$	4	$x^2 + 4x + 4 = (x + 2)^2$
$x^2 + 12x$	36	$x^2 + 12x + 36 = (x + 6)^2$
$x^2 - 6x$	9	$x^2 - 6x + 9 = (x - 3)^2$
$x^2 + x$	$\frac{1}{4}$	$x^2 + x + \frac{1}{4} = \left(x + \frac{1}{2}\right)^2$

Do you see the pattern? Provided that the coefficient of x^2 is 1, we complete the square by adding the square of $\frac{1}{2}$ of the coefficient of x.

Procedure for completing a square

Start	Add	Result
$x^2 + mx$	$\left(\frac{m}{2}\right)^2$	$x^2 + mx + \left(\frac{m}{2}\right)^2 = \left(x + \frac{m}{2}\right)^2$

NOW WORK PROBLEM 73.

The next example illustrates how the procedure of completing the square can be used to solve a quadratic equation.

4 **EXAMPLE 6 Solving a Quadratic Equation by Completing the Square**

Solve by completing the square: $2x^2 - 12x - 5 = 0$

SOLUTION First, we rewrite the equation as follows:

$$2x^2 - 12x - 5 = 0$$
$$2x^2 - 12x = 5$$

Next, we divide both sides by 2 so that the coefficient of x^2 is 1. (This enables us to complete the square at the next step.)

$$x^2 - 6x = \frac{5}{2}$$

Now complete the square by adding 9 to both sides.

$$x^2 - 6x + 9 = \frac{5}{2} + 9$$
$$(x - 3)^2 = \frac{23}{2}$$
$$x - 3 = \pm\sqrt{\frac{23}{2}} \qquad \text{Use the Square Root Method.}$$
$$x - 3 = \pm\frac{\sqrt{46}}{2} \qquad \sqrt{\frac{23}{2}} = \frac{\sqrt{23}}{\sqrt{2}} \cdot \frac{\sqrt{2}}{\sqrt{2}} = \frac{\sqrt{46}}{2}$$
$$x = 3 \pm \frac{\sqrt{46}}{2}$$

The solution set is $\left\{3 - \frac{\sqrt{46}}{2}, 3 + \frac{\sqrt{46}}{2}\right\}$

NOTE: If we wanted an approximation, say rounded to two decimal places, of these solutions, we would use a calculator to get $\{-0.39, 6.39\}$.

NOW WORK PROBLEM 79.

Solve a quadratic equation using the quadratic formula

5 **The Quadratic Formula**

We can use the method of completing the square to obtain a general formula for solving the quadratic equation.

$$ax^2 + bx + c = 0, \qquad a \neq 0$$

NOTE: There is no loss in generality to assume that $a > 0$, since if $a < 0$ we can multiply by -1 to obtain an equivalent equation with a positive leading coefficient.

As in Example 6, we rearrange the terms as

$$ax^2 + bx = -c, \qquad a > 0$$

Since $a > 0$, we can divide both sides by a to get

$$x^2 + \frac{b}{a}x = -\frac{c}{a}$$

Now the coefficient of x^2 is 1. To complete the square on the left side, add the square of $\frac{1}{2}$ of the coefficient of x; that is, add

$$\left(\frac{1}{2} \cdot \frac{b}{a}\right)^2 = \frac{b^2}{4a^2}$$

to both sides. Then

$$x^2 + \frac{b}{a}x + \frac{b^2}{4a^2} = \frac{b^2}{4a^2} - \frac{c}{a}$$

$$\left(x + \frac{b}{2a}\right)^2 = \frac{b^2 - 4ac}{4a^2} \qquad \frac{b^2}{4a^2} - \frac{c}{a} = \frac{b^2}{4a^2} - \frac{4ac}{4a^2} = \frac{b^2 - 4ac}{4a^2} \qquad (3)$$

Provided that $b^2 - 4ac \geq 0$, we now can use the Square Root Method to get

$$x + \frac{b}{2a} = \pm\sqrt{\frac{b^2 - 4ac}{4a^2}}$$ Square Root Method.

$$x + \frac{b}{2a} = \frac{\pm\sqrt{b^2 - 4ac}}{2a}$$ The square root of a quotient equals the quotient of the square roots. Also, $\sqrt{4a^2} = 2a$ since $a > 0$.

$$x = -\frac{b}{2a} \pm \frac{\sqrt{b^2 - 4ac}}{2a}$$ Add $-\frac{b}{2a}$ to both sides.

$$x = \frac{-b \pm \sqrt{b^2 - 4ac}}{2a}$$ Combine the quotients on the right.

What if $b^2 - 4ac$ is negative? Then equation (3) states that the left expression (a real number squared) equals the right expression (a negative number). Since this occurrence is impossible for real numbers, we conclude that if $b^2 - 4ac < 0$ the quadratic equation has no *real* solution.

We now state the *quadratic formula.*

Quadratic Formula

Consider the quadratic equation

$$ax^2 + bx + c = 0, \qquad a \neq 0$$

If $b^2 - 4ac < 0$, this equation has no real solution.

If $b^2 - 4ac \geq 0$, the real solution(s) of this equation is (are) given by the **quadratic formula.**

$$x = \frac{-b \pm \sqrt{b^2 - 4ac}}{2a} \qquad (4)$$

The quantity $b^2 - 4ac$ is called the **discriminant** of the quadratic equation, because its value tells us whether the equation has real solutions. In fact, it also tells us how many solutions to expect.

Discriminant of a Quadratic Equation

For a quadratic equation $ax^2 + bx + c = 0$:

1. If $b^2 - 4ac > 0$, there are two unequal real solutions.
2. If $b^2 - 4ac = 0$, there is a repeated solution, a root of multiplicity 2.
3. If $b^2 - 4ac < 0$, there is no real solution.

When asked to find the real solutions, if any, of a quadratic equation, always evaluate the discriminant first to see how many real solutions there are.

EXAMPLE 7 Solving a Quadratic Equation Using the Quadratic Formula

Use the quadratic formula to find the real solutions, if any, of the equation

$$3x^2 - 5x + 1 = 0$$

SOLUTION The equation is in standard form, so we compare it to $ax^2 + bx + c = 0$ to find a, b, and c.

$$\begin{aligned} 3x^2 - 5x + 1 &= 0 \\ ax^2 + bx + c &= 0 \qquad a = 3, b = -5, c = 1 \end{aligned}$$

With $a = 3$, $b = -5$, and $c = 1$, we evaluate the discriminant $b^2 - 4ac$.

$$b^2 - 4ac = (-5)^2 - 4(3)(1) = 25 - 12 = 13$$

Since $b^2 - 4ac > 0$, there are two real solutions, which can be found using the quadratic formula.

$$x = \frac{-b \pm \sqrt{b^2 - 4ac}}{2a} = \frac{-(-5) \pm \sqrt{13}}{2(3)} = \frac{5 \pm \sqrt{13}}{6}$$

The solution set is $\left\{\frac{5 - \sqrt{13}}{6}, \frac{5 + \sqrt{13}}{6}\right\}$.

EXAMPLE 8 Solving a Quadratic Equation Using the Quadratic Formula

Use the quadratic formula to find the real solutions, if any, of the equation

$$3x^2 + 2 = 4x$$

SOLUTION The equation, as given, is not in standard form.

$$\begin{aligned} 3x^2 + 2 &= 4x \\ 3x^2 - 4x + 2 &= 0 \qquad \text{Put in standard form.} \\ ax^2 + bx + c &= 0 \qquad \text{Compare to standard form.} \end{aligned}$$

With $a = 3$, $b = -4$, and $c = 2$, we find

$$b^2 - 4ac = (-4)^2 - 4(3)(2) = 16 - 24 = -8$$

Since $b^2 - 4ac < 0$, the equation has no real solution.

NOW WORK PROBLEMS 85 AND 91.

SUMMARY

Procedure for Solving a Quadratic Equation

To solve a quadratic equation, first put it in standard form:

$$ax^2 + bx + c = 0$$

Then:

Step 1: Identify a, b, and c.
Step 2: Evaluate the discriminant, $b^2 - 4ac$.
Step 3: (a) If the discriminant is negative, the equation has no real solution.
(b) If the discriminant is zero, the equation has one real solution, a repeated root.
(c) If the discriminant is positive, the equation has two distinct real solutions.

For conditions (b) and (c), if you can easily spot factors, use the factoring method to solve the equation. Otherwise, use the quadratic formula or the method of completing the square.

EXERCISE A.6 Answers to Odd-Numbered Problems Begin on Page AN-106.

In Problems 1–66, solve each equation.

1. $3x = 21$
2. $3x = -24$
3. $5x + 15 = 0$
4. $3x + 18 = 0$
5. $2x - 3 = 5$
6. $3x + 4 = -8$
7. $\frac{1}{3}x = \frac{5}{12}$
8. $\frac{2}{3}x = \frac{9}{2}$
9. $6 - x = 2x + 9$
10. $3 - 2x = 2 - x$
11. $2(3 + 2x) = 3(x - 4)$
12. $3(2 - x) = 2x - 1$
13. $8x - (2x + 1) = 3x - 10$
14. $5 - (2x - 1) = 10$
15. $\frac{1}{2}x - 4 = \frac{3}{4}x$
16. $1 - \frac{1}{2}x = 5$
17. $0.9t = 0.4 + 0.1t$
18. $0.9t = 1 + t$
19. $\frac{2}{y} + \frac{4}{y} = 3$
20. $\frac{4}{y} - 5 = \frac{5}{2y}$
21. $(x + 7)(x - 1) = (x + 1)^2$
22. $(x + 2)(x - 3) = (x - 3)^2$
23. $z(z^2 + 1) = 3 + z^3$
24. $w(4 - w^2) = 8 - w^3$
25. $x^2 = 9x$
26. $x^3 = x^2$
27. $t^3 - 9t^2 = 0$
28. $4z^3 - 8z^2 = 0$
29. $\frac{3}{2x - 3} = \frac{2}{x + 5}$
30. $\frac{-2}{x + 4} = \frac{-3}{x + 1}$
31. $(x + 2)(3x) = (x + 2)(6)$
32. $(x - 5)(2x) = (x - 5)(4)$
33. $\frac{2}{x - 2} = \frac{3}{x + 5} + \frac{10}{(x + 5)(x - 2)}$
34. $\frac{1}{2x + 3} + \frac{1}{x - 1} = \frac{1}{(2x + 3)(x - 1)}$
35. $|2x| = 6$
36. $|3x| = 12$
37. $|2x + 3| = 5$
38. $|3x - 1| = 2$
39. $|1 - 4t| = 5$
40. $|1 - 2z| = 3$
41. $|-2x| = 8$
42. $|-x| = 1$
43. $|-2|x = 4$
44. $|3|x = 9$
45. $|x - 2| = -\frac{1}{2}$
46. $|2 - x| = -1$

47. $|x^2 - 4| = 0$

48. $|x^2 - 9| = 0$

49. $|x^2 - 2x| = 3$

50. $|x^2 + x| = 12$

51. $|x^2 + x - 1| = 1$

52. $|x^2 + 3x - 2| = 2$

53. $x^2 = 4x$

54. $x^2 = -8x$

55. $z^2 + 4z - 12 = 0$

56. $v^2 + 7v + 12 = 0$

57. $2x^2 - 5x - 3 = 0$

58. $3x^2 + 5x + 2 = 0$

59. $x(x - 7) + 12 = 0$

60. $x(x + 1) = 12$

61. $4x^2 + 9 = 12x$

62. $25x^2 + 16 = 40x$

63. $6x - 5 = \frac{6}{x}$

64. $x + \frac{12}{x} = 7$

65. $\frac{4(x-2)}{x-3} + \frac{3}{x} = \frac{-3}{x(x-3)}$

66. $\frac{5}{x+4} = 4 + \frac{3}{x-2}$

In Problems 67–72, solve each equation by the Square Root Method.

67. $x^2 = 25$

68. $x^2 = 36$

69. $(x - 1)^2 = 4$

70. $(x + 2)^2 = 1$

71. $(2x + 3)^2 = 9$

72. $(3x - 2)^2 = 4$

In Problems 73–78, what number should be added to complete the square of each expression?

73. $x^2 + 8x$

74. $x^2 - 4x$

75. $x^2 + \frac{1}{2}x$

76. $x^2 - \frac{1}{3}x$

77. $x^2 - \frac{2}{3}x$

78. $x^2 - \frac{2}{5}x$

In Problems 79–84, solve each equation by completing the square.

79. $x^2 + 4x = 21$

80. $x^2 - 6x = 13$

81. $x^2 - \frac{1}{2}x - \frac{3}{16} = 0$

82. $x^2 + \frac{2}{3}x - \frac{1}{3} = 0$

83. $3x^2 + x - \frac{1}{2} = 0$

84. $2x^2 - 3x - 1 = 0$

In Problems 85–96, find the real solutions, if any, of each equation. Use the quadratic formula.

85. $x^2 - 4x + 2 = 0$

86. $x^2 + 4x + 2 = 0$

87. $x^2 - 5x - 1 = 0$

88. $x^2 + 5x + 3 = 0$

89. $2x^2 - 5x + 3 = 0$

90. $2x^2 + 5x + 3 = 0$

91. $4y^2 - y + 2 = 0$

92. $4t^2 + t + 1 = 0$

93. $4x^2 = 1 - 2x$

94. $2x^2 = 1 - 2x$

95. $x^2 + \sqrt{3}x - 3 = 0$

96. $x^2 + \sqrt{2}x - 2 = 0$

In Problems 97–102, use the discriminant to determine whether each quadratic equation has two unequal real solutions, a repeated real solution, or no real solution, without solving the equation.

97. $x^2 - 5x + 7 = 0$

98. $x^2 + 5x + 7 = 0$

99. $9x^2 - 30x + 25 = 0$

100. $25x^2 - 20x + 4 = 0$

101. $3x^2 + 5x - 8 = 0$

102. $2x^2 - 3x - 4 = 0$

In Problems 103–108, solve each equation. The letters a, b, and c are constants.

103. $ax - b = c, \quad a \neq 0$

104. $1 - ax = b, \quad a \neq 0$

105. $\frac{x}{a} + \frac{x}{b} = c, \quad a \neq 0, b \neq 0, a \neq -b$

106. $\frac{a}{x} + \frac{b}{x} = c, \quad c \neq 0$

107. $\frac{1}{x-a} + \frac{1}{x+a} = \frac{2}{x-1}, \quad a \neq 1$

108. $\frac{b+c}{x+a} = \frac{b-c}{x-a}, \quad c \neq 0, a \neq 0$

Problems 109–114 list some formulas that occur in applications. Solve each formula for the indicated variable.

109. Electricity $\frac{1}{R} = \frac{1}{R_1} + \frac{1}{R_2}$ for R

110. Finance $A = P(1 + rt)$ for r

111. Mechanics $F = \frac{mv^2}{R}$ for R

112. Chemistry $PV = nRT$ for T

113. Mathematics $S = \frac{a}{1 - r}$ for r

114. Mechanics $v = -gt + v_0$ for t

115. Show that the sum of the roots of a quadratic equation is $-\frac{b}{a}$.

116. Show that the product of the roots of a quadratic equation is $\frac{c}{a}$.

117. Find k such that the equation $kx^2 + x + k = 0$ has a repeated real solution.

118. Find k such that the equation $x^2 - kx + 4 = 0$ has a repeated real solution.

119. Show that the real solutions of the equation $ax^2 + bx + c = 0$ are the negatives of the real solutions of the equation $ax^2 - bx + c = 0$. Assume that $b^2 - 4ac \geq 0$.

120. Show that the real solutions of the equation $ax^2 + bx + c = 0$ are the reciprocals of the real solutions of the equation $cx^2 + bx + a = 0$. Assume that $b^2 - 4ac \geq 0$.

121. Which of the following pairs of equations are equivalent? Explain.

(a) $x^2 = 9$; $x = 3$
(b) $x = \sqrt{9}$; $x = 3$
(c) $(x - 1)(x - 2) = (x - 1)^2$; $x - 2 = x - 1$

122. The equation

$$\frac{5}{x + 3} + 3 = \frac{8 + x}{x + 3}$$

has no solution, yet when we go through the process of solving it we obtain $x = -3$. Write a brief paragraph to explain what causes this to happen.

123. Make up an equation that has no solution and give it to a fellow student to solve. Ask the fellow student to write a critique of your equation.

124. Describe three ways you might solve a quadratic equation. State your preferred method; explain why you chose it.

125. Explain the benefits of evaluating the discriminant of a quadratic equation before attempting to solve it.

126. Make up three quadratic equations: one having two distinct solutions, one having no real solution, and one having exactly one real solution.

127. The word *quadratic* seems to imply four (*quad*), yet a quadratic equation is an equation that involves a polynomial of degree 2. Investigate the origin of the term *quadratic* as it is used in the expression *quadratic equation.* Write a brief essay on your findings.

A.7 Intervals; Solving Inequalities

PREPARING FOR THIS SECTION ***Before getting started, review the following:***

> Real Number Line, Inequalities, Absolute Value (Appendix A, Section A.2, pp. A-15–A-18)

OBJECTIVES
1. Use interval notation
2. Use properties of inequalities
3. Solve inequalities
4. Solve combined inequalities
5. Solve polynomial and rational inequalities

Suppose that a and b are two real numbers and $a < b$. We shall use the notation $a < x < b$ to mean that x is a number *between* a and b. The expression $a < x < b$ is equivalent to the two inequalities $a < x$ and $x < b$. Similarly, the expression $a \leq x \leq b$ is equivalent to the two inequalities $a \leq x$ and $x \leq b$. The remaining two possibilities, $a \leq x < b$ and $a < x \leq b$, are defined similarly.

Although it is acceptable to write $3 \geq x \geq 2$, it is preferable to reverse the inequality symbols and write instead $2 \leq x \leq 3$ so that, as you read from left to right, the values go from smaller to larger.

A statement such as $2 \le x \le 1$ is false because there is no number x for which $2 \le x$ and $x \le 1$. Finally, we never mix inequality symbols, as in $2 \le x \ge 3$.

Intervals

Use internal notation 1

Let a and b represent two real numbers with $a < b$.

A **closed interval,** denoted by $[a, b]$, consists of all real numbers x for which $a \le x \le b$.

An **open interval,** denoted by (a, b), consists of all real numbers x for which $a < x < b$.

The **half-open,** or **half-closed, intervals** are $(a, b]$, consisting of all real numbers x for which $a < x \le b$, and $[a, b)$, consisting of all real numbers x for which $a \le x < b$.

In each of these definitions, a is called the **left endpoint** and b the **right endpoint** of the interval.

The symbol ∞ (read as "infinity") is not a real number, but a notational device used to indicate unboundedness in the positive direction. The symbol $-\infty$ (read as "negative infinity") also is not a real number, but a notational device used to indicate unboundedness in the negative direction. Using the symbols ∞ and $-\infty$, we can define five other kinds of intervals:

$[a, \infty)$	consists of all real numbers x for which $x \ge a$ $(a \le x < \infty)$
(a, ∞)	consists of all real numbers x for which $x > a$ $(a < x < \infty)$
$(-\infty, a]$	consists of all real numbers x for which $x \le a$ $(-\infty < x \le a)$
$(-\infty, a)$	consists of all real numbers x for which $x < a$ $(-\infty < x < a)$
$(-\infty, \infty)$	consists of all real numbers x $(-\infty < x < \infty)$

Note that ∞ and $-\infty$ are never included as endpoints, since neither is a real number.

Table 2 summarizes interval notation, corresponding inequality notation, and their graphs.

TABLE 2

Interval	Inequality	Graph
The open interval (a, b)	$a < x < b$	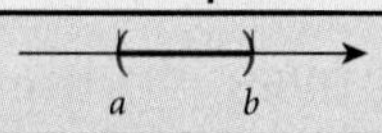
The closed interval $[a, b]$	$a \le x \le b$	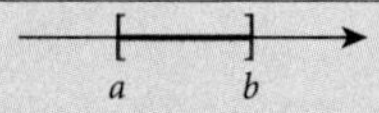
The half-open interval $[a, b)$	$a \le x < b$	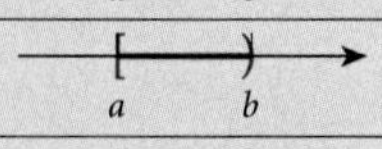
The half-open interval $(a, b]$	$a < x \le b$	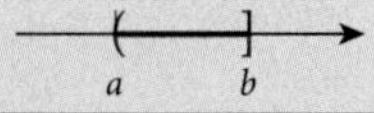
The interval $[a, \infty)$	$x \ge a$	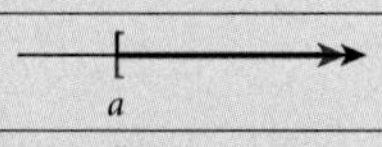
The interval (a, ∞)	$x > a$	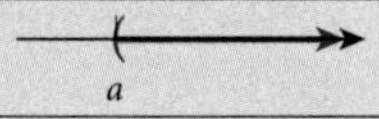
The interval $(-\infty, a]$	$x \le a$	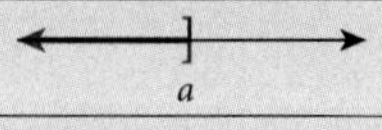
The interval $(-\infty, a)$	$x < a$	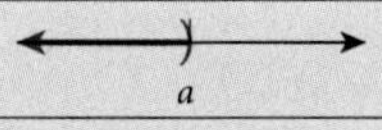
The interval $(-\infty, \infty)$	All real numbers	

EXAMPLE 1 **Writing Inequalities Using Interval Notation**

Write each inequality using interval notation.

(a) $1 \leq x \leq 3$ **(b)** $-4 < x < 0$ **(c)** $x > 5$ **(d)** $x \leq 1$

SOLUTION **(a)** $1 \leq x \leq 3$ describes all numbers x between 1 and 3, inclusive. In interval notation, we write $[1, 3]$.

(b) In interval notation, $-4 < x < 0$ is written $(-4, 0)$.

(c) $x > 5$ consists of all numbers x greater than 5. In interval notation, we write $(5, \infty)$.

(d) In interval notation, $x \leq 1$ is written $(-\infty, 1]$.

EXAMPLE 2 **Writing Intervals Using Inequality Notation**

Write each interval as an inequality involving x.

(a) $[1, 4)$ **(b)** $(2, \infty)$ **(c)** $[2, 3]$ **(d)** $(-\infty, -3]$

SOLUTION **(a)** $[1, 4)$ consists of all numbers x for which $1 \leq x < 4$.

(b) $(2, \infty)$ consists of all numbers x for which $x > 2$ $(2 < x < \infty)$.

(c) $[2, 3]$ consists of all numbers x for which $2 \leq x \leq 3$.

(d) $(-\infty, -3]$ consists of all numbers x for which $x \leq -3$ $(-\infty < x \leq -3)$.

NOW WORK PROBLEMS 1, 7, AND 15.

Properties of Inequalities

Use properties of inequalities **2**

The product of two positive real numbers is positive, the product of two negative real numbers is positive, and the product of 0 and 0 is 0. For any real number a, the value of a^2 is 0 or positive; that is, a^2 is nonnegative. This is called the **nonnegative property.**

For any real number a, we have the following:

Nonnegative Property

$$a^2 \geq 0 \tag{1}$$

If we add the same number to both sides of an inequality, we obtain an equivalent inequality. For example, since $3 < 5$, then $3 + 4 < 5 + 4$ or $7 < 9$. This is called the **addition property** of inequalities.

Addition Property of Inequalities

$$\text{If } a < b, \text{ then } a + c < b + c. \tag{2a}$$

$$\text{If } a > b, \text{ then } a + c > b + c. \tag{2b}$$

The addition property states that the sense, or direction, of an inequality remains unchanged if the same number is added to each side. Now let's see what happens if we multiply each side of an inequality by a non-zero number.

Begin with $3 < 7$ and multiply each side by 2. The numbers 6 and 14 that result yield the inequality $6 < 14$.

Begin with $9 > 2$ and multiply each side by -4. The numbers -36 and -8 that result yield the inequality $-36 < -8$.

Note that the effect of multiplying both sides of $9 > 2$ by the negative number -4 is that the direction of the inequality symbol is reversed. We are led to the following general **multiplication properties** for inequalities:

Multiplication Properties for Inequalities

If $a < b$ and if $c > 0$, then $ac < bc$.
If $a < b$ and if $c < 0$, then $ac > bc$. **(3a)**
If $a > b$ and if $c > 0$, then $ac > bc$.
If $a > b$ and if $c < 0$, then $ac < bc$. **(3b)**

The multiplication properties state that the sense, or direction, of an inequality *remains the same* if each side is multiplied by a *positive* real number, whereas the direction is *reversed* if each side is multiplied by a *negative* real number.

NOW WORK PROBLEMS 29 AND 35.

The **reciprocal property** states that the reciprocal of a positive real number is positive and that the reciprocal of a negative real number is negative.

Reciprocal Property for Inequalities

If $a > 0$, then $\dfrac{1}{a} > 0$. **(4a)**

If $a < 0$, then $\dfrac{1}{a} < 0$. **(4b)**

Solving Inequalities

An **inequality in one variable** is a statement involving two expressions, at least one containing the variable, separated by one of the inequality symbols $<, \leq, >$, or $\geq$. To **solve an inequality** means to find all values of the variable for which the statement is true. These values are called **solutions** of the inequality.

For example, the following are all inequalities involving one variable, x:

$$x + 5 < 8 \qquad 2x - 3 \geq 4 \qquad x^2 - 1 \leq 3 \qquad \frac{x + 1}{x - 2} > 0$$

Two inequalities having exactly the same solution set are called **equivalent inequalities.** As with equations, one method for solving an inequality is to replace it by a series of equivalent inequalities until an inequality with an obvious solution, such as $x < 3$, is obtained. We obtain equivalent inequalities by applying some of the same properties as those used to find equivalent equations. The addition property and the multiplication properties form the basis for the following procedures.

Procedures That Leave the Inequality Symbol Unchanged

1. Simplify both sides of the inequality by combining like terms and eliminating parentheses:

   Replace $(x + 2) + 6 > 2x + 5(x + 1)$
   by $x + 8 > 7x + 5$

2. Add or subtract the same expression on both sides of the inequality:

   Replace $3x - 5 < 4$
   by $(3x - 5) + 5 < 4 + 5$

3. Multiply or divide both sides of the inequality by the same positive expression:

   Replace $4x > 16$ by $\frac{4x}{4} > \frac{16}{4}$

Procedures That Reverse the Sense or Direction of the Inequality Symbol

1. Interchange the two sides of the inequality:

   Replace $3 < x$ by $x > 3$

2. Multiply or divide both sides of the inequality by the same *negative* expression.

   Replace $-2x > 6$ by $\frac{-2x}{-2} < \frac{6}{-2}$

As the examples that follow illustrate, we solve inequalities using many of the same steps that we would use to solve equations. In writing the solution of an inequality, we may use either set notation or interval notation, whichever is more convenient.

3 **EXAMPLE 3 Solving an Inequality**

Solve the inequality: $4x + 7 \geq 2x - 3$
Graph the solution set.

SOLUTION

$$4x + 7 \geq 2x - 3$$

$$4x + 7 - 7 \geq 2x - 3 - 7 \qquad \text{Subtract 7 from both sides.}$$

$$4x \geq 2x - 10 \qquad \text{Simplify.}$$

$$4x - 2x \geq 2x - 10 - 2x \qquad \text{Subtract } 2x \text{ from both sides.}$$

$$2x \geq -10 \qquad \text{Simplify.}$$

$$\frac{2x}{2} \geq \frac{-10}{2} \qquad \text{Divide both sides by 2. (The sense of the inequality symbol is unchanged.)}$$

$$x \geq -5 \qquad \text{Simplify.}$$

FIGURE 12

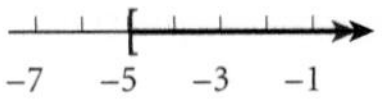

The solution set is $\{x \mid x \geq -5\}$ or, using interval notation, all numbers in the interval $[-5, \infty)$. See Figure 12 for the graph.

NOW WORK PROBLEM 43.

4 **EXAMPLE 4** **Solving Combined Inequalities**

Solve the inequality: $-5 < 3x - 2 < 1$
Graph the solution set.

SOLUTION Recall that the inequality

$$-5 < 3x - 2 < 1$$

is equivalent to the two inequalities

$$-5 < 3x - 2 \quad \text{and} \quad 3x - 2 < 1$$

We will solve each of these inequalities separately.

$-5 < 3x - 2$		$3x - 2 < 1$
$-5 + 2 < 3x - 2 + 2$	Add 2 to both sides.	$3x - 2 + 2 < 1 + 2$
$-3 < 3x$	Simplify.	$3x < 3$
$\frac{-3}{3} < \frac{3x}{3}$	Divide both sides by 3.	$\frac{3x}{3} < \frac{3}{3}$
$-1 < x$	Simplify.	$x < 1$

The solution set of the original pair of inequalities consists of all x for which

$$-1 < x \quad \text{and} \quad x < 1$$

FIGURE 13

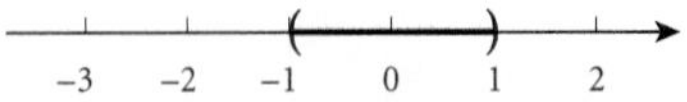

This may be written more compactly as $\{x \mid -1 < x < 1\}$. In interval notation, the solution is $(-1, 1)$. See Figure 13 for the graph.

NOW WORK PROBLEM 63.

EXAMPLE 5 **Using the Reciprocal Property to Solve an Inequality**

Solve the inequality: $(4x - 1)^{-1} > 0$
Graph the solution set.

SOLUTION Since $(4x - 1)^{-1} = \frac{1}{4x - 1}$ and since the Reciprocal Property states that when $\frac{1}{a} > 0$ then $a > 0$, we have

$$(4x - 1)^{-1} > 0$$

$$\frac{1}{4x - 1} > 0$$

$$4x - 1 > 0 \quad \text{Reciprocal Property.}$$

$$4x > 1$$

$$x > \frac{1}{4}$$

FIGURE 14

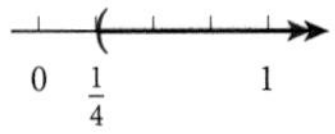

The solution set is $\left\{x \middle| x > \frac{1}{4}\right\}$; that is, all x in the interval $\left(\frac{1}{4}, \infty\right)$. Figure 14 illustrates the graph.

NOW WORK PROBLEM 67.

Polynomial and Rational Inequalities

Solve polynomial and rational inequalities **5**

The next four examples deal with polynomial and rational inequalities which are important for solving certain types of problems in calculus.

EXAMPLE 6 Solving a Quadratic Inequality

Solve the inequality $x^2 + x - 12 > 0$, and graph the solution set.

SOLUTION We factor the left side, obtaining

$$x^2 + x - 12 > 0$$
$$(x + 4)(x - 3) > 0$$

We then construct a graph that uses the solutions to the equation

$$x^2 + x - 12 = (x + 4)(x - 3) = 0$$

namely, $x = -4$ and $x = 3$. These numbers separate the real number line into three parts:

$$-\infty < x < -4 \qquad -4 < x < 3 \qquad 3 < x < \infty$$

or, in interval notation, into $(-\infty, -4)$, $(-4, 3)$, and $(3, \infty)$. See Figure 15(a).

Now if $x < -4$, then $x + 4 < 0$. We indicate this fact about the expression $x + 4$ by placing minus signs $(- - -)$ to the left of -4. If $x > -4$, then $x + 4 > 0$. We indicate this fact about $x + 4$ by placing plus signs $(+ + +)$ to the right of -4.

Similarly, if $x < 3$, then $x - 3 < 0$. We indicate this fact about $x - 3$ by placing minus signs to the left of 3. If $x > 3$, then $x - 3 > 0$. We indicate this fact about $x - 3$ by placing plus signs to the right of 3. See Figure 15 (b).

FIGURE 15

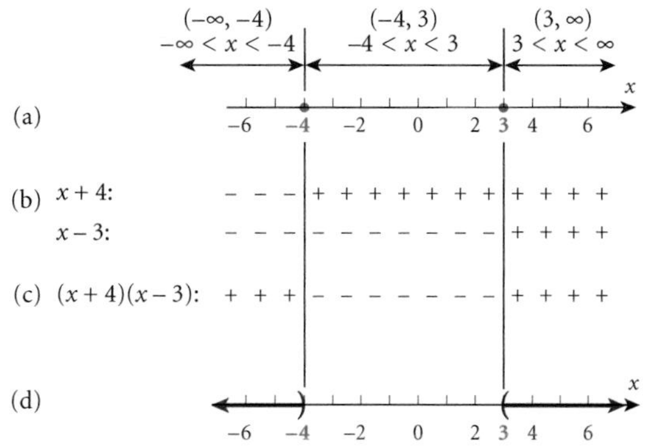

Now we prepare Figure 15(c). Since we know that the expressions $x + 4$ and $x - 3$ are both negative for $x < -4$, it follows that their product will be positive for $x < -4$. Since we know that $x + 4$ is positive and $x - 3$ is negative for $-4 < x < 3$, it follows that their product is negative for $-4 < x < 3$. Finally, since both expressions are positive for $x > 3$, their product is positive for $x > 3$.

We conclude that the product $(x + 4)(x - 3) = x^2 + x - 12$ is positive when $x < -4$ or when $x > 3$. The solution set is $\{x | x < -4 \text{ or } x > 3\}$. In interval notation the solution consists of the intervals $(-\infty, -4)$ or $(3, \infty)$. See Figure 15(d).

The preceding discussion demonstrates that the sign of each factor of the expression and, consequently, the sign of the expression itself, is the same on each interval that the real number line was divided into. An alternative, and simpler, approach to obtaining

Figure 15(c) would be to select a **test number** in each interval and use it to evaluate the expression to see if it is positive or negative. You may choose any number in the interval as a test number. See Figure 16.

FIGURE 16

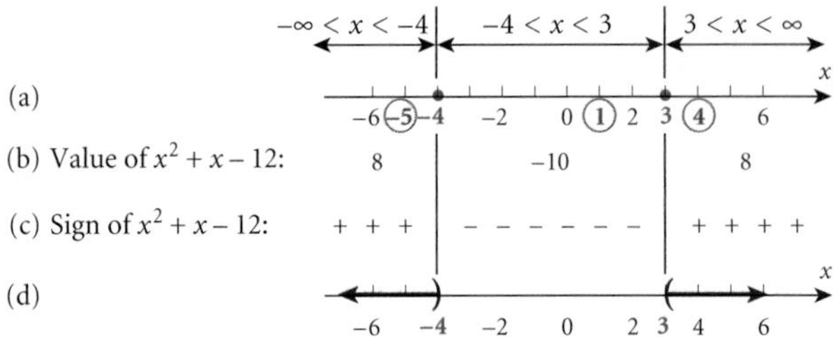

In Figure 16(a) the test numbers we selected, -5, 1, 4, have been circled. We evaluate the expression $x^2 + x - 12$ at each test number.

For $x = -5$: $(-5)^2 + (-5) - 12 = 8$, a positive number.
For $x = 1$: $1^2 + 1 - 12 = -10$, a negative number.
Fox $x = 4$: $4^2 + 4 - 12 = 8$, a positive number.

See Figure 16(b) and 16(c).

The rest of Figure 16 is obtained as before.

We shall employ the method of using a test number to solve inequalities. Here is another example showing all the details.

EXAMPLE 7 Solving a Quadratic Inequality

Solve the inequality $x^2 \leq 4x + 12$, and graph the solution set.

SOLUTION First, we rearrange the inequality so that 0 is on the right side:

$$x^2 \leq 4x + 12$$

$$x^2 - 4x - 12 \leq 0 \qquad \text{Subtract } 4x \text{ and 12 from each side.}$$

This inequality is equivalent to the one we seek to solve.

Next, we set the left side equal to 0 and solve the resulting equation:

$$x^2 - 4x - 12 = 0$$

$$(x + 2)(x - 6) = 0 \qquad \text{Factor.}$$

$$x + 2 = 0 \quad \text{or} \quad x - 6 = 0 \qquad \text{Zero-Product Property.}$$

$$x = -2 \quad \text{or} \quad x = 6$$

The solutions of the equation are -2 and 6, and they separate the real number line into three parts:

$$-\infty < x < -2 \qquad -2 < x < 6 \qquad 6 < x < \infty$$

See Figure 17.

In each part, select a test number. We will choose -3, 1, and 8, which are circled. See Figure 17(a).

Next we evaluate the expression $x^2 - 4x - 12$ at each test number.

For $x = -3$: $(-3)^2 - 4(-3) - 12 = 9$, a positive number.
For $x = 1$: $1^2 - 4(1) - 12 = -15$, a negative number.
For $x = 8$: $8^2 - 4(8) - 12 = 20$, a positive number.

See Figure 17(b) and 17(c).

The expression $x^2 - 4x - 12 < 0$ for $-2 < x < 6$. However, because the inequality we wish to solve is nonstrict, numbers x that satisfy the equation $x^2 - 4x + 12 = 0$ are also solutions of the inequality $x^2 - 4x - 12 \le 0$. We include -2 and 6, and the solution set of the given inequality is $\{x \mid -2 \le x \le 6\}$; that is, all x in the interval $[-2, 6]$. See Figure 17(d).

FIGURE 17

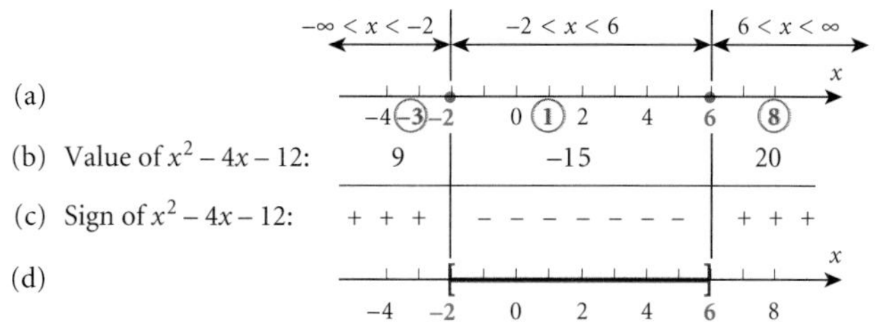

NOW WORK PROBLEM 75.

EXAMPLE 8 Solving a Polynomial Inequality

Solve the inequality $x^4 < x$, and graph the solution set.

SOLUTION We rewrite the inequality so that 0 is on the right side:

$$x^4 < x$$

$$x^4 - x < 0 \quad \text{Subtract } x \text{ from both sides.}$$

This inequality is equivalent to the one we wish to solve.

We proceed to solve the equation $x^4 - x = 0$ using factoring.

$$x^4 - x = 0$$

$$x(x^3 - 1) = 0 \quad \text{Factor out } x.$$

$$x(x - 1)(x^2 + x + 1) = 0 \quad \text{Difference of two cubes.}$$

$$x = 0 \quad \text{or} \quad x - 1 = 0 \quad \text{or} \quad x^2 + x + 1 = 0 \quad \text{Zero-Product Property.}$$

The solutions are 0 and 1, since the equation $x^2 + x + 1 = 0$ has no real solution.

Next we use 0 and 1 to separate the real number line into three parts:

$$-\infty < x < 0 \qquad 0 < x < 1 \qquad 1 < x < \infty$$

In each part, select a test number. We will choose $-1, \frac{1}{2}$, and 2. See Figure 18.

FIGURE 18

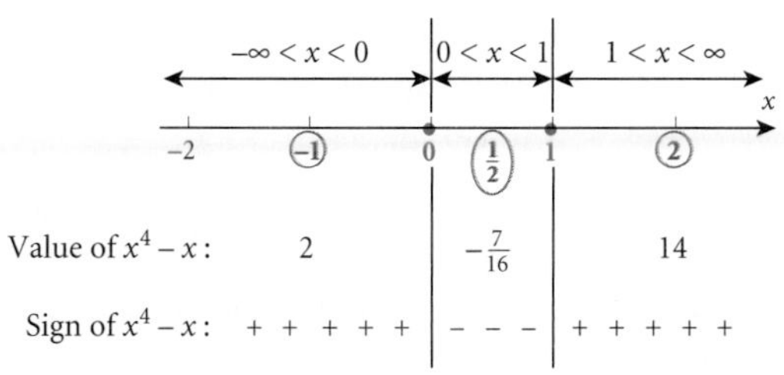

We evaluate the expression $x^4 - x$ at each test number.

For $x = -1$: $(-1)^4 - (-1) = 2$, a positive number.

For $x = \frac{1}{2}$: $\left(\frac{1}{2}\right)^4 - \left(\frac{1}{2}\right) = -\frac{7}{16}$, a negative number.

For $x = 2$: $2^4 - 2 = 14$, a positive number.

FIGURE 19

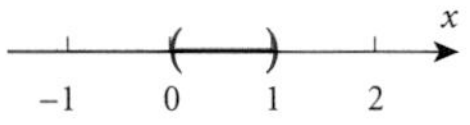

The expression $x^4 - x < 0$ for $0 < x < 1$. The solution set is $\{x | 0 < x < 1\}$; that is, all x in the interval $(0, 1)$. See Figure 19.

We have been solving inequalities by rearranging the inequality so that 0 is on the right side, setting the left side equal to 0, and solving the resulting equation. The solutions are then used to separate the real number line into intervals. But what if the resulting equation has no real solution? In this case we rely on the following result.

Theorem

If a polynomial equation has no real solutions, the polynomial is either always positive or always negative.

For example, the equation

$$x^2 + 5x + 8 = 0$$

has no real solutions. (Do you see why? Its discriminant, $b^2 - 4ac = 25 - 32 = -7$, is negative.) The value of $x^2 + 5x + 8$ is therefore always positive or always negative. To see which is true, we test its value at some number (0 is the easiest). Because $0^2 + 5(0) + 8 = 8$ is positive, we conclude that $x^2 + 5x + 8 > 0$ for all x.

NOW WORK PROBLEM 79.

Next we solve a rational inequality.

EXAMPLE 9 Solving a Rational Inequality

Solve the inequality $\frac{4x + 5}{x + 2} \geq 3$, and graph the solution set.

SOLUTION We first note that the domain of the variable consists of all real numbers except -2. We rearrange terms so that 0 is on the right side:

$$\frac{4x + 5}{x + 2} \geq 3$$

$$\frac{4x + 5}{x + 2} - 3 \geq 0 \quad \text{Subtract 3 from both sides.}$$

$$\frac{4x + 5 - 3(x + 2)}{x + 2} \geq 0 \quad \text{Rewrite using } x + 2 \text{ as the denominator.}$$

$$\frac{x - 1}{x + 2} \geq 0 \quad \text{Simplify.}$$

This inequality is equivalent to the one we wish to solve.

For rational expressions we set both the numerator and the denominator equal to 0 to determine the numbers to use to separate the number line. For this example we use -2 and 1 to divide the number line into three parts:

$$-\infty < x < -2 \qquad -2 < x < 1 \qquad 1 < x < \infty$$

Construct Figure 20, using $-3, 0,$ and 2 as test numbers.

FIGURE 20

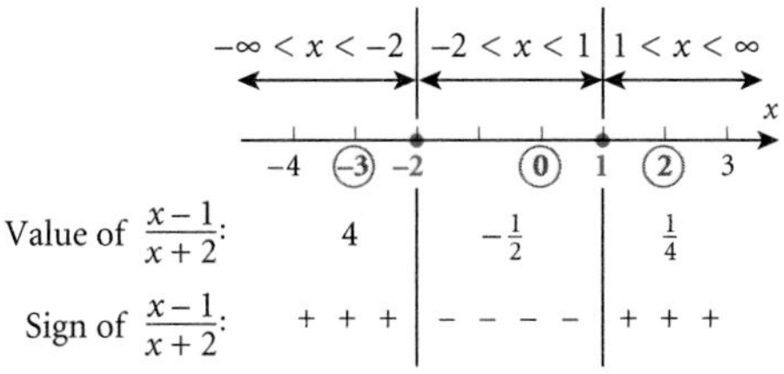

We conclude that $\dfrac{x-1}{x+2} > 0$ for $x < -2$ or for $x > 1$.

FIGURE 21

However, we want to know where the expression $\dfrac{x-1}{x+2}$ is positive or 0. Since $\dfrac{x-1}{x+2} = 0$ only if $x = 1$, we conclude that the solution set is $\{x \mid x < -2 \text{ or } x \geq 1\}$; that is, all x in the intervals $(-\infty, -2)$ or $[1, \infty)$. See Figure 21.

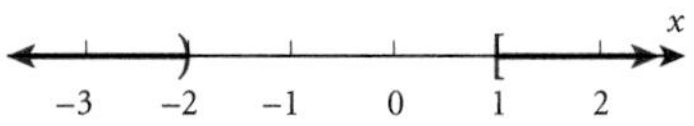

In Example 9 you may wonder why we did not first multiply both sides of the inequality by $x + 2$ to clear the denominator. The reason is that we do not know whether $x + 2$ is positive or negative and, as a result, we do not know whether to reverse the sense of the inequality symbol after multiplying by $x + 2$. However, there is nothing to prevent us from multiplying both sides by $(x + 2)^2$, which is always positive, since $x \neq -2$. (Do you see why?)

$$\frac{4x + 5}{x + 2} \geq 3 \qquad\qquad x \neq -2$$

$$\frac{4x + 5}{x + 2}(x + 2)^2 \geq 3(x + 2)^2$$

$$(4x + 5)(x + 2) \geq 3(x^2 + 4x + 4)$$

$$4x^2 + 13x + 10 \geq 3x^2 + 12x + 12$$

$$x^2 + x - 2 \geq 0$$

$$(x + 2)(x - 1) \geq 0 \qquad\qquad x \neq -2$$

This last expression leads to the same solution set obtained in Example 9.

 NOW WORK PROBLEM 91.

EXERCISE A.7   Answers to Odd-Numbered Problems Begin on Page AN-107.

In Problems 1–6, express the graph shown in color using interval notation. Also express each as an inequality involving x.

1.

   -1 0 1 2 3

2.

   -1 0 1 2 3

3.

   -2 -1 0 1 2

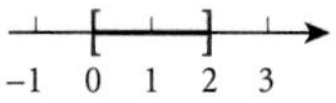

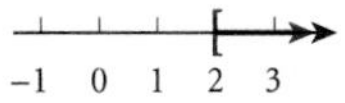

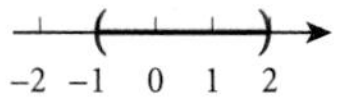

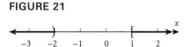

4. 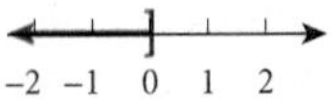

5. 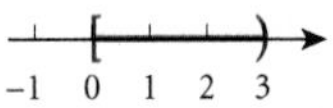

6. 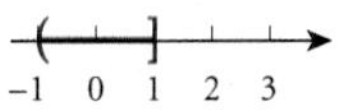

In Problems 7–14, write each inequality using interval notation, and graph each inequality using the real number line.

7. $0 \leq x \leq 4$
8. $-1 < x < 5$
9. $4 \leq x < 6$
10. $-2 < x < 0$
11. $x \geq 4$
12. $x \leq 5$
13. $x < -4$
14. $x > 1$

In Problems 15–22, write each interval as an inequality involving x, and graph each inequality using the real number line.

15. $[2, 5]$
16. $(1, 2)$
17. $(-3, -2)$
18. $[0, 1)$
19. $[4, \infty)$
20. $(-\infty, 2]$
21. $(-\infty, -3)$
22. $(-8, \infty)$

In Problems 23–28, an inequality is given. Write the inequality obtained by:
(a) Adding 3 to each side of the given inequality.
(b) Subtracting 5 from each side of the given inequality.
(c) Multiplying each side of the given inequality by 3.
(d) Multiplying each side of the given inequality by −2.

23. $3 < 5$
24. $2 > 1$
25. $4 > -3$
26. $-3 > -5$
27. $2x + 1 < 2$
28. $1 - 2x > 5$

In Problems 29–42, fill in the blank with the correct inequality symbol.

29. If $x < 5$, then $x - 5$ ________ 0.
30. If $x < -4$, then $x + 4$ ________ 0.
31. If $x > -4$, then $x + 4$ ________ 0.
32. If $x > 6$, then $x - 6$ ________ 0.
33. If $x \geq -4$, then $3x$ ________ −12.
34. If $x \leq 3$, then $2x$ ________ 6.
35. If $x > 6$, then $-2x$ ________ −12.
36. If $x > -2$, then $-4x$ ________ 8.
37. If $x \geq 5$, then $-4x$ ________ −20.
38. If $x \leq -4$, then $-3x$ ________ 12.
39. If $2x > 6$, then x ________ 3.
40. If $3x \leq 12$, then x ________ 4.
41. If $-\frac{1}{2}x \leq 3$, then x ________ −6.
42. If $-\frac{1}{4}x > 1$, then x ________ −4.

In Problems 43–100, solve each inequality. Express your answer using set notation or interval notation. Graph the solution set.

43. $x + 1 < 5$
44. $x - 6 < 1$
45. $1 - 2x \leq 3$
46. $2 - 3x \leq 5$
47. $3x - 7 > 2$
48. $2x + 5 > 1$
49. $3x - 1 \geq 3 + x$
50. $2x - 2 \geq 3 + x$
51. $-2(x + 3) < 8$
52. $-3(1 - x) < 12$
53. $4 - 3(1 - x) \leq 3$
54. $8 - 4(2 - x) \leq -2x$
55. $\frac{1}{2}(x - 4) > x + 8$
56. $3x + 4 > \frac{1}{3}(x - 2)$
57. $\frac{x}{2} \geq 1 - \frac{x}{4}$
58. $\frac{x}{3} \geq 2 + \frac{x}{6}$
59. $0 \leq 2x - 6 \leq 4$
60. $4 \leq 2x + 2 \leq 10$
61. $-5 \leq 4 - 3x \leq 2$
62. $-3 \leq 3 - 2x \leq 9$
63. $-3 < \frac{2x - 1}{4} < 0$

64. $0 < \frac{3x + 2}{2} < 4$

65. $1 < 1 - \frac{1}{2}x < 4$

66. $0 < 1 - \frac{1}{3}x < 1$

67. $(4x + 2)^{-1} < 0$

68. $(2x - 1)^{-1} > 0$

69. $0 < \frac{2}{x} < \frac{3}{5}$

70. $0 < \frac{4}{x} < \frac{2}{3}$

71. $(x - 3)(x + 1) < 0$

72. $(x - 1)(x + 2) < 0$

73. $-x^2 + 9 > 0$

74. $-x^2 + 1 > 0$

75. $x^2 + x > 12$

76. $x^2 + 7x < -12$

77. $x(x - 7) > -12$

78. $x(x + 1) > 12$

79. $4x^2 + 9 < 6x$

80. $25x^2 + 16 < 40x$

81. $(x - 1)(x^2 + x + 1) > 0$

82. $(x + 2)(x^2 - x + 1) > 0$

83. $(x - 1)(x - 2)(x - 3) < 0$

84. $(x + 1)(x + 2)(x + 3) < 0$

85. $-x^3 + 2x^2 + 8x < 0$

86. $-x^3 - 2x^2 + 8x < 0$

87. $x^3 > x$

88. $x^3 < 4x$

89. $x^3 > x^2$

90. $x^3 < 3x^2$

91. $\frac{x + 1}{1 - x} < 0$

92. $\frac{3 - x}{x + 1} < 0$

93. $\frac{(x - 1)(x + 1)}{x} < 0$

94. $\frac{(x - 3)(x + 2)}{x - 1} < 0$

95. $\frac{x - 2}{x^2 - 1} \geq 0$

96. $\frac{x + 5}{x^2 - 4} \geq 0$

97. $\frac{x + 4}{x - 2} \leq 1$

98. $\frac{x + 2}{x - 4} \geq 1$

99. $\frac{2x + 5}{x + 1} > \frac{x + 1}{x - 1}$

100. $\frac{1}{x + 2} > \frac{3}{x + 1}$

101. Computing Grades In your Economics 101 class, you have scores of 68, 82, 87, and 89 on the first four of five tests. To get a grade of B, the average of the first five test scores must be greater than or equal to 80 and less than 90. Solve an inequality to find the range of the score that you need on the last test to get a B.

102. General Chemistry For a certain ideal gas, the volume V (in cubic centimeters) equals 20 times the temperature T (in degrees Celsius). If the temperature varies from 80° to 120°C inclusive, what is the corresponding range of the volume of the gas?

103. Real Estate A real estate agent agrees to sell a large apartment complex according to the following commission schedule: \$45,000 plus 25% of the selling price in excess of \$900,000. Assuming that the complex will sell at some price between \$900,000 and \$1,100,000 inclusive, over what range does the agent's commission vary? How does the commission vary as a percent of selling price?

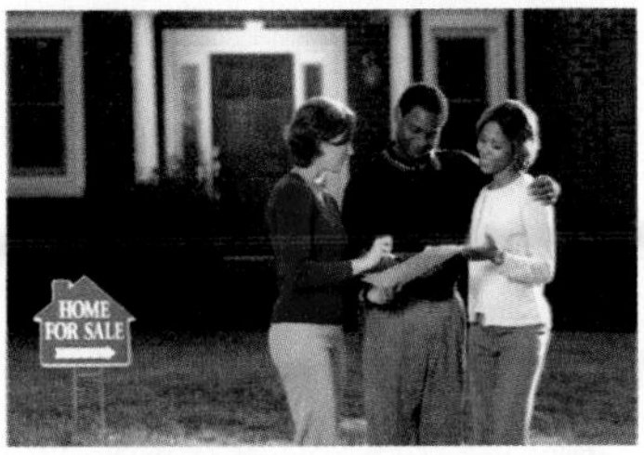

104. Sales Commission A used car salesperson is paid a commission of \$25 plus 40% of the selling price in excess of owner's cost. The owner claims that used cars typically sell for at least owner's cost plus \$70 and at most owner's cost plus \$300. For each sale made, over what range can the salesperson expect the commission to vary?

105. Federal Tax Withholding The percentage method of withholding for federal income tax (2003) states that a single person whose weekly wages, after subtracting withholding allowances, are over \$592, but not over \$1317, shall have \$74.35 plus 25% of the excess over \$592 withheld. Over what range does the amount withheld vary if the weekly wages vary from \$600 to \$800 inclusive?

Source: Internal Revenue Service, 2003.

106. Federal Tax Withholding Rework Problem 105 if the weekly wages vary from \$800 to \$1000 inclusive.

107. Electricity Rates Commonwealth Edison Company's charge for electricity in May 2003 is 8.275¢ per kilowatt-hour. In addition, each monthly bill contains a customer charge of \$7.58. If last summer's bills ranged from a low of \$63.47 to a high of \$214.53, over what range did usage vary (in kilowatt-hours)?

Source: Commonwealth Edison Co., Chicago, Illinois, 2003.

108. Water Bills The Village of Oak Lawn charges homeowners \$27.18 per quarter-year plus \$1.90 per 1000 gallons for water usage in excess of 12,000 gallons. In 2003, one homeowner's quarterly bill ranged from a high of \$76.52 to a low of \$34.78. Over what range did water usage vary?

Source: Village of Oak Lawn, Illinois, 2003.

109. Markup of a New Car The sticker price of a new car ranges from 12% to 18% higher than the dealer's cost. If the sticker price is \$8800, over what range will the dealer's cost vary?

110. IQ Tests A standard intelligence test has an average score of 100. According to statistical theory, of the people who take the test, the 2.5% with the highest scores will have scores of more than 1.96σ above the average, where σ (sigma, a number called the **standard deviation**) depends on the nature of the test. If $\sigma = 12$ for this test and there is (in principle) no upper limit to the score possible on the test, write the interval of possible test scores of the people in the top 2.5%.

111. Make up an inequality that has no solution. Make up one that has exactly one solution.

112. The inequality $x^2 + 1 < -5$ has no solution. Explain why.

113. Do you prefer to use inequality notation or interval notation to express the solution to an inequality? Give your reasons. Are there particular circumstances when you prefer one to the other? Cite examples.

114. How would you explain to a fellow student the underlying reason for the multiplication properties for inequalities. That is, the sense or direction of an inequality remains the same if each side is multiplied by a positive real number, whereas the direction is reversed if each side is multiplied by a negative real number.

A.8 *n*th Roots; Rational Exponents

PREPARING FOR THIS SECTION *Before getting started, review the following:*

> Exponents, Square Roots (Appendix A, Section A.2, pp. A-19–A-23)

OBJECTIVES
1. Work with *n*th roots
2. Simplify radicals
3. Rationalize denominators
4. Solve radical equations
5. Simplify expressions with rational exponents

*n*th ROOTS

The **principal *n*th root of a number *a*,** symbolized by $\sqrt[n]{a}$, where $n \geq 2$ is an integer, is defined as follows:

$$\sqrt[n]{a} = b \quad \text{means} \quad a = b^n$$

where $a \geq 0$ and $b \geq 0$ if $n \geq 2$ is even, and a, b are any real numbers if $n \geq 3$ is odd.

Notice that if a is negative and n is even then $\sqrt[n]{a}$ is not defined. When it is defined, the principal nth root of a number is unique.

Work with *n*th roots 1

The symbol $\sqrt[n]{a}$ for the principal nth root of a is sometimes called a **radical;** the integer n is called the **index,** and a is called the **radicand.** If the index of a radical is 2, we call $\sqrt[2]{a}$ the **square root** of a and omit the index 2 by simply writing $\sqrt{a}$. If the index is 3, we call $\sqrt[3]{a}$ the **cube root** of a.

EXAMPLE 1 Evaluating Principal *n*th Roots

(a) $\sqrt[3]{8} = \sqrt[3]{2^3} = 2$ **(b)** $\sqrt[3]{-64} = \sqrt[3]{(-4)^3} = -4$

(c) $\sqrt[4]{\frac{1}{16}} = \sqrt[4]{\left(\frac{1}{2}\right)^4} = \frac{1}{2}$ **(d)** $\sqrt[6]{(-2)^6} = |-2| = 2$

These are examples of **perfect roots,** since each simplifies to a rational number. Notice the absolute value in Example 1(d). If n is even, the principal nth root must be nonnegative.

In general, if $n \geq 2$ is a positive integer and a is a real number, we have

$$\sqrt[n]{a^n} = a, \quad \text{if } n \geq 3 \text{ is odd} \tag{1a}$$

$$\sqrt[n]{a^n} = |a|, \quad \text{if } n \geq 2 \text{ is even} \tag{1b}$$

NOW WORK PROBLEM 1.

Properties of Radicals

Let $n \geq 2$ and $m \geq 2$ denote positive integers, and let a and b represent real numbers. Assuming that all radicals are defined, we have the following properties:

$$\sqrt[n]{ab} = \sqrt[n]{a}\sqrt[n]{b} \tag{2a}$$

$$\sqrt[n]{\frac{a}{b}} = \frac{\sqrt[n]{a}}{\sqrt[n]{b}} \tag{2b}$$

$$\sqrt[n]{a^m} = (\sqrt[n]{a})^m \tag{2c}$$

When used in reference to radicals, the direction to "simplify" will mean to remove from the radicals any perfect roots that occur as factors. Let's look at some examples of how the preceding rules are applied to simplify radicals.

2 **EXAMPLE 2 Simplifying Radicals**

(a) $\sqrt{32} = \sqrt{16 \cdot 2} = \sqrt{16} \cdot \sqrt{2} = 4\sqrt{2}$

↑ 16 is a perfect square. ↑ (2a)

(b) $\sqrt[3]{16} = \sqrt[3]{8 \cdot 2} = \sqrt[3]{8} \cdot \sqrt[3]{2} = 2\sqrt[3]{2}$

↑ 8 is a perfect cube. ↑ (2a)

(c) $\sqrt[3]{-16x^4} = \sqrt[3]{-8 \cdot 2 \cdot x^3 \cdot x} = \sqrt[3]{(-8x^3)(2x)}$

↑ Factor perfect cubes inside radical. ↑ Combine perfect cubes.

$= \sqrt[3]{(-2x)^3 \cdot 2x} = \sqrt[3]{(-2x)^3} \cdot \sqrt[3]{2x}$

↑

$= -2x \cdot \sqrt[3]{2x}$ (2a)

NOW WORK PROBLEM 7.

EXAMPLE 3 Combining Like Radicals

(a) $-8\sqrt{12} + \sqrt{3} = -8\sqrt{4 \cdot 3} + \sqrt{3} = -8 \cdot \sqrt{4}\sqrt{3} + \sqrt{3} = -8 \cdot 2\sqrt{3} + \sqrt{3}$
$= -16\sqrt{3} + \sqrt{3} = -15\sqrt{3}$

(b) $\sqrt[3]{8x^4} + \sqrt[3]{-x} + 4\sqrt[3]{27x} = \sqrt[3]{2^3x^3x} + \sqrt[3]{-1 \cdot x} + 4\sqrt[3]{3^3x}$
$= \sqrt[3]{(2x)^3} \cdot \sqrt[3]{x} + \sqrt[3]{-1} \cdot \sqrt[3]{x} + 4\sqrt[3]{3^3} \cdot \sqrt[3]{x}$
$= 2x \cdot \sqrt[3]{x} - 1 \cdot \sqrt[3]{x} + 12 \cdot \sqrt[3]{x}$
$= (2x + 11)\sqrt[3]{x}$

NOW WORK PROBLEM 25.

Rationalizing

When radicals occur in quotients, it is customary to rewrite the quotient so that the denominator contains no radicals. This process is referred to as **rationalizing the denominator.**

The idea is to multiply by an appropriate expression so that the new denominator contains no radicals. For example:

If Denominator Contains the Factor	*Multiply By*	*To Obtain Denominator Free of Radicals*
$\sqrt{3}$	$\sqrt{3}$	$(\sqrt{3})^2 = 3$
$\sqrt{3} + 1$	$\sqrt{3} - 1$	$(\sqrt{3})^2 - 1^2 = 3 - 1 = 2$
$\sqrt{2} - 3$	$\sqrt{2} + 3$	$(\sqrt{2})^2 - 3^2 = 2 - 9 = -7$
$\sqrt{5} - \sqrt{3}$	$\sqrt{5} + \sqrt{3}$	$(\sqrt{5})^2 - (\sqrt{3})^2 = 5 - 3 = 2$
$\sqrt[3]{4}$	$\sqrt[3]{2}$	$\sqrt[3]{4} \cdot \sqrt[3]{2} = \sqrt[3]{8} = 2$

In rationalizing the denominator of a quotient, be sure to multiply both the numerator and the denominator by the same expression.

3

EXAMPLE 4 Rationalizing Denominators

Rationalize the denominator of each expression.

(a) $\dfrac{4}{\sqrt{2}}$ **(b)** $\dfrac{\sqrt{3}}{\sqrt[3]{2}}$ **(c)** $\dfrac{\sqrt{x} - 2}{\sqrt{x} + 2}$, $x \geq 0$

SOLUTION **(a)** $\dfrac{4}{\sqrt{2}} = \dfrac{4}{\sqrt{2}} \cdot \dfrac{\sqrt{2}}{\sqrt{2}} = \dfrac{4\sqrt{2}}{(\sqrt{2})^2} = \dfrac{4\sqrt{2}}{2} = 2\sqrt{2}$

↑ Multiply by $\dfrac{\sqrt{2}}{\sqrt{2}}$.

(b) $\dfrac{\sqrt{3}}{\sqrt[3]{2}} = \dfrac{\sqrt{3}}{\sqrt[3]{2}} \cdot \dfrac{\sqrt[3]{4}}{\sqrt[3]{4}} = \dfrac{\sqrt{3}\,\sqrt[3]{4}}{\sqrt[3]{8}} = \dfrac{\sqrt{3}\,\sqrt[3]{4}}{2}$

↑ Multiply by $\dfrac{\sqrt[3]{4}}{\sqrt[3]{4}}$.

(c) $\dfrac{\sqrt{x}-2}{\sqrt{x}+2} = \dfrac{\sqrt{x}-2}{\sqrt{x}+2} \cdot \dfrac{\sqrt{x}-2}{\sqrt{x}-2} = \dfrac{(\sqrt{x}-2)^2}{(\sqrt{x})^2-2^2}$

$$= \frac{(\sqrt{x})^2 - 4\sqrt{x} + 4}{x-4} = \frac{x - 4\sqrt{x} + 4}{x-4}$$

NOW WORK PROBLEM 33.

Equations Containing Radicals

When the variable in an equation occurs in a square root, cube root, and so on, that is, when it occurs under a radical, the equation is called a **radical equation.** Sometimes a suitable operation will change a radical equation to one that is linear or quadratic. The most commonly used procedure is to isolate the most complicated radical on one side of the equation and then eliminate it by raising each side to a power equal to the index of the radical. Care must be taken, because extraneous solutions may result. Thus, when working with radical equations, we always check apparent solutions. Let's look at an example.

4 **EXAMPLE 5 Solving Radical Equations**

Solve the equation: $\sqrt[3]{2x-4} - 2 = 0$

SOLUTION The equation contains a radical whose index is 3. We isolate it on the left side.

$$\sqrt[3]{2x-4} - 2 = 0$$

$$\sqrt[3]{2x-4} = 2 \quad \text{Add 2 to both sides.}$$

Now raise each side to the third power (since the index of the radical is 3) and solve.

$$\left(\sqrt[3]{2x-4}\right)^3 = 2^3 \quad \text{Raise each side to the 3rd power.}$$

$$2x - 4 = 8 \quad \text{Simplify.}$$

$$2x = 12 \quad \text{Solve for } x.$$

$$x = 6$$

Check: $x = 6$: $\sqrt[3]{2(6)-4} - 2 = \sqrt[3]{12-4} - 2 = \sqrt[3]{8} - 2 = 2 - 2 = 0.$

The solution is $x = 6$.

EXAMPLE 6 Solving Radical Equations

Solve the equation: $\sqrt{3x+4} = x$

SOLUTION The index of a square root is 2, so we square both sides

$$\sqrt{3x+4} = x$$

$$\left(\sqrt{3x+4}\right)^2 = x^2 \quad \text{Square both sides.}$$

$$3x + 4 = x^2 \quad \text{Simplify.}$$

$$x^2 - 3x - 4 = 0 \quad \text{Place in standard form.}$$

$$(x+1)(x-4)=0 \quad \text{Factor.}$$
$$x+1=0 \quad \text{or} \quad x-4=0 \quad \text{Zero-Product Property.}$$
$$x=-1 \quad \text{or} \quad x=4$$

There are two apparent solutions that need to be checked.

Check: $x=-1$: $\sqrt{3x+4}=\sqrt{3(-1)+4}=\sqrt{1}=1\neq -1$ $x=-1$ is extraneous

$x=4$: $\sqrt{3x-4}=\sqrt{3(4)+4}=\sqrt{16}=4$

The only solution is $x=4$.

NOW WORK PROBLEM 41.

Rational Exponents

Radicals are used to define rational exponents.

> If a is a real number and $n \geq 2$ is an integer, then
>
> $$\boxed{a^{1/n}=\sqrt[n]{a}} \qquad (3)$$
>
> provided that $\sqrt[n]{a}$ exists.

Note that if n is even and $a<0$, then $\sqrt[n]{a}$ and $a^{1/n}$ do not exist.

5 **EXAMPLE 7** **Using Equation (3)**

(a) $4^{1/2}=\sqrt{4}=2$
(b) $(-27)^{1/3}=\sqrt[3]{-27}=-3$
(c) $8^{1/2}=\sqrt{8}=2\sqrt{2}$
(d) $16^{1/3}=\sqrt[3]{16}=2\sqrt[3]{2}$

> If a is a real number and m and n are integers containing no common factors with $n \geq 2$, then
>
> $$\boxed{a^{m/n}=\sqrt[n]{a^m}=(\sqrt[n]{a})^m} \qquad (4)$$
>
> provided that $\sqrt[n]{a}$ exists.

We have two comments about equation (4):

1. The exponent m/n must be in lowest terms and n must be positive.
2. In simplifying $a^{m/n}$, either $\sqrt[n]{a^m}$ or $(\sqrt[n]{a})^m$ may be used. Generally, taking the root first, as in $(\sqrt[n]{a})^m$, is easier.

EXAMPLE 8 **Using Equation (4)**

(a) $4^{3/2}=(\sqrt{4})^3=2^3=8$
(b) $(-8)^{4/3}=(\sqrt[3]{-8})^4=(-2)^4=16$
(c) $(32)^{-2/5}=(\sqrt[5]{32})^{-2}=2^{-2}=\dfrac{1}{4}$

NOW WORK PROBLEM 45.

It can be shown that the laws of exponents hold for rational exponents. We use the laws of exponent in the next example.

EXAMPLE 9 Simplifying Expressions with Rational Exponents

Simplify each expression. Express your answer so that only positive exponents occur. Assume that the variables are positive.

(a) $\left(\dfrac{2x^{1/3}}{y^{2/3}}\right)^{-3}$ **(b)** $(x^{2/3}y)(x^{-2}y)^{1/2}$

SOLUTION **(a)** $$\left(\frac{2x^{1/3}}{y^{2/3}}\right)^{-3} = \left(\frac{y^{2/3}}{2x^{1/3}}\right)^{3} = \frac{(y^{2/3})^3}{(2x^{1/3})^3} = \frac{y^2}{2^3(x^{1/3})^3} = \frac{y^2}{8x}$$

(b) $$\begin{aligned}(x^{2/3}y)(x^{-2}y)^{1/2} &= (x^{2/3}y)[(x^{-2})^{1/2}y^{1/2}] \\ &= x^{2/3}yx^{-1}y^{1/2} = (x^{2/3}x^{-1})(y \cdot y^{1/2}) \\ &= x^{-1/3}y^{3/2} = \frac{y^{3/2}}{x^{1/3}}\end{aligned}$$

NOW WORK PROBLEM 61.

The next two examples illustrate some algebra that you will need to know for certain calculus problems.

EXAMPLE 10 Writing an Expression as a Single Quotient

Write the following expression as a single quotient in which only positive exponents appear.

$$(x^2 + 1)^{1/2} + x \cdot \frac{1}{2}(x^2 + 1)^{-1/2} \cdot 2x$$

SOLUTION

$$\begin{aligned}(x^2 + 1)^{1/2} + x \cdot \frac{1}{2}(x^2 + 1)^{-1/2} \cdot 2x &= (x^2 + 1)^{1/2} + \frac{x^2}{(x^2 + 1)^{1/2}} \\ &= \frac{(x^2 + 1)^{1/2}(x^2 + 1)^{1/2} + x^2}{(x^2 + 1)^{1/2}} \\ &= \frac{(x^2 + 1) + x^2}{(x^2 + 1)^{1/2}} \\ &= \frac{2x^2 + 1}{(x^2 + 1)^{1/2}}\end{aligned}$$

NOW WORK PROBLEM 65.

EXAMPLE 11 Factoring an Expression Containing Rational Exponents

Factor: $4x^{1/3}(2x + 1) + 2x^{4/3}$

SOLUTION We begin by looking for factors that are common to the two terms. Notice that 2 and $x^{1/3}$ are common factors. Then,

$$\begin{aligned}4x^{1/3}(2x + 1) + 2x^{4/3} &= 2x^{1/3}[2(2x + 1) + x] \\ &= 2x^{1/3}(5x + 2)\end{aligned}$$

NOW WORK PROBLEM 79.

EXERCISE A.8 Answers to Odd-Numbered Problems Begin on Page AN-108.

In Problems 1–28, simplify each expression. Assume that all variables are positive when they appear.

1. $\sqrt[3]{27}$
2. $\sqrt[4]{16}$
3. $\sqrt[3]{-8}$
4. $\sqrt[3]{-1}$
5. $\sqrt{8}$
6. $\sqrt[3]{54}$
7. $\sqrt[3]{-8x^4}$
8. $\sqrt[4]{48x^5}$
9. $\sqrt[4]{x^{12}y^8}$
10. $\sqrt[5]{x^{10}y^5}$
11. $\sqrt[4]{\dfrac{x^9y^7}{xy^3}}$
12. $\sqrt[3]{\dfrac{3xy^2}{81x^4y^2}}$
13. $\sqrt{36x}$
14. $\sqrt{9x^5}$
15. $\sqrt{3x^2}\sqrt{12x}$
16. $\sqrt{5x}\,\sqrt{20x^3}$
17. $(\sqrt{5}\,\sqrt[3]{9})^2$
18. $(\sqrt[3]{3}\,\sqrt{10})^4$
19. $(3\sqrt{6})(2\sqrt{2})$
20. $(5\sqrt{8})(-3\sqrt{3})$
21. $(\sqrt{3} + 3)(\sqrt{3} - 1)$
22. $(\sqrt{5} - 2)(\sqrt{5} + 3)$
23. $(\sqrt{x} - 1)^2$
24. $(\sqrt{x} + \sqrt{5})^2$
25. $3\sqrt{2} - 4\sqrt{8}$
26. $\sqrt[3]{-x^4} + \sqrt[3]{8x}$
27. $\sqrt[3]{16x^4} - \sqrt[3]{2x}$
28. $\sqrt[4]{32x} + \sqrt[4]{2x^5}$

In Problems 29–40, rationalize the denominator of each expression. Assume that all variables are positive when they appear.

29. $\dfrac{1}{\sqrt{2}}$
30. $\dfrac{6}{\sqrt[3]{4}}$
31. $\dfrac{-\sqrt{3}}{\sqrt{5}}$
32. $\dfrac{-\sqrt[3]{3}}{\sqrt{8}}$
33. $\dfrac{\sqrt{3}}{5 - \sqrt{2}}$
34. $\dfrac{\sqrt{2}}{\sqrt{7} + 2}$
35. $\dfrac{2 - \sqrt{5}}{2 + 3\sqrt{5}}$
36. $\dfrac{\sqrt{3} - 1}{2\sqrt{3} + 3}$
37. $\dfrac{5}{\sqrt[3]{2}}$
38. $\dfrac{-2}{\sqrt[3]{9}}$
39. $\dfrac{\sqrt{x+h} - \sqrt{x}}{\sqrt{x+h} + \sqrt{x}}$
40. $\dfrac{\sqrt{x+h} + \sqrt{x-h}}{\sqrt{x+h} - \sqrt{x-h}}$

In Problems 41–44, solve each equation.

41. $\sqrt[3]{2t - 1} = 2$
42. $\sqrt[3]{3t + 1} = -2$
43. $\sqrt{15 - 2x} = x$
44. $\sqrt{12 - x} = x$

In Problems 45–56, simplify each expression.

45. $8^{2/3}$
46. $4^{3/2}$
47. $(-27)^{1/3}$
48. $16^{3/4}$
49. $16^{3/2}$
50. $64^{3/2}$
51. $9^{-3/2}$
52. $25^{-5/2}$
53. $\left(\dfrac{9}{8}\right)^{3/2}$
54. $\left(\dfrac{27}{8}\right)^{2/3}$
55. $\left(\dfrac{8}{9}\right)^{-3/2}$
56. $\left(\dfrac{8}{27}\right)^{-2/3}$

In Problems 57–64 simplify each expression. Express your answer so that only positive exponents occur. Assume that the variables are positive.

57. $x^{3/4}x^{1/3}x^{-1/2}$
58. $x^{2/3}x^{1/2}x^{-1/4}$
59. $(x^3y^6)^{1/3}$
60. $(x^4y^8)^{3/4}$
61. $(x^2y)^{1/3}(xy^2)^{2/3}$
62. $(xy)^{1/4}(x^2y^2)^{1/2}$
63. $(16x^2y^{-1/3})^{3/4}$
64. $(4x^{-1}y^{1/3})^{3/2}$

In Problems 65–78, expressions that occur in calculus are given. Write each expression as a single quotient in which only positive exponents and/or radicals appear.

65. $\dfrac{x}{(1 + x)^{1/2}} + 2(1 + x)^{1/2}, \quad x > -1$
66. $\dfrac{1 + x}{2x^{1/2}} + x^{1/2}, \quad x > 0$
67. $2x(x^2 + 1)^{1/2} + x^2 \cdot \dfrac{1}{2}(x^2 + 1)^{-1/2} \cdot 2x$
68. $(x + 1)^{1/3} + x \cdot \dfrac{1}{3}(x + 1)^{-2/3}, \quad x \neq -1$

69. $\sqrt{4x+3}\cdot\dfrac{1}{2\sqrt{x-5}}+\sqrt{x-5}\cdot\dfrac{1}{5\sqrt{4x+3}},\quad x>5$

70. $\dfrac{\sqrt[3]{8x+1}}{3\sqrt[3]{(x-2)^2}}+\dfrac{\sqrt[3]{x-2}}{24\sqrt[3]{(8x+1)^2}},\quad x\neq 2, x\neq -\dfrac{1}{8}$

71. $\dfrac{\sqrt{1+x}-x\cdot\dfrac{1}{2\sqrt{1+x}}}{1+x},\quad x>-1$

72. $\dfrac{\sqrt{x^2+1}-x\cdot\dfrac{2x}{2\sqrt{x^2+1}}}{x^2+1}$

73. $\dfrac{(x+4)^{1/2}-2x(x+4)^{-1/2}}{x+4},\quad x>-4$

74. $\dfrac{(9-x^2)^{1/2}+x^2(9-x^2)^{-1/2}}{9-x^2},\quad -3<x<3$

75. $\dfrac{\dfrac{x^2}{(x^2-1)^{1/2}}-(x^2-1)^{1/2}}{x^2},\quad x<-1 \text{ or } x>1$

76. $\dfrac{(x^2+4)^{1/2}-x^2(x^2+4)^{-1/2}}{x^2+4}$

77. $\dfrac{\dfrac{1+x^2}{2\sqrt{x}}-2x\sqrt{x}}{(1+x^2)^2},\quad x>0$

78. $\dfrac{2x(1-x^2)^{1/3}+\dfrac{2}{3}x^3(1-x^2)^{-2/3}}{(1-x^2)^{2/3}},\quad x\neq -1, x\neq 1$

In Problems 79–90, expressions that occur in calculus are given. Factor each expression. Express your answer so that only positive exponents occur.

79. $(x+1)^{3/2}+x\cdot\dfrac{3}{2}(x+1)^{1/2},\quad x\geq -1$

80. $(x^2+4)^{4/3}+x\cdot\dfrac{4}{3}(x^2+4)^{1/3}\cdot 2x$

81. $6x^{1/2}(x^2+x)-8x^{3/2}-8x^{1/2},\quad x\geq 0$

82. $6x^{1/2}(2x+3)+x^{3/2}\cdot 8,\quad x\geq 0$

83. $3(x^2+4)^{4/3}+x\cdot 4(x^2+4)^{1/3}\cdot 2x$

84. $2x(3x+4)^{4/3}+x^2\cdot 4(3x+4)^{1/3}$

85. $4(3x+5)^{1/3}(2x+3)^{3/2}+3(3x+5)^{4/3}(2x+3)^{1/2},\quad x\geq -\dfrac{3}{2}$

86. $6(6x+1)^{1/3}(4x-3)^{3/2}+6(6x+1)^{4/3}(4x-3)^{1/2},\quad x\geq \dfrac{3}{4}$

87. $3x^{-1/2}+\dfrac{3}{2}x^{1/2},\quad x>0$

88. $8x^{1/3}-4x^{-2/3},\quad x\neq 0$

89. $x\left(\dfrac{1}{2}\right)(8-x^2)^{-1/2}(-2x)+(8-x^2)^{1/2}$

90. $2x(1-x^2)^{3/2}+x^2\left(\dfrac{3}{2}\right)(1-x^2)^{1/2}(-2x)$

A.9 Geometry Review

OBJECTIVES
1 Use the Pythagorean Theorem and its converse
2 Know geometry formulas

In this section we review some topics studied in geometry that we shall need for calculus.

Pythagorean Theorem

Use the Pythagorean Theorem and its converse 1

The *Pythagorean Theorem* is a statement about *right triangles.* A **right triangle** is one that contains a **right angle,** that is, an angle of 90°. The side of the triangle opposite the 90° angle is called the **hypotenuse;** the remaining two sides are called **legs.** In Figure 22 we have used c to represent the length of the hypotenuse and a and b to represent the lengths of the legs. Notice the use of the symbol ⌜ to show the 90° angle. We now state the Pythagorean Theorem.

FIGURE 22

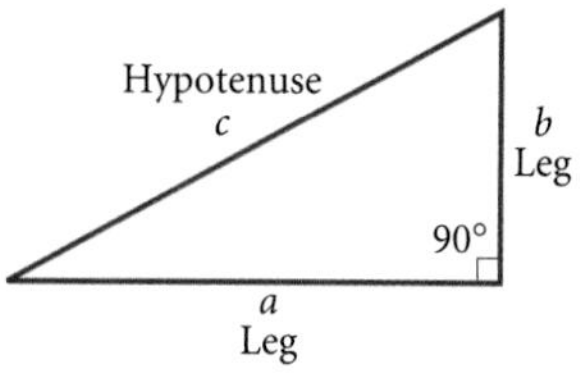

Pythagorean Theorem

In a right triangle, the square of the length of the hypotenuse is equal to the sum of the squares of the lengths of the legs. That is, in the right triangle shown in Figure 22,

$$\boxed{c^2 = a^2 + b^2} \tag{1}$$

EXAMPLE 1 Finding the Hypotenuse of a Right Triangle

In a right triangle, one leg is of length 4 and the other is of length 3. What is the length of the hypotenuse?

SOLUTION Since the triangle is a right triangle, we use the Pythagorean Theorem with $a = 4$ and $b = 3$ to find the length c of the hypotenuse. From equation (1), we have

$$c^2 = a^2 + b^2$$
$$c^2 = 4^2 + 3^2 = 16 + 9 = 25$$
$$c = \sqrt{25} = 5$$

NOW WORK PROBLEM 3.

The converse of the Pythagorean Theorem is also true.

Converse of the Pythagorean Theorem

In a triangle, if the square of the length of one side equals the sum of the squares of the lengths of the other two sides, then the triangle is a right triangle. The 90° angle is opposite the longest side.

EXAMPLE 2 Verifying That a Triangle Is a Right Triangle

FIGURE 23

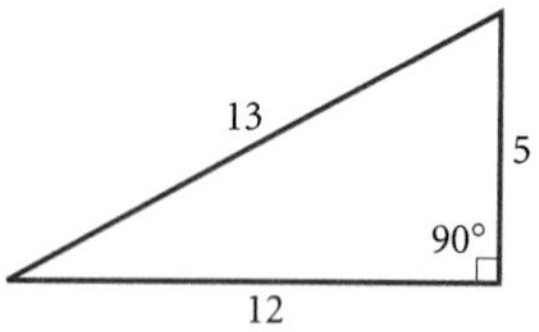

Show that a triangle whose sides are of lengths 5, 12, and 13 is a right triangle. Identify the hypotenuse.

SOLUTION We square the lengths of the sides.

$$5^2 = 25, \qquad 12^2 = 144, \qquad 13^2 = 169$$

Notice that the sum of the first two squares (25 and 144) equals the third square (169). So the triangle is a right triangle. The longest side, 13, is the hypotenuse. See Figure 23.

NOW WORK PROBLEM 11.

EXAMPLE 3 Applying the Pythagorean Theorem

The tallest inhabited building in the world is the Sears Tower in Chicago. If the observation tower is 1450 feet above ground level, how far can a person standing in the observation tower see (with the aid of a telescope)? Use 3960 miles for the radius of Earth. See Figure 24.

FIGURE 24

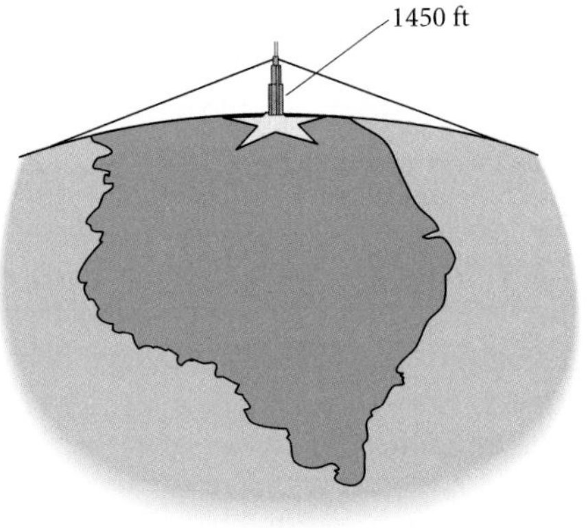

Source: Council on Tall Buildings and Urban Habitat (1997): Sears Tower No. 1 for tallest roof (1450 ft) and tallest occupied floor (1431 ft).

SOLUTION From the center of Earth, draw two radii: one through the Sears Tower and the other to the farthest point a person can see from the tower. See Figure 25. Apply the Pythagorean Theorem to the right triangle.

FIGURE 25

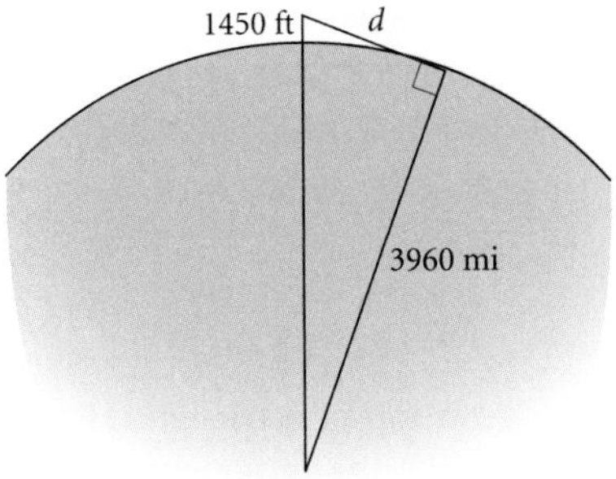

Since 1 mile = 5280 feet, 1450 feet $= \dfrac{1450}{5280}$ miles. Then we have

$$d^2 + (3960)^2 = \left(3960 + \frac{1450}{5280}\right)^2$$

$$d^2 = \left(3960 + \frac{1450}{5280}\right)^2 - (3960)^2 \approx 2175.08$$

$$d \approx 46.64$$

A person can see about 47 miles from the observation tower.

NOW WORK PROBLEM 37.

Geometry Formulas

Know geometry formulas 2

Certain formulas from geometry are useful in solving calculus problems. We list some of these formulas next.

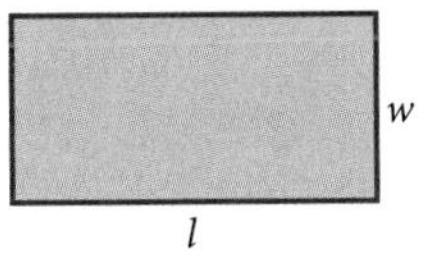

For a rectangle of length l and width w,

$$\text{Area} = lw \qquad \text{Perimeter} = 2l + 2w$$

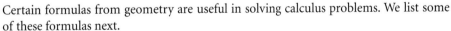

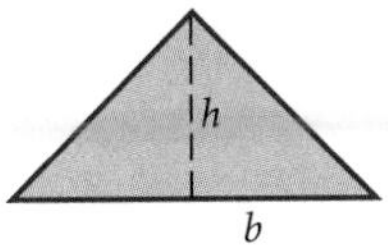

For a triangle with base b and altitude h,

$$\text{Area} = \frac{1}{2}bh$$

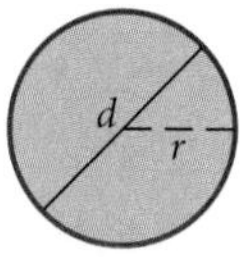

For a circle of radius r (diameter $d = 2r$),

$$\text{Area} = \pi r^2 \qquad \text{Circumference} = 2\pi r = \pi d$$

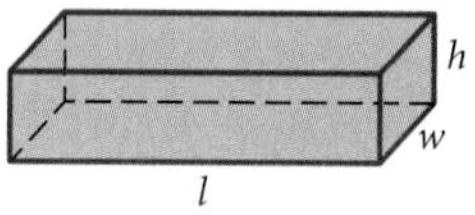

For a closed rectangular box of length l, width w, and height h,

$$\text{Volume} = lwh \qquad \text{Surface area} = 2lw + 2lh + 2wh$$

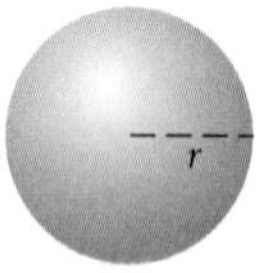

For a sphere of radius r,

$$\text{Volume} = \frac{4}{3}\pi r^3 \qquad \text{Surface area} = 4\pi r^2$$

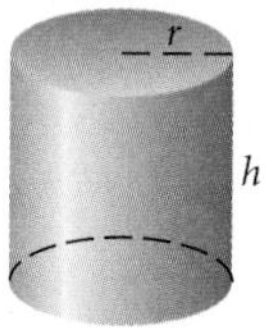

For a right circular cylinder of height h and radius r,

$$\text{Volume} = \pi r^2 h \qquad \text{Surface area} = 2\pi r^2 + 2\pi r h$$

NOW WORK PROBLEM 19.

EXAMPLE 4 Using Geometry Formulas

A Christmas tree ornament is in the shape of a semicircle on top of a triangle. How many square centimeters (cm) of copper are required to make the ornament if the height of the triangle is 6 cm and the base is 4 cm?

FIGURE 26

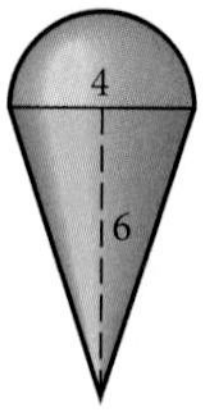

SOLUTION See Figure 26. The amount of copper required equals the shaded area. This area is the sum of the area of the triangle and the semicircle. The triangle has height $h = 6$ and base $b = 4$. The semicircle has diameter $d = 4$, so its radius is $r = 2$.

$$\begin{aligned}\text{Area} &= \text{Area of triangle} + \text{Area of semicircle}\\ &= \frac{1}{2}bh + \frac{1}{2}\pi r^2 = \frac{1}{2}(4)(6) + \frac{1}{2}\pi \cdot 2^2 \qquad b = 4;\, h = 6;\, r = 2.\\ &= 12 + 2\pi \approx 18.28\ \text{cm}^2\end{aligned}$$

About 18.28 cm^2 of copper are required.

NOW WORK PROBLEM 33.

EXERCISE A.9 Answers to Odd-Numbered Problems Begin on Page AN-109.

In Problems 1–6, the lengths of the legs of a right triangle are given. Find the hypotenuse.

1. $a = 5, \quad b = 12$

2. $a = 6, \quad b = 8$

3. $a = 10, \quad b = 24$

4. $a = 4, \quad b = 3$

5. $a = 7, \quad b = 24$

6. $a = 14, \quad b = 48$

In Problems 7–14, the lengths of the sides of a triangle are given. Determine which are right triangles. For those that are, identify the hypotenuse.

7. 3, 4, 5

8. 6, 8, 10

9. 4, 5, 6

10. 2, 2, 3

11. 7, 24, 25

12. 10, 24, 26

13. 6, 4, 3

14. 5, 4, 7

15. Find the area A of a rectangle with length 4 inches and width 2 inches.

16. Find the area A of a rectangle with length 9 centimeters and width 4 centimeters.

17. Find the area A of a triangle with height 4 inches and base 2 inches.

18. Find the area A of a triangle with height 9 centimeters and base 4 centimeters.

19. Find the area A and circumference C of a circle of radius 5 meters.

20. Find the area A and circumference C of a circle of radius 2 feet.

21. Find the volume V and surface area S of a rectangular box with length 8 feet, width 4 feet, and height 7 feet.

22. Find the volume V and surface area S of a rectangular box with length 9 inches, width 4 inches, and height 8 inches.

23. Find the volume V and surface area S of a sphere of radius 4 centimeters.

24. Find the volume V and surface area S of a sphere of radius 3 feet.

25. Find the volume V and surface area S of a right circular cylinder with radius 9 inches and height 8 inches.

26. Find the volume V and surface area S of a right circular cylinder with radius 8 inches and height 9 inches.

In Problems 27–30, find the area of the shaded region.

27.

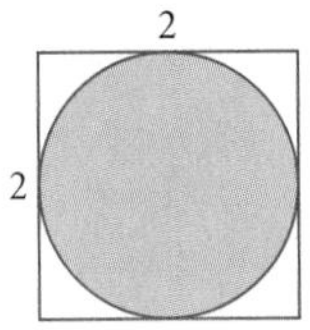

28.

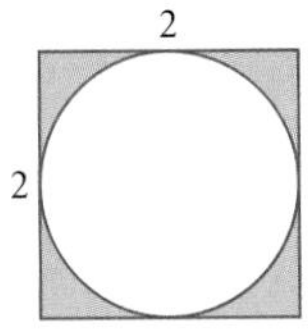

29.

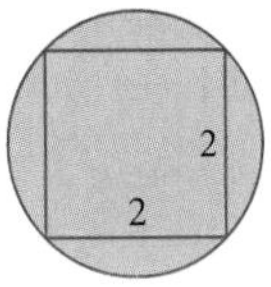

30.

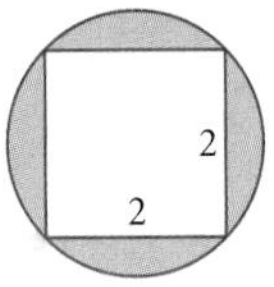

31. How many feet does a wheel with a diameter of 16 inches travel after four revolutions?

32. How many revolutions will a circular disk with a diameter of 4 feet have completed after it has rolled 20 feet?

33. In the figure shown, $ABCD$ is a square, with each side of length 6 feet. The width of the border (shaded portion) between the outer square $EFGH$ and $ABCD$ is 2 feet. Find the area of the border.

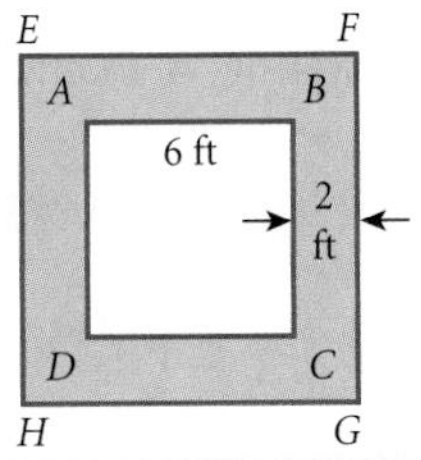

34. Refer to the figure above Problem 35. Square $ABCD$ has an area of 100 square feet; square $BEFG$ has an area of 16 square feet. What is the area of the triangle CGF?

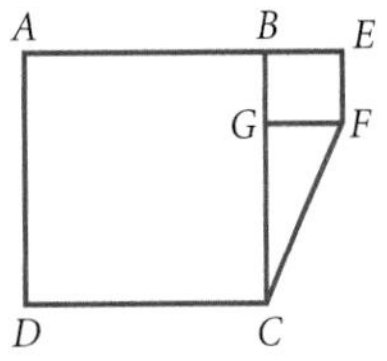

35. Architecture A Norman window consists of a rectangle surmounted by a semicircle. Find the area of the Norman window shown in the illustration. How much wood frame is needed to enclose the window?

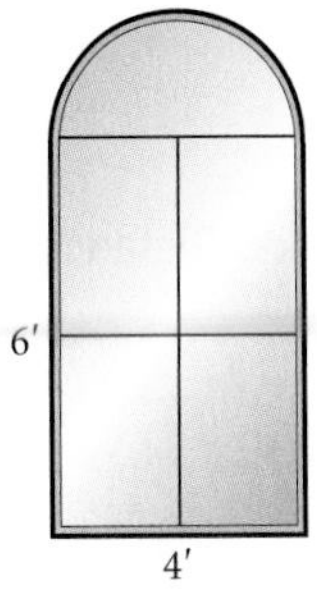

36. Construction A circular swimming pool, 20 feet in diameter, is enclosed by a wooden deck that is 3 feet wide. What is the area of the deck? How much fence is required to enclose the deck?

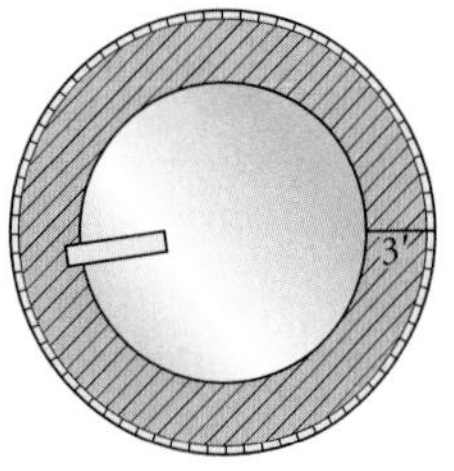

In Problems 37–39, use the facts that the radius of Earth is 3960 miles and 1 mile = 5280 feet.

37. **How Far Can You See?** The conning tower of the U.S.S. *Silversides,* a World War II submarine now permanently stationed in Muskegon, Michigan, is approximately 20 feet above sea level. How far can you see from the conning tower?

38. **How Far Can You See?** A person who is 6 feet tall is standing on the beach in Fort Lauderdale, Florida, and looks out onto the Atlantic Ocean. Suddenly, a ship appears on the horizon. How far is the ship from shore?

39. **How Far Can You See?** The deck of a destroyer is 100 feet above sea level. How far can a person see from the deck? How far can a person see from the bridge, which is 150 feet above sea level?

40. Suppose that m and n are positive integers with $m > n$. If $a = m^2 - n^2$, $b = 2mn$, and $c = m^2 + n^2$, show that a, b, and c are the lengths of the sides of a right triangle. (This formula can be used to find the sides of a right triangle that are integers, such as 3, 4, 5; 5, 12, 13; and so on. Such triplets of integers are called **Pythagorean triples.**)

41. You have 1000 feet of flexible pool siding and wish to construct a swimming pool. Experiment with rectangular-shaped pools with perimeters of 1000 feet. How do their areas vary? What is the shape of the rectangle with the largest area? Now compute the area enclosed by a circular pool with a perimeter (circumference) of 1000 feet. What would be your choice of shape for the pool? If rectangular, what is your preference for dimensions? Justify your choice. If your only consideration is to have a pool that encloses the most area, what shape should you use?

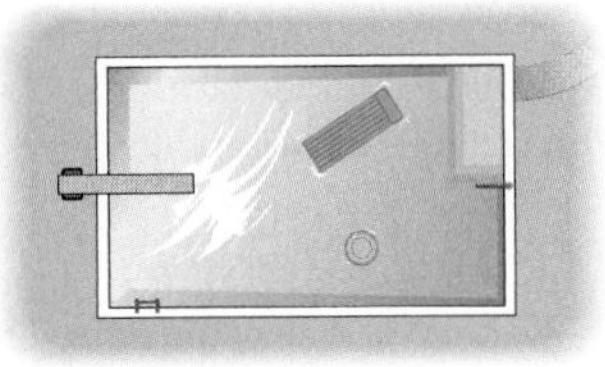

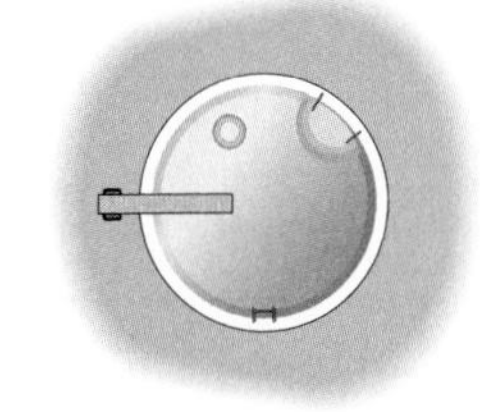

A.10 The Distance Formula

PREPARING FOR THIS SECTION *Before getting started, review the following:*

- Algebra Review (Appendix A, Section A.2, pp. A-15–A-23)
- Geometry Review (Appendix A, Section A.9, pp. A-80–A-82
- Rectangular Coordinates (Chapter 1, Section 1.1, pp. 2–3

OBJECTIVE 1 Use the Distance Formula

Use the distance formula 1 If the same units of measurement, such as inches or centimeters, are used for both the x-axis and the y-axis, then all distances in the xy-plane can be measured using this unit of measurement.

EXAMPLE 1 Finding the Distance between Two Points

Find the distance d between the points $(1, 3)$ and $(5, 6)$.

SOLUTION First we plot the points $(1, 3)$ and $(5, 6)$ as shown in Figure 27(a). Then we draw a horizontal line from $(1, 3)$ to $(5, 3)$ and a vertical line from $(5, 3)$ to $(5, 6)$, forming a right triangle, as in Figure 27(b). One leg of the triangle is of length 4 and the other is of length 3. By the Pythagorean Theorem, the square of the distance d that we seek is

$$d^2 = 4^2 + 3^2 = 16 + 9 = 25$$
$$d = 5$$

FIGURE 27

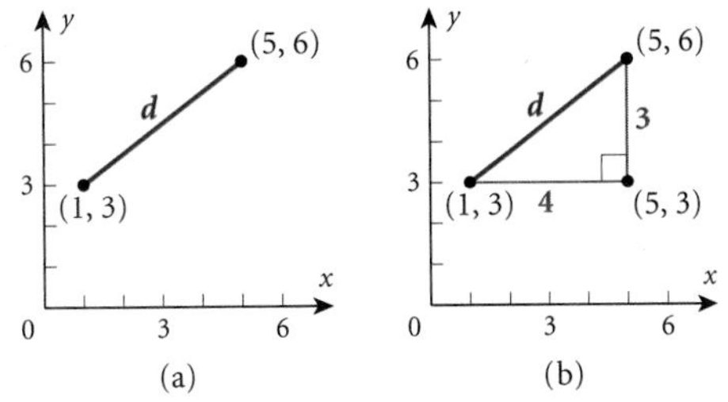

The **distance formula** provides a straightforward method for computing the distance between two points.

Distance Formula

The distance between two points $P_1 = (x_1, y_1)$ and $P_2 = (x_2, y_2)$, denoted by $d(P_1, P_2)$, is

$$d(P_1, P_2) = \sqrt{(x_2 - x_1)^2 + (y_2 - y_1)^2} \quad (1)$$

That is, to compute the distance between two points, find the difference of the x-coordinates, square it, and add this to the square of the difference of the y-coordinates. The square root of this sum is the distance. See Figure 28.

FIGURE 28

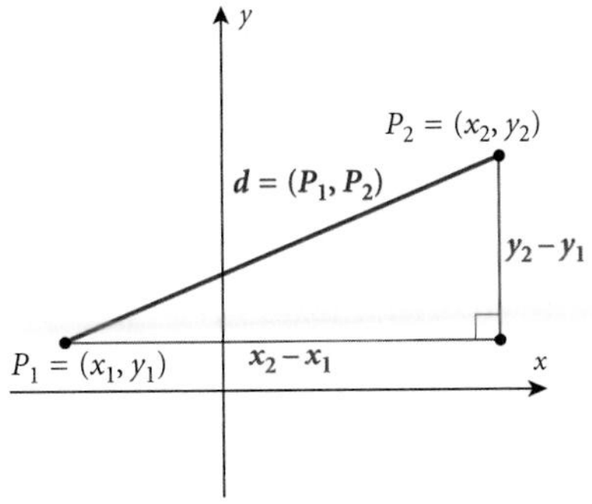

EXAMPLE 2 Finding the Distance between Two Points

Find the distance d between the points $(-4, 5)$ and $(3, 2)$.

SOLUTION Using the distance formula (1), the solution is obtained as follows:

$$d = \sqrt{[3 - (-4)]^2 + (2 - 5)^2} = \sqrt{7^2 + (-3)^2}$$
$$= \sqrt{49 + 9} = \sqrt{58} \approx 7.62$$

NOW WORK PROBLEMS 5 AND 9.

The distance between two points $P_1 = (x_1, y_1)$ and $P_2 = (x_2, y_2)$ is never a negative number. Furthermore, the distance between two points is 0 only when the points are identical, that is, when $x_1 = x_2$ and $y_1 = y_2$. Also, because $(x_2 - x_1)^2 = (x_1 - x_2)^2$ and $(y_2 - y_1)^2 = (y_1 - y_2)^2$, it makes no difference whether the distance is computed from P_1 to P_2 or from P_2 to P_1; that is, $d(P_1, P_2) = d(P_2, P_1)$.

Rectangular coordinates enable us to translate geometry problems into algebra problems, and vice versa. The next example shows how algebra (the distance formula) can be used to solve geometry problems.

EXAMPLE 3 Using Algebra to Solve Geometry Problems

Consider the three points $A = (-2, 1)$, $B = (2, 3)$, and $C = (3, 1)$.

(a) Plot each point and form the triangle ABC.
(b) Find the length of each side of the triangle.
(c) Verify that the triangle is a right triangle.
(d) Find the area of the triangle.

SOLUTION **(a)** Points A, B, and C and triangle ABC are plotted in Figure 29.

(b) $d(A, B) = \sqrt{[2 - (-2)]^2 + (3 - 1)^2} = \sqrt{16 + 4} = \sqrt{20} = 2\sqrt{5}$

$d(B, C) = \sqrt{(3 - 2)^2 + (1 - 3)^2} = \sqrt{1 + 4} = \sqrt{5}$

$d(A, C) = \sqrt{[3 - (-2)]^2 + (1 - 1)^2} = \sqrt{25 + 0} = 5$

FIGURE 29

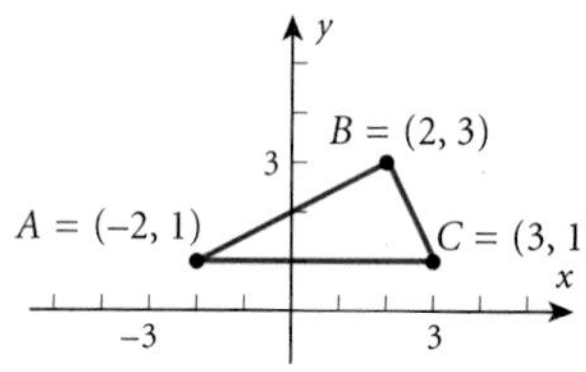

(c) To show that the triangle is a right triangle, we need to show that the sum of the squares of the lengths of two of the sides equals the square of the length of the third side. (Why is this sufficient?) Looking at Figure 29, it seems reasonable to conjecture that the right angle is at vertex B. To verify, we check to see whether

$$[d(A, B)]^2 + [d(B, C)]^2 = [d(A, C)]^2$$

We find that

$$[d(A, B)]^2 + [d(B, C)]^2 = (2\sqrt{5})^2 + (\sqrt{5})^2$$
$$= 20 + 5 = 25 = [d(A, C)]^2$$

so it follows from the converse of the Pythagorean Theorem that triangle ABC is a right triangle.

(d) Because the right angle is at B, the sides AB and BC form the base and altitude of the triangle. Its area is therefore

$$\text{Area} = \frac{1}{2}(\text{Base})(\text{Altitude}) = \frac{1}{2}(2\sqrt{5})(\sqrt{5}) = 5 \text{ square units}$$

NOW WORK PROBLEM 19.

EXERCISE A.10 Answers to Odd-Numbered Problems Begin on Page AN-109.

In Problems 1–14, find the distance $d(P_1, P_2)$ between the points P_1 and P_2.

1.
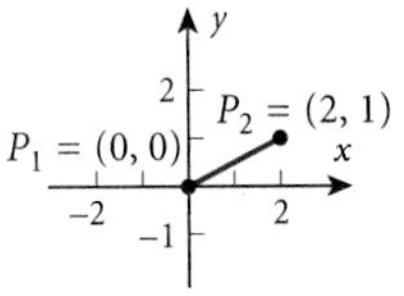

2.
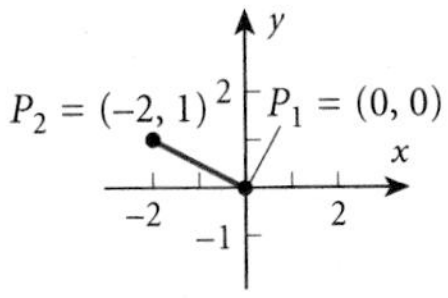

3.
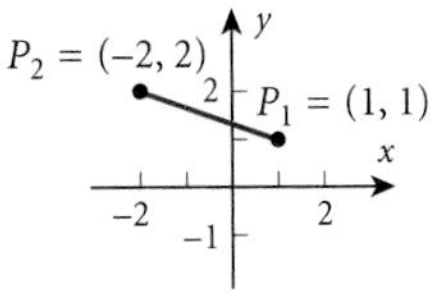

4.
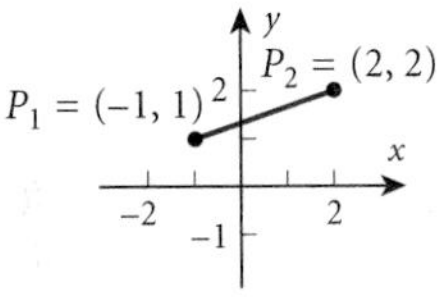

5. $P_1 = (3, -4);\ P_2 = (5, 4)$
6. $P_1 = (-1, 0);\ P_2 = (2, 4)$
7. $P_1 = (-3, 2);\ P_2 = (6, 0)$
8. $P_1 = (2, -3);\ P_2 = (4, 2)$
9. $P_1 = (4, -3);\ P_2 = (6, 4)$
10. $P_1 = (-4, -3);\ P_2 = (6, 2)$
11. $P_1 = (-0.2, 0.3);\ P_2 = (2.3, 1.1)$
12. $P_1 = (1.2, 2.3);\ P_2 = (-0.3, 1.1)$
13. $P_1 = (a, b);\ P_2 = (0, 0)$
14. $P_1 = (a, a);\ P_2 = (0, 0)$

In Problems 15–20, plot each point and form the triangle ABC. Verify that the triangle is a right triangle. Find its area.

15. $A = (-2, 5);\ B = (1, 3);\ C = (-1, 0)$
16. $A = (-2, 5);\ B = (12, 3);\ C = (10, -11)$
17. $A = (-5, 3);\ B = (6, 0);\ C = (5, 5)$
18. $A = (-6, 3);\ B = (3, -5);\ C = (-1, 5)$
19. $A = (4, -3);\ B = (0, -3);\ C = (4, 2)$
20. $A = (4, -3);\ B = (4, 1);\ C = (2, 1)$
21. Find all points having an x-coordinate of 2 whose distance from the point $(-2, -1)$ is 5.
22. Find all points having a y-coordinate of -3 whose distance from the point $(1, 2)$ is 13.
23. Find all points on the x-axis that are 5 units from the point $(4, -3)$.
24. Find all points on the y-axis that are 5 units from the point $(4, 4)$.

In Problems 25–28, find the length of the line segment. Assume that the endpoints of each line segment have integer coordinates.

25.
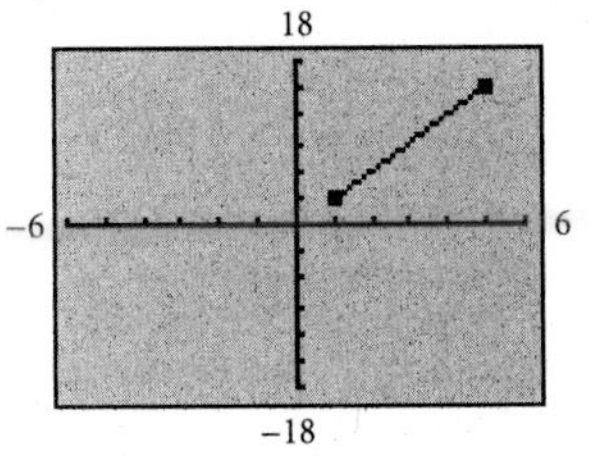

26.
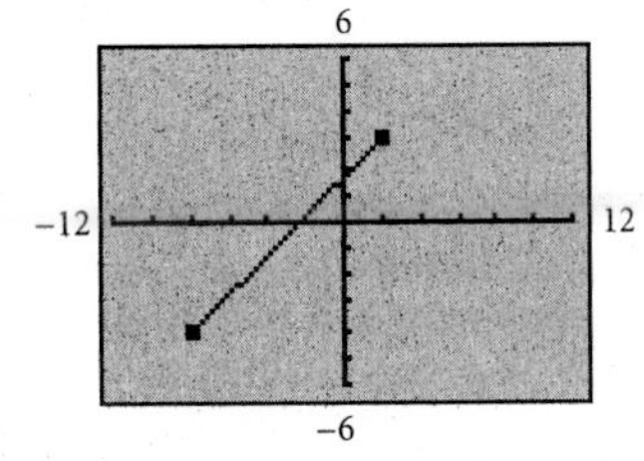

27.
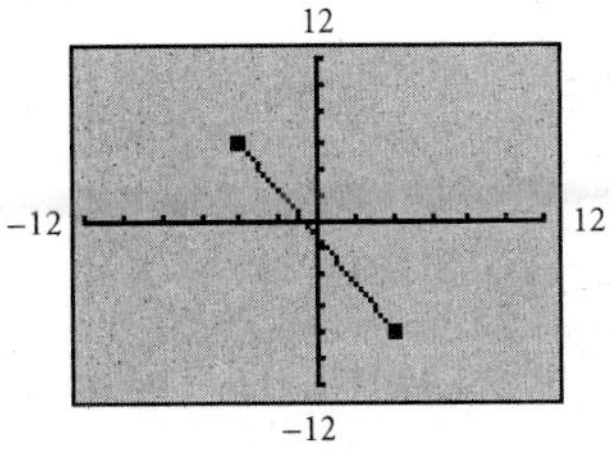

28.
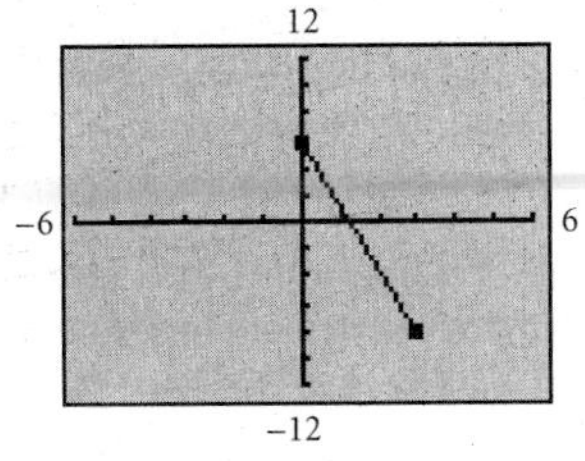

29. Baseball A major league baseball "diamond" is actually a square, 90 feet on a side (see the figure). What is the distance directly from home plate to second base (the diagonal of the square)?

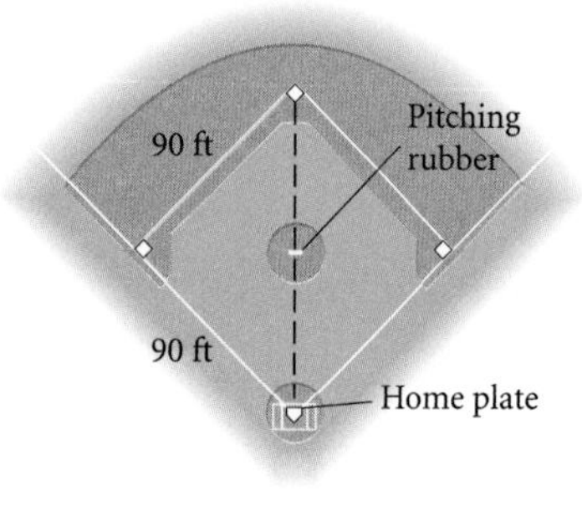

30. Little League Baseball The layout of a Little League playing field is a square, 60 feet on a side. How far is it directly from home plate to second base (the diagonal of the square)?

Source: Little League Baseball, Official Regulations and Playing Rules, 2003.

31. Baseball Refer to Problem 33. Overlay a rectangular coordinate system on a major league baseball diamond so that the origin is at home plate, the positive x-axis lies in the direction from home plate to first base, and the positive y-axis lies in the direction from home plate to third base.

(a) What are the coordinates of first base, second base, and third base? Use feet as the unit of measurement.
(b) If the right fielder is located at (310, 15), how far is it from the right fielder to second base?
(c) If the center fielder is located at (300, 300), how far is it from the center fielder to third base?

32. Little League Baseball Refer to Problem 34. Overlay a rectangular coordinate system on a Little League baseball diamond so that the origin is at home plate, the positive x-axis lies in the direction from home plate to first base, and the positive y-axis lies in the direction from home plate to third base.

(a) What are the coordinates of first base, second base, and third base? Use feet as the unit of measurement.
(b) If the right fielder is located at (180, 20), how far is it from the right fielder to second base?
(c) If the center fielder is located at (220, 220), how far is it from the center fielder to third base?

33. A Dodge Intrepid and a Mack truck leave an intersection at the same time. The Intrepid heads east at an average speed of 30 miles per hour, while the truck heads south at an average speed of 40 miles per hour. Find an expression for their distance apart d (in miles) at the end of t hours.

34. A hot-air balloon, headed due east at an average speed of 15 miles per hour and at a constant altitude of 100 feet, passes over an intersection (see the figure). Find an expression for the distance d (measured in feet) from the balloon to the intersection t seconds later

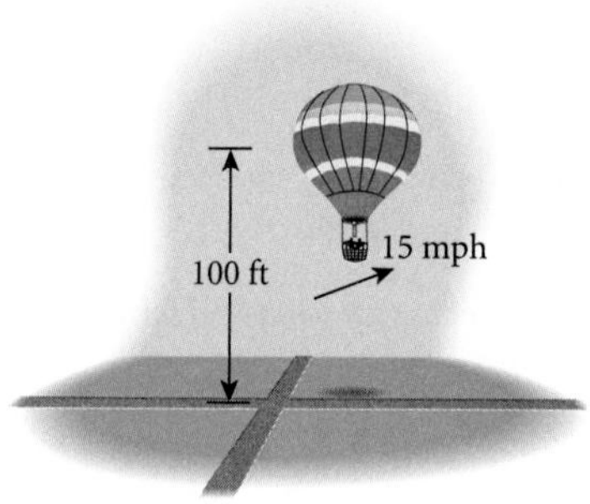

APPENDIX B

Using LINDO to Solve Linear Programming Problems

A number of available software packages can be used to solve linear programming problems. Because of the popularity of LINDO, we have chosen to include a brief section illustrating how it can be used to solve some linear programming problems. LINDO was designed to allow users to do simple problems in an easy, cost efficient manner. At the other extreme, LINDO has been used to solve real industrial programs involving more than 10,000 rows and several thousand variables. In solving such problems, LINDO uses a *revised* simplex method—one that exploits the special character of large linear programming problems. As a result, the intermediate tableaus obtained using LINDO may differ from those obtained manually. The latest version of LINDO, 6.1, is available for downloading at www.lindo.com/LND61@exe.

EXAMPLE 1 Using LINDO

Use LINDO to maximize

$$P = 3x_1 + 4x_2$$

subject to

$$\begin{aligned} 2x_1 + 4x_2 &\le 120 \\ 2x_1 + 2x_2 &\le 80 \\ x_1 \ge 0 \qquad x_2 &\ge 0 \end{aligned}$$

LINDO SOLUTION

The objective function

The constraints:

```
MAX       3 X1 + 4 X2
SUBJECT TO
                X1 + 4 X2 <=  120
                X1 + 2 X2 <=  80
END
```

Instruction to begin

```
: tabl
```

The initial tableau. SLK 2 is the slack variable associated with constraint number 2. ROW 1 ART is the objective row. The column for P is not written since pivoting does not affect it. The column on the far right is RHS. The bottom-row ART is added to LINDO to represent the pivot objective of minimizing the sum of the infeasibilities.

```
THE TABLEAU
     ROW  (BASIS)        X1        X2    SLK    2   SLK    3
       1 ART          -3.000    -4.000       .000       .000      .000
       2 SLK     2     2.000     4.000      1.000       .000   120.000
       3 SLK     3     2.000     2.000       .000      1.000    80.000
ART    3 ART          -3.000    -4.000       .000       .000      .000
```

Instruction to pivot The pivot element is in row 3, column X1. X1 will become a basic variable.

```
: piv
      X1 ENTERS AT VALUE  40.000  IN  ROW  3  OBJ. VALUE=  120.00
```

After pivoting, SLK 2 and X1 are basic variables. $P = 120$ at this stage; SLK 2 = 40; X1 = 40; SLK 3 = 0; X2 = 0.

```
: tabl
THE TABLEAU
     ROW  (BASIS)        X1        X2    SLK   2     SLK   3
      1 ART            .000    -1.000        .000     1.500    120.000
      2 SLK    2       .000     2.000       1.000    -1.000     40.000
      3          X1   1.000     1.000        .000      .500     40.000
```

The pivot element is in row 2, column X2. After pivoting, X2 and X1 are basic variables.

```
: piv
      X2 ENTERS AT VALUE  20.000  IN   ROW  2  OBJ. VALUE=  140.00
: tabl
THE TABLEAU
     ROW  (BASIS)        X1        X2    SLK   2    SLK   3
      1 ART            .000      .000        .500     1.000    140.000
      2          X2    .000     1.000        .500     -.500     20.000
      3          X1   1.000      .000       -.500     1.000     20.000
```

This is a final tableau. The solution is $P = 140$, X1 = 20, X2 = 20.

```
: piv
LP OPTIMUM FOUND AT STEP      2

        OBJECTIVE FUNCTION VALUE
        1)      140.00000

 VARIABLE         VALUE          REDUCED COST
       X1        20.000000           .000000
       X2        20.000000           .000000
```

The value of the slack variables are SLK 2 = 0, SLK 3 = 0.

```
        ROW   SLACK OR SURPLUS     DUAL PRICES
         2)          .000000         .500000
         3)          .000000        1.000000
```

Two pivots were used.

```
NO. ITERATIONS=       2
```

Compare the steps in Example 1, using LINDO, with the steps given in Example 1, Section 4.2.

EXAMPLE 2 Using LINDO

Use LINDO to maximize

$$P = 20x_1 + 15x_2$$

subject to

$$\begin{aligned} x_1 + x_2 &\geq 7 \\ 9x_1 + 5x_2 &\leq 45 \\ 2x_1 + x_2 &\geq 8 \\ x_1 \geq 0 \quad x_2 &\geq 0 \end{aligned}$$

SOLUTION

```
  MAX      20 X1 + 15 X2
  SUBJECT TO
            X1 + X2 >=    7
            9 X1 + 5 X2 <=    45
            2 X1 + X2 >=    8
  END
: tabl
THE TABLEAU
     ROW  (BASIS)          X1          X2    SLK   2    SLK   3    SLK   4
      1  ART          -20.000     -15.000      .000      .000      .000      .000
      2  SLK   2       -1.000      -1.000     1.000      .000      .000    -7.000
      3  SLK   3        9.000       5.000      .000     1.000      .000    45.000
      4  SLK   4       -2.000      -1.000      .000      .000     1.000    -8.000
ART   4  ART           -3.000      -2.000     1.000      .000     1.000   -15.000
```

```
: piv
        X1 ENTERS AT VALUE    5.0000      IN ROW    3 OBJ. VALUE =   100.00

: tabl
THE TABLEAU

     ROW  (BASIS)          X1       X2      SLK   2      SLK   3     SLK   4
       1 ART              .000   -3.889       .000       2.222        .000   100.000
       2 SLK    2         .000    -.444      1.000        .111        .000    -2.000
       3        X1       1.000     .556       .000        .111        .000     5.000
       4 SLK    4         .000     .111       .000        .222       1.000     2.000
ART    4 ART              .000    -.444      1.000        .111        .000    -2.000

: piv
        X2 ENTERS AT VALUE    4.5000      IN ROW    2 OBJ. VALUE =   117.50

: tabl
THE TABLEAU

     ROW  (BASIS)          X1       X2     SLK   2     SLK   3      SLK   4
       1 ART              .000     .000    -8.750      1.250        .000   117.500
       2        X2        .000    1.000    -2.250      -.250        .000     4.500
       3        X1       1.000     .000     1.250       .250        .000     2.500
       4 SLK    4         .000     .000      .250       .250       1.000     1.500
ART    4 ART              .000     .000    -8.750      1.250

: piv
SLK    2 ENTERS AT VALUE      2.0000      IN ROW   3 OBJ. VALUE =   117.50

: tabl
THE TABLEAU

     ROW  (BASIS)          X1       X2     SLK   2     SLK   3      SLK   4
       1 ART             7.000     .000      .000      3.000        .000   135.000
       2        X2       1.800    1.000      .000       .200        .000     9.000
       3 SLK    2         .800     .000     1.000       .200        .000     2.000
       4 SLK    4        -.200     .000      .000       .200       1.000     1.000

: piv
 LP OPTIMUM FOUND AT STEP        3

      OBJECTIVE FUNCTION VALUE
      1)       135.00000

 VARIABLE         VALUE          REDUCED COST
      X1          .000000          7.000000
      X2         9.000000           .000000

      ROW    SLACK OR SURPLUS     DUAL PRICES
      2)        2.000000           .000000
      3)         .000000          3.000000
      4)        1.000000           .000000

NO. ITERATIONS =        3
```

Compare the final tableau found using LINDO with the final tableau of Example 1, Section 4.4, namely,

$$
\begin{array}{c}
\\ s_1 \\ x_2 \\ s_3 \\ P
\end{array}
\begin{array}{c}
\begin{array}{cccccccc} P & x_1 & x_2 & s_1 & s_2 & s_3 & \text{RHS} \end{array} \\
\left[\begin{array}{cccccc|c}
0 & \frac{4}{5} & 0 & 1 & \frac{1}{5} & 0 & 2 \\
0 & \frac{9}{5} & 1 & 0 & \frac{1}{5} & 0 & 9 \\
0 & -\frac{1}{5} & 0 & 0 & \frac{1}{5} & 1 & 1 \\
\hline
1 & 7 & 0 & 0 & 3 & 0 & 135
\end{array}\right]
\end{array}
$$

The solution, in both cases, is $P = 135$, $x_2 = 9$, $x_1 = 0$, $s_1 = \text{SLK } 2 = 2$, $s_2 = \text{SLK } 3 = 0$, $s_3 = \text{SLK } 4 = 1$. Notice also that the intermediate tableaus are different. This is because LINDO uses a *revised* simplex method that requires additional computations not easily performed manually.

EXAMPLE 3 Using LINDO

Use LINDO to minimize

$$z = 5x_1 + 6x_2$$

subject to

$$\begin{aligned} x_1 + x_2 &\le 10 \\ x_1 + 2x_2 &\ge 12 \\ 2x_1 + x_2 &\ge 12 \\ x_1 &\ge 3 \\ x_1 \ge 0 \quad x_2 &\ge 0 \end{aligned}$$

SOLUTION

```
MIN      5 X1 + 6 X2
SUBJECT TO
            X1 + X2 <= 10
            X1 + 2 X2 >= 12
            2 X1 + X2 >= 12
            X1 >= 3
END
: tabl
THE TABLEAU
      ROW  (BASIS)        X1        X2     SLK  2     SLK  3     SLK  4     SLK  5
       1  ART          5.000     6.000       .000       .000       .000       .000       .000
       2  SLK    2     1.000     1.000      1.000       .000       .000       .000     10.000
       3  SLK    3    -1.000    -2.000       .000      1.000       .000       .000    -12.000
       4  SLK    4    -2.000    -1.000       .000       .000      1.000       .000    -12.000
       5  SLK    5    -1.000      .000       .000       .000       .000      1.000     -3.000
ART    5  ART         -4.000    -3.000       .000      1.000      1.000      1.000    -27.000
: piv
      X1 ENTERS AT VALUE   6.0000    IN ROW   4 OBJ. VALUE= -30.000
: tabl
THE TABLEAU
      ROW  (BASIS)        X1        X2     SLK  2     SLK  3     SLK  4     SLK  5
       1  ART           .000     3.500       .000       .000      2.500       .000    -30.000
       2  SLK    2      .000      .500      1.000       .000       .500       .000      4.000
       3  SLK    3      .000    -1.500       .000      1.000      -.500       .000     -6.000
       4        X1     1.000      .500       .000       .000      -.500       .000      6.000
       5  SLK    5      .000      .500       .000       .000      -.500      1.000      3.000
ART    5  ART           .000    -1.500       .000      1.000      -.500       .000     -6.000
: piv
      X2 ENTERS AT VALUE   4.0000    IN ROW   3 OBJ. VALUE= -44.000
: tabl
THE TABLEAU
      ROW  (BASIS)        X1        X2     SLK  2     SLK  3     SLK  4     SLK  5
       1  ART           .000      .000       .000      2.333      1.333       .000    -44.000
       2  SLK    2      .000      .000      1.000       .333       .333       .000      2.000
       3        X2      .000     1.000       .000      -.667       .333       .000      4.000
       4        X1     1.000      .000       .000       .333      -.667       .000      4.000
       5  SLK    5      .000      .000       .000       .333      -.667      1.000      1.000
ART    5  ART           .000      .000       .000      2.333      1.333
```

```
: piv
LP OPTIMUM FOUND AT STEP        2
OBJECTIVE FUNCTION VALUE
        1)      44.000000

 VARIABLE          VALUE          REDUCED COST
       X1         4.000000            .000000
       X2         4.000000            .000000
      ROW    SLACK OR SURPLUS      DUAL PRICES
       2)         2.000000            .000000
       3)          .000000          -2.333333
       4)          .000000          -1.333333
       5)         1.000000            .000000

NO. ITERATIONS =        2
```

Compare this result, using LINDO, to the result obtained in Example 2, Section 4.4. Note again that the same solution is found: $z = 44$, $x_1 = 4$, $x_2 = 4$, $s_1 = \text{SLK } 2 = 2$, $s_2 = \text{SLK } 3 = 0$, $s_3 = \text{SLK } 4 = 0$, $s_4 = \text{SLK } 5 = 1$. Also, note again the different intermediate tableaus.

EXERCISE B Answers to Odd-Numbered Problems Begin on Page AN-110.

In Problems 1–24, use LINDO (or any other software package) to solve each linear programming problem.

1. Maximize

$$P = 3x_1 + 2x_2 + x_3$$

subject to

$$\begin{aligned} 3x_1 + x_2 + x_3 &\le 30 \\ 5x_1 + 2x_2 + x_3 &\le 24 \\ x_1 + x_2 + 4x_3 &\le 20 \end{aligned}$$
$$x_1 \ge 0 \quad x_2 \ge 0 \quad x_3 \ge 0$$

2. Maximize

$$P = x_1 + 4x_2 + 3x_3 + x_4$$

subject to

$$\begin{aligned} 2x_1 + x_2 &\le 10 \\ 3x_1 + x_2 + x_3 + 2x_4 &\le 18 \\ x_1 + x_2 + x_3 + x_4 &\le 14 \end{aligned}$$
$$x_1 \ge 0 \quad x_2 \ge 0 \quad x_3 \ge 0 \quad x_4 \ge 0$$

3. Maximize

$$P = 3x_1 + x_2 + x_3$$

subject to

$$\begin{aligned} x_1 + x_2 + x_3 &\le 6 \\ 2x_1 + 3x_2 + 4x_3 &\le 10 \end{aligned}$$
$$x_1 \ge 0 \quad x_2 \ge 0 \quad x_3 \ge 0$$

4. Maximize

$$P = 3x_1 + x_2 + x_3$$

subject to

$$\begin{aligned} x_1 + x_2 + x_3 &\le 8 \\ 2x_1 + x_2 + 4x_3 &\ge 6 \end{aligned}$$
$$x_1 \ge 0 \quad x_2 \ge 0$$

5. Maximize

$$P = 2x_1 + x_2 + 3x_3$$

subject to

$$\begin{aligned} x_1 + x_2 - x_3 &\le 10 \\ x_2 + x_3 &\le 4 \end{aligned}$$
$$x_1 \ge 0 \quad x_2 \ge 0 \quad x_3 \ge 0$$

6. Maximize

$$P = 2x_1 + 2x_2 + 3x_3$$

subject to

$$\begin{aligned} x_1 - x_2 + x_3 &\le 6 \\ x_1 &\le 4 \end{aligned}$$
$$x_1 \ge 0 \quad x_2 \ge 0 \quad x_3 \ge 0$$

7. Maximize

$$P = x_1 + x_2 + x_3$$

subject to

$$\begin{aligned} x_1 + x_2 + x_3 &\le 6 \\ 4x_1 + x_2 &\ge 12 \end{aligned}$$
$$x_1 \ge 0 \quad x_2 \ge 0 \quad x_3 \ge 0$$

8. Maximize

$$P = 2x_1 + x_2 + 3x_3$$

subject to

$$\begin{aligned} -x_1 + x_2 + x_3 &\ge -6 \\ 2x_1 - 3x_2 &\ge -12 \end{aligned}$$
$$x_1 \ge 0 \quad x_2 \ge 0 \quad x_3 \ge 0$$

9. Maximize

$$P = 2x_1 + x_2 + 3x_3$$

subject to

$$\begin{aligned} 5x_1 + 2x_2 + x_3 &\le 20 \\ 6x_1 + x_2 + 4x_3 &\le 24 \\ x_1 + x_2 + 4x_3 &\le 16 \\ x_1 \ge 0 \quad x_2 \ge 0 \quad x_3 &\ge 0 \end{aligned}$$

10. Maximize

$$P = 3x_1 + 2x_2 + x_3$$

subject to

$$\begin{aligned} 3x_1 + 2x_2 - x_3 &\le 10 \\ x_1 - x_2 + 3x_3 &\le 12 \\ 2x_1 + x_2 + x_3 &\le 6 \\ x_1 \ge 0 \quad x_2 \ge 0 \quad x_3 &\ge 0 \end{aligned}$$

11. Maximize

$$P = 2x_1 + 3x_2 + x_3$$

subject to

$$\begin{aligned} x_1 + x_2 + x_3 &\le 50 \\ 3x_1 + 2x_2 + x_3 &\le 10 \\ x_1 \ge 0 \quad x_2 \ge 0 \quad x_3 &\ge 0 \end{aligned}$$

12. Maximize

$$P = 4x_1 + 4x_2 + 2x_3$$

subject to

$$\begin{aligned} 3x_1 + x_2 + x_3 &\le 10 \\ x_1 + x_2 + 3x_3 &\le 5 \\ x_1 \ge 0 \quad x_2 \ge 0 \quad x_3 &\ge 0 \end{aligned}$$

13. Maximize

$$P = 2x_1 + x_2 + x_3$$

subject to

$$\begin{aligned} -2x_1 + x_2 - 2x_3 &\le 4 \\ x_1 - 2x_2 + x_3 &\le 2 \\ x_1 \ge 0 \quad x_2 \ge 0 \quad x_3 &\ge 0 \end{aligned}$$

14. Maximize

$$P = 4x_1 + 2x_2 + 5x_3$$

subject to

$$\begin{aligned} x_1 + 3x_2 + 2x_3 &\le 30 \\ 2x_1 + x_2 + 3x_3 &\le 12 \\ x_1 \ge 0 \quad x_2 \ge 0 \quad x_3 &\ge 0 \end{aligned}$$

15. Maximize

$$P = 2x_1 + x_2 + 3x_3$$

subject to

$$\begin{aligned} x_1 + 2x_2 + x_3 &\le 25 \\ 3x_1 + 2x_2 + 3x_3 &\le 30 \\ x_1 \ge 0 \quad x_2 \ge 0 \quad x_3 &\ge 0 \end{aligned}$$

16. Maximize

$$P = 6x_1 + 3x_2 + 2x_3$$

subject to

$$\begin{aligned} 2x_1 + 2x_2 + 3x_3 &\le 30 \\ 2x_1 + 2x_2 + x_3 &\le 12 \\ x_1 \ge 0 \quad x_2 \ge 0 \quad x_3 &\ge 0 \end{aligned}$$

17. Maximize

$$P = 2x_1 + 4x_2 + x_3 + x_4$$

subject to

$$\begin{aligned} 2x_1 + x_2 + 2x_3 + 3x_4 &\le 12 \\ 2x_2 + x_3 + 2x_4 &\le 20 \\ 2x_1 + x_2 + 4x_3 &\le 16 \\ x_1 \ge 0 \quad x_2 \ge 0 \quad x_3 \ge 0 \quad x_4 &\ge 0 \end{aligned}$$

18. Maximize

$$P = 2x_1 + 4x_2 + x_3$$

subject to

$$\begin{aligned} -x_1 + 2x_2 + 3x_3 &\le 6 \\ -x_1 + 4x_2 + 5x_3 &\le 5 \\ -x_1 + 5x_2 + 7x_3 &\le 7 \\ x_1 \ge 0 \quad x_2 \ge 0 \quad x_3 &\ge 0 \end{aligned}$$

19. Maximize

$$P = 2x_1 + x_2 + x_3$$

subject to

$$\begin{aligned} x_1 + 2x_2 + 4x_3 &\le 20 \\ 2x_1 + 4x_2 + 4x_3 &\le 60 \\ 3x_1 + 4x_2 + x_3 &\le 90 \\ x_1 \ge 0 \quad x_2 \ge 0 \quad x_3 &\ge 0 \end{aligned}$$

20. Maximize

$$P = x_1 + 2x_2 + 4x_3$$

subject to

$$\begin{aligned} 8x_1 + 5x_2 - 4x_3 &\le 30 \\ -2x_1 + 6x_2 + x_3 &\le 5 \\ -2x_1 + 2x_2 + x_3 &\le 15 \\ x_1 \ge 0 \quad x_2 \ge 0 \quad x_3 &\ge 0 \end{aligned}$$

21. Maximize

$$P = x_1 + 2x_2 + 4x_3 - x_4$$

subject to

$$\begin{aligned} 5x_1 \qquad\;\; + 4x_3 + 6x_4 &\le 20 \\ 4x_1 + 2x_2 + 2x_3 + 8x_4 &\le 40 \end{aligned}$$

$$x_1 \ge 0 \qquad x_2 \ge 0 \qquad x_3 \ge 0 \qquad x_4 \ge 0$$

22. Maximize

$$P = x_1 + 2x_2 - x_3 + 3x_4$$

subject to

$$\begin{aligned} 2x_1 + 4x_2 + 5x_3 + 6x_4 &\le 24 \\ 4x_1 + 4x_2 + 2x_3 + 2x_4 &\le 4 \end{aligned}$$

$$x_1 \ge 0 \qquad x_2 \ge 0 \qquad x_3 \ge 0 \qquad x_4 \ge 0$$

23. Minimize

$$z = x_1 + x_2 + x_3 + x_4 + x_5 + x_6 + x_7$$

subject to

$$\begin{aligned} 4x_1 + 2x_2 + x_3 \qquad + 2x_5 + x_6 \qquad &\ge 75 \\ x_2 + 2x_3 + 3x_4 \qquad + x_6 \qquad &\ge 110 \\ x_5 + x_6 + 2x_7 &\ge 50 \end{aligned}$$

$$x_1 \ge 0 \qquad x_2 \ge 0 \qquad x_3 \ge 0 \qquad x_4 \ge 0$$

$$x_5 \ge 0 \qquad x_6 \ge 0 \qquad x_7 \ge 0$$

24. Minimize

$$z = x_1 + x_2 + x_3 + x_4 + x_5 + x_6 + x_7$$

subject to

$$\begin{aligned} 4x_1 + x_2 + 2x_3 \qquad + 2x_5 \qquad + x_7 &\ge 75 \\ 2x_2 + x_3 + 3x_4 \qquad\qquad + x_7 &\ge 180 \\ x_5 + 2x_6 + x_7 &\ge 50 \end{aligned}$$

$$x_1 \ge 0 \qquad x_2 \ge 0 \qquad x_3 \ge 0 \qquad x_4 \ge 0$$

$$x_5 \ge 0 \qquad x_6 \ge 0 \qquad x_7 \ge 0$$

APPENDIX C

Graphing Utilities

OUTLINE

C.1 The Viewing Rectangle

All graphing utilities, that is, all graphing calculators and all computer software graphing packages, graph equations by plotting points on a screen. The screen itself actually consists of small rectangles, called **pixels.** The more pixels the screen has, the better the resolution. Most graphing calculators have 48 pixels per square inch; most computer screens have 32 to 108 pixels per square inch. When a point to be plotted lies inside a pixel, the pixel is turned on (lights up). The graph of an equation is a collection of pixels. Figure 1 shows how the graph of $y = 2x$ looks on a TI-83 Plus graphing calculator.

FIGURE 1 $y = 2x$

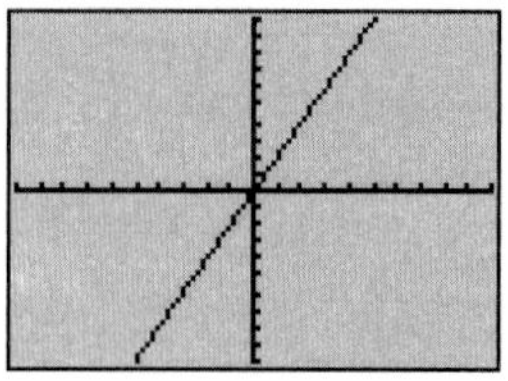

The screen of a graphing utility will display the coordinate axes of a rectangular coordinate system. However, you must set the scale on each axis. You must also include the smallest and largest values of x and y that you want included in the graph. This is called **setting the viewing rectangle** or **viewing window.** Figure 2 illustrates a typical viewing window.

FIGURE 2

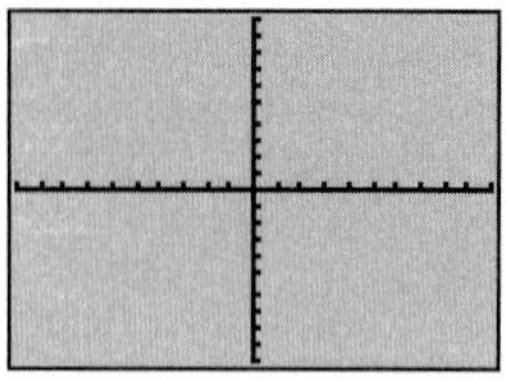

To select the viewing window, we must give values to the following expressions:

X min:	the smallest value of x
X max:	the largest value of x
X scl:	the number of units per tick mark on the x-axis
Y min:	the smallest value of y
Y max:	the largest value of y
Y scl:	the number of units per tick mark on the y-axis

Figure 3 illustrates these settings and their relation to the Cartesian coordinate system.

FIGURE 3

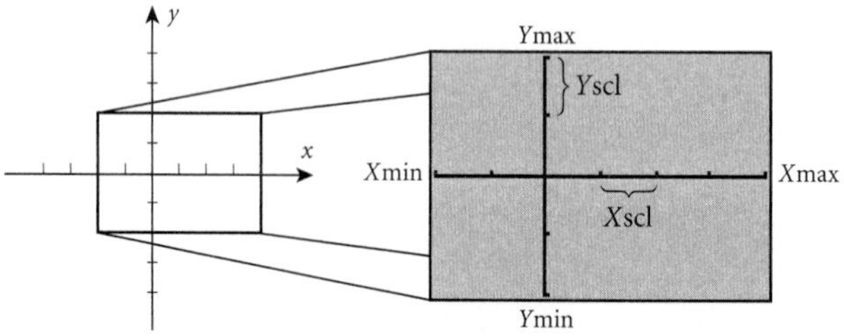

If the scale used on each axis is known, we can determine the minimum and maximum values of x and y shown on the screen by counting the tick marks. Look again at Figure 2. For a scale of 1 on each axis, the minimum and maximum values of x are -10 and 10, respectively; the minimum and maximum values of y are also -10 and 10. If the scale is 2 on each axis, then the minimum and maximum values of x are -20 and 20, respectively; and the minimum and maximum values of y are -20 and 20, respectively.

Conversely, if we know the minimum and maximum values of x and y, we can determine the scales being used by counting the tick marks displayed. We shall follow the practice of showing the minimum and maximum values of x and y in our illustrations so that you will know how the viewing window was set. See Figure 4.

FIGURE 4

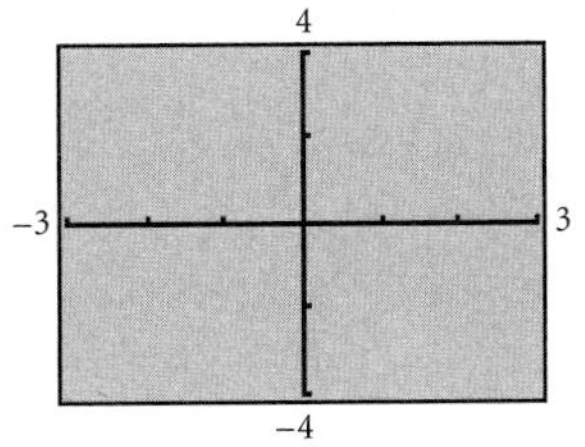

means

X min $= -3$	Y min $= -4$
X max $= 3$	Y max $= 4$
X scl $= 1$	Y scl $= 2$

EXAMPLE 1 Finding the Coordinates of a Point Shown on a Graphing Utility Screen

FIGURE 5

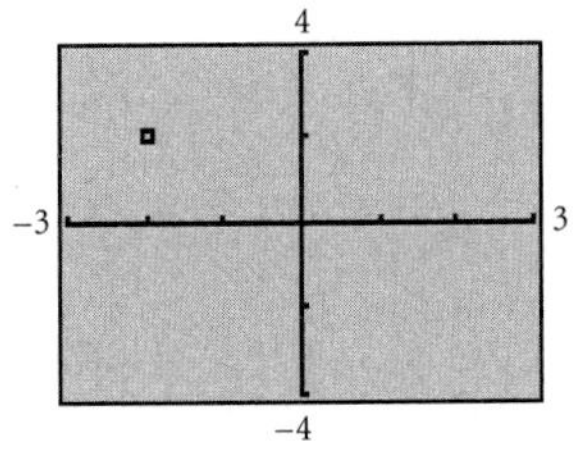

Find the coordinates of the point shown in Figure 5. Assume that the coordinates are integers.

SOLUTION First we note that the viewing window used in Figure 5 is

X min $= -3$	Y min $= -4$
X max $= 3$	Y max $= 4$
X scl $= 1$	Y scl $= 2$

The point shown is 2 tick units to the left on the horizontal axis (scale $= 1$) and 1 tick up on the vertical axis (scale $= 2$). The coordinates of the point shown are $(-2, 2)$.

EXERCISE C.1 Answers to Odd-Numbered Problems Begin on Page AN-110.

In Problems 1–4, determine the coordinates of the points shown. Tell in which quadrant each point lies. Assume that the coordinates are integers.

1.

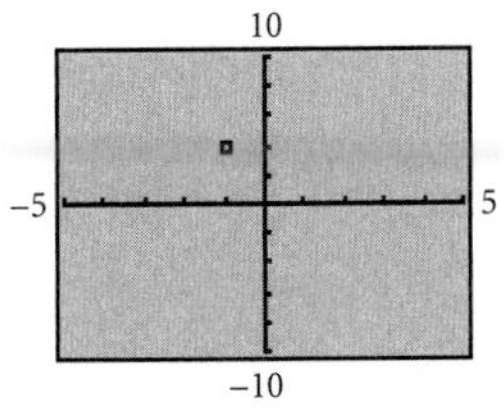

2.

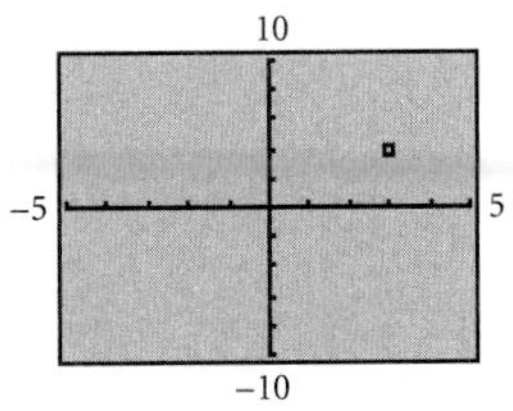

3.

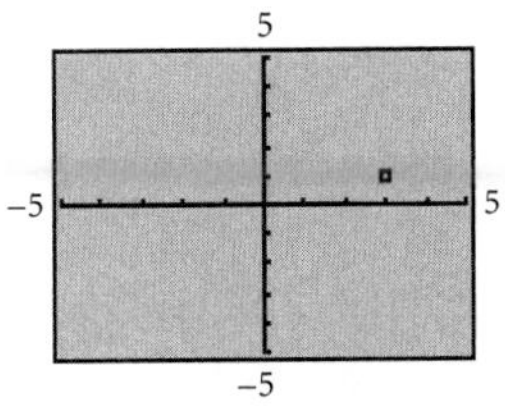

4.

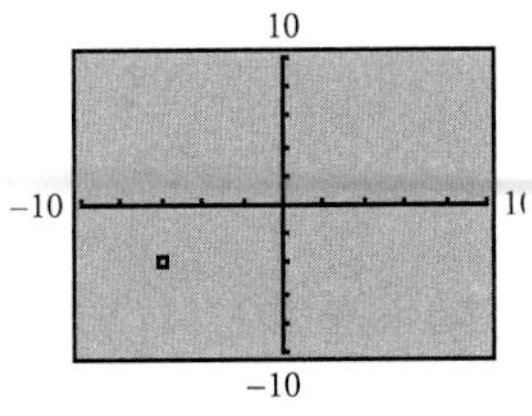

In Problems 5–10, determine the viewing window used.

5.

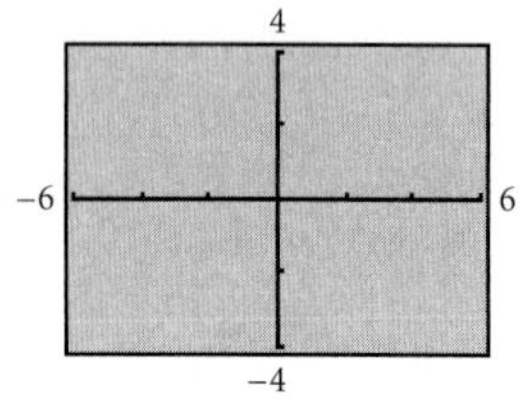

6.

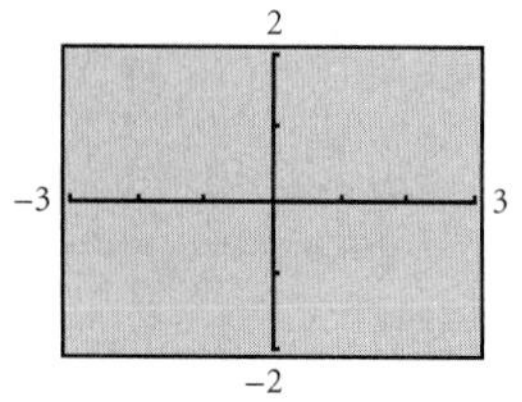

7.

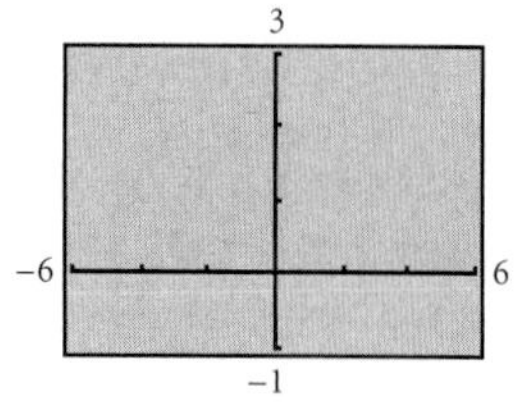

8.

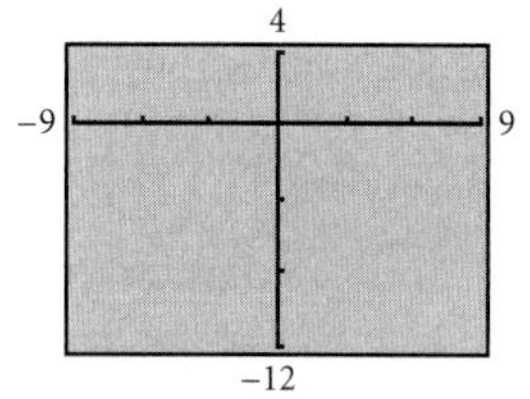

9.

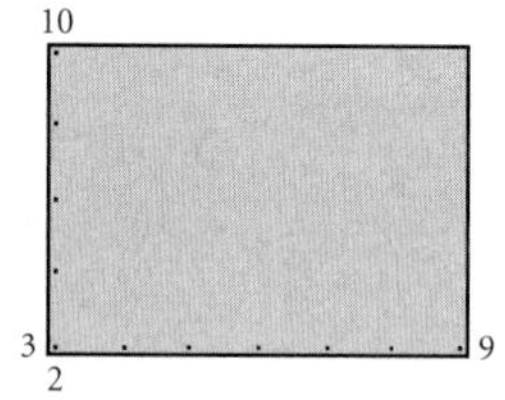

10.

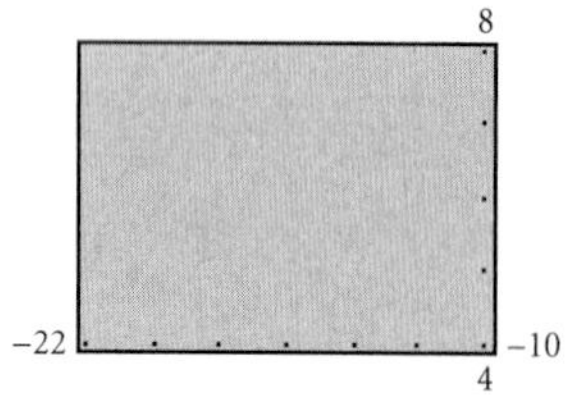

In Problems 11–16, select a setting so that each of the given points will lie within the viewing rectangle.

11. $(-10, 5)$, $(3, -2)$, $(4, -1)$

12. $(5, 0)$, $(6, 8)$, $(-2, -3)$

13. $(40, 20)$, $(-20, -80)$, $(10, 40)$

14. $(-80, 60)$, $(20, -30)$, $(-20, -40)$

15. $(0, 0)$, $(100, 5)$, $(5, 150)$

16. $(0, -1)$, $(100, 50)$, $(-10, 30)$

C.2 Using a Graphing Utility to Graph Equations

The graph of an equation in two variables can usually be obtained by plotting points in a rectangular coordinate system and connecting them. Graphing utilities perform these same steps when graphing an equation. For example, the T1-83 Plus determines 95 evenly spaced input values,* uses the equation to determine the output values, plots these points on the screen, and finally (if in the connected mode) draws a line between consecutive points.

To graph an equation in two variables x and y using a graphing utility requires that the equation be written in the form $y = \{\text{expression in } x\}$. If the original equation is not in this form, replace it by equivalent equations until the form $y = \{\text{expression in } x\}$ is obtained. In general, there are four ways to obtain equivalent equations.

**These input values depend on the values of X min and X max. For example, if X min = −10 and X max = 10, then the first input value will be −10 and the next input value will be* $-10 + \dfrac{10-(-10)}{94} = -9.7872$, *and so on.*

Procedures that Result in Equivalent Equations

1. Interchange the two sides of the equation:

$$\text{Replace} \quad 3x + 5 = y \quad \text{by} \quad y = 3x + 5$$

2. Simplify the sides of the equation by combining like terms, eliminating parentheses, and so on:

$$\text{Replace} \quad (2y + 2) + 6 = 2x + 5(x + 1)$$
$$\text{by} \quad 2y + 8 = 7x + 5$$

3. Add or subtract the same expression on both sides of the equation:

$$\text{Replace} \quad y + 3x - 5 = 4$$
$$\text{by} \quad y + 3x - 5 + 5 = 4 + 5$$

4. Multiply or divide both sides of the equation by the same nonzero expression:

$$\text{Replace} \quad 3y = 6 - 2x$$
$$\text{by} \quad \frac{1}{3} \cdot 3y = \frac{1}{3}(6 - 2x)$$

EXAMPLE 1 **Expressing an Equation in the Form $y = \{\text{expression in } x\}$**

Solve for y: $2y + 3x - 5 = 4$

SOLUTION We replace the original equation by a succession of equivalent equations.

$$2y + 3x - 5 = 4$$
$$2y + 3x - 5 + 5 = 4 + 5 \qquad \text{Add 5 to both sides.}$$
$$2y + 3x = 9 \qquad \text{Simplify.}$$
$$2y + 3x - 3x = 9 - 3x \qquad \text{Subtract } 3x \text{ from both sides.}$$
$$2y = 9 - 3x \qquad \text{Simplify.}$$
$$\frac{2y}{2} = \frac{9 - 3x}{2} \qquad \text{Divide both sides by 2.}$$
$$y = \frac{9 - 3x}{2} \qquad \text{Simplify.}$$

Now we are ready to graph equations using a graphing utility. Most graphing utilities require the following steps:

Steps for Graphing an Equation Using a Graphing Utility

STEP 1: Solve the equation for y in terms of x.

STEP 2: Get into the graphing mode of your graphing utility. The screen will usually display $y =$, prompting you to enter the expression involving x that you found in Step 1. (Consult your manual for the correct way to enter the expression; for example, $y = x^2$ might be entered as $x \wedge 2$ or as x^*x or as $x\ x^y\ 2$.)

STEP 3: Select the viewing window. Without prior knowledge about the behavior of the graph of the equation, it is common to select the **standard viewing window*** initially. The viewing window is then adjusted based on the graph that appears. In this text the standard viewing window will be

$$X\text{min} = -10 \qquad Y\text{min} = -10$$
$$X\text{max} = 10 \qquad Y\text{max} = 10$$
$$X\text{scl} = 1 \qquad Y\text{scl} = 1$$

STEP 4: Graph.

STEP 5: Adjust the viewing window until a complete graph is obtained.

EXAMPLE 2 Graphing an Equation on a Graphing Utility

Graph the equation: $6x^2 + 3y = 36$

SOLUTION **STEP 1:** We solve for y in terms of x.

$$6x^2 + 3y = 36$$
$$3y = -6x^2 + 36 \quad \text{Subtract } 6x^2 \text{ from both sides of the equation.}$$
$$y = -2x^2 + 12 \quad \text{Divide both sides of the equation by 3 and simplify.}$$

STEP 2: From the graphing mode, enter the expression $-2x^2 + 12$ after the prompt $y =$.

STEP 3: Set the viewing window to the standard viewing window.

STEP 4: Graph. The screen should look like Figure 6.

STEP 5: The graph of $y = -2x^2 + 12$ is not complete. The value of Ymax must be increased so that the top portion of the graph is visible. After increasing the value of Ymax to 12, we obtain the graph in Figure 7. The graph is now complete.

FIGURE 6

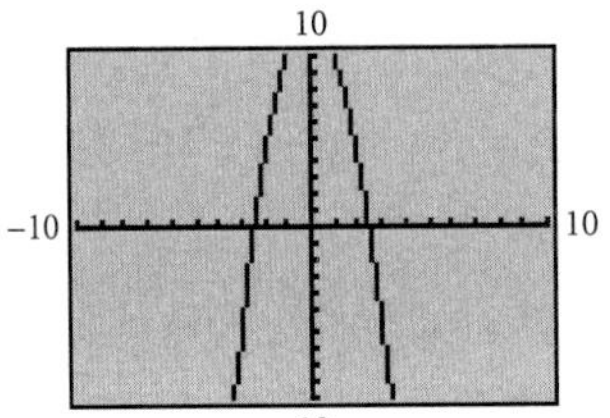

FIGURE 7

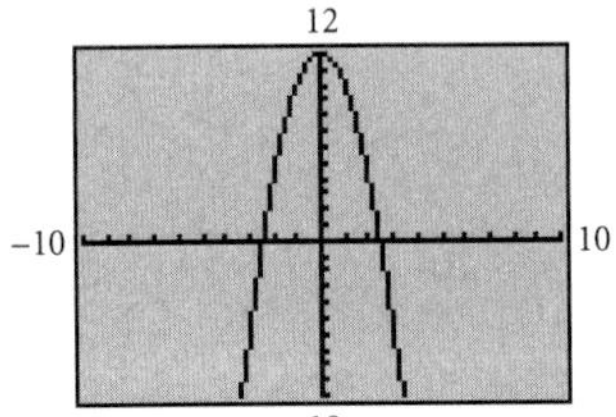

Some graphing utilities have a ZOOM-STANDARD feature that automatically sets the viewing window to the standard viewing window and graphs the equation.

Look again at Figure 7. Although a complete graph is shown, the graph might be improved by adjusting the values of Xmin and Xmax. Figure 8 shows the graph of $y = -2x^2 + 12$ using Xmin $= -4$ and Xmax $= 4$. Do you think this is a better choice for the viewing window?

FIGURE 8

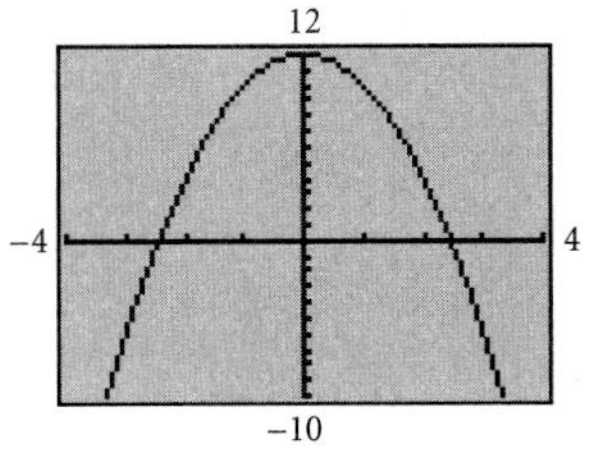

EXAMPLE 3   Creating a Table and Graphing an Equation

Create a table and graph the equation: $y = x^3$

SOLUTION   Most graphing utilities have the capability of creating a table of values for an equation. (Check your manual to see if your graphing utility has this capability.) Table 1 illustrates a table of values for $y = x^3$ on a TI-83 Plus. See Figure 9 for the graph.

TABLE 1

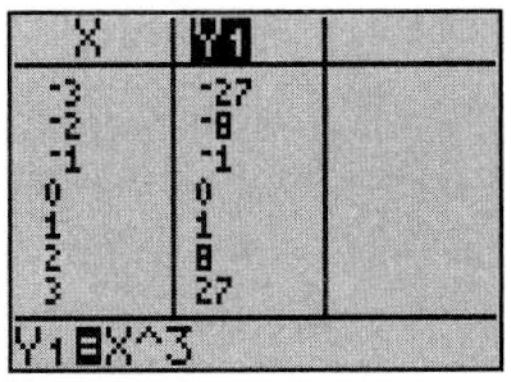

FIGURE 9

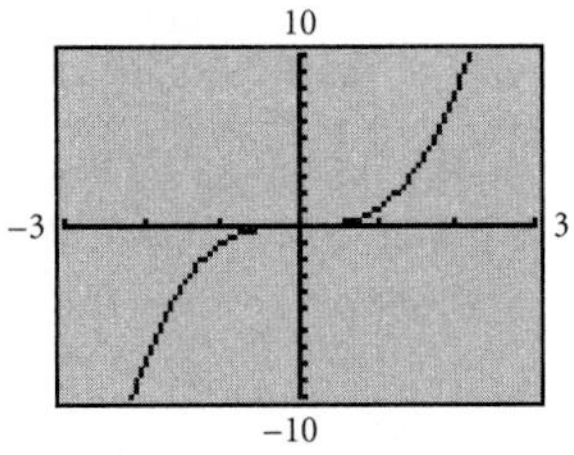

EXERCISE C.2   Answers to Odd-Numbered Problems Begin on Page AN-110.

In Problems 1–16, graph each equation using the following viewing windows:

(a) Xmin $= -5$	(b) Xmin $= -10$	(c) Xmax $= -10$	(d) Xmax $= -5$
Xmax $= 5$	Xmax $= 10$	Xmax $= 10$	Xmax $= 5$
Xscl $= 1$	Xscl $= 1$	Xscl $= 2$	Xscl $= 1$
Ymin $= -4$	Ymin $= -8$	Ymin $= -8$	Ymin $= -20$
Ymax $= 4$	Ymax $= 8$	Ymax $= 8$	Ymax $= 20$
Yscl $= 1$	Yscl $= 1$	Yscl $= 2$	Yscl $= 5$

1. $y = x + 2$

2. $y = x - 2$

3. $y = -x + 2$

4. $y = -x - 2$

5. $y = 2x + 2$

6. $y = 2x - 2$

7. $y = -2x + 2$

8. $y = -2x - 2$

9. $y = x^2 + 2$

10. $y = x^2 - 2$

11. $y = -x^2 + 2$

12. $y = -x^2 - 2$

13. $3x + 2y = 6$

14. $3x - 2y = 6$

15. $-3x + 2y = 6$

16. $-3x - 2y = 6$

17.–32. *For each of the above equations, create a table, $-3 \leq x \leq 3$, and list points on the graph.*

C.3 Square Screens

FIGURE 10

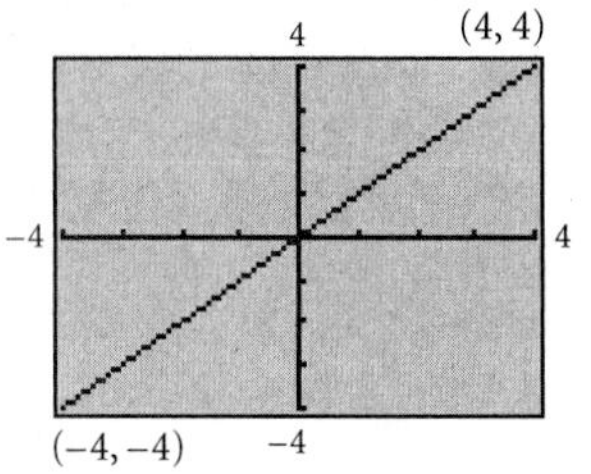

Most graphing utilities have a rectangular screen. Because of this, using the same settings for both x and y will result in a distorted view. For example, Figure 10 shows the graph of the line $y = x$ connecting the points $(-4, -4)$ and $(4, 4)$.

We expect the line to bisect the first and third quadrants, but it doesn't. We need to adjust the selections for *X*min, *X*max, *Y*min, and *Y*max so that a **square screen** results. On most graphing utilities, this is accomplished by setting the ratio of x to y at 3 : 2.* In other words,

$$2(X\max - X\min) = 3(Y\max - Y\min)$$

EXAMPLE 1 Examples of Viewing Rectangles That Result in Square Screens

(a) $X\min = -3$, $X\max = 3$, $X\text{scl} = 1$, $Y\min = -2$, $Y\max = 2$, $Y\text{scl} = 1$

(b) $X\min = -6$, $X\max = 6$, $X\text{scl} = 1$, $Y\min = -4$, $Y\max = 4$, $Y\text{scl} = 1$

(c) $X\min = -6$, $X\max = 6$, $X\text{scl} = 2$, $Y\min = -4$, $Y\max = 4$, $Y\text{scl} = 1$

FIGURE 11

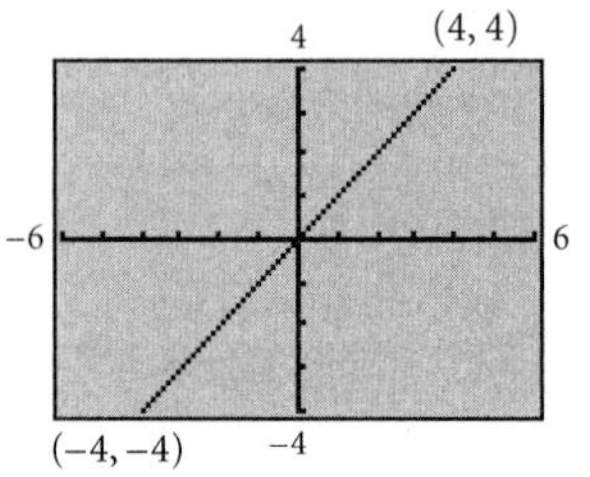

Figure 11 shows the graph of the line $y = x$ on a square screen using the viewing rectangle given in Example 1(b). Notice that the line now bisects the first and third quadrants. Compare this illustration to Figure 10.

**Some graphing utilities have a built-in function that automatically squares the screen. For example, the TI-85 has a ZSQR function that does this. Some graphing utilities require a ratio other than 3:2 to square the screen. For example, the HP 48G requires the ratio of x to y to be 2:1 for a square screen. Consult your manual.*

EXERCISE C.3 Answers to Odd-Numbered Problems Begin on Page AN-112.

In Problems 1–8, determine which of the given viewing rectangles result in a square screen.

1. $X\min = -3$, $X\max = 3$, $X\text{scl} = 2$, $Y\min = -2$, $Y\max = 2$, $Y\text{scl} = 2$

2. $X\min = -5$, $X\max = 5$, $X\text{scl} = 1$, $Y\min = -4$, $Y\max = 4$, $Y\text{scl} = 1$

3. $X\min = 0$, $X\max = 9$, $X\text{scl} = 3$, $Y\min = -2$, $Y\max = 4$, $Y\text{scl} = 2$

4. $X\min = -6$, $X\max = 6$, $X\text{scl} = 1$, $Y\min = -4$, $Y\max = 4$, $Y\text{scl} = 2$

5. $X\min = -6$, $X\max = 6$, $X\text{scl} = 1$, $Y\min = -2$, $Y\max = 2$, $Y\text{scl} = 0.5$

6. $X\min = -6$, $X\max = 6$, $X\text{scl} = 2$, $Y\min = -4$, $Y\max = 4$, $Y\text{scl} = 2$

7. $X\min = 0$, $X\max = 9$, $X\text{scl} = 1$, $Y\min = -2$, $Y\max = 4$, $Y\text{scl} = 1$

8. $X\min = -6$, $X\max = 6$, $X\text{scl} = 2$, $Y\min = -4$, $Y\max = 4$, $Y\text{scl} = 2$

9. If *X*min $= -4$, *X*max $= 8$, and *X*scl $= 1$, how should *Y*min, *Y*max, and *Y*scl be selected so that the viewing rectangle contains the point $(4, 8)$ and the screen is square?

10. If *X*min $= -6$, *X*max $= 12$, and *X*scl $= 2$, how should *Y*min, *Y*max, and *Y*scl be selected so that the viewing rectangle contains the point $(4, 8)$ and the screen is square?

C.4 Using a Graphing Utility to Graph Inequalities

It is easiest to begin with an example.

EXAMPLE 1 Graphing an Inequality Using a Graphing Utility

Use a graphing utility to graph: $3x + y - 6 \le 0$

SOLUTION We begin by graphing the equation $3x + y - 6 = 0$ ($Y_1 = -3x + 6$). See Figure 12.

FIGURE 12

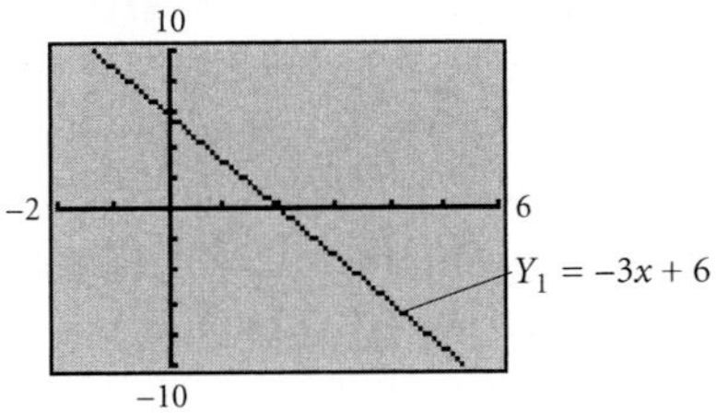

As with graphing by hand, we need to test points selected from each region and determine whether they satisfy the inequality. To test the point $(-1, 2)$, for example, enter $3(-1) + 2 - 6 \le 0$. See Figure 13(a). The 1 that appears indicates that the statement entered (the inequality) is true. When the point $(5, 5)$ is tested, a 0 appears, indicating that the statement entered is false. So, $(-1, 2)$ is a part of the graph of the inequality and $(5, 5)$ is not. Figure 13(b) shows the graph of the inequality on a TI-83 Plus.*

FIGURE 13

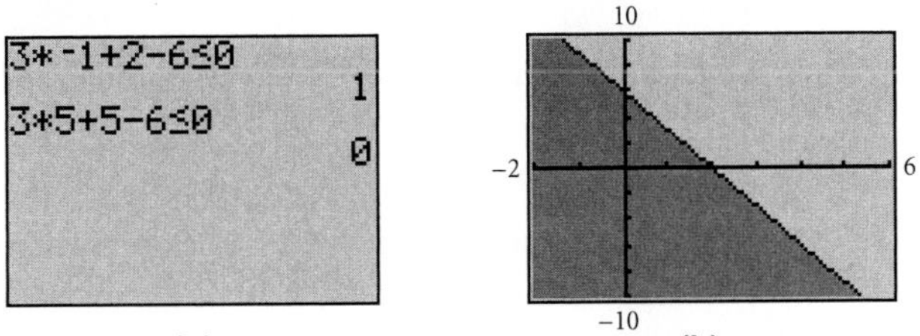

(a) (b)

The steps to follow to graph an inequality using a graphing utility are given next.

*Consult your owner's manual for shading techniques.

Steps for Graphing an Inequality Using a Graphing Utility

STEP 1 Replace the inequality symbol by an equal sign and graph the resulting equation, $y = \quad$.

STEP 2 In each of the regions, select a test point P.
(a) Use a graphing utility to determine if the test point P satisfies the inequality. If the test point satisfies the inequality, then so do all the points in the region. Indicate this by using the graphing utility to shade the region.
(b) If the coordinates of P do not satisfy the inequality, then none of the points in that region do.

C.5 Using a Graphing Utility to Locate Intercepts and Check for Symmetry

Value and Zero (or Root)

Most graphing utilities have an eVALUEate feature that, given a value of x, determines the value of y for an equation. We can use this feature to evaluate an equation at $x = 0$ to determine the y-intercept. Most graphing utilities also have a ZERO (or ROOT) feature that can be used to determine the x-intercept(s) of an equation.

EXAMPLE 1 Finding Intercepts Using a Graphing Utility

Use a graphing utility to find the intercepts of the equation $y = x^3 - 8$.

SOLUTION Figure 14(a) shows the graph of $y = x^3 - 8$.

FIGURE 14

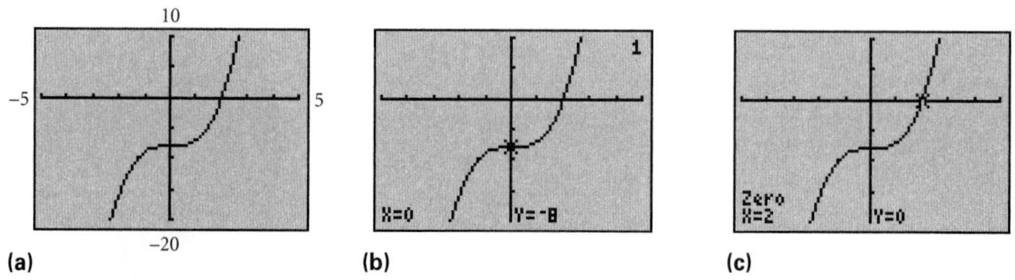

The eVALUEate feature of a TI-83 Plus graphing calculator accepts as input a value of x and determines the value of y. If we let $x = 0$, we find $y = -8$ so that the y-intercept is $(0, -8)$. See Figure 14(b).

The ZERO feature of a TI-83 Plus is used to find the x-intercept(s). See Figure 14(c). The x-intercept is $(2, 0)$.

TRACE

Most graphing utilities allow you to move from point to point along the graph, displaying on the screen the coordinates of each point. This feature is called TRACE.

EXAMPLE 2 Using TRACE to Locate Intercepts

Graph the equation $y = x^3 - 8$. Use TRACE to locate the intercepts.

SOLUTION Figure 15 shows the graph of $y = x^3 - 8$.

FIGURE 15

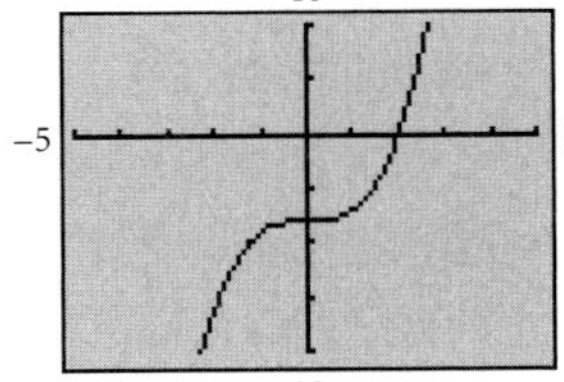

Activate the TRACE feature. As you move the cursor along the graph, you will see the coordinates of each point displayed. When the cursor is on the y-axis, we find that the y-intercept is -8. See Figure 16.

FIGURE 16

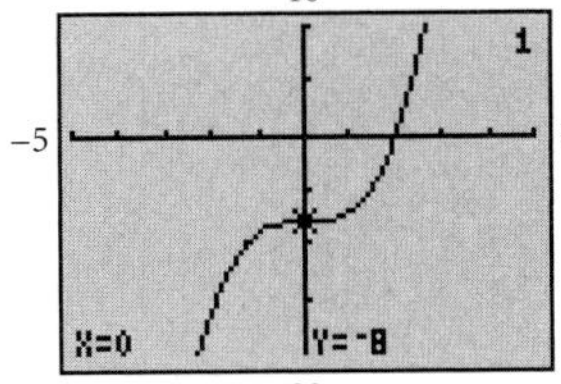

Continue moving the cursor along the graph. Just before you get to the x-axis, the display will look like the one in Figure 17(a). (Due to differences in graphing utilities, your display may be slightly different from the one shown here.)

FIGURE 17

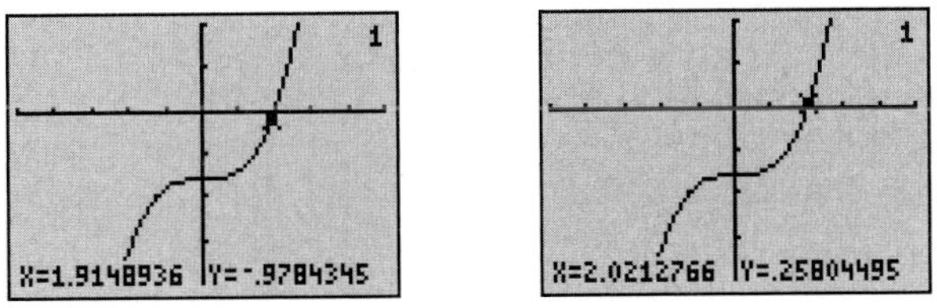

(a) (b)

In Figure 17(a), the negative value of the y-coordinate indicates that we are still below the x-axis. The next position of the cursor is shown in Figure 17(b). The positive value of the y-coordinate indicates that we are now above the x-axis. This means that between these two points the x-axis was crossed. The x-intercept lies between 1.9148936 and 2.0212766.

EXAMPLE 3 Graphing the Equation $y = \frac{1}{x}$

FIGURE 18

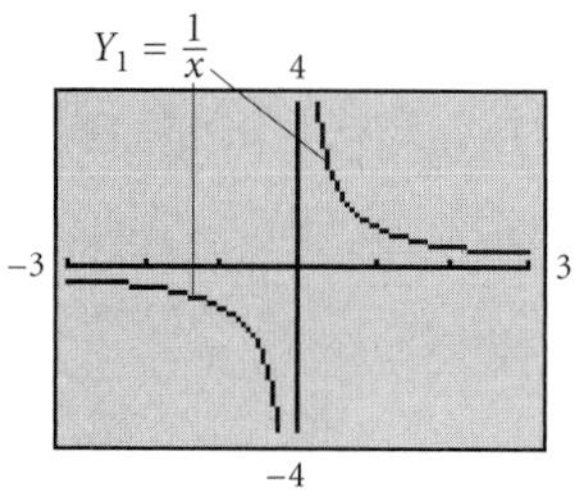

Graph the equation: $y = \frac{1}{x}$

with the viewing window set as

$$X\min = -3 \qquad Y\min = -4$$
$$X\max = 3 \qquad Y\max = 4$$
$$X\text{scl} = 1 \qquad Y\text{scl} = 1$$

Use TRACE to infer information about the intercepts and symmetry.

SOLUTION Figure 18 illustrates the graph. We infer from the graph that there are no intercepts; we may also infer that symmetry with respect to the origin is a possibility. The TRACE feature on a graphing utility can provide further evidence of symmetry with respect to the origin. Using TRACE, we observe that for any ordered pair (x, y) the ordered pair $(-x, -y)$ is also a point on the graph. For example, the points $(0.95744681, 1.0444444)$ and $(-0.95744681, -1.0444444)$ both lie on the graph.

EXERCISE C.5 Answers to Odd-Numbered Problems Begin on Page AN-112.

In Problems 1–6, use ZERO (or ROOT) to approximate the smaller of the two x-intercepts of each equation. Express the answer rounded to two decimal places.

1. $y = x^2 + 4x + 2$

2. $y = x^2 + 4x - 3$

3. $y = 2x^2 + 4x + 1$

4. $y = 3x^2 + 5x + 1$

5. $y = 2x^2 - 3x - 1$

6. $y = 2x^2 - 4x - 1$

In Problems 7–14, use ZERO (or ROOT) to approximate the ***positive*** *x-intercepts of each equation. Express each answer rounded to two decimal places.*

7. $y = x^3 + 3.2x^2 - 16.83x - 5.31$

8. $y = x^3 + 3.2x^2 - 7.25x - 6.3$

9. $y = x^4 - 1.4x^3 - 33.71x^2 + 23.94x + 292.41$

10. $y = x^4 + 1.2x^3 - 7.46x^2 - 4.692x + 15.2881$

11. $y = \pi x^3 - (8.88\pi + 1)x^2 - (42.066\pi - 8.88)x + 42.066$

12. $y = \pi x^3 - (5.63\pi + 2)x^2 - (108.392\pi - 11.26)x + 216.784$

13. $y = x^3 + 19.5x^2 - 1021x + 1000.5$

14. $y = x^3 + 14.2x^2 - 4.8x - 12.4$

In Problems 15–18, the graph of an equation is given.
(a) List the intercepts of the graph.
(b) Based on the graph, tell whether the graph is symmetric with respect to the x-axis, y-axis, and/or origin.

15.

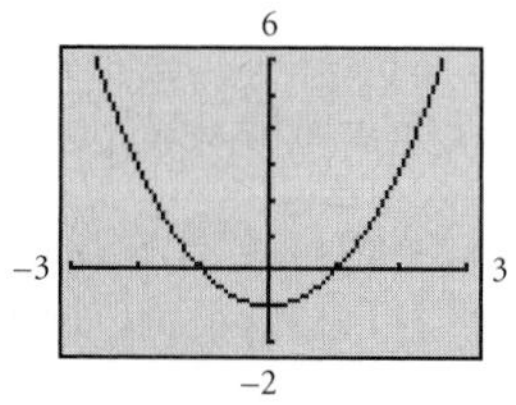

16.

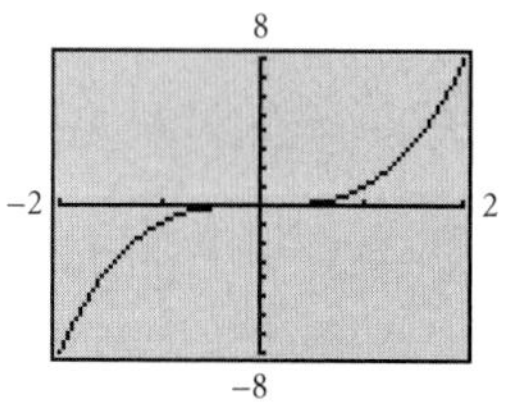

17.

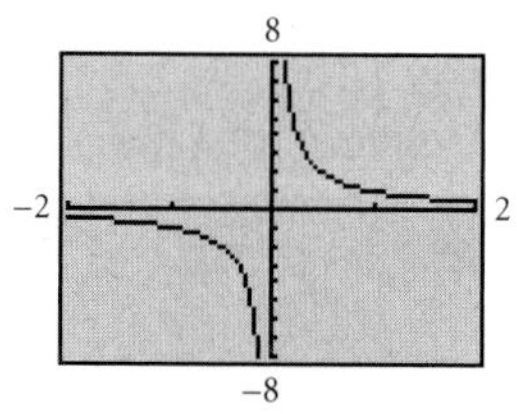

18.

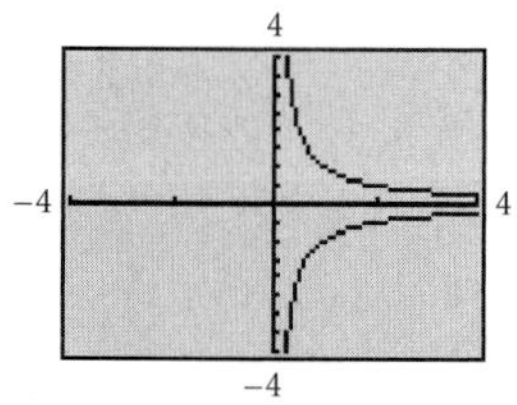

C.6 Using a Graphing Utility to Solve Equations

For many equations, there are no algebraic techniques that lead to a solution. For such equations, a graphing utility can often be used to investigate possible solutions. When a graphing utility is used to solve an equation, usually *approximate* solutions are obtained. Unless otherwise stated, we shall follow the practice of giving approximate solutions *rounded to two decimal places.*

The ZERO (or ROOT) feature of a graphing utility can be used to find the solutions of an equation when one side of the equation is 0. In using this feature to solve equations, we make use of the fact that the x-intercepts (or zeros) of the graph of an equation are found by letting $y = 0$ and solving the equation for x.

EXAMPLE 1 Using ZERO (or ROOT) to Approximate Solutions of an Equation

Find the solution(s) of the equation $x^2 - 6x + 7 = 0$. Round answers to two decimal places.

SOLUTION The solutions of the equation $x^2 - 6x + 7 = 0$ are the same as the x-intercepts of the graph of $Y_1 = x^2 - 6x + 7$. We begin by graphing the equation. See Figure 19(a).

FIGURE 19

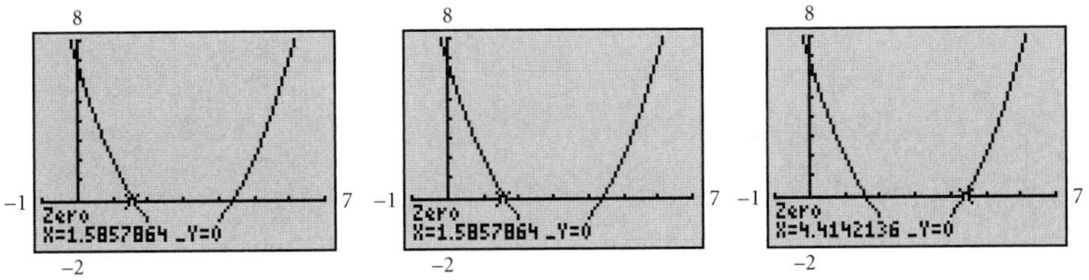

(a) (b) (c)

From the graph there appear to be two x-intercepts (solutions to the equation): one between 1 and 2, the other between 4 and 5.

Using the ZERO (or ROOT) feature of our graphing utility, we determine that the x-intercepts, and so the solutions to the equation, are $x = 1.59$ and $x = 4.41$, rounded to two decimal places. See Figures 19(b) and (c).

A second method for solving equations using a graphing utility involves the INTERSECT feature of the graphing utility. This feature is used most effectively when one side of the equation is not 0.

EXAMPLE 2 Using INTERSECT to Approximate Solutions of an Equation

Find the solution(s) to the equation $3(x - 2) = 5(x - 1)$.

SOLUTION We begin by graphing each side of the equation as follows: graph $Y_1 = 3(x - 2)$ and $Y_2 = 5(x - 1)$. See Figure 20(a).

FIGURE 20

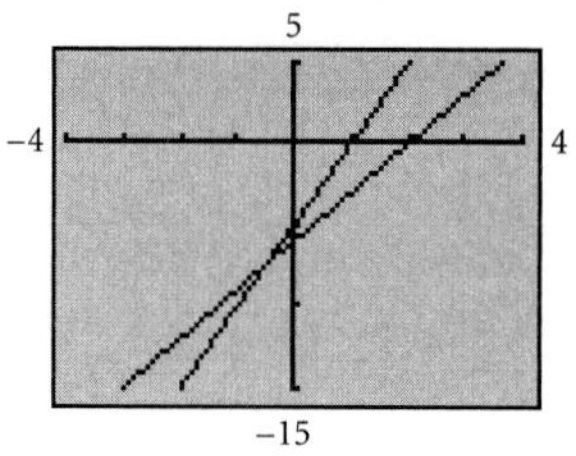

(a)   (b)

At the point of intersection of the graphs, the value of the y-coordinate is the same. We conclude that the x-coordinate of the point of intersection represents the solution to the equation. Do you see why? The INTERSECT feature on a graphing utility determines the point of intersection of the graphs. Using this feature, we find that the graphs intersect at $(-0.5, -7.5)$. See Figure 20(b). The solution of the equation is therefore $x = -0.5$. ▶

CHECK: We can verify our solution by evaluating each side of the equation with -0.5 STOred in x. See Figure 21. Since the left side of the equation equals the right side of the equation, the solution checks.

FIGURE 21

```
-.5→X
           -.5
3(X-2)
          -7.5
5(X-1)
          -7.5
```

▶

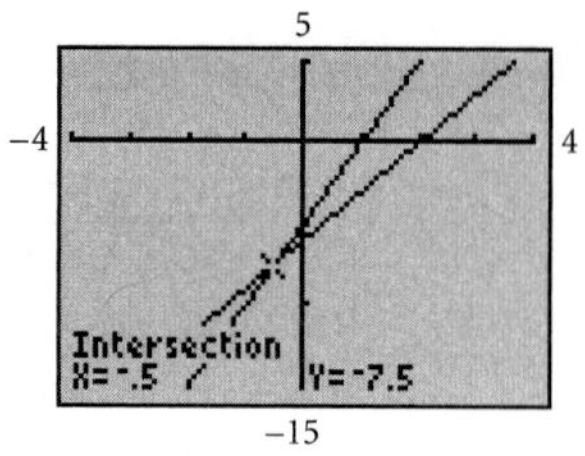

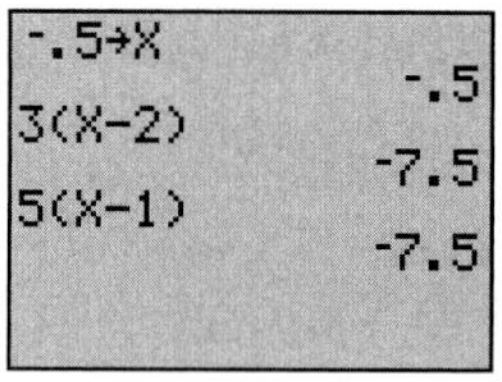

SUMMARY   The steps to follow for approximating solutions of equations are given next.

Steps for Approximating Solutions of Equations Using ZERO (or ROOT)

STEP 1: Write the equation in the form {expression in x} = 0.
STEP 2: Graph Y_1 = {expression in x}.
STEP 3: Use ZERO (or ROOT) to determine each x-intercept of the graph.

Steps for Approximating Solutions of Equations Using INTERSECT

STEP 1: Graph Y_1 = {expression in x on left side of equation}.
         Graph Y_2 = {expression in x on right side of equation}.
STEP 2: Use INTERSECT to determine each x-coordinate of the point(s) of intersection, if any.

EXAMPLE 3 Solving a Radical Equation

Find the real solutions of the equation $\sqrt[3]{2x - 4} - 2 = 0$.

SOLUTION Figure 22 shows the graph of the equation $Y_1 = \sqrt[3]{2x - 4} - 2$. From the graph, we see one x-intercept near 6. Using ZERO (or ROOT), we find that the x-intercept is (6, 0). The only solution is $x = 6$.

FIGURE 22

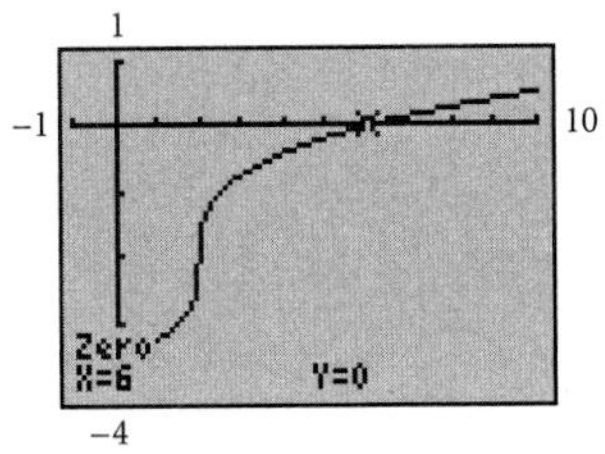

Answers to Odd-Numbered Problems

CHAPTER 1 Linear Equations

Exercise 1.1 (p. 15)

1. $A = (4, 2)$; $B = (6, 2)$; $C = (5, 3)$; $D = (-2, 1)$; $E = (-2, -3)$; $F = (3, -2)$; $G = (6, -2)$; $H = (5, 0)$

3. 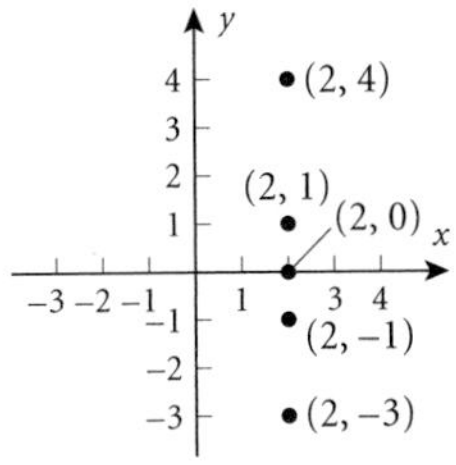

The set of points of the form $(2, y)$, where y is a real number, is a vertical line passing through $(2, 0)$ on the x-axis.

5.

x	0	-2	2	-2	4	-4
y	4	0	8	0	12	-4

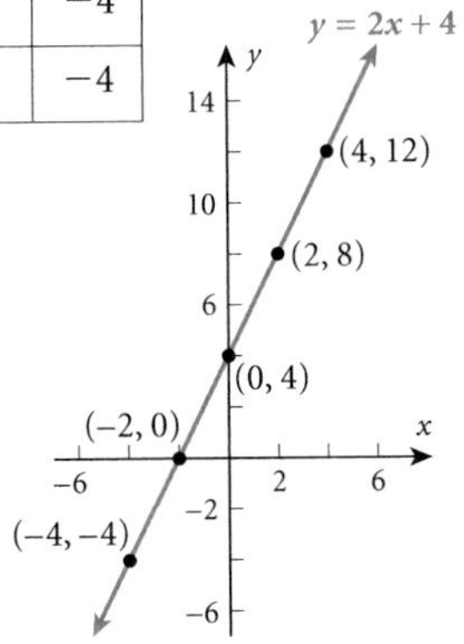

7.

x	0	3	2	-2	4	-4
y	-6	0	-2	-10	2	-14

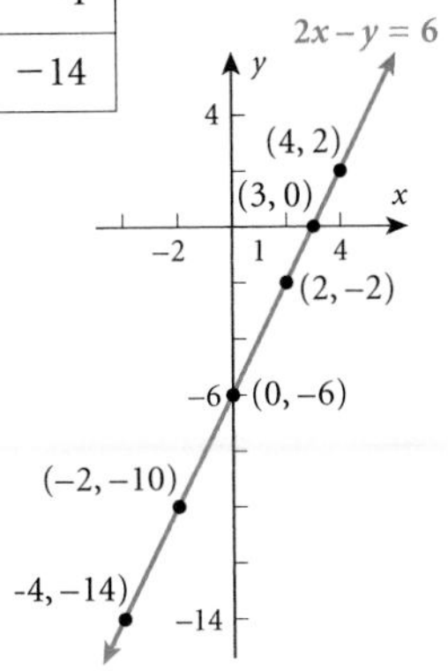

9. (a) Vertical line: $x = 2$
(b) Horizontal line: $y = -3$

11. (a) Vertical line: $x = -4$
(b) Horizontal line: $y = 1$

13. $m = \frac{1}{2}$ A slope of $\frac{1}{2}$ means that for every 2 unit change in x, y changes 1 unit.

15. $m = -1$ A slope of -1 means that for every 1 unit change in x, y changes by (-1) units.

17. $m = 3$ A slope of 3 means that for every 1 unit change in x, y will change 3 units.

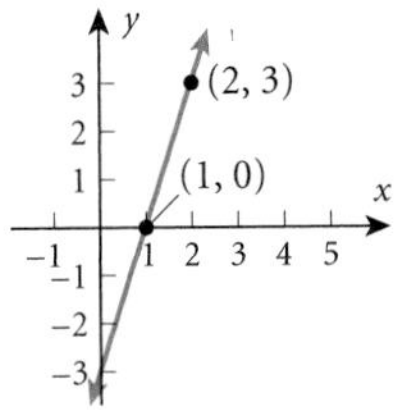

19. $m = -\frac{1}{2}$ A slope of $-\frac{1}{2}$ means that for every 2 unit change in x, y will change (-1) unit.

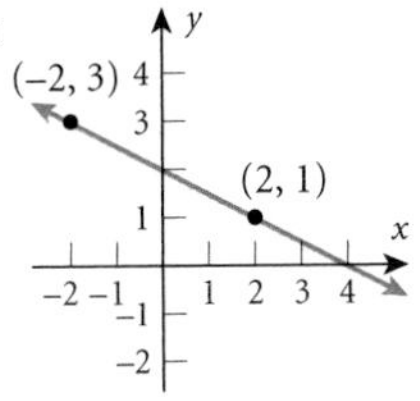

21. $m = 0$ A slope of 0 means that regardless how x changes, y remains constant.

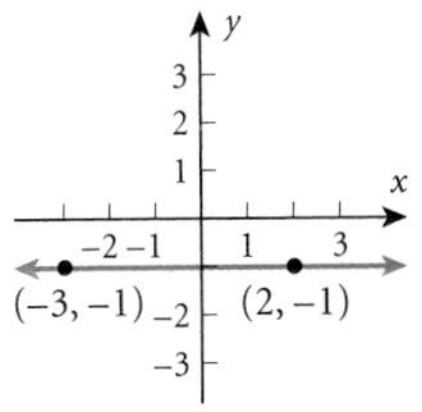

23. The slope is not defined.

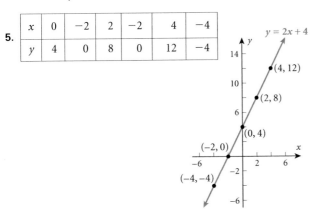

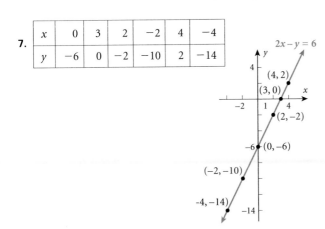

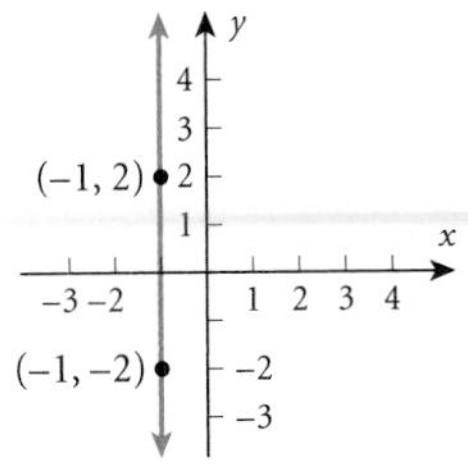

25.

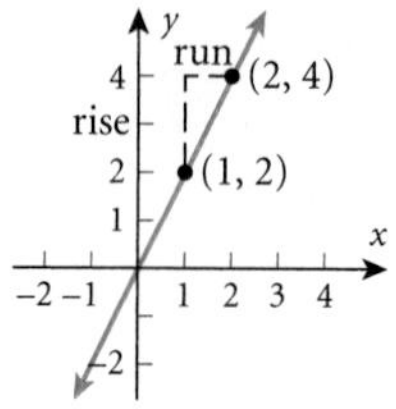

27.

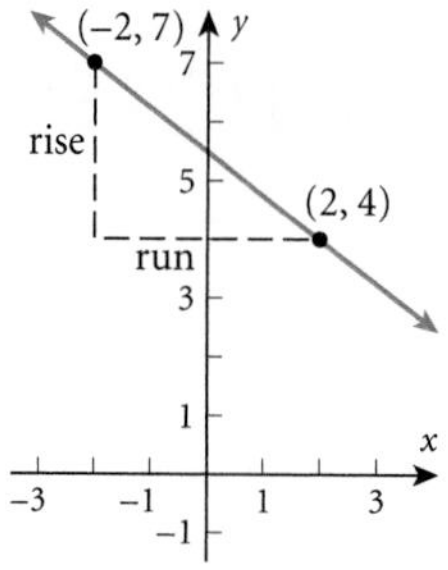

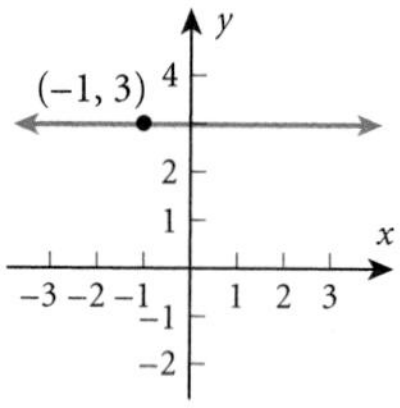

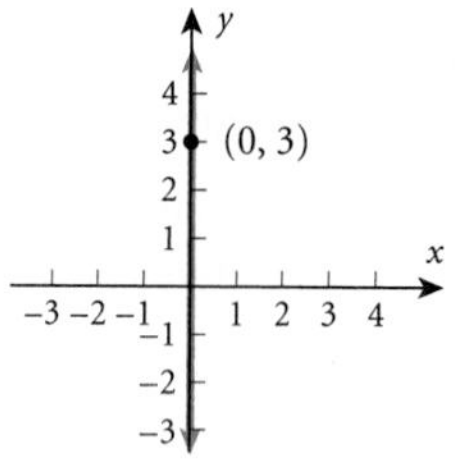

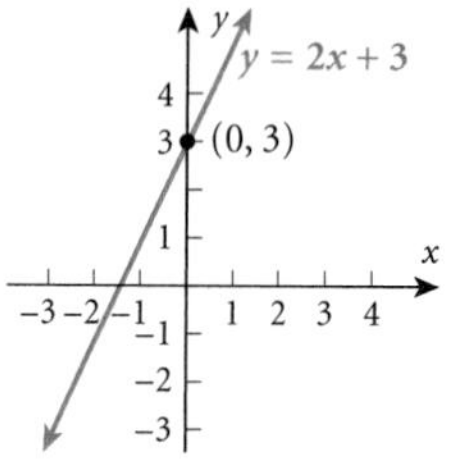

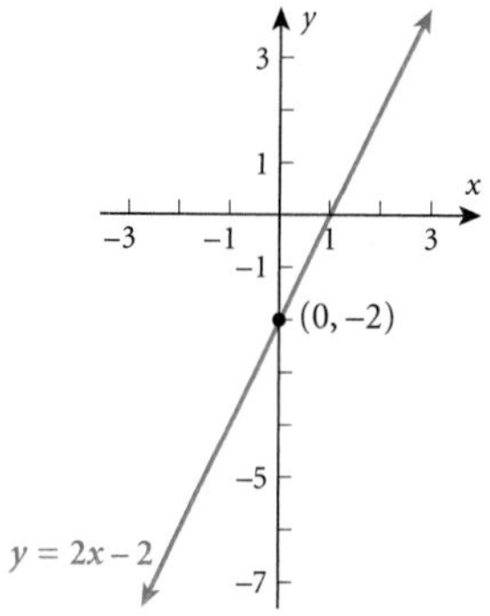

59. slope: $m = \frac{2}{3}$; y-intercept: $(0, -2)$

29.

31.

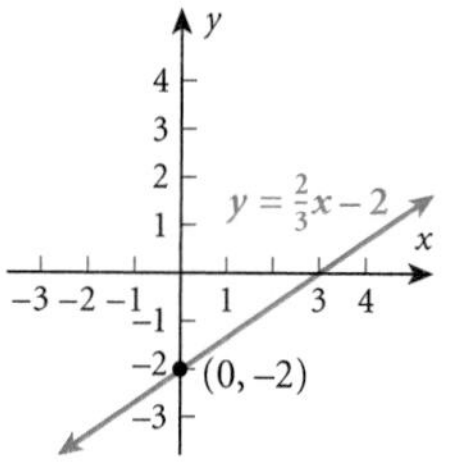

61. slope: $m = -1$; y-intercept: $(0, 1)$

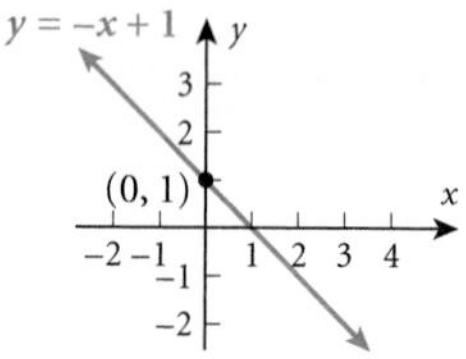

63. Slope is not defined; there is no y-intercept.

33. $x - 2y = 0$

35. $x + y = 2$

37. $2x - y = -9$

39. $2x + 3y = -1$

41. $x - 2y = -5$

43. $2x + y = 3$

45. $3x - y = -12$

47. $4x - 5y = 0$

49. $x - 2y = 2$

51. $x = 1$

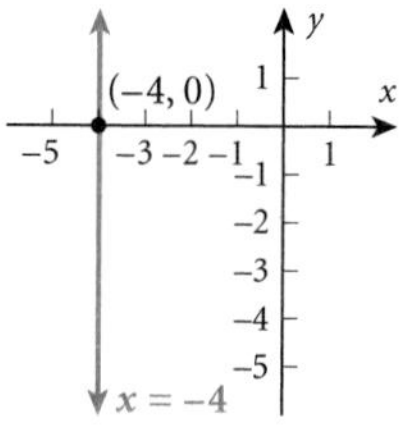

65. slope: $m = 0$; y-intercept: $(0, 5)$

53. $y = 4$

55. slope: $m = 2$; y-intercept: $(0, 3)$

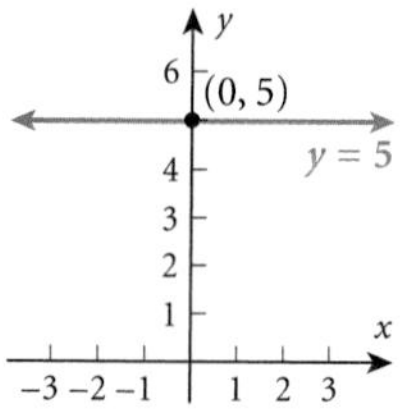

67. slope: $m = 1$; y-intercept: $(0, 0)$

57. slope: $m = 2$; y-intercept: $(0, -2)$

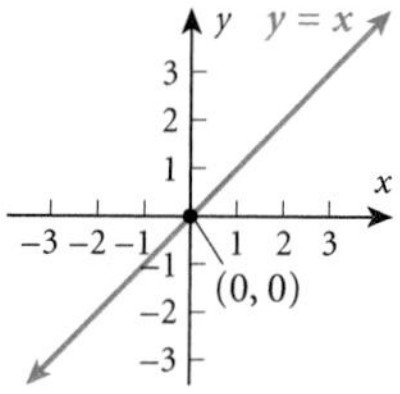

69. slope: $m = \frac{3}{2}$; y-intercept: $(0, 0)$

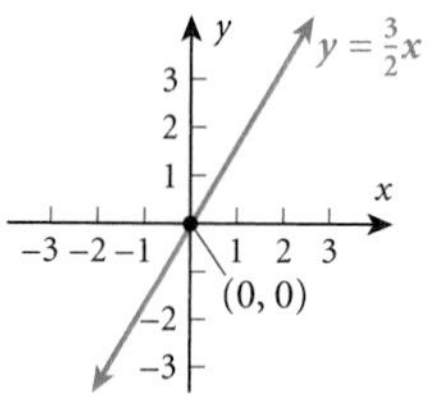

71. $y = -3$

73. $C = 0.122x$

75. (a) $C = 0.08275x + 7.58, 0 \leq x \leq 400$

(b)

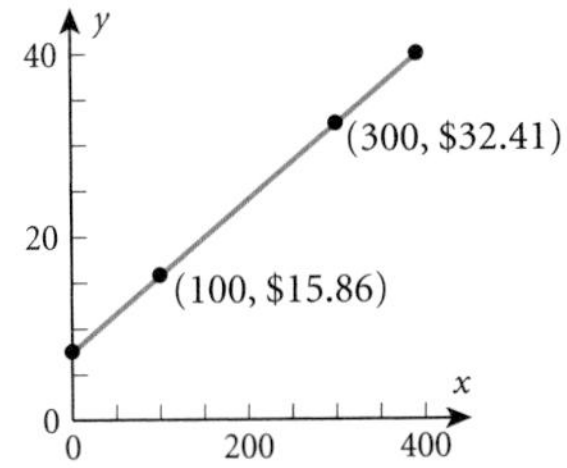

(c) The monthly charge for using 100 KWH is $15.86.

(d) The monthly charge for using 300 KWH is $32.41.

(e) The slope indicates that for every extra KWH used (up to 400 KWH), the electric bill increases by 8.275 cents.

77. $w = 4h - 129$

79. $C = 0.53x + 1{,}070{,}000$

81. (a) $C = \frac{5}{9}(F - 32)$ (b) $C = 20°$

83. (a) $y = -\frac{1}{75}t + 53.007$ (b) $y = 52.74$ billion gallons

(c) The slope tells us that the reservoir loses 1 billion gallons of water every 75 days.

(d) $y = 52.594$ billion gallons

(e) In 10.89 years the reservoir will be empty.

85. Window: X min $= -10$; X max $= 10$
Y min $= -10$; Y max $= 10$

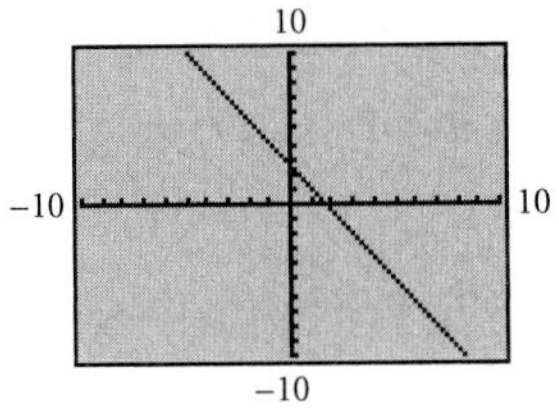

x-intercept: (1.67, 0); y-intercept: (0, 2.50)

87. Window: X min $= -10$; X max $= 10$
Y min $= -10$; Y max $= 10$

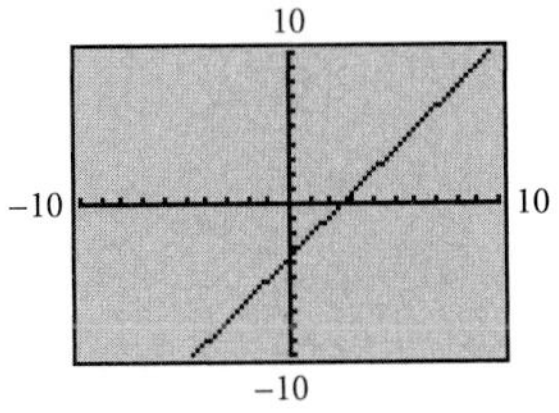

x-intercept: (2.52, 0); y-intercept: (0, −3.53)

89. Window: X min $= -10$; X max $= 10$
Y min $= -10$; Y max $= 10$

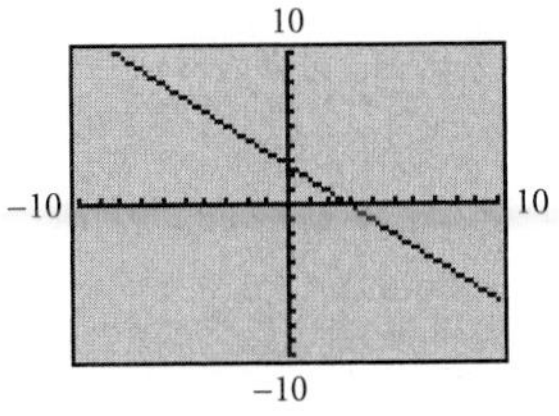

x-intercept: (2.83, 0); y-intercept: (0, 2.56)

91. Window: X min $= -10$; X max $= 10$;
Y min $= -10$; Y max $= 10$

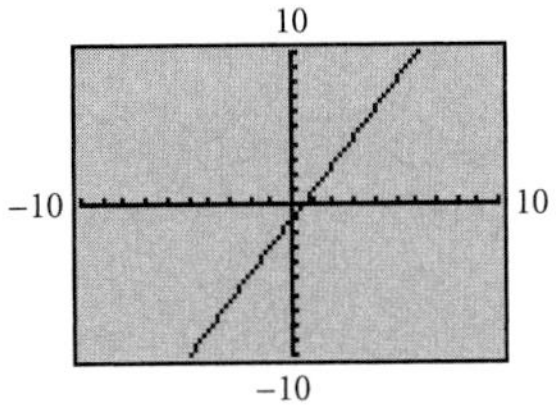

x-intercept: (0.78, 0); y-intercept: (0, −1.41)

93. (b)

95. (d)

97. $y = x + 2$ or $x - y = -2$

99. $y = -\frac{1}{3}x + 1$ or $x + 3y = 3$

101. (b), (c), (e), (g)

103. $y = 0$

105. Answers vary.

107. No; no.

109. The lines are identical.

111. Two lines can have the same y-intercept and the same x-intercept but different slopes only if their y-intercept is the point (0, 0).

Exercise 1.2 (p. 25)

1. parallel **3.** intersecting **5.** coincident

7. parallel **9.** intersecting **11.** intersecting

13. (3, 2)

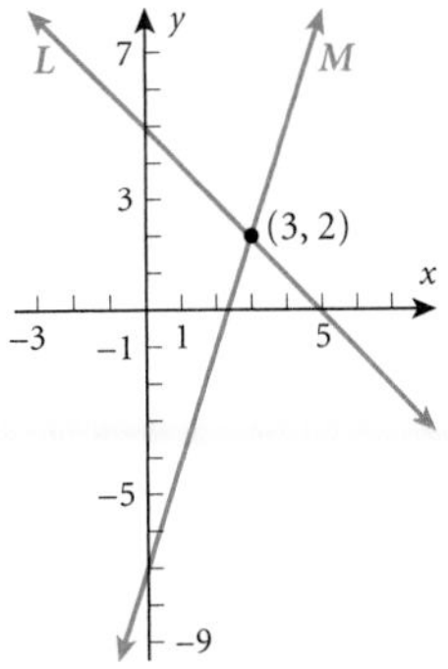

15. (3, 1)

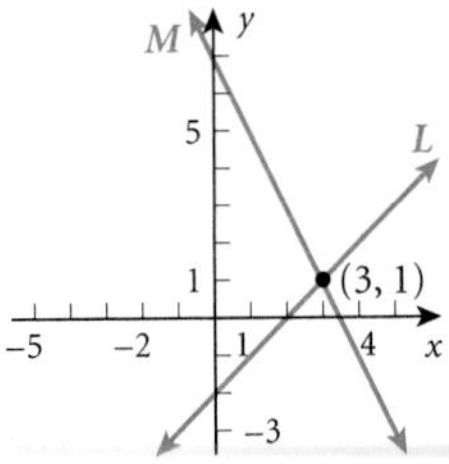

17. $(1, 0)$

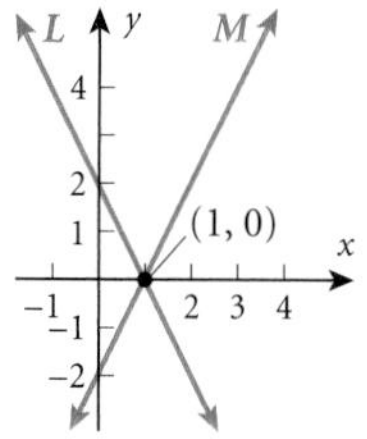

19. $(2, 1)$

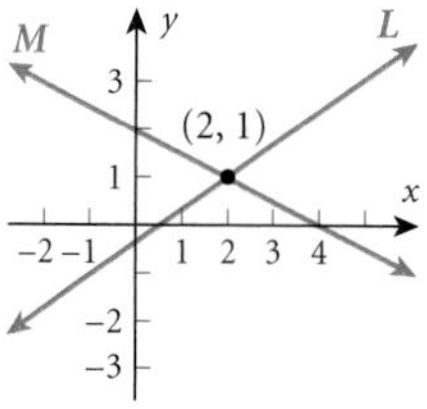

21. $(-1, 1)$

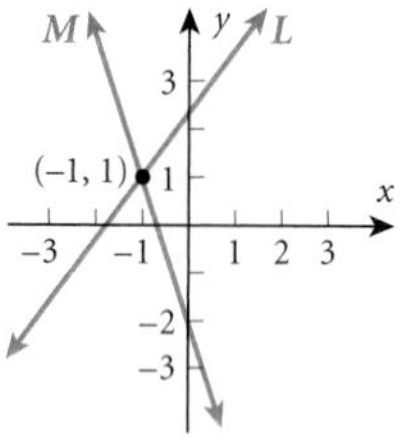

23. $(4, -2)$

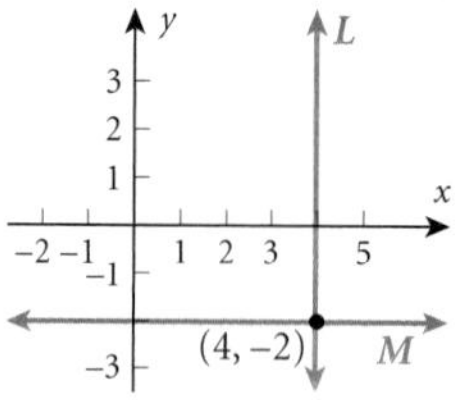

25. $m_1m_2 = \frac{1}{3} \cdot (-3) = -1$ **27.** $m_1m_2 = -\frac{1}{2} \cdot (2) = -1$

29. $m_1m_2 = -\frac{1}{4} \cdot (4) = -1$ **31.** $y = 2x - 3$ or $2x - y = 3$

33. $y = -\frac{1}{2}x + \frac{9}{2}$ or $x + 2y = 9$

35. $y = 4x + 6$ or $4x - y = -6$

37. $y = 2x$ or $2x - y = 0$ **39.** $x = 4$

41. $y = -\frac{1}{2}x - \frac{5}{2}$ or $x + 2y = -5$

43. $y = -\frac{1}{2}x + \frac{19}{30}$ or $15x + 30y = 19$

45. $t = 4$ **47.** $y = \frac{5}{19}x - \frac{85}{19}$ **49.** (c)

Exercise 1.3 (p. 33)

1. (a) $S = \$80{,}000$
(b) $S = \$95{,}000$
(c) $S = \$105{,}000$
(d) $S = \$120{,}000$

3. (a) $y = 650x - 1{,}287{,}500$
(b) The average cost of a compact car is predicted to be \$15,750.
(c) The slope can be interpreted as the average yearly increase in price of a compact car.

5. (a) $S = \frac{10}{7}t - \frac{16{,}594}{7}$
(b) The predicted average SAT score will be 492.

7. (a) $P = 0.53t - 1034.4$
(b) 27.7%
(c) The slope is the annual percentage increase of the population over 25 who hold bachelor's degrees or higher.

9. The break even point is $x = 30$.

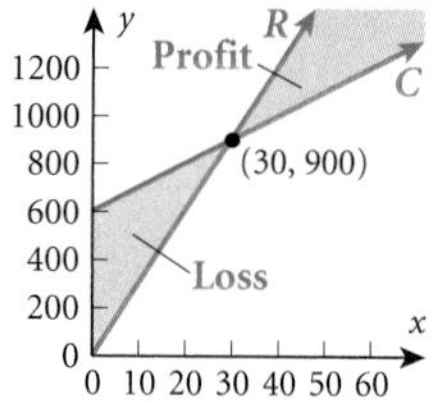

11. The break even point is $x = 500$.

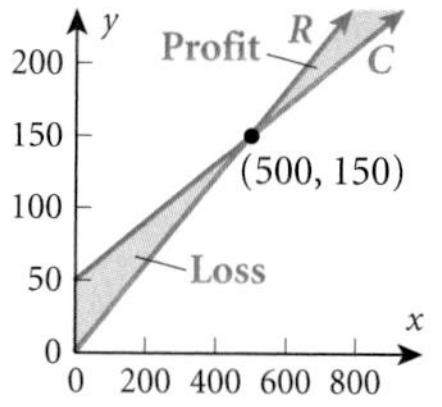

13. The break even point occurs when $x = 1200$ items are produced and sold.

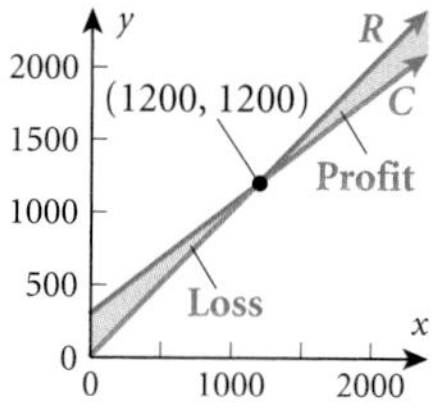

15. (a) The revenue from delivering x newspapers is given by $R = \$1.79x$
(b) The cost of delivering x newspapers is given by $C = \$0.53x + \$1{,}070{,}000$
(c) The profit from delivering x newspapers is given by $P = 1.26x - 1{,}070{,}000$
(d) (849207, 1520080) The company breaks even when 849,207 newspapers are delivered.
(e)

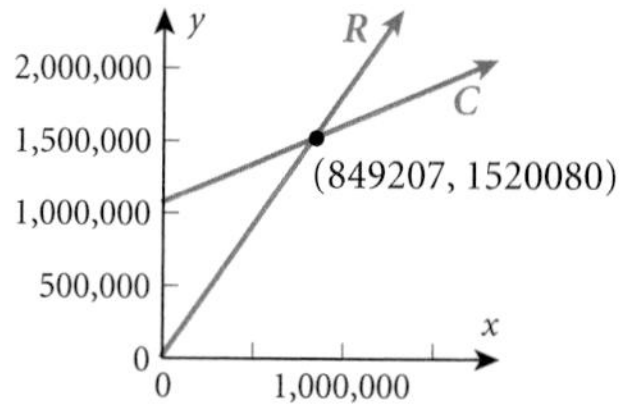

(f)

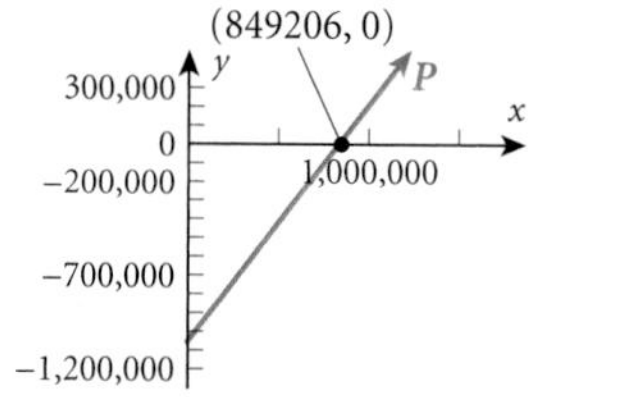

17. To break even put 20 caramels and 30 creams into each box; increase the number of caramels to obtain a profit.

19. Mr. Nicholson should invest \$50,000 in AA Bonds and \$100,000 in Savings and Loan Certificates.

21. Mix 25 pounds of Kona coffee with 75 pounds of Colombian coffee to obtain a blend worth \$10.80 per pound.

23. Mix 30 cubic centimeters of the 15% solution with 70 cubic centimeters of the 5% solution to obtain a solution that is 8% acid.

25. The market price is \$1.00.

27. The market price is \$10.00.

29. (a) Market price: \$1.00
(b) Supply demanded at market price: 1.1 units
(c)

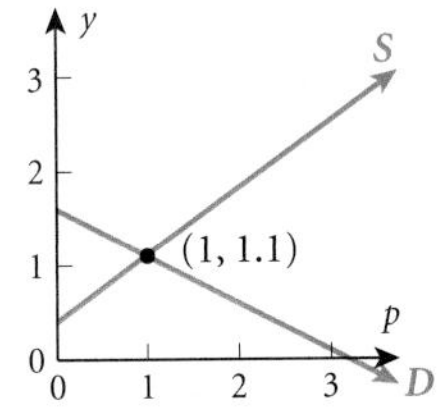

31. $D = -4p + 23$

Exercise 1.4 (p. 40)

1. A relation exists, and it appears to be linear.

3. A relation exists, and it appears to be linear.

5. No relation exists.

7. (a)

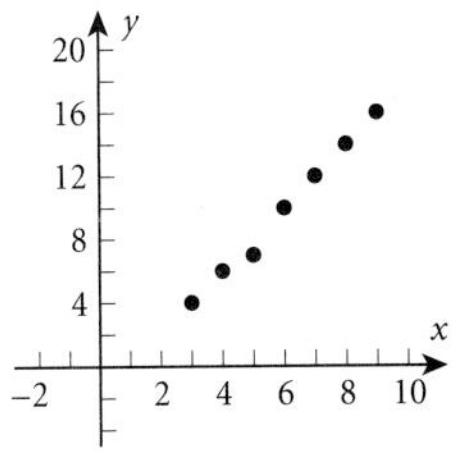

(b) $y = 2x - 2$

(c)

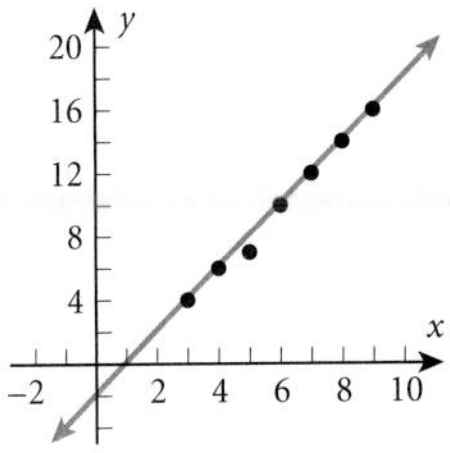

(d) Window: X min $= -2$; X max $= 10$
Y min $= -3$; Y max $= 20$

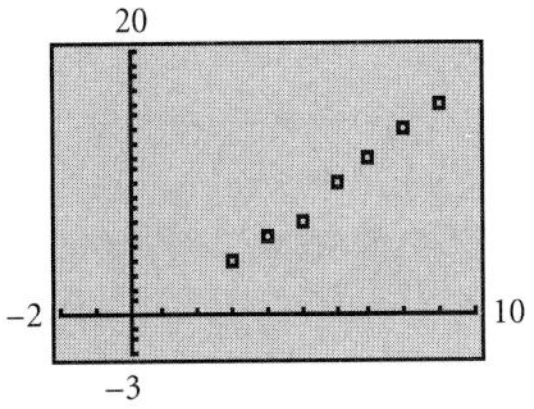

(e) Using the LinReg program, the line of best fit is:
$y = 2.0357x - 2.3571$.

(f)

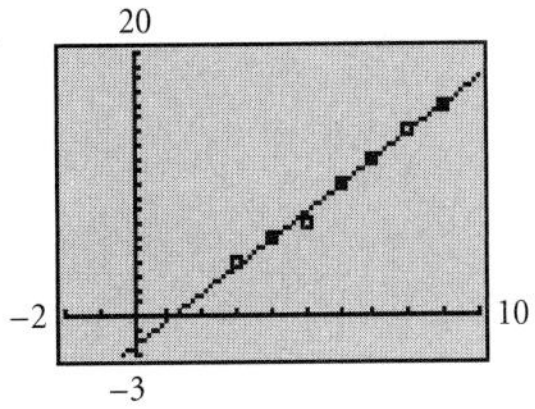

9. (a)

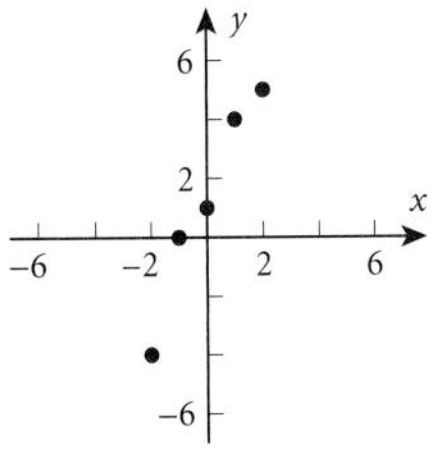

(b) $y = \frac{9}{4}x + \frac{1}{2}$

(c)

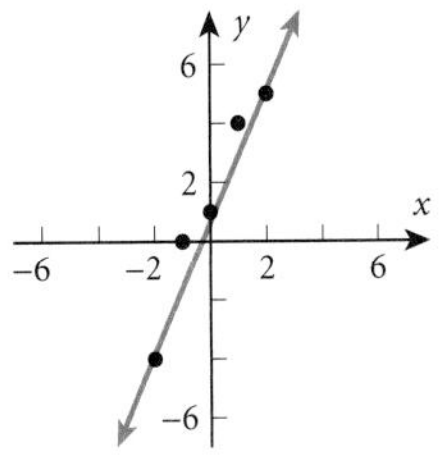

(d) Window: X min $= -6$; X max $= 6$
Y min $= -6$; Y max $= 7$

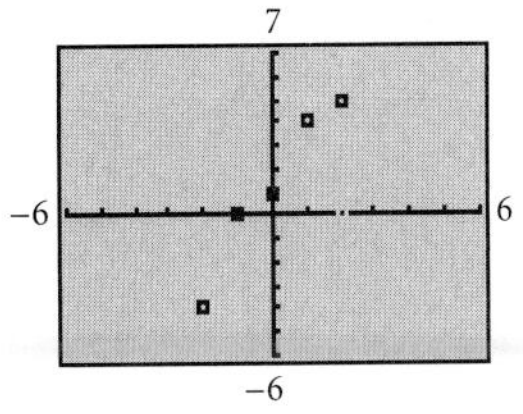

(e) Using the LinReg program, the line of best fit is:
$y = 2.2x + 1.2$.

(f)

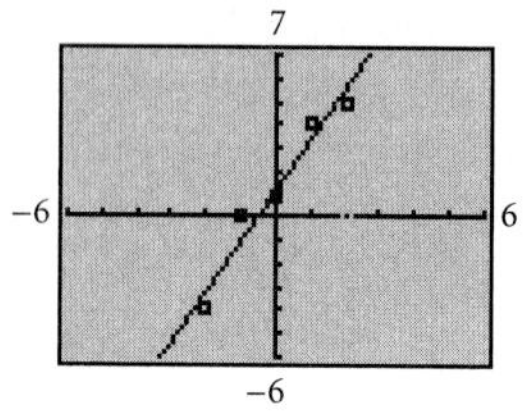

11. (a)

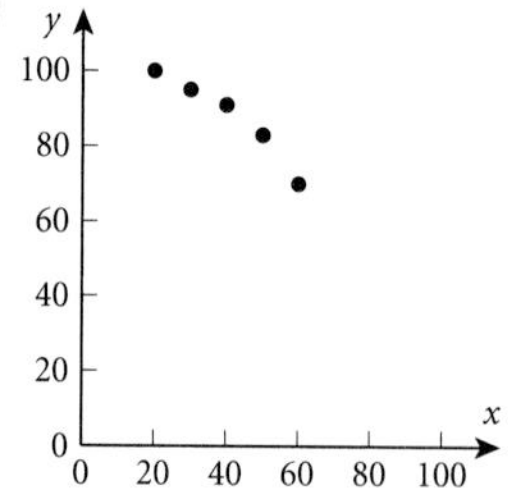

(b) $y = -\frac{3}{4}x + 115$

(c)

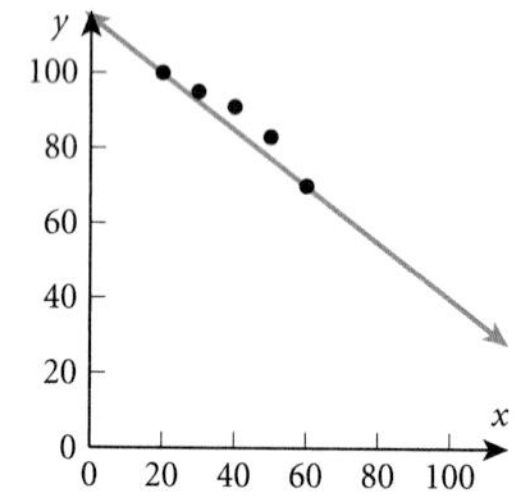

(d) Window: X min = 0; X max = 100
Y min = 1; Y max = 120

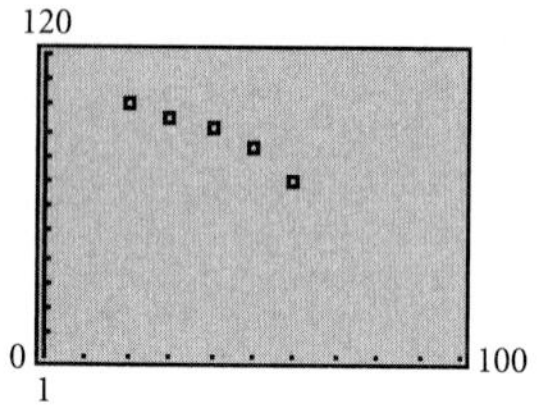

(e) Using the LinReg function the line of best fit is:
$y = -0.72x + 116.6$.

(f)

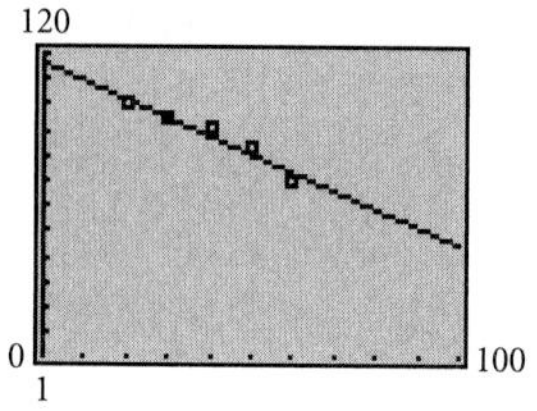

13. (a)

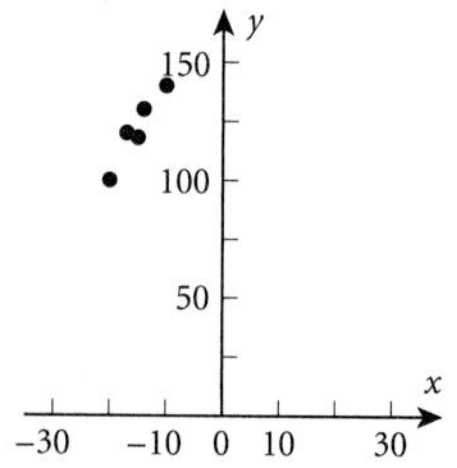

(b) $y = 4x + 180$

(c)

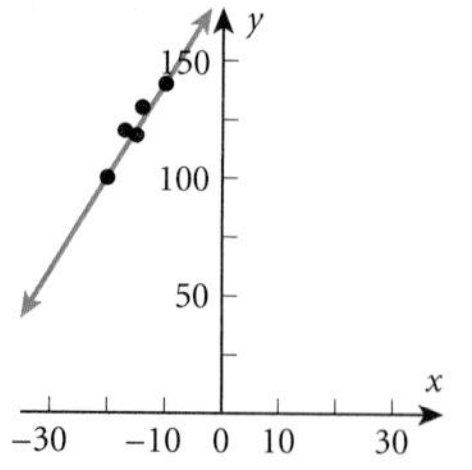

(d) Window: X min = −30; X max = 10
Y min = 0; Y max = 160

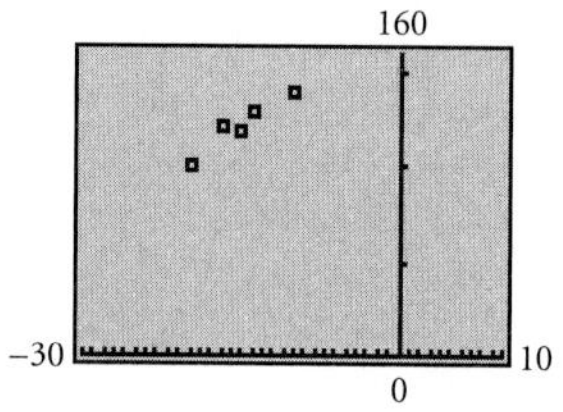

(e) Using the LinReg function, the line of best fit is:
$y = 3.86131x + 180.29197$.

(f)

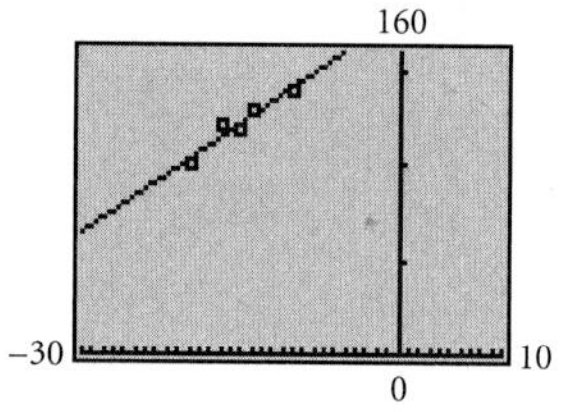

15. (a)

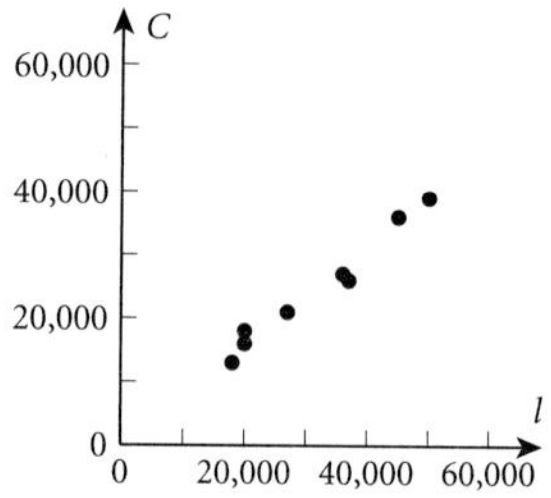

(b) $C = \frac{23}{30}I + \frac{2}{3}$

(c) The slope of this line indicates that a family will spend \$23 of every extra \$30 of disposable income.

(d) A family with a disposable income of \$42,000 is predicted to consume \$32,867 worth of goods.

(e) Using the LinReg function on the graphing utility, the line of best fit is: $y = 0.75489x + 0.62663$

17. (a) Window: X min = 0; X max = 75.5 (thousand)
Y min = 0; Y max = 236.5 (thousand)

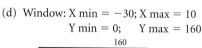

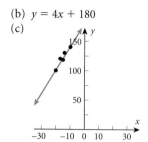

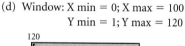

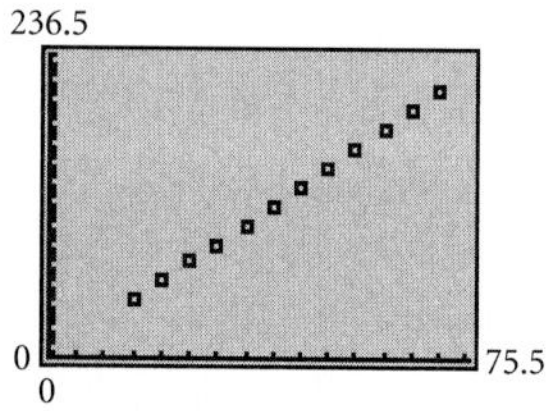

(b) Using the LinReg function, the line of best fit is: $y = 2.98140x - 0.07611$.

(c)

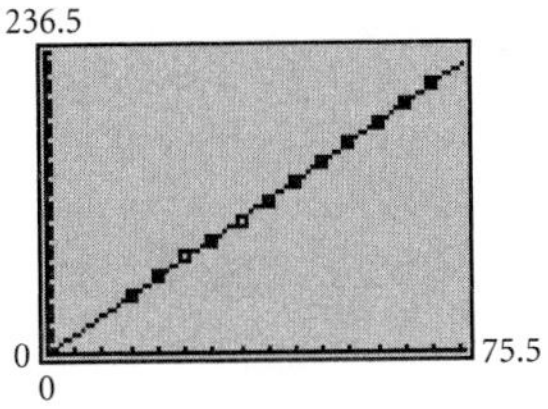

(d) The slope indicates that a person can borrow an additional \$2.98 for each additional dollar of income.

(e) An individual with an annual income of \$42,000 would qualify for a \$125,143 loan.

19. (a) Window: X min $= -10$; X max $= 110$
Y min $= 50$; Y max $= 70$

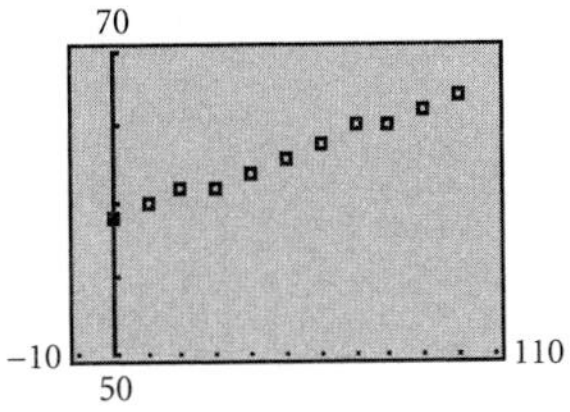

(b) Using the LinReg function, the line of best fit is: $y = 0.07818x + 59.0909$.

(c)

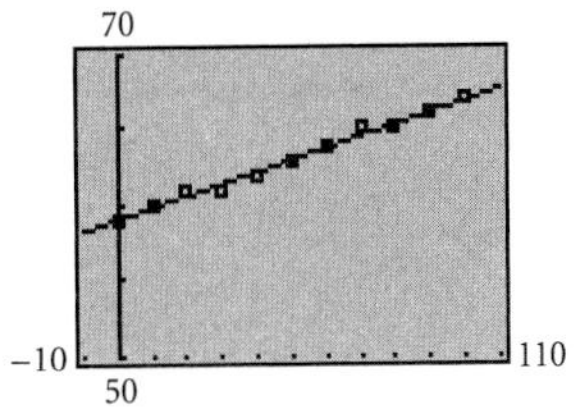

(d) The slope indicates the apparent change in temperature in a 65°F room for every percent increase in relative humidity.

(e) The apparent temperature of a room with an actual temperature of 65°F remains at 65° when the relative humidity is 75%.

CHAPTER 1 Review

True–False Items (p. 43)

1. F **2.** T **3.** T **4.** F

5. F **6.** T **7.** F **8.** T

9. F **10.** F

Fill in the Blanks (p. 43)

1. abscissa; ordinate, or x-coordinate; y-coordinate

2. Undefined; zero **3.** negative

4. parallel **5.** coincident

6. perpendicular **7.** intersecting

Review Exercises (p. 44)

1.

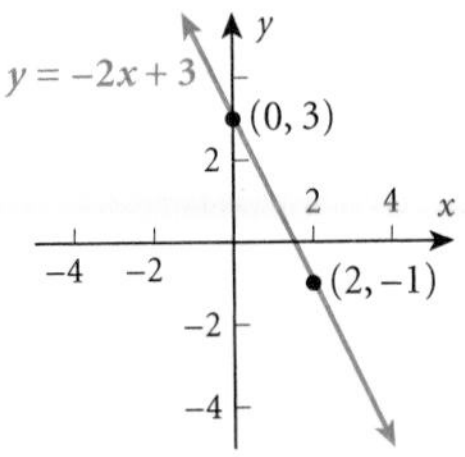

3.

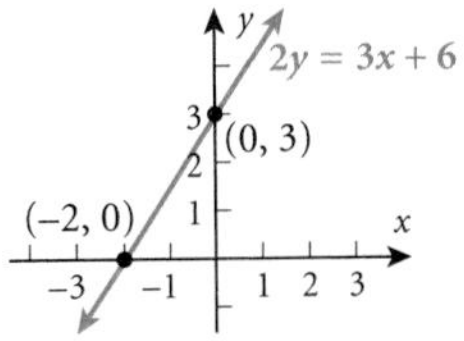

5. (a) $m = -\frac{1}{2}$; A slope $= -\frac{1}{2}$ means that for every 2 units x moves to the right, y moves down 1 unit.

(b) $y = -\frac{1}{2}x + \frac{5}{2}$ or $x + 2y = 5$

(c)

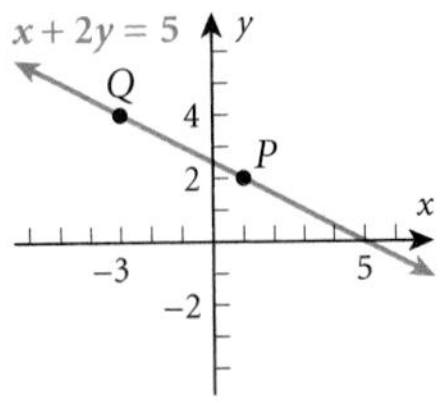

7. (a) $m = 2$; A slope $= 2$ means that for every 1 unit change in x, y changes 2 units.

(b) $y = 2x + 7$ or $2x - y = -7$

(c)

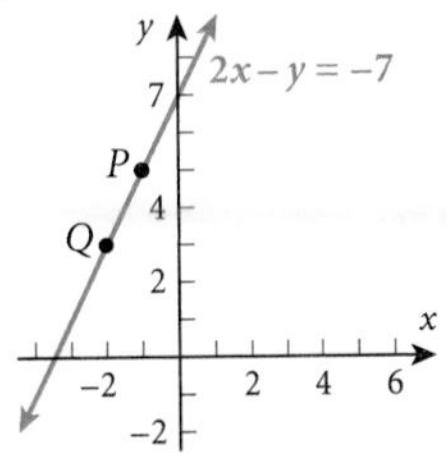

9. $y = -3x + 5$ or $3x + y = 5$

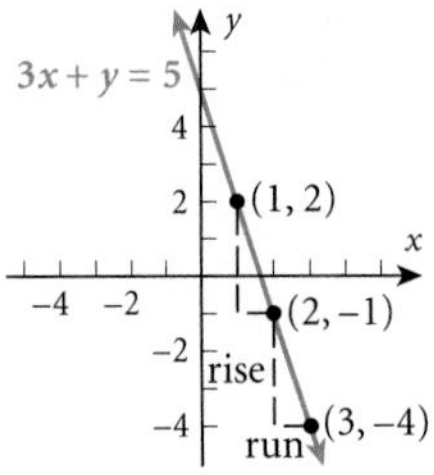

11. $y = 4$

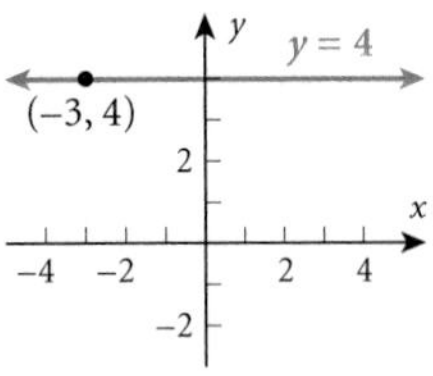

13. $x = 8$

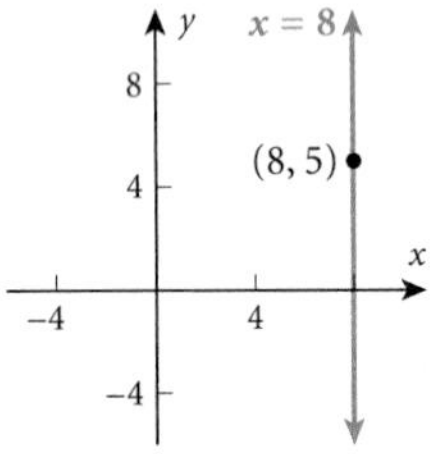

15. $y = -\frac{5}{2}x + 5$ or $5x + 2y = 10$

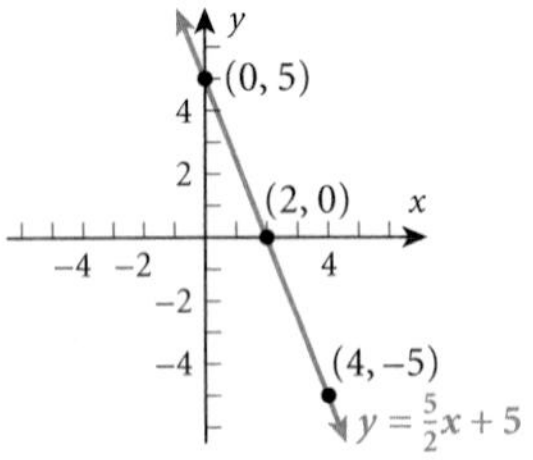

17. $y = -\frac{4}{3}x - 4$ or $4x + 3y = -12$

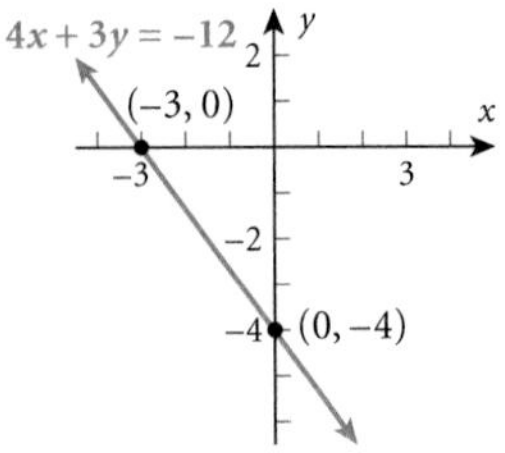

19. $y = -\frac{2}{3}x - \frac{1}{3}$ or $2x + 3y = -1$

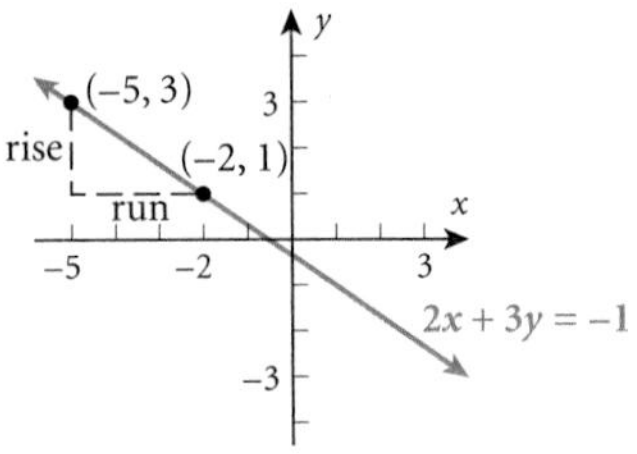

21. $y = \frac{3}{2}x + \frac{21}{2}$ or $3x - 2y = -21$

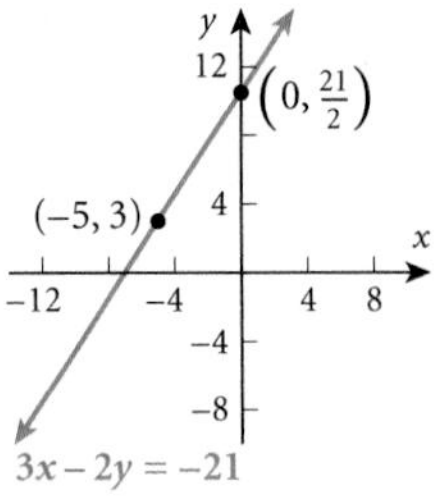

23. $m = -\frac{9}{2}$, y-intercept $(0, 9)$

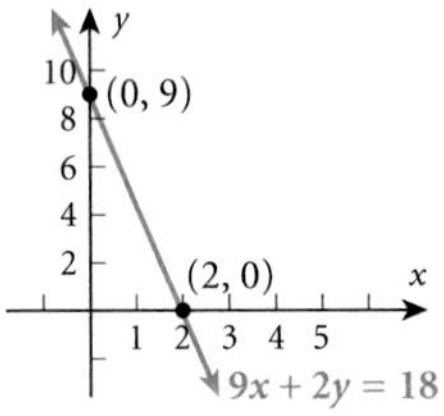

25. $m = -2$, y-intercept: $(0, \frac{9}{2})$

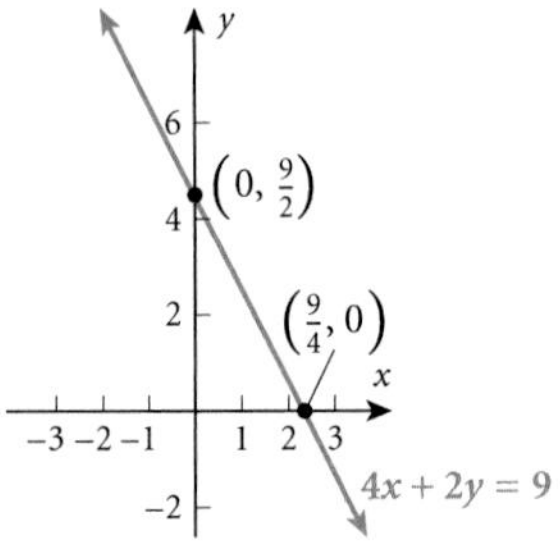

27. parallel **29.** intersecting **31.** coincident

33. $(5, 1)$

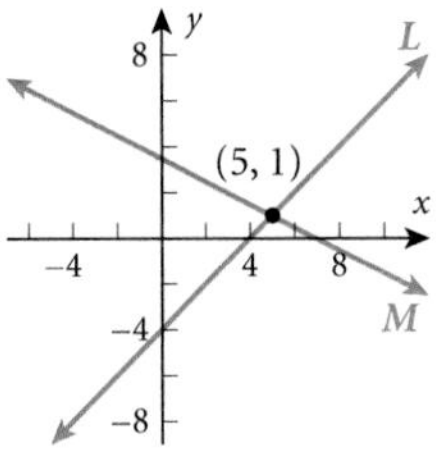

35. $(1, 3)$

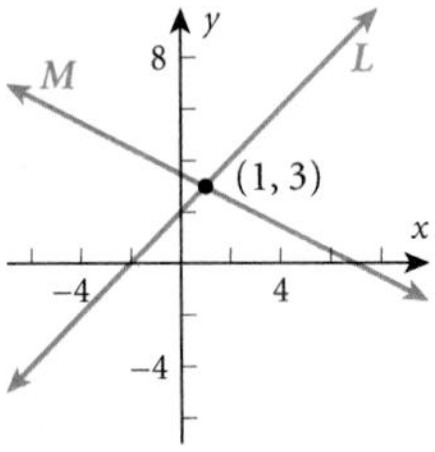

37. $(-2, 1)$

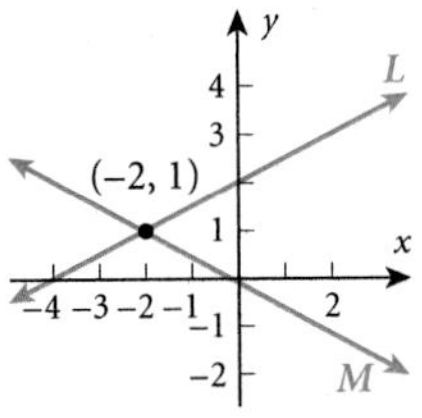

39. Invest \$78,571.43 in B-rated bonds and \$11,428.57 in the well-known bank.

41. (a) 120 people need to attend for the group to break even.
(b) 300 people need to attend to achieve the \$900 profit.
(c) If tickets are sold for \$12.00 each 86 people must attend to break even, and 215 people must attend to achieve a profit of \$900.

43. Relation does not appear to be linear.
(a)

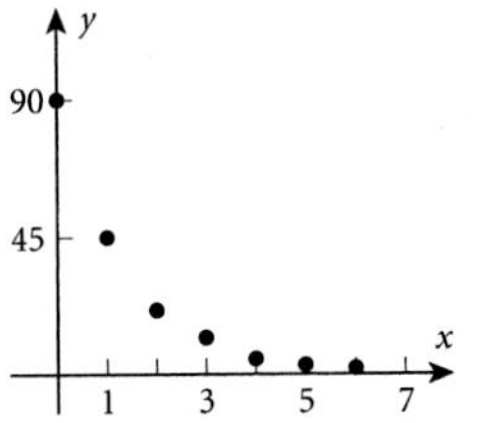

45. (a)

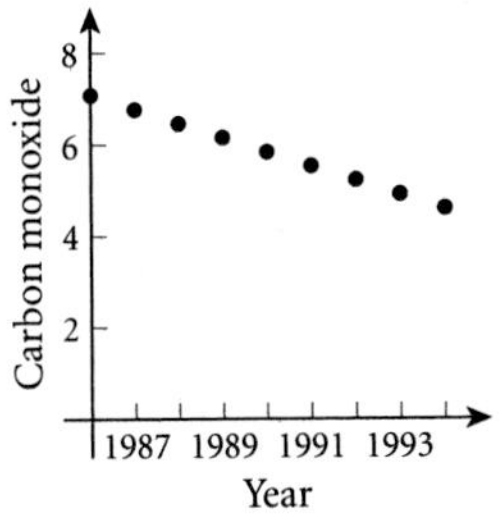

(b) $m = -0.2733$

(c) The slope indicates the average annual decrease in concentration of carbon monoxide between 1987 and 1990.
(d) $m = -0.33$
(e) The slope indicates the average annual decrease in concentration of carbon monoxide between 1990 and 1993.
(f) $m = -0.308$
(g) The slope indicates the average annual decrease in the concentration of carbon monoxide.
(i) The trend indicates that the average level of carbon monoxide is decreasing.

47. (a)

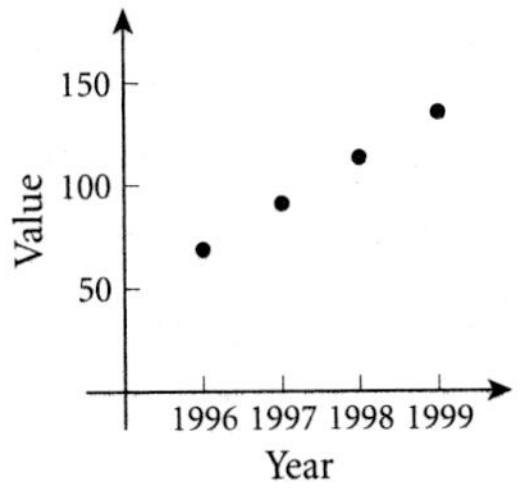

(b) The relation appears to be linear.
(c) $m = 22.05$
(d) The slope represents the average annual increase in value of a share of the Vanguard 500 Index Fund from 1996 to 1999.
(e) $y = 22.236x + 46.54$
(f) \$179.96

Mathematical Questions from Professional Exams

1. b **2.** d **3.** d **4.** d

5. c **6.** c **7.** b **8.** b

CHAPTER 2 Systems of Linear Equations; Matrices

Exercise 2.1 (p. 62)

1. Yes **3.** No **5.** Yes **7.** Yes **9.** Yes

11. The solution of the system is $x = 6$ and $y = 2$.

13. The solution of the system is $x = 3$ and $y = 2$.

15. The solution of the system is $x = 8$ and $y = -4$.

17. The solution of the system is $x = \frac{1}{3}$ and $y = -\frac{1}{6}$.

19. The system is inconsistent.

21. The solution of the system is $x = 1$ and $y = 2$.

23. The solutions of the system are $y = -\frac{1}{2}x + 2$ and x where x is any real number, or as $x = -2y + 4$ and y, where y is any real number.

25. The solution of the system is $x = 1$ and $y = 1$.

27. The solution of the system is $x = \frac{3}{2}$ and $y = 1$.

29. The solution of the system is $x = 4$ and $y = 3$.

31. The solution of the system is $x = \frac{4}{3}$ and $y = \frac{1}{5}$.

33. The solution of the system is $x = 8$, $y = 2$, and $z = 0$.

35. The solution of the system is $x = 2$, $y = -1$, and $z = 1$.

37. The system is inconsistent.

39. The solutions of the system are $x = 5z - 2$, $y = 4z - 3$, and z, where z is any real number.

41. The system is inconsistent.

43. The solution of the system is $x = 1$, $y = 3$, and $z = -2$.

45. The solution of the system is $x = -3$, $y = \frac{1}{2}$, and $z = 1$.

47. The dimensions of the floor should be 30 ft × 15 ft.

49. They should plant 219.8 acres of corn and 225.2 acres of soybeans.

51. 22.5 pounds of cashews should be mixed with the peanuts.

53. A bowl of noodles costs 571 yen and a carton of milk costs 220 yen.

55. The amount of the refund should be \$5.56.

57. 50 mg of the first liquid (20% vitamin C and 30% vitamin D) should be mixed with 75 mg of the second liquid.

59. Use 9.16 pounds of rolled oats and 8.73 pounds of molasse.

61. The theater has 100 orchestra seats, 210 main seats, and 190 balcony seats.

63. Kelly should invest \$8000 in treasury bills, \$7000 in treasury bonds, and \$5000 in corporate bonds.

Exercise 2.2 (p. 78)

1. $\left[\begin{array}{cc|c} 2 & -3 & 5 \\ 1 & -1 & 3 \end{array}\right]$

3. $\left[\begin{array}{cc|c} 2 & 1 & -6 \\ 3 & 1 & -1 \end{array}\right]$

5. $\left[\begin{array}{ccc|c} 2 & -1 & -1 & 0 \\ 1 & -1 & 1 & 1 \\ 3 & -1 & 0 & 2 \end{array}\right]$

7. $\left[\begin{array}{ccc|c} 2 & -3 & 1 & 7 \\ 1 & 1 & -1 & 1 \\ 2 & 2 & -3 & -4 \end{array}\right]$

9. $\left[\begin{array}{cccc|c} 4 & -1 & 2 & -1 & 4 \\ 1 & 1 & 0 & 0 & -6 \\ 0 & 2 & -1 & 1 & 5 \end{array}\right]$

11. $\left[\begin{array}{cccc|c} 1 & -1 & 1 & -1 & 0 \\ 2 & 3 & -1 & 4 & 5 \end{array}\right]$

13. $\left[\begin{array}{cc|c} 1 & -3 & -2 \\ 0 & 1 & 9 \end{array}\right]$

15. (a) $\left[\begin{array}{ccc|c} 1 & -3 & 4 & 3 \\ 0 & 1 & -2 & 0 \\ -3 & 3 & 4 & 6 \end{array}\right]$ (b) $\left[\begin{array}{ccc|c} 1 & -3 & 4 & 3 \\ 2 & 5 & 6 & 6 \\ 0 & -6 & 16 & 15 \end{array}\right]$

17. (a) $\left[\begin{array}{ccc|c} 1 & -3 & 2 & -6 \\ 0 & 1 & -1 & 8 \\ -3 & -6 & 2 & 6 \end{array}\right]$ (b) $\left[\begin{array}{ccc|c} 1 & -3 & 2 & -6 \\ 2 & -5 & 3 & -4 \\ 0 & -15 & 8 & -12 \end{array}\right]$

19. (a) $\left[\begin{array}{ccc|c} 1 & -3 & 1 & -2 \\ 0 & 1 & 4 & 2 \\ -3 & 1 & 4 & 6 \end{array}\right]$ (b) $\left[\begin{array}{ccc|c} 1 & -3 & 1 & -2 \\ 2 & -5 & 6 & -2 \\ 0 & -8 & 7 & 0 \end{array}\right]$

21. $\begin{cases} x + 2y = 5 \\ y = -1 \end{cases}$

The system is consistent and the solution is $x = 7$ and $y = -1$.

23. $\begin{cases} x + 2y + 3z = 1 \\ y + 4z = 2 \\ 0 = 3 \end{cases}$

The system is inconsistent.

25. $\begin{cases} x + 2z = -1 \\ y - 4z = -2 \\ 0 = 0 \end{cases}$

The system is consistent and has an infinite number of solutions. The solutions are $x = -2z - 1$, $y = 4z - 2$, and z, where z is any real number.

27. $\begin{cases} x_1 + 2x_2 - x_3 + x_4 = 1 \\ x_2 + 4x_3 + x_4 = 2 \\ x_3 + 2x_4 = 3 \end{cases}$

The system is consistent. The solutions are $x_1 = -17x_4 + 24$, $x_2 = 7x_4 - 10$, $x_3 = -2x_4 + 3$, and x_4, where x_4 is any real number.

29. $\begin{cases} x_1 + 2x_2 + 4x_4 = 2 \\ x_2 + x_3 + 3x_4 = 3 \\ 0 = 0 \end{cases}$

The system is consistent. The solutions are $x_1 = 2x_3 + 2x_4 - 4$, $x_2 = -x_3 - 3x_4 + 3$, x_3 and x_4, where x_3 and x_4 are any real numbers.

31. $\begin{cases} x_1 - 2x_2 + x_4 = -2 \\ x_2 - 3x_3 + 2x_4 = 2 \\ x_3 - x_4 = 0 \\ 0 = 0 \end{cases}$

The system is consistent. The solutions are $x_1 = x_4 + 2$, $x_2 = x_4 + 2$, $x_3 = x_4$, and x_4, where x_4 is any real number.

33. The solution of the system is $x = 2$ and $y = 4$.

35. The solution of the system is $x = 2$ and $y = 1$.

37. The solution of the system is $x = 2$ and $y = 1$.

39. The system is inconsistent.

41. The solution of the system is $x = \frac{1}{2}$ and $y = \frac{1}{3}$.

43. The solutions of the system are $y = -\frac{1}{3}x + \frac{2}{3}$ and x, where x is any real number, or $x = -3y + 2$ and y where y is any real number.

45. The solution of the system is $x = 2$ and $y = 3$.

47. The solution of the system is $x = \frac{2}{3}$ and $y = \frac{1}{3}$.

49. The solution of the system is $x = 1$, $y = 4$, and $z = 0$.

51. The solution of the system is $x = \frac{8}{5}$, $y = -\frac{12}{5}$, and $z = 6$.

53. The solution of the system is $x = 2$, $y = -1$, and $z = 1$.

55. The system is inconsistent.

57. The solution of the system is $x = \frac{1}{3}$, $y = \frac{2}{3}$, and $z = 1$.

59. The solution of the system is $x = \frac{2}{9}$, $y = -\frac{2}{3}$, and $z = \frac{2}{9}$.

61. The solution of the system is $x = 2$, $y = -1$, and $z = 3$.

63. The solution of the system is $x_1 = 20$, $x_2 = -13$, $x_3 = 17$, and $x_4 = -4$.

65. A mezzanine ticket costs \$54, a lower balcony ticket costs \$38, and a middle balcony ticket costs \$30.

67. There are 10 work stations set up for 2 students and 6 work stations set up for 3 students.

69. Carletta should invest \$4000 in treasury bills, \$4000 in treasury bonds, and \$2000 in corporate bonds.

71. The meal should consist of 1 serving of chicken, 1 serving of potatoes, and 2.5 servings of spinach.

73. 20 cases of orange juice, 12 cases of tomato juice, and 6 cases of pineapple juice are prepared.

75. The teacher should order 2 of the first package (20 white, 15 blue, 1 red), 10 of the second package (3 blue, 1 red), and 4 of the third package.

77. The recreation center should purchase 4 assorted cartons, 8 mixed cartons, and 5 single cartons.

79. To fill the order use 2 large cans, 1 mammoth can, and 4 giant cans.

83. Answers will vary.

Exercise 2.3 (p. 90)

1. No, the second row contains all zeros; it should be at the bottom.

3. No, there is a 1 above the leftmost 1 in the 2nd row. It should be a 0.

5. No, the leftmost 1 in the 2nd row is not to the right of the leftmost 1 in the 1st row.

7. Yes **9.** Yes **11.** Yes

13. Infinitely many solutions, $x = -y + 1$ and y, where y is any real number.

15. One solution, $x = 4$ and $y = 5$.

17. Infinitely many solutions, $x = 2z + 6$, and $y = -3z + 1$, and z, where z is any real number.

19. Infinitely many solutions, $x = -2y + 1$, y, and $z = 2$, where y is any real number.

21. Infinitely many solutions, $x = 1$, $y = 2$, and z, where z, is any real number.

23. One solution, $x = -1$, $y = 3$, and $z = 4$.

25. Infinitely many solutions, $x = z + 1$, $y = -2z + 1$, and z, where z is any real number.

27. Infinitely many solutions, $x_1 = x_4 + 4$, $x_2 = -2x_3 - 3x_4$, x_3, and x_4, where x_3 and x_4 are any real numbers.

29. $\left[\begin{array}{cc|c} 1 & 0 & 2 \\ 0 & 1 & 1 \end{array}\right]$ The solution of the system is $x = 2$ and $y = 1$.

31. $\left[\begin{array}{cc|c} 1 & 0 & 0 \\ 0 & 1 & 0 \\ 0 & 0 & 1 \end{array}\right]$ The system is inconsistent.

33. $\left[\begin{array}{cc|c} 1 & -2 & 4 \\ 0 & 0 & 0 \\ 0 & 0 & 0 \end{array}\right]$ The solutions of the system are $x = 2y + 4$, where y is any real number.

35. $\left[\begin{array}{ccc|c} 1 & 0 & 0 & -3 \\ 0 & 1 & 3 & 5 \\ 0 & 0 & 0 & 0 \end{array}\right]$ The solutions of the system are $x = -3$, $y = -3z + 5$, and z, where z is any real number.

37. $\left[\begin{array}{ccc|c} 1 & 0 & 0 & 3 \\ 0 & 1 & 0 & 2 \\ 0 & 0 & 1 & -4 \end{array}\right]$ The solution of the system is $x = 3$, $y = 2$, and $z = -4$.

39. $\left[\begin{array}{cccc|c} 1 & 0 & 0 & 0 & -17 \\ 0 & 1 & 0 & 0 & 24 \\ 0 & 0 & 1 & 0 & 33 \\ 0 & 0 & 0 & 1 & 14 \end{array}\right]$ The solution of the system is $x_1 = -17$, $x_2 = 24$, $x_3 = 33$, and $x_4 = 14$.

41. $\left[\begin{array}{cccc|c} 1 & 0 & 0 & -\frac{1}{3} & \frac{4}{3} \\ 0 & 1 & 0 & -\frac{11}{15} & \frac{14}{15} \\ 0 & 0 & 1 & \frac{4}{15} & -\frac{16}{15} \end{array}\right]$ The solution of the system is $x_1 = \frac{1}{3}x_4 + \frac{4}{3}$, $x_2 = \frac{11}{15}x_4 + \frac{14}{15}$, $x_3 = -\frac{4}{15}x_4 - \frac{16}{15}$, and x_4, where x_4 is any real number.

43. $\left[\begin{array}{ccc|c} 1 & -1 & 1 & 0 \\ 0 & 0 & 0 & 1 \end{array}\right]$ The system is inconsistent.

45. $\left[\begin{array}{ccc|c} 1 & 0 & \frac{7}{12} & 0 \\ 0 & 1 & -\frac{1}{4} & 0 \\ 0 & 0 & 0 & 1 \end{array}\right]$ The system is inconsistent.

47. $\left[\begin{array}{cccc|c} 1 & 0 & 0 & 0 & 1 \\ 0 & 1 & 0 & 0 & 2 \\ 0 & 0 & 1 & 0 & 0 \\ 0 & 0 & 0 & 1 & 1 \end{array}\right]$ The solution of the system is $x_1 = 1$, $x_2 = 2$, $x_3 = 0$, and $x_4 = 1$.

49. $\left[\begin{array}{ccc|c} 1 & 0 & 0 & 0 \\ 0 & 1 & 1 & 0 \\ 0 & 0 & 0 & 1 \end{array}\right]$ The system is inconsistent.

51. $\left[\begin{array}{ccc|c} 1 & 0 & 0 & 0 \\ 0 & 1 & -1 & -6 \\ 0 & 0 & 0 & 0 \end{array}\right]$ The solutions of the system are $x = 0$, $y = z - 6$, and z, where z is any real number.

53.

Amount in EE/E	Amount in I	Amount in HH/H
$ 0	$17,000	$8,000
$ 3,125	$14,875	$7,000
$ 6,250	$12,750	$6,000
$ 9,375	$10,625	$5,000
$12,500	$ 8,500	$4,000
$15,625	$ 6,375	$3,000
$18,750	$ 4,250	$2,000
$21,875	$ 2,125	$1,000
$25,000	$ 0	$ 0

55. Yes, the couple can still maintain their goals.

Amount in EE/E	Amount in I	Amount in HH/H
$ 0	$6,250	$18,750
$2,000	$4,550	$18,450
$4,000	$2,850	$18,150
$6,000	$1,150	$17,850
$7,353	$ 0	$17,647

57. There is insufficient information to determine the price of each food item.

Hamburger	Large Fries	Large Cola
$2.15	$0.88	$0.60
$2.10	$0.90	$0.65
$2.05	$0.92	$0.70
$2.00	$0.93	$0.75
$1.95	$0.95	$0.80
$1.90	$0.97	$0.85
$1.85	$0.98	$0.90

59. (a) The couple invests $20,000.

Treasury Bills	Corporate Bonds	Junk Bonds
$ 0	$10,000	$10,000
$1,000	$ 8,000	$11,000
$2,000	$ 6,000	$12,000
$3,000	$ 4,000	$13,000
$4,000	$ 2,000	$14,000
$5,000	$ 0	$15,000

(b) The couple invests $25,000.

Treasury Bills	Corporate Bonds	Junk Bonds
$12,500	$12,500	$ 0
$13,500	$10,500	$1,000
$14,500	$ 8,500	$2,000
$15,500	$ 6,500	$3,000
$16,500	$ 4,500	$4,000
$17,500	$ 2,500	$5,000
$18,500	$ 500	$6,000
$18,750	$ 0	$6,250

(c) Even if all $30,000 are invested in treasury bills (the investment with the lowest return), the interest income is $2100.

61.

Number of mg of 20% C, 30% D liquid	Number of mg of 40% C, 20% D liquid	Number of mg of 30% C, 50% D liquid
50	75	0
41.25	75.625	5
32.5	76.25	10
23.75	76.875	15
15	77.5	20
6.25	78.125	25
0	78.571	28.571

63. Answers will vary.

Exercise 2.4 (p. 103)

1. 2×2, a square matrix

3. 2×3 **5.** 3×2 **7.** 3×2

9. 2×1, a column matrix

11. 1×1, a column matrix, a row matrix, and a square matrix

13. False; the dimensions must be the same for two matrices to be equal.

15. True   **17.** True   **19.** True   **21.** True

23. False; the dimensions must be the same for two matrices to be equal.

25. $\begin{bmatrix} 1 & 1 \\ 6 & 7 \end{bmatrix}$

27. $\begin{bmatrix} 6 & 18 & 0 \\ 12 & -6 & 3 \end{bmatrix}$

29. $\begin{bmatrix} 2 & 4 & -8 \\ 1 & -4 & -1 \end{bmatrix}$

31. $\begin{bmatrix} 13a & 54 \\ -2b & -7 \\ -2c & -6 \end{bmatrix}$

33. $\begin{bmatrix} 1 & -1 & 4 \\ -5 & 1 & -1 \end{bmatrix}$

35. $\begin{bmatrix} 13 & -6 & -7 \\ -6 & 1 & -7 \end{bmatrix}$

37. $\begin{bmatrix} 9 & -5 & -6 \\ 1 & 1 & -3 \end{bmatrix}$

39. $\begin{bmatrix} -2 & -17 & 32 \\ 28 & 14 & 23 \end{bmatrix}$

41. $\begin{bmatrix} 5 & -2 & 3 \\ -12 & 1 & -5 \end{bmatrix}$

43. $\begin{bmatrix} 23 & -7 & -18 \\ -17 & -1 & -17 \end{bmatrix}$

45. $A + B = \begin{bmatrix} 3 & -5 & 4 \\ 5 & 3 & 3 \end{bmatrix} = B + A$

47. $A + (-A) = \begin{bmatrix} 0 & 0 & 0 \\ 0 & 0 & 0 \end{bmatrix} = 0$

49. $2B + 3B = \begin{bmatrix} 5 & -10 & 0 \\ 25 & 5 & 10 \end{bmatrix} = 5B$

51. $x = -4$ and $z = 4$

53. $x = 5$ and $y = 1$

55. $x = 4, y = -6,$ and $z = 6$

57. $\begin{bmatrix} -2 & 1 & 7 & 5 \\ 4 & 6 & 7 & 5 \\ -3.5 & 8 & -4 & 13 \\ 12 & -1 & 7 & 6 \end{bmatrix}$

59. $\begin{bmatrix} 19 & -11 & -14 & -15 \\ -12 & -13 & -21 & -17 \\ 15.5 & -24 & 19 & -39 \\ -29 & 10 & -14 & -11 \end{bmatrix}$

61. $\begin{bmatrix} 37 & -17 & 30 & 15 \\ 4 & 9 & 13 & 1 \\ 20.5 & 16 & -3 & 31 \\ 29 & 18 & 22 & 43 \end{bmatrix}$

63.

	Local	State	Fed
Male	542,978	1,154,869	135,237
Female	70,556	81,607	10,179

65.

	Associate's	Bachelor's	Master's	Doctoral
Male	218,000	573,000	181,000	26,700
Female	364,008	714,000	261,000	20,400

Yes, a 4×2 matrix could represent the situation.

67.

	Democrats	Republicans	Independents
Under $25,000	351	271	73
Over $25,000	203	215	55

69.

	LAS	ENG	EDUC
Male	250	225	80
Female	250	75	120

Exercise 2.5 (p. 114)

1. $[14]$ **3.** $[4]$ **5.** $[18 \quad -8]$

7. $\begin{bmatrix} 4 & 2 \\ 2 & 8 \end{bmatrix}$ **9.** $[4 \quad 6]$ **11.** $\begin{bmatrix} 4 & -14 \\ 12 & -32 \end{bmatrix}$

13. $\begin{bmatrix} 4 & 2 \\ 2 & 8 \\ 9 & 8 \end{bmatrix}$ **15.** $\begin{bmatrix} 9 & 1 \\ 5 & 4 \\ 11 & 7 \end{bmatrix}$

17. BA is defined and is a 3×4 matrix.

19. AB is not defined.

21. $(BA)C$ is not defined.

23. $BA + A$ is defined and is a 3×4 matrix.

25. $DC + B$ is defined and is a 3×3 matrix.

27. $\begin{bmatrix} -1 & 10 & -1 \\ -4 & 16 & -8 \end{bmatrix}$ **29.** $\begin{bmatrix} 11 & 5 \\ 13 & -9 \end{bmatrix}$

31. $\begin{bmatrix} 6 & 10 \\ 8 & 2 \\ -4 & 5 \end{bmatrix}$

33. $\begin{bmatrix} 3 & -1 \\ 4 & 2 \end{bmatrix}$ **35.** $\begin{bmatrix} 8 & 4 & 22 \\ 4 & 32 & 16 \end{bmatrix}$

37. $\begin{bmatrix} -14 & 7 \\ -20 & -6 \end{bmatrix}$ **39.** $\begin{bmatrix} 10 & 30 & 37 \\ 15 & 16 & 50 \\ -6 & 20 & -8 \end{bmatrix}$

41. $D(CB) = \begin{bmatrix} -6 & 42 & -9 \\ 1 & 20 & 6 \\ -7 & 4 & -18 \end{bmatrix} = (DC)B$

43. $\begin{bmatrix} 0.5 & 16 & -30 & 25 \\ 21 & 14 & 28 & 64 \\ 19.5 & 8 & -23 & 33 \\ -9.5 & 45 & -9 & 83 \end{bmatrix}$

45. $\begin{bmatrix} 31.5 & 251 & -31.5 & 143 \\ 861 & 350 & 791 & 420 \\ 369.5 & 115 & 206.5 & 215 \\ 412.5 & 882 & 451.5 & 491 \end{bmatrix}$

47. $\begin{bmatrix} 66 & 74 & 94 & 38 \\ 71 & 13 & 106 & 28 \\ 165 & 124.5 & 79 & 52 \\ 158 & -3 & 152 & 46 \end{bmatrix}$

49. $\begin{bmatrix} -5 & 23 & -102 & 44 \\ -108 & -56 & -70 & 122 \\ 108 & -152 & -67 & 36 \\ -346 & 279 & -249 & 187 \end{bmatrix}$

51. $AB = \begin{bmatrix} 5 & -2 \\ 6 & 4 \end{bmatrix} \neq \begin{bmatrix} 7 & -3 \\ 6 & 2 \end{bmatrix} = BA$

53. $x = 1$ or $x = \frac{1}{2}$

55. $a = d$ and $b = -c$

57. $A^2 = \begin{bmatrix} a & 1-a \\ 1+a & -a \end{bmatrix} \cdot \begin{bmatrix} a & 1-a \\ 1+a & -a \end{bmatrix} = \begin{bmatrix} 1 & 0 \\ 0 & 1 \end{bmatrix}$

59. Lee spent \$372 while Chan spent \$257.

61. $A^2 = \begin{bmatrix} 1 & 0 \\ 9 & 4 \end{bmatrix}, A^3 = \begin{bmatrix} 1 & 0 \\ 21 & 8 \end{bmatrix}$, and $A^4 = \begin{bmatrix} 1 & 0 \\ 45 & 16 \end{bmatrix}$

63. $A^2 = \begin{bmatrix} 1 & 0 \\ 0 & 1 \end{bmatrix}, A^3 = \begin{bmatrix} 1 & 0 \\ 0 & 1 \end{bmatrix}$, and $A^4 = \begin{bmatrix} 1 & 0 \\ 0 & 1 \end{bmatrix}$

67. $A^2 = \begin{bmatrix} -2.95 & -0.5 & -1.21 & 1.6 & -0.49 \\ -2.8 & 2.56 & 2.54 & 1.4 & 1.44 \\ 5.2 & 8.6 & 1.11 & -2.6 & -0.26 \\ -1.5 & -1 & 0.3 & 1 & 0.3 \\ 3.6 & -1.8 & -0.63 & -1.8 & -0.18 \end{bmatrix}$

$$A^{10} = \begin{bmatrix} 433.097 & -1583.562 & -369.817 & -216.548 & -141.638 \\ 1207.695 & 5998.438 & 2563.227 & -603.847 & 1045.705 \\ -1423.023 & 8065.070 & 4271.798 & 711.511 & 2023.364 \\ 479.476 & -246.538 & -231.543 & -239.737 & -140.557 \\ -899.341 & -3064.313 & -1627.502 & 449.670 & -698.609 \end{bmatrix}$$

$$A^{15} = \begin{bmatrix} -2247.845 & -118449.318 & -69122.364 & 1123.922 & -32134.304 \\ -11094.809 & 542433.513 & 249928.402 & 5547.405 & 110820.210 \\ 100366.783 & 831934.563 & 386083.791 & -50183.392 & 165597.136 \\ -18782.297 & -38432.094 & -18654.086 & 9391.148 & -7395.960 \\ -1633.665 & -306291.022 & -130154.569 & 816.833 & -56015.454 \end{bmatrix}$$

69. $A^2 = \begin{bmatrix} 1 & 0 & 1 \\ 1 & 2 & 1 \\ 2 & 2 & 2 \end{bmatrix}, A^{10} = \begin{bmatrix} 171 & 170 & 171 \\ 341 & 342 & 341 \\ 512 & 512 & 512 \end{bmatrix},$

$$A^{15} = \begin{bmatrix} 5461 & 5462 & 5461 \\ 10923 & 10922 & 10923 \\ 16384 & 16384 & 16384 \end{bmatrix}$$

71. Answers will vary.

Exercise 2.6 (p. 126)

1. $\begin{bmatrix} 1 & 2 \\ 2 & 3 \end{bmatrix}\begin{bmatrix} -3 & 2 \\ 2 & -1 \end{bmatrix} = \begin{bmatrix} 1 & 0 \\ 0 & 1 \end{bmatrix}$

3. $\begin{bmatrix} -1 & -2 \\ 3 & 4 \end{bmatrix}\begin{bmatrix} 2 & 1 \\ -\frac{3}{2} & -\frac{1}{2} \end{bmatrix} = \begin{bmatrix} 1 & 0 \\ 0 & 1 \end{bmatrix}$

5. $\begin{bmatrix} 1 & 2 & 3 \\ 2 & 3 & 4 \\ 1 & 2 & 1 \end{bmatrix}\begin{bmatrix} -\frac{5}{2} & 2 & -\frac{1}{2} \\ 1 & -1 & 1 \\ \frac{1}{2} & 0 & -\frac{1}{2} \end{bmatrix} = \begin{bmatrix} 1 & 0 & 0 \\ 0 & 1 & 0 \\ 0 & 0 & 1 \end{bmatrix}$

7. $\begin{bmatrix} 5 & -7 \\ -2 & 3 \end{bmatrix}$

9. $\begin{bmatrix} 4 & -1 \\ 3 & -1 \end{bmatrix}$

11. $\begin{bmatrix} \frac{3}{2} & -\frac{1}{2} \\ -2 & 1 \end{bmatrix}$

13. $\begin{bmatrix} 0 & 0 & 1 \\ 0 & 1 & 0 \\ 1 & 0 & 0 \end{bmatrix}$

15. $\begin{bmatrix} \frac{4}{9} & \frac{1}{9} & \frac{1}{9} \\ \frac{4}{3} & -\frac{2}{3} & \frac{1}{3} \\ \frac{7}{9} & -\frac{5}{9} & \frac{4}{9} \end{bmatrix}$

17. $\begin{bmatrix} 1 & -1 & 2 \\ -1 & 2 & -3 \\ -1 & 1 & -1 \end{bmatrix}$

19. $\begin{bmatrix} 2 & -1 & -1 & -2 \\ -1 & 1 & 1 & 2 \\ -2 & 1 & 2 & 3 \\ -1 & 1 & 1 & 1 \end{bmatrix}$

21. $\left[\begin{array}{cc|cc} 4 & 6 & 1 & 0 \\ 2 & 3 & 0 & 1 \end{array}\right] \Rightarrow \left[\begin{array}{cc|cc} 1 & \frac{3}{2} & \frac{1}{4} & 0 \\ 0 & 0 & -\frac{1}{2} & 1 \end{array}\right]$

23. $\left[\begin{array}{cc|cc} -8 & 4 & 1 & 0 \\ -4 & 2 & 0 & 1 \end{array}\right] \Rightarrow \left[\begin{array}{cc|cc} 1 & -\frac{1}{2} & -\frac{1}{8} & 0 \\ 0 & 0 & -\frac{1}{2} & 1 \end{array}\right]$

25. $\left[\begin{array}{ccc|ccc} 1 & 1 & 1 & 1 & 0 & 0 \\ 3 & -4 & 2 & 0 & 1 & 0 \\ 0 & 0 & 0 & 0 & 0 & 1 \end{array}\right]$

27. $\begin{bmatrix} 2 & -1 \\ -1 & 1 \end{bmatrix}$

29. $\begin{bmatrix} \frac{1}{3} & \frac{1}{6} \\ 0 & \frac{1}{4} \end{bmatrix}$

31. Inverse does not exist.

33. $\begin{bmatrix} \frac{7}{2} & -1 & -\frac{3}{2} \\ 1 & -1 & -1 \\ \frac{1}{2} & 1 & \frac{1}{2} \end{bmatrix}$

35. $\begin{bmatrix} \frac{1}{5} & \frac{2}{5} \\ \frac{2}{5} & -\frac{1}{5} \end{bmatrix} - \begin{bmatrix} -\frac{1}{5} & \frac{3}{5} \\ \frac{2}{5} & -\frac{1}{5} \end{bmatrix} = \begin{bmatrix} \frac{2}{5} & -\frac{1}{5} \\ 0 & 0 \end{bmatrix}$

37. $\begin{bmatrix} 42 & -18 & -5 \\ -9 & 4 & 1 \\ -7 & 3 & 1 \end{bmatrix}\begin{bmatrix} 2 \\ 1 \\ 3 \end{bmatrix} = \begin{bmatrix} 51 \\ -11 \\ -8 \end{bmatrix}$

39. $x = 36, y = -14$

41. $x = 2, y = 1$

43. $x = 88, y = -36$

45. $x = \frac{14}{9}, y = \frac{26}{3}, z = \frac{65}{9}$

47. $x = \frac{20}{3}, y = 24, z = \frac{56}{3}$

49. $x = -\frac{14}{9}, y = \frac{10}{3}, z = \frac{16}{9}$

51. $\begin{bmatrix} 0.0054 & 0.0509 & -0.0066 \\ 0.0104 & -0.0186 & 0.0095 \\ -0.0193 & 0.0116 & 0.0344 \end{bmatrix}$

53. $\begin{bmatrix} 0.0249 & -0.0360 & -0.0057 & 0.0059 \\ -0.0171 & 0.0521 & 0.0292 & -0.0305 \\ 0.0206 & 0.0082 & -0.0421 & 0.0005 \\ -0.0175 & 0.0570 & 0.0657 & 0.0619 \end{bmatrix}$

55. $\begin{bmatrix} \frac{1}{4} & -\frac{1}{16} & -\frac{9}{32} & \frac{5}{16} & \frac{3}{32} \\ -\frac{3}{2} & \frac{1}{8} & \frac{49}{16} & -\frac{21}{8} & \frac{5}{16} \\ \frac{7}{4} & -\frac{3}{16} & -\frac{91}{32} & \frac{47}{16} & -\frac{23}{32} \\ -\frac{1}{2} & \frac{3}{8} & \frac{11}{16} & -\frac{7}{8} & \frac{7}{16} \\ -\frac{5}{4} & \frac{5}{16} & \frac{77}{32} & -\frac{41}{16} & \frac{17}{32} \end{bmatrix}$

57. $x = 4.567, y = -6.444, z = -24.075$

59. $x = -1.187, y = 2.457, z = 8.265$

61. $AA^{-1} = \begin{bmatrix} a & b \\ c & d \end{bmatrix}\begin{bmatrix} \frac{d}{\Delta} & \frac{-b}{\Delta} \\ \frac{-c}{\Delta} & \frac{a}{\Delta} \end{bmatrix} = \begin{bmatrix} 1 & 0 \\ 0 & 1 \end{bmatrix}$

63. $\begin{bmatrix} 0 & \frac{1}{2} \\ \frac{1}{5} & -\frac{1}{10} \end{bmatrix}$

65. $\begin{bmatrix} -15 & 2 \\ 8 & -1 \end{bmatrix}$

Exercise 2.7: Application 1 (p. 136)

1. A's wages = C's wages = \$30,000; B's wages = \$22,500

3. A's wages = \$12,000; B's wages = \$22,000; C's wages = \$30,000

5. $X = \begin{bmatrix} 203.282 \\ 166.977 \\ 137.847 \end{bmatrix}$

7. Farmer's wages = \$20,000;

   Builder's wages = \$18,000;

   Tailor's wages = \$12,000;

   Rancher's wages = \$25,000

9. $X = \begin{bmatrix} 160 \\ 75.385 \end{bmatrix}$

Exercise 2.7: Application 2 (p. 140)

1. a. THIS IS KILLING MY GPA
   b. EVERYONE LOVES MICKEY

3. WHATS YOUR EMAIL ADDRESS

5. a. 21 11 47 27 57 33 49 28 85 50 35 22 80 47 85 55
   b. 61 33 101 61 54 30 64 36 52 33 70 45
   c. 39 22 97 58 59 37 42 27 48 25 31 20 49 28

Exercise 2.7: Application 3 (p. 143)

1.

Department	Total Costs Dollars	Direct Costs, Dollars	Indirect Costs for Services from Departments Dollars	
			S_1	S_2
S_1	3109.10	2000	345.46	763.63
S_2	2290.90	1000	1036.37	254.54
P_1	3354.55	2500	345.46	509.09
P_2	2790.91	1500	1036.37	254.54
P_3	3854.55	3000	354.46	509.09
Totals	15,400.01	10,000	3109.12	2290.89

Total of the service charges allocated to P_1, P_2, and P_3: \$3000.01
Sum of the direct costs of the service departments, S_1, and S_2: \$3000

Exercise 2.7: Application 4 (p. 149)

1. $\begin{bmatrix} 4 & 3 \\ 1 & 1 \\ 2 & 0 \end{bmatrix}$

3. $\begin{bmatrix} 1 & 0 & 1 \\ 11 & 12 & 4 \end{bmatrix}$

5. $[8 \quad 6 \quad 3]$

7. a. $y = \frac{54}{35}x + \frac{27}{5}$
   b. 17,743 units will be supplied.

9. $y = 1.498x + 36.135$

11. a. Not symmetric
    b. Symmetric
    c. Not symmetric
    Yes, for two matrices to be equal, they must have the same dimensions.

CHAPTER 2 Review

True–False Items (p. 151)

1. T 2. F 3. F 4. T 5. F 6. F 7. F

Fill in the Blanks (p. 151)

1. 3 × 2 2. one; infinitely many 3. rows; columns
4. inverse 5. 3 × 3 6. 5 × 5

Review Exercises (p .151)

1. $x = 2, y = -1$

3. $x = 2, y = -1$

5. No solution, the system is inconsistent.

7. $x = -1, y = 2, z = -3$

9. $x = \frac{7}{4}z - \frac{39}{4}, y = \frac{9}{8}z + \frac{69}{8}$, and z, where z is any real number.

11. $\begin{cases} 3x + 2y = 8 \\ x + 4y = -1 \end{cases}$

13. $\begin{cases} x = 4 \\ y = 6 \\ z = -1 \end{cases}$

15. $x = \frac{14}{9}, y = \frac{26}{9}$

17. $x = -103, y = 32, z = 9$

19. $x = 29, y = -10, z = -1$

21. No solution, the system is inconsistent.

23. $x = 9, y = -\frac{56}{3}, z = -\frac{37}{3}$

25. $x = 29, y = 8, z = -24$

27. $x = \frac{3}{7}z + \frac{10}{7}, y = \frac{5}{7}z - \frac{9}{7}$, and z, where z is any real number. Answers will vary. Three possible solutions are $x = \frac{10}{7}, y = -\frac{9}{7}$, when $z = 0$; $x = \frac{13}{7}, y = -\frac{4}{7}$, when $x = 1$; and $x = 1, y = -2$, when $z = -1$.

29. $x = -\frac{3}{5}z + 1, y = \frac{4}{5}z + 2$, and z, where z is any real number. Answers will vary.

31. No solution, the system is inconsistent.

33. One solution, $x = 11, y = -1$, and $z = -1$.

35. Infinitely many solutions, $x_1 = -2x_4 + 1, x_2 = -2x_4 - 1, x_3 = 3$, and x_4, where x_4 is any real number.

37. $\begin{bmatrix} 4 & -4 \\ 3 & 9 \\ 4 & 0 \end{bmatrix}$ The dimensions are 3×2.

39. $\begin{bmatrix} 6 & 0 \\ 12 & 24 \\ -6 & 12 \end{bmatrix}$ The dimensions are 3×2.

41. $\begin{bmatrix} 4 & -3 & 0 \\ 12 & -2 & -8 \\ -2 & 5 & -4 \end{bmatrix}$ The dimensions are 3×3.

43. $\begin{bmatrix} 8 & -13 & 8 \\ 9 & 2 & -10 \\ 18 & -17 & 4 \end{bmatrix}$ The dimensions are 3×3.

45. $\begin{bmatrix} 12 & -16 & 8 \\ 21 & 0 & -18 \\ 16 & -12 & 0 \end{bmatrix}$ The dimensions are 3×3.

47. $\begin{bmatrix} 1 & 2 & -1 \\ 0 & 4 & 2 \end{bmatrix}$ The dimensions are 2×3.

49. $\begin{bmatrix} 3 & -4 \\ 1 & 5 \\ 5 & -2 \end{bmatrix} = C$. The dimensions are 3×2.

51. $\begin{bmatrix} \frac{1}{3} & 0 \\ \frac{2}{3} & 1 \end{bmatrix}$

53. Inverse does not exist.

55. $\begin{bmatrix} \frac{1}{16} & \frac{1}{32} & \frac{1}{4} \\ \frac{3}{16} & \frac{3}{32} & -\frac{1}{4} \\ -\frac{3}{16} & \frac{13}{32} & \frac{1}{4} \end{bmatrix}$

57. $AB = BA$ when $x = w$ and $y = -z$.

59. Each box should contain 20 caramels and 30 creams. To obtain a profit, increase the number of caramels in each box, (decreasing the number of creams.)

61.

Almonds	Cashews	Peanuts
5	60	35
20	40	40
35	20	45
50	0	50

63. a. To attain \$2500 per year in income:

Treasury Bills	Corporate Bonds	Junk Bonds
\$35,000	\$5,000	\$ 0
\$35,500	\$4,000	\$ 500
\$36,000	\$3,000	\$1,000
\$36,500	\$2,000	\$1,500
\$37,000	\$1,000	\$2,000
\$37,500	\$ 0	\$2,500

b. To attain \$3000 per year in income:

Treasury Bills	Corporate Bonds	Junk Bonds
\$10,000	\$30,000	\$ 0
\$13,000	\$24,000	\$ 3,000
\$16,000	\$18,000	\$ 6,000
\$19,000	\$12,000	\$ 9,000
\$22,000	\$ 6,000	\$12,000
\$25,000	\$ 0	\$15,000

c. To attain $3500 per year in income:

Treasury Bills	Corporate Bonds	Junk Bonds
$ 0	$25,000	$15,000
$ 2,500	$20,000	$17,500
$ 5,000	$15,000	$20,000
$ 7,500	$10,000	$22,500
$10,000	$ 5,000	$25,000
$12,500	$ 0	$27,500

65. a.

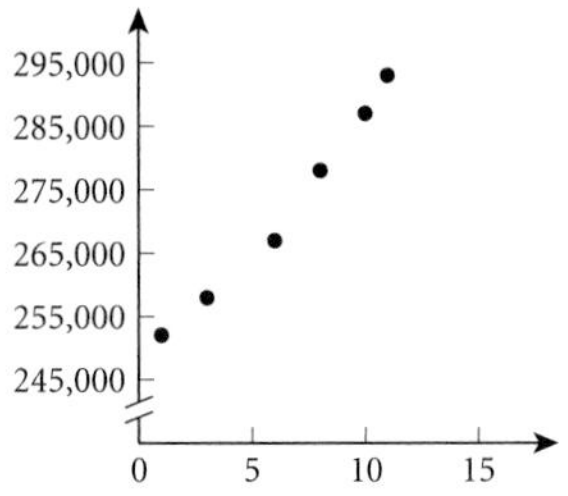

b. $y = 4096.774x + 245{,}870.968$

c. The number of people that emigrate is increasing by 4097 (rounding to the nearest person) people each year.

d. An estimated 307,323 people will emigrate in 2005.

e. An estimated 266,355 people emigrated in 1995.

f. $y = 4096.774x + 245{,}870.968$

67. 275.564 units of A, 98.86 units of B, and 179.548 units of C should be produced.

69.

Department	Total Costs (in Dollars)	Direct Cost In Dollars	Indirect Costs for Services (in Dollars)	
S_1	1745.45	800	349.09	596.36
S_2	5963.64	4000	174.55	1789.09
P_1	2445.45	1500	349.09	596.36
P_2	2216.37	500	523.64	1192.73
P_3	3338.18	1200	349.09	1789.09
Totals	15,709.09	8000	1745.46	5963.63

Mathematical Questions from Professional Exams (p. 156)

1. b **2.** b **3.** d **4.** c

CHAPTER 3 Linear Programming: Geometric Approach

Exercise 3.1 (p. 168)

1.

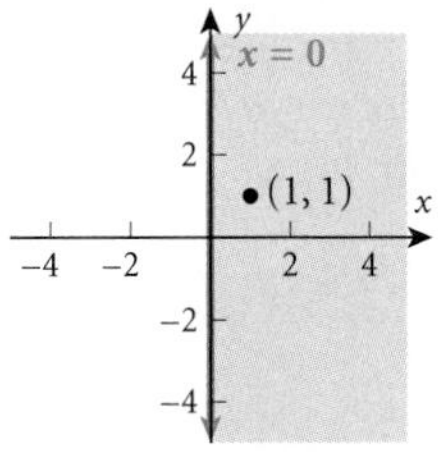

3.

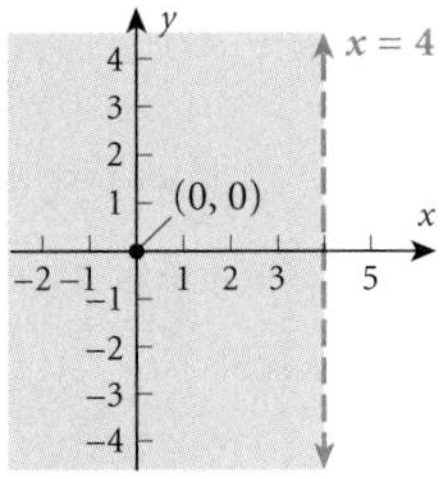

5.

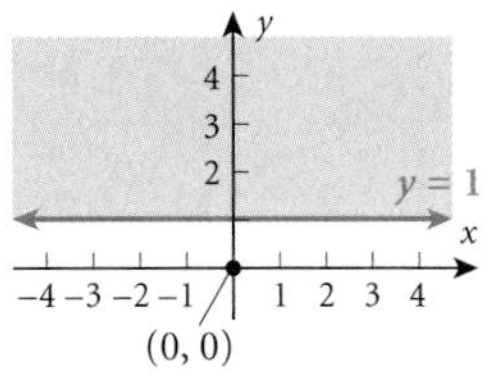

7.

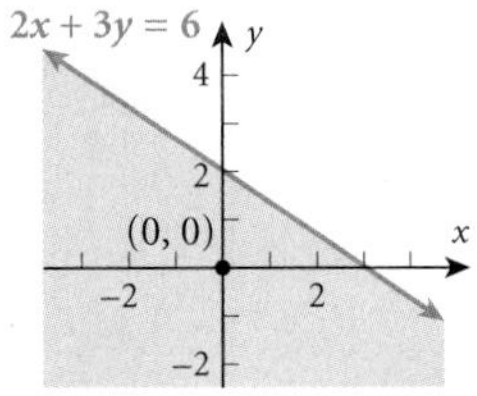

9.

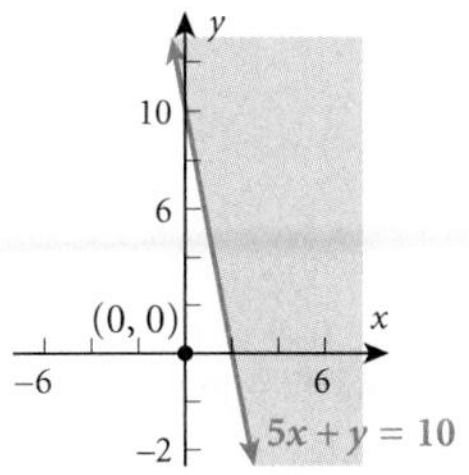

11.

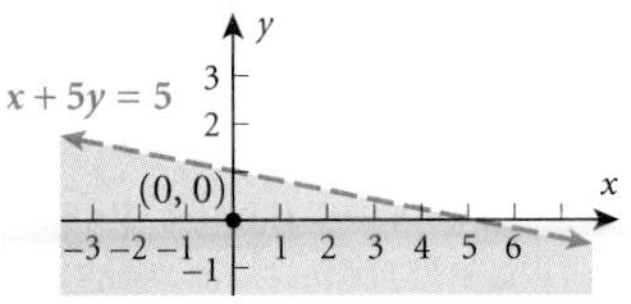

13. P_1 is part of the graph of the system, P_2 and P_3 are not.

15. P_1 and P_3 are part of the graph of the system, P_2 is not.

17. b **19.** c **21.** d **23.** c

25. Unbounded. Corner points: (2, 0), (0, 2).

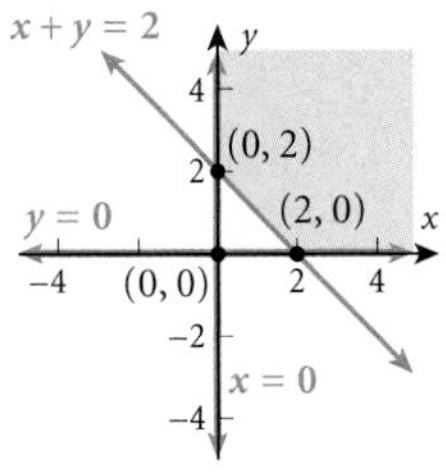

27. Bounded. Corner points: (2, 0), (3, 0), (0, 2).

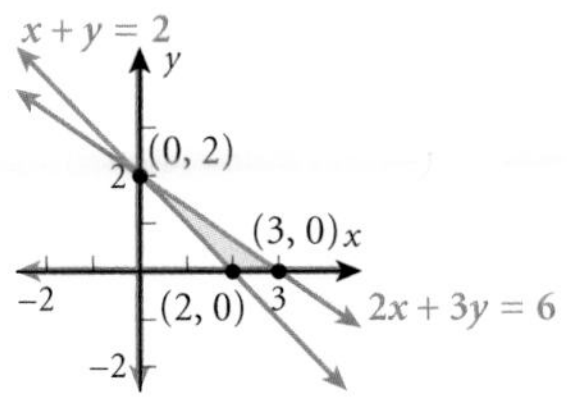

29. Bounded. Corner points: (2, 0), (5, 0), (2, 6), (0, 8), (0, 2).

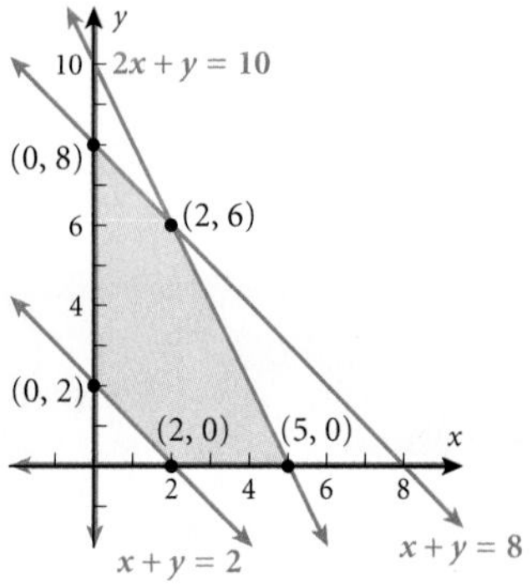

31. Bounded. Corner points: (2, 0), (4, 0), $(\frac{24}{7}, \frac{12}{7})$, (0, 4), (0, 2).

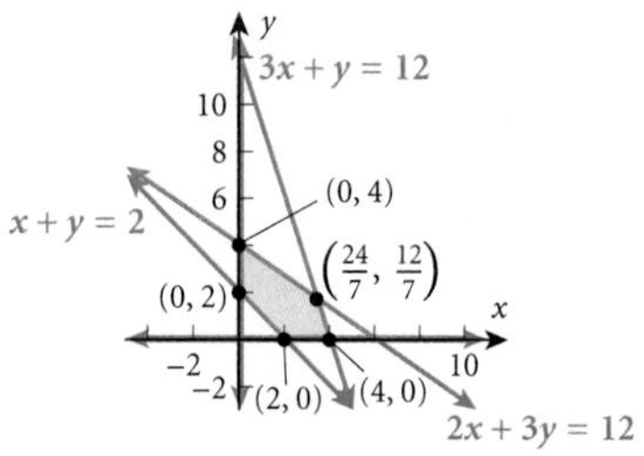

33. Unbounded. Corner points: (1, 0), $(0, \frac{1}{2})$, (0, 4).

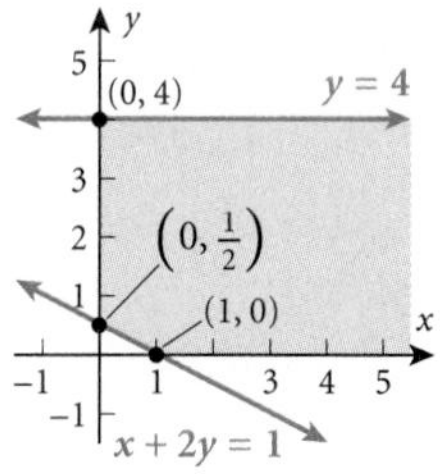

35. Bounded. Corner points: $(\frac{4}{5}, \frac{3}{5})$, (0, 3), (0, 1).

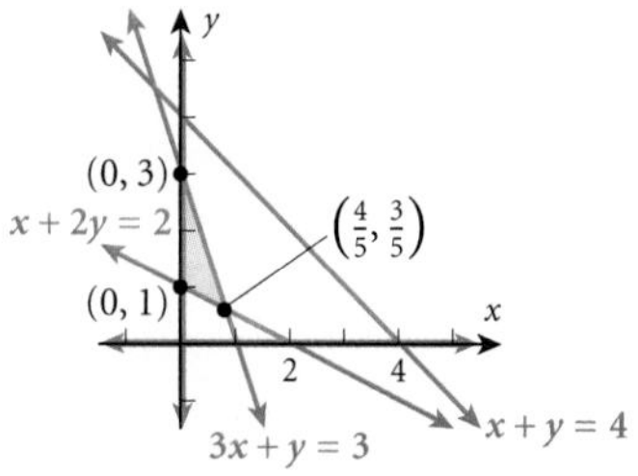

37. $\begin{cases} 4x + 8y \le 960 \\ 12x + 8y \le 1440 \\ x \ge 0 \\ y \ge 0 \end{cases}$

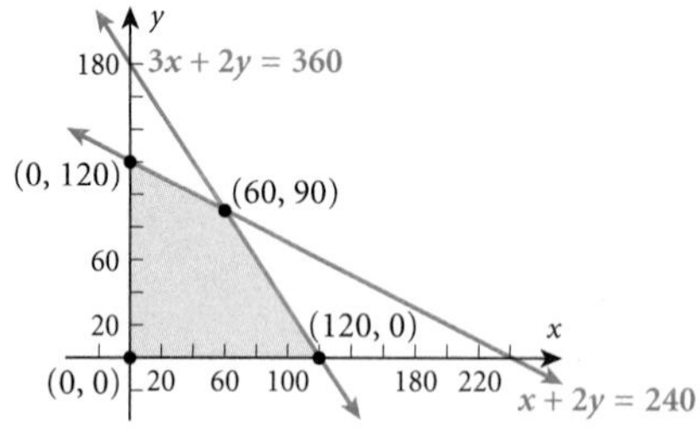

Corner points: (0, 0), (120, 0), (60, 90), (0, 120).

39. (a) $\begin{cases} 3x + 2y \le 80 \\ 4x + 3y \le 120 \\ y \ge 0 \\ x \ge 0 \end{cases}$

(b)

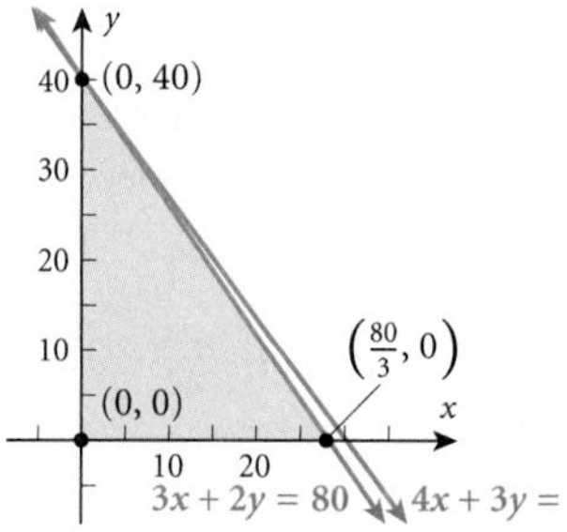

$3x + 2y = 80 \quad 4x + 3y = 120$
Corner points: (0, 0), $(\frac{80}{3}, 0)$, (0, 40).

41. (a) $\begin{cases} x + y \le 25{,}000 \\ x \ge 15{,}000 \\ y \le 10{,}000 \\ y \ge 0 \\ x \ge 0 \end{cases}$

(b)

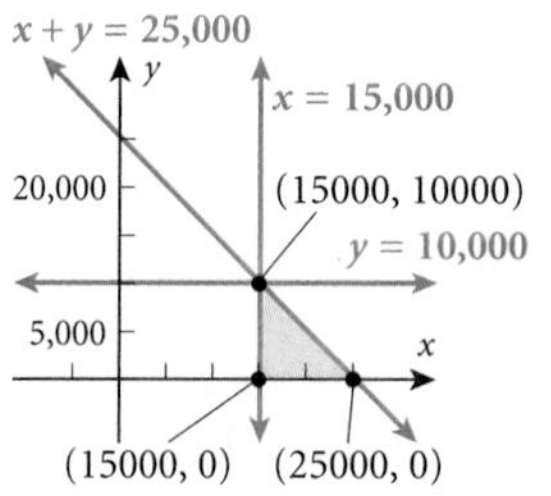

Corner points: (15000, 0), (25000, 0), (15000, 10000).

(c) (15000, 0) represents investing \$15,000 in treasury bills.
(25000, 0) represents investing \$25,000 in treasury bills.
(15000, 10000) represents investing \$15,000 in treasury bills and \$10,000 in corporate bonds.

43. (a) $\begin{cases} x + 2y \geq 5 \\ 5x + y \geq 16 \\ x \geq 0 \\ y \geq 0 \end{cases}$

(b)

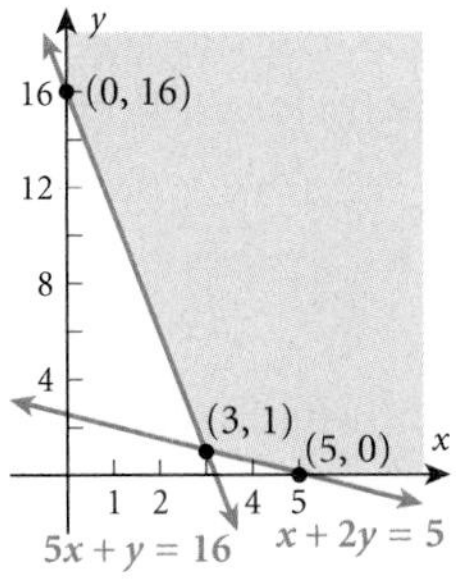

Corner points: $(0, 16), (3, 1), (5, 0)$.

45. (a) $\begin{cases} 5x + 4y \geq 85 \\ 3x + 3y \geq 70 \\ 2x + 3y \geq 50 \\ x \geq 0 \\ y \geq 0 \end{cases}$

(b)

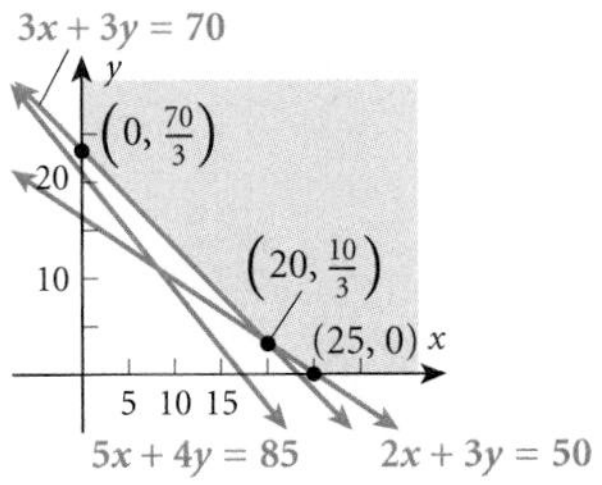

Corner points: $(0, \frac{70}{3}), (20, \frac{10}{3}), (25, 0)$.

47. Answers will vary.

Exercise 3.2 (p. 178)

1. Maximum of 38 at $(7, 8)$. Minimum of 10 at $(2, 2)$.

3. Maximum of 15 at $(7, 8)$. Minimum of 4 at $(2, 2)$.

5. Maximum of 55 at $(7, 8)$. Minimum of 14 at any point on the line segment between $(2, 2)$ and $(8, 1)$.

7. Maximum of 53 at $(7, 8)$. Minimum of 14 at $(2, 2)$.

9. Maximum of 81 at $(8, 1)$. Minimum of 22 at $(2, 2)$.

11. Corner points: $(0, 4), (3, 0), (13, 0)$.

13. Corner points: $(0, 0), (15, 0), (5, 10), (0, 10)$.

15. Corner points: $(3, 0), (10, 0), (10, 8), (0, 8), (0, 4)$.

17. Maximum of 14 at $(0, 2)$.

19. Maximum of 15 at $(3, 0)$.

21. Maximum of 56 at $(0, 8)$.

23. Maximum of 58 at $(6, 4)$.

25. Minimum of 0 at $(0, 0)$.

27. Minimum of 4 at $(2, 0)$.

29. Minimum of 4 at $(2, 0)$.

31. Minimum of $\frac{3}{2}$ at $(0, \frac{1}{2})$.

33. Maximum of 10 at any point on the line $x + y = 10$ between $(0, 10)$ and $(10, 0)$. Minimum of $\frac{20}{3}$ at $(\frac{10}{3}, \frac{10}{3})$.

35. Maximum of 50 at $(10, 0)$. Minimum of 20 at $(0, 10)$.

37. Maximum of 40 at $(0, 10)$. Minimum of $\frac{70}{3}$ at $(\frac{10}{3}, \frac{10}{3})$.

39. Maximum of 100 at $(10, 0)$. Minimum of 10 at $(0, 10)$.

41. Maximum of 192 at $(4, 4)$. Minimum of 54 at $(3, 0)$.

43. Maximun of 58 at $(4, 5)$. Minimum of 12 at $(0, 2)$.

45. Maximum of 240 at $(3, 10)$.

47. Maximum of 216 at $(2, 10)$.

49. Produce 90 low-grade packages and 105 high-grade packages for a maximum profit of \$69.

Exercise 3.3 (p. 184)

1. The farmer should plant x acres of soybeans, where $24 \leq x \leq 30$, and $y = 60 - 2x$ acres of corn for a maximum profit of \$9000.

3. She should invest \$12,000 in type A bonds and \$8000 in type B bonds for a maximum return of \$2400.

5. Manufacture 500,000 of each vitamin for a maximum profit of \$75,000.

7. The store should sell 30 microwaves and 20 stoves for a maximum revenue of \$13,000.

9. Purchase 2025 shares of Duke Energy Corp. and 555 shares of Eastman Kodak for a maximum annual yield of \$3267.

11. A child should have 1 of a serving of Gerber Banana Plum Granola and 0.7 of a serving of Gerber Mixed Fruit Carrot Juice for a minimum cost of \$1.44.

13. Run the ad for 20 months at the AOL website and 10 months at the Yahoo! website to reach a maximum of 2190 million people.

15. Manufacture 25 rolls of low-grade carpet (and no rolls of high-grade carpet) for a maximum income of \$2500.

CHAPTER 3 Review

True–False Items (p. 187)

1. T **2.** F **3.** T **4.** T **5.** T **6.** F

Fill in the Blanks (p. 187)

1. half plane **2.** objective **3.** feasible

4. bounded **5.** corner point

Review Exercises (p. 187)

1.

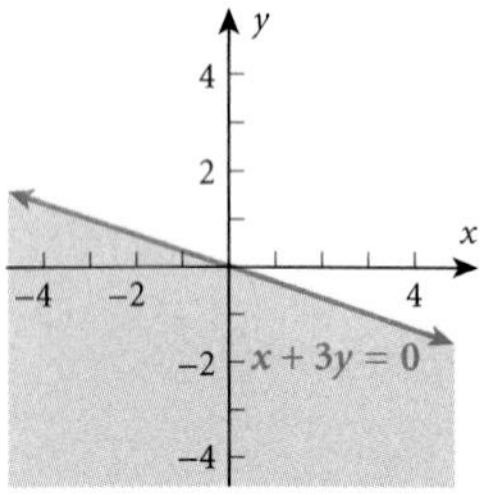

3.

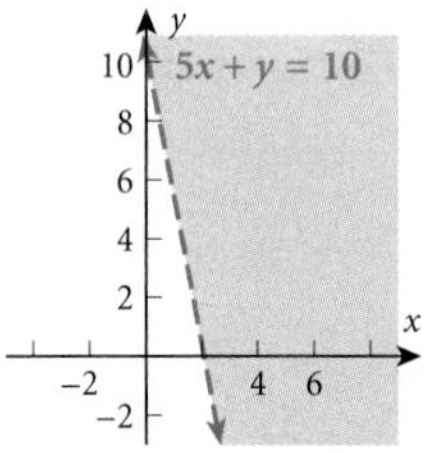

5. All 3 points P_1, P_2, and P_3 are part of the graph of the system.

7. a

9. Bounded.

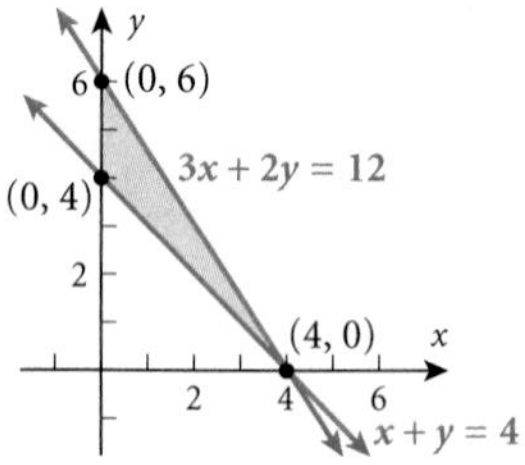

Corner points: (4, 0), (0, 4), (0, 6)

11. Bounded.

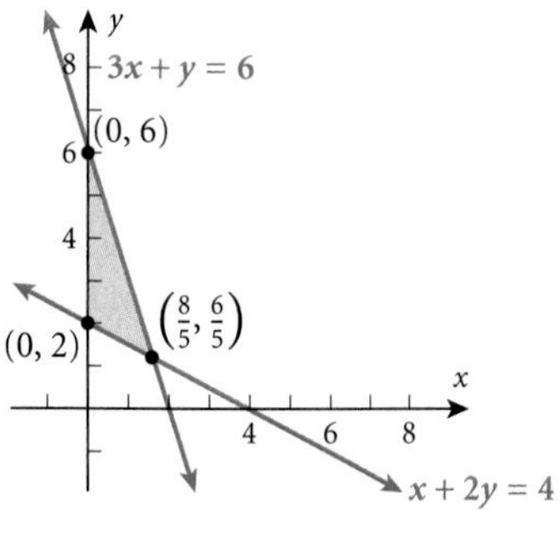

Corner points: (0, 2), $(\frac{8}{5}, \frac{6}{5})$, (0, 6)

13. Bounded.

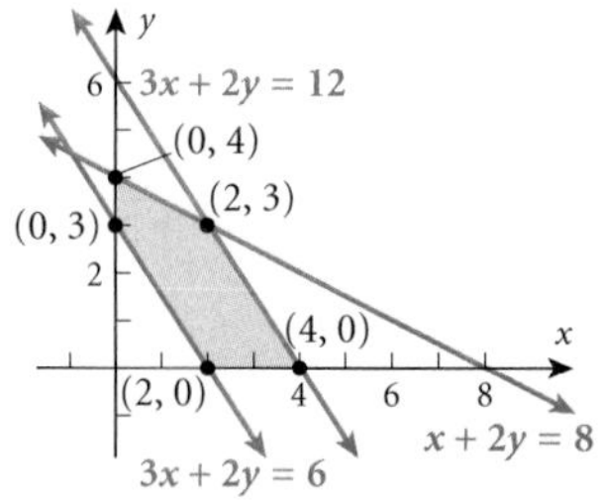

Corner points: (2, 0), (4, 0), (2, 3), (0, 3), (0, 4)

15. Maximum of $\frac{80}{3}$ at $\left(\frac{40}{3}, \frac{40}{3}\right)$.

17. Minimum of 20 at (0, 10).

19. Maximum of 40 at (20, 0), $(\frac{40}{3}, \frac{40}{3})$, and at any point on the line segment $2x + y = 40$ connecting them.

21. Minimum of 20 at (10, 0).

23. Maximum of 235 at (5, 8). Minimum of 60 at (4, 0), (0, 3) and at all points at points on the line segment $3x + 4y = 12$, connecting them.

25. Maximum of 155 at (5, 4). Minimum of 0 at (0, 0).

27. Maximum of 42 at (9, 8).

29. Maximum of 24 at (8, 8).

31. Minimum of $\frac{48}{5}$ at $\left(\frac{4}{5}, \frac{18}{5}\right)$.

33. Produce 8 pairs of downhill skis and 24 pairs of cross-country skis for a maximum profit of \$1760.

35. Katy should buy 7.5 lb of food A and 11.25 lb of food B for a minimum cost of \$18.75.

37. Give the child 1.58 servings of Gerber Banana Oatmeal and Peach and no Gerber Mixed Fruit Juice for a minimum cost of \$1.25.

39. Manufacture 550,000 high-potency vitamins and 250,000 thousand calcium-enriched vitamins for a maximum profit of \$67,500.

Mathematical Questions from Professional Exams

1. b **2.** a **3.** c **4.** c **5.** d **6.** c **7.** c

8. b **9.** b **10.** a **11.** b **12.** e **13.** c **14.** b

CHAPTER 4 Linear Programming: Simplex Method

Section 4.1 (p. 205)

1. Standard
3. Nonstandard
5. Nonstandard
7. Nonstandard
9. Standard
11. Cannot be modified
13. Cannot be modified

15. Constraints
$$\begin{aligned} &x_1 - x_2 - x_3 \le 6 \\ &-2x_1 + 3x_2 \le 12 \\ &x_3 \le 2 \\ &x_1 \ge 0, x_2 \ge 0, x_3 \ge 0 \end{aligned}$$

17. Slack variables
$$\begin{aligned} 5x_1 - 2x_2 + x_3 + s_1 &= 20 \\ 6x_1 + x_2 + 4x_3 + s_2 &= 24 \\ x_1 + x_2 + 4x_3 + s_3 &= 16 \end{aligned}$$

and initial simplex tableau

BV	P	x_1	x_2	x_3	s_1	s_2	s_3	RHS
s_1	0	5	−2	1	1	0	0	20
s_2	0	6	1	4	0	1	0	24
s_3	0	1	1	4	0	0	1	16
P	1	−2	−1	−3	0	0	0	0

19.
$$\begin{aligned} P - 3x_1 - 5x_2 &= 0 \\ 2.2x_1 - 1.8x_2 + s_1 &= 5 \\ 0.8x_1 + 1.2x_2 + s_2 &= 2.5 \\ x_1 + x_2 + s_3 &= 0.1 \end{aligned}$$
$x_1 \ge 0, x_2 \ge 0, x_3 \ge 0$
$s_1 \ge 0,\ s_2 \ge 0,\ s_3 \ge 0$

The initial simplex tableau is:

BV	P	x_1	x_2	s_1	s_2	s_3	RHS
s_1	0	2.2	−1.8	1	0	0	5
s_2	0	0.8	1.2	0	1	0	2.5
s_3	0	1	1	0	0	1	0.1
P	1	−3	−5	0	0	0	0

21.
$$\begin{aligned} &p - 2x_1 - 3x_2 - x_3 \\ &x_1 + x_2 + x_3 + s_1 = 50 \\ &3x_1 + 2x_2 + x_3 + s_2 = 10 \end{aligned}$$
$x_1 \ge 0, x_2 \ge 0, x_3 \ge 0, s_1 \ge 0, s_2 \ge 0$

The initial simplex tableau is:

BV	P	x_1	x_2	x_3	s_1	s_2	RHS
s_1	0	1	1	1	1	0	50
s_2	0	3	2	1	0	1	10
P	1	−2	−3	−1	0	0	0

23.
$$\begin{aligned} P - 3x_1 - 4x_2 - 2x_3 &= 0 \\ 3x_1 - x_2 + 4x_3 + s_1 &= 5 \\ x_1 + x_2 + s_2 &= 5 \\ x_1 - x_2 + x3 + s_3 &= 6 \end{aligned}$$
$x_1 \ge 0, x_2 \ge 0, x_3 \ge 0, s_1 \ge 0, s_2 \ge 0, s_3 \ge 0$

The initial tableau is:

BV	P	x_1	x_2	x_3	s_1	s_2	s_3	RHS
s_1	0	3	−1	4	1	0	0	5
s_2	0	1	1	0	0	1	0	5
s_3	0	2	−1	1	0	0	1	6
P	1	−3	−4	−2	0	0	0	0

25. Maximize $P = x_1 + 2x_2 + 5x_3$
Subject to the constraints
$$\begin{aligned} &x_1 - 2x_2 - 3x_3 \le 10 \\ &-3x_1 - x_2 + x_3 \le 12 \\ &x_1 \ge 0, x_2 \ge 0\ x_3 \ge 0 \end{aligned}$$

System with slack variables:
$$\begin{aligned} P - x_1 - 2x_2 - 5x_3 &= 0 \\ x_1 - 2x_2 - 3x_3 + s_1 &= 10 \\ -3x_1 - x_2 + x_3 + s_2 &= 12 \end{aligned}$$
$x_1 \ge 0, x_2 \ge 0, x_3 \ge 0, s_1 \ge 0, s_2 \ge 0$

Initial tableau:

BV	P	x_1	x_2	x_3	s_1	s_2	RHS
s_1	0	1	−2	−3	1	0	10
s_2	0	−3	−1	1	0	1	12
P	1	−1	−2	−5	0	0	0

27. Maximize $P = 2x_1 + 3x_2 + x_3 + 6x_4$
Subject to the constraints
$$\begin{aligned} &-x_1 + x_2 + 2x_3 + x_4 \le 10 \\ &-x_1 + x_2 - x_3 + x_4 \le 8 \\ &x_1 + x_2 + x_3 + x_4 \le 9 \\ &x_1 \ge 0, x_2 \ge 0, x_3 \ge 0, x_4 \ge 0 \end{aligned}$$

System with slack variables
$$\begin{aligned} P - 2x_1 - 3x_2 - x_3 - 6x_4 &= 0 \\ -x_1 + x_2 + 2x_3 + x_4 + s_1 &= 10 \\ -x_1 + x_2 - x_3 + x_4 + s_2 &= 8 \\ x_1 + x_2 + x_3 + x_4 + s_3 &= 9 \end{aligned}$$
$x_1 \ge 0, x_2 \ge 0, x_3 \ge 0, x_4 \ge 0$
$s_1 \ge 0, s_2 \ge 0, s_3 \ge 0$

BV	P	x_1	x_2	x_3	x_4	s_1	s_2	s_3	RHS
s_1	0	−1	1	2	1	1	0	0	10
s_2	0	−1	1	−1	1	0	1	0	8
s_3	0	1	1	1	1	0	0	1	9
P	1	−2	−3	−1	−6	0	0	0	0

29. New tableau:

BV	P	x_1	x_2	s_1	s_2	RHS
x_2	0	$\frac{1}{2}$	1	$\frac{1}{2}$	0	150
s_2	0	2	0	−1	1	180
P	1	0	0	1	0	300

New system:
$$\begin{aligned} x_2 &= -\tfrac{1}{2}x_1 - \tfrac{1}{2}s_1 + 150 \\ s_2 &= -2x_1 + s_1 + 180 \\ P &= -s_1 + 300 \end{aligned}$$

Current values: $P = 300, x_2 = 150, s_2 = 180$

31. New tableau:

BV	P	x_1	x_2	x_3	s_1	s_2	s_3	RHS
s_1	0	-2	0	0	1	0	-1	6
s_2	0	$\frac{5}{4}$	$-\frac{3}{2}$	0	0	1	$-\frac{1}{4}$	$\frac{55}{2}$
x_3	0	$\frac{3}{4}$	$\frac{1}{2}$	1	0	0	$\frac{1}{4}$	$\frac{9}{2}$
P	1	$\frac{5}{4}$	$-\frac{1}{2}$	0	0	0	$\frac{3}{4}$	$\frac{27}{2}$

New system: $s_1 = 2x_1 + s_3 + 6$

$$s_2 = -\tfrac{5}{4}x_1 + \tfrac{3}{2}x_2 + \tfrac{1}{4}s_3 + \tfrac{55}{2}$$
$$x_3 = -\tfrac{3}{4}x_1 - \tfrac{1}{2}x_2 - \tfrac{1}{4}s_3 + \tfrac{9}{2}$$
$$P = -\tfrac{5}{4}x_1 + \tfrac{1}{2}x_2 - \tfrac{3}{4}s_3 + \tfrac{27}{2}$$

Current values: $P = \frac{27}{2}, s_1 = 6, s_2 = \frac{55}{2}, x_3 = \frac{9}{2}$

33. New tableau:

BV	P	x_1	x_2	x_3	x_4	s_1	s_2	s_3	s_4	RHS
s_1	0	-3	0	1	0	1	0	0	0	20
x_4	0	2	0	0	1	0	1	0	0	24
s_3	0	0	-3	1	0	0	0	1	0	28
s_4	0	-2	-3	0	0	0	-1	0	1	0
P	1	7	-2	-3	0	0	4	0	0	96

New system:
$$s_1 = 3x_1 - x_3 + 20$$
$$x_4 = -2x_1 - s_2 + 24$$
$$s_3 = 3x_2 - x_3 + 28$$
$$s_4 = 2x_1 + 2x_3 + s_2$$
$$P = -7x_1 + 2x_2 + 3x_3 - 4s_2 + 96$$

Current values: $P = 96,\ s_1 = 20, x_4 = 24, s_3 = 28,$ $s_4 = 0$

Exercise 4.2 (p. 224)

1. (b); the pivot element is 1 in row 1, column 2

3. (a); the solution is $P = \frac{256}{7}, x_1 = \frac{32}{7}, x_2 = 0$ **5.** (c)

7. (b); the pivot element is 1 in row 3, column 4

9. The maximum is $P = \frac{204}{7} = 29\frac{1}{7}$ when $x_1 = \frac{24}{7}, x_2 = \frac{12}{7},$

11. The maximum is $P = 8$ when $x_1 = \frac{2}{3}, x_2 = \frac{2}{3},$

13. The maximum is $P = 6$ when $x_1 = 2, x_2 = 0.$

15. There is no maximum for P; the feasible region is unbounded.

17. The maximum is $P = 30$ when $x_1 = 0, x_2 = 0, x_3 = 10.$

19. The maximum is $P = 42$ when $x_1 = 1, x_2 = 10, x_3 = 0, x_4 = 0.$

21. The maximum is $P = 40$ when $x_1 = 20, x_2 = 0, x_3 = 0.$

23. The maximum is $P = 50$ when $x_1 = 0, x_2 = 15, x_3 = 5, x_4 = 0.$

25. The maximum profit is $1500 when the manufacturer makes 400 of Jean I, 0 of Jean II, and 50 of Jean III.

27. The maximum profit is $190 from the sale of 0 of product A, 40 of product B, and 75 of product C.

29. The maximum revenue is $275,000 when 200,000 gallons of regular, 0 gallons of premium, and 25,000 gallons of super premium are mixed.

31. The maximum return is $7830 when she invests $45,000 in stocks, $31,500 in corporate bonds, and $13,500 in municipal bonds.

33. The maximum profit is $14,400 when 180 acres of crop A, 20 acres of crop B, and 0 acres of crop C are planted.

35. The maximum revenue is $2800 for 50 cans of can I, no cans of can II, and 70 cans of can III.

37. The maximum profit is $12,000 from 1200 television cabinets and no sterco or radio cabinets.

39. The maximum profit is $30,000 when no TV are shipped from Chicago, 375 TVs are shipped from New York, and no TVs are shipped from Denver.

Section 4.3 (p. 236)

1. Standard form **3.** Not in standard form

5. Not in standard form

7. Maximize $P = 2y_1 + 6y_2$
subject to
$$y_1 + 2y_2 \le 2$$
$$y_1 + 3y_2 \le 3$$
$$y_1 \ge 0 \quad y_2 \ge 0$$
$$y_1 + 2y_2 \le 3$$

9. Maximize $P = 5y_1 + 4y_2$
subject to
$$y_1 + 2y_2 \le 3$$
$$y_1 + y_2 \le 1$$
$$y_1 \le 1$$
$$y_1 \ge 0 \quad y_2 \ge 0$$

11. Maximize $P = 2y_1 + 6y_2$
subject to
$$y_1 + 3y_2 \le 3$$
$$y_1 + y_2 \le 4$$
$$y_1 \ge 0 \quad y_2 \ge 0$$

13. The minimum is $C = 6$ when $x_1 = 0, x_2 = 2$

15. The minimum is $C = 12$ when $x_1 = 0, x_2 = 4$

17. The minimum is $C = \frac{21}{5}$ when $x_1 = \frac{8}{5}, x_2 = 0, x_3 = \frac{13}{5}$

19. The minimum is $C = 5$ when $x_1 = 1, x_2 = 1, x_3 = 0, x_4 = 0$

21. Mr Jones minimizes his cost at $ 0.22 when he adds 2 of pill P and 4 of pill Q to his diet.

23. Argus company has a minimum cost of $290 by producing 20 units of A, 30 units of B and 150 units of C.

25. Mrs Mintz minimizes her cost at $65.20 by purchasing 4 of Lunch $1, 3 of Lunch $2, and 2 of Lunch $3.

Section 4.4 (p. 250)

1. The maximum is $P = 44$ when $x_1 = 4, x_2 = 8$

3. The maximum is $P = 27$ when $x_1 = 9, x_2 = 0, x_3 = 0$

5. The maximum is $P = 7$ when $x_1 = 1, x_2 = 2$

7. $x_1 = 0, x_2 = 0, x_3 = \frac{20}{3}, z = \frac{20}{3}$

9. $M_1 \to A_1 : 100, M_2 \to A_1 : 400, M_1 \to A_2 : 300, M_2 \to A_2 : 0,$
$C = \$150,000$

CHAPTER 4   Review

True–False Items (p. 252)

1. T   **2.** F   **3.** T   **4.** F   **5.** T   **6.** F

Fill in the Blanks (p. 252)

1. Slack variables   **2.** column   **3.** $\geq$

4. Van Neuman Duality Principle

Review Exercises (p. 253)

1. In standard form

3. In standard form

5. Not in standard form

7. Not in standard form

9.

BV	P	x_1	x_2	x_3	s_1	s_2	s_3	RHS
s_1	0	2	5	1	1	0	0	100
s_2	0	1	3	1	0	1	0	80
s_3	0	2	3	3	0	0	1	120
P	1	-2	-1	-3	0	0	0	0

11.

BV	P	x_1	x_2	s_1	s_2	s_3	RHS
s_1	0	1	5	1	0	0	200
s_2	0	5	3	0	1	0	450
s_3	0	1	1	0	0	1	120
P	1	-6	-3	0	0	0	0

13.

BV	P	x_1	x_2	x_3	x_4	s_1	s_2	RHS
s_1	0	1	3	1	2	1	0	20
s_2	0	4	1	1	6	0	1	80
P	1	-1	-2	-1	-4	0	0	0

15. (a) The pivot element 2 is found in row s_2, column x_2. The new tableau after pivoting is

BV	P	x_1	x_2	s_1	s_2	RHS
x_1	0	1	0	-4	$-\frac{5}{2}$	15
x_2	0	0	1	1	$\frac{1}{2}$	5
P	1	0	0	1	$\frac{7}{2}$	125

11. $x_I = \frac{5}{8}, x_{II} = \frac{25}{4}, x_{III} = 0, C = \$7.50$

13. Ship 55 sets from the first warehouse to the first retailer, and 75 sets from the second to the second, for a minimum cost of $965.

15. The company should send 10 representatives from New York and 5 from San Francisco to Dallas and any combination of representatives to Chicago (10 from S. F; 0 from N4, or 9 from SF, 1 from NY or 8 from SF, 2 from N4) for a minimum cost of $6300

    (b) The resulting system of equation is
$$x_1 = 15 + 4s_1 + \tfrac{5}{2}s_2$$
$$x_2 = 5 - s_1 - \tfrac{1}{2}s_2$$
$$P = 125 - s_1 - \tfrac{7}{2}s_2$$
    (c) The new tableau is the final tableau. The solution is maximum $P = 125$ when $x_1 = 15$ and $x_2 = 5$.

17. (a) The pivot element 1 is found in row s_2, column x_3. The new tableau after pivoting is

BV	P	x_1	x_2	x_3	s_1	s_2	RHS
s_1	0	1	2	0	1	1	14
x_3	0	0	1	1	0	1	4
P	1	-2	2	0	0	3	12

    (b) The resulting system of equations is
$$s_1 = 14 - x_1 - 2x_2 - s_2$$
$$x_3 = 4 - x_2 - s_2$$
$$P = 12 + 2x_1 - 2x_2 - 3s_2$$
    (c) The problem requires additional pivoting. The new pivot 1 is found in row s_1, column x_1.

19. (a) The pivot element 0.5 is found in row s_1, column x_1. The tableau after pivoting is

BV	P	x_1	x_2	s_1	s_2	RHS
x_1	0	1	1	2	0	2
s_2	0	0	0.5	-2	1	1
P	1	0	0.5	5	0	5

    (b) The resulting system of equations is
$$x_1 = 2 - x_2 - s_1$$
$$s_2 = 1 - 0.5x_2 + 2s_1$$
$$P = 5 - 0.5x_2 - 5s_1$$
    (c) The new tableau is the final tableau. The maximum is $P = 5$ when $x_1 = 2$ and $x_2 = 0$.

21. (a) The pivot element 1 is found in row s_3, column x_1. The tableau after pivoting is

BV	P	x_1	x_2	x_3	s_1	s_2	s_3	RHS
s_1	0	0	0	4	1	-6	1	10
x_2	0	0	1	8	0	-4	1	8
x_1	0	1	0	3	0	-5	1	3
P	1	0	0	13	0	-15	3	14

(b) The resulting system of equations is

$$\begin{aligned} s_1 &= 10 - 4x_3 + 6s_2 - s_3 \\ x_2 &= 8 - 8x_3 + 4s_2 - s_3 \\ x_1 &= 3 - 3x_3 + 5s_2 - s_3 \\ P &= 14 - 13x_3 + 15s_2 - 3s_3 \end{aligned}$$

(c) No solution exists since in the pivot column (s_2) all 3 entries are negative.

23. The maximum is $P = 22{,}500$ when $x_1 = 0, x_2 = 100, x_3 = 50$

25. The maximum is $P = 352$ when $x_1 = 0, x_2 = \frac{6}{5}, x_3 = \frac{28}{5}$

27. In standard form

29. Not in standard form, constraints are not written as greater than or equal to inequalities

31. Not in standard form, constrains are not written as greater than or equal to inequalities

33. Maximize $P = 8y_1 + 2y_2$ subject to the constraints

$$\begin{aligned} 2y_1 + y_2 &\le 2 \\ 2y_1 - y_2 &\le 1 \\ y_1 \ge 0 \quad y_2 &\ge 0 \end{aligned}$$

35. Maximize $P = 100y_1 + 50y_2$ subject to the constraints

$$\begin{aligned} y_1 + 2y_2 &\le 5 \\ y_1 + y_2 &\le 4 \\ y_1 &\le 3 \\ y_1 \ge 0 \quad y_2 &\ge 0 \end{aligned}$$

37. Manimum is $C = 7$ when $x_1 = 3, x_2 = 1$

39. Manimum is $C = 350$ when $x_1 = 50, x_2 = 50$

41. Maximum is $P = 20$ when $x_1 = 0, x_2 = 4$

43. Minimum is $C = 6$ when $x_1 = 6, x_2 = 0$

45. Maximum is $P = 12{,}250$ when $x_1 = 0, x_2 = 5, x_3 = 25$

47. The brewer should brew no lite beer, 180 vats of regular beer, and 30 vats of dark beer to attain a maximum profit of $4500.

49. The manufacturer should ship no cars to dealer 1 and 25 cars to dealer 2 from warehouse 1, and ship 40 cars to dealer 1 and none to dealer 2 from warehouse 2. The minimum cost is $10,150.

51. The farmer will realize a maximum profit of $5714.29 if 0 acrses of corn, 0 acrses of wheat, and 142.85 acrses of soy are planted.

53. The pension fund should purchase 1071.4 shares of Duke Energy, 666.67 shares of Easteran Kodak, 294.12 shares of general motors 158.73 shares of H.J. Hieng to attain a maximum value of $ 3250.

CHAPTER 5 Finance

Exercise 5.1 (p. 267)

1. 60%

3. 110%

5. 6%

7. 0.25%

9. 0.25

11. 1.00

13. 0.065

15. 0.734

17. 150

19. 18

21. 105

23. 5%

25. 160%

27. 250

29. $\frac{1000}{3}$

31. $10

33. $45

35. $150

37. 10%

39. 33.3%

41. 13.3%

43. $1140

45. $1680

47. $1263.16; 10.52%

49. $2380.95; 9.52%

51. $489.00

53. $3\frac{1}{4}$ years

55. Take the discounted loan at 9% per annum.

57. Take the simple interest loan at 6.3% per annum.

59. Choose the simple interest loan at 12.3%.

61. $995,000

63. $62.20 interest received; interest rate is 1.24%

65. $4.40 interest received; interest rate is 1.06%

Exercise 5.2 (p. 276)

1. $1127.27

3. $578.81

5. $826.74

7. $98.02

9. $466.20

11. (a) $1124.86; interest earned: $124.86
(b) $1126.16; interest earned: $126.16
(c) $1126.82; interest earned: $126.82
(d) $1127.27; interest earned: $127.27

13. (a) $1126.49
(b) $1195.62
(c) $1268.99

15. (a) \$4438.55
(b) \$3940.16

17. (a) 8.16%
(b) 4.07%

19. 26.0%

21. 11.5 years

23. Choose b.

25. \$917.43; \$841.68

27. 5.35%

29. 6.82%

31. $6\frac{1}{4}$% compounded annually

33. 9% compounded monthly

35. \$109,400

37. \$656.07

39. \$10,420

41. \$29,137.83

43. \$42,640.10

45. yes

47. \$18,508.09

49. 3.67%; \$8.246 trillion

51. \$10,810.76

53. 9.1%

55. 15.3 years

57. \$940.90

59. \$858.73

61. 22.8 years

63. 11.2 years

Exercise 5.3 (p. 288)

1. \$1593.74

3. \$5073.00

5. \$7867.22

7. \$6977.00

9. \$113,201.03

11. \$147.05 per month

13. \$1868.68 per quarter

15. \$4119.18 per month

17. \$200.46 per month

19. \$2088.11 per year

21. \$62,822.56

23. \$9126.56

25. \$524.04 per month

27. \$22,192.08 per year

Payment \$	Sinking Fund Deposit \$	Cumulative Deposits	Accumulated Interest	Total \$
1	22,192.08	22,192.08	0	22,192.08
2	22,192.08	44,384.16	1775.37	46,159.53
3	22,192.08	66,576.24	5,468.13	72,044.37
4	22,192.08	88,768.32	11,231.68	100,000.00

29. \$205,367.98

31. \$16,555.75 per quarter

33. (a) \$180,611.12
(b) \$2395.33 semiannually

35. 34 years

37. Projected cost (2007): \$ 22,997.82
Projected cost including tax: \$ 25,125.11
Monthly sinking fund payment: \$ 495.78

39. Projected tuition 2012-2013: \$ 6361.34
Projected tuition 2013-2014: \$ 6685.13
Projected tuition 2014-2015: \$ 7025.41
Projected tuition 2015-2016: \$ 7383.00
Monthly sinking fund payment: \$ 521.29

Exercise 5.4 (p. 299)

1. \$15,495.62

3. \$856.60

5. \$85,135.64

7. \$229,100; \$25,239.51

9. \$470.73 per month

11. \$1719.43 per month

13. \$2008.18 per year

15. \$25,906.15

17. The monthly payment for the 9% loan is \$1342.71 and the monthly payment for the 8% loan is \$1338.30. The monthly payment for the 9% loan is larger. The total interest paid is larger for the 9% loan, \$242,813 compared to \$161,192. After 10 years the equity from the 8% loan is larger, \$89,695.33 compared to \$67,617.64.

19. \$55.82 per month

21. (a) \$1207.64
(b) \$36.24 per week

23. (a) \$40,000 down payment
(b) \$160,000 amount of loan
(c) \$1287.40 per month
(d) \$303,464
(e) 22 years, 4 months
(f) \$211,823.20

25. \$332.79 per month

27. \$474.01 per month; \$4,752.48 in interest paid

29. 30 year: monthly payment—\$966.37; total interest—\$235,893.20
15 year: monthly payment—\$1189.89; total interest—\$102,180.20

31. about 4 years and 8 months

33. Monthly payments are reduced by \$128.07. They will pay \$38,420.63 less in interest.

35. 62 months; \$50,870.77 in interest saved

Exercise 5.5 (p. 304)

1. (a) \$2930
(b) 14 payments
(c) 36 payments
(d) \$584.62

3. (a) 2162 trout
(b) 26 months

5. (a) $A_0 = 0, A_n = 1.02A_{n-1} + 500$
(b) 80 quarters, or 20 years
(c) \$159,738

7. (a) $A_0 = 150000, A_n = \left(1 + \frac{0.06}{12}\right)A_{n-1} - 899.33$
(b) $149,850.67
(c)

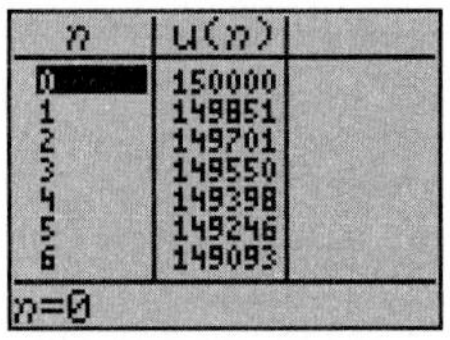

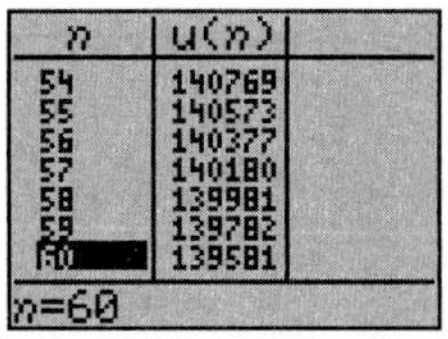

(d) After 4 years and 10 months (58 payments)
(e) With the 360th payment (30 years) the loan is paid.
(f) $173,758.80
(g) a. $A_0 = 150000, A_n = \left(1 + \frac{0.06}{12}\right)A_{n-1} - 999.33$
b. $149,750.67
c.

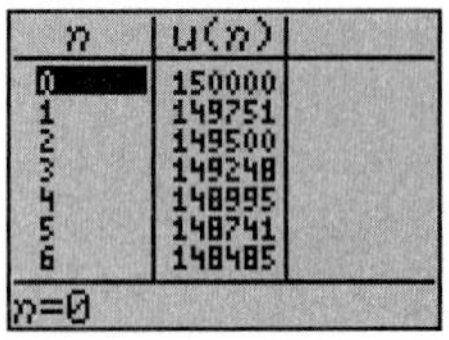

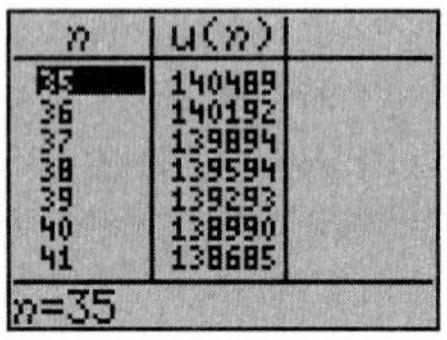

d. After 3 years and 1 month (37 payments)
e. With the 279th payment (23 yrs. 3 mos.) the loan is paid.
f. $128,167.43

Exercise 5.6 (p. 309)

1. The lease option is preferable.

3. Machine A is preferable.

5. $1086.46

7. $992.26

9. 4.38%

CHAPTER 5 Review

True–False Items (p. 311)

1. T **2.** T **3.** F **4.** F

Fill in the Blanks (p. 311)

1. proceeds **2.** present value **3.** annuity **4.** amortized

Review Exercises (p. 311)

1. 15 **2.** 350 **5.** $21\frac{3}{7}\%$ **7.** 80

9. 2200 **11.** Dan paid $19.80 in sales tax

13. $52.50 interest is charged. Dan must repay $552.50.

15. Warren must pay $19,736.84 to settle his debt.

17. $125.12

19. (c) The 10% per annum compounded monthly loan will cost Mike less.

21. $71.36 **23.** 5.95%

25. 5.87% compounded quarterly

27. The Corey's should save $1619.25 per month.

29. (a) The Ostedt's monthly payments are $2726.10.
(b) They will pay $517,830 in interest.
(c) After 5 years their equity is $117,508.93.

31. The monthly payments are $1,049.00. The equity after 10 years is $21,576.

33. Mr. Graff should pay $119,431.51 for the mine.

35. The investor should pay $108,003.59 for the well.

37. Mr. Jones will have saved $10,078.44.

39. Bill must deposit $37.98 every quarter.

41. The monthly payments are $141.22.

43. The effective rate of interest is 9.38%.

45. John will receive $1156.60 every 6 mos. for 15 years.

47. After 30 months there will be $2087.09 in the fund.

49. The student's monthly payment is $330.74.

51. Purchasing the trucks is the better choice.

53. $9845.94

Mathematical Questions from Professional Exams

1. b **2.** c

3. b **4.** b

5. d **6.** a

7. c

CHAPTER 6 Sets: Counting Techniques

Exercise 6.1 (p. 325)

1. true

3. false

5. false

7. true

9. true

11. $\{2, 3\}$

13. $\{1, 2, 3, 4, 5\}$

15. $\emptyset$

17. $\{a, b, d, e, f, q\}$

19. (a) $\{0, 1, 2, 3, 5, 7, 8\}$
(b) $\{5\}$
(c) $\{5\}$
(d) $\{0, 1, 2, 3, 4, 6, 7, 8, 9\}$
(e) $\{4, 6, 9\}$
(f) $\{0, 1, 5, 7\}$
(g) $\emptyset$
(h) $\{5\}$

21. (a) $\{b, c, d, e, f, g\}$
(b) $\{c\}$
(c) $\{a, h, i, j, k, l, m, n, o, p, q, r, s, t, u, v, w, x, y, z\}$
(d) $\{a, b, d, e, f, g, h, i, j, k, l, m, n, o, p, q, r, s, t, u, v, w, x, y, z\}$

23. (a)

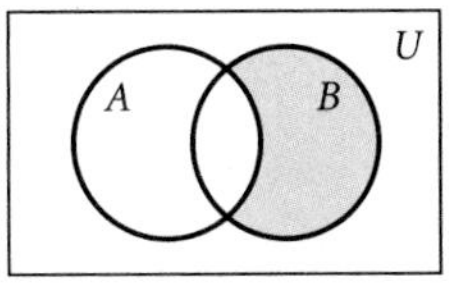

$\bar{A} \cap B$

(b)

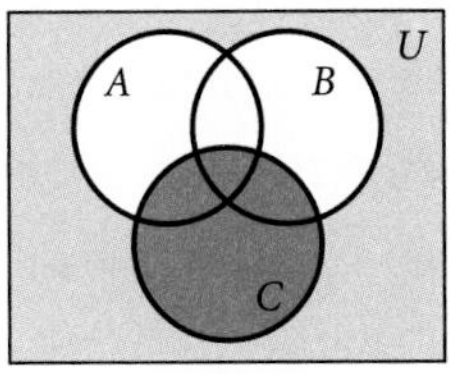

$(\bar{A} \cap \bar{B}) \cup C$

(c)

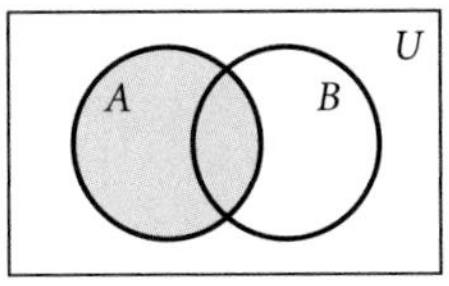

$A \cap (A \cup B)$

(d)

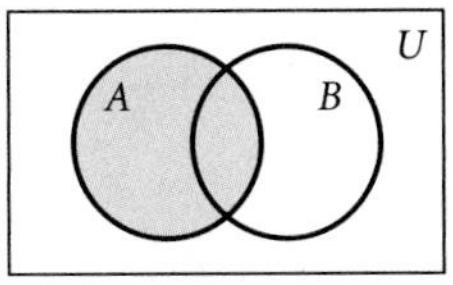

$A \cup (A \cap B)$

(e)

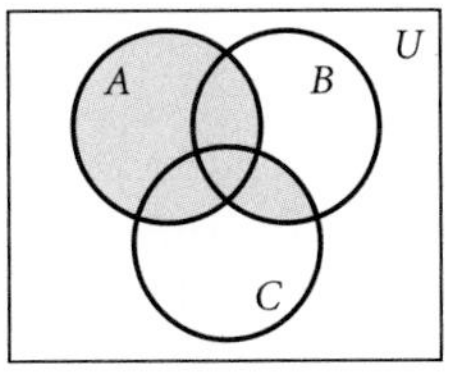

$(A \cup B) \cap (A \cup C)$

(f)

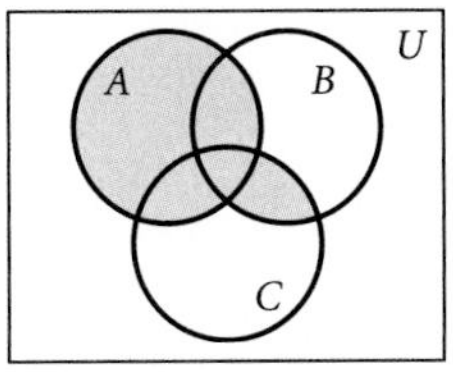

$A \cup (B \cap C)$

(g)

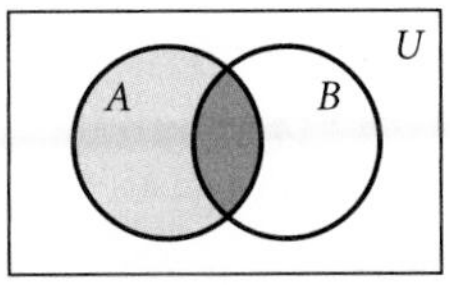

$A = (A \cap B) \cup (A \cap \bar{B})$

(h)

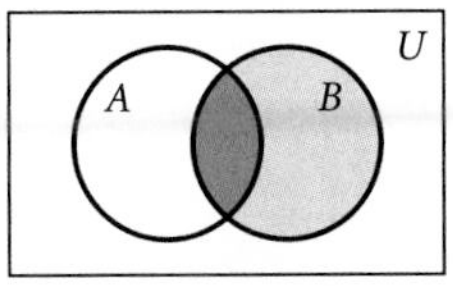

$B = (A \cap B) \cup (\bar{A} \cap B)$

25. $A \cap B$ is the set of members of the board of directors of IBM who are also customers of IBM.

27. $A \cup D$ is the set of all customers and/or stockholders of IBM.

29. $\bar{A} \cap D$ is the set of all members of the board of directors of IBM who are not customers of IBM.

31. $M \cap S$ is the set of all male students who smoke.

33. $\bar{M} \cup \bar{F}$ is the set of sophomores, juniors, and seniors together with the female freshmen.

35. $F \cap S \cap M$ is the set of male freshmen who smoke.

37. The subsets of $\{a, b, c\}$ are $\emptyset$, $\{a\}$, $\{b\}$, $\{c\}$, $\{a, b\}$, $\{a, c\}$, $\{b, c\}$, $\{a, b, c\}$.

Exercise 6.2 (p. 330)

1. $c(A) = 6$

3. $c(A \cap B) = 3$

5. $c[(A \cap B) \cup A] = 6$

7. $c(A \cup B) = 5$

9. $c(A \cap B) = 2$

11. $c(A) = 10$

13. 452 cars

15. $c(A) = 24$

17. $c(A \cup B) = 34$

19. $c(A \cup \bar{B}) = 15$

21. $c(A \cup B \cup C) = 54$

23. $c(A \cap B \cap C) = 3$

25. (a) 536 voters are Catholic or Republican.
(b) 317 voters are Catholic or over 54.
(c) 134 voters are Democtratic below 34 or over 54.

27. (a) 259 were seniors.
(b) 455 were women.
(c) 227 were on the dean's list.
(d) 76 seniors were on the deans list.
(e) 118 seniors were female.
(f) 93 women were on the dean's list.
(g) 912 students were in the college.

29. (a) 40 cars had power steering and air conditioning.
(b) 35 cars had automatic transmission and air conditioning.
(c) 40 cars had neither power steering nor automatic transmission.
(d) 205 cars were sold in July.
(e) 155 cars were sold with automatic transmission or air conditioning or both.

31. 8;

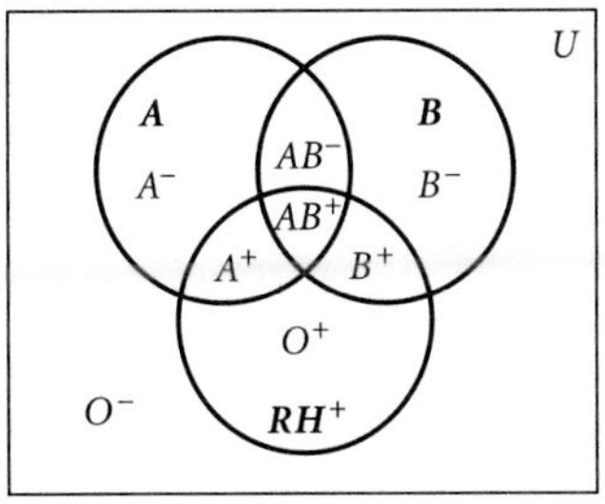

33. 46 use only one of the three brands.

35. $\varnothing$, $\{a\}$, $\{b\}$, $\{c\}$, $\{d\}$, $\{a, b\}$, $\{a, c\}$, $\{a, d\}$, $\{b, c\}$, $\{b, d\}$, $\{c, d\}$, $\{a, b, c\}$, $\{a, b, d\}$, $\{a, c, d\}$, $\{b, c, d\}$, $\{a, b, c, d\}$. There are 16 subsets of $\{a, b, c, d\}$.

Exercise 6.3 (p. 335)

1. 8 routes

3. 24 models

5. 864 outfits

7. 67,600 license plates

9. 120 arrangements

11. $26^4 \cdot 10^4 = 4{,}569{,}760{,}000$ user names

13. $26 \cdot 25^3 \cdot 10 \cdot 9^3 = 2{,}961{,}562{,}500$ user names

15. 360 words without repeated letters; 1296 words

17. 5040 rankings

19. $4^{10} \cdot 2^{15} = 2^{35} = 34{,}359{,}738{,}368 = 3.436 \times 10^{10}$ ways

21. (a) 6,760,000 different license plates
(b) 3,407,040 license plates without repeated digits
(c) 3,276,000 license plates with no repeated letters or digits

23. 16 distinguishable car types are produced.

25. 60 types of homes can be built.

27. 256 different numbers can be formed.

29. 125,000 different lock combinations are possible.

31. 8 different paths through the maze.

Exercise 6.4 (p. 342)

1. 60

3. 120

5. 90

7. 9

9. 28

11. 42

13. 40,320

15. 1

17. 56

19. 1

21. The ordered arrangements of length 3 formed from the letters a, b, c, d, and e are:
abc, abd, abe, acb, acd, ace, adb, adc, ade, aeb, aec, aed,
bac, bad, bae, bca, bcd, bce, bda, bdc, bde, bea, bec, bed,
cab, cad, cae, cba, cbd, cbe, cda, cdb, cde, cea, ceb, ced,
dab, dac, dae, dba, dbc, dbe, dca, dcb, dce, dea, deb, dec,
eab, eac, ead, eba, ebc, ebd, eca, ecb, ecd, eda, edb, edc

$P(5, 3) = 60$

23. 123, 124, 132, 134, 142, 143, 213, 214, 231, 234, 241, 242, 243, 312, 314, 321, 324, 341, 342, 412, 413, 421, 423, 431, 432; $P(4, 3) = 24$

25. 16 two-letter codes

27. 8 three-digit numbers

29. 24 ways

31. 60 three-letter codes

33. 6720 ways

35. 18,278 companies can be on the NYSE.

37. 132,860 ways

39. (a) 720 arrangements.
(b) 120 arrangements if S comes first.
(c) 24 arrangements if S must come first and Y last.

41. 19,958,400 ways

43. 3,368,253,000 ways

45. 32,760 ways officers can be chosen.

Exercise 6.5 (p. 349)

1. 15

3. 21

5. 5

7. 28

9. *abc, abd, abe, acd, ace, ade, bcd, bce, bde, cde*; $C(5, 3) = 10$

11. 123, 124, 134, 234; $C(4, 3) = 4$

13. 35 ways

15. 2380 ways

17. 1140 ways

19. 56 8-bit strings

21. 33,649

23. 90,720 different arrangements

25. 27,720 ways

27. 336 different committees

29. 27,720 ways

31. $\dfrac{100!}{22! \cdot 13! \cdot 10! \cdot 5! \cdot 16! \cdot 17! \cdot 17!} = 1.157 \times 10^{76}$

33. 1950 different collections

35. 1,217,566,350 ways to select.

37. 60 differents ways

39. 10,626 different samples

41. $P(50,15) = 2.943 \times 10^{24}$ different guesses.

43. $P(10,4) = 5040$ different 4 digit numbers (no repeated digits).

Section 6.6 (p. 357)

1. $x^5 + 5x^4y + 10x^3y^2 + 10x^2y^3 + xy^4 + y^5$

3. $x^3 + 9x^2y + 27xy^2 + 27y^3$

5. $16x^4 - 32x^3y + 24x^2y^2 - 8xy^3 + y^4$

7. 10

9. 405

11. 256 different subsets

13. 1023 non-empty subsets

15. 512 subsets

17. $$\binom{10}{7} = \binom{9}{7} + \binom{9}{6} = \left[\binom{8}{7} + \binom{8}{6}\right] + \binom{9}{6}$$
$$= \left[\binom{7}{7} + \binom{7}{6}\right] + \binom{8}{6} + \binom{9}{6} = \binom{6}{6} + \binom{7}{6} + \binom{8}{6} + \binom{9}{6}$$

19. $\binom{12}{6}$

21. $$k \cdot \binom{n}{k} = k \cdot \frac{n!}{k!(n-k)!} = \frac{n \cdot (n-1)!}{(k-1)!(n-k)!}$$
$$= n \cdot \frac{(n-1)!}{(k-1)!((n-1)-(k-1))!} = n \cdot \binom{n-1}{k-1}$$

CHAPTER 6 Review

True–False Items (p. 358)

1. true **2.** true **3.** false **4.** false

5. true **6.** false **7.** false

Fill in the Blanks (p. 359)

1. disjoint **2.** permutation

3. combination **4.** Pascal

5. binomial coefficients **6.** binomial theorem

7. 40

Review Exercises (p. 359)

1. $\subset, \subseteq$ **3.** none of these

5. none of these **7.** $\subset, \subseteq$

9. $\subseteq, =$ **11.** $\subseteq, =$

13. $\subset, \subseteq$ **15.** $\subseteq, =$

17. (a) $\{3, 6, 8, 9\}$ (b) $\{6\}$ (c) B
(d) B (e) $\emptyset$ (f) $\{1, 2, 3, 5, 6, 7, 8, 9\}$

19. (a)

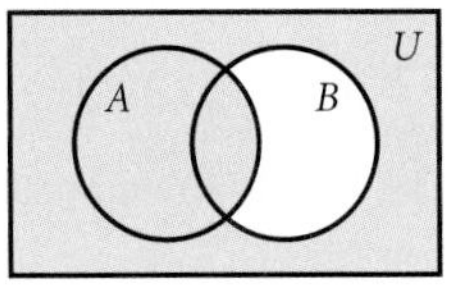

$A \cup \overline{B}$

(b)

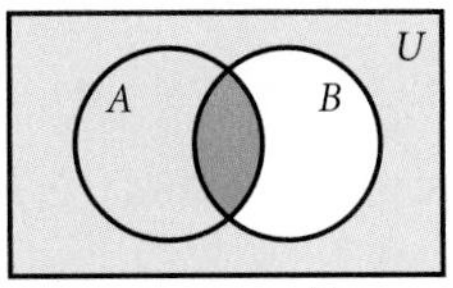

$(A \cap B) \cup \overline{B}$

(c)

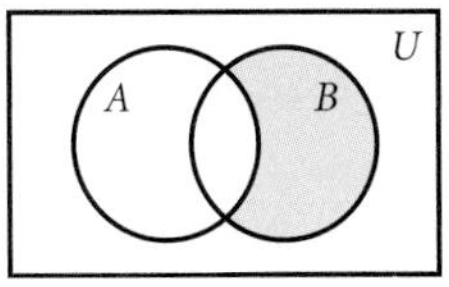

$B \cap \overline{A}$

(d)

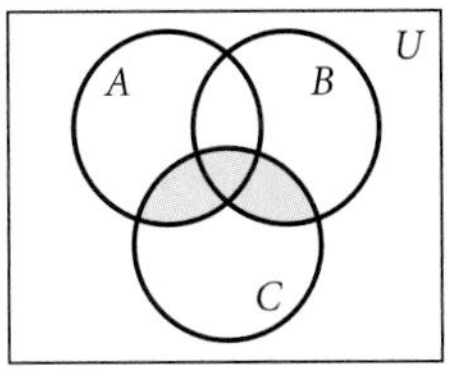

$(A \cup B) \cap C$

(e)

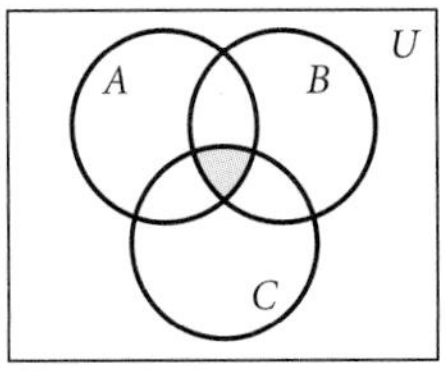

$(A \cap B) \cap (C)$

(f)

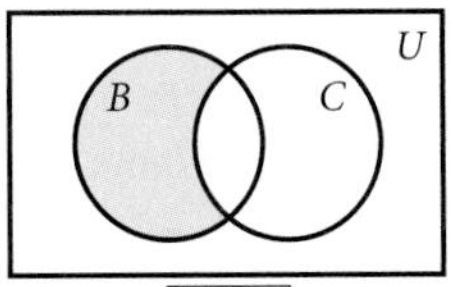

$\overline{(\overline{B} \cup C)}$

21.

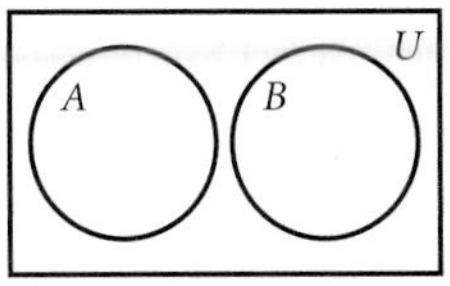

23. The set of all states whose names begin with A or which end with a vowel.

25. The set of all states whose names end with a vowel and lie east of the Mississippi River.

27. The set of all states whose names start with an A or end with a vowel and which lie east of the Mississippi River.

29. $c(A \cap B) = 3$

31. (a) $c(A \cup B) = 20$
(b) A and B are disjoint.

33. $c(A \cap B) = 1$; $c(A \cup B) = 5$

35. (a) 160 cars were sold in June.
(b) 35 cars had only power steering.

37. $\{1\}, \{1, 2\}, \{1, 3\}, \{1, 2, 3\}$ **39.** 1

41. 210 **43.** 12

45. 20 **47.** 12

49. 9900 **51.** 45

53. 1 **55.** 35

57. 9 **59.** 10 ways

61. 6 ways **63.** 72 different styles

65. 1024 different ways to answer

67. (a) 120 words
(b) 20 words if order is not important.

69. (a) 525 different committees
(b) 1715 different committees

71. 12,441,600 ways **73.** 20,790 different committees

75. 240 ways **77.** 24 ways

79. The speakers can be ordered 24 ways.

81. (a) 4845 samples will contain only good plums.
(b) 5700 samples will contain 3 good plums and 1 rotten plum.
(c) 7805 samples will contain one or more rotten plums.

83. 360 words can be made.

85. 302, 400 words can be made.

87. $x^4 + 8x^3 + 24x^2 + 32x + 16$

89. 560

CHAPTER 7 Probability

Exercise 7.1 (p. 373)

1. The outcomes from tossing a coin are H (heads) and T (tails). The sample space is {HH, HT, TH, TT}.

3. The outcomes from tossing each coin are H (heads) and T (tails). The sample space is {HHH, HHT, HTH, HTT, THH, THT, TTH, TTT}.

5. The outcomes from tossing each coin are H (heads) and T (tails), and the outcomes from tossing a die are the numbers 1, 2, 3, 4, 5, and 6. The sample space is {HH1, HT1, TH1, TT1, HH2, HT2, TH2, TT2, HH3, HT3, TH3, TT3, HH4, HT4, TH4, TT4, HH5, HT5, TH5, TT5, HH6, HT6, TH6, TT6}.

7. The sample space, S = {RA, RB, RC, GA, GB, GC} where R = Red and G = Green.

9. The sample space, S = {AA, AB, AC, BA, BB, BC, CA, CB, CC}

11. The sample space, S = {AA1, AB1, AC1, BA1, BB1, BC1, CA1, CB1, CC1, AA2, AB2, AC2, BA2, BB2, BC2, CA2, CB2, CC2, AA3, AB3, AC3, BA3, BB3, BC3, CA3, CB3, CC3, AA4, AB4, AC4, BA4, BB4, BC4, CA4, CB4, CC4}

13. The sample space, S = {RA1, RB1, RC1, GA1, GB1, GC1, RA2, RB2, RC2, GA2, GB2, GC2, RA3, RB3, RC3, GA3, GB3, GC3, RA4, RB4, RC4, GA4, GB4, GC4} where R = Red and G = Green.

15. $c(S) = 16$ 17. $c(S) = 216$ 19. $c(S) = 1326$

21. $c(S) = 676$ 23. Valid assignments are: 1, 2, 3, 6

25. Assignment 2 should be used. 27. $P(H) = \frac{3}{4}, P(T) = \frac{1}{4}$

29. $P(1) = \frac{2}{9}, P(2) = \frac{1}{9}, P(3) = \frac{2}{9}, P(4) = \frac{1}{9}, P(5) = \frac{2}{9}, P(6) = \frac{1}{9}$

31. Define W: A white ball is picked. $P(W) = \frac{3}{23}$

33. Define G: A green ball is picked. $P(G) = \frac{7}{23}$

35. Define R: A red ball is picked. $P(W \cup R) = \frac{8}{23}$

37. $P(\overline{G \cup R}) = \frac{11}{23}$ 39. $P(A) = \frac{1}{18}$

41. $P(E) = \frac{1}{9}$ 43. $P(E) = \frac{1}{6}$

45. Define A: an ace is drawn; H: a heart is drawn. $P(A \cap H) = \frac{1}{52}$

47. Define S: a spade is drawn. $P(S) = \frac{1}{4}$

49. Define F: a picture card is drawn. $P(F) = \frac{3}{13}$

51. Define E: a card with a number less than 6 is drawn. $P(E) = \frac{5}{13}$

53. Define A: an ace is drawn. $P(\overline{A}) = \frac{12}{13}$

55. Define H: A randomly selected person has health insurance. $P(H) = 0.845$

57. $P(H) = 0.5$, $P(T) = 0.5$, answers will vary but the results should be fairly close to the actual probabilities.

59. $P(H) = 0.75$, $P(T) = 0.25$, answers will vary but the results should be fairly close to the actual probabilities.

61. $P(\text{Red}) = \frac{1}{3}$, $P(\text{Yellow}) = \frac{2}{15}$, $P(\text{White}) = \frac{8}{15}$, answers will vary but the results should be fairly close to the actual probabilities.

Exercise 7.2 (p. 384)

1. Define E: A person selected at random has Rh-positive blood. $P(E) = 0.84$

3. Define E: A person selected at random has blood that contains the A antigen. $P(E) = 0.44$

5. Define E: A randomly selected ISP subscriber subscribes to America Online; F: A randomly selected ISP subscriber subscribes to MSN. $P(E \cup F) = 24.0\%$

7. Define E: A randomly selected ISP subscriber subscribes to an ISP other than one of the top 10. $P(E) = 60.1\%$

9. $P(\overline{A}) = 0.75$ 11. $P(A \cup B) = 0.65$ 13. $P(A \cup B) = 0.5$

15. Define E: Sum is 2; F: Sum is 12. E and F are mutually exclusive. $P(E \cup F) = \frac{1}{18}$

17. Define E: The Bears win; F: The Bears tie. $P(E \cup F) = 0.70$; $P(\overline{E \cup F}) = 0.30$

19. Define M: Jenny passes mathematics; E: Jenny passes English. $P(M \cap E) = 0.2$

21. (a) $P(A \cup B) = 0.7$ (b) $P(A \cap B) = 0.3$ (c) $P(B \cap \overline{A}) = 0.2$ (d) $P(\overline{A \cup B}) = 0.3$

23. Define T: Car needs a tune-up; B: Car needs a brake job.
(a) $P(T \cup B) = 0.68$
(b) $P(T \cap \overline{B}) = 0.58$
(c) $P(\overline{T \cup B}) = 0.32$

25. (a) $P(1 \text{ or } 2) = 0.57$
(b) $P(1 \text{ or more}) = 0.95$
(c) $P(0, 1, 2 \text{ or } 3) = 0.83$
(d) $P(3 \text{ or more}) = 0.38$
(e) $P(0 \text{ or } 1) = 0.29$
(f) $P(0) = 0.05$
(g) $P(1, 2 \text{ or } 3) = 0.78$
(h) $P(2 \text{ or more}) = 0.71$

27. $P(E) = \frac{3}{4}$ 29. $P(E) = \frac{5}{12}$ 31. $P(E) = \frac{1}{2}$

33. The odds for E: 3 to 2; The odds against E: 2 to 3

35. The odds for F: 3 to 1; The odds against F: 1 to 3

37. (a) The odds for E are 1 to 5.
(b) The odds for F are 1 to 17.
(c) The odds for $E \cup F$ are 2 to 7.

39. The odds against getting the interview are 23 to 27.

41. $P(A \cup B) = \frac{11}{15}$; the odds for A or B winning are 11 to 4.

Exercise 7.3 (p. 391)

1. Probability all 5 are defective $= 2.83 \times 10^{-4}$; probability at least 2 are defective $= 0.103$.

3. (a) $P(3H) = \frac{5}{16}$ (b) $P(0H) = \frac{1}{32}$

5. (a) Probability 3 sevens is 0.005.
(b) Probability of at least 2 sums of 7 or 11 is 0.126.

7. Probability of a repeated digit is 0.940.

9. Probability of no repeated letters is 0.006.

11. Probability the lists match is $\frac{1}{24}$.

13. Probability of least 2 of 6 people are born in the same month is 0.777.

15. Probability at least 2 senators have the same birthday is almost 1.

17. Probability L precedes E is $\frac{1}{2}$.

19. Probability the word begins with L is $\frac{1}{5}$.

21. Probability both the Giants and Dodgers are in the playoffs is $\frac{2}{65}$.

23. Probability the wild card team is from the Central division is $\frac{5}{13}$.

25. Probability of the given bridge hand is 0.005.

27. Probability no two passengers exit on the same floor is $\frac{105}{512}$.

Exercise 7.4 (p. 399)

1. $P(E) = 0.5$ 3. $P(E|F) = 0.429$ 5. $P(E \cap F) = 0.3$

7. $P(\overline{E}) = 0.5$ 9. $P(E|F) = 0.25, P(F|E) = 0.5$

11. $P(F) = 0.5$ 13. $P(E \cap F) = \frac{4}{13}$

15. (a) $P(E) = \frac{1}{2}$ (b) $P(F) = \frac{2}{3}$ 17. $P(C) = 0.69$

19. $P(C|A) = 0.9$ 21. $P(C|B) = 0.2$ 23. $P(E \cap F) = 0.1$

25. $P(F|E) = 0.2$ 27. $P(E|\overline{F}) = 0.667$

29. Probability exactly 2 girls, given 1st child is a girl is $\frac{1}{2}$.

31. $P(4H) = \frac{1}{16}$; Yes, if we know the 2nd throw is a head $P(4H|H) = \frac{1}{8}$.

33. Probability of drawing a heart and then a red is $\frac{25}{204}$; the probability of drawing a red and then a heart is $\frac{25}{204}$.

35. Probability of drawing 1 white and 1 yellow ball is $\frac{1}{5}$.

37. (a) Probability of drawing a red ace is $\frac{1}{26}$.
(b) Probability of drawing a red ace given an ace was drawn is $\frac{1}{2}$.
(c) Probability of drawing a red ace given a red was drawn is $\frac{1}{13}$.

39. Probability family has more than 2 children given it has at least one child is 0.625.

41. $P(E) = 0.4$ 43. $P(H) = 0.24$ 45. $P(E \cap H) = 0.1$

47. $P(G \cap H) = 0.08$ 49. $P(E|H) = 0.417$

51. $P(G|H) = 0.333$ 53. $P(E|F) = 0.107, P(F|E) = 0.818$

55. (a) $P(M) = \frac{724}{1009}$ (e) $P(A|M) = \frac{171}{724}$
(b) $P(A) = \frac{333}{1009}$ (f) $P(F|A \cup E) = \frac{213}{596}$
(c) $P(F \cap B) = \frac{72}{1009}$ (g) $P(M \cap \overline{B}) = \frac{383}{1009}$
(d) $P(F|E) = \frac{51}{263}$ (h) $P(F|\overline{E}) = \frac{117}{373}$

57. $P(E|F) = \frac{2}{23}$

59. Define E: Person is a Republican; F: a person voted for the Democrat. $P(E) = \frac{3}{4}$; $P(E|F) = \frac{11}{20}$.

61. $P(S) = 0.467$ 63. $P(W|E) = \frac{5}{11}$

Exercise 7.5 (p. 409)

1. $P(E \cap F) = 0.24$ 3. $P(F) = 0.125$

5. No, $P(E \cap F) = \frac{2}{9} \neq \frac{1}{9} = P(E)P(F)$.

7. (a) $P(E|F) = 0.2$ (c) $P(E \cap F) = 0.08$
(b) $P(F|E) = 0.4$ (d) $P(E \cup F) = 0.52$

9. $P(E \cap F \cap G) = \frac{4}{147}$

11. $P(E|F) = 0.5$ No, $P(E \cap F) = 0.1 \neq 0.06 = P(E)P(F)$.

13. No, $P(E \cap F) = \frac{1}{6} \neq \frac{1}{4} = P(E)P(F)$.

15. (a) Yes, $P(E \cap F) = \frac{1}{4} = P(E)P(F)$.
(b) No, $P(E \cap F) = \frac{1}{4} \neq \frac{5}{24} = P(E)P(F)$.

17. $P(H) = \frac{1}{2}$ 19. $P(4) = \frac{1}{6}$ 21. $P(\overline{4}) = \frac{5}{6}$

23. $F(5 \cup 6) = \frac{1}{3}$ 25. $P[(4 \cup 5 \cup 6) \cap H] = \frac{1}{4}$

27. $P(\text{RRR}) = \frac{1}{12}, P(\text{RRL}) = \frac{2}{12}, P(\text{RLR}) = \frac{1}{12}, P(\text{LRR}) = \frac{1}{12}$,
$P(\text{RLL}) = \frac{2}{12}, P(\text{LRL}) = \frac{2}{12}, P(\text{LLR}) = \frac{1}{12}, P(\text{LLL}) = \frac{2}{12}$
(a) $P(E) = \frac{1}{4}$ (b) $P(F) = \frac{2}{12}$ (c) $P(G) = \frac{1}{2}$ (d) $P(H) = \frac{1}{2}$

29. $P(E \cap F) = \frac{1}{4} = P(E)P(F)$

31. (a) Probability both children have heart disease is $\frac{9}{16}$.
(b) Probability neither child is diseased is $\frac{1}{16}$.
(c) Probability exactly 1 has disease is $\frac{3}{8}$.

33. (a) $P(3T) = \frac{27}{64}$ (b) $P(2H \text{ and } 1T) = \frac{9}{64}$

35. (a) Probability all recover is 0.656.
(b) Probability 2 recover is 0.049.
(c) Probability at least 2 recover is 0.996.

37. (a) Probability both are red is $\frac{9}{25}$.
(c) Probability one is red is $\frac{12}{25}$.

39. (a) $P(A|U) = \frac{8}{65}$ (b) $P(A|\overline{U}) = \frac{1}{103}$
(c) No, $P(U \cap A) = \frac{1}{21} \neq \frac{65}{3136} = P(U)P(A)$.

(d) No, $P(U \cap \overline{A}) = \frac{19}{56} \neq \frac{3445}{9408} = P(U)P(\overline{A})$.

(e) No, $P(\overline{U} \cap A) = \frac{1}{168} \neq \frac{103}{3136} = P(\overline{U})P(A)$.

(f) No, $P(\overline{U} \cap \overline{A}) = \frac{17}{28} \neq \frac{5459}{9408} = P(\overline{U})P(\overline{A})$.

41. (a) Probability both vote for the candidate is $\frac{4}{9}$.
(b) Probability neither vote for the candidate is $\frac{1}{9}$.
(c) Probability one votes for the candidate is $\frac{4}{9}$.

CHAPTER 7 Review

True–False Items (p. 413)

1. T **2.** F **3.** F **4.** F

5. T **6.** F **7.** T **8.** T

Fill in the Blanks (p. 413)

1. $\frac{1}{2}$ **2.** 32 **3.** 1, 0 **4.** 0.8

5. in favor of **6.** equally likely **7.** mutually exclusive

Review Exercises (p. 414)

1. $S = \{0, 1, 2, 3, 4, 5\}$

3. The outcomes for each child are boy (B) and girl (G). The sample space is {BB, BG, GB, GG}.

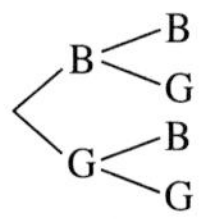

5. $P(\text{penny}) = \frac{4}{15}$, $P(\text{dime}) = \frac{1}{3}$, $P(\text{quarter}) = \frac{2}{5}$

7. $P(1) = \frac{1}{8}$, $P(2) = \frac{1}{4}$, $P(3) = \frac{1}{8}$, $P(4) = \frac{1}{8}$, $P(5) = \frac{1}{4}$, $P(6) = \frac{1}{8}$

9. (a) Let $X =$ the number of girls in a family of 4 children

X	$P(X)$
0	0.0625
1	0.25
2	0.375
3	0.25
4	0.0625

(b) (i) $P(0) = 0.0625$ (iii) $P(1) = 0.25$
(ii) $P(2) = 0.375$ (iv) $1 - P(4) = 0.9375$

11. (a) Let $X =$ the number of tails observed on three tosses of a coin

X	$P(X)$
0	$\frac{1}{8}$
1	$\frac{3}{8}$
2	$\frac{3}{8}$
3	$\frac{1}{8}$

(b) (i) $P(3) = \frac{1}{8}$ (iii) $P(2) = \frac{3}{8}$
(ii) $P(0) = \frac{1}{8}$ (iv) $P(2 \cup 3) = \frac{1}{2}$

13. (a) Probability both are blue is $\frac{10}{91}$.
(b) Probability 1 is blue is $\frac{45}{91}$.
(c) Probability at least 1 is blue is $\frac{55}{91}$.

15. (a) $P(A \cup B) = 0.6$ (c) $P(\overline{A \cup B}) = 0.4$
(b) $P(\overline{A}) = 0.7$ (d) $P(\overline{A} \cup B) = 0.8$

17. (a) $P(3) = \frac{21}{100}$ (b) $P(5) = \frac{23}{100}$ (c) $P(6) = \frac{73}{400}$

19. (a) $P(\overline{E}) = 0.35$ (c) No, because $P(E \cap F) = 0.3 \neq 0$.
(b) $P(E \cup F) = 0.75$

21. (a) $P(\overline{E}) = \frac{1}{2}$ (b) $P(F) = \frac{11}{24}$ (c) $P(\overline{F}) = \frac{13}{24}$

23. (a) $P(\overline{E}) = 0.70$ (e) $P(\overline{E \cap F}) = 1$
(b) $P(\overline{F}) = 0.55$ (f) $P(\overline{E \cup F}) = 0.25$
(c) $P(E \cap F) = 0$ (g) $P(\overline{E} \cup \overline{F}) = 1$
(d) $P(E \cup F) = 0.75$ (h) $P(\overline{E} \cap \overline{F}) = 0.25$

25. $P(E \cup P) = \frac{1}{2}$

27. (a) No, since $P(0) = \frac{1}{2} \neq P(1) = \frac{1}{8} \neq P(2) = \frac{3}{8}$.
(b) Outcome 0 has the highest probability.
(c) $P(F) = \frac{9}{64}$

29. $P(5 \cup 7 \cup 9) = \frac{5}{12}$

31. Odds in favor of throwing a 5 are 1 to 5.

33. Probability Bears win is $\frac{7}{13}$.

35. $P(E \cap F) = \frac{3}{16} = P(E)P(F)$

37. Define E: A student fails mathematics, and F: A student fails physics. (a) $P(E|F) = 0.333$; (b) $P(F|E) = 0.237$; (c) $P(E \cup F) = 0.560$

39. The probability is $\frac{5}{216}$.

41. Define E: A person has blue eyes, F: A person has brown eyes, G: A person is left handed.
(a) $P(E \cap G) = 0.025$ (b) $P(G) = 0.0625$ (c) $P(E|G) = 0.4$

43. (a) $P(F|E) = \frac{2}{5}$
(b) $P(E|F) = \frac{1}{5}$
(c) Let $E =$ scored over 80% and $F =$ took form A.
Yes, because $P(E \cap F) = 0.08 = P(E)P(F)$.
(d) Let $E =$ scored over 80% and $F =$ took form B.
Yes, because $P(E \cap F) = 0.12 = P(E)P(F)$.

45. Probability at least one matched is $\frac{2}{3}$.

47. (a) Probability all are underweight is 0.0002.
(b) Probability 2 are underweight is 0.083.
(c) Probability at most 1 is underweight is 0.910.

49. Probability all are born on different days is 0.017.

51. $P(E|F) = 0.2$

53. (a) Probability misses the 1st and gets the next 3 is 0.1029.
(b) Probability misses 10 in a row is 0.0282.

55. Probability the car is black is $\frac{1}{4}$.

Mathematical Questions from Professional Exams (p. 419)

1. b **2.** e **3.** b **4.** d **5.** c
6. d **7.** b **8.** a **9.** c **10.** b

CHAPTER 8 Additional Probability Topics

Exercise 8.1 (p. 430)

1. $P(E|A) = 0.4$ **3.** $P(E|B) = 0.2$ **5.** $P(E|C) = 0.7$

7. $P(E) = 0.31$ **9.** $P(A|E) = \frac{12}{31}$ **11.** $P(C|E) = \frac{7}{31}$

13. $P(B|E) = \frac{12}{31}$ **15.** $P(E) = 0.024$ **17.** $P(E) = 0.016$

19. $P(A_1|E) = 0.5; P(A_2|E) = 0.5$

21. $P(A_1|E) = 0.375; P(A_2|E) = 0.375; P(A_3|E) = 0.25$

23. $P(A_1|E) = 0.276; P(A_2|E) = 0.690; P(A_3|E) = 0.034$

25. $P(A_2|E) = 0; P(A_3|E) = 0.065; P(A_4|E) = 0;$
$P(A_5|E) = 0.065$

27. $P(U_I|E) = \frac{5}{15}; P(U_{II}|E) = \frac{3}{15} = 0.2; P(U_{III}|E) = \frac{7}{15}$

29. $P(M|CB) = 0.953$

31. $P(D|V) = 0.385; P(R|V) = 0.39; P(I|V) = 0.225$

33. $P(\text{Rock}|\text{positive}) = 0.385; P(\text{Clay}|\text{positive}) = 0.209;$
$P(\text{Sand}|\text{positive}) = 0.405$

35. $P(R) = 0.466; P(N|R) = 0.343$

37. The probability the nurse forgot is $\frac{9}{11}$.

39. Probability a student majors in engineering given she is female is 0.217.

41. (a) Probability a person is diseased given a positive test 0.858.
(b) Probability a patient has the disease given a positive test is 0.503.
(c) Probability a patient has the disease given two positive tests is 0.961.

Exercise 8.2 (p. 440)

1. $b(7, 4; .20) = 0.0287$ **3.** $b(15, 8; .30) = 0.0138$

5. $b(15, 10; \frac{1}{2}) = \frac{3003}{32768} = 0.0916$ **7.** 0.2969

9. $b(3, 2; \frac{1}{3}) = \frac{2}{9}$ **11.** $b(3, 0; \frac{1}{6}) = \frac{125}{216}$

13. $b(5, 3; \frac{2}{3}) = \frac{80}{243}$ **15.** $b(10, 6; .3) = 0.0368$

17. $b(12, 9; .8) = 0.2362$ **19.** $P(\text{At least 5 successes}) = 0.0580$

21. $b(8, 1; \frac{1}{2}) = \frac{1}{32}$ **23.** $P(\text{at least 5 tails}) = \frac{93}{256}$

25. $P(2H/\text{at least 1H occurs}) = \frac{28}{255}$ **27.** $b(5, 2; \frac{1}{6}) = \frac{625}{3888}$

29. (a) $b(8, 1; .05) = 0.2793$ (c) $1 - P(8, 0; .05) = 0.3366$
(b) $b(8, 2; .05) = 0.0515$ (d) $P(\text{Fewer than 3 defective}) = 0.9942$

31. $b(6, 3; .5) = \frac{5}{16} = 0.3125$

33. (a)

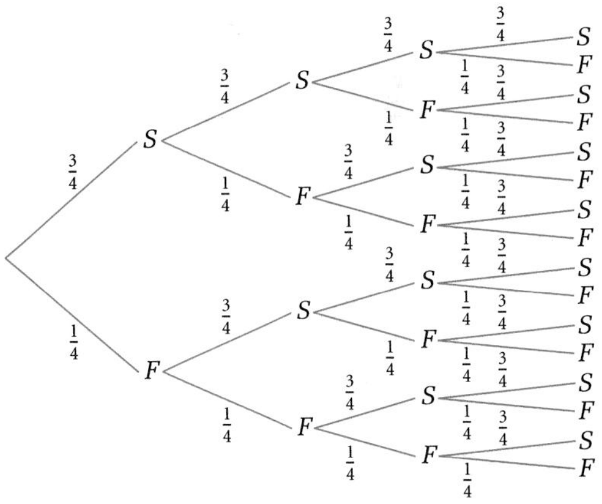

(b) $P(\text{Exactly 2 successes}) = \frac{54}{256}$

(c) $b\left(4, 2; \frac{1}{4}\right) = \frac{54}{256}$

35. $P(\text{At least 2 hits}) = 0.9996$

37. $P(\text{At least 10 correct}) = \frac{309}{2048}$
$P(\text{At least 12 correct}) = 0.6482$

39. $b(8, 8; .60) = 0.0007$ **41.** $b(10, 4; .23) = 0.1225$

43. (a) $P(\text{At least 5 correct}) = \frac{7}{64}$
(b) $P(\text{At least 5 correct}) = 0.3446$

45. (a) $b(10, 4; .124) = 0.0224$
(b) $b(10, 0; .124) = 0.2661$
(c) $P(\text{At most 5 are over 65}) = 0.9995$

47.

k	Actual Value of P(k)
0	0.0625
1	0.25
2	0.375
3	0.25
4	0.0625

49. $b(8, 3; .5) = 0.21875$

Exercise 8.3 (p. 450)

1. $E = 1.2$ **3.** $E = 50{,}800$ fans

5. Mary should pay 80 cents per game.

7. Dave should pay $1.67 to play.

9. The price exceeds the expected value by $0.75.

11. (a) $E = \$0.75$ (b) No, the game is not fair.
(c) To make the game fair, a player should lose $2.00 if 1 tail is thrown.

13. It is not fair; the expected loss is $0.42.

15. The expected loss is 1.2 cents, so Sarah should not play.

17. He should bet $7 to make the game fair.

19. Management should choose the second location.

21. $E = \frac{2000}{6} = 333\frac{1}{3}$ times **23.** $E = 10$ light bulbs

25. $E = 1$ person to have an unfavorable reaction.

27. $E = 2.734$ tosses.

29. The airline should schedule aircraft A to maximize expected profit.

Exercise 8.4 (p. 457)

1. Expected number of customers is 9; optimal number of cars is 9 for an expected daily profit of $168.

3.

Group size	2	3	4	5	6	7
Expected Tests per Saved Component	0.403	0.524	0.565	0.574	0.568	0.556

The optimal group size is 5.

5. (a) $E(X) = \$75{,}000 - 75000(.05^x) - 500x$ where x is the number of divers hired.
(b) Hiring two divers maximizes the net gain.

7. The probability the message is correctly received is 0.9647.

Exercise 8.5 (p. 459)

1. $P(X = 0) = \frac{1}{4}$; $P(X = 1) = \frac{1}{2}$; $P(X = 2) = \frac{1}{4}$

3. $P(X = 0) = \frac{1}{8}$; $P(X = 1) = \frac{3}{8}$; $P(X = 2) = \frac{3}{8}$; $P(X = 3) = \frac{1}{8}$

5. $P(X = 0) = \frac{1}{6}$; $P(X = 1) = \frac{1}{2}$; $P(X = 2) = \frac{3}{10}$; $P(X = 3) = \frac{1}{30}$

7. $E(X) = 1.2$

9. Actual probabilities: $P(X = 1) = \frac{1}{6}$, $P(X = 2) = \frac{1}{6}$, $P(X = 3) = \frac{1}{6}$, $P(X = 4) = \frac{1}{6}$, $P(X = 5) = \frac{1}{6}$; $P(X = 6) = \frac{1}{6}$.

11. $P(0.1 \le X \le 0.3) = 0.2$

13. $P(X = 2) = \frac{1}{2}$

CHAPTER 8 Review

True–False Items (p. 461)

1. T **2.** F **3.** F **4.** T

5. F **6.** T

Fill in the Blanks (p. 462)

1. Bayes' formula **2.** independent; the same

3. expected value of the experiment **4.** a real number

5. expected value

Review Exercises (p. 462)

1. $P(E|A) = 0.82$ **3.** $P(E|B) = 0.10$

5. $P(A|E) = 0.9866$ **7.** $P(B|E) = 0.0134$

9. $P(E|A) = 0.5$ **11.** $P(E|B) = 0.4$

13. $P(E|C) = 0.3$ **15.** $P(E) = 0.43$

17. $P(A|E) = 0.4651$ **19.** $P(B|E) = 0.4651$

21. $P(C|E) = 0.0698$

23. (a) $P(E|G) = \frac{9}{13}$ (e) $P(F|G) = \frac{3}{13}$
(b) $P(G|E) = \frac{12}{23}$ (f) $P(G|F) = \frac{2}{7}$
(c) $P(H|E) = \frac{22}{64}$ (g) $P(H|F) = \frac{17}{42}$
(d) $P(K|E) = \frac{11}{69}$ (h) $P(K|F) = \frac{13}{42}$

25. $P(C) = 0.163$

27. (a) $b(5, 0; .20) = 0.3277$ (b) $b(5, 3; .20) = 0.0512$

29. (a) $b(12, 12; \frac{1}{2}) = \frac{1}{2^{12}} \approx 0$

(b) The probability the student passes is $\frac{793}{2048}$.

(c) The odds in favor of passing are 793 to 1255.

31. $P(\text{At least 3 11's}) = .0016$

33. $E = 1.5$ **35.** $E = \$29.52$

37. (a) 18 cents would be a fair price for a ticket.

(b) Alice paid 56 cents extra for the eight tickets.

39. $E = 9.75$ **41.** (a) The expected profit is \$713,500

43. $E = \frac{500}{6} \approx 83.33$

45. The probability the code is correctly received is 0.9349.

47. (a) Probability pooled test is positive is $1 - (1 - p)^{30}$. The expected number of tests in a pooled sample is $E = 31 - 30\,(1 - p)^{30}$.

49. (a) $X = \{0, 1, 2, 3, 4, 5\}$
(b) $P(X = 0) = 0.5838; P(X = 1) = 0.3394;$ $P(X = 2) = 0.0702; P(X = 3) = 0.0064;$ $P(X = 4) = 0.0003; P(X = 5) \approx 0$
(c) $E(X) = 0.5$

51. (a) $X = \{0, 1, 2, 3\}$
(b) $P(X = 0) = .2917; P(X = 1) = .525; P(X = 2) = .175;$ $P(X = 3) = .0083$
(c) $E(X) = .8999$

51. $X = \{0, 1, 2, 3, 4, 5\}; P(X = 0) = .5158; P(X = 1) = .3651;$ $P(X = 2) = .1034; P(X = 3) = .0146; P(X = 4) = .0010;$ $P(X = 5) = .00002; E(X) = .62$

CHAPTER 9 Statistics

Exercise 9.1 (p. 472)

1. discrete **3.** continuous **5.** continuous **7.** discrete

9. discrete **11.** continuous

13. A poll should be taken either door-to-door or by means of the telephone of a cross section of people from different parts of the country.

15, 17. Answers will vary. All answers should include a method to choose a sample in which each member of the population has an equal chance of being selected.

19. Answers will vary. **21.** Answers will vary.

Exercise 9.2 (p. 478)

1. (a)

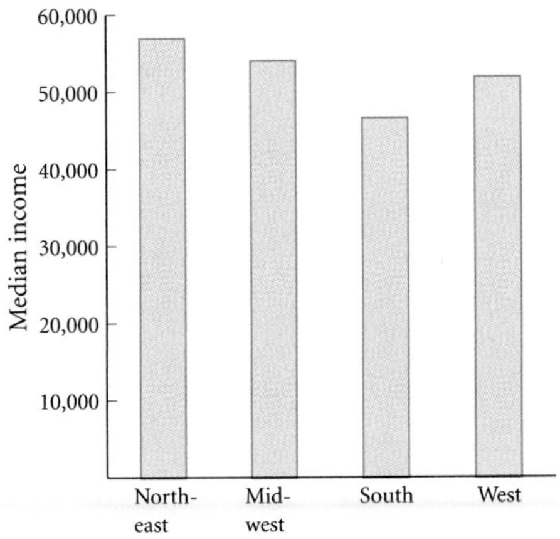

(b) Northeast has the highest median income.
(c) South has the lowest median income.

3. (a)

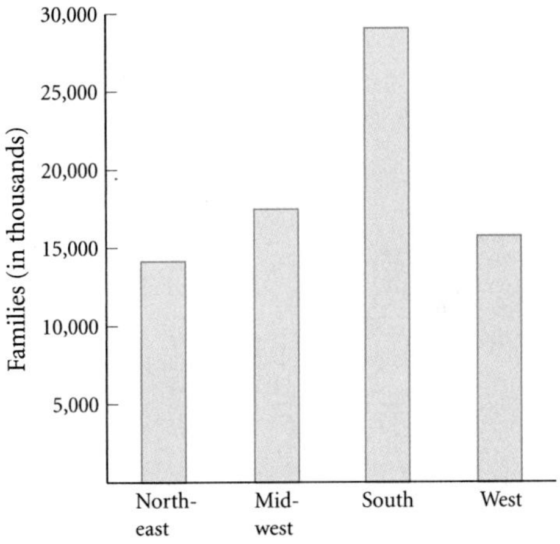

(b)

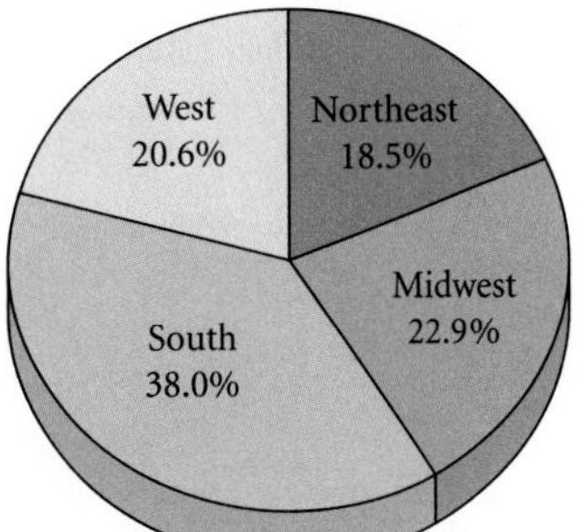

(d) The South has the most families.
(e) The Northeast has the fewest families.

5. (a)

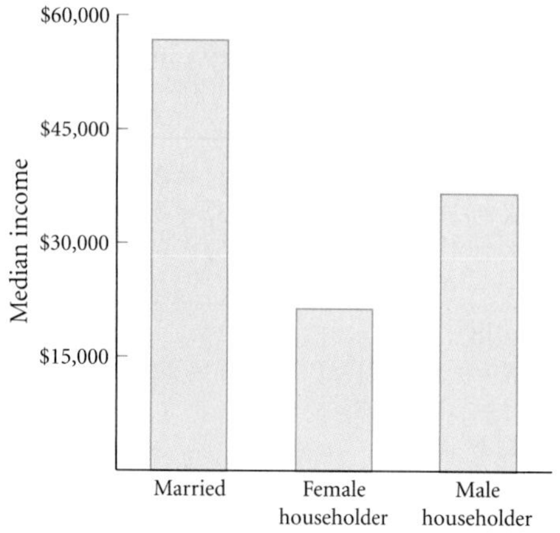

(b) Married couple families have the highest median income.
(c) Female householder—no spouse families have the lowest median income.

7. (a)

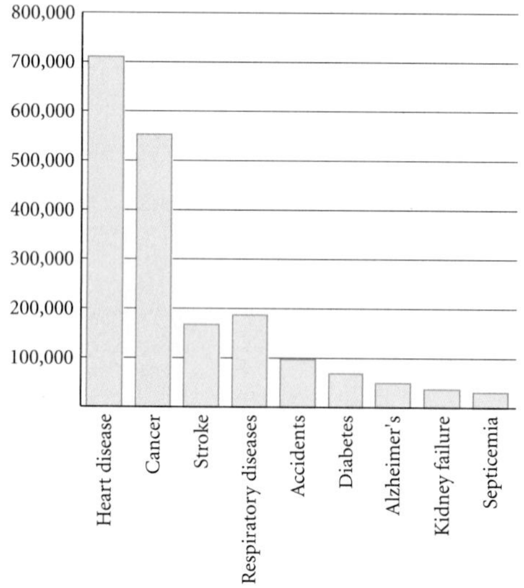

(b)

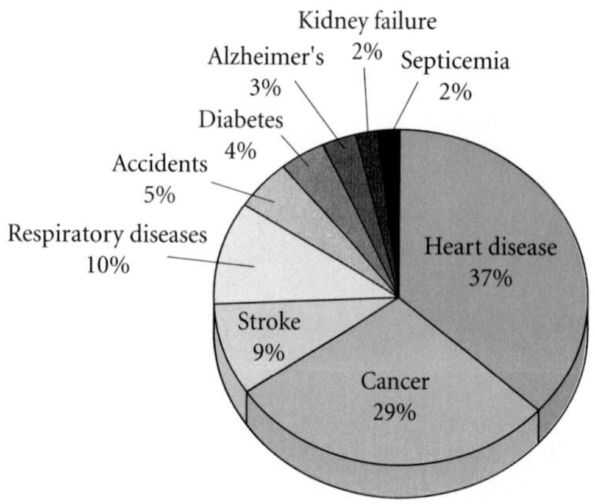

(d) Heart disease was the leading cause of death in 2000.

9. (a) Sky West had the highest percentage of on-time flights.
(b) Atlantic Coast had the lowest percentage of on-time flights.
(c) 84.3% of United Airline's flights were on time.

11. (a) Housing, fuel and utilities are the largest component of the CPI.
(b) Miscellaneous goods and services form the smallest component of the CPI.

Exercise 9.3 (p. 491)

1. (a)

Score	Frequency	Score	Frequency
25	1	41	5
26	1	42	3
28	1	43	1
29	1	44	2
30	3	45	1
31	2	46	2
32	1	47	1
33	2	48	3
34	2	49	1
35	1	50	1
36	2	51	1
37	4	52	3
38	1	53	2
39	1	54	2
40	1	55	1

(b)

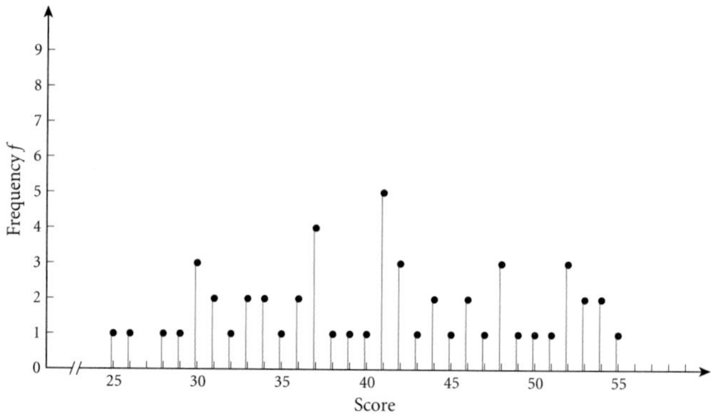

(c)

Class	Frequency	Class	Frequency	Class	Frequency	Class	Frequency
24–25.9	1	32–33.9	3	40–41.9	6	48–49.9	4
26–27.9	1	34–35.9	3	42–43.9	4	50–51.9	2
28–29.9	2	36–37.9	6	44–45.9	3	52–53.9	5
30–31.9	5	38–39.9	2	46–47.9	3	54–55.9	3

(d)

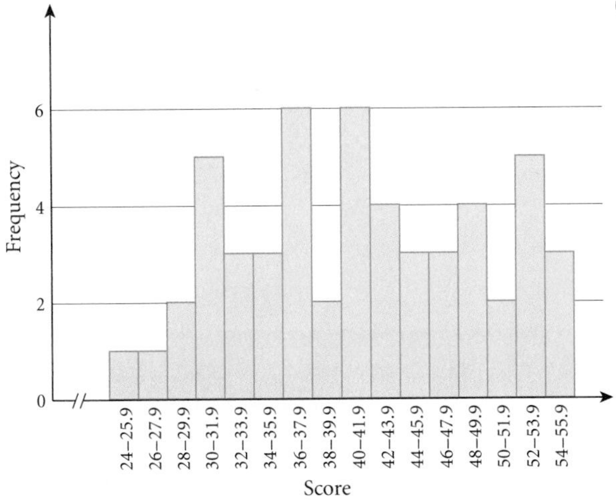

(e)

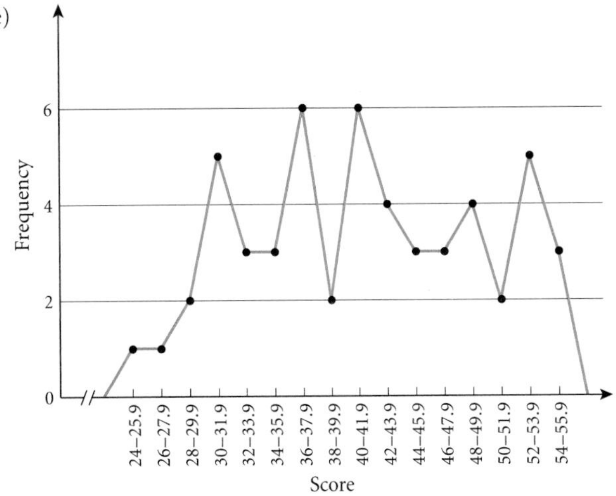

(f)

Class	Cumulative Frequency	Class	Cumulative Frequency
24–25.9	1	40–41.9	29
26–27.9	2	42–43.9	33
28–29.9	4	44–45.9	36
30–31.9	9	46–47.9	39
32–33.9	12	48–49.9	43
34–35.9	15	50–51.9	45
36–37.9	21	52–53.9	50
38–39.9	23	54–55.9	53

(g)

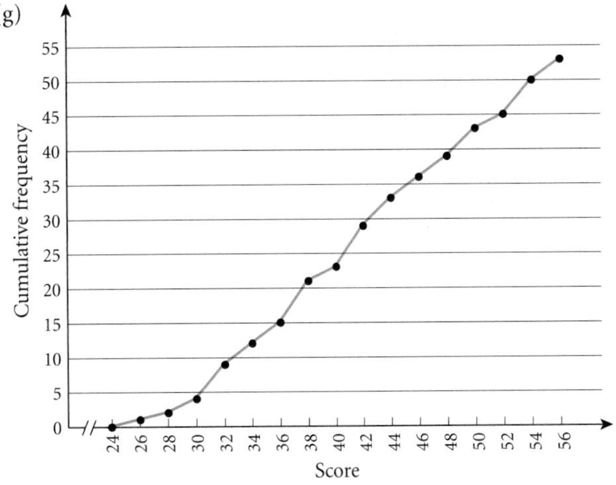

3. (a)

Class	Frequency	Class	Frequency	Class	Frequency	Class	Frequency
50–54.9	1	70–74.9	8	90–94.9	12	110–115.9	0
55–59.9	6	75–79.9	11	95–99.9	2	115–119.9	2
60–64.9	3	80–84.9	2	100–104.9	2		
65–69.9	6	85–89.9	12	105–109.9	4		

(b)

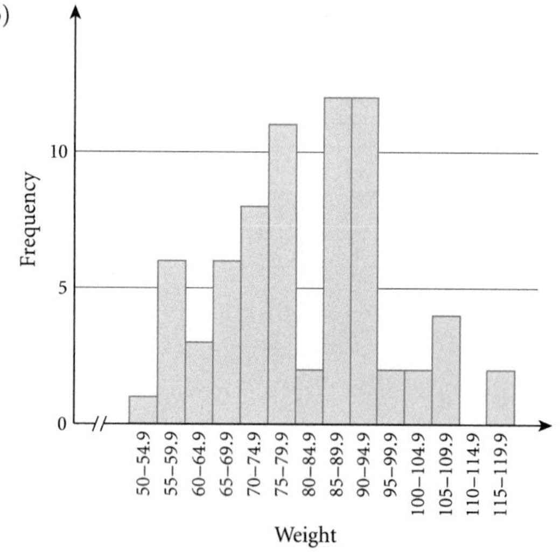

(c)

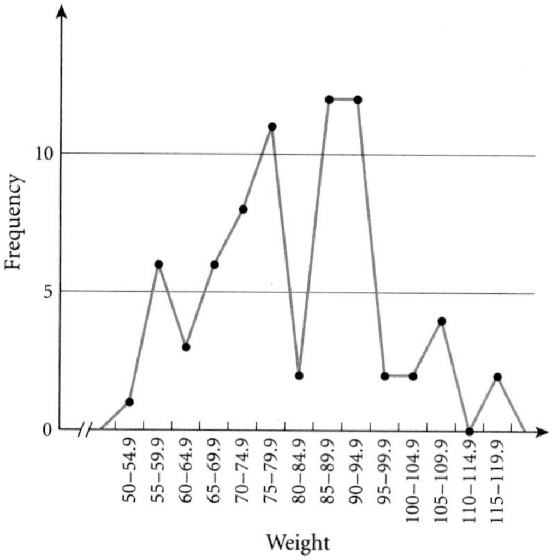

(d)

Class	Cumulative Frequency	Class	Cumulative Frequency	Class	Cumulative Frequency	Class	Cumulative Frequency
50–54.9	1	70–74.9	24	90–94.9	61	110–115.9	69
55–59.9	7	75–79.9	35	95–99.9	63	115–119.9	71
60–64.9	10	80–84.9	37	100–104.9	65		
65–69.9	16	85–89.9	49	105–109.9	69		

(e)

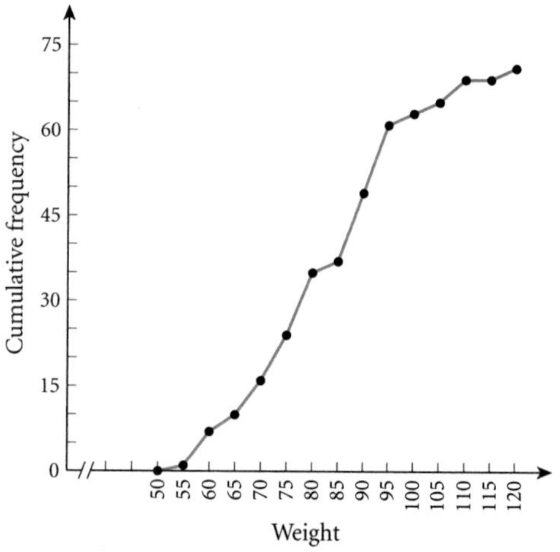

5. (a) There are 13 class intervals.

(b) The lower class limit of the 1st class interval is 20 years, the upper class limit is 24 years.

(c) The class width is 5 years.

(d) There are 1,300,000 drivers between the ages of 70 and 84.

(e) The interval 30–34 years has the most drivers.

(f) The interval 80–84 years has the fewest drivers.

(g)

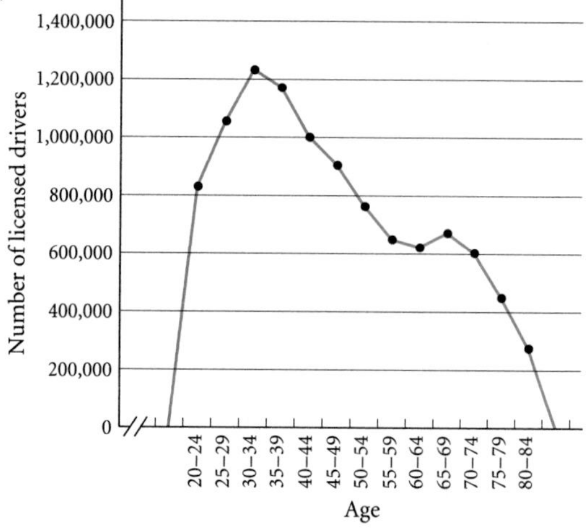

7.

Class	Frequency	Class	Frequency	Class	Frequency	Class	Frequency
20–29	1,860,000	40–49	1,870,000	60–69	1,230,000	80–89	220,000
30–39	2,400,000	50–59	1,420,000	70–79	1,030,000		

(a) There are 7 class intervals

(b) The lower class limit for the last interval is 80 years, the upper class limit is 89 years.

(c)

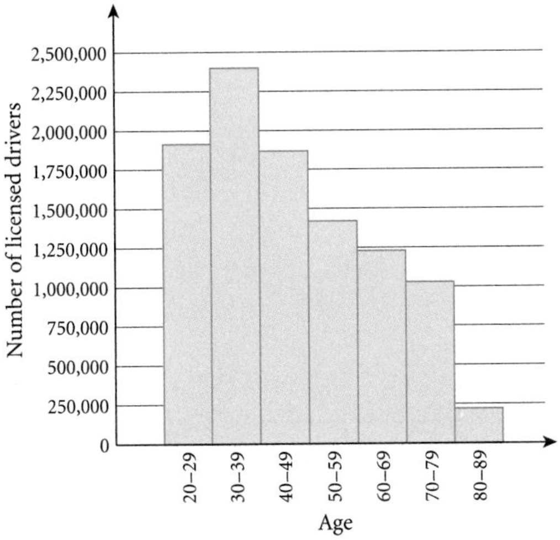

(d)

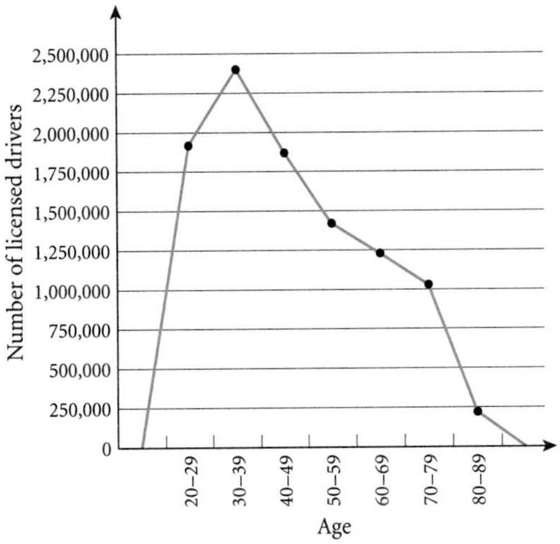

(e)

Class	Cumulative Frequency	Class	Cumulative Frequency	Class	Cumulative Frequency	Class	Cumulative Frequency
20–29	1,860,000	40–49	6,130,000	60–69	8,780,000	80–89	10,030,000
30–39	4,260,000	50–59	7,500,000	70–79	9,810,000		

(f)

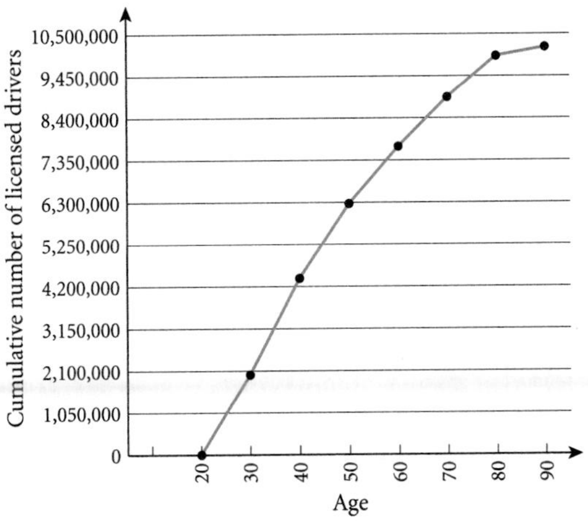

9. (a) There are 13 class intervals.

(b) The lower class limit of the 1st class interval is 20 years, the upper class limit is 24 years.

(c) The class width is 5 years.

(d)

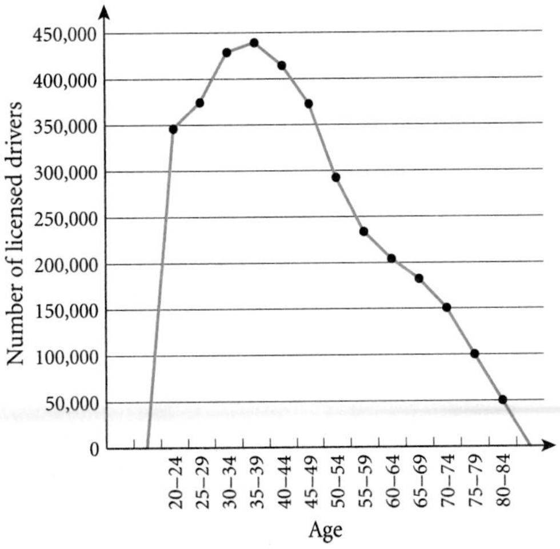

(e)

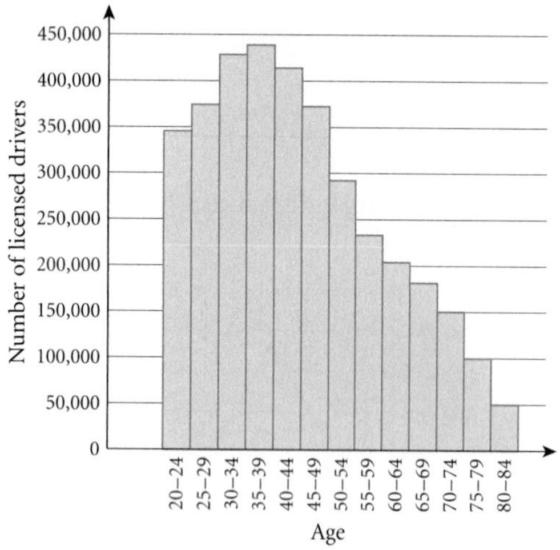

(f) Class interval 35–39 has the most licensed drivers.
(g) Class interval 80–84 has the fewest licensed drivers.

11. (a) There are 15 class intervals.
(b) The lower class limit for the first interval is 0, the upper limit is $999.
(c) The class width is $1000.
(d)

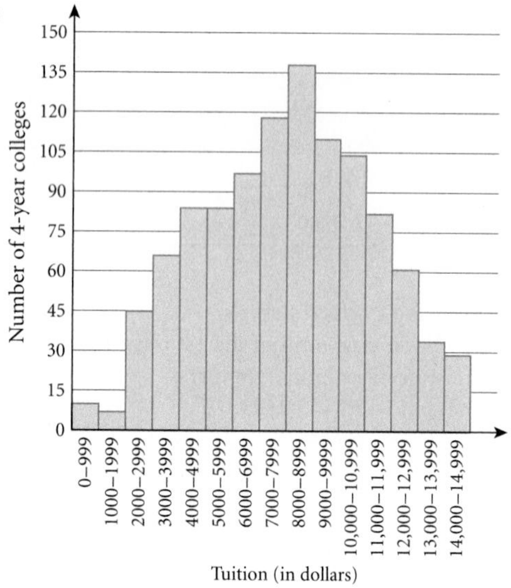

(e)

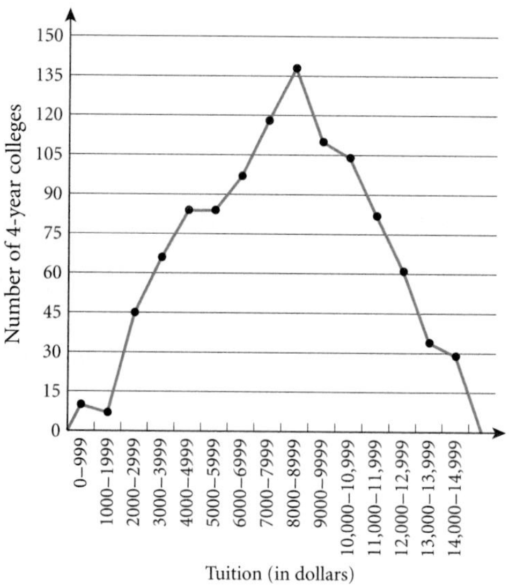

(f) Tuition between $8000 and $8999 occurs most frequently.

13. (a)

Class	Frequency	Class	Frequency	Class	Frequency	Class	Frequency
11.0–11.9	0	13.0–13.9	5	15.0–15.9	2	17.0–17.9	1
12.0–12.9	3	14.0–14.9	6	16.0–16.9	3	18.0–18.9	0

(b)

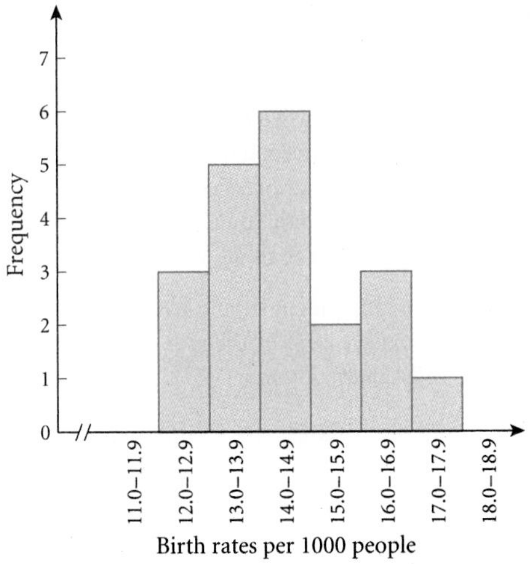

(c)

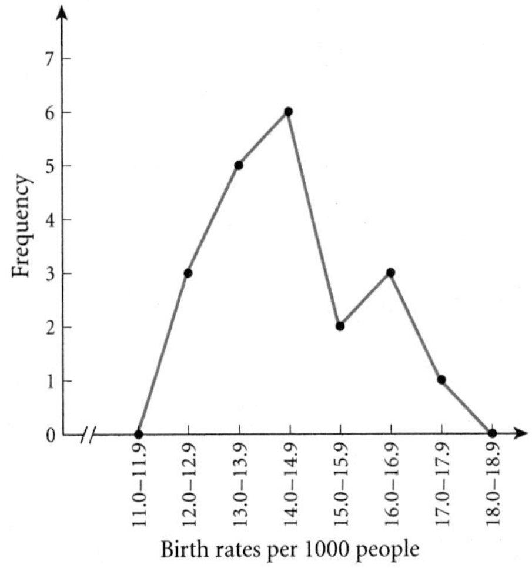

15. (a)

Class	Frequency	Class	Frequency	Class	Frequency	Class	Frequency
0–1.9	6	4–5.9	4	8–9.9	0	12–13.9	1
2–3.9	5	6–7.9	1	10–11.9	3		

(b)

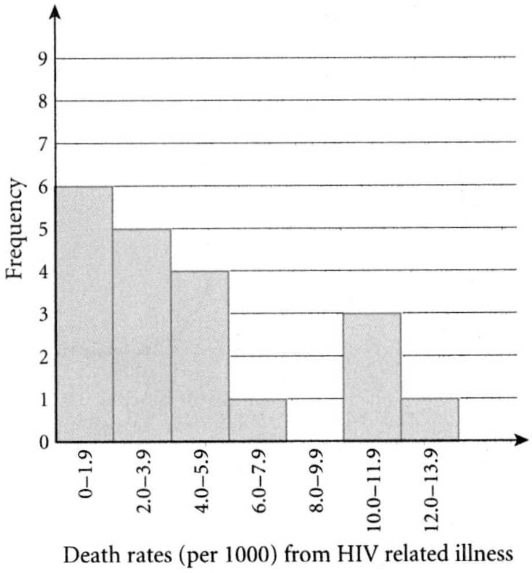

(c)

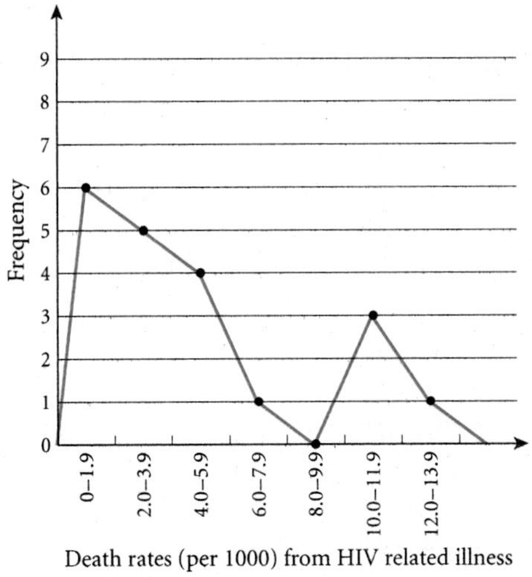

Exercise 9.4 (p. 502)

1. (a) mean: 31.25 (b) median: 30.5 (c) no mode

3. (a) mean: 70.4 (b) median: 70 (c) mode: 55

5. (a) mean: 78.8 (b) median: 82 (c) mode: 82

7. (a) mean: 73.33 (b) median: 77.5 (c) mode: 80

9. (a) mean: 31.29 years of age (b) median: 32 years of age
(c) mode: 32 years of age

11. The mean cost per share is $109.40.

13. (a) The mean age of a new mother in 2000 was approximately 27.668 years.
(b) The median age of a new mother in 2000 was approximately 27.45 years.

15. (a) The mean sales are approximately $119,966.
(b) The median sales are approximately $120,000.

17. (a) The mean age of a licensed driver was approximately 44.709 years.
(b) The median age of a licenced driver was approximately 42.5 years.

19. The mean tuition in 1992–93 was approximately \$8053.79.

21. (a) mean \$41,300; median \$36,000
(b) The median describes the 4 clustered salaries well.

Exercise 9.5 (p. 511)

1. $s = 7.058$ **3.** $s = 6$ **5.** $s = 13.946$

7. mean: $\bar{x} = 31.878$; standard deviation: $s = 7.921$.

9. mean 885.333 hours, standard deviation 69.681 hours.

11. (a) Range: 17 years (b) $s = 4.5548$ years
(c) $\sigma = 4.4807$ years

13. (a) Population; we have all of the mothers represented.
(b) The standard deviation is 6.367 years.

15. (a) Population data; all recorded earthquakes are included.
(b) The mean magnitude of the earthquakes is 3.278.
(c) The standard deviation of the magnitudes of the earthquakes recorded in 1998 is 1.382.

17. (a) $s = 15.693285$ years. (b) $\sigma = 15.693282$ years.
(c) Answers will vary.

19. (a) Population; all colleges of the kind are represented.
(b) The standard deviation of the tuition is \$3175.58.

21. (a) We expect at least 75% of the outcomes to be between 19 and 31.
(b) We expect at least 64% of the outcomes to be between 20 and 30.
(c) We expect at least 88.88% of the outcomes to be between 16 and 34.
(d) We expect at most 25% of the outcomes to be less than 19 or more than 31.
(e) We expect at most 11.11% of the outcomes to be less than 16 or greater than 34.

23. We expect at least 889 boxes to have between 0 and 12 defective watches.

25. (a) Population
(b) The mean number of births was 3,939,476.83.
(c) The standard deviation of births was 58,187.82.
(d) Exact; the data are not grouped.

Exercise 9.6 (p. 523)

1. $\mu = 8, \sigma = 1$ **3.** $\mu = 18, \sigma = 1$

5. (a) $z = -0.66$ (d) $z = 1.71$
(b) $z = -0.44$ (e) $z = 2.57$
(c) $z = -0.01$ (f) $z = 3$

7. (a) $A = 0.3133$ (d) $A = 0.3888$
(b) $A = 0.3642$ (e) $A = 0.4893$
(c) $A = 0.4989$ (f) $A = 0.2734$

9. A–5.48%; B–21.95%; C–34.37%; D–30.13%; F–8.08%

11. $A = 0.3085$ **13.** $A = 0.8181$

15. (a) 1365 women are between 62 and 66 inches.
(b) 1909 women are between 60 and 68 inches.
(c) 1995 women are between 58 and 70 inches.
(d) 3 women are taller than 70 inches.
(e) 3 women are shorter than 58 inches.

17. (a) Approximately 1 student should weigh at least 142 pounds.
(b) We would expect 70% of the students to weigh between 124.61 and 135.39 pounds.

19. 57.05% of the clothing can be expected to last between 28 and 42 months.

21. (a) Attendance lower than 10,525 will be in the lowest 70% of the figures.
(b) Approximately 77.46% of the attendance figures are between 8500 and 11,000 persons.
(c) Approximately 13.36% of the attendance figures differ from the mean by at least 1500 persons.

23. Kathleen had the highest relative standing.

25. (a)

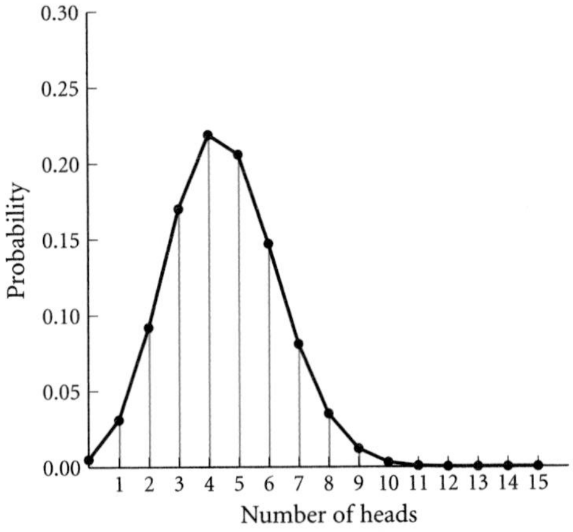

(b) Answers vary.
(c) mean 4.5, standard deviation 1.775

27. The approximate probability that there are between 285 and 315 successes is 0.7372.

29. The approximate probability of obtaining 300 or more successes is 0.5.

31. The approximate probability of obtaining 325 or more successes is 0.0307.

33. (a) The approximate probability of having at least 80 but no more than 90 hits is 0.2286.
(b) The approximate probability of having 85 or more hits is 0.0918.

35. The approximate probability of selecting at least 10 unsealed packages is 0.0116.

37.

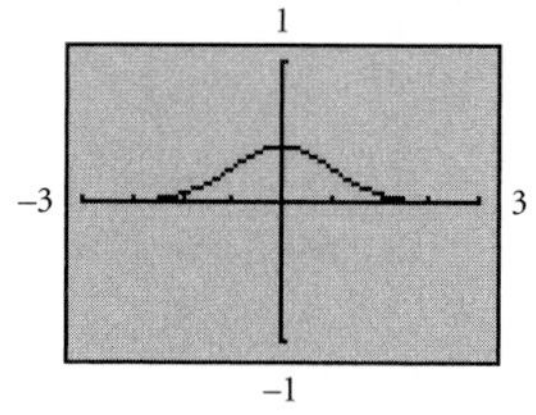

The graph assumes its maximum at $x = 0$.

39. $P[X \geq 10] \approx 0.0311$;

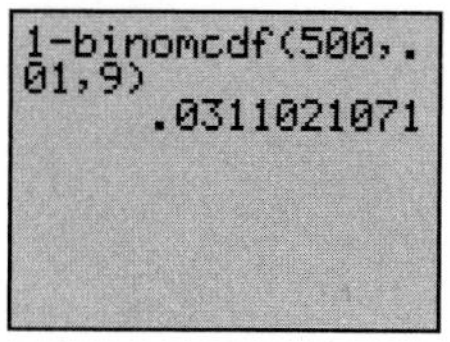

CHAPTER 9 Review

True–False Items (p. 527)

1. F **2.** T **3.** F **4.** T **5.** F

Fill in the Blanks (p. 527)

1. (a) mean (b) median (c) mode

2. the standard deviation **3.** bell **4.** Z-score

5. $\mu - k$; $\mu + k$

Review Exercises (p. 527)

1. circumference, continuous **3.** number of people, discrete

5. number of defective parts, discrete

7. Answers will vary. All answers should include a method to choose a sample of 100 students from the population in which each student has an equal chance of being chosen.

9. Answers will vary. All answers should give examples of possible bias.

11. (a)

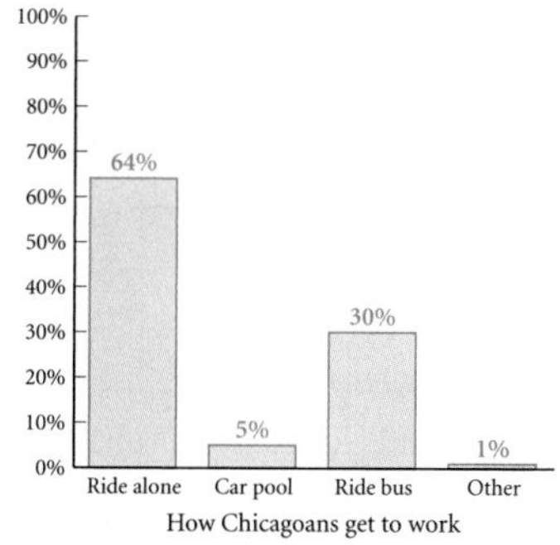

(b)

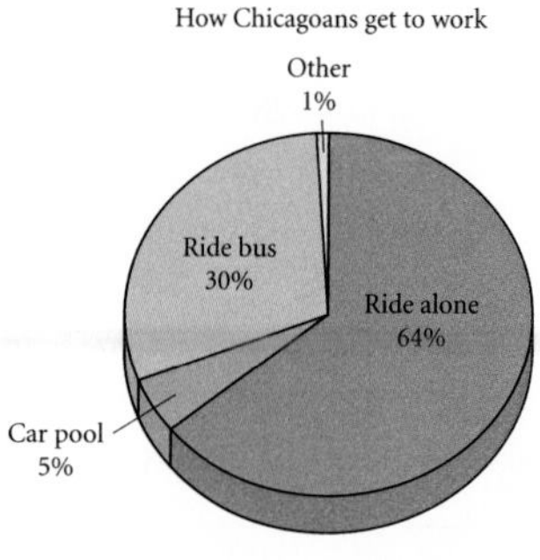

13. (a)

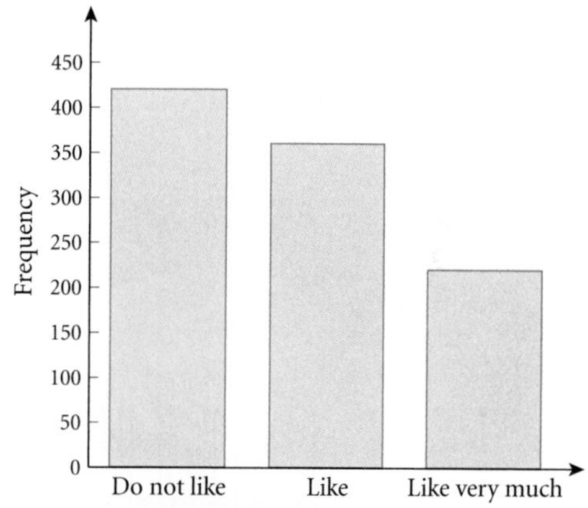

(b)

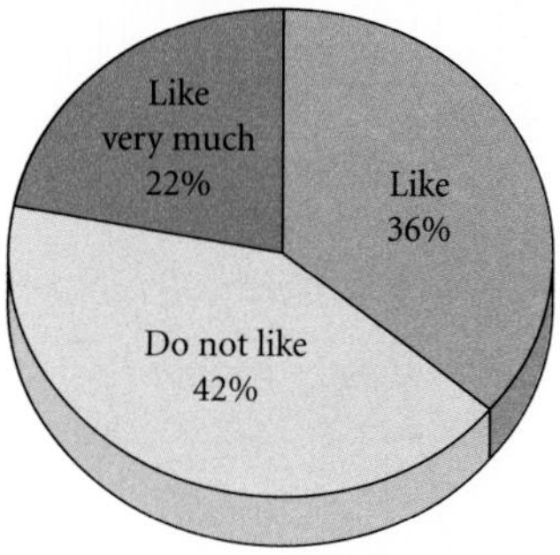

15. (a) American Indians made up the smallest percentage of 4-year college enrollment.
(b) Asian-Americans were overrepresented in 4-year colleges in 1997.
(c) Approximately 533,820 Hispanic students were enrolled in 4-year colleges in 1997.

17. (a) High school diploma represents the highest level of educational attainment of most Americans in 2000.
(b) Approximately 45,000,000 Americans have at least a bachelor's degree.
(c) Approximately 32,000,000 Americans do not have a high school diploma.
(d) Approximately 48,000,000 Americans have gone to college but do not have a bachelor's degree.

19. (a)

Score	Frequency	Score	Frequency	Score	Frequency	Score	Frequency
21	2	62	1	74	1	87	2
33	1	63	2	75	1	89	1
41	2	66	2	77	1	90	2
42	1	68	1	78	2	91	1
44	1	69	1	80	4	92	1
48	1	70	2	82	1	95	1
52	2	71	1	83	1	99	1
55	1	72	2	85	2	100	2
60	1	73	2				

The range is 79.

(b)

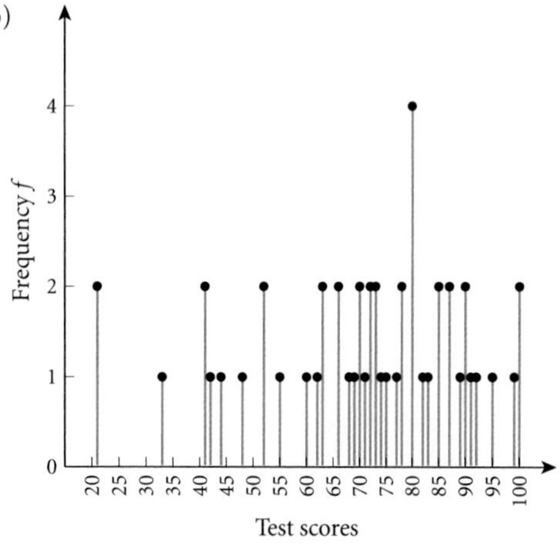

(c)

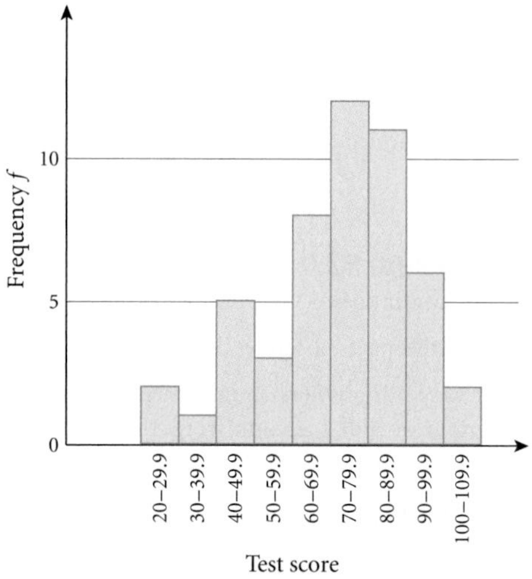

(d)

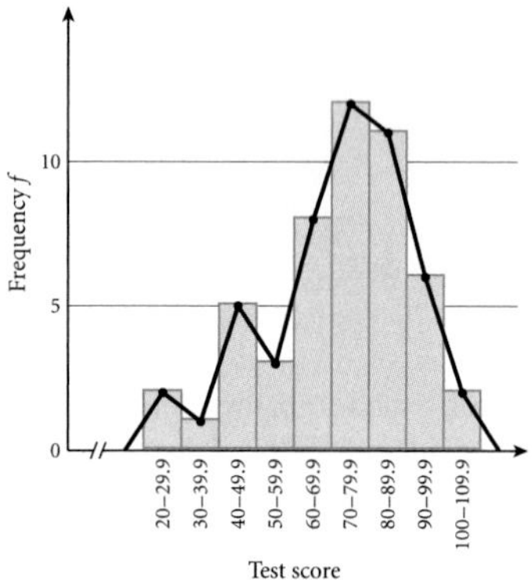

(f)

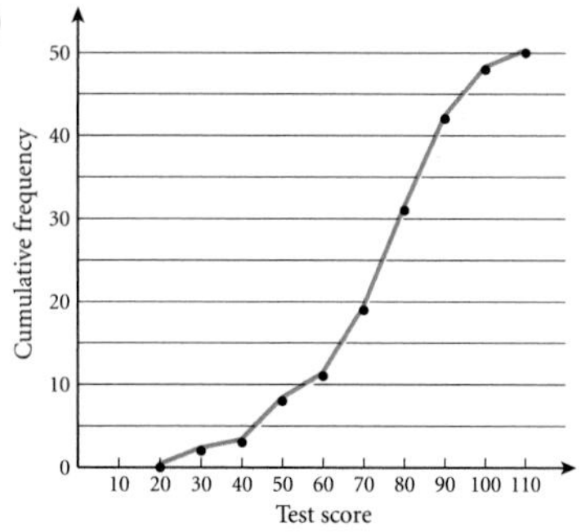

(e)

Score	Cumulative Frequency	Score	Cumulative Frequency
21	2	73	26
33	3	74	27
41	5	75	28
42	6	77	29
44	7	78	31
48	8	80	35
52	10	82	36
55	11	83	37
60	12	85	39
62	13	87	41
63	15	89	42
66	17	90	44
68	18	91	45
69	19	92	46
70	21	95	47
71	22	99	48
72	24	100	5

21. (a)

Time	Frequency	Time	Frequency	Time	Frequency	Time	Frequency	Time	Frequency
4′12″	1	4′46″	2	5′08″	1	5′43″	1	6′12″	1
4′15″	1	4′50″	1	5′12″	2	5′48″	1	6′30″	1
4′22″	1	4′52″	1	5′18″	1	5′50″	1	6′32″	1
4′30″	2	4′56″	1	5′20″	3	5′55″	1	6′40″	1
4′36″	1	5′01″	1	5′31″	2	6′01″	1	7′05″	1
4′39″	1	5′02″	1	5′37″	1	6′02″	1	7′15″	1
4′40″	1	5′06″	2	5′40″	2	6′10″	1		

The range is 3 minutes, 3 seconds.

(b)

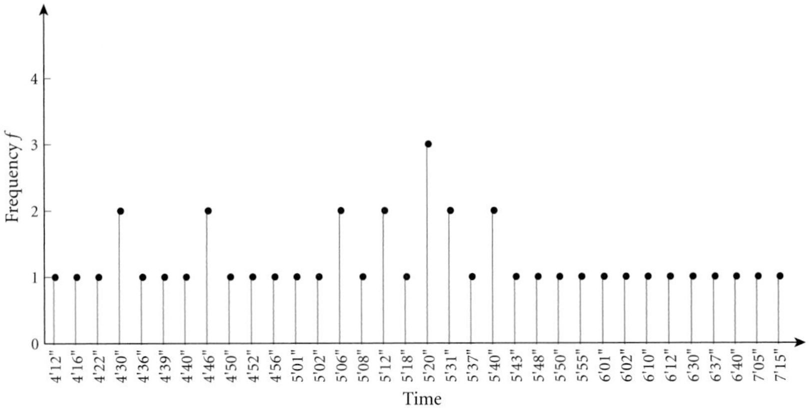

(c)

Class Interval	Frequency f_i
4′00″–4′29″	3
4′30″–4′59″	10
5′00″–5′29″	11
5′30″–5′59″	9
6′00″–6′29″	4
6′30″–6′59″	3
7′00″–7′29″	2

(d)

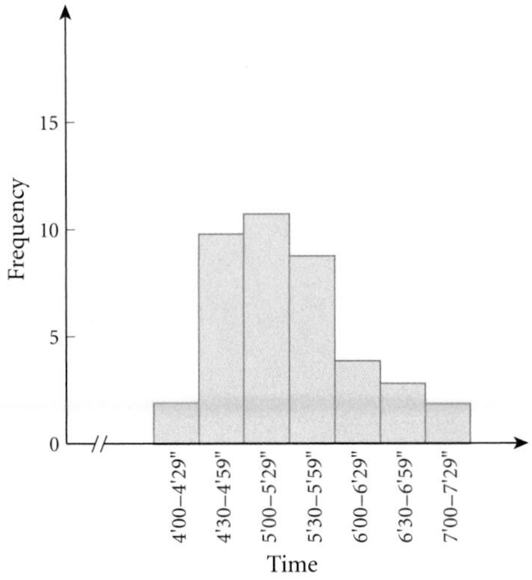

(e)

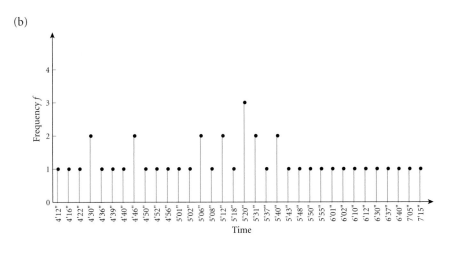

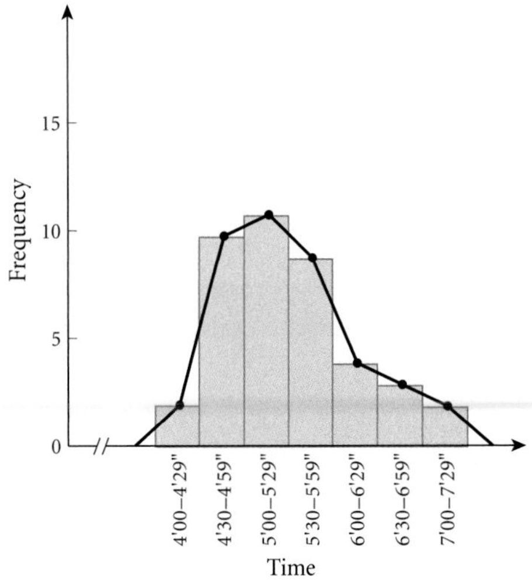

(f)

Class Interval	Cummulative Frequency
4′00″–4′29″	3
4′30″–4′59″	13
5′00″–5′29″	24
5′30″–5′59″	33
6′00″–6′29″	37
6′30″–6′59″	40
7′00″–7′29″	42

(g)

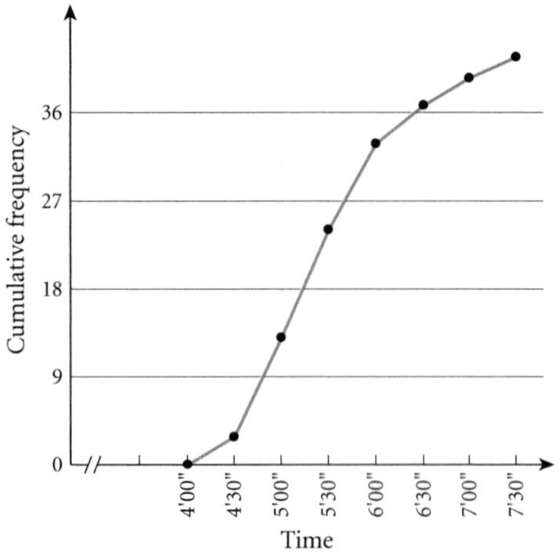

23. (a)

Age	Frequency	Age	Frequency
24	5	34	1
27	1	35	2
28	1	36	1
29	2	37	1
30	3	38	1
31	2	40	1
32	8	41	1
33	1		

(b)

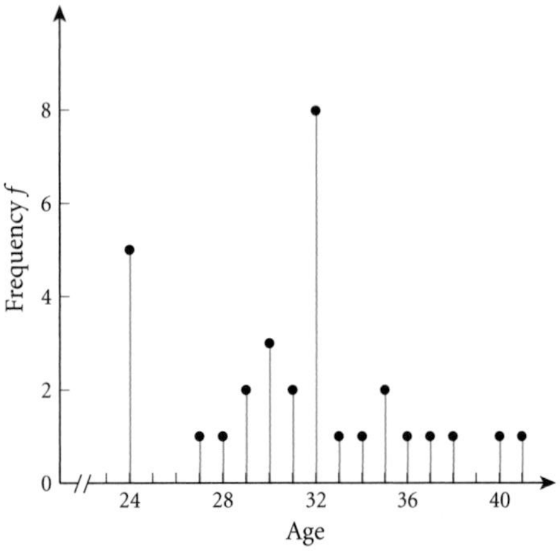

(c)

Class	Frequency	Class	Frequency
20.0–24.9	5	35.0–39.9	5
25.0–29.9	4	40.0–44.9	2
30.0–34.9	15		

There are 5 class intervals.

(d)

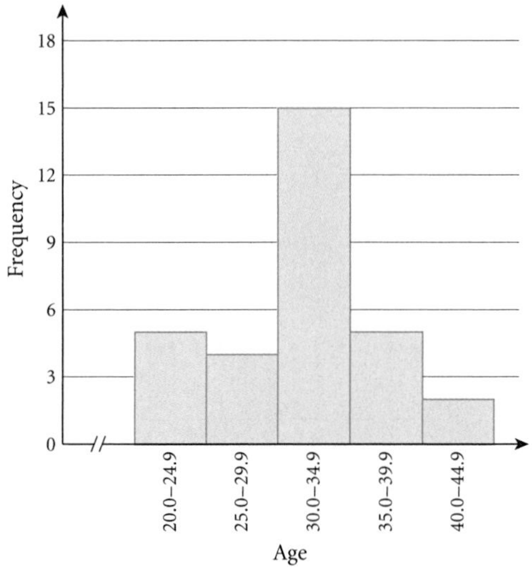

(e)

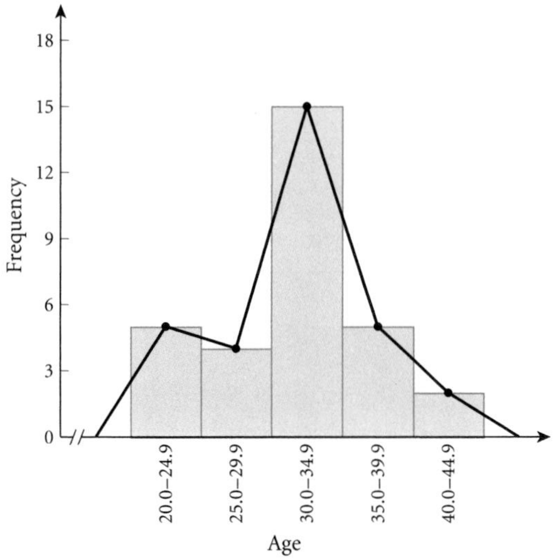

(f)

Class	Cumulative Frequency	Class	Cumulative Frequency
20.0–24.9	5	35.0–39.9	29
25.0–29.9	9	40.0–44.9	31
30.0–34.9	24		

(g)

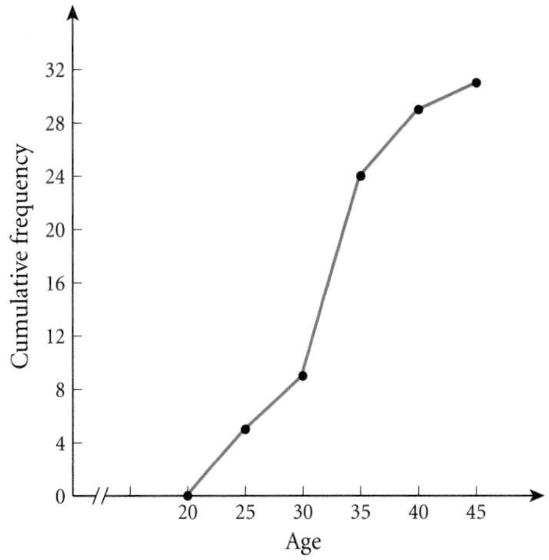

25. (a)

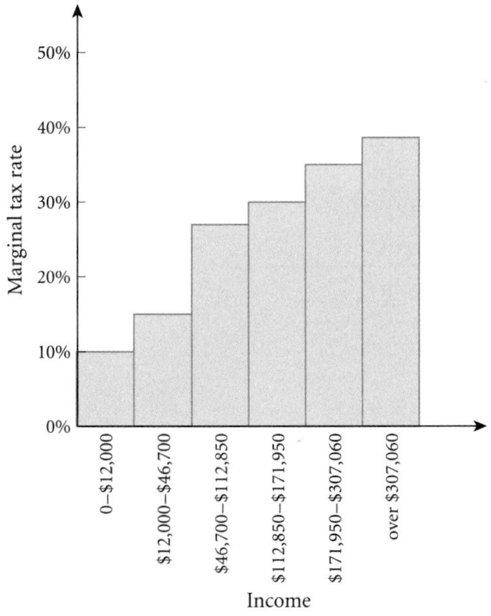

(b)

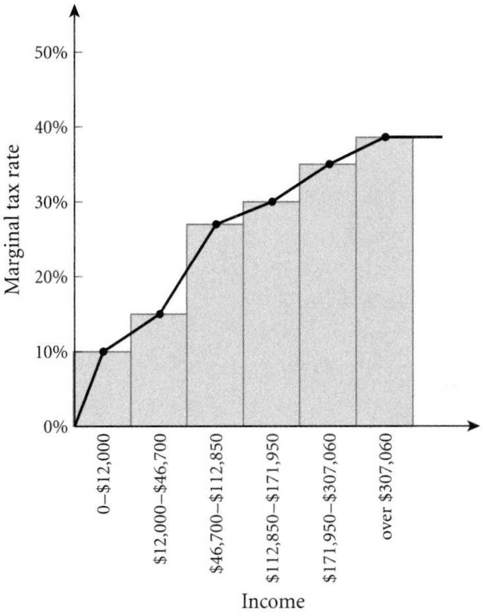

27. (a) Mean: 5.7273 (b) Median: 5 (c) Mode: 0, 4, 8 and 10
(d) Range: 12 (e) Standard deviation: 4.149

29. (a) Mean: 16.2 (b) Median: 7 (c) Mode: 7
(d) Range: 98 (e) Standard deviation: 29.5515

31. (a) Mean: 7 (b) Median: 7 (c) Mode: 7
(d) Range: 11 (e) Standard deviation: 3.6515

37. (a) Answers may vary. We assume they are a sample of Joe's scores calculating parts (b) and (c).
(b) Joe's mean score is 75.57.
(c) The standard deviation of Joe's scores is 2.99.

39. (a) The approximate mean age of a female in 2000 was 37.8 years.
(b) The approximate median age of a female in 2000 was 37 years.
(c) The approximate standard deviation of the ages of females in 2000 was 23.065 years.

41. We expect at least 75%, or 750 jars, to have between 11.9 and 12.1 ounces of jam.

43. The probability a bag weighs less than 9.5 or more than 10.5 pounds is less than 0.25.

45. $z = -0.667$

47. $z = 1.4$

49. $z = 1.667$

51. $A = 0.0855$

53. $A = 0.7555$

55. (a) 68.26% of the scores are between 20 and 30.
(b) 2.28% of the scores are above 35.

57. 0.17% of the dogs will die before reaching the age of 10 years, 4 months.

59. Bob scored equally well on both exams.

61. There is a probability of 0.9544 that this week's production will lie between 30 and 50.

63. The probability of obtaining more than 160 positive results is approximately 0.001.

65. The probability that in a group of 200 test-takers between 110 and 125 pass the test is 0.6893.

CHAPTER 10 Functions and Their Graphs

Exercise 10.1 (p. 547)

1. (a) $(3, -4)$ (b) $(-3, 4)$ (c) $(-3, -4)$

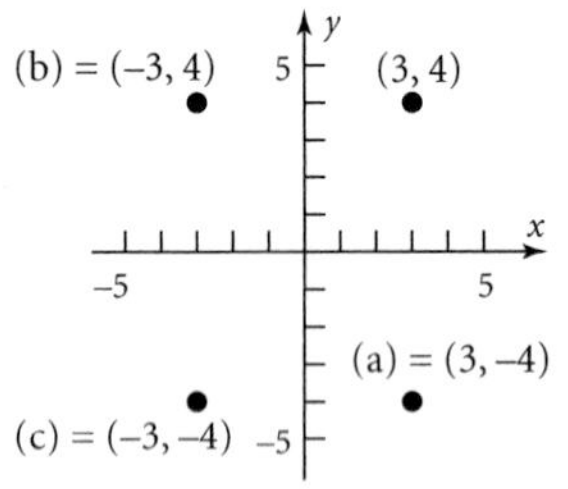

3. (a) $(-2, -1)$ (b) $(2, 1)$ (c) $(2, -1)$

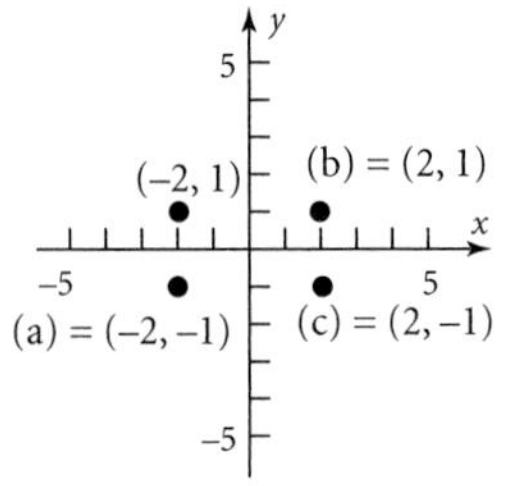

5. (a) $(1, -1)$ (b) $(-1, 1)$ (c) $(-1, -1)$

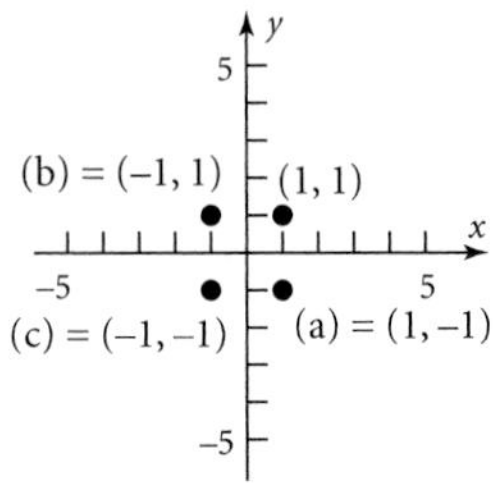

7. (a) $(-3, 4)$ (b) $(3, -4)$ (c) $(3, 4)$

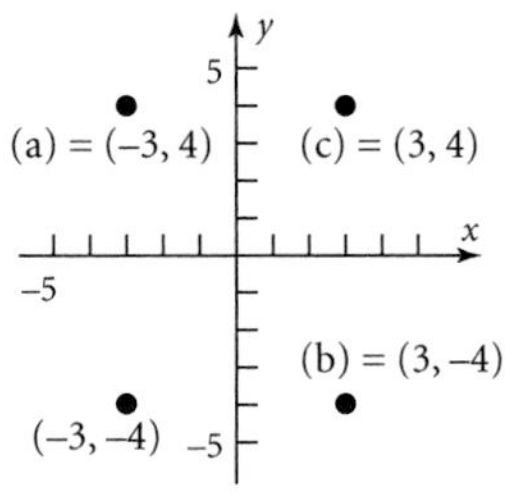

9. (a) $(0, 3)$ (b) $(0, -3)$ (c) $(0, 3)$

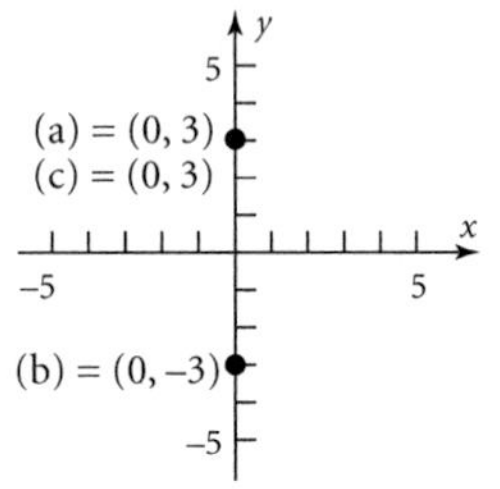

11. (a) The x-intercepts are $(-1, 0)$ and $(1, 0)$. There is no y-intercept.
(b) The graph is symmetric with respect to the x-axis, y-axis, and the origin.

13. (a) The x-intercepts are $\left(-\frac{\pi}{2}, 0\right)$ and $\left(\frac{\pi}{2}, 0\right)$. The y-intercept is $(0, 1)$.
(b) The graph is symmetric with respect to the y-axis.

15. (a) The x-intercept and the y-intercept are both $(0, 0)$.
(b) The graph is symmetric with respect to the x-axis.

17. (a) The x-intercept is $(1, 0)$. There is no y-intercept.
(b) The graph is not symmetric with respect to either axis or to the origin.

19. (a) The x-intercepts are $(-1, 0)$ and $(1, 0)$. The y-intercept is $(0, -1)$.
(b) The graph is symmetric with respect to the y-axis.

21. (a) There are no intercepts.
(b) The graph is symmetric with respect to the origin.

23. The point $(0, 0)$ is on the graph. The points $(1, 1)$ and $(-1, 0)$ are not on the graph.

25. The point $(0, 3)$ is on the graph. The points $(3, 0)$ and $(-3, 0)$ are not on the graph.

27. The points $(0, 2)$ and $(\sqrt{2}, \sqrt{2})$ are on the graph. The point $(-2, 2)$ is not on the graph.

29. The x-intercept and the y-intercept are $(0, 0)$. The graph is symmetric with respect to the y-axis.

31. The x-intercept and the y-intercept are $(0, 0)$. The graph is symmetric with respect to the origin.

33. The x-intercepts are $(-3, 0)$ and $(3, 0)$. The y-intercept is $(0, 9)$. The graph is symmetric with respect to the y-axis.

35. The x-intercepts are $(-2, 0)$ and $(2, 0)$. The y-intercepts are $(0, -3)$ and $(0, 3)$. The graph is symmetric with respect to the x-axis, the y-axis, and the origin.

37. The x-intercept is $(3, 0)$. The y-intercept is $(0, -27)$. The graph is not symmetric with respect to either axis or to the origin.

39. The x-intercepts are $(-1, 0)$ and $(4, 0)$. The y-intercept is $(0, -4)$. The graph is not symmetric with respect to either axis or to the origin.

41. The x-intercept and the y-intercept are both $(0, 0)$. The graph is symmetric with respect to the origin.

43. The x-intercept and the y-intercept are both $(0, 0)$. The graph is symmetric with respect to the y-axis.

45.

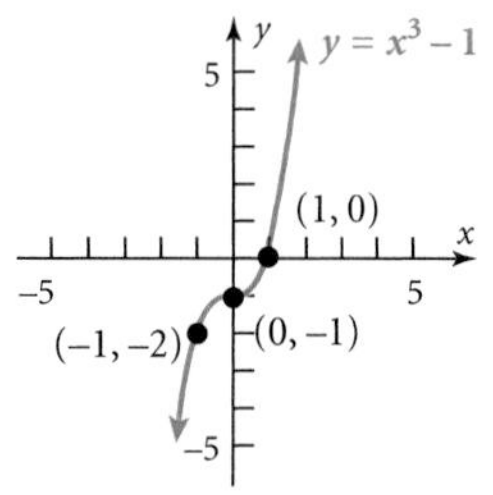

47.

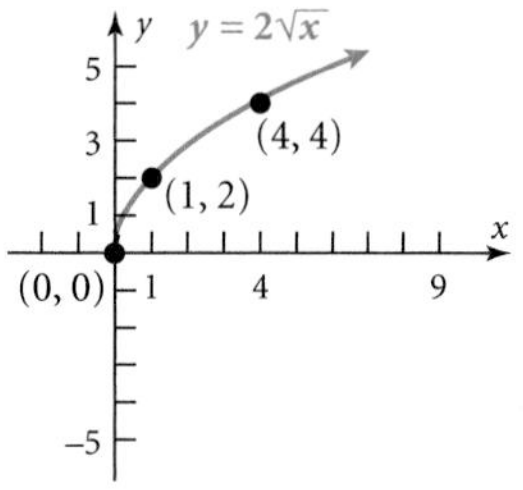

49. $a = -1$

51. $b = -\frac{2}{3}a + 2$

53. (a)

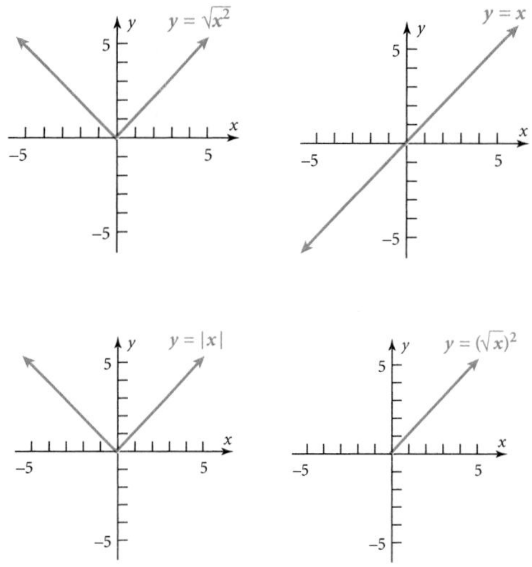

The graphs of $y = \sqrt{x^2}$ and $y = |x|$ are the same.

Exercise 10.2 (p. 558)

1. (a) $f(0) = -4$ (b) $f(1) = 1$
(c) $f(-1) = -3$ (d) $f(-x) = 3x^2 - 2x - 4$
(e) $-f(x) = -3x^2 - 2x + 4$

(f) $f(x+1) = 3x^2 + 8x + 1$

(g) $f(2x) = 12x^2 + 4x - 4$

(h) $f(x+h) = 3x^2 + 6xh + 2x + 3h^2 + 2h - 4$

3. (a) $f(0) = 0$ (b) $f(1) = \frac{1}{2}$

(c) $f(-1) = -\frac{1}{2}$ (d) $f(-x) = -\frac{x}{x^2+1}$

(e) $-f(x) = -\frac{x}{x^2+1}$ (f) $f(x+1) = \frac{x+1}{x^2+2x+2}$

(g) $f(2x) = \frac{2x}{4x^2+1}$

(h) $f(x+h) = \frac{x+h}{x^2+2hx+h^2+1}$

5. (a) $f(0) = 4$ (b) $f(1) = 5$

(c) $f(-1) = 5$ (d) $f(-x) = |x| + 4$

(e) $-f(x) = -|x| - 4$ (f) $f(x+1) = |x+1| + 4$

(g) $f(2x) = 2|x| + 4$ (h) $f(x+h) = |x+h| + 4$

7. (a) $f(0) = -\frac{1}{5}$ (b) $f(1) = -\frac{3}{2}$

(c) $f(-1) = \frac{1}{8}$ (d) $f(-x) = \frac{2x-1}{3x+5}$

(e) $-f(x) = \frac{2x+1}{5-3x}$ (f) $f(x+1) = \frac{2x+3}{3x-2}$

(g) $f(2x) = \frac{4x+1}{6x-5}$ (h) $f(x+h) = \frac{2x+2h+1}{3x+3h-5}$

9. 4 **11.** $2x + h - 1$

13. $3x^2 + 3xh + h^2$ **15.** $4x^3 + 6hx^2 + 4h^2x + h^3$

17. function **19.** function

21. not a function **23.** not a function

25. function **27.** not a function

29. all real numbers **31.** all real numbers

33. $\{x|x \neq -4, x \neq 4\}$ **35.** $\{x|x \neq 0\}$

37. $\{x|x \geq 4\}$ or the interval $[4, \infty)$

39. $\{x|x > 9\}$ or the interval $(9, \infty)$

41. $\{x|x > 1\}$ or the interval $(1, \infty)$

43. $A = -\frac{7}{2}$

45. $A = -4$, f is undefined at $x = -2$

47. $A = 8$, f is undefined at $x = 3$

49. $G(x) = 10x$ **51.** $R(x) = -\frac{1}{5}x^2 + 100x$

53. $R(x) = -\frac{1}{20}x^2 + 5x$

55. 28,027 thousand acres of wheat will be planted in 2010.

57. The expected Mathematics SAT score would be 456 in 2010.

59. (a) The cost per passenger is \$222.
(b) The cost per passenger is \$225.
(c) The cost per passenger is \$220.
(d) The cost per passenger is \$230.

61. (a) Yes (b) No (c) No (d) No

Exercise 10.3 (p. 572)

1. This is not the graph of a function.

3. This is the graph of a function.
(a) The domain is $\{x|-\pi \leq x \leq \pi\}$ or the interval $[-\pi, \pi]$. The range is $\{y|-1 \leq y \leq 1\}$ or the interval $[-1, 1]$.
(b) The x-intercepts are $\left(-\frac{\pi}{2}, 0\right)$ and $\left(\frac{\pi}{2}, 0\right)$. The y-intercept is $(0, 1)$.
(c) The graph is symmetric with respect to the y-axis.

5. This is not the graph of a function.

7. This is the graph of a function.
(a) The domain is $\{x|x > 0\}$ or the interval $(0, \infty)$. The range is all real numbers or the interval $(-\infty, \infty)$.
(b) The x-intercept is $(1, 0)$. There is no y-intercept.
(c) This graph does not have symmetry with respect to the x-axis, y-axis, or the origin.

9. This is the graph of a function.
(a) The domain is all real numbers or the interval $(-\infty,\infty)$. The range is $\{y|y \leq 2\}$ or the interval $(-\infty, 2]$.
(b) The x-intercepts are $(-3, 0)$ and $(3, 0)$. The y-intercept is $(0, 2)$.
(c) The graph is symmetric with respect to the y-axis.

11. This is the graph of a function.
(a) The domain is all real numbers or the interval $(-\infty,\infty)$. The range is $\{y|y \geq -3\}$ or the interval $[-3,\infty)$.
(b) The x-intercepts are $(1, 0)$ and $(3, 0)$. The y-intercept is $(0, 9)$.
(c) This graph does not have symmetry with respect to the x-axis, y-axis, or the origin.

13. (a) $f(0) = 3$, $f(-6) = -3$ (b) $f(6) = 0$, $f(11) = 1$
(c) $f(3)$ is positive. (d) $f(-4)$ is negative.
(e) $x = -3$, $x = 6$, and $x = 10$
(f) $f(x) > 0$ on the intervals $(-3, 6)$ and $(10, 11]$.
(g) The domain of f is $\{x|-6 \leq x \leq 11\}$ or the interval $[-6, 11]$.
(h) The range of f is $\{y|-3 \leq y \leq 4\}$ or the interval $[-3, 4]$.

(i) The x-intercepts are $(-3, 0)$, $(6, 0)$, and $(10, 0)$.
(j) The y-intercept is $(0, 3)$.
(k) The line $y = \frac{1}{2}$ intersects the graph 3 times.
(l) The line $x = 5$ intersects the graph once.
(m) $f(x) = 3$ when $x = 0$ and $x = 4$.
(n) $f(x) = -2$ when $x = -5$ and $x = 8$.

15. (a) Yes
(b) $f(-2) = 9$; the point $(-2, 9)$ is on the graph of f.
(c) $x = 0$ or $x = \frac{1}{2}$; the points $(0, -1)$ and $\left(\frac{1}{2}, -1\right)$ are on the graph of f.
(d) The domain of f is all real numbers or the interval $(-\infty, \infty)$.
(e) The x-intercepts are $\left(-\frac{1}{2}, 0\right)$ and $(1, 0)$.
(f) The y-intercept is $(0, -1)$

17. (a) No
(b) $f(4) = -3$; the point $(4, -3)$ is on the graph of f.
(c) $x = 14$; the point $(14, 2)$ is on the graph of f.
(d) The domain is the set $\{x \mid x \neq 6\}$.
(e) The x-intercept is $(-2, 0)$.
(f) The y-intercept is $\left(0, -\frac{1}{3}\right)$.

19. (a) Yes
(b) $f(2) = \frac{8}{17}$; the point $\left(2, \frac{8}{17}\right)$ is on the graph of f.
(c) $x = -1$ or $x = 1$; the points $(-1, 1)$ and $(1, 1)$ are on the graph of f.
(d) The domain is the set of all real numbers or the interval $(-\infty, \infty)$.
(e) The x-intercept is $(0, 0)$.
(f) The y-intercept is $(0, 0)$.

21. Yes **23.** No

25. f is increasing on the intervals $(-8, -2)$, $(0, 2)$, and $(5, \infty)$ or for $-8 < x < -2$, $0 < x < 2$ and $x > 5$.

27. There is a local maximum at $x = 2$. The local maximum is $f(2) = 10$.

29. f has local maxima at $x = -2$ and $x = 2$. The local maxima are $f(-2) = 6$, and $f(2) = 10$.

31. (a) The x-intercepts are $(-2, 0)$ and $(2, 0)$. The y-intercept is $(0, 3)$.
(b) The domain is $\{x | -4 \leq x \leq 4\}$or the interval $[-4, 4]$. The range is $\{y | 0 \leq y \leq 3\}$ or $[0, 3]$.
(c) The function is increasing on $(-2, 0)$ and $(2, 4)$ or for $-2 < x < 0$ and $2 < x < 4$. The function is decreasing on $(-4, -2)$ and $(0, 2)$ or for $-4 < x < -2$ and $0 < x < 2$.
(d) The function is even.

33. (a) The y-intercept is $(0, 1)$. There is no x-intercept.
(b) The domain is the set of all real numbers. The range is set of positive numbers or $\{y | y > 0\}$ or the interval $(0, \infty)$.
(c) The function is increasing on $(-\infty, \infty)$ or for all real numbers.
(d) The function is neither even nor odd.

35. (a) The x-intercepts are $(-\pi, 0)$, $(0, 0)$ and $(\pi, 0)$. The y-intercept is $(0, 0)$.
(b) The domain is $[-\pi, \pi]$. The range is $[-1, 1]$.
(c) The function is increasing on $\left(-\frac{\pi}{2}, \frac{\pi}{2}\right)$ or for $-\frac{\pi}{2} < x < \frac{\pi}{2}$. The function is decreasing on $\left(-\pi, -\frac{\pi}{2}\right)$ and $\left(\frac{\pi}{2}, \pi\right)$ or for $-\pi < x < -\frac{\pi}{2}$ and $\frac{\pi}{2} < x < \pi$.
(d) The function is odd.

37. (a) The x-intercepts are $\left(\frac{1}{2}, 0\right)$ and $\left(\frac{5}{2}, 0\right)$. The y-intercept is $\left(0, \frac{1}{2}\right)$.
(b) The domain is $\{x | -3 \leq x \leq 3\}$ or the interval $[-3, 3]$. The range is $\{y | -1 \leq y \leq 2\}$ or the inerval $[-1, 2]$.
(c) The function is increasing on $(2, 3)$ or for $2 < x < 3$. The function is decreasing on $(-1, 1)$ or for $-1 < x < 1$. The function is constant on $(-3, -1)$ and $(1, 2)$ or for $-3 < x < 1$ and $1 < x < 2$.
(d) The function is neither even nor odd.

39. (a) The function has a local maximum of 3 at $x = 0$.
(b) The function has local minima of 0 at $x = -2$ and $x = 2$.

41. (a) The function has a local maximum of 1 at $x = \frac{\pi}{2}$.
(b) The function has a local minimum of -1 at $x = -\frac{\pi}{2}$.

43. (a) $\frac{\Delta y}{\Delta x} = -4$ (b) $\frac{\Delta y}{\Delta x} = -8$
(c) $\frac{\Delta y}{\Delta x} = -10$

45. (a) $\frac{\Delta y}{\Delta x} = 5$ (b) $\frac{\Delta y}{\Delta x} = 5 = m_{\text{sec}}$
(c) $y = 5x$

47. (a) $\frac{\Delta y}{\Delta x} = -3$ (b) $\frac{\Delta y}{\Delta x} = -3 = m_{\text{sec}}$
(c) $y = -3x + 1$

49. (a) $\frac{\Delta y}{\Delta x} = x - 1$ (b) $\frac{\Delta y}{\Delta x} = 1 = m_{\text{sec}}$
(c) $y = x - 2$

51. (a) $\frac{\Delta y}{\Delta x} = x^2 + x$ (b) $\frac{\Delta y}{\Delta x} = 6 = m_{\text{sec}}$
(c) $y = 6x - 6$

53. (a) $\frac{\Delta y}{\Delta x} = -\frac{1}{1 + x}$ (b) $\frac{\Delta y}{\Delta x} = -\frac{1}{3} = m_{\text{sec}}$
(c) $y = -\frac{1}{3}x + \frac{4}{3}$

55. (a) $\frac{\Delta y}{\Delta x} = \frac{\sqrt{x} - 1}{x - 1}$ (b) $\frac{\Delta y}{\Delta x} = \sqrt{2} - 1 = m_{\text{sec}}$
(c) $y = (\sqrt{2} - 1)x + 2 - \sqrt{2}$

57. odd **59.** even **61.** odd

63. neither even nor odd **65.** even **67.** odd

69.

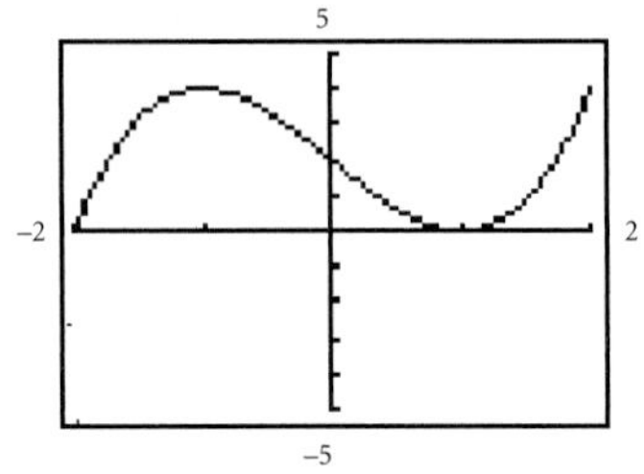

The function has a local minimum of 0 at $x = 1$ and a local maximum of 4 at $x = -1$. The function is increasing on $(-2, -1)$ and $(1, 2)$ and is decreasing on $(-1, 1)$.

71.

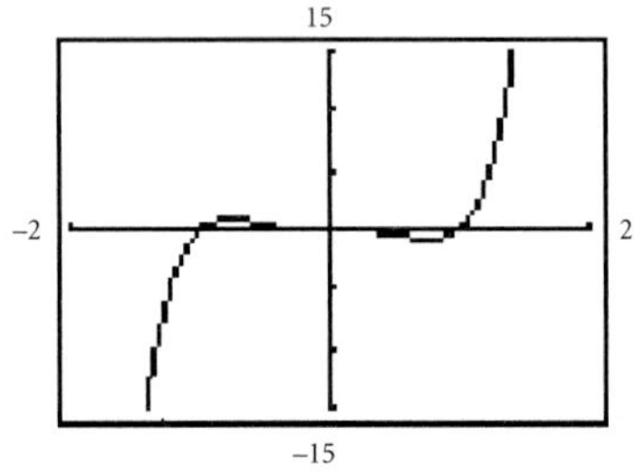

The function has a local minimum of -0.19 at $x = 0.77$ and a local maximum of 0.19 at $x = -0.77$. The function is increasing on $(-2, -0.77)$ and $(0.77, 2)$ and is decreasing on $(-0.77, 0.77)$.

73.

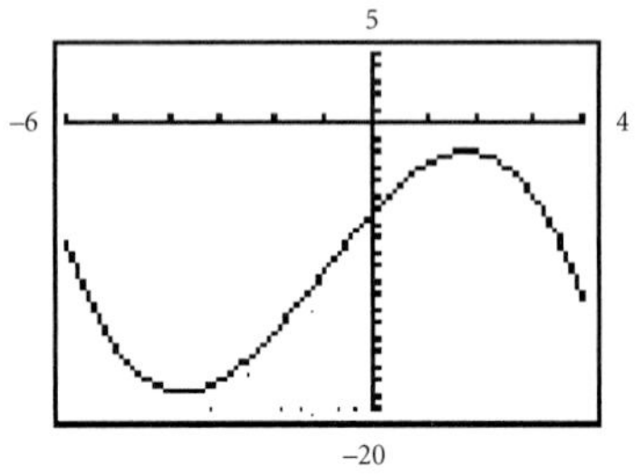

The function has a local minimum of -18.89 at $x = -3.77$ and a local maximum of -1.91 at $x = 1.77$. The function is increasing on $(-3.77, 1.77)$ and it is decreasing on $(-6, -3.77)$ and $(1.77, 4)$.

75.

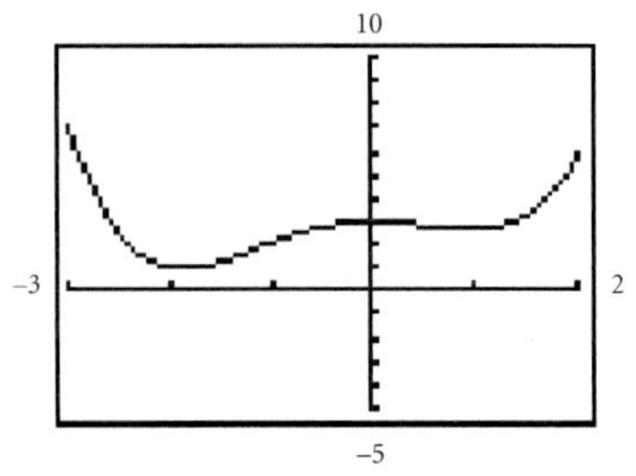

The function has a local minimum of 0.95 at $x = -1.87$, a local maximum of 3 at $x = 0$, and a local minimum of 2.65 at $x = 0.97$. The function is increasing on $(-1.87, 0)$ and $(0.97, 2)$ and is decreasing on $(-3, -1.87)$ and $(0, 0.97)$.

77. (a) $\dfrac{\Delta y}{\Delta x} = 1$ (b) $\dfrac{\Delta y}{\Delta x} = 0.5$ (c) $\dfrac{\Delta y}{\Delta x} = 0.1$

(d) $\dfrac{\Delta y}{\Delta x} = 0.01$ (e) $\dfrac{\Delta y}{\Delta x} = 0.001$

(f)

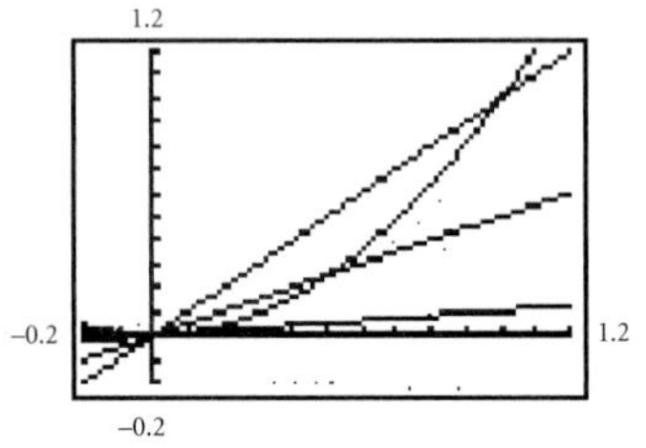

79. (a) 81.07 ft. (b) 129.59 ft.

(c) 26.63 ft. (d) The golf ball was hit 528.13 feet.

(e)

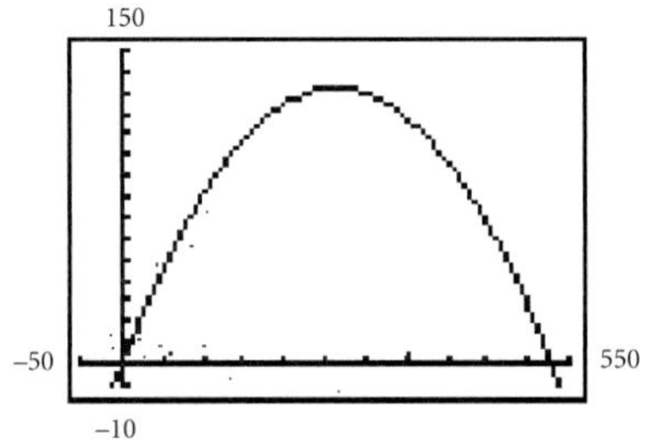

(f) The ball is at a height of 90 feet when it has traveled 115.07 feet and 413.05 feet.

(g)

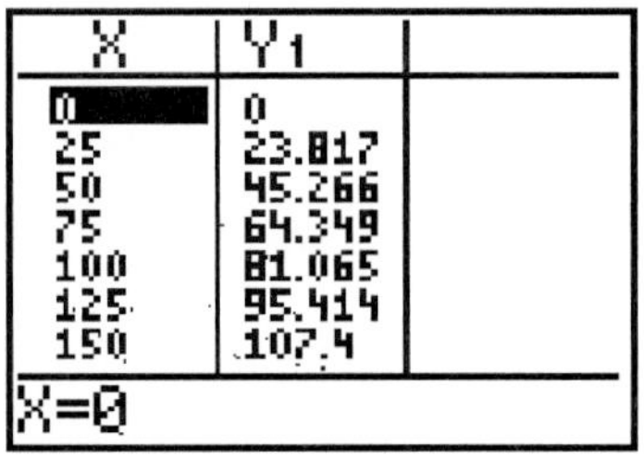

X	Y1
0	0
25	23.817
50	45.266
75	64.349
100	81.065
125	95.414
150	107.4

X=0

(h) The ball travels about 275 feet before reaching its maximum height. The maximum height of the ball is 132 feet.

(i) The ball travels 264 feet before reaching its maximum height.

81. (a) $V(x) = x(24 - 2x)^2 = 4x^3 - 96x^2 + 576x$

(b) The volume is 972 in^3.

(c) The volume is 160 in^3.

(d)

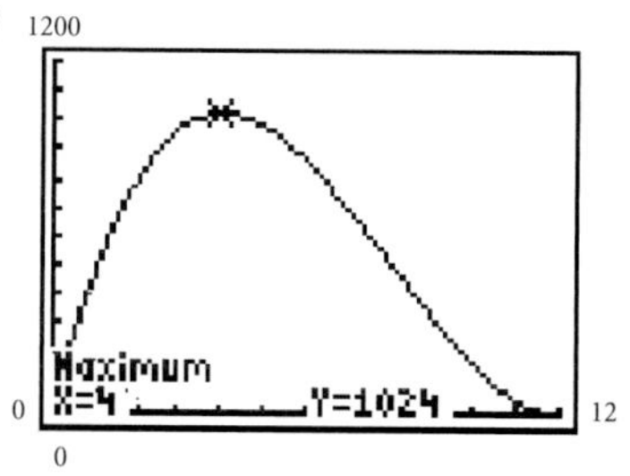

The volume V is largest when $x = 4$ in.

83. (a)

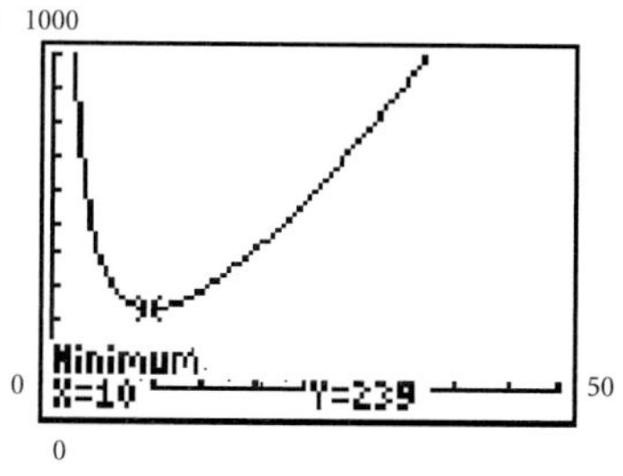

(b) Producing 10 riding lawn mowers minimizes average cost.

(c) The minimum average cost is \$239.

Exercise 10.4 (p. 585)

1. C. **3.** E. **5.** B. **7.** F.

9.

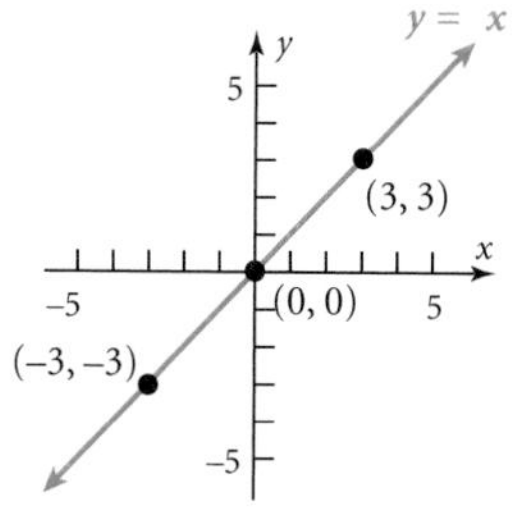

11.

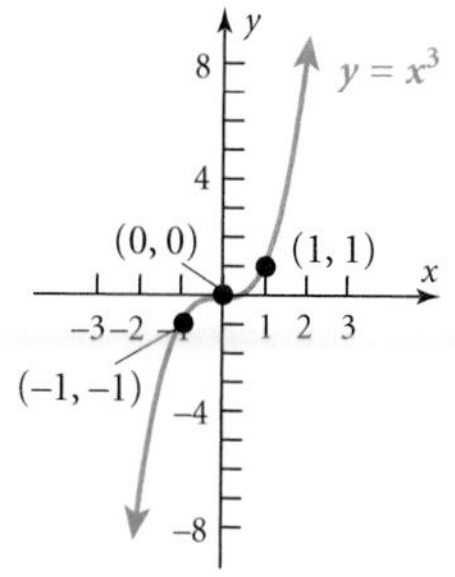

13.

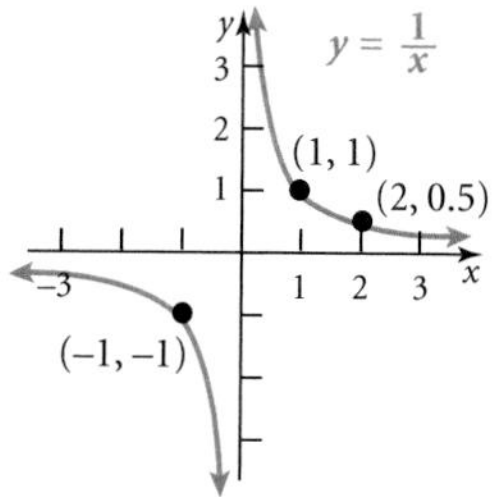

15.

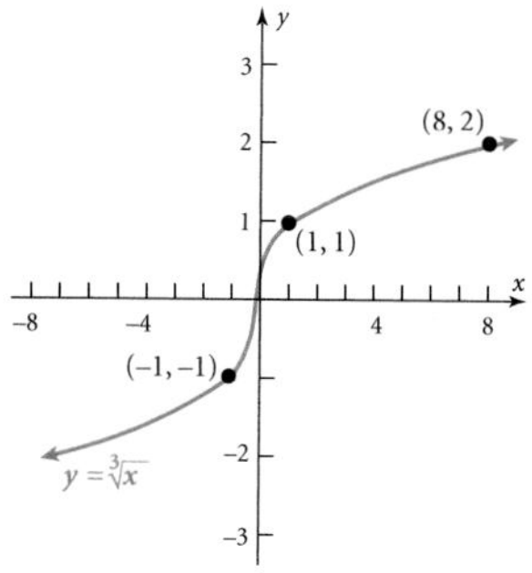

17. (a) $f(-2) = 4$ (b) $f(0) = 2$ (c) $f(2) = 5$

19. (a) $f(1.2) = 2$ (b) $f(1.6) = 3$ (c) $f(-1.8) = -4$.

21. (a) The domain is all real numbers, or the interval $(-\infty, \infty)$.

(b) There is no x-intercept. The y-intercept is $(0, 1)$.

(c)

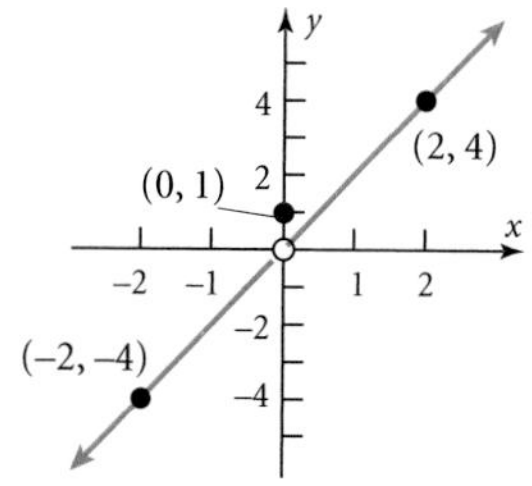

(d) The range is $\{y|y \neq 0\}$ or the intervals $(-\infty, 0)$ and $(0, \infty)$.

23. (a) The domain is all real numbers.

(b) There is no x-intercept. The y-intercept is $(0, 3)$.

(c)

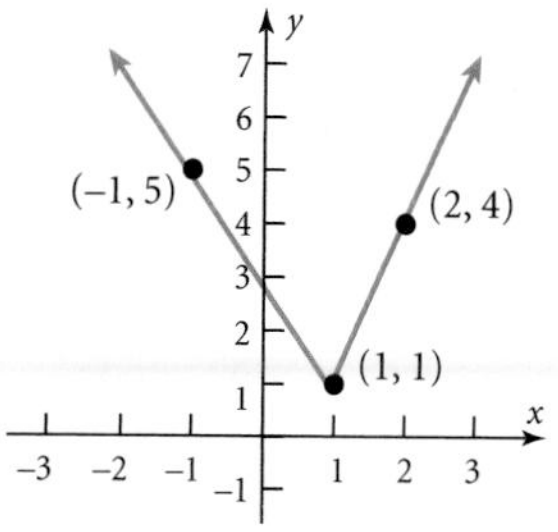

(d) The range is $\{y|y \geq 1\}$ or the interval $[1, \infty)$.

25. (a) The domain is the set of real numbers greater than or equal to -2, $\{x|x \geq -2\}$ or $[-2, \infty)$.
(b) The x-intercept is $(2, 0)$. The y-intercept is $(0, 3)$.
(c)

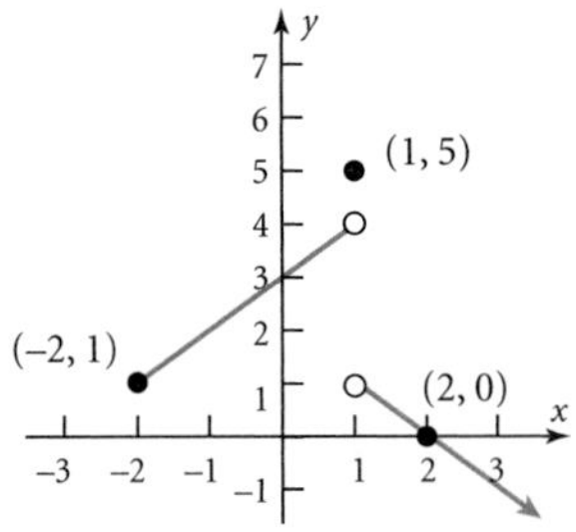

(d) $\{y|y = 5 \text{ or } y < 4\}$

27. (a) The domain is set of all real numbers or the interval $(-\infty, \infty)$.
(b) The x-intercepts are $(-1, 0)$ and $(0, 0)$. The y-intercept is $(0, 0)$.
(c)

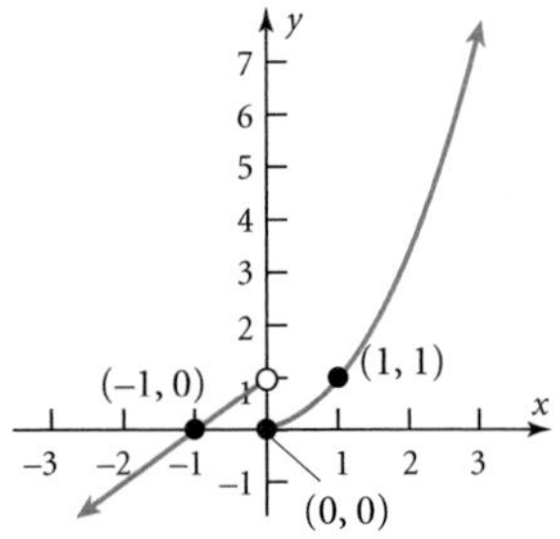

(d) The range is all real numbers or the interval $(-\infty,\infty)$.

29. (a) The domain is $\{x|x \geq -2\}$ or the interval $[-2, \infty)$.
(b) There is no x-intercept. The y-intercept is $(0, 1)$.
(c)

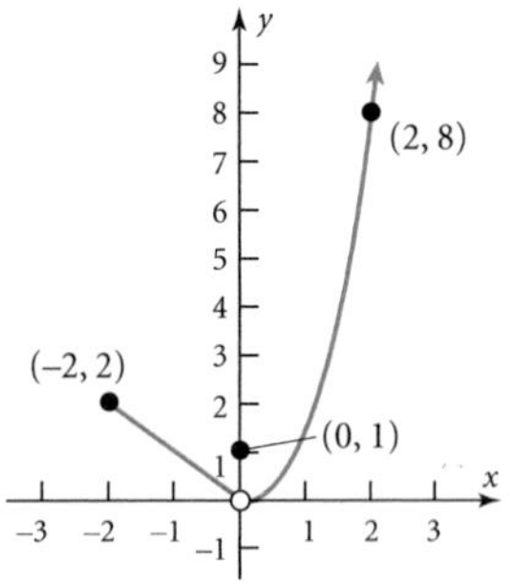

(d) The range is $\{y|y > 0\}$ or the interval $(0, \infty)$.

31. (a) The domain is all real numbers.
(b) The x-intercepts lie in the interval $[0, 1)$. The y-intercept is $(0, 0)$.
(c)

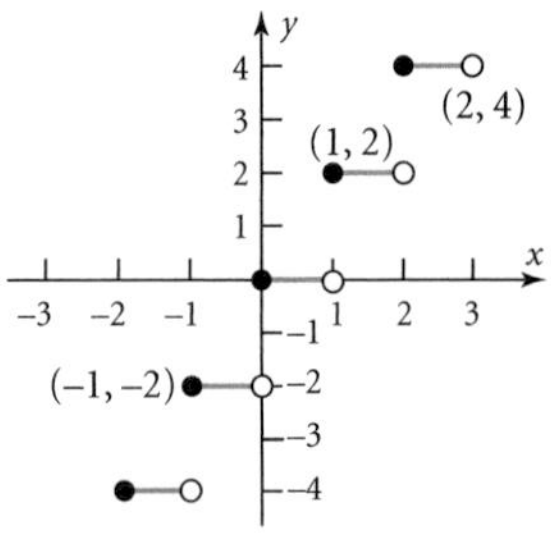

(d) The range is the set of even integers.

33. $f(x) = \begin{cases} -x & \text{if } -1 \leq x \leq 0 \\ \frac{1}{2}x & \text{if } 0 < x \leq 2 \end{cases}$

35. $f(x) = \begin{cases} -x & \text{if } x \leq 0 \\ 2 - x & \text{if } 0 < x \leq 2 \end{cases}$

37. (a) \$39.99 (b) \$43.74 (c) \$40.24

39. (a) \$59.33 (b) \$396.04
(c) $C(x) = \begin{cases} 0.99755x + 9.45 & \text{if } 0 \leq x \leq 50 \\ 0.74825x + 21.915 & \text{if } x > 50 \end{cases}$
(d)

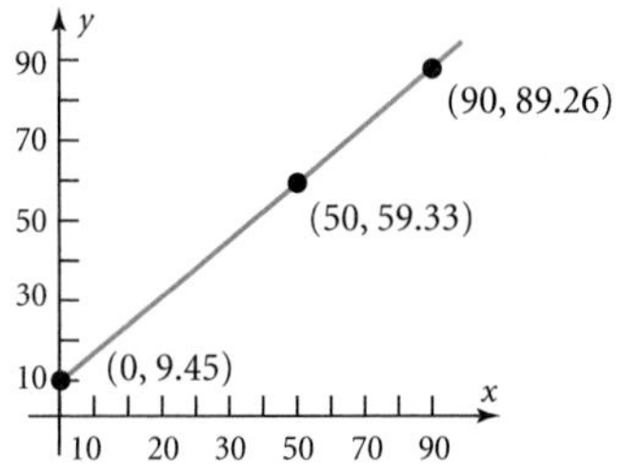

41. (a) 10°C. (b) 3.98°C. (c) −2.67°C.
(d) −3.70°C.
(e) For wind speeds under 1.79 m/sec, the wind chill factor is simply the air temperature.
(f) For wind speeds above 20 m/sec, the wind chill factor is a function of the air temperature.

43. $y = \begin{cases} 0.10x & \text{if } 0 < x \leq 7000 \\ 700 + 0.15(x - 7000) & \text{if } 7000 < x \leq 28{,}400 \\ 3910 + 0.25(x - 28{,}400) & \text{if } 28{,}400 < x \leq 68{,}800 \\ 34{,}926 + 0.33(x - 143{,}500) & \text{if } 143{,}500 < x \leq 311{,}950 \\ 90{,}514.5 + 0.35(x - 311{,}950) & \text{if } x > 311{,}950 \end{cases}$

45.

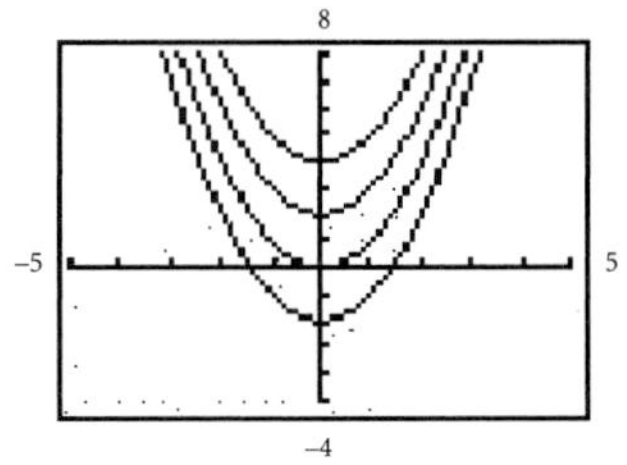

If $a > 0$, the graph of $y = x^2 + a$ is a vertical shift of the graph of $y = x^2$ up a units. If $a < 0$, the graph of

$y = x^2 + a$ is a vertical shift of the graph of $y = x^2$ down a units. The graph of $y = x^2 - 4$ is a vertical shift of the graph of $y = x^2$ down 4 units. The graph of $y = x^2 + 5$ is a vertical shift of the graph of $y = x^2$ up 5 units.

47.

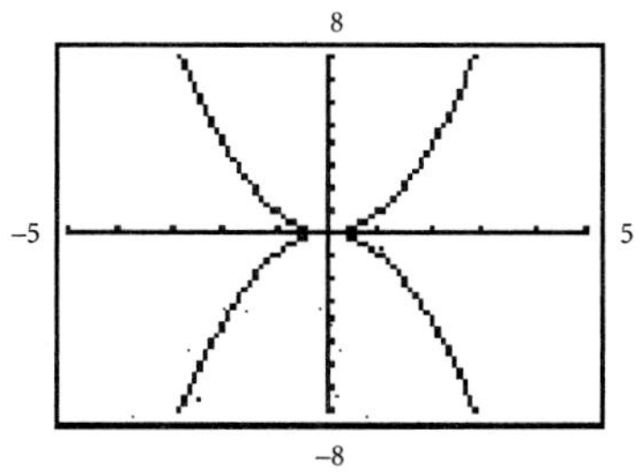

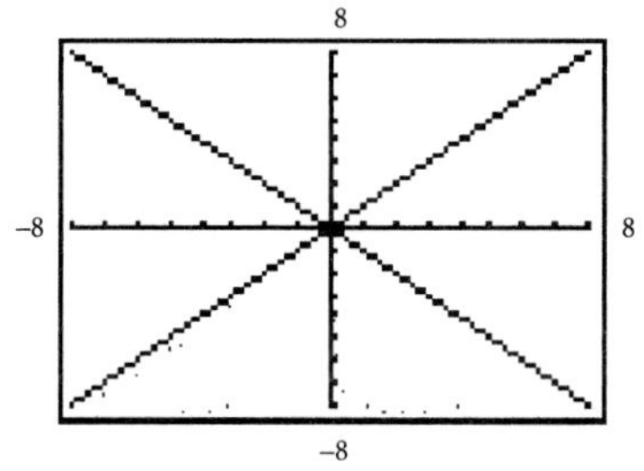

The graph of $y = -x^2$ is the reflection of $y = x^2$ about the x-axis. The graph of $y = -|x|$ is the reflection of $y = |x|$ about the x-axis.

49.

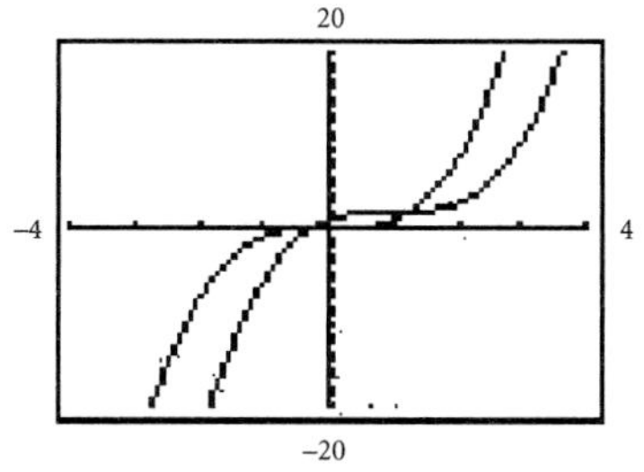

The graph of $y = (x - h)^3 + k$ is obtained by shifting the graph of $y = x^3$ h units to the right and k units up.

51.

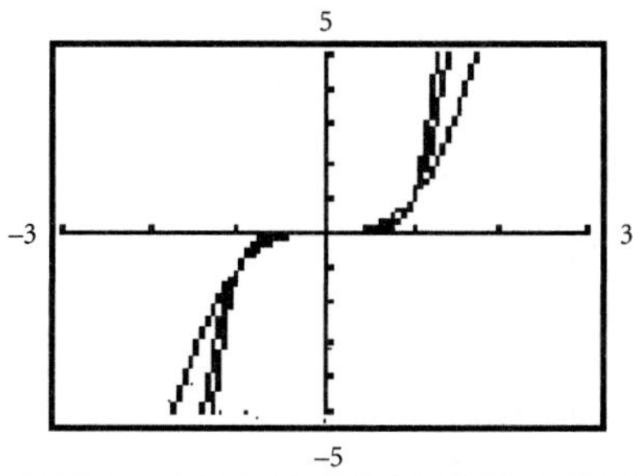

Each of the graphs increases in height as x increases. As the exponent increases the graphs become increasingly flatter near (0, 0).

Exercise 10.5 (p. 594)

1. B. **3.** H. **5.** A. **7.** F.

9. $y = (x - 4)^3$ **11.** $y = x^3 + 4$

13. $y = -x^3$ **15.** $y = -\sqrt{-x} - 2$

17. $y = -\sqrt{x + 3} + 2$ **19.** (c) (3, 0)

21.

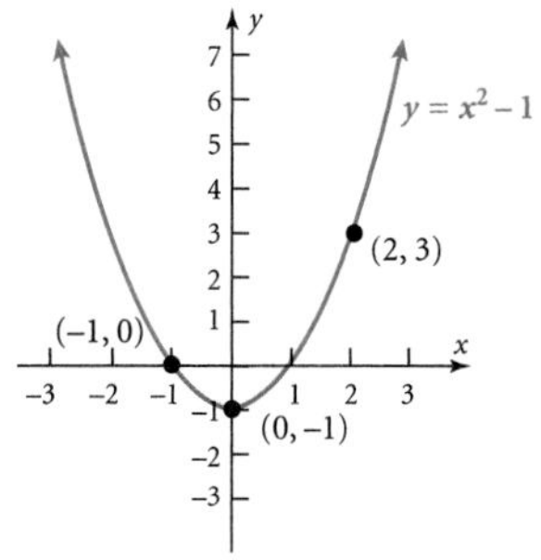

23.

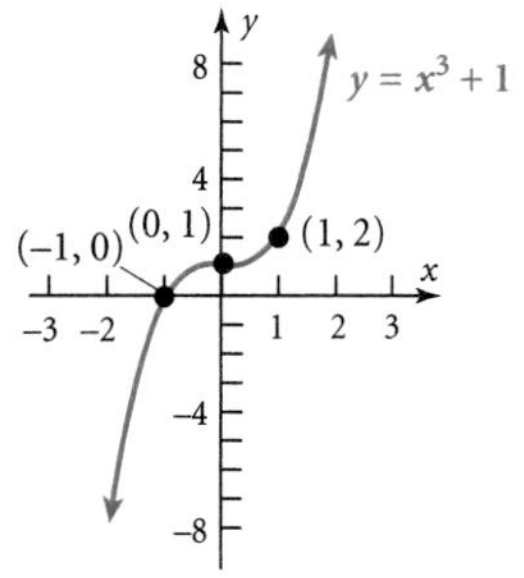

25.

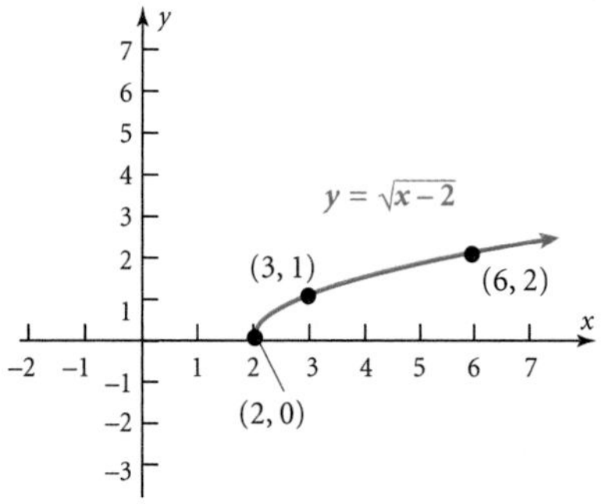

27.

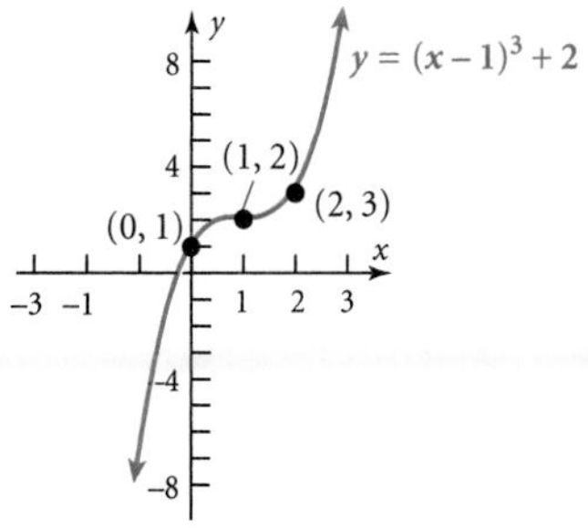

29.

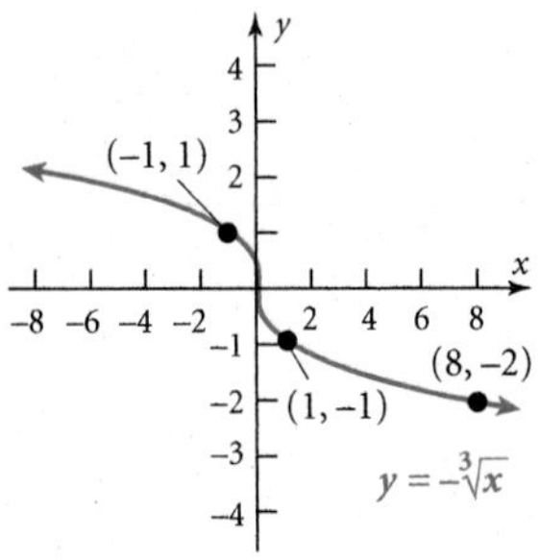

31.

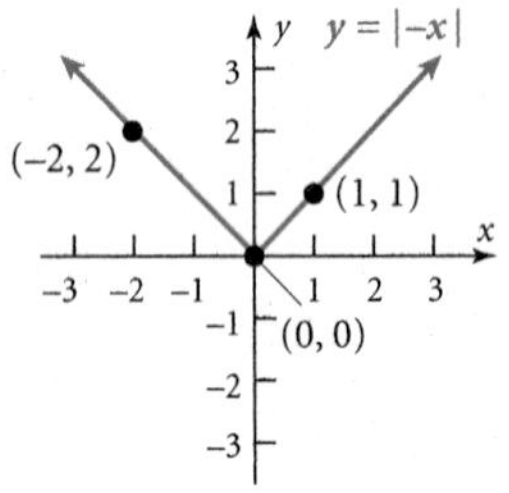

33.

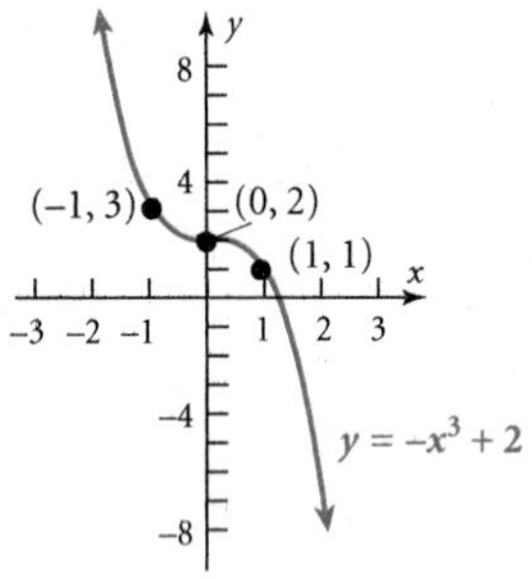

35.

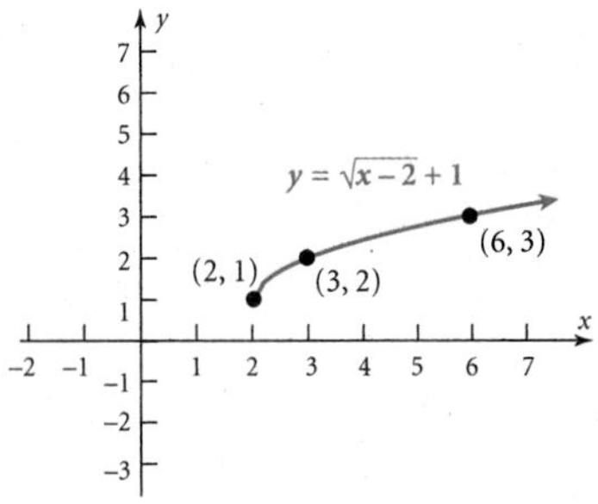

37.

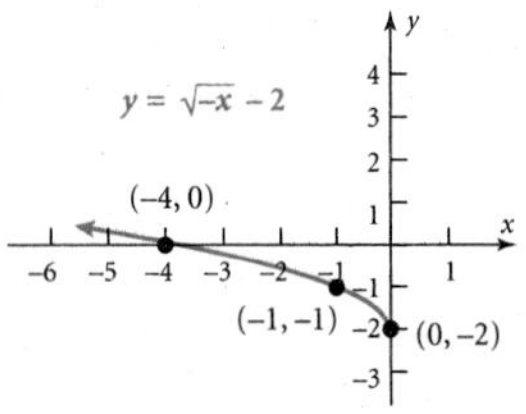

39. (a)

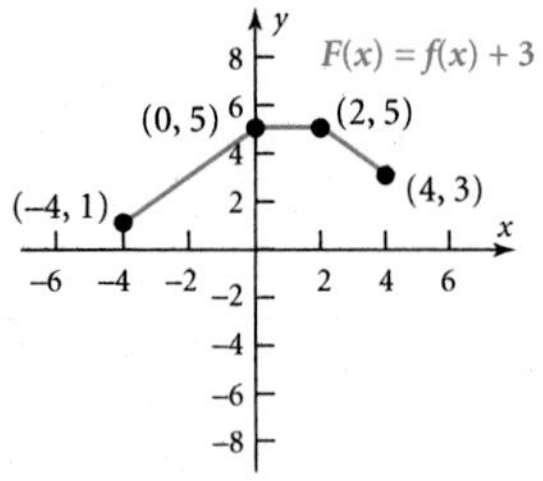

(b)

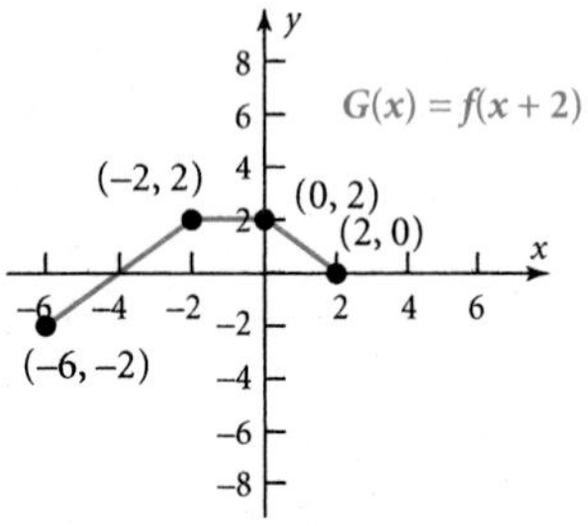

(c)

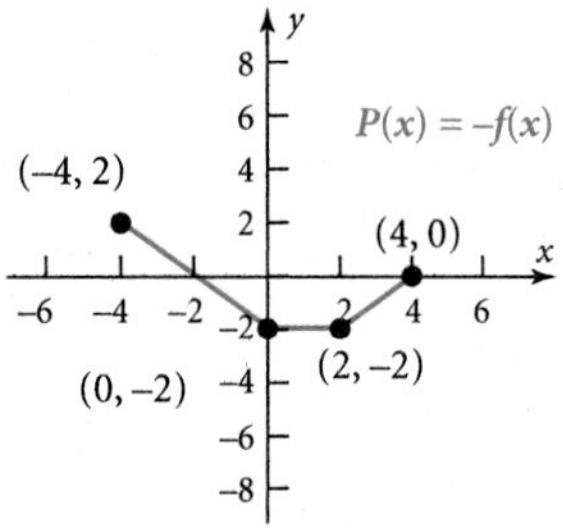

(d)

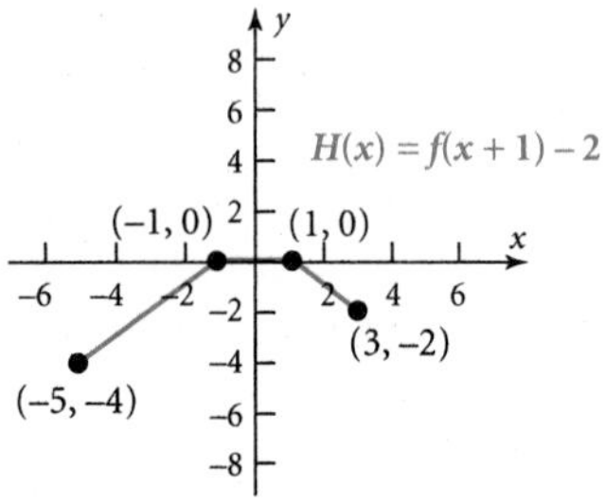

(e)

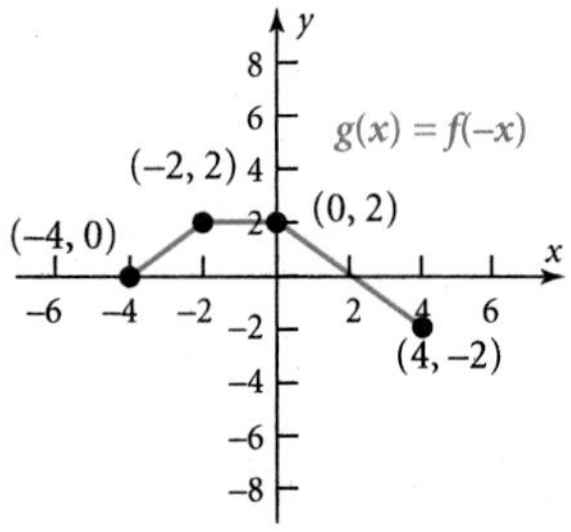

41. (a)

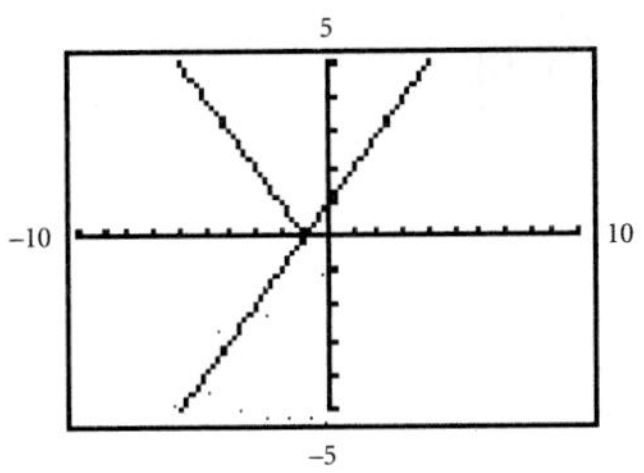

(b)

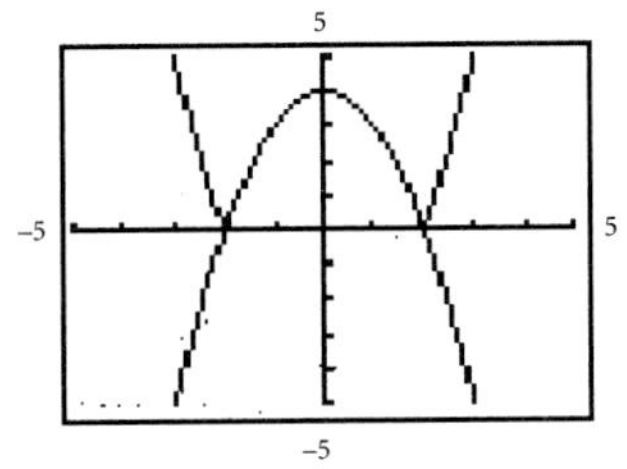

(c)

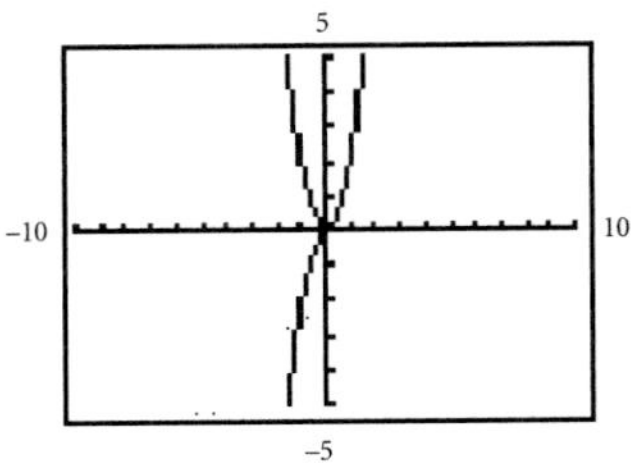

(d) The graph of $y = |f(x)|$ is obtained from the graph of $y = f(x)$ by reflecting about the x-axis those portions of the graph of $y = f(x)$ below the x-axis.

43. (a)

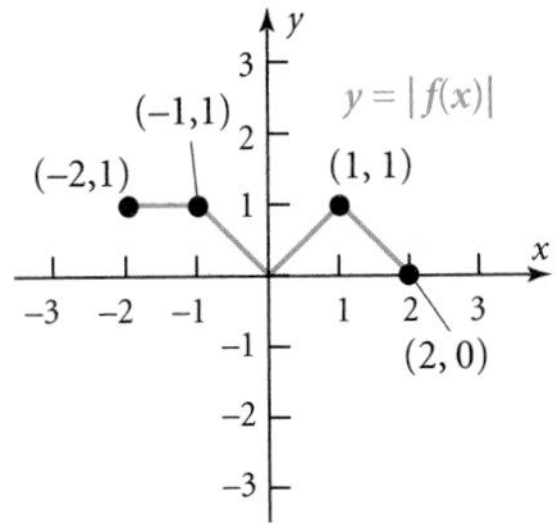

(b)

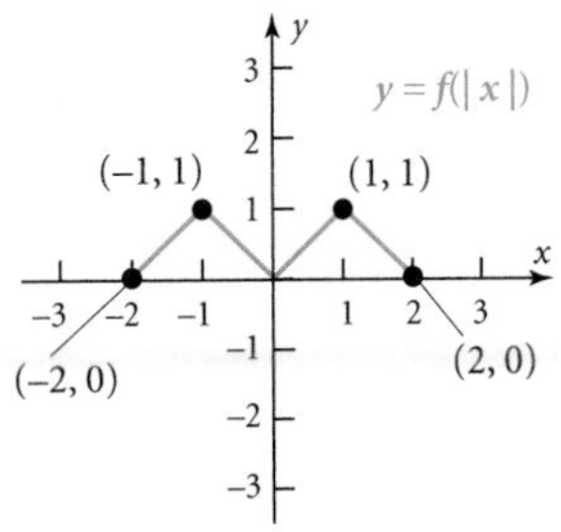

CHAPTER 10 Review

True – False Items (p. 598)

1. False **2.** False **3.** False
4. False **5.** True **6.** False

Fill in the Blanks (p. 598)

1. independent, dependent **2.** vertical
3. 5, −3 **4.** $a = -2$ **5.** $(-5, 0), (-2, 0), (2, 0)$

Review Exercises (p. 598)

1.

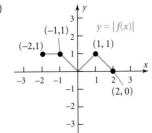

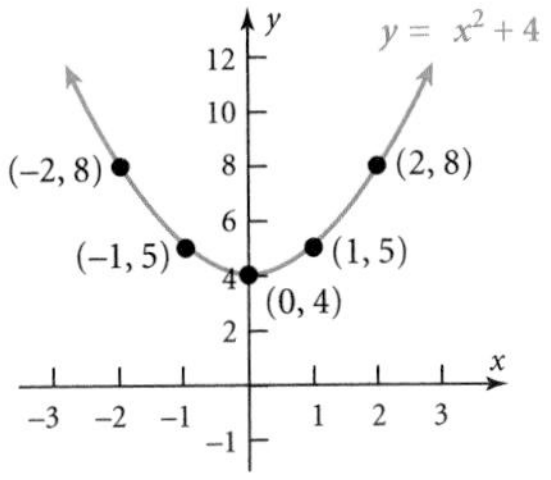

3. The x-intercept and the y-intercept are both $(0, 0)$. The graph is symmetric with respect to the x-axis.

5. The x-intercepts are $(-4, 0)$ and $(4, 0)$. The y-intercepts are $(0, -2)$ and $(0, 2)$. The graph is symmetric with respect to the x-axis, the y-axis, and the origin.

7. There is no x-intercept. The y-intercept is $(0, 1)$. The graph is symmetric with respect to the y-axis.

9. The x-intercepts are $(-1, 0)$ and $(0, 0)$. The y-intercepts are $(0, 0)$ and $(0, -2)$. The graph is not symmetric with respect to either axis nor the origin.

11. (a) $f(2) = 2$ (b) $f(-2) = -2$

(c) $f(-x) = -\dfrac{3x}{x^2 - 1}$ (d) $-f(x) = -\dfrac{3x}{x^2 - 1}$

(e) $f(x - 2) = \dfrac{3x - 6}{x^2 - 4x + 3}$ (f) $f(2x) = \dfrac{6x}{4x^2 - 1}$

13. (a) $f(2) = 0$ (b) $f(-2) = 0$

(c) $f(-x) = \sqrt{x^2 - 4}$ (d) $-f(x) = -\sqrt{x^2 - 4}$

(e) $f(x - 2) = \sqrt{x^2 - 4x}$ (f) $f(2x) = 2\sqrt{x^2 - 1}$

15. (a) $f(2) = 0$ (b) $f(-2) = 0$

(c) $f(-x) = \dfrac{x^2 - 4}{x^2}$ (d) $-f(x) = -\dfrac{x^2 - 4}{x^2}$

(e) $f(x - 2) = \dfrac{x^2 - 4x}{x^2 - 4x + 4}$ (f) $f(2x) = \dfrac{x^2 - 1}{x^2}$

17. $\{x | x \neq -3, x \neq 3\}$

19. $\{x|x \leq 2\}$ or the interval $(-\infty, 2]$ **21.** $\{x|x > 0\}$

23. $\{x|x \neq -3, x \neq 1\}$ **25.** $-4x - 2h + 1$

27. (a) Domain is $\{y|-4 \leq x \leq 3\}$ or the interval $[-4, 3]$, range is $\{y|-3 \leq y \leq 3\}$ or the interval $[-3, 3]$.
(b) The x-intercept and the y-intercept are both $(0, 0)$.
(c) $f(-2) = -1$ (d) $x = -4$
(e) The interval $(0, 3]$ or $\{x|0 < x \leq 3\}$

29. (a) Domain is the interval $(-\infty, \infty)$, range is the interval $(-\infty, 1]$.
(b) The function f is increasing on the intervals $(-\infty, -1)$ and $(3, 4)$ and is decreasing on the intervals $(-1, 3)$ and $(4, \infty)$.
(c) The function has a local maximum of 1 at $x = -1$ and a local maximum of 0 at $x = 4$. The function has a local minimum of -3 at $x = 3$.
(d) The graph is not symmetric with respect to either axis nor the origin.
(e) The function is neither even nor odd.
(f) The x-intercepts are $(-2, 0)$, $(0, 0)$, and $(4, 0)$. The y-intercept is $(0, 0)$.

31. odd **33.** even **35.** neither even nor odd

37. odd

39.

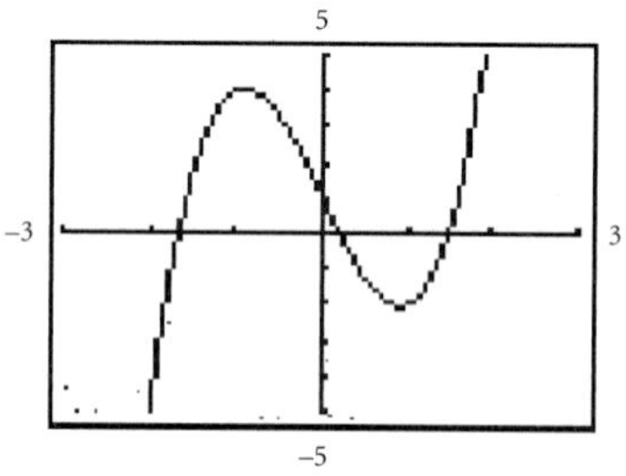

The function has a local maximum of 4.043 at $x = -0.913$ and a local minimum of -2.043 at $x = 0.913$. The function is increasing on the intervals $(-3, -0.913)$ and $(0.913, 3)$ and is decreasing on the interval $(-0.913, 0.913)$.

41.

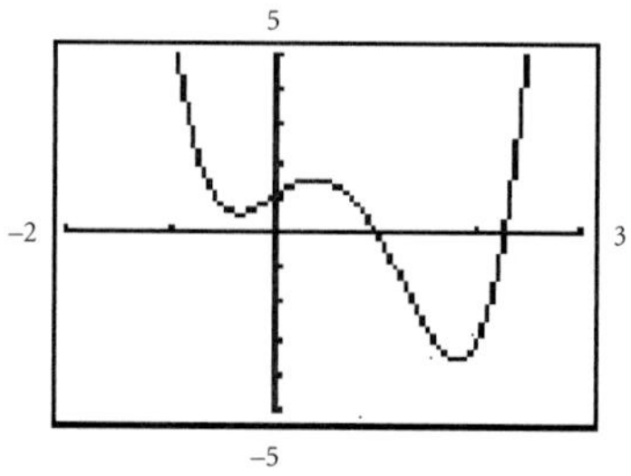

The function has local minima $f(-0.336) = 0.543$ and $f(1.798) = -3.565$, and a local maximum $f(0.414) = 1.532$. The function is increasing on the intervals $(-0.336, 0.414)$ and $(1.798, 3)$ and is decreasing on the intervals $(-2, -0.336)$ and $(0.414, 1.798)$.

43. (a) $\dfrac{\Delta y}{\Delta x} = 23$ (b) $\dfrac{\Delta y}{\Delta x} = 7$
(c) $\dfrac{\Delta y}{\Delta x} = 47$

45. $\dfrac{\Delta y}{\Delta x} = -5$ **47.** $\dfrac{\Delta y}{\Delta x} = -4x - 5$

49. Graphs (b), (c), (d), and (e) are graphs of functions.

51.

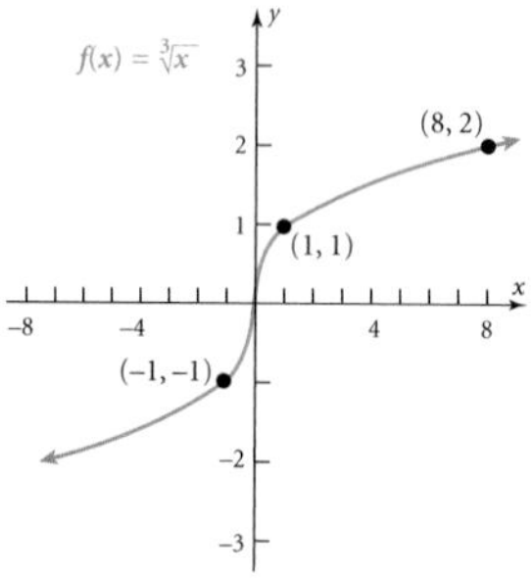

53. (a) The domain is $\{x|x > -2\}$ or the interval $(-2, \infty)$.
(b) The x-intercept and y-intercept are both $(0, 0)$.
(c)

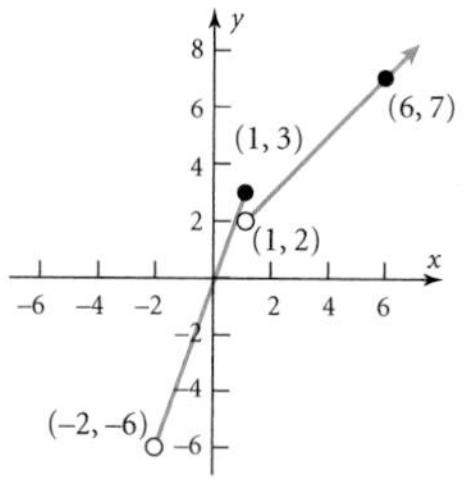

(d) The range is $\{y|y > -6\}$ or the interval $(-6, \infty)$.

55. (a) The domain is $\{x|x \geq -4\}$ or the interval $[-4, \infty)$.
(b) The y-intercept is $(0, 1)$. There is no x-intercept.
(c)

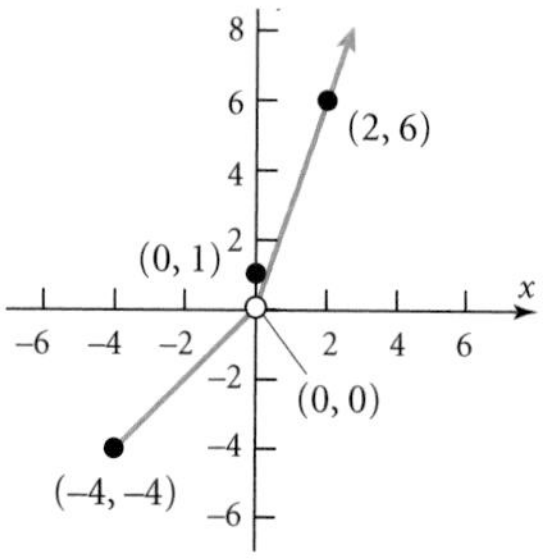

(d) The range is the intervals $[-4, 0)$ and $(0, \infty)$.

57.

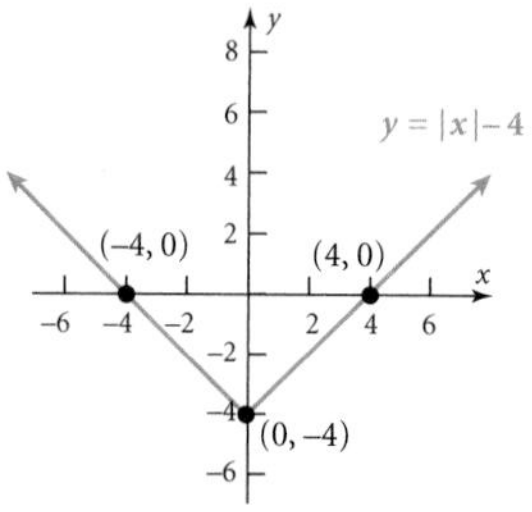

The x-intercepts are $(-4, 0)$ and $(4, 0)$. The y-intercept is $(0, -4)$. The domain is all real numbers or the interval $(-\infty, \infty)$, and the range is $\{y \mid y \geq -4\}$ or the interval $[-4, \infty)$.

59.

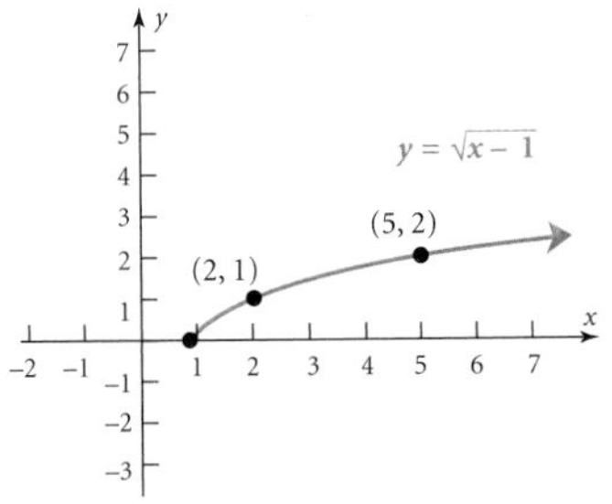

The x-intercept is $(1, 0)$. There is no y-intercept. The domain is $\{x \mid x \geq 1\}$ or $[1, \infty)$, and the range is $\{y \mid y \geq 0\}$ or the interval $[0, \infty)$.

61.

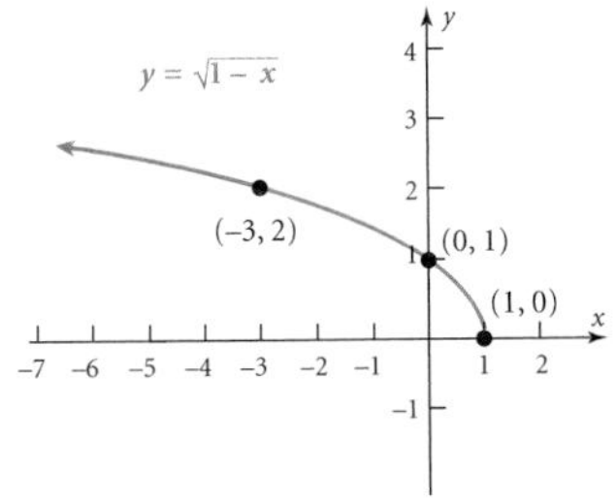

The x-intercept is $(1, 0)$. The y-intercept is $(0, 1)$. The domain is $\{x \mid x \leq 1\}$ or the interval $(-\infty, 1]$, and the range is $\{y \mid y \geq 0\}$ or the interval $[0, \infty)$.

63.

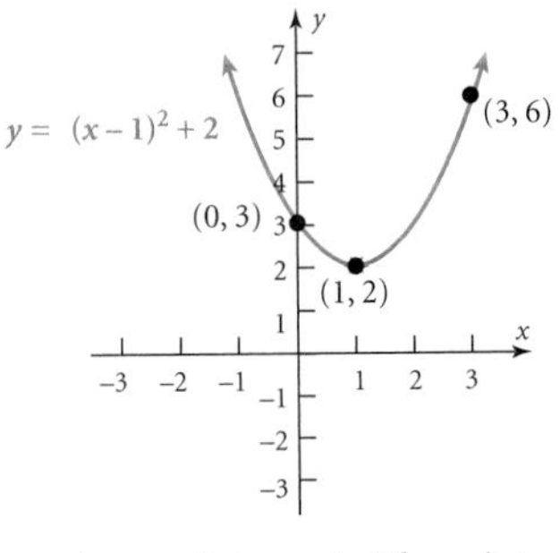

There is no x-intercept. The y-intercept is $(0, 3)$. The domain is $(-\infty, \infty)$, and the range is $\{y \mid y \geq 2\}$ or $[2, \infty)$.

65. (a)

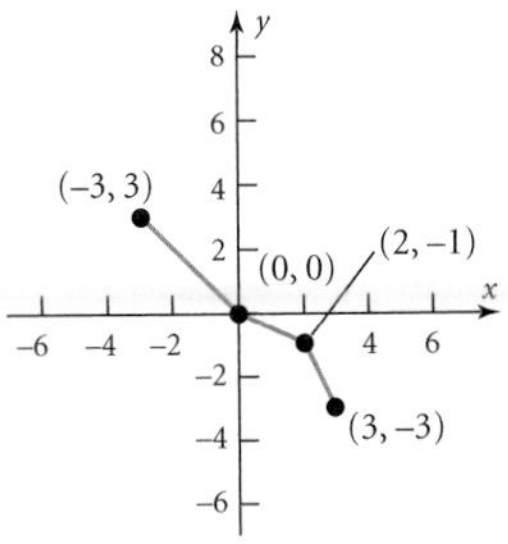

(b)

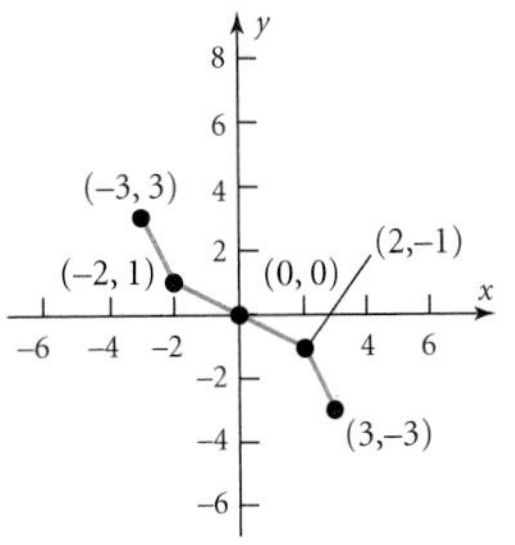

(c)

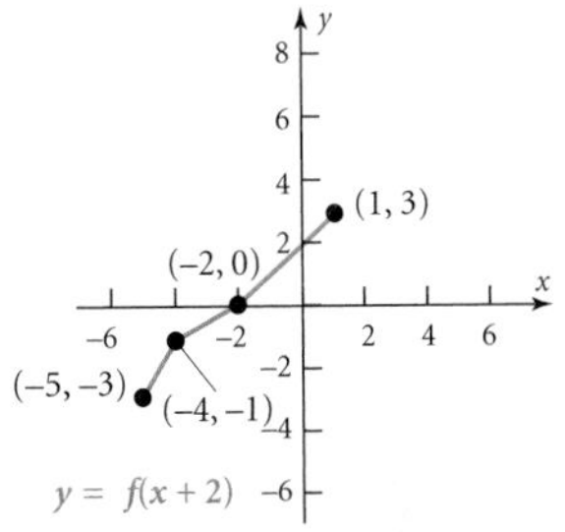

(d)

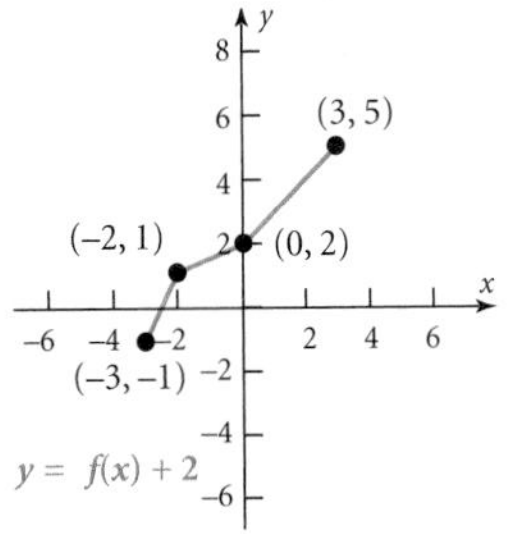

67. $f(x) = -2x + 3$

69. $A = 11$

71. $V(r) = 2\pi r^3$

73. (a) $R(x) = -\frac{1}{6}x^2 + 100x;\ 0 \leq x \leq 600$

(b) The revenue is \$13,333.33.

75. (a) $R(x) = -\frac{1}{5}x^2 + 20x,\ 0 \leq x \leq 100$

(b) The revenue is \$255.

77. (a) The total cost C in dollars is given by $C(r) = 0.12\pi r^2 + \frac{40}{r}$.

(b) The cost is \$16.03.

(c) The cost is \$29.13.

(d)

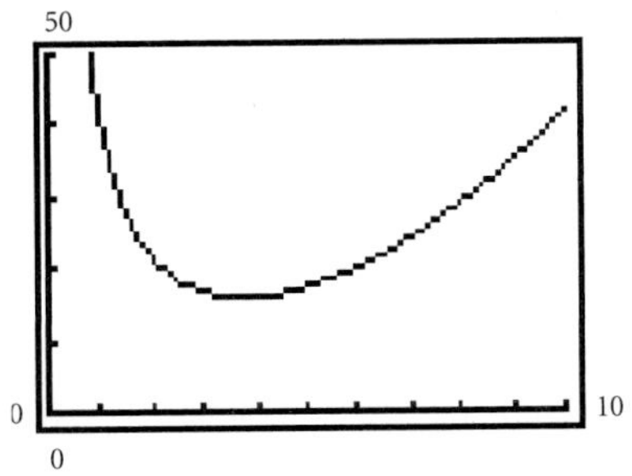

The cost is least when the radius is 3.758 cm.

Mathematical Questions from Professional Exams (p. 602)

1. (d) $0 \leq y < 4$

2. (d) 87

3. (c) $[-1, 0] \cup [1, \infty)$

CHAPTER 11 Classes of Functions

Exercise 11.1 (p. 614)

1. C. **3.** F. **5.** G. **7.** H.

9.

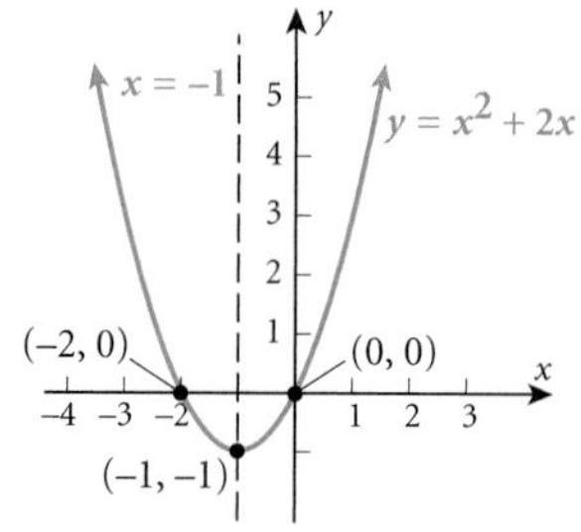

The graph opens upward. The vertex is located at the point $(-1, -1)$. The axis of symmetry is the line $x = -1$. The y-intercept is $(0, 0)$. The x-intercepts are $(-2, 0)$ and $(0, 0)$. The domain is the interval $(-\infty, \infty)$, and the range is the interval $[-1, \infty)$. The function is increasing on the interval $(-1, \infty)$, and is decreasing on the interval $(-\infty, 1)$.

11.

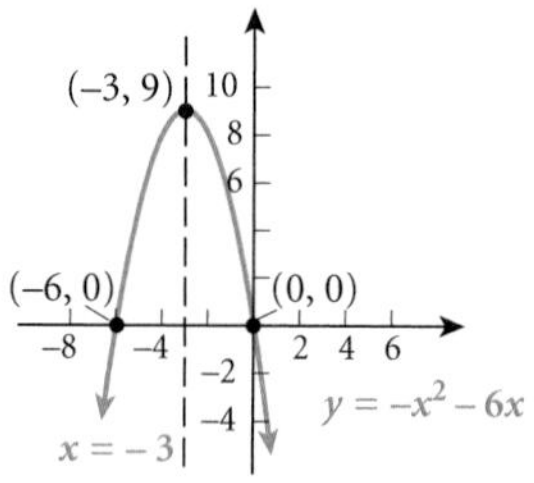

The graph opens downward. The vertex is located at the point $(-3, 9)$. The axis of symmetry is the line $x = -3$. The y-intercept is $(0, 0)$. The x-intercepts are $(-6, 0)$ and $(0, 0)$.The domain is the interval $(-\infty, \infty)$, and the range is the interval $(-\infty, 9]$. The function is increasing on the interval $(-\infty, -3)$, and is decreasing on the interval $(-3, \infty)$.

13.

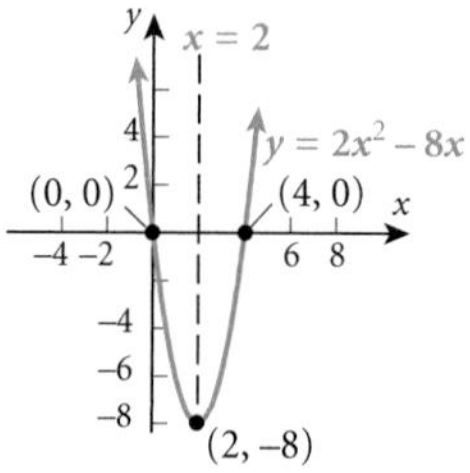

The graph opens upward. The vertex is located at the point $(2, -8)$. The axis of symmetry is the line $x = 2$. The y-intercept is $(0, 0)$. The x-intercepts are $(0, 0)$ and $(4, 0)$. The domain is the interval $(-\infty, \infty)$, and the range is the interval $[-8, \infty)$. The function is increasing on the interval $(2, \infty)$, and is decreasing on the interval $(-\infty, 2)$.

15.

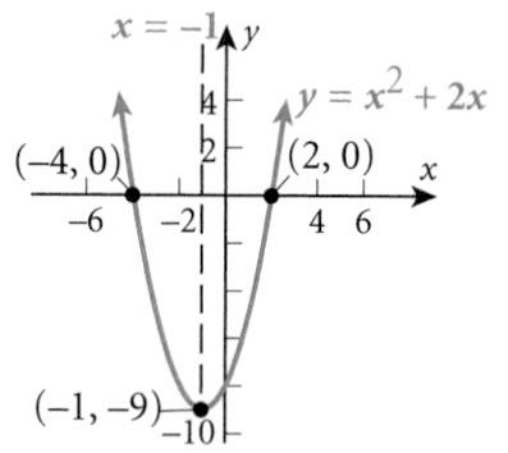

The graph opens upward. The vertex is located at the point $(-1, -9)$. The axis of symmetry is the line $x = -1$. The y-intercept is $(0, -8)$. The x-intercepts are $(-4, 0)$ and $(2, 0)$. The domain is the interval $(-\infty, \infty)$, and the range is the interval $[-9, \infty)$. The function is increasing on the interval $(-1, \infty)$, and is decreasing on the interval $(-\infty, -1)$.

17.

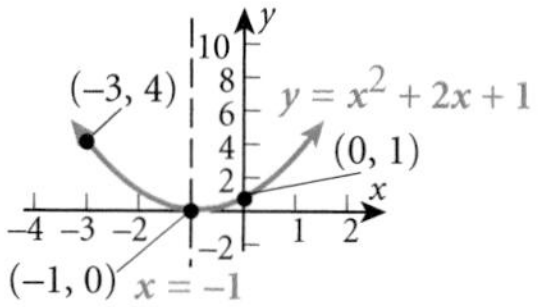

The graph opens upward. The vertex is located at the point $(-1, 0)$. The axis of symmetry is the line $x = -1$. The y-intercept is $(0, 1)$. The x-intercept is $(-1, 0)$. The domain is the interval $(-\infty, \infty)$, and the range is the interval $[0, \infty)$. The function is increasing on the interval $(-1, \infty)$, and is decreasing on the interval $(-\infty, -1)$.

19.

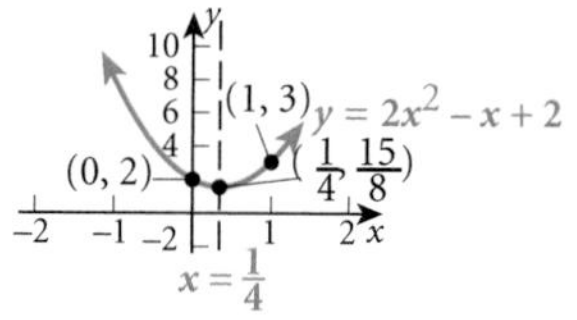

The graph opens upward. The vertex is located at the point $\left(\frac{1}{4}, \frac{15}{8}\right)$. The axis of symmetry is the line $x = \frac{1}{4}$. The y-intercept is $(0, 2)$. There is no x-intercept. The domain is the interval $(-\infty, \infty)$, and the range is the interval $\left[\frac{15}{8}, \infty\right)$. The function is increasing on the interval $\left(\frac{1}{4}, \infty\right)$, and is decreasing on the interval $\left(-\infty, \frac{1}{4}\right)$.

21.

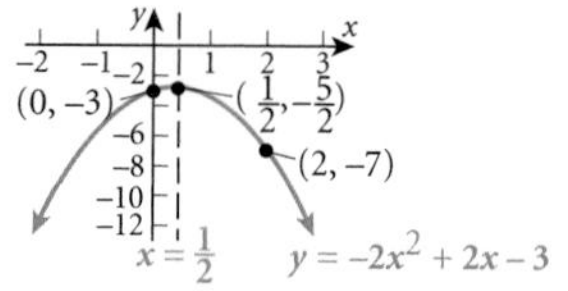

The graph opens downward. The vertex is located at the point $\left(\frac{1}{2}, -\frac{5}{2}\right)$. The axis of symmetry is the line $x = \frac{1}{2}$. The y-intercept is $(0, -3)$. There is no x-intercept. The domain is the interval $(-\infty, \infty)$, and the range is the interval $\left(-\infty, -\frac{5}{2}\right]$. The function is increasing on the interval $\left(-\infty, \frac{1}{2}\right)$, and is decreasing on the interval $\left(\frac{1}{2}, \infty\right)$.

23.

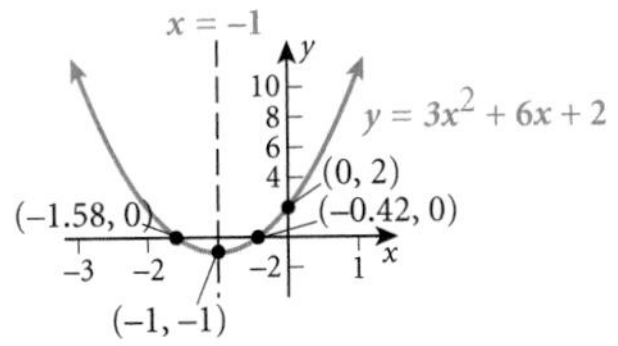

The graph opens up. The vertex is located at the point $(-1, -1)$. The axis of symmetry is the line $x = -1$. The y-intercept is $(0, 2)$. The x-intercepts are $(\frac{-\sqrt{3}-3}{3}, 0)$ $\approx (-1.58, 0)$ and $(\frac{\sqrt{3}-3}{3}, 0)$ $\approx (-0.42, 0)$. The domain is the interval $(-\infty, \infty)$, and the range is the interval $[-1, \infty)$. The function is increasing on the interval $(-1, \infty)$, and is decreasing on the interval $(-\infty, -1)$.

25.

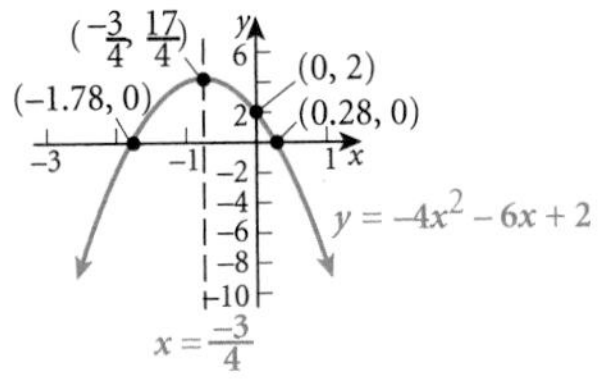

The graphs opens down. The vertex is located at the point $(-\frac{3}{4}, \frac{17}{4})$. The axis of symmetry is the line $x = -\frac{3}{4}$. The y-intercept is $(0, 2)$. The x-intercepts are $(\frac{-\sqrt{17}-3}{4}, 0)$ $\approx (-1.78, 0)$ and $(\frac{\sqrt{17}-3}{4}, 0)$ $\approx (0.28, 0)$. The domain is the interval $(-\infty, \infty)$, and the range is the interval $(-\infty, \frac{17}{4}]$. The function is increasing on the interval $(-\infty\ , -\frac{3}{4})$, and is decreasing on the interval $(-\frac{3}{4}, \infty)$.

27. The function has a minimum value of $f(-3) = -18$.

29. The function has a minimum value of $f(-3) = -21$.

31. The function has a maximum value of $f(5) = 21$.

33. The function has a maximum value of $f(2) = 13$.

35. (a) $f(x) = (x + 3)(x - 1)$, $f(x) = 2(x + 3)(x - 1)$, $f(x) = -2(x + 3)(x - 1)$, $f(x) = 5(x + 3)(x - 1)$

37. A unit price of \$500 should be established to maximize revenue. The maximum revenue is \$1, 000, 000.

39. (a) $R(x) = -\frac{1}{6}x^2 + 100x$

(b) The revenue is \$13, 333.33.

(c) A quantity of 300 units maximizes revenue. The maximum revenue is \$15, 000.

(d) The company should charge \$50.

41. (a) $R(x) = -\frac{1}{5}x^2 + 20x$

(b) The revenue is \$255.

(c) A quantity of 50 units maximizes revenue. The maximum revenue is \$500.

(d) The company should charge \$10.

43. (a) $A(x) = 200x - x^2$

(b) $x = 100$

(c) The maximum area is 10, 000 square yards.

45. The largest area is 2,000,000 m^2.

47. (a) The projectile is 39.0625 feet horizontally from the base of the cliff when it achieves is maximum height.

(b) The maximum height is 219.53125 feet above the water.

(c) The projectile strikes the water 170.024 feet from the base of the cliff.

(d)

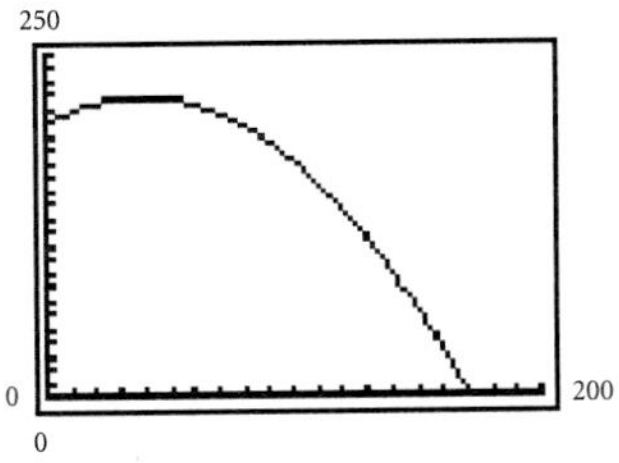

(e) The projectile is 135.698 feet from the cliff.

49. A depth of 3 inches will provide the maximum cross-sectional area.

51. 100 meters by $\frac{100}{2\pi}$ meters $\approx$ 100 meters by 63.66 meters

53. (a) There are the most hunters at the income level of \$56,584. At this income level, there are about 3685 hunters.

(b)

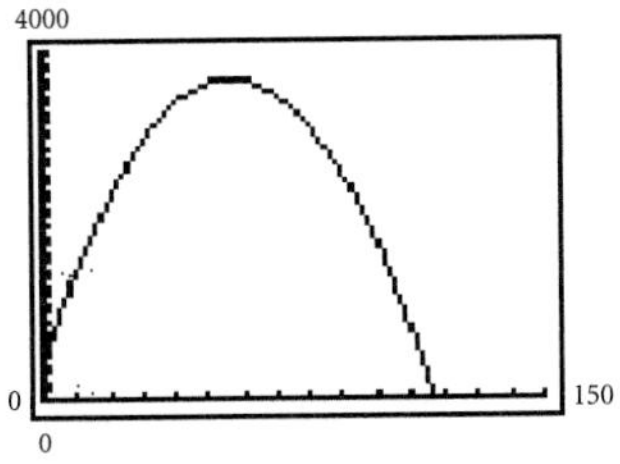

The number of hunters is increasing between the \$20,000 and \$40,000 income levels.

55. (a) The number of 23-year old male murder victims is 1795.

(b) The number of male murder victims at age 28 years is about 1456.

(c)

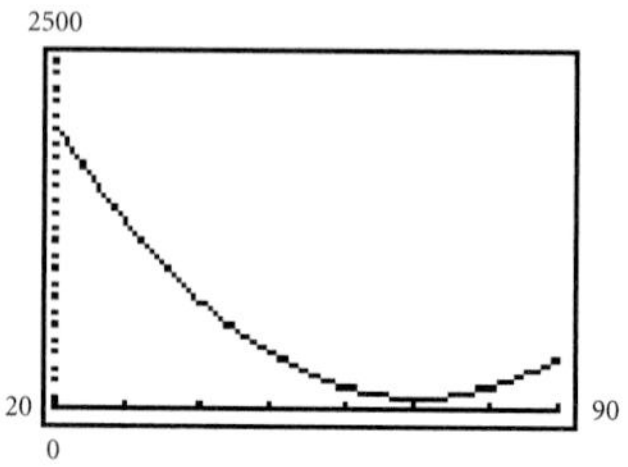

(d) The number of male murder victims decreases with age until age 70 and then begins to increase.

57. $x = \frac{a}{2}$; that is, when $\frac{1}{2}$ of the original amount is present.

59. Area $= \frac{38}{3}$ units^2

61. Area $= \frac{248}{3}$ units^2 **63.** $A = 25$ units^2

65.

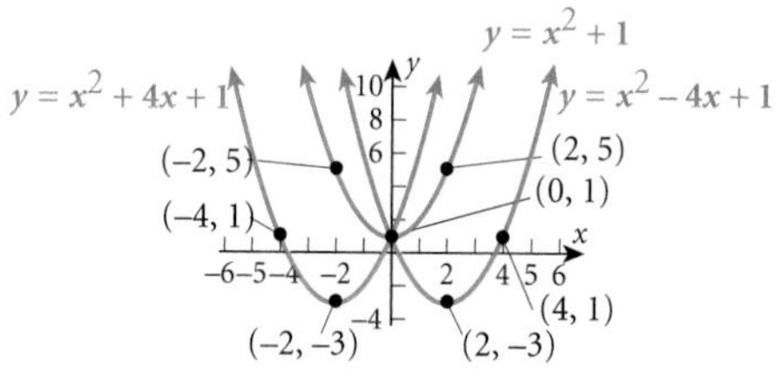

Exercise 11.2 (p. 625)

1. Answers will vary. Possible answers include $(-1, -1)$, $(0, 0)$ and $(1, 1)$.

3. origin

5.

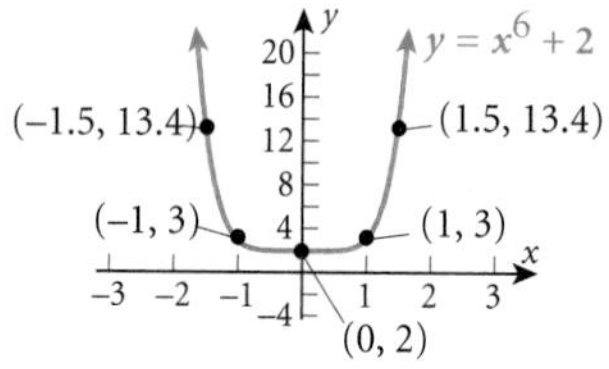

7.

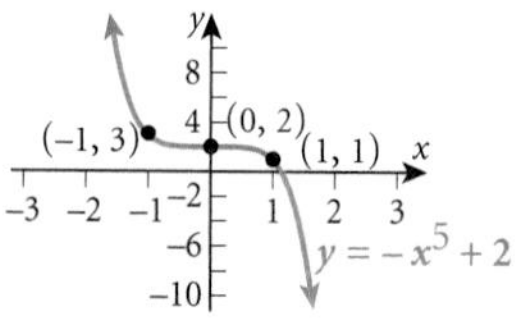

9.

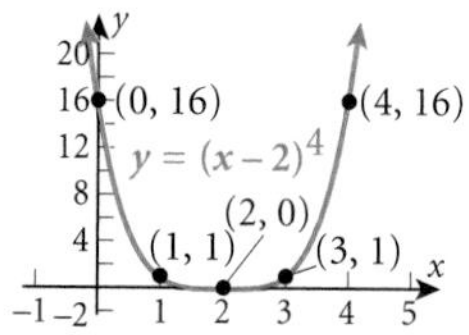

11. This is a polynomial function of degree 3.

13. This is a polynomial function of degree 2.

15. This is not a polynomial functon because the term $\frac{1}{x}$ has the monomial x in the denominator.

17. This is not a polynomial function because the exponent of the term $x^{3/2}$ is not a nonnegative integer.

19. This is a polynomial function of degree 4.

21. This is a polynomial function of degree 4.

23. $y = 3x^4$ **25.** $y = -2x^5$ **27.** $y = 5x^3$

29. $\{x|x \neq 3\}$ **31.** $\{x|x \neq 2, x \neq -4\}$

33. $\{x|x \neq -\frac{1}{2}, x \neq 3\}$ **35.** $\{x|x \neq 2\}$

37. $(-\infty, \infty)$ **39.** $\{x|x \neq -3, x \neq 3\}$

41. (a) The percentage of union membership in the labor force in 2000 was 17.04%.

(b) $u(75) = 18.02$. The percentage of union membership in the labor force in 2005 will be 18.02%.

Exercise 11.3 (p. 637)

1. (a) 11.2116 (b) 11.5873 (c) 11.6639 (d) 11.6648

3. (a) 8.8152 (b) 8.8214 (c) 8.8244 (d) 8.8250

5. (a) 21.2166 (b) 22.2167 (c) 22.4404 (d) 22.4592

7. 3.3201 **9.** 0.4274

11. This is not an exponential function.

13. This is an exponential function with base $a = 4$.

15. This is an exponential function with base $a = 2$.

17. This is not an exponential function.

19. B. **21.** D. **23.** A. **25.** E.

27.

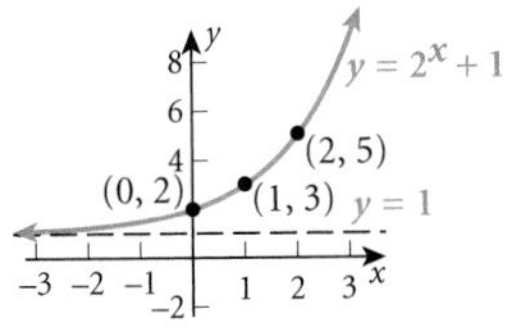

Domain = $(-\infty, \infty)$, Range = $(1, \infty)$; the horizontal asymptote is $y = 1$.

29.

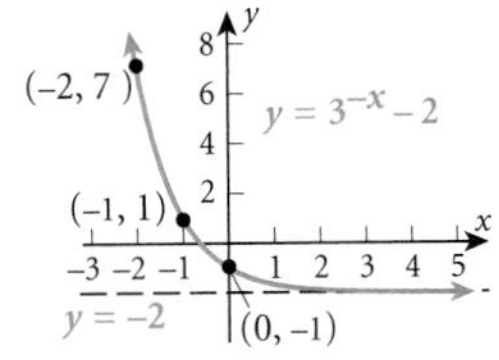

Domain = $(-\infty, \infty)$, Range = $(-2, \infty)$; the horizontal asymptote is $y = -2$.

31.

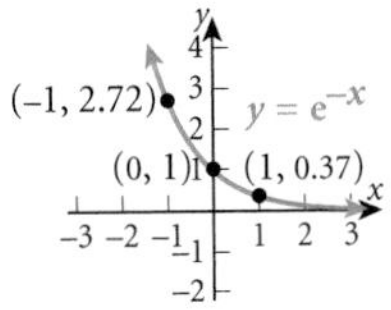

Domain = $(-\infty, \infty)$, Range = $(0, \infty)$; the horizontal asymptote is $y = 0$.

33.

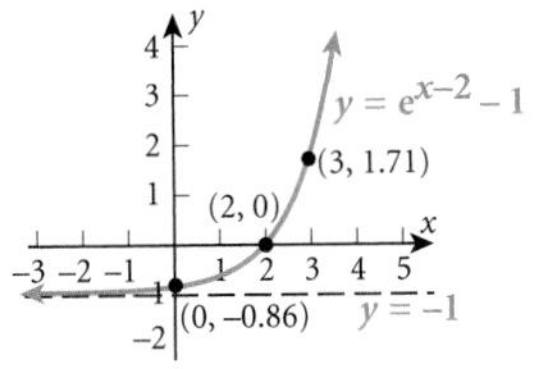

Domain = $(-\infty, \infty)$, Range = $(-1, \infty)$; the horizontal asymptote is $y = -1$.

35. $\{\frac{1}{2}\}$ **37.** $\{-\sqrt{2}, 0, \sqrt{2}\}$ **39.** $\{\frac{3-\sqrt{6}}{3}, \frac{3+\sqrt{6}}{3}\}$

41. $\{0\}$ **43.** $\{4\}$ **45.** $\{\frac{3}{2}\}$

47. $\{1, 2\}$ **49.** $\frac{1}{49}$ **51.** $\frac{1}{4}$

53. $y = 3^x$ **55.** $y = -6^x$

57. (a) 74.1% of light will pass through 10 panes.
(b) 47.2% of light will pass through 25 panes.

59. (a) There will be 44.35 watts after 30 days.
(b) There will be 11.61 watts after one year.

61. There will be 3.35 mg of the drug in the bloodstream after 1 hour. There will be 0.45 mg of the drug in the bloodstream after 6 hours.

63. (a) The probability that a car will arrive within 10 minutes of 12:00 PM is 0.632.
(b) The probability that a car will arrive within 40 minutes of 12:00 PM is 0.982.
(c) $F(t)$ approaches 1 as t becomes unbounded in the positive direction.
(d)

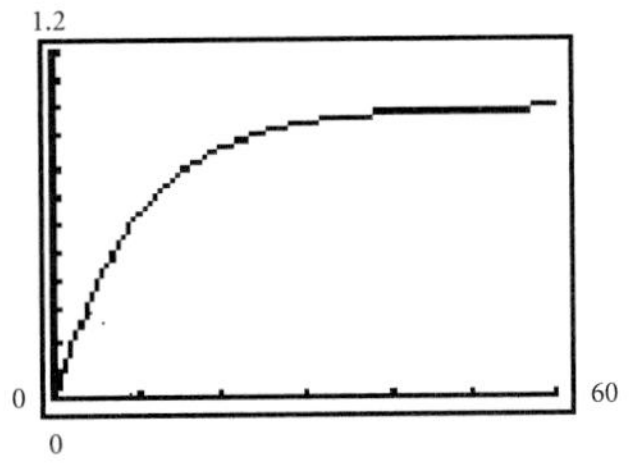

(e) About 6.931 minutes are needed for the probability to reach 50%.

65. (a) The probability that 15 cars will arrive between 5:00 PM and 6:00 PM is 5.2%.
(b) The probability that 20 cars will arrive between 5:00 PM and 6:00 PM is 8.9%.

67. (a) A 3-year old Civic DX Sedan costs \$12,123.27.
(b) A 9-year old Civic DX Sedan costs \$6442.80.

69. (a) 5.4 amperes, 7.6 amperes, 10.4 amperes
(b)

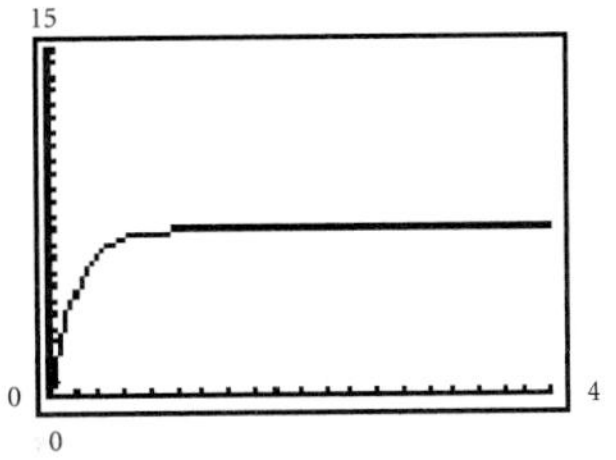

(c) The maximum current is 12 amperes.
(d) 3.3 amperes, 5.3 amperes, 9.4 amperes
(e)

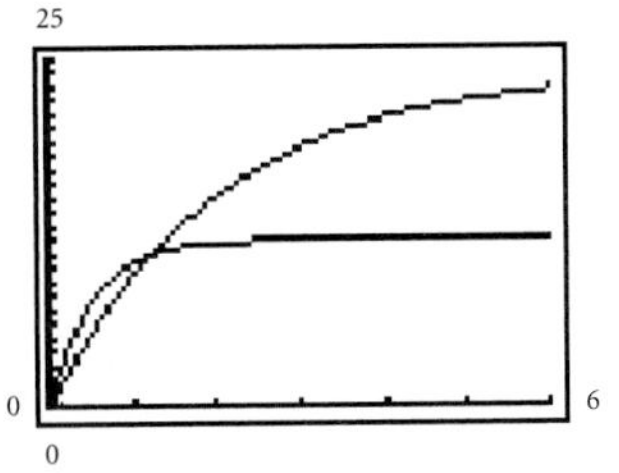

(f) The maximum current is 24 amperes.

71.

n	$2 + \frac{1}{2!} + \frac{1}{3!} + \cdots + \frac{1}{n!}$	Difference
4	2.7083333333	0.0099484951
6	2.7180555556	0.0002262729
8	2.7182787698	0.0000030586
10	2.7182818011	0.0000000273

73.
$$\frac{f(x+h) - f(x)}{h} = \frac{a^{x+h} - a^x}{h} = \frac{a^x a^h - a^x}{h} = a^x\left(\frac{a^h - 1}{h}\right)$$

75. $f(-x) = a^{-x}$

$= (a^x)^{-1}$

$= \dfrac{1}{a^x}$

$= \dfrac{1}{f(x)}$

77. (a) The relative humidity is 71%.
(b) The relative humidity is 73%.
(c) The relative humidity is 100%.

79. (a) $\sinh(-x) = \dfrac{e^{-x} - e^{-(-x)}}{2}$

$= \dfrac{e^{-x} - e^{x}}{2}$

$= -\dfrac{e^{x} - e^{-x}}{2}$

$= -\sinh x$

(b)

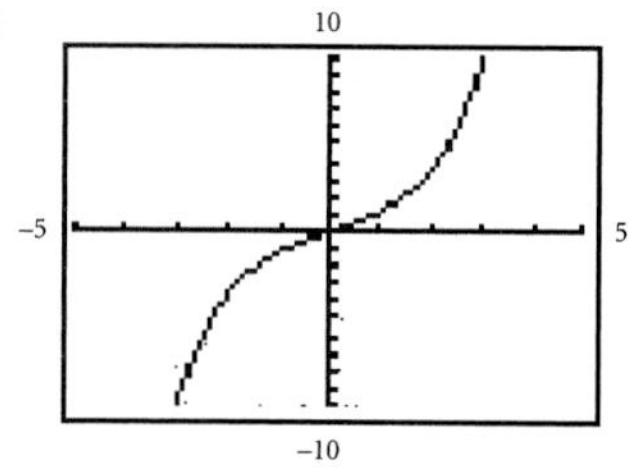

Exercise 11.4 (p. 650)

1. $\log_3 9 = 2$ **3.** $\log_a 1.6 = 2$ **5.** $\log_{1.1} M = 2$

7. $\log_2 7.2 = x$ **9.** $\log_x \pi = \sqrt{2}$ **11.** $\ln 8 = x$

13. $2^3 = 8$ **15.** $a^6 = 3$ **17.** $3^x = 2$

19. $2^{1.3} = M$ **21.** $(\sqrt{2})^x = \pi$ **23.** $e^x = 4$

25. 0 **27.** 2 **29.** −4

31. $\frac{1}{2}$ **33.** 4 **35.** $\frac{1}{2}$

37. $\{x \mid x > 3\}$ **39.** $\{x \mid x \neq 0\}$ **41.** $\{x \mid x > 0\}$

43. $\{x \mid x \geq 1\}$ **45.** 0.511 **47.** 30.099

49. $a = \sqrt{2}$

51.

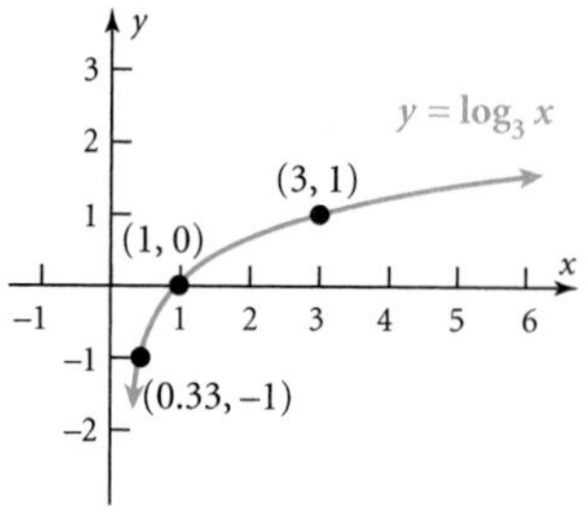

53.

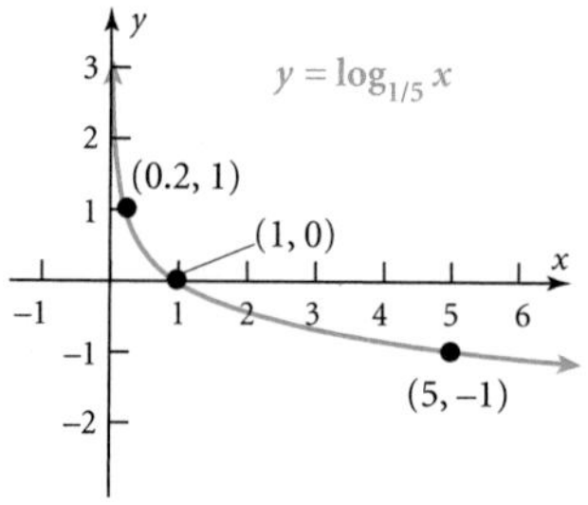

55. B. **57.** D. **59.** A.

61. E.

63.

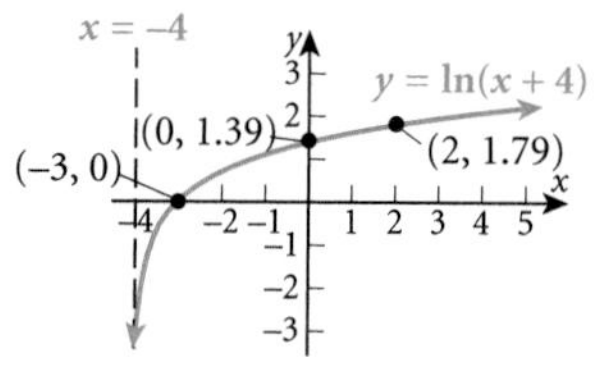

Domain $= (-4, \infty)$, Range $= (-\infty, \infty)$; the vertical asymptote is $x = -4$.

65.

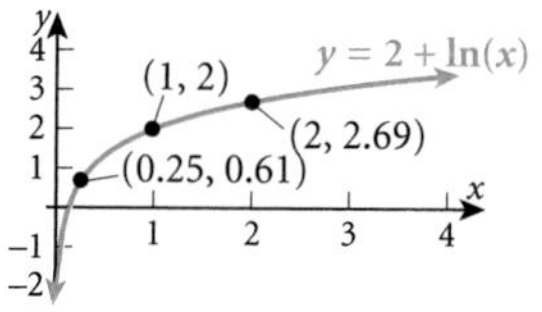

Domain $= (0, \infty)$, Range $= (-\infty, \infty)$; the vertical asymptote is $x = 0$.

67.

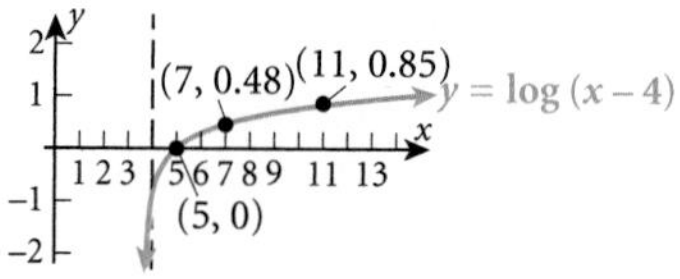

Domain $= (4, \infty)$, Range $= (-\infty, \infty)$; the vertical asymptote is $x = 4$.

69.

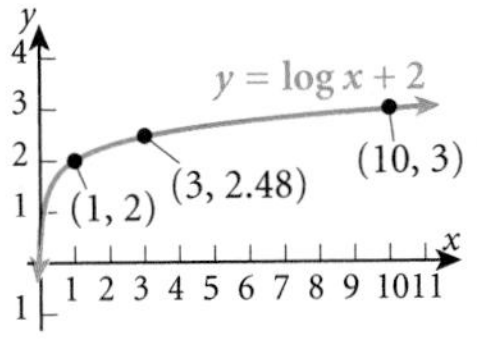

Domain = $(0, \infty)$, Range = $(-\infty, \infty)$; the vertical asymptote is $x = 0$.

71. $\{9\}$ **73.** $\left\{\frac{7}{2}\right\}$ **75.** $\{2\}$

77. $\{5\}$ **79.** $\{3\}$ **81.** $\{2\}$

83. $\left\{\frac{\ln 10}{3}\right\}$ **85.** $\left\{\frac{\ln 8 - 5}{2}\right\}$ **87.** $\{-2\sqrt{2}, 2\sqrt{2}\}$

89. $\{-1\}$

91. (a) 1 (b) 2 (c) 3
(d) The pH increases as the hydrogen ion concentration decreases.
(e) $[H^+] = 0.000316$
(f) $[H^+] = 3.981 \times 10^{-8}$

93. (a) The aircraft is 5.965 km above sea level.
(b) The height of the mountain is 0.900 km above sea level.

95. (a) It will take 6.931 minutes for the probability to reach 50%.
(b) It will take 16.094 minutes for the probability to reach 80%.

97. The time between injections is about 2.29 hours (2 hours, 17 minutes).

99. It takes 0.269 seconds to achieve a current of 0.5 ampere and 0.896 seconds to achieve a current of 1.0 ampere.

101. The population will be 309,124,000 people.

103. 50 decibels **105.** 110 decibels

107. The magnitude of the earthquake was 8.1 on the Richter scale.

109. (a) $k = \frac{50}{3} \ln\left(\frac{10}{3}\right) \approx 20.066$
(b) The risk is 90.9%.
(c) A blood alcohol concentration of 0.175 corresponds to a risk of 100%.
(d) A driver should be arrested with a blood alcohol concentration of 0.08 or greater.

Exercise 11.5 (p. 661)

1. 71 **3.** -4 **5.** 7

7. 1 **9.** 1 **11.** 2

13. $\frac{5}{4}$ **15.** 4 **17.** $a + b$

19. $b - a$ **21.** $3a$ **23.** $\frac{a + b}{5}$

25. $2 + \log_5 x$ **27.** $3 \log_2 z$

29. $1 + \ln x$ **31.** $\ln x + x$ **33.** $2 \log_a u + 3 \log_a v$

35. $2 \ln x + \frac{1}{2} \ln(1 - x)$

37. $3 \log_2 x - \log_2(x - 3)$

39. $\log x + \log(x + 2) - 2 \log(x + 3)$

41. $\frac{1}{3} \ln(x - 2) + \frac{1}{3} \ln(x + 1) - \frac{2}{3} \ln(x + 4)$

43. $\ln 5 + \ln x + \frac{1}{2} \ln(1 + 3x) - 3 \ln(x - 4)$

45. $\log_5(u^3 v^4)$ **47.** $-\log_3 x^{5/2}$

49. $\log_4 \frac{x - 1}{(x + 1)^4}$ **51.** $-2 \ln(x - 1)$

53. $\log_2[x(3x - 2)^4]$ **55.** $\log_a\left(\frac{25x^6}{\sqrt{2x + 3}}\right)$

57. $\log_2\left(\frac{(x + 1)^2}{x + 2x - 3}\right)$ **59.** 2.771

61. -3.880 **63.** 5.615 **65.** 0.874

67.

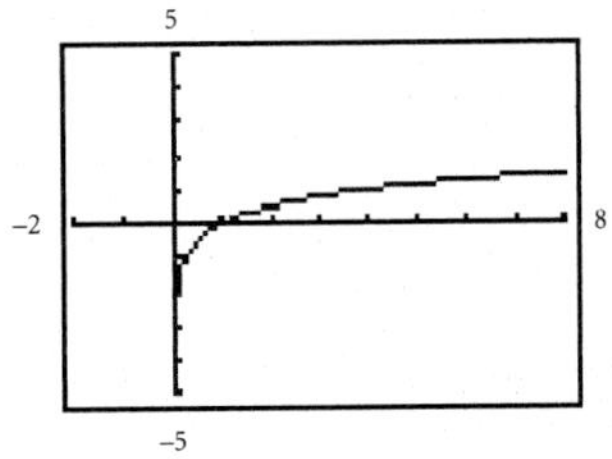

69.

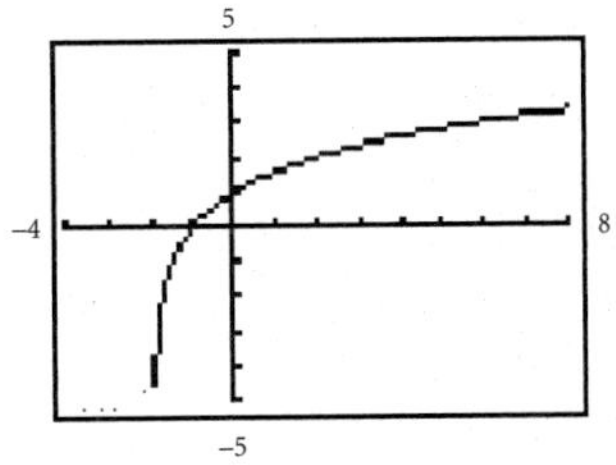

71.

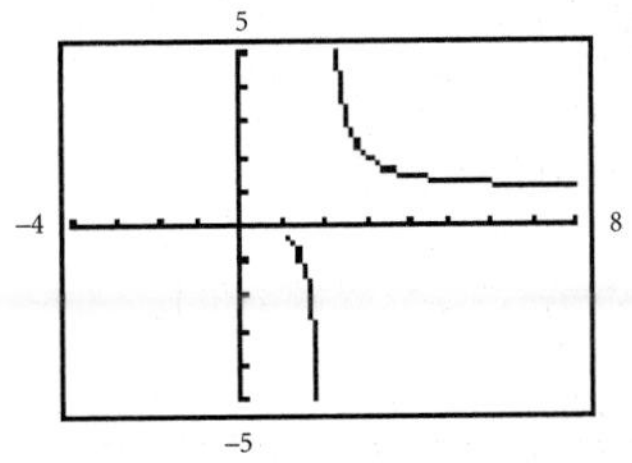

73. $y = Cx$ **75.** $y = C(x^2 + x)$

77. $y = Ce^{3x}$ **79.** $y = Ce^{-4x} + 3$

81. $y = \left(\frac{C\sqrt{2x+1}}{(x+4)^{1/3}}\right)^{1/3}$ **83.** 3 **85.** 1

87. $\log_a(x + \sqrt{x^2 - 1}) + \log_a(x - \sqrt{x^2 - 1})$

$= \log_a[(x + \sqrt{x^2-1})(x - \sqrt{x^2-1})]$

$= \log_a[x^2 - (x^2 - 1)]$

$= \log_a(1)$

$= 0$

89. $\ln(1 + e^{2x}) = \ln(e^{2x}(e^{-2x} + 1))$

$= \ln e^{2x} + \ln(1 + e^{-2x})$

$= 2x + \ln(1 + e^{-2x})$

91. $-f(x) = -\log_a x$

$= -\frac{\ln x}{\ln a}$

$= \frac{\ln x}{-\ln a}$

$= \frac{\ln x}{\ln a^{-1}}$

$= \frac{\ln x}{\ln(1/a)}$

$= \log_{1/a} x$

93. $f\left(\frac{1}{x}\right) = \log_a\left(\frac{1}{x}\right)$

$= \log_a x^{-1}$

$= -\log_a x$

$= -f(x)$

95. $\log_a\left(\frac{M}{N}\right) = \log_a(MN^{-1})$

$= \log_a M + \log_a N^{-1}$

$= \log_a M - \log_a N$

Exercise 11.6 (p. 666)

1. \$1127.50 **3.** \$580.92 **5.** \$98.02

7. \$466.20

9. The amount is \$1020.20, and the interest is \$20.20.

11. (a) \$4434.60 (b) \$3933.14

13. A 23.1% interest rate is required.

15. It will take approximately 11 years for the investment to triple.

17. \$913.93 is needed to get \$1000 in 1 year. \$835.27 is needed to get \$1000 in 2 years.

19. They should invest \$35,476.82.

21. A 22.0% interest rate is required.

23. (a) The Rule of 70 approximation is 70 years, which is greater than the actual solution of 69.3147 years by about 0.685 year.
(b) The Rule of 70 approximation is 14 years, which is greater than the actual solution of 13.8629 years by about 0.137 year.
(c) The Rule of 70 approximation is 7 years, which is greater than the actual solution of 6.9315 years by about 0.069 year.

CHAPTER 11 Review

True – False Items (p. 670)

1. True **2.** False **3.** True
4. True **5.** False **6.** False
7. False **8.** True **9.** False

Fill in the Blanks (p. 670)

1. parabola **2.** axis of symmentry
3. $-\frac{b}{2a}$ **4.** $(0, 1), (1, a), (-1, \frac{1}{a})$
5. 1 **6.** 4 **7.** $(0, \infty)$
8. $(1, 0), (a, 1), (\frac{1}{a}, -1)$ **9.** 1
10. $r \log_a M$

Review Exercises (p. 670)

1.

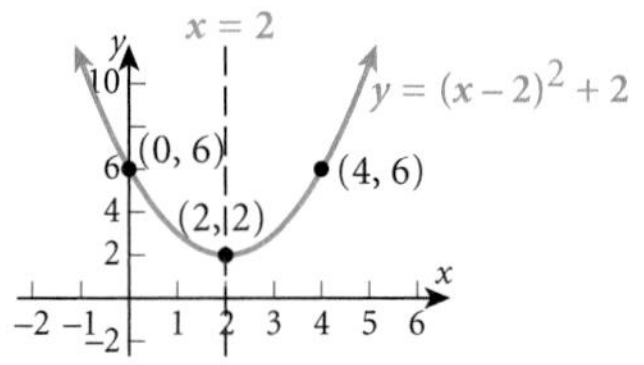

The graph opens up. The vertex is (2, 2). The axis of symmetry is the line $x = 2$. The y-intercept is (0, 6). There is no x-intercept.

3.

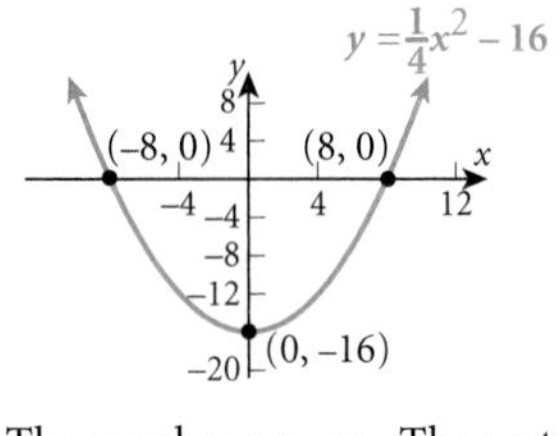

The graph opens up. The vertex is $(0, -16)$. The axis of symmetry is the line $x = 0$. The y-intercept is $(0, -16)$. The x-intercepts are $(-8, 0)$ and $(8, 0)$.

5.

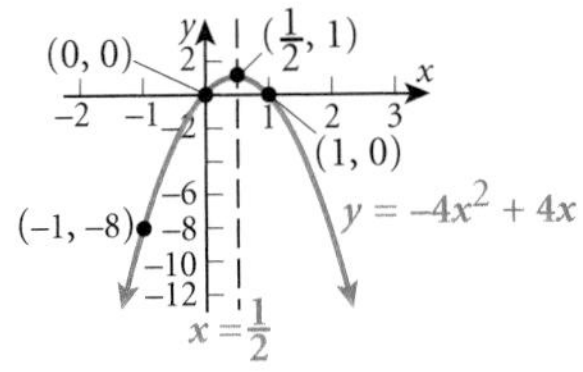

The graph opens down. The vertex is $\left(\frac{1}{2}, 1\right)$. The axis of symmetry is the line $x = \frac{1}{2}$. The y-intercept is $(0, 0)$. The x-intercepts are $(0, 0)$ and $(1, 0)$.

7.

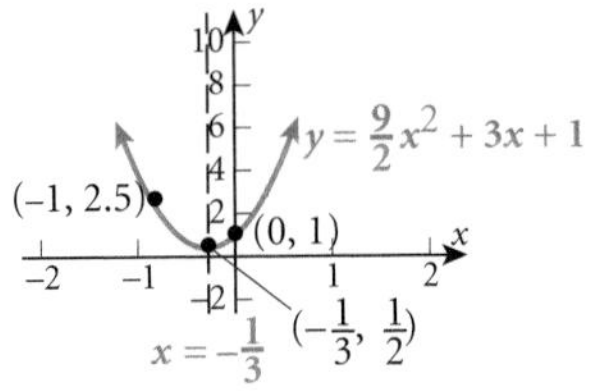

The graph opens up. The vertex is $\left(-\frac{1}{3}, \frac{1}{2}\right)$. The axis of symmetry is the line $x = -\frac{1}{3}$. The y-intercept is $(0, 1)$. There is no x-intercept.

9.

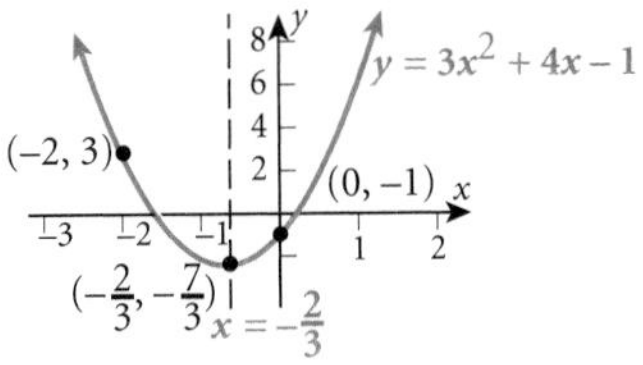

The graph opens up. The vertex is $\left(-\frac{2}{3}, -\frac{7}{3}\right)$. The axis of symmetry is the line $x = -\frac{2}{3}$. The y-intercept is $(0, -1)$. The x-intercepts are $\left(\frac{-2 - \sqrt{7}}{3}, 0\right)$ and $\left(\frac{-2 + \sqrt{7}}{3}, 0\right)$ or approximately $(-1.55, 0)$ and $(0.22, 0)$.

11. Minimum value = 1 **13.** Maximum value = 12

15. Maximum value = 16

17. Answers will vary. Possibilities include $(-1, 1)$, $(0, 0)$, and $(1, 1)$.

19.

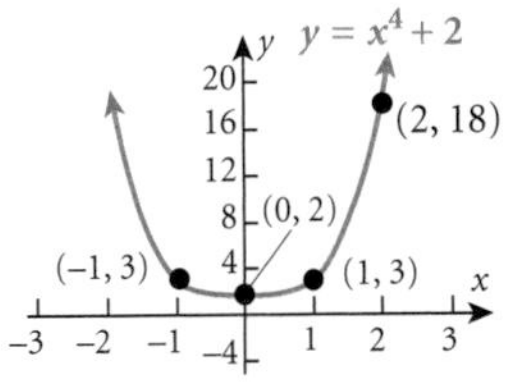

21.

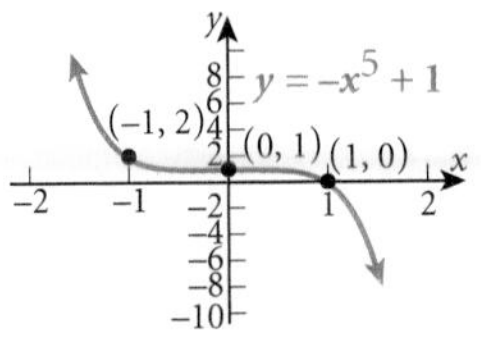

23. This is a polynomial function of degree 5.

25. This is not a polynomial function because the exponent in the term $5x^{1/2}$ is not a nonnegative integer.

27. $y = -2x^4$

29. The domain is $\{x|x \neq -3, x \neq 3\}$.

31. The domain is $\{x|x \neq -2\}$.

33. (a) 81 (b) 2 (c) $\frac{1}{9}$ (d) -3

35. $\log_5 z = 2$ **37.** $5^{13} = u$ **39.** $\{x|x > \frac{2}{3}\}$

41. $\{x|x < \frac{2}{3}\}$ **43.** -3 **45.** 4

47. 2 **49.** $\sqrt{2}$ **51.** 0.4

53. $\log_3 u + 2\log_3 v - \log_3 w$ **55.** $2\log x + \frac{1}{2}\log(x^3 + 1)$

57. $\ln x + \frac{1}{3}\ln(x^2 + 1) - \ln(x - 3)$ **59.** $\log_4 x^{25/4}$

61. $-2\ln(x + 1)$ **63.** $\ln\left(\frac{4x^3}{\sqrt{x^2 + x - 6}}\right)$

65. 2.124

67.

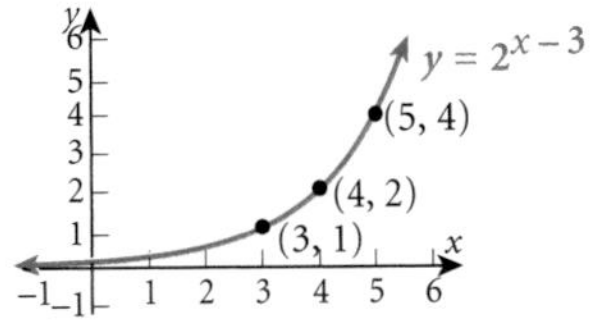

Domain = $(-\infty, \infty)$, Range = $(0, \infty)$; the x-axis is a horizontal asymptote.

69.

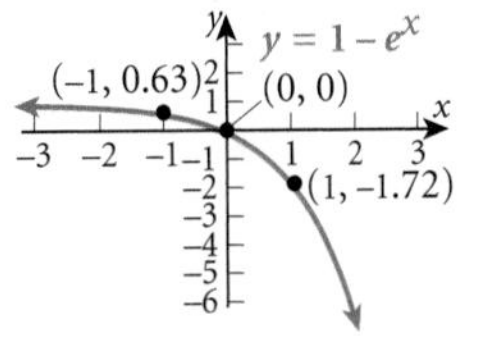

Domain = $(-\infty, \infty)$, Range = $(-\infty, 1)$; the line $y = 1$ is a horizontal asymptote.

71.

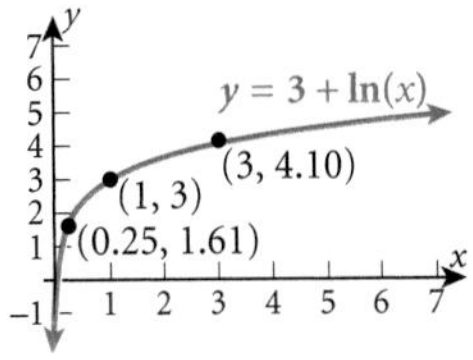

Domain = $(0, \infty)$, Range = $(-\infty, \infty)$; the y-axis is a vertical asymptote.

73. $\left\{\frac{1}{4}\right\}$ **75.** $\left\{\frac{-1 - \sqrt{3}}{2}, \frac{-1 + \sqrt{3}}{2}\right\}$

77. $\left\{\frac{1}{4}\right\}$ **79.** $\left\{\frac{12}{5}\right\}$ **81.** $\{11\}$

83. \$125.23 **85.** \$923.12

87. It will take almost 11.6 years to double.

89. The Piper Cub is 3229.5 meters above sea level.

91. 50 feet by 50 feet

93. 25 feet by $\frac{50}{\pi}$ feet $\approx$ 25 feet by 15.92 feet

95. (a) The limiting magnitude is 11.77.
(b) A diameter of 9.56 inches is required.

97. (a) The annual interest rate was 10.436%.
(b) The actual value will be $32,249.24.

99. (a) 63 clubs should be manufactured.
(b) The marginal cost is $151.90.

Mathematical Questions from Professional Exams (p. 675)

1. (a) or (e) **2.** (e) I, II, and III

3. (c) $\ln(y + \sqrt{y^2 + 1})$ **4.** (e) 125

5. (b) 1 **6.** (a) 2

7. (a) 3

CHAPTER 12 The Limit of a Function

Exercise 12.1 (p. 681)

1.

x	0.9	0.99	0.999
$f(x) = 2x$	1.8	1.98	1.998
x	1.1	1.01	1.001
$f(x) = 2x$	2.2	2.02	2.002

$\lim_{x \to 1} f(x) = 2$

3.

x	-0.1	-0.01	-0.001
$f(x) = x^2 + x$	2.01	2.0001	2.000001
x	0.1	0.01	0.001
$f(x) = x^2 + x$	2.01	2.0001	2.000001

$\lim_{x \to 0} f(x) = 2$

5.

x	-2.1	-2.01	-2.001
$f(x) = \dfrac{x^2 - 4}{x + 2}$	-4.1	-4.01	-4.001
x	-1.9	-1.99	-1.999
$f(x) = \dfrac{x^2 - 4}{x + 2}$	-3.9	-3.99	-3.999

$\lim_{x \to -2} f(x) = -4$

7.

x	-1.1	-1.01	-1.001
$f(x) = \dfrac{x^3 + 1}{x + 1}$	3.31	3.0301	3.003001
x	-0.9	-0.99	-0.999
$f(x) = \dfrac{x^3 + 1}{x + 1}$	2.71	2.9701	2.997001

$\lim_{x \to -1} f(x) = 3$

9. 32 **11.** 1 **13.** 4

15. 2 **17.** 3 **19.** 4

21. The limit does not exist.

23.

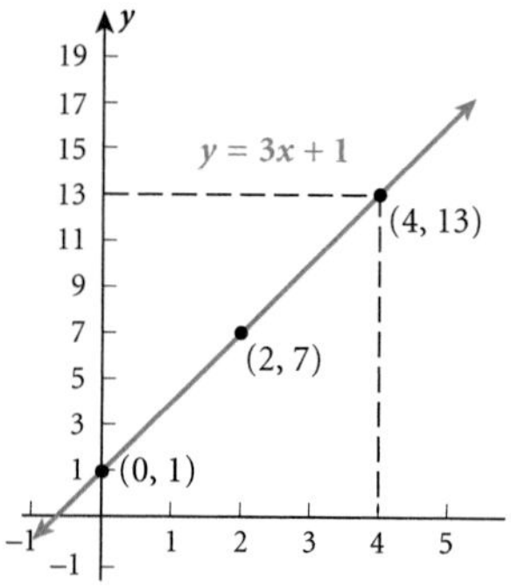

$\lim_{x \to 4} f(x) = 13$

25.

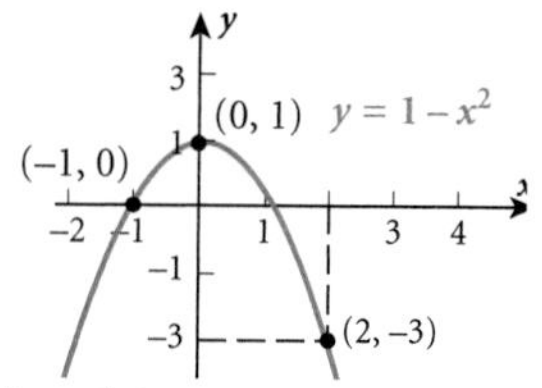

$\lim_{x \to 2} f(x) = -3$

27.

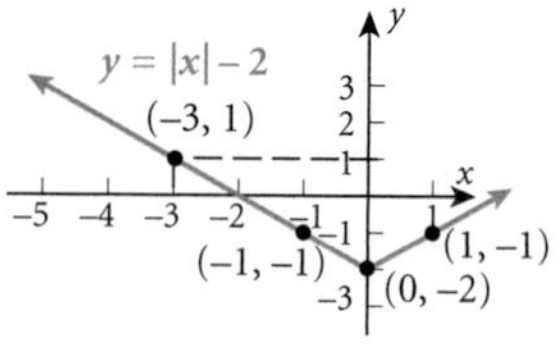

$\lim_{x \to -3} f(x) = 1$

29.

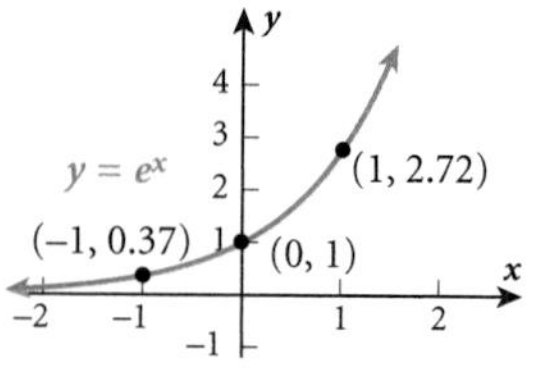

$\lim_{x \to 0} f(x) = 1$

31.

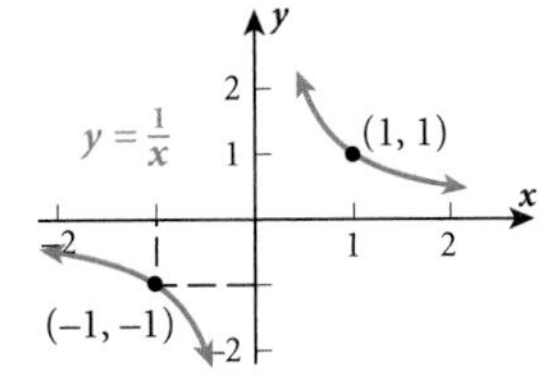

$\lim_{x \to -1} f(x) = -1$

33.

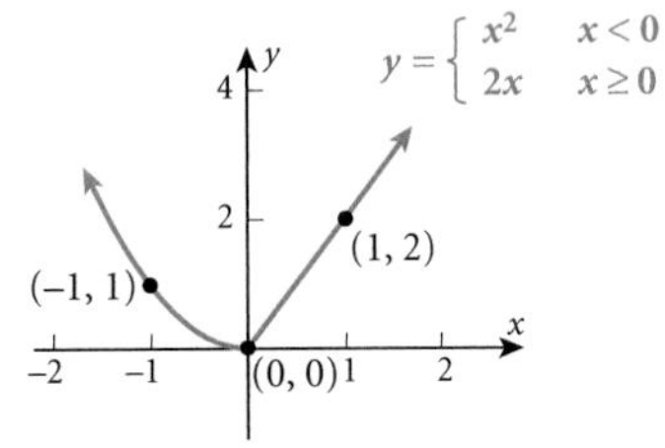

$\lim_{x \to 0} f(x) = 0$

35.

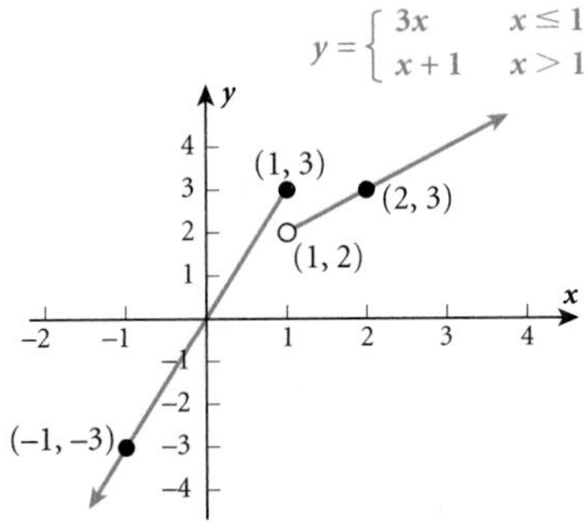

The limit does not exist.

37.

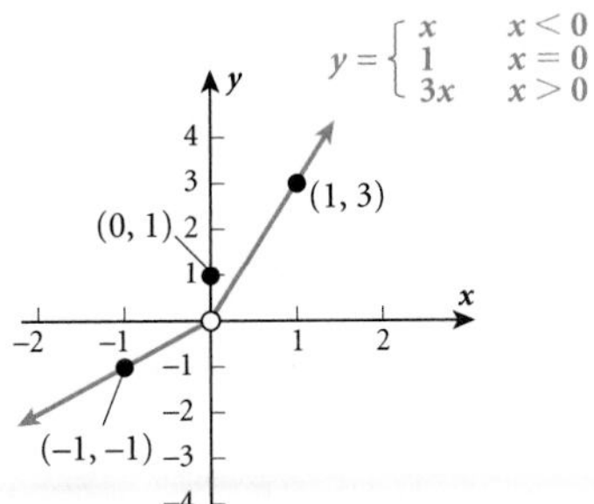

$\lim_{x \to 0} f(x) = 0$

39.

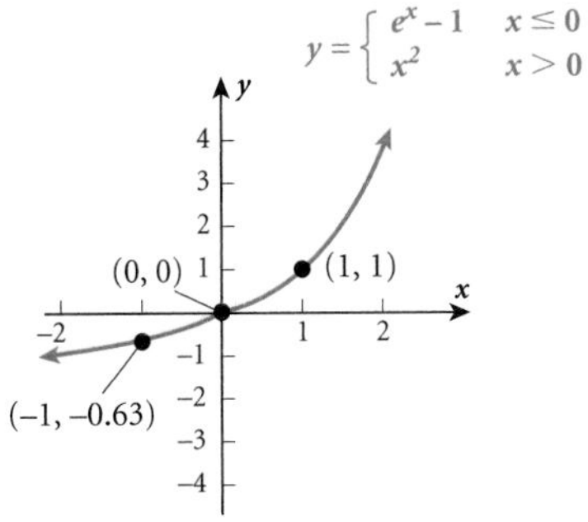

$\lim_{x \to 0} f(x) = 0$

41. 0.67 **43.** 1.6 **45.** 0

Exercise 12.2 (p. 690)

1. 5 **3.** 4 **5.** 8
7. 8 **9.** -1 **11.** 8
13. 3 **15.** -1 **17.** 32
19. 2 **21.** $\frac{7}{6}$ **23.** 3
25. 0 **27.** $\frac{2}{3}$ **29.** $\frac{8}{5}$
31. 0 **33.** 5 **35.** 6
37. -1 **39.** 0 **41.** -1
43. $\frac{1}{4}$ **45.** 10 **47.** 8
49. $\frac{4}{5}$ **51.** 10

Exercise 12.3 (p. 696)

1. $[-8, -3)$ or $(-3, 4)$ or $(4, 6]$ **3.** $(-8, 0), (-5, 0)$
5. $f(-8) = 0, f(-4) = 2$ **7.** 3
9. 2 **11.** 1
13. Yes, $\lim_{x \to 4} f(x) = 0$ **15.** No. **17.** Yes.
19. No. **21.** 5 **23.** 7
25. 1 **27.** 4 **29.** $-\frac{2}{3}$
31. $\frac{3}{2}$ **33.** Continuous **35.** Continuous
37. Discontinuous **39.** Discontinuous **41.** Discontinuous
43. Continuous **45.** Discontinuous **47.** Continuous
49. f is continuous on the interval $(-\infty, \infty)$.
51. f is continuous on the interval $(-\infty, \infty)$.
53. f is continuous on the interval $(0, \infty)$. f is not discontinuous for any numbers in its domain.
55. f is continuous on the interval $(-\infty, \infty)$.

57. f is continuous for all numbers in the set $\{x|x \neq -2, x \neq 2\}$. f is discontinuous at $x = -2$ and $x = 2$.

59. f is continuous on the intervals $(0, 1)$ and $(1, \infty)$. f is discontinuous at $x = 1$.

61. f is continuous for all number in the set $\{x|x \neq 0\}$. f is discontinuous at $x = 0$.

63. (a) $\lim_{x \to 350^-} C(x) = 39.99$
(b) $\lim_{x \to 350^+} C(x) = 39.99$
(c) The function $C(x)$ is continuous at $x = 350$.

65. (a) $W(v) =$

$$\begin{cases} 10 & 0 \le v < 1.79 \\ 33 - \dfrac{23(10.45 + 10\sqrt{v} - v)}{22.04} & 1.79 \le v \le 20 \\ -3.7034 & v > 20 \end{cases}$$

(b) $\lim_{v \to 0^+} W(v) = 10$

(c) $\lim_{v \to 1.79^-} W(v) = 10$

(d) $\lim_{v \to 1.79^+} W(v) = \dfrac{26{,}407}{1102} - \dfrac{575\sqrt{179}}{551} \approx 10.00095$

(e) $W(1.79) = \dfrac{26{,}407}{1102} - \dfrac{575\sqrt{179}}{551} \approx 10.00095$

(f) W is not continuous at $v = 1.79$ since $10 \neq 10.00095$.

(g) $\lim_{v \to 1.79^-} W(v) = 10.00$; $\lim_{v \to 1.79^+} W(v) = 10.00$; $W(1.79) = 10.00$; W is continuous at $v = 1.79$.

(i) $\lim_{v \to 20^-} W(v) = \dfrac{94{,}697}{2204} - \dfrac{11{,}500\sqrt{5}}{551} \approx -3.70332$

(j) $\lim_{v \to 20^+} W(v) = -3.7034$

(k) $W(20) = \dfrac{94{,}697}{2204} - \dfrac{11{,}500\sqrt{5}}{551} \approx -3.70332$

(l) W is not continuous at $v = 20$ since $-3.70332 \neq -3.7034$.

(m) $\lim_{v \to 20^-} W(v) = -3.70$; $\lim_{v \to 20^+} W(v) = -3.70$; $W(20) = -3.70$; W is continuous at $v = 20$.

Exercise 12.4 (p. 705)

1. 1 **3.** 2 **5.** 3

7. 0 **9.** ∞ **11.** $-\infty$

13. ∞ **15.** ∞ **17.** ∞

19. ∞

21. The horizontal asymptote is $y = 3$. The vertical asymptote is $x = 0$.

23. The horizontal asymptote is $y = 2$. The vertical asymptote is $x = 1$.

25. The horizontal asymptote is $y = 1$. The vertical asymptotes are $x = -2$ and $x = 2$.

27. (a) $\{x|x \neq 6\}$
(b) $[0, \infty)$
(c) The x-intercepts are $(-4, 0)$, and $(0, 0)$. The y-intercept is $(0, 0)$.
(d) $f(-2) = 2$ (e) $x = 4$ or $x = 8$
(f) f is discontinuous at $x = 6$.
(g) $x = 6$ (h) $y = 4$
(i) There is a local maximum of 2 at $x = -2$.
(j) There are local minima of 0 at $x = -4$ and $x = 0$ and a local minimum of 4 at $x = 8$.
(k) f is increasing on $(-4, -2)$ or $(0, 6)$ or $(4, \infty)$
(l) f is decreasing on $(-\infty, -4)$ or $(-2, 0)$ or $(6, 8)$
(m) 4 (n) ∞
(o) ∞ (p) ∞

29. $\lim_{x \to -1^-} R(x) = -\infty$, and $\lim_{x \to -1^+} R(x) = \infty$, so the graph of R will have a vertical asymptote at $x = -1$. $\lim_{x \to 1} R(x) = \frac{1}{2}$, but R is not defined at $x = 1$, so the graph of R will have a hole at $(1, \frac{1}{2})$.

31. $\lim_{x \to -1} R(x) = \frac{1}{2}$, but R is not defined at $x = -1$, so the graph of R will have a hole at $(-1, \frac{1}{2})$. $\lim_{x \to 1^-} R(x) = -\infty$ and $\lim_{x \to 1^+} R(x) = \infty$, so the graph of R will have a vertical asymptote at $x = 1$.

33. $R(x)$ is undefined at $x = 1$, where a hole appears at $(1, \frac{2}{9})$, and at $x = -2$, where a vertical asymptote occurs.

35. $R(x)$ is undefined at $x = 2$, where a hole appears at $(2, \frac{8}{5})$, and at $x = -3$, where a vertical asymptote occurs.

37. $R(x)$ is undefined at $x = -1$, where a hole appears at $(-1, 1)$.

39. (a) $C(x) = 79{,}000 + 10x$
(b) $\{x|x \ge 0\}$
(c) $\overline{C}(x) = \dfrac{79{,}000}{x} + 10$
(d) $\{x|x > 0\}$
(e) $\lim_{x \to 0^+} \overline{C}(x) = \infty$; the average cost of producing a few calculators is very high due to the fixed costs. Therefore, the average cost of producing no calculators is unbounded.
(f) $\lim_{x \to \infty} \overline{C}(x) = 10$; the more calculators that are produced, the closer the average cost gets to \$10 per calculator.

41. (a) $\lim_{x \to 100^-} C(x) = \infty$

CHAPTER 12   Review

True – False Items (p. 709)

1. True   **2.** False   **3.** True

4. True   **5.** True   **6.** True

7. True

Fill in the Blanks (p. 709)

1. $\lim\limits_{x \to c} f(x) = N$ **2.** equals   **3.** not exist

4. continuous   **5.** $\neq$ **6.** equals

7. $y = 2$, horizontal

Review Exercises (p. 709)

1. 12

3.

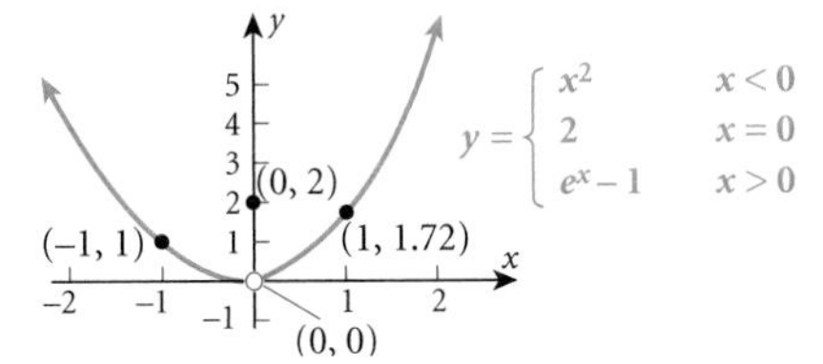

$$y = \begin{cases} x^2 & x < 0 \\ 2 & x = 0 \\ e^x - 1 & x > 0 \end{cases}$$

$\lim\limits_{x \to 0} f(x) = 0$

5. 9   **7.** 25   **9.** 4

11. 0   **13.** 64   **15.** -16

17. $\frac{1}{3}$ **19.** $\frac{6}{7}$ **21.** 0

23. $\frac{3}{2}$ **25.** $\frac{28}{11}$ **27.** $\frac{5}{3}$

29. $-\infty$ **31.** $-\infty$ **33.** ∞

35. Continuous   **37.** Discontinuous   **39.** Discontinuous

41. Continuous

43. The line $y = 0$ is a horizontal asymptote. The lines $x = -1$ and $x = 1$ are vertical asymptotes.

45. The $y = 5$ is a horizontal asymptote. The line $x = -2$ is a vertical asymptote.

47. (a) $(-\infty, 2)$ or $(2, 5)$ or $(5, \infty)$
(b) $(-\infty, \infty)$

(c) $(-2, 0), (0, 0), (1, 0), (6, 0)$

(d) $(0, 0)$

(e) $f(-6) = 2, f(-4) = 1$

(f) $f(-2) = 0, f(6) = 0$

(g) $\lim\limits_{x \to -4^-} f(x) = 4, \lim\limits_{x \to -4^+} f(x) = -2$

(h) $\lim\limits_{x \to -2^-} f(x) = -2, \lim\limits_{x \to -2^+} f(x) = 2$

(i) $\lim\limits_{x \to 5^-} f(x) = 2, \lim\limits_{x \to -5^+} f(x) = 2$

(j) $\lim\limits_{x \to 0} f(x)$ does not exist.

(k) $\lim\limits_{x \to 2} f(x)$ does not exist.

(l) No   (m) No   (n) No

(o) No   (p) Yes   (q) No

(r) $(-6, -4)$ or $(-2, 0)$ or $(6, \infty)$

(s) $(-\infty, -6)$ or $(0, 2)$ or $(2, 5)$ or $(5, 6)$

(t) $\lim\limits_{x \to -\infty} f(x) = \infty, \lim\limits_{x \to \infty} f(x) = 2$

(u) There is no local maximum. There is a local minimum of 2 at $x = -6$, a local minimum of 0 at $x = 0$, and a local minimum of 0 at $x = 6$.

(v) The line $y = 2$ is a horizontal asymptote. The line $x = 2$ is a vertical asymptote.

49. -11 **51.** $-\dfrac{1}{4}$

53. The graph has a hole at $(-4, -\frac{1}{8})$ because $\lim\limits_{x \to -4} R(x) = -\frac{1}{8}$, and R is not defined at $x = -4$. The graph has a vertical asymptote at $x = 4$ because $\lim\limits_{x \to 4^-} R(x) = -\infty$ and $\lim\limits_{x \to 4^+} R(x) = \infty$.

55. $R(x)$ is undefined at $x = 2$ and $x = 9$. A hole appears at $(2, -\frac{8}{7})$, and a vertical asymptote appears at $x = 9$.

59. (a) $\lim\limits_{x \to \infty} S(x) = \frac{4000}{7} \approx 571.4$

(b) For larger and larger advertising expenditures, the sales level will eventually level off at 571 units.

Mathematical Questions from Professional Exams (p. 712)

1. (b) $\frac{5}{6}$ **2.** (e) II, III   **3.** (d) $\dfrac{\sqrt{2}}{4}$

4. (e) $x \neq -1$

CHAPTER 13 The Derivative of a Function

Exercise 13.1 (p. 723)

1. $m_{tan} = 3, y = 3x + 5$

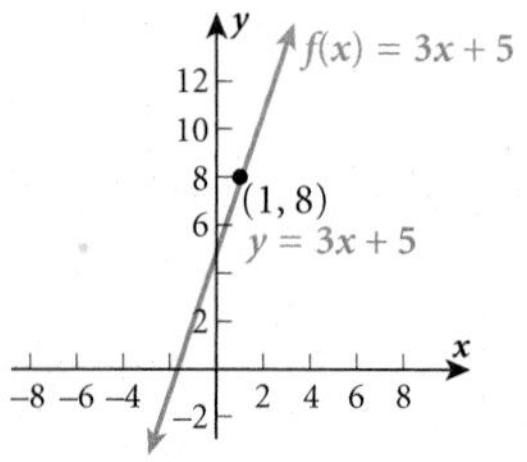

3. $m_{tan} = -2, y = -2x + 1$

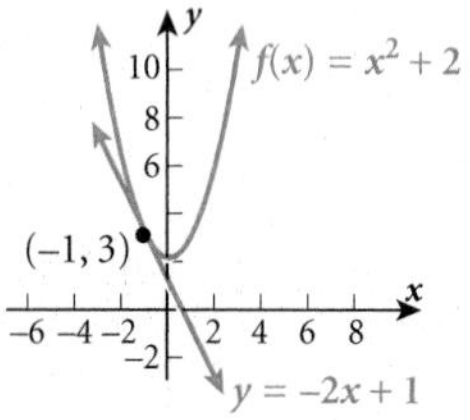

5. $m_{tan} = 12, y = 12x - 12$

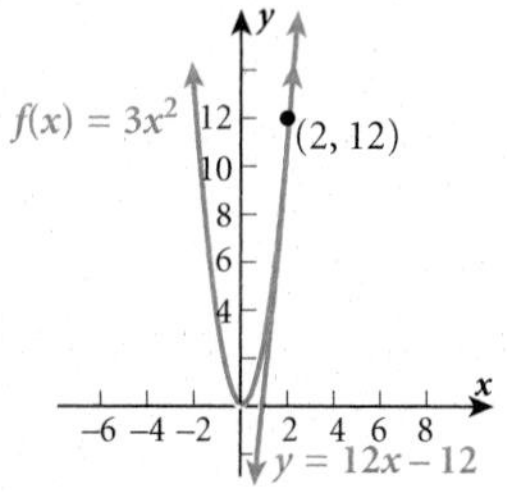

7. $m_{tan} = 5, y = 5x - 2$

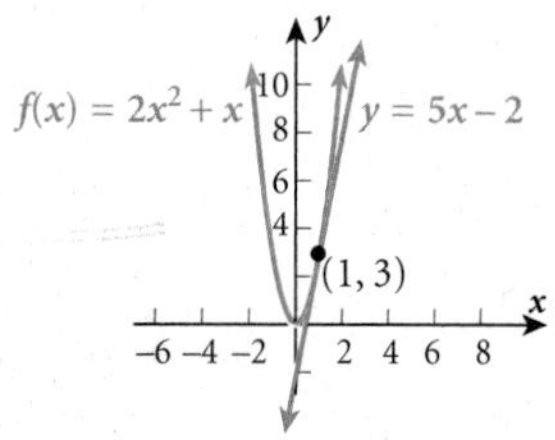

9. $m_{tan} = -4, y = -4x + 2$

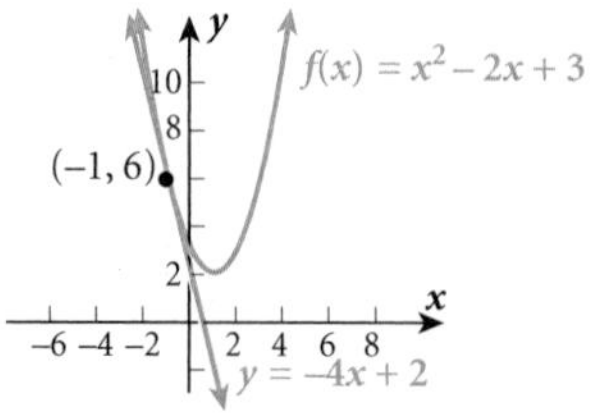

11. $m_{tan} = 1, y = x + 1$

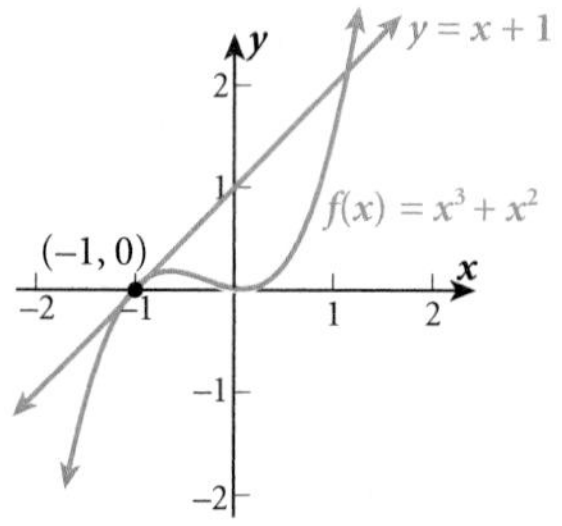

13. $f'(3) = -4$
15. $f'(0) = 0$
17. $f'(1) = 7$
19. $f'(0) = 4$
21. $f'(1) = 3$
23. $f'(1) = -1$
25. $f'(x) = 2$
27. $f'(x) = -2$
29. $f'(x) = 2x$
31. $f'(x) = 6x - 2$
33. $f'(x) = 3x^2$
35. $f'(x) = m$
37. (a) The average rate of change is 3.
 (b) The instantaneous rate of change at $x = 1$ is 3.
39. (a) The average rate of change is 12.
 (b) The instantaneous rate of change at $x = 1$ is 6.
41. (a) The average rate of change is 6.
 (b) The instantaneous rate of change at $x = 1$ is 4.
43. (a) The average rate of change is 7.
 (b) The instantaneous rate of change at $x = 1$ is 3.
45. $f'(-2) = 60$
47. $f'(8) = -\frac{3527}{4107} \approx -0.859$
49. $f'(0) = 1$
51. $f'(1) = 3e \approx 8.155$
53. $f'(1) = 0$
55. No.
57. The pilot should release the bomb at the point (2, 4).
59. (a) The average rate of change in sales is 74 tickets per day.
 (b) The average rate of change in sales is 94 tickets per day.
 (c) The average rate of change in sales is 110 tickets per day.
 (d) The instantaneous rate of change of sales on day 5 is 90 tickets per day.
 (e) The instantaneous rate of change of sales on day 10 is 130 tickets per day.

61. (a) The farmer is willing to supply 4500 crates for \$10 per crate.
(b) The farmer is willing to supply 7800 crates for \$13 per crate.
(c) The average rate of change in supply is 1100 crates per dollar.
(d) The instantaneous rate of change is 950 crates per dollar.

63. (a) $R'(x) = 8 - 2x$
(b) $C'(x) = 2$
(c) The break-even points are $x = 1$ and $x = 5$.
(d) $x = 3$
(e)

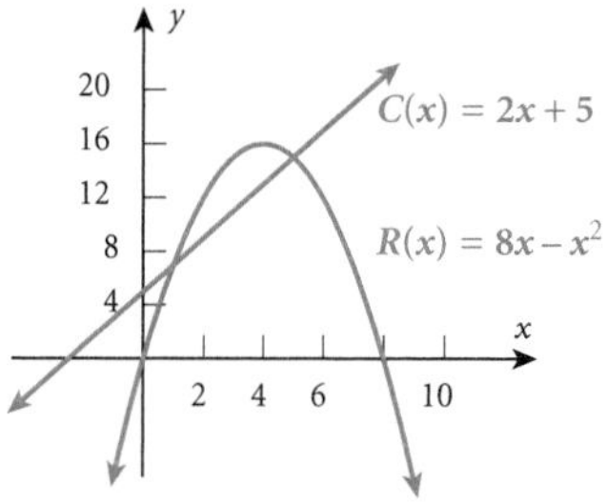

65. (a) $R(x) = -10x^2 + 2000x$
(b) $R'(x) = -20x + 2000$
(c) $R'(100) = \$0$
(d) The average rate of change in revenue is -10 dollars per ton.

67. (a) $R(x) = 90x - 0.02x^2$
(b) $R'(x) = 90 - 0.04x$
(c) $C'(x) = 10$
(d) The break-even points are $x = 0$ and $x = 4000$.
(e) The marginal revenue equals marginal cost at a production level of 2000 units.

69. The instantaneous rate of change of the volume with respect to the radius at $r = 3$ is $18\pi \approx 56.55$ cubic feet per foot.

Exercise 13.2 (p. 732)

1. $f'(x) = 0$ **3.** $f'(x) = 3x^2$ **5.** $f'(x) = 12x$

7. $f'(t) = t^3$ **9.** $f'(t) = 2x + 1$

11. $f'(x) = 3x^2 - 2x$ **13.** $f'(t) = 4t - 1$

15. $f'(x) = 4x^7 + 3$ **17.** $f'(x) = \frac{5}{3}x^4$

19. $f'(x) = 2ax + b$

21. $\frac{d}{dx}(-6x^2 + x + 4) = -12x + 1$

23. $\frac{d}{dt}(-16t^2 + 80t) = -32t + 80$

25. $\frac{dA}{dr} = 2\pi r$ **27.** $\frac{dV}{dr} = 4\pi r^2$

29. $f'(-3) = -24$ **31.** $f'(4) = 15$ **33.** $f'(3) = -4$

35. $f'(1) = 1$ **37.** $f'\left(-\frac{b}{2a}\right) = 0$ **39.** $\frac{dy}{dx} = 4$

41. $\frac{dy}{dx} = 8$ **43.** $\frac{dy}{dx} = -7$ **45.** $\frac{dy}{dx} = 1$

47. $\frac{dy}{dx} = 10$ **49.** $m_{\tan} = 3, y = 3x - 1$

51. $x = 2$ **53.** $x = -1, x = 1$

55. No.

57. $\left(-\frac{1}{3}, -\frac{1}{3}\right), \left(\frac{1}{3}, \frac{1}{3}\right)$ **59.** $y = -4x + 1, y = 4x - 7$

61. (a) The average cost is \$45.00.
(b) $C'(x) = 0.4x + 3$
(c) The marginal cost of production level of 100 pairs of eyeglasses is \$43.00
(d) We can interpret $C'(100)$ to be the cost of producing the 101st pair of eyeglasses.

63. (a) $V'(R) = 4kR^3$
(b) $V'(0.3) = 0.108k$ cm³/cm
(c) $V'(0.4) = 0.256k$ cm³/cm
(d) The amount of blood flowing through the artery increases by about $0.0175k$ cm³.

65. (a) The daily cost of producing 40 microwave ovens is \$3920.
(b) The marginal cost function is $C'(x) = 50 - 0.1x$.
(c) $C'(40) = 46$. The marginal cost of producing 40 microwave ovens may be interpreted as the cost to produce the 41st microwave oven.
(d) The cost of producing 41 microwave ovens is approximately \$3966.
(e) The actual cost of producing 41 microwave ovens is \$3965.95. The actual cost is \$0.05 less than the estimated cost.
(f) The actual cost of producing the 41st microwave oven is \$45.95.
(g) The average cost function is $\overline{C}(x) = \frac{2000}{x} + 50 - 0.05x$.
(h) The average cost of producing 41 microwave ovens is \$96.73.
(i) The average cost of producing 41 microwave ovens is \$50.78 greater than the actual cost of producing the 41st microwave oven, and \$50.73 greater than the estimated cost of producing the 41st microwave oven. Explanations will vary.

67. (a) The marginal price of beans in 1995 was $-\$2.431$.
(b) The marginal price of beans in 2002 was $-\$9.634$.

69. $V'(2) = 16\pi \approx 50.27$ cubic feet per foot

71. $A'(t) = 3a_3t^2 + 2a_2t + a_1$

73. Let $f(x) = x^n$.

$$f'(x) = \lim_{h \to 0} \frac{f(x+h) - f(x)}{h} = \lim_{h \to 0} \frac{(x+h)^n - x^n}{h}$$

$$= \lim_{h \to 0} \frac{x^n + nx^{n-1}h + \frac{n(n-1)}{2}x^{n-2}h^2 + \cdots + h^n - x^n}{h}$$

$$= \lim_{h \to 0} \frac{nx^{n-1}h + h^2(\frac{n(n-1)}{2})x^{n-2} + \text{terms involving } x \text{ and } h)}{h}$$

$$= \lim_{h \to 0} \left[nx^{n-1} + h(\frac{n(n-1)}{2})x^{n-2} + \text{terms involving } x \text{ and } h) \right]$$

$$= nx^{n-1}$$

Exercise 13.3 (p. 741)

1. $f'(x) = 16x - 2$

3. $f'(t) = 4t^3 - 6t$

5. $f'(x) = 18x^2 - 20x + 3$

7. $f'(x) = 24x^7 + 40x^4 + 9x^2$

9. $f'(x) = \dfrac{1}{(1 + x)^2}$

11. $f'(x) = -\dfrac{11}{(2x - 1)^2}$

13. $f'(x) = \dfrac{x^2 - 8x}{(x - 4)^2}$

15. $f'(x) = -\dfrac{6x^2 + 6x - 8}{(3x^2 + 4)^2}$

17. $f'(t) = \dfrac{4}{t^3}$

19. $f'(x) = -\dfrac{1}{x^2} - \dfrac{2}{x^3}$

21. $m_{\text{tan}} = 3, y = 3x - 1$

23. $m_{\text{tan}} = \dfrac{5}{4}, y = \dfrac{5}{4}x - \dfrac{3}{4}$

25. $x = -\dfrac{2}{3}, x = 1$

27. $x = -2, x = 0$

29. $y' = 9x^2 - 4x$

31. $y' = 32x - \dfrac{6}{x^3}$

33. $y' = \dfrac{1}{(3x + 5)^2}$

35. $y' = -\dfrac{8x}{(x^2 - 4)^2}$

37. $y' = \dfrac{12x^2 + 12x + 23}{(2x + 1)^2}$

39. $y' = -\dfrac{4x^2(x^2 + 12)}{(x^2 + 4)^2}$

41.

$V(t) = \dfrac{10,000}{t} + 6,000$

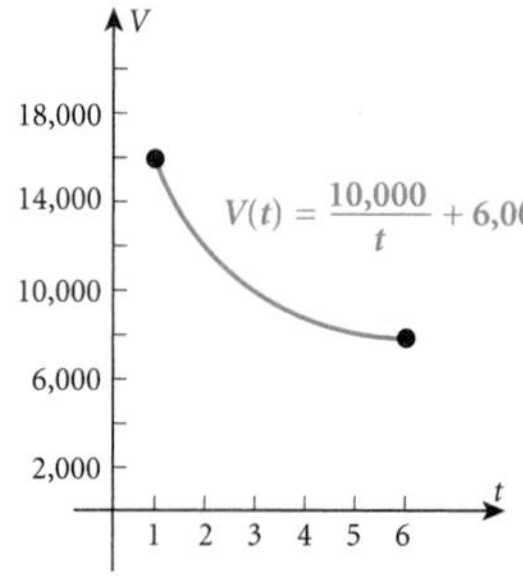

(a) The average rate of change is $-\$1000$ per year.

(b) $V'(t) = -\dfrac{10,000}{t^2}$ dollars per year.

(c) The instantaneous rate of change after two years is $-\$2500$ per year.

(d) The instantaneous rate of change after five years is $-\$400$ per year.

(e) Answers will vary.

43. (a) $R(x) = 10x + 40$ \qquad (b) $R'(x) = 10$
\qquad (c) $R'(4) = 10$ \qquad (d) $R'(6) = 10$

45. (a) $D'(p) = \dfrac{-1,000,000 - 200,000p}{(p^2 + 10p + 50)^2}$

(b) $D'(5) = -128, D'(10) = -48,$

$D'(15) = -\dfrac{6400}{289} \approx -22.15$

47. (a) The population is growing at a rate of 38.820 bacteria per hour.

(b) The population is growing at a rate of 35.503 bacteria per hour.

(c) The population is growing at a rate of 30.637 bacteria per hour.

(d) The population is growing at a rate of 24.970 bacteria per hour.

49. The rate of change of the intensity is -2 units per meter.

51. (a) $C'(x) = \dfrac{1}{10} - \dfrac{36,000}{x^2}$

(b) The marginal cost is $-\$0.044$ dollars per mph.

(c) The marginal cost is $-\$0.019$ dollars per mph.

(d) The marginal cost is $-\$0.078$ dollars per mph.

53. (a) $S'(r) = \dfrac{ag}{(g - r)^2}$

Exercise 13.4 (p. 747)

1. $f'(x) = 8(2x - 3)^3$

3. $f'(x) = 6x(x^2 + 4)^2$

5. $f'(x) = 12x(3x^2 + 4)$

7. $f'(x) = (4x + 1)(x + 1)^2$

9. $f'(x) = 8x(6x + 1)(2x + 1)^3$

11. $f'(x) = 3x^2(2x - 1)(x - 1)^2$

13. $f'(x) = \dfrac{-6}{(3x - 1)^3}$

15. $f'(x) = -\dfrac{8x}{(x^2 + 4)^2}$

17. $f'(x) = \dfrac{24x}{(x^2 - 9)^4}$

19. $f'(x) = \dfrac{3x^2}{(x + 1)^4}$

21. $f'(x) = \dfrac{(4x - 2)(2x + 1)^3}{3x^3}$

23. $f'(x) = \dfrac{(x^2 + 1)^2(5x^2 - 1)}{x^2}$

25. $f'(x) = \dfrac{3(x^2 - 1)(x^2 + 1)^2}{x^4}$

27. $f'(x) = \dfrac{6x(1 - x^2)}{(x^2 + 1)^3}$

29. (a) The car is depreciating at a rate of $7733.33 per year.

(b) The car is depreciating at a rate of $4793.39 per year.

(c) The car is depreciating at a rate of $3017.69 per year.

(d) The car is depreciating at a rate of $1972.79 per year.

31. (a) $p'(x) = -\dfrac{2000}{(x + 20)^2}$ \qquad (b) $R(x) = \dfrac{10,000x}{5x + 100} - 5x$

(c) $R'(x) = \dfrac{40,000}{(x + 20)^2} - 5$

(d) $R'(10) = 39.44$ dollars per pound, $R'(40) = 6.11$ dollars per pound

33. (a) The average rate of change in mass is -3.5 grams per hour.
(b) $M'(0) = -7$ grams per hour

Exercise 13.5 (p. 757)

1. $f'(x) = 3x^2 - e^x$

3. $f'(x) = e^x(x^2 + 2x)$

5. $f'(x) = \frac{e^x(x-2)}{x^3}$

7. $f'(x) = \frac{8x - 4x^2}{e^x}$

9. $y = (x^3 + 1)^5 \quad \frac{dy}{dx} = 15x^2(x^3 + 1)^4$

11. $y = \frac{x^2 + 1}{x^2 + 2} \quad \frac{dy}{dx} = \frac{2x}{(x^2 + 2)^2}$

13. $y = \left(\frac{1}{x} + 1\right)^2 \quad \frac{dy}{dx} = -\frac{2(x+1)}{x^3}$

15. $y = \left(\frac{1}{x^6} - 1\right)^5 \quad \frac{dy}{dx} = \frac{-30}{x^7}\left(\frac{1}{x^6} - 1\right)^4$

17. $y = e^{3x} \quad \frac{dy}{dx} = 3e^{3x}$

19. $y = e^{x^3} \quad \frac{dy}{dx} = 3x^2e^{x^3}$

21. $y' = 6x^2(x^3 + 1) = 6x^5 + 6x^2$

23. $f'(x) = 5e^{5x}$

25. $f'(x) = -16xe^{-x^2}$

27. $f'(x) = e^{x^2}(2x^3 + 2x)$

29. $f'(x) = 15e^{3x}$

31. $f'(x) = \frac{2x - x^2}{e^x}$

33. $f'(x) = \frac{e^{2x}(2x-1)}{x^2}$

35. $f'(x) = 2x - \frac{3}{x}$

37. $f'(x) = x + 2x \ln x$

39. $f'(x) = \frac{3}{x}$

41. $f'(x) = \ln(x^2 + 1) + \frac{2x^2}{x^2 + 1}$

43. $f'(x) = 1 + \frac{8}{x}$

45. $f'(x) = \frac{24(\ln x)^2}{x}$

47. $f'(x) = \frac{1}{x \ln 3}$

49. $f'(x) = \frac{x + 2x \ln x}{\ln 2}$

51. $f'(x) = (\ln 3)3^x$

53. $f'(x) = 2^x(2x + x^2 \ln 2)$

55. $y = 3x + 1$

57. $y = x - 1$

59. $y = 3x - 1$

61. $y = x - 1$

63. $y = x + 1$

65. (a) The reaction rate is 1.1 per unit.
(b) The reaction rate is 0.55 per unit.

67. The rate of change of the pressure with respect to the height is -1.130 kilograms per square meter per meter at a height of 500 meters and is -1.103 kilograms per square meter per meter at a height of 700 meters.

69. (a) $A'(t) = 18.9e^{-0.21t}$ percent of the market per year
(b) $A'(5) = 6.614$ percent of the market per year. In the sixth year, DVD players will penetrate approximately an additional 6.614 percent of the market.
(c) $A'(10) = 2.314$ percent of the market per year. In the eleventh year, DVD players will penetrate approximately an additional 2.314 percent of the market.
(d) $A'(30) = 0.035$ percent of the market per year. In the thirty-first year, DVD players will penetrate approximately an additional 0.035 percent of the market.

71. (a) $S'(x) = \frac{400{,}000}{x}$ dollars in sales per thousands of dollars of advertising cost
(b) $S'(10) = 40{,}000$ dollars in sales per thousands of dollars of advertising cost
(c) $S'(20) = 20{,}000$ dollars in sales per thousands of dollars of advertising cost
(d) Answers will vary.

73. (a) 1000 t-shirts can be sold at \$40.41.
(b) 5000 t-shirts can be sold at \$34.27.
(c) $p'(1000) = -\$0.0036$. This means that another t-shirt will be demanded if the price were reduced by \$0.0036.
(d) $p'(5000) = -\$0.00078$. This means that another t-shirt will be demanded if the price were reduced by \$0.00078.
(e) $R(x) = 50x - 4x \ln\left(\frac{x}{100} + 1\right)$
(f) $R'(1000) = \$36.77$. The revenue received for selling the 1001st t-shirt is \$36.77.
(g) $R'(5000) = \$30.35$. The revenue received for selling the 5001st t-shirt is \$30.35.
(h) $P(x) = 46x - 4x \ln\left(\frac{x}{100} + 1\right)$
(i) $P(1000) = \$36{,}408.42$
(j) $P(5000) = \$151{,}363.49$
(k) For $x = 3{,}631{,}550$, the profit is the greatest.
(l) A price of \$8.00 should be charged to maximize profit.

75. (a) $p'(t) = \frac{0.026}{t}$ dollars per year
(b) $p'(5) = 0.0052$ dollars per year
(c) $p'(10) = 0.0026$ dollars per year

77. Let $y = \ln u$ and $u = g(x)$. Then $y = \ln(g(x))$ and

$$\frac{dy}{dx} = \frac{dy}{du}\frac{du}{dx} = \frac{1}{u(x)}\frac{d}{dx}g(x) = \frac{\frac{d}{dx}g(x)}{g(x)}.$$

Exercise 13.6 (p. 766)

1. $f'(x) = 2 \quad f''(x) = 0$

3. $f'(x) = 6x + 1 \quad f''(x) = 6$

5. $f'(x) = -12x^3 + 4x \quad f''(x) = -36x^2 + 4$

7. $f'(x) = -\frac{1}{x^2} \quad f''(x) = \frac{2}{x^3}$

9. $f'(x) = 1 - \frac{1}{x^2} \quad f''(x) = \frac{2}{x^3}$

11. $f'(x) = \frac{1}{(x+1)^2}$ $f''(x) = -\frac{2}{(x+1)^3}$

13. $f'(x) = e^x$ $f''(x) = e^x$

15. $f'(x) = 6x(x^2+4)^2$ $f''(x) = 6(x^2+4)(5x^2+4)$

17. $f'(x) = \frac{1}{x}$ $f''(x) = -\frac{1}{x^2}$

19. $f'(x) = e^x(x+1)$ $f''(x) = e^x(x+2)$

21. $f'(x) = 2e^{2x}$ $f''(x) = 4e^{2x}$

23. $f'(x) = -\frac{1}{x(\ln x)^2}$ $f''(x) = \frac{2+\ln(x)}{x^2(\ln x)^3}$

25. (a) The domain is the interval $(-\infty, \infty)$.
(b) $f'(x) = 2x$
(c) The domain is the interval $(-\infty, \infty)$.
(d) $x = 0$
(e) The derivative $f'(x)$ exists for all values of x.
(f) $f''(x) = 2$
(g) The domain is the interval $(-\infty, \infty)$.

27. (a) The domain is the interval $(-\infty, \infty)$.
(b) $f'(x) = 3x^2 - 18x + 27$
(c) The domain is the interval $(-\infty, \infty)$.
(d) $x = 3$
(e) The derivative $f'(x)$ exists for all values of x.
(f) $f''(x) = 6x - 18$
(g) The domain is the interval $(-\infty, \infty)$.

29. (a) The domain is the interval $(-\infty, \infty)$.
(b) $f'(x) = 12x^3 - 36x^2$
(c) The domain is the interval $(-\infty, \infty)$.
(d) $x = 0, x = 3$
(e) The derivative $f'(x)$ exists for all values of x.
(f) $f''(x) = 36x^2 - 72x$
(g) The domain is the interval $(-\infty, \infty)$.

31. (a) The domain is the set $\{x | x \neq -2, x \neq 2\}$.
(b) $f'(x) = -\frac{x^2+4}{(x^2-4)^2}$
(c) The domain is the set $\{x | x \neq -2, x \neq 2\}$.
(d) The derivative $f'(x)$ is never zero.
(e) The derivative $f'(x)$ does not exist for $x = -2$ and $x = 2$.
(f) $f''(x) = \frac{2x(x^2+12)}{(x^2-4)^3}$
(g) The domain is the set $\{x | x \neq -2, x \neq 2\}$.

33. $f^{(4)}(x) = 0$ **35.** $f^{(20)}(x) = 0$ **37.** $f^{(8)}(x) = 5040$

39. $v = 32t + 20, a = 32$ **41.** $v = 9.8t + 4, a = 9.8$

43. $f^{(n)}(x) = e^x$

45. $f^{(n)}(x) = \frac{(-1)^{n+1}(n-1)!}{x^n}$

47. $f'(x) = 1 + \ln x$ $f^{(n)}(x) = \frac{(-1)^n(n-2)!}{x^{n-1}}$ for $n > 1$

49. $f^{(n)}(x) = n!\,2^n$ **51.** $f^{(n)}(x) = a^n e^{ax}$

53. $f^{(n)}(x) = \frac{(-1)^{n+1}(n-1)!}{x^n}$ **55.** $y'' - 4y = 0$

57. $f''(x) = x^2g''(x) + 4xg'(x) + 2g(x)$

59. (a) The velocity is 16 feet per second.
(b) The ball will reach its maximum height 2.5 seconds after it is thrown.
(c) The maximum height of the ball is 106 feet.
(d) The acceleration is -32 feet per second per second.
(e) The ball is in the for air 5.074 seconds.
(f) The velocity of the ball is -82.365 feet per second upon impact.
(g) The total distance traveled by the ball is 206 feet.

61. $v(1) = 3$ meters per second, $a(t) = 6t - 12$ meters per second.

63. (a) It takes 4.24 seconds for the rock to hit the ground.
(b) The average velocity is -20.8 meters per second.
(c) The average velocity is -14.7 meters per second.
(d) The velocity is -41.6 meters per second when the rock hits the ground.

Exercise 13.7 (p. 773)

1. $\frac{dy}{dx} = -\frac{x}{y}$ if $y \neq 0$

3. $\frac{dy}{dx} = -\frac{2y}{x}$ if $x \neq 0$

5. $\frac{dy}{dx} = \frac{2x-y}{x-2y}$ if $x - 2y \neq 0$

7. $\frac{dy}{dx} = \frac{2x+4y}{1-4x-2y}$ if $1 - 4x - 2y \neq 0$

9. $\frac{dy}{dx} = -\frac{2x}{y^2}$ if $y \neq 0$

11. $\frac{dy}{dx} = \frac{x-6x^2}{3y^2}$ if $y \neq 0$

13. $\frac{dy}{dx} = \frac{y^3}{x^3}$ if $x \neq 0$

15. $\frac{dy}{dx} = -\frac{y^2}{x^2}$ if $x \neq 0$

17. $\frac{dy}{dx} = \frac{ye^x - 2x}{2y - e^x}$ if $2y - e^x \neq 0$

19. $\frac{dy}{dx} = \frac{6x^2y^2e^x + y^3 - x^2y}{xy^2 - x^3}$ if $xy^2 - x^3 \neq 0$

21. $\frac{dy}{dx} = \frac{2x^2 - y^2}{2xy\ln x}$ if $y \neq 0, x \neq 1$

23. $\frac{dy}{dx} = \frac{-3x-6y}{6x+8y}$ if $6x + 8y \neq 0$

25. $\frac{dy}{dx} = \frac{3x^2 - 6xy + 3y^2 - 4x^3 - 4xy^2}{3x^2 - 6xy + 3y^2 + 4x^2y + 4y^3}$ if $3x^2 - 6xy + 3y^2 + 4x^2y + 4y^3 \neq 0$

27. $\frac{dy}{dx} = \frac{xy^2 - 3x^5 - 3x^2y^3}{3x^3y^2 + 3y^5 - x^2y}$ if $3x^3y^2 + 3y^5 - x^2y \neq 0$

29. $\frac{dy}{dx} = \frac{2xe^{x^2+y^2}}{1 - 2ye^{x^2+y^2}}$ if $1 - 2ye^{x^2+y^2} \neq 0$

31. $y' = -\frac{x}{y}$ $y'' = -\frac{x^2+y^2}{y^3}$ if $y \neq 0$

33. $y' = -\frac{2xy+y}{x^2+x}$ if $x \neq 0, x \neq -1$ $y'' = \frac{6x^2y + 6xy + 2y}{x^4 + 2x^3 + x^2}$

35. $m_{\tan} = -\frac{1}{2}$ $y = -\frac{1}{2}x + \frac{5}{2}$

37. The tangent line at $(1, 0)$ is $y = x - 1$.

39. $(0, 22)$ and $(0, 2)$ **41.** $(0, 24)$ and $(0, 4)$

43. (a) $m_{\tan} = \frac{dy}{dx} = -\frac{y+1}{x+4y}$ (b) $y = -\frac{1}{3}x + \frac{5}{3}$
(c) $(2, 1)$; $(6, -3)$

45. $\dfrac{dV}{dp} = \dfrac{V^3(V-b)^2}{2a(V-b)^2 - cV^3}$

47. (a) $\dfrac{dN}{dt} = \dfrac{430{,}163t^4 + 1{,}720{,}649t^2 + 1{,}720{,}658}{430{,}163t^5 + 1{,}720{,}655t^3 + 1{,}720{,}658t}$

(b) $N'(2) = \frac{2{,}580{,}977}{5{,}161{,}962} \approx 0.500$

$N'(4) = \frac{23{,}228{,}795}{92{,}915{,}244} \approx 0.250$

Exercise 13.8 (p. 778)

1. $f'(x) = \dfrac{4}{3}x^{1/3}$
3. $f'(x) = \dfrac{2}{3x^{1/3}}$
5. $f'(x) = -\dfrac{1}{2x^{3/2}}$
7. $f'(x) = 3(2x+3)^{1/2}$
9. $f'(x) = 3x(x^2+4)^{1/2}$
11. $f'(x) = \dfrac{1}{\sqrt{2x+3}}$
13. $f'(x) = \dfrac{9x}{\sqrt{9x^2+1}}$
15. $f'(x) = 5x^{2/3} - \dfrac{2}{x^{2/3}}$
17. $f'(x) = \dfrac{7}{3}x^{4/3} - \dfrac{4}{3x^{2/3}}$
19. $f'(x) = -\dfrac{4}{(x^2-4)^{3/2}}$
21. $f'(x) = \dfrac{e^{x/2}}{2}$
23. $f'(x) = \dfrac{1}{2x\sqrt{\ln x}}$
25. $f'(x) = \dfrac{e^{\sqrt[3]{x}}}{3x^{2/3}}$
27. $f'(x) = \dfrac{1}{3x(\ln x)^{2/3}}$
29. $f'(x) = e^x\left(\sqrt{x} + \dfrac{1}{2\sqrt{x}}\right)$
31. $f'(x) = \dfrac{e^{2x}(2x^2+x+2)}{\sqrt{x^2+1}}$
33. $\dfrac{dy}{dx} = -\dfrac{\sqrt{y}}{\sqrt{x}}$
35. $\dfrac{dy}{dx} = \dfrac{\sqrt{x^2+y^2} - x}{y}$
37. $\dfrac{dy}{dx} = \dfrac{y^{2/3}}{x^{2/3}}$
39. $\dfrac{dy}{dx} = -\dfrac{\sqrt{y}e^{\sqrt{x}-\sqrt{y}}}{\sqrt{x}}$

41. (a) The domain is the interval $[0, \infty)$.
(b) $f'(x) = \dfrac{1}{2\sqrt{x}}$
(c) The domain is the interval $(0, \infty)$.
(d) The derivative $f'(x)$ is nonzero on its domain.
(e) $x = 0$
(f) $f''(x) = -\dfrac{1}{4x^{3/2}}$
(g) The domain is the interval $(0, \infty)$.

43. (a) The domain is the interval $(-\infty, \infty)$.
(b) $f'(x) = \dfrac{2}{3x^{1/3}}$
(c) The domain is the set $\{x | x \neq 0\}$.
(d) The derivative $f'(x)$ is nonzero on its domain.
(e) $x = 0$
(f) $f''(x) = -\dfrac{2}{9x^{4/3}}$
(g) The domain is the set $\{x | x \neq 0\}$.

45. (a) The domain is the interval $(-\infty, \infty)$.
(b) $f'(x) = \dfrac{2}{3x^{1/3}} + \dfrac{2}{3x^{2/3}}$
(c) The domain is the set $\{x | x \neq 0\}$.
(d) $x = -1$
(e) $x = 0$
(f) $f''(x) = -\dfrac{2}{9x^{4/3}} - \dfrac{4}{9x^{5/3}}$
(g) The domain is the set $\{x | x \neq 0\}$.

47. (a) The domain is the interval $(-\infty, \infty)$.
(b) $f'(x) = \dfrac{4x}{3(x^2-1)^{1/3}}$
(c) The domain is the set $\{x | x \neq -1, x \neq 1\}$.
(d) $x = 0$
(e) $x = -1, x = 1$
(f) $f''(x) = \dfrac{4(x^2-3)}{9(x^2-1)^{4/3}}$
(g) The domain is the set $\{x | x \neq -1, x \neq 1\}$.

49. (a) The domain is the interval $[-1, 1]$.
(b) $f'(x) = \dfrac{1-2x^2}{\sqrt{1-x^2}}$
(c) The domain is the interval $(-1, 1)$.
(d) $x = -\dfrac{\sqrt{2}}{2}, x = \dfrac{\sqrt{2}}{2}$
(e) $x = -1, x = 1$
(f) $f''(x) = \dfrac{x(2x^2-3)}{(1-x^2)^{3/2}}$
(g) The domain is the interval $(-1, 1)$.

51. (a) $N'(t) = \dfrac{500}{(1+0.1t)^{3/2}}$
(b) Student enrollment will be increasing at the rate of 176.78 students per year.

53. $\dfrac{dz}{dx} = \dfrac{d}{dx}(x^{0.5}y^{0.4})$

$$0 = 0.5x^{-0.5}y^{0.4} + 0.4x^{0.5}y^{-0.6}\frac{dy}{dx}$$

$$\frac{dy}{dx} = -\frac{0.5x^{-0.5}y^{0.4}}{0.4x^{0.5}y^{-0.6}} = -\frac{5y}{4x}$$

55. (a) $A'(t) = \dfrac{3(t^{1/4}+3)^2}{4t^{3/4}}$ units per year
(b) The instantaneous rate of change of pollution is 2.34 units per year.

57. After 1 second, the child's velocity is 1.5 feet per second. The child strikes the ground with a velocity of 3 feet per second.

CHAPTER 13 Review

True – False Items (p. 781)

1. True 2. False 3. True
4. True 5. False 6. False
7. True

Fill in the Blanks (p. 782)

1. tangent 2. marginal cost
3. Chain Rule; Power Rule 4. velocity
5. 0 6. implicit

Review Exercises (p. 782)

1. $f'(2) = 2$ 3. $f'(2) = 4$ 5. $f'(1) = 0$
7. $f'(0) = 3$ 9. $f'(x) = 4$ 11. $f'(x) = 4x$
13. $f'(x) = 5x^4$ 15. $f'(x) = x^3$
17. $f'(x) = 4x - 3$ 19. $f'(x) = 14x$
21. $f'(x) = 15(x^2 - 6x + 6)$
23. $f'(x) = 24(16x^3 + 3x^2 - 5x + 1)$
25. $f'(x) = -\dfrac{16}{(5x-3)^2}$ 27. $f'(x) = -\dfrac{24}{x^{13}}$
29. $f'(x) = -\dfrac{3}{x^2} - \dfrac{8}{x^3}$ 31. $f'(x) = \dfrac{17}{(x+5)^2}$
33. $f'(x) = 10x^4(3x-1)(3x-2)^4$
35. $f'(x) = 7(x+1)^3(5x+1)$
37. $f'(x) = -\dfrac{2(x+1)}{(3x+2)^3}$ 39. $f'(x) = -\dfrac{42x^2}{(x^3+4)^3}$
41. $f'(x) = \dfrac{3(3x^2-4)(3x^2+4)^2}{x^4}$
43. $f'(x) = 3e^x + 2x$ 45. $f'(x) = 3e^{3x+1}$
47. $f'(x) = e^x(2x^2 + 11x + 7)$ 49. $f'(x) = -\dfrac{x}{e^x}$
51. $f'(x) = \dfrac{2e^{2x}(x-1)}{9x^3}$ 53. $f'(x) = \dfrac{1}{x}$
55. $f'(x) = x + 2x \ln x$ 57. $f'(x) = \dfrac{6x^2}{2x^3+1}$
59. $f'(x) = (\ln 2)2^x + 2x$ 61. $f'(x) = 1 + \dfrac{1}{x \ln 10}$
63. $f'(x) = \dfrac{1}{2\sqrt{x}}$ 65. $f'(x) = 5x^{2/3}$
67. $f'(x) = \dfrac{2x-3}{2\sqrt{x^2-3x}}$ 69. $f'(x) = \dfrac{x+9}{2(x+5)^{3/2}}$
71. $f'(x) = \dfrac{e^{x/2}(x+3)}{2}$ 73. $f'(x) = \dfrac{2+\ln x}{2\sqrt{x}}$
75. $f'(x) = 3x^2 \quad f''(x) = 6x$
77. $f'(x) = -3e^{-3x} \quad f''(x) = 9e^{-3x}$
79. $f'(x) = \dfrac{1}{(2x+1)^2} \quad f''(x) = -\dfrac{4}{(2x+1)^3}$
81. $\dfrac{dy}{dx} = \dfrac{10-y}{x+6y}$ if $x + 6y \neq 0$ 83. $\dfrac{dy}{dx} = \dfrac{8x - e^y}{xe^y}$ if $x \neq 0$
85. $m_{\tan} = -1, y = -x - 9$ 87. $m_{\tan} = 1, y = x + 1$
89. (a) The average rate of change is 7.
(b) The instantaneous rate of change at 2 is $f'(2) = 15$.
91. (a) 2.5 seconds elapse before the stone hits the water.
(b) The average velocity is −40 feet per second.
(c) The instantaneous velocity is −80 feet per second.
93. (a) The ball reaches its maximum height 4 seconds after it is thrown.
(b) The maximum height that the ball reaches is 262 feet.
(c) The total distance the ball travels is 518 feet.
(d) The velocity of the ball at any time t is $v(t) = 128 - 32t$ feet per second.
(e) The velocity of the ball is zero at $t = 4$ seconds, which is the time at which the ball's velocity changes from upward motion to downward motion.
(f) The ball is in the air for 8.0 seconds.
(g) The velocity of ball is −129.5 feet per second when it hits the ground.
(h) The acceleration at any time t is −32 feet per second per second.
(i) The velocity of ball is 64 feet per second when it has been in the air for 2 seconds and is −64 feet per second when it has been in the air for 6 seconds.
95. (a) $R(x) = -0.50x^2 + 75x$
(b) $R'(x) = -1.00x + 75$
(c) $C'(x) = 15$
(d) The break even points are $x = 10$ and $x = 110$.
(e) The marginal revenue and the marginal cost are equal when $x = 60$ units are produced.

Mathematical Questions from Professional Exams (p. 786)

1. (e) $2x^3e^{x^2} + 2xe^{x^2}$ 2. (d) $x \neq -3$
3. (d) $2bxe^{c^2+x^2}$ 4. (a) $2xe^{x^2}$
5. (e) 15 6. (c) $t = 2$

CHAPTER 14   Applications: Graphing Functions; Optimization

Exercise 14.1 (p. 793)

1. There is a horizontal tangent line at $(2, 0)$.

3. There is a horizontal tangent line at $(4, 16)$.

5. There is a horizontal tangent line at $(2, 9)$.

7. There is a vertical tangent line at $(0, 1)$.

9. There are horizontal tangent lines at $(-1, -1)$ and $(1, 3)$.

11. There is a vertical tangent line at $(0, -2)$.

13. There are horizontal tangent lines at $(0, 0)$ and $(8, -8192)$.

15. There is a horizontal tangent line at $(0, -1)$.

17. There is a horizontal tangent line at $(-1, -1)$, and there is a vertical tangent line at $(0, 0)$.

19. There is a horizontal tangent line at $(4, -12\sqrt[3]{2})$, and there is a vertical tangent line at $(0, 0)$.

21. There are horizontal tangent lines at $(-2, -12\sqrt[3]{4})$ and $(2, -12\sqrt[3]{4})$, and there is a vertical tangent line at $(0, 0)$.

23. There is a horizontal tangent line at $\left(-4, -\frac{\sqrt[3]{2}}{3}\right)$, and there is a vertical tangent line at $(0, 0)$.

25. There is a horizontal tangent line at $\left(-\frac{1}{2}, \frac{\sqrt[3]{4}}{3}\right)$, and a vertical tangent line at $(0, 0)$.

27. (a) Yes   (b) No   (c) vertical tangent line at $(0, 0)$

29. (a) No   (b) No   (c) f is not continuous at $x = 1$.

31. (a) Yes   (b) No   (c) no tangent line at $(0, 0)$
(d) $f(x) = \begin{cases} 3x & \text{if } x < 0 \\ x^2 & \text{if } x \geq 0 \end{cases}$

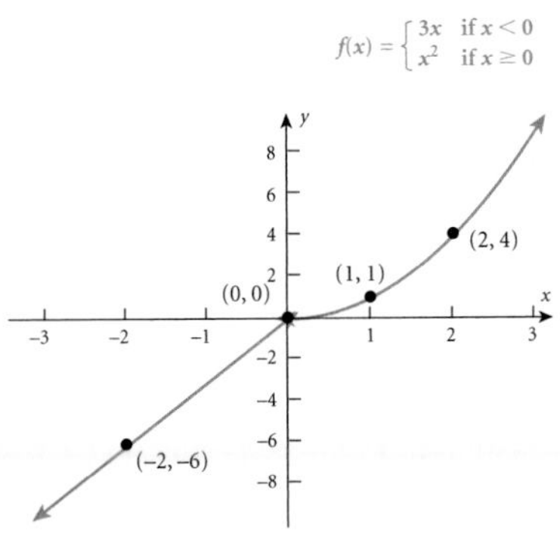

33. (a) No   (b) No   (c) f is not continuous at $x = 2$.

(d)

$f(x) = \begin{cases} 4x & \text{if } x \leq 2 \\ x^2 & \text{if } x > 2 \end{cases}$

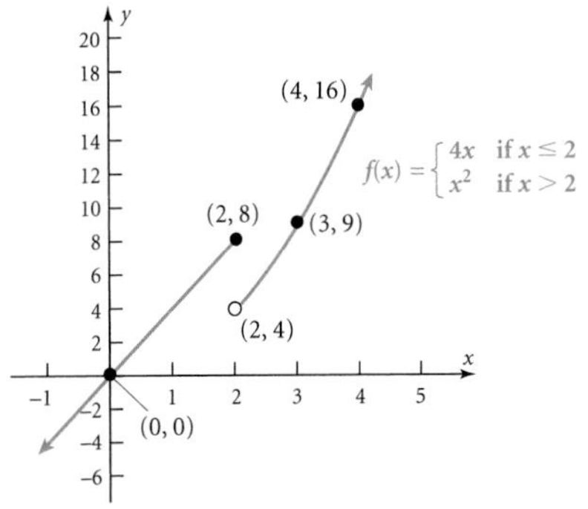

35. (a) Yes   (b) Yes, $f'(0) = 0$ (c) $f'(0)$ does exist.
(d)

$f(x) = \begin{cases} x^2 & \text{if } x < 0 \\ x^3 & \text{if } x \geq 0 \end{cases}$

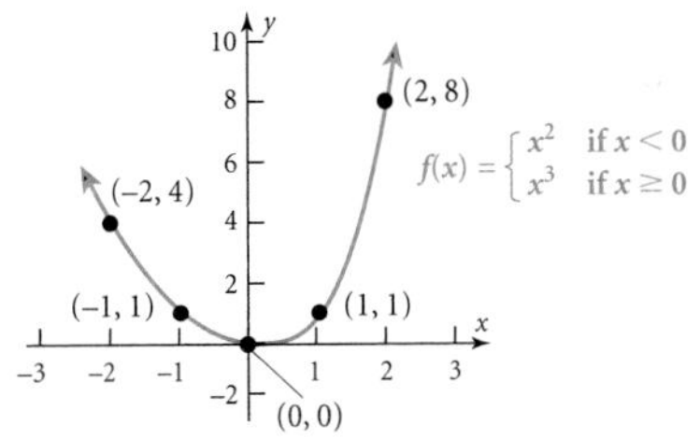

Exercise 14.2 (p. 803)

1. The domain is the interval $[x_1, x_9]$

3. The graph is increasing on the intervals (x_1, x_4), (x_5, x_7), and (x_8, x_9).

5. $x = x_4$, $x = x_6$, $x = x_7$, and $x = x_8$

7. f has a local maximum at (x_4, y_4) and at (x_7, y_7).

9. Step 1: The domain is the interval $(-\infty, \infty)$.
Step 2: The x-intercept is $(1, 0)$, and the y-intercept is $(0, -2)$.
Step 3: The graph is increasing on the interval $(-\infty, 1)$ and is decreasing on the interval $(1, \infty)$.
Step 4: There is a local maximum at $(1, 0)$.
Step 5: The tangent line is horizontal at $(1, 0)$.
Step 6: The end behavior is $y = -2x^2$.

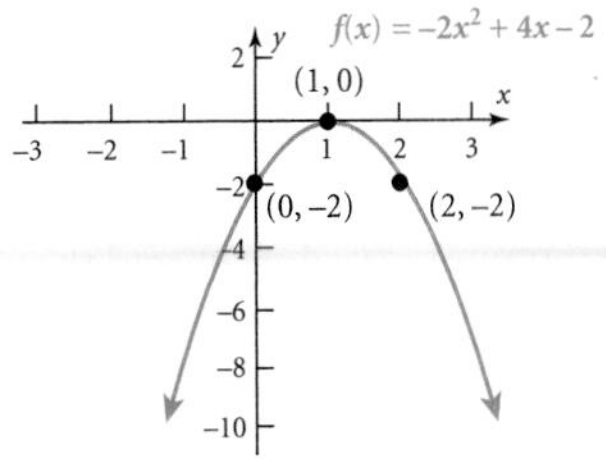

11. Step 1: The domain is the interval $(-\infty, \infty)$.
Step 2: The x-intercept is $(3, 0)$, and the y-intercept is $(0, -27)$.
Step 3: The graph is increasing on the interval $(-\infty, \infty)$.
Step 4: There are no local extrema.
Step 5: The tangent line is horizontal at $(3, 0)$.
Step 6: The end behavior is $y = x^3$.

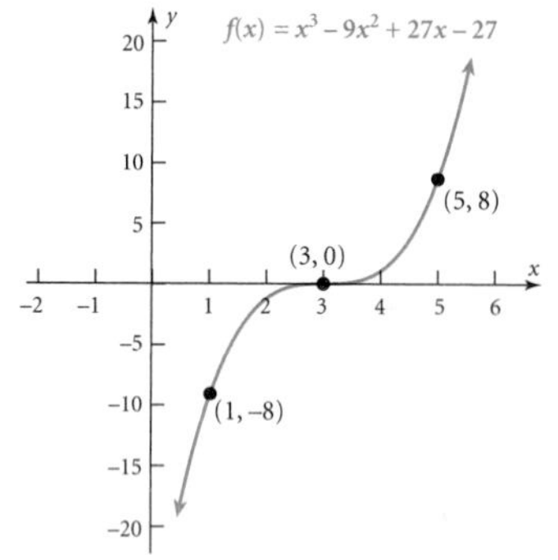

13. Step 1: The domain is the interval $(-\infty, \infty)$.
Step 2: The x-intercept and the y-intercept are both $(0, 0)$.
Step 3: The graph is increasing on the intervals $(-\infty, 2)$ and $(3, \infty)$ and is decreasing on the interval $(2, 3)$.
Step 4: There is a local maximum at $(2, 28)$, and there is a local minimum at $(3, 27)$.
Step 5: The tangent line is horizontal at the points $(2, 28)$ and $(3, 27)$.
Step 6: The end behavior is $y = 2x^3$.

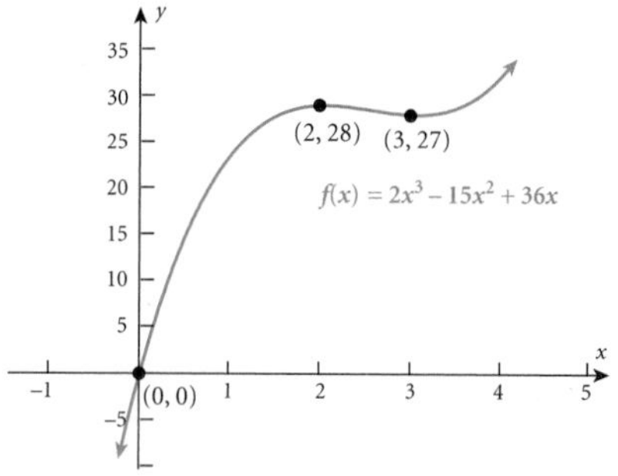

15. Step 1: The domain is the interval $(-\infty, \infty)$.
Step 2: The y-intercept is $(0, -1)$.
Step 3: The graph is increasing on the interval $(-1, 1)$ and is decreasing on the intervals $(-\infty, -1)$ and $(1, \infty)$.
Step 4: There is a local minimum at $(-1, -3)$, and there is a local maximum at $(1, 1)$.
Step 5: The tangent line is horizontal at the points $(-1, -3)$ and $(1, 1)$.
Step 6: The end behavior is $y = -x^3$.

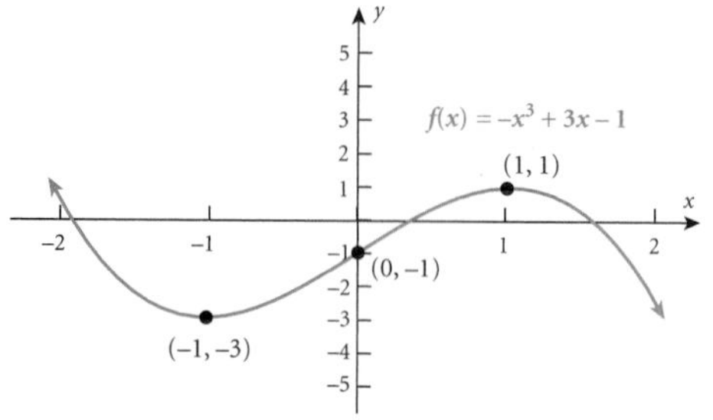

17. Step 1: The domain is the interval $(-\infty, \infty)$.
Step 2: The y-intercept is $(0, 2)$.
Step 3: The graph is increasing on the interval $(3, \infty)$ and is decreasing on the interval $(-\infty, 3)$.
Step 4: There is a local minimum at $(3, -79)$.
Step 5: The tangent line is horizontal at the points $(0, 2)$ and $(3, -79)$.
Step 6: The end behavior is $y = 3x^4$.

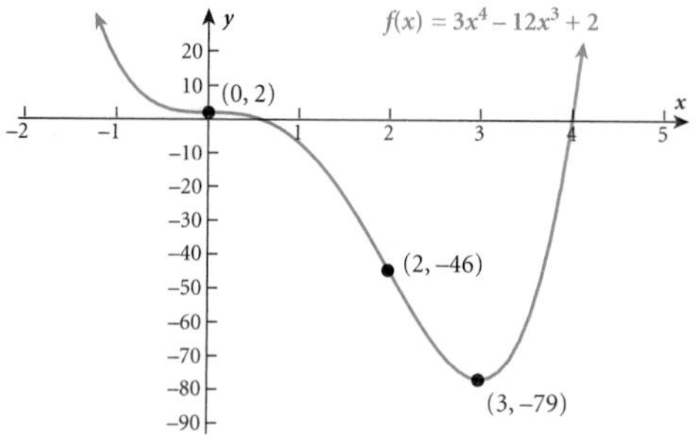

19. Step 1: The domain is the interval $(-\infty, \infty)$.
Step 2: The y-intercept is $(0, 1)$.
Step 3: The graph is increasing on the intervals $(-\infty, -1)$ and $(1, \infty)$, and is decreasing on the interval $(-1, 1)$.
Step 4: There is a local maximum at $(-1, 5)$, and there is a local minimum at $(1, -3)$.
Step 5: The tangent line is horizontal at $(-1, 5)$ and $(1, -3)$.
Step 6: The end behavior is $y = x^5$.

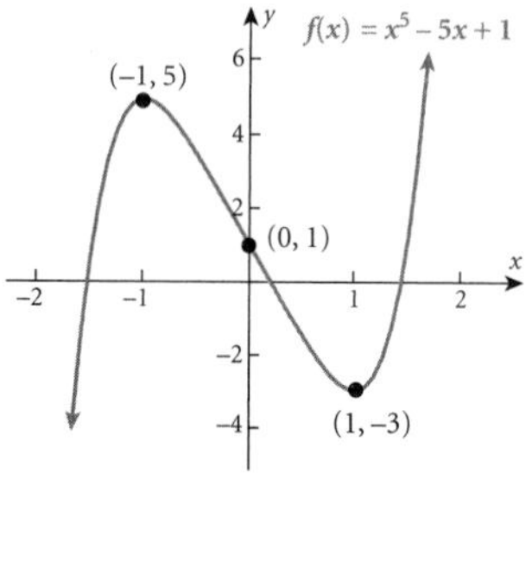

21. Step 1: The domain is the interval $(-\infty, \infty)$.
Step 2: The y-intercept is $(0, 1)$.
Step 3: The graph is increasing on the intervals $(-\infty, -2)$ and $(2, \infty)$ and is decreasing on $(-2, 2)$.
Step 4: There is a local maximum at $(-2, 65)$, and there is a local minimum at $(2, -63)$.
Step 5: The tangent line is horizontal at $(-2, 65)$, $(0, 1)$, and $(2, -63)$.
Step 6: The end behavior is $y = 3x^5$.

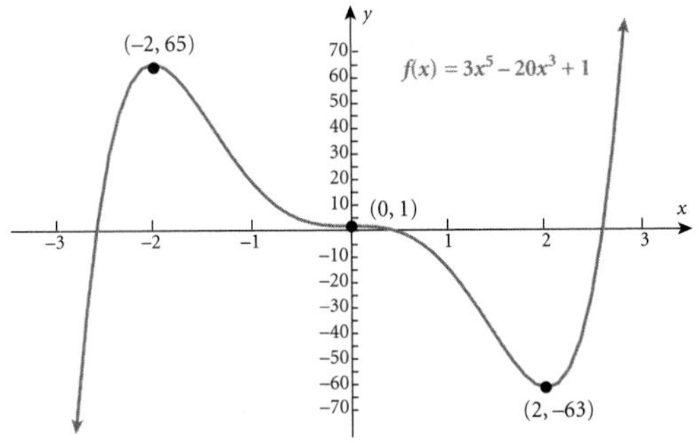

23. Step 1: The domain is the interval $(-\infty, \infty)$.
Step 2: The x-intercepts are $(0, 0)$ and $(0, -8)$, and the y-intercept is $(0, 0)$.
Step 3: The graph is increasing on the interval $(-1, \infty)$, and is decreasing on the interval $(-\infty, -1)$.
Step 4: There is a local minimum at $(-1, -1)$.
Step 5: The tangent line is vertical at $(0, 0)$ and horizontal at $(-1, -1)$.
Step 6: The end behavior is $y = x^{2/3}$.

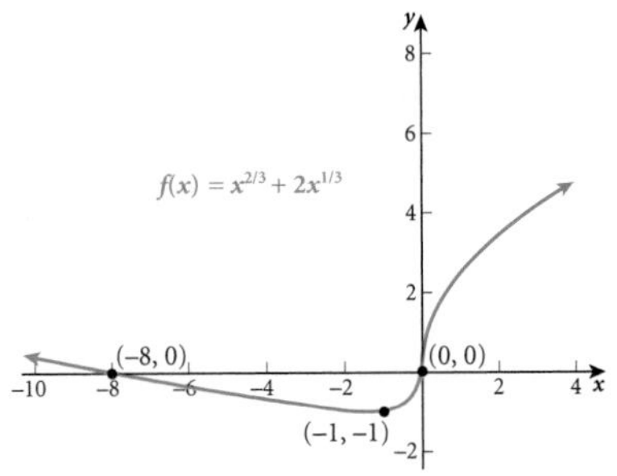

25. Step 1: The domain is the interval $(-\infty, \infty)$.
Step 2: The x-intercepts are $(-1, 0)$ and $(1, 0)$, and the y-intercept is $(0, 1)$.
Step 3: The graph is increasing on the intervals $(-1, 0)$ and $(1, \infty)$ and is decreasing on the intervals $(-\infty, -1)$ and $(0, 1)$.
Step 4: There is a local maximum at $(0, 1)$, and there are local minima at $(-1, 0)$ and $(1, 0)$.
Step 5: The tangent line is horizontal at $(0, 1)$ and is vertical at $(-1, 0)$ and $(1, 0)$.
Step 6: The end behavior is $y = x^{4/3}$.

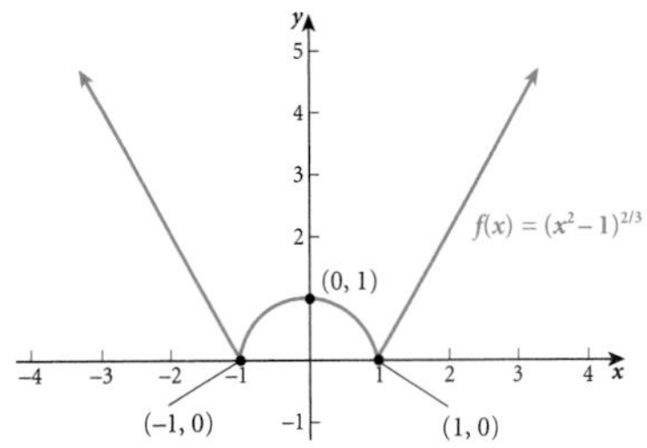

27. Step 1: The domain is the set $\{x \mid x \neq -4, x \neq 4\}$.
Step 2: The y-intercept is $(0, -\frac{1}{2})$.
Step 3: The graph is increasing on the intervals $(-\infty, -4)$ and $(-4, 0)$ and is decreasing on the intervals $(0, 4)$ and $(4, \infty)$.
Step 4: There is a local maximum at $(0, -\frac{1}{2})$.
Step 5: The tangent line is horizontal at $(0, -\frac{1}{2})$.
Step 6: The end behavior is the horizontal asymptote $y = 0$. The lines $x = -4$ and $x = 4$ are vertical asymptotes.

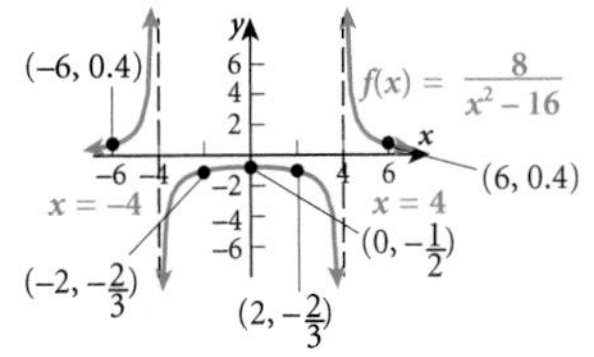

29. Step 1: The domain is the set $\{x \mid x \neq -3, x \neq 3\}$.
Step 2: The x-intercept and the y-intercept are both $(0, 0)$.
Step 3: The graph is decreasing on the intervals $(-\infty, -3)$, $(-3, 3)$, and $(3, \infty)$.
Step 4: There are no local extrema.
Step 5: There are no horizontal tangents or vertical tangents.
Step 6: The end behavior is the horizontal asymptote $y = 0$. The lines $x = -3$ and $x = 3$ are vertical asymptotes.

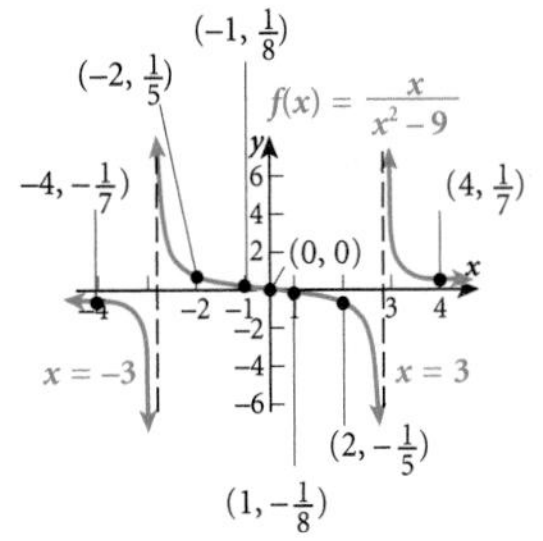

31. Step 1: The domain is the set $\{x \mid x \neq -2, x \neq 2\}$.
Step 2: The x-intercept and y-intercept are both $(0, 0)$.
Step 3: The graph is increasing on the intervals $(-\infty, -2)$ and $(2, 0)$, and is decreasing on the intervals $(0, 2)$ and $(2, \infty)$.
Step 4: There is a local maximum at $(0, 0)$.
Step 5: There is a horizontal tangent at $(0, 0)$.

Step 6: The end behavior is the horizontal asymptote $y = 1$. The lines $x = -2$ and $x = 2$ are vertical asymptotes.

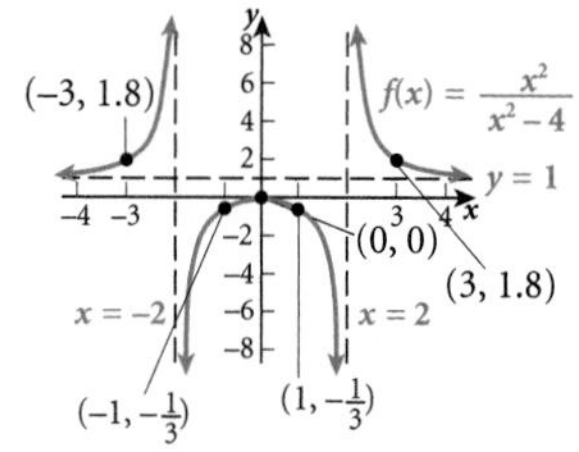

33. Step 1: The domain is the set $\{x | x > 0\}$.
Step 2: The x-intercept is $(1, 0)$.
Step 3: The graph is increasing on the interval $(0.37, \infty)$ and is decreasing on the interval $(0, 0.37)$.
Step 4: There is a local minimum at $(0.37, -0.37)$.
Step 5: There is a horizontal tangent at $(0.37, -0.37)$.
Step 6: $\lim_{x \to 0+} x \ln x = 0$ and $\lim_{x \to \infty} x \ln x = \infty$

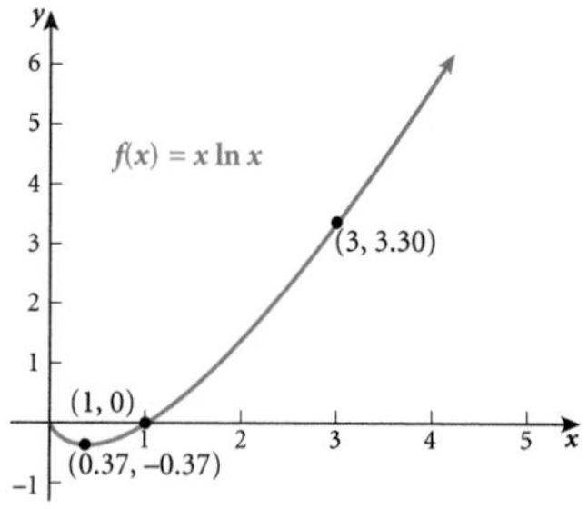

35. $S'(x) = 8x + 50 > 0$ for $1 \le x \le 10$. S is an increasing function.

37. (a) The graph of R is increasing on $(0, 2000)$ and is decreasing on $(2000, \infty)$.
(b) 2000 trucks need to be sold.
(c) The maximum revenue is \$20,000.
(d)

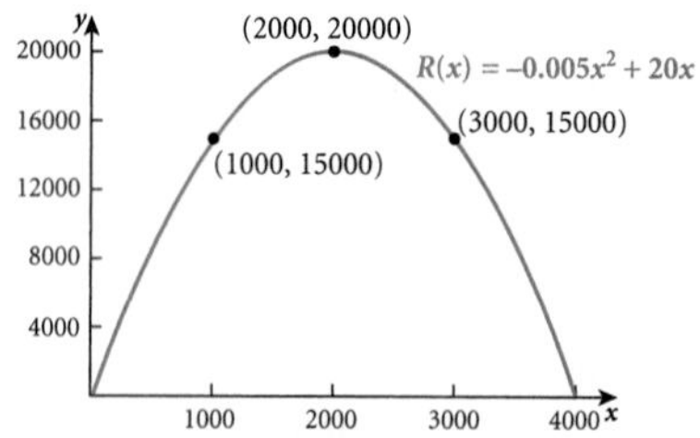

39. (a) The function is increasing on the interval $(0, 0.476)$
(b) The acreage of wheat planted from 2004 to 2008 will be decreasing.

41. (a) The yield will be increasing for amounts of nitrogen between 0 units and $\frac{20{,}164}{19{,}321} \approx 1.044$ units.
(b) The yield will be decreasing for amounts of nitrogen greater than $\frac{20{,}164}{19{,}321} \approx 1.044$ units.

43. $c = \frac{1}{2}$

45. $c = 0$

47. $c = \frac{3}{2}$

49. $c = \frac{2}{\sqrt[3]{3}} = \frac{2\sqrt[3]{9}}{3}$

Exercise 14.3 (p. 818)

1. The domain is $[x_1, x_4)$ or $(x_4, x_7]$

3. The graph of f is increasing on the intervals (x_1, x_3), $(0, x_4)$, and (x_4, x_6).

5. $x = 0$ and $x = x_6$

7. (x_3, y_3), (x_6, y_6)

9. The graph of f is concave up on the intervals (x_1, x_3) and (x_3, x_4).

11. The line $x = 4$ is a vertical asymptote.

13. The graph is concave down on $(-\infty, 2)$ and is concave up on $(2, \infty)$. The point $(2, -15)$ is the only inflection point.

15. The graph is concave down on $(0, 1)$ and is concave up on $(-\infty, 0)$ and $(1, \infty)$. The points $(0, -1)$ and $(1, 4)$ are the inflection points.

17. The graph is concave down on $(-\infty, 1)$ and is concave up on $(1, \infty)$. The point $(1, 68)$ is the only inflection point.

19. The graph is concave down on $(-\infty, -1)$ and $(0, 1)$ and is concave up on $(-1, 0)$ and $(1, \infty)$. The points $(-1, 7)$, $(0, 10)$, and $(1, 13)$ are the inflection points.

21. The graph is concave down on $(-\infty, 1)$ and is concave up on $(1, \infty)$. The point $(1, -5)$ is the only inflection point.

23. The graph is concave down on $(0, \infty)$ and is concave up on $(-\infty, 0)$. The point $(0, 2)$ is the only inflection point.

25. The graph is concave down on $(-\infty, -2)$ and is concave up on $(-2, 0)$ and $(0, \infty)$. The point $(-2, -12\sqrt[3]{4})$ is the inflection point.

27. The graph is concave up on $(-\infty, 0)$ and $(0, \infty)$. There are no inflection points.

29. Step 1: The domain is the interval $(-\infty, \infty)$.
Step 2: The y-intercept is $(0, 1)$.
Step 3: The graph is increasing on the intervals $(-\infty, 0)$ and $(4, \infty)$ and is decreasing on the interval $(0, 4)$.
Step 4: There is a local maximum at $(0, 1)$, and there is a local minimum at $(4, -31)$.
Step 5: The tangent line is horizontal at $(0, 1)$ and $(4, -31)$.
Step 6: The end behavior is $y = x^3$.
Step 7: The graph is concave up on the interval $(2, \infty)$ and is

concave down on the interval $(-\infty, 2)$. The point $(2, -15)$ is the only inflection point.

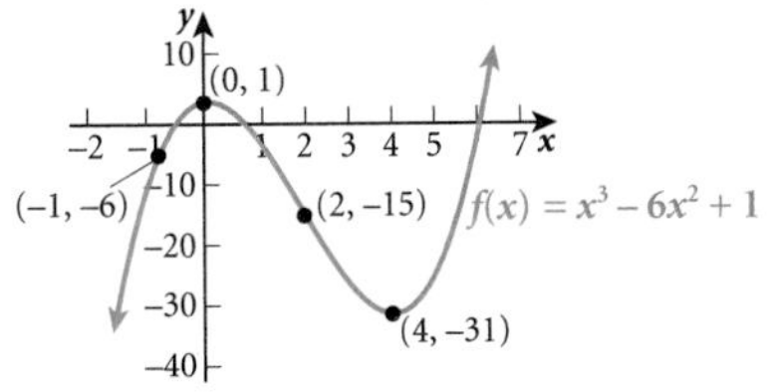

31. Step 1: The domain is the interval $(-\infty, \infty)$.
Step 2: The x-intercepts are $(-1, 0)$ and $(1, 0)$, and the y-intercept is $(0, 1)$.
Step 3: The graph is increasing on the intervals $(-1, 0)$ and $(1, \infty)$ and is decreasing on the intervals $(-\infty, -1)$ and $(0, 1)$.
Step 4: There is a local maximum at $(0, 1)$, and there are local minima at $(-1, 0)$ and $(1, 0)$.
Step 5: The tangent line is horizontal at $(-1, 0)$, $(0, 1)$, and $(1, 0)$.
Step 6: The end behavior is $y = x^4$.
Step 7: The graph is concave up on the intervals $\left(-\infty, -\frac{\sqrt{3}}{3}\right)$ and $\left(\frac{\sqrt{3}}{3}, \infty\right)$ and is concave down on the interval $\left(-\frac{\sqrt{3}}{3}, \frac{\sqrt{3}}{3}\right)$. The points $\left(-\frac{\sqrt{3}}{3}, \frac{4}{9}\right)$ and $\left(\frac{\sqrt{3}}{3}, \frac{4}{9}\right)$ are the inflection points.

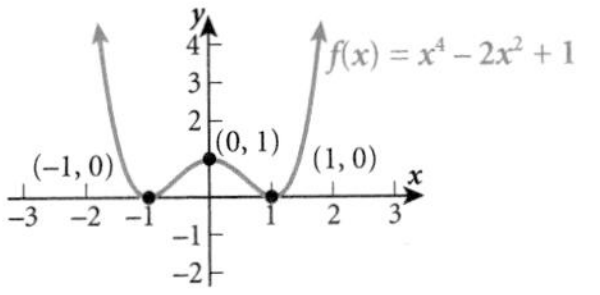

33. Step 1: The domain is the interval $(-\infty, \infty)$.
Step 2: The x-intercepts are $(10, 0)$ and $(0, 0)$, and the y-intercept is $(0, 0)$.
Step 3: The graph is increasing on the intervals $(-\infty, 0)$ and $(8, \infty)$ and is decreasing on $(0, 8)$.
Step 4: There is a local maximum at $(0, 0)$, and there is a local minimum at $(8, -8192)$.
Step 5: The tangent line is horizontal at $(0, 0)$ and $(8, -8192)$.
Step 6: The end behavior is $y = x^5$.
Step 7: The graph is concave up on the interval $(6, \infty)$ and is concave down on the interval $(-\infty, 6)$. The point $(6, -5184)$ is the only inflection point.

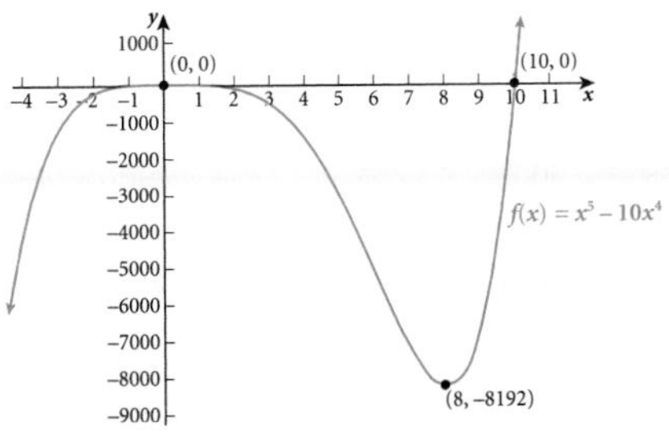

35. Step 1: The domain is the interval $(-\infty, \infty)$.
Step 2: The x-intercepts are $(3, 0)$ and $(0, 0)$, and the y-intercept is $(0, 0)$.
Step 3: The graph is increasing on the interval $\left(\frac{5}{2}, \infty\right)$ and is decreasing on the intervals $(-\infty, 0)$ and $\left(0, \frac{5}{2}\right)$.
Step 4: There is a local minimum at $\left(\frac{5}{2}, -\frac{3125}{64}\right)$.
Step 5: The tangent line is horizontal at $(0, 0)$ and $\left(\frac{5}{2}, -\frac{3125}{64}\right)$.
Step 6: The end behavior is $y = x^6$.
Step 7: The graph is concave up on the intervals $(-\infty, 0)$, and $(2, \infty)$ and is concave down on the interval $(0, 2)$. The points $(0, 0)$ and $(2, -32)$ are the inflection points.

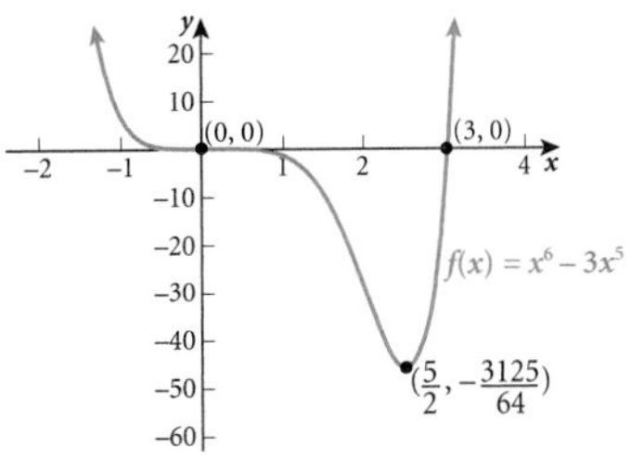

37. Step 1: The domain is the interval $(-\infty, \infty)$.
Step 2: The x-intercepts are $(4, 0)$ and $(0, 0)$, and the y-intercept is $(0, 0)$.
Step 3: The graph is increasing on the interval $(3, \infty)$ and is decreasing on the intervals $(-\infty, 0)$ and $(0, 3)$.
Step 4: There is a local minimum at $(3, -81)$.
Step 5: The tangent line is horizontal at $(0, 0)$ and $(3, -81)$.
Step 6: The end behavior is $y = 3x^4$.
Step 7: The graph is concave up on the intervals $(-\infty, 0)$ and $(2, \infty)$, and is concave down on the interval $(0, 2)$. The points $(0, 0)$ and $(2, -48)$ are the inflection points.

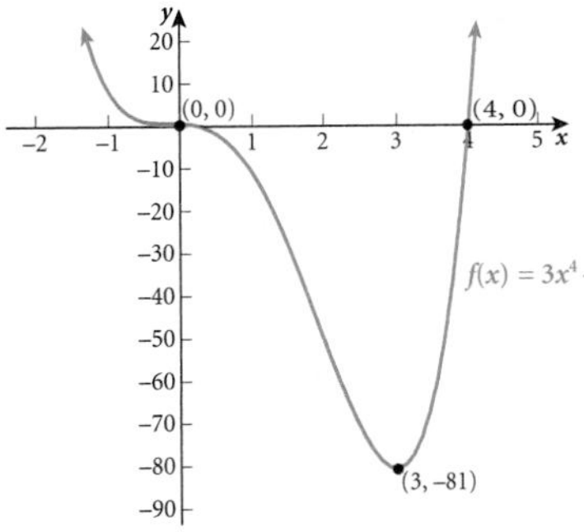

39. Step 1: The domain is the interval $(-\infty, \infty)$.
Step 2: The y-intercept is $(0, 4)$.
Step 3: The graph is increasing on the intervals $(-\infty, 0)$ and $(\sqrt[3]{4}, \infty)$ and is decreasing on the interval $(0, \sqrt[3]{4})$.
Step 4: There is a local maximum at $(0, 4)$, and there is a local minimum at $(\sqrt[3]{4}, 4-12\sqrt[3]{2})$.

Step 5: The tangent line is horizontal at $(0, 4)$ and $(\sqrt[3]{4}, 4-12\sqrt[3]{2})$.

Step 6: The end behavior is $y = x^5$.

Step 7: The graph is concave up on the interval $(1, \infty)$ and is concave down on the interval $(-\infty, 1)$. The point $(1, -5)$ is the only inflection point.

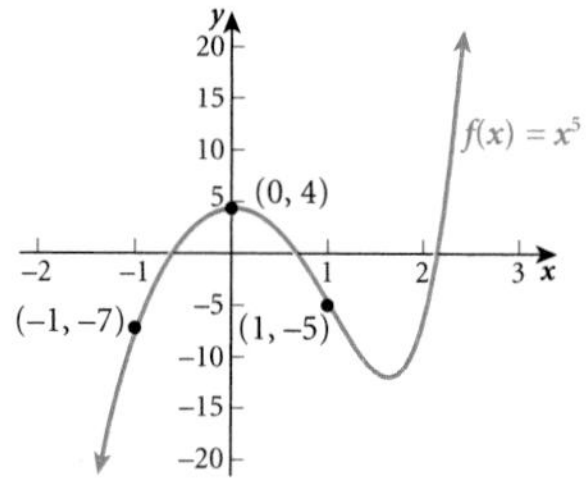

41. Step 1: The domain is the interval $(-\infty, \infty)$.

Step 2: The x-intercepts are $(10, 0)$ and $(0, 0)$, and the y-intercept is $(0, 0)$.

Step 3: The graph is increasing on the intervals $(-\infty, 0)$ and $(4, \infty)$ and is decreasing on the interval $(0, 4)$.

Step 4: There is a local maximum at $(0, 0)$, and there is a local minimum at $(4, -6\sqrt[3]{16})$.

Step 5: The tangent line is horizontal at $(4, -6\sqrt[3]{16})$ and is vertical at $(0, 0)$.

Step 6: The end behavior is $y = x^{5/3}$.

Step 7: The graph is concave up on the intervals $(-2, 0)$, $(0, \infty)$ and is concave down on the interval $(-\infty, -2)$. The point $(-2, -12\sqrt[3]{4})$ is the only inflection point.

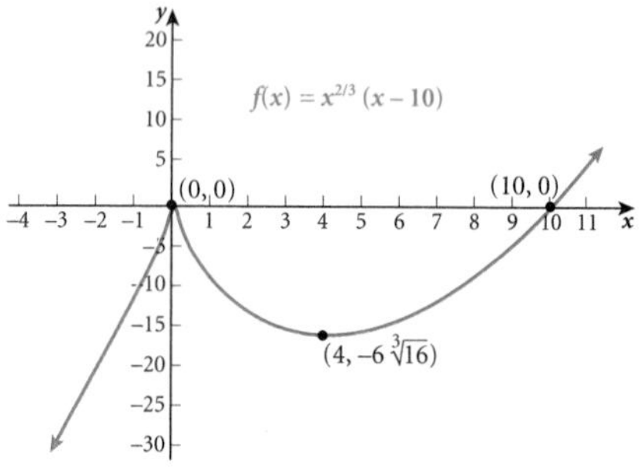

43. Step 1: The domain is the interval $(-\infty, \infty)$.

Step 2: The x-intercepts are $(-4, 0)$, $(4, 0)$, and $(0, 0)$, and the y-intercept is $(0, 0)$.

Step 3: The graph is increasing on the intervals $(-2, 0)$ and $(2, \infty)$ and is decreasing on the intervals $(-\infty, -2)$ and $(0, 2)$.

Step 4: There is a local maximum at $(0, 0)$, and there are local minima at $(-2, -12\sqrt[3]{4})$ and $(2, -12\sqrt[3]{4})$.

Step 5: The tangent line is horizontal at $(-2, -12\sqrt[3]{4})$ and $(2, -12\sqrt[3]{4})$ and is vertical at $(0, 0)$.

Step 6: The end behavior is $y = x^{8/3}$.

Step 7: The graph is concave up on the intervals $(-\infty, 0)$ and $(0, \infty)$. There is no inflection point.

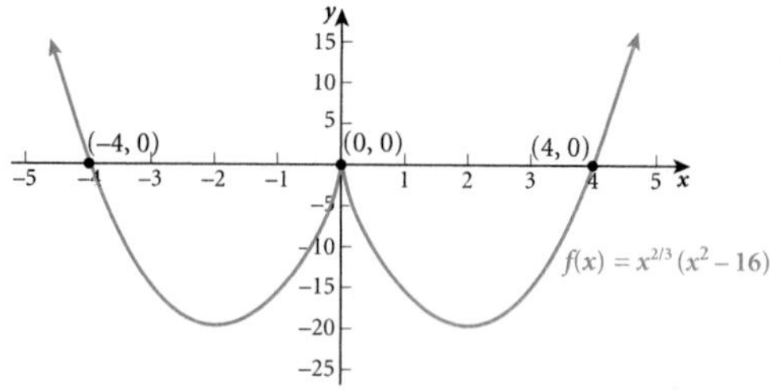

45. Step 1: The domain is the interval $(-\infty, \infty)$.

Step 2: The x-intercept and the y-intercept are both $(0, 0)$.

Step 3: The graph is increasing on the interval $(-1, \infty)$ and is decreasing on the interval $(-\infty, -1)$.

Step 4: There is a local minimum at $\left(-1, -\frac{1}{e}\right)$.

Step 5: The tangent line is horizontal at $\left(-1, -\frac{1}{e}\right)$.

Step 6: $\lim_{x\to\infty} xe^x = \infty$ and $\lim_{x\to-\infty} xe^x = 0$. The line $y = 0$ is a horizontal asymptote.

Step 7: The graph is concave up on the interval $(-2, \infty)$ and is concave down on the interval $(-\infty, -2)$. The point $\left(-2, -\frac{2}{e^2}\right)$ is the only inflection point.

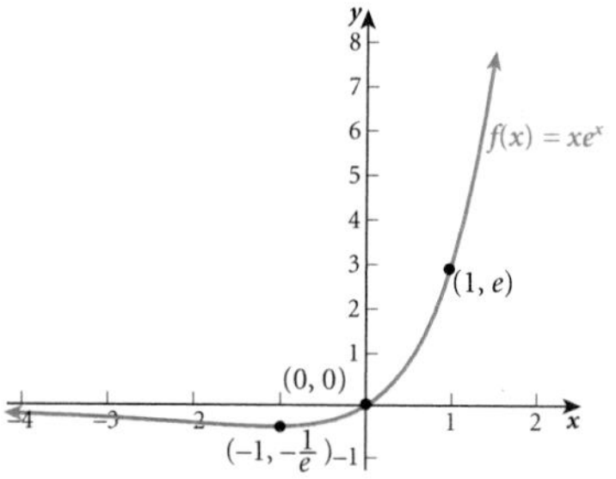

47. There is a local maximum at $(-1, 4)$ and a local minimum at $(1, 0)$.

49. There is a local minimum at $(-1, -4)$.

51. There is a local maximum at $(0, 2)$ and a local minimum at $(4, -254)$.

53. There is a local maximum at $(-1, -2)$ and a local minimum at $(1, 2)$.

59. $a = -3, b = 9$

61. (a) $\overline{C}(x) = 2x + \frac{50}{x}$

(b) The minimum average cost per item is \$20.

(c) $C'(x) = 4x$

(d)

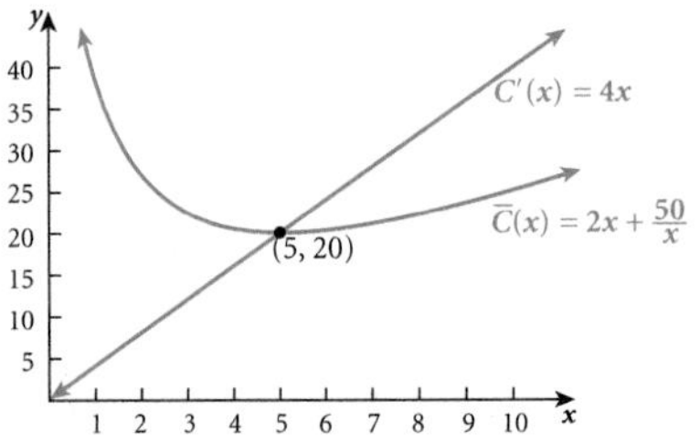

Point of intersection is (5, 20).

(e) The minimum average costs occurs at the production level for which the average cost equals the marginal cost.

63. (a) $\overline{C}(x) = \frac{500}{x} + 10 + \frac{x}{500}$

(b) The minimum average cost per item is \$12.

(c) $C'(x) = 10 + \frac{x}{250}$

(d)

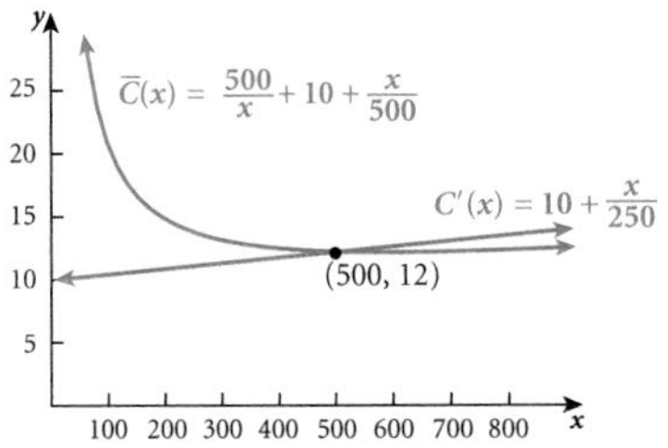

Point of intersection is (500, 12).

(e) The minimum average costs occurs at the production level for which the average cost equals the marginal cost.

65. (a) The domain of N is the interval $[0, \infty)$.

(b) The N-intercept is (0, 1). There is no t-intercept.

(c) N is increasing on the interval $(0, \infty)$.

(d) N is concave up on the interval $(0, \ln(49, 999))$ and is concave down on the interval $(\ln(49, 999), \infty)$.

(e) The inflection point is $(\ln(49, 999), 25, 000))$.

(f)

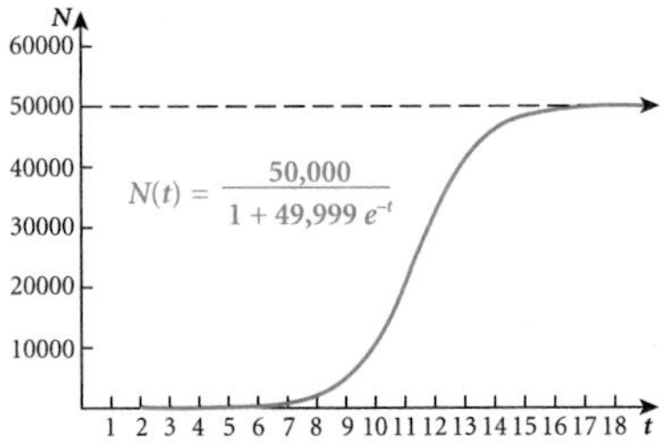

(g) The rumor is spreading at its greatest rate at $t = \ln(49, 999) \approx 10.82$ units of time since the rumor began to spread.

67. (a) $p'(t) = \dfrac{72, 000e^t}{(e^t + 9)^2}$

(b) The maximum growth rate occurs at $t = \ln(9) \approx$ 2.20 days.

(c) The equilibrium population is 8000.

(d)

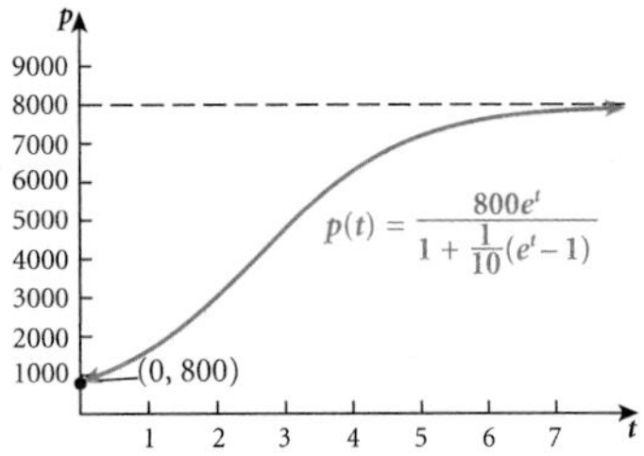

69. (a) The sales rate is a maximum at $x = \ln(50) \approx$ 3.91 months.

(b)

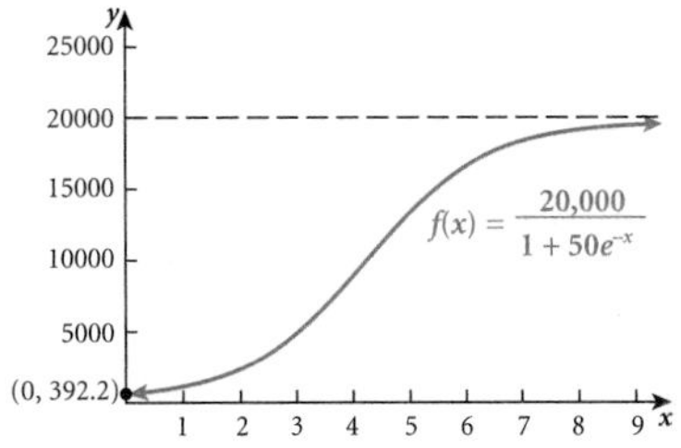

Exercise 14.4 (p. 831)

1. The absolute maximum is 15, and the absolute minimum is -1.

3. The absolute maximum is 1, and the absolute minimum is -39.

5. The absolute maximum is 16, and the absolute minimum is -4.

7. The absolute maximum is 1, and the absolute minimum is 0.

9. The absolute maximum is 1, and the absolute minimum is 0.

11. The absolute maximum is 4, and the absolute minimum is 2.

13. The absolute maximum is $\frac{1}{2}$, and the absolute minimum is $-\frac{1}{2}$.

15. The absolute maximum is 0, and the absolute minimum is $-\frac{1}{2}$.

17. The absolute maximum is $98\sqrt[3]{2}$, and the absolute minimum is 0.

19. The absolute maximum is $\frac{\sqrt[3]{12}}{9}$, and the absolute minimum is $-\sqrt[3]{2}$.

21. The absolute maximum is $10e^{10}$, and the absolute minimum is $-\frac{1}{e}$.

23. The absolute maximum is $\frac{1}{e}$, and the absolute minimum is 0.

25. The dimensions are 2 cm $\times$ 8 cm $\times$ 8 cm.

27. The dimensions are $20\sqrt[3]{2}$ cm $\times$ $20\sqrt[3]{2}$ cm $\times$ $10\sqrt[3]{2}$ cm $\approx$ 25.2 cm $\times$ 25.2 cm $\times$ 12.6 cm.

29. The radius is $\frac{4\sqrt[3]{25\pi^2}}{\pi} \approx 7.99$ cm, and the height is $\frac{10\sqrt[3]{25\pi^2}}{\pi} \approx 19.96$ cm.

31. The company should connect the telephone line 1.98 km from the box.

33. The most economical speed is 40 miles per hour.

35. The dimensions are 7 inches by 14 inches.

37.

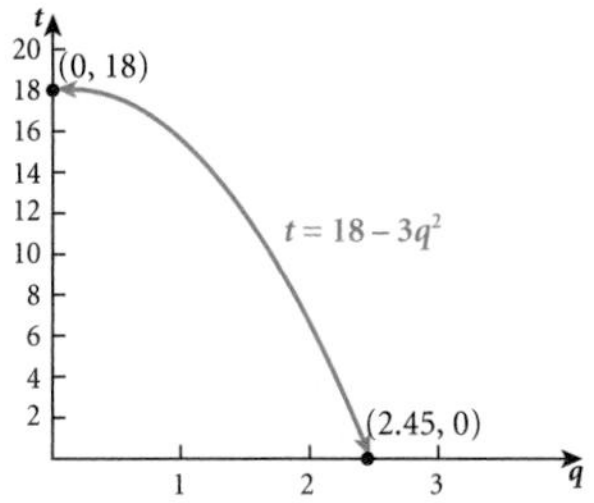

Demand for the product decreases as the tax rate increases. The optimal tax rate is 12%, and the revenue generated by this tax rate is 16.97 monetary units.

39. Let r be the radius, h be the height, and S be the surface area of the cylinder of volume V. We know that $V = \pi r^2 h$, so $h = \frac{V}{\pi r^2}$. The formula for S is
$S = 2\pi r^2 + 2\pi r h = 2\pi r^2 + \frac{2V}{r}$. Now $S'(r) = 4\pi r - \frac{2V}{r^2}$.

The only critical number of S is $r_c = \left(\frac{V}{2\pi}\right)^{1/3}$.
Furthermore, $S'(r) < 0$ if $0 < r < r_c$ and $S'(r) > 0$ if $r > r_c$. By the First Derivative Test, the surface area is least when $r = r_c = \left(\frac{V}{2\pi}\right)^{1/3}$. The height of the cylinder when $r = r_c$ is $h = \frac{V}{\pi r_c^2} = \left(\frac{V}{\pi}\right)\left(\frac{2\pi}{V}\right)^{2/3} = 2\left(\frac{V}{2\pi}\right)^{1/3} = 2r_c$.

41. The concentration is greatest two hours after the injection.

Exercise 14.5 (p. 836)

1. (a) $x = 4000 - 100p$ (b) $E(p) = \frac{p}{p - 40}$

(c) $E(5) = -0.143$, The demand decreases by approximately 1.43%.

(d) $E(15) = -0.6$, The demand decreases by approximately 6%.

(e) $E(20) = -1$, The demand decreases by approximately 10%.

3. (a) $x = 10{,}000 - 200p$ (b) $E(p) = \frac{p}{p - 50}$

(c) $E(10) = -0.25$, The demand decreases by approximately 1.25%.

(d) $E(25) = -1$, The demand decreases by approximately 5%.

(e) $E(35) = -2.333$, The demand decreases by approximately 11.665%.

5. $E(p) = \frac{p}{p - 200}$ $E(50) = -0.333$ The demand is inelastic.

7. $E(p) = -\frac{p}{p + 4}$ $E(10) = -0.714$ The demand is inelastic.

9. $E(p) = \frac{2p^2}{p^2 - 1000}$ $E(10) = -0.222$ The demand is inelastic.

11. $E(p) = \frac{p}{2p - 200}$ $E(10) = -0.056$ The demand is inelastic.

13. $E(p) = \frac{3p}{p - 4}$ $E(2) = -3$ The demand is elastic.

15. $E(p) = \frac{3\sqrt{p}}{6\sqrt{p} - 40}$ $E(4) = -0.214$ The demand is inelastic.

17. $E(4) = -1.23$ **19.** $E(5) = -1.18$

21. $E(5) = -39$ **23.** $E(2) = -0.125$

25. $E(100) = -3$

27. (a) The demand is elastic.
(b) The revenue will decrease.

29. (a) The demand is inelastic.
(b) The revenue will decrease.

31. (a) $x = 3000 - \frac{200}{3}p$
(b) $E(18) = -0.667$
(c) The demand will decrease by approximately 3.33%.
(d) The revenue will increase.

Exercise 14.6 (p. 844)

1. $\frac{dx}{dt} = -3$ **3.** $\frac{dx}{dt} = -\frac{8}{9}$

5. $\frac{dV}{dt} = 40$ **7.** $\frac{dV}{dt} = 5\pi^2$

9. The volume is increasing at a rate of 900 cubic centimeters per second.

11. The length of the leg of side length y is decreasing at a rate of $-\frac{8\sqrt{2009}}{2009} \approx 0.178$ centimeters per minute.

13. The surface area is shrinking at a rate of -0.75 square meters per minute.

15. The water level is rising at a rate of $\frac{1}{30}$ meter per minute.

17. The area of the spill is increasing at a rate of 316.67 square feet per minute.

19. (a) The rate of change in daily cost is \$500 per day.
(b) The rate of change in daily revenues is \$1480 per day.
(c) Revenue is increasing.
(d) $P(x) = -\dfrac{x^2}{10{,}000} + 10x - 5000$
(e) The rate of change in daily profit is \$980 per day.

21. Revenues are increasing at a rate of \$260,000 per year.

Exercise 14.7 (p. 851)

1. $dy = (3x^2 - 2)dx$

3. $dy = \dfrac{-x^2 + 2x - 6}{(x^2 + 2x - 8)^2}dx$

5. $\dfrac{dy}{dx} = -\dfrac{y}{x}\ \dfrac{dx}{dy} = -\dfrac{x}{y}$

7. $\dfrac{dy}{dx} = -\dfrac{x}{y}\ \dfrac{dx}{dy} = -\dfrac{y}{x}$

9. $\dfrac{dy}{dx} = \dfrac{2xy - x^2}{y^2 - x^2}\ \dfrac{dx}{dy} = \dfrac{y^2 - x^2}{2xy - y^2}$

11. $d(\sqrt{x - 2}) = \dfrac{1}{2\sqrt{x - 2}}dx$

13. $d(x^3 - x - 4) = (3x^2 - 1)dx$

15. $y = 2x - 3$

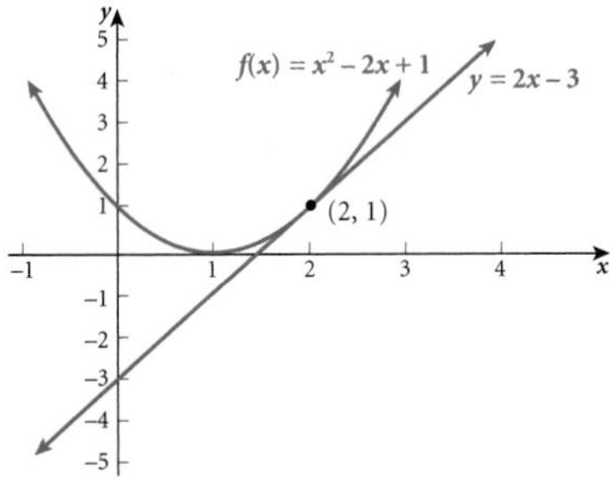

17. $y = \dfrac{1}{4}x + 1$

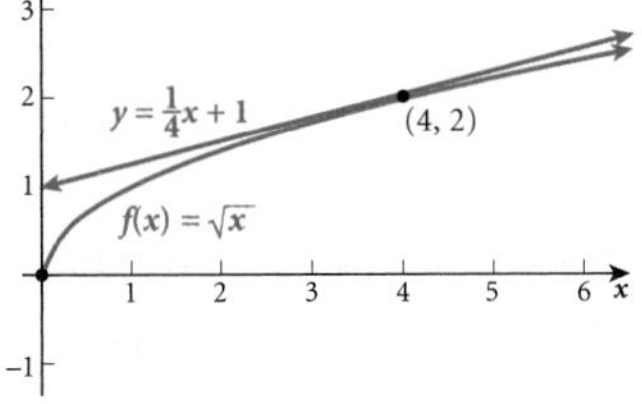

19. $y = x + 1$

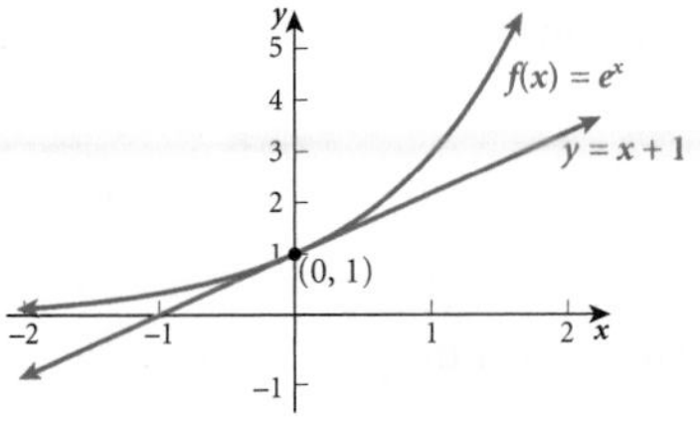

21. (a) The change is approximately 0.006.
(b) The change is approximately 0.00125.

23. The approximate increase in surface area is $2\pi \approx$ 6.28 square centimeters.

25. The approximate increase in volume is $3.6\pi \approx$ 11.31 cubic meters.

27. The percentage error is 6%.

29. The approximate loss is $0.8\pi \approx 25.13$ cubic centimeters.

31. The estimated height is 30 meters, and the percentage error of this estimate is 0.9%.

33. The clock will lose 70.28 minutes each day.

CHAPTER 14 Review

True – False Items (p. 854)

1. False **2.** True **3.** False
4. False **5.** True **6.** False

Fill in the Blanks (p. 854)

1. decreasing
2. decreasing, increasing
3. concave up
4. horizontal
5. concavity
6. $f'(x)\,dx$
7. linear approximation

Review Exercises (p. 854)

1. The graph has no horizontal or vertical tangent lines.

3. The graph has a horizontal tangent line at $\left(2, \dfrac{\sqrt[3]{2}}{6}\right)$ and a vertical tangent line at $(0, 0)$.

5. (a) Yes (b) No (c) vertical tangent line

7. (a) Yes (b) No (c) no tangent line

9. (a) The graph is increasing on the intervals $(-\infty, -2)$ and $(2, \infty)$ and is decreasing on the interval $(-2, 2)$.
(b) There is a local maximum at $\left(-2, \dfrac{48}{5}\right)$ and a local minimum at $\left(2, -\dfrac{48}{5}\right)$.

11. (a) The graph is increasing on the intervals $(-\infty, -2\sqrt{2})$ and $(-2\sqrt{2}, 0)$ and is decreasing on the intervals $(0, 2\sqrt{2})$ and $(2\sqrt{2}, \infty)$.
(b) The graph has a local maximum at $(0, 0)$.

13. (a) The graph is decreasing on $(-\infty, \infty)$.
(b) The graph has no local maxima or minima.

15. (a) The domain is the interval $(-\infty, \infty)$.
(b) The x-intercept is $(1, 0)$, and the y-intercept is $(0, -1)$.
(c) The graph is increasing on the interval $(-\infty, \infty)$.
(d) The graph has no local maxima or minima.

(e) The graph has a horizontal tangent at $(1, 0)$.
(f) The end behavior is $y = x^3$.
(g) The graph is concave down on the interval $(-\infty, 1)$ and is concave up on the interval $(1, \infty)$.The point $(1, 0)$ is the only inflection point.

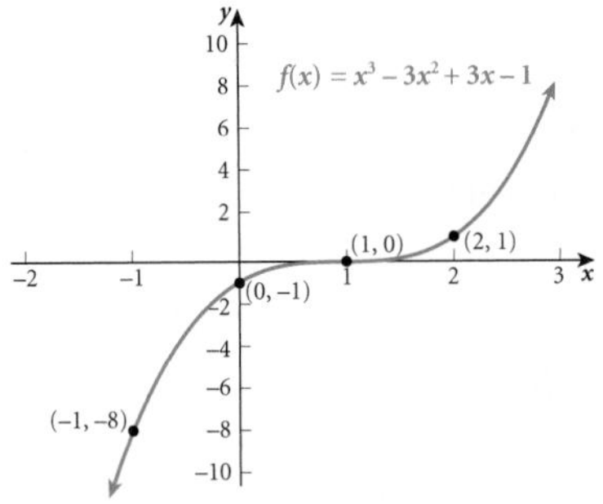

17. (a) The domain is the interval $(-\infty, \infty)$.
(b) The x-intercepts are $(-\sqrt[4]{5}, 0)$, $(0, 0)$, and $(-\sqrt[4]{5}, 0)$, and the y-intercept is $(0, 0)$.
(c) The graph is increasing on the intervals $(-\infty, -1)$ and $(1, \infty)$ and is decreasing on the interval $(-1, 1)$.
(d) The graph has a local maximum at $(-1, 4)$ and a local minimum at $(1, -4)$.
(e) The tangent lines to the graph are horizontal at $(-1, 4)$ and $(1, -4)$.
(f) The end behavior is $y = x^5$.
(g) The graph is concave down on the interval $(-\infty, 0)$ and is concave up on the interval $(0, \infty)$. The point $(0, 0)$ is the only inflection point.

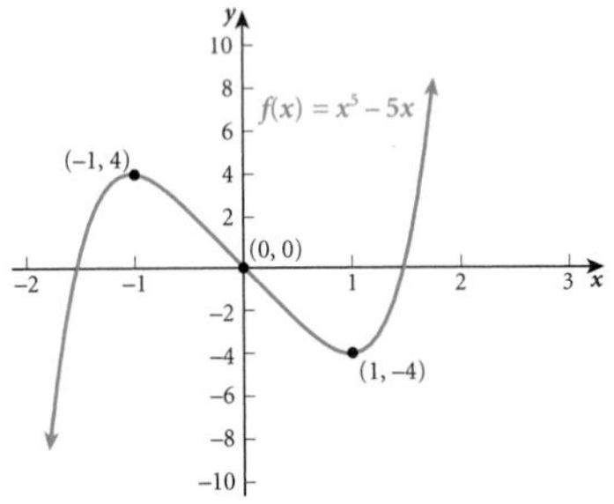

19. (a) The domain is the interval $(-\infty, \infty)$.
(b) The x-intercepts are $(-4, 0)$ and $(0, 0)$. The y-intercept is $(0, 0)$.
(c) The graph is increasing on the interval $(-1, \infty)$ and is decreasing on the interval $(-\infty, -1)$.
(d) The graph has a local minimum at $(-1, -3)$.
(e) The graph has a horizontal tangent at $(-1, -3)$ and a vertical tangent at $(0, 0)$.
(f) The end behavior is $y = x^{4/3}$.
(g) The graph is concave up on the intervals $(-\infty, 0)$ and $(2, \infty)$ and is concave down on the interval $(0, 2)$.

The points $(0, 0)$ and $(2, 6\sqrt[3]{2})$ are the inflection points.

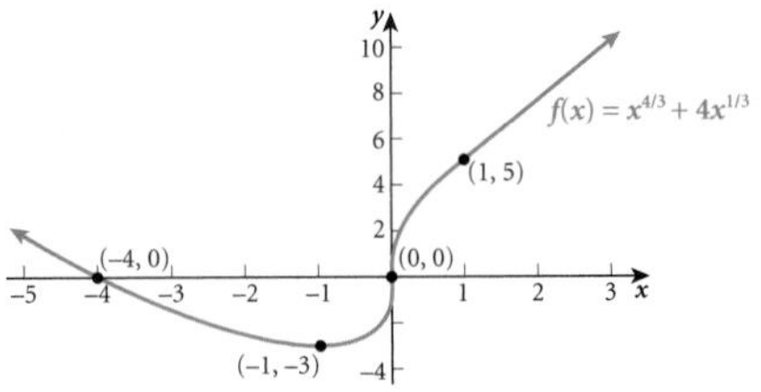

21. (a) The domain is the interval $(-\infty, \infty)$.
(b) The x-intercept and the y-intercept are both $(0, 0)$.
(c) The graph is increasing on the interval $(-1, 1)$ and is decreasing on the intervals $(-\infty, -1)$ and $(1, \infty)$.
(d) The graph has a local maximum at $(1, 1)$ and a local minimum at $(-1, -1)$.
(e) The graph has horizontal tangent lines at $(-1, -1)$ and $(1, 1)$.
(f) The end behavior is $y = 0$, which is a horizontal asymptote.
(g) The graph is concave down on the intervals $(-\infty, -\sqrt{3})$ and $(0, \sqrt{3})$ and is concave up on the intervals $(-\sqrt{3}, 0)$ and $(\sqrt{3}, \infty)$. The points $\left(-\sqrt{3}, -\frac{\sqrt{3}}{2}\right)$, $(0, 0)$, and $\left(\sqrt{3}, \frac{\sqrt{3}}{2}\right)$ are the inflection points.

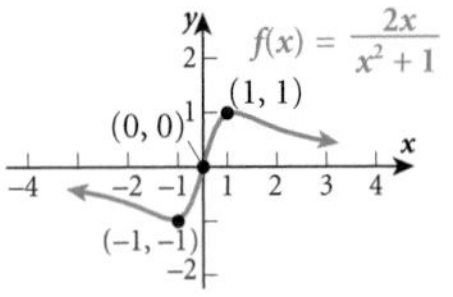

23. There is a local maximum at $\left(-\frac{1}{2}, 1\right)$ and a local minimum at $\left(\frac{1}{2}, -1\right)$.

25. There is a local maximum at $(0, 0)$ and local minima at $(-1, -1)$ and $(1, -1)$.

27. There is a local minimum at $\left(-1, -\frac{1}{e}\right)$.

29. The absolute maximum is 8, and the absolute minimum is –1.

31. The absolute maximum is 9, and the absolute minimum is 0.

33. The absolute maximum is 8, and the absolute minimum is –3.

35. $E(p) = \frac{2p^2}{p^2 - 500}$, The demand is elastic at $p = 20$.

37. $E(p) = \frac{p^2}{p^2 - 500}$, The demand is inelastic at $p = 10$.

39. (a) The demand function is elastic.
(b) Revenue will decrease if the price is raised.

41. $dy = (12x^3 - 6x^2 + 1)dx$ **43.** $dy = -\frac{5}{(x+1)^2}dx$

45. $y = 6x - 18$

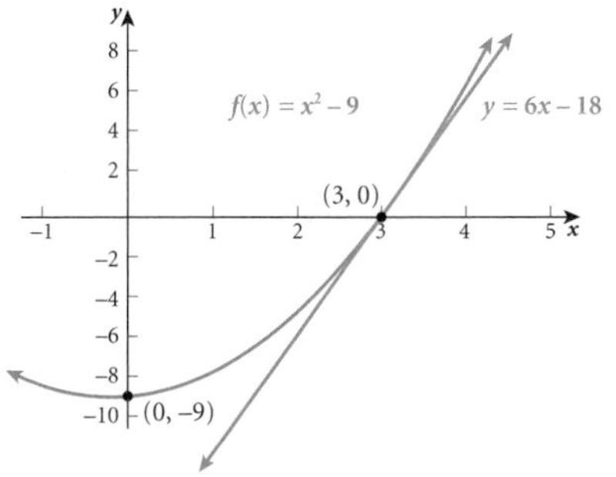

47. $\frac{dx}{dt} = -3$ **49.** $\frac{dy}{dt} = -\frac{9}{29}$

51. The surface area is increasing at a rate of $\frac{20}{3}$ square meters per minute.

53. (a) $\bar{C}(x) = 5x + \frac{1125}{x}$

(b) The minimum average cost is \$150.

(c) $C'(x) = 10x$.

(d)

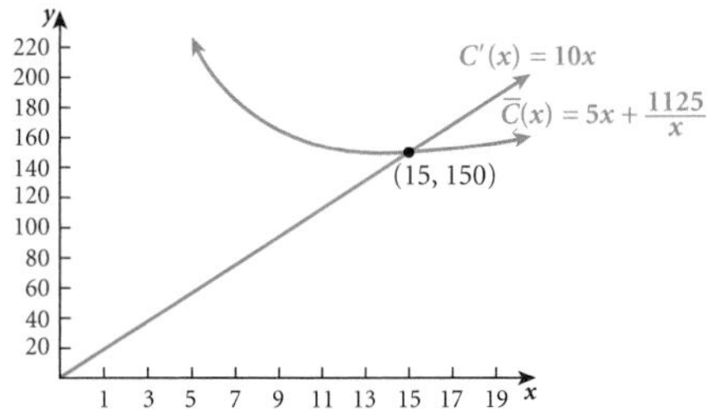

Intersection point is (15,150).

(e) The minimum average cost occurs at the production level where the average cost equals the marginal cost.

55. 98 units need to be sold.

57. The radius is $\frac{5\sqrt[3]{2\pi^2}}{\pi} \approx 4.30$ centimeters, and the height is $\frac{10\sqrt[3]{2\pi^2}}{\pi} \approx 8.60$ centimeters.

59. The decrease in area is approximately $40\pi \approx 125.7$ square millimeters.

61. (a) The demand decreases by approximately 1800 pounds.
(b) The demand decreases by approximately 9000 pounds.

63. (a) The concentration increases by approximately 0.0071 units.
(b) The concentration decreases by approximately 0.0208 units.

65. (b)

Mathematical Questions from Professional Exams (p. 858)

1. (b) $6X^2 + 8X + 3$ **2.** (d) $4X^2 + 6X + 2 + \frac{10}{X}$

3. (a) The average cost function multiplied by X.

4. (c) Substitute the solution(s) in the second derivative equation, and a positive solution indicates a minimum.

5. (a) $y = x^3e^{x^2}$ **6.** (b) Dec, 96

7. (d) 8,645 per year **8.** (d) 1, −2

9. (c) (0, 2) **10.** (c) $\frac{1}{3}$

11. (c) $\frac{7e^3}{4}$ **12.** (b) $-\frac{1}{2}$

13. (a) $\frac{1}{2\pi}$ **14.** (b) 0

15. (d) 54 **16.** (b) $-\frac{16}{5}$

CHAPTER 15 The Integral of a Function and Applications

Exercise 15.1 (p. 870)

1. $F(x) = \frac{x^4}{4} + K$ **3.** $F(x) = x^2 + 3x + K$

5. $F(x) = 4 \ln |x| + K$ **7.** $F(x) = \frac{2\sqrt{3}}{3}x^{3/2} + K$

9. $3x + K$ **11.** $\frac{x^2}{2} + K$

13. $\frac{3x^{4/3}}{4} + K$ **15.** $-\frac{1}{x} + K$

17. $2x^{1/2} + K$ **19.** $\frac{x^4}{2} + \frac{5x^3}{2} + K$

21. $\frac{x^3}{3} + 2e^x + K$ **23.** $\frac{x^4}{4} - \frac{2x^3}{3} + \frac{x^2}{2} - x + K$

25. $x - \ln |x| + K$ **27.** $2e^x - 3 \ln |x| + K$

29. $3x + 2\sqrt{x} + K$ **31.** $\frac{x^2}{2} - 2x + K$

33. $\frac{x^3}{3} - \frac{x^2}{2} + K$ **35.** $\frac{3x^5}{5} + 2 \ln |x| + K$

37. $4x + e^x + K$ **39.** $R(x) = 600x$

41. $R(x) = 10x^2 + 5x$

43. $C(x) = 7x^2 - 2800x + 4300$. The cost is a minimum when $x = 200$.

45. $C(x) = 10x^2 - 8000x + 500$. The cost is a minimum when $x = 400$.

47. (a) $C(x) = 9000 + 1000x - 10x^2 + \dfrac{x^3}{3}$

(b) $R(x) = 3400x$

(c) $P(x) = -9000 + 2400x + 10x^2 - \dfrac{x^3}{3}$

(d) A sales volume of 60 units yields maximum profit.

(e) The profit is $99,000.

(f)

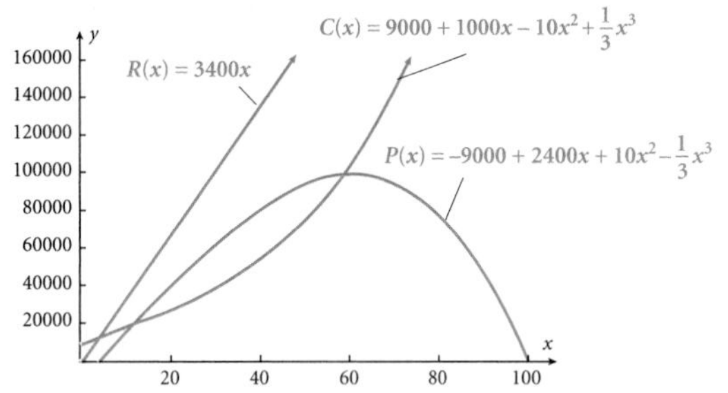

49. There will be 1,142,462 inmates in 2008.

51. The population will be 20,055 people in 10 months.

53. There will be 22,700 voting citizens in 3 years.

55. $2 - 2\ln 2 \approx 0.614$ milligrams were produced.

57. The reservoir will be empty after 500.28 days.

Exercise 15.2 (p. 876)

1. $\dfrac{(2x+1)^6}{12} + K$

3. $\dfrac{e^{2x-3}}{2} + K$

5. $\dfrac{1}{6-4x} + K$

7. $\dfrac{(x^2+4)^3}{6} + K$

9. $\dfrac{e^{x^3+1}}{3} + K$

11. $\frac{1}{2}(e^{x^2} - e^{-x^2}) + K$

13. $\dfrac{(x^4+3)^7}{21} + K$

15. $\dfrac{3(1+x^2)^{2/3}}{4} + K$

17. $\dfrac{2(x+3)^{5/2}}{5} - 2(x+3)^{3/2} + K$

19. $\ln(e^x + 1) + K$

21. $2e^{\sqrt{x}} + K$

23. $\dfrac{3(x^{1/3}-1)^7}{7} + K$

25. $\dfrac{-1}{2(x^2+2x+3)} + K$

27. $-\dfrac{2}{3(1+\sqrt{x})^4} + K$

29. $\dfrac{\ln|2x+3|}{2} + K$

31. $\dfrac{\ln(4x^2+1)}{8} + K$

33. $\dfrac{\ln(x^2+2x+2)}{2} + K$

35. The value of the car after two years is $19,414.55. The value of the car after four years is $16,624.02.

37. (a) $B(t) = 68.6e^{0.025t}$

(b) The budget will exceed $100 billion when $t = 15.08$, which is during January, 2016.

39. (a) $N(t) = 2000\,e^{0.01t} - 1600$

(b) The work force will reach 800 employees in 18.23 years.

41. Let $u = ax + b$. Then $du = a\,dx$, so $\frac{1}{a}\,du = dx$. Substituting, we have

$$\begin{aligned}\int (ax+b)^n\,dx &= \int u^n \cdot \frac{1}{a}\,du\\ &= \frac{1}{a}\int u^n\,du\\ &= \frac{1}{a}\frac{1}{n+1}u^{n+1} + K\\ &= \frac{(ax+b)^{n+1}}{a(n+1)} + K\end{aligned}$$

Exercise 15.3 (p. 881)

1. $\dfrac{xe^{4x}}{4} - \dfrac{e^{4x}}{16} + K$

3. $\dfrac{xe^{2x}}{2} - \dfrac{e^{2x}}{4} + K$

5. $-(x^2 + 2x + 2)e^{-x} + K$

7. $\dfrac{2x^{3/2}\ln x}{3} - \dfrac{4x^{3/2}}{9} + K$

9. $x(\ln x)^2 - 2x\ln x + 2x + K$

11. $\dfrac{x^3 \ln 3x}{3} - \dfrac{x^3}{9} + K$

13. $\dfrac{x^3(\ln x)^2}{s} - \dfrac{2x^3\ln x}{9} + \dfrac{2x^3}{27} + K$

15. $\dfrac{(\ln x)^2}{2} + K$

17. The population is 5389 ants after four days and is 6012 ants after one week.

19. The car is worth $14,061.64 after 2 years. The car is worth $9,640.02 after 4 years.

Exercise 15.4 (p. 889)

1. $\dfrac{7}{2}$

3. e

5. $\dfrac{2}{3}$

7. $-\dfrac{1}{15}$

9. $\dfrac{5}{3}$

11. $\dfrac{5}{24}$

13. $3\ln 2 + \dfrac{95}{3}$

15. $\dfrac{20}{3}$

17. 0

19. $\dfrac{8}{3}$

21. $\dfrac{e^2 - 3}{2}$

23. $1 - \dfrac{1}{e}$

25. $\ln 2$

27. $\dfrac{2\ln 2}{3}$

29. $\dfrac{5e^6}{4} - \dfrac{e^2}{4}$

31. $\dfrac{4}{9e^3} - \dfrac{7}{9e^6}$

33. $5\ln 5 - 4$

35. 0

37. 0

39. 2

41. 64

43. 36

45. 12

47. The cost increase is $567,000.

49. (a) The total number of labor-hours needed is 15,921.

(b) The total number of labor-hours needed is 35,089.

(c) The quantity k represents the difficulty of learning a new skill. The closer k is to 0, the longer the time required to master the new skill.

51. The projected deficit is $197 billion.

53. The total sales during the first year were $13,450.01.

55. The total number of labor-hours needed is 3660.

57. (a) $\dfrac{2}{3}$

(b) $\dfrac{16}{15}$

Exercise 15.5 (p. 902)

1. Area = 56

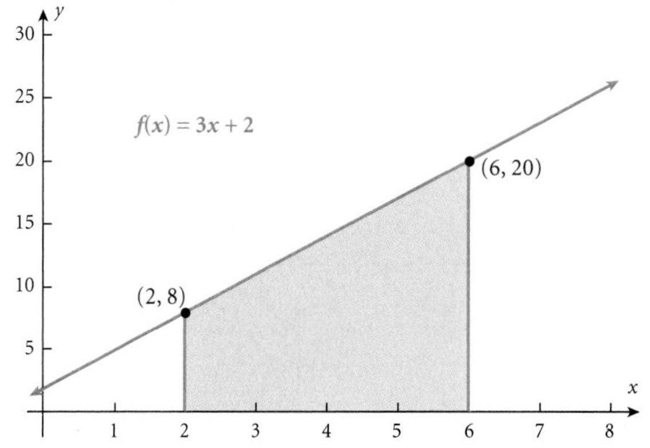

3. Area $= \frac{8}{3}$

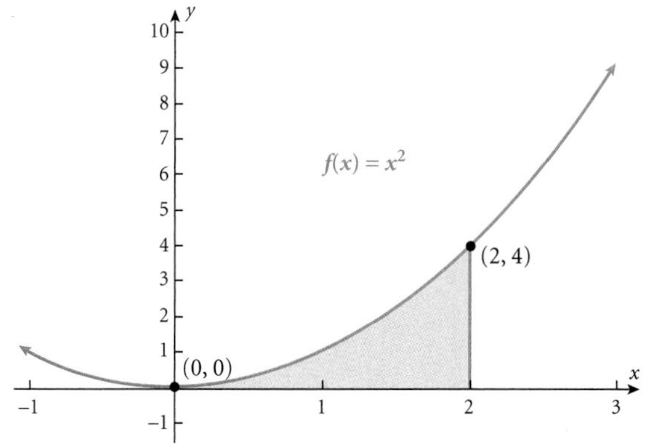

5. Area $= \frac{8}{3}$

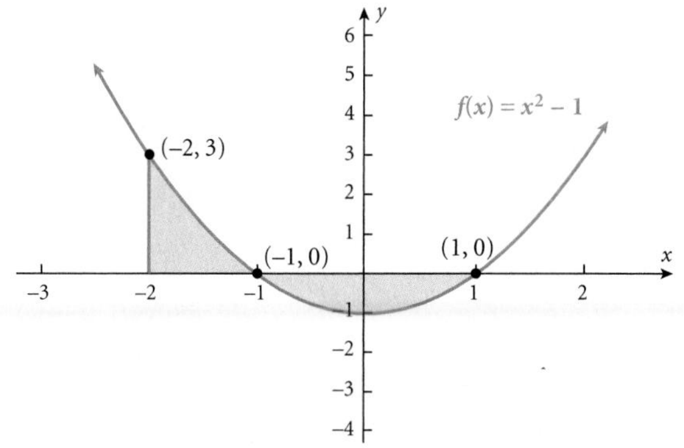

7. Area $= \frac{51}{4}$

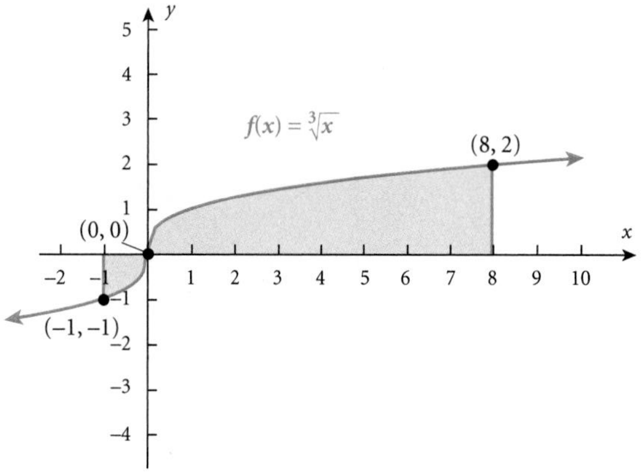

9. Area $= e - 1$

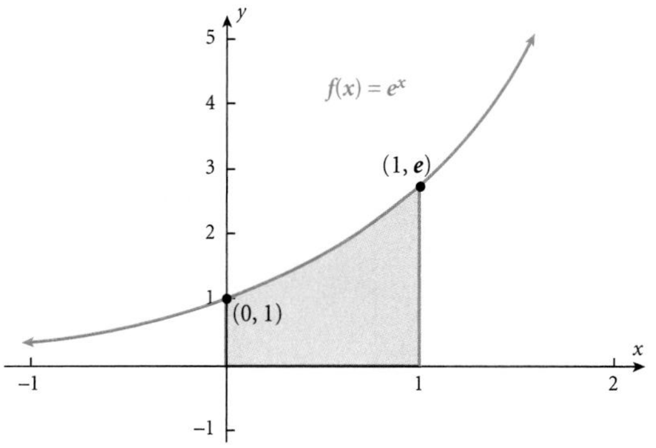

11. Area $= \frac{1}{2}$

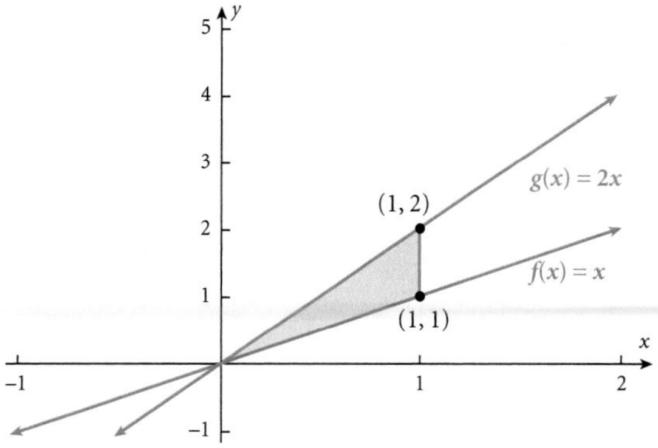

13. Area $= \frac{1}{6}$

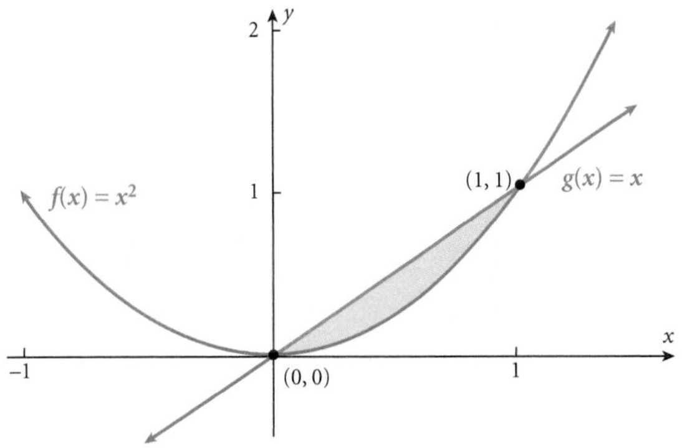

15. Area $= \frac{1}{6}$

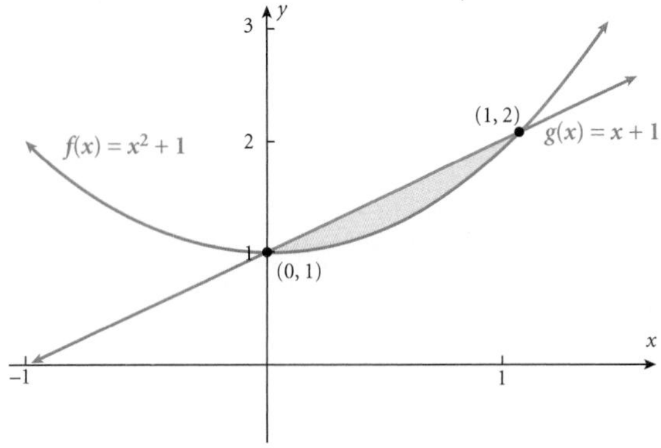

17. Area $= \frac{5}{12}$

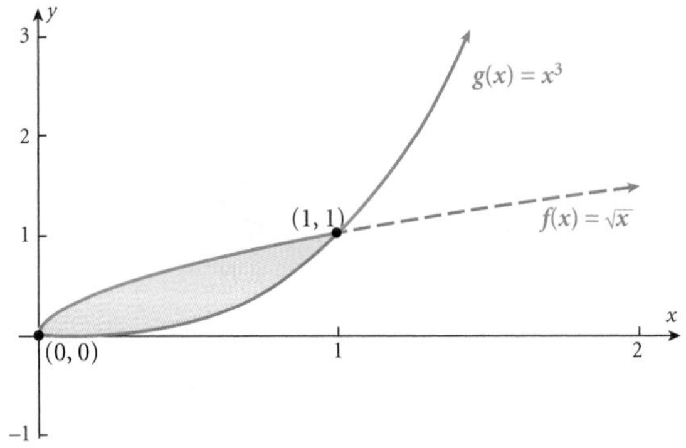

19. Area $= \frac{4}{15}$

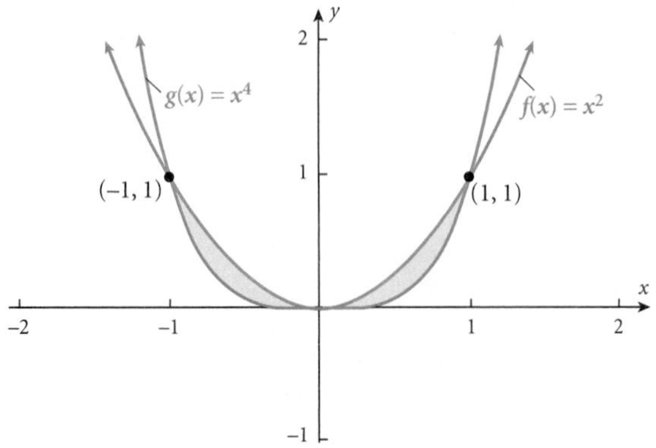

21. Area $= \frac{8}{3}$

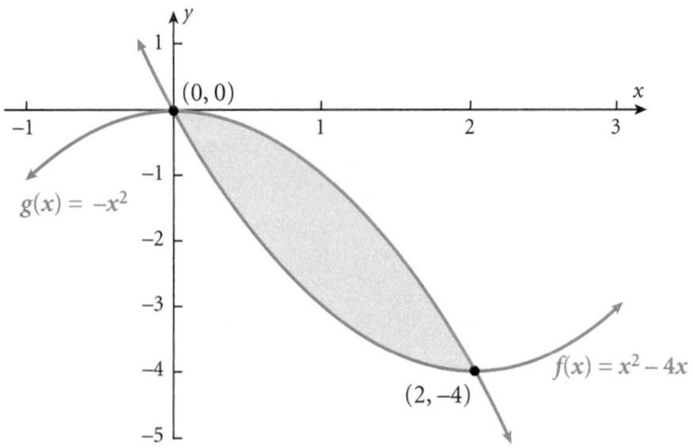

23. Area $= \frac{9}{2}$

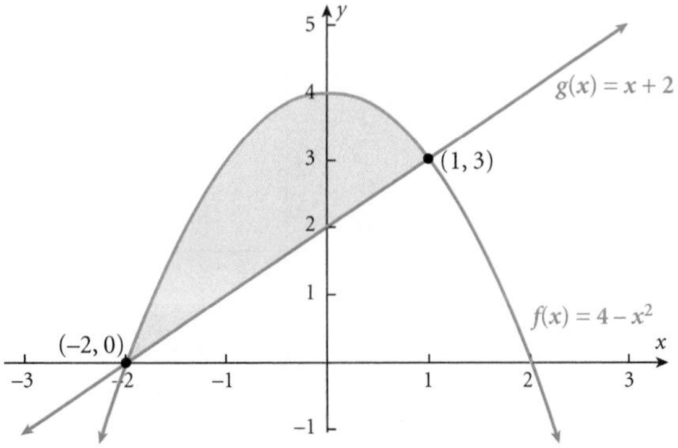

25. Area = 8

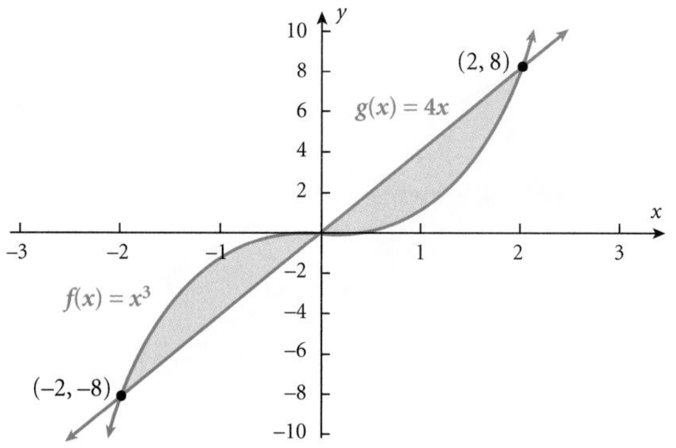

27. Area = $\frac{1}{3}$

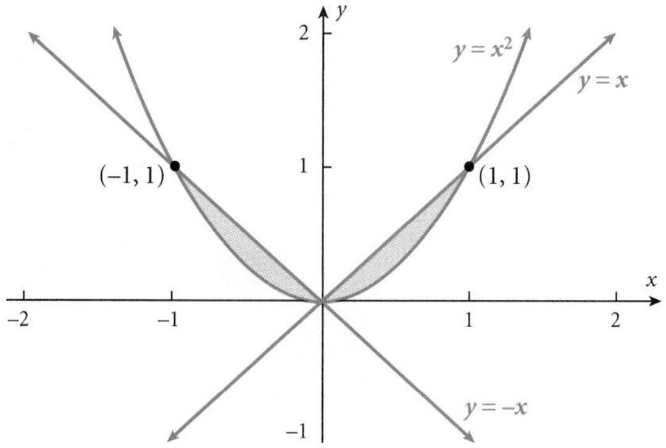

29. (a) The integral represents area below the line $y = 3x + 1$, above the the x-axis, and between the vertical lines $x = 0$ and $x = 4$.

(b)

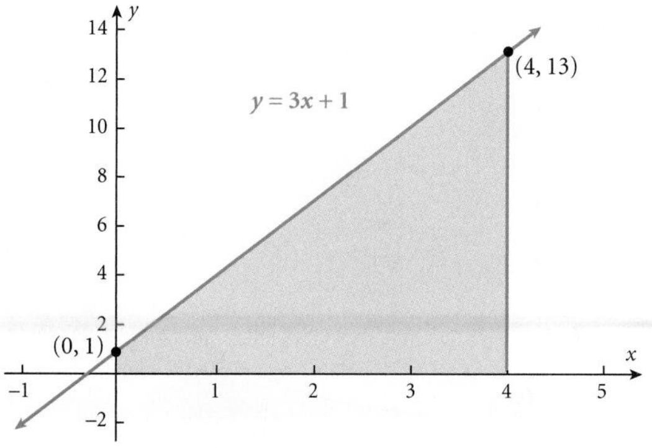

(c) 28

31. (a) The integral represents area below the graph of $y = x^2 - 1$, above the the x-axis, and between the vertical lines $x = 2$ and $x = 5$.

(b)

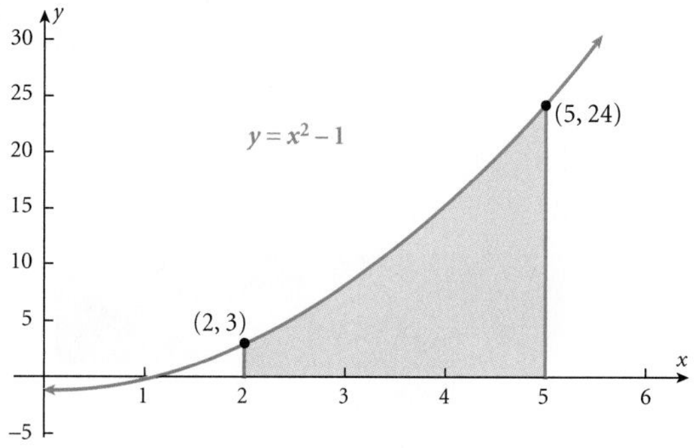

(c) 36

33. (a) The integral represents area below the graph of $y = e^x$, above the the x-axis, and between the vertical lines $x = 0$ and $x = 2$.

(b)

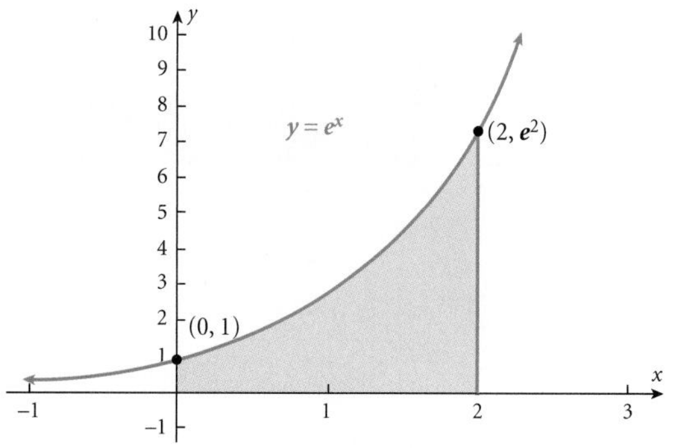

(c) $e^2 - 1$

35. The consumer's surplus is $\$\frac{40}{9} \approx \$4.44$, and the producer's surplus is $\$\frac{32}{9} \approx \$3.56$.

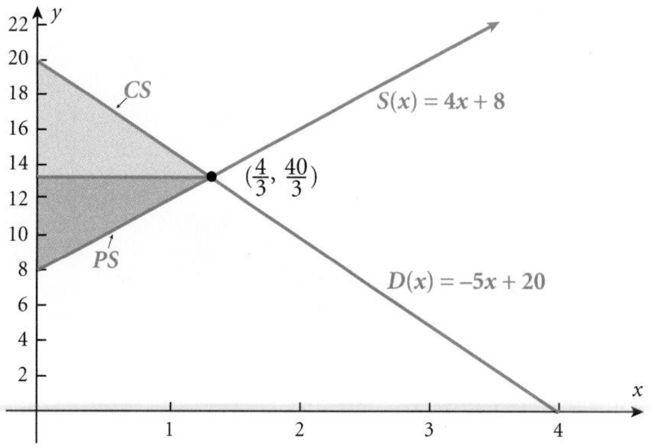

37. The operation should continue for 16 years. The profit that can be generated during this period is \$85.33 million.

39. (a) $c = \frac{\sqrt{3}}{3}$

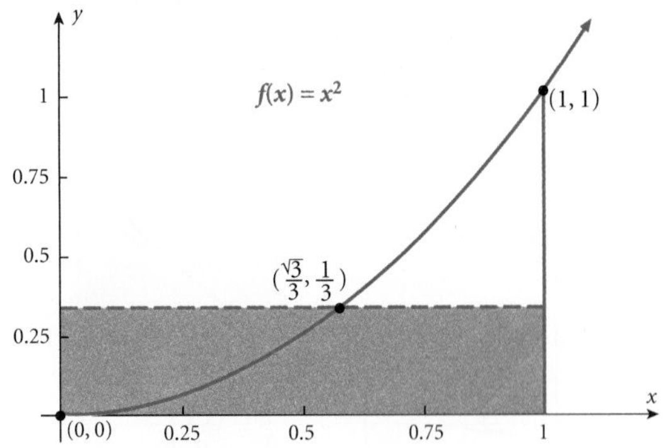

(b) $c = 2$

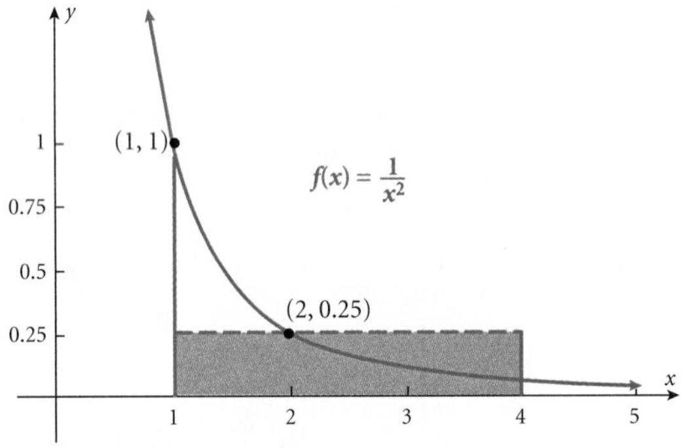

41. (a) x^2

(b) $\sqrt{x^2 - 2}$

(c) $\sqrt{x^x + 2x}$

Exercise 15.6 (p. 908)

1. $A \approx 3$ **3.** $A \approx 56$

5. (a)

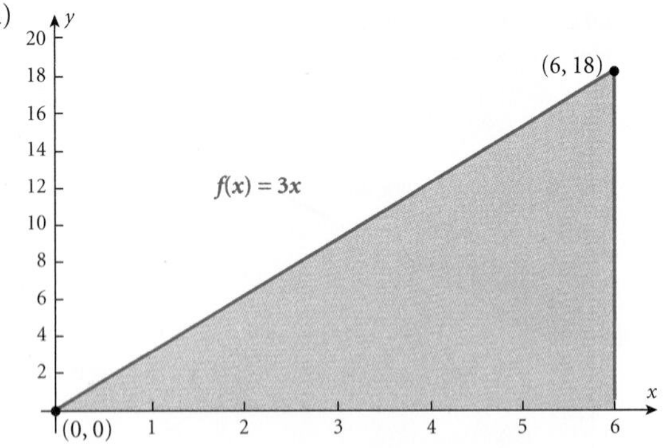

(b) $A \approx 36$

(c) $A \approx 72$

(d) $A \approx 45$

(e) $A \approx 63$

(f) $A = 54$

7. (a)

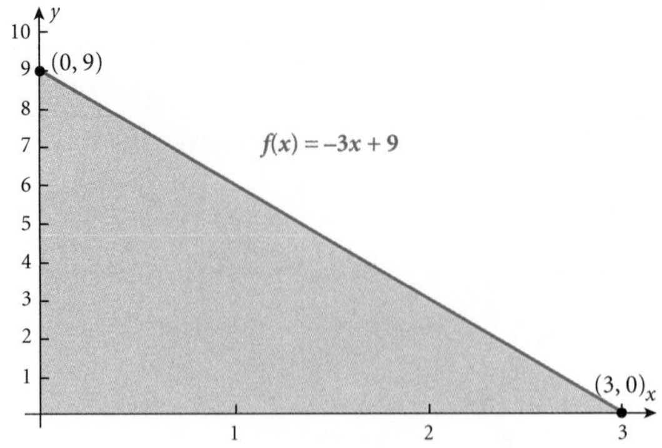

(b) $A \approx 18$

(c) $A \approx 9$

(d) $A \approx \frac{63}{4}$

(e) $A \approx \frac{45}{4}$

(f) $A = \frac{27}{2}$

9. (a)

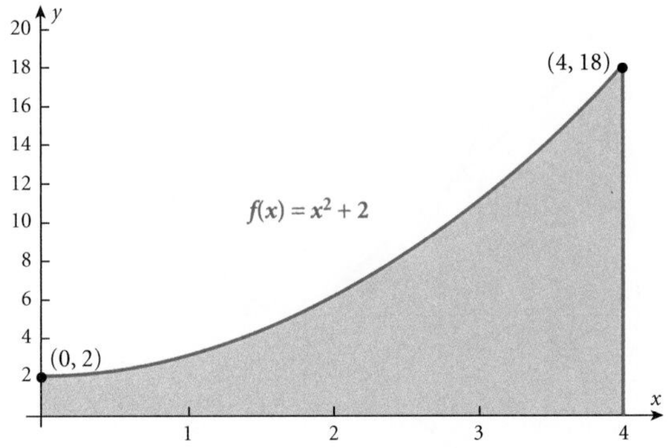

(b) $A \approx 22$

(c) $A \approx \frac{51}{2}$

(d) $A = \int_0^4 (x^2 + 2)dx$

(e) $A = \frac{88}{3}$

11. (a)

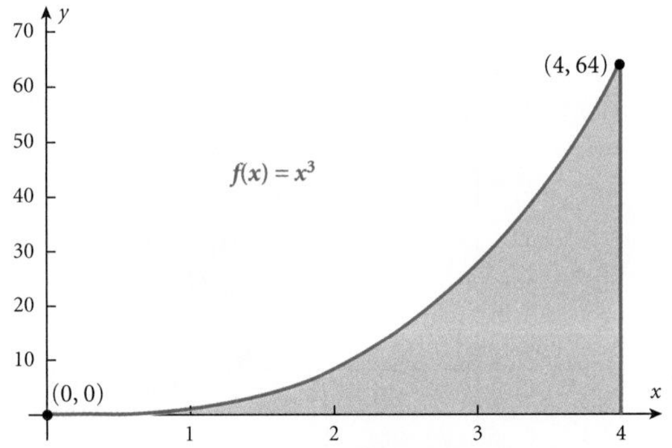

(b) $A \approx 36$

(c) $A \approx 49$

(d) $A = \int_0^4 x^3 dx$

(e) $A = 64$

13. (a)

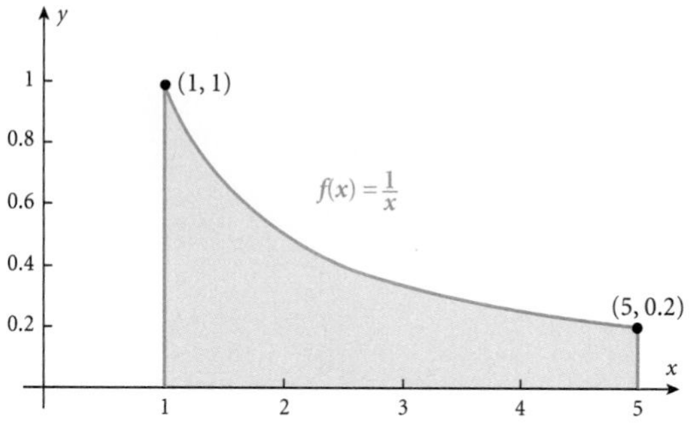

(b) $A \approx \frac{25}{12}$

(c) $A \approx \frac{4609}{2520} \approx 1.829$

(d) $A = \int_1^5 \frac{1}{x} dx$

(e) $A = \ln 5$

15. (a)

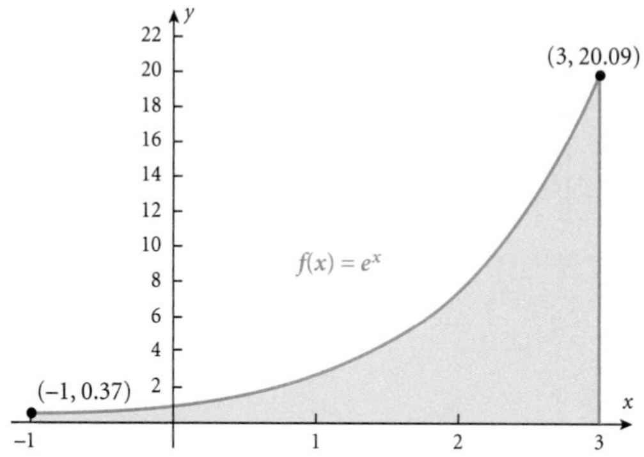

(b) $A \approx 11.475$

(c) $A \approx 15.197$

(d) $A = \int_{-1}^3 e^x dx$

(e) $A = e^3 - \frac{1}{e}$

17. 1.46

19. 38.29

21. (a)

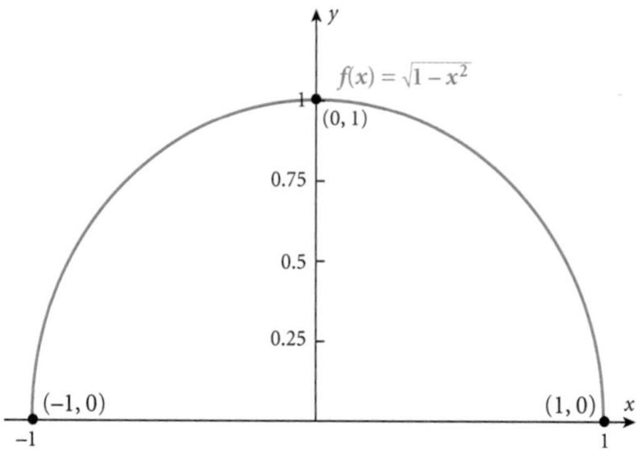

(b) $A \approx 1.42$

(c) $A \approx 1.52$

(d) $A = \int_{-1}^1 \sqrt{1 - x^2}\, dx$

(e) $A = 1.57$

(f) The actual area is $\frac{\pi}{2}$.

Exercise 15.7 (p. 916)

1. $y = \frac{x^3}{3} - x$

3. $y = \frac{x^3}{3} - \frac{x^2}{2} - \frac{3}{2}$

5. $y = \frac{x^4}{4} - \frac{x^2}{2} + 2x + 3$

7. $y = e^x + 3$

9. $y = \frac{x^2}{2} + x + \ln x - \frac{3}{2}$

11. There are 12,975 bacteria after one hour. There are 147,789 bacteria after 90 minutes. It will take 113.58 minutes to reach 1,000,000 bacteria.

13. There will be 7.68 grams of radium

15. The tree died 9727 years ago.

17. The population size is 6944 mosquitoes.

19. There are 55,418 bacteria.

21. (a) $P(t) = 10{,}000\ 5^{t/10}$ bacteria

(b) There were 250,000 bacteria.

(c) There were 20,000 bacteria after $t_1 = \frac{10 \ln 2}{\ln 5} \approx 4.31$ minutes.

23. The half-life is 4,620,981 years.

25. (a) $p(x) = 300 \left(\frac{1}{2}\right)^{x/200}$

(b) A price of \$106.07 should be charged.

(c) A price of \$89.19 should be charged.

CHAPTER 15 Review

True – False Items (p. 918)

1. True
2. False
3. False
3. False
4. False
5. False
6. True
7. True
8. True
9. False
10. True

Fill in the Blanks (p. 919)

1. $F'(x) = f(x)$
2. $\int f(x)dx$
3. integration by parts
4. lower, upper limits, integration
5. 0
6. $F(b) - F(a)$
7. $\int_0^2 \sqrt{x^2 + 1}\, dx$

Review Exercises (p. 919)

1. $F(x) = x^6 + K$
3. $F(x) = \frac{x^4}{4} + \frac{x^2}{2} + K$
5. $F(x) = 2\sqrt{x} + K$
7. $7x + K$
9. $\frac{5x^4}{4} + 2x + K$
11. $\frac{x^5}{5} - x^3 + 6x + K$
13. $3 \ln |x| + K$
15. $\ln |x^2 - 1| + K$
17. $\frac{e^{3x}}{3} + K$
19. $\frac{(x^3 + 3x)^6}{18} + K$
21. $\frac{2x^3}{3} - 3x^2 + K$
23. $e^{3x^2 + x} + K$
25. $\frac{2(x-5)^{5/2}}{5} + \frac{10(x-5)^{3/2}}{3} + K$
27. $\frac{xe^{4x}}{4} - \frac{e^{4x}}{16} + K$
29. $-\frac{\ln(2x)}{x} - \frac{1}{x} + K$
31. $R(x) = \frac{5x^2}{2} + 2x$
33. $C(x) = \frac{5x^2}{2} + 120{,}000x + 7500$; the cost is minimum at a production level of zero.
35. (a) $R(x) = 500x - 0.005x^2$
(b) 50,000 televisions need to be sold.
(c) The maximum revenue is \$12,500,000.
(d) The increase in revenue is \$625,000.
37. $-\frac{9}{2}$
39. $\frac{304}{3}$
41. $e + \frac{1}{e} - 2$
43. $\frac{1}{7}$
45. $\frac{e^6 - e^{-6}}{3}$
47. $3 - \frac{4}{e}$
49. 1
51. -10
53. 15

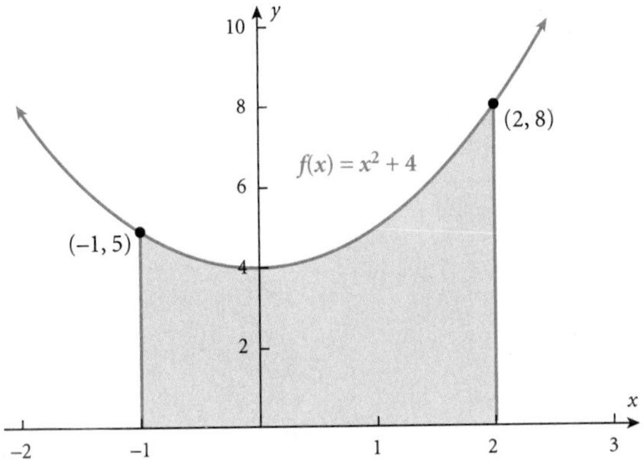

55. $e - \frac{1}{2}$

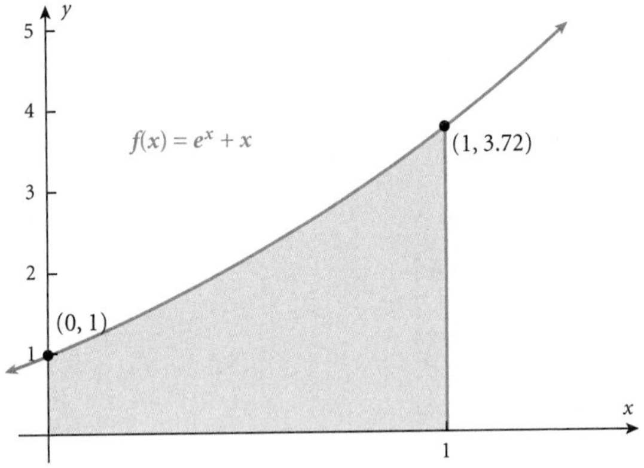

57. $\frac{31}{6}$

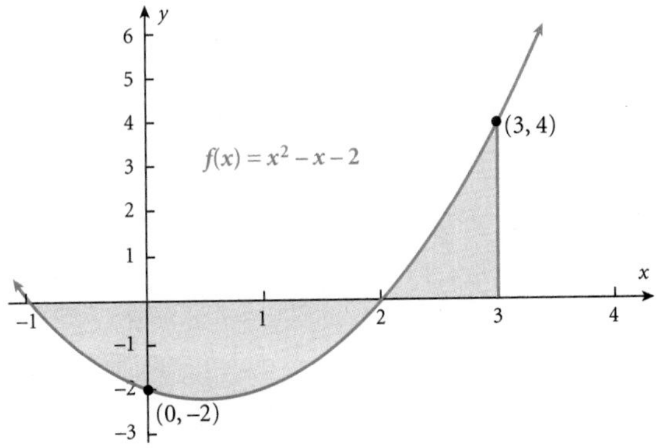

59. $\frac{125}{6}$

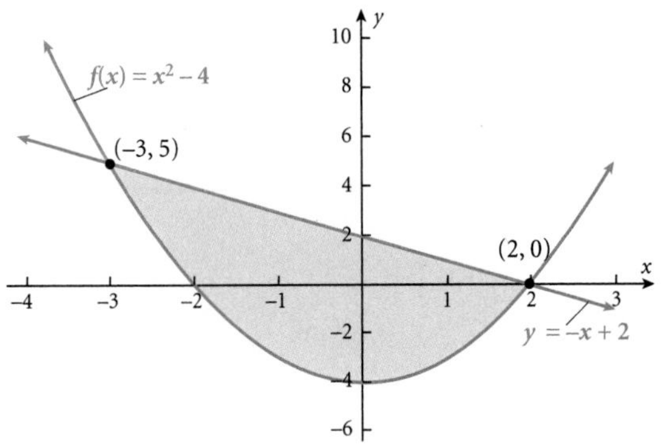

61. 8

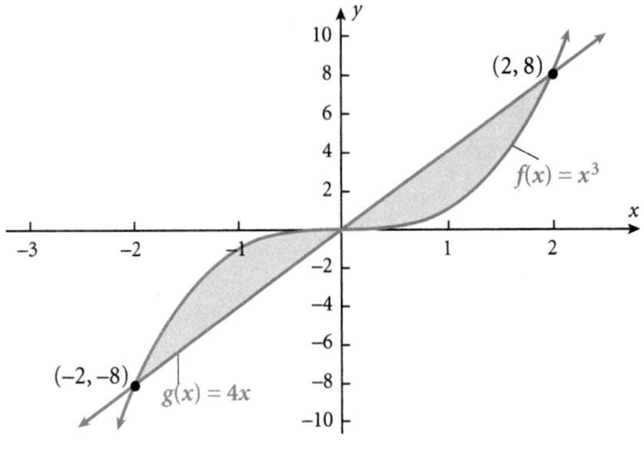

63. The monthly profit stays the same. The increase in profit is 0.

65. (a) 1 time unit

(b) 6 monetary units

67. (a)

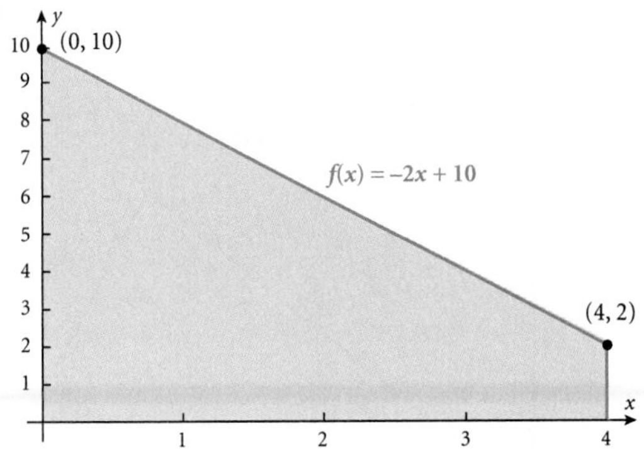

(b) $A \approx 28$

(c) $A \approx 20$

(d) $A \approx 26$

(e) $A \approx 22$

(f) $A = \int_0^4 (-2x + 10)dx$

(g) $A = 24$

69. 90.38

71. $y = \frac{x^3}{3} + \frac{5x^2}{2} - 10x + 1$

73. $y = \frac{e^{2x}}{2} - \frac{x^2}{2} + \frac{5}{2}$

75. $y = e^{10x}$

77. There will be 23,689 bacteria.

79. The burial ground is 7403 years old.

81. (a) $E(t) = \frac{t^3}{150} + \frac{t^2}{2} + 5$

(b) The economy will total \$18.33 million.

83. Margo should allow for 5432 labor-hours.

85. (a) The market price is \$10, and the demand level is 100 units.

(b) The consumer's surplus is \$100, and the producer's surplus is \$250.

(c)

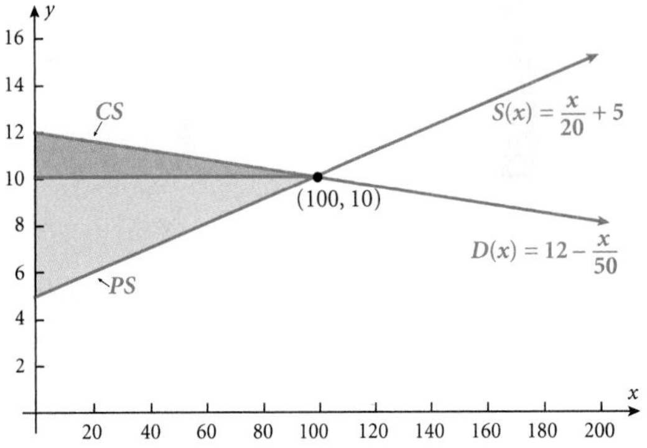

87. (a) $p(x) = 800\left(\frac{3}{4}\right)^{x/80}$

(b) A price of \$389.71 should be charged.

(c) A price of \$325.58 should be charged.

Mathematical Questions from Professional Exams (p. 924)

1. (c) 1

2. (d) $-\frac{1}{4}$

3. (c) $x^2e^x + 2xe^x$

4. (e) None of the above

5. (c) 62,208

6. (c) $(\ln x)^2 + C$

CHAPTER 16 Other Applications and Extensions of the Integral

Exercise 16.1 (p. 929)

1. This integral is improper because the upper limit of integration is infinite.

3. This integral is improper because the integrand is discontinuous at the lower limit of integration.

5. This integral is improper because the integrand is discontinuous at the lower limit of integration.

7. $\dfrac{e^{-4}}{4}$

9. The improper integral has no value.

11. $-\dfrac{5}{4}$

13. The improper integral has no value.

15. Area $=$ 2 square units

17. The capital value of the apartment is \$102,480.

19. (a) The area is equal to the integral, $\int_0^\infty r(t)\,dt$, which is the total reaction.

(b) This total reaction is $\frac{1}{2}$ units.

21. (a) The improper integral has no value.

(b) $\dfrac{3(\sqrt[3]{18} - \sqrt[3]{2}}{2}$

Exercise 16.2 (p. 932)

1. $\dfrac{1}{3}$ **3.** $\dfrac{2}{3}$ **5.** 9

7. -26 **9.** $e - 1$

11. The average population would be $8.22 \cdot 10^9$ people.

13. The average temperature is 37.5 degrees Celsius.

15. The average speed is 12 meters per second.

17. The average annual revenue is \$207.32 billion.

19. The average rainfall is 0.188118 inches.

Exercise 16.3 (p. 941)

1. $f(x) = \dfrac{1}{2} > 0$ on $[0, 2]$. $\displaystyle\int_0^2 f(x)dx = 1.$

3. $f(x) = 2x \geq 0$ on $[0, 1]$. $\displaystyle\int_0^1 f(x)dx = 1.$

5. $f(x) = \dfrac{3}{250}(10x - x^2) \geq 0$ on $[0, 5]$. $\displaystyle\int_0^5 f(x)dx = 1.$

7. $f(x) = \dfrac{1}{x} > 0$ on $[1, e]$. $\displaystyle\int_1^e f(x)dx = 1.$

9. $k = \dfrac{1}{3}$ **11.** $k = \dfrac{1}{2}$ **13.** $k = \dfrac{3}{250}$

15. $k = \dfrac{1}{\ln 2}$ **17.** $E(x) = 1$ **19.** $E(x) = \dfrac{2}{3}$

21. $E(x) = \dfrac{25}{8}$ **23.** $E(x) = e - 1$

25. The probability is $\dfrac{2}{5}$.

27. The probability is $\dfrac{1}{e^3} \approx 0.0498$.

29. The probability is 0.865.

31. (a) The probability is 0.0737.

(b) The probability is 0.2865.

33. (a) The probability is 0.5276.

(b) The probability is 0.0490.

(c) The probability is 0.3867.

35. The probability is 0.2865.

37. (a) $f(x) = \dfrac{1}{10}e^{-x/10}$

(b) The probability is 0.1954.

(c) The probability is 0.2231.

(d) The probability is 0.7769.

39. The expected waiting time is 9 minutes.

41. The contractor can be expected to be off by 2.29% on average.

43. (a) The probability is 0.242.

(b) The probability is 0.516.

45. (a) Let x be the cost of a new car, in thousands of dollars.

$$f(x) = \begin{cases} \dfrac{1}{35}x - \dfrac{2}{7} & \text{if } 10 \leq x \leq 17 \\ -\dfrac{1}{15}x + \dfrac{4}{3} & \text{if } 17 < x \leq 20 \end{cases}$$

(b) The probability is $\dfrac{5}{14} \approx 0.3571$.

(c)

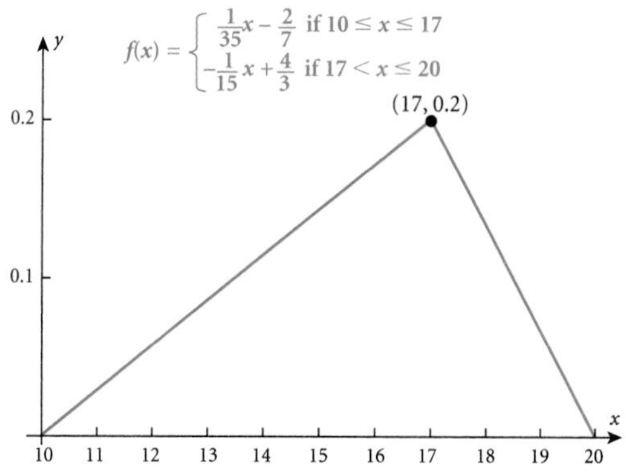

(d) The expected price is about $16,000.

(e) The expected price is $15,666.67.

47. (a) Answers will vary.

(b) $\Pr(0.6 \le X < 0.9) = 0.3$.

49. Let $f(x) = \frac{1}{b-a}$ on $[a, b]$. Then

$$\begin{aligned}\sigma^2 &= \int_a^b x^2\left(\frac{1}{b-a}\right)dx - [E(x)]^2\\ &= \int_a^b x^2\left(\frac{1}{b-a}\right)dx - \left[\int_a^b x\left(\frac{1}{b-a}\right)dx\right]^2\\ &= \frac{x^3}{3(b-a)}\bigg|_a^b - \left[\frac{x^2}{2(b-a)}\bigg|_a^b\right]^2\\ &= \frac{b^2+ab+a^2}{3} - \frac{b^2+2ab+a^2}{4}\\ &= \frac{4b^2+4ab+4a^2-3b^2-6ab-3a^2}{12}\\ &= \frac{(b-a)^2}{12}.\end{aligned}$$

CHAPTER 16 Review

True – False Items (p. 946)

1. False **2.** True **3.** False

Fill in the Blanks(p. 946)

1. average value

2. random variable

3. probability density

4. $\lim_{b \to \infty} \int_2^b f(x)\,dx$

5. $\lim_{b \to 2^-} \int_0^b f(x)\,dx$

Review Exercises (p. 946)

1. 1 **3.** 6

5. The improper integral has no value.

7. Area = 1 square unit. **9.** 5

11. $\frac{64}{3}$ **13.** 4

15. (a) $f(x) = \frac{8}{9}x \ge 0$ on $\left[0, \frac{3}{2}\right]$. $\int_0^{3/2} f(x)dx = 1$.

(b) $E(x) = 1$

17. (a) $f(x) = 12x^3(1 - x^2) \ge 0$ on $[0, 1]$.

$\int_0^1 f(x)\,dx = 1$.

(b) $E(x) = \frac{24}{35}$

19. The average yearly sales is 1255 units.

21. The average price is $14.47

23. (a) The probability is $\frac{1}{4}$.

(b) The probability is $\frac{5}{12}$.

(c) $E(X) = 4$

25. (a) $f(x) = \frac{3}{635{,}840}(x^2 - 28x + 196) \ge 0$ on $[20, 100]$.

$\int_{20}^{100} f(x)\,dx = 1$.

(b) The probability is $\frac{217}{7948} \approx 0.0273$.

(c) The probability is $\frac{607}{3974} \approx 0.153$.

(d) The man's expected age of death is 78.52 years.

27. (a) $f(x) = \frac{1}{2}x > 0$ on $[0, 2]$. $\int_0^2 f(x)\,dx = 1$.

(b) The probability is $\frac{1}{4} = 0.25$.

(c) The probability is $\frac{5}{16} = 0.3125$.

(d) The probability is $\frac{7}{16} = 0.4375$.

(e) $E(x) = \frac{4}{3}$.

29. (a) The probability is $\frac{1}{5} = 0.2$.

(b) The probability is $\frac{1}{3} \approx 0.3333$.

(c) The expected wait time is 7.5 minutes.

31. The probability is 0.0821.

33. (a) The probability is 0.1055.

(b) The probability is 0.3189.

35. The probability is 0.5134.

Mathematical Questions from Professional Exams (p. 949)

1. (d) $-\frac{1}{4}$ **2.** (e) 0 **3.** (b) $\frac{1}{4}$

4. (b) $\frac{7}{16}$ **5.** (d) $\frac{1}{4}$

CHAPTER 17 Calculus of Functions of Two or More Variables

Exercise 17.1 (p. 955)

1.

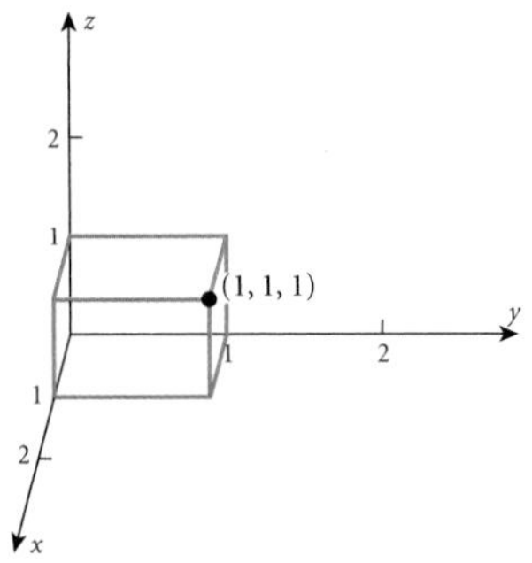

3.

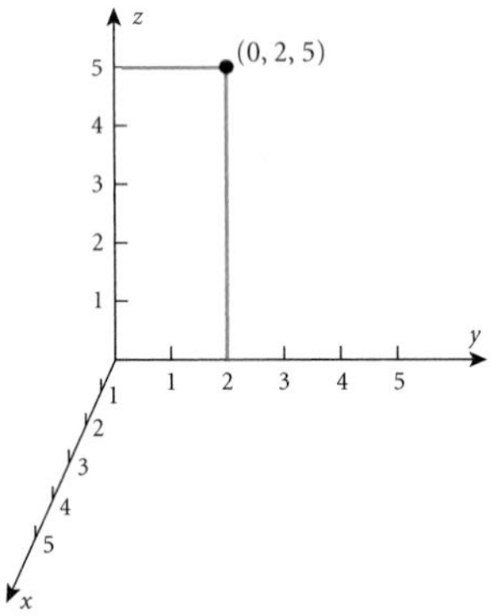

5.

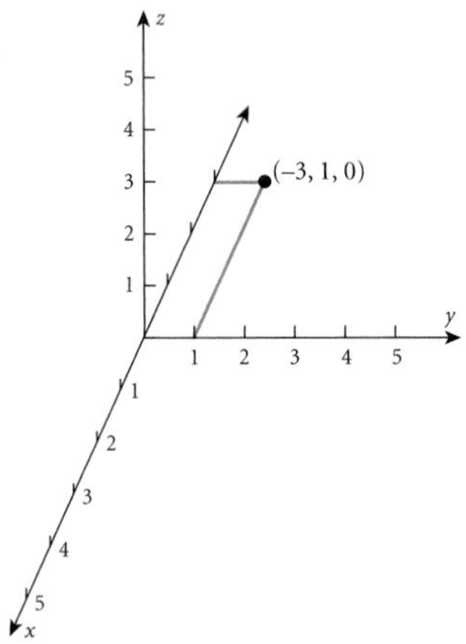

7. (0, 0, 3), (0, 1, 0), (0, 1, 3), (2, 0, 0), (2, 0, 3), (2, 1, 0)

9. (1, 2, 5), (1, 4, 3), (1, 4, 5), (3, 2, 3), (3, 2, 5), (3, 4, 3)

11. (−1, 0, 5), (−1, 2, 2), (−1, 2, 5), (4, 0, 2), (4, 0, 5), (4, 2, 2)

13. The plane through the point (0, 3, 0) that is parallel to the xz-plane

15. The yz-plane.

17. The plane through the point (0, 0, 5) that is parallel to the xy-plane.

19. $\sqrt{17}$ units **21.** $\sqrt{57}$ units **23.** $\sqrt{26}$ units

25. $(x-3)^2+(y-1)^2+(z-1)^2=1$

27. $(x+1)^2+(y-1)^2+(z-2)^2=9$

29. Center $=(-1,1,0)$, radius $=2$

31. Center $=(-2,-2,-1)$, radius $=3$

33. Center $=(2,0,-1)$, radius $=2$

35. $x^2+(y-3)^2+(z-6)^2=17$

Exercise 17.2 (p. 960)

1. $f(2,1)=5$ **3.** $f(2,1)=\sqrt{2}$ **5.** $f(2,1)=\frac{1}{5}$

7. $f(2,1)=3$ **9.** $f(2,1)=0$

11. (a) $f(1,0)=3$ (b) $f(0,1)=2$ (c) $f(2,1)=10$
(d) $f(x+\Delta x,y)=3x+3\Delta x+2y+xy+\Delta xy$
(e) $f(x,y+\Delta y)=3x+2y+2\Delta y+xy+x\Delta y$

13. (a) $f(0,0)=0$ (b) $f(0,1)=0$
(c) $f(a^2,t^2)=at+a^2$
(d) $f(x+\Delta x,y)=\sqrt{xy+\Delta xy}+x+\Delta x$
(e) $f(x,y+\Delta y)=\sqrt{xy+x\Delta y}+x$

15. (a) $f(1,2,3)=14$ (b) $f(0,1,2)=2$
(c) $f(-1,-2,-3)=-14$

17. The domain is the set $\{(x,y)\mid x\geq 0 \text{ and } y\geq 0\}$. This set is the first quadrant together with its border.

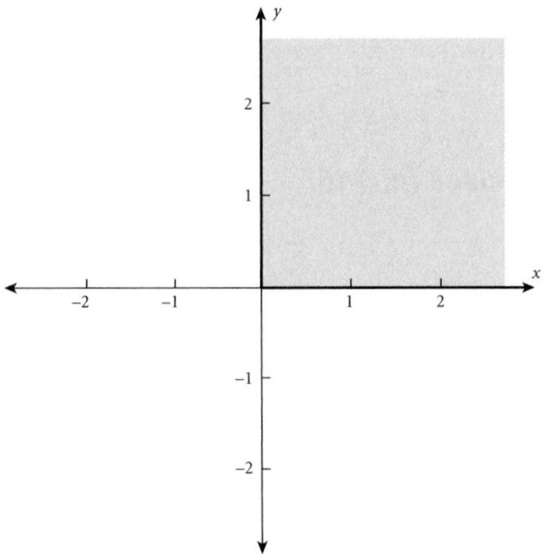

19. The domain is the set $\{(x, y) \mid x^2 + y^2 \le 9\}$. This set is the circle of radius 3 centered at the origin and its interior.

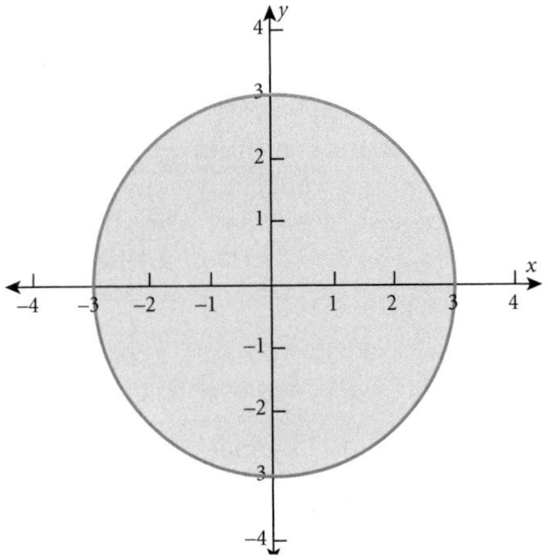

21. The domain is the set $\{(x, y) \mid x > 0 \text{ and } y > 0 \text{ and } y \ne 1\}$. This set is the first quadrant less the line $y = 1$.

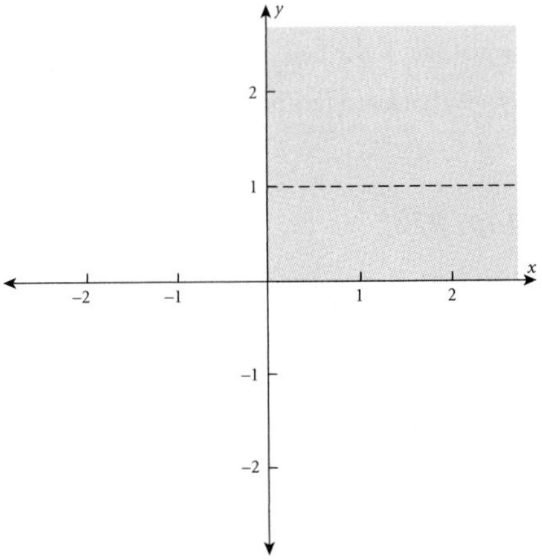

23. The domain is the set $\{(x, y) \mid x^2 + y^2 \ne 4\}$. This set is the union of the regions inside and outside of the circle of radius 2 centered at the origin.

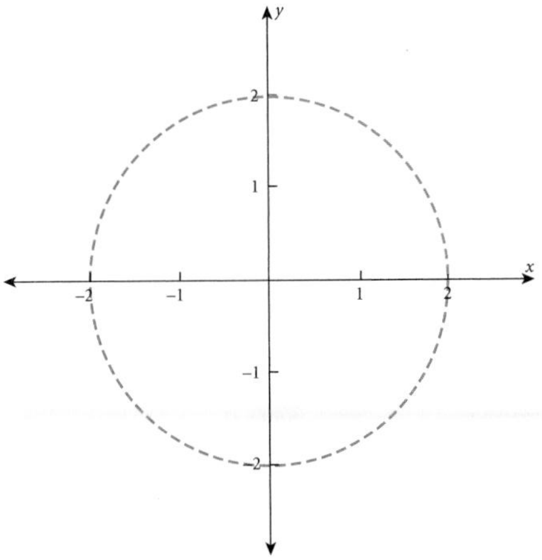

25. The domain is the set $\{(x, y) \mid (x, y) \ne (0, 0)\}$. This set is the entire *xy*-plane except for the origin.

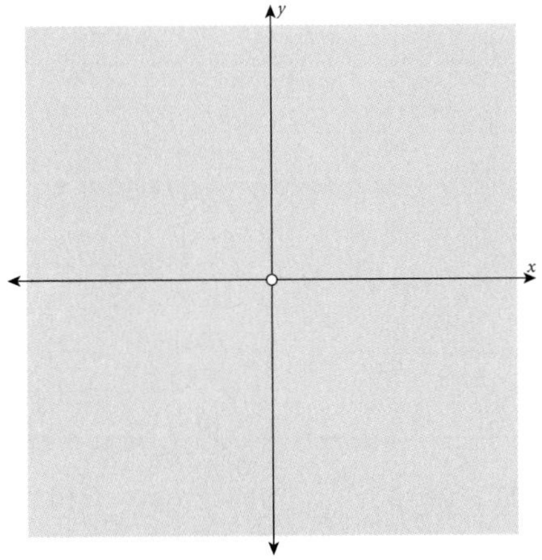

27. The domain is the set $\{(x, y, z) \mid x^2 + y^2 + z^2 \ge 16\}$. This set is the region on and outside of the sphere of radius 4 centered at the origin.

29. The domain is the set $\{(x, y, z) \mid x^2 + y^2 + z^2 \ne 0\}$. This set is the set of all points in space except for the origin (0, 0, 0).

31. (a) $f(x + \Delta x, y) = 3x + 3\Delta x + 4y$

(b) $f(x, y + \Delta y) = 3x + 4y + 4\Delta y$

(c) $\dfrac{f(x + \Delta x, y) - f(x, y)}{\Delta x} = 3$

(d) $\lim_{\Delta x \to 0} \dfrac{f(x + \Delta x, y) - f(x, y)}{\Delta x} = 3$

33. $C(r, h) = 600\pi r^2 + 1000\pi rh$ dollars

35. (a) 6.75 (b) 18 (c) 2 (d) 1.5

37. The total monthly bill is \$159.99.

39. (a) The heat index is 105.2°F.
(b) The relative humidity is 43%.
(c) The relative humidity is 55%.

Exercise 17.3 (p. 969)

1. $f_x(x, y) = 3$ $f_y(x, y) = -2 + 9y^2$
$f_x(2, -1) = 3$ $f_y(-2, 3) = 79$

3. $f_x(x, y) = 2x - 2y$ $f_y(x, y) = 2y - 2x$
$f_x(2, -1) = 6$ $f_y(-2, 3) = 10$

5. $f_x(x, y) = \dfrac{x}{\sqrt{x^2 + y^2}}$ $f_y(x, y) = \dfrac{y}{\sqrt{x^2 + y^2}}$

$f_x(2, -1) = \dfrac{2\sqrt{5}}{5}$ $f_y(-2, 3) = \dfrac{3\sqrt{13}}{13}$

7. $f_x(x, y) = -24x - 2y$ $f_y(x, y) = -2x + 3y^2 + 2y$
$f_{xx}(x, y) = -24$ $f_{yy}(x, y) = 6y + 2$
$f_{yx}(x, y) = -2$ $f_{xy}(x, y) = -2$

9. $f_x(x, y) = ye^x + e^y + 1$ $f_y(x, y) = e^x + xe^y$ $f_{xx}(x, y) = ye^x$ $f_{yy}(x, y) = xe^y$ $f_{yx}(x, y) = e^x + e^y$ $f_{xy}(x, y) = e^x + e^y$

11. $f_x(x, y) = \frac{1}{y}$ $f_y(x, y) = -\frac{x}{y^2}$ $f_{xx}(x, y) = 0$ $f_{yy}(x,y) = \frac{2x}{y^3}$

$f_{yx}(x, y) = -\frac{1}{y^2}$ $f_{xy}(x, y) = -\frac{1}{y^2}$

13. $f_x(x, y) = \frac{2x}{x^2 + y^2}$ $f_y(x, y) = \frac{2y}{x^2 + y^2}$ $f_{xx}(x, y) = \frac{2y^2 - 2x^2}{(x^2 + y^2)^2}$

$f_{yy}(x, y) = \frac{2x^2 - 2y^2}{(x^2 + y^2)^2}$ $f_{yx}(x, y) = -\frac{4xy}{(x^2 + y^2)^2}$

$f_{xy}(x, y) = -\frac{4xy}{(x^2 + y^2)^2}$

15. $f_x(x, y) = -\frac{2y + 10}{x^2y}$ $f_y(x, y) = \frac{x - 10}{xy^2}$ $f_{xx}(x, y) = \frac{8y + 20}{x^3y}$

$f_{yy}(x, y) = \frac{20 - 2x}{xy^3}$ $f_{yx}(x, y) = \frac{10}{x^2y^2}$ $f_{xy}(x, y) = \frac{10}{x^2y^2}$

17. $f_{xy} = 0 = f_{yx}$ **19.** $f_{xy} = 24x^3y + 14x = f_{yx}$

21. $f_{xy} = -\frac{2}{x^3} = f_{yx}$

23. $f_x(x, y, z) = 2xy - 3yz$ $f_y(x, y, z) = x^2 - 3xz$ $f_z(x, y, z) = 3z^2 - 3xy$

25. $f_x(x, y, z) = e^y$ $f_y(x, y, z) = xe^y + e^z$ $f_z(x, y, z) = ye^z$

27. $f_x(x, y, z) = \ln(yz) + \frac{y}{x}$ $f_y(x, y, z) = \ln(xz) + \frac{x}{y}$, $f_z(x, y, z) = \frac{x + y}{z}$

29. $f_x(x, y, z) = \frac{2x}{x^2 + y^2 + z^2}$ $f_y(x, y, z) = \frac{2y}{x^2 + y^2 + z^2}$

$f_z(x, y, z) = \frac{2z}{x^2 + y^2 + z^2}$

31. The slope is 20. **33.** The slope is $-\frac{2\sqrt{11}}{11}$.

35. The slope is 1. **37.** The slope is 1.

39. We have that $\frac{\partial z}{\partial x} = 2x$ and $\frac{\partial z}{\partial y} = 8y$. Now

$$x\frac{\partial z}{\partial x} + y\frac{\partial z}{\partial y} = x(2x) + y(8y) = 2(x^2 + 4y^2) = 2z.$$

41. We have that $\frac{\partial z}{\partial x} = \frac{x}{x^2 + y^2}$ and $\frac{\partial z}{\partial y} = \frac{x}{x^2 + y^2}$.

Then $\frac{\partial^2 z}{\partial x^2} = \frac{y^2 - x^2}{(x^2 + y^2)^2}$ and $\frac{\partial^2 z}{\partial y^2} = \frac{x^2 - y^2}{(x^2 + y^2)^2}$. Now

$$\frac{\partial^2 z}{\partial x^2} + \frac{\partial^2 z}{\partial y^2} = \frac{y^2 - x^2}{(x^2 + y^2)^2} + \frac{x^2 - y^2}{(x^2 + y^2)^2} = \frac{y^2 - x^2 + x^2 - y^2}{(x^2 + y^2)^2} = 0.$$

43. (a) $\frac{\partial z}{\partial x} = -20$, $\frac{\partial z}{\partial y} = -50$
(b) If the average price per pound of margarine remains fixed and the average price per pound of butter is increased by \$1, $\frac{\partial z}{\partial x}$ is the change in demand for butter. If the average price per pound of butter remains fixed and the average price per pound of margarine is increased by \$1, $\frac{\partial z}{\partial y}$ is the change in demand for margarine.

45. (a) $\frac{\partial A}{\partial N} = \frac{9}{I}$, $\frac{\partial A}{\partial I} = -\frac{9N}{I^2}$
(b) $\frac{\partial A}{\partial N}(78, 217) = 0.041$ $\frac{\partial A}{\partial I}(78, 217) = -0.015$.
(c) If he pitched 217 innings and he gave up 79 earned runs, his earned run average would increase by 0.041. If he pitched 218 innings and gave up 78 earned runs, his earned run average would decrease by 0.015.

47. (a) $\frac{\partial H}{\partial T} = 2.04901523 - 0.2247554r - 0.013667566t + 0.00245748tr + 0.00085282r^2 - 0.00000398tr^2$
(b) $\frac{\partial H}{\partial t}$ is the change in the heat index with respect to temperature, given a fixed humidity.
(c) $\frac{\partial H}{\partial r} = 10.141333127 - 0.2247554t - 0.10963434r + 0.00122874t^2 + 0.00170564tr - 0.00000398t^2r$
(d) $\frac{\partial H}{\partial t}$ is the change in the heat index with respect to humidity, given a fixed temperature.

49. No.

Exercise 17.4 (p. 976)

1. $(-1, 0), (0, 0), (1, 0)$ **3.** $(-1, -1), (0, 0), (1, 1)$

5. $(0, 0)$

7. The point $(0, 0)$ is a local minimum.

9. The point $\left(\frac{3}{2}, 0\right)$ is a local minimum.

11. The point $(-2, 4)$ is a saddle point.

13. The point $(2, -1)$ is a local minimum.

15. The point $(4, -2)$ is a local minimum.

17. The point $(0, 0)$ is a saddle point.

19. The point $(0, 0)$ is a saddle point, and the point $(2, 2)$ is a local minimum.

21. The point $(0, 0)$ is a critical point that is neither a saddle point nor a local extremum, and the point $\left(3, -\frac{9}{2}\right)$ is a saddle point.

23. The funtion has no critical points.

25. The maximize profits, the quantities sold should be $x \approx \frac{20{,}000}{7} \approx 2857$ units and $y = \frac{2000}{7} \approx 286$ units. The corresponding prices are $P = \frac{64{,}000}{7} \approx \$9{,}143$ and $q = \frac{54{,}000}{7} \approx \$7{,}714$. The maximum profit is $P\left(\frac{20}{7}, \frac{2}{7}\right) = \frac{128{,}000}{7} \approx \$18{,}286$

27. The manufacturer should produce 15,250 tons of grade A and 4100 tons of grade B to maximize profit.

29. For a fixed amount of the first drug, an amount of $\frac{2b}{3}$ of the second drug maximizes the reaction. For a fixed amount of the second drug, an amount of $\frac{2a}{3}$ of the first drug maximizes the reaction. If the amounts of both drugs are variable, $\frac{2a}{3}$ units of the first drug and $\frac{2b}{3}$ units of the second drug maximize the reaction.

31. There are no values of x and t that will maximize y.

33. (a) The dimensions are $43\frac{1}{3}$ inches by $21\frac{2}{3}$ inches by $21\frac{2}{3}$ inches.

(b) For the cylinder of maximum volume, the radius is $\frac{130}{3\pi} \approx 13.79$ inches and the height is $43\frac{1}{3}$ inches.

35. $x = 50$ tons, $y = 1$ ton per week

Exercise 17.5 (p. 985)

1. The maximum value is 15.
3. The minimum value is $\frac{1}{2}$.
5. The maximum value is 528.
7. The minimum value is 612.
9. The maximum value is 16,000.
11. The minimum value is $\frac{233}{12}$.
13. The two numbers are both 50.
15. The three numbers are all $\frac{5\sqrt{3}}{3}$.
17. To minimize cost, the factory should produce 18 units of type x and 36 units of type y.
19. The dimensions are $20\frac{2}{3}$ inches by $20\frac{2}{3}$ inches by $20\frac{2}{3}$.
21. (a) $178\frac{4}{7}$ units of capital and 750 units of labor will maximize the total production.
    (b) The maximum number of units of production is 529.14 units.
23. The dimensions are $\sqrt[3]{175}$ feet by $\sqrt[3]{175}$ feet by $\sqrt[3]{175}$ feet, which is approximately 5.593 feet by 5.593 feet by 5.593 feet.
25. The dimensions are $\sqrt[3]{12}$ feet by $\sqrt[3]{12}$ feet by $\frac{3\sqrt[3]{12}}{2}$ feet, which is approximately 2.29 feet by 2.29 feet by 3.43 feet.

Exercise 17.6 (p. 991)

1. $2y^3 + \frac{8}{3}$ 3. $18x^2 + 4x$ 5. $3y + \frac{5}{2}$

7. $8x - 22$ 9. $\frac{1}{3\sqrt{1 + y^2}}$ 11. $e^y(e^2 - 1)$

13. $(e^4 - 1)\,e^{-4y}$ 15. $\frac{2}{\sqrt{y + 6}}$ 17. 8

19. $\frac{35}{3}$ 21. $\frac{17}{6}$ 23. 22

25. 12 27. 24 29. 21

31. Volume $= \frac{35}{2}$ cubic units

CHAPTER 17 Review

True – False Items (p. 993)

1. True 2. False 3. False 4. False

Fill in the Blanks (p. 993)

1. surface 2. $2 - \sqrt{2}$ 3. $x = x_0$ 4. saddle point

Review Exercises (p. 993)

1. 3 units 3. $\sqrt{69}$ units 5. 5 units

7. The radius is 3 units.

9. $(x + 6)^2 + (y - 3)^2 + (z - 1)^2 = 4$

11. The center is the point $(1, -3, -8)$, and the radius is 5.

13. (a) $(x - 1)^2 + (y + 4)^2 + (z - 3)^2 = 36$
    (b) The center is the point $(1, -4, 3)$, and the radius is 6.

15. (a) $f(1, -3) = 11$ (b) $f(4, -2) = -8$

17. (a) $f(1, -3) = -\frac{1}{2}$ (b) $f(4, -2) = 0$

19. The domain is the entire xy-plane.

21. The domain is the set $\{(x, y) \mid y > x^2 + 4\}$, which is the set of points above the parabola $y = x^2 + 4$.

23. The domain is the set $\{(x, y) \mid (x + 2)^2 + y^2 \geq 9\}$, which is the set of points on or outside of the circle of radius 3 centered at the point $(-2, 0)$.

25. $f_x(x, y) = 2xy + 4$ $f_y(x, y) = x^2$ $f_{xx}(x, y) = 2y$,
$f_{xy}(x, y) = 2x$ $f_{yx}(x, y) = 2x$ $f_{yy}(x, y) = 0$

27. $f_x(x, y) = y^2e^x + \ln y$ $f_y(x, y) = 2ye^x + \frac{x}{y}$ $f_{xx}(x, y) = y^2e^x$,
$f_{xy}(x, y) = 2ye^x + \frac{1}{y}$ $f_{yx}(x, y) = 2ye^x + \frac{1}{y}$,
$f_{yy}(x, y) = 2e^x - \frac{x}{y^2}$

29. $f_x(x, y) = \frac{x}{\sqrt{x^2 + y^2}}$ $f_y(x, y) = \frac{y}{\sqrt{x^2 + y^2}}$,
$f_{xx}(x, y) = \frac{y^2}{(x^2 + y^2)^{3/2}}$ $f_{xy}(x, y) = -\frac{xy}{(x^2 + y^2)^{3/2}}$,
$f_{yx}(x, y) = -\frac{xy}{(x^2 + y^2)^{3/2}}$ $f_{yy}(x, y) = \frac{x^2}{(x^2 + y^2)^{3/2}}$

31. $f_x(x, y) = e^x\left(\ln(5x + 2y) + \frac{5}{5x + 2y}\right)$ $f_y(x, y) = \frac{2e^x}{5x + 2y}$,

$f_{xx}(x, y) = e^x\left(\ln(5x + 2y) + \frac{50x + 20y - 25}{(5x + 2y)^2}\right)$

$f_{xy}(x, y) = \frac{e^x(10x + 4y - 10)}{(5x + 2y)^2}$

$f_{yx}(x, y) = \frac{e^x(10x + 4y - 10)}{(5x + 2y)^2}$

$f_{yy}(x, y) = -\frac{4e^x}{(5x + 2y)^2}$

33. $f_x(x, y, z) = 3e^y + ye^z - 24xy$
$f_y(x, y, z) = 3xe^y + xe^z - 12x^2$
$f_z(x, y, z) = xye^z$

35. The slope is 12.

37. The slope is 1.

39. (a) The only critical point is $(-4, -2)$.
(b) The point $(-4, -2)$ is a local maximum.

41. (a) The only critical point is $(1, 2)$.
(b) The point $(1, 2)$ is a local maximum.

43. (a) The only critical point is $\left(0, \frac{9}{2}\right)$.
(b) The point $\left(0, \frac{9}{2}\right)$ is a local minimum.

45. The maximum value is $\frac{5900}{19}$.

47. The minimum value is $\frac{16}{5}$.

49. $-\frac{40y}{3}$

51. $24x^2 + 8$

53. 51

55. 448

57. 32

59. $\frac{27}{2}$

61. The volume is 672 cubic units.

63. (a) $\frac{\partial z}{\partial K} = 20\left(\frac{L}{K}\right)^{3/4}$ $\frac{\partial z}{\partial L} = 60\left(\frac{K}{L}\right)^{1/4}$
(b) $\frac{\partial z}{\partial K} = 1.257$ $\frac{\partial z}{\partial L} = 150.892$
(c) The factory should increase the use of labor. Explanations will vary.

65. $C_x(x, y) = 40$, $C_y(x, y) = 45$, If the number of deluxe vacuum cleaners produced remains fixed, increasing the production of standard vacuum cleaners by one will increase cost by \$40. If the number of standard vacuum cleaners produced remains fixed, increasing the production of deluxe vacuum cleaners by one will increase cost by \$45.

67. (a) $R(x, y) = -6x^2 + 3xy - 8y^2 + 350x + 400y$
(b) $R_x(x, y) = -12x + 3y + 350$,
$R_y(x, y) = 3x - 16y + 400$, If the demand for deluxe vacuum cleaners produced remains fixed, an increase of one in the demand for standard vacuum cleaners will change revenue by R_x dollars. If the demand for standard vacuum cleaners produced remains fixed, an increase of one in the demand for deluxe vacuum cleaners will change revenue by R_y dollars.

69. (a) $P(x, y) = -6x^2 + 3xy - 8y^2 + 310x + 355y - 1050$
(b) $P_x(50, 30) = -160$, $P_y(50, 30) = 70$, If the demand for deluxe vacuum cleaners produced remains at 30 vacuum cleaners, increasing the demand of standard vacuum cleaners from 50 to 51 will decrease profit by \$160. If the demand for standard vacuum cleaners produced remains at 50 vacuum cleaners, increasing the demand of deluxe vacuum cleaners from 30 to 31 will increase profit by \$70.

71. (a) 4000 units of brand x at a price of \$4,000 and 5000 units of brand y at a price of \$11,000 will maximize profit.
(b) The maximum profit is a loss of \$159,000.

73. (a) \$15,300 should be allocated to capital, and \$35,700 should be allocated to labor.
(b) The maximum number of units is 3409 units.

APPENDIX A Review

Exercise Appendix A.1 (p. A-14)

1. (a) 2 and 5 are natural numbers.
(b) −6, 2, and 5 are integers.
(c) $-6, \frac{1}{2}, -1.333\ldots, 2$, and 5 are rational numbers.
(d) π is an irrational number.
(e) All the numbers are real numbers.

3. (a) 1 is a natural number.
(b) 0 and 1 are integers.
(c) All the numbers are rational numbers.
(d) There are no irrational numbers in the set C.
(e) All the numbers are real numbers.

5. (a) There are no natural numbers in the set E.
(b) There are no integers in the set E.
(c) There are no rational numbers in the set E.
(d) All the numbers are irrational.
(e) All the numbers are real numbers.

7. (a) 18.953
(b) 18.952

9. (a) 28.653
(b) 28.653

11. (a) 0.063
(b) 0.062

13. (a) 9.999
(b) 9.998

15. (a) 0.429
(b) 0.428

17. (a) 34.733
(b) 34.733

19. $3 + 2 = 5$ **21.** $x + 2 = 3 \cdot 4$

23. $3y = 1 + 2$ **25.** $x - 2 = 6$

27. $\frac{x}{2} = 6$ **29.** 7 **31.** 6

33. 1 **35.** $\frac{13}{3}$ **37.** -11

39. 11 **41.** -4 **43.** 1

45. 6 **47.** $\frac{2}{7}$ **49.** $\frac{4}{45}$

51. $\frac{23}{20}$ **53.** $\frac{79}{30}$ **55.** $\frac{13}{36}$

57. $-\frac{16}{45}$ **59.** $\frac{1}{60}$ **61.** $\frac{15}{22}$

63. $6x + 24$ **65.** $x^2 - 4x$

67. $x^2 + 6x + 8$ **69.** $x^2 - x - 2$

71. $x^2 - 10x + 16$ **73.** $x^2 - 4$

79. Subtraction is not commutative.

81. Division is not commutative.

83. This is true by the symmetric property of real numbers.

85. All real numbers are either rational or irrational; no real number is both.

87. $0.9999\ldots = 1$

Exercise Appendix A.2 (p. A-25)

1.

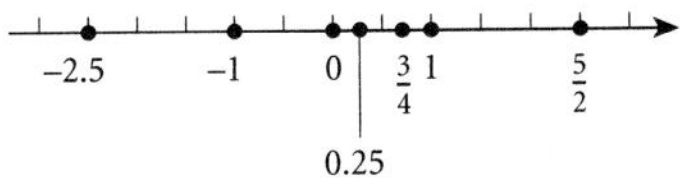

3. $>$ **5.** $>$ **7.** $>$ **9.** $=$

11. $<$ **13.** $x > 0$ **15.** $x < 2$ **17.** $x \leq 1$

19.

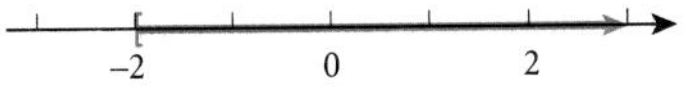

21.

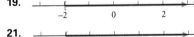

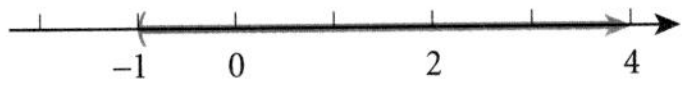

23. 1 **25.** 2 **27.** 6 **29.** 4

31. -28 **33.** $\frac{4}{5}$ **35.** 0 **37.** 1

39. 5 **41.** 1 **43.** 22 **45.** 2

47. (c) $x = 0$ **49.** (a) $x = 3$ **51.** none

53. (b) $x = 1$, (c) $x = 0$, (d) $x = -1$

55. $\{x \mid x \neq 5\}$ **57.** $\{x \mid x \neq -4\}$

59. $C = 0°$ **61.** $C = 25°$ **63.** 16 **65.** $\frac{1}{16}$

67. $\frac{1}{9}$ **69.** 9 **71.** 5 **73.** 4

75. $64x^6$ **77.** $\frac{x^4}{y^2}$ **79.** $\frac{x}{y}$ **81.** $-\frac{8x^3z}{9y}$

83. $\frac{16x^2}{9y^2}$ **85.** -4 **87.** 5 **89.** 4

91. 2 **93.** $\sqrt{5}$ **95.** $\frac{1}{2}$ **97.** 10; 0

99. 81 **101.** 304,006.671 **103.** 0.004

105. 481.890 **107.** 0.000

109. 4.542×10^2 **111.** 1.3×10^{-2}

113. 3.2155×10^4 **115.** 4.23×10^{-4}

117. 61,500 **119.** 0.001214

121. 110,000,000 **123.** 0.081

125. $A = l \cdot w$; $\{l|l > 0\}$, $\{w|w > 0\}$, $\{A|A > 0\}$

127. $C = \pi \cdot d$; $\{d|d > 0\}$, $\{C|C > 0\}$

129. $A = \frac{\sqrt{3}}{4}x^2$; $\{x|x > 0\}$, $\{A|A > 0\}$

131. $V = \frac{4}{3}\pi r^3$; $\{r|r > 0\}$, $\{V|V > 0\}$

133. $V = x^3$; $\{x|x > 0\}$, $\{V|V > 0\}$

135. (a) It costs \$6,000 to produce 1000 watches.
(b) It costs \$8,000 to produce 2000 watches.

137. (a) $|113 - 115| = |-2| = 2 \leq 5$
(b) $|109 - 115| = |-6| = 6 > 5$

139. (a) Yes, $|2.999 - 3| = |-0.001| = 0.001 \leq 0.01$.
(b) No, $|2.89 - 3| = |-0.11| = 0.11 > 0.01$.

141. No, $\frac{1}{3}$ is larger by 0.000333....

143. No.

Exercise Appendix A.3 (p. A-32)

1. 64 **3.** $\frac{1}{8}$ **5.** 1

7. 4 **9.** 3 **11.** 2

13. 4 **15.** $\frac{1}{64}$ **17.** $\frac{1}{4}$

19. (a) 11.2116
(b) 11.5873
(c) 11.6639
(d) 11.6648

21. (a) 8.8152
(b) 8.8214
(c) 8.8244
(d) 8.8250

23. (a) 21.2166
(b) 22.2167
(c) 22.4404
(d) 22.4592

25. 3 **27.** -1 **29.** 8

31. $\frac{1}{3}$ **33.** 2 **35.** 3

37. $x = \log_2 5$ **39.** $t = \log_{1.1} 10$ **41.** 55.590

43. 1385.002 **45.** 1499.364 **47.** 12,432.323

49. 2074.642

Exercise Appendix A.4 (p. A-38)

1. 1, 2, 3, 4, 5 **3.** $\frac{1}{2}, \frac{2}{3}, \frac{3}{4}, \frac{4}{5}, \frac{5}{6}$

5. 1, −4, 9, −16, 25 **7.** $\frac{1}{2}, \frac{2}{5}, \frac{2}{7}, \frac{8}{41}, \frac{8}{61}$

9. $-\frac{1}{6}, \frac{1}{12}, -\frac{1}{20}, \frac{1}{30}, -\frac{1}{42}$ **11.** $\frac{1}{e}, \frac{2}{e^2}, \frac{3}{e^3}, \frac{4}{e^4}, \frac{5}{e^5}$

13. 1, 3, 5, 7, 9 **15.** −2, −1, 1, 4, 8

17. 5, 10, 20, 40, 80 **19.** $3, 3, \frac{3}{2}, \frac{1}{2}, \frac{1}{8}$

21. 1, 2, 2, 4, 8

23. $A, A + d, A + 2d, A + 3d, A + 4d$

25. $\sqrt{2}, \sqrt{2 + \sqrt{2}}, \sqrt{2 + \sqrt{2 + \sqrt{2}}}, \sqrt{2 + \sqrt{2 + \sqrt{2 + \sqrt{2}}}}, \sqrt{2 + \sqrt{2 + \sqrt{2 + \sqrt{2 + \sqrt{2}}}}}$

27. (a) $a_1 = 2, r = 2$ (b) 2, 4, 8, 16 (c) 30

29. (a) $a_1 = -\frac{3}{2}, r = \frac{1}{2}$ (b) $-\frac{3}{2}, -\frac{3}{4}, -\frac{3}{8}, -\frac{3}{16}$ (c) $-\frac{45}{16}$

31. (a) $a_1 = \frac{1}{4}, r = 2$ (b) $\frac{1}{4}, \frac{1}{2}, 1, 2$ (c) $\frac{15}{4}$

33. (a) $a_1 = \sqrt[3]{2}, r = \sqrt[3]{2}$ (b) $\sqrt[3]{2}, \sqrt[3]{4}, 2, 2\sqrt[3]{2}$ (c) $2 + 3\sqrt[3]{2} + \sqrt[3]{4}$

35. (a) $a_1 = \frac{1}{2}, r = \frac{3}{2}$ (b) $\frac{1}{2}, \frac{3}{4}, \frac{9}{8}, \frac{27}{16}$ (c) $\frac{65}{16}$

Exercise Appendix A.5 (p. A-48)

1. $10x^5 + 3x^3 - 10x^2 + 6$ **3.** $2ax + a^2$

5. $2x^2 + 17x + 8$ **7.** $x^4 - x^2 + 2x - 1$

9. $6x^2 + 2$ **11.** $(x - 6)(x + 6)$

13. $(1 - 2x)(1 + 2x)$ **15.** $(x + 2)(x + 5)$

17. prime **19.** prime

21. $(x + 3)(5 - x)$ **23.** $3(x - 6)(x + 2)$

25. $y^2(y + 5)(y + 6)$ **27.** $(2x + 3)^2$

29. $(3x + 1)(x + 1)$ **31.** $(x - 3)(x + 3)(x^2 + 9)$

33. $(x - 1)^2(x^2 + x + 1)^2$ **35.** $x^5(x - 1)(x + 1)$

37. $(4x + 1)(5 - 4x)$ **39.** $(2y - 3)(2y - 5)$

41. $(x^2 + 1)(3x + 1)(1 - 3x)$ **43.** $(x - 6)(x + 3)$

45. $(x + 2)(x - 3)$ **47.** $3x(x - 2)^3(5x - 4)$

49. $(x - 1)(x + 1)(x + 2)$

51. $(x - 1)(x + 1)(x^2 - x + 1)$

53. $\frac{3(x - 3)}{5x}$ **55.** $\frac{x(2x - 1)}{x + 4}$

57. $\frac{5x}{(x - 6)(x - 1)(x + 4)}$ **59.** $\frac{2(x + 4)}{(x - 2)(x + 2)(x + 3)}$

61. $\frac{x^3 - 2x^2 + 4x + 3}{x^2(x - 1)(x + 1)}$ **63.** $-\frac{1}{x(x + h)}$

65. $2(3x + 4)(9x + 13)$ **67.** $2x(3x + 5)$

69. $5(x + 3)(x + 1)(x - 2)^2$ **71.** $3(4x - 1)(4x - 3)$

73. $6(3x - 5)(5x - 4)(2x + 1)^2$ **75.** $\frac{19}{(3x - 5)^2}$

77. $\frac{(x - 1)(x + 1)}{(x^2 + 1)^2}$ **79.** $\frac{x(3x + 2)}{(3x + 1)^2}$

81. $\frac{(x + 3)(1 - 3x)}{(x^2 + 1)^2}$

Exercise Appendix A.6 (p. A-58)

1. {7} **3.** {−3} **5.** {4} **7.** $\left\{\frac{5}{4}\right\}$

9. {−1} **11.** {−18} **13.** {−3} **15.** {−16}

17. {0.5} **19.** {2} **21.** {2} **23.** {3}

25. {0, 9} **27.** {0, 9} **29.** {21} **31.** {−2, 2}

33. {6} **35.** {−3, 3} **37.** {−4, 1} **39.** $\left\{-1, \frac{3}{2}\right\}$

41. {−4, 4} **43.** {2}

45. The equation has no solution. **47.** {−2, 2} **49.** {−1, 3}

51. {−2, −1, 0, 1} **53.** {0, 4} **55.** {−6, 2}

57. $\left\{-\frac{1}{2}, 3\right\}$ **59.** {3, 4} **61.** $\left\{\frac{3}{2}\right\}$

63. $\left\{-\frac{2}{3}, \frac{3}{2}\right\}$ **65.** $\left\{-\frac{3}{4}, 2\right\}$ **67.** {−5, 5} **69.** {−1, 3}

71. {−3, 0} **73.** 16 **75.** $\frac{1}{16}$ **77.** $\frac{1}{9}$

79. {−7, 3} **81.** $\left\{-\frac{1}{4}, \frac{3}{4}\right\}$ **83.** $\left\{\frac{-1 - \sqrt{7}}{6}, \frac{-1 + \sqrt{7}}{6}\right\}$

85. $\{2 - \sqrt{2}, 2 + \sqrt{2}\}$ **87.** $\left\{\frac{5 - \sqrt{29}}{2}, \frac{5 + \sqrt{29}}{2}\right\}$

89. $\left\{1, \frac{3}{2}\right\}$ **91.** The equation has no real solution.

93. $\left\{\frac{-1 - \sqrt{5}}{4}, \frac{-1 + \sqrt{5}}{4}\right\}$

95. $\left\{-\frac{\sqrt{3} - \sqrt{15}}{2}, -\frac{\sqrt{3} + \sqrt{15}}{2}\right\}$

97. The equation has no real solution.

99. The equation has a repeated real solution.

101. The equation has two unequal real solutions.

103. $\left\{\frac{b+c}{a}\right\}$ **105.** $\left\{\frac{abc}{a+b}\right\}$ **107.** $\{a^2\}$

109. $R = \frac{R_1R_2}{R_1 + R_2}$ **111.** $R = \frac{mv^2}{F}$ **113.** $r = \frac{S-a}{S}$

115. The solution set to the quadratic equation $ax^2 + bx + c = 0, a \neq 0$ is $\left\{\frac{-b-\sqrt{b^2-4ac}}{2a}, \frac{-b+\sqrt{b^2-4ac}}{2a}\right\}$. Adding the two solutions, we obtain

$$\frac{-b-\sqrt{b^2-4ac}}{2a} + \frac{-b+\sqrt{b^2-4ac}}{2a} =$$

$$\frac{-b-b-\sqrt{b^2-4ac}+\sqrt{b^2-4ac}}{2a} = -\frac{2b}{2a} = -\frac{b}{a}.$$

117. $k = -\frac{1}{2}$ or $k = \frac{1}{2}$

119. Because $b^2 - 4ac = (-b)^2 - 4ac \geq 0$, both equations $ax^2 + bx + c = 0$ and $ax^2 - bx + c = 0$ have real solutions. The solutions to the first equation are

$$x = \frac{-b-\sqrt{b^2-4ac}}{2a} \text{ and } x = \frac{-b+\sqrt{b^2-4ac}}{2a},$$

and the solutions to the second equation are

$$x = \frac{b-\sqrt{b^2-4ac}}{2a} = -\left(\frac{-b+\sqrt{b^2-4ac}}{2a}\right) \text{ and}$$

$$x = \frac{b+\sqrt{b^2-4ac}}{2a} = -\left(\frac{b+\sqrt{b^2-4ac}}{2a}\right).$$

121. The equations in (b) are equivalent, because $\sqrt{9} = 3$. In (a), -3 is a solution of $x^2 = 9$ but not of $x = 3$. In (c), 1 is a solution of $(x-1)(x-2) = (x-1)^2$ but not of $x - 2 = x - 1$.

Exercise Appendix A.7 (p. A-70)

1. $[0, 2], 0 \leq x \leq 2$ **3.** $(-1, 2), -1 < x < 2$

5. $[0, 3], 0 \leq x < 3$

7. $[0, 4]$

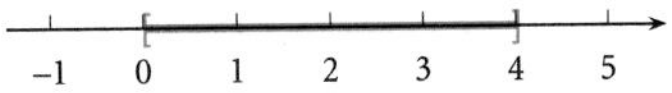

9. $[4, 6)$

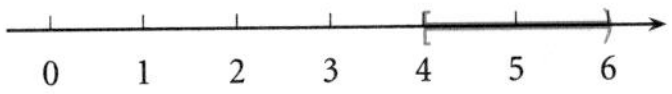

11. $[4, \infty)$

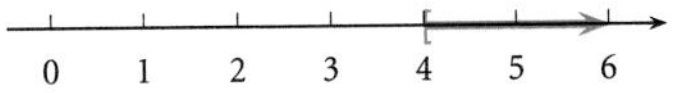

13. $(-\infty, -4)$

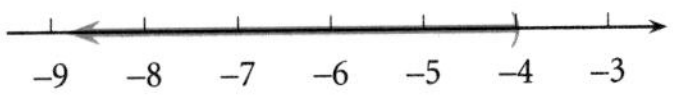

15. $2 \leq x \leq 5$

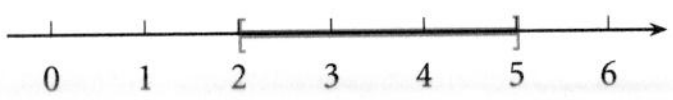

17. $-3 < x < 2$

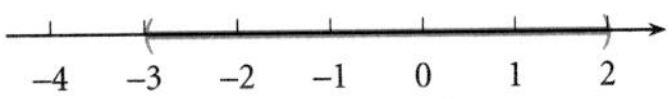

19. $x \geq 4$

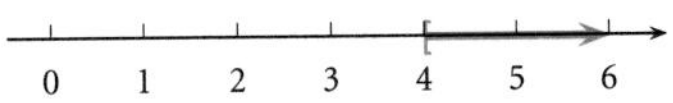

21. $x < -3$

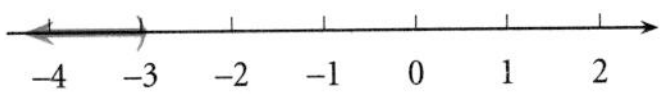

23. (a) $6 < 8$
(b) $-2 < 0$
(c) $9 < 15$
(d) $-6 > -10$

25. (a) $7 > 0$
(b) $-1 > -8$
(c) $12 > -9$
(d) $-8 < 6$

27. (a) $2x + 4 < 5$
(b) $2x - 4 < -3$
(c) $6x + 3 < 6$
(d) $-4x - 2 > -4$

29. $<$ **31.** $>$ **33.** $\geq$ **35.** $<$

37. $\leq$ **39.** $>$ **41.** $\geq$

43. $(-\infty, 4)$; $\{x|x < 4\}$

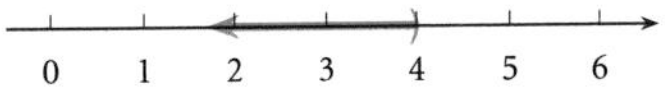

45. $[-1, \infty)$; $\{x|x \geq -1\}$

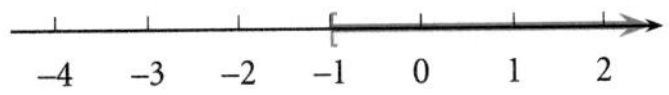

47. $(3, \infty)$; $\{x|x > 3\}$

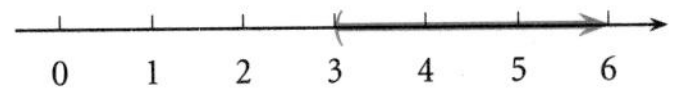

49. $[2, \infty)$; $\{x|x \geq 2\}$

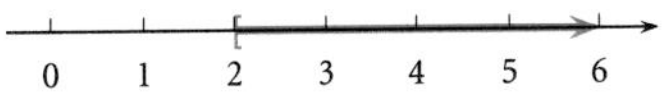

51. $(-7, \infty)$; $\{x|x > -7\}$

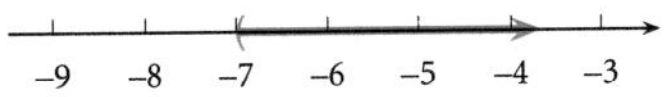

53. $(-\infty, \frac{2}{3}]$; $\{x|x \leq \frac{2}{3}\}$

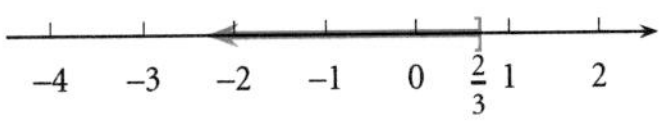

55. $(-\infty, -20)$; $\{x|x < -20\}$

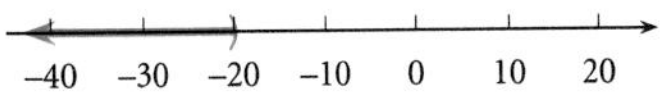

57. $[\frac{4}{3}, \infty)$; $\{x|x \geq \frac{4}{3}\}$

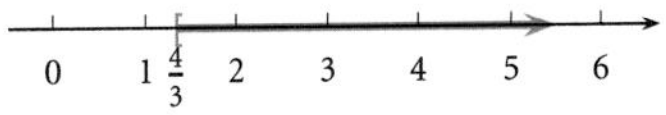

59. $[3, 5]$; $\{x|3 \leq x \leq 5\}$

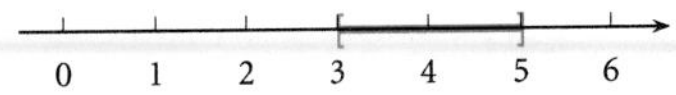

61. $[\frac{2}{3}, 3]$; $\{x|\frac{2}{3} \leq x \leq 3\}$

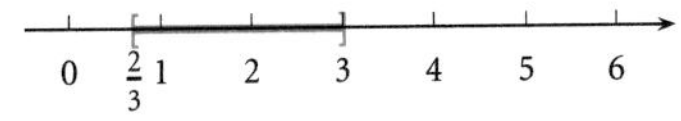

63. $\left(-\frac{11}{2}, \frac{1}{2}\right)$; $\{x \mid -\frac{11}{2} < x < \frac{1}{2}\}$

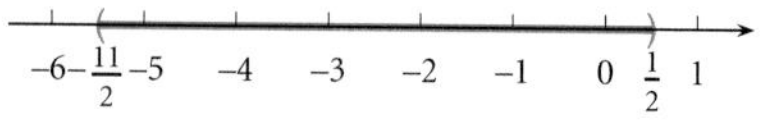

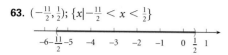

65. $(-6, 0)$; $\{x \mid -6 < x < 0\}$

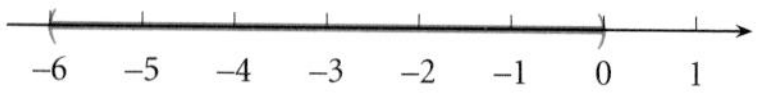

67. $\left(-\infty, -\frac{1}{2}\right)$; $\{x \mid x < -\frac{1}{2}\}$

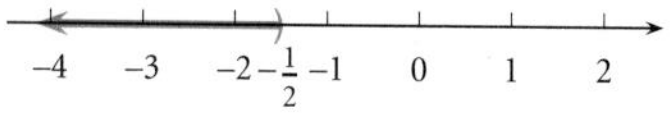

69. $\left(\frac{10}{3}, \infty\right)$; $\{x \mid x > \frac{10}{3}\}$

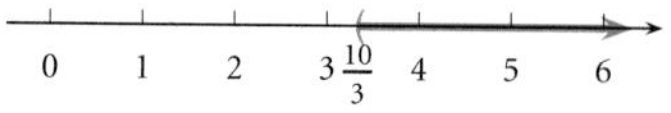

71. $(-1, 3)$; $\{x \mid -1 < x < 3\}$

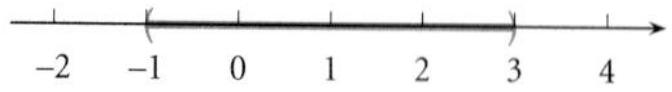

73. $(-3, 3)$; $\{x \mid -3 < x < 3\}$

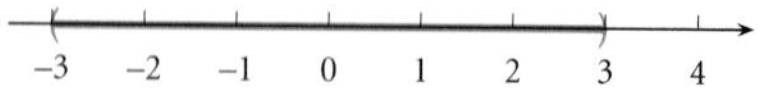

75. $(-\infty, -4)$ or $(3, \infty)$; $\{x \mid x < -4 \text{ or } x > 3\}$

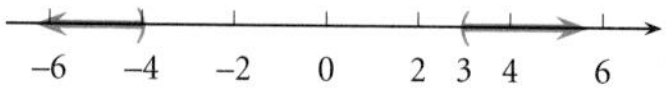

77. $(-\infty, 3)$ or $(4, \infty)$; $\{x \mid x < 3 \text{ or } x > 4\}$

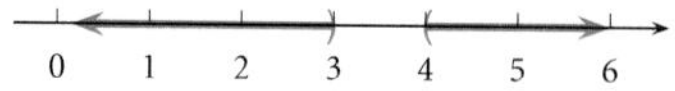

79. $\varnothing$

81. $(1, \infty)$; $\{x \mid x > 1\}$

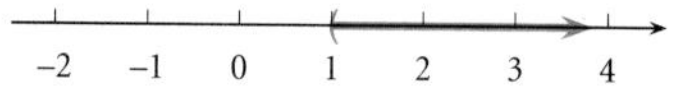

83. $(-\infty, 1)$ or $(2, 3)$; $\{x \mid x < 1 \text{ or } 2 < x < 3\}$

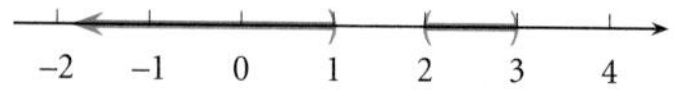

85. $(-2, 0)$ or $(4, \infty)$; $\{x \mid -2 < x < 0 \text{ or } x > 4\}$

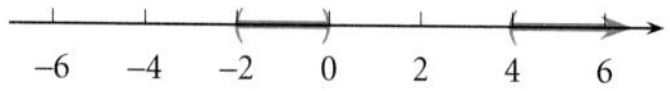

87. $(-1, 0)$ or $(1, \infty)$; $\{x \mid -1 < x < 0 \text{ or } x > 1\}$

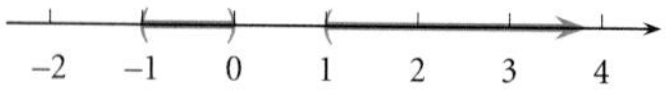

89. $(1, \infty)$; $\{x \mid x > 1\}$

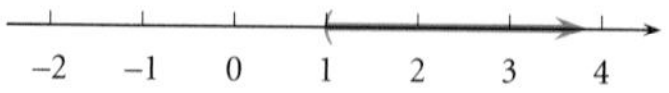

91. $(-\infty, -1)$ or $(1, \infty)$; $\{x \mid x < -1 \text{ or } x > 1\}$

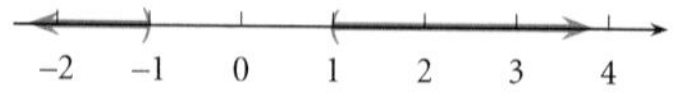

93. $(-\infty, -1)$ or $(0, 1)$; $\{x \mid x < -1 \text{ or } 0 < x < 1\}$

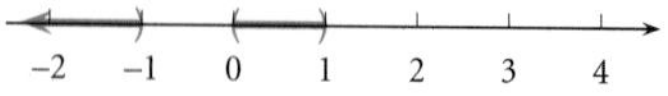

95. $(-1, 1)$ or $[2, \infty)$; $\{x \mid -1 < x < 1 \text{ or } x \geq 2\}$

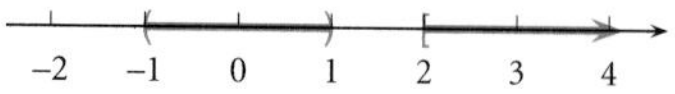

97. $(-\infty, 2)$; $\{x \mid x < 2\}$

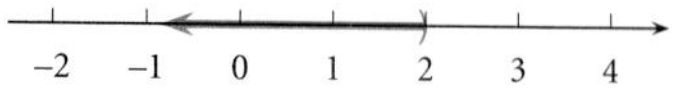

99. $(-\infty, -3)$ or $(-1, 1)$ or $(2, \infty)$; $\{x \mid x < -3 \text{ or } -1 < x < 1 \text{ or } x > 2\}$

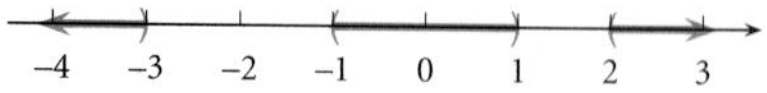

101. The solution is $74 \leq x < 124$, but assuming that the highest possible test score is 100, the range of possible exam scores that will enable you to earn a B is from 74 to 100.

103. The range of possible commissions is \$45, 000 to \$95, 000. The commission varies from 5% of the selling price to 8.6% of the selling price.

105. The amount withheld varies from \$81.35 to \$131.35.

107. Usage ranged from 657.41 kilowatt-hours to 2500.91 kilowatt-hours.

109. The dealer's cost range from \$7457.63 to \$7857.14.

Exercise Appendix A.8 (p. A-79)

1. 3 **3.** -2 **5.** $2\sqrt{2}$

7. $-2x\sqrt[3]{x}$ **9.** x^3y^2 **11.** x^2y

13. $6\sqrt{x}$ **15.** $6x\sqrt{x}$ **17.** $15\sqrt[3]{3}$

19. $12\sqrt{3}$ **21.** $2\sqrt{3}$ **23.** $x - 2\sqrt{x} + 1$

25. $-5\sqrt{2}$ **27.** $(2x - 1)\sqrt[3]{2x}$ **29.** $\frac{\sqrt{2}}{2}$

31. $-\frac{\sqrt{15}}{5}$ **33.** $\frac{5\sqrt{3} + \sqrt{6}}{23}$ **35.** $\frac{8\sqrt{5} - 19}{41}$

37. $\frac{5\sqrt[3]{4}}{2}$ **39.** $\frac{2x + h - 2\sqrt{x^2 + xh}}{h}$

41. $\left\{\frac{9}{2}\right\}$ **43.** $\{3\}$ **45.** 4

47. -3 **49.** 64 **51.** $\frac{1}{27}$

53. $\frac{27\sqrt{2}}{32}$ **55.** $\frac{27\sqrt{2}}{32}$ **57.** $x^{7/12}$

59. xy^2 **61.** $x^{4/3}y^{5/3}$ **63.** $\frac{8x^{3/2}}{y^{1/4}}$

65. $\frac{3x + 2}{(x + 1)^{1/2}} = \frac{3x + 2}{\sqrt{x + 1}}$

67. $\frac{x(3x^2 + 2)}{(x^2 + 1)^{1/2}} = \frac{x(3x^2 + 2)}{\sqrt{x^2 + 1}}$

69. $\dfrac{22x + 5}{10(4x^2 - 17x - 15)^{1/2}} = \dfrac{22x + 5}{10\sqrt{4x^2 - 17x - 15}}$

71. $\dfrac{x + 2}{2(x + 1)^{3/2}}$

73. $\dfrac{4 - x}{(x + 4)^{3/2}}$

75. $\dfrac{1}{x^2(x^2 - 1)^{1/2}} = \dfrac{1}{x^2\sqrt{x^2 - 1}}$

77. $\dfrac{1 - 3x^2}{2x^{1/2}(x^2 + 1)^2} = \dfrac{1 - 3x^2}{2\sqrt{x}(x^2 + 1)^2}$

79. $\dfrac{(5x + 2)(x + 1)^{1/2}}{2}$

81. $2x^{1/2}(x + 1)(3x - 4)$

83. $(x^2 + 4)^{1/3}(11x^2 + 12)$

85. $(2x + 3)^{1/2}(3x + 5)^{1/3}(17x + 27)$

87. $\dfrac{3(x + 2)}{2x^{1/2}}$

89. $\dfrac{2(2 - x)(2 + x)}{(8 - x^2)^{1/2}} = \dfrac{2(2 - x)(2 + x)}{\sqrt{8 - x^2}}$

Exercise Appendix A.9 (p. A-83)

1. $c = 13$ 3. $c = 26$ 5. $c = 25$

7. This is a right triangle. The hypotenuse is the side of length 5.

9. This is not a right triangle.

11. This is a right triangle. The hypotenuse is the side of length 25.

13. This is not a right triangle. 15. $A = 8$ inches^2

17. $A = 4$ inches^2 19. $A = 25\pi$ m^2, $C = 10$ m

21. $V = 224$ ft^3, $S = 232$ ft^2 23. $V = \frac{256\pi}{3}$ cm^3, $S = 64\pi$ cm^2

25. $V = 648\pi$ inches^3, $S = 306\pi$ inches^2

27. The area is π units^2. 29. The area is 2π units^2.

31. The wheel travels 64π inches after four revolutions.

33. The area is 64 ft^2.

35. The area of the window is $24 + 2\pi$ ft$^2 \approx 30.28$ ft^2. $16 + 2\pi \approx 22.28$ ft of wood frame are needed to enclose the window.

37. You can see a distance of 28,920 ft, which is about 5.478 miles.

39. A person can see 64,667 ft or 12.248 miles from the deck. A person can see 79,200 ft or 15.0 miles from the bridge.

41. The areas of the rectangular pools vary from 0 ft^2 to 62,500 ft^2. The shape of the rectangle of largest area is a square with side length 250 ft. The area of the circular pool is $\frac{250\,000}{\pi}$ ft$^2 \approx 79,577$ ft^2. The best choice for a pool of largest area would be the circular pool.

Exercise Appendix A.10 (p. A-88)

1. $\sqrt{5}$ 3. $\sqrt{10}$ 5. $2\sqrt{17}$

7. $\sqrt{85}$ 9. $\sqrt{53}$ 11. $\sqrt{6.89} \approx 2.625$

13. $\sqrt{a^2 + b^2}$

15. $d(A, B) = \sqrt{13}$
$d(B, C) = \sqrt{13}$
$d(A, C) = \sqrt{26}$
$(\sqrt{13})^2 + (\sqrt{13})^2 = (\sqrt{26})^2$
Area $= \frac{13}{2}$ square units

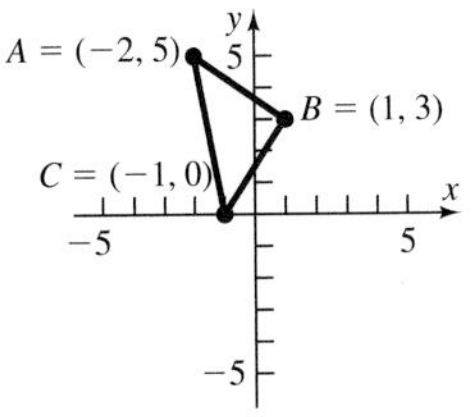

17. $d(A, B) = \sqrt{130}$
$d(B, C) = \sqrt{26}$
$d(A, C) = 2\sqrt{26}$
$(\sqrt{26})^2 + (2\sqrt{26})^2 = (\sqrt{130})^2$
Area $= 26$ square units

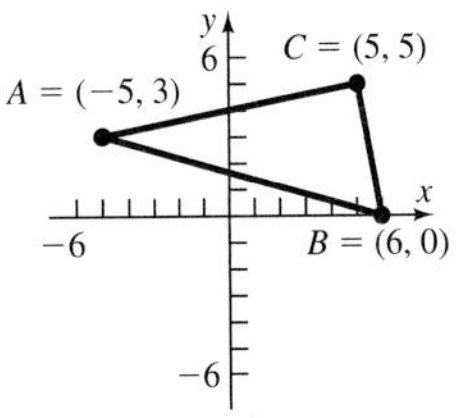

19. $d(A, B) = 4$
$d(B, C) = \sqrt{41}$
$d(A, C) = 5$
$4^2 + 5^2 = (\sqrt{41})^2$
Area $= 10$ square units

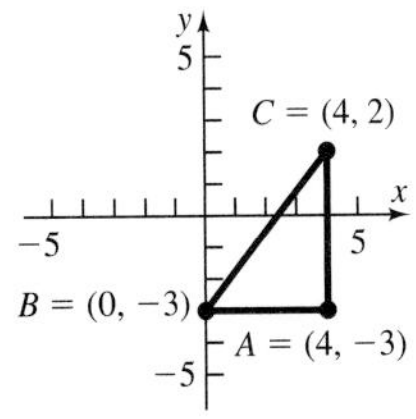

21. $(2, 2)$; $(2, -4)$ 23. $(0, 0)$; $(8, 0)$

25. $4\sqrt{10}$ 27. $2\sqrt{65}$

29. $90\sqrt{2} \approx 127.28$ ft

31. (a) First base $(90, 0)$; second base $(90, 90)$; third base $(0, 90)$
(b) $5\sqrt{2161}$ ft ≈ 232.4 ft
(c) $30\sqrt{149}$ ft ≈ 366.2 ft

33. $d = 50\,t$

APPENDIX B Using Lindo to Solve Linear Programming Problems

Exercise Appendix B (p. B-5)

1. Maximum of $P = 24$ when $x_1 = 0, x_2 = 12$, and $x_3 = 0$.
3. Maximum of $P = 15$ when $x_1 = 5, x_2 = 0$, and $x_3 = 0$.
5. Maximum of $P = 40$ when $x_1 = 14, x_2 = 0$, and $x_3 = 4$.
7. Maximum of $P = 6$ when $x_1 = 6, x_2 = 0$, and $x_3 = 0$.
9. Maximum of $P = 15.2$ when $x_1 = 1.6, x_2 = 4.8$, and $x_3 = 2.4$.
11. Maximum of $P = 15$ when $x_1 = 0, x_2 = 5$, and $x_3 = 0$.
13. No maximum value.
15. Maximum of $P = 30$ when $x_1 = 0, x_2 = 0$, and $x_3 = 10$.
17. Maximum of $P = 42$ when $x_1 = 1$, $x_2 = 10$, $x_3 = 0$, and $x_4 = 0$.
19. Maximum of $P = 40$ when $x_1 = 20, x_2 = 0$, and $x_3 = 0$.
21. Maximum of $P = 50$ when $x_1 = 0$, $x_2 = 15$, $x_3 = 5$, and $x_4 = 0$.
23. Minimum of $P = 76.25$ when $x_1 = 6.25, x_2 = 0, x_3 = 0, x_4 = 20, x_5 = 0, x_6 = 50$, and $x_7 = 0$.

APPENDIX C Graphing Utilities

Exercise Appendix C.1 (p. C-2)

1. $(-1, 4)$ quadrant II
3. $(3, 1)$ quadrant I
5. $X\min = -6$
   $X\max = 6$
   $X\text{scl} = 2$
   $Y\min = -4$
   $Y\max = 4$
   $Y\text{scl} = 2$
7. $X\min = -6$
   $X\max = 6$
   $X\text{scl} = 2$
   $Y\min = -1$
   $Y\max = 3$
   $Y\text{scl} = 1$
9. $X\min = 3$
   $X\max = 9$
   $X\text{scl} = 1$
   $Y\min = 2$
   $Y\max = 10$
   $Y\text{scl} = 2$
11. $X\min = -12$
    $X\max = 6$
    $X\text{scl} = 1$
    $Y\min = -4$
    $Y\max = 8$
    $Y\text{scl} = 1$
13. $X\min = -30$
    $X\max = 50$
    $X\text{scl} = 10$
    $Y\min = -100$
    $Y\max = 50$
    $Y\text{scl} = 10$
15. $X\min = -10$
    $X\max = 110$
    $X\text{scl} = 10$
    $Y\min = -20$
    $Y\max = 180$
    $Y\text{scl} = 20$

Exercise Appendix C.2 (p. C-6)

1. (a)
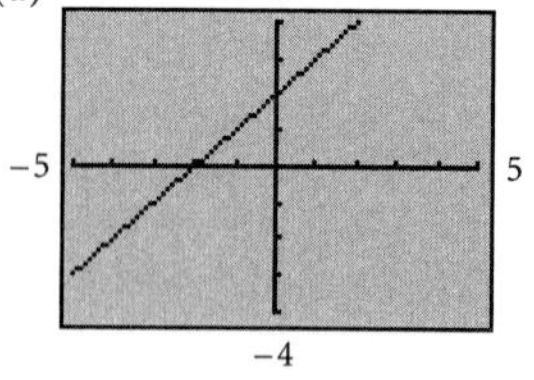

(b)
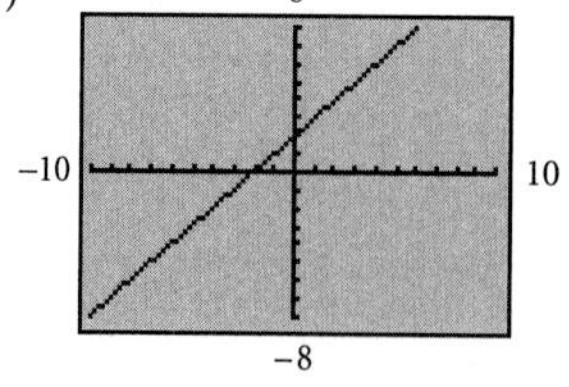

(c)
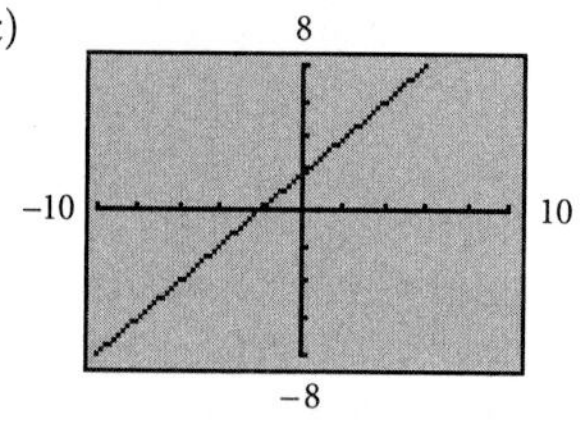

(d)
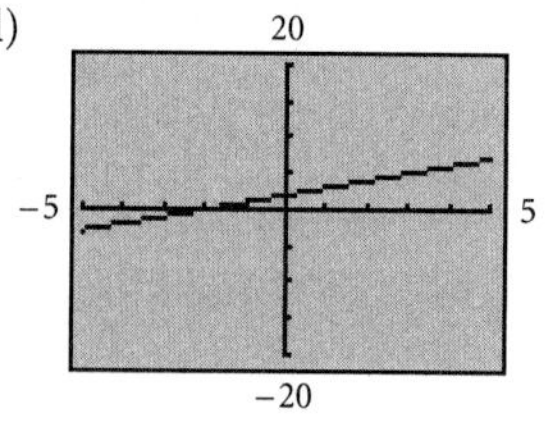

3. (a)
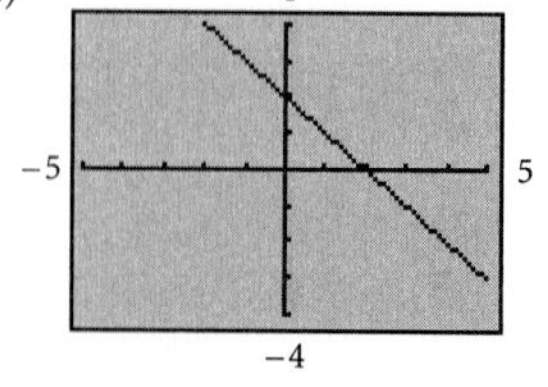

(b)
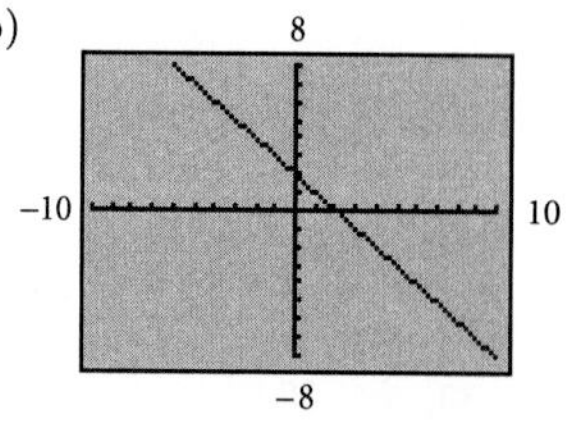

(c)
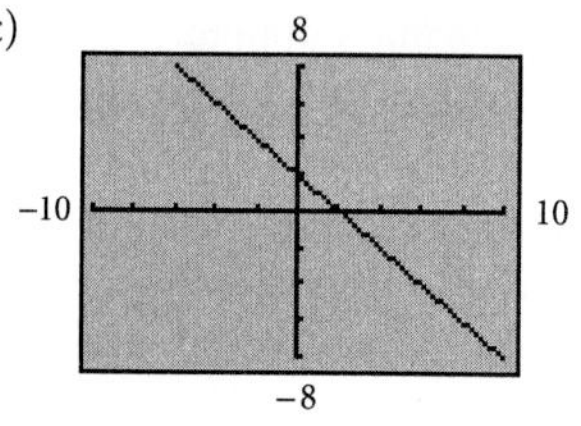

(d)
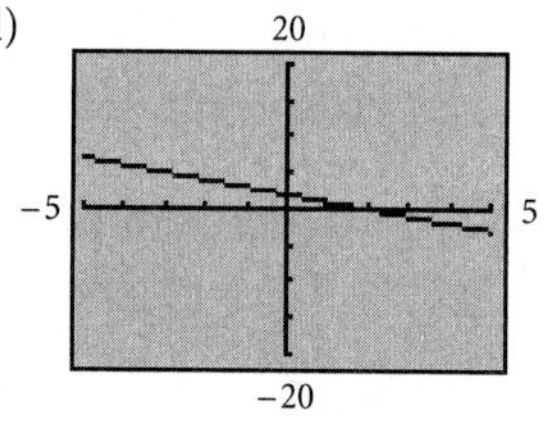

5. (a) (b) (c) (d)

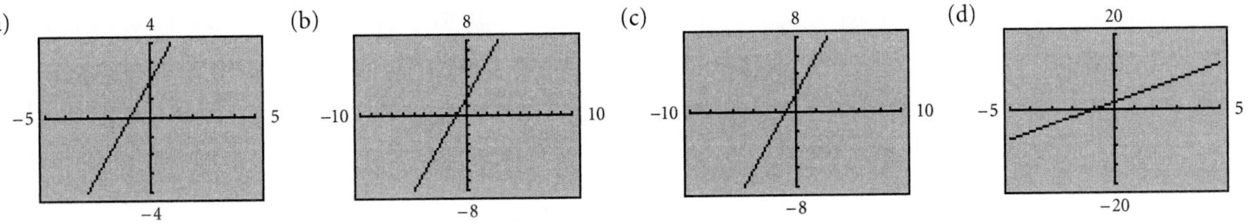

7. (a) (b) (c) (d)

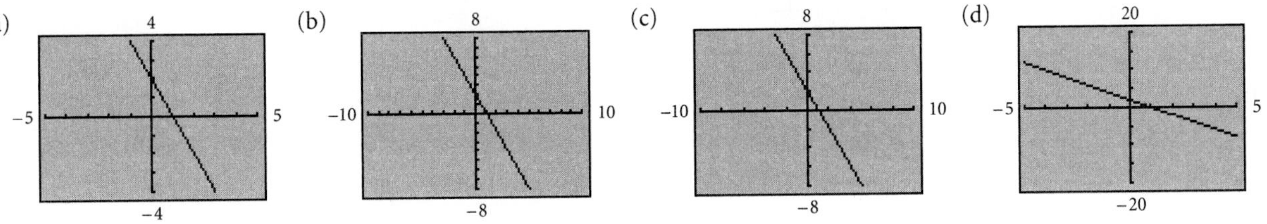

9. (a) (b) (c) (d)

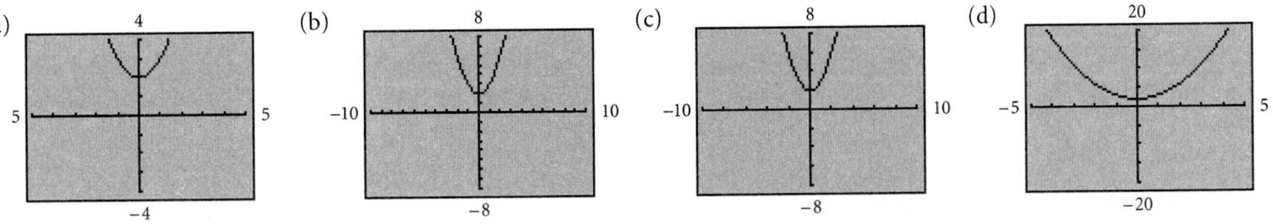

11. (a) (b) (c) (d)

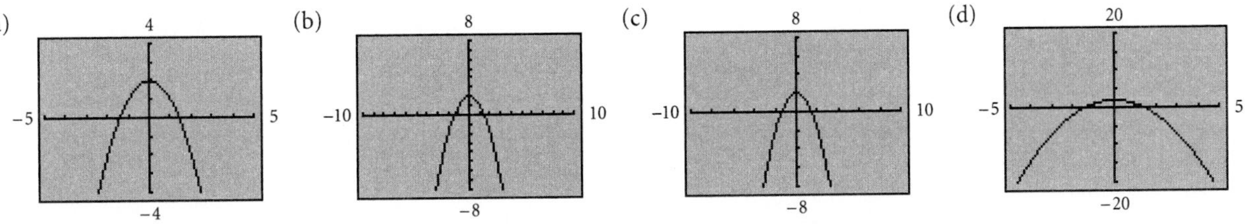

13. (a) (b) (c) (d)

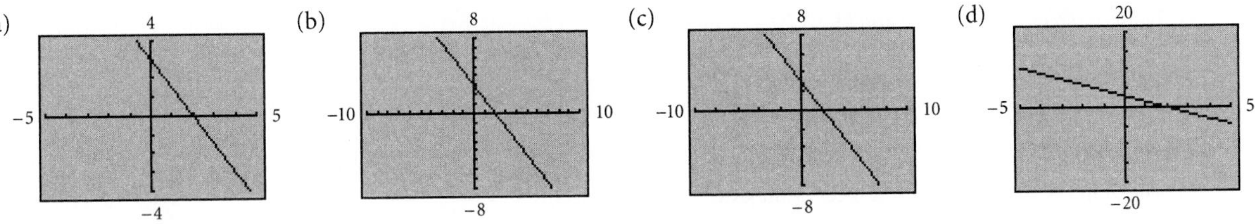

15. (a) (b) (c) (d)

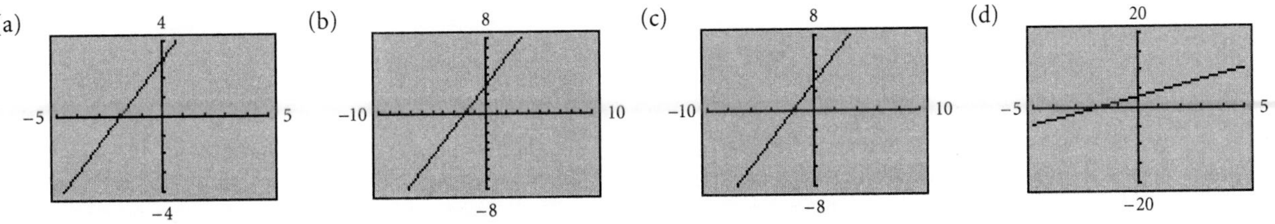

17.

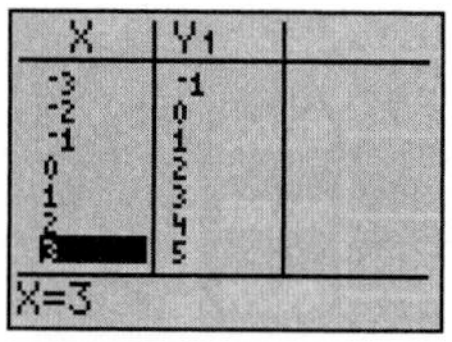

X	Y1
-3	-1
-2	0
-1	1
0	2
1	3
2	4
3	5

X=3

19.

X	Y1
-3	5
-2	4
-1	3
0	2
1	1
2	0
3	-1

X=3

21.

X	Y1
-3	-4
-2	-2
-1	0
0	2
1	4
2	6
3	8

X=3

23.

X	Y1
-3	8
-2	6
-1	4
0	2
1	0
2	-2
3	-4

X=3

25.

X	Y1
-3	11
-2	6
-1	3
0	2
1	3
2	6
3	11

X=3

27.

X	Y1
-3	-7
-2	-2
-1	1
0	2
1	1
2	-2
3	-7

X=3

29.

X	Y1
-3	7.5
-2	6
-1	4.5
0	3
1	1.5
2	0
3	-1.5

X=-3

31.

X	Y1
-3	-1.5
-2	0
-1	1.5
0	3
1	4.5
2	6
3	7.5

X=-3

Exercise Appendix C.3 (p. C-7)

1. Yes   **3.** Yes   **5.** No   **7.** Yes

9. $Y \min = 1$
$Y \max = 9$
$Y \operatorname{scl} = 1$

Exercise Appendix C.5 (p. C-11)

1. The smaller of the two x-intercepts is -3.41.

3. The smaller of the two x-intercepts is -1.71.

5. The smaller of the two x-intercepts is -0.28.

7. The positive x-intercept is $(3, 0)$.

9. The positive x-intercept is $(4.5, 0)$.

11. The positive x-intercepts are $(0.32, 0)$ and $(12.3, 0)$.

13. The positive x-intercepts are $(1, 0)$ and $(23, 0)$.

15. (a) The x-intercepts are $(-1, 0)$ and $(1, 0)$. The y-intercept is $(0, -1)$.

   (b) The graph is symmetric with respect to the y-axis.

17. (a) The graph has no intercepts.

   (b) The graph is symmetric with respect to the origin.

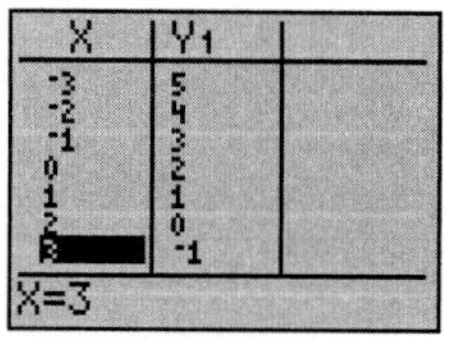

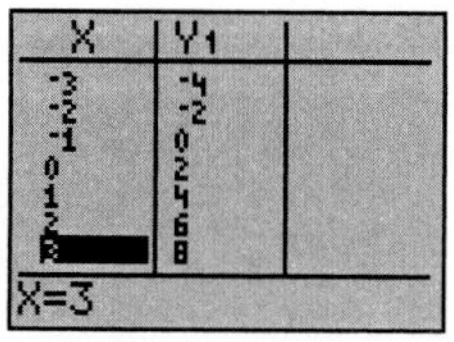

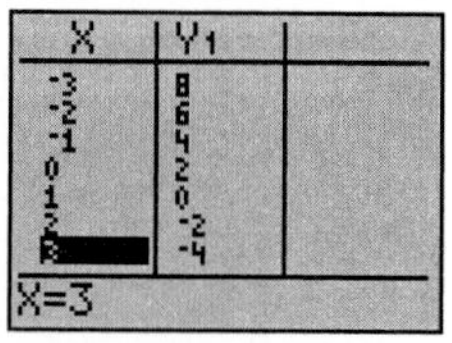

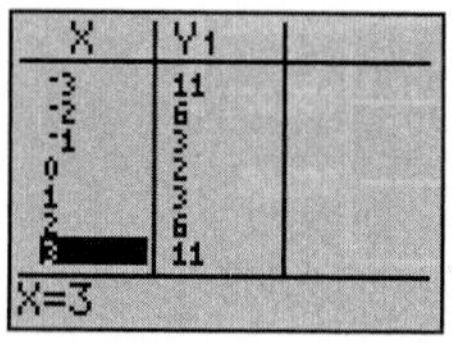

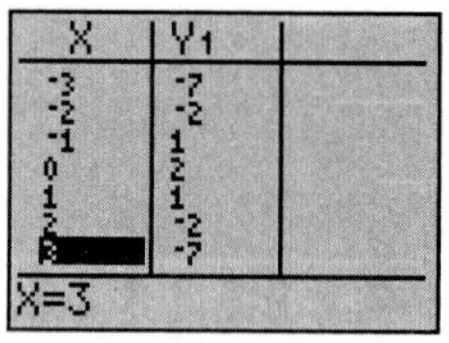

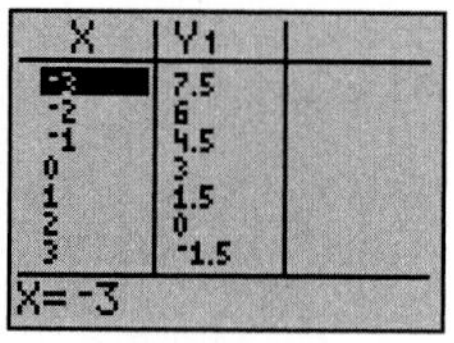

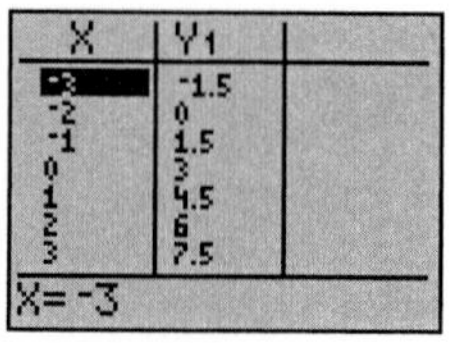

Photo Credits

Chapter 1

Page 1: Doug Menuez/PhotoDisc, Inc./Getty Images. Page 14: PhotoDisc, Inc./Getty Images. Page 27: Ryan McVay/PhotoDisc, Inc./Getty Images.

Chapter 2

Page 48: Arthur Tilley/Taxi/Getty Images. Page 56: Nik Wheeler/Corbis Images. Page 63: Peter Pearson/Stone/Getty Images. Page 129: Bill Lai/The Image Works. Page 154: Jacobs Stock Photography/PhotoDisc, Inc./Getty Images.

Chapter 3

Page 158: Image State. Page 167: Day Williams/Photo Researchers. Page 183: Allan H. Shoemake/Taxi/Getty Images.

Chapter 4

Page 194: B & C Alexander/Photo Researchers. Page 215: Plastock//Photonica 2003. Page 237: PhotoDisc, Inc./Getty Images. Page 251: Hans Peter Marten/Stone/Getty Images.

Chapter 5

Page 261: Ryan McVay/PhotoDisc, Inc./Getty Images. Page 268: Don Smetzer/Stone/Getty Images. Page 293: Mark Antman/The Image Works. Page 305: Corbis Digital Stock.

Chapter 6

Page 316: Phil Degginger/Stone/Getty Images. Page 327: Mark Lewis/Stone/Getty Images. Page 335: PhotoDisc, Inc./Getty Images. Page 341: Sean Justice/The Image Bank/Getty Images. Page 342: PhotoDisc, Inc./Getty Images.

Chapter 7

Page 364: Rob Crandall/The Image Works. Page 389: PhotoDisc, Inc./Getty Images. Page 402: Tannen Maury/The Image Works. Page 408: David Frazier/Photo Researchers.

Chapter 8

Page 420: D. Ducros/Photo Researchers. Page 423: Tom Raymond/Stone/Getty Images. Page 438: PhotoDisc, Inc./Getty Images. Page 439: Michael Melford/The Image Bank/Getty Images.

Chapter 9

Page 469: Corbis Digital Stock. Page 473: Corbis Digital Stock. Page 513: Francois Gohier/Photo Researchers. Page 519: Bob Daemmrich/The Image Works.

Chapter 10

Page 97: Digital Vision. Page 115: Simon Fraser/SPLPhoto Researchers Inc. Page 118: Corbis Digital Stock. Page 135: Monika Graff/The Image Works. Page 144: Lester Lefkowitz/Stone/Getty Images.

Chapter 11

Page 163: John Henley/Corbis Images. Page 201: Corbis Digital Stock. Page 209: Corbis Digital Stock. Page 227: PhotoDisc, Inc./Getty Images.

Chapter 12

Page 236: Syracuse Newspapers/Dick Blume/The Image Works. Page 257: Corbis Digital Stock. Page 264: PhotoDisc Red/Getty Images. Page 271: PhotoDisc Green/Getty Images.

Chapter 13

Page 273: PhotoDisc, Inc./Getty Images. Page 284: ©AP/Wide World Photos. Page 318: Corbis Digital Stock. Page 339: Corbis Digital Stock.

Chapter 14

Page 380: Digital Vision/Getty Images. Page 392 (left): Francesco Reginato/The Image Bank/Getty Images. Page 392 (right): Image State. Page 419: HIRB//Index Stock.

Chapter 15

Page 420: ©AP/Wide World Photos. Page 447: Comstock Images/Getty Images. Page 459: PhotoDisc, Inc./Getty Images. Page 473: SciMAT/Photo Researchers, Inc.

Chapter 16

Page 485: Kim Kulish/Corbis Images. Page 490: PhotoDisc, Inc./Getty Images. Page 499: PhotoDisc, Inc./Getty Images. Page 500: Monika Graff/The Image Works.

Chapter 17

Page 511: Stone/Getty Images. Page 521: Brendan Byrne/Digital Vision. Page 541: Rich La Salle/Index Stock.

Appendix A

Page 1: PhotoDisc, Inc./Getty Images. Page 63: ThinkStock LLC//Index Stock.

Index

DIFFERENTIAL CALCULUS

Derivative of f at a Number c	$f'(c) = \lim_{x \to c} \frac{f(x) - f(c)}{x - c}$	$f'(c)$ is the slope of the tangent line to the graph of f at the point $(c, f(c))$
Derivative of f at x	$f'(x) = \lim_{h \to 0} \frac{f(x + h) - f(x)}{h}$	$f'(x)$ is the instantaneous rate of change of $y = f(x)$ with respect to x
Derivative of a Constant Function	$\frac{d}{dx} b = 0$	b is a constant
Derivative of a Power Function	$\frac{d}{dx} x^n = nx^{n-1}$	n is a rational number
Derivative of $f(x) = e^x$	$\frac{d}{dx} e^x = e^x$	
Derivative of $f(x) = \ln x$	$\frac{d}{dx} \ln x = \frac{1}{x}$	
Derivative of a Sum	$\frac{d}{dx} [f(x) + g(x)] = \frac{d}{dx} f(x) + \frac{d}{dx} g(x)$	
Derivative of a Difference	$\frac{d}{dx} [f(x) - g(x)] = \frac{d}{dx} f(x) - \frac{d}{dx} g(x)$	
Derivative of a Constant Times a Function	$\frac{d}{dx} [c f(x)] = c \frac{d}{dx} f(x)$	c is a constant
Derivative of a Product	$\frac{d}{dx} [f(x) \cdot g(x)] = f(x) \frac{d}{dx} g(x) + g(x) \frac{d}{dx} f(x)$	
Derivative of a Quotient	$\frac{d}{dx} \frac{f(x)}{g(x)} = \frac{g(x) \frac{d}{dx} f(x) - f(x) \frac{d}{dx} g(x)}{[g(x)]^2}$	
Power Rule	$\frac{d}{dx} [g(x)]^n = n[g(x)]^{n-1} \frac{d}{dx} g(x)$	n is a rational number
Chain Rule	If $y = f(u)$ and $u = g(x)$, then $y = f(g(x))$ and $\frac{dy}{dx} = \frac{dy}{du} \cdot \frac{du}{dx}$	
The Differential of y	$dy = f'(x)\, dx$	

REVENUE, COST, AND PROFIT FUNCTIONS

Demand equation	$p = d(x)$	p is the price when x units are demanded (sold).
Revenue function	$R(x) = xp$	R is the revenue when x units are sold at the price p.
Cost function	$C = C(x)$	C is the cost of producing x units.
Average cost	$\overline{C}(x) = \frac{C(x)}{x}$	The average cost per unit when x units are produced.
Marginal revenue	$R'(x) = \frac{d}{dx} R(x)$	The revenue received from selling one additional unit.
Marginal cost	$C'(x) = \frac{d}{dx} C(x)$	The cost of producing one additional unit.
Profit function	$P(x) = R(x) - C(x)$	The profit derived from producing and selling x units.
Maximum Profit	$R'(x) = C'(x)$	Achieved when marginal revenue equals marginal cost.
Elasticity of Demand	$\lvert E(p) \rvert = \left\lvert \frac{p \cdot f'(p)}{f(p)} \right\rvert$	A measure of how quantity demanded changes with respect to a change in price. [$f(p)$ is the quantity demanded at the price p.]

STEPS FOR GRAPHING FUNCTIONS

STEP 1 Find the domain of f.

STEP 2 Locate the intercepts of f. (Skip the x-intercepts if they are too hard to find.)

STEP 3 Determine where the graph of f is increasing or decreasing.
A function f is increasing on an interval I if $f'(x) > 0$ on I.
A function f is decreasing on an interval I if $f'(x) < 0$ on I.

STEP 4 Find any local maxima or local minima of f by using the First Derivative Test or the Second Derivative Test.
If c is in the domain of a continuous function f and if f is decreasing for $x < c$ and is increasing for $x > c$, then at c there is a local minimum. $f(c)$ is the local minimum.
If c is in the domain of a continuous function f and if f is increasing for $x < c$ and is decreasing for $x > c$, then at c there is a local maximum. $f(c)$ is the local maximum.
If $f'(c) = 0$ and if $f''(c) > 0$, then at c there is a local minimum.
If $f'(c) = 0$ and if $f''(c) < 0$, then at c there is a local maximum.

STEP 5 Locate all points on the graph of f at which the tangent line is either horizontal or vertical.

STEP 6 Determine the end behavior and locate any asymptotes.

STEP 7 Locate the inflection points, if any, of the graph by determining the concavity of the graph.
If $f''(x) > 0$ on an interval I, the f is concave up on I.
If $f''(x) < 0$ on an interval I, the f is concave down on I.
Any point on the graph of f where the concavity changes is an inflection point.

INTEGRAL CALCULUS

Indefinite integral	$\int f(x)\,dx = F(x) + K$	$F'(x) = f(x)$; K is a constant
Definite integral	$\int_a^b f(x)\,dx = F(b) - F(a)$	$F'(x) = f(x)$
Integration by parts	$\int u\,dv = uv - \int v\,du$	
Area *A* under the graph of *f* from *a* to *b*	$A = \int_a^b f(x)\,dx$	$f(x) \geq 0$ on the interval $[a, b]$

PARTIAL DERIVATIVES

$$f_x(x, y) = \lim_{\Delta x \to 0} \frac{f(x + \Delta x, y) - f(x, y)}{\Delta x} \qquad f_y(x, y) = \lim_{\Delta y \to 0} \frac{f(x, y + \Delta y) - f(x, y)}{\Delta y}$$